ANNUAL REVIEW OF BIOCHEMISTRY

EDITORIAL COMMITTEE (1986)

ANNUAL REVIEW OF BIOCHEMISTRY

VOLUME 55, 1986

CHARLES C. RICHARDSON, *Editor*

Harvard Medical School

PAUL D. BOYER, *Associate Editor*

University of California, Los Angeles

IGOR B. DAWID, *Associate Editor*

National Institutes of Health

ALTON MEISTER, *Associate Editor*

Cornell University Medical College

ANNUAL REVIEWS, INC.   4139 EL CAMINO WAY   PALO ALTO, CALIFORNIA 94306 USA

℞ ANNUAL REVIEWS INC.
Palo Alto, California, USA

International Standard Serial Number: 0066–4309
International Standard Book Number: 0–8243–0855-7
Library of Congress Catalog Card Number: 50–13143

Typesetting by Kachina Typesetting Inc., Tempe, Arizona; John Olson, President Typesetting coordinator, Janis Hoffman

PRINTED AND BOUND IN THE UNITED STATES OF AMERICA

Annual Review of Biochemistry
Volume 55, 1986

CONTENTS

(continued)   v

(continued)

SOME RELATED ARTICLES IN OTHER *ANNUAL REVIEWS*

From the *Annual Review of Cell Biology*, Volume 2 (1986)

Microtubule-Associated Proteins, J. B. Olmsted

Mechanisms of Protein Translocation Across the Endoplasmic Reticulum, Peter Walter and Vishwanath Lingappa

Cotranslational and Posttranslational Translocation in Prokaryotic Systems, Jon Beckwith and Catharine Lee

The Role of Protein Kinase C in Transmembrane Signalling, Ushio Kikkawa and Yasutomi Nishizuka

Structure and Function of Nuclear and Cytoplasmic Ribonucleoprotein Particles, Gideon Dreyfuss

Chromatin Structure, D. S. Pederson, F. Thoma, and R. T. Simpson

Surface Proteins in Embryonic Development, Gerald M. Edelman

From the *Annual Review of Genetics*, Volume 19 (1985)

The Structure and Function of Yeast Centromeres, Louise Clarke and John Carbon

HPRT: Gene Structure, Expression, and Mutation, J. Timothy Stout and C. Thomas Caskey

Steroid Receptor Regulated Transcription of Specific Genes and Gene Networks, Keith R. Yamamoto

The Regulation of Transcription Initiation in Bacteria, William S. Reznikoff, Deborah A. Siegele, Deborah W. Cowing, and Carol A. Gross

In Vitro Mutagenesis, Michael Smith

Structure and Evolution of the Insulin Gene, Donald F. Steiner, S. J. Chan, J. M. Welsh, and S. C. M. Kwok

Selected Topics in Chromatin Structure, Joel C. Eissenberg, Iain L. Cartwright, Graham H. Thomas, and Sarah C. R. Elgin

From the *Annual Review of Immunology*, Volume 4 (1986)

Structural Correlates of Idiotopes, Joseph M. Davie, Michael V. Seiden, Neil S. Greenspan, Charles T. Lutz, Timothy L. Bartholow, and Brian L. Clevinger

The Molecular Genetics of the T-Cell Antigen Receptor and T-Cell Antigen Recognition, Mitchell Kronenberg, Gerald Siu, Leroy E. Hood, Nilabh Shastri

Regulation of the Assembly and Expression of Variable Region Genes, George D. Yancopoulos and Frederick W. Alt

Murine Major Histocompatibility Complex Class-I Mutants: Molecular Analysis and Structure-Function Implications, Stanley G. Nathenson, Jan Geliebter, Gertrude M. Pfaffenbach, and Richard A. Zeff

The Human Interleukin-2 Receptor, Warner C. Greene and Warren J. Leonard

From the *Annual Review of Medicine,* Volume 37 (1986)

> *Interleukins,* Charles A. Dinarello and James W. Mier
> *DNA Analysis in Genetic Disorders,* C. A. Francomano and H. H. Kazazian, Jr.

From the *Annual Review of Microbiology,* Volume 40 (1986)

> *Organization of the Genes for Nitrogen Fixation in Photosynthetic Bacteria and Cyanobacteria,* Robert Haselkorn
> *Recent Advances in Bacterial Ion Transport,* Barry P. Rosen

From the *Annual Review of Neuroscience,* Volume 9 (1986)

> *Genetics and Molecular Biology of Ionic Channels in* Drosophila, Mark A. Tanouye, C. A. Kamb, Linda E. Iverson, and Lawrence Salkoff
> *Inactivation and Metabolism of Neuropeptides,* Jeffrey F. McKelvy and Shmaryahu Blumberg

From the *Annual Review of Pharmacology and Toxicology,* Volume 26 (1986)

> *Molecular Pharmacology of Botulinum Toxin and Tetanus Toxin,* Lance L. Simpson

From the *Annual Review of Physiology,* Volume 48 (1986)

> *Inositol Phospholipid Metabolism in the Kidney,* D. A. Troyer, D. W. Schwertz, J. I. Kreisberg, and M. A. Venkatachalam

ANNUAL REVIEWS INC. is a nonprofit scientific publisher established to promote the advancement of the sciences. Beginning in 1932 with the *Annual Review of Biochemistry,* the Company has pursued as its principal function the publication of high quality, reasonably priced *Annual Review* volumes. The volumes are organized by Editors and Editorial Committees who invite qualified authors to contribute critical articles reviewing significant developments within each major discipline. The Editor-in-Chief invites those interested in serving as future Editorial Committee members to communicate directly with him. Annual Reviews Inc. is administered by a Board of Directors, whose members serve without compensation.

ANNUAL REVIEWS OF		SPECIAL PUBLICATIONS
Anthropology	Medicine	Annual Reviews Reprints:
Astronomy and Astrophysics	Microbiology	Cell Membranes, 1975–1977
Biochemistry	Neuroscience	Cell Membranes, 1978–1980
Biophysics and Biophysical Chemistry	Nuclear and Particle Science	Immunology, 1977–1979
Cell Biology	Nutrition	
Earth and Planetary Sciences	Pharmacology and Toxicology	Excitement and Fascination
Ecology and Systematics	Physical Chemistry	of Science, Vols. 1 and 2
Energy	Physiology	
Entomology	Phytopathology	History of Entomology
Fluid Mechanics	Plant Physiology	Intelligence and Affectivity,
Genetics	Psychology	by Jean Piaget
Immunology	Public Health	
Materials Science	Sociology	Telescopes for the 1980s

For the convenience of readers, a detachable order form/envelope is bound into the back of this volume.

Martin D. Kamen

Ann. Rev. Biochem. 1986. 55:1–34

A CUPFUL OF LUCK, A PINCH OF SAGACITY

Martin D. Kamen

University of California, San Diego, La Jolla, California 92093

CONTENTS

This discovery indeed is almost of that kind which I call *serendipity,* a very expressive word, which as I have nothing better to tell you, I shall endeavour to explain to you: you will understand it better by the derivation than by the definition. I once read a silly fairy tale, called *The Three Princes of Serendip:* as their highnesses travelled, they were always making discoveries, by accidents and sagacity, of things which they were not in quest of: for instance, one of them discovered that a mule blind of the right eye had travelled the same road lately, because the grass was eaten only on the left side, where it was worse than on the right—now do you understand *serendipity?*

Horace Walpole. 1754. (1a,b)

As the Twig is Bent . . .

The mention of the word "serendipity" to describe the combination of chance and sagacity as a crucial determinant of scientific creativity (2) is a well-worn cliché, but I can think of no better word to apply to the history of my half-century in research.

1

0066-4154/86/0701-0001$02.00

That I even got into research at all, and especially biochemical research, is amazing considering the circumstances of my upbringing. My parents were poor Russian Jewish immigrants who met and married in Toronto, where I was born August 27, 1913. My father had learned photography in Canada and opened a studio in Chicago. It was located in Hyde Park, near the campus of the University of Chicago, and we saw many of the faculty who came for portraits, as well as some distinguished visitors like E. Benes, the former prime minister of Czechoslovakia, and Fanny Bloomfield Zeisler, a leading pianist of the time. I showed considerable musical talent and got into the fix of being acclaimed as a child prodigy. The traumas attendant on this experience left me with a permanent aversion to center stage and conditioned me all through my scientific career to downplay achievement.

Throughout my youth, I had many friends who displayed a fervid interest in manipulating toy steam engines, crystal sets, and the like, but I took no interest in things mechanical. Indeed, I thought those who did a bit bizarre. As an occasional helper in our studio darkroom, I encountered some of the smells and arcane processes of a kind of chemistry but other than this had no experience with what might be called science. Entering the University of Chicago as a freshman in the spring quarter of 1930, I declared a major in English literature, probably because it was as good as any subject to expand my considerable background of readings in the classics.

The shock of the Great Depression took effect on the family fortune that summer and suddenly there was need to cast about for a means whereby I could make a living after graduation. My father suggested I try chemistry and as a dutiful son, I somewhat grudgingly agreed. That fall, I switched my major. So much for vocational guidance!

The vagaries of academic organization had assigned an eminent organic chemist, J. W. E. Glattfeld, to be my advisor the previous quarter, and he had labored mightily to fashion a curriculum that disposed of the general science requirement expeditiously. Now he was somewhat taken aback to find his ward opting for a science major. He would have been astounded to be told that inside ten years I would be writing him for advice on the chemistry of C_4-saccharinic acids, his specialty, in connection with my efforts to devise fast efficient syntheses of erythronic acids labeled with the short-lived carbon isotope, ^{11}C.

My skills as a musician helped support my studies in science. I received a partial scholarship for playing first viola in the university orchestra, and picked up bits of cash here and there as a jazz fiddler in the numerous havens of alcoholic refreshment to be found in the university area. Meanwhile I was discovering the excitement and fascination of science. By my junior year, I had become a complete convert. However, I retained a deep love for music and this was to be a source of happiness in later years, particularly as I had the good fortune to enjoy musical activity without having to make a living from it.

A stubborn tendency to impracticality persisted and resulted in my accumulating almost as many credits in mathematics and physics as in chemistry. I remained resistive to the arguments that the only practical areas were synthetic organic or inorganic chemistry and graduated with honors in physical chemistry after a hectic three years of undergraduate study. The research project in which I had participated to earn the "cum laude" was a determination of ammonia gas emission spectra excited by electrodeless discharge—part of a general program under the aegis of the senior professor, Dr. W. D. Harkins, to study products of ergosterol using radiations other than ultraviolet. The routine lab operations, as well as the implementation of the professor's research program—in the two completely disparate areas of nuclear physics and surface chemistry—were entrusted to a hardworking, imaginative, and much exploited young assistant professor, Dr. D. M. Gans. I became closely attached to him and even somewhat resentful of the pressures that he endured in his efforts to achieve tenure. I carried away from this experience a conviction never to be repressive or demanding of graduate students and fellows who might be under my supervision, if ever I attained the lofty status of a senior professor.

With the Depression still raging in the early 1930s, I had little incentive to leave the university and eagerly accepted Gans's suggestion I stay on and work for a doctorate with Harkins. Inheriting a cloud chamber apparatus that had been built by my predecessor, Henry W. Newson, and by Gans, I dreamed up an overly ambitious thesis research on the angular distribution of protons after collision with fast neutrons. I hoped from the results obtained to deduce something about the nature of intranuclear forces. The means at hand were inadequate, as it turned out. For one thing, the neutron sources were too weak, being made up of a few millicuries of thoria mixed with beryllium powder, However, three years of arduous effort allowed the accumulation of data on some 700 recoils which I used to support a thesis that earned me a PhD in the winter of 1936. A paper was written and published with, as customary, the professor as senior author (3).

Conditions in the outer world were still grim. Gans suggested I spend some time at the famed Radiation Laboratory in Berkeley, California, where the charismatic Ernest O. Lawrence had developed the cyclotron as the most powerful tool of nuclear physics. Harkins intended to build a cyclotron in Chicago, and I could expect to be taken on as a research assistant provided I could claim some expertise as a cyclotroneer. The Radiation Laboratory was Mecca for all aspiring young nuclear physicists, and I was intrigued at the prospect of going there. Working as a photographer, and saving proceeds of various engagements as a jazz fiddler over the Christmas season, I left for Berkeley just as the new year of 1937 was dawning.

Nuclear Euphoria

Although I had not written for permission to work with Lawrence and his group, I received a most hospitable welcome and was particularly happy to see that Franz N. D. Kurie, who had published results (4) similar to mine prior to the appearance of our paper, was on the laboratory staff and eager to press forward on the problem of neutron-proton scattering. I was stunned to see such a profusion of neutrons as emerged from the cyclotron. Each expansion of the cloud chamber Kurie had built in Berkeley showed literally thousands of recoil protons, whereas I had been used to seeing one or two every ten expansions. However, our joy was short-lived. We could not cope with the sheer abundance of neutrons that came from every direction and made accurate definition of the scattering angles impossible. However, serendipity began its entry into my life at this point.

We noticed when the chamber was filled with N_2 gas that almost every expansion showed one or two stubby tracks of recoil protons with a knob on the end. They were homogeneous in range (~ 1 cm), so that they could only come from neutrons moving so slowly that the recoils produced took up only the reaction energy, there being no excess kinetic energy imparted by the neutrons. There was good evidence (5) that slow neutrons were captured by nitrogen nuclei (^{14}N) to produce a carbon isotope, presumably of mass 14. We could assume that the knobs at the beginning of the stubby tracks were the relatively heavy and slow ^{14}C recoils and that we were observing the reaction: $^{14}N + {}^{1}n \rightarrow {}^{14}C + {}^{1}H$. I filed this observation away in the back of my head along with a mass of data I would be accumulating as the laboratory radiochemist on integral yields of radioisotope prediction. It would be crucial in later encouragement to proceed with the search for long-lived radiocarbon.

Serendipity continued its sway, as events moved forward. I joined with E. M. ("Ed") McMillan, a young laboratory staff member who was already a leader in the field of nuclear physics, to straighten out some anomalous results Lawrence had reported on platinum transmutation by deuterons. We welcomed a young graduate student from the department of chemistry, Samuel Ruben, to help with the difficult platinum metal radiochemistry needed to sort out the activities observed when platinum was bombarded by highly energetic deuterons (up to 8 MeV). We soon found that the anomalies in yields of radioactivity were the results of light element contamination. However, the authentic radioactivities that we discovered provided fresh anomalies in that there were more radioactivities observed than there were isotopes available for assignment. We had to conclude that we were making many nuclear isomers, thereby establishing for the first time the phenomenon of nuclear isomerism as a general occurrence in the heavy elements of the periodic system (6). The serendipitous aspect was not only this discovery but also the circumstance that

Ruben and I were thrown together. We became aware that we could form a fruitful partnership based on a strong mutual respect and friendship and of the fact we would have access to both the considerable facilities of the Radiation Laboratory and the department of chemistry. We reached an understanding to collaborate, with seniority for production and characterization of cyclotron products assigned to me and that for chemical applications to Ruben.

The particular benefit in this arrangement for me lay in that my time in the Radiation Laboratory was largely taken up in monitoring cyclotron productions of radioactive materials, and such productions were assuming increasing magnitude as Lawrence pressed his campaign to encourage biologists and medical researchers to use the cyclotron and its products. By working with Ruben, I could gain more time for our mutually conceived researches. For Sam, the advantages of ready access to the cyclotron were obvious. For him, too, there were growing commitments of time for departmental teaching assignments.

Being at the center of Radiation Laboratory isotope distribution, I was becoming well known to many researchers worldwide, some of whom became good friends later. One was George Hevesy, to whom on one occasion during World War II I sent ^{32}P in three separate batches to minimize loss at sea because of the submarine menace. He wrote a charming letter in reply in which he remarked that if any of the samples failed to reach Denmark, it would be necessary to advise workers on the radioactive content of seawater to be careful with the figures obtained. With others I entered into active collaboration, an example being research with Dr. G. A. Whipple and his group at Rochester on excretion of iron in dogs maintained in various stages of iron depletion.

Photosynthesis and CO_2 Fixation

The fruits of our partnership materialized swiftly. Sam had pressed vigorously his own campaign to interest biologists on campus in using radioisotopes. It should be noted that biochemists, and biologists in general, were low in the pecking order on the Berkeley campus and it took considerable courage for a young graduate student hoping to advance to faculty status in chemistry to risk his future on biochemical research. My first intimation of Sam's campaign came in a request from Lawrence to help I. L. Chaikoff, a young assistant professor in physiology, obtain samples of CO_2 labeled with 21 min ^{11}C. When I inquired what he wanted to do with such short-lived material, Lawrence told me of Chaikoff's ingenious plan to synthesize D-glucose from $^{11}CO_2$ expeditiously and specifically, using the photosynthetic ability of green plants, which everyone supposed produced D-glucose from CO_2. With the ^{11}C–uniformly labeled glucose so obtained, Chaikoff would study glucose metabolism in rats in short-term experiments. He had calculated he would have sufficient radioactive glucose to do a significant series of studies on glucose assimilation and excretion as CO_2.

When I told Sam about this scheme, he reacted angrily. He claimed it had been his idea, which he had convinced Chaikoff would work. The idea certainly had all the boldness and imagination one had come to expect from Sam. I learned that Dr. W. Zev Hassid had been recruited by Chaikoff to help in the isolation and purification of the labeled glucose expected to be formed by feeding $^{11}CO_2$ to barley, wheat, and sunflower plants. I already knew Zev because I played in chamber music sessions at his house organized by his wife, Leila, who was an accomplished violinist.

After many weeks of fruitless effort in which appreciable quantities of labeled glucose were not formed, Sam suddenly realized that it might be possible to solve the age-old problem of the primary CO_2 fixation product in green plant photosynthesis. He proposed I join him and Zev to mount a strong effort to isolate and characterize the labeled material actually formed from $^{11}CO_2$ in short-term experiments. It seemed simple enough. After exposure to the labeled CO_2, the plants could be cut up and placed in acidified boiling water to which carrier unlabeled bicarbonate was added to remove unreacted labeled CO_2. The water-soluble radioactive material could then be examined chemically to see in what compounds it had appeared. The identification could be performed adding carrier for the various compounds in which we guessed the radioactivity was contained. I agreed enthusiastically to give the project my full-time attention for the few months we optimistically expected would be required to solve the problem.

Some hundreds of experiments and three years later, we were still in the dark about the nature of the CO_2 fixation product, but we had learned much about the photosynthetic process. Elsewhere I have summarized the results we obtained (7), which were published in several papers in 1940 (8–10). The main results were to establish fixation in (a) compound(s) which contained hydroxyl and carboxyl moieties, the process being initiated by a reversible carboxylation of some CO_2 acceptor in the dark, followed by photoreduction to carbohydrate and other plant products. A fundamental mystery that was posed by our findings was how the plants obtained the free energy needed to drive the carboxylation reaction. All model reactions we knew of in vitro were strongly endothermic ($\Delta G_0' \sim 10$ kcal), whereas the photosynthetic "dark" carboxylation was exothermic ($\Delta G_0'$ estimated as ~ -2 kcal). Neither we nor anyone else then knew about phosphate, or more relevantly, ATP, or about photophosphorylation, which would be discovered over a decade later.

Our announcement of a dark CO_2 fixation met with complete disbelief among our biochemical friends in the Life Sciences building across the campus. Remarks like "you can't have photosynthesis without light" were heard. We persisted, generalizing the CO_2 uptake to a variety of living systems, including bacteria, fungi, protozoa, plant roots, and chopped rat liver. We observed, using a procedure based on thermal decarboxylation of the precipitates produced by addition of barium ions to carrier amounts of organic acids, that carboxylation was a general phenomenon (11). The reaction to our results at a center of microbiological research, the Hopkins Marine Station of Stanford University, was diametrically opposed to that of our colleagues in Berkeley. C. B. van Niel, who had been preaching the doctrine of CO_2 as a general metabolite for years, was delighted with our results. Some of his associates and disciples, such as H. A. Barker, S. F. Carson, and J. W. Foster, came to work with us. Together we decisively demonstrated authentic CO_2 fixation in anaerobic fermentations of methane bacteria (12), and oxidative metabolism of propionic acid bacteria and fungi. These researches extended and confirmed the discovery made by Harland G. Wood just a few years earlier, of heterotrophic CO_2 fixation in glycerol fermentations by propionic acid bacteria (13). We summarized our findings in a position paper, written in collaboration with our associates at Pacific Grove (14).

Being out of the mainstream of biochemistry, we were unaware that some biochemists, notably Hans Krebs, had become interested in pursuing experiments like ours with $^{11}CO_2$. Among the group working with A. Baird Hastings at Harvard on incorporation of ^{11}C-labeled lactate into glycogen in animal systems, there was Birgit Vennesland, who guessed correctly that CO_2 might be incorporated, and prevailed on Hastings to include control experiments with $^{11}CO_2$ that demonstrated CO_2 fixation (an excellent summary of the early history of CO_2 fixation is given in Ref. 15). The decisive data were reported by Earl A. Evans, Jr. and Louis Slotin at Chicago using the newly built cyclotron there. They showed that pigeon liver preparations incubated with $^{11}CO_2$ fixed radioactivity in the carboxyl groups of alpha-ketoglutarate (16). (Ironically, Slotin had gone to Chicago and accepted the appointment I had thought I might eventually fill.)

Undoubtedly, our discovery of general CO_2 fixation was one of the major positive contributions of the early work with isotopic tracers. Another important finding from our laboratory was the demonstration, using the rare stable oxygen isotope ^{18}O, that the oxygen evolved in photosynthesis originated from water oxygen and not from CO_2 (17). Our claims depended on the nonparticipation of exchange reactions that equilibrated oxygen from CO_2, molecular oxygen, and water, and that were favored at low pH. Although our experiments were performed at very alkaline pH (~ 10), there was no certainty that the internal cellular pH at the chloroplast was not in the acid range. It was also known that atmospheric oxygen reflected the ^{18}O content of atmospheric CO_2, rather than water (the so-called "Dole" effect), but in an article published a few years later, Barker and I discussed this anomaly and showed that it did not refute the conclusion that all atmospheric oxygen originated by photosynthetic photooxidation of water (18). Still later, our results incurred the ire of none other than Otto Warburg, assuredly one of the great figures in biochemistry in this century, who stubbornly clung to the notion that photosynthetic oxygen came from CO_2 (19). In this, as in his persistent claim that photosynthesis was nearly 100% efficient, he was wrong. More work from other laboratories decisively substantiated our conclusions. Thus, for example, Holt & French found (20), using isolated chloroplasts, and Dole & Jencks showed, using the small difference in ^{18}O content of naturally isotopically equilibrated water and oxygen (21), results consistent with those we reported originally.

Isotopy and Long-Lived Carbon (^{14}C)

The serendipitous route to ^{14}C was not at all clear in the winter of 1939 as Sam and I sat gloomily in the "Rat House," an old frame shack of a building in the chemistry complex where we did our photosynthesis experiments, assessing our chances of solving our original problem—the chemical nature of the first CO_2 fixation product in photosynthesis. Clearly we needed a long-lived carbon isotope. I recalled the stubby tracks I had seen a year previously in the nitrogen disintegration experiments with Franz Kurie. I knew that ^{14}C existed but that in all likelihood it was not long-lived. Nowadays, when ^{14}C is so common an item in nuclear catalogs, it may be difficult to imagine a time when its existence as a long-lived isotope was in doubt.

Ed McMillan had tried making some ^{14}C by exposing a pound bottle of ammonium nitrate crystals to the stray neutrons of the 37-inch cyclotron, but the bottle had been accidentally broken and he had not attempted to repeat the trial. Moreover, Sam Ruben had looked at some graphite targets that had received many microampere-hours of deuteron bombardment in connection with some experiments we had performed using 10 min radionitrogen (^{13}N) and found no activity that could be released by combustion to CO_2. Hence the reaction of deuteron capture in ^{13}C, followed by proton ejection to make ^{14}C, had been ineffectual, or else the half-life of the ^{14}C expected would have had to be greater than several years to escape detection. Theoretically, with ^{14}C having a nuclear spin of zero and ^{14}N, its product in beta decay, a spin of one, the degree of spin forbiddenness was simply not great enough to give ^{14}C a half-life of more than a few days, according to calculations made by J. R. Oppenheimer's group. I knew from my original observations in the cloud chamber that the energy upper limit for the postulated decay of ^{14}C to ^{14}N was no more than a few hundred kilovolt-equivalents. So all indications were that with this energy and the failure to see ^{14}C in the bombarded graphite, the radioactivity to be expected was probably quite short-lived, on the order of seconds.

However, a new development in cyclotron technology had occurred. Robert R. Wilson had confirmed a prediction of Lawrence that huge currents of high energy deuterons circulated in the interior of the cyclotron that were not being collected at the external target. Working with Wilson in the early morning hours from midnight to six AM, I had shown that highly infusible materials like ferrous phosphide could be soldered to the copper backing of probes Wilson had designed and these could be cooled well enough with water to withstand the great power input of the internal beams (on the order of kilowatts). I had been able to produce large quantities of ^{32}P and ^{59}Fe using such internal targets while the external beam was essentially untouched. Thus, the cyclotron could begin to keep up with the drastically increased demands for fast neutrons needed in radiotherapy and radioactive isotopes required for tracer research. In a paper that Wilson and I published on the internal target technique (22), I wrote prophetically, "The method of internal targets should find its most important application in the preparation of radioisotopes which are long-lived and difficult of activation, as well as in the demonstration of the existence of many radioisotopes as yet undiscovered."

I explained to Sam that it would be marvelous if I could reserve the cyclotron for a week or so to try the bombardment of graphite on an internal target, using all the beam available circulating inside the cyclotron. Such a procedure would raise the total activation several orders of magnitude above those ever tried before. But it was foolish to suppose I could ever get such a dispensation, particularly in view of the pessimism regarding the existence of ^{14}C as a long-lived isotope.

I had barely left Sam and returned to the Radiation Laboratory when I was called to Lawrence's office and told to begin an intensive search for ^{14}C or any other possible long-lived isotope among the elements of primary biological importance. It appeared that Harold Urey had been promoting vigorously the use of the rare stable isotopes (^{2}H, ^{13}C, ^{15}N, and ^{18}O) as tracers in biological research for which no adequate competing long-lived unstable isotopes were available.

The rest of the story is history (23). In a very short time, ^{14}C was discovered (24), as well as ^{3}H (25). As the account I wrote of the dramatic circumstances attending the birth of ^{14}C is buried in an obscure journal (26), I exhume it to quote:

> The weather in Berkeley during the winter months can be rugged. February of 1940 was no exception—as I was painfully aware while sitting at the controls of the ailing 37-inch cyclotron in the old Radiation Laboratory on the University campus. I had been there more or less continuously for three days and nights. As the operation drew to an end in the early morning hours of February 15, there was an extraordinary fanfare of driving rain on the tin roof, punctuated by the blasts of high voltage discharges in the bowels of the machine. Added to the general cacophony were occasional howls, screams and gutteral growls emanating from some recordings of French who-dunnits—a consequence of the activities of language classes which occasionally occupied the lab mezzanine in the upper reaches of the building. Bone-tired and red-eyed, I shut down the machine, rescued the remaining fragments of carbon target, which resembled so many bits of intensely radioactive bird gravel, and

shambled over to the ramshackle hut in which Dr. Samuel Ruben, my collaborator, worked and would be appearing shortly. These precious bits of discouraged graphite hopefully contained evidence for the existence of a long-lived radioactive form of carbon.

Indeed they did! Thus, the most valuable single tool in the nuclear armamentarium, ^{14}C, was revealed. It would contribute immeasurably to the study of life processes, as well as to those of death (as elaborated in the ^{14}C-dating technique invented by Willard C. Libby). Its full potential was realized when the uranium piles produced neutrons in weighable amounts (27a). In addition to its impact on all natural science, as well as archaeology, it provided direct proof (27b) through its anomalously long half-life (5700 years), apparently unique among beta-ray emitters, of the participation in nuclear beta decay of mesons and other particles smaller than neutrons and protons.

As the fall of 1940 neared, our prospects could not have been brighter. Plans were being made to create a future company under Research Corporation auspices to construct and operate a battery of cyclotrons dedicated to ^{14}C production, with an expansion of the whole radioisotope program centered at the Laboratory. In chemistry, Sam's position was growing stronger as his talents were being better perceived; and even some appreciation, however dim, of the worth of his collaborations with biologists was growing. Graduate students were eager to work with him on isotope tracer studies of organic reaction mechanisms and publications in this area were emerging from his group.

But our luck changed as the international situation worsened. Many months before Pearl Harbor, the laboratory had to go on a war footing. All non–war related research was halted. The ammonium nitrate tanks that had been installed to make ^{14}C were removed. I was assigned to head a program of isotope production and Sam was drawn off to work on various projects in chemical warfare. Our expectations of using ^{14}C to solve at last the problem that had started our researches in photosynthesis some three years previously were not to be realized. The minute quantities of ^{14}C remaining from our initial discovery were available for use in some pilot studies on the primary fixation product, as well as in a few researches on exchange reactions and mechanisms of fatty acid oxidation, a few of which were reported. Incidental to my full-time pre-occupations with the wartime isotope program, I crowded in a research on the production and radiochemistry (28) of the long-lived sulfur isotope, ^{35}S, in which isotopically enriched materials (namely AgCl with excess amounts of the ^{35}Cl isotope) were used for the first time to make an isotopic assignment of the radioactivity, in this case to the isotope ^{35}S.

Beset by the frustration of not being able to get on with the photosynthesis research, now that we had the tool, ^{14}C, at hand, we were wholly unprepared for the catastrophes that befell us next. Sam was killed in a laboratory accident in late 1943, and shortly thereafter in early 1944, I was separated from the Manhattan Project as an alleged security risk. This action taken by Army Intelligence, as I discovered many years later, was predicated solely on the need to prevent possible leaks of information incidental to my extensive social contacts resulting from my musical activities. Army pressures were exerted to close all opportunities for my employment as a scientist in the San Francisco Bay Area, but I managed to hang on working as a marine inspector at the Kaiser shipyard during the day. At night, I contrived to do some research with H. Albert ("Nook") Barker, who graciously offered me a haven in his laboratory. Nook, aided by Tom Norris, a former graduate student of Sam's, had obtained a small quantity (a few microcuries-equivalent) of ^{14}C from the ammonium nitrate tanks still sitting in the "Rat House" and proposed we examine the fermentations of glucose and xylose by the thermophile *Clostridium thermoaceticum,* which was anomalous in producing acetic acid in excess of the expected two moles per mole of carbohydrate dissimilated. It was possible that the excess acetic acid arose by complete synthesis from two moles of CO_2 as in a reaction scheme:

$$C_6H_{12}O_6 + 2H_2O \rightarrow 2CH_3COOH + 2CO_2 + \text{``8H''}$$

$$2CO_2 + \text{``8H''} \rightarrow CH_3COOH + 2H_2O$$

This suggestion was eminently testable by tracer procedures. Conducting the fermentation in the presence of $^{14}CO_2$, we showed that the appearance of labeled carbon in the acetate and dilution of the pool of $^{14}CO_2$ occurred in the amount expected, indicating that the two carbons of CO_2 could be used in the synthesis of one mole of acetic acid (29). We also performed other experiments on the agenda of the program we had previously considered and which had been prevented by the war effort. Thus, we examined an anaerobic fermentation of lactate and found a conversion of CO_2 to acetate and butyrate (30), as well as a process in *Clostridium kluyveri* that revealed that during conversion of ethanol and acetate to butyrate and caproate there was an intermediary participation of acetate in the production of the C_4 and C_6 fatty acids with addition of a C_2 unit to the carboxyl carbon of butyrate formed initially (31). These experiments, performed with weak samples of ^{14}C-labeled barium carbonate (a few microcuries–equivalent total in 30–800 milligrams), were the first biological experiments with ^{14}C reported. A major consequence of the work on *C. thermoaceticum* was the initiation of three decades of intensive investigations by Harland Wood, L. C. Ljungdahl, and their associates (32a,b), which elaborated elegantly the existence of a new pathway for synthesis of acetate from CO_2 through catalysis by the correnoid system of B-12.

Early in 1945, the efforts of the Radiation Laboratory administration to have me placed somewhere outside the Bay Area were successful. I was offered an appointment in the Washington University Medical School in St. Louis. The opportunity to begin a new career was welcomed eagerly, although I left Berkeley most reluctantly.

Good Science and Bad Politics

My duties in my new assignment required me to supervise the operation of the Washington University cyclotron which, with the diversion of the Berkeley cyclotrons to war-related research, had become the major source of radioisotope production in the country. It was the property of the Mallinckrodt Institute of Radiology, but its avuncular director, Dr. Sherwood Moore, was happy to see it used for the support of a general program of isotope research at the medical school and on the campus. For help in cyclotron maintenance and in building a mass spectrometer to extend researches to the use of rare, stable isotopes, I could rely on the cooperation of Professor Arthur L. Hughes, the chairman of the physics department, and two excellent technical associates, A. Schulke and H. Huth. A number of investigators in the medical school were anxious to begin or expand projects involving isotopic tracers and materials in basic biomedical research and clinical applications. I had only an empty laboratory to start with but had been allocated ample space in the institute at the medical school and had a modest budget to begin accumulation of laboratory facilities and the employment of technical help.

There was no possibility, at least for some time, of renewing the researches Sam and I had expected to continue. Moreover, two major research groups were already engaged in such an effort, one at Berkeley where Melvin Calvin was taking up the continuation of our original researches in photosynthesis, with Andrew A. Benson providing some continuity, and the other at Chicago under the direction of Hans Gaffron. While I was slowly building my own research effort, I had ample opportunity to collaborate in numerous projects at the medical school. A group, led by Dr. E. V. Cowdry in the department of anatomy, was embarking on an ambitious program in cancer research at the associated Barnard Skin and Cancer Hospital. It received the first shipment of ^{14}C to be made at Oak Ridge. I was a coauthor of a number of papers with these investigators, including studies in calcium metabolism in rotifers, as well as on the effects of phosphate metabolism in human tissues dosed with carcinogens.

Another research of more basic character involved the study of blood dyscrasias in collaboration with Dr. Carl V. Moore in hematology. Dr. Moise Grinstein, a noted porphyrin chemist visiting from Argentina, proposed to use ^{14}C-glycine to study hemoglobin synthesis in various clinical applications. David Shemin and David Rittenberg had shown that ^{15}N-glycine was an efficient precursor of heme and had suggested that the whole molecule was incorporated into the porphyrin moiety as well as into the globin (33). Accordingly, I proposed that we use ^{14}C-carboxyl-labeled glycine, available from Dr. R. B. Loftfield at the Massachusetts Institute of Technology, for these studies. In preliminary experiments with a dog and a rat, we were surprised to find that carboxyl carbon of glycine entered the globin but not the heme (34). This finding was a key to the eventual elegant unraveling of the details of heme biosynthesis by Shemin (35).

One day, J. W. ("Joe") Kennedy, the newly arrived chairman of the chemistry department, phoned me to inquire if anyone was interested in how viruses multiply. I replied that a free trip, lodging, and a state dinner in Stockholm awaited the successful solver of this problem, whereupon Kennedy proposed an experiment he thought might provide the answer. It involved labeling with high specific ^{32}P activity, the phosphorus of tobacco mosaic virus, or some other suitable virus, using it to infect tobacco leaves, and assaying the viral progeny for ^{32}P content. If none was found, this would indicate the parent virus molecule acted simply as a template. If the ^{32}P were distributed at random through the progeny, it would follow that the parent was fragmented into conserved structures used to resynthesize new virus. I suggested that a better system was *Escherichia coli* bacteriophage because it was not possible to quantify infective units in the case of tobacco mosaic virus and also because we had in residence an outstanding expert on the physiology and manipulation of bacteriophage. He was Alfred V. Hershey in the department of bacteriology. The experiments that resulted included the participation of my first graduate student, Howard Gest, and produced data that were clearly interpretable as a consequence of the fragmentation hypothesis. Moreover, they showed that the recoil process occurring in the beta decay of ^{32}P to ^{32}S denatured the viral molecule, a kind of "suicide." The paper we published (36) was the basis for the development in later years of a thriving area of bacteriophage research based on the "suicide" phenomenon.

Most of my collaborative efforts were expended in a series of productive researches with Sol Spiegelman, then at the beginning of his brilliant career. We studied phosphorus metabolism in yeast. We found differing pool sizes for various yeast phosphates and nucleotides, including a very metabolically active fraction of metaphosphate not in equilibrium with the major metaphosphate pool. We designed experiments to obtain information on phosphate flux in and out of the various intracellular phosphate components. Information so obtained could be related to mechanisms of enzyme formation with particular reference to energy requirements. The results we reported (37) were suggestive with respect to the genetic regulation of enzyme and protein synthesis. We proposed a concept of gene action in which partial replicas of genes were continually produced in the nucleus and exported to the cytoplasm, there to exist as self-duplicating units in a competition for protein which eventually determined the enzymic composition of the cell. We used the term "plasmagene," coined previously by others, for these self-duplicating units. This term might be said to have prefigured "messenger RNA." (This article got us a lot of notoriety because *Time* magazine wrote a report based on it, titled "Tempest in the Cells.")

The tempo of tracer activity nationwide was accelerating. I found myself attending numerous conferences and seminars as well as writing review articles. The first of these, which provided a respite from the dreaded St. Louis summer in the air-conditioned facilities of the medical school library, was a response to a request from Professor G. Evelyn Hutchison at Yale to assess the status of isotope biochemistry in geochemistry, particularly with respect to the suggestion by the reknowned Russian, W. I. Vernadsky, that anomalies in isotope ratios of various elements in the biosphere could result mainly from intervention by living systems. I had to review all the data then existent on isotope effects and provide a critique of the experimental findings. The article I wrote (38) was a pioneer effort in biogeochemistry. I concluded that natural physicochemical effects could account for all the variations in isotope ratios reported.

The next summer I attended the symposium at Cold Spring Harbor on cytoplasmic factors in heredity and stayed on to write a full-scale text on radioactive tracers in biology for Academic Press (39). I was working under extreme pressure to finish the book before returning to the trenches in St. Louis, and my failure to appear in the outdoors occasioned much comment and concern among the symposium participants. One day I did emerge, blinking in the sun, like a troglodyte coming out of a cave and clad only in swimming trunks, to be waylaid by Max Delbruck and his wife Manny, who led me to the pier where I was deposited in their boat and moved out a half-mile onto the Sound. As I was lolling languidly in the bow, Max suddenly turned and ordered me into the water, saying the swim back would do me good as I needed the exercise. This experience helped my resolve to stay indoors!

As Kurt Jacoby, the fatherly publisher of Academic Press, had surmised, the timing for such a book was perfect. A large, pent-up demand for a modern text on tracer isotope methodology existed and sales were phenomenal for such a technical treatise. At the Press, they called it "Our *Gone with the Wind*." I was particularly gratified with the favorable reviews it received, especially from one of my early heros, Fritz Paneth. Hevesy also wrote a glowing letter, as did another idol, A. J. Kluyver. The only sour note was a brief review in Russia which complimented the translator but complained there were too few references to Russian work. The book went into three editions and many reprintings before the increasing burden of keeping it current forced me to discontinue further production.

Shortly after the arrival in 1947 of my first graduate student, Howard Gest, who had heard about me from friends at Oak Ridge, like Charley Coryell, Harrison Davies, and Waldo Cohn, I acquired two more graduate students, Jack Siegel—also from Oak Ridge—and Herta Bregoff. I had been brooding about the many discussions Sam Ruben and I had had concerning the possibility that phosphorylation in some way made ATP in the light and could be the driving force in the photosynthetic dark CO_2 uptake. Sam had actually published a note (40) on some of his speculations, basing a model scheme on the Warburg-Christian enol phosphopyruvate reaction. My level of sophistication about phosphate metabolism had risen markedly since then, as I sweated at the bench learning to be a biochemist by isolating and purifying enzymes and phosphate esters of the glycolytic pathway and synthesizing various phosphorylated substrates needed in the studies on yeast metabolism.

I had suggested to Howard that he do his thesis work on the effects in algae of illumination on phosphorus turnover. To aid his efforts, I had arranged to have him attend van Niel's famous summer course. He came back enthusiastic about phototrophic bacteria as additional systems for study. I became similarly affected. Siegel was assigned as a thesis subject a reinvestigation, using isotopic carbon, of the claim by J. W. Foster that certain strains of *Pseudomonas* effected a photooxidation of isopropanol that involved simple dehydrogenation of substrate to acetone with simultaneous reduction of CO_2 to cell material, according to the reaction:

$$\begin{array}{ccc} CH_3 & & CH_3 \\ | & & | \\ 2\ H\text{-}C\text{-}OH + CO_2 \rightarrow 2 & C{=}O + (CH_2O) + H_2O \\ | & & | \\ CH_3 & & CH_3 \end{array}$$

It soon became apparent that the phototrophic bacteria were potentially better objects than algae for study, primarily because they might be more amenable to genetic manipulation, if needed, and because they lacked the oxygen-evolving capacity of algae and so might provide an easier access to isolation of the precursors of oxygen. An amazing sequence of serendipitous happenings, beginning with the inadvertent use of glutamate rather than ammonia as a nitrogen source in growth experiments, related in detail elsewhere (41), led Gest and me to the important discoveries of hydrogen photoevolution (42) and nitrogen fixation (43) in the nonsulfur, purple photosynthetic bacteria. These phenomena underscored a basic relation between hydrogen and nitrogen metabolism that opened new areas of research of significance in later years relative to enzymic mechanisms of nitrogen fixation and storage of bioenergy. As an example, a whole field of investigation has developed centered on the mechanisms whereby nitrogenase effects hydrogen evolution.

The demonstration that so large and varied a family of microorganisms as the photosynthetic bacteria included nitrogen fixers shocked the workers in the field and thereby introduced us to new friends among the microbiologists, especially Perry Wilson and his associates, as well as Robert Burris at the University of Wisconsin—a profitable connection as it turned out very shortly.

During his summer in van Niel's laboratory, Siegel found that the original Foster strains no longer existed. He obtained other strains by enrichment cultures, which could perform the dehydrogenation of isopropanol to acetone and beyond. He also extended our observations on photohydrogen production, showing it to be a general phenomenon among phototrophic bacteria and not confined to *Rhodospirillum rubrum,* the original strain investigated. Herta Bregoff established the stoichiometry of the reaction with malate and other substrates and showed that hydrogen evolution began only after complete exhaustion of ammonia in the culture medium.

All of this frenetic and highly productive activity was superimposed on an intense social and musical life that brought me into intimate contact with leading intellectual and musical circles in the city. Unfortunately, it also stirred Security, now the F.B.I., to renewed surveillance because of my many new friends whose liberal politics reinforced suspicion.

Soon after my arrival in St. Louis, Chancellor Arthur H. Compton had taken office and among his first acts was an audience with me in which he attempted to allay my fears of further reprisals stemming from the affair in Berkeley. He had been consulted at the time of my dismissal and had regretfully agreed to it solely on the grounds that it was a precautionary action. He was determined that no further injustice be done, a decision which laid a considerable burden on him in the next decade as he withstood pressures brought to bear by trustees and some others swayed by the campaign of vilification which attended the F.B.I. investigations. From my own reading of history and from the early news reports, it was clear that there was no lack of those who would exploit the mindless hysteria in the wake of the Cold War and attempt to turn prophylaxis into radical surgery.

My fears were soon realized. In response to my application for a passport to attend a symposium in France to which I had been invited, as well as to give lectures in Israel and Australia, federal agents raided the travel office in New York and seized the passport that had been issued a few days previously. For the next seven years I was embroiled in a struggle with the passport division of the Department of State, as well as in a libel action against the *Chicago Tribune* and its associated paper, the *Washington Times-Herald.* One is reminded of Cesare Sterbini's famous lines in the celebrated Calumny aria from Rossini's *Barber of Seville* wherein the elderly Don Bartolo is instructed in the gentle art of calumny by the cynical music master, Don Basilio:

. . . Oh! calumny is like the sigh
Of gentlest zephyrs breathing by;
How softly sweet, along the ground
Its first shrill voice is heard around:
So soft, that, sighing 'mid the bowers,
It scarcely fans the drooping flowers.

Thus will the voice of calumny,
More subtle than the plaintive sigh,
In many a serpent-writhing, find
Its secret passage to the mind,—
The heart's most inmost feelings gain,
Bedim the sense, and fire the brain.

Then passing on from tongue to tongue,
It gains new strength, it sweeps along
In giddier whirl from place to place,
And gains fresh vigor in its race;

Till, like the sounds of tempests deep,
That through the woods in murmurs sweep
And howl amid their caverns drear,
It shakes the trembling soul with fear.

At length the fury of the storm
Assumes its wildest, fiercest form,—
In one loud crash of thunder roars,
And, like an earthquake, rocks the shores.
While all the frowning vault of heaven,
With many a fiery bolt is riven.

Thus calumny, a simple breath,
Engenders ruin, wreck and death;
And sinks the wretched man forlorn,
Beneath the lash of slander torn,
The victim of public scorn.

The storm first broke early in the summer of 1948 in the middle of the exciting discoveries of photohydrogen evolution. Summoned by subpoena to Washington to testify at hearings of the House UnAmerican Activities Committee (HUAC), I turned to Edward U. Condon for help. Condon, the head of the Bureau of Standards, was a prime target for HUAC in its attempt to exploit the Cold War hysteria to bring discredit on the memory of Franklin D. Roosevelt. Condon, writing to me at the time, remarked wryly, "They have the verdict; now all they need is the evidence," and arranged legal help from the office of Arnold, Porter, and Fortas, which was already submerged in appeals for aid from a legion of victims in and out of the government employ.

The press in St. Louis rallied to my cause, encouraged by strong statements of support from Chancellor Compton. The storm passed, not without considerable prejudice to my chances of academic advancement as the notoriety associated with me because of the libel and the passport difficulties effectively choked off offers I would have otherwise received as my career in science prospered. A new and much more formidable threat than the HUAC, Ruth B. Shipley, the chief of the passport division, remained as a barrier to be overcome. The resulting battle to establish the rights of free citizens to a passport that could not be withheld by an arbitrary and capricious bureaucrat consumed another four years. I was fortunate to find legal counsel, ethically motivated and supremely competent and unafraid of professional reprisals, in the persons of Nathan H. David and Alexander E. Boskoff. Taking on the McCormick enterprises, they helped me win the war with a unanimous jury verdict against the *Washington Times-Herald*, upholding our contention I had been libeled, and awarding damages. In a most fortunate development, the McCormick organization, through its powerful connections among various federal agencies, had attempted to bolster its allegation by producing hitherto highly secret files that contained obvious lies, misstatements, and the raw unevaluated data that had formed the basis for the actions taken by the Army in 1945. Without those efforts by Col. McCormick's informants we would have never known the true character of the fabrications that had caused me such trouble for a decade. They would have remained safely concealed, always available to foment continued persecution and harassment.

A Major Diversion

While the grim counterpoint of federal harassment to the main tune of scientific progress continued as the 1950s began, my three graduate students finished and left, leaving no prospects of new recruits because I still had no formal appointment in an academic unit empowered to engage in graduate studies. I was an employee of the department of radiology and had been given the dispensation of supervising doctoral thesis work only because the students involved could qualify for support by the departments of bacteriology or chemistry. It seemed most expedient to follow the practice of other investigators in the medical school and rely on help from postdoctoral fellows. Never having had one before, I wrote for advice to Burris and Wilson at Wisconsin and through them learned that Leo Vernon was available. He had finished work on the characterization of mitochondrial cytochrome c reductase and had the experience with redox enzymes I thought essential to implement ideas I had generated on the mechanism of energy bioconversion in photosynthetic bacteria. I had corresponded with Fritz Lipmann and thereby had sharpened my own perceptions of how ATP might be involved. A beginning required elaborating the enzyme composition of the bacterial chromatophores. Influenced by van Niel's general theory of photosynthesis, I thought the chromatophores should exhibit a reaction with free energy storage analogous to the Hill reaction in chloroplasts. The difficulty in documenting such a process lay in the fact that the oxidizing component of the primary photosynthetic act (customarily schematized as the photoproduction of "OH" from water) was not eliminated by evolution as oxygen, as in green plant photosynthesis. Rather, it would accumulate and back-react with the reductants produced simultaneously, so that only heat would result with no storage of isolated reductants and oxidants. It was clear that enzymic mechanisms prevented such back-reactions in vivo, so I reasoned that by determining what these enzymes were one could devise suitable reagent species to show the Hill analogue of the chloroplast reaction in the anaerobic chromatophores.

Meanwhile, the main line of endeavor remained the elucidation of the light effects on metabolism in the phototrophic bacteria. Research fellows came to aid in this effort—John Glover and Stanley Ranson from England, H. Van Genderen from Holland, and R. Collet from Switzerland—as did Sam Ajl, a newly arrived junior faculty member in the department of bacteriology. The major finding was that dark metabolism of substrates followed the course of the Krebs cycle but that light suppressed cycling of substrate carbon and diverted it almost completely to cell synthesis. As an example, the methyl carbon of acetate appeared quickly in CO_2 in the dark, but not in the light. Equilibration of acetate carbon through the Krebs cycle could be demonstrated by analysis of the label content in cycle intermediates.

On Leo's arrival, we discussed what enzymes to start looking for, and Leo suggested we begin by examining extracts for redox components of the cytochrome system. I was dubious because I was still under the influence of the dogma, then current, that cytochromes were present only in respiratory systems. Leo reminded me he had experience only with the aerobic enzymic components of mitochondrial respiration and coupled ATP synthesis. On his first trial, he found to our surprise that there were relatively large amounts of a heme pigment practically identical spectroscopically with mitochondrial cytochrome c!

I wondered how such a feature of R. rubrum extracts had escaped previous observation in all the years (at least a half century) this bacterium had been studied as the prototype of bacterial photosynthesis. Writing to Roger Stanier, an authority on bacterial metabolism and general microbiology as well as an old friend, I inquired whether he had ever noticed anything peculiar about R. rubrum extracts, particularly the pink color. He replied that it was well known that such extracts were pink, probably owing to the presence of bacteriochlorophyll decomposition products. It appeared no one had ever looked at the spectrum!

It may well be asked why photohydrogen evolution and nitrogen fixation, two such remarkable metabolic features of $R.$ $rubrum,$ had also eluded detection through the years. The answer lay in the circumstance that the accepted nitrogen source had always been ammonia, an inhibitor of photohydrogen evolution. Nitrogen fixation had thereby been missed. An irony of history can be remarked here. The noted Dutch microbiologist, M. W. Beijerinck, had originally thought that the free-living nitrogen-fixer, *Azotobacter,* should be called *Para-chromatium,* in view of its strong morphological resemblance to *Chromatium,* the phototrophic sulfur bacterium, but had decided against this because he thought *Chromatium* did not fix nitrogen. Half a century later, we showed that Beijerinck's intuition had been correct!

The news that $R.$ $rubrum$ chromatophores contained cytochrome c met with the expected disbelief in many quarters. In England, Sidney Elsden repeated our experiments and convinced himself that indeed $R.$ $rubrum$ harbored a cytochrome c. He went further and did some experiments we had postponed until the fall because Leo and I were to be in van Niel's laboratory that summer. Elsden found that the new cytochrome c differed from its mitochondrial analogue in having an isoelectric point near neutral pH and being totally unreactive with the mitochondrial cytochrome c oxidase, although, as we had found, it reacted well with the reductase. We jointly published a short note (44), announcing the existence of the new cytochrome, termed later "cytochrome c_2" as the names "c" and "c_1" were already assigned to mitochondrial cytochromes. This unexpected diversion from the main objective of our research—from definition of the mechanisms of conversion of light energy in chromatophores to a study of structure and function of cytochromes c—eventually developed over the next quarter century into the new field of comparative biochemistry of iron proteins.

Further studies on cytochrome content and associated enzymology in chromatophores revealed a new kind of C-type cytochrome, which exhibited the anomaly of high-spin spectra like those of hemoglobins and catalases, while maintaining the characteristic covalent binding of the heme moiety through thioether linkage to cysteinyl residues as in cytochromes c and c_2 (45a).

Adapting an approach by Alan Mehler using oxygen as a "Hill oxidant," we were also able to demonstrate photoreactions in chromatophores and chloroplasts sequestering oxidants and reductants. We observed stoichiometric yields of oxygen uptake and acetaldehyde formation consistent with the oxidation of ascorbate through 2,6-dichlorophenyl indophenol coupled photochemically with catalase-mediated peroxidation of ethanol. In both spinach chloroplasts and $R.$ $rubrum$ chromatophores, we could conclude that a Hill system, with molecular oxygen as a Hill oxidant, was operative (45b).

Our next problem was to show that a Hill system was generated under anaerobic conditions. While we were struggling to find an uphill process such as simultaneous reduction of a low-potential H-donor, like NAD, and oxidation of a high-potential acceptor, like cytochrome c, I received a letter from Albert Frenkel, who had worked with Sam Ruben and me as a graduate student in plant physiology back in Berkeley on magnesium ion exchange reactions in chlorophylls and chlorophyll derivatives. He had been on a sabbatical fellowship in Fritz Lipmann's laboratory. Following a suggestion by Lipmann that he look at the reactions in chromatophores that were our interest in St. Louis, he had proceeded to do the simple and straightforward experiment of incubating *R. rubrum* chromatophores anaerobically in the light with ADP and inorganic phosphate. Incredulously, I read that he had found ATP formation in amounts far beyond what might have resulted from utilization of trace amounts of oxygen that might have been present. He had discovered photophosphorylation in chromatophores, the reaction needed to complete the scheme of coupled photochemical energy conversion (46). Independently, Arnon and his associates in Berkeley found the same reaction in green plant chloroplasts (47). Vernon and I could note ruefully that we had missed this discovery by too much cogitation when the phenomenon was readily demonstrable with a simple, direct experiment.

Our work on chromatophore photooxidations and cytochrome composition had been supported partially by a fellowship grant from the National Institutes of Health (NIH) to Vernon. The Institute officials had been quite eager to make this award, but—suddenly reneging on an implied commitment to continue support—they terminated the award, with no time allowed for us to find other employment opportunities for Vernon, who was in no way at fault. This was but one example of many that characterized the political interference in research stemming from Cold War pressures. Fortunately, Leo found employment soon at Brigham Young University and went on to a distinguished career, including many years as research director at the C. F. Kettering Foundation.

With funds from the National Science Foundation, thanks to William Consolazio and Estella Engel, and the Kettering Foundation, as well as from my modest budget, I found it possible to bring Jack Newton to St. Louis. He confirmed an observation I had made that a C-type cytochrome existed in extracts from the photoanaerobic *Chromatium,* characterized it further, and went on to show such extracts could carry out photophosphorylation (48). Meanwhile, I spent much time in the trenches prosecuting the war with the Establishment. Its successful outcome, with victory in the libel action in 1955, the subsequent issuance of the passport, and the decision of the Justice Department to clear my record, brought a resumption of funding from the NIH.

Some Relevant Asides

In the next quarter century, our researches on the metabolism of phototrophic bacteria intensified. At steady state, the number of investigators in our group was well over two dozen, distributed more or less evenly among postdoctoral fellows, graduate students, and technical support personnel, creating a heavy administrative burden for me. Our efforts involved collaborations with a half dozen research organizations in this country and abroad.

Serendipity continued to play its role in a substantial record of achievement, which included not only opening new research areas, such as the comparative biochemistry of cytochromes and structure-function relations in bacterial iron proteins, but also adventures in academe, as I participated with Nate Kaplan in the creation of a new department of graduate biochemistry at Brandeis University, and with others in founding a new campus of the University of California at La Jolla, and a new academic unit in molecular biology at the University of Southern California. However, I have given most attention in this prefatory chapter to the first quarter century of my career, because it is the story that uniquely records the effect of societal impact.

Recall the legend of Damocles fated to sit at a sumptuous banquet with a sword over his head. I, too, had a scientific banquet spread before me, owing to the fantastic metabolic versatility of the phototrophic bacteria, but the sword of the Federal Establishment dangled overhead. Fortunately, the sword was eventually removed and I could partake of the banquet freely, as I have done since. Only a very brief summary of my experiences, wholly inadequate in chronicling the individual contributions of many colleagues and students, can be offered in the following section.

Big Questions and Small Answers

The phototrophic bacteria exhibit the ability to grow utilizing every known metabolic pattern except oxygenic photosynthesis. Their complexity of cellular organization is probably unsurpassed in living organisms. For example, both anaerobic photophosphorylation and aerobic phosphorylation often exist side by side in their membrane systems (chromatophores) whereas in green algae and plants they are neatly separated and packaged in cellular organelles (chloroplasts and mitochondria). I can cite an interesting case in point—the primitive animal or protozoan *Euglena gracilis*—in which the functional C-type cytochrome ("cytochrome c-552") is localized in the chloroplast, whereas a mitochondrial C-type cytochrome ("cytochrome c-556") functions as the substrate of a cytochrome c-oxidase outside the chloroplast in the respiratory apparatus (49).

I first became interested in this aspect of bacterial metabolism in following up the original observations of H. Nakamura on inhibition of oxygen uptake by light (50). Many investigations on the respiratory systems of *R. rubrum* and related bacteria in our laboratory (see Refs. 51–53 as examples) and elsewhere have failed to resolve wholly the fundamental mechanism of this phenomenon, but serendipity has entered often in the discovery of new heme proteins and their functional variability.

The ability to travel freely broadened my base of expertise on structure-function relationships of redox enzymes and systems. I worked in Sweden, Germany, England, Japan, and Australia with outstanding researchers like A. H. Theorell, G. Drews, R. Hill, K. Okunuki, F. Egami, and R. Lemberg and their associates. I still remember one exciting afternoon spent with David Keilin and Robin Hill in Cambridge, when Hill and I stood at the bench feeding reagents and protein samples to Keilin who, peering into his hand visual spectroscope, explored the ligand behavior of the heme protein Vernon and I had just discovered and which came to be called cytochrome c'. At the end of the day, having learned more cytochrome chemistry than I had known previously, I had the satisfaction of hearing Keilin remark that he would not have believed such a heme protein could exist.

Indeed, he was not alone in this belief. The anomalous behavior of cytochrome c' has intrigued and mystified us and others for several decades. It exhibits absorption spectra and magnetic properties usually associated with "unsaturated" high-spin heme proteins, like myoglobins and peroxidases, but a redox behavior, prosthetic group, and mode of binding like that of the "saturated" cytochromes c. Its ligand reactions (54) are not typical of either of these in that it reacts only with the small neutral ligands, CO in the reduced state, and NO in both the reduced and oxidized states. It is highly autooxidizable, encouraging the notion it might function as an oxidase, but no convincing data in support of this idea were ever obtained (54, 55). The oxidase function in *R. rubrum* and other phototrophic bacteria is served mainly by an ubiquitous B-type "cytochrome o" (54).

The magnetic properties of cytochrome c' posed more difficulties. Originally, Anders Ehrenberg and I found (56) a magnetic moment for the oxidized *Chromatium* protein much lower than expected for a pure high-spin heme compound. We suggested the protein might exist as a thermal mixture of high- and low-spin states. An alternative view, based on later EPR studies, proposed cytochromes c' to be examples of intermediate spin states with quantum mechanical admixtures of high-spin (57). A trail of investigation eventually led me to the campus of the University of Southern California where P. J. Stephens had a unique magnetic circular dichroism determination apparatus wherewith this question could be probed further. We concluded (58) that the data available could be rationalized by the assumption that cytochrome c' was essentially a high-spin heme protein, a conclusion also reached by others (59). The question still remains open. Very probably the explanation is that the ligand field strength that stabilizes the intermediate spin state is so close to that of the upper state that very small changes in the ligand environment can change resultant spin moments (60).

Attempts to establish the function of this ubiquitous protein have been fruitless so far. It does not act as an oxygen-transport heme protein, as in myoglobins. Cyril Appleby and I looked at spectra of cytochrome c' down to extremely low oxygen pressures permitted by "NAK" alloy without seeing formation of an oxygen adduct. No evidence for peroxidase activity has been found.

It seemed possible that cytochrome c' might be an artifact of isolation procedures and that it existed in vivo as a cytochrome of the B- or C-type. This notion was encouraged by finding of a stable low-spin cytochrome c in a number of phototrophic bacteria ("cytochromes c-556") homologous with cytochromes c'. However, this rationalization seems to have been laid to rest decisively by the recent demonstration of F. R. Salemme, P. Weber, and their associates that the three-dimensional structure of *Rhodospirillum molischianum* resembles not that of a mitochondrial-type cytochrome c but those of cytochrome b-562, hemerythrin, apoferritin monomer, and tobacco mosaic coat protein (61). These resemblances to so many functionally dissimilar proteins provide no clues at all to functions. The tertiary structure provides a basis for understanding most of the anomalous properties of cytochromes c', including the kinetics of CO binding, studied by M. A. Cusanovich and Q. H. Gibson (62), which indicate, as originally noted (63), the reactions to be complex and with limited ligand accessibility.

Relief from the troubles occasioned by cytochromes c' was afforded by the family of mitochondrial-type cytochromes c_2, which we found to be distributed widely among phototrophic bacteria, as well as in at least one nonphotosynthetic bacterium, the nitrate-reducer, *Paracoccus denitrificans*. These cytochromes, structurally and functionally, comprise a group that overlaps continuously in homology with the eukaryotic mitochondrial cytochromes c. The first sequence determination of a bacterial cytochrome was accomplished in our laboratory by Karl Dus and Knut Sleiten—that of the prototype cytochrome c_2, from *R. rubrum* (64)—and was followed by many others in the laboratory of R. P. Ambler.

It is amusing to reflect on the fact that this first sequence determination required an expensive, complex, and cumbersome setup that cost the American taxpayer about $1000 per residue, whereas the rapid development of the Sequenator by Per Edman and advances in peptide analysis soon reduced this cost by three orders of magnitude!

Many studies on the physicochemical properties of these heme proteins have been reported but I note just one—that on pH dependence of redox potentials of cytochrome c_2 (65)—because to my knowledge the paper in which we reported our results was the only one among several hundred I submitted over the years in which the reviewers not only recommended its acceptance without revision but urged accelerated publication!

Our interest in functional expression paralleled researches on structure. Kinetic experiments on the structural basis for the inability of cytochromes c_2 to react well with eukaryotic mitochondrial cytochrome c oxidase, while exhibiting good reactivity with the reductase—a phenomenon, it will be recalled, which delayed our original recognition of the unique character of the bacterial cytochrome c_2—have promoted insights into the comparative biochemistry of cytochromes c. B. Errede in our laboratory performed extensive kinetic analyses of the reactions with mitochondrial oxidase using a selected group of cytochromes c_2. She found that the lysine residues effective in promoting binding and reactivity with the oxidase were the same (66) as those determined by direct chemical modifications carried out by E. Margoliash and his associates at Northwestern University and by F. Millett and his group at the University of Arkansas. A finding—provocative from the evolutionary standpoint—was the demonstration that the eukaryotic cytochrome c from the primitive protozoan, *Tetrahymena pyriformis,* reacted as predicted hardly at all with the beef heart oxidase because it lacked all the lysines required for productive binding. It is pertinent that this primitive animal possesses a prokaryotic D-type oxidase, rather than a eukaryotic cytochrome c oxidase.

A special project to probe structure-function relationships in cytochromes c was the basis for a thesis proposed by Arthur Robinson. He set up a Merrifield machine to make selected heme peptides that could be used to establish parameters for study of the intact protein. One might also use them to study ligand reactions as compared with those of cytochrome c. The main outcome was a refinement by John Sharp and Robinson of synthetic techniques, which were calibrated by use in producing an active lysozymelike polypeptide in association with John Rupley and his group at the University of Arizona. The final preparation gave quite a good specific activity, nearly 2–3% of the native protein (67). Later, Robinson used his apparatus to develop the suggestion that controlled deamidation of peptides in proteins might be a basis for biological timing in the aging process (68).

The three-dimensional structure of the *R. rubrum* cytochrome c_2 determined by F. R. Salemme in Professor J. Kraut's laboratory was among the earliest for cytochromes c and provided the basis for deductions on structure-function relationships (69), a basis further elaborated by Gary Smith using NMR spectroscopy (70). A serendipitous aspect of this work was his discovery with G. W. Pettigrew of a new N-terminal blocking group derived from *N,N*-dimethylproline in the mitochondrial cytochrome c of the insect parasite *Crithidia oncopelti* (71). This amino acid had not been reported as existing in proteins but only in the free form.

The researches on the bacterial cytochromes resulted in wholly new perspectives on potential variabilities in structure and function of the heme group in relation to protein. We showed that at least 12 subgroups of the cytochromes *c* exist, based mainly on homologies determined by R. P. Ambler and his associates, using samples supplied largely by R. G. Bartsch and T. E. Meyer in our laboratory. A summary of the present states of knowledge can be found in recent reviews (72, 73).

Another round of serendipity heralded the discovery of yet another kind of bacterial iron protein. Robert Bartsch, just beginning his fruitful career, turned his attention to a reexamination of *Chromatium* extracts. He used DEAE columns, rather than the carboxymethyl cellulose resins we had used previously. Under the conditions required for DEAE chromatography, there appeared a new, readily eluted fraction that we had missed. It contained a greenish-brown protein that proved reversibly oxidizable and exhibited a high positive redox potential. The absorption spectra of both redox forms showed little structure, only a general absorption with a maximum in the near ultraviolet somewhat more pronounced in the reduced form (74). Karl Dus determined its primary structure (75). Later, Henk de Klerk found a similar protein in *Rhodopseudomas gelatinosa,* thereby showing it was distributed not only in the sulfur, but also nonsulfur, purple phototrophic bacteria (76). Subsequently its occurrence in other purple phototrophic bacteria as well as in at least one nonphotosynthetic microorganism has been established. Amino acid sequences from at least seven examples at this time, as determined by S. Tedro and T. E. Meyer, support classification in a group separate from other nonheme proteins, such as ferredoxins and rubredoxin. The sobriquet "HIPIP" (acronym for "High Potential Iron Protein") has been applied, referring to its major distinguishing characteristic, the high positive redox potential, although recently some examples of intermediate potential have been found.

Although ferredoxins and HIPIPs have the same 4Fe-4S cluster as a prosthetic group, the range of redox potentials determined by the protein component extends over 800 mV. An attempt to rationalize this fact has been made (77) based on the proposal that the identical 4Fe-4S clusters are in different diastereomeric environments and also exploiting suggestions on hydrogen bonding deduced from high resolution X-ray structure determinations (78a). These effect nonidentical interactions of a tyrosine residue (Tyr-19) with the inorganic sulfurs in the cluster. However, the expected correlations between circular dichroism spectra and redox potentials are not found (78b). Furthermore, the tyrosine originally supposed to be conserved is not present in *Rhodospirillum tenue* (79). An understanding of the structural basis for HIPIP redox potentials still eludes us. Its attainment should help provide a rationale for structure-function relationships in important enzymes, like hydrogenase and nitrogenase.

In 1957 the first international symposium on enzymes held in Japan afforded the opportunity to meet Professors Okunuki and Egami and their associates and students. Thereby, a long series of important collaborations began with T. Horio, S. Taniguchi, T. Yamanaka, J. Yamashita, and T. Kakuno, all of whom contributed to the successful prosecution of many researches too numerous to record here. In the course of these collaborations, I became familiar with the extensive Japanese literature on heme protein research and microbiology, and found that H. Iwasaki in the laboratory of Professor T. Mori in Nagoya had anticipated our discovery of cytochrome c' (80) as had K. Hori in the same laboratory with HIPIP (81). These observations could not be carried further at Nagoya where neither the procedures for bulk production of pure bacterial cultures nor the collaborating laboratories needed for X-ray structure determinations or for application of the many existent spectroscopies were available. Such requirements were fully met during my tenure at Brandeis and at La Jolla, so that the full significance of the qualitative observations at Nagoya could be appreciated.

While our interest in structure and function of iron proteins took center stage, our experiments were always interdigitated with researches on the photosynthetic apparatus in bacteria. We examined model systems such as one in which Richard Kassner showed that in the obligatory presence of coproporphyrin, ferredoxin could be photoreduced by ascorbate or other electron donors whereas the back-oxidation with oxygen did not require such a catalyst (82).

A major question, stemming from the early work on phosphate nutrition, was the minimal composition of the photosynthetic system in bacteria. In 1967 I was asked to direct a new biochemical unit at the laboratory of photosynthesis in Gif-sur-Yvette where J. Lavorel had just become the new director of research. This opportunity, largely made possible through the efforts of Jacques Monod, to exploit a nutritional approach to the definition of a minimal photosynthetic apparatus, resulted in the creation of a small but vigorously active group under the able direction of Françoise Reiss-Husson. I also hoped to begin a search for a good transducing phage to facilitate the use of powerful methods of nuclear bacterial genetics and molecular biology.

The demonstration by Kassner and me that manganese was not essential in bacterial photosynthetic growth (83), however important it might be in oxygenic photosynthesis, influenced our decision to focus on the iron requirement. At Gif, we used a special strain of *Rhodopseudomonas spheroides* to establish that with minimal iron there were correlations at steady state growth between low levels of ATP and photopigment synthesis, with some indications that low ATP levels preceded onset of bacteriochlorophyll appearance (84, 85). It was certain that photopigment synthesis accompanied low levels of ATP not only in *Rps. spheroides,* but also in *Rhodospirillum rubrum,* as Jurgen Oelze and I found in Freiburg (86, 87), and Greg Schmidt and I found with *Chromatium vinosum* in La Jolla (88). At Gif, another group led by J. Maroc showed that in the strictly anaerobic sulfate-reducers, vitamins K were normal components, as well as in the phototrophic bacteria (90). Other researches at Gif, reflecting the biophysical approach there, probed fluorescence yields, both variable and constant, to establish correlations with aging (91a). Oelze and I also showed that photoanaerobic and dark aerobic transport chains had common compartments in *R. rubrum*, but that the inhibitor, 2-hydroxy diphenyl, which acted between the ubiquinone pool and cytochrome *b*, could distinguish the two pathways, inhibiting maximally only the phototrophic NADH-dependent reactions in light-grown bacteria, as compared with the succinate-dependent processes (87).

Some forays into the sulfur metabolism of both phototrophes and sulfate reducers should be mentioned, as they present some unfinished business. A collaboration with Kassner and G. C. Wagner established redox potentials for the couples postulated as involved in sulfite reduction in *Desulfovibrio* strains (91b). One of these couples appeared capable of mediating reductive disproportionation of trithionite to sulfite and thiosulfite with a menaquinone (MK-6). We suggested looking for the associated enzyme. Later, Harry Peck came to La Jolla and collaborated with Siv Tedro and me in demonstrating the general existence of a soluble sulfite reductase in phototrophes (91c). Harry, it should be recalled, has had a distinguished career, beginning as a graduate student with Howard Gest.

The molecular biology program was not developed during my time at Gif largely because of difficulties in recruiting investigators, who showed understandable reluctance to leave the safe confines of the *E. coli* bacteriophage system to embark on the uncertain search for transduction in new systems, even though a strong center of molecular genetics, headed by Boris Ephrussi and Piotr Slonimski, was close by in the next building. We did some preliminary experiments at La Jolla to answer the question of whether a special messenger RNA was required to effect transitions from dark aerobic nonpigment synthesis to photosynthetic anaerobic bacteriochlorophyll production. In *Rps apheroides* we saw no difference in RNA composition comparing bleached cells grown under high oxygen peressure and those grown under normal anaerobic light conditions (92). In other experiments in *R. rubrum* (93) using RNA pulse-labeled with tritiated uracil, no evidence for control of photosynthetic protein synthesis at the transcriptional level was found.

It is apparent that the modern biochemical genetic approach is coming into its own. Reports from the laboratories of G. Drews, H. Gest, B. Marrs, S. Kaplan, and D. I. Friedman already describe substantial progress in isolation of various gene transfer agents and temperate phages. Very recently I have heard of researches by Dr. Fevzi Daldal, who has constructed photosynthetically active mutants of phototrophic bacteria that lack the gene for synthesis of cytochrome c_2. My own experience in relatively recent times has been confined to some interesting collaborative efforts with A. F. Garcia and G. Drews using mutant strains of *Rps. capsulata*. In these experiments, membranes capable only of oxidative phosphorylation were isolated from a strain lacking the ability to synthesize photopigments. These were incubated with reaction centers from a photosynthetically revertant strain. A complete reconstitution of a reaction system was achieved, functional in that it could, on actinic illumination, couple ATP synthesis to photosynthetic electron transport as evidenced by concurrent photoreduction of endogenous cytochrome b, photooxidation of cytochrome c, and bleaching of reaction center bacteriochlorophyll (94, 95). These experimental results, wholly consistent with our early observations as well as those of others (96), suggest that the two electron transport chains in phototrophic bacteria—one associated with anaerobic photophosphorylation, the other with aerobic oxidative phosphorylation—have at least some shared components.

An important set of graduate thesis experiments to define light-induced electron transport in *Chromatium* sp. was performed by M. A. Cusanovich using chromatophores in which photoinduced changes in cytochrome content were studied as a function of imposed redox potential (97, 98). Two pathways appeared to be functional, one involving a pigment-heme complex coupled to a "cytochrome c-555," and another coupled to a low-potential component and a "cytochrome c-552." An optimum potential for coupling of photoinduced electron transport to phosphorylation occurred at 50–100 mV, much like the value found earlier (99) for *R. rubrum*. These results were further extended in thesis work by S. Kennel (100), who also noted that the functional cytochrome c complex was a membrane-bound "cytochrome c-556,552" with one component at high potential, the other at low, and that the solubilized protein showed an amino acid composition totally different from the soluble cytochrome c-553 previously isolated (101). These results point the way to future investigations, emphasizing the fact that soluble components, as isolated, may not be the truly functional proteins in photoinduced bioenergy conversion. In this connection, F. Daldal's experiments, cited above, should be noted. Such considerations also bear obviously on speculations about the evolution of bioenergetic mechanisms.

A diversion in recent times, inspired by the energy crisis of the late 1970s, brought our attention back to an old interest, the hydrogenases. In collaboration with N. O. Kaplan, John Benemann, and Jeffrey Berenson, we demonstrated that spinach chloroplast preparations could be coupled photochemically to clostridial hydrogenase to produce hydrogen and oxygen by a splitting of water (102). This much-quoted experiment indicated that if the problems of chloroplast and hydrogenase stability could be solved, a reactor to use cheap solar energy could be devised. Our further studies, based largely on the ideas and considerable skills of Alexander Klibanov, were directed toward simple means of stabilizing the most reactive as well as least air-stable hydrogenase, that from *C. kluyveri*. An attractive and elegant solution involving simple adsorption of the enzyme on solid supports treated with polycations was demonstrated (103). We wrote a review of our studies indicating there is no substantial difficulty in production of stable, active hydrogenases (104a).

Our results on the bacterial cytochromes c have led to proposals in some quarters that "phylogenetic trees" intended to show divergence of eukaryotic mitochondrial cytochromes c might be extended to include the prokaryotic cytochromes c. However, in addition to the difficulties introduced by functional heterogeneity of the prokaryotic proteins, some showing multifunctional behavior as in *Rps. spheroides* cytochrome c_2, and others with no known functionality, as in other cytochromes from the same microorganism, there occurs a "saturation" phenomenon, by which is meant at a certain limit of change there is no further divergence. Instead convergence or back-mutation may become dominant. This is probably the case for cytochromes c_2, which overlap in homology with the most extremely divergent cytochromes c from fungi and protozoa in eukaryotes. Hence available data do not as yet permit construction of all-inclusive phylogenetic trees, although they may be quite valid for the unique case of the eukaryotic cytochromes c, constrained by the mitochondrial organization to the single function of mediation of coupled electron transport between the two terminal complexes of the respiratory chain. Terry Meyer and I have had a paper on this subject written for some time with the happy title "Only God Can Make a Tree." It is now in press (104b).

As I have mentioned, my childhood experiences conditioned me to avoid crowds, with a consequent disinclination for "hot" areas of research. I was most happy opening doors through which not only I but others could pass and prosper. This was reflected in a Festschrift (105) consisting of papers published at a symposium to mark my survival to age 65. Success in organizing the meeting and seeing the proceedings through to publication, a colossal task, was due to the devoted efforts of Nathan and Goldie Kaplan and Arthur Robinson.

L'Envoi

The banquet has only begun. Many more discoveries await the investigator hardy enough to meet the challenges posed by the phototrophic bacteria. Just recently, a remarkable repeat of history has been reported, from the laboratory of Howard Gest who, it will be recalled, inadvertently substituted glutamate for ammonia in a culture medium for *R. rubrum,* thereby setting the stage for the discoveries of photoevolution of hydrogen and nitrogen fixation in phototrophic bacteria. He has published a report thirty-five years later in which similar inadvertence, this time a substitution of ammonium sulfate for ammonium chloride, created conditions for the emergence of a new genus. Gest has called it "Heliobacterium chlorum" (106)—a fastidious phototrophic microorganism harboring a new bacteriochlorophyll (107). It could grow because in the soil samples used the enrichment cultures included sulfate-reducing bacteria that produced sulfide, thereby discouraging the overgrowth of more commonly encountered phototrophic forms. This episode underscores the importance of greater sophistication in future researches about nutrition factors, especially manipulations of nutrients. The metabolic versatility of phototrophic bacteria requires close control of environmental conditions and dictates more concern about culture history than has been exercised in the past in some quarters. The variety of cellular responses to small changes in growth parameters, like temperature, pH, trace metal composition, etc, can be well appreciated from a reading of the doctoral thesis of Paul Weaver (108). [See also a review by Drews & Oelze (109).]

As I enter my sixth decade of research activity, I hope to remain a guest at the banquet table, joining in the excitement of chance discovery whether by others or myself and hoping I can even display some sagacity. For the opportunities to experience the fascination of research fun and games in these last five decades I note the generous support given by the National Institutes of Health and the National Science Foundation. In early times, the officials responsible included especially Dr. William Consolazio and Dr. George Meader. Later, Dr. Marvin Cassman helped at a crucial time, while Ms. Estella K. Engel was a constant source of encouragement and support. When the increasing burdens of academic commitment and grant administration drew me away from the bench, Dr. R. G. Bartsch filled the gap, giving added guidance and counsel to many generations of graduate students. Dr. T. E. Meyer is owed special mention for years of arduous and demanding effort in advancing our studies on the many new forms of bacterial iron proteins, as well as helping to organize the inchoate masses of data accumulated.

I am indebted to Profesor Z. Gromet-Elhanan, whose invitation to participate as a consultant and coinvestigator on a US Israel Binational Grant has kept me au courant on continuing progress in the understanding of reaction center biochemistry. In this connection I should not neglect to mention the stimulus given me by the researches of George Feher, Mel Okumura, and their associates at La Jolla. It is possible a most important contribution on my part was enticing George Feher to apply his considerable skills and expertise in solid-state science to the study of primary photoreactions in bacterial photometabolism.

Finally I record my gratitude to the members of the Editorial Board for the recognition implicit in their invitation to write this prefatory chapter.

ACKNOWLEDGMENTS

Leaders of laboratories who collaborated either personally or by sending postdoctoral fellows may be listed: R. P. Ambler, H. and M. Baltscheffsky, B. Chance, R. E. Dickerson, G. Drews, A. Ehrenberg, F. Egami, H. B. Gray, J. Falk, T. Flatmark, Q. H. Gibson, Y. Hatefi, R. Hill, T. Horio, R. J. Kassner, J. Kraut, N. O. Kaplan, J. Lavorel, R. Lemberg, R. K. Morton, K. Okunuki, S. Sano, R. Sato, J. Schiff, P. Slonimski, and A. H. Theorell. Valuable technical help was provided, particularly by Ms. S. Tedro, and by Mrs. H. Smith, Mrs. R. Smith, S. Lindroth, and R. Pabst.

Literature Cited

1a. Remer, T. C., ed. 1965. *Serendipity and the Three Princes,* p. 6. Norman, Okla: Univ. Okla. Press

1b. Rosenau, M. J. 1935. *J. Bacteriol.* 29:91

2. Cannon, W. B. 1945. *The Way of the Investigator.* New York: W. W. Publisher

3. Harkins, W. D., Gans, D. M., Kamen, M. D., Newson, H. W. 1935. *Phys. Rev.* 47:52

4. Kurie, F. N. D. 1933. *Phys. Rev.* 44:461

5. Burcham, W. E., Goldhaber, M. 1936. *Proc. Cambridge Philos. Soc.* 32:632

6. McMillan, E. M., Kamen, M. D., Ruben, S. 1937. *Phys. Rev.* 52:375

7. Kamen, M. D. 1949. *Photosynthesis in Plants,* a symposium, ed. J. Franck, W. E. Loonis, p. 365. Ames: Iowa State College Press

8. Ruben, S., Kamen, M. D., Hassid, W. Z. 1940. *J. Am. Chem. Soc.* 62:3443

9. Ruben, S., Kamen, M. D., Perry, L. H. 1940. *J. Am. Chem. Soc.* 62:3450

10. Ruben, S., Kamen, M. D. 1940. *J. Am. Chem. Soc.* 62:3451

11. Ruben, S., Kamen, M. D. 1940. *Proc. Natl. Acad. Sci. USA* 26:418

12. Barker, H. A., Ruben, S., Kamen, M. D. 1940. *Proc. Natl. Acad. Sci. USA* 26:426

13. Wood, H. G., Werkman, C. H. 1936. *Biochem. J.* 30:48

14. van Niel, C. B., Ruben, S., Carson, S. F., Kamen, M. D., Foster, J. W. 1942. *Proc. Natl. Acad. Sci. USA* 28:8

15. See 1974. *Mol. Cell Biol.* 5:79

16. Evans, E. A. Jr., Slotin, L. 1940. *J. Biol. Chem.* 136:301

17. Ruben, S., Randall, M., Kamen, M. D., Hyde, J. L. 1941. *J. Am. Chem. Soc.* 63:877

18. Kamen, M. D., Barker, H. A. 1945. *Proc. Natl. Acad. Sci. USA* 31:8

19. Warburg, O. 1958. *Science* 128:68

20. Holt, A. S., French, C. S. 1948. *Arch. Biochem. Biophys.* 19:429

21. Dole, M., Jencks, G. 1944. *Science* 100:409

22. Wilson, R. R., Kamen, M. D. 1938. *Phys. Rev.* 54:1031

23. Kamen, M. D. 1963. *Science* 140:584

24. Ruben, S., Kamen, M. D. 1940. *Phys. Rev.* 57:549

25. Alvarez, L., Cornog, R. 1939. *Phys. Rev.* 56:613

26. Kamen, M. D. 1972. *Environ. South West* 448:11

27a. Snell, A. H. 1976. *Oak Ridge Natl. Lab. Rev.* Fall:45

27b. Primakoff, H. 1982. See Ref. 105, p. 45

28. Kamen, M. D. 1941. *Phys. Rev.* 60:537; 1942. 62:303

29. Barker, H. A., Kamen, M. D. 1945. *Proc. Natl. Acad. Sci. USA* 31:219

30. Barker, H. A., Kamen, M. D., Haas, V. 1945. *Proc. Natl. Acad. Sci. USA* 31:355

31. Barker, H. A., Kamen, M. D., Bornstein, B. T. 1945. *Proc. Natl. Acad. Sci. USA* 31:373

32a. Wood, H. G., Drake, H. L., Hu, S.-I. *Biochemistry Symposium,* ed. E. E. Snell, p. 29. Palo Alto, Calif: Annual Reviews

32b. Wood, H. G. 1982. See Ref. 105, p. 99

33. Shemin, D., Rittenberg, D. 1946. *J. Biol. Chem.* 166:621

34. Grinstein, M., Kamen, M. D., Moore, C. V. 1948. *J. Biol. Chem.* 174:767

35. Shemin, D. 1982. See Ref. 105, p. 117

36. Hershey, A. V., Kamen, M. D., Kennedy, J. W., Gest, H. 1951. *J. Gen. Physiol.* 34:305

37. Spiegelman, S., Kamen, M. D. 1946. *Science* 104:581

38. Kamen, M. D. 1946. *Bull. Am. Mus. Nat. Hist.* 87:107

39. Kamen, M. D. 1947. *Radioactive Tracers in Biology.* New York: Academic. 281 pp.

40. Ruben, S. 1943. *J. Am. Chem. Soc.* 65:279

41. Kamen, M. D., Gest, H. 1952. *Phosphorus Metabolism,* ed. W. D. McElroy, B. Glass. Baltimore, Md: Johns Hopkins

42. Gest, H., Kamen, M. D. 1948. *Science* 109:558

43. Kamen, M. D., Gest, H. 1948. *Science* 109:560

44. Elsden, S. R., Kamen, M. D., Vernon, L. P. 1953. *J. Am. Chem. Soc.* 43:492

45a. Vernon, L. P., Kamen, M. D. 1954. *J. Biol. Chem.* 211:643, 663

45b. Vernon, L. P., Kamen, M. D. 1954. *Arch. Biochem. Biophys.* 51:122

46. Frenkel, A. 1954. *J. Am. Chem. Soc.* 76:5568

47. Arnon, D. I., Whatley, F. R., Allen, M. B. 1954. *J. Am. Chem. Soc.* 76:6724

48. Newton, J., Kamen, M. D. 1956. *Biochim. Biophys. Acta* 21:74; 1957. 25:462

49. Perini, F., Kamen, M. D., Schiff, J. A. 1964. *Biochim. Biophys. Acta* 88:88, 91

50. Nakamura, H. 1937. *Acta Phytochim. Tokyo* 9:189

51. Horio, T., Kamen, M. D. 1962. *Biochemistry* 1:1141

52. Taniguchi, S., Kamen, M. D. 1963. *Biochim. Biophys. Acta* 74:438

53. Yamashita, J., Kamen, M. D., Horio, T. 1969. *Arch. Mikrobiol.* 60:304
54. Taniguchi, S., Kamen, M. D. 1963. *Biochim. Biophys. Acta* 112:1
55. Chance, B., Horio, T., Kamen, M. D., Taniguchi, S. 1966. *Biochim. Biophys. Acta* 112:1
56. Ehrenberg, A., Kamen, M. D. 1965. *Biochim. Biophys. Acta* 102:333
57. Maltempo, M. M., Moss, T. H. 1976. *Q. Rev. Biophys.* 9:181
58. Rawlings, J., Stephens, P. J., Nafie, L. A., Kamen, M. D. 1977. *Biochemistry* 16:1725
59. Emptage, M. H., Zimmerman, R., Que, L. Jr., Munck, E., Hamilton, W. D., Orme-Johnston, W. H. 1977. *Biochim. Biophys. Acta* 495:12
60. Williams, R. J. P. 1971. *Cold Spring Harbor Symp. Quant. Biol.* 36:53
61. Weber, P. C., Bartsch, R. G., Cusanovich, M. A., Hamlin, R. C., Howard, A., et al. 1980. *Nature* 286:302
62. Cusanovich, M. A., Gibson, Q. H. 1973. *J. Biol. Chem.* 248:822
63. Gibson, Q. H., Kamen, M. D. 1966. *J. Biol. Chem.* 241:1969
64. Dus, K., Sletten, K., Kamen, M. D. 1968. *J. Biol. Chem.* 243:5507
65. Pettigrew, G. W., Meyer, T. E., Bartsch, R. G., Kamen, M. D. 1975. *Biochim. Biophys. Acta* 430:197
66. Errede, B., Kamen, M. D. 1978. *Biochemistry* 17:1015
67. Sharp, J. J., Robinson, A. B., Kamen, M. D. 1975. *J. Am. Chem. Soc.* 95:6097
68. Robinson, A. B., McKerrow, J. H., Cary, P. 1970. *Proc. Natl. Acad. Sci. USA* 66:753
69. Salemme, F. R., Kraut, J., Kamen, M. D. 1973. *J. Biol. Chem.* 248:7701
70. Smith, G. M. 1979. *Biochemistry* 18:1628
71. Pettigrew, G. W., Smith, G. M. 1977. *Nature* 265:661
72. Meyer, T. E., Kamen, M. D. 1982. *Adv. Protein Chem.* 35:105
73. Kamen, M. D. 1983. *Fed. Proc.* 42:2815
74. Bartsch, R. G. 1963. *Bacterial Photosynthesis,* ed. H. Gest, A. San Pietro, L. Vernon, p. 315. Yellow Springs, Ohio: Antioch
75. Dus, K. 1973. *J. Biol. Chem.* 248:7318
76. de Klerk, H., Kamen, M. D. 1966. *Biochim. Biophys. Acta* 112:175
77. Carter, C. W. Jr. 1977. *J. Biol. Chem.* 252:7802
78a. Carter, C. W. Jr., Kraut, J., Freer, S. T., Xuong, Ng., Alden, R. A., Bartsch, R. G. 1974. *J. Biol. Chem.* 249:4212
78b. Przysiecki, C. T., Meyer, T. E., Cusanovich, M. A. 1985. *Biochemistry* In press

79. Tedro, S. M., Meyer, T. E., Kamen, M. D. 1977. *J. Biol. Chem.* 254:1495
80. Iwasaki, H. 1960. *J. Biochem. (Tokyo)* 47:174
81. Hori, K. 1961. *J. Biochem. (Tokyo)* 50:481
82. Kassner, R. J., Kamen, M. D. 1967. *Proc. Natl. Acad. Sci. USA* 58:2445
83. Kassner, R. J., Kamen, M. D. 1968. *Biochim. Biophys. Acta* 153:270
84. Jolchine, G., Reiss-Husson, F., Kamen, M. D. 1969. *Proc. Natl. Acad. Sci. USA* 64:650
85. Fanica-Gaignier, M., Clement-Metral, J., Kamen, M. D. 1971. *Biochim. Biophys. Acta* 226:135
86. Oelze, J., Kamen, M. D. 1971. *Biochim. Biophys. Acta* 234:137
87. Oelze, J., Kamen, M. D. 1975. *Biochim. Biophys. Acta* 287:1
88. Schmidt, G. L., Kamen, M. D. 1970. *Arch. Microbiol.* 76:151
89. Maroc, J., de Klerk, H., Kamen, M. D. 1968. *Biochim. Biophys. Acta* 102:621
90. Maroc, J., Azerad, R., Kamen, M. D., Le Gall, J. 1970. *Biochim. Biophys. Acta* 197:87
91a. de Klerk, H., Govindjee, Kamen, M. D., Lavorel, J. 1969. *Proc. Natl. Acad. Sci. USA* 62:972
91b. Wagner, G. C., Kassner, R. J., Kamen, M. D. 1974. *Proc. Natl. Acad. Sci. USA* 71:253
91c. Peck, H. D. Jr., Tedro, S., Kamen, M. D. 1974. *Proc. Natl. Acad. Sci. USA* 71:2404
92. Borda, L., Green, M. H., Kamen, M. D. 1969. *J. Gen. Microbiol.* 56:345
93. Yamashita, J., Kamen, M. D. 1968. *Biochim. Biophys. Acta* 161:162; 1969. 182:322
94. Garcia, A. F., Drews, G., Kamen, M. D. 1974. *Proc. Natl. Acad. Sci. USA* 71:4213
95. Garcia, A. F., Drews, G., Kamen, M. D. 1974. *Biochim. Biophys. Acta* 387:129
96. Thore, A., Keister, D. L., San Pietro, A. 1969. *Arch. Mikrobiol.* 67:378
97. Cusanovich, M. A., Bartsch, R. G., Kamen, M. D. 1968. *Biochim. Biophys. Acta* 153:397
98. Cusanovich, M. A., Kamen, M. D. 1968. *Biochim. Biophys. Acta* 153:376, 418
99. Horio, T., Kamen, M. D. 1962. *Biochemistry* 1:144
100. Kennel, S. J., Bartsch, R. G., Kamen, M. D. 1972. *Biophys. J.* 12:802
101. Kennel, S. J., Kamen, M. D. 1971. *Biochim. Biophys. Acta* 234:485; 253:153
102. Benemann, J. R., Berenson, J. A., Kap-

lan, N. O., Kamen, M. D. 1973. *Proc. Natl. Acad. Sci. USA* 70:2317

103. Klibanov, A. M., Kaplan, N. O., Kamen, M. D. 1978. *Proc. Natl. Acad. Sci. USA* 75:3640

104a. Klibanov, A. M., Kaplan, N. O., Kamen, M. D. 1980. *Enzyme Engineering,* ed. H. Weetal, G. P. Royer, 5:135 Plenum

104b. Meyer, T., Kamen, M. 1985. *Proc. Natl. Acad. Sci. USA.* In press

105. Kaplan, N. O., Robinson, A. B., eds. 1982. *From Cyclotrons to Cytochromes.* New York: Academic

106. Gest, H., Favinger, J. L. 1983. *Arch. Microbiol.* 136:11

107. Brockman, H. J., Lipinski, A. 1983. *Arch. Microbiol.* 136:17

108. Weaver, P. 1975. *Environmental and mutational variations in the photosynthetic apparatus of* Rhodospirillum rubrum. PhD thesis. Univ. Calif., San Diego

109. Drews, G., Oelze, J. 1981. *Advances in Microbial Physiology,* ed. A. H. Rose, J. G. Morris, 22:1–92. New York: Academic

Ann. Rev. Biochem. 1986. 55:35–67

LECTINS AS MOLECULES AND AS TOOLS[1]

Halina Lis and Nathan Sharon

Department of Biophysics, The Weizmann Institute of Science, Rehovoth, Israel

CONTENTS

[1]The literature survey was completed in June 1985.

0066-4154/86/0701-0035$02.00

PERSPECTIVES AND SUMMARY

In 1973, when the first review on lectins appeared in *Annual Review of Biochemistry* (1), interest in these sugar-specific, cell-agglutinating proteins (2, 3) was just starting to gain momentum and the literature on the subject was limited in volume and in scope. Since then, over 15,000 articles dealing with lectins have been published. The ubiquitous occurrence of lectins—in plants, animals, and microorganisms—has been firmly established. The number of purified lectins has increased from a dozen to well over 100. Whereas ten years ago only the amino acid sequence of concanavalin A had been determined, at present close to ten complete lectin sequences are known, some derived from the cloned lectin genes, as well as numerous partial sequences. Studies of the molecular genetics of lectins have provided examples of novel controls of gene expression. The extensive homologies observed between lectins from taxonomically related sources demonstrate that these proteins have been conserved throughout evolution and argue strongly that they must have an important role in nature. Still, despite efforts to understand the physiological function of lectins, we know little with certainty about this intriguing question, in striking contrast to our detailed knowledge of the molecular properties of a large number of lectins. Much effort has been invested in the search for applications for lectins. As a result, lectins have become invaluable tools in biological and medical research, in areas as diverse as bacterial typing and bone marrow transplantation. In addition they continue to be the focus of intense interest in their own right. In short, as the one hundredth anniversary of the discovery of lectins approaches (4), the field is thriving and attracting increased attention.

In this article we concentrate on selected advances in lectin research that have occurred since the subject was last reviewed in this series (5). Primarily, we deal with plant lectins, still the largest and best-characterized group. We discuss in some detail a new group of lectins, those of bacterial origin, and survey briefly invertebrate and vertebrate lectins. Finally, we present some examples of applications of lectins for the study of glycoconjugates in solution and on cell surfaces. For more detailed coverage, the reader is referred to the recently published treatise on the properties, functions, and applications of lectins (6), and to other recent reviews (7–9), symposia proceedings, and books (10–14).

PLANT LECTINS

Molecular Properties

Lectins, whether from plants or other sources, are routinely detected and quantitated by their ability to agglutinate erythrocytes and are readily purified by affinity chromatography on immobilized carbohydrates (7, 15). They are classified into a small number of specificity groups (mannose[2], galactose, N-acetylglucosamine, N-acetylgalactosamine, L-fucose, and N-acetylneuraminic acid) according to the monosaccharide that is the most effective inhibitor of the agglutination of erythrocytes or precipitation of carbohydrate-containing polymers by the lectin. Although found primarily in seeds, lectins are also present in other plant tissues (8, 9, 16). In some plant families, such as the Leguminoseae (17) or the Gramineae (18), lectins are present in many species, whereas in others, such as the Euphorbiaeceae (19), they have been found in a few species only. Usually, a particular source contains (a) lectin(s) belonging to a single specificity group, but in an increasing number of cases, two (or more) lectins that differ in their specificity are found in the same plant (see Table 1). Individual lectins frequently occur as a group of closely related proteins, designated as isolectins, the synthesis of which is under direct genetic control (9). Analysis of isolectin patterns in different wheat species has indicated that each genome directs the synthesis of a distinct subunit species of wheat germ agglutinin (26). In diploids, a single subunit species is present giving rise to only one molecular form of the lectin. In polyploids, however, the different polypeptide chains, coded for by the different genomes, combine randomly with identical or nonidentical partners, forming both homomeric and heteromeric dimers. Under suitable conditions, the subunits of different wheat germ agglutinin isolectins can also be interchanged in vitro (7). Moreover, the subunits of lectins from different cereal species (wheat, rye, and barley) can be exchanged to form intergeneric, heteromeric lectins (27). The lectins from these three sources are indeed similar in their physicochemical properties and carbohydrate specificity (28).

As a rule, there is one sugar-binding site per subunit. This has been demonstrated also for the lectins from soybean (29), peanut (30), and lima bean (31), previously reported to contain two binding sites per molecule consisting of four subunits. Two combining sites have, however, been found in the four-subunit lectins from *Datura stramonium* (32) and from jackfruit *(Artocarpus integrifolia)* (33). On the other hand, two combining sites are present in each subunit of wheat germ agglutinin (20), and in one of the subunits (the B chain) of ricin (34, 35). The subunits of the same lectin usually have the same sugar specificity, and the binding sites are equivalent. Exceptions to this generalization are the isolectins of phytohemagglutinin (PHA), the lectin of the red kidney bean *(Phaseolus vulgaris),* composed of varying proportions of the E and L subunits that form the tetrameric E_4-PHA and L_4-PHA, and the *Griffonia simplicifolia* I isolectins made up of varying proportions of subunits A and B (cf Table 1). Another example is the tetrameric lectin from *Dolichos biflorus* made up of two closely related subunits, only one of which appears to bind sugars (36).

---

[2]All sugars are in the D-pyranose form unless otherwise noted.

Table 1   Plants that contain multiple lectins that differ in their specificity

Source	Lectin	Specificity Carbohydrate	Specificity Blood type	Refere
Griffonia simplicifolia	I-A$_4$	GalNAc	A	2(
	I-B$_4$	Gal	B	2(
	II	GlcNAc	[a]	2(
	IV	Galβ3GlcNAcβ3Galβ4Glc ↑α2 ↑α4 L-Fuc L-Fuc	Lewis b	2▮
Laburnum alpinum	I	(GlcNAc)$_{2-3}$	O	2ᦗ
	II	Gal	O	
Phaseolus vulgaris	E-PHA	[b]	—	2(
	L-PHA	[b]	[a]	
Ulex europaeus	I	L-Fuc	O	2▮
	II	GlcNAc	O	
Vicia cracca	I	GalNAc	A	2ᦔ
	II	Man, Glc	—	
Vicia villosa	A$_4$	GalNAc	A$_1$	2⒋
	B$_4$	[b]	[a]	
Viscum album	I	Gal	—	2ᦕ
	II	GalNAc	—	

[a] Does not agglutinate human erythrocytes of blood types O, A, B.
[b] See Table 2.

Primary Sequences

In addition to concanavalin A (37), the lectins from fava bean (38), lentil (39), soybean (40), *Onobrychis viciifolia* (41), *Dioclea grandiflora* (42), and *Lathyrus ochrus* (43, 44) have been completely sequenced by conventional techniques. Several other legume lectins—for example those from peanut (45), *Vicia cracca* (46), and *Lathyrus odoratus* (47)—have been incompletely sequenced. The primary structures of soybean agglutinin (48), pea lectin (49), L-PHA, and E-PHA (50) have been deduced from the nucleotide sequence of cDNA reversely transcribed from the specific mRNAs.

Structurally, two classes of legume lectins have been recognized: (*a*) those comprised of either identical or nearly identical subunits of M_r 25,000–30,000 (one-chain lectins); and (*b*) those made up of two different subunits, the light α chain and the heavy β chain (two-chain lectins). The latter lectins occur only in the Vicieae tribe. Nevertheless, all these lectins exhibit extensive homologies when properly aligned, i.e. by placing the β chains of the two-chain lectins along the NH$_2$-terminal sequences of the one-chain lectins, followed by the α chains (51). Concanavalin A and the recently isolated lectin from *D. grandiflora* (42), both belonging to the Diocleae tribe, occupy a special position. In these cases homology with the other one-chain legume lectins is obtained by aligning, with a few deletions, the amino ends of the legume lectins with residue

123 of the Diocleae lectins, proceeding to the COOH terminal of the latter lectins, and continuing along their NH_2-terminal region (40). This has been referred to as circular homology (see Figure 1). In all legume lectins examined, several positions are invariant (e.g. Phe-6 and Phe-11) or highly conserved; in a number of other positions, only conservative substitutions occur (51). These conserved amino acids include two that correspond to residues previously identified in concanavalin A as being important in the binding of Ca^{2+} and Mn^{2+}. Similarly, the amino acids contributing to the three-dimensional structure of the hydrophobic binding cavity of concanavalin A (see *Carbohydrate Specificity* section below) are highly conserved in homologous positions in all of the lectins investigated.

The first complete amino acid sequence of a cereal (Gramineae) lectin, wheat germ agglutinin, has been established (52). The NH_2-terminal residue is pyroglutamyl (53). A high degree of sequence homology is found between several parts of the molecule (52), as would be expected since the subunits of this lectin consist of four isostructural domains, each stabilized by four interlocking and homologously placed disulfide bonds.

Carbohydrate Units of Glycoprotein Lectins

Most lectins contain covalently bound carbohydrate. Two types of glycoprotein lectins have been discerned: (*a*) those containing primarily mannose and *N*-

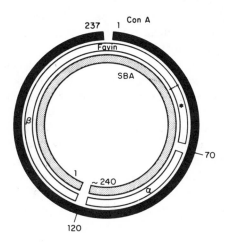

Figure 1   Schematic drawing of the alignment of fava bean lectin (favin) α and β chains (open bars), soybean agglutinin (SBA) (stippled bar), and concanavalin A (dark bar) showing the circular permutation that gives maximum homology among the sequences of these lectins. The dashed line transecting the favin β chain indicates the region where the homologous chain of the lentil lectin ends. The location of the covalently bound carbohydrate in favin is indicated by the asterisk. [From Hemperly & Cunningham (40).] The location of the carbohydrate of soybean agglutinin is unknown.

acetylglucosamine; and (*b*) those containing L-arabinose and galactose. Soybean agglutinin is the only lectin of the first group for which the structure of the carbohydrate unit has been unequivocally established (54). The structure is identical to that of the N-linked $Man_9GlcNAc_2$ oligosaccharide found in several animal glycoproteins, and nearly identical to the core portion of yeast mannans (55). This provides strong evidence for the evolutionary conservation of such structures and concurs with the finding of a common biosynthetic pathway of protein *N*-glycosylation in animals, plants, and yeasts.

Several lectins of the first type also contain L-fucose and xylose (7); the latter monosaccharide has not been found with mannose and *N*-acetylglucosamine in the same carbohydrate unit of animal or yeast glycoproteins (55). The structure of xylose-containing units has been investigated to a limited extent. The lectin from Tora bean contains the pentasaccharide core Manα6(Manα3)Manβ4GlcNAcβ4GlcNAc common to all *N*-glycoproteins, with xylose attached by a β→2 linkage to the β-linked mannose (56). The same position and linkage were reported earlier for the xylose residue of bromelain (57); in this glycoprotein, however, the core oligosaccharide appears to lack the α→3 linked mannose. Oligosaccharides with core regions like that in Tora bean lectin or like that in bromelain are present in *Vicia graminea* lectin (58). On the basis of immunological studies, a structure similar to that of the core region of bromelain has been proposed for the carbohydrate units of the lectins from *Sophora japonica* and *Wistaria floribunda* (59).

Glycoprotein lectins of the second type have been found only in plants of the Solanaceae family (7). Both the potato and the *D. stramonium* lectins contain the tri- and tetra-arabinofuranosides (L-Ara*f*α3)$_{0–1}$L-Ara*f*β2L-Ara*f*β2L-Ara*f* β-linked to hydroxyproline. In addition these lectins contain galactose, and in the case of the *D. stramonium* lectin also Gal1→3Gal, both α-linked to serine (59a). Neither of these linking groups, nor arabinose, has been found in animal glycoproteins.

Chemically deglycosylated potato (60) and tomato (61) lectins retained their hemagglutinating activity and carbohydrate specificity, although with the tomato lectin the product exhibited lower agglutinating activity than the parent compound. These findings support earlier contentions, based on chemical and enzymatic modification of the carbohydrate chains of several legume lectins, that the carbohydrate moieties are not required for biological activity (8).

Biosynthesis

Experiments hybridizing soybean agglutinin cDNA with soybean leaf DNA have shown the presence in soybean plants of two related lectin genes, L1 and L2 (62). The L1 gene codes for the seed lectin, whereas the function of the L2 gene, which is expressed at only low levels, is unknown. Both genes are present in a lectin-deficient soybean line, but the L1 gene contains a large (3.4-

kilobase) insertion within its coding region (62). It has been suggested that the insertion sequence, with the structural features of a transposable element, blocks the transcription of the L1 gene in lectin-deficient soybean lines. In the cotyledons of *Phaseolus vulgaris* (63), soybean (62), and pea (64), transcripts hybridizing with lectin cDNA begin to accumulate during midmaturation and then decrease during late maturation. The increase in mRNA coincides with the period of maximal production of the seed lectins and indicates that lectin accumulation in the seeds is regulated at the transcriptional level.

The intracellular sites of synthesis and protein modification, and the transport pathways of plant lectins, as demonstrated in studies with tissue slices and cell-free systems, are generally not different from those of animal glycoproteins. *N*-Glycosylation proceeds via the dolichol phosphate cycle, and it is inhibited by tunicamycin (65, 66). The primary translation product of the fava bean lectin mRNA is a polypeptide of M_r 29,000 (67). The NH_2-terminal signal sequence of 29 residues is followed by the β chain, and indirect evidence was obtained for the presence of the α chain in this polypeptide. The lectin thus appears to be synthesized as a single polypeptide precursor with the structure NH_2-signal-β-chain-α-chain-COOH. Pea lectin, too, is synthesized as a single polypeptide precursor of M_r 25,000 with the same orientation as that of fava bean lectin (64). A cotranslational peptide cleavage in the endoplasmic reticulum removes the signal sequence and the product is then transported to the protein bodies where it is cleaved to yield the β and α chains (M_r 17,000 and 7,000, respectively) of the mature pea lectin. These studies provide convincing evidence for the hypothesis, formulated on the basis of the sequence homologies between the two-chain and one-chain legume lectins, that the former are derived from a single polypeptide precursor (68).

The circular homology observed between concanavalin A and other legume lectins (see *Primary Sequences* section above) has been postulated to be due to rearrangements within a common ancestral genome, possibly involving gene duplication (51). According to a recent report, however, the amino acid sequence derived from cDNA complementary to the mRNA for concanavalin A has a direct homology, not a circular one, with other legume lectins (69). The cDNA contained a region corresponding to 29 residues of the signal sequence, followed by a coding region corresponding to amino acids 119–237 of concanavalin A, a region coding for 15 amino acids not found in the mature lectin, and finally a region corresponding to amino acids 1–118 of the lectin. Although differences from the sequence of concanavalin A have been noted in 12 positions, it has been concluded that the sequence derived from the cDNA represents a precursor of concanavalin A and not of a closely related polypeptide. It has been suggested that during the formation of mature concanavalin A there must be a transposition and ligation (between residues 118 and 119) of two peptides produced from the precursor polypeptide. This type of modification has not been reported previously for eukaryotes, although it has been postulated to occur in bacteriophage λ (70). The presence in concanavalin A of a large proportion of subunits consisting of two fragments, which correspond to residues 1–118 and 119–237, may thus be due to incomplete ligation and not to the proteolytic cleavage of the intact subunits, as originally assumed (37).

Biosynthetic experiments in vivo provided evidence that the A and B chains of ricin are encoded by a single mRNA species (71). This has been confirmed by sequencing of the cloned DNA complementary to ricin-precursor mRNA (72). The coding region of the cDNA corresponds to an NH_2-terminal sequence of 24 amino acids, followed by the A chain amino acids, which are joined to the B chain via a linking region of 12 amino acids. Translation in the presence of dog pancreatic microsomes results in N-glycosylation of the ricin precursor (66). Intracellular transport of the glycosylated precursor from the endoplasmic reticulum to a denser vesicle fraction, tentatively identified as the Golgi apparatus, is accompanied by modifications of the oligosaccharide units, including addition of L-fucose.

The genes encoding the subunits of E-PHA and L-PHA are located on the same chromosome, 4 kilobases apart, and contain regions encoding signal peptides of 21 and 20 amino acids, respectively. The mRNA coding regions of the two genes are 90% homologous, suggesting that they are derived from the duplication of an ancestral gene (50). E-PHA contains three potential glycosylation sites, whereas two such sites are present in L-PHA. Earlier evidence seemed to indicate that in each of the isolectins two carbohydrate chains are present per subunit, one comprised of mannose and N-acetylglucosamine, and the other containing xylose and L-fucose as well (73). The latter carbohydrate unit appears to be derived from an oligomannose unit by modifications that occur after the protein moves from the endoplasmic reticulum to the Golgi (74). Whereas the incorporation of L-fucose has been directly demonstrated, nothing is known about the location and mechanism of xylose incorporation. In the Golgi, the modified oligosaccharide also acquires one or more outer N-acetylglucosamine residues which are subsequently removed when PHA arrives at its destination in the protein bodies (75). This is reminiscent of the transient attachment of N-acetylglucosamine during the biosynthesis of lysosomal hydrolases of animal cells (76, 77). In the latter system, however, this N-acetylglucosamine is incorporated in the form of GlcNAc-1-P, to generate the Man-6-P recognition marker, whereas phosphate is not incorporated into the oligosaccharide of PHA. The sorting signals and posttranslational processing steps for glycoproteins that are transported to the lysosomes (76, 77) and their plant equivalent, the protein bodies, seem thus to be different.

The pathway of biosynthesis of Gramineae lectins is unusual in that removal of the signal sequence from 23,000-dalton precursor proteins to give the

18,000-dalton mature lectin subunits is essentially a posttranslational event (78, 79). In the developing rice embryos, maturation involves an additional posttranslational modification that splits the bulk of the 18,000-dalton subunits into two fragments (79).

Carbohydrate Specificity

Detailed information has become available on the carbohydrate specificity of many plant lectins, in particular for oligosaccharides and glycopeptides (20, 80). It is based on inhibition of hemagglutination or polysaccharide precipitation, and also on other techniques, e.g. affinity chromatography on Sepharose-bound lectins (81–85) and affinity electrophoresis (86, 87). Application of physicochemical methods, such as those used in the investigation of combining sites of enzymes and antibodies, is providing thermodynamic and kinetic data on the binding process between lectins and carbohydrates (88, 89). Association constants for the interaction of a lectin with a series of carbohydrates correlate well with the relative inhibitory activity of the same sugars (8). Rather unexpectedly, sugars that are good lectin inhibitors in solution do not always bind to immobilized lectins. With concanavalin A (81, 82) and *Ricinus communis* agglutinin (83), only carbohydrates with association constants in solution greater than 4×10^6 M^{-1} bind to the immobilized lectins.

Earlier work has shown that certain lectins react poorly, if at all, with monosaccharides, although they combine readily with oligosaccharides, and that among lectins that interact with monosaccharides, several exhibit pronounced preference for di-, tri-, and tetrasaccharides (20, 80). Recent studies have defined the specificity of some lectins of the former group and provided additional information on lectins of the latter group (see Table 2). The lectin from *D. stramonium,* previously shown to be specific for β1→4 linked oligomers of *N*-acetylglucosamine, also binds disaccharides such as Galβ4GlcNAc (*N*-acetyllactosamine) and Manβ4GlcNAc as strongly as it binds GlcNAcβ4-GlcNAc (32). It exhibits high affinity for oligosaccharides with repeating *N*-acetyllactosamine sequences, as well as for oligosaccharides containing a core mannose substituted at C-2 and C-6 by *N*-acetyllactosamine (98). Physicochemical measurements of the interaction of the lectins from peanut (99) and from *Erythrina cristagalli* (100) with a variety of carbohydrates have shown that the increased affinity for the disaccharides Galβ3GalNAc and Galβ4GlcNAc, respectively, as compared with galactose, is due to enthalpic effects, strongly suggesting that these lectins have extended binding sites.

Many lectins recognize terminal nonreducing saccharides, while others also recognize internal sugar sequences. A few lectins recognize carbohydrate sequences together with the amino acid to which they are linked, as first demonstrated with the lectin from the mushroom *Agaricus bisporus* (101). The lectin of *V. graminea,* specific for human blood group N, requires a cluster of

Table 2 Lectins with pronounced specificity for oligosaccharides

| | | Specificity | | |
| | | Oligosaccharide | | |
Source of lectin	Monosaccharide	Structure	RIA[a]	Reference
Datura stramonium	GlcNAc	GlcNAcβ4GlcNAcβ4GlcNAc	550	32
Dolichos biflorus	GalNAc	GalNAcα3GalNAc	36	90
Erythrina cristagalli	Gal	Galβ4GlcNAc	30–50	91, 92
Peanut	Gal	Galβ3GalNAc	50	20
Phaseolus vulgaris (E-PHA)	—	Galβ4GlcNAcβ2Manα6⟍ GlcNAcβ4—Manβ4-R[b] GlcNAcβ2Manα3⟋		93
Phaseolus vulgaris (L-PHA)	—	Galβ4GlcNAcβ2⟍ Man Galβ4GlcNAcβ6⟋		94

Vicia graminea	—	NH$_2$-Leu | (Galβ3GalNAcα)-Ser | (Galβ3GlcNAcα)-Thr | (Galβ3GalNAcα)-Thr | Glu-COOH		95, 96
Vicia villosa	GalNAc	NH$_2$ | (GalNAcα)-Ser | (Pro)$_2$ | Gly | (Ala)$_2$ | (GalNAcα)-Thr-COOH	120	97
Wheat germ	GlcNAc	GlcNAcβ4GlcNAcβ4GlcNAc	3000	20

[a] Relative inhibitory activity with the inhibitory activity of the monosaccharide arbitrarily set as 1.
[b] R = GlcNAcβ4GlcNAc.

Galβ3GalNAc units linked to serine (or threonine) in close vicinity to an NH_2-terminal leucine residue, as present in asialoglycophorin from type N human erythrocytes (95, 96). Peptides that contain one or more N-acetyl-galactosaminyl-serine units are about 10–15 times more effective as inhibitors of hemagglutination by the *Vicia villosa* lectin B_4 than is N-acetylgalactosamine alone, whereas a heptapeptide that contains two N-acetylgalacto samine residues, one bound to the NH_2-terminal serine and the other to the COOH-terminal threonine, is over 100 times more inhibitory (97). It appears that complex-type oligosaccharides must be linked to asparagine in order to interact with columns of Sepharose-bound pea and fava bean lectins (85).

Evidence has been presented on the importance of molecular shape in determining lectin-carbohydrate interactions. For example, *Ulex europaeus* lectin I binds both L-Fucα2Galβ4GlcNAc and L-Fucα2Galβ3GlcNH$_2$ which, although structurally different, have similar topographies for the binding by the lectin, as inferred from theoretical molecular modeling (102). Acetylation of the amine of the latter trisaccharide dramatically changes the relevant topography, and the product L-Fucα2Galβ3GlcNAc is not bound by the lectin.

Considerable information on the three-dimensional structure of N-linked oligosaccharides is emerging from high resolution NMR studies (103, 104). The NMR experiments show that the oligosaccharides°have well-defined conformations with limited flexibility around the glycosidic bonds, except for the possibility of the formation of two rotational isomers about the C5–C6 bond in the α1→6 linked mannose of the pentasaccharide core. The prevalence of either of the two isomers depends on the type of substitution of the mannose residues of the core. In particular, substitution of the β-linked mannose by N-acetylglucosamine linked β1→4 ("bisecting" N-acetylglucosamine) fixes the orientation of the Manα6Man arm into one of the two possible conformations. Oligosaccharides in this fixed conformation appear to bind poorly to concanavalin A, since earlier studies have shown that introduction of a bisecting N-acetylglucosamine into compounds of the general structure R'-2Manα6-(R''-2Manα3)Manβ4R (where either R' or R'' is β1→2-linked N-acetylglucosamine) decreased binding to the lectin by a factor of 10 (81). This conformation also prevents the interaction with four enzymes involved in the intracellular conversion of oligomannose saccharides into complex and hybrid units, namely the two N-acetylglucosaminetransferases II and III, an αL-fucosyltransferase, and mannosidase II (105, 106).

Model building based on X-ray data for the lectin and NMR data for branched mannose oligosaccharides have shown that concanavalin A possesses an extended binding site complementary to the trisaccharide Manα3(Manα6)Man (107). Supporting the above conclusion is the finding that the relative affinity of Manα3(Manα6)Man to concanavalin A is 130-fold greater than that of methyl α-mannoside (107a). It is also in agreement with the results of frontal affinity chromatography of mannose-containing glycopeptides on immobilized concanavalin A (82). The trisaccharide binds to the lectin with Manα6 occupying the subsite to which free mannose or methyl α-mannoside binds (107).

Using fast-reaction techniques it was found that the association rate constants for the binding of 4-methylumbelliferyl β-N-acetylgalactosamine to soybean agglutinin (29) and of N-dansylgalactosamine to the *E. cristagalli* lectin (108) (1.1×10^4 M^{-1}s^{-1} and 4.8×10^4 M^{-1}s^{-1}, respectively) are about 10^5 times smaller than expected for a diffusion-controlled process. Similar low values have been obtained for the binding of sugars to other lectins examined (e.g. concanavalin A and peanut agglutinin) with the exception of wheat germ agglutinin, for which it is about 1.5×10^6 M^{-1}s^{-1} (109). The dissociation of lectin-carbohydrate complexes is also very slow, as first demonstrated for concanavalin A (110), and more recently for soybean agglutinin with 4-methylumbelliferyl β-N-acetylgalactosamine (0.36 s^{-1}, Ref. 29), and for *E. cristagalli* lectin with N-dansylgalactosamine (about 0.5 s^{-1}, Ref. 108). The dissociation of 4-methylumbelliferyl βGalβ1→3GalNAc from its complex with peanut agglutinin is about 100 times slower than for other mono- and disaccharides (109a), which is very suggestive of site-ligand complementarity. Even more pronounced is the exceedingly slow dissociation rate (0.003 s^{-1}) for *Lens culinaris* lectin bound to erythrocytes (109b), although here additional factors are probably involved. Unlike enzymes, which for efficient catalysis require rapid dissociation, the slow kinetics of carbohydrate binding may facilitate the interaction of lectins with cells. It allows time for multivalent interactions of different receptors with a lectin molecule together with topological reorganization of the cross-linked receptor complexes in the viscous cell membrane.

Despite considerable effort, the contribution of hydrophobic forces and of hydrogen bonds to the energy of interaction of carbohydrates with lectins is not known. From theoretical considerations it was inferred that the preference of concanavalin A for methyl α-mannoside over methyl β-mannoside is due mainly to hydrophobic interactions of the α-methoxy group with Leu 99 or Tyr 100, but that the ability of the α-isomer to form more and stronger hydrogen bonds with the protein is a contributing factor (111). The model-building studies mentioned above have suggested that the binding of Manα3(Manα6)-Man to concanavalin A involves both hydrogen bonds and extensive van der Waals contacts. On the other hand, based mainly on studies of the interaction of *U. europaeus* lectin I and *G. simplicifolia* lectin IV with a wide range of synthetic trisaccharides, it has been proposed that binding involves the interaction of a large, essentially nonpolar, surface of the oligosaccharide, which terminates in a tight cluster of only two or three hydroxyl groups (112, 112a). This polar grouping appears to participate in a highly specific complexation reaction with a polar grouping at or near the periphery of the combining site that is essential for the subsequent complex formation with the lectin. The polar grouping provided by the protein thus acts as a "gate" to the combining site while the interactions between the carbohydrate and the amino acids of the combining site are mainly hydrophobic.

Hydrophobic Binding Sites

Lectins frequently bind hydrophobic glycosides, or other hydrophobic sugar derivatives, more strongly than analogous nonhydrophobic derivatives. A recent example is *N*-dansylgalactosamine, which binds to *E. cristagalli* lectin 60 times more strongly than *N*-acetylgalactosamine (100). Such findings suggest the existence of hydrophobic regions near the carbohydrate binding sites of the lectins. In addition, several lectins bind hydrophobic compounds devoid of sugar moieties. Whenever tested, such binding is not inhibited by specific sugars, indicating that the hydrophobic ligands bind to lectins at sites distinct from the carbohydrate binding sites. In lima bean lectin the distance between the hydrophobic and carbohydrate binding sites has been calculated from fluorescence energy transfer studies to be 28 Å (113). This value is close to that found earlier in concanavalin A (35 Å) by crystallographic studies (20).

Among the hydrophobic compounds that interact with lectins are indoleacetic acid, 1,8-anilinonaphthalenesulfonic acid, 2,6-toluidonylnaphthalenesulfonic acid, as well as adenine and its *N*-6-derivatives (114). The association constants for binding of indoleacetic acid are low; those for 1,8-anilinonaphthalenesulfonic acid are $10^3-10^4 \, M^{-1}$ (115), and for adenine and its derivatives $10^5-10^6 \, M^{-1}$ (116). The latter values are similar to or higher than the association constants for the binding of mono- and disaccharides to lectins (7, 20). Some of the hydrophobic compounds mentioned above are phytohormones, which are believed to play an important role in controlling plant growth. It has been suggested that lectins may function not only by virtue of their ability to bind carbohydrates but also by serving as binding proteins for biologically active hydrophobic ligands (114).

VERTEBRATE LECTINS

Vertebrate lectins are divided into two classes: (*a*) integral membrane lectins that require detergents for their extraction; and (*b*) soluble lectins (5, 117). The first group consists of lectins that differ in their sugar specificities (mannose, L-fucose, mannose-6-phosphate, *N*-acetylgalactosamine) and physicochemical properties (118, 119). Among the best-characterized lectins of this class are the galactose and mannose/*N*-acetylglucosamine–specific receptors of mammalian and avian hepatocytes, respectively (120), and the receptor for mannose-6-

phosphate present in a variety of cells (77). Recently, an endogenous lectin specific for thiodigalactoside and N-acetylgalactosamine has been isolated from microsomal fractions of baby hamster kidney cells (121), and a galactose-specific lectin has been purified from extracts of metastatic B16 melanoma cells (122). Membrane lectins seem to be present on the surfaces of various other cells, such as mouse and human lymphocytes, 3LL Lewis lung carcinoma, and mouse leukemia L1210 (123, 124); none of these lectins has been isolated.

Of the soluble vertebrate lectins, the first to be purified was the β-galactoside-specific lectin from the electric organ of the eel *Electricus electricus* (125, 126). Subsequently, lectins with the same specificity were found in many organs of teleosts, amphibia, avians, and mammals, and in embryos at the very early stages of their development (117). The concentration of these lectins in tissues is often ontogenically regulated, reaching a maximum at a given stage of development. However, in the thymus and spleen of mouse, calf, and chicken, the lectin titer remains high throughout adulthood. As in plants, isolectins have also been found in animals (127). The soluble β-galactoside-specific vertebrate lectins are of a similar molecular size, consisting each of two subunits of M_r 13,500–16,500 (117). They exhibit immunological cross-reactivity, which can be correlated to the phylogenetic distance separating the species from which the lectins are derived (128). All β-galactoside-specific lectins require a reducing agent to maintain their carbohydrate binding activity. In the case of the eel lectin, the reducing agent is required to prevent the oxidation of a tryptophan residue in the binding site (126).

Vertebrates also contain β-galactoside binding proteins that have been isolated as monomers (M_r 13,000–14,000). Of these, only the lactose binding protein from chicken (designated chicken lactose-lectin II) has been well characterized (117, 129). It differs from the dimeric lactose-specific chicken lectin I in subunit molecular weight, isoelectric point, and peptide map, as well as immunologically. Since chicken lactose-lectin II agglutinates erythrocytes, it must either have two carbohydrate binding sites per monomer or form dimers or oligomers when interacting with cells.

Not all soluble vertebrate lectins are β-galactoside specific (117). Those from the serum of the eel *Anguilla rostrata* and from *Xenopus laevis* oocytes, for example, are specific for L-fucose and both α- and β-galactosides, respectively. They also differ from the β-galactoside-specific lectins in that they each consist of 12 subunits.

INVERTEBRATE LECTINS

Lectins are found in practically all of the approximately 30 phyla and the various classes and subclasses of invertebrates (14, 130), mainly in the hemolymph and sexual organs, e.g. albumin glands and eggs (131, 132). They appear also to be present on the membranes of hemocytes (133, 134), cells that function as primitive and rather unspecific immunological protectors. Only a few of these lectins have been purified and characterized—for example the lectins of the snail *Helix pomatia* (20), of the crabs *Limulus polyphemus* (135) and *Carcinoscopus rotunda cauda* (136), of the lobster *Homarus americanus* (137), of the clam *Tridacna maxima* (138), and of the slug *Limax flavus* (139). With the exception of the blood-type-specific *H. pomatia* lectin, the other lectins mentioned above, as well as most of the hemolymph lectins, are specific for sialic acid. Recently a lectin that exhibits pronounced specificity for 9-*O*- and 4-*O*-acetylated sialic acids has been purified from the hemolymph of the crab *Cancer antennarius* (140).

BACTERIAL SURFACE LECTINS

Many intact bacteria possess the ability to bind to and agglutinate erythrocytes and other types of cells (12, 141). These activities are frequently inhibited by sugars, suggesting the presence of lectins on the bacterial surfaces as proposed almost ten years ago (142). Such lectins often exist in the form of filamentous appendages known as fimbriae or pili. The best-characterized bacterial lectins are type 1 fimbriae of *Escherichia coli* specific for mannose, and type P fimbriae specific for Galα4Gal. Different types of fimbriae may be present in the same bacterial culture (143–145). Lectinlike substances, i.e. preparations that bind sugars but do not agglutinate cells, have been obtained from several organisms—e.g. *Myxococcus xanthus* (specific for certain galactose derivatives), oral actinomyces (specific for galactose), and Mycoplasma spp. (specific for sialic acid).

Escherichia Coli *Type 1*

Most, or perhaps all, *E. coli* strains (and many *Salmonellae* species) carry type 1 fimbriae when grown under suitable conditions (146, 147). Such bacteria may spontaneously shift from a fimbriated phase to a nonfimbriated one and back, a phenomenon known as phase variation. This phase variation occurs at a relatively high frequency—approximately one per thousand bacteria per generation, which is several orders of magnitude higher than the rate of mutation and is genetically controlled at the transcriptional level (148).

Type 1 fimbriae are easily detached from the bacterial cell by shearing and purified by differential centrifugation. They consist of a stable array of identical protein subunits of about 170 amino acids (M_r 15,700–17,000) (146). The fimbriae are highly resistant to disruption, but can be dissociated reversibly into subunits by treatment with saturated guanidine hydrochloride (149). Type 1 fimbriae from different strains of *E. coli* have similar amino acid composition and molecular size but differ from those of type 1 fimbriae of *Klebsiella pneumoniae* and *Salmonella typhimurium*. The complete amino acid sequence of the type 1 fimbrial subunit of *E. coli* has been derived from the nucleotide sequence of the *fim* A gene coding for these fimbriae (150). It shows pronounced homologies, in particular in the NH_2-terminal and COOH-terminal regions, with the amino acid sequence of the major subunit of type P fimbriae, and with K99 fimbriae, although none of them exhibits immunological kinship, nor do they share sugar specificity.

Five genes encoded in an 11.2-kilobase segment of bacterial DNA are associated with the expression of type 1 fimbriae (151). Four of these genes code for proteins that control the biosynthesis of the fimbrial subunits or their assembly, and the fifth codes for a protein with a molecular weight of 19,000 which is the precursor of the fimbrial subunit (152). After synthesis, the nascent chains are processed and translocated into or through the inner membrane, and the subunits are then assembled into intact fimbriae (153).

Hapten inhibition studies of the agglutination of yeasts by *E. coli* and by isolated type 1 fimbriae show that the branched oligosaccharides Manα6 (Manα3)Manα6(Manα3)ManαOMe and Manα6(Manα3)Manα6-(Manα2-Manα3)ManαOMe, the trisaccharide Manα3Manβ4GlcNAc, and the aromatic glycoside *p*-nitrophenyl α-mannoside, are strong inhibitors (154). The combining site of the lectin of *E. coli* probably corresponds to that of a trisaccharide and may be in the form of a cavity on the surface of the lectin. There may be a hydrophobic region in the vicinity of the combining site.

The specificity of *K. pneumoniae* is similar to that of *E. coli*, whereas that of *Salmonella* is different, since with all six *Salmonellae* species examined *p*-nitrophenyl α-mannoside, as well as the trisaccharide Manα3Manβ4GlcNAc, is a weak inhibitor, less effective than methyl α-mannoside (155). Although classified under the general term "mannose-specific," the type 1 fimbrial lectins of different genera may thus differ in their sugar specificity. Within a given genus, however, all species and strains tested exhibit a similar specificity.

In general, it appears that mannose-specific bacterial lectins preferentially bind structures found in short oligomannose chains and in hybrid units of N-linked glycoproteins. Indeed, mammalian cells treated with swainsonine (an inhibitor of the processing of N-linked oligosaccharide units of glycoproteins), or ricin-resistant mutants of baby hamster kidney cells (which express increased levels of N-linked oligomannose or hybrid type oligosaccharides) bind increased numbers of mannose-specific *E. coli* (156, 157).

Mannose-specific bacteria bind avidly through their surface lectins to various types of phagocytic cell, such as mouse peritoneal macrophages and human neutrophils (158, 159). Binding elicits a burst of metabolic activity in the phagocytes, including induction of chemiluminescence (160) and of protein iodination (161). It is often followed by ingestion and death of the bacteria, a sequence of steps characteristic for the phagocytosis of opsonized bacteria (162, 163). None of these events occurs in the presence of methyl α-mannoside or mannose, sugars which minimally affect immune phagocytosis (158), or when the bacteria are devoid of type 1 fimbriae (162).

Escherichia Coli *Type P*

Agglutination and adherence reactions of many *E. coli* strains are not inhibited by mannose. Some of these agglutinate human erythrocytes carrying the P blood group antigen, but not those from individuals lacking this antigen (i.e. of the p⁻ phenotype). Such bacteria produce fimbriae that are designated as "type P fimbriae" or "Pap pili" (pili associated with pyelonephritis) (164). Bacteria with type P fimbriae also undergo phase variation (143, 165). Purified type P fimbriae retain the hemagglutinating activity and binding specificity expressed by the intact bacteria (166). They are morphologically similar to type 1 fimbriae, but are immunologically and structurally distinct from the latter. The Pap genes encoding formation and expression of the type P fimbriae of *E. coli* J96 were localized to an 8.5-kilobase chromosomal DNA fragment, which codes for at least 8 proteins (M_r 13,000–81,000) (167, 168). The amino acid sequence of the major type P fimbrial subunit of M_r 19,500 has been deduced from the sequence of the corresponding gene (169). Recent evidence suggests that a minor component of type P fimbriae, and not the major one, may carry the carbohydrate binding site of the fimbriae (170).

The best inhibitors of agglutination of human erythrocytes by type P fimbriated *E. coli* and of adhesion of these bacteria to cells are compounds containing the disaccharide Galα4Gal (164, 171, 171a). The receptor for these bacteria on epithelial cells is presumably globotetraosylceramide, a glycolipid containing the Galα4Gal group (164, 166, 172).

Other Escherichia Coli

The receptor for the mannose-resistant *E. coli* K99 fimbrial lectin has been isolated from equine erythrocytes and characterized as Neu5Gcα2→3Galβ-4Glcβ1-ceramide (173). From hemagglutination inhibition studies using human erythrocytes, it was concluded that several strains of *E. coli* are specific for *N*-acetylneuraminyl α2→3 galactosides (174).

The pyelonephritigenic *E. coli* strain 1H 11165 specifically agglutinates human erythrocytes carrying the M blood group antigen, which is located in the NH₂-terminal region of glycophorin A (175). Glycophorin A from type M cells, but not from type N cells, bound to the bacteria. The most efficient inhibitor of binding is the NH₂-terminal glycooctapeptide from glycophorin A^M. This peptide contains the essential serine residue and the alkali-labile oligosaccharides, both of which are recognized by the bacterium.

Actinomyces

Most human strains of *Actinomyces viscosus* and of *Actinomyces naeslundii* carry fimbriae designated as type 2 (176, 177). Such strains coaggregate with certain streptococci and agglutinate sialidase-treated erythrocytes, both reactions being inhibited by galactose, methyl β-galactoside, and lactose (178). Although isolated type 2 fimbriae alone do not agglutinate sialidase-treated erythrocytes, lactose-specific agglutination occurs when these cells are incubated with multivalent complexes, formed by cross-linking the fimbriae with the specific antibody. Type 2 fimbriae should therefore be considered as lectinlike substances.

The most effective disaccharide inhibitor of *A. viscosus* is Galβ3GalNAc (at least 10-fold more active than lactose) (179). This finding, in conjunction with experiments on the effects of lectins on the binding of actinomyces to sialidase-treated monolayers of cultured epithelial cells, suggests that the receptor for actinomyces on the epithelial cells is most likely the O-linked disaccharide Galβ3GalNAc (180). The nature of the receptor for the actinomyces on the streptococcal cells remains to be elucidated.

Mycoplasma

Several mycoplasma strains agglutinate erythrocytes and bind to surfaces of eukaryotic cells via receptors that are sensitive to sialidase (181, 182). Attempts to isolate lectins from these organisms have not been successful.

Glycophorin binds readily to *Mycoplasma gallisepticum,* while asialoglycophorin binds poorly (181, 183). Monosaccharides, including *N*-acetylneuraminic acid, are not inhibitory in any of the systems tested. Clustering of sialic acid residues linked either α2→3 to galactose or α2→6 to *N*-acetylgalactosamine, as found in glycophorin, has been proposed to be required for effective binding to *M. gallisepticum.*

Mycoplasma pneumoniae is specific for *N*- acetylneuraminic acid attached by α2→3 linkage to the terminal galactose residue of the poly-*N*-acetyllactosamine sequence of blood type I/i antigen (184, 185). This conclusion is based in part on measurements of the adherence to sheet cultures of the organism of sialidase-treated human erythrocytes that have been resialylated by specific sialyltransferases (184). Highest levels of adherence (70-fold higher compared to desialylated erythrocytes) are observed with erythrocytes having the sequence NeuAcα2→3Galβ1→4(or3)GlcNAc on their surfaces. Also, oligosaccharides containing sialic acid linked α2→3 to galactose are more inhibitory than the corresponding α2→6 isomers. Submaxillary mucins known to contain sialic acid linked α2→6 to *N*-acetylgalactosamine are not inhibitory. In its preference for sialic acid linked α2→3 to galactose residues, *M. pneumoniae* differs from *M. gallisepticum.* Minor sialylated oligosaccharides of the type carried by glycoprotein bands 3 and 4.5 and by glycolipids appear to be the main receptors for *M. pneumoniae* on human erythrocytes.

BIOLOGICAL ROLES

There is considerable support, but little solid evidence, for the belief that lectins function primarily as recognition molecules. This function may be expressed differently in different organisms and also in different organs or tissues of the same organism (see Table 3).

Plant Lectins

In plants, two proposed functions of lectins are currently attracting most attention: (*a*) as mediators of symbiosis between plants and microorganisms; and (*b*) in protection of plants against phytopathogens. The hypothesis that recognition of rhizobia by lectins of the plant host may account for the specificity in the initiation of nitrogen-fixing symbiosis (190, 191) has stimulated numerous investigations (16, 192, 193); however, most of the supporting evidence is indirect, and the hypothesis continues to be the subject of debate. The controversy centers mainly on two points: first, whether lectins and their appropriate receptors are present at the same time and place at which nodulation by the bacteria occurs; and second, whether there is a correlation between the ability of the host lectin to recognize a specific strain of Rhizobium and the nodulating ability of the strain. At present the only lectin isolated from roots that can bind to a specific nodulating strain of Rhizobium is trifoliin of white clover (194). This lectin has been suggested to reversibly cross-bridge receptors on the root hair cell wall with bacterial capsular polysaccharides and/or lipopolysaccharides as a prelude to nodulation (192).

Table 3 Roles of lectins in nature[a]

Plants	Attachment of nitrogen-fixing bacteria to legumes
	Protection against phytopathogens
Animals	Endocytosis and intracellular translocation of glycoproteins
	Regulation of cell migration and adhesion
	Recognition determinants in nonimmune phagocytosis (159)
	Binding of bacteria to epithelial cells (186)
Microorganisms	Attachment of bacteria and parasites [e.g. amoeba (187, 188) and *Plasmodium* (189)] to host cells
	Recognition determinants in nonimmune phagocytosis
	Recognition determinants in cell adhesion of slime molds

[a]References are given only for roles not discussed in text.

The proposal that lectins may be involved in the defense of plants against fungal, bacterial, and viral pathogens during germination and early growth of the seedlings is supported primarily by two lines of evidence: (*a*) the binding of lectins to various fungi and their ability to inhibit fungal growth and germination (195, 196); and (*b*) the presence of lectins at the potential site of invasion by the infectious agents (197). It is less clear whether lectins serve in a similar defense capacity in mature plant tissues.

Animal Lectins

Membrane lectins are thought to mediate the binding of soluble extracellular and intracellular glycoproteins as well as of cells. The classical examples are the binding of asialoglycoproteins by a galactose-specific lectin (receptor) on mammalian liver cells, and of asialo-agalactoglycoproteins by mannose/*N*-acetylglucosamine–specific lectins (receptors) on avian hepatocytes; both interactions are probably key steps in the removal of these glycoproteins from the circulatory system (118). Another example is the pinocytosis of glycoproteins with terminal nonreducing mannose and/or *N*-acetylglucosamine residues by macrophages (198). This uptake is mediated by a macrophage surface lectin specific for mannose and *N*-acetylglucosamine. It is likely that the same lectin is also responsible for the binding and phagocytosis of yeasts and zymosan (159). The mannose-6-phosphate–specific lectin mediates the targeting of hydrolytic enzymes to the lysosomes (77, 199). Galactose-specific lectins present on various human and murine tumors (122, 200, 201) were suggested to influence the pathogenesis of cancer metastasis by promoting the formation of tumor cell aggregates (emboli) in the circulation and their adhesion to the endothelial layer of the capillaries.

A common function proposed for the soluble vertebrate lectins is to bind to the complementary glycoconjugates on and around the cells that release them (117). This proposal is based on the finding that these proteins, which are initially concentrated inside cells, are ultimately found extracellularly (202). For example, chicken lactose lectin I, which is concentrated intracellularly in developing muscle, becomes extracellular with maturation (203). Likewise, a rat β-galactoside binding lectin in lung is concentrated in elastic fibers, a specialized form of extracellular matrix (204).

As mentioned, humoral lectins are ubiquitous within invertebrates; since these taxa lack immunoglobulins, the possibility has been raised that the humoral lectins might be their functional analogs. Although there is some evidence that invertebrate hemolymph lectins might function as opsonins (205), this is far from being certain (206). The main argument against the possible involvement of lectins in defense functions is the apparent lack of diversity of their carbohydrate specificities, which are directed mainly towards sialic acids.

Lectins of Microorganisms

Bacterial surface lectins and lectinlike substances seem to be involved in the initiation of infection by mediating bacterial adherence to epithelial cells—for example in the urinary and gastrointestinal tracts (141, 207). This has been best documented for *E. coli* carrying type 1 or type P fimbriae, and for type 1 fimbriated *K. pneumoniae*. Thus, type 1 fimbriated strains are considerably more infective than their isogenic nonfimbriated counterparts (208–210). Sugars that inhibit binding of the bacteria to epithelial cells in vitro (208, 211, 212), as well as antibodies to the lectins (162, 213–215) or to the lectin receptors (215), decrease significantly the rate of urinary tract infection in experimental animals. The pattern of distribution on oral epithelial surfaces of actinomyces carrying the galactose (lactose)-specific lectins supports the assumption that these lectins are the principal mediators of adherence, colonization, and establishment of specific microbial communities in oral cavities by binding to streptococci and to epithelial cells (216).

Nonimmune phagocytosis mediated by bacterial surface lectins (see *Escherichia Coli Type 1* section above) may be of clinical relevance in nonimmune hosts and in tissues, such as the renal medulla, where opsonic activity is poor (159, 162).

In the case of slime molds, recent evidence indicates that discoidin I, an endogenous lectin from *Dictyostelium discoideum* synthesized as the mold aggregates, functions to promote cell-substratum attachment and ordered cell migration during morphogenesis (217), rather than cell-cell adhesion as originally proposed. Thus, two mutants of *D. discoideum* with low levels of discoidin I failed to form streams and to spread on plastic but were only partially impaired in their ability to form aggregates.

APPLICATIONS

The earliest applications of lectins, still in wide use, were for blood typing and for mitogenic stimulation of lymphocytes (218, 219). The ready availability of a large number of lectins with different carbohydrate specificities has led to their extensive utilization as reagents for the study of simple and complex carbohydrates in solution and on cell surfaces (7, 218), for the identification and separation of cells (220), and for the selection of lectin-resistant mutants of animal cells with altered glycosylation patterns (221, 222). With a few exceptions, in which lectins of nonplant origin were employed [e.g. those of *Helix pomatia* (223), *Limulus polyphemus* (224), *Homarus americanus* (225), and *Limax flavus* (226, 227)], such studies are done with plant lectins.

Glycoconjugates in Solution

Radioactively labeled lectins and various lectin conjugates (e.g. with enzymes, biotin, fluorescent dyes, or colloidal gold) serve as specific and sensitive reagents for the detection of glycoproteins separated on polyacrylamide gels, either directly or after blotting (228–230). The usefulness of this method can be greatly increased by in situ chemical or enzymatic modifications of the separated glycoproteins (228, 231). Protocols for the isolation of glycoproteins now routinely include affinity chromatography on immobilized lectins as a major purification step. Recent examples of lectin-purified glycoproteins are: the receptors for insulin (232) and epidermal growth factor (233, 234); glycocalicin, the predominant glycoprotein of the human platelet membrane (235); C2, the second component of human complement (236); the sulfated glycoproteins of calf thyroid plasma membrane (237); and the major plasma membrane glycoprotein of Ehrlich ascites cells (238).

The high resolving power of lectins permits separation of closely related compounds, such as variants of glycoproteins that differ in their glycosylation pattern, or of glycopeptides and oligosaccharides that differ in their structure to a small extent only. Chromatography on concanavalin A, in addition to serving as a standard procedure to demonstrate the presence of oligomannose units, is employed to separate compounds containing dibranched complex oligosaccharides from those with more highly branched structures (81). Behavior on columns of L-PHA provides information relative to the branching pattern of tribranched complex oligosaccharides (84), while chromatography on lentil or pea lectin serves to demonstrate the presence of L-fucose linked to the chitobiose unit of the core (239). Terminal nonreducing α-linked galactose can be detected by *G. simplicifolia* I-B$_4$ lectin (238).

Two distinct subpopulations of the mammalian β-adrenergic receptors, one containing oligomannose units and the other with complex type carbohydrates, have been obtained by affinity chromatography on concanavalin A and wheat germ agglutinin, respectively (240). Affinity chromatography on a series of lectins of the common α-subunit of the pituitary glycoprotein hormones, isolated from normal human serum and from patients with pituitary tumors, revealed core fucosylation and increased branching in the tumor glycoproteins (241). Ovalbumin, which has a single carbohydrate attachment site at Asn 239, has been separated by successive affinity chromatography on concanavalin A and wheat germ agglutinin into eight subfractions; three of these were homogenous with respect to their carbohydrate and did not exhibit the microheterogeneity characteristic of glycoproteins (242). In a related application, two biosynthetic intermediates of ovalbumin, which differ in the extent of carbohydrate processing, were separated by affinity chromatography on concanavalin A (243).

Lectin chromatography of glycopeptides and of oligosaccharides has been employed to analyze changes in glycan branching and sialylation of the Thy-1 antigen during normal differentiation of mouse lymphocytes (244); to demonstrate structural changes in the carbohydrate chains of human thyroglobulin upon malignant transformation of the human thyroid gland (245); and to study the carbohydrate moieties of various glycoproteins, such as the epidermal growth factor (246) and low-density lipoprotein receptors (247), the murine major histocompatibility antigen (248), band 3 from adult human erythrocytes (249), human amniotic fluid fibronectin (250), and surface glycoproteins of a ricin-resistant mutant of baby hamster kidney cells (251). The effect of bisecting N-acetylglucosamine on the enzymatic transfer of galactose to various biantennary glycopeptides was examined by separation of the products on immobilized lectins (252). Sugar nucleotides have also been fractionated on lectins, as illustrated by the separation of UDP-Gal from UDP-Glc and of UDP-GalNAc from UDP-GlcNAc by high performance liquid chromatography on immobilized *Ricinus communis* agglutinin (253).

Glycoconjugates on Cells and Organelles

There are numerous reports on the application of lectins in histochemical and cytochemical studies (218, 254–256). Changes in lectin binding patterns have been observed during embryonic differentiation (257, 258), cell maturation (259), aging (260), metaplastic alterations (261, 262), malignant transformation (263–267), and many other pathological conditions, such as lysosomal storage diseases (268), inflammatory bowel disease (269), psoriasis (270), and pneumococcal meningitis (271).

Staining with lectins is of use in the identification of various classes of cell. In particular, peanut agglutinin serves for the identification of immature thymocytes of mice and man (220), of germinal center cells (272), and of human monocytes (273). The same lectin may be used as a histochemical marker of the epidermis in the early amphibian embryo (274) and of human granular cells regardless of their location (275). *U. europaeus* lectin I binds specifically to vascular endothelium (276) and can thus facilitate the detection of vascular invasion by tumor cells. Lectins differentiate between cultured tumor cells of high and low metastatic potential (277). They may also serve as an aid in the investigation and classification of lymphocytic proliferative diseases (278).

Experiments with lectins have helped to localize the intracellular sites of protein glycosylation. Using concanavalin A, it was found that the lipid-linked oligosaccharides $Man_{3-5}GlcNAc_2$ are located on the cytoplasmic side of microsomes from cultured fibroblasts, while the lipid linked $Man_{6-9}GlcNAc_2$ and $Glc_{1-3}Man_9GlcNAc_2$ are facing the lumen of the endoplasmic reticulum (279). The $Man_5GlcNAc_2$-pyrophosphoryl-dolichol seems therefore to be synthesized on the cytoplasmic side of the endoplasmic reticulum membrane and then translocated to the luminal side, where it is converted to $Glc_3Man_9GlcNAc_2$-pyrophosphoryl-dolichol which serves as the donor in peptide glycosylation. Staining of IgM-secreting myeloma cells with concanavalin A and wheat germ agglutinin led to the conclusion that the stack of Golgi cisternae is composed of two subcompartments—a proximal one rich in immature N-glycosyl units (of the oligomannose type) and a distal one, containing mature oligosaccharides (of the N-acetyllactosamine type)—and that the axis of trans-Golgi transport runs from the proximal to the distal face (280). These results support the hypothesis that the maturation of N-glycoproteins in the Golgi is a vectorial process (281).

Labeling of *Saccharomyces cerevisiae* protoplast membranes with ferritin derivatives of concanavalin A and wheat germ agglutinin has shown that the yeast chitin synthase receives N-acetylglucosamine residues from UDP-GlcNAc at the cytoplasmic face of the membrane and transfers them vectorially to a growing chain of chitin that is concomitantly extruded from the protoplasts (282).

PHA was shown to bind to the T lymphocyte antigen receptor on human T leukemia cell lines and tonsil lymphocytes (283). Concanavalin A and wheat germ agglutinin probably also bind to the same antigen receptor, whereas *H. pomatia* agglutinin, which is not mitogenic, does not. These results are in line with earlier suggestions that all lymphocyte mitogens share a common receptor (284).

Mapping Neuronal Pathways

Lectins conjugated to horseradish peroxidase have proved to be useful markers in mapping central neuronal pathways, since the conjugates are taken up by neurons and transported within the axons (285). In the case of wheat germ agglutinin, transport is in both anterograde and retrograde directions, as well as transneuronal, while L-PHA (286, 287) and ricin (288) conjugates are only transported in the anterograde and retrograde directions, respectively. Neither E-PHA nor peanut agglutinin is transported effectively in either direction (286). Uptake of the lectin conjugates is apparently receptor-mediated, since in the few cases tested, injection of the conjugate together with the specific sugar or free lectin diminished the subsequent labeling of neuronal projections (289–291). When ricin conjugates were used as tracers, morphological lesions and cell death in neurons were observed (288, 292). Such "suicide transport" offers a new approach for tackling neurobiological problems, for example by denervating target organs in the peripheral nervous system.

Typing of Bacteria

In recent years, lectins have been shown to distinguish between microbial species (293). For example, *Neisseria gonnorrhoeae* can be differentiated from other *Neisseriae* and related bacteria by its agglutination with wheat germ agglutinin (294), and *Bacillus anthracis* can be identified with the aid of soybean agglutinin (295). Ten out of 28 serovars of *Bacillus thurginensis* were individualized according to their ability to interact with various lectins (296). Lectins discriminate between pathogenic and nonpathogenic strains of *Trypanosoma cruzi* and between different morphological stages, i.e. amastigotes and promastigotes, of *Leishmania donovani* (297).

Cell Separation

The selective interaction of peanut agglutinin with immature thymocytes is the basis of the widely used method for the separation of these cells from mature thymocytes by selective agglutination and other techniques (e.g. affinity chromatography) (298). Recently this lectin has been employed to separate germ cells from somatic cells in mouse testis (299). Soybean agglutinin is used for the separation of mouse T and B splenocytes (300). Sequential agglutination of mouse splenocytes with peanut agglutinin and soybean agglutinin yielded a stem-cell-enriched fraction devoid of graft versus host activity and capable of reconstituting lethally irradiated allogeneic mice (301). An important outgrowth of this finding is the use of soybean agglutinin for the removal from human bone marrow of mature T cells and the isolation of a population enriched in hemopoietic stem cells, suitable for transplantation into haploidentical recipients (302). The method has been successfully employed in the treatment of more than 50 children born with severe combined immune deficiency (303–305) and is now being tested in cases of leukemia (306, 307).

In a different approach, hybrid molecules, made by covalent linking of ricin to monoclonal anti–T cell antibodies, were used to eliminate mature T cells from human bone marrow (308). The remaining cells were successfully transplanted into suitably conditioned, HLA-matched recipients with acute lymphocytic leukemia.

ACKNOWLEDGMENT

We wish to thank Mrs. Dvorah Ochert for her invaluable editorial assistance.

Literature Cited

1. Lis, H., Sharon, N. 1973. *Ann. Rev. Biochem.* 42:541–74
2. Sharon, N., Lis, H. 1972. *Science* 177:949–59
3. Goldstein, I. J., Hughes, R. C., Monsigny, M., Osawa, T., Sharon, N. 1980. *Nature* 285:66
4. Stillmark, H. 1888. *Inaug. Diss., Dorpat*
5. Barondes, S. H. 1981. *Ann. Rev. Biochem.* 50:207–31
6. Liener, I. E., Sharon, N., Goldstein, I. J., eds. 1986. *The Lectins: Properties, Functions and Applications in Biology*

and Medicine. New York: Academic. In press
7. Lis, H., Sharon, N. 1984. In *Biology of Carbohydrates,* ed. V. Ginsburg, P. W. Robbins, Vol. 2, pp. 1–85. New York: Wiley
8. Lis, H., Sharon, N. 1981. In *The Biochemistry of Plants: A Comprehensive Treatise,* Vol. VI: *Proteins and Nucleic Acids,* ed. A. Marcus, pp. 371–447. Academic
9. Etzler, M. E. 1985. *Ann. Rev. Plant Physiol.* 36:209–34

10. Goldstein, I. J., Etzler, M. E., eds. 1983. *Chemical Taxonomy, Molecular Biology and Function of Plant Lectins. Progress in Clinical and Biological Research,* Vol. 138. New York: Liss. 298 pp.
11. Bøg-Hansen, T. C., ed. 1981–1985. *Lectins: Biology, Biochemistry, Clinical Biochemistry,* Vol. 1 (1981); Vol. 2 (1982); Vol. 3 (1983); Vol. 4 (1985). Berlin/New York: de Gruyter
12. Mirelman, D., ed. 1986. *Microbial Lectins and Agglutinins.* New York: Wiley. In press
13. Olden, K., Parent, J. B., eds. 1986. *Vertebrate Lectins: Recent Research.* New York: Van Nostrand Reinhold. In press
14. Cohen, E., ed. 1984. *Recognition Proteins, Receptors and Probes: Invertebrates. Progress in Clinical and Biological Research,* Vol. 157. New York: Liss. 207 pp.
15. Lis, H., Sharon, N. 1981. *J. Chromatogr.* 215:361–72
16. Etzler, M. E. 1986. See Ref. 6
17. Pusztai, A., Croy, R. R. D., Grant, G., Stewart, J. C. 1983. In *Seed Proteins,* ed. J. Doussant, J. Morre, J. Vaughan, pp. 53–82. New York: Academic
18. Stinissen, H. M., Peumans, W. J., Carlier, A. R. 1983. *Planta* 159:105–11
19. Lalonde, L., Fountain, D. W., Kermode, A., Ouellette, F. B., Scott, K., et al. 1984. *Can. J. Bot.* 62:1671–77
20. Goldstein, I. J., Hayes, C. E. 1978. *Adv. Carbohydr. Chem. Biochem.* 35:127–340
21. Shibata, S., Goldstein, I. J., Baker, D. A. 1982. *J. Biol. Chem.* 257:9324–29
22. Konami, Y., Yamamoto, K., Tsuji, T., Matsumoto, I., Osawa, T. 1983. *Hoppe-Seyler's Z. Physiol. Chem.* 364:397–405
23. Baumann, C., Rüdiger, H., Strosberg, A. D. 1979. *FEBS Lett.* 102:216–18
24. Tollefsen, S. E., Kornfeld, R. 1983. *J. Biol. Chem.* 258:5165–71
25. Franz, H., Ziska, P., Kindt, A. 1981. *Biochem. J.* 195:481–84
26. Peumans, W. J., Stinissen, H. M., Carlier, A. R. 1982. *Planta* 154:562–67
27. Peumans, W. J., Stinissen, H. M., Carlier, A. R. 1982. *Planta* 154:568–72
28. Peumans, W. J., Stinissen, H. M. 1983. See Ref. 10, pp. 99–116
29. De Boeck, H., Lis, H., van Tilbeurgh, H., Sharon, N., Loontiens, F. G. 1984. *J. Biol. Chem.* 259:7067–74
30. Neurohr, K. J., Young, N. M., Mantsch, H. H. 1980. *J. Biol. Chem.* 255:9205–9
31. Roberts, D. D., Goldstein, I. J. 1984. *Arch. Biochem. Biophys.* 230:316–20
32. Crowley, J. F., Goldstein, I. J., Arnarp, J., Lönngren, J. 1984. *Arch. Biochem. Biophys.* 231:524–33
33. Appukuttan, P. S., Basu, D. 1985. *FEBS Lett.* 180:331–34
34. Zentz, C., Frenoy, J.-P., Bourrillon, R. 1979. *Biochimie* 61:1–6
35. Houston, L. L., Dooley, T. P. 1982. *J. Biol. Chem.* 257:4147–51
36. Etzler, M. E., Gupta, S., Borrebaeck, C. 1981. *J. Biol. Chem.* 256:2367–70
37. Cunningham, B. A. 1975. *Pure Appl. Chem.* 41:31–46
38. Hopp, T. P., Hemperly, J. J., Cunningham, B. A. 1982. *J. Biol. Chem.* 257:4473–83
39. Foriers, A., Lebrun, E., Van Rapenbusch, R., De Neve, R., Strosberg, A. D. 1981. *J. Biol. Chem.* 256:5550–60
40. Hemperly, J. J., Cunningham, B. A. 1983. *Trends. Biochem. Sci.* 8:100–2
41. Kouchalakos, R. N., Bates, O. J., Bradshaw, R. A., Hapner, K. D. 1984. *Biochemistry* 23:1824–30
42. Richardson, M., Campos, F. D. A. P., Moreira, R. A., Ainouz, I. L., Begbie, R., et al. 1984. *Eur. J. Biochem.* 144:101–11
43. Richardson, M., Rougé, P., Sousa-Cavada, B., Yarwood, A. 1984. *FEBS Lett.* 175:76–81
44. Yarwood, A., Richardson, M., Sousa-Cavada, B., Rougé, P. 1985. *FEBS Lett.* 184:104–9
45. Lauwereys, M., Foriers, A., Sharon, N., Strosberg, A. D. 1985. *FEBS Lett.* 181:241–44
46. Baumann, C. M., Strosberg, A. D., Rüdiger, H. 1982. *Eur. J. Biochem.* 122:105–10
47. Kolberg, J., Michaelsen, T. E., Sletten, K. 1980. *FEBS Lett.* 117:281–83
48. Vodkin, L. O., Rhodes, P. R., Goldberg, R. B. 1983. *Cell* 34:1023–31
49. Higgins, T. J. V., Chandler, P. M., Zurawski, G., Button, S. C., Spencer, D. 1983. *J. Biol. Chem.* 258:9544–49
50. Hoffman, L. M., Donaldson, D. D. 1985. *EMBO J.* 4:883–89
51. Strosberg, A. D., Lauwereys, M., Foriers, A. 1983. See Ref. 10, pp. 7–20
52. Wright, C. S., Gavilanes, F., Peterson, D. L. 1984. *Biochemistry* 23:280–87
53. Nagata, Y. 1985. *Agric. Biol. Chem.* 49:535–36
54. Dorland, L., van Halbeek, H., Vliegenthart, J. F. G., Lis, H., Sharon, N. 1981. *J. Biol. Chem.* 256:7708–11
55. Sharon, N., Lis, H. 1982. In *The Proteins,* ed. H. Neurath, R. L. Hill. 5:1–144. New York: Academic. 3rd ed.
56. Ohtani, K., Misaki, A. 1980. *Carbohydr. Res.* 87:275–85
57. Ishihara, H., Takahashi, N., Oguri, S., Tejima, S. 1979. *J. Biol. Chem.* 254:10715–19

58. Prigent, M.-J., Montreuil, J., Strecker, G. 1984. *Carbohydr. Res.* 131:83–92
59. Kaladas, P. M., Goldberg, R., Poretz, R. D. 1983. *Mol. Immunol.* 20:727–35
59a. Ashford, D., Desai, N. N., Allen, A. K., Neuberger, A., O'Neil, M. A., Selvendran, R. L. 1982. *Biochem. J.* 201:199–208
60. Desai, N. N., Allen, A. K., Neuberger, A. 1983. *Biochem. J.* 211:273–76
61. Kilpatrick, D. C., Graham, C., Urbaniak, S. J., Jeffree, C. E., Allen, A. K. 1984. *Biochem. J.* 220:843–47
62. Goldberg, R. B., Hoschek, G., Vodkin, L. O. 1983. *Cell* 33:465–75
63. Hoffman, L. M., Ma, Y., Barker, R. F. 1982. *Nucleic Acids Res.* 10:7819–28
64. Higgins, T. J. V., Chrispeels, M. J., Chandler, P. M., Spencer, D. 1983. *J. Biol. Chem.* 258:9550–52
65. Chrispeels, M. J. 1983. *Planta* 157:454–61
66. Lord, J. M. 1985. *Eur. J. Biochem.* 146:411–16
67. Hemperly, J. J., Mostov, K. E., Cunningham, B. A. 1982. *J. Biol. Chem.* 257:7903–9
68. Foriers, A., Wuilmart, C., Sharon, N., Strosberg, A. D. 1977. *Biochem. Biophys. Res. Commun.* 75:980–86
69. Carrington, D. M., Auffret, A., Hanke, D. E. 1985. *Nature* 313:64–67
70. Hendrix, R. W., Casjens, S. R. 1974. *Proc. Natl. Acad. Sci. USA* 71:1451–55
71. Butterworth, A. G., Lord, J. M. 1983. *Eur. J. Biochem.* 137:57–65
72. Lamb, F. I., Roberts, L. M., Lord, J. M. 1985. *Eur. J. Biochem.* 148:265–70
73. Vitale, A., Warner, T. G., Chrispeels, M. J. 1983. *Planta* 160:256–63
74. Vitale, A., Ceriotti, A., Bollini, R., Chrispeels, M. J. 1984. *Eur. J. Biochem.* 141:97–104
75. Vitale, A., Chrispeels, M. J. 1984. *J. Cell Biol.* 99:133–40
76. Kornfeld, R., Kornfeld, S. 1985. *Ann. Rev. Biochem.* 54:631–64
77. Hasilik, A., von Figura, K. 1986. *Ann. Rev. Biochem.* 55:
78. Stinissen, H. M., Peumans, W. J., Carlier, A. R. 1982. *Plant Mol. Biol.* 1:277–90
79. Stinissen, H. M., Peumans, W. J., Carlier, A. R. 1983. *Plant Mol. Biol.* 2:33–40
80. Goldstein, I. J., Poretz, R. D. 1986. See Ref. 6
81. Baenziger, J. U., Fiete, D. 1979. *J. Biol. Chem.* 254:2400–7
82. Ohyama, Y., Kasai, K., Nomoto, H., Inoue, Y. 1985. *J. Biol. Chem.* 260:6882–87
83. Baenziger, J. U., Fiete, D. 1979. *J. Biol. Chem.* 254:9795–99
84. Cummings, R. D., Kornfeld, S. 1982. *J. Biol. Chem.* 257:11230–34
85. Katagiri, Y., Yamamoto, K., Tsuji, T., Osawa, T. 1984. *Carbohydr. Res.* 129:257–65
86. Bøg-Hansen, T. C. 1983. In *Solid Phase Biochemistry: Analytical and Synthetic Aspects*, ed. W. H. Scouten, pp. 223–51. New York: Wiley
87. Hořejší, V. 1981. *Anal. Biochem.* 112:1–8
88. Loontiens, F. G., Clegg, R. M., Van Landschoot, A. 1983. *J. Biosci.* 5(Suppl. 1):105
89. Loontiens, F. G., De Boeck, H., Clegg, R. M. 1985. *J. Biosci.* 8:426–36
90. Baker, D. A., Sugii, S., Kabat, E. A., Ratcliffe, R. M., Hermentin, P., Lemieux, R. U. 1983. *Biochemistry* 22:2741–50
91. Iglesias, J. L., Lis, H., Sharon, N. 1982. *Eur. J. Biochem.* 123:247–52
92. Kaladas, P. M., Kabat, E. A., Iglesias, J. L., Lis, H., Sharon, N. 1982. *Arch. Biochem. Biophys.* 217:624–37
93. Yamashita, K., Hitoi, A., Kobata, A. 1983. *J. Biol. Chem.* 258:14753–55
94. Hammarström, S., Hammarström, M.-L., Sundblad, G., Arnarp, J., Lönngren, J. 1982. *Proc. Natl. Acad. Sci. USA* 79:1611–15
95. Duk, M., Lisowska, E., Kordowicz, M., Wasniowska, K. 1982. *Eur. J. Biochem.* 123:105–12
96. Prigent, M. J., Verezbencomo, V., Sinaÿ, P., Cartron, J. P. 1984. *Glycoconjugate J.* 1:73–80
97. Tollefsen, S. E., Kornfeld, R. 1983. *J. Biol. Chem.* 258:5172–76
98. Cummings, R. D., Kornfeld, S. 1984. *J. Biol. Chem.* 259:6253–60
99. De Boeck, H., Matta, K. L., Claeyssens, M., Sharon, N., Loontiens, F. G. 1983. *Eur. J. Biochem.* 131:453–60
100. De Boeck, H., Loontiens, F. G., Lis, H., Sharon, N. 1984. *Arch. Biochem. Biophys.* 234:297–304
101. Presant, C. A., Kornfeld, S. 1972. *J. Biol. Chem.* 247:6937–45
102. Hindsgaul, O., Norberg, T., Le Pendu, J., Lemieux, R. U. 1982. *Carbohydr. Res.* 109:109–42
103. Brisson, J.-R., Carver, J. P. 1983. *Can. J. Biochem. Cell Biol.* 61:1067–78
104. Carver, J. P., Brisson, J.-R. 1984. In *Biology of Carbohydrates*, ed. V. Ginsburg, P. W. Robbins, Vol. 2, pp. 289–331. New York: Wiley
105. Schachter, H., Narasimhan, S., Gleeson, P., Vella, G. 1983. *Can. J. Biochem. Cell Biol.* 61:1049–66

106. Brisson, J.-R., Carver, J. P. 1983. *Biochemistry* 22:3671–80, 3680–86
107. Carver, J. P., Mackenzie, A. E., Hardman, K. D. 1985. *Biopolymers* 24:49–63
107a. Brewer, F., Bhattacharyya, L., Brown, R. D. III, Koenig, S. H. 1985. *Biochem. Biophys. Res. Commun.* 127:1066–71
108. De Boeck, H., MacGregor, R. B., Clegg, R. M., Sharon, N., Loontiens, F. G. 1985. *Eur. J. Biochem.* 149:141–45
109. Clegg, R. M., Loontiens, F. G., Sharon, N., Jovin, T. M. 1983. *Biochemistry* 22:4797–804
109a. Loontiens, F. G. 1983. *FEBS Lett.* 162:193–96
109b. Hoebeke, J., Foriers, A., Schreiber, A. B., Strosberg, A. D. 1978. *Biochemistry* 17:5000–5
110. Gray, R. D., Glew, R. H. 1973. *J. Biol. Chem.* 248:7547–51
111. Sekharudu, Y. C., Rao, V. S. R. 1984. *Int. J. Biol. Macromol.* 6:337–47
112. Spohr, U., Hindsgaul, O., Lemieux, R. U. 1985. *Can J. Chem.* 63:2644–52
112a. Hindsgaul, O., Khare, D. P., Bach, M., Lemieux, R. U. 1985. *Can. J. Chem.* 63:2653–58
113. Kella, N. K. D., Roberts, D. D., Shafer, J. A., Goldstein, I. J. 1984. *J. Biol. Chem.* 259:4777–81
114. Roberts, D. D., Goldstein, I. J. 1983. See Ref. 10, pp. 131–41
115. Roberts, D. D., Goldstein, I. J. 1983. *Arch. Biochem. Biophys.* 224:479–84
116. Roberts, D. D., Goldstein, I. J. 1983. *J. Biol. Chem.* 258:13820–24
117. Barondes, S. H. 1984. *Science* 223:1259–64
118. Ashwell, G., Harford, J. 1982. *Ann. Rev. Biochem.* 51:531–54
119. Monsigny, M., Kieda, C., Roche, A. C. 1983. *Biol. Cell* 47:95–110
120. Drickamer, K. 1985. See Ref. 13
121. Stojanovic, D., Hughes, R. C. 1984. *Biol. Cell* 51:197–206
122. Raz, A., Lotan, R. 1981. *Cancer Res.* 41:3642–47
123. Roche, A.-C., Barzilay, M., Midoux, P., Junqua, S., Sharon, N., Monsigny, M. 1983. *J. Cell. Biochem.* 22:131–40
124. Monsigny, M., Roche, A.-C., Midoux, P. 1984. *Biol. Cell* 51:187–96
125. Teichberg, V. I., Silman, I., Beitsch, D. D., Resheff, G. 1975. *Proc. Natl. Acad. Sci. USA* 72:1383–87
126. Levi, G., Teichberg, V. I. 1981. *J. Biol. Chem.* 256:5735–40
127. Fitzgerald, J. E., Catt, J. W., Harrison, F. L. 1984. *Eur. J. Biochem.* 140:137–41
128. Levi, G., Teichberg, V. I. 1982. *FEBS Lett.* 148:145–48
129. Beyer, E. C., Zweig, S. E., Barondes, S. H. 1980. *J. Biol. Chem.* 255:4236–39
130. Gold, E. R., Balding, P. 1975. *Receptor-specific Proteins: Plant and Animal Lectins.* Amsterdam: Excerpta Medica. 440 pp.
131. Yeaton, R. W. 1981. *Dev. Comp. Immunol.* 5:391–402
132. Gilboa-Garber, N., Susswein, A. J., Mizrahi, L., Avichezer, D. 1985. *FEBS Lett.* 181:267–70
133. Cheng, T. C., Marchalonis, J. J., Vasta, G. R. 1984. See Ref. 14, pp. 1–15
134. Vasta, G. R., Cheng, T. C., Marchalonis, J. J. 1984. *Cell. Immunol.* 88:475–88
135. Barondes, S. H., Nowak, T. P. 1978. *Methods Enzymol.* 50:302–5
136. Dorai, D. T., Bachhawat, B. K., Bishayee, S., Kannan, K., Rao, D. R. 1981. *Arch. Biochem. Biophys.* 209:325–33
137. Abel, C. A., Campbell, P. A., Vander-Wall, J., Hartman, A. L. 1984. See Ref. 14, pp. 103–14
138. Baldo, B. A., Sawyer, W. H., Stick, R. V., Uhlenbruck, G. 1978. *Biochem. J.* 175:467–77
139. Miller, R. L., Collawn, J. F. Jr., Fish, W. W. 1982. *J. Biol. Chem.* 257:7574–80
140. Ravindranath, M. H., Higa, H. H., Cooper, E. L., Paulson, J. C. 1985. *J. Biol. Chem.* 260:8850–56
141. Sharon, N. 1986. See Ref. 6
142. Ofek, I., Mirelman, D., Sharon, N. 1977. *Nature* 265:623–25
143. Nowicki, B., Rhen, M., Väisänen-Rhen, V., Pere, A., Korhonen, T. K. 1984. *J. Bacteriol.* 160:691–95
144. Väisänen-Rhen, V. 1984. *Infect. Immun.* 46:401–7
145. Karch, H., Leying, H., Büscher, K.-H., Kroll, H.-P., Opferkuchi, W. 1985. *Infect. Immun.* 47:549–54
146. Brinton, C. C. Jr. 1965. *Trans. NY Acad. Sci.* 27:1003–54
147. Duguid, J. P., Old, D. C. 1980. See Ref. 207, pp. 184–217
148. Eisenstein, B. I. 1981. *Science* 214:337–39
149. Eshdat, Y., Silverblatt, F. J., Sharon, N. 1981. *J. Bacteriol.* 148:308–14
150. Klemm, P. 1984. *Eur. J. Biochem.* 143:395–99
151. Orndorff, P. E., Falkow, S. 1984. *J. Bacteriol.* 160:61–66
152. Dodd, D. C., Bassford, P. J. Jr., Eisenstein, B. I. 1984. *J. Bacteriol.* 159:1077–79
153. Dodd, D. C., Eisenstein, B. I. 1984. *J. Bacteriol.* 160:227–32

154. Firon, N., Ofek, I., Sharon, N. 1983. *Carbohydr. Res.* 120:235–49
155. Firon, N., Ofek, I., Sharon, N. 1984. *Infect. Immun.* 43:1088–90
156. Elbein, A. D., Pan, Y. T., Solf, R., Vosbeck, K. 1983. *J. Cell. Physiol.* 115:265–75
157. Firon, N., Duksin, D., Sharon, N. 1985. *FEMS Microbiol. Lett.* 27:161–65
158. Bar-Shavit, Z., Goldman, R., Ofek, I., Sharon, N., Mirelman, D. 1980. *Infect. Immun.* 29:417–24
159. Sharon, N. 1984. *Immunol. Today* 5: 143–47
160. Blumenstock, E., Jann, K. 1982. *Infect. Immun.* 35:264–69
161. Perry, A., Ofek, I., Silverblatt, F. J. 1983. *Infect. Immun.* 39:1334–45
162. Silverblatt, F. J., Cohen, L. S. 1979. *J. Clin. Invest.* 64:333–36
163. Öhman, L., Hed, J., Stendahl, O. 1982. *J. Infect. Dis.* 146:751–57
164. Svanborg-Eden, C., Leffler, H. 1986. See Ref. 12
165. Rhen, M., Mäkelä, P. H., Korhonen, T. K. 1983. *FEMS Microbiol. Lett.* 19:267–71
166. Korhonen, T. K., Vaisanen, V., Saxen, H., Hultberg, H., Svenson, S. B. 1982. *Infect. Immun.* 37:286–91
167. Hull, R. A., Gill, R. E., Hsu, P., Minshew, B. H., Falkow, S. 1981. *Infect. Immun.* 33:933–38
168. Normark, S., Lark, D., Hull, R., Norgren, M., Baga, M., et al. 1983. *Infect. Immun.* 41:942–49
169. Baga, M., Normark, S., Hardy, J., O'Hanley, P., Lark, D., et al. 1984. *J. Bacteriol.* 157:330–33
170. Uhlin, P. E., Norgren, M., Baga, M., Normark, S. 1985. *Proc. Natl. Acad. Sci. USA* 82:1800–4
171. Svanborg-Edén, C., Freter, R., Hagberg, L., Hull, R., Hull, S., et al. 1982. *Nature* 298:560–62
171a. Bock, K., Breimer, M. E., Brignole, A., Hansson, G. C., Karlsson, K.-A., et al. 1985. *J. Biol. Chem.* 260:8545–51
172. Leffler, H., Svanborg-Edén, C. 1981. *Infect. Immun.* 34:920–29
173. Smit, H., Gaastra, W., Kamerling, J. P., Vliegenthart, J. F. G., de Graaf, F. K. 1984. *Infect. Immun.* 46:578–84
174. Parkkinen, J., Finne, J., Achtman, M., Väisänen, V., Korhonen, T. K. 1983. *Biochem. Biophys. Res. Commun.* 111: 456–61
175. Jokinen, M., Ehnholm, C., Väisänen-Rhen, V., Korhonen, T., Pipkorn, R., et al. 1985. *Eur. J. Biochem.* 147:47–52
176. Cisar, J. O. 1982. In *Host-Parasite Interactions in Periodontal Diseases*, ed. R. J. Genco, S. E. Mergenhagen, pp. 121–31. Am. Soc. Microbiol.
177. Cisar, J. O. 1986. See Ref. 12
178. McIntire, F. C., Vatter, A. E., Baros, J., Arnold, J. 1978. *Infect. Immun.* 21:978–88
179. McIntire, F. C., Crosby, L. K., Barlow, J. J., Matta, K. L. 1983. *Infect. Immun.* 41:848–50
180. Brennan, M. J., Cisar, J. O., Vatter, A. E., Sandberg, A. L. 1984. *Infect. Immun.* 46:459–64
181. Kahane, I., Banai, M., Razin, S., Feldner, J. 1982. *Rev. Infect. Dis.* 4:S185–92
182. Razin, S. 1986. See Ref. 12
183. Glasgow, L. R., Hill, R. L. 1980. *Infect. Immun.* 30:353–61
184. Loomes, L. M., Uemura, K., Childs, R. A., Paulson, J. C., Rogers, G. N., et al. 1984. *Nature* 307:560–63
185. Loomes, L. M., Uemura, K., Feizi, T. 1985. *Infect. Immun.* 47:15–20
186. Ashkenazi, S., Mirelman, D. 1984. *Pediatr. Res.* 18:1366–71
187. Kobiler, D., Mirelman, D. 1980. *Infect. Immun.* 29:221–25
188. Ravdin, J. I., Murphy, C. F., Salata, R. A., Guerrant, R. L., Hewlett, E. L. 1985. *J. Infect. Dis.* 151:804–15
189. Vanderberg, J. P., Gupta, S. K., Schulman, S., Oppenheim, J. D., Furthmayr, H. 1985. *Infect. Immun.* 47:201–10
190. Hamblin, J., Kent, S. P. 1973. *Nature New Biol.* 245:28–30
191. Bohlool, B. B., Schmidt, E. L. 1974. *Science* 185:269–71
192. Dazzo, F. B., Sherwood, J. E. 1983. See Ref. 10, pp. 209–23
193. Pueppke, S. G. 1983. See Ref. 10, pp. 225–36
194. Dazzo, F. B., Truchet, G. L. 1983. *J. Membr. Biol.* 73:1–16
195. Mirelman, D., Galun, E., Sharon, N., Lotan, R. 1975. *Nature* 256:414–16
196. Barkai-Golan, R., Mirelman, D., Sharon, N. 1978. *Arch. Microbiol.* 116:119–24
197. Mishkind, M., Raikhel, N. V., Palevitz, B. A., Keegstra, K. 1982. *J. Cell Biol.* 92:753–64
198. Stahl, P. D., Wileman, T. E., Diment, S., Shepherd, V. L. 1984. *Biol. Cell* 51:215–18
199. Sahagian, G. G. 1984. *Biol. Cell* 51: 207–14
200. Lotan, R., Lotan, D., Raz, A. 1985. *Cancer Res.* 45:4349–53
201. Gabius, H.-J., Engelhardt, R., Cramer, F., Bätge, R., Nagel, G. A. 1985. *Cancer Res.* 45:253–57
202. Barondes, S. H., Cerra, R. F., Cooper,

D. N. W., Haywood-Reid, P. L., Roberson, M. M. 1984. *Biol. Cell* 51:165–72
203. Barondes, S. H., Haywood-Reid, P. L. 1981. *J. Cell Biol.* 91:568–72
204. Cerra, R. F., Haywood-Reid, P. L., Barondes, S. H. 1984. *J. Cell Biol.* 98:1580–89
205. Renwrantz, L., Stahmer, A. 1983. *J. Comp. Physiol.* 149:535
206. Bayne, C. J., Moore, M. N., Carefoot, T. H., Thompson, R. J. 1979. *J. Invertebr. Pathol.* 34:1
207. Beachey, E. H., ed. 1980. *Bacterial Adherence. Receptors and Recognition, Ser. B*, Vol. 6. London: Chapman & Hall
208. Fader, R. C., Davis, C. P. 1980. *Infect. Immun.* 30:554–61
209. Iwahi, T., Abe, Y., Nako, M., Imada, A., Tsuchiya, K. 1983. *Infect. Immun.* 39:1307–15
210. Hagberg, L., Engberg, I., Freter, R., Lam, J., Olling, S., Svanborg-Edén, C. 1983. *Infect. Immun.* 40:273–83
211. Aronson, M., Medalia, O., Schori, L., Mirelman, D., Sharon, N., Ofek, I. 1979. *J. Infect. Dis.* 139:329–32
212. Roberts, J. A., Kaack, B., Källenius, G., Möllby, R., Winberg, J., Svenson, S. B. 1984. *J. Urology* 131:163
213. Silverblatt, F. J., Weinstein, R., Rene, P. 1982. *Scand. J. Infect. Dis. Suppl.* 33:79
214. O'Hanley, P., Lark, D., Falkow, S., Schoolnik, G. 1985. *J. Clin. Invest.* 75:347–60
215. Abraham, S. N., Babu, J. P., Giampapa, C. S., Hasty, D. L., Simpson, W. A., Beachey, E. H. 1985. *Infect. Immun.* 48:625–28
216. Cisar, J. O., Brennan, M. J., Sandberg, A. L. 1985. In *Molecular Basis for Oral Microbial Adhesion*, ed. B. R. Rosan, S. E. Mergenhagen, pp. 159–69. Washington, DC: Am. Soc. Microbiol.
217. Springer, W. R., Cooper, D. N. W., Barondes, S. H. 1984. *Cell* 39:557–64
218. Lis, H., Sharon, N. 1986. See Ref. 6
219. Lis, H., Sharon, N. 1977. In *The Antigens*, ed. M. Sela, 4:429–529. New York: Academic
220. Sharon, N. 1983. *Adv. Immunol.* 34:213–98
221. Briles, E. B. 1982. *Int. Rev. Cytol.* 75:101–65
222. Stanley, P. 1983. *Methods Enzymol.* 96:157–84
223. Hellström, U., Hammarström, M.-L., Hammarström, S., Perlmann, P. 1984. *Methods Enzymol.* 106:153–68
224. Muresan, V., Sarras, M. P. Jr., Jamieson, J. D. 1982. *J. Histochem. Cytochem.* 30:947–55
225. Herron, L. R., Abel, C. A., VanderWall, J., Campbell, P. A. 1983. *Eur. J. Immunol.* 13:73–78
226. Roth, J., Lucocq, J. M., Charest, P. M. 1984. *J. Histochem. Cytochem.* 32:1167–76
227. Schulte, B. A., Spicer, S. S., Miller, R. L. 1984. *Histochem. J.* 16:1125–32
228. Gershoni, J. M. 1985. *Trends Biochem. Sci.* 10:103–6
229. Kijimoto-Ochiai, S., Katagiri, Y. J., Ochiai, H. 1985. *Anal. Biochem.* 147:222–29
230. Rohringer, R., Holden, D. W. 1985. *Anal. Biochem.* 144:118–27
231. Nicolson, G. L., Irimura, T. 1984. *Biol. Cell* 51:157–64
232. Hedo, J. A., Harrison, L. C., Roth, J. 1981. *Biochemistry* 20:3385–93
233. Cohen, S., Fava, R. A., Sawyer, S. T. 1982. *Proc. Natl. Acad. Sci. USA* 79:6237–41
234. Hock, R. A., Nexø, E., Hollenberg, M. D. 1980. *J. Biol. Chem.* 255:10737–43
235. Tsuji, T., Tsunehisa, S., Watanabe, Y., Yamamoto, K., Tohyama, H., Osawa, T. 1983. *J. Biol. Chem.* 258:6335–39
236. Schultz, D. R., Arnold, P. I. 1984. *Acta Pathol. Microbiol. Immunol. Scand.* 92(Suppl. 284):59–66
237. Edge, A. S. B., Spiro, R. G. 1984. *J. Biol. Chem.* 259:4710–13
238. Eckhardt, A. E., Goldstein, I. J. 1983. *Biochemistry* 22:5280–89
239. Kornfeld, K., Reitman, M. L., Kornfeld, R. 1981. *J. Biol. Chem.* 256:6633–40
240. Stiles, G. L., Benovic, J. L., Caron, M. G., Lefkowitz, R. J. 1984. *J. Biol. Chem.* 259:8655–63
241. Chapman, A. J., Gallagher, J. T., Beardwell, C. G., Shalet, S. M. 1984. *J. Endocrinol.* 103:111–16
242. Kato, Y., Iwase, H., Hotta, K. 1984. *Anal. Biochem.* 138:437–44
243. Kato, Y., Iwase, H., Hotta, K. 1984. *J. Biochem.* 95:455–63
244. Carlsson, S. R. 1985. *Biochem. J.* 226:519–25
245. Yamamoto, K., Tsuji, T., Tarutani, O., Osawa, T. 1984. *Eur. J. Biochem.* 143:133–44
246. Childs, R. A., Gregoriou, M., Scudder, P., Thorpe, S. J., Rees, A. R., Feizi, T. 1984. *EMBO J.* 3:2227–33
247. Cummings, R. D., Kornfeld, S., Schneider, W. J., Hobgood, K. K., Tolleshaug, H., et al. 1983. *J. Biol. Chem.* 258:15261–73
248. Swiedler, S. J., Freed, J. H., Tarentino, A. L., Plummer, T. H. Jr., Hart, G. W. 1985. *J. Biol. Chem.* 260:4046–54

249. Fukuda, M., Dell, A., Oates, J. E., Fukuda, M. N. 1984. *J. Biol. Chem.* 259:8260–73
250. Krusius, T., Fukuda, M., Dell, A., Ruoslahti, E. 1985. *J. Biol. Chem.* 260:4110–16
251. Gleeson, P. A., Feeney, J., Hughes, R. C. 1985. *Biochemistry* 24:493–503
252. Narasimhan, S., Freed, J. C., Schachter, H. 1985. *Biochemistry* 24:1694–700
253. Tokuda, M., Kamei, M., Yui, S., Koyama, F. 1985. *J. Chromatogr.* 323: 434–38
254. Schrevel, J., Gros, D., Monsigny, M., eds. 1981. *Prog. Histochem. Cytochem.* 14(2):1–269
255. Leathem, A. J. C., Atkins, N. J. 1983. In *Techniques in Immunocytochemistry*, ed. G. R. Bullock, P. Petrusz, 2:39–70. London: Academic
256. Roth, J. 1978. *Exp. Pathol.* (Suppl. 3):1–186
257. Watanabe, M., Muramatsu, T., Shirane, H., Ugai, K. 1981. *J. Histochem. Cytochem.* 29:779–90
258. Maylié-Pfenninger, M., Jamieson, J. D. 1980. *J. Cell Biol.* 86:96–103
259. Zieske, J. D., Bernstein, I. A. 1982. *J. Cell Biol.* 95:626–31
260. Bischof, W., Aumuller, G. 1982. *Prostate* 3:507–13
261. Orgad, U., Alroy, J., Ucci, A. A., Gavris, V. 1983. *Lab. Invest.* 48:65A
262. Wells, M., Taylor, M. J., Dixon, M. F. 1984. *J. Pathol.* 143:A3
263. Cooper, H. S. 1984. *Hum. Pathol.* 15:904–6
264. Fischer, J., Klein, P. J., Vierbuchen, M., Skutta, B., Uhlenbruck, G., Fischer, R. 1984. *J. Histochem. Cytochem.* 32: 681–89
265. Freeman, H. J. 1983. *J. Histochem. Cytochem.* 31:1241–45
266. Kluskens, L. F., Kluskens, J. L., Bibbo, M. 1984. *Am. J. Clin. Pathol.* 82:259–66
267. Walker, R. A. 1984. *J. Pathol.* 144:109–18
268. Alroy, J., Orgad, U., Ucci, A. A., Pereira, M. E. A. 1984. *J. Histochem. Cytochem.* 32:1280–84
269. Jacobs, L. R., Huber, P. W. 1985. *J. Clin. Invest.* 75:112–18
270. Kariniemi, A.-L., Holthöfer, H., Miettinen, A., Virtanen, I. 1983. *Br. J. Dermatol.* 109:523–29
271. Vierbuchen, M., Klein, P. J. 1984. *Lab. Invest.* 48:181–86
272. Reichert, R. A., Gallatin, W. M., Weismann, I. L., Butcher, E. C. 1983. *J. Exp. Med.* 157:813
273. Rosenberg, M., Chitayat, D., Tzehoval, E., Waxdal, M. J., Sharon, N. 1985. *J. Immunol. Methods* 81:7–13
274. Slack, J. M. W. 1985. *Cell* 41:237–47
275. Schwechheimer, K., Moller, P., Schnabel, P., Waldherr, R. 1983. *Virchows Arch. Pathol. Anat. Physiol.* 399:289–97
276. Miettinen, M., Holthöfer, H., Lehto, V.-P., Miettinen, A., Virtanen, I. 1983. *Am. J. Clin. Pathol.* 79:32–36
277. Fogel, M., Altevogt, P., Schirrmacher, V. 1983. *J. Exp. Med.* 157:371–76
278. Strauchen, J. A. 1984. *Am. J. Pathol.* 116:370–76
279. Snider, M. D., Rogers, O. C. 1984. *Cell* 36:753–61
280. Tartakoff, A. M., Vassalli, P. 1983. *J. Cell Biol.* 97:1243–48
281. Snider, M. D. 1984. In *Biology of Carbohydrates*, ed. V. Ginsburg, P. W. Robbins, 2:163–98. New York: Wiley
282. Cabib, E., Bowers, B., Roberts, R. L. 1983. *Proc. Natl. Acad. Sci. USA* 80: 3318–21
283. Chilson, O. P., Boylston, A. W., Crumpton, M. J. 1984. *EMBO J.* 3: 3239–45
284. Dillner-Centerlind, M., Axelsson, B., Hammarström, S., Hellström, U., Perlmann, P. 1980. *Eur. J. Immunol.* 10:434–42
285. Mesulam, M., ed. 1982. *Tracing Neural Connections with Horseradish Peroxidase*. New York: Wiley. 251 pp.
286. Gerfen, C. R., Sawchenko, P. E. 1984. *Brain Res.* 290:219–38
287. Luiten, P. G. M., terHorst, G. J., Karst, H., Steffens, A. B. 1985. *Brain Res.* 329:374–78
288. Harper, C. G., Gonatas, J. O., Mizutani, T., Gonatas, N. K. 1980. *Lab. Invest.* 42:396–404
289. Gonatas, N. K., Harper, C., Mizutani, T., Gonatas, J. O. 1979. *J. Histochem. Cytochem.* 27:728–34
290. Trojanowski, J. Q., Gonatas, J. O., Gonatas, N. K. 1981. *J. Neurocytol.* 10:441–56
291. Trojanowski, J. Q., Gonatas, J. O., Gonatas, N. K. 1982. *Brain Res.* 231: 33–50
292. Wiley, R. G., Blessing, W. W., Reis, D. J. 1982. *Science* 216:889–90
293. Doyle, R. J., Keller, K. F. 1984. *Eur. J. Clin. Microbiol.* 3:4–9
294. Doyle, R. J., Nedjat-Haiem, F., Keller, K. F., Frasch, C. E. 1984. *J. Clin. Microbiol.* 19:383–87
295. Graham, K., Keller, K., Ezzell, J., Doyle, R. 1984. *Eur. J. Clin. Microbiol.* 3:210–12
296. DeLucca, A. J. II. 1984. *Can. J. Microbiol.* 30:1100–4
297. De Miranda, S. I. K., Pereira, M. E. A. 1984. *Am. J. Trop. Med. Hyg.* 33:839–44

298. Reisner, Y., Sharon, N. 1984. *Methods Enzymol.* 108:168–79
299. Maekawa, M., Nishimune, Y. 1985. *Biol. Reprod.* 32:419–25
300. Reisner, Y., Ravid, A., Sharon, N. 1976. *Biochem. Biophys. Res. Commun.* 72:1585–91
301. Reisner, Y., Itzicovitch, L., Meshorer, A., Sharon, N. 1978. *Proc. Natl. Acad. Sci. USA* 75:2933–36
302. Reisner, Y., Kapoor, N., Kirkpatrick, D., Pollack, M. S., Cunningham-Rundles, S., et al. 1983. *Blood* 61:341–48
303. Friedrich, W., Vetter, U., Heymer, B., Reisner, Y., Goldmann, S. F., et al. 1984. *Lancet* 1:761–65
304. Friedrich, W., Goldmann, S. F., Ebell, W., Blütters-Sawatzki, R., Gaedicke, G., et al. 1985. *Eur. J. Pediatr.* 144:125–30
305. Zepp, F., Mannhardt, W., Düber, J., Gehler, J., Beetz, R., Schulte-Wissermann, H. 1984. *Exp. Hematol.* 12(Suppl. 15):89–90
306. Slavin, S., Waldmann, H., Or, R., Cividalli, G., Naparstek, E., et al. 1985. *Transplant. Proc.* 17:465–67
307. O'Reilly, R. J., Collins, N. H., Kernan, N., Brochstein, J., Dinsmore, R., et al. 1985. *Transplant. Proc.* 17:455–59
308. Filipovich, A. H., Youle, R. J., Neville, D. M., Vallera, D. A., Quinones, R. R., Kersey, J. H. 1984. *Lancet* 1:469–72

Ann. Rev. Biochem. 1986. 55:69–102

ARACHIDONIC ACID METABOLISM[1]

Philip Needleman, John Turk[2], Barbara A. Jakschik, Aubrey R. Morrison[2], and James B. Lefkowith

Department of Pharmacology, Washington University School of Medicine, 660 South Euclid Avenue, St. Louis, Missouri 63110

CONTENTS

[1]Investigations by the authors in areas discussed in this chapter have been supported in part by grants from the National Institutes of Health: (JT) AM34388; (BAJ) HL21874 and HL31922; (JL) NIH Clinical Investigator Award HL01313; (ARM) AM30542 and (PN) 5-P50-HL17646, HL14397, and HL20787.
[2]Department of Medicine

0066-4154/86/0701-0069$02.00

The synthesis and function of arachidonic acid metabolites remains the focus of extensive investigation. This subject has previously been reviewed in the *Annual Review of Biochemistry* in 1975, 1978, and 1983. The current effort is organized into a brief consideration of the products and function of the major synthetic enzymes, namely cyclooxygenase, the endoperoxide-dependent enzymes, 12-lipoxygenase, 15-lipoxygenase, 5-lipoxygenase, leukotriene synthetic enzymes, and cytochrome p450. In addition, attention is directed at the role and function of arachidonic acid metabolism in certain models of pathophysiological conditions. Finally, recent advances in the study of essential fatty acid deficiency, especially as related to arachidonic acid metabolism, warrant consideration.

CYCLOOXYGENASE PRODUCTS

Prostaglandin Endoperoxide Synthase

Arachidonic acid (1, 2) and certain other polyunsaturated fatty acids (3–5) may be transformed into prostaglandins (PG) by the enzyme prostaglandin endoperoxide synthase (PES) (1, 2, 6). The cyclooxygenase activity of PES inserts two molecules of oxygen into arachidonate to yield a 15-hydroperoxy-9,11-endoperoxide with a substituted cyclopentane ring (PGG_2). A peroxidase activity of PES reduces PGG_2 to its 15-hydroxy analogue (PGH_2). The cyclooxygenase and peroxidase activities of PES reside in a single protein (2, 6, 7). PES is membrane-associated but has been detergent-solubilized and purified to homogeneity (7, 8). The subunit molecular weight of the purified enzyme is about 72,000 (7, 8). Each subunit requires one molecule of heme for maximal catalytic activity (8). As recently reviewed (6), the specific activity of the enzyme ranges between 2.4 and 23.4 μmoles/min/mg at 24°C, and the K_m for both arachidonate and O_2 is about 5 μM. Immunocytochemical studies indicate that PES is contained in endoplasmic reticulum and nuclear membrane but not plasma membrane or mitochondrial membrane of cultured fibroblasts (9).

Cyclooxygenase-catalyzed fatty acid oxidation occurs slowly initially and later accelerates (10). Exogenous hydroperoxides eliminate this kinetic lag phase at concentrations (10^{-7} to 10^{-8} M) far below the K_m (10^{-5} M) of the peroxidase activity of PES (11). A continuous requirement for activator hydroperoxide is implied by termination of PES-catalyzed substrate oxidation upon addition of glutathione and glutathione peroxidase (10). Cyclooxygenase-catalyzed oxygen consumption declines to zero before complete consumption of fatty acid substrate, and a second burst of oxygen consumption occurs upon addition of fresh enzyme (10). Such self-deactivation of the cyclooxygenase appears to occur in intact cells as well as with purified enzyme preparations and may limit in vivo prostaglandin biosynthesis (12, 13).

The peroxidase activity of PES exhibits a K_m of about 20 μM for PGG_1, catalyzes the reduction of other lipid hydroperoxides and H_2O_2, and also undergoes self-deactivation (14, 15). Various reducing cosubstrates increase the number of turnovers before deactivation, and these cosubstrates may be covalently modified during this process (16, 17). Cosubstrate oxidation by PES has been postulated to activate certain procarcinogenic substances (17, 18).

Nonsteroidal anti-inflammatory agents inhibit the cyclooxygenase but not the hydroperoxidase activity of PES (19). As recently reviewed (6), aspirin acetylates the enzyme resulting in irreversible inhibition; certain acetylenic fatty acids, such as eicosa-5,8,11,14-tetrynoic acid (ETYA), also inactivate the cyclooxygenase, possibly by acting as suicide substrates.

Metabolism of the Prostaglandin Endoperoxides

THROMBOXANES   In blood platelets the principal metabolite of PGG_2 and PGH_2 is thromboxane A_2 (TxA_2). The oxetane ring of TxA_2 spontaneously hydrolyzes ($t_{1/2} = 30$ sec) to the hemiacetal thromboxane B_2 (TxB_2) (2). TxA_2 but not TxB_2 contracts vascular smooth muscle and induces platelet aggregation and serotonin release at concentrations below 20 nM.

Platelet thromboxane synthase activity is associated with dense tubular membranes and catalyzes formation of 12-hydroxy-heptadeca-trienoic acid (HHT) as well as TxA_2 from PGH_2. Some HHT must also arise from the nonenzymatic breakdown of PGH_2. A 750-fold purification of the enzyme has been achieved by affinity chromatography with an immobilized inhibitor of thromboxane synthase (20). Similar enrichment for thromboxane synthase activity and for cytochrome p450 optical properties during the purification suggest that thromboxane synthase may be a cytochrome p450–type hemoprotein (20).

Inhibitors of thromboxane synthase include imidazole (21), certain substituted imidazoles (such as dazoxiben) (22), certain pyridine derivatives (such as OKY-1581) (23), prostaglandin analogues containing a pyridine rather than a cyclopentane ring (24), and PGH_2 analogues such as 15-deoxy-9,11-azo-PGH_2 (25). The compound 9,11-azo-PGH_2 mimics the action of thromboxane as does 9,11-methanoepoxy-PGH_2 (U46619) (26).

Thromboxane synthase inhibitors prevent TxA_2 formation but not aggregation by washed platelets in response to arachidonate (21). Coupled with the TxA_2-mimetic properties of stable endoperoxide analogues such as U44619, this has suggested that TxA_2 and PGH_2 may activate platelets by a common receptor (27). Certain analogues of PGH_2 (e.g. 15-deoxy-9,11-epoxyimino-PGH_2) and of TxA_2 (e.g. pinane-TxA_2) prevent PGH_2-induced platelet aggregation at concentrations that do not inhibit thromboxane synthase (26, 28). Such compounds may represent antagonists at a putative TxA_2/PGH_2 receptor (26–28).

Thromboxane B_2 infused intravenously into man is rapidly metabolized to a variety of compounds that are excreted into the urine, and the most abundant of these is 2,3-dinor-TxB$_2$ (29). No circulating plasma metabolite has been identified, and peripheral blood TxB$_2$ derives largely from platelet activation during blood collection (30).

PROSTACYCLIN   Vascular endothelial cells convert PGH$_2$ to prostacyclin (PGI$_2$), an enol-ether which spontaneously hydrolyzes ($t_{1/2}$ = 10 min at pH 7.4, 24°C) to 6-keto-PGF$_{1\alpha}$ (31). These compounds were first identified from rat stomach (32), and PGI$_2$ was subsequently shown to be responsible for arachidonate-induced coronary vasodilation observed by Needleman and coworkers (33, 34) and for the antiaggregatory influence on platelets of medium from incubation of PGH$_2$ and aortic microsomes observed by Vane and coworkers (35). Vascular and nonvascular smooth muscle cells also synthesize PGI$_2$ (36, 37), although endothelial cells have a higher synthetic capacity due to a higher cyclooxygenase content than myocytes (37). Not all endothelial cells synthesize PGI$_2$ (38).

PGI$_2$ synthase has been purified to homogeneity by affinity chromatography using immobilized monoclonal antibody to PGI$_2$ synthase activity (39) or immobilized endoperoxide substrate analogue (40). Immunofluorescence studies indicate that the enzyme is located in the plasma membrane and nuclear membrane of a wide variety of smooth muscle cells (36). The purified enzyme exhibits a subunit molecular weight of about 50,000 and a specific activity of 1000–2000 μmole PGI$_2$/min/mg protein (39, 40). The K_m for PGH$_2$ is about 5 μM (41) and self-deactivation occurs during catalysis (39). The optical behavior of the purified enzyme suggests that it may be a cytochrome p450–type hemoprotein (42). PGI$_2$ synthase is inactivated by a variety of lipid hydroperoxides (41, 43), a process that is partially prevented by reducing compounds that are radical scavengers (43).

Thrombin stimulates PGI$_2$ synthesis by cultured endothelial cells (44), and ADP stimulates PGI$_2$ synthesis by isolated rabbit aorta (45). TxA$_2$-mimetic PGH$_2$ analogues stimulate PGI$_2$ synthesis by cultured vascular smooth muscle cells (46). Platelet-derived PGH$_2$ also appears to be utilized by endothelial cells for PGI$_2$ synthesis (47, 48), and this effect is magnified by thromboxane synthase inhibitors (49). Augmentation of vascular cell PGI$_2$ production by platelet activators and platelet release products may serve to limit the area of platelet deposition about a site of vascular injury. After endothelial cells have been stimulated to produce PGI$_2$, they become unresponsive to a second stimulus, apparently due to deactivation of the cyclooxygenase (12, 13). Vascular PGI$_2$ synthesis may therefore occur in phasic bursts which must be actively initiated rather than as a continuous process.

The vasodepressor substances histamine (50) and bradykinin (51) also stimulate PGI_2 synthesis by cultured endothelial cells, and this response could mediate a component of the depressor effects of these compounds in vivo. PGI_2 production by cultured endothelial cells declines as the cells reach confluence and increases when the cells are seeded into fresh medium at lower density (52). Similar phenomena occur with cultured 3T3-L1 preadipocytes (53), and exogenous PGI_2 retards while indomethacin promotes the differentiation of these cells into mature adipocytes (54). PGI_2 may therefore modulate the proliferative and differentiation state of some cells.

PGI_2 elevates cAMP levels in platelets (55), vascular smooth muscle cells (56), and toad bladder epithelium (57), and this biochemical action appears to mediate the antiaggregatory effects of PGI_2 on platelets (55), the activation of myocyte cholesterol ester hydrolase (56), and increase in toad bladder short circuit current (57). Activation of adenylate cyclase via a PGI_2 receptor may represent a general mechanism for the effects of PGI_2 on responsive cells. Membranes from a cultured, PGI_2-responsive cell line have been reported to contain a specific protein receptor for PGI_2 with a maximum binding capacity of 350 fmol/mg membrane protein and a molecular weight of 82,000 (58).

PGI_2 and 6-keto-$PGF_{1\alpha}$ infused intravenously into man are rapidly metabolized to a variety of compounds, and the principal urinary metabolite is 2,3-dinor-6-keto-$PGF_{1\alpha}$ (59). Circulating levels of PGI_2 (measured as 6-keto-$PGF_{1\alpha}$) in human plasma are less than 3 pg/ml, which is insufficient to exert any recognized biologic activity (60). Circulating metabolites of PGI_2 include 2,3-dinor-13,14-dihydro-6,15-diketo $PGF_{1\alpha}$ and its 20-carboxy analogue (61). The compound 6-keto-PGE_1 is produced from 6-keto-$PGF_{1\alpha}$ upon perfusion through rabbit liver (62), and this compound inhibits platelet aggregation, although less potently than PGI_2 (63).

PROSTAGLANDIN D_2 Isomerization of PGH_2 to PGD_2 can be catalyzed by serum albumin (K_m 6 μM, specific activity 0.87 mol PGH_2/min/mol albumin at 37°C) (64). PGH to PGD isomerase activity has been purified to electrophoretic homogeneity from the cytosol of rat brain (65) and rat spleen (66). The brain enzyme has a molecular weight of 80,000–85,000, a K_m for arachidonate of 8 μM, and requires glutathione. The spleen enzyme does not require glutathione and has a molecular weight of 26,000–34,000. Both enzymes have a specific activity two orders of magnitude higher than that of albumin, and both are inactivated by sulfhydryl-modifying reagents. A systematic survey of rat tissues indicated that PGH to PGD isomerase activity is higher in brain and spinal cord than in other tissues (67). PGD_2 is the principal cyclooxygenase product of rat and human mast cells (68).

PGD_2 inhibits platelet aggregation (69), increases platelet cAMP content, and has a platelet membrane receptor distinct from that for PGI_2 (70). Intravenously infused PGD_2 is a peripheral vasodilator, pulmonary vasoconstrictor, and bronchoconstrictor in the dog (71). Inhaled PGD_2 is a bronchoconstrictor in man (72).

The high concentrations of PGH to PGD isomerase in central nervous system suggest that PGD_2 may have neuromodulatory actions (67). PGD_2 augments cAMP content of cultured neuroblastoma cells (67) and induces depolarization of neuroblastoma-glioma hybrid cells (73). PGD_2 appears to decrease norepinephrine release from adrenergic nerve terminals in the cat (74). Instillation of PGD_2 into the third ventricle of rats increases slow-wave sleep (75).

PGD_2 intravenously infused into primates is metabolized to compounds reflecting various combinations of 11-keto reduction, dehydrogenation of the 15-hydroxyl, reduction of the delta^{13} double bond, β-oxidation, and ω-oxidation (76). Urinary excretion of markedly increased amounts of two metabolites of PGD_2 has been demonstrated in several patients with systemic mastocytosis, and mass spectrometric quantitation of one of these metabolites has been used as a diagnostic tool in this disorder (77, 78). Overproduction of PGD_2 may be involved in hypotensive episodes of patients with mastocytosis; chronic aspirin therapy reduces such attacks (77).

A novel circulating PGD_2 metabolite has recently been identified in plasma of a patient with mastocytosis and urine of a normal subject (79). This metabolite is a diastereoisomer of $PGF_{2\alpha}$, which has a beta configuration of the C-9 hydroxyl group and which is equipotent to $PGF_{2\alpha}$ as a pressor agent in the rat. The compound is formed from PGD_2 by an NADPH-requiring, cytosolic activity in liver.

Albumin catalyzes transformation of PGD_2 to 9-deoxy-delta^{9,12}-PGD_2, 15-deoxy-delta^{12,14}-PGD_2, and 9,15-dideoxy-delta^{9,12,14}-PGD_2 (80). In protein-free buffers, PGD_2 decomposes to several compounds including 9-deoxy-delta^9-PGD_2, and this compound suppresses the proliferation of cultured leukemia cells in vitro (81). Formation of this compound in vivo or by living cells has apparently not yet been demonstrated.

PROSTAGLANDIN E_2 Isomerization of PGH to PGE is catalyzed by a microsomal enzyme partially purified from sheep vesicular gland which exhibits a K_m of 10 μM for PGH_1 and a specific activity of 1–2 μmol PGE_2/min/mg protein (82). PGH to PGE isomerase activity from kidney is markedly stimulated by glutathione (83) and inactivated by sulfhydryl-modifying reagents (84).

As recently reviewed (85), PGE_2 is the predominant arachidonate metabolite from the kidney of many species, appears to antagonize ADH-induced water reabsorption (86, 87), and may participate in the modulation of renin release along with PGI_2 (88). PGE_2 is also produced by macrophages and may mediate some effects of macrophages on neighboring cells as well as influencing the functional state of the macrophage itself, as discussed in recent reviews (85, 89, 90). The possible participation of PGE_2 and other cyclooxygenase products in a variety of other processes has also recently been reviewed (91–94).

PRODUCTS OF THE 12-LIPOXYGENASE

The 12-Lipoxygenase Enzyme

Lipoxygenases catalyze incorporation of one oxygen molecule into polyunsaturated fatty acids containing a 1,4-*cis,cis*-pentadiene system to yield a 1-hydroperoxy-2,4-*trans,cis*-pentadiene product. The regional specificity of the lipoxygenase is designated by the number of the product carbon bearing the hydroperoxy group. The platelet 12-lipoxygenase converts arachidonate to 12-S-hydroperoxy-eicosa-5,8,10,14-tetraenoic acid (12-S-HPETE). The initial step in catalysis is stereospecific removal of the pro-S-hydrogen atom at C-10 with subsequent antarafacial insertion of molecular oxygen at C-12 (95).

Platelet 12-lipoxygenase activity is contained in both cytosolic and membrane fractions, and the cytosolic activity is an aggregated, lipid-containing form (96, 97). The kinetic properties, ionic sensitivity, and regional specificity of the cytosolic and membrane-associated activities are not distinguishable (98). The 12-lipoxygenase may therefore be a loosely membrane-associated enzyme and appears to be preferentially associated with internal platelet membranes rather than plasma or granule membranes (97, 98). Human platelet cytosolic 12-lipoxygenase activity elutes from size-exclusion columns in bands with approximate molecular weights of 100,000 and 160,000 and only the higher-molecular-weight species exhibits associated peroxidase activity (99). The 12-lipoxygenase activity from rat lung exhibits no peroxidase activity (100).

The platelet 12-lipoxygenase exhibits a kinetic lag period and is activated by trace concentrations of hydroperoxides (97, 101). Platelet 12-HETE synthesis continues for at least two hours after collagen stimulation, although TxB_2 synthesis ceases after 3 minutes (101). The 12-lipoxygenase K_m for arachidonate is 5 μM for the rat lung enzyme (100) and 3.4 μM for the human platelet enzyme (97). Platelet 12-lipoxygenase has no known cofactor requirements and is fully active without exogenous calcium, although a rat basophilic leukemia cell 12-lipoxygenase is stimulated fivefold by 1 mM Ca^{2+} (102).

The platelet 12-lipoxygenase is not inhibited by nonsteroidal anti-inflammatory agents but is inhibited by ETYA, nordihydroguaiaretic acid (NDGA), BW755C, and a number of other compounds (97, 103, 104).

Metabolism of 12-HPETE

Reduction of 12-HPETE to its hydroxy analogue 12-HETE has been attributed to a peroxidase activity associated with the 12-lipoxygenase which is inhibited by nonsteroidal anti-inflammatory agents (99), but this observation needs substantiation. More recent evidence indicates that the selenium form of glutathione peroxidase plays a major role in reducing 12-HPETE to 12-HETE (105–107). Arachidonate stimulates platelet oxidation of [1-^{14}C]-glucose to

$^{14}CO_2$ via the hexose monophosphate shunt, and this effect is diminished by lipoxygenase inhibitors (105). Reduction of 12-HPETE in disrupted platelets is dependent on GSH (106). Platelets from selenium-deficient rats have impaired ability to reduce 12-HPETE (107).

The hydroxy-epoxide compounds 10-hydroxy-11,12-epoxy-eicosa-5,8,14-trienoic acid and 8,11,12-trihydroxy-eicosa-5,9,14-trienoic acid are produced by platelets from arachidonic acid via 12-HPETE (108, 109). Hematin catalyzes intramolecular rearrangement of the hydroperoxy group of 12-HPETE to yield 8-hydroxy- and 10-hydroxy-11,12-epoxy-eicosatrienoic acids (110, 111). Similar catalytic activity is observed with rat lung homogenate but is not heat-labile and probably not enzymatic (112). Stereochemical analysis of the optical center at carbon 8 of the 8-hydroxy isomer also suggests a nonenzymatic origin (113). In contrast, epoxide hydrolase activity from rat lung homogenate, which converts the hydroxy-epoxy acids to trihydroxy acids, is heat-labile and likely enzymatic (112).

Platelet-derived 12-HETE may be subjected to a second oxygenation by the neutrophil 5-lipoxygenase and reduced to yield 5(S),12(S)-dihydroxy-eicosa-6(trans),8(cis),10(trans), 14(cis)-tetraenoic acid (114). Neutrophils also convert 12-HETE to a 12,20-dihydroxy-eicosatetraenoic acid (115, 116).

Actions of 12-Lipoxygenase Products

Although 12-lipoxygenase activity is present in platelets of all species so far examined, the role of 12-lipoxygenase products in platelet function is not yet known (97). Exogenous 12-HPETE inhibits collagen-induced platelet aggregation with an IC_{50} of 2–6 μM, and higher concentrations (IC_{50} = 15–25 μM) inhibit the platelet cyclooxygenase (117). A subset of patients with myeloproliferative disorders has virtually absent platelet 12-lipoxygenase activity and a markedly increased incidence of hemorrhagic events, but a causal relationship between these phenomena is not established (118).

The migration of aortic smooth muscle cells in vitro is stimulated by intact platelets and by exogenous 12-HETE at concentrations as low as one femtomolar (119). The effect with intact platelets is abolished by ETYA but enhanced by indomethacin. Exogenous 12-HETE is at least five orders of magnitude more potent than other HETE isomers in this system (120).

Platelets markedly stimulate neutrophil 5-lipoxygenase product generation, and this effect may be due to platelet-derived 12-HPETE because it is not observed with ETYA-treated platelets (121). Exogenous 12-HPETE also produces the effect (EC_{50} = 3 μM). This suggests that 12-HPETE may amplify production of LTB_4 and other 5-lipoxygenase products at inflammatory loci.

Isolated pancreatic islets from the rat synthesize 12-HETE (122–127). Concentrations of glucose that stimulate insulin secretion also stimulate islet 12-HETE production (123, 126), and several structurally distinct 12-lipoxygenase inhibitors suppress glucose-induced insulin secretion (123–125, 127). The concentration of the lipoxygenase inhibitors required to suppress insulin secretion corresponds closely to the concentration required to inhibit 12-HETE synthesis (124). Selective cyclooxygenase inhibitors do not suppress glucose-induced insulin secretion by isolated islets (123, 127), and intact, isolated islets do not appear to generate significant amounts of 5-lipoxygenase, 15-lipoxygenase, or cytochrome p450 monooxygenase products from arachidonate (122, 124, 125). Isolated islets convert 12-HPETE to compounds including 8-hydroxy-11,12-epoxy-eicosa-5,9,14-trienoic acid, and the latter compound has recently been reported to augment glucose-induced insulin secretion threefold from isolated islets at a concentration of 2 µM (128).

Both 12-HETE and 5-HETE induce secretion of neutrophil-specific granules and augment IgE-mediated release of mast cell granules at concentrations between 1 and 10 µM (129, 130). Exogenous, radiolabeled 12-HETE and 5-HETE are incorporated into neutrophil and mast cell phospholipids, and it has been suggested that this may modify the tendency of granule and plasma membranes to fuse (129, 130).

PRODUCTS OF THE 15-LIPOXYGENASE

The 15-Lipoxygenase Enzyme

The 15-lipoxygenase of rabbit neutrophils is primarily cytosolic and has been partially purified (131). The crude but not the purified enzyme is stimulated by divalent metal cations. Activity elutes with an approximate molecular weight of 68,000 on gel filtration chromatography. The K_m for arachidonate is 28 µM. The enzyme is inhibited by sulfhydryl modifying reagents, ETYA (IC_{50} = 1.7 µM), and BW755C (IC_{50} = 75 µM), and activity is slightly enhanced by 560-µM indomethacin. Nonenzymatic decomposition of 15-HPETE results in formation of 15-HETE; its 15-keto analogue; 13-hydroxy-14,15-epoxy-eicosa-5,8,11-trienoic acid; and 11,14,15-trihydroxy-eicosa-5,8,12-trienoic acid (131).

A 15-lipoxygenase in rabbit reticulocytes converts arachidonate to 12-HPETE as well as 15-HPETE (132) and undergoes self-deactivation associated with oxidation of a single methionine residue (133). The reticulocyte enzyme has a molecular weight of 78,000, and contains 5% neutral sugars and one nonheme iron atom per molecule (134). Immunoprecipitation studies indicate that this enzyme is contained only in reticulocytes and not in other tissues with lipoxygenase activity (135). The enzyme oxygenates fatty acid in phospholipid as well as unesterified substrate, and mitochondrial membranes are more readily oxidized than plasma membranes (135).

Metabolism of 15-HPETE

Either 15-HPETE or its reduction product 15-HETE may be oxygenated by the 5-lipoxygenase and then reduced to yield 5,15-dihydroxy-eicosa-

6,13(*trans*),8,11(*cis*)-tetraenoic acid (5,15-diHETE) (136). This compound may also be formed by the action of the 15-lipoxygenase on 5-HETE. The compound is formed from arachidonate by rat peritoneal leukocytes (136), human eosinophils and neutrophils (136, 137), murine eosinophils (138), and probably other cells. In porcine and human blood granulocytes, 15-HPETE may be dehydrated to a 14,15-oxido-eicosa-5,8,10,12-tetraenoic acid (14,15-LTA$_4$) analogous to leukotriene A$_4$ (139, 140). Formation of this epoxide intermediate is inferred from the isolation of its hydrolysis products, which are two diastereoisomeric 8,15-dihydroxy compounds with a conjugated triene (delta^{9,11,13}) structure in which the C-15 oxygen atom derives from molecular oxygen and the C-8 oxygen atom derives from water (139). Two additional 8,15-dihydroxy-conjugated triene compounds and two 14,15-dihydroxy-conjugated triene compounds are also formed in these systems, and both hydroxyl oxygen atoms derive from molecular oxygen in these four compounds. The formation of those compounds may be rationalized by the action of a 12-lipoxygenase-like enzyme on 15-HPETE with initial abstraction of the pro-S hydrogen atom at C-10 (141). A [1,5] radical migration to C-14 could then produce 14,15-LTA$_4$ upon reaction with the 15-hydroperoxy group or a 14,15-dihydroperoxide upon reaction with molecular oxygen. A [1,3] radical migration to C-8 could produce an 8,15-dihydroperoxide upon reaction with molecular oxygen. Experiments with 15-S-HPETE stereospecifically labeled with either a pro-S or a pro-R ^{3}H-atom at C-10 support this mechanism and indicate that the formation of 14,15-LTA$_4$ is enzyme-catalyzed and differs in stereochemistry from nonenzymatic formation of 14,15-LTA$_4$ from 15-S-HPETE catalyzed by hemoglobin (141).

The known hydrolysis products of 14,15-LTA$_4$ are epimeric at C-8 (139), are also formed from synthetic 14,15-LTA$_4$ by boiled leukocytes (142), and are therefore likely generated by nonenzymatic hydrolysis of 14,15-LTA$_4$. Glutathione adducts of 14,15-LTA$_4$ have not been identified from intact cells incubated with arachidonate, but such compounds are formed from synthetic 14,15-LTA$_4$ by rat basophilic leukemia cells and have spasmogenic activity on guinea pig ileum (143).

The dominant monohydroxylated arachidonate metabolite from human eosinophils incubated with ionophore A23187 and exogenous arachidonate is 15-HETE (137, 144, 145), and these cells also synthesize 15-series conjugated triene diols (137, 144–146). Under these conditions, 5-HETE is the major monohydroxylated arachidonate from human neutrophils (144, 146, 153), and far smaller amounts of 15-series conjugated triene diols are produced by neutrophils (137, 144, 145). Both murine (138) and porcine (147) eosinophils also synthesize 15-HETE and 15-series conjugated triene diols when stimulated with A23187 and exogenous arachidonate. In the absence of exogenous arachidonate smaller amounts of 15-lipoxygenase products are generated by human eosinophils (145, 146), and under all stimulation conditions LTC_4 is a major eosinophil product (144–146). The amplification of 15-lipoxygenase product generation by exogenous arachidonate may reflect the high K_m (28 μM) of the 15-lipoxygenase for this substrate (131) or arachidonate-induced cytolysis.

Human neutrophils incubated with 15-HPETE and A23187 form at least two trihydroxylated compounds with a conjugated tetraene structure: 5,6,15(L)-trihydroxy-eicosa-7,9,11,13-tetraenoic acid (Lipoxin A) and 5(D),14,15(L)-trihydroxy-eicosa-6,8,10,12-tetraenoic acid (Lipoxin B) (148, 149). Analogous products are formed from the 15-hydroperoxy derivative of eicosapentaenoic acid (150). It appears likely that the 5-lipoxygenase participates in the formation of these compounds, and the configuration of the optical center at C-5 is compatible with this possibility (149). Whether triol formation involves a 5,6- or a 14,15-oxido intermediate is uncertain. An alternative mechanism involves a third dioxygenation with a 12-lipoxygenase-like hydrogen atom abstraction at C-10 followed by [1,5] radical migration to C-6 (for lipoxin A) or to C-14 (for lipoxin B) followed by reaction with molecular oxygen and reduction. This mechanism appears to explain formation of trihydroxy tetraenes from 5,15-diHETE by the purified reticulocyte lipoxygenase (151).

Actions of 15-Lipoxygenase Products

The neutrophil 5-lipoxygenase and platelet 12-lipoxygenase are inhibited by 15-HETE (IC_{50} = 1–5 μM), although cyclooxygenase activity is not (152). Conversely, production of 5-HETE and LTB_4 by a cultured mast/basophil cell line has been reported to be stimulated by 15-HETE (EC_{50} = 3 μM) (153). It has been suggested that 15-HETE may participate in a complex network of regulatory interactions between various lipoxygenase pathways (152, 153).

Two of the 15-series conjugated triene diols have been reported to have biologic activity. The compound 8(S), 15(S)-dihydroxy-eicosa-5,11(*cis*), 9,13(*trans*)-tetraenoic acid has been reported to exert chemotactic activity comparable to LTB_4 on human neutrophils (154). Others observe little activity of this compound on human neutrophil chemokinesis or aggregation (155). The compound (erythro-14(R), 15(S))-dihydroxy-eicosa-5,8(*cis*), 10,12(*trans*)-tetraenoic acid has been reported to inhibit (IC_{50} = 10^{-7} to 10^{-8}M) LTB_4-induced superoxide anion generation by human neutrophils (156). Lipoxin A has been reported to induce superoxide anion generation from human neutrophils at submicromolar concentrations (149).

LEUKOTRIENES

Leukotrienes are potent biologically active compounds that are synthesized by way of the 5-lipoxygenase pathway in neutrophils, eosinophils, monocytes, mast cells, and keratinocytes (157–162) as well as lung, spleen, brain, and heart (163–166). Prior to their structural elucidation, leukotrienes (LT) were recognized in perfusates of lung, as slow reacting substances (SRS) after stimulation with cobra venom (167), and as slow reacting substances of anaphylaxis (SRS-A) after immunological challenge (168). The structure of SRS-A was elucidated in 1979 (169, 170) and found to be a mixture of the peptidoleukotrienes LTC_4, LTD_4, and LTE_4. The dihydroxyleukotriene, LTB_4, was first isolated from polymorphonuclear leukocytes (157). The characterization of the structure of each product in the 5-lipoxygenase-leukotriene pathway has been extensively reviewed (171). The name leukotrienes was derived from their original discovery in leukocytes and from the presence of a conjugated triene in the molecule. The conjugated triene is responsible for the characteristic UV absorption spectrum (157, 172). The subscripts denote the number of double bonds present. Therefore, leukotrienes derived from arachidonic acid (n-6) belong to the 4-series, those from eicosapentanoic acid (n-3) to the 5 series, and those from eicosatrienoic (Mead) acid (n-9) to the 3-series (173).

Biosynthesis

Arachidonic acid is oxygenated to 5-hydroperoxy-6,8,11,14-eicosatetraenoic acid (5-HPETE) by the 5-lipoxygenase. The labile allylic epoxide (LTA) is formed by abstraction of the pro-R hydrogen atom at C10 and elimination of a hydroxyl moiety from the hydroperoxy group (174). LTA_4 can be hydrolyzed enzymatically to LTB_4 (5S, 12R-dihydroxy-6,8,10,14-eicosatetraenoic acid) and nonenzymatically to dihydroxyeicosatetraenoic acids that are isomers of LTB_4 (175). LTA_4 can also be converted to LTC_4 by the addition of glutathione at C6 by a glutathione transferase (176). Removal of glutamic acid from LTC_4 by γ-glutamyl transpeptidase generates LTD_4 (177). Further removal of a glycine residue by a dipeptidase is responsible for the formation of LTE_4 (178). The action of γ-glutamyl transpeptidase on LTE_4 has been shown to cause the production of LTF_4 by the reincorporation of a glutamyl moiety (179, 180). However, the generation of LTF_4 by tissues or cells has not been demonstrated. The enzymatic mechanism of the individual steps in the pathway has been reviewed earlier (171).

5-LIPOXYGENASE STIMULATION BY CALCIUM   Studies with a cell-free enzyme system obtained from rat basophilic leukemia (RBL-1) cells showed that the 5-lipoxygenase was specifically stimulated by calcium in a concentration-dependent fashion (181–183). The enhancement by calcium clarifies the mechanism by which leukotriene formation by RBL-1 cells (184, 185), polymorphonuclear leukocytes (186), and macrophages (187) from exogenously added arachidonic acid was potentiated by the addition of the calcium ionophore, A23187. The 5-lipoxygenase seems to be the only enzyme in this pathway that is stimulated by calcium. The addition of A23187 is not required for leukotriene synthesis from 5-HPETE by whole cells (188), and LTC_4 and LTB_4 can be formed from LTA_4 by the cell-free enzyme system from RBL-1 cells without the addition of calcium and in the presence of EDTA (189–191).

KINETIC STUDIES   The kinetics of the 5-lipoxygenase-leukotriene pathway are quite different from those of the cyclooxygenase pathway, where the first step, the cyclooxygenase itself, is rate limiting. In leukotriene biosynthesis, the rate limiting steps occur at the LTA-hydrolase and glutathione transferase. Kinetic studies were performed with the cell-free enzyme system from RBL-1 cells in the absence of glutathione (i.e. no peptidoleukotriene formation was possible). When the concentration of arachidonic acid was varied from 1 to 300 μM, LTB_4 formation reached a plateau at 10 μM arachidonate, and LTA_4 and 5-HETE formation peaked at 100 μM respectively. At 100 μM arachidonic acid, seven times as much LTA_4 as LTB_4 was synthesized, and at 300 μM arachidonic acid three times as much 5-HETE as LTA_4 was synthesized (190). Similar results were obtained with human neutrophils that were stimulated with 10 μM A23187 in the presence of exogenous arachidonic acid (1 to 300 μM) (192). An excess of LTA_4 formation was also observed when homogenate of RBL-1 cells was incubated with arachidonic acid and glutathione, and LTC_4 and LTD_4 were generated (191). The relatively small conversion of LTA_4 to LTB_4 and LTC_4 resulted in nonenzymatic breakdown products of LTA_4, 6-*trans*-LTB_4 and 5,6-diHETE. The formation of 6-*trans*-LTB_4 is not confined to the cell-free enzyme system, but has also been observed with immunologic challenge of bone marrow–derived mouse mast cells (193). These experiments suggest that this enzyme system has a much larger capacity to synthesize 5-HETE and LTA_4 than LTB_4. So far no biological activity has been reported for the 6-*trans* isomers of LTB_4, except when used in high concentration (194). These findings raise the question whether LTA_4 has some intracellular function that has not yet been discovered.

Kinetic studies with partially purified 5-lipoxygenase from RBL-1 cells showed a K_m of 17 μM for arachidonic acid (195). The partially purified enzyme remained calcium-dependent (183, 196) and was also enhanced by ATP. The latter action was shared by other nucleotides, but to a lesser degree (183). The K_m for LTA_4 of the purified LTA-hydrolase is 20–30 μM (197).

The 5-lipoxygenase and the dehydrase are quite prone to inactivation by peroxides. This has been observed with the cell-free system from RBL-1 cells (190, 198), polymorphonuclear leukocytes stimulated with A23187 (192), and after immunologic challenge of bone marrow–derived mouse mast cells or exposure of macrophages to zymosan. When the mast cells or macrophages were first treated with catalase and then stimulated, larger amounts of 5-lipoxygenase products were observed. A similar enhancement was observed with treatment of superoxide dismutase, suggesting that superoxide and possibly hydroxyl radicals may also inactivate the 5-lipoxygenase-leukotriene pathway (193).

SUBCELLULAR LOCALIZATION OF THE ENZYMES   Production of 5-HETE, LTA_4, and LTB_4 were observed only in the $100,000 \times g$ supernatant of RBL-1 cell homogenates (190), indicating that the 5-lipoxygenase, dehydrase, and LTA-hydrolase are soluble enzymes. In murine keratinocytes the enzymes necessary for LTB_4 synthesis were also found in the cytosolic fraction (199). The LTA-hydrolase has been purified to apparent homogeneity from the cytosolic portion of human leukocytes. It is a monomeric protein with a molecular weight of 68,000–70,000. The amino acid content and N-terminal sequence has been determined (197). The glutathione transferase, γ-glutamyl transpeptidase, and dipeptidase are particulate enzymes (191). The γ-glutamyl transpeptidase and dipeptidase have been further localized in the plasma membranes (200). The highest concentration of the LTC-forming glutathione transferase activity in RBL-1 cells was found in a fraction that is markedly enriched in granules (189). Others have reported activity in all particulate fractions of RBL-1 cell homogenate after differential centrifugation (201). The difference may be due to the methods employed for cell lysis. These studies indicate that the glutathione transferase involved in LTC_4 synthesis differs from conventional glutathione transferases, which are predominantly soluble enzymes.

The subcellular localization of the leukotriene-forming enzymes differs from the microsomal enzymes responsible for prostaglandin biosynthesis. The enzymes required for LTB_4 synthesis are cytosolic, while the enzymes involved in peptidoleukotriene formation from LTA_4 are particulate. Therefore, LTA_4, which is synthesized by a soluble enzyme, has to reach the particulate enzyme in the granules or some other subcellular fraction to be converted to LTC_4. Since the LTD-forming enzyme is in the plasma membrane, LTC_4 is most likely converted to LTD_4 as it passes through the plasma membrane to the outside of the cell.

It has been reported that plasma of various mammalian species, including human, can convert LTA_4 to LTB_4. The activity appeared to be enzymatic since it was destroyed by heating to 56°C or digestion with pronase. The enzyme did not seem to originate from leukocyte lysis during the isolation of plasma. Lysosomal enzyme activity was not detectable in most plasma samples and there was no correlation between leukocyte degranulation and LTB_4 formation (202). Human erythrocytes do not have the total 5-lipoxygenase-leukotriene pathway, but they contain a cytosolic hydrolase which converts LTA_4 to LTB_4 (203). These findings suggest that any LTA_4 that might escape from leukocytes could be converted by plasma and/or erythrocytes into the potent chemotactic factor LTB_4.

SUBSTRATE SPECIFICITY   The substrate specificity for the 5-lipoxygenase-leukotriene pathway is quite stringent. Incubation of a variety of polyenoic fatty acids with the cell-free enzyme system from RBL-1 cells demonstrated that the double bond at C5 is critical since the fatty acids 20:5(5,8,11,14,17), 20:3(5,8,11), 19:4(5,8,11,14), and 18:4(5,8,11,14) were readily converted to LTA. The K_m with partially purified 5-lipoxygenase for 5,8,11,14,17-eicosapentaenoic acid was 24 μM, for 5,8,11-eicosatrienoic acid was 32 μM, and for arachidonic acid was 17 μM (195). Fatty acids with their initial double bond at C4 (i.e. carbon number 4 of the fatty acid), C6, C7, or C8 are not converted by 5-lipoxygenase to leukotrienes (204–206). A very limited conversion of 8,11,14-eicosatetraenoic acid to LTC_3 by mouse mastocytoma cells (207) and to 8,15-LTB_3 by polymorphonuclear leukocytes (208) has been reported. The formation of 8,15-LTB_3 was not enhanced by the calcium ionophore, A23187 (207). It is therefore questionable that this leukotriene is a 5-lipoxygenase product. With the partially purified 5-lipoxygenase from RBL-1 cells, the conversions of 8,11,14-eicosatetraenoic acid and 4,7,10,13,16,19-docosahexaenoic acid were insignificant (195).

Incubations of 5,11,14-, 5,8,14-, and 5,8,11-eicosatetraenoic acid with the cell-free enzyme system from RBL-1 cells indicated that the 5-lipoxygenase requires double bonds at C5 and C8 for catalysis to occur, and the dehydrase (LTA synthesis) requires double bonds at C5, C8, and C11. 5,11,14-eicosatetraenoic acid was not oxygenated by the 5-lipoxygenase, and 5,8,14-eicosatetraenoic acid was converted to 5-HETE but not to the allylic epoxide (209). The double bond at C14 is essential for LTB formation since 5,8,11-eicosatrienoic (Mead) acid was readily converted to LTA_3, but only insignificant amounts of LTB_3 were formed (210). The findings with the RBL-1 cell system were confirmed in experiments with neutrophils from essential fatty acid–deficient rats that were enriched in 5,8,11-eicosatrienoic acid (211). When these cells were stimulated with A23187, LTA_3 was formed but very little or no LTB_3 was formed. LTA_4 production was reduced by 64% as compared to neutrophils from normal rats, and LTB_4 by 87%. A greater reduction in LTB_4 generation as compared to LTA_4 was also observed when neutrophils from normal rats were exposed to A23187 in the presence of exogenous 5,8,11-eicosatrienoic acid. This inhibition of LTB_4 synthesis was due to a metabolite of 5,8,11-eicosatrienoic acid and not to the acid itself (211).

An additional double bond at C17 (5,8,11,14,17-eicosapentaenoic acid) did not significantly alter LTB formation as compared to arachidonate by the cell-free enzyme system from RBL-1 cells (205). Eicosapentaenoic acid was also converted to LTB_5 by murine mastocytoma cells (161, 212) and neutrophils (213). However, the efficiency of conversion in neutrophils was only one fourth that of arachidonic acid (213).

In contrast to the stringent structural requirements of a polyenoic fatty acid to be converted to LTB, a specific double bond configuration is not necessary for the glutathione transferase. It seems that any fatty acid that is converted to LTA is further converted to LTC (204, 205, 214). This low specificity for the fatty acid portion was further illustrated in a study with a series of acetylenic fatty acids. Any of the 33 acetylenic fatty acids (varying in position and number of triple bonds as well as in chain length) tested (1–100 μM) inhibited the glutathione transferase while only some inhibited the 5-lipoxygenase (215). On the other hand, the glutathione transferase involved in LTC_4 synthesis does not act upon adducts of halogenated nitrobenzene, which are used routinely for the measurement of conventional glutathione transferases in liver and other tissues (201). Some soluble glutathione transferases from liver can form LTC_4 from LTA_4 (216, 217). With regard to the peptide portion, the glutathione transferase has more restricted requirements since only glutathione seems to serve as a substrate, but not cysteine or cysteinylglycine (182, 205). The studies with the particulate LTC-forming glutathione transferase show that this enzyme differs from conventional glutathione transferases not only by its subcellular localization but also by its substrate requirement.

Metabolism

The metabolism of peptidoleukotrienes in whole animals and by cells has been reviewed in detail (171). LTB_4 can be further converted by ω-oxidation to 20-hydroxy-LTB_4 and subsequently to 20-carboxy-LTB_4 by polymorphonuclear leukocytes (218). The ω-oxidation of LTB_4 to 20-hydroxy-LTB_4 by polymorphonuclear leukocytes is catalyzed by a NADPH-dependent, cytochrome p450-like enzyme (219, 220). When [$^3H_8$]LTB_4 was administered to monkeys or rabbits, 25% of the ^3H-activity was recovered in urine. The major nonvolatile urinary metabolite was 20-hydroxy-LTB_4 (0.8%). Analysis of plasma and urine showed that in contrast to previously studied eicosanoids, more than 70% of the injected [$^3H_8$]LTB_4 was recovered in tritiated water, suggesting that LTB_4 was extensively degraded by β-oxidation (221).

Receptors and Action

Leukotrienes have potent biological actions. Comprehensive reviews of the biological effects of the leukotrienes have been published (222, 223). Very briefly, the peptidoleukotrienes contract respiratory, vascular, and intestinal smooth muscles. The action of peptidoleukotrienes on the vasculature include arteriolar constriction, dilation of venules, and plasma exudation. They are coronary vasoconstrictors, have a negative inotropic effect on the heart, and are arrhythmogenic. The effects on the respiratory system include constriction of bronchi, especially the smaller airways, and increased mucus secretion. Among the peptidoleukotrienes LTE_4 is generally less potent than LTC_4 and LTD_4. LTB_4 is a potent chemotactic agent for neutrophils and eosinophils. It can cause plasma exudation, but is less potent than the peptidoleukotrienes and requires the presence of polymorphonuclear leukocytes. These actions have implicated the leukotrienes as mediators in asthma, immediate hypersensitivity reactions, inflammatory reactions, and myocardial infarction.

Leukotrienes appear to bind to specific receptors. The first evidence for multiple receptors for peptidoleukotrienes came from the observation that FPL55712 blocked the action of LTD_4 on guinea pig lung parenchyma strip but not that of LTC_4 (224, 225). The first evidence for multiple receptors for peptidoleukotrienes came from the observation that FPL55712, which was considered a selective SRS-A antagonist, blocked the action of LTD_4 on guinea pig lung parenchyma strip but not that of LTC_4 (224, 225). Using crude SRS-A, it was demonstrated that FPL55712 is a more potent antagonist on guinea pig ileum than on lung parenchymal strip (226). Results obtained from binding studies with radiolabeled ligands provide further evidence for multiple receptors for peptidoleukotrienes. $[^3H]LTC_4$ was found to bind specifically to uterine membranes with saturation kinetics. In the presence of calcium, binding was achieved in the 5 nM range, but in the absence of calcium the affinity decreased approximately 14-fold. $[^3H]LTD_4$ did not bind to the same membrane preparation. Binding was also enhanced by serine borate, which inhibits the conversion of LTC_4 to LTD_4 (227). A smooth muscle cell line derived from Syrian hamster vas deferens has also been shown to have specific binding sites for LTC_4 with a K_d of 5 nM and a receptor density of 250,000 per cell. The K_d for LTD_4 and LTE_4 was 100–1000 times greater than for LTC_4. Competition studies with various structural analogs of LTC_4 showed that compounds that had smooth muscle contractile activity effectively displaced $[^3H]LTC_4$. In contrast, an inactive analogue, lacking a free amino acid group, did not compete with $[^3H]LTC_4$ (228). Receptors for LTC_4 on rat lung membranes with a K_d of 40 nM have been observed. LTD_4 and LTE_4 displaced the LTC_4 with a K_d of only 4 μM and 10 μM, respectively (229). In guinea pig lung, receptors for LTD_4 have been described with binding or pharmacological studies that also display high affinity for LTE_4, but not for LTC_4 (230, 231). In addition, guanine nucleotides inhibit the binding of LTD_4 to guinea pig lung but not that of LTC_4 (232). Plasma membrane fractions of guinea pig ileum smooth muscle exhibited saturable binding sites for $[^3H]LTD_4$ with a K_d of 2.2 nM and for $[^3H]LTC_4$ with a K_d of 8.5 nM (233). The ratio of these two K_d values is comparable to the molar ratio of LTD_4 and LTC_4 initiating similar contractile responses on the intact tissue (234).

The binding sites for LTB_4 appear to be distinct from those of peptidoleukotrienes. LTB_4 did not cross-react with the receptors for peptidoleukotrienes (227, 229, 230). Binding studies with LTB_4 and polymorphonuclear leukocytes are problematic since these cells readily metabolize LTB_4 to 20-hydroxy-LTB_4. Specific saturable binding of LTB_4 to human polymorphonuclear leukocytes at 4°C has been observed (235–238). The 6-*trans* isomers of LTB_4 cross-reacted 30–50% (235, 236). High-and low-affinity receptors, with K_ds of 0.39 nM and 61 nM respectively, have been described that were thought to be associated with the chemotactic and secretagogue activity of LTB_4, respectively (237). Binding studies with LTB_4 comparing human and rat polymorphonuclear leukocytes showed marked species difference. Human polymorphonuclear leukocytes displayed high- and low-affinity binding sites with a K_d of 14 nM and 56,000 sites per cell, and a K_d of 280 nM and 469,000 sites per cell, respectively. Rat polymorphonuclear leukocytes exhibited only high-affinity binding with a K_d of 4.5 nM and 6400 sites per cell and did not metabolize LTB_4. Unlike human polymorphonuclear leukocytes, the cells from the rat also did not respond to LTB_4 by chemotaxis. Cells from both species aggregated to LTB_4 with similar specificity and magnitude (238). These studies suggest different receptors may be associated with the various functions of LTB_4.

CYTOCHROME p450

The cytochrome p450 enzymes are those heme proteins possessing a sulfur ligated to the iron that forms carbon monoxide complexes exhibiting a major absorption band at 450 nm. The enzymes are widely distributed in animal tissues, plants, and microorganisms, and appear to be a multigene family (239–242) whose main function is the monooxygenation of lipophilic substances. The cytochrome p450 enzymes exhibit a wide substrate heterogeneity as evidenced by differential inhibition with various monoclonal antibodies (243). A typical cytochrome p450–catalyzed oxygenation reaction proceeds as follows: $RH + O_2 + NAD(P)H + H^+ \rightarrow ROH + H_2O + NAD(P^+)$, where RH represents the substrate. In liver microsomes a flavoprotein (NADPH-cytochrome p450 reductase) containing both flavin mononucleotide and flavin adenine dinucleotide catalyzes electron transfer from reduced pyridine nucleotide to the cytochrome. The mechanisms involved in oxygen activation by cytochrome p450 have been previously described (244). In addition the enzyme can, independent of O_2 and NADPH, act as a peroxidase using organic hydroperoxides in the following reaction: $ROOH + R'H \rightarrow ROH + R'OH$ (244, 245).

Many drugs, chemicals, and endogenous compounds are oxygenated by cytochrome p450 in mammalian tissues and in some instances potentially toxic or carcinogenic epoxides are formed (246, 247). In addition to metabolism by the cyclooxygenase and lipoxygenase, naturally occurring olefins are oxygenated by microsomal cytochrome p450 enzymes (246–252).

Arachidonic Acid Metabolism

ω and ω-1 oxidation of arachidonic acid by cytochrome p450 was catalyzed by microsomal preparations of rabbit renal cortex (253, 254), purified renal cytochrome p450 (255), and microsomal preparations of rabbit liver (255–257). These metabolic transformations result in the formation of 19-OH and 19-oxo eicosatetraenoic acid (by ω-1 oxidation) and 20-OH-eicosatetraenoic and eicosatetraen-1,20-dioic acids (by ω oxidation). This (these) p450 enzyme(s) appear(s) to be inducible by β-napthoflavone (257). Other arachidonate metabolites have been demonstrated to be produced by a phenobarbital-inducible hepatic and renal microsomal cytochrome p450, including a series of products formed by epoxidation of the double bonds of the fatty acid, namely 14(15) oxido-, 11(12)oxido-, 8(9) oxido-, and 5(6) oxido- eicosatrienoic acid. The epoxides can then be further converted to vic-diols by hepatic and renal cortical microsomal and cytosolic fractions and by purified microsomal epoxide hydrolase (253, 257). In addition, both types of oxidations can occur in concert to produce trihydroxy fatty acids (258). Recently, a cytochrome p450 enzyme metabolizing arachidonic acid has been localized to the cells isolated from the thick ascending limb of the loop of Henlé of rabbit (259). Incubation of the cells with [^{14}C]-arachidonic acid produced major products that were chromatographically different from known lipoxygenase products and their synthesis was not inhibited by indomethacin (i.e. cyclooxygenase-independent) but was partially inhibited by SKF-525A and carbon monoxide, suggesting that they were produced by a cytochrome p450. Figure 1 indicates the potential metabolic transformations of arachidonic acid by microsomal p450.

Metabolism of Cyclooxygenase and Lipoxygenase Products

Early studies demonstrated that guinea pig kidney cortical microsomes could metabolize PGA$_1$ to the ω and ω-1 hydroxylation products 19 OH- and 20-OH-PGA$_1$ (260, 261). Later studies (262, 263) also demonstrated ω,ω-1 hydroxylation of PGE$_1$ by a guinea pig kidney cortical microsomal enzyme. Monooxygenase enzymes catalyzing ω ($K_m = 17$ μM) and ω-1 ($K_m = 130$ μM) hydroxylations have also been found in the liver (264). Enzyme mechanisms were studied in a reconstituted rabbit liver microsomal enzyme system containing a phenobarbital-inducible isozyme or a 5,6-benzoflavone-inducible isozyme of p450, NADPH-cytochrome p450 reductase, phosphatidylcholine, and NADPH (265). Significant metabolism of prostaglandins by the phenobarbital-inducible isozyme occurred only in the presence of cytochrome b$_5$. Under these conditions, PGE$_1$ hydroxylation was linear for 45 min and maximal rates were obtained with a 1:1:2 molar ratio of reductase:cytcochrome b$_5$:p450 isozyme. The phenobarbital-induced cytochrome p450 catalyzed the conversion of PGE$_1$, PGE$_2$, and PGA$_1$ to their respective 19- and 20-hydroxy metabolites in a ratio of 5:1. The benzoflavin-induced isozyme, which is also inducible by many carcinogens (266), differed from the phenobarbital-induced isozyme in that it catalyzed prostaglandin hydroxylation at substantial rates in the absence of cytochrome b$_5$, and was specific for position 19 of all three prostaglandins and had an order of activity PGA$_1 >$ PGE$_1 >$ PGE$_2$. However, addition of

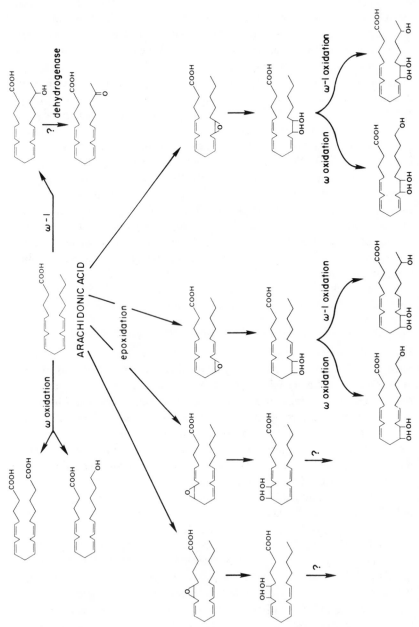

Figure 1 Metabolic pathways for arachidonic acid through ω, ω-1 oxidation and epoxygenase pathways.

cytochrome b_5 increased 19 hydroxylation 1.6- to 3-fold depending on substrate. Finally, it has recently been demonstrated that 5(6) epoxyeicosatrienoic acid is metabolized by sheep seminal vesicle microsomes to 5(6) epoxy-PGE_1, 5(6) epoxy-PGG_1, and 5(6) epoxy-PGH_1 and to the (5S,6S) and (5R,6R) isomers of 5-hydroxy-PGI_1 (267–269). The relative importance of cytochrome p450 metabolism compared to PG hydroxydehydrogenase is currently unknown.

Leukotrienes are further metabolized by polymorphonuclear leukocytes by ω oxidation to 5S,12R,20-trihydroxy-6-cis,8,10,-$trans$,14-cis-eicosatetraenoic acid (20-OH LTB_4) and to a dicarboxylic acid (20-COOH LTB_4) (218, 270). More recently, it has been demonstrated that the ω oxidation of LTB_4 in polymorphonuclear leukocytes is specific for LTB_4 and 5S,12S-dihydroxy-6,8,10,14 eicosatetraenoic acid (271). Furthermore, a sonicate of human neutrophils catalyzed an NADPH-dependent conversion of leukotriene B_4 to a more polar metabolite (220). The conversion did not occur in anaerobic conditions and was inhibited by carbon monoxide and p-chloromercuribenzoate but not by KCN, NaN_3, catalase, mannitol, or superoxide dismutase. However, the activity was observed in the 100,000 $\times$ g supernatant from the homogenate in which the cytochrome p450 was not detected.

Biological Effects

Incubations of anterior pituitary cells with the synthetically produced epoxygenase metabolites of arachidonic acid caused an increase in the release of luteinizing hormone (LH) (272). ETYA (10 μM) completely inhibited LHRH-stimulated release of LH from rat anterior pituitary cell suspensions while indomethacin was ineffective and nordihydroguiaretic acid was only marginally effective. The synthetic epoxyeicosatrienoic acids (273, 274) were used to stimulate LH secretion. The most potent metabolite was 5(6) epoxyeicosatrienoic acid, which exhibited an ED_{50} of 30–80 nM and a maximal response at 1 μM. It is unclear whether the effect is attributable to intact 5(6) EET or to stimulation of the formation of a cyclooxygenase metabolite. These studies should be repeated in the presence of indomethacin. The other epoxides were active but (11,12>8,9>14,15-epoxyeicosatrienoic acid) less active than 5(6) epoxyeicosatrienoic acid. Anterior pituitary cell microsomes were demonstrated to metabolize arachidonic acid to the four cis epoxyeicosatrienoic acids (5(6)-, 8(9)-, 11(12)-, and 14(15)- epoxyeicosatrienoic acids) (275).

Arachidonic acid was shown to stimulate in vitro release of somatostatin from rat hypothalamic median eminence (276). This effect is inhibited by ETYA but not by indomethacin. The microsomal fractions from the rat hypothalamus catalyzed the NADPH-dependent metabolism of arachidonic acid to the major products of 5,6-epoxyeicosatrienoic acid (ED_{50} = 10^{-8}M) and its hydration product 5,6-dihydroxyeicosatrienoic acid (ED_{50} = 5 $\times$

10^{-7}M), both of which were capable of stimulating somatostatin release. In a mouse pituitary tumor cell line (AtT-20), inhibitors of cytochrome p450 enzymes, SKF-525A and piperonyl butoxide (277), were demonstrated to suppress secretagogue-induced ACTH release (278), suggesting a role for the epoxides in pituitary function. In isolated rat pancreatic islets, 5(6) epoxyeicosatrienoic acid stimulates insulin release selectively while 14(15) > 11(12) > 8(9) epoxyeicosatrienoic acids release glucagon (279). These experiments need to be confirmed in other laboratories.

The role of the cytochrome p450 pathway in renal function is still not clearly defined. However, it has been reported that 5(6) epoxyeicosatrienoic acid or possibly one of its metabolites inhibits sodium absorption and potassium secretion in rabbit cortical collecting tubule (280). Furthermore, the arachidonic acid metabolite formed by the cytochrome p450 of medullary thick ascending limb of loop of Henlé inhibited vascular Na^+, K^+ ATPase activity (281).

ARACHIDONIC ACID METABOLISM IN TISSUE INJURY

The intrinsic capability to metabolize arachidonic acid is often dramatically altered in pathophysiologic states. In tissue injury, various inflammatory cells (e.g. neutrophils or monocytes), each of which is capable of metabolizing arachidonate to bioactive eicosanoids (282, 283), invade a tissue and interact with the resident cells (e.g. fibroblasts, smooth muscle cells, etc). This interaction often leads to a quantitative, and sometimes qualitative, change in the arachidonic acid metabolic capability of a tissue. Hydronephrosis (284), myocardial infarction (166), glomerulonephritis (285), renal vein constriction (286), pulmonary fibrosis (287), rheumatoid arthritis (288), and ulcerative colitis (289) all represent states of tissue injury in which the arachidonic acid metabolism of the involved tissue is markedly altered.

The hydronephrotic kidney (HNK) from rabbits with unilateral ureteral obstruction is a useful model for understanding the mechanisms of the enhanced arachidonate metabolism (284, 290) seen in inflammatory disorders. Histologically, the HNK exhibits changes characteristic of chronic inflammation: there is an invasion of mononuclear cells into the renal cortex and a proliferation of the fibroblast-like interstitial cells (284). Simultaneously, there is a dramatic increase in renal cyclooxygenase activity, especially in the cortex: the V_{max} of the cortical cyclooxygenase increases twentyfold (284, 291). Additionally, thromboxane synthetase activity becomes manifest in the HNK whereas no activity is present in the normal kidney (292).

These biochemical changes also occur in the ex vivo perfused HNK, which exhibits a time-dependent increase in PGE_2 and thromboxane A_2 production in response to the inflammatory peptide agonist bradykinin (290, 293). Normal kidneys produce a much more modest amount of PGE_2 and no thromboxane in response to peptide agonists, and show no change in metabolism during perfusion. The time-dependent increase in the production of arachidonate metabolites by the HNK can be blocked by inhibitors of protein synthesis (i.e. cycloheximide, or actinomycin D) indicating that the HNK actively synthesizes cyclooxygenase (290).

There is an intimate relationship between the inflammatory cell infiltrate seen in hydronephrosis and the increase in renal arachidonate metabolism as indicated by experiments utilizing the leukocyte agonist, endotoxin, and by ablation studies using nitrogen mustard. Endotoxin stimulates macrophages in culture, causing these inflammatory cells to produce PGE_2 and to become cytolytic (294). When endotoxin is injected into the ex vivo perfused HNK an immediate and sustained release of arachidonate metabolism is seen (295). Normal kidneys, which contain no inflammatory cells, do not respond to endotoxin. Administration of endotoxin to animals in vivo also markedly increases the production of arachidonate metabolites by the HNK but not the normal kidney (293). Treating hydronephrotic animals with nitrogen mustard renders animals leukopenic and prevents the invasion of macrophages into inflammatory sites. This treatment suppresses the peptide-stimulated arachidonate metabolism of the HNK and abolishes the response of the HNK to in vivo endotoxin (293). The macrophage thus appears to be a critical component of the enhanced arachidonic acid metabolism seen in the rabbit HNK.

The studies with nitrogen mustard additionally suggest that the macrophage is the cellular location of the thromboxane synthetase in this pathophysiologic state. Cortical microsomes from the HNK from nitrogen mustard–treated animals exhibit an attenuated thromboxane synthetase activity relative to the control HNK (293). Thus, the presence of thromboxane synthetase can be correlated with the presence of this inflammatory cell.

Macrophages are able to affect dramatically the proliferation and arachidonate metabolism of fibroblasts. Depletion of monocytes by antimacrophage serum and steroids delays the appearance of fibroblasts at sites of inflammatory injury and suppresses their proliferation (296). Macrophages are also able to elaborate a factor that stimulates fibroblast proliferation (297) and PGE_2 synthesis (298, 299). The interaction of macrophages and fibroblasts in the HNK, and how this interaction leads to an enhancement in arachidonate metabolism, is elucidated by culturing explants of cortical tissue. HNK explants yield a monolayer of both fibroblasts and monocytes, whereas normal kidney explants yield only fibroblasts (300). The cultures from the HNK exhibit an enhanced release of PGE_2 in response to the peptide agonist bradykinin and endotoxin relative to cultures from normal kidneys. Passaging cultures from the HNK leads to a depletion of the mononuclear cells, a decreased response to bradykinin, and a loss of the endotoxin response. Furthermore, conditioned medium from adherent rabbit peripheral blood mononuclear cells stimulates PGE_2

synthesis in both the HNK and normal kidney cultures. These studies show that macrophages in the HNK cause an increase in fibroblast arachidonate metabolism through an elaborated factor.

The nature of this factor (which has been termed MNCF for mononuclear cell factor) and its effects on fibroblasts have been widely investigated. MNCF results in a dramatic increase in PGE_2 synthesis from dermal fibroblasts (301), gingival fibroblasts (299), synovial cells (298), and lung fibroblasts (287). MNCF enhances both basal and bradykinin-induced fibroblast PGE_2 release (301). Studies with microsomes demonstrate that MNCF increases the V_{max} of the fibroblast cyclooxygenase (K_m is unchanged) (301). Both the enhanced basal and agonist-stimulated release in MNCF-pretreated cells are dependent on new protein synthesis (301). Thus, one can draw an analogy between these in vitro data and the intact perfused HNK in which the time-dependent increase in arachidonate metabolism is also dependent on new protein synthesis.

Partial purification of macrophage-conditioned medium yields a monokine of 11,000–13,000 daltons which enhances fibroblast proliferation (302) and prostaglandin synthesis (303). This factor comigrates with interleukin-1 in several chromatographic systems (302). Exogenous interleukin-1 is also able to stimulate fibroblast prostaglandin production (303, 304). Therefore, interleukin-1 may be an important constituent of MNCF and may mediate the communication between mononuclear cells and fibroblasts in inflammatory states.

In summary, the enhanced arachidonate metabolism in experimental hydronephrosis results from the invasion of macrophages and their interaction with the resident fibroblast-like interstitial cells. Ureteral obstruction triggers a regional inflammatory response resulting in the invasion of blood-borne monocytes which become tissue macrophages. These cells are capable of releasing (a) factor(s) (possibly interleukin-1) which stimulate(s) fibroblast proliferation, cortical microsomal cyclooxygenase activity, and PGE_2 release (i.e. intrinsic arachidonate metabolism). The enhanced thromboxane synthetase levels and thromboxane A_2 appear to come directly from the macrophage. This model of inflammatory injury thus serves to elucidate the general principles of how inflammatory cells interact with resident tissue cells to produce dramatic alterations in tissue arachidonate metabolism.

MANIPULATION OF ARACHIDONIC ACID METABOLISM BY ESSENTIAL FATTY ACID DEFICIENCY

Mammals were first shown to have an absolute requirement for dietary (n-6) fatty acids (i.e. linoleate and arachidonate) in the 1930s (305). A state of essential fatty acid (EFA) deficiency was induced by feeding animals a fat-free diet and a variety of pathophysiologic effects were noted: dermatitis, reproductive inefficiency, and papillary necrosis. This state of EFA deficiency is characterized biochemically by a decrease in (n-6) and (n-3) fatty acids and by an accumulation of (n-9) fatty acids, notably 20:3 (n-9) (termed Mead acid), which is not a constituent of normal tissues. The ratio of the Mead acid to arachidonate is used to define the deficiency state biochemically with a ratio of greater than 0.4 used as the dividing line between normal and EFA-deficient animals.

Since there is, in general, a decrease in tissue levels of arachidonate, EFA deficiency is a useful tool in probing the role of arachidonic acid and its metabolites (i.e. the prostaglandins and leukotrienes) in various physiologic and pathologic states. The accumulated triene, 20:3 (n-9), cannot be metabolized to prostaglandins (306) or leukotriene B_3 (210), although it can be metabolized to leukotriene C_3 (204). Using the deficiency state has advantages over using pharmacologic inhibitors of arachidonate metabolism (e.g. aspirin or indomethacin) in that one can perform long-term studies without running into problems with drug toxicity or the nonspecific effects of this class of drugs. Furthermore, using selected fatty acid realimentation (e.g. (n-3) fatty acids such as eicosapentaenoic acid), one can essentially restructure membranes.

The application of EFA deficiency, however, entails a detailed appreciation of its effects on the fatty acid content of individual tissues as well as on the fatty acid content of individual phospholipids within a tissue. The effects of EFA deprivation on the arachidonate content and phospholipid composition of different tissues are quite diverse (307). Hepatic lipids, for example, are readily depleted of arachidonate. In contrast, the renal cortex tenaciously retains arachidonate, whereas the heart can actually double its content of arachidonate even in the face of a maximal restriction of essential fatty acids. This increase in cardiac arachidonate is due solely to a fourfold increase in arachidonyl-phosphatidylethanolamine. The renal cortex, however, shows preservation of the arachidonate content in phosphatidylethanolamine, phosphatidylserine, and phosphatidylcholine. Phosphatidylinositol, on the other hand, shows a depletion of arachidonate in liver, heart, and kidney.

Experiments on the mechanism behind these tissue-specific changes in arachidonate using labeled fatty acids have elucidated a general mechanism for the conservation of arachidonic acid (307). Labeled arachidonate initially localizes to the liver following an intraperitoneal injection. Over a period of several days, however, the liver is depleted of label whereas the heart and renal cortex accumulate labeled arachidonate. The EFA-deficient heart and renal cortex accumulate three to five times the activity of the control heart and renal cortex, respectively. The EFA-deficient heart accumulates arachidonate selectively in phosphatidylethanolamine (eight times greater than control), whereas the EFA-deficient renal cortex accumulates arachidonate in phosphatidylethanolamine, phosphatidylserine, and phosphatidylcholine. This enhanced uptake is specific for arachidonate over 20 : 3 (n-9). These studies suggest that the liver serves to supply other tissues with arachidonate in EFA deficiency, and that the heart and renal cortex both contain mechanisms to accumulate arachidonate selectively in certain phospholipids.

One notable effect of EFA deficiency is the ability of this state to abrogate the inflammatory nephritis seen in a murine model of the human disease systemic lupus erythematosus (308). The mechanism of the salutary effect of EFA deficiency may in part be explained by the effects of EFA depletion on inflammatory cells. EFA deficiency leads to a modest depletion of membrane arachidonate in neutrophils and its replacement by the Mead acid (211). These neutrophils, however, form virtually no leukotriene B_4 (211), which may be an important chemotactic stimulus in inflammation (309). Little or no leukotriene B_3 is formed, either (211). The decrease in LTB_4 production appears to be due to inhibition by a lipoxygenase metabolite of the Mead acid (211). Exogenous Mead acid inhibits the conversion of LTA_4 to LTB_4 by rat basophilic leukemic cell homogenates, and this inhibition is blocked by nordihydroguaiaretic acid, a lipoxygenase inhibitor.

Diets rich in (n-3) fatty acids can also protect against the inflammatory nephritis in murine lupus (310). Again the beneficial effects of (n-3) fatty acid substitution may be explainable by the effects of the (n-3) fatty acid enrichment on inflammatory cells. Macrophages from mice with murine lupus fed a fish-oil diet synthesize less PGE_2 and TxB_2, and exhibit a decreased degree of Ia positivity (311). Furthermore, neutrophils can incorporate eicosapentaenoic acid and metabolize this fatty acid to LTB_5 (213). This leukotriene has only one tenth the potency of LTB_4 in inducing chemotaxis or neutrophil aggregation (312).

Another important effect of EFA deficiency is its ability to cause a scaly dermatitis and an increase in transepidermal water loss (305). Studies using the novel (n-6) fatty acid, columbinic acid (5E, 9Z, 12Z-octadecatrienoic acid), have elucidated the role of (n-6) fatty acids in maintaining epidermal integrity. Columbinic acid is metabolized via cyclooxygenase to a 9-hydroxyoctadecatri-enoic acid, and is also metabolized via lipoxygenase to a 13-hydroxyoctadecat-rienoic acid (313). This latter product is the primary columbinate metabolite produced by epidermal homogenates. No cyclized (i.e. prostaglandinlike) products are formed. Both native columbinic acid and its 13-hydroxy metabo-lite are capable of reversing the scaly dermatitis of EFA deficiency, whereas the 9-hydroxy metabolite is inactive (313). Neither hydroxy metabolite, however, is able to reverse the increase in transepidermal water loss; only the parent compound is active (314). It is of note that epidermis lacks the ability to desaturate and chain elongate eighteen-carbon fatty acids (315). These studies show that epidermal integrity is a complex function of (n-6) fatty acids and their lipoxygenase metabolites, and that prostaglandins are not essential for epidermal integrity.

A recently described effect of EFA deficiency is its ability to potentiate the anesthetic effect of volatile anesthetics (316). EFA-deficient animals are significantly more sensitive to the action of halothane, isoflurane, and cyclopropane than normal animals. Barbiturate sleeping times are unaffected by EFA deficiency. This increase in anesthetic potency cannot by reversed by (n-3) fatty acid supplementation and is not a general function of membrane unsaturation, but can be specifically reversed by realimentation with linoleate. When brain fatty acid composition is analyzed it is seen that the reversal induced by linoleate is accompanied by a selective normalization of the percentage of arachidonate in phosphatidylinositol. Arachidonate percentage in the other brain phospholipids changes modestly or not at all. These data suggest that arachidonyl-phosphatidylinositol is important to the action of volatile anesthetics.

In summary, EFA deficiency is a useful probe for elucidating the role of arachidonate and its metabolites in both physiologic and pathologic situations. The deficiency state has proven valuable in helping to understand the importance of eicosanoids in inflammation and epidermal function, and may help clarify the mechanism of action of volatile anesthetics.

Literature Cited

1. Samuelsson, B., Granström, E., Green, K., Hamberg, M., Hammarström, S. 1975. *Ann. Rev. Biochem.* 44:669–95
2. Samuelsson, B., Goldyne, M., Granström, E., Hamberg, M., Hammarström, S., Malmsten, C. 1978. *Ann. Rev. Biochem.* 47:997–1029
3. Needleman, P., Raz, A., Minkes, M. S., Ferendelli, J. A., Sprecher, H. 1979. *Proc. Natl. Acad. Sci. USA* 76:944–48
4. Granström, E. 1984. *Inflammation* 8: S15–S25 (Suppl.)
5. Sprecher, H., Van Rollins, M., Sun, F., Wyche, A., Needleman, P. 1982. *Proc. Natl. Acad. Sci. USA* 257:3912–18
6. Pace-Asciak, C. R., Smith, W. L. 1983. *Enzymes* 16:543–603
7. Pagels, W. R., Sachs, R. J., Marnett, L. J., DeWitt, D. L., Day, J. S., Smith, W. L. 1983. *J. Biol. Chem.* 258:6517–23
8. Roth, G. J., Machuga, E., Strittmatter, P. 1981. *J. Biol. Chem.* 256:10018–22
9. Rollins, T. E., Smith, W. L. 1980. *J. Biol. Chem.* 255:4872–75
10. Hemler, M. E., Lands, W. E. M. 1980. *J. Biol. Chem.* 255:6253–61
11. Lands, W. E. M. 1984. *Prostaglandins Leukotrienes Med.* 13:35–46
12. Kent, R. S., Diedrich, S. L., Whorton, A. R. 1983. *J. Clin. Invest.* 72:455–65
13. Brotherton, A. F. A., Hoak, J. C. 1983. *J. Clin. Invest.* 72:1255–61
14. Ohki, S., Ogino, N., Yamamoto, S., Hayaishi, O. 1979. *J. Biol. Chem.* 254:829–36
15. Egan, R. W., Gale, P. H., Kuehl, F. A. 1979. *J. Biol. Chem.* 254:3295–302
16. Reed, G. A., Brooks, E. A., Eling, T. E. 1984. *J. Biol. Chem.* 259:5591–95
17. Josephy, P. D., Eling, T. E., Mason, R. P. 1983. *J. Biol. Chem.* 258:5561–69
18. Zenser, T. V., Mattammal, M. B., Wise, R. W., Rice, J. R., Davis, B. B. 1983. *J. Pharmacol. Exp. Ther.* 227:545–50
19. Mizuno, K., Yamamoto, S., Lands, W. E. M. 1982. *Prostaglandins* 23:743–57
20. Ullrich, V., Haurand, R. 1983. *Adv. Prostaglandin Thromboxane Leukotriene Res.* 11:105–10
21. Needleman, P., Raz, A., Ferrendelli, J. A., Minkes, M. 1977. *Proc. Natl. Acad. Sci. USA* 74:1716–20
22. FitzGerald, G. A., Brash, A. R., Oates, J. A., Pedersen, A. R. 1983. *J. Clin. Invest.* 71:1336–43
23. Ito, T., Ogawa, K., Sakai, K., Watanabe, J., Satake, T., et al. 1983. *Adv. Prostaglandin Thromboxane Leukotriene Res.* 11:245–51
24. Corey, E. J., Pyne, S. J., Schafer, A. I. 1983. *Tetrahedron Lett.* 24:3291–94
25. Gorman, R. R. 1983. *Adv. Prostaglandin Thromboxane Leukotriene Res.* 11:235–40
26. Nicolaou, K. C., Magolda, R. L., Smith, J. B., Aharony, D., Smith, E. F., et al.

1979. *Proc. Natl. Acad. Sci. USA*
76:2566–70
27. LeBreton, G. C., Venton, D. L., Enke,
S. E., Halushka, P. V. 1979. *Proc. Natl.
Acad. Sci. USA* 76:4097–101
28. Fitzpatrick, F. A., Bundy, G. L., Gorman, R. R., Honohan, T. 1978. *Nature*
275:764–66
29. Roberts, L. J., Sweetman, B. J., Oates,
J. A. 1981. *J. Biol. Chem.* 256:8384–93
30. FitzGerald, G. A., Pedersen, A. K., Patrono, C. 1983. *Circulation* 67:1174–77
31. Johnson, R. A., Morton, D. R., Kinner,
J. A., Gorman, R. R., McGuire, J. C., et
al. 1976. *Prostaglandins* 12:915–28
32. Pace-Asciak, C. R., Wolfe, L. S. 1971.
Biochemistry 10:3657–64
33. Kulkarni, P. S., Roberts, R., Needleman, P. 1976. *Prostaglandins* 12:337–53
34. Isakson, P. C., Raz, A., Denny, S. E.,
Pure, E., Needleman, P. 1977. *Proc.
Natl. Acad. Sci. USA* 74:101–5
35. Gryglewski, R. J., Bunting, S., Moncada, S., Flower, R. J., Vane, J. R. 1976.
Prostaglandins 12:685–713
36. Smith, W. L., DeWitt, D. L., Allen, M.
L. 1983. *J. Biol. Chem.* 258:5922–26
37. DeWitt, D. L., Day, J. S., Sonnenberg,
W. K., Smith, W. L. 1983. *J. Clin. Invest.* 72:1882–88
38. Charo, I. F., Shak, S., Karasek, M. A.,
Davison, P. A., Goldstein, I. M. 1984. *J.
Clin. Invest.* 74:914–19
39. DeWitt, D. L., Smith, W. L. 1983. *J.
Biol. Chem.* 258:3285–93
40. Ullrich, V., Castle, L., Weber, P. 1981.
Biochem. Pharmacol. 30:2033–36
41. Salmon, J. A., Smith, D. R., Flower, R.
J., Moncada, S., Vane, J. R. 1978.
Biochim. Biophys. Acta 523:250–62
42. Graf, H., Ruf, H.-H., Ullrich, V. 1983.
Angew. Chem. Int. Ed. Engl. 22:487–88
43. Ham, E. A., Egan, R. W., Soderman, D.
D., Gale, P. H., Kuehl, F. A. 1979. *J.
Biol. Chem.* 254:2191–94
44. Weksler, B., Ley, C. W., Jaffe, E. A.
1978. *J. Clin. Invest.* 62:923–30
45. Boeynaems, J. M., Galand, N. 1983.
Biochem. Biophys. Res. Commun. 112:
290–96
46. Hassid, A. 1984. *Biochem. Biophys.
Res. Commun.* 123:21–26
47. Marcus, A. J., Weksler, B. B., Jaffe, E.
A., Broekman, M. J. 1980. *J. Clin. Invest.* 66:979–86
48. Schafer, A. I., Crawford, D. D., Gim-
brone, M. A. 1984. *J. Clin. Invest.*
73:1105–12
49. Needleman, P., Wyche, A., Raz, A.
1979. *J. Clin. Invest.* 63:345–49
50. Baenziger, N. L., Fogerty, F. J., Mertz,
L. F., Chernuta, L. F. 1981. *Cell*
24:915–23

51. Hong, S. L., Deykin, D. 1982. *J. Biol.
Chem.* 257:7151–54
52. Evans, C. E., Billington, D., McEvoy,
F. A. 1984. *Prostaglandins Leukotrienes
Med.* 14:255–66
53. Hyman, B. T., Stoll, L. L., Spector, A.
A. 1982. *Biochim. Biophys. Acta* 713:
375–85
54. Hopkins, N. K., Gorman, R. R. 1981.
Biochim. Biophys. Acta 663:457–66
55. Gorman, R. R., Bunting, S., Miller, O.
V. 1977. *Prostaglandins* 13:377–88
56. Hajjar, D. P., Weksler, B. B., Falcone,
D. J., Hefton, J. M., Tack-Goldman, K.,
et al. 1982. *J. Clin. Invest.* 70:479–88
57. Pohlman, T., Yates, J., Needleman, P.,
Klahr, S. 1983. *Am. J. Physiol.* 244:
F270–77
58. Leigh, P. J., Cramp, W. A., MacDer-
mot, J. 1984. *J. Biol. Chem.* 259:12431–
36
59. Rosenkranz, B., Fischer, C., Weimer,
K. E., Frolich, J. C. 1980. *J. Biol.
Chem.* 255:10194–98
60. Blair, I. A., Barrow, S. E., Waddell, K.
A., Lewis, P. J., Dollery, C. T. 1982.
Prostaglandins 23:579–89
61. Rosenkranz, B., Fischer, C., Frolich, J.
C. 1981. *Clin. Pharmacol. Ther.* 29:
420–24
62. Wong, P. Y-K., Malik, K. U., Desiderio, D. M., McGiff, J. C., Sun, F. F.
1980. *Biochem. Biophys. Res. Commun.*
93:486–94
63. Miller, O. V., Aiken, J. W., Shebuski,
R. J., Gorman, R. R. 1980. *Prostaglandins* 20:391–99
64. Watanabe, T., Narumiya, S., Shimizu,
T., Hayaishi, O. 1982. *J. Biol. Chem.*
257:14847–53
65. Shimizu, T., Yamamoto, S., Hayaishi,
O. 1979. *J. Biol. Chem.* 259:5222–28
66. Crist-Hazelhof, E., Nugteren, D. H.
1979. *Biochim. Biophys. Acta* 572:43–51
67. Shimizu, T., Mizuno, M., Amano, T.,
Hayaishi, O. 1979. *Proc. Natl. Acad.
Sci. USA* 76:6231
68. Lewis, R. A., Soter, N. A., Diamond, P.
T., Austen, K. F., Oates, J. A., et al.
1982. *J. Immunol.* 129:1627–31
69. Watanabe, T., Shimizu, T., Narumiya,
S., Hayaishi, O. 1982. *Arch. Biochem.
Biophys.* 216:372–79
70. Schafer, A. I., Cooper, B., O'Hara, D.,
Handin, R. I. 1979. *J. Biol. Chem.*
254:2914–17
71. Wasserman, M. A., DuCharme, D. W.,
Griffin, R. L., DeGraaf, G. L., Robinson, F. G. 1977. *Prostaglandins* 13:255–
69
72. Hardy, C. C., Robinson, C., Tatters-
field, A. E., Holgate, S. T. 1984. *N.
Engl. J. Med.* 311:209–13

73. Kondo, K., Shimizu, T., Hayaishi, O. 1981. *Biochem. Biophys. Res. Commun.* 98:648–55
74. Hemker, D. P., Aiken, J. W. 1980. *Prostaglandins* 20:321–32
75. Ueno, R., Honda, K., Inoue, S., Hayaishi, O. 1983. *Proc. Natl. Acad. Sci. USA* 80:1735–37
76. Ellis, C. K., Smigel, M. D., Oates, J. A., Oelz, O., Sweetman, B. J. 1979. *J. Biol. Chem.* 254:4152–63
77. Roberts, L. J., Sweetman, B. J., Lewis, R. A., Austen, K. F., Oates, J. A. 1980. *N. Engl. J. Med.* 303:1400–4
78. Roberts, L. J. 1982. *Methods Enzymol.* 86:559–70
79. Liston, T. E., Roberts, L. J. 1985. *Proc. Natl. Acad. Sci. USA.* 82:6030–34
80. Fitzpatrick, F. A., Wynalda, M. A. 1983. *J. Biol. Chem.* 258:11713–18
81. Fukushima, M., Kato, T., Ota, K., Arai, Y., Narumiya, S., et al. 1982. *Biochem. Biophys. Res. Commun.* 109:626–33
82. Ogino, N., Miyamoto, T., Yamamoto, S., Hayaishi, O. 1977. *J. Biol. Chem.* 252:890–95
83. Sheng, W. Y., Lysz, T. A., Wyche, A., Needleman, P. 1983. *J. Biol. Chem.* 258:2188–92
84. Sheng, W. Y., Wyche, A., Lysz, T., Needleman, P. 1982. *J. Biol. Chem.* 257:14632–34
85. Currie, M. G., Needleman, P. 1984. *Ann. Rev. Physiol.* 46:327–41
86. Sato, M., Dunn, M. J. 1984. *Am. J. Physiol.* 247:F423–33
87. Garcia-Perez, A., Smith, W. L. 1984. *J. Clin. Invest.* 74:63–74
88. Freeman, R. H., Davis, J. O., Villarreal, D. 1984. *Circ. Res.* 54:1–9
89. Bonney, R. J., Humes, J. L. 1984. *J. Leukocyte. Biol.* 35:1–10
90. Chensue, S. W., Kunkel, S. L. 1983. *Clin. Lab. Med.* 3:677–94
91. Goodwin, J. S., Ceuppens, J. 1983. *J. Clin. Immunol.* 3:295–315
92. Davies, P., Bailey, P. J., Goldenberg, M. M. 1984. *Ann. Rev. Immunol.* 2:335–57
93. Ruzicka, T., Printz, M. P. 1984. *Rev. Physiol. Biochem. Pharmacol.* 100:122–60
94. Stenson, W. F., Parker, C. N. 1983. *Clin. Rev. Allergy* 1:369–84
95. Hamberg, M., Hamberg, G. 1980. *Biochem. Biophys. Res. Commun.* 95:1090–97
96. Nugteren, D. H. 1982. *Methods Enzymol.* 86:49–54
97. Aharony, D., Smith, J. B., Silver, M. J. 1984. In *The Leukotrienes*, ed. L. W. Chakrin, D. M. Bailey, pp. 103–23. Orlando, Fla: Academic. 308 pp.
98. Lagarde, M., Croset, M., Authi, K. S., Crawford, M. 1984. *Biochem. J.* 222:495–500
99. Siegel, M. I., McConnell, R. T., Porter, N. A., Cuatrecasas, P. 1980. *Proc. Natl. Acad. Sci. USA* 77:308–12
100. Yokoyama, C., Mizuno, K., Mitachi, H., Yoshimoto, T., Yamamoto, S., et al. 1983. *Biochim. Biophys. Acta.* 750:237–43
101. Hwang, D. H. 1982. *Lipids* 17:845–47
102. Hamasaki, Y., Tai, H.-H. 1984. *Biochim. Biophys. Acta* 793:393–98
103. Salari, H., Braquet, P., Borgeat, P. 1984. *Prostaglandins Leukotrienes Med.* 13:53–60
104. Van Wauwe, J., Goossens, J. 1983. *Prostaglandins* 26:725–30
105. Bryant, R. W., Simon, T. C., Bailey, J. M. 1982. *J. Biol. Chem.* 257:14937–43
106. Chang, W.-C., Nakao, J., Orimo, H., Murota, S. 1982. *Biochem. J.* 202:771–76
107. Bryant, R. W., Simon, T. C., Bailey, J. M. 1983. *Biochem. Biophys. Res. Commun.* 117:183–89
108. Walker, I. C., Jones, R. L., Wilson, N. H. 1979. *Prostaglandins* 18:173–78
109. Bryant, R. W., Bailey, J. M. 1979. *Prostaglandins* 17:9–18
110. Pace-Asciak, C. R. 1984. *Biochim. Biophys. Acta* 793:485–88
111. Pace-Asciak, C. R. 1984. *J. Biol. Chem.* 259:8332–37
112. Pace-Asciak, C. R., Mizuno, K., Yamamoto, S. 1983. *Prostaglandins* 25:79–84
113. Pace-Asciak, C. R., Granström, E., Samuelsson, B. 1983. *J. Biol. Chem.* 258:6835–40
114. Borgeat, P., Picard, S., Vallerand, P. 1981. *Prostaglandins Leukotrienes Med.* 6:557–80
115. Wong, P. Y.-K., Westlund, P., Hamberg, M., Granström, E., Chao, P. H.-W., et al. 1984. *J. Biol. Chem.* 259:2683–86
116. Marcus, A. J., Safier, L. B., Ullman, H. L., Broekman, M. J., Islam, N., et al. 1984. *Proc. Natl. Acad. Sci. USA* 81:903–7
117. Aharony, D., Smith, J. B., Silver, M. J. 1982. *Biochim. Biophys. Acta* 718:193–200
118. Schafer, A. I. 1982. *N. Engl. J. Med.* 306:381–86
119. Nakao, J., Ooyama, T., Chang, W.-C., Murota, S., Orimo, H. 1982. *Atherosclerosis* 43:143–50
120. Nakao, J., Ooyama, T., Ito, H., Chang, W.-C., Murota, S. I. 1982. *Atherosclerosis* 44:339–42
121. Maclouf, J., Fruteau de Laclos, B.,

Borgeat, P. 1982. *Proc. Natl. Acad. Sci. USA* 79:6042–46
122. Turk, J., Colca, J. R., Kotagal, N., McDaniel, M. L. 1984. *Biochim. Biophys. Acta* 794:110–24
123. Turk, J., Colca, J. R., Kotagal, N., McDaniel, M. L. 1984. *Biochim. Biophys. Acta* 794:125–36
124. Turk, J., Colca, J. R., McDaniel, M. L. 1985. *Biochim. Biophys. Acta* 834:23–36
125. Turk, J., Wolf, B. A., Comens, P. C., Colca, J., Jakschik, B., et al. 1985. *Biochim. Biophys. Acta* 835:1–17
126. Metz, S. A. 1985. *Proc. Natl. Acad. Sci. USA* 82:198–202
127. Yamamoto, S., Ishii, M., Nakadate, T., Nakaki, T., Kato, R. 1983. *J. Biol. Chem.* 258:12149–52
128. Pace-Asciak, C. R., Martin, J. M. 1984. *Prostaglandins Leukotrienes Med.* 16:173–80
129. Stenson, W. F., Parker, C. W., Sullivan, T. J. 1980. *Biochem. Biophys. Res. Commun.* 96:1045–52
130. Stenson, W. F., Parker, C. W. 1980. *J. Immunol.* 124:2100–4
131. Narumiya, S., Salmon, J. A., Cottee, F. H., Weatherley, B. C., Flower, R. J. 1981. *J. Biol. Chem.* 256:9583–92
132. Bryant, R. W., Bailey, J. M., Schewe, T., Rapoport, S. M. 1982. *J. Biol. Chem.* 257:6050–55
133. Rapoport, S., Hartel, B., Hausdorf, G. 1984. *Eur. J. Biochem.* 139:573–76
134. Wiesner, R., Hausdorf, G., Anton, M., Rapoport, S. 1983. *Biomed. Biochim. Acta* 42:431–36
135. Rapoport, S. M., Schewe, T., Wiesner, R., Halangk, W., Ludwig, P., et al. 1979. *Eur. J. Biochem.* 96:545–61
136. Maas, R. L., Turk, J., Oates, J. A., Brash, A. R. 1982. *J. Biol. Chem.* 257:7056–67
137. Turk, J., Maas, R. L., Brash, A. R., Roberts, L. J., Oates, J. A. 1982. *J. Biol. Chem.* 257:7068–76
138. Turk, J., Rand, T. H., Maas, R. L., Lawson, J. A., Brash, A. R., et al. 1983. *Biochim. Biophys. Acta* 750:78–90
139. Maas, R. L., Brash, A. R., Oates, J. A. 1981. *Proc. Natl. Acad. Sci. USA* 78:5523–27
140. Jubiz, W., Radmark, O., Lindgren, J. A., Malmsten, C., Samuelsson, B. 1981. *Biochem. Biophys. Res. Commun.* 99:976–86
141. Maas, R. L., Brash, A. R. 1983. *Proc. Natl. Acad. Sci. USA* 80:2884–88
142. Sok, D.-E., Han, C.-O., Shieh, W.-R., Zhou, B.-N., Sih, C. J. 1982. *Biochem. Biophys. Res. Commun.* 104:1363–70
143. Sok, D.-E., Chung, T., Sih, C. J. 1983.

Biochem. Biophys. Res. Commun. 110:273–79
144. Verhagen, J., Bruynzeel, P. L. B., Koedam, J. A., Wassink, G. A., de Boer, M., et al. 1984. *FEBS Lett.* 168:23–28
145. Henderson, W. R., Harley, J. B., Fauci, A. S. 1984. *Immunology* 51:679–86
146. Borgeat, P., Fruteau de Laclos, B., Rabinovitch, H., Picard, S., Braquet, P., et al. 1984. *J. Allergy Clin. Immunol.* 74:310–15
147. Turk, J., Maas, R. L., Roberts, L. J., Brash, A. R., Oates, J. A. 1983. *Adv. Prostaglandin Thromboxane Leukotriene Res.* 11:123–31
148. Serhan, C. N., Hamberg, M., Samuelsson, B. 1984. *Biochem. Biophys. Res. Commun.* 118:943–49
149. Serhan, C. N., Hamberg, M., Samuelsson, B. 1984. *Proc. Natl. Acad. Sci. USA* 81:5335–39
150. Wong, P. Y.-K., Hughes, R., Lam, B. 1985. *Biochem. Biophys. Res. Commun.* 126:763–72
151. Kuhn, H., Wiesner, R., Stender, H. 1984. *FEBS Lett.* 177:255–59
152. Vanderhoek, J. Y., Bryant, R. W., Bailey, J. M. 1982. *Biochem. Pharm.* 31:3463–67
153. Vanderhoek, J. Y., Tare, N. S., Bailey, J. M., Goldstein, A. L., Pluznik, D. H. 1982. *J. Biol. Chem.* 257:12191–95
154. Shak, S., Perez, D. H., Goldstein, I. M. 1983. *J. Biol. Chem.* 258:14948–53
155. Evans, J., Ford-Hutchinson, A. W., Fitzsimmons, B., Rokach, J. 1984. *Prostaglandins* 28:435–38
156. Radmark, O., Serhan, C., Hamberg, M., Lundberg, U., Ennis, M. D., et al. 1984. *J. Biol. Chem.* 259:13011–16
157. Borgeat, P., Samuelsson, B. 1979. *J. Biol. Chem.* 254:7865–69
158. Jorg, A., Henderson, W. R., Murphy, R. C., Klebanoff, S. J. 1982. *J. Exp. Med.* 155:390–402
159. Rouzer, C. A., Scott, W. A., Hamill, A. L., Cohn, Z. A. 1980. *J. Exp. Med.* 152:1236–47
160. Razin, E., Menzia-Huerta, J. M., Stevens, R. L., Lewis, R. A., Liu, F. T., et al. 1983. *J. Exp. Med.* 157:189–201
161. Mencia-Huerta, J. M., Razin, E., Ringle, E. W., Corey, E. J., Hoover, D., et al. 1983. *J. Immunol.* 130:1885–90
162. Ziboh, V. A., Marcelo, C. L., Voorhees, J. J. 1981. *J. Invest. Dermatol.* 76:307
163. Morris, R. H., Taylor, G. W., Piper, P. J., Tippins, J. R. 1980. *Nature* 285:104–6
164. Malik, K. U., Wong, P. Y.-K. 1981.

Biochem. Biophys. Res. Commun. 103: 511–20
165. Lindgren, L. A., Hökfelt, T., Dahlen, S. E., Patrono, C., Samuelsson, B. 1984. *Proc. Natl. Acad. Sci. USA* 81:6212–16
166. Evers, A. S., Murphree, S., Saffitz, J. E., Jakschik, B. A., Needleman, P. 1985. *J. Clin. Invest.* 75:992–99
167. Feldberg, W., Kellaway, C. H. 1938. *J. Physiol.* 94:187–226
168. Kellaway, C. H., Trethewie, E. F. 1940. *Q. J. Exp. Physiol. Cogn. Med. Sci.* 30:121–45
169. Murphy, R. C., Hammarström, S., Samuelsson, B. 1979. *Proc. Natl. Acad. Sci. USA* 76:4275–79
170. Hammarström, S., Murphy, R. C., Samuelsson, B., Clark, D. A., Mioskowski, C., Corey, E. J. 1979. *Biochem. Biophys. Res. Commun.* 91:1266–72
171. Hammarström, S. 1983. *Ann. Rev. Biochem.* 52:355–77
172. Morris, H. R., Taylor, G. W., Piper, P. J., Sirois, P., Tippins, J. R. 1978. *FEBS Lett.* 87:203–6
173. Samuelsson, B., Hammarström, S. 1980. *Prostaglandins* 19:645–48
174. Panossian, A., Hamberg, M., Samuelsson, B. 1982. *FEBS Lett.* 150:511–13
175. Borgeat, P., Samuelsson, B. 1979. *Proc. Natl. Acad. Sci. USA* 76:3213–17
176. Radmark, O., Malmsten, C., Samuelsson, B. 1980. *Biochem. Biophys. Res. Commun.* 96:1679–87
177. Orning, L., Hammarström, S., Samuelsson, B. 1980. *Proc. Natl. Acad. Sci. USA* 77:2014–17
178. Sok, D.-E., Pai, J.-K., Atrache, V., Kang, Y.-C., Sih, C. 1981. *Biochem. Biophys. Res. Commun.* 101:222–29
179. Anderson, M. A., Allison, R. D., Meister, A. 1982. *Proc. Natl. Acad. Sci. USA* 79:1088–91
180. Bernstrom, K., Hammarström, S. 1982. *Biochem. Biophys. Res. Commun.* 109: 800–4
181. Jakschik, B. A., Sun, F. F., Steinhoff, M. M. 1980. *Biochem. Biophys. Res. Commun.* 95:103–10
182. Jakschik, B. A., Lee, L. H. 1980. *Nature* 287:51–52
183. Ochi, K., Yoshimoto, T., Yamamoto, S. 1983. *J. Biol. Chem.* 258:5754–58
184. Jakschik, B. A., Falkenhein, S., Parker, C. W. 1977. *Proc. Natl. Acad. Sci. USA* 74:4577–81
185. Jakschik, B. A., Lee, L. H., Shuffer, G., Parker, C. W. 1978. *Prostaglandins* 16:733–48
186. Borgeat, P., Samuelsson, B. 1979. *Proc. Natl. Acad. Sci. USA* 76:2148–52
187. Tripp, C. S., Mahoney, M., Needleman, P. 1985. *J. Biol. Chem.* 260:5895–98
188. Parker, C. W., Koch, D., Huber, M. M., Falkenhein, S. F. 1980. *Biochem. Biophys. Res. Commun.* 94:1037–43
189. Jakschik, B. A., Kuo, C. G. 1983. *Adv. Prostaglandin Thromboxane Leukotriene Res.* 11:141–45
190. Jakschik, B. A., Kuo, C. G. 1983. *Prostaglandins* 25:767–82
191. Jakschik, B. A., Harper, T., Murphy, R. C. 1982. *J. Biol. Chem.* 257:5346–49
192. Sun, F. F., McGuire, J. C. 1984. *Biochim, Biophys. Acta* 794:56–64
193. Wei, Y. F., Jakschik, B. A. 1985. *Fed. Proc.* 44:986
194. Lewis, R. A., Goetzl, E. J., Drazen, J. M., Soter, N. A., Austen, K. F., Corey, E. J. 1981. *J. Exp. Med.* 154:1243–48
195. Furukawa, M., Yoshimoto, T., Ochi, K., Yamamoto, S. 1984. *Biochim. Biophys. Acta* 795:458–65
196. Parker, C. W., Aykent, S. 1982. *Biochem. Biophys. Res. Commun.* 109: 1011–16
197. Radmark, O., Shimikzu, T., Jornvall, H., Samuelsson, B. 1984. *J. Biol. Chem.* 259:12339–45
198. Egan, R. W., Tischler, A. N., Baptista, E. M., Ham, E. A., Soderman, D. D., Gale, P. H. 1983. *Adv. Prostaglandin Thromboxane Leukotriene Res.* 11:151–57
199. Ziboh, V. A., Casebolt, T. L., Marcelo, C. L., Voorhees, J. J. 1984. *J. Invest. Dermatol.* 83:248–51
200. Kuo, C. G., Lewis, M. T., Jakschik, B. A. 1984. *Prostaglandins* 28:929–38
201. Bach, M. K., Brashler, J. R., Morton, D. R. 1984. *Arch. Biochem. Biophys.* 230:455–65
202. Fitzpatrick, F., Haeggstrom, J., Granström, E., Samuelsson, B. 1983. *Proc. Natl. Acad. Sci. USA* 80:5425–29
203. Fitzpatrick, F., Ligget, W., McGee, J., Bunting, S., Morton, D., Samuelsson, B. 1984. *J. Biol. Chem.* 259:11403–7
204. Jakschik, B. A., Sams, A. R., Sprecher, H., Needleman, P. 1980. *Prostaglandins* 20:401–10
205. Jakschik, B. A., Kuo, C. G., Wei, Y. F. 1985. In *Prostaglandins, Leukotrienes and Cancer,* Vol. 1, *Basic Biochemical Processes,* ed. K. Horn, L. Marnett, pp. 51–75. Boston, Mass: Nijhoff
206. Corey, E. J., Shih, C., Cashman, J. R. 1983. *Proc. Natl. Acad. Sci. USA* 80:3581–84
207. Hammarström, S. 1981. *J. Biol. Chem.* 256:7712–14
208. Jubiz, W., Nolan, G. 1983. *Biochem. Biophys. Res. Commun.* 114:855–62
209. Wei, Y. F., Evans, R. W., Morrison, A. R., Sprecher, H., Jakschik, B. A. 1985. *Prostaglandins* 29:537–45

210. Jakschik, B. A., Morrison, A. R., Sprecher, H. 1983. *J. Biol. Chem.* 258:12797–800
211. Stenson, W. F., Prescott, S. M., Sprecher, H. 1984. *J. Biol. Chem.* 259:11784–89
212. Murphy, R. C., Pickett, W. C., Culp, B. R., Lands, W. E. M. 1981. *Prostaglandins* 22:613–22
213. Prescott, S. M. 1984. *J. Biol. Chem.* 259:7615–21
214. Hammarström, S. 1980. *J. Biol. Chem.* 255:7093–94
215. Jakschik, B. A., DiSantis, D. M., Sankarappa, S. K., Sprecher, H. 1982. *Adv. Prostaglandin Thromboxane Leukotriene Res.* 9:127–35
216. Bach, M. K., Brashler, J. R., Morton, D. R., Steel, L. K., Kaliner, M. A., Hugli, T. E. 1982. *Adv. Prostaglandin Thromboxane Leukotriene Res.* 9:103–14
217. Mannervik, B., Jensson, H., Alin, P., Hammarström, S. 1984. *FEBS Lett.* 174:289–93
218. Hansson, G., Lindgren, J. A., Dahlen, S. E., Hedqvist, P., Samuelsson, B. 1981. *FEBS Lett.* 130:107–12
219. Shak, S., Goldstein, I. M. 1984. *Biochem. Biophys. Res. Commun.* 123:475–81
220. Sumimoto, H., Takeshige, K., Sakai, H., Minakami, F. 1984. *Biochem. Biophys. Res. Commun.* 125:615–21
221. Serafin, W. E., Oates, J. A., Hubbard, W. C. 1984. *Prostaglandins* 27:899–911
222. Piper, P. 1984. *Physiol. Rev.* 64:744–61
223. Stjernschantz, J. 1984. *Med. Biol.* 62:215–30
224. Drazen, J. M., Austen, K. F., Lewis, R. A., Clark, D. A., Goto, G., et al. 1980. *Proc. Natl. Acad. Sci. USA* 77:4354–58
225. Krell, R. D., Osborn, R., Vickery, L., Falcone, K., O'Donnell, M., Gleason, J., et al. 1981. *Prostaglandins* 22:387–409
226. Fleisch, J. H., Haisch, K. D., Spaethe, S. M. 1982. *J. Pharmacol. Exp. Ther.* 221:146–57
227. Levinson, L. 1984. *Prostaglandins* 28:229–40
228. Krilis, S., Lewis, R. A., Corey, E. J., Austen, K. F. 1983. *J. Clin. Invest.* 72:1516–19
229. Pong, S.-S., DeHaven, R. N., Kuehl, F. A. Jr., Egan, R. W. 1983. *J. Biol. Chem.* 258:9616–19
230. Pong, S.-S., DeHaven, R. N. 1983. *Proc. Natl. Acad. Sci. USA* 80:7415–19
231. Krell, R. D., Tsai, B. S., Berdoulay, A., Barone, M., Giles, R. 1983. *Prostaglandins* 25:171–78
232. Hogaboom, G. K., Mong, S., Wu, H.

L., Crooke, S. T. 1983. *Biochem. Biophys. Res. Commun.* 116:1136–43
233. Krilis, S., Lewis, R. A., Corey, E. J., Austen, K. F. 1984. *Proc. Natl. Acad. Sci. USA* 81:4529–33
234. Lewis, R. A., Drazen, J. M., Austen, K. F., Clark, D. A., Corey, E. J. 1980. *Biochem. Biophys. Res. Commun.* 96:271–77
235. Goldman, D. W., Goetzl, E. J. 1982. *J. Immunol.* 129:1600–4
236. Kreisle, R. A., Parker, C. W. 1983. *J. Exp. Med.* 157:628–41
237. Goldman, D. W., Gifford, L. A., Goetzl, E. J. 1984. *Fed. Proc.* 43:845
238. Kreisle, R. A., Parker, C. W., Griffin, G. L., Senior, R. M., Stenson, W. F. 1985. *J. Immunol.* 134:3356–63
239. Lu, A. Y. H., West, S. B. 1980. *Pharmacol. Rev.* 31:277–96
240. Coon, M. J., Black, S. C., Koop, D. R., Morgan, E. T., Tarr, G. E. 1982. In *Microsomes, Drug Oxidations and Drug Toxicity,* ed. R. Sato, R. Kato, pp. 138–46. Japan Sci. Soc.
241. Nebert, D. W. 1979. *Mol. Cell. Biochem.* 27:27–46
242. Nebert, D. W., Eisen, H. J., Negishi, M., Lang, M. A., Hjelmeland, L. M. 1981. *Ann. Rev. Pharmacol. Toxicol.* 21:431–62
243. Flyino, T., Park, S. S., West, D., Gelboin, H. V. 1982. *Proc. Natl. Acad. Sci. USA* 79:3682–86
244. White, R. E., Coon, M. J. 1980. *Ann. Rev. Biochem.* 49:315–56
245. Wheeler, E. L. 1983. *Biochem. Biophys. Res. Commun.* 110:646–53
246. Wislocki, P. G., Miwa, G. T., Lu, A. Y. H. 1980. In *Enzymatic Basis of Detoxication,* ed. W. B. Jacoby, 1:135–82. New York: Academic
247. Robbins, E. L. 1961. *Fed. Proc.* 20:272–76
248. Wada, F., Shibata, H., Goto, M., Sakamoto, Y. 1968. *Biochem. Biophys. Acta* 162:518–24
249. Ellin, A., Orrenius, S. 1975. *Mol. Cell. Biochem.* 8:69–70
250. Ellin, A., Jakobsson, S. V., Schenkman, J. B., Orrenius, S. 1972. *Arch. Biochem. Biophys.* 150:64–71
251. Gibson, G. G., Cinti, D. L., Sligar, S. G., Schenkman, J. B. 1980. *J. Biol. Chem.* 255:1867–73
252. Cooper, D. Y. 1973. *Ann. NY Acad. Sci.* 1212:1–467
253. Morrison, A. R., Pascoe, N. 1981. *Proc. Natl. Acad. Sci. USA* 78:7375–78
254. Oliw, E. H., Lawson, J. A., Brash, A. R., Oates, J. A. 1981. *J. Biol. Chem.* 256:9924–31
255. Capdevila, J., Parkhill, L., Chacos, N.,

Okita, R., Masters, B. S., Estabrook, R. 1981. *Biochem. Biophys. Res. Commun.* 101:1357–63

256. Capdevila, J., Chacos, N., Werringloer, J., Prough, R. A., Estabrook, R. 1981. *Proc. Natl. Acad. Sci. USA* 78:5362–66

257. Oliw, E. H., Guengerich, P., Oates, J. A. 1982. *J. Biol. Chem.* 2157:3771–81

258. Oliw, E. H., Oates, J. A. 1981. *Prostaglandins* 22:863–71

259. Ferreri, N. R., Schwartzman, M., Ibrahim, N. G., Chandler, P. N., McGiff, J. C. 1984. *J. Pharmacol. Exp. Ther.* 231:441–48

260. Israelsson, U., Hamberg, M., Samuelsson, B. 1968. *Eur. J. Biochem.* 11:390–94

261. Samuelsson, B., Granström, E., Green, K., Hamberg, M. 1971. *Ann. NY Acad. Sci.* 180:138–63

262. Navarro, J., Piccolo, D. E., Kupfer, D. 1978. *Arch. Biochem. Biophys.* 191:125–33

263. Kupfer, D., Navarro, J., Piccolo, D. E. 1978. *J. Biol. Chem.* 253:2804–17

264. Theoharides, A. D., Kupfer, D. 1981. *J. Biol. Chem.* 256:2168–75

265. Vatsis, K. P., Theoharides, A. D., Kupfer, D., Coon, M. J. 1982. *J. Biol. Chem.* 257:11221–29

266. Haugen, D. A., Coon, M. J. 1976. *J. Biol. Chem.* 251:7929–39

267. Oliw, E. H. 1984. *J. Biol. Chem.* 259:2716–21

268. Oliw, E. H. 1984. *Biochem. Biophys. Acta* 793:408–15

269. Oliw, E. H. 1983. *FEBS Lett.* 172:279–83

270. Lindgren, J. A., Hansson, G., Samuelsson, B. 1981. *FEBS Lett.* 128:329–35

271. Shak, S., Goldstein, I. M. 1984. *J. Biol. Chem.* 259:10181–87

272. Snyder, G. D., Capdevila, J., Chacos, N., Manna, S., Falck, J. R. 1983. *Proc. Natl. Acad. Sci. USA* 80:3504–7

273. Falck, J. R., Manna, S. 1982. *Tetrahedron Lett.* 23:1755–56

274. Corey, E. J., Miwa, H., Falck, J. R. 1979. *J. Am. Chem. Soc.* 101:1586–87

275. Capdevila, J., Snyder, G. D., Falck, J. R. 1984. *FEBS Lett.* 178:319–22

276. Capdevila, J., Chacos, N., Falck, J. R., Manna, S., Negro-vilar A., Ojeda, S. 1983. *Endocrinology* 113:421–23

277. Testa, B., Jenner, P. 1981. *Drug Metab. Rev.* 12:1–117

278. Luini, A. G., Axelrod, J. 1985. *Proc. Natl. Acad. Sci. USA* 82:1012–14

279. Falck, J. R., Manna, S., Moltz, J., Chacos, N., Capdevila, J. 1983. *Biochem. Biophys. Res. Commun.* 114:743–49

280. Jacobson, H. R., Corona, S., Capdevila, J., Chacos, N., Manna, S. et al. 1983.

16th Ann. Meet. Am. Soc. Nephrology 193A

281. Schwartzman, M., Ferreri, N. R., Carroll, M. A., Songu-Mize, E., McGiff, J. C. 1985. *Nature* 314:620–22

282. Stenson, W. F., Parker, C. W. 1979. *J. Clin. Invest.* 64:1457–65

283. Humes, J. L., Sadowski, S., Galavage, M., Goldenberg, M., Subers, E., et al. 1982. *J. Biol. Chem.* 257:1591–94

284. Okegawa, T., Jonas, P. E., DeSchryver, K., Kawasaki, A., Needleman, P. 1983. *J. Clin. Invest.* 71:81–90

285. Lianos, E. A., Andres, G. A., Dunn, M. J. 1983. *J. Clin. Invest.* 72:1439–48

286. Zipser, R., Myers, S., Needleman, P. 1980. *Circ. Res.* 47:231–37

287. Clark, J. G., Kostal, K. M., Marino, B. A. 1983. *J. Clin. Invest.* 72:2082–91

288. McGuire, M. K., Meats, J. E., Ebsworth, N. M., Harvey, L., Murphy, G., et al. 1982. *Rheumatol. Int.* 2:113–20

289. Zipser, R. D., Patterson, J. B., Kao, H. W., Hauser, C. J., Locke, R. 1985. *Am. J. Physiol.* 249:6457–63

290. Needleman, P., Wyche, A., Bronson, S. D., Holmberg, S., Morrison, A. R. 1979. *J. Biol. Chem.* 254:9772–77

291. Sheng, W. Y., Lysz, T. A., Wyche, A., Needleman, P. 1983. *J. Biol. Chem.* 258:2188–92

292. Morrison, A. R., Nishikawa, K., Needleman, P. 1977. *Nature* 267:259–60

293. Lefkowith, J. B., Okegawa, T., DeSchryver-Kecskemeti, K., Needleman, P. 1984. *Kidney Int.* 26:10–17

294. Taffet, S. M., Russell, S. W. 1981. *J. Immunol.* 126:424–27

295. Okegawa, T., DeSchryver-Kecskemeti, K., Needleman, P. 1983. *J. Pharmacol. Exp. Ther.* 225:213–18

296. Leibovich, S. J., Ross, R. 1975. *Am. J. Pathol.* 78:71–91

297. Leibovich, S. J., Ross, R. 1976. *Am. J. Pathol.* 84:501–14

298. Dayer, J-M., Robinson, D. R., Krane, S. M. 1977. *J. Exp. Med.* 145:1399–404

299. D'Souza, S. M., Englis, D. J., Clark, A., Russell, R. G. G. 1981. *Biochem. J.* 198:391–96

300. Jonas, P. E., Leahy, K. M., DeSchryver-Kecskemeti, K., Needleman, P. 1984. *J. Leukocyte Biol.* 35:55–64

301. Whitely, P. J., Needleman, P. 1984. *J. Clin. Invest.* 74:2249–53

302. Schmidt, J. A., Oliver, C. N., Lepe-Zuniga, J. L., Green, I., Gery, I. 1984. *J. Clin. Invest.* 73:1462–72

303. Albrightson, C. R., Baenziger, N. L., Needleman, P. 1985. *J. Immunol.* 135:1872–77

304. Mizel, S. B., Dayer, J-M., Krane, S. M.,

Mergenhagen, S. E. 1981. *Proc. Natl. Acad. Sci. USA* 78:2474–77

305. Holman, R. T. 1969. *Prog. Chem. Fats Other Lipids* 9:275–348
306. Struijk, C. B., Beerthuis, R. K., Pabson, H. J. J., Van Dorp, D. A. 1966. *Rec. Trav. Chim. Pays-Bas* 85:1233–50
307. Lefkowith, J. B., Flippo, V., Sprecher, H., Needleman, P. 1985. *J. Biol. Chem.* In press
308. Hurd, E. R., Johnston, J. M., Okita, J. R., MacDonald, P. C., Ziff, M., Gilliam, J. N. 1981. *J. Clin. Invest.* 67:476–85
309. Palmblad, J., Malmsten, C. L., Uden, A. M., Radmark, O., Engstedt, L., Samuelsson, B. 1981. *Blood* 58:658–61
310. Prickett, J. D., Robinson, D. R., Steinberg, A. D. 1981. *J. Clin. Invest.* 68:556–59

311. Kelley, V. E., Ferretti, A., Izui, S., Strom, T. B. 1985. *J. Immunol.* 134:1914–19
312. Lee, T. H., Mencia-Huerta, J. M., Shih, C., Corey, E. J., Lewis, R. A., Austen, K. F. 1984. *J. Biol. Chem.* 259:2383–89
313. Elliott, W. J., Morrison, A. R., Sprecher, H. W., Needleman, P. 1985. *J. Biol. Chem.* 260:987–92
314. Elliott, W. J., Sprecher, H. W., Needleman, P. 1985. *Biochim. Biophys. Acta.* 835:158–60
315. Chapkin, R. S., Ziboh, V. A. 1984. *Biochem. Biophys. Res. Commun.* 124:784–92
316. Evers, A. E., Elliott, W. J., Lefkowith, J. B., Needleman, P. 1985. *J. Clin. Invest.* In press

Ann. Rev. Biochem. 1986. 55:103–36

SINGLE-STRANDED DNA BINDING PROTEINS REQUIRED FOR DNA REPLICATION

John W. Chase

Department of Molecular Biology, Albert Einstein College of Medicine, Bronx, New York 10461

Kenneth R. Williams

Department of Molecular Biophysics and Biochemistry, Yale University, New Haven, Connecticut 06510

CONTENTS

103

PERSPECTIVES AND SUMMARY

This review will focus on a class of proteins that have a very high affinity for single-stranded DNA (ssDNA), bind with no sequence specificity, and partici- pate in DNA metabolic reactions. As expected from their proposed mode of action, these proteins are required in amounts that are stoichiometric, rather than catalytic, with respect to the amount of ssDNA present, as are polymerases, helicases, etc. Many bind cooperatively to ssDNA and stimulate their cognate DNA polymerase. Often an apparently simple subject becomes increasingly complex as more is learned about it until order is restored by better understanding. So it is with ssDNA binding proteins and their cellular func- tions. Perhaps the most confusing issue recently raised is their identification and role in eukaryotes. We will review here the best-studied prokaryotic ssDNA binding proteins, the product of gene 32 of phage T4 (gp32) and the *Escherichia coli* SSB (the product of the *ssb* gene) in an attempt to establish general guidelines for this class of "DNA binding proteins." We will then review two of the best-studied eukaryotic ssDNA binding proteins, adenovirus DBP and calf thymus UP1, in order to compare them to the prokaryotic proteins. Whether ssDNA binding proteins with similar modes of action exist in both prokaryotes and eukaryotes is a question we cannot yet answer with certainty. We hope this presentation will aid investigators in thinking about the problem and ultimately advancing the field of DNA metabolism in all systems. No attempt has been made to review completely all ssDNA binding proteins and we apologize to our colleagues who work on important proteins of this type in other systems. Several earlier reviews pertinent to this subject are listed (1–12).

NOMENCLATURE

Nomenclature in this field has, from the beginning, been a problem. The first prokaryotic proteins of this type to be described (T4 gp32 and *E. coli* SSB) were originally thought of as DNA-unwinding proteins (13, 14) and DNA-melting proteins (15). Later the term "helix-destabilizing protein" (HDP) was suggested (1) and still later the more general term "single-stranded DNA binding protein" became extensively used (16, 17). None of these terms seems completely satisfactory. "DNA-unwinding protein" and "helix-destabilizing protein" im- ply a function that neither the gp32 nor *E. coli* SSB possesses, that is, neither protein by itself can denature native double-stranded DNA (dsDNA) (13–15, 18). "Single-stranded DNA binding protein," on the other hand, implies too little about the function of these proteins. RNA polymerase, RecA protein, and lactate dehydrogenase, for example, all bind to single-stranded DNA to varying degrees, but these are not the types of protein we will consider here. The more that is learned about *E. coli* SSB and gp32, the more likely it appears that "stabilization" of single-stranded DNA is at least one important aspect of the mechanism of action of these proteins. Perhaps the term "ssDNA stabilizing" protein would have been more appropriate; however, we will make no recommendation here for an entirely new system of nomenclature since the terms now in use are firmly entrenched. The general term "SSB" appears more widely accepted than "HDP" and so it would seem reasonable to use it as a class designation. Considering the number of such proteins that are now known from diverse groups of prokaryotic and eukaryotic systems, it is unlikely that unique and descriptive names can be given to each. Rather, it seems more reasonable to agree on such a class designation in the way polymerases, ligases, and topoisomerases, etc, have been used, and to clearly indicate the source of the protein when using these designations [e.g. F factor SSB (19)]. We suggest that *ssb* should be used as the general designation for the genes encoding any of these proteins with appropriate description to make it completely clear which genome is being described (e.g. the F factor *ssb* gene or the yeast *ssb* gene). In cases where more than one protein and gene are found in a particular organism, appropriate extensions can be added such as *ssbA, ssbB,* etc, in analogy to *polA, polB,* etc. Again, in analogy to the established systems in prokaryotes, we suggest that multiple SSBs from one system be assigned Roman numerals to distinguish them (e.g. yeast SSB I, yeast SSB II, etc in analogy to DNA polymerase I and DNA polymerase II, etc).

CRITERIA FOR ssDNA BINDING PROTEINS DISCUSSED IN THIS REVIEW

If we agree that the class designation "SSB" should describe proteins from a variety of systems with similar modes of action to *E. coli* SSB and gp32, then various criteria should be applied to a newly isolated protein before it is named an "SSB." The most meaningful studies in this regard would be combined biochemical and genetic analyses; however, in the case of most eukaryotes, a genetic analysis is usually difficult. From a biochemical standpoint, the following guidelines might be useful for categorizing proteins that bind to ssDNA.

DNA Binding

The most obvious criterion by which to evaluate a DNA binding protein is the strength with which it binds to various nucleic acids. Proteins of the type we consider here have a great preference for single- compared to double-stranded nucleic acids. These preferences can be demonstrated by the behavior of the protein on columns of single- and double-stranded DNA cellulose. In low salt the protein would be expected to flow through the latter and bind to the former. In practice ssDNA cellulose chromatography is usually the most advantageous step in their purification and the concentration of salt necessary to remove the bound protein can be a useful preliminary indication of the binding affinity. Although no absolute numbers can be given, in our experience proteins that bind weakly, requiring an ionic strength equivalent to less than 0.4 M NaCl for elution, should probably be looked on with some skepticism. For example, one protein that was studied in detail as an ssDNA binding protein was recently shown to be lactate dehydrogenase (20). Both gp32 and SSB require greater than 1.0 M NaCl for complete elution from ssDNA cellulose.

Most of these proteins have a strong affinity for ssRNA as well as ssDNA so that the relative binding affinity to these nucleic acids is not a generally useful property for defining SSBs, although substantial differences can sometimes be found, particularly when specific homopolymers are compared.

Stimulation of DNA Polymerase

One enzymatic property that is valuable in determining the functional significance of ssDNA binding proteins is their ability to specifically stimulate their cognate DNA polymerase. Many DNA polymerases, particularly those responsible for DNA replication, are stimulated by a specific ssDNA binding protein. However, when considering this property, careful consideration should be given to the template. Stimulation on nicked duplex DNA may not be seen unless accessory proteins are present to facilitate strand displacement and therefore primed ssDNA templates should also be included in any study. Such templates could take the form of (*a*) gapped duplex DNA, (*b*) a synthetic template consisting of a homopolymer and a complementary oligonucleotide, or (*c*) single-stranded phage DNA with a primer. In practice, several of these should be tried. The ability to stimulate the activity of DNA polymerase sets these proteins apart from other single-stranded nucleic acid binding proteins such as the bacteriophage fd gene 8 protein or the murine leukemia virus protein, p10, both of which are structural proteins that can be isolated from the intact virions, or the fd gene 5 protein, which actually prevents replicative DNA synthesis.

Stoichiometry

The binding protein should be required in stoichiometric quantities with respect to the template rather than in catalytic amounts. For this reason we do not consider the C_1C_2 polymerase α cofactors as members of this "class," although they do stimulate HeLa cell polymerase α (21).

Cooperativity

Eco SSB, gp32, and adenovirus DBP bind cooperatively to ssDNA; however, calf thymus UP1 does not. Additional work is needed to evaluate the general importance of this parameter.

In addition to these general properties, both gp32 and SSB share several additional features that have been found in at least some other functionally homologous proteins and that might be of further use in classifying single-stranded nucleic acid binding proteins. Both proteins facilitate the renaturation of DNA, and have an acidic COOH terminus that is more exposed to proteolysis when the protein is bound to ssDNA then when it is free in solution. Finally, it would seem advisable to obtain an amino acid composition and some amino acid sequence data on any new ssDNA binding protein. The amino acid composition would readily detect the high glycine content and characteristic dimethylarginine residues of the numerous proteins associated with heterogeneous nuclear RNA (hnRNA). A computer search of a short stretch of sequence corresponding to even six amino acids is sufficient to rapidly identify high-mobility group proteins (HMGs), histones, prokaryotic ribosomal proteins, dehydrogenases, and numerous other proteins that might fortuitously bind ssDNA. Just such a search of the National Biomedical Research Foundations Protein Data Base succeeded in identifying a rat liver ssDNA binding protein as being lactate dehydrogenase (20).

PHAGE T4 gp32

Biological Role

Studies on conditional lethal mutations provide a striking demonstration of the importance of the gene 32 protein (gp32) in T4 DNA metabolism. Within two minutes after cells infected with T4 phage carrying the tsP7 mutation in gene 32 are shifted to a nonpermissive temperature, all T4 DNA replication stops (22, 23) and the normal 600–1000S intracellular T4 DNA is converted to small DNA segments (34–80S) that range from one fourth to two genomes in length (23). This latter result suggests that gp32 normally protects ssDNA from nuclease digestion (24), thus preventing these regions from becoming lethal dsDNA cuts that cannot be repaired. In addition to DNA replication and repair, gp32 also plays a key role in T4 DNA recombination (25, 26). Studies with the amA453 mutant demonstrate that gp32 is essential for the formation of the "joint" DNA molecules that are required for recombination (25). The genetic experiments of Mosig et al (27) suggest that gp32 may actually be an integral part of a complex of proteins that catalyzes T4 DNA recombination and is associated with the membrane.

Many of the conclusions derived from these in vivo studies have now been confirmed in vitro. This is particularly true with respect to the effect of gp32 on T4 DNA replication. On a primed ssDNA template the addition of gp32 results in a 5–10-fold increase in the rate of synthesis by the gp43 DNA polymerase (28). This rate enhancement appears to result primarily from the ability of gp32 to melt-out double-helical hairpin structures in the ssDNA template (29). In the absence of gp32 these stable secondary structures impede the progress of T4 DNA polymerase thus decreasing the rate of DNA synthesis. A similar increase in the rate as well as the processivity of DNA synthesis (30, 31) is seen when gp32 is added to a system containing gp43 as well as the three T4 DNA polymerase "accessory" proteins (gp44, gp45, gp62). The ability of gp32 to stimulate the rate of DNA replication is even more apparent when a nicked dsDNA template is used in place of the primed ssDNA template. In the absence of gp32, the gp43 T4 DNA polymerase is essentially unable to use a nicked dsDNA template. The addition of gp32 alone allows some limited DNA synthesis to occur in A-T-rich regions (32); however, the rate of synthesis is less than 0.2% of that seen under optimum in vitro conditions (33). Reasonable rates of synthesis and full-length copying of dsDNA templates require the presence of the three polymerase accessory proteins. When gp32 is added to a system containing a nicked dsDNA template, gp43, and its three accessory proteins, the rate of DNA synthesis is increased 100–200-fold (34, 35). Since the rate of fork movement in this "five protein system" increases almost linearly with the gp32 concentration, it appears that gp32 must make a significant contribution towards destabilizing the dsDNA in front of an advancing DNA replication complex (35). These observations as well as previous in vivo studies (36) also clearly establish that gp32 plays a stoichiometric rather than catalytic role in T4 DNA replication. Consistent with this conclusion, it has been estimated that a T4-infected cell contains as many as 10,000 copies of gp32 (37).

Helix-destabilization per se cannot be the only function of gp32 in DNA replication, because the E. coli SSB protein, which is also a strong helix-destabilizing protein, cannot substitute for gp32 in these reactions (14, 38). This result suggests that gp32 and SSB may impose quite different conformations onto the ssDNA. It is also possible that gp32 but not the E. coli SSB protein can interact directly with other proteins in the T4 DNA replication complex, and that these protein: protein interactions are an essential aspect of gp32 function. Support for the latter idea derives both from genetic studies (27) and from the results of protein-affinity chromatography. Formosa et al (40) have shown that when crude lysates of T4-infected E. coli are applied to an affinity column containing immobilized gp32, at least 10 T4-encoded proteins are specifically retained. Nine of these proteins have been identified as being involved in either T4 DNA replication or recombination.

There is also in vitro data that supports the involvement of gp32 in DNA repair and recombination. Curtis & Alberts (23) have shown that gp32 can protect ssDNA from deoxyribonuclease I digestion. The ability of gp32 to catalyze the renaturation of complementary single-strands (13), the formation of heteroduplex joints between suitable segments of DNA (41), and, in the presence of RecA protein, the formation of D-loops between dsDNA and homologous single-strands (42), could all be envisioned to play essential roles in DNA recombination.

While cooperative binding of gp32 to ssDNA is undoubtedly crucial to most of its functions in DNA metabolism, the ability of gp32 to also bind ssRNA allows it to control its own rate of synthesis at the level of translation (43–45). Based on both in vivo and in vitro data, a model for gp32 autogenous regulation that requires that during synthesis gp32 bind first to all the ssDNA present has been proposed (46–49). At this point the "free" gp32 concentration rises until it reaches a threshold level necessary for gp32 to bind specifically to its own mRNA and to prevent further gp32 synthesis. The 10^1–10^4 higher affinity [depending on the particular homopolymer tested, see Ref. (50)] of gp32 for ssDNA over ssRNA ensures that all of the intracellular ssDNA will be saturated with gp32 before binding occurs on mRNA. The specificity for gp32 mRNA (as opposed to other T4 mRNAs) seems to result from the presence of a uniquely unstructured region that spans about 50 nucleotides near the presumed ribosome binding site of gp32 mRNA (47, 48). Since increasing the gp32 concentration by an additional three- to fourfold above that normally found in vivo will shut off the synthesis of other T4 proteins (51), it is obvious that the concentration of "free" gp32 must be tightly controlled (49).

Protein:Nucleic Acid Interactions

Formation of the gp32:single-stranded nucleic acid complex significantly alters several physicochemical properties of both the protein and the nucleic acid. With respect to the optical properties of the nucleic acid, gp32 binding results in a hyperchromic change in synthetic polynucleotides such as poly-(dA), which normally contain intrachain base stacking (52), that exceeds the change that can be induced by heating (15). In addition, gp32 binding results in a significant reduction in the DNA ellipticity above about 250 nm (15, 53) and a twofold increase in the fluorescence emission of poly(1,N^6-ethenoadenylic acid) (54). All of these changes have been attributed to the ability of gp32 to totally unstack the bases of both ribose- and deoxyribose-containing polynucleotides (3, 15, 54). This unstacking may result in part from the highly expanded conformation that gp32 imposes onto the ssDNA. Even though the mass of the gp32:fd ssDNA complex is 13 times greater than that for free fd ssDNA, sucrose gradient centrifugation reveals that the complex sediments only about 1.3 times faster than free fd ssDNA (13). This result implies that gp32 binding leads to a sixfold increase in the frictional coefficient of fd ssDNA (13). That gp32 binding significantly extends the length of the polynucleotide backbone is evident from electron microscopy studies that demonstrate that gp32 binding leads to about a 50% increase in the distance between neighboring nucleotide bases (55). Electron microscopy and sucrose gradient centrifugation studies carried out on gp32:fd ssDNA complexes prepared in the presence of excess ssDNA reveal that gp32 tends to bind in long clusters indicative of strong gp32:gp32 cooperative protein:protein interactions (13, 55).

Other changes that are evident in the physicochemical properties of gp32 as a result of binding to ssDNA are chemical shifts in the ^{1}H-NMR resonances of several aromatic amino acids (8, 56), and a marked decrease in the intrinsic fluorescence of gp32 (57). The number of aromatic protons involved in the ^{1}H-NMR shifts increases with the length of the oligonucleotide until no further changes are seen when the oligonucleotide length exceeds about eight nucleotide bases (56). These data suggest that the binding groove of gp32 accommodates about eight bases (56). Although previous estimates of the binding site size of gp32 range from five (58, 59) to eleven (53) nucleotides, most values reported (13, 15, 50, 60, 61) are between these two extremes and are therefore in reasonably good agreement with the NMR data (56). The chemical shifts that are induced in the aromatic region of the ^{1}H-NMR spectrum of gp32 as a result of binding to oligonucleotides are thought to arise from hydrophobic interactions between the side chains of aromatic amino acids and the bases of the ssDNA (56). These interactions may impose a regular "ladder" pattern on the single-stranded nucleotides that may be a general structural feature of single-stranded DNA binding proteins (56). The extent of gp32 fluorescence-quenching increases from about 3% for a dinucleotide to about 35% for d(pt)$_8$ (57); this increase, like the NMR study (56), also suggests the involvement of additional aromatic amino acids in gp32 as the length of the oligonucleotide is increased over this range. No further increase in the extent of fluorescence-quenching is seen upon going to d(pT)$_{16}$ or denatured calf thymus DNA (57).

Fluorescence-quenching is the physicochemical probe that has been most extensively used to measure the affinity of gp32 for single-stranded nucleic acids. These studies (54, 58, 60) demonstrate that gp32 has an affinity that varies from $2 \times 10^5 \, M^{-1}$ to $6 \times 10^5 \, M^{-1}$ for oligonucleotides that contain from two to eight nucleotides. Surprisingly, the affinity of gp32 for these oligonucleotides is not significantly increased by the presence of a 5'-terminal phosphate (58) or by increasing the length of the oligonucleotide from two to eight bases (54, 58). In addition, this interaction shows very little dependence upon salt concentration, base composition, or sugar type so that the affinity of gp32 for an oligodeoxyribonucleotide is only about twofold greater than that for the corresponding oligoribonucleotide (54, 58). In contrast to oligonucleotides, binding of gp32 to polynucleotides is so sensitive to ionic strength that a tenfold increase in the NaCl concentration results in almost a 10^7-fold decrease in apparent affinity (54). Also in contrast to oligonucleotides, there is a marked effect of base composition on the affinity of gp32 for different polynucleotides (49, 50, 61, 62). Under ionic conditions that are thought to approach those found in vivo, the affinity of gp32 for deoxypolynucleotides increases from about $2 \times 10^7 \, M^{-1}$ to about $2 \times 10^9 \, M^{-1}$ in the following order: poly(dA) < poly(dU) < T4 ssDNA < poly(dC) < poly(dG) < poly(dT) (49). With the exception of poly(dC), the affinity of gp32 for the corresponding ribo-homopolynucleotide is about 10^{-1} times that for the deoxyribo-homopolynucleotide (49). Poly(dC) is unusual in that the apparent binding constant of this polynucleotide to gp32 is almost 10^4 times larger than that for poly(C) (49).

Most of the increase in the affinity of gp32 for polynucleotides compared to oligonucleotides results from the fact that the former are long enough to permit cooperative binding. The apparent binding constant is now equal to $K\omega$ where K is an intrinsic affinity resulting from the gp32:single-stranded nucleic acid interaction and ω is a unitless cooperativity parameter that is equal to about 2×10^3 (50, 54), and reflects protein:protein interactions between adjacent molecules of gp32 bound to the polynucleotide lattice (54). Since ω is not dependent upon ionic strength or base composition (50, 54), it has been proposed that gp32 can exist in two distinct conformations corresponding respectively to when gp32 is bound noncooperatively to an isolated oligonucleotide and to when it is bound cooperatively to a polynucleotide (54). Based on the salt concentration dependence of the binding of gp32, the polynucleotide binding mode allows approximately three positively charged amino acids that had previously been shielded to now come in contact with negatively charged phosphates of the polynucleotide backbone. These additional charge:charge interactions would therefore account for the large increase in salt sensitivity upon going from the oligonucleotide to the polynucleotide binding mode. This model (54) is consistent with previous limited proteolysis studies (63, 64) that suggest that cooperative binding induces a conformational change in gp32 that increases the exposure of the acidic, COOH-terminal "A" region and decreases the exposure of the NH_2-terminal "B" region to cleavage by several different proteases. Oligonucleotides that are too short to permit cooperative binding have no effect on the rate of proteolysis of either the "A" or "B" region (63). In terms of a conceptual model for gp32 binding to polynucleotides, the "A" region is thought to form an "arm" or "flap" that normally covers (and perhaps actually interacts with) at least three basic residues on the gp32 surface (54). The conformational change that occurs upon cooperative binding displaces this flap, which makes it more accessible to proteolysis (and presumably for interacting with other T4 DNA replication proteins), and allows these basic amino acids to contact the phosphodiester backbone of single-stranded nucleic acids.

In addition to binding single-stranded nucleic acids, gp32 also has a measurable affinity for double-stranded nucleic acids. Using a sedimentation velocity binding technique, Jensen et al (15) found that gp32 has an affinity of 8×10^3 M^{-1} for native calf thymus DNA in 0.05 M salt. The primary reason that binding of gp32 to ssDNA is 10^4–10^5 times stronger than it is to dsDNA is that in the latter case binding is noncooperative so that ω is about equal to 1 rather than 2×10^3 (15). Because of this large difference in affinity for ssDNA versus dsDNA, gp32 should be an effective equilibrium destabilizer of double-helical DNA and RNA. This potential is realized in vitro with synthetic polynucleotides. In 10-mM KCl and 10-mM $MgSO_4$, the addition of a stoichiometric amount of gp32 decreases the T_m of poly[d(A-T)] from 65°C to 25° (13). Surprisingly, gp32 does not destabilize naturally occurring dsDNA (13). Under conditions where gp32 should decrease the T_m of T7 dsDNA by more than 60°C, it actually has no effect (15). There appears to be a "kinetic block" that prevents the attainment of equilibrium and the expected destabilization (15). Since ssDNA binding proteins melt dsDNA by trapping ssDNA loops that occur spontaneously as a result of "breathing" rather than by first binding to the dsDNA and then actually forcing the strands apart (2) in a manner similar to that of a helicase, it is possible that the loops that form in naturally occurring dsDNA are too short and too transient to permit gp32 to nucleate cooperative binding (15). Since in the noncooperative binding mode gp32 binds almost equally to both single-stranded and double-stranded nucleic acids (15, 50), only when cooperative binding is possible is gp32 able to act as a helix-destabilizing protein. Even without this "kinetic block," binding measurements done under physiologically relevant conditions indicate that on thermodynamic grounds alone gp32 by itself should not be able to melt any fully duplex DNA or RNA structures in vivo including those containing only A·T or A·U base pairs (49). At the estimated in vivo concentration of free gp32 (2–3 µM), all ssDNA and ssRNA sequences should, however, be saturated with gp32. In addition, gp32 should be able to melt most adventitious stem-loop (hairpin) structures that form in ssDNA, but because of the lower affinity of gp32 for ssRNA, similar structures in mRNA would be stable to gp32 (49).

Although the exact nature of the kinetic block that prevents gp32 from destabilizing dsDNA in vitro is not completely understood, recent experiments are beginning to delineate the kinetics of gp32 binding to ssDNA. Under conditions of excess nucleic acid, gp32 binding is at least a three-step process: 1. preequilibrium formation of noncooperatively bound protein, 2. association of free gp32 to singly contiguous sites, hence forming the first cooperatively bound clusters, and finally 3. redistribution of the growing clusters to form the final equilibrium distribution (65). The second step in this process apparently involves a facilitated translocation of noncooperatively bound gp32 molecules along the ssDNA lattice until they find an adjacent gp32 molecule and engage in cooperative binding. This ability of gp32 to "slide" or "hop" along the ssDNA lattice could allow it to more quickly saturate transiently single-stranded regions as well as to keep pace with a replication complex moving at 1,000 nucleotides·sec^{-1} (65). The kinetics of gp32 binding in vivo may be somewhat different from these in vitro experiments in that there would not be such a large excess of available single-stranded nucleic acid binding sites and the free gp32 concentration would be about an order of magnitude higher than used in this study (65), that is 2–3 μM (49). At these higher protein concentrations gp32 would exist in the form of linear aggregates (66) that may be able to bind as a unit to ssDNA. The kinetics of dissociation of gp32 from ssDNA have been studied by the use of ionic strength jumps (67, 68). The data are consistent with a mechanism involving only the dissociation of single gp32 molecules from the ends of cooperatively bound clusters (67, 68).

Based on both the equilibrium and kinetic studies, gp32 would be expected to play a "passive" role in DNA metabolism in vivo; i.e. while it should be able to quickly saturate single-stranded sequences, gp32 by itself should not be able to melt natural dsDNA or even hairpin loops in ssDNA that are held together by more than five or six base pairs (2, 49). In concert with other proteins at the replication fork, gp32 may contribute to helix-destabilization (35), but in general its function seems to be to provide a polynucleotide : protein complex that is resistant to nucleases and is conformationally appropriate to serve as a substrate for other enzymes and proteins involved in DNA metabolism. In view of these findings it is surprising that more attention has not been given to the ability of gp32, under physiologically relevant conditions, to catalyze the renaturation of complementary ssDNA. The addition of gp32 results in a more than 1000-fold increase in the rate of renaturation of single-stranded T4 DNA (13). The primary effect of gp32 in this reaction is to accelerate the rate of the bimolecular nucleation step, which is normally the "slow step" in renaturation. The normally fast "zippering" process is not appreciably affected by gp32. While most of its supposed functions can be accounted for on the basis of the gp32 : nucleic acid interaction, recent protein affinity chromatography (40) and genetic studies (27) suggest that gp32 may play a larger role than was first supposed in maintaining the topological stability of the multiprotein complexes involved in DNA replication and recombination.

Structural and Functional Domains

Based on the amino acid sequence of gp32 (69, 70) and the nucleotide sequence of its gene (48), gp32 contains 301 amino acids and has a molecular weight of 33,488. Although gp32 should have a net charge of -10 at pH 7, which is in agreement with an experimentally determined isoelectric point of 5.0 (64), the charge distribution is quite asymmetric. While the NH_2-terminal half of gp32 has a net charge of $+10$, the COOH-terminal half has a net charge of -20. The monomeric gp32 behaves hydrodynamically, as though it were a prolate ellipsoid with an axial ratio of 4 : 1 and an overall length approaching 12 nm (13). Electron microscopy studies on crystals of a large proteolytic fragment of gp32 confirm these dimensions (71). Since in its complex with ssDNA each gp32 molecule may only span a distance of about 3.3 nm (assuming a binding site size of 7) and an internucleotide spacing of about 0.47 nm (55), adjacent molecules of gp32 probably overlap one another to some extent. This overlap presumably contributes to the cooperativity that is observed when gp32 binds ssDNA (13, 15, 55). Although gp32 exists as a monomer in dilute solution [<0.5 μM, Ref. (66)], it undergoes indefinite self-association so that at the estimated free gp32 concentration in vivo [~3.0 μM, Ref. (49)] it would have an apparent molecular weight corresponding to that of a dimer or trimer (66). While these gp32 : gp32 interactions that occur in solution are stronger and more salt-sensitive than those that give rise to cooperative binding of gp32 to ssDNA (8, 50, 58, 66), it nonetheless seems likely that the two processes are related.

The finding that limited proteolysis can be used to generate functionally active fragments of gp32 (53, 72, 73) has facilitated more detailed structure/function studies. The regions spanning residues 9–21 and 253–275 in gp32 are particularly susceptible to cleavage by a wide variety of different proteinases (63, 64, 69, 73–75). Clearly these two regions are exposed on the surface of gp32. Cleavage at any point between residues 9 and 21 removes the basic NH_2-terminal "B" region and produces gp32* − B. Cleavage at any point between residues 253 and 275 removes the acidic, COOH-terminal "A" region and produces gp32* − A. Cleavage at both sites results in a 26,000-dalton fragment called gp32* − (A+B). Since the in vitro properties of these individual cleavage products appear to be identical irrespective of the enzyme used in their preparation, it would appear that the essential residues within the "A" and "B" domains must be contained within residues 276–301 and 1–9, respectively (69, 70).

Determination of the function of the NH_2-terminal "B" region, which has the sequence Met-Phe-Lys-Arg-Lys-Ser-Thr-Ala-Glu-, has been approached both by comparing the properties of gp32* − A with those of gp32* − (A+B), and more recently, by characterizing gp32* − B directly. Regardless of which approach is taken, the conclusion is the same: the NH_2-terminal "B" region is essential for gp32 : gp32 protein interactions whether they occur between "free" gp32 molecules in solution or between adjacent gp32 molecules bound to an ssDNA lattice. Hence, removal of the "B" region results in a gp32 molecule that binds noncooperatively to ssDNA [ω = 1; Refs. (9, 60, 76, 76a)] and that no longer forms linear self-aggregates in solution (9, 75). That both kinds of gp32 : gp32 interactions can be eliminated by removing only 9 of the first 301 amino acids in gp32 supports the notion that these two processes are closely related. As expected from the loss of cooperative gp32 : gp32 interactions, gp32* − (A+B) has an approximately equal affinity of about $10^6 M^{-1}$ for both $d(pT)_8$ and poly(dT)(60). Despite the loss of cooperative protein : protein interactions, gp32* − (A+B) binding to ssDNA brings about polynucleotide lattice deformation comparable to that observed with gp32 (76). Along with the fact that both gp32 and gp32* − (A+B) have similar affinities for $d(pT)_8$ (60, 76), these data suggest that removal of the "A" and "B" regions from gp32 has only a minimal effect on its intrinsic interaction with nucleic acids. Although these studies leave no doubt that the NH_2-terminal B region is essential for both cooperative binding and indefinite self-aggregation, it has not yet been demonstrated that this domain is directly involved in either process.

Removal of the COOH-terminal "A" region to produce gp32* − A has relatively little effect on the equilibrium binding properties of this protein (60, 76). The only major difference is that oligonucleotide binding by gp32* − A is more salt-dependent (and therefore in this respect more closely resembles polynucleotide binding by gp32) than in the case of gp32 (76). This result is most easily interpreted in terms of the proteolytic cleavage removing the "arm" or "flap" of gp32 that has been postulated to prevent approximately three basic amino acids on gp32 from interacting with phosphates on ssDNA when gp32 is in the oligonucleotide binding mode (76). Hence, gp32* − A may in effect be in a conformation that more closely resembles the normal gp32 "polynucleotide binding mode" before gp32* − A even binds ssDNA. If so, the lack of a requirement for gp32* − A to undergo a conformational change from the oligonucleotide to the polynucleotide binding mode might be important in understanding why removal of the COOH-terminal "A" region enables gp32 to denature naturally occurring dsDNA in vitro. Although the affinity of gp32 for polynucleotides is increased 2–3-fold by removal of the "A" region, this small thermodynamic difference between the two proteins is not enough to account for the observation that under conditions where gp32 has no effect, gp32* − A can decrease the T_m of T4 dsDNA by 70°C (60, 64, 76). It seems likely therefore that melting of native dsDNA by gp32* − A must follow a different kinetic pathway from that used by gp32 (76).

In addition to increasing the ability of gp32 to melt native DNA, proteolytic removal of the COOH-terminal A region results in a protein that can no longer catalyze the renaturation of native dsDNA (64) and significantly alters the T4 in vitro replication complex (38). While gp32* − A can substitute for gp32 in leading-strand synthesis, there is no lagging-strand synthesis in the presence of gp32* − A (38). The absence of RNA primer extension (lagging-strand synthesis) appears to be due to destabilization of the 3'-hydroxy chain terminus by gp32* − A. At least in vitro, the "A" region of gp32 appears to be essential to limit the helix-destabilizing ability of gp32. This control might be modulated by protein:protein interactions between gp32 and either the T4 DNA polymerase, 43P, or the T4 RNA priming protein, 61P, since both of these interactions have been shown to require an intact "A" region. The COOH-terminal A region may not be as indispensable in vivo to gp32 function as suggested by these in vitro DNA replication experiments. Minegawa et al (77) have isolated a mutant phage that induces a gp32 lacking approximately 30 amino acids at the COOH terminus that is nonetheless viable. This model for the function of the COOH terminus of gp32 is analogous to that proposed for the high-mobility group protein, HMG-1. As in the case of gp32, proteolytic removal of a very acidic COOH-terminal region results in increasing the helix-destabilizing ability of HMG-1 (78); this effect may result from this region otherwise partially occluding a positively charged ssDNA binding site (79). Instead of interacting with other replication proteins as in the case of gp32, the acidic COOH terminus of HMG-1 is thought to interact with basic histones (80).

Since removal of both the "A" and "B" regions of gp32 has only a minimal effect on the intrinsic interaction between gp32 and ssDNA, it is reasonable to conclude that the ssDNA binding domain is contained within the region spanning residues 22–253. This conclusion is further supported by direct ultraviolet-light–induced cross-linking of gp32 to ssDNA (9), as well as by an unusual clustering of missense mutations that map between residues 36 and 125 (81). Several approaches have been used to further define the molecular details of the gp32:ssDNA interaction. Based on studies (54) of the salt-sensitivity of the gp32:ssDNA interaction it has been proposed that binding involves at least three ionic interactions between positively charged amino acids on the surface of gp32 and negatively charged phosphates on the DNA. Chemical modification experiments with tetranitromethane suggest that four or five of the eight tyrosines in gp32 are essential for binding to ssDNA (53). Several independent approaches suggest that 1–2 tryptophyl residues are also essential for gp32 binding (82–85). Studies with model peptides (86) suggest that these crucial tryptophyl residue(s) are involved in stacking interactions with the nucleic acid bases. These conclusions relating to the involvement of aromatic amino acids in gp32 binding are in excellent agreement with ^{1}H-NMR studies that demonstrate that the resonances of at least one phenylalanine, one tryptophan, and five tyrosine residues are involved in the chemical shift changes observed upon nucleotide binding (56). Many of the aromatic proton NMR shifts observed on oligonucleotide complex formation are similar to those previously observed with gene 5 protein of bacteriophage fd (87). In the case of gene 5 protein, complex formation is known to involve intercalation with one phenylalanine and three tyrosine residues (87). Radiolysis experiments (83) suggest that at least one cysteine is also located in the binding site of gp32.

When these physicochemical and genetic studies are taken together with the known gp32 sequence it is indeed tempting to speculate that the tyrosine-rich region that spans residues 72–116 must make an important contribution to ssDNA binding (9, 69, 70). This region is unusual in that it contains 6 of the 8 tyrosines in gp32 and these residues are approximately equally spaced, each separated by 6–10 amino acids. In addition, this region contains 3 of the 4 cysteines and 2 of the 5 tryptophans in gp32 (69, 70). There is also a Lys-Arg-Lys sequence spanning residues 110–112 that could be important for ionic interactions with ssDNA. The tyrosine-rich region of gp32 is further intriguing in that if this region is turned around so as to read from the COOH- to the NH$_2$-terminal end, then it would contain the putative binding domain of retroviral nucleic acid binding proteins (88). This domain occurs once in nucleic acid binding proteins of murine leukemia viruses (88), and is repeated twice in the homologous structural proteins from avian sarcoma virus (89) and bovine leukemia virus (90). It has the general structure (90):

Cys-(X)$_2$-Cys-(X)$_3$-Gly-His-(X)$_4$-Cys.

After turning the tyrosine-rich region of gp32 around, the only difference would be that the Gly-His sequence would be one residue further away from the middle cysteine in gp32 than on other retroviral proteins (90). We hope that the recent crystallization of both gp32* − A (71) and gp32* − (A+B) (91) will inspire further work on crystallization and X-ray diffraction analysis of gp32 : oligonucleotide complexes that ultimately should provide a more definitive understanding of the molecular basis for the interaction of gp32 with nucleic acids.

ESCHERICHIA COLI SSB (Eco SSB)

Biological Role

Eco SSB was first described by Sigal et al (14), and shortly thereafter studies were published by Weiner et al (92) and Molineux et al (93). In view of what was already known at the time about gp32 from phage T4, the importance of SSB in DNA metabolism was anticipated. Subsequently the requirement for SSB in various in vitro phage DNA replication systems was discovered (94), thus strengthening the expectation that the protein played a central role in E. coli DNA replication. However, it was not until mutations in the SSB structural gene were identified that these expectations were confirmed (95). Although at least eight[1] mutations in Eco SSB are known, only two have been extensively characterized in vivo and their mutant gene products studied in vitro. These are ssb-1 and ssb-113 (formerly lexC113). In general, SSB-deficient mutants are similar: 1. They confer temperature-sensitive lethality underlining the essential cellular role of SSB (99, 100). 2. They exhibit increased UV-sensitivity (99, 100). 3. They are defective in their ability to amplify the synthesis of RecA protein as a response to DNA damage (101). 4. They are defective in SOS repair (102, 103) and λ prophage induction (100). 5. Finally,

---

[1]Four mutations have been produced by oligonucleotide mutagenesis (J. Chase, J. Murphy, and K. Williams, unpublished). The ssb-2 mutation has not been well-characterized (96). The ssb-3 mutation has only recently been reported (C. Schmellik-Sandage and E. Tessman, unpublished). The lexC113 (ssb-113)- and ssb-114- (97) encoded proteins each contain a pro→ser substitution at the penultimate amino acid residue and their tryptic peptide fingerprints are identical (98). However differences between the effects of these mutations have been reported (97).

certain abnormalities in recombination have been shown (99, 104). There are, however, significant differences in the effects of temperature on the expression of these defects in *ssb-1* and *ssb-113* mutant strains. Both mutations lead to a temperature-sensitive defect in DNA replication and hence growth, but at the temperature permissive for growth *ssb-1* mutant strains are relatively normal in all functions tested, expressing the defects cited above only at the temperature restrictive for growth. In contrast, *ssb-113* mutant strains demonstrate most of these deficiencies even at the permissive temperature for growth. In addition, there are differences in postirradiation DNA degradation between strains containing these mutations (103). Thus the *ssb-1* mutation appears to alter a function of SSB generally required for DNA replication, recombination, and repair, whereas the effects of the *ssb-113* mutation are more complex. In vitro biochemical studies of the purified mutant proteins can now explain certain of these observations; however, important questions remain to be answered, especially relating to the *ssb-113*-encoded protein.

The stoichiometric requirement for *Eco* SSB in DNA replication is well-documented (94). Even though there are only an estimated 300–350 copies of SSB (tetramer) per cell (92, 105) as compared to 10,000 copies of gp32 in a T4-infected *E. coli* cell (13), when normalized to the expected number of replication forks, these two proteins are present in nearly equivalent amounts. There is sufficient SSB or gp32 to cover about 1400 nucleotides of ssDNA per replication fork. In vitro, SSB increases the fidelity of DNA replication by tenfold (106) and stimulates *E. coli* DNA polymerases II and III* but not polymerase I or T4 DNA polymerase (14, 92, 93, 107). Pausing by DNA polymerase III assemblies has been correlated with regions of secondary structure and has been shown to be relieved by SSB (108). *Eco* SSB is also necessary for primosome assembly (94) and is required in an in vitro system for replication of *oriC*-containing minichromosomes (109).

In vitro *Eco* SSB is required for methyl-directed repair of DNA mismatches (110). In vivo *ssb-113* enhances Tn10 precise excision by 15-fold (111). The mechanism by which *Eco* SSB functions in these processes is not yet known.

A number of observations suggest that there exists some specific relationship between *Eco* SSB and RecA protein (12). While it is tempting to speculate that protein:protein interaction may govern this relationship, no such evidence yet exists. In vitro studies of RecA-mediated reactions suggest that SSB affects ssDNA by removing secondary structures, probably allowing RecA protein easier and more direct access to the DNA (112, 112a). The possible involvement of SSB in activating RecA protease has not yet been investigated.

The question of the regulation of *Eco* SSB synthesis is not yet clear but does not appear to be analogous to that described for gp32. The *ssb* gene is located adjacent to the *uvrA* gene which is part of the SOS system. They are transcribed in opposite directions. Although SSB is involved in the SOS response, two studies conclude that it is not amplified under conditions of SOS induction (113) and not repressed by the *lexA* gene product (114), while one study concludes that both the *ssb* and *uvrA* genes are controlled by a common LexA binding site (115).

Recently an entire class of SSB-like proteins has been identified in transmissible plasmids [(116, 117), P. Ruvolo and J. Chase, unpublished]. The protein encoded by F plasmid (FSSB) has been sequenced (19). This protein has extensive homology to *Eco* SSB within the DNA binding domain and diverges extensively in the COOH-terminal region, although the five COOH-terminal amino acid residues are identical. It is interesting to remember that the *ssb-113* mutation of *Eco* SSB occurs at the penultimate residue. The biological role of these plasmid-encoded proteins is not yet known.

Protein:Nucleic Acid Interactions

While *E. coli* SSB and T4 gp32 appear to be functionally homologous, they do not share any obvious sequence homology and moreover, there appear to be fundamental differences in the topology of their ssDNA:protein complexes. These differences probably arise from the observation that SSB is a tetramer composed of four identical subunits while, at least at low protein concentrations, gp32 is a monomer. At higher protein concentrations, gp32 does undergo indefinite aggregation. However, whether it is the monomer or these linear clusters of gp32 that actually bind ssDNA, the final complex can still be conceptualized in a two-dimensional sense as monomers of gp32 aligned along a linear ssDNA lattice. SSB, on the other hand, exists in solution as an extremely stable asymmetric tetramer that shows no tendency to undergo further aggregation (92, 93, 119). The unique features of the interaction of SSB with ssDNA are related to the arrangement of ssDNA around this tetramer.

SSB appears similar to gp32 in that binding results in an expansion of circular fd ssDNA (14) and an unstacking of the nucleotide bases (53). However, whereas gp32 binding increases the internucleotide spacing by 50% (55), SSB binding decreases the spacing by 35% (14). Since the estimated internucleotide spacing in the SSB:ssDNA complex is too small for a linear ssDNA chain (14), the ssDNA is probably wrapped around a core of either one (120) or two (121) SSB tetramers. The tetrameric structure of SSB is preserved when SSB binds oligonucleotides (16, 120, 122), and presumably the SSB tetramer is the species that also interacts with longer ssDNA molecules. Consistent with this interpretation, electron microscopy of SSB:ssDNA complexes formed at low protein to nucleic acid ratios reveals a striking beaded appearance (14, 121, 123) that suggests that SSB organizes ssDNA in a manner similar to the organization of duplex DNA by histones (121). Nuclease digestion and equilibrium density banding of SSB:ssDNA complexes indicates that each of these beads contains 145 nucleotides of ssDNA wrapped around two SSB tetramers

(121). Between each nucleosomal-like bead is a stretch of 30 bases of protein-free ssDNA that might be available to interact with other proteins such as RecA (121, 123). This model is consistent with a previous fluorescence anisotropy experiment that indicates that the SSB:ssDNA complex has significant local flexibility that probably results from "small gaps of free DNA between protein stretches" (16). When similar complexes are formed at higher protein to nucleic acid ratios the ssDNA is extended into a smooth-contoured nucleoprotein filament that does not show any breaks in the protein sheath (123).

These two kinds of complexes may correspond to SSB binding to ssDNA with an apparent binding site size of 33 bases (smooth-contoured) versus 65 bases (beaded structure) respectively (124). Fluorescence-quenching experiments indicate that these two binding modes, which are characterized by different final extents of fluorescence-quenching, are in slow equilibrium, and that below 0.01 M NaCl SSB has a site size of 33 while above 0.20 M NaCl SSB has a site size of 65 (124). At intermediate NaCl concentrations, a mixture of these two binding modes exists, which explains previous large discrepancies in the literature concerning the binding site size of SSB. While the exact topology of these two apparently quite distinct SSB:ssDNA complexes is still in doubt, it may be significant that the transition between these two binding modes is modulated by NaCl concentrations in the physiological range. Intuitively it would also seem likely that the smooth-contoured structure, where each SSB tetramer might contact two adjacent SSB tetramers, would be characterized by a higher degree of cooperativity than the beaded structure. In the latter structure each SSB tetramer may only be able to contact one adjacent SSB tetramer. The ability to form this nucleosomal-like structure might be a key to a heretofore-unrecognized aspect of SSB function.

One of the major questions that remains concerning the mechanism of SSB binding to ssDNA is the number of SSB monomers that actually contact the ssDNA in either of its binding modes. Based on sedimentation experiments, SSB can bind four $d(pT)_8$, two $d(pT)_{16}$, and one $d(pT)_{30-40}$ molecule (120). Equilibrium binding studies on the length dependence of oligonucleotide binding demonstrate approximately a 10-fold increase in affinity in going from $d(pCpT)_2$ to $d(pCpT)_3$, but only about a twofold increase in going from $d(pCpT)_3$ to $d(pCpT)_4$ (122). These results suggest that each SSB monomer can interact with 6–8 bases. When SSB is purified without resorting to the use of denaturing conditions, it has an affinity of about 2×10^6 M^{-1} for $d(pT)_8$ compared to about 10^8 M^{-1} for $d(pT)_{16}$ (125, 126). If $d(pT)_{16}$ can span two SSB monomers then it might be predicted to have an affinity approaching 4×10^{12} M^{-1}. Similarly, if $d(pT)_{30-40}$ can bind in such a way as to perfectly span all four SSB monomers, then its affinity might approach 1.6×10^{25} M^{-1}. Electron microscopy (14, 121, 122), filter binding (126), density gradient centrifugation (14), and nuclease digestion experiments (126) all indicate that

SSB binding to ssDNA is highly cooperative, so SSB might be expected to have an even higher affinity for poly(dT). Actually, the affinity of SSB for $d(pT)_{30-40}$ is estimated to be only about 5×10^8 M^{-1} (120) and that for poly(dT) or ssDNA to be in the range $10^8–10^{10}$ M^{-1} (124, 125). Obviously, these values are too low to permit SSB binding to poly(dT) to be viewed as simply the sum of the free energy of binding of four $d(pT)_8$ molecules plus a cooperativity parameter. The fivefold increase in the extent of fluorescence-quenching observed when SSB is complexed with poly(dT) as compared to four $d(pT)_8$ molecules suggests there may be significant differences in the conformation of SSB in these two complexes (16, 18). Indeed, this notion is supported by partial proteolysis studies that reveal that cooperative binding induces a conformational change in SSB that is similar to that seen with gp32 (18). Lower than expected binding affinities for $d(pT)_{16}$ and $d(pT)_{30-40}$ could also easily be explained by the fact that it is unlikely that each $d(pT)_{16}$ or $d(pT)_{30-40}$ molecule would bind in such a way as to perfectly span two or four SSB binding sites respectively. Similarly, it is quite possible that in either or both the "smooth-contoured" and "beaded structures" that have been observed, all four SSB subunits do not equally contact the ssDNA lattice. It is, however, useful to keep these "theoretical maximum" values for SSB binding affinity in mind when thinking about models for the interaction of SSB with ssDNA.

While it is generally agreed that SSB binds cooperatively to ssDNA, there is considerable doubt concerning the actual value for this parameter. Based on SSB cluster length measurements from electron microscopy as well as the results of nuclease digestion experiments, ω may be as high as $10^4–10^5$ for SSB binding to ssDNA (122, 126). If in the poly(dT):SSB complex only two monomers actually contact the ssDNA lattice, then a reasonable estimate of the cooperativity parameter might be obtained by comparing the affinity of SSB for poly(dT) with that for $d(pT)_{16}$. Unfortunately, fluorescence-quenching experiments provide only a minimum value for the affinity of SSB for poly(dT). A filter binding assay, however, suggests that SSB has an affinity of about 6×10^{10} M^{-1} for ssDNA. When compared to the affinity of SSB for $d(pT)_{16}$, the former affinity leads to an approximate value of about 10^3 for the cooperativity parameter (18), which would be close to that reported for gp32. Obviously, more detailed studies need to be done to better evaluate the cooperativity of SSB binding in both the "nucleosomal" and "smooth-contoured" binding modes.

In addition to binding ssDNA cooperatively, the available data indicate that SSB and gp32 share several other binding characteristics. Like gp32, SSB binds poly(dT) the tightest of any polynucleotide tested (92, 120, 125), and in general binds about 10 times better to ssDNA than to ssRNA (14, 92, 125). SSB binds ssDNA without regard to sequence (122) and with an affinity that is at least 10^3 times greater than for dsDNA (18, 125). Conditions have not been found where SSB can denature native dsDNA (92, 18), but in the presence of

2-mM spermidine at pH 7, SSB does catalyze dsDNA renaturation (128). Kinetic experiments indicate that the dissociation rate [about 1 molecule sec^{-1}, Ref. (120, 129)] of even noncooperatively bound SSB is much too slow to keep pace with a replication fork moving at 500–1,000 nucleotides sec^{-1}, so, as in the case of gp32, some kind of facilitated transport must be invoked. In support of this idea, NMR studies suggest that SSB can translocate along an ssDNA lattice without completely dissociating from it (129). In addition there is evidence for direct transfer of cooperatively bound clusters of SSB molecules from one ssDNA strand to another (126), which could provide for a "hopping" as opposed to a simple sliding mechanism of linear transport.

In contrast to these similarities, the SSB : ssDNA complex differs substantially from the gp32 : ssDNA complex with respect to its stability to increasing NaCl concentration. The reason may be that this parameter greatly affects the mode of SSB binding to ssDNA (121, 123, 124). Increasing the NaCl concentration has a large and predictable effect on the gp32 : ssDNA complex, which can be understood over the range spanning at least from 0.1 to 1.0 M NaCl in terms of counter-ion release consequent to formation of the protein : nucleic acid complex (50, 54). The binding of gp32 to ssDNA is so sensitive to NaCl concentration that a 10-fold increase leads to a 10^7-fold decrease in affinity (54). In contrast to these results, increasing the NaCl concentration from 0.04 to 0.15 M has little effect on SSB binding to ssDNA (122). However, a further increase to 0.29 M NaCl results in a precipitous decline in both the cooperativity of binding and in the intrinsic SSB : ssDNA interaction (122). Increasing the NaCl concentration above 0.29 M NaCl once again has little effect on binding so that even in 1–5 M NaCl SSB still has an affinity of 10^7–10^8 M^{-1} for poly(dT) (18). In general, a tenfold increase in NaCl concentration decreases the affinity of SSB for ssDNA by less than 100-fold (18, 120, 125). From these preliminary data it appears that the overall binding of SSB to ssDNA is less sensitive to ionic strength and therefore probably involves fewer ionic interactions than the gp32 : ssDNA complex. The unusually abrupt decrease in binding affinity as the ionic strength is increased through the physiological range probably reflects alterations in the structure and topology of the SSB : ssDNA complex (121, 123, 124), rather than simply being a direct effect on SSB : ssDNA ionic interactions.

Structural and Functional Domains

The *E. coli ssb* gene encodes a protein of 177 amino acid residues of M_r = 18,873 (130), which has a pI of 6.0, a Stokes radius of 38 Å, and a frictional coefficient of 1.36 (92). The native form of the protein is a tetramer that is very stable in vitro, at least above 0.05 μM SSB, at 25°C, and is most likely the active species in vivo (16, 18, 92, 122). In contrast to gp32, SSB does not form aggregates even at protein concentrations as high as 5 μM (119). While SSB has been reported to be stable to boiling for 2 min (92), a more recent study (131) indicates that this treatment may lead to a change in apparent binding site size. Coupled with the observation that SSB undergoes an irreversible thermal transition at 71°C (9), it would seem ill-advised to unnecessarily expose SSB to elevated temperatures.

The secondary structure of *Eco* SSB predicted on the basis of its amino acid sequence suggested several functional domains (130). These have been further defined by partial proteolysis studies (18) and the analysis of purified mutant SSB proteins [(98, 119) and J. Chase, J. Murphy, and K. Williams, unpublished].

The amino-terminal two thirds of the protein (residues 1–105) is predicted to be highly ordered containing primarily α-helix and β-pleated sheet. This region of SSB contains its DNA binding domain since a proteolytic fragment containing only these residues binds to ssDNA at least as well as intact SSB. This NH_2-terminal region also includes most of the residues important for monomer:monomer interactions within the SSB tetramer and for cooperative SSB binding to ssDNA.

The region containing residues 106–165 is predicted to be a random coil and lacks any charged amino acids. This entire domain appears to be more exposed to proteolysis when SSB is bound to ssDNA, which suggests that the region would be available to interact with other proteins. The COOH-terminal region of SSB, containing residues 166–177, is striking because of its apparent similarity to the COOH-terminal region of gp32. Both proteins contain an acidic region at their COOH terminus that is susceptible to partial proteolysis (18, 63, 64). In both cases cooperative binding to DNA results in a conformational change that increases the exposure of these COOH-terminal domains to proteolysis. Oligonucleotides that are too short to allow cooperative protein:protein interactions fail to influence the rate of proteolysis of *Eco* SSB as is also the case with gp32. Finally, in both cases proteolytic removal of the COOH terminus alters the kinetics of DNA binding and increases the helix-destabilizing ability of the protein. By analogy with gp32, the COOH-terminal portion of SSB may be important for interacting with other proteins involved in DNA metabolism and perhaps in some way modulating the ability of SSB to interact with DNA.

Based on a 270-MHz ^{1}H-NMR study (129), it was concluded that SSB binding does not involve "extensive stacking interactions between the nucleotide bases and aromatic amino acid side chains of SSB." Preliminary results at 500 MHz, however, demonstrate perturbations in the aromatic region of the NMR spectrum that are induced by oligonucleotide binding (R. Prigodich, J. Coleman, J. Chase, K. Williams, unpublished), and that are qualitatively similar to those seen with gp32. In the case of gp32, aromatic amino acid side chains are thought to form an array of hydrophobic pockets that loosely intercalate with nucleic acid bases (56).

Other physicochemical studies also indicate that at least some phenylalanine and tryptophan residues are important for SSB binding. Studies of protein: ssDNA cross-linking induced by ultraviolet irradiation have shown that only one covalent linkage, involving phenylalanine 60, forms between SSB and ssDNA (132). That this residue is actually at the interface of the SSB: ssDNA complex is supported by recent in vitro mutagenesis studies (J. Chase, J. Murphy, K. Williams, unpublished). While in 30-mM NaCl SSB can lower the T_m of poly[d(A-T)] by 33°C, a mutant SSB containing alanine in place of phenylalanine 60 can only decrease the T_m of this polynucleotide by 21°C (J. Chase, J. Murphy, K. Williams, unpublished). In addition, both fluorescence (16) and optically detected magnetic resonance spectroscopy (133) suggest that at least one SSB tryptophan residue is involved in stacking interactions with ssDNA. In contrast to gp32, chemical modification experiments (16, 53) suggest that surface tyrosine residues are not involved in SSB binding to ssDNA.

Studies on a temperature-sensitive SSB mutant protein suggest that the single histidine residue is essential for tetramer formation. The cellular effects of the *ssb-1* mutation and the effects of increased expression of the *ssb-1* gene product can reasonably be explained as a defect in the stability of the SSB tetramer. In the mutant protein tyrosine is substituted for histidine at residue 55, within a putative α-helical region contained in the DNA binding domain (residues 4–115). The *ssb-1* mutation profoundly decreases the stability of the tetrameric structure of the protein, which can be reversed in vitro by increasing the SSB-1 protein concentration (119). As expected from these data, increased production of SSB-1 mutant protein in vivo reverses the cellular effects of the *ssb-1* mutation (134). The explanation for the conditional lethality of *ssb-1* and its suppression by increased production of SSB-1 protein appears to be a simple matter of chemical equilibrium: the tetrameric structure of SSB-1 is unstable at low protein concentrations and the SSB-1 monomer is thermally unstable but the tetramer is not. Increased production of SSB-1 shifts equilibrium in favor of tetramer and therefore enhances the thermal stability of SSB-1.

The *ssb-113* mutation results in the replacement of the penultimate amino acid residue, proline-176, by serine and occurs within a region of the protein not required for DNA binding. Nevertheless the cellular effects of the mutation are in some respects more severe than those of the *ssb-1* mutation. In vitro, the *ssb-113*-encoded protein binds to ssDNA as well as the wild-type protein. In fact, studies of its ability to lower the thermal melting transition of poly[d(A-T)] show it to be a better helix-destabilizing protein than wild-type SSB (98). Binding affinities calculated from thermal melting transition measurements are 2–5 times greater than for wild-type SSB. These measurements agree with electron microscopic measurements that show that *ssb-113*-encoded protein can melt-out ssDNA secondary structure at ionic strengths where wild-type

SSB is ineffective (J. Griffith, personal communication). It is interesting that the single amino acid replacement resulting from the *ssb-113* mutation has in vitro effects (though less severe) that qualitatively parallel complete loss of the entire COOH-terminal portion of the protein. These observations agree with the notion that the COOH-terminal portion of the protein may modulate DNA binding affinity, perhaps affecting the ability of the bound protein to dissociate from the DNA. An alternative explanation may be that the *ssb-113*-encoded protein is defective in a protein:protein interaction. Although only a small amount of direct evidence now exists to support the latter idea, it would not be surprising (especially in light of gp32 results described above) if a number of such interactions are found. One such interaction has been reported with protein n which forms an isolatable complex with SSB whose stability is further increased if protein n' is present (135). Also, genetic data related to certain *rep* mutants can be interpreted as suggesting a specific interaction between the *rep* gene product (3'→5' helicase) and SSB (136). The latter observation should be pursued since an intimate relationship between SSB and a helicase would seem propitious.

EUKARYOTIC SSBs?

Having reviewed the fundamental properties of the best-characterized pro-karyotic SSBs, we ask whether proteins with fundamentally similar modes of action participate in eukaryotic DNA metabolism. The answer is at best hazy. Shortly after SSBs were discovered in prokaryotes they were sought in eu-karyotes. Apparently similar proteins (i.e. proteins that bind to ssDNA and stimulate cognate DNA polymerase) were found (137, 138). However, recent studies cause us to question the similarity between these proteins and those from prokaryotes, particularly those from higher eukaryotes.

Adenovirus DBP and SSBs

The DNA replication mechanisms of a number of simple eukaryotic viruses (parvoviruses and adenoviruses) proceed through ssDNA intermediates reminiscent of prokaryotic phages (10, 94). Here it seems most likely that proteins analogous to SSBs will be found to participate in DNA replication. The best-studied of these viruses, adenovirus, does in fact produce a DNA binding protein (Ad DBP—which we would prefer to call Ad SSB) of approximately M_r = 59,000 with many similarities to prokaryotic SSBs. (Ad DBP is still commonly referred to as 72K owing to its anomalous electrophoretic morility.) In vivo studies demonstrate that this protein is essential for adenovirus DNA replication and many thermosensitive mutants have been isolated containing DBP mutations (139). In vivo and in vitro studies of DNA replication show that DBP is essential for elongation (140, 141) and may also be involved in initiation, although host factors can substitute for it in initiation of DNA replication (140–142). DBP is the most abundant early viral protein (5 × 10^6 molecules per cell) (143). There is a stoichiometric requirement for DBP in vitro (144). Its binding site size is approximately seven nucleotides (145). The purified protein binds tightly and cooperatively to ssDNA (143, 145), and puts this DNA into an extended conformation (145). DBP is highly asymmetric in solution (f/f$_0$ = 1.82) (145). Self-association of DBP occurs in vitro and is dependent on protein concentration and ionic strength (143). DBP partially protects ssDNA against hydrolysis by several nucleases (145, 146). Phosphorylation of serine and threonine residues occurs in DBP (147), but there is disagreement on whether it affects DNA binding (147, 148). However, the possible significance of even slight modulation of DNA binding by such a mechanism should not be overlooked (consider the COOH terminus of gp32 and *Eco* SSB and the effect of the *ssb-113* mutation discussed above).

Partial proteolysis products of DBP have been isolated (148). The COOH-terminal portion (about two thirds) of the protein binds to DNA (in prokaryotic SSBs it is the NH$_2$-terminal portion that binds to DNA) and is active in DNA replication (149), while the NH$_2$-terminal portion (about one third of the protein), which contains most or all of the phosphorylation sites, does not bind to DNA (147, 148). Although quantitative measurements of the binding affinity of these proteolysis products compared to intact DBP have not yet been made, the COOH-terminal peptide binds more tightly to ssDNA than intact DBP as judged by its elution from an ssDNA cellulose column at higher ionic strength (148). Thus the NH$_2$-terminal domain of DBP could serve a role in modulating DNA binding as we speculate the COOH-terminal domain does in gp32 and *Eco* SSB. The DBPs encoded by Ad2 and Ad5 are highly conserved. Both contain 529 amino acid residues and only 9 residue changes occur, all within the NH$_2$-terminal domain (150). The DBPs encoded by Ad5 and Ad12 are not as highly conserved and it is interesting to note that their COOH-terminal DNA binding domains are much more highly conserved (80%) than are the NH$_2$-terminal portions of these proteins (45%) (151). These observations bear a striking resemblance to similar comparisons of *Eco* SSB and prokaryotic plasmid-encoded SSBs [(19); P. Ruvolo and J. Chase, unpublished].

The COOH-terminal portion of DBP complements the H5*ts*125 mutant (pro→ser substitution at position 413) at nonpermissive temperatures in an in vitro DNA replication system (149). This together with the observation that DBP mutants that affect the host range map in the NH$_2$-terminal portion of the protein (150, 152) lends support to the idea that DBP contains at least two functional domains.

Despite these similarities to prokaryotic SSBs several interesting differences exist. DBP does not lower the thermal melting transition of poly[d(A-T)] but rather seems to stabilize it and DBP binds to the termini of duplex DNA (153).

Perhaps the latter processes are related. Although no catalytic activities are known to be associated with Ad DBP, there is evidence that it does affect host range (152, 154), the NH$_2$-terminal domain has been implicated in late gene expression (154), and the COOH-terminal domain has been implicated in early gene expression (155) in addition to DNA replication (149). The fact that cells infected with the H5ts125 DBP mutant produce higher levels of early virus mRNA at the nonpermissive temperature than cells infected with wild-type virus (156, 157) is caused by increased stability of message (158). Thus DBP autoregulates its own synthesis (159). In view of the known autoregulation of gp32 it is interesting to speculate that a similar function of DBP may have evolved into a function able to participate in a wider variety of processes in adenovirus.

Despite the large emphasis on the role of DBP in adenovirus DNA replication, detailed physical studies of the interaction of DBP with DNA and RNA have not been performed. Considering the fact that reasonable quantities of DBP can be obtained, there are no great technical problems preventing such analysis and that of proteolysis products as well.

There is still a question of whether duplex adenovirus DNA is complexed with cellular histones (10). If it is complexed with nonhistone cellular proteins or viral proteins during replication, this together with the basic mechanism of adenovirus DNA replication may suggest an even closer relationship to prokaryotic DNA replication than to that of higher eukaryotes. DNA of papovaviruses (e.g. SV40 and polyoma virus) is condensed by cellular histones and replication proceeds bidirectionally utilizing Okazaki fragment intermediates for lagging-strand synthesis (10). SV40 does not encode a protein functionally analogous to prokaryotic SSBs or similar to Ad DBP. The question of whether such a protein is provided by the host is unanswered and leads to the question of whether such proteins even exist in higher eukaryotes.

Yeast SSBs

Above the simple eukaryotic viruses, yeast plasmids currently represent the best system for studies of eukaryotic DNA replication. Three single-stranded DNA binding proteins have now been isolated from yeast. SSB I (M_r = 45,000) specifically stimulates yeast DNA polymerase I and increases its processivity (160, 160a). It is estimated to be present at approximately 20,000 molecules per cell (160). The gene encoding SSB I has been cloned and recent work has demonstrated that it is nonessential (161). SSB II appears to bind more tightly to ssDNA than SSB I, has an M_r = 50,000, and is also present at a high level (160). SSB III (mtSSB) is mitochondrial and has an M_r = 20,000 (160). Considering the utility of yeast genetics and molecular biology, it would seem that the functional role of these proteins in yeast DNA metabolism can be quickly determined. This coupled with careful in vitro physical studies of the proteins should soon allow us to ascertain the functional relationship between these proteins and prokaryotic SSBs.

Calf Thymus UP1

When Herrick & Alberts (137) subjected extracts from calf thymus to a procedure similar to that used to purify gp32 and *E. coli* SSB, they isolated three major ssDNA binding proteins. The predominant protein, the 24,000-dalton UP1, appeared similar in many respects to the prokaryotic ssDNA binding proteins. Thus, UP1 binds tightly (K_{app} _~ 10^7 M^{-1} for denatured *Clostridium perfringens* DNA) but without apparent sequence specificity to ssDNA (162). UP1 binds slightly better to ssDNA than to ssRNA and its 1500-fold greater affinity for ssDNA than for dsDNA makes it a very effective helix-destabilizing protein (162). In 35-mM NaCl the stoichiometric addition (assuming a binding site size of about 7 phosphates) of UP1 decreased the T_m of poly[d(A-T)] by about 45°C (162). UP1 binding disrupts base-stacking interactions and imposes a rigid, extended conformation onto the ssDNA (162, 163). Most importantly, UP1 stimulated the activity of DNA polymerase alpha by 10-fold and this stimulation was specific in that gp32 could not substitute for UP1 (163). While this stimulation of DNA polymerase alpha was presumed to reflect the in vivo role of UP1, there were, nonetheless, several differences noted in the properties of UP1 compared to other prokaryotic ssDNA binding proteins that were known to play essential roles in DNA replication. In particular, UP1 is heterogeneous with respect to size and charge. Individual UP1 species range in size from about 22,500 to 25,500 daltons and have isoelectric points that range from about 6.5 to 7.5 (137). In addition, UP1 exists in solution as a stable monomer (while *Eco* SSB is a tetramer) that does not seem to have any tendency to aggregate as do gp32 and Ad DBP. The importance of larger molecular forms to the interaction of these proteins with DNA is not clear but it is interesting to note the difference. With respect to its interactions with nucleic acid, UP1 is unusual in that it appears to bind noncooperatively and unlike gp32 or *E. coli* SSB, UP1 can destabilize natural dsDNA (162). A further difference with respect to UP1 is that this protein does not catalyze the renaturation of dsDNA.

These differences between the properties of UP1 and other prokaryotic ssDNA binding proteins may arise from the fact that UP1 might play a somewhat different role in vivo than was originally supposed. Recent immunological (164) and HPLC peptide mapping studies (K. Williams, B. Merrill, K. Stone, A. Kumar, W. Szer, M. Pandolfo, S. Riva, unpublished) demonstrate that UP1 is actually a proteolytic fragment corresponding to the first 195 amino acid residues in the 32,000-dalton A_1 hnRNP protein. Immediately following transcription, heterogeneous nuclear RNA (hnRNA) becomes associated with a group of at least six proteins to form a ribonucleoprotein complex that is believed to function in the stabilization, transport, and posttranscriptional processing of unique sequence transcripts (165). If care is taken to inhibit ribonuclease activity, hnRNP complexes can be isolated by centrifugation as 200–300S heterodisperse structures that can be converted upon mild ribonuclease digestion to a relatively monodisperse 40S species. Electron microscopy of the 200–300S complexes suggests a "beads on a string" model in which the 40S beads are separated by less protein-rich filamentous "spacers" that appear to be susceptible to ribonuclease. SDS polyacrylamide gel electrophoresis (SDS PAGE) of the 40S complexes reveals three closely spaced doublets corresponding to the A_1/A_2, B_1/B_2, and C_1/C_2 proteins respectively (165). That UP1 is derived from the A_1 hnRNP protein has now been confirmed by direct sequencing of several peptides from A_1. The amino acid sequence of all A_1 tryptic peptides so far sequenced, corresponding to a total of 45 residues, matches exactly the previously published sequence for UP1 (166, 167). Since the A_1 protein was isolated from HeLa cells and the UP1 was derived from calf thymus, it is evident that the primary structure of at least this hnRNP is highly conserved. Using synthetic oligonucleotide probes, whose sequences were based on the amino acid sequence of calf thymus UP1 (166, 167), a cDNA clone has been isolated from a cDNA library in λgt11 constructed with total poly(A) mRNA of newborn rat brain that corresponds to the A_1 protein (168). The amino acid sequence of UP1 (166, 167) matches exactly the first 195 amino acids predicted from the DNA sequence of this clone. Based on the DNA sequence, the A_1 hnRNP protein contains another 124 amino acids that constitute a very glycine-rich (about 40%) domain at the COOH-terminus of this protein (F. Cobianchi, K. Williams, and S. Wilson, unpublished).

One of the intriguing questions posed by these recent data concerns the physiological significance of this apparent proteolysis. Most of the available data suggest that the UP1 protein may simply be an artifact resulting from proteolysis of A_1 that occurs during the UP1 purification. Overnight incubation at 30°C of partially purified hnRNP proteins results in their extensive conversion to lower-molecular-weight species that co-migrate with UP1 on SDS PAGE (169). This conversion is not slowed by conventional protease inhibitors and is catalyzed by at least one trypsin-like enzyme that appears to purify together with hnRNP proteins (169). Similarly, highly purified A_1 protein can be quantitatively converted by partial trypsin digestion to a stable, lower-molecular-weight protein that exactly corresponds to UP1 protein after SDS PAGE (A. Kumar, W. Szer, K. Williams, unpublished). Presumably, trypsin cleaves the A_1 protein at arginine 195 which is the usual COOH-terminus of UP1 (166, 167). Several other UP1-like proteins that range in size from about 25,000 to 28,000 (170) daltons can also be readily converted by partial proteolysis to a 24,500 species (B. Merrill and K. Williams, unpublished), thus accounting for at least some of the microheterogeneity observed in previous preparations of UP1 (137). When crude HeLa cell extracts are screened by Western blot analysis with polyclonal antibodies directed against UP1, only proteins that correspond in molecular weight to the "core" hnRNP proteins are detected (164). While similar experiments on crude extracts from calf thymus do reveal some UP1 protein, most of the cross-reactive species present again correspond in molecular weight to known hnRNP proteins (164). Thus while it is still possible that some UP1 may be present in vivo, it is clear that most of the UP1 purified from these tissues actually results from in vitro proteolysis of the A_1 hnRNP protein. Previous estimates of the relative abundance of the UP1 protein in vivo are therefore clearly too high (137). The major piece of evidence that argues against UP1 simply being an in vitro artifact is that this apparently quite specific proteolysis endows the protein with a unique functional property, the ability to stimulate DNA polymerase alpha. Preliminary experiments suggest that partially purified A_1/A_2 proteins cannot substitute for UP1 in this reaction. As expected, the A_1/A_2 proteins do stimulate DNA polymerase alpha if they are first converted by partial proteolysis to proteins that correspond in molecular weight to UP1 (S. Riva, unpublished). Although the idea appears to be highly speculative, it is nonetheless intriguing to contemplate that limited in vivo proteolysis might be used to alter the function of a protein that is normally involved in hnRNA metabolism to one that might play a role in DNA metabolism. It is also still possible that alternative explanations involving partial gene duplication or differential splicing could also account for the appearance of small amounts of UP1 in vivo.

Even if UP1 turns out to be an artifact of the isolation procedure, it is interesting to note that there is actually a reasonably good analogy between the functions of prokaryotic ssDNA binding proteins and eukaryotic hnRNP proteins. Both classes of proteins play a structural role that is essential to transient stabilization of single-stranded nucleic acids. In both cases, binding of these proteins results in the removal of competing secondary structures and intrachain base-stacking as well as the imposition of an extended structure that is suitable to then serve as a substrate for other enzymes involved in nucleic acid metabolism. Based on their similar functions and effects on nucleic acid structure, it seems likely that both prokaryotic ssDNA binding proteins and eukaryotic hnRNP proteins will be found to share similar mechanisms for binding single-stranded nucleic acids. Indeed, there is already indirect evidence that suggests that like the binding of gp32, aromatic amino acids may also be involved in the binding of UP1 to single-stranded nucleic acids. The amino acid sequence of UP1 revealed the presence of a 91-residue internal repeat such that when residues 3–93 are aligned with 94–184, 33% of the amino acids are identical and an additional 39% of the residues can be accounted for by single-base changes (167). The importance of aromatic amino acids in UP1 is indicated by the high degree of homology of these amino acids, in particular in the A (residues 1–93) and B (residues 94–184) domains of UP1. When the A and B domains are aligned, the homology with respect to the position of aromatic amino acids is greater than 80%. There is one stretch, in particular, in which there are three aromatic and three positively charged residues in identical positions (residues 51–62 and 142–152), which we have previously speculated represent two independent ssDNA binding domains (167). This idea is supported by photoinduced cross-linking experiments that demonstrate that following exposure of the UP1 : $[^{32}P]d(pT)_8$ complex to ultraviolet light, both the "A" and "B" domains become independently linked to $[^{32}P]d(pT)_8$ (B. Merrill, K. Stone and K. Williams, unpublished). The presence of multiple nucleic acid binding domains may actually prove to be a general feature of eukaryotic single-stranded nucleic acid binding proteins. In addition to hnRNP proteins, multiple binding domains have previously been postulated to exist in high-molecular-weight members of a class of nonhistone chromosomal proteins referred to as HMG or high-mobility group proteins (171) as well as in the S1 ribosomal protein (172) and in structural proteins from avian sarcoma (89) and bovine leukemia virus (90).

From a practical standpoint, the finding that UP1 is probably derived from limited proteolysis of the A_1 hnRNP protein poses a serious obstacle to the isolation of other, perhaps less abundant, eukaryotic ssDNA binding proteins that might be more analogous than is UP1 to their prokaryotic counterparts. It is now clear that all major hnRNP proteins as well as their in vitro proteolysis products bind tightly to ssDNA cellulose and elute over a broad range of salt concentrations extending at least from 0.6 M to 0.9 M NaCl (169). The abundance of these proteins in vivo and the fact that they can be readily isolated by conventional chromatographic procedures (169) that were originally designed to purify eukaryotic ssDNA binding proteins (137) requires that any new eukaryotic "ssDNA binding protein" be extensively characterized before ascribing any functional significance to its ability to bind ssDNA cellulose. There are, however, a number of characteristic properties that can be used to rule out the possibility that the protein is actually an hnRNP protein. The hnRNP proteins have an unusual amino acid composition, including high glycine (25%) and the modified basic residue identified as N^G,N^G-dimethyl-arginine. The most abundant of these proteins are basic with isoelectric points of 8.4–9.2, have blocked NH_2-termini, and have a size of 32,000–44,000 daltons (165). Since antibodies directed against UP1 cross-react with all major hnRNP proteins (164), it appears likely that this class of proteins will be found to share at least some homologous sequences in common. The availability of polyclonal antibodies to hnRNP proteins (164), and the completion of the amino acid sequence of the most abundant hnRNP protein, the A_1 protein (168), should therefore allow more definitive tests to be done to determine if a ssDNA binding protein is actually an hnRNP protein. In this regard it has recently been shown that two ssDNA binding proteins from calf thymus with molecular weights of 48,000 and 61,000 (174) do not cross-react with antibodies to UP1 (S. Riva, unpublished). These proteins stimulate the activity of DNA polymerase alpha by as much as 100-fold (174). The stimulation appears to arise from the ability of these proteins to block nonproductive polymerase-binding sites on ssDNA sequences, thereby increasing the probability for an association of polymerase with 3'-OH primer ends. It is not yet known if these two proteins are actually required for any step in eukaryotic DNA replication (174). In fact, except for the adenovirus DNA binding protein, convincing evidence is not yet available for the involvement of SSB proteins in eukaryotic DNA replication.

CONCLUSIONS

The bacteriophage T4 gp32, *E. coli* SSB, and the adenovirus DBP proteins provide three excellent prototypes for a class of ssDNA-stabilizing proteins that play an essential role in DNA replication. While these three proteins do not share any obvious sequence homology, they nonetheless are similar with respect to their interactions with nucleic acids. Each of these proteins has a region (COOH-terminal for *Eco* SSB and gp32 and NH_2-terminal for Ad DBP) that seems to be important for modulating the helix-destabilizing activity of the protein and perhaps also for interacting with other proteins involved in DNA metabolism. The existence of multiple binding modes and the increasing evidence that indicates that the ability to interact with other DNA metabolizing proteins is an essential aspect of gp32 and SSB function suggests that prokaryotic ssDNA binding proteins may fulfill a more sophisticated and dynamic role in vivo than implied by their name.

Even though the calf thymus UP1 protein may not prove to be the long-sought eukaryotic analogue of gp32 and *Eco* SSB, studies on this protein have already revealed another distinctive structural motif, multiple nucleic acid binding sites, that seems to be a common feature of several single-stranded nucleic acid binding proteins. Preliminary data also points to the involvement of aromatic amino acids in the binding of UP1 to single-stranded nucleic acids. If more detailed physicochemical studies support this idea, then it would imply that numerous prokaryotic and eukaryotic ssDNA and ssRNA binding proteins might share a common mechanism of binding and of recognizing single-stranded nucleic acids.

The general features of DNA replication in prokaryotes and eukaryotes are clearly similar and therefore it is likely that proteins exist in eukaryotes that stabilize ssDNA at replication forks and perhaps confer unique structure to that ssDNA of importance for the interaction of other DNA metabolic proteins with the DNA. We question, however, whether many of the eukaryotic proteins that have been described to date and suggested as performing these functions are the correct proteins. Alternatively, the subtleties of eukaryotic DNA metabolism may be such that basic differences exist between these proteins in prokaryotes and eukaryotes. Progress in answering these questions will no doubt be directly related to the ability to perform genetic analysis in the system under study. Clear answers will first emerge in lower eukaryotic systems such as yeast and *Drosophila* where sophisticated genetic analysis can be performed. Much additional work (and less direct approaches) will be required in higher eukaryotic systems where the necessary genetic manipulations cannot yet be done.

ACKNOWLEDGMENTS

Research from the author's laboratories was supported by NIH Research Grants GM11301, GM23451, GM34697, and CA13330 to J.W.C. and GM31539 and NSF Grant PCM83–02908 to K.R.W.

We are grateful to Dr. William Konigsberg, Mary LoPresti, Barbara Merrill, Janet Murphy, and Kathy Stone for their participation in our research and to Vivian Gradus for her patience and skill in preparing the manuscript.

Literature Cited

1. Alberts, B., Sternglanz, R. 1977. *Nature* 269:655–61
2. von Hippel, P. H., Jensen, D. E., Kelly, R. C., McGhee, J. D. 1977. In *Nucleic Acid-Protein Recognition*, ed. H. J. Vogel, pp. 65–90. New York: Academic
3. Coleman, J. E., Oakley, J. L. 1980. *CRC Crit. Rev. Biochem.* 7:247–89
4. DePamphilis, M. L., Wassarman, P. M. 1980. *Ann. Rev. Biochem.* 49:627–66
5. Falaschi, A., Cobianchi, F., Riva, S. 1980. *Trends Biochem. Sci.* 5:154–57
6. Duguet, M. 1981. *Biochimie* 63:649–69
7. Geider, K., Hoffmann-Berling, H. 1981. *Ann. Rev. Biochem.* 50:233–60
8. Kowalczykowski, S. D., Bear, S. D., von Hippel, P. H. 1981. *Enzymes* 14a:373–444
9. Williams, K. R., Konigsberg, W. H. 1981. In *Gene Amplification and Analysis*, ed. J. G. Chirikjian, T. S. Papas, 2:475–508. North-Holland/Amsterdam: Elsevier
10. Challberg, M. D., Kelly, T. J. 1982. *Ann. Rev. Biochem.* 51:901–34
11. Williams, K. R., Konigsberg, W. H. 1983. In *Bacteriophage T4*, ed. C. K. Mathews, E. M. Kutter, G. Mosig, P. B. Berget, pp. 82–89. Washington, DC: Am. Soc. Microbiol.
12. Chase, J. W. 1984. *BioEssays* 1:218–22
13. Alberts, B. M., Frey, L. 1970. *Nature* 227:1313–18
14. Sigal, N., Delius, H., Kornberg, T., Gefter, M. L., Alberts, B. M. 1972. *Proc. Natl. Acad. Sci. USA* 69:3537–41
15. Jensen, D. E., Kelly, R. C., von Hippel, P. H. 1976. *J. Biol. Chem.* 251:7215–28
16. Bandyopadhyay, P. K., Wu, C.-W. 1978. *Biochemistry* 17:4078–85
17. Kornberg, A. 1980. In *DNA Replication*. San Francisco: Freeman
18. Williams, K. R., Spicer, E. K., LoPresti, M. B., Guggenheimer, R. A., Chase, J. W. 1983. *J. Biol. Chem.* 258:3346–55
19. Chase, J. W., Merrill, B. M., Williams, K. R. 1983. *Proc. Natl. Acad. Sci. USA* 80:5480–84
20. Williams, K. R., Reddigari, S., Patel, G. 1985. *Proc. Natl. Acad. Sci. USA* 82:5260–64
21. Pritchard, C. G., DePamphilis, M. L. 1983. *J. Biol. Chem.* 258:9801–9
22. Riva, S., Cascino, A., Geiduschek, E. P. 1970. *J. Mol. Biol.* 54:85–102
23. Curtis, M. J., Alberts, B. 1976. *J. Mol. Biol.* 102:793–816
24. Wu, J.-R., Yeh, Y.-C. 1973. *J. Virol.* 12:758–65
25. Tomizawa, J.-I., Anraku, N., Iwama, Y. 1966. *J. Mol. Biol.* 21:247–53
26. Berger, H., Warren, A., Fry, K. 1969. *J. Virol.* 3:171–75
27. Mosig, G., Luder, A., Garcia, G., Dannenberg, R., Bock, S. 1979. *Cold Spring Harbor Symp. Quant. Biol.* 43:501–15

28. Huberman, J. A., Kornberg, A., Alberts, B. M. 1971. *J. Mol. Biol.* 62:39–52
29. Huang, C.-C., Hearst, J. E. 1980. *Anal. Biochem.* 103:127–39
30. Huang, C.-C., Hearst, J. E. 1981. *J. Biol. Chem.* 256:4087–94
31. Newport, J. W., Kowalczykowski, S. C., Lonberg, N., Paul, L. S., von Hippel, P. 1980. In *Mechanistic Studies on DNA Replication and Genetic Recombination. ICN-UCLA Symp. Mol. Cell. Biol.*, ed. B. M. Alberts, C. F. Fox, 19:485–505. New York: Academic
32. Nossal, N. G. 1974. *J. Biol. Chem.* 249:5668–76
33. Sinha, N. K., Morris, C. F., Alberts, B. M. 1980. *J. Biol. Chem.* 255:4290–303
34. Nossal, N. G., Peterlin, B. M. 1979. *J. Biol. Chem.* 254:6032–37
35. Alberts, B. M., Barry, J., Bedinger, P., Burke, R. L., Hibner, U., Liu, C. C. et al. 1980. See Ref. 31, pp. 449–71
36. Sinha, N. K., Snustad, D. P. 1971. *J. Mol. Biol.* 62:267–71
37. Alberts, B. M. 1970. *Fed. Proc.* 29:1154
38. Burke, R. L., Alberts, B. M., Hosoda, J. 1980. *J. Biol. Chem.* 255:11484–93
39. Deleted in proof
40. Formosa, T., Burke, R. L., Alberts, B. M. 1983. *Proc. Natl. Acad. Sci. USA* 80:2442–46
41. Wackernagel, W., Radding, C. M. 1974. *Proc. Natl. Acad. Sci. USA* 71:431–35
42. Shibata, T., DasGupta, C., Cunningham, R. P., Radding, C. M. 1980. *Proc. Natl. Acad. Sci. USA* 77:2606–10
43. Krisch, H. M., Bolle, A., Epstein, R. H. 1974. *J. Mol. Biol.* 88:89–104
44. Gold, L., O'Farrell, P. Z., Russel, M. 1976. *J. Biol. Chem.* 251:7251–62
45. Russel, M., Gold, L., Morrissett, H., O'Farrell, P. Z. 1976. *J. Biol. Chem.* 251:7263–70
46. Gold, L., Lemaire, G., Martin, C., Morrissett, H., O'Conner, P., et al. See Ref. 2, pp. 91–113
47. Krisch, H. M., Duvoisin, R. M., Allet, B., Epstein, R. H. 1980. See Ref. 31, pp. 517–26
48. Krisch, H. M., Allet, B. 1982. *Proc. Natl. Acad. Sci. USA* 79:4937–41
49. von Hippel, P. H., Kowalczykowski, S. C., Lonberg, N., Newport, J. W., Paul, L. S., et al. 1982. *J. Mol. Biol.* 162:795–818
50. Newport, J. W., Lonberg, N., Kowalczykowski, S. C., von Hippel, P. 1981. *J. Mol. Biol.* 145:105–21
51. Lemaire, G., Gold, L., Yarus, M. 1978. *J. Mol. Biol.* 126:73–90
52. Riley, M., Maling, B., Chamberlin, M. J. 1966. *J. Mol. Biol.* 20:359–89
53. Anderson, R. A., Coleman, J. E. 1975. *Biochemistry* 14:5485–91
54. Kowalczykowski, S. C., Lonberg, N., Newport, J. W., von Hippel, P. H. 1981. *J. Mol. Biol.* 145:75–104
55. Delius, H., Mantell, N. J., Alberts, B. 1972. *J. Mol. Biol.* 67:341–50
56. Prigodich, R. V., Casas-Finet, J., Williams, K. R., Konigsberg, W., Coleman, J. E. 1984. *Biochemistry* 23:522–59
57. Kelly, R. C., von Hippel, P. H. 1976. *J. Biol. Chem.* 251:7229–39
58. Kelly, R. C., Jensen, D. E., von Hippel, P. H. 1976. *J. Biol. Chem.* 251:7240–50
59. Suau, P., Toulme, J. J., Helene, C. 1980. *Nucleic Acids Res.* 8:1357–72
60. Spicer, E. K., Williams, K. R., Konigsberg, W. H. 1979. *J. Biol. Chem.* 254:6433–36
61. Bobst, A. M., Langemeier, P. W., Warwick-Koochaki, P., Bobst, E., Ireland, J. C. 1982. *J. Biol. Chem.* 257:6184–93
62. Bobst, A. M., Pan, Y.-C. 1975. *Biochem. Biophys. Res. Commun.* 67:562–70
63. Williams, K. R., Konigsberg, W. H. 1978. *J. Biol. Chem.* 253:2463–70
64. Hosoda, J., Moise, H. 1978. *J. Biol. Chem.* 253:7547–55
65. Lohman, T. M., Kowalczykowski, S. C. 1981. *J. Mol. Biol.* 152:67–109
66. Carroll, R. B., Neet, K., Goldthwait, D. A. 1975. *J. Mol. Biol.* 91:275–91
67. Lohman, T. M. 1984. *Biochemistry* 23:4656–65
68. Lohman, T. M. 1984. *Biochemistry* 23:4665–75
69. Williams, K. R., LoPresti, M. B., Setoguchi, M., Konigsberg, W. H. 1980. *Proc. Natl. Acad. Sci. USA* 77:4614–17
70. Williams, K. R., LoPresti, M. B., Setoguchi, M. 1981. *J. Biol. Chem.* 256:1754–62
71. Cohen, H. A., Chiu, W., Hosoda, J. 1983. *J. Mol. Biol.* 169:235–48
72. Hosoda, J., Takacs, B., Brack, C. 1974. *FEBS Lett.* 47:338–42
73. Moise, H., Hosoda, J. 1976. *Nature* 259:455–58
74. Tsugita, A., Hosoda, J. 1978. *J. Mol. Biol.* 122:255–58
75. Hosoda, J., Burke, R. L., Moise, H., Kubota, I., Tsugita, A. 1980. See Ref. 31, pp. 505–13
76. Lonberg, N., Kowalczykowski, S. C., Paul, L. S., von Hippel, P. H. 1981. *J. Mol. Biol.* 145:123–38
76a. Williams, K. R., Sillerud, L. O., Schafer, D. E., Konigsberg, W. H. 1979. *J. Biol. Chem.* 254:6426–32
77. Minegawa, T., Yonesaki, T., Fujisawa, H. 1983. *Plant Cell Physiol.* 24(3):403–9

78. Isackson, P. J., Reeck, G. R. 1982. *Biochim. Biophys. Acta* 697:378–80
79. Carballo, M., Puigdomenech, P., Tancredi, T., Palau, J. 1984. *EMBO J.* 3:1255–61
80. Reeck, G. R., Isackson, P. J., Teller, D. C. 1982. *Nature* 300:76–78
81. Doherty, D. H., Gauss, P., Gold, L. 1982. *Mol. Gen. Genet.* 188:77–90
82. Toulme, J.-J., Helene, C. 1980. *Biochim. Biophys. Acta* 606:95–104
83. Casas-Finet, J., Toulme, J.-J., Cazenave, C., Santus, R. 1984. *Biochemistry* 23:1208–13
84. Toulme, J.-J., LeDoan, T., Helene, C. 1984. *Biochemistry* 23:1195–1201
85. LeDoan, T., Toulme, J.-J., Helene, C. 1984. *Biochemistry* 23:1202–7
86. Helene, C., Toulme, J.-J., LeDoan, T. 1979. *Nucleic Acids Res.* 7:1945–54
87. O'Connor, T. P., Coleman, J. E. 1983. *Biochemistry* 22:3375–81
88. Henderson, L. E., Copeland, T. D., Sowder, R. C., Smythers, G. W., Oroszlan, S. 1981. *J. Biol. Chem.* 256:8400–6
89. Misono, K. S., Sharief, F. S., Leis, J. 1980. *Fed. Proc.* 39:1611
90. Copeland, T. D., Morgan, M. A., Oroszlan, S. 1983. *FEBS Lett.* 156:37–40
91. McKay, D., Williams, K. R. 1982. *J. Mol. Biol.* 160:659–61
92. Weiner, J. H., Bertsch, L. L., Kornberg, A. 1975. *J. Biol. Chem.* 250:1972–80
93. Molineux, I. J., Friedman, S., Gefter, M. L. 1974. *J. Biol. Chem.* 249:6090–98
94. Marians, K. J. 1984. *CRC Crit. Rev. Biochem.* 17:153–215
95. Meyer, R. R., Glassberg, J., Kornberg, A. 1979. *Proc. Natl. Acad. Sci. USA* 76:1702–5
96. Auerbach, J. I., Howard-Flanders, P. 1981. *J. Bacteriol.* 146:713–17
97. Johnson, B. F. 1984. *Arch. Microbiol.* 138:106–12
98. Chase, J. W., L'Italien, J. J., Murphy, J. B., Spicer, E. K., Williams, K. R. 1984. *J. Biol. Chem.* 259:805–14
99. Glassberg, J., Meyer, R. R., Kornberg, A. 1979. *J. Bacteriol.* 140:14–19
100. Vales, L., Chase, J. W., Murphy, J. B. 1980. *J. Bacteriol.* 143:887–896
101. Baluch, J., Chase, J. W., Sussman, R. 1980. *J. Bacteriol.* 144:489–98
102. Lieberman, H. B., Witkin, E. M. 1983. *Mol. Gen. Genet.* 190:92–100
103. Whittier, R. F., Chase, J. W. 1983. *Mol. Gen. Genet.* 190:101–11
104. Golub, E. I., Low, K. B. 1983. *Proc. Natl. Acad. Sci. USA* 80:1401–5
105. Cuozzo, M., Silverman, P. 1985. *J. Biol. Chem.* Submitted for publication
106. Kunkel, T. A., Meyer, R. R., Loeb, L.

A. 1979. *Proc. Natl. Acad. Sci. USA* 76:6331–35
107. Fay, P. J., Johanson, K. O., McHenry, C. S., Bambara, R. A. 1982. *J. Biol. Chem.* 257:5692–99
108. LaDuca, R. J., Fay, P. J., Chuang, C., McHenry, C. S., Bambara, R. A. 1983. *Biochemistry* 22:5177–87
109. Fuller, R., Kaguni, J., Kornberg, A. 1981. *Proc. Natl. Acad. Sci. USA* 78:7370–74
110. Lu, A.-L., Welsh, K., Clark, S., Su, S.-S., Modrich, P. 1984. *Cold Spring Harbor Symp. Quant. Biol.* 49:589–96
111. Lundblad, V., Kleckner, N. 1985. *Genetics* 109:3–19
112. Muniyappa, K., Shaner, S. L., Tsang, S. S., Radding, C. M. 1984. *Proc. Natl. Acad. Sci. USA* 81:2757–61
112a. Tsang, S. S., Muniyappa, K., Azhderian, E., Gonda, D. K., Radding, C. M., et al. 1985. *J. Mol. Biol.* In press
113. Salles, B., Paoletti, C., Villani, G. 1983. *Mol. Gen. Genet.* 189:175–77
114. Alazard, R. J. 1983. *Mutat. Res.* 109:155–68
115. Brandsma, J. A., Bosch, D., Brackendorf, C., van de Putte, P. 1983. *Nature* 305:243–45
116. Kolodkin, A. L., Capage, M. A., Golub, E. I., Low, K. B. 1983. *Proc. Natl. Acad. Sci. USA* 80:4422–26
117. Golub, E. I., Low, K. B. 1985. *J. Bacteriol.* 162:235–41
118. Deleted in proof
119. Williams, K. R., Murphy, J. B., Chase, J. W. 1984. *J. Biol. Chem.* 259:11804–11
120. Krauss, G., Sindermann, H., Schomburg, U., Maass, G. 1981. *Biochemistry* 20:5346–52
121. Chrysogelos, S., Griffith, J. 1982. *Proc. Natl. Acad. Sci. USA* 79:5803–7
122. Ruyechan, W. T., Wetmur, J. G. 1976. *Biochemistry* 15:5057–64
123. Griffith, J. D., Harris, L. D., Register, J. 1984. *Cold Spring Harbor Symp. Quant. Biol.* 49:553–59
124. Lohman, T. M., Overman, L. B. 1985. *J. Biol. Chem.* 260:3594–603
125. Molineux, I. J., Pauli, A., Gefter, M. L. 1975. *Nucleic Acids Res.* 2:1821–37
126. Schneider, R. J., Wetmur, J. G. 1982. *Biochemistry* 21:608–15
127. Deleted in proof
128. Christiansen, C., Baldwin, R. L. 1977. *J. Mol. Biol.* 115:441–54
129. Romer, R., Schomburg, U., Krauss, G., Maass, G. 1984. *Biochemistry* 23:6132–37
130. Sancar, A., Williams, K. R., Chase, J.

W., Rupp, W. D. 1981. *Proc. Natl. Acad. Sci. USA* 78:4274–78
131. Bobst, E. V., Bobst, A. M., Perrino, F. W., Meyer, R. R., Rein, D. C. 1985. *FEBS Lett.* 181:133–37
132. Merrill, B. M., Williams, K. R., Chase, J. W. 1984. *J. Biol. Chem.* 259:10850–56
133. Cha, T. A., Maki, A. H. 1984. *J. Biol. Chem.* 259:1105–9
134. Chase, J. W., Murphy, J. B., Whittier, R. F., Lorensen, E., Sninsky, J. J. 1983. *J. Mol. Biol.* 64:193–211
135. Low, R. L., Shlomai, J., Kornberg, A. 1982. *J. Biol. Chem.* 257:6242–50
136. Tessman, E. S., Peterson, P. K. 1982. *J. Bacteriol.* 152:572–86
137. Herrick, G., Alberts, B. 1976. *J. Biol. Chem.* 251:2124–32
138. van der Vliet, P. C., Levine, A. J. 1973. *Nature New Biol.* 246:170–74
139. Young, C. S. H., Shenk, T., Ginsberg, H. S. 1984. In *The Adenoviruses*, ed. H. S. Ginsberg, pp. 125–72. New York: Plenum
140. Friefeld, B. R., Krevolin, M. D., Horwitz, M. S. 1983. *Virology* 124:380–89
141. van Bergen, B. G., van der Vliet, P. C. 1983. *J. Virol.* 46:642–48
142. Friefeld, B. R., Lichy, J. H., Field, J., Gronostajski, R. M., Guggenheimer, R. A., et al. 1984. *Curr. Top. Microbiol. Immunol.* 110:221–55
143. Schechter, N. M., Davies, W., Anderson, C. W. 1980. *Biochemistry* 19:2802–10
144. Kaplan, L. M., Ariga, H., Hurwitz, J., Horwitz, M. S. 1979. *Proc. Natl. Acad. Sci. USA* 76:5534–38
145. van der Vliet, P. C., Keegstra, W., Jansz, H. S. 1978. *Eur. J. Biochem.* 86:389–98
146. Nass, K., Frenkel, G. D. 1980. *J. Virol.* 35:314–19
147. Linne, T., Philipson, L. 1980. *Eur. J. Biochem.* 103:259–70
148. Klein, H., Maltzman, W., Levine, A. J. 1979. *J. Biol. Chem.* 254:11051–60
149. Ariga, H., Klein, H., Levine, A. J., Horwitz, M. S. 1980. *Virology* 101:307–10
150. Kruijer, W., van Schaik, F. M. A., Sussenbach, J. S. 1982. *Nucleic Acids Res.* 10:4493–500
151. Kruijer, W., van Schaik, F. M. A., Speijer, J. G., Sussenbach, J. S. 1981. *Virology* 128:140–53
152. Kruijer, W., van Schaik, F. M. A., Sus-

senbach, J. S. 1981. *Nucleic Acids Res.* 9:4439–57
153. Fowlkes, D. M., Lord, S. T., Linne, T., Pettersson, U., Philipson, L. 1979. *J. Mol. Biol.* 132:163–80
154. Klessig, D. F., Grodzicker, T. 1979. *Cell* 17:957–66
155. Nevins, J. R., Winkler, J. J. 1980. *Proc. Natl. Acad. Sci. USA* 77:1893–97
156. Carter, T. H., Blanton, R. A. 1978. *J. Virol.* 28:450–56
157. Blanton, R. A., Carter, T. H. 1979. *J. Virol.* 29:458–65
158. Babich, A., Nevins, J. R. 1981. *Cell* 26:371–79
159. Nicolas, J. C., Suarez, F., Levine, A. J., Girard, M. 1981. *Virology* 108:521–24
160. Jong, A. Y. S., Aebersold, R., Campbell, J. 1985. *J. Biol. Chem.* In press
160a. LaBonne, S. G., Dumas, L. B. 1983. *Biochemistry* 22:3214–19
161. Jong, A. Y. S., Campbell, J. 1986. *Proc. Natl. Acad. Sci. USA*. In press
162. Herrick, G., Alberts, B. 1976. *J. Biol. Chem.* 251:2133–41
163. Herrick, G., Delius, H., Alberts, B. 1976. *J. Biol. Chem.* 251:2142–46
164. Valentini, O., Biamonti, G., Pandolfo, M., Morandi, C., Riva, S. 1985. *Nucleic Acids Res.* 13:337–46
165. Beyer, A., Christensen, M. E., Walker, B. W., LeStourgeon, W. M. 1977. *Cell* 11:127–38
166. Williams, K. R., Stone, K. L., LoPresti, M. B., Merrill, B. M., Planck, S. R. 1985. *Proc. Natl. Acad. Sci. USA* 82:5666–70
167. Merrill, B. M., LoPresti, M. B., Stone, K. L., Williams, K. R. 1985. *J. Biol. Chem.* 261: In press
168. Cobianchi, F., Williams, K. R., Wilson, S. H. 1985. *Fed. Proc.* 44:856
169. Pandolfo, M., Valentini, O., Biamonti, G., Morandi, C., Riva, S. 1985. *Nucleic Acids Res.* 13:6577–90
170. Valentini, O., Biamonti, G., Mastromei, G., Riva, S. 1984. *Biochim. Biophys. Acta* 782:147–55
171. Reeck, G. R., Teller, D. C. 1985. In *Progress in Non-Histone Protein Research*, Vol. 2, ed. I. Bekhor. Boca Raton, Fla: CRC
172. Thomas, J. O., Szer, W. 1982. *Prog. Nucleic Acids Res.* 27:157–87
173. Deleted in proof
174. Sapp, M., Konig, H., Riedel, H. D.; Richter, A., Knippers, R. 1985. *J. Biol. Chem.* 260:1550–56

Ann. Rev. Biochem. 1986. 55:137–166

REACTIVE OXYGEN INTERMEDIATES IN BIOCHEMISTRY[1]

Ali Naqui and Britton Chance

D501 Richards Building, Department of Biochemistry and Biophysics, University of Pennsylvania, Philadelphia, Pennsylvania 19104

and

Institute for Structural and Functional Studies, University City Science Center, 3401 Market Street, Philadelphia, Pennsylvania 19104

Enrique Cadenas

Institute fur Physiologische Chemie, Universitat Dusseldorf, Moorenstraße 5 D-4000 Dusseldorf 1, West Germany

CONTENTS

[1]Abbreviations used: EPR, electron paramagnetic resonance; CD, circular dichroism; MCD, magnetic circular dichroism; IR, infrared; RR, resonance Raman; NMR, nuclear magnetic resonance; ENDOR, electron nuclear double resonance; EXAFS, extended X-ray absorption fine structure; SOD, superoxide dismutase; HRP, horseradish peroxidase; PMS, phenazine methosulfate; ATP, adenosine triphosphate; PG, prostaglandin.

137

0066-4154/86/0701-0137$02.00

PERSPECTIVES AND SUMMARY

Reaction of molecular oxygen with proteins, which either transport it or reduce it to varying degrees as a result of electron transfer, has been and remains an exciting and productive field. Reactive oxygen intermediates in biochemistry is an extremely important field, particularly for membrane proteins that are capable of transmembrane charge separations. It has practical implications also, because while oxygen is an essential component for living organisms, certain intermediates of O_2 can be highly damaging to the intracellular environment.

Instead of looking into various biochemical and enzymatic systems, we have concentrated mainly on a single enzyme (cytochrome oxidase) for several reasons (see below). This is one of the few enzymes that reduces molecular O_2 all the way to H_2O, thus giving an opportunity to follow all the known intermediate states of O_2 reduction. In the earlier stages of investigation, mainly optical and EPR spectroscopy were used to study the reaction of cytochrome oxidase. In recent years the whole arsenal of biochemical and biophysical techniques, including magnetic susceptibility, circular dichroism (CD), magnetic circular dichroism (MCD), infrared (IR), resonance Raman (RR), Mossbauer absorption, nuclear magnetic resonance (NMR), electron nuclear double resonance (ENDOR), and extended X-ray absorption fine structure spectroscopy (EXAFS), have been used. Thus cytochrome oxidase has become, in addition to one of the most important enzymes, one of the most studied enzymes.

An intriguing property of cytochrome a_3, probably due to its structural plasticity, is that it can assume different structures for different roles, such as of an electron (like other cytochromes), oxygen (like hemoglobin), and peroxide acceptor (like peroxidases).

Recently the reaction of cytochrome oxidase and O_2 at room temperature has been reinvestigated using time-resolved biophysical techniques, and these studies found a striking similarity between their results and those found at low temperatures a decade ago. Recent renewed interest in cytochrome oxidase and its reaction with H_2O_2 using different techniques has been extremely useful.

Simpler bacterial terminal oxidases of aa_3 type showed in their reaction with O_2 a surprising similarity to their mammalian counterparts.

It is now clear that the reaction mechanisms of oxygen reduction at low and room temperatures are very similar. The first step in O_2 reduction is binding of O_2 to the active site followed by electron transfer. While this may proceed via a concerted two-electron transfer, the one-electron transfer cannot be ruled out. Whether the two-electron transfer is followed by a two-electron transfer or two one-electron transfers is not clear.

However, it is quite surprising that the "third substrate" of cytochrome oxidase (a proton, the first two being cytochrome c and O_2) has received very little experimental attention, i.e. it is unclear at what stage H^+ is taken up or released. The importance of these studies cannot be overestimated in view of the fact that cytochrome oxidase also acts as a proton pump. We think that in the near future this question will receive more attention partly because Mitchell has now accepted and proposed his mechanism for a proton pump in cytochrome oxidase.

We have also discussed in limited space the generation of singlet O_2 in biochemical systems, because there has been a recent surge of interest in this area. Until recently, generation of singlet O_2 was mainly hypothetical and conclusions were based upon indirect methods. In the last few years, however, there has been direct evidence of singlet O_2 generation in enzyme systems, especially peroxidases, and indeed the respiratory chain as well.

INTRODUCTION

The presence of enzymes catalyzing the reduction of molecular oxygen was detected by Bertrand (1) and Yoshida (2) at the end of the last century, and Bertrand introduced the term "oxidase" to distinguish this class of enzymes (3).

Although Spitzer (4, 5) first suggested that iron is involved in oxidizing enzymes, it was Warburg (6–8) who first clearly formulated this concept as a scientific theory and proposed that oxygen is activated in the cell by an iron-containing enzyme—the *Atmungsferment*. He also showed that carbon monoxide inhibits cellular respiration and that this inhibition is reversed by bright light. This provided the evidence that his respiratory enzyme was a heme compound.

In 1925, Keilin (9) showed that the four-banded absorption spectra observed by MacMunn (10) in 1884 (and termed as "myo- and histo-haematins") are composed of superimposed alpha and beta bands of three distinct hemechrome compounds which he termed as *a, b,* and *c*. Keilin also then provided the concept of a respiratory chain that transported electrons from succinate to O_2 via a chain of cytochromes with a terminal oxidase. However, Warburg and Keilin could not reconcile many differences between them [for historical details see (11–13)]. Keilin & Hartree's finding in 1939 that the spectrum of cytochrome $a_3 \cdot CO$ is close to that predicted by Warburg did reconcile most of the problem but apparently the compound of a_3 and CO was found to be insensitive to bright light (14).

In 1953, Chance (15–19) succeeded in demonstrating the light sensitivity of the $a_3 \cdot CO$ compound. He showed that the photodissociation difference spectrum is identical to the static difference spectrum and to Warburg's photochemical action spectrum. This result, which is now generally accepted, together with earlier results of Keilin & Warburg, made clear that Warburg's respiratory enzyme, *Atmungsferment*, cytochrome oxidase (the cyanide- and CO-sensitive enzyme responsible for the oxidation of the classical cytochrome system), and cytochrome a_3 are identical, and thus the stage was set for further investigation.

Scope of the Review

Reactive oxygen intermediates in biochemistry is a very wide subject. It is impossible to discuss all aspects of this problem in any review. In 1978, Keevil & Mason (20) had compiled what they termed a "complete catalog of 220 oxidases" (by "oxidases" they meant all enzymes that catalyze reactions of molecular O_2, rather than any specific type of reaction). So instead of listing all relevant enzymes, we will concentrate on the reaction of O_2 and its reductive intermediates (particularly H_2O_2) with cytochrome c oxidase and will briefly discuss singlet oxygen in enzyme systems.

Cytochrome oxidase has been chosen for several reasons. The first is the authors' scientific interest. Second, it is probably one of the most important enzymes due to its central role in the mitochondrial respiratory chain. Third, it is one of the most studied enzymes. Finally, it may be termed as an "ultimate" or "super" enzyme because cytochrome oxidase not only contains different metal ions, but also possesses varied reactivity [its main role is, of course, of an oxidase, but apart from that it possesses catalase (21), peroxidase (22–25), superoxide dismutase (26; A. Naqui, B. Chance, manuscript in preparation), and probably carbonmonoxy oxygenase activity (27–29)].

The reaction of molecular oxygen and its reduction intermediates with cytochrome oxidase includes almost all of the stages of the reductive cycle of O_2. It also couples two substrates, one of which is a one-electron donor (reduced cytochrome c) and the other (O_2) a four-electron acceptor. This in itself is an intriguing problem.

In the discussion of the reaction of cytochrome oxidase with molecular oxygen and its reductive intermediates, the results of recently (post-1981) published papers will be stressed. Apart from the cytochrome oxidase–O_2 reaction, we will also discuss the question of singlet oxygen in enzyme systems. Although singlet oxygen generation in chemical systems is well-defined [e.g. see (30)], only recently was it detected directly in biosystems.

The Chemistry of Molecular Oxygen

The chemistry of molecular oxygen has been discussed several times, and some excellent review articles are available (31–36). However, a short overview stressing some points seems essential.

Before we go to the thermodynamic questions of O_2 reduction, it may be interesting to note why O_2 has been "chosen" by nature to act as the terminal oxidant of the respiratory chain. This question has been extensively discussed by George, who stated, ". . . the kinetic reactivity of the halogens makes them unsuitable as biological oxidants, and nitrogen is too poor an oxidizing agent. Oxygen is thus the only element in the most appropriate physical state, with a satisfactory solubility in water and with desirable combinations of kinetic and thermodynamic properties" (36).

The kinetic inertness of dioxygen in solution can be explained by its electronic structure (37–40). Although molecular oxygen contains an even number of electrons, it has two unpaired electrons in its molecular orbitals (Figure 1), and is said to be in a triplet ground state. These two electrons have the same spin quantum number, and if O_2 attempts to oxidize another atom or molecule by accepting a pair of electrons from it, both new electrons must be of parallel spin so as to fit into the vacant spaces of the orbitals. Usually, a pair of electrons in an atomic or molecular orbital would have antiparallel spins. This imposes a restriction on oxidation by O_2. The other problem lies in the fact that the one-electron reduction of molecular O_2 is thermodynamically unfavorable [the redox potential of reaction $O_2 + e^- \rightarrow O_2^-$ is -0.33 V, see Table 1; the thermodynamic values have been adopted from (36–43)]. It has been proposed

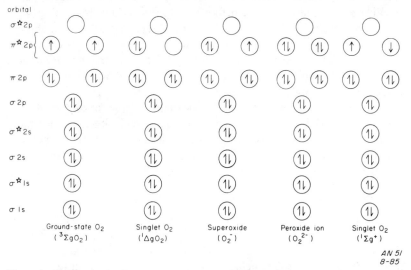

AN 51
8-85

Figure 1 Bonding in the diatomic oxygen molecule. Redrawn from (40) with permission.

Table 1   Thermodynamic parameters for the reduction of molecular O_2 to H_2O

Reaction(s)			E_0 (pH 7), V
Four-electron transfer			
	$1/4\ O_2 + H^+ + e^-$	$\rightleftharpoons 1/2\ H_2O$	+ 0.82
Two two-electron transfers			
	$1/2\ O_2 + H^+ + e^-$	$\rightleftharpoons 1/2\ H_2O_2$	+ 0.3
	$1/2\ H_2O_2 + H^+ + e^-$	$\rightleftharpoons H_2O$	+ 1.35
Four one-electron transfers			
	$O_2 + e^-$	$\rightleftharpoons O_2^-$	− 0.33
	$O_2^- + e^- + 2\ H^+$	$\rightleftharpoons H_2O_2$	+ 0.94
	$H_2O_2 + H^+ + e^-$	$\rightleftharpoons OH + H_2O$	+ 0.38
	$OH + H^+ + e^-$	$\rightleftharpoons H_2O$	+ 2.33

(37, 39) that for this thermodynamic restriction a concerted two-electron transfer is the likely reductive pathway in the mechanisms of enzymes reducing O_2 to H_2O (but see below). However, in biological systems, the thermodynamic barriers can be lowered owing to tight binding of the intermediates to the enzyme.

To overcome the spin restriction discussed above, the oxygen molecule can interact with another paramagnetic center to participate in exchange coupling. The transition metal ions frequently have unpaired electrons and are excellent catalysts of O_2 reduction (39). Another way of increasing the reactivity of O_2 is to "move" one of the unpaired electrons in a way that alleviates the spin restriction. The singlet O_2 has no unpaired electrons [the other singlet O_2 (Σg^+) usually decays to the ($^1\Delta g$) state before it has time to react with anything (40)]. On the basis of chemical models, Malmstrom (37) concluded that in the absence of a large driving force for electron transfer from the metal ion to O_2, only singlet O_2 will bind. However, if the driving force of electron transfer is large and if the redox potential is "right," molecular O_2 will also bind, as clearly evidenced by hemoglobin.

It appears that the high affinity of cytochrome oxidase for oxygen at room temperature (compared to at low temperature) is due to a kinetic rather than a thermodynamic property. The very rapid electron transfer at the higher temperature, reducing oxygen to a variety of compounds (see below), creates a trap that affords the key to the physiological function of cytochrome oxidase (44).

Cytochrome Oxidase

As stated earlier, cytochrome c oxidase (ferrocytochrome c: O_2 oxidoreductase E.C.1.9.3.1) is the terminal oxidase of the mitochondrial respiratory chain. Different aspects of cytochrome oxidase have been reviewed and recently a conference was held solely dedicated to this enzyme [for recent reviews and materials of that conference see (37, 45–52)]. However, a short introduction is necessary for those who are not familiar with this particular enzyme, as it will often be referred to in the following sections. Cytochrome oxidase catalyzes the electron transfer from reduced cytochrome c to molecular O_2, thus reducing it to H_2O in a four-electron transfer process. This electron transfer process produces a proton gradient across the membrane, which in turn drives the production of ATP. (In this review, however, our attention will be concentrated only on the electron transport to O_2.)

Cytochrome oxidase contains at least four metal atoms per functional unit: two hemes (cytochrome a and a_3) and two associated copper atoms (Cu_a and Cu_{a_3}). Two of the copper and iron atoms are EPR-visible (Cu_a and Fe_a), and the other two are antiferromagnetically coupled to form a binuclear center (Fe_{a_3}–Cu_{a_3}) which is EPR-undetectable. It is generally accepted that both copper ions are divalent [but see (53, 54)] and both cytochromes are in the Fe^{III} state [but see (55)] in the resting (as isolated) state. Reduced cytochrome c, the physiological substrate of cytochrome oxidase, transfers electrons to the "Fe_a–Cu_a site," which acts as an electron pool. The Fe_{a_3}–Cu_{a_3} binuclear center binds O_2 and reduces it to H_2O.

This membrane protein has at least 7 (50, 51) [up to 13 (56)] subunits and spans the membrane several times. Probably all of the redox metal centers are located in the two largest subunits. Subunit three has been implicated in the proton translocating activity [(45, 57–60), but see also (61)].

Recently, it has been reported that cytochrome oxidase contains zinc and magnesium as well as Cu and Fe (62, 63). The Zn and Mg content is half that of the copper and iron, i.e. the metal ion ratio is proposed to be $Fe : Cu : Zn : Mg = 2 : 2 : 1 : 1$.

A schematic introductory diagram of cytochrome oxidase is shown in Figure 2. The values of the distances quoted are the minimum and maximum values found in the literature (64–71); other parameters are those that the authors prefer. Alternate views are also included in parentheses.

OXYGEN INTERMEDIATES

The "Oxy" Compound

As cytochrome oxidase is the terminal oxidase in the respiratory chain and responsible for the utilization of about 90% of all O_2 consumed by living organisms (45), the detection of an "oxy"–cytochrome oxidase compound eluded many investigators. In 1958, Okunuki et al (72) reported that when O_2 is added to the reduced cytochrome oxidase, shifts in the Soret region peak to higher wavelengths were observed. The new 428-nm absorbing species was termed in an "oxygenated" state, and was believed to be an oxygen compound of cytochrome oxidase, similar to oxymyoglobin (72, 73). But the sluggish

144   NAQUI, CHANCE, & CADENAS

CYTOCHROME c OXIDASE

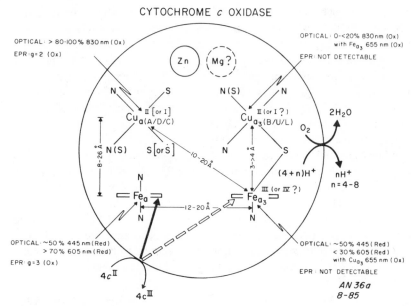

Figure 2   Schematic diagram of cytochrome oxidase.

behavior of this putative oxygenated compound ruled against such a conclusion.

Lemberg, however, advanced arguments based on chemical approaches against the presence of O_2 in the oxygenated compound and instead proposed a peroxidatic activity of the "oxygenated" enzyme, but made no special effort to change the nomenclature (74–78).

Gibson & Greenwood (79–81) studied the reaction of reduced cytochrome oxidase with O_2 and found that the lifetime is too short to observe a stable "oxy" compound at room temperature. The first oxygen concentration–dependent step had a second order rate constant of $\sim 10^8 M^{-1}s^{-1}$, which was followed by a first order rate constant of $3 \times 10^4 \ s^{-1}$. Erecinska & Chance (82) studied the same reaction at a subzero temperature $(-30°C)$, and found an active intermediate in the cytochrome oxidase-O_2 reaction absorbing at 428 nm. A few years later, due to limitations of the flow apparatus, Chance et al (44, 83–88) introduced the now well-known triple trapping method and for the first time trapped the cytochrome oxidase–O_2 compound (compound A in Chance's terminology). They showed that compound A is stable only at around $-130°C$. Since then the triple trapping method has been extensively used by various investigators for different systems. Clore et al (89–93) repeated the experiments of Chance et al using both optical and EPR spectroscopy and essentially reconfirmed the earlier results. However, they proposed that the reaction proceeds via at least three intermediates (I, II, and III when starting from the fully reduced enzyme and I_M, II_M, and III_M starting from the mixed valence state) as opposed to two intermediates observed by Chance et al. They also assigned slightly different structural configurations for the low-temperature intermediates. However, all their data were based on purified cytochrome oxidase, whereas Chance used mainly the membrane-bound oxidase, which appears to form intermediates that are fully occupied.

While Chance et al (83) preferred a Fe $_{a_3}^{2+}$ · O_2 Cu $_{a_3}^{1+}$ structural configuration for their compound A, Clore et al (93) proposed that electron delocalization occur in their compound I. This delocalization can proceed to different degrees, as far as Fe $_{a_3}^{3+}$ · O_2^{2-} Cu $_{a_3}^{2+}$ (94). There is no structural information to prefer one over the other. The situation is somewhat similar to the oxy-hemoglobin and oxy-myoglobin studies where delocalized electrons on oxygen have been proposed to be the state of bound O_2 [e.g. see (95, 96) for a discussion]. In the absence of any further structural information, we will for the time being continue to assign the original proposed structure by Chance et al to compound A. A detailed structural study of Compound A could be useful.

The "oxy" compound has optical spectra similar to CO-difference spectra but is not photolabile (83, but see below). A similar type of "oxy" compound also forms from the mixed valence state (83) (that is, when the cytochrome oxidase is partially oxidized, presumably the Fe_a and Cu_a are oxidized, Fe_{a_3} and Cu_{a_3} are reduced, and CO is bound to Fe $_{a_3}^{2+}$).

Recently, the reaction of cytochrome oxidase with O_2 has been reinvestigated by several authors, both by traditional optical methods (97–100), and by rapid scanning optical spectroscopy (101) and time-resolved resonance Raman (RR) spectroscopy (102, 103). These experiments are extremely important not only because they all support the idea that the cytochrome oxidase–O_2 reaction proceeds via an "oxy" intermediate which is extremely unstable at room temperature, but also because they demonstrate the similarity between the low-temperature and room-temperature experiments (97–103).

Babcock et al (102, 103), using the time-resolved RR spectroscopy, studied the reaction of fully and partially reduced cytochrome oxidase and O_2. They also concluded that the oxygen reduction proceeds via an "oxy" compound which has characteristics similar to oxyhemoglobin and oxymyoglobin. However, they concluded that the "oxy" intermediate is photolabile. The puzzling result of their experiments is that there is no apparent change in the RR spectra during the first 40 μsec when fully reduced cytochrome oxidase is used or during the first 75 μsec when the mixed valence state is the starting material. The authors concluded that this is due to the extreme photolability of the "oxy" compound. This contradicts the well-established time course of the cytochrome oxidase–O_2 reaction at room temperature followed by optical spectroscopy and studied in several laboratories (79–81, 97–101). To reconcile this discrepancy, we offer an alternate explanation of the RR results. At 40 μsec after the initiation of the reaction, there is no change in the RR spectrum when a higher-power and focused laser is used, while there is a change if a defocused and less-intense laser power is used. This result can be explained as a rereduction of the heme on the focused and intense laser, whereas in the defocused and less-intense laser power, the reaction proceeds without rereduction [a rough calculation shows that the $t_{1/2}$ for formation of the "oxy" compound is ≈ 10 μsec and 23 μsec for the next step (a_3 oxidation; formation of compound B?) under the conditions of the experiment]. If this is true, then the RR difference spectral peak at 1378 cm^{-1} and 1588 cm^{-1} can be attributed to the formation of compound B–type intermediates and not the "oxy" compound. (We would like to note that the difference spectra shown in Figure 2 of (102) is noisy, so the peak assignments, we think, are tentative.) The oxidation state marker band is at 1378 cm^{-1} and is moved substantially from 1371 cm^{-1} observed for the Fe^{III}. It is possible that the 1378 cm^{-1} band is due to a ferryl heme (Fe^{IV}). We will return to this question later.

Similar arguments can be forwarded for the mixed valence experiments. Thus, it seems that there is not much discrepancy between the time course of the reaction followed by optical and RR spectroscopy.

Orii used a rapid scanning spectrophotometer and repeated the flow-flash experiments (101). He, like others, also observed an "oxy" compound as the primary reaction product. The spectral characteristics of the first compound observed between 5 μsec and 50 μsec are similar to the CO difference spectra. But their amplitudes are different, suggesting either the occupancy of "oxy" compound is low [Chance found the occupancy of the "oxy" compound to be very low even at subzero temperatures (82)], or compound A has substantially decayed by the time the difference spectrum was recorded. Both possibilities are the result of very fast electron transfer to the "oxy" compound. Hill & Greenwood (97–99), repeating the flow-flash experiments, concluded that the reaction proceeds via any "oxy" intermediate but proposed a branched mechanism.

From these experiments, we can conclude that (a) the reaction of cytochrome oxidase with O_2 proceeds via an "oxy" compound which is followed by fast electron transfer to the O_2 molecule, and (b) although there may be subtle differences in the detailed reaction mechanism between low-temperature and room-temperature experiments, the main features are conserved.

Is Superoxide Generated in Cytochrome Oxidase–O_2 Reaction?

Although cytochrome oxidase has been extensively studied, the reaction mechanism is not yet fully understood. Whether the reduction of oxygen proceeds via a sequential electron transfer or synchronized multiple electron transfer is still unresolved. It has been reported that the SOD (superoxide dismutase) does not inhibit the reaction of cytochrome c oxidation by cytochrome oxidase (104). Greenwood et al (105) performed a flow-flash experiment and reported that SOD does not effect the decay of the "oxygenated oxidase" to oxidized enzyme. These experiments indicate that either superoxide production is bypassed by a two-electron transfer, or if produced, the superoxide anion is not released from the active site of the enzyme. The synchronized two-electron transfer from the bimetallic center (Fe_{a_3}–Cu_{a_3}) to O_2 is favored for thermodynamic considerations as pointed out earlier (37). Superoxide ion is also potentially hazardous [(106), but see (107)]. However, one must note that the redox potentials are given for solutions and will be different in a protein environment when these intermediates are tightly bound to the enzyme active site. As we discussed earlier, this is one way to overcome the spin restriction!

There are several reports in the literature that can be interpreted as evidence for the formation of a superoxide ion during the reaction of cytochrome oxidase and oxygen. For example, Fridovich & Handler (108) reported that the sulfite oxidation rate is greatly enhanced when both cytochrome c and cytochrome oxidase are present. This enhanced rate is largely diminished in the presence of cyanide or when either cytochrome oxidase or cytochrome c is absent. From these results, they concluded that a superoxide ion is produced during the reaction.

A radical signal with an EPR value close to $g = 2$ has been reported from turnover experiments (109). Especially interesting is the fact that this radical signal is larger in amplitude during "high-turnover conditions." In another experiment (110), a $g = 2$ signal has also been seen (see below). A superoxide ion, if produced and tightly bound to the active site, may be largely EPR-undetectable due to the spin coupling with $Fe_{a_3}^{3+}$, therefore making it difficult to detect its generation even by EPR. Clore et al have suggested that the first detectable cytochrome oxidase–O_2 compound (compound I) is an equivalent of the superoxide cytochrome oxidase compound because of the electron delocalization from the Fe_{a_3}–Cu_{a_3} center to the oxygen molecule (93). Greenwood et al also favored sequential one-electron transfer (105).

It has been reported earlier that cytochrome oxidase in the resting state possesses an SOD activity (26, 111). We have recently shown that the SOD activity is greatly enhanced (about eightfold) in the pulsed state (A. Naqui, B. Chance, manuscript in preparation). These observations suggest that cytochrome oxidase is a superoxide scavenger by itself and has a "built-in defense mechanism" against the potentially hazardous O_2^-, whether it is produced by accident or as an intermediate. So, even if produced it will be largely undetected in solution.

The "Peroxy" Intermediate

The result of the two-electron transfer to molecular oxygen is a peroxy state (O_2^{2-} or H_2O_2 if protonated).

In cytochrome oxidase there are two types of peroxy intermediates: 1. those formed upon addition of H_2O_2 or alkylhydroperoxides; 2. those formed at low or room temperature as a result of the reaction of fully reduced or mixed valence cytochrome oxidase and O_2. We will address these two aspects separately.

REACTION OF CYTOCHROME OXIDASE WITH H_2O_2 Okunuki & Orii were the first to study the reaction of cytochrome oxidase with H_2O_2 in detail (21). Okunuki et al (72, 112, 113) discovered the so-called "oxygenated" oxidase (see above). Later, Lemberg and his colleagues studied oxygenated oxidase and its reaction with H_2O_2 and proposed a peroxidatic role for oxygenated oxidase (74–78). Orii & King studied the decay process of the "oxygenated oxidase" and found that it decays in two distinct steps (114, 115). More recently, Bickar et al, unaware of the earlier work of Okunuki & Orii, reinvestigated the reaction of cytochrome oxidase with H_2O_2 (116). They found that resting oxidase binds some H_2O_2, whereas the pulsed state (reduced and reoxidized) binds H_2O_2 more efficiently and more quickly. We attribute this to the heterogeneity of the resting oxidase as described by Brudvig et al (117) and Naqui et al (118). The K_D, determined from the Scatchard plot, was found to be <10 μM. Although they discussed the possibility of a ferryl iron for Fe_{a_3}, they preferred a $Fe_{a_3} \cdot H_2O_2$ in ferric state. They also found that reduced cytochrome c was readily oxidized by the cytochrome oxidase–H_2O_2 complex in the absence of O_2. It should be pointed out that Lemberg & Gilmour (78) have studied the reaction of "oxygenated" oxidase with a variety of electron donors and found that many of them were able to react with the oxygenated oxidase efficiently, including reduced cytochrome c.

Orii, independently, proposed a peroxidatic cycle, and studied in detail the reaction of cytochrome c with cytochrome oxidase–H_2O_2 (22). In the absence of O_2 the second order rate constant is $4 \times 10^6 \, M^{-1} s^{-1}$ and the K_m (with respect to cytochrome c) is 2×10^{-6} M for the peroxidatic cycle. The second order rate constant is close to that found for the oxidase activity. Apparent K_m with respect to H_2O_2 was found to be 0.18 mM (about 20 times larger than the K_D; the value was determined by Bickar et al, and is consistent with the idea that generally $K_m \neq K_D$). These results, together with the earlier results, demonstrate that the peroxidatic cycle for cytochrome oxidase is similar to peroxidases. This peroxidatic cycle is inhibited by cyanide and azide. However, the kinetic curves are quite different from the oxidase activity.

Kumar et al (119, 120) very recently reexamined the reaction of H_2O_2 with cytochrome oxidase and the relationship between pulsed and oxygenated oxidases. In the course of this study, they discovered a new form of cytochrome oxidase and termed it as the "420-nm form." The 420-nm form is produced when the reduced cytochrome oxidase is reoxidized in the absence of H_2O_2. When H_2O_2 was added to this 420-nm form, the Soret peak moved to 428 nm establishing clearly that "oxygenated" oxidase is indeed a peroxide adduct of cytochrome oxidase (119). They also assigned the optical and EPR signals observed at various times by various groups upon reoxidation of reduced cytochrome oxidase either to the 420-nm or 428-nm absorbing forms. The 420-nm form retains the 655-nm band, the 605-nm band is enhanced, and only this form has the $g = 5$ EPR signal first discovered by Shaw et al (121). Upon peroxide addition the 655-nm band and the $g = 5$ EPR signal is abolished and the 580-nm band is developed. Thus the well-known changes in the alpha region of the oxygenated oxidase have been separated into two components. This observation also agrees with Wikstrom's earlier contention that the 580-nm band is due to peroxide binding to the active site (122).

Antonini & Brunori (123–130) approached the relationship between the resting and the reoxidized state from a different perspective. They showed that the cytochrome c oxidation rate is several fold faster in the reoxidized form than in the resting form under certain conditions and coined the term "pulsed."

Different authors have referred to the reduced and reoxidized form of cytochrome oxidase in different terms: oxygenated (21, 72, 113–115), pulsed (123–130), oxygen-pulsed (131), oxy-ferri (27), peroxidatic or non–sulfur bridged (25), etc.

We propose to simplify this confusing terminology. It is now apparent that Okunuki's original proposal for an "oxy" compound is no longer valid, and a true "oxy" compound has been found to exist as an extremely labile intermediate at room temperature and can only be trapped at low temperatures. In order to make a clean sweep of a somewhat messy and confusing nomenclature, we propose in this review serious consideration of the renaming of the pulsed peroxide or oxygenated species as a *peroxide compound,* and the unligated species, the presumed pulsed compound which has recently been shown to absorb at 420 nm, now appears to be a *peroxidatic form* of cytochrome oxidase. Some may wish to continue to use the term "pulsed."

We will now address the question on possible structure of these two states. The simplest explanation of the available data is that cytochrome oxidase in the resting state is in some sort of a "closed" conformation (69). The reduction and reoxidation produces a more open conformation in the 420-nm form. This form is more reactive than the resting state as shown by its higher reactivity towards cyanide (118, 132), H_2O_2, and cytochrome c oxidation. This open conformation probably is five coordinate at the Fe_{a_3} site and accepts H_2O_2 as a sixth ligand at Fe_{a_3} to produce the 428-nm form (119, 120).

Beinert has searched for and found no evidence for retention of oxygen on cytochrome a_3 in the $g = 5$ form (121, 131, 133). Our finding of only small spectroscopic changes during the resting to 420-nm form transition also argues against any major changes in the valence state of the redox center of cytochrome oxidase during the transition. This explanation is further strengthened by the recent observation, using RR spectroscopy, that the 420-nm is similar to the resting state except that it is much more photoreducable than the resting oxidase under the same conditions (134). This indicates a conformational change. There is, however, no substantial change in the spin state and redox state marker bands. An alternate possibility that the 420-nm and 428-nm forms are both peroxide derivatives of oxidase, while not ruled out, is less attractive at this moment because (a) the 420-nm form is produced under conditions where no peroxide should be present (119, 120), (b) O_2 has not been detected as a ligand of a_3 in the $g = 5$ form (121, 131, 133), and (c) the RR spectrum does not show any sharp change in the spin state marker bands (134).

However, in a recent paper, Wriggleswroth (135) has extended the earlier observation on the binding of peroxide to cytochrome oxidase and observed that the 605-nm and the 580-nm bands are formed at different peroxide concentrations and long incubation times, suggesting that peroxide itself may be able, in some cases, to induce the transition to the pulsed form.

X-ray absorption spectroscopy studies from this laboratory have delineated structural differences between the resting and pulsed forms of cytochrome oxidase (25). A sulfur atom bridging the active-site iron and copper in the resting state has been suggested to be linked only to the copper atom in the pulsed state, thus rendering the Fe_{a_3} in the pulsed state more open. However, at that time, the 420-nm form was not yet found and it is possible that the form used for the EXAFS studies was a mixture of 420-nm and 428-nm forms. More recent EXAFS study of the pulsed-peroxide form shows that the Fe_{a_3} in the pulsed-peroxide form may be a ferryl iron with a short (1.71 Å) Fe-O distance (136). The results are shown in Figure 3. The possibility is quite intriguing in view of the fact that pulsed oxidase is similar to the peroxidases (25, 137) [and peroxidases are known to form ferryl iron; see e.g. (138)] and a peroxidatic function of the pulsed oxidase is quite clear. We searched for a ferryl type of iron in pulsed peroxide form using RR spectroscopy (134). Although we observed a high- to low-spin transition from resting to pulsed-peroxide, we were unable to detect a ferryl iron as judged from the oxidation state marker region even at a very low laser power. It is possible that if formed, the ferryl iron is photoreduced even at that low power similar to compound I of horseradish peroxidase (HRP) (139, 140). One interesting observation is that the oxidation state marker band shifts from 1371 cm^{-1} (for resting) to 1374 cm^{-1} (for the pulsed-peroxide), while a ferryl iron is expected to have the band $\sim$1380 cm^{-1} (141). The same observation has been made by Carter et al for resting oxidase + H_2O_2 and compound C (142). One may hypothesize that the pulsed-peroxide may be a mixture of ferric and ferryl form and if there is any endogenous electron available the $(Fe_{a_3}O)^{3+}$ may receive an electron and become $(Fe_{a_3}O)^{2+}$ according to the following scheme:

$$ Fe^{3+}_{a_3} + H_2O_2 \rightleftharpoons Fe^{3+}_{a_3} \cdot H_2O_2 \xrightarrow{-H_2O} (Fe_{a_3}O)^{3+} \xrightarrow{\overline{e}} (Fe_{a_3}O)^{2+} \qquad 1. $$

$$ \longleftarrow \text{EPR detectable } Cu_{a_3}? \longrightarrow $$

COMPOUNDS B AND C   Chance and coworkers, using the triple trapping method, found that the first "oxy" compound is transformed into compound B (from the fully reduced state) or C (from the mixed valence state) as a result of electron transfer from cytochrome oxidase to O_2 (83–88). Clore et al have also observed several phases in the low-temperature experiments indicative of electron transfer (89–93). Both of the groups have discussed in detail the possible structures for these low-temperature compounds.

The reactions of compound B with reducing agents were studied later by Chance et al (143). Those experiments were carried out to test the hypothesis that compound B is indeed a peroxy intermediate. They studied the reaction of compound B with an organic electron donor, phenozine methosulfate (PMS), and found that the second order rate constant of the reaction was $\sim$10^3 M^{-1}s^{-1}, which is quite a fast rate at $-60°$C! These results point once again to the fact that compound B is a peroxy intermediate similar to peroxidase compounds with peroxides. However, it should be noted that compound B could be heterogeneous due to electron transfer from Cu_a and Fe_a to the $(a_3 \, O_2 \, Cu_{a_3})$ site. Both Chance et al (83) and Clore et al (93) detected this phenomenon. Similar experiments with compound C would be interesting.

Greenwood & Gibson (79–81), and later Greenwood and others (97–99, 105), studying the reaction of cytochrome oxidase and oxygen, have delineated three phases. The first is the formation of the oxycompound (compound A), and the two later phases consist of intramolecular electron transfer to O_2. Although no specific intermediate could be detected at room temperature, it is clear that a peroxy intermediate is formed. Indeed, Greenwood et al (105), using the flow-flash experiments with the mixed valence CO compound, concluded that the first product observed after the oxygen-dependent fast phase is not oxidized enzyme as prepared but is similar to the oxygenated form of the enzyme. Carter et al, studying compound C with RR spectroscopy, concluded that compound C is very similar to a peroxy compound of cytochrome oxidase (142).

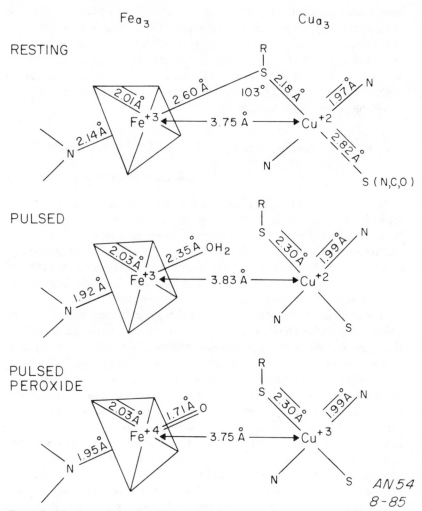

AN 54
8-85

Figure 3 Pictorial representation of X-ray absorption results on the binuclear active site structures. Adopted from (136) with permission.

Erecinska & Chance (82) have studied the reaction of cytochrome oxidase and O_2 at subzero temperatures ($-30°C$) using a cooled stop-flow apparatus and observed an intermediate with an absorption band at 428 nm. At that time they proposed this early intermediate to be an "oxy" compound. Our later (119, 120) studies tentatively reassigned this compound to be a peroxy intermediate in the reaction cycle.

The flow-flash experiments, coupled with RR spectroscopy, have been carried out by Babcock et al (102, 103) (see above). As we have discussed earlier, their experimental data can be explained as evidence of a low-spin intermediate (band at 1588 cm^{-1}) where ferryl iron formation is a possibility (1378 cm^{-1}). Babcock et al (102, 103) do not discuss this possibility at all and prefer that the intermediate having the different RR spectral peaks at 1378 cm^{-1} and 1588 cm^{-1} is an "oxy" compound. This intermediate can actually be a compound B–type intermediate at room temperature similar to peroxide compounds of peroxidases (138).

This brings us to the question of relationships between the peroxidases and cytochrome oxidase. Reaction of cytochrome oxidase with H_2O_2 and its fast decay in the presence of electron donors such as reduced cytochrome c indicates a peroxidase-type reaction mechanism. Chance & Powers have studied a series of peroxidases and their H_2O_2 compounds, and have compared them with the pulsed cytochrome oxidase and its peroxide adduct. They found striking similarities between the peroxidases and pulsed oxidase and their peroxide compounds (25, 137). On the basis of these results, they proposed a mechanism of cytochrome oxidase reaction where pulsed oxidase acts as a peroxidase and a "scavenger" of any H_2O_2 produced during the reaction (Figure 4).

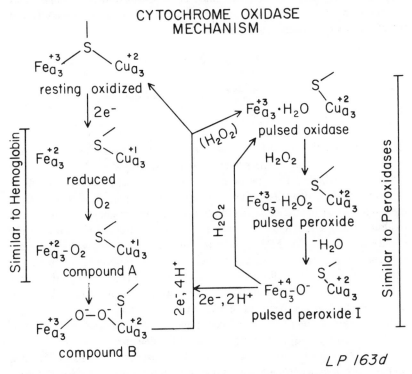

Figure 4 Mechanism of oxygen reduction by cytochrome oxidase including a role for pulsed oxidase as a peroxidase proposed by Chance et al. Reproduced from (25) with permission.

Product of Three-Electron Transfer to O_2

The product of the third electron transfer to O_2 and the state of cytochrome oxidase is not unequivocably established. However, there are a few reports in the literature shedding some light on this aspect, and this question is closely related to the detection of the EPR signal of Cu_{a_3}.

We pointed out in an earlier section that Cu_{a_3} is antiferromagnetically coupled to the cytochrome a_3 in the resting state and that is why it is EPR-undetectable. One would expect to detect the $Cu_{a_3}^{2+}$ signal during turnover because at a certain stage the coupling between Fe_{a_3} and Cu_{a_3} could be broken, rendering Cu_{a_3} detectable. This remained a puzzle for a long time as no EPR signal from Cu_{a_3} could be detected. It was in 1980 that Reinhammar et al (144) reported a new Cu^{2+} EPR signal and assigned it to the Cu_{a_3} in cytochrome oxidase. Subsequently, Karlsson & Andreasson (145) also detected a Cu_{a_3} signal during the oxidation of the fully reduced enzyme. Later Karlsson et al (146) detected the Cu_{a_3} signal at low temperature using a method similar to the triple trapping method. This copper signal has an unusual shape and relaxation properties. They proposed that these unusual properties were due to an interaction with a nearby paramagnetic ion and could be a result of the three-electron transfer to oxygen from cytochrome oxidase:

$$\underset{I}{Fe_{a_3}^{3+} \diagdown \overset{O^-}{\underset{O^-}{\diagup}} \diagdown Cu_{a_3}^{2+}} \xrightarrow[\underset{e^-}{}]{\overset{2H^+ \quad H_2O}{\diagdown \diagup}} \underset{II}{Fe_{a_3}^{4+}=O \quad Cu_{a_3}^{2+}} \qquad 2.$$

AN 53
8-85

[It is important to note the difference of opinion between the two groups regarding the structure of the primary compound (see above)].

However, whether Cu_{a_3} becomes EPR-detectable upon transfer of three electrons to O_2 or not needs further comment. For example, Cline et al (110), in their study to define the coordination of the Cu_{a_3} environment in cytochrome oxidase, used the ENDOR (electron nuclear double resonance) method. They noted that when the Cu_{a_3} EPR signal was visible, ". . . the signals of $g = 3$ and 6 (assigned to cytochrome a and a_3 respectively) are more or less completely absent." Under these conditions some Cu_a is also oxidized, although the signal is considerably decreased compared to the resting state. It could well be that only one electron has been transferred to O_2 from Cu_{a_3}, thus generating a superoxide ion giving rise to the radical signal seen by these authors [Figure 1b of (110), upper curve] and rendering Cu_{a_3} EPR-detectable.

Brunori & Gibson (100) repeated the flow-flash experiments at 2°C at a higher oxygen concentration and monitored the reaction in the Soret region. They observed that after the initial very rapid phase ($t_{1/2} \approx 30$ μs), there is a kinetic component which is manifested by a plateau before the oxidation step of Fe_a. The plateau lasts ~400 μsec and is less distinguishable with a low O_2 concentration. They concluded that this plateau with a rate constant of ~5000 s^{-1} at 2°C may lead to a species in which a third electron (from Cu_a^{1+}?) is added to the peroxy complex leading to the breakage of the oxygen–oxygen bond and expulsion of the first water molecule, having a structure similar to $Fe_{a_3}^{4+} = O\ Cu_{a_3}^{2+}$.

Although these reports suggested that the result of the three-electron transfer to O_2 is a quadrivalent iron and EPR-detectable Cu_{a_3}, the picture is not entirely clear. It is unclear whether at "the plateau" Cu_{a_3} could be detected by EPR. It is worthy of note that Babcock's observed species and "the plateau" have a similar time course taking the difference in conditions (temperature and O_2 concentration) into account. Further, Cu_{a_3} has been detected under very different conditions, where the state of other metal sites seem to be different. More experimental data are necessary before a detailed picture can be drawn.

Fully Reoxidized States of Cytochrome Oxidase

Clearly the result of a four-electron transfer to O_2 is the production of H_2O but it is not quite clear what the state of cytochrome oxidase is once it has been reduced and reoxidized either from the fully reduced or mixed valence states.

Gibson & Greenwood have suggested that the result of full oxidation of the reduced oxidase is similar to the resting oxidase (147). Later Greenwood et al suggested that the oxidation product of the mixed valence state is similar to oxygenated oxidase (105). Using Chance's triple trapping method, Denis (148) has shown that the 655-nm band is present when starting from the fully reduced state but when starting with the mixed valence state, the 655-nm band is abolished. These results together with our finding that the 655-nm band is abolished by the addition of H_2O_2 to the pulsed oxidase shows that the 428-nm form is produced when the experiment is initiated from the mixed valence state. Using the triple trapping method, Beinert has observed the $g = 5$ EPR signal after the formation and decay of compounds A and B (149). We find that the $g = 5$ EPR signal is a property of the 420-nm pulsed state that is abolished upon peroxide addition. Probably that is why traditionally prepared oxygenated oxidase lacked the $g = 5$ EPR signal (150). Together with Beinert's results this finding shows that the oxidation end product of the fully reduced cytochrome oxidase is the 420-nm form. The end product also binds cyanide much faster [about 1000-fold faster than the resting form at −16°C (A. Naqui, unpublished results)]. However, on the basis of the EXAFS result, it was concluded that the end product is similar to the sulfur-bridged resting oxidase (136). At this moment we are unable to reconcile this difference with the EXAFS results.

It seems that due to the very small spectroscopic difference between the pulsed (420-nm) and the resting state, Gibson & Greenwood (147) concluded that the reoxidized state is similar to the resting state. Later, Rosen et al (151) found that the reoxidized state is somewhat different from the resting state, as did Orii (101), who concluded that the intermediate III is produced as a result of oxidation of cytochrome oxidase. (Orii's terminology is not to be confused with Clore's. Orii's intermediate III (114, 115) is the decay product of the "oxygenated" oxidase and is similar, but not identical, to the resting oxidase.)

On the basis of these results, we propose a simpler scheme which adequately explains the results:

$$3.$$

SINGLET OXYGEN

The metabolism of hydroperoxides by several hemoproteins or halide-dependent enzymatic reactions can proceed, at least formally, as a hydroperoxide disproportionation with formation of dioxygen in an excited state, singlet molecular oxygen ($^1O_2$):

$$2\ ROOH \rightarrow 2\ ROH + {}^1O_2 \qquad\qquad 4.$$

The generation of $^1O_2$ as a consequence of an enzymatic reaction could represent a key step in the metabolism of hydroperoxides. Examples are given by the metabolism of hydroperoxides by cytochrome P-450 (in the absence of electron donors and hydroxylatable substrates) (152), the prostaglandinendoperoxide synthase reaction (153), and the halide-mediated H_2O_2 decomposition by lactoperoxidase, chloroperoxidase, and myeloperoxidase (154–160). In the latter case, free $^1O_2$ is generated. These possibilities are summarized in Figure 5. A short survey on the molecular mechanisms, the requirements to observe photoemission, and the physiological role for $^1O_2$ formation are given below; more detailed discussions are found in the respective references. $^1O_2$ formation was monitored as emission of light accompanying the enzymatic reaction, either in the red ($^1O_2$ dimol emission) or in the near-infrared ($^1O_2$ monomol emission) region.

Cytochrome P-450

The generation of excited states during the metabolism of hydroperoxides by microsomal fractions takes into account the oxene transferase activity of cytochrome P-450, which allows the transfer of an O_2 atom from the oxene donors like hydroperoxides to substrates of the monooxygenase system.

ENZYMATIC GENERATION OF $^1O_2$

P-450 (Fe^{3+})-catalyzed heterolytic
scission of hydroperoxides

$$Fe^{3+} + ROOH \longrightarrow [FeO]^{3+} + ROH$$

$$[FeO]^{3+} + ROOH \longrightarrow Fe^{3+} + ROH + {}^1O_2$$

$$2 \ ROOH \longrightarrow 2 \ ROH + {}^1O_2$$

PG-hydroperoxidase-catalyzed
reduction of PGG2

$$Fe^{3+} + PGG_2 \longrightarrow [FeO]^{3+} + PGH_2$$

$$[FeO]^{3+} + PGG_2 \longrightarrow Fe^{3+} + PGH_2 + {}^1O_2$$

$$PGG_2 + PGG_2 \longrightarrow 2 \ PGH_2 + {}^1O_2$$

H_2O_2 disproportionation catalyzed by
lactoperoxidase, myeloperoxidase, or chloroperoxidase

$$H_2O_2 + Br^- + H^+ \longrightarrow HOBr + H_2O$$

$$H_2O_2 + HOBr \longrightarrow H^+ + Br^- + H_2O + {}^1O_2$$

$$H_2O_2 + H_2O_2 \longrightarrow 2 \ H_2O + {}^1O_2$$

AN 55 (E.C.)
8-85

Figure 5 Enzymatic generation of singlet oxygen.

The heterolytic cleavage of the O–O bond of the hydroperoxide proceeds under formation of a transient $(FeO)^{3+}$ species (see 161, 162). This active O_2 complex can either hydroxylate or epoxidize a monooxygenase substrate or may react with a second molecule of hydroperoxide, yielding, in part, $^1O_2$ (Figure 5). An analogous reaction occurs with chloroperoxidase and H_2O_2 through a scrambling mechanism: the evolving dioxygen was derived from two different H_2O_2 molecules (163). Chloroperoxidase closely resembles cytochrome P-450 and probably also contains an axial thiolate ligand as postulated for cytochrome P-450 (164).

The low-level chemiluminescence originating during the metabolism of oxene donors by microsomes or isolated cytochrome P-450 under anaerobic conditions was explained by this mechanism, and the excited state responsible for chemiluminescence was identified, in terms of dimol emission spectral criteria, as $^1O_2$ (152). Different oxene donors elicited chemiluminescence and the following order of efficiency was observed: iodosobenzene (an oxene donor that has only one O_2 atom)—cumene hydroperoxide-t-butyl hydroperoxide—H_2O_2. The requirements to observe photoemission during hydroperoxide disproportionation rely on the absence of electron donors and monooxygenase substrates to cytochrome P-450. The fact that the oxene donor–supported chemiluminescence of cytochrome P-450 proceeds under anaerobic conditions indicates that the photoemissive species formed upon interaction of a microsomal component with a hydroperoxide could originate from the O_2 atom present in the hydroperoxide molecule. Moreover, the chemiluminescence in anaerobiosis excludes an important possible component of photoemission, that is, the formation of excited states during lipid peroxidation.

In short, the chemiluminescence arising from hydroperoxide breakdown catalyzed by cytochrome P-450 can involve either a heterolytic or a homolytic scission of the hydroperoxide. In the former mechanism, the generation of excited states occurs in a primary fashion in a reaction that can proceed under anaerobic conditions. In the latter mechanism the generation of excited states occurs in a secondary fashion as a consequence of free radical interactions initiated by RO and O_2 and is required to support lipid peroxidation. Of course it remains to be established whether these are general and significant reactions in O_2 toxicity with a promoting effect on lipid peroxidation. However, the generation of $^1O_2$—a secondary diffusible oxidizing species ("secondary" to the oxenoid intermediary)—by cytochrome P-450 might be involved in the oxidative activation of xenobiotics (165–167).

Prostaglandin-Endoperoxide Synthase

The oxidation of arachidonic acid by prostaglandin (PG)-endoperoxide synthase involves cyclooxygenase and hydroperoxidase activities in a single protein. The reaction is associated with the generation of a potent oxidant (168) produced during the peroxidase-catalyzed conversion of prostaglandin G_2 (PGG_2) to prostaglandin H_2 (PGH_2). This oxidizing equivalent, with the oxidizing capacity of $HO^{\cdot}$, would be potent enough to account for the rapid loss of activity of the enzymes involved in the initial steps of the prostaglandin biosynthesis cascade, and for the oxidation of chemicals. Under appropriate conditions, it might be predicted to have pathological significance.

Regarding the possible mechanism underlying the generation of the oxidizing equivalent, several hypotheses or lines of evidence were published. The analogy between some of these enzymatic reactions and the chemistry of $^1O_2$ was pointed out, and moreover, it was suggested that an enzyme-bound or metal-bound form of $^1O_2$ might be the reactive species (169). Recent evidence was provided for the generation of $^1O_2$ during the enzymatic conversion of PGG_2 to PGH_2. Evidence was based on a spectral distribution of chemiluminescence resembling $^1O_2$ dimol emission and enhancement and quenching of photoemission by 1,4-diazabicyclooctane and beta-carotene, respectively.

The function of $^1O_2$ in cooxygenation systems, however, remains uncertain. Since the oxygenation is an important step in the activation of chemicals to toxic and/or carcinogenic intermediates, the possibility exists that PG synthase, like the cytochrome P-450–dependent monooxygenase, might participate in the oxidative activation of xenobiotics. The pathological aspects of the generation of such a powerful oxidant during PG hydroperoxidase reaction was related to inflammation [see (170)], because of the effect of certain antiinflammatory agents, which act as free radical scavengers. A possible relationship between atherosclerosis and hypertension, on the one hand, and this oxidizing species, on the other, was also pointed out.

Myelo-, Chloro-, and Lactoperoxidases

Monomol emission of $^1O_2$ was observed during the lactoperoxidase-, myeloperoxidase-, and chloroperoxidase-H_2O_2-halide systems (154–160), and hypothetically, during the decomposition of H_2O_2 by catalase (157). The former group of enzymatic reactions is associated with emission at 1268 nm and free $^1O_2$ would be responsible for the observed chemiluminescence. The catalase-catalyzed decomposition of H_2O_2, a reaction that proceeds in the absence of halides or cofactors, yields light emission at 1642 nm, and possibly a bound or microenvironmentally perturbed form of $^1O_2$ would be generated. Another interpretation for the 1642-nm emission of catalase could originate from an energy transfer from $^1O_2$ to the Fe-heme coordination complex at the active site (157). [For an alternative view see (171)].

These experimental observations were possible because of the recent development of sensitive infrared spectroscopic techniques to detect $^1O_2$ in solution (154, 159), and are understood as the logical consequence following the original discovery of $^1O_2$ generation by Khan & Kasha in 1963 (172).

The identification of a 1270-nm emission band, along with isotope effects, lends strong support to the generation of $^1O_2$ during the above-described enzymatic reactions because apart from $^1O_2$, there is no other molecular electronic state capable of emitting in the 1270-nm region. The mechanism for the generation of $^1O_2$ during these peroxidase reactions proceeds in a two-step reaction as a disproportionation of H_2O_2; the requirement of halide serves as a trigger of H_2O_2 decomposition with intermediate formation of a hypohalous acid; a second molecule of H_2O_2 is decomposed by the latter with generation of $^1O_2$ (Figure 5). A scrambling mechanism as that proposed for chloroperoxidase (163) does not apply to catalase, since the O_2 evolved originates from a single H_2O_2 molecule. These peroxidase-catalyzed reactions requiring halogen ion cofactors are suggestive of the classic inorganic H_2O_2/OCl^- reaction, where the separation of Cl^- from $ClOO^-$ leads to the generation of O_2 in singlet excited state (172). Formation of $^1O_2$ is near stoichiometric in both the lactoperoxidase H_2O_2–Br^- systems, efficient in the myeloperoxidase-H_2O_2–Br^- system, and modest in the myeloperoxidase H_2O_2–Cl^- reaction (154–156).

The biological relevance of the generation of $^1O_2$ by these enzymes is explained for the myeloperoxidase activity in terms of the microbicidal and cytotoxic activity of $^1O_2$, thus establishing the proposed $^1O_2$-mediated model of killing of polymorphonuclear leukocytes (173). However, under physiological conditions, neutrophils stimulated with phorbol myristate acetate and suspended in Cl^--containing buffer did not yield any chemiluminescence at 1270 nm (indicative of $^1O_2$ monomol emission). Under the conditions existing in the phagocytic vacuole of the neutrophil, the production of $^1O_2$ is very low; therefore it is not a major product of the neutrophil respiratory burst (174).

CONCLUDING REMARKS

Relationship Between Low- and Room-Temperature Studies

It is particularly interesting to note that the similarity between the low-temperature and room-temperature pathways in the cytochrome oxidase–oxygen reaction has become increasingly clear. Although Chance et al (83) have proposed a sequential mechanism at low temperature, Clore et al (93), repeating the same experiments, preferred a branched mechanism. Hill & Greenwood (97) also preferred a branched mechanism, while others (101–103) proposed a sequential mechanism at room temperature. The main experimental evidence for a branched mechanism is that some redox centers are oxidized heterogeneously. It is, however, not clear whether this heterogeneity is real [see (45) for an explanation] or another manifestation of the well-established heterogeneity of the cytochrome oxidase preparations (117, 118). Whatever the case, it is clear that the reaction sequences are very similar to both at low and room temperatures particularly as Chance finds no discontinuity of Arrhenius plot from $-100°C$ to $-23°C$ (83).

Another point is that some optical bands and EPR signals seen at room temperature for various species can be correlated to the bands and signals observed at low temperature. This concerns the discovery of the 420-nm form, and assigning the 655-nm, 580-nm, and slightly enhanced 605-nm bands of the optical spectra and the $g = 5$ EPR signal to either the 420-nm form or the peroxide adduct of the 420-nm form (119, 120). These assignments helped explain the $g = 5$ signal observed by Beinert (121, 143), the 655-nm band observed by Denis (148) at low temperatures, and the 580-nm band seen by Wikstrom (122).

The similarity between the low- and room-temperature reaction mechanisms of cytochrome oxidase once again demonstrates the validity of the cryoenzymological methods (175), and helps us to observe the fascinating reaction pathways in "slow motion" so that the reaction mechanism can be better understood and some of the states analyzed in detail by freezing certain slow motion pictures into "snapshots."

Future Directions

The most interesting results of the structural studies of cytochrome oxidase are the homologies of cytochrome oxidase with other well-known and structurally defined forms of hemoproteins which apparently cytochrome oxidase can borrow by means of the structural plasticity of the multisubunit macromolecule. Cytochrome a_3 serves as 1. an electron acceptor similar in axial ligands to cytochrome a, cytochrome c, cytochrome c_1, and possibly cytochrome b. 2. an oxygen acceptor similar in structural parameters to hemoglobin and having a second order rate constant of the same very high magnitude. 3. peroxidase where the axial ligands reaction center dimensions and reactivities are similar to those of the peroxidases. Whether such plasticity of reaction center enables cytochrome oxidase to acquire other activities such as superoxide dismutase has yet to be tested from the structural standpoint.

As a membrane protein, cytochrome oxidase has defied the attempts of biochemists to crystalize it. However, mammalian cytochrome oxidase has been crystalized (176), and everybody in the field is awaiting the crystal structure of this important protein, as in the case of the photosynthetic reaction center (177). Other structural tools such as RR, EXAFS, etc will continue to be very informative because of their ability to use samples in liquid states and study the dynamics of the process rather than static states. RR and EXFAS are also useful in studying the binuclear metal site.

When it seemed that the identity of the metal ions (heme and copper atoms) were more or less clear, Caughey's discovery (62, 63) of Zn and Mg in cytochrome oxidase stirred new interest in metal ions and their roles. Although a proton donating (to O_2) role of Zn and proton translocating role of Mg have been proposed (62, 63), it is far from clear what role Zn and Mg play.

Due to the simplicity and striking resemblance to the more complex mammalian oxidases, bacterial oxidases [for a review see (178)] have received great attention during the past few years and are likely to be more vigorously investigated. The comparative studies are likely to shed more light on the mechanism of oxygen reduction by cytochrome oxidase and related systems.

Another aspect of the oxygen reduction, which has received very little attention, is the steps where protons are taken up and released. Recently, we have studied the pH dependence of the low-temperature intermediates and found that the conversion of compound B_1 to B_2 occurs with release of protons by the enzyme and probably a proton is taken up in the step between compound A and B_1 (179, and J. C. Vincent et al, manuscript in preparation). This question will receive more attention in the future since at present unanimity has been established on the previously diverging views on whether cytochrome oxidase separates charges by means of a proton pump. Even now Mitchell agrees with this viewpoint (180). This adds new and important dimensions to cytochrome oxidase, which incorporates a proton pump, and a highly essential protolytic amino acid group and gating mechanisms which it may share with the well-known transmembrane protein bacterial rhodopsin and possibly the ATP synthetase as well. Thus new directions will certainly seriously take into account the properties of charge separation and how this is coupled to the active site, problems on which we ourselves speculated in the past (181).

In the 1970s, the generation of singlet oxygen was postulated by using indirect methods and suffered from several drawbacks. Recently, it became possible to observe its generation directly in the IR region using ultrasensitive detectors. So it seems that singlet oxygen in enzyme systems will certainly receive more attention in the future.

ACKNOWLEDGMENTS

The authors are indebted to Drs. G. T. Babcock, R. A. Copeland, T. G. Frey, B. Hoffman, C. Kumar, J. S. Leigh, Jr., Y. Li, M. Lundeen, T. Ohnishi, L. Powers, D. Rousseau, T. G. Spiro, and T. Yonetani for discussions and comments; to Drs. A. U. Khan and D. Rousseau for providing manuscripts prior to publication. This research was partially supported by NIH Grants HL 31909, HL 18708, GM 31992, and RR 01633.

Literature Cited

1. Bertrand, G. 1894. *C.R. Acad. Sci. Paris* 118:1215–18
2. Yoshida, H. 1883. *J. Chem. Soc.* 43: 472–86
3. Bertrand, G. 1897. *Ann. Chim. (Phys.)* 12:115–40
4. Rohmann, F., Spitzer, W. 1895. *Ber. Dtsch. Chem. Ges.* 28:567–72
5. Spitzer, W. 1897. *Pfluegers Arch. Gesamte Physiol. Menschen Tiere* 67: 615–56
6. Warburg, O. 1924. *Biochem. Z.* 152: 479–94
7. Warburg, O. 1925. *Ber. Dtsch. Chem. Ges.* 58:1001–11
8. Warburg, O. 1926. *Biochem. Z.* 177: 471–86
9. Keilin, D. 1925. *Proc. R. Soc. London, Ser. B* 98:312–39
10. MacMunn, C. A. 1886. *Philos. Trans. R. Soc. London* 177:267–98
11. Keilin, D. 1966. *The History of Cell Respiration and Cytochrome.* Cambridge: Cambridge Univ. Press
12. Nicholls, P., Chance, B. 1974. In *Molecular Mechanisms of Oxygen Activation,* ed. O. Hayaishi, pp. 479–534. New York: Academic
13. Slater, E. C., Van Gelder, B. F., Minnaert, K. 1965. In *Oxidase and Redox Related Systems,* ed. T. E. King, H. S. Mason, M. Morrison, pp. 667–706. New York: Wiley
14. Keilin, D., Hartree, E. F. 1939. *Proc. R. Soc. London Ser. B* 127:167–91
15. Chance, B. 1953. *J. Biol. Chem.* 202: 383–96
16. Chance, B. 1953. *J. Biol. Chem.* 202: 397–406
17. Chance, B. 1953. *J. Biol. Chem.* 202: 407–16
18. Chance, B., Smith, L., Castor, L. N. 1953. *Biochim. Biophys. Acta* 12:289–98
19. Castor, L. N., Chance, B. 1955. *J. Biol. Chem.* 217:453–65
20. Keevil, T., Mason, H. S. 1978. *Methods Enzymol.* 52:3–40
21. Orii, Y., Okunuki, K. 1963. *J. Biochem.* 54:207–13
22. Orii, Y. 1982. *J. Biol. Chem.* 257: 9246–48
23. Lemberg, R., Gilmour, M. B. 1967. *Biochim. Biophys. Acta* 143:500–17
24. Chance, B., Waring, A., Saronio, C. 1978. In *Energy Conservation in Biological Membranes,* ed. G. Schafer, M. Klingenberg, pp. 56–73. Berlin: Springer-Verlag
25. Chance, B., Kumar, C., Powers, L., Ching, Y. C. 1983. *Biophys. J.* 44:353–63
26. Markossian, K. A., Poghossian, A. A., Paitian, N. A., Nalbandyan, R. M. 1978. *Biochem. Biophys. Res. Commun.* 81: 1336–43
27. Nicholls, P., Chanady, G. A. 1981. *Biochim. Biophys. Acta* 634:256–65
28. Bickar, D., Bonaventura, C., Bonaventura, J. 1984. *J. Biol. Chem.* 259:10777–83
29. Brzezinski, P., Malmstrom, B. G. 1985. *FEBS Lett.* 187:111–14
30. Khan, A. U. 1976. *J. Phys. Chem.* 80:2219–28
31. George, P., Griffith, J. S. 1953. In *The*

Enzymes, ed. P. D. Boyer, H. Lardy, K. Myrback, 1:347–89. New York: Academic

32. Taube, H. 1965. *J. Gen. Physiol.* 49:29–50

33. Bennett, L. E. 1973. *Prog. Inorg. Chem.* 18:1–76

34. Hamilton, G. A. 1974. See Ref. 12, pp. 405–51

35. Valentine, J. S. 1973. *Chem. Rev.* 73:235–45

36. George, P. 1965. See Ref. 13, pp. 3–36

37. Malmstrom, B. G. 1982. *Ann. Rev. Biochem.* 51:21–59

38. Green, M. J., Hill, H. A. O. 1984. *Methods Enzymol.* 105:3–22

39. Caughey, W. S., Wallace, W. J., Volpe, J. A., Yoshikawa, S. 1976. In *The Enzymes,* ed. P. D. Boyer, 3:299–344. New York: Academic

40. Halliwell, B., Gutteridge, J. M. C. 1984. *Biochem. J.* 219:1–14

41. Ilan, Y. A., Czapski, G., Meisel, D. 1976. *Biochim. Biophys. Acta* 430:209–24

42. Wood, P. M. 1974. *FEBS Lett.* 44:22–24

43. Malkin, R., Malmstrom, B. G. 1970. *Adv. Enzymol.* 33:177–244

44. Chance, B., Saronio, C., Leigh, J. S. Jr. 1975. *Proc. Natl. Acad. Sci. USA* 72:1635–40

45. Wikstrom, M., Krab, K., Saraste, M. 1981. *Cytochrome Oxidase. A Synthesis.* New York: Academic

46. *J. Inorg. Biochem.* 1985. 23:143–393. Contains papers presented at the Rome-Caprarola Meet. on Cytochrome Oxidase, Oct. 1984

47. Brunori, M., Rotilio, G. 1984. *Methods Enzymol.* 105:22–35

48. Hatefi, Y. 1985. *Ann. Rev. Biochem.* 54:1015–69

49. Brunori, M., Wilson, M. T. 1982. *Trends Biochem. Sci.* 7:295–99

50. Capaldi, R. A. 1982. *Biochim. Biophys. Acta* 694:291–306

51. Capaldi, R. A., Malatesta, F., Darley-Usmar, V. M. 1983. *Biochim. Biophys. Acta* 726:135–48

52. Wikstrom, M., Saraste, M. 1984. In *Bioenergetics,* ed. L. Ernster, pp. 49–94. Amsterdam: Elsevier

53. Chan, S. I., Bocian, D. F., Brudvig, G. W., Morse, R. H., Stevens, T. H. 1979. In *Cytochrome Oxidase,* ed. T. E. King, Y. Orii, B. Chance, K. Okunuki, pp. 177–88. Amsterdam: Elsevier

54. Hu, V. W., Chan, S. I., Brown, G. S. 1977. *Proc. Natl. Acad. Sci. USA* 74:3821–25

55. Seiter, C. H. A., Angelos, S. G. 1980. *Proc. Natl. Acad. Sci. USA* 77:1806–8

56. Kadenbach, B., Ungibauer, M.,

Jarausch, J., Buge, U., Kuhn-Nentivig, L. 1983. *Trends Biochem. Sci.* 8:398–400

57. Chan, S. H. P., Freedman, J. A. 1983. *FEBS Lett.* 162:344–48

58. Azzi, A., Casey, R. P., Nalecz, M. J. 1984. *Biochim. Biophys. Acta* 768:209–26

59. Penttila, T. 1983. *Eur. J. Biochem.* 133:355–61

60. Thelen, M., O'Shea, P. S., Petrone, G., Azzi, A. 1985. *J. Biol. Chem.* 260:3626–31

61. Puettner, I., Carafoli, E., Malatesta, F. 1985. *J. Biol. Chem.* 260:3719–23

62. Einarsdottir, O., Caughey, W. S. 1984. *Biochem. Biophys. Res. Commun.* 124:836–42

63. Einarsdottir, O., Caughey, W. S. 1985. *Fed. Proc.* 44:1780 (Abstr.)

64. Brudvig, G. W., Blair, D. F., Chan, S. I. 1984. *J. Biol. Chem.* 259:11001–9

65. Goodman, G., Leigh, J. S. Jr. 1985. *Biochemistry* 24:2310–17

66. Scholes, C. P., Janakiraman, R., Taylor, H., King, T. E. 1984. *Biophys. J.* 45:1027–30

67. Ohnishi, T., LoBrutto, R., Salerno, J. C., Bruckner, R. C., Frey, T. G. 1982. *J. Biol. Chem.* 257:14821–25

68. Mascarhenar, R., Wei, Y. H., Scholes, C. P., King, T. E. 1983. *J. Biol. Chem.* 258:5348–51

69. Powers, L., Chance, B., Ching, Y., Angiolillo, P. 1981. *Biophys. J.* 34:465–98

70. Scott, R. A., Cramer, S. P., Beinert, H. 1982. *SSRL Activity Rep.* 82/01, VIII–89

71. Scott, R. A., Schwartz, J. R., Cramer, S. 1984. In *EXAFS and Near Edge Structure III,* ed. K. O. Hodgdon, B. Hedman, J. E. Penner-Hahn, pp. 111–16. Berlin: Springer-Verlag

72. Okunuki, K., Hagihara, B., Sekuzu, I., Horio, T. 1958. In *Proc. Int. Symp. Enzyme Chem.,* pp. 264–72. Tokyo: Maruzen

73. Wainio, W. W. 1965. See Ref. 13, pp. 622–38

74. Williams, G. R., Lemberg, R., Cutler, M. E. 1968. *Can. J. Biochem.* 46:1371–79

75. Lemberg, R., Mansley, G. E. 1966. *Biochim. Biophys. Acta* 118:19–35

76. Lemberg, R., Stranbury, J. 1967. *Biochim. Biophys. Acta* 143:37–51

77. Lemberg, R., Cutler, M. E. 1970. *Biochim. Biophys. Acta* 197:1–10

78. Lemberg, R., Gilmour, M. V. 1967. *Biochim. Biophys. Acta* 143:500–17

79. Gibson, Q. H., Greenwood, C. 1963. *Biochem. J.* 86:541–54

80. Gibson, Q. H., Greenwood, C. 1965. *J. Biol. Chem.* 240:2694–98

81. Greenwood, C., Gibson, Q. H. 1967. *J. Biol. Chem.* 242:1782–87
82. Erecinska, M., Chance, B. 1972. *Arch. Biochem. Biophys.* 151:304–15
83. Chance, B., Saronio, C., Leigh, J. S. Jr. 1975. *J. Biol. Chem.* 250:9226–37
84. Chance, B., Graham, N., Legallais, V. 1975. *Anal. Biochem.* 67:552–79
85. Chance, B., Leigh, J. S. Jr. 1977. *Proc. Natl. Acad. Sci. USA* 74:4777–80
86. Chance, B., Saronio, C., Waring, A., Leigh, J. S. Jr. 1978. *Biochim. Biophys. Acta* 503:37–55
87. Chance, B., Saronio, C., Leigh, J. S. Jr., Ingledew, W. J., King, T. E. 1978. *Biochem. J.* 171:787–98
88. Chance, B., Saronio, C., Leigh, J. S. Jr. 1979. *Biochem. J.* 177:931–41
89. Clore, G. M., Chance, E. M. 1979. *Biochem. J.* 177:613–21
90. Clore, G. M., Chance, E. M. 1978. *Biochem. J.* 173:799–810
91. Clore, G. M., Chance, E. M. 1978. *Biochem. J.* 173:811–20
92. Clore, G. M., Andreasson, L. E., Karlsson, B., Aasa, R., Malmstrom, B. G. 1980. *Biochem. J.* 185:155–67
93. Clore, G. M., Andreasson, L. E., Karlsson, B., Aasa, R., Malmstrom, B. G. 1980. *Biochem. J.* 185:139–54
94. Karlsson, B., Aasa, R., Vanngard, T., Malmstrom, B. G. 1981. *FEBS Lett.* 131:186–88
95. Caughey, W. S., ed. 1979. *Biochemical and Clinical Aspects of Oxygen.* New York: Academic. 866 pp.
96. Caughey, W. S., Barlow, C. H., Maxwell, J. C., Volpe, J. A., Wallace, W. J. 1975. *Ann. NY Acad. Sci.* 244:1–9
97. Hill, B. C., Greenwood, C. 1984. *Biochem. J.* 218:913–21
98. Hill, B. C., Greenwood, C. 1983. *Biochem. J.* 215:659–67
99. Hill, B. C., Greenwood, C. 1984. *FEBS Lett.* 166:362–66
100. Brunori, M., Gibson, Q. H. 1983. *EMBO J.* 2:2025–26
101. Orii, Y. 1984. *J. Biol. Chem.* 259:7187–90
102. Babcock, G. T., Jean, J. M., Johnston, L. N., Palmer, G., Woodruff, W. H. 1984. *J. Am. Chem. Soc.* 106:8305–6
103. Babcock, G. T., Jean, J. M., Johnston, L. N., Woodruff, W. H., Palmer, G. 1985. *J. Inorg. Biochem.* 23:243–51
104. Markossian, K. A., Nalbandyan, R. M. 1975. *Biochem. Biophys. Res. Commun.* 67:870–76
105. Greenwood, C., Wilson, M. T., Brunori, M. 1974. *Biochem. J.* 137:205–15
106. Fridovich, I. 1974. *Adv. Enzymol.* 41:35–97

107. Fee, J. A. 1980. In *Metal Activation of Dioxygen,* ed. T. G. Spiro, pp. 209–37. New York: Wiley
108. Fridovich, I., Handler, P. 1961. *J. Biol. Chem.* 236:1836–40
109. Wilson, M. T., Jensen, P., Aasa, R., Malmstrom, B. G., Vanngard, T. 1982. *Biochem. J.* 203:483–92
110. Cline, J., Reinhammar, B., Jensen, P., Venters, R., Hoffman, B. M. 1983. *J. Biol. Chem.* 258:5124–28
111. Markosyan, K. A., Nalbandyan, R. M. 1978. *Biokhimiya* 43:1143–49
112. Sekuzu, I., Takemori, S., Yonetani, T., Okunuki, K. 1959. *J. Biochem.* 46:43–49
113. Orii, Y., Okunuki, K. 1963. *J. Biochem.* 53:489–99
114. Orii, Y., King, T. E. 1972. *FEBS Lett.* 21:199–202
115. Orii, Y., King, T. E. 1976. *J. Biol. Chem.* 251:7487–93
116. Bickar, D., Bonaventura, J., Bonaventura, C. 1982. *Biochemistry* 21:2661–66
117. Brudvig, G. W., Stevens, T. H., Morse, R. H., Chan, S. I. 1981. *Biochemistry* 20:3912–21
118. Naqui, A., Kumar, C., Ching, Y. C., Powers, L., Chance, B. 1984. *Biochemistry* 23:6222–27
119. Kumar, C., Naqui, A., Chance, B. 1984. *J. Biol. Chem.* 259:2073–76
120. Kumar, C., Naqui, A., Chance, B. 1984. *J. Biol. Chem.* 259:11668–71
121. Shaw, R. W., Hansen, R. E., Beinert, H. 1978. *J. Biol. Chem.* 253:6637–40
122. Wikstrom, M. 1981. *Proc. Natl. Acad. Sci. USA* 78:4051–54
123. Antonini, E., Brunori, M., Colosimo A., Greenwood, C., Wilson, M. T. 1977. *Proc. Natl. Acad. Sci. USA* 74:3128–32
124. Bonaventura, C., Bonaventura, J., Brunori, M., Wilson, M. T. 1978. *FEBS Lett.* 85:30–34
125. Colosimo, A., Brunori, M., Sarti, P., Antonini, E., Wilson, M. T. 1981. *Isr. J. Chem.* 21:30–33
126. Brunori, M., Colosimo, A., Rainoni, G., Wilson, M. T., Antonini, E. 1979. *J. Biol. Chem.* 254:10769–75
127. Brunori, M., Colosimo, A., Sarti, P., Antonini, E., Wilson, M. T. 1981. *FEBS Lett.* 126:195–98
128. Brunori, M., Colosimo, A., Wilson, M. T., Sarti, P., Antonini, E. 1983. *FEBS Lett.* 152:75–78
129. Wilson, M. T., Peterson, J., Antonini, E., Brunori, M., Colosimo, A., Wyman, J. 1981. *Proc. Natl. Acad. Sci. USA* 78:7115–18

130. Sarti, P., Colosimo, A., Brunori, M., Wilson, M. T., Antonini, E. 1983. *Biochem. J.* 209:81–89
131. Armstrong, F., Shaw, R. W., Beinert, H. 1983. *Biochim. Biophys. Acta* 722:61–71
132. Brittain, T., Greenwood, C. 1976. *Biochem. J.* 155:453–55
133. Shaw, R. W., Rife, J. E., O'Leary, M. H., Beinert, H. 1981. *J. Biol. Chem.* 256:1105–7
134. Copeland, R. A., Naqui, A., Chance, B., Spiro, T. G. 1985. *FEBS Lett.* 182:375–79
135. Wrigglesworth, J. M. 1984. *Biochem. J.* 217:715–19
136. Powers, L., Chance, B. 1985. *J. Inorg. Biochem.* 23:207–17
137. Chance, B., Powers, L., Ching, Y., Poulos, T., Schonbaum, G. R., et al. 1984. *Arch. Biochem. Biophys.* 235:596–611
138. Chance, B. 1951. *Adv. Enzymol.* 12:153–90
139. Van Wart, H. E., Zimmer, J. 1985. *J. Am. Chem. Soc.* 107:3379–81
140. Teraoka, J., Ogura, T., Kitagawa, T. 1982. *J. Am. Chem. Soc.* 104:7354–56
141. Rakhit, G., Spiro, T. G., Uyeda, M. 1976. *Biochem. Biophys. Res. Commun.* 71:803–8
142. Carter, K. R., Antalis, T. M., Palmer, G., Ferris, N. S., Woodruff, W. H. 1981. *Proc. Natl. Acad. Sci. USA* 78:1652–55
143. Chance, B., O'Connor, P., Yang, E. 1982. In *Oxidases and Related Redox Systems,* ed. T. E. King, H. S. Mason, M. Morrison, pp. 1019–35. Oxford: Pergammon
144. Reinhammar, B., Malkin, R., Jensen, P., Karlsson, B., Andreasson, L. E., et al. 1980. *J. Biol. Chem.* 255:5000–3
145. Karlsson, B., Andreasson, L. E. 1981. *Biochim. Biophys. Acta* 635:73–80
146. Karlsson, B., Aasa, R., Vanngard, T., Malmstrom, B. G. 1981. *FEBS Lett.* 131:186–88
147. Gibson, Q. H., Greenwood, C. 1964. *J. Biol. Chem.* 239:586–90
148. Denis, M. 1977. *FEBS Lett.* 84:296–98
149. Shaw, R. W., Hansen, R. E., Beinert, H. 1979. *Biochim. Biophys. Acta* 548:386–96
150. Muijsers, A. O., Tiesjema, R. H., Van Gelder, B. F. 1971. *Biochim. Biophys. Acta* 234:481–92
151. Rosen, S., Branden, R., Vanngard, T., Malmstrom, B. G. 1977. *FEBS Lett.* 74:25–30
152. Cadenas, E., Sies, H., Graf, H., Ullrich, V. 1983. *Eur. J. Biochem.* 130:117–21
153. Cadenas, E., Sies, H., Nastainczyk, W., Ullrich, V. 1983. *Hoppe-Seyler's Z. Physiol. Chem.* 364:519–28
154. Kanofsky, J. R. 1983. *J. Biol. Chem.* 258:5991–93
155. Kanofsky, J. R. 1984. *J. Photochem.* 25:105–13
156. Kanofsky, J. R. 1984. *J. Biol. Chem.* 259:5596–600
157. Khan, A. U. 1983. *J. Am. Chem. Soc.* 105:7195–97
158. Khan, A. U. 1984. *Biochem. Biophys. Res. Commun.* 122:668–75
159. Khan, A. U. 1984. *J. Photochem.* 25:327–34
160. Khan, A. U., Gebauer, P., Hager, L. P. 1983. *Proc. Natl. Acad. Sci. USA* 80:5195–97
161. Ullrich, V. 1977. In *Microsomes and Drug Oxidations,* ed. V. Ullrich, I. Roots, A. Hildebrandt, R. O. Estabrook, A. H. Conney, pp. 119–210. New York: Plenum
162. Ullrich, V. 1984. In *Oxygen Radicals in Chemistry and Biology,* ed. W. Bors, M. Saran, D. Tait, pp. 391–404. Berlin: de Gruyter
163. Hager, L. P., Doubek, D. L., Silverstein, R. M., Hargis, J. H., Martin, J. C. 1972. *J. Am. Chem. Soc.* 94:4364–66
164. Hollenberg, P. F., Hager, L. P. 1974. *J. Biol. Chem.* 248:2630–33
165. Ullrich, V., Castle, L., Weber, P. 1981. *Biochem. Pharmacol.* 30:2033–36
166. White, R. E., Coon, M. J. 1980. *Ann. Rev. Biochem.* 49:315–56
167. Cadenas, E., Sies, H. 1982. *Eur. J. Biochem.* 124:349–56
168. Marnett, L. J., Reed, G. A. 1979. *Biochemistry* 18:2923–29
169. Foote, C. S. 1968. *Acc. Chem. Res.* 1:104–10
170. Kuehl, F. A., Humes, J. L., Ham, E. A., Egan, R. W., Dougherty, H. W. 1980. In *Advances in Prostaglandin and Thromboxane Research,* ed. B. Samuelsson, P. V. Ramwell, R. Paoletti, pp. 77–86. New York: Raven
171. Kanofsky, J. R. 1984. *J. Am. Chem. Soc.* 106:4277–78
172. Khan, A. U., Kasha, M. 1963. *J. Chem. Phys.* 39:2105–6
173. Allen, R. C. 1975. *Biochem. Biophys. Res. Commun.* 63:675–83
174. Kanofsky, J. R., Tauber, A. I. 1983. *Blood* 62(Suppl. 1):82a
175. Douzou, P. 1977. *Cryobiochemistry. An Introduction.* London: Academic
176. Ozawa, T., Tanaka, M., Wakabayashi,

T. 1982. *Proc. Natl. Acad. Sci. USA* 79:7175–79
177. Deisenhofer, J., Epp, O., MiKi, K., Huber, R., Michel, H. 1984. *J. Mol. Biol.* 180:385–98
178. Poole, R. K. 1983. *Biochim. Biophys. Acta* 726:205–43
179. Vincent, J. C., Kumar, C., Naqui, A., Chance, B. 1983. *Biophys. J.* 41:321a (Abstr.)

180. Mitchell, P., Mitchell, R., Moody, A. J., West, I., Baum, H., Wrigglesworth, J. 1985. *FEBS Lett.* 188:1–8
181. Chance, B., Powers, L., Ching, Y. 1981. In *Mitochondria and Microsomes: In Honor of Lars Ernster,* ed. C. P. Lee, G. Schatz, G. Dallner, pp. 271–92. Reading, MA: Addison-Wesley

Ann. Rev. Biochem. 1986. 55:167–193

LYSOSOMAL ENZYMES AND THEIR RECEPTORS

Kurt von Figura and Andrej Hasilik

Physiologisch-Chemisches Institut, Westfälische Wilhelms-Universität, Waldeyerstr. 15, D-4400 Münster, West Germany

CONTENTS

PERSPECTIVES AND SUMMARY

Interest in lysosomes and lysosomal enzymes was stimulated by the existence of some 30 inherited lysosomal storage disorders in man. The enzyme defects involved in most of these disorders were identified in the 1970s; see review by Neufeld, Lim, & Shapiro in this series in 1975 (1). Presently, these mutations are being characterized at the level of DNA and RNA.

Targeting of lysosomal enzymes is part of the more general question: how do eukaryotic cells transport proteins synthesized in the rough endoplasmic reticulum to diverse destinations? Hickman & Neufeld discovered, in 1972, that the multiple deficiency of lysosomal enzymes in I-cell disease results from a

167

0066-4154/86/0701-0167$02.00

deficiency in a recognition marker that is common to lysosomal enzymes and required for targeting the enzymes to lysosomes (2). This observation provided the basis for many subsequent studies that eventually led to the identification of the recognition marker and its receptor. A 215-kd receptor, which recognizes mannose 6-phosphate residues in lysosomal enzymes, has been identified as an essential component of a system that in many cells allows for specific transport of lysosomal enzymes to lysosomes. It was originally identified as a cell-surface receptor binding exogenous lysosomal enzymes and mediating their transfer to lysosomes along the pathway of receptor-mediated endocytosis. We now know that this receptor functions also in transport of endogenous lysosomal enzymes and that its presence in organelles that constitute elements of the secretory pathway is important for that function. The combined application of biochemical and cytological methods has significantly contributed to the present knowledge of lysosomal enzyme transport. Further, the current application of recombinant DNA methods to the study of lysosomal enzymes and their receptors is expected to provide answers to many unresolved questions.

This review is limited to discussion of synthesis and transport of lysosomal enzymes in mammalian tissues. We focus on the processing of the oligosaccharides in lysosomal enzymes, on the mannose 6-phosphate–specific receptor (MPR), and on its role in the transport of lysosomal enzymes. The transport of lysosomal enzymes and the function of mannose 6-phosphate–specific receptors have been the subject of recent reviews (3–5).

BIOSYNTHESIS OF LYSOSOMAL ENZYMES

Synthesis and Modifications in Endoplasmic Reticulum

Most of the proteins that are localized partially or fully to the extracytosolic side of the endoplasmic reticulum, Golgi complex, lysosomes, and nuclear and plasma membranes have in common a signal sequence that directs the ribosomes engaged in their synthesis to the rough endoplasmic reticulum [reviewed in (6)]. This common signal sequence is a stretch of 15–30 mainly hydrophobic amino acids that is localized to the N-terminus. When protruding from the ribosome this stretch first forms a complex with a cytosolic ribonucleoprotein called signal recognition particle (7). Subsequently, this complex binds to a receptor at the surface of the rough endoplasmic reticulum that is called the SRP receptor or the docking protein (8, 9), and the nascent protein is transferred into the lumen of the rough endoplasmic reticulum. Most of the soluble proteins that are released into the endoplasmic reticulum lose the signal peptide before the synthesis of their polypeptide is completed.

A difference in size between polypeptides synthesized in vitro (nonglycosylated and possibly containing the signal peptide) and those synthesized in cultured cells treated with tunicamycin, an inhibitor of glycosylation (nonglycosylated products lacking cleavable signal peptide), is the commonly accepted evidence for the existence of a signal peptide in a protein. Using this technique, it has been shown that porcine cathepsin D (10), mouse cathepsin D, and rat β-glucuronidase (11), as well as α and β chains of human β-hexosaminidase (12), transiently contain signal peptides. Direct evidence for the presence of a signal sequence in porcine cathepsin D has been obtained by Erickson et al (13), who determined partial N-terminal sequences of porcine cathepsin D synthesized in vitro both in the presence and in the absence of membranes. The translation of cathepsin D is also regulated by the signal recognition particle (14).

As has been demonstrated for several (nonlysosomal) glycoproteins, a glucosylated oligosaccharide is transferred to certain asparagine residues in the polypeptide intruding into the lumen of the rough endoplasmic reticulum (15–17). Usually this transfer takes place prior to folding of the protein backbone. The oligosaccharide transferred to asparagine residues is preassembled in an "activated" form as a derivative of dolichol pyrophosphate [(18), see Figure 1]. The asparagine-linked oligosaccharides in lysosomal

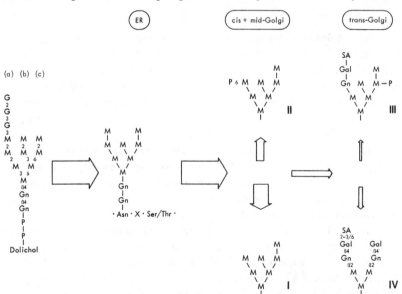

Figure 1   Main stages in the processing of oligosaccharides in lysosomal enzymes. The organelles to which the processing is localized are indicated at the top of the figure. Four typical structures that have been found in lysosomal enzymes as formed in different parts of the Golgi complex are shown: I = high-mannose, II = phosphorylated high-mannose, III = phosphorylated hybrid, IV = complex oligosaccharide. Hybrid oligosaccharides without phosphate groups have been found also. The arrows indicate the approximate relative abundance of the four oligosaccharide types. The symbols are: G = glucose, Gal = galactose, Gn = *N*-acetylglucosamine, M = mannose, SA = *N*-acetyl neuraminic acid, P = phosphate. Single numbers indicate the positions of α-anomerically linked sugars.

enzymes are subject to processing that follows the principles elucidated in the past decade for secretory and membrane glycoproteins. These have been reviewed in the previous issue of this series by Kornfeld & Kornfeld (18) and will be only briefly mentioned. We will focus on the reactions that have been studied in lysosomal enzymes and which in part are specific for them.

Processing of the oligosaccharide is initiated by "trimming" reactions. The removal of the first glucose residue is effected by glucosidase I, takes place within a few minutes of the transfer of the oligosaccharide, and may even precede the completion of the polypeptide synthesis (19). Removal of the two other glucoses by glucosidase II is a much slower process (20). Recently, removal of outer glucose residues within 1 min of synthesis has been observed in cathepsin D in human fibroblasts (21). The trimming in the rough endoplasmic reticulum also involves a specific α-mannosidase (22), and in general seems to yield octamannosyl chains. Specific removal of a mannose residue from the terminus of the middle branch is likely to facilitate further processing in the Golgi complex (23).

Transport to the Golgi

Several cytological observations suggest that the transport is mediated by smooth vesicles formed in "transitional" elements of the endoplasmic reticulum (24, 25). The transit times from the reticulum to the Golgi complex vary among different products (26–29), and it is not known whether this results from characteristic interactions of the individual products with other components remaining in or leaving the reticulum. In the case of membrane-associated histocompatibility antigens, the transport depends on the availability of β_2-microglobulin (30). Lodish et al (26, 31) postulated that a receptor protein in the endoplasmic reticulum membrane regulates the selective transport of secretory proteins into transport vesicles en route to the Golgi. 1-Deoxynojirimycin, an inhibitor of the trimming glucosidases (32, 33), has been shown to inhibit secretion of α_1-proteinase inhibitor in rat hepatocytes (34) and HepG2 cells (31), and the transport of lysosomal enzymes into lysosomes in fibroblasts (21). In the presence of 1-deoxynojirimycin, these glycoproteins were retarded in the endoplasmic reticulum. Upon subcellular fractionation of cells treated with the drug, the retarded glycoproteins were found in the microsomal fraction and their carbohydrates did not show the characteristics of processing in the Golgi complex (21, 31).

In rat hepatocytes and HepG2 cells, the selectivity of the effect of 1-deoxynojirimycin was indicated by the fact that it did not inhibit the secretion of albumin (31, 34) or of the glycoprotein C3 and transferrin (31). In a mixed population of hybridoma cells, the drug inhibited the secretion of IgD and not of IgM (35). This differential inhibition corresponded well to a differential effect on the formation of complex oligosaccharides. In the case of the membrane-associated glycoprotein v-erbB, the transport to the plasma membrane was not affected, although the processing was blocked (36). It appears that the inhibition of the processing in the presence of 1-deoxynojirimycin interferes with the transport of certain soluble glycoproteins, including lysosomal enzymes. The inhibition of transport of affected glycoproteins was incomplete and molecules that eventually reached their normal extracellular or lysosomal destinations contained at least some normally processed oligosaccharides (21, 34). Under normal conditions, the transport of proteins between organelles may be differentially influenced by interactions with other components of the system. A well-known example is the retention of β-glucuronidase in microsomal organelles containing a protein called egasyn (37).

Common Modifications in the Golgi

TRANSPORT   The Golgi complex is an elaborate membrane system, in which a number of modification reactions, in particular the synthesis of the carbohydrate portion of the various glycoconjugates, are accomplished [reviewed in (38)]. The complex consists of a stack of flat or fenestrated cisternae with associated tubules and vesicles with polar orientation. The so-called *cis* part receives the product of biosynthesis from the rough endoplasmic reticulum (24) and the other pole, the *trans* part, is marked in secretory cells by condensing vacuoles and secretory vesicles (25). Functionally, it is useful to consider three regions within the stack: *cis, mid,* and *trans,* as suggested by Griffith et al (39) and Rothman et al (40). Most studies on the intracellular transport deal with membrane proteins. The various parts of the Golgi complex, though not rigidly separated, are involved in different carbohydrate modification reactions (see below). Lipids, membrane-associated proteins and soluble proteins that are produced in the endoplasmic reticulum are subject to a vectorial (*cis* to *trans*) flow through the Golgi complex. The complex's own constituents behave as a stationary phase and their relative distribution through the cisternae may be maintained through a counterflow resembling the distillation process (41). It has been pointed out by Slot & Geuze (42) that small vesicles with a large membrane/volume ratio may efficiently accomplish the transport of membrane components. Indeed, small vesicles in the vicinity of the Golgi complex are enriched in various receptors (42–44), and it should be of interest to test the possibility that some vesicles are enriched in the membrane constituents of various parts of the Golgi complex and serve to reflux these constituents between the neighboring organelles. As far as transport between the *cis* and *mid* parts of the Golgi complex is concerned, it has been suggested by Rothman and coworkers that it is the biosynthetic product that is passed over in small vesicles. This suggestion is based on studies of the transport of the membrane glycoprotein G of the vesicular stomatitis virus in an in vitro system, in which the budding of vesicles from Golgi cisternae was observed (45).

As judged from the modifications of the carbohydrates, lysosomal enzymes can pass through all three parts of the complex. It is a matter of debate, however, whether their passage through the *trans*-Golgi is obligatory. This question is related to the localization of the sorting step and will be discussed below.

COMMON MODIFICATIONS   Depending on the protein moiety, the outer-chain mannose residues (see Figure 1) in glycoproteins entering *cis*-Golgi are subjected to further trimming. This is accomplished by at least two α-mannosidases in an ordered sequence, and also involves a specific *N*-acetylglucosaminyl transferase. The first enzyme in this sequence, mannosidase I, is defined as an enzyme hydrolyzing high-mannose oligosaccharides to yield $Man_5 GlcNAc_2$ (46), and is an accepted marker for *cis*-Golgi (47). Its product may subsequently be processed by *N*-acetylglucosaminyl transferase I to yield $GlcNAc Man_5 GlcNAc_2$ (48, 49).

Recently, immunocytochemical evidence has been presented for localization of this enzyme in the *mid*-cisternae of the Golgi complex (50). Mannosidase II is highly specific for the product of *N*-acetylglucosaminyl transferase I and converts it to $GlcNAc Man_3 GlcNAc_2$. Mannosidase II seems to be rather broadly distributed through the Golgi complex, with a maximum activity in the middle portion of the complex (51). With the aid of specific antibodies directed to mannosidase II, a rather uniform distribution of the enzyme in the elements of the Golgi complex (52) has been firmly established. If mannosidase II does not act on its substrate, addition of galactose and sialic acid results in formation of hybrid oligosaccharides. The action of mannosidase II is prevented e.g. by the presence of the so-called bisecting *N*-acetylglucosamine linked to β-mannose (53, 54). The product of mannosidase II is the preferred substrate of *N*-acetylglucosaminyl transferase II. This reaction opens the pathway for the synthesis of an array of complex oligosaccharides with two or more antennas (18, 55). The final steps in this synthesis take place in *trans*-Golgi cisternae, where galactosyl transferase is localized as has been demonstrated by Roth & Berger (56).

In earlier studies, ample indirect evidence has been obtained of the presence of hybrid or complex oligosaccharides in lysosomal enzymes. This evidence is based on sensitivity to neuraminidase, binding to immobilized and cellular lectins, and carbohydrate analyses, in which fucose, galactose, and sialic acid were detected. Structural data on complex oligosaccharides in soluble lysosomal enzymes is available for β-glucuronidase from human spleen (57). This enzyme contains a small amount of complex oligosaccharides with two antennas, of which one is incomplete. Bi- and triantennary oligosaccharides comprising about 80% of the total oligosaccharides were found in human β-glucocerebrosidase (58). It should be pointed out that this hydrolase is an example of a membrane-associated lysosomal enzyme (59). The observation, in lysosomal enzymes, of short oligosaccharides containing only four (57, 60) or just a single sugar residue (61), points to intralysosomal degradation. This may in part explain the low amounts of typical complex oligosaccharide structures found in lysosomal enzymes. As determined by lectin binding and resistance to endoglucosaminidase H, an enzyme cleaving most of the high-mannose and hybrid oligosaccharides (62), complex oligosaccharides have been found in metabolically labeled mouse β-glucuronidase (63), human fibroblasts cathepsin D, β-hexosaminidase and arylsulfatase B (64, 65), and in Chang liver α-galactosidase (66), which contains some tri- or tetraantennary oligosaccharides. In newly synthesized cathepsin D, oligosaccharides resistant to endoglucosaminidase H and containing galactose comprise almost 30% of the total oligosaccharide population (64, 67).

While the content of complex oligosaccharides in soluble lysosomal enzymes of normal fibroblasts is generally low, it is elevated in I-cell and mucolipidosis III fibroblasts, predominantly in their secretions (64, 68–71). This striking change results from a defect in these mutant cells in the phosphorylation of high-mannose oligosaccharides (see below).

Sialylated hybrid oligosaccharides have been found in cathepsin D and β-hexosaminidase of human fibroblasts and contained 5–10% of the radioactivity in anionic oligosaccharides released by endoglucosaminidase H (72). A group of hybrid oligosaccharides that contain phosphate in addition to sialic acid was characterized by Varki & Kornfeld (73) and will be discussed below.

Formation of Mannose 6-Phosphate Residues

REACTIONS   Observations on the efficiency and saturability of the endocytosis of lysosomal enzymes led Hickman & Neufeld to propose, in 1972 (2), that lysosomal enzymes carry a specific recognition marker. By 1980, it became apparent from a series of observations contributed from several laboratories that the recognition marker in lysosomal enzymes is a carbohydrate (74), that it is related to mannose (75), sensitive to alkaline phosphatase, and probably represented by mannose 6-phosphate residues (76–79). Mannose 6-phosphate was identified in lysosomal enzymes as a component of oligosaccharides cleavable by endoglucosaminidase H (80–83). Phosphate has been found in the carbohydrate of all soluble lysosomal enzymes tested including β-glucuronidase (82), β-hexosaminidase, α-glucosidase and cathepsin D (83), α-L iduronidase (84), arylsulfatase A (85), arylsulfatase B (65), myeloperoxidase (86), acid phosphatase (87), and cathepsin C (88). Subsequent to the identification of the phosphorylated residue, it was observed that some of the phosphorylated oligosaccharides contain N-acetyl-glucosaminyl-1-phospho-6-mannose diester groups (72, 89–91). The anomeric configuration of the N-acetylglucosaminyl residue was found to be α (72, 89, 90, 92). Previously, phosphodiester compounds of sugars or sugar alcohols had been found in cell walls of various microorganisms. In analogy to the synthesis of phosphomannan in yeast (93), two carbohydrate modification reactions, which form the phosphorylated recognition marker, were identified: transfer of N-acetylglucosaminyl 1-phosphate from the UDP-N-acetylglucosamine to the C-6 hydroxyl of a mannose residue, and hydrolysis of the covering N-acetylglucosamine residue (see Figure 2).

The enzyme catalyzing the first reaction, UDP-N-acetylglucosamine:lysosomal enzyme N-acetylglucosaminyl-1-phosphotransferase (referred to as transferase), was first demonstrated in membranes from rat liver, fibroblasts, and Chinese hamster ovary cells (94–96), and was partially purified from rat liver Golgi preparations (97). [β-^{32}P]UDP-N-acetylglucosamine is a suitable substrate for determination of transferase activity and can be prepared from [γ-^{32}P]ATP using commercially available enzymes (98). N-acetylglucosamine 1-phosphate can be transferred to the C-6 position of mannose residues in precursor and mature forms of lysosomal enzymes, high-mannose oligosaccharides, and α-methylmannoside (96, 97). However, lysosomal enzymes are phosphorylated at least 100 times more efficiently than the oligosaccharides or α-methylmannoside. Phosphorylation of high-mannose oligosaccharides in nonlysosomal glycoproteins is not detectable (97), or only barely so (96). The acceptor activity of lysosomal enzymes is destroyed by denaturation (96). When lysosomal enzymes are deglucosylated with endoglucosaminidase H, they become specific inhibitors of the N-acetylglucosaminyl-1-phosphate transfer to lysosomal enzymes (99). Apparently, lysosomal enzymes contain a unique, denaturable structure that is distinct from the acceptor oligosaccharide, common to all lysosomal enzymes, and recognized by the transferase. The dual recognition of lysosomal enzymes by the transferase is strongly supported by the failure of deglycosylated lysosomal enzymes to inhibit phosphorylation of α-methyl mannoside (99), and the characterization of a mutant that phosphorylates high-mannose oligosacchrides and α-methyl mannoside but not lysosomal enzymes (100).

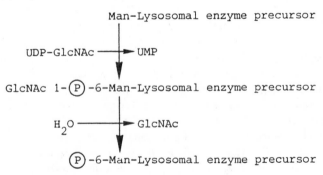

Figure 2   Two-step biosynthesis of mannose 6-phosphate residues in lysosomal enzymes.

Primary structures of several lysosomal enzymes have been determined. No homologies are found when the primary sequences in the vicinity of the two glycosylation sites in porcine cathepsin D (101), one in rat cathepsin B (102), and four potential glycosylation sites found in the established partial sequence of human α-fucosidase (103) are compared. Therefore, the signal for the phosphorylation is not represented (in different species) by a common sequence adjacent to the glycosylation sites. Phosphorylation at different glycosylation sites within a single enzyme is random (63, 64). Thus, it is likely that the signal is present in the lysosomal enzyme in a single copy. Because the signal is sensitive to treatment with heat, sodium dodecylsulfate, or trypsin (99), it is probably part of the protein and dependent on the tertiary rather than primary structure.

In the cell, the precursor forms of the lysosomal enzymes serve as acceptors for the transferase, which is localized to the Golgi complex (104, 105). Within the Golgi complex, transferase activity decreases from *cis* to *trans* (51, 106, 107). At the time of phosphorylation, the bulk of oligosaccharides contain six to nine mannose residues (63). Analyses of mutants showed that phosphorylation of glucosylated high-mannose oligosaccharides is possible in branches that do not contain glucose (108), and that truncated oligosaccharides with only five mannose residues can be phosphorylated (108, 109). Mutants provide some indication of whether a single transferase is responsible for phosphorylation. In I-cell patients, in fibroblasts, and in all tissues examined, the transferase activity is absent (94, 110–113), and lysosomal enzymes do not contain phosphate (83, 85, 87, 114), pointing to the existence of a single transferase. In a milder form of the disease, mucolipidosis III, transferase activity (100, 111, 115) and phosphorylation of lysosomal enzymes (85, 116) are markedly reduced. The residual phosphorylation leads to the formation of oligosaccharides with one as well as with two phosphate groups (115). This further supports the concept of a single transferase.

The second enzyme in the specific pathway, N-acetylglucosamine 1-phosphodiester α-N-acetylglucosaminidase, has been partially purified from rat liver (117, 118) and human placenta (119). The enzyme is immunologically and catalytically distinct from the lysosomal α-N-acetylglucosaminidase. It hydrolyzes UDP-N-acetylglucosamine, but not the corresponding arylglycoside. The mechanism of hydrolysis is that of a glycosidase and not a phosphodiesterase (120). It fractionates with Golgi membranes (104, 117, 118) and in fractionated Golgi membranes, it is distributed at slightly higher densities than the transferase (51, 105–107), which suggests a *mid*-Golgi localization.

STRUCTURE AND DISTRIBUTION OF PHOSPHORYLATED OLIGOSACCHARIDES IN LYSOSOMAL ENZYMES   In Figure 1, examples of the main types of oligosaccharides found in lysosomal enzymes are shown. The structures of the phosphorylated oligosaccharides were studied in purified lysosomal enzymes as well as in total cellular or secreted glycoproteins. Only a minor portion of the oligosaccharides in lysosomal enzymes becomes phosphorylated. This portion does not exceed 30% in β-glucuronidase synthesized in mouse lymphoma cells (63). Only about 25% of the radioactive mannose was released as phosphorylated oligosaccharides from cathepsin D and β-hexosaminidase that were isolated from NH_4Cl-induced secretions of human fibroblasts and therefore not subjected to degradation in lysosomes (72). In lysosomal enzymes isolated from tissues, the portion of phosphorylated oligosaccharides is even lower. In a β-glucuronidase preparation from human spleen containing forms enriched in mannose 6-phosphate, phosphorylated oligosaccharides accounted for 10% of total oligosaccharides (57, 92), in β-glucuronidase from rat liver lysosomes phosphorylated oligosaccharides were not detectable (91), and in cathepsin D from porcine spleen they accounted for less than 4% of total oligosaccharides (121). Pulse-chase labeling studies indicate the presence of secondary modifications to the oligosaccharides in lysosomal enzymes including removal of blocking N-acetylglucosamine residues (63, 89) and removal of phosphate (63). Therefore, the amount of phosphorylated oligosaccharides and the relative amounts of the different species depend greatly on the source and subcellular location of a lysosomal enzyme.

In all studies, oligosaccharides with one phosphate group were found to be two to five times more frequent than oligosaccharides with two phosphate groups (63, 72, 73, 90, 92). Primarily, the phosphate is found in the diester form. In the cathepsin D and β-hexosaminidase preparations from human skin fibroblasts mentioned above (72), more than 80% of the phosphorylated oligosaccharides contained one or two covered phosphates. A similar portion was found in phosphorylated oligosaccharides in β-glucuronidase and the total cellular glycoproteins isolated from mouse lymphoma cells after a three-hour labeling period (89, 90). In the lysosomal enzymes isolated from tissues, phosphorylated oligosaccharides are found to contain either exclusively monoesters [cathepsin D from porcine spleen, (121)] exclusively diesters [microsomal β-glucuronidase from rat liver, (91)], or predominantly diesters [β-glucuronidase from human spleen, (92)]. Oligosaccharides that contain both a phosphomonoester and phosphodiester group represent in all preparations a minor species (73, 92). The underlying oligosaccharides contain four to nine mannose residues, but species with six to eight mannose residues are the most frequent forms (63, 72, 73, 89–92, 121).

Varki & Kornfeld (90) analyzed in detail the position of the phosphate groups in cellular glycoproteins of mouse lymphoma cells. According to their analysis, the mannose residues that can be phosphorylated are represented by the three residues at the nonreducing termini of branches a through c and the penultimate residues of branches a and c in the Man_8 $GlcNAc_2$ oligosaccharide (shown in

Figure 1). This oligosaccharide is probably the predominant product of the endoplasmic reticulum α-mannosidase (18, 22). Phosphorylation of these five mannose residues is not random and phosphorylated mannoses are most commonly found in branches c and a. During the processing covering N-acetylglucosamine residues and outer nonphosphorylated mannose residues are removed. As a result, the oligosaccharides underlying phosphomonoesters are in general smaller than those carrying phosphodiesters, which suggests that the cleavage of the covering N-acetylglucosamine residue is followed by removal of mannoses (63, 73, 90). In oligosaccharides with two phosphate groups, the phosphates always occur on two different branches.

Recently, phosphorylated oligosaccharides that contain one or two sialic acid residues were identified by Varki & Kornfeld (73) in P388D$_1$ macrophage–like cells. These hybrid oligosaccharides contain only a single phosphate as mono- or diester on branch c and lack a bisecting N-acetylglucosamine on the β-linked mannose. Hybrid oligosaccharides containing phosphodiester groups also were found in thyroglobulin secreted by malignant thyroid tissue (122).

Lysosomal enzymes usually contain two or more oligosaccharides per subunit. This holds true for human spleen β-glucuronidase (63, 57), cathepsin D, the α and β chain of β-hexosaminidase, and arylsulfatase B from human skin fibroblasts (64, 65). Any of the oligosaccharides in mouse β-glucuronidase (63) and human cathepsin D (64) may become phosphorylated. From the extent of phosphorylation, it appears that on the average one or less than one oligosaccharide per subunit is phosphorylated. The percentage of phosphate groups that becomes uncovered in the Golgi complex is unknown. The phosphate groups in less than 20% of the phosphorylated oligosaccharides in cathepsin D and β-hexosaminidase become uncovered (72), provided that NH_4Cl does not interfere with the action of phosphodiester α-N-acetylglucosaminidase. Under these conditions, less than 10% of cathepsin D and β-hexosaminidase subunits leave the Golgi complex with high-mannose oligosaccharides containing phosphomonoesters.

Analyses of the oligosaccharides in β-glucuronidase from human spleen (57, 92), mouse lymphoma cells (89), rat liver microsomes and lysosomes (91), and cathepsin D from porcine spleen (121, 123) indicate that neutral high-mannose structures are the prevailing oligosaccharides in these enzymes. The sizes of these oligosaccharides depend on the enzyme sources. In β-glucuronidase from human spleen and from mouse lymphoma and P388D$_1$ cells, the high-mannose oligosaccharides with nine and eight mannose residues are most frequent (57, 63, 89, 92), whereas in β-glucuronidase from rat liver lysosomes, forms with five mannose residues prevail (91). The presence of neutral high-mannose oligosaccharides depends largely on phosphorylation. This is indicated by the paucity of high-mannose oligosaccharides in lysosomal enzymes from I-cell fibroblasts (64, 68, 69).

MANNOSE 6-PHOSPHATE–DEPENDENT TRANSPORT IN VIVO

It appears that mannose 6-phosphate residues participate in the transport of lysosomal enzymes in two physiologically significant ways. First, in certain cells, they are indispensable for targeting endogenous lysosomal enzymes to lysosomes. In 1972, Hickman & Neufeld (2) reported the important finding that lysosomal enzymes in I-cell fibroblasts lack the recognition marker required for receptor-mediated endocytosis. They proposed that the inability of I-cell fibroblasts to equip their enzymes with this recognition marker caused the intracellular deficiency and extracellular accumulation of many lysosomal enzymes that is observed in I-cell fibroblasts. The subsequent demonstration that the inability of I-cell fibroblasts to synthesize mannose 6-phosphate residues in lysosomal enzymes (83, 114) was the result of a deficiency in N-acetylglucosamine 1-phosphotransferase (94, 110) further identified mannose 6-phosphate residues as indispensable for transport of lysosomal enzymes in fibroblasts. The general importance of these residues is indicated by the excessive levels of lysosomal enzymes in all body fluids of I-cell patients and the morphological alterations in many tissues that result from deficiency in intracellular lysosomal enzymes (124).

The second function of mannose 6-phosphate residues is to mediate intercellular exchange of lysosomal enzymes. This function is exemplified in female carriers of Hunter disease. Hunter disease is an X-linked lysosomal disorder that is characterized by a deficiency in iduronate sulfatase. The female carriers have two populations of cells, one of which is deficient in iduronate sulfatase due to inactivation of the X-chromosome carrying the nonaffected gene. Yet, the phenotype of carriers as well as the morphology of their tissues is normal (125). This is explained by transfer of iduronate sulfatase from the normal to the deficient cell population. This phenomenon of cross-correction has been studied by Neufeld and coworkers (1) in fibroblasts and shown to depend on the mannose 6-phosphate–containing recognition marker mentioned above.

The functions of the phosphorylated recognition markers depend on their interaction with specific mannose 6-phosphate receptors. Therefore, the two functions also apply to at least one of the receptors, which will be discussed in the following section.

MANNOSE 6-PHOSPHATE–SPECIFIC RECEPTORS

Two Distinct Receptors

Two mannose 6-phosphate–specific receptors are known. These receptors can be differentiated by dependence on divalent cations. The cation-independent receptor has been extensively characterized with regard to structure, biosynthesis, turnover, subcellular location, and function and will be discussed in detail below and be referred to as MPR. The existence of a second cation-dependent receptor is suggested by a recent report by Hoflack & Kornfeld (126). These authors found that membranes from mouse P388D$_1$ macrophage–like cells bind lysosomal enzymes in a saturable and mannose-6 phosphate–dependent manner. More importantly, binding is strictly dependent on divalent cations. P388D$_1$ macrophages do not synthesize detectable amounts of the cation-independent receptor (127). In these cells, the transport of lysosomal enzymes to lysosomes is probably mediated by the cation-dependent receptors.

The Cation-Independent Receptor (MPR)

ISOLATION OF MPR   MPR is a membrane protein that can be solubilized in the presence of nonionic detergents without loss of its binding properties. The first purification of MPR was achieved by Sahagian et al (128). The authors isolated the MPR from bovine liver using the lysosomal enzyme β-galactosidase immobilized to Sepharose 4B. Because of easier accessibility, yeast phosphomannan and secretions from *Dictyostelium discoideum* became the preferred materials for preparation of affinity matrixes (73, 129, 130). The receptors have been isolated from a variety of tissues and cells including bovine liver, human fibroblasts, rat hepatocytes, Chinese hamster ovary cells, and rat chondrosarcoma (128, 131). In several cells of human origin, including normal fibroblasts, HepG2, U937, and HL-60 cells, the receptor constitutes 0.1–0.5% of total membrane protein (our unpublished results).

STRUCTURAL FEATURES   In SDS-polyacrylamide gel electrophoresis, MPR from different sources behaves as a glycoprotein with an M_r of about 215,000. It contains fucose and terminal sialic acid residues (127, 131), phosphoserine groups, and intrachain disulfide linkages (132). The latter are indicated by a decrease in electrophoretic mobility after reduction. The receptors span the membrane. A small portion of about 10 kd protrudes at the cytosolic side of membrane and harbors the C-terminus of the receptor. The greater portion of the receptor polypeptide is exposed at the luminal side of organelles and the outer side of the plasma membrane. This portion contains the mannose 6-phosphate binding site(s) (133, 134).

BIOSYNTHESIS AND TURNOVER OF MPR   The receptors undergo a series of posttranslational modifications, some of which cause minor changes in the mobility in SDS polyacrylamide electrophoresis (127, 132). The high-mannose oligosaccharides in the receptor are converted predominantly to complex-type structures. The processing of the oligosaccharides is remarkably slow and not complete even two to three hours after synthesis (127, 132). Phosphorylation of the receptor on serine residues is restricted to the pool of mature receptors. The phosphate has a half-life about seven times shorter than the protein in MPR (132). Additional posttranslational modifications are indicated by sequential acquistion of immunoreactivity and of binding activity in Chinese hamster ovary cells (132). In fibroblasts and adherent Chinese hamster ovary cells, the half-life of the protein moiety ranges from 10 to 29 hours (132, 135, 136). The turnover is not affected if saturating amounts of exogenous ligands are added to cells or if endogeneous ligands are lacking (135). Exposure to NH_4Cl or leupeptin, two inhibitors of lysosomal proteolysis, also has no significant effect (135). Sahagian (137) found under specific conditions that the medium of Chinese hamster ovary cells contained immunoreactive fragments, supposedly representing degradation products of MPR. The author proposed a nonlysosomal mechanism for receptor degradation that depends on binding of secreted lysosomal enzymes to MPR.

BINDING SPECIFICITY   Isolated receptors incorporated into liposomes (130, 131, 138) or immobilized on various matrixes (73, 130, 131, 139) bind lysosomal enzymes in a saturable and mannose 6-phosphate–dependent manner. The binding characteristics of the purified receptors (73, 128, 130, 139) are comparable to receptors in isolated membranes (140–143) or at the cell surface (144–146). The lysosomal enzymes bind with K_D's in the nanomolar range. The binding is competitively inhibited by mannose 6-phosphate ($K_i = 0.05$–2 mM) and lysosomal enzymes ($K_i = 2$–40 nM). It does not depend on divalent cations and drops precipitously between pH 6 and pH 5.

By studying the binding of oligosaccharides to immobilized receptors, Fischer et al (130) and Varki & Kornfeld (73) could show that phosphomonoester groups in oligosaccharides are essential for binding. Oligosaccharides with one phosphomonoester residue were retarded on receptor-substituted columns and oligosaccharides with two phosphates in monoester linkage bound firmly to the column. Neutral oligosaccharides or oligosaccharides with one or two phosphates in diester linkage to N-acetylglucosamine did not bind. The affinity to immobilized receptors compared well with the ability of oligosaccharides to become internalized (147, 148). The K_{uptake} of oligosaccharides with one or two phosphates in monoester linkage were 3.2×10^{-7} and 3.9×10^{-8} M (148). Neutral oligosaccharides and those with one phosphate in diester linkage were not internalized. The 10-fold lower K_{uptake} of oligosaccharides with two phosphates in monoester linkages argues for an extended binding site in the receptor. Earlier observations of Sly and coworkers (149, 150), who compared the uptake and inhibitor activity of phosphomannan fragments containing multiple phosphomonoesters to that of pentamannosyl monophosphate, suggested a cooperativity in binding. The polyvalent fragment was a potent inhibitor and was internalized, whereas the monovalent pentasaccharide was not internalized and was no more inhibitory than mannose 6-phosphate. Recognition of carbohydrates neighboring the mannose 6-phosphate residues contributes to binding. This is evident from the following three observations. First, the K_{uptake} of monophosphorylated oligosaccharides is 100-fold lower than the K_i of mannose 6-phosphate (148). Second, presence of outer mannose residues accessible to jack bean α-mannosidase can lower the affinity of oligosaccharides with phosphomonoester groups for the receptor. Third, the positions of phosphomonoester groups on the oligosaccharides influence the affinity for the receptors (73). Gabel et al (151) analyzed the oligosaccharides in endogenous glycoproteins bound to the receptors in mouse lymphoma cells. The receptor-bound material was highly enriched in oligosaccharides bearing one or two phosphomonoesters, whereas phosphorylated oligosaccharides in nonreceptor-bound material were enriched in phosphodiesters.

Lysosomal enzymes from *Dictyostelium discoideum* have been widely used as tools to measure uptake via the MPR and to isolate the receptors. Their binding properties (145, 146, 152) are indistinguishable from those of lysosomal enzymes. Yet, the recognition of the slime mold enzymes by the receptors is resistant to exhaustive treatment with alkaline phosphatase (153). Gabel et al (154) recently identified these alkaline phosphatase–resistant groups as methylphosphomannose residues. Thus, mannose 6-phosphate residues covered with methyl groups are recognized by the MPR, whereas those covered with *N*-acetylglucosamine, as occurs in lysosomal enzymes of mammalian origin, do not bind to the MPR.

DISTRIBUTION OF MPR IN CELLULAR MEMBRANES   Early studies on binding of lysosomal enzymes suggested that most of the receptor is localized to intracellular membranes (143). In subcellular fractionation studies, the receptor was found in the plasma membrane (128), in Golgi membranes (155), in coated vesicles (134, 156) and in endosomes (157), but not in lysosomes (132, 136). In various immunocytochemical investigations (158–161), the consensus finding is that MPR is present in the plasma membrane (including coated pits), in coated vesicles present in the vicinity of the plasma membrane, in organelles with tubular extensions that have been defined as CURL (162) [the vesicular elements of CURL are equivalent to endosomes (163) or receptosomes (164)], and in the Golgi area including the nearby coated vesicles, although in certain tissues it may be difficult to detect MPR in the plasma membrane (161). However, the findings with regard to the distribution of MPR in intracellular membranes are strikingly divergent. Thus, in rat hepatocytes and some other cells Brown & Farquhar (159) find MPR to be localized specifically to one or two cisternae in *cis*-Golgi, whereas Geuze and coworkers (160) observe MPR well distributed through all of the Golgi complex and the *trans*-Golgi reticulum in rat hepatocytes. In HepG2 cells, more MPR is present in the *trans*-Golgi reticulum than in the Golgi itself (161). The binding of the specific antibodies used in these studies was visualized with an immunoperoxidase-conjugated second antibody (159) or with protein A–coated gold particles (160, 161). In these studies, little MPR was found in the endoplasmic reticulum. In contrast, in Chinese hamster ovary cells endoplasmic reticulum was reported to be rich in this receptor (158). Finally, extensive (44, 159) or limited (158, 160) labeling of MPR was observed in the membranes of lysosomal organelles. While it is presently impossible to reconcile these conflicting findings, it is possible that the use of different visualization techniques, the application of different criteria for the identification of the organelles, and the study of different cell types all may have contributed to the apparent differences in these observations.

ITINERARIES OF LYSOSOMAL ENZYMES AND OF THE CATION-INDEPENDENT RECEPTOR (MPR)

Although the role of MPR in targeting endogenous lysosomal enzyme is firmly established, no conclusive answers are available about where receptors and ligands combine (binding site), where the receptor-ligand complexes are segregated from the secretory pathway, where the ligands dissociate from the receptors, where receptors and ligands are separated, and along which pathway receptors recycle.

BINDING SITE   Since the binding property of a lysosomal enzyme depends on formation of, at minimum, one uncovered mannose 6-phosphate residue, its interaction with MPR would be possible at or *trans* to the location of the uncovering enzyme. Localization of MPR to probably all membranes of the Golgi complex and of the uncovering α-*N*-acetylglucosaminidase to *mid*-Golgi suggests that *mid*-Golgi is the most proximal site where the binding of lysosomal enzymes can occur.

SEGREGATION SITE   The binding to receptors is thought to be followed by the selective packaging into vesicles. The purpose of this packaging is to separate newly synthesized lysosomal enzymes from other products that traverse the Golgi complex and are destined for secretion and certain cellular membranes. The most proximal site where segregation could occur is the binding site itself. The restriction of MPR to the *cis*-Golgi cisternae and neighboring membranes that Brown & Farquhar (159) observed led them to propose that the lysosomal enzyme-receptor complexes are segregrated in the *cis*-Golgi into clathrin-coated vesicles. This hypothesis is difficult to reconcile with the presence of sialylated, phosphorylated hybrid and complex type oligosaccharides in lysosomal enzymes, which are bound to the receptors (151), en route to lysosomes (67), or present in lysosomes (64, 165). Considering the *trans*-Golgi localization of the terminal glycosyltransferases that add galactose and sialic acid residues to hybrid and complex type oligosaccharides (41, 56), lysosomal enzymes are expected to pass the *trans*-Golgi cisternae. As processing appears more likely to occur in nonreceptor than in receptor-bound form, a significant fraction of lysosomal enzyme is likely not to bind before reaching *trans*-Golgi cisternae.

The prevalence of neutral high-mannose oligosaccharides in lysosomal enzymes (see above) also has been utilized as an argument to propose that segregation takes place predominantly in *cis*-Golgi. Lysosomal enzyme precursors secreted in the presence of NH_4Cl retain their neutral high-mannose oligosaccharides (72). These findings, together with the preferred occurrence of complex oligosaccharides in lysosomal enzymes from I-cell fibroblasts (see above), indicate that the exemption of neutral high-mannose oligosaccharides from processing to complex structures is due to the presence of phosphorylated oligosaccharides rather than to binding to MPR or to a bypass of *trans*-Golgi cisternae.

In double-labeling experiments with immunogold, Geuze and coworkers (160, 161) colocalized MPR, lysosomal enzymes, and albumin to the Golgi cisternae and the *trans*-Golgi reticulum. In HepG2 cells, albumin was colocalized with MPR and lysosomal enzymes even at the level of packaging in the coated buds of the *trans*-Golgi reticulum (161).

Theoretically, the most distal site at which segregation of lysosomal enzymes could occur is the plasma membrane. Available data on lysosomal enzyme transport neither contradict nor favor a pathway via the plasma membrane. Such a pathway has been suggested to involve transport of MPR-ligand complexes along the secretory route to the plasma membrane and internalization of the complexes along the pathway of receptor-mediated endocytosis. The pool of lysosomal enzymes bound to MPR detectable by immunofluorescence or immunogold at the plasma membrane of fibroblasts (166) or HepG2 cells (161) may represent the newly synthesized endogenous lysosomal enzymes moving via the plasma membrane. If transport via the plasma membrane constitutes a major pathway, the enzymes at the plasma membrane must be protected from displacement by mannose 6-phosphate (167). However, the lysosomal enzymes at the plasma membrane also may represent enzymes that are reinternalized following secretion. Although this secretion-recapture pathway (168) is of importance in certain conditions such as the Hunter-carrier, it appears to constitute only a minor pathway for delivery of lysosomal enzymes to lysosomes. If cells are grown in the presence of mannose 6-phosphate and related compounds, which block MPR-dependent uptake of lysosomal enzymes, little if any accumulation of lysosomal enzymes in the medium and no intracellular reduction were found (167, 169–171). A portion of the lysosomal enzymes present at the cell surface may function in the degradation of the components of the extracellular matrix (172). Further studies may well show that the location of the segregation site depends on the cell type and that within a single cell segregation may be a smooth process occurring along an extended section of the secretory pathway.

COATED MEMBRANES AND VESICLES   Ultrastructural studies localized MPR and lysosomal enzymes to coated membranes and coated vesicles neighboring the Golgi complex or *trans*-Golgi reticulum and at the plasma membrane. Willingham et al (173) identified the latter as being part of the endocytosis pathway for exogenous lysosomal enzymes. Highly purified preparations of coated vesicles from brain, liver, and human placenta contain lysosomal enzymes. In such preparations, the enzymes are enriched in precursor forms and at least in part bound to MPR (156, 174). Coated membranes immunoselected from fibroblasts, and analyzed for the lysosomal enzyme cathepsin D, contained exclusively the precursor forms of this enzyme (175). It was estimated that during transport the precursors of cathepsin D are associated with coated membranes only for a short period not exceeding a few minutes. This agrees with the view that coated vesicles rapidly lose their coats. The observation that precursors of cathepsin D contain oligosaccharides of the complex type (our unpublished results) supports the notion that coated membranes are involved in lysosomal transport distal to *trans*-Golgi.

DISSOCIATION OF MPR-LYSOSOMAL ENZYME COMPLEXES   Like many other receptors involved in transport of ligands to lysosomal or prelysosomal structures, MPR participates in many rounds of transport. The amount of internalized ligand is far in excess of the number of MPR binding sites when the protein synthesis is blocked by cycloheximide (141, 144). Reutilization implies dissociation and separation of receptors and ligands and transport of unoccupied receptors back to the binding site. Receptor-ligand complexes dissociate below pH 5.7 (128, 141). In vivo the same mechanism appears to be utilized for dissociation of receptor-ligand complexes. This is illustrated by the effects of conditions that prevent acidification of intracellular organelles. (Acidification of organelles is reviewed by Helenius & Mellman in this volume.) When dissociation is inhibited, cells should rapidly become depleted in unoccupied receptors. As a consequence, receptor-dependent functions (i.e. sorting of endogenous and endocytosis of exogenous lysosomal enzymes) should be blocked. This is exactly what is observed in cells exposed to weak bases (83, 84, 116, 141, 169, 176, 177) and the ionophore monensin (177–179). Weak bases and ionophores raise the pH of acidic organelles above 6 (180). Among the acidic subcellular compartments (181–184), lysosomes, CURL, and coated vesicles are candidates for the dissociation site. In the laboratories of Robbins (185, 186) and Sly (187, 188), several mutants of Chinese hamster ovary cells were characterized that are defective in ATP-dependent acidification of endosomes but not of lysosomes. Such mutations result in an enhancement in the secretion of lysosomal enzymes and a decrease in the endocytosis. These observations define the dissociation of receptor-enzyme complexes clearly as a prelysosomal event.

SEPARATION OF MPR AND LYSOSOMAL ENZYMES   Reutilization of receptors implies that following dissociation, receptors and ligands are separated. Recent morphologic studies by Geuze and coworkers (43, 160, 161) shed some light on where receptors might be separated from lysosomal enzymes. A membrane reticulum containing tubular and vesicular structures extending in hepatocytes from the peripheral cytoplasm down to the *trans*-Golgi area was described by Geuze et al (162) as the compartment where, following endocytosis, the asialoglycoprotein receptors are separated from their ligands. The receptors concentrate in the tubular membranes. Free clathrin is frequently observed adjacent to these tubules, suggesting that coated structures mediate retrieval of receptors from the tubules. The ligands concentrate in the lumina of the smooth vesicular structures, which may develop into or fuse with secondary lysosomes. In rat hepatocytes, MPR colocalizes in CURL with the asialoglycoprotein receptor (43). In HepG2 cells, MPR accumulates in the tubules of CURL, while most of the lysosomal enzymes are found in the lumina of the smooth vesicular structures (161). Thus, CURL is a likely candidate as the site where the receptors are physically separated from lysosomal enzymes. The absence or paucity of receptors in lysosome-enriched fractions from Chinese hamster ovary cells (132) and fibroblasts (133) supports the view that separation of MPR and its ligands occurs prelysosomally.

CYCLING OF MPR   The pathway(s) along which receptors are transported for reutilization are unknown. Models for transport of MPR have to take into account that the receptors functioning in transport of exogenous and endogenous lysosomal enzymes share at least one organelle, in which they mix. This is indicated by the inhibition of segregation and of endocytosis of lysosomal enzymes (189, 190), and the rapid tagging of all cellular receptors with antibodies (133, 137, 190) in cells exposed to antireceptor antibodies. CURL is the most likely candidate for the organelle where receptors coming from the Golgi and the plasma membrane mix. Separate pathways may exist for their return to the Golgi and the plasma membrane. We favor a simple view, in which all receptors are transported from the CURL-tubules to the Golgi complex, where they mix with newly synthesized receptors. Unoccupied MPR following the secretory route would replenish the pool of plasma membrane receptors. As discussed above, MPR occupied with endogneous ligands may follow the same pathway or diverge somewhere before reaching the plasma membrane. In the latter case, MPR ferrying endogenous ligands would gain access to CURL along a route different from the pathway of receptor-mediated endocytosis, and mix in CURL with MPR coming from the cell surface.

Conditions perturbing the transport of MPR to or from CURL, as well as those inhibiting the dissociation of ligands, may be expected to change the subcellular distribution of MPR. Gonzalez-Noriega et al (141) observed a depletion of MPR binding sites at the cell surface with use of chloroquine-treated fibroblasts, and proposed this depletion resulted from inhibition of receptor recycling to the cell surface. When measured as antigenic sites, the number of cell surface receptors is also decreased. However, the uptake of iodinated anti-MPR antibodies is largely resistant to chloroquine and related weak bases (our unpublished results). This indicates that chloroquine alters the steady-state concentration in membranes, but allows for recycling of receptors. Since the ligands do not dissociate in cells exposed to weak bases, ligand-occupied and hence functionally inactive receptors may recycle in such cells. A significant portion of lysosomal enzymes internalized by Chinese hamster ovary cells is returned to the cell surface, indicating that ligand occupation per se does not prevent the backflow of receptors (4).

Farquhar and coworkers (44, 191, 192) proposed that the movements of MPR are induced upon its occupation with or dissociation from ligand. In cells that do not synthesize ligands for MPR (I-cell fibroblasts or hepatocytes exposed to tunicamycin), receptors accumulated in the cis-Golgi cisternae and in adjacent coated vesicles, whereas they were steadily removed from endosomes/lysosomes. An opposite distribution was observed in chloroquine-treated hepatocytes, where ligands cannot be separated from receptors. The authors proposed that MPR shuttles nonconstitutively with movement from the cis-Golgi to endosomes/lysosomes being triggered by occupation with ligand, and movement from endosomes/lysosomes to the cis-Golgi area by dissociation of ligands. At present, this attractive model is difficult to accommodate with the observation that cycloheximide (which inhibits the synthesis of endogenous ligands) is without apparent effect on receptor distribution in hepatocytes (193), and on uptake of antireceptor antibodies (190). Furthermore, cycloheximide does not affect the exchange of receptors between cell surfaces and intracelluar membranes, and the kinetics of receptor exchange between intracellular membranes and cell surfaces are similar in I-cell and control fibroblasts and are not significantly affected by weak bases and monensin (our unpublished results). Thus, the ultrastructural and biochemical analyses indicate that dissipation of pH gradients within cells (e.g. by chloroquine or monensin) may alter the relative rates of the movement of the receptor to and from a compartment such that the overall cycling rate is not appreciably changed.

DELIVERY OF LYSOSOMAL ENZYMES INTO DENSE LYSOSOMES   The route of the transport of lysosomal enzymes from CURL to lysosomes is not known.

It may involve packaging into specific transport vesicles, which fuse with existing lysosomes. A more likely possibility is a gradual transition of (light) CURL elements to (dense) lysosomes (163, 194). Vesicular elements of the CURL may lose their tubular, receptor-dense connections and detach from the continuous network of CURL. These vesicles, which may be considered as new (light) lysosomes, gradually acquire the high density characteristic of lysosomes. The traditional view that secondary lysosomes are the organelles in which lysosomal enzymes first encounter their substrates and initiate degradation has to be modified. Ligands that are internalized by receptor-mediated endocytosis are separated from their receptors in CURL (162, 193) and mix in the lumen of vesicular CURL elements with lysosomal enzymes. Since CURL is an acidic organelle (182), degradation may well be initiated therein. Some recent experiments designed to trace the compartments in which endocytosed ligands are subject to proteolysis indicate that the degradation is initiated in light organelles that cofractionate with endosomal markers [(195), and personal communication from S. Diment and P. Stahl]. At present, these organelles cannot unequivocally be identified with CURL (vesicles still connected with tubules and subject to membrane recycling) or with light organelles defined above as new lysosomes. New lysosomes should share many properties with CURL, such as composition of the fluid content, but the former should be distinguished by the lack of membrane components, which are subject to recycling. Since internalized lysosomal enzyme can induce degradation of material stored in secondary lysosomes (196), the possibility has to be considered that new lysosomes can fuse with existing secondary lysosomes.

Lysosomal enzymes are converted by a series of proteolytic steps into mature forms, which are typically represented by the forms isolated from tissues. For cathepsin D, the conversion is initiated in light organelles and completed in dense lysosomes (67). In the light organelles the precursor is converted into an intermediate independent of ATP-driven acidification (our unpublished results). Generation of mature forms from the intermediate depends on ATP-driven acidification and thiolproteinases and is localized in lysosomes (197). The processing is likely to be initiated in light organelles discontinuous with CURL, such as new (light) lysosomes according to the definition given above. This view is based on the absence of proteolytically processed forms of cathepsin D in the medium of cultured fibroblasts (169), in spite of the exocytosis of fluid from CURL (198).

For details of proteolytic maturation of lysosomal enzymes and their fate in lysosomes the reader is referred to recent reviews (37, 199).

MODEL FOR MPR-DEPENDENT TRANSPORT OF LYSOSOMAL ENZYMES   In this section we propose a speculative model (Figure 3), which attempts to reduce the number of signals required for transport to a minimum. Nascent lysosomal enzymes are sorted into the endoplasmic reticulum with the aid of amino-terminal signal peptides. A yet undefined structure common to all lysosomal enzymes serves as a signal for the phosphorylation by N-acetylglucosaminyl-1-phosphotransferase. This process, in turn, initiates the generation of mannose 6-phosphate residues, which serve as signals for the MPR. Binding to MPR somewhere in Golgi complex removes lysosomal enzymes from the fluid content that is destined for secretion. By utilizing the secretory route, the complexes can reach the plasma membrane, where a signal residing in MPR allows for collection in coated pits. Along the pathway of receptor-mediated endocytosis, the complexes are ferried to CURL, where receptors and ligands are separated. The same signal allowing for concentration in coated pits at the plasma membrane may be utilized in CURL for collecting the receptors into vesicles that move to the Golgi complex.

The proposed model might reflect a primitive pathway, from which more sophisticated pathways evolved. The latter may allow for a shorter path between the binding site and the segregation site and avoid transport via the plasma membrane. Necessarily, the evolution of these pathways would require the formation of additional signals. Functionally analogous signals have been postulated to direct transport of Golgi-derived products to distinct destinations, such as different domains of the plasma membrane (200, 201) or elements of either a constitutive or a regulated secretory pathway (202). Such signals may trigger packaging of receptor-ligand complexes into specific vesicles that mature into CURL or guide the return of receptors from CURL to either Golgi or plasma membrane. In Figure 3 such shorter pathways are indicated by dashed lines.

MANNOSE 6-PHOSPHATE–INDEPENDENT TRANSPORT

Several lines of evidence indicate that lysosomal enzymes can be ferried in a manner that is independent of mannose 6-phosphate–specific receptors. In I-cell fibroblasts, variable residual amounts of lysosomal enzymes are found within the lysosomes (69). The fraction of newly synthesized enzyme targeted to lysosomes may be as high as 20–50% for α-glucosidase and cathepsin D (87,

Figure 3   A model of the transport of lyso-somal enzymes. The thin lines refer to path-ways of the enzymes and the bold lines to those of MPR. For further explanation see text.

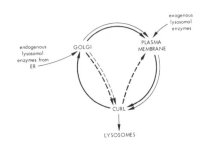

169), whereas for other enzymes, such as β-hexosaminidase, α-L-iduronidase, and arylsulfatase A it is below the limit of detection (84, 87, 169, 203). The activity of acid phosphatase is normal in I-cell fibroblasts. This has to be attributed to secondary effects, since only one third of the newly synthesized acid phosphatase polypeptides are targeted to lysosomes (87). The normal activities in I-cell fibroblasts of β-glucocerebrosidase and acetyl CoA:α-glucosaminide N-acetyltransferase, two integral membrane enzymes, suggest that membrane proteins find their way independent of mannose 6-phosphate residues. This agrees with the absence of mannose 6-phosphate residues in membrane proteins of the lysosomal membrane (our unpublished results), including β-glucocerebrosidase (204).

In liver, spleen, kidney, and brain from I-cell patients, the activities of many lysosomal enzymes are normal, although these tissues are deficient in N-acetylglucosaminyl 1-phosphotransferase (112, 113). As the residual lysosomal enzymes in I-cell fibroblasts, these enzymes must use mannose 6-phosphate–independent mechanisms for transport into lysosomes. Therefore, we must consider that mannose 6-phosphate–independent mechanisms also contribute to targeting of lysosomal enzymes in normal tissues.

ACKNOWLEDGMENTS

Thanks are due to Dr. J. Conary for critical reading of the manuscript and Mrs. R. Rumpff for typing the manuscript. Our research on lysosomal enzymes is sponsored by the Deutsche Forschungsgemeinschaft and Fonds der Chemischen Industrie.

Literature Cited

1. Neufeld, E. F., Lim, T. W., Shapiro, L. J. 1975. *Ann. Rev. Biochem.* 44:357–76
2. Hickman, S., Neufeld, E. F. 1972. *Biochem. Biophys. Res. Commun.* 49:992–99
3. Creek, K. E., Sly, W. S. 1984. In *Lysosomes in Biology and Pathology*, ed. J. T. Dingle, R. T. Dean, W. Sly, pp. 63–82. Amsterdam Elsevier
4. Sahagian, G. G. 1985. In *Recent Research on Vertebrate Lectins. Advanced Cell Biology Monographs*, ed. B. Parent, K. Olden. New York: van Nostrand In press
5. Marchase, R. B., Koro, L. A., Hiller, A. M. 1984. In *The Receptors*, ed. P. M. Conn, pp. 261–311. Orlando: Academic
6. Walter, P., Gilmore, R., Blobel, G. 1984. *Cell* 38:5–8
7. Walter, P., Blobel, G. 1982. *Nature* 299:691–98
8. Gilmore, R., Blobel, G., Walter, P. 1982. *J. Cell Biol.* 95:463–69
9. Meyer, D. I., Krause, E., Dobberstein, B. 1982. *Nature* 297:647–50
10. Erickson, A. H., Blobel, G. 1979. *J. Biol. Chem.* 254:11771–74
11. Rosenfeld, M. G., Kreibich, G., Popov, D., Kato, K., Sabatini, D. D. 1982. *J. Cell Biol.* 93:135–43
12. Proia, R., Neufeld, E. F. 1982. *Proc. Natl. Acad. Sci. USA* 79:6360–64
13. Erickson, A. H., Conner, G. E., Blobel, G. 1981. *J. Biol. Chem.* 256:11224–31
14. Erickson, A. H., Walter, P., Blobel, G. 1983. *Biochem. Biophys. Res. Commun.* 115:275–80
15. Kiely, M., McKnight, G. S., Schimke, R. 1976. *J. Biol. Chem.* 251:5490–95
16. Bergman, L. W., Kuehl, W. M. 1978. *Biochemistry* 17:5174–80
17. Rothman, J. E., Katz, F. N., Lodish, H. F. 1978. *Cell* 15:1447–54
18. Kornfeld, R., Kornfeld, S. 1985. *Ann. Rev. Biochem.* 54:631–64
19. Atkinson, P. H., Lee, J. T. 1984. *J. Cell Biol.* 98:2245–49
20. Hubbard, S. C., Robbins, P. W. 1979. *J. Biol. Chem.* 254:4568–76

21. Lemansky, P., Gieselmann, V., Hasilik, A., von Figura, K. 1984. *J. Biol. Chem.* 259:10129–35
22. Bischoff, J., Kornfeld, R. 1983. *J. Biol. Chem.* 258:7907–10
23. Tabas, I., Kornfeld, S. 1979. *J. Biol. Chem.* 254:11655–63
24. Jamieson, J. D., Palade, G. E. 1968. *J. Cell Biol.* 39:589–603
25. Farquhar, M. G., Palade, G. E. 1981. *J. Cell Biol.* 91:77s–103s
26. Lodish, H. F., Kong, N., Snider, M., Strous, G. J. A. M. 1983. *Nature* 304:80–83
27. Fries, E., Gustafsson, L., Peterson, P. A. 1984. *EMBO J.* 3:147–52
28. Scheele, G., Tartakoff, A. 1985. *J. Biol. Chem.* 260:926–31
29. Gabel, C. A., Kornfeld, S. 1984. *J. Cell Biol.* 99:296–305
30. Sege, K., Rask, L., Peterson, P. A. 1981. *Biochemistry* 20:4523–30
31. Lodish, H. F., Kong, N. 1984. *J. Cell Biol.* 98:1720–29
32. Saunier, B., Kilker, R. D. Jr., Tkacz, J. S., Quaroni, A., Herscovics, A. 1982. *J. Biol. Chem.* 257:14155–61
33. Romero, P. A., Saunier, B., Herscovics, A. 1985. *Biochem. J.* 226:733–40
34. Gross, V., Andus, T., Tran-Thi, T.-A., Schwarz, R. T., Decker, K., Heinrich, P. C. 1983. *J. Biol. Chem.* 258:12203–9
35. Peyrieras, N., Bause, E., Legler, G., Vasilov, R., Claesson, L., et al. 1983. *EMBO J.* 2:823–32
36. Schmidt, J. A., Beug, H., Hayman, M. J. 1985. *EMBO J.* 4:105–12
37. Skudlarek, M. D., Novak, E., Swank, R. T. 1984. See Ref. 3, pp. 17–34
38. Tartakoff, A. M. 1983. *Int. Rev. Cytol.* 85:221–52
39. Griffith, G., Quinn, P., Warren, G. 1983. *J. Cell Biol.* 96:835–50
40. Rothman, J. E., Miller, R. L., Urbani, L. J. 1984. *J. Cell Biol.* 99:260–71
41. Rothman, J. E. 1981. *Science* 213:1212–19
42. Slot, J. W., Geuze, H. J. 1983. *J. Histochem. Cytochem.* 31:1049–56
43. Geuze, H. J., Slot, J. W., Strous, G. J. A. M., Peppard, J., von Figura, K., et al. 1984. *Cell* 37:195–204
44. Brown, W. J., Farquhar, M. G. 1984. *Proc. Natl. Acad. Sci. USA* 81:5135–39
45. Balch, W. E., Glick, B. S., Rothman, J. E. 1984. *Cell* 39:525–36
46. Tulsiani, D. R. P., Hubbard, S. C., Robbins, P. W., Touster, O. 1982. *J. Biol. Chem.* 257:3660–68
47. Dunphy, W. G., Rothman, J. E. 1983. *J. Cell Biol.* 97:270–75
48. Harpaz, N., Schachter, H. 1980. *J. Biol. Chem.* 255:4894–902
49. Oppenheimer, C. L., Hill, R. L. 1981. *J. Biol. Chem.* 256:799–804
50. Dunphy, W. G., Brands, R., Rothman, J. E. 1985. *Cell* 40:463–72
51. Goldberg, D., Kornfeld, S. 1983. *J. Biol. Chem.* 258:3159–65
52. Novikoff, P. M., Tulsiani, D. R. P., Touster, O., Yam, A., Novikoff, A. B. 1983. *Proc. Natl. Acad. Sci. USA* 80:4364–68
53. Narasimhan, S. 1982. *J. Biol. Chem.* 257:10235–42
54. Allen, S. D., Tsai, D., Schachter, H. 1984. *J. Biol. Chem.* 259:6984–90
55. Schachter, H., Narasimhan, S., Gleeson, P., Vella, G. 1983. *Can. J. Biochem. Cell Biol.* 61:1049–66
56. Roth, J., Berger, E. 1982. *J. Cell Biol.* 92:223–29
57. Howard, D. R., Natowicz, M., Baenziger, J. U. 1982. *J. Biol. Chem.* 257:10861–68
58. Takasaki, S., Murray, G. J., Furbish, F. S., Brady, R. O., Barranger, J. A., Kobata, A. 1984. *J. Biol. Chem.* 259:10112–17
59. Erickson, A., Ginns, E. I., Barranger, J. A. 1985. *J. Biol. Chem.* In press
60. Taniguchi, T., Mizuochi, T., Towatari, T., Katunuma, N., Kobata, A. 1985. *J. Biochem.* 97:973–76
61. Takahashi, T., Dehdrani, A. H., Schmidt, P. G., Tang, J. 1984. *J. Biol. Chem.* 259:9874–82
62. Kobata, A. 1979. *Anal. Biochem.* 100:1–14
63. Goldberg, D. E., Kornfeld, S. 1981. *J. Biol. Chem.* 256:13060–67
64. Hasilik, A., von Figura, K. 1981. *Eur. J. Biochem.* 121:125–29
65. Steckel, F., Hasilik, A., von Figura, K. 1983. *J. Biol. Chem.* 258:14322–26
66. Sweeley, C. C. 1983. *Arch. Biochem. Biophys.* 224:186–95
67. Gieselmann, V., Pohlmann, R., Hasilik, A., von Figura, K. 1983. *J. Cell Biol.* 97:1–5
68. Kress, B., Freeze, H., Herd, J. K., Alhadeff, J. A., Miller, A. 1980. *J. Biol Chem.* 255:955–61
69. Miller, A. L., Kress, B. C., Stein, R., Kinnon, C., Kern, H., et al. 1981. *J. Biol. Chem.* 256:9352–62
70. Vladutiu, G. D., Rattazzi, M. C. 1975. *Biochem. Biophys. Res. Commun.* 67:956–64
71. Kress, B. C., Hirani, S., Freeze, H. H., Little, L., Miller, A. L. 1982. *Biochem. J.* 207:421–28
72. Hasilik, A., Klein, U., Waheed, A.,

Strecker, G., von Figura, K. 1980. *Proc. Natl. Acad. Sci. USA* 77:7074–78

73. Varki, A., Kornfeld, S. 1983. *J. Biol. Chem.* 258:2808–18

74. Hickman, S., Shapiro, L. J., Neufeld, E. F. 1974. *Biochem. Biophys. Res. Commun.* 57:55–61

75. Hieber, V., Distler, J., Myerowitz, R., Schmickel, R. D., Jourdian, G. W. 1976. *Biochem. Biophys. Res. Commun.* 73:710–17

76. Kaplan, A., Achord, D. T., Sly, W. S. 1977. *Proc. Natl. Acad. Sci. USA* 74:2026–30

77. Kaplan, A., Fischer, D., Achord, D., Sly, W. S. 1977. *J. Clin. Invest.* 60:1088–93

78. Sando, G. N., Neufeld, E. F. 1977. *Cell* 12:619–27

79. Ullrich, K., Mersmann, G., Weber, E., von Figura, K. 1978. *Biochem. J.* 170:643–50

80. von Figura, K., Klein, U. 1979. *Eur. J. Biochem.* 94:347–54

81. Distler, J., Hieber, V., Sahagian, G., Schmickel, R., Jourdian, G. W. 1979. *Proc. Natl. Acad. Sci. USA* 76:4235–39

82. Natowicz, M. R., Chi, M. M.-Y., Lowry, O. H., Sly, W. S. 1979. *Proc. Natl. Acad. Sci. USA* 76:4322–26

83. Hasilik, A., Neufeld, E. F. 1980. *J. Biol. Chem.* 255:4946–50

84. Myerowitz, R., Neufeld, E. F. 1981. *J. Biol. Chem.* 256:3044–48

85. Waheed, A., Hasilik, A., Cantz, M., von Figura, K. 1982. *Eur. J. Biochem.* 122:119–23

86. Hasilik, A., Pohlmann, R., Olsen, R. L., von Figura, K. 1984. *EMBO J.* 3:267–76

87. Lemansky, P., Gieselmann, V., Hasilik, A., von Figura, K. 1985. *J. Biol. Chem.* 260:9023–30

88. Mainferme, F., Wattiaux, R., von Figura, K. 1985. *Eur. J. Biochem.* 153:211–16

89. Tabas, I., Kornfeld, S. 1980. *J. Biol. Chem.* 255:6633–39

90. Varki, A., Kornfeld, S. 1980. *J. Biol. Chem.* 1980. 255:10847–58

91. Mizuochi, T., Nishimura, J., Kato, K., Kobata, A. 1981. *Arch. Biochem. Biophys.* 209:298–303

92. Natowicz, M., Baenziger, J. U., Sly, W. S. 1982. *J. Biol. Chem.* 257:4412–20

93. Karson, E. M., Ballou, C. E. 1978. *J. Biol. Chem.* 253:6484–92

94. Hasilik, A., Waheed, A., von Figura, K. 1981. *Biochem. Biophys. Res. Commun.* 98:761–67

95. Reitman, M. L., Kornfeld, S. 1981. *J. Biol. Chem.* 256:4275–81

96. Reitman, M. L., Kornfeld, S. 1981. *J. Biol. Chem.* 256:11977–80

97. Waheed, A., Hasilik, A., von Figura, K. 1982. *J. Biol. Chem.* 257:12322–31

98. Lang, L., Kornfeld, S. 1984. *Anal. Biochem.* 140:264–69

99. Lang, L., Reitman, M., Tang, J., Roberts, R. M., Kornfeld, S. 1984. *J. Biol. Chem.* 259:14663–71

100. Varki, A., Reitman, M. L., Kornfeld, S. 1981. *Proc. Natl. Acad. Sci. USA* 78:7773–77

101. Shewale, J. G., Tang, J. 1984. *Proc. Natl. Acad. Sci. USA* 81:3703–7

102. Takio, K., Towatari, T., Katunuma, N., Teller, D. C., Titani, K. 1983. *Proc. Natl. Acad. Sci. USA* 80:3660–70

103. de Wet, J. R., Fukushima, H., Dewji, N. N., Wilcox, E., O'Brien, J. S., Helinski, D. R. 1984. *DNA* 3:437–47

104. Waheed, A., Pohlmann, R., Hasilik, A., von Figura, K. 1981. *J. Biol. Chem.* 256:4150–52

105. Deutscher, S. L., Creek, K. E., Merion, M., Hirschberg, C. B. 1983. *Proc. Natl. Acad. Sci. USA* 80:3938–42

106. Pohlmann, R., Waheed, A., Hasilik, A., von Figura, K. 1982. *J. Biol. Chem.* 257:5723–25

107. Morré, D. J., Minnifield, N., Creek, K. E., Navas, P. 1985. *Special FEBS Meet.*, p. 144 (Abstr.) Algarve, Portugal: printed in Dept. Zool. Univ. Coimbra

108. Gabel, C. A., Kornfeld, S. 1982. *J. Biol. Chem.* 257:10605–12

109. Stahl, J., Robbins, A. R., Krag, S. 1982. *Proc. Natl. Acad. Sci. USA* 79:2296–300

110. Reitman, M. L., Varki, A., Kornfeld, S. 1981. *J. Clin. Invest.* 67:1574–79

111. Varki, A., Reitman, M. L., Vannier, A., Kornfeld, S., Grubb, J. H., Sly, W. S. 1982. *Am. J. Hum. Genet.* 34:717–29

112. Waheed, A., Pohlmann, R., Hasilik, A., von Figura, K., van Elsen, A., Leroy, J. G. 1982. *Biochem. Biophys. Res. Commun.* 105:1052–58

113. Owada, M., Neufeld, E. F. 1982. *Biochem. Biophys. Res. Commun.* 105:814–20

114. Bach, G., Bargal, R., Cantz, M. 1979. *Biochem. Biophys. Res. Commun.* 91:976–81

115. Waheed, A., Hasilik, A., Cantz, M., von Figura, K. 1982. *Hoppe Seyler's Z. Physiol. Chem.* 363:169–78

116. Robey, P., Neufeld, E. F. 1982. *Arch. Biochem. Biophys.* 213:251–57

117. Varki, A., Kornfeld, S. 1980. *J. Biol. Chem.* 255:8398–401

118. Varki, A., Kornfeld, S. 1981. *J. Biol. Chem.* 256:9937–43

119. Waheed, A., Hasilik, A., von Figura, K. 1981. *J. Biol. Chem.* 256:5717–21

120. Varki, A., Sherman, N., Kornfeld, S.

1983. *Arch. Biochem. Biophys.* 222: 145–49
121. Nakao, Y., Kozutsumi, Y., Kawasaki, T., Yamashina, I., van Halbeek, H., Vliegenthart, J. F. G. 1984. *Arch. Biochem. Biophys.* 229:43–54
122. Yamamoto, K., Tsuji, T., Tarutani, O., Osawa, T. 1985. *Biochim. Biophys. Acta* 838:84–91
123. Takahashi, T., Schmidt, P. G., Tang, J. 1983. *J. Biol. Chem.* 258:2819–30
124. Neufeld, E. F., McKusick, V. A. 1983. In *The Metabolic Basis of Inherited Disease*, ed. J. B. Stanbury, J. B. Wyngaarden, D. S. Frederickson, J. L. Goldstein, M. S. Brown, pp. 751–77. New York: McGraw-Hill. 5th ed.
125. McKusick, V. A., Neufeld, E. F. 1983. See Ref. 124, pp. 778–87
126. Hoflack, B., Kornfeld, S. 1985. *Proc. Natl. Acad. Sci. USA* 82:4428–32
127. Goldberg, D. E., Gabel, C. A., Kornfeld, S. 1983. *J. Cell Biol.* 97:1700–6
128. Sahagian, G. G., Distler, J., Jourdian, G. W. 1981. *Proc. Natl. Acad. Sci. USA* 78:4289–93
129. Sahagian, G. G., Distler, J. J., Jourdian, G. W. 1982. *Methods Enzymol.* 83:392–96
130. Fischer, H. D., Creek, K. E., Sly, W. S. 1982. *J. Biol. Chem.* 257:9938–43
131. Steiner, A. W., Rome, L. H. 1982. *Arch. Biochem. Biophys.* 214:681–87
132. Sahagian, G. G., Neufeld, E. F. 1983. *J. Biol. Chem.* 258:7121–28
133. von Figura, K., Gieselmann, V., Hasilik, A. 1985. *Biochem. J.* 225:543–47
134. Sahagian, G. G., Steer, C. J. 1985. *J. Biol. Chem.* 260:9838–42
135. Creek, K. E., Sly, W. S. 1983. *Biochem. J.* 214:353–60
136. von Figura, K., Gieselmann, V., Hasilik, A. 1984. *EMBO J.* 3:1281–86
137. Sahagian, G. G. 1984. *Biol. Cell* 51:207–14
138. Campbell, C. H., Miller, A. C., Rome, L. H. 1983. *Biochem. J.* 214:413–19
139. Talkad, V., Sly, W. S. 1983. *J. Biol. Chem.* 258:7345–51
140. Rome, L. H., Miller, J. 1980. *Biochem. Biophys. Res. Commun.* 92:986–93
141. Gonzalez-Noriega, A., Grubb, J. H., Talkad, V., Sly, W. S. 1980. *J. Cell Biol.* 85:839–52
142. Fischer, H. D., Gonzalez-Noriega, A., Sly, W. S. 1980. *J. Biol. Chem.* 255:5069–74
143. Fischer, H. D., Gonzalez-Noriega, A., Sly, W. S., Morré, J. D. 1980. *J. Biol. Chem.* 255:9608–15
144. Rome, L. H., Weissmann, B., Neufeld, E. F. 1979. *Proc. Natl. Acad. Sci. USA* 76:2331–34
145. Freeze, H. H., Yeh, R. Y., Miller, A. L. 1983. *J. Biol. Chem.* 258:8928–33
146. Shepherd, V. L., Freeze, H. H., Miller, A. L., Stahl, P. D. 1984. *J. Biol. Chem.* 259:2257–61
147. Creek, K. E., Sly, W. S. 1982. *J. Biol. Chem.* 257:9931–37
148. Natowicz, M., Hallett, D. W., Frier, C., Chi, M., Schlesinger, P. H., Baenziger, J. U. 1983. *J. Cell Biol.* 96:915–19
149. Fischer, H. D., Natowicz, M., Sly, W. S., Bretthauer, R. K. 1980. *J. Cell Biol.* 84:77–86
150. Kaplan, A., Fischer, D., Sly, W. S. 1978. *J. Biol. Chem.* 253:647–50
151. Gabel, C. A., Goldberg, D. E., Kornfeld, S. 1982. *J. Cell Biol.* 95:536–42
152. Freeze, H. H., Miller, A. L., Kaplan, A. 1980. *J. Biol. Chem.* 255:11081–84
153. Cladaras, M. H., Graham, T., Kaplan, A. 1983. *Biochem. Biophys. Res. Commun.* 116:541–46
154. Gabel, C. A., Costello, C. E., Reinhold, V. N., Kurz, L., Kornfeld, S. 1984. *J. Biol. Chem.* 259:13762–69
155. Brown, W. J., Farquhar, M. G. 1985. *J. Cell Biol.* 101:138a (Abstr.)
156. Campbell, C. H., Rome, L. H. 1983. *J. Biol. Chem.* 258:13347–52
157. Dickson, R. B., Beguinot, L., Hanover, J. A., Richert, N. D., Willingham, M. C., Pastan, I. 1983. *Proc. Natl. Acad. Sci. USA* 80:5335–39
158. Willingham, M. C., Pastan, I. H., Sahagian, G. G. 1983. *J. Histochem. Cytochem.* 31:1–11
159. Brown, W. J., Farquhar, M. G. 1984. *Cell* 36:295–307
160. Geuze, H. J., Slot, J. W., Strous, G. J. A. M., Hasilik, A., von Figura, K. 1984. *J. Cell Biol.* 98:2047–54
161. Geuze, H. J., Slot, J. W., Strous, G. J. A. M., Hasilik, A., von Figura, K. 1985. *J. Cell Biol.* In press
162. Geuze, H. J., Slot, J. W., Strous, G. J. A. M., Lodish, H. F., Schwartz, A. L. 1983. *Cell* 32:277–87
163. Helenius, A., Mellman, I., Wall, D., Hubbard, A. 1983. *Trends Biochem. Sci.* 8:245–50
164. Pastan, I., Willingham, M. C. 1983. *Trends Biochem. Sci.* 8:250–54
165. Vladutiu, G. D. 1983. *Biochim. Biophys. Acta* 760:363–70
166. von Figura, K., Voss, B. 1979. *Exp. Cell Res.* 121:267–76
167. von Figura, K., Weber, E. 1978. *Biochem. J.* 176:943–50
168. Neufeld, E. F., Sando, G. N., Garvin, A. J., Rome, L. H. 1977. *J. Supramol. Struct.* 6:95–101
169. Hasilik, A., Neufeld, E. F. 1980. *J. Biol. Chem.* 256:4937–45

170. Vladutiu, G. D., Rattazzi, M. 1979. *Clin. Invest.* 63:595–601
171. Sly, W. S., Natowicz, M., Gonzalez-Noriega, A., Grubb, J. H., Fischer, H. D. 1981. In *Lysosomes and Lysosomal Storage Diseases,* ed. J. W. Callahan, J. A. Lowden, pp. 131–43. New-York: Raven
172. Roff, C. F., Wang, J. L. 1983. In *Glycoconjugates Proc. 7th Int. Symp. Glycoconjugates,* ed. M. A. Chester, D. Heinegard, A. Lundblad, S. Svensson, pp. 520–21. Lund:Rahms
173. Willingham, M. C., Pastan, I. H., Sahagian, G. G., Jourdian, G. W., Neufeld, E. F. 1981. *Proc. Natl. Acad. Sci. USA* 78:6967–71
174. Tümmers, S., Zühlsdorf, M., Robenek, H., Hasilik, A., von Figura, K. 1983. *Hoppe-Seyler's Z. Physiol. Chem.* 364:1287–95
175. Schulze-Lohoff, E., Hasilik, A., von Figura, K. 1985. *J. Cell Biol.* 101:824–29
176. Sando, G. N., Titus-Dillon, P., Hall, C. W., Neufeld, E. F. 1979. *Exp. Cell Res.* 119:359–64
177. Merion, M., Sly, W. S. 1983. *J. Cell Biol.* 96:644–50
178. Pohlmann, R., Krüger, S., Hasilik, A., von Figura, K. 1984. *Biochem. J.* 217:649–58
179. Vladutiu, G. D., Rattazzi, M. G. 1980. *Biochem. J.* 192:813–20
180. Ohkuma, S., Poole, B. 1978. *Proc. Natl. Acad. Sci. USA* 75:3327–31
181. Schneider, D. L. 1981. *J. Biol. Chem.* 256:3858–64
182. Tycko, B., Maxfield, F. R. 1982. *Cell* 28:643–51
183. Galloway, C. J., Dean, G. E., Marsh, M., Rudnick, G., Mellman, I. 1983. *Proc. Natl. Acad. Sci. USA* 80:3334–38
184. Forgac, M., Cantley, L., Wiedenmann, B., Altstiel, L., Branton, D. 1983. *Proc. Natl. Acad. Sci. USA* 80:1300–3
185. Robbins, A. R., Peng, S. R., Marshall, J. L. 1983. *J. Cell Biol.* 96:1064–71
186. Robbins, A. R., Oliver, C., Bateman, J. L., Krag, S. S., Galloway, C. J., Mellman, I. 1984. *J. Cell Biol.* 99:1296–1308

187. Merion, M., Schlesinger, P., Brooks, R. M., Moehring, J. M., Moehring, T. J., Sly, W. S. 1983. *Proc. Natl. Acad. Sci. USA* 80:5315–19
188. Sly, W. S., Merion, M., Schlesinger, P., Moehring, J. M., Moehring, T. J. 1983. In *Protein Synthesis,* ed. A. K. Abraham, T. S. Eikhom, I. F. Pryme, pp. 239–51. Clifton, NJ: Humana
189. Creek, K. E., Grubb, J. H., Sly, W. S. 1983. *J. Cell Biol.* 97:253a (Abstr.)
190. Gartung, C., Braulke, T., Hasilik, A., von Figura, K. 1985. *EMBO J.* 4:1725–30
191. Brown, W. J., Constantinescu, E., Farquhar, M. G. 1984. *J. Cell Biol.* 99:320–26
192. Brown, W. J., Farquhar, M. G. 1984. *J. Cell Biol.* 99:231a (Abstr.)
193. Geuze, H. J., Slot, J. W., Strous, G. J., Luzio, J. P., Schwartz, A. L. 1984. *EMBO J.* 3:2677–85
194. Brown, M. S., Anderson, R. G. W., Goldstein, J. L. 1983. *Cell* 32:663–67
195. Berg, T., Ford, T., Kindberg, G., Blomhoff, R., Drevon, C. 1985. *Exp. Cell Res.* 156:570–74
196. Rome, L. H., Garvin, A. J., Alietta, M. M., Neufeld, E. F. 1979. *Cell* 17:143–53
197. Gieselmann, V., Hasilik, A., von Figura, K. 1985. *J. Biol. Chem.* 260:3215–20
198. Besterman, J. M., Airhart, J. A., Woodworth, R. C., Low, R. B. 1981. *J. Cell Biol.* 91:716–27
199. Hasilik, A., von Figura, K. 1984. See Ref. 3, pp. 3–16
200. Rindler, M. J., Ivanov, I. E., Plesken, H., Rodriguez-Boulan, E., Sabatini, D. D. 1984. *J. Cell Biol.* 98:1304–19
201. Matlin, K. S., Simons, K. 1984. *J. Cell Biol.* 98:2131–39
202. Gumbiner, B., Kelly, R. B. 1982. *Cell* 28:51–59
203. Waheed, A., Hasilik, A., von Figura, K. 1982. *Hoppe-Seyler's Z. Physiol. Chem.* 363:425–30
204. Murray, G. J., Jonsson, L. V., Sorrell, S. H., Ginns, E. I., Tager, J. M., et al. 1985. *Fed. Proc.* 44:709 (Abstr.)

Ann. Rev. Biochem. 1986. 55:195–224

TRANSMEMBRANE TRANSPORT OF DIPHTHERIA TOXIN, RELATED TOXINS, AND COLICINS[1]

David M. Neville, Jr. and Thomas H. Hudson

Laboratory of Molecular Biology, National Institute of Mental Health, Bethesda, Maryland 20205

CONTENTS

PERSPECTIVES AND SUMMARY

Toxins of plant and bacterial origin have enormous importance in the medical and veterinary sciences. These protein toxins, acting in highly specific ways to disrupt crucial biochemical pathways, continue to provide biochemists with

---

[1]The US Government has the right to retain a nonexclusive royalty-free license in and to any copyright covering this paper.

novel insights regarding the pathways themselves. There are many known toxins and certainly more to be discovered. Among these, there is diversity in host cell susceptibility, biochemical mechanisms of toxin action and, more recently appreciated, mechanisms by which the toxins reach the cellular compartments wherein their toxic biochemical actions occur. The last source of toxin diversity accentuates the foremost feature common to all: the toxins must be inserted through or transported across the host cell's plasma membrane (processes termed translocation) in order to effect their toxic actions. This review focuses on the membrane translocation event.

The greatest amount of detailed knowledge relating to the translocation event comes from the study of diphtheria toxin (DT) action on eukaryotic cells and bacteriocin action on prokaryotes. These two toxins are discussed separately with their many parallels emphasized. Using these systems as models, data from other toxin systems are introduced.

Structure/function relationships are mapped for several toxins using mutant or chemically modified toxins. In general these analyses demonstrate specific protein domains which are responsible for the various functions needed for intoxication, i.e. binding, translocation, and biochemical action. Structure similarities within these domains are examined between toxins with the eye to determining critical requirements for functioning.

The kinetics of cellular inactivation by several toxins are presented in detail. It is in this respect that the similarities between toxins are most striking. After the addition of toxin a lag period is observed in which no cellular inactivation occurs. The length of the lag is inversely dependent on the toxin concentration, and is followed by cellular inactivation which exhibits first order kinetics. Evidence is presented for DT and bacteriocins showing that the first order rate is the result of single-hit inactivation kinetics. The significance of this model for the study of toxin translocation is discussed in detail.

In addition to providing details concerning the molecular pathogenesis of a variety of diseases, the study of toxin translocation provides insights in two major areas of research in eukaryotes. A great deal of evidence has accumulated indicating that protein toxins utilize receptor-mediated endocytotic pathways to gain entrance to the cell prior to the translocation process. Toxins, then, are being used to study the uptake and processing of proteins by the vesicular systems of the cell. Finally, attempts have been made to couple the remarkable potency of the protein toxins with specific ligands (such as monoclonal antibodies) in order to direct the toxin to specific, targeted cells. One drawback to the use of these hybrid toxins (immunotoxins) has been the relative inefficiency of their translocation processes. By better understanding the events involved in the efficient translocation of intact toxins, more efficient immunotoxins may be designed.

DIPHTHERIA TOXIN

Overview

Diphtheria toxin is a 58,342-dalton protein secreted by *Corynebacterium diphtheriae* infected with the β *tox* phage. The toxin is the agent responsible for the devastating morbidity and mortality associated with diphtheria toxin epidemics down through the ages and the inactivated toxin is the immunogenic active agent responsible for protection by immunization. We consider diphtheria toxin to be the prototype of the two-chain or multi-domain bacterial and plant toxins in which one chain, the B chain, binds to the external plasma membrane receptors and the other chain, the A chain, is an enzyme whose substrate lies within the cell cytosol. These toxins operate by a minimum of three sequential steps: 1. binding to plasma membrane cell surface receptors; 2. transport to the cytosol compartment; and 3. enzymatic alteration of the intracellular substrate leading to a change in function of the cell (1). We will consider diphtheria toxin as a prototype for these toxins because it is the most extensively studied toxin. This is largely due to the interest of A. M. Pappenheimer, Jr. and his colleagues in elucidating the mechanisms of diphtheria toxin action, an interest which has already spanned over three decades of productive research.

Several recent reviews of diphtheria toxin (2), ADP-ribosylating toxins (3), and toxins in general (4, 5) are available. The transport steps remain a puzzle and are the focus of this review. How does a cell transport a high-molecular-weight, highly charged, hydrophilic protein across the lipid bilayer to the cytosol compartment? This same question is being asked in other related systems such as the cotranslational transport of newly synthesized proteins across the endoplasmic reticulum membrane on a pathway that leads to external secretion (6), the transport of proteins synthesized in the cytosol that are destined for the interior of the mitochondrion (7), and entry of a wide variety of viruses into eukaryotic cells (8). Receptor-mediated entry of bacteriocins into bacteria (9) and phage nucleic acids into bacteria (9) are other systems in which macromolecular transport across membranes has been investigated.

Molecular Genetics

Diphtheria toxin is secreted by *Corynebacterium diphtheriae* carrying the *tox* structural gene, which resides in a group of closely related corynebacterio-phages, β *tox*[+], γ *tox*[−], and ω *tox*[+]. These phages are lysogenic; and nontoxigenic diphtheria strains can be converted to toxigenic strains by lysogenization with *tox*[+] phages (1). Toxin production occurs either during lytic replication or as a lysogen replication. Toxin production in wild-type strains only occurs under limiting iron conditions (10). The control of synthesis by iron appears to be at the level of transcription and involves bacterial host factors. The mechanism of *tox* regulation by iron is not completely understood; however, the simplest model appears to be that of negative regulation at the toxin promoter (10, 11). The β phage chromosome has been mapped and the map resembles that of the coliphage lambda; the genes involved in phage head formation and assembly are in one cluster and those involved in tail formation in another (12). A variety of mutant toxins exhibiting altered binding, transport, and enzymatic activities have been isolated and have provided considerable insight into the interrelationships between toxin structure and function. These are discussed in the following section. Because *Corynebacteria* are not obviously susceptible to conjugation, transduction, transformation, or transfection, the utility of the various mutants for genetic analysis has been hampered. Consequently several groups have cloned nontoxic fragments of the *tox* gene (13). An 831-base-pair segment of β *tox*-45 genome encoding fragment A of the toxin, the *tox* promoter and signal sequence has been cloned in *Escherichia coli* K12. The ADP-ribosyl transferase activity was secreted into the periplasmic space (13). Evidence indicates that a TAGGAT "−10" sequence is most likely to confer *tox* promoter activity in *E. coli* (14).

Structure

Diphtheria toxin (DT) is secreted from the bacterium as a single polypeptide chain 535 residues in length. Most of the protein sequence and its relation to functional domains were determined by classical genetic and protein chemistry techniques and more recently by molecular biological approaches that provided the DNA sequence (15, 16). The DNA sequence indicates that the protein contains a 25-residue leader sequence. Interesting features of the toxin structure are shown in Figure 1. The growth medium of late cultures *Corynebacterium diphtheriae* contains tryptic-like enzymes capable of cleaving between arginine 193 and serine 194, providing a two-chain structure held together by an interchain disulfide bond between cystine 186 and cystine 201. Reduction followed by size exclusion chromatography achieved separation and purification of both chains (17). Early work showed that the smaller chain, known as the A chain, contained all the ribosyl transferase activity (17). This chain was relatively nontoxic to cells, requiring at least a 10^5-fold increase in A chain concentration over intact toxin concentration for equal toxicity (1, 4).

Five toxin-related proteins obtained by mutation with nitrosoguanidine have proved useful in elucidating toxin structural and functional interrelationships (18a). These mutant toxins, known as CRM materials (cross-reacting materials), are depicted in Figure 1 along with other structural details of DT. CRM 197 contains a single A chain substitution of glycine in wild-type to glutamic acid at position 52 (16). It binds to the diphtheria toxin receptor and competes toxicity but is nontoxic and lacks ADP-ribosyl transferase activity (18a). CRM

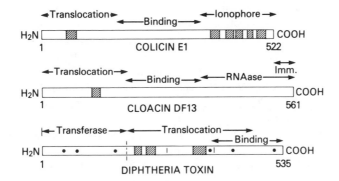

Figure 1 Structure-function relationship in diphtheria toxin, cloacin DF13, and colicin E1 molecules. Shaded areas represent sites of possible alpha helical amphipathic domains. In the DT map: the vertical dashed line represents the cleavage point between arg 193 and ser 194; the dots represent the loci of the point mutations detailed in text; and the vertical solid lines represent the early termination points of CRM 30 and CRM 45.

228 contains five amino acid replacements (18b). Two are located in the A chain at positions 79 and 162 and result in the loss of ADP-ribosyl transferase activity. It appears then that the active site of the A chain spans residues 52 to 162. Since NAD binding proteins show a nucleotide fold, it has been suggested that this fold resides between residue 52 and 162 (16). CRM 228 exhibits B chain substitutions at residues 378 and 431 and is fivefold less active in competing diphtheria toxin toxicity than CRM 197 (18a). Two CRMs containing early termination signals were found: CRM 30, which terminates at residue 281, and CRM 45, which terminates at residue 386, the numbered name referring to the approximate molecular weight of the mutant protein in kilodaltons. Both contain full A chain activity but have considerably reduced toxicity. However, on a weight basis, CRM 45 is approximately 10-fold more toxic than A chain for both sensitive and resistant cells (18a, 19). CRM 45 exhibits a 1,000-fold reduction in toxicity compared to wild-type toxin, a normal A chain activity, and does not compete wild-type DT toxicity (18a), indicating that it lacks an essential binding or transport function or both. It has been proposed that the transport or translocation function of diphtheria toxin B chain is located in a hydrophobic lipid-binding domain of CRM 45 (20). The sequence analysis shows that there are two helical amphipathic spans in the early portions of the B chain. These regions are between 207 and 225 and between 242 and 264 (21). These regions have high hydrophilicity. Their negative and positive charges are separated so that they present a large hydrophilic face containing ion pairs and a narrow hydrophobic face. This is characteristic of serum lipoproteins and is considered to be the structure responsible for lipoproteins binding to phospholipid head groups. Both of these

alpha helical amphipathic domains exhibit 50% homology with apolipoprotein A1 (21). A third amphipathic helical region exists between residues 346 and 372. This helical region exhibits a high degree of hydrophobicity, and in contrast to the previous regions, presents a large hydrophobic face and a narrow hydrophilic face. This region resembles the transmembrane helical portions of intrinsic membrane proteins (21). The carboxyl terminal region of the B chain is believed to contain the toxin receptor binding site since this region is missing from CRM 45 (18a).

Recently a new CRM has been described, CRM 1001, in which a tyrosine was substituted for cysteine 471. This CRM has full enzymatic activity, is nontoxic, and competes wild-type toxin toxicity (22). The presumption is that either a transport function has been eliminated by a conformation change resulting from loss of the interchain B chain disulfide, or that the cysteine 471 is directly involved in transport. It is interesting to note that the translocation function placement is within the region believed to be the B chain binding site. A similar close association between the ricin B chain binding site and the ricin transport function has been noted (23). The available evidence that interrelates structure and function indicates that the individual functional domains are spread over rather broad structural domains and in the case of the B chain are not limited to a single contiguous span for either the binding function or the translocation function.

Isolated fragment A of DT contains a single NAD^+ binding site ($K_d = 8.3$ μM), which is involved in both the ADP-ribose transfer to EF-2 and the glycohydrolase reaction (24). Exposure of DT fragment A and NAD^+ mixtures to UV light results in the covalent linkage of the nicotinamide moiety of NAD^+ to the glutamate at residue 148 (25). The cross-linking is highly efficient (approaching 1 mol/mol) and is not observed in CRM 197, suggesting that the Glu-148 residue is at or near the catalytic center (25).

DT, as isolated from bacterial filtrates, is found either complexed with nucleotide or free of bound nucleotide (26). The endogenous nucleotide is largely adenyl-(3',5') uridine 3'-monophosphate (ApUp), which structurally resembles NAD^+ (27). NAD, adenine, and nicotinamide competitively inhibit ApUp binding to nucleotide-free DT (28). Values of K_d of ApUp for intact, nicked DT range between 0.2 nM and 1.8 nM, whereas intact, nicked DT binds NAD^+ with a K_d of about 10 μM (29). The relative order of affinities of isolated DT A chain for ApUp and NAD^+ are reversed with intact DT (29). ApUp blocks both the binding and hydrolysis of NAD^+ by DT, and CRM 197 binds ApUp approximately 5,000-fold less tightly than does wild-type DT (28).

A variety of phosphate-containing compounds have been reported to block DT action on cells (26, 30, 31). These compounds include nucleoside triphosphates and inositol hexaphosphate, and act by directly binding to the toxin, thereby blocking toxin-receptor interactions (26, 32). Affinity labeling studies have localized the cationic COOH terminal portion of the B chain as the site of toxin interaction with polyphosphates [termed the P-site (33)]. The model that has emerged from these studies suggests that the endogenous nucleotide (ApUp) binds to and blocks both the nucleotide binding site on the A chain of DT and the P-site on the B chain. This model predicts a close three-dimensional proximity of the two toxin sites (28, 29). DT has been crystalized and X-ray crystallographic studies have been undertaken (34, 35).

When diphtheria toxin is purified on either DEAE or sizing columns, as much as one third of the material can be found to exist as a dimer depending on the previous history (T. H. Hudson, unpublished observations). Peaks corresponding to trimers and tetramers are also noted in lower amounts. Since both dimer and monomer toxins exist in bound and free nucleotide species, unfractionated preparations of diphtheria toxin are mixtures of at least four different species.

Enzymatic Activity

The A chain of diphtheria toxin is an enzyme that catalyzes the transfer of the ADP-ribosyl moiety of endogenous NAD^+ (36) to a modified histidine residue in the N-terminal region of elongation factor 2 (EF-2) (37). ADP-ribosylation of this residue, diphthamide, renders EF-2 inactive in protein synthesis (36, 37). When the kinetics of inactivation of EF-2 and protein synthesis are observed in the same cells, both fall exponentially following a dose-dependent lag period (38, 39). ADP ribosylated EF-2 has a greatly weakened binding to the ribosome (40). The region around diphthamide is probably associated with binding since ribosomal-bound EF-2 is resistant to ADP-ribosylation catalyzed by diphtheria toxin (40). Cycloheximide protects EF-2 from ADP-ribosylation by diphtheria toxin, presumably by freezing it on the ribosome (41). Cycloheximide protection is prevented in the presence of puromycin (42).

The structure of ADP-ribosylated diphthamide has been determined (43). Overhauser enhancement of NMR spectra indicates that the configuration of the N-glycoside bond is inverted as is the case with the ribosyl transferase found in pertussis, cholera, and pseudomonas exotoxin A and the enzyme induced by bacteriophage T4 (37). This mechanism differs from NAD glycohydrolase activity present in endoplasmic reticulum and the inner membrane of mitochondria which, like snake venom NADases, can utilize histidine and alcohols as acceptors with maintenance of the β configuration found in NAD^+ (37, 44, 45). The ADP-ribosyl-diphthamide bond is acid- and alkaline-stable, and formation releases 5.2 kcal per mol at pH 7.0 (46). The reaction is easily reversible using a pH of 5, an excess of nicotinamide, and DT A chain (38, 46).

Cells having a volume of about 2 picoliters contain approximately 10^6 molecules of EF-2, which corresponds to a concentration of 2 μM (39). The intracellular concentration of NAD^+ is 200 μM (39), and under intracellular conditions the enzyme is not saturated with respect to EF-2 (39) but is with respect to NAD^+ (46). Under these conditions and in the presence of DT A chain, the conversion of EF-2 to EF-2-ADPR is first order and the rate constant is linearly related to the DT A chain concentration (39). By extrapolation of these data, an A chain concentration of 10^{-12} M or approximately 1 molecule per cell will inactivate 37% (e^{-1}) of a cell's EF-2 in 24 hours.

The selection of DT-resistant cells from sensitive cell populations has resulted in the establishment of cell lines altered in toxin binding and/or translocation to the cytosol (see following section) as well as the loss in sensitivity of EF-2 to A chain–catalyzed ADP-ribosylation (47, 48). The last group is further classified as putative mutants in EF-2 or in the posttranslational modification of EF-2 (47, 48). Resistant cells altered in the EF-2 modification process when exposed to 10^{-10} M DT exhibit an 80% reduction in plating efficiency (47). This suggests that the diphtheria toxin may have another effect on cells besides the ADP-ribosylation of EF-2. Such an effect can be due to the occupancy of DT receptors or to membrane insertion phenomena that are believed to take place during the translocation step. Similar mutants have been generated in *Saccharomyces cerevisiae* and the lesion localized to the posttranslational modification process (49, 50). Mutants in the EF-2 structural gene act codominantly following somatic cell hybridization with wild-type, sensitive cells (47). Exposure of these hybrids to DT results in the loss of 50% of control protein synthesis and a 50% decrease in growth rates. When toxin is removed, these cells maintain the 50% reduction in growth rate for a period of up to 6–7 days. This indicates that DT is long-lived and active after gaining entrance to the cytosol compartment (47).

Diphthamide has not been localized to any other protein except EF-2 (51). The tryptic peptide containing diphthamide is not ribosylated by DT A chain and NAD^+, indicating that neighboring structural domains provide recognition features (52). Conservation of diphthamide throughout all eukaryotic cells and the archaebacteria (53) suggests an important function yet to be elucidated. Recently a mono-ADP-ribosyl transferase that copurified with crude EF-2 preparations with activity similar to DT yet inhibitable by histamine has been reported in a transformed baby hamster kidney (BHK) cell line (54).

Binding and Uptake

The properties of receptors and receptor-mediated processes can be studied phenomenologically, that is, the response of the system to a variety of agonists and competitive antagonists can be quantitated. Rank order of potency relationships among the agonists and antagonists can be established. Alternatively, receptors can be identified by binding studies performed with labeled agonists and competition with unlabeled material can generate competitive displacement curves. Through Scatchard analysis, affinity and receptor number can be determined. To relate the functional receptors identified in phenomenological studies with the receptors identified in binding studies requires a sufficiently diverse panel of agonists and antagonists to make a correlation between the two types of studies (55). Alternatively, a variety of responsive systems that vary in receptor number and affinity can be used to provide the correlations between binding studies and dose-response curves. The latter procedure requires close coupling between receptor occupancy and the cellular response. That is, a linear (or hyperbolic) range of response with receptor occupancy is required. The correlation of toxin receptors identified through binding studies with functional receptors has proven difficult for DT and other toxins because of the lack of close coupling and because of a general lack of sufficient numbers of mutants exhibiting differing binding affinities. In addition, some resistant cell lines appeared to show both receptor and transport defects, complicating the analysis (56).

Before receptors were identified on cells by binding studies, the receptor-mediated nature of the DT intoxication process was demonstrated by the saturation of the rate constant of protein synthesis inactivation with increasing toxin doses (18a). Competition of toxicity with the nontoxic mutant CRM 197 was also observed. Kinetic studies indicated that CRM 197 competed the dose-dependent lag period as well as the rate constant of inactivation of protein synthesis. From the saturation of the inactivation rate constant of protein synthesis the saturable process was estimated to have an equilibrium dissociation constant, K_d, in the vicinity of 5×10^{-7}. Curiously, the lag period continued to shorten at concentrations 100-fold above this value. The same phenomenon has been noted with ricin (57). A variety of cell types having widely differing sensitivities to DT were also observed to have roughly the same dissociation constant (K_d) in the range of $1-2 \times 10^{-9}$ M (58, 59). Receptor number varied between 600 and 200,000 per cell. Cell lines derived from kidney, such as Vero, have the highest number of receptors and the highest sensitivity to DT. Binding studies using ^{125}I-labeled DT on HeLa S-3 cells indicate that the product of receptor affinity and number is too low to measure (58). The studies mentioned above were all performed by employing a 100–500-fold excess of cold over labeled DT to estimate the nonspecific binding. The tracer range extended out to about 20×10^{-9} M. Nonsaturable binding (estimated by taking the final limiting slope of the total binding curve and extrapolating through the ordinate) was subtracted to obtain specific binding. Scatchard analysis using this method indicates a single class of high-affinity sites.

Diphtheria toxin binding studies employing isolated cell membranes, a fixed amount of tracer, and competing this tracer with cold toxin up to 5×10^{-5} M revealed heterogeneous binding with the apparent affinities detected in the area of 5×10^7 M^{-1} down to 10^6 M^{-1} (60). Studies performed in this manner and studies in which wash times are minimized will detect lower affinity binding. In the study cited above no change between membranes derived from toxin-sensitive cells and toxin-resistant cells (mouse and rat) were noted, yet the binding to both types of cell membranes required divalent cation and was competed by ATP in agreement with cellular toxicity studies (60).

Recently, binding studies performed with nicked and unnicked CRM 197 revealed that nicked CRM 197 had a 3-fold higher equilibrium affinity constant, K_a, than unnicked CRM 197 (61). In the same study, a hybrid toxin composed of CRM 228 B chain and wild-type A chain was constructed and its binding studied. Wild-type DT binding and protein synthesis inhibition was competed with nicked and unnicked CRM 197, CRM 228, and the hybrid toxin, CRM 228A–CRM197B. The rank orders of competition for DT binding and protein synthesis inhibition were the same. These studies indicate: 1. Certain domains of the A chain influence the binding functions heretofore solely localized to the B chain, 2. The DT binding observed was directed, at least in part, towards the functional receptor, i.e. resulted in intoxication.

Kinetic studies of the inhibition of protein synthesis of murine L cells, which are 10,000-fold more resistant to diphtheria toxin than sensitive cells, show an approach to saturation of the first order rate constant of protein synthesis inactivation (K_i) with increasing dose (56). The curves are similar to those previously reported in HeLa cells with the exception that the dose-dependent lag period is much less prominent and the approach to saturation occurs at 5×10^{-5} M instead of 10^{-8} M. The saturable nature of this process indicates the presence of a receptor. Intoxication of murine L cells by diphtheria toxin in some way involves the toxin B chain since equal toxicity occurs at a 10-fold lower dose of intact toxin compared to A chain (56). Monoclonal antibodies generated against diphtheria toxin that inhibit the toxicity of sensitive cells do not inhibit the toxicity of diphtheria toxin towards murine L cells (62). This indicates that the receptor noted in the kinetic studies of L cell intoxication is different from the receptor mediating toxicity in sensitive cells. Whether this low-affinity receptor is also present on the external surface of sensitive cells remains to be demonstrated, although a receptor with the correct affinity was identified in cell membranes (60). Interestingly, the low-affinity receptor identified in these studies was competable with ATP in both the murine and the sensitive species. A glycoprotein capable of binding diphtheria toxin has been solubilized and isolated from toxin-sensitive cells (63a). DT binding activity was undetectable in preparations of the one insensitive (mouse) cell line tested. The binding of nucleotide-free DT was competable with ATP whereas DT complexed with ApUp did not bind (63b). The exact relationship between this glycoprotein and the functional receptor has yet to be determined.

Attempts have also been made to correlate uptake of labeled diphtheria toxin with toxicity. By performing the uptake experiments in the presence of methylamine, which blocks degradation of internalized toxin, uptake studies can be used to quantitate the presence of receptors mediating internalization below the range accessible to binding studies (64). The correlation of uptake of DT in the presence of methylamine with the sensitivity of the cell lines to DT reveals a good correlation in rank order and a rather weak correlation when absolute uptake is correlated with the concentration of toxin required to achieve a 50% reduction in protein synthesis. For example, Vero cells and human embryonic lung cells have the same sensitivity to DT yet differ in uptake by a factor of 9 (64). In this study murine L cells fail to demonstrate uptake. The toxin concentration in this study was 1.7×10^{-9} M. When receptor-mediated uptake of rhodamine-labeled DT was assayed by fluorescence microscopy, no differences could be noted between sensitive cells and murine 3T3 cells (65). However, the toxin concentration for these studies was in the range of 3–5 $\times$ 10^{-7} M. The difference in these two uptake studies of DT by murine cells may reflect the fact that the functional receptor present on murine cells has a much lower affinity than that present on the sensitive cell types.

In a series of studies designed to isolate a variety of EF-2 mutants from cells susceptible to genetic manipulation, functional diphtheria toxin receptors were identified in *Saccharomyces cerevisiae* spheroplasts (66). After a 2-hour exposure to DT, cell viability was reduced 1.5 logs as was protein synthesis. Kinetic studies on protein synthesis revealed what appears to be a dose-dependent lag period and an increasing inactivation rate constant of protein synthesis with increasing toxin concentration in the range between 8×10^{-9} and 8×10^{-7} M. Toxicity was competed by CRM 197 and ATP as is the case for the functional diphtheria toxin receptor. Whether the receptor-mediated process in *S. cerevisiae* resembles that of the sensitive vertebrate cells or the more resistant murine cells remains to be elucidated. Binding studies have not been reported.

As discussed in the section on enzymatic activity, a number of mutants in EF-2 have been described that exhibit resistance to DT and these have been labeled as class 2 mutants (47). DT-resistant mutants having normal EF-2 have been described by a number of groups (47, 67a, 67b, 67c). Binding and uptake studies already mentioned indicate that certain cells contain varying amounts of the high-affinity diphtheria toxin receptor (58, 59). In addition, murine cells are believed to constitute a transport type defect since under conditions giving normal values of uptake and near saturation of receptors, the inactivation rate constant of protein synthesis is only one tenth that of sensitive intoxicated cells (56). Toxin-resistant mutants whose affinity for DT is decreased have also been described (68). Another interesting mutant, classed as a transport mutant, is characterized by a lesion in vesicle acidification due to malfunctioning of the ATPase-driven proton pump (67b). These mutants have a decreased sensitivity for DT toxin and pseudomonas toxin and an increased sensitivity towards ricin. Several enveloped viruses requiring a pH-dependent processing step exhibit diminished infectivity for this mutant. Because of the complexity of the transport process as outlined in the next section it is likely that a variety of different types of transport mutants will be isolated.

Kinetics of Intoxication

SINGLE-HIT KILLING PROCESS   Early work on diphtheria toxin intoxication indicated that only a small fraction of endocytosed toxin was productive of toxicity (39, 58, 69). Therefore, investigations of the uptake route productive of toxicity rather than the degradation pathway required subtler techniques. For this reason kinetic studies utilizing cell survival or inhibition of protein synthesis or inactivation of EF-2 as their end-points have been relied on to provide insights into the intoxication process. Figure 2 *(left)* shows an early kinetic study on the intoxication process (70). HeLa cells were exposed to diphtheria toxin in the presence of ammonia which was known to block transport of toxin to the cytosol but not affect the absorption of toxin to the cell. At various times cells were washed free of toxin and ammonia and plated in complete medium. After four days viable cells were scored. The logarithmic inactivation of surviving cells as a function of exposure time indicates a single-hit process. There is a finite period of time required to absorb a lethal dose of toxin. The steeper curve at a high dose indicates that this time is inversely related to the toxin external dose.

The first order process in Figure 2 *(left)* can be described by a single exponential which gives a constant probability with exposure time that the lethal event will occur in any live cell sometime during the four-day assay period. Because of the probabilistic nature of the process, lethal events accumu-

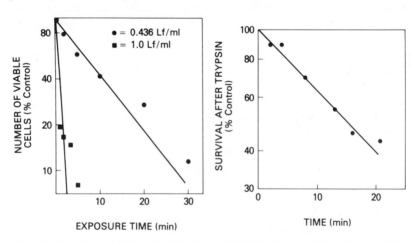

Figure 2   Left: Rate of delivery of a lethal dose of diphtheria toxin to HeLa cells at two toxin concentrations (Lf, flocculating units) [redrawn from Ref. (70)]. *Right:* Rate of delivery of a lethal dose of colicin K to *E. coli* cells [redrawn from Ref. (105)].

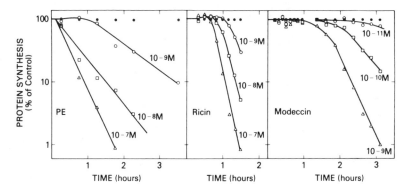

Figure 3   Protein synthesis inhibition in Green monkey kidney (Vero) cells by: *pseudomonas aeruginosa* exotoxin A, *(left);* ricin, *(center);* and modeccin toxin, *(right)*. Cells were exposed to the indicated toxin concentrations at time zero and the rate of protein synthesis was determined by a 20-minute pulse of ^{14}C-l-leucine at the indicated times (T. H. Hudson, unpublished).

late with time at a constant percentage of the remaining viable cells. A lethal single-hit process, such as the one shown in Figure 2, does not specify the number of offending molecules responsible for the lethal event. One molecule per event is an attractive model but not the only model.

FIRST ORDER INACTIVATION OF PROTEIN SYNTHESIS   When protein synthesis inactivation is measured as an end-point of toxicity and plotted as a function of time, first order inactivation kinetics are seen for pseudomonas exotoxin A, ricin, and modeccin, illustrated in Figure 3, and diphtheria toxin, illustrated in the right panel of Figure 4. Several other notable features are seen which are similar for all of the toxins. There is a dose-dependent lag period which precedes the first order inactivation of protein synthesis, and there is a post-lag curvature to the plots as if a portion of the lag period were not an absolute lag but rather represented a prior obligatory reaction which had to be completed before first order inactivation of protein synthesis could proceed at its maximum rate (constant) (57). A second prominent feature is the fact that the slopes of the inactivation of protein synthesis, K_i (first order rate constant of inactivation), are not linearly proportional to the input toxin concentration. In cases where receptor occupancy has been determined, first order rate constants of ricin (57) and DT (18a) are more nearly proportional to the square root of receptor occupancy.

For diphtheria toxin, the inactivation of EF-2 via ADP-ribosylation has also been followed kinetically and is found to obey roughly the same kinetics as protein synthesis inactivation including the dose-dependent lag period (38, 39). Using conditions similar to those encountered in vivo with respect to NAD^+ concentration and EF-2 concentration, the first order inactivation of EF-2

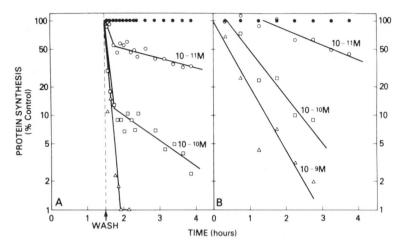

Figure 4   Left: Protein synthesis inhibition of Vero cells intoxicated with DT in the presence of cycloheximide. Both cycloheximide and DT are removed from the cells at the arrow. *Right:* Protein synthesis inhibition of Vero cells intoxicated with DT in the absence of cycloheximide. [redrawn from Ref. (42)].

observed in Vero cells at varying external toxin concentrations could be related to an in vitro toxin concentration giving a similar value of K_i. At toxin concentrations approaching receptor saturation (10^{-9} M), the rates of inactivation indicated that approximately 200 molecules of diphtheria toxin A chain were present within each cell (39). This calculation seems at variance with a model that explains the single-hit kinetics as the introduction of a single molecule of DT A chain into the cytosol. Rather, the lethal event could be ascribed to the accumulation of several hundred DT A chain molecules within the cytosol of most of the cells within the population in the very short period of time between the end of the dose-dependent lag period and the establishment of a first order inactivation of EF-2.

THE RATE-LIMITING STEP IN DT INTOXICATION   The considerations detailed above raise the question of what constitutes the rate-limiting step in diphtheria toxin intoxication. Was it the ADP-ribosylation of EF-2 or the entry of the toxin A chain into the cytosol compartment? The experiment shown in the left panel in Figure 4 indicates that translocation to the cytosol and not ADP-ribosylation is the rate-limiting step (42). In Figure 4 these two events are dissociated by treating cells by cycloheximide before and during the application of toxin. Cycloheximide has been shown to block the ADP-ribosylation of EF-2 by diphtheria toxin. Cycloheximide does this by freezing EF-2 on the ribosome in a state inactive to the toxin A chain. (This protection is blocked by another ribosomal binding agent puromycin.) Cells to which diphtheria toxin is pre-bound in the cold in the presence of cycloheximide are first washed free of unbound toxin and then warmed up for a period of 1.5 hours still in the presence of cycloheximide. The cycloheximide is then washed out and the rate of protein synthesis is measured as a function of time by treating the cells with short pulses of ^{14}C-L-leucine. Immediately following wash out, the K_i values are 5- to 8-fold higher than those observed in cells treated with the same concentration of DT as shown in the right panel of Figure 4. This high rate of inactivation of protein synthesis only lasts about 10 minutes, then the K_i values abruptly change to the values seen without cycloheximide.

THE CASE FOR CONCERTED ENTRY   An interpretation of the above data is that during the 1.5-hour incubation with DT in the presence of cycloheximide, toxin A chain gained entrance to the cytosol of a certain fraction of the cells. Removal of cycloheximide allowed ADP-ribosylation to occur at a high rate due to the accumulated cytosolic A chain activity. This rate was reflected in the initial rapid inactivation of protein synthesis after cycloheximide removal. The second, slower, K_i resulted from the inactivation of EF-2 by A chain undergoing the translocation process after the cycloheximide washout. It was concluded that the rate-limiting step was toxin translocation to the cytosol. The maximal rate of ADP-ribosylation was only observed after a lengthy accumulation of A chain in the cytosol. Therefore, the overall rate of the intoxication process was governed by the slower first step of translocation. The inflection points for the slope changes seen in Figure 4, panel A, at 55%, 13%, and < 2% of the control protein synthesis represent the percentages of EF-2 not made accessible to A chain during the cycloheximide incubation and initial period of rapid EF-2 inactivation. These represent the fractions of cells for each DT concentration that did not experience the lethal event of toxin translocation.

The number of cytosolic DT A chain molecules present in each cell at the time of cycloheximide wash out can be estimated by using a turnover number of EF-2 ADP-ribosylation by DT A chain (39) and adjusting to cytosol pH (36). In a quantal model, these A chain concentrations represent the weighted average of rate constants $K_1, K_2, K_3 \ldots$ associated with fractions of the cell population $F_1, F_2, F_3, \ldots$ which have experienced 1, 2, 3 . . . quantal events during the cycloheximide incubation. Poisson analysis of these kinetic data allows the calculation of the quantal size (number of DT A chain molecules that translocate during one entry event). For 10^{-11} M, 10^{-10} M, and 10^{-9} M DT the quantal size is 83, 88, and 77 A chain molecules, respectively. The quantal size of DT translocation is, therefore, constant over the range where receptor occupancy changes by 50-fold (42).

TWO-POPULATION STATES OF INTOXICATED CELLS   The above interpretation of the intoxication process predicts a two-population state model of intoxicated cells. One state synthesizes protein at the control rate and the other has virtually no protein synthesis. Cells are rapidly converted from one state to another and this probability constant is the rate-limiting step observed in protein synthesis inactivation of large numbers of cells assayed as an aggregate and constitutes the lethal event. The lethal event is the entry of a bolus of toxin into the cytosol compartment. Kinetics indicate that the bolus contains approximately 80 molecules. Thus the single-hit kinetics do not correspond to one molecule gaining entrance to the cytosol but rather to an event that has a certain probability with time of occurring and delivers a large number of molecules. Since the bolus size stays constant and since at higher receptor occupancy the probability that the entry event occurs is greater, it is postulated that with higher receptor occupancy of toxin more toxin packages of constant size are made, each having a constant probability of entry (42).

When the rate of protein synthesis in single cells is quantitated by autoradiographic techniques at early times in the intoxication process, two populations of cells are noted for diphtheria toxin (42): a cell population synthesizing protein at the control rate and cells essentially having no synthesis of protein. With time an increase in the population not synthesizing protein is observed balanced by a decrease in the population synthesizing protein at the control rate.

Ricin, Modeccin, and Pseudomonas Exotoxin A. The Rate-Limiting Step

It has been noted that the kinetics of protein synthesis inactivation are remarkably similar for the different toxins that utilize different receptors and apparently different entry mechanisms (57–59) (see Figure 3). When protein synthesis on individual cells was determined by autoradiographic techniques following treatment with ricin, modeccin, and pseudomonas exotoxin A, the results were similar to those observed with diphtheria toxin. There are two population states, and intoxication consists of a gradual interconversion of cells synthesizing protein at the control rate to cells synthesizing no protein at all. When graded doses of cycloheximide are exposed to the same type of cell, protein synthesis in individual cells decreases gradually and a multipopulation state is observed (T. H. Hudson, M. A. G. Kimak, and D. M. Neville, Jr., submitted for publication). Earlier models of diphtheria toxin intoxication of cells explain the first order inactivation kinetics as being the result of a steady state concentration of diphtheria toxin A chain in the cytosol (72). The rate-limiting step in this model is the ADP-ribosylation of EF-2. This model has difficulty explaining the simultaneous presence of single-hit inactivation kinetics observed in Figures 2 and 3 with the large number of toxin A chain molecules present per cell as derived from in vitro EF-2 ribosylation with graded A chain concentrations (39). Moreover, the model requires some mechanism to maintain this steady state within the cell, which in the face of continual toxin input would require a continual removal or degradation. However, the available evidence indicates that intracellular DT A chain is stable for days (47, 73).

Dose-Dependent Lag Period

The dose-dependent lag period noted in all the toxins has been widely interpreted as representing some form of processing step that must be accomplished before toxin gains entrance to the cytosol (20, 57). In the case of diphtheria toxin this processing step is acid-dependent (74, 77). Mutants that have a defect in vesicle acidification are highly resistant to diphtheria and pseudomonas toxin as well as several enveloped viruses which also require acidification (67a, b, c). The prolongation of the lag period by lipid-soluble weak bases is generally considered to be due to a rise in pH of the endocytotic vesicle by concentration of the protonated base within the vesicle following passive diffusion of the free base. At high concentrations of DT the lag period can be shortened to within minutes by a brief exposure to pH 5.5 (74). Carboxylic acid ionophores such as monensin, which collapse the cation and proton gradient within the vesicle, also protect cells against DT (76). It appears that if the dose-dependent portion of the lag period represents a processing event, this event has the opposite pH dependence for ricin whose dose-dependent lag period is lengthened by exposure of the medium to acid and shortened by the exposure of medium to alkaline conditions (57). As previously mentioned, mutants deficient in vesicle acidification have enhanced sensitivity towards ricin toxicity (67a). A point mutation in diphtheria toxin produced by the reversal of a nonsense mutation with a suppressor strain has resulted in a toxin mutant that has a decreased lag period in Chinese hamster ovary (CHO) cells but not in Vero cells (78).

The hydrophobicity of the NH_2-terminal portion of the diphtheria toxin B chain and the interaction of CRM 45 with detergents has suggested that this portion of DT inserts into the membrane (20). Lipid bilayers exposed to DT at low pH demonstrate the formation of voltage-dependent channels (79, 81). Because low pH is a requirement for this event as well as the processing event for DT, these two may be related.

The protein synthesis inactivation curves shown in Figures 3 and 4 indicate that although inactivation is first order with time, inactivation is not linearly proportional to dose. This has previously been noted with ricin where the dependence of K_i values varies with the square root of the receptor occupancy (57). One possible mechanistic interpretation of this finding is that the toxin dimerizes and the toxin dimers are utilized as channels for the entry of monomeric toxin. Since dimers are inactive, that is that they can never gain entrance to the cytosol, any concentration greater than the dimerization constant would result in a square root dependence of K_i on receptor occupancy (57).

We have carried out a similar calculation done on data of HeLa cell intoxication with diphtheria toxin which reveals that the K_i values depend on receptor occupancy to the 0.57 power. Occupancy was varied by competition with CRM 197. These data show that CRM 197 also competes the dose-dependent lag period (18a). When the log of the dose-dependent lag period is plotted vs the log of receptor occupancy by wild-type DT the lag also varies by the 0.57 power of wild-type toxin occupancy. This indicates that the process responsible for the blunted dependence of entry on receptor occupancy is occurring during the putative processing event.

The ability of CRM 197 to compete the dose-dependent lag period indicates that if this lag period represents a processing step, the A chain point mutation in CRM 197 renders this mutant incapable of performing the processing event. This raises the question as to whether toxin-bound nucleotide could be involved in the processing event. Ricin B chain, which has the same tracer binding and cold displacement profile to cells as ricin, also competes the ricin dose-dependent lag period and therefore cannot perform the putative processing step (57).

Modeling the Kinetic Data

For certain toxins, such as DT and ricin, there may be a portion of the lag period that is not dose-dependent and that represents the time taken to route the toxin following endocytosis to a compartment where the processing event can take place. If the processing event is considered to be an obligatory first order reaction preceding a membrane entry event that is also a first order process, then the curved portion of the inactivation curve represents various half-lives of the initial first order reaction (57).

The rate constant governing the first order processing may be a lumped constant consisting of a pH-dependent term and a membrane concentration–dependent term to fit the observed pH and dose dependency of the "lag" period. With this modeling the "log linear" portion of the inactivation curve is not strictly log linear but approaches linearity to the extent of the number of half-lives completed for the initial reaction (57). This scheme is illustrated in Figure 5.

The model shown above has the advantage that in experiments utilizing prebinding followed by washing away free toxin, various inhibitors may be applied and removed at certain times with some possibility that inhibitor effects can be said to be effecting different sequential steps of the multi-step process: binding, endocytosis, routing, processing, and transmembrane transport. (The last step, protein synthesis inactivation, is not rate-limiting at any time and cannot be independently observed without inhibitors as in Figure 4 or by monitoring single cells.)

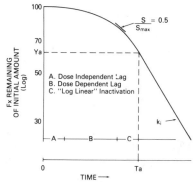

Figure 5   Model of the kinetics of intoxication.

An Energy Requirement for Toxin Transmembrane Transport

Utilizing the scheme diagrammed in Figure 5, cells were deprived of energy sources during the "log linear" or membrane translocation portion of the intoxication process elicited with either diphtheria toxin or ricin. This was accomplished by a one-hour cooling to 0°C or by exposure to a mixture of deoxyglucose and azide (T. H. Hudson, M. A. G. Kimak, and D. M. Neville, Jr., submitted for publication). Following warming or inhibitor wash out, the kinetics of protein synthesis inactivation were again followed. The level of protein synthesis was unchanged during energy deprivation and post–wash out proceeded with the same slope (unlike) the cycloheximide wash out experiment of Figure 4). This result indicates that cells did not accumulate toxin in their cytosol compartment during energy deprivation (71). Treatment of DT-intoxicated cells with NH_4Cl during this period permitted a continual drop in protein synthesis, suggesting that the necessary energy source is not a proton gradient (42).

COLICINS AND CLOACINS

Utilization of Bacterial Receptor-Mediated Transport Systems

Bacteriocins are protein antibiotics secreted by one strain of bacteria and toxic to the same or similar strains of bacteria. Several recent reviews are available (9, 82, 83). In general the proteins are approximately 60 kilodaltons in size and are encoded on plasmids. The best-studied bacteriocins are made by and intoxicate *E. coli* and are called colicins. The kinetics of colicin inactivation as a function of dose exhibit saturability. Genetic studies indicate that a number of different colicins utilize the same bacterial outer membrane protein receptors. These are generally given the same letter, thus colicins E1, E2, and E3 utilize the same receptor. Colicin E1 achieves its toxin affect by collapsing the membrane potential gradient without inhibiting respiration (84). Colicin E2 is an endonuclease that inflicts single and double strand breaks on the bacterial genome (85), and colicin E3 is a specific ribosomal ribonuclease that cleaves a single phosphodiester bond of the 16S RNA of the 30S ribosomal subunit, approximately 50 nucleotides in from the 3' end (86, 87). Colicins exhibit immunity in that they are nontoxic for the bacteria making them. For colicin E2 and E3 (and cloacin DF13) this has been shown to be due to the fact that a small acidic protein of approximately 10 kilodaltons is bound tightly to the enzymatically active region of the colicin by noncovalent forces inactivating the enzymatic activity. This binding protein is made in excess by the bacterium synthesizing the colicin and is in some way removed by the bacterium experiencing the toxic effects (88).

Obviously colicin E2 and E3 must be transported across the bacterial outer and inner membranes for their effects to take place. Colicin E1 must in some fashion be inserted into the inner plasma membrane to achieve a permanent collapse of the electrical membrane gradient. However, membrane channel effects and cytosolic enzymatic effects are not mutually exclusive. A bacteriocin from *Enterobacter cloacae*, cloacin DF13, exhibits the ribosomal nuclease activity of E3 (89), yet also induces K^+ efflux and Na^+ influx following a lag period (90). Colicin E2 also causes membrane changes but only if the host *E. coli* is lysogenized with λ (91). The effects are K^+ efflux and inhibition of the proton-driven transport of proline and β-D-galactosides. The *rex* gene of λ is responsible for these effects and mediates a variety of other membrane effects independent of colicin E2 (9). The chemistry of colicin transport into the membrane and across the membrane is of particular interest. The colicins utilize transport systems already present in the bacterial membrane (9, 92a, 92b). For example, colicin M utilizes an outer membrane receptor known as Ton A which is also involved in ferrichrome uptake. Colicin M also utilizes a general chelated iron transport system present in the inner membrane known as Ton B. Both Ton A and Ton B are required for T1 phage infection. Chelated iron transport is driven by the proton-motive force. Other examples of shared transport systems between low-molecular-weight *E. coli* transport systems, phage transport systems, and colicin transport systems have been noted (92c).

Binding

Several binding studies of bacteriocins to cells have been reported. Colicin Ia labeled with ^{125}I and retaining 85% of its original bioactivity binds to *E. coli* in a saturable manner. Scatchard analysis reveals two affinities, 2500 sites per cell at 10^{10} M^{-1} and 2000 sites per cell at 10^9 M^{-1}. A killing unit, defined as the amount of colicin required to reduce survivors by a factor of e^{-1}, corresponded to 10 molecules bound per cell for a majority of the cells and 200 molecules per cell for a small fraction of the cells (93).

Cloacin DF13 purified from its immunity complex has only 8% of the complex toxicity (94). Similar findings have been reported for colicins E2 and E3 (88). Binding to intact cells or vesicles derived from spheroplasts were performed using biosynthetically labeled DF13. Cloacin or cloacin complex binding to intact cells was saturable with $K_a = 6 \times 10^7 \, M^{-1}$ and 425 sites per cell, while vesicles exhibited higher affinity, $8 \times 10^8 \, M^{-1}$ and 10^5 sites per vesicle (94). Since no differences were detected between cloacin and its immunity complex, it was concluded that the immunity protein plays an essential role in the transmembrane transport process in intact cells (94). Unfortunately, toxicity to intact cells and spheroplasts was not compared. Colicin E1 can intoxicate wild-type spheroplasts and spheroplasts from resistant cells. Higher multiplicities are required to elicit effects of spheroplasts (95, 96). The nature of the interaction with the spheroplast membrane has not been determined.

Structure

The relationship between structural and functional domains for cloacin DF13 was deduced by studying a variety of mutants and localizing them to unique proteolytic fragments (97). The nucleotide sequences of the cloacin gene has now been determined and the 561-amino-acid sequence thus derived. The NH_2-terminal region up to residue 200 is rich in hydrophobic residues and is involved in membrane translocation (Figure 1). Helicity as determined by the analysis of Chow & Fasman is absent from the translocating domain until residue 119 (98). We note considerable amphipathic character of the helix over the next 20 residues (cross-hatched areas in Figure 1). The receptor binding domain spans residues 200–420, the RNAase domain 360–531, and the immunity binding region 525–561. A similar domain structure has been observed for colicins E3, E2, and E1 (99, 100).

The nucleotide sequence for the colicin gene E1 has also been determined (101). This membrane-acting colicin in common with Ia and Ib has an axial ratio 20 giving a prolate elipsoid of 370 Å × 19 Å. There are three predicted alphahelical stretches of 40, 100, and 35 residues consecutively spaced between residues 50 and 255. We note considerable amphipathic character over residues 57–81. The ability of E1 to exhibit ionophore activity and to form voltage-dependent ion channels has been localized to a COOH terminal 20-K_d proteolytic fragment (100, 102). This region is rich in hydrophobic residues. It has been suggested that there is a series of five membrane-spanning amphipathic helixes in this domain responsible for ion channel properties of E1 (102), and it is further proposed that a single or double amino acid protonation in this region is responsible for the pH-dependent insertion of E1 into artificial membranes and the formation of voltage-dependent channels (102). Groupings of membrane-spanning amphipathic helixes have been proposed to explain ion conduction through acetylcholine-activated sodium channels (103, 104a) and voltage-dependent sodium channels (see Catterall, this volume).

Single-Hit Inactivation Kinetics

The inactivation of *E. coli* by a variety of colicins follows single-hit inactivation kinetics. When bacteria are exposed to colicins for varying amounts of time and then diluted out and plated and assayed for colony forming units, the number of survivors falls in a log linear fashion with exposure time (104b). These data are remarkably similar to the curves previously shown for diphtheria toxin in Figure 2, left panel. The effective exposure time to colicin can also be varied by adding trypsin at various times following colicin absorption. Survivors also fall in a log linear fashion although the slope is reduced by the trypsin treatment (discussed below) (see Figure 2, right panel). When the in vivo kinetics of protein synthesis inactivation by the ribosomal nuclease colicin E3 or cloacin DF13 are observed (see Figure 6), the parallels with the toxins affecting eukaryotic protein synthesis (Figure 3) are striking. An equally striking comparison is the kinetics of the loss of preloaded Rb^+ following treatment with colicin K (96) shown in Figure 7. The fall off in each case is log linear, the linear

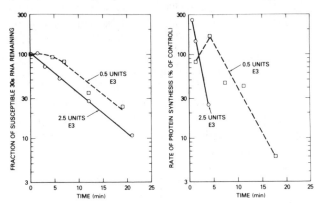

Figure 6   Kinetics of colicin E3–induced fragmentation of 30S ribosomal RNA *(left)* and synthesis inhibition *(right)* [redrawn from Ref. (86)]. *Left:* The susceptible 30S RNA was set at 110% of the highest value attained experimentally. From this and all other experimental points the fragmentation found in the absence of colicin E3 was subtracted. The experimental values after exposure to E3 for the indicated times are expressed as the percent of susceptible 30S RNA. *Right:* The rate of protein synthesis in control cells was taken as the slope of [³H] leucine incorporation with time. The rates of protein synthesis in E3-treated cells was determined by determining the amount of [³H]leucine incorporated between experimentally determined points, dividing by the time interval between those points, and plotting (as the percent of the control rate) at the mid-time between the two points. Because the ordinate scaling is arbitrary, the K_i values cannot be compared between the left and right panels.

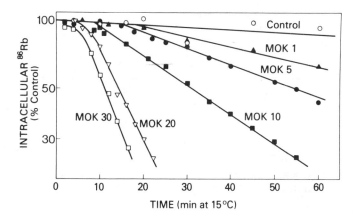

Figure 7   Kinetics of colicin K action on *E. coli* K12 cells as measured by the release of [⁸⁶Rb] from preloaded cells. Colicin concentration given as MOK (multiplicity of killing) [redrawn from Ref. (96)].

portion of the curve is related to the dose of colicins, and this phase is preceded by a dose-dependent lag period.

The log linear decrease in *E. coli* survivals caused by colicins K1, E1, E2, and E3 can be reversed by trypsin treatment; however, the rescue window size decreases with time as shown in Figure 2, right panel (105). It has been shown that the plating efficiency after trypsin rescue of colicin K exposure correlates with the fractional leakage of β galactosides. Thus permeabilized cells are dead cells and with time there is a constant probability of converting a cell from a live impermeable state to a permeable dead state. This probability is directly proportional to the colicin K multiplicity (105). These results were explained by the assumption that each killing unit of colicin absorbed has a constant probability per unit time of undergoing the transition from a state that is potentially lethal and trypsin-sensitive to a state that is damaged and trypsin-resistant. One killing unit of colicin is defined as that amount of colicin that decreases the survivors by a factor of e^{-1}. Further studies with a variety of radioactively labeled colicins showed that a killing unit corresponded to generally about 10 molecules of colicin specifically bound to the cell receptors (93, 106).

For colicins E1, E2, and E3, the decrease in rescuable cells by trypsin with time could be prevented by exposing the cells to energy blockers such as 2,4-dinitrophenol (105). The picture then that emerged in colicin intoxication was that of a population of cells that initially absorbed a potentially lethal dose of colicin. Subsequently individual cells undergo a transition to cells manifesting damage. The probability constant of this transition, and thus the first order rate constant, is proportional to the mean number of potentially damaging molecules absorbed per cell (105). Thus for colicin K the rate-limiting step defined by the first order rate constant appeared to be the transition from a trypsin-sensitive receptor-bound state to a trypsin-insensitive membrane-inserted state producing an ion channel. The first order inactivation kinetics exhibited by colicin E3 and cloacin DF13 also suggested that the rate-limiting step was entry of nuclease activity to the cytosol, particularly since the onset of leakage of potassium for DF13 was closely correlated with the onset of protein synthesis inhibition (110). If the rate-limiting step is the transport step of the colicin to the compartment where it is active, then the second step, collapse of the membrane potential for the membrane-acting colicins or the enzymatic inactivation of ribosomes for colicin E2 and DF13, must be faster than the rate-limiting step. In the case of the membrane-active colicins, it has been demonstrated that these proteins open up voltage-dependent channels in lipid bilayers and ion conducting channels in liposomes (107a). The lipid bilayer channels have a conductance of 3–10 picosiemans and a pore radius corresponding to 3.1 Å (107b). It has been calculated that one such channel can within several minutes collapse the bacterial membrane potential (107b). Thus the rate-limiting single-hit process could be the insertion of a single colicin-induced channel per bacterium. However, there is some question about the relevance of the bilayer work or similar experiments done in liposomes with respect to the physiological situation. In cells the membrane-active colicins collapse the membrane potential, but the proton gradient is maintained (84), and transport systems capable of being energized by the proton gradient in the absence of a membrane potential remain functional (84, 108). Cells treated with colicin E1, Ia, or K actually have a stimulated phosphoenolpyruvate-dependent phosphotransferase systems and can continue to accumulate α-methylglucoside against a concentration gradient (9). Therefore the colicin-induced channel is highly specific. However the colicin-induced ion channel conductivity in artificial bilayers or liposomes is of low specificity exhibiting little discrimination between cations (109).

The ribonuclease from colicin DF13 has been purified and its enzymatic constants have been determined in the absence of any immunity protein. The K_m for the reaction of DF13 with ribosomes is 17.7×10^{-6} M under conditions containing all the supplements necessary for protein synthesis (89). From the extrapolated value of V_{max}, a dissociation constant of the enzyme-substrate complex was calculated to be 8 s^{-1}. The authors conclude that a few DF13 molecules can inactivate most of the bacterial ribosomes in 20 min. Since the concentration of ribosomes (4×10^{-5} M) is in the neighborhood of the K_m, a velocity of half V_{max} would be in order and we calculate that it would take 50 min for one molecule of DF13 in a single bacterium to inactivate one half of the ribosomes. The kinetics for inactivation of protein synthesis at a saturating multiplicity with DF13 indicates that after a 15-min lag period protein synthesis falls from the control rate at least 0.1-fold in 15 min (110). Michaelis constants are not available on colicin E3 or E2. However, from the available data in the literature (87, 88), we calculate (extrapolated from high enzyme low substrate to low enzyme high substrate concentration) that 1 molecule of E3, free of immunity protein, probably requires somewhere between 40 and 16 min to inactivate one half of the *E. coli* ribosomes. Kinetic data on protein synthesis inactivation obtained with intact cells indicates that after a 10–15-min lag period, protein synthesis almost totally ceases within a 3-min period (86). These considerations raise the possibility that the enzymatic colicins and perhaps even the membrane-acting colicins may undergo their transmembrane transport step in a concerted fashion delivering more than one active molecule to the necessary compartment. If the concerted entry is the rate-limiting step, single-hit kinetics would still apply as in the case previously discussed with diphtheria toxin.

Dose-Dependent Lag Period

Prior to the log linear entry process, the enzymatic and membrane-active colicins exhibit a dose-dependent lag period (Figures 6 and 7), which is lengthened at lower temperatures (96). Our view is that this is best explained by a temperature-dependent concerted obligatory processing event, and is analogous to the putative processing event already described for diphtheria toxin. The DT lag period is also temperature-dependent (111, 112).

Parallels with Bacteriophage DNA Transport

We have previously mentioned that various bacteriophages utilized low-molecular-weight iron transport systems for the transport of the DNA into the cytosol in some manner as yet not completely understood. Studies with a wide variety of phages including T1, T4, T5, and T7 indicate that a membrane potential is required for the transport of DNA across the membrane [reviewed in (9)]. Prior depolarization of the membrane will inhibit DNA injection. For T4 at a multiplicity of 0.5, the infection rate falls precipitously as the membrane potential is reduced below -100 mV. Similarly, the injection of phage DNA causes a depolarization of the membrane potential (113). Another parallel with the colicin is that in the case of T4 there is an activation of transport systems that are coupled to the phosphoenolpyruvate phosphotransferase system. The process of T1 DNA injection across the membrane results in a transient depolarization of the membrane potential. However this depolarization is highly specific (114). There is a large outflow of potassium and apparently a large influx of protons. However, neutral low-molecular-weight molecules such as glucosides did not show leakage, indicating that if a pore is formed it is a highly specific pore. When T4 phage injection is slowed down, by lowering the temperature to 10°C, and the kinetics studied by means of a blender interruption experiment, a lag period can be demonstrated (113). T5 phage shows an interesting feature in that the initial transport event is a transport of 8% of the DNA across the membrane. This material is transcribed and translated at which point the remainder of the DNA is transported (9). This is an interesting process since it represents a transport spanning the membrane with the deliverance of an active portion of a macromolecule into the cytosol compartment. It has been proposed that the transport of negatively charged DNA is accomplished by means of the proton-motive force gradient and that the depolarization observed is caused by an influx of protons neutralizing the DNA negative charges with respect to the negative membrane potential (115). This proton influx is balanced by K^+ efflux. Further, it has been proposed that the Ton B chelated iron uptake system (driven by the proton-motive force) is involved in mediating this vectorially coupled process (115).

GENERAL MODELS OF TOXIN TRANSMEMBRANE TRANSPORT, FUTURE PERSPECTIVES

The many similarities between diphtheria toxin and the colicins have been detailed: the dose-dependent lag period, the probabilistic nature of the transport event, the requirement of an energy source, the multi-domain structure with localized hydrophobic regions, the suggestion of membrane-spanning amphipathic helixes, and the property of forming voltage-dependent channels in lipid bilayers and ion-conducting channels in liposomes. Since certain colicins can elicit both membrane depolarization and cytosol nuclease activity, the minimum transport in these cases would consist of producing a membrane-spanning insertion and placing the active enzymatic site within the cytosol. The transmembrane insertion may or may not have specific channel properties depending on other host factors controlling the membrane. Moreover, the transmembrane configuration is not accessible to trypsin. We think that the available evidence indicates that this model could serve equally well for diphtheria toxin and related toxin membrane transport.

The relationships between phage DNA transport systems and colicins suggest that the energetics may have a common basis. For phage nucleic acid the energy requirement is related to the maintenance of the membrane potential and the movement of counter ions during DNA transport with a resultant transient membrane depolarization. How the maintenance of the bacterial membrane potential is needed for colicin membrane insertion should be a fruitful area for study. Similarly, the coupling between the energy requirement for DT and ricin transmembrane transport and the transport event should yield new insights into the overall process. For diphtheria toxin the concerted entry of 80 or so molecules from a vesicle would seem to require a process that can rapidly spread throughout the vesicle. The gating of a single ion channel can produce cascades leading to membrane depolarization. Perhaps the transmembrane transport of one toxin molecule could promote the concerted entry of the others within the same vesicle.

TOXIN CONJUGATES AND IMMUNOTOXINS

A considerable amount of the current toxin literature is concerned with linking toxins to other binding proteins or monoclonal antibodies in the case of immunotoxins. Early work in this area regarded the toxins as drugs to be targeted by a targeting reagent (116). A more recent formulation attempts to use the inherent translocation properties of multi-domain toxins and to change the toxin binding properties, i.e. to construct functional analogues of toxins with altered binding specificities (117–119). Progress in this area has been recently reviewed (120). Highly specific reagents are available for in vitro use (121, 122), and some are undergoing clinical trials as adjuncts to bone marrow transplantation (123). However, to date it has been impossible to separate the toxin translocation domain from the toxin receptor binding domain (present on nontarget as well as target cells) (23). This has prevented the construction of high-efficacy immunotoxins with the requisite specificity for in vivo use (57, 120).

Literature Cited

1. Pappenheimer, A. M. Jr. 1977. *Ann. Rev. Biochem.* 46:69–94
2. Uchida, T. 1982. *Molecular Action of Toxins and Viruses*, pp. 1–31, London/ New York: Elsevier, 369 pp.
3. Foster, J. W., Kinney, D. M. 1985. *CRC Crit. Rev. Microbiol.* 11:273–98
4. Middlebrook, J. L., Dorland, R. B. 1984. *Microbiol. Rev.* 48:199–221
5. Eidels, L., Proia, R. L., Hart, D. A. 1983. *Microbiol. Rev.* 47:596–620
6. Kaderbhai, M. A., Ridd, D. H., Robinson, A., Austen, B. M. 1984. *Biochem. Soc. Trans.* 12:917–20
7. Reid, G. A. 1985. *Curr. Top. Membr. Transp.* 24:295–336
8. Marsh, M., Kielian, M. C., Helenius, A. 1984. *Biochem. Soc. Trans.* 12:981–83
9. Schweiger, M., Hirsch-Kauffman, M. 1982. See Ref. 2, pp. 191–217
10. Welkos, S. L., Holmes, R. K. 1981. *J. Virol.* 37:936–45
11. Murphy, J. R., Bacha, P. 1979. *Microbiology*, ed. D. Schlessinger, pp. 181–86, Washington, DC: Am. Soc. Microbiol. 356 pp.
12. Buck, G. A., Groman, N. B. 1981. *J. Bacteriol.* 148:143–52
13. Leong, D., Coleman, K. D., Murphy, J. R. 1983. *Science* 220:515–17
14. Leong, D., Murphy, J. R. 1985. *J. Bacteriol.* In press
15. Greenfield, L., Bjorn, M. J., Horn, G., Fong, D., Buck, G. A., et al. 1983. *Proc. Natl. Acad. Sci. USA* 80:6853–57
16. Giannini, G., Rappuoli, R., Ratti, G. 1984. *Nucleic Acids Res.* 12:4063–69
17. Drazin, R., Kandel, J., Collier, R. J. 1971. *J. Biol. Chem.* 246:1504–10
18a. Uchida, T., Pappenheimer, A. M. Jr., Harper, A. A. 1973. *J. Biol. Chem.* 248:3845–50
18b. Kaczorek, M., Delpeyroux, F., Chenciner, N., Streeck, R. E., Murphy, J. R. et al. 1983. *Science* 221:855–58
19. Pappenheimer, A. M. Jr., Harper, A. A., Moynihan, M., Brockes, J. P. 1982. *J. Infect. Dis.* 145:94–102
20. Boquet, P., Silverman, M. S., Pappenheimer, A. M. Jr., Vernon, W. B. 1976. *Proc. Natl. Acad. Sci. USA* 73:4449–53
21. Lambotte, P., Falmagne, P., Capiau, C., Zanen, J., Ruysschaert, J.-M., Dirkx, J. 1980. *J. Cell Biol.* 87:837–40
22. Rappuoli, R., Ratti, G., Giannini, G., Perugini, M., Murphy, J. R. 1985. *Mol. Biol. Microb. Pathog. Meet.*, Lülea. Abstr. In press

23. Youle, R. J., Murray, G. J., Neville, D. M. Jr. 1981. *Cell* 23:551–59
24. Kandel, J., Collier, R. J., Chung, D. W. 1974. *J. Biol. Chem.* 249:2088–97
25. Carroll, S. F., Collier, R. J. 1984. *Proc. Natl. Acad. Sci. USA* 81:3307–11
26. Lory, S., Collier, R. J. 1980. *Proc. Natl. Acad. Sci. USA* 77:267–71
27. Barbieri, J. T., Carroll, S. F., Collier, R. J., McCloskey, J. A. 1981. *J. Biol. Chem.* 256:12247–51
28. Collins, C. M., Collier, R. J. 1984. *J. Biol. Chem.* 259:15159–62
29. Collins, C. M., Barbieri, J. T., Collier, R. J. 1984. *J. Biol. Chem.* 259:15154–58
30. Middlebrook, J. L., Dorland, R. B., Leppla, S. H. 1978. *J. Biol. Chem.* 253:7325–30
31. Middlebrook, J. L., Dorland, R. B. 1979. *Can. J. Microbiol.* 25:285–90
32. Chang, T., Neville, D. M. Jr. 1978. *J. Biol. Chem.* 253:6866–71
33. Proia, R. L., Wray, S. K., Hart, D. A., Eidels, L. 1980. *J. Biol. Chem.* 255:12025–33
34. Collier, R. J., Westbrook, E. M., McKay, D. B., Eisenberg, D. 1982. *J. Biol. Chem.* 257:5283–85
35. McKeever, B., Sarma, R. 1982. *J. Biol. Chem.* 257:6923–25
36. Honjo, T., Nishizuka, Y., Hayaishi, O. 1969. *Cold Spring Harbor Symp. Quant. Biol.* 34:603–8
37. Oppenheimer, N. J., Bodley, J. W. 1981. *J. Biol. Chem.* 256:8579–81
38. Youle, R. J., Neville, D. M. Jr. 1979. *J. Biol. Chem.* 254:11089–96
39. Moynihan, M. R., Pappenheimer, A. M. Jr. 1981. *Infect. Immun.* 32:575–82
40. Gill, D. M., Pappenheimer, A. M. Jr., Baseman, J. B. 1969. *Cold Spring Harbor Symp. Quant. Biol.* 34:595–602
41. Gale, E. F., Cundlittle, E., Reynolds, P. E., Richmond, M. H., Waring, M. J. 1981. *The Molecular Basis of Antibiotic Action,* pp. 402–547. New York: Wiley. 547 pp.
42. Hudson, T. H., Neville, D. M. Jr. 1985. *J. Biol. Chem.* 260:2675–80
43. Bodley, J. W., Upham, R., Crow, F. W., Tomer, K. B., Gross, M. L. 1984. *Arch. Biochem. Biophys.* 230:590–93
44. Richter, C., Winterhalter, K. H., Baumhüter, S., Lötscher, H.-R., Moser, B. 1983. *Proc. Natl. Acad. Sci. USA* 80:3188–92
45. Yost, D. A., Anderson, B. M. 1983. *J. Biol. Chem.* 258:3075–80
46. Honjo, T., Nishizuka, Y., Kato, I., Hayaishi, O. 1971. *J. Biol. Chem.* 246:4251–60
47. Draper, R. K., Chin, D., Eurey-Owens, D., Scheffler, I. E., Simon, M. I. 1979. *J. Cell Biol.* 83:116–25
48. Moehring, J. M., Moehring, T. J., Danley, D. E. 1980. *Proc. Natl. Acad. Sci. USA* 77:1010–14
49. Chen, J.-Y. C., Bodley, J. W., Livingston, D. M. 1985. *Mol. Cell. Biol.* In press
50. Chen, J.-Y. C., Bodley, J. W. 1985. *Fed. Proc.* 44:124 (Abstr.)
51. Dunlop, P. C., Bodley, J. W. 1983. *J. Biol. Chem.* 258:4754–58
52. Van Ness, B. G., Barrowclough, B., Bodley, J. W. 1980. *FEBS Lett.* 120:4–6
53. Pappenheimer, A. M. Jr., Dunlop, P. C., Adolph, K. W., Bodley, J. W. 1983. *J. Bacteriol.* 153:1342–47
54. Lee, H., Iglewski, W. J. 1984. *Proc. Natl. Acad. Sci. USA* 81:2703–7
55. Freychet, P., Roth, J., Neville, D. M. Jr. 1971. *Proc. Natl. Acad. Sci. USA* 68:1833–37
56. Heagy, W. E., Neville, D. M. Jr. 1981. *J. Biol. Chem.* 256:12788–92
57. Esworthy, R. S., Neville, D. M. Jr. 1984. *J. Biol. Chem.* 259:11496–504
58. Middlebrook, J. L., Dorland, R. B., Leppla, S. H. 1978. *J. Biol. Chem.* 253:7325–30
59. Didsbury, J. R., Moehring, J. M., Moehring, T. J. 1983. *Mol. Cell. Biol.* 3:1283–94
60. Chang, T., Neville, D. M. Jr. 1978. *J. Biol. Chem.* 253:6866–71
61. Mekada, E., Uchida, T. 1985. *J. Biol. Chem.* 260:12148–207
62. Hayakawa, S., Uchida, T., Mekada, E., Moynihan, M. R., Okada, Y. 1983. *J. Biol. Chem.* 258:4311–17
63a. Proia, R. L., Hart, D. A., Holmes, R. K., Holmes, K. V., Eidels, L. 1979. *Proc. Natl. Acad. Sci. USA* 76:685–89
63b. Proia, R. L., Eidels, L., Hart, D. A. 1981. *J. Biol. Chem.* 256:4991–97
64. Mekada, E., Kohno, K., Ishiura, M., Uchida, T., Okada, Y. 1982. *Biochem. Biophys. Res. Commun.* 109:792–99
65. Keen, J. H., Maxfield, F. R., Hardegree, M. C., Habig, W. H. 1982. *Proc. Natl. Acad. Sci. USA* 79:2912–16
66. Murakami, S., Bodley, J. W., Livingston, D. M. 1982. *Mol. Cell. Biol.* 2:588–92
67a. Moehring, J. M., Moehring, T. J. 1983. *Infect. Immun.* 41:998–1009
67b. Robbins, A. R., Peng, S. S., Marshall, J. L. 1983. *J. Cell Biol.* 96:1064–71
67c. Marnell, M. H., Mathis, L. S., Stookey, M., Shia, S., Stone, D. K., Draper, R. K. 1984. *J. Cell Biol.* 99:1907–16
68. Draper, R. K., Chin, D., Stubbs, L.,

Simon, M. I. 1978. *J. Supramol. Struct.* 9:47–55

69. Bonventre, P. F., Saelinger, C. B., Ivins, B., Woscinski, C., Amorini, M. 1975. *Infect. Immun.* 11:675–84

70. Duncan, J. L., Groman, N. B. 1969. *J. Bacteriol.* 98:963–69

71. Hudson, T. H., Kimak, M. A. G., Neville, D. M. Jr. 1985. Submitted for publication

72. Pappenheimer, A. M. Jr., Gill, D. M. 1973. *Science* 182:353–58

73. Yamaizumi, M., Uchida, T., Takamatsu, K., Okada, Y. 1982. *Proc. Natl. Acad. Sci. USA* 79:461–65

74. Sandvig, K., Olsnes, S. 1981. *J. Biol. Chem.* 256:9068–76

75. Sandvig, K., Olsnes, S. 1980. *J. Cell Biol.* 87:828–32

76. Marnell, M. H., Stookey, M., Draper, R. K. 1982. *J. Cell Biol.* 93:57–62

77. Draper, R. K., Simon, M. I. 1980. *J. Cell Biol.* 87:849–54

78. Bacha, P., Murphy, J. R., Moynihan, M. 1980. *J. Biol. Chem.* 255:10658–62

79. Donovan, J. J., Simon, M. I., Montal, M. 1982. *Nature* 298:669–72

80. Kagan, B. L., Finkelstein, A., Colombini, M. 1981. *Proc. Natl. Acad. Sci. USA* 78:4950–54

81. Misler, S. 1983. *Proc. Natl. Acad. Sci. USA* 80:4320–24

82. Jakes, K. S. 1982. See Ref. 2, pp. 131–67

83. Konisky, J. 1978. *The Bacteria*, ed. L. N. Ornston, J. R. Sokatch, 6:71–136. New York: Academic. 603 pp.

84. Konisky, J., Tokuda, H. 1978. *Proc. Natl. Acad. Sci. USA* 75:2579–83

85. Schaller, K., Nomura, M. 1976. *Proc. Natl. Acad. Sci. USA* 73:3989–93

86. Samson, A. C. R., Senior, B. W., Holland, I. B. 1972. *J. Supramol. Struct.* 1:135–44

87. Boon, T. 1971. *Proc. Natl. Acad. Sci. USA* 68:2421–25

88. Jakes, K. S., Zinder, N. D. 1974. *Proc. Natl. Acad. Sci. USA* 71:3380–84

89. Oudega, B., de Graaf, F. K. 1976. *Biochim. Biophys. Acta* 425:296–304

90. de Graaf, F. K. 1973. *Antonie van Leeuwenhoek* 39:109–19

91. Beppu, T., Yamamoto, H., Arima, K. 1975. *Antimicrob. Agents Chemother.* 8:617–26

92a. Davies, J. K., Reeves, P. 1975. *J. Bacteriol.* 123:96–101

92b. Davies, J. K., Reeves, P. 1975. *J. Bacteriol.* 123:102–7

92c. Neville, D. M. Jr., Chang, T.-M. 1978. *Current Topics in Membranes and Transport*, ed. F. Bonner, A. Kleinzeller, 10:65–150. New York: Academic 367 pp.

93. Konisky, J., Cowell, B. S. 1972. *J. Biol. Chem.* 247:6524–29

94. Oudega, B., Klaasen-Boor, P., Sneeuwloper, G., de Graaf, F. K. 1977. *Eur. J. Biochem.* 78:445–53

95. Bhattacharyya, P., Wendt, L., Whitney, E., Silver, S. 1970. *Science* 168:998–1000

96. Wendt, L. 1970. *J. Bacteriol.* 104:1236–41

97. Gaastra, W., Oudega, B., de Graaf, F. K. 1978. *Biochim. Biophys. Acta* 540:301–12

98. van den Elzen, P. J. M., Walters, H. H. B., Veltkamp, E., Nijkamp, H. J. J. 1983. *Nucleic Acids Res.* 11:2465–77

99. Ohno-Iwashita, Y., Imahori, K. 1980. *Biochemistry* 19:652–59

100. Ohno-Iwashita, Y., Imahori, K. 1982. *J. Biol. Chem.* 257:6446–51

101. Yamada, M., Ebina, Y., Miyata, T., Nakazawa, T., Nakazawa, A. 1982. *Proc. Natl. Acad. Sci. USA* 79:2827–31

102. Davidson, V. L., Brunden, K. R., Cramer, W. A., Cohen, F. S. 1984. *J. Membr. Biol.* 79:105–18

103. Stevens, C. F. 1985. *Trends Neurosci.* 8:1–2

104a. Finer-Moore, J., Stroud, R. M. 1984. *Proc. Natl. Acad. Sci. USA* 81:155–59

104b. Nomura, M. 1964. *Proc. Natl. Acad. Sci. USA* 52:1514–21

105. Plate, C. A., Luria, S. E. 1972. *Proc. Natl. Acad. Sci. USA* 69:2030–34

106. Farid-Sabet, S. 1982. *J. Bacteriol.* 150:1383–90

107a. Kayalar, C., Erdheim, G. R., Shanafelt, A., Goldman, K. 1984. *Curr. Top. Cell. Regul.* 24:301–12

107b. Schein, S. J., Kagan, B. L., Finkelstein, A. 1978. *Nature* 276:159–63

108. Ramos, S., Kaback, H. R. 1977. *Biochemistry* 16:854–59

109. Konisky, J., Tokuda, H. 1979. *Proc. Natl. Acad. Sci. USA* 76:6167–71

110. de Graaf, F. K., Planta, R. J., Stouthamer, A. H. 1971. *Biochim. Biophys. Acta* 240:122–36

111. Kato, I., Pappenheimer, A. M. Jr. 1960. *J. Exp. Med.* 112:329–49

112. Strauss, N., Hendee, E. D. 1959. *J. Exp. Med.* 109:144–63

113. Kalasauskaite, E. V., Kadisaite, D. L., Daugelavicius, R. J., Grinius, L. L., Jasaitis, A. A. 1983. *Eur. J. Biochem.* 130:123–30

114. Wagner, E. F., Ponta, H., Schweiger, M. 1980. *J. Biol. Chem.* 255:534–39

115. Wagner, E. F., Schweiger, M. 1980. *J. Biol. Chem.* 255:540–42

116. Moolten, F. L., Cooperband, S. R. 1970. *Science* 169:68–70

117. Chang, T. M., Neville, D. M. Jr. 1977. *J. Biol. Chem.* 252:1505–14
118. Chang, T. M., Dazord, A., Neville, D. M. Jr. 1977. *J. Biol. Chem.* 252:1515–22
119. Youle, R. J., Neville, D. M. Jr. 1980. *Proc. Natl. Acad. Sci. USA* 77:5483–86
120. Neville, D. M. Jr. 1986. *CRC Critical Reviews in Therapeutic Drug Carrier Systems,* Vol. 2, Boca Raton, FL: CRC. 2:In press
121. Vallera, D. A., Ash, R. C., Zanjani, E. D., Kersey, J. H., LeBein, T. W., et al. 1983. *Science* 222:512–14
122. Casellas, P., Canat, X., Fauser, A. A., Gros, O., Laurent, G., et al. 1985. *Blood* 65:289–97
123. Filipovich, A. H., Vallera, D. A., Youle, R. J., Neville, D. M. Jr., Kersey, J. H. 1985. *Transplant Proc.* 17:442–44

Ann. Rev. Biochem. 1986. 55:225–48

MOLECULAR ASPECTS OF SUGAR:ION COTRANSPORT

J. Keith Wright, Robert Seckler, and Peter Overath

Max-Planck-Institut für Biologie, Corrensstrasse 38, D7400 Tübingen, West Germany

CONTENTS

INTRODUCTION

Kinetic investigations over many years have established that the transport of nutrients into cells is mediated by specific carriers localized within the cell membrane. In its simplest form, transport was viewed as a reversible, cyclic process involving the binding of a solute to the carrier at the outer face of the membrane, movement of the complex to the inner face, release of substrate to the cytoplasm, and, finally, reorientation of the empty carrier back to the outer membrane face (1–3). An essential feature of this kinetic model was that a single binding site be alternately exposed on the outer or inner side of the membrane. However, there was considerable freedom as to how such a process might occur on the molecular level. The later view of membrane structure as a fluid phospholipid bilayer with proteins embedded or attached led to the proposal that carriers are membrane-spanning proteins, unlikely capable of translational movement across the membrane (4). The structural equivalent of

225

the kinetic cycle is now considered to be steps for binding and release of substrates, connected by conformational changes inside a protein or a complex of several proteins. Although the three-dimensional structure and the detailed molecular mechanism are not yet known for any transport system, biochemical, biophysical, and genetic investigations in the past 10 years have provided ample evidence that transport is mediated by a class of membrane-spanning proteins conferred with high internal flexibility.

Transport of sugars occurs by four distinct mechanisms. First, many mammalian cells contain a *hexose transporter*[1], studied in most detail in erythrocytes (5, 6), which merely equilibrates sugars across the membrane and, therefore, conforms in its kinetic properties most closely to a classical carrier. The second class of transport systems, which is the subject of this review, includes systems that effect the simultaneous translocation of sugars and cations (H^+ or Na^+). Therefore, the proteins involved are called *sugar:ion cotransporters*. In the presence of an electrochemical potential gradient of Na^+ or H^+ they bring about the active transport of sugars. Similar transporters are involved in the active uptake of amino acids. The remaining two classes have thus far only been found in bacteria. The *binding-protein-dependent transport systems,* of which maltose transport in *Escherichia coli* is a typical example, mediate the active transport of sugars, most likely at the expense of ATP hydrolysis. The binding proteins are situated in the periplasm and a complex of several proteins is located in the cytoplasmic membrane (7, 8). The last type of transport system, called the *phosphoenol pyruvate:sugar phosphotransferase system* differs principally from the three others in that the sugar is simultaneously phosphorylated while translocated through the bacterial cytoplasmic membrane (9).

Active transport by sugar:ion cotransport is a ubiquitous means by which cells concentrate such essential nutrients. These systems make use of the omnipresent, unequal distribution of ions across cytoplasmic membranes. Cotransporters occur in bacteria (10), yeasts (11, 12), algae (13), plants (14, 15), fungi (16), parasitic protozoa (17), and higher eukaryotes (18, 19). The following is an attempt to summarize present knowledge on the systems that have been studied in detail in the past years.

COTRANSPORTERS

Early experiments demonstrated that the uptake of D-glucose from the lumen of kidney and small intestine across the brush-border membrane was intimately

---

[1]There is at present no accepted nomenclature for transport proteins. Designations like carrier, permease, transporter, or translocator are synonymous and their preferred use for a given system frequently has historical reasons. We choose to use the noncommittal term *transporter* whenever possible.

associated with sodium ions (18, 20). Likewise, the transport of lactose into *Escherichia coli* was found to be associated with the movement of protons (21). These and a host of other observations have fortified the concept, formulated in the theories of gradient coupling for Na^+ (18) and of chemiosmotic coupling for H^+ (22), that the transport of sugars against their concentration gradients is coupled to the movement of ions down the gradients of their electrochemical potentials. The gradients of the electrochemical potential of ions in turn are created by primary active transport systems such as the Na^+/K^+-ATPase or the respiratory chain.

The brush-border membranes of kidney tubuli and of the small intestine contain two Na^+-dependent pathways for D-glucose. Presently, it is unclear whether in the intestine these two pathways represent the activities of two different cotransporters (23) or of a single cotransporter operating in two modes (24). However, in rabbit kidney, two distinct cotransporters have been clearly identified, one in the outer cortex and the other in the outer medulla (25, 26). The aromatic β-glucoside phlorizin is frequently employed as an inhibitor of high affinity.

Lactose uptake in *E. coli* is catalyzed by an H^+ : galactoside cotransporter located in the cytoplasmic membrane. The protein, also called lactose permease (27), is the product of the *lacY* gene of the *lac* operon. This cotransporter recognizes a large number of α- and β-galactosides (28). In addition to lactose, β-D-galactosyl-1-thio-β-D-galactoside (GalSGal), methyl 1-thio-β-D-galactoside, and melibiose (O-α-D-galactoside-(1→6)-α-D-glucose) are used in transport studies. Because of their high affinities for the cotransporter, GalSGal and *p*-nitrophenyl α-D-galactoside (NpαGal) are employed in binding and transport studies.

The Na^+-dependent active transport of melibiose in *E. coli* strains B and K12 (29) and in *Salmonella typhimurium* (30) involves another cotransporter encoded by the *melB* gene. The substrates methyl 1-O-α-D-galactoside, methyl 1-thio-β-D-galactoside, and GalSGal are frequently employed in transport studies. The high-affinity substrate NpαGal is used for binding and transport measurements. The cotransporter accepts H^+ and Li^+ in addition to Na^+ for the transport of some galactosides. Therefore, this cotransporter might be related to an ancestral transport protein important in the evolution of Na^+-dependent cotransporters from the H^+-dependent cotransporters (31).

COTRANSPORT IN THEORY

A basic scheme of cotransport is shown in Figure 1. In the absence of substrate (S) and ion (ION) the transporter can exist in two conformations, C' and C'', exposing the binding sites for S and ION alternately to the outer (') or inner face ('') of the cytoplasmic membrane. Addition of S and ION to the medium leads to the formation of a ternary complex or intermediate C·S·ION'. The ternary intermediate reorients effecting the exposure of the binding sites to the cytoplasm (C·S·ION' to C·S·ION''). Dissociation of C·S·ION'' via C·S'' or C·ION'' leads to the formation of the unloaded transporter in the C'' conformation which completes the cycle by reorientation to the C' state. This cyclic process catalyzes the coupled transport of S and ION into the cell (influx), characterized by the translocation rate constants k_c^+ and k_o^-. If the cycle operates in the reverse direction, characterized by rate constants k_o^+ and k_c^-, S and ION move from the cytoplasm to the surrounding medium (efflux). If S and ION are present on both sides of the membrane, the transporter transfers cosubstrates between the two compartments via the ternary complexes C·S·-ION' and C·S·ION'' (exchange). For additional definitions see the legend to Figure 1.

When the fluxes of two species across a membrane via a transporter are coupled, the gradients of their electrochemical potentials are also coupled. Therefore, ion:sugar cotransporters are biological energy transducers that convert the energy stored in the transmembrane gradient of the electrochemical potential of the cotransported ion ($\Delta \widetilde{\mu}_{ION}$) into a gradient of the chemical

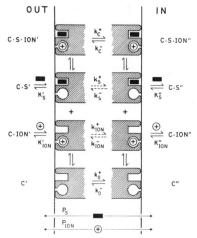

Figure 1   The cotransport cycle. The cotransporter (C) possesses one binding site for the sugar (S) and one binding site for the ion (ION), both of which can be exposed concertedly on the outer (') or inner face ('') of the membrane. Ternary intermediates C·S·ION' or C·S·ION'' are formed from C' or C'' via binary intermediates C·ION' and C·S' or C·ION'' and C·S'' characterized by binding constants K_{ION}' and K_S' or K_{ION}'' and K_S'', respectively. Strictly coupled cotransport occurs by reorientation exclusively via the ternary intermediates as determined by the rate constants k_c^+ and k_c^- and the reorientation of the unloaded transporter characterized by the rate constants k_o^+ and k_o^-. Reorientations of binary intermediates indicated by rate constants k_{ION}^+ and k_{ION}^- or k_S^+ and k_S^- effect the uncoupled movement of ION or S. Such processes are called internal leaks or slips (11, 12). External leaks are caused by passive diffusion of the cosubstrates and are characterized by permeability coefficients P_S and P_{ION}.

potential of the sugar by coupling the transmembrane fluxes of ion and sugar. The free energy available from a monovalent cation crossing the membrane under reversible conditions is

$$\Delta\widetilde{\mu}_{ION} = RT \ln \frac{[ION]''}{[ION]'} + F\Delta\psi \qquad\qquad 1.$$

where R is 8.314 J mol $^{-1}K^{-1}$, T the absolute temperature, F is 96, 490 J Amp $^{-1}s^{-1}$, and $\Delta\Psi$ is the difference in the electrical potential $\Psi'' - \Psi'$. The energy required for the transport of a mole of a neutral sugar under reversible conditions is

$$\Delta\mu_S = RT \ln \frac{[S]''}{[S]'} . \qquad\qquad 2.$$

In the absence of any internal or external leaks (cf Figure 1) the cotransporter inversely couples the free-energy gradients of ion and the sugar so that

$$\Delta\mu_S = - n_{ION} \cdot \Delta \widetilde{\mu}_{ION}, \qquad\qquad 3.$$

where n_{ION} is the symport stoichiometry, i.e. the number of ions translocated per molecule of sugar.

In the course of active transport, cotransporters are exposed to three gradients: the concentration gradient of the ion $\Delta\mu_{ION}$, the concentration gradient of the sugar $\Delta\mu_S$, and the electrical-potential gradient $\Delta\Psi$. At the equilibrium of Equation 3, the rates of sugar influx and efflux are equal. From this requirement, a relation between the parameters of the model and the driving gradient can be derived:

$$\frac{[S]''}{[S]'} = \frac{[ION]'}{[ION]''} \frac{k_c^+ k_o^- K_{ION}'' K_S''}{k_c^- k_o^+ K_{ION}' K_S'} = \frac{[ION]'}{[ION]''} \cdot e \frac{-F\Delta\psi}{RT} . \qquad 4.$$

This restriction follows from the cyclic nature of the transport process. The significance of this equation can be illustrated by calculating the distribution of the eight cotransport intermediates (Figure 1) as a function of the prevailing electrochemical gradients of ion and sugar (32, 33).

An example of the distribution of transport intermediates during sugar accumulation by an H^+-coupled transporter is given in Figure 2. In the absence of ΔpH and $\Delta\Psi$, $[C\cdot H]'$ and $[C\cdot H]''$ are equal (A). Imposition of $\Delta pH = 2$ yields $[C\cdot H]'/[C\cdot H]'' = 100$, exactly determined by $[H]'/[H]''$ (B). Addition of sugar to the external compartment to $S' = K_S'$ perturbs the previous distribution of intermediates (C). This nonequilibrium distribution drives the cycling of cotransporters and the influx of sugar (see arrows above CHS' and C''). At an intermediate stage, [S]'and [S]'' are equal and the distribution of intermediates is different (D). Finally, when [CS]''/[CS]' = 100, corresponding to [S]''/[S]' = 100, no net transport of sugar can occur (E). At equilibrium (Equation 4), the concentration gradients of ion and sugar are exactly reflected in the concentration ratios of certain cotransport intermediates:

$$\frac{[C \cdot H]'}{[C \cdot H]''} = \frac{[H^+]'}{[H^+]''} \quad \text{and} \quad \frac{[C \cdot S]''}{[C \cdot S]'} = \frac{[S]''}{[S]'} . \qquad 5.$$

Therefore, the mass action of ion and sugar causes redistributions of intermediates, which result in an accumulation of sugar according to Equation 3. Thermodynamic considerations do not require changes in rate or dissociation constants for solute accumulation with ΔpH. If ΔpH-dependent changes in rate or dissociation constants occur, they are internally compensated due to the restriction given by Equation 4.

Active transport with ΔΨ is also associated with redistributions of cotransporters. The imposition of ΔΨ may cause increases in k_c^+, k_o^-, K_{ION}'', and K_S'' or decreases in k_c^-, k_o^+, K_{ION}' and K_S' (see Equation 4 for [ION]' = [ION]''). However these effects may be distributed among the parameters of the cycle, the product of all changes, analyzed by Equation 4, is exactly

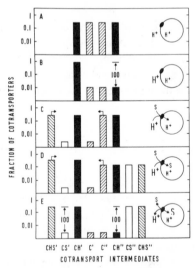

Figure 2   Active transport of solute with ΔpH. The concentrations of outward- and inward-oriented intermediates are calculated for $k_c^+ = k_c^- = k_o^+ = k_o^-$, $K_H' = K_H''$, and $K_S' = K_S''$, as is Equation 5 in the text. Panels A to D represent different times at which S' = S'' = 0, ΔpH = 0 (panel A); S' = S'' = 0, ΔpH = 2 (H' = 100 × K_H' and H'' = K_H'', panel B); S' = K_S, S'' = 0, ΔpH = 2 (panel C); S' = S'' = K_S, ΔpH = 2 (panel D); or S' = K_S, S'' = 100 K_S, ΔpH = 2 (equilibrium, panel E). Taken from Ref. 33.

determined by the magnitude of $\Delta\Psi$. In any step in the cotransport cycle where the ion or charged groups of the protein move through $\Delta\Psi$, electrical work is performed. The equilibrium constant for this step is $\Delta\Psi$-dependent. For example, if the entire electrical work is performed in the interconversion of the unloaded cotransporter affecting k_o^+ and k_o^-, then the equilibrium in the presence of $\Delta\Psi$ is described by

$$\frac{[C]'}{[C]''} = \left(\frac{k_o^-}{k_o^+}\right)_{\Delta\psi=0} \cdot e \frac{-F\Delta\psi}{RT} .$$ 6.

The displacement of the equilibrium of the unloaded cotransporter by $\Delta\Psi$ also causes an equivalent redistribution of the binary complexes C·ION' and C·-ION'' (cf 33). As in the case discussed above (Figure 2B), the asymmetric distribution of binary complexes causes net inward flow of solute. Simulations involving $\Delta\Psi$-induced affinity changes, e.g. for $(K_s'')_{\Delta\Psi<0} >> (K_s'')_{\Delta\Psi=0}$, likewise demonstrate changes in the distribution of certain intermediates (33).

In summary, transmembrane gradients of the sugar, the ion, and the electrical potential are reflected in the distributions of the cotransport intermediates. The cotransport equilibrium (Equation 4) has a counterpart in the magnitude of redistribution of cotransporters due to the mass action of sugar (MA_S) or ion (MA_{ION}) and the product of the individual redistributions caused by the performance of electrical work (EW) so that $MA_S = MA_{ION} \cdot EW$.

COTRANSPORT IN PRACTICE

Real transport systems can be described by the model presented here. The minimal requirements for cotransport are binding sites for ion and sugar and a conformational change enabling them to be presented concertedly to each of two compartments. There is sufficient experimental evidence to allow Na^+ : D-glucose cotransport in the intestine (19, 24, 34–36) and H^+ : galactoside cotransport in *E. coli* (37–41) to be examined in detail.

The intestinal *Na^+ : D-glucose cotransporter* contains a binding site for D-glucose and one or possibly two for Na^+. As expected, binding sites are accessible from both sides of the membrane. There appears to be no fixed order of binding or release. However, the equilibrium between C' and C'' in the absence of $\Delta\tilde{\mu}_{Na^+}$ lies on the side of C''. Therefore, the externally added glucoside phlorizin binds weakly. When the external concentration of Na^+ is raised or an externally positive $\Delta\Psi$ is imposed the glucoside binds readily to the cotransporter. This finding illustrates clearly that redistributions of cotransporters have occurred. The binding of external Na^+ to C' displaces the equilibrium between C·Na' and C·Na'' by mass action. The performance of electrical work in the interconversion of C' and C'' likewise increases the number of external phlorizin binding sites. The effect of $\Delta\Psi$ can be interpreted as the acceleration of the outward movement of a negatively charged portion of the protein functioning as a gate. Tentatively, these charges have been identified with carboxylate groups which may also serve as the binding site(s) for Na^+. The half-saturation constant for the binding of Na^+ to this site is about 40 mM. Therefore, under physiological conditions of about 0.1 M Na^+, the Na^+ binding site is saturated. The cotransporter is poised as C·Na' to bind glucose and to transport ion and sugar to the internal compartment. The turnover number for cotransport in brush-border membrane vesicles is estimated at $1-20s^{-1}$ (s^{-1} is equivalent to mol substrate/s/mol cotransporter; 24, 42–45). The binary intermediates C·Na and C·S do not reorient. Internal Na^+ and glucose inhibit influx. Thus, k_S^- and k_{ION}^- must be negligible or zero, and there should be strict coupling of Na^+ and glucose fluxes.

Estimates of the stoichiometry of Na^+:D-glucose cotransport in intestinal cells vary between 1 and 3 in different preparations (23, 24, 35, 45–50). In the kidney, two separate cotransporters can be identified (25, 26, 43). The cotransporter in the outer cortex with a low affinity for glucose catalyzes 1:1 cotransport. The second protein in the outer medulla with a higher affinity for glucose catalyzes active transport with 2 Na^+ per glucose thus permitting the efficient removal of glucose present at low concentration in the filtrate. Some of the problems of determining stoichiometries in brush-border membrane vesicles are discussed in Refs. 19, 47, and 51.

The H^+ : *galactoside cotransporter* of E. coli contains a binding site for H^+ and a binding site for α- or β-galactosides (28, 52). The binding sites are accessible from both sides of the membrane, as influx and efflux both exhibit saturation (27, 39, 40, 53, 54). The affinity of the cotransporter for H^+ is high with a pK_H between 8.3 and 10.3 (54–57). As in the previous example, the ion binding site is saturated under physiological conditions. The protonated intermediates will predominate.

Quantitative studies indicate the presence of only a single binding site for galactoside per polypeptide chain. Every substrate tested competitively inhibits the binding and transport of all other galactosides. Therefore, all substrates are recognized by the common galactosyl residue.

Examination of the transport of GalSGal in the absence of $\Delta\widetilde{\mu}_{H^+}$, where S' = S'' at equilibrium, cf Equation 4, discloses that this substrate binds with the same affinity to the cotransporter on both sides of the membrane ($K_S' = K_S'' = 80$ μM, in cells). Furthermore, this symmetry extends to the rate constants for the reorientation of the binding sites, $k_o^- = k_o^+ = 20\,s^{-1}$ and $k_c^+ = k_c^- = 20\,s^{-1}$. Interestingly, the occupation of the binding sites does not hinder or accelerate the rate of the conformational change in this case. Thus, there is a rapid equilibrium between C' and C'' and between C·S·H' and C·S·H''. In the absence of $\Delta\widetilde{\mu}_{H^+}$, there are nearly equal concentrations of C' and C'' in contrast to the predominance of C'' in the Na^+ : glucose cotransporter at $\Delta \widetilde{\mu}_{Na^+} = 0$.

The dependence of influx, efflux, and exchange on $\Delta\Psi$ indicates that the membrane potential increases k_o^- and decreases k_o^+ (38, 53, 54). Therefore, as in the case of Na^+ : glucose cotransport, the equilibrium between C' and C'' is displaced towards C' by the performance of electrical work. In contrast, ΔpH causes a displacement of binary intermediates by mass action alone (cf Figure 2). In the presence of $\Delta\Psi$ and ΔpH, hydrophilic substrates like GalSGal and lactose can be accumulated nearly 1000-fold. On the other hand, galactosides exhibiting a significant passive permeability across the membrane (P_s in Figure 1) are accumulated to markedly lower levels (58).

The turnover number for GalSGal is about 25 s^{-1} in cells and 1 s^{-1} in right-side-out vesicles (52, 59). Because of the inherent reversibility of all steps in the cotransport cycle, active transport should be possible in right-side-out and in inside-out vesicles as long as $\Delta \widetilde{\mu}_{H^+}$ is made internally negative. The same should be true in reconstituted proteoliposomes. Indeed, active transport can be observed in all three preparations (59–62). In addition, because of the functional symmetry of the cotransporter, nearly the same kinetic parameters are observed in the three preparations.

A stoichiometry of 1 H^+ per lactose or GalSGal can be measured under many conditions in cells, vesicles, and proteoliposomes (21, 32, 59, 63–67). However, under some conditions values between 0 and 2 are estimated (68–70). One possible explanation is that the stoichiometry of cotransport is variable. This would inevitably entail inner leaks and constant dissipation of $\Delta \widetilde{\mu}_{H^+}$. Alternatively, the apparent deviations of the stoichiometry from unity could reflect the difficulty in performing such measurements. If inner leaks were present, their magnitudes could be increased by mutation. Two mutants have been isolated under conditions where a low accumulation of a poisonous galactoside could be caused by uncoupling H^+ and galactoside fluxes. Both mutants exhibit similar kinetic properties and grow poorly on lactose and not at all on melibiose (71). Further analysis of one mutant demonstrates that transport is merely slow but not uncoupled (72). Thus, these mutants do not support the concept of inner leaks as significant pathways in H^+ : galactoside cotransport, analogous to the absence of inner leaks in Na^+ : D-glucose cotransport.

The essential features of the model described above can be summarized. The cotransporter operates in a functionally symmetrical way in the absence of $\Delta \widetilde{\mu}_{H^+}$. Active transport occurs when equilibria among cotransport intermediates are displaced. The imposition of $\Delta\Psi$ increases k_o^- and decreases k_o^+, and ΔpH causes redistribution of binary intermediates by mass action. There is much evidence to support this model, although some aspects remain controversial. Two effects not covered by this description remain to be clarified. These are the $\Delta \widetilde{\mu}_{H^+}$-dependent affinity of some galactosides for the cotransporter (32, 39, 52, 53) and galactoside accumulation lower than that predicted by Equation 3 at high galactoside concentrations (27).

For GalSGal and a number of other substrates (class I) the dissociation constant K_S ($\Delta \tilde{\mu}_{H^+} = 0$) and the half-saturation constant for active transport K_T ($\Delta \tilde{\mu}_{H^+} < 0$) are similar (28). Thus, the imposition of $\Delta \tilde{\mu}_{H^+}$ does not reduce K'_S. For lactose and some other substrates (class II), the imposition of $\Delta \tilde{\mu}_{H^+}$, or at least $\Delta \Psi$, causes the affinity of various substrates to increase to different extents: $K_S/K_T = 160$ for lactose to 3 for NpαGal (52). Because the electrical work for active transport is performed in the steps k_o^+ and k_o^- and because both lactose and GalSGal are accumulated about 1000-fold at low external concentrations, the shift from a high K_S to a low K_T must be a secondary effect. The restriction of Equation 4 requires that such an ancillary change be balanced by a change of equal magnitude elsewhere in the catalytic cycle. Kinetic measurements suggest that the affinity of the cotransporter for lactose is high on both sides of the membrane in the presence of $\Delta \Psi$ (54, 59, 73). Therefore, the ratio K''_S/K'_S in Equation 4 appears to be about 1 for all galactosides, even though for some galactosides like lactose, K'_S and K''_S depend on $\Delta \Psi$. One possible explanation for the $K_S \rightarrow K_T$ shift is that $\Delta \Psi$ causes a structural change in part of the sugar-binding site. The binding of class I substrates is not influenced by this change, whereas the binding of class II substrates is altered. There is other evidence that class I and class II substrates, while occupying a common recognition site for the galactosyl moiety in the cotransporter, interact with the protein in slightly different ways (72, 74).

The second effect is the attainment of a maximal internal concentration of galactoside with increasing external concentrations. For hydrophilic disaccharides this effect cannot be explained by inner or outer leaks (75, 76). Examination of the kinetics of GalSGal transport reveals, however, that the cotransporter is kinetically so designed that net fluxes of galactosides at high concentrations are not possible, because the affinities for H^+ and galactoside are high on both sides of the membrane and k_c^+ and k_c^- are nearly equal (40). Such a kinetic design is an interesting form of self-regulation of cotransporters by which the cell is protected against the toxic effects of high internal levels of a solute without resorting to dissipative leaks.

One final effect on cotransport deserves consideration. Lowering the temperature through the midpoint of the phase transition of the membrane phospholipids inhibits transport (77). For example, in cotransporter-overproducing fatty-acid auxotrophs, a decrease in the temperature of 20°C causes a 40-fold decrease in the rate of galactoside transport. Over the same temperature range, the affinity of the carrier for galactoside increases slightly and the number of accessible binding sites is unchanged (52). Thus, the phase transition of the membrane phospholipids affects the reorientation steps k_o and k_c and not the number of active cotransporters or their affinity for substrate.

Few systems have been studied in as much detail as the two discussed here. However, the following summary of observations made on other ion : sugar cotransporters suggests there may be functional similarities among these diverse systems.

The K_T for sugar influx decreases with increasing concentration of the ion. This behavior has been observed for the two cotransporters discussed above and has been documented for Na^+/Li^+ : galactoside cotransport in $E.$ $coli$ and $S.$ $typhimurium$ (78–80), H^+ : D-glucose transport in $Chlorella$ (13), for H^+ : hexose transport in yeast (81), and for H^+ : sucrose cotransport in bean leaf discs (82). On a qualitative level, this effect is due to the formation of a ternary (or higher) intermediate containing ion and sugar and can be rationalized by either ordered or random binding of the cosubstrates. Important is that from the dependence of the change on the ion concentration, a rough estimate of the affinity of the ion for the cotransporter can be made. Assuming ambient external concentrations of 0.1 M Na^+, 10^{-7} M H^+ for $E.$ $coli,$ or 10^{-5} M H^+ for yeast, the ion-binding sites are nearly saturated in every case.

The performance of electrical work by which $\Delta\Psi$ drives active transport has been localized to the reorientation steps for three additional systems, Na^+ : glucose cotransport in porcine cell lines (45) and Na^+ : melibiose cotransport in $S.$ $typhimurium$ and $E.$ $coli$ (78–80).

Stoichiometries near unity have been reported for other sugar : H^+ cotransporters. In yeasts, maltose (11), lactose (83), D-fucose (81), and D-xylose (84) are cotransported with a single H^+. Hexose : H^+ cotransport in sugar cane cells also proceeds with a stoichiometry near unity (85). While the cotransport of sugar and H^+ has been widely demonstrated in plant cells, technical problems such as rapid re-extrusion of H^+ have hindered accurate measurements of the stoichiometry (14, 15). Galactoside : H^+ cotransport in $Streptococcus$ $lactis$ evinces a 1 : 1 stoichiometry (86). The transport of D-glucose and its analogues in $Neurospora$ $crassa$ is accompanied by the movement of 0.5 to 1 H^+ (16). In the alga $Chlorella,$ the uptake of glucose and similar sugars involves concomitant movement of 1 to 2 H^+ at pH 5.3–5.7 (13).

DETECTION, PURIFICATION, AND RECONSTITUTION

Because ion : sugar cotransporters amount to only 1% or less of all proteins in their respective cytoplasmic membranes, their detection, purification, and functional reconstitution into phospholipid vesicles are difficult. For every system, a thorough analysis of the kinetics and substrate specificity in cells and derived cytoplasmic membrane vesicles as well as a systematic search for suitable inhibitors is a prerequisite for a subsequent biochemical characterization. Bacterial systems offer the enormous advantage that mutants lacking the transport activity can be isolated and serve as controls in every stage of analysis.

Moreover, the synthesis of many bacterial transport systems is inducible by the transported substrate or suitable analogues. Therefore, noninduced cells provide a further convenient control. Ideally, the purified transporter should be reconstituted into large (>1 μm) lipid vesicles in such a way that all molecules have the same orientation as in cells. The turnover number of the transporter for both the sugar and the ion as well as the affinity on either face of the membrane should be the same as in the cell from which the protein is derived. At present, all reconstituted systems fall short of this goal.

Continued attempts over the past ten years to detect, isolate, and reconstitute the *Na$^+$:D-glucose cotransporters* from intestinal or renal brush-border membrane vesicles have been recently critically reviewed (19, 87). Phlorizin binding to membrane vesicles in the presence of an inward-directed Na$^+$ gradient leads to estimates of 0.011–0.032 nmol binding sites/mg protein (corresponding to 0.1–0.2% for a protein of $M_r = 75,000$; $K_D = 4$–8 μM) for the intestinal transporter (34, 88) and 0.03–0.17 nmol/mg for the renal transporter ($K_D = 0.3$–1 μM, Refs. 89–92). Labeling experiments with 4-azidophlorizin (93), iodoacetamidophlorizin (94), fluoresceinisothiocyanate (34, 95), or 10-*N*-(bromoacetyl)amino-1-decyl-β-D-glucopyranoside (96) suggest that (part of) the transporter from both tissues is a protein of $M_{SDS}{}^2$ around 72,000. Because some of these labels are fairly specific, they should be useful in tracing the transporter during purification. A protein of similar size has been isolated by an affinity column from detergent extracts of small-intestinal membranes using a monoclonal antibody recognizing the glucose/phlorizin binding site of the transporter (97). Reconstitution experiments have relied almost entirely on Na$^+$-coupled glucose transport as an assay. Although glucose active uptake in proteoliposomes has been demonstrated repeatedly (90, 98–102), quantitative evaluation of the data suggests either only a small fraction of the transporters has been successfully reconstituted or that the reconstituted transporters have a turnover number much lower than in brush-border membrane vesicles (19, 87). Quantitation of transporters by rate measurements in proteoliposomes is notoriously difficult because of their small and heterogenous size and their generally undefined passive permeability. Observed transport rates may be heavily weighted in terms of a small portion of the total population of cotransporters present in the larger vesicles, because these vesicles have more stable driving gradients and fill up more slowly. A high passive permeability causes the rapid dissipation of artificially imposed $\Delta \tilde{\mu}_{ION}$, while a small permeability leads to the build-up of a counterdirected $\Delta \tilde{\mu}_{ION}$ during cotransport.

Recently, several bacterial ion:sugar transporters have been reconstituted. Work on the *galactoside:H$^+$ cotransporter* of *E. coli* is discussed in some

---

[2] M_{SDS} is the apparent molecular weight of a protein as determined by polyacrylamide–gel electrophoresis in the presence of sodium dodecyl sulfate.

detail because the *lacY* gene product is the only ion : sugar cotransporter which has been purified to homogeneity. The transporter was first identified as a membrane protein (M_{SDS} = 31,000) by labeling a thiol group with *N*-ethylmaleimide. This thiol group is protected by substrates against modification (103, 104). In cytoplasmic membranes with amplified transporter levels, labeling is highly specific (105, 106). An equally specific, covalent labeling of the transporter is achieved by using 4-nitro [2-^{3}H] phenyl α-D-galactopyrano-side as a photoaffinity reagent (107, 108). Furthermore, the amount of transporter in membranes can be quantitatively estimated by differentially labeling cells with radioactive amino acids or by equilibrium binding measurements using a high-affinity substrate (52, 58, 106, 109, 110). Comparison of these estimates leads to the conclusion that the transporter binds one galactoside per polypeptide chain (58, 111). In cytoplasmic membranes from induced cells carrying only the chromosomal *lacY* gene, the transporter amounts to about 1% of the protein, while membranes from cells carrying the *lacY* gene on a multicopy plasmid contain up to 16% (105, 111). Such membranes provide suitable material for purification. In one procedure (62, 73, 112–114), *E. coli* cell envelopes (inner and outer membranes) are first extracted with urea and sodium cholate. Membrane proteins are then solubilized in the presence of additional *E. coli* phospholipids in octylglucoside, and the extract is passed over a DEAE sepharose column. The eluate containing essentially pure transporter is used for the formation of proteoliposomes by a dilution regimen. Another method (59, 106, 115) uses cytoplasmic membranes pre-extracted with 5-sulfosalicylate. Solubilization in dodecyl-O-β-D-maltoside, a detergent in which the transporter retains the ability to bind substrate, passage over an anion exchange column, and removal of residual impurities by anti-impurity antibodies lead to a product of at least 95% purity. All transporter molecules retain the ability to bind substrate. After addition of *E. coli* phospholipids the detergent is removed by polystyrene beads. Both in this and the previously described method the initially formed proteoliposomes are small; their size can be increased by freeze-thaw cycles. About half of the carrier molecules in the proteoliposomes have the same orientation as in cells while the other half are inverted (116). Because the transporter behaves kinetically symmetrically these two populations cannot readily be distinguished by transport measurements. By a number of criteria, most importantly the K_D of substrate binding, the K_T for active transport, and the galactoside : H$^+$ cotransport stoichiometry, the purified transporter has the same properties as in cells or membrane vesicles (59, 62, 73). Also, the turnover number for influx of galactosides in the presence of $\Delta \widetilde{\mu}_{H^+}$ is very similar to that in right-side-out cytoplasmic membrane vesicles. This rate is 10 times smaller than in cells. Although this discrepancy remains to be explained, there is at present little doubt that the *lacY* gene product is the only protein required for galactoside : H$^+$ symport. The cotransporter can be reconstituted in a variety of phospholipids (117); however, the observed transport activity is apparently lower than in *E. coli* phospholipids.

Four other bacterial ion : sugar transporters have been identified and reconstituted into proteoliposomes. The *Na*$^+$ *: melibiose cotransporter* has been identified as the product of the *melB* gene (M_{SDS} = 30,000) in hybrid-plasmid bearing maxicells or in proteoliposomes obtained by dilution of octylglucoside-solubilized membrane proteins (118). The reconstituted transporter evinces melibiose active transport or melibiose countertransport coupled to Na$^+$ or H$^+$ (119, 120). The *proton-coupled transporters for galactose* (*galP* gene product, M_{SDS} = 34,000–39,000, Ref. 121), *arabinose* (*araE* gene product, M_{SDS} = 39,000, Ref. 122) and *xylose* (*xylE* gene product, M_{SDS} = 39,000, Ref. 123) have been identified and reconstituted (123, 124) into proteoliposomes along lines similar to those described for the lactose transporter. Because all four transporters occur in low abundance, their purification will likely require the development of overproducing strains by gene amplification and by addition of more efficient promoters to the respective structural genes.

STRUCTURE

The elucidation of the structure of cation : sugar cotransporters has been limited to methods such as predictions for the folding of polypeptide chains from amino acid sequences, the analysis of mutations affecting the interaction with substrates, the use of antibodies or proteases as high-molecular-weight, membrane-impermeable probes, and chemical modification by group-specific reagents. Protection of a transporter by a substrate against inactivation by a group-specific reagent has generally been taken as an indication for the location of certain amino acids at or near the binding site (e.g. Refs. 125–127). This conclusion could be premature because substrate binding may trigger a conformational change in the protein leading to the protection of a group distant from the binding site. Nevertheless, the selective modification of a transporter is not only useful for the identification of transport proteins (95, 103, 121–123) but also for the introduction of fluorescent monitoring groups (34, 95, 128, 129).

While neither Na$^+$ nor glucose or phlorizin appears to protect the *Na*$^+$ *: glucose cotransporter* against inactivation by thiol reagents (130), substrate protection against amino- and phenol-group specific reagents has been observed (95, 126, 127). These results suggest the presence of a lysine residue at the binding site for glucose (95, 126) and a tyrosine residue at the sodium site (127). Protection by Na$^+$ against inactivation by phenol reagents has also been observed for the Na$^+$: glucose cotransporter from calf kidney cortex (125). A fluorescein label introduced at the protectable amino group of the intestinal transporter has been used to follow conformational changes of the protein upon Na^+ binding (34).

The structural asymmetry of the intestinal Na^+ : glucose cotransporter has been studied using thiol-specific reagents of different membrane permeabilities and by comparing the accessibility of the transporter towards a probe in intact brush-border membrane vesicles and after permeabilization of the membranes by detergents such as deoxycholate. First, in intact brush-border membrane vesicles both the inactivation of the transporter by organo-mercurials and the restoration of phlorizin binding by thiols occurs faster with permeable than impermeable reagents. This differential effect is not observed with deoxycholate-treated vesicles (131). Second, the glucose transporter is inactivated by papain, trypsin, or chymotrypsin in deoxycholate-disrupted membranes but not in intact brush-border membrane vesicles (132). These experiments suggest that the cotransporter has one or more thiol groups and protease-sensitive sites on the cytosolic face of the membrane. Therefore, the protein is structurally asymmetric with respect to the plane of the membrane.

The interpretation of estimates for the functional size of the Na^+ : glucose cotransporter from rabbit or calf kidney by radiation inactivation (92, 133–135) is not straightforward because of differences in the assay conditions used and the presence of two cotransport systems in the kidney. In addition, there may be differences between species, and the very high target size of $M_{app} \approx 10^6$ observed under some conditions may indicate aggregation of integral proteins in the brush-border membrane (92, 135). Smaller target sizes ranging from $M_{app} = 110,000$ for phlorizin binding in the rabbit kidney (92) to $M_{app} = 345,000$ for tracer equilibrium exchange in calf kidney brush-border membrane vesicles (134) or $M_{app} = 343,000$ for Na^+-dependent glucose transport reconstituted from rabbit kidney (135) are taken as evidence for an oligomeric structure of the transporter. As in other cases (136), radiation inactivation appears to be a questionable method for estimating the size of integral membrane proteins.

The primary structure of the *galactoside : H^+ cotransporter* from *E. coli* has been deduced from the nucleotide sequence of the *lacY* gene (137) and by sequencing the N- and C-termini of the protein (138, 139). Furthermore, the transporter reacts with antibodies raised against the chemically synthesized C-terminus (140, 141). The *lacY* gene codes for a protein of $M_r = 46,500$ compared to $M_{SDS} \approx 30,000$ determined by gel electrophoresis in the presence of sodium dodecylsulfate (103, 105, 138). This result can be explained by the retention of a compact structure in the presence of sodium dodecylsulfate (142). On the other hand, complete unfolding apparently occurs in hexamethylphosphoric triamide ($M_{app} = 47,000$, Ref. 143).

The following observations suggest that the galactoside transporter functions as a monomer. First, the transporter binds one molecule of galactoside per polypeptide chain both in the membrane and in a dodecylmaltoside micelle (59). Second, the transporter purified in this detergent is a monomer as shown by sedimentation equilibrium (144). Third, the time-dependent phosphorescence anisotropy of the eosin-labeled transporter reconstituted into vesicles of dimyristoylphosphatidylcholine yields a rotational correlation time of 25 µs corresponding to $M_r = 47,200$, implying that the protein is a monomer. The monomeric state is maintained in the presence of a membrane potential (129). This last result is at variance with a target-size analysis by radiation inactivation, which suggests a $\Delta \tilde{\mu}_{H^+}$-induced change from $M_{app} = 45,000–50,000$ to $M_{app} = 85,000–100,000$ (145). Finally, an oligomeric structure was proposed by the existence of trans-dominant $lacY$ mutants (146), one of which exhibits a change in amino acid residue 262 from Gly to Asp (147). However, the negatively dominant phenotype of $lacY^-/lacY^+$-heterodiploid strains could not be confirmed by analysis of transport rates (144).

Because a monomer is the functional unit, the amino acid sequence of the galactoside transporter must be folded in such a way as to provide binding sites for a hydrophilic sugar and a proton and a pathway through the hydrophobic interior of the lipid bilayer. Both circular dichroism (148) and Raman scattering (142) indicate a high α-helix content of the transporter. The latter method yields an α-helix content of about 70%, a β-strand content below 10%, and β-turns contributing 15%. About one third of the residues in α-helices and most other residues are exposed to water. In combination with structural predictions, these data lead to the model shown in Figure 3 (32, 40, 142; compare also Refs. 137, 139, 148, 149). Eight hydrophobic, 1–6, 13, and 14, and two amphipathic membrane-spanning α-helices, 10 and 11, are predicted. These helices may form a ring, possibly elliptical, in contact with the phospholipids (129). The remaining more hydrophilic segments, 8, 9, 12, and 13, are presumed to form the inner core of the protein and are likely involved in the formation of the sugar-binding site.

The following amino acid residues are expected to be located near the sugar-binding site. Alkylation of Cys_{148} (150) prohibits sugar binding, most likely by steric hindrance. Replacement of this amino acid by either a glycine (151, 152) or a serine residue (153) still permits galactoside active transport, albeit with decreased apparent affinity and rate. Cys_{148} is suggested to be located within the protein at a depth corresponding to carbon atom C-5 of the phospholipid acyl chains (128). Mutants with altered sugar specificity have been isolated (146, 147, 149, 154) by screening for growth on maltose which is not transported by the wild type. Exchanges in codons 177 (Ala→Val or Thr), 236 (Tyr→Phe, Asn, Ser, or His), or 266 (Thr→Ile) imply that Ala_{177}, Tyr_{236}, and Thr_{266} are involved in sugar recognition (cf assignments in Figure 3). In the model for the tertiary structure in Ref. 149, these residues and residue Cys_{148},

PERIPLASM

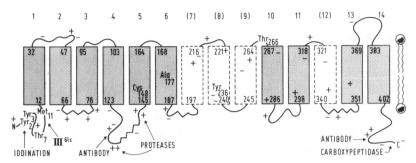

Figure 3   Model for the transmembrane arrangement of the polypeptide chain of the galactoside : H^+ transporter from *E. coli* (40, 142). The model predicts 10 membrane-spanning α-helixes (shaded boxes) of which helixes 1–6 and 13 and 14 are hydrophobic while helixes 10 and 11 are amphipathic, i.e. they are hydrophobic on one side only. Segments 7, 8, 9, and 12 are more hydrophilic and are not assumed to be membrane-spanning; they are nevertheless predicted to be largely α-helical. The N-terminus faces the cytoplasmic side of the transporter as inferred from iodination studies (139) and the interaction of the transporter with factor III^{glc} of the phosphotransferase system (167). The C-terminus and the loop around amino acid residue 135 also have been located on the cytoplasmic side using antipeptide antibodies and proteases (140, 141, 168). Two epitopes recognized by monoclonal antibodies are located on the periplasmic face of the protein (169, 170). See text for further details.

located in membrane-spanning α-helixes, are juxtaposed in the center of the protein.

The Na^+/H^+ : *melibiose cotransporter* of *E. coli* is also a very hydrophobic protein (M_r = 52,000, Ref. 155). Although there is no apparent homology between this transporter and the galactoside transporter, putative membrane-spanning segments are likewise most evident near the N- and C-termini. Thus, the overall three-dimensional organization of the two proteins may be similar.

Mutants with altered cation recognition have been described for the Na^+/H^+ : melibiose cotransporter (156–159) which are incapable of H^+-coupled transport. One class accepts only Na^+ as the cotransported ion (158); a second type with a $Pro_{122} \rightarrow Ser$ exchange engages in Li^+ : melibiose cotransport, while retaining the ability to use Na^+ (157–159).

INTRICACIES AND GENERALIZATIONS

Many experimental and theoretical results have led to considerable refinement in our understanding of how the active transport of sugars occurs. On the one hand, the energy sources for many such processes are found to be transmembrane free-energy gradients, and the concentrative transport of many sugars is simultaneously coupled to the movement of H^+ or Na^+ down the gradients of their electrochemical potentials. On the other hand, the concepts of mobile or rotating carriers have been abandoned in favor of the two-conformational-state model with alternately accessible binding sites. In reviewing the present state of knowledge on the kinetics, purification, and structure of cotransporters, several more general problems have become apparent which will be discussed in this section. Also, the comparison of the various systems allows certain generalizations to be made which suggest common functional and structural features of the proteins mediating sugar:ion cotransport.

Kinetic descriptions of sugar:ion cotransport are complex and of considerable flexibility, because up to thirteen parameters determine fluxes at any instant. Although nonmathematical rules employed in the evaluation of enzyme kinetics (160) are also applicable to cotransport, treatment of experimental results within the framework of a mathematical model is highly desirable. The development of a quantitative and consistent model requires the collection of sufficient and accurate data. Discrepancies in the kinetic description of cotransporters can frequently be traced to an inadequate experimental design or interpretation of transport measurements. For example, measuring the actual value of $\Delta \widetilde{\mu}_{ION}$ under the ambient conditions during kinetic studies is difficult. Also, the experimental difficulties connected with determining the *stoichiometry* of cotransport are considerable and agreement on the precise values of stoichiometries does not appear to be forthcoming. Indeed, because the present methods for determining stoichiometries seem to have reached the limit of refinement, the question of integral vs nonintegral (i.e. variable) stoichiometries may remain moot for some time, pending the development of new techniques.

In spite of the experimental difficulties associated with the accurate determination of the individual parameters of the cotransport cycle, some qualitative features appear to be shared by many cotransporters. The rate of cotransport is slow. The *turnover numbers* for sugar transport are $1-10^2$ s^{-1}. The rates of ion transport by channels and mobile carriers, like valinomycin, are notably larger at 10^6-10^8 s^{-1} and 10^2-10^5 s^{-1}, respectively (161). Many enzymes have turnover numbers in the range 10^2-10^5 s^{-1}; in some cases, catalysis appears to be limited by the rate of a conformational change in the protein with a rate of $1-10^4$ s^{-1} (162). The conformational change causing the binding sites for ion and sugar to change their accessibility may determine the rate of cotransport. To maintain strict coupling, the energy barriers for the reorientation of the binary intermediates must be higher than those for the reorientation of unloaded cotransporters and ternary intermediates. The slowness of cotransport may reflect the height of the barriers required to distinguish between binary and other intermediates.

The external *binding sites for ions* in the cotransporters examined here appear to be saturated under physiological conditions. Thus, transport can occur as soon as the sugar is bound. The external medium can be effectively scavenged for sugars. The affinity of the H^+ : galactoside cotransporter of $E.$ $coli$ is so high (pK up to 10.3) that how the cotransporter becomes deprotonated at neutral pH is unclear. There have been several suggestions that H^+ involved in cotransport may not rapidly equilibrate with the H^+ in the bulk phase (163–165). Rapid one- or two-dimensional diffusion of H^+ along or within the membrane could make deprotonation of high-affinity sites possible. Na^+ may not be able to use such pathways as well owing to the special role of hydrogen bonding in constructing many such pathways. Lastly, the saturation of the external ion binding site would imply that k_{ION}^+ in Figure 1 is small; otherwise constant dissipation of $\Delta \tilde{\mu}_{ION}$ would result.

The $electrical$ $work$ appears to be performed in the steps corresponding to the reorientation of the loaded or unloaded transporter (k_c and k_o in Figure 1). The active transport of sugar has thus far not been demonstrated to occur by the thermodynamically equally admissible possibilities of increasing $K_S^{''}$ or $K_{ION}^{''}$ or of decreasing $K_S^{'}$ or $K_{ION}^{'}$(Equation 4). If this observation extends to other cotransporters, then some general principles of protein architecture may be the reason. The performance of electrical work in a binding step requires a continuously variable affinity of the form $K_{(\Delta\Psi<0)} = K_{(\Delta\Psi=0)} \exp (\pm n_{ION} F\Delta\Psi/ RT)$. Perhaps such a potential variation is at odds with structural requirements for specifity, determined by precise protein-substrate contacts. This problem does not arise if the electrical work is performed in the reorientation steps.

As discussed for the galactoside : H^+ cotransporter from $E.$ $coli,$ these transport systems may be subject to a form of $self-regulation$ also exhibited by the Ca^{2+}-pumping ATPase of sarcoplasmic reticulum (166). At a low external sugar concentration, thermodynamic efficiency is high and Equation 3 obtains. This suggests that cotransporters are designed like enzymes to avoid wasting metabolic energy, e.g. kinases do not hydrolyze ATP in the absence of an acceptor. However, as the internal concentration of one or both cosubstrates rises, the internally oriented cotransporters become saturated, and even a strictly coupled cotransporter may be unable to achieve equilibrium within a finite time because net transport of sugar becomes slow. Thus, kinetic inefficiency prohibits the rapid attainment of thermodynamic equilibrium, and thereby protects the cell from high internal levels of sugar without resorting to leaks.

The $purification$ and $reconstitution$ of cotransporters still pose severe problems. First, cotransporters are integral membrane proteins of low abundance. At least in bacterial systems, the problem of low abundance can be overcome by cloning techniques. Membrane proteins must be solubilized as protein-detergent micelles to apply purification techniques developed for water-soluble proteins. An alternative method, exploited in only a few instances, is the use of organic solvents. This approach to purification might facilitate removal of lipids and might open new possibilities for crystallization. Second, solubilizing a cotransporter prohibits measuring its physiological activity, transport. Suitable alternatives for monitoring a purification are specific labeling of the protein, immunochemical tests, or substrate binding assays. Third, once the cotransporter is purified, the protein must be reincorporated into a membrane, and its activity must be accurately measured to determine if the activity present in the cells is obtained. Monitoring reconstitution attempts initially by a stoichiometric assay, such as substrate binding, appears more reliable than transport measurements, because binding assays do not depend upon properties of proteoliposomes unrelated to the intrinsic ability of the protein to perform active transport.

Structural information about cotransporters is at present indirectly obtained. Sequencing the structural genes for cotransporters from prokaryotes and eukaryotes will likely yield more primary sequences in the near future. Although the phospholipid and aqueous phases impose restrictions upon the folding of integral membrane proteins, there is no way to predict the three-dimensional structure from the primary sequences. This information is contained in the crystal structures of unloaded and fully occupied cotransporters in both conformations. However, the formation of crystals is hampered by the use of detergents and the paucity of protein-protein contact possible. The hydrophobic surfaces are covered with detergent molecules, and the hydrophilic faces may not project far into the aqueous phase.

A growing understanding of how membrane proteins are structurally designed suggests the presence of several motifs in the architecture of cotransporters. On the periphery of the protein a hydrophobic external surface is required. The structure of bacteriorhodopsin suggests a palisade of hydrophobic α-helixes as a possible solution. The binding sites for sugar and ion could be located in a hydrophilic core separated from the lipids by this structure. Ionophores may give some ideas of what structures are involved in binding ions. The sugar-binding sites in enzymes and other proteins, whose structures have been determined by X-ray diffraction, might hold clues as to how similar sites in cotransporters may be constructed. There is no recognized protein model for the gating mechanism responsible for the alternate exposure of the binding sites because this is the unique feature of transporters.

Cotransporters are proteins with remarkable properties. On the one hand, a conformational change causes binding sites for sugar and ion to become accessible to two different compartments. This conformational change must entail considerable spatial reorganization of parts of the protein and has often been compared to the movement of a gate within a pore. On the other hand, the structure of the transporter must be so compact that solutes as small as H^+ do not diffuse across the membrane through the protein matrix. Kinetic models depict cotransport as a series of steps occurring on a time scale of 10–100 ms. In contrast, structural models, although providing more detail on the molecular level, are static. Further elucidation of the molecular mechanism of ion : sugar cotransport will require the melding of dynamic and static models. Application of the entire arsenal of physical, biochemical, and genetic techniques will be required to achieve this goal.

ACKNOWLEDGMENT

The views of the authors have profited from many discussions with the members of the research group in Tübingen. We thank I. Riede and K. Dornmair for reading and U. Hieke for typing the manuscript. Many colleagues have helpfully provided us with preprints of their work.

Literature Cited

1. Widdas, W. F. 1952. *J. Physiol.* 118:23–39
2. Rosenberg, T., Wilbrandt, W. 1957. *J. Gen. Physiol.* 41:289–96
3. LeFevre, P. G. 1975. *Curr. Top. Membr. Transp.* 7:109–215
4. Singer, S. J. 1974. *Ann. Rev. Biochem.* 43:805–33
5. Jones, M. N., Nickson, J. K. 1981. *Biochim. Biophys. Acta* 650:1–20
6. Carruthers, A. 1984. *Prog. Biophys. Mol. Biol.* 43:33–69
7. Hengge, R., Boos, W. 1983. *Biochim. Biophys. Acta* 737:443–78
8. Shuman, H. A., Treptow, N. A. 1985. *The Enzymes of Biological Membranes*, ed. A. N. Martonosi, 3:561–75. New York: Plenum. 2nd ed.
9. Postma, P., Lengeler, J. 1985. *Microbiol. Rev.* 48:232–69
10. Harold, F. M. 1977. *Curr. Top. Bioenerg.* 6:84–149
11. Eddy, A. A. 1982. *Adv. Microbiol. Physiol.* 23:1–78
12. Kotyk, A. 1983. *J. Bioenerg. Biomembranes* 15:307–19
13. Komor, E., Tanner, W. 1974. *J. Gen. Physiol.* 64:568–81
14. Giaquinta, R. T. 1983. *Ann. Rev. Plant Physiol.* 34:347–87
15. Reinhold, L., Kaplan, A. 1984. *Ann. Rev. Plant Physiol.* 35:45–83
16. Slayman, C. L., Slayman, C. W. 1974. *Proc. Natl. Acad. Sci. USA* 71:1935–39
17. Zilberstein, D., Dwyer, D. M. 1985. *Proc. Natl. Acad. Sci. USA* 82:1716–20
18. Crane, R. K. 1977. *Rev. Physiol. Biochem. Pharmacol.* 78:101–59
19. Semenza, G., Kessler, M., Hosang, M., Weber, J., Schmidt, U. 1984. *Biochim. Biophys. Acta* 779:343–79
20. Riklis, E., Quastel, J. H. 1958. *Can. J. Biochem.* 36:347–62

21. West, I. C. 1970. *Biochem. Biophys. Res. Commun.* 41:655–61
22. Mitchell, P. 1973. *J. Bioenerg.* 4:63–91
23. Kaunitz, J. D., Wright, E. M. 1984. *J. Membr. Biol.* 79:41–51
24. Dorando, F. C., Crane, R. K. 1984. *Biochim. Biophys. Acta* 772:273–87
25. Barfuss, D. W., Schafer, J. A. 1981. *Am. J. Physiol.* 241:F322–32
26. Turner, R. J., Moran, A. 1982. *Am. J. Physiol.* 242:F406–14
27. Rickenberg, H. V., Cohen, G. N., Buttin, G., Monod, J. 1956. *Ann. Inst. Pasteur* 91:829–57
28. Kennedy, E. P. 1970. In *The Lactose Operon*, ed. J. R. Beckwith, D. Zipser, pp. 49–92. New York: Cold Spring Harbor Lab. 437 pp.
29. Prestidge, L. S., Pardee, A. B. 1965. *Biochim. Biophys. Acta* 100:591–93
30. Stock, J., Roseman, S. 1971. *Biochem. Biophys. Res. Commun.* 44:132–38
31. Maloney, P. C., Wilson, T. H. 1985. *Bioscience* 35:43–48
32. Overath, P., Wright, J. K. 1983. *Trends Biochem. Sci.* 8:404–8
33. Wright, J. K. 1986. Submitted for publication
34. Peerce, B. E., Wright, E. M. 1984. *J. Biol. Chem.* 259:14105–12
35. Hopfer, U., Groseclose, R. 1980. *J. Biol. Chem.* 255:4453–62
36. Aronson, P. S. 1978. *J. Membr. Biol.* 42:81–98
37. Kaback, H. R. 1983. *J. Membr. Biol.* 76:95–112
38. Ghazi, A., Shechter, E. 1981. *Biochim. Biophys. Acta* 645:305–15
39. Overath, P., Wright, J. K. 1980. *Ann. NY Acad. Sci.* 358:292–306
40. Wright, J. K., Dornmair, K., Mitaku, S., Möröy, T., Neuhaus, J.-M., et al. 1985. *Ann. NY Acad. Sci.* In press

41. Kaczorowski, G. J., Kaback, H. R. 1979. *Biochemistry* 18:3691–97
42. Toggenburger, G., Kessler, M., Semenza, G. 1982. *Biochim. Biophys. Acta* 688:557–71
43. Turner, R. J., Moran, A. 1982. *J. Membr. Biol.* 70:37–45
44. Wright, E. M., Peerce, B. E. 1985. *Ann. NY Acad. Sci.* In press
45. Lever, J. E. 1984. *Biochemistry* 23:4697–702
46. Kaunitz, J. D., Gunther, R., Wright, E. M. 1982. *Proc. Natl. Acad. Sci. USA* 79:2315–18
47. Kimmich, G., Randles, J. 1980. *Biochim. Biophys. Acta* 596:439–44
48. Goldner, A. M., Schultz, S. G., Curran, P.F. 1969. *J. Gen. Physiol.* 53:362–83
49. Kessler, M., Semenza, G. 1979. *FEBS Lett.* 108:205–8
50. Beck, J. C., Saktor, B. 1978. *J. Biol. Chem.* 253:5531–35
51. Turner, R. J. 1983. *J. Membr. Biol.* 76:1–15
52. Wright, J. K., Riede, I., Overath, P. 1981. *Biochemistry* 20:6404–15
53. Kaczorowski, G. J., Robertson, D. E., Kaback, H. R. 1979. *Biochemistry* 18:3697–704
54. Wright, J. K. 1986. *Biochim. Biophys. Acta* In press
55. Yamato, I., Rosenbusch, J. P. 1983. *FEBS Lett.* 151:102–4
56. Bentaboulet, M., Kepes, A. 1981. *Eur. J. Biochem.* 117:233–38
57. Page, M. G. P., West, I. C. 1981. *Biochem. J.* 196:721–31
58. Overath, P., Teather, R. M., Simoni, R. D., Aichele, G., Wilhelm, U. 1979. *Biochemistry* 18:1–11
59. Wright, J. K., Overath P. 1984. *Eur. J. Biochem.* 138:497–508
60. Teather, R. M., Hamelin, O., Schwarz, H., Overath, P. 1977. *Biochim. Biophys. Acta* 467:386–95
61. Lancaster, J. R., Hinkle, P. C. 1977. *J. Biol. Chem.* 252:7657–61
62. Viitanen, P. V., Garcia, M. L., Kaback, H. R. 1984. *Proc. Natl. Acad. Sci. USA* 81:1629–33
63. West, I. C., Mitchell, P. 1973. *Biochem. J.* 132:587–92
64. Zilberstein, D., Schuldiner, S., Padan, E. 1979. *Biochemistry* 18:669–73
65. Patel, L., Garcia, M. L., Kaback, H. R. 1982. *Biochemistry* 21:5805–10
66. Booth, I. R., Mitchell, W. J., Hamilton, W. A. 1979. *Biochem. J.* 182:687–96
67. Ahmed, S., Booth, I. R. 1981. *Biochem. J.* 200:573–81
68. ten Brink, B., Lolkema, J. S., Hellingwerf, K. J., Konings, W. N. 1981. *FEMS Microbiol. Lett.* 12:237–40
69. Ramos, S., Kaback, H. R. 1977. *Biochemistry* 16:4271–75
70. Ahmed, S., Booth, I. R. 1981. *Biochem. J.* 200:583–89
71. Wilson, T. H., Seto-Young, D., Bedu, S., Putzrath, R. M., Müller-Hill, B. 1985. *Curr. Top. Membr. Transp.* 23:121–34
72. Wright, J. K., Seckler, R. 1985. *Biochem. J.* 227:287–89
73. Garcia, M. L., Viitanen, P. V., Forster, D. L., Kaback, H. R. 1983. *Biochemistry* 22:2524–31
74. Thérisod, H., Ghazi, A., Houssin, C., Shechter, E. 1985. *Biochim. Biophys. Acta* 814:68–76
75. Booth, I. R., Hamilton, W. A. 1980. *Biochem. J.* 188:467–73
76. Heinz, E. 1978. *Mechanics and Energetics of Biological Transport*, pp. 98–119. Berlin:Springer-Verlag. 159 pp.
77. Overath, P., Thilo, L. 1978. *Int. Rev. Biochem.* 19:1–44
78. Tokuda, H., Kaback, H. R. 1977. *Biochemistry* 16:2130–36
79. Bassilana, M., Damiano-Forano, E., Leblanc, G. 1985. *Biochem. Biophys. Res. Commun.* 129:626–31
80. Lopilato, J., Tsuchiya, T., Wilson, T. H. 1978. *J. Bacteriol.* 134:147–56
81. van den Broek, P. J. A., van Steveninck, J. 1982. *Biochim. Biophys. Acta* 693:213–20
82. Delrot, S., Bonnemain, J.-L. 1981. *Plant Physiol.* 67:560–64
83. Barnett, J. A., Sims, A. P. 1982. *J. Gen. Microbiol.* 128:2303–10
84. Hauer, R., Höfer, M. 1982. *Biochem. J.* 208:459–64
85. Komor, E., Thom, M., Maretzki, A. 1981. *Planta* 153:181–92
86. Kashket, E. R., Wilson, T. H. 1974. *Biochem. Biophys. Res. Commun.* 59:879–86
87. Koepsell, H. 1986. *Rev. Physiol. Biochem. Pharmacol.* 104
88. Tannenbaum, C., Toggenburger, G., Kessler, M., Rothstein, A., Semenza, G. 1977. *J. Supramol. Struct.* 6:519–33
89. Kinne, R. 1976. *Curr. Top. Membr. Transp.* 8:209–67
90. Koepsell, H., Menuhr, H., Ducis, I., Wissmüller, T. F. 1983. *J. Biol. Chem.* 258:1888–94
91. Lin, J. T., Hahn, K. D. 1983. *Anal. Biochem.* 129:337–44
92. Turner, R. J., Kempner, E. S. 1982. *J. Biol. Chem.* 257:10794–97
93. Hosang, M., Gibbs, E. M., Diedrich, D. F., Semenza, G. 1981. *FEBS Lett.* 130:244–48
94. Lin, J. T. 1985. *Ann. NY Acad. Sci.* In press

95. Peerce, B. E., Wright, E. M. 1984. *Proc. Natl. Acad. Sci. USA* 81:2223–26
96. Neeb, M., Fasold, H., Koepsell, H. 1985. *FEBS Lett.* 182:139–44
97. Schmidt, U. M., Eddy, B., Fraser, C. M., Venter, J. C., Semenza, G. 1983. *FEBS Lett.* 161:279–83
98. Crane, R. K., Malathi, P., Preiser, H. 1976. *FEBS Lett.* 67:214–16
99. Fairclough, P., Malathi, P., Preiser, H., Crane, R. K. 1979. *Biochim. Biophys. Acta* 553:295–306
100. Im, W. B., Ling, K. Y., Faust, R. G. 1982. *J. Membr. Biol.* 65:131–37
101. Lin, J. T., DaCruz, M. E. M., Riedel, S., Kinne, R. 1981. *Biochim. Biophys. Acta* 640:43–54
102. Ducis, I., Koepsell, H. 1983. *Biochim. Biophys. Acta* 730:119–29
103. Fox, C. F., Kennedy, E. P. 1965. *Proc. Natl. Acad. Sci. USA* 54:891–99
104. Fox, C. F., Carter, J. R., Kennedy, E. P. 1967. *Proc. Natl. Acad. Sci. USA* 57:698–705
105. Teather, R. M., Müller-Hill, B., Abrutsch, U., Aichele, G., Overath, P. 1978. *Mol. Gen. Genet.* 159:239–48
106. Wright, J. K., Teather, R. M., Overath, P. 1983. *Methods Enzymol.* 97:158–75
107. Kaczorowski, G. J., LeBlanc, G., Kaback, H. R. 1980. *Proc. Natl. Acad. Sci. USA* 77:6319–23
108. Goldkorn, T., Rimon, G., Kaback, H. R. 1983. *Proc. Natl. Acad. Sci. USA* 80:3323–26
109. Jones, T. H. D., Kennedy, E. P. 1969. *J. Biol. Chem.* 244:5981–87
110. Kennedy, E. P., Rumley, M. K., Armstrong, J. B. 1974. *J. Biol. Chem.* 249:33–37
111. Teather, R. M., Bramhall, J., Riede, I., Wright, J. K., Fürst, M., et al. 1980. *Eur. J. Biochem.* 108:223–31
112. Newman, M. J., Wilson, T. H. 1980. *J. Biol. Chem.* 255:10583–86
113. Newman, M. J., Foster, D. L., Wilson, T. H., Kaback, H. R. 1981. *J. Biol. Chem.* 256:11804–8
114. Foster, D. I., Garcia, M. L., Newman, M. J., Patel, L., Kaback, H. R. 1982. *Biochemistry* 21:5634–38
115. Wright, J. K., Schwarz, H., Straub, E., Overath, P., Bieseler, B., Beyreuther, K. 1982. *Eur. J. Biochem.* 124:545–52
116. Seckler, R., Wright, J. K. 1984. *Eur. J. Biochem.* 142:269–79
117. Chen, C. C., Wilson, T. H. 1984. *J. Biol. Chem.* 259:10150–58
118. Hanatani, M., Yazyu, H., Shiota-Niiya, S., Moriyama, Y., Kanazawa, H., et al. 1984. *J. Biol. Chem.* 259:1807–12
119. Tsuchiya, T., Ottina, K., Moriyama, Y.,
   Newman, M. J., Wilson, T. H. 1982. *J. Biol. Chem.* 257:5125–28
120. Wilson, D. M., Ottina, K., Newman, M. J., Tsuchiya, T., Ito, S., Wilson, T. H. 1985. *Membr. Biochem.* 5:269–90
121. Macpherson, A. J. S., Jones-Mortimer, M. C., Horne, P., Henderson, P. J. F. 1983. *J. Biol. Chem.* 258:4390–96
122. Macpherson, A. J. S., Jones-Mortimer, M. C., Henderson, P. J. F. 1981. *Biochem. J.* 196:269–83
123. Henderson, P. J. F., Macpherson, A. J. S. 1985. *Methods Enzymol.* 104(IIB): In press
124. Henderson, P. J. F., Kagawa, Y., Hirata, H. 1983. *Biochim. Biophys. Acta* 732:204–9
125. Lin, J. T., Stroh, A., Kinne, R. 1982. *Biochim. Biophys. Acta* 692:210–17
126. Weber, J., Semenza, G. 1983. *Biochim. Biophys. Acta* 731:437–47
127. Peerce, B. E., Wright, E. M. 1985. *J. Biol. Chem.* 260:6026–31
128. Mitaku, S., Wright, J. K., Best, L., Jähnig, F. 1984. *Biochim. Biophys. Acta* 776:247–58
129. Dornmair, K., Corin, A. F., Wright, J. K., Jähnig, F. 1985. *EMBO J.* 4:3633–38
130. Biber, J., Weber, J., Semenza, G. 1983. *Biochim. Biophys. Acta* 728:429–37
131. Klip, A., Grinstein, S., Semenza, G. 1979. *Biochim. Biophys. Acta* 558:233–45
132. Klip, A., Grinstein, S., Semenza, G. 1979. *J. Membr. Biol.* 51:47–73
133. Kinne, R., DaCruz, M. E. M., Lin, J. T. 1984. *Curr. Top. Membr. Transp.* 20:245–58
134. Lin, J. T., Szwarc, K., Kinne, R., Jung, C. Y. 1984. *Biochim. Biophys. Acta* 777:201–8
135. Takahashi, M., Malathi, P., Preiser, H., Jung, C. Y. 1985. *J. Biol. Chem.* 260:10551–56
136. Jørgensen, P. L. 1982. *Biochim. Biophys. Acta* 694:27–68
137. Büchel, D. E., Gronenborn, B., Müller-Hill, B. 1980. *Nature* 283:541–45
138. Ehring, R., Beyreuther, K., Wright, J. K., Overath, P. 1980. *Nature* 283:537–40
139. Bieseler, B., Prinz, H., Beyreuther, K. 1985. *Ann. NY Acad. Sci.* In press
140. Seckler, R., Wright, J. K., Overath, P. 1983. *J. Biol. Chem.* 258:10817–20
141. Carrasco, N., Herzlinger, D., Mitchell, R., DeChiara, S., Danho, W., et al. 1984. *Proc. Natl. Acad. Sci. USA* 81:4672–76
142. Vogel, H., Wright, J. K., Jähnig, F. 1985. *EMBO J.* 4:3625–31

143. König, B., Sandermann, H. 1982. *FEBS Lett.* 147:31–34
144. Wright, J. K., Weigel, U., Lustig, A., Bocklage, H., Mieschendahl, M., et al. 1983. *FEBS Lett.* 162:11–15
145. Goldkorn, T., Rimon, G., Kempner, E. S., Kaback, H. R. 1984. *Proc. Natl. Acad. Sci. USA* 81:1021–25
146. Mieschendahl, M., Büchel, D., Bocklage, H., Müller-Hill, B. 1981. *Proc. Natl. Acad. Sci. USA* 78:7652–56
147. Markgraf, M., Bocklage, H., Müller-Hill, B. 1985. *Mol. Gen. Genet.* 198: 473–75
148. Foster, D. L., Boublik, M., Kaback, H. R. 1983. *J. Biol. Chem.* 258:31–34
149. Brooker, R. J., Wilson, T. H. 1985. *Proc. Natl. Acad. Sci. USA* 82:3959–63
150. Beyreuther, K., Bieseler, B., Ehring, R., Müller-Hill, B. 1982. In *Methods in Protein Sequence Analysis,* ed. M. Elzinga, pp. 139–48. Clifton, NJ: Humana
151. Trumble, W. R., Viitanen, P. V., Sarkar, H. K., Poonian, M. S., Kaback, H. R. 1984. *Biochem. Biophys. Res. Commun.* 119:860–67
152. Viitanen, P. V., Menick, D. R., Sarkar, H. K., Trumble, W. R., Kaback, H. R. 1985. *Biochemistry.* In press
153. Neuhaus, J. M., Soppa, J., Wright, J. K., Riede, I., Blöcker, H., et al. 1985. *FEBS Lett.* 185:83–88
154. Shuman, H. A., Beckwith, J. 1979. *J. Bacteriol.* 137:365–73
155. Yazyu, H., Shiota-Niiya, S., Shimamoto, T., Kanazawa, H., Futai, M., Tsuchiya, T. 1984. *J. Biol. Chem.* 259:4320–26
156. Niiya, S., Yamasaki, K., Wilson, T. H.,
Tsuchiya, T. 1982. *J. Biol. Chem.* 257:8902–6
157. Tsuchiya, T., Oho, M., Shiota-Niiya, S. 1983. *J. Biol. Chem.* 258:12765–67
158. Shiota, S., Yamane, Y., Futai, M., Tsuchiya, T. 1985. *J. Bacteriol.* 162: 106–9
159. Yazyu, H., Shiota, S., Futai, M., Tsuchiya, T. 1985. *J. Bacteriol.* 162: 933–37
160. Cleland, W. W. 1963. *Biochim. Biophys. Acta* 67:104–37
161. Läuger, P., Benz, R., Stark, G., Bamberg, E., Jordan, P. C., et al. 1981. *Q. Rev. Biophys.* 14:513–98
162. Hammes, G. G., Schimmel, P. R. 1970. In *The Enzymes,* ed. P. D. Boyer, 2:67–114. New York:Academic. 584 pp.
163. Williams, R. J. P. 1978. *Biochim. Biophys. Acta* 505:1–44
164. Kell, D. B. 1979. *Biochim. Biophys. Acta* 549:55–99
165. Westerhoff, H. V., Melandri, B. A., Venturoli, G., Azzone, G. F., Kell, D. B. 1984. *Biochim. Biophys. Acta* 768: 257–92
166. De Meis, L., Inesi, G. 1985. *FEBS Lett.* 185:135–38
167. Nelson, S. D., Wright, J. K., Postma, P. W. 1983. *EMBO J.* 2:715–20
168. Seckler, R., Möröy, T., Wright, J. K., Overath, P. 1986. *Biochemistry* In press
169. Carrasco, N., Tahara, S. M., Patel, L., Goldkorn, T., Kaback, H. R. 1982. *Proc. Natl. Acad. Sci. USA* 79:6894–98
170. Herzlinger, D., Viitanen, P. V., Carrasco, N., Kaback, H. R. 1984. *Biochemistry* 23:3688–93

Ann. Rev. Biochem. 1986. 55:249–85
Copyright © 1986 by Annual Reviews Inc. All rights reserved

GENETICS OF MITOCHONDRIAL BIOGENESIS

Alexander Tzagoloff and Alan M. Myers

Department of Biological Sciences, Columbia University, New York, NY 10027

CONTENTS

INTRODUCTION

In this review we will attempt to summarize what is currently known about the comparative roles of mitochondrial and nuclear DNA in directing the biogenesis of respiratory-competent mitochondria. Rather than being exhaustive, the discussion will strive to frame the available information around

249

0066-4154/86/0701-0249$02.00

three questions: 1. What is the degree of genetic autonomy of mitochondria? 2. How is the expression of mitochondrial DNA regulated by nuclear gene products? and, 3. How are cytoplasmically synthesized proteins transported into mitochondria and can the transport process in itself account for the functional and structural continuity of mitochondria? For more detailed discussions of these topics there are several excellent reviews on the genetic organization of mitochondrial DNA by Dujon (1) and Wallace (2) and a comprehensive treatment of protein import into mitochondria by Hay et al (3). Also recommended are the proceedings of several recent symposia on mitochondrial genes (4) and nucleomitochondrial interactions (5).

THE GENETIC FUNCTION OF MITOCHONDRIAL DNA

Mitochondria (along with chloroplasts) occupy a unique position among cellular organelles because of their possession of a separate genome and all the enzymatic machinery for transcribing and translating the genetic information into functional proteins. This circumstance has been responsible for the great deal of interest in defining the degree of genetic autonomy of the organelle and in exploiting the simplicity of the mitochondrial system in answering basic questions of eukaryotic gene expression.

Earlier observations that mitochondria synthesize certain constituent polypeptides of cytochrome oxidase (6–8), coenzyme QH_2-cytochrome c reductase (8–10), and of ATPase (8, 11) suggested that these proteins were also likely to be encoded in mitochondrial DNA (mtDNA). This has been borne out, first by genetic analysis of yeast mtDNA (1) and subsequently by direct sequencing of the genome from fungal (1, 12–15), animal (16–21), and plant sources (22, 23).

In *Saccharomyces cerevisiae,* it is possible to induce point mutations or small deletions in mtDNA that result in a respiratory deficiency (24). Such strains called *mit⁻* mutants have the same gross phenotype as the classical ρ^- mutant (25). They are unable to utilize nonfermentable substrates and consequently form small colonies when cultivated on limiting concentrations of glucose or other fermentable sugars. Unlike ρ^- mutants, however, *mit⁻* mutants retain a fully functional system of mitochondrial protein synthesis and therefore do not exhibit the pleiotropic absence of cytochromes a, a_3, and b (24, 26, 27). Most *mit⁻* strains fall into one of four different phenotypic classes. These are defined by mutants deficient in 1. cytochrome oxidase, 2. coenzyme QH_2-cytochrome c reductase, 3. cytochrome oxidase and coenzyme QH_2-cytochrome c reductase, and 4. oligomycin-sensitive ATPase. Biochemical analyses of each mutant class provided the first evidence that the *mit⁻* phenotypes were the result of lesions in structural components of the above three inner-membrane enzymes. For example, mutations leading to a loss of cytochrome oxidase (cytochromes a, a_3) were mapped to three separate loci on mtDNA designated *oxi1*, *oxi2*, and *oxi3* (26). Based on the patterns of cytochrome oxidase–specific polypeptides in various *mit⁻* mutants it was concluded that these three genes code for subunit 1 *(oxi3)*, subunit 2 *(oxi1)*, and subunit 3 *(oxi2)* of the enzyme (26, 28, 29). Similar data suggested that the mitochondrial genome of yeast also codes for cytochrome b (30) and for several subunits of the oligomycin-sensitive ATPase complex (31–33). The genes and their known products are listed in Table 1 for the yeast, human, and *Neurospora crassa* genomes.

In addition to coding for components of the respiratory chain and the coupling ATPase, yeast mtDNA houses the genes for a full set of tRNAs (79, 80) and the two rRNAs (81) of mitochondrial ribosomes. The rRNAs and tRNAs were the first recognized products of mtDNA not only in yeast but also in mammalian mitochondria (82, 83). In fact, many of the tRNA genes of yeast mtDNA had been mapped relative to one another prior to any information about the physical map of this genome (84). Mitochondrial genes that function in protein synthesis have been termed *syn⁻* genes. They include the 15S and 21S rRNAs, 25 tRNAs, and a single protein subunit of mitochondrial ribosomes, var1 (55). Genetic markers for *syn* genes include mutations conferring antibiotic resistance (85–88) and *syn⁻* mutations that result in loss of function of var1 (89), tRNAs (90), or ribosomes (91). *syn⁻* mutants are phenotypically similar to ρ^- mutants. They do not synthesize the normal complement of mitochondrially encoded proteins and consequently are multiply deficient in respiratory enzymes and in the ATPase.

To what extent is the genetic composition of the yeast genome representative of other mtDNAs? A number of mtDNAs of different phylogenetic groups have been studied during the past several years (12–23, 92–94). It is now clear that there exist substantive differences in the coding properties of the fungal, animal, and plant genomes. Even though the diversity in the size of mtDNA of different yeasts is considerable (2, 92, 95, 96), their genetic compositions are fairly constant. Higher fungi such as *N. crassa* and *Aspergillus nidulans* share a set of genes in common with yeast but have additional genes that are not present in the yeast genome (70, 73, 97, 98). This is also true of animal mtDNAs (16–21). All the fungal and animal mtDNAs studied to date code for the two rRNAs, a variable but complete set of tRNAs sufficient to recognize all the codons of the mitochondrial genetic code, and the mRNAs for subunits 1, 2, and 3 of cytochrome oxidase, the cytochrome b component of coenzyme QH_2-cytochrome c reductase, and subunits 6 and 8 of the ATPase complex. By inference the same genes are probably also present in the genomes of plants. This has already been confirmed for the rRNAs (99), several subunits of cytochrome oxidase, and cytochrome b (22, 23).

The sequence analysis of animal mitochondrial DNAs has revealed in addition to the common set of genes, seven other reading frames (URF for un-

Table 1 Genes of fungal and human mitochondrial DNAs[a]

Gene product	Saccharomyces cerevisiae			Homo sapiens		Neurospora crassa	
	Gene	Markers	Ref.	Gene	Ref.	Gene	Ref.
sb 1 of cyt. ox.	oxi3	mit⁻	12,26,34	COI	16	COI	64
sb 2 of cyt. ox.	oxi1	mit⁻	26,35,36	COII	16	COII	65
sb 3 of cyt. ox	oxi2	mit⁻	26,37	COIII	16	COIII	66
cytochrome b	cob/box	mit⁻, ant^r	38,39	cyt b	16	cyt b	67
sb 6 of ATPase	oli2/pho1	mit⁻, ant^r	40,41	ATPase 6	16	oli2	68
sb 8 of ATPase	aap1	mit⁻	32,42	URF A6L	16	URF$_x$	68
sb 9 of ATPase	oli1/pho2	mit⁻, ant^r	43,44	—	—	mal	69
sb of Q reductase		—		URF 1	16,63	URF 1	70
sb of Q reductase		—		URF 2	16,63	URF 2	71
sb of Q reductase		—		URF 3	16,63	URF 3	71
sb of Q reductase		—		URF 4	16,63	?	
sb of Q reductase		—		URF 4L	16,63	?	
sb of Q reductase		—		URF 5	16,63	URF 5	72
?				URF 6	16	?	
?						URF N	73
?						URF U	73

Gene product	Designation	Phenotype	Refs	Designation	Refs	Designation	Refs
?	[RF1]	none	45				—
?	[RF2]	none	40				—
intron transposition factor	*fit1*	—	46–48				—
maturase	[bI2]	mit^-	39				—
maturase	[bI3]		49				—
maturase	[bI4]	mit^-	38,50,51				—
maturase	[aI1]	mit^-	12,52,53				—
maturase	[aI2]	mit^-	12,52				—
maturase?	[aI3]		12,52				—
maturase	[aI4]	*mim*	12,52,54				—
maturase?	[aI5]		34				—
maturase?	[aI6]		34				—
?	—					[ATPase 6 intron]	68
?	—					[cyt b intron]	67
ribosomal protein?	*var1*	syn^-	55			[21S rRNA intron]	74,75
ribosomal protein	—	syn^-,ant^r	46,56	16S RNA	16	lg rRNA	74,75
21S rRNA	lg rRNA	ant^r	57,58	12S RNA	16	sm rRNA	75
15S rRNA	sm rRNA	syn^-	59,60	D, etc		asp, etc	76–78
tRNAs	asp, etc		61,62				—
9S RNA	*tsl*						—

ᵃThe names used to designate genes are not always consistent in the literature. The most commonly used names have been chosen here. Intron-encoded reading frames and putative coding sequences in the yeast genome have not yet been named according to the usual conventions for gene designations. They have been bracketed to indicate that the descriptive and shorthand notations are used only as a matter of convenience.

identified reading frame) with matching transcripts and translation products (100, 101). These genes are not detected in yeast mtDNA; however, some URFs with potential products homologous to those of animal mitochondria are present in *N. crassa* (70–72) and in *A. nidulans* (97, 98). The proteins encoded by the URFs have hydrophobic compositions and are therefore probably located in the membrane. The identities and possible functions of the URF products are still not clear although there is now evidence that most of them may be component polypeptides of NADH-coenzyme Q reductase. This important member of the respiratory chain (102) is composed of more than 20 different polypeptides (103, 104) and is one of the most complex enzymes of the respiratory chain. Studies with HeLa cells indicate six of the URF products cofractionate with NADH-coenzyme Q reductase (63). Additional evidence that the URF products are subunits of the reductase comes from a reexamination of the mitochondrially synthesized proteins in *N. crassa* (105). Experiments utilizing specific inhibitors of mitochondrial and cytoplasmic ribosomes indicate that four polypeptides associated with purified NADH-coenzyme Q reductase are translated on mitochondrial ribosomes (105). These proteins constitute minor components of the total mitochondrial translation products and for this reason were probably overlooked in earlier studies.

The synthesis of NADH-coenzyme Q reductase of *S. cerevisiae* is not blocked by chloramphenicol (106), indicating that it occurs entirely on cytoplasmic ribosomes. It is not clear at present whether the genes corresponding to the URFs have been transferred to nuclear DNA or are absent in yeast. The NADH-coenzyme Q complex of *S. cerevisiae* differs from the mammalian enzyme in several important respects. Thus, some of the iron-sulfur centers of the mammalian enzyme have been reported to be absent in yeast mitochondria (107). Even more important from a functional standpoint is the lack of phosphorylation in the NADH to coenzyme Q segment of the yeast respiratory chain (108). These earlier observations may be the biochemical correlates of more recent genetic data.

Another intriguing difference between yeast and animal mtDNA is the absence in the latter of the gene for subunit 9 [dicyclohexylcarbodimide (DCCD) binding protein] responsible for the proton translocating activity of the ATPase (109). In higher eukaryotes, this subunit of the ATPase is therefore encoded by a nuclear gene. This is also true in *N. crassa,* although in this organism a copy of the gene is conserved in mtDNA (69, 110). The proteins encoded by the nuclear and mitochondrial genes are very homologous but only the nuclear gene appears to be expressed (110, 111). The mitochondrial copy in *N. crassa* may be an example of a mitochondrial pseudogene. It is still not excluded, however, that the mitochondrial gene may be expressed during a specific stage of the cell cycle or under physiological conditions that have yet to be recognized.

While *S. cerevisiae* lacks the set of URFs now thought to code for part of the NADH-coenzyme Q reductase, it has other putative coding sequences whose functions are only now beginning to be unraveled. Most of these are located within intervening sequences of the genes for cytochrome *b* (38, 39), subunit 1 of cytochrome oxidase (12, 34), and the 21S rRNA (46). In the case of the cytochrome *b* introns, there is strong evidence that the reading frames are expressed into products ("maturases") required for RNA splicing [(39, 50, 51), see discussion below]. A similar function has been ascribed to the reading frames found within some introns of the *oxi3* gene (52, 53). Recently, Michel & Lang (112) have noted significant homology between the protein sequences encoded by two of the *oxi3* introns and certain regions of viral reverse transcriptases. This important observation points to the possibility that the mitochondrial introns may have a more diversified function than was originally suspected.

The reading frame located in the intron of the yeast 21S rRNA gene is a clear instance of an intron-encoded protein serving a function other than RNA splicing. This intron is dispensable (not all strains of *S. cerevisiae* have a split 21S rRNA gene) and is transmitted with an unusually high bias in heteroallelic crosses (113). Transmission of the intron has been correlated with the introduction of a double-stranded break at the intron boundary of the recipient DNA (114) similar to what has been observed in other gene conversion events in yeast (115–117). Mitochondrial protein synthesis is required for the intron transmission and mutations within the intron reading frame prevent both transmission and cleavage of the recipient genome at the acceptor site (47, 48). These results indicate that gene conversion in mitochondria depends on expression of the intron-encoded protein. Whether the intron codes for the endonuclease responsible for cleaving the double-stranded DNA or has some other function remains to be clarified.

The variation seen in the protein products encoded in different mtDNAs probably reflects divergent evolutionary trends that may have been established soon after the introduction of this DNA in eukaryotes. The simplest and most highly specialized mitochondrial genome is found in animals. It codes for the two rRNAs, the minimum number of tRNAs needed to decode the endogenous messages, and a small number of constitutively expressed proteins all of which function in electron transport and in oxidative phosphorylation. This genetic information is organized in a highly economical way with a conspicuous absence of what might be recognized as regulatory elements in the DNA sequence. This genome therefore appears to have reached an evolutionary cul-de-sac with little opportunity left for change, except perhaps by further reduction of its genetic information. From an evolutionary perspective the situation seems to be potentially more dynamic in fungi. As indicated there is now evidence that in addition to the set of respiratory components, yeast mtDNA also codes for proteins that facilitate the acquisition or loss of genetic information by transposition or gene conversion. The differences seen in the organization and structure of mitochondrial genes among different strains of *S. cerevisiae* (2) suggest that in this yeast the capability for shuffling of genetic information has not been lost. The detection of mitochondrial sequences in yeast nuclear DNA (118) and the already-mentioned evidence for the existence of what appears to be a nonfunctional copy of the ATPase subunit 9 gene in *N. crassa* mtDNA (69) support the idea that transfer of genetic information between mitochondria and the nucleus has occurred relatively recently.

ORGANIZATION OF MITOCHONDRIAL DNA

The size of mitochondrial DNA ranges from approximately 16 kb in animals to more than 100 kb in plants (2). Not only is there a heterogeneity of genome size but more importantly also of the way in which the genetic material is packaged. The genomes of *S. cerevisiae* and of animal mitochondria represent the extremes in the evolution of this DNA. They have been studied in the most detail and for these reasons will be reviewed here.

Structure of Mammalian mtDNA

The genomes of man (16), mouse (18), cow (17), and frog (21) mitochondria have been completely sequenced. They are nearly identical in size and have the same organization and complement of genes that code for very homologous products. The locations of the genes and identity of the products are shown in Figure 1 for the human genome. Most of the protein coding genes are present in one strand of the circular duplex molecule; the tRNA genes, however, are distributed in both strands. The most remarkable organizational feature of this DNA and one that has important consequences for the production of the different mitochondrial RNAs is the absence of intergenic sequences and the almost general presence of one or more tRNA genes between the rRNA and the mRNA coding regions. All the genes are flush-ended thereby leaving no room for extraneous sequences. The frugal use of coding material is also evident in the absence of intervening sequences and the unusually small sizes of the rRNAs and tRNAs, but not of the reading frames. The proteins encoded in animal mtDNA are of the same size as their homologues in other organisms. The overall impression is of an extremely specialized genome with little opportunity for further simplification.

Transcription and Processing of RNA in Animal Mitochondria

It was known from studies of HeLa cell mtDNA that both strands of this genome are transcribed from single promoters giving rise to primary transcripts that contain all the information encoded in each of the two strands (119, 120). This

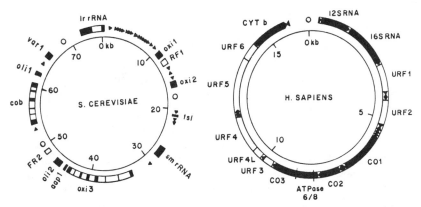

Figure 1 Maps of the yeast and human mitochondrial genomes. *S. cerevisiae:* The positions of the genes with known products are depicted by the solid bars. The introns of *oxi3, cob* and large (lr) rRNA genes as well as two putative coding sequences are shown by the open bars. The solid triangles indicate the locations of the tRNA genes and point in the directions of their transcription. The open circles show some of the origins of replication (136, 137). All the genes except one tRNA are transcribed clockwise. The physical map of the DNA is based on studies reported in Ref. 138. The present map has been adapted from Ref. 139. The gene products are identified in Table 1. *H. sapiens:* The genes of human mtDNA with known products are shown as solid bars. The URFs are indicated by the open bars. Subunit 8 of the ATPase is located adjacent and upstream of the subunit 6 gene. The tRNA genes are shown as solid triangles. All the genes except URF 6 and eight tRNAs are transcribed clockwise. The origin of replication of the H strand is depicted by the open circle. The gene products are identified in Table 1.

together with the organizational features of the DNA has led to a simple and elegant model of how the mature rRNAs, tRNAs, and mRNAs are generated (121, 122). The maturation of the three classes of transcripts requires only three different types of RNA processing enzymes. The most important are the endonucleases responsible for the excision of the tRNAs that demarcate the starts and ends of the rRNAs and mRNAs. An RNase P-type endonuclease has been purified from human mitochondria (123). This enzyme probably catalyzes cleavage of the precursor RNA on the 5' side of the tRNA sequences (123). The smaller transcripts generated as a result of the endonucleolytic removal of the tRNAs are further matured by base modification of the tRNAs (124) and the rRNAs (125) and polyadenylation of the mRNA precursors (126).

Another consequence of the structure of the mammalian mitochondrial genes is the absence of 5' leaders in the messages. This has been confirmed by direct sequence analysis of a number of different mRNAs whose 5' termini start either with the AUG or AUA initiation codons predicted by the DNA sequence (122). The mechanism by which mitochondrial ribosomes interact with leaderless messages is not known. Mammalian mitochondrial ribosomes are themselves unusual in terms of the subunit sizes and protein content relative to RNA (127, 128).

The continuous transcription of mtDNA in animals cannot account for the 10–30-fold higher concentrations of rRNA relative to mRNAs. A transcriptional attenuation mechanism has been proposed to explain the more efficient production of the rRNA. Support for the model comes from the sequences of the 16S rRNA which show ragged 3' ends (129) suggesting that the mature molecules can arise from transcription termination as well as precise endonucleolytic cleavage of a large precursor RNA (122). Two distinct transcription start sites separated by less than 100 bases on the H strand were identified by labeling the 5' di- and triphosphate termini with GTP in the presence of guanylyl transferase and mapping the "capped" ends on mtDNA (130). Kinetic studies indicated that one of the H-strand sites produces transcripts at the high rates seen for rRNA while transcripts initiated at the second site are produced at lower rates more typical of mRNA synthesis (130). These results have been interpreted to indicate that transcription of the rRNA cistron occurs predominantly by initiation from the more active site. Transcription is attenuated by termination near the 3' end of the 16S rRNA gene. The transcript issued from the less active start site overrides the stop signal and serves as the precursor for the mRNAs and tRNAs. Other studies, however, indicate a single transcriptional start site (131). Exactly how the site of initiation of transcription can influence activity of the RNA polymerase complex at an attenuation region some 2700 base pairs farther removed still needs to be explained.

At present there is no evidence for transcriptional attenuation of the protein-coding regions of the genome. The mRNAs are therefore probably synthesized in a fixed stoichiometry and whatever differences exist in the steady-state abundance of this class of transcripts can be assumed to be a function of their inherent stabilities.

The absence of separate promoters for the mitochondrial genes in animal genomes, as well as of 5' leaders in the mature mRNAs for modulating the efficiency of translation, argues against the existence of any elaborate mechanism for regulating the synthesis of this set of proteins. The proteins encoded in the genome of animal mitochondria are probably without exception constituents of enzymes that make up the terminal respiratory chain. These enzymes exist in a fixed stoichiometry that need not be changed either during the cell cycle or stage of growth of the organism. The invariant character of the respiratory chain in higher animals could explain the seeming absence of means for individually regulating the expression of mitochondrial gene products.

Structure of Yeast mtDNA

The mitochondrial genome of *S. cerevisiae* consists of 70–75 kb. About 70% of this DNA has been sequenced, including most of the genetically important regions. Superficially there is little similarity between yeast mtDNA and the more compact genome of animal mitochondria. With the exception of a single tRNA, all the genes are transcribed from the same strand of DNA (Figure 1). The exception is a threonine tRNA (132) that probably arose by duplication of another tRNA gene. This aberrant mitochondrial threonine tRNA is responsible for the recognition of the CUN family of leucine codons as threonine. Most of the yeast mitochondrial genes are separated by long stretches of A/T-rich sequences that are randomly interspersed with short G/C clusters (133, 134). The functions of the A/T spacers and of the G/C clusters are not known presently. The earlier suggestion that the G/C clusters might act as signals for RNA processing (135) has not been borne out by more recent data. As mentioned already yeast mtDNA shares a common set of genes with the mammalian genome but does not have the genes corresponding to the URFs. The order in which the yeast genes are arranged is also completely different from what is seen in animal mtDNAs. For example in *S. cerevisiae,* the two rRNAs are separated by tRNA as well as two cytochrome oxidase genes (Figure 1). Although most of the tRNA genes occur in one main cluster, others are dispersed throughout the genome. The scattered tRNA genes may be important in processing of the primary transcripts. Thus almost all of the protein coding regions as well as the two rRNAs are flanked either on the 5' or 3' side by a tRNA gene. In most cases these tRNAs are cotranscribed with the rRNA or protein coding genes and their excision is therefore an essential aspect of mRNA maturation (140–143).

The large size of the yeast genome is due not only to the presence of intergenic spacers but also to the long intervening sequences found in the genes for cytochrome *b* (38, 39), subunit 1 of cytochrome oxidase (12, 34), and of the 21S rRNA (46). The number of introns in these genes is strain-dependent in *S. cerevisiae.* For example, while most commonly used laboratory strains have five introns in the cytochrome *b* gene (long gene), there are other strains in which the gene has only two introns (144). Similar variations have been noted in the intron compositions of the cytochrome oxidase subunit 1 gene (144, 145). *S. cerevisiae* has one of the largest and most complex mitochondrial genomes among yeasts, some of which have genome sizes almost as small as those of animal mitochondria (95) and others intermediate between the two (92, 96). The smaller genomes are indicative of shorter intergenic regions and less frequent occurrence of intervening sequences.

Transcription and Processing of Mitochondrial RNA in Yeast

Several approaches have been used to assess the number of promoters on the mitochondrial genome of *S. cerevisiae.* The use of guanylyl transferase to cap the 5' termini of the primary transcripts has been exploited in several laboratories to map the transcriptional initiation sites in the *S. cerevisiae* genome (146–148). Some 20 different primary transcripts have been identified and of these 13 have been characterized and localized on mtDNA (148). The sequence determination of the 5' ends of these transcripts has led to the recognition of a common nine-nucleotide-long sequence (5'-ATATAAGTA-3') which marks the initiation of transcription in yeast mitochondria (148, 149). All the transcripts start with the terminal A of the nonanucleotide sequence (148, 149). Three primary transcripts have 5' termini in regions of the genome proposed to be origins of replication (137), a finding consistent with the idea that mtDNA replication in yeast requires an RNA primer (149). Unlike the mammalian genome, whose replication has been studied extensively (150), almost nothing is known about the mechanism of replication of yeast mtDNA.

Mitochondrial promoters have also been localized by examining transcription of the genomes in ρ^- mutants with different deletion endpoints in their mtDNA. Such strains are capable of generating wild-type transcripts provided the promoter regions have not been deleted (140–143). Finally, substantial progress has been made in reconstituting a mitochondria-free transcription system. Although the polymerase has not been purified to homogeneity, active preparations have been obtained that are capable of correct initiation of transcription with ρ^- genomes or mtDNA sequences in plasmids (151, 152). While the in vitro studies have confirmed the requirement of the nonanucleotide sequence for initiation of transcription, they also indicate mitochondrial promoters are more complex structures with additional sequence elements that remain to be identified. At present there is also no information concerning the signals for termination of transcription. The 3' termini of the primary transcripts are therefore not known.

Based on the locations of the promoters and analysis of the transcripts emanating from different regions of the yeast genome, the known primary transcripts are all composed of two or more coding sequences, one of which is often a tRNA (140–142, 153, 154). There are no rules about the relative positions of the genes or the types of processing steps that will give rise to the mature transcripts. The tRNA sequences can be located either 3' or 5' of the protein coding regions (140, 154). Furthermore, excision of the tRNA does not necessarily generate the mature 5' or 3' termini of the mRNAs. Almost always additional endonucleolytic cleavages are necessary to complete the processing of the mRNAs.

There are probably three different types of endonucleases that effect processing of mitochondrial transcripts. One is an RNase P-type nuclease that cleaves on the 5' side of the tRNA (61). In analogy with the bacterial enzyme, the mitochondrial nuclease requires a small RNA cofactor whose gene has been mapped near the methionine tRNA gene (62). Whether the RNA can catalyze the cleavage on its own or acts in conjunction with a protein has not been established. Since the RNase P activity is specific for the 5' side of the tRNA, there must be another enzyme that cleaves on the 3' side of the tRNA. Most of the mature messages terminate near a common sequence (5'-AUUCUUA-3') that has been proposed to act as a signal for a general 3' processing endonucle-ase (153, 154). In some polycistronic transcripts the 3' processing signal occurs more than once, which suggests that processing may involve a number of independent cleavages (153). In addition to these general 5' and 3' processing activities there is evidence for the existence of other processing enzymes that act on specific transcripts. A case in point is a protein required for 5'-end processing of the cytochrome *b* pre-RNA (155).

Processing of mitochondrial transcripts is much more elaborate in *S. cerevisiae* than in animal systems. Nonetheless, there are motifs suggestive of a common ancestry. Despite the fact that the mitochondrial genome of *S. cerevisiae* is transcribed from different promoters, the primary transcriptional units are polycistronic RNAs requiring the action of RNAses to generate the mature mRNAs. As in the animal system, excision of tRNA sequences is important for maturation of the rRNAs and mRNAs. Unlike the mammalian messages, those of yeast and other fungi are not polyadenylated and have 5' untranslated leaders of variable lengths. Transcription and processing of mtDNA in other yeasts show features intermediate between *S. cerevisiae* and animal mitochondria. The organization of genes in the mitochondrial genome of *Schizosaccharomyces pombe* (95) is very similar to that of mammalian mtDNA. Intergenic spacers are extremely short and with one exception all the protein coding sequences are separated by tRNA genes. Although there are no published data on transcription and processing in this yeast, the prediction is that the mechanism of RNA maturation will be essentially the same as in animals. *S. cerevisiae* and other yeasts with smaller genomes represent an evolutionary spectrum illustrative of the general tendency of this mtDNA to simplify by specialization.

Mitochondrial RNA Splicing

Recent investigations of mitochondrial RNA splicing have focused on two related findings. The first is the well-documented occurrence of intron open reading frames (ORFs) contiguous with preceding exons. Many of the intron ORFs encode proteins termed maturases required for removal of the intervening sequences in which they are located. The second observation is the similarity of some mitochondrial introns to the rRNA intron of *Tetrahymena,* which is capable of self-splicing in the absence of protein (156, 157).

Table 2 presents a comparison of mitochondrial introns from different organisms. Both genetic and biochemical evidence strongly support the idea that splicing of the cytochrome *b* pre-mRNA in yeast is contingent on expression of the proteins encoded in the second and fourth introns of the gene (39, 50, 51). Thus, mutations resulting in premature termination codons within the intron reading frames act in *trans* and prevent splicing of the RNA (39, 165). Analysis of mitochondrial translation products in such mutants shows the

Table 2   Classification of mitochondrial introns[a]

Organism	Gene	Intron	ORF	Type	Evidence for maturase	Ref.
Saccharomyces cerevisiae	oxi3	aI1	+[e]	II	yes	12,34,145
	oxi3	aI2	+[e]	II	yes	
	oxi3	aI3	+	I		
	oxi3	aI4	+	I	yes	
	oxi3	aI5	+	I		
	oxi3	aI6	+	I		
	oxi3	aI7	−	II		
	oxi3	aI8	−	?		
	cob	bI1	−	II		
	cob	bI2	+	I	yes	40,80
	cob	bI3	+	I		
	cob	bI4	+	I	yes	
	cob	bI5	−	I		
	1r rRNA	1	+	I		46
Neurospora crassa	cyt b	bI1	−	I		67,158
	cyt b	bI2	+	I		
	oli2	pI1	−	?		68
	oli2	pI2	+	I		
	1r RNA		+	I		74
Aspergillus nidulans	oxiA[b]	aI1	−	I		159
	oxiA[b]	aI2	+	I		
	oxiA[b]	aI3	+	I		
	cob	bI1	+	I		160
	1r rRNA	1	+	I		13
Schizosaccharomyces pombe	coxI[c]	aI1	+	I		161
	coxI	aI2	+	I		
	cob	bI1	+[e]	II		15
Podospora anserina	coxI[c]	aI1	+[e]	II		162
Kluveromyces thermotolerans	1r rRNA	1	+	I		47
Zea mays	cox2[d]	1	−	II		22
Triticum aestivum	cox2[d]	1	−	II		163
O. sativa	cox2[d]	1	−	II		164

[a]Mitochondrial introns whose sequences are known have been grouped into two classes as described in the text. Open reading frames (ORF) found in both types of introns have been shown to code for maturases. Examples of both Group I and Group II introns are known to lack open reading frames.

[b]gene coding for cytochrome oxidase subunit 3.

[c]genes coding for cytochrome oxidase subunit 1.

[d]genes coding for cytochrome oxidase subunit 2.

[e]ORFs potentially coding for proteins with homology to retroviral reverse transcriptases.

accumulation of novel polypeptides, which based on their sizes and antigenic properties have been interpreted to be hybrid proteins with sequences derived from the upstream exon (cytochrome *b*) and intron reading frames (51, 166). Similar analyses have shown that the first and second introns of the cytochrome oxidase subunit 1 pre-mRNA also encode maturases (52, 53). Mutations within the ORF of intron 4 of the cytochrome *b* gene are suppressed by specific missense mutations in intron 4 of the cytochrome oxidase subunit 1 gene, implying that the latter intron encodes a latent maturase that can be mutationally activated (54). In addition to these mitochondrially encoded maturases there are also nuclearly encoded proteins with maturase activity (167–169).

Mitochondrial introns along with certain introns of nuclei and chloroplasts do not conform to the GU· · ·AG rule typical of eukaryotic pre-mRNAs (170). The intron sequences can be classified into two distinct types based on the presence of short conserved sequence elements that predict regions of similar secondary structure (159, 171–173). Thus the self-splicing intron of *Tetrahymena* rRNA is a member of Group I introns along with many other mitochondrial introns (Table 2). This observation has prompted investigation into the possibility of self-splicing in mitochondria. Self-splicing of the intron in the 21S rRNA gene of yeast mitochondria has been implicated by detection of intermediates (174) similar to those found during self-splicing of the ribosomal precursor RNA from *Tetrahymena* nuclei (157). Self-splicing of this intron and of two introns encoded in the gene for subunit 1 of cytochrome oxidase (introns aI3 and aI5) has been detected in vitro (175). Similar processing intermediates of the large ribosomal RNA intron have been detected in *N. crassa* (176). This intron as well as the first intron of the cytochrome *b* pre-mRNA is capable of self-splicing in vitro (177).

Whether self-splicing of Group I introns occurs in mitochondria is unclear since many are known to require *trans* acting factors for excision. In addition to the nuclear and mitochondrially encoded maturases of yeast at least three different nuclear genes have been reported to promote splicing of the *N. crassa* large rRNA (178–180). Mutations in one of these genes simultaneously induces splicing defects in at least two different Group I introns (180). A possible explanation for these observations is that the *trans* acting factors in some way facilitate formation or stabilization of particular secondary structures thus enabling the self-catalyzed splicing of Group I introns. A common mechanism of splicing Group I introns is suggested by the finding that many of the ORFs contained within these sequences predict two separate decapeptide sequences that are highly conserved in maturases of different species (67, 158, 161, 181).

The splicing mechanism involved in removal of Group II introns from mitochondrial pre-RNAs is likely to be different from that of the Group I introns. Four Group II introns are known to form circular molecules after excision (182, 183) which may be "lariat" intermediates found in splicing of nuclear pre-mRNAs (184). An interesting feature of Group II introns is that all four ORFs located within these sequences potentially encode proteins that share homology with each other as well as with retroviral reverse transcriptases (112).

Multisubunit splicing complexes have been implicated in the processing of introns in nuclear pre-mRNAs, thus predicting that multiple genes are required for removal of intron sequences. In yeast, however, only a relatively small number of mutant genes have been found to affect mitochondrial splicing. For example, only a single nuclear gene was found to be required for splicing of the last intron in the cytochrome b pre-mRNA (167) among over 200 nuclear complementation groups required for mitochondrial function (see below). None of these nuclear genes were found to be required for removal of the fourth intron of this gene, which suggests that the intron 4 maturase alone may be sufficient for splicing. Although based on negative evidence, the available data support the conclusion that splicing of mitochondrial introns is not accomplished by large multisubunit complexes.

The evolution of mitochondrial introns is particularly interesting in light of the known functions of these elements. Intron-encoded maturases regulate expression of their host genes at the level of RNA splicing. In the case of the maturase encoded by the fourth intron of the cytochrome b gene there is coordinate regulation of two respiratory enzymes, cytochrome oxidase and coenzyme QH_2-cytochrome c reductase (185). Despite these regulatory functions the introns do not confer an obvious selective advantage to the cell, since mitochondrial genomes lacking specific introns of the cytochrome b gene are fully capable of supporting respiratory metabolism during vegetative growth (186, 187). The wide range of intron compositions in homologous genes of closely related yeast species (Table 2) suggests that mitochondrial introns originally may have spread to different genomes by transposition or gene conversion–like mechanisms, possibly involving sequence elements capable in some way of promoting their own propagation.

DEPENDENCE OF MITOCHONDRIAL BIOGENESIS ON NUCLEAR GENES

It has been apparent for some time that despite the importance of mtDNA for the maintenance and propagation of respiratory functional mitochondria, the most fundamental processes involved in biogenesis of this organelle are in fact dependent on the expression of genes located in nuclear DNA. This is most clearly illustrated by the long-standing observation that ρ^0 mutants of yeast lacking intact mtDNA nonetheless have organelles morphologically and functionally related to mitochondria. Such respiratory-deficient mitochondria have most of the enzymatic capabilities of wild-type mitochondria except for the few respiratory and ATPase proteins encoded in mtDNA (188).

The perpetuation of functional mitochondria during cell growth and division therefore requires not only an understanding of the role of mtDNA, but equally if not more importantly, of the structural and regulatory genes located in nuclear DNA. Until recently these have received little attention mainly because of past emphasis on the part of those working in this field to decipher the genetic contribution of the much simpler mitochondrial genome. More recently a number of laboratories have turned their attention to the nuclear genome particularly from the standpoint of understanding how the expression of the two compartmentally separate groups of genes are coordinately regulated and how the large number of proteins synthesized in the cytoplasm are imported into mitochondria. The number of genes required to specify an organelle as complex as the mitochondrion can be anticipated to comprise a respectable fraction of the total genetic information of a cell. Given the ease with which mutants of *S. cerevisiae* can be isolated as well as the means for gene cloning in this organism, the goal of arriving at a complete description of this set of genes and of their products is not beyond attainment.

PET *Genes and* pet *Mutants*

PET genes are by definition nuclear genes whose expression is required for the morphogenesis of respiratory-competent mitochondria. These genes may code for products that have a direct function in mitochondrial respiration and oxidative phosphorylation or they may affect oxidative metabolism of mitochondria indirectly. For example a gene coding for a mitochondrial ribosomal protein qualifies as a *PET* gene since a defect in mitochondrial protein synthesis would lead to a respiratory-deficient phenotype. Mutations in *PET* genes (*pet* mutants) were first studied by Sherman (189). Like ρ^- mutants, *pet* mutants cannot grow on nonfermentable substrates even though they grow normally on glucose.

Several laboratories are currently engaged in attempting to mutationally saturate the *PET* genes of the *S. cerevisiae* genome. Reasoning that some *PET* products may have functions shared by the rest of the cell, some laboratories have chosen to isolate temperature-sensitive *pet* strains (152, 169, 190). Although this has the obvious advantage of including a greater assortment of mutants, it also has some shortcomings. The most serious of these is that the mutant phenotype is not always clearly differentiated at the two temperatures. This can be a serious problem if the mutant is to be used to clone the gene by complementation.

The collection obtained in our laboratory consists of nonconditional mutants selected for their inability to grow on the nonfermentable substrate glycerol. At present there are several thousand independent isolates, which based on complementation tests represent approximately 200 complementation groups defining different *PET* genes. The number of mutants per complementation group varies from 1 to as many as 50. Since mutants within a given complementation group generally have the same phenotype, the biochemical characterizations have been confined to a single representative from each group. These have been studied for the intactness of the respiratory chain by both enzyme assays and spectral measurements of cytochromes. The mutants have also been analyzed for their ability to express mitochondrial genes at the level of the mRNAs and of the translation products. Based on these assays most of the *pet* mutants can be grouped into five phenotypic classes (Table 3); some groups, however, with more exotic properties, defy this simple classification. The phenotypes do not in any sense pinpoint the nature of the precise biochemical lesion; they are nonetheless useful for further biochemical screening. For example, mutants with lesions in components of the mitochondrial translation machinery or RNA polymerase would be expected to have a pleiotropic phenotype. Similarly, mutations in catalytic subunits of specific enzymes such as cytochrome oxidase or coenzyme QH_2-cytochrome *c* reductase should express specific deficiencies in these enzymes. The assignment of mutants to the somewhat general groups described above is a time-saving measure in the search for more specific types of mutants.

A mutant collection representative of a particular set of genes offers the added advantage of being a tool for testing specific hypotheses. For example, the occurrence of separate promoters for different genes in yeast mtDNA affords the possibility of individually regulating their expression by transcriptional rates. This could be achieved by means of *trans* acting protein factors encoded in nuclear DNA. Mutations in such gene products would be expected to cause a deficiency of a specific respiratory enzyme (e.g. cytochrome oxidase) due to lowered rates of transcription of the mitochondrial gene(s). The existence of such transcriptional factors can be tested by analysis of mitochondrial mRNAs in mutants exhibiting specific deficiencies in respiratory enzymes. To date mutants of this type have not been detected among the existent collection.

In seeking to mutationally saturate a group of genes it is important to have a means for estimating the total number of complementation groups that will express a particular phenotype, in the present case loss of ability to grow on nonfermentable substrates. In principle it should be possible to predict the total number of genes from the average number of mutants per complementation group by the Poisson formula. For the collection of nonconditional *pet* mutants obtained in our laboratory the average is 7–8 members per complementation group. This implies nearly complete saturation of all the genes obtainable by the selection method used and further predicts a bell-shaped distribution with the largest number of groups having 7–8 mutants. In reality, however, the largest

Table 3 Phenotypes of *S. cerevisiae pet* mutants

		Enzyme deficiency			
	None	Pleiotropic	ATPase	Cyt. Ox.	Co. QH$_2$-c red
Number of complementation groups	80	45–50	3–4	40	30
Gene products identified	1. enzymes of CoQ biosynthesis 2. expression of cytochrome c	1. aminoacyl-tRNA synthetases 2. ribosomal proteins 3. elongation factor	1. α subunit of F$_1$ 2. β subunit of F$_1$	1. subunit 4 2. subunit 5 3. subunit 6 4. maturases	1. 5'-end processing of cyt. b pre-mRNA 2. maturase 3. translation of cyt. b

number of complementation groups are those having only one mutant. This nonideal behavior cannot be explained by the size heterogeneity of the target or the general complexity of the system but rather argues for some special circumstance by which the selection protocol excludes certain types of *pet* mutants. The loss of a special class of *pet* mutants can be explained by the recent observation that mutations in genes coding for components of the mitochondrial protein synthesizing system lead to a secondary instability in mtDNA that can result in a quantitative conversion to ρ^- and ρ^0 mutants (191). Unstable mutants include strains with lesions in mitochondrial ribosomal proteins, aminoacyl-tRNA synthetases, and other protein factors required for translation. Strains with stringent mutations in this class of genes would be scored as mitochondrial mutants and would therefore be discarded during the isolation procedure. Despite this caveat, in practice *pet* strains defective in mitochondrial protein synthesis can be isolated as nonconditional mutants albeit they are un-derrepresented in the collection. As might be predicted they tend to have leaky phenotypes and are found in complementation groups with relatively few members.

Characterization of PET *Genes*

The methods developed for cloning of yeast genes by transformation with recombinant plasmid libraries (192) can also be used to clone *PET* genes for which appropriate mutants are available (193). Approximately one third of the genes represented by our *pet* collection have been cloned by this method. Of these some 20 genes have been sequenced and their products tentatively identified. A list of the genes for which the products have been identified is presented in Table 4.

Additional *PET* genes have been cloned by physical methods. Several of the genes were isolated from libraries of nuclear DNA through the use of synthetic (202, 207, 208) or heterologous probes (204, 209). Different methods for enrichment of cytoplasmic mRNAs coding for mitochondrial proteins were also employed to select *PET* genes from genomic libraries by the technique of competitive RNA hybridization (212, 213). The observation that polyribo-somes engaged in the synthesis of mitochondrial proteins are found to be associated with the outer membrane of yeast mitochondria (210, 211) has been exploited successfully to isolate the genes for citrate synthetase and a num-ber of outer-membrane proteins (212). Enrichment of glucose-repressible mRNAs coding for specific mitochondrial proteins by velocity gradient sedi-mentation was used to obtain the genes for nuclearly encoded subunits of coenzyme QH_2-cytochrome c reductase (213, 214) and of cytochrome oxidase (200).

Table 4 *pet* genes and their protein products

Complementation group	Gene name	Product	Gene sequence	Ref.
G1	*ATP2*	β subunit of F1	yes	194
G26	*MSY*	tyrosyl-tRNA synthetase	yes	a
G30	—	aminoacyl-tRNA synth?	yes	a
G36	*CBP2*	cyt. *b* maturase	yes	167
G43	*OP1*	ADP/ATP carrier		195
G46	*COX5*	subunit 5 of cyt. ox.	yes	196–198
G50	*ATP1*	α subunit of F1	yes	b
G60	*CBP1*	5'-end processing of cyt. *b* pre-mRNA	yes	199
G66	*COX4*	subunit 4 of cyt. ox.	yes	200
G69	*MST1*	threonyl-tRNA synth.	yes	201
G71	*COX6*	subunit 6 of cyt. ox.	yes	202
G79	*MSS116*	splicing of sb 1 of cyt. ox. pre-mRNA		203
G82	—	ribosomal protein?	yes	a
G89	*MRP1*	ribosomal protein	yes	a
G94	*MSD*	aspartyl-tRNA synth.	yes	a
G96	*MSS51*	splicing of sb 1 of cyt. ox.	yes	168
G100	—	aminoacyl-tRNA synth.	yes	a
G120	*MSF1*	α subunit of phenylalanyl-tRNA synth.	yes	a
G127	—	ribosomal protein?	yes	a
G130	*tufM*	elongation factor	yes	204
G133	—	ribosomal protein	yes	a
G151	CBP3	cyt. *b* translation factor?	yes	a
G154	CBP6	cyt. *b* translation factor	yes	205
G181	MSW	tryptophanyl-tRNA synth.	yes	206
G195	—	ribosomal protein?	yes	a

[a]M. Crivellone, A. Gampel, J. Hill, T. J. Koerner, I. Muroff, A. Myers, unpublished studies.
[b]M. Douglas, personal communication.

NUCLEAR-MITOCHONDRIAL INTERACTIONS

The previous discussion has centered around the question of the relative genetic contributions of mitochondrial and nuclear DNA towards the biogenesis of mitochondria. The very fact that genetic information is distributed among two spatially separated compartments implies the existence of mechanisms for ensuring a coordinate expression of the proteins and/or RNAs encoded in the two genomes. The regulatory signals responsible for coupling the output of the two sets of genes are now just beginning to be investigated and a discussion of this topic must therefore be largely conjectural.

Growth and division of eukaryotic cells is accompanied by a concomitant increase in mitochondrial mass. This is accomplished not by de novo formation of mitochondria but rather by the addition of newly synthesized lipids and proteins to preexisting organelles (215, 216) such that functional constancy is preserved. The latter condition depends on balanced rates of synthesis of the catalytic and structural constituents by the cytoplasmic and mitochondrial systems. Another aspect of regulation is seen in organisms such as yeast with the capability of adjusting their respiratory potential to the metabolic needs of the cell. The enzymatic composition of mitochondria in *S. cerevisiae* can differ in accordance with the carbon source or in response to physiological conditions (217, 218). Whether this type of regulation also exists in higher eukaryotes is not clear at present. There are, however, considerable tissue-specific compositional and functional differences in mitochondria that arise from the expression or lack thereof of developmentally controlled genes.

Do Mitochondria Affect the Expression of Nuclear DNA?

There are four well-documented examples of mitochondrial enzymes with subunit polypeptides derived from both nuclear and mitochondrial genes (219). They are the two respiratory complexes cytochrome oxidase and coenzyme QH_2-cytochrome *c* reductase, the oligomycin-sensitive ATPase, and mitochondrial ribosomes. In animal mitochondria and fungi such as *N. crassa*, NADH-coenzyme Q reductase can be added to this list (104, 105). Each of these oligomeric enzymes contains nuclearly encoded subunits whose functions are not always well understood but are known to be essential for catalytic activity. Even though this group of proteins interacts with subunits encoded in mtDNA, their synthesis or for that matter the synthesis of *PET* products in general does not depend on the presence of an intact mitochondrial genome. This is supported by the fact that mitochondrial constituents such as cytochrome *c* (25), F_1 ATPase (220), and ribosomal proteins (221) are not significantly different in wild-type and in ρ^- or ρ^0 mutants, the latter having no mtDNA. Combined with the observation that mitochondria of wild-type and of ρ^- mutants have essentially identical protein compositions [except for the proteins translated on mitochondrial ribosomes, (188)], these data make a cogent argument against the involvement of mitochondria in regulating either transcription of *PET* genes or translation of the cytoplasmic mRNAs. Whether this is true in other organisms is difficult to assess since their mtDNA cannot be readily deleted in the way that it can in *S. cerevisiae*. In the absence of evidence to the contrary, it is probably a safe assumption that mtDNA in general exerts little if any influence on nuclear genes.

Requirement of Nuclear Gene Products for Expression of mtDNA

Nuclearly encoded enzymes and proteins play an important role in transcription of yeast mtDNA, in processing of precursor RNAs, and in translation of mitochondrial messages. The RNA polymerase and virtually all the proteins constituting the translational machinery of mitochondria are specified by nuclear genes. Similarly, many of the RNA processing reactions are catalyzed by enzymes imported from the cytoplasm. Together these enzymes are needed for the synthesis of the entire complement of mitochondrially encoded proteins. There are other nuclear genes that function in the expression of individual mitochondrial products. Synthesis of cytochrome b, for example, depends on a group of nuclear genes that code for proteins involved in 5'-end processing (155) and intron excision of the pre-mRNA (167, 169) as well as translation of the mature message (205). A similar requirement of transcript-specific processing and translation factors has been reported for subunit 1 (168) and subunit 2 (222) of cytochrome oxidase. The existence of such an elaborate assortment of different proteins with functions confined to single transcripts is difficult to rationalize in terms of a simple constitutive expression of the target genes. More likely, they may have secondary functions in regulating the rates at which individual proteins such as cytochrome b are produced under different growth conditions or different stages of the cell cycle.

Mitochondrial genes could be regulated during transcription, processing of the primary transcripts, and/or translation. At present there is no evidence for transcriptional regulation either of the genome as a whole or of single genes. Despite the numerous yeast *pet* mutants in existence, none have been demonstrated to be affected in transcription of mtDNA. Negative evidence by itself is not conclusive, but in this case it does suggest that transcriptional factors, if they exist, are not numerous and therefore are unlikely to operate at the level of the individual genes. There are additional considerations mitigating against stringent control of mtDNA transcription. Mitochondria code for both rRNAs, the tRNAs, and the var1 proteins of the small ribosomal subunit, the latter having been shown to be essential for proper ribosome assembly (223). Lack of transcription of such genes would result in the loss of the protein synthetic capacity of mitochondria. Since this activity has been found to be a prerequisite for the maintenance of the wild-type mitochondrial genome (191, 224), a corollary prediction is that any regulatory mechanism involving substantially decreased rates of mitochondrial transcription would lead to an instability in mtDNA paralleled by an induction of ρ^- mutants. Nevertheless, changes in transcription of mtDNA can probably be tolerated within some range without affecting the genotype of the cell.

The respiratory activity of yeast mitochondria undergoes dramatic changes in response to two environmental factors, oxygen and glucose. Under anaerobic conditions or when grown on glucose as the carbon source, synthesis of the respiratory complexes and of the mitochondrial ATPase is repressed and as a result mitochondria are almost totally devoid of the ability to oxidize nonfermentable substrates and to generate ATP by oxidative phosphorylation (218). The effect of glucose on the synthesis of key respiratory carriers offers a favorable experimental situation for probing the regulation of nuclear and mitochondrial genes. The steady-state levels of the cytoplasmic mRNAs for cytochrome c (225), of certain subunits of coenzyme QH_2-cytochrome c reductase (213), of cytochrome oxidase (226), and of the F_1 ATPase (227) are reduced during fermentative growth. In the case of cytochrome c, the difference in mRNA concentration is as much as several orders of magnitude in cells grown on glucose and on the nonfermentable substrate lactate (228). Recent studies indicate the presence of different cis acting sequence elements or upstream activation sites (UAS) in the 5' regions of the structural gene for cytochrome c (228, 229). These sequences have been shown to affect the transcriptional efficiency as a function of glucose (down regulation) and of heme (up regulation). Upstream activation sites probably represent a general mechanism by which transcription of certain PET genes is suppressed in yeast growing on glucose.

The aforementioned observation that the stability of wild-type mtDNA is contingent on mitochondrial protein synthesis precludes stringent control of genes coding for aminoacyl-tRNA synthetases, ribosomal proteins, and other enzymes that are involved in a general way in mitochondrial DNA and RNA metabolism. The latter would include RNA polymerase and processing enzymes responsible for the maturation of the rRNAs and tRNAs. Nuclear genes coding for proteins that process specific mitochondrial RNAs or facilitate translation of specific mRNAs could be under glucose control and provide a mechanism for regulating the production of structural subunits encoded in the two different genomes. The synthesis of cytochrome b, for example, could be limited by suppressing transcription of $CBP6$, a nuclear gene coding for a basic protein that stimulates translation of cytochrome b mRNA (205). Thus the expression of nuclear and mitochondrial genes could be coordinated by regulating transcription not only of PET genes coding for structural subunits of enzymes such as coenzyme QH_2-cytochrome c reductase and cytochrome oxidase but also of other nuclear genes whose products are needed for the synthesis of the mitochondrial partner proteins.

The extent to which glucose and oxygen affect production of mitochondrial gene products is still not clear. Some studies indicate that certain transcripts such as those encoding cytochrome b are present at reduced levels in glucose-repressed cells (230). In a separate study, however, the levels of all mitochondrial transcripts examined were found to be identical in cells grown on glucose, lactate, or the nonrepressing sugar galactose (C. L. Dieckmann, personal communication). The efficiency with which mitochondrial mRNAs are translated as a function of glucose repression has been difficult to estimate since this activity is measured under very artificial conditions (in the presence of cycloheximide to block cytoplasmic protein synthesis) which may give an erroneous impression of the normal protein synthetic rates.

In summary, this remains one of the least-well-understood aspects of mitochondrial biogenesis. While there are suggestions that nuclear gene products do exert some control over mtDNA, the contention that mitochondrial genes are expressed in a constitutive manner cannot be refuted by concrete evidence. Also lacking at present are data bearing on the metabolic adaptability of mitochondria in higher eukaryotes.

TRANSPORT OF PROTEINS INTO MITOCHONDRIA

The majority of mitochondrial proteins are synthesized on cytoplasmic ribosomes and therefore need to be transported to one of the four internal compartments of the organelle. Not only is it necessary for this group of proteins to be targeted specifically to mitochondria, but once internalized each species must be delivered selectively to either one of the two membranes or two soluble spaces. The mechanisms by which externally synthesized proteins are sorted out internally is probably the single most important problem in understanding how mitochondria are assembled. The viewpoint taken here is that proper assembly of lipid and protein components as well as their correct topological arrangement can only be achieved by a process of growth of existing mitochondria. In other words mitochondria and perhaps other cellular organelles as well represent a template principle in whose absence assembly would not be possible even if all the needed genetic information were available to the cell. The validity of this assertion can be supported by several compelling arguments. First, there is experimental evidence for the existence in the outer mitochondrial membrane of receptor proteins that recognize among the large number of different cytoplasmic proteins the subset destined to become part of mitochondria (231–234). The segregation of newly synthesized proteins to the appropriate cellular membrane systems assumes an already existing condition in which the correct distribution of receptors has been established. Secondly, energy-coupled processes such as oxidative phosphorylation and transport are catalyzed by enzymes with defined membrane orientations. The topology of membrane proteins is probably determined to a large extent by their locations at the time they integrate into the lipid bilayer. Location with respect to the two sides of the membrane is obviously also of paramount importance for extrinsic membrane proteins. For example the orientation of the ATPase complex of the inner membrane is such that the F_1 component faces the matrix side (235). The five different subunit polypeptides of F_1 are synthesized on cytoplasmic ribosomes (220) and are transported individually to the matrix where they associate to form the oligomeric F_1 molecule (236, 237). F_1 in turn is bound to a group of more hydrophobic proteins constituting F_0, an integral component of the inner membrane (238). It is self-evident that orientation of the entire complex with respect to the external and internal phases is crucially dependent on the manner in which the F_0 proteins are physically disposed in the membrane and the location of the F_1 subunits at the time they interact with each other and with the F_0 units. Orientation is therefore contingent on the transport of proteins to the appropriate cellular compartment which in turn presupposes the existence of a membrane.

Posttranslational versus Cotranslational Transport

The transfer of proteins from the cytoplasm to mitochondria is not obligatorily coupled to their synthesis as is the transport of secretory proteins into the lumen space of endoplasmic reticulum (239). Many studies have shown that mitochondrial precursor proteins translated in cell-free systems can be transported into the proper compartments upon incubation with isolated mitochondria (3). Posttranslational transport is also supported by the detection of sizable pools of precursors in the cytosol, and by in vivo pulse-chase studies, which indicate accumulation of extramitochondrial precursors that can be chased into mitochondria (240, 241). This is not to say, however, that the transport process cannot be initiated at the nascent polypeptide stage for some proteins. Yeast mitochondria have been found to have bound cytoplasmic polysomes engaged in a preferential synthesis of mitochondrial proteins (242). Suissa & Schatz (243) were able to extract mRNA enriched for mitochondrial proteins from polysomes bound to the outer membrane of yeast mitochondria. These authors found, however, that not all mitochondria-specific mRNAs fractionated with the membrane-bound polysomes (243). It is conceivable that the recognition of proteins by mitochondrial outer-membrane receptors could occur at different stages of completion of the polypeptide chain depending on the protein. Thus, some proteins could become attached to the receptors during chain elongation while others might have to be completed in order to be recognized as a mitochondrial protein.

The mechanism of insertion of the endogenously synthesized group of proteins into the inner membrane has not been studied due to the difficulty in reconstituting an in vitro mitochondrial translation system. Yeast mitochondrial ribosomes are tightly bound to the inner membrane (244, 245) suggesting that membrane insertion of the hydrophobic proteins occurs cotranslationally.

Primary Translation Products

Most but not all mitochondrial proteins are synthesized on cytoplasmic ribosomes as precursors with amino-terminal extensions usually ranging from 20–50 amino acids (3). The precursors are proteolytically processed during transport in a single step by a specific protease located in the matrix (246, 247). Some mitochondrial precursors require two consecutive proteolytic cleavages to form the mature products (248).

Even though there is currently a fairly long list of mitochondrial precursors with known sequences, the primary or tertiary structural features that determine a signal sequence have not been deduced. This is emphasized by the fact that the proteolytic cleavage sites cannot be predicted from the known sequence. Suggestive homologies have been noticed between some precursor proteins (249), but a survey of a broader group of precursor proteins fails to reveal any common primary sequence feature except that they tend to have a net positive charge due to an excess of arginine and lysine residues. Even this rule has its exception in the case of a subunit of the coenzyme QH_2-cytochrome c reductase with a high proportion of acidic residues in its amino-terminal region (214).

The factors that determine whether a protein is synthesized as a precursor have been speculated on but are still an open issue. Some presequences appear to have all the information needed to direct foreign proteins to mitochondria (250). Other reports suggest, however, that the amino-terminal extension is necessary but not sufficient (251). There are many examples of mitochondrial proteins synthesized without presequences (231, 252–255). The addressing signals can therefore be encoded in the primary and/or tertiary structure of proteins independent of the presence of presequences.

In addition to acting as signals for sequestration in mitochondria, amino-terminal extensions could also play a role in directing precursor proteins to particular compartments of the organelle. The notion more exactly would be that regardless of their ultimate destinations, proteins with amino-terminal extensions must at some stage either travel through or have part of their amino-terminal sequence exposed to the matrix in order to be processed. Most protein constituents of the inner membrane (236, 256–258) and of the matrix (212, 259–261) are synthesized as larger precursors. On the other hand, many outer-membrane and intermembrane constituents lack presequences (253–255, 262). There are exceptions. The intermembrane enzyme cytochrome c peroxidase is made as a precursor (263). Similarly, some inner-membrane proteins such as the ATP/ADP carrier (252) and cytochrome c (264) are transported without cleavage of their polypeptide chains. Cytochrome c, however, is a water-soluble extrinsic protein bound to the outer surface of the inner membrane; its transport is essentially to the intermembrane space from where it can interact with the membrane. The ATP/ADP carrier and other examples that seemingly contradict a role of the presequence in internal assortment could also form a group of proteins that, like cytochrome c, are integrated into the inner membrane from the outer surface.

Transport of proteins across membranes appears to be an irreversible process. Once the protein is internalized and delivered to its residence, it becomes permanently fixed. This implies that mature proteins are no longer recognized by the transport system. This is easy to understand when the protein in question is cleaved. Other mechanisms must exist that achieve the same end without changing the primary structure. One such mechanism involves chemical modification of the protein with an attendant change in tertiary structure. Neupert and colleagues have shown that the covalent attachment of heme to apocytochrome c occurs in the intermembrane space (232). The addition of heme is a rapid event which results in a conformational change in the protein molecule. The outer-membrane receptor implicated in the transport of this respiratory carrier has been found to interact only with the apocytochrome c. The addition of heme therefore fixes cytochrome c in the interior of the mitochondrion. Conceivably, other proteins lacking presequences undergo changes in tertiary structure also, perhaps as a result of their associations with membrane lipids or other proteins when they are part of oligomeric enzymes.

Mitochondrial Receptors

The import of proteins into mitochondria is thought to be initiated by receptor proteins capable of discriminating mitochondrial from other cytoplasmic proteins. The receptors could be part of a larger transport complex or they could be responsible for both recognition and translocation. The earliest evidence supporting the existence of outer-membrane receptors came from studies on the transport of cytochrome c (231–233). This evidence can be summarized as follows. The outer membrane of $N.$ $crassa$ mitochondria has a protease-sensitive component that binds apocytochrome c but not the functional cytochrome (232). In vitro studies with isolated mitochondria have shown that transport of apocytochrome c results in a rapid conversion to cytochrome c (232). This process can be experimentally arrested at an intermediate stage of transport by addition of deuteroheme. In the presence of this heme analogue apocytochrome c becomes bound to the membrane but is prevented from being translocated to the interior. Removal of deuteroheme allows the apocytochrome c to be transported to the intermembrane compartment and chemically modified by attachment of heme (232). These experiments indicate that translocation is not obligatorily coupled to receptor binding but internalization (fixation in the interior) is dependent on the enzymatic attachment of heme to the apoprotein. The number of other proteins recognized by the cytochrome c receptor is not known. The transport of precursors to ATPase subunit 9 and cytochrome c_1 (231, 265) is not inhibited by excess apocytochrome c indicating a lack of involvement of the cytochrome c receptor. Mitochondrial precursors synthesized in vitro have also been shown to bind to outer-membrane vesicles in a manner indicative of specific interaction with receptors (234). An initial binding of proteins to surface-located receptors is probably a general feature of protein import by mitochondria.

Certain proteins need to traverse both membranes of mitochondria before reaching their normal location. This is true of matrix enzymes and surprisingly, also some intermembrane constituents that follow a somewhat circuitous transport route (248, 266). How does transport across the outer and inner membrane differ? Does the inner membrane have its own set of receptors that recognize a subset of proteins in the intermembrane compartment? At least one important difference has been noted in the mechanism of transport of proteins to the inner membrane and to the matrix. Schatz and collaborators have amassed an impressive body of experimental data showing that whereas transport across the outer membrane is an energy-independent process, integration of precursors into the inner membrane or their transport into the matrix has an absolute requirement for energy (3, 267, 268). The delivery of proteins to the inner membrane or the matrix is inhibited by uncouplers (267) and by reagents that prevent formation of an electrochemical potential without affecting the matrix pool of ATP (268, 269). These observations indicate that the Mitchelian potential is the driving force for transport of this class of proteins.

At present there are no experimental data bearing on the question of inner membrane receptors. The favored models picture points of contact between the outer and inner membranes. Presumably, regions of membrane contact (fusion ?) contain receptors that recognize and mediate a direct transfer of proteins across both membranes. This model is based in part on electron microscopic observations of frequent regions of close apposition of the two membranes (211).

The resolution of the mitochondrial transport machinery is currently being attempted by genetic (270) and traditional biochemical approaches (234). These studies hold the promise of filling in much of the detailed information on which our understanding of mitochondrial biogenesis ultimately rests.

Transport Routes

Studies on the transport of selected proteins into mitochondria have uncovered a variety of different transport routes. Their salient features are summarized below.

1. Outer-membrane proteins made without amino-terminal extensions are recognized by receptors and inserted directly into the outer membrane in the absence of any energy requirement.
2. Two different transport routes have been described for proteins of the intermembrane space. The first involves a direct energy-independent transfer across the outer membrane. The second mechanism entails an initial transfer of the precursor protein to the inner membrane. This phase is accompanied by proteolytic cleavage of part of the amino-terminal sequence by the matrix protease. The partially processed and still membrane-bound intermediate is then released into the intermembrane space as the mature protein by a second proteolytic cleavage at the amino-terminal end. This cleavage is believed to be catalyzed by a protease located in the intermembrane space.

3. The transport of matrix proteins shows an energy requirement. It is still an open question whether the two membranes are crossed in a single concerted step or in two separate steps. Most of the matrix proteins are synthesized as precursors and are probably processed by the same matrix protease.

4. Transport of proteins to the inner membrane has also been shown to follow at least two different routes. Extrinsic proteins attached to the outer surface of the membrane are transported to the intermembrane space from whence they interact with the membrane. Other extrinsic proteins, such as the F_1 ATPase, that interact with components facing the inside, are transported to the matrix. Transport of integral proteins is thought to occur by the same mechanism that transfers proteins to the matrix except that these proteins are locked in the lipid bilayer by stop-transfer sequences. These proteins can be synthesized either with or without presequences. Another mechanism illustrated by the transport of cytochrome c_1 also involves two consecutive proteolytic cleavages. Insertion of this protein into the inner membrane in the first step is accompanied by the first cleavage which removes part of the amino-terminal extension. This is followed by attachment of heme and a second proteolytic cleavage. This mechanism is very similar to that operating in the transport of cytochrome b_2 to the intermembrane space except that release of the mature cytochrome c_1 into the intermembrane compartment is prevented by a hydrophobic membrane-anchoring sequence.

ASSEMBLY OF MITOCHONDRIAL ENZYMES

Most of the oxidative metabolism of mitochondria is catalyzed by enzymes that are located in the matrix and inner membrane (271). These enzymes should be viewed not only as units of function but also of structure. This is particularly true of the inner-membrane respiratory and ATPase complexes which together with the lipids account for the physical and chemical properties of the membrane. Studies of protein import have provided the basic outline of how cytoplasmically synthesized proteins of mitochondria are recognized and transferred to the various internal compartments of this organelle (3). Out of this information have come important clues about the assembly of mitochondrial enzymes. For example, the fact that the constituent polypeptides of multisubunit enzymes are transported as individual precursors (3, 236) clearly establishes that assembly of the holoenzymes occurs within mitochondria.

In regard to membrane enzymes, assembly of the subunits must occur in a manner that ensures not only a proper quaternary structure but also a correct orientation of the final product in the membrane. Several laboratories are currently addressing these issues by using the ATPase complex as a model mitochondrial membrane enzyme (237, 272). The ATPase consists of 9–10 nonidentical polypeptides of which 6–7 are translated on cytoplasmic ribosomes (273) and 3 are products of mitochondrial genes (32, 40–44). Five of the cytoplasmically synthesized subunits constitute the F_1 ATPase. The precursors to the F_1 subunits are transported to the matrix where they are processed and assembled into the oligomer. Recent studies have demonstrated both transport and assembly of the F_1 complex in isolated mitochondria (237). This system offers the promise of reconstituting ATPase assembly in an in vitro situation.

In the native ATPase complex, F_1 is associated with a set of hydrophobic proteins (F_0) located in the inner membrane (238). Binding of F_1 to F_0 is governed by protein-protein interactions and in vivo probably occurs spontaneously once synthesis of a functional F_0 unit is completed. This aspect of ATPase biogenesis has been investigated by examining assembly of the complex in mit^- mutants with lesions in the F_0 polypeptides (272). These studies suggest that assembly of a functional F_0 unit is initiated by the synthesis of subunit 9 and is followed by a sequential integration and assembly of subunit 8 and subunit 6 (272). All three F_0 subunits are synthesized on mitochondrial ribosomes and their insertion into the inner membrane is presumed to occur cotranslationally.

As alluded to earlier in this article, the orientation of the ATPase must be determined by the directions from which insertion of the subunits takes place. In the case of the mitochondrially translated polypeptides, insertion occurs from the matrix side. This probably confers an orientation such that the binding site on F_0 that recognizes F_1 is exposed to the matrix for final assembly of the complex.

Literature Cited

1. Dujon, B. 1981. In *Molecular Biology of the Yeast Saccharomcyes. Life Cycle and Inheritance*, ed. J. N. Strathern, E. W. Jones, J. R. Broach, pp. 505–635. Cold Spring Harbor, NY: Cold Spring Harbor
2. Wallace, D. C. 1982. *Microbiol. Rev.* 46:208–40
3. Hay, R., Bohni, P., Gasser, S. 1984. *Biochim. Biophys. Acta* 779:65–87
4. Slonimski, P. P., Borst, P., Attardi, G., eds. 1982. *Mitochondrial Genes*. Cold Spring Harbor, NY: Cold Spring Harbor
5. Schweyen, R. J., Wolf, K., Kaudewitz, F., eds. 1983. *Mitochondria 1983. Nucleo-Mitochondrial Interactions*. Berlin: de Gruyter
6. Weiss, H., Sebald, W., Bucher, Th. 1971. *Eur. J. Biochem.* 22:19–26
7. Mason, T. L., Poyton, R. O., Wharton, D. C., Schatz, G. 1973. *J. Biol. Chem.* 248:1346–54
8. Koch, G. 1976. *J. Biol. Chem.* 251: 6097–107
9. Weiss, H., Ziganke, B. 1974. *Eur. J. Biochem.* 41:63–71
10. Katan, M. B., Van Harten-Loosbroek, N., Groot, G. S. P. 1976. *Eur. J. Biochem.* 70:409–17
11. Tzagoloff, A., Meagher, P. 1971. *J. Biol. Chem.* 246:7328–36
12. Bonitz, S. G., Coruzzi, G., Thalenfeld, B. E., Tzagoloff, A., Macino, G. 1980. *J. Biol. Chem.* 255:11927–41
13. Netzker, R., Kochel, H. G., Kunzel, H. 1982. *Nucleic Acids Res.* 10:4783–94
14. Citterich, M. H., Morelli, G., Macino, G. 1983. *EMBO J.* 2:1235–42
15. Lang, B. F., Ahne, F., Bonen, L. 1985. *J. Mol. Biol.* 184:353–66
16. Anderson, S., Bankier, A. T., Barrell, B. G., de Bruijn, M. H. L., Coulson, A. R., et al. 1981. *Nature* 290:457–65
17. Anderson, S., de Bruijn, M. H. L., Coulson, A. R., Eperon, I. C., Sanger, F., Young, I. G. 1982. *J. Mol. Biol.* 156:683–717
18. Bibb, M. J., Van Etten, R. A., Wright, C. T., Walberg, M. W., Clayton, D. A. 1981. *Cell* 26:167–80
19. Saccone, C., Cantatore, P., Gadaleta,

G., Gallerani, R., Lanave, C., et al. 1981. *Nucleic Acids Res.* 9:4139–48
20. Clary, D. O., Wolstenholme, D. R. 1984. *Nucleic Acids Res.* 12:2367–78
21. Roe, B. A., Ma, D.-P., Wilson, R. K., Wong, J. F.-H. 1985. *J. Biol. Chem.* 260:9759–74
22. Fox, T. D., Leaver, C. J. 1981. *Cell* 26:315–23
23. Isaac, P. G., Jones, V. P., Leaver, C. J. 1985. *EMBO J.* 4:1617–23
24. Tzagoloff, A., Akai, A., Needleman, R. B., Zulch, G. 1975. *J. Biol. Chem.* 250:8236–42
25. Slonimski, P. P. 1953. *Formation des Enzymes Respiratoire Chez la Levure.* Paris: Mason (In French)
26. Slonimski, P. P., Tzagoloff, A. 1976. *Eur. J. Biochem.* 61:27–41
27. Slonimski, P. P., Pajot, P., Jacq, C., Foucher, M., Perrodin, G., et al. 1978. In *Biochemistry and Genetics of Yeast,* ed. M. Bacila, B. L. Horecker, A. O. M. Stoppani, pp. 339–68. New York: Academic
28. Cabral, F., Solioz, M., Rudin, Y., Schatz, G., Clavilier, L., Slonimski, P. P. 1978. *J. Biol. Chem.* 253:297–304
29. Weiss-Brummer, B., Guba, R., Haid, A., Schweyen, R. J. 1979. *Curr. Genet.* 1:75–83
30. Claisse, M. L., Spyridakis, A., Slonimski, P. P. 1977. In *Mitochondria 1977, Genetics and Biogenesis of Mitochondria,* ed. W. Bandlow, R. J. Schweyen, K. Wolf, F. Kaudewitz, pp. 337–44. Berlin: de Gruyter
31. Sebald, W., Wachter, E., Tzagoloff, A. 1979. *Eur. J. Biochem.* 100:599–606
32. Macreadie, I. G., Choo, W. M., Novitski, C. E., Marzuki, S., Nagley, P., et al. 1982. *Biochem. Int.* 5:129–36
33. Roberts, H., Choo, W. M., Murphy, M., Marzuki, S., Lukins, H. B., Linnane, A. 1979. *FEBS Lett.* 108:501–4
34. Hensgens, L. A. M., de Haan, L. B. M., van der Horst, G., Grivell, L. A. 1983. *Cell* 32:379–89
35. Fox, T. D. 1979. *Proc. Natl. Acad. Sci. USA* 76:6534–38
36. Coruzzi, G., Tzagoloff, A. 1979. *J. Biol. Chem.* 254:9324–30
37. Thalenfeld, B. E., Tzagoloff, A. 1980. *J. Biol. Chem.* 255:6173–80
38. Nobrega, F. G., Tzagoloff, A. 1980. *J. Biol. Chem.* 255:9828–37
39. Lazowska, J., Jacq, C., Slonimski, P. P. 1980. *Cell* 22:333–48
40. Macino, G., Tzagoloff, A. 1980. *Cell* 20:507–17
41. Novitski, C. E., Macreadie, I. G., Maxwell, R. J., Lukins, H. B., Linnane, A. W., Nagley, P. 1983. In *Manipulation and Expression of Genes in Eucaryotes,* ed. P. Nagley, A. W. Linnane, W. J. Peacock, J. A. Pateman, pp. 297–304. New York: Academic
42. Macreadie, I. G., Novitski, C. E., Maxwell, R. J., John, U., Ooi, B. G., et al. 1983. *Nucleic Acids Res.* 11:4435–51
43. Macino, G., Tzagoloff, A. 1979. *J. Biol. Chem.* 254:4617–23
44. Hensgens, L. A. M., Grivell, L. A., Borst, P., Bos, J. L. 1979. *Proc. Natl. Acad. Sci. USA* 76:1663–67
45. Coruzzi, G., Bonitz, S. G., Thalenfeld, B. E., Tzagoloff, A. 1980. *J. Biol. Chem.* 256:12780–87
46. Dujon, B. 1980. *Cell* 20:185–97
47. Jacquier, A., Dujon, B. 1985. *Cell* 41: 383–94
48. Macreadie, I. G., Scott, R. M., Zinn, A. R., Butow, R. A. 1985. *Cell* 41:395–402
49. Gargouri, A., Lazowska, J., Slonimski, P. P. 1983. See Ref. 5, pp. 259–68
50. Dhwahle, S., Hanson, D. K., Alexander, N. J., Perlman, P. S., Mahler, H. R. 1981. *Proc. Natl. Acad. Sci. USA* 78: 1778–82
51. Weiss-Brummer, B., Rodel, G., Schweyen, R. J., Kaudewitz, F. 1982. *Cell* 29:527–36
52. Hensgens, L. A. M., van der Horst, G., Vos, H. L., Grivell, L. A. 1984. *Curr. Genet.* 8:457–65
53. Carignani, G., Groudinsky, O., Frezza, D., Schiavon, E., Bergantino, E., Slonimski, P. P. 1983. *Cell* 35:733–42
54. Dujardin, G., Jacq, C., Slonimski, P. P. 1982. *Nature* 298:628–32
55. Hudspeth, M. E. S., Ainly, W. M., Shamard, D. S., Butow, R. A., Grossman, L. I. 1982. *Cell* 30:617–26
56. Sor, F., Fukuhara, H. 1983. *Nucleic Acids Res.* 11:339–48
57. Sor, F., Fukuhara, H. 1980. *C. R. Acad. Sci. Ser. D* 291:933–36
58. Li, M., Tzagoloff, A., Underbrink-Lyon, K., Martin, N. C. 1982. *J. Biol. Chem.* 257:5921–28
59. Bonitz, S. G., Tzagoloff, A. 1980. *J. Biol. Chem.* 255:9075–81
60. Newman, D., Pham, H. D., Underbrink-Lyon, K., Martin, N. C. 1980. *Nucleic Acids Res.* 8:5007–16
61. Martin, N. C., Underbrink-Lyon, K. 1981. *Proc. Natl. Acad. Sci. USA* 78:4743–47
62. Miller, D. L., Martin, N. C. 1983. *Cell* 34:911–17
63. Chomyn, A., Mariottini, P., Cleeter, M. W. J., Ragan, C. I., Matsuno-Yagi, A., et al. 1985. *Nature* 314:592–97
64. Burger, G., Scriven, C., Machleidt, W., Werner, S. 1982. *EMBO J.* 1:1385–91

65. De Jonge, J., de Vries, H. 1983. *Curr. Genet.* 7:21–28
66. Browning, K. S., RajBhandary, U. L. 1982. *J. Biol. Chem.* 257:5253–56
67. Citterich, M. H., Morelli, G., Macino, G. 1983. *EMBO J.* 2:1235–42
68. Citterich, M. H., Morelli, G., Nelson, M. A., Macino, G. 1983. See Ref. 5, pp. 355–69
69. Van den Boogaart, P., Samallo, J., Agsteribbe, E. 1982. *Nature* 298:187–89
70. Burger, G., Werner, S. 1985. *J. Mol. Biol.* In press
71. de Vries, H., de Jonge, J. C., Schrage-Tabak, C. 1986. In *Nuclear-Mitochondrial Interactions*, ed. C. Saccone, A. M. Kroon. Amsterdam: North Holland. In press
72. Nelson, M. A., Macino, G. 1986. See Ref. 71
73. Burger, G., Werner, S. 1983. See Ref. 5, pp. 330–42
74. Burke, J. M., RajBhandary, U. L. 1982. *Cell* 31:509–20
75. Heckman, J. E., RajBhandary, U. L. 1979. *Cell* 17:583–95
76. Heckman, J. E., Yin, S., Alzner-DeWeerd, B., RajBhandary, U. L. 1979. *J. Biol. Chem.* 254:12694–700
77. Heckman, J. E., Sarnoff, J., Alzner-DeWeerd, B., Yin, S., RajBhandary, U. L. 1980. *Proc. Natl. Acad. Sci. USA* 77:3159–63
78. DeVries, H., De Jonge, J. C., Bakker, H., Meurs, H., Kroon, A. 1979. *Nucleic Acids Res.* 6:1791–803
79. Wesolowski, M., Fukuhara, H. 1979. *Mol. Gen. Genet.* 170:261–75
80. Bonitz, S. G., Berlani, R., Coruzzi, G., Li, M., Macino, G., et al. 1980. *Proc. Natl. Acad. Sci. USA* 77:3167–70
81. Sanders, J. P. M., Heyting, C., Borst, P. 1975. *Biochem. Biophys. Res. Commun.* 65:699–707
82. Borst, P. 1972. *Ann. Rev. Biochem.* 41:333–76
83. Dawid, I. B. 1972. *Dev. Biol.* 29:139–51
84. Fukuhara, H., Bolotin-Fukuhara, M., Hsu, H. J., Rabinowitz, M. 1976. *Mol. Gen. Genet.* 145:7–17
85. Coen, D., Deutsch, J., Netter, P., Petrochilo, E., Slonimski, P. P. 1970. In *Control of Organelle Development*, pp. 449–96. Cambridge: University Press
86. Linnane, A. W., Lamb, A. J., Christodoulou, C., Lukins, H. B. 1968. *Proc. Natl. Acad. Sci. USA* 59:1288–93
87. Wolf, K., Dujon, B., Slonimski, P. P. 1973. *Mol. Gen. Genet.* 125:53–90
88. Blanc, H., Wright, C. T., Bibb, M. J., Wallace, D. C., Clayton, D. A. 1981. *Proc. Natl. Acad. Sci. USA* 78:3789–93
89. Zassenhaus, H. P., Perlman, P. S. 1982. *Curr. Genet.* 6:179–88
90. Berlani, R. E., Pentella, C., Macino, G., Tzagoloff, A. 1980. *J. Bacteriol.* 141:1086–97
91. Bolotin-Fukuhara, M., Sor, F., Fukuhara, H. 1983. See Ref. 5, pp. 455–67
92. Clark-Walker, G. D., Sriprakash, K. S. 1983. *EMBO J.* 2:1465–72
93. Eperon, I. C., Janssen, J. W. G., Hoeijmakers, J. H. J., Borst, P. 1983. *Nucleic Acids Res.* 11:105–25
94. Boer, P., Bonen, L., Lee, R. W., Gray, M. W. 1985. *Proc. Natl. Acad. Sci. USA* 82:3340–44
95. Lang, B. F., Ahne, F., Distler, S., Trinkl, H., Kaudewitz, F. 1983. See Ref. 5, pp. 313–29
96. Clark-Walker, G. D., McArthur, C. R., Sriprakash, K. S. 1983. *J. Mol. Evol.* 19:333–41
97. Brown, T. A., Davies, R. W., Ray, J. A., Waring, R. B., Scazzocchio, C. 1983. *EMBO J.* 3:427–35
98. Scazzocchio, C., Brown, T. A., Waring, R. B., Ray, J. A., Davies, R. W. 1983. See Ref. 5, pp. 303–11
99. Stern, D. B., Dyer, T. A., Lonsdale, D. M. 1982. *Nucleic Acids Res.* 11:3333–37
100. Chomyn, A., Hunkapiller, M. W., Attardi, G. 1981. *Nucleic Acids Res.* 9:867–70
101. Chomyn, A., Mariottini, P., Gonzalez-Cadavid, N., Attardi, G., Strand, D. D., et al. 1983. *Proc. Natl. Acad. Sci. USA* 80:5535–39
102. Hatefi, Y., Haavik, A. G., Griffiths, D. E. 1962. *J. Biol. Chem.* 237:1676–80
103. Heron, C., Smith, S., Ragan, C. I. 1979. *Biochem. J.* 181:435–43
104. Hatefi, Y., Galante, Y. M., Stigall, D. L., Ragan, C. I. 1979. *Methods Enzymol.* 56:577–602
105. Ise, W., Haiker, H., Weiss, H. 1985. *EMBO J.* 4:2075–80
106. Tzagoloff, A. 1982. *Mitochondria.* New York: Plenum
107. Ohnishi, T., Asakura, T., Yonetani, T., Chance, B. 1971. *J. Biol. Chem.* 246:5960–66
108. Ohnishi, T., Kawaguchi, K., Hagihara, B. 1966. *J. Biol. Chem.* 241:1797–803
109. Senior, A. E., Wise, J. G. 1983. *J. Membr. Biol.* 73:105–24
110. Sebald, W., Hoppe, J., Wachter, E. 1979. In *Function and Molecular Aspects of Biomembrane Transport*, ed. E. Quagliariello, pp. 63–74. Amsterdam: Elsevier
111. De Vries, H., De Jonge, J. C., Arnberg, A., Peijnenburg, A. C. M., Agsteribbe, E. 1983. See Ref. 5, pp. 343–56
112. Michel, F., Lang, B. F. 1985. *Nature* 316:641–43

113. Zinn, A. R., Butow, R. A. 1984. *Cold Spring Harbor Symp. Quant. Biol.* 49: 115–21
114. Zinn, A. R., Butow, R. A. 1985. *Cell* 40:887–95
115. Kostriken, R., Strathern, J. N., Klar, A. J. S., Hicks, J. B., Heffron, F. 1983. *Cell* 35:167–74
116. Hicks, J. B., Strathern, J. N., Klar, A. J. S. 1979. *Nature* 282:478–83
117. Szostak, J. W., Orr-Weaver, T. L., Rothstein, R. J., Stahl, F. W. 1983. *Cell* 33:25–35
118. Farrelly, F., Butow, R. A. 1983. *Nature* 301:296–301
119. Clayton, D. A. 1984. *Ann. Rev. Biochem.* 53:573–94
120. Murphy, W. I., Attardi, B., Tu, C., Attardi, G. 1975. *J. Mol. Biol.* 99:809–14
121. Barrell, B. G., Anderson, S., Bankier, A. T., De Bruijn, M. H. L., Chen, E., et al. 1980. In *Biological Chemistry of Organelle Formation*, ed. Th. Bucher, W. Sebald, H. Weiss, pp. 9–25. Berlin: Springer-Verlag
122. Ojala, D., Montoya, J., Attardi, G. 1981. *Nature* 290:470–74
123. Doerson, C. J., Guerrier-Takada, C., Altman, S., Attardi, G. 1985. *J. Biol. Chem.* 260:5942–49
124. Roe, B. A., Wong, J. F. H., Chen, E. Y., Armstrong, P. W., Stankiewicz, A., et al. 1982. See Ref. 4, pp. 45–49
125. Dubin, D. T., Taylor, R. H. 1978. *J. Mol. Biol.* 121:523–40
126. Ojala, D., Attardi, G. 1974. *J. Mol. Biol.* 82:151–74
127. Boynton, J. E., Gillham, N. W., Lambowitz, A. M. 1980. In *Ribosomes: Structure, Function and Genetics*, ed. G. Chambliss, G. Craven, J. Davies, K. Davis, L. Kahan, M. Nomura, pp. 903–50. Baltimore: University Park
128. Matthews, D. E., Hessler, R. A., Denslow, N. D., Edwards, J. E., O'Brien, T. W. 1982. *J. Biol. Chem.* 257:8788–94
129. Dubin, D. T., Montoya, J., Timko, K. D., Attardi, G. 1982. *J. Mol. Biol.* 157:1–19
130. Montoya, J., Christianson, T., Levens, D., Rabinowitz, M., Attardi, G. 1982. *Proc. Natl. Acad. Sci. USA* 79:7195–99
131. Chang, D. D., Clayton, D. A. 1984. *Cell* 36:635–43
132. Li, M., Tzagoloff, A. 1979. *Cell* 18:47–53
133. Bernardi, G. 1982. See Ref. 4, pp. 269–78
134. Bernardi, G. 1980. *FEBS Lett.* 115: 159–62
135. Tzagoloff, A., Nobrega, M., Akai, A.,

136. Faugeron-Fonty, G., Le Van Kim, C., de Zamaroczy, M., Goursot, R., Bernardi, G. 1984. *Gene* 32:459–73
137. de Zamaroczy, M., Faugeron-Fonty, G., Baldacci, G., Goursot, R., Bernardi, G. 1984. *Gene* 32:439–57
138. Morimoto, R., Rabinowitz, M. 1979. *Mol. Gen. Genet.* 170:25–48
139. Dujon, B. 1983. See Ref. 5, pp. 1–24
140. Christianson, T., Edwards, J. C., Mueller, D. M., Rabinowitz, M. 1983. *Proc. Natl. Acad. Sci. USA* 80:5564–68
141. Thalenfeld, B. E., Hill, J., Tzagoloff, A. 1983. *J. Biol. Chem.* 258:610–15
142. Locker, J., Rabinowitz, M. 1981. *Plasmid* 6:302–14
143. Frontali, L., Palleschi, C., Francisci, S. 1982. *Nucleic Acids Res.* 10:7283–93
144. Borst, P. 1980. See Ref. 121, pp. 27–41
145. Hensgens, L. A. M., Arnberg, A. C., Roosendaal, E., van der Horst, G., Van der Veen, R., et al. 1983. *J. Mol. Biol.* 164:35–58
146. Edwards, J. C., Osinga, K. A., Christianson, T., Hensgens, L. A. M., Janssens, P. M., et al. 1983. *Nucleic Acids Res.* 11:8269–82
147. Levens, D., Tichio, B., Ackerman, E., Rabinowitz, M. 1981. *J. Biol. Chem.* 256:5226–32
148. Christianson, T., Rabinowitz, M. 1983. *J. Biol. Chem.* 258:14025–33
149. Osinga, K. A., De Haan, M., Christianson, T., Tabak, H. F. 1982. *Nucleic Acids Res.* 10:7993–8006
150. Clayton, D. A. 1982. *Cell* 28:693–705
151. Edwards, J. C., Levens, D., Rabinowitz, M. 1982. *Cell* 31:337–46
152. Edwards, J. C., Christianson, T., Mueller, D., Biswas, T. K., Levens, D., et al. 1983. See Ref. 5, pp. 69–78
153. Thalenfeld, B. E., Bonitz, S. G., Nobrega, F. G., Macino, G., Tzagoloff, A. 1983. *J. Biol. Chem.* 258:14065–68
154. Zassenhaus, H. P., Martin, N. C., Butow, R. A. 1984. *J. Biol. Chem.* 259:6019–27
155. Dieckmann, C. L., Koerner, T. J., Tzagoloff, A. 1984. *J. Biol. Chem.* 259: 4722–31
156. Kruger, K., Grabowski, P. J., Zaug, A. J., Sands, J., Gottschling, D. E., Cech, T. R. 1982. *Cell* 31,147–57
157. Zaug, A. J., Grabowksi, P. J., Cech, T. R. 1983. *Nature* 301:578–83
158. Burke, J. M., Breitenberger, C., Heckman, J. E., Dujon, B., RajBhandary, U. L. 1984. *J. Biol. Chem.* 259:504–11
159. Davies, R. W., Waring, R. B., Ray, J. A., Brown, T. A., Scazzocchio, C. 1982. *Nature* 300:719–24

160. Waring, R. B., Davies, R. W., Lee, S., Grisi, E., McPhail-Berks, M., Scazzocchio, C. 1981. *Cell* 27:4–11
161. Lang, B. F. 1984. *EMBO J.* 3:2129–36
162. Osiewacz, H. D., Esser, K. 1984. *Curr. Genet.* 8:299–305
163. Bonen, L., Boer, P. H., Gray, M. W. 1984. *EMBO J.* 3:2531–36
164. Kao, T.-H., Moon, E., Wu, R. 1984. *Nucleic Acids Res.* 12:7305–15
165. Anziano, P. Q., Hanson, D. K., Mahler, H. R., Perlman, P. S. 1982. *Cell* 30:925–32
166. Claisse, M. L., Slonimski, P. P., Johnson, J., Mahler, H. R. 1980. *Mol. Gen. Genet.* 177:375–87
167. McGraw, P., Tzagoloff, A. 1983. *J. Biol. Chem.* 258:9459–68
168. Faye, G., Simon, M. 1983. *Cell* 32:77–87
169. Pillar, T., Lang, B. F., Steinberger, I., Vogt, B., Kaudewitz, F. 1983. *J. Biol. Chem.* 258:7954–59
170. Breathnach, R., Benoist, C., O'Hare, K., Gannon, F., Chambon, P. 1978. *Proc. Natl. Acad. Sci. USA* 75:4853–57
171. Cech, T. R., Kyle-Tanner, N., Tinoco, I. Jr., Weil, B. R., Zuker, M., Perlman, P. S. 1983. *Proc. Natl. Acad. Sci. USA* 80:3903–7
172. Michel, F., Dujon, B. 1983. *EMBO J.* 2:33–38
173. Michel, F., Jacquier, A., Dujon, B. 1982. *Biochimie* 64:867–81
174. Tabak, H. F., Van der Horst, G., Osinga, K. A., Arnberg, A. C. 1984. *Cell* 39:623–29
175. Van der Horst, G., Tabak, H. F. 1985. *Cell* 40:759–66
176. Garriga, G., Lambowitz, A. M. 1984. *Cell* 39:631–41
177. Garriga, G., Lambowitz, A. M. 1983. *J. Biol. Chem.* 258:14745–48
178. Mannella, C. A., Collins, R. A., Green, M. R., Lambowitz, A. M. 1979. *Proc. Natl. Acad. Sci. USA* 76:2635–39
179. Bertrand, H., Bridge, P., Collins, R. A., Garriga, G., Lambowitz, A. M. 1982. *Cell* 29:517–26
180. Collins, R. A., Lambowitz, A. M. 1985. *J. Mol. Biol.* 184:413–28
181. Waring, R. B., Davies, R. W., Scazzocchio, C., Brown, T. A. 1982. *Proc. Natl. Acad. Sci. USA* 79:6332–36
182. Arnberg, A. C., Van Ommen, G. J. B., Grivell, L. A., Van Bruggen, E. F. J., Borst, P. 1980. *Cell* 19:313–19
183. Halbreich, A., Pajot, P., Foucher, M., Grandchamp, C., Slonimski, P. P. 1980. *Cell* 19:321–29
184. Ruskin, B., Krainer, A. R., Maniatis, T., Green, M. R. 1984. *Cell* 38:317–31
185. De La Salle, H., Jacq, C., Slonimski, P. P. 1982. *Cell* 28:721–32
186. Labouesse, M., Slonimski, P. P. 1982. *EMBO J.* 2:269–76
187. Hill, J., McGraw, P., Tzagoloff, A. 1985. *J. Biol. Chem.* 260:3235–38
188. Schatz, G., Groot, G. S. P., Mason, T. L., Rouslin, W., Wharton, D. C., Saltzgaber, J. 1972. *Fed. Proc.* 31:21–29
189. Sherman, F. 1963. *Genetics* 48:375–85
190. Michaelis, G., Mannhaupt, G., Pratje, E., Fischer, E., Naggert, J., Schweizer, E. 1982. See Ref. 4, pp. 311–21
191. Myers, A. M., Pape, L. K., Tzagoloff, A. 1985. *EMBO J.* 4:2087–92
192. Struhl, K., Stinchcomb, D. T., Scherer, S., Davis, R. W. 1979. *Proc. Natl. Acad. Sci. USA* 76:1035–39
193. Dieckmann, C. L., Tzagoloff, A. 1983. *Methods Enzymol.* 97:355–60
194. Saltzgaber-Muller, J., Kunapuli, S. P., Douglas, M. G. 1983. *J. Biol. Chem.* 258:11465–70
195. O'Malley, K., Pratt, P., Robertson, J., Lilly, M., Douglas, M. G. 1982. *J. Biol. Chem.* 257:2097–103
196. Koerner, T. J., Hill, J., Tzagoloff, A. 1985. *J. Biol. Chem.* 260:9613–15
197. Seraphin, B., Simon, M., Faye, G. 1985. *Curr. Genet.* 9:435–39
198. Cumsky, M. G., Ko, C., Trueblood, C. E., Poyton, R. O. 1985. *Proc. Natl. Acad. Sci. USA* 82:2235–39
199. Dieckmann, C. L., Homison, G., Tzagoloff, A. 1984. *J. Biol. Chem.* 259:4732–38
200. Maarse, A. C., van Loon, A. P. G. M., Riezman, H., Gregor, I., Schatz, G., Grivell, L. A. 1984. *EMBO J.* 3:2831–37
201. Pape, L. K., Koerner, T. J., Tzagoloff, A. 1985. *J. Biol. Chem.* 260:15362–70
202. Wright, R. M., Ko, C., Cumsky, M. G., Poyton, R. O. 1984. *J. Biol. Chem.* 259:15401–7
203. Faye, G., Simon, M. 1983. See Ref. 5, pp. 433–39
204. Nagata, S., Tsunetsugo-Yokota, Y., Naito, A., Kaziro, Y. 1983. *Proc. Natl. Acad. Sci. USA* 80:6192–96
205. Dieckmann, C. L., Tzagoloff, A. 1985. *J. Biol. Chem.* 260:1513–20
206. Myers, A. M., Tzagoloff, A. 1985. *J. Biol. Chem.* 260:15371–77
207. Montgomery, D. L., Hall, B. D., Gillam, S., Smith, M. 1978. *Cell* 14:673–80
208. Lomax, M. I., Bachman, N. J., Nasoff, M. S., Caruthers, M. H., Grossman, L. I. 1984. *Proc. Natl. Acad. Sci. USA* 81:6295–99
209. Montgomery, D. L., Leung, D. W., Smith, M., Shalit, P., Faye, G., Hall, B.

D. 1980. *Proc. Natl. Acad. Sci. USA* 77:541–45
210. Kellems, R. E., Butow, R. A. 1972. *J. Biol. Chem.* 247:8043–50
211. Kellems, R. E., Allison, V. F., Butow, R. A. 1975. *J. Cell Biol.* 65:1–14
212. Suissa, M., Suda, K., Schatz, G. 1984. *EMBO J.* 3:1773–81
213. van Loon, A. P. G. M., de Groot, R. J., van Eyk, E., van der Horst, G. T. J., Grivell, L. A. 1982. *Gene* 20:323–27
214. van Loon, A. P. G. M., de Groot, R. J., de Haan, M., Dekker, A., Grivell, L. A. 1984. *EMBO J.* 3:1039–43
215. Luck, D. J. L. 1965. *J. Cell Biol.* 24:461–70
216. Plattner, H., Salpeter, M. M., Saltzgaber, J., Schatz, G. 1970. *Proc. Natl. Acad. Sci. USA* 66:1252–59
217. Criddle, R. S., Schatz, G. 1969. *Biochemistry* 8:322–34
218. Mahler, H. R., Bastos, R. N., Feldman, F., Flury, U., Lin, C. C., et al. 1975. In *Membrane Biogenesis,* ed. A. Tzagoloff, pp. 15–61. New York: Plenum
219. Schatz, G., Mason, T. L. 1974. *Ann. Rev. Biochem.* 43:51–87
220. Schatz, G. 1968. *J. Biol. Chem.* 243:2192–99
221. Faye, G. 1976. *FEBS Lett.* 69:167–70
222. Muller, P. P., Reif, M. K., Zonghou, S., Sengstag, C., Mason, T. L., Fox, T. D. 1984. *J. Mol. Biol.* 175:431–52
223. Terpstra, P., Butow, R. A. 1979. *J. Biol. Chem.* 254:12662–69
224. Williamson, D. H., Marudas, N. G., Wilkie, D. 1971. *Mol. Gen. Genet.* 111:209–23
225. Zitomer, R. S., Montgomery, D. L., Nichols, D. L., Hall, B. D. 1979. *Proc. Natl. Acad. Sci. USA* 76:3627–31
226. Lustig, A., Padmanaban, G., Rabinowitz, M. 1982. *Biochemistry* 21:309–16
227. Szekely, E., Montgomery, D. L. 1984. *Mol. Cell. Biol.* 4:939–46
228. Guarente, L., Mason, T. 1983. *Cell* 32:1279–86
229. Guarente, L., Lalonde, B., Gifford, P., Alani, E. 1984. *Cell* 36:503–11
230. Zennaro, E., Grimaldi, L., Baldacci, G., Frontali, L. 1985. *Eur. J. Biochem.* 147:191–96
231. Zimmerman, R., Hennig, B., Neupert, W. 1981. *Eur. J. Biochem.* 116:455–60
232. Hennig, B., Neupert, W. 1981. *Eur. J. Biochem.* 121:203–13
233. Hennig, B., Koehler, H., Neupert, W. 1983. *Proc. Natl. Acad. Sci. USA* 80:4963–67
234. Riezman, H., Hay, R., Gasser, S., Daum, G., Schneider, G., et al. 1983. *EMBO J.* 2:1105–11

235. Kagawa, Y., Racker, E. 1966. *J. Biol. Chem.* 241:2475–82
236. Lewin, A. S., Gregor, I., Mason, T. L., Nelson, N., Schatz, G. 1980. *Proc. Natl. Acad. Sci. USA* 77:3998–4002
237. Lewin, A. S., Norman, D. 1983. *J. Biol. Chem.* 258:6750–55
238. Kagawa, Y., Racker, E. 1966. *J. Biol. Chem.* 241:2461–66
239. Blobel, G., Dobberstein, B. 1975. *J. Cell Biol.* 67:835–51
240. Hallermayer, G., Zimmerman, R., Neupert, W. 1977. *Eur. J. Biochem.* 81:523–32
241. Reid, G. A., Schatz, G. 1982. *J. Biol. Chem.* 257:13062–67
242. Ades, I. Z., Butow, R. A. 1980. *J. Biol. Chem.* 255:9918–24
243. Suissa, M., Schatz, G. 1982. *J. Biol. Chem.* 257:13048–55
244. Grivell, L. A., Reijinders, L., Borst, P. 1971. *Biochim. Biophys. Acta* 247:91–103
245. Spithill, T. W., Trembath, M. K., Lukins, H. B., Linnane, A. W. 1978. *Mol. Gen. Genet.* 164:155–62
246. Cerletti, N., Bohni, P. C., Suda, K. 1983. *J. Biol. Chem.* 258:4944–49
247. McAda, P. C., Douglas, M. G. 1982. *J. Biol. Chem.* 257:3177–82
248. Gasser, S. M., Ohashi, A., Daum, G., Bohni, P. C., Gibson, J., et al. 1982. *Proc. Natl. Acad. Sci. USA* 79:267–71
249. Horwich, A. L., Fenton, W. A., Williams, K. R., Kalousek, F., Kraus, J. P., et al. 1984. *Science* 224:1068–74
250. Hurt, E. C., Pesshold-Hurt, B., Schatz, G. 1984. *FEBS Lett.* 178:306–10
251. Geller, B. L., Britten, M. L., Biggs, C. M., Douglas, M. G. 1983. See Ref. 5, pp. 607–18
252. Zimmerman, R., Neupert, W. 1980. *Eur. J. Biochem.* 109:217–29
253. Sagara, Y., Ito, A. 1982. *Biochim. Biophys. Res. Commun.* 109:1102–7
254. Freitag, H., Janes, M., Neupert, W. 1982. *Eur. J. Biochem.* 126:197–202
255. Gasser, S. M., Schatz, G. 1983. *J. Biol. Chem.* 258:3427–30
256. van Loon, A. P. G. M., Kreike, J., De Ronde, A., van der Horst, G. T., Gasser, S. M., Grivell, L. A. 1983. *Eur. J. Biochem.* 135:457–63
257. Michel, R., Wachter, E., Sebald, W. 1979. *FEBS Lett.* 101:373–76
258. Mihara, K., Omura, T., Harano, T., Brenner, S., Fleischer, S., Rajagopalan, K. V., Blobel, G. 1982. *J. Biol. Chem.* 257:3355–58
259. Conboy, J. G., Rosenberg, L. E. 1981. *Proc. Natl. Acad. Sci. USA* 78:3073–77

260. Raymond, Y., Shore, G. C. 1981. *J. Biol. Chem.* 256:2087–90
261. Mueckler, M. M., Himeno, M., Pitot, H. C. 1982. *J. Biol. Chem.* 257:7178–80
262. Watanabe, K., Kubo, S. 1982. *Eur. J. Biochem.* 123:587–92
263. Kaput, J., Goltz, S., Blobel, G. 1982. *J. Biol. Chem.* 257:15054–58
264. Zimmerman, R., Paluch, W., Neupert, W. 1979. *FEBS Lett.* 108:141–46
265. Teintze, M., Slaughter, M., Weiss, H., Neupert, W. *J. Biol. Chem.* 257:10364–71
266. Daum, G., Gasser, S. M., Schatz, G. 1982. *J. Biol. Chem.* 257:13075–80
267. Nelson, N., Schatz, G. 1979. *Proc. Natl. Acad. Sci. USA* 76:4365–69
268. Gasser, S. M., Daum, G., Schatz, G. 1982. *J. Biol. Chem.* 13034–41
269. Schleyer, M., Schmidt, B., Neupert, W. 1982. *Eur. J. Biochem.* 125:109–16
270. Yafe, M. P., Schatz, G. 1984. *Proc. Natl. Acad. Sci. USA* 8:4819–23
271. Lehninger, A. L. 1965. *The Mitochondrion. Molecular Basis of Structure and Function.* New York: Benjamin
272. Marzuki, S., Hadikusumo, R. G., Chou, W. M., Watkins, L., Lukins, H. B., Linnane, A. W. 1983. See Ref. 5, pp. 535–49
273. Todd, R. D., Griesenbeck, T. A., Douglas, M. G. 1980. *J. Biol. Chem.* 255:5461–67

Ann. Rev. Biochem. 1986. 55:287–315
Copyright © 1986 by Annual Reviews Inc. All rights reserved

CARBOHYDRATE-BINDING PROTEINS: TERTIARY STRUCTURES AND PROTEIN-SUGAR INTERACTIONS

Florante A. Quiocho

Department of Biochemistry, Rice University, Houston, Texas 77251

CONTENTS

PERSPECTIVES AND SUMMARY

Proteins and enzymes that bind carbohydrates are present in large numbers in all living cells and are involved in a myriad of important biological functions. This abundance is primarily due to the fact that carbohydrates derived from carbon dioxide fixation constitute the bulk of the organic matter on the earth, and together with their various derivatives and polymeric forms, are utilized for many essential purposes in the cell. Sugars are the central energy source for mechanical work and chemical reactions in living cells. Phosphate derivatives of monosaccharides are important in these energy transformations, and several sugar derivatives, particularly adenosine triphosphate, serve key roles in the

287

0066-4154/86/0701-0287$02.00

storage and transfer of energy. The uptake of sugars into the cell and their subsequent utilization require transport proteins and a whole assemblage of enzymes. In addition, carbohydrates in the form of polysaccharides are used for fuel storage and as structural elements. Nucleic acids, which control the biosynthesis of proteins and the transfer of genetic information, are also polymers with carbohydrate constituents. Polysaccharides associated with proteins (glycoproteins) are found in blood and various secretions, cell membranes, and connective tissues. The continuing upsurge of interest in protein-saccharide interactions is fueled by their essential role in biological recognition and adhesion processes. Glycoproteins on the cell surface have been implicated in these processes.

The present review is not intended to describe comprehensively the functions of carbohydrate-binding proteins or the nature of the carbohydrates and polysaccharides or glycoproteins that are associated with these functions. Rather this essay, which relies mainly on the results of X-ray crystallographic studies, is devoted to two related topics of current interest—the three-dimensional structures of the proteins and enzymes that bind carbohydrates and the molecular basis of the interactions between proteins and saccharide ligands.

TERTIARY STRUCTURES

It is evident from the list in Table 1 that there are many enzymes and proteins that bind carbohydrates whose tertiary structures have been or are being determined. These proteins vary in size, subunit composition, and function. Moreover, there is wide diversity in the three-dimensional structures of these proteins.

Lysozyme from hen egg white has the distinction of being both the first enzyme molecule and the first saccharide-binding protein whose structure was determined (1, 2). It catalyzes the hydrolysis of $\beta(1-4)$-linked N-acetyl sugars of the polysaccharide component of bacterial cell walls. Egg white lysozyme is the smallest of the known saccharide-binding protein structures. It is roughly ellipsoidal and is composed of two domains with an extended groove between them that serves as the polysaccharide binding site. Lysozyme is an example of the $\alpha + \beta$ class of proteins; proteins belonging to this class have α-helixes and β-strands that do not intertwine, but instead tend to segregate along the polypeptide chain (3). The N-terminal domain is a compact globular region with a hydrophobic core. The second domain is composed primarily of β-sheets and is more hydrophilic than the N-terminal domain.

Glycogen phosphorylase and the debranching enzyme are required for the first two steps in glycogen degradation. Phosphorylase catalyzes the first step by sequential phosphorylation of $\alpha(1-4)$-linked glucosyl sugars to yield α-D-glucose-1-phosphate. The debranching enzyme hydrolyzes the $\alpha(1-6)$ linkages

Table 1   Structure determination of carbohydrate-binding proteins

Enzymes/proteins	M_r of subunit	Oligomeric structure	Resolution (Å)	Ref.
Lysozyme	14,600	Monomer	1.5	48
Hexokinase A isozyme	50,000	Monomer	2.1	67
Phosphofructokinase	33,900	Tetramer	2.4	11
2-Keto-3-deoxy-6-phosphoglu- conate aldolase	24,000	Trimer	2.8	19
Glycogen phosphorylase				
a	97,500	Dimer	2.1	4
b	97,500	Dimer	2.0	5
Neuraminidase	50,000	Tetramer	2.8	27
Xylose isomerase	41,300	Tetramer	4.0	17
Glucose-6-phosphate isomerase	66,000	Dimer	2.6	18
Phosphoglucomutase[a]	64,000	Monomer	2.7	
Taka-amylase A	54,000	Monomer	3.0	20
α-Amylase	53,000	Monomer	5.0	72
Debranching enzyme	170,000	Monomer	crystal	6
Glyceraldehyde-3-phosphate de- hydrogenase	36,000	Tetramer	2.7	12
Triose phosphate isomerase	27,000	Dimer	2.5	16
Phosphoglycerate kinase	45,000	Monomer	2.5	10
Phosphoglycerate mutase (yeast)	25,000	Tetramer	2.8	73
Arabinose-binding protein	33,170	Monomer	1.7	31
Galactose-binding protein[b]	33,300	Monomer	2.0	
Maltose-binding protein[c]	40,500	Monomer	2.8	
Hemagglutinin	75,000	Trimer	3.0	26
Wheat germ agglutinin	17,000	Dimer	2.2	21
Concanavalin A	26,000	Tetramer	2.0	22
Pea lectin				
α	6,000	Tetramer $(\alpha\beta)_2$	6.0	25
β	17,500			
Ricin				
A Chain	32,000	Dimer (AB)	4.0	74
B Chain	32,000			
Glycoprotein from *trypanosoma brucei*	62,500	Dimer	6.0	75

[a]M. Rossmann, W. R. Ray, manuscript in preparation.
[b]M. N. Vyas, N. K. Vyas, F. A. Quiocho, unpublished data.
[c]J. C. Spurlino, F. A. Quiocho, unpublished data.

at the branch point of glycogen. Phosphorylase has the largest subunit structure determined thus far (4, 5), but the monomeric debranching enzyme is larger and its structure should be elucidated soon (6). The crystal structures of rabbit muscle phosphorylase *a* and *b* have been determined at 2.1 and 2 Å resolutions, respectively (4, 5). Despite its large size (841 amino acid residues), the phosphorylase subunit is compact and made up of only two large domains. Both domains exhibit α/β topology. The first approximately 480 residues fold into the N-terminal domain. Within this domain is a subdomain which contains the glycogen storage site. The remainder of the molecule constitutes the C-terminal domain. The N-terminal domain contains several binding sites that control various regulatory functions, whereas the C-terminal domain contains the covalently linked coenzyme pyridoxal phosphate and many of the residues required for enzymatic activity. An unusual feature of glycogen phosphorylase is its large number of ligand binding sites, some of which are at relatively large distances from each other.

Several of the carbohydrate-binding proteins, though often completely unrelated in their functions, have very similar structures that are characterized by the presence of two distinct globular domains with a deep cleft between them. Many ligand binding sites occur in the cleft where the domains come together. The proteins that have this feature include the sugar-binding proteins that serve as initial receptors for bacterial active transport and chemotaxis (7, 8), and the many enzymes that are involved in carbohydrate metabolism, such as the kinases [e.g. hexokinase (9), phosphoglycerate kinase (10), phosphofructokinase (11)], and dehydrogenases [e.g. glyceraldehyde-3-phosphate dehydrogenase (12)]. The domains usually have a central β-pleated sheet, consisting mostly of parallel strands, sandwiched by helixes. The α-helixes and β-sheet strands approximately alternate along the chain, producing an α/β domain topology. At least one of the domains has a topology similar to the nucleotide fold. In all of these domains the parallel strands and the N-terminals of helixes point towards the binding-site cleft between the domains.

In many proteins and enzymes with multidomain structures, each distinct domain is formed from one continuous polypeptide segment with only one stretch of polypeptide connecting any two domains (13). There are, however, some exceptions in the carbohydrate-binding proteins. For example, there are three peptide segments that connect the two domains of phosphofructokinase (11), hexokinase (9), and remarkably, the transport/chemotaxis binding proteins specific for L-arabinose (7), D-galactose (8), Leu, Ile, or Val (14), and sulfate (15). The overall structures of the four binding proteins are very similar, but the three connecting segments are not necessarily located at the same region of the polypeptide chain. These proteins also lack significant sequence homology. The existence of multiple segments connecting domains creates an additional complication in our understanding of protein folding.

There are several structures, commonly observed in a group of enzymes that catalyzes the isomerization of carbohydrates and sugar-phosphates, that are very similar to each other, but altogether different from those described thus far. The monomers of triose phosphate isomerase (16), and the larger domain of xylose isomerase (17) and glucose-6-phosphate isomerase (18), are all comprised of a core of eight parallel β-strands twisted into a cylinder (or β-barrel) with eight helixes packed around the outside. The same structural feature is also exhibited by 2-keto-3-deoxy-6-phosphogluconate aldolase (19) and the larger domain of taka-amylase A (20). The smaller domain of taka-amylase A, which is linked to the larger domain by a single-peptide segment, exhibits a totally different domain structure; it has an eight-stranded antiparallel β-sandwich structure.

Concanavalin A and wheat germ agglutinin represent two different plant lectins whose complete molecular structures have been determined (21–23). Plant lectins are a group of proteins that bind carbohydrates. Although the specific functions of these lectins are unknown, the numerous biological effects of plant lectins on animal cells, for example the stimulation of mitosis in lymphocytes and the modulation of cell surface components, have been studied widely (24). Plant lectins have also been valuable reagents both for affinity chromatographic isolation of glycoproteins and as probes of cell surfaces. These proteins are generally composed of multiple subunits; wheat germ agglutinin has two identical subunits, and concanavalin A has four identical subunits at neutral pH that dissociate into dimers below pH 6. Though relatively small in size and lacking in secondary structure, the wheat germ agglutinin monomer folds into an assembly of four isostructural domains, each stabilized by four interlocking homologically placed disulfide bonds (21). On the other hand, the concanavalin A subunit is much larger and is distinguished by a structure made up almost entirely of two large pleated sheets (22, 23). One sheet extends through the center of the monomer and provides gross structure to the subunit; the other sheet forms a flat surface at the back of the molecule and participates in both monomer-monomer and dimer-dimer interactions. Concanavalin A is an example of an all β type protein. Since pea lectin and concanavalin A belong to the same class of lectin and show sequence similarities, it is not surprising that the low-resolution structure of the former looks very much like the latter (25).

Although hemagglutinin and neuraminidase are both integral membrane glycoproteins of influenza virus with specificity for carbohydrates, they have structures that are not only complex but completely different from each other (26–28). Of the proteins with specificity for saccharides, the structure of the hemagglutinin is one of the most complex and intriguing. The hemagglutinin molecule is composed of three identical subunits of 550 amino acids and containing 25% carbohydrate by weight. It recognizes sialic acid found on glycoproteins or glycolipids on the target cell surface and mediates subsequent fusion of viral and host cell membranes. The trimeric glycoprotein consists of three major domains—the main component extracellular to the membrane, a small hydrophobic peptide that spans the membrane, and a small hydrophilic domain on the interior side of the membrane. It is the main component, cleaved from the viral membrane by proteolytic digestion, whose structure has been determined. The hemagglutinin trimer is 135 Å long with a triangular cross-section varying in radius from 15 to 40 Å. The entire trimer structure consists of two structurally distinct regions—a triple-stranded coiled-coil of α-helixes that extends 76 Å from the membrane, and a globular region of antiparallel β-sheets that contains the sugar-binding site and the variable antigenic sites. The monomer folds essentially into two domains, a globular head distal to the membrane end and a fibrous stem proximal to the membrane end.

Neuraminidase catalyzes cleavage of the α-ketosidic linkage between the terminal sialic acid and an adjacent sugar residue of cell surface oligosaccharide of hemagglutinin. During infection, this reaction permits transport of the virus through mucin and destroys the hemagglutinin receptor on the host cell, thus allowing the elution of progeny virus particles from the infected cell. In contrast with the hemagglutinin, neuraminidase is a tetramer with a box-shaped head, $100 \times 100 \times 60$ Å, attached to a slender stalk. The head, which is the main extracellular component, can be cleaved off by pronase and crystalized. The extracellular components are also different as the structure of the neuraminidase head is globular in shape and each monomer is composed of six topologically identical β-sheets arranged in a propeller formation (27).

The foregoing brief discussion makes abundantly clear the considerable variability in the structures of proteins and enzymes with specificity for carbohydrates. Though it has been suggested that complex tertiary structures may be governed by a set of relatively simple principles (29), one would not have predicted the great diversity in these structures. Undoubtedly there are other structural motifs that remain to be unraveled.

PROTEIN-SUGAR INTERACTIONS

The study of protein-ligand interactions had its beginning almost a century ago when Emil Fischer, referring to the selectivity exhibited by certain 'carbohydrase' enzymes, stated: ". . . I would like to say that the enzyme and the glucoside have to fit each other like a lock and key" (30). As will become evident in this essay, the essence of this simple view of protein-sugar interaction remains intact.

There has been considerable advancement in our detailed understanding of protein-sugar interactions at the molecular level, mainly due to several crystallographic studies, beginning twenty years ago with the difference Fourier analysis of the structure of lysozyme and its complexes with sugar inhibitors (2). Since then several complexes involving other carbohydrate-binding proteins have been similarly analyzed, some (including those of lysozyme) to at least 2 Å resolution. So far, only the 1.7 Å resolution structure of the complex of L-arabinose-binding protein with L-arabinose has been extensively refined (31). These accomplishments over the span of twenty years, albeit gained from the studies of only a limited class of proteins, mirror the considerable advancement in the field of X-ray crystallography of biological macromolecules.

The importance of crystallographic refinement, especially in understanding the precise molecular details of the interactions between proteins and their ligands, cannot be overemphasized. Refinement obviates the lack of accuracy and the interpretative bias inherent in the use of difference Fourier synthesis to study these interactions. Moreover, when very high resolution data are available, refinement gives useful information about the location of bound water or solvent molecules and the mobility of parts of the structure.

This discussion of protein-sugar interactions draws heavily on the results of the highly refined 1.7 Å structure of the complex between L-arabinose-binding protein and L-arabinose. (Most of these results are shown in Figures 1 and 2 and Tables 2 and 3). These data provide the only precise structure of a sugar-protein complex, and we believe that the complex contains the main essential features of sugar-protein interactions. This survey also takes into account results of analysis of other complexes (see Table 4 for a list). A variety of monosaccharides and oligosaccharides that act as competitive inhibitors and a transition state analogue have been used to characterize the saccharide-binding site in lysozyme at 2 Å resolution (32–35). One of these complexes is shown in Figure 3. Figure 4 shows the complex of phosphofructokinase with fructose-6-phosphate that has been partially refined at 2.4 Å resolution (11). Complexes of wheat germ agglutinin orginally analyzed at 2.8 Å resolution (36, 37) have been recently reinterpreted in light of the complete molecular structure of the native protein determined at 1.8 Å resolution (21). One of these complexes is shown in Figure 5. On the basis of the partial 2.1 Å structure of hexokinase, it has been possible to identify tentatively the residues (but not the specific groups) within hydrogen-binding distance of o-toluoylglucosamine, an analogue of the substrate glucose (38). Several difference Fourier syntheses have been computed for phosphorylase complexed with glucose and several analogues (4, 5, 39, 40). All of these compounds bind with their glucopyranose moieties in essentially the same position.

Information about the mode of binding of oligosaccharides has also been obtained from difference Fourier maps that show the binding of maltose to part of the active site of taka-amylase A (20) and the interaction of a number of oligosaccharides with the glycogen storage site of phosphorylase, which is distinct from the catalytic site (4, 5, 41–43). Although it will not be reviewed here, several X-ray studies have been conducted on complexes of enzymes involved in sugar metabolism with their triose phosphate substrates [e.g. (10, 12, 13)]. These triose phosphate substrates are bound mainly via salt-link with the phosphate group.

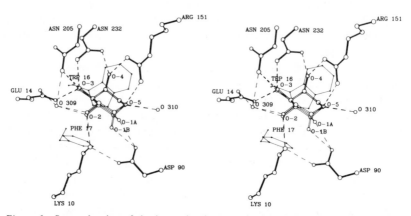

Figure 1   Stereo drawing of the interaction between the bound α or β anomeric form of *L*-arabinose and *L*-arabinose-binding protein. Dashed lines indicate hydrogen bonds. The α-anomeric hydroxyl is labeled O-1A and the β-hydroxyl O-1B. Note that the refined positions of both sugar anomers are not exactly coincident [see Ref. (31)] and that Trp 17 is partially overlapped with C-3, C-4, and C-5 of the sugar.

Locations of Sugar-Binding Sites in Proteins

The locations of the binding sites in proteins vary, but for monosaccharides they are usually found in clefts formed between domains. Enzymes and proteins with bilobal structures constitute a large proportion of carbohydrate-binding proteins. The sugar-binding sites of enzymes with the α/β barrel domain structure are located in a shallow cavity at the carboxyl end of the parallel strands and/or the N-terminus of the α-helixes of the barrel. A pocket at the distal tip of the influenza virus hemagglutinin and neuraminidase serves as the sugar receptor site. The sugar sites for these membrane surface proteins are entirely consistent with the role of these proteins in virus-host interactions. The extended binding site for oligosaccharide substrates in lysozyme (32, 33) and taka-amylase A (20) are located in a groove along each protein surface. Glycogen phosphorylase has two saccharide-binding sites per subunit, separated by about 30 Å, with very different features: the site for catalysis of glycogen breakdown is located in a deep cleft between the N- and C-terminal domains, whereas the glycogen storage site is located on a surface consisting primarily of two antiparallel helices in a subdomain of the N-terminal domain (41–43).

Wheat germ agglutinin has unique sugar-binding sites. There are two binding sites per subunit, which are noncooperative, spatially distinct, and formed by amino acid residues of both protomers (21). This unique feature manifests itself in the high degree of internal pseudosymmetry observed between domains in the dimer molecule, and suggests that dimerization and formation of sugar-binding sites may have occurred in the early evolution of this molecule (44).

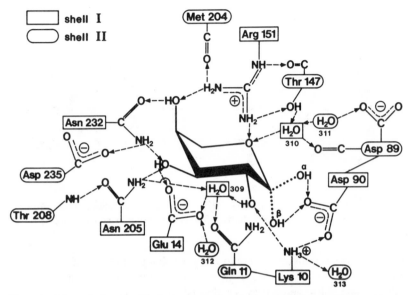

Figure 2   Schematic diagram of the intricate networks of hydrogen bonds formed in the arabinose-binding protein-sugar complex. Shell I represents the essential residues H-bonded to the sugars and to the adjacent second shell of residues. Adapted from Ref. 31 and reproduced by permission from *Nature*.

In many enzymes with more than one substrate, such as those involved in glycolysis (e.g. kinases, dehydrogenases), the number of domains is commensurate with the number of ligands required for the catalysis. The various domains, each of which binds one specific substrate or a cofactor, meet together precisely in a cleft between the domains to consummate the catalytic reaction.

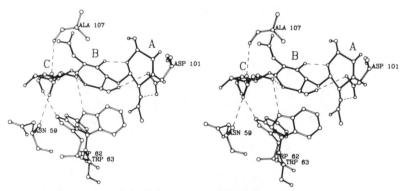

Figure 3   Stereo view of the hydrogen bonds formed between lysozyme and tri-*N*-acetyl-glucosamine. The coordinates used to produce this drawing were obtained from Ref. 33 for the trisaccharide molecule and from the Protein Data Bank for the hen egg white lysozyme.

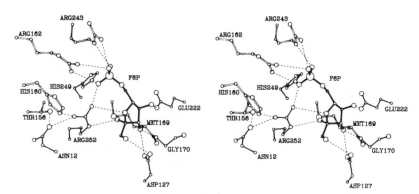

Figure 4   Stereo view of the current model of the hydrogen-bonding between phosphofructoki-nase and fructose-6-phosphate (F6P) (See also Ref. 11). Photograph kindly provided by Dr. P. Evans.

This process often involves a conformational change. In contrast, in proteins such as lysozyme, taka-amylase A, and arabinose-binding protein, which act only on one substrate and do not depend on cofactors, both domains are extensively utilized in binding.

A clear relationship has been established between the α/β topology of the domain structure of many proteins (including some of those listed in Table 1) and the position and geometry of the ligand-binding sites (45). Because of the right-handedness of the βαβ connection in these domains and the reversal of strand order, there is a pocket where the loops that connect strands with their respective helices are on opposite sides of the sheet at the carboxy ends of the

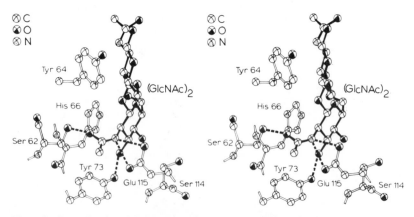

Figure 5   Stereo drawing of the complex of wheat germ agglutinin with di-*N*-acetyl-glucosamine. Reproduced with permission (21). Photograph courtesy of Dr. C. Wright.

Table 2 Hydrogen bonds between L-arabinose-binding protein and α/β-L-arabinose[a]

Donors (X)	Acceptors (Y)	α-L-arabinose		β-L-arabinose	
		$X\cdots Y$ (Å)	$X-\hat{H}\cdots Y$ (°)	$X\cdots Y$ (Å)	$X--\hat{H}\cdots Y$ (°)
O-1	Asp 90 OD2	2.77	159	2.74	166
O-2	Wat 309 O	2.61	155	2.60	156
O-3	Glu 14 OE2	2.77	176	2.66	173
O-4	Asn 232 OD1	2.62	178	2.62	176
Lys 10 NZ	O-2	2.73	146	2.86	151
Asn 205 ND2	O-3	3.03	169	3.09	167
Asn 232 ND2	O-3	2.97	162	2.98	159
Arg 151 NH2	O-4	2.82	167	2.81	168
Arg 151 NH1	O-5	3.05	161	2.99	160
Wat 310 O	O-5	2.80	—	2.74	—
	Overall mean =	2.82 (0.15)	164 (10)	2.81 (0.16)	164 (8)

[a]Adapted from Ref. 31 and reproduced by permission of *Nature*. The N-terminal domain supplies Lys 10, Glu 14, and Asp 90, whereas the C-terminal domain provides Arg 151, Asn 205, and Asn 232.

strands. Differences in orientation, length, and amino acid sequence of these loops can lead to formation of a binding crevice without requiring a change in the basic folding pattern of the domain.

Many carbohydrate substrates that are bound in clefts are buried to varying degrees. This observation provides evidence for ligand-induced conformational changes. The nature and significance of these changes are discussed in another section below.

Table 3 van der Waals interactions between L-arabinose-binding protein and L-arabinose[a]

Sugar atoms	No. of contacts <4 Å	Mean distance (Å)
C-1	5	3.70
C-2	3	3.83
C-3	3	3.73
C-4	7	3.70
C-5	6	3.74
O-1 (α)	7	3.47
O-1 (β)	6	3.60
O-2	3	3.63
O-3	6	3.60
O-4	5	3.57
O-5	3	3.68
Overall mean =		3.66 (0.10)

[a]Summary of data to be published, N. K. Vyas and F. A. Quiocho.

Table 4 Hydrogen bond interactions in protein-saccharide complexes[a]

Protein	Ligand	Sugar						Other groups
		O-1	O-2	O-3	O-4	O-5	O-6	
L-Arabinose-binding protein (31)	L-arabinose	Asp90 OD2	Lys10 NZ Wat309 O	Glu14 OE2 Asn205 ND2 Asn232 ND2	Asn232 OD1 Arg151 NH2	Arg151 NH2 Wat310 O	—	—
Wheat germ agglutinin (21)	(GlcNAc)$_2$[b] (site 1)	—	—	Tyr173 OH	0	0	0	2-Acetamido 2-H-bonds
Phosphorylase b (5)	Heptenitol	Leu136 N	Asn284 ND2 Tyr572 OH	Glu671 OE1 Ser673 N	Asn483 ND2 Gly674 N	His376 ND1	His376 ND1	—
Hexokinase (38)	O-Toluoyl glucosamine	Gln277 XE	—	Asn188 XD Asn245 XD	Asp189 OD2 Asn188 XD Asn215 XD	0	Asp189 OD1	0
Phosphofructose kinase (11)	Fructose-6 phosphate	Arg252 NE	Asp127 OD1	Asp127 OD2 Gly170 N	Glu222 OE1	Arg252 NH	—	6-Phosphate 3 Salt-links

Enzyme	Substrate	Subsite							2-Acetamido
Lysozyme (32,33)	(GlcNAc)₃	A	—	—	—	0	0	0	2-Acetamido 1 H-bond
		B	—	0	—	0	0	Asp101 OD1	0
		C	—	0	Trp63 NE	0	0	Trp62 NE	2-Acetamido 2 H-bonds
Taka-amylase (20)	Maltose	3	0	0	Arg344 NH	0	0	Trp83 NE	—
		4	Asp297 OD	0	His296 NE2	—	0	His122 NE2	—
Phosphorylase a (4,41)ᶜ	Malto-heptaose	G4	0	0	Ser428 OG	—	0	0	—
		G5	Glu432 OE1, Glu432 OE2, Lys436 NZ	—	Val430 O, Glu432 OE2	—	Asn406 ND2	Tyr403 O, Asn406 ND2, Gln407 NE2	—
		G6	0	—	Lys436 NZ	—	Gln407 NE2	Gln407 OE1	—

ᵃThe data were adapted from the references enclosed in parentheses. Dashes (—) indicate hydroxyl groups that are involved in glycosidic bonds, substituted by other groups, or not present. Zeros indicate no H-bonds.

ᵇGlcNAc = N-acetyl glucosamine.

ᶜAdditional data from Drs. E. Goldsmith and R. J. Fletterick, Private communication.

Hydrogen Bonds and van der Waals Contacts

Despite the wide diversity in the tertiary structures and binding-site topologies of carbohydrate-binding proteins, there emerge certain definite patterns and common basic features of the interactions between proteins and their carbohydrate substrates. The crystallographic refinement of the structure of the liganded form of *L*-arabinose-binding protein has provided the most precise detailed analysis of the mode of binding of a sugar substrate to a protein. Because this analysis demonstrates the many essential aspects of protein-sugar interactions and may provide a useful framework within which to study other complexes, it will be the major focus of the discussion.

The *L*-arabinose-binding protein (ABP) is one of the six different binding proteins, including those specific for *D*-galactose, maltose, Leu/Ile/Val, sulfate, and phosphate, whose structures are currently under investigation in our laboratory (8, 14, 15, 31). Structures of complexes of several of these proteins with their ligands have provided a wealth of new information on a variety of protein-ligand interactions. These periplasmic proteins are members of a large class of proteins that serve as initial components of high-affinity active transport systems for a large variety of carbohydrates, amino acids, and ions in gram-negative bacteria. Several of the sugar-binding proteins also act as initial receptors for chemotaxis. Protein components embedded in the cytoplasmic membrane, distinct for either chemotaxis or transport, are further required for both processes. These membrane-bound components appear to recognize the liganded form of the binding proteins in preference to the unliganded protein.

The crystal structure of the ABP complex was extensively refined at 1.7 Å resolution to a crystallographic R-factor of 13.7% (31) using the restrained least squares method developed by Hendrickson & Konnert (46). The final refined structure differed from ideal bond lengths and angle distances by overall root-mean-square deviations of 0.016 Å and 0.044 Å, respectively, and the probable mean error in coordinates was 0.14 Å. This order of precision and resolution is required to make more definitive interpretation and to come to meaningful conclusions concerning protein-ligand interactions. Furthermore, as the binding-site region with the bound *L*-arabinose has one of the lowest average isotropic *B*-factors and root-mean-square coordinate shifts, this region of the molecule is more precise than the overall mean and thus more highly ordered. The markedly low mean-square amplitudes of vibration of the binding site must result from the presence of the sugar substrate. That the protein-sugar interaction as revealed by structure refinement is far more complex and intricate than initially deduced from an excellent 2.4 Å Fourier map calculated with multiple isomorphous replacement phases (47) attests to the need for structure refinement.

Details of the hydrogen-bonding between ABP and *L*-arabinose are shown in Figures 1 and 2. The data in Tables 2 and 3 provide a quantitative description of the ABP-sugar interaction. Although the types of hydrogen bonds are not strictly comparable, the parameters listed in Table 2 are within the range observed for hydrogen bonds in highly refined protein structures [e.g. see Ref. (48)] and in crystal structures of sugars (49). Furthermore, these data may serve as an initial yardstick with which to examine interactions in other protein-carbohydrate complexes.

The most notable feature of the ABP-arabinose complex is the unique binding-site geometry that accommodates either the α or β anomeric form of the L-arabinose substrate. The key to this novel stereospecificity is the precise alignment of atom OD2 of residue Asp 90, which enables it to form a hydrogen bond with either the α (equatorial) or the β (axial) anomeric hydroxyl. The remaining interactions, such as hydrogen bonds and van der Waals contacts, are all essentially identical for both sugar anomers. It is also noteworthy that all sugar atoms are involved in these interactions.

As will be amply demonstrated, hydrogen bonding is the dominant interaction in protein-carbohydrate complex formation. Because sugars are highly polar organic molecules, they are highly solvated in aqueous solution. Thus, in the complex formation, they exchange their solvation shell of water for the polar groups that make hydrogen bond interactions in the binding site of the protein. Concomitantly, water molecules hydrogen-bonded to the polar groups of the protein are displaced.

The hydrogen bonding pattern between ABP and L-arabinose has several important underlying features. With the exception of the anomeric hydroxyl that participates solely as a hydrogen bond donor, all the hydroxyl groups of the sugar serve simultaneously as hydrogen bond donors and acceptors. This is consistent with two important concepts arising from the analysis of simple sugars (49, 50): (a) the anomeric hydroxyl is a stronger-than-average hydrogen bond donor and weaker-than-average hydrogen bond acceptor and (b) as a result of the "cooperative effect," the simultaneous participation of the nonanomeric hydroxyl groups as donor and acceptor groups generally leads to stronger-than-average hydrogen bonds.

Remarkably, the cooperative hydrogen bonding system in the complex involving all of the nonanomeric hydroxyls has the simple type:

$$(NH)_n \rightarrow OH \rightarrow O \qquad\qquad 1.$$

where NH and O are hydrogen bond donor and acceptor groups, respectively, OH is a nonanomeric sugar hydroxyl, and $n = 1$ or 2 (see Figures 1 and 2).

Each of the hydroxyl L-arabinose groups acts as a hydrogen bond donor to only one acceptor group from the protein or an isolated water molecule. This pattern is also predominantly found in the crystal structures of sugars, although it has been observed occasionally that an OH group can participate in bifurcated hydrogen bonds (in which the "active" proton of the OH is H-bonded to two basic groups) (49). Bifurcated hydrogen bonds are weaker than the predominant type.

Very favorable hydrogen bond geometry, resulting in strong hydrogen bond formation, is achieved by the simultaneous involvement of the nonanomeric hydroxyls as donor and acceptor groups (see Figure 1). For example, the OH-3 of the sugar donates one hydrogen bond and accepts two; this leaves it fully coordinated in an essentially tetrahedral arrangement, including the sugar -C-O bond. On the other hand, each of the hydrogen bonds donated to or by O-2 and O-4 is shared by both lone pairs of electrons of the acceptor group. These favorable hydrogen-bondings are evidenced by the formation of two planar arrangements of atoms, one consisting of C-2 and O-2 of the sugar, Lys 10 NZ, and Wat 309 O, and the other consisting of C-4 and O-4, Arg 151 NH2, and Asn 232 OD1 (see Figure 1).

Extensive networks of hydrogen bonds are formed, radiating from the sugar molecule and extending to at least three shells of residues around the vicinity of the sugar. The networks, which encompass the sugar, the first shell of essential residues, and the second shell, are shown in Figure 2. This structure provides the first clear mapping of hydrogen bond networks in a protein-sugar complex.

With the exception of Lys 10, all of the residues hydrogen-bonded to the sugar have planar side chains with two or more functional groups capable of forming multiple hydrogen bonds (e.g. Glu 14, Asp 90, Asn 205, Asn 232, and Arg 151). As indicated in Figure 2, two of these residues (Arg 151 and Asn 232) make bidentate hydrogen bonds with the sugar, and all participate in intricate networks of hydrogen bonds within the binding-site region. As a consequence, all the functional groups of the essential residues and every one of the potential hydrogen bond donor groups of these residues and of the *L*-arabinose are fully utilized. The formation of the bidentate hydrogen bonds with Arg 151 or Asn 232 is such that the atoms OD1 and ND2 of Asn 232 and O-3 and O-4 of the sugar are coplanar, as are atoms NH1 and NH2 of Arg 151 and Ara O-5 and O-4. Lys 10 is also engaged in multiple interactions crucial to ligand binding. It is in an excellent position to donate a hydrogen bond to O-2, to stabilize the alignment of Asp 90 via a salt-link, and to make van der Waals contact with either anomeric hydroxyl. Although each domain contributes three essential residues, the C-terminal domain makes two more H-bonds with the sugar than the N-terminal domain (Table 2).

The sugar ring oxygen (O-5) is almost tetrahedrally coordinated, with each of the two ring oxygen *p* orbitals directed at a hydrogen bond donor.

Two sequestered water molecules play dual roles: (*a*) they mediate the interactions between sugars and ligand site residues, and (*b*) they provide additional noncovalent linkages between the two domains at the opening of the cleft. These linkages, in addition to those formed via the bound sugars, stabilize the closed or liganded form of ABP.

Finally, the bound *L*-arabinose (see Figure 6) and essentially all of the hydrogen-bonding residues are buried and inaccessible to the bulk solvent. The solvent-accessible surface area of free *L*-arabinose is reduced by 98% upon binding. The salt bridge between Lys 10 and Asp 90 is also largely inaccessible to the bulk solvent. Furthermore it is noteworthy that the charged residues Glu 14 and Arg 151 do not have counterions.

As summarized in Table 3, there are numerous van der Waals contacts to the bound sugar anomers that provide additional stability to the complex. A cluster of these contacts, which is of interest as it has been shown that the binding of ligand causes a fluorescence change (51, 52), occurs between a nonpolar patch of the arabinose (consisting of C-3, C-4, and C-5) and Trp 16 and Phe 17 (see Figure 1). It is important to note that although the number of van der Waals interactions to both anomeric hydroxyls is almost the same (Table 3), several of these contacts are not entirely the same, owing to the different positions of the anomeric hydroxyls and to the asymmetry of the protein.

Despite the almost complete entrapment of the sugar and the extensive interactions as a result of the complex formation, both anomers of *L*-arabinose are bound in the normal *C1* full chair conformation commonly observed for simple monopyranosides. There is no evidence for sugar ring distortions.

As previously discussed (31), the molecular details of the interaction between arabinose-binding protein and its substrate have provided an excellent basis for essentially all previous results from sugar-binding studies in solution, especially those relating to sugar specificity, tight affinity, sugar-induced fluorescence change, pH effect, anomeric effect, etc. We have also utilized the known structure of the ABP-arabinose complex and the postulated sequence homology between ABP and the *lac* repressor to predict the structure of the sugar-inducer binding site in the repressor protein (53).

The structural analysis of ABP provides the first direct view of a binding site exactly complementary to both α and β anomeric forms of a sugar substrate. Due to the differences in the electronic structure of the bonds to the anomeric carbon atom (54), sugar molecules with an equatorial anomeric substituent are

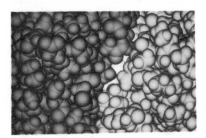

Figure 6   A space-filling model, based on the 1.7 Å refined structure, of the sugar-binding cleft of arabinose-binding protein showing two domains (grey and black) and the *L*-arabinose molecule (white) buried in the cleft. Figure courtesy of Drs. N. K. Vyas and J. S. Sack.

less stable than those with an axial substituent; this relative stability is commonly referred to as the *anomeric effect* (55, 56). Moreover, since the anomeric effect is enhanced by an environment with a low dielectric constant, the binding site of ABP would be expected to further destabilize the α anomer of *L*-arabinose (with an equatorial anomeric hydroxyl) or would favor the β sugar substrate (with an axial hydroxyl). Consequently, our discovery that almost equal amounts of both anomers are bound to ABP is unexpected (31) and raises the question of how the anomeric effect is minimized or neutralized. We have proposed that neutralization is accomplished by the formation of two hydrogen bonds between the sugar-ring oxygen and the guanidinium group of Arg 151 and the water molecule 310 (see Figures 1 and 2) (31). Though both hydrogen bonds are formed with both anomers, the ring oxygen of the equatorial anomer should delocalize electron density toward these hydrogen bonds more readily than does the axial anomer.

It is important to indicate that the lack of anomeric specificity of ABP correlates well with the transport function of the binding protein: the open-chain, aldehyde form of *L*-arabinose derived from both anomeric sugars is utilized in the biosynthesis of pentose phosphates (57). Many of the enzymes of glucose metabolism (e.g. hexokinase, glucokinase, glucose-6-phosphatase) can also act on both anomeric forms of their respective sugar substrates.

Although many of the hydrogen bonds listed in Table 4 are not yet firmly established, two common features of hydrogen bond interactions in many of the complexes have emerged: (*a*) the simultaneous participation of nonanomeric sugar hydroxyl groups as donor and acceptor groups, discussed previously, and (*b*) the extensive involvement of polar residues with planar side chains that are capable of forming multiple hydrogen bonds.

There are 54 hydrogen bonds involving only sugar hydroxyls and ring oxygens listed in Table 4. Of the 41 residues and water molecules engaged in hydrogen-bonding, 25 (or 61%) are formed with residues with planar multi-functional polar side chains, including 15 with charged groups (mostly from carboxylate residues) and 10 with amide side chains (mostly from Asn residues). The rest of the hydrogen bonds involve 9 monofunctional side chains (mostly Tyr and Trp residues), 5 peptide groups, and 2 water molecules. If the several additional hydrogen bonds predicted to occur upon binding of oligosaccharide substrates to lysozyme [(33); see Table 5] and taka-amylase A (Figure 7) are included, the H-bond distributions do not change significantly. Hexokinase is an extreme case in which only planar polar side chains are utilized in hydrogen-bonding.

The two common features reflect the widespread involvement of sugar hydroxyls in multivalent hydrogen bonds, primarily of the type exclusively encountered in the arabinose-binding protein complex (see Equation 1 above). Moreover, in light of the extensive networks of hydrogen bonds found upon

Table 5 Summary of data for some protein-carbohydrate interactions[a]

Protein	Ligand	K_d (M)	Total no. of H bonds	Total no. of van der Waals < 4.0 Å	Ligand exposure	ΔG^0 (kcal·mole⁻¹)	ΔH^0 (kcal·mole⁻¹)	ΔS^0 (cal·K⁻¹·mole⁻¹)
Lysozyme (32,33,76,77)	GlcNAc	2.6×10^{-2}	3	—	Partial	-2.2	-5.6	-11.4
	(GlcNAc)$_2$	1.9×10^{-4}	—	—	Partial	-5.1	-10.0	-16.7
	(GlcNAc)$_3$	6.6×10^{-6}	6	48	Partial	-7.0	-12.8	-19.7
	(GlcNAc)$_6$	2.3×10^{-6}	15	141	Partial	-7.7	-11.5	-12.6
Arabinose-binding protein (31,78)	α/β-L-arabinose	4.1×10^{-7}	10	47/48	Buried	-8.7	-15.3	-22.1
	α/β-D-galactose	5.8×10^{-7}	11[b]	53/54[b]	Buried	-8.5	-15.0	-21.8
Wheat germ agglutinin (21,36,79)	(GlcNAc)$_2$ (site 1)	2.5×10^{-4}	4	~8	Partial	-4.9	—	—
	α-(2-3)-neuraminyl-lactose	1.6×10^{-3}	—	—	—	-3.8	-13.3	-31.9
Hexokinase (38,67,80)	D-glucose	1.8×10^{-4}	~7	—	Buried	-5.1	-0.73	+14.7
Phosphorylase (5)	Heptenitol	3×10^{-3}	9	~21	Buried	—	—	—

[a]The numbers in parentheses indicate the references from which data were obtained.
[b]D. Wilson, N. K. Vyas, F. A. Quiocho, unpublished data.

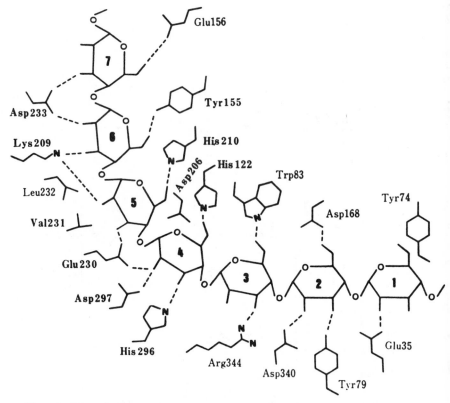

Figure 7   Schematic diagram of the proposed complex of taka-amylase A with a maltoheptaose substrate in a ribbonlike conformation. Adapted from Ref. 20.

complete analysis of the binding protein-arabinose complex, the characteristics of hydrogen bond interactions observed indicate that similar networks are likely to occur in many of these complexes. The networks in the other complexes remain to be fully mapped. As clearly demonstrated in the arabinose-binding protein complex, the basic finding is that all potential hydrogen bond donor groups, particularly those in low dielectric constant environments, must be utilized.

There are at least 13 instances in which sugar hydroxyls participate in multivalent hydrogen bonds or simultaneously serve as donors and acceptors (Table 4). Moreover, there are also eight examples of bidentate hydrogen bonds: two in arabinose-binding protein, phosphofructokinase and phosphorylase, and one in hexokinase and lysozyme.

An unusual finding, based on the data in Table 4, is the widespread use of charged residues (His residues included) in sugar binding, accounting for 41%

of those listed. All of the complexes analyzed thus far make use of charged residues, which vary in quantity from one in the complexes of lysozyme (Figure 3) and wheat germ agglutinin (Figure 5) to four in the arabinose-binding protein complex (Figure 1; see also Table 4). As many as twelve charged residues are utilized in the proposed taka-amylase A complex (Figure 7). This finding is significant since virtually all of these complexes involve uncharged sugar substrates. Furthermore, there are charged residues used in sugar binding in the *L*-arabinose-binding protein (Figure 2), wheat germ agglutinin (Figure 5), lysozyme, phosphorylase, and taka-amylase that are uncompensated for (see also Table 4). It is also noteworthy that the isolated binding-site residues Glu 14 and Arg 151 of the arabinose-binding protein are also buried and inaccessible to the bulk solvent. The side chain of Glu 14 lies at the first turn of a helix, whereas the guanidinium group of Arg 151 occupies the fifth residue of another helix of four turns and only extends very close to the N-terminus of the helix. Hydrogen bond dipoles, together with the fractional positive charge arising from the helix macrodipole, could stabilize the buried charge of Glu 14. As the helix macrodipole would have a destabilizing effect, the charge of Arg 151 is neutralized mainly by hydrogen bond dipoles induced by it. The occurrence of uncompensated buried charged residues may not be so unusual after all, as it has been recently discovered that the sulfate substrate that is completely sequestered in a deep cleft of the *Escherichia coli* sulfate-binding protein is held in place mainly by seven hydrogen bonds; there are no countercharges within van der Waals distance of the sulfate dianion (15). These experimental findings of far-reaching importance are prime evidence that isolated charge(s) buried in a solvent-inaccessible site within a protein can be stabilized by interactions other than salt-links.

While the comparison of the complexes listed in Table 4 is presently limited to the more general properties, it is likely that many of the detailed features and geometries of these complexes will be similar to those solidly established in the binding protein-arabinose complex. However, more refined structures are necessary to make meaningful comparisons. Such highly refined structures could also serve another purpose; some structures for protein-ligand complexes have resulted in hydrogen bonds that must be viewed as questionable or very tentative. For example it is indicated that O-1 and O-4 of the heptenitol bound in phosphorylase, O-6 of the sugar bound in subsite C of lysozyme, and O-6 of the maltose bound in taka-amylase A participate solely as recipients of hydrogen bonds (see Table 4). These preliminary findings are contrary to the hydrogen-bonding usually observed in water or sugar structures—the hydroxyl groups may serve simultaneously as hydrogen bond donor and acceptor groups or strictly as donors. One seldom, if ever, finds hydroxyl groups acting solely as a hydrogen bond acceptors, leaving free the "active" hydrogen atom. The hydrogen bond proposed between the anomeric hydroxyl of monopyranosides (e.g.

heptenitol, glucose) and the donor peptide NH group of Leu 136 of phosphorylase is probably nonexistent or at best a very weak bond [see Table 4 and Refs.
(5, 39)], particularly when the anomeric effect is also taken into consideration.
Of course it is possible that the "active" hydrogen atoms of the sugar hydroxyls
in question are donated to water molecules.

Structure refinement at high resolution is necessary to establish clearly which
of the functional groups of Asn and Gln residues participate in hydrogen-
bonding sugar substrates. [A problematic case is the complex of hexokinase
(Table 4)]. However, reasonable assignments can still be made from complexes
whose structures cannot be refined at better than 2 Å by taking into consideration the entire hydrogen-bonding pattern of the complex (including H-bond
networks) and the hydrogen bond geometry and by making use of the guidelines
derived from results of refined structures of complexes such as that of the
arabinose-binding protein.

Many van der Waals contacts, less readily described than hydrogen bonds,
are also undoubtedly formed in protein-sugar complexes. This frequency
notwithstanding, there is little data identifying these contacts in protein-sugar
complexes (see Table 5). This deficiency is related to the lack of several refined
protein-sugar structures that are required in order to establish confidently these
contacts between nonhydrogen atoms. Only the contacts in the binding protein-
sugar complex have been firmly established (Table 3). The contacts in the
lysozyme-trisaccharide inhibitor complex were obtained from 2 Å resolution
difference Fourier maps and the additional ones in the proposed hexasaccharide
binding were deduced from model-building experiments (33). The van der
Waals interactions in the complex of wheat germ agglutinin-disaccharide are
tentative (21), as are those in the phosphorylase-heptenitol complex (5).

Depending on which L-arabinose anomer is bound, there are either 47 or 48
van der Waals interactions of <4 Å in the binding protein complex (Table 3),
none of them shorter than 3.19 Å. Of special significance is the finding that
some of these contacts are confined to a cluster of nonpolar atoms within the
sugar. In the L-arabinose molecule, the disposition of both the equatorial OH-3
and the axial OH-4 to one side of the ring creates a cluster of nonpolar atoms
(C-3, C-4, and C-5). This nonpolar cluster partially overlaps with residue Trp
16 (Figure 1). This specific type of nonpolar interaction, which is certain to
contribute to the specificity, may be more common than heretofore anticipated.
A similar type of interaction is formed between Trp 183 of E. coli galactose-
binding protein and the bound glucose (N. K. Vyas, M. N. Vyas, F. A.
Quiocho, unpublished data). It is this author's view that the van der Waals
contacts involving Trp 62 of lysozyme (Figure 3) and Tyr 64 and His 66 of
wheat germ agglutinin (Figure 5) are likely of this type. In all cases, the
aromatic residue is partially stacked or face-to-face with the pyranose ring.

Interestingly, there are about the same number of hydrogen bonds and nonpolar contacts in the complex of arabinose-binding protein with monosaccharide as there are in the complex of lysozyme with the tri-*N*-acetylglucosamine (Table 5). This finding is in part consistent with the observation that the monosaccharide is almost completely enclosed in the binding site, while the trisaccharide located in subsites A, B, and C in lysozyme is partially exposed to the bulk solvent. The occupation of the three additional sugar subsites (D, E, and F) in lysozyme, as proposed in the model-building experiment, triples the number of van der Waals interactions, most of the contacts occurring in subsites D and E [(33); see Table 5].

One can conclude on the basis of the structural results that hydrogen bonds and van der Waals interactions are extensively utilized in protein-sugar interactions. The available thermodynamic data of sugar-binding studies in solution are consistent with this conclusion (Table 5). With the exception of the hexokinase complex, all the thermodynamic parameters, especially ΔH^0 and ΔG^0, have negative values, which are expected for complexes that are stabilized mainly by hydrogen bond and van der Waals interactions (58, 59). These structural and thermodynamic data do not support the proposal that hydrophobic effects, which are expected to exhibit $\Delta S^0 > 0$ and $\Delta H^0 \approx 0$, are the dominant forces stabilizing protein-ligand complexes (60, 61) or more specifically protein-sugar complexes (62).

The thermodynamic data are, however, difficult to correlate with the other binding data summarized in Table 5. There are significant differences in the ligand affinity and in the number of hydrogen bonds and van der Waals contacts formed in these complexes that are not reflected in the overall thermodynamic data. The only unique feature of the binding protein-arabinose complex that could account for it having a much higher affinity is that the bound monopyranoside is the most buried in any of the complexes studied thus far and is engaged in very extensive networks of hydrogen bonds. It is also difficult to rationalize the thermodynamic data for the hexokinase complex, particularly in light of the fact that the buried sugar, as in the other complexes, is bound by hydrogen bonds and nonpolar contacts. These absences of anticipated correlations illustrate the ubiquitous difficulty in understanding thermodynamic data in structural terms.

The extent of the contributions of hydrogen bonds and van der Waals interactions to the overall stability of protein-sugar complexes is somewhat difficult to ascertain. Using estimates for the energies of hydrogen bonds and van der Waals interactions without due regard to other factors such as polarization and induced dipoles, one would expect that the former would provide the major contributions to binding. Preliminary results of empirical energy minimization calculation of the refined structure of the complex between the binding protein and *L*-arabinose using AMBER (63) are consistent with this conclusion (H. Luecke, F. A. Quiocho, unpublished data).

As hydrogen bonds are highly directional, they are mainly responsible for conferring stereospecificity on the sugar-binding site and ensuring correctness of fit for substrates. Moreover, as hydrogen bonds, especially those that are part of networks, have fewer degrees of freedom than water, they offer a more stable solvation shell for the bound sugars. An additional feature of hydrogen bonds that is important to enzymes and transport protein components is that these bonds are stable enough to provide significant ligand binding but are of sufficiently low strength to allow rapid ligand dissociation. Ligand affinity and the kinetics of ligand binding are fundamentally related to the functions of binding proteins in transport and chemotaxis (15, 31, 52). The bimolecular association rate is a measure of the minimum response time for chemotaxis, and the rate of ligand dissociation defines the maximum possible velocity for the corresponding transport system. The ratio of the uptake and release rates (viz. the affinity) determines the sensitivity of the two physiological functions. The proposed mechanism for the binding protein–dependent transport system entails translocation of solute via protein components with differing ligand site affinities, high in the binding protein on the periplasmic (uptake) side and low in the protein component(s) confined to the plasma membrane (at the discharge side), in synchrony with conformational changes that alter sites accessibility (15, 31). A gradient of ligand sites affinities can be easily achieved by decreases primarily in the number of hydrogen bonds that will be formed in each site.

Oligosaccharide Binding

Our present views of the binding of sugar polymers to proteins come principally from studies of active sites of lysozyme (32, 33) and taka-amylase A (20) and the glycogen storage site of phosphorylase (42, 43).

Both lysozyme and taka-amylase A catalyze the hydrolysis of 1–4 glycosidic bonds of polysaccharides, and each consists of two domains separated by a groove sufficiently large to bind a hexasaccharide and a heptasaccharide, respectively. However, as already discussed above, the topologies of the two enzymes, and as well as the four domains, are entirely different. Despite total lack of structural homology, there are strong similarities in the mode of binding of the oligosaccharide substrates to both enzymes. Notably, these substrates are predicted to bind in a ribbonlike configuration [with the O-5 of one sugar hydrogen-bonded to O-3 of the next sugar (e.g. see Figure 3)], which follows the natural curvature of the binding-site groove. A combination of X-ray analysis with a model-building experiment has identified six sugar subsites (A–F) in the groove of lysozyme (32, 33) and seven in taka-amylase A (Figure 7). In both enzymes, two acidic residues, one in a more hydrophilic environment and the other in a more hydrophobic environment, are close to the scissile glycosidic bond and these two residues are essential in the acid-base catalytic mechanism proposed for glycosidic bond hydrolysis (20, 32).

One distinguishing feature of the proposed ligand binding and catalytic mechanism of lysozyme involves a distortion of the sugar bound in subsite D to a sofa conformation (32, 33). This distortion was proposed as a way to relieve the close contacts or overcrowding between the enzyme and the C(6)–O(6) atoms of the sugar. It would be of interest to determine if a similar substrate distortion is applicable to taka-amylase A. However, theoretical calculations indicate that the sugar unit is bound in its usual full chair conformation in subsite D of lysozyme (64).

The glycogen storage site of phosphorylase, having two distinguishing features, presents an entirely different view of an oligosaccharide binding site [Figure 8; see also Ref. (41)]. First, the storage site is located in a shallow groove formed mainly by two antiparallel α-helixes in the subdomain of the N-terminal domain, rather than between two domains, which is characteristic of the catalytic site. The groove is also much shallower than those of lysozyme and taka-amylase A. Second, the storage site accommodates the oligosaccharide ligand in its preferred left-handed helical conformation (65, 66) with the 2-hydroxyl of one glucose unit hydrogen-bonded to the 3-hydroxyl of the adjacent sugar. The structural analysis of the binding of maltoheptaose has revealed at least four subsites in the storage site, each site occupied by a glucose unit at positions 3, 4, 5, and 6 along the maltoheptaose (see Table 4) (41). The most extensive hydrogen bond and nonpolar interactions occur at subsite 5.

The basic features of the interactions between these enzymes and their

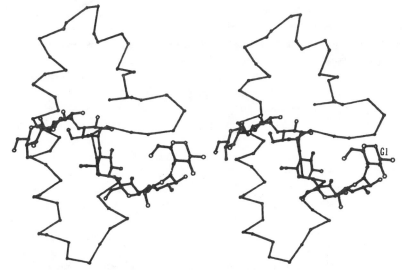

Figure 8   The glycogen storage site of phosphorylase with a bound left-handed helical maltoheptaose. G1 represents the first glucose unit. Photograph kindly provided by Drs. E. Goldsmith and R. Fletterick.

oligosaccharide ligands are all essentially the same, and do not differ from those already discussed. The extensive use of residues with planar polar side chains in sugar binding is clearly evident in the proposed taka-amylase A-oligosaccharide complex (see Figure 7). In these protein-oligosaccharide complexes, it is apparent that the energy difference between the preferred helical conformation for oligosaccharides and the ribbonlike configuration is such that either conformation can adapt to a given binding site.

Sugar-Induced Conformational Change

Changes in the protein conformation may occur upon sugar binding. These changes can be divided into two categories: local changes, affecting only residues in and around the binding site region, and large changes, often requiring relative movement between domains of bilobal proteins and enzymes. Small changes were first detected in difference Fourier analysis of the binding of monosaccharide and oligosaccharide inhibitors to lysozyme (32). This local conformation change comprises a 1 Å shift of residue Trp 62 which results in the narrowing of the binding-site cleft and thereby produces better contacts between enzyme and sugar. Sugar-induced local changes have also been observed in other proteins (e.g. the catalytic and storage sites of phosphorylase (5, 41, 43).

Large changes in conformation were first seen following independent determination of the structures of the unliganded and liganded forms of the bilobal hexokinase: the unliganded form has a much wider binding-site cleft than the enzyme with bound glucose (9, 67). The mechanism proposed to account for the two different structures basically entails a movement of one domain relative to another about a hinge between the two domains. This movement results in the closing of the binding-site cleft located between the two lobes (68). This motion buries the sugar molecule and presumably excludes water molecules from the binding site. A similar hinge-bending motion appears to occur in other bilobal proteins such as phosphoglycerate kinase (10, 69) and arabinose-binding protein (47, 70). X-ray scattering measurements indicate a 1 Å decrease in the radius of gyration of these proteins upon ligand binding. The refined structure of the complex of the arabinose-binding protein with L-arabinose clearly shows an almost enclosed substrate molecule (Figure 6) and the bound sugar cannot leave without a conformational change. The fact that the two domains in hexokinase and arabinose-binding protein are connected by three separate peptide segments suggests considerable restriction in the hinge-bending motion. Furthermore, theoretical conformational energy calculations of the arabinose-binding protein show that large changes in the width of the binding-site cleft involve only modest changes in the protein's internal energy (71).

Conformational changes concomitant with carbohydrate binding serve several functions. They allow essential residues to be properly oriented for binding and/or catalysis. A hinge-bending motion between domains can result in closure of the bilobal structure around the substrate, thereby shielding it from water. Exclusion of water near the catalytic site of phosphorylase and hexokinase should facilitate phosphorylysis and phosphate transfer, respectively. The low dielectric constant environment resulting from enclosure of sugar substrates should also increase the strength of hydrogen bonds in the complexes. A ligand-induced conformational change is a feature believed to be crucial to the function of the periplasmic binding proteins in both transport and chemotactic processes. It generates the appropriate stereochemistry for the specific interaction of the liganded protein with the membrane-bound protein components, in preference to the unliganded form, thus initiating the translocation process or flagella motion.

CONCLUDING REMARKS

Carbohydrate-binding proteins have widely diverse three-dimensional structures and binding site topologies. Nevertheless, the features of protein-saccharide interactions are very similar. A great deal of our understanding of these features, particularly the highly refined structure of the arabinose-binding protein complex, has come from high resolution crystallographic studies. Hydrogen bonds and van der Waals contacts are the dominant forces that stabilize these interactions. The interactions are precise, and stereospecificity is achieved mainly by the orientation of hydrogen-bonding residues (which are themselves fixed by intricate hydrogen bond networks with other residues within the binding sites). The majority of the residues in the binding-site region have planar polar side chains with two or more functional groups engaged in multiple hydrogen-bonding. In several complexes, additional specificity and stability is conferred by partial stacking of aromatic residues with the sugar-ring structure. Exact complementarity between proteins and substrates is often achieved even when several residues from different domains are involved in forming the binding-site cleft and a conformational change is required to juxtapose these residues and modulate access to/from the binding site. Such complexity and precision in the formation of the protein-sugar complexes strongly indicate that replacements of binding-site residues by site-directed mutagenesis will offer very little further detailed understanding of sugar-protein interactions and, in most cases, will be unlikely to lead to "better" enzymes or proteins. On the other hand, binding, crystallographic, and theoretical studies of the interaction with modified sugar substrates (by systematic substitution of each of the hydroxyl groups by hydrogen or fluorine) should prove extremely fruitful, especially in dissecting the energetics of sugar binding.

ACKNOWLEDGMENTS

I am grateful to the following past and present coworkers for their invaluable contributions to this project: N. K. Vyas, G. L. Gilliland, M. E. Newcomer, D. M. Miller, and G. N. Phillips. I also thank B. Kubena, L. Rodseth, and J. S. Sack for assistance in preparing the manuscript. The author's research was supported by grants from the National Institutes of Health (GM-21371), the Welch Foundation (C-581), and the National Science Foundation (PCM82-14057).

Literature Cited

1. Blake, C. C. F., Koenig, D. F., Mair, G. A., North, A. C. T., Phillips, D. C., Sarma, V. R. 1965. *Nature* 206:757–61
2. Johnson, L. N., Phillips, D. C. 1965. *Nature* 206:761–63
3. Levitt, M., Chothia, C. 1976. *Nature* 261:552–58
4. Fletterick, R. J. 1983. *Proc. Robert A. Welch Found. Conf. Chem. Res.* 27:173–214
5. McLaughlin, P. J., Stuart, D. I., Klein, H. W., Oikonomakos, N. G., Johnson, L. N. 1984. *Biochemistry* 23:5862–73
6. Osterlund, B. R., Hayakawa, K., Madsen, N. B., James, M. N. G. 1984. *J. Mol. Biol.* 174:557–59
7. Gilliland, G. L., Quiocho, F. A. 1981. *J. Mol. Biol.* 146:341–62
8. Vyas, N. K., Vyas, M. N., Quiocho, F. A. 1983. *Proc. Natl. Acad. Sci. USA* 80:1782–96
9. Anderson, C. M., Stenkamp, R. E., Steitz, T. A. 1978. *J. Mol. Biol.* 123:15–33
10. Blake, C. C. F., Rice, D. W. 1981. *Philos. Trans. R. Soc. London Ser. B* 293:93–104
11. Evans, P. R., Farrants, G. W., Hudson, P. J. 1981. *Philos. Trans. R. Soc. London Ser. B* 293:53–62
12. Biesecker, G., Harris, J. I., Thierry, J.-C., Walker, J. E., Wonacott, A. J. 1977. *Nature* 266:328–33
13. Schulz, G. E., Schirmer, R. H. 1979. *Principles of Protein Structure*, pp. 84–92. New York: Springer-Verlag. 314 pp.
14. Saper, M. A., Quiocho, F. A. 1983. *J. Biol. Chem.* 258:11057–62
15. Pflugrath, J. W., Quiocho, F. A. 1985. *Nature* 314:257–60
16. Alber, T., Banner, D. W., Bloomer, A. C., Petsko, G. A., Phillips, D., et al.

1981. *Philos. Trans. R. Soc. London Ser. B* 293:159–71
17. Carrell, H. L., Rubin, B. H., Hurley, T. J., Glusker, J. P. 1984. *J. Biol. Chem.* 259:3230–36
18. Achari, A., Marshall, S. E., Muirhead, H., Palmieri, R. H., Noltmann, E. A. 1981. *Philos. Trans. R. Soc. London Ser. B* 293:145–57
19. Mavridis, I. M., Hatada, M. H., Tulinsky, A., Lebioda, L. 1982. *J. Mol. Biol.* 152:419–44
20. Matsuura, Y., Kusunoki, M., Harada, W., Kakudo, M. 1984. *J. Biochem.* 95:697–702
21. Wright, C. S. 1984. *J. Mol. Biol.* 178:91–104
22. Edelman, G. M., Cunningham, B. A., Reeke, G. N. Jr., Becker, J. W., Waxdal, M. J., Wang, J. L. 1972. *Proc. Natl. Acad. Sci. USA* 69:2580–84
23. Hardman, K. D., Ainsworth, C. F. 1972. *Biochemistry* 11:4910–19
24. Sharon, N. 1984. *Adv. Immunol.* 34:213–98
25. Meehan, E. J., McDuffie, J., Einspahr, H., Bugg, C. E., Suddath, F. L. 1982. *J. Biol. Chem.* 257:13278–82
26. Wilson, K. A., Skehel, J. J., Wiley, D. C. 1981. *Nature* 289:366–73
27. Varghese, J. N., Laver, W. G., Colman, P. M. 1983. *Nature* 303:35–40
28. Colman, P. M., Varghese, J. N., Laver, W. G. 1983. *Nature* 303:41–44
29. Chothia, C. 1984. *Ann. Rev. Biochem.* 53:537–72
30. Fischer, E. 1984. *Ber. Dtsch. Chem. Ges.* 27:2985–93
31. Quiocho, F. A., Vyas, N. K. 1984. *Nature* 310:381–86
32. Blake, C. C. F., Johnson, L. N., Mair, G. A., North, A. C. T., Phillips, D. C.,

Sarma, V. R. 1967. *Proc. R. Soc. London Ser. B* 167:378–88

33. Imoto, T., Johnson, L. N., North, A. C. T., Phillips, D. C., Rupley, J. A. 1972. *The Enzymes*, ed. P. Boyer, 7:665–868. New York: Academic. 959 pp.

34. Ford, L. O., Johnson, L. N., Machin, P. A., Phillips, D. C., Tjian, R. 1974. *J. Mol. Biol.* 88:349–71

35. Kelly, J. A., Sielecki, A. R., Sykes, B. D., James, M. N. G., Phillips, D. C. 1979. *Nature* 282:875–78

36. Wright, C. S. 1980. *J. Mol. Biol.* 141:267–91

37. Wright, C. S. 1980. *J. Mol. Biol.* 139:53–60

38. Anderson, C. M., Stenkamp, R. E., McDonald, R. C., Steitz, T. A. 1978. *J. Mol. Biol.* 123:207–19

39. Sprang, S. R., Goldsmith, E. J., Fletterick, R. J., Withers, S. G., Madsen, N. B. 1982. *Biochemistry* 21:5364–71

40. Johnson, L. N., Jenkins, J. A., Wilson, K. S., Stura, E. A., Zanotti, G. 1980. *J. Mol. Biol.* 140:565–80

41. Goldsmith, E., Fletterick, R. J. 1983. *Pure Appl. Chem.* 55:577–88

42. Goldsmith, E., Sprang, S., Fletterick, R. 1982. *J. Mol. Biol.* 156:411–27

43. Johnson, L. N., Stura, E. A., Sansom, M. S. P., Babu, Y. S. 1983. *Biochem. Soc. Trans.* 11:142–44

44. Wright, H. T., Brooks, D. M., Wright, C. S. 1985. *J. Mol. Evol.* 21:133–38

45. Branden, C.-I. 1980. *Q. Rev. Biophys.* 13:317–38

46. Konnert, J. H., Hendrickson, W. A. 1980. *Acta Crystallogr. A* 36:344–50

47. Newcomer, M. E., Gilliland, G. L., Quiocho, F. A. 1981. *J. Biol. Chem.* 256:13213–17

48. Artymiuk, P. J., Blake, C. C. F. 1981. *J. Mol. Biol.* 152:737–62

49. Jeffrey, G. A., Takagi, S. 1978. *Acc. Chem. Res.* 11:264–70

50. Jeffrey, G. A., Lewis, L. 1978. *Carbohydr. Res.* 60:179–82

51. Parsons, R. G., Hogg, R. W. 1974. *J. Biol. Chem.* 249:3602–7

52. Miller, D. M. III, Olson, J. S., Pflugrath, J. W., Quiocho, F. A. 1983. *J. Biol. Chem.* 258:13665–72

53. Sams, C., Vyas, N. K., Quiocho, F. A., Matthews, K. S. 1984. *Nature* 310:429–30

54. Jeffrey, G. A., Pople, J. A., Binkley, J. S., Vishveshwara, S. 1978. *J. Am. Chem. Soc.* 100:373–79

55. Edward, J. T. 1955. *Chem. Ind.* 1102–4

56. Lemieux, R. U. 1963. *Moleular Rearrangements*, pp. 713–68. New York: Wiley-Interscience

57. Englesberg, E. 1971. *Metabolic Pathways: Metabolic Regulation*, ed. H. J. Vogel, 5:257–96. New York: Academic. 576 pp.

58. Ross, P. D., Subramanian, S. 1981. *Biochemistry* 20:3096–102

59. Pimentel, G. C., McClellan, A. L. 1971. *Ann. Rev. Phys. Chem.* 22:347–95

60. Chothia, C., Janin, J. 1975. *Nature* 256:705–8

61. Janin, J., Chothia, C. 1978. *Biochemistry* 17:2943–48

62. Kabat, E. A., Liao, J., Burzynska, M. H., Wong, T. C., Thorgersen, H., Lemieux, R. U. 1981. *Mol. Immunol.* 18:873–81

63. Weiner, P. K., Kollman, P. A. 1981. *J. Comp. Chem.* 2:287–303

64. Warshel, A., Levitt, M. 1976. *J. Mol. Biol.* 103:227–48

65. Rees, D. A., Smith, P. J. C. 1975. *J. Chem. Soc. Perkin Trans. 2* 1975:836–40

66. Quigley, G. A., Sarko, A., Marchesault, R. H. 1970. *J. Am. Chem. Soc.* 92:5834–39

67. Bennett, W. S., Steitz, T. A. 1980. *J. Mol. Biol.* 140:183–209

68. Anderson, C. M., Zucker, F. H., Steitz, T. A. 1979. *Science* 204:375–80

69. Pickover, C. A., McKay, D. B., Engelman, D. M., Steitz, T. A. 1979. *J. Biol. Chem.* 254:1323–29

70. Newcomer, M. E., Lewis, B. A., Quiocho, F. A. 1981. *J. Biol. Chem.* 256:13218–22

71. Mao, B., Pear, M. R., McCammon, J. A., Quiocho, F. A. 1982. *J. Biol. Chem.* 257:1131–33

72. Payan, F., Haser, R., Pierrot, M., Frey, M., Astier, J. P. 1980. *Acta Crystallogr. B* 36:416–21

73. Winn, S. I., Watson, H. C., Harkins, R. N., Fothergill, L. A. 1981. *Philos. Trans. R. Soc. London Ser. B* 293:121–30

74. Villafranca, J. E., Robertus, J. D. 1981. *J. Biol. Chem.* 256:167–69

75. Freymann, D. M., Metcalf, P., Turner, M., Wiley, D. C. 1984. *Nature* 311:167–69

76. Bjurulf, C., Laynez, J., Wadso, I. 1970. *J. Eur. Biochem.* 14:47–52

77. Banerjee, S. K., Holler, E., Hess, G. P., Rupley, J. A. 1975. *J. Biol. Chem.* 250:4355–67

78. Fukada, H., Sturtevant, J. M., Quiocho, F. A. 1983. *J. Biol. Chem.* 258:13193–98

79. Kronis, K. A., Carver, J. P. 1985. *Biochemistry* 24:834–40

80. Takahashi, K., Casey, J. L., Sturtevant, J. M. 1981. *Biochemistry* 20:4693–97

Ann. Rev. Biochem. 1986. 55:317–37

MUTANTS IN GLUCOSE METABOLISM

Dan G. Fraenkel

Department of Microbiology and Molecular Genetics, Harvard Medical School, Boston, Massachusetts 02115

CONTENTS

Perspectives and Summary

This review is about mutant studies in microbial glucose catabolism (Figure 1). It is largely restricted to the reactions of the glycolytic and pentose-phosphate pathways as far as pyruvate, principally for *Escherichia coli* and the yeast *Saccharomyces cerevisiae*. There is also mention of the same mutants in oxidative bacteria, where the reactions are employed in a somewhat different pattern. Emphasis is on contributions to understanding of metabolism, i.e. the degree to which mutants are informative about use and control of the reactions in vivo. Genetic studies are mentioned only in passing, but there is citation of cloning and sequencing because of their use in enzymology, particularly now that reverse genetics is allowing deliberate in vivo residue alteration.

Examples of particular interest to the reviewer, such as phosphofructokinase, are emphasized, but the subject is not a large one and there is some attempt at brief comprehensiveness so as to update previous reviews (1, 2).

0066-4154/86/0701-0317$02.00

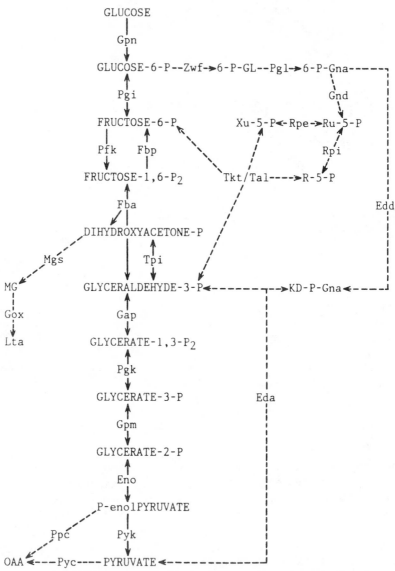

Figure 1 Reactions in *Escherichia coli* and/or *Saccharomyces cerevisiae* mentioned in this article. Substrates and products are not all shown; optically active compounds are all *D*. Substrate abbreviations: 6-P-GL, 6-phosphogluconolactone; 6-P-Gna, 6-phosphogluconate; Ru-5-P, Ribulose-5-P; Xu-5-P, Xylulose-5-P; R-5-P, ribose-5-P; KD-P-Gna, 2-keto-3-deoxy-6-phosphogluconate; MG, methylglyoxal; Lta, lactate; OAA, oxalacetate. Abbreviations for the reactions are: Gpn, glucose phosphorylation; Pgi, phosphoglucose isomerase; Pfk, fructose-6-P 1-kinase; Fbp, fructose-1,6-P$_2$ phosphatase; Fba, fructose-1,6-P$_2$ aldolase; Tpi, triose-phosphate isomerase; Gap, glyceraldehyde-3-phosphate dehydrogenase; Pgk, phosphoglycerate kinase;

Glucose Kinase

E. COLI   In enteric bacteria growing on glucose phosphorylation is concomitant with uptake by the phosphoenolpyruvate phosphotransferase system (PTS). Although enterics also contain glucokinase (3), PTS mutants grow very slowly on glucose (4–6) unless other uptake systems are available (7). Conversely, glucokinase mutants of *E. coli* (4, 5), and fructokinase mutants of *Aerobacter aerogenes* (8), grow well on glucose and fructose, respectively, employing the PTS. The role of the kinases may be in phosphorylation of hexoses produced internally, e.g. from lactose in *E. coli* and sucrose in *A. aerogenes*. Nonetheless, mutants lacking the kinases employ the PTS for internally generated sugars also (4, 5, 8), as discussed (10). Cloning (11) and purification (12) of *E. coli* glucokinase have been reported.

YEAST   There are three kinases, the hexokinases P-I (or A) and P-II (or B) (13, 14), and glucokinase (15). The structural genes were identified by deficiency mutants, *hxk1, hxk2,* and *glk1,* respectively (16, 17), and have been cloned (18–20).

To the first degree, properties of the mutants reflect properties of the enzymes. Thus, P-I and P-II, but not glucokinase, also phosphorylate fructose, and mutants lacking either one grow on fructose while the double mutant does not; similarly, double mutants lacking any two of the three enzymes grow on glucose while the triple mutant does not (21). Thus the individual enzymes are adequate for growth. However, the actual growth rates on glucose of strains with a single functioning kinase gene were somewhat lower than wild-type and rates of glucose utilization differed considerably (21). [It is known that in some wild-type strains P-II is the predominant isoenzyme expressed in growth on glucose while P-I is carbon catabolite repressed, appearing only after glucose exhaustion or in growth on ethanol (22–24).]

In addition to metabolic phosphorylation of their substrates, two other roles for the kinases have been proposed. One is in carbon catabolite repression. Selections for mutants expressing, in the presence of glucose, classes of otherwise repressed enzymes yielded *hex1* (25) and *glr1* (26), allelic to *HXK2,* and known *hxk2* alleles also showed the derepressed phenotype (27, 28). When

←——————————————————————————————

Gpm, phosphoglycerate mutase; Eno, enolase; Pyk, pyruvate kinase; Zwf, glucose-6-phosphate dehydrogenase; Pgl, 6-phosphogluconolactonase; Gnd, 6-phosphogluconate dehydrogenase; Rpe, ribulose-5-phosphate epimerase; Rpi, ribose-5-phosphate isomerase; Tkt, transketolase; Tal, transaldolase; Edd, 6-P-Gna dehydrase; Eda, KD-P-Gna aldolase; Mgs, methylglyoxal synthase; Gox, glyoxylase. (Most, but not all, of the corresponding genes are named by the same three-letter abbreviations.) Most of the reactions also occur in pseudomonads. Their main route to 6-P-GNA, via 2-ketogluconate (see text), is not shown, but the Entner-Doudoroff pathway (reactions Edd and Eda), which is also found in *E. coli,* is shown.

P-II was restored to such mutants by introduction of cloned *HXK2*, conferring very high enzyme level, repression was restored (19), while introduction of cloned *HXK1* did not have the same effect (20). A variety of measurements [e.g. (29)] including assayed glucose phosphorylation activity, growth rate, metabolite levels, and ethanol formation, have not clearly indicated that the derepressed phenotype of *hxk2* strains is just a secondary consequence of somewhat impaired glucose metabolism. Rather, such results might implicate for isoenzyme P-II (but not P-I) a regulatory role in carbon catabolite repression, perhaps via a special domain (28–30).

Another role for the glucose phosphorylating enzymes in yeast may be in transport. Uptake of glucose and fructose by wild-type strains seems to show at least two K_m values, e.g. for glucose "low", ~ 1 mM and "high", ~ 20 mM. In a mutant lacking isoenzymes P-I and P-II fructose uptake showed only the high K_m component while glucose uptake showed both, while in a mutant lacking all three kinases glucose also had only high K_m uptake. Low K_m uptake was restored by introduction of plasmids for the missing kinases (31). While secondary effects depending on phosphorylation and metabolism of the sugar might account for such results, a similar pattern of low and high K_m uptake was observed for the nonmetabolized analogue 6-deoxyglucose (32), and these results have led us to speculate that each of the three phosphorylating enzymes might even have a direct role in uptake, perhaps through binding to the carriers or through some metabolic action other than hexose phosphorylation. Or, since low K_m glucose uptake appears to be both derepressible and subject to inactivation (33), one might also imagine those processes being affected in the mutants.

Phosphoglucose Isomerase

E. COLI   In enteric bacteria, phosphoglucose isomerase (*pgi*) mutants grow more slowly than the wild-type strain on glucose and other sugars whose catabolism is via glucose-6-P, and this growth depends on the pentose-phosphate pathway as a route from glucose-6-P to fructose-6-P. In growth on other substances they are defective in content of glucose and its derivatives (1). Nonsense and Mu-phage insertion mutants were more recently obtained (34) as well as deletions of the entire gene (this lab, unpublished). Thus it seems possible that neither the enzyme, nor glucose itself, is essential in *E. coli* (in laboratory cultivation). A portion of phosphoglucose isomerase in *E. coli* was located to the periplasm (35), and a minor form of the enzyme, with altered charge, was also identified (36), as was a clone (37).

YEAST   In contrast to the *E. coli pgi* mutants, yeast phosphoglucose isomerase mutants (*pgi1*) do not grow on glucose (38–42). The several reports have indicated somewhat different permissive conditions for yeast *pgi* mutants, perhaps reflecting different degrees of leakiness, and glucose auxotrophy was reported (41, 43). (Glucose itself may be essential in yeast, composing cell wall glucan.) *PGI1* is probably the structural gene (44), and a clone was reported (45).

Phosphofructokinase and Fructose Bisphosphatase

E. COLI

Phosphofructokinase   The major phosphofructokinase (fructose-6-P 1-kinase) of *E. coli*, Pfk-1, accounting for ~90% of the activity in the K12 wild-type strain, is an allosteric protein showing cooperativity for fructose-6-P, activation by nucleoside diphosphates, and inhibition by phosphoenolpyruvate (46–48). Its structural gene, *pfkA* (49), has been sequenced (50), and the protein resembles closely in sequence and key residues the enzyme from *Bacillus stereothermophilus* and the two portions of the rabbit muscle enzyme (50). A variety of deficiency mutants have been reported (51), including deletions of the entire gene (52, 53). A nonsense mutation, presumably close to the C-terminus, gave a mutant protein still sensitive to effectors but with less affinity for fructose-6-P (51); limited C-terminal proteolysis, on the other hand, gave an enzyme of reduced cooperativity for fructose-6-P and insensitive to effectors (54).

As expected, *pfkA* mutants are impaired in their growth on substances whose catabolism is via fructose-6-P. The degree of impairment is particularly severe for mannitol and other substrates of the PTS, as if its function is somehow impaired. This impairment is partially relieved by constitutivity of isocitrate lyase (55), but the mechanism may be complex (56).

On non-PTS substrates, like lactose and glucose-6-P, *pfkA* mutants grow at about one-third of the wild-type rate (55, 56). Their metabolism does not solely depend on the pentose-phosphate pathway, for an additional block in glucose-6-P deghydrogenase only partially reduces growth (55). On the other hand, their loss of the minor phosphofructokinase, Pfk-2, almost completely blocks growth (57, 58).

Pfk-2, the minor isoenzyme, normally present at about the 10% level, was found through an unlinked suppressor mutation, *pfkB1*, which restored growth of a *pfkA* mutant (59) and governs a phosphofructokinase activity different from Pfk-1 (60). Other *pfkB* mutations abolish Pfk-2 activity (57, 58). *pfkB1* causes about a 25-fold increase in the amount of the minor isoenzyme without changing its kinetic characteristics (61). The *pfkB* gene, specifying Pfk-2, has been sequenced (62), and *pfkB1* is a promoter mutation increasing its transcription (63).

Pfk-2 has been characterized (47, 61, 64–68). Initial studies indicated a nonallosteric enzyme, noncooperative for fructose-6-P and having neither nucleoside diphosphates nor phosphoenolpyruvate as effectors (60, 61, 64).

However, later work showed, in contrast to the relative insensitivity of Pfk-1 (46), inhibition by ATP (65, 68) and also by fructose-1,6-P_2 (66). Pfk-2 was found as a dimer (65) or tetramer (61), depending on ligands; Pfk-1 is a tetramer (47). Pfk-1 and Pfk-2 do not share antigenic determinants (64) and the sequences are not homologous (50).

Many enteric bacteria contain material cross-reactive with Pfk-2, and in some *E. coli* clinical isolates there was indication for high-level activity (47). Nonetheless, the normal function of Pfk-2 in *E. coli* K12 is not known, and the low level of activity from the wild-type allele, *pfkB*$^+$, may be removed (*pfkB* mutation) without obvious effects on growth, provided the strain is still *pfkA*$^+$ so as to have Pfk-1 (58). Physiological conditions do not seem to much affect the level of Pfk-2, which is low in *pfkB*$^+$ strains and high in *pfkB1* strains (63, 69, 70). Pfk-1, on the other hand, is found in about a 3-fold higher level anaerobically than aerobically (70, 71). The level of Pfk-2 also does not depend on *pfkA* (59–61, 64, 69).

As mentioned, when the amount of Pfk-2 is increased to about the Pfk-1 range, it can substitute for Pfk-1. Several isogenic strains, including the wild-type (high-level Pfk-1, low-level Pfk-2) and a strain with no Pfk-1 but a high level of Pfk-2, were compared (72). Yields were similar but growth rate of the latter strain was somewhat less on glucose (aerobically or anaerobically) and on glucose-6-P (aerobically), and a larger difference was seen for anaerobic growth on glucose-6-P. All strains grew similarly on glycerol (aerobically) or glycerol plus fumarate (anaerobically). Strains with a high level of both enzymes also grew normally.

However, another strain originally constructed as carrying *pfkB1* was defective in growth on glycerol (57). This defect was later associated with the *pfkB* allele, which had changed. The new allele, *pfkB1**, with a secondary mutation in the structural gene, was associated with a high level of a mutant version of Pfk-2, called Pfk-2* (58), which differs from Pfk-2 in several interesting ways. First, they seem to have different kinetic mechanisms, both ordered Bi Bi, F-6-P/ATP/ADP/F-1,6-P_2 in Pfk-2 but ATP/F-6-P/F-1,6-P_2/ADP in Pfk-2* (67). This difference may account for Pfk-2* being less sensitive to inhibition by fructose-1,6-P_2 (66). Second, they differ in their inhibitability (Pfk-2) or not (Pfk-2*) by MgATP, probably acting at an allosteric site (68). Third, Pfk-2* is always dimer (J. Babul, personal communication).

In vivo, Pfk-2* in high level was found to allow somewhat more rapid growth on glucose of mutants without Pfk-1 than did Pfk-2 itself (possibly accounting for unintentional selection of the new allele). Restored growth by Pfk-2* was associated with normal levels of fructose-6-P, but with Pfk-2 this substrate remained high (66). Thus, it may be that ATP inhibition of Pfk-2 prevents complete restoration of high flux and normal growth even with a large amount of the enzyme, and it is the loss of inhibition in Pfk-2* that significantly improves growth and lowers substrate concentration.

The mechanism of growth impairment on gluconeogenic carbon sources in strains carrying *pfkB1** is not understood. The phenomenon is somewhat dependent on strain background (58), occurs regardless of the presence of Pfk-1 (66), and depends on high-level Pfk-2*, for its loss restores growth (57, 58). The weak growth on glycerol is associated with exceptionally low fructose-6-P and an excess of fructose-1,6-P_2 (66), as if reconversion of fructose-6-P to fructose-1,6-P_2 by uninhibited Pfk-2* causes fructose-6-P starvation. However, assessment of phosphofructokinase in vivo during gluconeogenic growth, comparing gamma-position labeling of ATP with 1-position labeling of fructose-1,6-P_2 in cultures growing on glycerol-3-P in the presence of $^{32}P_i$, showed no higher than marginal relative labeling in any of several strains, including *pfkB1** (73). This result seems to exclude rapid futile cycling between fructose-1,6-P_2 and fructose-6-P as causing the growth impairment.

Fructose bisphosphatase   Fructose bisphosphatase mutants of *E. coli* fail to grow on substances such as glycerol, succinate, and acetate but grow on most sugars; the enzyme is expressed constitutively and, like other fructose bisphosphatases, is sensitive to inhibition by AMP (1). Through cloning, the locus, *fbp*, was shown to be the structural gene and a deletion obtained on the chromosome (74). The enzyme was characterized from a clone with high level (75).

A mutant fructose bisphosphatase insensitive to inhibition by AMP has also been obtained. When it replaces the wild-type enzyme in normal amount growth is not impaired on a variety of sugars and gluconeogenic carbon sources. There are no evident growth abnormalities with the normal enzyme in very high level either (200).

YEAST

Phosphofructokinase   Phosphofructokinase is the largest glycolytic enzyme in yeast, and a variety of measurements, as recently evaluated (76), indicate a native molecular weight of ~800,000 and an $\alpha_4\beta_4$ structure with subunits of similar sizes but somewhat different compositions (77).

The enzyme is sensitive to a variety of effectors, including AMP, fructose-2,6-P_2, and P_i as activators and ATP as inhibitor. Allosteric models, supported by binding, kinetic, and X-ray work, have been proposed (78–81). It has also been suggested from binding stoichiometry and fructose-6-P protection of SH groups that α and β might be functionally different, regulatory and catalytic, respectively (82). However, there have been no reports of catalytic activity being recovered from the separated subunits, of ligand binding to the separated subunits, or of reconstitution of activity from dissociated enzyme.

Two main types of mutants have been reported. In the first type kinetic characteristics of the enzyme are altered. Thus, there are cases of decreased affinity for fructose-6-P (83) and increased ATP inhibitability (84), and of loss of cooperativity for fructose-6-P, of ATP inhibition, or of fructose-2,6-P_2 stimulation (85, 86). The mutations map to one or the other of two unlinked genes, *PFK1* and *PFK2* [unfortunately assigned different numbers in different laboratories, (87)]. Most of the mutations affecting "regulatory" properties are in *PFK1* [as used in Ref. (83)], and cloning and other experiments show *PFK1* to govern α (83). Mutants with altered phosphofructokinase activity contained both α and β subunits (83); perhaps the soluble enzyme requires both.

The second type of mutation in the same genes causes complete loss of phosphofructokinase activity as measured in extracts (42, 83, 84, 87). This class includes the *pfk1-1* mutant, which lacks α (83), and a *pfk2* mutant that lacks β (this lab, unpublished). Surprisingly, even these presumptive null mutants grew well on glucose, although with lags when respiration was impaired (85, 87). On the other hand, *pfk1 pfk2* double mutants do not grow on glucose (83, 86, 87). We have shown, for one such pair of mutations, that the single-gene mutants contained high levels of fructose-6-P in their growth on glucose, consistent with at least a partial block in phosphofructokinase, and the double mutant, completely impaired, accumulated even more. These results, together with experiments with labeled glucose and with mutants carrying additional glycolytic blocks, show that although the single-gene mutants lack the soluble enzyme they still likely possess a functional phosphofructokinase activity depending on the other gene (83). A similar conclusion may be drawn from other studies (42, 86, 87), although a different interpretation has been made [major use of the pentose-phosphate pathway, (88)].

Evidence has been reported for a particulate phosphofructokinase activity in *pfk2* mutants, found in cells treated with toluene and dependent on *PFK1* (89, 90). A similar activity is readily observed in a *pfk2* mutant strain carrying *PFK1* on a plasmid and, hence, having excess α subunit (this lab, unpublished). The expected *PFK2*-dependent activity in *pfk1* mutants, however, remains elusive.

Fructose bisphosphatase   Fructose bisphosphatase in *S. cerevisiae* is subject to control both of activity—inhibition by both AMP and fructose-2,6-P_2 (91, 92)—and of amount, being at a negligible level in growth on glucose but much higher in gluconeogenic growth and in the derepressed phase of growth after glucose exhaustion from a culture (93). Amount is governed both at the level of mRNA (94), and by protein turnover (95), exemplified by its catabolite inactivation, the loss of activity and antigen after glucose addition to a culture. This process includes a rapid and reversible partial inactivation (91, 92) by phosphorylation of the enzyme (96), likely governed by cAMP-dependent protein kinase and fructose-2,6-P_2 (97, 98). The enzyme from *Kluyveromyces fragilis* and its phosphorylation in vitro have also been studied (99).

The structural gene *FBP1* has been cloned in both *S. cerevisiae* [through its expression in *E. coli,* (94)] and in *Schizosaccharomyces pombe* (100). The size of the protein subunit in *S. cerevisiae* is not clear, with recent estimates differing widely (92, 94, 101, 102).

In *S. cerevisiae,* a null mutation in the structural gene has been constructed; the mutant grew normally on glucose and required hexose supplementation for growth on respiratory substrates (94). Another mutant has been reported (103), as well as one in *Schizosaccharomyces pombe* (104).

Functions of the various controls of activity in yeast have yet to be explained, in particular, the significance of catabolite inactivation. The process does not occur in *Schizosaccharomyces pombe* (104), and futile cycling between fructose-6-P and fructose-1,6-P_2 was not observed even in a mutant strain of *S. cerevisiae* with delayed inactivation (105).

Fructose-2,6-P_2 in yeast   As mentioned, fructose-2,6-P_2 is an activator of phosphofructokinase (106–108). Its level is increased by glucose (109) and by anaerobiosis (110), and its presence seems to help explain the fact of rapid flux through phosphofructokinase in spite of relatively low fructose-6-P (111). The higher levels of fructose-2,6-P_2 in the presence of glucose are also in accord with its inhibition of fructose bisphosphatase (91, 92) and possible role in catabolite inactivation. However, considering the sensitivity of the enzymes, it is not clear whether the levels in wild-type strains in gluconeogenic growth are noninhibitory to fructose bisphosphatase and insufficient to activate phosphofructokinase.

Fructose-6-P 2-kinase (112, 113) is present in higher activity in growth on glucose (111) and is also higher anaerobically (110). Rapid increase in activity after glucose addition (111) involves its cAMP-dependent phosphorylation (112, 114); as emphasized (112), activation by phosphorylation is opposite to the response in liver. Even very low fructose 6-P 2-kinase activities in yeast would likely be sufficient for the synthesis of the cofactor, particularly if it does not turn over rapidly, and relative stability of fructose-2,6-P_2 level was seen in several strains after glucose exhaustion (111). Fructose-2,6-P_2 2-phosphatase activity in yeast has not yet been reported. Mutants in the 2-kinase have not been described; the activity is present in strains without the 1-kinase (111).

Aldolase and Triose-Phosphate Isomerase

E. COLI   *E. coli* has a type 2 aldolase (115), and mutants are unable to grow on hexoses and derivatives (116–118). Growth on gluconate was restored by secondary loss of 6-phosphogluconate dehydrogenase (*gnd*), allowing use of the Entner-Doudoroff pathway; further selection for growth on glucose gave an additional block at phosphoglucose isomerase (119).

The growth of *fda* mutants, even nonconditional mutants, on substances such as glycerol—where aldolase should also be required—is not understood. Residual aldolase activity in the mutants might depend on a second aldolase, and a type 1 aldolase has been described in the Crookes' strain of *E. coli* (120). Evidently this enzyme is not induced or is of insufficient activity to allow growth of *fda* mutants on hexoses. The two *E. coli* aldolases were found subject to metabolic effectors (115, 120). In *E. coli* the aldolase reaction is not at equilibrium in vivo, for C-b from glucose did not label equally the 1- and 6-positions of fructose-1,6-P_2 (121, 122).

Triose-phosphate isomerase activity should be needed in the gluconeogenic direction, and *E. coli tpi* mutants required ribose (123) or glycerol (118) supplementation. The failure of *tpi* mutants to grow on sugars is harder to explain. Further metabolism of the glyceraldehyde-3-P might be adequate, but methylglyoxal enzymatically formed from dihydroxyacetone-P is toxic [reviewed in (124)]. Methylglyoxal can be converted to D-lactate by glyoxylase, and there is evidence for some use of this pathway in *tpi* mutants (124). It is not clear whether it serves a normal role, e.g. for growth or at least catabolism in conditions of limited P_i, or for D-lactate formation (124), and methylglyoxal synthase mutants have not been reported.

The *E. coli fda* and *tpi* genes have been cloned (37), and *tpi* has been sequenced (125).

YEAST   Aldolase mutants, *fba1*, have been identified among mutants growing in enriched medium with ethanol as a carbon source but not glucose (126). Assay showed both cleavage and condensation reactions to be reduced by ~95%, and accumulation was observed of fructose-1,6-P_2 from glucose, and of dihydroxyacetone-P from ethanol. The mutants contained cross-reacting material.

A yeast triose-phosphate isomerase mutant, *tpi1*, grew on enriched medium with ethanol plus glycerol but not glucose. Dihydroxyacetone-P levels were abnormally high, and increased by glucose (42). The gene has been cloned (45), sequenced (127), and had site-specific mutagenesis performed (128).

In yeast there is the possibility of glycerol formation from dihydroxyacetone-P (129), but there is no information for *tpi1* mutants. A route in yeasts to D-lactate via methylglyoxal was proposed (130), but there is uncertainty about methylglyoxal synthase in *S. cerevisiae* (130, 131). A glyoxylase I mutant had been reported as unable to grow on glycerol, and it had been suggested that some methylglyoxal is made from glyceraldehyde-3-P in normal glycerol metabolism (131).

Glyceraldehyde-3-Phosphate to Phosphoenolpyruvate

E. COLI   Single-gene mutants lacking glyceraldehyde-3-P dehydrogenase (*gap*), phosphoglycerate kinase (*pgk*), or enolase (*eno*) have been reported (118, 132, 133); like *tpi* mutants they grow on succinate plus glycerol but on neither substance alone. Metabolite accumulations and specific nutritional differences among the mutants were reported, and a titration of the glycerol requirement showed triose-P derivatives to serve about 1/20 of the total metabolic need (118). The sensitivity of these mutants to glucose and other sugars was found to involve several factors, including severe catabolite repression and osmotically preventable lysis but not methylglyoxal formation (118). *gap* mutants were assessed for several enzyme functions (antigenicity, NAD binding, etc) and it was concluded that the enzyme is unlikely to serve any essential noncatalytic role, such as buffering of ligand concentration (134). *pgk* and *eno* (37), and *gap* (135), have been cloned.

YEAST   Single-gene phosphoglycerate kinase (*pgk1*) and phosphoglycerate mutase mutants (*gpm1*) were obtained among isolates unable to grow with glucose but still growing on enriched medium with glycerol plus ethanol (42, 136). Metabolite accumulations were reported (42). Both genes have been cloned (45, 137) and *PGK1* has been sequenced (138).

One might assume that glyceraldehyde-3-P dehydrogenase and enolase mutants would have also appeared in such selections, but cloning and sequencing, proceeding from enriched mRNA, has revealed multiple genes, three (*GAP*) for glyceraldehyde-3-P dehydrogenase (139) and two (*ENO*) for enolase (140). All these genes are expressed as mRNA (139, 141), and for enolase differential expression of the two genes as mRNA is reflected in the amounts of their products, the slightly different two subunits forming three dimeric isoenzymes, one, the *ENO2* product, being induced by glucose (142). An *eno1* mutant, constructed by reverse genetics, was not notably defective in growth (142), and the Eno-1 isoenzyme has been identified as a heat-shock protein (143).

Pyruvate Kinase and Oxalacetate Formation

E. COLI   There are two isoenzymes, Pyk-F, activated by fructose-1,6-P_2, and Pyk-A, activated by AMP and other metabolites. They differ structurally (144) and genetically (145). Conditions influencing their individual expressions have been reported (69, 146). Mutants, *pykF* or *pykA*, lacking either one grow adequately on glucose and other sugars (145, 147) while double mutants grow feebly, except aerobically on substrates of the PTS, where pyruvate formation in uptake may bypass the block (147).

Phosphoenolpyruvate carboxylase is used in *E. coli* to form 4C from 3C compounds, and *ppc* mutants require appropriate supplementation for growth on glucose (148). The enzyme is subject to a variety of effectors [e.g. (149)], and two approaches have been taken to assessing their individual functions. First, metabolite levels have been determined for a variety of growth conditions and employed with the enzyme in vitro (150, 151). As expected, the activity was higher with metabolite levels from conditions where the enzyme was needed, but this method gave V/V_{max} values of only 2% even for growth on glucose. A second approach is genetic, and mutants in effector response have been described (152, 153). The gene has been cloned (154).

YEAST   *pyk1* mutants, readily isolated, do not grow on glucose but do grow gluconeogenically (2). The gene has been cloned (45, 155) and sequenced (155). The enzyme is activated by fructose-1,6-P_2, and a mutant with enzyme not requiring this effector was reported (156).

In yeast, the route to dicarboxylic acids is probably pyruvate carboxylase (2).

Pentose-Phosphate Pathway

E. COLI   Mutants in the oxidative branch, lacking glucose-6-P dehydrogenase (*zwf*), 6-phosphogluconolactonase (*pgl*), or 6-phosphogluconate dehydrogenase (*gnd*), and the genetics of *zwf* were discussed earlier (1). Plasmids [*zwf, gnd,* (37)], or phages [*pgl,* (157); *gnd,* (158)] have been identified. There is considerable work on *gnd,* including its sequencing (159), cell-free enzyme synthesis (160), use of high-level strains for enzyme purification (161, 162), as well as studies on the mechanism of growth rate dependence of its expression (163–166).

In *E. coli,* single-gene mutants lacking these enzymes are relatively unimpaired in growth on glucose (1). Their synthesis of ribose evidently employs the nonoxidative pentose-phosphate pathway (1). Experiments with tritiated glucose (167) showed some labeling of amino acids from the 3-position, consistent with the partial origin of NADPH from the 6-phosphogluconate dehydrogenase reaction in the wild-type strain, and no labeling in a *zwf* mutant, as expected; however, labeling from the 1- and 6-positions, particularly anaerobically, was not greatly affected, and it was concluded that a major route to NADPH might still be unknown. Increased origin of NADPH from the two dehydrogenases of the pentose-phosphate pathway, however, was clearly seen in a mutant lacking phosphoglucose isomerase.

There has been little mutant study of the nonoxidative branch of the pentose-phosphate pathway. Transketolase mutants (*tkt*) are unable to grow on pentoses and are defective in aromatic amino acid and heptose biosynthesis (1). Like the single-gene mutants in the oxidative branch, they do not require ribose supplementation, although double mutants, *gnd tkt,* do (168). Thus, the mutants demonstrate that either the oxidative or the nonoxidative route is adequate for ribose synthesis, as confirmed by labeling (1, 168, 169). The only other reaction with reported mutants is ribose-phosphate isomerase. There are two isoenzymes, A and B. *rpiA* mutants are ribose auxotrophs (170). Normal function of isoenzyme B is unclear, but the enzyme is expressed for ribose catabolism in *rpiA* mutants and, mutationally derepressed, can substitute for isoenzyme A (171).

Flux determination in the pentose-phosphate pathway of *E. coli* is not understood. Labeling experiments (168, 169, 172) suggest reduced use of the oxidative branch anaerobically, *pgi* mutants do not grow anaerobically on glucose (nor *edd* mutants on gluconate), and limitation in NADPH reoxidation may be an explanation. The idea that flux might be limited in the same way aerobically was not supported by a study of pyridine nucleotide levels in wild-type and in a mutant with elevated glucose-6-P dehydrogenase (173); also, loss of transhydrogenase made no difference to growth of a *pgi* mutant on glucose (174). An attempt to reconcile the kinetic characteristics of 6-phosphogluconate dehydrogenase, metabolite levels, and in vivo use of the reaction in a variety of strains and growth conditions was likewise of limited success (161).

It should be mentioned that most estimates of glucose metabolism [including (161)] via the pentose-phosphate pathway neglect recycling of fructose-6-P, and exact description of the fluxes has rarely been attempted (175). Nonetheless, values in the 25% range are common and not greatly out of line with the rate of glucose use in a *pgi* mutant (1)—as if the pathway is not very expandable. Flux through the nonoxidative pentose-phosphate pathway is unlikely to be limiting, because growth on pentoses is quite rapid (1).

YEAST   6-Phosphogluconate dehydrogenase mutants (*gnd1*) lacked 80% of wild-type enzyme level, and *gnd2* mutants lost the rest. Strains carrying *gnd1* did not grow on glucose, possibly because of 6-phosphogluconate accumulation, secondary loss of glucose-6-P dehydrogenase (*zwf1* mutation) restoring growth. *GND1* and *ZWF1* are likely the structural genes, and single-gene *zwf1* mutants were unaffected in their growth on glucose (176).

Mutants in Oxidative Bacteria

The reactions of carbohydrate intermediary metabolism chosen for discussion above are, with exceptions, found in almost all cells, higher and lower, including oxidative bacteria such as the pseudomonads where there has also been considerable mutant study [reviewed by Lessie & Phibbs, (177)]. Their pattern of sugar metabolism differs from *E. coli* in that phosphofructokinase is absent and the Entner-Doudoroff pathway predominates. [In *E. coli* the Entner-Doudoroff pathway is the main route for gluconate metabolism but is employed marginally with glucose, (1).] Thus, phosphoglucose isomerase mutants in *Rhizobium* and *Pseudomonas* (178–180), grow normally on glucose and gluconate but not on substances catabolized via fructose-6-P, such as mannitol. Mutants blocked in the Entner-Doudoroff pathway are blocked in growth on gluconate and glucose as well (180–186). Even when present, 6-phosphogluconate dehydrogenase and the nonoxidative pentose-phosphate pathway are generally insufficient for growth of the latter mutants (177, 183).

Mutants of oxidative bacteria lacking glucose-6-P dehydrogenase, accordingly, would be expected to have a phenotype like *pgi* mutants but additionally to be blocked in growth on glucose, and this is the case in *Rhizobium meliloti* (187). In many pseudomonads, however, glucose metabolism is primarily by its oxidation, possibly periplasmic, to 2-ketogluconate, using membrane-bound and membrane-energizing dehydrogenases; the 2-ketogluconate is taken up, phosphorylated and reduced to 6-phosphogluconate (177). Uptake of glucose itself and metabolism via glucose-6-P is also possible, and the latter route predominates with low glucose concentrations (188), with glucose produced internally from disaccharides (188), and in anaerobic growth with nitrate (189). Accordingly, *Pseudomonas aeruginosa* mutants lacking glucose-6-P dehydrogenase did not grow on glucose anaerobically with nitrate (189) but did grow aerobically on glucose (179, 180) although a lag was noted (179). Glucose dehydrogenase mutants may (188) or may not (181, 190) grow on glucose, but problems with induction complicate interpretation (177). Mutants lacking 2-ketogluconokinase or 2-ketogluconate-6-P reductase were cited as growing very slowly on glucose or gluconate (181). These results generally support the primacy, in aerobic growth, of the route via gluconate. Preferential labeling of alginic acid in *P. aeruginosa* from the 6-C compared to 1-C of glucose (191, 192), and its synthesis in mutants (180, 192), could be interpreted in the same way.

The enzymes of the glycolytic pathway between fructose-1,6-P_2 and pyruvate are present in pseudomonads, but the mode of their employment is not entirely clear. Many species use a PTS route for fructose so it enters general metabolism at fructose-1,6-P_2 (193). Nonetheless, experiments with 1-labeled fructose fit with its metabolism principally using the Entner-Doudoroff pathway (193), and phosphoglucose isomerase mutants are impaired in growth on fructose (179). Glycerol, entering general metabolism at dihydroxyacetone-P (185), might use either the Entner-Doudoroff pathway or lower glycolysis, since mutants blocked either in the first step of the Entner-Doudoroff pathway (182–185), or in phosphoglycerate kinase [*pgk*, (180, 194)] or glyceraldehyde-3-P dehydrogenase [*gap*, (180)], still grew on glycerol.

pgk and *gap* mutants were not reported as impaired in growth on glucose either (180, 194), which raises the interesting possibility of recycling of glyceraldehyde-3-P back into the Entner-Doudoroff pathway (177). Or, a route via methylglyoxal (124, 195) might be reconsidered, which would be consistent with a report of the glucose 6-position being catabolized via lactate (196).

Finally, as shown by a mutant study in *Pseudomonas aeruginosa* (197), the pathway to dicarboxylic acids employs pyruvate carboxylase.

Conclusions and Qualifications

In many of the mutant studies loss of a major enzyme of central intermediary metabolism has the simplest expected consequence: a blocked pathway, substrate accumulation, and inability to grow on substances entering metabolism before the block. Such results serve as genetic demonstration of the role of the enzyme. There are interesting metabolic problems even with mutants of this type. For example, factors governing substrate accumulations sometimes are little known. And a secondary consequence of certain blocks, the inhibition of growth on otherwise permissive carbon sources by carbon sources whose metabolism is blocked, is generally not understood either. In many cases metabolite accumulations are associated in some way with the inhibition, for it is released by earlier blocks, and study of inhibitions may reveal normal control mechanisms. For example (1), an *E. coli* aldolase mutant accumulated fructose-1,6-P_2 from glucose, growth on glycerol was inhibited, and inhibition of glycerol kinase by this metabolite was established. And yet, even in this case it is likely that the mechanism of growth inhibition is more complex, for it occurred on other carbon sources also. It is also possible that unusually high concentrations of metabolites might interfere with reactions in nonphysiological ways.

Of equal interest are cases where unusually high metabolite concentrations do not seem to have adverse effects, as seen in the growth of a variety of yeast glycolysis mutants even in their permissive media (42). A striking example is gluconeogenic growth occurring in the presence of high levels of fructose-2,6-P_2 (111).

A second class of result with deficiency mutants is where loss of an enzyme does not have much effect on growth. Sometimes, (*a*), this result accords with the fact of known isoenzymes, and is informative about their potential function. Thus, any one of the yeast hexokinases, or either of the *E. coli* pyruvate kinases, is adequate for growth on glucose. In other cases, (*b*), minor isoenzymes have only been recognized through studies of a mutant lacking the major one (e.g. *E. coli* phosphofructokinase-2), or remain to be clearly established as explaining the leakiness of certain mutants (e.g. *E. coli* aldolase). A different explanation, (*c*), is suggested for yeast phosphofructokinase single-gene mutants, where lack of the soluble enzyme reveals activities apparently dependent on the remaining subunit.

In the examples above the mutations are to loss of functional gene product, and could even be deletions. A third and entirely different type of mutant was also mentioned, where the gene product retains catalytic activity but has different kinetic or regulatory characteristics. There are few examples yet: *E. coli* phosphofructokinase-2*, *E. coli* fructose bisphosphatase insensitive to inhibition by AMP, yeast phosphofructokinase insensitive to ATP inhibition, and mutants in *E. coli* phosphoenolpyruvate carboxylase were mentioned. Such mutants are of considerable potential interest in evaluating the in vivo roles of allosteric controls. The general impression from limited studies is that this type of enzyme alteration does not necessarily always greatly perturb metabolism.

One important qualification (1) to cases where loss or alteration of an enzyme seems not to have much effect on growth needs reemphasis. In general, a limited number of situations have been examined and those often only in a qualitative way. In other conditions growth or metabolism might be profoundly affected. Furthermore, small effects on growth rate can be large effects in the context of competition with wild-type strains. For example, *E. coli* *zwf* and *gnd* mutants were found to grow at about normal rate on glucose in batch culture (1), but in mixed culture with wild-type in chemostats a *gnd* mutant was rapidly counterselected (198) while a *zwf* mutant was not (199). However, one looks for large effects before small ones.

The question of sensitivity of detection of differences applies also to estimations of metabolic fluxes (and their perturbations in mutants), as analyzed (175) for the pentose-phosphate pathway. The several cases where labeling regimes did not clearly reveal cycling between fructose-6-P and fructose-1,6-P_2 (73, 105, 122) are subject to this caution.

Finally, two general qualifications need be made with respect to drawing conclusions about metabolism in wild-type strains from studies of mutants. The first one refers to the common situation of a mutational block preventing growth. This result may, but does not necessarily, reflect that the pathway in question is the sole one normally employed. For three examples, consider (*a*) a yeast phosphoglucose isomerase mutant not growing on glucose, and, (*b*) and (*c*), *E. coli* mutants lacking the aldolase of glycolysis (*fda*) or of the Entner-Doudoroff pathway (*eda*), respectively, not growing on gluconate. In case (*a*), a conclusion that the pentose-phosphate pathway has a low flux in the wild-type strain would likely be correct, for labeling data suggests the same thing (83). In (*b*) and (*c*), on the other hand, a variety of experiments, both with other mutants [growth being restored in (*b*) by a block in 6-phosphogluconate dehydrogenase and in (*c*) by a block in the previous reaction of the Entner-Doudoroff pathway] and with labeling (161) show that gluconate metabolism normally employs two pathways.

Thus, gluconate negativity of the single-gene mutants of examples (*b*) and (*c*) likely reflects problems of toxicity. Sometimes, however, available data are insufficient. For example, as mentioned, an *E. coli* mutant completely blocked at phosphofructokinase (*pfkA pfkB*) grows on neither hexoses nor pentoses (58), showing that the pentose-phosphate pathway does not function as an adequate sole pathway for glucose metabolism—in the mutant. The extent of such cyclic metabolism in the wild-type strain is another question.

The second qualification about the relevance to normal metabolism of studies with blocked mutants is for those cases where growth is unimpaired. The variety of examples were discussed mainly in terms of accounting for growth of the mutants. Yet the availability of other pathways in the mutants may or may not reflect their normal function in the wild-type strain. Thus, the adequacy of yeast hexokinase P-I in a mutant lacking the other two kinases involves its expression in a situation where it is normally repressed, and the question of its function in the wild-type strain is unresolved. Equally interesting are those cases [e.g. *E. coli* pyruvate kinase, (144)] where two isoenzymes are normally present at an adequate level. Sometimes the normality of mutant growth seems to indicate the presence in the wild-type strain of a reaction or pathway not yet clearly established [e.g. for NADP reduction in *E. coli,* (167)]. Or, the result with a mutant demands consideration of somewhat unexpected use of a known route, as in the growth on glucose of a *Pseudomonas* phosphoglycerate kinase mutant, which in part led to the suggestion (177) of recycling of glyceraldehyde-3-P back into the Entner-Doudoroff pathway even in wild-type strains. Thus, one use of mutants is their pointing to aspects of normal metabolism that are unclear.

ACKNOWLEDGMENT

I thank Joan Stone for her dedicated assistance.

Literature Cited

1. Fraenkel, D. G., Vinopal, R. T. 1973. *Ann. Rev. Microbiol.* 27:69–100
2. Fraenkel, D. G. 1982. In *The Molecular Biology of the Yeast Saccharomyces,* ed. J. N. Strathern, E. W. Jones, J. R. Broach, pp. 1–37. New York: Cold Spring Harbor Lab.
3. Kamel, M. Y., Allison, D. P., Anderson, R. L. 1966. *J. Biol. Chem.* 241:690–94
4. Fraenkel, D. G., Falcoz-Kelly, F., Horecker, B. L. 1964. *Proc. Natl. Acad. Sci. USA* 52:1207–13
5. Curtis, S. J., Epstein, W. 1975. *J. Bacteriol.* 122:1189–99
6. Simoni, R. D., Levinthal, M., Kundig, F. D., Kundig, W., Anderson, B., et al. 1967. *Proc. Natl. Acad. Sci. USA* 58:1963–70
7. Postma, P. W., Stock, J. B. 1980. *J. Bacteriol.* 141:476–84
8. Kelker, N. E., Anderson, R. L. 1970. *J. Bacteriol.* 112:1441–43
9. Deleted in proof
10. Thompson, J., Chassy, B. M. 1985. *J. Bacteriol.* 162:224–34
11. Fukada, Y., Shotaro, Y., Shimosaka, M., Kuosaku, M., Kimura, A. 1983. *J. Bacteriol.* 156:922–25
12. Fukuda, Y., Yamaguchi, S., Shimosaka, M., Murata, K., Kimura, A. 1984. *Agric. Biol. Chem.* 48:2541–48
13. Colowick, S. P. 1973. In *The Enzymes,* ed. P. D. Boyer, 9:1–48. New York: Academic. 3rd ed.
14. Barnard, E. A. 1975. *Methods Enzymol.* 42:6–20
15. Maitra, P. K. 1975. *Methods Enzymol.* 42:25–30
16. Lobo, Z., Maitra, P. K. 1977. *Genetics* 86:727–44
17. Maitra, P. K., Lobo, Z. 1983. *Genetics* 105:501–15
18. Walsh, R. B., Kawasaki, G., Fraenkel, D. G. 1983. *J Bacteriol.* 154:1002–4
19. Fröhlich, K.-U., Entian, K.-D., Mecke, D.1984. *Mol. Gen. Genet.* 194:144–48
20. Entian, K.-D., Kopetzki, E., Fröhlich, K.-U., Mecke, D. 1984. *Mol. Gen. Genet.* 198:50–54
21. Lobo, Z., Maitra, P. K. 1977. *Arch. Biochem. Biophys.* 182:637–43

22. Kopperschläger, G., Hofmann, E. 1969. *Eur. J. Biochem.* 9:419–23
23. Gancedo, J.-M., Clifton, D., Fraenkel, D. G. 1977. *J. Biol. Chem.* 252:4443–44
24. Muratsubaki, H., Katsume, T. 1979. *Biochem. Biophys. Res. Commun.* 86: 1030–36
25. Entian, K.-D., Zimmermann, F. K., Scheel, I. 1977. *Mol. Gen. Genet.* 156: 99–105
26. Michels, C. A., Romanowski, A. 1980. *J. Bacteriol.* 143:674–79
27. Entian, K.-D., Mecke, D. 1982. *J. Biol. Chem.* 257:870–74
28. Michels, C. A., Hahnenberger, K. M., Sylvestre, Y. 1983. *J. Bacteriol.* 153: 574–78
29. Entian, K.-D., Fröhlich, K.-U. 1984. *J. Bacteriol.* 158:29–35
30. Kopetzki, E., Entian, K.-D. 1985. *Eur. J. Biochem.* 146:657–62
31. Bisson, L. F., Fraenkel, D. G. 1983. *Proc. Natl. Acad. Sci. USA* 80:1730–34
32. Bisson, L. F., Fraenkel, D. G. 1983. *J. Bacteriol.* 155:995–1000
33. Bisson, L. F., Fraenkel, D. G. 1984. *J. Bacteriol.* 159:1013–17
34. Vinopal, R. T., Hillman, J. D., Schulman, H., Reznikoff, W. S., Fraenkel, D. G. 1975. *J. Bacteriol.* 122:1172–74
35. Friedberg, I. 1972. *J. Bacteriol.* 112: 1201–5
36. Schreyer, R., Böck, A. 1973. *J. Bacteriol.* 115:268–76
37. Thomson, J., Gerstenberger, P. D., Goldberg, D. E., Gociar, E., Orozco de Silva, A., Fraenkel, D. G. 1979. *J. Bacteriol.* 137:502–6
38. Maitra, P. K. 1971. *J. Bacteriol.* 107:759–69
39. Clifton, D., Weinstock, S. B., Fraenkel, D. G. 1978. *Genetics* 88:1–11
40. Herrera, L. S., Pascual, C. 1978. *J. Gen. Microbiol.* 108:305–10
41. Navon, G., Shulman, R. G., Yamane, T., Eccleshall, T. R., Lam, K.-B., et al. 1979. *Biochemistry* 18:4487–99
42. Ciriacy, M., Breitenbach, I. 1979. *J. Bacteriol.* 139:152–60
43. Pascual, C., Alonso, A., Perez, C., Herrera, L. S. 1979. *Arch. Microbiol.* 121:17–21
44. Maitra, P. K., Lobo, Z. 1977. *Mol. Gen. Genet.* 56:55–60
45. Kawasaki, G., Fraenkel, D. G. 1982. *Biochem. Biophys. Res. Commun.* 108: 1107–12
46. Blangy, D., Buc, H., Monod, J. 1968. *J. Mol. Biol.* 31:13–35
47. Kotlarz, D., Buc, H. 1982. *Methods Enzymol.* 90:60–70
48. Martel, A., Garel, J.-R. 1984. *J. Biol. Chem.* 259:4917–21
49. Thomson, J. 1977. *Gene* 1:347–56
50. Hellinga, H. W., Evans, P. R. 1985. *Eur. J. Biochem.* 149:363–73
51. Vinopal, R. T., Clifton, D., Fraenkel, D. G. 1975. *J. Bacteriol.* 122:1162–71
52. Pahel, G., Bloom, F. R., Tyler, B. M. 1979. *J. Bacteriol.* 138:653–56
53. Albin, R., Silverman, P. M. 1984. *Mol. Gen. Genet.* 197:261–71
54. Le Bras, G., Garel, J.-R. 1982. *Biochemistry* 21:6656–60
55. Vinopal, R. T., Fraenkel, D. G. 1974. *J. Bacteriol.* 118:1090–100
56. Roehl, R. A., Vinopal, R. T. 1976. *J. Bacteriol.* 126:852–60
57. Vinopal, R. T., Fraenkel, D. G. 1975. *J. Bacteriol.* 122:1153–61
58. Daldal, F., Fraenkel, D. G. 1981. *J. Bacteriol.* 147:935–43
59. Morrissey, A. T. E., Fraenkel, D. G. 1972. *J. Bacteriol.* 112:183–87
60. Fraenkel, D. G., Kotlarz, D., Buc, H. 1973. *J. Biol. Chem.* 248:4865–66
61. Babul, J. 1978. *J. Biol. Chem.* 253: 4350–55
62. Daldal, F. 1984. *Gene* 28:337–42
63. Daldal, F. 1983. *J. Mol. Biol.* 168:285–305
64. Kotlarz, D., Buc, H. 1977. *Biochem. Biophys. Acta* 484:35–48
65. Kotlarz, D., Buc, H. 1981. *Eur. J. Biochem.* 117:569–74
66. Daldal, F., Babul, J., Guixé, V., Fraenkel, D. G. 1982. *Eur. J. Biochem.* 126: 373–79
67. Campos, G., Guixé, V., Babul, J. 1984. *J. Biol. Chem.* 259:6147–52
68. Guixé, V., Babul, J. 1985. *J. Biol. Chem.* 260:11001–5
69. Kotlarz, D., Garreau, H., Buc, H. 1975. *Biochim. Biophys. Acta* 381:257–68
70. Babul, J., Robinson, J. P., Fraenkel, D. G. 1977. *Eur. J. Biochem.* 74:533–37
71. Reichelt, J. L., Doelle, H. W. 1971. *Antonie van Leeuwenhoek. J. Microbiol. Serol.* 37:497–506
72. Robinson, J. P., Fraenkel, D. G. 1978. *Biochem. Biophys. Res. Commun.* 81: 858–63
73. Daldal, F., Fraenkel, D. G. 1983. *J. Bacteriol.* 153:390–94
74. Sedivy, J. M., Daldal, F., Fraenkel, D. G. 1984. *J. Bacteriol.* 158:1048–53
75. Babul, J., Guixé, V. 1983. *Arch. Biochem. Biophys.* 225:944–49
76. Chaffotte, A. F., Laurent, M., Tijane, M., Tardieu, A., Roucous, C., et al. 1984. *Biochemie* 66:49–58
77. Tijane, M. N., Chaffotte, A. F., Yon, J. M., Laurent, M. 1982. *FEBS Lett.* 148:267–70
78. Laurent, M., Seydoux, F., Dessen, P. 1979. *J. Biol. Chem.* 7515–20

79. Laurent, M., Tijane, M. N., Roucous, C., Seydoux, F. J., Tardieu, A. 1984. *J. Biol. Chem.* 259:3124–26

80. Nissler, K., Kessler, R., Schellenberger, W., Hofmann, E. 1979. *Biochem. Biophys. Res. Commun.* 91:1462–67

81. Kessler, R., Nissler, K., Schellenberger, W., Hofmann, E. 1982. *Biochem. Biophys. Res. Commun.* 107:506–10

82. Tijane, M. N., Chaffotte, A. F., Seydoux, F. J., Roucous, C., Laurent, M. 1980. *J. Biol. Chem.* 255:10188–93

83. Clifton, D., Fraenkel, D. G. 1982. *Biochemistry* 21:1935–42

84. Nadkarni, M., Parmar, L., Lobo, Z., Maitra, P. K. 1984. *FEBS Lett.* 175:294–98

85. Lobo, Z., Maitra, P. K. 1983. *FEBS Lett.* 139:93–96

86. Lobo, Z., Maitra, P. K. 1983. *J. Biol. Chem.* 258:1444–49

87. Breitenbach-Schmitt, I., Heinisch, J., Schmitt, H. D., Zimmermann, F. K. 1984. *Mol. Gen. Genet.* 195:530–35

88. Breitenbach-Schmitt, I., Schmitt, H. D., Heinisch, J., Zimmermann, F. K. 1984. *Mol. Gen. Genet.* 195:536–40

89. Lobo, Z., Maitra, P. K. 1982. *FEBS Lett.* 137:279–82

90. Nadkarni, M., Lobo, Z., Maitra, P. K. 1982. *FEBS Lett.* 147:251–55

91. Noda, T., Hoffschulte, H., Holzer, H. 1984. *J. Biol. Chem.* 259:7191–97

92. Gancedo, J.-M., Maźon, M. J., Gancedo, C. 1982. *Arch. Biochem. Biophys.* 218:478–82

93. Foy, J. J., Bhattacharjee, J. K. 1978. *J. Bacteriol.* 136:647–56

94. Sedivy, J. M., Fraenkel, D. G. 1985. *J. Mol. Biol.* In press

95. Funayama, S., Gancedo, J. M., Gancedo, C. 1980. *Eur. J. Biochem.* 109:61–66

96. Mazon, M. J., Gancedo, J. M., Gancedo, C. 1982. *J. Biol. Chem.* 257:1128–30

97. Pohlig, G., Wingender-Drissen, R., Noda, T., Holzer, H. 1983. *Biochem. Biophys. Res. Commun.* 115:317–24

98. Gancedo, J. M., Mazon, M. J., Gancedo, C. 1983. *J. Biol. Chem.* 258:5998–99

99. Toyoda, Y., Sy, J. 1984. *J. Biol. Chem.* 259:8718–723

100. Vassarotti, A., Friesen, J. D. 1985. *J. Biol. Chem.* 260:6348–53

101. Funayama, S., Molano, J., Gancedo, C. 1979. *Arch. Biochem. Biophys.* 197:170–77

102. Rittenhouse, J., Harrsch, P. B., Marcus, F. 1984. *Biochem. Biophys. Res. Commun.* 120:467–73

103. Gancedo, C., Delgado, M. A. 1984. *Eur. J. Biochem.* 139:651–55

104. Vassarotti, A., Boutry, M., Colson, A. M. 1982. *Arch. Microbiol.* 133:131–36

105. Bañuelos, M., Fraenkel, D. G. 1982. *Mol. Cell. Biol.* 2:921–29

106. Avigad, G. 1981. *Biochem. Biophys. Res. Commun.* 102:985–91

107. Bartrons, R., Van Schaftingen, E., Vissers, S., Hers, H.-G. 1982. *FEBS Lett.* 143:137–40

108. Nissler, K., Otto, A., Schellenberger, W., Hofmann, E. 1983. *Biochem. Biophys. Res. Commun.* 111:294–300

109. Lederer, B., Vissers, S., Van Schaftingen, E., Hers, H.-G. 1981. *Biochem. Biophys. Res. Commun.* 103:1281–87

110. Furuya, E., Kotaniguchi, H., Hagihara, B. 1982. *Biochem. Biophys. Res. Commun.* 105:1519–23

111. Clifton, D., Fraenkel, D. G. 1983. *J. Biol. Chem.* 258:9245–49

112. François, J., Van Schaftingen, E., Hers, H. G. 1984. *Eur. J. Biochem.* 145:187–93

113. Yamashoji, S., Hess, B. 1984. *FEBS Lett.* 172:51–54

114. Yamashoji, S., Hess, B. 1984. *FEBS Lett.* 178:253–56

115. Baldwin, S. A., Perham, R. N., Stribling, D. 1978. *Biochem. J.* 169:633–41

116. Böck, A., Neidhardt, F. C. 1966. *J. Bacteriol.* 92:470–76

117. Cooper, R. A. 1969. *FEBS Symp.* 19:99–106

118. Irani, M. H., Maitra, P. K. 1977. *J. Bacteriol.* 132:398–410

119. Schreyer, R., Böck, A. 1973. *J. Bacteriol.* 115:268–76

120. Baldwin, S. A., Perham, R. N. 1978. *Biochem. J.* 169:643–52

121. Shulman, R. G., Brown, T. R., Ugurbil, K., Ogawa, S., Cohen, S. M., Den Hollander, J. A. 1979. *Science* 205:160–66

122. Chambost, J.-P., Fraenkel, D. G. 1980. *J. Biol. Chem.* 55:2867–69

123. Anderson, A., Cooper, R. A. 1969. *FEBS Lett.* 4:19–20

124. Cooper, R. A. 1984. *Ann. Rev. Microbiol.* 38:49–68

125. Pichersky, E., Gottlieb, L. D., Hess, J. F. 1984. *Mol. Gen. Genet.* 195:314–20

126. Lobo, Z. 1984. *J. Bacteriol.* 160:222–26

127. Alber, T., Kawasaki, G. 1982. *J. Mol. Appl. Genet.* 1:419–34

128. Petsko, G. A. Jr., Davenport, R. C., Frankel, D., Rajbhandary, U. L. 1984. *Biochem. Soc. Trans.* 12:229–32

129. Gancedo, C., Gancedo, J.-M., Sols, A. 1968. *Eur. J. Biochem.* 5:165–72

130. Babel, W., Hofmann, K. H. 1981. *FEMS Microbiol. Lett.* 10:133–36

131. Penninckx, M. J., Jaspers, C. J., Legrain, M. J. 1983. *J. Biol. Chem.* 258:6030–36

336 FRAENKEL

132. Hillman, J. D., Fraenkel, D. G. 1975. *J. Bacteriol.* 122:1175–79
133. Irani, M. H., Maitra, P. K. 1976. *Mol. Gen. Genet.* 145:65–71
134. Hillman, J. D. 1979. *Biochem. J.* 179:99–107
135. Branlant, G., Flesch, G., Branlant, C. 1983. *Gene* 25:1–7
136. Lam, K.-B., Marmur, J. 1977. *J. Bacteriol.* 130:746–49
137. Hitzeman, R. A., Clarke, L., Carbon, J. 1980. *J. Biol. Chem.* 255:12073–80
138. Hitzeman, R. A., Hagie, F. E., Chen, C. Y., Seeburg, P. H., Derynck, R. 1982. *Nucleic Acids Res.* 10:7791–808
139. Holland, J. P., Labieniec, L., Swimmer, C., Holland, M. J. 1983. *J. Biol. Chem.* 258:5291–99
140. Holland, M. J., Holland, J. P., Thill, G. P., Jackson, K. A. 1981. *J. Biol. Chem.* 256:1385–95
141. Musti, A. M., Zehner, Z., Bostian, K. A., Paterson, B. M., Kramer, R. A. 1983. *Gene* 25:133–43
142. McAlister, L., Holland, M. J. 1982. *J. Biol. Chem.* 257:7181–88
143. Iida, H., Yahara, I. 1985. *Nature* 315:688–90
144. Malcovati, M., Valentini, G. 1982. *Methods Enzymol.* 90:170–79
145. Garrido-Pertierra, A. G., Cooper, R. A. 1983. *FEBS Lett.* 162:420–22
146. Kornberg, H. L., Malcovati, M. 1973. *FEBS Lett.* 32:257–59
147. Garrido-Pertierra, A. G., Cooper, R. A. 1977. *J. Bacteriol.* 129:1208–14
148. Kornberg, H. L. 1966. *Biochem. J.* 99:1–11
149. Smith, T. E., Balasubramanian, K. A., Beezley, A. 1980. *J. Biol. Chem.* 255:1635–42
150. Morikawa, M., Izui, K., Taguchi, M., Katsuki, H. 1980. *J. Biochem.* 87:441–49
151. Izui, K., Taguchi, M., Morikawa, M., Katsuki, H. 1981. *J. Biochem.* 90:1321–31
152. Morikawa, M., Izui, K., Katsuki, H. 1977. *J. Biochem.* 81:1473–85
153. McAlister, L. E., Evans, E. L., Smith, T. E. 1981. *J. Bacteriol.* 146:200–8
154. Sabe, H., Miwa, T., Kodaki, T., Izui, K., Hiraga, S., Katsuki, H. 1984. *Gene* 31:279–83
155. Burke, R. L., Tekamp-Olson, P., Najarian, R. 1983. *J. Biol. Chem.* 258:2193–201
156. Maitra, P. K., Lobo, Z. 1977. *Eur. J. Biochem.* 78:353–60
157. Kupor, S. R., Fraenkel, D. G. 1969. *J. Bacteriol.* 100:1296–301
158. Wolf, R. E. Jr., Fraenkel, D. G. 1974. *J. Bacteriol.* 117:468–76
159. Nasoff, M. S., Baker, H. V. II, Wolf, R. E. Jr. 1984. *Gene* 27:253–64
160. Isturiz, T., Wolf, R. E. Jr. 1975. *Proc. Natl. Acad. Sci. USA* 72:4381–84
161. Orozco de Silva, A., Fraenkel, D. G. 1979. *J. Biol. Chem.* 254:10237–42
162. Wolf, R. E. Jr., Shea, F. M. 1979. *J. Bacteriol.* 138:171–75
163. Wolf, R. E. Jr., Prather, D. M., Shea, F. M. 1979. *J. Bacteriol.* 139:1093–96
164. Farrish, E. E., Baker, H. V. II, Wolf, R. E. Jr. 1982. *J. Bacteriol.* 152:584–94
165. Baker, H. V. II, Wolf, R. E. Jr. 1983. *J. Bacteriol.* 153:771–81
166. Baker, H. V. II, Wolf, R. E. Jr. 1984. *Proc. Natl. Acad. Sci. USA* 81:7669–73
167. Csonka, L. N., Fraenkel, D. G. 1977. *J. Biol. Chem.* 252:3382–91
168. Josephson, B. L., Fraenkel, D. G. 1974. *J. Bacteriol.* 118:1082–89
169. Johnson, R., Krasna, A. I., Rittenberg, D. 1973. *Biochemistry* 12:1969–77
170. Skinner, A. J., Cooper, R. A. 1971. *FEBS Lett.* 12:293–96
171. Essenberg, M. K., Cooper, R. A. 1975. *Eur. J. Biochem.* 55:323–32
172. Caprioli, G., Rittenberg, D. 1969. *Biochemistry* 8:3375–84
173. Orthner, C. L., Pizer, L. I. 1974. *J. Biol. Chem.* 249:3750–55
174. Hanson, R. L., Rose, C. 1979. *J. Bacteriol.* 38:783–87
175. Katz, J., Rognstad, R. 1967. *Biochemistry* 6:2227–47
176. Lobo, Z., Maitra, P. K. 1982. *Mol. Gen. Genet.* 185:367–68
177. Lessie, T. G., Phibbs, P. V. Jr. 1984. *Ann. Rev. Microbiol.* 38:359–87
178. Arias, A., Cerveñansky, C., Gardiol, A., Martinez-Drets, G. 1979. *J. Bacteriol.* 137:409–14
179. Phibbs, P. V. Jr., McCowen, S. M., Feary, T. W., Blevins, W. T. 1978. *J. Bacteriol.* 133:717–28
180. Banerjee, P. C., Vanags, R. I., Chakrabarty, A. M., Maitra, P. K. 1983. *J. Bacteriol.* 155:238–45
181. de Torrontegui, G., Diaz, R., Wheelis, M. L., Ćanovas, J. L. 1976. *Mol. Gen. Genet.* 144:307–11
182. Blevins, W. T., Feary, T. W., Phibbs, P. V. Jr. 1975. *J. Bacteriol.* 121:942–49
183. Allenza, P., Lessie, T. L. 1982. *J. Bacteriol.* 150:1340–47
184. Roehl, R. A., Feary, T. W., Phibbs, P. V. Jr. 1983. *J. Bacteriol.* 156:1123–29
185. Heath, H. E., Gaudy, E. T. 1978. *J. Bacteriol.* 136:638–46
186. Ronson, C. A., Primrose, S. B. 1979. *J. Gen. Microbiol.* 112:77–88
187. Cerveñanský, C., Arias, A. 1984. *J. Bacteriol.* 160:1027–30

188. Berka, T. R., Allenza, P., Lessie, T. G. 1984. *Curr. Microbiol.* 11:143–48
189. Hunt, J. C., Phibbs, P. V. Jr. 1983. *J. Bacteriol.* 154:793–802
190. Quay, S. C., Friedman, S. B., Eisenberg, R. C. 1972. *J. Bacteriol.* 112:291–98
191. Lynn, A. R., Sokatch, J. R. 1984. *J. Bacteriol.* 158:1161–62
192. Banerjee, P. C., Vanags, R. I., Chakrabarty, A. M., Maitra, P. K. 1985. *J. Bacteriol.* 161:458–60
193. Sawyer, M. H., Baumann, P., Baumann, L., Berman, S. M., Ćanovas, J. L., Berman, R. H. 1977. *Arch. Microbiol.* 112:49–55
194. Aparicio, M. L., Ruiz-Amil, M., Vicente, M., Ćanovas, J. L. 1971. *FEBS Lett.* 14:326–28
195. Cooper, R. A. 1974. *Eur. J. Biochem.* 44:81–86
196. Rizza, V., Hu, A. S. L. 1973. *Biochem. Biophys. Res. Commun.* 54:168–75
197. Phibbs, P. V. Jr., Feary, T. W., Blevins, W. T. 1974. *J. Bacteriol.* 118:999–1009
198. Hartl, D. L., Dykhuizen, D. E. 1981. *Proc. Natl. Acad. Sci. USA* 78:6344–48
199. Dykhuizen, D. E., Hartl, D. L. 1983. *Genetics* 105:1–18
200. Sedivy, J. M., Babul, J., Fraenkel, D. G. 1986. *Proc. Natl. Acad. Sci. USA.* In press

Ann. Rev. Biochem. 1986. 55:339–72

TRANSCRIPTION TERMINATION AND THE REGULATION OF GENE EXPRESSION

Terry Platt

Department of Biochemistry, University of Rochester Medical Center, 601 Elmwood Avenue, Rochester, New York 14642

CONTENTS

INTRODUCTION

The transcriptional apparatus of the cell utilizes both simple and sophisticated mechanisms to control mRNA biosynthesis, and events related to termination of transcription at the 3' ends of genes or gene clusters are of major importance. The efficiency of termination, and its modulation by attenuator sites, antitermination factors, and processing/modification signals, can have profound effects on gene expression. The vast prokaryotic literature of the past 15 years is covered by numerous reviews (1–5), and parallel developments in eukaryotic systems are gaining momentum as well (6–8). From 1981 to the present, over

339

0066-4154/86/0701-0339$02.00

300 articles have been published that are concerned with some aspect of transcription termination, and nearly one third of these involve eukaryotic systems. The desirability of focusing on these recent developments has necessitated limiting direct citations in this review to work of the past 5 years. The primary emphasis has turned out to be biochemical, but with an implicit acknowledgment of the contributions from genetic approaches that have laid the foundations for these endeavors.

By analogy with promoters, terminators may be considered "constitutive" or "regulatable." There is a general correlation of the latter with factor-dependence, either to compensate for features lacking in the nucleic acid sequence itself, or to provide greater versatility than can be encoded in a termination signal alone. The simplest signals capable of specifying termination by core RNA polymerase molecules are straightforward: an RNA hairpin followed by multiple uridines will suffice, and in eukaryotes, polymerase III stops upon encountering (some) uridine-encoding stretches. Though phenomena such as attenuation now seem comprehensible, the complexity of events occurring at other types of termination sites is proving to be much more extensive than previously envisioned. The simple sites in some instances have multiple functions and in others do not behave as expected. In more elaborate situations, such as termination with rho in *Escherichia coli* or polymerase II in higher cells, the sequence or structure determinants remain elusive. Additional complications are provided by the identification of numerous termination regions whose properties cannot be accommodated in the simple models.

Two questions might frame our attempt to integrate observations in both prokaryotic and eukaryotic systems with general considerations about mechanism and regulation: (*a*) What nucleic acid sequence(s) or structure, if any, constitute a signal for transcription termination, and (*b*) How does the transcriptional apparatus recognize and utilize the termination signal? We shall see that the answers to these questions are still dismayingly vague. What should become clear, however, is that a remarkable number of common themes unite the variety that exists in both prokaryotic and eukaryotic systems. Indeed, it now appears that the utilization, regulation, and integration of different termination mechanisms depends less on the type of organism than on the complex metabolic needs of the cell itself.

STRUCTURAL FEATURES OF TERMINATION SITES

This section will compare and contrast different classes of termination sites at the structural level, and later sections will emphasize their functional requirements and responses. The classifications used are somewhat arbitrary, and since responses at termination sites in vivo may differ profoundly from characteristics in vitro, interpretations and predictions should be offered only with proper caution and qualification.

Prokaryotes

SIMPLE TERMINATORS   Many bacterial transcripts have a series of uridine residues at the 3' end, preceded by a GC-rich region of dyad symmetry in the DNA, and such sites are generally efficient terminators (9). Characterization of synthetic structures confirms that self-complementary regions and a stretch of uridines are sufficient for termination (without factors), though not always at high efficiency in vitro (10); to what extent the loop or flanking sequences also play a role has not been determined. Crucial signals reside in the RNA (11), and the significant participation of DNA-DNA interactions in the termination event has been ruled out (12, 13). Base analogue substitutions or point mutations in the inverted repeat that strengthen the potential RNA hairpin encoded by the dyad symmetry increase termination (14–16); alterations that reduce the number of terminal uridines also weaken the termination response (10, 16). Most evidence thus supports the model that (*a*) the dyad symmetry permits the formation of intramolecular hairpin structures in the RNA transcript, eliciting a pause by RNA polymerase, and (*b*) the uridines facilitate dissociation of the transcript from the template, since RNA-DNA hybrids with rU-dA pairing are exceptionally unstable.

Some terminators have a run of adenines (preceding the GC-rich region) that can provide a symmetric counterpart to the uridine-encoding region, and should thus function in both directions. This has been demonstrated for the terminators of the *rho* gene (17) and for the *rrnB* operon (S. Li, C. L. Squires, personal communication); sites with similar structure have also been found at the end of *rpoBC* (18), *dnaG* and *rpoD* (19), the alpha operon (20), and the M1 RNA of RNase P (21). Such bidirectional termination sites could be used to provide important genes with a protective barrier to transcription from convergent promoters, as has now been documented in vivo for *tonB* in *E. coli* (22), *his* in *Salmonella typhimurium* (23), and for transposon Tn10 (24). Currently unsolved questions concern the importance (if any) of sequences in the loop, and why some good hairpin terminators have so few consecutive uridines immediately adjacent; nuclease sensitivity suggests that cruciform structures in the DNA may play a role in termination at certain sites (25).

ATTENUATORS   A modulated termination signal encountered by the transcription complex as it moves towards a gene or through an operon provides a more subtle and sensitive way of economically controlling gene expression. Such "attenuation" of transcription is best understood in the amino acid biosynthetic operons, such as histidine, tryptophan, phenylalanine, threonine, leucine, and isoleucine (3). In these polycistronic gene clusters, a "leader region" 100–200 nucleotides long is transcribed before RNA polymerase reaches the initiation codon of the first structural gene. In the simplest case, the leader transcript can form two mutually exclusive RNA secondary structures whose formation, determined by interaction with the translational apparatus and indirectly by the intracellular level(s) of controlling molecules, governs directly the fate of the RNA polymerase molecule (Figure 1). One structure allows continuation of transcription into the operon, while the other causes termination of the transcript prior to the coding region of the downstream structural gene.

The attenuator site is usually defined as the region where the leader RNA ends, with local structural features strongly resembling rho-independent transcription terminators, though frequency of termination is determined by interactions between several regions of the transcript. In *trp*, for example (Figure 1), the termination event requires only the 3:4 RNA structure and the uridine residues (16), but efficiency is modulated by the ability of the transcript to form structures 1:2 or 2:3, and by translation of the leader peptide (26). When the ribosome follows polymerase closely, the early structures are sterically blocked and the 3:4 hairpin predominates; if the ribosome stalls prematurely, owing to lack of charged tRNA(s), 2:3 can form, preventing 3:4 formation and thus termination. However, in the complete absence of translation, the termination structure forms in almost every transcript and functions as

a.

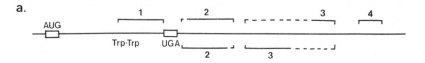

b.

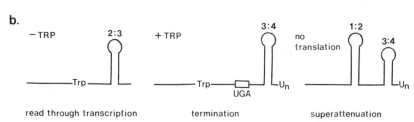

Figure 1   Attenuation in the tryptophan operon. (*a*) The tryptophan leader transcript is represented indicating the translational starts codon (AUG), the regulatory tryptophan codons (Trp), and the translational stop codon (UGA). The solid portions of the brackets indicate the extent of possible base pairing between regions 1, 2, 3, and 4. The RNA structures involved in termination, 1:2 and 3:4, are shown above the transcript. The RNA structure involved in readthrough transcription, 2:3, is shown below. (*b*) The model for attenuation. In the absence of tryptophan, the ribosome stalls at the Trp codons, allowing formation of structure 2:3 and readthrough transcription. Under conditions of excess tryptophan, the ribosome reaches the translational stop codon, structure 3:4 forms, and termination occurs within the run of uridine residues. In the absence of translation, structures 1:2 and 3:4 form, causing transcription to terminate. (From Ref. 5, adapted from Ref. 3).

a superattenuator (27). Mutations that remove the uridine residues distal to structure 3:4 prevent termination, but nevertheless allow pausing of the polymerase (28). Both pausing and termination in the *trp* attenuator are affected similarly by point mutations in the β subunit of RNA polymerase in vivo and in vitro, suggesting that both effects operate via a common active site in RNA polymerase (29).

The *his* operon leader region displays an even more extravagant structure than those of other biosynthetic operons: only the first and last of the four potential structures is expected to be required for the attenuation response, and a structural similarity in this leader region to tRNA^{His} suggests that attenuation control by cognate tRNA-binding proteins may be mediated in other ways as well (30). Similarly, a tRNA^{Met} gene is in the leader region of the *nusA* operon (31), but whether such other structures serve as additional latent pausing sites to couple translation to transcription (discussed later in detail), as a means to open the region to other cellular controls, or to mediate termination at attenuator sites in some other way, remains to be answered.

There are now many examples of attenuator sites between individual genes within transcriptional units, most notably in every operon that encodes an RNA polymerase subunit: sigma (19), β and β' (32), alpha (20), and even *nusA* (31, 33). The mechanisms involved in these instances are different than those used in the biosynthetic operons, and in a more general sense, attenuation has come to include other kinds of control mediated by RNA structure. For example, the ribosome's ability to initiate protein synthesis can be prevented by specific secondary structures to control expression of *lamB* (34), *ermC* (35), and *rplJ* (36), but discussion of this "translational attenuation" is beyond the scope of this article.

RHO-DEPENDENT TERMINATORS   Attempts to compare and contrast the known rho-dependent terminators are complicated by several factors. Relatively few examples of rho-dependent terminators have been identified at the sequence level, and of these, only two, lambda t_{R1} and *trp t'*, have been studied extensively. A third site, in the *tyrT* locus encoding tRNA^{Tyr}, also exhibits rho-dependence as part of a complex transcription unit (36a), but little is known in general for other tRNA genes. The apparent involvement of rho factor in some aspects of attenuation is also not understood, and will not be discussed here (see 1–5). For known sites, the character of the terminator varies considerably with respect to the type of gene being regulated and the organism in which the terminator originated. The initial and simplest assumption that all rho-dependent terminators should have similar features has not revealed any striking sequence or structural homologies [see (2, 37, 38)] and characterization of more rho-dependent terminators is necessary before any generalities emerge.

In the lambda rightward operon, up to five tandem rho-dependent termination sites have been identified in vitro: all exhibit some dyad symmetry, but each site responds differently to salt conditions (39, 40), and in its sensitivity to rho and nusA proteins (39). Comparison of t_{R1} II with other rho-dependent terminators, such as tRNA^{Tyr}, lambda t_{R0} and ColE1, revealed some common features, but the most promising conserved sequence turned out to serve inefficiently as a transcriptional pause site and failed to function as a rho-dependent terminator in vitro (41). Deletion analysis into upstream sequences showed that successively larger deletions eliminated rho-dependent termination at t_{R1}, yet spaced mutations across this region indicated that no single alteration was capable of eliminating rho-dependent termination (42), suggesting that, as with *trp t'* (see below), sequences required for rho-dependent termination are upstream of the termination endpoints.

In the *trp* operon, mutations increasing readthrough transcription in vivo led to the initial discovery of the tandem structure of the *trp* termination sites (43). The second of these sites, *trp t'* (located 250 bp downstream of the last structural gene in an AT-rich region with little secondary structural potential), is very efficient, but requires rho factor both in vivo and in vitro, and a substantial intact segment of mRNA upstream from the points of termination, as judged by deletions that eliminate termination (5, 11). This upstream region alone appears sufficient to specify rho-dependent termination 50–100 nucleotides downstream at a position that does not depend on a particular sequence (44).

Cumulative evidence from many laboratories thus supports some general characteristics of rho-dependent terminators: (*a*) sequence contained in the termination region itself cannot alone be responsible for rho-dependent termination, (*b*) sequences considerably upstream of the point(s) where transcription ceases convey the signal for termination, presumably via interactions between rho factor and the mRNA, (*c*) a region spanning about 80 bp encodes most of the signals required for termination, (*d*) an unstructured region of RNA is not enough to cause termination, and (*e*) the termination endpoints do not appear to correspond to any specific sequence requirements. The RNA region upstream of the endpoints is characteristically unstructured and low in G content, though around the termination endpoints, weak secondary structure in the RNA may confer specificity. These features alone are not enough to specify termination, and a requirement for cytidine residues within this region seems likely. Little else is known, and attempts to find a consensus sequence have been frustrating (37, 41), though an extensive list of such sites has been compiled (38). Deletion analyses have also not implicated specific sequences, but instead indicates that a relatively large region of RNA is needed (5, 42), in accord with estimates for a rho-binding site size of 80–100 nucleotides (45, 46). It is not clear whether there is a minimum transcript length, as well over 250 nucleotides are necessary in some cases, but this may reflect the specific transcript and terminator being tested [see (2, 40) and references therein]. Models in which RNA wraps around the rho hexamer to allow activation of the rho NTPase activity, as shown in Figure 2, are consistent with the available evidence (2, 4, 47), but other mechanisms have not been ruled out.

Rho factor has been implicated in the phenomenon of mutational polarity (reduction of distal expression in a polycistronic operon due to a translational stop codon in a proximal gene) by causing premature transcription termination [see(2) for discussion]. This invites a comparison of rho-dependent terminators with the polar site identified in the *his* operon of *S. typhimurium* (48). A local 100-nucleotide segment spanning the polar site resembles *trp t'* in its low G composition (15%), though it is flanked by short (30-nucleotide) regions containing 50% G residues. A curiously regular spacing of cytidines (12 ± 1) is evident, and a 28-bp deletion relieving polarity removes 3 of these (48, S. Ciampi, J. Roth, personal communication) as shown in Figure 3; similar spacing occurs in *trp t'* and the leader region of the *rho* gene. We would predict by analogy to known rho-dependent sites that regions of polarity may be AT-rich and/or possess little secondary structure, and remain hopeful that a molecular understanding of this phenomenon is not far off.

An intriguing possibility is suggested by three facts: (*a*) rho ATPase activation requires pyrimidines in the RNA, (*b*) pyrimidines dominate the third position in amino acid codon usage, and (*c*) a spacing of 12 (precisely 4 codons) is consistent with the rho monomer "site size." In *hisG*, with one exception, the C's are all in third positions (see Figure 3), and in similar fashion rho could recognize many untranslated regions of a transcript, owing to this regular

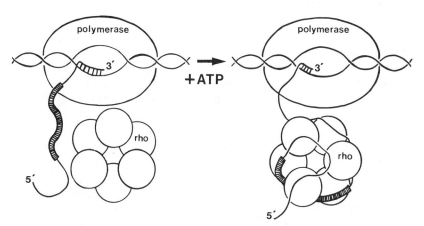

Figure 2   A model illustrating rho-dependent termination. In the absence of translation, a rho-recognition site in the nascent RNA transcript is accessible to rho protein, which binds to the RNA as a hexamer. A large portion of the transcript is probably wrapped around the hexamer, and additional interactions may occur at the secondary subunit site. Upon hydrolysis of nucleoside triphosphates, rho can catalyze release of the transcript (see text for details). From Ref. 5.

Are C's Involved in "Nucleating" Rho-dependent Termination?

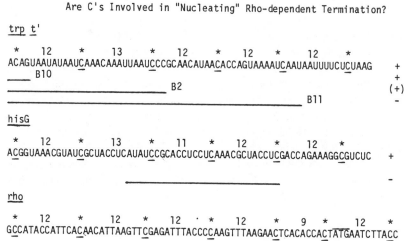

Figure 3 Are C's involved in nucleating rho binding to its recognition site? Sequences from the end of an operon (*trp t'*), the beginning of a gene (rho), and a polar site (*hisG*) are shown; bars indicate deletions affecting function. See text for discussion.

occurrence of pyrimidines that is at least somewhat independent of coding information. This hypothesis has the virtue of unifying polarity with extragenic termination via rho recognition by "nucleation" at cytidine bases and specific binding to each of the six subunits of rho. Such a mechanism could utilize the lone essential sulfhydryl at cys_{202} via a transient covalent intermediate as is seen in other RNA binding proteins (49, 50).

Eukaryotes

A hierarchy in complexity is also seen in eukaryotic termination sites, though actually defining such a site is complicated by the likelihood in many cases that termination, per se, is far removed from the location corresponding to the mature 3' end, which is generated by posttranscriptional cleavage and processing.

RNA POLYMERASE III   The simplest sites specifying termination for RNA polymerase III have clusters of four or more T residues embedded (preferably) in a GC-rich region (51). No requirement for dyad symmetry, or any secondary structure in the transcript, is apparent for the Xenopus enzyme, nor do factors other than the elongating polymerase III molecule seem to be involved in termination (52). Similar requirements apply to the calf thymus and HeLa cell enzymes (53), and removal of the "oligo T" region beyond a human tRNA gene causes efficient readthrough transcription (54). The reverse effect can occur by introducing point mutations to cause premature termination: both yeast and

Xenopus enzymes respond to these mutations, suggesting that the simple signal is evolutionarily conserved (55), though yeast requires a longer T-stretch than Xenopus, presumably to avoid premature termination in those genes with four consecutive internal U's (55a). Furthermore, alterations of the T-cluster in yeast that diminish termination also result in loss of correct processing of the longer precursor (55a).

The commonly observed T cluster may be sufficient for termination, but it is now apparent that pol III can utilize other kinds of signals as well. Analysis of the mouse M4 5S gene demonstrates that it lacks the expected thymine cluster just past the coding region, although accurate transcription and termination occur in vitro: instead, this gene has a run of 23 A's (suggesting that a mechanism other than rU-dA instability must be involved) and no other features of the usual pol III site (56). Unusual features are also evident in one case of an *Alu* family transcript: 80% of the RNA produced in vitro is terminated within a run of 9 A's at the end of the *Alu* repeat, immediately past an imperfect hairpin—the remaining readthrough extends to a conventional pol III T-cluster 150 bp downstream, which is the type of site usually used with *Alu* family transcripts (57). This and the example of M4 5S may represent a new class of pol III terminators, characterized by a run of A's and some resemblance to the rho-dependent sites in prokaryotes. Yet another class may be exemplified by the lysine and phenylalanine tRNA genes of Xenopus, where T_4 clusters are only partially effective—there is a surprising dependence on surrounding sequence (e.g. TTTTA changed to TTTTC results in readthrough) but no structure (S. Clarkson, personal communication), which possibly reflects the different transcription factor requirements of tRNA genes compared to 5S.

RNA POLYMERASE I   Transcription of ribosomal RNA, by RNA polymerase I, probably produces the precursor 3' end directly by a termination event, and until recently, it was generally accepted that the 3' end of the pre-rRNA was coincident, or nearly so, with the mature end of 28S RNA (58). It now appears, however, that these results must have been due to the first in a series of rapid processing events of a precursor species that is several hundred nucleotides longer [see (59)]. In yeast, though the region necessary for 3' end formation is confined to sequences between -36 and $+74$ (60), termination itself is thought to occur near $+210$, close to the upstream enhancing element of the next rDNA repeat (R. Planta, personal communication). In the frog, transcription proceeds about 235 nucleotides beyond the 28S endpoint, just past a 12-bp palindrome (P. Labhart, R. Reeder, personal communication); curiously, transcription from within the "*Bam* islands" of the (usually nontranscribed) spacer region also seems to encounter a "spacer-transcript" terminator just upstream from the next 40S initiation site, with some homology to the proposed 40S terminator at $+235$ (60a, T. Moss, personal communication). Yeast and frog may thus have two different mechanisms utilizing termination of transcription to maximize polymerase activity, bringing molecules that have just completed one transcript to a position where they are poised for activation at the next repeat unit. In mouse, readthrough transcription continues to a major endpoint about 565 nucleotides past the mature end of 28S at a T_6 cluster (61–63), and competition experiments suggest that a DNA-binding protein may be involved (62). Mutations within a specific palindromic octanucleotide sequence (GGTCGACC) demonstrate that it is required for function, while the 30% activity of an 18-mer containing this sequence but positioned elsewhere indicates that other elements, such as the stretches of pyrmidines that normally surround this sequence, are also necessary (I. Grummt, personal communication).

RNA POLYMERASE II   The enzyme responsible for synthesizing messenger RNA precursors and small nuclear RNAs exhibits remarkable variety in its apparent progress beyond the coding regions of genes. It is imperative to acknowledge that it has not been formally demonstrated in any eukaryotic system that there is no transcription occurring beyond the observed 3' end(s) of the mRNA, hence "termination" will be used here to refer to an event that must be occurring (i.e. pol II ceasing transcription), without specifying where or how. Studies in yeast have provided some of the most specific and revealing insights into these mysteries. A small deletion just downstream from the *CYC1* structural gene results in a 90% reduction of detectable mRNA and all of the remainder is longer than normal (several species, with up to 1000 additional nucleotides, can be found) and fully polyadenylated (64). Of special interest is the identification of extragenic *SUT* loci (suppressors of termination) presumably encoding nuclear proteins involved in termination (65). Surprisingly, the presence of different 3' noncoding sequences appears to increase *CYC1* mRNA stability and affect translational efficiency (66). Some homology with the deleted region from *CYC1* has been identified as critical for accurate termination in the *Drosophila* ADE8 gene introduced into yeast (67), and limited homologies with some other yeast genes such as *ILV2* (68) make it plausible that a general signal is present. At least partial function is confined to a 21-bp region that contains the sequence TTTTTATA, common to some other yeast "terminators," and additional downstream sequences may participate in governing its efficiency (69).

A bidirectional terminator seems to exist upstream of the URA3 promoter, though it exhibits no homology with other yeast "terminators" and its function in vivo is unknown (70). An 80-bp fragment from the *CYC1* terminator inserted into the intron of actin seems partially functional, resulting in reasonable yield of a short polyadenylated mRNA in the forward orientation, but not in reverse (H. Ruohola, R. Parker, B. Osborne, T. Platt, unpublished). Within a coding region, this same fragment seems to function in both orientations (B. Osborne,

L. Guarente, unpublished) and it, as well as the *ADH1* terminator, inhibits expression when placed within the untranscribed UAS region of *GAL1* (71), but it remains to be seen what parameters account for these curious effects. Analysis of transcripts from the 2μ circle indicates that the polyadenylated 3' ends (with one exception) lie between 20 and 60 nucleotides of the translation stop codon (71a). Each region in this instance has some homology to the tripartite signal proposed previously (64), and deletions into the region increase readthrough transcription and differentially affect mRNA stability (71a). Similarly, analyses of yeast *GAL1-7-10* (72) and *URA3* (73) transcripts have located multiple mRNA endpoints with reasonable accuracy, yet specific signals that can cause pol II termination in yeast have not been identified.

In higher eukaryotes, the components directing 3' end formation of the nonpolyadenylated histone gene transcripts are among the best understood [see (8)]. They have obvious features reminiscent of prokaryotic rho-independent termination sites, but the RNA hairpin and the highly conserved nearby distal sequence (CAAGAAAGA) are essential for the processing events that generate the mature 3' end (74), and their removal does not prevent (normal) termination within the next 200 nucleotides (75). The lack of these features in the basal level (replication-independent) histone genes, whose transcripts are polyadenylated (76), supports the interpretation that they are involved in maturation rather than termination. As discussed in a subsequent section, the other category of nonpolyadenylated pol II transcripts, the U-series of small nuclear RNAs, also have signals clearly required for correct 3' end formation. Yet despite the observation that (for U1) transcription does not extend farther than 60 nucleotides beyond the mature end (77), specific signals for termination per se have not been identified experimentally—suggestions for possible pol II sites distal to U1 (78) or U2 (79) by sequence homology remain to be confirmed.

For that vast majority of pol II transcripts that are polyadenylated, many of which are over 100 kb long, the multiplicity of observed 3' ends plus the variability in distance beyond the coding region renders analysis difficult. Although eliminating the AATAAA polyadenylation signal results in apparent readthrough transcription in many instances, there is no direct evidence that it is a necessary component of the termination signal, and distal sequences must certainly be involved as well [see (8), and references therein]. The 3' termination point of ovalbumin transcription coincides with a short region nearly 1 kb past the end of the structural gene, but no specific signals are apparent (80). As with the yeast *URA3* gene (73), multiple 3' ends are seen in *Drosophila* tropomyosin (81), chicken pro-collagen (82) and, in the mouse, alpha-amylase (83), and dihydrofolate reductase (84). In the latter two cases, transcription proceeds between 1 and 4 kb beyond the last poly(A) site, as it also does for the mouse β-globin gene, which has only a single polyadenylation site (85). In another context, this globin terminator functions (only) in one orientation, still

1.4 kb beyond the poly(A) site, and the signal must be encoded several hundred bases prior to the 3' end (86).

A compilation of nearly 100 mammalian genes does reveal a consensus sequence YGTGTTYY downstream from the AATAAA polyadenylation signal in two thirds of the cases examined (87), and in SV40 an important signal is embedded in the sequence AGGTTTTTT, located about 60 bases beyond the AATAAA (87a). Generally, such elements occur as GT clusters [see (8)] and are essential for the formation of 3' ends, but whether they are necessary for cleavage or polyadenylation rather than termination has not been determined. Substantial conservation among the 3' untranslated regions of numerous vertebrate genes (primarily actins) is also tantalizing, though the function may be translational rather than transcriptional (88). Conclusions based on examining primary structure should take a cautious cue from factor-dependent termination in prokaryotic systems, where the critical processes are still far from understood. The places where transcription truly seems to stop are heterogeneous and nonspecific, and we discuss below the possibility that this reflects a loss of processivity rather than a response to a specific signal.

FACTORS THAT REGULATE TERMINATION EFFICIENCY AND LOCATION

As with simple terminators, the functional signals for factor-dependent termination appear to reside in the RNA, are located somewhat upstream from the point of termination, and may involve secondary structure (or lack thereof) in the transcript, but generally do not have the obvious features of symmetry and uridine-encoding regions described above. As examination of such sequences has not revealed extensive common features of primary or secondary structure among these complex terminators (38, 45), either the relevant regions have not been compared (they may lie far upstream from the point of termination), or specific sequences are not necessarily required for recognition. Possibly some termination sites share an absence of particular features (specific forbidden nucleotides, or secondary structures), and we surmise that the factors themselves can somehow mediate the essential information by their interactions with the transcription complex.

Prokaryotes

RHO AND OTHER TERMINATION FACTORS   The *rho* gene has been cloned (89), and its coding sequence directs the synthesis of a 419-amino-acid protein (17), active as a hexamer (90). The gene is unusual in having a long leader region (250 bp) of unknown function, perhaps for the autoregulation suggested by previous observations (89, 91, 92), since a rho-dependent termination site overlaps the coding region (5). In the presence of RNA and NTPs, the protein is configured in two distinct domains: the isolated amino-terminal domain (31 kd) can bind RNA, but connection to the carboxyterminal domain is required for the RNA-dependent NTP hydrolytic activity (93). Mutations affecting RNA binding are common (94–96), but more exotic variants, such as "super-rho" proteins, have also been isolated (91). Crystals of rho protein show promise for elucidating the detailed structure by X-ray diffraction (L. Lim, S. Mathews, R. Grant, T. Platt, unpublished), and several amino acid changes in altered rho proteins have been identified (R. Grant, K. Sullivan, J. Ackerson, T. Platt, unpublished). Overproducing vectors (97, 98) will greatly simplify biochemical characterization—induction by nalidixic acid, for example, can yield about 25 mg of rho per gram of cells, or 40% of the soluble protein (98).

Figure 2 illustrates one model for rho-mediated transcription termination that is consistent with the known evidence on the action of rho (2, 4, 47). As the polymerase complex transcribes untranslated regions, nascent RNA is exposed, and destabilizing conditions enhance rho recognition (45, 99). When these regions contain the proper signal(s) for rho, rho seems to interact with the RNA simultaneously at two distinct sites (100). In the presence of ATP, activation of the primary site by a polyribonucleotide is coordinated with interactions at the small secondary RNA binding site to activate the rho NTPase, which is essential for termination. The *rho115* mutation, in particular, seems to affect rho at this primary polynucleotide binding site (94, 101), but the mechanisms coupling ATP hydrolysis to termination remain obscure. Just how the interactions between rho and elongation complex cause termination at some distance from the recognition site is also unknown. Genetic and biochemical evidence for protein-protein interactions between rho, nus proteins, and RNA polymerase have been discussed previously [see (2)], and in vitro observations now support the direct participation of nusA protein (discussed below) in mediating rho action as well (102, L. Peritz, J. Greenblatt, personal communication). A new factor, tau, has also been identified as a termination factor in *E. coli,* but is not yet well characterized (103, 103a).

RIBOSOMES: POLARITY AND ATTENUATION   In the interior of gene clusters, termination can be regulated by more complex mechanisms. Two of these, polarity and attenuation, effectively couple transcription to translation via ribosomal masking or exposure of signals in the RNA transcript. Halting translation within an early gene of a polycistronic operon can reduce distal gene expression, but mutations in rho factor can alleviate this "polar" effect [see (94)]: it is as if RNA polymerase encounters previously latent termination sites that become active owing to the lack of translation. Current models for polarity are based on the early proposals of Adhya, Gottesman, and coworkers [see (2) for discussion] and incorporate two premises: (*a*) there are rho-dependent sites of transcription termination within operons, and (*b*) these sites are not susceptible to rho as long as the mRNA is concurrently translated. Polar effects can be suppressed not only by mutations in rho factor, but by the antitermination factors N and Q of bacteriophage lambda (104–106).

Coupling between transcription and translation is also manifested in the phenomenon of attenuation, where translation of a leader peptide containing cognate amino acids governs formation of a distal hairpin terminator, but also requires that the progress of polymerase be coordinated with that of the ribosome. For example, in the *trp* leader region (Figure 1), the early hairpin structure 1:2 does not function as a terminator in vitro, but as a pause site (28, 107) that is enhanced by nusA protein (108); similar nusA enhancement is seen in the *ilvB* and *ilvGEDA* attenuator regions (109). Because translation can activate a paused complex to restore transcriptional activity (110), this pause site may help couple transcription to translation in vivo by slowing the polymerase and permitting the translating ribosome to stay close behind (110a). Both pausing and termination in the *trp* attenuator are affected similarly by point mutations in the β subunit of RNA polymerase in vivo and in vitro, suggesting that some feature is being recognized by a common active site (29). Additional factors are not required for transcript release (111) or for the attenuator to function in vitro (112). In vivo the response is closely tied to cellular physiology: moderate typtophan depletion causes derepression of the *trp* promoter to increase RNA initiation; more severe starvation leads to a further increase in *trp* mRNA by the relief of attenuation (113). The *nusA1* mutation appears to enhance distal *trp* gene expression at the attenuator (114), and overproduction of tryptophanyl-tRNA synthetase has a similar effect, presumably by binding much of the trp-tRNA, mimicking starvation (115).

Attenuators that function in other operons control transcription termination in several ways. The *rplKAJL-rpoBC* operon, which encodes four ribosomal proteins and the RNA polymerase subunits β and β', contains a terminator that is about 70% efficient in vivo in reducing transcription distal to *rplL* (32). Attenuation in the *rrnB* ribosomal RNA operon in vitro requires nusA protein and ppGpp to mediate pausing and termination (116). Transcription of the S10 ribosomal protein operon is regulated autogenously by the L4 protein, which acts to attenuate transcription in the leader region of the S10 mRNA, possibly by direct interaction of L4 with the terminator (117, 118). In the phenylalanine-tRNA synthetase operon, a leader peptide contains five phenylalanine codons, and alternate RNA structures could govern the extent of termination (119).

Interesting developments in a number of other systems bear mentioning. Conversion of the four leucine codons in the *S. typhimurium leu* operon leader region into threonine codons abolishes regulation by leucine, with new control by threonine (120). In the *pyrBI* operon, depletion of UTP (rather than any particular amino acid) governs attenuation, with readthrough transcription occurring after polymerase pauses at multiple uridines in the transcript, because the caught-up ribosome can then prevent formation of the attenuator RNA hairpin (121–122); support for this model also suggests that a second regulatory mechanism may exist (123). In *pyrE,* a similar effect can be elicited by replacing rare codons in the leader peptide region by common ones, though the difference is only evident at high UTP concentrations (124). These many results all indicate that it is the relative rates of transcription and translation that determine the extent of attenuation. In more elaborate coupling between parallel operons, *ilvB(N)* expression is regulated not only by valine and leucine in "end-product attenuation," as expected from the sequence (125), but by alanine and threonine in "substrate attenuation," to help coordinate carbon flow through the *ilvGEDA* operon pathway (126). Several reviews offer a more extensive discussion of attenuation than can be covered here (3, 127).

THE NUS PRODUCTS AND ANTITERMINATION   Specific protein factors can also reduce or prevent termination by modifying either the transcribing polymerase or sequestering the termination signal (e.g. "antitermination" occurs at attenuators when formation of the RNA hairpin is prevented). The best-characterized systems involve two antitermination proteins of bacteriophage lambda, N and Q, which turn on early and late gene expression during lytic growth. The properties of N protein in lambda and related phages have been extensively reviewed (104, 129, 130) and N-modified polymerase has been shown to cause readthrough transcription at both rho-dependent and rho-independent terminators (131) as well as relieve polarity of IS insertion elements. Additional requirements include a cis-acting N-utilization *(nut)* site, the host nus proteins (132), and other as-yet-unidentified components (133, 134).

Two specific regions of homology are involved in the response at *nut* sites. There is evidence for species-specific recognition of the Box A sequence CGCTCCTA: a 1-bp difference in *S. typhimurium* prevents heterologous function with the respective nusA protein from *E. coli* (135). Possibly Box A is only required at elevated temperatures (136), while Box B, having a region of dyad symmetry but with less defined sequence, appears to be essential for antitermination under most conditions (129), with other sequences and/or the spacing of these regions also mediating the response (137). Since translation of the upstream mRNA influences recognition, Box A must be recognized as an RNA signal rather than DNA (138); this has been confirmed by the demonstration in vitro that nusA protein binds to RNA immediately upstream of the Box A sequence (138a). Together, N and a *nut* site can antiterminate transcription from other promoters (137), and may be involved in the combined recognition of Box A and Box B (L. Peritz, J. Greenblatt, personal communication).

The nusA protein, first identified genetically [see (132)] and proven biochemically to be "L-factor" (139), binds specifically to N protein (140) and to core polymerase (141), and is involved in transcriptional pausing and termination (9, 39, 142). Evidence for the participation of nusA as a general transcription factor has been summarized previously (133, 139). Though extracts from mutant *nusA1* cells cannot support N function, activity is restored by adding purified nusA, directly implicating nusA protein in N antitermination (134, 143). Possibly nusA acts as an "adapter" to couple N and polymerase in the formation of an antitermination complex (140). *NusB* mutations can restore N activity to *nusA1* mutants (144), indicating a possible nusA-nusB interaction, and it may be associated with the transcription termination activity of polymerase since some *nusB* mutants partially relieve polarity [see (114, 145)].

Antitermination is also defective in S30 extracts prepared from strains mutant in *nusB,* or *nusE* (encoding ribosomal protein S10), but N function in the absence of translation and ribosomes still requires the S10 protein or the 30S particle (134, 143, 146). Mutations in the β subunit of polymerase reduce or eliminate N function and support a model where polymerase is modified by N protein or by N protein plus a host factor (129); evidence that nusA stabilizes an alternate conformation of RNA polymerase is provided by cross-linking experiments (J. Greenblatt, J. Li, personal communication). In addition, some rho mutations appear to inhibit N activity in vivo by suppressing antitermination at rho-dependent sites (148). A two-stage process may be involved, since stoichiometric amounts of N, nusA, and nusB are supplemented by a number of other previously unidentified *E. coli* proteins (J. Greenblatt, personal communication). A still open question is how N-modified transcription can be terminated, as it appears to occur over a region, perhaps involving a series of N-unresponsive terminators [(149, 150), A. Podhajska, W. Szybalski, personal communication].

Antitermination can also be seen in vitro with purified Q protein of lambda, RNA polymerase, DNA containing the Q-utilization site (*qut,* located close to a Box A sequence), and nusA (151). After initiation, polymerase pauses at nucleotide 16, and Q addition allows transcriptional readthrough; with Q and nusA present, this pause is barely detectable and RNA is quickly chased into longer transcripts, suggesting that Q may function in concert with nusA by accelerating polymerase through the pause, preventing the first step in termination from occurring (152).

Antitermination control of ribosomal RNA (rRNA) operon expression in *E. coli* is more complex [see (153)]. Insertion of IS elements into *rrnC* shows no polar effect (154), and readthrough of a polar site in 16S rDNA requires a short segment of DNA just distal to the P2 promoter, in a region that contains homologies to Boxes A, B, and C of the lambda *nut* sites [(155, 156), M. Cashel, personal communication]. Indeed, overproduction of *nut* transcription can inhibit rRNA synthesis (157), and antitermination is defective in a *nusB5* host (158), implying that some elements of lambda regulation are shared.

However, both promoter and leader regions of *rrn* operons seem able to confer antitermination, which in detail may involve different specific mechanisms than those used by N and Q (159). A temperature-sensitive *E. coli* mutant affecting the elongation of rRNA transcripts has been described (160), and integration host factor probably plays some role in antitermination (161), but details of host protein involvement remain obscure. Recent evidence does suggest that a specific protein factor may mediate antitermination in the tryptophanase operon via a BoxA-like sequence (161a).

Efficient transcription termination and proper maturation of the *rrnB* transcript require 76 bp of sequence distal to the 5S transcript, which contains two dyad symmetry structures, T1 and T2 (162, 163). The similarly structured *rrnE* operon reveals an interesting homology distal to the first RNA structure that may be involved in proper termination of the antitermination complex (164), and mutagenesis of T1 allows readthrough transcription, as does a mutation between T1 and T2 (M. Cashel, personal communication). Further studies acquire special importance because antiterminated *rrn* transcriptional complexes would most likely be detrimental to the cell if they continued transcription beyond the *rrn* operons.

PROCESSING AND RETROREGULATION   The characteristics of a termination region can also influence expression of the upstream gene(s) encoded by the transcript. In bacteriophage lambda, the *sib* site, located distal to the *int* gene, serves two functions that control mRNA half-life. With RNA initiation at the *int* promoter, the hairpin encoded by the dyad symmetry of the *sib* site causes polymerase to terminate, generating a complete and stable *int* transcript, but initiation at pL (modified by N protein) does not respond to the encoded termination signal, and a longer readthrough transcript results (165, 166). Added RNA sequences adjacent to the sib "terminator" structure permit formation of an RNAse III site, resulting in endonucleolytic cleavage that destroys the hairpin structure and renders the *int* mRNA highly labile, presumably due to 3' exonucleolytic degradation (167).

At the end of the *trp* operon, the tandem *trp t* and *trp t'* sites both appear to be involved in transcription termination, though in vivo, the 3' end of *trp* mRNA coincides with the RNA hairpin of *trp t*. Paradoxically, readthrough transcription at *trp t* is observed in vivo in *rho⁻* strains, or in strains with deletions of the *trp t'* region. We have proposed that most *trp* mRNA must be rho-terminated at the distal *trp t'* site, producing a long unstructured trailer region, with subsequent degradation by a 3'-endonuclease back to the RNA hairpin (corresponding to *trp t*) yielding the observed 3' ends (2). Indeed, in vitro, transcripts terminating at the distal rho-dependent *trp t'* site can be trimmed by the 3' exonuclease RNase II into shorter species similar to those terminating at the proximal *trp t* site, and with vectors carrying *trp t'* alone to distal *galK,* there is

3-fold less *galK* activity than with both sites present (168). The RNA sequences upstream of *trp t'* are presumably required for binding and/or activation of the NTPase activity of rho factor essential for its ability to catalyze termination (at the *trp t'* site). Since a high level of this RNA-dependent NTPase activity would probably be deleterious to the cell, posttranscriptional degradation of this region may be essential. Both terminators in this context have a dual function: *trp t* as a minor terminator and a protective barrier to further degradation of upstream structural mRNA, and *trp t'* as the rho-recognition site and a substrate for 3' exonucleases.

In these instances, an important region of RNA lies downstream of the RNA hairpin: an RNase III cleavage site for *sib* (167), and the rho-recognition region for *trp,* without which termination at *trp t'* would not occur (2, 5). However, the *trp* gene product levels are not linked to the terminator response in any obvious way, while with the *sib* site, the specific response determines upstream *int* expression. Similarly, with *glyA,* an insertion between the structural gene and the hairpin terminator reduces expression to 30% of normal (169), and *dnaG* expression also seems to be retroregulated by an attenuatorlike site (19). Termination signals at the ends of operons involved in sugar catabolism, such as *lac* (170), *ara* (171), and *gal* (172), bear some resemblance to other sites, but have enough difference that they are not fully interpretable, possibly because 3' processing events also participate in regulation. Segmental stabilization of the 5' portion of the polycistronic *rcxA* transcript in *Rhodopseudomonas capsulata* also appears owing to two strong hairpin structures, and the differential stability is maintained when this transcript is made in *E. coli* (173). Repetitive extragenic palindromic (REP) sequences, encoding large stable hairpins of unknown function, could also be involved in retroregulation and mRNA processing (169, 174, 175); one unusual hairpin with no adjacent uridines is found just distal to the *serB* gene (175a). Such schemes provide for subtle levels of regulation linking requirements for operon expression to the configuration of the termination site(s), and the metabolic needs of the cell.

Eukaryotes

PARAMETERS AFFECTING POLYMERASES   Analysis of factor-dependent versus factor-independent termination in eukaryotes is less clear-cut than in prokaryotes, owing to the complexities of the polymerases themselves. The definition of an "external" factor is not always easy to pin down, but the ability to isolate transcription complexes for all three enzymes offers the promise of better biochemical accessibility (176). Polymerase III is able to terminate 5S transcription with whatever proteins it carries during elongation, though what it needs for tRNA or *Alu* family transcription is not clear. Polymerase I may require a DNA binding factor(s), as judged by titration experiments (62), but supportive evidence is lacking. For polymerase II, there is likely to be a functional, though not necessarily obligate, coupling of posttranscriptional events to whatever mechanisms cause termination. In yeast, recognition of a consensus sequence TTTTTATA by a rho-like factor has been proposed (69), which could correspond to one class of the trans-acting *SUT* mutations that have been identified (65). Interactions with particular regions in the transcription unit must also be governed by factors reflecting cell type. In the mouse H4 histone genes, these sequences are localized around the 3' end (177), and termination between the μ and δ immunoglobin genes occurs only upon maturation of B-lymphocyte secretor cells (178).

Other parameters may influence termination. For example, though attenuation as defined in prokaryotes cannot exist in the eukaryotic domain because transcription and translation occur in different subcellular compartments, conceptually similar situations occur. In the leader region of the yeast *LEU2* gene, an open reading frame for a leucine-rich peptide, followed by a possible hairpin structure, suggests a model for regulation via an inducer complex that, in response to leucine deprivation, can stabilize one of the alternative RNA structures in the nucleus and thereby control premature termination (179). In SV40, after initiation of late transcription, polymerase II also terminates prematurely at a hairpin site (180). Like prokaryotic attenuation, production of this short aborted RNA probably depends on alternative conformations at the 5' end, which in turn are determined by the presence or absence of the agnoprotein; these elements may also regulate the production of the other two capsid proteins (181).

In controlling transcription, higher-order structures may also play a role: altered chromatin conformation, as probed by hypersensitivity to nuclease, correlates with the 3' ends of the yeast *GAL* genes, whether induced or glucose-repressed (182), and with the 3' ends of the lysozyme (182a) and globin (182b) genes. Chromatin changes are seen with the transcriptional state for globin (183) and sea urchin histones (184)—perhaps the conformational accessibility of the DNA is more significant for polymerase II than the other polymerases (185).

PROCESSING AND POLYADENYLATION   In yeast, where all mRNAs are polyadenylated, it has been suggested that processing and poly(A) addition are tightly coupled to termination of transcription (64, 66). No convincing signals specifying poly(A) addition have been identified however [see (71a)], and a simple possibility not yet ruled out is that every mRNA 3'-OH becomes polyadenylated. In higher eukaryotes, the consensus AAUAAA site (186) is important but not sufficient for 3' end formation and/or polyadenylation. Introduction of a single base change in this sequence shows that polyadenylation can still occur efficiently, even though the mutation renders processing inefficient (187). Other experiments suggest that polyadenylation is inefficient under conditions where termination itself is inefficient (188), possibly because the poly(A) polymerase complex must interact with a specific recognition signal near the 3' end of mRNA precursors to catalyze subsequent polyadenylation (189). In the absence of AAUAAA, a distal minor site (AAUACA) can be used with high efficiency (190), and clearly transcription can sometimes proceed efficiently through multiple polyadenylation sites without diminution (83, 84, 190a), hence termination is not invariably tied to processing and poly(A) addition. Sequences or signals downstream from the processing/polyadenylation site(s) are commonly required for generating correct 3' ends of pol II transcripts, as in adenovirus (191), β-globin (191a), SV40 (87a, 192, 193), bovine growth hormone (194), HSV thymidine kinase (195), and both U1 (196) and U2 (197) snRNAs. Alteration or deletion of these signals generally reduces 3' end formation, and may result in detectable levels of larger read-through transcripts. Stage- and tissue-specific factors also influence the relative use of alternative processing or polyadenylation sites (83, 198), and changes in chromatin structure may reflect (or cause) some of these events (184).

With the polymerase II transcripts of the sea urchin H2A and H3 histone genes, precursor mRNAs with terminals 100–200 nucleotides downstream from the terminal-inverted repeat of the mature transcript are subsequently processed back to the hairpin region to generate the mature 3' end (75, 199). Though normally unpolyadenylated, the H2A transcript contains poly(A) addition signals in its 3' noncoding region that can be utilized in COS cells (200). For H3, this processing requires the participation of a small ribonucleoprotein particle, whose specificity resides with its RNA component, U7, a 60-nucleotide-long species with base complementarity to the conserved precursor ends of histone mRNAs (201).

Studies in vitro illuminate the details of these processes even further. For *Drosophila* histone H3, an endonuclease has been partially purified that appears to generate the 3' end directly (202), and in a HeLa extract the endonucleolytic cleavage event can be separated from subsequent polyadenylation when certain ATP analogues are used (203, 204). Alterations in the AAUAAA sequence can prevent the cleavage (205, 206), but surprisingly, in one instance, the AAUAAA signal appears sufficient to direct polyadenylation at a distant 3' end even in the absence of an endonucleolytic cut (207). It seems likely that in addition to a suitable hexanucleotide (of which AAUAAA is the best) and downstream elements, other factors must also play a role. The possible complexity may be presaged by the example of sea urchin H2A transcription in HeLa cells, where a tripartite structure is implicated in causing polymerase to stop, apparently independent of processing: it includes part of the H2A coding region, and a pair of sequence blocks in the 3' flanking region (M. Johnson, C. Norman, M. Reeve, J. Scully, N. Proudfoot, personal communication). Overall, three points deserve emphasis. First, GT-rich signals distal to the polyadenylation point are necessary for efficient and correct processing (8, 87, 87a). Second, evidence suggests that the small nuclear RNAs U1 or U4 may be involved in site selection or utilization for generating the 3' ends (87a, 203, 207a) whether for processing or termination. Third, the elements may vary in sequence and spatial proximity to one another, and specify a variety of effects that link termination, processing, and polyadenylation.

MECHANISTIC AND STRUCTURAL CONSIDERATIONS

The Dynamics of Termination

ELONGATION, PAUSING, AND RNA RELEASE   RNA polymerase does not elongate RNA chains at a constant rate, and hesitates at specific sites in vitro, and it has been generally accepted that a transient pause is a prerequisite for termination in vivo. The common feature of many of these sites is significant dyad symmetry, permitting hairpin formation in the transcript. Even hairpins other than those normally involved in termination can elicit a pause, such as the early (1:2) hairpin in the *trp* leader region (28, 107), the analogous hairpin in the *thr* leader region (14), and several sites in the cII region of the lambda genome (208). As the hairpin is strengthened, through analogue incorporation into the RNA or by increasing the extent of the dyad symmetry in the DNA, the pause increases, suggesting that the strength of the RNA-RNA interactions in a hairpin controls the length of the pause. However, observations that not all hairpins elicit pausing (116, 208, 209) belie a simple interpretation. In one very clear instance, on MVD-1 cDNA (MVD-1 RNA is a QB replicase template), RNA polymerase pauses significantly at some non-hairpin sites, and shows no sign of doing so at strong nearby hairpins (209). Indications are that at least two different kinds of transcriptional pause site exist (some influenced by upstream regions and others not), which differ considerably from the replication pauses on essentially the identical template (209).

In the absence of rho, each of the sites of the lambda t_{R1} terminator region functions as a pause site for the transcribing complex (208, 210). The ability to elicit a pause does not correlate with the relative strength of the terminator, and although not every pause site acts as a termination site, each termination site acts as a pause site. RNA polymerase also hesitates at other rho-dependent termination sites in the absence of rho (208, 211, 212), but rho does not affect the kinetics of pausing, functioning only to release the nascent transcript (208, 210, 213). In contrast, nusA does enhance pausing, and also lowers the initial rate of transcript release (210). The effects of nusA on elongation have been thoroughly characterized (133, 214), and an excellent discussion of transcription kinetics can be found in von Hippel et al (4).

In eukaryotic systems, pausing has also been shown to occur in certain cases.

Polymerase II pauses in vitro on prokaryotic phage templates of both T7 (215) and lambda (216); in the latter instance termination occurs about 35% of the time as well. Temporally regulated pausing and premature termination seems to be involved in the control of adenovirus expression, at a site resembling the lambda t_{R1} site (217, 218). The participation of a renaturase activity that facilitates pol II transcription is likely to be essential for correct and efficient termination and RNA release (219).

The final step of transcription itself, after the polymerase ceases copying the template, is dissociation of the ternary transcription complex into the component molecules. The extent to which this is a concerted event may vary from one situation to another, and obtaining clearly interpretable data is not always easy. At the *trp* attenuator (111) and at the early rho-independent site in phage T7 (213) in vitro, it now appears that the ternary complex spontaneously dissociates at the hairpin site. At certain other terminator sites, the nusA protein seems able to mediate efficient termination and release (219a).

At a rho-dependent site of T7, these terminal events appear to occur essentially at the same time, but whether they are obligatorily coupled has not yet been determined (213). On a lambda template, rho-catalyzed release of the transcript seems to depend on obligate pausing by polymerase, either natural or induced by 3'-O-methyl nucleotide incorporation (220). Moreover, the length of the RNA, and its particular affinity for rho (substitution of inosine for guanosine increases this affinity and shortens the minimal RNA) both have significant effects on the extent of termination (40, 46, 100, 220, 221).

THE PROCESSIVITY OF POLYMERASES   One of the secrets to understanding termination of transcription may lie in considerations of the ability of RNA polymerase molecules to copy, with extremely high fidelity, extraordinarily long stretches of DNA, over 100 kb for some eukaryotic genes. Any factors that relieve the elongation complex from the constraints of this processivity will de facto increase the probability of terminating transcription. In prokaryotes, we know of two general mechanisms that seem to be employed: (*a*) at rho-independent sites, pausing at RNA hairpins that are followed by multiple uridines results in termination, and (*b*) at other sites, protein factors (such as rho or the ribosome) interacting with the RNA, and subsequently polymerase, decrease the latter's ability to continue elongation. Antitermination factors may be viewed as contributing the reverse—fixing the elongation complex in a highly processive form that is thus resistant to termination signals encoded in the nucleotide sequence or mediated by other factors. Curiously, the interaction of nusA with the elongation complex does not seem to be processive, at least in vitro (214).

In eukaryotic cells, where polymerase II must be very processive and is cut off from direct regulation by translation, termination signals (as specific nucleotide sequences where transcription stops) may not exist, except in very specific cases, e.g. SV40 (180, 181) and adenovirus (217, 218). It could suffice to remove the processivity constraint (perhaps by dissociation of a stabilizing protein factor, analogous to nusA, associated with the elongating polymerase) in concert with the endonucleolytic cleavage that must accompany polyadenylation, for example. Subsequently, a less stable polymerase complex would dissociate over a distance at positions that were thermodynamically most favorable. The AT-richness of intergenic regions would be expected to enhance the probability of termination, and the total absence of processivity of wheat germ pol II in vitro on a poly[d(A-T)] template (222) is consistent with this idea.

Mechanisms: Similarities and Differences

COMMON FEATURES OF RNA POLYMERASES AND COMPARATIVE STUDIES   RNA polymerases both fulfill the basic necessity for faithfully copying specific portions of the DNA template and are the focus for almost all transcriptional regulation, from initial binding to a promoter site through release of the completed transcript. Though the variety of responses that can occur at termination sites alone is staggering, a remarkable similarity among these complex molecules is now emerging. Sequence analysis of the large subunit of the yeast RNA polymerase II and RNA polymerase III reveals six blocks of homology at the amino acid level, shared to a high degree with the *E. coli* β' subunit (223); the analogous subunits of *Drosophila* (224) and mammalian (J. Ahearn, J. Corden, personal communication; J. Ingles, personal communication) polymerase II have similar homologies.

Some tests of the ubiquity of the termination response have been carried out. The yeast SUP4 termination region (normally utilized by polymerase III) exhibits no detectable response when transcribed by the bacterial enzyme in vitro (J. Mott, R. Rothstein, T. Platt, unpublished), and a polymerase II site (the CYC-1 terminator) introduced into *E. coli* between the *gal* promoter and *galK* displays no response in vivo or in vitro (R. Gallagher, K. Zaret, F. Sherman, T. Platt, unpublished). However, wheat germ polymerase II can recognize the *oop* terminator of phage lambda in vitro and stop with 35% efficiency, though the obvious hairpin structure per se is not crucial for function [(216), J. Sharp, W. Hatfield, personal communication]. Since punctuation signals are already maintained across a wide variety of bacterial orders [see (9)], some conservation between prokaryotes and eukaryotes would not be unexpected, though it may occur more at the mechanistic than at the structural level.

EVOLUTION AND INTEGRATION OF TERMINATION MECHANISMS   The structure and signals resident in the RNA are the pivot for control of termination, but are subject to the influence of many other factors. The utility and versatility of this mode of transcriptional regulation depends to a large extent on coordination with other processes in the cell, far beyond the prevention of run-on transcription around the genome, which would itself be inhibitory (225). We may approach our initial two questions about the characteristics and functional properties of termination sites from another vantage point: where did they come from and where have they gone? Though it is almost axiomatic that termination is an essential function, how such signals developed is unclear, as anyone who has tried to devise genetic selections based on terminator function is aware.

Some insights arise from the unusual viewpoint (analogous to the argument that bacteria are more highly evolved than humans) that in bacteria, rho-dependent sites may have evolved before rho-independent sites. Any protein factor with rho-like properties would confer a selective advantage to the cell, because lack of translation would be coupled to the cessation of transcription, via recognition of a stretch of untranslated RNA with relatively little secondary structure (to distinguish mRNA from tRNA and rRNA). If the ribosome encountered a nonsense codon, this factor would communicate to RNA polymerase that further transcription was fruitless, and release it from the DNA to initiate elsewhere. Such properties are consistent with those of rho factor and rho-dependent sites as we know them, and with the involvement of rho in mutational polarity as well as in transcription termination at the ends of genes or operons. Possibly hairpin terminators are the evolutionary latecomers, and arose first as a way of protecting genes from exonuclease digestion (that was required, in turn, to remove the stimulatory effects on NTPase activity of the RNA immediately preceding a rho-dependent termination site). In time, due to the association of hairpins with the ends of a majority of genes, and because of pausing induced by these structures, RNA polymerase may have gained the ability to recognize hairpins as termination signals in their own right, when followed by a suitable uridine-encoding region.

In eukaryotic cells, the three polymerases would have developed in parallel. Termination of the stable RNAs, it may be argued, requires little in the way of control because of the obligate maturation steps that follow precursor biosynthesis [see (226)], and hence these genes can utilize "constitutive" termination signals recognized directly by the enzyme. The 10- to 1000-fold greater size of polymerase II products, and their complexity in terms of cellular requirements for expression at specific times in the cell cycle, specific stages in development, or in particular tissues of the organism, demand sophisticated control over biosynthesis, processing, and targeting. Mechanisms such as attenuation, anti-termination, processing, and polyadenylation may all play a role, as we have seen here.

Several examples of what we know about the production of 3' ends are schematically illustrated in Figure 4, where a termination site is thus most properly considered in relation to its related function(s). Such regions may be composed of a number of subelements, not all of which are present in any one terminator, but which can enhance its utility to the cell (Figure 5). In the majority of bacterial genes (excepting tRNA and rRNA), we generally presume that the mature end of the RNA transcript *(mer)* corresponds to that resulting from termination itself *(ter),* since hairpins plus uridines can evoke the same response in vitro (Figure 4*a*). In more complex cases, such as the end of the tryptophan operon, (Figure 4*b*), *ter* = *mer* only 25% of the time, and the rest of *mer* is generated as the consequence of rho binding to the factor entry region *(fer)* causing polymerase to stop some distance downstream in the *ter* region. The lack of structure plus the free 3' end provides accessibility to exoribonucle-ase action that halts at the hairpin barrier *(bar)* to yield the remaining 75% of *mer,* which is coincidentally identical to that produced by termination directly. In eukaryotic cells, the site of endonuclease cleavage *(sec)* operationally replaces *ter* in generating the 3' end(s), entirely disconnecting the requirement for termination per se from the processing events that produce mature mRNA. For the cell-cycle-specific histone genes (Figure 4*c*), *bar* is still a hairpin (coincidentally an essential part of the *fer* region), but this function may be replaced in the case of other pol II transcripts by polyadenylation (Figure 4*d*). The presence of the *bar* site can thus be a locus for control, providing for variation in the half-life of different RNA species. In eukaryotes, particularly where the primary transcript is far removed from its final form and destination, even more complex mechanisms may be involved in coupling termination (to the extent that it occurs as a specific event) to polyadenylation, cleavage, splicing, and export from nucleus to cytoplasm.

SUMMARY, CONCLUSIONS, AND QUESTIONS FOR THE FUTURE

Our ability to determine nucleotide sequences has progressed from infancy to a full-fledged technology in a scant two decades. It was widely believed in the early days that common function required common sequence, and that the key to understanding gene expression and regulation lay in simply determining the nucleic acid sequences. In fact, the linear sequences of DNA and RNA have told us tantalizingly little about how the signals are decoded, and deciphering the functional properties of the encoded signals has necessitated consideration of the secondary, tertiary, and quaternary conformations of both nucleic acids and proteins, and the contacts between them. The premise that termination sites, regardless of their function in regulation, would still have certain com-mon structural features required for the basic response of RNA polymerase, has proved to be true only in the simplest cases. Far more striking than the similarities are the astonishing variety and complexity of the signals residing in

HOW 3′ ENDS CAN BE MADE AND PRESERVED

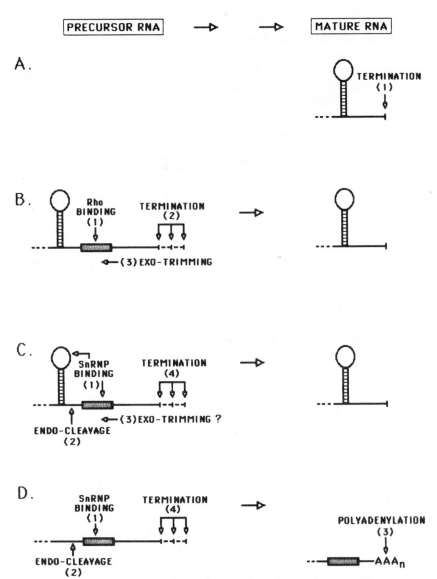

Figure 4 Mechanisms for generating the 3′ ends of transcripts. In prokaryotes, mRNA 3′ ends are generated directly by termination (*A*) or may result from nuclease processing, as appears to occur at the end of the tryptophan operon after rho-dependent termination (168) or with tRNA maturation [see (226)] (*B*). In eukaryotes, RNA hairpins and some distal sequences are essential for correct production of the 3′ ends in histones, probably by endo-cleavage, though participation of exo-trimming has not been ruled out, (*C*), while in all other known mRNAs either endonucleolytic cleavage or termination probably generates the site for 3′ poly(A) addition with participation of distal sequences (*D*); the poly(A) may serve a controllable protective function like that of the hairpins.

SUBELEMENTS ASSOCIATED WITH TERMINATOR REGIONS

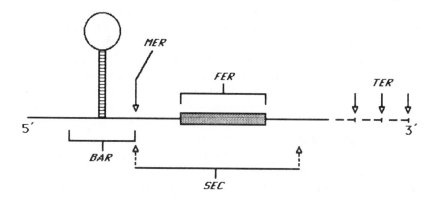

FER = FACTOR ENTRY REGION

TER = REGION WHERE POLYMERASE STOPS TRANSCRIPTION

SEC = SITE OF ENDONUCLEASE CLEAVAGE

BAR = BARRIER TO ACTION OF RIBONUCLEASE

MER = MATURE END OF RNA

Figure 5   Subelements associated with terminator regions. See text for discussion.

the nucleic acid sequence while the polymerase molecules themselves are proving to exhibit remarkable structural homologies.

A general feature of prokaryotic signals specifying termination of transcription is that although they are encoded in the DNA, recognition is mediated through the RNA transcript, generally by using specific structural features rather than particular sequences. In the factor-independent case, this takes the form of a tight self-complementary hairpin loop, which must be identified by some component of the elongation complex. Usually there is no reason to doubt the assumed purpose of hairpins as termination sites, but in the larger context of controlling gene expression, the ability to function as a barrier to exonuclease degradation may be not only important, but dominant, as the example of *trp t* and its relationship to *trp t'* suggests. Clearly, the dual function, providing for the additional protection of mRNA from degradation, may prove to be a common feature of genetic control elements.

In other cases, such as rho-dependent termination, the signal may be (necessarily) more vague, not discernible by searches for consensus sequence or structure. It is not known what recognition elements govern behavior of rho-dependent sites, nor do we understand clearly the function of even the best-studied accessory factors in detail. Examination of some 150 prokaryotic termination sites of all types reveals only a few possible consensus sequences both upstream (CGGGC/G) and downstream (TCTG) of the RNA endpoints

(38), though it should be pointed out that dyad symmetry, for example, would not emerge from a strict consensus search.

Expression of different genes and operons can be independently modified by controlling the extent of termination with several different mechanisms. Therefore, it is not surprising that termination signals exhibit such diversity, despite sharing several characteristics. Auxilliary proteins such as N and Q are used to modify normal termination under special circumstances, and undoubtedly the number of factors and structures involved will increase as the sensitive control that transcription termination exerts over gene expression becomes better defined. With the eukaryotic genome, though impressive work has been done in vivo, and with isolated nuclei in vitro, not enough data has accumulated to allow the construction of detailed models for termination. It should be emphasized that no direct evidence exists that termination, rather than processing, generates a specific 3' end in any pol II transcript. Indeed, even at the conceptual level, it is an extraordinary challenge to conceive of, much less execute, an experiment that can truly discriminate between these possibilities without the availability of reliable in vitro systems.

The parallels between prokaryotes and eukaryotes are nevertheless striking. Tandem terminators, for example, may represent precursor forms of multiply repetitive elements in eukaryotes. The facets of prokaryotic mRNA biosynthesis and degradation are remarkably similar to the interrelationships between processing and termination in the histone transcripts and their precursors. The utility of hairpin regions as pause sites, factor loading regions, termination signals, and as protective barriers to exonucleases as well as to convergent transcription, is astonishing. The RNA binding requirement or rho for termination activity makes it a sort of rudimentary ribonucleoprotein particle, and a small imaginative hop leads to a particle that has sequestered its own RNA permanently, to facilitate recognition of more specific sequences by complementarity. Such a mechanism in turn permits "action at a distance," as appears to occur with the termination component of both rho-dependent and pol II activities. If a loss of processivity by the enzyme itself is primarily responsible for termination, such a loss explains why the set of 3' ends may occur over such a broad region in both cases. It would also not be surprising if antitermination mechanisms are commonly used in long eukaryotic transcription units, especially those with giant introns, to provide the extraordinary processivity that is required to maintain transcription through the final exon.

Thus, though remarkable progress has been made in the past five years, vistas of even greater complexity have appeared. Examples where termination participates in controlling gene expression are ever more widespread, yet the molecular mechanisms that actually influence the behavior of the transcription complex are still virtually unknown. The structural homologies of polymerases, the functional similarities of processing and attenuation mechanisms, and the subelements and their interrelationships at the ends of genes

(Figure 5) provide additional hope that comparisons between prokaryotic and eukaryotic systems will continue to provide insights that further our general understanding of transcription termination and its control. Future work will require the combined efforts of geneticists and biochemists to elucidate the molecular details of these events in prokaryotes, and help determine whether the similarities with eukaryotic systems indicate a general conservation of molecular regulatory strategies. As often indicated first by genetic approaches in vivo, studies in vitro of protein-protein interactions, in addition to those between protein and nucleic acid, will be of paramount importance. The variety and complexity of these multifactor interactions both in vivo and in vitro raises the challenge of considering new approaches and explanations that will lead to an increased knowledge of the mechanisms of transcription termination.

ACKNOWLEDGMENTS

I thank my many colleagues for their thoughtful and constructive comments, particularly Sankar Adhya, Dave Bear, Jack Greenblatt, and Ken Zaret. Special recognition should be accorded the pioneers in this field whose pre-1981 work could not be cited here, including S. Adhya, M. Chamberlin, D. Court, B. deCrombrugghe, A. Das, D. Friedman, M. Gottesman, J. Richardson, J. Roberts, M. Rosenberg, W. Szybalski, C. Yanofsky, and their coworkers. I am indebted to the members of my laboratory for their intellectual contributions and interactions over the past 10 years, and am especially grateful to Jill Galloway, Ray Grant, John Mott, and Hannele Ruohola for their enthusiastic and understanding support during the writing of this manuscript. Contributions to the artwork by Jill Galloway and Shari Harwell are gratefully acknowledged. Additional support was provided by grants from the NIH (GM-35658) and NSF (PCM-8411035).

Literature Cited

1. Kolter, R., Yanofsky, C. 1982. *Ann. Rev. Genet.* 16:113–34
2. Platt, T., Bear, D. G. 1983. *Gene Function in Prokaryotes,* ed. J. Beckwith, J. Davies, J. A. Gallant, pp. 123–61. New York: Cold Spring Harbor Lab.
3. Bauer, C. E., Carey, J., Kasper, L. M., Lynn, S. P., Waechter, D. A., Gardner, J. F. 1983. See Ref. 2, pp. 65–89
4. von Hippel, P. H., Bear, D. G., Morgan, W. D., McSwiggen, J. A. 1984. *Ann. Rev. Biochem.* 53:389–446
5. Galloway, J. L., Platt, T. 1986. *Regulation of Gene Expression—25 Years On,* ed. C. F. Higgins, I. R. Booth. New York: Cambridge Univ. Press. pp. 155–78
6. Sentenac, A., Hall, B. 1982. *The Molecular Biology of the Yeast Saccharomyces: Metabolism and Gene Expression,* ed. J. N. Strathern, E. W. Jones, J. R. Broach, pp. 561–606. New York: Cold Spring Harbor Lab.
7. Nevins, J. R. 1983. *Ann. Rev. Biochem.* 52:441–66
8. Birnstiel, M. L., Busslinger, M., Strub, K. 1985. *Cell* 41:349–59
9. Platt, T. 1981. *Cell* 24:10–23
10. Christie, G. E., Farnham, P. J., Platt, T. 1981. *Proc. Natl. Acad. Sci. USA* 78:4180–84
11. Sharp, J. A., Platt, T. 1984. *J. Biol. Chem.* 259:2268–73
12. Farnham, P. J., Platt, T. 1982. *Proc. Natl. Acad. Sci. USA* 79:998–1002
13. Ryan, T., Chamberlin, M. J. 1983. *J. Biol. Chem.* 258:4690–93
14. Gardner, J. F. 1982. *J. Biol. Chem.* 257:3896–904
15. Rosenberg, M., Chepelinsky, A. B.,

McKenney, K. 1983. *Science* 222:734–39

16. Stroynowski, I., Kuroda, M., Yanofsky, C. 1983. *Proc. Natl. Acad. Sci. USA* 80:2206–10
17. Pinkham, J. L., Platt, T. 1983. *Nucleic Acids Res.* 11:3531–45
18. Squires, C., Krainer, A., Barry, G., Shen, W. F., Squires, C. L. 1981. *Nucleic Acids Res.* 9:6827–39
19. Burton, Z. F., Gross, C. A., Watanabe, K. K., Burgess, R. 1983. *Cell* 32:335–49
20. Bedwell, D., Davis, G., Gosink, M., Post, L., Nomura, M., et al. 1985. *Nucleic Acids Res.* 13:3891–903
21. Reed, R. E., Baer, M. F., Guerrier-Takada, C., Donis-Keller, H., Altman, S. 1982. *Cell* 30:627–36
22. Postle, K., Good, R. F. 1985. *Cell* 41:577–85
23. Carlomagno, M. S., Riccio, A., Bruni, C. B. 1985. *J. Bacteriol.* 162:362–68
24. Schollmeier, K., Gartner, D., Hillen, W. 1985. *Nucleic Acids Res.* 13:4227–37
25. Sheflin, L. G., Kowalski, D. 1985. *Nucleic Acids Res.* 13:6137–54
26. Yanofsky, C., Das, A., Fisher, R., Kolter, R., Berlin, V. 1983. *Gene Expression. UCLA Symp. Mol. Cell. Biol.*, ed. D. H. Hamer, M. J. Rosenberg, 8:295–310. New York: Liss
27. Stroynowski, I., van Cleemput, M., Yanofsky, C. 1982. *Nature* 298:38–41
28. Farnham, P. J., Platt, T. 1981. *Nucleic Acids Res.* 9:563–77
29. Fisher, R. F., Yanofsky, C. 1983. *J. Biol. Chem.* 258:8146–50
30. Ames, B. N., Tsang, T. H., Buck, M., Christman, M. F. 1983. *Proc. Natl. Acad. Sci. USA* 80:5240–42
31. Ishii, S., Kuroki, K., Imamoto, F. 1984. *Proc. Natl. Acad. Sci. USA* 81:409–13
32. Ralling, G., Linn, T. 1984. *J. Bacteriol.* 158:279–85
33. Nakamura, Y., Mizusawa, S. 1985. *EMBO J.* 4:527–32
34. Hall, M. N., Gabay, J., Debarbouille, M., Schwartz, M. 1982. *Nature* 295:616–18
35. Dubnau, D. 1985. *EMBO J.* 4:533–37
36. Christensen, T., Johnsen, M., Fiil, N. P., Friesen, J. D. 1984. *EMBO J.* 3:1609–12
36a. Rossi, J., Egan, J., Hudson, L., Landy, A. 1981. *Cell* 26:305–14
37. Morgan, W. D., Bear, D. G., Litchman, B. L., von Hippel, P. H. 1985. *Nucleic Acids Res.* 13:3739–54
38. Brendel, V., Hamm, G. H., Trifonov, E. N. 1986. *J. Biomolec. Struct. Dyn.* 3
39. Lau, L. F., Roberts, J. W., Wu, R. 1982. *Proc. Natl. Acad. Sci. USA* 79:6171–75
40. Morgan, W. D., Bear, D. G., von Hippel, P. H. 1983. *J. Biol. Chem.* 258:9553–64
41. Lau, L. F., Roberts, J. W., Wu, R., Georges, F., Narang, S. A. 1984. *Nucleic Acids Res.* 12:1287–99
42. Lau, L. F., Roberts, J. W. 1985. *J. Biol. Chem.* 260:574–84
43. Wu, A. M., Christie, G. E., Platt, T. 1981. *Proc. Natl. Acad. Sci. USA* 78:2913–17
44. Platt, T., Mott, J. E., Galloway, J. L., Grant, R. A. 1985. *Sequence Specificity in Transcription and Translation. UCLA Symp. Mol. Cell. Biol.* (NS), Vol. 30, ed. R. Calendar, L. Gold, pp. 151–60. New York: Liss
45. Bear, D. G., McSwiggen, J. A., Morgan, W. D., von Hippel, P. H. 1985. See Ref. 44, pp. 137–50
46. Ceruzzi, M. A. F., Bektesh, S. L., Richardson, J. P. 1985. *J. Biol. Chem.* 260:9412–18
47. Richardson, J. P. 1983. *Microbiology—1983*, ed. D. Schlessinger, pp. 31–34. Washington, DC: Am. Soc. Microbiol.
48. Ciampi, M. S., Schmid, M. B., Roth, J. R. 1982. *Proc. Natl. Acad. Sci. USA* 79:5016–20
49. Santi, D. V., Garrett, C. E., Barr, P. J. 1983. *Cell* 33:9–10
50. Uhlenbeck, O. C., Carey, J., Romanuik, P. J., Lowary, P. T., Beckett, D. 1983. *J. Biomol. Struct. Dyn.* 1:539–52
51. Bogenhagen, D. F., Brown, D. D. 1981. *Cell* 24:261–70
52. Cozzarelli, N. R., Gerrard, S. P., Schlissel, M., Brown, D. D., Bogenhagen, D. F. 1983. *Cell* 34:829–35
53. Watson, J. B., Chandler, D. W., Gralla, J. D. 1984. *Nucleic Acids Res.* 12:5369–84
54. Adeniyi-Jones, S., Romeo, P. H., Zasloff, M. 1984. *Nucleic Acids Res.* 12:1101–15
55. Koski, R. A., Allison, D. S., Worthington, M., Hall, B. D. 1982. *Nucleic Acids Res.* 10:8127–43
55a. Allison, D. S., Hall, B. D. 1985. *EMBO J.* 4:2657–64
56. Emerson, B. M., Roeder, R. G. 1984. *J. Biol. Chem.* 259:7926–35
57. Hess, J., Perez-Stable, C., Wu, G. J., Weir, B., Tinoco, I., Shen, C.-K. J. 1985. *J. Mol. Biol.* 185:7–21
58. Sollner-Webb, B., Tower, J. 1986. *Ann. Rev. Biochem.* 55
59. Gurney, T. 1985. *Nucleic Acids Res.* 13:4905–19
60. Kempers-Veenstra, A. E., van Heerikhuizen, H., Musters, W., Klootwijk, J., Planta, R. J. 1984. *EMBO J.* 3:1377–82
60a. Moss, T. 1983. *Nature* 302:223–30

61. Grummt, I., Sorbaz, H., Hofmann, A., Roth, E. 1985. *Nucleic Acids Res.* 13: 2293–304
62. Grummt, I., Öhrlein, A., Maier, U., Hassouna, N., Bachellerie, J.-P. 1985. *Cell* 43:801–10
63. Miwa, T., Kominami, R., Muramatsu, M. 1986. *Mol. Cell Biol.* Submitted
64. Zaret, K. S., Sherman, F. 1982. *Cell* 28:563–73
65. Kotval, J., Zaret, K. S., Consaul, S., Sherman, F. 1983. *Genetics* 103:367–88
66. Zaret, K. S., Sherman, F. 1984. *J. Mol. Biol.* 76:107–35
67. Henikoff, S., Kelly, J. D., Cohen, E. H. 1983. *Cell* 33:607–14
68. Falco, S. C., Dumas, K. S., Livak, K. J. 1985. *Nucleic Acids Res.* 13:4011–27
69. Henikoff, S., Cohen, E. H. 1984. *Mol. Cell Biol.* 4:1515–20
70. Yarger, J. G., Armilei, G., Gorman, M. C. 1986. *Mol. Cell Biol.* 6: In press
71. Brent, R., Ptashne, M. 1984. *Nature* 312:612–15
71a. Sutton, A., Broach, J. R. 1985. *Mol. Cell Biol.* 5:2770–80
72. Nogi, Y., Fukasawa, T. 1983. *Nucleic Acids Res.* 11:8555–68
73. Buckholz, R. G., Cooper, T. G. 1983. *Mol. Cell Biol.* 3:1889–97
74. Birchmeier, C., Folk, W., Birnstiel, M. L. 1983. *Cell* 35:433–40
75. Birchmeier, C., Schümperli, D., Sconzo, G., Birnstiel, M. L. 1984. *Proc. Natl. Acad. Sci. USA* 81:1057–61
76. Wells, D., Kedes, L. 1985. *Proc. Natl. Acad. Sci. USA* 82:2834–38
77. Kunkel, G. R., Pederson, T. 1985. *Mol. Cell Biol.* 5:2332–40
78. Manser, T., Gesteland, R. F. 1982. *Cell* 29:257–64
79. Tani, T., Watanabe-Nagasu, N., Okada, N., Ohshima, Y. 1983. *J. Mol. Biol.* 168:579–94
80. LeMeur, M. A., Galliot, B., Gerlinger, P. 1984. *EMBO J.* 3:2779–86
81. Boardman, M., Basi, G. S., Storti, R. V. 1985. *Nucleic Acids Res.* 13:1763–76
82. Aho, S., Tate, V., Boedtker, H. 1983. *Nucleic Acids Res.* 11:5443–50
83. Hagenbüchle, O., Wellauer, P. K., Cribbs, D. L., Schibler, U. 1984. *Cell* 38:737–44
84. Frayne, E. G., Leys, E. J., Crouse, G. F., Hook, A. G., Kellems, R. E. 1984. *Mol. Cell Biol.* 4:2921–24
85. Citron, B., Falck-Pedersen, E., Salditt-Georgieff, M., Darnell, J. E. Jr. 1984. *Nucleic Acids Res.* 12:8723–31
86. Falck-Pedersen, E., Logan, J., Shenk, T., Darnell, J. E. Jr. 1985. *Cell* 40:897–905

87. McLauchlan, J., Gaffney, D., Whitton, J. L., Clements, J. B. 1985. *Nucleic Acids Res.* 13:1347–68
87a. Sadofsky, M., Connelly, S., Manley, J. L., Alwine, J. C. 1985. *Mol. Cell Biol.* 5:2713–19
88. Yaffe, D., Nudel, U., Mayer, Y., Neuman, S. 1985. *Nucleic Acids Res.* 13:3723–37
89. Brown, S., Albrechtsen, B., Pedersen, S., Klemm, P. 1982. *J. Mol. Biol.* 162: 283–98
90. Finger, L. R., Richardson, J. P. 1982. *J. Mol. Biol.* 156:203–19
91. Tsurushita, N., Hirano, M., Shigesada, K., Imai, M. 1984. *Mol. Gen. Genet.* 196:458–64
92. Barik, S., Bhattacharya, P., Das, A. 1985. *J. Mol. Biol.* 182:495–508
93. Bear, D. G., Andrews, C. L., Singer, J. D., Morgan, W. D., Grant, R. A., et al. 1985. *Proc. Natl. Acad. Sci. USA* 82:1911–15
94. Richardson, J. P., Carey, J. L. 1982. *J. Biol. Chem.* 257:5767–71
95. Housley, P. R., Whitfield, H. J. 1982. *J. Biol. Chem.* 257:2569–77
96. Kent, R. B., Guterman, S. K. 1982. *Mol. Gen. Genet.* 187:330–34
97. Shigesada, K., Tsurushita, N., Matsumoto, Y., Imai, M. 1984. *Gene* 29:199–209
98. Mott, J. E., Grant, R. A., Ho, Y.-S., Platt, T. 1985. *Proc. Natl. Acad. Sci. USA* 82:88–92
99. Richardson, J. P., Macy, M. R. 1981. *Biochemistry* 20:1133–39
100. Richardson, J. P. 1982. *J. Biol. Chem.* 257:5760–66
101. Sharp, J. A., Guterman, S. K., Platt, T. 1986 *J. Biol. Chem.* 261:2524–28
102. Schmidt, M. C., Chamberlin, M. J. 1984. *J. Biol. Chem.* 259:15000–2
103. Briat, J.-F., Chamberlin, M. J. 1984. *Proc. Natl. Acad. Sci. USA* 81:7373–77
103a. Briat, J. F., Bollag, G., Kearney, C. A., Molineaux, I., Chamberlin, M. J. 1986 Submitted for publication
104. Greenblatt, J. 1981. *Cell* 24:8–9
105. Ward, D. F., Gottesman, M. E. 1982. *Science* 216:946–51
106. Forbes, D., Herskowitz, I. 1982. *J. Mol. Biol.* 160:549–69
107. Winkler, M. E., Yanofsky, C. 1981. *Biochemistry* 20:3738–44
108. Fisher, R., Yanofsky, C. 1983. *J. Biol. Chem.* 258:9208–12
109. Hauser, C. A., Sharp, J. A., Hatfield, L. K., Hatfield, G. W. 1985. *J. Biol. Chem.* 260:1765–70
110. Landick, R., Carey, J., Yanofsky, C. 1985. *Proc. Natl. Acad. Sci. USA* 82:4663–67

110a. Fisher, R. F., Das, A., Kolter, R., Winkler, M. E., Yanofsky, C. 1985. *J. Mol. Biol.* 182:397–409
111. Berlin, V., Yanofsky, C. 1983. *J. Biol. Chem.* 258:1714–19
112. Das, A., Crawford, I. P., Yanofsky, C. 1982. *J. Biol. Chem.* 257:8795–98
113. Yanofsky, C., Kelley, R. L., Horn, V. 1984. *J. Bacteriol.* 158:1018–24
114. Ward, D. F., Gottesman, M. E. 1981. *Nature* 292:212–15
115. Das, A., Yanofsky, C. 1984. *J. Bacteriol.* 160:805–7
116. Kingston, R. E., Chamberlin, M. J. 1981. *Cell* 27:523–31
117. Lindahl, L., Archer, R., Zengel, M. J. 1983. *Cell* 33:241–48
118. Zengel, J. M., Lindahl, L. 1985. *J. Bacteriol.* 163:140–47
119. Springer, M., Mayaux, J.-F., Fayat, G., Plumbridge, J. A., Graffe, M., et al. 1985. *J. Mol. Biol.* 181:467–78
120. Carter, P. W., Weiss, D. L., Weith, H. L., Calvo, J. M. 1985. *J. Bacteriol.* 162:943–49
121. Turnbough, C. L. Jr., Hicks, K. L., Donahue, J. P. 1983. *Proc. Natl. Acad. Sci. USA* 80:368–72
121a. Roland, K. L., Powell, F. E., Turnbough, C. L. Jr. 1985. *J. Bacteriol.* 163:991–99
122. Navre, M., Schachman, H. K. 1983. *Proc. Natl. Acad. Sci. USA* 80:1207–11
123. Levin, H. L., Schachman, H. K. 1985. *Proc. Natl. Acad. Sci. USA* 82:4643–47
124. Bonekamp, F., Andersen, H. D., Christensen, T., Jensen, K. F. 1985. *Nucleic Acids Res.* 13:4113–23
125. Friden, P., Newman, T., Freundlich, M. 1982. *Proc. Natl. Acad. Sci. USA* 79:6156–60
126. Hauser, C., Hatfield, G. W. 1984. *Proc. Natl. Acad. Sci. USA* 81:76–79
127. Yanofsky, C. 1984. *Mol. Biol. Evol.* 1(2):143–61
128. Deleted in proof
129. Friedman, D. I., Gottesman, M. 1983. *Lambda II*, ed. R. W. Hendrix, J. W. Roberts, F. W. Stahl, R. A. Weisberg, pp. 21–52. New York: Cold Spring Harbor Lab.
130. Franklin, N. C. 1985. *J. Mol. Biol.* 181:75–84
131. Szybalski, W., Drahos, D., Luk, K.-C., Somasekhar, G. 1983. See Ref. 47, pp. 35–38
132. Friedman, D. I., Schauer, A. T., Mashni, E. J., Olson, E. R., Baumann, M. F. 1983. See Ref. 47, pp. 39–42
133. Greenblatt, J. 1984. *Can. J. Biochem. Cell Biol.* 62:79–88
134. Goda, Y., Greenblatt, J. 1985. *Nucleic Acids Res.* 13:2569–82
135. Friedman, D. I., Olson, E. R. 1983. *Cell* 34:143–49
136. Peltz, S. W., Brown, A. L., Hasan, N., Podhajska, A. J., Szybalski, W. 1985. *Science* 228:91–93
137. Drahos, D., Galluppi, G. R., Caruthers, M., Szybalski, W. 1982. *Gene* 18:373–54
138. Olson, E. R., Tomich, C.-S. C., Friedman, D. I. 1984. *J. Mol. Biol.* 180:1053–63
138a. Tsugawa, A., Kurihara, T., Zuber, M., Court, D., Nakamura, Y. 1985. *EMBO J.* 4:2337–42
139. Kung, H.-F. 1983. See Ref. 47, pp. 53–56
140. Greenblatt, J., Li, J. 1981. *J. Mol. Biol.* 147:11–23
141. Greenblatt, J., Li, J. 1981. *Cell* 24:421–28
142. Greenblatt, J., McLimont, M., Hanly, S. 1981. *Nature* 292:215–20
143. Das, A., Wolska, K. 1984. *Cell* 38:165–73
144. Ward, D. F., DeLong, A., Gottesman, M. E. 1983. *J. Mol. Biol.* 168:73–85
145. Swindle, J., Ajioka, J., Georgopoulos, C. 1983. See Ref. 47, pp. 63–65
146. Das, A., Ghosh, B., Barik, S., Wolska, K. 1985. *Proc. Natl. Acad. Sci. USA* 82:4070–74
147. Deleted in proof
148. Das, A., Gottesman, M. E., Wardwell, J., Trisler, P., Gottesman, S. 1983. *Proc. Natl. Acad. Sci. USA* 80:5530–34
149. Honigman, A. 1981. *Gene* 13:299–309
150. Burt, D. W., Brammar, W. J. 1982. *Mol. Gen. Genet.* 185:462–67
151. Grayhack, E. J., Roberts, J. W. 1982. *Cell* 30:637–48
152. Grayhack, E. J., Yang, X., Lau, L. F., Roberts, J. W. 1985. *Cell* 42:259–69
153. Lamond, A. 1985. *Trends Biochem. Sci.* 10:271–74
154. Siehnel, R. J., Morgan, E. A. 1983. *J. Bacteriol.* 153:672–84
155. Li, S. C., Squires, C. L., Squires, C. 1984. *Cell* 38:851–60
156. Aksoy, S., Squires, C. L., Squires, C. 1984. *J. Bacteriol.* 159:260–64
157. Sharrock, R. A., Gourse, R. L., Nomura, M. 1985. *J. Bacteriol.* 163:704–8
158. Sharrock, R. A., Gourse, R. L., Nomura, M. 1985. *Proc. Natl. Acad. Sci. USA* 82:5275–79
159. Holben, W. E., Prasad, S. M., Morgan, E. A. 1985. *Proc. Natl. Acad. Sci. USA* 82:5073–77
160. Liebke, H. H., Speyer, J. F. 1983. *Mol. Gen. Genet.* 189:314–20
161. Peacock, S., Weissbach, H. J., Nash, H.

A. 1984. *Proc. Natl. Acad. Sci. USA* 81:6009–13

161a. Stewart, V., Yanofsky, C. 1985. *J. Bacteriol.* 164:731–40

162. Elford, R. M., Holmes, W. M. 1983. *J. Mol. Biol.* 168:557–61

163. Szeberenyi, J., Apirion, D. 1983. *J. Mol. Biol.* 168:525–61

164. Liebke, H., Hatfull, G. 1985. *Nucleic Acids Res.* 13:5515–25

165. Schindler, D., Echols, H. 1981. *Proc. Natl. Acad. Sci. USA* 78:4475–79

166. Guarneros, G., Montanez, C., Hernandez, T., Court, D. 1982. *Proc. Natl. Acad. Sci. USA* 79:238–42

167. Schmeissner, U., McKenney, K., Rosenberg, M., Court, D. 1984. *J. Mol. Biol.* 176:39–53

168. Mott, J. E., Galloway, J. L., Platt, T. 1985. *EMBO J.* 4:1887–91

169. Plamann, M. D., Stauffer, G. V. 1985. *J. Bacteriol.* 161:650–54

170. Hediger, M. A., Johnson, D. F., Nierlich, D. P., Zabin, I. 1985. *Proc. Natl. Acad. Sci. USA* 82:6414–18

171. Lin, H.-C., Lei, S.-P., Studnicka, G., Wilcox, G. 1985. *Gene* 34:129–34

172. Debouck, C., Riccio, A., Schumperli, D., McKenney, K., Jeffers, J., et al. 1985. *Nucleic Acids Res.* 13:1841–53

173. Belasco, J. G., Beatty, J. T., Adams, C. W., von Gabain, A., Cohen, S. N. 1985. *Cell* 40:171–81

174. Higgins, C. F., Ames, G. F.-L., Barnes, W. M., Clement, J. M., Hofnung, M. 1982. *Nature* 298:760–62

175. Stern, M. J., Ames, G. F.-L., Smith, N. H., Robinson, E. C., Higgins, C. F. 1984. *Cell* 37:1015–26

175a. Neuwald, A. F., Stauffer, G. V. 1985. *Nucleic Acids Res.* 13:7025–39

176. Culotta, V. C., Wides, R. J., Sollner-Webb, B. 1985. *Mol. Cell Biol.* 5:1582–90

177. Lüscher, B., Stauber, C., Schindler, R., Schümperli, D. 1985. *Proc. Natl. Acad. Sci. USA* 82:4389–93

178. Mather, E. L., Nelson, K. J., Haimovich, J., Perry, R. P. 1984. *Cell* 36:329–38

179. Andreadis, A., Hsu, Y.-P., Kohlhaw, G. B., Schimmel, P. 1982. *Cell* 31:319–25

180. Hay, N., Skolnik-David, H., Aloni, Y. 1982. *Cell* 29:183–93

181. Hay, N., Aloni, Y. 1985. *Mol. Cell Biol.* 5:1327–34

182. Proffitt, J. H. 1985. *Mol. Cell Biol.* 5:1522–24

182a. Fritton, H. P., Sippel, A. E., Igo-Kemenes, T. 1983. *Nucleic Acids Res.* 11:3467–85

182b. Groudine, M., Kohwi-Shigematsu, T., Gelvias, R., Stamatoyannopoulos, G.,

Papayannopolou, T. 1983. *Proc. Natl. Acad. Sci. USA* 80:7551–55

183. Hofer, E., Hofer-Warbinek, R., Darnell, J. E. 1982. *Cell* 29:887–93

184. Bryan, P. N., Olah, J., Birnstiel, M. L. 1983. *Cell* 33:843–48

185. Miller, T. J., Mertz, J. E. 1982. *Mol. Cell Biol.* 2:1595–607

186. Fitzgerald, M., Shenk, T. 1981. *Cell* 24:251–60

187. Montell, C., Fisher, E. F., Caruthers, M. H., Berk, A. J. 1983. *Nature* 305:600–5

188. Miller, T. J., Stephens, D. L., Mertz, J. E. 1982. *Mol. Cell Biol.* 2:1581–94

189. Manley, J. L. 1983. *Cell* 33:595–605

190. Mason, P. J., Jones, M. B., Elkington, J. A., Williams, J. G. 1985. *EMBO J.* 4:205–11

190a. Amara, S. G., Evans, R. M., Rosenfeld, M. G. 1984. *Mol. Cell Biol.* 4:2150–60

191. McDevitt, M. A., Imperiale, M. J., Ali, H., Nevins, J. R. 1984. *Cell* 37:993–99

191a. Gil, A., Proudfoot, N. J. 1984. *Nature* 312:473–74

192. Sadofsky, M., Alwine, J. C. 1984. *Mol. Cell Biol.* 4:1460–68

193. Conway, L., Wickens, M. 1985. *Proc. Natl. Acad. Sci. USA* 82:3949–53

194. Woychik, R. P., Lyons, R. H., Post, L., Rottman, F. M. 1984. *Proc. Natl. Acad. Sci. USA* 81:3944–48

195. Cole, C. N., Stacy, T. P. 1985. *Mol. Cell Biol.* 5:2104–13

196. Hernandez, N. 1985. *EMBO J.* 4:1827–37

197. Yuo, C.-Y., Ares, M. Jr., Weiner, A. M. 1985. *Cell* 42:193–202

198. Henikoff, S., Sloan, J. S., Kelly, J. D. 1983. *Cell* 34:405–14

199. Krieg, P. A., Melton, D. A. 1984. *Nature* 308:203–6

200. Nordstrom, J. L., Hall, S. L., Kessler, M. M. 1985. *Proc. Natl. Acad. Sci. USA* 82:1094–98

201. Strub, K., Galli, G., Busslinger, M., Birnstiel, M. L. 1984. *EMBO J.* 3:2801–7

202. Price, D. H., Parker, C. S. 1984. *Cell* 38:423–29

203. Moore, C. L., Sharp, P. A. 1984. *Cell* 36:581–91

204. Moore, C. L., Sharp, P. A. 1985. *Cell* 41:845–55

205. Montell, C., Fisher, E. F., Caruthers, M. H., Berk, A. J. 1983. *Nature* 305:600–5

206. Wickens, M., Stephenson, P. 1984. *Science* 226:1045–51

207. Manley, J. L., Yu, H., Ryner, L. 1985. *Mol. Cell Biol.* 5:373–79

207a. Berget, S. M. 1984. *Nature* 309:179–82

208. Morgan, W. D., Bear, D. G., von Hip-

pel, P. H. 1983. *J. Biol. Chem.* 258: 9565–74

209. LaFlamme, S. E., Kramer, F. R., Mills, D. R. 1985. *Nucleic Acids Res.* 13:8425–40

210. Lau, L. F., Roberts, J. W., Wu, R. 1983. *J. Biol. Chem.* 258:9391–97

211. Kassavetis, G. A., Chamberlin, M. J. 1981. *J. Biol. Chem.* 256:2777–86

212. Farnham, P. J., Greenblatt, J., Platt, T. 1982. *Cell* 29:945–51

213. Andrews, C., Richardson, J. P. 1985. *J. Biol. Chem.* 260:5826–31

214. Schmidt, M. C., Chamberlin, M. J. 1984. *Biochemistry* 23:197–203

215. Kadesch, T. R., Chamberlin, M. J. 1982. *J. Biol. Chem.* 257:5286–95

216. Hatfield, G. W., Sharp, J. A., Rosenberg, M. 1983. *Mol. Cell Biol.* 3:1687–93

217. Maderious, A., Chen-Kiang, S. 1984. *Proc. Natl. Acad. Sci. USA* 81:5931–35

218. Mok, M., Maderious, A., Chen-Kiang, S. 1984. *Mol. Cell Biol.* 4:2031–40

219. Kane, C. M., Chamberlin, M. J. 1985. *Biochemistry* 24:2254–62

219a. Schmidt, M. C., Chamberlin, M. J. 1986. Submitted for publication

220. Morgan, W. D., Bear, D. G., von Hippel, P. H. 1984. *J. Biol. Chem.* 259: 8664–71

221. Ceruzzi, M., Richardson, J. P. 1985. See Ref. 44, p. 161–70

222. Durand, R., Job, C., Teissere, M., Job, D. 1982. *FEBS Lett.* 150:477–81

223. Allison, L. A., Moyle, M., Shales, M., Ingles, C. J. 1985. *Cell* 42:599–610

224. Biggs, J., Searles, L. L., Greenleaf, A. L. 1985. *Cell* 42:611–21

225. Adhya, S., Gottesman, M. 1982. *Cell* 29:939–44

226. Apirion, D. 1984. *Processing of RNA*, ed. D. Apirion. Boca Raton, Fla: CRC

Ann. Rev. Biochem. 1986. 55:373–95

DOUBLE-STRANDED RNA REPLICATION IN YEAST: THE KILLER SYSTEM[1]

Reed B. Wickner

Section on Genetics of Simple Eukaryotes, Laboratory of Biochemical Pharmacology, National Institute of Arthritis, Diabetes, and Digestive and Kidney Diseases, National Institutes of Health, Building 4, Room 116, Bethesda, Maryland 20892

CONTENTS

PERSPECTIVES AND SUMMARY

Strains of the yeast *Saccharomyces cerevisiae* carry as many as five nonhomologous species of double-stranded RNA (dsRNA), called L-A, L-BC, T, W, and M(1–4). All show non-Mendelian inheritance. Of these, L-A (3, 5, 6), L-BC (3), and M (7–10) are known to be families of natural variants, different strains carrying genetically and/or physically distinct molecules of the given family. L-A, L-BC, and M are found in intracellular viruslike particles

---

(3, 11). While these particles are not infectious, they can be introduced into spheroplasts by polyethylene glycol-induced fusion. Among transformants selected for a DNA plasmid present at the same time, a large proportion of clones have incorporated the viruslike particles (M. El-Sherbeini, K. A. Bostian, personal communication). Each M dsRNA encodes a secreted protein toxin, called a "killer toxin," and immunity to the homologous toxin (12). Strains carrying M_1 dsRNA, for example, are said to be K_1 killers. K_1 killers kill most wild-type strains lacking M, as well as M_2-carrying K_2 killers, M_3-carrying K_3 killers, etc. M dsRNA and the resultant killer phenotype show cytoplasmic inheritance (13), that is, transmission to all progeny in meiotic crosses ($4+:0$ segregation) (13), and transmission by cytoplasmic mixing (cytoduction). Recently, chromosomally determined killer phenomena in strains lacking M or lacking all dsRNAs have been reported (10).

This review will focus on the dsRNAs and killer phenomena of *Saccharomyces* as have other recent reviews (14–17). dsRNAs and killer phenomena are also widespread in other yeasts (8, 18) and filamentous fungi [reviewed in (19)]. Extensive studies of the biochemical and physical properties of these "mycoviruses" have been carried out. A linear dsDNA plasmid encoding a toxin-immunity system in *Kluyveromyces lactis* has also been studied extensively (20). The early descriptions of antiviral factors from *Penicillium stoloniferum* led to the discovery of interferon and the dsRNAs that were inducing its production [reviewed in (21)]. Killer phenomena have also been described in *Ustilago, Torulopsis, Debaromyces, Hansenula, Kluyveromyces, Candida, Pichia,* and *Cryptococcus* (8, 18, 22).

Motivations for studying the *Saccharomyces* killer systems include the following:

1. This is a model eukaryotic virus for which the host genetics and molecular biology can be readily studied. Animal virus studies have, of necessity, focused on the viral genome and have often ignored the role of host cellular functions in viral replication because of the difficulty of carrying out animal or animal cell genetics. As will be detailed below, over 40 chromosomal genes affecting yeast dsRNA virus replication in various ways have been described.
2. The processing and secretion of the killer toxin have been extensively studied as a model for protein processing and secretion in general, and the toxin action may serve as a model for hormone action or intercellular communication.
3. A secretion vector based on the toxin precursor has been developed (23). Further, since methods of amplifying M dsRNAs to thousands of copies per cell are known, they have the potential to be used as vectors.
4. Fermentation processes carried out with sensitive strains are often spoiled by contaminating killer strains (usually K_2 killers) introduced with the

grapes, rice, barley, etc. Methods have been developed to introduce M dsRNAs into industrial fermentation strains, both to provide immunity to some killer contaminants and to kill other sensitive contaminants (24–28).

5. Recently, strains of *Candida* have been typed using sensitivity to a battery of killer strains, in analogy to phage typing of *Salmonella,* as an aid to studying the epidemiology of infections with this human pathogen (29).

The reaction of many (including yeast geneticists) on reading about the killer system is that it is very complex. To assist the reader, a table is included (Table 1) summarizing all the elements of the system.

DOUBLE-STRANDED RNAs IN *SACCHAROMYCES CEREVISIAE*

M dsRNAs: The Toxin-Immunity Precursor Protein

Six groups of killer strains have been found among *Saccharomyces* strains, with strains assigned to different groups if they kill each other (7–10, 13). The K_1 phenotype is widely distributed among laboratory strains and wild-type strains (30), while K_2 has been found almost exclusively among fermentation contaminants (7). The K_3 group consists of strains derived from a single wine yeast (9). Recently Hara and coworkers have found three new killer groups, of which one is based on dsRNA while the other two seem to be chromosomally determined (10).

The K_1 and K_2 toxins are encoded by M_1 and M_2 dsRNAs, respectively (12). Nearly the entire sequence of M_1 dsRNA is known as a combination of direct RNA sequencing and sequencing of cDNA clones of the region encoding the toxin precursor (Figure 1) (31–35). Recently some 200 nucleotides at each of the ends of M_2 have been sequenced directly, enabling comparisons with M_1 (36).

As shown in Figure 1, the preprotoxin-immunity sequence comprises bases 14–964 from the left-hand end of M_1. This is followed by a region of about 200 bp of all A residues on the message strand and all U residues on the antimessage strand (31, 37). Most of the right-hand end of M_1 has been sequenced and does not contain any sizable open reading frames (34, 35). Deletion mutants (37) lacking all but bases 1–230 on the left in Figure 1 and the rightmost approximately 500 bp (35, 37) are transcribed (35), packaged (38), and replicated, and inhibit the replication of wild-type M_1 dsRNA (84, 85) in analogy with defective-interfering particles of animal cells. Thus the signals for transcription, packaging, and replication must be within the retained regions of these mutants.

The size of M_1 and M_2 dsRNAs varies rapidly by up to 300 bp without affecting copy number, toxin production, or immunity (39). The site of this

Table 1   Components of the *Saccharomyces cerevisiae* killer systems

dsRNAs	
M_1	a 1.8-kilobase dsRNA that encodes the K_1 preprotoxin-immunity protein. Strains carrying M_1 have the genotype [KIL-k_1]. Secretion of an active K_1 toxin is the K_1^+ phenotype. Resistance to the K_1 toxin is the R_1^+ phenotype.
M_2	a 1.5-kilobase dsRNA that codes for a second toxin which kills M-o- and M_1-containing cells. M_2, [KIL-k_2], K_2^+, and R_2^+ are analogous to M_1, [KIL-k_1], K_1^+, and R_1^+.
L-A	a 4.5-kilobase dsRNA that is noninfectious and stably maintained like a plasmid, but encapsidated like a virus. It encodes its major capsid protein (81 kilodaltons). The genes [EXL], [NEX], and [HOK] (see below) are present in various combinations on various forms of L-A, denoted L-A-E ([EXL] alone), L-A-HN ([HOK] and [NEX]; found in all K_1 killers), L-A-HE ([HOK] and [EXL]), or L-A-H ([HOK] only). L-A depends on MAK3, MAK10, and PET18 for replication.
L-B and L-C	4.5-kilobase dsRNAs unrelated to L-A and present in VLPs with different major proteins. L-B and L-A or L-C and L-A are compatible. L-B and L-C show some sequence homology. L-(BC) means an L dsRNA related to L-B and L-C type.
T and W	2.76-kilobase and 2.25-kilobase minor dsRNAs. They do not cross-hybridize with each other, other dsRNAs, or cell DNA. They are cytoplasmically inherited. The copy number of T and W is induced 10-fold by growth at 37°C.
Cytoplasmic genes	
[HOK]	*h*elper *o*f *k*iller. This non-Mendelian gene supplies the helper function needed by M_1 for replication in a wild-type strain. It is located on certain forms of L-A dsRNA, namely, L-A-HN, L-A-HE, or L-A-H. To test for [HOK], the combination of L-A-E + M_1 is introduced from a *ski2* strain into the *SKI*$^+$ M-o strain to be tested. L-A-E cannot support M_1 replication in a *SKI*$^+$ host, but if [HOK] is present, M_1 will be maintained.
[EXL]	*ex*c*l*uder of M_2 dsRNA. This non-Mendelian trait prevents the replication of M_2 if [NEX] is absent, but not if [NEX] is present. [EXL] is located on certain forms of L-A, namely, L-A-E and L-A-HE. L-A-E acts by lowering the copy number of the L-A-H in the M_2 strain.
[NEX]	M_2 *n*on*ex*cludable by [EXL], but does not prevent exclusion of M_2 by strains carrying M_1. [NEX] is located on L-A-HN, the form of L-A found in wild-type K_1 killer strains.
[B]	a cytoplasmic gene which allows M_1 to dispense with certain of the *MAK* gene products and elevates the copy number of M_1, making the strain a superkiller.
[KIL-d]	a mutant form of [KIL-k_1] in which haploid strains unstably maintain and incompletely express M_1, but diploid strains stably maintain and normally express M_1.
[KIL-n]	a mutant form of M_1 dsRNA in which no active toxin is produced, but cells are immune to toxin action.
[KIL-s]	deletion mutants of M_1 which prevent the replication of (suppress) the parent M_1 dsRNA. These mutants are analogous to defective interfering particles of animal viruses.

Table 1 *(continued)*

Chromosomal genes	
MAK	*ma*intenance of [*K*IL-k$_1$]. *MAK* genes comprise at least 32 chromosomal genes necessary to maintain M$_1$ dsRNA. Mutants carrying the recessive alleles, *mak*, are K$^-$R$^-$M-o. At least some of the genes are also required for the maintenance of M$_2$. *MAK3, MAK10*, and *PET18* are needed by all forms of L-A. *MAK27* is needed by L-A-E.
clo	a complex chromosomal defect resulting in loss of L-B or L-C.
SKI	*s*uper*ki*ller. Mutants carrying the recessive alleles of *ski2, ski3, ski4, ski6, ski7*, or *ski8* produce more killer toxin and have increased copy number of M$_1$, M$_2$, L-A, and L-(BC). *ski$^-$* mutations eliminate M's need for [HOK] and for some of the *MAK* genes and prevent the exclusion of M$_2$ by L-A-HN in *mkt* strains. The *ski5* mutant has none of these traits and lacks a cell surface protease that normally degrades the toxin.
MKT	*m*aintenance of [KIL-*k*$_2$] in the presence of [NEX]. Strains having a recessive allele, *mkt* (about 80% of laboratory strains), cannot maintain [KIL-k$_2$] at 30°C if [NEX] (L-A-HN) is present.
MKS	*mkt s*uppressor. Like most *ski$^-$* mutations, mutations in *mks1, mks2*, or *MKS50* prevent exclusion of M$_2$ by L-A-HN in *mkt1* hosts, but do not show other characteristics of *ski$^-$* mutations.
KEX	*k*iller *ex*pression. Two chromosomal genes needed to process the toxin precursor. Mutants are K$^-$R$^+$. The *kex* mutants also do not properly process the α-hormone precursor, and many other secreted proteins are affected. The *KEX2* product is a protease that cleaves on the COOH side of a pair of basic residues.
REX	*r*esistance *ex*pression. One chromosomal gene needed to express M$_1$-determined resistance to the toxin.
KRE	*k*iller *r*esistant. Three chromosomal genes needed for normal toxin action on sensitive cells. *KRE1* affects β(1-6)glucan, the normal toxin cell wall receptor. *KRE2* also affects toxin binding, but *KRE3* does not.
SEC	*sec*retion. Chromosomal genes for general protein secretion; mutants in these genes are conditional lethal and also affect killer toxin secretion.

variation is probably the poly rA · poly rU region (E. M. Hannig, M. J. Leibowitz, personal communication).

Recently, Hannig & Leibowitz (36) have directly sequenced about 210 bp at each end of the 1550-bp M$_2$ dsRNA and, in addition, showed that this molecule, like M$_1$, has a substantial internal poly rA · poly rU region. While there is no cross-hybridization between M$_1$ and M$_2$, there are several short homologies, the most striking of which is the sequence 3' . . . ACACACGAGU . . . 5', which begins at base 105 from the 3' end of the message strand in both M$_1$ and M$_2$ dsRNAs (see H in Figure 1). While other short homologies were noted, this is the longest and is one of the few cases where the location is the same in M$_1$ and M$_2$ dsRNAs, suggesting that this region could be a signal for some process such as transcription, replication, packaging, etc. While little homology of

M_1 ds RNA

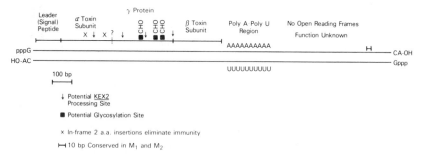

Figure 1   The top strand is the message or "+" strand. Sequence data are available for all of M_1 except for the area immediately to the right of the poly rA · poly rU region.

primary structure was found, the secondary structure of the 5' end of the message strand seems to be similar. An AUG at positions 7–9 of the message strand begins an open reading frame continuing as far as the authors have sequenced (68 amino acids). Also, in vitro translation of isolated message strand produces a main species of 28 kd and a minor 38-kd species, either of which could be the K_2 preprotoxin.

The K_1 killer toxin is made as a 34-kd precursor protein which is glycosylated at asparagine residues to form a 42-kd glycoprotein (40). The 42-kd protein is then processed to yield the mature K_1 toxin consisting of a 9.5-kd α subunit and a 9.0-kd β subunit (33), neither of which is glycosylated (41). These subunits have been located in the preprotoxin sequence (33). An amino-terminal hydrophobic signal peptide of 44 amino acids, called δ, is followed by the α toxin subunit, a region called γ, which is not incorporated into the toxin, and finally the carboxy-terminal β toxin subunit. The α-γ border is not precisely known but is estimated based on the size of α. Since all of the potential glycosylation sites are in the γ region, it has been suggested that the γ protein might remain in the cell membrane as the immunity protein after α and β are cut out to form the toxin (33). Vernet et al. (42) have recently obtained several lines of evidence that this is not the case and that the toxin precursor itself may be the immunity protein (see below).

Studies of the sequence of events in the processing and secretion of the K_1 toxin have produced important information about these processes. The hydrophobic leader (signal) peptide of 44 residues is longer than usual, and it is not clear when it is cleaved from the toxin precursor protein. Recently, Skipper et al (23) have fused the preprotoxin leader sequence (52 amino acids) to a bacterial carboxymethylcellulase gene and obtained dramatic stimulation of secretion of cellulase activity by yeast cells. Until the toxin is released into the medium, the precursor protein seems to be membrane-bound, not just inside membranes, as judged by the severe conditions required to extract it (40). Bussey and coworkers (40, 43) have used chromosomal mutants in the general secretion process, called *sec* mutants (44), chromosomal mutants with defects relatively specific for killer toxin secretion, called *kex* mutants (45, 46), and M_1 dsRNA mutants defective in toxin secretion, to examine the processing of the protoxin and secretion of toxin. Mutants in *sec18* (accumulates endoplasmic reticulum), *sec7* (accumulates golgi), and *sec1* (accumulates secretory vesicles) each are capable of making the 43-kd glycosylated protoxin. The same is true of *kex1* and *kex2* mutants. The protoxin is relatively stable in *kex1* and *sec18* mutants but is unstable in *sec7*, *sec1*, and *kex2* mutants, and *sec1* mutants even secrete a small amount of toxin. *KEX2* encodes a membrane-bound protease that cuts on the carboxy side of a pair of basic residues (47). The cut forming the amino terminus of β is such a site (33), and three other such potential *KEX2* cutting sites are shown in Figure 1. The fact that *sec18^{ts}*, *sec7^{ts}*, and *sec1^{ts}* mutants are defective in secretion of toxin at the nonpermissive temperature shows that the protoxin traverses the normal secretion pathway. That *kex1* and *kex2* mutants do not secrete toxin but do grow normally shows that these gene products are somewhat specific for certain secreted products. The *kex2* mutants also fail to secrete normal α factor (46), the pheromone produced by α mating-type cells to prepare *a* mating type cells for mating. Instead they secrete a highly glycosylated form of prepro-α factor (47).

Secretion of active toxin also requires a TPCK-sensitive protease step (43). The *KEX2* protease is not sensitive to TPCK (47), so this must be a distinct processing step from that catalyzed by *KEX2*. TPCK, a chymotrypsin inhibitor, stabilizes protoxin in wild-type cells as well as in *sec7* and *sec1* mutants, but not in *kex2* mutants (43).

Among mutants of M_1 dsRNA failing to secrete any toxin protein was one whose protoxin showed greater than normal stability. This mutant may define a recognition site for one of the processing enzymes (40).

Using a cDNA clone of the preprotoxin sequence, it has been shown that removing nine amino-terminal amino acids and changing three others at the amino terminus results in a protoxin that is not detectably glycosylated, but this protein is nonetheless processed to yield active toxin and makes the cell immune (48). This indicates that the carbohydrate portion of the protoxin is probably not essential for either immunity function or for processing.

It is not yet possible to make a detailed model of protoxin processing, but it seems likely that protoxin follows the standard secretion pathway and that most proteolytic events occur in the golgi or secretory vacuoles or both. The *kex2* mutants and TPCK inhibition define two of the enzymes involved.

After the toxin has been secreted from the plasma membrane, there are two processes that reduce the amount that appears in the medium. The *SKI5* gene controls a phenyl methylsulfonyl fluoride-sensitive protease, located on the cell surface, which degrades a substantial proportion of secreted toxin molecules (93). Thus *ski5* mutants have the superkiller phenotype. As discussed below, $\beta(1–6)$D-glucan linkages in the cell walls are an essential element in the toxin's binding to sensitive cells (94). Killer cells are immune but nonetheless have these linkages. Some secreted toxin is bound by $\beta(1–6)$D-glucan linkages and does not appear in the medium. The *kre1* mutants (95), which largely lack this linkage, secrete an increased amount of toxin and are superkillers (93).

IMMUNITY PROTEIN   Immunity to the killer toxin is encoded in the pre-protoxin gene as shown by the fact that the cDNA clone carrying only this sequence gives immunity to cells (48). Since the α and β regions are part of the mature toxin molecule, it was suggested that γ might be the immunity function (33). Several lines of evidence suggest that this is not the case and that the protoxin itself may be the immunity protein (42). Mutants of M_1 dsRNA that produce apparently unprocessed protoxin do not secrete toxin but remain immune to the toxin (40). Similarly, the *kex1* and *kex2* mutants produce no toxin but make protoxin and are immune (43).

Site-specific mutations introduced into various points in a cDNA clone of preprotoxin have been used to determine sites required for immunity function (42). An in-frame deletion of the middle one third of the γ region from the *Acc*I to *Hinf*I sites (Figure 1) eliminates toxin production and results, of course, in a shorter protoxin but does not affect immunity. However, in-frame insertions of Leu-Glu into either of the two hydrophobic domains of α (after Val-85 or after Val-116) eliminate immunity and substantially reduce toxin production. Immunity was also reduced by the deletion of the leader with replacement of the first seven amino acids of α with Met-Gly-Asp-Glu. Deletion of the β region also does not affect immunity. Vernet et al (42) attempted to detect the putative γ protein using antibodies to synthetic peptides of γ. These antibodies could detect toxin precursor, but no γ protein was detected. These results, taken together, suggest that the protoxin, and not γ, is the immunity factor and that the α domain, and conceivably the amino-terminal one third of γ, are important for this activity. Vernet et al suggest that the protoxin could, for example, bind to toxin receptors on the cell membrane much more tightly than toxin itself but could be unable to carry out the toxin action and would thus prevent the active toxin from binding and acting.

K_1 TOXIN ACTION   The receptor for the K_1 killer toxin is the $\beta(1–6)$D-glucan linkages in the cell wall (94, 96). This structure specifically binds the K_1 toxin and has been used in a purification procedure for the K_1 toxin (94). Mutants in the *kre1* or *kre2* genes (95) have a substantial reduction in the amount of this linkage in their glucan and as a result are resistant to toxin action even though they carry no M_1 genome (94). The K_1 toxin causes rapid inhibition of net proton extrusion and potassium uptake by sensitive cells (97, 98). Soon the inside-outside difference in pH decreases and, later, potassium and ATP efflux are observed along with cell death and cessation of macromolecular synthesis (99, 100). The action of the toxin involves the *KRE3* gene product (95), but the point at which this gene is involved is not yet defined. Likewise, the mechanism of M_1-determined immunity, which requires the chromosomal *REX1* gene (101), is not yet known, but Vernet et al (42) have speculated that it may involve binding of the protoxin to the toxin receptor without toxin action (see above).

The L-A Family: [HOK], [NEX], [EXL], and Viruslike Particle Coat Protein

Most *Saccharomyces* strains, whether killers or not, have a 4.5-kb linear dsRNA called L (1, 2). Recently it was recognized that L dsRNA comprised two unrelated families of molecules, called the L-A family and the L-BC family (3, 6, 49). Some strains have a member of only one family, some have only a member of the other, while many have one of each, and a few have neither. L-A encodes the major coat protein of the viruslike particles in which both it and M dsRNA are encapsidated, a protein of 81,000 daltons (3, 50). There are several natural variants of L-A which, while related to varying degrees in sequence, show markedly different genetic activity (3, 5, 6, 49, 51) (Table 2). The cytoplasmic genes [HOK], [NEX], and [EXL] (described below) are found in various combinations on various natural variants of L-A. For example, L-A-HN has [HOK] and [NEX], while L-A-E has only [EXL].

[HOK] is a function of L-A that is necessary for the maintenance of M_1 or M_2 in a wild-type host (6, 52). This function is, presumably, the ability of L-A to

Table 2 Comparison of forms of L-A dsRNA

	Ability to maintain			Ability to protect M_2 against L-A-E	Copy number	Comments
	M_1		M_2			
	SKI^+ host	ski^- host	SKI^+ host			
L-A-HN	+	+	+	+	High	Found in all wild-type K_1 killers and many nonkillers
L-A-HE	+	+	−		High	Found in strains 200 and 201
L-A-H	+	+	+	−	Medium	Found in some K_2 killer strains
L-A-E	−	+	−		Low	Found in strain AN33

provide the major viruslike particle (VLP) coat protein, although there are other possibilities. As will be discussed further below, in a *ski⁻* host, [HOK] is not needed for M replication or maintenance (6), so L-A's role for M replication may be to antagonize the negative effects of *SKI* products. L-A-E, like the other L-A variants, encodes a VLP protein of about 81,000 daltons (3), but does not provide the [HOK] function essential for M_1 or M_2. This may be because the coat protein is qualitatively or quantitatively inadequate to encapsidate M dsRNA, or because the copy number of L-A-E is lower than that of L-A species having [HOK].

[EXL] is defined as the ability of certain L-A's to exclude M_2 from a strain not carrying [NEX] (5). [NEX] was at first defined as a cytoplasmic gene preventing [EXL] action (5). Later, a second action of [NEX] became apparent, namely, the exclusion, at 30°C or above, of M_2 from strains carrying a recessive allele of the chromosomal *mkt1* or *mkt2* genes (5, 51). M_1 is not so excluded by [NEX] in an *mkt⁻* host, hence the name of these genes (*m*aintenance of *K*-*t*wo).

Natural variants of L-A include those with [EXL] only (L-A-E), [HOK] only (L-A-H), [HOK] and [NEX] (L-A-HN), and [HOK] and [EXL] (L-A-HE). All wild-type K_1 killers examined so far carry L-A-HN, while the lone wild-type K_2 killer studied had L-A-H (3, 51, 53). The L-A-HN originating from a K_1 strain can maintain M_2, while L-A-H, from a K_2 strain, can also maintain M_1 (53). Thus, the previous use of the names L_1 and L_2 (54) or L_{1A} and L_{2A} (14) to designate the L-A's derived from K_1 and the single K_2 strain studied is confusing. While the structural differences observed are substantial, they may be due to genetic drift or selective factors other than the strain's carrying M_1 or M_2.

A number of partial cDNA clones of L-A have recently been isolated (55) and sequenced (56). These reveal the presence of a 170-bp inverted repeat separated by 328 bp of nonrepeated DNA located about 2.3–2.5 kb from the end of L-A at which transcription starts. The 170-bp inverted repeat has stop codons in all reading frames in one direction (56). Since yeast ribosomes usually translate only the first open reading frame from the 5' end of the message, the location of these stop codons suggests that, barring splicing or other complexities, the coat protein gene is located at the 5' end of the message strand. As in the case of M_1, where the right-hand end of the molecule has no large open reading frames, the right-hand end of L-A may not have a protein coding function.

L-B and L-C dsRNA

Most *Saccharomyces* strains, whether killers or not, carry a species of dsRNA essentially the same size as L-A, but with no apparent homology to L-A as judged by blotting-hybridization experiments (3). Two of these, called L-B and L-C, yield dramatically different fingerprint patterns after ribonuclease T_1 digestion, but nonetheless have significant homology by hybridization both to each other and to the non-L-A L-sized dsRNAs present in many other strains (3). This indicates that L-B and L-C are members of another family of related molecules. In this case, the functional aspects of the heterogeneity are unknown, and it is not clear whether this L-BC family has any functional relation to the killer phenomena. Since some K_1 killer strains lack L-BC entirely (3), it is clear that M_1 does not need L-BC for its maintenance or replication.

L-B and L-C, like L-A and M, are found in intracellular VLPs. These particles, like those encapsidating L-A and M, have an RNA polymerase activity that produces single-stranded transcripts, but the major coat proteins of the particles are smaller, about 77,000 and 73,000 daltons (3). These coat proteins are encoded by the L-BC molecule (57).

L-A VLPs and L-BC VLPs are each found in two sizes, the large sedimenting at about 160S, and the smaller sedimenting at about 80S (3). Both large and small particles have RNA polymerase activity (3). In strains carrying both, L-A is primarily in large particles, while L-BC is primarily in the small particles (38).

T and W dsRNAs

Recently, two new species of dsRNA called T (2.7 kb) and W (2.25 kb) have been found in many *Saccharomyces* strains (4). T and W have no homology, by Northern blots, with each other or with L-A, L-BC, M_1, or M_2 dsRNAs or with cellular DNA. They are present in very low copy number in cells grown at 20–30°C, but their copy number, in many strains, increases by 10- to 20-fold when cells are grown at 37°C (4). There are other strains in which T and W are not heat-inducible, and crosses show that this heat inducibility is dominant and controlled by a nonchromosomal genetic element. This is just the opposite of M and L-A dsRNAs, which are readily cured by growth of cells at 37–39°C (49, 58). T and W are inherited as cytoplasmic factors but do not seem related to previously described non-Mendelian genes or molecules. While no functional relation has yet been established with the killer systems, all killer strains examined to date carry W dsRNA, suggesting a possible relationship (4).

As described below, there are many genetic and physiologic factors that modify the copy numbers of the various dsRNA species. However, a typical K_1 killer strain grown to stationary phase on rich medium at 30°C could have about 1000 copies of L-A, 150 copies of M_1, 100 copies of L-BC, 20 copies of T, and 20 copies of W per cell.

TRANSCRIPTION OF DOUBLE-STRANDED RNA

The VLPs encapsidating L-A and M dsRNAs have an RNA polymerase activity that synthesizes a full-length plus strand (message strand) copy of each dsRNA (59–63). These single-stranded RNAs (ssRNAs) are released from the VLPs in vitro (60). In vivo both full-length and smaller plus strand copies of L-A and M_1 have been observed, and each of these species has message activity in vitro (64, 73). It is not clear whether the partial-length in vivo molecules are primary transcripts, breakdown products, replication intermediates, or pause products in the synthesis of full-length molecules. Both the full- and partial-length in vivo single-stranded M_1 molecules have poly A regions (64), probably due to copying of the poly A · poly U region of the M_1 dsRNA (65).

The in vivo role of the VLP RNA polymerase producing single-stranded L-A, L-BC, and M copies could be to produce transcripts for translation purposes, to make replication intermediates, or both. In the case of L-A, pulse-chase experiments and strand-separation gels show that the minus (anti-message) strand of L-A dsRNA is labeled 10 minutes earlier than the plus (message) strand (66). This means that cold plus strands synthesized before the pulse are incorporated into L-A dsRNA molecules before those being labeled during the pulse. This indicates that the L-A plus strand is a replication intermediate in vivo (66).

It is not yet clear what gene(s) encode the VLP RNA polymerase(s), and this is now one of the most exciting questions in this system. Candidates include the dsRNAs themselves and the large number of chromosomal genes known to be involved in their replication and maintenance.

REPLICATION OF DOUBLE-STRANDED RNA

L-A and M_1 show an apparently conservative mode of replication (67, 73) like reovirus (69) and unlike the semiconservative replication of *Penicillium stoloniferum* dsRNA virus (70, 102). In the case of L-A, it has been shown that the plus strand is made before the minus strand (66). Presumably, the plus strands produced by the VLP RNA polymerase conservatively copying the double-stranded template then serve as a template for minus-strand synthesis.

Calculations, based on particle sedimentation rate, radius, and density, such as those of Bruenn (71) following Luria et al (72), show that L-A VLPs are about 23% RNA and have a molecular mass of 1.22×10^7 daltons. This means that there are 2.8×10^6 daltons of L-A per particle or, since L-A has a molecular mass of 3.0×10^6, just one L-A dsRNA molecule per particle.

Recently, evidence for structural and functional heterogeneity of VLPs containing M_1 has been found (R. Esteban, R. B. Wickner, submitted for publication). CsCl density gradient centrifugation shows two peaks of M_1 VLPs. The dense (heavy) peak (M_1-H) VLPs have two M_1 dsRNA molecules per particle, while the lighter peak (M_1-L VLPs) have only one M_1 dsRNA per VLP. The M_1-H, M_1-L, and L-A VLPs are about the same size by electron microscopy.

The M_1-H VLP RNA polymerase produces only ssRNA in vitro, all of which is extruded from the particles, while the M_1-L VLPs produce some extruded ssRNA and some RNA that remains within the particles. The product remaining in the particles is a mixture of single- and double-stranded material, the dsRNA being the size of M_1 (R. Esteban, R. B. Wickner, submitted for publication). The peak of M_1 dsRNA synthesis is in particles lighter than M_1-L VLPs, presumably those carrying only a single (+) M_1 strand.

L-A VLPs have only one L-A dsRNA per particle. No particles are found with two L-A dsRNAs per particle. All of the ssRNA polymerase product of L-A VLPs is extruded into the reaction mixture. Unlike in M_1-L VLPs, none is retained within the particles. We suggest that this difference reflects simply the inherent capacity of the particle for dsRNA—a sort of "head-full" hypothesis. We suggest that the L-A-encoded major coat protein forms a structure primarily adapted to hold one L-A dsRNA and that it can hold no more, not even another L-A ssRNA transcript. M_1 is less than half the length of L-A and so two M_1 dsRNA molecules will fit comfortably inside one VLP, but the new (+) ssRNA transcripts are all extruded for lack of space. Particles with only one M_1 dsRNA molecule are not crowded and so retain some of their (+) ssRNA transcripts which can then be copied to make a second M_1 dsRNA molecule inside the same particle (R. Esteban, R. B. Wickner, submitted for publication).

Recently, VLPs synthesizing a complete L-A (−) strand to form L-A dsRNA in vitro have been detected and separated by CsCl gradients, from VLPs producing (+) single strands (T. Fujimura, R. B. Wickner, in preparation). The (−) strand-producing particles contain only a (+) single strand. After they have synthesized a (−) strand, they then begin producing (+) single strands which are extruded from the particles. This indicates that a single particle has the RNA polymerase activities making both (+) and (−) strands. Whether these are the same or different proteins remains to be seen.

L-BC VLPs synthesizing L-BC dsRNA have likewise been isolated and separated from those making full-length ssRNA transcripts (T. Fujimura, R. B. Wickner, in preparation).

A disagreement exists as to whether L dsRNA (probably L-A) replication is influenced by the cell cycle. One study (67) indicates no influence of time in the cell cycle on the rate of L dsRNA synthesis, while a second study (68) concludes that L is synthesized in G1 but not during S phase. Whether the disagreement is due to methodological problems, a strain difference, or some other factor is unclear. Cell cycle arrest has a marked effect on the amount of single-stranded L and M RNA in the cell (73). Arrest of cells at G_o using the pheromone α factor results in a 3-fold increase of L and M free full-length plus strands, but arrest in G1 produced by shifting a $cdc7^{ts}$ mutant to the nonpermissive temperature produced a further 8-fold accumulation of L plus strands and a further 280-fold accumulation of M_1 plus strands (73).

A large number of chromosomal genes are involved in the replication or maintenance of specific dsRNAs (Table 3). At least 30 genes (called *MAK* genes) are needed for M_1 replication (74–78). Among these, *MAK3, MAK10,* and *PET18* are required for L-A (3, 6), but none of these 30 are needed for T, W, or L-BC [(3, 4) and R. B. Wickner, unpublished observations]. *MAK8* encodes ribosomal protein L3 (79), and *MAK1* is the topoisomerase I gene (80). The polyamines spermidine or spermine are needed for M_1 and M_2; putrescine does not suffice (81, 82). L-A-E needs polyamines, but in this case putrescine is sufficient (82).

Each of the three independent *pet18* mutants studied have three phenotypes, namely, temperature-dependent loss of L-A and M dsRNA, loss of mitochondrial DNA, and temperature-sensitive cell growth (30, 74, 83). When L-A VLPs isolated from wild-type cells are compared with those isolated from *pet18* strains grown at the permissive temperature for L-A replication, those from the mutant were relatively temperature-unstable with release of L-A dsRNA at the elevated temperature (T. Fujimura, R. B. Wickner, in press). M_1-containing VLPs were not so affected, showing the same stability whether from mutant or wild-type cells. This is the first case in which evidence has been obtained for the nature of the role of a chromosomal gene in dsRNA replication. It also suggests that M and L-A VLPs, while sharing the L-A-encoded major VLP protein, must have important structural differences.

In order to investigate the pleiotropic effects of *pet18* mutations, Toh-e & Sahashi (92) have cloned and sequenced a DNA segment that complements the temperature-sensitive cell growth defect and the defect in M and L-A maintenance of *pet18* mutants. The smallest clone complementing the temperature-sensitive and *mak⁻* phenotypes had four long open reading frames. Complementation of the cell growth and dsRNA replication defects of *pet18* mutants

Table 3 Dependence of dsRNAs on chromosomal genes

dsRNA	MAK3, MAK10, PET18[a]	MAK1 (TOP1), MAK8 (TCM1), SPE2, many other MAK genes	MAK27, SPE10	MKT1, MKT2	CLO
M_1	+[b]	+	+	−	−
M_2	+	+	+	+	−
L-A-HN	+	−	−	−	−
L-A-E	+	−	+	−	−
L-BC	−[b]	−	−	−	+
T	−	−	−	−	−
W	−	−	−	−	−

[a]*PET18* is also needed for maintenance of mitochondrial DNA, but this is a comlpex locus with several genes (see text).

[b]"+" means the indicated chromosomal gene(s) is (are) needed for replication or maintenance by the indicated dsRNA. "−" means there is no such requirement.

both required at least two of these genes (open reading frames), which they named *mak31* and *mak32*. Each was needed for both cell growth and M_1 dsRNA replication. Furthermore, the several *pet18* mutants studied are in fact all large deletions missing more than 3 kb, including the entire *mak31* and *mak32* sequences (92). The loss of mitochondrial DNA is due to a completely different gene, also presumably deleted in the same mutants (92). I. Chou and G. R. Fink (personal communication) have also cloned this region and conclude that the *pet18* deletion is due to recombination between nearby repeated Ty1 elements. Thus, the *MAK31* and *MAK32* products could be regarded as genetically defined heat shock proteins, that is, proteins needed for cell growth only at high temperature. It is not yet known whether their synthesis increases at elevated temperature.

The results of Toh-e & Sahashi (92) also have important implications concerning the VLP stability experiments discussed above. Usually a thermolabile structure obtained from a temperature-sensitive mutant would be interpreted to mean that the mutant gene product was present in the structure. In this case, a complete deletion is producing a temperature-sensitive phenotype and thermolabile particle structure. Clearly, further work should be directed at locating the *MAK31* and *MAK32* gene products and at determining which deficiency—*mak31* or *mak32*—is responsible for the thermolabile VLPs.

Cloning and sequence analysis of other *MAK* genes is also proceeding. *MAK8 (TCM1)* (104, 105), *MAK1* (DNA topoisomerase I) (106), *SPE2* (S. K. Taneja, C. W. Tabor, H. Tabor, personal communication), *SPE10* (107), *MAK16* (R. B. Wickner, J. Crowley, D. Kaback, unpublished observations), *MAK11* (T. Icho, R. B. Wickner, unpublished observations), *MAK18* (H.-S. Lee, R. B. Wickner, unpublished observations), and *MKT1* (R. B. Wickner, T. J. Koh, unpublished observations) have all been cloned and in some cases sequenced.

Many of the *mak⁻* mutations result in defects in cell growth in addition to their defects in M replication. The *mak16-1* mutation (77), *mak30-1* mutation [(74) and A. Sugino, unpublished observations], and the *pet18* mutation (83) result in temperature-sensitive cell growth, while mutation of *mak6* (= *lts5*) results in cold sensitivity for growth (52). Mutations in *mak13, mak15, mak17, mak18, mak20, mak22,* and *mak27* result in slow cell growth at any temperature (77).

Although no *MAK* genes have been found necessary for L-BC, T, or W replication, a complex chromosomal defect called *clo⁻* which results in loss of L-BC has been described (4). As yet, no chromosomal genes involved in T or W replication have been described, probably because there is as yet no phenotype associated with these molecules that could be used to find mutants affecting their replication. As described previously, L-A-HN ([NEX]) excludes M_2 in strains carrying the recessive allele of either *MKT1* or *MKT2* (5, 51). However, M_1 does not appear to require the *MKT* gene products.

In analogy with the defective interfering deletion mutants of animal viruses, deletion mutants of M_1 dsRNA have been described and were discussed above. Insofar as has been tested, they require the same chromosomal *MAK* gene products as does the parental M_1 dsRNA (85).

Although L-A, L-BC, M, T, and W dsRNAs are all linear molecules, they seem to have different requirements for chromosomal genes. This is perhaps reminiscent of bacterial phage and plasmid systems in which replicons that are structurally grossly similar use different enzyme systems for replication.

REGULATION OF DOUBLE-STRANDED RNA REPLICATION

The factors determining the copy number of the various dsRNAs include the chromosomal *MAK, SKI, PET18, CLO, MKT, SPE, MKS*, and *KRB* genes and the cytoplasmic factors [EXL], [NEX], [HOK], and [B] (formerly [KIL-b]).

The *ski⁻* mutants were first isolated based on their superkiller phenotype (86). Afterward, they were found to have a number of other phenotypes:

1. Increased copy number of M_1, M_2, L-A, and L-BC dsRNAs. This means that the *SKI* products repress replication of these dsRNAs (52, 86, 87).

2. Suppression of many of the *mak⁻* mutations (88). This could mean that these *MAK* products are repressors of the *SKI* products—which are themselves repressors of dsRNA replication. This would imply that the *mak⁻* mutants would affect the copy numbers of M, L-A, and L-BC as do the *SKI* products. However, insofar as is known, no *MAK* genes affect L-BC copy number (87). This also suggests that *ski⁻* and *mak⁻* mutants are not simply different types of mutations in the same genes. The *mak16-1* mutation is epistatic to all *ski⁻* mutations tested.

3. Making M_1 maintenance and replication independent of [HOK], the function of L-A that is normally essential for M_1 replication (6, 52). In *SKI⁺* strains, L-A-E cannot maintain M_1, but it can do so in *ski⁻* strains (52). Other strains have been isolated having no [HOK], [NEX], or [EXL] activity and only very low copy number of L-A dsRNA in which M_1 is stable only if the strain is *ski⁻* [(52, 53) and T. Fujimura, R. B. Wickner, unpublished observations]. The role of L-A in M replication may be to protect M from repression by the *SKI* products, perhaps by the L-A-encoded coat protein directly shielding M from *SKI* repression. Alternatively, perhaps the more abundant L-A soaks up *SKI* products so they cannot act on M.

4. M_2 normally needs *MKT1* if L-A-HN ([NEX] function) is present, but not if the host is *ski⁻* (52). Using this property, 65 *ski⁻* mutants were isolated. Complementation tests (52) showed that these comprised six genes (*ski2*,

ski3, ski4, ski6, ski7, and *ski8*) which have now all been mapped [(86) and R. B. Wickner, unpublished observations]. Other mutations showing property 4, but not the other properties of *ski⁻* mutants, comprised dominant mutations at one locus *(MKS50)* and recessive mutations at two other loci *(mks1* and *mks2)* (52).

5. Cold sensitivity for cell growth (at 8°C) if and only if an M replicon is present (52). This effect is seen as well with M_1, M_2, or with S_3, a deletion mutant lacking nearly all of the preprotoxin coding sequence (see above). This means that the cold sensitivity for growth is not due to overproduction of the toxin or of immunity protein. Rather, it seems to be due to elevated levels of an M replicon (52). We suspect that some gene product critical for cell growth at low temperatures and for M replication (e.g. *MAK6 = LTS5*) is used up by the high copy M replicon, leaving none for the cell to use for growth. Elimination of M results in derepression of L-A replication (see below) and thus much higher levels of total dsRNA than with M present. Nevertheless, such *ski⁻* L-A M-o strains are not detectably cold sensitive for growth. Thus the total dsRNA load is not critical, just the load of M replicons.

The *SKI8* gene has recently been cloned (S. S. Sommer, R. B. Wickner, in preparation), and it is expected that analysis of this clone will shed light on the nature of the *SKI8* protein.

Another dsRNA repression effect is the lowering of L-A copy number by M dsRNA (87). Elimination of M results in a 5- to 10-fold increase of L-A dsRNA. However, M does not repress or otherwise alter the level of L-BC. This effect, too, is probably an effect of an M replicon rather than of an M-encoded product, since suppressive mutants, such as S_3, lacking nearly the entire preprotoxin gene—the only known large open reading frame—show the same extent of repression of L-A as their parent M_1 dsRNA (87).

In a wild-type killer strain there are thus two factors repressing L-A, namely *SKI* products and M dsRNA. Eliminating the *SKI* products (by making the strain *ski⁻*) eliminates a repressor of L-A, but, because the copy number of M increases, the copy number of L-A actually decreases about 10-fold. This indicates that M repression of L-A is more powerful than *SKI* repression of L-A (87).

Another factor repressing L-A is growth in glucose, which produced a 2- to 3-fold lowering of L copy number (probably L-A) in a strain lacking M (89). L-BC apparently does not repress L-A (T. J. Koh, R. B. Wickner, unpublished observations).

The *KRB1* mutation (90) is an interesting genetic puzzle. Introduction of M_1 into a *mak7-1* mutant generally results in loss of M_1 as the cells grow, but, with surprising frequency, dominant suppressor mutations arise. These mutations are called *KRB1* (for *k*iller *r*eplication *b*ypass). (Note that, following standard nomenclature for *Saccharomyces,* the dominant allele, which is the mutant in this case, is written in capital letters.) Crosses of the type *KRB1 mak7-1* K^+ × *krb1 mak7-1* K^- produce uniform 2 K^+ : 2 K^- segregation with such tight centromere linkage that it is not clear that any second division segregations have been observed in over a thousand tetrads examined. It was unlinked to the known chromosomal centromeres and defines a new centromere (90, 91). By genetic criteria it is a new (17th) chromosome (91).

The large number of yeast genes mapped make it somewhat surprising to find a new chromosome. M dsRNA can be eliminated (cured) from killer strains either by growth in the presence of sublethal concentrations of cycloheximide or by growth at elevated temperatures (39–40°C) (see above). Surprisingly, elimination of M_1 from *mak7-1 KRB1* K_1^+ strains by either of these means appears to eliminate the *KRB1* mutation (90). The latter finding would suggest that *KRB1* is a mutation of M_1 or of M_1 and a chromosomal gene, but a normal *MAK7*-dependent M_1 is present in *krb1* segregants of a cross of the type *mak7 KRB1* K_1^+ × *MAK*$^+$ *krb*$^+$ M-o. No DNA copy of M_1 was detected in *KRB1* or wild-type strains. The solution of this puzzle probably awaits a more detailed knowledge of M_1 replication and the role of the *MAK7* gene in this process.

M_2 dsRNA can be excluded (eliminated) from cells by three different mechanisms involving three different dsRNAs (51):

1. M_1 excludes M_2 (5, 7, 51). This presumably is due to the competition between M_1 and M_2 for replication factors, such as the many *MAK* gene products they both use.
2. In an *mkt1* or *mkt2* strain, L-A-HN (the [NEX] function) excludes M_2 (3, 5, 49). Here again, one may hypothesize that L-A-HN and M_2 compete for the *MKT* gene products and that M_2 loses this competition. The fact that M_1 and M_2 lower L-A copy number (87) also suggests competition for a factor that both require. However, the *MKT* alleles available do not appear to affect L-A-HN copy number. A deletion mutant of *mkt1* can be constructed, as the *MKT1* gene has recently been cloned (R. B. Wickner, T. J. Koh, unpublished).
3. L-A-E excludes M_2 (3, 5, 51, 53). This exclusion occurs even in an *MKT*$^+$ strain and so differs from the L-A-HN exclusion of M_2 which is observed only in *mkt*$^-$ strains. In addition, L-A-E only excludes M_2 from a strain carrying M_2 and L-A-H, not from one carrying M_2 and L-A-HN. When mutants resistant to L-A-E action were selected from an M_2 L-A-H-carrying strain, they were found to have an altered L-A (called L-A-HR), suggesting that exclusion by L-A-E involves its action on L-A-H (51). Further studies showed that L-A-E dramatically lowers the copy number of L-A-H (without actually excluding it), but not that of L-A-HR, the mutant molecule resistant to the action of L-A-E (53). L-A-E is also unable to lower the L-A-HN copy

number. Thus, it seems that L-A-E excludes M_2 by lowering the L-A-H copy number. L-A-E cannot provide the helper function ([HOK]) for M_1 or M_2 (see above), but L-A-H or L-A-HN can. By lowering L-A-H copy number, L-A-E prevents L-A-H's providing M_2 with sufficient [HOK] function and M_2 is lost. Consistent with this explanation is the finding that L-A-E can exclude M_1 from a specially constructed strain carrying M_1 and L-A-H, but not from wild-type K_1 killers which have M_1 and L-A-HN (53).

Thus far, all the cytoplasmic factors affecting M replication have been located on either M itself or on some form of L-A. However, a cytoplasmic factor, called [B] (originally [KIL-b]), has been found that makes the products of *MAK11, MAK4, MAK7,* and *MAK17* dispensable for M_1 replication (88). This cytoplasmic factor also increases the copy number of M_1 about 4-fold (87) and makes the strain a superkiller. Recent experiments have shown that this factor is not located on M_1 or on L-A (R. Esteban, R. B. Wickner, unpublished observations).

The pattern of suppression of *mak$^-$* mutations by [B] and by *ski$^-$* mutations can be used to classify *mak$^-$* mutations. The *mak16-1* mutation is not suppressed by any of these factors and, among genes needed for M_1, is so far unique in this regard. The *mak3, mak10,* and *pet18* mutations, needed for both L-A and M replication, are suppressed by *ski1-1,* but not by other *ski$^-$* mutations. The other *mak$^-$* mutations are suppressed by all *ski$^-$* mutations. [B] suppresses all the *mak$^-$* mutations tested except *mak16-1, mak3, mak10, pet18, mak18, mak19, mak21, mak26,* and *mak27* mutations [(87, 88) and R. B. Wickner, unpublished data]. The biochemical significance of this phenomenon is not yet known.

CONCLUDING REMARKS

Figure 2 shows a heuristic model that summarizes some aspects of the genetic control of dsRNA replication. The *SKI* genes protect the cells from the deleterious effects of high levels of M replicons. But why are even higher levels of L-A apparently harmless? We suppose this is because the rate-limiting host proteins used by L-A for its replication are not essential to the host, but one or more of those used by M is essential to the host and limited. This makes control of M copy number most critical and, indeed, *SKI* products repress M copy number more dramatically than they do L-A.

According to our current models for the replication of dsRNAs, there is, in each case, conservative transcription of dsRNA to yield ($+$) strands which are used as a template to make ($-$) strands yielding a new dsRNA. For M, L-A, and L-BC, both of these steps have been observed to occur in vitro in VLPs. In view of the dramatically different genetic control of M, L-A, L-BC, T, and W

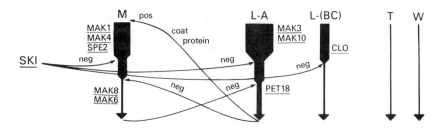

REPLICATED dsRNAs

Figure 2 Model of dsRNA replication control. The attenuation of dsRNA copy number by *SKI* products or other dsRNAs is indicated schematically by the thickness of the corresponding arrow. Only a few of the many *MAK* genes affecting M replication are shown on the diagram. Current information does not yet allow distinction between those directly involved and those indirectly involved through regulatory effects, for example.

replication, it would indeed be surprising if their mechanisms of replication, maintenance, and/or segregation were not significantly different. It should be pointed out that the genes encoding the various VLP RNA polymerase activities discussed above are completely unknown.

In view of the genetic interactions described here, we can imagine that M's interactions with the host evolved by a series of steps. Early M molecules arose from an mRNA or by infection from outside and developed more and more replication ability (i.e. higher and higher copy number) using a few host proteins (like *MAK16*) to get themselves duplicated. Selection was favored by increased toxin production, and spreading was by cell mating. The high copy number, however, caused cell death at low temperatures, and cells were selected that had evolved a means (*SKI* genes) to control or block M replication. The *SKI* products eliminated M from the cells, except for those M molecules clever enough to learn to use various *MAK* genes and L-A to partially antagonize the *SKI* products. Modern yeast cells have reached an equilibrium M copy number based on the cost-benefit balance between the pathology and energy load M replication and toxin production place on the cell and the advantage of being able to kill other strains of the species competing for an ecological niche.

The current state of knowledge of the killer system dsRNA replication includes (*a*) VLP enzyme systems apparently capable of carrying out all the steps of M, L-A, and L-BC replication and (*b*) extensive genetics that has defined a large number of chromosomal and cytoplasmic genes controlling the replication processes. Many of these gene products will undoubtedly be part of the nuts-and-bolts of replication, while others will be controlling the process in various ways.

Using mutants and the in vitro systems, a definition of the roles of the nuts-and-bolts gene products should be possible. Several clones of the chromosomal genes affecting the various dsRNAs are now in hand and should make possible elucidation of the mechanisms of the interactions among these genes. For example, suppression by ski^- mutations of many of the mak^- mutations could be due to MAK products repressing SKI products, which, in turn, repress M replication. This can now be tested directly.

Yeast dsRNAs provide a rich system for the study of viral-host interactions and other problems, as outlined in the Perspectives and Summary section above. A combined genetic, enzymologic, and molecular cloning approach is beginning to yield important insights.

ACKNOWLEDGMENTS

The author wishes to thank his coworkers and collaborators, past and present, particularly Mike Leibowitz, Akio Toh-e, Susan Porter Ridley, Steve Sommer, Steven Ball, Micheline Wésolowski, François Hilger, Françoise Boutelet, Murray Cohn, Anil Tyagi, Celia Tabor, Herb Tabor, Ernie Hannig, Tsutomu Fujimura, Rosa Esteban, Tateo Icho, and Hyun-Sook Lee.

Literature Cited

1. Bevan, E. A., Herring, A. J., Mitchell, D. J. 1973. *Nature* 245:81–86
2. Vodkin, M., Katterman, F., Fink, G. R. 1974. *J. Bacteriol.* 117:681–86
3. Sommer, S. S., Wickner, R. B. 1982. *Cell* 31:429–41
4. Wésolowski, M., Wickner, R. B. 1984. *Mol. Cell. Biol.* 4:181–87
5. Wickner, R. B. 1980. *Cell* 21:217–26
6. Wickner, R. B., Toh-e, A. 1982. *Genetics* 100:159–74
7. Naumova, T. L., Naumov, G. I. 1973. *Genetika* 9:85–90
8. Young, T. W., Yagiu, M. 1978. *Antonie van Leeuwenhoek J. Microbiol. Serol.* 44:59–77
9. Extremera, A. L., Martin, I., Montoya, E. 1982. *Curr. Genet.* 5:17–19
10. Hara, S., Iimura, Y., Otsuka, K. 1983. *J. Jpn. Soc. Agric. Chem.* 57:897 (From Japanese)
11. Herring, A. J., Bevan, E. A. 1974. *J. Gen. Virol.* 22:387–94
12. Bostian, K. A., Hopper, J. E., Rogers, D. T., Tipper, D. J. 1980. *Cell* 19:403–14
13. Somers, J. M., Bevan, E. A. 1969. *Genet. Res.* 13:71–83
14. Tipper, D. J., Bostian, K. A. 1984. *Microbiol. Rev.* 48:125–56
15. Wickner, R. B. 1985. *Cell Technol.* 4:312–21 (From Japanese)
16. Wickner, R. B. 1985. *Current Topics in Medical Mycology,* ed. R. McGinnis, 1:286–312. New York: Springer-Verlag
17. Bussey, H. 1984. *Microbiol. Sci.* 1:62–66
18. Stumm, C., Hermans, J. M., Middelbeek, E. J., Croes, A. F., de Vries, G. J. M. L. 1977. *Antonie van Leeuwenhoek J. Microbiol. Serol.* 43:125–28
19. Lemke, P. A., ed. 1979. *Viruses and Plasmids in Fungi.* New York: Dekker. 653 pp.
20. Gunge, N. 1983. *Ann. Rev. Microbiol.* 37:253–76
21. Buck, K. W. 1979. See Ref. 19, pp. 1–42
22. Kandel, J. S., Stern, T. A. 1979. *Antimicrob. Agents Chemother.* 15:568–71
23. Skipper, N., Sutherland, M., Davies, R. W., Kilburn, D., Miller, R. C., et al. 1985. *Science.* 230:958–60
24. Ouchi, K., Akiyama, H. 1976. *J. Ferment. Technol.* 54:615–23
25. Ouchi, K., Wickner, R. B., Toh-e, A., Akiyama, H. 1979. *J. Ferment. Technol.* 57:483–87
26. Hara, S., Iimura, Y., Otsuka, K. 1980. *Am. J. Enol. Vitic.* 31:28–33
27. Hara, S., Iimura, Y., Oyama, H., Kozeki, T., Kitano, K., Otsuka, K. 1981. *Agric. Biol. Chem.* 45:1327–34
28. Young, T. W. 1981. *J. Inst. Brew.* 87:292
29. Polonelli, L., Archibusacci, C., Sestito, M., Morace, G. 1983. *J. Clin. Microbiol.* 17:774–80
30. Fink, G. R., Styles, C. A. 1972. *Proc. Natl. Acad. Sci. USA* 69:2846–49
31. Skipper, N. 1983. *Biochem. Biophys. Res. Commun.* 114:518–25
32. Skipper, N., Thomas, D. Y., Lau, P. C. K. 1984. *EMBO J.* 3:107–11
33. Bostian, K. A., Elliott, Q., Bussey, H.,

Burn, V., Smith, A., Tipper, D. J. 1984. *Cell* 36:741–51

34. Thiele, D. J., Wang, R. W., Leibowitz, M. J. 1982. *Nucleic Acids Res.* 10:1661–78

35. Thiele, D. J., Hannig, E. M., Leibowitz, M. J. 1984. *Virology* 137:20–31

36. Hannig, E. M., Leibowitz, M. J. 1985. *Nucleic Acids Res.* 13:4379–400

37. Fried, H. M., Fink, G. R. 1978. *Proc. Natl. Acad. Sci. USA* 75:4224–28

38. Thiele, D. J., Hannig, E. M., Leibowitz, M. J. 1984. *Mol. Cell. Biol.* 4:92–100

39. Sommer, S. S., Wickner, R. B. 1984. *Mol. Cell. Biol.* 4:1747–53

40. Bussey, H., Sacks, W., Galley, D., Saville, D. 1982. *Mol. Cell. Biol.* 2:346–54

41. Palfree, R., Bussey, H. 1979. *Eur. J. Biochem.* 93:487–93

42. Vernet, T., Boone, C., Greene, D., Thomas, D. Y., Bussey, H. 1986. *Natural Antimicrobial Systems Symposium*, ed. R. G. Board. Bath, England: Bath Univ. Press. In press

43. Bussey, H., Saville, D., Greene, D., Tipper, D. J., Bostian, K. A. 1983. *Mol. Cell. Biol.* 3:1362–70

44. Novick, P., Ferro, S., Schekman, R. 1981. *Cell* 25:561–69

45. Wickner, R. B., Leibowitz, M. J. 1976. *Genetics* 82:429–42

46. Leibowitz, M. J., Wickner, R. B. 1976. *Proc. Natl. Acad. Sci. USA* 73:2061–65

47. Julius, D. J., Brake, A., Blair, L., Kunisawa, R., Thorner, J. 1984. *Cell* 37:1075–89

48. Lolle, S., Skipper, N., Bussey, H., Thomas, D. Y. 1984. *EMBO J.* 3:1383–87

49. Sommer, S. S., Wickner, R. B. 1982. *J. Bacteriol.* 150:545–51

50. Hopper, J., Bostian, K. A., Rowe, L. B., Tipper, D. J. 1977. *J. Biol. Chem.* 252:9010–17

51. Wickner, R. B. 1983. *Mol. Cell. Biol.* 3:654–61

52. Ridley, S. P., Sommer, S. S., Wickner, R. B. 1984. *Mol. Cell. Biol.* 4:761–70

53. Hannig, E. M., Leibowitz, M. J., Wickner, R. B. 1985. *Yeast* 1:57–65

54. Field, L. J., Bobek, L. A., Brennan, V. E., Reilly, J. D., Bruenn, J. A. 1982. *Cell* 31:193–200

55. Bobek, L. A., Bruenn, J. A., Field, L. J., Gross, K. W. 1982. *Gene* 19:225–30

56. Bruenn, J. A., Madura, K., Siegel, A., Miner, Z., Lee, M. 1985. *Nucleic Acids Res.* 13:1575–91

57. El-Sherbeini, M., Tipper, D. J., Mitchell, D. J., Bostian, K. A. 1984. *Mol. Cell. Biol.* 4:2818–27

58. Wickner, R. B. 1974. *J. Bacteriol.* 117:1356–57

59. Herring, A. J., Bevan, E. A. 1977. *Nature* 268:464–66

60. Welsh, D. J., Leibowitz, M. J., Wickner, R. B. 1980. *Nucleic Acids Res.* 8:2349–63

61. Welsh, D. J., Leibowitz, M. J. 1980. *Nucleic Acids Res.* 8:2365–75

62. Bruenn, J., Bobek, L., Brennan, V., Held, W. 1980. *Nucleic Acids Res.* 8:2985–97

63. Welsh, D. J., Leibowitz, M. J. 1982. *Proc. Natl. Acad. Sci. USA* 79:786–89

64. Bostian, K. A., Burn, V. E., Jayachandran, S., Tipper, D. J. 1983. *Nucleic Acids Res.* 11:1077–97

65. Hannig, E. M., Thiele, D. J., Leibowitz, M. J. 1984. *Mol. Cell. Biol.* 4:101–9

66. Newman, A. M., McLaughlin, C. S. 1984. *12th Int. Conf. Yeast Genet. Mol. Biol.*, Abstr. D8

67. Newman, A. M., Elliott, S. G., McLaughlin, C. S., Sutherland, P. A., Warner, R. C. 1981. *J. Virol.* 38:263–71

68. Zakian, V., Wagner, D. W., Fangman, W. L. 1981. *Mol. Cell. Biol.* 1:673–79

69. Joklik, W. K., ed. 1981. *The Reoviruses.* New York: Plenum. 571 pp.

70. Buck, K. W. 1979. See Ref. 19, pp. 93–160

71. Bruenn, J. A. 1980. *Ann. Rev. Microbiol.* 34:49–68

72. Luria, S. E., Darnell, J. E., Baltimore, D., Campbell, A. 1978. *General Virology.* New York: Wiley. 578 pp. 3rd ed.

73. Sclafani, R. A., Fangman, W. L. 1984. *Mol. Cell. Biol.* 4:1618–26

74. Wickner, R. B., Leibowitz, M. J. 1976. *J. Mol. Biol.* 105:427–43

75. Wickner, R. B. 1978. *Genetics* 88:419–25

76. Wickner, R. B. 1979. *Genetics* 92:803–21

77. Wickner, R. B., Leibowitz, M. J. 1979. *J. Bacteriol.* 140:154–60

78. Guerry-Kopecko, P., Wickner, R. B. 1980. *J. Bacteriol.* 144:1113–18

79. Wickner, R. B., Ridley, S. P., Fried, H. M., Ball, S. G. 1982. *Proc. Natl. Acad. Sci. USA* 79:4706–8

80. Thrash, C., Voelkel, K., Dinardo, S., Sternglanz, R. 1984. *J. Biol. Chem.* 259:1375–77

81. Cohn, M. S., Tabor, C. W., Tabor, H. 1978. *J. Biol. Chem.* 253:5225–27

82. Tyagi, A. K., Wickner, R. B., Tabor, C. W., Tabor, H. 1984. *Proc. Natl. Acad. Sci. USA* 81:1149–53

83. Leibowitz, M. J., Wickner, R. B. 1978. *Mol. Gen. Genet.* 165:115–21

84. Somers, J. M. 1973. *Genetics* 74:571–79

85. Ridley, S. P., Wickner, R. B. 1983. *J. Virol.* 45:800–12

86. Toh-e, A., Guerry, P., Wickner, R. B. 1978. *J. Bacteriol.* 136:1002–7
87. Ball, S. G., Tirtiaux, C., Wickner, R. B. 1984. *Genetics* 107:199–217
88. Toh-e, A., Wickner, R. B. 1980. *Proc. Natl. Acad. Sci. USA* 77:527–30
89. Oliver, S. G., McCready, S. J., Holm, C., Sutherland, P. A., McLaughlin, C. S., Cox, B. S. 1977. *J. Bacteriol.* 130: 1303–9
90. Wickner, R. B., Leibowitz, M. J. 1977. *Genetics* 87:453–69
91. Wickner, R. B., Boutelet, F., Hilger, F. 1983. *Mol. Cell. Biol.* 3:415–20
92. Toh-e, A., Sahashi, Y. 1985. *Yeast.* 1:159–72
93. Bussey, H., Steinmetz, O., Saville, D. 1983. *Curr. Genet.* 7:449–56
94. Hutchins, K., Bussey, H. 1983. *J. Bacteriol.* 154:161–69
95. Al-Aidroos, K., Bussey, H. 1978. *Can. J. Microbiol.* 24:228–37
96. Bussey, H., Saville, D., Hutchins, K., Palfree, R. G. E. 1979. *J. Bacteriol.* 140:888–92
97. de la Pena, P., Barros, F., Gascon, S., Ramos, S., Lazo, P. S. 1980. *Biochem. Biophys. Res. Commun.* 96:544–50
98. de la Pena, P., Barros, F., Gascon, S., Lazo, P. S., Ramos, S. 1981. *J. Biol. Chem.* 256:10420–25
99. Bussey, H., Sherman, D. 1973. *Biochim. Biophys. Acta* 298:868–75
100. Skipper, N., Bussey, H. 1977. *J. Bacteriol.* 129:668–77
101. Wickner, R. B. 1974. *Genetics* 76:423–32
102. Buck, K. W. 1980. *The Eukaryotic Microbial Cell, Soc. Gen. Microbiol. Symp., 30th,* ed. G. W. Gooday, D. Lloyd, A. P. J. Trinci, pp. 330–75. Cambridge, England: Cambridge Univ. Press
103. Bevan, E. A., Herring, A. J. 1976. *Genetics, Biogenesis, and Bioenergetics of Mitochondria,* ed. W. Bandlow, R. J. Schweyen, D. Y. Thomas, K. Wolf, F. Kaudewitz, pp. 153–62. Amsterdam: de Gruyter
104. Fried, H. M., Warner, J. R. 1981. *Proc. Natl. Acad. Sci. USA* 78:238–42
105. Schultz, L. D., Friesen, J. D. 1983. *J. Bacteriol.* 155:8–14
106. Thrash, C., Bankier, A. T., Barrell, B. G., Sternglanz, R. 1985. *Proc. Natl. Acad. Sci. USA* 82:4374–78
107. Fonzi, W. A., Sypherd, P. S. 1985. *Mol. Cell. Biol.* 5:161–66

Ann. Rev. Biochem. 1986. 55:397–425
Copyright © 1986 by Annual Reviews Inc. All rights reserved

BACTERIAL PERIPLASMIC TRANSPORT SYSTEMS: STRUCTURE, MECHANISM, AND EVOLUTION

Giovanna Ferro-Luzzi Ames

Department of Biochemistry, University of California, Berkeley, California 94720

CONTENTS

OVERVIEW

Gram-negative bacteria have a complex cell surface, consisting of three layers: an outer membrane, composed of protein, phospholipid, and lipopolysaccharide (1); the cell wall proper or peptidoglycan; and an inner, or cytoplasmic membrane, composed of protein and phospholipid. Nutrients, therefore, have to pass through this rather formidable architectural and protective structure. The outer membrane constitutes an important barrier against a variety of assaults, such as enzymatic and detergent attack in the digestive tract, and

397

0066-4154/86/0701-0397$02.00

antibiotic entry (1). Limited permeability to small solutes is available through the outer membrane by way of proteinaceous channels, constituted of a special class of proteins which can be substrate-nonspecific (the proteins in this case are named *porins*) or substrate-specific (1). The cell wall proper is commonly regarded as an entirely permeable layer, conferring rigidity to the cell while forming a widely open network through which nutrient diffusion occurs readily. The cytoplasmic membrane, on the other hand, is impermeable to almost every solute unless a special transport system is provided.

Transport systems can be categorized according to a variety of principles. For the purposes of this review these systems are classified according to their response to a physical treatment, osmotic shock (2), into shock-sensitive and shock-resistant permeases. The osmotic shock procedure consists of plasmolyzing the cells in sucrose followed by a rapid dilution in cold distilled water (with or without a small amount of magnesium added). The plasmolysis step is performed in the presence of EDTA and Tris buffer, which are essential for slightly "weakening" the outer membrane; the membrane is then partially ruptured during the rapid influx of water upon dilution. The mechanics of this process, along with a wealth of information on the nature and properties of the outer membrane, are illustrated in an excellent recent review by Nikaido & Vaara (1). Here it will suffice to say that during osmotic shock a special class of proteins is released into the medium, the *periplasmic proteins*, so called because they are thought to reside in a special cell compartment, the *periplasm*, located between the inner and the outer membranes. The term "periplasmic protein" should be taken more as an operational definition than as an indication of a physical location since a protein such as EF-Tu (elongation factor Tu), which must be functioning well inside the cell, actually is released by a treatment very similar to osmotic shock (3, 4), while a number of phosphatases, which are typical periplasmic enzymes as far as their function is concerned, are poorly released by osmotic shock (5).

Shock-sensitive permeases, also referred to as *periplasmic permeases,* then are those systems that are inactivated during osmotic shock because of the loss of an essential protein component which is referred to as the periplasmic component. In all cases studied, the periplasmic component is a protein that binds the transported solute with high affinity. The first of these periplasmic transport proteins to be shown to be released by osmotic shock was a sulfate-binding protein, part of a sulfate permease in *Salmonella typhimurium* (6). Subsequently numerous permeases have been shown to belong to this class and several have been extensively characterized (7, 7a). Shock-resistant permeases, on the other hand, are those that retain all of their activity [and therefore, their component(s)] upon osmotic shock. A classic representative of this class is the β-galactoside permease which has been intensively studied (8, 9). This permease is composed of a single protein, the product of the *lacY* gene, which is tightly membrane-bound. No other bacterial shock-resistant system has yet been characterized sufficiently for the generalization to be made that all shock-resistant permeases are constituted of a single protein component.

An additional characteristic subsequently used to distinguish between these two classes of permeases has been the nature of the mechanism of energy coupling. Shock-resistant permeases are powered by the proton-motive force (8, 9), while energy coupling in shock-sensitive permeases has been postulated to depend on substrate-level phosphorylation energy (10, 11); however, this mechanism is presently controversial, as will be discussed later.

In this review I will discuss the present status of knowledge concerning periplasmic permeases[1], and I will try to generalize the findings so as to produce a coherent picture of transport through these systems. I will at the end attempt to foresee which directions of research might be most promising. This review is not meant to be an encyclopedic collection of everybody's data. Thus, salient examples that represent specific aspects will be chosen throughout the review. I apologize to those researchers who might feel slighted or forgotten; many references have been left out since they appear already in recent reviews covering many detailed aspects of periplasmic transport (7, 12, 13) and because of lack of space, rather than lack of relevance.

GENERAL COMPOSITION OF PERIPLASMIC PERMEASES

It is important to realize that all transport assays involve whole cells or, at best, membrane vesicles. Thus, in the past, the use of such complex assay systems did not lead to an understanding of the composition of the individual transport systems. It was only when careful genetic analysis was introduced into this field that it became possible to unravel the complexities of these permeases. With the availability in recent years of recombinant DNA technology, a number of periplasmic permeases have been analyzed in great detail and a general picture is emerging. At the moment, the most thoroughly characterized systems are those for histidine, maltose, branched-chain amino acids, oligopeptides, β-methyl galactoside, ribose, arabinose, and phosphate. Many others are in various stages of development.

Initial evidence that shock-sensitive systems have multiple components was obtained for the histidine permease, the first genetically characterized permease, by showing that a class of transport-deficient mutants could be obtained that had an intact periplasmic histidine-binding protein (14), thus suggesting that one or more additional proteins were necessary for transport (14–16). A

---

[1]Data obtained from *Salmonella typhimurium* and *Escherichia coli* will be pooled, since these organisms are very closely related.

complementary experiment demonstrated that mutants defective in the histidine-binding protein were defective in transport. However, these results still left open the possibility that the mutants were unable to transport not because of the absence of the periplasmic binding protein, but because of the lack of another protein regulated by, or simultaneously with, the binding protein. Incontrovertible proof was obtained by correlating the temperature sensitivity of a mutant histidine-binding protein to temperature sensitivity of transport (17). Today the important proof that a periplasmic binding protein is a component of a transport system can also be obtained by reconstituting the transport system (see below), i.e. by adding back the periplasmic component to cells deprived of it either mutationally or by physical manipulation, and observing recovery of transport activity.

Accumulated evidence shows that periplasmic permeases are typically composed of one periplasmic substrate-binding protein and three membrane-bound components. This has been found to be true for seven permeases: histidine (18), maltose (19), branched-chain amino acids (12), oligopeptides (20), ribose (21), β-methyl galactoside (22–24), and phosphate (25a). In additional cases there is strong evidence of a similar complexity (e.g. glutamine and arabinose). The histidine and the maltose permeases, which will be referred to most extensively in this review, are schematically represented in Figure 1, both from the genetic and architectural points of view. The outer membrane is represented as containing pores which allow entrance of the substrate into the periplasm. The membrane-bound components of periplasmic permeases are represented as forming a complex within the cytoplasmic membrane, without a specific arrangement since their interactions with each other and with the periplasmic component are not known. As will be discussed later, indirect evidence suggests an interaction between the histidine-binding protein J and the membrane-bound protein P in the case of the histidine permease, and between the maltose-binding protein and both the MalG and MalF proteins, and the MalK and the MalG proteins in the case of the maltose permease. The genes coding for the components of the histidine permease constitute a single operon, while those of the maltose permease constitute two divergent operons.

Figure 2 shows the genetic structure of the other most extensively characterized permeases. In all of these cases a single operon contains all known transport components. In at least two cases (histidine and the branched-chain amino acids) there is an additional, separately regulated gene coding for a second periplasmic binding protein of different specificity (*argT* and *livJ*, respectively), which will be discussed later. In some cases the region contains additional genes whose function is either unknown at present [e.g. *livL* in the branched-chain amino acid permease (12)] or is involved in further catabolism of the transported substrates (e.g. *rbsK* codes for a ribose kinase, 26). It is of interest to realize that while it became clear very early that the periplasmic

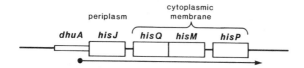

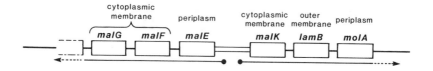

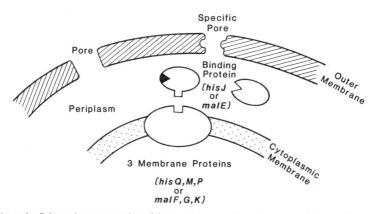

Figure 1   Schematic representation of the genetic composition *(top)* and architecture *(bottom)* of the histidine and maltose permeases. The histidine permease genes constitute an operon under control of the regulatory locus *dhuA*. The maltose permease genes constitute two divergent transcription units. The arrows indicate direction of transcription. The *lamB* gene codes for the outer membrane protein producing maltodextrin-specific channels. Gene *molA* codes for a periplasmic protein of unknown function. Downstream from *malG* there is an open reading frame also of unknown function. The *hisJ* and the *malE* genes code for the histidine- and maltose-binding proteins, respectively. The three membrane-bound proteins are represented as forming a complex of an unspecified nature. The periplasmic binding protein is shown in two different conformations, with and without bound substrate (▲), respectively. The possibility of an interaction between the liganded binding protein and the membrane-bound complex is represented by a protuberance and matching indentation, respectively. Substrates penetrate the outer membrane either through nonspecific hydrophilic pores (most commonly) or through pores that exhibit substrate specificity.

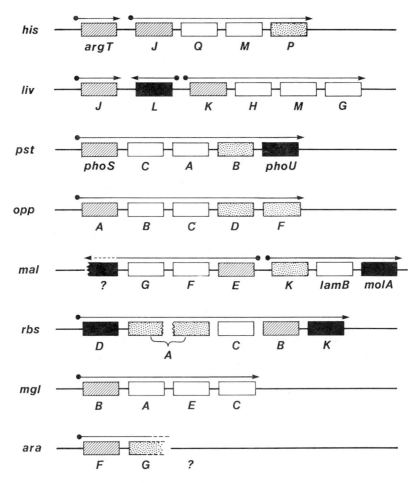

Figure 2 Genetic structure of periplasmic transport operons. Each operon codes for at least one binding protein (hatched boxes) and usually three membrane-bound components involved in transport (blank and stippled boxes). Black boxes represent genes that are not or not known to be involved in transport. Some of the genes represented as black boxes may later turn out to be transport genes: e.g. *rbsD*, which is likely to be a membrane-bound transport protein (M. Hermodson, S. Buckel, personal communication). Stippled boxes represent genes coding for a family of homologous membrane-bound components. Gene *rbsA* is composed of two homologous halves (represented as two broken halves): these probably resulted from a gene duplication (21). The sequence of the *liv, mgl,* and *ara* systems is not yet complete, and their structure has been established primarily by genetic means; either *livH* or *livM* or *livG* might eventually be shown to belong to the "conserved" family. The arrows indicate direction of transcription, but are not accurately located at the start and termination sites since these have not been established in most cases. These are known for the *his* system (15, and G. Ames Ferro-Luzzi, K. Storm, unpublished results). The genes are not drawn to scale. Regulatory sites are not represented since they are not well-characterized in most cases.

binding protein was not the only component of these permeases (14), the existence of more than one membrane-bound protein did not become evident until the identification of the first membrane-bound component was achieved (27) and the first complete sequence of a transport operon was obtained (18). The data available on the arabinose permease indicate the existence of only one membrane-bound component (28); when the complete DNA sequence of this system becomes available, it is likely that it will also show the presence of additional components. Standard genetic techniques (e.g. 16, 29) such as were used in the past for the genetic characterization of these periplasmic systems should be easily applicable to analysis of any newly discovered system. Recombinant DNA techniques can greatly facilitate the analysis.

It seems then, that complexity is the norm for this particular mode of transport. In trying to understand why this should be so, we can only speculate. These systems typically concentrate substrates inside the cell against a very large gradient (e.g. 10^5-fold in the case of maltose, 30); perhaps achieving and maintaining such large concentration gradients requires a complex mechanism, possibly in relation to energy coupling. Alternatively, the high efficiency of transport that these systems usually display requires a complex structure. They are in fact able to scavenge solutes off very low concentrations: e.g. the apparent K_m for uptake is 0.01 μM for the histidine permease (14), 1 μM for the maltose permease (30), and 0.2 μM for the phosphate permease (31). By comparison, the apparent K_m for the transport of lactose through the shock-resistant, monocomponent, β-galactoside permease *(lacY)* is 190 μM (32). The very high efficiency of these permeases may be a necessity for the cells, at least where amino acid transport is concerned, since "leakage" of biosynthetically produced amino acids seems to occur and these high-affinity permeases recapture and concentrate the lost amino acid (15). This recapture may constitute an important evolutionary advantage, thus justifying the existence of complex, multicomponent permeases, since amino acid biosynthesis is expensive: e.g. it has been calculated that 41 ATP molecules are being sacrificed for each histidine molecule made (33). Evidently, a lost or an externally encountered histidine molecule is very valuable!

The very high substrate affinity of amino acid permeases explains why bioassays are possible, i.e. assays that allow the determination of very low levels of amino acids (or vitamins) by the use of bacteria requiring those amino acids. An interesting parameter in this respect is the *limit concentration* (15, 34), which defines the lowest external concentration required by the permease in order to supply the cell with enough amino acid to carry on protein synthesis without the need for its biosynthesis: at lower concentrations the bacteria supplement the external supply with biosynthetically produced amino acid by relieving whichever regulatory mechanism is operating (feedback and/or repression). The limit concentration for histidine is 0.15 μM in *S. typhimurium*.

At external histidine concentrations below this value the permease becomes limiting. This finding has led to the development of a transport assay that is particularly convenient and that has the added advantage that it does not disturb the cells in any nonphysiological fashion (34).

An interesting aspect of some periplasmic systems is the fact that the membrane-bound components are multifunctional, i.e. they are needed for transport of additional substrates and for that purpose are utilized by more than one binding protein. For example, in the case of the histidine permease, the membrane-bound proteins are also essential for transport of arginine under conditions of nitrogen starvation, via the lysine-arginine-ornithine-binding protein (LAO protein, coded for by the *argT* gene; see Figure 2) (35). This means that whatever mechanism is deduced from the available data, it will have to include the alternating interaction with and removal of each of the different periplasmic components from contact with the same set of membrane-bound components. The existence of alternative periplasmic binding proteins (LivJ and LivK) utilizing the same set of membrane-bound components has been shown also for the branched-chain amino acid permease (12, 36). It is possible that this occurrence will turn out to be more frequent than we presently believe, since the existence of additional transport activities through a set of common membrane-bound proteins has not been investigated exhaustively in most cases.[2]

In this general introduction to the architectural makeup of these permeases, mention should be made of the difficulty that large-size substrates may encounter in crossing the outer membrane. In the case of the maltose permease, it has been shown that this system is also needed for transporting higher polymers of glucose (maltodextrins), up to six or seven glucose residues long (37). Permeation of these large molecules through the outer membrane would not be possible were it not for the presence of a substrate-specific channel-forming protein, the product of the *lamB* gene (which also functions as receptor for phage λ) which is part of the maltose transport operon (Figure 1). Since permeability through the outer membrane usually seems to be unencumbered for substrates of periplasmic permeases, and since the properties of the LamB protein have been reviewed recently (1, 19, 38), this subject will not be discussed further.

CHARACTERISTICS OF INDIVIDUAL COMPONENTS

The Periplasmic Substrate-Binding Protein

The binding proteins are the most thoroughly analyzed of these transport components for obvious reasons. Their purification is easy (since they are

[2]In the maltose operon, the *molA* gene codes for a periplasmic protein (M. Hofnung, personal communication) that might be a binding protein that utilizes the *malK, malF*, and *malG* components. This may be a case analogous to that of *hisJ/argT* and *livJ/livK*.

soluble and easily assayable proteins), they can be obtained in large quantities, and most of them are remarkably stable. A summary of the properties they share as a group is as follows (reviewed in 7; see references for individual cases therein). They are monomeric proteins with molecular weights varying from about 25,000 to about 56,000 [the latter is the largest known binding protein; it is an oligopeptide-binding protein recently identified (39) and characterized (40)]; several are stable to heat (e.g. the histidine-binding protein maintains 70% of its activity after 10 min at 100°C, 17); they have high affinity for their substrates; they undergo a conformational change upon binding of substrate; and they have two functionally and genetically separable active domains. The latter three properties, discussed below, are the most interesting with respect to the function of these proteins.

The binding affinity (K_D) is between 0.1 and 1 μM for sugar substrates and around 0.1 μM for amino acids. Presumably because of this high affinity, some of these proteins have been purified with tightly bound substrates, which in some cases has led to controversy and confusion. Thus, the galactose-binding protein binds tightly one molecule of glucose (which is an even better substrate than galactose, 41) per molecule of protein (42, 43); this led to biphasic Scatchard plots variously interpreted as being due to the presence of two binding sites on the molecule (44) or to the presence of an impurity in the radioactive galactose utilized as substrate (41). Similarly, purified arabinose- and histidine-binding proteins contain bound substrate (45). The substrate, which can be easily removed by reversible denaturation with guanidine-HCl (45), causes no problem for kinetic measurements if these are performed by equilibrium dialysis against a large volume of ligand rather than against a volume comparable to that occupied by the protein (42, 46). A thorough analysis by stopped-flow rapid-mixing techniques of the kinetics of ligand-binding to the arabinose-, galactose- and maltose-binding proteins revealed that in all cases the variation in the affinity constants for several different substrates was due primarily to differences in the dissociation rate constants (which varied by 100-fold), while the association rate constants were similar for all substrates (45). The rate of ligand dissociation would then be the determining factor defining the maximum possible velocity for the corresponding transport system. Computations showed that in all cases the observed V_{max} values for transport are equal to or smaller than the values obtained assuming that ligand dissociation is limiting (45), thus indicating that ligand dissociation is not the limiting factor in transport. The point should be made here that comparison of the affinity as measured in vitro with the affinity of the transport system in vivo should always be made cautiously, especially when this correlation is used as evidence that the binding protein is a component of that transport system. Such correlation may be coincidental, since the binding protein, once removed from its in vivo location, might present a K_D for binding that might be totally unrelated (either better or worse) to its K_D in vivo and to the apparent K_m of the entire transport system. Its almost certain interaction with the membrane might drastically affect its kinetic properties.

Binding proteins undergo conformational change upon binding of substrate as has been measured in the case of the histidine-, maltose-, arabinose-, ribose-, galactose-, glutamine-, and leucine-isoleucine-valine-binding proteins by a variety of methods (see 7 for additional references): fluorescence spectroscopy (45, 47–53), nuclear magnetic resonance (54–57), scanning calorimetry (58), and immunology (59). The use of a distant reporter group method, which was applied to the histidine-, galactose-, and maltose-binding proteins (48, 49, 60), has the particular advantage that it can distinguish between movement of a residue located right at the ligand-binding site and changes reflecting movement in regions of the molecule removed from the actual ligand-binding site. The demonstration that a substrate-induced conformational change occurs in all binding proteins analyzed indicates that this is an essential aspect of the mechanism of action of these proteins.

It is commonly postulated that binding proteins interact with the membrane-bound components. However, direct evidence for this hypothesis is lacking. In the case of the histidine-binding protein, such a hypothesis is based on four separate sets of data. The first consists of the characterization of a mutant J protein that cannot function in transport despite the fact that it has an intact histidine-binding site (61). This suggests the existence of a region of the protein that is essential for transport but not necessary for binding histidine. By using the distant reporter group method (48) and by nuclear magnetic resonance investigations (62), it was shown that this mutant J protein is unable to undergo a normal ligand-induced conformational change, thus adding further evidence that the specific conformational change is intimately involved in the functioning of binding proteins in transport.

The second piece of evidence depends on an entirely genetic argument (63): a *hisP* mutation has been characterized that suppresses the described mutation in the second domain of the J protein; this can be interpreted most easily, but not only, as a result of an interaction between different components of a structure, in which the J defect is corrected by a corresponding alteration in the interacting P protein. The possibility that the mutated P protein might be functioning altogether without the aid of the J protein was discarded by isolating a derivative strain that carried the mutated P protein but lacked entirely the J protein: this strain was unable to transport histidine.

The third piece of evidence is deduced from comparing the sequence of the J protein with that of the closely related lysine-arginine-ornithine-binding protein (LAO, the *argT* gene product): these two proteins have an overall homology of 70%, but two portions of the molecules are better than 90% homologous (64). Since both the J and the LAO proteins require the Q, M, and P proteins for function and since the mutation in the P-interaction site of the J protein is located in one of these two highly homologous stretches, this suggests that the two proteins are involved in an identical function, presumably the interaction with a common membrane component (64, 65), besides each of them having a specific substrate-binding site.

The fourth piece of evidence comes from the preliminary characterization of *hisJ* mutants that interfere with the proper functioning of the membrane components, such as would be expected from a mutant-binding protein that binds irreversibly to one of them (G. Ames Ferro-Luzzi, unpublished results). Genetic evidence was also obtained for an interaction occurring between the maltose-binding protein and the *malF* and *malG* gene products. Mutants are available in these genes that allow transport of maltose in the absence of the binding protein (see below); the introduction into these mutants of a wild-type *malE* gene completely inhibits maltose uptake (66). These data have been explained by postulating an interaction between the MalF and MalG proteins and the maltose-binding protein: presumably the function of the mutated membrane components is inhibited by the presence of wild-type binding protein because of a nonproductive and interfering interaction with the altered membrane components. Evidence that the maltose-binding protein has separate domains, for binding maltose and for interacting with the MalF and MalG proteins, has not yet been obtained. Indirect evidence for the existence of two sites was also obtained for the glutamine permease by chemically altering the glutamine-binding protein in such a way that its binding activity was unaffected while its ability to participate in transport was lost (67). Even though all of these data can be explained in terms of an interaction between the periplasmic and the membrane-bound components, this hypothesis will have to be supported finally by biochemical data.

Studies on the maltose-binding protein have shown that this protein interacts with the outer membrane. A column carrying the maltose-binding protein as a covalently bound affinity ligand retained the outer membrane protein, LamB (68), indicating that the maltose-binding protein, besides its maltose-binding site, must have at least one additional functional site responsible for interacting with the outer membrane. The relationship between these two sites has not been established. Genetic evidence also suggests that the maltose-binding protein interacts with the LamB protein (37).

A great deal of important structural information has been derived from the X-ray crystallographic studies of several binding proteins, mostly from F. Quiocho's laboratory. The most advanced structural analysis has been achieved on the arabinose-binding protein and a summary of these results will be given here. Analysis of several other binding proteins is at various levels of sophistication and gives results that are entirely compatible with a general picture based on the structure of the arabinose-binding protein. This protein molecule is arranged in two globular domains (lobes) forming a cleft and connected by a flexible hinge: the overall shape resembles a kidney bean (69). The binding site for the substrate is located in the concave region between the two lobes. The molecule is flexible, the cleft closing down somewhat upon binding of substrate (70, 71), thus "trapping" the substrate deep within the protein (from which the name of "Venus's flytrap" for this type of binding mechanism). These data agree with the evidence that binding proteins in general undergo a conformational change upon binding of substrate. Since there is a direct relationship between higher affinity and lower rate of dissociation of a substrate from the closed form of the protein, it is likely that this closed, more stable, liganded form of the protein is "encouraged" to release the substrate by an external stimulus, presumably by interaction with the membrane-bound transport components (see below for model). Other binding proteins have been analyzed by X-ray crystallography, mostly by the Quiocho group. It seems evident that an overall bilobate shape is the norm for these proteins: the leucine-isoleucine-valine-binding protein (72), the galactose-binding protein from *E. coli* (73), the galactose-binding protein from *S. typhimurium* (74), and the sulfate-binding protein (75).

Among recent findings was an interesting one obtained by nuclear magnetic resonance spectroscopy which showed that formation of hydrogen bonds is involved in the protein-ligand interaction in the case of the glutamine-binding protein (57), in particular involving the amide bond of glutamine. It is also interesting that with the solution of the structure of the sulfate-binding protein of *S. typhimurium* with bound substrate at 2.0 Å resolution, it became apparent that the charged oxygen atoms of the sulfate molecule are stabilized by hydrogen bonds with the protein, rather than by salt-bridges (75). There has been no evidence that binding proteins undergo covalent modification except for a report that an arginine-binding protein can be phosphorylated by ATP while transporting arginine (76). No evidence of phosphorylation of the histidine-binding protein was obtainable (G. Ames Ferro-Luzzi, unpublished results; 77).

The Membrane-Bound Components

Membrane-bound components are the hardest to study. They are often present in very small amounts, they need to be solubilized before they can be subjected to biochemical studies, and no activity assay is as yet available for any of them. Their number has been determined for several permeases by genetic or recombinant DNA techniques, and in all cases in which exhaustive data are available they are present in the number of three (Figure 2; however, see later comment on *oppF*). This uniformity indicates that something in the basic makeup of these permeases requires such architectural and functional composition. Two of the components are highly hydrophobic (18, 20, 21, 25, 25a, 78–80). The third one has an amino acid sequence that is not recognizably hydrophobic (21a, 25, 25a, 78, 81), despite the fact that it is membrane-bound; however, the strength of the membrane attachment might vary from one system to the other. For example, the MalK protein is thought to be a peripheral membrane protein because it is released upon sonication (38), and because it can be recovered in the soluble protein fraction in mutants defective in *malG*, indicating that it might be anchored to the membrane via the MalG protein (82). The HisP protein, on the other hand, is not released either by sonication or in *hisQ* or *hisM* mutants (G. Ames Ferro-Luzzi, K. Nikaido, unpublished results). The HisP, HisQ, and HisM proteins have been shown to be located in the cytoplasmic membrane (27; R. Jaskot, G. Ames Ferro-Luzzi, unpublished results). Similarly, the products of the *malF* and *malK* genes have been localized in the cytoplasmic membrane (83, 84). An interesting observation is that the sequence of all of these membrane-bound components lacks a signal peptide and, therefore, they are not processed (the latter, however, has been shown only for the HisP protein; 18). This is in agreement with findings obtained for other cytoplasmic membrane proteins: thus, it is fair to generalize that proteins targeted for the cytoplasmic membrane do not require a signal peptide and processing.

It had been previously observed that HisP and MalK are homologous to each other (78); as DNA sequences of additional permeases were obtained, it became clear that each periplasmic permease contains a member of a family of homologous proteins. I will refer to each member of this family as the "conserved" component. Figure 2 highlights the genes coding for the conserved components in each transport system as stippled boxes: i.e. *hisP, pstB, oppD, oppF, malK,* and *rbsA.* The latter codes for a large protein composed of two homologous halves (21a). A possible exception is apparent in the *opp* system, which contains genes coding for two conserved components, *oppD* and *oppF.* However, it is not known whether both are essential for transport (I. Hiles, C. Higgins, personal communication). Preliminary results for the arabinose system indicate that *araG* also belongs to this homologous family (R. Hogg, personal communication). Figure 3 shows an alignment of the sequences of these conserved proteins. Several areas of strong conservation are marked as sites A to F: presumably these protein regions are involved in the performance of functions that are common to the entire family. Higgins et al (81) have noted that some of the conserved regions share homology with several proteins known to have ATP-binding sites such as the α and β-subunits of the proton-translocating ATPase, myosin, adenylate kinase, RecA protein, and others (85). These are the sites labeled as A and B in Figure 3 and in Refs. 81 and 85.

Recently it has been found that both HisP and MalK carry a nucleotide-binding site by obtaining covalent derivatives between these proteins and the photoaffinity label 8-azido-ATP (86). Competition of 8-azido-ATP labeling with a variety of nucleotide-containing compounds suggested that ATP and/or

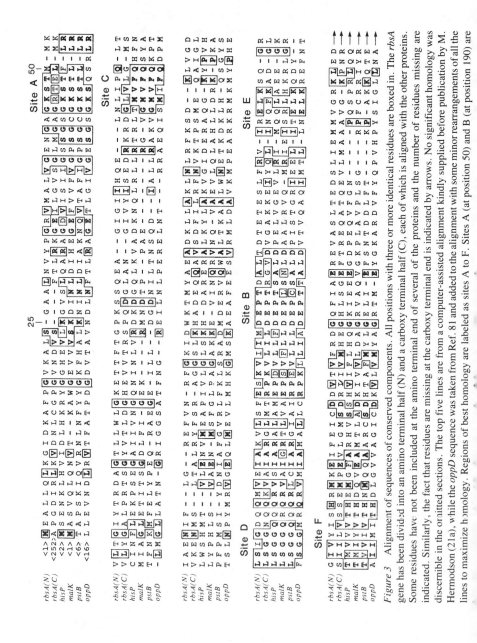

Figure 3   Alignment of sequences of conserved components. All positions with three or more identical residues are boxed in. The *rbsA* gene has been divided into an amino terminal half (N) and a carboxy terminal half (C), each of which is aligned with the other proteins. Some residues have not been included at the amino terminal end of several of the proteins and the number of residues missing are indicated. Similarly, the fact that residues are missing at the carboxy terminal end is indicated by arrows. No significant homology was discernible in the omitted sections. The top five lines are from a computer-assisted alignment kindly supplied before publication by M. Hermodson (21a), while the *oppD* sequence was taken from Ref. 81 and added to the alignment with some minor rearrangements of all the lines to maximize homology. Regions of best homology are labeled as sites A to F. Sites A (at position 50) and B (at position 190) are

GTP are the natural substrates of the HisP protein (86). This protein was also shown to bind to several dye columns (86; R. Jaskot, G. Ames Ferro-Luzzi, unpublished results) and to bind ATP in partially purified preparations (D. Speiser, G. Ames Ferro-Luzzi, unpublished results). The MalK protein has been purified and has been shown to bind ATP with an affinity constant of 10^{-4} M (K. Nikaido, H. Nikaido, personal communication). A fusion product between the *oppD* gene product and β-galactosidase was also shown to react with the ATP analogue 5'-*p*-fluorosulfonylbenzoyladenosine (81). The discovery of a nucleotide-binding site in the conserved components is an exciting advance in this field. Since nucleotide-binding has been demonstrated for several of these proteins, it is reasonable to generalize that all conserved membrane components bind a nucleotide and that such function is essential for all periplasmic systems. The possibility that such interaction might be involved in the energy-coupling mechanism has been suggested (81, 86). However, attempts to demonstrate ATP hydrolysis have failed, despite numerous variations in reaction conditions: no evidence of exchange of either inorganic phosphate or pyrophosphate into ATP, nor of hydrolysis of GTP or UTP, was obtained (G. Ames Ferro-Luzzi, D. Speiser, unpublished results; and K. Nikaido, H. Nikaido, personal communication). However, in no case was ATP hydrolysis assayed while substrate transport (histidine or maltose) was occurring concomitantly. This crucial experiment, which involves the availability of an in vitro reconstituted system that actively transports, would give a better understanding of the involvement of the possible nucleotide-binding site in energy coupling. However, it is possible that the bound nucleotide does not undergo hydrolysis and that it serves an entirely regulatory function.

None of the other membrane-bound components from any other system has been purified, or shown to have a functional site. The *hisQ* and *hisM* genes bear strong homology to each other (87) indicating that they originated by a gene duplication. A hydropathicity plot of *hisQ* and *hisM* indicates very similar patterns which, together with their sequence homology, suggested the possibility that these two proteins form a pseudodimer in the membrane (87). The *hisQ* protein can be labeled with 8-azido ATP (86; in this publication the label was erroneously thought to be located in the M protein), thus it is possible that it also has a nucleotide-binding site. However, the extent of labeling is erratic and might be due to the proximity of the P protein to the Q protein, rather than to an independent binding site. A clear homology is also evident between proteins PstC and PstA (G. Ames Ferro-Luzzi, R. Doolittle, unpublished results). Homology has also been detected between MalF and MalG (80). However, none was evident between OppB and OppC.

Genetic data obtained in the maltose and histidine systems have indicated that some of the membrane components might carry a substrate-binding site. The best evidence comes from mutants in *malF* and *malG* which can be obtained in a strain deleted for *malE* (i.e. completely lacking the binding protein) and which are able to transport maltose (66, 88). The apparent affinity for maltose in these mutants is essentially identical and about 2000-fold larger than in the wild type (in presence of maltose-binding protein). However, their V_{max} varies, indicating that the mutant proteins may have acquired the ability to expose their binding site, rather than having acquired sites with improved affinity (66). It is important to establish whether the MalK protein is still an essential component of transport in these mutants. Spheroplasts prepared from such mutants are able to transport maltose, thus indicating that the mutation has altered a membrane component so that the system is able to function without the aid of a periplasmic binding protein (which is lost during the vesicle preparation). Other evidence is derived from the existence of *hisQ* and *hisM* mutants that have an altered specificity of transport (18, 89): an alteration in a substrate-binding site is the most likely explanation of the properties of these mutants.

TRANSPORT MODELS

From all the available information, and especially making strong use of the details known for histidine and maltose transport, a model derived from the general model of Figure 1 with some added details is presented in Figure 4. The substrate crosses the outer membrane (through specific or unspecific channels) and encounters the binding protein, by which it is bound reversibly. The effect of the binding event on the concentration of substrate within the periplasm can be most easily understood by imagining the situation as equivalent to that of a small dialysis bag containing binding protein immersed in a large volume of solution containing the substrate (actually these are the conditions for the large-volume dialysis binding assay mentioned earlier; 46): the concentration of free substrate is the same inside and outside of the dialysis bag (or periplasm), but the total concentration (bound plus free) is higher inside and dependent on the binding protein concentration (fully induced maltose-binding protein has been estimated to be 1 mM in the periplasm; 90) and on its affinity for the substrate. It is thus irrelevant to speak of a function of binding proteins as increasing the free concentration of the substrate within the periplasm; but it is likely that the ability of the highly concentrated binding proteins to trap substrate within the periplasm and present it in a concentrated form to the next transport component is an important aspect of their function. It is also possible that binding of substrate may not be as freely and fully reversible as we picture it to be in vitro: the binding protein may actually maintain a higher level of bound substrate than expected from simple in vitro kinetics, possibly by a "cage" effect similar to that thought to be responsible for maintaining a DNA-DNA-binding protein complex intact through gel electrophoresis (91). In either case, it seems reasonable that the binding protein with substrate bound to it is the species that performs the next step in transport.

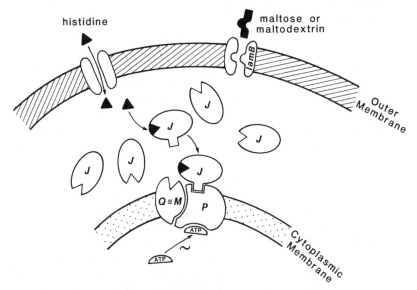

Figure 4 Transport model. The proteins represented are those of the histidine operon. The equivalent components for maltose would be: the maltose-binding protein (MalE) equivalent to J; the MalK protein, equivalent to P; the MalF and MalG proteins, equivalent to Q and M. The squiggle (∼) suggests an involvement of ATP in energy coupling in an unknown way. The J protein is shown as interacting with the P protein on the basis of genetic evidence. In the maltose system available evidence suggests interaction between the binding protein and the MalF and MalG proteins. Possibly, the binding protein interacts with all three of the membrane-bound components. See text for details.

Figure 4 also shows that the liganded binding protein undergoes a conformational change which allows its interaction with one of the membrane-bound components. Conformational changes in the membrane-bound apparatus triggered by the interaction with the liganded binding protein elicit both the release of substrate from the binding protein and the appearance of binding site(s) on the membrane-bound component(s), which allow passage of the substrate from one binding site to the other, into the inside of the cell. It is also necessary to postulate that the membrane components do not have an accessible (or active) binding site unless the loaded binding protein has interacted with the membrane, because no transport has been detected in spheroplasts or membrane vesicles deprived of the binding protein, or in mutants lacking the binding protein.

The loaded site of the binding protein must be very close (juxtaposed) to the next site, otherwise a released substrate molecule would be no different than any other substrate molecule that wandered into the periplasm, which would then negate any special function to the binding protein itself. Since a concentration gradient is established, energy must be consumed at some stage of transport. Energy coupling might occur, in an unspecified fashion, via the nucleotide-binding site. While the initial steps—binding of substrate to the periplasmic component and conformational change of the latter—are well established, all of the rest is based on indirect evidence. Interaction might be with one of the hydrophobic membrane-bound components and/or with the "conserved" protein element (possibly in order to properly position the periplasmic protein relative to the hydrophobic membrane components). The existence of a substrate-binding site on the hydrophobic component(s) is likely, but not proven. And the involvement of the "conserved" component with the energy coupling process is entirely theoretical. If the newly discovered nucleotide-binding site is involved in energy coupling, then ATP hydrolysis presumably occurs concomitant with transport. This hypothesis is far from proven and it is still entirely possible that energy is derived otherwise than from ATP (see below). In this case, the nucleotide-binding site might have a regulatory function.

An analogous model utilizing the opening and closing of a pore through the membrane-bound permease components, instead of binding sites, is a viable alternative to the binding site model depicted in Figure 4 (18). In this case the liganded periplasmic component triggers opening of the pore and the substrate diffuses through the pore, which closes up again once the free binding protein is released from the membrane. This ensures one-way entry into the cytoplasm. It is hard to reconcile with this model the results obtained with mutants in the membrane-bound components that cause a change in specificity or eliminate the need for the binding protein.

A brief comment should be made here concerning the relative amounts of binding protein and membrane components. In the case of the histidine permease it was shown that the histidine-binding protein is in excess over the membrane-bound components (30-fold excess of J over P; 27, 92) and is limiting in transport (92). The level of the membrane-bound components is critical in histidine transport, since a decrease in the level of any one of them yielded corresponding decreases in rates of uptake (92). On the other hand, in the case of the maltose permease a decrease in the level of the maltose-binding protein to 20% of its fully induced level, with presumably unchanged levels in the membrane-bound components, is not accompanied by a decrease in the rate of uptake (93), indicating that the amount of maltose-binding protein is not limiting for maltose transport. This result will have to be explained in terms of the finding that the MalK protein is present in amounts comparable to those of the maltose-binding protein: ($>10^4$ molecules per cell (84 and 93, respectively), thus suggesting that its level might be carefully poised to take fuller advantage of the entire complement of binding protein. This particular arrangement may be necessary for a permease, such as the maltose one, which transports very large amounts of substrate: the V_{max} for maltose is about 40 μmol/min/g dry weight as opposed to 1 μmol/min/g dry weight for the histidine permease. However, it should be remembered that estimates on the amount of the MalK protein differ widely (compare 84 with 82). It also should be noted that careful quantitation of the two hydrophobic components has not been achieved in any system; thus it is premature to draw any hypothesis specifically based on the level of these proteins.

In order to prove the correctness of various aspects of the model of Figure 4 it will be necessary to isolate, purify, and characterize each of the membrane-bound components, and to reconstitute the entire system in liposome vesicles. This is the hardest area of investigation. However, it should at least be possible to confirm biochemically in the near future the interaction of the periplasmic component with one or more of the membrane-bound components, since overproduction of these proteins by genetic engineering has been achieved in several systems, thus allowing efficient visualization of reaction products. New tools, such as direct visualization of protein location by electron microscopy of ultrathin sections of frozen cells (94), might also prove very useful towards this end.

ENERGY COUPLING

In 1973 Berger & Heppel presented data that were interpreted to indicate that ATP or some form of phosphate bond energy is responsible for powering periplasmic systems, as opposed to shock-resistant systems which would be powered directly by the proton-motive force (10, 11). Data were obtained with cells that had been starved to eliminate endogenous energy sources and that, in addition, were either treated with a variety of metabolic poisons or were defective in the proton-translocating ATPase. The treatments were intended to affect differentially and specifically various metabolic steps, either lowering the ATP level or the proton-motive force. The results can be summarized as follows: arsenate inhibits periplasmic systems much more strongly than shock-resistant ones; in an ATPase mutant periplasmic systems cannot be energized by D-lactate or phenazine methosulfate/ascorbate, while shock-resistant ones can; finally, in an ATPase mutant periplasmic systems are slightly more resistant to proton uncouplers than shock-resistant systems. However, since any one of the treatments that was used could have affected multiple metabolic pathways, either directly or indirectly, and since the presumed changes were rarely monitored by direct assay, interpretation of these data should be made with caution. For example, since good assays of the proton-motive force were not available, the energized membrane state was assumed to have been low or high under a certain set of conditions and thus to explain a certain set of results.

Today we are aware of the complicated interrelationships between the proton-motive force and a variety of cell functions, therefore we do not expect that a simple relationship exists between addition of an inhibitor and its unique effect on the proton-motive force. In light of later results, it is interesting that under several conditions then tested the two classes of transport systems behaved very similarly with relatively small variations in the extent of the response. For example, treatment with cyanide or with the uncoupler FCCP dramatically inhibited transport of both glutamine and proline (representative of shock-sensitive and shock-resistant permeases respectively); differences in the response between the two classes were seen only at low cyanide concentration and were minor (in either wild-type or *unc* cells) (11). These data could be interpreted to say equally well either that the two classes of permeases respond similarly or that they respond differently to proton-motive-force inhibitors. A troublesome ambiguity also was evident in the correlation between levels of ATP and extent of transport. Upon addition of cyanide to wild-type cells, actual measurement of the ATP level indicated that it was either unchanged or increased; however, glutamine transport was drastically decreased. This suggests that ATP is not the only factor energizing periplasmic systems. Additional results from two other laboratories contribute further complications to the problem. Plate clearly demonstrated that under conditions where the ATP level of cells is unchanged, but the proton-motive force is decreased, glutamine transport is also decreased, thus indicating 1. that ATP is not sufficient to power transport, and 2. that the proton-motive force plays a role (directly or indirectly) in periplasmic transport (95). In agreement with Plate's results, Singh & Bragg showed that periplasmic systems are functional only under conditions in which a proton-motive force is expected to be generated (96). Earlier experiments indicating that ATP is directly involved in energizing both periplasmic and shock-resistant systems should be reinterpreted today taking into consideration the fact that a proton-motive force could have been built up by way of the ATPase activity (97). Studies with the maltose permease were interpreted as indicating that the level of intracellular ATP is not the direct energy source and that the proton-motive force is not important in powering maltose transport. These latter results may also need reinterpretation since no assays of the ATP levels or of the proton-motive force were performed, in particular since severe inhibition of transport by uncouplers was clearly seen in the wild-type strain (98).

An interesting set of experiments implicating acetylphosphate as the energy source for periplasmic systems was presented by Hong and his collaborators (99). These experiments suffered from the same sources of ambiguity as the above experiments, since measurements of acetylphosphate were not performed and the correlation between transport and acetylphosphate was therefore indirect. Later this laboratory presented preliminary data that indicated that under appropriate conditions in which arsenate inhibited glutamine transport very strongly the levels of ATP and acetylphosphate were normal or only slightly decreased. Therefore, it was suggested that not acetylphosphate, but a compound derived from it, was the true energy-coupling factor (100). This same laboratory also presented data obtained in membrane vesicles reconstituted with glutamine-binding protein (see below), which are basically in agreement with those obtained in whole cells, in that they suggest that the proton-motive force is required, but is not the only requirement and that a factor that is synthesized only in the presence of the enzymes involved in the biosynthesis of acetylphosphate, acetylkinase, and/or phosphotransacetylase, is also needed (101).

It should be pointed out that whatever the effect of the proton-motive force may be, it probably is not the coupling of transport to movement of protons because it has been impossible to show proton translocation during transport through several periplasmic systems (102, 103). However, it should be remembered that these determinations were performed under conditions that are not known to encourage vigorous transport to occur, i.e. very high bacterial and substrate concentrations and anaerobiosis, and these nonphysiological conditions may be responsible for artifactual results. A tentative conclusion might be formed at this point that the proton-motive force plus at least another factor, probably not ATP itself, are important in the functioning of periplasmic permeases. It is possible that the function of the proton-motive force is to maintain the "proper" configuration of the transport complex rather than to supply directly the energy needed for accumulation. No attempts have been made to distinguish between these two possibilities.

Critical discussions of theories and experiments concerning energy coupling in shock-sensitive systems have appeared fairly recently (19, 104), and the reader is referred to these reviews for additional details. Here it suffices to say that since it is unclear which of the myriad of metabolic intermediates are present under any one set of circumstances, it is advisable at this moment to restrict generalizations to a simple statement that the two classes of transport systems can behave fundamentally differently with respect to energization, without specifying the primary differences any further. Clearly, this aspect of research is just taking its first steps and further work will be necessary in order to understand energy coupling in these permeases.

RECONSTITUTION

An essential step in the understanding of the mechanism of action of periplasmic permeases is the ability to have them function in a much simpler system, such as membrane vesicles. Ideally, study of transport in liposomes is the final goal of transportologists. While the latter is probably in the distant future as far as periplasmic systems are concerned, reconstitution of binding protein–dependent active transport in membrane vesicles has been recently obtained

(105, reviewed in 106). Briefly, the method consists of a standard preparation of membrane vesicles from lysozyme-generated spheroplasts. Addition to these vesicles of high concentrations of binding protein (about 2.0 mg per ml for the glutamine-binding protein) and of an energy source allows transport of glutamine into the vesicles. A crucial factor seems to be that the vesicles be prepared from a mutant strain lacking the binding protein (101); failure to do so yields vesicles with high residual levels of transport that cannot be increased by the addition of binding protein. This result may explain numerous previous failures to obtain reconstitution of transport in membrane vesicles. In this respect it may be interesting that membrane vesicles prepared from wild-type strains contain a complement of binding protein that is not liberated by the spheroplasting procedure and that seems to behave differently from reconstituted transport: namely, the residual protein cannot be washed away from the vesicles, residual transport occurs efficiently without addition of an energy source and is not sensitive to arsenate, and the apparent K_m for transport is about 10 times greater than that obtained in reconstituted vesicles or intact cells (101). These results may indicate that the residual protein is bound in a special form in the vesicles that is not its "normal" state. Similar results were obtained with the β-methylgalactosidepermease (106a).

One of the important uses of such reconstituted systems is the ability to investigate energy coupling factors, since it is to be hoped that in the much simplified environment of a membrane vesicle it should be possible to investigate individual energy candidates without excessive interference from general metabolism. These hopes have been somewhat frustrated by the finding that vesicles are capable of substantial metabolism (101). Up to now results confirm 1. the importance of the proton-motive force since uncouplers and inhibitors of respiration interfere with reconstituted transport; 2. the noninvolvement (at least in a unique way) of either ATP or acetylphosphate as energy-coupling factors since transport can be abolished under conditions where both compounds are still present in the membrane vesicles (101). However, since these experiments did not involve quantitation of ATP and acetylphosphate and their correlation with various levels of transport, these conclusions should be considered tentative. A serious problem in this kind of reconstitution experiments is the fact that vesicle preparations vary greatly in their ability to utilize external or trapped energy sources, depending on the exact methodology of their preparation (note added in press in Ref. 101). It is to be hoped that as conditions of vesicle preparation are standardized and their metabolic capabilities are thoroughly characterized, these problems will be eliminated. Another source of concern is the very low level of reconstitution achieved; no experiments have been published to address specifically this question. From published data it can be calculated that reconstituted transport is of the order of 1% of that of a corresponding suspension of intact cells. Finally, it is also important that reconstituted vesicles be shown to be effective for a variety of periplasmic systems. At present the method has been shown to be effective for the glutamine permease (101), for the β-methylgalactoside-permease (106a), and for the histidine permease (A. Gee, G. Ames Ferro-Luzzi, unpublished results).

Another reconstitution method utilizes spheroplasts instead of membrane vesicles (107). This reconstitution procedure has been shown to be effective in the case of the *pst* permease for inorganic phosphate (107), ribose (108), glutamine (109), glutamate (110), and histidine (A. Gee, G. Ames Ferro-Luzzi, unpublished results); and the procedure can be used to demonstrate that a particular protein is truly a component of transport by showing transport recovery upon addition of the binding protein. It can also be useful to study the requirements for the interaction between the binding protein and the membrane components. It would not give much of an advantage over whole cells for the study of energy coupling since the entire complement of cytoplasmic proteins is presumably present inside the spheroplasts. The effectiveness of recovery of transport by this method is very good: between approximately 30 and 80% of that of untreated cells (e.g. 107).

A method for introducing binding proteins into whole cells treated with calcium phosphate to permeabilize the outer membrane has been recently developed (111 and reviewed in Ref. 112). The method is based on the permeabilization of the outer membrane by high osmotic pressure in the presence of calcium phosphate, at 0°C. The permeabilized outer membrane allows entry of entire proteins in a manner possibly similar to that allowing DNA entry during transformation (noteworthy, however, is the fact that for proteins there is no need of a temperature shock). Again, this reconstitution method, like the one using spheroplasts, can be used for the direct demonstration of the involvement of a binding protein in transport. The procedure is simpler than the one involving spheroplasts. Like the latter method, it also could be used to determine the nature of the interaction of the binding protein with the other components of transport. However, it should be kept in mind that reconstitution in this case is into a presumably intact periplasmic environment, which, while having the advantage of being more "natural," also has the disadvantage of being more complicated, as compared to a presumably un-encumbered surface in spheroplasts or vesicles.

The latter reconstitution method has been used to show reconstitution of transport and chemotaxis by the addition of a binding protein, and to study the mobility of maltose-binding protein within the periplasm by bleaching, with a laser beam, fluorescently labeled binding protein that had been reconstituted into calcium phosphate–treated cells (113). Interestingly, this latter experiment indicated that the mobility of the maltose-binding protein is extremely low, about 100-fold lower than expected for a protein of the same size in the cytoplasm. Three hypotheses were put forward to explain these results: (*a*) the protein is bound to immobile receptors (it was excluded that such receptors are the inner membrane components of transport or the chemotactic receptor MCP-II by showing that low mobility was unaltered in mutants lacking these proteins); (*b*) the periplasm is compartmentalized; (*c*) the periplasm might be extremely viscous. An additional possibility, that the protein forms multimers, cannot explain this low rate of diffusion, since aggregates of extraordinary size would have to form before the mobility of a protein is sufficiently slowed (C. Higgins, personel communication).

It is to be hoped that use of a combination of these reconstitution methods will bring forth a variety of useful data concerning the architecture of the transporting complex, its mechanism of action, and its energy-coupling reactions.

EVOLUTIONARY ASPECTS

A comparison of the characteristics of the known periplasmic systems suggests that the underlying mechanism is the same for all of them. All the permeases that have been studied extensively have a similar composition, requiring one (or more, see below) periplasmic component and three membrane-bound components. The genes coding for the components are closely linked on the chromosome, probably forming an operon in all cases (though the maltose permease is coded by two divergent operons). An interaction may occur between the periplasmic binding protein and the membrane-bound components. An additional interesting complexity, which also seems to be shared by several periplasmic systems, is the duplication and divergent evolution of the gene coding for the periplasmic component. This has been shown clearly for the *hisJ* gene, which is highly homologous (70%) to the adjacent gene *argT*, coding for the LAO protein (Figure 2; 64). A remarkably analogous situation exists for the branched-chain amino acid permease, where two homologous (80%) periplasmic proteins are found (36): one is specific for leucine (LS protein, coded for by the *livJ* gene), and the other binds leucine, isoleucine, and valine (LIV protein, coded for by the *livK* gene; 12, 114). In each case the two proteins require the same membrane-bound proteins for their function and it is likely that their genes, which are closely linked on the chromosome, originated by duplication of a single ancestral gene (64). Scattered information supports the notion that duplication of the periplasmic component is a common characteristic of these transport systems. The galactose- and the arabinose-binding proteins may also have originated from a duplication since they are antigenically related, they share a small but definite homology (115), and their genes are located quite close to each other on the chromosome (116). A gene with unknown function[2] but coding for a periplasmic protein is part of the

---

[2]In the maltose operon, the *molA* gene codes for a periplasmic protein (M. Hofnung, personal communication) that might be a binding protein that utilizes the *malK*, *malF*, and *malG* components. This may be a case analogous to that of *hisJ/argT* and *livJ/livK*.

maltose transport operon (117, 118; M. Hofnung, personal communication). A malate-inducible periplasmic protein is immunologically related to a citrate-binding protein (G. Sweet, W. Kay, personal communication). Besides the duplication yielding two related periplasmic proteins, evidence for an additional duplication within the histidine and phosphate transport operons can be drawn from the definite homology existing between *hisQ* and *hisM* (87), and between *pstC* and *pstA* (G. Ames Ferro-Luzzi, R. Doolittle, unpublished data; M. Hermodson, S. Buckel, personal communication).

Considering their similarity despite the complexity of their organization, it has been hypothesized that all the periplasmic systems have originated by duplication and divergence from a single ancestral system, perhaps already containing a duplication both of the periplasmic component and of one of the membrane components (87). Each system would have evolved a different specificity while retaining the same basic architecture. A search for homologies among parallel components of all these systems could answer this question. We have already seen that strong homology exists between one of the membrane-bound components of each of six different permeases (Figures 2 and 3). A comparison among the hydrophobic membrane components revealed significant homologies between RbsC and PstA and PstC and between MalF and PstC, and weak homology between MalF and HisM (G. Ames Ferro-Luzzi, R. Doolittle, unpublished data). Comparison of all the available sequences of periplasmic components has shown that the galactose-, arabinose-, and ribose-binding proteins are significantly homologous to each other (119, 120), but that no significant homologies exist between several completely unrelated binding proteins (G. Ames Ferro-Luzzi, T. Farrah, M. Johnson, R. Doolittle, unpublished data). In favor of a structural and functional relationship between the binding proteins are the extensive studies of Quiocho and his collaborators on the X-ray structure of several such proteins, which showed in all cases strongly similar two-domain structures, even though the proteins transported substrates that were quite different from each other (leucine-isoleucine-valine, arabinose, and sulfate) (69). Thus, there is reasonable evidence that a complex ancestral system would have spawned the present multiplicity of periplasmic permeases.

If all periplasmic systems are derived from an ancestral system that already contained a duplication of the periplasmic component, why is the homology between the two periplasmic components of one permease greater than the homology between those same components and their duplicated versions as they appear in a permease of completely different specificity? In other words, why would LAO resemble J more than either LAO or J resembles LS or LIV? A reasonable explanation is that LAO and J are constrained in their evolution by their need to interact with the same membrane component, whereas LS and LIV presumably need to interact with their own set of specialized membrane components. This hypothesis is supported by the fact that within the J and LAO proteins there are segments that are much more highly conserved (>90%

homology) than the overall protein sequence (70% homology). These conserved regions are believed to be involved in forming that domain of each protein molecule that is believed to be responsible for the interaction with the membrane components since mutational alteration in one of these highly conserved regions of the J protein results in a loss of the ability of J to interact properly with P (63, 64). Therefore, this system, and independently each of the other systems, would have evolved as a "package."

CONCLUSION

Work on the structure and function of periplasmic systems has reached a stage where essential generalizations can be made because research into numerous systems has started yielding abundant results and, most important, revealing through their similarities the existence of a common composition and organization. The four basic aims in the study of these transport systems are at various levels of development.

First, the characterization of the protein composition and genetic organization of a periplasmic permease has been essentially accomplished, since one transport operon (histidine) has been completely sequenced and all gene products have been identified and overproduced. Several other permeases (oligopeptide, ribose, maltose, phosphate, leucine-isoleucine-valine) are getting very close to complete characterization and showing great similarity to each other, and many others are in various stages of characterization.

Second, the investigation of the molecular mechanism of action and the architecture of these systems is beginning to take off, with the first attempts at purifying the protein components and determining their possible enzymatic functions. Antibodies against a number of these proteins have been obtained and more should be available soon. The state of the field is such that very precise questions can now be asked concerning the existence and nature of protein-protein interaction sites by altering specific amino acid residues through the use of recombinant DNA technology and as dictated by the available X-ray structure determination of several of these proteins. By similar techniques the nature of the nucleotide-binding site on the conserved components and of possible substrate-binding sites on the membrane components can also be explored. With the availability of these tools, in conjunction with the ever essential genetic approach, it should be possible to elucidate the molecular architecture of these permeases in the near future.

Third, the least advanced of the three areas of research is the study of the mechanism of energy coupling. Because of the unavailability of a well-characterized in vitro system, the complexities arising from doing this research in whole cells have caused the results to be ambiguous. However, the results are tantalizing because they clearly point out differences with the energy-coupling mechanism of shock-resistant permeases. This area will undoubtedly require a great deal of dedication, but it will be very rewarding if successful.

Finally, an intriguing aspect emerging more and more clearly as more information becomes available is the clear-cut sequence homology between several components from independent permeases. Since we can guess that a cell may contain a few dozen periplasmic permeases, it will be an interesting speculative problem to trace their genealogy as more are being discovered and compared to each other. In addition, some important clues concerning their mechanism of action might emerge from such studies. The next few years should bring understanding and excitement to the field of periplasmic transport with all its varied approaches and corps of dedicated and enthusiastic followers.

ACKNOWLEDGMENTS

The author's experimental work cited in this review was supported by a research grant from the National Institutes of Health (AM-12121). I would like to thank the many authors who generously sent preprints of their manuscripts.

Literature Cited

1. Nikaido, H., Vaara, M. 1985. *Microbiol. Rev.* 49:1–32
2. Neu, H. C., Heppel, L. A. 1965. *J. Biol. Chem.* 240:3685–92
3. Jacobson, G. R., Takacs, B. J., Rosenbusch, J. P. 1976. *Biochemistry* 15:2297–303
4. Ames, G. F.-L., Nikaido, K. 1979. *J. Biol. Chem.* 254:9947–50
5. Kier, L. D., Weppelman, R., Ames, B. N. 1977. *J. Bacteriol.* 130:399–410
6. Pardee, A. B., Prestidge, L. S., Whipple, M. B., Dreyfuss, J. 1966. *J. Biol. Chem.* 241:3962–69
7. Furlong, C. E. 1986. In *Escherichia coli and Salmonella typhimurium: Cellular and Molecular Biology*, ed. F. C. Neidhardt. Washington, DC: ASM. In press
7a. Furlong, C. E. 1986. *Methods Enzymol.* 125:279–88
8. Kaback, H. R. 1983. *J. Membr. Biol.* 76:95–112
9. Overath, P., Wright, J. K. 1983. *Trends Biochem. Sci.* 8:404–8
10. Berger, E. A. 1973. *Proc. Natl. Acad. Sci. USA* 70:1514–18
11. Berger, E. A., Heppel, L. A. 1974. *J. Biol. Chem.* 249:7747–55
12. Landick, R., Oxender, D. L., Ames, G. F.-L. 1985. In *The Enzymes of Biological Membranes*, Vol. 3, ed. A. N. Martonosi, pp. 577–615. New York: Plenum
13. Antonucci, T. K., Oxender, D. L. 1986. *Adv. Microbiol. Physiol.* 27: In press
14. Ames, G. F.-L., Lever, J. 1970. *Proc. Natl. Acad. Sci. USA* 66:1096–103
15. Ames, G. F.-L. 1972. In *Biological Membranes. ICN-UCLA Symp. Mol. Biol.* ed. C. F. Fox, pp. 409–26. New York: Academic. 501 pp.
16. Ames, G. F.-L., Noel, K. D., Taber, H., Spudich, E. N., Nikaido, K., et al. 1977. *J. Bacteriol.* 129:1289–97
17. Ames, G. F.-L., Lever, J. E. 1972. *J. Biol. Chem.* 247:4309–16
18. Higgins, C. F., Haag, P. D., Nikaido, K., Ardeshir, F., Garcia, G., Ames, G. F.-L. 1982. *Nature* 298:723–27
19. Hengge, R., Boos, W. 1983. *Biochem. Biophys. Acta* 737:443–78
20. Hiles, I. D., Higgins, C. F. Submitted for publication
21. Bell, A. W., Buckel, S. D., Groarke, J. M., Hope, J. N., Kingsley, D. H., Hermodson, M. A. Submitted for publication
21a. Buckel, S. D., Bell, A. W., Mohana Rao, J. K., Hermodson, M. A. Submitted for publication
22. Harayama, S., Bollinger, J., Iino, T., Hazelbauer, G. L. 1983. *J. Bacteriol.* 153:408–15
23. Robbins, A. R., Guzman, R., Rotman, B. 1976. *J. Biol. Chem.* 251:3112–16
24. Muller, N., Heine, H.-G., Boos, W. 1985. *J. Bacteriol.* In press
25. Surin, B. P., Rosenberg, H., Cox, G. B. 1985. *J. Bacteriol.* 161:189–98
25a. Amemura, M., Makino, K., Shinagawa, H., Kobayashi, A., Nakata, A. 1985. *J. Mol. Biol.* 184:241–50

26. Hope, J. N., Bell, A. W., Hermodson, M. A., Groarke, J. M. Submitted for publication
27. Ames, G. F.-L., Nikaido, K. 1978. *Proc. Natl. Acad. Sci. USA* 75:5447–51
28. Hogg, R. W. 1984. In *Microbiology-1984*, ed. L. Leive, D. Schlessinger, pp. 38–41. Washington, DC: Am. Soc. Microbiol.
29. Ames, G. F.-L. 1974. In *Methods in Enzymology: Biomembranes*, Vol. 32, ed. S. Fleischer, L. Packer, R. W. Estabrook, pp. 849–56. New York: Academic
30. Szmelcman, S., Schwartz, M., Silhavy, T. J., Boos, W. 1976. *Eur. J. Biochem.* 65:13–19
31. Rosenberg, H., Gerdes, R. G., Chegwidden, K. 1977. *J. Bacteriol.* 131:505–11
32. Winkler, H. H., Wilson, T. H. 1966. *J. Biol. Chem.* 241:2200–11
33. Brenner, M., Ames, B. N. 1971. In *Metabolic Regulation. Metabolic Pathways*, Vol. 5, ed. H. J. Vogel, pp. 349–87. New York: Academic
34. Ames, G. F.-L. 1984. *Arch. Biochem. Biophys.* 104:1–18
35. Kustu, S. G., Ames, G. F.-L. 1973. *J. Bacteriol.* 116:107–13
36. Landick, R., Oxender, D. L. 1985. *J. Biol. Chem.* 260:8257–61
37. Wandersman, C., Schwartz, M., Ferenci, T. 1979. *J. Bacteriol.* 140:1–13
38. Shuman, H. A., Treptow, N. A. 1985. See Ref. 12, pp. 561–75
39. Higgins, C. F., Hardie, M. M. 1983. *J. Bacteriol.* 155:1434–38
40. Guyer, C. A., Morgan, D. G., Osheroff, N., Staros, J. V. 1985. *J. Biol. Chem.* 260:10812–18
41. Zukin, R. S., Strange, P. G., Heavey, L. R., Koshland, D. E. Jr. 1977. *Biochemistry* 16:381–86
42. Richarme, G., Kepes, A. 1974. *Eur. J. Biochem.* 45:127–33
43. Miller, D. M. III, Olson, J. S., Quiocho, F. A. 1980. *J. Biol. Chem.* 255:2465–71
44. Boos, W., Gordon, A. S., Hall, R. E., Price, H. D. 1972. *J. Biol. Chem.* 247:917–24
45. Miller, D. M. III, Olson, J. S., Pflugrath, J. W., Quiocho, F. A. 1983. *J. Biol. Chem.* 258:13665–72
46. Lever, J. E. 1972. *Anal. Biochem.* 50:73–83
47. Boos, W., Gordon, A. S., Hall, R. E., Price, H. D. 1972. *J. Biol. Chem.* 247:916–24
48. Zukin, R. S., Clos, M. F., Hirsch, R. E. 1986. *Biophys. J.* In press
49. Zukin, R. S., Steinman, H. M., Hirsch, R. E. 1984. See Ref. 28, pp. 53–56

50. Zukin, R. S. 1979. *Biochemistry* 18:2139–45
51. Penrose, W. R., Zand, R., Oxender, D. L. 1970. *J. Biol. Chem.* 245:1432–37
52. Weiner, J. H., Heppel, L. A. 1971. *J. Biol. Chem.* 246:6933–41
53. Zukin, R. S., Hartig, P. R., Koshland, D. E. Jr. 1979. *Biochemistry* 18:5599–605
54. Kreishman, G. P., Robertson, D. E., Ho, C. 1973. *Biochem. Biophys. Res. Commun.* 53:18–23
55. Robertson, D. E., Kroon, P. A., Ho, C. 1977. *Biochemistry* 16:1443–51
56. Cedel, T. E., Cottam, P. F., Meadows, M. D., Ho, C. 1984. *Biophys. Chem.* 19:279–87
57. Shen, Q., Simplacenau, V., Cottam, P. F., Ho, C. 1985. *Biophys. J.* 47:88a
58. Gaudin, C., Marty, B., Ragot, M., Sari, J. C., Belaich, J. P. 1980. *Biochimie* 62:741–46
59. Rotman, B., Ellis, J. H. 1972. *J. Bacteriol.* 111:791–96
60. Zukin, R. S., Hartig, P. R., Koshland, D. E. Jr. 1977. *Proc. Natl. Acad. Sci. USA* 74:1932–36
61. Kustu, S. G., Ames, G. F.-L. 1974. *J. Biol. Chem.* 249:6976–83
62. Manuck, B. A., Ho, C. 1979. *Biochemistry* 18:566–73
63. Ames, G. F.-L., Spudich, E. N. 1976. *Proc. Natl. Acad. Sci. USA* 73:1877–81
64. Higgins, C. F., Ames, G. F.-L. 1981. *Proc. Natl. Acad. Sci. USA* 78:6038–42
65. Ames, G. F.-L., Higgins, C. F. 1983. *Trends Biochem. Sci.* 8:97–100
66. Treptow, N. A., Shuman, H. A. 1985. *J. Bacteriol.* 163:654–60
67. Hunt, A. G., Hong, J.-S. 1983. *Biochemistry* 22:851–54
68. Bavoil, P., Nikaido, H. 1981. *J. Biol. Chem.* 256:11385–88
69. Gilliland, G. L., Quiocho, F. A. 1981. *J. Mol. Biol.* 146:341–62
70. Newcomer, M. E., Lewis, B. A., Quiocho, F. A. 1981. *J. Biol. Chem.* 256:13218–22
71. Mao, B., Pear, M. P., McCammon, J. A., Quiocho, F. A. 1982. *J. Biol. Chem.* 257:1131–33
72. Saper, M. A., Quiocho, F. A. 1983. *J. Biol. Chem.* 258:11057–62
73. Vyas, N. K., Vyas, M. N., Quiocho, F. A. 1983. *Proc. Natl. Acad. Sci. USA* 80:1792–96
74. Mowbray, S. L., Petsko, G. A. 1983. *J. Biol. Chem.* 258:7991–97
75. Pflugrath, J. W., Quiocho, F. A. 1985. *Nature* 314:257–60
76. Celis, R. T. F. 1984. *Eur. J. Biochem.* 145:403–11

77. Ames, G. F.-L., Nikaido, K. 1981. *Eur. J. Biochem.* 115:525–31
78. Gilson, E., Higgins, C. F., Hofnung, M., Ames, G. F.-L., Nikaido, H. 1982. *J. Biol. Chem.* 257:9915–18
79. Froshauer, S., Beckwith, J. 1984. *J. Biol. Chem.* 259:10896–903
80. Dassa, E., Hofnung, M. 1985. *EMBO J.* 4:2287–93
81. Higgins, C. F., Hiles, I. D., Whalley, K., Jamieson, D. K. 1985. *EMBO J.* 4:1033–40
82. Shuman, H. A., Silhavy, T. J. 1981. *J. Biol. Chem.* 256:560–62
83. Shuman, H. A., Silhavy, T. J., Beckwith, J. 1980. *J. Biol. Chem.* 255:168–74
84. Bavoil, P., Hofnung, M., Nikaido, H. 1980. *J. Biol. Chem.* 255:8366–69
85. Walker, J. E., Saraste, M., Runswick, M. J., Gay, N. J. 1982. *EMBO J.* 1:945–51
86. Hobson, A. C., Weatherwax, R., Ames, G. F.-L. 1984. *Proc. Natl. Acad. Sci. USA* 81:7333–37
87. Ames, G. F.-L. 1985. *Curr. Top. Membr. Transp.* 23:103–19
88. Shuman, H. A. 1982. *J. Biol. Chem.* 252:5455–61
89. Payne, G., Spudich, E. N., Ames, G. F.-L. 1985. *Mol. Gen. Genet.* 200:493–96
90. Dietzel, I., Kolb, V., Boos, W. 1978. *Arch. Microbiol.* 118:207–18
91. Fried, M., Crothers, D. 1981. *Nucleic Acids Res.* 9:6505–25
92. Stern, M. J. 1986. PhD thesis. Univ. Calif., Berkeley
93. Manson, M. D., Boos, W., Bassford, P. J., Rasmussen, B. A. 1985. *J. Biol. Chem.* In press
94. Bernadac, A., Lazdunski, C. 1985. *Biol. Cell.* 41:211–16
95. Plate, C. A. 1979. *J. Bacteriol.* 137:221–25
96. Singh, A. P., Bragg, P. D. 1977. *J. Supramol. Struct.* 6:389–98
97. Singh, A. P., Bragg, P. D. 1976. *Biochem. Biophys. Acta* 438:450–61
98. Ferenci, T., Boos, W., Schwartz, M., Szmelcman, S. 1977. *Eur. J. Biochem.* 75:187–93
99. Hong, J.-S., Hunt, A. G., Masters, P. S., Lieberman, M. A. 1979. *Proc. Natl. Acad. Sci. USA* 76:1213–17
100. Hong, J.-S., Hunt, A. G. 1980. *J. Supramol. Struct.* 4:77
101. Hunt, A. G., Hong, J.-S. 1983. *Biochemistry* 22:844–50
102. Henderson, P. J. E., Giddens, R. A., Jones-Mortimer, M. C. 1977. *Biochem. J.* 162:309–20
103. Daruwalle, K. R., Paxton, A. T., Henderson, P. J. F. 1981. *Biochem. J.* 200:611–27
104. Hunt, A. G., Hong, J.-S. 1982. In *Membranes and Transport,* Vol. II, ed. A. N. Martonosi. New York: Plenum
105. Hunt, A. G., Hong, J.-S. 1981. *J. Biol. Chem.* 256:11988–91
106. Hong, J.-S. 1986. *Methods Enzymol.* 125:302–9
106a. Rotman, B., Guzman, R. 1984. See Ref. 28, pp. 57–60
107. Gerdes, R. G., Strickland, K. P., Rosenberg, H. 1977. *J. Bacteriol.* 131:512–18
108. Robb, F. T., Furlong, C. E. 1980. *J. Supramol. Struct.* 13:183–90
109. Masters, P. S., Hong, J.-S. 1981. *Biochemistry* 20:4900–4
110. Barash, H., Halpern, Y. S. 1971. *Biochem. Biophys. Res. Commun.* 45:681–88
111. Brass, J. M., Boos, W., Hengge, R. 1981. *J. Bacteriol.* 146:10–17
112. Brass, J. M. 1986. *Methods Enzymol.* 125:289–301
113. Higgins, C. F., Brass, J. M., Foley, M., Rugman, P. A., Birmingham, J., Garland, P. B. 1986. *J. Bacteriol.* 165 (3) In press
114. Landick, R., Anderson, J. J., Mayo, M. M., Gunsalus, R. P., Mavromara, P., et al. 1980. *J. Supramol. Struct.* 14:527–37
115. Mahoney, W. C., Hogg, R. W., Hermodson, M. A. 1981. *J. Biol. Chem.* 256:4350–56
116. Clark, A. F., Hogg, R. W. 1981. *J. Bacteriol.* 147:920–24
117. Duplay, P., Bedouelle, H., Charbit, A., Clement, J. M., Dassa, E., et al. 1984. See Ref. 92, pp. 29–33
118. Clement, J. H., Hofnung, M. 1981. *Cell* 27:507–14
119. Argos, P., Mahoney, W. C., Hermodson, M. A., Haney, M. 1981. *J. Biol. Chem.* 256:1131–33
120. Groarke, J. M., Mohana Rao, J. K., Hermodson, M. A. Submitted for publication.

Ann. Rev. Biochem. 1986. 55:427–53

TAURINE: BIOLOGICAL UPDATE

Charles E. Wright, Harris H. Tallan, and Yong Y. Lin

Office of Mental Retardation and Developmental Disabilities of the State of New York, Department of Developmental Biochemistry, New York State Institute for Basic Research in Developmental Disabilities, Staten Island, New York 10314

Gerald E. Gaull

Department of Nutrition and Medical Affairs, Searle Food Resources, Inc., Skokie, Illinois 60076

CONTENTS

PERSPECTIVES AND SUMMARY

Ten years ago, taurine was thought to be mainly the biochemically inert end product of methionine and cysteine metabolism, along with inorganic sulfate. Free taurine, however, is found in millimolar concentrations, especially in the tissues that are excitable (1), are rich in membranes (1), and that generate oxidants (cf 2, 3).

0066-4154/86/0701-0427$02.00

What is taurine doing? New data from various laboratories are focusing on a central mechanism, namely: the role of taurine as a nutrient is to protect cell membranes by attenuating toxic compounds. Taurine is most abundant where reduced oxygen molecules are generated and where other potentially toxic substances, such as bile acids, retinoids, and xenobiotics are found. The chemical reactivity of the amino group allows taurine to perform this function, and it is probably the sulfonic acid end of the molecule that is critical in its performance.

A need for taurine in nutrition was first identified in early 1975 with publication of the preliminary communication on retinal degeneration in the taurine-deficient cat (4) and in work showing the failure of formula-fed preterm infants to maintain normal plasma and urine taurine concentrations (5). Since then interest in taurine has burgeoned.

The classical literature on taurine has been comprehensively reviewed from its discovery in 1827 (6). Six symposium proceedings published between 1976 and 1985 (7–12) provide an archive of the recent renaissance in interest in this molecule. Its probable neurochemical role has just been authoritatively reviewed (13). Therefore, we have not attempted to be encyclopedic.

The purpose of this review is to focus on the areas of emerging current interest and on biochemical data pointing to the more general biological function mentioned above. The outlook will necessarily be that of the authors, who worked together for a number of years on this interesting molecule.

BASIC BIOCHEMISTRY

Chemistry

Taurine, 2-aminoethanesulfonic acid, is a β-amino acid with a molecular mass of 125 daltons. It is a colorless compound, crystalizing as tetragonal needles; m.p. 328°C (decomp.) (14). It is soluble in water, 10.48 g/100 ml at 25°C (1, 14). The dielectric constant at pH 2.8 is 41–53 (15, 16).

Chemical analysis of taurine has been carried out by colorimetric (17–22), fluorometric (23, 24), radiometric (25), and enzymatic (26) methods. Chromatographic procedures by amino acid analyzer (27–33), high performance liquid chromatography (34–38), and gas chromatography (39–45) are now the methods of choice. Gas chromatograph–mass spectrometry (GCMS) has also been utilized to study plasma taurine kinetics in the rhesus monkey (42, 43). Natural abundance [13C]taurine nuclear magnetic resonance (NMR) has been used to record the presence of taurine in whole tissue from a number of marine mollusks (46). More recently, chemically enriched [13C]taurine NMR has been used to study the complexation of calcium by taurine; the direct metal ion interaction with taurine appears to be small (see below) (47). Enriched 13C NMR has also been used to study the dynamic state of taurine that has been taken up by cultured human lymphoblastoid cells. The results indicated that extracellular taurine taken up by human lymphoblastoid cells is incorporated into a mobile large pool of intracellular taurine and has no observable interaction with cellular macromolecular components or with metal ions (48).

Like other amino acids, taurine behaves as an amphoteric electrolyte; however, some physicochemical differences in taurine are expected as a result of the presence of the sulfonate ion. Taurine possesses a more acidic acid function (pK_1' 1.5) as well as a more acidic ammonium function (pK_2' 8.74) than other amino acids (49). For example, its structural analogues, β-alanine and α-alanine, have a pK_1' and pK_2' of 3.6 and 10.19, and 2.34 and 9.67, respectively (50).

The capacity of amino acid dipole ions to form a metal complex is an important feature of their biological activity. Taurine forms less stable metal complexes with various transition metals than do other amino acids, as indicated by the small stability constants (log K_S) shown in Table 1.

Despite the significant biological effect of taurine on calcium and zinc binding in living systems, a contribution by direct interaction between taurine and metals appears to be minimal. Dolara and coworkers (53, 54), using natural abundance ^{13}C NMR, found the stability constant for taurine-calcium complex formation to be low. They estimated that 8.2% of myocardial calcium is bound to taurine. Irving and coworkers (47) refined this approach by using ^{13}C-enriched taurine, and calculated the stability constants to be even lower. They estimated that only 1% of myocardial calcium complexed with taurine. The low stability constants of taurine-metal complexes are the result of the sulfonate ion, which is a poor ligand. Thus, it fails to form a stable five-membered ring complex of metal and amino acid in which both the amino and the acidic groups participate in chelate formation. The direct interaction between taurine and metal ions is mainly attributable to the electric association between metal cations and the sulfonate anion or to the interaction between metal ions and the nitrogen's unshared pair of electrons.

The amino group of taurine can react with carboxylic acids to form amide linkages. Conjugation of bile acids with taurine occurs in most vertebrates. Certain xenobiotics react in similar fashion (46). Retinoic acid reacts with taurine to form retinotaurine in biliary metabolites (55). Taurine also forms peptide bonds in some low-molecular-weight acidic peptides in calf brain synaptosomal and vesicular preparations (56, 57). Formation of taurine chloramines by chlorination of the amino group with hypochlorous acid has recently been implicated in biological systems (2, 58). Deamination of taurine to isethionic acid (2-hydroxyethanesulfonic acid) presumably proceeds through the aldehyde intermediate, 2-oxoethanesulfonic acid (59).

Many taurine derivatives have been synthesized chemically for study of their pharmacological activities. They are synthesized by modification of the amino group and/or formation of sulfonamides; the compounds include a series of

Table 1   Stability constants as log K_s of the complexes of divalent amino acids with metals (51, 52)

Amino acids	Cu^{2+}	Ni^{2+}	Zn^{2+}	Co^{2+}	Cd^{2+}	Fe^{2+}	Mn^{2+}	Mg^{2+}
Glycine	15.4	11.0	9.3	8.9	8.1	7.8	5.5	4
L-Proline	16.8	11.3	10.2	9.3	8.7	8.3	5.5	<4
DL-Tryptophan	15.9	10.2	9.3	8.5	8.1	7.6	5	<4
L-Asparagine	14.9	10.6	8.7	8.4	6.8	6.5	4.5	4
DL-Alanine	15.1	—	—	8.4	—	7.3	—	—
β-Alanine	12.9	4.6	4.1	7	3.7	4	—	—
Taurine	8	3.1	4.6	4	3.1	0	3	—

2-acylamidoethanesulfonamides (60), 2-guanidoethanesulfonic acid (taurocyamine) (61), 2-ureidoethanesulfonic acid (carbamyltaurine) (62), and 2-ethanethiosulfonic acid (thiotaurine) (63).

The sulfinic acid group in hypotaurine, 2-aminoethanesulfinic acid, is quite different in chemical reactivity from the sulfonic acid function in taurine. Unlike sulfonic acids, sulfinic acids retain a lone electron pair on the sulfur atom and hence can behave as a nucleophile contributing to a high chemical reactivity. Sulfinic acids are quite acidic as compared with carboxylic acids but are less acidic than sulfonic acids. Sulfinic acids are known to be readily oxidized by various oxidizing agents that may be present in biological systems. The sulfinic acids also undergo facile autooxidation to give sulfonic acids, and the autooxidation is often accelerated by the presence of chloride ion or metal ions (64). Disproportionation of sulfinic acids takes place readily and gives the corresponding thiosulfonic and sulfonic acids (65). As in disproportionation, reaction of sulfinic acids with disulfides also gives the corresponding thiosulfonic acids (66, 67). The reaction may have considerable interest analogous to the possible reaction of hypotaurine with disulfide bonds in proteins.

General Metabolism

BIOSYNTHESIS   There are a number of possible synthetic routes by which taurine could be derived from precursor cysteine (13): (a) oxidation of cysteine to 3-sulfinoalanine (cysteine sulfinic acid) and then to cysteic acid, which is decarboxylated to taurine; (b) oxidation of cysteine to 3-sulfinoalanine and decarboxylation of the latter to hypotaurine, which is oxidized to taurine; (c) reaction of cysteine with phosphopantothenate to form phosphopantothenoyl cysteine, which is decarboxylated to phosphopantethein, from which cysteamine is split off and oxidized to hypotaurine. The enzymatic apparatus to carry out these transformations is in place [for detailed description, see (13)], except for an uncertainty as to the mechanism of oxidation of hypotaurine to taurine. Evidence that this, too, is enzymic in nature has been presented (68), but the enzyme has yet to be isolated and characterized. A fourth biosynthetic route that has been proposed utilizes inorganic sulfate and serine to form cysteic acid via the sequence $SO_4 + ATP \rightarrow$ adenosine 5'-phosphosulfate $\rightarrow$ 3'-phosphoadenosine 5'-phosphosulfate, which reacts with α-aminoacrylate (derived from serine) to yield cysteic acid. Regulation of the 3-sulfinoalanine pathways could occur by removal of the compound via transamination with 2-oxoglutarate, a reaction catalyzed by aspartate aminotransferase (69). In the mouse, about 85% of administered 3-sulfinoalanine is decarboxylated, about 15% transaminated (70).

Earlier studies (cf 13) indicated that the dominant route varies among species and among tissues. The cysteamine pathway is the main provider of taurine in the heart and possibly in rat brain, the 3-sulfinoalanine pathway being most important in rat and dog liver (71). However, recent work by Kuriyama et al (72) with mice failed to detect any cysteamine or cystamine in brain, suggesting absence of the cysteamine pathway in that organ; the compounds were present in kidney and heart. Liver perfusion experiments (rat and cat) failed to detect the formation of any labeled taurine from [^{35}S]sulfate (73), nor could such formation be demonstrated in feeding experiments with rats (74). It was suggested that the sulfate pathway might be operative only when other pathways were absent (13).

The question remains whether 3-sulfinoalanine is preferentially oxidized, yielding taurine via cysteic acid, or decarboxylated to yield taurine via hypotaurine. This may simply be a matter of the relative activities and affinities of the cysteine dioxygenase and of 3-sulfinoalanine decarboxylase, both routes being active simultaneously. The choice of route is not without consequence, since hypotaurine is a strong competitive inhibitor of taurine uptake into cells, whereas cysteic acid has no effect on taurine transport. It may be for this reason that hypotaurine, if formed, is not permitted to accumulate, and generally is not found in mammalian tissues. The accumulation of hypotaurine in regenerating rat and guinea pig liver (75) may reflect a temporary shift from oxidation of 3-sulfinoalanine to decarboxylation, coupled with an inability to remove hypotaurine as fast as formed in these circumstances.

ENDOGENOUS VS EXOGENOUS TAURINE   The capacity to synthesize taurine, by whatever route, varies widely among species, rendering some, as the cat, totally dependent on dietary sources. How much taurine actually is made by the biosynthetic apparatus has been determined experimentally in a few instances. Huxtable & Lippincott (76, 77) studied rats, which have a high biosynthetic capacity of some 50 μmol/day from dietary methionine. The rat utilizes dietary taurine when available, gradually displacing biosynthesized taurine as excess taurine is eliminated in the urine. On a taurine-deficient diet, less taurine is excreted, and tissue taurine pools continue to consist mostly of biosynthetic taurine. Thus, after 87 days on a diet containing [^{3}H]taurine and

[^{35}S]methionine, 58% of the total carcass taurine was derived from the diet, 29% had been biosynthesized, and 13% was unlabeled taurine from the body pool at the start of the experiment; on a taurine-free diet, 54% of the taurine had been biosynthesized and 46% was unlabeled. That the rate of biosynthesis is constant was shown by summing the [^{35}S]taurine excreted in the urine and remaining in the carcass after 63 days on taurine-free and taurine-containing diets; these totals were similar. Since the percent contribution of diet and of biosynthesis to the total taurine pool did not vary much from one tissue to another, it was suggested that the rate of exchange of taurine from tissue to tissue is greater than the rate of excretion into the urine. Similar results were obtained with mice (78).

The derivation of body taurine is of particular interest in the case of the fetus and the neonate, which have high concentrations of taurine in brain (cf 79). Sturman (80) has determined that at birth 59% of rat brain taurine came from the mother and 41% had been biosynthesized by the fetus during gestation. The proportions for the whole rat are essentially the same. After birth, labeled body taurine continues to be mobilized by the brain and at 12 days constitutes 21% of the increased content of taurine in that organ. By weaning, the origin of whole body taurine was as follows: 6% from mother in utero, 7% from mother's milk, 87% biosynthesized by pup (pre- and post-natally). This proportion is similar to the 80% biosynthesized: 20% dietary taurine in the adult rat.

The effects of pregnancy and lactation on taurine concentrations in the dam have been described by Stipanuk et al (81). Most striking was an 88% increase in liver taurine concentration just before delivery, followed by a drop to 30% of control as lactation commenced, seemingly a mobilization of taurine to provide the large amounts present in early milk.

The actions of guanidinoethanesulfonic acid (GES) on tissue concentrations of taurine indicate that it affects dietary taurine and biosynthesized taurine differently (82). After administration of GES to rats, the ratio of dietary to biosynthetic taurine increased in all tissues, although the actual content of taurine was decreased in tissues other than brain. The mechanism of this effect is not clear; it was suggested that in addition to inhibition of transport there was also inhibition of biosynthesis.

EXCRETION OF TAURINE IN URINE AND BILE   As noted above, under conditions of inadequate availability of taurine, tissue stores are conserved by a decrease in the excretion of the compound. This renal adaptation is accomplished by changes in the rate of reuptake of taurine by the brush border membrane of the renal cortex (83, 84). Kinetic analysis of taurine uptake by brush border membrane vesicles from adult and from 28-day-old rats showed alterations in the V_{max} of uptake, the apparent K_m (50 μM) being unchanged (85, 86). The V_{max} was greater than the normal value of 150 pmol/mg protein/

min in vesicles from animals fed a diet low in sulfur amino acids (345 pmol/mg protein/min); it was less than normal in vesicles from animals fed a high taurine diet (120 pmol/mg protein/min). Age-related differences were slight (85, 87). Similar results were obtained with collagenase-isolated tubules, which have a low-affinity (apparent K_m, 2.56 mM) high capacity uptake system in addition to the high-affinity (apparent K_m, 84 µM) low capacity system (85, 86). For both systems, V_{max} values were affected by the differences in diet, as well as by a 72-hour period of fasting (85). Renal adaptation to diet appeared to be "blunted" by the fast, which tended to return the kinetic parameters to normal (85). A detailed study has been made of the properties of the high-affinity uptake system of rat brush-border membrane vesicles (87).

The conjugation of taurine with bile acids is carried out by most animal species, although glycine conjugates in varying amounts are produced in some cases. The enzymic basis of these species differences has been elucidated by Vessey (88).

TURNOVER   Recent experiments by Huxtable (77) in rats show that the half-lives of taurine turnover for various tissues and organs are quite similar: on a taurine-enriched diet, 4.8 ± 1.1 days for all visceral organs, 5.5 ± 1.1 days for all brain areas, 11.4 days for whole body; on a taurine-free diet, 8.7 ± 2.0 days for visceral organs, 6.7 days for brain areas, 15.0 days for whole body. If there is "exchange" of taurine among the tissues and organs, as suggested, reuptake of labeled taurine would increase the apparent half-lives.

In mice, whole body turnover of taurine has a half-life of 18.6 days (78).

Metabolism in Particular Organs

PLATELETS   Together with the other formed elements of the blood, platelets are a readily accessible human material and have been the subject of much investigation, particularly as a potential model of the aminergic neurons of the nervous system. The high taurine content of platelets (89) prompted study of the taurine uptake. Ahtee et al (90) found a saturable, temperature- and Na^+-dependent system with an apparent K_m of 25.4 µM. Dinitrophenol and iodoacetic acid were inhibitory. The authors noted obtaining aberrant results when platelets were studied in the presence of plasma, as was often done. Subsequent studies indicated the presence of three transport systems (91, 92). These comprised a high-affinity uptake with apparent K_m of 2.9 µM; a medium-affinity uptake, apparent K_m 100 µM; and a low-affinity uptake, apparent K_m 360 µM. Later workers reported finding two systems with apparent K_m values of 8 and 200 µM, respectively (93), corresponding to the previously reported high- and medium-affinity uptakes. It is of interest that the uptake of taurine by platelets is greater than the uptake of other amino acids, 5-fold that of GABA, 3-fold that of β-alanine and L-glutamate, and 1.4-fold that of D-aspartate (94).

The function of taurine in the platelet remains unknown. There have been, however, many attempts to connect changes in platelet taurine content or uptake with various conditions and inherited disorders. In particular, a defect has been sought in platelets from persons with retinitis pigmentosa, a variably inherited group of disorders in which retinal changes occur similar to those suffered by the taurine-deficient cat. However, a consistent deficiency in platelet taurine content or uptake has not been demonstrated (93, 95, 96).

Alterations in platelet taurine content or uptake have been reported for a varied group of dysfunctions. Platelets from patients with degenerative progressive myoclonus epilepsy showed consistently lower taurine uptake than that in control platelets, although uptake of GABA, serotonin, and dopamine was normal (97). Apparent K_m values for taurine uptake were about the same for patient and control groups (36 μM); V_{max} was 166 $\pm$ 18 (9 patients) and 202 $\pm$ 24 (9 controls) nmol/min/10^8 platelets. The difference is slight. Similarly, the uptake of taurine into platelets from seizure-susceptible rats (289 $\pm$ 25 nmol/sec/kg protein) was lower than in platelets from seizure-resistant rats (383 $\pm$ 14), $P < 0.05$ (98). Uptake into many brain areas was markedly reduced as well. Platelets from patients with Down syndrome also had a reduced taurine uptake (99): V_{max} of 30.5 nmol/min/ml packed cells, compared with 59.1 for normal platelets ($P < 0.01$). Apparent K_m values, determined in plasma, are all high, and do not differ significantly.

The taurine content of platelets from patients with congestive heart disease tended to be higher than control, but not significantly (100). Uptake of taurine by such platelets was significantly less than control ($P < 0.05$). Platelets from hypertensive patients were not affected. On the other hand, platelets from 24-week-old spontaneously hypertensive rats with elevated blood pressures had increased taurine and increased uptake (100). Rats made hypertensive by subtotal nephrectomy and administration of NaCl had a lower taurine content in platelets and a lower uptake (100). As the authors note, it is not clear that either animal model reflects the human situation.

Finally, Baskin et al have found that the taurine content per platelet in platelets from hypothyroid patients was lower than that in control platelets, whereas platelets from hyperthyroid patients had a greater taurine content than the control value (101). Platelets from subjects experiencing a migraine headache had a higher concentration of taurine than platelets taken during the postictal phase (102).

LYMPHOBLASTOID CELLS   Lymphocytes and cultured lymphoblastoid cells represent another source of easily accessible human material in which to examine the possible biological role of taurine. In both types of cells, over 60% of the free amino acid pool is taurine (89, 103). Human lymphoblastoid cell lines can be grown on a taurine-deficient, biochemically defined culture medium (104). The large taurine pool found in these cells when they are grown on serum-supplemented medium is totally depleted after about one week in biochemically defined medium. Although such cells replicate generation after generation under the latter condition, there is a decreased percent of viable cells in the culture. The addition of taurine to the biochemically defined medium results in increased viability, which occurs in a dose-dependent manner (104).

Optimum growth of these cells is maintained at 100 μM taurine, the approximate normal concentration in human plasma. The apparent K_m for the uptake of taurine is about 15–20 μM (105), which is similar to that of a number of other cells that have been examined. The fact that the apparent K_m of these cells is far below the average plasma concentration suggests that the transport sites for the high-affinity low-capacity uptake system allow the concentration of taurine to decrease considerably before the cellular uptake of taurine is limited by availability of substrate. Although it takes several days for the cells to be depleted of taurine, it takes only a few minutes for them to replenish their pools when taurine is made available in the medium.

Taurine uptake by cultured human lymphoblastoid cells is inhibited by a number of diverse compounds (105). Structural analogues, such as hypotaurine (2-aminoethanesulfinic acid) and β-alanine, are competitive inhibitors and provide insight into the structural requirements of the taurine receptor. Most of the inhibitors studied (105), however, are uncompetitive in nature. These compounds presumably affect the taurine receptor or transport system by perturbing neighboring components of the cell membrane. Thus, substances of varied structure and function, although acting on different systems, could all be inhibiting taurine uptake or enhancing taurine efflux nonspecifically, thereby producing similar pharmacological effects based on deprivation of taurine. In the animal, such a pharmacological effect could initiate or aggravate pathological changes in the retina, for that tissue is particularly vulnerable to taurine deficiency.

Among the compounds tested for possible effects on the taurine transport system of cultured human lymphoblastoid cells were two drugs, chlorpromazine and chloroquine (106). Chlorpromazine, a phenothiazine derivative, is a tranquilizer and antiemetic. It inhibits strongly the uptake of taurine by a synaptosomal preparation from rat cerebral cortex (107), and has a similar effect on the lymphoblastoid cells (105). Chloroquine, an antimalarial and anti-inflammatory agent, has a multiplicity of effects on lysosomes (cf 107) and rapidly depletes cystinotic fibroblasts of stored cystine (108). It is a moderately strong inhibitor of taurine uptake by lymphoblastoid cells, comparable in effectiveness to β-alanine and guanidinoethanesulfonate (105). Both chlorpromazine and chloroquine, though differing in therapeutic application, have long been known to cause retinal damage upon prolonged or excessive dosage

(109, 110). The possibility is thus raised that the common property of inhibition of taurine uptake underlies the common production of retinal damage.

RETINA   Taurine is exceptionally abundant as a free amino acid in the vertebrate retina (111–113). Taurine concentrations range from 10 mM in the frog retina to approximately 50 mM in the rat and rabbit retina (114). Intermediate amounts are found in the retina of other species, including man (114, 116; J. A. Sturman, personal communication). Along with the pineal gland (cf 117–123), the retina possesses the highest concentration of taurine of any region derived from the central nervous system. Taurine is concentrated in the retinal pigmented epithelium and photoreceptor cell layer, with maximal concentrations (50–80 mM) present in the inner segments and outer nuclear layer (124–126).

Taurine may be supplied to the retina via the vitreous humor (127–129) or the circulating plasma (130), or may be synthesized in situ. Activities for cysteine oxidase, 3-sulfinoalanine decarboxylase, and hypotaurine oxidase have been observed in retinal tissues from several animal species (131–137). The taurine-synthesizing enzyme, 3-sulfinoalanine decarboxylase, has been localized by immunohistochemical methods to be prominent in the inner nuclear layer and the ganglion cell layer of the rabbit retina (137) and the inner plexiform layer and inner nuclear layer of the rat retina (138, 139). In addition, in vitro evidence has been obtained for the synthesis of taurine from cysteine in the baboon, guinea pig, and bull frog retina (114, 140, 141). Although it is apparent that the retinas of some animals can synthesize taurine, the millimolar concentrations present in the retina can be accounted for by the ability of the retinal pigmented epithelium to accumulate taurine (142–150). A temperature-sensitive, sodium-dependent, energy-dependent, high-affinity transport system (114, 142, 150) allows the retinal pigmented epithelium to rapidly concentrate taurine against a 400- to 500-fold gradient (114). This occurs prior to being slowly transferred to the photoreceptors and neural retina (151, 129). This slow passage of taurine is not well understood, but probably reflects the action of the blood-retinal barrier which excludes free entry of taurine and other neuroactive amino acids into the subretinal space (150, 152, 153). The presence in the retina of two binding proteins specific for taurine (154–156) may be important in the accumulation and transport of taurine into the inner retina.

Although a function for taurine in the retina has not been clarified, in recent years a definite connection has been demonstrated between the availability of taurine to the retina and the integrity of that organ (cf 157), for depletion of taurine under various circumstances has been found to lead to abnormal electroretinograms or to degeneration of photoreceptor cells. This has been demonstrated in a number of species: initially in cats fed a taurine-free casein diet (4, 158), later in infant rhesus monkeys fed commercial human infant formula almost devoid of taurine (159), in rats receiving the taurine transport inhibitor guanidinoethanesulfonic acid (160–162), and in human children receiving taurine-free total parenteral nutrition (163, 164). Retinal abnormalities also have been found in patients with intestinal disease and bacterial overgrowth and in rats with bacterial overgrowth in surgically prepared blind loops; in these conditions, bacterial catabolism of taurine depletes the host (165). In most cases, supplementation with taurine reverses or prevents the retinal disturbance (4, 159, 163, 164, 166, 167).

Several possible functions for taurine in the retina have been suggested: 1. Taurine may act in some species as an inhibitory neural transmitter (111). This is supported by the observations that photoreceptor taurine is released after photoexcitation (168–171) and intravitreal administration of taurine reversibly inhibits the in vitro b-wave of the electroretinogram (119). 2. Taurine may regulate osmotic pressure in the eye (172, 173), since taurine exposure inhibits the swelling and disorganization of isolated rod outer segments induced by prolonged illumination (100). 3. Taurine may regulate retinal Ca^{2+} homeostasis. At low Ca^{2+} concentrations, taurine stimulates an ATP-dependent uptake of Ca^{2+} (174–180); however, it decreases Ca^{2+} uptake at high Ca^{2+} concentrations (175, 181, 182). 4. Taurine inhibits the phosphorylation of specific retinal membrane proteins (180). Although the mechanism(s) is (are) not completely understood, taurine may be influencing a kinase or altering the ability of the retinal membrane to be phosphorylated via direct membrane interaction. 5. Taurine reverses melatonin- and serotonin-induced inhibition of phagocytosis in the retinal pigmented epithelium (183). Thus, taurine may be involved in the normal phagocytic function of the retina. 6. Taurine may stabilize retinal membranes either directly in maintaining structure after retinoid exposure (184) or by preventing increased lipid peroxidation after exposure to oxidant generating systems (168, 185). Finally, 7. Taurine may scavenge hypochlorous acid which is generated by a peroxidase present in retinal pigmented epithelial cells (186, 187). This antioxident property of taurine may attenuate the biocidal capability of retinal pigmented epithelium during the phagocytosis of shed rod outer segments.

BRAIN   Most aspects of the neurochemistry of taurine have been reviewed by Oja & Kontro (13). A new development of great interest is here described.

Sturman et al (188, 189) have shown that in full or almost complete taurine deficiency, achieved in kittens by deprivation of taurine both pre- and post-natally, physical and morphological dysfunctions occur, including cellular aberrations in the brain. Kittens born to deficient mothers and nursed on such mothers' taurine-limited milk are smaller than normal, at 8 weeks (weaning) being 60% of normal weight. Similar kittens given oral supplements of taurine reach full size, demonstrating directly that taurine is necessary for growth. The deficient kittens display other abnormalities as well: abnormal development of the hind legs, a peculiar gait with excessive abduction and paresis, and thoracic kyphosis. These symptoms are indicative of cerebellar dysfunction.

It had been found earlier (190) that, in addition to the retinal degeneration in taurine-deficient cats, there is disorganization and disruption of the cells of the tapetum lucidum, a reflective layer behind the retina of cats. Study of the cerebellum then disclosed an alteration in its morphology. Many cells of the external granule cell layer were still in evidence at 8 weeks in cerebellum from deficient kittens, but had largely migrated to the internal molecular layer in normal cerebellum, as expected. Mitotic figures in many cells of the deficient cerebellar external granule cell layer indicated that cell division was still going on, although this process normally is complete by about 3 weeks of age. No mitotic figures were seen in the normal tissue. Supplementation of deficient kittens with taurine resulted in an increase in the cerebellar taurine content to normal and the occurrence of normal migration of cells. The delay in cell migration in the absence of adequate taurine must upset the strict timetable of cerebellar development, resulting in a failure to form many synaptic connections.

Recently, it has been found that the visual cortex of deficient kittens also is affected by lack of taurine (191). At birth, neuroblasts at both the ventricular and pial zones have failed to complete their differentiation and have not migrated into the molecular layer; subsequent arborization is poor and much organization must be lost.

TAURINE IN NUTRITION AND DISEASE

Taurine in Human Nutrition

In studies on feeding the preterm infant it was noted that in infants fed various formulas, compared with those fed pooled human milk, most plasma and urine free amino acids were present in higher concentration in the former group than in the latter. Taurine in contrast was present in lower concentration in the plasma and urine of the formula-fed infants. This evidence suggests that a major source of taurine is the diet. This hypothesis is consistent with the low activity of 3-sulfinoalanine decarboxylase known to be present in mature human liver (1); low activity was found both in mature and immature human liver (192).

The milk of many species contains taurine (28), but the amount varies greatly. Furthermore, the concentration of taurine in milk decreases during lactation. The milk of the dairy cow, a chronically lactating mammal, has very little taurine. The early milk that the cow feeds its calf, however, has a concentration of taurine like that of human milk. These results explain why the taurine concentration in plasma and urine of the formula-fed preterm infant is lower than that of an infant fed human milk, and why standard infant formulas contained so little taurine.

Full-term infants have a similar nutritional response to formulas low in taurine (193, 194). When taurine is added to formulas in concentrations the same as that found in human milk, their plasma and urine concentrations of taurine are the same as those in infants fed human milk (195, 196). Taurine, thus, is easily bioavailable. Further, when adult man is chronically fed synthetic liquid diets devoid of taurine, he excretes no taurine in the urine (E. Ahrens, S. Moore, personal communication). These data, in sum, suggest that the limited ability of the human infant to synthesize taurine is probably a function of species rather than of age. The work of Klein and coworkers (198) using ^{18}O in man is consistent with the belief that, even in adult man, the ability to synthesize taurine is present, albeit limited. It would be interesting to know how far man can expand his ability to synthesize taurine, if he can, and the circumstances under which such expansion occurs.

Y. Nishihara (personal communication) has compared blood and milk obtained simultaneously from eight lactating women. In each the taurine concentration of milk was far higher than that in blood. Therefore, taurine is actively secreted or transferred, rather than merely spilled, into human milk.

Taurine is rapidly transferred from the rat dam to the pup via the milk (200). Although the rat has a great ability to synthesize taurine, a significant proportion of the taurine in the pup's brain is derived from the milk. Detailed studies (80, 201) confirmed this basic finding. Sturman contends that the mother contributes as much taurine to the pup from birth to weaning as she does during gestation, about 40% of the taurine of each pup. Huxtable and Lippincott suggest that 13% of the taurine content of the pup is obtained from the milk. Mouse pups nursed by dams treated with guanidinoethylsulfonate, an inhibitor of taurine uptake, have about half the retinal taurine concentration of the pups whose mothers were not so treated (202). Radioactivity from [^{35}S]taurine injected into the lactating dams was reduced in the milk, providing further evidence of active transfer by the dam rather than synthesis by the pup.

Nutritionists have been mainly concerned with the identification of essential nutrients. The assumption has been that we could identify all essential nutrients, estimate their dietary requirements, and understand their metabolic function. Adequate nutrition was defined as having enough, and in some cases not too much, of these essential nutrients. Nutrition is now concerned with food-related problems that have little to do with essential nutrients, e.g. obesity, hypertension, arteriosclerosis, and dental caries. These diseases are closely related to diet. However, there is little reason to believe that they are related to essential nutrients in the old sense.

A new concept helps bridge the gap between nutritional "old think" and nutritional "new think," i.e. the conditionally essential nutrient (203). Such nutrients are usually synthesized or acquired and stored in amounts adequate to maintain health. Nutritional inadequacy of such nutrients occurs in patients who have various inherited or acquired diseases that alter the metabolism of such nutrients or who are subjected to extreme conditions that deplete nutrient pools. In some cases, there is inadequate storage, such as in the premature infant. In other cases, such as in taurine-free total parenteral nutrition, the deficiency may be iatrogenic.

The first examples of the nutrient role of taurine involved such extreme conditions. In the taurine-deficient cat, the events were the feeding of an abnormal diet plus the inability of the cat to conserve taurine and to synthesize it (4). In the human infant the events were premature birth, making breast-feeding difficult, followed by the feeding of an unnatural diet (infant formula) because of this difficulty, plus the inability to synthesize taurine in amounts adequate to maintain normal body pools (192–196). These interactions allowed the identification of a conditionally essential nutrient in each case.

One of the main objectives of the experiments in which infant formulas were supplemented with taurine (195, 196) was to determine whether or not such supplementation would alter bile acid conjugation (195), fat absorption (204), or bile acid kinetics (205). The increased availability of taurine in formulas maintains its predominance in intraduodenal bile acid conjugates, a situation similar to that found in preterm infants fed pooled human milk (195). Taurine predominance of bile acid conjugation neither alters bile acid kinetics (205) nor increases fat absorption under the conditions of that experiment (204). It does, however, in children with cystic fibrosis (206). This latter effect, perhaps, is another example of conditional essentiality.

The effects of taurine deficiency on the retina are not limited to the cat; it has been identified both in nonhuman primates and man. Identification of such functional impairment or structural damage, however, has required the imposition of iatrogenic factors, i.e. it is conditionally essential. After two years of feeding taurine-deficient infant formulas to the monkey, there is considerable damage to the cone outer segments (207). In man, abnormalities in retinal function, as measured by electroretinograph, and pigment epithelium defects, as measured with fluorescein angiography, occur in the "blind-loop" syndrome (165), in which intestinal stasis results in an abnormal flora. This situation results in an abnormally low concentration of plasma taurine. The bacterial flora in the "blind loop" evidently can do what man cannot, i.e. actively metabolize taurine, resulting in deficiency and in retinal abnormalities in the host.

Children maintained on long-term total parenteral nutrition also develop low plasma taurine concentrations. No clear signs of retinal degeneration appear by ophthalmoscopy (163, 164), but electroretinography and visual evoked potentials in these patients identify clear-cut functional impairment. These abnormalities are reversed by the administration of taurine.

Retinitis Pigmentosa

The degenerative changes that occur in the retina of taurine-deficient cats resemble those found in the retina of patients with retinitis pigmentosa, suggesting that this genetically and clinically heterogeneous group of human diseases might be related to a defect in taurine availability or utilization (150, 167, 209). However, the concentration of taurine in the plasma of patients with retinitis pigmentosa was found to be normal, both in a varied group mostly with recessive inheritance (208) and in a group consisting mostly of "simplex" cases (no affected relatives) (209). Recently, it was reported from India (210) that among patients of unspecified inheritance type, some had normal plasma taurine concentrations, but others had a mean plasma taurine concentration only 4% of the normal mean. The reason for this difference is not evident, but may involve variations in type of inheritance of in diet (vegetarian diets common in India may be low in taurine). Measurement of whole blood free amino acids (211) showed normal taurine concentrations in a large group of patients, one third of whom had a recessive mode of inheritance and one third of whom were "simplex"; the taurine values thus obtained actually represent the contents of the platelets and leukocytes, which together contain 90% of the taurine in whole blood (89). A direct analysis of platelets from mainly "simplex" cases also revealed normal taurine concentrations (209). Thus, it is evident that retinitis pigmentosa occurs in persons with normal concentrations of taurine in plasma, platelets, and probably leukocytes.

The possibility remains that it is uptake of taurine into retina that is defective in persons with retinitis pigmentosa. Since retinal tissue is not readily available, uptake measurements have been made with isolated human platelets, which have multiple active uptake systems, (90–92, 99) including one similar to that in retina (212). In one study (95), a 20% decreased uptake of taurine was found in platelets of 12 patients of unspecified type, but apparently mostly "simplex." In another investigation (213), platelets from "simplex" patients had normal uptake, but there was about a 20% increased uptake in platelets from "multiplex" cases (two sibs affected) and from X-linked hemizygous males. Other types had normal platelet uptake (213, 214). Thus, there was no consistent evidence from study of platelets that altered taurine uptake mechanisms are involved in the etiology of retinitis pigmentosa.

Platelets, however, are only fragments of somatic cells and might not express a possible defect in taurine transport. Recently it has been shown that cultured human lymphoblastoid cells also possess a high-affinity active uptake system for taurine (105). These cells provide two advantages over platelets. First, they are intact somatic cells, rather than cell fragments. Second, any inherited defect in taurine uptake can be propagated in culture, thereby providing additional evidence that such a defect is genetic. Furthermore, their use would allow continued study of any defect identified.

Taurine uptake in cells from patients with retinitis pigmentosa (RP) differed significantly from that of control cell lines (215, 216). The mean V_{max} for affected cell lines is lower than the mean of control cell lines, whereas the mean apparent K_m for the affected cell lines is greater than that for control cell lines. Individual affected cell lines differ from the controls in having a low V_{max} or a high apparent K_m, or in some cases both. These findings indicate that a defective taurine uptake could be a contributory factor in the manifestation of the RP group of disorders. The defect may be in taurine uptake per se, or may be a deficiency in some component of the cell membrane.

This recent study suggests that a defective taurine uptake system is one component of multifactorial disturbances that lead to pigmentary degeneration of the retina, whatever the mode of inheritance of this degeneration. In the retina itself, it may be that certain cell types are especially vulnerable to a relatively mild taurine deficiency or that particular cells have a more severely affected uptake system. How taurine could act in maintaining the viability and functioning of such cells remains to be ascertained. It would appear, however, that defective uptake of taurine under conditions in which this amino acid is limiting is a biochemical defect manifest in nonretinal cells that can be propagated in culture. More extensive biochemical studies of taurine uptake are needed in families in which the mode of inheritance and the functional attributes (217) are carefully defined. Clinical disease in the different genetic forms of retinitis pigmentosa might be the result of any of a number of possible defects in taurine transport, all leading to a relative unavailability of taurine at a site where its presence is essential for function. Nutritional deficiency severe enough to result in reduced plasma concentrations might then be an exacerbating factor.

TOWARD A BIOLOGICAL ROLE

Membrane Protection

There are now six lines of evidence suggesting that taurine plays an important role in the stabilization of membranes. 1. The original work of Hayes et al (4) demonstrates the disruption of the membranes of the photoreceptor cell in the taurine-deficient cat. This work has been extended to include effects on the tapetum lucidum of the taurine-deficient cat (190) and on the cones of the photoreceptor layer of the taurine-deficient monkey (207). 2. Taurine counteracts damage caused by external agents to skeletal muscle, intracellular membrane (218), and heart muscle (219). 3. Taurine protects isolated rod outer segments from frog retina against structural damage induced by illumination and by oxidants (184). 4. Taurine protects cultured cells against retinol-induced and ferrous sulfate–induced damage and/or swelling (220, 221). 5. Taurine has an antioxidant effect in rabbit spermatozoa (222). This raises the possibility that the protective effects mentioned above might be mediated through an ability to reduce lipid peroxidation. 6. Taurine protects the hepatic cell against damage to the membrane by carbon-tetrachloride (223).

The membrane-stabilizing effect of taurine has been examined in cultured human lymphoblastoid cells (220, 221). Retinoids were used as toxic agents because of their known membrane destabilizing action and antiproliferative effects (224, 225) and because retinol is critical to the function of the photoreceptor cells. Retinol and retinoic acid gave a time- and dose-dependent decrease in cell viability. Taurine (5–20 mM) increased the viability of the cells treated with 10 µM retinol. Adding zinc alone at 50 or 100 µM also provided protection. Taurine and zinc together are synergistic, but still do not give complete protection. This was afforded by the addition of α-tocopherol (known to protect against retinol).

Cells exposed to retinol were found to have increased in size and become spherical. The presence of taurine and zinc or of α-tocopherol maintained a cell size similar to that of controls. The effect was not the result of lipid peroxidation, since retinol did not increase malonaldehyde (MDA) production (220). It is significant that taurine and zinc, two water-soluble substances, and α-tocopherol, a lipid-soluble substance, have additive effects. This observation suggests that the mechanisms are different and complementary.

Exposure of lymphoblastoid cells to ferrous sulfate and ascorbate [a particularly effective oxidant-generating system (226)] decreased cell viability and increased cell size. The addition of 1–20 mM taurine afforded virtually complete protection, in terms of both cell viability and of cell swelling. β-Alanine, glycine, and hypotaurine also partially protected, as did well-known antioxidants, i.e. α-tocopherol, albumin, and mannitol.

In contrast to the retinoids, iron-ascorbate increased MDA formation (221), as would be expected with a powerful oxidant-generating system. Neither taurine, α-tocopherol, nor mannitol are able to counteract MDA formation, although albumin does.

Because of the cell swelling seen both in the retinoid toxicity model (220) and in the oxidation model (221), changes in the major ions of the incubation media were examined. The omission or replacement of the major osmotic determinants of the incubation medium decreased cell injury. This suggests that damage to the membrane results in increased permeability and transfer of water. Thus, taurine protects the cells from damage caused by iron-ascorbate, but it does not do so through a reduction in the extent of lipid peroxidation. Rather, the protective effect seems to be related to an action on permeability to ions and water.

These differences in taurine's protective effects may be related to differences in the reduced oxygen molecules involved. The oxidative effects may be exerted on components other than polyunsaturated fatty acids. Such oxidative effects would not be detected by the thiobarbituric reaction. It is possible also that the peroxidative effects and the nonperoxidative effects both result in permeability changes, which in turn, result in osmotic changes due to shifts in water and ions, with resulting cell swelling and damage (227). Whether the protective effect of taurine on membranes is direct or indirect remains to be determined.

Other evidence also suggests that the protective effect of taurine is mediated through an action on membrane permeability and ion transport. Taurine modifies calcium fluxes in sarcolemmal, synaptosomal, and retinal subcellular fractions (181, 228, 229). It induces changes in transmembrane potentials consistent with modifications in potassium permeability (230). It has been related to changes in sodium gradients in cultured cells (231) and to osmolar changes in marine invertebrates (232).

Therefore, taurine seems to counteract a variety of agents that result in leaky membranes. Taurine may have a dual role in this regard. First, it may react with these deleterious agents via its amino group to detoxify them. Second, it may also exert a direct protective effect in preventing the ionic and water shifts that result in cellular damage and death.

Detoxification

Lithocholic acid and its conjugates are known to induce intrahepatic cholestasis in a variety of experimental animals (233). Cholestasis results even though lithocholate is normally only a minor component of bile that undergoes sulfation, a process thought to reduce its potential toxicity. It has been known for some time that certain secondary bile acids in man have pyrogenic and inflammatory properties. The prototype is lithocholate, and these two adverse effects are separate in man (234).

Lithocholic acid sulfate injected into guinea pigs, which predominantly conjugate bile acids with glycine, quickly decreases bile flow and then induces pathological changes in the liver (235). Lithocholic acid induces cholestasis by directly affecting the structure and function of the bile canalicular membrane (236).

Administration of taurine (0.5%) in the drinking water one, two, or five days before intravenous injection of lithocholic acid sulfate prevents the cholestasis and the morphological changes that result from administration of this secondary bile acid. Although sulfation increases the solubility of lithocholic acid and of taurolithocholic acid, it does not change that of glycolithocholic acid. A high glycine-to-taurine ratio would result in the formation of the cholestatic glycolithocholic acid sulfate. These authors (235) suggest there may be a theoretical advantage to man in maintaining a low glycine-to-taurine ratio by the supplementation of formulas and parenteral solutions with taurine. The advantage, of course, is in detoxifying the secondary bile acids rather than increasing the lipid absorption characteristic of the primary bile acids. Viewed from this perspective, conjugation results in a smaller secondary bile acid pool, and taurine conjugation ameliorates the hepatoxicity of the secondary bile acids.

Only about 1% of taurine in man is in the bile acid pool (237). Nonetheless, bile is a major route of excretion of taurine, and the conjugation of taurine with bile acids has been thought to be the only biochemical reaction in which taurine has a role as substrate. There are a number of theoretical reasons to suggest that taurine is even more effective than glycine in allowing conjugated bile acids to abet the absorption of lipids (238).

Since present evidence suggests that taurine conjugation of bile acids apparently has no effect on fat absorption except under the pathological conditions found in cystic fibrosis (206), one might ask what its major role could be. Free bile salts are not normally present in human bile (238). The majority of bile salts are the primary bile salts, which are either trihydroxylated or dihydroxylated. The monohydroxylated bile salts, the secondary bile salts, are formed by degradation of the primary bile acids by the bacterial flora of the gut. Thus, during the enterohepatic circulation of bile acids, some of the major primary bile acids, cholic and chenodeoxycholic acids, undergo bacterial degradation to yield deoxycholic and lithocholic acids, respectively. One advantage of conjugation of bile acids is that it results in a smaller total bile acid pool size, which in turn, results in a smaller secondary bile acid pool size (239), i.e. in less potentially toxic material.

Recent evidence suggests that the amino group of taurine will react with the hydroxyl group of retinol to form retinotaurine (55, 240). Since it has been found only in rat bile, it seems likely to be an excretory product. Although well-characterized chemically, its physiological significance is uncertain. Certainly, retinotaurine is another example of the amino group of taurine reacting with a potentially toxic compound.

Taurine conjugates with certain xenobiotics. These are organic carboxylic acids and form peptide linkages with the amino group of taurine (241). Not only are endogenously produced toxins, such as secondary bile acids, detoxified with taurine, but exogenous toxins can be conjugated as well.

Antioxidation

Although taurine conjugates with bile acids (237, 238) and xenobiotics (241), these reactions consume only a small percent of the total taurine (242). Its ubiquity among a variety of animal species (cf 1), abundance in the free amino acid pools (cf 1, 89, 103), inability to be translationally incorporated into proteins, and efficient mechanisms of cellular uptake (105) and tissue conservation all suggest that taurine may have other and possibly more fundamental biochemical functions.

Recent in vitro experiments have demonstrated that taurine may be a necessary and metabolically active amino acid. In a cell culture system where taurine concentrations can be controlled, taurine enhances the proliferation of human lymphoblastoid cells in a concentration-dependent manner (104). This action of taurine could result from its counteracting the antiproliferative effects of oxidative and peroxidative products, which may cause extensive damage and subsequent cellular death. This interpretation is supported by the fact that many known antioxidants also enhance the proliferation of cultured cells in biochemically defined cell culture systems (cf 243). Furthermore, taurine concentrations are high in cells and tissues that possess considerable potential for producing oxidants. For example, there is a high concentration of taurine in the retina (111, 125), a tissue where various oxidants are generated photolytically and enzymatically. There is an exceptionally high concentration of taurine in neutrophils (103), cells that enzymatically produce oxidants during the phagocytic process (cf 244). Taurine is also a radioprotectant (245), and since radiation is a potent generator of oxidizing molecules, taurine may be interacting with these oxidants.

In biological systems, the one-electron reduction of molecular oxygen produces a variety of oxidants. For protection, cells possess a battery of low-molecular-weight scavengers and several enzymes that scavenge these oxidants. One such enzyme scavenger is superoxide dismutase which combines two superoxide anions to form hydrogen peroxide. Although superoxide anion is a weak base, it will spontaneously dismutate at low pH. Other scavengers include catalase and specific peroxidases that reduce hydrogen peroxide to water. During the peroxidase-catalyzed reaction, the myeloperoxidase of neutrophils (246–249) and juvenile monocytes (250) and the eosinophil peroxidase (251–253) catalyze the formation of hypochlorous acid (HOCl) from hydrogen peroxide and chloride anion.

Hypochlorous acid is a potent oxidizing agent; its biocidal properties are well known, especially for the disinfection of water. It can directly oxidize a variety of biologically significant substances such as carbohydrates, nucleic acids, peptides, and amino acids. It can also form secondary chlorinating agents such as molecular chlorine, N-chloramides, and N-chloramines (254–256).

When hypochlorous acid reacts with any organic or inorganic amine, the amino group can acquire either one or two chlorine atoms. This reaction, however, is dependent upon the concentration and ionic strength of the environment. Above pH 6, the amino group acquires one chlorine atom, forming a monochloramine; below pH 6, the amino group acquires two chlorine atoms, resulting in a dichloramine.

Most chlorinated α-amino acids are rather unstable (2, 257). They spontaneously deaminate, decarboxylate, and dechlorinate to form respective aldehydes which are extremely toxic. This oxidation reaction can also occur on peptide linkages, resulting in the degradation of proteins. However, amino acids that possess amino groups at other carbon positions react with HOCl to form relatively stable N-chloramines (258). Taurine is a β-amino acid that can be chlorinated by the myeloperoxidase/H_2O_2/Cl^- system, thus forming a stable taurine chloramine (58, 254, 255, 258, 259). Early studies (258) have shown that taurine is a competitive inhibitor of the decarboxylation and deamination of amino acids by myeloperoxidase. Once formed, taurine chloramine participates in few reactions, oxidizing reduced sulfhydryls and thioether bonds (246, 247). Taurine chloramine has also been shown to inhibit myeloperoxidase-catalyzed reactions (249, 259).

Taurine chloramine can be synthesized by the reaction between taurine and sodium hypochlorite. This reaction proceeds rapidly, with one mole of taurine reacting with one mole of hypochlorite anion (Y. Y. Lin, C. E. Wright, unpublished results). The reaction occurs with equal stochiometry within the pH range 3.5–11.0. Under acid pH conditions taurine dichloramine is the predominant form produced. Taurine dichloramine probably has little significance in a biological system due to its instability at physiological pH. Unlike other amino acid chloramines examined, taurine chloramine is rather stable, although not at low pH. Experiments examining the stability of this molecule (Y. Y. Lin, C. E. Wright, unpublished results) indicate a half-life of approximately 2.5 days. This is exceptionally long. Most chlorinated α-amino acids decompose within minutes after formation. A few, such as glycine chloramine and α-alanine chloramine, have half-lives between 15 and 20 hours.

Taurine chloramine is an oxidant, but it reacts with fewer organic molecules than does hypochlorous acid. Thus, taurine chloramine formation attenuates the biocidal activity of hypochlorous acid generated via enzymatic, chemical, or photochemical processes. This property of taurine may protect the cells from autolysis.

Neutrophils and eosinophils have been shown to release chloramines (58, 250, 259). Different types of cultured cells have been examined for their ability to release chloramines into the extracellular milieu (2). Cultured monkey neutrophils, mouse olfactory epithelial cells, human myeloid leukemia cells, human lymphoblastoid cells, and bovine retinal pigmented epithelial cells all release chloramines (2). Chloramine formation was not observed in human lymphoblastoid cells growing in a biochemically defined medium devoid of exogenous sources of taurine. It was detected only after taurine was added back into the cell culture medium. In cultured cells that can form taurine chloramine, e.g. human lymphoblastoid cells, formation is linear with time and the number of cells present in the incubation assay (2). These results suggest that the major chloramine being produced is taurine chloramine. Taurine chloramine production was not observed in cultured human primary fibroblasts, amniocytes, or mouse adenocarcinoma cells. These cell have no detectable peroxidase activity and demonstrate low phagocytic capabilities. Thus it appears that the formation of taurine chloramine by cultured cells depends on their in vivo capabilities to phagocytize and on having a peroxidase activity which may be similar to myeloperoxidase.

The formation of taurine chloramine in several different cell lines suggests that taurine is not merely a metabolically inert end product of methionine catabolism. Rather, taurine may function in biological systems as a general detoxifier (220), eliminating excessive cholates, removing xenobiotics, and scavenging chlorine oxidants. Taurine may thereby specifically protect cells from self-destruction during processes that generate oxidants (221).

Literature Cited

1. Jacobsen, J. G., Smith, L. N. 1968. *Physiol. Rev.* 48:424–511
2. Wright, C. E., Lin, T., Lin, Y. Y., Sturman, J. A., Gaull, G. E. 1985. See Ref. 12, pp. 137–47
3. Wright, C. E., Lin, T., Lin, Y. Y., Sturman, J. A. 1984. *Fed. Proc.* 43:616
4. Hayes, K. C., Carey, R. E., Schmidt, S. Y. 1975. *Science* 188:949–51
5. Räihä, N., Heinonen, K., Rassin, D. K., Gaull, G. E. 1975. *Pediatrics* 57:659–74
6. Tiedemann, F., Gmelin, L. 1827. *Ann. Physik. Chem.* 9:326–37
7. Huxtable, R., Barbeau, A., eds. 1976. *Taurine*. New York: Raven
8. Barbeau, A., Huxtable, R., eds. 1978. *Taurine and Neurological Disorders*. New York: Raven
9. Schaffer, S. W., Baskin, S. I., Kocsis, J. J., eds. 1981. *The Effects of Taurine on Excitable Tissues*. New York: Spectrum
10. Huxtable, R. J., Pasantes-Morales, H., eds. 1982. *Taurine in Nutrition and Neurology*. New York: Plenum
11. Kuriyama, K., Huxtable, R. J., Iwata, H., eds. 1983. *Sulfur Amino Acids: Biochemical and Clinical Aspects*. New York: Liss
12. Oja, S. S., Ahtee, L., Kontro, P., Paasonen, M. K., eds. 1985. *Taurine: Biological Actions and Clinical Perspectives*. New York: Liss
13. Oja, S. S., Kontro, P. 1983. In *Handbook of Neurochemistry*, ed. A. Lajtha, 3:501–33. New York: Plenum. 2nd ed.
14. Ansell, G. B. 1959. In *Data for Biochemical Research*, ed. R. M. C. Dawson, W. H. Elliott, K. M. Jones, pp. 2–28. Oxford: Clarendon
15. Carr, W., Shull, W. 1939. *Trans. Faraday Soc.* 35:579–87
16. Devoto, G. 1932. *Atti. Reale Accad. Naz. Lincei Rend. Cl. Sci. Fis. Mat. Nat.* 15:471–73
17. Pentz, E. I., Davenport, C. H., Glover, W., Smith, D. D. 1957. *J. Biol. Chem.* 288:433–45
18. Ling, N. R. 1957. *J. Clin. Pathol.* 10:100
19. Garvin, J. E. 1960. *Arch. Biochem. Biophys.* 91:219–25
20. Sorbo, B. 1961. *Clin. Chim. Acta* 6:87–90
21. Gaitonde, M. K., Short, R. A. 1971. *Analyst* (London) 96:274–80
22. Anzano, M. A., Naewbanij, J. O., Lamb, A. J. 1978. *Clin. Chem.* 24:321–25
23. Yoshikawa, K., Kuriyama, K. 1976. *Jpn. J. Pharmacol.* 26:649–54
24. Yoneda, Y., Takashima, S., Hirai, K., Kuriyama, E., Yukawa, Y., et al. 1977. *Jpn. J. Pharmacol.* 27:881–88
25. Karlsson, A., Fonnum, F., Malthe-Soressen, D., Storm-Mathisen, J. 1974. *Biochem. Pharmacol.* 23:3053–61
26. Lombardini, J. B. 1975. *J. Pharmacol. Exp. Ther.* 193:301–8
27. Tachiki, K. H., Hendrie, H. C., Kellams, J., Aprison, M. H. 1977. *Clin. Chim. Acta* 75:455–65
28. Rassin, D. K., Sturman, J. A., Gaull, G. E. 1978. *Early Human Dev.* 2:1–13
29. Tachiki, K. H., Baxter, C. F. 1979. *J. Neurochem.* 33:1125–29
30. Connolly, B. M., Goodman, H. O. 1980. *Clin. Chem.* 26:508–10
31. James, L. B. 1981. *J. Chromatogr.* 209: 479–83
32. Gurusiddaiah, S., Brosemer, R. W. 1981. *J. Chromatogr.* 223:179–81
33. Erbersdobler, H. F., Greulich, H. G.,

Trautwein, E. 1983. *J. Chromatogr.* 254:332–34
34. Shihabi, Z. K., White, J. P. 1979. *Clin. Chem.* 25:1368–69
35. Stuart, J. D., Wilson, T. D., Hill, D. W., Walters, F. H., Feng, S. Y. 1979. *J. Liquid Chromatogr.* 2:809–21
36. Larsen, B. R., Grosso, D. S., Chang, S. Y. 1980. *J. Chromatogr.* 18:233–36
37. Stabler, T. V., Siegel, A. L. 1981. *Clin. Chem.* 27:1771
38. Wheler, G. H. T., Russell, J. T. 1981. *J. Liquid Chromatogr.* 4:1281–91
39. Caldwell, K. A., Tappel, A. L. 1968. *J. Chromatogr.* 26:635–40
40. Shahrokhi, F., Gehrke, C. W. 1968. *J. Chromatogr.* 36:31–41
41. Horning, E. C., VandenHeuvel, W. J. A., Creech, B. G. 1963. *Methods Biochem. Anal.* 11:69–147
42. Irving, C. S., Klein, P. D. 1980. *Anal. Biochem.* 107:251–59
43. Matsubara, Y., Lin, Y. Y., Sturman, J. A., Gaull, G. E., Marks, L. M., Irving, C. S. 1985. *Life Sci.* 36:1933–40
44. Kataoka, H., Yamamoto, S., Makita, M. 1984. *J. Chromatogr.* 306:61–68
45. Kataoka, H., Ohnishi, N., Makita, M. 1985. *J. Chromatogr.* 339:370–74
46. Norton, R. S. 1979. *Comp. Biochem. Physiol. B* 63:67–72
47. Irving, C. S., Hammer, B. E., Danyluk, S. S., Klein, P. D. 1980. *J. Inorg. Biochem.* 13:137–50
48. Lin, Y. Y., Wright, C. E., Gaull, G. E., Zagorski, M., Gonnella, N. C., Nakanishi, K. 1985. Submitted for publication
49. Andrew, S., Schmidt, C. L. A. 1927. *J. Biol. Chem.* 73:651–54
50. Greenstein, J. P., Winitz, M. 1961. *Chemistry of the Amino Acids,* Vol. 1, pp. 486–91. New York: Wiley
51. Albert, A. 1950. *Biochem. J.* 47:531
52. Sakurai, H., Takeshima, S. 1983. See Ref. 11, pp. 398–99
53. Dolara, P., Franconi, F., Giotti, A., Basosi, R., Valensin, G. 1978. *Biochem. Pharmacol.* 27:803–4
54. Dolara, P., Ledda, F., Mugelli, A., Mantelli, L., Zilletti, L., et al. 1978. See Ref. 8, pp. 151–59
55. Skare, K. L., Schnoes, H. K., DeLuca, H. F. 1982. *Biochemistry* 21:3308–17
56. Lähdesmäki, P., Marnela, K. M. 1985. See Ref. 12, pp. 105–14
57. Marnela, K. M., Timonen, M., Lähdesmäki, P. 1984. *J. Neurochem.* 43:1650–53
58. Weiss, S. J., Klein, R., Slivka, A., Wei, M. 1982. *J. Clin. Invest.* 70:598–607
59. Read, W. O., Welty, J. D. 1962. *J. Biol. Chem.* 237:1521–22
60. Andersen, L., Sundman, L. O., Linden,

I. B., Kontro, P., Oja, S. S. 1983. *J. Pharm. Sci.* 73:106–8
61. Ditrich, E. 1878. *J. Prakt. Chem. Neue Folge* 18:63–78
62. Salkowski, E. 1873. *Ber. Dtsch. Chem. Ges.* 6:1191–93
63. Sorbo, B. 1958. *Bull. Soc. Chim. Biol.* 40:1859–64
64. Bredereck, H., Wagner, A., Blaschke, R., Demetriades, G., Kottenhahn, K. G. 1959. *Chem. Ber.* 92:2628–36
65. Otto, R. 1868. *Ann. Chem. Pharm.* 145:317–29
66. Kice, J. L., Bowers, K. W. 1962. *J. Am. Chem. Soc.* 84:2384–89
67. Kice, J. L., Morkved, E. H. 1964. *J. Am. Chem. Soc.* 86:2270–78
68. Oja, S. S., Kontro, P. 1981. *Biochim. Biophys. Acta* 677:350–57
69. Recasens, M., Benezra, R. 1981. In *Amino Acid Neurotransmitters,* ed. F. V. DeFeudis, P. Mandel, pp. 545–50. New York: Raven
70. Griffith, O. W. 1983. *J. Biol. Chem.* 258:1591–98
71. Scandurra, R., Politi, L., Dupré, S., Moriggi, M., Barra, D., Cavallini, D. 1977. *Bull. Mol. Biol. Med.* 2:172–77
72. Kuriyama, K., Ida, S., Ohkuma, S. 1984. *J. Neurochem.* 42:1600–6
73. Hardison, W. G. M., Wood, C. A., Proffitt, J. H. 1977. *Proc. Soc. Exp. Biol. Med.* 155:55–58
74. Lombardini, J. B., Medina, E. V. 1978. *J. Nutr.* 108:428–33
75. Sturman, J. A. 1980. *Life Sci.* 26:267–72
76. Huxtable, R. J., Lippincott, S. E. 1982. *Drug-Nutrient Interac.* 1:153–68
77. Huxtable, R. J. 1981. *J. Nutr.* 111:1275–86
78. Huxtable, R. J., Lippincott, S. E. 1982. *J. Nutr.* 112:1003–10
79. Sturman, J. A., Hayes, K. C. 1980. *Adv. Nutr. Res.* 3:231–99
80. Sturman, J. A. 1981. *Dev. Brain Res.* 2:111–28
81. Stipanuk, M. H., Kuo, S. M., Hirschberger, L. L. 1984. *Life Sci.* 35:1149–55
82. Huxtable, R. J. 1982. *J. Nutr.* 112:2293–300
83. Rozen, R., Scriver, C. R. 1982. *Proc. Natl. Acad. Sci. USA* 79:2101–5
84. Friedman, A., Albright, P. W., Chesney, R. W. 1981. *Life Sci.* 29:2415–19
85. Chesney, R. W., Friedman, A. L., Albright, P. W., Gusowski, N. 1982. *Proc. Soc. Exp. Biol. Med.* 170:493–501
86. Chesney, R. W., Gusowski, N., Theissen, M. 1984. *Pediatr. Res.* 18:611–18
87. Chesney, R. W., Gusowski, N., Dabbagh, S., Theissen, M., Padilla, M., Diehl, A. 1985. *Biochim. Biophys. Acta* 812:702–12

88. Vessey, D. A. 1978. *Biochem. J.* 174: 621–26
89. Soupart, P. 1962. In *Amino Acid Pools*, ed. J. T. Holden, pp. 220–62. New York: Elsevier
90. Ahtee, L., Boullin, D. J., Paasonen, M. K. 1974. *Br. J. Pharmacol.* 52:245–51
91. Gaut, Z. N., Nauss, C. B. 1976. See Ref. 7, pp. 91–98
92. Nauss-Karol, C., VanderWende, C. 1981. See Ref. 9, pp. 81–91
93. Voaden, M. J., Hussain, A. A., Chan, I. P. R. 1982. *Br. J. Ophthalmol.* 66:771–75
94. Hambley, J. W., Johnston, G. A. R. 1985. *Life Sci.* 36:2053–62
95. Airaksinen, E. M., Sihvola, P., Airaksinen, M. M., Sihvola, M., Tuovinen, E. 1979. *Lancet* 1:474–75
96. Airaksinen, E. M., Airaksinen, M. M., Sihvola, P., Marnela, K. M. 1981. *Metab. Pediatr. Ophthalmol.* 5:45–48
97. Airaksinen, E. M. 1979. *Epilepsia* 20: 503–10
98. Bonhaus, D. W., Huxtable, R. J. 1983. *Neurochem. Int.* 5:413–19
99. Boullin, D. J., Airaksinen, E. M., Paasonen, M. K. 1975. *Med. Biol.* 53: 184–86
100. Paasonen, M. K., Himberg, J.-J., Penttila, O., Solatunturi, E., Ylitalo, P. 1980. In *Pharmacological Control of Heart and Circulation*, ed. L. Tardos, L. Szekeres, J. G. Papp, pp. 221–25. New York: Pergamon
101. Baskin, S. I., Klekotka, S. J., Kendrick, Z. V., Bartuska, D. G. 1979. *J. Endocrinol. Invest.* 2:245–49
102. Dhopesh, V. P., Baskin, S. I. 1982. *Headache* 22:165–66
103. Fukuda, K., Hirai, Y., Yoshida, H., Nakajima, T., Usui, T. 1982. *Clin. Chem.* 29:1758–61
104. Gaull, G. E., Wright, C. E., Tallan, H. H. 1983. See Ref. 11, pp. 297–304
105. Tallan, H. H., Jacobson, E., Wright, C. E., Schneidman, K., Gaull, G. E. 1983. *Life Sci.* 33:1853–60
106. Tallan, H. H., Schneidman, K. 1984. *Fed. Proc.* 43:1779
107. Schmid, R., Sieghart, W., Karobath, M. 1975. *J Neurochem.* 25:5–9
108. States, B., Lee, J., Segal, S. 1983. *Metabolism* 32:272–78
109. Bond, W. S., Yee, G. C. 1980. *Am. J. Hosp. Pharm.* 37:74–78
110. Olansky, A. 1982. *J. Am. Acad. Dermatol.* 6:19–23
111. Pasantes-Morales, H., Klethi, J., Ledig, M., Mandel, P. 1972. *Brain Res.* 41: 494–97
112. Cohen, A. I., McDaniel, M., Orr, H. T. 1973. *Invest. Ophthalmol.* 12:686–93
113. Macaione, S., Ruggieri, P., DeLuca, F., Tucci, G. 1974. *J. Neurochem.* 22:881–91
114. Voaden, M. J., Oradeu, A. C. I., Marshall, J., Lake, N. 1981. See Ref. 9, pp. 145–60
115. Deleted in proof
116. Schmidt, S. Y., Aguirre, G. D. 1985. *Invest. Ophthalmol. Visual Sci.* 26:679–83
117. Green, J. P., Day, M., Robinson, J. D. 1962. *Biochem. Pharmacol.* 11:957–58
118. Vellan, E. J., Gjessing, L. R., Stalsber, H. 1970. *J. Neurochem.* 17:699–701
119. Guidotti, A., Badiami, G., Pepeau, G. 1972. *J. Neurochem.* 19:431–35
120. Crabai, F., Sitzer, A., Pepeau, G. 1974. *J. Neurochem.* 23:1091–92
121. Lombardini, J. B. 1976. See Ref. 7, pp. 311–26
122. Grosso, D. A., Bressler, R., Benson, B. 1978. *Life Sci.* 22:1789–98
123. Klein, D. C., Wheler, G. H. T., Weller, J. L. 1983. See Ref. 11, pp. 169–81
124. Kennedy, A. J., Voaden, M. J. 1974. *J. Neurochem.* 23:1093–98
125. Orr, H. T., Cohen, A. I., Lowry, O. H. 1976. *J. Neurochem.* 26:609–11
126. Voaden, M. J., Lake, N., Marshall, J., Morjaria, B. 1977. *Exp. Eye Res.* 25: 249–57
127. Bito, L. Z. 1985. In *Handbook of Neurochemistry*, ed. A. Lajtha, 9:477–506. New York: Plenum. 2nd ed.
128. Pasantes-Morales, H., Bonaventure, N., Wioland, N., Mandel, P. 1973. *Int. J. Neurosci.* 5:235–41
129. Pourcho, R. G. 1977. *Exp. Eye Res.* 25:119–27
130. Steinberg, R. H., Miller, S. S. 1979. In *The Retinal Pigment Epithelium*, ed. K. M. Zinn, M. F. Marmor, pp. 205–25. Cambridge: Harvard Univ. Press
131. Macaione, S., Tucci, G., De Luca, G., Di Giorgio, R. M. 1976. *J. Neurochem.* 27:1411–15
132. Mathur, R. L., Klethi, J., Ledig, M., Mandel, P. 1976. *Life Sci.* 18:75–80
133. Pasantes-Morales, H., Lopez-Colome, A. M., Salceda, R., Mandel, P. 1976. *J. Neurochem.* 27:1103–6
134. Di Giorgio, R. M., Tucci, G., Macaione, S. 1975. *Life Sci.* 16:429–36
135. Di Giorgio, R. M., Macaione, S., De Luca, G. 1977. *Life Sci.* 20:1657–62
136. Lin, C. T., Lei, H. Z., Wu, J. Y. 1982. *Invest. Ophthalmol. Visual Sci. Suppl.* 22:114
137. Su, Y. Y. T., Lin, C. T., Song, G. X., Wu, J. Y. 1983. *Trans. Am. Soc. Neurochem.* 14:131
138. Lin, C. T., Song, G. X., Wu, J. Y. 1985. *Brain Res.* 337:293–98

139. Lin, C. T., Song, G. X., Wu, J. Y. 1985. *Brain Res.* 331:71–80
140. Nishimura, C., Ida, S., Kuriyama, K. 1983. *J. Neurosci. Res.* 9:59–67
141. Nishimura, C., Ida, S., Kuriyama, K. 1983. See Ref. 11, pp. 233–50
142. Schmidt, S. Y. 1980. *Exp. Eye Res.* 31:373–79
143. Starr, M. S. 1978. *Brain Res.* 151:604–8
144. Starr, M. S., Voaden, M. J. 1972. *Vision Res.* 12:1261–69
145. Salceda, R. 1980. *Neurochem. Res.* 5:561–72
146. Schmidt, S. Y. 1981. See Ref. 9, pp. 177–85
147. Pourcho, R. G. 1981. *Exp. Eye Res.* 32:11–20
148. Adler, R. 1983. *J. Neurosci. Res.* 10:369–79
149. Schmidt, S. Y., Berson, E. L. 1978. *Exp. Eye Res.* 27:191–98
150. Voaden, M. J. 1982. In *Problems of Normal and Genetically Abnormal Retinas,* ed. R. M. Clayton, J. Haywood, H. W. Reading, A. Wright, pp. 353–62. New York: Academic
151. Young, R. W. 1969. In *The Retina: Morphology, Function and Clinical Characteristics,* ed. B. R. Straatsma, M. O. Hall, R. A. Allen, F. Crescitelli, pp. 172–210. Los Angeles: Univ. Calif. Press
152. Miller, S. S., Steinberg, R. H. 1979. *J. Gen. Physiol.* 14:237–59
153. Miller, S. S., Steinberg, R. H. 1976. *Exp. Eye Res.* 23:177–90
154. Lopez-Colome, A. M., Pasantes-Morales, H. 1980. *J. Neurochem.* 34:1047–52
155. Lombardini, J. B. 1981. *Soc. Neurosci.* 7:321
156. Lombardini, J. B., Priem, S. D. 1983. *Exp. Eye Res.* 37:239–50
157. Sturman, J. A., Hayes, K. C. 1980. In *Advances in Nutritional Research,* ed. H. H. Draper, 3:231–99. New York: Plenum
158. Schmidt, S. Y., Berson, E. L., Hayes, K. C. 1976. *Invest. Ophthalmol.* 15:47–52
159. Neuringer, M., Sturman, J. A., Wen, G. Y., Wisniewski, H. M. 1985. See Ref. 12, pp. 53–62
160. Lake, N. 1981. *Life Sci.* 29:445–48
161. Lake, N. 1982. *Neurochem. Res.* 7:1385–90
162. Pasantes-Morales, H., Quesada, O., Carabez, A., Huxtable, R. J. 1983. *J. Neurosci. Res.* 9:135–43
163. Geggel, H. S., Ament, M. E., Heckenlively, J. R., Kopple, J. D. 1982. *Clin. Res.* 30:486A
164. Geggel, H. S., Ament, M. E., Heckenlively, J. R., Martin, D. A., Kopple, J. D. 1985. *N. Engl. J. Med.* 312:142–46
165. Sheikh, K. 1981. *Gastroenterology* 80:1363
166. Hayes, K. C., Rabin, A. R., Berson, E. L. 1975. *Am. J. Pathol.* 78:505–24
167. Berson, E. L., Hayes, K. C., Rabin, A. R., Schmidt, S. Y., Watson, G. 1976. *Invest. Ophthalmol.* 15:52–58
168. Pasantes-Morales, H., Ademe, R. M., Quesada, O. 1981. *J. Neurosci. Res.* 6:337–48
169. Pasantes-Morales, H., Quesada, O., Carabez, A. 1981. *J. Neurochem.* 36:1583–86
170. Salceda, R., Lopez-Colome, A. M., Pasantes-Morales, H. 1977. *Brain Res.* 135:186–91
171. Pasantes-Morales, H., Urban, P. F., Klethi, J., Mandel, P. 1973. *Brain Res.* 51:375–78
172. Hoffman, E. K., Hendil, K. B. 1976. *J. Comp. Physiol.* 108:279–86
173. Thurston, J. H., Hauhart, R. E., Dirgo, J. A. 1980. *Life Sci.* 26:1561–68
174. Kuo, C. H., Miki, N. 1980. *Biochem. Biophys. Res. Commun.* 94:646–51
175. Lopez-Colome, A. M., Pasantes-Morales, H. 1981. *Exp. Eye Res.* 32:771–80
176. Pasantes-Morales, H. 1982. *Vision Res.* 22:1487–93
177. Pasantes-Morales, H., Ordonez, A. 1982. *Neurochem. Res.* 7:317–28
178. Lombardini, J. B. 1983. *J. Neurochem.* 40:402–6
179. Lombardini, J. B. 1983. See Ref. 11, pp. 251–62
180. Lombardini, J. B. 1985. *J. Neurochem.* 45:268–75
181. Pasantes-Morales, H., Ademe, R. M., Lopez-Colome, A. M. 1979. *Brain Res.* 172:131–38
182. Pasantes-Morales, H., Cruz, C. 1983. See Ref. 11, pp. 263–76
183. Ogino, N., Matsumura, M., Shirakawa, H., Tsukahara, I. 1983. *Ophthalmic Res.* 15:72–89
184. Cruz, C., Pasantes-Morales, H. 1983. *J. Neurochem.* 41:Suppl.S134
185. Pasantes-Morales, H., Cruz, C. 1984. *J. Neurochem. Res.* 11:303–12
186. Armstrong, D., Connole, E., Feeney, L., Berman, E. R. 1978. *J. Neurochem.* 31:761–69
187. Armstrong, D., Santangelo, G., Connole, E. 1981. *Curr. Eye Res.* 1:225–42
188. Sturman, J. A., Moretz, R. C., French, J. H., Wisniewski, H. M. 1985. *J. Neurosci. Res.* 13:405–16
189. Sturman, J. A., Moretz, R. C., French, J. H., Wisniewski, H. M. 1985. *J. Neurosci. Res.* 13:521–28

190. Wen, G. Y., Sturman, J. A., Wisniewski, H. M., Lidsky, A. A., Cornwell, A. G., Hayes, K. C. 1979. *Invest. Ophthalmol. Visual Sci.* 18:1201–6

191. Palackal, T., Sturman, J. A., Moretz, R. C., Wisniewski, H. M. 1985. *Trans. Am. Soc. Neurochem.* 16:186

192. Rassin, D. K., Gaull, G. E., Heinonen, K., Räihä, N. C. R. 1977. *Pediatrics* 59:407–22

193. Järvenpää, A. L., Räihä, N. C. R., Rassin, D. K., Gaull, G. E. 1982. *Pediatrics* 70:214–20

194. Järvenpää, A. L., Rassin, D. K., Räihä, N. C. R., Gaull, G. E. 1982. *Pediatrics* 70:221–30

195. Järvenpää, A. L., Rassin, D. K., Kuitunen, P., Gaull, G. E., Räihä, N. C. R. 1983. *Pediatrics* 72:677–83

196. Rassin, D. K., Gaull, G. E., Järvenpää, A. L., Räihä, N. C. R. 1983. *Pediatrics* 71:179–86

197. Deleted in proof

198. Klein, P. D., James, W. P., Wong, W. W., Irving, C. S., Murgatroyd, P. R., et al. 1984. *Hum. Nutr. Clin. Nutr.* 38:95–106

199. Deleted in proof

200. Sturman, J. A., Rassin, D. K., Gaull, G. E. 1977. *Pediatr. Res.* 11:28–33

201. Huxtable, R. J., Lippincott, S. E. 1983. *Ann. Nutr. Metab.* 27:107–116

202. Lake, N. 1983. *Neurochem. Res.* 8:881–87

203. Chipponi, J. X., Bleier, J. C., Santi, M. T., Rudman, D. 1982. *Am. J. Clin. Nutr.* 35:1112–16

204. Järvenpää, A. L., Räihä, N. C. R., Rassin, D. K., Gaull, G. E. 1983. *Pediatrics* 71:171–78

205. Watkins, J. B., Järvenpää, A. L., Szcepanik-Van Leeuwen, S., Klein, P. D., Rassin, D. K., et al. 1983. *Gastroenterology* 85:793–800

206. Roy, C. C., Weber, A. M., Morin, C. L., Lepage, G., Brisson, G., et al. 1982. *J. Pediatr. Gastroenterol. Nutr.* 1:469–78

207. Sturman, J. A., Wen, G. Y., Wisniewski, H. M., Neuringer, M. D. 1984. *Int. J. Dev. Neurosci.* 2:121–29

208. Berson, E. L., Schmidt, S. Y., Rabin, A. R. 1976. *Br. J. Ophthalmol.* 60:142–47

209. Airaksinen, E. M., Oja, S. S., Marnela, K. M., Sihvola, P. 1980. *Ann. Clin. Res.* 12:52–54

210. Uma, S. M., Satapathy, M., Sitaramayya, A. 1983. *Biochem. Med.* 30:49–52

211. Arshinoff, S. A., McCulloch, J. C., Macrae, W., Stein, A. N., Marliss, E. B. 1981. *Br. J. Ophthalmol.* 65:626–30

212. Schmidt, S. Y., Berson, E. L. 1980. *Invest. Ophthalmol. Visual Sci.* 19:1274–80

213. Voaden, M. J., Hussain, A. A., Chan, I. P. R. 1982. *Br. J. Ophthalmol.* 66:771–75

214. Airaksinen, E. M., Airaksinen, M. M., Sihvola, P., Marnela, K. M. 1981. *Metab. Pediatr. Ophthalmol.* 5:45–48

215. Wright, C. E., Tallan, H. H., Gaull, G. E. 1985. *Invest. Ophthalmol. Visual Sci.* 26:Suppl.132

216. Wright, C. E., Tallan, H. H., Gaull, G. E., Nussbaum, R., Carr, R., et al. 1986. Submitted for publication

217. Bird, A. C. 1982. In *Problems of Normal and Genetically Abnormal Retinas*, ed. R. M. Clayton, J. Haywood, H. W. Reading, A. Wright, pp. 327–32. New York: Academic

218. Huxtable, R. J., Bressler, R. 1973. *Biochim. Biophys. Acta* 323:573–83

219. Kramer, J. H., Chovan, J. P., Schaffer, S. W. 1981. *Am. J. Physiol.* 240: H238–46

220. Pasantes-Morales, H., Wright, C. E., Gaull, G. E. 1984. *J. Nutr.* 114:2256–61

221. Pasantes-Morales, H., Wright, C. E., Gaull, G. E. 1985. *Biochem. Pharmacol.* 34:2205–7

222. Alvarez, J. G., Storey, B. T. 1984. *Biol. Reprod.* 29:548–55

223. Nakashima, T., Takino, T., Kuriyama, K. 1983. See Ref. 11, pp. 449–59

224. Gery, I. 1980. *Invest. Ophthalmol. Visual Sci.* 19:751–59

225. Stillwell, W., Bryant, L. 1983. *Biochim. Biophys. Acta* 731:483–86

226. Lewis, D. A. 1984. *Biochem. Pharmacol.* 33:1705–14

227. VanGelder, N. M., Barbeau, A. 1985. See Ref. 12, pp. 149–63

228. Azari, J., Huxtable, R. J. 1980. *Eur. J. Pharmacol.* 61:217–23

229. Pasantes-Morales, H., Gamboa, A. 1980. *J. Neurochem.* 34:244–46

230. Gruener, R., Markovitz, D., Huxtable, R., Bressler, R. 1975. *J. Neurol. Sci.* 24:351–60

231. Kurzinger, K., Hamprecht, B. 1981. *J. Neurochem.* 37:956–67

232. Allen, J. A., Garrett, M. R. 1971. *Adv. Mar. Biol.* 9:205–9

233. Yousef, I. M., Tuchweber, B., Vonk, R. J., Masse, D., Audet, M., Roy, C. C. 1981. *Gastroenterology* 80:233–41

234. Palmer, R. H., Glickman, P. B., Kappas, A. 1962. *J. Clin. Invest.* 41:1573–77

235. Dorvil, N. P., Yousef, I. M., Tuchweber, B., Roy, C. C. 1983. *Am. J. Clin. Nutr.* 37:221–32

236. Kakis, G., Yousef, I. M. 1978. *Gastroenterology* 75:595–607
237. Sturman, J. A., Hepner, G. W., Hofmann, A. F., Thomas, P. J. 1975. *J. Nutr.* 105:1206–14
238. Hofmann, A. F., Small, D. M. 1967. *Ann. Rev. Med.* 18:333–76
239. O'Maille, E. R. L., Richards, T. G., Short, A. H. 1965. *J. Physiol.* 180:67–79
240. Skare, K. L., Sietsema, W. K., De Luca, H. F. 1982. *J. Nutr.* 112:1626–30
241. Emudianughe, T. S., Caldwell, J., Smith, R. L. 1983. *Xenobiotica* 13:133–38
242. Hepner, G. W., Sturman, J. A., Hofmann, A. F., Thomas, P. J. 1973. *J. Clin. Invest.* 52:433–40
243. Barnes, D., Sato, G. 1980. *Cell* 22:649–55
244. Klebanoff, S. J., Clark, R. A. 1978. *The Neutrophil: Function and Clinical Disorders.* Amsterdam: North Holland. 810 pp.
245. Dilley, J. V. 1972. *Radiat. Res.* 50:191–96
246. Naskalski, J. W. 1977. *Biochim. Biophys. Acta* 485:291–300
247. Zgliczynski, J. M., Selvaraj, R. J., Paul, B. B., Stelmaszynska, T., Poskitt, P. K.

F., Sbarra, A. J. 1977. *Proc. Soc. Exp. Biol. Med.* 154:418–22
248. Test, S. T., Weiss, S. J. 1984. *J. Biol. Chem.* 259:399–405
249. Grisham, M. B., Jefferson, M. M., Thomas, E. L. 1984. *J. Biol. Chem.* 259:6757–65
250. Lampert, M. B., Weiss, S. J. 1983. *Blood* 62:645–51
251. Wever, R., Plat, H., Hamers, M. N. 1981. *FEBS Lett.* 123:327–31
252. Buys, J., Wever, R., Ruttenberg, E. J. 1984. *Immunology* 51:601–7
253. Bolscher, B. G. J. M., Plat, H., Wever, R. 1984. *Biochim. Biophys. Acta* 784:177–86
254. Thomas, E. L. 1979. *Infect. Immun.* 23:522–31
255. Thomas, E. L. 1979. *Infect. Immun.* 25:110–16
256. Scully, F. E., Bempong, M. A. 1982. *Environ. Health Perspect.* 46:111–16
257. Evans, O. M. 1982. *Anal. Chem.* 54:1579–82
258. Zgliczynski, J. M., Stelmaszynska, T., Dumanski, J., Ostrowski, W. 1971. *Biochim. Biophys. Acta* 235:419–24
259. Thomas, E. L., Grisham, M. B., Jefferson, M. M. 1983. *J. Clin. Invest.* 72:441–54

Ann. Rev. Biochem. 1986. 55:455–81

EXTRALYSOSOMAL PROTEIN DEGRADATION[1]

S. Pontremoli and E. Melloni

Institute of Biological Chemistry, University of Genoa, 16132 Genoa, Italy

CONTENTS

[1]Abbreviations used: CANP, calcium-dependent neural proteinase; μM CANP, low-calcium-requiring form of CANP; mM CANP, high-calcium-requiring form of CANP; PKC, protein kinase C; ER, endoplasmic reticulum; PMA, phorbol 12-myristate 13-acetate; f-Met-Leu-Phe, *N*-formyl-L-methionyl-L-leucyl-L-phenylalanine; AMC, 7-amino-4-methyl-coumarin.

0066-4154/86/0701-0455$02.00

1. PERSPECTIVES AND SUMMARY

The existence in a variety of cells of nonlysosomal systems for protein degradation is now generally accepted. This subject has been covered in reviews (1–4) and symposia (5–7).

For the purposes of the present discussion we will define two distinct types of extralysosomal proteolytic systems, each related to a specific physiological function. One type (A) catalyzes selective degradation of proteins to small peptides and amino acids. The second (B) is characterized by the ability to catalyze selective but limited proteolytic modification of proteins.

Proteolytic enzymes belonging to type A probably play a major role in: 1. Degradation of short-lived proteins (3, 8–11), abnormal proteins, and proteins produced in excessive amounts (12). 2. Degradation of cellular organelles, and/or soluble proteins during cell maturation (13). 3. Selective protein degradation that accounts for the differences in half-lives of intracellular proteins and for the changes in enzyme levels in response to modifications of metabolic requirements (2–4).

Proteolytic enzymes belonging to type B seem to be involved primarily in: 1. Posttranslational maturation of proteins, as required for their transport, secretion, or control of activity [see Section 2; for review see (14)]. 2. Partial degradation of protein components of the cytoskeletal network and the contractile apparatus (see Section 3). 3. Limited proteolytic modifications of membrane-bound proteins including cell-surface receptors (see Section 3). 4. Limited selective proteolysis of regulatory enzymes resulting in their activation or inactivation or of soluble proteins as an initial signal for their subsequent complete degradation (see Section 2.2).

Since these proteolytic processes occur in the nonlysosomal compartment, in each case mechanism(s) must exist that provide for selective recognition of the substrate(s). In the degradative systems (type A) indiscriminate proteolysis seems to be prevented by a requirement for an ATP-dependent conjugation of target proteins with ubiquitin (12, 15, 16). An additional system, still to be defined with respect to selectivity, appears to be enhanced by ATP and by digestible substrates possessing an appropriate conformation (17–19). Other examples of recognition signals are peroxidation of membrane lipids (20) associated with spontaneous activation of a proteolytic system linked to cell maturation. Targeting of protein substrates may also be accomplished by mixed function oxidation (21, 22); phosphorylation (23–26); mixed disulfide formation (27–30); or oxidation of SH groups (31).

In those systems involved in limited proteolysis (type B) two main classes of proteinases may be considered: (a) signal and processing proteinases and (b) a novel class of thiol endopeptidases identified as Ca^{2+}-dependent neutral proteinases (hereinafter called CANP). Signal peptides in the precursor protein, defined as pre- or prosequences, direct the proteinases of the first class, while variations in the intracellular concentration of Ca^{2+} and changes in intracellular localization appear to regulate the activity of the CANP class of proteinases. Since intracellular mobilization of Ca^{2+} appears to be the primary signal for these proteinases, their activation may be related to cell responses following surface stimulation by external ligands.

For proteinases herewith classified under type B, targeting of proteins may direct selective recognition of substrates. This has recently been shown for liver microsomal HMG-CoA reductase which in its phosphorylated form (32) becomes more susceptible to degradation by CANP into active 52–56-kd fragments (33, 34). As will be clear from the following discussion, although several problems remain to be elucidated, significant progress has been made in the understanding of the mechanisms of extralysosomal protein degradation. The aim of this review is to describe and critically evaluate the most recent developments in this field.

The discussion has been restricted to those mammalian proteolytic systems that catalyze limited and highly specific protein degradations. This chapter considers the properties and mechanism of action of signal and processing peptidases in relation to their physiological functions. In addition, special attention will be devoted to the structural and catalytic properties of Ca^{2+}-dependent thiol endopeptidases (CANPs) in an effort to identify common physiological functions for these proteinases in the expression of various cell activities. Other aspects of nonlysosomal protein degradation have been reviewed (2–4, 7, 12).

2. SIGNAL AND PROCESSING PROTEINASES

This class of proteinases catalyzes the limited proteolysis of target proteins, converting larger precursors designated as pre- or proproteins into the mature, biologically active proteins, having precise cellular localization (35).

The presequence, also designated as the signal peptide (36), is an extension at the NH_2 terminus, and is required for the delivery of proteins to membrane organelles (37) or for their translocation into mitochondria (38, 39). The enzymes that cleave these presequences are called "signal peptidases."

The prosequence, which may occur in any part of the protein molecule, ensures the correct folding of the protein and prevents expression of its biological activity (35). The enzymes cleaving the prosequences are here defined as "processing proteinases."

In addition to pre- and prosequences, in other proteins as-yet-undefined segments are removed by limited proteolysis. This process affects certain enzymes and is promoted by a heterogeneous and incompletely defined class of processing proteinases producing molecular forms with altered catalytic activity.

The following discussion is restricted to a brief survey of the known properties of each group of proteinases with special emphasis on homologies and
specificities. The mechanism by which single proteins are delivered to
organelles, mitochondria, or membranes is not included in the present chapter.

2.1. Signal Peptidases of the Endoplasmic Reticulum (ER)

Experimental data have established that ER signal peptidases are localized on
the outer surface of the membranes (40–44). Very little is known concerning the
properties of these enzymes, and most of the experiments have been carried out
only with crude preparations. All signal peptidases thus far investigated can be
characterized as Zn^{2+}-endopeptidases that remove the entire signal peptide
without intermediate steps (45). Further degradation of the signal peptide,
observed with crude preparations or with intact microsomes, has been attributed to the action of proteinases on the cleavage product (46).

The site of cleavage, deduced from the residues present at the junction
between the presegment and the protein (47) and from the hydrolysis of the
synthetic substrate succinyl-Ala-Ala-Phe-AMC (43), contains amino acids
with small side chains (Ala, Gly, Ser, Cys) at the carbonyl position.

2.2. Mitochondrial Signal Peptidases

Despite many reports concerning the identification of mitochondrial precursor
proteins [for review see (39)] and the characterization of the molecular weights
of the precursor and product forms, little information is available about the
molecular and catalytic properties of the proteinases involved in the maturation
processes.

Several reports describing the properties of mitochondrial signal peptidases
indicate that these enzymes, having molecular masses of approximately 108 kd,
are metallo (Zn) endopeptidases with neutral pH optima, that may be located
either in the matrix or in the inner membranes of mitochondria (48–53). In the
case of ornithine carbamoyltransferase (OTC) signal peptidase, this dual
localization may be artifactual, or related to the presence of two enzymes (51),
one located in the matrix and responsible for the formation of mature OTC, and
the other located in both compartments, converting pre-OTC to a form of
intermediate size. The problems of distinct enzymes involved in the processing
of OTC and the formation of an intermediate during the maturation of the
precursor remain unsolved questions as recently reviewed by Hay et al (39).

2.3. Specificity of ER and Mitochondrial Signal Peptidases

Recent studies have provided some indication concerning the specificities of
the two classes of signal peptidases.

A first general type of specificity has been deduced from the observation that
mitochondrial and ER precursors are processed only by the homologous enzymes (39). Comparative analyses of sequences deduced from cloned com-

plementary DNA (54, 55) have indicated that the signal peptides of mitochondrial precursors from different species are characterized by two significant homologies: one concerns the peptide sequence, the other the nature of the amino acids (Gly, Ala, Ser, Cys) present at the site of cleavage. The only exception is at present pre-OTC in which a different amino acid (glutamine) has been identified at the sensitive peptide bond (53).

In the ER precursors of different eukaryotes the signal peptides are also characterized by a high degree of sequence homology and by the presence of small aliphatic side chains of amino acids at the site of cleavage (37). Thus mitochondrial and ER signal peptidases appear to be directed by different signal peptides sequences to similar sites of cleavage.

An interesting consideration is that such sequences contain information that not only directs the proteolytic cleavage but also relates to the transport mechanism. This is indicated by the high content of positively charged amino acid residues (56) in mitochondrial precursors, and by the high content of hydrophobic residues in ER signal peptides (37).

2.4. Processing Proteinases

2.4.1 CONVERSION OF PROHORMONES   The conversion of pre-prohormones to hormones involves in sequential steps: 1. removal of the presegment by the signal peptidases; 2. intermediate or final processing of the resulting prohormones by characteristic endopeptidases; 3. formation of the biologically active peptide by specific amino- or carboxypeptidases.

Step 1 is catalyzed by a group of proteinases previously discussed (see Section 2.1). Step 2 is promoted by a class of "processing proteinases" identified in different tissues, but very poorly characterized in their molecular and catalytic properties.

A pertinent example of processing proteinase is an acid thiol "trypsinlike" peptidase, present in rat pituitary lobes and involved in processing of pro-opiomelanocortin (57–59), and provasopressin (64). In adrenomedullary chromaffin granules trypsinlike activities have been shown to be involved in processing of proenkephalin (60–63). A unique endopeptidase, which specifically cleaves between Lys-Arg sequences, has recently been purified from bovine adrenomedullary chromaffin cells (65). The enzyme has neutral pH activity and has been tentatively identified as a serine protease.

In brain cortex a thiol proteinase of molecular mass 90 kd with neutral pH optima has been identified and found to be responsible for the conversion of somatostatin 28 to somatostatin 14 (66).

The pro-β–nerve growth factor (pro-β-NGF) is converted to the mature form by a still-undefined proteinase isolated in association with the pro-β-NGF in a high-molecular-weight complex (67). Proteinases similar in properties to those described above have been shown to be involved in the maturation of many precursors, e.g. proinsulin, proglucagon, prosomatostatin, prorelaxin, and others in mammalian cells, in yeast, and in anglerfish islet (35, 68–71).

In spite of these differences in the catalytic properties, it seems well established that processing proteinases are directed in their specific site of cleavage by pairs of basic amino acid residues (lysine and arginine) located in defined regions of the precursors and usually absent in the mature forms (72–78). Cleavage can remove these residues thereby generating directly the mature hormones. Alternatively cleavage can take place in a position that leaves one or both basic residues extending at either the COOH or the NH_2 terminus end of the still immature peptide. In these cases, a carboxypeptidase or an aminopeptidase (79), or both in sequence (with the function of processing enzymes), remove these residues and promote formation of the active peptide hormone.

The catalytic and molecular properties of these carboxypeptidases, in contrast to those of the aminopeptidases, have been extensively studied and well characterized in bovine pituitary and brain (80–85), in bovine secretory granules from adrenal medulla (80, 81, 84, 86–88), and in rat neural lobes (89). These carboxypeptidases are similar in many of their enzymic properties including thiol and metal (zinc) dependence and therefore distinct from other known carboxypeptidases (82–84, 86–89).

2.4.2 OTHER PROTEOLYTIC MODIFICATIONS   Mature proteins undergo other proteolytic modifications but in many cases it remains to be established that these occur in vivo and that they are related to important physiological functions. A few pertinent examples taken from the most recent literature are illustrated below.

Rat liver guanylate cyclase is activated in its plasma membrane–bound form by a proteolytic (90) mechanism apparently involved in the physiological regulation of this enzyme. Activation by limited proteolysis has also been demonstrated for Ca^{2+}-ATPase of chloroplast coupling factor 1 (91) and for rat brain phosphatidylinositol phosphodiesterase (92). The activity of cystathionine beta-synthase in rat liver appears to be regulated by limited proteolysis that results in 30-fold greater affinity of the enzyme for homocysteine (93). The processing of active forms of methionyl-tRNA synthetase has been shown to proceed through the formation of multiple forms of the enzyme produced by endogenous proteolysis (94). Removal of a 1–2-kd peptide from the M_1 type of pyruvate kinase changes its kinetic properties (95).

Additional roles in the limited proteolysis of cytosolic enzymes may be assigned to some lysosomal cathepsins. The recently described rat liver cathepsin T (96, 97) and rabbit liver cathepsin M (98) have been shown to catalyze limited proteolysis of tyrosine aminotransferase (99) and aldolase (100–102), respectively. Cathepsin M, which is localized on the outer surface of lysosomes, has direct accessibility to cytosolic substrates (103, 104), and may be responsible for the accumulation of an inactive digested form of aldolase in the livers of fasted rabbits (101).

In yeast, the participation of proteolysis in inactivation or activation of various enzymes is established by studies with mutants lacking specific proteinases (105–107).

Finally, it must be emphasized that although very little is known about the specificity of these proteinases, it is conceivable that some recognition signals must be present to direct their activity. These signals presumably involve selective cellular locations, targeting by covalent modifications, the presence of highly conserved regions (108), and undoubtedly involve other considerations.

2.5. General Comments on the Properties of Signal and Processing Proteinases

In contrast to the Ca^{2+}-dependent neutral proteinases, the signal and processing proteinases have well-defined functions in intracellular protein processing and transport, but are poorly understood in terms of enzyme structure and catalytic properties.

In common with other nonlysosomal proteinases, signal and processing proteinases must be even more selectively directed to specific recognition of the site of cleavage. The signal peptide hypothesis has been recently confirmed by the observation that binding to signal peptidases of signal sequence analogues inhibits processing and translocation of proproteins such as prolipoproteins (109). Whether the specificity is due to a single enzyme with a different intracellular location or to the existence of different proteinases is still an open question. The involvement of one or two proteinases in the maturation of the precursors would imply the formation of intermediate products and the requirement for additional site(s) of recognition. Most important is the possibility that processing proteinases may be Ca^{2+}-dependent like endopeptidases (79), which would provide an elaborate mechanism for their activation; this would add to their function the determination of the rate and extent of protein maturation. Future elucidation of the enzymic properties of these endopeptidases may contribute to the precise understanding of their physiological role; at present they are considered to function in the terminal step of a complex mechanism.

3. Ca^{2+}-ACTIVATED NEUTRAL PROTEINASES

3.1. Erythrocyte CANP

3.1.1 LOCALIZATION AND SUBSTRATE SPECIFICITY   In human (110–112) and porcine erythrocytes (113) only one type of CANP is present. The enzyme shows intermediate requirements of Ca^{2+} with $K_{0.5}$ of 40–50 μM; thus it is closer in its properties to the mM form of CANP. In rat erythrocytes, CANP appears to be present in both the cytosol and in the plasma membrane (114), while in human erythrocytes, lysed in the presence of chelating agents, CANP is recovered exclusively in the soluble fraction (110, 112, 115, 116). Degradation of membrane proteins in Ca^{2+}-enriched erythrocytes (117, 118) and the presence of immunoreactive trace amounts (2%) of CANP in plasma membranes (119) are consistent with our recent data indicating that the cytosolic proteinase is transiently translocated by μM Ca^{2+} to the inner face of the plasma membranes (120).

Erythrocyte CANP actively degrades globin chains that have been deprived of the heme group, and also with less efficiency isolated hemoglobin chains, but native tetrameric hemoglobin is completely resistant (121). The proteinase shows a high degree of specificity, cleaving a single peptide bond between Lys_{11}-Ala_{12} and Lys_8-Ser_9 in alpha chains and beta chains, respectively (121). In intact erythrocytes or in inside-out erythrocyte vesicles CANP appears to selectively degrade the membrane cytoskeletal proteins designated band 2.1 and band 4.1 (120, 122).

3.1.2 MOLECULAR STRUCTURE AND ACTIVATION MECHANISMS   CANP is present in human erythrocytes as an inactive proenzyme of molecular mass 110 kd composed of two polypeptide chains (112), one 80 kd containing the catalytic site (115, 120–124), the other 30 kd whose function remains unknown. High (millimolar) concentrations of Ca^{2+} convert the proenzyme to active CANP by a two-step mechanism (116, 123, 124). The first step, dissociation of the heterodimeric proenzyme, is followed by the autocatalytic conversion of the 80-kd inactive subunit to a 75-kd proteinase that is fully active in the presence of micromolar concentrations of Ca^{2+}. In the presence of a digestible substrate the same conversion of the inactive 110-kd proenzyme is promoted by micromolar concentrations of Ca^{2+}. The mechanism has the characteristics of a self-terminating process since it is interrupted after the substrate has undergone limited digestion (123).

An alternative mechanism for the formation of active CANP occurs in the presence of micromolar concentrations of Ca^{2+}. These low concentrations of Ca^{2+} promote binding of the inactive 80-kd subunit to the erythrocyte membrane and conversion of the bound subunit to the 75-kd subunit (120) which is then released as fully active enzyme (125). Binding and activation are not observed with the 110-kd heterodimer. Membrane phospholipids have been identified as the natural compounds responsible for such activation (126). Their effect has been attributed to a large decrease in the concentration of Ca^{2+} required for the conversion of the 80-kd to the 75-kd active subunit. This activation mechanism resembles that proposed for activation of protein kinase C (127).

Another mechanism for inducing CANP activity is based on the synergistic effects of Mn^{2+} and Ca^{2+}. This mechanism involves changes in the conformation of the proteinase without dissociation of the heterodimer or autoproteolytic conversion of the catalytic subunit (128). The concentration of Mn^{2+} present in red cells supports this hypothesis, but its physiological significance remains to be established. Finally it must be emphasized that in the presence of high concentrations of Ca^{2+} activation of CANP can proceed rapidly and according to the mechanism herewith first described. However, the high intracellular concentrations of Ca^{2+} are restricted to pathophysiological conditions (129–132).

The model, reported in Figure 1, illustrates the possible mechanisms responsible for the irreversible activation of human erythrocytes that CANP promoted by calcium ions.

3.1.3 FUNCTION OF ERYTHROCYTE CANP   In human and rat erythrocytes CANP catalyzes the degradation of submembraneous cytoskeleton proteins or transmembrane proteins (114, 119, 122, 133–135). The physiological significance of these limited degradation processes may be related to changes in cell shape, deformability, and lateral movement of transmembrane proteins that are correlated with a number of well-known functions, including removal of senescent erythrocytes with alteration in antigen exposure (136, 137). In intact

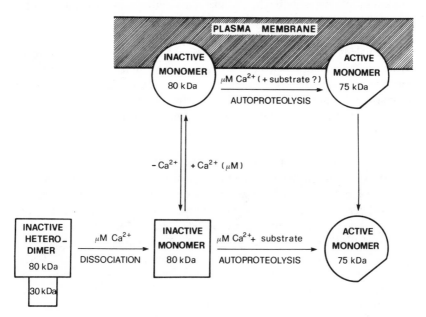

Figure 1   Proposed model for the activation of human erythrocyte CANP. For details see text.

erythrocytes CANP is also involved in the degradation of excess alpha or beta hemoglobin chains (138). This degradation appears to be a three-step process: 1. CANP catalyzes the hydrolysis of a single peptide bond in the hemoglobin chains (118). 2. The modified chains become attached to the inner surface of the erythrocyte membrane, but only after modification of the membrane by CANP (122). 3. The modified chains are degraded to small fragments by the action of intrinsic membrane-bound acid proteinases (122, 139, 140). Thus, degradation of excess globin chains in intact erythrocytes (138), a process which we have also demonstrated in reconstructed systems (122), represents the result of an integrated function of CANP and membrane proteinases and provides a physiological significance to the specificity of the proteinase toward soluble and membrane-bound proteins.

3.2. Human Neutrophil CANP

3.2.1 PROPERTIES AND ACTIVATION MECHANISM(S)   Human neutrophils contain only an mM form of CANP which is localized in the cytosolic fraction (141, 142). The enzyme is a monomer with a molecular mass of 85 kd having a $K_{0.5}$ for Ca^{2+} of 50 μM, with negligible activity at 1–5 μM (142). The neutrophil CANP does not undergo, under any known conditions, auto-proteolytic conversion to a μM form. However we have presented evidence that proteolytic activity of CANP is expressed in the presence of 1–5 μM Ca^{2+}; such expression promotes its binding to the neutrophil plasma membrane. In the bound form, CANP becomes active at these physiological concentrations of Ca^{2+} (142, 143). This activation is fully reversible since the enzyme returns to its original high-Ca^{2+}-requiring form (mM CANP) when it is dissociated from neutrophil membranes by EDTA. As in the case of human erythrocytes (126), membrane phospholipids have been identified as the natural compounds responsible for the activation of the proteinase (142). The model reported in Figure 2 summarizes the activation mechanism of human neutrophil CANP as described above.

3.2.2 FUNCTION OF NEUTROPHIL CANP   In confirmation of previous observations from Nishizuka's laboratory on rat brain enzyme (144), neutrophil CANP has been shown to convert the phospholipid-Ca^{2+}-dependent form of protein kinase C (PKC) to a form that is fully active in the absence of Ca^{2+} and phospholipids. As identical proteolytic modification of PKC has been observed by us (143) in a reconstructed system containing neutrophil membranes and physiological concentrations of Ca^{2+} following binding of PKC and of CANP to the membranes. This proteolytic conversion of PKC also occurs in intact neutrophils stimulated with PMA or f-Met-Phe-Leu (145). These results suggest that following a localized intracellular mobilization of Ca^{2+} induced by external stimuli (146–148), cytosolic PKC and CANP are translocated to the plasma membranes where active CANP promotes proteolytic conversion of

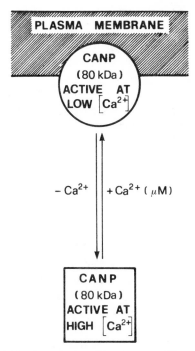

Figure 2   Proposed model for the activation of human neutrophil CANP. For details see text.

PKC. Activated PKC is then released from the membranes, and able to reach other cell compartments, where phosphorylation of target substrates may be related to well-known cell responses (149–152) including neutrophil degranulation (S. Pontremoli, E. Melloni, personal unpublished observations) and increased mobility. We suggest that CANP may be involved in the modulation of cell responses to activation by promoting the membrane to cytosol translocation of active PKC known to play a crucial role in the transduction of extracellular signals (153, 154). This specific role of CANP appears to be related to its site of activation and would explain, on a physiological basis, the presence of a single nonconvertible mM form of CANP in human neutrophils. These considerations do not exclude the participation of CANP, in other basic neutrophil functions, by selective degradation of membrane or cytoskeletal proteins, independent of modification of PKC. These ideas should also be kept in mind when considering the role of CANP in other cell types.

3.3. Platelet CANP

3.3.1 LOCALIZATION AND SUBSTRATE SPECIFICITY   Following earlier reports describing the presence of a single mM form (155–157), it has recently been established that bovine and human platelets contain both types of CANP (μM and mM) (158, 159). The two forms appear to be structurally independent, and no indication for autoproteolytic conversion of the mM to the μM form is yet available.

The two types of CANP have similar molecular masses of 105–110 kd and each is composed of two subunits: one of 80 kd and one of 25–30 kd (156, 157, 159).

Although both types of CANP have been isolated from the cytosol fraction, in platelets the proteinase also undergoes association to plasma membrane following stimulation with thrombin (160). This idea of functional localization receives support from the observation that in activated platelets, membrane glycoprotein receptors (157, 161–166) and PKC (167, 168) are selectively digested by the proteinase.

CANP also degrades specific polypeptides of the contractile system of platelets, including alpha-actinin (169), actin-binding protein (ABP) (170), P-235 protein (170), and calmodulin-binding proteins (171).

3.3.2 FUNCTION OF PLATELET CANP   The involvement of CANP in platelet functions has also been postulated on the basis of observations showing that many agents that induce platelet activation also elevate the intracellular concentration of Ca^{2+} (172–181). Based also on the observation that active CANP is present in stimulated platelets (182), it can be suggested that the proteinase is involved in the reorganization of the cytoskeleton, in changes in the cell shape, as well as in the disruption of proteins affecting polymerization of actin filaments (183, 184). Another function, related to the localization of CANP in activated platelets (164), involves degradation of membrane receptors (157, 164) and thereby modulation of the receptor-ligand interactions at the cell surface. Finally, as in other cells, CANP is responsible for the proteolytic degradation of membrane-bound PKC (167, 168) followed by intracellular release of the active kinase. Recent reports have indicated that, in conditions that would favor activation of CANP, proteolytic modification of Von Wille-brand Factor occurs (185). No indication is at present available concerning changes in the functional properties of the modified factor.

Interesting analogies exist between the function of CANP in human platelets and in neutrophils. In both types of cells the mM form of CANP can presumably be activated by association to the plasma membrane where specific target substrates are localized. Association appears to occur following stimulation of both cells and is mediated by intracellular mobilization of Ca^{2+}, suggesting the existence of a common linkage between activation of CANP and regulation of the stimulus-induced response in these cells.

3.4. Brain CANP

3.4.1 LOCALIZATION AND SUBSTRATE SPECIFICITY   The Ca^{2+}-activated proteinase was first isolated from this tissue by Guroff (186). More recently brain has been shown to contain multiple forms of the proteinase. An mM

CANP was first identified (187, 188) in rat and in monkey brain (185), while μM forms of the proteinase have been described in bovine brain and rat neurons (189–196). The three structurally distinct types of CANP, identified in rat brain by Zimmerman & Schlaepfer (197) and characterized by different sensitivities to Ca^{2+}, are thought to arise from a common high-molecular-weight precursor by autoproteolysis induced by calcium ions. This is consistent with the finding that the mM CANP (193) purified from synaptosomal membranes can be converted to the μM form. The number and type of CANPs present in nerve cells, as in other cells, remains an open question, since the presence of some forms may be due to the use of Ca^{2+}-buffers during the isolation procedure (193).

According to many reports, both mM and μM forms of CANP are present in the soluble fraction of the cell (186, 189–191, 193). Different cellular localizations are suggested by the observations of Siman et al (193) and Ishizaki et al (198), who reported the presence of CANP in association with synaptosomal plasma membrane or with the neurofilament-enriched cytoskeleton. More recently Ishizaki et al have reported the purification of the cytosolic and of the cytoskeleton-bound forms of CANP from bovine spinal cord (199). Both enzymes have identical molecular weight and subunit composition but apparently have different Ca^{2+} requirements for degradation of neurofilament polypeptides. It appears that proper localization of CANP confers to the enzyme a preference for the type of substrate to be digested. Based on the evidence that binding of CANP to the erythrocyte membrane leads to its activation at physiological concentrations of Ca^{2+} (120), association of the brain proteinase with the synaptosomal membrane or the cytoskeleton may also represent part of the activation process of this enzyme in a specific site of the cell.

The subunit composition and molecular weight of brain CANP differ from those in other tissues not only with respect to molecular weight heterogeneity, but particularly in the absence of the 30-kd subunit. It is however still controversial if the 30-kd subunit is lost during the purification, particularly if buffers are employed containing high concentrations of Ca^{2+} which are known to produce rapid dissociation of the heterodimeric structure of the proteinase (116, 124). Malik et al (196) have reported the isolation of an 18-kd protein which copurifies with CANP and have suggested a possible regulatory role for this polypeptide. It is not yet clear, however, if this protein is a contaminant or a true subunit of the enzyme.

The molecular mass of the catalytic subunit of brain CANPs varies between 70 kd and 150 kd (189, 191, 193, 197). This heterogeneity may result from the conversion of a common high-molecular-weight precursor into different low-molecular-weight active species (197). Further studies are necessary to establish if these proteinases are single gene products or as shown for muscle separate gene products (200).

In brain, CANP(s) has been shown to degrade specifically cytoskeleton components including neurofilaments and microtubule-associated proteins (196–198) in addition to the spectrinlike protein fodrin (193, 195). CANP has recently been shown to degrade, with different specificities, various neuropeptides (189), such as dynorphin, neurotensin, neoendorphin, and others. In rat brain the native form of PKC was first demonstrated to be converted by CANP in a fully active phospholipid-Ca^{2+}-independent form (190); this conversion is probably related, as discussed earlier, to the intracellular localization of active PKC.

In terms of peptide bond specificity, limited proteolysis of the various neuropeptides (189) has revealed that brain CANP selectively hydrolyzes those in which either Lys or Tyr is present at the P_1 position and amino acids with large aliphatic side chains are present at the P_2 position.

3.4.2 FUNCTION OF BRAIN CANP   Selective degradation of fodrin catalyzed by membrane-associated CANP has been proposed as the initial event in persistent changes in synaptic chemistry and ultrastructure, which lead to an increase in the number of glutamate receptors (194, 201, 202) and associated modifications of the functional properties of neuronal circuits that are detected after learning (203). These structural-functional modifications, which probably require mobilization of surface receptors and variation in the cell shape, are facilitated by CANP-dependent proteolysis of neurofilament and microtubule proteins (194, 195, 197).

Another important function of CANP may be the posttranslational modification of filament proteins, possibly required for the migration of the neurofilaments and for the filament-membrane interaction within the axon (194, 197).

The processing of neuropeptides has also been proposed as a physiological role for CANP (189). Indeed a hormone convertase activity has recently been suggested as a major physiological function for proteinases belonging to the CANP family (79).

3.5. Kidney CANP

In rat and porcine kidney an mM and a μM form of CANP have been identified and shown to be localized intracellularly and diffused in the cytoplasm in proximal and distal tubules and collecting ducts, suggesting a function in active transport and reabsorption (204–206).

Both forms are heterodimers with subunits of 80–83 kd and 25–29 kd. On the basis of peptide mapping and distinct molecular masses it has been concluded that the two catalytic subunits are structurally different (205).

3.6. Muscle CANP

3.6.1 LOCALIZATION AND SUBSTRATE SPECIFICITY   In mammalian and chicken gizzard smooth muscle (207, 208), and also in human and chicken skeletal muscle (209–212), a single mM form of CANP has been detected. On the other hand, multiple forms of the proteinase with the properties of mM and μM CANP have been identified in pig (213–217), rabbit (218–222), and rat skeletal muscle (197), and in heart muscle of several species (197, 223–227). Conversion of the mM to the μM forms (207, 208, 222, 224, 228–231) promoted by Ca^{2+} has also been reported for the muscle proteinase. This conversion is accompanied by the generation of low-molecular-weight forms of the large and of the small enzyme subunits. These forms produced by auto-proteolysis appear to be different, on the basis of immunological properties and differences in peptide maps (216, 217, 220, 221, 227), from the native μM forms of CANP. It has recently been observed that in the presence of phospho-lipids the requirement of Ca^{2+} for the autoproteolytic conversion of the mM form of CANP, isolated from bovine vascular smooth muscle, is significantly reduced (208). This mechanism of activation has been discussed in detail in Section 3.1.2.

Although several reports have indicated the presence of CANP in Z-disk of myofibrils (231, 232), or in association with the plasma membrane (231–233), its localization remains controversial, particularly since recent reports have indicated that the proteinase is present throughout the cytoplasm with no preferential location on either plasma membrane or Z-disk (234). Data obtained by Ishiura (235) estimate that less than 4% of total cellular CANP is bound to Z-disk.

According to various reports the molecular mass of dimeric CANP is approx-imately 100–110 kd, made up of polypeptide chains of 76–82 kd for the larger subunit and 28–30 kd for the smaller one (207–209, 213–215, 218, 220, 221, 225–227, 230, 236). The single catalytic subunit of monomeric CANP has molecular mass of approximately 80 kd (206). It must be emphasized that the isolation of monomeric forms of CANP in muscle does not exclude the possibil-ity that this finding may result from the loss of the smaller subunit during the purification procedures (200). It must be recalled however that monomeric forms of CANP have been found in other cell types (42, 196).

The complete primary structure of the 80-kd subunit of chicken skeletal muscle deduced from the nucleotide sequence of the cloned complementary DNA (237) has suggested an architectural organization of the protease in four distinct domains. Those identified as II and IV were considered responsible for protease activity and for liganding of Ca^{2+} because of sequences similar to those of thiol proteinases and of calmodulinlike calcium-binding proteins, respectively. Domains I and II were suggested to be involved in Ca^{2+} depen-dence and in the binding of the inhibitor or of the activator, respectively.

Also of interest are preliminary DNA sequence analyses of the kex2 yeast gene (79, 238); this encodes a Ca^{2+}-dependent endopeptidase which reveals differences in sequence homologies as compared to the muscle CANP gene product (237). Yeast proteinase, in addition to hydrophobic domains responsible for anchoring to the bilayer, contains one region with cysteine residues presumably involved in catalysis and a negatively charged stretch for liganding of Ca^{2+}. This segment is not strictly homologous to the characteristic sequences present in calmodulin-binding proteins. These differences in sequences of Ca^{2+}-binding domains may be pertinent to the fact that skeletal muscle CANP corresponds to an mM type in contrast to the yeast Kex2 endopeptidase that is a μM form.

Although the different forms of CANP have distinct catalytic subunits, they all appear to have similar substrate specificity. Soluble enzymes, such as phosphorylase kinase and phosphorylase phosphatase (239–243), have been identified as specific cytosolic substrates, while proteins belonging to the contractile system or to the cytoskeleton represent the principal structural target molecules (224).

3.6.2 FUNCTION OF MUSCLE CANP   A commonly proposed function of CANP in muscle is for the disassembly of the three-dimensional architecture of myofibrils initiated by removal of Z-disks followed by release of individual myofibrils proteins which are further degraded by CANP. This sequence of events is considered to occur during metabolic turnover of myofibrillar structures (182, 200, 214, 225, 244–247).

Association of CANP to the cell membrane (231–233) may be related to its control of the assembly of the microfilament system which occurs through depolymerization-polymerization cycles.

In rat uteri (248) CANP has been postulated to be involved in tissue degradation and resorption resulting in loss of intracellular and extracellular proteins during the involution period.

An enhanced activity of CANP has been postulated to be responsible for those structural changes occurring under pathological conditions and is supported by the following observations: (a) in atrophying muscle tissue, the levels of the proteinase are largely increased (249); (b) in experimentally induced myocardial infarction a significant elevation in Ca^{2+} concentration occurs; and (c) specific CANP inhibitors reduce the size of acute myocardial infarctions (250, 251). Further support of this correlation is provided by experimental data showing (252) that increase in the Ca^{2+} influx, occurring in experimentally induced myopathy, is accompanied by dissolution of the Z-lines.

3.7. Tumor Cell CANP

Ehrlich-Ascites tumor cells (EAT) contain an mM CANP mainly localized in the soluble fraction (253) and with minor quantities associated with the detergent-resistant cytoskeleton (254). The cytosolic proteinase is composed of two subunits of molecular masses of 72 kd and 29 kd, respectively (253–255). Based on the high affinity of CANP for arginine methyl ester-Sepharose resin, it was suggested that the proteinase has a peptide bond specificity similar to that established for other CANPs (121, 189). It selectively degrades vimentin and desmin, and has therefore been implicated in the reorganization of the filament network during various stages of the cell cycle. Since the intermediate filament proteins lose their nucleic acid binding capacity after digestion with CANP (256, 257), the action of this proteinase has been tentatively related to gene expression.

3.8. Eye Tissue CANP

3.8.1 LENS   Bovine lens contains an mM CANP (258) that is localized in the soluble fraction of the cortex, and is totally absent in the epithelium (259). The enzyme is a dimer made up of 80-kd and 29-kd subunits. It degrades the intermediate filament proteins vimentin and the A and B chains of alpha crystallin and has been implicated in the disappearance of intermediate filaments during terminal differentiation of lens epithelial cells. The degradation of alpha-crystallin suggests that CANP may also play a role in cataractogenesis, a condition in which both free and bound Ca^{2+} levels in lens tissue are increased to 0.4–4 mM (260).

3.8.2 RETINA   The two forms of CANP identified in porcine retina (261) have molecular structures resembling those of the lens enzyme. The large amount of these CANP enzymes in retina and their ability to preferentially degrade the microtubule-associated proteins 1 and 2 suggest CANP's participation in microtubule polymerization-depolymerization cycles.

3.9. Lung CANP

Lung homogenate contains an mM CANP (262). The enzyme appears to be localized in the cytosol, and shows a molecular weight of 110,000 in gel-chromatography. Identical enzyme forms are present in lung from neonatal and adult mice.

Lung CANP digests the regulatory (R) subunit of cAMP-dependent protein kinase forming a smaller fragment (37 kd) that no longer inhibits the catalytic subunit. The varying amounts of modified R subunits present in the lungs of neonatal and adult mice suggest that CANP may operate in vivo, possibly to prolong the activation of the cAMP-dependent kinase (262).

3.10. Liver CANP

Rat liver contains both mM and μM CANPs (263, 264), which are indistinguishable on the basis of molecular weight, and contains 80-kd and 28-kd subunits (264, 265). The removal of the smaller subunit from the mM type, by treatment with high concentrations of Ca^{2+}, does not change the activity of the isolated catalytic subunit.

In rabbit liver, two forms of mM CANP are present (266–268), one having a molecular mass of 150 kd, the other of 200 kd. Each enzyme is composed of two subunits having molecular masses of 80 kd and 95 kd, respectively. The two mM types of CANP are converted to μM forms when exposed to either high levels of Ca^{2+} or to physiological concentrations of Ca^{2+} in the presence of a digestible substrate (267, 268). These results indicate that in liver the activation of CANP involves mechanisms similar to those operating in human erythrocytes (see Section 3.1.2) despite the differences in subunit structure. A specific function for CANP in liver has not yet been identified.

3.11. General Comments on the Properties and Physiological Functions of Ca^{2+}-Dependent Neutral Proteinases

3.11.1 PHYSIOLOGICAL FUNCTIONS OF CANPS   From the preceding discussion it seems clear that a most interesting question, anticipated in the introduction, concerns possible common functions of the Ca^{2+}-dependent neutral proteinases in different cell types. To answer this question it is important to review the properties of these proteinases. In fact, in spite of its cytosolic localization, CANP degrades specific substrates located in three distinct cellular compartments. These have been identified in the previous discussion and correspond to: the cytoplasm, the cytoskeleton network, and the plasma membrane. Finally, in almost all cells CANP is considered to be present in an inactive form due to the low intracellular concentrations of Ca^{2+}. Thus on the basis of these properties it is difficult to visualize how the proteinase can fulfill its physiological roles.

A first answer is provided by the consideration that the best known signal that promotes intracellular activation of the proteinase is the mobilization of Ca^{2+} from intracellular stores. This signal is probably not the only one but is certainly that more extensively characterized and occurs in many cells as an early event in response to stimulation by extracellular ligands (146–148, 172–181).

This provides a first common correlation between cell stimulation and activation of the proteinase.

A second point to be considered is that under physiological conditions the signal provided by Ca^{2+} can promote activity of the soluble proteinases whose requirements are satisfied by low concentrations of the metal ion. Alternatively Ca^{2+} can promote activation of the mM forms of CANP by inducing a translocation of the proteinases from the cytosol to the plasma membrane, followed by conversion of CANP to a μM form which is subsequently released as a soluble, physiologically active proteinase (120, 125). In other tissues Ca^{2+} can promote translocation to the membrane of the mM CANP, which in the bound form is active at μM Ca^{2+} but returns to the original high Ca^{2+} requirements when it is released from the membrane (142, 143).

Thus Ca^{2+} either activates CANP or promotes its translocation in the active form to different cell compartments where it acts on endogenous substrates.

This differential localization and activation must depend on specific cell requirements and be related to the presence, in a cell, of either different molecular species of CANP or of a single mM form of the proteinase.

Based on these observations the following functions may be considered for CANP: 1. In a stimulated cell activated CANP may degrade cytoskeletal or submembraneous cytoskeletal proteins thus participating in the spatial organization of the cytoplasmic network and thereby in the determination of cell morphology, cell locomotion, and intracellular movement of organelles. 2. CANP may be involved in the amplification of the cellular response to activation by promoting a limited proteolytic degradation of membrane-bound growth factors (269) or receptors (164). 3. Modification of protein kinase C by CANP may result in the production of soluble Ca^{2+} phospholipid-independent forms of the kinase which may translocate to the cytosol with enhanced phosphorylating activity (143, 167, 168). 4. Finally, in response to intracellular or extracellular stimuli CANP may be activated and promote the selective limited proteolysis of soluble proteins which may be required to initiate their subsequent complete degradation (138). Limited proteolysis of cytosolic hormone receptors may also be part of the mechanism for receptor degradation or intracellular translocation (270, 271).

We may therefore conclude that by acting on proteins located in three different cellular compartments, CANP participates as a key enzyme in the transduction and amplification of external stimuli. Utilizing Ca^{2+} as a common signal for activation or for translocation and activation, using phospholipids (126, 141) and/or an appropriate substrate (123) to increase its sensitivity to Ca^{2+}, and acting on a variety of substrates in different cell compartments, CANP can accomplish common functions in different cell types.

3.11.2 CALCIUM SENSITIVITY OF DIFFERENT MOLECULAR SPECIES OF CANP   The physiological significance of the presence of two structurally unrelated forms of CANP, one requiring high, the other low concentrations of Ca^{2+}, has puzzled many investigators in recent years. We believe that at present a more precise understanding of this problem is available.

It now appears that many of the high-Ca^{2+}-requiring forms, defined as mM CANP, are convertible to μM forms (120, 123, 193, 207, 208, 222, 224, 228–231); suggesting that in many cases, regardless of the type present in the cell, all Ca^{2+}-dependent proteinases express their catalytic activity at μM concentrations of Ca^{2+}. Without autoproteolytic conversion of the mM to the μM active form, physiological activity could not be expressed. In addition such mechanism would be essential to promote CANP activity in cells in which only a single mM type is detected.

Two main objections to the physiological significance of this proenzyme-active enzyme conversion were based on the fact that this process also required concentrations of Ca^{2+} as high as those necessary for activity and on the observation that, in some cases, the converted forms still require 0.1–0.2 mM Ca^{2+} (228–230).

Recent observations, however, have shown that the autoproteolytic conversion of the mM to the μM form can occur at physiological concentrations of Ca^{2+} provided the presence of a digestible substrate (123), or following binding of the still-inactive proteinase form to the phospholipid bilayer of the plasma membrane (120) in a translocation process mediated by μM Ca^{2+}. This second mechanism is probably the only one involved in the activation of nonconvertible mM forms of CANP (142), and an alternative one for complete activation of those converted forms with still relatively high Ca^{2+} requirements.

It thus becomes apparent that the physiological roles of the two distinct molecular species of CANP are not explained simply on the basis of different Ca^{2+} requirements, but on the specific cell functions and the particular role of CANP in promoting these specific functions.

The latter mechanism is illustrated in neutrophils (142) and in platelets (160) where the activation of CANP through cytosol-membrane translocation may promote selective degradation of membrane-bound protein kinase C (143, 167, 168) or of surface receptors (157, 164), thereby amplifying the signal of an external ligand.

Since the mM CANP cannot be converted into the μM form, the activity of CANP in these cells can presumably be expressed only when the proteinase is in the membrane-bound form.

Similar considerations may apply to cells in which the response to external stimuli requires degradation of submembraneous cytoskeleton proteins.

In conclusion we can postulate that the presence of one or two types of CANP carrying eventually different substrate specificities must be related to a specific cell function, and to the localization of target substrates, since the activation of the μM and of the mM forms occurs presumably by different mechanisms and in different cellular compartments.

3.11.3 MECHANISMS OF REGULATION OF CANPS   Several mechanisms have been proposed for the modulation of CANP activity. A first one is obviously related to the concentration of Ca^{2+} and thereby to the existence of a tight coupling between Ca^{2+} mobilization and modulation of the intracellular activity of CANP.

An additional regulatory property has been ascribed to a specific natural inhibitor(s) (112, 272–277) that coexists with CANP in the cytosol of nearly all cells examined, but whose exact role in the modulation of the proteinase activity is still unclear. The formation of the proteinase inhibitor complex occurs only in the presence of high concentrations of Ca^{2+}; thus the effect of the inhibitor on the activity of the µM form of CANP cannot be determined (116).

We have postulated that the function of the inhibitor may be to prevent uncontrolled and fast activation of CANP in conditions in which a rapid elevation in the intracellular concentration of Ca^{2+} may occur (129–132). We have also shown that the inhibitor prevents, through the formation of the enzyme-inhibitor complex (112), association of CANP to the membranes (125) and thereby its sequential activation.

This last effect of the inhibitor may represent its most relevant physiological function in the modulation of CANP activity. A highly sensitive control mechanism can be accomplished, as previously indicated, by the substrate on the basis of its susceptibility to digestion and of its concentration (123). In addition this mechanism of regulation is characterized by a self-terminating signal which interrupts activation once the substrate has undergone proteolytic modification (123).

Low-molecular-weight proteins (17–20 kd) present in calf (196) and bovine (278) brain and capable of stimulating 2–3-fold the activity of the proteinase without change in the affinity for Ca^{2+} have also been proposed as natural activators of CANP.

A possible regulatory mechanism may be suggested by observations from several laboratories (128, 159, 196, 199, 214, 224, 226, 227, 279, 280) indicating that Mn^{2+}, Mg^{2+}, or other metal ions reduce the Ca^{2+} requirement of mM CANP which thereby becomes active at µM concentrations of Ca^{2+} (128). Due to its reversibility this activation mechanism may be of great physiological importance.

Finally, it has recently been shown that a kinase activity, which copurifies with CANP from rat skeletal muscle and brain, promotes phosphorylation of the proteinase (32, 281) and of specific substrate proteins of the neurofilament (200). The effects of this phosphorylation on the functional properties of CANP are still unknown.

ACKNOWLEDGMENTS

Our own studies have been supported by the Italian Consiglio Nazionale delle Ricerche, Progetto Finalizzato "Ingegneria Genetica e Basi Molecolari delle Malattie Ereditarie," and Progetto Finalizzato "Oncologia," Sottoprogetto "Biologia Cellulare."

Literature Cited

1. Barrett, A. J., ed. 1977. *Proteinases in Mammalian Cells and Tissues.* Amsterdam: North Holland
2. Holzer, H., Heinrich, P. C. 1980. *Ann. Rev. Biochem.* 49:69–91
3. Amenta, J. S., Brocher, S. C. 1981. *Life Sci.* 28:1195–208
4. Grisolia, S., Wheatley, D. N. 1984. In *Life Chemistry Reports*, Vol. 2, pp. 257–97. London: Harwood Academic
5. Cohen, G. N., Holzer, H., eds. 1979. *Limited Proteolysis in Microorganisms.* DHEW Publ. No(NIH). Washington, DC: USGPO
6. Katunuma, N., Umezawa, H., eds. 1983. *Proteinase Inhibitors, Medical and Biological Aspects.* Tokyo: Japan Sci. Soc. Press/Berlin: Springer-Verlag
7. Khairallah, E. A., Bond, J. S., Bird, J. W. C., eds. 1985. *Intracellular Protein Catabolism.* New York: Liss
8. Knowles, S. P., Ballard, F. J. 1976. *Biochem. J.* 156:609–17
9. Neff, N. T., DeMartino, G. N., Goldberg, A. L. 1979. *J. Cell. Phys.* 101:439–58
10. Wildenthal, K., Wakeland, J. R., Ord, J. M., Stull, J. T. 1980. *Biochem. Biophys. Res. Commun.* 96:793–98
11. Rote, K. V., Rechsteiner, M. 1983. *J. Cell. Phys.* 116:103–10
12. Hershko, A., Ciechanover, A. 1982. *Ann. Rev. Biochem.* 51:335–64
13. Müller, M., Dubiel, W., Rothmann, J., Rapoport, S. 1980. *Eur. J. Biochem.* 109:405–10
14. Gething, M. J. ed. 1985. *Protein Transport and Secretion.* Cold Spring Harbor, NY: Cold Spring Harbor Lab.
15. Ciechanover, A., Finley, D., Varshavsky, A. 1984. *Cell* 37:57–65
16. Hershko, A. 1985. See Ref. 7, pp. 11–16
17. Chung, C. H., Goldberg, A. L. 1982. *Proc. Natl. Acad. Sci. USA* 78:4931–35
18. Chung, C. H., Waxman, L., Goldberg, A. L. 1983. *J. Biol. Chem.* 258:215–21
19. Goldberg, A. L., Voellmy, R., Chung, C. H., Menon, A. S., Desautels, M., et al. 1985. See Ref. 7, pp. 33–45
20. Rapoport, S. M., Shewe, T., Wiesner, R., Halangk, W., Ludwig, P., et al. 1979. *Eur. J. Biochem.* 96:545–61
21. Rivett, A. J., Roseman, J. E., Oliver, C. N., Levine, R. L., Stadtman, E. R. 1985. See Ref. 7, pp. 317–28
22. Herrath, M., Holzer, H. 1985. See Ref. 7, pp. 329–40
23. Bergström, G., Ekman, P., Humble, E., Engström, L. 1978. *Biochim. Biophys. Acta* 532:259–67

24. Hall, E. R., McCully, V., Cottam, G. L. 1979. *Arch. Biochem. Biophys.* 195:315–24
25. Holzer, H. 1984. In *Enzyme Regulation by Reversible Phosphorylation. Further Advances,* ed. P. Cohen, pp. 143–52. Amsterdam: Elsevier
26. Noda, T., Hoffschulte, H., Holzer, H. 1984. *J. Biol. Chem.* 259:7191–96
27. Brown, P. A., Khairallah, E. A. 1982. *Fed. Proc.* 41:868
28. Tischler, M. E., Fagan, J. M. 1982. *Arch. Biochem. Biophys.* 217:191–96
29. Khairallah, E. A., Bartolone, J., Brown, P., Bruno, M. K., Makowski, G., Wood, S. 1985. See Ref. 7, pp. 373–83
30. Tischler, M. E., Fagan, J. M., Allen, D. 1985. See Ref. 7, pp. 363–72
31. McKay, M. J., Bond, J. S. 1985. See Ref. 7, pp. 351–61
32. Zimmerman, U.-J. P., Schlaepfer, W. W. 1984. *Biochem. Biophys. Res. Commun.* 120:767–74
33. Liscum, L., Cummings, R. D., Anderson, R. G. W., De Martino, G. N., Goldstein, J. L., Brown, M. S. 1983. *Proc. Natl. Acad. Sci. USA* 80:7165–69
34. Parker, R. A., Miller, S. J., Gibson, D. M. 1984. *Biochem. Biophys. Res. Commun.* 125:629–35
35. Docherty, K., Steiner, D. F. 1982. *Ann. Rev. Physiol.* 44:625–38
36. Blobel, G., Dobberstein, B. 1975. *J. Cell. Biol.* 67:852–62
37. Sabatini, D. D., Kreibich, G., Morimoto, T., Adesnik, M. 1982. *J. Cell. Biol.* 92:1–22
38. Schatz, G., Butow, R. A. 1983. *Cell* 32:316–18
39. Hay, R., Böhni, P., Gasser, S. 1984. *Biochim. Biophys. Acta* 779:65–87
40. Jackson, R. C., Blobel, G. 1977. *Proc. Natl. Acad. Sci. USA* 74:5598–602
41. Shields, D., Blobel, G. 1978. *J. Biol. Chem.* 253:3753–56
42. Kaschnitz, R., Kreil, G. 1978. *Biochem. Biophys. Res. Commun.* 83:901–7
43. Strauss, A. W., Zimmerman, M., Boime, I., Ashe, B., Mumford, R. A., Alberts, A. W. 1979. *Proc. Natl. Acad. Sci. USA* 76:4225–29
44. Fujimoto, Y., Watanabe, Y., Uchida, M., Ozaki, M. 1984. *J. Biochem.* 96:1125–31
45. Stern, J. B., Jackson, R. C. 1985. *Arch. Biochem. Biophys.* 237:244–52
46. Haebner, J. F., Rosenblatt, M., Dee, P. C., Potts, J. T. Jr. 1979. *J. Biol. Chem.* 254:10596–99

47. Davis, B. D., Tai, P.-C. 1980. *Nature* 283:433–38
48. Mori, M., Miura, S., Tatibana, M., Cohen, P. P. 1980. *Proc. Natl. Acad. Sci. USA* 77:7044–48
49. Miura, S., Mori, M., Amaya, Y., Tatibana, M. 1982. *Eur. J. Biochem.* 122: 641–47
50. Morita, T., Miura, S., Mori, M., Tatibana, M. 1982. *Eur. J. Biochem.* 122:501–9
51. Conboy, J. G., Fenton, W. A., Rosenberg, L. E. 1982. *Biochem. Biophys. Res. Commun.* 105:1–7
52. McAda, P. C., Douglas, M. G. 1982. *J. Biol. Chem.* 257:3177–82
53. Fenton, W. A., Hack, A. M., Helfgott, D., Rosenberg, L. E. 1984. *J. Biol. Chem.* 259:6616–21
54. Morohashi, K., Fujii-Kuriyama, Y., Okada, Y., Sogawa, K., Hirose, T., et al. 1984. *Proc. Natl. Acad. Sci. USA* 81:4647–51
55. Horwich, A. L., Fenton, W. A., Williams, K. R., Kalousek, F., Kraus, J. P., et al. 1984. *Science* 224:1068–74
56. Horwich, A. L., Fenton, W. A., Firgaira, F. A., Fox, J. E., Kolansky, D., et al. 1985. *J. Cell. Biol.* 100:1515–21
57. Chang, T. L., Loh, Y. P. 1983. *Endocrinology* 112:1832–38
58. Loh, Y. P., Chang, T. L. 1982. *FEBS Lett.* 137:57–62
59. Chang, T. L., Loh, Y. P. 1984. *Endocrinology* 114:2092–99
60. Lindberg, I., Yang, H. Y. T., Costa, E. 1982. *Biochem. Biophys. Res. Commun.* 106:186–93
61. Lindberg, I., Yang, H. Y. T., Costa, E. 1984. *J. Neurochem.* 42:1411–19
62. Evangelista, R., Ray, P., Lewis, R. V. 1982. *Biochem. Biophys. Res. Commun.* 106:895–902
63. Mizuno, K., Miyata, A., Kangawa, K., Matsuo, H. 1982. *Biochem. Biophys. Res. Commun.* 108:1235–42
64. Chang, T. L., Gainer, H., Russell, J. T., Loh, Y. P. 1982. *Endocrinology* 111: 1607–14
65. Mizuno, K., Kojima, M., Matsuo, H. 1985. *Biochem. Biophys. Res. Commun.* 128:884–91
66. Gluschankof, P., Morel, A., Gomez, S., Nicolas, P., Fahy, C., Cohen, P. 1984. *Proc. Natl. Acad. Sci. USA* 81:6662–66
67. Berger, E. D., Shooter, E. M. 1977. *Proc. Natl. Acad. Sci. USA* 74:3647–51
68. Hudson, P., Haley, J., Cronk, M., Shine, J., Niall, H. 1981. *Nature* 291: 127–31
69. Fletcher, D. J., Quigley, J. P., Bauer, E., Noe, B. D. 1981. *J. Cell Biol.* 90:312–22
70. Noe, B. D., Debo, G., Spiess, J. 1984. *J. Cell Biol.* 99:578–87
71. Noe, B. D., Moran, M. N. 1984. *J. Cell Biol.* 99:418–24
72. Comb, M., Seeburg, P. H., Adelman, J., Eiden, L., Herbert, E. 1982. *Nature* 295:663–66
73. Gubler, U., Seeburg, P., Hoffman, B. J., Gage, L. P., Udenfriend, S. 1982. *Nature* 295:206–8
74. Noda, M., Furutani, Y., Takahashi, H., Toyosato, M., Hirose, T., et al. 1982. *Nature* 295:202–6
75. Nakanishi, S., Inoue, A., Kita, T., Nakamura, M., Chang, A. C. Y., et al. 1979. *Nature* 278:423–27
76. Land, H., Schütz, G., Schmale, H., Richter, D. 1982. *Nature* 295:299–303
77. Jacobs, J. W., Goodman, R. H., Chiu, W. W., Dee, P. C., Habener, J. F., et al. 1981. *Science* 213:457–59
78. Craig, R. K., Hall, L., Edbrooke, M. R., Allison, J., McIntyre, I. 1982. *Nature* 295:345–47
79. Fuller, R. S., Brake, A. J., Blair, L., Julius, D. I., Thorner, J. 1985. See Ref. 14, pp. 97–102
80. Fricker, L. D., Supattapone, S., Snyder, S. H. 1982. *Life Sci.* 31:1841–44
81. Supattapone, S., Fricker, L. D., Snyder, S. H. 1984. *J. Neurochem.* 42:1017–23
82. Fricker, L. D., Snyder, S. H. 1983. *J. Biol. Chem.* 258:10950–55
83. Supattapone, S., Fricker, L. D., Snyder, S. H. 1983. *J. Neurochem.* 42:1017–23
84. Hook, V. Y. H., Eiden, L. E., Brownstein, M. 1982. *Nature* 295:341–42
85. Docherty, K., Carroll, R. J., Steiner, D. F. 1982. *Proc. Natl. Acad. Sci. USA* 79:4613–17
86. Fricker, L. D., Snyder, S. H. 1982. *Proc. Natl. Acad. Sci. USA* 79:3886–90
87. Hook, V. Y. H. 1984. *Neuropeptides* 4:117–26
88. Hook, V. Y. H., Eiden, L. E. 1984. *FEBS Lett.* 172:212–18
89. Hook, V. Y. H., Loh, P. Y. 1984. *Proc. Natl. Acad. Sci. USA* 81:2776–80
90. Lacombe, M.-L., Hanoune, J. 1979. *J. Biol. Chem.* 254:3697–99
91. Moroney, J. V., McCarty, R. E. 1982. *J. Biol. Chem.* 257:5910–14
92. Hirasawa, K., Irvine, R. F., Dawson, R. M. C. 1982. *Biochem. J.* 206:675–78
93. Skovby, F., Kraus, J. P., Rosenberg, L. E. 1984. *J. Biol. Chem.* 259:588–93
94. Siddiqui, F. A., Yang, D. C. H. 1985. *Biochim. Biophys. Acta* 828:177–87
95. Fujii, Y., Kobashi, K., Nakai, N. 1984. *Arch. Biochem. Biophys.* 233:310–13
96. Gohda, E., Pitot, H. C. 1980. *J. Biol. Chem.* 255:7371–79

97. Gohda, E., Pitot, H. C. 1981. *J. Biol. Chem.* 256:2567–72
98. Pontremoli, S., Melloni, E., Salamino, F., Sparatore, B., Michetti, M., Horecker, B. L. 1982. *Arch. Biochem. Biophys.* 214:376–85
99. Hargrove, J. L., Granner, D. K. 1981. *J. Biol. Chem.* 256:8012–17
100. Pontremoli, S., Melloni, E., Michetti, M., Salamino, F., Sparatore, B., Horecker, B. L. 1982. *Proc. Natl. Acad. Sci. USA* 79:2451–54
101. Pontremoli, S., Melloni, E., Michetti, M., Salamino, F., Sparatore, B., Horecker, B. L. 1982. *Proc. Natl. Acad. Sci. USA* 79:5194–96
102. Horecker, B. L., Erickson-Viitanen, S., Melloni, E., Pontremoli, S. 1985. *Curr. Top. Cell Regul.* 25:77–89
103. Pontremoli, S., Melloni, E., Michetti, M., Salamino, F., Sparatore, B., Horecker, B. L. 1982. *Biochem. Biophys. Res. Commun.* 106:903–9
104. Pontremoli, S., Melloni, E., Damiani, G., Michetti, M., Salamino, F., et al. 1984. *Arch. Biochem. Biophys.* 233:267–71
105. Wolf, D. H. 1982. *Trends Biochem. Sci.* 7:35–37
106. Achstetter, T., Ehmann, C., Osaki, A., Wolf, D. H. 1984. *J. Biol. Chem.* 259:13344–48
107. Achstetter, T., Emter, O., Ehmann, C., Wolf, D. H. 1984. *J. Biol. Chem.* 259:13334–43
108. MacGregor, J. S., Hannappel, E., Xu, G.-J., Pontremoli, S., Horecker, B. L. 1982. *Arch. Biochem. Biophys.* 217:652–64
109. Dev, I. K., Harvey, R. J., Ray, P. H. 1985. *J. Biol. Chem.* 260:5891–94
110. Pontremoli, S., Melloni, E., Salamino, F., Sparatore, B., Michetti, M., et al. 1980. *Eur. J. Biochem.* 110:421–30
111. Murakami, T., Hatanaka, M., Murachi, T. 1981. *J. Biochem.* 90:1809–16
112. Melloni, E., Sparatore, B., Salamino, F., Michetti, M., Pontremoli, S. 1982. *Biochem. Biophys. Res. Commun.* 106:731–40
113. Kikuchi, T., Yumoto, N., Sasaki, T., Murachi, T. 1984. *Arch. Biochem. Biophys.* 234:639–45
114. Pant, H. C., Virmani, M., Gallant, P. E. 1983. *Biochem. Biophys. Res. Commun.* 117:372–77
115. Melloni, E., Sparatore, B., Salamino, F., Michetti, M., Pontremoli, S. 1982. *Biochem. Biophys. Res. Commun.* 107:1053–59
116. Melloni, E., Salamino, F., Sparatore, B., Michetti, M., Pontremoli, S. 1984. *Biochem. Int.* 8:477–89
117. Allen, D. W., Cadman, S. 1979. *Biochim. Biophys. Acta* 551:1–9
118. Lorand, L., Bjerrum, O. J., Hawkins, M., Lowe-Krentz, L., Siefring, G. E. 1983. *J. Biol. Chem.* 258:5300–5
119. Hatanaka, M., Yoshimura, N., Murakami, T., Kannagi, R., Murachi, T. 1984. *Biochemistry* 23:3272–76
120. Pontremoli, S., Melloni, E., Sparatore, B., Salamino, F., Michetti, M., Sacco, O., Horecker, B. L. 1985. *Biochem. Biophys. Res. Commun.* 128:331–38
121. Melloni, E., Salamino, F., Sparatore, B., Michetti, M., Pontremoli, S. 1984. *Biochim. Biophys. Acta* 788:11–16
122. Pontremoli, S., Melloni, E., Sparatore, B., Michetti, M., Horecker, B. L. 1984. *Proc. Natl. Acad. Sci. USA* 81:6714–17
123. Pontremoli, S., Sparatore, B., Melloni, E., Michetti, M., Horecker, B. L. 1984. *Biochim. Biophys. Res. Commun.* 123:331–37
124. Pontremoli, S., Melloni, E. 1986. *Calcium and Cell Function.* Vol. VI: 159–83
125. Pontremoli, S., Salamino, F., Sparatore, B., Michetti, M., Sacco, O., Melloni, E. 1985. *Biochim. Biophys. Acta.* 831:335–39
126. Pontremoli, S., Melloni, E., Sparatore, B., Salamino, F., Michetti, M., et al. 1985. *Biochem. Biophys. Res. Commun.* 129:389–95
127. Kaibuchi, K., Takai, Y., Nishizuka, Y. 1981. *J. Biol. Chem.* 256:7146–49
128. Pontremoli, S., Sparatore, B., Salamino, F., Michetti, M., Melloni, E. 1985. *Arch. Biochem. Biophys.* 239:517–22
129. La Celle, P. L., Kirkpatrick, F. H., Udkow, M. P., Arkin, B. 1983. In *Red Cell Shape*, ed. M. Bessis, R. I. Weed, P. F. Leblond, pp. 69–80. New York: Springer-Verlag
130. Eaton, J. W., Shelton, T. D., Swofford, H. L., Kolpin, C. E., Jacob, H. S. 1973. *Nature* 246:105–6
131. Palek, J. 1973. *Blood* 42:988
132. La Celle, P. K. 1971. In *Red Cell Membrane*, ed. R. I. Weed, E. R. Jaffe, P. A. Miesher, pp. 10–25. New York: Grune & Stratton
133. Triplett, R. B., Wingate, J. M., Carraway, K. L. 1972. *Biochem. Biophys. Res. Commun.* 49:1014–20
134. King, L. E., Morrison, M. 1977. *Biochem. Biophys. Acta* 471:162–68
135. Lang, R. D. A., Wickenden, C., Wynne, J., Lucy, J. A. 1984. *Biochem. J.* 218:295–305
136. Lutz, H. V., Flepp, R., Stringaro-Wipf, G. 1984. *J. Immunol.* 133:2610–18
137. Kay, M. M. B., Goodman, S. R., Sorensin, K., Whitfield, C. F., Wong, P., et al. 1983. *Proc. Natl. Acad. Sci. USA* 80:1631–35

138. Melloni, E., Salamino, F., Sparatore, B., Michetti, M., Pontremoli, S. 1982. *Arch. Biochem. Biophys.* 216:495–502

139. Pontremoli, S., Salamino, F., Sparatore, B., Melloni, E., Morelli, A., et al. 1979. *Biochem. J.* 181:559–68

140. Melloni, E., Sparatore, B., Salamino, F., Michetti, M., Pontremoli, S. 1982. *Arch. Biochem. Biophys.* 218:579–84

141. Legendre, J. L., Jones, H. P. 1983. *J. Reticuloendothel. Soc.* 34:89–97

142. Pontremoli, S., Sparatore, B., Salamino, F., Michetti, M., Sacco, O., Melloni, E. 1985. *Biochem. Int.* 11:35–44

143. Melloni, E., Pontremoli, S., Michetti, M., Sacco, O., Sparatore, B., et al. 1985. *Proc. Natl. Acad. Sci. USA* 82:6435–39

144. Kishimoto, A., Kajikawa, N., Shiota, M., Nishizuka, Y. 1983. *J. Biol. Chem.* 258:1156–64

145. Melloni, E., Pontremoli, S., Michetti, M., Sacco, O., Sparatore, B., Horecker, B. L. 1986. *J. Biol. Chem.* In press

146. Mottola, C., Romeo, D. 1982. *J. Cell. Biol.* 93:129–34

147. Pozzan, T., Lew, P. D., Wollheim, C. B., Tsien, R. Y. 1983. *Science* 221: 1413–15

148. Lagast, H., Pozzan, T., Waldvogel, F. A., Lew, P. D. 1984. *J. Clin. Invest.* 73:878–83

149. Snyderman, R., Goetzl, E. J. 1981. *Science* 213:830–37

150. Andrews, P. C., Babior, B. M. 1983. *Blood* 61:333–40

151. White, J. R., Huang, C.-K., Hill, J. M., Naccache, P. H., Becker, E. L., Sha'afi, R. I. 1984. *J. Biol. Chem.* 259:8605–11

152. Andrews, P. C., Babior, B. M. 1984. *Blood* 64:883–90

153. Nishizuka, Y. 1984. *Nature* 308:693–98

154. Nishizuka, Y., Takai, Y., Kishimoto, A., Kikkawa, U., Kaibuchi, K. 1984. In *Recent Progress in Hormone Research*, ed. R. O. Greep, 40:301–41. New York: Academic

155. Phillips, D. R., Jakabova, M. 1977. *J. Biol. Chem.* 252:5602–5

156. Truglia, J. A., Stracher, A. 1981. *Biochem. Biophys. Res. Commun.* 100:814–22

157. Yoshida, N., Weksler, B., Nachman, R. 1983. *J. Biol. Chem.* 258:7168–74

158. Sakon, M., Kambayashi, J., Ohno, H., Kosaki, G. 1981. *Thromb. Res.* 24:207–14

159. Tsujinaka, T., Shiba, E., Kambayashi, J., Kosaki, G. 1983. *Biochem. Int.* 6:71–80

160. Lucas, R. C., Lawrence, J. J., Stracher, A. 1979. *J. Cell. Biol.* 83:77a

161. Nachman, R. L., Jaffe, E. A., Weksler, B. B. 1977. *J. Clin. Invest.* 52:2745–56

162. Caen, J. P., Nurden, A. T., Jeanneau, C., Michel, H., Tobelem, G., et al. 1976. *J. Lab. Clin. Med.* 87:586–96

163. Phillips, D. R. 1980. *Prog. Hemostasis Thromb.* 5:81–109

164. McGowan, E. B., Yeo, K.-T., Detwiler, T. C. 1983. *Arch. Biochem. Biophys.* 227:287–301

165. Nurden, A. T., Caen, J. P. 1976. *Thromb. Haemostasis* 35:139–50

166. Berndt, M. C., Phillips, D. R. 1981. *Thromb. Haemostasis* 46:75

167. Tapley, P. M., Murray, A. W. 1984. *Biochem. Biophys. Res. Commun.* 118: 835–41

168. Tapley, P. M., Murray, A. W. 1984. *Biochem. Biophys. Res. Commun.* 122: 158–64

169. Gache, Y., Landon, F., Touitou, H., Olomucki, A. 1984. *Biochem. Biophys. Res. Commun.* 124:877–81

170. Fox, J. E. B., Goll, D. E., Reynolds, C. C., Phillips, D. R. 1985. *J. Biol. Chem.* 260:1060–66

171. Kosaki, G., Tsujinaka, T., Kambayashi, J., Morimoto, K., Yamamoto, K., et al. 1983. *Biochem. Int.* 6:767–75

172. Feinman, R. D., Detwiler, T. C. 1974. *Nature* 249:172–73

173. Detwiler, T. C., Charo, I. F., Feinman, R. D. 1978. *Thromb. Haemostasis* 40:207–11

174. Le Breton, G. C., Dinerstein, R. J., Roth, L. J., Feinberg, H. 1976. *Biochem. Biophys. Res. Commun.* 71:362–70

175. Feinstein, M. B., Fraser, C. 1975. *J. Gen. Physiol.* 66:561–81

176. Feinstein, M. B. 1980. *Biochem. Biophys. Res. Commun.* 93:593–600

177. Massini, P., Naf, U. 1980. *Biochim. Biophys. Acta* 598:575–87

178. Feinstein, M. B., Rodan, G. A., Cutler, L. S. 1981. In *Platelets in Biology and Pathology*, ed. J. L. Gordon, 2:76–82. Amsterdam/New York: Elsevier Biomedical

179. Rink, T. J., Smith, S. W., Tsien, R. Y. 1982. *FEBS Lett.* 148:21–26

180. Käser-Glanzmann, R., Jakabova, M., George, J. N., Lüscher, E. F. 1977. *Biochim. Biophys. Acta* 466:429–40

181. White, G. C. 1980. *Biochim. Biophys. Acta* 631:130–38

182. Fox, J. E. B., Reynolds, C. C., Phillips, D. R. 1983. *J. Biol. Chem.* 258:9973–81

183. Dayton, W. R., Goll, D. E., Stromer, M. H., Reville, W. J., Zeece, M. G., Robson, R. M. 1975. *Cold Spring Harbor Conf. Cell Prolif.* 2:551–77

184. Hartwig, J. H., Tyler, J., Stossel, T. P. 1980. *J. Cell. Biol.* 87:841–48

185. Kunicki, T. J., Montgomery, R. R., Schullek, J. 1985. *Blood* 65:352–56
186. Guroff, G. 1964. *J. Biol. Chem.* 239: 149–55
187. Inoue, M., Kishimoto, A., Takai, Y., Nishizuka, Y. 1976. *J. Biol. Chem.* 251: 4476–78
188. Inoue, M., Kishimoto, A., Takai, Y., Nishizuka, Y. 1977. *J. Biol. Chem.* 252: 7610–16
189. Hirao, T., Takahashi, K. 1984. *J. Biochem.* 96:775–84
190. Kishimoto, A., Kajikawa, N., Shiota, M., Nishizuka, Y. 1983. *J. Biol. Chem.* 258:1156–64
191. Klein, I., Lehotay, D., Gondek, M. 1981. *Arch. Biochem. Biophys.* 208: 520–27
192. Zimmerman, U.-J. P., Schlaepfer, W. W. 1982. *Biochemistry* 21:3977–83
193. Siman, R., Baudry, M., Lynch, G. 1983. *J. Neurochem.* 41:950–55
194. Baudry, M., Bundman, M. C., Smith, E. K., Lynch, G. S. 1981. *Science* 212: 937–38
195. Siman, R., Baudry, M., Lynch, G. 1984. *Proc. Natl. Acad. Sci. USA* 81:3572–76
196. Malik, M. N., Fenko, M. D., Iqbal, K., Wisniewski, H. M. 1983. *J. Biol. Chem.* 258:8955–62
197. Zimmerman, U.-J. P., Schlaepfer, W. W. 1984. *J. Biol. Chem.* 259:3210–18
198. Ishizaki, Y., Tashiro, T., Kurokawa, M. 1983. *Eur. J. Biochem.* 131:41–45
199. Ishizaki, Y., Kurokawa, M., Takahashi, K. 1985. *Eur. J. Biochem.* 146:331–37
200. Ishiura, S. 1981. *Life Sci.* 29:1079–87
201. Baudry, M., Lynch, G. 1980. *Proc. Natl. Acad. Sci. USA* 77:2298–302
202. Siman, R., Baudry, M., Lynch, G. 1985. *Nature* 313:225–28
203. Lynch, G., Baudry, M. 1984. *Science* 224:1057–63
204. Yoshimura, N., Kikuchi, T., Sasaki, T., Kitahara, A., Hatanaka, M., Murachi, T. 1983. *J. Biol. Chem.* 258:8883–89
205. Kitahara, A., Sasaki, T., Kikuchi, T., Yumoto, N., Yoshimura, N., et al. 1984. *J. Biochem.* 95:1759–66
206. Yoshimura, N., Hatanaka, M., Kitahara, A., Kawaguchi, N., Murachi, T. 1984. *J. Biol. Chem.* 259:9847–52
207. Hataway, D. R., Werth, D. K., Haeberle, J. R. 1982. *J. Biol. Chem.* 257:9072–77
208. Coolican, S. A., Hathaway, D. R. 1984. *J. Biol. Chem.* 259:11627–30
209. Suzuki, K., Ishiura, S., Tsuji, S., Katamoto, T., Sugita, M., Imahori, K. 1979. *FEBS Lett.* 104:355–58
210. Ishiura, S., Murofushi, H., Suzuki, K., Imahori, K. 1978. *J. Biochem.* 84:225–30
211. Sugita, H., Ishiura, S., Suzuki, K., Imahori, K. 1980. *J. Biochem.* 87:339–41
212. Suzuki, K., Tsuji, S., Ishiura, S. 1981. *FEBS Lett.* 136:119–22
213. Dayton, W. R., Goll, D. E., Zeece, M. G., Robson, R. M. 1976. *Biochemistry* 15:2150–58
214. Dayton, W. R., Reville, W. J., Goll, D. E., Stromer, M. H. 1976. *Biochemistry* 15:2159–67
215. Dayton, W. R., Schollmeyer, J. V., Lepley, R. A., Cortés, L. R. 1981. *Biochim. Biophys. Acta* 659:48–61
216. Dayton, W. R. 1982. *Biochim. Biophys. Acta* 709:166–72
217. Wheelock, M. J. 1982. *J. Biol. Chem.* 257:12471–74
218. Tsuji, S., Imahori, K. 1981. *J. Biochem.* 90:233–40
219. Tsuji, S., Ishiura, S., Takanashi-Nakamura, M., Katamoto, T., Suzuki, K., Imahori, K. 1981. *J. Biochem.* 90:1405–11
220. Kubota, S., Suzuki, K. 1982. *Biomed. Res.* 3:699–702
221. Inomata, M., Hayashi, M., Nakamura, M., Imahori, K., Kawashima, S. 1983. *J. Biochem.* 93:291–94
222. Kubota, S., Suzuki, K., Yamahori, K. 1981. *Biochem. Biophys. Res. Commun.* 100:1189–94
223. Mellgren, R. L. 1980. *FEBS Lett.* 109: 129–33
224. Croall, D. E., De Martino, G. 1983. *J. Biol. Chem.* 258:5660–65
225. Otsuka, Y., Tanaka, H. 1983. *Biochem. Biophys. Res. Commun.* 111:700–9
226. Hara, K., Ichihara, Y., Takahashi, K. 1983. *J. Biochem.* 93:1435–45
227. Croall, D. E., De Martino, G. 1984. *Biochim. Biophys. Acta* 788:348–55
228. Suzuki, K., Tsuji, S., Kubota, S., Kimura, Y., Imahori, K. 1981. *J. Biochem.* 90:275–78
229. Suzuki, K., Tsuji, S., Ishiura, S., Kimura, Y., Kubota, S., Imahori, K. 1981. *J. Biochem.* 90:1787–93
230. Mellgren, R. L., Repetti, A., Muck, T. C., Easly, J. 1982. *J. Biol. Chem.* 257:7203–9
231. Dayton, W. R., Schollmeyer, J. V. 1980. *J. Cell Biol.* 87:267a
232. Dayton, W. R., Schollmeyer, J. V. 1981. *Exp. Cell Res.* 136:423–33
233. Barth, R., Elce, J. S. 1981. *Am. J. Physiol.* 240:E493–98
234. Kleese, W. C., Goll, D. E. 1980. *J. Cell Biol.* 87:84a
235. Ishiura, S., Sugita, H., Nonaka, I., Imahori, K. 1980. *J. Biochem.* 87:343–46
236. Goll, E. G., Edmunds, T., Kleese, W. C., Sathe, S. K., Shannon, J. D. 1985. See Ref. 7, pp. 151–64

237. Ohno, S., Emori, Y., Imajoh, S., Kawasaki, H., Kisaragi, M., Suzuki, K. 1984. *Nature* 312:566–70
238. Julius, D. I., Brake, A. J., Blair, L., Kumisawa, R., Thorner, J. 1984. *Cell* 37:1075–81
239. Meyer, W. L., Fisher, E. H., Krebs, E. G. 1964. *Biochemistry* 3:1033–39
240. Drummond, G. I., Duncan, L. 1966. *J. Biol. Chem.* 241:3097–103
241. Huston, R. B., Krebs, E. G. 1968. *Biochemistry* 7:2116–22
242. Drummond, G. I., Duncan, L. 1968. *J. Biol. Chem.* 243:5532–38
243. Mellgren, R. L., Aylward, J. H., Killilea, S. D., Lee, E. Y. C. 1979. *J. Biol. Chem.* 254:648–52
244. Reddy, M. K., Rabinowitz, M., Zak, R. 1983. *Biochem. J.* 209:635–41
245. Azanza, J.-L., Raymond, J., Robin, J.-M., Cottin, P., Ducastaing, A. 1979. *Biochem. J.* 183:339–47
246. Reddy, M. K., Etlinger, J. D., Rabinowitz, M., Fishman, D. A., Zak, R. 1975. *J. Biol. Chem.* 250:4278–84
247. Hathaway, D. R., Werth, D. K., Haeberle, J. R. 1982. *Clin. Res.* 30:192A
248. Elce, J. S., Baenziger, J. E., Young, D. C. R. 1984. *Biochem. J.* 220:507–12
249. Okitani, A., Goll, D. E., Stromer, M. H., Robson, R. M. 1976. *Fed. Proc.* 35:1746
250. Toyo-oka, T., Masaki, T. 1979. *J. Mol. Cell. Cardiol.* 11:769–86
251. Toyo-oka, T., Kamishiro, T., Masaki, M., Masaki, T. 1982. *Jpn. Heart J.* 23:829–34
252. Leonard, J. P., Salpeter, M. M. 1979. *J. Cell Biol.* 82:811–19
253. Traub, P. 1984. *Arch. Biochem. Biophys.* 228:120–32
254. Nelson, W. J., Traub, P. 1981. *Eur. J. Biochem.* 116:51–57
255. Nelson, W. J., Traub, P. 1982. *J. Biol. Chem.* 257:5544–53
256. Traub, P., Nelson, W. J. 1982. *Mol. Biol. Rep.* 8:239–47
257. Nelson, W. J., Traub, P. 1983. *Mol. Cell. Biol.* 3:1146–56
258. Yoshida, H., Murachi, T., Tsukahara, I. 1984. *Biochim. Biophys. Acta* 798:252–59
259. Roy, D., Chiesa, R., Spector, A. 1983. *Biochem. Biophys. Res. Commun.* 116:204–9
260. Hightower, K. R., Reddy, V. N. 1982. *Exp. Eye Res.* 34:413–21
261. Yoshimura, N., Tsukahara, I., Murachi, T. 1984. *Biochem. J.* 223:47–51
262. Beer, D. G., Butley, M. S., Malkinson, A. M. 1984. *Arch. Biochem. Biophys.* 228:207–19
263. De Martino, G. N. 1982. *Biochem. Biophys. Res. Commun.* 108:1325–30
264. De Martino, G. N. 1981. *Arch. Biochem. Biophys.* 211:253–57
265. De Martino, G. N., Croall, D. E. 1985. See Ref. 7, pp. 117–26
266. Pontremoli, S., Melloni, E., Salamino, F., Sparatore, B., Michetti, M., Horecker, B. L. 1984. *Proc. Natl. Acad. Sci. USA* 81:53–56
267. Melloni, E., Pontremoli, S., Salamino, F., Sparatore, B., Michetti, M., Horecker, B. L. 1984. *Arch. Biochem. Biophys.* 232:505–12
268. Melloni, E., Salamino, F., Sparatore, B., Michetti, M., Pontremoli, S., Horecker, B. L. 1984. *Arch. Biochem. Biophys.* 232:513–19
269. Cohen, S., Ushiro, H., Stoscheck, C., Chinkers, M. 1982. *J. Biol. Chem.* 257:1523–31
270. Puca, G. A., Nola, E., Sica, V., Bresciani, F. 1977. *J. Biol. Chem.* 252:1358–66
271. Vedeckis, W. V., Freeman, M. R., Schrader, W., O'Malley, B. W. 1980. *Biochemistry* 19:335–43
272. Waxman, L. 1978. In *Protein Turnover and Lysosome Function*, ed. H. L. Segal, D. J. Doyle, pp. 363–77. New York: Academic
273. Nishiura, I., Tanaka, K., Yamato, S., Murachi, T. 1978. *J. Biochem.* 84:1657–59
274. Takahashi-Nakamura, M., Tsuji, S., Suzuki, K., Imahori, K. 1981. *J. Biochem.* 90:1583–89
275. Mellgren, R. L., Carr, T. 1983. *Arch. Biochem. Biophys.* 225:779–86
276. Nakamura, M., Inomata, M., Hayashi, M., Imahori, K., Kawashima, S. 1984. *J. Biochem.* 96:1399–407
277. Takano, E., Yumoto, N., Kannagi, R., Murachi, T. 1984. *Biochem. Biophys. Res. Commun.* 122:912–17
278. DeMartino, G. N., Blumenthal, D. K. 1982. *Biochemistry* 21:4297–303
279. Kawashima, S., Nomoto, M., Hayashi, M., Inomata, M., Nakamura, M., Imahori, K. 1984. *J. Biochem.* 95:95–101
280. Suzuki, K., Ishiura, S. 1983. *J. Biochem.* 93:1463–71
281. Zimmerman, U.-J. P., Schlaepfer, W. W. 1985. *Biochem. Biophys. Res. Commun.* 129:804–11

Ann. Rev. Biochem. 1986. 55:483–509
Copyright © 1986 by Annual Reviews Inc. All rights reserved

PLATELET ACTIVATING FACTOR: A BIOLOGICALLY ACTIVE PHOSPHOGLYCERIDE

Donald J. Hanahan

Department of Biochemistry, University of Texas Health Science Center, San Antonio, Texas 78284-7760

CONTENTS

PERSPECTIVES AND SUMMARY

In 1971, Henson (1) first proposed that there was an interaction between leukocytes and platelets whereby a soluble, "fluid phase mediator" was re-

483

leased from the leukocytes of sensitized immunized rabbits. This mediator(s) then activated platelets releasing vasoactive amines. Subsequently two groups of investigators, Sirganian & Osler (2) and Benveniste, Henson & Cochrane (3) independently reported a confirmation of that earlier observation. The latter group also coined the term, platelet activating factor, and this led to further studies by Benveniste and collaborators on the chemical identification of the mediator. During this period the Paris group gave substance to the fact that platelet activating factor was potentially a lipid but did not precisely define its chemical nature. However, in 1979, our laboratory in San Antonio (4) reported the semisynthesis of an alkylacetylglycerophosphocholine (AGEPC) which mimicked exactly the biological behavior of the naturally occurring material (5), and very shortly thereafter Benveniste and colleagues (6) reported on an alternate semisynthetic pathway to the same compound. At the same time, Snyder and collaborators in Tennessee (7) published a semisynthetic approach to AGEPC which they considered to be the hypotensive agent under active study in their laboratory. It is now well established that the hypotensive agent and the platelet activating factor are one and the same compound. However it was not until some nine months later that our group in San Antonio was successful in elucidating the structure of the naturally occurring platelet activating factor generated from sensitized rabbit basophils (8). Its chemical structure is particularly novel as shown in Figure 1.

This compound is unique in several ways: 1. It represents the first bona fide example of a biologically active phosphoglyceride. 2. It possesses an O-alkyl ether residue at the sn-1 position and a short chain acyl moiety, i.e. acetyl, at the sn-2 position. At the sn-3 position, the polar head group in all naturally formed platelet activating factors is that of an O-phosphocholine group. Although the configuration of the naturally occurring material is presumed to be that of the sn-3 type by reference to the activity of synthetic model compounds, in actual fact the amounts of material isolated in the usual instance are too small to allow a strict confirmation of its stereochemistry.

To date, it has been shown that 1-O-alkyl derivative is some 300-fold more active than the analogous fatty acyl derivative (4). Further, all reports on the

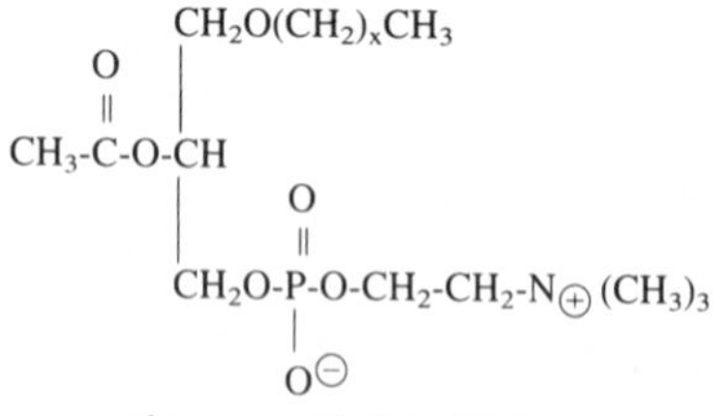

$$CH_2O(CH_2)_xCH_3$$
$$\underset{\|}{\overset{O}{}}\quad|$$
$$CH_3\text{-}C\text{-}O\text{-}CH$$
$$|\qquad\overset{O}{\underset{\|}{}}$$
$$CH_2O\text{-}P\text{-}O\text{-}CH_2\text{-}CH_2\text{-}N_{\oplus}(CH_3)_3$$
$$|$$
$$O^{\ominus}$$

where $x = 13:0$ to $17:1$

Figure 1 Structural formula of platelet activating factor (1-O-alkyl-2-acetyl-sn-glycero-3-phosphocholine).

Table 1 Biological activity of synthetic 1-O-alkyl-2-acyl-sn-glycero 3-phosphocholines[a,b]

Substituent at sn-2 position		Biological activity
Acetyl		1×10^{-11} M
Propionyl		5×10^{-10} M
Butyryl		1×10^{-7} M
Hexanoyl	no activity at	1×10^{-6} M
Palmitoyl		1×10^{-6} M

[a]Unpublished observations, D. Hanahan et al
[b]Activity based on ability of derivatives to cause 50% secretion of serotonin from washed rabbit platelets

chemical structure of naturally derived platelet activating factor have until recently demonstrated primarily the structure given above. Variations in the substituent on the sn-2 position pf the AGEPC molecule as related to chain length show a striking influence on biological activity (see Table 1). Simple removal of the acyl moiety renders the resulting lyso derivative without any detectable biological activity.

It is evident then that alkylacetylglycerophosphocholine (AGEPC) is a very potent lipid chemical mediator and the nature of the substituents at the sn-2 position are of particular importance in expressing biological activity. A number of laboratories have shown the importance of structural variations on biological activity (9–11).

Starting with research groups in San Antonio, Oak Ridge, and Paris as the nidus of research on the biochemistry and biology of this unique compound, a quite spectacular interest has emerged that is evident in the active publication rate in this area of study. In fact it is interesting to note the many diverse systems that are sensitive to platelet activating factor and/or appear to synthesize it, as shown in Table 2.

In developing the format for this review, the decision was made to center attention on specific areas of high interest at this point in this field and to describe selectively certain, but not all, papers in a particular area. To this end the review considers such diverse topics as structure of naturally occurring platelet activating factor produced in several different cells, the nature of the binding of this molecule to cells and the subsequent biochemical and biological reactions, a brief excursion into the biosynthetic and metabolic pathway, and a broad consideration of the biological behavior of the molecule.

Table 2 Cells or tissues sensitive to and/or capable of synthesizing platelet activating factor

Platelets	Neutrophils
Alveolar macrophages	Smooth muscle
Basophils	Liver

Several in-depth, excellent review articles on the chemistry and biochemistry of platelet activating factor have appeared in the past few years and one is urged to read them to obtain a proper perspective of the field. To explore the topic of the chemistry/biochemistry of ether-containing lipids in more specific detail, see the review articles by Snyder et al (12), O'Flaherty & Wykle (13), and the book edited by Mangold & Paltauf (14).

CHEMICAL CHARACTERISTICS OF NATURALLY PRODUCED PLATELET ACTIVATING FACTOR

In 1980, the chemical structure of naturally produced platelet activating factor from IgE (rabbit) basophils was elucidated and shown to be alkylacetylglycerophosphocholine (AGEPC, see formula above) (8). To date, all reports on the platelet activating factor activity reported to be present or secreted by many cell types have confirmed this basic structure. However, as should have been expected, the possibility of the concomitant formation of a long-chain fatty acyl acetyl glycerophosphocholine could not be excluded and as a matter of fact has now been reported and will be discussed later.

Of considerable interest and importance, it is now possible to use a combination of sophisticated techniques not only to detect amounts of platelet activating factor in the picogram range but also to provide a structural analysis of the active principle. Though one cannot assign a stereochemical conformation by the elegant mass spectrometry procedures in use, careful use of phospholipases with stereochemical preferences, e.g. the sn-3 configuration, will allow a reasonable decision regarding optical characteristics of the material. While structural analysis of naturally produced platelet activating factor has progressed exceedingly well, questions must be raised as to the considerable variation reported in the several laboratories on the chemical nature of the 1-O-alkyl chain, i.e. length and degree of unsaturation. The composition has run the gamut from platelet activating factor with a single chain length to one with as many as five to six species. Though there can certainly be selectivity in the synthetic mechanism within cells, there are concerns regarding possible technical problems during structural determination and in the subsequent interpretation of the results. In the papers reviewed below, the basic idea was to illustrate the sophistication of the techniques employed and to reflect on the technical facets of the approach.

In a study of the biosynthesis of platelet activating factor by human neutrophils, Clay et al (15) employed human neutrophils subjected to activation by opsonized zymosan or calcium ionophore (A23187). Production of platelet activating factor under the experimental conditions used occurred within 10 minutes for the zymosan challenge and at two minutes for the ionophore treatment. The total lipids extracted from these cells were separated by silica gel

G thin layer chromatography. The plate was sectioned and each area extracted with methanol and assayed for biologic activity. The high-activity fractions were combined and subjected to reverse-phase HPLC separation. In addition to biological assays and high-performance liquid chromatography these investigators employed gas chromatography/mass spectrometry to identify structures. Certain facets of the results are intriguing. First, an average of 84% of tritiated AGEPC standard and 100% of the biologic activity were recovered from the thin layer plate. This type of result is unusual since recoveries from thin layer plates are commonly in a much lower range and highly variable. Further, only a single molecular species was identified, namely 1-O-hexadecyl-2-acetyl glycero-3-phosphocholine. The mass spectrometric assay system involved analysis of the "diglyceride" formed by hydrolysis of the phospholipid with 50% HF at 4°C. Controls were run to ensure that the HF treatment did not cause significant degradation to unidentified products but no recoveries are given.

Oda and collaborators (16) detected two chemical types, namely 1-O-hexadecyl- and 1-O-octadecyl-2-acetyl-glycero-3-phosphocholine present in a ratio of 4:1, produced by A23187 stimulation of neutrophils. Experimentally human neutrophils were stimulated first with cytochalasin B and then with A23187. Subsequently the lipids, to which an internal standard of 16:0 d$_3$ AGEPC was added, were purified by silicic acid, thin layer, and aluminum oxide chromatography. A final purification was achieved by silica gel H thin layer chromatography using a solvent system of methanol-water, 2:1, v/v. The platelet activating factor–containing fractions were subjected to phospholipase C treatment, which removed the polar head group, and resulting "diglycerides" converted to the tert-butyldimethylsilyl ether derivatives. The latter compounds were then chromatographed on silica gel H thin layer plates, using a solvent system of hexane/diethyl ether, 9:1, which effectively separated the 1-O-alkyl-2 acetyl species from the 1-O-fatty acyl-2-acetyl species. These t-butyl derivatives were then analyzed by electron impact mass spectrometry using selected ion monitoring techniques. The ions at m/z 245, 415, and 443 corresponded to the 16:0 and 18:0 alkyl chain lengths. Insignificant amounts of the 18:1 species were found. On the basis of an internal standard, Oda et al (16) calculated that cells without stimulation contain ~1 picogram 16:0 and 18:0 AGEPC, respectively, whereas the cells stimulated with A23187 produced 16:0 AGEPC of the order of 64 picomoles and 18:0 AGEPC of the order of 16 picomoles. Interestingly, the molar ratio of ether-containing choline phosphoglycerides to AGEPC in these stimulated cells was 138. These authors noted that there was a long fatty acyl containing 1-O-long chain fatty acyl-2-acetyl glycero-3-phosphocholine present in the neutrophils. This latter substance must be removed by thin layer chromatography, with a solvent system of hexane/diethyl ether, 9:1, or else it could be wrongly identified as an even-numbered alkyl ether derivative. This was an important observation, but unfortunately no clear evidence was presented as to its composition. Also since the fatty acyl derivative also has biological activity (though much less than the alkyl ether type), it may play an important role in cellular metabolism.

In a study comparable to those cited above, Pinckard and associates (17) subjected human neutrophilic polymorphonuclear leukocytes to sequential stimulation with cytochalasin B and N-formyl methionyl leucyl phenylalanine (FMLP). The lipids were isolated from the pelleted cells and supernates, purified by thin layer chromatography, and those fractions containing platelet activating factor activity were analyzed by normal and reverse-phase HPLC and fast-atom-bombardment mass spectrometry. On the basis of relative retention times and mass spectral data, these investigators concluded that the major species was apparently a 16 : 0 alkyl ether but also noted that there were at least four other molecular species present. Though no quantitative data were presented, casual examination of the HPLC patterns would suggest that four minor components comprised no more than 5% of the total sample. No reference was made to the possible presence of an analogous fatty acyl derivative.

In 1983, Billah & Johnston (18) first reported the presence of platelet activating factor in human amniotic fluid. A combination of thin layer chromatographic behavior, aggregating activity towards platelets, and certain chemical treatments strongly suggested that the active principal was indeed an alkyl acetyl GEPC. Recently, Nishihara and collaborators (19) provided a more definitive answer on the chemical structure of the platelet activating factor–like material found in human amniotic fluid during labor. In addition to thin layer chromatography, and platelet aggregating ability, these investigators employed gas chromatography/mass spectrometry to confirm the presence of an alkyl ether–linked choline phosphoglyceride. The platelet activating factor fraction isolated from a thin layer chromatographic plate was hydrolyzed with phospholipase C and any acyl esters were removed by alkaline hydrolysis. The resulting alkyl glycerols were converted to the trimethylsilyl ether derivative and assayed by combined gas chromatography/mass spectrometry. Although five peaks emerged on gas chromatography, Nishihara et al could only identify one, a 1-O-octadecyl derivative. They concluded that this was the only species of platelet activating factor released in the amniotic fluid. However, until the other peaks found on the gas chromatogram are identified, the conclusion that only an 18 : 0 type is present is premature.

The question raised above about the presence of a long-chain fatty acyl analogue of the more commonly studied 1-O-alkyl derivative, which had been largely overlooked by most investigators except Mueller and coworkers (20), has now been answered in a most convincing manner by Satouchi and colleagues (21). Using rabbit polymorphonuclear neutrophils preincubated with cytochalasin B and then stimulated with A23187, these investigators isolated the platelet activating factor–rich fraction by procedures described in an earlier paper. Subsequent to phospholipase C treatment, the resulting "diglycerides" derived from the alkyl acetylglycerophosphocholine and the acyl acetyl-glycerophosphocholine were converted to the tBDMS derivatives and separated on thin layer chromatography. A solvent system of hexane/diethyl ether, 9 : 1, v/v, allowed a satisfactory separation of these two classes of derivatives which were then analyzed by gas liquid chromatography/mass spectrometry coulped with selected ion monitoring. Those ions selected for identification were those representative of molecular weight, acetyl, and long-chain fatty acyl groups. Interestingly, three fatty acyl species were detected, namely palmitic, stearic, and oleic acids. Of considerable interest and importance, the analytical data on the levels of the alkyl acetyl and the acyl acetyl forms showed an almost equal (and apparently concomitant) production of each, though the palmitic acid derivative was the most abundant. These results raise the possibility that the enzymes responsible for the deacylation of the choline phosphoglycerides and the ensuing reacetylation of the lyso derivates may not have an absolute requirement for a 1-O-alkyl ether linkage at the C-1 position. In an almost identical study in which human neutrophils were stimulated with A23187 and the resulting platelet activating factor analyzed by fast-atom-bombardment mass spectrometry (22), the presence of an acyl acetyl type platelet activating factor [in addition to the alkyl derivatives (16:0, 17:0, 18:0, 18:1)] was apparently dismissed as an apparent contaminant. This decision is difficult to rationalize since the acyl acetyl derivative is present, and though it has less biological activity towards platelets, nevertheless it does exert an effect. It may be that this derivative acts on other systems in cell and that compartmentalization of species occurs.

Finally, an interesting communication by Mallet & Cunningham (23) showed that platelet activating factor could be isolated from the lesional scale of psoriatic patients. Inasmuch as nanogram quantities of platelet activating factor could be recovered from 100 milligrams of human lesional scale, structural identification could be undertaken. A combination of thin layer chromatography, hydrolysis by phospholipase C, and then HPLC led to the isolation of the 1-O-alkyl glycerols. The latter were derivativized to the tert-butyldimethyl silyl ethers and analyzed by gas chromatography/mass spectrometry. Both 16 : 0 and 18 : 0 alkyl residues, in a ratio of 3 : 1, were identified. However no mention was made of the possible presence of a 1-O fatty acyl analogue.

BIOSYNTHETIC ROUTE TO PLATELET ACTIVATING FACTOR

The stimulation of certain cells, mainly of the phagocytic class, can lead to the formation of the biologically active phosphoglyceride, 1-O-alkyl-2-acetyl-glycero-3-phosphocholine (AGEPC; platelet activating factor). Of high importance then is a delineation of the reaction pathway leading to this potent chemical mediator. Inasmuch as this subject has been covered very effectively in a recent review by Snyder et al (12), it is necessary only to comment on two important points. It is evident that a major biosynthetic pathway involves the enzyme, acetyl-CoA : 1-O-alkyl-2-lyso-glycero-3-phosphocholine acetyl transferase. Figure 2 depicts the reaction catalyzed by this enzyme.

The source of I probably derives from the 2-fatty acyl derivative found in very high concentration in the choline-containing phosphoglycerides of phagocytic cells (24). An alternate pathway has been shown to involve the conversion of 1-O-alkyl-2-acetyl-glycerol to II through the cytidine diphosphocholine pathway (25). However, this reaction scheme, if viewed as a de novo scheme in the cell per se, would demand first the probable formation of 1-O-alkyl-fatty glycerol by phospholipase C attack on the parent phosphoglycerides, replacement of the long-chain fatty acyl group with an acetyl residue, and concomitant formation of the phosphocholine derivative. It would appear that this would be an inefficient pathway given the ability of these cells to form a lyso derivative by attack of phospholipase A_2 on the parent phosphoglycerides followed by acetylation by the transferase pathway.

The formation of platelet activating factor in platelets has been the subject of some debate but it is now clear that the platelets of rabbits as well as humans can produce this compound, albeit in much lesser amounts than in phagocytic cells. In 1983, Alam et al (26) reported that human and rabbit platelets, stimulated with calcium ionophore, could form platelet activating factor. This activity was isolated (from a total lipid extract) by HPLC and its behavior mimicked that of a synthetic (and radioactive) standard. On the basis of the aggregation profile on rabbit platelets, these authors calculated that the stimulated human cells produced in the range of 1.8 picomoles per 5×10^8 cells and the rabbit cells of the order of 3 picomoles per 5×10^8 cells. In this study, the entire platelet incubate was extracted for lipid and hence no information was provided on the amount of

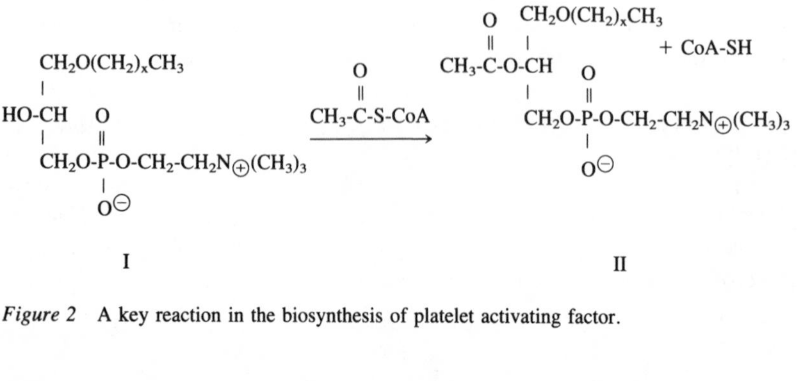

Figure 2   A key reaction in the biosynthesis of platelet activating factor.

platelet activating factor secreted by these cells. A later study by McKean & Silver (27) directed towards an understanding of phospholipid biosynthesis in platelets, showed that acylation of lyso platelet 1-O-alkyl-2-(lyso)-glycero-3-phosphocholine by long-chain unsaturated fatty acyl CoA residues could be accomplished through use of the platelet microsomes. The acyl-CoA sn-glycero-3-phosphocholine acyl transferase was membrane bound, had a pH optimum of 7.5, and was insensitive to magnesium but inhibited by calcium ions. The 1-O-alkyl lyso derivative was acylated some three to fourteen times slower than the analogous 1-O-long-chain fatty acyl lyso compound. Interestingly, the preferred fatty acyl donors were the CoA esters of linoleic and arachidonic acids.

Chignard et al (28) observed that rabbit platelets contained sufficient ether-linked glycerophospholipid to provide the substrate for platelet activating factor biosynthesis. The conclusion was reached that rabbit platelets, upon stimulation with thrombin, could form platelet activating factor at a level of 5 nanograms per 5×10^8 cells. These authors further concluded that rabbit platelets, upon stimulation with thrombin, converted approximately 10% of the ether-linked choline phosphoglycerides to the lyso form and approximately 0.2% to platelet activating factor.

While the pathway by which platelet activating factor can be synthesized in cells appears quite well defined, the ultimate metabolic fate of the newly produced mediator is less clear. A new and exciting dimension to this story has emerged with the demonstration that relatively little of the platelet activating factor formed by cells under stimulatory conditions is released. This is dramatically illustrated in a study by McIntyre et al (29) in which it was shown that cultured human endothelial cells, when stimulated with histamine, bradykinin, or ATP, could form platelet activating factor for a period up to 45 minutes. Concomitantly prostacyclin (PGI$_2$) was produced but only up to 7.5 minutes after stimulation with the same agents. However the most significant observation was that PGI$_2$ was released from the cell monolayer, whereas the platelet activating factor remained cell bound with no detectable release. They proposed that this lipid chemical mediator may influence some of the inflammatory properties of histamine and bradykinin as well as affecting the vascular surface and ultimately cell-cell interaction (with platelets, or neutrophils, for example). This is a particularly fascinating observation since the intercellular reactions of platelet activating factor may be of utmost importance in its potential pathophysiologic role and in effecting cell-cell interactions.

In an investigation designed to explore the modulation of synthesis and release of platelet activating factor from human polymorphonuclear leukocytes, Ludwig and associates (30) found that using stimulants such as cytochalasin B, FMLP, and calcium, only 30–40% of the total activity produced by these cells was released. A pattern emerged in which there was a distinct lag phase between biosynthesis and release.

Mueller et al (20), in a well-designed investigation, presented data on the molecular species of platelet activating factor biosynthesized by stimulated rabbit peritoneal and human peripheral neutrophils. These cells were treated with ionophore, opsonized zymosan, or FMLP (including a pretreatment with cytochalasin B), the platelet activating factor isolated by thin layer chromatography and then subjected to resolution by reverse-phase high performance liquid chromatography. A heterogeneous display of alkyl side chains (chain length and unsaturation) was obtained. The platelet activating factor derived from rabbit neutrophils contained alkyl chains of 15:0, 16:0, 18:0, and 18:1, whereas the human neutrophil platelet activating factor contained alkyl chains of 16:0, 17:0 (two isomers), 18:0, and 18:1. They concluded that the compositions of the alkyl chains of the platelet activating factors generated from these two cell sources were significantly different from that of the presumed precursor pools of 1-O-alkyl-2- long-chain fatty acyl-glycerophosphocholine. They further concluded that the nature of the alkyl chains on the newly synthesized platelet activating factor did not depend on the stimulus used.

A debatable issue in the study of the metabolism of platelet activating factor has been the question of the specificity of the fatty acyl group inserted into the lyso derivative of platelet activating factor formed on interaction of this agonist with cells such as the alveolar macrophage. In a particularly interesting article, Robinson and collaborators (31) investigated the coenzyme A–dependent and coenzyme A–independent specificities of the acylation of lysophospholipids. These investigators concluded that there were three separate mechanisms operative in this process: the first, a coenzyme A–independent transacylation; the second, a coenzyme A–dependent transacylation; and the third, a acyl coenzyme A–dependent acylation. In the absence of coenzyme A, ATP, and magnesium ions, the acylation process exhibited a high preference for arachidonic acid. In the presence of these cofactors, there was no specific preference for arachidonic acid. These provocative results shed new light on the pathway by which the lyso form of platelet activating factor is metabolized in cells. The question remains as to the possible compartmentation of these acylation reactions and whether the substrate and enzyme system are ever in a milieu devoid of the coenzyme A, ATP, and magnesium ions in the intact cell. These questions were addressed in a thoughtful manner by Robinson et al (31).

The role of Ca^{2+} in platelet activating factor formation in rat peritoneal macrophages was investigated by Gomez-Cambronero et al (32). The activation of the enzyme, lyso (platelet activating factor):acetyl CoA acetyl transferase, was sensitive to the level of extracellular Ca^{2+}, but through the use of TMB-8 (an intracellular calcium transport antagonist), it was shown that there was no inhibition of the activation of the enzyme induced by zymosan in the presence of extracellular Ca^{2+}. Zymosan induced a rapid increase in Ca^{2+} entry into the cell and the authors concluded that this was a crucial event in the activation of the acetyl transferase and in the formation of platelet activating factor.

In a study designed to investigate the formation of platelet activating factor in the eosinophils from patients with eosinophilia, Lee and colleagues (33) noted that this biosynthetic pathway was stimulated in a dose-, time- and Ca^{2+}/Mg^{2+}-dependent manner by a number of agonists. These included eosinophil chemotactic factor, C5a, formyl-methionylleucylphenylalanine, or A23187 (ionophore). The latter stimulant maintained an elevated level of the enzyme, 1-O-alkyl-2-(lyso)-sn-glycero-3-phosphocholine acetyl CoA : acetyl transferase, for at least fifteen minutes, whereas the other agonists showed one- to three-minute peak periods. The basic message in this paper was that chemotactic stimulation of eosinophils was intimately involved in the biosynthesis and release of platelet activating factor and this may relate to the function of eosinophils in inflammatory and allergic reactions.

Kramer et al (34) reported that the metabolic handling of platelet activating factor in human platelets involves two enzyme systems. The first is a Ca^{2+}-independent acetyl hydrolyase, present in the cytosol, that deacetylates alkylacetylglycerophosphocholine to the lyso form, and the second a coenzyme A–independent N-ethyl maleimide–sensitive transacylase. The latter enzyme is localized in the platelet membrane and incorporates a long-chain fatty acid into the lyso form to yield an alkyl acylglycerophosphocholine. The alkyl acyl derivative possesses no activating factor activity and is considered the major metabolic product resulting from the interaction of platelet activating factor with the platelet. Kramer and associates further provided convincing evidence supporting the exclusive incorporation of arachidonic acid into the lyso platelet activating factor. These investigators (as well as several others) believe that this arachidonoyl-containing ether phosphocholineglyceride is the primary substrate for attack during agonist action on a cell such as the platelet, yielding lysoplatelet activating factor which then can be acetylated to yield a biologically active derivative.

Touqui and colleagues (35) have proposed that the long-chain fatty acyl derivative of lysoplatelet activating factor is an important precursor of platelet activating factor in the stimulated platelet rabbit platelet. In the latter instance thrombin action on rabbit platelets led to the release of platelet activating factor and lysoplatelet activating factor into the medium. At the same time there was a significant reduction in the level of alkylacylglycerophosphocholine in the cell. It was proposed that an alkylacetylglycerophosphocholine cycle was operative in which platelet activating factor was first converted to the long-chain fatty acyl derivative which then can yield platelet activating factor when the cell is stimulated. Part of this cycle has been established with considerable certainty, namely the formation of a long-chain acyl derivative, but the latter portion of the proposed cycle needs further careful kinetic analysis to provide a definitive answer.

Earlier Albert & Snyder (36) had reported that a deacylation-reacylation cycle was of prime importance in the metabolism of arachidonic acid in alveolar macrophages and involves the deacylation of the arachidonic acid–rich ether phosphocholine fraction upon stimulation. Presumably the resulting lysoderivative could be acetylated to the corresponding biologically activating factor.

INTERACTION OF PLATELET ACTIVATING FACTOR WITH CELLS

Binding Characteristics

A particularly active and important area of research on this mediator has centered on developments of the concept of a receptor-mediated process for its activation of cells. The experimental approach on this problem is fraught with difficulties since the very high biological potency of this factor makes it mandatory to use a synthetic, radiolabeled sample of very high specific activity. To date the only labeled platelet activating factor with sufficiently high specific activity is that which is labeled in the 1-O-alkyl moiety with tritium. The specific activity should be greater than 30–40 Ci/mmole to allow a reasonable estimation of the high-affinity, low-capacity sites, which are considered to be the primary receptor sites. The latter may number from 100 to 500 and hence examination of the binding characteristics by a Scatchard plot can involve significant errors due to the low levels of radioactivity encountered in this region of the binding profile.

Perhaps one of the earliest investigations on the binding of (chemically) well-characterized platelet activating factor was that of Valone et al (37). Using human platelets and semisynthetic platelet activating factor, labeled with tritium in the 1-O-alkyl position, they showed that this material, containing a mixture of the 1-O-hexadecyl and 1-O-octadecyl-2 acetyl-sn-glycero-3-phosphocholine species, was bound to platelets in a consistent manner at 20°C. A Scatchard analysis of the binding profile showed two distinct binding sites, one of which was considered a high-affinity, low-capacity site with nearly 1400 molecules bound per platelet, with a K_D of 37 ± 3 nM. The other site showed an almost infinite capacity for binding this labeled material. A number of analogues of platelet activating factor, e.g. the lyso derivative and the 2-benzoyl analogue, were used in an attempt to block the binding but in general did not affect the reaction unless 50–5000-fold greater concentrations (than in platelet activating factor) were employed. Treatment of the platelets with the active factor for five minutes at 37°C caused irreversible binding and desensitization to further AGEPC addition. Even at two minutes there was full deactivation (in the absence of added calcium ions), even though 50% of the specific sites, i.e. those of high affinity and low capacity, were still available. In a subsequent report, Valone (38) described the isolation of a human platelet membrane protein that bound 1-O-hexadecyl-2-acetyl-sn-glycero-3-phosphocholine. The procedure involved isolation of the AGEPC binding proteins by affinity chromatography on a column of AGEPC noncovalently bound to human serum albumin coupled to AH Sepharose. Solubilized platelet proteins were passed through this column and the unabsorbed protein discarded. Then washing with a high concentration of AGEPC removed a protein (single-component, 180,000 daltons) which was considered to be the AGEPC binding protein. No further proof that this protein preparation was specific for AGEPC or had binding characteristics comparable to that of the intact platelet was presented.

Kloprogge & Akkerman (39), in a study of the binding kinetics of tritiated AGEPC to human platelet, observed a single class of specific receptors numbering near 250. Prior treatment of the platelets with nonlabeled platelet activating factor rendered the platelet incapable of binding tritiated platelet activating factor. This desensitization appeared to result from a loss of available binding sites (undoubtedly of the high-affinity, low-capacity type). Chesney et al (40) showed that pre-exposure of human platelets to low aggregating levels, 8 $\times$ 10^{-11} M, of platelet activating factor led to desensitization or lack of response of the platelets to aggregating concentrations, 8 $\times$ 10^{-10} M, of platelet activating factor. These investigators concluded that this effect was caused by a decrease in the affinity of the high-affinity site rather than a loss of binding sites. However under the conditions of their experimental procedures and given the very low levels of radioactivity encountered, a decision with regard to decreased affinity of high-affinity sites, which are present in very low numbers, must be approached with great caution. Hwang and associates (41) published an interesting article on specific receptors for tritiated 1-O-octadecyl/ hexadecyl-2-acetyl-sn-glycero-3-phosphocholine on membranes derived from rabbit platelets and guinea pig smooth muscle. Their careful study showed that there were 150–300 receptors. Further a series of platelet activating factor analogues showed binding characteristics comparable to their relative ability to induce aggregation in intact platelets. Specific receptor sites were found in purified plasma membranes from several smooth muscle sources as well as from neutrophils. Apparently cells unresponsive to platelet activating factor such as the erythrocyte and alveolar macrophages did not exhibit any binding sites.

An investigation similar to those mentioned above for platelets has been conducted on polymorphonuclear leukocytes (neutrophils). Valone & Goetzl (42) claimed that there were two distinct AGEPC binding sites in these cells, one a high-affinity, low-capacity type which had a K_D of 0.11 nM (compared to 37 nM in platelets) and bound some 5 $\times$ 10^6 molecules per cell comparable to an estimated 1400 molecules per platelet. The other site, exhibiting a considerably lower affinity and greater capacity, was considered to be a nonspecific site(s) and not associated with any receptor-mediated process. Interestingly, no stereospecific requirements were noted either with sn-1 or sn-3 AGEPC, since both bound to the neutrophils. This is in contradistinction to high stereospecificity noted with the platelets. In a similar study, Bussolino et al (43) found the binding pattern of tritiated platelet activating factor to human polymorphonuclear leukocytes to exhibit two distinct binding sites, one saturable with a K_D of 45 nM (approximately 28,000 sites) and the other a nonsaturable site. These investigators also claimed that the K_D for AGEPC binding to human platelets was 20 nM (with approximately 1600 sites), but unfortunately did not provide any experimental details on the latter topic.

In earlier studies from the laboratory of Sanchez Crespo (44), evidence was presented that the rat platelet unlike the rabbit was insensitive to platelet activating factor. This suggested that the rat platelet did not possess specific receptors for this lipid chemical mediator. In a subsequent paper from the same laboratory, Iñarrea et al (45) provided additional support for this observation through a study of the binding characteristics of tritiated platelet activating factor to platelets isolated from human, rabbit, and rat. Rat platelets showed only nonspecific binding, whereas, as expected, the rabbit and human cells possessed specific receptor sites. Two different sites were identified in the human and rabbit platelet, one of which was saturable and the other of infinite binding capacity. The high-affinity sites associated with humans and rabbits showed K_D values of 1.6 nM and 0.9 nM, respectively. While this report provides a strong reason why thrombocytopenia cannot be induced in rats, it does not exclude interaction of this agonist with other tissues and cells as shall be shown later in this review.

Inhibition of Induced Reactions

Pertinent to an understanding of the mechanism of the receptor-mediated action of AGEPC on cells, it is important to have antagonists to the binding site and/or site of action of this potent lipid chemical mediator.

To this end, a significant advance in stimulating research in the area was the report by Terashita and colleagues (46) in 1983 on the behavior of a synthetic analogue of platelet activating factor, termed CV3988, which they regarded as a specific antagonist of this agonist's action toward rabbit platelets. The structural formula of this compound is shown in Figure 3.

Chemically this compound is rac-3-(N-n-octadecyl-carbamoyloxy)-2-methoxypropyl-2-thiazolioethyl phosphate. It had no effect on aggregation induced by ADP, arachidonic acid, collagen, or calcium ionophore. At 3×10^{-6} to 3×10^{-5} M it inhibited aggregation induced by AGEPC at 3×10^{-8} M. In 1985, Valone (47), using washed human platelets, showed that CV3988

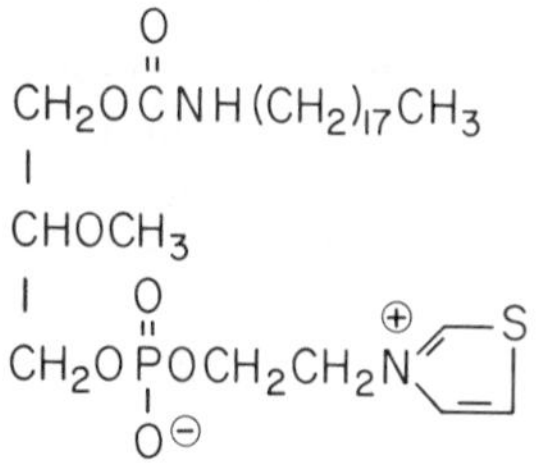

$$\text{CH}_2\text{O}\overset{\displaystyle\text{O}}{\overset{\|}{\text{C}}}\text{NH(CH}_2)_{17}\text{CH}_3$$

$$|$$

$$\text{CHOCH}_3$$

$$\text{CH}_2\text{O}\overset{\displaystyle\text{O}}{\overset{\|}{\text{P}}}\text{OCH}_2\text{CH}_2\text{N}\overset{\oplus}{\underset{}{}}\text{S}$$

$$\overset{|}{\text{O}}{}^{\ominus}$$

CV-3988

Figure 3 CV3988, a potent inhibitor of platelet activating factor.

inhibited AGEPC-induced aggregation in a dose-dependent manner, yielding an IC_{50} of 2.9×10^{-8} M. Concomitantly it was observed that this antagonist inhibited the binding of tritiated AGEPC in a concentration-dependent manner with an IC_{50} near 7×10^{-8} M. These data suggested that CV3988 reacts by inhibiting the binding of AGEPC to its receptor on the platelet.

Recently Terashita et al (48) presented detailed evidence on the inhibition by CV3988 (see structural formula above) of tritiated platelet activating factor binding to washed rabbit platelets. They also described results obtained with human and guinea pig platelets. The specific binding of ^{3}H platelet activating factor was inhibited in a dose-dependent manner by unlabeled platelet activating factor and CV3988. The IC_{50} values for unlabeled platelet activating factor and for CV3988 inhibition were 1.2×10^{-9} M and 8×10^{-8} M, respectively. This inhibition affected only the specific site receptor with little or no observable effect on the nonspecific binding. On the basis of the binding assay results, a total of 568 ± 50 specific receptors per platelet was calculated. CV3988 inhibited specific binding of tritiated platelet activating factor to human and guinea pig platelets with an IC_{50} value of 1.6×10^{-7} M and 1.8×10^{-7} M, respectively. This latter antagonist also inhibited platelet activating factor–induced aggregation in rabbit, guinea pig, and human platelets. On the other hand, rat and mouse platelets showed no specific receptors for platelet activating factor. Only one point of concern need be raised on this study and this centers on the possible desensitization of the platelets under the assay conditions. Although these investigators showed that the six-minute preincubation with tritiated platelet activating factor in the receptor binding assay led to reversal by addition of CV3988, it was not clear whether these platelets after the six-minute preincubation period with the agonist would undergo normal aggregation.

Increased interest in exploration of antagonists to AGEPC binding to cells has resulted in a series of papers in which a group of chemically diverse, rather novel compounds have been explored as antagonists of AGEPC-induced activation of cells. Shen and coworkers (49) isolated and characterized from a

Chinese herbal plant (haifenteng) a unique substance and showed it to have the chemical structure shown in Figure 4. In addition this compound was orally active and inhibited in vivo induced vascular permeability. Using a receptor preparation isolated from rabbit platelet membranes, they showed that kadsurenone inhibited PAF binding with a K_i of 5.8×10^{-8} M (using AGEPC at 6.3×10^{-9} M). An assay system involving rabbit platelets in plasma indicated that this compound was an effective inhibitor of PAF-induced changes in these cells.

In a novel approach to the study of antagonists of PAF action, Chesney and her associates (50) examined the inhibitory capacity of α-adrenergic antagonists such as phenoxybenzamine and phentolamine on the binding of PAF to human gel filtered platelets. These compounds specifically inhibited platelet activating factor–induced platelet and neutrophil aggregation and degranulation, whereas yohimbine and prazosin showed little or no activity. In addition these authors noted that gel-filtered platelets exhibited a second wave of aggregation and secreted alpha granules and dense granules upon exposure to platelet activating factor. These cells appeared to have high-affinity and low-affinity binding sites for this agonist. The high-affinity site was calculated to bind some 400 molecules per platelet and the low-affinity site showed an uptake of near 9000 molecules per platelet. Both phenoxybenzamine and phentolamine, α-adrenergic antagonists, elicited competitive inhibition of specific binding at an IC_{50} of 50 μM and 240 μM, respectively. A similar inhibitory profile was noted for these compounds in the aggregation assay. These investigators concluded on the basis of these and other observations that the high-affinity platelet activating factor receptor is not a typical $α_2$-receptor but is affected in some manner by these particular α-adrenergic antagonists. Interestingly, verapamil competitively inhibited the platelet activating factor binding and activity. It may be that these effects are localized in or near a calcium channel but this possibility remains to be clarified.

Recently in a provocative article, Kornecki et al (51) reported that two pyschotropic triazolobenzodiazepine drugs, alprazolam and triazolam, inhibited platelet activating factor–induced shape, aggregation, and secretion patterns in human (washed) platelets. This agonist at 7 nM induced 50% of the maximum initial velocity of aggregation of gel-filtered platelets, but if 10 μM

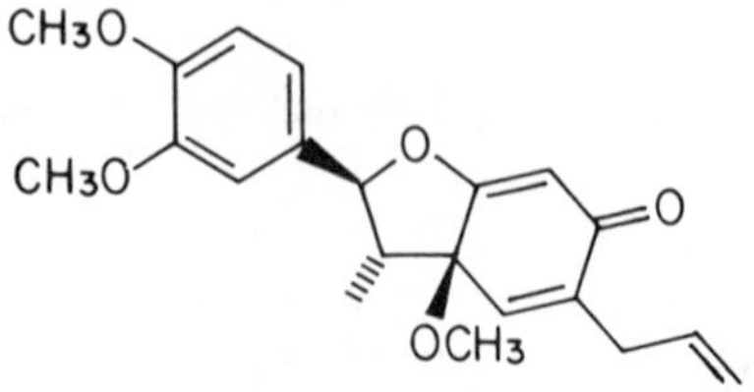

Figure 4 Kadsurenone, a novel antagonist of platelet activating factor activity.

alprazolam were present, this value for PAF activity was changed to 35 nM. Apparently these drugs are specific inhibitors of platelet activating factor–induced activation of human platelets and did not affect such agonists as ADP, thrombin, epinephrine, A23187, collagen, or arachidonic acid. The particularly challenging facet of this study is that platelet activating factor may play a role in neuronal function and this may explain the mechanism by which these pyschotropic drugs act on cells.

Utilizing a different approach on the types of inhibitors suitable for characterization of the platelet activating factor binding site, Tokumura and associates (52) explored the behavior of several closely related (chemically) analogues of platelet activating factor. These derivatives, which bear modification in the polar head, are shown in Figure 5. A comparison of these compounds with CV3988 was undertaken and showed that I was the most potent inhibitor with an IC_{50} of 4.1×10^{-8} M against a challenge of 1×10^{-10} M 1-O-hexadecyl-2-acetyl-sn-glycero-3-phosphocholine in the serotonin secretion assay. These results on serotonin secretion are provided in Table 3.

In addition compound I was the most effective in completely blocking the degradation of inositol phospholipids and the production of phosphatidic acid.

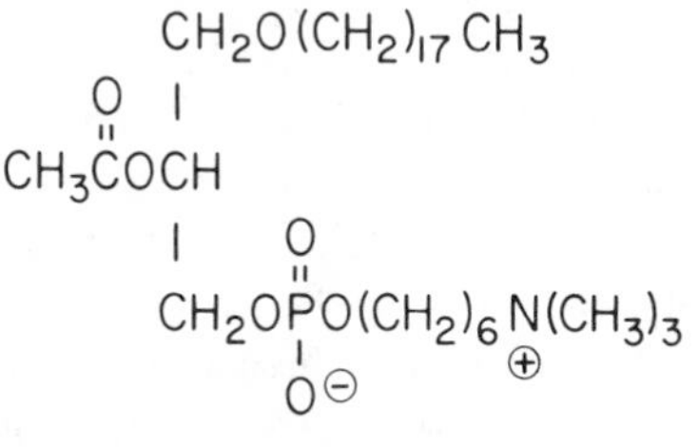

$$CH_2O(CH_2)_{17}CH_3$$
$$\overset{O}{\underset{\|}{CH_3C}}OCH \ |$$
$$CH_2O\overset{O}{\underset{\underset{O^\ominus}{|}}{\overset{\|}{P}}}O(CH_2)_6 \underset{\oplus}{N(CH_3)_3}$$

I

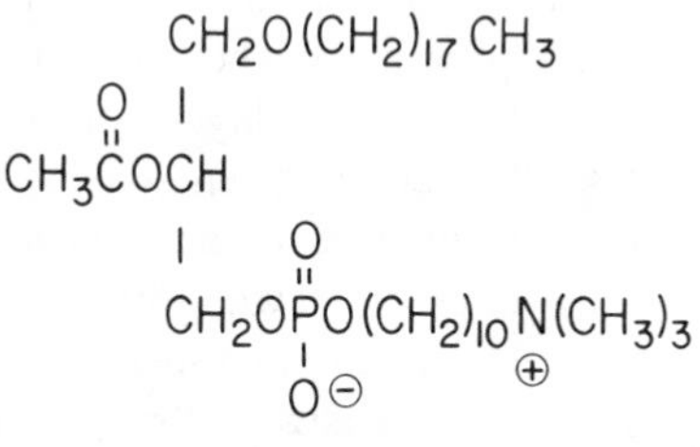

$$CH_2O(CH_2)_{17}CH_3$$
$$\overset{O}{\underset{\|}{CH_3C}}OCH \ |$$
$$CH_2O\overset{O}{\underset{\underset{O^\ominus}{|}}{\overset{\|}{P}}}O(CH_2)_{10} \underset{\oplus}{N(CH_3)_3}$$

II

$$CH_2O\overset{O}{\overset{\|}{C}}NH(CH_2)_{14}CH_3$$
$$|$$
$$CHOCH_3$$
$$| \quad O$$
$$CH_2O\overset{\|}{\underset{\underset{O^\ominus}{|}}{P}}OCH_2CH_2\underset{\oplus}{N(CH_3)_3}$$

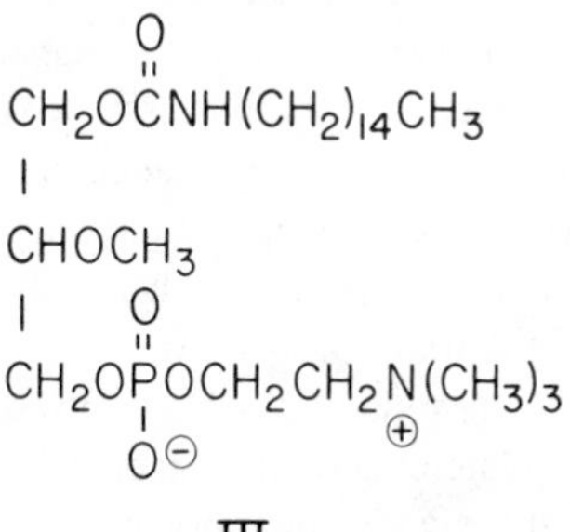

III

Figure 5 Structural analogues of 1-O-alkyl-2-acetyl-sn-glycero-3-phosphocholine (platelet activating factor) with inhibitory properties.

Table 3 Inhibitory activity of structural analogues of AGEPC on serotonin release from washed rabbit platelets stimulated with AGEPC[a]

Compound	IC_{50} (M) AGEPC, at 1×10^{-10} M
I	4.1×10^{-8}
II	2.0×10^{-7}
III	7.0×10^{-7}
CV3988	5.9×10^{-7}

[a]Taken from Tokumura et al (52).

The latter are the very early indicators (usually within 5 seconds) of the interaction of platelet activating factor with the platelet. It was evident from these studies that each inhibitor occupied the same receptor site as the agonist. It appears possible, through use of such closely related structural analogues, to define more precisely the putative receptor for platelet activating factor on the platelet.

As will be described in a later section, platelet activating factor induced a dramatic release of glucose (glycogenolysis) in the perfused (fed) rat liver (53). Subsequent studies by Buxton et al (54) demonstrated that this agonist will at the same time cause vasoconstriction of the hepatic vasculature. The stimulation of hepatic glycogenolysis and vasoconstriction was shown to be inhibited by two structural analogues of platelet activating factor, i.e. CV 3988 and U66985 (structures given above). When platelet activating factor was coinfused with either one of these two antagonists a greater inhibition of the agonist effects was observed with U66985 than that noted with CV3988. Evidence was presented in support of a specific receptor site inhibition by these antagonists.

Lad and associates (55) proposed that a pertussis toxin–sensitive site(s) on human neutrophils is closely associated with the action of platelet activating factor on these cells. Pretreatment of human neutrophils with pertussis toxin inhibited the platelet activating factor–induced chemotaxis, superoxide generation, aggregation, and release of lysozyme. It was suggested that a GTP binding protein (termed "Ni") is the target for the pertussis toxin and that inhibition ensues from this interaction. It will be of importance to confirm this inhibiting capability of pertussis toxin on platelets which apparently do not contain a GTP binding protein. Another important aspect of such a study would be to establish with certainty the purity of the toxin preparation.

An interesting observation was reported by Vigo (56) who demonstrated that C-reactive protein inhibited platelet activating factor–induced human platelet aggregation and stabilized the platelet membrane against the lytic action of lysophosphatidyl choline. C-reactive protein is described as an acute phase reactant whose serum levels rise significantly as a result of trauma or tissue injury. The liver is an important source of this protein which appears to have a high affinity for phosphocholine-containing lipids or polysaccharides in the presence of Ca^{2+}. It appeared from the data provided by Vigo (56) that the C-reactive protein inhibits platelet activating factor by reacting directly with it and the membrane. Simultaneous addition of C-reactive protein and platelet activating factor did not cause inhibition. If platelets were incubated with C-reactive protein thirty minutes prior to addition of platelet activating factor, inhibition was of the order of 72%, whereas preincubation of platelet activating factor and C-reactive protein for thirty minutes led to a 90% inhibition of activity. Arachidonic acid release was inhibited by C-reactive protein and this suggested inhibition of phospholipase in the platelet. In an isolated system, it was shown that this protein could inhibit phospholipase A_2 action towards dipalmitoyl phosphatidyl choline bilayers.

OTHER BIOLOGICAL SYSTEMS OR CONDITIONS AFFECTED BY/OR SENSITIVE TO PLATELET ACTIVATING FACTOR

In the preceding sections, attention has been focused primarily on the platelet as the target cell for studies on platelet activating factor, with secondary emphasis on the neutrophil and alveolar macrophage. As this field has developed, interest has expanded in exploring the biological activity of this potent lipid chemical mediator in many diverse systems. A sampling of these studies reveals the almost ubiquitous influence of this agonist in the intact animal, in the isolated cell, and in its cellular components. A few examples of the biological spectrum of platelet activating factor action are outlined below.

Liver

Inasmuch as one mode of action of platelet activating factor is to initiate a significant turnover of the inositol phosphoglycerides in washed rabbit platelets (59) and as various hormones and α-adrenergic agonists exhibit a similar behavior (58, 59), a study of the metabolic effects of platelet activating factor on rat liver revealed some dramatic alterations. In an initial report in this area, Shukla et al (53) observed that 1-*O*-hexdecyl-2-acetyl-*sn*-glycero-3-phospho-choline had a pronounced effect on the turnover of the phosphoinositides in isolated hepatocytes and a particular stimulatory action on glycogenolysis in the intact perfused (fed) rat liver. The latter behavior is clearly shown in Figure 6.

Not only was there a most positive effect of this agonist on the release of glucose, but there was also a stereospecific requirement, with the *sn*-1 con-figuration of platelet activating factor exhibiting no activity. In addition the deacylated form of this agonist, or lyso platelet activating factor, had no

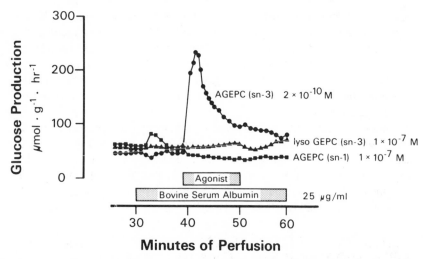

Figure 6   Effects of *sn*-3-AGEPC, *sn*-1-AGEPC, and *sn*-3-lyso-GEPC on glucose release in perfused rat liver. (Reproduced with permission from authors.)

influence on the process. As shown in a subsequent paper by Buxton and colleagues (60), this agonist exerted its action in a dose-dependent, calcium-dependent manner. Characteristic of the response was an increase in lactate and acetoacetate output and a decreased oxygen consumption. Through repeated infusion of the platelet activating factor, a homologous desensitization could be evoked, with no apparent subsequent effect on the response of the liver to agonists such as phenylephrine and glucagon. This system provides a new approach to understanding the biological activity of platelet activating factor wherein a well-established biochemical parameter, namely glycogenolysis, can be used as a guideline.

Ischemic Bowel Necrosis

Gonzalez-Crussi & Hsueh (61) reported an interesting finding in which rats were subjected to treatment with platelet activating factor and/or lipopolysac-charide (LPS) via the abdominal aorta. After a period of five minutes to three hours, the animals were examined for lesions in the abdominal and intrathoracic organs using histological techniques. LPS alone in doses of 10–40 μg caused no bowel necrosis whereas platelet activating factor alone at 2 μg caused de-monstrable lesions. However a combination of 1 μg platelet activating factor plus 20 μg LPS caused the reproducible (and dramatic) formation of necrotiz-ing lesions. There was a significant morphologic similarity of these alterations to those noted in human necrotizing enterocolitis. These authors concluded that platelet activating factor and lipopolysaccharide act synergistically in develop-ment of bowel necrosis syndrome. These most interesting observations provide a framework for further investigation of the cooperative action of platelet activating factor and other agents in cellular functions and events.

Exocrine Secretory Glands

The potential for platelet activating factor to behave in an acetylcholinelike fashion was pursued by Söling et al (62). Using pancreatic lobules from the exocrine pancreas or the parotid glands of guinea pig, they studied the effect of this agonist and certain of its analogues on exocytosis (amylase release), calcium uptake, and inositol phosphoglyceride metabolism and compared these results with those obtained with carbamoylcholine. There was a dose-dependent effect, in the 5 nM range, of the platelet activating factor on exocytosis and a dose-dependent stimulation of calcium uptake and inositol phospholipid turnover. Of considerable interest, the 2-O-methyl analogue as well as the sn-1 isomer of platelet activating factor were active in this biological system. Söling et al concluded that the alterations induced by platelet activating factor were similar to those observed with acetylcholine.

Endotoxin-Induced Hypotension

Recently Doebber and colleagues (63) explored the involvement of platelet activating factor in endotoxin-induced hypotension in rats. Infusion of endotoxin into rats evoked a time-dependent appearance of platelet activating factor in blood and also induced hypotension. Chemical proof for platelet activating factor, at least the 1-O-alkyl ether derivative, was lacking, but infusion of semisynthetic platelet activating factor produced a hypotension comparable to that seen with endotoxin. Interestingly, the infusion of kadsurenone, described in an earlier section as an antagonist of platelet activating factor, decreased the hypotension induced by platelet activating factor to a significant level and caused a comparable level reversal of the endotoxin-related hypotension. Terashita et al (64) also have reported similar results with the antagonist, CV3988.

Lungs

When platelet activating factor was injected into a plateletfree, perfused (isolated) guinea pig lung, Hamasaki et al (65) noted a significant increase in airway and pulmonary edema. At the same time there was synthesis of thromboxane A_2 and PGI_2.

Hypoxia

Prevost and colleagues (66) reported that hypobaric hypoxia in rats caused the release of platelet activating factor into the lung alveolar lavage. The level of platelet activating factor was estimated by an aggregation assay. Limited chemical proof was offered, though, that 1-O-alkyl-2-acetyl-sn-glycero-3-phosphocholine was indeed the agent present in the lavage fluid.

Cardiac Anaphylaxis

Levi et al (67) established that platelet activating factor was a putative mediator of cardiac anaphylaxis in the guinea pig. During anaphylaxis in the isolated guinea pig heart induced by guinea pig antidinitrophenyl bovine-α-globulin IgG_1, platelet activating factor was released into the coronary effluent. This released agonist may contribute partly to cardiac dysfunction during anaphylaxis in vitro. In addition, administration of platelet activating factor to nonsensitized isolated hearts mimicked several of the functional changes noted during cardiac anaphylaxis. These included a decrease in contractility and coronary flow and decreased atrioventricular function.

Vascular Permeability

In a detailed study, Humphrey et al (68) characterized by histological techniques the vasoactive and leukotactic properties of platelet activating factor. The target tissues were the rat cremaster muscle and the skin. Not only did platelet activating factor exhibit a potent vasoactive effect, it stimulated leukocyte emigration in postcapillary venules. Interestingly the vasoactive properties of this agonist appeared to be neutrophil-independent and also independent of mast cell and platelet stimulation. The data suggested that platelet activating factor may act upon the microvasculature by direct stimulation of venular endothelial cells.

Using human subjects, Archer and coworkers (69) evaluated the histological changes in subjects given intradermal injection of platelet activating factor at a level of 200 picomoles per site. Platelet activating factor elicited the expected wheal and flare reaction within a 15-minute period, with no response noted with the lyso derivative or a saline control. At four to twelve hours after injection of platelet activating factor, there was intravascular accumulation of neutrophils, together with a perivascular mixed cellular infiltrate and at twenty four hours there was less intense cellular infiltration, with the main component being lymphocytes and histiocytes. Archer et al proposed that platelet activating factor, being able to mimic many facets of acute and chronic inflammation, may be involved in the pathogenesis of psoriasis.

Leukemic Cells

The cytotoxic ability of some eleven different alkyl phosphoglyceride analogues of platelet activating factor was tested on human leukemia cells (HL-60), human polymorphonuclear neutrophils, and Detroit 551 human skin fibroblasts (70). Three derivatives, specifically 1-O-alkyl-2-acetamido-glycero-3 phosphocholine, 1-O-alkyl-2-methoxy-glycero-3-phosphocholine, and 3-O-alkyl-2-acetyl-sn-glycero-1-phosphocholine, were found to have selective toxic action against HL-60 cells as compared to normal human skin fibroblasts and human neutrophils. The mechanism of action remains unknown but it was suggested that these cytotoxic compounds are not easily metabolized by these cells due to lack of an ether cleavage enzyme and they could interfere with normal phosphoglyceride metabolism involving specific phospholipases.

Macrophages

In a particularly interesting study, Hayashi et al (71) showed that platelet activating factor and certain of its analogues exhibited potent activity towards guinea pig peritonal macrophages. The index of activity was stimulation of glucose consumption by the macrophages. Though the effect was less on the macrophages (8.5×10^{-6} M for 50% activation) than for guinea pig platelets (2.9×10^{-10} M for 50% activation), the overall effects with specific analogues were most significant. Specifically, 1-O-octadecyl-2-O-methyl-sn-glycero-3-phosphocholine exhibited activity higher than that of 1-O-octadecyl-2-acetyl-sn-glycero-3-phosphocholine (platelet activating factor). Stereochemical specificity was noted with the sn-3 form showing activity whereas the sn-1 isomer did not. Further a lyso derivative gave no response in the system. Thus, as Hayashi et al observed, the requirements for activity in the macrophage system paralleled that of the platelets, namely a 1-O-alkyl group in the sn-1 position, a small, not bulky group on the sn-2 position, and a polar head group containing a phosphocholine at the sn-3 position of the glycerol backbone. Inhibition of activation was achieved with CV3988. These investigators concluded that platelet activating factor and its analogues bound to a common cell surface receptor on the macrophage.

CONCLUDING REMARKS

In the relatively short period of time dating from the proof in 1979 that "platelet activating factor" was indeed a unique chemical entity and further substantiation in 1980 of the chemical structure of the naturally produced material, developments in this field have been extremely rapid. Given that this factor is an important lipid chemical mediator and is involved in inflammatory reactions, an even more important facet of this substance is that it is a biologically active phosphoglyceride. To those interested in the biochemistry of phospholipids over the years there has always been the question of why there were so many diverse types of these compounds in cell membranes and whether the individual classes of phospholipids in particular would play any role other than that of a chemical component of the membrane providing a semipermeable physical barrier. Recently this question was addressed in a review article (72). Examples were cited wherein the phospholipids behaved as "co-factors" in certain enzymatic reactions, for example in amplification of the interaction of Factor Xa, calcium, and Factor V to form a lipoprotein that attacks prothrombin and converts it to thrombin. No chemical change in the phospholipids was noted even though the stimulatory behavior of these lipids was very significant. Another example centered on the turnover of phosphatidyl-myoinositol 4,5 bis phosphate in a cell membrane subsequent to an agonist interaction. One of the resulting products, inositol triphosphate, is considered by many investigators to behave as a second messenger for the intracellular movement of calcium. The third example was that of the formation of platelet activating factor. At this point in time it appears that this biologically active phosphoglyceride interacts with a specific receptor site in a cell and imparts a signal perhaps via a specific phosphorylation event. However, evidence supporting any specific metabolic alteration to the platelet activating factor during its interaction at the high-affinity, low capacity site has not been forthcoming. However there are qualifications to the last statement, particularly as it concerns the potential detectability of any metabolic products of such an interaction. This problem is brought clearly into focus by examination of the amounts of this lipid chemical mediator required for initiating the biological reaction. It is not unusual in the platelet activating factor assay system using washed rabbit platelets to observe significant agonist activity at a level of 1×10^{-12} M. In the usual test system, 1×10^9 platelets are used, and from studies on the binding of AGEPC to platelets there are 100–500 high-affinity, low-capacity binding sites that are the sites associated with the biological activity. Then indeed the number of molecules per cell associated with the active sites in such a reaction mixture is very small.

Pursuant to the above observations on the biologically active component of the interaction of platelet activating factor with the cell it is important to point out that the majority of the platelet activating factor is metabolized via a noncalcium–dependent pathway to an inactive product in which the acetyl component is removed and a long-chain fatty acyl moiety, believed to be primarily arachidonic acid, is inserted. This particular species is then considered to be the main source of arachidonic acid released during agonist attack on the cell, and the resulting lyso derivative is the substrate for formation of platelet activating factor. This reaction sequence is shown in Figure 7.

As has been evident from the literature cited here, interest has centered on the alkyl ether phosphoglycerides as the important backbone of the platelet activating factor produced upon stimulation of cells. Further, considerable interest has developed on the phagocytic cells as the primary source of platelet activating factor since they contain very high concentrations of the alkyl ether phosphocholine derivative. However, it is quite evident that liver can produce platelet activating factor under suitable experimental conditions and it is well documented that the liver contains a low level of 1-O-alkyl ethers. Hence, a question to be answered is that of the basic level of a glyceryl ether precursor needed for production of platelet activating factor. If one considers the high potency of this mediator, then actually only very small amounts would be needed and hence in this sense the phagocytic cells would possess no significant biosynthetic advantage. Finally, it is now well established that stimulated cell can produce 1-O-fatty acyl-2-acetyl-glycero-phosphocholine as well as 1-O-alkyl-2-acetyl-glycerophosphocholine. Though the fatty acyl derivative is

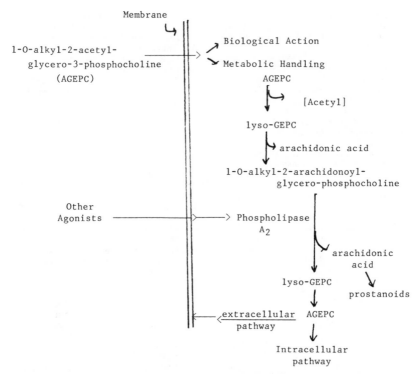

Figure 7 A proposed scheme for the metabolic handling of platelet activating factor by cells.

some 300-fold less active than the 1-*O*-alkyl analogue, it does not mean that the fatty acyl–containing components are not metabotically important. It simply may affect different reactions and pathways than noted with the more widely acclaimed 1-*O*-alkyl derivative. Certainly it bears serious consideration as an endogenously generated lipid chemical mediator.

Literature Cited

1. Henson, P. M. 1970. *J. Exp. Med.* 131:287–304
2. Sirganian, R. P., Osler, A. G. 1971. *Immunology* 106:1244–51
3. Benveniste, J., Henson, P. M., Cochrane, C. G. 1972. *J. Exp. Med.* 136:1356–77
4. Demopoulos, C. A., Pinckard, R. N., Hanahan, D. J. 1979. *J. Biol. Chem.* 254:9355–58
5. Pinckard, R. N., Farr, R. S., Hanahan, D. J. 1979. *J. Immunol.* 123:1847–57
6. Benveniste, J., Tence, M., Varenne, P., Bidault, J., Boullet, C., Polonsky, J. 1979. *C. R. Acad. Sci. Ser. D* 289:1037–40

7. Blank, M. L., Snyder, F., Byers, L. W., Brooks, B., Muirhead, E. E. 1979. *Biochem. Biophys. Res. Commun.* 90: 1194–200
8. Hanahan, D. J., Demopoulos, C. A., Liehr, J., Pinckard, R. N. 1980. *J. Biol. Chem.* 255:5514–16
9. Snyder, F. 1985. *Med. Res. Rev.* 5:107–40
10. O'Flaherty, J. T., Salzer, W. L., Cousart, S., McCall, C. E., Piantadosi, C., et al. 1983. *Res. Commun. Chem. Pathol. Pharmacol.* 39:291–309
11. Satouchi, K., Pinckard, R. N., McManus, L. M., Hanahan, D. J. 1981. *J. Biol. Chem.* 256:4425–32

12. Snyder, F., Lee, T.-C., Wykle, R. L. 1985. In *The Enzymes of Biological Membranes*, ed. A. Martonosi, 2:1–58. New York: Plenum

13. O'Flaherty, J. T., Wykle, R. L. 1983. *Clin. Rev. Allergy* 1:353–67

14. Mangold, H. K., Paltauf, F., eds. 1983. *Ether Lipids, Biochemical and Biomedical Aspects.* New York: Academic

15. Clay, K. L., Murphy, R. C., Andres, J. L., Lynch, J., Henson, P. M. 1984. *Biochem. Biophys. Res. Commun.* 121: 815–25

16. Oda, M., Satouchi, K., Yasunaga, K., Saito, K. 1985. *J. Immunol.* 134:1090–93

17. Pinckard, R. N., Jackson, E. M., Hoppens, C., Weintraub, S. T., Ludwig, J. C., et al. 1984. *Biochem. Biophys. Res. Commun.* 122:325–32

18. Billah, M. M., Johnston, J. M. 1983. *Biochem. Biophys. Res. Commun.* 113: 51–58

19. Nishihara, J., Ishibashi, T., Imai, Y., Muramatsu, T. 1984. *Lipids* 19:907–10

20. Mueller, H. W., O'Flaherty, J. T., Wykle, R. L. 1984. *J. Biol. Chem.* 259:14554–59

21. Satouchi, K., Oda, M., Yasunaga, K., Saito, K. 1985. *Biochem. Biophys. Res. Commun.* 128:1409–17

22. Weintraub, S. T., Ludwig, J. C., Mott, G. E., McManus, L. M., Lear, C., Pinckard, R. N. 1985. *Biochem. Biophys. Res. Commun.* 129:868–87

23. Mallet, A. I., Cunningham, F. M. 1985. *Biochem. Biophys. Res. Commun.* 126: 192–98

24. Nakagawa, Y., Sugiara, T., Waku, K. 1985. *Biochim. Biophys. Acta* 833:323–29

25. Satouchi, K., Oda, M., Saito, K., Hanahan, D. J. 1984. *Arch. Biochem. Biophys.* 234:318–21

26. Alam, I., Smith, J. B., Silver, M. J. 1983. *Thromb. Res.* 30:71–79

27. McKean, M. L., Silver, M. J. 1985. *Biochem. J.* 225:723–29

28. Chignard, M., Le Couedic, J.-P., Coeffier, E., Benveniste, J. 1984. *Biochem. Biophys. Res. Commun.* 124:637–43

29. McIntyre, T. M., Zimmerman, G. A., Satoh, K., Prescott, S. M. 1985. *J. Clin. Invest.* 76:271–80

30. Ludwig, J. C., McManus, L. M., Clark, P. O., Hanahan, D. J., Pinckard, R. N. 1984. *Arch. Biochem. Biophys.* 232: 102–10

31. Robinson, M., Blank, M. L., Snyder, F. 1985. *J. Biol. Chem.* 260:7889–95

32. Gomez-Cambronero, J., Iñarrea, P., Alonso, F., Sanchez Crespo, M. 1984. *Biochem. J.* 219:419–24

33. Lee, T.-C., Lenihan, D. J., Malone, B., Roddy, L. L., Wasserman, S. I. 1984. *J. Biol. Chem.* 259:5526–30

34. Kramer, R. M., Patton, G. M., Pritzker, C. R., Deykin, D. 1984. *J. Biol. Chem.* 259:13316–20

35. Touqui, L., Jacquemin, C., Dumarey, C., Vargaftig, B. B. 1985. *Biochim. Biophys. Acta* 833:111–18

36. Albert, D. H., Snyder, F. 1984. *Biochim. Biophys. Acta* 796:92–101

37. Valone, F. H., Coles, E., Reinhold, V. R., Goetzl, E. J. 1982. *J. Immunol.* 129:1637–41

38. Valone, F. H. 1984. *Immunology* 52: 169–74

39. Kloprogge, E., Akkerman, J. W. N. 1984. *Biochem. J.* 223:901–9

40. Chesney, C. M., Pifer, D. D., Huch, K. M. 1983. *Inserm. Symp. No. 23*, ed. J. Benveniste, B. Arnoux, pp. 177–83. New York: Elsevier

41. Hwang, S.-B., Lee, C.-S., Cheah, M. J., Shen, T. Y. 1983. *Biochemistry* 22: 4756–63

42. Valone, F. H., Goetzl, E. J. 1983. *Immunology* 48:141–49

43. Bussolino, F., Tetta, C., Camussi, G. 1984. *Agents Actions* 15:15–17

44. Sanchez Crespo, M., Alonso, F., Iñarrea, P., Egido, J. 1981. *Agents Actions* 11:599

45. Iñarrea, P., Gomez-Cambronero, J., Nieto, M., Sanchez Crespo, M. 1984. *Eur. J. Pharmacol.* 105:309–15

46. Terashita, Z., Tsushima, S., Yoshioka, Y., Nomura, H., Inada, Y., Nishikawa, K. 1983. *Life Sci.* 32:1975–82

47. Valone, F. H. 1985. *Biochim. Biophys. Res. Commun.* 126:502–8

48. Terashita, Z., Imura, Y., Nishikawa, K. 1985. *Biochem. Pharmacol.* 34:1491–95

49. Shen, T. Y., Hwang, S.-B., Chang, M. N., Doebber, T. W., Lam, M.-H., et al. 1985. *Proc. Natl. Acad. Sci. USA* 82: 672–76

50. Chesney, C. M., Pifer, D. D., Huch, K. M. 1985. *Biochem. Biophys. Res. Commun.* 127:24–30

51. Kornecki, E., Ehrlich, Y. H., Lenox, R. H. 1984. *Science* 226:1454–56

52. Tokumura, A., Homma, H., Hanahan, D. J. 1985. *J. Biol. Chem.* 260:12710–14

53. Shukla, S. D., Buxton, D. B., Olson, M. S., Hanahan, D. J. 1983. *J. Biol. Chem.* 10212–14

54. Buxton, D. B., Hanahan, D. J., Olson, M. S. 1985. *Biochem. Pharmacol.* In press

55. Lad, P. M., Olson, C. V., Grewal, I. S. 1985. *Biochem. Biophys. Res. Commun.* 129:632–38

56. Vigo, C. 1985. *J. Biol. Chem.* 260: 3418–22
57. Shukla, S. D., Hanahan, D. J. 1982. *Biochem. Biophys. Res. Commun.* 106: 697–703
58. Prpic, V., Blackmore, P. F., Exton, J. H. 1982. *J. Biol. Chem.* 257:11323–31
59. Downes, C. P., Michell, R. H. 1982. *Cell Calcium* 3:467–502
60. Buxton, D. B., Shukla, S. D., Hanahan, D. J., Olson, M. S. 1984. *J. Biol. Chem.* 259:1468–71
61. Gonzalez-Crussi, F., Hseuh, W. 1983. *Am. J. Pathol.* 112:127–35
62. Söling, H. D., Eibl, H., Fest, W. 1984. *Eur. J. Biochem.* 144:65–72
63. Doebber, T. W., Wu, M. S., Robbins, J. C., Choy, B. M., Chang, M. N., Shen, T. Y. 1985. *Biochem. Biophys. Res. Commun.* 127:799–808
64. Terashita, Z., Imura, Y., Nishikawa, K., Sumida, S. 1984. *Kyoto Conf. Prostaglandins*. New York: Raven. In press

65. Hamasaki, Y., Mojarad, M., Saga, T., Tai, H.-H., Said, S. I. 1984. *Am. Rev. Respir. Dis.* 129:742–46
66. Prevost, M. C., Cariven, C., Simon, M. F., Chap, H., Douste-Blazy, L. 1984. *Biochem. Biophys. Res. Commun.* 119: 58–63
67. Levi, R., Burke, J. A., Guo, Z.-G., Hattori, Y., Hoppens, C. M., et al. 1984. *Circulation Res.* 54:117–24
68. Humphrey, D. M., McManus, L. M., Hanahan, D. J., Pinckard, R. N. 1984. *Lab. Invest.* 50:16–25
69. Archer, C. B., Page, C. P., Morley, J., MacDonald, D. M. 1985. *Br. J. Dermatol.* 112:285–90
70. Hoffman, D. R., Hajdu, J., Snyder, F. 1984. *Blood* 63:545–52
71. Hayashi, H., Kudo, I., Inoue, K., Onozaki, K., Tsushima, S., et al. 1985. *J. Biochem.* 97:1737–45
72. Hanahan, D. J., Nelson, D. R. 1984. *J. Lipid Res.* 25:1528–35

Ann. Rev. Biochem. 1986. 55:511–38

STRUCTURAL ASPECTS OF THE RED CELL ANION EXCHANGE PROTEIN[1]

Daniel Jay

Department of Biochemistry and Molecular Biology, Harvard University, Cambridge, Massachusetts 02138

Lewis Cantley

Department of Physiology, Tufts University School of Medicine, Boston, Massachusetts 02111

CONTENTS

---

[1]Abbreviations used: BIDS, 4-benzamido-4'-isothiocyanostilbene-2,2'-disulfonate; DIDS, 4,4'-Diisothiocyanostilbene-2,2'-disulfonate; NAP-taurine, *N*-(4-azido-2-nitrophenyl)-2-aminoethane sulfonate; PITC, phenylisothiocyanate. The abbreviations for fragments of band 3 are explained in the legend of Figure 1.

511

0066-4154/86/0701-0511$02.00

PERSPECTIVES AND SUMMARY

The erythrocyte anion exchange protein, band 3, is arguably the best understood mammalian transport system. This protein was originally named for its migration position upon sodium dodecyl sulfate gel electrophoretic separation of human red cell plasma membrane proteins. It is the third largest of the major proteins and is the most abundant protein in the red cell membrane (25% of total membrane protein, 1.2×10^6 copies per cell; Fairbanks et al, Ref. 1). Interest in this protein was stimulated when it was established that specific inhibitors of red cell anion exchange such as the stilbene disulfonates covalently label band 3 (2, 3). The ease of purification of red cell plasma membranes coupled with the relative abundance of band 3 has facilitated studies of both the structure and mechanism of action of this transport system.

Band 3 is an integral membrane glycoprotein which appears as a diffuse band with a molecular weight of 90,000–100,000 when analyzed by sodium dodecyl sulfate polyacrylamide gel electrophoresis. Red cell plasma membranes can be enriched for band 3 by high pH extraction of peripheral proteins such as spectrin; after such extraction band 3 makes up 70% of the membrane protein and is still capable of carrying out anion exchange. Further purification of Triton X-100–solubilized band 3 to homogeneity can be achieved by chromatographic techniques including anion exchange (4, 5), *p*-chloromercuribenzamido ethyl agarose (6), or Concanavalin A sepharose (7). The detergent-solubilized and purified protein retains specific binding sites for the stilbene disulfonates indicating retention of a native state (8).

The ability to prepare inside-out and right-side-out red cell membrane vesicles (9) has simplified the analysis of sidedness of band 3 with respect to the membrane. Band 3 has been successfully reconstituted into phospholipid vesicles with partial recovery of activity as judged by external anion-dependent anion efflux which is inhibited by stilbene disulfonates (6, 10). It should be noted that under these conditions band 3 shows a marked reduction in binding of stilbene disulfonates (11). This effect may be due to alterations in band 3 by the high levels of octyl glucoside used in reconstitution (11a). The protein can be roughly divided into two domains that independently carry out separate functions; an amino terminal cytoplasmic domain and a carboxy terminal membrane-associated domain. The membrane-associated domain facilitates the 1 : 1

exchange of chloride for bicarbonate across the erythrocyte membrane. This rapid exchange allows the red cell to function as a carrier of the bicarbonate derived from respiratory CO_2 from the tissues to the lungs. In addition it has been proposed that this exchange is important for regulating the postcapillary pH of the red cell during gas exchange (12, 13). The transport activity appears to require only the membrane domain since the cytoplasmic domain can be proteolytically removed without detectably altering anion exchange activity (14). The cytoplasmic domain is responsible for two functions. The amino-terminal portion of this domain binds several glycolytic enzymes (5) as well as hemoglobin (15). The physiological role of this binding remains unclear. The cytoplasmic domain also possesses the binding site of ankyrin which in turn interacts with spectrin and thus aids in anchoring the cytoskeleton to the membrane (16).

This review summarizes the current knowledge of the structure of band 3. It is not intended to be comprehensive, but is meant to assimilate recent findings, notably the determination of the sequence of murine band 3 (17). This information is used to construct a model suggesting how the band 3 protein folds to form an anion transport site. For more comprehensive treatment of the subject the reader is directed to excellent general reviews (18–21) or specific reviews on chemical inhibitors of band 3 (11, 22) or oligomeric structure (23).

STRUCTURE OF THE BAND 3 PROTEIN

Proteolytic Cleavage Sites

For the purpose of clarity and convenience a new nomenclature for proteolytic fragments is suggested and will be used throughout the review. Fragments will be named by a three-unit code in which the first unit represents the reagent used to generate the N terminus of the fragment or is N if the N terminus is the same as that of the intact protein (Ch, chymotrypsin; Cn, Cyanogen bromide; Pa, papain; Pe, pepsin; Sc, S-cyanylation; Tr, trypsin). Similarly the second unit represents the protease that gave rise to the C terminus of the fragment or is C if the C terminus is the same as the intact protein. The third unit is the size of the fragment in kilodaltons. An example of this is TrCh-17 which corresponds to a 17,000-dalton fragment derived from a tryptic cleavage at the N terminus and a chymotryptic cleavage at the C terminus. The major proteolytic cleavage sites are illustrated in Figure 1.

The band 3 polypeptide can be cleaved on the cytoplasmic face of the membrane by trypsin to form fragments NTr-40 and TrC-55, which are respectively the cytoplasmic and membrane-associated domains (5, 24). The NTr-40 fragment can be further cleaved by trypsin to form fragments NTr-23 and TrTr-20. Alternatively the N-terminal 202 amino acids can be removed in a single fragment NSc-22 by S-cyanylation. The membrane-associated domain

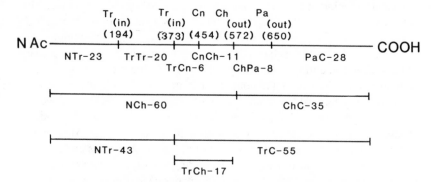

Figure 1 Proteolytic fragmentation of band 3. The nomenclature for proteolytic fragments is a three-subunit code. The first unit defines the amino terminus of the fragment; the second unit defines the carboxy terminus of the fragment; the third unit is the size of the fragment in kilodaltons. Abbreviations used are: C, original carboxy terminus of band 3; Ch, chymotryptic cleavage site; Cn, cyanogen bromide cleavage site; N, original amino terminus of band 3; Pa, papain cleavage site; Tr, trypsin cleavage site. Numbers in parentheses refer to the residue number of the cleavage site based on the murine band 3 sequence.

TrC-55 can be further subdivided by an exofacial chymotryptic cleavage to form fragments TrCh-17 and ChC-35 (24, 25). Cleavage at this site does not appreciably alter anion exchange activity (14), and the fragments retain a stable interaction in the bilayer or in nonionic detergent (26). By chemical modifications discussed later it is evident that both TrCh-17 and ChC-35 are required for anion exchange.

Addition of chymotrypsin to intact erythrocytes allows cleavage at the exofacial site to generate fragments NCh-60 and ChC-35. Fragment ChC-35 can be further cleaved by extracellular papain to form fragments ChPa-8 and PaC-28 (27). It should be noted that the amino terminus of ChC-35 is also cleaved by papain so that the ChPa-8 fragment is more accurately called a PaPa-8 fragment. This cleavage is known to inactivate anion exchange. Since additional papain-sensitive sites exist near the exofacial chymotryptic cleavage site it is not possible to produce the fragment NPa-68 by exofacial papain treatment.

Sequence

The primary structure of human band 3 has been difficult to determine due to long hydrophobic stretches in the membrane domain that are intractable to classical protein sequencing techniques. Kaul et al (28) have sequenced the first 200 amino acids of the protein while others have sequenced smaller stretches of the membrane domain, notably the N terminal 37 amino acids of an 11,000-dalton cyanogen bromide fragment derived from the C terminus of TrCh-17 (29) and the 72 amino acids of a papain subfragment of the fragment ChPa-8 (30).

The advent of recombinant DNA technology and DNA sequencing allowed for rapid determination of the primary sequence of proteins, circumventing the problems associated with conventional protein sequencing. A major difficulty in applying these techniques to human band 3 was the inability to obtain significant amounts of messenger RNA encoding for the band 3 polypeptide. Although band 3 protein exists in high copy number in the mature erythrocyte, the RNA is present only early in erythroid development. Since such precursor cells do not circulate in the human circulatory system, obtaining sizable amounts of human band 3 messenger RNA was a profound difficulty.

Braell & Lodish (31, 32) circumvented this problem by utilizing anemic mouse spleen as their source. This tissue contained a high percentage of erythroid precursors and was readily available. They demonstrated the presence of murine band 3 mRNA by specific immunoprecipitation of the polypeptide from in vitro translation. Braell & Lodish also demonstrated that murine band 3 was structurally similar to the human protein as judged by similar proteolysis patterns of the two native species. Using this mRNA, Kopito & Lodish (17) constructed a cDNA library in the expression vector lambda-gt11 and isolated clones encoding murine band 3 by antibody screening. The complete sequence of mouse band 3 is presented in Figure 2.

The mouse protein is 929 amino acids long. Homology with known sequences of the human band 3 varies from good to almost identical. Interestingly the region of the known human band 3 sequence that is least homologous is the 50 amino acids at the amino terminus. This region of the human sequence includes the glycolytic enzyme-binding domain (33), and consistent with the divergence in sequence, mouse band 3 does not appear to bind glycolytic enzymes (17). It has been noted that for human band 3, 16 of the first 31 residues are acidic while none are basic (33, 34), suggesting that the interaction with glycolytic enzymes may be with a positively charged region of the proteins. The amino terminus of mouse band 3 is slightly less negatively charged, with 11 of the first 31 residues acidic, and with two basic residues.

The protein sequences established for two hydrophobic fragments of the membrane domain of human band 3 (corresponding to sequences 456–492 and 579–641 of murine band 3; Figure 2) are almost identical with the respective sequences of the mouse protein. This is somewhat surprising since if the sole function of these regions is to traverse the bilayer as alpha helixes, there should be no serious constraints on the identity of the side chains provided they are hydrophobic and allow alpha helical structure. This is in contrast with the histocompatibility antigens of mouse and human (H-2 and HLA respectively), where the transmembrane segment is not conserved between these two species. The lack of variation between the mouse and human sequences may suggest that the side chains play some additional role in specific helical packing or transport site structure.

MetGlyAspMetArgAspHisGluGluValLeuGluIleProAspArgAspSerGluGluGluLeuGluAsnIle 25
IleGlyGlnIleAlaTyrArgAspLeuThrIleProValThrGluMetGlnAspProGluAlaLeuProThrGlu 50
GlnThrAlaThrAspTyrValProSerSerThrSerThrProHisProSerSerGlyGlnValTyrValGluLeu 75
GlnGluLeuMetMetAspGlnArgAsnGlnGluGluGlnTrpValGluAlaAlaAlaHisTrpIleGlyLeuGluGlu 100
AsnLeuArgGluGlyGlyValTrpGlyArgProHisLeuSerTyrLeuThrPheTrpSerLeuLeuGluLeuGln 125
LysValPheSerLysGlyThrPheLeuLeuGlyLeuAlaGluThrSerLeuAlaGlyValAlaAsnHisLeuLeu 150
AspCysPheIleTyrGluAspGlnIleArgProAsnArgGluGluLeuLeuArgAlaLeuLeuLysArg 175
SerHisAlaGluAspLeuGlyAsnLeuGluGlyValLysProAlaValLeuThrArgSerGlyGlyAlaSerGlu 200
ProLeuLeuProHisGlnProSerLeuGluThrGlnLeuTyrCysGlyGlnAlaGluGlyGlySerGluGlyPro 225
SerThrSerGlyThrLeuLysIleProProAspSerGluThrThrLeuValLeuIleGlyArgAlaAsnPheLeu 250
GluLysProValLeuGlyPheValArgLeuLysGluAlaAlaValProLeuGluAspLeuValLeuProGluProVal 275
GlyPheLeuLeuValLeuLeuGlyProGluAlaProHisValAspTyrThrGlnLeuGlyArgAlaAlaAlaAlaThr 300
LeuMetThrGluArgValPheArgIleThrAlaSerMetAlaHisAsnArgGluGluLeuLeuArgSerLeuGlu 325
SerPheLeuAspCysSerLeuValLeuProProThrAspAlaProSerGluLysAlaLeuLeuAsnLeuValPro 350
ValGlnLysGlyLeuLeuArgArgArgTyrLeuProSerProAlaLysProAspProAsnLeuTyrAsnThrLeu 375
AspLeuAsnGlyGlyLysGlyGlyProGlyAspGluAspAspProLeuArgArgThrGlyArgIlePheGlyGly 400
LeuIleArgAspIleArgArgArgTyrProTyrTyrLeuSerAspIleThrAspAlaLeuSerProGlnValLeu 425
AlaAlaValIlePheIleTyrPheAlaAlaLeuSerProAlaValThrPheGlyGlyLeuLeuGlyGluLysThr 450
ArgAsnLeuMetGlyValSerGluLeuLeuIleSerThrAlaValGlnGlyIleLeuPheAlaLeuLeuGlyAla 475
GlnProLeuLeuValLeuGlyPheSerGlyProLeuLeuValPheGluGluAlaPhePheSerPheCysGluSer 500
AsnAsnLeuGluTyrIleValGlyArgAlaTrpIleGlyPheTrpLeuIleLeuLeuValMetLeuValValAla 525
PheGluGlySerPheLeuValGlnTyrIleSerArgTyrThrGlnGluIlePheSerPheLeuIleSerLeuIle 550
PheIleTyrGluThrPheSerLysLeuIleLysIlePheGlnAspTyrProLeuGlnGlnThrTyrAlaProVal 575
ValMetLysProLysProGlnGlyProValProAsnThrAlaLeuPheSerLeuValLeuMetAlaGlyThrPhe 600
LeuLeuAlaMetThrLeuArgLysPheLysAsnSerThrTyrPheProGlyLysLeuArgArgValIleGlyAsp 625
PheGlyValProIleSerIleLeuIleMetValLeuValAspSerPheIleLysGlyThrTyrThrGlnLysLeu 650
SerValProAspGlyLeuLysValSerAsnSerSerAlaArgGlyTrpValIleHisProLeuGlyLeuLeuTyrArg 675
LeuPheProThrTrpMetMetPheAlaSerValLeuProAlaLeuLeuValPheIleLeuIlePheLeuGluSer 700
GlnIleThrThrLeuIleValSerLysProGluArgLysMetIleLysGlySerGlyPheHisLeuAspLeuLeu 725
LeuValValGlyMetGlyGlyValAlaAlaLeuPheGlyMetProTrpLeuSerAlaThrThrValArgSerVal 750
ThrHisAlaAsnAlaLeuThrValMetGlyLysAlaSerGlyProGlyAlaAlaAlaGlnIleGlnGluValLys 775
GluGlnArgIleSerGlyLeuLeuValSerValLeuValGlyLeuSerIleLeuMetGluProIleLeuSerArg 800
IleProLeuAlaValLeuPheGlyIlePheLeuTyrMetGlyValThrSerLeuSerGlyIleGlnLeuPheAsp 825
ArgIleLeuLeuLeuPheLysProProLysTyrHisProAspValProPheValLysArgValLysThrTrpArg 850
MetHisLeuPheThrGlyIleGlnIleIleCysLeuAlaValLeuTrpValValLysSerThrProAlaSerLeu 875
AlaLeuProPheValLeuIleLeuThrValProLeuArgArgLeuIleLeuProLeuIlePheArgGluLeuGlu 900
LeuGluCysLeuAspGlyAspAspAlaLysValThrPheAspGluGluAsnGlyLeuAspGluTyrAspGluVal 925
ProMetProVal 929

Figure 2   Complete sequence of murine band 3. The protein sequence is derived from the sequence of the clone isolated from a murine spleen cDNA library. The hydrophobic regions are underlined. Adapted from Kopito & Lodish, 1985 (17).

A comparison of murine band 3 sequence with sequences in the Dayhoff data base revealed no significant homology with any known protein sequence. Moreover, analysis of the putative membrane-spanning regions (as judged by hydrophobicity index) compared with other multihelical integral membrane proteins (such as bacteriorhodopsin, rhodopsin, bacterial K-ATPase, and lac permease) also revealed no significant homology.

Although the homology of the murine band 3 protein with those fragments of human band 3 that have been sequenced is relatively good, there are certainly differences between the two as judged by amino acid composition. One notable case is observed by comparing the amino acid composition of human TrCh-17 with that of the corresponding section of the murine protein. While the composition of the human polypeptide indicates two histidines in TrCh-17 (35), the murine sequence has none (17). Thus direct comparison of the murine band 3 sequence with the chemical and structural information obtained for the human protein is made with the caveat that there may exist significant differences between the two proteins. It should now be a considerably easier task to clone and sequence the human band 3 using fragments of the mouse clone as probes. Further comparisons may shed light on what domains must be conserved for the transport function. Also it would be of interest to sequence the clones of avian band 3 (36). Although anion exchange activity of the avian band 3 is similar to that of the human protein, the avian erythrocyte contains two distinct but similar band 3 proteins, and like the murine protein these proteins fail to bind glycolytic enzymes (37).

Chemical Modifications

Several covalent inhibitors of anion exchange have been used for structural studies, the most useful being the stilbene disulfonates (38). They have been shown to react exclusively with band 3 when added to intact erythrocytes at neutral pH, and when bound at a 1:1 stoichiometry they completely inhibit anion exchange (39). Two covalent stilbene disulfonate reagents have been used, the bifunctional reagent, 4,4'-diisothiocyanostilbene-2,2'disulfonate (DIDS), and the monofunctional reagent, 4-benzamido-4'-isothiocyanostilbene-2,2'disulfonate (BIDS). Both molecules can be tritiated by catalytic reduction with $[^3H_2]$ (40). The reactive site on band 3 is a lysine residue in the TrCh-17 fragment (14) and this corresponds to either lysine 558 or 561 in the murine sequence. Incubation with DIDS at pH 9.5 allows for cross-linking of TrCh-17 and ChC-35 fragments (41). The site of cross-linking has not been identified but it has been further localized to the PaC-28 fragment (71). Reductive methylation of the reactive lysine on the PaC-28 fragment reduces chloride transport by 75% (42), suggesting that it may have functional importance (see *Kinetic Model* section below).

The photoreactive reagent NAP-taurine is a substrate for transport by band 3

(43) and has been used as an affinity label. This molecule can inhibit competitively from the cytoplasmic side of the membrane ($K_I = 20$ μM). Using [^{35}S]-NAP-taurine, Knauf et al (44) mapped the site of covalent labeling to the TrCh-17 fragment. This labeling is greatly reduced by pretreatment with DIDS suggesting mutually exclusive sites. This conclusion is supported by the studies of Macara & Cantley (45), which showed competition between stilbene disulfonates and NAP-taurine for equilibrium binding to band 3.

Three other residues can be modified to inhibit activity and may be involved in the structure of the active site. Modification with a water-soluble carbodiimide has been shown to inactivate anion exchange and the modified carboxylates have been mapped to the fragment ChPa-8 (46). There are two aspartates in the sequence of the corresponding mouse fragment (Asp 625 and Asp 639). These two residues are near the two ends of a hydrophobic stretch that is known to span the bilayer (see *Active Site Residues* section below). Asp 639 is 10 residues from the exofacial papain cleavage site and is the most likely candidate for modification. It should be noted that there is an additional aspartate in the human sequence corresponding to residue 644 (30).

Phenylisothiocyanate (PITC) has also been shown to specifically inactivate anion transport (47, 48), and the site of covalent modification has been mapped to a lysine residue near the center of the P5 fragment sequenced by Brock et al (30). This site corresponds to lysine 608 in the mouse protein sequence and maps to the inside surface of the protein (see below), which makes it the only residue located on the cytoplasmic side whose modification has been demonstrated to inhibit function.

Reaction of band 3 with pyridoxal phosphate plus sodium borohydride inhibits anion transport (48a, 48b). The label was localized to TCh-17 and to ChC-35 fragments; however cells having band 3 labeled in only the TCh-17 fragment had normal anion transport (48b), arguing that the ChC-35 fragment must participate in anion exchange.

Wieth and coworkers (49), using phenylglyoxal, and Zaki (50), using 1,2-cyclohexanedione, have demonstrated that two different arginine residues are essential for anion exchange. This conclusion is supported by the pH profile showing titratable groups with pK$_A$s of approximately 11 (51). These arginine residues have been localized to the ChC-35 fragment (51a).

Although cysteine residues have been shown to be unnecessary for activity (52), studies of their location have been useful in determining the topology (52) and oligomeric structure (53, 54). Human band 3 contains six cysteine residues as judged by amino acid composition (28) and this number is in agreement with the predicted murine sequence. Five of these residues can be labeled with *N*-ethylmaleimide; three in the cytoplasmic domain and two in the membrane domain (52). The two in the membrane domain are localized to the ChC-35 fragment, and by further dissection of this fragment by trypsin they have been shown to be contained in a 9,000-dalton fragment (54a). Ramjeesingh et al (55) have located the sixth cysteine that does not react with *N*-ethylmaleimide in the TrCh-17 fragment approximately 40 residues from the chymotrypsin cleavage site.

These approximate locations for the cysteine residues in human band 3 are consistent with the cysteine locations in the murine sequence. The two cysteine residues in the ChC-35 fragment are located at positions 861 and 903 of the murine sequence, 42 residues apart, and the cysteine residue in the region of TrCh-17 is located 79 residues from the proposed chymotrypsin cleavage site. The suggestion by Ramjeesingh et al (55) that the two cysteine residues in ChC-35 are located at the N terminus of this fragment and that the glycosylation site is located at the C terminus of the PaC-28 fragment is inconsistent with the locations of these sites in the murine sequence and is also inconsistent with the location of the glycosylation site in human band 3 determined by a novel end-labeling technique (56) (see *Glycosylation* section below).

The three cysteine residues associated with the cytoplasmic domain of human band 3 have all been localized to the TrTr-20 fragment. One has been mapped to the N terminus of this fragment while another is within 50 residues of the C terminus (54). These locations are not consistent with the mouse sequence, which has cysteine residues at positions 152 (within the NTr-23 fragment), 215 (20 residues from the N terminus of TrTr-20), and 330 (76 residues from the C terminus of this fragment). These discrepancies are likely to be differences between species rather than errors in mapping since there is no cysteine residue in the first 201 residues of human band 3 (28).

Topology of Membrane Domain

The structural studies on human band 3 in combination with the hydropathy maps from the murine sequence allow a prediction of the tertiary structure of band 3. Chemical labeling and proteolytic digestion experiments demonstrate a minimum of four membrane-spanning regions while the hydrophobicity index of murine band 3 suggests the possibility of a maximum of 12 (17). On the basis of the evidence discussed below we propose eight helical membrane-spanning regions and a hairpin loop into the membrane as illustrated in Figure 3. This model is similar to the model proposed by Kopito & Lodish (17) with some differences at the C terminus. Those regions for which the location of the peptide with respect to the membrane is strongly supported by labeling and proteolytic digestion studies are indicated by solid lines while ambiguous assignments are indicated by dotted lines. Additional labeling studies with membrane-impermeant agents will be necessary to further test this model. Demonstrating the actual number of membrane-spanning helixes may only be definitively obtained by crystallography.

Strong evidence exists supporting the assignment of transmembrane region 1

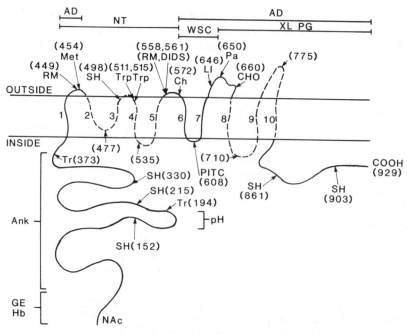

Figure 3 Topology and sites of chemical labeling and proteolytic cleavage of band 3. The model for the topology of band 3 is based on the established locations of sites of chemical labeling and proteolytic cleavage of human band 3 and on the hydropathy plot of the murine band 3 sequence. Solid lines crossing the bilayer are used for regions whose topology is established by experiment while dotted lines are used for regions that are proposed to cross the bilayer on the basis of hydrophobicity of the sequence. The cytoplasmic domain is drawn entirely with a solid line but no attempt is made to accurately represent the topology of this domain. Numbers in parentheses refer to the postulated residue number of a site based on the murine band 3 sequence. Sites whose locations have not been determined exactly are represented by straight lines over the region to which the location of the site has been limited.

Abbreviations used are: AD, adamantane diazirine reactive sites; Ank, ankyrin binding site; Ch, chymotryptic cleavage site; CHO, glycosylation site; COOH, carboxy terminus; DIDS, covalent stilbene disulfonate labeling site; GE, glycolytic enzyme binding site; Hb, hemoglobin binding site; LI, lactoperoxidase-catalyzed radioiodination site; Met, methionine; NAc, *N*-acetylated methionine at the amino terminus; NT, NAP-taurine labeling site; Pa, papain cleavage site; pH, pH-sensitive hinge region; PITC, phenyl-isothiocyanate labeling site; RM, site of reductive methylation of lysine; SH, cysteine; Tr, tryptic cleavage site; Trp, tryptophan; WSC, water-soluble carbodiimide reactive site; XL, site of DIDS cross-linking.

in Figure 3. The tryptic cleavage site that gives rise to fragment TrC-55 is located on the cytoplasmic surface of the cell since this site is only accessible from the inside (5, 24). By comparing the protein sequence determined at the tryptic cleavage site of human band 3 (29) with the murine sequence (17), the tryptic site corresponds to residue 373 in the murine sequence. It should be noted that in the murine sequence residue 373 is an asparagine while in the human sequence it is a lysine. A lysine located in a cyanogen bromide fragment TrCn-6 derived from the N terminus of TrCh-17 (corresponding to murine Lys 449) was shown to be exofacial by reductive methylation under conditions that favored extracellular labeling (57). In addition, using a lipophilic photoactivated cross-linking reagent, adamantane diazirine, Goldman found significant labeling of the TrCn-6 fragment implying that this fragment is in contact with the lipid bilayer (58). These data indicate that TrCn-6 must span the membrane. Consistent with this conclusion for the human protein, the murine band 3 has a stretch of 23 uncharged residues with only two polar side chains in this region (Val 424–Gly 447) suggesting a transmembrane alpha helix (see *Transport Site Conformational Changes* section below for discussion of the transmembrane alpha helix).

The topology of the region between Met 454 and Tyr 572 is more difficult to determine. Jennings & Nicknish (57) have demonstrated exofacial reductive methylation of a lysine residue on the C-terminal fragment derived from cyanogen bromide digestion of TrCh-17 to CnCh-11. In addition the C terminus of this fragment is known to face the outside of the cell since this is the location of the exofacial chymotryptic and one of the exofacial papain cleavage sites. External labeling by stilbene disulfonates also may occur here (at lysine 558 or 561). Since the two ends of CnCh-11 are outside the cell this fragment must span the membrane an even number of times or not at all. The human CnCh-11 fragment has been partially sequenced at its N terminus (29) and out of 37 residues there is only a single conservative difference from the corresponding murine sequence (17). The extreme hydrophobicity of CnCh-11 in both species is suggestive though not conclusive of a transmembrane region. Interestingly, adamantane diazirine did not significantly label CnCh-11 (58), suggesting that this putative transmembrane region is either in the bilayer but inaccessible to the lipid environment owing to helix packing or that the entire region bounded by CnCh-11 is outside the cell. Although the data are consistent with the latter possibility, the resistance of this portion of the polypeptide to extracellular proteases supports the former possibility.

In Figure 3 we suggest that CnCh-11 spans the bilayer twice (hydrophobic stretches 4 and 5). The region corresponding to hydrophobic stretches 2 and 3 has 32 sequential uncharged residues (Leu 459–Phe 490) with only 5 polar side chains. This stretch is not sufficiently long to span the bilayer twice in an alpha helical configuration and we have proposed a loop into the bilayer with a turn at Pro 477 near the middle of the hydrophobic stretch. The proposed transmembrane stretches 4 and 5 have charged groups in regions that would be within the bilayer if arranged in alpha helical transmembrane configurations. The possible involvement of these charged groups in anion transport is discussed below.

The assignment of transmembrane regions 6 and 7 is well supported. This region of the protein corresponds to fragment ChPa-8, which is generated by extracellular cleavage by chymotrypsin at the amino terminus and extracellular cleavage by papain at the carboxy terminus (24, 27). Brock et al (30) sequenced the 72 amino acids that coincide approximately with this fragment. Exofacial labeling with lactoperoxidase catalyzed radioiodination of a tyrosine residue near the carboxy terminus of this fragment (murine Tyr 646) but not the tyrosine in the middle of this fragment (murine Tyr 614). The latter tyrosine residue is flanked by three basic residues on each side, and the failure to label suggests a cytosolic location. As discussed above, phenylisothiocyanate labels a lysine residue corresponding to Lys 608 in the murine sequence and this site is only six residues from the cytosolic tyrosine residue (30). Thus ChPa-8 must cross the bilayer twice. The transmembrane regions indicated by stretches 6 and 7 (Figure 3) correspond to the hydrophobic sequences 589–606 and 622–642 of the murine protein. The hydrophobic stretch designated 7 has two Asp residues that are implicated in the transport site (see below).

The topography of PaC-28 is much more difficult to predict. The N terminus of this fragment is clearly outside the cell by the criteria of the presence of an exofacial papain cleavage site (27) and the location of the carbohydrate moiety within 10 residues of this site on both the human (56) and murine (17) proteins. The location of two cysteine residues can be postulated to be within 40 residues of the C terminus of human band 3 on the basis of the murine sequence. Rao (52) has shown that these residues are accessible only from the inside of the cell, implying that this portion of the sequence is cytosolic. These data demonstrate that the PaC-28 fragment spans the bilayer at least once while the hydrophobicity index of the murine sequence suggests the possibility of a maximum of five transmembrane regions.

We have proposed three transmembrane stretches in PaC-28. Since Cys 903 is inside the cell and the residues between 903 and the C terminus are mostly hydrophilic, it is likely that the C terminus (residues 903–929) is inside the cell. Reithmeier has shown that band 3 can be digested with carboxypeptidase Y in ghosts but not in intact cells consistent with this hypothesis (R. Reithmeier, personal communication). Also since the nearest cysteine (residue 861) is also inside the cell and the 42 residues between the cysteines are unlikely to form two helical spanning regions, it is likely that the region from cysteine 861 to the C terminus is all inside the cell. Thus, PaC-28 must cross the bilayer an odd number of times. In Figure 3 we suggest the three transmembrane stretches indicated by 8, 9, and 10. Stretch 8 corresponds to 23 uncharged residues with only two polar side groups in the murine sequence (676–698). Stretch 9 contains 19 consecutive nonpolar residues (724–742) and stretch 10 contains 24 uncharged residues with only four polar groups (801–824). There is no conclusive evidence that the loop between 8 and 9 is cytosolic or that the loop between 9 and 10 is exofacial.

Topology of Cytoplasmic Domain

Since there are no restraints imposed by transmembrane regions in the amino-terminal cytoplasmic domain of band 3, little is known about its tertiary structure. However, Low et al (59) proposed a model for the cytoplasmic domain and elements of this model are used in Figure 3. Based on electron microscopy studies (60) and hydrodynamic studies (61), the cytoplasmic domain is highly extended with an axial ratio of greater than 10. There appears to be a pH-regulated hinge region around the trypsin and S-cyanylation site (59). While it is clear that hemoglobin binds at the very N-terminal portion of the protein (62), and that the glycolytic enzymes also bind at or quite near this segment (33), the location of the ankyrin binding is still unresolved. Since antibodies against the cytoplasmic domain seem to partially compete with ankyrin for binding to band 3 (63) but not with hemoglobin (59), it is likely that ankyrin binding occurs C terminal to the glycolytic enzyme binding region.

POSTTRANSLATIONAL MODIFICATIONS

Glycosylation

Approximately 8% of the molecular weight of band 3 is carbohydrate and almost all of the carbohydrate is associated with a single asparagine-linked oligosaccharide chain (34, 64). The diffuse appearance of band 3 on SDS PAGE has been attributed to heterogeneity in the carbohydrate (65). The oligosaccharide chain can be cleaved from the protein by endoglycosidases such as endo B galactosidase (66) and Peptide-N-Glycosidase (G. J. Chin, D. G. Jay, unpublished results).

By cleavage with hydrazine into a large and small oligosaccharide chain, Tsuji et al (67, 68) were able to sequence part of the large-chain and all of the small-chain oligosaccharide. By using various endoglycosidases and analysis with high-atom-density mass spectroscopy, Fukuda et al (69, 70) elucidated the structure of the entire oligosaccharide chain for both adult and fetal forms of band 3.

The carbohydrate moiety was localized to the ChC-35 fragment (34) and more recently Jennings et al (71) further located it to the PaC-28 subfragment of ChC-35. These studies were done by proteolytic dissection of band 3 labeled on the carbohydrate by tritiated borohydride following galactose oxidase treatment. Recently the location of the glycosylation site has been mapped to 280 ± 30 residues from the C terminus of band 3 by a novel procedure for mapping sites of interest in the sequence of proteins (56). The procedure exploits the ability to end-label proteins (72). End-labeled TrC-55 and ChC-35 fragments were subjected to partial proteolysis and peptides containing the glycosyl moiety were isolated by lectin sepharose chromatography. The smallest band observed by SDS PAGE and autoradiography contains both the radiolabeled N terminus of the original fragment and the glycosyl moiety; the size of the peptide defines the location in the sequence of the glycosylation site. This site is located just C terminal to the exofacial papain cleavage site. As correlative evidence, Kopito & Lodish (17) have identified two potential glycosylation sites in the sequence of murine band 3. One of these is located in ChPa-8 and has been shown not to be glycosylated in human band 3 (30). The other potential site is just C terminal to the exofacial papain cleavage site in good agreement with the empirical result determined by mapping the glycosylation site of human band 3.

Phosphorylation

Drickamer (73) determined that phosphorylation of band 3 occurred primarily on the N-terminal cytoplasmic domain although some label in the membrane domain was reported. More recently Waxman (74) reported that phosphorylation was exclusively at the amino-terminal cytosolic domain. A major site of phosphorylation is at Tyr 8 (75). It is at present not known whether phosphorylation plays any physiological role. Macara & Cantley (20) suggested the possibility that because of its location in the glycolytic enzyme-binding domain, phosphorylation may play some role in regulating association with these enzymes. It may be noteworthy that in mouse band 3, which fails to associate with glycolytic enzymes, no potential tyrosine phosphorylation site is present in this region of the molecule.

Methylation

Band 3 can be specifically methylated by freeze-thawing cells in the presence of S-adenosyl-[methyl-H^3] methionine (76). This methylation appears to occur on internal aspartates and one site of methylation is the hinge region between membrane and cytoplasmic domains. The physiological significance of this covalent modification is unclear.

BIOSYNTHESIS AND ASSEMBLY

The biosynthesis of human band 3 has been difficult to study because band 3 is not actively synthesized in mature erythrocytes and large quantities of synchronous human erythroid precursors cannot be obtained. Work on nonhuman erythroid precursors, however, has yielded much information. Working with phenylhydrazine-treated anemic rabbits, Koch et al (77, 78) demonstrated that red cell membrane proteins were asynchronously synthesized, the largest proteins being made during the earliest stages of erythroid differentiation. Using an avian erythroid system, Weiss & Chan (79) could demonstrate band 3 synthesis in polychromatophilic erythroblasts. By immunochemical techniques Foxwell & Tanner (80) showed that band 3 was inserted in the plasma membrane between the polychromatophilic normoblast and reticulocyte stages of differentiation. Murine band 3 biosynthesis was investigated by the time course of incorporation of [S^{35}]-methionine (injected in vivo) into band 3 (81). Band 3 was synthesized after the bulk of spectrin and actin, and its biosynthesis was complete before reticulocyte proteins were synthesized.

Braell & Lodish (32) demonstrated that in vivo murine band 3 is inserted into the endoplasmic reticulum cotranslationally in its mature orientation as judged by proteolytic specificity. This species contains a carbohydrate moiety that is sensitive to endo-glycosidase H implying a high-mannose core structure. Within 30 minutes this form is converted to an endo H–resistant form and after 40 minutes the protein is found at the plasma membrane. By isolating mRNA from Friend erythroleukemia cells (82) or from anemic mouse spleen (83), Braell & Lodish (32) have studied membrane translocation of murine band 3 in vitro. They established that band 3 is synthesized on membrane-bound polysomes and that there is no cleavable signal sequence. By determining that dog pancreas microsomes need not be added until the entire cytoplasmic domain is synthesized, Braell & Lodish inferred that there exists an internal signal sequence near the N terminus of the membrane-associated domain.

TRANSPORT SITE STRUCTURE: A MODEL

The location of the residues in the primary sequence that interact with covalent inhibitors of anion exchange along with the predicted topography of the polypeptide chain places certain restrictions on the arrangement of these residues to form the anion transport site. An additional restriction can be imposed if one assumes that the transmembrane regions are in an alpha helical configuration. The side groups of sequential residues of an alpha helix project from the helical axis at intervals of 100° angles and are displaced 1.5 Å along the direction of the axis. Thus in the folded protein, side groups that are at intervals of 3–4 residues apart in the sequence are in close contact at the same side of the helix.

The assumption that the transmembrane regions of band 3 are in an alpha helical configuration is supported by circular dichroism spectra (4, 84). The limit pepsin and papain fragments of band 3 that remain associated with the membrane have maximum molecular weights of approximately 4000 and are 86–94% alpha helical as judged by circular dichroism studies (84). In order for a peptide fragment to span the lipid bilayer of a cell membrane in an alpha helical configuration the helix must be approximately 45–50 Å in length (i.e. 30–35 residues or about 4000 daltons). The data of Oikawa et al (84) are consistent with this structure. In addition, since the middle 30 Å of the width of the bilayer is composed of hydrophobic fatty acid chains, the side groups of the middle 20 residues of the helix should be nonpolar. The two ends of the hydrophobic region are expected to contain clusters of polar residues for interaction with the lipid head groups. These criteria are certainly satisfied by the proposed transmembrane regions 1, 6, 8, 9, and 10 (Figure 3), which have stretches of 20–24 mostly nonpolar residues with no charged side groups framed by clusters of polar residues. As discussed above, the region indicated by 2 and 3 is suggested to loop into the membrane because of its unusually long stretch of 32 uncharged and mostly nonpolar residues. On the other hand, the regions indicated by 4, 5, and 7 all have long runs of hydrophobic residues but with charged side groups fewer than 20 residues apart near the two ends of the hydrophobic stretches. Although the charged groups suggest that these are not true transmembrane regions, the evidence that 7 spans the bilayer is quite strong. The argument for assigning regions 4 and 5 as transmembrane alpha helices is discussed below.

Active Site Residues

A model for the arrangement of the transmembrane stretches to form an anion transport site is proposed in Figure 4. The left half of the figure is a view of the model as seen from the cytosolic side of the membrane and the right half is a view from the outside of the cell. The cylinders represent transmembrane alpha helices with the same numbering system as specified in Figure 3, and the heavy solid line represents the portions of the polypeptide that loop out of the membrane. The arrows indicate the direction from the N to the C terminus. The amino-terminal half of the protein including stretches 1, 2, and 3 has been omitted and the last 70 residues on the cytoplasmic side at the carboxy terminus have been omitted. The relative angles of projection from the helical axis for the side groups in each transmembrane section were calculated assuming a 100° turn per residue along the full length of the helix from cytosolic to extracellular side. The positively charged residues (Arg, Lys, and His) are shaded and negatively charged residues (Glu and Asp) are unshaded.

Although the order of the helices in Figure 4 is somewhat arbitrary, the arrangement presented was chosen on the basis of the projection of charged groups from the helical axes. Of particular interest are the side groups on helixes 4, 5, 6, and 7. In all four cases, the polar and charged residues map to one face of the helix. For example, if region 4 is arranged as an alpha helix, Arg 509, which is near the extracellular side of the hydrophobic stretch, protrudes from the axis at the exact same angle as Glu 527, which is 24 Å away near the cytosolic side of the hydrophobic stretch. Thus, both of these residues could face the same transport cavity, one at each side of the membrane.

The argument for region 5 being transmembrane relies very heavily on the observation that the charged groups all protrude from the same face of the helix.

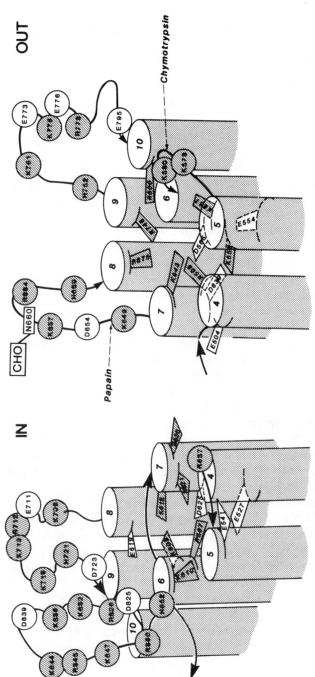

Figure 4 A model for the arrangement of transmembrane regions of band 3. The left half of the figure is a view of band 3 as seen from the cytosolic side of the membrane and the right half is a view from the outside of the cell. The cylinders represent transmembrane alpha helixes with the same numbering system as specified in Figure 3 and the heavy solid line represents the portions of the polypeptide that loop out of the membrane. The arrows indicate the direction from N to C terminus. The amino-terminal half of the protein including stretches 1, 2, and 3 has been omitted and the last 70 residues on the cytoplasmic side at the carboxy terminus have been omitted. The location of all the charged amino acids are indicated by their single-letter codes (D=Asp, E=Glu, H=His, K=Lys, R=Arg) and a number indicating the position in the sequence. Those residues whose side groups are proposed to project from the transmembrane helixes are indicated by rectangles. The relative angles of projection from the helical axis for the side groups in each transmembrane section were calculated assuming a 100° turn per residue along the full length of the helix from cytosolic to extracellular side. The positively charged residues (R, K, and H) are shaded and negatively charged residues (E and D) are unshaded.

The distance from the cluster of polar molecules ending with Glu 541 to the intermittent series of charged residues beginning with Glu 554 is only 19.5 Å along the helical axis, less than the width of the hydrophobic region of a cell membrane. However, as pointed out by Kopito & Lodish (17), the series of charged groups near the carboxy-terminal end of this region, Glu 554, Lys 558, Lys 561, and Asp 565, all map within an arc of 60° from the helical axis and are within 10.5 Å of each other along the length of the axis. In addition, almost all the other polar but uncharged side groups in region 5 protrude from the same face of the helix as this charged cluster (not shown in Figure 4). Thus, although region 5 is relatively polar suggesting an extramembrane location, when folded into an alpha helix, this region can be placed in the membrane such that the charged residues all face a central hydrophilic cavity produced from the polar side groups of helixes 4, 6, and 7 (Figure 4). These polar residues are thus shielded from the hydrophobic milieu of the lipid bilayer.

An additional reason for placing region 5 in the membrane is the location of the DIDS reactive site in the charged cluster near the carboxy terminus of this region (Lys 558 or Lys 561). Fluorescence resonance energy transfer studies of Rao et al (8) indicate that this site must protrude into the membrane since the distance from probes located at the cytoplasmic cysteine residues to the stilbene disulfonate binding site was determined to be less than the width of the bilayer (34–42 Å). In addition, Falke et al (85) showed that band 3, which had been extensively digested with papain such that the largest remaining fragments were in the range of 4000 daltons (i.e. which had conditions similar to those of Oikawa et al, Ref. 84 discussed above), retained the stilbene disulfonate–inhibitable Cl^- binding site, although the binding affinity was reduced. Thus, the residues that make up the anion transport site (and stilbene disulfonate binding site) must be buried in a region inaccessible to the nonspecific protease, papain. The relative shortness of the stretch of hydrophilic residues that loop out of the membrane to form the bend between helix 4 and helix 5 suggests that these helixes are close together.

Region 6 is proposed to make up part of the transport site since modification of Lys 608 by PITC inhibits anion exchange (30). This residue is in a positively charged cluster at the cytosolic surface, and when arranged in a helical configuration, the side groups of Arg 607, Lys 608, and Lys 610 all protrude from the same face of the helix. In the model presented in Figure 4 this cluster is proposed also to face the central hydrophilic cavity.

Helix 7 is of particular interest since it is an established transmembrane region which contains a carboxylate group whose modification inhibits anion exchange (46). Only two carboxylate groups exist in this region, Asp 625 on the cytosolic half of the hydrophobic stretch and Asp 639, near the extracellular surface. These two residues map to the same face of helix 7 within a 40° arc and 21 Å apart along the length of the helix. This distance is less than the width of the hydrophobic region of a lipid bilayer indicating that one or both of these groups is buried in the membrane. In addition, the other charged groups of this region, Lys 618, Arg 620, and Arg 621 on the cytosolic side and Lys 643 on the extracellular side, also map to the same face of helix 7. Thus, in the model presented in Figure 4, the side of helix 7 containing these charged residues is portrayed as facing the central hydrophilic cavity. It seems likely that the essential carboxylate that is modified by the extracellularly added water-soluble carbodiimide is Asp 639.

The arrangement of regions 8, 9, and 10 is more arbitrary. The hydrophobic stretches of these three helices do not have charged residues that would map to sites deep in the membrane. However, in all three cases the end of the hydrophobic stretch that maps to the extracellular surface contains an Arg residue. An arginine residue essential for anion exchange has been located in the region of band 3 containing these three stretches (51) and the arrangement in Figure 4 suggests that Arg 675, Arg 748, and/or Arg 800 might contribute to the hydrophilic cavity.

No evidence for an essential role for the residues in the regions indicated by 1, 2, and 3 has thus far been reported and this section of the protein has been omitted from the model in Figure 4. These regions may aid in shielding the polar groups of helices 4, 5, and 7 from the surrounding lipid (see OLIGOMERIC STRUCTURE section below).

The location of tryptophan residues in band 3 by fluorescence techniques is also consistent with the model in Figure 4. By resonance energy transfer to fluorescent fatty acids, Kleinfeld et al (86) showed that two major clusters of tryptophan residues exist: one cluster extends 10 Å into the cytosol and the other is located near the extracellular surface. In the models portrayed in Figures 3 and 4, six tryptophan residues are located on the cytosolic side of band 3 (four between residues 89 and 119 of the N-terminal cytoplasmic domain and two at 849 and 866 of the C-terminal cytoplasmic region), four are located in the transmembrane region, and one is on the extracellular loop between helices 7 and 8. The four residues in the transmembrane regions are on the extracellular half of helices 4, 8, and 9, about 10–15 Å from the extracellular surface in the model of Figure 4. These locations are consistent with the predictions of Kleinfeld et al (86). This arrangement is also consistent with the observations of Macara et al (87) that hydrophilic collisional quenchers have little effect on tryptophan fluorescence when added to the extracellular surface but cause significant quenching when added to the cytosolic side. The fact that stilbene disulfonates quench 40–50% of the tryptophan fluorescence of band 3 (8) further argues that the DIDS binding site protrudes into the membrane allowing a close approach between DIDS and the transmembrane tryptophan residues.

Positively Charged Funnel

The striking feature of the model in Figure 4 is that when viewed from either side of the membrane, the protein appears as a positively charged funnel that filters into a small cluster of 2 or 3 negatively charged residues. The loops out of the membrane on the extracellular surface between helixes 5–6, 7–8, and 9–10 are all positively charged and may contribute to repulsion of cations and selection for anions. Similarly, the loops into the cytosol between helixes 8–9 and at the carboxy terminus of 10 have a high density of positive charge. The lip of the hydrophilic cavity on the extracellular side is surrounded by the positively charged residues Arg 509, Lys 558, Lys 561, Lys 643, Arg 675, and Arg 748 and Arg 800. These residues should act as a further selection for anions. Deeper into the cavity, the negative charge of residues Asp 639 and Glu 554 are proposed to block transport and act as part of the gating mechanism (see *Kinetic Model* section below). When viewed from the cytosolic side a very similar lip of positively charged residues (Arg 607, Lys 608, Lys 610, Lys 618, Arg 620, and Arg 621) is seen, and deeper into the cavity are the negatively charged residues Glu 527 and Asp 625. Thus as one might expect for an exchange mechanism in which anions are sequentially transported in opposite directions across the membrane, the structures of the transport sites approached from opposite sides of the membrane are quite similar.

Kinetic Model

Kinetic data are overwhelmingly in support of a ping pong mechanism for anion exchange (88; for review see Ref. 20). In such a model, an anion enters the transport site from the extracellular surface and is transported and released on the cytosolic side. Then a cytosolic anion binds to the vacated site and is transported to the outside of the cell. When this anion is released, a transport site is available on the outside to begin a new cycle. No net charge movement occurs since a second anion cannot enter the cell until an anion exits the cell to regenerate the extracellular transport site. Consistent with this model, net anion movement through band 3 is at least 10^4 times slower than anion exchange (22), and this slow net transport is thought to be due to leakage of anions through the gate rather than the return of empty transport sites (see Ref. 20 for discussion). Also consistent with this model is the ^{35}Cl NMR data of Falke & Chan (89) and Falke et al (90). They showed by direct equilibrium binding studies that a single Cl^- binding site can be approached from either side of the membrane and that stilbene disulfonates added to the extracellular side prevent Cl^- binding from both sides of the membrane. Furthermore, they showed that Cl^- binding to and release from the transport site from either side of the membrane was fast compared to the rate of the transport step.

The surprising prediction of the ping pong model is that the transport site of band 3 must be capable of existing in two different stable conformations that do not interconvert in the absence of a transportable anion (at least on the time scale of transport experiments). In order to explain the hysteretic nature of the transport site and other features of the exchange kinetics, a gating mechanism involving formation of two alternative salt bonds between a carboxylate group and two positively charged residues at the transport site was proposed (20, 91). The essence of the model is that an anion entering the transport site from the extracellular side would break the salt bond by competing with the carboxylate for binding to the cationic group and that the carboxylate would reform a salt bond with a second nearby cationic group. This conformational change would allow movement of the anion through the transport system to the cytosolic side of the membrane but would block access to the transport site by any additional external anions. The original conformation could only be restored if an anion entered the transport site from the cytosolic side and broke the newly formed salt bond to allow the original salt bond to reform (i.e. the reverse of the first reaction). During this step, transport to the outside of the cell would occur and approach to the transport site from the cytosolic side would be blocked. A similar somewhat more complex model involving alternating salt bond formation to explain the ping pong kinetics was proposed by Wieth & Brahm (92).

The salt bond models are consistent with the location of aspartate and glutamate residues deep in the hydrophilic cavities of both the extracellular and cytosolic surfaces of band 3 (Figure 4). It is difficult to predict what positively charged residues might be involved in these proposed salt bonds. Since no lysine or arginine residues exist in the center of the hydrophobic stretches, it seems likely that the intracellular and extracellular cavities form independent salt bonds. pH titration data suggests the involvement of extracellular carboxylates and arginine residues (49, 51) in the transport mechanism. Involvement of a lysine residue with an unusually high pK_A (>11) due to participation in a salt bond is also possible. As discussed above, Asp 639 is likely to be the essential carboxylate modified by the water-soluble carbodiimide (46), and either Arg 675, Arg 748, or Arg 800 is likely to be the essential residue detected by labeling with phenylglyoxal (51). These residues are good candidates for formation of salt bonds at the extracellular cavity. From equilibrium binding measurements detected by ^{35}Cl NMR, Falke & Chan (93) show that phenylglyoxal blocks Cl^- binding at the transport site but argue that 1,2-cyclohexanedione reacts with a different arginine residue in a cavity through which Cl^- must pass to access the transport site.

The two lysine residues that form the DIDS cross-link between NCh-60 and ChC-35 are also of potential interest. The residue on helix 5 that readily reacts with DIDS (Lys 558 or Lys 561) can undergo reductive methylation without inhibition of anion exchange (42). It is possible that the methylated lysine might still function in formation of a salt bond. However, methylation of the lysine on PaC-28 that forms the other end of the cross-link does inhibit anion exchange. This site only reacts with DIDS at high pH and its location in the sequence is unknown. The rather large size of the DIDS molecule (>20 Å) along with the length of the lysine side groups indicates that the distance between the protein backbones that form the two ends of this cross-link may be quite large (> 30 Å). Three lysine residues are present in the extracellular domain of the murine PaC-28 fragment in the model of Figure 4. All three are in regions proposed to loop out of the membrane. It remains to be determined whether the DIDS-reactive lysine in the PaC-28 fragment is conserved between human and murine species. Further location of this site in the human band 3 sequence and verification of its essential role in anion exchange will provide important information about the transport site structure. Three residues on the extracellular surface in the model in Figure 4 (Lys 580, Lys 643, and Lys 649) could potentially form this bond. Since DIDS analogues lacking the lysine-reactive isothiocyanate groups are equally potent inhibitors, it is possible that none of the DIDS-reactive sites functions directly in the transport mechanism.

Less is known about the charged residues on the cytosolic surface. As discussed above, only Lys 608 has been implicated as essential for transport due to chemical labeling. However, a cytosolic group with a pK_A of 6.1 must be deprotonated for transport activity (94). This could be a carboxylate in an unusually hydrophobic environment such as Glu 527 or Asp 625. In the model in Figure 4, Arg 607 and Arg 621 appear to be good candidates for formation of salt bonds with either Glu 527 or Asp 625 to regulate transport.

Transport Site Conformational Changes

Because of the relatively rapid rate of anion exchange (10^5 sec^{-1} at 37°), it is unlikely that the transport cycle involves a conformational change in which a major movement in the protein backbone occurs. One attraction of the alternating salt bond model is that little if any movement in the protein backbone is necessary. It has been difficult to assess protein conformational changes associated with Cl^- movement because of the rapid rate of transport. However, using temperature jump techniques with a fluorescent stilbene disulfonate probe for the transport site, it has been possible to observe a rapid binding step followed by a protein conformational change (95). The conformational change was much slower than the Cl^- transport rate and could possibly reflect an attempt to transport the much bulkier fluorescent anion. Interestingly, when eosin maleimide was reacted covalently with the extracellular transport site its fluorescence was shielded from quenching by extracellular Cs^+ but was quenched by Cs^+ added to the cytosolic surface, suggesting a partial transport of this bulky anion (87). Addition of stilbene disulfonates to band 3 had no detectable effect on the circular dichroism spectra (84), and only a small effect on the tryptophan fluorescence once the quenching due to energy transfer was eliminated (87). These results also argue against a major protein conformational change.

OLIGOMERIC STRUCTURE

Band 3 is known by sedimentation analysis to exist as a dimer in nonionic detergents (4, 96). This interaction can be mediated by the membrane domain since the cytoplasmic domain can be proteolytically removed with retention of the dimeric state (26). Although the predominance of the dimeric species in the membrane is still subject to question, there exists a great deal of circumstantial evidence for such predominance. Freeze fracture analysis of red cell membranes has shown the existence of intramembranous particles (60), and these have been quantitated to approximately 4×10^5 particles per cell (96). These particles are thought to be made up largely of band 3 protein, and since the number of band 3 molecules (1.2×10^6/cell) exceeds the number of particles, it appears that each particle contains more than one band 3 molecule. Steck (53) has demonstrated that band 3 can be cross-linked to dimers by oxidation with Cu^{2+} o-phenanthroline but he also observed trimers and tetramers. Wang & Richards, using DTBP (97), and Reithmeier & Rao, using Cu^{2+} o-phenanthroline at 0 degrees C (98), showed cross-linking of band 3 exclusively to dimers. To argue against the likelihood of dimer formation by random collision, Khiem & Ji (99) utilized a photoactivatable cross-linker with a half-life in the millisecond timescale and also demonstrated dimer cross-linking. Fluorescence resonance energy transfer measurements between probes at the ion transport sites have verified the dimeric association and have also suggested the presence of tetrameric species in the membrane (91). Evidence that a tetrameric species forms a water channel through planar lipid bilayers was presented by Benz et al (100) supporting other evidence from Solomon's laboratory that band 3 forms a water channel in the red cell membrane (101).

The locations of cross-linked residues have been studied. Reithmeier & Rao (98) demonstrated that all three cysteine residues in the cytoplasmic domain can randomly participate in the Cu^{2+} o-phenanthroline cross-linking, suggesting that all three cysteines are at the dimer interface. Using impermeant cross-linking reagents Staros & Khakkad (102) have provided evidence that the TrCh-17 fragment of one subunit is in close contact with the ChC-35 fragment of the other. More recently Jennings & Nicknish (103) localized a site of intermolecular cross-linking with bis-sulfosuccinimidyl suberate, another impermeant reagent, near the exofacial chymotryptic cleavage site, suggesting that this region of the membrane domain is at the dimer subunit interface.

Whether or not the dimeric association is essential for the transport function of band 3 has not been determined. Small anions are clearly transported by the two halves of the dimer independently (20, 45). In any event, the dimeric association is extremely tight even in solutions of nonionic detergents: batches of solubilized band 3 dimers separately tagged with different fluorescent reporter groups failed to randomize upon mixing for several days as judged by resonance energy transfer measurements (unpublished results of I. Macara, J. Lytton, L. Cantley).

A model for the possible arrangement of transmembrane helixes with respect to the dimer interface is proposed in Figure 5. The monomer structure is the same as in Figure 4. The model takes into account the locations of cross-linking sites discussed above. The major region of dimer interaction is proposed to be helixes 5 and 6. The extracellular loop between these two transmembrane regions is the site of extracellular chymotrypsin cleavage and is also the site where bis-sulfosuccinimidyl suberate covalently cross-links the two subunits (103). The model is also consistent with the observation of Staros & Khakkad (102) that TrCh-17 of one subunit cross-links to ChC-35 of the other. The extracellular loops at the ends of transmembrane regions 4 and 5 (located in TrCh-17) are close to the extracellular loop between 9 and 10 (located in ChC-35). Regions 1, 2, and 3 have been omitted (these regions are in TrCh-17) but may also contribute to the cross-linking.

One attraction of the arrangement in Figure 5 is that the extracellular loops between 7–8 and 9–10 from the two subunits may be viewed as forming a single large positively charged funnel allowing access to both transport sites. Such an arrangement has been suggested on the basis of the apparent steric interactions that occur when stilbene disulfonates with bulky groups attached at the 4 and 4' ends of the stilbene moiety bind to adjacent sites of a dimer (91). Such an interaction is compatible with the relatively close proximity of the stilbene disulfonate binding sites of adjacent subunits (28–45 Å; Ref. 91). Furthermore, immobilized monomers of band 3 failed to bind stilbene disulfonates unless dimerization was allowed to proceed, suggesting that interactions between two halves of the dimer are essential for formation of the stilbene disulfonate binding site (104). Thus, it seems plausible that the two transport sites per dimer are close to the dimer interface and even possible that residues from both dimers contribute to these sites. In the model presented in the right half of Figure 5, the two rectangles indicate the approximate sizes of the DIDS molecules (20 Å) compared with the diameters of the cross-sections of the transmembrane alpha helixes (5 Å). One end of the DIDS molecule is placed near the reactive lysine residues of helix 5. The other end is likely to be near the loop between helix 7 and 8 since papain cleavage in this loop affects which lysine residue is involved in cross-linking (see discussion above).

When viewed from the cytosolic side (left half of Figure 5), one can also envision a structure in which the extramembranous loops from the two subunits of the dimer cooperate to form a positively charged funnel to the transport sites. Information from two-dimensional crystals of band 3 would be extremely valuable in assessing the proposed dimeric cavity model. Thus far no crystals

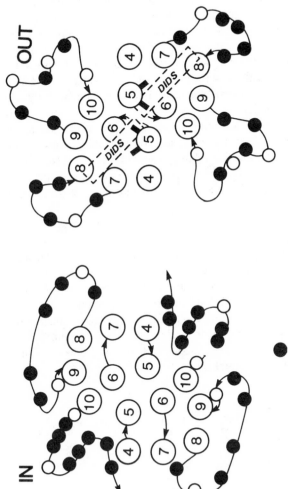

Figure 5   A model for the dimeric association of band 3. The left half is the dimer viewed from the cytosolic side of the membrane and the right half is the view from the extracellular side. The large circles with numbers represent the transmembrane alpha helixes with the same numbering as in Figures 3 and 4. The small circles represent the approximate locations of positively charged (shaded) and negatively charged (unshaded) residues on the extramembrane loops. The charged groups on the helixes have been omitted with the exception of Lys 558 and Lys 561 of helix 5 where DIDS reacts. The dimer interface is between transmembrane regions 5 and 6 of adjacent subunits. The proposed locations of the DIDS binding sites are indicated by the two rectangles. The size of the rectangles indicates the approximate dimensions of the DIDS molecule relative to the cross-sectional diameters of the helixes. See the text for further explanation.

are available. For a more detailed discussion of the oligomeric structure, see review by Jennings (23).

FUTURE DIRECTIONS

Now that the cDNA for band 3 has been cloned, exciting new information should be obtainable by expression of genes that have undergone site-specific mutagenesis. The model in Figure 4 makes predictions as to what residues are likely to be important in transport. It is possible, for example, that mutation of the Asp and Glu residues that are in the hydrophobic sections of helixes 4, 5, and 7 to small uncharged residues will convert band 3 into an anion channel. The effect of mutations on the insertion of band 3 into the membrane could also reveal important information about this poorly understood phenomenon. Further comparisons of residues conserved between species might also add to our understanding of the transport sites. Finally, additional labeling studies are still needed to conclusively determine the folding pattern of band 3 with respect to the membrane, and crystallization of the protein in either two or three dimensions would be invaluable for assessing the structure of the transmembrane helixes.

ACKNOWLEDGMENTS

This work was supported by grant GM35912 awarded to LC from the National Institutes of Health. Daniel Jay is a Junior Fellow in the Harvard Society of Fellows and LC is an Established Investigator of the American Heart Association. We wish to thank Guido Guidotti and Reinhart Reithmeier for many helpful discussions. We thank Ron Kopito for a copy of his manuscript containing the murine sequence prior to acceptance. We also thank M. Jennings, J. Kyte, J. Falke, T. Osawa, H. Passow, J. Brahm, A. Rothstein, and R. Reithmeier for preprints and unpublished information.

Literature Cited

1. Fairbanks, G., Steck, T. L., Wallach, D. F. H. 1971. *Biochemistry* 10:2606–17
2. Cabantchik, Z. I., Rothstein, A. 1974. *J. Membr. Biol.* 15:207–26
3. Ho, M. K., Guidotti, G. 1975. *J. Biol. Chem.* 250:675–83
4. Yu, J., Steck, T. L. 1975. *J. Biol. Chem.* 250:9170–75
5. Yu, J., Steck, T. L. 1975. *J. Biol. Chem.* 250:9176–84
6. Lukacovic, M. F., Feinstein, M. B., Sha'afi, R. I., Perrie, S. 1981. *Biochemistry* 20:3145–51
7. Findlay, J. B. C. 1974. *J. Biol. Chem.* 249:4398–403
8. Rao, A., Martin, P., Reithmeier, R. A. F., Cantley, L. C. 1979. *Biochemistry* 18:4505–16
9. Steck, T. L., Weinstein, R. S., Straus, J. H., Wallach, D. F. H. 1970. *Science* 168:255–57
10. Ross, A. H., McConnell, H. M. 1978. *J. Biol. Chem.* 253:4777–82
11. Cabantchik, Z. I., Volsky, D. J., Ginsburg, H., Loyter, A. 1980. *Ann. NY Acad. Sci.* 341:444–54
11a. Lieberman, D. M., Reithmeier, R. A. F. 1983. *Biochemistry* 22:4028–33
12. Crandall, E. D., Bidani, A. 1981. *J. Appl. Physiol.* 50:265–71

13. Crandall, E. D., Mathew, S. J., Fleischer, R. S., Winter, H. I., Bidani, A. 1981. *J. Clin. Invest.* 68:853–62
14. Grinstein, S., Ship, S., Rothstein, A. 1978. *Biochim. Biophys. Acta* 507:294–304
15. Shaklai, N., Yguerabide, J., Ranney, H. M. 1977. *Biochemistry* 16:5593–97
16. Hargreaves, W. R., Giedd, K. N., Verkleij, A., Branton, D. 1980. *J. Biol. Chem.* 255:11965–72
17. Kopito, R. R., Lodish, H. F. 1985. *Nature.* 316:234–38
18. Guidotti, G. 1980. In *Membrane Transport in Erythrocytes, Alfred Benzon Symp. 14*, ed. U. V. Lassen, H. H. Ussing, J. O. Wieth, pp. 300–8. Copenhagen: Munksgaard
19. Steck, T. L. 1978. *J. Supramol. Struct.* 8:311–24
20. Macara, I. G., Cantley, L. C. 1983. In *Cell Membranes, Methods and Reviews,* ed. E. Elson, W. Frazier, L. Glaser, 1:41–87. New York: Plenum
21. Jennings, M. L. 1985. *Ann. Rev. Physiol.* 47:519–33
22. Knauf, P. A. 1979. *Curr. Top. Membr. Transp.* 912:249–363
23. Jennings, M. L. 1984. *J. Membr. Biol.* 80:105–17
24. Drickamer, L. K. 1976. *J. Biol. Chem.* 251:5115–23
25. Jenkins, R. E., Tanner, M. J. A. 1977. *Biochem. J.* 161:139–47
26. Reithmeier, R. A. F. 1979. *J. Biol. Chem.* 254:3054–60
27. Jennings, M. L., Adams, M. F. 1981. *Biochemistry* 20:7118–23
28. Kaul, R. K., Murthy, S. N. A., Reddy, A. G., Steck, T. L., Kohler, H. 1983. *J. Biol. Chem.* 258:7981–90
29. Mawby, W. J., Findlay, J. B. C. 1983. *Biochem. J.* 205:465–75
30. Brock, C. J., Tanner, M. J. A., Kempf, C. 1983. *Biochem. J.* 213:577–86
31. Braell, W. A., Lodish, H. F. 1981. *J. Biol. Chem.* 256:11337–44
32. Braell, W. A., Lodish, H. F. 1982. *Cell* 28:23–31
33. Murthy, P. S. N., Liu, T., Kaul, R. K., Kohler, H., Steck, T. L. 1981. *J. Biol. Chem.* 256:11203–8
34. Drickamer, L. K. 1978. *J. Biol. Chem.* 253:7242–48
35. Steck, T. L., Koziarz, J. J., Singh, M. K., Reddy, G., Kohler, H. 1978. *Biochemistry* 17:1216–22
36. Cox, J. V., Moon, R. T., Lazarides, E. 1985. *J. Cell Biol.* 100:1548–57
37. Jay, D. G. 1983. *J. Biol. Chem.* 258:9431–36
38. Cabantchik, Z. I., Knauf, P. A., Rothstein, A. 1978. *Biochim. Biophys. Acta* 515:239–303
39. Lepke, S., Fasold, H., Pring, M., Passow, H. 1976. *J. Membr. Biol.* 29:147–77
40. Ship, S., Shami, Y., Breuer, W., Rothstein, A. 1977. *J. Membr. Biol.* 33:311–24
41. Jennings, M. L., Passow, H. 1979. *Biochim. Biophys. Acta* 554:498–519
42. Jennings, M. L. 1982. *J. Biol. Chem.* 257:7554–59
43. Cabantchik, Z. I., Knauf, P. A., Ostwald, T., Markus, H., Davidson, L., et al. 1976. *Biochim. Biophys. Acta* 455:526–37
44. Knauf, P. A., Breuer, W., McCulloch, L., Rothstein, A. 1978. *J. Gen. Physiol.* 72:631–49
45. Macara, I. G., Cantley, L. C. 1981. *Biochemistry* 20:5695–701
46. Bjerrum, P. J. 1983. In *Structure and Function of Membrane Proteins,* ed. E. Quagliariello, F. Palmieri, pp. 107–15. Amsterdam: Elsevier
47. Sigrist, H., Kempf, C., Zahler, P. 1980. *Biochim. Biophys. Acta* 597:137–44
48. Kempf, C., Brock, C., Sigrist, H., Tanner, M. J. A., Zahler, P. 1981. *Biochim. Biophys. Acta* 641:88–98
48a. Cabantchik, I. Z., Balshin, M., Breuer, W., Rothstein, A. 1975. *J. Biol. Chem.* 250:5130–36
48b. Nanri, H., Hamasaki, N., Minakami, S. 1983. *J. Biol. Chem.* 258:5985–89
49. Wieth, J. O., Bjerrum, P. J., Borders, C. L. 1982. *J. Gen. Physiol.* 79:283–312
50. Zaki, L. 1981. *Biochem. Biophys. Res. Commun.* 99:243–51
51. Wieth, J. O., Bjerrum, P. J. 1982. *J. Gen. Physiol.* 79:253–82
51a. Bjerrum, P. J., Wieth, J. O., Borders, C. L. 1983. *J. Gen. Physiol.* 81:453–84
52. Rao, A. 1979. *J. Biol. Chem.* 254:3503–11
53. Steck, T. L. 1972. *J. Mol. Biol.* 66:295–305
54. Rao, A., Reithmeier, R. A. F. 1979. *J. Biol. Chem.* 254:6144–50
54a. Ramjeesingh, M., Gaarn, A., Rothstein, A. 1981. *J. Bioenerg. Biomembr.* 13:411–23
55. Ramjeesingh, M., Gaarn, A., Rothstein, A. 1983. *Biochim. Biophys. Acta* 729:150–60
56. Jay, D. G. 1986. *Biochemistry.* In press
57. Jennings, M. L., Nicknish, J. S. 1984. *Biochemistry* 23:6432–36
58. Goldman, D. 1980. *Identification of the intramembranous regions of transport proteins.* PhD thesis. Harvard Univ., Cambridge, Mass. 33 pp.
59. Low, P. S., Westfall, M. A., Allen, D.

P., Appell, K. C. 1984. *J. Biol. Chem.* 259:13070–76
60. Weinstein, R. S., Khodadad, J. K., Steck, T. L. 1978. *J. Supramol. Struct.* 8:325–35
61. Appell, K. C., Low, P. S. 1981. *J. Biol. Chem.* 256:11104–11
62. Walder, J. A., Chatterjee, R., Steck, T. L., Low, P. S., Musso, G. F., et al. 1984. *J. Biol. Chem.* 259:10238–46
63. Bennett, V., Stenbuck, P. J. 1980. *J. Biol. Chem.* 255:6424–32
64. Fukuda, M., Eshdat, Y., Tarone, G., Marchesi, V. T. 1978. *J. Biol. Chem.* 253:2419–28
65. Markowitz, S., Marchesi, V. T. 1981. *J. Biol. Chem.* 256:6463–68
66. Fukuda, M. N., Fukuda, M., Hakomori, S. 1979. *J. Biol. Chem.* 254:5458–65
67. Tsuji, T., Irimura, T., Osawa, T. 1980. *Biochem. J.* 187:677–86
68. Tsuji, T., Irimura, T., Osawa, T. 1981. *J. Biol. Chem.* 256:10497–502
69. Fukuda, M., Dell, A., Fukuda, M. N. 1984. *J. Biol. Chem.* 259:4782–91
70. Fukuda, M., Dell, A., Oates, J. E., Fukuda, M. N. 1984. *J. Biol. Chem.* 259:8260–73
71. Jennings, M. L., Adams-Lackey, M., Denny, G. H. 1984. *J. Biol. Chem.* 259: 4652–60
72. Jay, D. G. 1984. *J. Biol. Chem.* 259: 15572–78
73. Drickamer, L. K. 1977. *J. Biol. Chem.* 252:6909–17
74. Waxman, L. 1979. *Arch. Biochem. Biophys.* 195:300–14
75. Dekowski, S. A., Rybicki, A., Drickamer, K. 1983. *J. Biol. Chem.* 258:2750–53
76. Terwilliger, T. L., Clarke, S. 1981. *J. Biol. Chem.* 256:3067–76
77. Koch, P. A., Gartrell, J. E., Gardner, F. H., Carter, J. R. 1975. *Biochim. Biophys. Acta* 389:162–76
78. Koch, P. A., Gardner, F. H., Gartrell, J. E., Carter, J. R. 1975. *Biochim. Biophys. Acta* 389:177–87
79. Weiss, M. J., Chan, L. N. L. 1978. *J. Biol. Chem.* 253:1892–97
80. Foxwell, B. M. J., Tanner, M. J. A. 1981. *Biochem. J.* 195:129–37
81. Chang, H., Langer, P. J., Lodish, H. F. 1976. *Proc. Natl. Acad. Sci. USA* 73:3206–10
82. Sabban, E. L., Sabatini, D. D., March-

esi, V. T., Adesnik, M. 1980. *J. Cell. Physiol.* 104:261–68
83. Sabban, E., Marchesi, V., Adesnik, M., Sabatini, D. D. 1981. *J. Cell Biol.* 91:637–46
84. Oikawa, K., Lieberman, D. M., Reithmeier, R. A. F. 1985. *J. Biol. Chem.* 260: In press
85. Falke, J. J., Kanes, K. J., Chan, S. I. 1985. *J. Biol. Chem.* 260:13294–303
86. Kleinfeld, A. M., Lukacovic, M., Matayoshi, E. D., Holloway, P. 1982. *Biophys. J.* 37:146a
87. Macara, I. G., Kuo, S., Cantley, L. C. 1983. *J. Biol. Chem.* 258:1785–92
88. Gunn, R. B., Frohlich, O. 1979. *J. Gen. Physiol.* 74:351–74
89. Falke, J. J., Chan, S. I. 1985. *J. Biol. Chem.* 260:9537–44
90. Falke, J. J., Kanes, K. J., Chan, S. I. 1985. *J. Biol. Chem.* 260:9545–51
91. Macara, I. G., Cantley, L. C. 1981. *Biochemistry* 20:5095–105
92. Deleted in proof
93. Falke, J. J., Chan, S. I. 1984. *Biophys. J.* 45:91–92
94. Wieth, J. O., Brahm, J., Fundler, J. 1980. *Ann. NY Acad. Sci.* 341:394–418
95. Dix, J. A., Verkman, A. S., Solomon, A. K., Cantley, L. C. 1979. *Nature* 282:520–22
96a. Weinstein, R. S., Khodadad, J. K., Steck, T. L. 1980. See Ref. 18, pp. 35–48
96b. Clarke, S. 1975. *J. Biol. Chem.* 250: 5459–69
97. Wang, K., Richards, F. M. 1975. *J. Biol. Chem.* 250:6622–26
98. Reithmeier, R. A. F., Rao, A. 1979. *J. Biol. Chem.* 254:6151–55
99. Khiem, D. J., Ji, T. H. 1977. *J. Biol. Chem.* 252:8524–31
100. Benz, R., Tosteson, M. T., Schubert, D. 1984. *Biochim. Biophys. Acta* 775:347–55
101. Lukacovic, M. F., Verkman, A. S., Dix, J. A., Solomon, A. K. 1984. *Biochim. Biophys. Acta* 778:253–59
102. Staros, J. V., Khakkad, B. P. 1983. *J. Membr. Biol.* 74:247–54
103. Jennings, M. L., Nicknish, J. S. 1985. *J. Biol. Chem.* 260:5472–79
104. Boodhoo, A., Reithmeier, R. A. F. 1984. *J. Biol. Chem.* 259:785–90

Ann. Rev. Biochem. 1986. 55:539–67

PROTEOGLYCAN CORE PROTEIN FAMILIES[1]

John R. Hassell[2], James H. Kimura[3], and Vincent C. Hascall[4]

CONTENTS

[1]The US Government has the right to retain a nonexclusive royalty-free license in and to any copyright covering this paper.

[2]Laboratory of Developmental Biology and Anomalies, [4]Bone Research Branch, National Institute of Dental Research, National Institutes of Health, Bethesda, Maryland 20892

[3]Departments of Biochemistry and Orthopedic Surgery, Rush-Presbyterian-St. Luke's Medical Center, Chicago, Illinois 60612

INTRODUCTION

Proteoglycans are complex glycoconjugates that contain one or more glyco-saminoglycan chains, such as chondroitin sulfate, heparan sulfate, or keratan sulfate, covalently bound to a protein [(1, 2) for reviews]. This protein is referred to throughout this review as the *core protein*. N- and O-linked oligosac-charides are also often covalently bound to the core protein. The exceptional diversity of proteoglycans is derived from both the number of different core proteins, and from the polydispersity produced by the large variety of posttrans-lational modifications required to construct the final macromolecule. Figure 1 outlines the general steps involved in proteoglycan synthesis. The initial translation product, referred to throughout this review as the *precursor protein*, is released into the rough endoplasmic reticulum, frequently having N-linked high-mannose oligosaccharides attached. The vast majority of posttranslational modifications occur when the precursor protein is processed in the Golgi complex. These include one or more of the following: 1. addition of the glycosaminoglycan chains onto appropriate serine and threonine residues (or asparagine residues for corneal keratan sulfate); 2. addition of O-linked oligo-saccharides onto appropriate serine and threonine residues; 3. conversion of high-mannose N-linked oligosaccharides to complex forms; and 4. possible processing of the protein by removal of portions of the polypeptide. In some cases phosphate may be added onto serine residues in the core protein (3, 4) and onto many of the xylose residues (4, 5) which link the glycosaminoglycan chains to the core protein, although where these phosphorylation steps occur in the cell remains to be determined.

Once completed, the proteoglycan can have a large variety of fates depend-ing upon the cell type and function of the macromolecule. It may enter a storage granule as for the heparin proteoglycan in mast cells (6) and the heparan sulfate proteoglycan in the synaptic vesicles of the electric organ of the electric eel (7); it may insert in the membrane of secretory vesicles and be deposited on the cell surface as an intercalated, integral membrane component, as for a chondroitin sulfate proteoglycan on human melanoma cells (8) and a hybrid proteoglycan with both heparan sulfate and chondroitin sulfate on mouse mammary epithelial cells (9); or it may be packaged in secretory vesicles and secreted into the

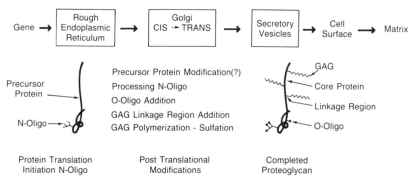

Figure 1   Proteoglycan biosynthesis

extracellular matrix as a structural component, as for the major hybrid proteoglycan of cartilage which contains both chondroitin sulfate and keratan sulfate chains (1, 2) and the heparan sulfate proteoglycan in glomerular basement membranes (10).

One important aspect for characterizing a proteoglycan is to determine the size of the core protein. The precursor protein can be isolated by using antibodies (Figure 2) to the proteoglycan and its size determined by SDS-PAGE, molecular sieve chromatography, or ultracentrifugation. Alternatively, the bulk of the glycosaminoglycan side chains can be removed from the proteoglycan by digestion with the appropriate enzyme(s) (chondroitinase, keratanase, heparitinase) to yield a *core preparation* (Figure 2) whose size can be likewise determined. Because precursor proteins and core preparations contain significant amounts of carbohydrate (oligosaccharides, and in the case of core preparations, residual glycosaminoglycan linkage regions as well), the molecular weight of the core protein will be smaller. The molecular weights of some core proteins have also been approximated from compositional analyses to estimate protein content combined with determined molecular weights for either intact proteoglycans or for core preparations.

At this time, little is known about the chemistry and structure of the core proteins. Classic protein chemistry and sequencing approaches are frustrated by the large amounts of complex carbohydrate structures on a core protein, or by the paucity of material available, or by its frequently very large size. Through the use of antibodies, however, the precursor protein for at least five proteoglycans [(see Refs. 8, 11, 12) for examples] and cell free translation products for the protein of two cartilage proteoglycans (13, 14) have been identified, and effective use of enzymatic digestion has permitted the size of core preparations from at least 10 proteoglycans to be determined (Table 1). Furthermore, recent studies using molecular biology techniques have de-

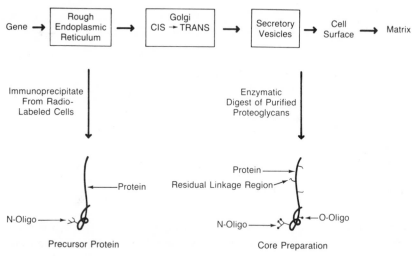

Figure 2   Core protein size determination

termined the entire amino acid sequence of one small proteoglycan (15). Such studies will make it possible to identify those proteoglycans that are distinctly different from those that are related and belong to a common gene family.

The proteoglycan field is rapidly expanding, making a comprehensive review of all the relevant recent literature beyond the scope of this article. Our objective is to focus attention on the characteristics of the precursor proteins and the core preparations because this portion of the proteoglycan is directly related to the primary gene product and will eventually provide the basis for defining the different gene families.

MAJOR PROTEOGLYCAN OF CARTILAGE

Structure and Properties

The predominant proteoglycan in cartilage is a large macromolecule with extreme polydispersity in size and composition [(1, 2, 16–19) for recent reviews]. A typical molecule, M_r = approximately 2,500,000, consists of a core protein (estimated M_r = 200,000–300,000) with about 100 chondroitin sulfate chains and up to 130 keratan sulfate chains. The keratan sulfate chains are bound to the protein by linkage structures closely related to the numerous O-linked oligosaccharides which are also found distributed along the core protein (20–25). Other substituents on the protein include about 10–15 N-linked oligosaccharides of the complex type (20, 21, 26), and phosphate esters on up to 80% of the xylose residues that initiate chondroitin sulfate chains (4, 5) as well as a few on serine residues (3, 4).

The properties of the glycosaminoglycan chains, which account for 80–90%

Table 1 Estimated molecular weights of various forms of the proteins of different proteoglycans

PG source	GAG type[a]	Precursor protein[b]	Core protein[b]	Core preparation[b]	Reference
mast cell	Hep,CS			10	196
L-2 cell	CS		10[f]	25	15,198
hepatocyte	HS			35	149
PG-I[c]	CD,DS			45	110
PG-II[d]	CS,DS	43,47	38[g]	43,47	12,110
cornea	KS			53	144,145
epithelial	CS,HS hybrid	56		53	162
epiphysis[e]	CS,DS			68	199
skin fibroblast	HS			90[i]	164
granulosa cell	DS,HS			240	156,158
melanoma	CS	220		250	8,152
smooth muscle cells	CS			320	T. N. Wight, M. Lark, personal communication
cartilage	CS,KS hybrid	370	200–300[h]	400	11,68
granulosa cell	DS			400	161
EHS, L-2	HS	400		400	178,184
skin fibroblast	CS			500	126

[a]Abbreviations used: Hep, heparin; HS, heparan sulfate; CS, chondroitin sulfate; DS, dermatan sulfate; KS, keratan sulfate
[b]molecular mass in kd
[c]cartilage, bone
[d]skin, bone, tendon, cartilage, cornea
[e]chick growth cartilage, also sterna
[f]determined from cDNA clone
[g]determined from precursor protein synthesized in presence of tunicamycin
[h]estimated from core preparations and protein contents
[i]exists as a 180-kd disulfide bonded dimer

of the mass, dominate the properties of the proteoglycan. In some cases, variation in glycosaminoglycan chain type, number, and length can account for all the size polydispersity of newly synthesized proteoglycans as in cultures of chondrosarcoma chondrocytes (27). In other cases, these parameters account for some, but not all, of the size polydispersity present in the proteoglycan population (28). For example, some studies of proteoglycans isolated from hyaline cartilages indicate that variations occur in the size of the core protein. This variation may be continuous (29–31), giving rise to polydispersity, or discontinuous (32), giving rise to heterogeneity (see below).

The distribution of substituents along the core protein reflects regional specialization of the polypeptide. A portion at the N-terminus with $M_r =$

60,000–70,000 (33, 34), referred to as the hyaluronic acid–binding region, contains: 1. few or no glycosaminoglycan chains, 2. most of the N-linked oligosaccharides (20), 3. a binding site that interacts specifically with hyaluronic acid (35–37) to permit the formation of large aggregates, and 4. a site that interacts with the link proteins involved in aggregate stabilization (37–39). An adjacent region of the core protein, M_r = approximately 30,000 and referred to as the keratan sulfate–attachment region, contains 60% or more of the keratan sulfate chains, while the remainder towards the carboxyl terminus, referred to as the chondroitin sulfate–attachment region, contains the majority of the chondroitin sulfate chains as well as a few of the keratan sulfate chains and O-linked oligosaccharides (40).

The ability of the core protein of this proteoglycan to interact with hyaluronic acid distinguishes it from other core protein types, more so than does its size or composition. The nature of the highly specific interactions involved in proteoglycan binding to hyaluronic acid and the importance of the link protein for stabilizing this interaction have been reviewed extensively [see (18, 41) in particular] and will not be discussed here. An additional observation not covered in these reviews is that neither the N-linked oligosaccharides on the proteoglycan nor those on the link protein appear to be critically involved in the aggregation process (42). Biosynthetic studies have established that aggregation occurs extracellularly (43, 44). In the chondrosarcoma chondrocyte system, aggregation occurred rapidly (44), and subsequent studies showed that the core protein precursor could bind into aggregates (5, 11), suggesting that no inactive precursor form of the hyaluronic acid–binding region existed. Recent developments using hyaline cartilage organ cultures, however, indicate that at least under some conditions, newly synthesized proteoglycans cannot bind to hyaluronic acid as well as the same molecules do after being allowed to "mature" in the tissue for a period of time, typically 24 hours (45–48). At high concentrations of hyaluronic acid, 8–10% compared with the usual concentration of 1–2%, the newly synthesized proteoglycans bind nearly to the same extent as the bulk of unlabeled proteoglycans isolated from the tissue, and the transition to the higher-affinity form was shown to be temperature-dependent, increasing with increases in temperature up to 60°C (45, 46). These results suggest that different conformational states of the hyaluronic acid–binding region with different affinity for hyaluronic acid may exist.

Electron Microscopy

Cartilage proteoglycans can be examined by electron microscopy, and several studies on both monomeric and aggregated molecules spread as monolayers in cytochrome c have visualized a number of the substructures originally described from biochemical studies (49–58). For aggregated molecules, evidence for variation in the core protein length was seen (51, 52, 54, 56) with some indication that this variation occurred primarily in the length of the chondroitin sulfate–attachment region of the protein (51). The core protein lengths for proteoglycans in aggregates isolated from fetal hyaline cartilage were significantly more uniform, however (57). Antibodies to the hyaluronic acid–binding region have been used in combination with electron microscopy of spread proteoglycan molecules to show that this region of the core is localized at one end (58). Cartilage proteoglycans have also been rotary shadowed after being sprayed onto freshly cleaved mica and examined in the electron microscope (59). Such preparations reveal that the monomer contains two adjacent globular domains at one end of the core protein, one of which appears to be the hyaluronic acid–binding region. Micrographs of aggregates reveal a striking beadlike structure along the central filament of hyaluronic acid representing the hyaluronic acid–binding region and the link proteins known to be involved in aggregate formation.

Gold-labeled polyclonal antibodies against the core protein have been used in immunolocalization for electron microscopy studies to quantitate proteoglycans in the extracellular matrix and the chondrocytes in thin sections of cartilage (60). These procedures were used in combination with ricin and chondroitin sulfate localization to demonstrate that glycosaminoglycan and oligosaccharide biosynthetic events occur primarily in the medial/trans Golgi cisternae of chondrocytes (61).

Core Protein Molecular Weight

The molecular weights of core proteins of cartilage proteoglycans have been estimated by several methods. Indirect measurements based on protein content and determinations of the molecular weight of intact proteoglycans from bovine hyaline cartilages give estimates of $M_r = 180,000–300,000$ (28, 62–64). Removal of chondroitin sulfate chains, but not keratan sulfate chains, from a bovine hyaline cartilage proteoglycan with chondroitinase digestion, yielded a core preparation which was about 45% protein with a $M_r = 460,000$ determined by ultracentrifugation (65). Chondroitinase digestion of the proteoglycan from a rat chondrosarcoma, which contains O-linked oligosaccharides but no keratan sulfate, yielded a core preparation with 60% protein that gave a single major band on SDS-PAGE with a $M_r = 400,000$ (11, 66). Assuming that the protein and residual sugar structures contribute equally, these values would correspond to core proteins with estimated M_r of 210,000 (bovine cartilage) and 240,000 (rat chondrosarcoma) respectively. Digestion of a chick cartilage proteoglycan with both chondroitinase and keratanase (an endo-β-galactosidase) also produced a core preparation giving a single band on SDS-PAGE with $M_r = $ approximately 400,000 (67, 68), indicative of a core protein molecular weight close to that estimated for the chondrosarcoma proteoglycan. Treatment of a rat chondrosarcoma proteoglycan preparation with anhydrous hydrogen fluoride removed 95% of the chondroitin sulfate and oligosaccharides (69), and yielded a core preparation with a single band at M_r = approximately 210,000 on SDS-PAGE (70). The general agreement of these determinations suggests that the core proteins in the population of completed proteoglycans resident in a cartilage matrix have molecular weights of 200,000–300,000.

Cell free translation of mRNA from cartilage followed by immunoprecipitation has yielded values on SDS-PAGE of 330,000–340,000 for the translation product from chick sternum mRNA (13, 14) and from rat chondrosarcoma mRNA (14), with the latter being slightly smaller. In both cases only one major band was observed. Cell free translation of bovine cartilage mRNA yielded an immunoprecipitation product for the core protein with M_r greater than 300,000 (71).

The precursor protein in chondrocytes from the rat chondrosarcoma was initially identified and isolated by immunoprecipitation and by its ability to bind to hyaluronic acid in aggregates (11). It migrated on SDS-PAGE as a single, major band with M_r = 370,000. The precursor protein exhibits a half-life in these cells of 60–100 minutes (11, 72) while somewhat shorter half-lives have been estimated in bovine cartilage chondrocytes (73, 74). Most of the intracellular lifetime of the precursor protein is spent in the rough endoplasmic reticulum (75). The size of the precursor protein pool can be increased by treating cells with the ionophore monensin (76–78) or with diethylcarbamazine (79), which block movement through the Golgi apparatus with a concomitant increase in material in the rough endoplasmic reticulum. A precursor protein of M_r = approximately 400,000 has also been immunoprecipitated from ^{35}S-methionine-labeled chick sternal chondrocytes (79a). Most steps for conversion of the precursor protein to the completed proteoglycan must occur rapidly in the Golgi apparatus (11, 61). Kinetic studies with the chondrosarcoma chondrocytes (11) have not revealed any intermediates between the precursor protein and the completed proteoglycan with a sufficiently long half-life for detection. Studies with chick cartilage chondrocytes (68, 70, 80), however, suggest that larger intermediates may occur; species of M_r = approximately 400,000 and 600,000 were immunoprecipitated from subcellular fractions which contained elements of the Golgi apparatus (70). Since both the translation product and the precursor protein appear significantly larger than the estimated value for the core protein in the completed proteoglycan, it is likely that some processing of the precursor protein occurs rapidly and nearly concurrent with conversion to the completed proteoglycan in the Golgi apparatus.

Heterogeneity of Core Proteins

Electrophoresis of proteoglycans from mature hyaline cartilages in large pore agarose-acrylamide gels reveals two, and occasionally more, distinct bands of high molecular size (81–83). This suggests that there are two distinct populations of proteoglycans in these tissues since it is difficult to reconcile a single polydisperse population with discrete banding in this technique. The two major bands were shown to differ significantly in their ratios of keratan sulfate to chondroitin sulfate with the faster migrating band having a higher ratio (82). There is a developmental transition in the banding pattern, since proteoglycans from fetal hyaline cartilage give only a single band which migrates at a somewhat higher molecular size than either of the two major bands found in mature hyaline cartilages (84). Interestingly, cartilage from mature individuals with the human genetic dwarfism pseudoachondroplasia lacks the faster migrating band (85), and chondrocytes in the tissue accumulate material in a highly distended rough endoplasmic reticulum, perhaps the protein precursor for the missing proteoglycan.

Recently, the aggregating proteoglycans from bovine nasal septum were subdivided into two populations, a major one representing 43% of the total proteoglycans with M_r = approximately 3,500,000, and a minor one representing 15% of the total with M_r = approximately 1,300,000 (32). The former population corresponded to the slower migrating band on the large pore gels and contained a lower ratio of keratan sulfate to chondroitin sulfate than the latter population which corresponded to the faster migrating band. The two populations had core proteins with estimated molecular weights that differed by a factor of two, the larger with M_r = approximately 210,000 and the smaller with M_r = approximately 94,000. After ^{131}I-iodination and trypsin digestion, the two populations gave similar two-dimensional peptide maps with only a few distinctly different peptides, and they showed some differences in immunoreactivity toward different antisera. Further, the smaller proteoglycan population contained a distinct keratan sulfate–attachment region whereas the larger did not. These data suggest that the smaller proteoglycan population is not a degradation product of the larger and that two distinct core proteins are present. However, biosynthetic studies with explants from mature human hyaline cartilage have indicated that only the larger proteoglycan is detected as a synthetic product even though both species are present in the tissue (86, 87, 87a). This suggests in this case that at least some of the smaller proteoglycan may be derived as a degradation product of the larger (See Refs. 87 and 87a for extensive discussion of these issues).

Chondroitinase digestion of purified aggregating proteoglycans from the chondrosarcoma followed by SDS-PAGE also showed a second, smaller core preparation as a minor constituent (66). However, aggregates in such samples frequently have trace amounts of several smaller species which contain the hyaluronic acid binding region, including the smallest form with M_r = approximately 65,000, which is devoid of glycosaminoglycans (33, 88, 89). Further, only one core preparation is seen in chondroitinase digests of newly synthesized proteoglycans isolated from cell cultures from this tumor, and only a single precursor protein is usually observed (5, 11, 33, 75). Therefore, in this case the smaller proteoglycan components appear to be degradation products of the largest form derived some time after the proteoglycans have been secreted into the matrix.

The evidence overall suggests that in mature hyaline cartilage there are at least two core proteins of different length but with similar hyaluronic acid–binding activity and immunological determinants, whereas there appears to be only one in the chondrosarcoma and in fetal or immature cartilage. However, the situation is made more complex by as yet poorly understood degradation processes that occur in the matrix. It would be of interest to see if more than one precursor protein can be identified in chondrocytes from mature hyaline cartilages.

Similar Proteoglycans in Noncartilaginous Tissues

Large aggregating proteoglycans with dermatan sulfate or chondroitin sulfate side chains are not restricted to cartilage. Proteoglycans with the ability to bind to hyaluronic acid have been found in smooth muscle cultures (90, 91), in aorta (92–96), and in glial cell cultures (97). Chondroitinase digests of this proteoglycan from smooth muscle cell cultures were recently shown to yield core preparations of M_r 320,000 (T. N. Wight, M. Lark, personal communication). They also appear to be present in tendon (98), fibroblasts (99, 100), and developing bone mesenchyme (101). Proteoglycans with cross-reactivity to antibodies against cartilage proteoglycans have been found in aorta (92, 102) and bone (101). Further, the large proteoglycans present in small amounts in bone and tendon yield core preparations after chondroitinase and keratanase digestion with M_r = approximately 400,000, similar to cartilage proteoglycans (L. W. Fisher, personal communication).

Diseases

There are two other known genetic anomalies in addition to the human pseudoachondroplasia mentioned above that involve defects in synthesis of the core protein of cartilage proteoglycans. Chondrocytes from cartilage of the nanomelic chick lack the core protein precursor (104, 105). The same defect appears to be present in the mouse cartilage matrix deficiency (cmd/cmd) (106). In both cases, the proteoglycan is absent in the cartilage although the chondrocytes make essentially normal amounts of link protein (105, 106).

SMALL PROTEOGLYCANS IN COLLAGENOUS CONNECTIVE TISSUES

General Characteristics

Most, perhaps all, fibrillar collagenous connective tissues contain one or more small proteoglycans, usually with dermatan sulfate as the glycosaminoglycan constituent. Members of this group typically have core proteins with estimated

M_r = 40,000–50,000, with one or two glycosaminoglycan chains attached. N-linked oligosaccharides, and in some cases, O-linked oligosaccharides are also present on the core protein. Recent work with these proteoglycans isolated from cartilage indicates that two distinctly different core proteins are probably present in this group. Additionally, a unique small proteoglycan with keratan sulfate chains has been isolated from cornea. Some features of these proteoglycans in different connective tissues are described in the following sections.

Cartilage

A low-buoyant-density, small proteoglycan has been isolated from bovine nasal cartilage and chemically characterized (107). Chondroitinase digestion of the macromolecule produced a core preparation of M_r = 45,000 when analyzed by SDS-PAGE. It contained one to two chondroitin sulfate chains of M_r = approximately 35,000 with no evidence for the presence of dermatan sulfate. The core protein was unusually rich in leucine. This proteoglycan had been observed in several cartilages from baboon (108), in bovine articular cartilage (109, 110), in chick growth plate cartilage (111), and as a biosynthetic product in organ culture of bovine articular cartilage (112, 113). Less of this proteoglycan is found in immature cartilage than in older cartilage (110, 114).

A similar population of proteoglycans has been isolated from mature bovine articular cartilage, which contains about 10 times more of these proteoglycans than does bovine fetal epiphyseal cartilage (110). In dissociative solvents, only one size class of proteoglycan was observed with M_r of 90,000–120,000. Additionally, after chondroitinase digestion, a core preparation with closely spaced bands near M_r = 45,000 on SDS-PAGE was obtained. The glycosaminoglycan chains contained significant amounts of iduronic acid and are therefore classified as dermatan sulfate. A subpopulation of the proteoglycans, referred to as dermatan sulfate PG-I, exhibited an unusual self-aggregation, migrating as a broad band with apparent M_r = 165,000–285,000 when electrophoresed on SDS-PAGE gels at high ionic strength; whereas a second subpopulation, referred to as dermatan sulfate PG-II, electrophoresed in the expected M_r = 90,000–120,000 range. Further analyses revealed that PG-I and PG-II have distinctly different antigenic properties; polyclonal antibodies against the mixed population detected only PG-II. After chondroitinase digestion of PG-II, followed by SDS-PAGE and immunotransfer, the antiserum stained two closely spaced bands at M_r = 43,000 and 47,000, similar to results observed for this family of proteoglycans isolated from a number of tissues (see below). Amino acid sequence analyses indicate that PG-II has an identical N-terminal sequence as that for the small proteoglycan from skin (115, 116, 116a) whereas PG-I does not have the same N-terminal sequence (L. C. Rosenberg, personal communication). Neither PG-I nor PG-II cross-reacts

with antisera against the larger, aggregating proteoglycan species from cartilage. The mechanism for the self-association of PG-I is not clear, although it
may be mediated in part by the self-association of the dermatan sulfate chains
(118). There were no clear differences in the chemical properties of the
dermatan sulfate chains for PG-I and PG-II.

Tendon

Investigations of the small proteoglycans from tendon provide several lines of
evidence that suggest that these proteoglycans are closely associated with
collagen fibrils. Almost 90% of the proteoglycans isolated from proximal
bovine flexor tendon, which is under essentially only tensile load, are in the
small proteoglycan family (98). As with the small proteoglycans from cartilage
described above, these tendon proteoglycans have a small proportion of PG-I
molecules as indicated by their ability to self-associate, although PG-II predominates (K. G. Vogel, personal communication). The glycosaminoglycan
chains are dermatan sulfate, with approximately 70% iduronic acid and with
average $M_r = 37,000$. After chondroitinase digestion, a core preparation with
an apparent $M_r = 48,000$ was produced. The intact proteoglycan as well as to a
lesser extent, the core preparation, can inhibit fibrillogenesis of type I and type
II collagen in an in vitro assay system, whereas the isolated dermatan sulfate
chains have no effect (120). Interestingly, the population of small proteoglycans isolated from either cartilage or aorta has no effect on fibrillogenesis. The results suggest that the core protein of the small dermatan sulfate
proteoglycan from tendon can bind selectively to collagen and inhibit fibrillogenesis, and differs in this respect from other proteoglycans in this family.
The small proteoglycan from tendon has the same N-terminal sequence as that
of the small proteoglycan from skin (K. G. Vogel, personal communication).
An antiserum to these proteoglycans recognizes cartilage PG-II, but not PG-I,
and cross-reacts with bovine bone PG-II (L. W. Fisher, personal communication; K. G. Vogel, personal communication).

Electron microscopic studies of developing and mature rat tendon (121, 122)
utilizing cuprinolinic blue as a selective stain for proteoglycans showed that the
proteoglycans are periodically arrayed on the collagen fibril in close association
with the d and e band. Further, there was a correlation between the dermatan
sulfate content of the tendon and the average diameter of the collagen fibrils in
tendons from animals of different age. Because the predominant proteoglycans
in this tissue are the small species, these results suggest that they have a function
in either tissue collagen morphogenesis, or tissue maintenance, or both. The
periodic association of proteoglycan with collagen has been seen in other
tissues, notably aorta where small matrix granules align with the collagen fibril
(1). Skin fibroblasts cultured on collagen gels synthesize a dermatan sulfate
proteoglycan which is localized primarily in the gel (122a).

Bone

A sequential extraction procedure, 4 M guanidine HCl followed by 4 M guanidine HCl with 0.5 M EDTA, was used to isolate a proteoglycan population in the second extract that was present in the mineral phase of fetal bovine bone (101). The predominant proteoglycans in this population were in the small proteoglycan family and contained chondroitin sulfate chains (no dermatan sulfate was found) of $M_r = 40,000$. On SDS-PAGE analyses, two broad proteoglycan bands were seen; the smaller (equivalent to cartilage PG-II) was suggested to have a single chondroitin sulfate chain, while the larger (equivalent to cartilage PG-I) was suggested to be the same core protein with two glycosaminoglycan chains. Chondroitinase digestion of the mixed proteoglycan population yielded core preparations with the characteristic two-band pattern around M_r = approximately 43,000. More refined analyses since then have shown that the core preparation derived from bone PG-I appears to give a single band at $M_r = 45,000$, while that from bone PG-II gives the doublet at $M_r = 43,000$ and $46,000$ (L. W. Fisher, personal communication). PG-I and PG-II have significantly different amino acid compositions (123). PG-I had a composition very similar to that of the low-buoyant-density proteoglycan isolated from bovine nasal cartilage (107), characterized by an unusually high leucine content. PG-II had lower leucine and higher amounts of aspartate/asparagine and glutamate/glutamine. Bone PG-I does not have dermatan sulfate, or has dermatan sulfate with low iduronic acid content. Antisera raised against bone PG-II cross-react with tendon (98) and cartilage (110) PG-II (K. G. Vogel, personal communication).

A recent extensive study of purified small proteoglycan populations from a number of bovine connective tissues (124), including adult bovine bone (125), distinguished two subpopulations on the basis of immunological cross-reactivity. The preparations from bone, tendon, sclera, and cornea showed extensive cross-reactivity, but not with preparations from either nasal cartilage or aorta. The latter two showed extensive cross-reactivity with each other. It is possible that the proteoglycan preparations from the first group of tissues contain primarily PG-II and generated antibodies predominantly against this population in agreement with the results discussed above, while the cartilage and aorta preparations contain primarily PG-I and generated antibodies predominantly against this population. The sample from bovine nasal cartilage had the high leucine content (107) characteristic of bone PG-I (123). This could also account for the apparent immunological differences between the cartilage samples derived from nasal septum (124) and mature articular cartilage (110).

Skin

Several investigators have reported that skin fibroblasts synthesize both small and large proteoglycans. The large proteoglycans (100, 126) have core preparations after chondroitinase with M_r = approximately 500,000 (126), and the small, predominant proteoglycans appear to be primarily PG-II. Recently radiolabeled precursors and antisera against the core protein of the small proteoglycan population were used to study their biosynthesis in the presence or absence of tunicamycin, which inhibits N-linked oligosaccharide synthesis (12). The results suggested that the characteristic doublet observed for core preparations at M_r = 43,000 and 47,000 result from the presence of different numbers, either two or three, of N-linked oligosaccharides on the core protein. They also suggested that this proteoglycan did not contain any O-linked oligosaccharides and that the core protein devoid of any N-linked oligosaccharides has a M_r = 39,000.

A small dermatan sulfate proteoglycan, isolated from porcine skin (127), has similar overall properties to those described above. This proteoglycan population was also isolated from bovine skin (115, 116, 116a, 128) and has been sequenced through the first 24 amino acids from the N-terminus (115). Position 4 of the sequence contains a serine residue substituted with a complex carbohydrate, quite likely a glycosaminoglycan chain (115, 116a). Interestingly, the core protein of a small proteoglycan isolated from human placental membrane has an identical amino acid sequence through the first nine residues, after which it differs (129).

Cornea

Intact proteoglycans have been isolated from bovine (130–134) and monkey (135–137) cornea by dissociative extraction and ion exchange chromatography. Two major proteoglycan populations of small molecular size were identified, one with dermatan sulfate chains and one with keratan sulfate chains. The hydrodynamic size distribution of these proteoglycans permit them to fit well within the space between the regular lattice of collagen fibrils in the corneal stroma (138), and it is presumed that the proteoglycans facilitate and maintain the orderly packing of the collagen fibrils required for the optical transparency of this tissue. In support of this, disorganization of the collagen fibril packing is observed during corneal wound healing where distinctly different and larger proteoglycans are synthesized (139, 140); and the human disease corneal macular dystrophy, which also involves disorganization of the matrix, is characterized by altered synthesis of the keratan sulfate proteoglycan (141–143).

The dermatan sulfate proteoglycan population is related to the PG-II proteoglycans discussed above as indicated by the production of the characteristic core preparation doublet at M_r = 43,000 and 47,000 after chondroitinase digestion (144), and by having similar immunological properties, amino acid composition, and peptide maps as the small proteoglycan populations from bone, tendon, and sclera (124). These corneal proteoglycans contain one to two dermatan sulfate chains of $M_r = 55,000$, distinguished by their low level of sulfation, as well as N-linked oligosaccharides (135).

The keratan sulfate proteoglycan is considered unique to the cornea. Two studies (144, 145) have used keratanase digestion of bovine corneal proteoglycans followed by SDS-PAGE to estimate the size of the core preparation for the keratan sulfate proteoglycans. In one (145), in which the proteoglycans were purified by ethanol precipitation and step elution from an ion exchange column, two bands with apparent $M_r = 40,000$ and 55,000, were observed. In the other, in which the proteoglycans were isolated by gradient elution from an ion exchange column, a major band with apparent $M_r = 53,000$ was observed with only a faint band at 40,000 (144). It remains to be determined whether the smaller core protein band is derived from the larger, possibly as a degradation product. The proteoglycan contains two to three branched oligosaccharide linkage regions with a structure related to complex N-linked oligosaccharides; each contains two keratan sulfate chains (146). High-mannose N-linked oligosaccharides are also present, one of which has an unusual four-mannose residue structure and which may be related to synthesis of the keratan sulfate linkage oligosaccharide region (25, 146). Antibodies against the core protein of the keratan sulfate proteoglycans from bovine (145) and monkey (136, 143) cornea do not cross-react with the corneal dermatan sulfate proteoglycan population. The difference in size of the core protein as well as in its immunological determinants makes it likely that the core protein of the keratan sulfate proteoglycan is a separate gene product.

Cytochemical localization of proteoglycans in cornea suggests that keratan sulfate proteoglycans were localized on the a and c bands while the dermatan sulfate proteoglycans were localized on the d and e bands (147).

Summary

Current evidence indicates that the core proteins of PG-I, PG-II, and the keratan sulfate proteoglycan of cornea are distinctly different. The degree of similarity of the PG-II proteoglycans from different tissues will be established once the amino acid sequences of their core proteins are known. Recently, cDNA clones have been developed for the core proteins of the bone PG-II (A. Day, M. Young, J. Termine, personal communication) and the placental membrane proteoglycan (E. Ruoslahti, T. Krusius, personal communication) respectively. The former recognizes two closely spaced mRNA in Northern blots of bone cell mRNA, and a doublet band around $M_r = 40,000$ was immunoprecipitated from cell free translations with antisera to PG-II from either bone or tendon. Thus, the basis for the characteristic doublet derived from PG-II may be more complicated than a simple difference in the number of N-linked oligosaccharides (12). The development of appropriate cDNA clones for PG-II from different tissue sources will clarify this problem as well as determine to what extent different members of this group constitute a gene family.

CELL SURFACE PROTEOGLYCANS

General Characteristics

Cell surface proteoglycans comprise a fascinating group containing a large diversity of structures and core proteins. Proteoglycans in this class are either directly intercalated in the lipid bilayer of the cell membrane as integral membrane components, presumably through a hydrophobic region of the core protein, or associated with the cell membrane by interaction with other cell surface macromolecules. A pericellular region of the core protein contains the glycosaminoglycan chains and oligosaccharides. In general, it appears that some of the intercalated proteoglycans are "shed" from the cell surface as processed forms that lack the hydrophobic region, while the rest are eventually internalized and degraded in lysosomes. The available data strongly suggest that there are several distinctly different core proteins represented in this group, and the following examples have been chosen to reflect these differences. A recent, more extensive review of cell-associated proteoglycans has been published in this journal (149).

Hepatocytes

Cell surface proteoglycans were initially isolated from liver and hepatocytes (150). Two forms of heparan sulfate proteoglycans were found, a peripheral, non-intercalated form with a core preparation after heparatinase digestion of M_r = 27,000, and an intercalated form with a core preparation of M_r = 35,000. Both forms contained four to five heparan sulfate chains of apparent M_r = 14,000, and the non-intercalated form was presumed to be derived from the intercalated form. Polyclonal antibodies against the intercalated proteoglycan stain the sinusoidal plasmalemma and some intracellular organelles of hepatocytes as well as other cell surfaces (151). The staining pattern indicates that this proteoglycan differs from two other proteoglycans with heparan sulfate chains, those from basement membranes and from mouse mammary epithelial cells (see below).

Human Melanoma Cells

A search for monoclonal antibodies that recognize epitopes specific for the cell surface of human melanoma cells uncovered one that reacts with the core protein of a chondroitin sulfate proteoglycan (8, 152–155). In pulse-chase studies, the antibody identified a precursor protein with M_r = 210,000–220,000 which increased in size to approximately 240,000 after a 15-minute chase followed shortly by the appearance of a 250,000-dalton glycoprotein and the proteoglycan, with the latter predominating (8, 152). Chondroitinase digestion of the cell-associated proteoglycan yielded a core preparation with M_r = approximately 250,000. The initial 210,000–220,000- and the 240,000-dalton products were sensitive to endoglycosidase H digestion indicating that they contained high-mannose N-linked oligosaccharides, whereas the 250,000-dalton species was not, indicating that the N-linked oligosaccharides were converted to the complex type. The 250,000-dalton form may contain O-linked oligosaccharides which would account for its larger size. The 250,000-dalton glycoprotein was originally thought to be an intermediate in the assembly of the proteoglycan, but recent studies (154) indicate that it is probably a separate product, representing a proportion of the protein precursor that bypasses the glycosaminoglycan assembly step. This was illustrated by treating cells with ammonium chloride at concentrations that did not inhibit protein synthesis but greatly diminished conversion of protein precursor to proteoglycan. In this case, the proportion of the 250,000-dalton glycoprotein increased at the expense of the proteoglycan. The 250,000-dalton glycoprotein gives rise to a "shed" form with $M_r = 170,000$, nearly the same as the core protein of the chondroitinase-digested "shed" form of the proteoglycan (J. R. Harper, personal communication).

Immunolocalization studies with the monoclonal antibody show that the proteoglycan is located in punctated foci and filamentous structures on the melanoma cell surface (153). The antibody stained only fetal melanocytes and neural cells as well as melanoma cells (153). It also interfered with melanoma cell adhesion and spreading, which suggests that this proteoglycan may have a role in cell attachment.

Rat Ovarian Granulosa Cells

Extensive studies of the metabolism of proteoglycans in cultures of rat ovarian granulosa cells have shown that these cells contain two cell surface proteoglycans, one with dermatan sulfate chains (156) and the other with heparan sulfate chains (157). The dermatan sulfate proteoglycan contains a core preparation after chondroitinase digestion with $M_r = $ approximately 240,000 (156), and the heparan sulfate proteoglycan yields a core preparation after heparatinase digestion with the same size (M. Yanagishita, personal communication). Both proteoglycans contain four to six glycosaminoglycan chains of $M_r = $ 40,000, as well as N-linked and O-linked oligosaccharides. The metabolism of both proteoglycans is very similar; about 35% of each are "shed" from the surface into the medium as slightly smaller proteoglycans, while the remainder are internalized and eventually entirely degraded in lysosomes via two degradation pathways (158, 159). The half-life of these proteoglycans on the cell surface is about four hours.

The same cells also synthesize a separate dermatan sulfate proteoglycan with a much larger core preparation after chondroitinase digestion, $M_r = 400,000$ on SDS-PAGE. The proteoglycan contains about 20 glycosaminoglycan chains, a few N-linked oligosaccharides, and more than 200 O-linked oligosaccharides

(156). Essentially all of this proteoglycan is secreted into the medium in cell cultures, and it is a major constituent of follicular fluid (160). The core protein lacks the hyaluronic acid binding region characteristic of the major cartilage proteoglycans, and therefore may not be related to this family, although it appears to have a similar core protein size.

Mouse Mammary Epithelial Cells

Mouse mammary epithelial cells synthesize an unusual cell surface proteoglycan in which individual molecules contain both chondroitin sulfate and heparan sulfate chains bound to the core protein (9, 161). The intact proteoglycan, M_r = 250,000, intercalates in liposomes. A proportion of the proteoglycans are processed to a smaller form that cannot intercalate and is "shed" into the medium (162). Trypsin treatment of the cells releases a species of similar size as the "shed" form. Digestion of the trypsin-released form of the proteoglycan with both chondroitinase and heparatinase generates a core preparation of apparent M_r = 53,000, whereas chondroitinase digestion alone yields products with sizes intermediate between the untreated proteoglycan and the 53,000-dalton core protein. The actual size of the core protein would be smaller when the contribution of the residual carbohydrate structures is taken into account. The "average" proteoglycan molecule, whether intact, trypsin-released, or "shed," contains one to two chondroitin sulfate chains of M_r = approximately 17,000 and up to four heparan sulfate chains of M_r = 36,000; a minor proportion may contain only heparan sulfate chains. The mechanisms by which the cell synthesizes both glycosaminoglycan chains on the same core are unknown.

A monoclonal antibody directed against the core protein has been developed (163). This antibody has been used to identify a precursor protein isolated from the cells with M_r = 56,000, significantly larger than the expected size of the protein in the 53,000-dalton species. How much of the precursor protein might be removed during conversion to the intact proteoglycan and how much during the "shedding" process is unknown. The antibody stains the surfaces of the cells, but not the basement membrane matrix produced by the cells (163).

Skin Fibroblasts

Another heparan sulfate proteoglycan, which is distinctly different from any described above, has been isolated from skin fibroblasts (164–166). The core protein in this case is identical to, or closely resembles in size and in biological properties, the transferin receptor, a cell membrane glycoprotein. Both the core preparation derived from heparatinase digestion of the proteoglycan and the transferin receptor contain two M_r = 90,000 subunits linked by disulfide bonds. The proteoglycan, however, contains four to six heparan sulfate chains of M_r = 20,000 bound to each 90,000-dalton subunit for an overall M_r = approximately 350,000. Like the transferin receptor, the proteoglycan forms a complex with transferin which was immunoprecipitated by antibodies specific for transferin. Various smaller heparan sulfate species that are "shed" into the medium could not bind to transferin. Further, a large proportion, probably most, of the transferin receptors do not contain glycosaminoglycan chains. If the core protein is identical to the transferin receptor, the case may be analogous to the situation with the melanoma chondroitin sulfate proteoglycan in which some of the core protein can be intercalated into the cell membrane in a form without glycosaminoglycan chains as described above.

Summary

The presently known cell surface proteoglycans have core proteins that range in M_r from less than 40,000 in hepatocytes to approximately 200,000 in melanoma cells and perhaps even larger in colon carcinoma cells (167). Antibodies to these proteoglycans (melanoma cells, hepatocytes, and epithelial cells) stain cell surfaces, indicating that they differ in this respect from matrix-associated proteoglycans. Further, the staining indicates that there is some tissue or cell specificity for some of these proteoglycans. While some of the core proteins may be related, such as the chondroitin sulfate proteoglycan on the melanoma cells and the dermatan sulfate and heparan sulfate proteoglycans on the granulosa cells, all of which have very similar core protein sizes, others may be quite distinct, such as the heparan sulfate proteoglycans on hepatocytes, skin fibroblasts, and granulosa cells, which have significantly different core protein sizes. It is clear that the final classification of these proteoglycans will depend on characterization of core protein structures since the nature of the glycosaminoglycan chains varies. Interestingly, core proteins of similar size on the granulosa cell surface are substituted with either dermatan sulfate chains or heparan sulfate chains whereas the core protein of the proteoglycan on the epithelial cell surface is substituted with both glycosaminoglycans to form hybrid molecules. The mechanisms used by the cell to distinguish whether a core protein is to be substituted with dermatan sulfate, heparan sulfate, or both are unknown, but it is possible that differences in the primary structures of the core proteins are involved.

BASEMENT MEMBRANE PROTEOGLYCANS

General Considerations

Basement membranes are thin sheets of extracellular matrix that separate endothelial cells, epithelial cells, muscle cells, fat cells, and neural tissue from adjacent connective tissue stroma. Heparan sulfate proteoglycans have been isolated from a variety of basement membranes of normal tissues or cells, such as kidney glomerular basement membrane (10, 168–174) or corneal endothelial cells (175), as well as from basement membrane–producing tumors and tumor cell lines (176–186a). Chondroitin sulfate proteoglycans have also been shown to be present in some basement membranes (10, 173, 174, 187), although less is known about their chemical structures. The basement membrane proteoglycans appear to be critically involved in such processes as tissue morphogenesis (188, 189) and determination of permselective properties through the membrane (168, 169, 186a, 189a). Immunocytochemical and histochemical localization shows that they are arrayed on the lamina densa which suggests that they interact with other basement membrane components through specific interactions involving the core protein and/or the glycosaminoglycan chains.

In contrast with the large cartilage proteoglycans, which are found to be in the same size class regardless of tissue source, basement membrane proteoglycans vary considerably in size and structure. However, they have some immunological determinants in common on their core proteins which also serve to distinguish them from the cell surface proteoglycans. The relationship of the various basement membrane proteoglycans to one another will be understood when their core proteins have been characterized and precursor proteins identified.

Glomerulus

Several studies indicate that the heparan sulfate proteoglycans isolated from glomeruli (173, 174) or from purified basement membrane (10) have $M_r =$ 130,000–200,000 with four to five heparan sulfate chains of $M_r = 25,000$–30,000 bound to a core protein of estimated $M_r = 30,000$. Thus, they have similar general structures to the cell surface heparan sulfate proteoglycans isolated from hepatocytes (described above). However, these two proteoglycans do not appear to be antigenically related; an antiserum to the glomerular basement membrane proteoglycan selectively binds to basement membranes (172), whereas one to the hepatocyte cell surface proteoglycan selectively binds to cell surfaces (151).

Antibodies to glomerulus proteoglycan immunoprecipitate the 400,000-dalton precursor protein to the EHS proteoglycan (M. A. J. Dickinson, P. E. Brenchley, J. R. Hassell, J. C. Anderson, personal communication). This indicates that these native and tumor proteoglycans, although different in size, have some common immunological determinants. In biosynthetic studies, variable amounts of a similar size chondroitin sulfate proteoglycan have been observed; glomerular basement membrane isolated from perfused rat kidney contained 10–15% of this species (10), and proteoglycans isolated from organ culture of isolated glomeruli from rabbit (173) and rat (174) kidney contained much larger proportions. The half-life of proteoglycans in rat glomerular basement membrane labeled in vivo is approximately one week (190) indicating that they turn over rapidly in this tissue.

Tumors and Tumor Cells

A number of tumors and tumor cell lines, such as the EHS (Englebreth-Holm-Swarm) tumor and the PYS-2 and L2 cells, make basement membrane components. Heparan sulfate proteoglycans produced by the EHS murine basement membrane tumor have been extensively studied (176–182). Two proteoglycans have been obtained: a low-buoyant-density proteoglycan with high molecular weight, M_r = approximately 750,000 by gel filtration (177) or 400,000–600,000 by sedimentation equilibrium centrifugation (182; J. Engel, M. Paulsson, R. Timpl, personal communication), and a large core preparation, M_r = 400,000 (177); and a high-buoyant-density proteoglycan with low molecular weight, M_r = 130,000–300,000, and a small core preparation, M_r = 35,000–135,000 (177, 182). Both proteoglycans contain three to five heparan sulfate chains with estimated M_r = 60,000 when compared with chondroitin sulfate standards by molecular sieve chromatography (176, 177, 182) or 29,000 for the high density form and 43,000 for the low density form when estimated by ultracentrifugation or from electron microscopy measurements (179, 182; Engel, M. Paulsson, R. Timple, personal communication). While some studies suggest that the smaller proteoglycans represent a separate population (182), kinetic experiments suggest that at least a portion of this species is derived from the larger proteoglycan (177), and the smaller species shares antigenic determinants with the larger (177, 181).

The L-2 cells also make a large (M_r = approximately 750,000) heparan sulfate proteoglycan (183, 184), while PYS (parietal yolk sac carcinoma) cells synthesize a smaller heparan sulfate proteoglycan of M_r = approximately 400,000 which contains four to five chains of M_r = 30,000 (185, 186). Polyclonal antibodies against the L-2 cell proteoglycan (184) as well as those against the EHS tumor proteoglycans react with basement membranes in a wide variety of normal tissues but do not recognize cell surfaces in immunohistochemical studies (176, 181, 183). These antibodies are indistinguishable in this regard from those directed against the glomerular basement membrane proteoglycan described above. Further, these antibodies have been used to identify a M_r = 400,000 protein precursor as a biosynthetic product in EHS tumor cells (178) as well as in embryonic muscle cells (180) and L-2 cells (184).

Both forms of the EHS tumor proteoglycans have been examined by electron microscopy either after spraying onto cleaved mica followed by rotary shadowing or by Kleinschmidt spreading techniques (179, 182). Preparations of the large proteoglycan show mainly molecules with long protein filaments, approximately 180 nm in length, with four to six side branches clustered at one end. Other investigators find a more compact structure for the core protein, possibly indicating the presence of domain structures (J. Engel, M. Paulsson, R. Timpl, personal communication). Preparations of the small proteoglycan show mainly a 42-nm core with four to six side branches distributed along its length. The asymmetrical location of the side chains on the larger proteoglycan has been confirmed by the isolation of a $M_r = 200,000$ fragment, which does not contain heparan sulfate, from trypsin digests of the proteoglycan (178).

Normal Cells

A monoclonal antibody prepared against frog tadpole tail connective tissue antigens has been isolated and shown to be directed against a heparan sulfate proteoglycan ($M_r = 500,000–600,000$) which yields a core preparation after heparatinase digestion of $M_r = 350,000–400,000$ (191). Like the antibodies to basement membranes (discussed above), this antibody localizes in a variety of basement membranes, including muscle, blood vessels, nerve sheath, and notochord; but it is especially concentrated in the specialized basal lamina of the neuromuscular junction. The similar size of the core preparations for this proteoglycan and the large proteoglycan from the EHS tumor suggest that the core proteins may be related.

INTRACELLULAR VESICLE PROTEOGLYCANS

A heparin proteoglycan present in secretory granules of mast cells was original-ly isolated from rat skin (192). Subsequent studies indicated that this pro-teoglycan had an unusual core protein of small size with a high proportion of serine and glycine residues (193). Pronase digestion left a core containing essentially only these two amino acids and it was suggested that at least a portion of the core protein contained a repeating ser-gly sequence with about 15 repeats required to accommodate the number of heparin side chains expected to be present. Isolation of the heparin proteoglycan from rat serosal mast cells gave a preparation that had an amino acid composition with more than 80% serine and glycine residues (6).

In separate studies with a mouse interleukin 3–dependent cell line derived from bone marrow mast cells, a chondroitin sulfate proteoglycan present in secretory granules was isolated (194, 195). This proteoglycan had an apparent $M_r = 200,000–250,000$, with multiple glycosaminoglycan chains and oligo-saccharides attached. The chondroitin sulfate was oversulfated with a high proportion of 4,6-disulfated galactosamine residues, a variant originally classi-fied as chondroitin sulfate E. The core preparation after chondroitinase diges-tion yielded a single band on SDS-PAGE with $M_r =$ approximately 10,000, and it was not degraded significantly by a number of proteases. Over 60% of the residues in the core protein were serine and glycine. It was suggested that the core proteins for the heparin and chondroitin sulfate E proteoglycans were homologous but probably distinct gene products based upon differences in amino acid compositions (195).

A similar proteoglycan was isolated from a rat basophilic leukemia cell line (197). The proteoglycans (M_r = 100,000–150,000) in this case appear to be hybrids with both heparan sulfate and dermatan sulfate chains (M_r = approximately 12,000) on individual core proteins. More than 50% of the amino acids are serine and glycine, and the core protein is resistant to a variety of proteases. The dermatan sulfate was oversulfated with the unusual disaccharide, iduronic acid-2-SO_4-N-acetylgalactosamine-4-SO_4, classifying the glycosaminoglycan as chondroitin sulfate B. The cells exocytosed the proteoglycan along with histamine and beta-hexosaminidase when stimulated to secrete. Since all of these storage granule proteoglycans contain oversulfated glycosaminoglycan chains and protease-resistant core proteins, they are thought to function in retention or storage of the cationic mediators and neutral proteases known to be present in the granules.

What appears to be a closely related chondroitin sulfate proteoglycan has been isolated from L-2 rat yolk sac tumor cells (197). The proteoglycan gave a core preparation of M_r = 25,000 after chondroitinase digestion. A cDNA clone which includes the sequence for the entire core protein of this proteoglycan has been isolated and characterized (15). The sequence contains 104 amino acids with a central region of 49 residues which contains alternating serines and glycines. Thus, the core protein would be approximately 10,000 in molecular weight and would contain the ser-gly repeating structure for initiating the glycosaminoglycan chains predicted from the chemical studies. The sequence contained no X-asparagine-Y-serine sequence, and the protein therefore would not contain N-linked oligosaccharides. It is not known whether this proteoglycan is related to secretory vesicles in this system.

Another small dermatan sulfate proteoglycan with M_r = approximately 250,000 has been isolated from PYS-2 cells (198) and has properties that suggest that its core protein may be related to that for the proteoglycan from L-2 cells. It has a large number of dermatan sulfate chains (M_r = approximately 15,000) that are clustered on a protease-resistant portion of the core protein. A core preparation with M_r = 27,000 and 54,000 bands was obtained after chondroitinase digestion.

A heparan sulfate proteoglycan has been isolated from the synaptic vesicles of the electroplaque organ of the electric eel (7). The proteoglycan has a core protein with estimated M_r = 70,000–140,000, and is associated with the inner membrane of the vesicles. It was suggested that these proteoglycans are deposited onto the cell surface in the synaptic cleft when the vesicles fuse with the membrane during neurotransmission activity. Thus, this proteoglycan may be involved in storage of transmitter molecules and may be related to the heparan sulfate proteoglycans concentrated in the neuromuscular junctions (described above).

TYPE IX COLLAGEN

It is now clear that a collagenous component in cartilage contains glycosaminoglycan side chains, and hence fits the definition of a proteoglycan. This molecule was, in fact, originally isolated from chick tibia as a dermatan sulfate proteoglycan designated PG-lt and shown to contain a collagenous core (199). Collagen investigators then isolated this molecule, designating it as p-M collagen (200), p-HMW collagen (202), and type IX collagen (203), its present designation. It has a $M_r = 300,000$ with three disulfide-bonded subunits of $M_r = 115,000$ 84,000, and 68,000 (compared with collagen molecular weight standards on SDS-PAGE). The 115,000-dalton subunit yields a 68,000-dalton core preparation after chondroitinase digestion, indicating that it contains one glycosaminoglycan chain. All three subunits react with antibodies to HMW collagen (202). Genetic clones have been obtained for the 84,000-dalton subunit (α-1 type IX collagen) and for the protein of the 115,000-dalton subunit (α-2 type IX collagen) (203). Sequences derived from these clones distinguish this molecule from other collagen types. It is not clear whether all type IX collagen molecules contain a glycosaminoglycan chain. This collagen has been postulated to play a role in stabilizing the network of collagenous fibers found in cartilage tissue (204).

CONCLUDING REMARKS

Proteoglycans are a diverse group of macromolecules having in common the presence of glycosaminoglycan side chains. Historically they have been classified according to the type of glycosaminoglycan chain or by the tissue of origin. Other connective tissue components such as the collagens or glycoproteins (link protein, fibronectin, laminin, osteonectin, etc) are primarily distinguished from one another by properties of their protein (size, immunological determinants, peptide maps, primary sequence, etc). In this review, we have attempted to begin to use similar criteria as a basis for proteoglycan classification. Several major families have emerged for those proteoglycans whose core proteins have been best characterized: 1. a group of large proteoglycans with hyaluronic acid–binding regions found primarily in cartilages but present in smaller amounts in aorta, bone, tendon, smooth muscle cells, etc; 2. small proteoglycans with a characteristic core preparation doublet at 43,000 and 47,000 daltons found in most connective tissues; and, 3. small proteoglycans characteristic of intracellular storage granules with a long repeating ser-gly region in the core protein. Other groups such as the basement membrane and cell surface proteoglycans appear more complex and cannot be sorted into recognizable families at present. In this regard, the production of cDNA clones for a wide variety of proteoglycan core proteins will be most useful in determining core protein structures and establishing the guidelines for classifying the expected gene families.

Literature Cited

1. Hascall, V. C., Hascall, G. K. 1981. In *Cell Biology of Extracellular Matrix*, ed. E. Hay, pp. 39–63. New York: Plenum
2. Hascall, V. C. 1981. In *Biology of Carbohydrates*, ed. V. Ginsburg, P. Robbins, 1:1–49. New York: Wiley
3. Anderson, R. S., Schwartz, E. R. 1984. *Arthritis Rheum.* 27:58–71
4. Oegema, T. R., Kraft, E. L., Jourdian, G. W., vanValen, T. R. 1984. *J. Biol. Chem.* 259:1720–26
5. Kimura, J. H., Lohmander, L. S., Hascall, V. C. 1984. *J. Cell. Biochem.* 26:261–78
6. Metcalfe, D. D., Smith, J. A., Austen, K. F., Silbert, J. E. 1980. *J. Biol. Chem.* 255:11753–58
7. Carlson, S. S., Kelley, R. B. 1983. *J. Biol. Chem.* 258:11082–91
8. Bumol, T. F., Reisfeld, R. A. 1982. *Proc. Natl. Acad. Sci. USA.* 79:1245–49
9. Rapraeger, A., Jalkanen, M., Endo, E., Koda, J., Bernfield, M. 1985. *J. Biol. Chem.* 260:11046–52
10. Kanwar, Y. S., Hascall, V. C., Farquhar, M. G. 1981. *J. Cell Biol.* 90:527–32
11. Kimura, J. H., Thonar, E. J.-M., Hascall, V. C., Reiner, A., Poole, A. R. 1981. *J. Biol. Chem.* 256:7890–97
12. Glossl, J., Beck, M., Kresse, H. 1984. *J. Biol. Chem.* 259:14144–50
13. Upholt, W. B., Vertel, B. M., Dorfman, A. 1979. *Proc. Natl. Acad. Sci. USA.* 76:4847–51
14. Vertel, B. M., Upholt, W. B., Dorfman, A. 1984. *Biochem. J.* 217:259–63
15. Bourdon, M. A., Oldberg, A., Pierschbacher, M., Ruoslahti, E. 1985. *Proc. Natl. Acad. Sci. USA.* 82:1321–25
16. Hardingham, T. E. 1984. In *Molecular Biophysics of the Extracellular Matrix*, ed. S. Arnott, D. Rees, E. Morris, pp. 1–19. Clifton, NJ: Humana
17. Heinegard, D., Paulsson, M. 1984. In *Extracellular Matrix Biochemistry*, ed. K. Piez, A. Reddi, pp. 277–328. New York: Elsevier
18. Muir, I. H. M. 1980. In *The Joints and Synovial Fluid*, ed. L. Sokolof, 2:27–94. New York: Academic
19. Roden, L. 1980. In *The Biochemistry of Glycoproteins and Proteoglycans*, ed.

W. Lennarz, pp. 267–371. New York: Plenum
20. Lohmander, L. S., De Luca, S., Nilsson, B., Hascall, V. C., Caputo, C. B., et al. 1980. *J. Biol. Chem.* 255:6084–91
21. Nilsson, B., De Luca, S., Lohmander, S., Hascall, V. C. 1982. *J. Biol. Chem.* 257:10920–27
22. Santer, V., White, R. J., Roughley, P. J. 1982. *Biochim. Biophys. Acta* 716:277–82
23. Thonar, E. J.-M., Sweet, M. B. E. 1979. *Biochim. Biophys. Acta* 584:353–57
24. Thonar, E. J.-M., Lohmander, L. S., Kimura, J. H., Fellini, S. A., Yanagishita, M., Hascall, V. C. 1983. *J. Biol. Chem.* 258:11564–70
25. Hascall, V. C. 1983. In *Limb Development and Regeneration, Part B*, ed. R. Kelley, P. Goetinck, J. MacCabe, pp. 3–15. New York: Liss
26. De Luca, S., Lohmander, L. S., Nilsson, B., Hascall, V. C., Caplan, A. I. 1980. *J. Biol. Chem.* 255:6077–83
27. Fellini, S. A., Kimura, J. H., Hascall, V. C. 1981. *J. Biol. Chem.* 256:7883–89
28. Hascall, V. C., Sajdera, S. W. 1970. *J. Biol. Chem.* 245:4920–30
29. Heinegard, D. 1977. *J. Biol. Chem.* 252: 1980–89
30. Hardingham, T. E., Ewins, R. J. F., Muir, H. M. 1976. *Biochem. J.* 157:127
31. Rosenberg, L. C., Wolfenstein-Todel, C., Margolis, R., Pal, S., Strider, W. 1976. *J. Biol. Chem.* 251:6439–44
32. Heinegard, D., Wieslander, J., Sheehan, J., Paulsson, M., Sommarin, Y. 1985. *Biochem. J.* 225:95–106
33. Stevens, J. W., Oike, Y., Handley, C., Hascall, V. C., Hampton, A., Caterson, B. 1984. *J. Cell. Biochem.* 26:247–59
34. Bonnet, F., Le Gledic, S., Perin, J.-P., Jolles, J., Jolles, P. 1983. *Biochim. Biophys. Acta* 743:82–90
35. Hardingham, T. E., Muir, H. 1972. *Biochim. Biophys. Acta* 279:401–5
36. Hardingham, T. E., Muir, H. 1973. *Biochem. J.* 135:905–8
37. Hascall, V. C., Heinegard, D. 1974. *J. Biol. Chem.* 249:4242–49
38. Heinegard, D., Hascall, V. C. 1979. *J. Biol. Chem.* 254:921–26
39. Bonnet, F., Dunham, D. G., Harding-

ham, T. E. 1985. *Biochem. J.* 228:77–85
40. Heinegard, D., Axelsson, I. 1977. *J. Biol. Chem.* 252:1971–79
41. Hascall, V. C. 1977. *J. Supramol. Struct.* 7:101–20
42. Lohmander, L. S., Fellini, S. A., Kimura, J. H., Stevens, R. L., Hascall, V. C. 1983. *J. Biol. Chem.* 258:12280–86
43. Bjornsson, S., Heinegard, D. 1981. *Biochem. J.* 199:17–29
44. Kimura, J. H., Hardingham, T. E., Hascall, V. C., Solursh, M. 1979. *J. Biol. Chem.* 254:2600–9
45. Bayliss, M. T., Ridgeway, G. D., Ali, S. Y. 1984. *Biosci. Rep.* 4:827–33
46. Bayliss, M. T., Ridgeway, G. D., Ali, S. Y. 1983. *Biochem. J.* 215:705–8
47. Oegema, T. R. 1980. *Nature* 288:583–85
48. Roughley, P. J., Killackey, B. 1984. *Trans. Orthop. Res.* 9:115
49. Rosenberg, L., Hellmann, W., Kleinschmidt, A. K. 1970. *J. Biol. Chem.* 245:4123–30
50. Thyberg, J., Lohmander, S., Heinegard, D. 1975. *Biochem. J.* 151:157–66
51. Buckwalter, J. A., Rosenberg, L. C. 1982. *J. Biol. Chem.* 257:9830–39
52. Kimura, J. H., Osdoby, P., Caplan, A. I., Hascall, V. C. 1978. *J. Biol. Chem.* 253:4721–29
53. Heinegard, D., Lohmander, S., Thyberg, J. 1978. *Biochem. J.* 175:913–19
54. Rosenberg, L., Margolis, R., Hellmann, W., Kleinschmidt, A. K. 1975. *J. Biol. Chem.* 250:1877–83
55. Hascall, G. K. 1980. *J. Ultrastruct. Res.* 70:369–75
56. Faltz, L. L., Reddi, A. H., Hascall, G. K., Martin, D., Pita, J. C., Hascall, V. C. 1979. *J. Biol. Chem.* 254:1375–80
57. Buckwalter, J. A., Rosenberg, L. C. 1983. *Col. Rel. Res.* 3:489–504
58. Buckwalter, J. A., Poole, A. R., Reiner, A., Rosenberg, L. C. 1982. *J. Biol. Chem.* 257:10529–32
59. Wiedemann, H., Paulsson, M., Timpl, R., Engel, J., Heinegard, D. 1984. *Biochem. J.* 224:331–33
60. Ratcliffe, A., Fryer, P. R., Hardingham, T. E. 1984. *J. Histochem. Cytochem.* 32:193–201
61. Ratcliffe, A., Fryer, P. R., Hardingham, T. E. 1985. *J. Cell Biol.* 101:2355–65
62. Pasternack, S. G., Veis, A., Bren, M. 1974. *J. Biol. Chem.* 249:2206–11
63. Reihanian, H., Jamieson, A. M., Tang, L.-H., Rosenberg, L. 1979. *Biopolymers* 18:1727–47
64. Sheehan, J. K., Nieduszynski, I. A.,

Phelps, C. F., Muir, H., Hardingham, T. E. 1978. *Biochem. J.* 171:109–14
65. Hascall, V. C., Riola, R. L. 1972. *J. Biol. Chem.* 247:4529–38
66. Kimata, K., Hascall, V. C., Kimura, J. H. 1982. *J. Biol. Chem.* 257:3827–32
67. Oike, Y., Kimata, K., Shinomura, T., Suzuki, S. 1982. *J. Biol. Chem.* 257:9751–58
68. Oike, Y., Kimata, K., Shinomura, T., Nakazawa, K., Suzuki, S. 1982. *Biochem. J.* 191:193–207
69. Olson, C. A., Krueger, R., Schwartz, N. B. 1985. *Anal. Biochem.* 146:232–37
70. Schwartz, N. B., Habib, G., Campbell, S., D'Elvlyn, D., Gartner, M., et al. 1985. *Fed. Proc.* 44:369–72
71. Treadwell, B. V., Mankin, D. P., Ho, P. K., Mankin, H. J. 1980. *Biochemistry* 19:2269–75
72. Mitchell, D., Hardingham, T. E. 1981. *Biochem. J.* 196:521–29
73. Cole, N. N., Lowther, D. A. 1969. *FEBS Lett.* 2:351–53
74. McQuillan, D. J., Handley, C. J., Robinson, H. C., Ng, K., Tzaicos, C., et al. 1984. *Biochem. J.* 224:977–88
75. Fellini, S. A., Kimura, J. H., Hascall, V. C. 1984. *J. Biol. Chem.* 259:4634–41
76. Mitchell, D., Hardingham, T. E. 1982. *Biochem. J.* 202:249–54
77. Nishimoto, S. K., Kajiwara, T., Tanzer, M. L. 1982. *J. Biol. Chem.* 257:10558–61
78. Burditt, L. J., Ratcliffe, A., Fryer, P. R., Hardingham, T. E. 1985. *Biochim. Biophys. Acta* 844:247–55
79. Stevens, R. L., Parsons, W. G., Austen, K. F., Hein, A., Caulfield, J. P. 1985. *J. Biol. Chem.* 260:5777–86
79a. Pacifici, M., Soltesz, R., Thal, G., Shanley, J. D., Boettiger, D., Holtzer, H. 1983. *J. Cell Biol.* 97:1724–36
80. Geetha-Habib, M., Campbell, S. C., Schwartz, N. B. 1984. *J. Biol. Chem.* 259:7300–10
81. McDevitt, C. A., Muir, H. 1971. *Anal. Biochem.* 44:612–22
82. Pearson, J. P., Mason, R. M. 1978. *Biochem. Soc. Trans.* 6:244–46
83. Stanescu, V., Maroteaux, P., Sobczak, E. 1977. *Biochem. J.* 163:103–9
84. Stanescu, V., Maroteaux, P., Stanescu, R., Sobczak, E. 1975. *Biol. Neonate* 27:361–67
85. Stanescu, V., Stanescu, R., Maroteaux, P. 1984. *J. Bone Jt. Surg.* 66-A:817–36
86. Oegema, T. E., Thompson, R. 1986. *Cartilage Biochemistry*, ed. K. Kuettner, R. Schleyerbach, V. Hascall. Raven. In press
87. Bayliss, M. See Ref. 86

87a. Thonar, E. J. M., Bjornsson, S., Kuettner, K. See Ref. 86
88. Caputo, C. B., MacCallum, D. K., Kimura, J. H., Schrode, J., Hascall, V. C. 1980. *Arch. Biochem. Biophys.* 204:220–33
89. Poole, A. R., Reiner, A., Mort, J. S., Reihanian, H., Jamieson, A. M., et al. 1984. *J. Biol. Chem.* 259:14849–56
90. Chang, Y., Yanagishita, M., Hascall, V. C., Wight, T. N. 1983. *J. Biol. Chem.* 258:5679–88
91. Wight, T. N., Hascall, V. C. 1983. *J. Cell Biol.* 96:167–76
92. Gardell, S., Baker, J., Caterson, B., Heinegard, D., Roden, L. 1980. *Biochem. Biophys. Res. Commun.* 95:1823–31
93. McMurtrey, J., Radhakrishnamurthy, B., Dalferes, E. R., Berenson, G. S., Gregory, J. D. 1979. *J. Biol. Chem.* 254:1621–26
94. Oegema, T. R., Hascall, V. C., Eisenstein, R. 1979. *J. Biol. Chem.* 254:1312–18
95. Vijayagopal, P., Radhakrishnamurthy, B., Srinivasan, S. R., Berenson, G. S. 1985. *Biochim. Biophys. Acta* 839:110–18
96. Wagner, W. D., Rowe, H. A., Connor, J. R. 1983. *J. Biol. Chem.* 258:11136–42
97. Norling, B., Glimelius, B., Westermark, B., Wasteson, A. 1978. *Biochem. Biophys. Res. Commun.* 84:914–21
98. Vogel, K. G., Heinegard, D. 1985. *J. Biol. Chem.* 260:9298–306
99. Schafer, I. A., Sitabkha, L., Pandy, M. 1984. *J. Biol. Chem.* 259:2321–30
100. Coster, L., Carlstedt, I., Malmstrom, A. 1979. *Biochem. J.* 183:669–81
101. Fisher, L. W., Termine, J. D., Dejter, S. W., Whitson, S. W., Yanagishita, M., et al. 1983. *J. Biol. Chem.* 258:6588–94
102. Mangkarnkanok-Mark, M., Eisenstein, R., Baker, R. M. 1981. *J. Histochem. Cytochem.* 29:547–52
103. Deleted in proof
104. Argraves, W. S., McKeown-Longo, P. J., Goetinck, P. F. 1981. *FEBS Lett.* 131:265–68
105. McKeown-Longo, P. J., Velleman, S. G., Goetinck, P. F. 1983. *J. Biol. Chem.* 258:10779–85.
106. Kimata, K., Barrach, H.-J., Brown, K. S., Pennypacker, J. P. 1981. *J. Biol. Chem.* 256:6961–68
107. Heinegard, D., Paulsson, M., Inerot, S., Carlstrom, C. 1981. *Biochem. J.* 197:355–66
108. Stanescu, V., Sweet, M. B. E. 1981. *Biochim. Biophys. Acta* 673:101–13
109. Swann, D. A., Garg, H. G., Sotman, S.

L., Hermann, H. 1983. *J. Biol. Chem.* 258:2683–88
110. Rosenberg, L. C., Choi, H. U., Tang, L.-H., Johnson, T. L., Pal, S., et al. 1985. *J. Biol. Chem.* 260:6304–13
111. Shinomura, T., Kimata, K., Oike, Y., Noro, A., Hirose, N., et al. 1983. *J. Biol. Chem.* 258:9314–22
112. Hascall, V. C., Handley, C. J., McQuillan, D. J., Hascall, G. K., Robinson, H. C., Lowther, D. A. 1983. *Arch. Biochem. Biophys.* 224:206–23
113. Campbell, M. A., Handley, C. J., Hascall, V. C., Campbell, R. A., Lowther, D. A. 1984. *Arch. Biochem. Biophys.* 234:275–89
114. Stanescu, V., Stanescu, R. 1983. *Biochim. Biophys. Acta* 757:377–81
115. Pearson, C. H., Winterbottom, N., Fackre, D. S., Scott, P. G., Carpenter, M. R. 1983. *J. Biol. Chem.* 258:15101–4
116. Pringle, G. A., Dodd, C. M., Osborn, J. W., Pearson, C. H., Mosmann, T. R. 1985. *Collagen Relat. Res.* 5:23–39
116a. Chopra, R. K., Pearson, H., Pringle, G. A., Fackre, D. S., Scott, P. G. 1986. *Biochem. J.* In press
117. Deleted in proof
118. Fransson, L.-A., Coster, L., Nieduszynski, I. A., Phelps, C. F., Sheehan, J. K. 1984. See Ref. 16, pp. 95–118
119. Deleted in proof
120. Vogel, K. G., Paulsson, M., Heinegard, D. K. 1984. *Biochem. J.* 223:587–97
121. Scott, J. E., Orford, C. R. 1981. *Biochem. J.* 197:213–16
122. Scott, J. E., Orford, C. R., Hughes, E. W. 1981. *Biochem. J.* 195:573–81
122a. Gallagher, J. T., Gasiunas, N., Schor, S. L. 1983. *Biochem. J.* 215:107–16
123. Fisher, L. W. 1985. In *The Chemistry and Biology of Mineralized Tissues*, ed. W. Butler, pp. 188–96. Birmingham, Ala: Ebsco Media
124. Heinegard, D. K., Bjorne-Persson, A., Coster, L., Franzen, A., Gardell, S., et al. 1985. *Biochem. J.* 230:181–94
125. Franzen, A., Heinegard, D. K. 1984. *Biochem. J.* 224:59–66
126. Habuchi, H., Kimata, K., Suzuki, S. 1986. *J. Biol. Chem.* 261:1031–40
127. Damle, S. P., Coster, L., Gregory, J. D. 1982. *J. Biol. Chem.* 257:5523–27
128. Pearson, C. H., Gibson, G. J. 1982. *Biochem. J.* 201:27–37
129. Brennan, M. J., Oldberg, A., Pierschbacher, M. D., Ruoslahti, E. 1984. *J. Biol. Chem.* 259:13742–50
130. Antonopoulos, C. A., Axelsson, I., Heinegard, D., Gardell, S. 1974. *Biochim. Biophys. Acta* 388:108–19

131. Axelsson, I., Heinegard, D. 1975. *Biochem. J.* 145:491–500
132. Axelsson, I., Heinegard, D. 1978. *Biochem. J.* 169:517–30
133. Keller, R., Stein, T., Stuhlsatz, H. W., Greiling, H., Ohst, E., et al. 1981. *Hoppe-Seyler's Z. Physiol. Chem.* 362:327–36
134. Gregory, J. D., Coster, L., Damle, S. P. 1982. *J. Biol. Chem.* 257:6965–70
135. Hassell, J. R., Newsome, D. A., Hascall, V. C. 1979. *J. Biol. Chem.* 254:12346–54
136. Nakazawa, K., Hassell, J. R., Hascall, V. C., Newsome, D. A. 1983. *Arch. Biochem. Biophys.* 222:105–16
137. Nakazawa, K., Newsome, D. A., Nilsson, B., Hascall, V. C., Hassell, J. R. 1983. *J. Biol. Chem.* 258:6051–55
138. Axelsson, I. 1977. *Keratan Sulfate Proteoglycans.* PhD thesis. University Lund, Sweden
139. Cintron, C., Hassinger, L. C., Kublin, C., Newsome, D. A. 1983. *J. Ultrastruct. Res.* 65:13–22
140. Hassell, J. R., Cintron, C., Kublin, C., Newsome, D. A. 1983. *Arch. Biochem. Biophys.* 222:362–69
141. Hassell, J. R., Newsome, D. A., Nakazawa, K., Rodrigues, M., Krachmer, J. 1982. In *Extracellular Matrix*, ed. S. Hawkes, J. Wang, pp. 397–406. New York: Academic
142. Klintworth, G. K., Smith, C. F. 1983. *Lab. Invest.* 48:603–12
143. Nakazawa, K., Hassell, J. R., Hascall, V. C., Lohmander, L. S., Newsome, D. A., Krachmer, J. 1984. *J. Biol. Chem.* 259:13751–57
144. Hassell, J. R., Hascall, V. C., Ledbetter, S., Caterson, B., Thonar, E. J.-M., et al. 1984. In *Proc. 8th Symp. Ocular and Visual Dev.*, ed. S. Hilfer, J. Sheffield, pp. 101–14. New York: Springer-Verlag
145. Conrad, G. W., Johnson-Ager, P., Woo, M. L. 1982. *J. Biol. Chem.* 257:464
146. Nilsson, B., Nakazawa, K., Hassell, J. R., Newsome, D. A., Hascall, V. C. 1983. *J. Biol. Chem.* 258:6056–63
147. Scott, J. E. 1986. *Ciba Found. Symp.*, ed. J. Whelan. New York: Wiley. In press
148. Deleted in proof
149. Hook, M., Kjellen, L., Johansson, S., Robinson, J. 1984. *Ann. Rev. Biochem.* 53:847–69
150. Kjellen, L., Pettersson, I., Hook, M. 1981. *Proc. Natl. Acad. Sci. USA* 78:5371–75
151. Stow, J. L., Kjellen, L., Unger, E., Hook, M., Farquhar, M. G. 1985. *J. Cell Biol.* 100:975–80
152. Bumol, T. F., Walker, L. E., Reisfeld, R. A. 1984. *J. Biol. Chem.* 259:12733–41
153. Harper, J. R., Bumol, T. F., Reisfeld, R. A. 1984. *J. Immunol.* 132:2096–103
154. Harper, J. R., Reisfeld, R. A., Quaranta, V. 1986. *J. Biol. Chem.* In press
155. Ross, A. H., Cossu, G., Herlyn, M., Bell, J. R., Steplewski, Z., Kaprowski, H. 1983. *Arch. Biochem. Biophys.* 225:370–83
156. Yanagishita, M., Hascall, V. C. 1984. *J. Biol. Chem.* 259:10260–69
157. Yanagishita, M., Hascall, V. C. 1983. *J. Biol. Chem.* 258:12857–64
158. Yanagishita, M., Hascall, V. C. 1984. *J. Biol. Chem.* 259:10270–83
159. Yanagishita, M., Hascall, V. C. 1984. *J. Biol. Chem.* 259:10260–69
160. Yanagishita, M., Rodbard, D., Hascall, V. C. 1979. *J. Biol. Chem.* 254:911–20
161. David, G., Van der Berghe, H. 1985. *J. Biol. Chem.* 260:11067–74
162. Rapraeger, A., Bernfield, M. 1985. *J. Biol. Chem.* 260:4103–9
163. Jalkanen, M., Nguyen, H., Rapraeger, A., Kurn, N., Bernfield, M. 1985. *J. Cell Biol.* 101:976–84
164. Fransson, L.-A., Carlstedt, I., Coster, L., Malmstrom, A. 1984. *Biochemistry* 81:5657–61
165. Carlstedt, I., Coster, L., Malmstrom, A., Fransson, L.-A. 1983. *J. Biol. Chem.* 258:11629–35
166. Coster, L., Malmstrom, A., Carlstedt, I., Fransson, L.-A. 1983. *Biochem. J.* 215:417–19
167. Iozzo, R. V. 1984. *J. Cell Biol.* 99:403–17
168. Farquhar, M. G. 1981. See Ref. 1
169. Kanwar, Y. S., Farquhar, M. G. 1979. *Proc. Natl. Acad. Sci. USA* 76:4493–97
170. Kanwar, Y. S., Veis, A., Kimura, J. H., Jakubowski, M. L. 1984. *Proc. Natl. Acad. Sci. USA* 81:762–66
171. Parthasarathy, N., Spiro, R. G. 1984. *J. Biol. Chem.* 259:12749–55
172. Stow, J. L., Sawada, H., Farquhar, M. G. 1985. *Proc. Natl. Acad. Sci. USA* 82:3296–300
173. Stow, J. L., Glasgow, E. F., Handley, C. J., Hascall, V. C. 1983. *Arch. Biochem. Biophys.* 225:950–57
174. Kobayashi, S., Oguri, K., Kobayashi, K., Okayama, M. 1983. *J. Biol. Chem.* 258:12051–57
175. Robinson, J., Gospodarowicz, D. 1984. *J. Biol. Chem.* 259:3818–24
176. Oldberg, Å., Robey, P. G., Barrach, H.-J., Wilczek, J., Rennard, S. I., Martin, G. R. 1980. *Proc. Natl. Acad. Sci. USA* 77:4494–98

177. Hassell, J. R., Leyshon, W. C., Ledbetter, S. R., Tyree, B., Suzuki, S., et al. 1985. *J. Biol. Chem.* 260:8098–105
178. Ledbetter, S. R., Tyree, B., Hassell, J. R., Horigan, E. A. 1985. *J. Biol. Chem.* 260:8106–13
179. Laurie, G. W., Hassell, J. R., Martin, G. R., Fisher, L., Lewis, M. S. 1986. Submitted for publication
180. Hassell, J. R., Horigan, E. A., Mosley, G. L., Ledbetter, S. R., Kleinman, H. K., Chandrasekhar, S. 1985. In *Progress in Clinical and Biological Research*, ed. J. Lash, K. Kratochivil, L. Saxen, 171:75–86. New York: Liss
181. Dziadek, M., Fujiwara, S., Paulsson, M., Timpl, R. 1985. *EMBO J.* 4:905–12
182. Fujiwara, S., Wiedeman, H., Timpl, R., Lustig, A., Engel, J. 1984. *Eur. J. Biochem.* 143:145–57
183. Fenger, M., Wewer, U., Albrechtsen, R. 1984. *FEBS Lett.* 173:75–79
184. Wewer, U. M., Albrechtsen, R., Hassell, J. R. 1985. *Differentiation.* 30:61–67
185. Oohira, A., Wight, T. N., McPherson, J., Bornstein, P. 1982. *J. Cell Biol.* 92:357–67
186. Oohira, A., Wight, T. N., Bornstein, P. 1983. *J. Biol. Chem.* 258:2014–21
186a. Rohrbach, D. H., Wagner, C. W., Star, V., Martin, G. R., Brown, K. S., Yoon, J.-W. 1983. *J. Biol. Chem.* 258:11671–77
187. Bernfield, M. R., Banerjee, S. D., Cohn, R. H. 1972. *J. Cell Biol.* 52:674
188. Gordon, J. R., Bernfield, M. R. 1980. *Dev. Biol.* 74:118
189. Thompson, H. A., Spooner, B. S. 1982. *Dev. Biol.* 89:417–24
189a. Mynderse, L. A., Hassell, J. R., Kleinman, H. K., Martin, G. R., Martinez-Hernandez, A. 1983. *Lab. Invest.* 48:292–302
190. Cohen, M. P., Surma, M. L. 1982. *Biochim. Biophys. Acta* 716:337–40
191. Anderson, M. J., Fambrough, D. M. 1983. *J. Cell Biol.* 97:1396–411
192. Horner, A. A. 1971. *J. Biol. Chem.* 246:231–39
193. Robinson, H. C., Horner, A. A., Hook, M., Ogren, S., Lindahl, U. 1978. *J. Biol. Chem.* 253:6687–93
194. Stevens, R. L., Austen, K. F. 1982. *J. Biol. Chem.* 257:253–59
195. Stevens, R. L., Otsu, K., Austen, K. F. 1985. *J. Biol. Chem.* 260:14194–200
196. Seldin, D. C., Austen, K. F., Stevens, R. L. 1985. *J. Biol. Chem.* 260:11131–39
197. Oldberg, A., Hayman, E. G., Ruoslahti, E. 1981. *J. Biol. Chem.* 256:10847–52
198. Couchman, J. R., Woods, A., Hook, M., Christner, J. E. 1985. *J. Biol. Chem.* 260:13755–62
199. Nora, A., Kimata, K., Oike, Y., Shinomura, T., Maeda, N., et al. 1983. *J. Biol. Chem.* 258:9323–31
200. von der Mark, K., von Merkel, M., Wiedemann, H. 1984. *Eur. J. Biochem.* 138:629–33
201. Bruckner, P., Mayne, R., Tuderman, L. 1983. *Eur. J. Biochem.* 136:333–39
202. Vaughan, L., Winterhalter, K. H., Bruckner, P. 1985. *J. Biol. Chem.* 260:4758–63
203. van der Rest, M., Mayne, R., Ninomiya, Y., Seidah, N. G., Chretien, M., Olsen, B. R. 1985. *J. Biol. Chem.* 260:220–25
204. Bruckner, P., Vaughan, L., Winterhalter, K. H. 1985. *Proc. Natl. Acad. Sci. USA* 82:2608–12

Ann. Rev. Biochem. 1986. 55:569–97
Copyright © 1986 by Annual Reviews Inc. All rights reserved

THE ROLE OF ANTISENSE RNA IN GENE REGULATION

Pamela J. Green¹, Ophry Pines, and Masayori Inouye

Department of Biochemistry, State University of New York at Stony Brook, Stony Brook, New York 11794

CONTENTS

PERSPECTIVES AND SUMMARY

Gene expression in both prokaryotes and eukaryotes is controlled by the products of regulatory genes. According to a great number of studies in the past 20 years, the products of such genes were determined to be proteins termed activators or repressors. Recently, naturally occurring regulatory genes have been discovered that direct the synthesis of RNA which can directly control gene expression. These newly discovered RNA repressors are highly specific inhibitors of gene expression. The regulatory RNA contains a sequence that is complementary to the target RNA, and binding of the two RNAs occurs by base

¹Present Address: Laboratory of Plant Molecular Biology, The Rockefeller University, 1230 York Avenue, New York, New York 10021

0066-4154/86/0701-0569$02.00

pairing. The term "antisense RNA" has been coined to designate this regulatory RNA. The genes that direct the synthesis of antisense RNA are designated antisense genes.

Antisense genes were initially discovered in prokaryotes. Within a relatively short period of time, antisense RNA has been identified in the regulation of diverse and complex phenomena in bacteria such as plasmid replication and incompatibility, Tn10 transposition, osmoregulation of porin expression, regulation of phage reproduction, and autoregulation of cAMP-receptor protein synthesis. At present, naturally occurring antisense genes in eukaryotes have not been identified.

Three lines of research have emerged following the discoveries of antisense genes: (a) searches for additional systems in which antisense RNA regulation is naturally involved, (b) the development of systems in which artificially constructed antisense genes regulate a cellular or viral gene of interest, and (c) studies probing the mechanisms by which antisense RNA affects gene expression.

The finding that antisense RNA can inhibit gene expression in natural systems led to the development of strategies to artificially regulate genes using antisense RNA. With relatively simple manipulations, antisense RNA complementary to a chosen mRNA can be synthesized in vivo and may be used to inhibit the expression of the respective target gene. The function of endogenous genes has now been suppressed both in prokaryotes and eukaryotes by artificial antisense genes. In eukaryotes, direct microinjection of antisense RNA (synthesized in vitro) into cells has also resulted in the specific inhibition of gene expression. The potential use of antisense RNAs, for not only basic research but also applied research, is demonstrated by the types of genes that have been the successful targets of antisense RNA–mediated inhibition. These include, for example, developmentally regulated genes and genes that code for products that may be harmful for the host such as oncogenes and genes that are required for virus production. Thus, it should be possible to use antisense RNAs not only for gene therapy but also for preventing viral infection.

As mentioned earlier, some of the requirements for antisense RNA regulation are understood. However, many aspects of the mechanism are still unclear. This review will attempt to cover the accumulated data and our current understanding of antisense RNA regulation.

PROKARYOTIC ANTISENSE REGULATORY SYSTEMS

Plasmid Replication

The regulatory role of small complementary RNAs first became apparent through the study of plasmid replication in *Escherichia coli*. For members of the ColE1 and FII Plasmid groups, replication is negatively controlled by a distinct untranslated RNA species approximately 100 nucleotides long (1, 2). In addition to regulating plasmid copy number, the small RNAs prohibit the stable maintenance of two similar plasmids in the same cell, a phenomenon known as incompatibility. Recently two small RNAs were shown to regulate the replication of a *Staphylococcus aureus* plasmid, pT181, in a similar way (3). Yet despite sharing a common purpose, the ColE1, FII, and pT181 regulatory RNAs have different modes of action.

ColE1   The negative control of replication of ColE1-type plasmids is accomplished by RNA I which is transcribed in the opposite direction from the same DNA that encodes the primer RNA for DNA replication at the origin (1, 4–6). Replication of ColE1 initiates when the primer precursor (RNA II) hybridizes to its DNA template and is cleaved by ribonuclease H at the origin (7). Using the cleaved RNA as a primer, DNA synthesis by DNA polymerase I can ensue. RNA I functions to block processing of the primer precursor by hybridizing with it (1, 4) to prevent the formation of a suitable ribonuclease H substrate (i.e. a DNA:RNA hybrid). RNA I is complementary to nucleotides −552 to −447 of RNA II (8) which initiates 555 nucleotides upstream (at −555) of the replication origin (7).

Many mutant studies, which have been reviewed previously (9, 10), and ribonuclease sensitivity experiments (8, 11, 12) indicate that RNA I exists in a tRNA-like cloverleaf structure containing three stem-and-loops. The three stem-and-loops (1, 6, 13) function together with the 5' end of RNA I (11, 14, 15) to achieve hybrid formation with RNA II. Detailed in vitro kinetic studies support an initial transient interaction between the loop regions of RNA I and RNA II followed by pairing beginning at the 5' end of RNA I as depicted in Figure 1A (11). These initial contacts then lead to complete hybrid formation.

An alternative model for RNA I secondary structure where RNA I forms a dimer (by pairing in an antiparallel fashion) has also been proposed (9). The RNA I dimer has a double-stranded rather than a single-stranded cloverleaf structure and contains three interior loops. Closure of the interior loops has been proposed as a mechanism by which mutations might inactivate RNA I. While this model provides an attractive explanation for the effects of several mutations (9), it is important to realize that a dimer of RNA I has not been observed thus far.

Maximal negative control of plasmid replication by RNA I requires a ColE1-encoded protein consisting of 63 amino acid residues (16–19), referred to as Rop (16) or Rom (17). A target site involved in conferring sensitivity to Rop-RNA I-mediated inhibition exists between 52 and 135 nucleotides downstream of the RNA II transcription start site (18, 19). Recent work indicates that Rop protein aids in the binding of RNA I to RNA II by specifically enhancing the initial reversible interaction between the two species (17, 20). Rop protein

A

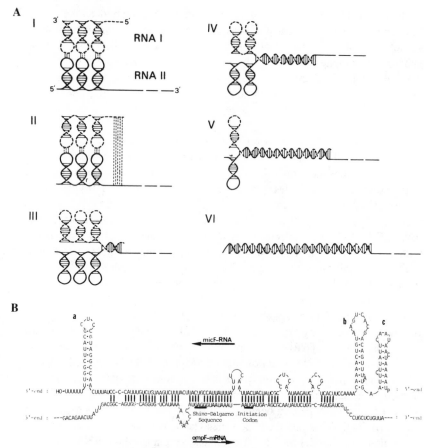

B

in the presence of RNA I may also elicit transcription termination or pausing as transcripts consistent with termination 200 (and possibly 110) nucleotides downstream of the primer start site have been observed in vitro (21). Such an effect has not been observed with ColE1 (17). Rop requires RNA I for activity, but RNA I can also function alone. The contribution of Rop varies depending upon the conditions or the parameter being assayed. For example, Rop has been observed to enhance binding of RNA I and RNA II by twofold (17), and decrease copy number (17) or in vitro replication (21) by threefold. In the absence of Rop approximately four times more RNA I is required to block pMB1 primer processing in vitro (21).

FII   Plasmids that belong to the FII incompatibility group are also regulated by a small untranslated RNA, but in contrast to the ColE1 situation, the regulatory

C

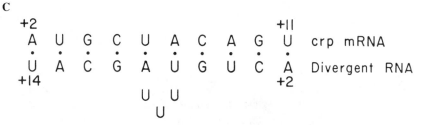

```
+2                        +II
A  U  G  C  U  A  C  A  G  U    crp mRNA
•  •  •  •  •  •  •  •  •  •
U  A  C  G  A  U  G  U  C  A    Divergent  RNA
+14                       +2
         U  U
            U
```

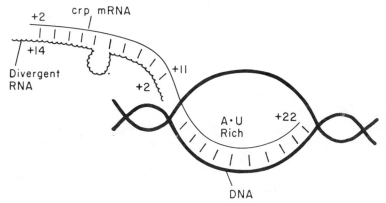

Figure 1 Proposed interactions of three prokaryotic antisense RNAs and their respective target RNAs.

(A) Model for the stepwise binding of ColE1 RNA I to RNA II (11). In step I RNA I and RNA II interact at the loops of their folded structures. This interaction facilitates pairing (step II) that starts at the 5' end of RNA I (step III). At this stage, the loop-to-loop contacts may be broken. Progressive pairing continues as the stem-and-loop structures unfold (steps IV and V). It should be noted that the three pairs of loops do not necessarily interact simultaneously, and RNA II may interact in an alternative folded structure (11). This figure was adapted from references 11 and 20.

(B) Model for hybrid formation between *micF* RNA and *ompF* mRNA (51). The complementary region of the *micF* RNA is sandwiched by two stable stem-and-loop structures: *(a)* $\Delta G = -12.5$ and *(b)* $\Delta G = -4.5$. Hybrid formation blocks translation of the *ompF* mRNA because the Shine-Dalgarno sequence and the AUG initiation codon are within the base-paired region. The formation of the hybrid may also promote rapid degradation of the *ompF* mRNA. This figure is adapted from Ref. 51.

(C) Model for the inhibition of *crp* transcription by antisense RNA (57). *Top:* structure of the hybrid between the 5' end of the *crp* mRNA and the first 14 nucleotides of the antisense RNA is shown. The RNA : RNA hybrid resembles the initial segment of rho-independent terminators (58). *Bottom:* the complete structure of the rho-independent terminator is shown. The structure consists of the *crp* mRNA-antisense RNA hybrid followed by an A-U–rich crp mRNA-DNA hybrid. Like in a rho-independent terminator, the RNA-RNA hybrid causes the polymerase to pause and the instability of the A-U–rich RNA-DNA hybrid leads to release of the *crp* transcript and termination of transcription. This figure was adapted from Ref. 57.

RNA [called RNA I (2), RNA E (22), or CopA RNA (23)] likely acts as a translational repressor (22, 24, 25). The target of RNA I is the mRNA for the essential plasmid replication protein, RepA1 (24, 26, 27). The repression mechanism is somewhat complicated by the fact that two RNA transcripts, RNA-CX and RNA-A (RNA-II) (24, 28–30), each contain the RepA1 coding sequence and a 91-nucleotide upstream region that is complementary to RNA I. RNA-CX and RNA-A are transcribed in the same direction from the same DNA but RNA-CX initiates about 380 bp upstream of RNA-A. In vivo, however, at or above the normal copy number, only RNA-CX appears to be a primary target for RNA I–mediated repression (31). The other RepA1 mRNA, RNA-A, is controlled by the presence of excess amounts of the CopB protein (30a) (RepA2) (31), a transcriptional repressor (32, 32a, 32b). The 5' region of RNA-CX (not included in RNA-A) contains the coding sequence for CopB; thus in the RNA-CX transcript, the RNA I complementary region is sandwiched between the RepA1 and CopB coding sequences (24, 33). It has been proposed that the hybridization between RNA I and RNA-CX induces a change in RNA-CX secondary structure in the region of the RepA1 ribosome binding site that blocks translation (24). The DNA sequences of several mutants indicate that a single-stranded loop in the RNA I secondary structure may be an important determinant for hybrid formation (27, 32a, 33, 34).

pT181   Like the ColE1 and FII plasmids discussed above, the *S. aureus* plasmid, pT181, has evolved a replication control mechanism whereby small complementary RNAs play a pivotal role. Two regulatory RNAs (RNA I and RNA II) (3) appear to repress the replication of pT181 by inhibiting the translation of the RepC protein, an essential replication factor (35–37). The RepC coding sequence is contained in both RNA III and RNA IV, which have different 5' ends (3). RNA I and RNA II differ at their 3' ends and are therefore complementary to the 5' untranslated regions of the two RepC mRNAs to different extents. The regulatory contributions of the individual pT181 transcripts have not been determined. RNA I and RNA II share no significant sequence homology with their ColE1 or FII group counterparts; however there may be some similarities at the level of RNA secondary structure. Again, the analysis of copy number mutants (38) supports the contention that the primary interaction between regulatory and target transcripts involves contacts between complementary single-stranded loops (3).

Tn10 Transposition

Detailed analysis of the transposon Tn10 has revealed that the activity of this element can also come under antisense RNA control. Transposition of Tn10 is mediated by functions specified primarily by IS10-Right (39, 40), one of two inverted repeat sequences that comprise the ends of the transposon. A long open reading frame encoding the transposase spans nearly the entire length of IS10-Right from the outside towards the inside of the element. Transcription of the transposase gene is directed by a weak promoter called pIN (41, 42). The much stronger pOUT promoter, located just inside pIN, specifies the synthesis of an antisense RNA (RNA-OUT) that is complementary to the first 36 nucleotides of the transposase mRNA (RNA-IN) transcribed from pIN. Recently a 70-nucleotide-long RNA was identified as the major in vivo RNA made from pOUT of IS10-Right (42a). The concentration of this RNA was estimated at 5–10 molecules per cell containing a single copy of Tn10.

Several observations indicate that RNA-OUT negatively regulates the synthesis of the IS10 transposase by forming a hybrid with the transposase translation initiation site region of RNA-IN (43, 44). First, single-copy Tn10 elements transpose at much lower frequencies in *E. coli* when IS10-Right is present on a multicopy plasmid (43). This phenomenon has been designated multicopy inhibition (MCI). Deletion and mutational analyses have limited the position of the determinants responsible for MCI to the outermost 180 bp of IS10-Right. This region contains the pOUT transcription unit (41, 42). Second, it has become clear that the mechanism of MCI involves inhibition of transposase gene expression. In addition, the mode of MCI is primarily posttranslational. MCI plasmids have been shown to decrease the expression of a transposase-*lacZ* gene fusion (i.e. a translational fusion) by more than 90% while operon fusions (i.e. transcriptional fusions) were inhibited by less than 50% (43).

Finally, pairing between RNA-IN and RNA-OUT has been demonstrated in vitro (44). Many mutant derivatives of RNA-OUT have been analyzed in this manner to determine the mechanism by which hybrid formation occurs. The results are most easily interpreted in terms of a potential stem-and-loop structure in RNA-OUT that begins near the 5' end of the molecule and includes the region capable of pairing with RNA-IN. This structure can be divided into two domains: a 21-nucleotide loop domain and a stem domain consisting of about 45 nucleotides. The 5' end region of RNA-IN is complementary to part of the RNA-OUT loop and one strand of the stem. Of all the mutations tested in an in vitro pairing assay, only two were found to decrease the ability of RNA-OUT to pair with wild-type RNA-IN. Both of the pairing-defective mutants clearly increase the base-pairing potential within the loop domain of RNA-OUT. RNA-OUT mutations that do not affect in vitro pairing between RNA-IN and RNA-OUT include one mutation in the loop that does not alter the base-pairing potential and several others that are located in the strand of the RNA-OUT stem that is not complementary to RNA-IN. While the latter set of mutations do not alter the efficiency of in vitro hybrid formation, several do decrease the calculated stability of the stem domain. Their lack of effect on the in vitro pairing reaction indicates that hybrid formation does not initiate at the base of the RNA-OUT stem. Based on these and other observations, a model has been proposed (J. D. Kittle, N. Kleckner, unpublished results) which predicts that pairing begins with a rate-limiting interaction between the 5' end of RNA-IN and the loop domain of RNA-OUT. After this initial pairing, one strand of the RNA-OUT stem is displaced as hybrid formation proceeds.

It has been demonstrated previously (45) that transposition frequencies in vivo are related to transposase levels. Thus, it is easy to understand how the pOUT transcript, when synthesized from a multicopy plasmid, might control Tn10 transposition frequency by inactivating the transposase mRNA. Tn10 also contains other overlapping transcripts such as *tetD* RNA and *tetC* RNA (46, 47). However, at present, the activities of these RNAs as well as antisense RNA species possibly encoded by other transposons (see Refs. cited in 42, 48, and 49) are more obscure.

Bacterial and Phage Gene Expression

Antisense RNAs are naturally encoded by both bacterial and phage genomes. The structural and functional analyses of these species and those described in the preceding two sections have led to a better understanding of the differing roles of antisense RNAs in complex regulatory schemes.

micF   The first natural *E. coli* antisense RNA to be identified was the micF RNA (50, 51). It was discovered during the characterization of *ompC*, a gene for one of the two major *E. coli* outer membrane porins. Surprisingly, several *ompC* promoter clones were found to repress the synthesis of the other major outer membrane porin, OmpF. The region responsible for this activity was localized to a 300-bp fragment upstream of the *ompC* promoter. Sequence analysis revealed that a stretch of approximately 80 bp was 70% homologous to the DNA encoding the 5' end of the *ompF* mRNA including the ribosome binding site and the *ompF* initiation codon. From S1 mapping and *lacZ* fusion studies, it became clear that the homologous DNA is transcribed in the opposite direction from *ompC* giving rise to a 174-nucleotide RNA that is complementary to a region of the *ompF* mRNA encompassing the translation initiation site. It was proposed that the complementary RNA, designated micRNA (*mRNA-interfering complimentary RNA), inhibits *ompF* expression by forming a hybrid with the *ompF* mRNA (50, 51) (see Figure 1B). This model predicted that artificial micRNAs could be used to regulate selected genes as discussed in the next section.

Figure 1B shows that 44 nucleotides of the 5' untranslated leader region including the Shine-Dalgarno sequence, and 28 nucleotides of the coding region of the *ompF* mRNA, are included in the hybrid. Several small bulges and internal loops are present in the hybrid due to the short regions of nonhomology between the *micF* and *ompF* genes. The base-paired region of the micRNA (*micF* RNA) is flanked by two stable stem-and-loops, a and b (see Figure 1B).

Based on this structure, it seems most likely that *micF* RNA inhibits OmpF synthesis by repressing translation.

Northern blot experiments with a *micF* probe have detected the *micF* RNA (50) and two other smaller RNA species in cells containing *micF* on a multicopy plasmid. The T1 digestion products of these RNAs indicate that smaller RNAs are derived from *micF* and that the major *micF* RNA species (4.5S) is missing about 80 nucleotides from the *micF* RNA 5' end (J. Andersen, N. Delihas, unpublished results). The physiological significance of the smaller RNAs and the events leading to their production are not yet known. In similar experiments, the production of the *micF* RNA (50) was associated with a greatly diminished level of *ompF* mRNA. This could be due to degradation or premature termination of the *ompF* message upon hybrid formation.

Transcription of *micF* appears to be coordinated with the complex induction pathway leading to *ompC* mRNA synthesis (50). Two regulatory proteins, OmpR and EnvZ, control the expression of *ompC* and *ompF* in response to the osmolarity of the culture medium (for a review see Ref. 52). While the total amount of OmpC and OmpF in the outer membrane remains constant, the proportions of the individual proteins vary depending on the culture conditions. For example, in high-osmolarity medium, the outer membrane contains mainly OmpC and very little OmpF. The simultaneous transcription of *ompC* and *micF* predicts that the role of *micF* RNA is to rapidly repress OmpF synthesis as OmpC induction proceeds. The phenotype of a chromosomal mutation deleting *ompC* and the region upstream including *micF* (53) is consistent with this assessment. The deletion mutant synthesizes OmpF constitutively but has little effect on β-galactosidase transcription from the *ompF* promoter. The translational regulatory function responsible for the OmpF constitutive phenotype of the deletion mutant is apparently present in the region upstream of *ompC* due to the fact that *ompF* is osmoregulated in *ompC* mutants.

Although the inhibitory activity of *micF* RNA can adequately explain these results, several new observations indicate that the picture is somewhat more complicated. As discussed above, the chromosomal mutation that deletes both these genes along with *ompC* gives rise to constitutive OmpF synthesis (53). Yet when *micF* was replaced with the gene for kanamycin resistance, no marked effect on the osmoregulation of *ompF* was detected (54). However, the *ompC* gene of this mutant was expressed even in low osmolarity (54). A detailed analysis of the transcriptional regulation of *micF* is clearly necessary to reconcile these results which were obtained using different strains grown under different conditions.

crp   Synthesis of the *E. coli* cAMP receptor protein (CRP) is autoregulated both in vivo (55) and in vitro (56). Recently, cAMP-CRP was shown to repress transcription of the *crp* gene indirectly, by inducing the synthesis of an antisense RNA (57). The transcription start sites of the antisense RNA and the *crp* mRNA are 3 bp apart and on opposite DNA strands. Although the sequence coding for the antisense RNA and the *crp* mRNA do not overlap, nucleotides 2–6 and 10–14 of the antisense RNA are complementary to nucleotides 2–11 of the *crp* mRNA (see Figure 1*C*). CRP activates transcription of the antisense RNA in vivo and in vitro in the presence of cAMP. In vitro synthesis of antisense RNA specifically inhibits transcription from the *crp* promoter. This was demonstrated in two ways. First, the antisense promoter was inactivated by a linker insertion that did not affect the binding of CRP to its nearby site. In the absence of antisense transcription, in vitro transcription of *crp* from the mutagenized template was no longer repressed by the addition of cAMP-CRP. Moreover, the addition of purified antisense RNA to the reaction specifically inhibited *crp* transcription from either the normal or the mutagenized template.

As shown in Figure 1*C*, the proposed hybrid formed between the antisense RNA and the *crp* mRNA is followed by an A–U–rich stretch. The resemblance of this structure to a rho-independent terminator (58) has led to the proposal that the antisense RNA acts by inducing premature termination of the *crp* mRNA. For this reason, the regulatory RNA has been designated antisense attenuator RNA. It is likely that antisense attenuator RNA will provide an additional approach to artificially regulate other genes of interest.

PHAGE λ   Antisense RNA also seems to be one of the regulatory factors involved in λ development. A promoter, designated P_{aQ}, has recently been shown to direct antisense transcription of part or all of the amino-terminal region of the λ*Q* gene coding sequence (59). Q protein is the antiterminator required for λ late gene transcription. The antisense transcript is thought to repress the synthesis of the Q protein by interfering with *Q* transcription or translation. Originally implied by its DNA sequence (D. Court, as cited in 59), the existence of P_{aQ} was recently confirmed by in vitro transcription and genetic analysis (59). Transcription initiation at P_{aQ} requires CII protein and a putative CII binding site has been identified. CII protein is also required to activate transcription of the integrase and λ repressor genes which are necessary for the establishment of lysogeny. A mutation in the CII binding site *(paq1)*, which does not effect Q function, resulted in the relief of inhibition of late gene expression (in a λ *cI857 cro20* phage) which is dependent upon CII. From these data it was concluded that P_{aQ} was responsible in part for this phenomenon which has been designated "CII dependent inhibition" (59). Furthermore, P_{aQ} transcription appears to favor the lysogenic over the lytic pathway because the *paq1* mutation causes λcI857 to form clear rather than turbid plaques at low temperature.

An additional antisense RNA which may play a role in λ development is the OOP RNA (λ 4S RNA) (60, 61). The role of OOP RNA is unknown at present but its structure has been well characterized (62, 63). The 3' end of the

77-nucleotide-long *oop* transcript is complementary to the last 17 amino acid codons of the CII mRNA coding sequence. The region of OOP RNA that could hybridize to 10 out of the 17 complementary CII codons is contained within the putative secondary structure thought to be the OOP RNA terminator. *oop* transcription initiates in the intercistronic region between the *O* and *cII* genes, just upstream of the Shine-Dalgarno sequence for the O protein. The *oop* gene is located near the λ *ori* region but it has been shown to be dispensable for the normal function of the origin (64) and therefore is unlikely to serve as the primer for the initiation of DNA replication. At present, there is no evidence that OOP RNA functions to repress the expression of *cII,* but the possibility could add an interesting twist to the emerging role of the antisense RNA transcribed from P_{aQ} which is induced by CII (see above).

PHAGE P22   Another intricate regulatory system involving a small antisense RNA has been identified in the *Salmonella* phage P22 (65). The *s*mall *a*ntisense *R*NA, called *sar* RNA, is complementary to the Shine-Dalgarno sequence of the mRNA for the P22 antirepressor protein, Ant, as well as the region between *ant* and the cotranscribed gene *arc*. The *sar* RNA contains no open reading frame and has been shown to inhibit antirepressor synthesis in trans. Mutations in the *sar* promoter that decrease promoter strength in vitro (66) give rise to higher levels of antirepressor synthesis in vivo. It has been proposed that *sar* RNA negatively regulates *ant* expression by forming a hybrid with the *ant* mRNA which represses translation (65).

Artificial Antisense RNA (micRNA)

The discovery of *micF* RNA prompted the development of a system designed to artificially regulate bacterial gene expression with antisense RNA (67). In addition to facilitating the controlled expression of various bacterial genes, the artificial "mic" system has provided insight as to the mechanism of antisense regulation and the characteristics of effective antisense RNAs. It was reasoned that an artificial *mic* gene[2] could be created by positioning a DNA fragment coding for a portion of an mRNA between a suitable promoter and transcription terminator, in the antisense orientation. The mic transcripts from such a construct would be complementary to the target mRNA over the region covered by the cloned DNA fragment.

The mic cloning vector pJDC402, shown in Figure 2A, has been used to efficiently construct and regulate a number of antisense genes in *E. coli* (67,

---

[2]The term "*mic* gene" has been used to define a gene capable of encoding an RNA complementary to all or part of a specific mRNA (67). The target gene for the micRNA is shown in parentheses. For example, transcription of a *mic (lpp)* gene will result in the production of RNA complementary to the *lpp* mRNA.

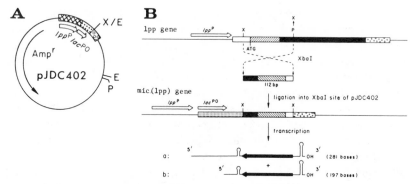

Figure 2   Structure of the mic vector pJDC402 and construction of a *mic(lpp)* gene

(*A*) Structure of pJDC402. *lpp*ᵖ (cross-hatches) and *lac*ᵖᵒ (solid dots) represent the lipoprotein promoter and the lactose promoter operator, respectively. The promoters are separated from the *lpp* transcription termination region (open dots) by a unique *Xba*I site (X). The *Pvu*II (P) and the EcoRI (E) sites are also indicated. Amp^r designates the ampicillin resistance gene.

(*B*) Construction of a *mic(lpp)* gene using pJDC402. Open arrows represent promoters. The open bar represents the region coding for the 5' nontranslated region of the *lpp* mRNA. The slashed bar and the solid bar correspond to the coding regions for the lipoprotein signal sequence and mature portion, respectively. The *Pvu*II site in the *lpp* gene was converted to an *Xba*I site by inserting an *Xba*I linker. To construct the *mic(lpp)* gene, the 112-bp *Xba*I fragment was inserted into the unique *Xba*I site of pJDC402 in the reverse (antisense) orientation. (*a*) and (*b*) show the mic(*lpp*) RNAs initiating at the *lpp* and *lac* promoters respectively. The solid arrows represent the portions of the *mic(lpp)* RNAs that are complementary to the *lpp* mRNA. This figure was adapted from Ref. 67.

68). pJDC402 contains the pIN-II (69) promoter system [which consists of the highly expressed lipoprotein *(lpp)* promoter (70) and the *lac* promoter-operator region in tandem] and the *lpp* transcription terminator separated by a unique *Xba*I site. This configuration ensures high-level transcription of DNA fragments inserted into the *Xba*I site in the presence of *lac* inducers such as IPTG (isopropyl-β-D-thiogalactopyranoside). Another feature incorporated into the mic vector is the presence of sequences with the potential to form stem-and-loop structures on either side of the *Xba*I site. These structures flank the antisense sequences of the artificial micRNAs and were fashioned after the predicted structure of the natural *micF* RNA (50, 51).

Figure 2*B* shows an example of *mic* gene construction using pJDC402 and the *E. coli lpp* gene. After converting the *Pvu*II site in the *lpp* gene to an *Xba*I site, a small fragment containing the Shine-Dalgarno sequence for ribosome binding and the coding region for the first 29 amino acid residues of pro-lipoprotein was excised. This fragment was then inserted into the *Xba*I site of pJDC402 in the opposite orientation. In the presence of IPTG, transcription of the resulting *mic(lpp)* gene yields antisense RNA that is complementary to the ribosome binding site region of the *lpp* mRNA. Lipoprotein synthesis has been

Table 1 Artificial prokaryotic antisense genes

Target mRNA	Sequences covered[a]	Level of inhibition[b]	Reference
lpp	L_{16}-SD-AA_{1-29}	High	67
	AA_{3-29}	Low	67
	AA_{43-63}	Low	67
ompC	L_{72}-SD-AA_{1-32}	High	67
	L_{20}-SD-AA_{1-32}	Moderate	67
ompA	L_{60}-SD-AA_{1-66}	Low	67
	L_{128}-SD	Low	67
	AA_{3-66}	Low	67
	L_{74}	Very Low	67
	L_{128}-SD-AA_{1-346}	Low	68
	L_{128}-SD-AA_{1-274}	Low	68
lacZ	AA_{8-1007}	High	75
	AA_{8-279}	Low	75
	$AA_{888-1007}$	Very Low	75
	AA_{7-147}	High	76

[a]Regions of the target mRNA covered by antisense sequences are abbreviated as follows: L, untranslated leader region; SD, Shine-Dalgarno sequence; AA, amino acid coding sequences. The number of leader nt or the amino acid codons involved are indicated as subscripts. Amino acid codons for the outer membrane proteins correspond to positions in the primary translation products prolipoprotein, proOmpC, and proOmpA.

[b]Approximate levels of inhibition in the presence of the various antisense genes are indicated: very low, $\leq 25\%$ decrease in expression; low, 26–75% decrease in expression; moderate, 3–6-fold decrease in expression; high, > 6-fold decrease in expression. Values obtained from different references [with the exception of (67) and (68)] cannot be compared directly due to differences in antisense gene structure and assay conditions.

shown to decrease 16-fold within five min after induction of this artificial *mic(lpp)* RNA (67). Since the functional half-life of the *lpp* mRNA is known to be 12 min (71), the primary influence of micRNA must be exerted after transcription has initiated. Yet, the synthesis of artificial antisense RNA was found to drastically decrease the cellular level of *lpp* mRNA, and *micF* RNA had an analogous effect on *ompF* mRNA levels. Therefore, the inhibitory effects of micRNA:mRNA hybrid formation extend beyond translation.

An almost complete inhibition of OmpC production was similarly achieved using a *mic(ompC)* construct covering most of the *ompC* mRNA leader, the Shine-Dalgarno sequence, and 32 amino acids of the coding sequence (see Table 1). A shorter antisense RNA lacking most of the sequences complementary to the 5' leader region of the *ompC* mRNA was less effective. This result points to the importance of 5' end complementarity and/or the length of the complementary region in antisense RNA activity (67). Further support for such a contention comes from the finding that cells engineered to produce micRNA complementary to the 5' region of the *tet* mRNA became approximately 50% more sensitive to tetracycline than control cultures (P. Roy, personal communication).

Another *E. coli* gene, the *ompA* gene, was found to be much less responsive to repression by antisense RNA (67, 68). As shown in Table 1, none of a series of seven *mic(ompA)* constructs inhibited OmpA synthesis efficiently. The explanation for this lack of response is unclear at present, but the *ompA* mRNA has the potential to form a high degree of secondary structure (52, 72–74), which may render it more resistant to hybridization with antisense RNAs. Curiously, *mic(ompA)* RNA complementary to only the coding sequence is as effective as the *mic(ompA)* RNA complementary to the ribosome binding site and the 5' leader region. This is clearly not the case for the *mic(lpp)* RNAs (see Table 1). The two micRNAs directed against coding segments of the *lpp* mRNA are about eightfold less effective than the *mic(lpp)* RNA covering the ribosome binding site shown in Figure 2B. However, recent experiments using the *E. coli lacZ* gene indicate that antisense RNA complementary to the coding portion of the message can efficiently block β-galactosidase synthesis (75, 76). The extent of antisense RNA length and amino-terminal complementarity appear to be correlated with high inhibition levels in this system as was observed for *mic(lpp)* and *mic(ompC)* (see Table 1). The use of extensive coding region complementarity may be particularly helpful in the development of highly specific antisense RNAs against some genes such as *ompC* and *lpp*, which share a common sequence (of about 70 bp which is 80% homologous) in the ribosome binding site region (67).

One approach to increase the level of micRNA-mediated repression has been to increase the dosage of micRNA in the cell. This is most easily achieved by constructing plasmids containing more than one *mic* gene. In the mic cloning vector pJDC402, the mic transcription unit is flanked by two *Hin*fI sites which allow the entire *mic* gene to be excised and reinserted into a unique restriction site in the plasmid containing the original *mic* gene. Using this strategy, a plasmid containing two copies of the *mic(lpp)* gene shown in Figure 2B was created. With this construct, it was found that doubling the dosage doubled the response, i.e. a 31-fold inhibition with two genes per plasmid versus a 16-fold inhibition with a plasmid containing a single *mic* gene. The expression of the *ompA* gene was similarly affected by increases in *mic(ompA)* gene dosage although the degree of inhibition was not proportional to the number of *mic* genes present (67, 68). The *ompA* studies also showed that the level of repression elicited by two different *mic(ompA)* genes on the same plasmid is greater than when either *mic* gene is present alone (68).

mic Immune System

The experiments described in the preceding sections clearly demonstrate that antisense RNA (micRNA) can be used to effectively block the expression of normal cellular genes. These findings imply that detrimental foreign genes that enter the cell should be susceptible to a similar mode of control. Recently, antisense RNAs have been directed against viral genes in order to test this exciting prediction as well as emphasize its importance as a novel alternative approach for the development of pathogen resistance (77).

The virus chosen for the model "mic immune system" (77) was the *E. coli* F-specific bacteriophage SP (78). The single-stranded (+) RNA genome of SP codes for a coat protein, a coat read-through protein, a replicase, and protein A required for phage maturation (Y. Inokuchi, A. Harashima, unpublished results). Based on the precedent that micRNAs complementary to the Shine-Dalgarno sequence and translation initiation site repress gene expression most effectively in bacteria (67), two restriction fragments, A and B, from an SP cDNA clone were isolated for *mic* gene construction. Fragment A contains the Shine-Dalgarno and initiation codon sequences from the SP coat protein gene and fragment B contains the same determinants from the replicase gene. The *mic* genes were constructed from these fragments (*micA,* and *micB*) and a third fragment (C) covering the 3' end of the SP genome *(micC),* by the same strategy shown in Figure 2*B* except the mic cloning vector pJDC406 (77) was used. This vector is a modified version of pJDC402 (Figure 2) which allows the convenient excision of IPTG-inducible *mic* transcription units with *Sma*I. Multiple *mic* genes can then be inserted into the unique *Pvu*II and *Aat*II sites of pJDC406 as desired.

A triple *mic* derivative of pJDC406 (pMIC-A:B:C) directing antisense transcription from fragments A, B, and C, was thus constructed and shown to confer upon *E. coli* resistance to phage SP in the presence of IPTG (77). While control cells containing pJDC406 lysed when infected with the virus at an m.o.i. (multiplicity of infection) of 15, cells containing pMIC-A:B:C did not, although a cessation of growth did occur. Plaque formation was also inhibited on cells containing the triple *mic* plasmid. The plating efficiency on pMIC-A:B:C cells induced with IPTG was one tenth that observed with the control and the size of the plaques that formed on the pMIC-A:B:C strain was greatly diminished. In another experiment, phage titers were determined at various times after addition of phage SP at an m.o.i. of 0.1. In the presence of IPTG, the number of infective centers decreased about fourfold in cells carrying pMIC-A:B:C and the titer 100 min after infection was about 5% that of the control. These results indicate that the inhibitory effect of micRNA production is present throughout infection.

The inhibition of phage protein production is presumed to be at least partially responsible for the resistance exhibited by the pMIC-A:B:C strain. The finding that coat protein production is greatly diminished after micRNA synthesis is induced (77) supports this viewpoint. This effect coupled with degradation of the phage RNA could explain the poor yield of phage from cells containing pMIC-A:B:C.

The largest contributions to the immunity conferred by pMIC-A:B:C appear to come from the *micA* and the *micB* genes. Plasmids containing the *micA* and/or the *micB* genes in either the single or double *mic* configuration inhibit SP production more than *micC* plasmids. While the *micC* RNA may bind to the 3' end of SP RNA and inhibit replication, *micA* and *micB* RNAs, which are complementary to the coat protein and replicase ribosome binding site regions, respectively, have more antiviral activity (77). It is interesting to note that a *mic* gene was recently constructed from the ribosome binding site portion of the SP protein A gene. Protein A is required for phage SP maturation. A plasmid containing one copy of the protein A *mic* gene was found to confer more resistance to SP (almost complete resistance) than any plasmid described above, including pMIC-A:B:C (A. Hirashima, M. Inouye, manuscript in preparation).

These experiments demonstrate that the mic immune system can effectively block the proliferation of a desired target virus. Another important feature of the system is that it is highly specific. Cells carrying pMIC-A:B:C are fully sensitive to related phages Qβ and GA.

ANTISENSE RNA IN EUKARYOTIC SYSTEMS

Antisense Genes

The capacity of a nucleic acid to act as a specific inhibitor of gene expression in eukaryotes was initially suggested by the early work with Rous sarcoma virus (RSV) (79). In that study, a short single-stranded oligodeoxyribonucleotide was added to chick embryo fibroblast tissue cultures that had been infected with RSV. The oligonucleotide, which was complementary to 13 nucleotides of the terminally redundant sequences of RSV, caused the inhibition of virus production. Subsequent in vitro experiments (80) showed that the same oligonucleotide could function to inhibit translation of the RSV 35S RNA in a wheat germ cell free translation system and could act as a primer for reverse transcriptase using RSV RNA as a template. The latter observation indicates that hybridization of an oligonucleotide to viral RNA can occur in vitro. Noteworthy is the fact that the previous work had shown that formation of a DNA-mRNA hybrid could arrest translation of eukaryotic mRNA in vitro (81). The question of whether inhibition of RSV production occurred by a mechanism involving translational inhibition in vivo was not pursued.

Experiments with the thymidine kinase (TK) gene from Herpes simplex virus (HSV) first demonstrated that an artificial antisense RNA gene could be used to regulate the expression of a selected gene in eukaryotes (82). For these studies TK DNA sequences were inserted in either orientation between the promoter of HSV-TK or the Murine sarcoma virus long terminal repeat and an SV40 polyadenylation signal. A mixture of plasmids containing the artificial sense and antisense HSV-TK genes at a ratio of 1 : 100, respectively, was microinjected into mouse LTK⁻ cells. This resulted in a significantly lower level of transient TK expression compared to that from cells that received the sense gene only. Inhibition of HSV-TK laid the groundwork for subsequent successful studies using chicken TK (83), mouse β-actin (83), RSV *env* (84), *Drosophila* hsp26 (85), *Dictyostelium discoidin* (85a), and human N-*ras* (Hall et al, personal communication) and the bacterial β-galactosidase (86) and chloramphenicol acetyl transferase (83), antisense genes (see Table 2).

In the initial HSV-TK experiments (82), a plasmid designed to direct the synthesis of a 1364-base antisense RNA complementary to the sense TK mRNA from +51 to +1415 (where +1 represents the transcription initiation site) was found to decrease transient TK gene expression by about 4–5-fold. Similar results were subsequently obtained when an antisense gene (from −80 to +343) covering the 5' end of the TK mRNA was used (83). With a hybrid HSV-chicken TK fusion gene construct, it was shown that a complementary RNA sequence of only 52 nucleotides from the 5' untranslated region could inhibit TK expression very efficiently; more efficiently in fact, than a long sequence complementary to most of the TK coding region including the initiation codon. Further studies with antisense TK genes (83) attest to the specificity of this type of inhibition. The anti-HSV-TK inhibited the expression of HSV-TK gene in mouse LTK⁻ cells but had no effect on the expression of the chicken-TK gene which bears no sequence homology with HSV-TK (83). Conversely, the anti-chicken-TK specifically inhibited the expression of the chicken TK gene without altering HSV-TK gene expression. Such specificity has also been observed with prokaryotic antisense systems (67).

The TK experiments described above were performed using technically demanding microinjection methodology. In parallel studies, DNA-mediated transformation has been found to be an expedient alternative. Transformation of LTK⁻ cells with sense and antisense genes (at a ratio of 1 : 100) was found to effectively reduce colony formation on selective medium. Similar to the microinjection experiments, the best results were obtained using the HSV-chicken TK gene fusion and the antisense construct covering 52 nucleotides of the 5' untranslated regions of the target mRNA. TK activity measurements revealed comparable inhibitory effects (83). An inducible TK antisense RNA gene controlled by the mouse mammary tumor virus (MMTV) promoter was also constructed. When the MMTV promoter was induced cells containing the inducible antisense gene exhibited a dose-dependent decrease in TK enzyme activity (83). An inhibition of approximately 90% was achieved. Furthermore, induced TK transformants grew slower in a medium containing a limiting concentration of thymidine.

In an independent study, transformation analyses have been carried out using a chimeric dihydrofolate reductase (DHFR) antisense TK construct that can be

Table 2 Antisense inhibition of gene expression in eukaryotic cells

A. Antisense RNA genes

Inhibited gene	Source of the gene	Host cells	Method of application into the cells	References
Thymidine kinase	Herpes simplex virus	Mouse L	microinjection + transformation	82,83,87
Thymidine kinase	Chicken	Mouse L	microinjection + transformation	83
β-Actin	Mouse	Mouse L	transformation	83
		Monkey BSC-1	microinjection	83
hsp26	Drosophila	Drosophila	transformation	85
N-ras	Human	Mouse 3T3	transformation	a
env	Rous sarcoma virus	Avian fibroblasts	transformation	84
Discoidin 1	Dictyostelium	Dictyostelium	transformation	85a
β-galactosidase	Enterobacteriaceae	Mouse 3T3	transformation	86
Chloramphenicol acetyl transferase	Enterobacteriaceae	Mouse L	transformation	83

B. Microinjected antisense RNA or complementary DNA

Inhibited gene	Inhibitor	Source of the gene	Host cells	Reference
β-Globin	RNA	Xenopus	Frog oocytes	89
Thimidine kinase	RNA	Herpes simplex virus	Frog oocytes	90
β-Actin	RNA	Mouse	Monkey BSC-1	83
Kruppel	RNA	Drosophila	Drosophila embryos	93
Chloramphenicol acetyl transferase	RNA	Enterobacteriaceae	Frog oocytes	90
Interleukin 2	DNA	Human	Frog oocytes	91
Interleukin 3	DNA	Human	Frog oocytes	91
Tumor necrosis factor	DNA	Human	Frog oocytes	92

[a]Hall et al, personal communication.

amplified in mouse L cells (87). Selection of cells resistant to increasing concentrations of methotrexate (a competitive inhibitor of DHFR) increased the production of an mRNA species containing both the DHFR coding region (in the sense orientation) and the TK antisense sequences (in the 3' noncoding segment). This amplification (up to $5-10 \times 10^3$ molecules/cell) resulted in progressive reduction in TK expression (10–15% activity remained). The residual TK activity is comparable to that observed in the study using the inducible MMTV TK antisense construct discussed above.

Both transformation and microinjection have been used to demonstrate that the chromosomal mouse actin gene is sensitive to antisense inhibition (83). Fewer colonies appeared following transformation with an antisense β-actin gene compared to control transformations. Furthermore, cells harboring micro-injected antisense β-actin genes showed a significant reduction in the number of intracellular actin cables. The functional inactivation of an endogenous human N-*ras* gene in mouse NIH/3T3 cells has been suggested (A. Hall, H. Patterson, C. Marshal; personal communication). In these DNA-mediated transformation experiments, an inducible antisense N-*ras* gene was used. Upon induction, cells containing the antisense *ras* gene appeared to revert to a more normal morphology but had a transformed phenotype in the absence of the inducer. An antisense gene has also been shown to inhibit the expression of an endogenous cellular gene in *Drosophila* (85). *Drosophila* tissue culture cells were stably transformed with a heat-inducible antisense hsp26 gene. Following heat induction, the cells produced much less (up to 89% inhibition) of the heat-shock protein hsp26 than untransformed cells. The inhibition was specific since closely related heat-shock proteins did not appear to be affected. An example of endogenous gene repression by antisense RNA has recently been demonstrated in the lower eukaryote *Dictyostelium* (85a). Using an antisense RNA construction of the discoidin gene that was transformed into *Dictyostelium*, the expression of three endogenous discoidin genes was successfully inhibited. Transformants exhibited a greater than 90% reduction in accumulated discoidin mRNA and protein, and exhibited a phenotype identical to discoidin-deficient mutants.

A potential antisense viral immune system (mic immune system) has recently been suggested using antisense Rous sarcoma virus (RSV) *env* genes (84). A quail cell line that was previously infected and transformed with RSV deleted in the *env* gene was used in this study. This cell line has been designated R(−)Q. It is known that plasmids containing the *env* gene are capable of efficiently rescuing infectious virus from R(−)Q cells (88). When a plasmid containing the *env* gene was cotransfected with plasmids containing an antisense *env* gene, significant inhibition of virus production was observed (up to 80% inhibition). Antisense genes corresponding to both the 3' coding region or the 5' end of the *env* gene appeared to be efficient inhibitors of virus production in the rescue system. Furthermore, cotransfection of the individual antisense constructs and a plasmid containing the whole RSV genome also resulted in significant inhibition of virus expression in R(−)Q cells.

The mechanism of action of antisense RNA in eukaryotes is not clear at present. This is mainly due to the limited data concerning the fate of the target mRNA and antisense RNA. In one study both sense and antisense HSV-TK RNAs were carefully followed in tissue culture cells (87). In cells expressing high levels of antisense TK RNA as part of a chimeric DHFR mRNA (see above), the TK mRNA was primarily found in the nucleus. In contrast, the distribution of antisense TK RNA was found to be 60% cytoplasmic and 40% nuclear. It appears that synthesis of antisense TK RNA changes the distribution of the TK mRNA since the latter is found primarily in the cytoplasm of the parental cells (which do not produce antisense TK RNA). Moreover, nearly all of the RNA:RNA hybrids containing the TK message were detected in the nucleus. A chimeric DHFR antisense TK gene corresponding to the 3' end of the TK gene repressed TK expression to approximately the same degree as a construction that also included a sequence complementary to the 5' end of the TK message. These results suggest that antisense RNA hybridizes to the RNA message in the nucleus and thereby inhibits its transport into the cytoplasm or renders it extremely labile upon arrival in the cytoplasm.

Given the apparent efficiency of antisense RNA covering the region of translation initiation in one study (83) and the nuclear effect in another (87), it is possible that both cytoplasmic and nuclear hybridization can occur in animal cells. The specific antisense gene construction and the conditions for the expression of such genes may favor one or the other of these mechanisms.

Microinjection of Antisense RNA

An alternative approach to exploit antisense RNA involves directly introducing specific antisense RNA into individual cells. This approach is advantageous for certain biological systems that lack methods for stable transformation, or promoters that can be used for efficient expression. This section describes studies in which antisense RNAs were synthesized and capped in vitro and then microinjected into various cells. Table 2 presents the genes that have been successfully inhibited by microinjected antisense RNA.

The first of such experiments reported described the direct microinjection of antisense β-globin RNA into frog oocytes (89). *Xenopus* β-globin mRNA and antisense RNA were synthesized in vitro from cDNA clones using the SP6 RNA polymerase system. Antisense RNAs complementary to various regions of the globin message were injected into frog oocytes and incubated a number of hours before the cells were injected with globin mRNA. Specific inhibition of globin production was observed while the expression of unrelated messages was unaffected. Globin synthesis was inhibited only when the antisense RNA included sequences complementary to the 5' end or the translation initiation site region of the globin message. Antisense RNAs complementary to the 3' translated or untranslated terminus of the message had no observable inhibitory effect on globin production. Of interest is the finding that oocytes injected with globin mRNA a number of hours prior to injection of globin antisense RNA can still exhibit significant inhibition of globin production. It has been suggested from these observations that globin antisense RNA inhibits globin mRNA translation even after the message is loaded onto polysomes; although this inhibition is apparently less efficient than the inhibition of translation initiation. When RNA was extracted from oocytes and treated with ribonucleases, to digest single-stranded RNA, protected duplex RNA was found only in cells containing both globin mRNA and antisense RNA indicating that hybridization does indeed occur in vivo. This hybridization was time-dependent because no duplex globin RNA was found immediately after oocytes (containing globin message) were injected with globin antisense RNA. Since both the inhibitory RNA and the target mRNA were microinjected into the cytoplasm, inhibition likely occurred in this subcellular compartment. This is in contrast to the nuclear inhibitory mode of the chimeric DHFR antisense TK-RNA gene. As discussed in the previous section, these two suggested mechanisms of antisense RNA action are not mutually exclusive and may occur at the same time in a single cell. In this regard, it is interesting to note that mRNA appears to be stable in the presence of antisense RNA whether the antisense RNA is derived from an antisense gene or is microinjected (82, 87, 89, 90). This is in marked contrast to the bacterial case, where the target mRNA appears extremely unstable (see the prokaryotic section).

As described in the previous section and Table 2 the synthesis of the HSV-TK and chloramphenicol acetyl transferase (CAT) has been successfully inhibited in mouse L cells using specific antisense genes (90). Similar results have been obtained by RNA microinjection experiments; the mRNAs for these enzymes were introduced into frog oocytes and were shown to be inhibited by the corresponding antisense RNAs (which were injected five hours earlier). In the case of the CAT antisense RNA, the RNAs complementary to the 5' end of the message were more effective. However, unlike the globin case, antisense RNAs complementary only to the 3' coding region were able to block CAT and TK synthesis to a certain extent. CAT expression from a plasmid microinjected into the nucleus was also inhibited by CAT antisense RNA that was injected into the cytoplasm.

A method for hybridization arrest of mRNA translation in *Xenopus* oocytes has been described (91). Using this method it has been possible to inhibit the expression of interleukin 2, interleukin 3 (91), and tumor necrosis factor (92) in oocytes by short synthetic deoxyribonucleotides (14–23 nucleotides) or single-stranded DNAs that are complementary to the corresponding mRNAs. When the complementary DNA was hybridized to the message prior to injection, a strong inhibition (up to 99%) of gene expression was observed. Substantial inhibition (up to 88%) of specific mRNA translation also occurred when oligonucleotides were injected into oocytes one hour before the mRNA. This indicates that oligonucleotide-mediated hybrid arrest can occur in vivo. The efficiency of hybridization and of DNA:RNA duplexes relative to RNA:RNA hybrids has not been tested in vivo.

Application of antisense RNA to a developmental system has been successfully shown with the Kruppel (Kr) gene of *Drosophila* (93). The Kr gene, which is transcribed and expressed at a certain stage in *Drosophila* embryo development, is known to play a role in controlling segmentation of the embryo. The severity of the Kruppel mutant phenotype depends on the residual level of Kr^+ activity. While heterozygous (Kr/+) embryos survive with small defects, homozygous mutant embryos die before hatching and exhibit deletions and sometimes duplications of various segments of the embryo. Wild-type *Drosophila* embryos injected with antisense Kr RNA were found to develop lethal Kr phenotypes at a high frequency. The phenotypes obtained by antisense RNA were consistent with only a partial inhibition of Kr gene expression. Experiments performed with heterozygous (Kr/+) embryos yielded similar results and also demonstrated that phenocopy severity is dose-dependent. This method can certainly be applied to many other developmentally regulated genes and to the study of how and when such genes function during development. The partial inhibition of Kr gene expression that was observed is similar to results obtained with other microinjected antisense RNAs (89, 90).

Possible Existence of Antisense RNA in Eukaryotic Cells

The existence of naturally occurring antisense RNA genes has not yet been demonstrated in eukaryotes. This section summarizes various observations concerning the roles of intramolecular base pairing, intermolecular base pairing, and RNAs affecting translation in eukaryotes, that may throw some light on what to expect of natural antisense RNA regulation in eukaryotes.

INTRAMOLECULAR BASE PAIRING   The possible base pairing within the c-myc mRNA has been suggested (94) to affect expression and explain, in part, a mechanism for activation of this oncogene. The gene consists of three exons. The first exon does not code for protein but the second contains the first ATG initiation codon. Exon one contains a sequence that is complementary to a region in exon two. It has been predicted that a stem-and-loop structure ($\Delta G° = -90$ kcal/mole) exists in the c-myc mRNA under physiological conditions (94). The first AUG is found within the postulated loop of this structure. Translocation of the c-myc gene in a non-Hodgkin-lymphoma, Manca, was shown to occur downstream of the first exon and results, therefore, in a c-myc mRNA devoid of the first exon. Loss of the first exon abolishes potential base pairing and formation of the above secondary structure and may cause the observed activation of translation. This interpretation is consistent with the fact that the formation of a secondary structure in a eukaryotic mRNA (95–97) or antisense RNA base pairing to mRNA can inhibit translation of a message. Furthermore, loss of exon one may be a common feature of human Burkitt lymphomas and murine plasmacytomas in which c-myc is rearranged (98, 99). Based on these observations, it seems that intramolecular base pairing may be a possible route to modulate gene expression in eukaryotes.

INTERMOLECULAR BASE PAIRING   Primary transcripts containing mRNA sequences are present both in the cytoplasm and the nucleus as ribonucleoproteins. When considering the possibility of antisense RNA regulation in eukaryotes, it must be realized that in addition to the RNA species involved, the proteins associated with these RNAs may play an important role. mRNA splicing is an example of a natural process involving RNA:RNA base pairing. The U1 RNA of the U1 snRNP (small nuclear ribonucleoprotein) contains at its 5' end a sequence that is complementary to the common 5' splice junction of an intron, and has been shown to bind 5' splice sites in vitro (100). The hypothesis that the U1 snRNP plays an important role in splicing was supported by reconstitution experiments in vitro (101, 102): (a) Removal of the eight complementary nucleotides from the 5' end of U1 snRNA by site-directed hydrolysis with ribonuclease H results in complete loss of splicing activity; (b) Antibodies directed against the U1 snRNP inhibit splicing at an early stage. In this case, the protein portion of the snRNP almost certainly plays a role in mediating the hybridization between U1 RNA and pre-mRNA.

RNAS AFFECTING TRANSLATION   Numerous reports in the literature suggest that various RNA species isolated from eukaryotes can modulate translation in vitro (103–106). Interesting examples of RNAs that can inhibit translation in vitro have been found in cytoplasmic fractions of chick embryo muscle. One class of these are small U-rich oligoribonucleotides (103, 104) referred to as translational control RNA (tcRNA). Of particular interest is a certain tcRNA that has been shown to inhibit translation of the myosin heavy chain (MHC) mRNA (107). This MHC tcRNA is associated with the MHC mRNA ribonucleoprotein complex and has been hypothesized to allow sequestering in the cytoplasm of inactive MHC mRNA which is unassociated with ribosomes. An analogous idea has been suggested regarding other mRNAs that are accumulated in an inactive form in the cytoplasm (108–110). Isolated MHC tcRNA has been sequenced (102 nucleotides) and shown to share short stretches of homology with U1 snRNA. It has been speculated that MHC tcRNA inhibits translation via hybridization to the MHC mRNA polyadenylation signal. Besides the finding that MHC tcRNA binds weakly to oligo dA cellulose columns, there is no additional experimental evidence to support this hypothesis.

A second class of RNAs termed iRNA from chick embryo muscle has been isolated in the form of a 10S ribonucleoprotein complex (iRNP) (105). These RNAs, which are not U-rich, are distinct from snRNAs and tcRNA described above. The iRNPs contain heterogeneous small RNAs in the 70–90-nucleotide range. Both the iRNA and the iRNP are potent inhibitors of in vitro translation. In vitro studies show that iRNA inhibits the binding of mRNA to 43S ribosomal complex (105) and is thus an inhibitor of translation initiation. Both the iRNA and tcRNAs may be considered a family of RNA molecules. It becomes important to ask if these RNAs are general inhibitors of translation or whether the different RNA components individually contain sequences that provide information for the inhibition of specific mRNAs.

It is possible that eukaryotic antisense RNA may be a small RNA in the form of a ribonucleoprotein and may have limited complementarity to the target mRNA similar to the U1 snRNA that is involved in splicing. In this regard, it is interesting that there are many small RNAs in eukaryotes whose functions have not been determined (111–114). An example of this is the small nuclear RNAs (180–240 nucleotides) that are transcribed from the 5' flanking region of the mouse dihydrofolate reductase gene (DHFR) (113). These small RNAs, which are not polyadenylated, and the DHFR mRNA are transcribed in opposite directions. Transcription of these small RNAs initiates just upstream of the DHFR promoter region. The small RNAs contain sequences complementary to the first 10 nucleotides of the DHFR transcript and also to a short region following the DHFR translation stop codon. The function of these small RNAs and the significance of their complementarity to the DHFR transcript is unknown. However, one may point out that there is a similarity between the organization of the DHFR genetic loci and some prokaryotic antisense RNA genes in which the antisense gene is adjacent to or overlaps the target gene (43, 57).

PROSPECTS

From the observations compiled in this review, it has become apparent that antisense RNA can function to repress the expression of endogenous bacterial and eukaryotic genes as well as genes originating from viruses, episomes, and mobile genetic elements. Repression can be accomplished with microinjected antisense RNA or RNA transcripts from a natural or artificial antisense gene. In either case, the effect is specific and is likely to be dependent upon the formation of an antisense RNA : mRNA hybrid structure. The level of inhibition becomes more pronounced as the concentration of antisense RNA is increased, but the structure of both RNAs is also an important determinant.

In prokaryotes, natural systems have been discovered where the antisense RNA directly blocks translation of the target mRNA by hybridizing to, and presumably obscuring, the ribosome binding site and initiation codon. The direct inhibition of transcription by a bacterial antisense RNA has also been reported. In most instances, it is not known whether these two mechanisms are mutually exclusive. Similar mechanistic questions exist concerning the action of artificial antisense RNAs in eukaryotic cells. Antisense RNA is known to exist in the cytoplasm, and it can likely inhibit translation in that cellular compartment. However, antisense RNA has also been demonstrated to form hybrids with target transcripts in the nucleus. Such base-paired structures may preclude processing and export events required for mRNA function. Future investigations into the behavior of antisense RNAs covering introns should help test these possibilities.

Many experiments have been aimed at determining the optimal antisense RNA structure. For both prokaryotic and eukaryotic regulatory systems, this seems to be dictated at least in part by the nature of the target mRNA. Yet, in bacteria, a clear trend is still evident. The most effective antisense RNAs are complementary to the 5' end region of the target mRNA which generally includes the translation initiation site (see Table 1). With eukaryotic antisense systems such complementarity does not always confer an advantage. Antisense RNAs covering other parts of the mRNA such as the 5' untranslated region or the 3' coding region have also been associated with efficient inhibition in eukaryotic cells. Some of the observed variability is expected to be related to subtleties inherent to the different physical approaches and cell types used. This makes defining a universal antisense RNA structure for achieving optimal repression particularly difficult. At present this must still be done on a case by case basis. The analyses performed thus far indicate that several parameters can be varied to maximize repression in a given system. These include 1. the region of the target mRNA covered by the antisense transcript; 2. the length of the potential target RNA-antisense RNA hybrid; 3. the organization of the antisense transcript; 4. the potential of the antisense transcript to form secondary structure; and 5. the nature of the promoter and other signals that control how the antisense transcript will be synthesized and treated inside the cell. Following the fate and localization of both the antisense and the target RNA species will help to determine the optimal characteristics and cellular mechanisms that contribute to effective antisense RNA action.

The discovery of natural prokaryotic antisense systems and the successful use of antisense RNA to regulate selected genes of interest should facilitate a number of important applications. The conditional inhibition of target gene expression using inducible antisense genes is one of the most significant offshoots of the preliminary work. This approach is now being used as a powerful tool to dissect the roles of specific gene products in simple and complex biological systems. Antisense RNA–induced phenocopies have distinct advantages over conventional conditional mutants because they can be created independent of temperature and without altering the nature of the target gene. This approach would particularly enhance the study of some diploid genetic systems where dominant mutations are difficult to isolate. In addition to known genes, unknown genes (perhaps isolated by virtue of their regulatory properties or sequence homology with other genes) can be repressed with specific antisense RNAs in order to identify their function or temporal expression requirements in defined model systems. The use of transformation and microinjection methodologies demonstrated that antisense RNA can be effectively introduced into a large number of different cell types.

The progress made by the development of the mic immune system emphasized that antisense RNA can confer resistance against pathogenes that introduce nucleic acids into bacterial cells during infection. Moreover, the RSV experiments demonstrate the promise of parallel approaches in eukaryotic cells. It is also possible to direct antisense RNAs against plasmid- or transposon-encoded resistance determinants to render infectious bacteria more sensitive to antibiotics.

The potential use of antisense RNA in gene therapy is an additional frontier that may be developed on the basis of existing technology. Genes such as oncogenes or defective mutant alleles should be susceptible to inhibition using antisense genes that are stably expressed in transgenic animals and plants. The expression of antisense genes in these systems can be controlled using inducible, constitutive, or tissue-specific promoters. If necessary, a normal copy of the gene can be introduced to replace the repressed version. In this case, the antisense RNA can be specifically directed against a region of the mutant gene that is not homologous to the normal copy. A similar strategy of differential inhibition can be used to elucidate the roles of individual members of multigene families in complex expression systems. These and other potential applications of antisense RNA in basic and applied research should allow studies of gene expression in development to proceed in exciting new directions.

One interesting avenue that might be pursued relates to the question of how natural antisense genes may have become fixed in different populations. A partial gene duplication may explain how the *E. coli micF* gene arose. Similarly, in eukaryotes, gratuitous antisense transcription of pseudogenes could result in regulatory systems that confer a selectable advantage.

ACKNOWLEDGMENTS

We thank our many colleagues for providing us with preprints and other unpublished results during the preparation of this manuscript. We also gratefully acknowledge Drs. Manuel Perucho and Stephen Pollitt for critically reading the manuscript and Peggy Yazulla for her secretarial assistance. This work was supported by grants from the National Institute of General Medical Sciences (GM19043) and from the American Cancer Society (NP387F). O.P. was supported by a Dr. Chaim Weizmann fellowship.

Literature Cited

1. Tomizawa, J. I., Itoh, T. 1981. *Proc. Natl. Acad. Sci. USA* 78:6096–100
2. Rosen, J., Ryder, T., Ohtsubo, H., Ohtsubo, E. 1981. *Nature* 290:794–99
3. Kumar, C. C., Novick, R. P. 1985. *Proc. Natl. Acad. Sci. USA* 82:638–42
4. Tomizawa, J., Itoh, T., Selzer, G., Som, T. 1981. *Proc. Natl. Acad. Sci. USA* 78:1421–25
5. Lacatena, R. M., Cesareni, G. 1981. *Nature* 294:623–26
6. Stuitje, A. R., Spelt, C. E., Veltkamp, E., Nijkamp, H. J. J. 1981. *Nature* 280:264–67
7. Itoh, T., Tomizawa, J. 1980. *Proc. Natl. Acad. Sci. USA* 77:2450–54
8. Morita, M., Oka, A. 1979. *Eur. J. Biochem* 97:435–43
9. Davison, J. 1984. *Gene* 28:1–15
10. Scott, J. R. 1984. *Microbiol. Rev.* 48:1–23
11. Tomizawa, J. 1984. *Cell* 38:861–70
12. Tamm, J., Polisky, B. 1983. *Nucleic Acids Res.* 11:6381–97
13. Lacatena, R. M., Cesareni, G. 1983. *J. Mol. Biol.* 170:635–50
14. Tamm, J., Polisky, B. 1985. *Proc. Natl. Acad. Sci. USA* 82:2257–61
15. Castagnoli, L., Lacatena, R. M., Cesareni, G. 1985. *Nucleic Acids Res.* 13:5353–67
16. Cesareni, G., Muesing, M. A., Polisky, B. 1982. *Proc. Natl. Acad. Sci. USA* 79:6313–17
17. Tomizawa, J., Som, T. 1984. *Cell* 38:871–78
18. Cesareni, G., Cornelissen, M., Lacatena, R. M., Castagnoli, L. 1984. *EMBO J.* 3:1365–69
19. Som, T., Tomizawa, J. 1983. *Proc. Natl. Acad. Sci. USA* 80:3232–36
20. Tomizawa, J. 1985. *Cell* 40:527–35
21. Lacatena, R. M., Banner, D. W., Castagnoli, L., Cesareni, G. 1984. *Cell* 37:1009–14
22. Easton, A. M., Rownd, R. H. 1982. *J. Bacteriol.* 152:829–39
23. Stougaard, P., Molin, S., Nordstrom, K. 1981. *Proc. Natl. Acad. Sci. USA* 78:6008–12
24. Womble, D. D., Dong, X. N., Wu, R. P., Luckow, V. A., Martinez, A. F., Rownd, R. H. 1984. *J. Bacteriol.* 160:28–35
25. Light, J., Molin, S. 1983. *EMBO J.* 2:93–98
26. Light, J., Molin, S. 1982. *Mol. Gen. Genet.* 187:486–93
27. Givskov, M., Molin, S. 1984. *Mol. Gen. Genet.* 194:286–92
28. Easton, A. M., Sampathkumar, P., Rownd, R. H. 1981. In *The Initiation of DNA Replication*, ed. D. S. Ray, pp. 125–41. New York: Academic
29. Lurz, R., Danbara, H., Ruckert, R., Timmis, K. N. 1981. *Mol. Gen. Genet.* 183:490–96
30. Rosen, J., Ryder, T., Ohtsubo, H., Ohtsubo, E. 1980. *Nature* 290:794–79
30a. Riise, E., Stovgaard, P., Bindslev, B., Nordstrom, K., Molin, S. 1982. *J. Bacteriol.* 151:1136–45
31. Rownd, R. H., Womble, D. D., Dong, X. N., Luckow, V. A., Wu, R. P. 1984. In *Plasmids in Bacteria*, ed. D. Helinski, S. N. Cohen, D. Clewell, D. Jackson, A. Hollaender, pp. 382–98. New York: Plenum
32. Light, J., Molin, S. 1981. *Mol. Gen. Genet.* 184:56–61
32a. Womble, D. D., Dong, X., Luckow, V. A., Wu, R. P., Rownd, R. H. 1985. *J. Bacteriol.* 161:534–43
32b. Dong, X., Womble, D. D., Luckow, V. A., Rownd, R. H. 1985. *J. Bacteriol.* 161:544–51
33. Nordstrom, K., Molin, S., Light, J. 1984. *Plasmid* 12:71–90
34. Brady, G., Frey, J., Danbara, H., Timmis, K. N. 1983. *J. Bacteriol.* 154:429–36
35. Khan, S. A., Carleton, S. M., Novick, R. P. 1981. *Proc. Natl. Acad. Sci. USA* 78:4902–6
36. Novick, R. P., Adler, G. K., Majumder, S., Khan, S. A., Carleton, S., Iordanescu, S. 1982. *Proc. Natl. Acad. Sci. USA* 79:4108–12
37. Novick, R. P., Adler, G. K., Projan, S. J., Carleton, S., Highlander, S. K., et al. 1984. *EMBO J.* 3:2300–405
38. Carleton, S. C., Projan, S., Highlander, S. K., Mogazeh, S., Novick, R. P. 1984. *EMBO J.* 3:2407–14
39. Foster, T. J., Davis, M. A., Roberts, D. E., Takeshita, K., Kleckner, N. 1981. *Cell* 23:201–13
40. Kleckner, N., Morisato, D., Roberts,

D., Bender, J. 1984. *Cold Spring Harbor Symp. Quant. Biol.* 49:235-44
41. Halling, S. M., Simons, R. W., Way, J. C., Walsh, R. B., Kleckner, N. 1982. *Proc. Natl. Acad. Sci. USA* 79:2608-12
42. Simons, R. W., Hoopes, B. C., McClure, W. R., Kleckner, N. 1983. *Cell* 34:673-82
42a. Lee, Y., Schmidt, F. J. 1985. *J. Bacteriol.* 164:556-62
43. Simons, R. W., Kleckner, N. 1983. *Cell* 34:683-91
44. Kittle, J. D., Kleckner, N. 1984. *Cold Spring Harbor Lab. Bacteriophage Meet.* 48 pp. (Abstr.)
45. Morisato, D., Way, J. C., Kim, H.-J., Kleckner, N. 1983. *Cell* 32:799-807
46. Schollmeier, K., Hillen, W. 1984. *J. Bacteriol.* 160:499-503
47. Braus, G., Argast, M., Back, C. F. 1984. *J. Bacteriol.* 160:504-9
48. Normark, S., Bergström, S., Edlund, T., Grundström, T., Jourin, B., et al. 1983. *Ann. Rev. Genet.* 17:499-525
49. Kleckner, N. 1981. *Ann. Rev. Genet.* 15:341-404
50. Mizuno, T., Chou, M.-Y., Inouye, M. 1984. *Proc. Natl. Acad. Sci. USA* 81:1966-70
51. Mizuno, T., Chou, M.-Y., Inouye, M. 1983. *Proc. Jpn. Acad. Sci.* 59:335-38
52. Green, P. J., DiRienzo, J. M., Yamagata, H., Inouye, M. 1985. In *Organization of Prokaryotic Cell Membranes*, ed. B. K. Ghosh, 2:45-104. Boca Raton, Fla: CRC
53. Schnaitman, C. A., McDonald, G. A. 1984. *J. Bacteriol.* 159:555-63
54. Matsuyama, S.-I., Mizushima, S. 1985. *J. Bacteriol.* 162:1196-202
55. Cossart, P., Gicquel-Sanzey, B. 1985. *J. Bacteriol.* 161:454-57
56. Aiba, H. 1983. *Cell* 32:141-49
57. Okamoto, K., Freundlich, M. Submitted for publication
58. Rosenberg, M., Court, D. 1979. *Ann. Rev. Genet.* 13:319-53
59. Hoopes, B. C., McClure, W. R. 1985. *Proc. Natl. Acad. Sci. USA* 82:3134-38
60. Wulff, D. L., Rosenberg, M. 1983. In *Lambda II*, ed. R. W. Hendrix, J. W. Roberts, F. W. Stahl, R. A. Weisberg, pp. 53-73. Cold Spring Harbor, NY: Cold Spring Harbor Lab.
61. Furth, M. E., Wickner, S. H. 1983. See Ref. 60, pp. 145-173
62. Schwartz, E., Scherer, G., Hobom, G., Kossel, H. 1978. *Nature* 272:410-14
63. Rosenberg, M., DeCrombrugghe, B., Musso, R. 1976. *Proc. Natl. Acad. Sci. USA* 73:717-21

64. Tsurimoto, T., Matsubara, K. 1982. *Proc. Natl. Acad. Sci. USA* 79:7639-43
65. Wu, T., Liao, S., McClure, W. R., Susskind, M. M. 1984. *Cold Spring Harbor Lab. Bacteriophage Meet.* 182 pp. (Abstr.)
66. Liao, S.-M., McClure, W. R. 1984. *Cold Spring Harbor Lab. Bacteriophage Meet.* 101 pp. (Abstr.)
67. Coleman, J., Green, P. J., Inouye, M. 1984. *Cell* 37:429-36
68. Green, P. J., Mitchell, C., Inouye, M. 1985. *Biochemie.* 67:763-67
69. Nakamura, K., Inouye, M. 1982. *EMBO J.* 1:771-15
70. Nakamura, K., Inouye, M. 1979. *Cell* 18:1109-17
71. Hirashima, A., Inouye, M. 1973. *Nature* 242:405-9
72. Green, P. J., Inouye, M. 1984. *J. Mol. Biol.* 176:431-42
73. Movva, N. R., Nakamura, K., Inouye, M. 1980. *J. Mol. Biol.* 143:317-28
74. Movva, N. R., Nakamura, K., Inouye, M. 1980. *Proc. Natl. Acad. Sci. USA* 77:3845-49
75. Ellison, M. J., Kelleher, R. J. III, Rich, A. 1985. *J. Biol. Chem.* 260:9085-87
76. Pestka, S., Daugherty, B. L., Jung, V., Hotta, K., Pestka, R. K. 1984. *Proc. Natl. Acad. Sci. USA* 81:7525-28
77. Coleman, J., Hirashima, A., Inokuchi, Y., Green, P. J., Inouye, M. 1985. *Nature* 315:601-3
78. Sakurai, T., Miyake, T., Watanabe, I. 1968. *Jpn. J. Microbiol.* 12:544-46
79. Zamecnik, P. C., Stephenson, M. L. 1978. *Proc. Natl. Acad. Sci. USA* 75:280-84
80. Stephenson, M. L., Zamecnik, P. C. 1978. *Proc. Natl. Acad. Sci. USA* 75:285-88
81. Paterson, B. M., Roberts, B. E., Kuff, E. L. 1977. *Proc. Natl. Acad. Sci. USA* 74:4370-74
82. Izant, J. G., Weintraub, H. 1984. *Cell* 36:1007-15
83. Izant, J. G., Weintraub, H. 1985. *Science* 229:345-51
84. Stoltzfus, C. M., Chang, L. 1985. *Mol. Cell. Biol.* 5:2341-48
85. McGarry, T. J., Lindquist, S. 1986. *Proc. Natl. Acad. Sci. USA.* 88:399-403
85a. Crowley, T. E., Nellew, W., Gomer, R. H., Firtel, R. A. 1985. *Cell.* 43:633-41
86. Rubenstein, J. L. R., Nicolas, J. F., Jacob, F. 1984. *C. R. Acad. Sci.* Ser. III 299:271-74
87. Kim, S. K., Wold, B. J. 1985. *Cell* 42:129-38

88. Stacey, D. W., Allfrey, V. G., Hanafusa, H. 1977. *Proc. Natl. Acad. Sci USA* 74:1614–18
89. Melton, D. A. 1985. *Proc. Natl. Acad. Sci. USA* 82:144–48
90. Harland, R., Weintraub, H. 1985. *J. Cell. Biol.* 101:1094–99
91. Kawasaki, E. S. 1985. *Nucleic Acids Res.* 13:4991–5004
92. Wang, A. M., Creasey, A. A., Ladner, M. B., Lin, L. S., Strickler, J., et al. 1985. *Science* 228:149–54
93. Rosenberg, U. B., Preiss, A., Seifert, E., Jackle, H., Knipple, D. C. 1985. *Nature* 313:703–6
94. Saito, H., Hayday, A. C., Wiman, K., Hayward, W. S., Tonegawa, S. 1983. *Proc. Natl. Acad. Sci. USA* 30:7476–80
95. Kozak, M. 1980. *Cell* 19:79–90
96. Pelletier, J., Sonenberg, N. 1985. *Cell* 40:515–26
97. Baim, S. B., Pietras, D. F., Eustice, D. C., Sherman, F. 1985. *Mol. Cell Biol.* 5:1839–46
98. Adams, J. M., Gerondakis, S., Webb, E., Corcoran, L. M., Cory, S. 1983. *Proc. Natl. Acad. Sci. USA* 80:1982–86
99. Stanton, L. W., Watt, R., Marcu, K. B. 1983. *Nature* 303:401–6
100. Mount, S. M., Pettersson, I., Hinterberger, M., Karmas, A., Steitz, J. A. 1983. *Cell* 33:509–18

101. Kramer, A., Keller, W., Appel, B., Luhrmann, R. 1984. *Cell* 38:299–307
102. Keller, W. 1984. *Cell* 39:423–25
103. Heywood, S. M., Kennedy, D. S., Bester, A. J. 1974. *Proc. Natl. Acad. Sci. USA* 71:2428–31
104. Bester, A. J., Kennedy, D. S., Heywood, S. M. 1975. *Proc. Natl. Acad. Sci. USA* 72:1523–27
105. Sarkar, S. 1984. *Prog. Nucleic Acid Res. Mol. Biol.* 31:267–93
106. Coppola, G., Bablanian, R. 1983. *Proc. Natl. Acad. Sci. USA.* 80:75–79
107. McCarthy, T. L., Siegel, E., Mroczkowski, B., Heywood, S. M. 1983. *Biochemistry* 22:935–41
108. Northemann, W., Schmelzer, E., Heinrich, P. C. 1980. *Eur. J. Biochem.* 112:451–59
109. Vincent, A., Civelli, O., Maundrel, K., Scherrer, K. 1980. *Eur. J. Biochem.* 112:617–33
110. Eller, M. S., Cullinan, R. E., McGuire, P. M. 1982. *Fed. Proc.* 41:1457 (Abstr.)
111. Reddy, R. 1985. *Nucleic Acids Res.* 13:r155–r161 (Suppl.)
112. Jelinek, W., Leinwand, L. 1978. *Cell* 15:205–14
113. Farnham, P. J., Abrams, J. M., Schimke, R. T. 1985. *Proc. Natl. Acad. Sci. USA* 82:3978–82
114. Wolin, S. L., Steitz, J. A. 1984. *Proc. Natl. Acad. Sci. USA* 81:1996–2000

Ann. Rev. Biochem. 1986. 55:599–629
Copyright © 1986 by Annual Reviews Inc. All rights reserved

BIOLOGICAL CATALYSIS BY RNA

Thomas R. Cech and Brenda L. Bass[1]

Department of Chemistry and Biochemistry, University of Colorado, Boulder, Colorado 80309

CONTENTS

PERSPECTIVES AND SUMMARY

Most of the chemical reactions that take place in biological systems are accelerated above the basal chemical reaction rate. There is more than one way in which such rate enhancement is achieved. Many reactions are catalyzed by enzymes, which can be defined as macromolecules that (*a*) accelerate the rate of a reaction by providing an alternative reaction pathway with a lower activation energy, (*b*) are highly specific with respect to the substrates they act upon and

[1]Brenda L. Bass is currently a postdoctoral fellow at the Hutchinson Cancer Research Center, 1124 Columbia Street, Seattle, Washington 98104

0066-4154/86/0701-0599$02.00

the products they generate, and (c) are not consumed in the reaction, so that one enzyme molecule can sequentially process many substrate molecules (1–3).

Other biochemical reactions are quasi-catalytic; they involve macromolecules that meet the first two criteria for an enzyme but not the third. Proteins such as type I restriction endonucleases (4, 5), poly(ADP-ribose) synthetase (6, 7), and the transmethylase for O^6-methylguanine (8) mediate very specific reactions with very large rate acceleration but are inactivated in the reactions. Proteins that mediate site-specific DNA recombination reactions appear to act stoichiometrically rather than catalytically (9–11). In other cases, reactions that occur at a specific site within a protein (12) or RNA molecule are greatly accelerated by the folded structure of the molecule. In such intramolecular catalysis (1, 3, 13), the macromolecule is physically changed in the reaction, so the third part of the definition of an enzyme is not met. Intramolecular catalysis and true catalysis both involve the acceleration of very specific reactions, so we have found it useful to discuss them together in this review.

Biological catalysis has long been considered to be exclusively the realm of proteins. In the past few years, however, several examples of RNA catalysis have emerged. Two RNA molecules, RNase P and the excised rRNA intervening sequence from *Tetrahymena,* can act as true enzymes. Intramolecular catalysis is involved in certain RNA splicing reactions, the cleavage of a phage T_4 RNA, and reactions that are part of the life cycle of certain RNA molecules that infect plants. The site-specific cleavage of tRNA by metal ions involves intramolecular catalysis by functional groups on the tRNA and true catalysis by the metal ion. In other cases, RNA-protein complexes are known to be involved in mediating biochemical reactions, but the respective contributions of the RNA and protein moieties have not yet been determined. These include the catalysis of protein synthesis by ribosomes and the involvement of small nuclear ribonucleoprotein particles in a variety of RNA processing reactions.

After describing specific examples of RNA catalysis and intramolecular catalysis, we will compare the intrinsic catalytic capability of RNA with that of protein. Clearly there are differences between these macromolecules. Proteins have a much greater diversity of functional groups than RNA, including groups that are well suited for general acid-base catalysis and for the formation of hydrophobic pockets. Nevertheless, when mechanisms of catalysis are considered, an overall similarity between protein and RNA catalysis emerges. The formation of a noncovalent complex between the substrate and a binding site within the RNA or protein is crucial to catalysis by both macromolecules. In addition to noncovalent binding, covalent intermediates are formed in some protein-catalyzed reactions and in reactions catalyzed by the *Tetrahymena* IVS RNA. Finally, both protein and RNA catalysts have been proposed to enhance the reactivity of the substrate by facilitating the formation of the transition state

(13–15). Thus, the two macromolecules appear to share major catalytic strategies. It remains possible that further study of RNA catalysis will reveal heretofore unknown modes of catalysis that are either specific to RNA or general to all enzymes.

SITE-SPECIFIC, METAL ION–CATALYZED CLEAVAGE OF tRNA

Several divalent metal ions catalyze hydrolysis of the sugar-phosphate backbone of tRNA^{Phe}, primarily at a site within the D loop. Cleavage by Mg^{2+} occurs at pH 9.5 but is negligible at pH 7 (16). At neutral pH efficient cleavage is promoted by Zn^{2+} and especially Pb^{2+} (17). Eu^{3+} also catalyzes hydrolysis, and the initial cleavage products are the same as those produced by Mg^{2+}, Zn^{2+}, and Pb^{2+} (18).

X-ray crystallographic studies have provided a structural basis for interpreting the Pb^{2+}-catalyzed cleavage reaction. Brown et al (19, 20) have analyzed the structure of the Pb^{2+}-tRNA^{Phe} complex at pH 7.4, where the tRNA is cleaved between residues D17 and G18, and at pH 5.0, where Pb^{2+} is bound to a similar site but no cleavage has occurred. Rubin & Sundaralingam (21) obtained data for a different crystal form of Pb^{2+}-tRNA^{Phe} at pH 6.5, where cleavage occurred. In both studies, three major Pb^{2+}-binding sites were observed; site 1 is in proximity to the major cleavage site. At site 1, the Pb^{2+} ion is bound mainly to O-4 of the base moiety of U59 and N-3 of the base moiety of C60. These residues are in the T loop, far from D17/G18 in the primary structure of tRNA but close in the tertiary structure. The Pb^{2+} displaces an Mg^{2+} bound near this site in the native structure.

Cleavage mechanisms have been proposed, based on the structural data, the pH dependence of the cleavage, and the nature of the reaction products (5' hydroxyl, 2',3'-cyclic phosphate). Brown et al (19, 20) propose that a Pb^{2+}-bound hydroxyl group abstracts the proton from the 2'-OH of D17. The resulting 2'-O$^-$ then attacks the phosphorus atom of the phosphate at D17/G18 in a reaction analogous to random alkaline hydrolysis of RNA. This mechanism is depicted in Figure 1. Sundaralingam et al (22) consider a similar mechanism and another in which the Pb^{2+} ionizes an intervening water molecule that in turn promotes ionization of the 2'-OH.

Three factors are thought to be critical (19, 20). First, the section of the molecule around D17 is less rigidly structured than the rest of the molecule, thereby permitting reactivity. Second, the bound Pb^{2+} ion is positioned such that a hydroxyl group in its coordination sphere is in a favorable orientation and distance to abstract the proton from the 2'-OH. Third, a Pb^{2+}-bound water molecule has $pK_a \simeq 7.7$ (Table 1); thus, the Pb^{2+} coordination sphere is expected to be occupied by a hydroxyl group at neutral pH. As shown in Table

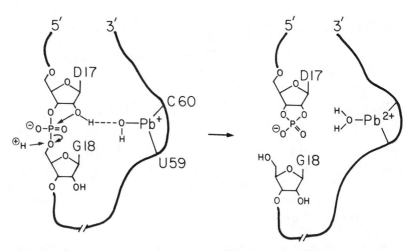

Figure 1   A proposed mechanism for Pb^{2+}-catalyzed site-specific cleavage of yeast tRNA^{Phe}. (*Left*) Prior to cleavage of the sugar-phosphate backbone, the aquo $Pb(OH)^+$ is depicted as being bound to and positioned by the tRNA^{Phe} molecule. (*Right*) Following cleavage, the Pb^{2+} moiety can be released by the tRNA to catalyze another cleavage reaction. Adapted with permission from (19, 22).

1, other divalent metal ions have considerably higher pK_a's; their coordination spheres are expected to be occupied by water rather than OH^- at $pH << pK_a$.

Thus, metal ion–catalyzed cleavage of tRNA^{Phe} can be considered to be an intramolecular version of a metallo-enzyme-catalyzed reaction (19). The D-loop acts as the substrate. The remainder of the tRNA acts as the enzyme; it binds the Pb^{2+} ion and, via tertiary interactions, binds the D-loop to position the bond between D17 and G18 for cleavage.

RIBONUCLEASE P

It has been known for some time that RNase P, the enzyme responsible for the cleavage that produces the mature 5' terminus of tRNA molecules, requires an RNA as well as a protein subunit for its in vivo activity (24–26). It is now clear that the active site resides within the RNA moiety (27, 28). Under certain conditions such as high concentrations of magnesium ion, the RNA component alone is able to catalyze the maturation of tRNA precursors in vitro. By itself the protein has no activity under any conditions tested. RNA catalysis has been shown to occur with the RNase P of *Escherichia coli* (27, 28), *Bacillus subtilis* (27, 29, 30), and *Salmonella typhimurium* (31). The presence of an RNA component in the RNase P of eukaryotes (32, 33) suggests that RNA catalysis may be ubiquitous in the maturation of 5' ends of tRNAs.

Table 1   Acid dissociation constants for hydrolysis of hydrated metal ions (M^{2+} + $H_2O \rightleftarrows MOH^+ + H^+$)[a]

M^{2+}	pK_a
Ba^{2+}	13.47
Ca^{2+}	12.85
Mg^{2+}	11.44
Mn^{2+}	10.59
Co^{2+}	10.2
Ni^{2+}	9.86
Fe^{2+}	9.5
Zn^{2+}	8.96
Pb^{2+}	7.71

[a]Values at 25°C (23).

To date most studies have been performed using the RNA subunit from the RNase P of *E. coli* (M1 RNA) or *B. subtilis* (P RNA). The sequences of the two RNA molecules are quite different, yet an active RNase P complex can be reconstituted with the protein subunit from one bacterium and the RNA of the other (27, 30). One interpretation of this result is that the RNA molecules have similar structures in solution, and it is the structure of the RNA that the protein recognizes.

The RNase P reaction is dependent on RNA structure, has a strict requirement for a divalent cation (Mg^{2+} or Mn^{2+}), does not require energy from the hydrolysis of nucleoside triphosphates, and yields reaction products with 3' hydroxyl and 5' phosphate termini (27). In these respects the RNase P reaction resembles the reactions of Group I introns which will be discussed later. Modification of the 3' hydroxyl of P RNA or M1 RNA does not affect its reactivity (34, 35). Thus, the reaction differs from those of Group I intervening sequences, which require the 3' hydroxyl of the RNA chain or the 3' hydroxyl of a free guanosine molecule.

It has been proposed that a hexacoordinated Mg^{2+} ion binds to the catalytic site on M1 RNA, and that a hydroxyl ligand of the Mg^{2+} ion acts as a general base to catalyze hydrolysis of the substrate RNA (35, 36). The rate of the M1 RNA-catalyzed cleavage of pre-tRNA does not increase between pH 5.5 and 9.5 (35). Although the lack of a pH dependence is not, at first glance, what one would expect for a reaction involving $[Mg(OH_2)_5OH]^+$ (pK_a for $Mg(OH_2)_6^{2+}$ = 11.44; Table 1), it has been noted that large pK_a perturbations can occur in protein enzymes and in analogy, perhaps in RNA catalysts as well (35). The cleavage rate of a synthetic pre-tRNA substrate by *B. subtilis* P RNA does depend on pH (34), although the dependence is not as great as that expected if OH$^-$ were the nucleophile (15).

Manganese is the only metal ion which can substitute for Mg^{2+}; Co^{2+}, Fe^{2+}, Ni^{2+}, Zn^{2+}, Ca^{2+}, and Sr^{2+} do not promote the reaction of M1 RNA (35). Only Zn^{2+}, Ca^{2+}, Mg^{2+}, and Mn^{2+} have been assayed with P RNA, and the results for these metals are similar to those with M1 RNA (29).

Mixed metal experiments in which the reaction is performed in the presence of magnesium and another metal indicate there are two types of metal binding sites: catalytic and structural sites (35). While only Mg^{2+} can function efficiently at the catalytic site, Mg^{2+}, Ca^{2+}, Sr^{2+}, and to a lesser extent Mn^{2+} can preserve the structural properties of the RNA molecule. It has been proposed that magnesium functions most efficiently at the catalytic site not only because of its size but also because as a hard acid it can complex with hard bases such as OH$^-$ and H_2O (35).

Although the RNA subunit of RNase P contains the catalytic active site, the protein subunit is clearly required for in vivo activity. RNase P provides an opportunity to learn how protein may aid in RNA-catalyzed reactions. Polyamines, high concentrations of Mg^{2+}, and polyethylene glycol can substitute for the protein of E. coli RNase P, suggesting that the protein has a role in electrostatic shielding (27, 29, 35). It has been proposed that the protein, in addition, aids in an essential conformational change (37).

Evidence for a structural change is derived from digestion of the M1 RNA with double-strand and single-strand specific nucleases in the presence or absence of the protein (37). Kinetic studies with the E. coli RNase P also implicate an essential conformational change. Catalysis by M1 RNA alone demonstrates a short lag time that disappears when M1 RNA is dialyzed out of 7 M urea (27). The lag may represent the time required for M1 RNA to assume a potentially active structure. The lag time is decreased when the pH is increased from 7 to 7.5 suggesting the putative conformational change is pH dependent (37a).

The salt-dependence of the P RNA–catalyzed reaction has been studied extensively (29). These data also suggest that the protein may provide electrostatic shielding. In addition, the character of the ion dependence, the inhibition by high concentrations of $SO_4{}^{2-}$, and the stimulation by ethanol and dimethyl sulfoxide suggest the B. subtilis protein aids in a conformational transition (29).

RNA SPLICING

Eukaryotic genes are often interrupted by stretches of noncoding DNA called intervening sequences (IVS) or introns. RNA polymerases transcribe both the exons (coding sequences) and the intervening sequences to give large precursor

RNAs. The intervening sequences are subsequently removed by a process known as RNA splicing (for review 38, 39, and Sharp et al, this volume). Like other site-specific nucleic acid recombination processes (e.g. phage λ integration, DNA transposition), RNA splicing is a complex process that requires sequence recognition, strand cleavage, and ligation to a different site. It must be a very accurate process; inaccuracy in pre-mRNA splicing would destroy the reading frame for protein synthesis, and inaccuracy in pre-rRNA and pre-tRNA splicing would be expected to produce nonfunctional ribosomes and tRNAs. In many cases RNA splicing involves the RNA molecule not only as a substrate, but also as a catalyst or quasi-catalyst.

Tetrahymena Pre-rRNA

SELF-SPLICING   In many species of *Tetrahymena,* a ciliated protozoan, every copy of the nuclear 26S rRNA gene is interrupted by a ~400 nucleotide IVS (40–43). RNA splicing is an early step in the processing of the pre-rRNA (41, 42). The IVS is excised as a discrete linear molecule (44–47), which is subsequently converted to a circular form (46, 48).

Excision of the IVS and ligation of the exons were found to occur in vitro when pre-rRNA was incubated in a solution containing guanosine, Mg^{2+}, and salt with no added protein (49). Self-splicing also occurred with a fragment of pre-rRNA transcribed in vitro from a recombinant DNA template using purified *E. coli* RNA polymerase, thereby showing that the reactivity was intrinsic to the RNA and not due to some tightly associated protein with splicing activity (50).

In the self-splicing and autocyclization reactions, covalent bonds within the RNA molecule are broken and rejoined. Thus, the RNA is not a true catalyst. Yet it is known that the activity within the IVS RNA that promotes these reactions is not consumed or inactivated in the reactions (51–53). Thus, terms such as "catalytic activity" are appropriate.

THREE CATEGORIES OF SELF-PROCESSING REACTIONS   The two-step mechanism proposed for *Tetrahymena* pre-rRNA splicing (54, 55) is shown in Figure 2a. Both steps in splicing occur by transesterification, an exchange of phosphate esters that produces no net change in the number of ester linkages. This mechanism explains how RNA ligation can take place without an external energy source as is often provided by ATP or GTP hydrolysis. In the first transesterification, it appears that the 3' hydroxyl of a free guanosine molecule (GTP, GDP, GMP, or guanosine) acts as the nucleophile, attacking the 5' splice site. This reaction leaves a 3' hydroxyl group at the end of the 5' exon, which can then act as the nucleophile for the second transesterification reaction, exon ligation. The proposed intermediate, in which the IVS and the 3' exon are still connected, has recently been isolated and shown to be active in the second step (65), thereby satisfying a major prediction of the model.

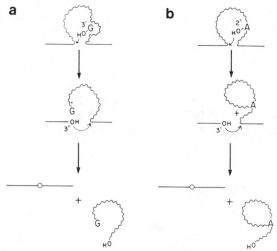

Figure 2 Transesterification mechanisms for RNA splicing (56). (*a*) Self-splicing of the *Tetrahymena* pre-rRNA and other precursor RNAs containing Group I IVSs (54, 55, 57–60). (*b*) Splicing of nuclear pre-mRNA (61, 62; see Sharp et al, this volume) and self-splicing of precursor RNAs containing Group II IVSs (63, 64). The IVS is excised as a lariat held together by a branch containing a 2'–5' phosphodiester bond. —, exons. ᴧᴧᴧ, IVS.

Cyclization of the excised IVS RNA also occurs by transesterification. The 3' hydroxyl of the 3'-terminal guanosine of the IVS RNA attacks a specific phosphate near the 5' end of the molecule, releasing an oligonucleotide of length 15 and producing a circular RNA covalently closed by a 5'–3' phosphodiester bond (55). Cyclization is a reversible reaction. Pyrimidine-rich oligonucleotides such as UpCpU-OH attack the cyclization junction, linearizing the molecule and restoring the 3'-terminal guanosine (52).

A number of additional IVS RNA-mediated reactions have been described (65–68). All of the reactions, including the two steps of splicing and the cyclization reactions described above, can be categorized into three groups: (*a*) transesterification by attack of guanosine at phosphates that follow two or three pyrimidine nucleotides (and, as a minor reaction, at phosphates that follow other guanosine residues); (*b*) transesterification by attack of oligopyrimidines (such as the 5' exon, which terminates with CUCUCU-OH) at the phosphate that follows the 3'-terminal guanosine residue of the IVS; and (*c*) specific hydrolysis at the splice sites (65).

HOW THE REACTIONS ARE CATALYZED   In a general sense, it is clear that catalysis requires the native structure of the RNA and 1–10 mM magnesium ion. The requirement for structure is shown by the loss of reactivity at high temperatures (51, 69) and at high concentrations of the denaturants formamide and urea (A. Zaug, unpublished), various intercalating dyes (70), and deoxyoligonucleotides that hybridize to various regions of the IVS RNA (T. Inoue, unpublished). Studies of the effects of site-specific mutations and small deletions within the IVS RNA show that reactivity is extremely sensitive to the structure of the molecule (68, 69, 71–73). Models of the secondary and tertiary structure of the IVS will be presented in the section regarding Group I IVSs.

The requirement for Mg^{2+} can be met by Mn^{2+} but not by Ca^{2+}, Zn^{2+}, Co^{2+}, or Pb^{2+} (A. Zaug, T. Inoue, B. Bass, T. Cech, unpublished). Polyamines or high concentrations of salt do not substitute for Mg^{2+}. The magnesium ion requirement is difficult to interpret. The secondary and tertiary structure of the IVS RNA is sensitive to the concentration of Mg^{2+}, as evidenced by chemical modification studies (74) and spectrophotometric melting analysis (D. Turner, A. Zaug, T. Cech, unpublished). Thus, while it is attractive to think of Mg^{2+} acting directly at the active site (15, 36), it is difficult to separate its possible direct role in catalysis from its indirect role in formation of the active structure of the RNA.

Three catalytic strategies of the IVS RNA have been identified. First, the RNA has a specific interaction with the guanosine nucleotide that is the attacking group for the first transesterification reaction of splicing. The K_m for the interaction is 32 μM (75, 76). Structure-reactivity studies with the guanosine substrate (75) led to a model in which the guanine base of guanosine is bound to the RNA via four hydrogen bonds (Figure 3). These interactions are seen as providing the specificity of the reaction for guanosine. It is likely that stacking interactions and perhaps other interactions also contribute to the free energy of binding. The recent finding that deoxyG and dideoxyG must be present at high concentrations to act as competitive inhibitors of splicing indicates that there are also significant binding interactions with the 2' OH and perhaps the 3' OH (76). These studies have also indicated that the 2' OH has importance in reactivity beyond its role in binding, because even very high concentrations of deoxyG do not promote the splicing reaction (76).

Second, the IVS RNA has a specific interaction with the stretch of pyrimidine residues at the end of the 5' exon that is the attacking group for the second transesterification reaction of splicing. Progressive deletion of the 5' exon and replacement with plasmid vector sequences has shown that the CUCUCU sequence preceding the 5' splice site is important for splicing (J. Price, J. Engberg, T. Cech, unpublished). Certain pyrimidine-rich oligonucleotides can function as 5' exons, cleaving the pre-rRNA at the 3' splice site and becoming covalently attached to the 3' exon (67). In this intermolecular exon-ligation reaction (or trans-splicing), the oligonucleotides follow the reactivity series CCCCC > UCU > CU > UUU > CC > UU (67). The reversal of IVS RNA cyclization (52) is analogous to exon ligation with respect to both the requirements for the attacking group (UCU-OH) and the phosphate being attacked (the phosphate that follows the 3' terminal guanosine of the IVS). Thus it has been

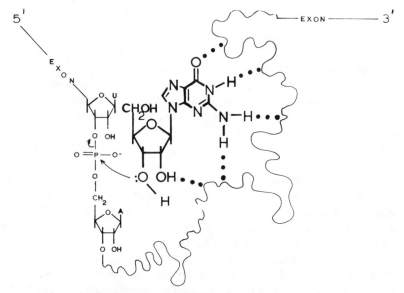

Figure 3   Model for the guanosine binding site in the *Tetrahymena* IVS RNA (75, 76). The proposed hydrogen bonds (dotted lines) could involve one or two adjacent nucleotides of the RNA or, at the other extreme, five different bases, sugars, or phosphates that are brought into close proximity by the RNA secondary and tertiary structure. Also shown is the nucleophilic attack by the 3' oxygen of guanosine at the phosphorus atom of the phosphate at the 5' splice site. The resulting transesterification has been proposed to be the first step of RNA self-splicing (49). Adapted with permission from (75).

proposed that selection of both the 5' splice site and the cyclization site could be determined by the same oligopyrimidine binding site within the IVS RNA (52, 65, 67). Mutagenesis of the major cyclization site in the IVS causes cyclization to shift to new sites. The major sites always follow three consecutive pyrimidine residues, showing that the sites are chosen primarily because of this sequence rather than their position (68). An oligopyrimidine-binding site also provides a reasonable explanation for these data.

The third catalytic strategy of the IVS RNA is to enhance the reactivity of phosphates at the sites of nucleophilic attack. The initial evidence that the molecule could strain or activate specific phosphates was the observation that the circular IVS RNA undergoes hydrolysis precisely at the bond formed during cyclization, giving a linear form with 5' phosphate and 3' hydroxyl termini (51). Hydrolysis is first order with respect to hydroxide ion concentration, consistent with direct attack (specific base catalysis) (15). Site-specific hydrolysis requires Mg^{2+} and the native structure of the RNA. The phosphate at the 3' splice site of the pre-rRNA is also subject to site-specific hydrolysis (65). This may be an identical reaction to the site-specific hydrolysis of the circular IVS RNA, because in both cases the labile phosphate ester follows the guano-

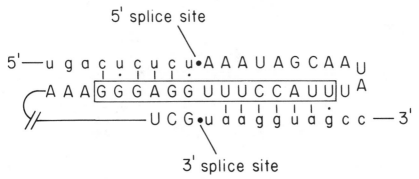

Figure 4   The internal guide sequence model of Davies et al (77, 78). Lower case letters, exons; capital letters, IVS. The internal guide sequence (boxed nucleotides) has been proposed to base-pair with both exons to form a precise alignment structure for RNA splicing. Sequences shown are for the *Tetrahymena thermophila* pre-rRNA (79). Similar structures can be drawn for most Group I IVSs (78, 80).

sine residue at the 3' end of the IVS. The phosphate at the 5' splice site also appears to undergo specific hydrolysis, although not to the extent of that at the 3' splice site (65). The interpretation of the hydrolysis reactions is that the phosphates are activated to facilitate nucleophilic attack by UCU-OH or G-OH; in the absence of the biologically relevant nucleophile, they are subject to attack by OH^-. Hydrolysis at the 3' splice site is not a productive pathway for RNA splicing. It is not a major reaction in vivo (46, 47), presumably because splicing occurs so rapidly in the presence of GTP.

SINGLE ACTIVE SITE AND TRANSLOCATION   During splicing, a free guanosine molecule attacks the 5' splice site; in an excised IVS RNA molecule with a blocked 3' end, a free guanosine molecule can attack the cyclization site (66). The guanosine at the 3' end of the IVS normally attacks the cyclization site; in a molecule in which the 5' exon is still attached to the IVS, the 3'-terminal guanosine can attack either the normal cyclization site or the 5' splice site (65). In all cases, the site of attack is preceded by a tripyrimidine sequence. It has therefore been postulated that all of the IVS RNA-mediated reactions might take place in the same active site (65).

Davies and coworkers have described a model for the structure of the *Tetrahymena* pre-rRNA (77) analogous to their models for various mitochondrial pre-RNAs (78). In this model, an internal guide sequence aligns the 5' and 3' exons for splicing (Figure 4). In the proposed pairing of the 5' exon with the internal guide sequence, the UCU preceding the 5' splice site is paired with a GGA in the internal guide sequence, entirely consistent with the biochemical data on splicing described above. The proposed pairing of the 3' exon with the internal guide sequence is hard to reconcile with the results of Been & Cech

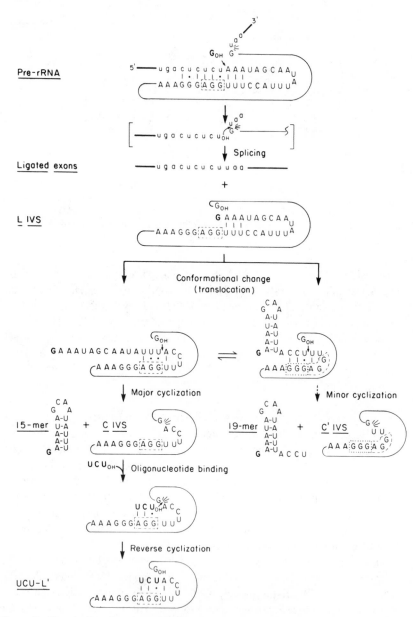

Figure 5 The single active site model for self-splicing and other self-processing reactions mediated by the *Tetrahymena* IVS RNA (52, 65, 67, 68). Lower case letters, exons; capital letters, IVS. The pre-rRNA is shown with the 5' exon paired with the internal guide sequence (77, 81). The critical oligopyrimidine binding site within the internal guide sequence is boxed. It is part of the active site for transesterification. The labile phosphodiester bond at the 3' splice site is depicted as a

(68), that showed that this portion of the internal guide sequence could be deleted without preventing splicing. It remains possible that this interaction is helpful in the wild-type structure but can be compensated by other structures (65).

The idea of a single active site and the available information about the mechanism of the various IVS RNA-mediated reactions are put in the context of a revised internal guide sequence model in Figure 5. A conformational change is required to bring different oligopyrimidine sequences into the active site for attack by guanosine. This translocation step bears some similarity to that involved in protein synthesis on ribosomes, in that interactions between triplets are involved.

Group I Precursor RNAs

CONSERVED SEQUENCES AND THE CORE STRUCTURE   Many fungal mitochondrial mRNA and rRNA genes are interrupted by IVSs. A large group of these have been defined "Group I" (81) based on a set of conserved sequence elements, each about ten nucleotides in length (78, 81, 82). The four conserved sequences have been called **A, B,** *box*9L (or element **9L**), and *box*2 (or element **2**) by Cech et al (83) and P, Q, R, and S by Davies et al (78). Another pair of sequences, **9R** and **9R'** [also called E' and E, respectively (78)] are not highly conserved in primary sequence, but are always complementary to each other. The six sequence elements always occur in the same polarity along the IVS, 5'-**9R'-A-B-9L-9R-2**-3'. In addition to these sequence elements, most Group I IVSs have a potential internal guide sequence (78; see previous section).

The same conserved sequence elements occur in the self-splicing *Tetrahymena* nuclear rRNA IVS (77, 83–85). Thus, the *Tetrahymena* IVS is a Group I IVS. Two Group I IVSs occur in the nuclear rRNA genes of *Physarum polycephalum* (86–89), while others occur in maize and bean chloroplast tRNA genes (90). The IVS that interrupts the phage T_4 thymidylate synthase gene (91, 92) has a sequence that is consistent with it belonging to Group I (93).

---

strained bond (65). Following splicing, the oligopyrimidine binding site is unoccupied, so a local conformational change can bring another tripyrimidine sequence into the binding site. The UUU preceding the major cyclization site is the only remaining pyrimidine that can bind to the GGA. The CCU preceding the minor cyclization site could presumably bind to a portion of the internal guide sequence adjacent to the GGA, as shown. In either case the 3' terminal guanosine residue of the IVS RNA can occupy the guanosine binding site and undergo transesterification to produce a circular IVS RNA (C IVS or C' IVS). Cyclization produces a strained GpA or GpU bond (15, 51, 52). Following cyclization, the oligopyrimidine binding site is again unoccupied and available for binding tripyrimidines such as UCU, which can attack the cyclization junction and linearize the IVS RNA (52).

These conserved sequences are thought to pair, **A** with **B**, **9L** with **2**, and **9R** with **9R'**, to fold the IVSs into the same core secondary and tertiary structure. This structure is illustrated for one of the mitochondrial mRNA IVSs and the *Tetrahymena* rRNA IVS in Figure 6. Evidence for these structures originally came from comparative sequence analysis of different IVSs and computer modeling (77, 78, 81, 85). In addition, there is strong genetic evidence for the **9R'/9R** and **9L/2** interactions in the case of cytochrome *b* IVS 4 (94, 101). For the *Tetrahymena* IVS, direct biochemical analysis of the structure using ribonucleases, chemical modifying agents, and methidiumpropyl-EDTA-Fe(II) (102) as probes has given support to a structure similar to that shown in Figure 6*b* (74, 83, 103).

Elements **9R'**, **9L**, **9R**, and **2** are the sites of cis-dominant, splicing-defective mutations in several of the mitochondrial IVSs (Figure 6*a*). There is mounting evidence that the analogous structural elements are required for *Tetrahymena* pre-rRNA self-splicing. In the *Tetrahymena* IVS, a single base change in element **2** destroys splicing activity in vitro as well as in vivo in *E. coli* (73). Changing two bases in element **9L** destroys self-splicing activity, but activity can be restored by a compensatory change in two bases of element **2** (69). This latter observation provides evidence that self-splicing activity requires pairing of elements **9L** and **2**. The particular nucleotide sequence also affects reactivity, perhaps because it affects the stability of the pairing interaction.

SELF-SPLICING   Several mitochondrial Group I IVSs have been found to be self-splicing in vitro. These include the first IVS of cytochrome *b* pre-mRNA from *Neurospora crassa* (58), the IVS of pre-rRNA from yeast (60), the third and fifth IVSs of cytochrome oxidase pre-mRNA from yeast (60), and some unidentified mitochondrial IVSs from *Podospora anserina* (104). In all cases guanosine is added to the 5' end of the IVS during splicing, and the splicing mechanism follows that shown for the *Tetrahymena* IVS in Figure 2*a*. The phage T_4 thymidylate synthase IVS undergoes self-excision in vitro (93), although the details of the mechanism have not yet been reported. In the case of the yeast mitochondrial rRNA IVS, the linear IVS RNA undergoes cyclization with release of four nucleotides from its 5' end (59), consistent with a transesterification mechanism. The fact that Group I IVSs have a common splicing mechanism is probably a reflection of their common ancestry, as indicated by the conserved sequence elements.

Some Group I IVSs are not self-splicing in vitro. One interpretation is that these IVSs require association with proteins to fold into the correct structure for self-splicing (58, 83, 105). Those IVSs that do self-splice in vitro may have their activity enhanced by the binding of proteins in vivo. Such proteins would provide a level of catalysis beyond that intrinsic to the RNA.

Group II Precursor RNAs

A small group of mitochondrial IVSs have conserved sequences and postulated secondary structures distinct from those of Group I IVSs, and have been designated Group II (81, 85). One of these, the last IVS in the yeast mitochondrial cytochrome oxidase pre-mRNA, has been found to undergo self-splicing in vitro (63, 64). Like self-splicing of Group I IVSs, the reaction requires Mg^{2+}. It differs in having no requirement for guanosine or any other free nucleotide as a substrate. The reaction requires spermidine, which presumably stabilizes the active conformation of the RNA. The IVS is excised as a "lariat," a branched RNA held together by a 2'–5' phosphodiester bond (63, 64). Similar lariat RNAs were previously characterized as intermediates in nuclear pre-mRNA splicing (Sharp et al, this volume). The mechanism for splicing of the cytochrome oxidase Group II pre-mRNA therefore appears to be analogous to that established for nuclear pre-mRNA splicing, as shown in Figure 2b.

The proposed mechanism of Group II RNA splicing also bears some similarity to that of Group I RNA splicing. Both processes are two-step transesterification reactions, the first step leading to cleavage at the 5' splice site and the second to cleavage at the 3' splice site and exon ligation. The second step is strictly analogous in the two systems. The first step differs; in Group I splicing, the nucleophile is the 3' hydroxyl of a guanosine bound to the RNA (75), while in Group II splicing the nucleophile is the 2' hydroxyl of a nucleotide within the RNA chain. It is the use of the 2' hydroxyl that generates the lariat structure.

How might a Group II IVS catalyze its own splicing? It seems reasonable that the key elements are Mg^{2+} ion and the folded structure of the RNA, as in the self-splicing of Group I IVSs. The structure of the RNA must bring the lariat-forming 2' hydroxyl group in proximity to the 5' splice site and catalyze transesterification. The nucleophilicity of a 2' hydroxyl within an RNA chain would be expected to be considerably less than that of the terminal 2' or 3' hydroxyl groups, which have a cis-diol configuration. The self-splicing of the Group II pre-mRNA shows that, in the context of a folded RNA molecule, even the chemically less reactive 2' hydroxyl can participate in a cleavage-ligation reaction. The accuracy of the reaction is presumably achieved by recognition of the sequence GUGCG which occurs immediately downstream from the 5' splice site within Group II IVSs (56, 63). It remains to be seen whether the structure of the RNA also activates the phosphate to be attacked, as in the case of Group I splicing.

Nuclear Pre-mRNA

Splicing of nuclear pre-mRNA, like the self-splicing of the Group II IVS described above, involves a lariat RNA intermediate (Figure 2b). Although nuclear pre-mRNAs do not undergo self-splicing in vitro, it is known that the IVS RNA has some role in directing the splicing reaction. The IVS contains

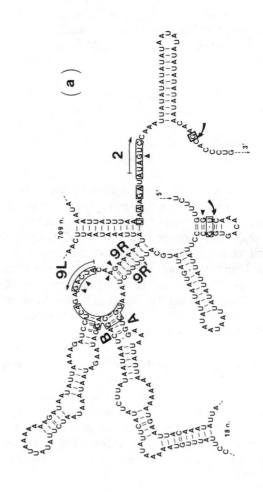

(a)

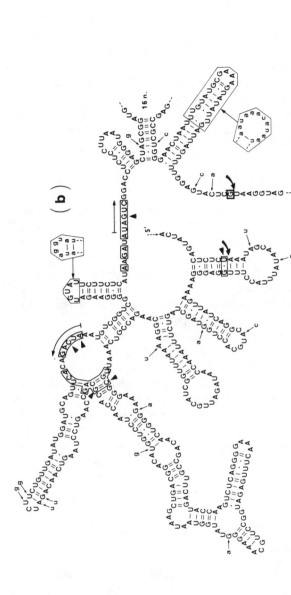

Figure 6 Secondary structure models of Group I IVSs according to Michel & Dujon (85). Large arrows designate splice sites. Heavily boxed residues are invariant in *Saccharomyces cerevisiae* mitochondrial Group I IVSs. The arrows above sequence elements **9L** and **2** indicate the established base-pairing interaction (69, 94), which in these models is at the level of tertiary structure. (*a*) The fourth IVS in the yeast mitochondrial cytochrome oxidase pre-mRNA (95). The fourth IVS in the closely related fourth IVS of the cytochrome *b* gene (96–99). (*b*) *T. thermophila* Solid triangles point to the sites of splicing-defective mutations in this IVS or the closely related fourth IVS of the cytochrome *b* gene (96–99). (*b*) *T. thermophila* (upper case letters) and *T. pigmentosa* (lower case letters) pre-rRNA IVSs (79, 100). Solid triangles indicate splicing-defective single-base or double-base mutations (69, 73; L. Barfod, T. Cech, unpublished). Adapted with permission from (85). Similar models have been proposed by others (74, 77, 83).

three conserved sequences, located at the 5' splice site, at the 3' splice site, and at the site of branch formation; their importance in the splicing reaction has been confirmed by mutational analysis. On the other hand, deletion of large regions of an IVS outside these three conserved sequences does not generally affect splicing (reviewed in detail by Sharp et al, this volume). This is in sharp contrast to the self-splicing IVSs, in which a large portion of the RNA structure is highly conserved and required for splicing activity (71–73). Thus, it is not surprising that nuclear pre-mRNA splicing requires trans-acting elements in addition to the conserved IVS sequences. The possibility of RNA catalysis arises because some of these trans-acting elements, the snRNPs (small nuclear ribonucleoprotein particles), contain RNA.

Small nuclear RNAs (snRNAs) are stable, relatively abundant RNA molecules found in the nuclei of a wide variety of eukaryotes (106, 107). The more abundant snRNAs range in size from about 90 to 220 nucleotides, and their sequences are highly conserved in evolution. Each snRNA exists in a complex with several polypeptides to give a discrete snRNP (108). The U1 snRNP (109, 110), the U2 snRNP (111, 112), and perhaps other snRNPs are required for nuclear pre-mRNA splicing in vitro. The 5' end of the U1 snRNA appears to be involved in recognition of the 5' splice site, probably by base-pairing with the conserved $GU_G^A AGU$ sequence at the 5' end of the IVS (113). The U2 snRNA has been proposed to interact with the sequence near the 3' end of the IVS that contains the lariat-forming 2'-hydroxyl group (111, 114, 115).

In addition to their role in sequence recognition, snRNPs could have a role in catalysis of pre-mRNA splicing. As speculated by Kruger et al (50), it is possible that nuclear pre-mRNA splicing is similar to self-splicing, with the catalytic RNA component being trans-acting (present on separate snRNP molecules) in the former case and cis-acting (the IVS itself) in the latter. SnRNPs appear to require their protein as well as their RNA components to be functional. As in the case of RNase P, however, it may be that the RNA contains the active site and the proteins have an auxiliary role. Such arguments now appear much more plausible with the finding that a Group II IVS undergoes self-splicing by the same mechanism used for nuclear pre-mRNA splicing (Figure 2*b*).

THE TETRAHYMENA IVS RNA AS A TRUE, REGENERATED ENZYME

The L-19 IVS RNA is missing the first 19 nt of the IVS, which contain the major and minor cyclization sites. Therefore it cannot undergo intramolecular reaction (51). The L-19 IVS still retains activity, however, and can catalyze cleavage-ligation reactions on other RNA molecules. The L-19 IVS RNA is regenerated after each of these intermolecular reactions (53). Thus, while the role of the IVS RNA in the cell is probably restricted to self-splicing, it can turn over as a true enzyme in vitro. Like many protein enzymes, it is stable for hours under reaction conditions.

A probable reaction pathway has been described (53). When certain oligo-ribonucleotides are incubated with the L-19 IVS RNA enzyme in the presence of Mg^{2+}, a transesterification reaction occurs, resulting in covalent attachment of a portion of the oligonucleotide substrate to the 3' end of the enzyme. This covalent enzyme-substrate intermediate reacts in one of two ways: (a) The intermediate can be attacked by an oligonucleotide with a free 3' hydroxyl terminus; transesterification occurs, a new ligated oligonucleotide is produced, and the L-19 IVS RNA enzyme is regenerated. (b) The intermediate can undergo hydrolysis, freeing the remaining fragment of the substrate RNA with 5'-phosphate and 3'-hydroxyl termini and regenerating the enzyme. More specifically, with pC_5-OH as a substrate, the net reactions are (a) $2 pC_5$-OH $\rightarrow$ pC_4-OH $+ pC_6$-OH and (b) pC_5-OH $\rightarrow pC_4$-OH $+ pC$-OH. The latter reaction is minor at pH 7.5 and enhanced at pH 9. While the reactions have been studied most extensively with short oligonucleotides, the enzyme also catalyzes cleavage-ligation of large RNA substrates.

In both the transesterification and the hydrolysis reactions, the enzyme is unreactive toward DNA substrates and shows considerable specificity for the sequence of its RNA substrate (53). Of the homopentanucleotides, C_5 is the only good substrate; the transesterification reaction at neutral pH and 42°C has $K_m = 40$ µM and $k_{cat} = 1.7$ min^{-1}. There is limited reaction with U_5 and none with A_5. DeoxyC_5 is a competitive inhibitor of the reaction ($K_i = 200$ µM). Based on such results, it was proposed that the active site of the L-19 IVS RNA enzyme is the same site that binds oligopyrimidines in the self-splicing and cyclization reactions, as shown in Figure 5.

Protein ribonucleases that are active on single-stranded RNA substrates have specificity only at the mononucleotide level (e.g. RNase T_1 cleaves after guanosine, RNase A after pyrimidines, and RNase U_2 after adenosine residues). Although substrate specificity studies are still in progress, it is already clear that the L-19 IVS RNA enzyme recognizes three or more nucleotides in choosing a reaction site. Thus, it has more base-sequence specificity for single-stranded RNA than any known protein ribonuclease.

REPLICATION OF PLANT INFECTIOUS AGENTS

Plant viroids, the smallest infectious agents known, are single-stranded, covalently closed circles of RNA (117, 118). The viroid replication cycle appears to involve the conversion of multimeric linear molecules to infectious monomeric circular forms, perhaps by way of monomeric linear intermediates (reviewed in Ref. 119). Some satellite RNAs, linear RNA molecules encapsidated during plant virus infections, undergo oligomerization and cleavage back to monomers during their replication cycle (Ref. 120 and papers cited therein).

In three cases there is experimental evidence that cleavage of these oligomeric RNAs to give monomers occurs by self-processing. Robertson et al (121) have shown that a dimeric viroid RNA, produced by transcription of a recombinant DNA template with phage RNA polymerase in vitro, is inefficiently cleaved to give linear monomer-size RNA when incubated in a Mg^{2+}-containing buffer at 39°C with no added protein. A portion of the reaction products can be circularized by treatment with wheat germ RNA ligase, evidence that these molecules are full-length and have 2',3'-cyclic phosphate termini (121). Prody et al (122) have found that dimeric and trimeric forms of the tobacco ringspot virus satellite are partially converted to monomeric linear RNA when incubated in vitro with no added protein. Partial nucleotide sequence analysis indicates that the monomer RNA produced by cleavage in vitro has the same sequence as authentic monomer RNA from virus particles. Furthermore, the monomer RNA produced in vitro has biological activity. Preliminary evidence has also been obtained for the reverse reaction, self-catalyzed dimerization of the linear monomers (122). As in the viroid RNA experiments, there is the suggestion that the reaction products have 2',3'-cyclic phosphate termini, but RNA fingerprinting data are not yet available (122). Once the nature of the termini and the sequence at the cleavage sites are determined for both the in vitro and in vivo reaction products, it will be easier to evaluate whether self-processing reactions could have a role in the replication cycle of these plant infectious agents. Efficient self-processing of dimeric RNA transcripts of avocado sunblotch viroid and of its complementary strand has recently been found (C. J. Hutchins, P. D. Rathjen, A. C. Forster, R. H. Symons, personal communication).

The *Tetrahymena* IVS RNA mediates its own oligomerization when exposed to a heating-cooling treatment in vitro, and the oligomers are subsequently converted to monomeric and oligomeric circles (66). The similarity of these reactions to steps in the life cycle of the plant infectious agents may be more than coincidence, considering that viroid RNAs and satellite RNAs contain sequences homologous to the conserved sequence elements that define Group I IVSs (120, 123, 124). The occurrence of these sequences suggests that the biologically active structure of the infectious RNA might resemble that of a Group I IVS, with folding of the RNA perhaps being facilitated by proteins in vivo. It has been proposed that self-catalyzed reactions might account for the conversion of oligomeric viroid RNA to monomeric circles (66, 119, 124) and for the oligomerization of the peanut stunt virus-associated satellite RNA (120). The 3' hydroxyl groups required for such schemes are different from the 2',3'-cyclic phosphates thought to be produced during in vitro cleavage of these RNAs, as described above.

OTHER SYSTEMS

An RNA molecule transcribed from bacteriophage T_4 undergoes site-specific self-cleavage in vitro, producing species 1 RNA as found in T_4-infected *E. coli* (125). Cleavage results in 5'-hydroxyl and 3'-phosphate termini. The reaction requires a monovalent cation and is stimulated by non-ionic detergents. It is not affected by the addition of 5 mM EDTA and is not stimulated by the addition of divalent cations; the apparent lack of a divalent cation requirement distinguishes this reaction from the other quasi-catalytic cleavage reactions discussed above. A dodecanucleotide that contains the major cleavage site is not cleaved under the same reaction conditions, providing evidence against involvement of a contaminating nuclease and evidence that the intact structure of the RNA is required (125).

Many biochemical processes in eukaryotes are mediated by small ribonucleoprotein particles. In the nucleus, the U7 snRNP is required for the correct processing of the 3' end of a histone poly A^- mRNA (126, 127), and U1 and perhaps other snRNPs are required for polyadenylation of other mRNAs (128, 129). In the cytoplasm, the interaction between ribosomes and the endoplasmic reticulum is mediated by the signal recognition particle, which contains a 295-nt RNA molecule (130, 131). The presence of both protein and RNA in such particles will make it difficult to determine whether the RNA plays a catalytic role in these reactions, or is simply binding to certain sequences on an RNA substrate and thereby "marking" them for processing by protein catalysts.

An even more complex situation is found with the ribosome, a ribonucleoprotein particle containing 3 or 4 RNA molecules and more than 50 polypeptides. The finding that the structure of rRNA is highly conserved in evolution (132) and the finding of drug-resistance mutations in the rRNA (reviewed in 133), have forced a reevaluation of the idea that the rRNAs simply provide the framework for assembly of a complex of protein enzymes. In addition, 16S and 23S rRNA form a binary complex (134) which, when supplemented with a limited number of ribosomal proteins, is capable of catalyzing many of the steps of protein synthesis to a small extent (135).

The proper positioning of the tRNA substrates and the mRNA template may play a major role in the catalysis of protein synthesis. This catalytic function may be best performed by other RNA molecules, the rRNAs. According to this view, the ribosomal proteins refine and tune the structure and activity of a translational apparatus that is basically an RNA machine (136).

COMPARISON OF RNA AND PROTEINS AS BIOLOGICAL CATALYSTS

It is commonly agreed that it is the rate and specificity of protein-catalyzed reactions that distinguishes them from uncatalyzed and nonenzymically catalyzed reactions (2, 137–139). The catalytic properties of a protein enzyme result from its ability to form a complex with its substrate. Using noncovalent interactions enzymes assume a precise secondary and tertiary structure, creating three-dimensional cavities or active sites. Noncovalent interactions are also responsible for binding the substrate to the enzyme active site.

The kinetic parameters for two RNA catalysts, the RNA subunit of RNase P and the IVS RNA of *Tetrahymena thermophila,* are listed in Table 2. The K_m values of the RNA catalysts are within the range of those observed for protein catalysts, while the k_{cat} values are on the low end of the range. However, the reactions certainly proceed at a rate and with a specificity more typical of protein enzymes than chemical catalysts (15, 75). In addition, at least three pieces of evidence suggest that RNA catalysis occurs by a mechanism similar to that of protein catalysis, involving the formation of a complex between the catalyst and the substrate. First, all of the RNA catalysts that have been studied are dependent on their native structure for catalytic activity (27, 49, 69, 71–73). Second, the RNase P RNA and IVS RNA demonstrate Michaelis-Menten kinetics (27, 30, 53, 75). Finally, both RNase P and the IVS RNA show high substrate specificity (27, 52, 53, 67, 75, 76). Thus, RNA and protein catalysts can provide similar rate enhancement and specificity and, in addition, seem to rely on the same principles to achieve their catalytic properties. We will now discuss more specifically the properties that allow each macromolecule to function as a biological catalyst.

Substrate Binding Sites

RNA and protein both have secondary and tertiary structures and thus have the capability to form three-dimensional substrate binding sites. There are differences between RNA and protein, however, that will most likely be reflected in the organization of binding sites and in the type of substrate each molecule may bind. The most obvious difference is the homogeneity of RNA side chains when compared to the side chains of a protein. A protein polymer forms from a pool of twenty different amino acid monomers, while RNA forms from a group of only four different monomers. In contrast to the range of chemical properties represented among the amino acids, the four bases contain similar functional groups and have similar hydrophilicity. Thus, although RNA can catalyze reactions very efficiently, it may be more limited in the type of reactions it can catalyze and the substrates upon which it can act.

An RNA polymer differs from a protein polymer in the way it utilizes noncovalent interactions to fold into a secondary and tertiary structure. The secondary structure of a protein is created by intramolecular hydrogen bonds between the repeating peptide bond backbone. The amino acid side chains are directed outward from the hydrogen bonding mesh. The hydrophobic interactions that occur between the aqueous solvent and the amino acid side

Table 2 Kinetic parameters of pre-rRNA self-splicing and selected enzyme-catalyzed reactions

Biological catalyst	Substrate	K_m (mM)	k_{cat} (min^{-1})	k_{cat}/K_m (M^{-1} min^{-1})	Ref.
IVS RNA[a]	guanosine	0.032	0.91	28×10^3	76
	U$_3$	0.03	0.4	13×10^3	52
L − 19 IVS RNA	C$_5$	0.04	1.7	42×10^3	53
RNase P	pre-tRNA^{Tyr}	0.0005	2	4×10^6	27
M1 RNA	pre-tRNA^{Tyr}	0.0005	1	2×10^6	27
P RNA	A$_3$C-tRNA$_F^{Met}$	0.0001	2.2	2.2×10^7	30
RNase A	CpC[b]	4.0	14400	3600×10^3	140
	C>p[c]	3.0	330	110×10^3	140
RNase T$_1$	GpA	0.055	5760	105×10^6	141

[a]Self-splicing of the *Tetrahymena* pre-rRNA in which the IVS RNA acts as an intramolecular catalyst. All other entries refer to true catalysts.
[b]Formation of 2',3' cyclic phosphate.
[c]Hydrolysis of 2',3' cyclic phosphate.

chains provide the driving force for the folding of the protein into a tertiary structure (139, 142). Enzymes are usually globular proteins rather than fibrous proteins, with extensive tertiary structure and very closely packed interiors (142, 143).

The secondary structure of a polynucleotide is characterized by the formation of specific hydrogen bonds between the four bases. In contrast to a protein, it is the side chains, rather than the backbone, that are located on the interior of the secondary structure and provide the hydrogen bonding network (144). Although tertiary interactions do exist, solubility is not a driving force for these interactions in RNA because the highly charged sugar-phosphate backbone is exposed to the solvent. RNA molecules are less compact than proteins (145).

Although more data are necessary before conclusive statements regarding RNA binding sites can be made, the unique features of RNA structure can be predicted to have major consequences for the substrate binding sites within an RNA molecule. For example, it seems likely that because the charged phosphates are directed outward from the hydrogen bonding network, most of an RNA molecule will be accessible to water and hydrophobic pockets will not be common. From their study of the effects of solvents on the formation of a double helix of A_7U_7, Hickey & Turner (146) conclude that hydrophobic bonding is not a major contributor to stabilization of the RNA helix. On the other hand, Porschke (147) finds that amides of hydrophobic amino acids such as Val, Ile, and Leu destabilize the RNA double helix and concludes that single-stranded RNA may form hydrophobic sites for binding hydrophobic ligands. Thus, it is premature to dismiss the possibility that a folded RNA molecule could form a hydrophobic binding site.

It is interesting to note that RNase T_1, a protein whose substrate is ribonucleic acid, does not have a distinct cleft for substrate binding (148). The enzyme-substrate complex involves specific hydrogen bonds from the peptide backbone to the guanine base, and a tyrosine residue is thought to stabilize the binding by stacking of its aromatic ring with the purine base. For an RNA catalyst that binds a nucleic acid substrate, base-stacking might fill the role that hydrophobic interactions play in protein catalysis.

The base-pairing properties of RNA may actually make it a better macromolecule than protein for binding single-stranded nucleic acid substrates. A protein enzyme, because of its extensive folding, constructs its binding sites so that residues from very different parts of the molecule are involved in binding a single substrate. The X-ray crystal structure of tRNA indicates that such long range interactions can exist within an RNA molecule (149, 150). Thus, it is possible that binding sites also involve nonadjacent residues. However, RNA has the additional capability to bind a nucleic acid substrate with adjacent residues, by Watson-Crick or other base-pairing interactions. In the model presented in Figure 5, this type of linear binding site is proposed to bind the pyrimidines that precede the 5' splice site and the major and minor cyclization sites of the *Tetrahymena* pre-rRNA.

It seems likely that RNA will frequently utilize base-pairing interactions to bind its substrate. However, the existing data in no way rule out the participation of other interactions in binding substrates to RNA. For example, the phosphate oxygens of RNA can act as hydrogen bond acceptors, and the 2' hydroxyls can function as hydrogen bond donors (152a,b). Optimal binding of guanosine to the IVS RNA requires the ribose hydroxyls and, thus, interactions in addition to base pairing are implicated (75, 76).

General Acid-Base Catalysis, Covalent Intermediates, and Bond Strain

In addition to an enzyme-substrate complex, other catalytic strategies contribute to enzyme catalysis. For example, in some cases functional groups at the active site act as general acid and base catalysts. Amino acid residues ionizable near neutral pH such as histidine ($pK_a \simeq 6$) and cysteine ($pK_a \simeq 8$) are especially well-suited for general acid-base catalysis. Nucleic acids do not contain many moieties that are ionizable near neutral pH (Table 3). Certain modified nucleosides such as 7-methylguanosine are ionizable near neutral pH, but neither RNase P nor the *Tetrahymena* IVS RNA contain modified nucleosides. A 5'-terminal phosphate has a $pK_a \simeq 6$ and has in fact been proposed as a possible proton donor-acceptor for catalysis of peptidyl transfer (153). Neither RNase P nor the *Tetrahymena* IVS RNA requires a 5'-terminal phosphate for activity.

In protein enzymes, interactions at the active site can dramatically shift a pK_a from its normal value (154, 155). By analogy, it is possible that in a folded RNA molecule, pK_a's could be shifted such that the nucleotides in RNA would be more useful in donating or accepting protons. In fact, *E. coli* 5S rRNA releases a proton when the pH is increased from 7 to 8 (156). Thus, RNA enzymes may also utilize general acid-base catalysis.

The first step of *Tetrahymena* pre-rRNA splicing occurs by a transesterification reaction in which the 3' hydroxyl of a guanosine becomes covalently linked to the 5' end of the IVS (Figures 2a, 3). During the reaction the hydrogen of the 3' hydroxyl must be lost. It seems likely that a general base catalyst at the active site mediates the proton transfer. The 2' hydroxyl of the guanosine is essential for the reaction; dG cannot substitute for G (75, 76). One possibility is that this functional group is the general base involved in the reaction. Cyclization of the excised IVS RNA is analogous to the first step of splicing in that the nucleophile is the 3' hydroxyl of guanosine; cyclization also requires proton transfer. However, the rate of cyclization is independent of pH in the range 4.7–9.0 (15). Thus, if general base catalysis is involved in the reaction, the functional group involved does not ionize in this pH range.

Another strategy used by some protein enzymes is to form a covalent intermediate with a substrate. A covalent intermediate has been observed in the

Table 3 Functional groups in RNA with potential for general acid-base catalysis[a]

Acid or base	pK$_a$
Terminal 2' or 3' hydroxyl	~12.5
N-3 of uridine	10.1
N-1 of guanosine	10.0
N-1 of 7-methylguanosine	7.8
5'-terminal phosphate	6.1– 6.4
N-3 of cytidine	4.5
N-1 of adenosine	3.9
N-7 of guanosine	2.4
Internal phosphate	~1.0

[a]Values for nucleoside 5' phosphates at 20°C extrapolated to zero ionic strength from (151), except 7-methylguanosine-5'-triphosphate from (152).

reactions of the L-19 IVS RNA enzyme with oligoribonucleotide substrates. The covalent linkage involves the 3' hydroxyl of the terminal guanosine of the L-19 RNA (53). So far, this is the only example of a covalent intermediate in an RNA-catalyzed reaction. However, covalent bond formation by a ribose 2'-OH is involved in the self-splicing of the Group II IVS RNA (Figure 2b), and ribose hydroxyls are commonly used to form covalent bonds in enzyme-catalyzed reactions that involve RNA (e.g. aminoacyl-tRNA synthetase and RNA polymerase). These examples emphasize that the ribose hydroxyls can be good nucleophiles that form stable linkages, giving RNA the potential for covalent catalysis.

Protein catalysts are thought to facilitate the formation of the transition state by providing active site groups that bind the transition state better than the unreacted substrate, or possibly by distorting the substrate upon binding (13, 14). A similar situation has been proposed to exist in an RNA-catalyzed reaction. The phosphodiester bond that follows the 3'-terminal guanosine of the *Tetrahymena* IVS RNA in the circular IVS RNA, in the pre-rRNA, or in the covalent IVS RNA enzyme-substrate intermediate is particularly susceptible to hydrolysis (51, 53, 65). The rate of hydrolysis is enhanced 12 orders of magnitude over that of a simple phosphate diester and ~ 10 orders of magnitude over that of an average phosphodiester bond in RNA (53). The lability of this bond has been interpreted in terms of the phosphate being strained or activated to promote nucleophilic attack (attack by the 3' hydroxyl of the 5' exon in the case of the self-splicing reaction and attack by the 3' hydroxyl of an oligonucleotide on the covalent intermediate in the case of the L-19 IVS RNA enzyme reaction). Thus, it appears likely that the folded structure facilitates the formation of the transition state.

Conformational Change

All current discussions of protein catalysis stress its dynamic nature; conformational change is considered essential to catalysis (157–159). Conformational changes could occur in an RNA molecule by making and breaking helixes, by stacking and unstacking helixes, or by altering tertiary interactions (133). There is evidence for conformational change in catalysis by the IVS RNA and by the RNA of RNase P from both *E. coli* and *B. subtilis* (27, 29, 69). It is not yet clear how these conformational changes contribute to catalysis.

In addition to gross conformational changes, intramolecular motion on the picosecond time scale also occurs in proteins (157, 160). It has been proposed that such fluctuations are a means of collimating thermal energy to produce high free energy events at the catalytic site (160, 161). Molecular dynamics calculations indicate that RNA molecules also have a great deal of internal motion (162, 163). It is possible that this motion acts to enhance the reactivity of the various phosphodiester bonds that are cleaved in the reactions of the *Tetrahymena* IVS RNA and RNase P.

CONCLUSION

In their discussions about the origin of the translation apparatus, Crick, Orgel, Woese, and others foresaw the possibility that RNA might have catalytic activity (136, 164, 165). The general consensus, however, was that RNA would not be a very good catalyst, and that its function would have been taken over by proteins during evolution.

The new findings about RNA catalysts give a quite different perspective. In self-processing reactions and in true enzymatic reactions with other polynucleotides as substrates, RNA can promote reactions with rate enhancement and specificity comparable to those of protein enzymes. RNase P recognizes the structure of tRNA with as much fidelity as any of the protein enzymes that use tRNA as a substrate. The *Tetrahymena* L-19 IVS RNA enzyme cleaves single-stranded RNA with more sequence specificity than any known protein ribonuclease, and it accelerates the rate of hydrolysis 12 orders of magnitude over the basal-level rate of hydrolysis of simple phosphate diesters. Thus, these RNA catalysts are not easily dismissed as molecular fossils, waiting to be replaced by better protein enzymes. It seems likely they have been retained because they are such good catalysts for reactions involving RNA substrates. It appears to be the limited versatility of RNA catalysts, rather than any deficit in catalytic efficiency or accuracy, that is responsible for the relatively restricted occurrence of RNA as a biological catalyst.

ACKNOWLEDGMENTS

We are grateful to our present and former laboratory colleagues who helped shape the ideas presented here, and to the many colleagues elsewhere who provided ideas and preprints of their work. We thank Alice Sirimarco and Jan Logan for their help in preparing the manuscript. Our research has been supported by grants from the National Institutes of Health and the American Cancer Society.

Literature Cited

1. Jencks, W. P. 1969. *Catalysis in Chemistry and Enzymology*. New York: McGraw-Hill
2. Walsh, C. 1979. *Enzymatic Reaction Mechanisms*. San Francisco: Freeman
3. Bender, M. L., Bergeron, R. J., Komiyama, M. 1984. *The Bioorganic Chemistry of Enzymatic Catalysis*. New York: Wiley
4. Eskin, B., Linn, S. 1972. *J. Biol. Chem.* 247:6183–91
5. Yuan, R., Heywood, J., Meselson, M. 1972. *Nature New Biol.* 240:42–43
6. Jump, D. B., Smulson, M. 1980. *Biochemistry* 19:1024–30
7. Ogata, N., Veda, K., Kawaichi, M., Hayaishi, O. 1981. *J. Biol. Chem.* 256:4135–37
8. Robins, P., Cairns, J. 1979. *Nature* 280:74–76
9. Nash, H. 1981. *Ann. Rev. Genet.* 15:143–67
10. Reed, R. R. 1981. *Cell* 25:713–19
11. Abremski, K., Hoess, R. 1984. *J. Biol. Chem.* 259:1509–14
12. Little, J. W. 1984. *Proc. Natl. Acad. Sci. USA* 81:1375–79
13. Fersht, A. 1977. *Enzyme Structure and Mechanism*. Reading/San Francisco: Freeman
14. Wells, T. N. C., Fersht, A. R. 1985. *Nature* 316:656–57
15. Zaug, A. J., Kent, J. R., Cech, T. R. 1985. *Biochemistry*. 24:6211–18
16. Wintermeyer, W., Zachau, H. G. 1973. *Biochim. Biophys. Acta* 299:82–90
17. Werner, C., Krebs, B., Keith, G., Dirheimer, G. 1976. *Biochim. Biophys. Acta* 432:161–75
18. Rordorf, B. F., Kearns, D. R. 1976. *Biopolymers* 15:1491–504
19. Brown, R. S., Hingerty, B. E., Dewan, J. C., Klug, A. 1983. *Nature* 303:543–46
20. Brown, R. S., Dewan, J. C., Klug, A. 1985. *Biochemistry*. 24:4785–801
21. Rubin, J. R., Sundaralingam, M. 1983. *J. Biomol. Struct. Dyn.* 1:639–46
22. Sundaralingam, M., Rubin, J. R., Cannon, J. F. 1984. *Int. J. Quantum Chem: Quantum Biol. Symp.* 11:355–66
23. Baes, C. F. Jr., Mesmer, R. E. 1976. *The Hydrolysis of Cations*. New York: Wiley
24. Kole, R., Baer, M. F., Stark, B. C., Altman, S. 1980. *Cell* 19:881–87
25. Ikemura, T., Shimura, Y., Sakano, H., Ozeki, H. 1975. *J. Mol. Biol.* 96:69–86
26. Altman, S., Guerrier-Takada, C., Frankfort, J., Robertson, H. D. 1982. In *Nucleases*, ed. S. Linn, R. Roberts, pp. 243–74. New York: Cold Spring Harbor Lab.
27. Guerrier-Takada, C., Gardiner, K., Marsh, T., Pace, N., Altman, S. 1983. *Cell* 35:849–57
28. Guerrier-Takada, C., Altman, S. 1984. *Science* 223:285–86
29. Gardiner, K. J., Marsh, T. L., Pace, N. R. 1985. *J. Biol. Chem.* 260:5415–19
30. Marsh, T. L., Pace, B., Reich, C., Gardiner, K., Pace, N. R. 1986. In *Sequence Specificity in Transcription and Translation*, ed. R. Calendar, L. Gold. *UCLA Symp. Mol. Cell. Biol. (NS)*, New York: Plenum
31. Baer, M., Altman, S. 1985. *Science* 228:999–1002
32. Akaboshi, E., Guerrier-Takada, C., Altman, S. 1980. *Biochem. Biophys. Res. Commun.* 96:831–37
33. Kline, L., Nishikawa, S., Soll, D. 1981. *J. Biol. Chem.* 256:5058–63
34. Marsh, T. L., Pace, N. R. 1985. *Science* 229:79–81
35. Guerrier-Takada, C., Haydock, K., Allen, L., Altman, S. 1986. *Biochemistry*. In press
36. Haydock, K., Allen, L. C. 1985. In *Progress in Clinical and Biological Research*, ed. R. Rein, 172A:87–98. New York: Liss
37. Guerrier-Takada, C., Altman, S. 1984. *Biochemistry* 23:6327–34
37a. Altman, S., Guerrier-Takada, C. 1986. Submitted for publication
38. Cech, T. R. 1983. *Cell* 34:713–16
39. Greer, C. L., Abelson, J. 1984. *Trends Biochem. Sci.* 9:139–41
40. Wild, M. A., Gall, J. G. 1979. *Cell* 16:565–73
41. Cech, T. R., Rio, D. C. 1979. *Proc. Natl. Acad. Sci. USA* 76:5051–55
42. Din, N., Engberg, J., Kaffenberger, W., Eckert, W. 1979. *Cell* 18:525–32
43. Din, N., Engberg, J. 1979. *J. Mol. Biol.* 134:555–74

44. Zaug, A. J., Cech, T. R. 1980. *Cell* 19:331–38
45. Carin, M., Jensen, B. F., Jentsch, K. D., Leer, J. C., Nielson, O. F., Westergaard, O. 1980. *Nucleic Acids Res.* 8:5551–66
46. Brehm, S. L., Cech, T. R. 1983. *Biochemistry* 22:2390–97
47. Kister, K.-P., Müller, B., Eckert, W. A. 1983. *Nucleic Acids Res.* 11:3487–502
48. Grabowski, P. J., Zaug, A. J., Cech, T. R. 1981. *Cell* 23:467–76
49. Cech, T. R., Zaug, A. J., Grabowski, P. J. 1981. *Cell* 27:487–96
50. Kruger, K., Grabowski, P. J., Zaug, A. J., Sands, J., Gottschling, D. E., Cech, T. R. 1982. *Cell* 31:147–57
51. Zaug, A. J., Kent, J. R., Cech, T. R. 1984. *Science* 224:574–78
52. Sullivan, F. X., Cech, T. R. 1985. *Cell* 42:639–48
53. Zaug, A. J., Cech, T. R. 1986. *Science* 231:470–75
54. Zaug, A. J., Cech, T. R. 1982. *Nucleic Acids Res.* 10:2823–38
55. Zaug, A. J., Grabowski, P. J., Cech, T. R. 1983. *Nature* 301:578–83
56. Cech, T. R. 1986. *Cell* 44:207–10
57. Garriga, G., Lambowitz, A. M. 1983. *J. Biol. Chem.* 258:14745–48
58. Garriga, G., Lambowitz, A. M. 1984. *Cell* 39:631–41
59. Tabak, H. F., Van der Horst, G., Osinga, K. A., Arnberg, A. C. 1984. *Cell* 39: 623–29
60. Van der Horst, G., Tabak, H. F. 1985. *Cell* 40:759–66
61. Padgett, R. A., Konarska, M. M., Grabowski, P. J., Hardy, S. F., Sharp, P. A. 1984. *Science* 225:898–903
62. Ruskin, B., Krainer, A. R., Maniatis, T., Green, M. R. 1984. *Cell* 38:317–31
63. Peebles, C. L., Mecklenburg, K. L., Perlman, P. S., Tabor, J., Cheng, H. L. 1986. *Cell* 44:213–23
64. Van der Veen, R., Arnberg, A. C., Van der Horst, G., Bonen, L., Tabak, H. F., Grivell, L. A. 1986. *Cell* 44:225–34
65. Inoue, T., Sullivan, F. X., Cech, T. R. 1986. *J. Mol. Biol.* In press
66. Zaug, A. J., Cech, T. R. 1985. *Science* 229:1060–64
67. Inoue, T., Sullivan, F. X., Cech, T. R. 1985. *Cell* 43:431–37
68. Been, M., Cech, T. R. 1985. *Nucleic Acids Res.* 13:8389–408
69. Burke, J. M., Kaneko, K. J., Irvine, K. D., Oettgen, A. B., Zaug, A. J., Cech, T. R. 1985. Submitted for publication
70. Tanner, N. K., Cech, T. R. 1985. *Nucleic Acids Res.* 13:7741–58
71. Price, J. V., Kieft, G. L., Kent, J. R.,

Sievers, E. L., Cech, T. R. 1985. *Nucleic Acids Res.* 13:1871–89
72. Price, J. V., Cech, T. R. 1985. *Science* 228:719–22
73. Waring, R. B., Ray, J. A., Edwards, S. W., Scazzocchio, C., Davies, R. W. 1985. *Cell* 40:371–80
74. Inoue, T., Cech, T. R. 1985. *Proc. Natl. Acad. Sci. USA* 82:648–52
75. Bass, B. L., Cech, T. R. 1984. *Nature* 308:820–26
76. Bass, B. L., Cech, T. R. 1986. Submitted for publication
77. Waring, R. B., Scazzocchio, C., Brown, T. A., Davies, R. W. 1983. *J. Mol. Biol.* 167:595–605
78. Davies, R. W., Waring, R. B., Ray, J. A., Brown, T. A., Scazzocchio, C. 1982. *Nature* 300:719–24
79. Kan, N. C., Gall, J. G. 1982. *Nucleic Acids Res.* 10:2809–22
80. Waring, R. B., Davies, R. W. 1984. *Gene* 28:277–91
81. Michel, F., Jacquier, A., Dujon, B. 1982. *Biochimie* 64:867–81
82. Waring, R. B., Davies, R. W., Scazzocchio, C., Brown, T. A. 1982. *Proc. Natl. Acad. Sci. USA* 79:6332–36
83. Cech, T. R., Tanner, N. K., Tinoco, I. Jr., Weir, B. R., Zuker, M., Perlman, P. S. 1983. *Proc. Natl. Acad. Sci. USA* 80:3903–7
84. Burke, J. R., RajBhandary, U. L. 1982. *Cell* 31:509–20
85. Michel, F., Dujon, B. 1983. *EMBO J.* 2:33–38
86. Gubler, U., Wyler, T., Braun, R. 1979. *FEBS Lett.* 100:347–50
87. Campbell, G. R., Littau, V. C., Melera, P. W., Allfrey, V. G., Johnson, E. M. 1979. *Nucleic Acids Res.* 6:1433–48
88. Nomiyama, H., Sakaki, Y., Takagi, Y. 1981. *Proc. Natl. Acad. Sci. USA* 78:1376–80
89. Nomiyama, H., Kuhara, S., Kukita, T., Otsuka, T., Sakaki, Y. 1981. *Nucleic Acids Res.* 9:5507–20
90. Bonnard, G., Michel, F., Weil, J. H., Steinmetz, A. 1984. *Mol. Gen. Genet.* 194:330–36
91. Chu, F. K., Maley, G. F., Maley, F., Belfort, M. 1984. *Proc. Natl. Acad. Sci. USA* 81:3049–53
92. Belfort, M., Pedersen-Lane, J., West, D., Ehrenman, K., Maley, G., et al. 1985. *Cell* 41:375–82
93. Chu, F. K., Maley, G. F., Belfort, M., Maley, F. 1985. *J. Biol. Chem.* 260: 10680–88
94. Holl, J., Rödel, G., Schweyen, R. J. 1985. *EMBO J.* 4:2081–85
95. Bonitz, S. G., Coruzzi, G., Thalenfeld,

B. E., Tzagoloff, A. 1980. *J. Biol. Chem.* 255:11927–41
96. Netter, P., Jacq, C., Carignani, G., Slonimski, P. P. 1982. *Cell* 28:733–38
97. De La Salle, H., Jacq, C., Slonimski, P. P. 1982. *Cell* 28:721–32
98. Weiss-Brummer, B., Rödel, G., Schweyen, R. J., Kaudewitz, F. 1982. *Cell* 29:527–36
99. Anziano, P. Q., Hanson, D. K., Mahler, H. R., Perlman, P. S. 1982. *Cell* 30:925–32
100. Wild, M. A., Sommer, R. 1980. *Nature* 283:693–94
101. Weiss-Brummer, B., Holl, J., Schweyen, R. J., Rödel, G., Kaudewitz, F. 1983. *Cell* 33:195–202
102. Hertzberg, R. P., Dervan, P. B. 1982. *J. Am. Chem. Soc.* 104:313–15
103. Tanner, N. K., Cech, T. R. 1985. *Nucleic Acids Res.* 13:7759–79
104. Michel, F., Cummings, D. J. 1985. *Curr. Genet.* 10:69–79
105. Wollenzien, P. L., Cantor, C. R., Grant, D. M., Lambowitz, A. M. 1983. *Cell* 32:397–407
106. Weinberg, R., Penman, S. 1968. *J. Mol. Biol.* 38:289–304
107. Reddy, R., Busch, H. 1981. In *The Cell Nucleus,* ed. H. Busch, 8:261–306. New York: Academic
108. Lerner, M. R., Steitz, J. A. 1979. *Proc. Natl. Acad. Sci. USA* 76:5495–99
109. Padgett, R. A., Mount, S. M., Steitz, J. A., Sharp, P. A. 1983. *Cell* 35:101–7
110. Krämer, A., Keller, W., Appel, B., Luhrmann, R. 1984. *Cell* 38:299–307
111. Black, D. L., Chabot, B., Steitz, J. A. 1985. *Cell* 42:737–50
112. Krainer, A. R., Maniatis, T. 1985. *Cell* 42:725–36
113. Mount, S. M., Pettersson, I., Hinterberger, M., Karmas, A., Steitz, J. A. 1983. *Cell* 33:509–18
114. Keller, E. B., Noon, W. A. 1984. *Proc. Natl. Acad. Sci. USA* 81:7417–20
115. Ruskin, B., Green, M. R. 1985. *Cell.* 43:131–42
116. Deleted in proof
117. Gross, H. J., Domdey, H., Lossow, C., Jank, P., Raba, M., et al. 1978. *Nature* 273:203–8
118. Deiner, T. O. 1979. *Viroids and Viroid Diseases.* New York: Wiley
119. Branch, A. D., Robertson, H. D. 1984. *Science* 223:450–55
120. Collmer, C. W., Hadidi, A., Kaper, J. M. 1985. *Proc. Natl. Acad. Sci. USA* 82:3110–14
121. Robertson, H. D., Rosen, D. L., Branch, A. D. 1985. *Virology* 142:441–47
122. Prody, G. A., Bakos, J. T., Buzayan, J.

M., Schneider, I. R., Bruening, G. 1985. *Science* In press
123. Dinter-Gottlieb, G., Cech, T. R. 1984. *Abstr. Meet. on RNA Processing,* p. 37. New York: Cold Spring Harbor Lab.
124. Dinter-Gottlieb, G. 1986. Submitted for publication
125. Watson, N., Gurevitz, M., Ford, J., Apirion, D. 1984. *J. Mol. Biol.* 172:301–23
126. Galli, G., Hofstetter, H., Stunnenberg, H. G., Birnstiel, M. L. 1983. *Cell* 34:823–28
127. Strub, K., Galli, G., Busslinger, M., Birnstiel, M. L. 1984. *EMBO J.* 3:2801–7
128. Moore, C. L., Sharp, P. A. 1984. *Cell* 36:581–91
129. Moore, C. L., Sharp, P. A. 1985. *Cell* 41:845–55
130. Walter, P., Blobel, G. 1982. *Nature* 299:691–98
131. Li, W.-Y., Reddy, R., Henning, D., Epstein, P., Busch, H. 1982. *J. Biol. Chem.* 257:5136–42
132. Noller, H. F., Woese, C. R. 1981. *Science* 212:403–10
133. Noller, H. F. 1984. *Ann. Rev. Biochem.* 53:119–62
134. Burma, D. P., Nag, B., Tewari, D. S. 1983. *Proc. Natl. Acad. Sci. USA* 80:4875–78
135. Burma, D. P., Tewari, D. S., Srivastava, A. K. 1985. *Arch. Biochem. Biophys.* 239:427–35
136. Woese, C. R. 1972. In *Exobiology,* ed. C. Ponnamperuma, pp. 301–41. Amsterdam: North Holland
137. Jencks, W. P. 1975. *Adv. Enzymol.* 43:219–410
138. Fersht, A. R. 1984. *Trends Biochem. Sci.* 9:145–47
139. Page, M. I. 1984. In *The Chemistry of Enzyme Action,* ed. M. I. Page, pp. 1–54. Amsterdam: Elsevier
140. Richards, F. M., Wyckoff, H. W. 1971. In *The Enzymes,* ed. P. D. Boyer, 4:647–806. New York: Academic
141. Osterman, H. L., Walz, F. G. Jr. 1978. *Biochemistry* 17:4124–30
142. Chothia, C. A. 1984. *Ann. Rev. Biochem.* 53:537–72
143. Klapper, M. H. 1971. *Biochim. Biophys. Acta* 229:557–66
144. Sigler, P. B. 1975. *Ann. Rev. Biophys. Bioeng.* 4:447–527
145. Cantor, C. R., Schimmel, P. R. 1980. *Biophysical Chemistry,* Vol. I, p. 190. San Francisco: Freeman
146. Hickey, D. R., Turner, D. H. 1985. *Biochemistry* 24:2086–94

147. Porschke, D. 1985. *J. Mol. Evol.* 21: 192–98
148. Heinemann, U., Saenger, W. 1982. *Nature* 299:27–31
149. Kim, S. H., Quigley, G. J., Suddath, F. L., McPherson, A., Sneden, D., et al. 1973. *Science* 179:285–88
150. Robertus, J. D., Ladner, J. E., Finch, J. T., Rhodes, D., Brown, R. S., et al. 1974. *Nature* 250:546–51
151. Ts'o, P. O. P. 1974. In *Basic Principles in Nucleic Acid Chemistry,* ed. P. O. P. Ts'o, 1:453–584. New York: Academic
152. Rhoads, R. E., Hellmann, G. M., Remy, P., Ebel, J.-P. 1983. *Biochemistry* 22:6084–88
152a. Quigley, G. J., Rich, A. 1976. *Science* 194:796–806
152b. Helene, C., Lancelot, G. 1982. *Prog. Biophys. Mol. Biol.* 39:1–68
153. Garrett, R. A., Wooley, P. 1982. *Trends Biochem. Sci.* 7:385–86
154. Hunkapiller, M. W., Smallcombe, S. J., Whitaker, D. R., Richards, J. H. 1973. *Biochemistry* 12:4732–43
155. Bull, P., Zaldivak, J., Venegas, A., Martial, J., Valenzuela, P. 1975. *Biochem. Biophys. Res. Commun.* 64:1152–59
156. Kao, T. H., Crothers, D. M. 1980. *Proc. Natl. Acad. Sci. USA* 77:3360–64
157. Karplus, M., McCammon, J. A. 1983. *Ann. Rev. Biochem.* 53:263–300
158. Ricard, J., Noat, G., Nari, J. 1984. *Eur. J. Biochem.* 145:311–18
159. Somogyi, B., Welch, G. R., Damjanovich, S. 1984. *Biochim. Biophys. Acta* 768:81–112
160. Kell, D. B. 1982. *Trends Biochem. Sci.* 7:351
161. McCammon, J. A., Wolynes, P. G., Karplus, M. 1979. *Biochemistry* 18:927–42
162. Rigler, R., Wintermeyer, W. 1983. *Ann. Rev. Biophys. Bioeng.* 12:475–505
163. Harvey, S. C., Prabhakaran, M., Mao, B., McCammon, J. A. 1984. *Science* 223:1189–91
164. Crick, F. H. C. 1968. *J. Mol. Biol.* 38:367–79
165. Orgel, L. E. 1968. *J. Mol. Biol.* 38:381–93

Ann. Rev. Biochem. 1986. 55:631–61
Copyright © 1986 by Annual Reviews Inc. All rights reserved

NONVIRAL RETROPOSONS: GENES, PSEUDOGENES, AND TRANSPOSABLE ELEMENTS GENERATED BY THE REVERSE FLOW OF GENETIC INFORMATION

Alan M. Weiner

Department of Molecular Biophysics and Biochemistry, Yale University School of Medicine, 333 Cedar Street, New Haven, Connecticut 06510

Prescott L. Deininger

Department of Biochemistry, Louisiana State University Medical Center, New Orleans, Louisiana 70112

Argiris Efstratiadis

Department of Human Genetics and Development, Columbia University, 701 West 168th Street, New York, NY 10032

CONTENTS

0066-4154/86/0701-0631$02.00

PERSPECTIVES AND SUMMARY

Movement of genetic information from one locus to another is known as transposition, and in principle, the information could be carried from the parental locus to the target locus as well by RNA as by DNA. Nonetheless, although DNA-mediated transposition has been documented in both pro-karyotes and eukaryotes, RNA-mediated transposition appears to be restricted to eukaryotes. In order to distinguish RNA-mediated transpositions from DNA-mediated events, the "reverse" flow of genetic information from RNA back into DNA has been termed "retroposition," and the transposed information is therefore known as a "retroposon" (1, 2). The known retroposons can be divided into viral and nonviral superfamilies based on common structural features (Table 1). The viral superfamily, for which the avian and rodent retroviruses serve as a prototype, has been extensively reviewed elsewhere (3–7).

Over the past few years, retroposition of nonviral cellular RNA species has emerged as a major evolutionary force contributing to the continuous sequence duplication, dispersion, and rearrangement that maintain the remarkable fluid-ity of eukaryotic genomes. Because nonviral retroposition has given rise to many large families of pseudogenes and transposable elements that confer no obvious advantage on the organism, Orgel & Crick (8) and Doolittle & Sapien-za (9) made the provocative proposal that nonviral retroposons could be thought of as molecular parasites that infest the genome but rarely confer a selective advantage (for a recent review, see Ref. 9a). This point of view is surely extreme, given the opportunistic ability of natural selection to find good use for even the most peculiar mutations. In fact, mobile elements in prokaryotes (bacteriophage lambda, mu, P1, and P2, as well as the insertion sequences IS10 and IS50) can confer a selective advantage on their hosts (reviewed in Ref. 10). Moreover, although the genome can be thought of at any instant as a mosaic of sequences that contribute to the organismal phenotype (genic sequences) and those that do not (nongenic sequences), the very fluidity of the genome guarantees a constant flow of sequences back and forth between these two abstract genetic compartments. Thus retroposition, which creates novel se-

Table 1 Retroposons[a]

Viral superfamily	Nonviral superfamily	
Retroviruses (all vertebrates examined)	RNA polymerase II transcripts	
Endogenous or exogenous	Functional semiprocessed retrogene[g]	
Nondefective or defective	Processed retropseudogenes (Table 2)	
Retrotransposons[b]	snRNA retropseudogenes (human and rodent)[h]	
Ty (yeast)	F family (*Drosophila*)[i]	⎤
Copia family (*Drosophila*)[c]	LINE1 family (human, monkey, mouse, rat)[i,j]	⎬ LINES
DIRS-1 (*Dictyostelium*)	RNA polymerase III transcripts	⎦
Bs1 (maize)[d]	7SL RNA retropseudogenes	⎤
IAP (rodents)	Unprocessed (RNA sequence intact or truncated)[k]	
THE1 repeats (human)[e]	"Processed" (internally deleted)[l]	
VL30 (rat and mouse)	B1 superfamily (rodents)[m]	
	Homologous composite of processed monomers	
	Dimeric Type I Alu family (primates)	
	Tetrameric Type I Alu family (human)	
	7SK RNA retropseudogenes	
	Unprocessed (RNA sequence intact or truncated)[k]	
	tRNA retropseudogenes	
	Unprocessed (3' truncated)[n]	
	"Processed" (highly mutated and evolved)[o]	⎬ SINES
	Monomer family (galago)	
	B2 superfamily (rodents)[p]	
	ID repeats (rat and mouse[q])	
	C repeats (rabbit)	
	C repeats (artiodactyls[r])	
	Heterologous composites[s]	
	tRNA with 7SL retropseudogenes	
	Type II Alu family (galago)[t]	
	Type I Alu with LINE1 (human)	
	7SK retropseudogene with Type I Alu family (human)	
	C with BCS (artiodactyls)	
	B2 with OBY (mouse)	⎦
	Polymerase unknown	
	BCS or A repeats (artiodactyls)	
	Type I and II rDNA insertions (*Drosophila* and *Bombyx*)[u]	
	RIME repeats (Trypanozoons)[i,v]	

Table 1 *(continued)*

Viral superfamily	Nonviral superfamily
	Distinguishing Hallmarks
Dispersed in genome	Dispersed in genome
Bounded by long terminal repeats (LTRs)	No terminal direct or inverted repeats
Transposition intermediate is RNA polymerase II transcript	Both RNA polymerase II and III transcripts serve as transposition intermediates
Active transposition (element presumed to encode reverse transcriptase and/or integrase)[f]	Passive transposition
Generate 4–6-bp target site duplications characteristic of the retroposon	Generate 7–21-bp target site duplications (occasionally shorter or longer)
No 3' terminal poly(A) tract	Often have 3' terminal poly(A) tract
May contain introns that are removed to generate subgenomic mRNAs	Intronless even when parental sequence contains introns (one known exception)

[a] References are given only for retroposons not discussed further in the text.

[b] Term proposed for virallike transposons with no obligatory extracellular phase (5).

[c] Includes copia, 297, 17.6, B104, gypsy, and HMS Beagle; copia itself appears to encode an *env* protein, and may be nondefective retrovirus (4).

[d] Tentatively classified as retrotransposon based on structural similarity to Ty and copia families (19).

[e] No homology detected with retroviral *pol* genes (20); could be progenitor or derivative of retrovirus.

[f] These activites could also be supplied in *trans* by another member of the viral superfamily.

[g] Rat and mouse preproinsulin I gene.

[h] Human U1 (21, 22), U2 (23, 24), U3 (25), U4 (26, 26a), U6 (27); mouse U6 (28); rat U3 (29, 30).

[i] Classified as RNA polymerase II because AATAAA polyadenylation signal precedes 3' poly(A) tract.

[j] Published experiments (31, 32) cannot distinguish RNA polymerase II readthrough transcription from transcripts initiating within the element.

[k] Included as SINEs because 7SL retropseudogenes with 5' truncations have given rise to "secondary" retropseudogenes with 5' substitutions (33), and 7SK retropseudogenes can generate mobile composites with the Type I Alu family (34).

[l] Provisionally classified as "processed," because the internal sequences may have been deleted at the RNA level (see text).

[m] Closely related to rodent 4.5S RNA; for a discussion see Ref. 2.

[n] See Ref. 35.

[o] Could be derived from tRNA gene or retrogene (see text).

[p] Closely related to rodent 4.5S_I RNA and rat R.dre.1 element; for a discussion see Ref. 2.

[q] J. G. Sutcliffe, personal communication.

[r] Goat and cow.

[s] Upstream element listed first.

[t] 5' Monomer family + 3' Alu right monomer.

[u] May not be retroposon; for discussion, see Refs. 22 and 22a.

[v] See Ref. 35a.

quence combinations through the duplicative dispersion of genetic information, can shape and reshape the eukaryotic genome in many different ways.

Nonviral retroposons have also been the subject of many previous reviews (1,2,11–18). We do not intend to belabor this information, but rather to update the earlier reviews. In particular, we discuss new data showing that many if not all SINEs [short mobile elements in the terminology of Singer, 1982 (13)] are retropseudogenes derived from known RNA polymerase III transcripts such as 7SL RNA and tRNA. We also review new data confirming that the major families of LINEs (long mobile elements) found in different mammals are closely related to each other. We then discuss the molecular mechanisms that have been proposed for nonviral retroposition. Finally, we emphasize that the contemporary families of SINEs and LINEs arose very recently in evolution, so that youth itself can explain why the families are relatively homogeneous within a species, but exhibit characteristic differences between species.

THE AMAZING VARIETY OF RETROPOSONS

The Viral Superfamily

Table 1 lists representative retroposons of the viral superfamily and all the known nonviral retroposons. The structure and life cycle of the prototypic vertebrate retroviruses have been reviewed elsewhere (3, 6, 7). While many members of the viral superfamily encode their own reverse transcriptase and integrase, other members may be defective in one or both of these activities, and require that the enzyme(s) be supplied in *trans* by a nondefective member of the viral superfamily residing in the same genome. For example, the human THE1 repeats are 2.3 kb long with 350-bp long terminal repeats (LTRs), are present in about 10,000 copies per genome, generate 5-bp duplications of the target sequence, and encode a 2.0-kb polyadenylated RNA. However, the DNA sequence of the retroposon appears to be unrelated to that of any known retrovirus (20). The intracisternal A-type particle (IAP) appears to be a retrovirus without an obligatory extracellular phase (36), and the viruslike 30S RNA sequence (VL30) may be similar (37).

Although it was immediately obvious that the Ty family of elements in yeast, and the copia family of elements in *Drosophila*, had most if not all of the distinguishing features of the viral superfamily, it was only recently that the extraordinary power of contemporary yeast genetics was harnessed to demonstrate that Ty elements must transpose through an RNA intermediate (5). A Ty element was constructed with a powerful inducible GAL1 promoter replacing part of the upstream LTR, and an intron derived from the rp51 ribosomal protein gene inserted at an innocuous site preceding the downstream LTR. Transposition of this element healed the upstream LTR and precisely eliminated the intron, thus verifying both transit through an RNA intermediate, and reverse transcription to generate a circular retrovirallike transposition intermediate with a solitary LTR that is a composite of the upstream and downstream LTRs (38). Interestingly, induction of the GAL1 promoter on the marked Ty element mobilized other unmarked Ty elements as well, implying that the reverse transcriptase can also work in *trans*. However, since yeast appears to lack retroposons of the nonviral superfamily, it may be that the Ty reverse transcriptase is specific for Ty transcripts, and cannot work efficiently on other cellular RNAs.

The complete DNA sequence of a representative copia element (4) revealed only weak homology with a number of retroviral proteins, including the reverse transcriptase, but surprisingly good homology with the part of the *pol* gene encoding the retroviral integrase. Moreover, copia elements appear more similar to yeast Ty than to several vertebrate retroviruses or even the *Drosophila* copia-like element 17.6. In fact, the 17.6 element appears to encode an *env* product, but copia and Ty do not, suggesting that the retrovirallike particles containing copia RNA and a reverse transcriptase may be viral pseudotypes in which copia RNA is packaged by the *env* product of a nondefective retroposon of the viral superfamily.

New members of the viral superfamily will undoubtedly be found as additional species are scrutinized. For example, the DIRS-1 element of *Dictyostelium discoideum* may belong to this family, although it has inverted nonidentical LTRs (ITRs). One of the open reading frames in DIRS-1 encodes a protein with significant homology to regions of reverse transcriptase (39).

The Nonviral Superfamily

With the curious exception of the four ribosomal RNA species (18S, 28S, and 5.8S rRNAs transcribed by RNA polymerase I, and the 5S rRNA transcribed by RNA polymerase III), all the other major classes of cellular RNA species are known to have given rise to retroposons (Table 1). Despite the extraordinary variety of nonviral retroposons, these elements share certain features. All the nonviral retroposons correspond to a partial or complete DNA copy of a cellular RNA species. Some of the parental RNA species are cytoplasmic (tRNA, 7SL RNA, mRNA), while others are predominantly or exclusively nuclear (snRNAs, Alu, and Alu-equivalent sequences). With only two exceptions (the functional semiprocessed rat preproinsulin I gene and the incompletely processed U2 snRNA pseudogene discussed below), nonviral retroposons are derived from fully processed RNAs. For example, retropseudogenes derived from processed mRNAs always have 3' poly(A) tails and lack introns. The sequence of the mature RNA is often intact (particularly in the case of processed mRNAs) but can be truncated at either the 5' end, the 3' end, or rarely at both ends (33). [The RNA information present in LINE1 specimens can also be scrambled relative to the prototype LINE1 sequence, but it is not clear whether such scrambling takes place during or after insertion into a new chromosomal site (see below).] Insertion of a nonviral retroposon usually generates a target site duplication of 7–21 bp, but shorter and longer direct repeats (40) or none at all (22, 25) have occasionally been observed. Interestingly, the length of the target site duplication is not characteristic of the element. Alu sequences, for example, have been found to make target site duplications varying from 9 to 21 bp (41).

Perhaps most surprising of all, the nonviral retroposons exhibit no consistent structural similarities. Nonviral retroposons vary in length from as little as 33 bp (the U2.1 pseudogene) to over 6 kb (the LINE1 prototype). Although nonviral retroposons usually have a 3' terminal poly(A) tract, most of the snRNA pseudogenes and all of the artiodactyl SINEs represent obvious exceptions. Finally, both RNA polymerase II and III transcripts can give rise to nonviral retroposons. Thus, neither the 5' terminal structure of the RNA (cap or triphosphate) nor the 3' terminal structure [usually poly(A) or oligo(U)] is a prerequisite for retroposition. As discussed in detail below, the amazing variety of nonviral retroposons strongly suggests that these elements transpose passively, i.e. they do not encode or even specify the enzymes responsible for their retroposition. However, the long open reading frame in the LINE1 prototype could be an exception to this generalization.

RETROPOSONS DERIVED FROM PROCESSED MESSAGES

Processed Retropseudogenes

Processed retropseudogenes (previously called "processed genes" in the terminology of Ref. 42) resemble a cDNA copy of a fully processed mRNA species (Tables 1 and 2). These retropseudogenes always include the 3' terminal poly(A) tract of the parental mRNA species, lack any introns present in the parental gene, and often extend to the normal 5' cap site (e.g. Refs. 63, 68, and 69); however, 5' truncations of processed mRNAs are not uncommon. In some cases, 5' truncation is caused by the insertion of another retroposon (53, 60), but in other cases the processed retropseudogene appears to be derived from an aberrant transcript generated by faulty splicing or by initiation downstream from the normal cap site. For example, the retropseudogene derived from the human lambda light chain is missing the V region (42), while the human epsilon heavy chain retropseudogene is missing the VDJ region (79, 80). Perhaps, genes that are subject to strict tissue-specific regulation in the soma usually give rise to retropseudogenes derived from aberrant germline transcripts. Of course we cannot exclude incomplete reverse transcription as the cause of 5' truncation; however, the abundance of full-length retropseudogenes (Table 2) makes incomplete reverse transcription a less likely explanation.

Since the essential promoter elements for RNA polymerase II lie upstream from the transcriptional initiation site, retroposition of a correctly initiated mRNA will almost always generate an inactive retropseudogene. For example, the processed human metallothionein II retropseudogene is inactive, despite the fact that the coding region remains intact (55–56a). Such inactive retropseudogenes degenerate by neutral drift, unless retroposition affected the function of adjacent DNA sequences. In contrast, when mRNA coding regions are duplicated by DNA-mediated events such as tandem duplication or translocation, the flanking regulatory elements (and introns) are preserved; thus the new locus is potentially active, and may be subject to positive or negative selection

Table 2 Retropseudogenes derived from processed messages

Species	Protein	Number of genes	Chromosomal location of gene(s)	Number of retropseudogenes	Chromosomal location of pseudogenes	mRNA sequence in pseudogene intact	5' truncated	Initiates upstream	Refs.
human	triosephosphate isomerase	1	12	5–6 [3]	≠12	+			43
human	argininosuccinate synthetase	1	9	14 [2]	6, 9, X, Y etc	+			44–46
human	phosphoglycerate kinase	1 somatic 1 testis-specific	X ?	1	6	+*			47
human	glyceraldehyde-3-phosphate dehydrogenase	1	12	~25	one on X	+			48–50
mouse	glyceraldehyde-3-phosphate dehydrogenase	1	6	200					48
human	lactate dehydrogenase	?	5	? [1]	≠5 (one on 3)				51
human	dihydrofolate reductase	1		~5 [4]			+ [2]		52, 52a, 53
human	metallothionein[a]	2 MT-I 1 MT-II	16	~11[1 MT-I] [1 MT-II]	1, 4, 18, 20 etc	+			54–58
mouse	cytochrome c	1 somatic 1 testis-specific		20–30 [3]		+			59
rat	cytochrome c	1 somatic		20–30 [7]		+ [3]	+ [1][b]	+ [3]	60, 61
mouse	ribosomal protein L32	1*		16–20 [3][c]		+			62, 63
human	ribosomal protein L32	1*		~20 [1][d]		+			64
mouse	ribosomal protein L30	1*		≥15[4]		+			65
mouse	ribosomal protein L7	1–2*		≥20 [0]					66

Species	Gene	Functional genes	Chrom.	Retropseudogenes [sequenced]	Chrom.				Ref.
mouse	ribosomal protein L18	?		≥8 [0]					67
rat	α-tubulin	2*		10–20 [1]		+			68
human	β-tubulin	2		15–20 [5]		+ [4]	+ [1]		69
human	β-actin	1*		~20 [2]		+	+		70, 71
mouse	cytoplasmic γ-actin	?		? [1]^c			+		72
mouse	myosin light chain	1		1 [1]			+		73
human	nonmuscle tropomyosin	1		≥3 [1]		+*			74
mouse	cytokeratin endo A	1		1 [0]					75
mouse	tumor antigen p53	1	11	1 [1]	14		+		76
mouse	α-globin	2	11	1 [1]	15			+	17
mouse	pro-opiomelanocortin	1	12	1 [1]	19*		+		77, 78
human	Ig ε heavy chain	1 Cε	14	1 [1]^f	9		+		79, 80
human	Ig λ light chain	6 Cλ	22	? [1]^g	≠22		+		42
human	c-Ha-ras	1	11	1 [1]	X	+*	+*		81
human	c-Ki-ras	1	12	1 [1]		+*			82
human	c-raf	1	3	1 [0]	4				83

Numbers in brackets indicate number of retropseudogenes currently sequenced; * denotes uncertainty

a Protein sequencing data suggests the existence of a third MT-I gene.
b Rat LINE1 sequence lies upstream from point of truncation.
c One retropseudogene is in the second intron of the DHFR gene.
d The sequenced retropseudogene is in the first intron of HLA-SBβ gene.
e Inserted into a mouse LINE1 sequence.
f Missing VDJ sequences.
g Missing V sequence.

in addition to neutral drift. Occasionally, however, a processed mRNA may insert fortuitously downstream from a foreign promoter, or may acquire a promoter after retroposition. The intronless chicken calmodulin gene could have arisen in this way, although the other hallmarks of retroposition have yet to be demonstrated (84). Finally, in rare instances, retroposition of an aberrant mRNA initiating upstream from the normal cap site may move the normal promoter along with the mRNA sequence (Table 2), and thus potentially be able to generate a functional processed retrogene. The rat preproinsulin I gene discussed below is an example of such a rare event, and is presumably derived from an aberrant germline transcript of this otherwise tissue-specific mRNA. However, retropseudogenes derived from aberrant upstream transcripts are not necessarily subject to positive selection, as the mouse alpha-globin retropseudogene (reviewed in Ref. 17) and the rat cytochrome c retropseudogenes (60) demonstrate.

Curiously, nonviral retroposition is most commonly found in mammals, even when the same gene family has been examined in a variety of organisms. For example, the *Drosophila* and chicken alpha- and beta-tubulin genes are no less abundant than those of mammals, but mammals have 20–30 tubulin retropseudogenes while *Drosophila* and chicken have none. Similarly, all of actin and cytochrome c loci in the *Drosophila* and chicken genomes are true genes, while most of the human beta-actin and rodent cytochrome c loci are retropseudogenes. Most dramatically, retropseudogenes for the mammalian glyceraldehyde-3-phosphate dehydrogenase are very abundant (about 25 copies in man, rabbit, guinea pig, and hamster, but more than 200 copies in rat and mouse) while the same gene is single copy in the chicken. However, the existence of the F family in *Drosophila* and the putative calmodulin retrogene in the chicken (discussed above) suggest that retroposition can occur in other organisms, although it is much more frequent in mammals.

A Functional Semiprocessed Retrogene

The rat and mouse preproinsulin I gene is for the moment the sole example of a functional retrogene (40). Rats and mice (as well as three species of fish) have two nonallelic insulin genes, designated preproinsulin I and II, which are almost equally expressed. Gene I contains a single small intron in the 5′ noncoding region, whereas gene II contains this same small intron as well as an additional larger intron within the coding region for the C peptide (85). Gene II most closely resembles the ancestral gene, because the unique chicken, dog, guinea pig, and human genes each have two introns. The precise deletion of the small intron from gene I suggested that this locus might be a semiprocessed retrogene derived from the parental gene II, but a deletion in the 5′ flanking region of the rat gene I obscured the upstream direct repeat. Comparison of the mouse gene I with the rat genes I and II was therefore required to establish that both the rat and mouse preproinsulin I genes are functional retrogenes derived from an aberrantly initiated and partially processed gene II transcript. The retroposed sequence is flanked by 41-nt direct repeats, and the polyadenylation signal is followed by an $ACCA_4$ tract in the rat and an $ACCA_8$ tract in the mouse. The aberrant transcript appears to have initiated at least 0.5 kb upstream from the normal cap site, so that the RNA intermediate for retroposition carried most if not all of the preproinsulin II promoter and regulatory sequences.

Although organisms can clearly survive with only one insulin gene, even when the product has diminished biological activity as is the case in guinea pig (86), fixation of the functional semiprocessed preproinsulin I gene appears to reflect positive selection. This conclusion is based on the independent preservation of two functional genes in two different species over sufficient evolutionary time for one of the genes to be inactivated by drift. Thus, replacement substitutions have accumulated in regions encoding the signal and C peptides (which are eliminated by protein processing), but such substitutions are completely absent from the regions encoding the A and B chains of the mature hormone. The existence of such strong negative selection on the A and B chains in gene I suggests that the two-copy state was initially subject to positive selection. This conclusion is strengthened by the observation that the sequences upstream from the promoter and within the single small intron (where rat gene I and II can be compared with mouse gene I) as well as the 3' flanking sequences (where the rat and mouse gene I can be compared) are evolving neutrally.

The murine preproinsulin gene I is unlikely to be the only functional mRNA retroposon. The intronless globin gene of Chironomus (87) might be another example. All but one of the 17–20 intronless actin genes in *Dictyostelium* are functional (88), and these might also be retrogenes; however, the extreme A-richness of the flanking sequences would obscure the hallmarks of retroposition. It is also possible that the overlapping subsets of introns in the actin multigene families of many species could be explained by invoking independent retropositions of different partially spliced transcripts.

LINES (LONG INTERSPERSED REPEATED SEQUENCES)

Mammalian genomes contain 20–50 thousand copies of a long (6–7-kb) interspersed element known as the LINE1 or L1 family (reviewed in Refs. 1, 2, 13–15). Many fragments of this family had been independently named and characterized in several mammalian species before they were recognized in 1983 as parts of a single family of retroposons. Now it is clear that each mammalian genome is inhabited by a related but generally species-specific L1 family. Each of these families can be thought of as an especially abundant and remarkably complex family of processed retropseudogenes, although we should keep in mind that some L1 elements may actually be functional retrogenes. The structural features of typical L1 specimens support this conclusion. Most L1 elements terminate at the 3' end with an A-rich tract, sometimes preceded by a polyadenylation signal. L1 elements clearly contain one or more open reading frames (ORFs; Refs. 15, 89–92). L1 elements are mobile, and usually make target site duplications at the site of insertion (93–98). Moreover, a variety of polymorphisms caused by L1 insertions (or deletions) suggest that at least some contemporary L1 elements are capable of retroposition (99–101; A. V. Furano, personal communication). Finally, the putative retroposition intermediate, a homogeneous 6.5-kb polyadenylated transcript, can be detected in the cytoplasm of relatively undifferentiated human NTera2 teratocarcinoma cells (102).

However, this simple summary is misleadingly neat:

No complete L1 element has yet been sequenced; instead, when the sequenced fragments of the human, monkey, dog, rat, and mouse L1 families are aligned with the rat L1 family, for which over 6.5 kb of composite sequence is available, a "consensus" sequence can be derived for the rat, mouse, and human L1 families (92). An independently derived consensus for the primate (human and monkey) L1 sequence covering approximately 6 kb without interruption has also been compiled (sequence unpublished but available upon request; Ref. 15). Unfortunately, parts of these consensus sequences are represented by only a single specimen, and there is no guarantee that this specimen is free from mutations affecting the provisional ORFs. Thus, the three major ORFs in the primate L1 consensus may be fused or extended upstream by minor changes in the current consensus sequence (15).

Many L1 elements are severely truncated at the 5' end, so that the sequences from the 3' end are present at more than fivefold higher copy number than the complete elements (93–95, 103–106). Such 5' truncation has usually been attributed to incomplete reverse transcription, but other explanations are possible (see below).

Some L1 elements are internally scrambled or deleted, lack a 3' terminal A-rich tract, and have inserted cleanly into the target site without duplication (91). These L1 elements lack the obvious marks of retroposition, and were probably dispersed by a DNA-mediated event such as recombination between extrachromosomal L1-containing circles and the chromosome (107–109a).

Although human NTera2 cells contain a 6.5-kb L1 transcript, L1-related transcripts in more highly differentiated cells are heterogeneous, confined primarily to the nucleus, and may reflect readthrough transcription by RNA polymerase II from promoters outside the element (96, 104, 110–112). In addition, readthrough transcription from promoters for RNA polymerase III lying outside the element may also contribute to the heterogeneity of L1-related transcripts (113).

Finally, although the primate and rodent L1 families exhibit about 60%

homology over a region of about 1,500 bp that includes one of the ORFs, no significant homology can be detected in the first 2 kb from the 5' end of the L1 elements from the two species (T. N. H. Lee, M. F. Singer, T. Fanning, unpublished results quoted in Ref. 15), and the presumed 3' untranslated regions (200 bp in primates and about 700 bp in rats and mice) are completely different. Such species-specific differences at the 5' and 3' ends of mammalian L1 families may reflect the formation of composite elements derived from an ancestral L1 sequence which was present only in low copy number after the mammalian radiation (see below for further discussion of composite elements). Indeed, two human composite L1 elements, each starting with an Alu sequence, have been described (97).

The abundance of the mammalian L1 family raises several obvious questions:

Do the 6–7-kb elements represent the complete L1 sequence, or do these elements arise by retroposition of processed mRNAs derived from a larger functional gene that is present in much lower copy number? Such a hypothetical parental element could have introns, or even belong to the retroviral superfamily; in either case, regulation of the parental element, and of the processed retrogenes derived from it, might be very different.

Can the many truncated and scrambled L1 sequences themselves serve as templates for further retropositions, or are these structurally diverse genomic L1 sequences "frozen" in place like other processed mRNAs? The answer to this question presumably depends on the details of L1 transcription. Do L1 sequences contain internal promoters for RNA polymerase II, or could read-through from random RNA polymerase II promoters upstream [stimulated perhaps by one or more internal enhancers within the L1 sequence (114)] supply transcripts for retroposition? If a 3' terminal poly(A) tract is required for retroposition, can the AAUAAA polyadenylation signal remaining in a processed L1 retrogene function without the recently identified downstream components of the polyadenylation signal (115–117)?

Are L1 sequences abundant because they are abundantly transcribed during the time in development when germline retroposition occurs most efficiently, or because the L1 element itself carries sequences which make it a particularly efficient substrate for retroposition? For example, retroposition could occur in cleavage-stage embryos where the 6.5-kb transcript may be abundant (102), or in oocytes during the prolonged lampbrush stage characteristic of mammalian oogenesis (see below). Alternatively, L1 transcripts might encode a *cis*-acting reverse transcriptase, or a karyophilic protein that can transport the mRNA or cDNA into the nucleus (90). However, the 3' terminal ORF present in most mammalian L1 specimens does not appear to be homologous to the conserved domains of retroviral reverse transcriptases (118).

Why are the structures of genomic L1 sequences so much more complex than those of other abundant retroposons such as the Alu family? The high frequency of 5' truncation might reflect blocks to reverse transcription caused by RNA secondary structure (90, 106, 119), or by RNA branch structures, nucleotide modification, or protein binding. Alternatively, the overrepresentation of sequences derived from the 3' end of L1 elements might reflect preferential nuclease attack at specific sites in genomic L1 sequences, followed by recombination and subsequent retroposition that is dependent on sequences in the 3' end of the element (92). This interpretation is consistent with the observation that some L1 specimens from different species share common sites of 5' (or even 3') truncation, regardless of whether they appear to have arisen by an RNA- or DNA-mediated process (92). Site-specific nuclease attack, followed by recombination, could also generate the discrete sizes of extrachromosomal L1-containing circles observed in monkey cells (107); subsequent integration of these circles followed by additional recombination events could then generate permuted and scrambled L1 sequences (91, 108). Still another possible explanation for scrambling of L1 sequences is the ability of retroviral reverse transcriptases to promote high levels of viral recombination, presumably because strand displacement is a fundamental property of the reverse transcriptase reaction (120); even a small amount of recombination between abundant L1 transcripts would quickly generate a diverse population of L1 sequences.

Although the LINE1 family is clearly most abundant in mammalian genomes, analogous LINEs are probably present in the genomes of lower eukaryotes. For example, the F elements of *Drosophila* are mobile, present in about 50 copies per genome, make 8–13-bp target site duplications upon insertion, and have a 3' poly(A) tract preceded by a AATAAA polyadenylation signal (121). Of five cloned F elements, three appear to be full-length 4.7-kb elements whereas the two others are truncated at the 5' end.

SINES (SHORT INTERSPERSED REPEATED SEQUENCES)

SINEs are short (approximately 70–300 bp) repetitive elements that are often present in over 100,000 copies per genome. Almost all known SINEs appear to be retroposons; only the bovine consensus sequence (BCS family; Refs. 122 and 123) and the mouse OBY family (2) are possible exceptions. Unlike members of LINE1 families, members of each SINE family generally have well-defined 5' and 3' ends, with variation between elements occurring primarily in the characteristic simple sequence [usually oligo(A)] found at the 3' end of the element. SINEs usually make target site duplications of 7–21 bp upon insertion. Most SINE families except for the BCS (123) and OBY families (2) have been shown to carry a functional internal promoter. As anticipated by

Jagadeeswaran et al (124) and first clearly documented by Rogers (2), SINEs with an internal RNA polymerase III promoter can cotranscribe adjacent chromosomal sequences and thereby mobilize them. In particular, the propensity of SINEs to insert into the 3' terminal simple sequence tail of other SINEs can generate mobile composite elements (Table 1; also, see below).

The Human Alu Family and the Rodent B1 Family

The human Alu family is the best studied of all SINEs, and will serve here as a paradigm for this class of retroposons. The 500,000 Alu elements in the human genome constitute a remarkable 5–6% of the genome by mass (125). The Alu, LINE1, and THE1 families, together with poly(CA), account for most of the highly repetitive DNA in the human genome (20, 126). Alu elements, which are approximately 300 bp long, are now known to be dimeric retropseudogenes, derived from 7SL RNA by one or more internal deletions of 7SL sequence followed by a dimerization (see below). The right monomer is 31 nucleotides longer than the left because the left monomer has sustained a more extensive internal deletion of 7SL sequence. Individual Alu elements diverge from the consensus by about 14% as a result of single base changes, insertions, and deletions (reviewed in Ref. 41).

The internal promoter defines the 5' end of the Alu element by directing the initiation of transcription by RNA polymerase III at a fixed distance upstream (131). This promoter exhibits a typical A and B block consensus (127). One or more A blocks lie close to the 5' end of the element (positions 5 and 31), and appear to increase both the strength (128, 129) and accuracy (129, 130) of initiation; however, only the B block between positions 70 and 100 appears to be essential for transcription (128, 129).

Curiously, no monomeric Alu elements have ever been found in the human genome, although the rodent equivalent of the Alu sequence (the B1 super-family; Table 1) is almost exclusively monomeric. The absence of monomeric human Alu elements may not be difficult to explain if only the left Alu monomer has a functional promoter; the right monomer may be inactive (128–130). Inactivity of the right monomer promoter is puzzling, because the right monomer is more homologous to the 7SL sequence than the left (133) and might have been expected to better preserve the 7SL promoter structure; however, even single base changes could in principle affect the promoter activity of either monomer. Although it is tempting to attribute the transcriptional inactivity of the right monomer to the 31 additional nucleotides that lie within the right monomer B block, we would then be at a loss to explain what sequences function as the B block for intact 7SL genes (33).

The ability of a newly retroposed Alu element to serve as a template for further retropositions will depend on whether the Alu promoter can function efficiently in the context of new 5' flanking sequences. However, the Alu promoter is derived from the 7SL promoter, which is known to require compatible upstream sequences for efficient transcription (134). Perhaps the many mutations in the left monomer relative to the right have rendered the left promoter less dependent on upstream sequences and therefore capable of more efficient retroposition.

A simple A-rich sequence defines the 3' end of Alu element and almost all other SINEs, except for artiodactyl SINEs, which lack an A-rich tail and usually have simple repeating sequences [e.g. $(AGC)_n$] instead (122, 123). The length of the A-rich sequence varies from 4 to more than 50 bp between individual Alu elements, and the A-rich sequence, though often a relatively pure homopolymer tract, is sometimes supplemented with or even replaced by simple sequences such as $(NA_x)_y$ (2, 12, 41).

Transcription by RNA polymerase III initiates at the 5' end of the element, transcribes through the entire element including the 3' terminal A-rich tract, and then terminates beyond the downstream direct repeat at one or more random oligo(dT) tracts in the adjacent chromosomal sequences (128–135). Reverse transcription is then primed on the 3' terminal A-rich tract, either by the 3' oligo(U) tract of the Alu transcript itself (124) or by another cellular RNA or DNA (21). In either case, the part of the Alu transcript derived from the 3' flanking chromosomal sequences will be lost upon reverse transcription, thus preserving the "anonymity" of the retroposed information. Although self-priming provides an attractive explanation for the high efficiency of Alu retroposition, one should keep in mind that processed mRNAs cannot possibly self-prime, and that artiodactyl SINEs do not terminate with self-complementary simple sequences (122, 123).

The abundance of Alu and other SINE families is usually attributed to the ability of many (and perhaps all) newly retroposed elements to serve as templates for further retroposition. But if reverse transcription is primed at random sites within the 3' A-rich tract, why doesn't the A-rich tract grow shorter with each successive round of retroposition? One possibility is that the first few bases of cDNA might slip back and prime again, thereby lengthening the tail and perhaps amplifying other simple sequences present in the parental A-rich tract (124). Another possibility is de novo addition of either a homopolymer or simple sequence at the staggered chromosomal break before insertion (2). A third very interesting possibility is that the A-rich region might be expanded at any time after retroposition by one or more of the mechanisms commonly invoked to explain the expansion of simple satellite sequences (1, 136–138). Perhaps the most compelling argument that a mechanism of this type may be at work on the 3' ends of Alu sequences is that similar expansions can also occur at the internal A-rich region following the left monomer (139, 140).

Alu and B1 Sequences are Processed 7SL Retropseudogenes

Human Alu sequences are dimeric, but the homologous rodent sequences (the B1 superfamily) are monomeric. How did these efficient retroposons arise? Today there is no doubt that both Alu and B1 sequences are derived from 7SL RNA, although the details of this process are still a matter of speculation (143).

The 7SL RNA is a component of the signal recognition particle, the cytoplasmic ribonucleoprotein particle required for cotranslational secretion of membrane and secretory proteins into the lumen of the rough endoplasmic reticulum (141). As expected for a component of the translational apparatus, 7SL RNA has been highly conserved throughout evolution. Thus, the finger-prints of chicken, mouse, and human 7SL RNA are similar. The sequences of rat and human 7SL RNA are identical, as are the 3' terminal sequences of dog and human 7SL RNA (literature reviewed in Ref. 142). Finally, the sequences of human and *Xenopus* 7SL RNA are very similar, and can be convincingly aligned with the *Drosophila* 7SL RNA sequence (143).

The sequence of the human Alu right monomer is almost identical to the 7SL RNA sequence, except for the deletion of 155 internal nucleotides from the 7SL sequence. The human Alu left monomer also lacks these 155 nt, as well as an adjacent 31 nt of internal 7SL sequence. The rodent B1 sequence resembles the human Alu right monomer, but the internal deletion of 7SL sequence extends 14 nt further downstream.

The belief that Alu and B1 sequences are derived from the 7SL sequence, rather than vice versa, rests primarily on the remarkable evolutionary conserva-tion of 7SL RNA. In addition, the existence of abundant "unprocessed" 7SL retropseudogenes in the human genome (5' and/or 3' truncated; Ref. 33) demonstrates that mechanisms are available for generating 7SL retropseu-dogenes, some of which could be "processed" (internally deleted). The deriva-tion of Alu and B1 sequences from 7SL is also consistent with the absence of Alu-like sequences (but the presence of 7SL genes) in the *Drosophila* genome (143, 144). Curiously, although both *Xenopus* and sea urchin contain sequ-ences that cross-hybridize with Alu probes, 7SL probes fail to detect a 7SL RNA in the urchin (144).

How, then, did the 7SL sequence give rise to the monomeric B1 and dimeric Alu sequences, and why are the rodent and human repeats structurally differ-ent? For simplicity, we will assume that 7SL sequences originally gave rise to an ancestral retroposon closely resembling the contemporary human Alu right monomer, and that subsequent rearrangements of this ancestral monomeric Alu then generated the dimeric human Alu and the monomeric rodent B1 sequences; however, it is equally possible that these two families arose by independent deletion events directly from 7SL RNA. As we have seen above, both B1 and Alu sequences retropose using an RNA polymerase III transcript as the intermediate. Both these SINEs have an internal promoter for RNA polymerase III (which defines the 5' end of the element) and a 3' terminal A-rich tract that serves as the priming site for reverse transcription (thereby defining the 3' end of the element). Thus, generation of the ancestral Alu right monomer from the 7SL sequence (itself an RNA polymerase III transcription unit) would require at least two changes: acquisition of an A-rich tract immediately following the 3' end of the mature RNA sequence, and deletion of 155 internal nt. We will argue that both these initial events occurred at the RNA level.

Addition of an A-rich tract precisely at the 3' end of the 7SL RNA coding region is very unlikely to have taken place at the DNA level, i.e. by deletion or recombination. In contrast, it is easy to imagine addition of a poly(A) tail to 7SL RNA. Although this kind of polyadenylation might be considered aberrant, the 7SL7 and 7SL23 retropseudogenes provide a powerful precedent (33). In both these pseudogenes, the 3' end of the 7SL RNA sequence almost precisely abuts a poly(A) tail, and flanking direct repeats confirm that the pseudogenes were generated by retroposition. Moreover, there are other examples of retroposons with poly(A) tails that are derived from RNAs that are not normally polyadenylated (e.g. the human U1 snRNA pseudogene U1.101 of Ref. 21 and the rat U3 snRNA pseudogene H3.3 of Ref. 30). In fact, a significant fraction of 7SL RNA is retained on an oligo(dT) cellulose column (E. Ullu, personal communication). Other unusual polyadenylated RNA polymerase III transcripts have also been reported (145).

We speculate that formation of a 7SL retropseudogene with a poly(A) tail created a retroposon that could preserve the 3' end of the 7SL RNA sequence through successive retropositions. Although the subsequent deletion of 155 nt from this retroposon could have occurred at the DNA level, a case can be made that this event also took place at the RNA level (143). First, the major sites of micrococcal nuclease attack on 7SL RNA within the intact signal recognition particle (146) correlate well with the boundaries of the deletion found in the human Alu right monomer (133). Second, in the probable secondary structure of 7SL RNA (142) the resulting 5' and 3' halves of the nascent Alu right monomer would be held together by strong secondary structure. These two observations suggest that the human Alu right monomer might have resulted from nuclease attack around positions 100 and 250 of 7SL RNA in the intact signal recognition particle. Ligation of the 5' and 3' halves of the sequence, followed by retroposition of the internally deleted RNA sequence, would then generate the first human Alu right monomer.

The left human Alu monomer could have arisen independently from a polyadenylated 7SL RNA, following nuclease attack and deletion of 186 (155 + 31) nt. This would be consistent with the presence of a micrococcal nuclease-sensitive site at position 76. Alternatively, the human Alu left monomer could have arisen from the ancestral human Alu right monomer by a deletion at the DNA level between the hyphenated short direct repeats (TGCAgtgAGC and

TGCActccAGC), which are present at the appropriate positions; such deletions frequently remove all of the upstream direct repeat and portions of the downstream direct repeat (145a).

Finally, the human Alu right and left monomers were fused to form the contemporary dimeric element. This step is not difficult to imagine, since retroposons frequently insert into the A-rich tail of other retroposons (2). In fact, two dimeric Alu elements can occasionally fuse to form a tetrameric Alu element; the resulting tetramer is flanked by one set of direct repeats, indicating that it has transposed as a unit (145b, c, d). Similarly, although the Type I Alu sequence of the prosimian galago closely resembles the dimeric Alu of higher primates (147), the galago Type II Alu sequence appears to be an independent composite of an Alu right monomer and a tRNA-derived SINE designated the Monomer family (148, 149; see below).

As mentioned above, the absence of monomeric Alu elements from the human genome can be ascribed in part to the inactivation of the promoter in the right monomer (128–130). Further experiments will be required to determine whether this inactivation is due to single base changes, or to the 155-nucleotide deletion of 7SL sequence; the possibility remains that the deletion of 31 additional nt in the left monomer activates the left promoter. The failure of the left Alu monomer to spawn monomeric elements suggests that the internal oligo(A) tract following this monomer is too short to serve as an efficient template for reverse transcription.

Most Other SINEs are tRNA Retropseudogenes

Just as the Alu family arose from 7SL RNA, so many other families of SINEs are derived from tRNA or tRNA genes (2, 149–151). The homology between the internal RNA polymerase III promoters of SINEs and tRNAs was immediately obvious (123, 148, 152–154), but more extensive homologies were found only recently. Such homology was first noted between the goat C family and a cysteine tRNA (2). Subsequently, the rat ID, mouse B2, rabbit C, and bovine 73-bp repeats (149–151) were also found to display strong homologies to tRNA, although sequence divergence within the repetitive elements has led to some disagreement regarding the exact parental tRNA species. A particularly convincing homology can be found between the galago Monomer family and an initiator methionine tRNA (149). Structures resembling typical tRNA stems and loops can also be drawn for the various SINE sequences, although the base-pairing is much poorer than that in tRNA. However, the consensus sequences of several SINE families are more homologous to the presumed parental tRNA and exhibit much better base pairing. This suggests that individual SINE elements are subject to neutral drift from the parental tRNA sequence without strong selection for tRNA-like structures (149).

How were these SINEs derived from tRNA? In each case, the tRNA homology lies at the 5' end of the SINE sequence as expected for an internal RNA polymerase III promoter; however, we should not discount the possibility that mutations have increased the efficiency of retroposition by rendering the promoter less dependent on compatible 5' flanking sequences (see above, and Ref. 134). Also, as for the ancestral Alu, addition of a 3' terminal oligo(A) or a simple repeating sequence was necessary to serve as an efficient priming site for reverse transcription. Since the 3' terminal poly(A) tract of the galago Monomer (149) and rat ID elements (151) very nearly abuts the tRNA sequence, just as the 3' terminal A-rich tract abuts the 7SL sequence in the Alu right monomer, these elements probably arose by aberrant polyadenylation of the parental tRNA. In contrast, the 3' terminal A-rich tract of the rodent B2 and rabbit C families lie far downstream from the tRNA homology (84 and 251 nt respectively; Ref. 151), suggesting that these SINEs arose by retroposition of a readthrough transcript derived from a tRNA gene, pseudogene, or retropseudogene.

We noted above that the tRNA homology within SINE sequences has diverged significantly from the parental tRNA sequence. Although much of this divergence can be attributed to neutral drift, another interesting interpretation is that the RNA sequence within SINEs must become nonfunctional before the element can evolve into an efficient retroposon. For the Alu family, the deletion of 155 nt from the 7SL sequence may have been sufficient in itself to alter function, for example by abolishing the ability of the RNA to bind the protein components of the signal recognition particle (141). For SINE families derived from tRNA sequences, an accumulation of point mutations may be necessary to abolish tRNA-like folding, RNA processing, or protein binding, before the retroposon can escape a strong negative selection. This may explain why the homologies between SINE families and tRNA seldom exceed 70%, while the homology between Alu and 7SL RNA exceeds 80% (133). Obviously, such mutations must occur prior to multiplication, and should therefore become obvious as the SINE consensus sequences improve; sequence divergence after multiplication to high copy number would be much less likely to affect the consensus sequence.

SINEs abound in all mammalian genomes. In addition to the human and rodent families already discussed, the rabbit C family (154) and the unrelated artiodactyl C family (122, 123, 155) display all the characteristics of short transposable elements. The rabbit C family terminates with the expected 3' terminal oligo(A) tract, but surprisingly, the artiodactyl C family terminates much more often with simple sequence repeats than with oligo(A). How reverse transcription could be primed on such a template is a mystery, although slippage of the initial cDNA and repriming provides an attractive possible mechanism which could also maintain the length of the 3' terminal A-rich tract in SINEs (124; also, see above).

SINEs may not be restricted to mammals. In vitro transcription of total genomic DNA from salmon, newt, and tortoise with a HeLa cell extract yields a small homogeneous RNA polymerase III transcript in each case (151, 156, 157). These transcripts are derived from highly repetitive sequences with tRNA-like structures, and may represent transposable elements (157). In addition, the mouse B1 sequence hybridizes well to DNA from a few species such as maize and chicken, but not to DNA from many other species (158). However, the abundant OAX transcript in *Xenopus* (159) and the CR1 sequence in chicken (160) now appear to lack the hallmarks of retroposition. Moreover, despite extensive studies on the structure and organization of genes encoding mRNA and snRNA in both *Xenopus* and chicken, no retropseudogenes have been definitively identified in either organism.

Can SINEs Serve as Tissue-Specific Markers?

Recently, the provocative proposal has twice been made that the presence of a particular mobile element within an mRNA transcript may serve to identify the tissue from which that mRNA is derived. The rat ID (or "identifier") sequence was proposed as a marker for brain-specific transcripts (153, 161), and the mouse "Set 1" (or B2) sequences were claimed to be a marker for mRNAs specific to both normal embryonic and oncogenically transformed cells (162, 163). How successfully have these claims been substantiated?

The 82-nucleotide rat ID sequence has all the distinguishing marks of a retroposon: flanking direct repeats and a 3' terminal A-rich region (164), an internal RNA polymerase III promoter (161), and high copy number in the genome of rats and mice (165). The element has been found in the 5' flanking region of the *v*-Ha-*ras* gene (166), in introns of mRNA precursors (164), and in the 3' untranslated regions of mature mRNA (153). The ID element is closely related to (and may in fact be a retroposon derived from) two small RNA polymerase III transcripts known as BC1 and BC2, which are found exclusively in neural tissue in the animal (161) but are present in many rat cell lines (J. G. Sutcliffe, personal communication). BC2 is about 160 nucleotides long, and may be a polyadenylated form of the 110-nucleotide BC1 RNA.

The bold hypothesis that ID elements could serve as a marker for neural-specific RNA polymerase II transcription units was based on two separate (and, in retrospect, unrelated) observations. First, the small BC1 and BC2 RNAs transcribed by RNA polymerase III are found in cytoplasmic RNA from brain, but not in RNA from kidney or liver (153). Second, the closely related ID retroposon sequences are found in introns of some brain-specific mRNAs (165). Thus it was conceivable that (*a*) the same factors that activate the absolutely neural-specific transcription of BC1 and BC2 by RNA polymerase III could also activate the transcription of ID sequences within a subset of neural-specific mRNA transcription units, and (*b*) the transcription of ID sequences within neural-specific genes by RNA polymerase III could in turn activate tissue-specific transcription of these mRNAs by RNA polymerase II. The ability of the ID sequence to function as a transcriptional enhancer in certain transient expression assays (167) increased the credibility of this model for transcriptional regulation by ID sequences.

The most serious objection to the ID hypothesis is that the basic outlines of mammalian embryogenesis and development clearly antedate the relatively recent multiplication of species-specific SINEs in mammalian genomes. Thus retroposons such as ID cannot play an obligatory role in mammalian gene regulation. Although this objection rules out the ID hypothesis in its original form, the possibility remained that neural-specific transcription of ID sequences by RNA polymerase III could increase transcription from adjacent RNA polymerase II promoters. This more modest hypothesis predicts that ID sequences should be overrepresented in neural-specific genes: random insertion of the ID retroposon into nonneural genes would be expected to cause inappropriate expression of those genes in neural tissue, and thus be subject to negative selection. In contrast, ID sequences could accumulate in neural-specific genes because these insertions would simply "reinforce" normal tissue-specific regulation, and thus be selectively neutral. Consistent with prediction, Sutcliffe et al (161) found that readthrough transcription of ID sequences by RNA polymerase II in purified nuclei was greatly increased when the nuclei were isolated from brain tissue.

However, subsequent examination of total in vivo nuclear RNA by Owens et al (167a) demonstrated beyond a doubt that the earlier results (161) were misleading, and that ID sequences are present in similar abundance in the nuclear RNA polymerase II transcripts of other organs. Thus ID sequences are unlikely to function as transcriptional regulatory elements. The function of ID sequences, if any, remains to be determined. One interesting possibility is that RNA polymerase III transcripts of ID sequences might hybridize in vivo to RNA polymerase II transcripts containing ID sequences in inverted orientation; such hybrids could influence posttranscriptional regulation of neural-specific transcripts.

Scott et al (169) originally cloned cDNAs derived from mRNA species that are specifically activated by SV40 transformation of BALB/c 3T3 cells. These cDNA clones fell into a small number of cross-hybridizing sets, one of which (Set 1) turned out to share a small dispersed repetitive element. Murphy et al (162) then demonstrated by Northern blotting that polyadenylated mRNAs reacting with the Set 1 probe peak during embryonic organogenesis and decline thereafter; in addition, mRNAs reacting with the Set 1 probe are much more prevalent in undifferentiated EC embryonal carcinoma cells and F9 teratocarcinoma cells than after differentiation. Thus, it was proposed that the Set 1 repetitive element might be a "general onco-fetal marker." Surprisingly, however, the Set 1 element was subsequently reported to be none other than a mouse B2 sequence, a typical SINE (163).

How could a mouse B2 sequence be mistaken for a general onco-fetal marker? Singh et al (170) have recently demonstrated that RNA polymerase III transcription of mouse B2 sequences is increased 5- to 20-fold by SV40 transformation of mouse NIH 3T3 cells; moreover, even in untransformed 3T3 cells, transcription of B2 sequences by RNA polymerase III is sensitive to growth conditions and virtually disappears at high cell densities. The careful observations of Singh et al (170) confirm previous observations that transcription of SINEs and LINEs is often derepressed in relatively undifferentiated tissues (102, 106, 162, 168). Although it is possible that B2 sequences might activate adjacent RNA polymerase II transcription units in undifferentiated cells (a version of the original ID hypothesis), the apparent enrichment for B2 sequences in mRNA from undifferentiated cells could be explained in many other ways. For example, SINEs might preferentially retropose into active chromatin, and the chromatin structure of germline cells might resemble that of undifferentiated somatic cells.

SPECULATIONS ON THE MECHANISM(S) OF RETROPOSITION

Despite considerable speculation, the mechanism of retroposition remains unknown. Here we attempt to clarify the major questions, and to classify the distinguishing features of each proposed mechanism as succinctly as possible. All authors agree that the structural characteristics of nonviral retroposons (Table 1) lead to the inescapable conclusion that RNA information has been inserted into a staggered chromosomal break; however, the agreement ends there. Given the extraordinary variety of RNA species that can serve as substrates for reverse transcription, as well as the variety of retrogene structures derived from them (Table 1), it is appropriate to ask whether one mechanism or many is responsible for retroposition. In the absence of proof to the contrary, we are tempted to believe that the known kinds of nonviral retroposition can be reconciled with a single mechanism, and our discussion reflects that prejudice.

Is the RNA copied into cDNA before insertion (21, 71, 124), or is the RNA inserted into the chromosome and then copied into cDNA in situ (2, 17, 26, 119)? Models of the second type are acceptable for retropseudogenes in which the 3' end of the RNA sequence overlaps the downstream direct repeat, but are inadequate for the many retropseudogenes where overlap is not observed. In addition, since incompletely processed nuclear RNA species might be expected to serve as substrates for reverse transcription in situ, models of the second type also suggest that unprocessed retropseudogenes should be relatively abundant. However, only two partially processed retrogenes are known out of the many documented examples: the functional rat preproinsulin I retrogene which retains a single intron (40), and a human U2 snRNA pseudogene which retains the 10 extra nucleotides at the 3' end characteristic of preU2 snRNA (23, 171). In fact, the absence of introns from processed genes may imply that cytoplasmic but not nuclear RNA usually serves as the template for reverse transcription. To explain how small nuclear RNAs such as U1 and U2 could give rise to large families of processed retropseudogenes, we suggest that the snRNA may be reverse transcribed after export to the cytoplasm, but before assembly into a karyophilic snRNP particle (172). Similarly, transcripts of most SINEs, LINEs, and viral retroposons, although concentrated in the nucleus, may not be entirely excluded from the cytoplasm (4, 5, 96, 104, 172a).

What primes synthesis of the cDNA? If the RNA is copied into cDNA before insertion, reverse transcription might be self-primed for RNAs whose 3' end is complementary to an internal RNA sequence (proposed for SINEs in Ref. 124 and for certain snRNAs in Ref. 25); however, bimolecular priming by a second cellular RNA or by a DNA fragment must be invoked for all other RNA species (173). If the RNA is reverse transcribed in situ, the chromosome itself would presumably serve as the primer (2, 17, 23, 119).

What is the source of the reverse transcriptase activity? This vexing question remains unanswered (2, 17, 21, 23, 119, 124) and further speculation is unlikely to clarify the issue. We have already mentioned that retropseudogenes are relatively rare in *Drosophila* (121) and apparently absent in yeast (Table 1), despite the presence of endogenous viral retroposons that encode reverse transcriptases (4, 5). These reverse transcriptases may be specific for the viral templates. We also mentioned that retrogenes are rare or absent in birds, although endogenous and exogenous retroviruses abound in these species today. Thus the mere presence of reverse transcriptase activity in an organism is not sufficient to generate retrogenes; the right kind of reverse transcriptase must be present in the right part of the right cells at the right time in development.

We are intrigued by the possibility that differences in gametogenesis could account for the abundance of nonviral retrogenes in mammals compared to chickens, amphibia, or *Drosophila*. Spermatogenesis is very similar in all these species, and is therefore unlikely to make a significant contribution to the observed frequency of retroposition. In keeping with this conclusion, the three known mouse cytochrome *c* retrogenes are not derived from the testis-specific isozyme, but rather from the somatic isozyme which is presumably expressed in oocytes (59). In contrast, mammalian oogenesis differs from that in other organisms by prolonging the lampbrush stage (the diplotene of meiotic prophase) from birth to ovulation. This state of relatively suspended animation may last for as long as 40 years in humans, but for only several months in amphibians and for less than three weeks in birds; the meroistic oocytes of *Drosophila* skip the lampbrush stage altogether because nurse cells assume the role of lampbrush chromosomes (174). If retroposition in mammals does occur predominantly (or even exclusively) in the female germline, retrogenes may be underrepresented on the Y chromosome.

Why does the staggered break usually vary in length from 7 to 21 bp, although the direct repeats can be as short as 0 bp (22, 25) and as long as 41 bp (40)? The variable length of the direct repeats confounds any simple argument that a single protein (or protein complex) could be responsible for integration (but see Ref. 24) as is thought to be the case for viral retroposons (7). However, although the direct repeats cannot be reconciled with a single consensus sequence, the strand corresponding to the RNA sequence is often rich in adenine (2, 26a) especially at the 5' end (17, 71, 175). Perhaps random nicking in A+T-rich regions produces staggered breaks with either 5' or 3' extended ends, the former serving as retroposon insertion sites, the latter as substrates for the DNA tailing reactions proposed by Rogers (1, 2) to account for "zero option" insertions. The preponderance of direct repeats with lengths between 7 and 21 bp could be explained if larger breaks were efficiently repaired, while smaller breaks were often fatal. Random nicking in A+T-rich regions would also explain why A-rich chromosomal sites often serve as hotspots for multiple insertions as first noted by Rogers (2). Thus the first retroposition event would not inactivate the target site and might in fact activate it, perhaps by introducing (or at least duplicating) oligo(dA) tracts. This, as Rogers (2) was the first to emphasize, must surely be the origin of the many mobile composite retroposons such as the human Alu, the galago Type II Alu (148), the artiodactyl C-BCS family (123), the mouse B2-OBY family (2), and the human 7SK-Type I Alu family (clone 11 of Ref. 34).

Why does the RNA sequence in many retrogenes exhibit significant overlap with the downstream direct repeat? This has been variously attributed to in situ reverse transcription of the RNA primed by an extended 3' end at the staggered break (2, 17, 23, 119), or to reverse transcription followed by hybridization of the 5' end of the cDNA to an extended 5' end at the staggered break (24, 71). However, the postulated hybridization cannot be obligatory in all cases, since the RNA sequence in many retrogenes fails to overlap the downstream direct repeat. Perhaps the 3' end of a cDNA (corresponding to the 5' end of the RNA sequence) is first attached to an extended 5' end on the upstream side of the staggered break; optional hybridization between the 5' end of the cDNA and the extended 5' end of the downstream side of the chromosomal break would then determine whether, and to what degree, the 3' end of the retrogene would be truncated (24). This would provide a natural explanation for the puzzling observation that retroposons with A-rich 3' tails (Alu elements and processed retropseudogenes derived from polyadenylated mRNAs) are rarely if ever truncated at the 3' end, whereas retropseudogenes derived from tRNA (35), 7SL RNA (33), 7SK RNA (34), and mature snRNAs (U1, U2, U3, U4, and U6) are almost always truncated at the 3' end, except in those rare instances where the pseudogene appears to be derived from an aberrantly polyadenylated molecule of the RNA (21–23, 28, 33). In contrast, occasional examples of 5' truncation (29, 30, 33, 42, 79, 80) could reflect incomplete reverse transcription of a normally initiated RNA, complete reverse transcription of an aberrantly initiated RNA, or aberrant splicing. Given the large number of processed retropseudogenes spanning the complete mRNA sequence, aberrant initiation or splicing appear to be the more likely explanations for the 5' truncation of tissue-specific immunoglobulin mRNAs (42, 79, 80).

A WORD ABOUT THE EVOLUTION OF SINES AND LINES

Why are Alu and L1 sequences relatively homogeneous within each species, but often characteristically different between species? One possibility is that these repetitive sequences are subject to extensive gene conversion, so that each family is constrained to coevolve as a unit (see Ref. 176 for a general discussion of coevolution). However, recent work (138, 177) suggests that none of the Alu sequences in the chimpanzee beta globin cluster has been converted since the separation of the chimpanzee and human lineages. This result agrees well with the general observation that gene conversions are rare in mammals, except at hotspots such as certain immunoglobulin and histocompatibility loci (e.g. see Ref. 178). Another possibility is that Alu and L1 sequences might coevolve by frequent reciprocal recombination; however, we can rule this out because the observed stability of mammalian karyotypes is not compatible with even a low level of chromosomal translocation. A third possibility is that Alu and L1 sequences in the human and mouse genomes have reached a steady state, so that old sequences are continually replaced by new ones. However, a steady state would require a mechanism for specifically removing these sequences from the genome, and there is no real evidence for this. A steady state would also imply that some Alu and L1 sequences should be far older than average, while in fact divergence within the Alu and L1 families is relatively monodisperse. Most Alu elements diverge from the consensus by 8–20%, and only a very few of the more than 50 available sequences diverge by as much as 28% (41; P. L. Deininger, unpublished calculations). Similarly, many mouse L1 sequences differ from each other by less than 5% (90, 179).

If Alu and L1 sequences do not coevolve by gene conversion or recombination, and are not reliably eliminated from the genome before they degenerate, what accounts for the low level of intraspecific divergence and the existence of characteristic interspecific differences? We believe the evidence favors a fourth possibility, namely, that the expansion of these repetitive sequences within the genome began quite recently in evolutionary time, and probably continues today. We therefore suggest that very few copies of the original Alu sequence were present in the ancestral primate some 55 Myr ago. The comparative youth of the family would then explain not only the relative lack of sequence divergence within it, but also the observed species-specific differences between the galago (prosimian) and human Alu families (147). Mutations occurring when there were few Alu sequences in the genome ("founder sequences") would be able to alter the consensus slightly in the two species. However, once the copy number of the Alu family increased, it would be difficult for subsequent mutations to alter the consensus. Thus, the Alu family consensus sequences for new world monkeys and humans do not differ significantly (147), suggesting that the copy number of the Alu family must have been high enough to stabilize the consensus before these species diverged. Similar arguments can be made for the mouse L1 family. Assuming that mouse L1 sequences were present at very low copy number 12 Myr ago, significant species-specific differences between the L1 families of *Mus caroli* and *Mus platythrix* (179) can be explained by expansion of different L1 "founder sequences" within each genome. The more subtle species-specific differences between the L1 families of *Mus domesticus* and *M. caroli* suggest that these closely related species diverged when the copy number of the L1 family was sufficiently high for the consensus to resist change.

Although the Alu and L1 families expanded quite recently in evolutionary time, we do not really know whether these families are continuing to expand today. We also do not know whether the Alu family has increased exponentially (as expected if each new Alu element can retropose as efficiently as its parent) or if only a small number of Alu elements can function as active retroposons (perhaps because 5' flanking regions strongly influence the efficiency of transcription; see Ref. 134). Thus the expansion of the Alu and L1 families may more nearly approximate a linear than an exponential function. Finally, we do not know whether there is a relatively sharp "saturation value," beyond which further expansion of a repetitive DNA sequence family would become subject to a disproportionately strong negative selection.

CONCLUDING REMARKS

Eukaryotic genomes are not as tidy as the genomes of prokaryotes. Introns, huge intergenic regions, satellite sequences, pseudogenes, and many families of transposable elements suggest that excess DNA is often not subject to a strong negative selection. Retroposition helps to maintain the complexity and fluidity of eukaryotic genomes by generating genes, pseudogenes, transposable elements, and novel combinations of DNA sequences. The resulting wealth of genetic variation serves as raw material for positive selection, as well as for negative selection and neutral drift. However, all retroposons are insertional mutagens, and transposable elements in particular can increase rapidly if unchecked by strong negative selection (92). Thus retroposition, like all other forms of genetic variation, is both "good" and "bad" for the organism. Before we can reach a more sophisticated understanding of the role of retroposition in evolution, we will need to know more about the detailed molecular mechanism(s) of retroposition.

Literature Cited

1. Rogers, J. 1983. *Nature* 305:101–2
2. Rogers, J. 1985. *Int. Rev. Cytol.* 93:187–279
3. Weiss, R., Teich, N., Varmus, H., Coffin, J., eds. 1985. *RNA Tumor Viruses.* Cold Spring Harbor, NY: Cold Spring Harbor Lab. 2nd. ed.
4. Mount, S. M., Rubin, G. M. 1985. *Mol. Cell. Biol.* 5:1630–38
5. Boeke, J. D., Garfinkel, D. J., Styles, C. A., Fink, G. R. 1985. *Cell* 40:491–500
6. Baltimore, D. 1985. *Cell* 40:481–82
7. Panganiban, A. T. 1985. *Cell* 42:5–6
8. Orgel, L. E., Crick, F. H. C. 1980. *Nature* 284:604–7
9. Doolittle, W. F., Sapienza, C. 1980. *Nature* 284:601–3
9a. Wichman, H. A., Potter, S. S., Pine, D. S. 1985. *Nature* 317:77–81
10. Hartl, D. L., Dykhuizen, D. E., Miller, R. D., Green, L., de Framond, J. 1983. *Cell* 35:503–10
11. Jelinek, W. R., Schmid, C. W. 1982. *Ann. Rev. Biochem.* 51:813–44
12. Schmid, C. W., Jelinek, W. R. 1982. *Science* 216:1065–70
13. Singer, M. F. 1982. *Cell* 28:433–34
14. Singer, M. F. 1982. *Int. Rev. Cytol.* 76:67–112
15. Singer, M. F., Skowronski, J. 1985. *Trends Biochem. Sci.* 10:119–22
16. Sharp, P. A. 1983. *Nature* 301:471–72
17. Vanin, E. F. 1984. *Biochem. Biophys. Acta* 782:231–41
18. Jeffreys, A. J., Harris, S. 1984. *Bioessays* 1:253–58
19. Johns, M. A., Mottinger, J., Freeling, M. 1985. *EMBO J.* 4:1093–102
20. Paulson, K. E., Deka, N., Schmid, C. W., Misra, R., Schindler, C. W., et al. 1985. *Nature* 316:359–61
21. Van Arsdell, S. W., Denison, R. A., Bernstein, L. B., Weiner, A. M., Manser, T., Gesteland, R. F. 1981. *Cell* 26:11–17
22. Denison, R. A., Weiner, A. M. 1982. *Mol. Cell. Biol.* 2:815–28
22a. Eickbush, T. H., Robins, B. 1985. *EMBO J.* 4:2281–85
23. Hammarstrom, K., Westin, G., Bark,

C., Zabielski, J., Pettersson, U. 1984. *J. Mol. Biol.* 179:157–69
24. Van Arsdell, S. W., Weiner, A. M. 1984. *Nucleic Acids Res.* 12:1463–71
25. Bernstein, L. B., Mount, S. M., Weiner, A. M. 1983. *Cell* 32:461–72
26. Hammarstrom, K., Westin, G., Pettersson, U. 1982. *EMBO J.* 1:737–39
26a. Bark, C., Hammarstrom, K., Westin, G., Pettersson, U. 1985. *Mol. Cell. Biol.* 5:943–48
27. Theissen, H., Rinke, J., Traver, C. N., Luhrmann, R., Appel, B. 1985. *Gene.* In press
28. Ohshima, Y., Okada, N., Tani, T., Itoh, Y., Itoh, M. 1981. *Nucleic Acids Res.* 9:5145–58
29. Reddy, R., Henning, D., Chirala, S., Rothblum, L., Wright, D., Busch, H. 1985. *J. Biol. Chem.* 260:5715–19
30. Stroke, I. L., Weiner, A. M. 1985. *J. Mol. Biol.* 184:183–93
31. Shafit-Zagardo, B., Brown, F. L., Zavodny, P. J., Maio, J. J. 1983. *Nature* 304:277–80
32. Heller, D., Jackson, M., Leinwand, L. 1984. *J. Mol. Biol.* 173:419–36
33. Ullu, E., Weiner, A. M. 1984. *EMBO J.* 3:3303–10
34. Murphy, S., Altruda, F., Ullu, E., Tripodi, M., Silengo, L., Melli, M. 1984. *J. Mol. Biol.* 177:575–90
35. Pratt, K., Eden, F. C., You, K. H., O'Neill, V. A., Hatfield, D. 1985. *Nucleic Acids Res.* 13:4765–75
35a. Hasan, G., Turner, M. J., Cordingley, J. S. 1984. *Cell* 37:333–41
36. Ono, M., Toh, H., Miyata, T., Awaya, T. 1985. *J. Virol.* 55:387–94
37. Stoye, J., Coffin, J. 1985. See Ref. 3, pp. 357–404
38. Varmus, H. E., Swanstrom, R. 1985. See Ref. 3, pp. 75–134
39. Cappello, J., Handelsman, K., Lodish, H. F. 1985. *Cell* 43:105–15
40. Soares, M. B., Schon, E., Henderson, A., Karathanasis, S. K., Cate, R., et al. 1985. *Mol. Cell. Biol.* 5:2090–103
41. Schmid, C. W., Shen, C.-K. J. 1986. *Molecular Evolutionary Genetics,* ed. R.

J. MacIntyre, pp. 323–58. New York: Plenum
42. Hollis, G. F., Hieter, P. A., McBride, O. W., Swan, D., Leder, P. 1982. *Nature* 296:321–25
43. Brown, J. R., Daar, I. O., Krug, J. R., Maquat, L. E. 1985. *Mol. Cell. Biol.* 5:1694–706
44. Freytag, S. O., Beaudet, A. L., Bock, H. G., O'Brien, W. E. 1984. *Mol. Cell. Biol.* 4:1978–84
45. Freytag, S. O., Bock, H. G., Beaudet, A. L., O'Brien, W. E. 1984. *J. Biol. Chem.* 259:3160–66
46. Su, T. S., Nussbaum, R. L., Airhart, S., Ledbetter, D. H., Mohandas, T., et al. 1984. *Am. J. Hum. Genet.* 36:954–64
47. Pani, K., Singer-Sam, J., Munns, M., Yoshida, A. 1985. *Gene* 35:11–18
48. Piechaczyk, M., Blanchard, J. M., Riaad-El-Sabouty, S., Dani, C., Marty, L., Jeanteur, P. 1984. *Nature* 312:469–71
49. Hanauer, A., Mandel, J. L. 1984. *EMBO J.* 3:2627–33
50. Benham, F. J., Hodgkinson, S., Davis, K. E. 1984. *EMBO J.* 3:2635–40
51. Tsujibo, H., Tiano, H. F., Li, S. S. 1985. *Eur. J. Biochem.* 147:9–15
52. Anagnou, N. P., O'Brien, S. J., Shimada, T., Nash, W. G., Chan, M. J., Nienhuis, A. W. 1984. *Proc. Natl. Acad. Sci. USA* 81:5170–74
52a. Maurer, B. J., Carlock, L., Wasmuth, J., Attardi, G. 1985. *Somatic Cell. Mol. Genet.* 11:79–85
53. Shimada, T., Chen, M. J., Nienhuis, A. W. 1984. *Gene* 31:1–8
54. LeBeau, M. M., Diaz, M. O., Karin, M., Rowley, J. D. 1985. *Nature* 313:709–11
55. Karin, M., Eddy, R. L., Henry, W. M., Haley, L. L., Byers, M. G., Shows, T. B. 1984. *Proc. Natl. Acad. Sci. USA* 81:5494–98
56. Karin, M., Richards, R. I. 1982. *Nature* 299:797–802
56a. Karin, M., Richards, R. I. 1984. *Environ. Health Perspect.* 54:111–15
57. Varshney, U., Gedamu, L. 1984. *Gene* 31:135–45
58. Schmidt, C. J., Hamer, D. H., McBride, O. W. 1984. *Science* 224:1104–6
59. Linbach, K. J., Wu, R. 1985. *Nucleic Acids Res.* 13:617–30
60. Scarpulla, R. C. 1984. *Mol. Cell. Biol.* 4:2279–88
61. Scarpulla, R. C. 1985. *Nucleic Acids Res.* 13:763–75
62. Feagin, J. E., Setzer, D. R., Schimke, R. P. 1983. *J. Biol. Chem.* 258:2480–87
63. Dudov, K. P., Perry, R. P. 1984. *Cell* 37:457–68

64. Trowsdale, J., Kelly, A., Lee, J., Carson, S., Austin, P., Travers, P. 1984. *Cell* 38:241–49
65. Wiedemann, L. M., Perry, R. P. 1984. *Mol. Cell. Biol.* 4:2518–28
66. Klein, A., Meyuhas, O. 1984. *Nucleic Acids Res.* 12:3763–76
67. Peled-Yalif, E., Cohen-Binder, I., Meyuhas, O. 1984. *Gene* 29:157–66
68. Lemischka, I., Sharp, P. A. 1982. *Nature* 300:330–35
69. Lee, M. G., Lewis, S. A., Wilde, C. D., Cowan, N. J. 1983. *Cell* 33:477–87
70. Leavitt, J., Gunning, P., Porreca, P., Ng, S.-Y., Lin, C.-S., Kedes, L. 1984. *Mol. Cell. Biol.* 4:1961–69
71. Moos, M., Gallwitz, D. 1983. *EMBO J.* 2:757–61
72. Tokunaga, K., Yoda, K., Sakiyama, S. 1985. *Nucleic Acids Res.* 13:3031–42
73. Robert, B., Daubas, P., Akimenko, M.-A., Cohen, A., Garner, I., et al. 1984. *Cell* 39:129–40
74. Maclead, A. R., Talbot, K. 1983. *J. Mol. Biol.* 167:523–37
75. Vasseur, M., Duprey, P., Brulet, P., Jacob, F. 1985. *Proc. Natl. Acad. Sci. USA* 82:1155–59
76. Zakut-Houri, R., Oren, M., Bienz, B., Lavie, V., Hazum, S., Givol, D. 1983. *Nature* 306:594–97
77. Takahashi, H., Mishina, M., Numa, S. 1983. *FEBS Lett.* 156:67–71
78. Uhler, M., Herbert, E., D'Eustachio, P., Ruddle, F. H. 1983. *J. Biol. Chem.* 258:9444–53
79. Battey, J., Max, E. E., McBride, W. O., Swan, D., Leder, P. 1982. *Proc. Natl. Acad. Sci. USA* 79:5956–60
80. Ueda, S., Nakai, S., Nishida, Y., Hisajima, H., Honjo, T. 1982. *EMBO J.* 1:1539–44
81. Miyoshi, J., Kagimoto, M., Soeda, E., Sakaki, Y. 1984. *Nucleic Acids Res.* 12:1821–28
82. McGrath, J. P., Capon, D. J., Smith, D. H., Chen, E. Y., Seeburg, P. H., et al. 1983. *Nature* 304:501–6
83. Bonner, T., O'Brien, S. J., Nash, W. G., Rapp, U. R., Morton, C. C., Leder, P. 1984. *Science* 223:71–74
84. Stein, J. P., Munjaal, R. P., Lagace, L., Lai, E. C., O'Malley, B. W., Means, A. R. 1983. *Proc. Natl. Acad. Sci. USA* 80:6485–89
85. Lomedico, P., Rosenthal, N., Efstratiadis, A., Gilbert, W., Kolodner, R., Tizard, R. 1979. *Cell* 18:545–58
86. Chan, S. J., Episkopou, V., Zeitlin, S., Karathanasis, S., MacKrell, A., et al. 1984. *Proc. Natl. Acad. Sci. USA* 81:5046–50

87. Antoine, M., Niessing, J. 1985. *Nature* 310:795–98
88. Romans, P., Firtel, R. A. 1985. *J. Mol. Biol.* 183:311–26
89. Manuelidis, L. 1982. *Nucleic Acids Res.* 10:3211–19
90. Martin, S. L., Voliva, C. F., Burton, F. H., Edgell, M. H., Hutchison, C. A. III. 1984. *Proc. Natl. Acad. Sci. USA* 81:2308–12
91. Potter, S. S. 1984. *Proc. Natl. Acad. Sci. USA* 81:1012–16
92. Soares, M. B., Schon, E., Efstratiadis, A. 1985. *J. Mol. Evol.* 22:117–33
93. Grimaldi, G., Skowronski, J., Singer, M. F. 1984. *EMBO J.* 3:1753–59
94. Lerman, M. I., Thayer, R. E., Singer, M. F. 1983. *Proc. Natl. Acad. Sci. USA* 80:3966–70
95. Thayer, R. E., Singer, M. F. 1983. *Mol. Cell. Biol.* 3:967–73
96. DiGiovanni, L., Haynes, S. R., Misra, R., Jelinek, W. R. 1983. *Proc. Natl. Acad. Sci. USA* 80:6533–37
97. Miyake, T., Migita, K., Sakaki, Y. 1983. *Nucleic Acids Res.* 11:6837–46
98. Nomiyama, H., Tsuzuki, T., Wasasugi, S., Fukado, M., Shumada, K. 1984. *Nucleic Acids Res.* 12:5225–34
99. Economou-Pachnis, A., Lohse, M. A., Furano, A. V., Tsichlis, P. N. 1985. *Proc. Natl. Acad. Sci. USA* 82:2857–61
100. Burton, F. H., Loeb, D. D., Chao, S. F., Hutchison, C. A. III, Edgell, M. H. 1985. *Nucleic Acids Res.* 13:5071–84
101. Shyman, S., Weaver, S. 1985. *Nucleic Acids Res.* 13:5085–93
102. Skowronski, J., Singer, M. F. 1985. *Proc. Natl. Acad. Sci. USA* 82:6050–54
103. Fanning, T. G. 1983. *Nucleic Acids Res.* 11:5073–91
104. Kole, L. B., Haynes, S. R., Jelinek, W. R. 1983. *J. Mol. Biol.* 165:257–86
105. Voliva, C. F., Jahn, C. L., Comer, M. B., Hutchison, C. A. III, Edgell, M. H. 1983. *Nucleic Acids Res.* 11:8847–59
106. Bennett, K. L., Hill, R. E., Pietras, D. F., Woodworth-Gutai, M., Kane-Haas, C., et al. 1984. *Mol. Cell. Biol.* 4:1561–71
107. Schindler, C. W., Rush, M. G. 1985. *J. Mol. Biol.* 131:161–73
108. Jones, R. S., Potter, S. S. 1985. *Proc. Natl. Acad. Sci. USA* 82:1989–93
109. Fujimoto, S., Tsuda, T., Toda, M., Yamagishi, H. 1985. *Proc. Natl. Acad. Sci. USA* 82:2072–76
109a. Riabowol, K., Shmookler Reis, R. J., Goldstein, S. 1985. *Nucleic Acids Res.* 13:5563–84
110. Whitney, F. R., Furano, A. V. 1984. *J. Biol. Chem.* 259:10:481–92
111. Schmeckpeper, B. J., Scott, A. F.,
Smith, K. D. 1984. *J. Biol. Chem.* 259:1218–25
112. Jackson, M., Heller, D., Leinwand, L. 1985. *Nucleic Acids Res.* 13:3389–403
113. Manley, J. L., Colozzo, M. T. 1982. *Nature* 300:376–79
114. Lueders, K. K., Fewell, J. W., Kuff, E. L., Koch, T. 1984. *Mol. Cell. Biol.* 4:2128–35
115. Gil, A., Proudfoot, N. J. 1984. *Nature* 312:473–74
116. McDevitt, M. A., Imperiale, M. J., Ali, H., Nevins, J. P. 1984. *Cell* 37:993–99
117. Sadofsky, M., Alwine, J. C. 1984. *Mol. Cell. Biol.* 4:1460–68
118. Patarca, R., Haseltine, W. A. 1984. *Nature* 309:728
119. Voliva, C. F., Martin, S. L., Hutchison, C. A., Edgell, M. H. 1984. *J. Mol. Biol.* 178:795–813
120. Junghans, R. P., Boone, L. R., Skalka, A. M. 1982. *Cell* 30:53–62
121. DiNocera, P. P., Digan, M. E., Dawid, I. B. 1983. *J. Mol. Biol.* 168:715–27
122. Watanabe, Y., Tsukada, T., Notake, M., Nakanishi, S., Numa, S. 1982. *Nucleic Acids Res.* 10:1459–92
123. Spence, S. E., Young, R. M., Garner, K. J., Lingrel, J. B. 1985. *Nucleic Acids Res.* 13:2171–86
124. Jagadeeswaran, P., Forget, B. G., Weissman, S. M. 1981. *Cell* 26:141–42
125. Rinehart, F. P., Ritch, T. G., Deininger, P. L., Schmid, C. W. 1981. *Biochemistry* 20:3003–10
126. Sun, L., Paulson, K. E., Schmid, C. W., Kadyk, L., Leinwand, L. 1984. *Nucleic Acids Res.* 12:2669–91
127. Fowlkes, D., Shenk, T. 1980. *Cell* 22:405–13
128. Fuhrman, S., Deininger, P., LaPorte, P., Friedmann, T., Geiduschek, E. P. 1981. *Nucleic Acids Res.* 9:6439–56
129. Perez-Stable, C., Ayres, T., Shen, C.-K. J. 1984. *Proc. Natl. Acad. Sci. USA* 81:5291–95
130. Paolella, G., Lucero, M. A., Murphy, M. H., Baralle, F. E. 1983. *EMBO J.* 2:691–96
131. Duncan, C. H., Jagadeeswaran, P., Wang, R. C., Weissman, S. M. 1981. *Gene* 13:185–96
132. Deleted in proof
133. Ullu, E., Murphy, S., Melli, M. 1982. *Cell* 29:195–202
134. Ullu, E., Weiner, A. M. 1985. *Nature* 318:371–74
135. Haynes, S. R., Jelinek, W. R. 1981. *Proc. Natl. Acad. Sci. USA* 78:6130–34
136. Smith, G. P. 1976. *Science* 191:528–35
137. Hamada, H., Petrino, M. G., Kakunga, T. 1982. *Proc. Natl. Acad. Sci. USA* 79:6465–69

138. Sawada, I., Willard, C., Shen, C.-K. J., Chapman, B., Wilson, A., Schmid, C. W. 1985. *J. Mol. Evol.* 22:316–22
139. Tsukada, T., Watanabe, Y., Nakai, Y., Imura, H., Nakanishi, S., Numa, S. 1982. *Nucleic Acids Res.* 10:1471–76
140. Saffer, J. D., Lerman, M. I. 1983. *Mol. Cell. Biol.* 3:960–64
141. Walter, P., Blobel, G. 1982. *Nature* 299:691–98
142. Gundelfinger, E. D., DiCarlo, M., Zopf, D., Melli, M. 1984. *EMBO J.* 3:2325–32
143. Ullu, E., Tschudi, C. 1984. *Nature* 312:171–72
144. Ullu, E., Esposito, V., Melli, M. 1982. *J. Mol. Biol.* 161:195–201
145. Carlson, D. P., Ross, J. 1983. *Cell* 34:857–64
145a. Efstratiadis, A., Posakony, J. W., Maniatis, T., Lawn, R. M., O'Connell, C., et al. 1980. *Cell* 21:653–68
145b. Degen, S. J. F., MacGillivray, R. T. A., Davie, E. W. 1983. *Biochemistry* 22:2087–97
145c. Lee, M. G.-S., Loomis, C., Cowan, N. 1984. *Nucleic Acids Res.* 12:5823–36
145d. Flemington, E., Traina-Dorge, V., Slagel, V., Bradshaw, H., Deininger, P. L. Submitted for publication
146. Gundelfinger, E. D., Krause, E., Melli, M., Dobberstein, B. 1983. *Nucleic Acids Res.* 11:7364–75
147. Daniels, G. R., Fox, G. M., Loewensteiner, D., Schmid, C. W., Deininger, P. L. 1983. *Nucleic Acids Res.* 11:7579–93
148. Daniels, G. R., Deininger, P. L. 1983. *Nucleic Acids Res.* 11:7595–10
149. Daniels, G. R., Deininger, P. L. 1985. *Nature* 317:819–22
150. Lawrence, C. B., McDonnell, D. P., Ramsey, W. J. 1985. *Nucleic Acids Res.* 13:4239–51
151. Sakamoto, K., Okada, N. 1985. *J. Mol. Evol.* 22:134–40
152. Krayev, A. S., Markusheva, T. V., Kramerov, D. A., Ryskov, A. P., Skryabin, K. G., et al. 1982. *Nucleic Acids Res.* 10:7461–75
153. Sutcliffe, J. G., Milner, R. J., Bloom, F. E. Lerner, R. A. 1982. *Proc. Natl. Acad. Sci. USA* 79:4942–46
154. Cheng, J.-F., Printz, R., Callaghan, T., Shuey, D., Hardison, R. C. 1984. *J. Mol. Biol.* 176:1–20
155. Schimenti, J. C., Duncan, C. H. 1984. *Nucleic Acids Res.* 12:1641–55
156. Matsumoto, K., Murakami, K., Okada, N. 1984. *Biochem. Biophys. Res. Commun.* 124:514–22
157. Endoh, H., Okada, N. 1986. *Proc. Natl. Acad. Sci. USA.* 83:251–55
158. Blin, N., Weber, T., Alonso, A. 1983. *Nucleic Acids Res.* 11:1375–88
159. Ackerman, E. J. 1983. *EMBO J.* 2:1417–22
160. Stumph, W. E., Baez, M., Beattie, W. G., Tsai, M.-J., O'Malley, B. W. 1983. *Biochemistry* 22:306–15
161. Sutcliffe, J. G., Milner, R. J., Gottesfeld, J. M., Lerner, R. A. 1984. *Nature* 308:237–41
162. Murphy, D., Brickell, P. M., Latchman, D. S., Willison, K., Rigby, P. W. J. 1983. *Cell* 35:865–71
163. Brickell, P. M., Latchman, D. S., Murphy, D., Willison, K., Rigby, P. W. J. 1983. *Nature* 306:756–60
164. Barta, A., Richards, R. I., Baxter, J. D., Shine, J. 1981. *Proc. Natl. Acad. Sci. USA* 78:4867–71
165. Milner, R. J., Bloom, F. E., Lai, C., Lerner, R. A., Sutcliffe, J. G. 1984. *Proc. Natl. Acad. Sci. USA* 81:713–17
166. Minarovits, J., Kovacs, Z., Foldes, I. 1984. *FEBS Lett.* 174:208–10
167. McKinnon, R. D., Shinnick, T. M., Sutcliffe, J. G. Submitted for publication
167a. Owens, G. P., Chaudhari, N., Hahn, W. E. 1985. *Science* 229:1263–65
168. Vasseur, M., Condamine, H., Duprey, P. 1985. *EMBO J.* 4:1749–53
169. Scott, M. R. D., Westphal, K.-H., Rigby, P. W. J. 1983. *Cell* 34:557–67
170. Singh, K., Carey, M., Saragosti, S., Botchan, M. 1985. *Nature* 314:553–56
171. Yuo, C.-Y., Ares, M. Jr., Weiner, A. M. 1985. *Cell* 42:193–202
172. Mattaj, I. W., DeRobertis, E. M. 1985. *Cell* 40:111–18
172a. Adeniyi-Jones, S., Zasloff, M. 1985. *Nature* 317:81–84
173. Chen, P.-J., Cywinski, A., Taylor, J. M. 1985. *J. Virol.* 54:278–84
174. Davidson, E. H. 1976. *Gene Activity in Early Development*, pp. 319–49. New York: Academic
175. Daniels, G. R., Deininger, P. L. 1985. *Nucleic Acids Res.* 13:8939–54
176. Dover, G. A., Flavell, R. B. 1984. *Cell* 38:622–23
177. Sawada, I., Beal, M. P., Shen, C.-K. J., Chapman, B., Wilson, A. C., Schmid, C. 1983. *Nucleic Acids Res.* 11:8087–101
178. Jeffreys, A. J., Wilson, V., Thein, S. L. 1985. *Nature* 316:76–79
179. Martin, S. L., Voliva, C. F., Hardies, S. C., Edgell, M. H., Hutchison, C. A. III. 1985. *Mol. Biol. Evol.* 2:127–40

Ann. Rev. Biochem. 1986. 55:663–700

ACIDIFICATION OF THE ENDOCYTIC AND EXOCYTIC PATHWAYS

Ira Mellman, Renate Fuchs, and Ari Helenius

Department of Cell Biology, Yale University School of Medicine, 333 Cedar Street, P.O. Box 3333, New Haven, Connecticut 06510

CONTENTS

0066-4154/86/0701-0663$02.00

SUMMARY AND PERSPECTIVES

Most of the membrane organelles in a typical eukaryotic cell belong to the elements of the exo- and endocytic pathways, referred to collectively as the vacuolar system. They include the endoplasmic reticulum, the Golgi complex, the secretory vacuoles, the endosomes, and the lysosomes and other organelles involved in biosynthesis, processing, transport, storage, release, and degradation of soluble and membrane-bound macromolecules. Both the inward- and the outward-directed pathways consist of a series of organelles through which material passes in an orderly and sequential fashion by means of specific transport vesicles and membrane fission and fusion reactions. Both pathways effect molecular sorting of soluble and membrane-bound components, and both pathways have in some cells the task of storage of molecules for subsequent mobilization. Mechanisms for membrane retrieval between the compartments are in place to ensure the balance in membrane flow. The pathways are, also, functionally interconnected at one or more levels.

An important similarity among most of the organelles of the vacuolar system is the presence of H^+-ATPases responsible for generating an internal acidic environment. The acidity of lysosomes and certain secretory vesicles such as chromaffin granules has long been known, but only recently has it become clear that many other vacuolar organelles such as endosomes, the Golgi complex, and coated vesicles are also acidic. In the endocytic pathway incoming material encounters progressively decreasing pH as it moves through the prelysosomal compartments towards the lysosomal compartment. A similar trend of progressively decreasing pH is seen in the exocytic pathways followed by many secretory products (i.e. from the endoplasmic reticulum to secretory vesicles). The properties of the proton pumps responsible for generating the proton gradients are similar in all these organelles; they are electrogenic, they show clear-cut substrate specificity for ATP, they are inhibited by low concentrations of N-ethylmaleimide, and judging by their insensitivity to vanadate they may not have a phosphorylated intermediate. The composite data indicates that the H^+-ATPases encountered in the vacuolar system are distinct from the mitochondrial F_1F_0-type ATPases as well as the plasma membrane-type H^+-ATPases found in the gastric mucosa, yeast, and other fungi. We call these pumps "vacuolar H^+-ATPases," and describe, in this review, their properties in some detail.

The pH in the various organelles differs, and the functions of acidity are complex and variable. In lysosomes, the low pH provides favorable conditions for enzymatic hydrolyses. In chromaffin granules, the proton gradient is used as an energy source for the coupled transport of biogenic amines. In receptor-mediated endocytosis the difference in pH between the endosome and the extracellular environment is used by the cell to provide asymmetry to the recycling circuit between the two compartments. The pH difference allows, in this case, incoming receptors, ligands, and fluid-phase components to display different properties in intra- and extracellular compartments. Exposure to the endosomal low pH induces conformational changes which in turn can lead to the dissociation of receptor-ligand complexes, changes in ligand solubility, activation of latent activities, etc. In many cases these alterations determine the subsequent sorting and fate of the incoming molecules. The principle of acid activation in the endocytic pathway is exploited by viruses and bacterial toxins whose penetration into the host cell cytoplasm is triggered by acid-induced conformational changes.

The major challenges are now to identify and characterize the vacuolar H^+-ATPases, to determine how they are targeted to the various organelles of residence, and to elucidate how the pH in these organelles is regulated. It will also be important to determine how receptors and other migratory molecules respond to acidity at a molecular level, and to determine what role the decreasing pH and possibly other differences in ionic composition have when considered from the point of view of entire pathways.

THE ORGANELLES OF THE EXO- AND ENDOCYTIC PATHWAYS

The Endocytic Pathway

Endocytosis is the general term used for the internalization of extracellular fluid or particles by invagination and pinching off of the plasma membrane (for reviews see 1–7). Endocytosis serves an important role in uptake of nutrients, scavenging of extracellular material, and in the internalization of receptor-bound ligands such as hormones, growth factors, lipoproteins, and antibodies. A distinction is usually made between internalization of large particles *(phagocytosis)* and small particles, solutes, and fluid *(pinocytosis)*. The molecular processes underlying phagocytosis and pinocytosis are different and can be operationally distinguished. Phagocytosis is particle-activated and cytochalasin-sensitive, and it is usually a property of professional phagocytic cells (amoebae, macrophages, polymorphonuclear leukocytes, etc). Pinocytic uptake, on the other hand, is continuous even in the absence of added ligands. It is cytochalasin-resistant, and it is observed in variable extent in nearly all cell types.

The organelles involved in phagocytosis are the plasma membrane, the prelysosomal phagocytic vacuoles—*the phagosomes*—, and the fusion product between a phagosome and one or more primary or secondary lysosomes—*the phagolysosomes* (Figure 1). Of these the phagolysosomes are definitely acidic as first demonstrated by Metchnikoff in 1893 (8), who fed litmus paper to protozoa and observed it changing color from blue to red. It is likely, however,

EXOCYTOSIS   PINOCYTOSIS   PHAGOCYTOSIS

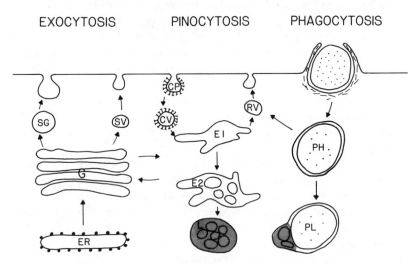

Figure 1   A schematic view of membrane traffic in the vacuolar system. Abbreviations used: ER, endoplasmic reticulum; SG, secretory granule; SV, secretory vesicle; L, lysosome; E1, peripheral endosome; E2, perinuclear endosome; CV, coated vesicle; RV, recycling vesicle; CP, coated pit; PL, phagolysosome; PH, phagosome. The exocytic pathway branches within the Golgi complex into a regulated route (SG) and a constitutive route (SV).

that phagosomes begin to become acidic even before they fuse with a lysosome (9–11).

The general pathways of pinocytosis are also schematically depicted in Figure 1. *Coated pits* and *coated vesicles* constitute major vehicles for pinocytic uptake of both fluid, membrane-bound ligands, and receptors (see 4, 5, 7, 12). Pathways involving uncoated vesicles may also exist (13–18), but information about their relative importance is lacking. Whereas isolated coated vesicle preparations have been found to contain proton ATPase of the vacuolar type (19, 20), it remains unclear whether the ATPase is present specifically in the endocytic coated vesicles (21–23).

The *endosomes* constitute the next station in the pathway (24–34). These relatively recently discovered prelysosomal organelles (26) represent a heterogeneous, complex set of vacuoles located in the peripheral and perinuclear cytoplasm. The endosomes (or subpopulations thereof) have also been called receptosomes (27), CURL (28), endocytic vesicles (30), etc. They consist of complex vacuolar and tubular elements, and many have the appearence of multivesicular bodies (30a). They constitute the main site for sorting of endocytosed material and membrane recycling in the cell. For recent reviews the reader is referred to Ref. 25, 31–34. Endosomes are acidic (35–42), and the acid pH is crucial for many of their functions.

The secondary *lysosomes*, the main intracellular digestive compartment, constitute the terminal compartment of the pathway. With a pH of 4.5–5.0, the lysosomes are the most acidic organelles in animal cell (43–45).

The Exocytic Pathways

The exocytic pathways are responsible for secretion, posttranslational modification, and transport of lipid and protein to the plasma membrane, the lysosomes, and the organelles of the vacuolar system. The main organelles are the *endoplasmic reticulum*, the *Golgi apparatus*, and *secretory granules* (Figure 1). The secretory granules are either part of a constitutive exocytic pathway, or they belong to regulated pathways where exocytosis is triggered by specific stimuli (45a). The pH of the endoplasmic reticulum is not known, but it is usually assumed to be close to neutrality. It is becoming increasingly clear that elements of the Golgi complex contain proton pumps, and are mildly acidic (22, 23, 46). Many exocytic granules of the triggered type contain vacuolar ATPases, and acidity is important for their proper function (47–50). Figure 1 does not depict the various coated and noncoated transport vesicles thought to function as carriers between organelles, some of which may also be acidic (19, 20, 23).

CHARACTERISTICS OF PROTON ATPases

There are three general categories of proton ATPases in eukaryotic cells: ATPases of the mitochondrial-bacterial F_1F_0 class, the plasma membrane proton ATPases, and the vacuolar ATPases. They are easily distinguished on the basis of their sensitivities to a variety of inhibitory agents as well as on the basis of their general functional characteristics. This information is summarized in Table 1.

Proton ATPases of the F_1F_0 class occur in the inner mitochondrial membrane, in chloroplasts, and (in prokaryotes) in the plasma membrane (for review, see Refs. 61, 69). While the precise subunit compositions of F_1F_0-type enzymes can vary, their general structural features are highly conserved. They consist of a water-soluble domain referred to as F_1, and a hydrophobic transmembrane domain, F_0. F_1 is composed of at least five polypeptide chains ranging in molecular mass from <10 kd to 50–60 kd (the total molecular mass is approximately 380 kd) (61, 70). In contrast, the F_0 domain consists of at least three small subunits (each <20 kd) and presumably represents the transmembrane proton channel (70). The F_1 complex is responsible for the ATPase and ATP-synthase activities. The reaction cycle of F_1F_0 ATPases does not appear to involve the formation of a phosphoenzyme intermediate.

Plasma membrane proton ATPases are typified by the proton pumps found in the plasma membranes of *Neurospora* and yeast, where their function is

Table 1   Characteristics of the proton ATPases

| Inhibitor[c] | Mitochondrial F_1F_0 | ATPase CLASS Plasma membrane | | Vacuolar |
		Gastric	Fungal	
Oligomycin	+	−	−	−
Efrapeptin	+	−	−	−
Azide	+	−	−	−
NBD-Cl⁻	+	−	−	+[a]
DCCD	+	−	−	+[b]
Vanadate	−	+	+	−
Ouabain	−	−	−	−
NEM	−	−	−	+
Tributyl tin	+	−	−	+
DIDS	?	−	−	+
Zn^{2+}	?	?	?	+
Nitrate	?	?	?	+

Properties

Molecular mass (kd)	320–390	94–115	100	80–116?
No. of subunits	8–10	1	1	?
Electrogenic	+	−	+	+
Anion stimulation	Cl^-	−	Cl^-	Cl^-
Cation stimulation	−	K^+	−	−
Mg^{2+} dependence	+	+	+	+
Phosphorylated intermediate	−	+	+	?
References	(54, 58, 61, 62, 70)	(54, 64, 66, 71–73)	(51, 52, 63)	(51, 53, 55, 57, 59, 60, 65, 68)

[a] Concentrations required for inhibition are at least 10-fold lower than those required to inhibit ATPases of the F_1F_0 class.

[b] Generally, DCCD concentrations needed for inhibition are at least 10-fold higher than those required to inhibit ATPases of the F_1F_0 class.

[c] Abbreviations used: NBD-Cl⁻, 4-chloro-7-nitrobenzo-2-oxa-1,3-diazole; DCCD, dicyclohexyl-carbodiimide; NEM, N-ethylmaleimide; DIDS, 4,4′-diisothiocyanostilbene-2,2-disulfonic acid.

analogous to that of the Na^+,K^+-ATPase of the mammalian plasma membrane: i.e. generation of a transmembrane electrical potential (52, 57, 63). In mammals, the only known proton ATPase of this type is the K^+,H^+-ATPase of the gastric parietal cell whose major function is the secretion of acid into the stomach (64, 66). Plasma membrane proton pumps form stable phosphorylated intermediates and are sensitive to inhibition by the transition-state analogue sodium orthovanadate (71, 72). Thus, these proton pumps are similar in reaction mechanism and inhibitor sensitivity to other well-studied ion transport ATPases, such as the Na^+,K^+-ATPase and sarcoplasmic reticulum Ca^{2+}-ATPase. While these enzymes can all be considered to belong to the same general class of ATPases, it should be pointed out that the proton ATPases (as well as the Ca^{2+}-ATPase) are insensitive to ouabain, a potent inhibitor of the Na^+,K^+-ATPase.

Both the fungal proton ATPase and the gastric mucosa K^+,H^+-ATPase have been isolated and found to consist of a major 95–115-kd polypeptide chain which contains the catalytic portion of these molecules and is phosphorylated during the reaction cycle (64, 68). A tryptic peptide containing the fluorescein isothiocyanate (FITC)–binding site—presumably the ATP-binding site—of the H^+,K^+-ATPase is closely related to the corresponding regions of the Na^+, K^+-, and Ca^{2+}-ATPases, whose complete amino acid sequences have recently been deduced from cloned cDNAs (73–75).

While the plasma membrane proton ATPases may have similar structures and reaction mechanisms, their ion transport characteristics differ. Proton transport by the H^+,K^+-ATPase across the parietal cell plasma membrane into the stomach lumen is molecularly coupled to the transport of K^+ from the stomach lumen into the parietal cell: i.e. ATP drives an obligatory electroneutral exchange of K^+ for H^+ (64, 66). Na^+ cannot substitute for K^+. In contrast, proton translocation via the fungal plasma membrane ATPase does not require direct coupling to another cation (or anion) and is thus definitely electrogenic (52, 63).

The vacuolar proton ATPases are distinguished from the other two classes by virtue of their inhibitor specificities, lack of coupling to counter-ion transport (Table 1), and intracellular distribution. They are not inhibited by oligomycin, efrapeptin, azide, aurovertin (mitochondrial inhibitors), nor by classical inhibitors of the plasma membrane—type ATPases, such as vanadate and ouabain. The only known "specific" inhibitors are the alkylating reagents such as N-ethylmaleimide (NEM) and 4-chloro-7-nitrobenzo-2-oxa-1,3-diazole (NBD-Cl) (19, 20, 36, 47, 60). While these agents will inhibit ATPases of other classes, they are effective against vacuolar ATPases at relatively low concentrations (e.g. complete inhibition by NEM can be obtained at 10 μM) (50). Conversely, dicyclohexylcarbodiimide (DCCD), which is a potent inhibitor of F_1F_0 ATPases at micromolar concentrations, is only effective against vacuolar ATPases in the millimolar range (19, 36, 76). Dauromycin, an antibiotic with some detergentlike properties, has also been reported to be a selective inhibitor of the vacuolar enzymes (77).

Vacuolar ATPases appear to be electrogenic, proton translocation proceeding without direct molecular coupling to other cations or anions (unlike the K^+,H^+-ATPase of the gastric mucosa, see above). As a result, acidification is accompanied by the generation of an interior positive membrane potential.

Since this potential difference prevents the continued translocation of protons into the vesicle lumen, it must be dissipated by the influx of external anions (negative charges) or the efflux of internal alkali cations (positive charges). A diagram illustrating the ion fluxes that may accompany vacuole acidification is given in Figure 2. A more detailed description of the proton ATPases found in endocytic and secretory organelles will be given below.

IDENTIFICATION OF ACIDIC VACUOLAR ORGANELLES

A variety of biochemical, functional, and morphological approaches have been used to identify acidic compartments and to measure vacuole acidification in intact cells and cell-free systems. Clearly, organelles for which all three types of data exist constitute the most convincing demonstrations. At present, these include lysosomes, endosomes, and certain secretory granules. In this section we summarize the general methods used to study organelle acidification in order to establish the criteria that have been used to identify vesicles with a low pH.

Measurement of Vacuole Acidification

LIPOPHILIC WEAK BASES   The most common and convenient method to study vacuolar acidification relies on the use of lipophilic weak bases. These agents are membrane-permeant when uncharged at neutral pH and relatively membrane-impermeable once protonated. Thus, if allowed to equilibrate with either intact cells or isolated organelles, weak bases will accumulate within membrane vesicles that have acidic internal pH. In quantitative terms, the degree of accumulation will depend on the magnitude of the transmembrane pH gradient and on the total internal volume of the acidic vesicles.

Biochemical determinations of acidification using lipophilic weak bases are

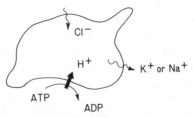

Figure 2   Mechanism of ATP-dependent acidification of endocytic and secretory organelles. A vacuolar ATPase catalyzes the electrogenic translocation of protons into the vacuole creating both a positive internal membrane potential and a low internal pH. The membrane potential is dissipated by passive influx of permeant cytoplasmic anions (Cl^-) or efflux of internal cations (K^+).

generally made in one of two ways. First, optical measurements are possible using dyes such as acridine orange or 9-amino acridine (19, 47, 67, 78). These dyes exhibit characteristic alterations in their absorbance spectra and fluorescence intensity as a function of concentration. Consequently, acidification can be followed by the quenching of acridine fluorescence which occurs as the dye becomes concentrated within acidic vesicles. The fluorescence quenching of these dyes results from the concentration-dependent increase in resonance energy transfer among individual acridine molecules. While neither fluorescence nor absorbance measurements can be converted into actual pH values, radiolabeled weak bases such as ^{14}C-methylamine can be used to obtain quantitative data (20, 76, 79, 80). Following equilibration with the labeled base, the vesicles are collected by centrifugation (or filtration) and the radioactivity determined. Since it is also possible to measure the internal volume of the vesicle population, one can obtain the intravesicular concentration of the pH probe and estimate the internal pH.

While rapid and convenient, these methods are limited by the fact that they are nonselective and thus only as reliable as the homogeneity of the organelle fraction used. Even a slight contamination could yield aberrant results if the contaminating membranes contained a proton pump and accounted for a disproportionately large fraction of the total intravesicular volume.

The ability of lipophilic weak bases to partition into any vesicle with a low internal pH has been exploited to allow the morphological identification of acidic organelles in intact cells. At the level of the fluorescence microscope, acridine orange has long been used for this purpose. This approach has been particularly useful in tissue culture cells where it is relatively specific for lysosomes and phagolysosomes, the most acidic intracellular compartments (81, 82). More recently weak bases such as the probes dinitro-amino-phenyl (DAMP) and primaquine (21–23) have been developed for use in conjunction with electron microscopy. High-affinity antibodies to dinitrophenol or primaquine (respectively) allow the localization of these probes by immunocytochemistry using the immunoperoxidase technique or using protein A–gold conjugates on ultrathin cryosections or Lowacryl sections.

SELECTIVE LABELING WITH ENDOCYTIC TRACERS   Specific methods have been developed to study the acidification of endocytic organelles by selectively introducing pH probes into endosomes or lysosomes by endocytosis. This approach was first introduced by Ohkuma & Poole (44), who found that fluorescein isothiocyanate–labeled dextran (FITC-dextran), a macromolecular marker of fluid-phase pinocytosis, could also be used as an intracellular pH probe. The intensity and excitation spectrum of fluorescein fluorescence are titratable functions of pH. Therefore, standard curves can be constructed relating fluorescence intensity (or excitation maxima) and pH. The relationship between fluorescence and pH is also independent of the concentration or ionic environment of the fluorochrome. By incubating cultured macrophages for several hr at 37°C in medium containing FITC-dextran, the marker was selectively localized in lysosomes. Since the dextran polymer is not a substrate for lysosomal hydrolases, the marker remained intralysosomal almost indefinitely. Extracellular FITC-dextran was removed, the cells transferred to a spectrofluorometer, and the intralysosomal pH estimated to be approximately 4.8 (44). Addition of weak bases (e.g. ammonium chloride, chloroquine) or the carboxylic ionophores nigericin or monensin rapidly (in <30 sec) increased the amount of intracellular fluorescence to a level close to the pH of the external medium, reflecting the dissipation by these agents of the pH gradient across the lysosomal membrane (see below). Removal of the inhibitors from the medium resulted in the restoration of the acidic pH, indicating that the lysosomes were capable of re-acidification, in a fashion that appeared to depend on cytoplasmic ATP concentration (83). Recently, this approach has been used to study acidification of endocytic vesicles in cell-free systems (see below).

FUNCTIONAL MEASUREMENTS OF VESICLE ACIDIFICATION   A third general approach to identify acidic compartments makes use of acid-sensitive properties of incoming ligands or the intrinsic enzyme activites present in the organelles of interest. For example, the known effect of low pH on the fusogenic activity of enveloped animal viruses has been exploited to identify endosomes as the first acidic compartment encountered by newly internalized virus particles (38) (see below). The pH in lysosomes has been assessed using amino acid methyl esters, which are hydrolyzed by the acid esterases and specifically accumulate in the lysosomes (84).

Functional assays for acidification have also been developed for a variety of acidic secretory granules, such as chromaffin granules and platelet dense granules (50, 85, 86). Each of these organelles has a transporter to facilitate the uptake and concentration of biogenic amines in response to a pH gradient (see below). By monitoring the accumulation of labeled serotonin or 5-hydroxytryptamine, one can obtain an indirect but accurate measure of granule acidification. It is important to note that even if the granule fraction is contaminated by other acidic organelles, the coupled assay ensures that only granule acidification will be measured, since only the granule membrane possesses the amine transporter.

ACIDIFICATION OF ENDOCYTIC AND SECRETORY VACUOLES

Lysosomes

A number of laboratories have shown that in intact cells, the intralysosomal environment is maintained at a pH of 4.6–5.0 (9, 35, 43, 44). It is now well-established that the low pH is not caused by a Donnan potential for protons, as originally thought (for detailed review see 45), but is due to the activity of a proton ATPase (56, 60, 76). Among the first and most convincing demonstrations was provided by Ohkuma and coworkers (60). Rats were given parenteral injections of FITC-dextran which accumulated in lysosomes following endocytosis. Enriched lysosome fractions from the liver were prepared and acidification determined using the intralysosomal FITC-dextran as a pH probe. The lysosomes retained an internal pH of 5.0 after isolation and were shown to have a finite but low intrinsic permeability to H^+, K^+, Na^+. Addition of Mg^{2+}-ATP led to a rapid decrease in internal pH which was insensitive to oligomycin, NaN_3, DCCD, and vanadate but could be completely blocked by 1 mM NEM. Substitution of Mg^{2+} by Cu^{2+} or Zn^{2+} also completely prevented acidification. GTP was found to support acidification activity nearly as well as ATP, although the possible role of nucleoside diphosphokinase activity which could interconvert GTP and ATP was not assessed. The external cation composition (Na^+, K^+, or choline) had no significant effect while permeant external anions potentiated ATP-dependent proton translocation. The anion conductance exhibited selectivity for $Br^- \geq Cl^- >$ phosphate $> F^-$, sulfate. Acidification was reduced in the absence of external anions.

While these results were compatible with an electrogenic mechanism for the proton ATPase (Figure 2), they did not exclude the possibility of electroneutral transport of protons and anions. Although initial reports suggested that protons may be cotransported with external phosphate anions (76), recent evidence has demonstrated that lysosomal acidification is electrogenic (56, 59, 87). Using the potential sensitive fluorescent dye Di-S-C_3 (5), Harikumar & Reeves (56) showed, for example, that rat kidney lysosomes developed an interior positive membrane potential after addition of ATP. The presence of a permeant external anion (Cl^-) abolished the effect. By preventing the development of an electrical gradient for protons, external anions facilitate the development of a greater chemical gradient for protons, i.e. a more acidic intralysosomal pH.

The lysosomal proton pump has not yet been isolated. However, Reggio et al (81) identified an antibody to lysosomal membranes that recognizes a 100-kd polypeptide cross-reactive on Western blots with a preparation of the 100-kd H^+/K^+-ATPase isolated from hog gastric mucosa. Whether this antigen is in fact the lysosomal proton ATPase remains to be determined. These results do suggest, however, the interesting possibility that although on mechanistic grounds the H^+/K^+-ATPase and the lysosomal proton ATPase fall into distinct classes of enzyme, at the molecular level they may be related.

Endosomes

The first direct evidence indicating that endosomes are acidic was provided by Tycko & Maxfield (35). FITC was coupled to alpha-2 macroglobulin and incubated with cultured fibroblasts at 37°C to permit its receptor-mediated endocytosis. Incubation times were brief in an attempt to ensure that most of the internalized ligand was selectively localized in endosomes. Individual cells were analyzed by fluorescence microscopy at two excitation wavelengths and the FITC-labeled alpha-2 macroglobulin found in acidic compartments within 15–20 min after internalization. Since little of the ligand was expected to have entered lysosomes at this time point, it appeared likely that acidification occurred in a prelysosomal compartment, i.e. endosomes.

Similar findings have been made using a variety of other FITC-conjugated receptor-bound ligands and markers of fluid phase endocytosis. In addition to single-cell measurements made by fluorescence microscopy (39–41, 88), endosome acidification has also been monitored in cell suspensions by fluorescence spectrophotometry (37, 41a) and flow cytofluorometry (42). Experiments using FITC-labeled transferrin have been particularly conclusive because this ligand enters the endosomal compartment but escapes transport to lysosomes (37) (see below).

While it is difficult to convert optical measurements of FITC fluorescence in intact cells to actual pH values (10), a consensus appears to have been reached that endosomes maintain an internal pH of between 5 and 6, i.e. somewhat less acidic than lysosomes. Part of the difficulty in establishing a more precise figure reflects the likelihood that endosomal pH may vary over a considerable range, even within the same cell. For example, internalized pH-sensitive markers often appear to be transferred from less acidic to progressivley more acidic endosomes as a function of time after endocytosis (10, 41a, 42, 88). The method and cell type used for pH measurements may also influence the results. For example, using spectrofluorometric measurements, FITC-transferrin was found to reach endosomes of pH 5–6 in K562 cells (37). However, in Chinese hamster ovary (CHO) cells, single-cell measurements made using fluorescence microscopy do not detect FITC-transferrin in vesicles of pH less than 6.5 (39).

Like lysosomes, endosomes lower their internal pH by an ATP-dependent mechanism. This was first demonstrated using crude fractions of endosomes which had been selectively labeled with FITC-dextran (36, 41a). Galloway et al (36) established that the endosomal proton pump was distinct from the F_1F_0- and plasma membrane-type proton ATPases, but very similar to the lysosomal ATPase. Its inhibitor sensitivity is typical for vacuolar ATPases (Table 1). Similar results have been obtained using fluorescent pH probes coupled to ligands for receptor-mediated endocytosis, such as alpha-2 macroglobulin (40), asialoorosomucoid, and transferrin, except that the endosomal membrane has a high permeability for protons (R. Fuchs, I. Mellman, in preparation). Thus, endosomes labeled by markers of fluid phase endocytosis and by receptor-mediated endocytosis are functionally equivalent with respect to their capacity for ATP-dependent acidification.

In recent work, we have further characterized the mechanism of endosome acidification using highly purified fractions of rat liver endosomes obtained by free flow electrophoresis and sucrose density gradient centrifugation (R. Fuchs, I. Mellman, in preparation; M. Marsh, S. Schmid, H. Kern, E. Harms, I. Mellman, A. Helenius, in preparation). The bioenergetics of ATP-dependent proton transport in endosomes were found to be very similar to those of lysosomes, except that the endosomal membrane seemed somewhat more permeable to protons and K^+, and less permeable to Cl^-. Unlike lysosomes, however, GTP did not substitute for ATP. Endosome acidification was also shown to be electrogenic.

Analogous results have been obtained for two other classes of nonlysosomal endocytic vesicles. ATP-dependent acidification of FITC-dextran-labeled vesicles isolated from rat kidney proximal tubule also appears to be electrogenic and largely inhibitable by NEM (65). In contrast to endosomes from rat liver, fibroblasts, and macrophages, the proton ATPase of these kidney vesicles may be directly stimulated by Cl^-. Similarly, "multivesicular bodies," endocytic organelles isolated from the livers of estradiol-treated rats, have been shown using nonselective pH probes to possess an electrogenic proton ATPase similar to the proton pumps of endosomes and lysosomes (67).

Taken together, these studies show that the ability of endosomes to establish and maintain a proton gradient of one-to-two orders of magnitude requires not only the activity of an NEM-sensitive, ATP-driven electrogenic proton pump, but also a regulated permeability of the endosomal membrane to certain anions and/or cations.

Coated Vesicles

Considerable interest has been generated by reports that, like endosomes and lysosomes, coated vesicles contain an ATP-driven proton pump. Coated vesicle fractions prepared by standard techniques from bovine brain (19, 20, 90) and from rat liver (89) have been examined. ATP-dependent acidification occurs via a typical vacuolar proton ATPase (Table 1). As described above for endosomes and lysosomes, proton translocation in coated vesicle fractions also appears to be electrogenic, with acidification being favored by the presence of external permeant anions (Cl^-, Br^-, but not gluconate) or by voltage clamp conditions (internal and external K^+ in the presence of valinomycin).

Racker and colleagues have recently been able to solubilize the active ATPase from bovine brain coated vesicle membranes using deoxycholate (91). Functional reconstitution into liposomes was also accomplished from the crude solubilizate; interestingly, the addition of phosphatidyl serine (PS) was found to stabilize the enzyme during solubilization. Partial purification of the deoxycholate-solubilized ATPase has indicated that the ATPase activity copurifies with a polypeptide of 116 kd (91). Unfortunately, it has not yet been possible to reconstitute the purified ATPase into liposomes. Therefore its final identification and the identification of other polypeptides that may be part of the proton pump must await further investigation.

While considerable information has been obtained describing coated vesicle acidification, some basic problems remain. First, the intracellular origin of the vesicles used is unknown. For example, it is not clear what fraction of the vesicles are endocytic or Golgi-derived. Coated vesicle fractions have recently been shown to consist of biochemically and immunologically distinct subpopulations (92, 93). Accordingly, the implication that proton pumps in the plasma membrane are internalized in coated vesicles and subsequently delivered to endosomes, although possible, is without experimental support. Indeed, evidence obtained using the cytochemical probes of intravesicular pH (DAMP, primaquine; see above) suggests that peripheral or endocytic coated vesicles do not have an acidic internal pH (21–23). In contrast, Golgi-region coated vesicles often label with these reagents. A second major problem concerns the purity of coated vesicle preparations. When purified by established techniques, they may be contaminated by proton pump–containing uncoated membranes such as submitochondrial particles (94) or endosomes.

Golgi Complex and Endoplasmic Reticulum

Since certain carboxylic ionophores (e.g. monensin) and in some instances lipophilic weak bases can disrupt normal Golgi function (see below), it has been suspected that Golgi vesicles and cisternae might maintain an acidic internal pH. Direct evidence that Golgi membranes are capable of ATP-dependent acidification first came from in vitro studies of rat liver Golgi fractions (47, 95). Glickman et al (47) showed that the Golgi-associated proton pump was a typical vacuolar ATPase and that Golgi membranes contained a discrete Cl^- conductance. Interestingly, similar data were obtained for fractions of rat liver rough and smooth endoplasmic reticulum (96).

Independent confirmation that elements of the Golgi may contain a proton pump was provided by electron microscopic immunocytochemistry. Using the pH-sensitive cytochemical probes DAMP and primaquine, the acidic internal pH of Golgi-associated vesicles and, occasionally, the *trans*-most Golgi cisterna, has been obtained in a number of cell types (21–23). Localization of these probes within elements of the endoplasmic reticulum has not been observed, suggesting that this organelle does not possess a proton pump or that the endoplasmic reticulum ATPase is used to establish electrical as opposed to pH gradients.

Secretory Granules

Most of the secretory granules studied thus far have been found to have an acidic pH and an ATP-driven proton pump. These include: chromaffin granules from adrenal medulla (97, 98), platelet dense granules (53) neurosecretory granules from the pituitary (78), and cholinergic synaptic vesicles (100, 101).

In contrast, the storage granules of the rat parotid gland are thought not to be acidic or to contain a proton ATPase (102). While the functions of low intragranular pH may vary (see below), it is clear that the proton ATPase of each of these organelles closely reflects the properties of the ATPases already described for lysosomes, endosomes, coated vesicles, and Golgi membranes (Table 1).

Acidification has been most extensively studied in the bovine adrenal chromaffin granule. While the ATPase of these granules was originally thought to be of the F_1F_0 class, the work of Nelson and colleagues (98, 103) has demonstrated that chromaffin granule ATPase is a typical vacuolar proton pump. Chromaffin granule acidification is NEM-sensitive, electrogenic, and attains an internal pH of ~5.7, as measured by ^{14}C-methylamine partitioning and ^{31}P-ATP nuclear magnetic resonance (104).

Chromaffin granules are a source for purification of the ATPase since inhibition of ATP-dependent proton translocation by NEM correlates well with the inhibition of up to 70% of the ATP hydrolytic activity associated with the purified membranes (103; G. Dean, P. Nelson, G. Rudnick, submitted for publication). These results suggest (but do not prove) that the major ATPase activity associated with chromaffin granules is due to the proton pump. The ATPase has been solubilized in cholate and $C_{12}E_9$ and functionally reconstituted into liposomes (103). As is the case with the coated vesicle ATPase, addition of phosphatidyl serine stabilizes the detergent-solubilized enzyme. Partial purification of the granule proton pump has been accomplished, and NEM-sensitive ATPase activity copurifies with four polypeptides ranging in molecular mass from 115 kd to 20 kd (105, 105a; G. Dean, C. Galloway, I. Mellman, G. Rudnick, unpublished results). A photoactivatable ATP analogue, 3'-O-(4-benzoyl)benzoyl-ATP, specifically labels only the 98-kd polypeptide (C. Galloway, G. Dean, G. Rudnick, I. Mellman, unpublished results). In contrast, NEM labels the 115-kd and 57-kd polypeptides. Final identification of the polypeptides that comprise the ATPase must await its complete purification and functional reconstitution.

One other acidic secretory vesicle that deserves note is found in various urinary epithelia, such as the amphibian bladder and the mammalian collecting duct (82, 106). These vesicles contain a vacuolar-type proton ATPase which can be inserted into the apical plasma membrane by exocytic fusion, resulting in urinary acidification. A potent stimulus for the secretory event appears to be increased carbon dioxide concentrations (107). Once inserted into the cell surface, these proton pumps can be retrieved by endocytosis and apparently stored in the intracellular vesicles. This system provides the only example of regulated translocation of proton ATPases from one compartment to another. It also constitutes the only direct evidence, by electrophysiological measurement, that vacuolar ATPases can be present on the plasma membrane.

In a preliminary report, the kidney ATPase has been partially purified from

CHAPS-solubilized membranes (107a). Functional reconstitution into liposomes was reported and proton pumping activity found to copurify with several polypeptides, the largest of which was 80 K.

Yeast and Plant Vacuoles

Many fungal and plant cells have intracellular vacuoles that maintain an acidic internal pH (for review, see Ref. 50). Best studied are the vacuoles of yeast and *Neurospora*, which have been isolated and found to contain an ATP-dependent proton pump (51, 57). It is now clear that the yeast vacuole is analogous to the mammalian cell's lysosome, since it contains an array of hydrolytic enzymes and continuously receives extracellular macromolecules internalized by endocytosis (108). Like mammalian lysosomes, ATP-driven acidification of isolated yeast vacuoles is sensitive to NEM, high concentrations of DCCD, Cu^{2+}, and Zn^{2+}.

Uchida et al (68) recently reported the isolation of the yeast vacuole ATPase. The purified enzyme consisted of three polypeptides of 89 kd, 64 kd, and 19.5 kd. While these polypeptides differ somewhat from those reported for the putative coated vesicle and chromaffin granule ATPases, the possibility of proteolysis by vacuole hydrolases has not yet been eliminated. Moreover, the purified yeast enzyme has not yet been reconstituted, so its identification remains tentative.

Partial purification of the vacuole ATPase from corn coleoptyles has also been accomplished (108a,b). In this case, however, functional proton pumping activity is associated with an 80 K polypeptide.

It is clear that there is no consensus regarding the structure of the proton ATPase from any single source. Thus, determining the structural relationships of proton pumps from different acidic organelles must await considerable additional information.

Regulation of Intravesicular pH

If the same or similar proton pumps are present in all endocytic and exocytic membranes, how is their internal pH regulated? As discussed above, there is ample evidence that different endosomal compartments may have different pH's, and that endosomes and secretory granules have a less acidic pH than lysosomes. Regulation could occur by controlling the number of ATPase molecules present per organelle, or by modulating the activity (e.g. by phosphorylation or some other covalent modification) of a constant number of ATPases. Alternatively, the observations that many organelles have defined permeabilities to anions, protons, and other cations suggests that net proton flux and accumulation can be controlled by regulating the ion permeability characteristics of each membrane. For example, rat liver endosomes appear more permeable to K^+ and less permeable to Cl^- than rat liver lysosomes. Therefore, the endosomes may be less acidic because they may be less able to accumulate protons (limited Cl^- permeability will favor the development of a membrane potential, limiting proton accumulation) and less able to retain protons (cytoplasmic K^+ can more easily exchange with internal H^+ if the conductance to both ions is significant).

INHIBITORS OF VACUOLAR ACIDIFICATION IN INTACT CELLS

In addition to their use as nonselective pH-probes in vitro, lysosomotropic, or more accurately "acidotropic" (109), weak bases are widely used to elevate the pH within acidic vacuoles of living cells. Agents such as ammonium chloride, chloroquine, and methylamine are relatively lipophilic in their unprotonated form and they penetrate the membranes of cells and vacuoles. Upon entering an acidic environment they become protonated and too polar to escape rapidly through the membrane. The equilibrium thus favors accumulation within the organelles (109, 110). The increase in vacuolar pH, which can amount to as much as 1–2 pH units and which occurs within a minute or two after addition of the agents to the medium, is due to the neutralization of protons by the weak base and possibly to the loss of the buffering capacity (45, 110). The effect of the acidotropic weak bases depends on the concentration of the drugs, acidity of the organelles, and on the pH of the medium. When more than one base is present the effects are additive.

The main cellular effects of acidotropic weak bases are related to the increase in vacuolar pH. Lysosomal degradation by acidic hydrolases is inhibited (109–113), molecular sorting and recycling in endosomes is affected (114–116), virus and toxin entry is blocked (117–119), and exocytosis in the regulated pathway is disturbed (49), just to mention a few examples. The acidotropic weak bases are unfortunately not free of undesirable side effects. The most important is the swelling of the acidic vacuoles leading to a dramatic vacuolization of the cell's cytoplasm (120–123). This effect may severely affect the pathways under study. Some of the amines also directly inhibit certain enzymes such as transglutaminase (124) and cathepsin D (121). It has, moreover, been suggested that intracellular fusion events may be inhibited (125, 126). For a recent review on the effects and side effects of these agents the reader is referred to Dean et al (126).

The second group of agents used to elevate vacuolar pH are the carboxylic ionophores such as monensin, nigericin, and X537A. They intercalate into membranes and mediate exchange of monovalent cations through the membrane (127). Given the high concentration of K^+ in the cytoplasm their major effect on the acidic organelles is to exchange protons for potassium ions, thereby effecting a rise in vacuolar pH. The overall effects on vacuolar function are similar to those of lysosomotropic agents (41, 128–133). But some differences are also evident: e.g. they do not affect pH as rapidly and their effects on Golgi function tend to be more marked at low concentrations than those on other acidic organelles (133). The reason for these differences is not known.

ROLE OF ACIDIFICATION IN THE ENDOCYTIC PATHWAYS

Material internalized by the phagocytosis and pinocytosis is usually exposed to decreasing pH very soon after internalization. It only takes five minutes or less before the pH of the compartment in which the material is located drops well below the extracellular level (9–11, 35, 37–42, 134–136).

Phagocytosis

It is likely that phagosomes become acidic prior to the onset of fusion with lysosomes, although acidification may be preceded by a transient alkalinization of the phagosome's interior (9–11). The nature of this acidification reaction has not been studied in detail. If it is caused by a vacuolar ATPase similar to that found in endosomes, which seems likely, the enzyme must either be derived from the plasma membrane as a component of the phagosomal membrane, or be delivered to the newly formed phagosome by fusion of ATPase-carrying vesicles. Although not usually considered intrinsic to the plasma membrane, vacuolar ATPases have been observed in the plasma membrane under some conditions. Gluck et al (137) found that urinary acidification can occur by insertion of a vacuolar type ATPase into the luminal membrane of the epithelial cells (see above); while in *Paramecium,* which depends on phagocytosis for uptake of nutrients, evidence has been obtained for the fusion of newly formed digestive vacuoles with smaller, nonlysosomal vacuoles ("acidosomes") around the time when the pH begins to decrease (138). This suggests that the ATPase might be delivered to the phagosome from an intracellular source. Both of these examples are from highly specialized cell types and may not reflect the situation in most mammalian cells.

It is not clear why phagosomes are acidic. It has been proposed that acid pH might facilitate the fusion of phagosomes with lysosomes (139), although it does not seem to be a prerequisite (82). Intracellular parasites such *Legionella pneumophilia* and *Toxoplasma gondii,* which multiply in phagosomes (140), may be able to inhibit phagosomal acidification and in this way avoid acid-inactivation or phagosome-lysosome fusion followed by digestion in phago-lysosomes (139, 141).

Receptor-Mediated Endocytosis

RECEPTOR RECYCLING   Whereas the same uncertainty applies to the origin of the proton pumps in endosomes, the functions of the acidic pH are much better understood than that of the phagosome. Low pH has emerged as an important factor regulating receptor traffic during receptor-mediated endocytosis and receptor recycling. In a general sense, the acidity of endosomes provides the feature of asymmetry which is required in any cyclic pathway involved in net transport. The receptors, ligands, or receptor-ligand complexes are differentially affected by the pH in the various compartments of the recycling circuit and their properties are changed accordingly. The elegance of using pH in this way is the ease by which receptors in the pathway can be individually modulated. Receptor systems vary greatly in their responses to acid pH. In this section, we focus on some of the general aspects that have emerged from the study of a variety of such systems. For recent reviews on the topic the reader is referred to (3, 17, 25, 31, 34, 142).

Receptors usually recycle between the plasma membrane and the endosome compartment. The intracellular portion of the recycling circuit takes as a rule 15 min or less, and the plasma membrane phase can be as short as a few minutes, depending on the properties of the receptor, and whether ligand is present (143–145). Receptor recycling provides a rapid and economical way to utilize receptors for multiple rounds of uptake. In many cases the receptor has been found to recycle at a finite basal rate even in the absence of added ligand. This has been suggested for at least three receptors, the Fc receptor (146, 153, 154), the LDL receptor (147), and the transferrin receptor (148–150). Indirect evidence for a large number of receptors is provided by the observation that they are depleted from the cell surface in the presence of acidotropic agents and monensin, again irrespective of whether ligands are added or not (114–116, 129, 151, 152). The basal recycling probably reflects the continuous nonspecific pinocytic uptake of membrane and fluid which is generally thought to involve the internalization of 50–200% of the plasma membrane surface area per hour (155, 156). The frequency at which a receptor molecule can be predicted to be internalized is thus once every 0.5–2 hours. The actual rates of basal recycling are, however, variable depending on whether the particular receptors are efficiently integrated into coated pits in their unoccupied state (7, 147, 150, 157–159). LDL receptors in fibroblasts are, for example, localized preferentially in coated pits even in the absence of ligand (7, 147). Many other receptors move rapidly into the coated microdomains of the cell surface only when associated with ligand, the addition of which apparently enhances the uptake rate of the receptor above the basal rate (157, 158).

Upon entering the endosomal compartment, the unoccupied receptors and receptor-ligand complexes encounter a pH of 5.0–6.5, a condition that alters properties of many complexes and profoundly affects their fate. The responses that the incoming receptor-ligand complexes and free receptors display fall roughly into three categories: those where the receptor (or receptor with bound ligand) is reversibly altered by acid pH, those where only the ligand is modified, and those where both ligand and receptor are unaffected by low pH. These receptor types will be discussed below.

RECEPTORS ALTERED BY LOW PH   Receptors are often found to bind their ligands with high or intermediate affinity at neutral pH (the pH of the extracellular medium) and only weakly in mildly acidic pH (the pH of the endosome). The receptor-ligand complexes formed on the plasma membrane therefore dissociate once they reach the endosome. The receptor is free to recycle to the plasma membrane, bind new ligand molecules, and continue through repeated cycles of uptake (for reviews see 25, 31, 33, 109, 159). The dissociated ligands become part of the fluid volume of the endosome, and usually get delivered to lysosomes. Only a fraction (one third or less) of ligand is typically secreted as a consequence of the fusion of recycling vesicles with the plasma membrane (160). The low pH in endosomes thus regulates the sorting of ligand from the receptor.

The molecular basis for the change in binding affinity is not clear. The limited data available suggests that it is primarily the receptor that reacts to low pH and not the ligand. Several of the receptors that have lectin activity (the asialoglycoprotein receptor, the mannose receptor, and the mannose-6-phosphate receptors) mediate uptake of a large variety of glycosylated ligands including "neoglycoproteins" with artificially attached carbohydrate side chains (161, 162). As it seems unlikely that all such ligands behave similarly when exposed to acid pH, the receptor is probably the main target for pH regulation. The recycling of unoccupied receptors is, moreover, often found to be inhibited by agents that elevate endosomal pH (see below). More direct evidence for acid-induced conformational changes in the asialoglycoprotein receptor and the epidermal growth factor (EGF) receptor have been obtained by DiPaola & Maxfield (163), who found that the receptors are differentially cleaved by trypsin, radioiodinated by lactoperoxidase, and labeled by hydrophilic photolabels at acidic and neutral pH. Much additional work is needed to clarify at the molecular level the changes that bring about the altered receptor properties.

When endosomal acidification is inhibited by the addition of acidotropic agents or carboxylic ionophores, intracellular dissociation of the receptor-ligand complex is fully or partially blocked (30, 144, 164). This prevents further binding of ligand to the receptor, and internalization of ligand may come to a halt after one round of uptake (114–116, 129, 151, 152, 165–169a). The actual fate of the nondissociated complexes is variable. Usually, the majority of receptor-ligand complexes accumulate in the highly swollen intracellular vacuoles, and the cell surface is correspondingly depleted of most of its receptor molecules. It is significant that acidotropic agents and carboxylic ionophores block not only the recycling of occupied receptors but also the constitutive recycling of unoccupied receptors. Treatment of cells with the drugs at 37°C in the absence of ligand thus usually results in a dramatic decrease of receptor number on the cell surface which may or may not be further accentuated by the

presence of ligand. The vacuoles in which receptors and receptor-ligand complexes accumulate are probably endosomal, having a low buoyant density and a
low content of lysosomal marker enzymes (170, 171). A definite identification
of them as bona fide endosomes is, however, somewhat arbitrary owing to the
drastic changes in morphology, size, and properties of acidic organelles in the
drug-treated cells mentioned above.

Why are the receptors and receptor-ligand complexes in the drug-treated
cells not recycled, and why are they not delivered to lysosomes? Studies with
receptors and fluid-phase markers suggest that the recycling pathway and the
pathway to lysosomes may only be partially blocked by the presence of
inhibitors; receptors apparently continue to be recycled and delivered to the
lysosomes albeit at a reduced rates (172–173a). The reasons for the block in the
traffic of most acid-sensitive receptors are probably quite complex. A major
contributing factor may be a trivial one: receptor accumulation in vacuoles may
simply reflect the massive redistribution of membrane into the cell and the
change in vacuolar structure induced by the drugs. As already mentioned above
it is known that more than half of the cell surface is internalized after addition of
acidotropic bases, and recent studies on primaquine effects on hepatocytes
suggest even more dramatic increases in vacuolar surface area at the expense of
the plasma membrane (123). This redistribution alone could explain most of the
internalization effect. The size of individual vacuoles, moreover, increases
drastically, which must have other functional consequences. The swelling of
the vacuoles may, for instance, prevent their movement in the cytoplasm,
making productive contacts with other organelles such as lysosomes more
difficult (173). It is interesting, in this context, that (hydroxymethyl)-
aminomethane (Tris) has been shown to inhibit insulin receptor recycling but
not its transport to lysosomes (159). Tris belongs to a group of acidotropic weak
bases whose vacuolating effects are less drastic than most of the more commonly used aciditropic bases (120).

Another possible explanation for the inhibition is that receptors in their
neutral pH conformation are somehow prevented from entering the recycling or
the lysosome-directed pathways in the endosome. According to this view,
endosomes would possess a mechanism for sorting receptors on the basis of
their conformation, aggregation state, charge, or other pH-dependent property.
It is quite possible that such a mechanism exists but the molecular principles
remain to be elucidated.

THE LIGAND IS SENSITIVE TO ACID PH   The prime example of this class is the
uptake system for transferrin, the major iron-transporting protein in serum. It
binds two ferric iron atoms with very high affinity and delivers them to cells that
carry functional transferrin receptors (see 174). The receptor is a disulfide-
linked dimer of two 90K transmembrane glycoproteins (see 175, 176). The

ligand-receptor complex is rapidly internalized through coated pits and recycled via endosomes to the cell surface (177–181). In the process, transferrin donates one or both of its iron atoms to ferritin at an intracellular site. After returning to the cell surface, the apotransferrin-receptor complex dissociates and both its components participate in subsequent rounds of iron transport.

Studies by Karin & Mintz (179), Klausner et al (182), Octave et al (183), and Dautry-Varsat et al (184) have shown that this elegant and economical pathway depends on carefully adjusted pH dependencies between the carrier protein and the receptor on the one hand, and between the iron atoms and the transferrin molecule on the other. While transferrin has quite a high affinity for its receptor at neutral pH, apotransferrin has high affinity at acidic pH but relatively low affinity at neutral pH. The iron, which is extremely tightly bound to transferrin at neutral pH, on the other hand, dissociates easily at mildly acidic pH (185, 186). These facts, in conjunction with the low endosomal pH, presumably suffice to explain why the transferrin is stripped of its iron, why the apotransferrin reappears in circulation, and why the majority of the receptor and transferrin escape degradation in lysosomes. Addition of lysosomotropic agents has a clear-cut inhibitory effect on iron accumulation by the cell, but recycling continues albeit at a reduced rate (187–190).

Other systems where the primary action of pH is on the ligand include viruses and toxins, which are described below in detail. The receptor-dependent recycling of high-density lipoprotein (HDL) in macrophages (191) may also belong in this category, although it is unclear what role the acid pH may have. The internalized lipoprotein is thought to bind cholesterol in the endosomes and then transport it out of the cell. This is the first case in which receptor-mediated endocytosis may remove material from cells.

LIGAND-RECEPTOR COMPLEXES ARE INSENSITIVE TO ACID PH   The best known examples of this class are the receptors for choriogonadotropin (172) and immunoglobulins (154, 192–195). Here, the receptor-ligand complexes are either destroyed in lysosomes or transported across epithelial cells. In most of these cases the receptors are "sacrificial," i.e. they are only used once before being either degraded in the lysosomes or proteolytically cleaved so that ligand-binding portion is removed.

The IgG Fc receptor (a 55-kd transmembrane glycoprotein) of macrophages is a well-studied example of pH-insensitive receptors (153, 153a, 154, 173a). The receptor binds to the Fc portion of antibodies present on soluble immune complexes and antibody-coated particles. Internalization occurs by coated vesicles or phagosomes, depending on the size of the particle, and the complexes become sequestered with the receptors into endosomes or phagosomes. The ligands do not dissociate at acid pH, and the receptor is therefore not liberated for recycling. Instead the entire complex is destroyed in the lysosomes or phagolysosome, and the receptor is thus effectively "down-regulated." Continued uptake requires new receptor synthesis.

Studies on the Fc receptor have provided a first clue as to the principles governing the sorting events in the endosomal compartment (153, 153a, 173a). It has been shown that the valency of the ligand that is bound alters the intracellular pathway. In unoccupied form, or when bound to a monovalent ligand, the receptor displays normal basal recycling into and out from endosomes. It is only when associated with a polyvalent ligand that the receptor-ligand complex is directed to the lysosomal route. Receptor traffic and "down regulation" is thus dependent on the valency of the ligand and presumably the aggregation state of the receptor-ligand complex. It is conceivable that pH-dependent aggregation/disaggregation determines the fate of other receptors in a similar way, and that clustered receptors are handled differently in endosomes than monomeric.

A different situation is encountered during transepithelial transport of IgG in the intestine of newborn rats via the IgG-receptor; this receptor has high affinity for IgG at neutral and acidic pH which allows the complex to form in the intestinal lumen and pass through endosomes undissociated on their way from the apical to the basal surface of the cells (192, 193). The relatively high pH in the interstitial fluid at the basolateral side is thought to induce dissociation of the transepithelially transported IgG. In this particular case, it is not yet known whether the receptor is reutilized.

GENERAL CONSIDERATIONS   The properties of receptor-mediated uptake and its dependence on endosomal pH may in specific receptor systems be much more complex than suggested by our simple division into three receptor categories. It is possible, and probably quite common, that receptor systems fit the criteria outlined above only partially. Only a fraction of the ligand may, for instance, dissociate during the time the complexes spend in endosomes. It is also important to recognize that the properties of a receptor uptake system may vary depending on cell type or physiological state, a point clearly illustrated by insulin receptor studies (see 159).

Another complication is that the pH is not uniform throughout the various elements of the endosomal compartment. Ligands encounter decreasing pH with time in the endosomal compartment (41a, 42). In most cells the average, initial endosomal pH is probably no lower than about pH 5.5–6.0. Within the next 30 minutes it drops to around 5, and finally reaches its minimum within lysosomes where values as low as 4.5–5.0 have been recorded. Judging by the kinetics of recycling (about 10–15 min for the intracellular phase), most of the acid-sensitive receptors do not remain in the endosomal compartment long enough to encounter the more acidic pH values. This is in agreement with the observation that acid-releasable ligands usually dissociate at mildly acidic pH values, i.e. at values close to pH 6.0. Given the kinetics of the recycling pathway, ligands that dissociate at pH values lower than this may not be able to dissociate effectively in time to be recycled. Such considerations also affect virus entry. We have shown that wild-type Semliki Forest viruses which penetrate at pH 6.0 do so only 4 min after internalization in BHK21 cells, whereas a mutant virus that is activated at pH 5.3 penetrates 40 minutes later (196). Such a delay, caused by the relatively low activation pH, may be important in the infection of cells such as nerve cells where the peripheral endosomes can be distant from the cell body and located in a region where penetration would not result in efficient infection (196a). Morphological localization studies have shown that endosomes with mildly acidic pH tend to be located in the peripheral cytoplasm and those with more acidic pH are perinuclear (88). The peripheral endosomes are the ones into which incoming material is initially channeled.

Virus Entry

The entry of enveloped animal virus provides one of the most dramatic and well-studied examples of the role of vacuolar acidity. It also provides the clearest examples of what changes membrane proteins can undergo when exposed to mildly acidic pH in endosomes.

A large number of viruses have been identified that depend on adsorbtive endocytosis and acid pH for their penetration (for reviews see 197–201). In the best-studied cases, Influenza virus and Semliki Forest virus, acid pH in endosomes induces a conformational change in the spike glycoproteins of the virus, which are thus activated to mediate the fusion of the viral membrane with the vacuolar membrane (202–207). As a result, the viral genome and other components inside the viral membrane are released into the cytoplasmic compartment. The group of viruses suspected to follow a similar pathway is rapidly expanding. It already includes enveloped viruses from the following families: toga viruses, rhabdoviruses, orthomyxoviruses, baculoviruses, and retroviruses. Evidence has also been obtained for endocytic, acid-triggered penetration of some non-enveloped viruses such as poliovirus, reovirus, and adenovirus (208–211). In general, viruses that follow this pathway of entry are distinguished by the fact that their entry is inhibited by acidotropic weak bases and carboxylic ionophores (212–214), and the enveloped ones frequently display membrane fusion activity when present in media of low pH in vitro (204, 207, 215, 216).

Semliki Forest virus and Influenza virus particles are internalized by adsorptive endocytosis in coated pits, and, in the case of Influenza, in small noncoated vesicles as well (202, 217, 218). The virus particles are next delivered to endosomes from which penetration usually occurs (203, 205). The fusion activity of Semliki Forest virus requires a pH of 6.1 or below, while the pH needed for Influenza is between pH 5.1 and 6.0 depending on the strain. The Semliki Forest virus spikes responsible for the fusion are heterotrimers containing two integral membrane glycoproteins (E1 and E2, 49K and 51K respectively) and a peripheral one (E3, 10K) (219). When exposed to acid, both E1 and E2 undergo irreversible conformational changes (220). The change in E1 depends in addition on the presence of cholesterol. The steroid requirement is particularly interesting because Semliki Forest virus can only fuse with membranes that contain cholesterol or other steroids with a 3-β-hydroxyl group (220, 221).

The Influenza hemagglutinin is a homotrimer where each unit is composed of two disulfide-bonded subunits, HA1 (56K) and HA2 (24K). The structure of the external domain of the hemagglutinin has been determined by X-ray crystallography (222), the major antigenic epitopes have been located on the molecule (223), and large amounts of structural and functional information is available. It is known that the HA1 subunits form the top domain of the spike and carry a binding site for sialic acid responsible for virus attachment to the cell surface. The stem of the 13.5-nm-long spike is mainly composed of the three HA2 subunits. The N-terminal of HA2 located in the stem region is hydrophobic and highly conserved, and several lines of evidence indicate that it plays a central role in the fusion reaction.

Studies using biochemical, immunochemical, genetic, and morphological techniques (224–232) have shown that an irreversible conformational change accompanies the acid-induced activation of HA. The change involves partial or total dissociation of the three subunits of HA leading to the opening of the molecule along its long axis. As the three subunits become separated from each other at the top of the molecule apolar moities previously hidden within the stem become accessible. The most favored view is that the exposure of the hydrophobic moiety, which includes the N-terminal peptides of the three HA2 subunits, allows the protein to become hydrophobically attached to the target membrane. The dual attachment in the viral and the target membrane may bring the membranes into close enough proximity to allow fusion.

While much remains to be learned about the mechanism of the fusion reaction it is already clear that it is exposure to low pH that induces the dramatic changes in the spike glycoproteins. Recent genetic studies show that single mutations in HA that affect intersubunit salt bridges, the cavity in which the hydrophobic N-terminal of HA2 is located, or the N-terminal of HA2 sequence itself, can change the efficiency of fusion and its pH dependence (224–226, 232). Mutants of Semliki Forest virus with altered pH dependence have been also isolated and partially characterized (196).

Protein Toxins

Many bacterial and plant toxins are thought to enter the vacuolar apparatus of animal cells by adsorbtive endocytosis, and some have been shown to depend on acidic pH for their penetration (234–236). The best characterized of these are diphtheria toxin, *Pseudomonas aeruginosa* exotoxin, and the plant toxin modeccin (234–240). Each consists of two subunits, A and B, of which the smaller (A) carries the toxic activity and the larger (B) is needed for translocation of the A chain into the cytoplasm. The A-chains inhibit protein synthesis either by ADP-ribosylating elongation factor 2 (diphtheria toxin and *Pseudomonas* endotoxin) or by inactivating the 60S ribosomal subunit (modeccin). Diphtheria toxin is best characterized in terms of structure and biological effect. After binding to yet unidentified receptors the toxin is internalized. Penetration occurs from endosomes and requires an acidic environment (pH < 5.5) (241, 242). Intoxication is blocked by agents that elevate vacuolar pH, and the block can be bypassed by lowering medium pH to 5.0 and below, which induces penetration directly through the plasma membrane (119, 241–245).

Several approaches have shown that diphtheria toxin undergoes an irreversible change in conformation and solubility properties when exposed to acidic pH. It binds nonionic detergent and it attaches firmly to artificial membranes (246–248). There is also evidence from model membrane studies suggesting that it can form transmembrane, voltage-dependent anion channels (249–252), and it has been assumed that these channels allow the passage of the A-chain through the membrane. The exact nature of the conformational change remains unclear, but it is assumed that some of the relatively long hydrophobic stretches in the sequence of the B subunit may become exposed, and that these insert into the endosomal membrane. Fluorescence spectroscopy suggests exposure of buried tryptophan-containing sites in the molecule (253). Only relatively small changes in secondary structure are, however, detected by circular dichroism measurements (254). Whereas the critical role of low pH seems relatively well-established the details involved in the actual penetration reaction remain unclear.

The Functions of Lysosomes

The lysosomes are the most acidic compartments of most cells. The low pH facilitates the activity of lysosomal hydrolases, most of which have pH optima in the 4.5–5.5 range. The low pH may be generally advantageous for hydrolytic reactions, and many substrates become acid-denatured and thus more susceptible to enzymatic degradation. A further advantage is the safety aspect; by confining most of their hydrolytic enzymes into lysosomes and adjusting their pH optima to extreme values, the cell diminishes the risks involved if there is "accidental" release of the enzymes into the cytoplasm and extracellular fluid.

Whether the pH of lysosomes has other functions is not clear. It is conceivable that the proton gradient could be used for coupled transport of degradation products out of the lysosomes. This has recently been proposed for the transport of cysteine (255). The low pH could also be useful in the uptake of weak bases of physiological importance, and facilitate the escape of weak acids into the cytosolic compartment.

ROLE OF ACIDITY IN THE EXOCYTIC PATHWAY

It is becoming increasingly apparent that low intravesicular pH plays important roles in various stages of the exocytic pathway. Thus far, the transport of certain small molecules into secretion granules, the efficient targeting and storage of secretory proteins, and the orderly flow of traffic along the secretory pathway have all been found to directly or indirectly depend on pH gradients established by vacuolar proton ATPases.

Coupled Transport of Biogenic Amines into Secretory Granules

Many acidic secretory granules use an electrochemical proton gradient to mediate the coupled transport of biogenic amines (for review, see Ref. 48, 50). The best-known examples include the adrenal chromaffin granule (epineph-rine), the platelet dense granule (serotonin), and the synaptic vesicle (catechol-amines, acetylcholine) (53, 86, 97, 98, 103, 104). In each case, generation of the electrochemical proton potential is accomplished by vacuolar-type proton ATPases described in detail above. While the precise details of amine transport remain to be determined, it is well-established that the granule membrane contains a reserpine-sensitive transporter that can mediate the exchange of cytoplasmic amines for internal protons. The transporter appears to be relative-ly nonspecific, and can accept a variety of substrates: e.g. epinephrine, nore-pinephrine, serotonin, and dopamine. Although both neutral and protonated amines can be transported into the granule, both cases result in the efflux of net positive charge (e.g. one H^+ per unprotonated amine, two H^+ per protonated amine). In this way, the granule can efficiently accumulate amines to internal concentrations as high as 0.1 M. In vitro, amine transport can be stimulated by ATP-driven proton pumping or by artificially imposed diffusion potentials (48, 50). Either an electrical or chemical proton gradient can be chemiosmotically coupled to amine accumulation in isolated granules.

Not all secretory granules utilize transmembrane proton gradients to drive the accumulation of biogenic amines. For example, secretory granules found in the neurohypophysis have a proton ATPase and acidic interiors, but serve to package and store polypeptide hormones (78–80). While the function of low pH in these organelles is not yet clear, it has been suggested that the acidic internal environment provides the optimal conditions for the proteolytic processing enzymes that cleave prohormones into mature species (46). In addition, low pH may play a role in the biogenesis of these organelles, by influencing the targeting of newly synthesized hormones into the nascent granule (see below).

Other secretory granules, such as those found in the rat parotid gland, seem to contain neither a proton pump nor an acidic internal pH (102). Therefore, an obligatory role for pH gradients in events such as the fusion of the granule and plasma membranes during exocytosis appears unlikely.

Targeting of Secretory Proteins

Exocytosis of secretory proteins by mammalian cells can occur along regulated or unregulated (constitutive) pathways (1, 2, 128) (Figure 1). The "regulated pathway" is characterized by the storage of a protein in secretory granules and released only after a cell receives the appropriate stimulus. In general, these granules form at the *trans* face of the Golgi apparatus and contain high concentrations of the secreted protein. This type of exocytosis is generally associated with specialized secretory cells, such as the pancreatic acinar cell or the neurosecretory cells of the pituitary gland. In contrast, the "constitutive pathway" is characterized by the continuous, unregulated release of secretory products directly following their synthesis. There is no significant storage or concentration of the secreted proteins in intracellular vesicles prior to release. The constitutive pathway is characteristic of most cell types, such as fibroblasts or plasma cells, which continuously secrete proteins such as fibronectin and immunoglobulin, respectively. The constitutive pathway also operates in many cells that are capable of regulated secretion (49, 128, 256). What determines whether a particular protein is secreted in a regulated or a constitutive manner is not yet known. However, there is evidence to suggest that acidic pH in organelles of the exocytic pathway plays an important regulatory role.

The AtT-20 cell line is derived from a mouse pituitary tumor and is capable of the regulated secretion of mature ACTH as well as the constitutive exocytosis of other secreted and membrane proteins (49, 256). A fraction of immature ACTH (i.e. proopiomelanocortin, POMC) is also released constitutively, suggesting that packaging in secretory granules is a prerequisite for maturation of the hormone precursor. Kelly and colleagues have been able to transfect AtT-20 cells with plasmids containing the genes for other regulated secretory proteins, such as insulin and trypsinogen, which are not normally expressed by pituitary cells (257, 258). The fact that these proteins are efficiently packaged and stored in cytoplasmic secretory granules and are subject to regulated exocytosis by the addition of cyclic AMP suggests the existence of a highly conserved and generalized mechanism which facilites the proper targeting of certain secretory proteins to storage granules. Conceivably this mechanism relies on some type of secretory protein receptor; e.g. a receptor that might mediate the selective transport of newly synthesized proteins from the Golgi to the forming granule.

While the existence of intracellular receptors for the transport of secreted proteins remains to be demonstrated, it is of interest that acidophilic agents such as chloroquine dramatically change the pattern of secretion in AtT-20 cells

(49). Chloroquine-treated cells can no longer efficiently package POMC in secretory granules. Instead, the hormone is diverted from the regulated pathway to the constitutive pathway and exocytosed continuously, without modulation by added cyclic AMP. While the mode of chloroquine action is not known, one can draw analogies with its effects on receptor-mediated endocytosis. Since chloroquine would be expected to elevate the internal pH of the Golgi, Golgi-associated vesicles, and the secretory granules, it is conceivable that the low-pH-facilitated discharge of POMC from its receptor is inhibited, tying up all available receptors and leading to the constitutive release of newly synthesized hormone. A similar series of events is believed to occur in the transport of newly synthesized lysosomal enzymes in chloroquine-treated cells (161). Enzyme release from the mannose-6-phosphate receptor is inhibited by the elevated pH, the pool of free receptors decreases, and the enzymes are released constitutively into the medium. Whatever the mechanism leading to the packaging of ACTH in secretory granules, it seems clear that acidic pH in one or more organelles plays a critical role in the sorting events that are known to occur.

pH and Membrane Traffic on the Exocytic Pathway

EFFECTS OF MONENSIN   Treatment of cells with carboxylic ionophores can lead to dramatic alterations in the processing and transport of membrane and secretory proteins on the exocytic pathway (for review, see Ref. 128). As first shown by Tartakoff & Vassalli (259), monensin arrests the secretion of immunoglobulin by plasma cells within the Golgi. Since the intracellular immunoglobulin fails to receive terminal galactose or sialic acid residues, it has been suggested that the block occurs at medial Golgi cisternae, prior to the compartment containing terminal transferase activities (260, 261). However, the effects of monensin on transport are not always predictable. For instance, in hepatoma cells, neither transport transferrin nor vesicular stomatitis virus G protein reaches the plasma membrane following monensin treatment but terminal glycosylation is blocked only for transferrin (262, 263). Interestingly, both proteins can be localized by immunocytochemistry to the same Golgi vesicles. Yet another variation is the mouse macrophage Fc receptor, which reaches the plasma membrane in monensin-treated cells but fails to be galactosylated (264). Some of these differences may be due to differences in the cell types used and to variations in the rate that individual proteins are transported through the Golgi. Nevertheless, it seems likely that monensin does not exert a single, well-defined block in the exocytic pathway.

Monensin facilitates the electroneutral exchange of Na^+, K^+, and H^+. As such, it would be expected to dissipate transmembrane pH gradients in Golgi compartments, an effect often invoked as its mechanism of inhibition. However, any possible K^+ and/or Na^+ gradients would also be disrupted. In addition, monensin may exert other effects on the cell that could indirectly affect Golgi function; one such effect is the hyperpolarization of the electrical potential across the plasma membrane (128).

Nevertheless, since monensin causes a rapid (within 30 min) and dramatic dilation of Golgi cisternae and Golgi-associated vesicles (128), it seem clear that the drug does directly affect the internal ionic environment of the Golgi apparatus. Why should the disruption of Golgi pH or cation gradients lead to dilation, inhibition of transport, and aberrant glycoprotein processing? One possibility is that monensin treatment inhibits membrane vesiculation, thereby slowing the rate at which membrane leaves the *trans* Golgi. Alternatively, the dissipation of transmembrane ion gradients may disrupt the transport of sugar nucleotide precursors or other metabolites into or out from the Golgi, conceivably leading to less efficient glycosyltransferase activity and possibly osmotic imbalance. However, the available evidence suggests that the transport of at least sugar nucleotide precursors is not coupled to transmembrane ion gradients (265).

EFFECTS OF ACIDOTROPIC AGENTS   Although the influences of acidophilic agents on the exocytic pathway have been less extensively studied, several examples are known in which the transport of membrane and secretory proteins is altered by amines. Agents such as chloroquine can cause dilation of Golgi cisternae and Golgi-associated vesicles in a manner at least superficially similar to monensin (121). In hepatoma cells, primaquine will greatly slow the secretion of albumin and transferrin, as well as prevent the budding of vesicular stomatitis virus at the plasma membrane (266). In contrast to the effects of monensin in the same cells, terminal glycosylation of G protein is not inhibited by primaquine treatment. This is consistent with the observation that the primaquine block seems to occur at the *trans*-most cisterna of the Golgi stack. In cultured kidney epithelial cells, transport of the Influenza virus spike glycoprotein, hemagglutinin, is also slowed in the presence of NH_4Cl (267).

Unlike monensin, acidotropic drugs would be expected only to dissipate transmembrane pH gradients. Therefore, all of these effects should be a direct or indirect reflection of an elevation of intravesicular pH. As discussed earlier, it is conceivable that this pH change affects the interaction of proteins on the exocytic pathway with a possible class of receptors responsible for transporting newly synthesized membrane and secretory proteins from the Golgi to the plasma membrane. In the AtT-20 cell experiments (see above), it was not determined whether chloroquine slowed the overall rate of constitutive secretion in addition to diverting ACTH from the regulated pathway.

GENETIC APPROACHES

The importance of intravesicular acidification in the endocytic and exocytic pathways is illustrated by mutant cell lines that exhibit acidification defects both in intact cells and in vitro (41a, 268–271). In general, these cell lines were isolated by selection for the ability to grow in the presence of agents (e.g. bacterial toxins, enveloped viruses) which require low intravesicular pH to penetrate the cytosol and kill their host cell. Both nonlethal and temperature-sensitive mutations have been isolated.

Among the best-characterized mutants is the series of CHO cell lines isolated by Robbins and colleagues (260, 271, 268). These cells exhibited defects in the ATP-dependent acidification of endosomes in vitro. In contrast, lysosome acidification—both in vitro and in vivo—appears normal. Genetic complementation analysis revealed that the mutants fall into two distinct complementation groups, designated *end-1* and *end-2;* thus, at least two genes may control endosome acidification in CHO cells.

Cells of both complementation groups are phenotypically similar and exhibit pleiotropic defects in the endocytic pathway consistent with defective endosome acidification: resistance to a variety of low-pH-requiring viruses and toxins; reduced uptake of exogenously added lysosomal enzymes via the mannose-6-phosphate receptor; failure to dicharge iron from internalized transferrin [also postulated as being the reason for the lethality of the temperature-sensitive phenotype (269)]; inefficient retention of newly synthesized lysosomal enzymes. Interestingly, these cells also exhibit aberrant Golgi-related functions, and, for example, defects in the terminal glycosylation of some viral spike glycoproteins and several endogenous secretory glycoproteins (e.g. fibronectin) (268; A. Robbins, in preparation). The wild-type phenotype is restored for each of these defects in complementing *end-1* × *end-2* hybrids, correlating with the restoration of endosome acidification activity. That the effects of these mutations is a reflection of altered acidification or ion transport activity is further supported by the fact that low concentrations of monensin closely mimic the mutant phenotype (268).

The nature of the molecular defect in either *end-1* or *end-2* cells is unknown. Nevertheless, these observations do suggest the possibility that endosomes and Golgi membranes share at least one gene product, related to ion transport, which is required for the normal function of both organelles. For example, the mutations could affect a subunit common to the endosomal and Golgi proton pumps. Alternatively, they may affect a shared anion or cation transporter whose activity indirectly controls intravesicular acidification. In any event, localization of the defects in these cells may permit the demonstration that the acidification mechanisms and ion permeabilities defined in cell-free systems also function in intact cells. They may also permit a definition of the biochemical and ontogenetic relationship between ion transport mechanisms in the endocytic and exocytic pathways.

Since the mutant gene products have not yet been identified, it is difficult to speculate on the significance of the observation that both the *end-1* and *end-2* mutations affect endosome acidification without affecting the acidification of lysosomes. This observation may indicate that there are differences between the acidification mechanisms of endosomes and lysosomes. On the other hand, they may simply reflect a quantitative difference in the concentration of a mutant protein in the two organelles. For example, if a mutation partially affects the V_{max} of a proton pump shared by endosomes and lysosomes, and if lysosomes contain many more copies of the pump than do endosomes, the observed phenotype might be expected to result. Therefore, it is not yet possible to conclude on the basis of the genetic findings that there exist fundamental differences in the acidification of endosomes and lysosomes.

CONCLUSIONS

It is clear that the acidification of intracellular organelles plays numerous and critical roles in maintaining the normal function of the endocytic and exocytic pathways. In every case studied thus far, acidification is mediated by a unique class of ATP-driven electrogenic proton pumps, which we have referred to as the "vacuolar H^+-ATPases." The characteristics of the ATPases found in all endocytic and exocytic organelles are strikingly similar, suggesting the possibility that the same proton pump is distributed throughout the vacuolar apparatus. Although the definitive identification and isolation of the ATPase of any organelle has yet to be reported, this suggestion is further supported by the genetic evidence, which shows that two distinct mutations affecting endosome acidification also affect Golgi functions in a manner at least superficially similar to the effects of monensin and acidophilic agents on the exocytic pathway. However, crucial questions remain as to the biogenesis, origin, and inter-organelle transit of the ATPase. While it is appealing to suggest that the proton pump exists on the plasma membrane, is internalized in coated vesicles, and is subsequently distributed throughout the vacuolar system, no information exists to support this scheme.

If the same (or greatly similar) proton pumps are present in all endocytic and exocytic membranes, how is their internal pH regulated? There is ample evidence that different endosomal compartments may have different pH's, and that endosomes and secretory granules have a less acidic pH than lysosomes. Regulation could occur by controlling the number of ATPase molecules present per organelle, or by modulating the activity (e.g. by phosphorylation or some other covalent modification) of a constant number of ATPases. Alternatively, the recent finding that many organelles have defined permeabilities to anions, protons, and other cations suggests that net proton flux and accumulation can be controlled by regulating the ion permeability characteristics of each membrane. For example, rat liver endosomes appear more permeable to K^+ and less permeable to Cl^- than rat liver lysosomes. Therefore, the endosomes may be less acidic because they may be less able to accumulate protons (limited Cl^-

permeability will favor the development of a membrane potential, limiting proton accumulation) and less able to retain protons (cytoplasmic K^+ can more easily exchange with internal H^+ if the conductances for both ions is significant).

Now that the basic role and mechanism of organelle acidification has been defined, future work must concentrate on the crucial questions of the regulation, biogenesis, and functions of acidification in the endocytic and exocytic pathways. It is already clear that such work will require the combined application of biochemical, morphological, and genetic approaches.

Literature Cited

1. Palade, G. E. 1975. *Science* 189:347–58
2. Farquhar, M. G. 1983. *Fed. Proc.* 42:2407–13
3. Silverstein, S. C., Steinman, R. M., Cohn, Z. A. 1977. *Ann. Rev. Biochem.* 46:669–722
4. Steinman, R. M., Mellman, I. S., Muller, W. A., Cohn, Z. A. 1983. *J. Cell Biol.* 96:1–27
5. Pearse, B. M. F., Bretscher, M. S. 1981. *Ann. Rev. Biochem.* 50:85–107
6. Edelson, P. J., Cohn, Z. A. 1978. In *Membrane Fusion*, ed. G. Poste, G. L. Nicholson, pp. 387–405. Amsterdam: Elsevier/North Holland
7. Goldstein, J. L., Anderson, R. G. W., Brown, M. S. 1979. *Nature* 279:679–85
8. Metchnikoff, E. 1893. *Lectures on the Comparative Pathology of Inflammation.* London: Kegan, Paul, Trench, Trübner & Co. Ltd.
9. Geisow, M. J., D'Arcy Hart, P., Young, M. R. 1981. *J. Cell Biol.* 89:645–52
10. Heiple, J. M., Taylor, D. L. 1982. *J. Cell Biol.* 94:143–49
11. McNeil, P. L., Tanasugarn, L., Meigs, J. B., Taylor, D. L. 1983. *J. Cell Biol.* 97:692–702
12. Marsh, M., Helenius, A. 1980. *J. Mol. Biol.* 142:439–54
13. Huet, C., Ash, J. F., Singer, S. J. 1980. *Cell* 21:429–38
14. Willingham, M. C., Yamada, S. S. 1978. *J. Cell Biol.* 78:480–87
15. Ghosh, D. C., Wellner, R. B., Cragoe, E. J. Jr., Wu, H. C. 1985. *J. Cell Biol.* 101:350–57
16. Moya, M., Dautry-Varsat, A., Goud, B., Louvard, D., Boquet, P. 1985. *J. Cell Biol.* 101:548–59
17. Pastan, I. H., Willingham, M. C. 1983. *Ann. Rev. Physiol.* 43:239–50
18. Carrière, D., Casellas, P., Richer, G., Gros, P., Jansen, F. K. 1985. *Exp. Cell Res.* 156:327–40
19. Stone, D. K., Xiao-Song, X., Racker, E. 1983. *J. Biol. Chem.* 258:4059–62
20. Forgac, M., Cantley, L., Wiedenmann, B., Altstiel, L., Branton, D. 1983. *Proc. Natl. Acad. Sci. USA* 80:1300–3
21. Anderson, R. W. G., Galck, J. R., Goldstein, J. L., Brown, M. S. 1984. *Proc. Natl. Acad. Sci. USA* 81:4838–42
22. Schwartz, A. L., Strous, G. J. A. M., Slot, J. W., Geuze, H. J. 1985. *EMBO J.* 4:899–940
23. Anderson, R. G. W., Pathak, R. K. 1985. *Cell* 40:635–43
24. Tolleshaug, H., Berg, T., Frolich, W., Norum, K. R. 1979. *Biochim. Biophys. Acta* 585:71–84
25. Helenius, A., Mellman, I., Wall, D., Hubbard, A. 1983. *Trends Biochem. Sci.* 8:245–50
26. Straus, W. 1964. *J. Cell Biol.* 20:497–507
27. Willingham, M. C., Pastan, I. H. 1980. *Cell* 21:67–77
28. Geuze, H. J., Slot, J. W., Strous, G. J. A. M., Lodish, H. F., Schwartz, A. L. 1983. *Cell* 32:277–87
29. McKanna, J. A., Haigler, H. T., Cohen, S. 1979. *Proc. Natl. Acad. Sci. USA* 76:5689–93
30. Wall, D., Wilson, G., Hubbard, A. 1980. *Cell* 21:79–93
30a. Marsh, M., Griffiths, G., Dean, G., Mellman, I., Helenius, A. 1986. *Proc. Natl. Acad. Sci. USA* In press
31. Bergeron, J. J. M., Cruz, J., Khan, M. N., Posner, B. I. 1985. *Ann. Rev. Physiol.* 47:383–403
32. Pastan, I. H., Willingham, M. C. 1983. *Trends Biochem. Sci.* 8:250–54
33. Brown, M. S., Anderson, R. G. W., Goldstein, J. L. 1983. *Cell* 32:663–67
34. Hopkins, C. R. 1983. *Nature* 305:684–85
35. Tycko, B., Maxfield, F. R. 1982. *Cell* 28:643–51

36. Galloway, C. J., Dean, G. E., Marsh, M., Rudnick, G., Mellman, I. 1983. *Proc. Natl. Acad. Sci. USA* 80:3334–38
37. Van Renswoude, J., Bridges, K. R., Harford, J. B., Klausner, R. D. 1982. *Proc. Natl. Acad. Sci. USA* 79:6186–90
38. Marsh, M., Bolzau, E., Helenius, A. 1983. *Cell* 32:931–40
39. Yamashiro, D. J., Fluss, S. R., Maxfield, F. R. 1983. *J. Cell Biol.* 97:929–34
40. Tycko, B., Keith, C. H., Maxfield, F. R. 1983. *J. Cell Biol.* 97:1762–76
41. Maxfield, F. R. 1982. *J. Cell Biol.* 95:676–81
41a. Merion, M., Schlesinger, P., Brooks, J. M., Moehring, J. M., Moehring, T. J., Sly, W. S. 1983. *Proc. Natl. Acad. Sci. USA* 80:5315–19
42. Murphy, R. F., Powers, S., Cantor, C. R. 1984. *J. Cell Biol.* 98:1757–62
43. Coffey, J. V., De Duve, C. 1968. *J. Biol. Chem.* 243:3255–63
44. Ohkuma, S., Poole, B. 1978. *Proc. Natl. Acad. Sci. USA* 75:3327–31
45. Reeves, J. B. 1984. In *Lysosomes in Biology and Pathology*, ed. J. T. Dingle, R. T. Dean, W. Sly, pp. 175–199. Amsterdam: Elsevier
45a. Kelly, R. B. 1985. *Science* 230:25–32
46. Loh, Y. P., Brownstein, M. J., Gainer, H. 1984. *Ann. Rev. Neurosci.* 7:189–222
47. Glickman, J., Croen, K., Kelly, S., Al-Awqati, Q. 1983. *J. Cell Biol.* 97:1303–8
48. Rudnick, G. 1986. *Ann. Rev. Physiol.* 48:In press
49. Moore, H.-P., Gumbiner, B., Kelly, R. B. 1983. *Nature* 302:434–36
50. Rudnick, G. 1985. In *Physiology of Membrane Disorders*, ed. T. E. Andreoli, D. D. Fanestil, J. H. Hoffman, S. G. Schultz. New York: Plenum. 2nd ed. In press
51. Bowman, E. J. 1983. *J. Biol. Chem.* 258:15238–44
52. Brooker, R. J., Slayman, C. W. 1983. *J. Biol. Chem.* 258:222–26
53. Dean, G. E., Fishkes, H., Nelson, P. J., Rudnick, G. 1984. *J. Biol. Chem.* 259:9569–74
54. Deleage, G., Penin, F., Godinot, C., Gautheron, D. C. 1983. *Biochim. Biophys. Acta* 725:464–71
55. Forgac, M., Cantley, L. 1984. *J. Biol. Chem.* 259:8101–5
56. Harikumar, P., Reeves, J. P. 1983. *J. Biol. Chem.* 258:10403–10
57. Kakinuma, Y., Ohsumi, Y., Anraku, Y. 1981. *J. Biol. Chem.* 256:10859–63
58. Linnett, P. E., Beechey, R. B. 1979. *Methods Enzymol.* 55:472–518
59. Moriyama, Y., Takano, T., Ohkuma, S. 1984. *J. Biochem.* 95:995–1007
60. Ohkuma, S., Moriyama, Y., Takano, T. 1982. *Proc. Natl. Acad. Sci. USA* 79:2758–62
61. Pedersen, P. L. 1982. *Ann. NY Acad. Sci.* 402:1–20
62. Penefsky, H. S. 1979. *Methods Enzymol.* 55:297–303
63. Perlin, D. S., Kasamo, K., Brooker, R. J., Slayman, C. W. 1984. *J. Biol. Chem.* 259:7884–92
64. Rabon, E., Gunther, R. D., Soumarmon, A., Bassilian, S., Lewin, M., Sachs, G. 1985. *J. Biol. Chem.* 260:10200–7
65. Sabolic, I., Haase, W., Burckhardt, G. 1985. *Am. J. Physiol.* 248:F836–F44
66. Sachs, G., Koelz, H. R., Berglindh, T., Rabon, E., Saccomani, G. 1982. In *Membranes and Transport*, ed. A. Martonosi, 2:633–43. New York: Plenum
67. Van Dyke, R. W., Hornick, C. A., Belcher, J., Scharschmidt, B. F., Havel, R. J. 1985. *J. Biol. Chem.* 260:11021–26
68. Uchida, E., Ohsumi, Y., Anraku, Y. 1985. *J. Biol. Chem.* 260:1090–95
69. Dunn, S. D., Heppel, L. A. 1981. *Arch. Biochem. Biophys.* 210:421–36
70. Schneider, E., Altendorf, K. 1984. *Trends Biochem. Sci.* 9:51–53
71. Faller, L., Jackson, R., Malinowska, D., Mukidjam, E., Rabon, E., et al. 1982. *Ann. NY Acad. Sci.* 402:146–63
72. Gustin, M. C., Goodman, D. B. P. 1982. *J. Biol. Chem.* 257:9629–33
73. Farley, R. A., Faller, L. D. 1985. *J. Biol. Chem.* 260:3899–901
74. Shull, G. E., Schwartz, A., Lingrel, J. B. 1985. *Nature* 316:691–95
75. MacLennan, D. H., Brandl, C. J., Korczak, B., Green, N. M. 1985. *Nature* 316:696–700
76. Schneider, D. L. 1983. *J. Biol. Chem.* 258:1833–38
77. Stone, D. K., Xie, X.-S., Racker, E. 1984. *J. Biol. Chem.* 259:2701–3
78. Russell, J. T. 1984. *J. Biol. Chem.* 259:9496–507
79. Scherman, D., Nordmann, J. J. 1982. *Proc. Natl. Acad. Sci. USA* 79:476–79
80. Russell, J. T., Holz, R. W. 1981. *J. Biol. Chem.* 256:5950–53
81. Reggio, H., Bainton, D., Harms, E., Coudrier, E., Louvard, D. 1984. *J. Cell Biol.* 99:1511–26
82. Kielian, M. C., Cohn, Z. A. 1980. *J. Cell Biol.* 85:754–65
83. Poole, B., Ohkuma, S. 1981. *J. Cell Biol.* 90:665–69
84. Reeves, J. P. 1979. *J. Biol. Chem.* 254:8914–21
85. Corcoran, J. J., Wilson, S. P., Kirshner, N. 1984. *J. Biol. Chem.* 259:6208–14
86. Fishkes, H., Rudnick, G. 1982. *J. Biol. Chem.* 257:5671–77

87. Dell'Antone, P. 1984. *FEBS Lett.* 168: 15–22
88. Tanasugarn, L., McNeil, P., Reynolds, G. T., Taylor, D. L. 1984. *J. Cell Biol.* 98:717–24
89. Van Dyke, R. W., Steer, C. J., Scharschmidt, B. F. 1984. *Proc. Natl. Acad. Sci. USA* 81:3108–12
90. Xie, X.-S., Stone, D. K., Racker, E. 1983. *J. Biol. Chem.* 258:14834–38
91. Stone, D. K., Xie, X.-S., Racker, E. 1985. *Methods Enzymol.* In press
92. Pfeffer, S., Kelly, R. B. 1985. *Cell.* 40:949–57
93. Wiedenmann, B., Lawley, K., Grund, C., Branton, D. 1985. *J. Cell Biol.* 101:12–18
94. Van Dyke, R. W., Scharscmidt, B. F., Steer, C. J. 1985. *Biochim. Biophys. Acta* 812:423–36
95. Zhang, F., Schneider, D. L. 1983. *Biochem. Biophys. Res. Commun.* 114: 620–25
96. Rees-Jones, R., Al-Awqati, Q. 1984. *Biochemistry* 23:2236–40
97. Johnson, R. G., Beers, M. F., Scarpa, A. 1982. *J. Cell Biol.* 257:10701–7
98. Cidon, S., Nelson, N. 1983. *J. Biol. Chem.* 258:2892–98
99. Deleted in proof
100. Toll, L., Howard, B. D. 1980. *J. Biol. Chem.* 255:1787–89
101. Anderson, D. C., King, S. C., Parsons, S. M. 1981. *Biochem. Biophys. Res. Commun.* 103:422–28
102. Arvan, P., Rudnick, G., Castle, J. D. 1984. *J. Biol. Chem.* 259:13567–72
103. Cidon, S., Ben-David, H., Nelson, N. 1983. *J. Biol. Chem.* 258:11684–88
104. Johnson, R. G., Carty, S. E., Scarpa, A. 1982. See Ref. 66, pp. 237–44
105. Nelson, N., Cidon, S. 1985. *Methods Enzymol.* In press
105a. Percy, J. M., Peyde, J. G., Apps, D. K. 1985. *Biochem. J.* 231:557–64
106. Gluck, S., Kelly, S., Al-Awqati, Q. 1982. *J. Cell Biol.* 257:9230–33
107. Cannon, C., van Adelsberg, J., Kelly, S., Al-Awqati, Q. 1985. *Nature* 314: 443–46
107a. Gluck, S. 1986. *Kidney Int.* (Abstr.) In press
108. Riezman, H. 1985. *Cell.* 90:1001–9
108a. Mandala, S., Taiz, L. 1985. *Plant Physiol.* 78:104–9
108b. Mandala, S., Taiz, L. 1985. *Plant Physiol.* 78:327–33
109. de Duve, C. 1983. *Eur. J. Biochem.* 137:391–97
110. de Duve, C., de Barsy, T., Poole, B., Trouet, A., Tulkens, P., van Hoof, F. 1974. *Biochem. Pharmacol.* 23:2495–531
111. Seglen, P. O., Grinde, B., Solheim, A. E. 1979. *Eur. J. Biochem.* 95:215–25
112. Lie, S. O., Schofield, B. 1973. *Biochem. Pharmacol.* 22:3109–14
113. Goldstein, J. L., Brunschede, G. Y., Brown, M. S. 1975. *J. Biol. Chem.* 250:7854–62
114. Sando, G. N., Titus-Dillon, P., Hall, C. W., Neufeld, E. F. 1979. *Exp. Cell Res.* 19:359–64
115. Tolleshaug, H., Berg, T. 1979. *Biochem. Pharmacol.* 28:2919–22
116. Tietze, C., Schlesinger, P., Stahl, P. 1980. *Biochem. Biophys. Res. Commun.* 93:1–8
117. Jensen, E. M., Liu, O. C. 1963. *Proc. Soc. Exp. Biol. Med.* 112:456–59
118. Helenius, A., Marsh, M., White, J. 1982. *J. Gen. Virol.* 58:47–61
119. Kim, K., Gorman, N. B. 1965. *J. Bacteriol.* 90:1552–56
120. Ohkuma, S., Poole, B. 1981. *J. Cell Biol.* 90:656–64
121. Wibo, M., Poole, B. 1974. *J. Cell Biol.* 63:430–40
122. Ose, L., Reinertsen, R., Berg, T. 1980. *Exp. Cell Res.* 126:109–19
123. Zÿderhand, J., Schwartz, A., Slot, J., Strous, G., Geuze, H. 1986. Submitted for publication
124. Davies, P. J. A., Davies, D. R., Levitszki, A., Maxfield, F. R., Milhaud, P., et al. 1980. *Nature* 283:162–66
125. Gordon, A. H., D'Arcy Hart, P., Young, M. R. 1980. *Nature* 286:79–80
126. Dean, R. T., Jessup, W., Roberts, C. R. 1984. *Biochem. J.* 217:27–40
127. Pressman, B. C. 1976. *Ann. Rev. Biochem.* 45:501–30
128. Tartakoff, A. M. 1983. *Cell* 32:1026–28
129. Basu, S. K., Goldstein, J. L., Anderson, R. G., Brown, M. S. 1981. *Cell* 24:493–502
130. Marnell, M. H., Stookey, M., Draper, R. K. 1982. *J. Cell Biol.* 93:57–62
131. Marsh, M., Wellsteed, J., Kern, H., Harms, E., Helenius, A. 1982. *Proc. Natl. Acad. Sci. USA* 79:5297–301
132. Smolyo, A. P., Garfield, R. E., Chacko, S., Smolyo, A. V. 1975. *J. Cell Biol.* 66:425–43
133. Tartakoff, A. M., Vasalli, P. 1977. *J. Exp. Med.* 146:1332–45
134. Jensen, M. S., Bainton, D. F. 1973. *J. Cell Biol.* 56:379–88
135. Fok, A. K., Lee, Y., Allen, R. D. 1982. *J. Protozool.* 29:409–14
136. Kielian, M. C., Cohn, Z. A. 1980. *J. Cell Biol.* 85:754–65
137. Gluck, S., Cannon, C., Al-Awqati, Q. 1982. *Proc. Natl. Acad. Sci. USA* 79:4327–31

138. Allen, R. D., Fok, A. K. 1983. *J. Cell Biol.* 97:566–70
139. Horwitz, M. A., Maxfield, F. R. 1984. *J. Cell Biol.* 99:1936–43
140. Gorman, M. B. A. 1977. *Rev. Microbiol.* 31:507–33
141. Sibley, L. D., Weidner, E., Krahenbuhl, J. L. 1985. *Nature* 315:416–18
142. Anderson, R. G. W., Kaplan, J. 1983. *Mod. Cell Biol.* 1:1–52
143. Schwartz, A. L., Fridovich, S. E., Lodish, H. F. 1982. *J. Biol. Chem.* 257: 4230–37
144. Harford, J., Bridges, K., Ashwell, G., Klausner, R. D. 1983. *J. Biol. Chem.* 258:3191–97
145. Bleil, J. D., Bretscher, M. S. 1982. *EMBO J.* 1:351–55
146. Mellman, I. S., Steinman, R. M., Unkeless, J. C., Cohn, Z. A. 1980. *J. Cell Biol.* 87:712–22
147. Anderson, R. G. W., Brown, M. S., Beisigel, U., Goldstein, J. L. 1982. *J. Cell Biol.* 90:523–31
148. Watts, C. 1985. *J. Cell Biol.* 100:633–37
149. Hopkins, C. R., Trowbridge, I. S. 1983. *J. Cell Biol.* 97:508–21
150. Hopkins, C. 1985. *Cell* 40:199–208
151. Tolleshaug, H., Berg, T. 1979. *Biochem. Pharmacol.* 28:2912–22
152. Kaplan, J., Keogh, E. A. 1981. *Cell* 24:925–32
153. Mellman, I., Plutner, H., Ukkonen, P. 1984. *J. Cell Biol.* 98:1163–69
153a. Mellman, I., Plutner, H. 1984. *J. Cell Biol.* 98:1170–77
154. Mellman, I. 1982. In *Membrane Recycling,* ed. D. Evered, pp. 35–58. *CIBA Found. Symp. 92.* London: Pitman
155. Steinman, R. M., Brodie, S. E., Cohn, Z. A. 1976. *J. Cell Biol.* 68:665–87
156. Thilo, L., Vogel, G. 1980. *Proc. Natl. Acad. Sci. USA* 77:1015–19
157. Schlessinger, J. 1980. *Trends Biochem. Sci.* 5:210–14
158. Helenius, A., Kartenbeck, J., Simons, K., Fries, E. 1980. *J. Cell Biol.* 84:404–20
159. Marshall, S. 1985. *J. Biol. Chem.* 260: 4136–44
160. Besterman, J. M., Airhart, J. A., Woodworth, R. C., Low, R. B. 1981. *J. Cell Biol.* 91:717–72
161. Shepherd, V. L., Lee, Y. C., Schlesinger, P. H., Stahl, P. D. 1981. *Proc. Natl. Acad. Sci. USA* 78:1019–1022
162. Hubbard, A. L. 1982. See Ref. 154, pp. 109–15
163. DiPaola, M., Maxfield, F. R. 1984. *J. Biol. Chem.* 259:9163–71
164. Wileman, T., Boshans, R. L., Schlesinger, P., Stahl, P. 1984. *Biochem. J.* 220:665–75
165. King, A. C., Hernaez-Davis, L., Cuatrecasas, P. 1980. *Proc. Natl. Acad. Sci. USA* 77:3283–87
166. Tietze, C., Schlesinger, P., Stahl, P. 1982. *J. Cell Biol.* 92:417–29
167. Sando, G. N., Titus-Dillon, P., Hall, C. W., Neufeld, E. F. 1979. *Exp. Cell Res.* 19:359–64
168. Gonzalez-Noriega, A., Grubb, J. H., Talked, V., Sly, W. S. 1980. *J. Cell Biol.* 85:839–52
169. Van Leuven, F., Cassiman, J. J., Van Den Berghe, H. 1980. *Cell* 20:37–43
169a. Schwartz, A. L., Bolognesi, A., Fridovich, S. E. 1984. *J. Cell Biol.* 98:732–38
170. Pastan, I. H., Willingham, M. C. 1981. *Science* 214:504–9
171. Merion, M., Sly, W. S. 1983. *J. Cell Biol.* 96:644–50
172. Berg, T., Tolleshaug, H. 1980. *Biochem. Pharmacol.* 29:917–26
172a. Ascoli, M. 1984. *J. Cell Biol.* 99:1242–50
173. Stein, B. S., Bensch, K. G., Sussman, H. H. 1984. *J. Biol. Chem.* 259:14762–72
173a. Ukkonen, P., Lewis, V., Marsh, M., Helenius, A., Mellman, I. 1986. *J. Exp. Med.* In press
174. Aisen, P., Listowsky, I. 1980. *Ann. Rev. Biochem.* 49:357–93
175. Newman, R., Schneider, C., Sutherland, R., Vodinelich, L., Greaves, M. 1982. *Trends Biochem. Sci.* 7:397–400
176. Kuhn, L. C., McClellan, A., Ruddle, F. H. 1984. *Cell* 37:95–103
177. Hemmaplardh, D., Morgan, E. H. 1977. *Br. J. Haematol.* 36:85–96
178. Octave, J.-N., Schneider, Y.-J., Trouet, A., Crichton, R. R. 1981. *Trends Biochem. Sci.* 8:217–20
179. Karin, M., Mintz, B. 1981. *J. Biol. Chem.* 258:3245–52
180. Van Renswoude, J., Bridges, K. R., Harford, J. B., Klausner, R. D. 1982. *Proc. Natl. Acad. Sci. USA* 79:6186–90
181. Harding, C., Heuser, J., Stahl, P. 1983. *J. Cell Biol.* 97:329–39
182. Klausner, R. D., Ashwell, G., van Renswoude, J., Harford, J. B., Bridges, K. R. 1983. *Proc. Natl. Acad. Sci. USA* 80:2263–66
183. Octave, J.-N., Schneider, Y.-J., Hoffmann, P., Trouet, A., Crichton, R. R. 1982. *Eur. J. Biochem.* 123:235–40
184. Dautry-Varsat, A., Ciechanover, A., Lodish, H. F. 1983. *Proc. Natl. Acad. Sci. USA* 80:2258–62

185. Princiotto, J. V., Zapolski, A. J. 1975. *Nature* 255:87–88
186. Lestas, A. N. 1976. *Br. J. Haematol.* 32:341–50
187. Morgan, E. H. 1981. *Biochim. Biophys. Acta* 642:119–34
188. Ciechanover, A., Schwartz, A. L., Dautry-Varsat, A., Lodish, H. F. 1983. *J. Biol. Chem.* 258:9681–89
189. Klausner, R. D., van Renswoude, J., Ashwell, G., Kempf, C., Schechter, A. N., et al. 1983. *J. Biol. Chem.* 218:4715–24
190. Harford, J., Wolkoff, A. W., Ashwell, G., Klausner, R. D. 1983. *J. Cell Biol.* 96:1824–28
191. Schmitz, G., Robenek, H., Lohmann, U., Assmann, G. 1985. *EMBO J.* 4:613–22
192. Abrahamson, D. R., Rodewald, R. 1981. *J. Cell Biol.* 91:270–80
193. Rodewald, R. 1980. *J. Cell Biol.* 71:666–70
194. Mostov, K. E., Kraehenbuhl, J. P., Blobel, G. 1980. *Proc. Natl. Acad. Sci. USA* 77:7257–61
195. Kuhn, L., Kraehenbuhl, J. P. 1982. *Trends Biochem. Sci.* 7:299–302
196. Kielian, M. C., Keränen, S., Kääriäinen, L., Helenius, A. 1984. *J. Cell Biol.* 98:139–45
196a. Helenius, A., Dons, R., White, J., Kielian, M. 1986. *Banbury Rep.* No. 22. Cold Spring Harbor, NY: Cold Spring Harbor Press. In press
197. Helenius, A., Marsh, M., White, J. 1980. *Trends Biochem. Sci.* 5:104–6
198. White, J., Kielian, M., Helenius, A. 1983. *Q. Rev. Biophys.* 16:151–95
199. Lenard, J., Miller, D. K. 1981. *Cell* 28:5–6
200. Kielian, M., Helenius, A. 1985. In *Toga Viruses*, ed. H. Fraenkel-Conrat, R. Wagner. In press
201. Marsh, M. 1984. *Biochem. J.* 218:1–10
202. Matlin, K., Reggio, H., Helenius, A., Simons, K. 1983. *J. Cell Biol.* 91:601–13
203. Marsh, M., Bolzau, E., White, J., Helenius, A. 1983. *J. Cell Biol.* 96:455–61
204. Maeda, T., Ohnishi, S. 1980. *FEBS Lett.* 122:283–87
205. Yoshimura, A., Ohnishi, S.-I. 1984. *J. Virol.* 51:497–504
206. Matlin, K., Reggio, H., Helenius, A., Simons, K. 1982. *J. Mol. Biol.* 156:609–31
207. White, J., Matlin, K., Helenius, A. 1981. *J. Cell Biol.* 89:674–79

208. Madshus, I. H., Olsnes, S., Sandvig, K. 1984. *J. Cell Biol.* 98:1194–200
209. Madshus, I. H., Olsnes, S., Sandvig, K. 1984. *EMBO J.* 3:1945–50
210. Silverstein, S. C., Dales, S. 1968. *J. Cell Biol.* 36:197–230
211. Seth, P., Pastan, I., Willingham, M. C. 1985. *J. Biol. Chem.* 260:9598–602
212. Jensen, E. M., Force, E. E., Unger, J. B. 1961. *Proc. Soc. Exp. Biol. Med.* 107:447–51
213. Helenius, A., Marsh, M., White, J. 1982. *J. Gen. Virol.* 58:47–61
214. Marsh, M., Wellsteed, J., Kern, H., Harms, E., Helenius, A. 1982. *Proc. Natl. Acad. Sci. USA* 79:5297–301
215. White, J., Helenius, A. 1980. *Proc. Natl. Acad. Sci. USA* 77:3273–77
216. Huang, R. T. C., Rott, R., Klenk, H. D. 1980. *Virology* 104:294–302
217. Helenius, A., Kartenbeck, J., Simons, K., Fries, E. 1980. *J. Cell Biol.* 84:404–20
218. Dourmashkin, R. R., Tyrell, D. A. J. 1974. *J. Gen. Virol.* 24:129–41
219. Garoff, H., Frischauf, A.-M., Simons, K., Lerach, H., Delius, H. 1980. *Nature* 288:236–41
220. Kielian, M., Helenius, A. 1985. *J. Cell Biol.* 101:2284–91
221. Kielian, M., Helenius, A. 1984. *J. Virol.* 52:281–83
222. Wilson, I. A., Skehel, J. J., Wiley, D. C. 1981. *Nature* 289:366–73
223. Wiley, D. C., Wilson, I. A., Skehel, J. J. 1981. *Nature* 289:373–78
224. Gething, M.-J., Doms, R., York, D., White, J. 1986. *J. Cell Biol.* In press
225. Doms, R. W., Gething, M.-J., Henneberry, J., White, J., Helenius, A. 1986. *J. Virol.* In press
226. Daniels, R. S., Downie, J. C., Hay, A. J., Knossow, M., Skehel, J. J., Wang, M. L., Wiley, D. C. 1985. *Cell* 40:431–39
227. White, J., Helenius, A., Kartenbeck, J. 1982. *EMBO J.* 1:217–22
228. Skehel, J. J., Bayley, P. M., Brown, E. B., Martin, S. R., Waterfield, M. D., et al. 1982. *Proc. Natl. Acad. Sci. USA* 79:968–72
229. Webster, R. G., Brown, L. E., Jackson, D. C. 1983. *Virology* 126:587–99
230. Yewdell, J. W., Gerhard, W., Bächi, T. 1983. *J. Virol.* 48:239–48
231. Daniels, R. S., Douglas, A. R., Skehel, J. J., Wiley, D. C. 1983. *J. Gen. Virol.* 64:1657–62
232. Rott, R., Orlich, M., Klenk, H.-D., Wang, M. L., Skehel, J. J., Wiley, D. C. 1984. *EMBO J.* 3:3329–32
233. Deleted in proof

234. Olsnes, S., Phil, A. 1982. In *Molecular Action of Toxins and Viruses*, ed. Cohen, van Heyningan, pp. 51–105. Amsterdam: Elsevier
235. Collier, R. J., Kaplan, D. A. 1984. *Sci. Am.* 251(1):56–64
236. Pappenheimer, A. M. Jr. 1977. *Ann. Rev. Biochem.* 46:69–94
237. Kaczorek, M., Delpeyroux, F., Chenciner, N., Streeck, R. E., Murphy, J. R., et al. 1983. *Science* 221:855–58
238. Greenfield, L., Bjorn, M. J., Horn, G., Fong, D., Buck, G. A., et al. 1983. *Proc. Natl. Acad. Sci. USA* 80:6853–67
239. Ratti, G., Rappuoli, R., Giannini, G. 1983. *Nucleic Acids Res.* 11:6589–95
240. Uchida, T. 1983. *Pharmacol. Ther.* 19:107–22
241. Draper, R. K., Simon, M. I. 1980. *J. Cell Biol.* 87:849–54
242. Sandvig, K., Olsnes, S. 1980. *J. Cell Biol.* 87:828–32
243. Sandvig, K., Olsnes, S. 1982. *J. Biol. Chem.* 257:7504–13
244. Draper, R. K., O'Keefe, D. O., Stookey, M., Graves, J. 1984. *J. Biol. Chem.* 259:4083–88
245. Marnell, M. H., Shia, S.-P., Stookey, M., Draper, R. K. 1984. *Infect. Immunol.* 44:145–50
246. Zalman, L. S., Wisnieski, B. J. 1984. *Proc. Natl. Acad. Sci. USA* 81:3341–45
247. Hu, V. H., Holmes, R. K. 1984. *J. Biol. Chem.* 259:12226–33
248. Blewitt, M., Zhao, J. M., Mc Keever, B., Sarma, R., London, E. 1984. *Biochem. Biophys. Res. Commun.* 120:286–90
249. Kagan, B. L., Finkelstein, A., Colombini, M. 1981. *Proc. Natl. Acad. Sci. USA* 78:4950–54
250. Donovan, J. J., Simon, M. I., Draper, R. K., Montal, M. 1981. *Proc. Natl. Acad. Sci. USA* 78:172–76
251. Misler, S. 1983. *Proc. Natl. Acad. Sci. USA* 80:4320–24
252. Misler, S. 1984. *Biophys. J.* 45:107–9

253. Blewitt, M. G., Chung, L. A., London, E. 1986. *Biochemistry.* In press
254. Iglewski, B. H., Kabat, D. 1975. *Proc. Natl. Acad. Sci. USA* 72:2284–88
255. Pisoni, R. L., Thoene, J. G., Christensen, H. N. 1985. *J. Biol. Chem.* 260:4791–98
256. Gumbiner, B., Kelly, R. B. 1982. *Cell* 28:51–59
257. Moore, H.-P., Walker, M. D., Lee, F., Kelly, R. B. 1983. *Cell* 35:531–38
258. Burgess, T. L., Craik, C. S., Kelly, R. B. 1985. *J. Cell Biol.* 101:639–45
259. Tartakoff, A. M., Vassalli, D. 1977. *J. Exp. Med.* 146:1332–45
260. Tartakoff, A. M., Vassalli, D. 1979. *J. Cell Biol.* 79:284–99
261. Griffiths, G., Quinn, P., Warren, G. 1983. *J. Cell Biol.* 96:835–50
262. Strous, G. J. A. M., Lodish, H. F. 1980. *Cell* 22:709–17
263. Strous, G. J. A. M., Willemsen, R., van Kerkhof, P., Slot, J. S., Geuze, H. J., Lodish, H. F. 1983. *J. Cell Biol.* 97:1815–22
264. Green, S. A., Plutner, H., Mellman, I. 1985. *J. Biol. Chem.* 260:9867–74
265. Capasso, J. M., Hirschberg, C. B. 1984. *Proc. Natl. Acad. Sci. USA* 81:7051–55
266. Strous, G. J., Du Main, A., Zijderhand-Bleekemolen, J. E., Slot, J. W., Schwartz, A. L. 1985. *J. Cell Biol.* 101:531–39
267. Matlin, K., Simons, K. 1983. *Cell* 34:233–43
268. Robbins, A. R., Oliver, C., Bateman, J. L., Krag, S. S., Galloway, C. J., Mellman, I. 1984. *J. Cell Biol.* 99:1296–308
269. Marnell, M. H., Mathis, L. S., Stookey, M., Shia, S.-P., Stone, D. K., Draper, R. K. 1984. *J. Cell Biol.* 99:1907–16
270. Robbins, A. R., Peng, S. S., Marshall, J. L. 1983. *J. Cell Biol.* 96:1064–71
271. Klausner, R. D., van Renswoude, J., Kempf, C., Rao, K., Bateman, J. L., Robbins, A. R. 1984. *J. Cell Biol.* 98:1098–101

Ann. Rev. Biochem. 1986. 55:701-32
Copyright © 1986 by Annual Reviews Inc. All rights reserved

DISCONTINUOUS TRANSCRIPTION AND ANTIGENIC VARIATION IN TRYPANOSOMES

Piet Borst

Division of Molecular Biology, The Netherlands Cancer Institute (Antoni van Leeuwenhoek Huis), Plesmanlaan 121, 1066 CX Amsterdam, The Netherlands

CONTENTS

0066-4154/86/0701-0701$02.00

1. SUMMARY

The main theme of this review is the discontinuous synthesis of mRNAs in trypanosomes. This novel process was discovered in the unicellular eukaryote *Trypanosoma brucei,* but it is probably a general feature of the order of Kinetoplastida, to which several other major human pathogens belong. Discontinuous RNA synthesis involves a sequence of 35 nucleotides (nt) found at the 5' end of all trypanosome mRNAs analyzed. The 35-nt sequence is encoded in arrays of 1.35-kb repeats that are clustered in the genome. The primary transcript of the 1.35-kb repeat is an RNA of 140 nt that carries the 35-nt sequence at its 5' end. The 35-nt sequence is transferred from the 140-nt precursor to pre-mRNAs made elsewhere in the genome. The process has not yet been reconstructed in vitro, and whether transfer involves priming of pre-mRNA synthesis, RNA-RNA ligation followed by splicing, trans-splicing, or more than one of these mechanisms, is still unknown. Circumstantial evidence makes priming the least likely of these alternatives. Why it is advantageous to trypanosomes to make their mRNAs in such an unusual fashion is unclear. As yet, there is no evidence for discontinuous synthesis of mRNAs in organisms other than kinetoplastid flagellates.

The mini-exon sequence was first found in mRNAs for Variant-specific Surface Glycoproteins (VSGs), and the control of the synthesis of these proteins is a second theme of this review. Silent VSG genes may be activated by their duplicative transposition to a telomeric expression site. The transposition process looks like a gene conversion, mediated by short blocks of sequence homology. Activation of the transposed gene is due to its insertion into an active transcription unit, i.e. to promoter addition. Telomeric VSG genes can also be activated without duplication. This can occur by a reciprocal translocation in which a silent telomeric gene exchanges position with a gene residing in an active expression site. A VSG gene may also be activated without detectable translocation, however, by the transcriptional activation of the silent expression site in which it is located. How this occurs is still unknown, because the transcription units are so long that the promoter for pre-mRNA synthesis has not yet been reached by chromosome walking. A simple mechanism in which a mobile promoter moves between telomeres has been rendered unlikely by the demonstration that two telomeric transcription units can be simultaneously active when one of them is interrupted by a large DNA insertion. Active VSG gene expression sites yield several minor mRNA-like RNAs, in addition to VSG mRNA. At least one of these RNAs could specify a novel (membrane ?) protein, unrelated to VSGs, but the actual synthesis and function of this protein remains to be demonstrated.

An unusual feature of the transcription of genes for variant surface glycoproteins is its insensitivity to high concentrations of α-amanitin; other trypanosomal protein genes, rRNA genes, and 5S RNA genes behave like their counterparts in other eukaryotes in this respect. Whether the variant surface glycoproteins are transcribed by an (rRNA-specific) RNA polymerase I or by a separate RNA polymerase II remains to be sorted out.

The analysis of housekeeping genes in trypanosomes has only recently begun and some of the available data are presented here together with a summary of possible sequences involved in the demarcation of transcription units. Homologous trypanosome transformation and in vitro transcription systems are urgently required to test the role of putative regulatory sequences and to dissect the process of discontinuous transcription.

2. GENERAL INTRODUCTION

Trypanosomes are unicellular eukaryotic flagellates and the causative agents of serious diseases in man (sleeping sickness, Chagas' disease) and his livestock. They have become popular as research objects because of their economic importance and the interesting biochemical mechanisms they have evolved to cope with the vicissitudes of parasitic life. In historic order the four major biochemical peculiarities (1) of trypanosomes are:

1. Antigenic variation, i.e. the ability to evade host antibodies by altering the antigenic composition of the surface coat. Antigenic variation is only found in the African trypanosomes and their close relatives, not in the South American *Trypanosoma cruzi*.
2. A kinetoplast, i.e. a part of the mitochondrion that contains the kinetoplast DNA, a highly unusual form of mitochondrial DNA, consisting of an enormous network of catenated circles. This is a characteristic shared by all representatives of the order Kinetoplastida, which includes *Leishmania* and *Crithidia*. The analysis of kinetoplast DNA is yielding a wealth of interesting results: diverse replication intermediates (2, 3); unusual mitochondrial genes (4–10); bent DNA (11–13); mini-exon rRNAs (5, 14–17); proteins starting with leucine (4, 7); and evidence for a most remarkable form of RNA editing resulting in the insertion of U-residues at 3 positions into an mRNA (4, 6, 7).
3. A glycosome, i.e. a microbody containing the glycolytic pathway from glucose to 3-P-glycerate (18, 19). The glycosome has been found in all Kinetoplastida analyzed and it may be another common characteristic of the order.
4. Discontinuous transcription of protein-coding genes, i.e. the synthesis of mRNAs in which RNA segments derived from two transcription units are joined. The limited evidence available suggests that this is also a common characteristic of Kinetoplastida.

This review will primarily deal with discontinuous transcription. For historic reasons this topic is closely linked to antigenic variation, and the genes for the Variant-specific Surface Glycoproteins (VSGs) remain the most intensively studied protein-coding genes in trypanosomes. Hence, recent developments in this area relevant to transcription are also summarized. Lack of space has precluded the covering of other interesting aspects of antigenic variation and trypanosome biochemistry in general, but I have included references to recent developments in Sections 3 and 6.

Trypanosome biochemistry has been discussed in the *Annual Reviews* series by Englund et al (20) and antigenic variation by Boothroyd (21). Many other recent reviews deal with antigenic variation in general (22–27), with special emphasis on the molecular genetics (28–31, 31a) or the structure of the surface coat (32). Most of these reviews contain more information about the biology of Kinetoplastida than the minimal information provided here and list the literature before 1984.

3. ANTIGENIC VARIATION AND DISCONTINUOUS TRANSCRIPTION: AN OVERVIEW

A. Antigenic Variation (see 22–31)

The antigenic determinants recognized on intact trypanosomes by host antibodies are located in the surface coat, which consists of a single protein species, the Variant-specific Surface Glycoprotein (VSG). [Other names are also used for this protein but I shall stick here to the name coined by Cross (33), who first isolated it.] The progeny of a single trypanosome can make more than a hundred different coats that have no exposed antigenic determinant in common. Switching from one coat to the next occurs at low frequency (about 10^{-6} per trypanosome division during chronic infections in laboratory animals, see 33a) and is not induced by antibody or other external stimuli. The trypanosome does not economize on the genetic information required to encode his large wardrobe of coats by assembling genes from gene segments. Each entire VSG is encoded by a separate gene and there are some 10^3 VSG genes (and pseudogenes) per trypanosome nucleus. A major route for VSG gene activation is the duplicative transposition of a gene copy to an expression site located at the end of a chromosome. Activation of the transposed gene is due to its insertion into an active transcription unit, i.e. to promoter addition (34), as illustrated in the upper half of Figure 1. This is not the only route for VSG gene activation, however, and in total four mechanisms are now known to be used by trypanosomes to activate silent VSG genes. These are schematically summarized in Figure 2:

1. A silent chromosome-internal gene may be activated by duplicative transposition to an active telomeric expression site. In all cases analyzed the

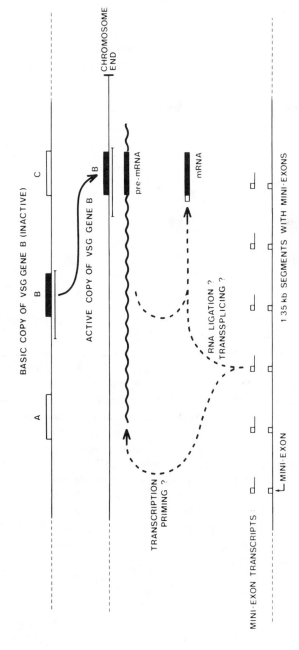

Figure 1   The synthesis of mRNA for variant-specific surface glycoprotein in trypanosomes. The upper part of the figure illustrates the activation of a silent gene by duplicative transposition to a telomeric expression site, where the gene is transcribed from a far upstream initiation site. The lower part of the scheme shows the synthesis of the 140-nt mini-exon-derived RNA on a separate chromosome. How this RNA donates its 35-nt terminal sequence to the mature mRNA is not yet known.

four modes of VSG gene activation

| duplicative transposition | telomere conversion | reciprocal translocation | in situ activation |

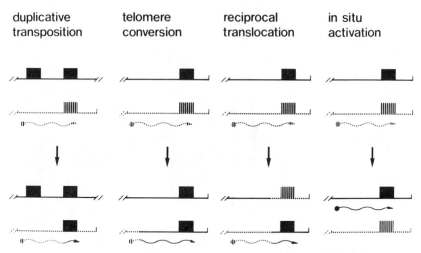

Figure 2 The four modes of VSG gene activation in trypanosomes.

transposed segment starts 1–2 kb upstream of the gene and ends within or shortly downstream of the gene (34–40).

2. A silent telomeric gene may be activated by duplicative transposition to an active telomeric expression site (38, 41–50). The segment transposed in this case is often much longer than 4 kb; it may involve up to 40 kb of upstream sequence and all traceable sequence downstream of the transposed VSG gene (43, 49, 50). The transposed segment may also cover only part of the gene (segmental gene conversion) (30, 31, 42, 43, 50a).

The duplicative transposition of a VSG gene to an expression site has the characteristics of a gene conversion in which the VSG gene residing in the site acts as acceptor and is lost (see 30, 31 for a detailed discussion). The suggestion that this conversion involves an error-prone DNA polymerase (51) was based on inadequate evidence and has been sunk (31, 52). Whether conversion is initiated by a specific endonuclease is unclear. The frequency of conversion is low and circumstantial evidence indicates that genes may also be translocated to silent telomeric sites (36, 48, 53). The cellular machinery for generalized recombination may therefore be sufficient to catalyze these events, given the fact that telomeres are also the site of frequent rearrangements in yeast (54).

3. An inactive telomeric VSG gene may be activated by exchanging position with a VSG gene that resides in an active telomeric expression site via a reciprocal translocation (55).

4. Inactive telomeric expression sites for VSG genes may be activated by an unknown mechanism. This will be discussed in Section 6.

B. Timing of VSG Gene Expression; the Metacyclic Repertoire

Antigenic variation as a strategy to evade the immune response does not only require the ability to change surface antigens. It also requires some programming of the order in which antigen genes are expressed. If the next VSG gene to be expressed were selected randomly in each trypanosome, the repertoire of antigens would be rapidly exhausted early in a chronic infection. That variants indeed appear in a loosely programmed order, characteristic of the stock analyzed, has been amply demonstrated, but how trypanosomes accomplish this is still largely unclear. The predominant early VSG genes, activated first when trypanosomes are transferred by syringe to another rodent, are invariably telomeric (36, 38, 39, 43–50, 56–61). The preferential activation of these genes appears to be due to two factors: first, some telomeric genes reside in an inactive expression site that can be reactivated; second, telomeric genes can enter an active expression site by telomere conversion and this process occurs more frequently than the duplicative transposition of chromosome-internal genes to an expression site (36, 45, 61). The preferential activation of telomeric genes is more a function of the telomere than of the VSG gene residing in it, because transfer of a late gene to a telomere that is preferentially activated can transform the gene into a predominant early gene (39, 60, 61). It seems likely that the degree of sequence homology between donor and acceptor gene will play a role in determining timing (39), but direct evidence for this is lacking. It also seems improbable that such homology would be the major determinant of timing, because this would lead to a vicious circle in which the same set of closely related genes would be activated again and again. Another factor that may affect timing is growth rate. Some variants grow faster than others and may therefore appear earlier in a chronic infection (see 62). A further selection might be imposed by events during coat switching. Esser & Schoenbechler (63) have recently shown that during switching trypanosomes contain a mixed coat. Mixing looks homogeneous at the light-microscopic level but has not been analyzed at the electron-microscopic or molecular level. It seems likely, however, that close contacts between two different VSGs occur during the switching period. Although the overall shape and properties of different VSGs are similar (32), it is conceivable that certain combinations would result in a poor coat. This would obviously introduce an additional selection against certain switches at the coat protein level. An even more theoretical possibility is an involvement of the idiotype network (64). The predominant early Variant Antigen Types induce a strong antibody response that can be expected to reverberate through the idiotype network. The anti-idiotype antibodies produced may prevent the outgrowth of some VATs. There are no data bearing on these conjectures.

A special form of programming of VSG gene expression occurs in the tsetse fly, when trypanosomes reacquire a coat in the salivary glands (see 65). Only a small and strain-specific set of 10–20 VSG genes is then activated and this metacyclic repertoire differs from the predominant early genes active in mammals. The metacyclic VSG genes studied are all telomeric (44, 66, 67, 67a), and some of them can also be used in the bloodstream repertoire (67). How the metacyclic genes are selectively activated in the fly is still unclear. The current working hypothesis is that there are one or more telomeric VSG gene expression sites that are specifically activated in the tsetse fly (66–68).

C. The Synthesis of Variant-Specific Surface Glycoproteins

VSGs are glycoproteins with about 450–500 amino acid residues. In the surface coat they are present as a tightly packed single layer of homo-dimers, with the amino-terminal domain exposed to host antibodies. The C-terminus is anchored in the membrane via a complex P-glycolipid, attached through an ethanolamine to the free COOH of the terminal amino acid (21, 26, 32, 69–79).

VSGs are made as precursor molecules containing both a conventional N-terminal signal peptide and a hydrophobic C-terminal extension of 17–23 amino acids. The C-terminal extension is replaced in a tunicamycin-insensitive step by the terminal P-glycolipid and this occurs within one minute after completion of the preprotein, presumably still in the endoplasmic reticulum (80, 81). Reports of later addition are attributable to the use of inadequate methods (see 81, 82). The rapidity of the addition is most simply explained if the glycolipid is preconstructed and added in one step.

The function of the unique terminal P-glycolipid is not known. Most procedures that damage trypanosomes release VSG molecules that contain the carbohydrate part of the P-glycolipid but not the hydrophobic moiety (32, 83). Release is catalyzed by an enzyme with the characteristics of a phospholipase C (84) and leaves the hydrophobic anchor 1,2-dimyristyl glycerol in the membrane (85, 86, 86a). It is therefore conceivable that the unusual P-glycolipid is used to allow rapid release of VSGs during coat-switching or antibody attack or when the trypanosome enters the tsetse fly (see 32, 86). When the trypanosome is taken up by the fly a metabolic upheaval occurs and VSG synthesis stops (87). The termination of VSG synthesis is not accompanied by detectable alterations in the silenced VSG gene (88–90) and can be induced in vitro by the simultaneous action of two signals: a decrease in temperature of 37°C to 27°C and addition of cisaconitate (87). The effects are reversible for up to 8 h. Eventually repression becomes irreversible and the coat is shed. Shedding is much slower than when the coat is released after trypanosome damage, however, and this makes it uncertain whether rapid release is an artifact of phospholipase leakage from lysosomes, rather than the induction of a physiological process. The simple axenic medium for the cultivation of bloodstream trypanosomes (92, 93) that has recently been developed should simplify the study of coat turnover.

D. Initial Experiments on Discontinuous Transcription

Historically the study of discontinuous transcription started with the analysis of VSG gene transcription, because the 35-nt sequence present at the 5' end of trypanosome mRNAs was first found in VSG mRNAs. Conventional analyses had not given any sign of introns in VSG genes (94, 95), but Bernards et al (96) happened to note in S1 protection experiments that the cDNA for VSG 118 protected about 10 nt more at the 5' end of the mRNA than the gene for VSG 118. This indicated the presence of a short sequence in the mRNA not adjacently encoded in the gene. Reverse transcription showed this "mini-exon sequence" on VSG 118 to be 35 nt long (97) and the same sequence was found at the 5' end of two other VSG mRNAs (98).

A search for mini-exon genes soon revealed a complex situation: there are about 200 mini-exon genes per trypanosome nucleus (52, 99); each gene is present in a 1.35-kb repeat and these repeats are clustered in the genome (99, 100). Some "orphon" mini-exon genes are present outside repeat arrays (89, 99, 100) and I shall return to these below. Mini-exon genes are bordered at their 3' end by a conventional eukaryotic mRNA splice donor site (99); the linkage of the mini-exon sequence with the body of the VSG 118 mRNA occurs at the position of a eukaryotic splice acceptor site, which has the canonical AG (97). This prompted a speculative model in which arrays of mini-exon genes are located upstream of a transcribed VSG gene and act as a multiple-promoter annex donor of 5' mRNA ends (99). This model was rapidly shown to be incorrect by further experiments: first, all attempts to demonstrate linkage of mini-exon genes to downstream VSG genes were unsuccessful. Second, the mini-exon sequence was shown to be present on many mRNAs, not only mRNAs for VSGs (101, 102). Third, a sequence related to the mini-exon sequence was found in mRNAs of kinetoplastid flagellates that do not show antigenic variation (103, 104). Fourth, the transcription of mini-exon genes turned out to yield only a 140-nt RNA (105–107), rather than the large RNAs predicted by the presence of mini-exon genes in one transcription unit with downstream VSG genes. Although the initial idea of a multiple-promoter annex mini-exon donor (99) is probably correct, we now know that the mini-exon sequence is joined to the body of mRNAs by an unconventional step, thus far only observed in kinetoplastid flagellates, i.e. discontinuous transcription.

4. THE EVIDENCE FOR DISCONTINUOUS TRANSCRIPTION

Discontinuous transcription is an unusual mode of mRNA production and the initial evidence for this "outlandish" process (24) was only tentative. Even today the evidence remains indirect, and the firm conviction among insiders that discontinuous transcription actually takes place in trypanosomes mainly rests on the combined results of five types of experiments:

1. All mRNAs analyzed thus far contain the mini-exon sequence These include 13 unknown mRNAs picked at random from a cDNA clone bank by De Lange et al (101); several mRNAs, including developmental stage-specific, stage-regulated, and constitutive mRNAs, analyzed by Parsons et al (102); and mRNAs for α- and β-tubulin (108–110) and calmodulin (111). Although all mRNAs in *T. brucei* may therefore contain the mini-exon sequence, as also indicated by quantitative hybridization experiments (101), this remains to be proven. For instance, to my knowledge no attempt has been made to analyze RNA enriched in sequences without mini-exon or to study protein synthesis in the presence of oligonucleotides blocking the translation of mRNAs containing mini-exons.

2. The primary transcription product of mini-exon repeats is a small RNA If many protein-coding genes yield mRNAs that contain the mini-exon sequence, and the genes for the mini-exon sequence are clustered, it becomes a problem to provide each protein-coding gene with its mini-exon. This could either be achieved by discontinuous transcription, as shown in Figure 1, or by clustering protein-coding genes downstream of mini-exon arrays and transcribing large segments of DNA into gigantic pre-mRNAs. To test the second alternative, three groups looked for transcripts of mini-exon repeats and each found a 140-nt RNA in total trypanosome RNA. This RNA has variously been called mini-exon transcript, mini-exon-derived RNA (medRNA), or small spliced leader RNA and been sized at 141, 139, 137, and 135 nt (105–107, 112). I shall use medRNA and 140 nts in this review.

The medRNA contains the 35-nt sequence at its 5' end. It is present in much lower concentration than the small rRNAs and this has complicated its analysis. The fact that medRNA (like trypanosome mRNAs) can only act as an acceptor for the guanylyl transferase of vaccinia virus (capping enzyme) only after chemical decapping, indicates that it contains a cap (112). The cap is not the trimethyl cap found on small nonribosomal nuclear RNAs in eukaryotes, because medRNA does not react with antibodies specifically directed against RNAs with trimethyl-caps (R. G. Nelson, N. Agabian, personal communication). A minor medRNA, 15 nucleotides shorter at the 3' end than the 140-nt transcript, has recently been detected (112). Whether this is a functional donor of the 35-nt sequence, an intermediate in all mRNA synthesis, or a side-product is not yet known.

The 1.35-kb repeat in which the mini-exon gene is contained has been sequenced in three *T. brucei* strains (103, 105, 113) and the corresponding repeats in several other kinetoplastid flagellates have been sequenced in part or completely (103, 107). Detailed speculations have been presented on the possible functions of sequence motifs in the *T. brucei* repeats (105, 113), but most of these seem unrealistic as these motifs are not conserved in the mini-exon repeats of other kinetoplastid flagellates. Conservation is largely limited to the mini-exon sequence itself (29 nt out of 35 in three other genera), an octamer directly upstream, the splice donor GT and some 13 bp downstream, and a T-rich stretch just downstream of the end of the medRNA-coding stretch (103, 107). Even the size of the medRNAs varies from about 140 nt in *T. brucei* to 105 in *T. cruzi* and 95 in *Leptomonas collosoma* (107).

Although the existence of medRNA suggests discontinuous transcription, it could be a side-product, because pre-mRNAs in trypanosomes are processed at a very high rate (see below), making some of the major primary transcripts difficult to find. Kooter et al (106) therefore set up a run-on assay to study synthesis of nascent RNA in isolated trypanosome nuclei. With this assay they established that the medRNA is made at a very high rate: in dot blots, the medRNA transcripts hybridize approximately as intensely as the rRNAs and much higher than major mRNAs, like the tubulin mRNA. This is compatible with medRNA being an intermediate in mRNA synthesis. Transcription of the medRNA gene is at least 700-fold higher than transcription of the remainder of the mini-exon repeat sequence (106). This shows that the medRNA is not made as part of a larger precursor.

3. The synthesis of medRNA is sensitive to α-amanitin, whereas the transcription of VSG genes is not   The availability of a nuclear run-on system for trypanosomes made it possible to study whether medRNA and pre-mRNAs are made by the same RNA polymerase as judged by sensitivity to inhibitors. Half-maximal inhibition by α-amanitin was found at 5 μg/ml for tubulin genes, 50 μg/ml for 5S RNA genes, and > 1 mg/ml for rRNA genes (112). This is similar to results obtained with other eukaryotes and suggests that *T. brucei* has at least three different RNA polymerases, polI for rRNA, polII for mRNA, and polIII for synthesis of (4S and) 5S RNA (see also 113a, 113b). The synthesis of medRNA was found to be about twofold less sensitive to α-amanitin than synthesis of tubulin RNA, but about 5-fold less sensitive than synthesis of 5S RNA (112). Since the medRNA is capped, a property unique to polII transcripts, Laird et al (112) and Lenardo et al (112a) tentatively conclude that synthesis of medRNA occurs by polII. Since medRNA is synthesized at a much higher rate than tubulin RNA, its synthesis might be subject to other rate-limiting factors than RNA polymerase synthetic capacity, resulting in a lower sensitivity to α-amanitin.

The most interesting outcome of the inhibitor experiments, however, was the remarkable behavior of VSG gene transcription (114). Initial experiments already showed that the synthesis of VSG gene transcripts was unusual since it depended on a rapid isolation procedure for nuclei. If nuclei were prepared by standard methods, involving a lengthy procedure for disrupting the trypano-somes by N_2 cavitation, VSG gene transcription was specifically lost. The effect of α-amanitin on VSG gene transcription was even more remarkable: hardly any inhibition was found with concentrations up to 1 mg/ml. If medRNA synthesis is sensitive to α-amanitin, whereas VSG pre-mRNA synthesis is not, it is highly unlikely that both RNAs are made by the same polymerase. This is another argument supporting discontinuous transcription of one class of pro-tein-coding genes.

4. Chromosomes containing no mini-exon sequence can yield mRNAs that do If discontinuous transcription occurs, it should be possible to link RNA segments encoded on different chromosomes. Trypanosomes do not condense their chromosomes, but a partial size fractionation of *T. brucei* chromosome-sized DNA molecules (further referred to as chromosomes) can be obtained by Pulsed-Field Gradient (PFG) gel electrophoresis. In such "molecular karyotypes," the mini-exon genes reside mainly near the gel slot and in the 2-megabase pair (Mb) area (see Section 6A). However, VSG genes are present in all chromosomal fractions and in three cases it has been shown (two tentatively, the other rigorously) that a VSG gene on a small chromosome yields an mRNA with a mini-exon sequence, even though the chromosome itself contains less than 0.5 mini-exon gene copy (108, 115, 116). These results establish that chromosomes without mini-exon genes, detectable by hybridiza-tion, can give rise to mRNAs that contain the mini-exon sequence.

5. The tandem arrays of tubulin genes contain no mini-exons Tubulin genes in *T. brucei* are arranged in two allelic clusters of tandem arrays of alternating α and β genes. Both genes yield mRNAs containing a 5' mini-exon sequence, but this sequence is neither present in the $\alpha\beta$ and $\beta\alpha$ intergenic regions, nor directly in front of the cluster (108–110, 117). This organization provides strong evidence for addition of the mini-exon sequence in trans. One could still argue that the cluster is transcribed into one gigantic pre-mRNA which starts with a mini-exon sequence and that each pre-mRNA would only yield a single mature mRNA. This would be excessively wasteful, however.

None of the arguments presented thus far proves that mRNAs in *T. brucei* are made discontinuously. Not finding long precursor transcripts of mini-exon repeats could be an artifact of isolated nuclei. The different α-amanitin sensitivities of mini-exon and VSG gene transcription could be misleading: since there are 200 mini-exon genes per nucleus, the experiments are not sensitive enough to exclude the possibility that a few mini-exon genes are transcribed in an α-amanitin-insensitive fashion. Hybridization experiments can exclude that a chromosome contains a complete mini-exon gene, but not that it contains a split mini-exon gene. Although valid, these alternatives are obviously farfetched and I find the sum total of the evidence now sufficiently strong to consider discontinuous transcription established in the remainder of this review. Attempts to get more direct evidence by in vitro reconstruction experiments have thus far failed. Even pulse-chase experiments in vivo remain to be done.

5. EVIDENCE FOR DISCONTINUOUS TRANSCRIPTION IN KINETOPLASTIDA OTHER THAN *T. BRUCEI*

Although the evidence is still rather patchy, it now looks as if discontinuous transcription of protein-coding genes is a characteristic of protein-coding genes in Kinetoplastida in general. Repeats of a sequence closely related to the mini-exon of *T. brucei* have been found in several other *Trypanosoma* genera (103, 104, 107) and in *Leptomonas collosoma* (104, 107); a sequence more distantly related to mini-exons has also been detected in *Herpetomonas* (103, 104), *Phytomonas* (103), *Leishmania* (104), and *Crithidia* (103). The length of the repeats in which the mini-exon is contained varies from 1.35 kb in *T. brucei* down to 0.4 kb in *Crithidia* (103), and more extensive homology is restricted to the mini-exon itself and adjacent areas (103, 104, 107). The mini-exon-like repeats of *T. cruzi* and *L. collosoma* are transcribed into small mini-exon-derived RNAs of 105 and 95 nucleotides, respectively (107). Moreover, total poly(A)$^+$ RNA from *T. cruzi* and *T. vivax* hybridizes in an analogous fashion to a mini-exon probe as *T. brucei* RNA, suggesting that many, if not all, mRNAs carry the mini-exon sequence. The only mRNAs studied thus far in detail are the tubulin mRNAs of *Leishmania* (118) and the mRNA for an unknown abundant *T. cruzi* protein (119). The latter was shown to have the *T. cruzi* mini-exon sequence at its 5' end. By reverse transcription the *Leishmania* tubulin mRNAs were found to have a sequence of about 35 nts at their 5' end that was not adjacently encoded in the genome (118). Presumably this is a mini-exon sequence.

6. CONTROL OF TRANSCRIPTION OF VSG GENES

A. *Mini-Chromosomes, Molecular Karyotype, and Ploidy of Kinetoplastida*

Pulsed-Field Gradient gel electrophoresis allows the separation of chromo-some-sized DNA molecules of *T. brucei* into four main fractions (115, 116, 120–125):

1. Large DNA, remaining close to the slot.
2. About five chromosomes in the megabasepair range. Initially these ran like a single band, but they have been resolved in recent experiments (122, 124, P. J. Johnson, personal communication).
3. A set of 5–7 intermediate chromosomes ranging in size between 200 and 700 kb.
4. Mini-chromosomes ranging from 30 to 150 kb.

The number and size of the chromosomes in categories 2–4 is highly variable and no two *T. brucei* strains have identical molecular karyograms (121, 122, 124). Whereas most lab strains studied have about a hundred mini-chromosomes of 50–150 kb, some *T. b. gambiense* strains have much fewer and also smaller (25–50 kb) mini-chromosomes (122).

Mini-chromosomes contain four types of sequences: telomeric sequences, VSG genes, 70-bp repeats, and a 177-bp satellite DNA, which is nearly exclusively present on mini-chromosomes and intermediate chromosomes (56, 120–122, 126, 127). All housekeeping genes analyzed are in large DNA or in the megabasepair area (120, 122). Mini-chromosomes have only been found in Kinetoplastida that undergo antigenic variation (121). The claim (128) that *T. cruzi* also contains mini-chromosomes, characterized by a 196-bp satellite (129), has been disproven (121, W. C. Gibson, unpublished observations). It is therefore likely that mini-chromosomes have evolved to increase the versatility of antigenic variation. Chromosomes smaller than 500 kb are not essential for antigenic variation, however, because *T. equiperdum* has few and *T. vivax* none (121).

Isoenzyme analyses and renaturation studies had suggested earlier that *T. brucei* is diploid for housekeeping genes and this has recently been confirmed by analysis of restriction site polymorphisms (130). Why then are VSG genes usually present in one copy per nucleus? It is possible that the small chromosomes that lack housekeeping genes are haploid, but the bulk of the VSG genes is in large DNA and in the megabasepair area (120), and it is unlikely that all of these genes are on separate chromosomes without housekeeping genes (122). The apparent haploid nature of these VSG genes may therefore be due to their rapid evolution and dispersal through the genome (131), leading to the presence of different genes at corresponding positions in homologous chromosomes (see 130).

The extensive differences in the molecular karyograms of different *T. brucei* strains can be explained by the frequent duplicative and nonduplicative translocation of telomeric VSG genes often accompanied by large stretches of upstream DNA (115, 132), but other DNA rearrangements, like transposition of RIME (see below), may contribute. Indeed, even homologous chromosomes may show substantial differences in mobility in PFG gels (122). The mobility of

DNA in these gels is not a simple function of size (123), however, and some of the differences observed may therefore be due to alterations in DNA structure rather than DNA size.

B. Anatomy of Trypanosome Telomeres

The four ways in which trypanosomes can activate a previously silent VSG gene are depicted in Figure 2. Active genes are invariably present in a telomeric expression site. The repetitive sequences present in telomeres have made them hard to clone as recombinant DNA and hence no complete telomere has been cloned, only segments. The following picture of the average trypanosome telomere is therefore based on patchy data and tentative.

The structure of the very end of the chromosome is not known; the most peripheral part analyzed consists of 5' CCCTAA 3' hexanucleotide tandem arrays which contain nicks or gaps that can be labeled by DNA polymerase I (133, 134). These hexanucleotide arrays vary in length and may be up to 20 kb long. Adjacent are CCCTAA-derived repeats of 29 bp, then an AT-rich stretch with many homopolymeric regions followed by a VSG gene (133). The VSG gene is accompanied by a stretch of upstream DNA of 1–2 kb of average base composition. Then follows a tandem array of imperfect repeats originally called 70-bp repeats by Liu et al (135) and later also called 76-bp repeats, (136), upstream repeat sequences (21), or V-sequences (40). The length of the repeats is highly variable because of an $(AAT)_n$ block in which n varies between 4 and 115 (40). The repeats bristle with structurally interesting sequences. They contain areas with potential for forming D-DNA, Z-DNA, or a cruciform; the $(T-G)_n$ is postulated to act as a hot spot for recombination and the TGTTG sequence, found in eukaryotic transposable elements (40, 135). Whether any of these sequences has physiological significance is not known. Probably the entire "barren" region in front of the VSG gene, not cut by most restriction enzymes, consists of 70-bp repeats. Only in one case in which this region is rather short (3 kb) is this assumption supported by sequence data (50).

At an early stage several groups noted that active VSG genes (not known then to be telomeric) reside in restriction fragments that vary in size in different trypanosome variant antigen types. We now know that this variation is the result of two types of sequence rearrangements. One type consists of the growth and contraction of telomeres, probably caused by terminal addition of hexamer units to the tip of the telomere and occasional deletion of large blocks of these repeats (133, 137, 138). The other type of rearrangement is strictly associated with antigenic switching (137), and is due to the variation in size of the 70-bp region. When a VSG gene enters the expression site by duplicative transposition, the crossover between the 70-bp repeats in front of the incoming gene and the 70-bp repeats of the expression site takes place in different repeats when the same gene is activated twice (34). This variable crossover provides a plausible explanation for the variable size of the array of 70-bp repeats upstream of the VSG gene in the expression site.

The structural analysis of regions upstream of the 70-bp repeats has long been thwarted by the instability of recombinant-DNA clones containing long stretches of 70-bp repeats. The repeats have recently been crossed, however, in two telomeres (50, 114, 139). In the 221 telomere of strain 427 up to 40 kb has been obtained in clone (50, 114, and J. Kooter, P. Borst, unpublished). This long area yields at least 5 minor transcripts that are retained on oligo(dT) columns and appear to contain a mini-exon sequence as judged by sandwich hybridization experiments. The upstream transcript closest to the 221 VSG gene corresponds to a pseudo-VSG gene (50). More interesting is the unassigned reading frame (329 amino acids) that Cully et al (139) have discovered further upstream in the 221 telomere and the 60% homologous reading frame (325 amino acids) directly upstream of the 70-bp repeats in another telomere, the 117 telomere. These expression-site–associated genes (ESAGs) yield mRNAs that contain a poly(A) tail and a 5' mini-exon sequence and code for proteins that have no homology with VSGs. These proteins contain an N-terminal sequence that looks like a signal sequence and a hydrophobic region immediately preceding their C-termini that could act as a membrane anchor. These ESAG proteins might therefore represent minor components of the surface coat. Another function is less likely, as ESAGs are not transcribed in culture form trypanosomes (114, 139). The fact that ESAG probes recognize 14–25 bands in Southern blots (139) is compatible with the idea that each functional expression site contains an ESAG. It should be noted, however, that the experiments required to test whether ESAGs actually give rise to proteins have not yet been completed. Whether the other minor mRNA-like RNAs coming from further upstream code for proteins is also not known.

The homology between telomeres extends far upstream of the 70-bp repeat area. The only telomere for which extensive upstream segments have been cloned and analyzed, the telomere that carries the VSG 221 gene in *T. brucei* 427, remains homologous to 20–30 other telomeres for at least 35 kb upstream of the 70-bp repeats (J. M. Kooter, personal communication). The homology is so high that the cross-hybridization is retained under the most stringent hybridization conditions. Since 200–250 telomeres are present, it is not clear why only 20–30 telomeres cross-hybridize. It is possible that this is a special set of telomeres that contain a VSG gene promoter.

C. Processing of VSG Pre-mRNA

Delineation of VSG gene transcription units is complicated by discontinuous transcription. One cannot simply walk upstream of an expressed VSG gene and look for a sequence corresponding to the start of the mRNA, because this sequence is part of another transcription unit. As an alternative, nascent transcripts made in isolated nuclei have been hybridized to cloned DNA fragments upstream of the expressed gene. This analysis has shown that all available upstream fragments of the two telomeres analyzed thus far are transcribed approximately at the same rate and, moreover, transcription is insensitive to α-amanitin, like transcription of the VSG gene itself (34, 50, 114). The simplest interpretation of these results is that the upstream area is part of one long VSG gene transcription unit. In the case of the telomere that contains VSG gene 221, the analysis has now been pushed 40 kb upstream of the gene, i.e. 35 kb upstream of the 70-bp repeat area, which is only 3 kb in this telomere, and still the end of the transcription unit is not in sight. No precursor coming even near the size of this transcription unit has been detected and this has been attributed to the rapid processing of the putative precursor. Some parts of the transcription unit are indeed represented at very low level in total RNA (50). Formally, however, the remote possibility remains that several short α-amanitin-resistant transcription units are adjacent and transcribed at the same rate. The promoter for VSG gene transcription could then be anywhere upstream of the VSG gene.

If the 40 kb upstream of VSG gene 221 consists of a single transcription unit, this unit must yield multiple mRNA-like RNAs containing a mini-exon sequence, the VSG 221 mRNA itself, the corresponding ESAG transcript, and four other minor transcripts (see preceding section). Each transcript could be derived from a separate pre-mRNA provided with a single mini-exon sequence by priming or ligation. This would be rather wasteful, however. A more plausible hypothesis is that a single pre-mRNA yields multiple mature transcripts. This could either occur by trans-splicing or by processing followed by ligation with (part of) medRNA. A choice between these alternatives is not yet possible. The final level of ESAG and VSG gene transcript differs by a factor 700 (139). Differential posttranscriptional control must be responsible for this, if both transcripts are cut from the same precursor RNA.

No intermediate in pre-mRNA processing has been fully characterized yet. Several minor transcripts have been found that specifically hybridize with the 1–2-kb stretch of DNA in front of a VSG gene that is cotransposed to the expression site (97, 140). Some of these transcripts run over the edge of the transposed segment and some of them have their 5' end at specific positions in the 70-bp repeats spanning this edge (34). All these transcripts are enriched in the poly(A)$^+$ RNA fraction, but whether they contain mini-exon sequences is not known. Further analysis of these minor RNAs may shed light on the processing pathway for VSG pre-mRNA, but it will be difficult to establish which RNAs are intermediates in VSG mRNA synthesis and which side-products of this process.

D. Nonduplicative Activation of VSG Genes

Already in the first DNA blots analyzed to study the mode of activation of VSG genes, some genes were observed that were activated without detectable duplication (94, 141). Such genes were later found to be invariably located next to chromosome ends (23, 56) and this led to the speculation (52) that such genes can enter a telomeric expression site by a reciprocal translocation, as schematically indicated in Figure 2. That this can actually occur in real life was recently demonstrated by Pays et al (55). The cross-over in this event seems to have taken place in the 70-bp repeats in front of the telomeric genes and was readily detected in standard Southern blots, because it occurred within 10 kb of the genes. A more complex exchange was analyzed by Van der Ploeg et al (115) and the gene inactivation in this event is now also interpreted to be a reciprocal translocation. In this case cross-overs are so far upstream that they cannot be detected by Southern blotting and the reciprocal translocation has been inferred from the movement of VSG genes and chromosome-specific repeats between chromosomes detected in PFG gels (L. H. T. Van der Ploeg, personal communication).

The fact that translocations occur that cannot be detected in Southern blots makes it rather difficult to prove rigorously that activation of VSG genes can occur without any movement at all. Several papers argue that this is the case, simply on the basis of Southern blots (38, 45–48, 48a, 58–60). Although the interpretation is plausible, proof is not rigorous. Even in the case described by Van der Ploeg et al (120), in which a switch was analyzed by PFG gels, proof is incomplete, because complex rearrangements in which several events occur in one switch cannot be excluded. In fact, in this case a telomeric gene thought to be activated in situ has acquired two extra 70-bp repeats in front that could not have arisen by a simple duplication of nearby repeats (50). Although the hypothesis that VSG genes can be activated in situ is reasonable and will be the basis for the following discussion, it should be kept in mind that trypanosomes have disregarded our cherished hypotheses before.

Analysis of the postulated in situ activation requires information on the transcription start of VSG pre-mRNA. Since the start of the transcription unit has not been reached, it is not yet known how the promoter is activated and inactivated in situ. Switching is apparently not induced by an external stimulus and it occurs at low rate (10^{-5} per division). These characteristics indicate that some kind of gene rearrangement is involved (142). The possibility that activation involves a movable promoter element hopping from telomere to telomere was tested by Cornelissen et al (143). They found a trypanosome variant in which the switch from telomere A to B had been accompanied by a 30-kb insert upstream of A. The region upstream of the insert was still transcribed at a high rate and in an α-amanitin-insensitive fashion but transcripts did not cross the insert and only mature mRNA B was produced. This variant shows that two telomeric expression sites can be simultaneously active and hence argues against a mobile promoter. No inserts have been found in other inactivated telomeric expression sites, but as long as the promoter for the VSG pre-mRNA has not been identified, one cannot exclude that this is so far upstream that most insertions/deletions remain undetected. Movable transcriptional terminators might therefore be involved in activation/inactivation of expression sites, but other possibilities can be envisaged (see 143).

The telomeric insert of Cornelissen et al (143) is not identical to the RIbosomal Mobile Element (RIME), a trypanosome genetic element, first detected in an rRNA gene. RIME is only 500 bp and it contains an A-rich stretch at one end, rather than (inverted) repeats, and it might therefore move by reverse transcription (144). The 200 or so copies of RIME are close to randomly distributed over the genome and do not seem to sit preferentially in telomeres (P. J. Johnson, personal communication).

The activity of telomeric VSG genes has been linked with two other phenomena, mini-exon orphons and telomere modification. The possible association of VSG gene transcription with a specific subset of mini-exon orphons (mini-exon repeats that are not part of the tandem arrays of mini-exon repeats), was raised by Parsons et al (89). A DNA fragment with one of these orphons was present in bloodstream trypanosomes but absent in the culture trypanosomes (which make no VSG) derived from them. More recent experiments have shown that the presence of this DNA fragment is due to modification of an EcoRV site in bloodstream trypanosomes but not in culture trypanosomes (R. G. Nelson, M. Parsons, N. Agabian, personal communication). There is no evidence that this orphon yields a unique medRNA and it is therefore unlikely that it could have any effect on the control of VSG gene expression.

A remarkable form of DNA modification has been observed in inactive telomeric VSG genes in bloodstream trypanosomes (145, 146). This modification renders PvuII and PstI restriction sites uncleavable and it is intimately linked with the shutoff of telomeric expression sites: only inactive telomeric VSG genes in bloodstream trypanosomes are modified; actively transcribed telomeric VSG genes, chromosome-internal VSG genes (which cannot be transcribed in situ), and telomeric VSG genes in culture form trypanosomes (in which the VSG gene transcription machinery is inactive) are not detectably modified. Modification has three other unusual features:

1. Modification of each restriction site is only partial in a clonal trypanosome population.
2. Modification is highest close to the end of the chromosome.
3. The level of modification increases with the size of the telomeric segment downstream of the VSG gene.

It is hard to believe that such a sophisticated modification system would not have an important role in VSG gene expression. Bernards et al (145) have even proposed an elaborate model in which modification controls the activity of the upstream VSG gene promoter. This model has lost some of its attraction with the finding that a large DNA insertion between promoter and VSG gene can lead to shutoff of VSG gene transcription and concomitant acquisition of modification without affecting the activity of the upstream promoter (143). Nevertheless, it remains improbable that modification is only a routine consequence of shutting off a VSG gene expression site.

The nature of the DNA modification that prevents cleavage of PvuII and PstI sites is unknown. Substitution of dC in telomeric DNA fractions of *T. equiperdum* by an unknown nucleoside X has been reported in 1983 (147). The nature of X has still not been identified and its amount seems too high to be explained by the limited telomeric modifications of only PvuII and PstI sites.

7. THE TRANSCRIPTION UNITS OF HOUSEKEEPING GENES

The transcription of several housekeeping genes in trypanosomes has recently been analyzed. The most informative data have come from the analysis of tubulin genes and genes for phosphoglycerate kinase (PGK) in *T. brucei* and the genes for an abundant protein of unknown function in *T. cruzi*. These will be discussed in turn.

Tubulin genes in *T. brucei* are arranged in long arrays of alternating α and β genes (108–110, 117). There are probably two allelic clusters (130), each containing 12 α, β doublets (110). The complete nucleotide sequence of the 3.7-kb repeat has recently been published (117). The repeats yield two transcripts, the α- and β-tubulin mRNAs. Both contain the mini-exon sequence. No evidence for the presence of a tubulin precursor transcript in total *T. brucei* RNA has been obtained in RNA blotting or S1 nuclease protection experiments. About 170 nts of the intergenic regions are not represented in mRNA (109, 110), which would be compatible with the independent synthesis of pre-α and pre-β mRNAs. The results with the PGK genes indicate, however, that this may be an oversimplification.

T. brucei contains three tandem PGK genes—A, B, and C—that are highly homologous (148). Gene B codes for cytosolic PGK and gene C for glycosomal PGK. Gene B yields a 1.8-kb, gene C a 2.5-kb mRNA; the difference is due to a long 3' untranslated region in C. The cytosolic protein is the major isoenzyme in culture trypanosomes, the glycosomal enzyme in bloodstream trypanosomes and the difference in isoenzyme ratio is reflected in mRNA ratio which is about 10 : 1 in culture form and 1 : 10 in bloodstream trypanosomes. Gene A codes for a third minor isoenzyme, not yet detected in extracts; it is specified by a 2.0-kb mRNA present in about equal (low) amounts in bloodstream and culture trypanosomes.

There are three indications that this gene cluster may be transcribed into one long precursor and that the interesting developmental regulation of the expression of these genes takes place at the posttranscriptional level (148 and B. W. Swinkels, W. C. Gibson, P. Borst, unpublished experiments). First, minor transcripts considerably larger than the mature PGK mRNAs are detectable in RNA blots. Second, S1 protection experiments using single-stranded DNA probes indicate that the intergenic regions between these genes, not represented in mature mRNAs, are nevertheless transcribed; also minor transcripts going further upstream than gene A can be detected in this way. Third, gene C (for which a fairly long gene-specific probe is available) is transcribed at high rate in isolated nuclei, irrespective of whether these nuclei are isolated from culture form (low mRNA level) or bloodstream (high mRNA level) trypanosomes. The simplest explanation for these results is that the whole PGK gene cluster is transcribed into one large precursor which is rapidly processed to yield the three gene-specific mRNAs. Processing by trans-splicing would be simplest, but cutting followed by ligation of medRNA would also explain the data.

Additional evidence for multigene precursor RNAs has recently been obtained for a set of genes coding for a very abundant *T. cruzi* protein of unknown function (119). There are about 30 genes for this protein, arranged in 0.9-kb tandem repeats. In addition to the mature mRNA of about 940 nt, which carries a mini-exon sequence at its 5' end, several minor RNAs were detected at the position expected for the dimer and trimer of mRNA. These molecules also hybridized with an "inter-exon" probe which contains the intergenic region not represented in mature mRNA. Gonzalez et al (119) present a speculative model which can account for the multiple bands by stepwise processing of a long precursor. Trans-splicing is envisaged to provide the final mRNA with a mini-exon, but I do not think that a "cut and ligate" alternative is excluded by the data. Cloning and sequence analysis of the putative precursor RNAs may shed more light on this.

In summary, the analysis of transcripts of the PGK genes and the array of *T. cruzi* genes underlines the possibility that trypanosome housekeeping genes are part of long transcription units, which argues against priming of pre-mRNA by medRNA as the major source of mini-exon sequences in mRNAs. A corollary is that posttranscriptional controls must play a major role in regulating steady-state mRNA levels.

8. DISCONTINUOUS TRANSCRIPTION; ADDITIONAL FACTS AND SPECULATIONS ABOUT MECHANISM

As indicated in Figure 1, three basic mechanisms can be envisaged for the linkage of the mini-exon sequence to the remainder of mRNAs: priming, ligation, and trans-splicing. There is insufficient evidence yet to choose between these alternatives. The analysis is hampered by the lack of success in setting up trypanosome transformation or in vitro mRNA synthesis and by difficulties in charting complete trypanosome pre-mRNA transcription units and the processing intermediates derived from these. I shall limit the discussion therefore to a recapitulation of the facts that need to be accounted for and a brief justification of my (present) bias against priming as the mechanism for mRNA synthesis.

In previous sections I have summarized some of the major facts about discontinuous transcription. These include the presence of the mini-exon sequence on a wide variety of mRNAs, possibly all mRNAs; the capped medRNA as the main transcript of mini-exon repeats; and the probable transcription of mini-exon and VSG gene by different polymerases. This last observation is important in terms of mechanism: it makes a "jumping polymerase" which starts out transcribing a mini-exon gene and then jumps to a protein-coding gene improbable. Other observations bearing on the mechanism of discontinuous transcription are:

1. There are two medRNAs which differ in 3' end, a major one of 140 nt and a minor one of 123 nt (112). This may indicate that the 3' end is not essential for function, but it remains to be shown that the minor (shorter) medRNA is an intermediate in mRNA synthesis.

2. The non-mini-exon part of medRNA is not detectable in RNA larger than 140 nt (105, 106), even in high-resolution experiments (112). This argues against any model in which medRNA is elongated (i.e. as primer) or is attached to longer RNA by bimolecular RNA ligation. This is a negative experiment, however, and if ligation is immediately followed by splicing, one might not detect the intermediate. The experiment argues against priming of RNA synthesis by the complete medRNA. Trans-splicing also predicts the existence of certain RNA intermediates, however: the free mini-exon sequence and various forms of the non-mini-exon part of medRNA. By analogy with pre-mRNA splicing in other eukaryotes, the non-mini-exon part could either remain free in linear form or as a lariat or circle, or be attached as branch to pre-mRNA. None of these have been found either.

3. Only limited homology exists between medRNA and sequences around the splice acceptor site of pre-mRNAs in the case of VSG genes, tubulin genes, PGK genes, the TIM gene, and calmodulin genes (see below). This shows that the synthesis of mature mRNAs does not depend on homology between medRNA and the areas of the pre-mRNA mentioned. As mentioned before, not a single promoter for a pre-mRNA has been identified in *T. brucei,* and whether any part of medRNA has homology to a promoter area is therefore unknown.

4. The mini-exon part of medRNA is flanked by a eukaryotic splice donor (99) and this is attached to the body of the mRNA following a eukaryotic splice

acceptor AG doublet in all mRNAs analyzed (49, 97, 98, 109–111). This indicates that the mini-exon sequence is attached to the body of the mRNA by splicing and it argues against the mini-exon part acting as primer for pre-mRNA synthesis in an analogous fashion as the influenza virus RNA replicase uses the start of host mRNAs as primers (cap snatching).

5. In addition to medRNA, several other small *T. brucei* RNAs can be enzymatically capped after prior decapping (112), suggesting the presence of trypanosome analogues of the small nuclear RNAs of other eukaryotes. Small RNAs containing a trimethyl-G cap have recently been detected with specific antibodies in *T. brucei* and *T. cruzi*, and one of these is partly homologous to human U1 RNA (R. G. Nelson, N. Agabian, personal communication). This indicates that trypanosomes contain analogues of the small RNAs required for splicing in other eukaryotes.

6. Knowledge of intermediates in pre-mRNA synthesis in *T. brucei* is still very limited. As detailed in previous sections, some transcription units are long and seem to yield several mature transcripts that contain a mini-exon sequence. This argues against priming, but the evidence is inconclusive. Trypanosomes may choose to waste a lot of RNA by making only one mRNA out of a precursor that can potentially yield many. Also the evidence that some transcription units are very long is seductive but circumstantial.

7. All trypanosome genes analyzed thus far do not contain introns (disregarding the mini-exon complication). The inventory includes several VSG genes, and the genes for α- and β-tubulins, calmodulin (3 genes), three forms of P-glycerate kinase, triose-P isomerase, and glyceraldehyde-P dehydrogenase. Admittedly, this is a small sample and inadequate to rule out that some trypanosome genes do have introns. In yeast, the intron-containing genes are also a small minority. It is therefore premature to conclude that mini-exon addition is the only form of RNA splicing employed in mRNA maturation in trypanosomes. Nevertheless, it makes the argument that medRNA only serves to assemble the splicing machinery somewhat tenuous.

8. MedRNA involvement appears to be limited to synthesis of mRNAs. The rRNAs and 5S RNA do not hybridize to a mini-exon probe (99, 100); 5S and 5.8S rRNAs do not contain even a segment of the mini-exon at their 5' end (124); and 5S RNA appears to be a primary transcript because it can be capped without prior decapping by vaccinia guanylyl transferase (112, 149).

This brief recapitulation of the main observations shows that none of the main three mechanisms for discontinuous transcription can be ruled out at present. Points 1 and 2 argue against priming by the complete medRNA. Priming by the mini-exon part of medRNA is made unlikely by point 4. A mechanism in which the first 50 nt or so would prime mRNA synthesis is not ruled out by any of the data, however, if priming does not involve extensive sequence homology detectable by hybridization and if priming does not involve a jumping polymerase (reinitiation). Point 7 makes priming clearly an unattractive mechanism, however.

Ligation of complete medRNA to complete or growing pre-mRNAs is not easily reconciled with points 2 and 6, but transfer of part of the medRNA would not be detected. The data are also fully compatible with a sequence in which the pre-mRNA is processed first, then (part of) medRNA is added rapidly followed by splicing.

Trans-splicing is the simplest alternative and this derives some respectability from the demonstration that trans-splicing can actually occur in mammalian extracts, admittedly under highly artificial conditions (150, 151). The mRNA precursors predicted by the trans-splicing alternative have not been detected, however, nor have the medRNA pieces (point 2). These are negative experiments that might be due to the short half-life of these intermediates or their preferential loss during isolation.

Both in trans-splicing and in RNA ligation the transfer machinery must somehow be able to distinguish pre-mRNA (pieces) from pre-rRNA. How this is done is not known. The pre-mRNA might be directly associated with a ribonucleoprotein particle analogous to the spliceosome of other eukaryotes (152, 153). There are no data on this yet for trypanosomes.

9. WHY DISCONTINUOUS TRANSCRIPTION?

Making mRNAs the way trypanosomes do cannot be the most efficient way of doing it. What advantage could this devious route for mRNA synthesis possibly have, or have had to the early ancestor of the Kinetoplastida that evolved it? Possible answers to this question are closely linked to the, as yet unknown, mechanism of discontinuous transcription. If discontinuous transcription occurs by priming, the medRNA detour could provide a way to overcome the well-known starting problem of RNA polymerases. A pool of medRNA, supplied by 200 genes, would allow the trypanosome to make rapidly large amounts of mRNA from some single-copy genes. This could be of special advantage to organisms with complex life cycles, requiring sudden adaptations to new environments. Although priming is biochemically sound and not without precedent, because it is used by some RNA viruses to make mRNA (see 154), I admit that *Escherichia coli* and yeast can also produce massive amounts of mRNA of single-copy genes in a short time, without resorting to discontinuous transcription. Why trypanosomes would have a less versatile polII than yeast remains unexplained.

If RNA-RNA ligation or trans-splicing is a step in discontinuous transcription, the mini-exon sequence might serve a specific function, e.g. capping, assembly of the splicing-processing-poly-adenylation machinery, transport from nucleus to cytoplasm, translation, etc. Why this would benefit from a separate mini-exon rather than a mini-exon gene as part of each transcription unit is not obvious. Economy does not seem a major consideration in the evolution of eukaryotic nuclear genomes. It might be easier to keep the mini-exon sequence uniform, however, when mini-exons were in long tandem arrays rather than scattered through the genome. It is also possible that the mini-exon sequence can only exert its special function if it is present in a short RNA, for instance during assembly of the splicing machinery. Finally, discontinuous synthesis might be used to overcome another type of starting problem. It is conceivable that efficient starting of mRNA synthesis is limited to sites near the chromosome scaffold and that such sites cannot be close together. For efficient use of DNA, short genes must then be assembled in multicistronic units. Since multicistronic mRNAs are not effectively read in eukaryotes, the cistrons must be cut apart at the RNA level and capped. Since capping is an early event in mRNA synthesis, it may only be possible to cap processed RNAs by the devious route of medRNA. The price is a heavy reliance on posttranscriptional controls for regulating mRNA levels. Why medRNA would be synthesized effectively in short transcription units and pre-mRNAs not remain unexplained.

In theory, discontinuous RNA synthesis could be used to solve other biological problems. It could serve, for instance, to generate diversity at the RNA level by allowing linkage of any of a set of splice donor sites to any of a set of splice acceptors. Discontinuous synthesis could also be used to reprogram protein synthesis by switching to a new set of leader sequences and modified ribosomes that only recognize the new leader. There is no evidence that such applications are actually used in trypanosomes or elsewhere in nature.

None of the possible reasons marshalled here for discontinuous transcription looks particularly compelling. It seems therefore unwise to prejudice the issue by opting for priming, as Krug (154) has done.

10. PROMOTERS, RNA PROCESSING, TRANSCRIPTION TERMINATION, AND POLYADENYLATION SIGNALS

Neither a homologous trypanosome DNA transformation system, nor an in vitro polII transcription system has been reported. Hence, no rigorous test system is available to analyze the significance of putative regulatory sequences in trypanosome DNA. Nevertheless, sequences are available and people have gazed at them. I shall briefly summarize the results, but critical readers have been warned that this section will mainly deal with flights of imagination.

No typical eukaryotic mRNA promoters, with CAAT box and TATA sequences at appropriate spacing, have been found in front of the start of medRNA, the only mRNA-like RNA for which the start is known (106, 154a). The speculation that a TATTTG sequence centered at position -27 could be the

TATA-box equivalent involved in medRNA synthesis (105) was not based on evidence and has been rendered unlikely by the observation that only the TTTTG part of this sequence is present in *T. cruzi* and *T. vivax* and moreover not at −27 but at −76 *(T. cruzi)* and −53 *(T. vivax)* (103).

The sequence around the mini-exon joining site is now available for several mRNAs and this has led to speculations about sequences involved in joining. Agabian and coworkers noted at an early stage that a sequence homologous to the mini-exon was present in the leader of α- but not β-tubulin mRNA (108, 109). This mini-exon-like sequence is part of the α-tubulin gene and its significance is unclear. It may be an evolutionary remnant from a time when individual genes had contiguous mini-exons, as suggested by Sather & Agabian (109), or it may have strayed in by accident (my bias). An element present directly 5' to the mini-exon joining area of α- and β-tubulin genes is TTTCT(G) sequence which is also present in the mini-exon. An even longer block of sequence homology is present in the same position in an unidentified gene (quoted in Ref. 109). Such an element could serve to align mini-exon and splice acceptor sequence.

The mammalian intron branch acceptor site is CTRAY (155), in which only the A seems to be essential. Such sequences can be found in the first 80 bp in front of all *T. brucei* splice acceptor sites, if the consensus is relaxed to YYRAY (P. Laird, personal communication) with the exception of the calmodulin genes. These data do not allow a stringent interpretation, in view of the nonstringent sequence requirement of this splice branch site in animals. Analyses on nascent RNA made in isolated trypanosome nuclei have recently shown that VSG gene primary transcripts extend between 100 and 450 nts beyond the site of poly(A) addition. (M. Timmers, T. De Lange, personal communication). The only sequence common to the termination areas of the nascent RNAs of three VSG genes was GCAGCT. This sequence also occurs within coding areas of genes, and the T-rich surroundings of the putative terminator box may therefore be essential for specific recognition.

Early sequence analyses of VSG mRNAs already showed that the AAUAAA polyadenylation signal used by higher eukaryotes is not found near the end of trypanosome mRNAs (see 23). The analysis of transcripts of the mobile element RIME then led to the suggestion that AAAAUUPyU might replace AAUAAA (144), but this signal is not present close to the poly(A) tail of tubulin mRNAs (109, 110). Rather, Sather & Agabian (109) have noted an AAAT(G)TGT sequence close to the poly(A) addition sites of tubulin mRNAs, whereas Kimmel et al (117) have pointed out a sequence of 22 bp, present with few changes both in the α- and β- intergenic region. This sequence starts with TATGT, a sequence implied in polyadenylation in yeast. The sequence downstream of the β gene is not close to the polyadenylation sites of the β-tubulin, however. Moreover, I do not see either the sequence of Kimmel et al or that of Sather & Agabian close to the polyadenylation site of published (23, 135) VSG mRNAs. At present there is therefore no good candidate for a polyadenylation signal.

11. THE DNA-DEPENDENT RNA POLYMERASES OF KINETOPLASTIDA

Experiments with isolated *T. brucei* nuclei have indicated the presence of four RNA-synthesizing activities, distinguishable by template specificity and sensitivity to α-amanitin (112, 114). The analysis of RNA polymerases from *T. brucei* does not match these results. Only a single peak of polymerase activity eluted from DEAE sephadex and this peak was inhibited by about 50% by 50 μg α-amanitin per ml (156). With CM-sephadex two polymerase species were subsequently resolved (156a, T. J. C. Beebee, personal communication), a salt-sensitive, α-amanitin-insensitive (PolI ?), and one half-maximally inhibited by about 50 μg α-amanitin per ml (PolIII ?). In contrast, RNA polymerase activities in *Crithidia,* another representative of the Kinetoplastida, were readily resolved in three peaks on DEAE sephadex, peak 1 insensitive, peaks 2 and 3 sensitive to α-amanitin (156). It seems unlikely to me that *T. brucei* and *Crithidia* would differ in such essentials as RNA polymerases I–III. The results with isolated trypanosome nuclei strongly indicate that PolI, II, and III should be present in *T. brucei* and possibly a fourth RNA polymerase. I therefore expect that the unusual result with *T. brucei* is an artifact of cell fractionation and that special precautions will be required to isolate an RNA polymerase type II from *T. brucei*.

12. PROSPECTS

This review has presented a field in flux. Discontinuous transcription seems established, but how and why is still unclear. Knowledge of the mechanism of antigenic variation is more advanced but several major problems remain unsolved and we are still far from a complete understanding of the overall operation of the system. These gaps determine to a major extent current research priorities.

Insight into the mechanism of discontinuous transcription will require homologous in vitro systems for RNA synthesis and processing. To define the sequences that control gene expression, a homologous trypanosome transformation system is essential. Further information on the extent to which discontinuous transcription is used among Kinetoplastida and related microorganisms may shed light on the origin and selective advantages of discontinuous transcription.

Central problems in the molecular genetics of antigenic variation are the nonduplicative activation/inactivation of VSG genes and the timing of VSG gene expression. Continued chromosome walking should soon answer the question of how transcription of VSG genes is controlled. The timing problem is more complex and directly related to the feasibility of vaccination against Trypanosomiasis. It is no problem to make VSGs in *E. coli* (157, 158), but if the vaccine has to protect against 10^3 different coats that keep changing, the outlook is grim. The possibility that the classical man-infective *T. gambiense* strains may share the same VSG gene repertoire (159, 160) makes the choice of early expressed VSG genes in these strains an interesting issue. The structure and biosynthesis of the unusual P-glycolipid that anchors VSGs in the membrane should also soon be settled, but the structure of the surface coat itself remains a daunting challenge for the coming years.

As I emphasized in the introduction, discontinuous transcription and antigenic variation are only two of several biochemical peculiarities of Kinetoplastida. Such peculiarities are of interest in their own right, but they can also provide illuminating insights into more general biochemical problems. The growth and contraction of chromosome ends are most spectacularly seen in the immodest trypanosome telomeres, but they could well be a feature of chromosomes in general. The mini-rRNAs of the mitochondria of Kinetoplastida are by far the smallest rRNAs in nature and should be able to provide information on the minimal rRNA sequences required for ribosome function. The glycosome is the only microbody in nature that contains glycolytic enzymes and the study of these enzymes is yielding information on the evolutionary origin of microbody enzymes and their delivery into the organelle. The discovery of trypanothione (161), a novel cofactor for glutathione reductase in trypanosomatids, represents a recent example of the metabolic originality of trypanosomes. These and other areas of trypanosome biochemistry may get their share of the limelight as antigenic variation runs into the mopping up phase.

Kinetoplastida are responsible for some of the major diseases of man and his domestic animals and effective chemotherapeutic agents against these parasites are few and toxic. This is remarkable in view of the fact that the biochemical differences between parasite and host are so extensive. It should be easier to make good drugs against trypanosomiasis than against flu or cancer. Eventually the wonderfully exotic biochemistry of trypanosomes should provide the basis for their downfall.

ACKNOWLEDGMENTS

I am indebted to colleagues in other labs for preprints and to Peter W. Laird, Jan M. Kooter, Bart W. Swinkels, Dr. Wendy C. Gibson, Dr. Patricia J. Johnson, Dr. Albert W. C. A. Cornelissen, Dr. Theodore C. White, and Dr. David R. Greaves for their critical comments and to Helga A. Woudt for typing many versions of the manuscript. The experimental work in my lab was supported by grants from the Netherlands Foundation for Chemical Research (SON) and from the UNDP/World Bank/WHO Special Programme for Research and Training in Tropical Diseases (T16/181/T7/34).

Literature Cited

1. Opperdoes, F. R. 1985. *Br. Med. Bull.* 41:130–36
2. Hajduk, S. L., Klein, V. A., Englund, P. T. 1984. *Cell* 36:484–92
3. Kitchin, P. A., Klein, V. A., Englund, P. T. 1985. *J. Biol. Chem.* 260:3844–51
4. Benne, R. 1985. *Trends Genet.* 1:117–21
5. Benne, R., Agostinelli, M., De Vries, B. F., Van den Burg, J., Klaver, B., et al. 1983. In *Mitochondria 1983: Nucleo-Mitochondrial Interactions,* ed. R. J. Schweyen, K. Wolf, F. Kaudewitz, pp. 285–302. Berlin: de Gruyter
6. Hensgens, L. A. M., Brakenhoff, J., De Vries, B. F., Sloof, P., Tromp, M. C., et al. 1984. *Nucleic Acids Res.* 12:7327–44
7. Benne, R., Van den Burg, J., Brakenhoff, J., De Vries, B. F., Nederlof, P., et al. 1985. In *Proc. of Int. Symp. on Achievements and Perspectives in Mitochondrial Research.* Amsterdam: Elsevier. In press
8. De la Cruz, V. F., Neckelman, N., Simpson, L. 1984. *J. Biol. Chem.* 259: 15136–47
9. Johnson, B. J. B., Hill, G. C., Donelson, J. E. 1984. *Mol. Biochem. Parasitol.* 13:135–46
10. Payne, M., Rothwell, V., Jasmer, D. P., Feagin, J. E., Stuart, K. 1985. *Mol. Biochem. Parasitol.* 15:159–70
11. Marini, J. C., Effron, P. N., Goodman, T. C., Singleton, C. K., Wells, R. D., et al. 1984. *J. Biol. Chem.* 259:8974–79
12. Ntambi, J. M., Marini, J. C., Bangs, J. D., Hajduk, S. L., Jimenez, H. E., et al. 1984. *Mol. Biochem. Parasitol.* 12:273–86
13. Wu, H. M., Crothers, D. M. 1984. *Nature* 308:509–13
14. Eperon, I. C., Janssen, J. W. G., Hoeijmakers, J. H. J., Borst, P. 1983. *Nucleic Acids Res.* 11:105–25
15. Sloof, P., Van den Burg, J., Voogd, A., Benne, R., Agostinelli, M., et al. 1985. *Nucleic Acids Res.* 13:4171–90
16. De la Cruz, V. F., Lake, J. A., Simpson, A. M., Simpson, L. 1985. *Proc. Natl. Acad. Sci. USA* 82:1401–5
17. De la Cruz, V. F., Simpson, A. M., Lake, J. A., Simpson, L. 1985. *Nucleic Acids Res.* 13:2337–56
18. Opperdoes, F. R., Borst, P. 1977. *FEBS Lett.* 80:360–64
19. Opperdoes, F. R., Baudhuin, P., Coppens, I., De Roe, C., Edwards, S. W., et al. 1984. *J. Cell. Biol.* 98:1178–84
20. Englund, P. T., Hajduk, S. L., Marini, J. C. 1982. *Ann. Rev. Biochem.* 51:695–726
21. Boothroyd, J. C. 1985. *Ann. Rev. Microbiol.* 39:475–502
22. Borst, P. 1983. In *Mobile Genetic Elements,* ed. J. A. Shapiro, pp. 621–59. New York: Academic
23. Borst, P., Cross, G. A. M. 1982. *Cell* 29:291–303
24. Borst, P., Bernards, A., Van der Ploeg, L. H. T., Michels, P. A. M., Liu, A. Y. C., et al. 1983. *Eur. J. Biochem.* 137:383–89
25. Parsons, M., Nelson, R. G., Agabian, N. 1984. *Immunol. Today* 5:43–50
26. Donelson, J. E., Turner, M. J. 1985. *Sci. Am.* 252:32–39
27. De Lange, T. 1985. *Int. Rev. Cytol.* (Suppl.). In press
28. Michels, P. A. M. 1984. In *Oxford Surveys on Eukaryotic Genes,* ed. N. MacLean, 1:145–68. Oxford: Oxford Univ. Press
29. Bernards, A. 1985. *Biochim. Biophys. Acta* 824:1–15
30. Pays, E. 1985. *Prog. Nucl. Acids Res. Mol. Biol.* 32:1–26
31. Pays, E. 1985. *Ann. Inst. Pasteur/Immunol.* 1366:25–39
31a. Donelson, J. E., Rice-Ficht, A. C. 1985. *Microbiol. Rev.* 49:107–25
32. Turner, M. J., Cardoso de Almeida, M. L., Gurnett, A. M., Raper, J., Ward, J. 1985. *Curr. Top. Microbiol. Immunol.* 117:23–55
33. Cross, G. A. M. 1975. *Parasitology* 71:393–417
33a. Lamont, G. S., Tucker, R. S., Cross, G. A. M. 1985. *Parasitology.* In press
34. De Lange, T., Kooter, J. M., Luirink, J., Borst, P. 1985. *EMBO J.* 4:3299–306
35. Michels, P. A. M., Liu, A. Y. C., Bernards, A., Sloof, P., Van der Bijl, M. M. W., et al. 1983. *J. Mol. Biol.* 166:537–56
36. Liu, A. Y. C., Michels, P. A. M., Bernards, A., Borst, P. 1985. *J. Mol. Biol.* 182:383–96
37. Donelson, J. E., Murphy, W. J., Brentano, S. T., Rice-Ficht, A. C., Cain, G. D. 1983. *J. Cell. Biochem.* 23:1–12
38. Myler, P. J., Allison, J., Agabian, N., Stuart, K. 1984. *Cell* 39:203–11
39. Laurent, M., Pays, E., Van der Werf, A., Aerts, D., Magnus, E., et al. 1984. *Nucleic Acids Res.* 12:8319–29
40. Aline, R. Jr., MacDonald, G., Brown, E., Allison, J., Myler, P., et al. 1985. *Nucleic Acids Res.* 13:3161–77
41. De Lange, T., Kooter, J. M., Michels, P. A. M., Borst, P. 1983. *Nucleic Acids Res.* 11:8149–65
42. Pays, E., Van Assel, S., Laurent, M.,

Darville, M., Vervoort, T., et al. 1983. *Cell* 34:371–81
43. Pays, E., Delauw, M. F., Van Assel, S., Laurent, M., Vervoort, T., et al. 1983. *Cell* 35:721–31
44. Laurent, M., Pays, E., Delinte, K., Magnus, E., Van Meirvenne, N. 1984. *Nature* 308:370–73
45. Myler, P., Nelson, R. G., Agabian, N., Stuart, K. 1984. *Nature* 309:282–85
46. Longacre, S., Hibner, U., Raibaud, A., Eisen, H., Baltz, T., et al. 1983. *Mol. Cell. Biol.* 3:399–409
47. Allison, J., Rothwell, V., Newport, G., Agabian, N., Stuart, K. 1984. *Nucleic Acids Res.* 12:9051–66
48. Young, J. R., Turner, M. J., Williams, R. O. 1984. *J. Cell. Biochem.* 24:287–95
48a. Young, J. R., Miller, E. N., Williams, R. O., Turner, M. J. 1983. *Nature* 306:196–98
49. Bernards, A., De Lange, T., Michels, P. A. M., Huisman, M. J., Borst, P. 1984. *Cell* 36:163–70
50. Bernards, A., Kooter, J. M., Borst, P. 1985. *Mol. Cell. Biol.* 5:545–53
50a. Pays, E., Houard, S., Pays, A., Van Assel, S., Dupont, F., et al. 1985. *Cell* 42:821–29
51. Rice-Ficht, A. C., Chen, K. K., Donelson, J. E. 1982. *Nature* 298:676–79
52. Borst, P., Bernards, A., Van der Ploeg, L. H. T., Michels, P. A. M., Liu, A. Y. C., et al. 1983. In *Genetic Rearrangement,* ed. K. F. Chater, C. A. Cullis, D. A. Hopwood, A. A. W. B. Johnston, H. W. Woolhouse, pp. 207–33. London: Croom Helm
53. Aline, R. F., Stuart, K. 1985. *Mol. Biochem. Parasitol.* 16:11–20
54. Horowitz, H., Thorburn, P., Haber, J. E. 1984. *Mol. Cell. Biol.* 4:2509–17
55. Pays, E., Guyaux, M., Aerts, D., Van Meirvenne, N., Steinert, M. 1985. *Nature* 316:562–64
56. Williams, R. O., Young, J. R., Majiwa, P. A. O. 1982. *Nature* 299:417–21
57. Young, J. R., Shah, J. S., Matthyssens, G., Williams, R. O. 1983. *Cell* 32:1149–59
58. Buck, G. A., Longacre, S., Raibaud, A., Hibner, U., Giroud, C., et al. 1984. *Nature* 307:563–66
59. Buck, G. A., Jacquemot, C., Baltz, T., Eisen, H. 1984. *Gene* 32:329–36
60. Michels, P. A. M., Van der Ploeg, L. H. T., Liu, A. Y. C., Borst, P. 1984. *EMBO J.* 3:1345–51
61. Aline, R. F. Jr., Scholler, J. K., Nelson, R. G., Agabian, N., Stuart, K. 1985. *Mol. Biochem. Parasitol.* 17:311–20
62. Myler, P. J., Allen, A. L., Agabian, N.,

Stuart, K. 1985. *Infect. Immun.* 47:684–90
63. Esser, K. M., Schoenbechler, M. J. 1985. *Science* 229:190–93
64. Jerne, N. K. 1985. *EMBO J.* 4:847–52
65. Tetley, L., Vickerman, K. 1985. *J. Cell Sci.* 74:1–19
66. Lenardo, M. J., Rice-Ficht, A. C., Kelly, G., Esser, K. M., Donelson, J. E. 1984. *Proc. Natl. Acad. Sci. USA* 81:6642–46
67. Cornelissen, A. W. C. A., Bakkeren, G. A. M., Barry, J. D., Michels, P. A. M., Borst, P. 1985. *Nucleic Acids Res.* 13:4661–76
67a. Delauw, M. F., Pays, E., Steinert, M., Aerts, D., Van Meirvenne, N., et al. 1985. *EMBO J.* 4:989–93
68. Turner, C. M. R., Barry, J. D., Vickerman, K. 1985. *Parasitology.* In press
69. Olafson, R. W., Clarke, M. W., Kielland, S. L., Pearson, T. W., Barbet, A. F., et al. 1984. *Mol. Biochem. Parasitol.* 12:287–98
70. Lalor, Th. M., Kjeldgaard, M., Shimamoto, G. T., Strickler, J. E., Konigsberg, W. H., et al. 1984. *Proc. Natl. Acad. Sci. USA* 81:998–1002
71. Freymann, D. M., Metcalf, P., Turner, M., Wiley, D. C. 1984. *Nature* 311:167–69
72. Barbet, A. F. 1985. *Mol. Biochem. Parasitol.* 14:175–85
73. Miller, E. N., Allan, L. M., Turner, M. J. 1984. *Mol. Biochem. Parasitol.* 13:67–81
74. Miller, E. N., Allan, L. M., Turner, M. J. 1984. *Mol. Biochem. Parasitol.* 13:309–22
75. Shapiro, S. Z., Pearson, T. W. 1984. In *Parasite Antigens, Receptors and the Immune Response,* ed. T. W. Pearson. New York/Basel: Dekker
76. Baltz, T., Giroud, Ch., Baltz, D., Duvillier, G., Degand, P., et al. 1982. *EMBO J.* 1:1393–98
77. Duvillier, G., Nouvelot, A., Richet, C., Baltz, T., Degand, P. 1983. *Biochem. Biophys. Res. Commun.* 114:119–25
78. Baltz, T., Duvillier, G., Giroud, Ch., Richet, C., Baltz, D., et al. 1983. *FEBS Lett.* 158:174–79
79. Rifkin, M. R., Fairlamb, A. H. 1985. *Mol. Biochem. Parasitol.* 15:245–56
80. Bangs, J. D., Hereld, D., Krakow, J. L., Hart, G. W., Englund, P. T. 1985. *Proc. Natl. Acad. Sci. USA* 82:3207–11
81. Ferguson, M. A. J., Duszenko, M., Lamont, G. S., Overath, P., Cross, G. A. M. 1985. *J. Biol. Chem.* In press
82. Grab, D. J., Webster, P., Verjee, Y.

1984. *Proc. Natl. Acad. Sci. USA* 81: 7703–7
83. Cardoso de Almeida, M. L., Turner, M. J. 1983. *Nature* 302:349–52
84. Jackson, D. G., Voorheis, H. P. 1985. *J. Biol. Chem.* 260:5179–83
85. Ferguson, M. A. J., Cross, G. A. M. 1984. *J. Biol. Chem.* 259:3011–15
86. Ferguson, M. A. J., Haldar, K., Cross, G. A. M. 1985. *J. Biol. Chem.* 260: 4963–68
86a. Ferguson, M. A. J., Low, M. G., Cross, G. A. M. 1985. *J. Biol. Chem.* In press
87. Czichos, J., Ehlers, B., Nonnengaesser, C., Overath, P. 1985. *Cell.* In press
88. Overath, P., Czichos, J., Stock, U., Nonnengaesser, C. 1983. *EMBO J.* 2: 1721–28
89. Parsons, M., Nelson, R. G., Stuart, K., Agabian, N. 1984. *Proc. Natl. Acad. Sci. USA* 81:684–88
90. Delauw, M. F., Pays, E., Steinert, M., Aerts, D., Van Meirvenne, N., et al. 1985. *EMBO J.* 4:989–93
91. Deleted in proof
92. Baltz, T., Baltz, D., Giroud, Ch., Crockett, J. 1985. *EMBO J.* 4:1273–77
93. Duszenko, M., Ferguson, M. A. J., Lamont, G. S., Rifkin, M. R., Cross, G. A. M. 1985. *J. Exp. Med.* 162:1256–63
94. Borst, P., Frasch, A. C. C., Bernards, A., Van der Ploeg, L. H. T., Hoeijmakers, J. H. J., et al. 1981. *Cold Spring Habor Symp. Quant. Biol.* 45:935–43
95. Boothroyd, J. C., Paynter, C. A., Coleman, S. L., Cross, G. A. M. 1982. *J. Mol. Biol.* 157:547–56
96. Bernards, A., Van der Ploeg, L. H. T., Frasch, A. C. C., Borst, P., Boothroyd, J. C., et al. 1981. *Cell* 27:497–505
97. Van der Ploeg, L. H. T., Liu, A. Y. C., Michels, P. A. M., De Lange, T., Borst, P. 1982. *Nucleic Acids Res.* 10:3591–604
98. Boothroyd, J. C., Cross, G. A. M. 1982. *Gene* 20:281–89
99. De Lange, T., Liu, A. Y. C., Van der Ploeg, L. H. T., Borst, P., Tromp, M. C., et al. 1983. *Cell* 34:891–900
100. Nelson, R. G., Parsons, M., Barr, P. J., Stuart, K., Selkirk, M. 1983. *Cell* 34:901–9
101. De Lange, T., Michels, P. A. M., Veerman, H. J. G., Cornelissen, A. W. C. A., Borst, P. 1984. *Nucleic Acids Res.* 12:3777–90
102. Parsons, M., Nelson, R. G., Watkins, K. P., Agabian, N. 1984. *Cell* 38:309–16
103. De Lange, T., Berkvens, T. M., Veerman, H. J. G., Frasch, A. C. C., Barry,

J. D., et al. 1984. *Nucleic Acids Res.* 12:4431–43
104. Nelson, R. G., Parsons, M., Selkirk, M., Newport, G., Barr, P. J., et al. 1984. *Nature* 308:665–67
105. Campbell, D. A., Thornton, D. A., Boothroyd, J. C. 1984. *Nature* 311:350–55
106. Kooter, J. M., De Lange, T., Borst, P. 1984. *EMBO J.* 3:2387–92
107. Milhausen, M., Nelson, R. G., Sather, S., Selkirk, M., Agabian, N. 1984. *Cell* 38:721–29
108. Agabian, N., Nelson, R. G., Parsons, M., Milhausen, M., Sather, S., et al. 1985. In *Genome Rearrangement,* pp. 153–72. Liss
109. Sather, S., Agabian, N. 1985. *Proc. Natl. Acad. Sci. USA.* 82:5695–99
110. Imboden, M., Blum, B., De Lange, T., Braun, R., Seebeck, T. 1985. *J. Mol. Biol.* In press
111. Tschudi, Ch., Young, A. S., Ruben, L., Patton, C. L., Richards, F. F. 1985. *Proc. Natl. Acad. Sci. USA* 82:3998–4002
112. Laird, P. W., Kooter, J. M., Loosbroek, N., Borst, P. 1985. *Nucleic Acids Res.* 13:4253–66
112a. Lenardo, M. J., Dorfman, D. M., Donelson, J. E. 1985. *Mol. Cell. Biol.* 5:2487–90
113. Dorfman, D. M., Donelson, J. E. 1984. *Nucleic Acids Res.* 12:4907–21
113a. Cordingley, J. S. 1985. *Mol. Biochem. Parasitol.* 17:321–30
113b. Lenardo, M. J., Dorfman, D. M., Reddy, L. V., Donelson, J. E. 1985. *Gene* 35:131–41
114. Kooter, J. M., Borst, P. 1984. *Nucleic Acids Res.* 12:9457–72
115. Van der Ploeg, L. H. T., Cornelissen, A. W. C. A., Michels, P. A. M., Borst, P. 1984. *Cell* 39:213–21
116. Guyaux, M., Cornelissen, A. W. C. A., Pays, E., Steinert, M., Borst, P. 1985. *EMBO J.* 4:995–98
117. Kimmel, B. E., Samson, S., Wu, J., Hirschberg, R., Yarbrough, L. R. 1985. *Gene* 35:237–48
118. Landfear, S. M., Wirth, D. F. 1985. *Mol. Biochem. Parasitol.* 15:61–82
119. Gonzalez, A., Lerner, T. J., Huecas, M., Sosa-Pineda, B., Nogueira, N., et al. 1985. *Nucleic Acids Res.* 13:5789–804
120. Van der Ploeg, L. H. T., Schwartz, D. C., Cantor, C. R., Borst, P. 1984. *Cell* 37:77–84
121. Van der Ploeg, L. H. T., Cornelissen, A. W. C. A., Barry, J. D., Borst, P. 1984. *EMBO J.* 3:3109–15
122. Gibson, W. C., Borst, P. 1985. *Mol. Biochem. Parasitol.* In press

123. Bernards, A., Kooter, J. M., Michels, P. A. M., Moberts, R. M. P., Borst, P. 1985. *Gene.* In press

124. Dorfman, D. M., Lenardo, M. J., Reddy, L. V., Van der Ploeg, L. H. T., Donelson, J. E. 1985. *Nucleic Acids Res.* 13:3533–49

125. Gibson, W. C. 1985. *Parasitol. Today.* 1:64–65

126. Sloof, P., Menke, H. H., Caspers, M. P. M., Borst, P. 1983. *Nucleic Acids Res.* 11:3889–3901

127. Rothwell, V., Aline, R. Jr., Parsons, M., Agabian, N., Stuart, K. 1985. *Nature* 313:595–97

128. Gonzalez, A., Prediger, E., Huecas, M. E., Nogueira, N., Lizardi, P. M. 1984. *Proc. Natl. Acad. Sci. USA* 81:3356–60

129. Sloof, P., Bos, J. L., Konings, A. F. J. M., Menke, H. H., Borst, P., et al. 1983. *J. Mol. Biol.* 167:1–21

130. Gibson, W. C., Osinga, K. A., Michels, P. A. M., Borst, P. 1985. *Mol. Biochem. Parasitol.* 16:231–42

131. Borst, P., Bernards, A., Van der Ploeg, L. H. T., Michels, P. A. M., Liu, A. Y. C., et al. 1984. In *Molecular Biology Host-Parasite Interactions,* ed. N. Agabian, H. Eisen, 13:205–17. New York: Liss

132. Van der Ploeg, L. H. T., Cornelissen, A. W. C. A. 1984. *Philos. Trans. R. Soc. London, Ser. B* 307:13–26

133. Van der Ploeg, L. H. T., Liu, A. Y. C., Borst, P. 1984. *Cell* 36:459–68

134. Blackburn, E. H., Challoner, P. B. 1984. *Cell* 36:447–57

135. Liu, A. Y. C., Van der Ploeg, L. H. T., Rijsewijk, F. A. M., Borst, P. 1983. *J. Mol. Biol.* 167:57–75

136. Campbell, D. A., Van Bree, M. P., Boothroyd, J. C. 1984. *Nucleic Acids Res.* 12:2759–74

137. Bernards, A., Michels, P. A. M., Lincke, C. R., Borst, P. 1983. *Nature* 303:592–97

138. Pays, E., Laurent, M., Delinte, K., Van Meirvenne, N., Steinert, M. 1983. *Nucleic Acids Res.* 23:8137–48

139. Cully, D. F., Ip, H. S., Cross, G. A. M. 1985. *Cell* 42:173–82

140. Pays, E., Lheureux, M., Steinert, M. 1982. *Nucleic Acids Res.* 10:3149–63

141. Williams, R. O., Young, J. R., Majiwa, P. A. O. 1979. *Nature* 299:847–49

142. Borst, P. 1986. *Biochem. Int.* In press

143. Cornelissen, A. W. C. A., Johnson, P. J., Kooter, J. M., Van der Ploeg, L. H. T., Borst, P. 1985. *Cell* 41:825–32

144. Hasan, G., Turner, M. J., Cordingley, J. S. 1984. *Cell* 37:333–41

145. Bernards, A., Loosbroek, N., Borst, P. 1984. *Nucleic Acids Res.* 12:4153–70

146. Pays, E., Delauw, M. F., Laurent, M., Steinert, M. 1984. *Nucleic Acids Res.* 12:5235–49

147. Raibaud, A., Gaillard, C., Longacre, S., Hibner, U., Buck, G., et al. 1983. *Proc. Natl. Acad. Sci. USA* 80:4306–10

148. Osinga, K. A., Swinkels, B. W., Gibson, W. C., Borst, P., Veeneman, G. H., et al. 1985. *EMBO J.* 4:3811–17

149. Lenardo, M. J., Dorfman, D. M., Reddy, L. V., Donelson, J. E. 1985. *Gene* 35:131–41

150. Solnick, D. 1985. *Cell* 42:157–64

151. Konarska, M. M., Padgett, R. A., Sharp, P. A. 1985. *Cell* 42:165–71

152. Grabowski, P. J., Seiler, S. R., Sharp, P. A. 1985. *Cell* 42:345–53

153. Frendewey, D., Keller, W. 1985. *Cell* 42:355–67

154. Krug, R. M. 1985. *Cell* 41:651–52

154a. Michiels, F., Muyldermans, S., Hamers, R., Matthyssens, G. 1985. *Gene* 36:263–70

155. Keller, E. B., Noon, W. A. 1984. *Proc. Natl. Acad. Sci. USA* 81:7417–20

156. Kitchin, P. A., Ryley, J. F., Gutteridge, W. E. 1984. *Comp. Biochem. Physiol.* 77:223–31

156a. Earnshaw, D. L., Beebee, T. J. C., Gutteridge, W. E. 1985. *Biochem. Biophys. Res. Commun.* 131:844–48

157. Parsons, M., Smit, J., Nelson, R. G., Stuart, K., Agabian, N. 1984. *Mol. Biochem. Parasitol.* 10:207–16

158. Lenardo, M. J., Brentano, S. T., Donelson, J. E. 1984. *Nucleic Acids Res.* 12:4637–52

159. Massamba, N. N., Williams, R. O. 1984. *Parasitology* 88:55–65

160. Paindavoine, P., Pays, E., Laurent, M., Geltmeyer, Y., Le Ray, D., et al. 1985. *Parasitology* In press

161. Fairlamb, A. H., Blackburn, P., Ulrich, P., Chait, B. T., Cerami, A. 1985. *Science* 227:1485–87

Ann. Rev. Biochem. 1986. 55:733–71

EUKARYOTIC DNA REPLICATION

Judith L. Campbell

Divisions of Biology and Chemistry, California Institute of Technology, Pasadena, California 91125

CONTENTS

PERSPECTIVES

What distinguishes prokaryotic and eukaryotic modes of DNA replication most clearly is that bacterial chromosomes form a single replicon copied from a single initiation point and eukaryotic chromosomes consist of multiple replicons that initiate at multiple points. Thus, eukaryotes have a significant problem in coordinating orderly replication of the genome that bacteria do not have. On the basis of the advances in our knowledge of eukaryotic systems that will be

733

0066-4154/86/0701-0733$02.00

reviewed here, however, it appears that there is a high degree of conservation of mechanism between bacteria and higher cells at the level of initiation within a given replicon. A model of replication initiation summarizing our current understanding of both prokaryotic and eukaryotic systems is shown in Figure 1.

Current evidence supports a role for RNA polymerase in initiation of replication in vivo. There are two stages at which RNA polymerase might function—

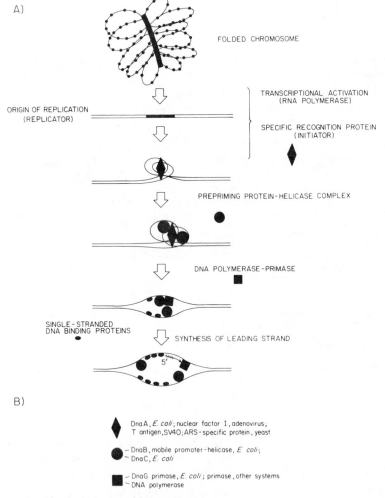

A)

FOLDED CHROMOSOME

ORIGIN OF REPLICATION
(REPLICATOR)

TRANSCRIPTIONAL ACTIVATION
(RNA POLYMERASE)

SPECIFIC RECOGNITION PROTEIN
(INITIATOR)

PREPRIMING PROTEIN-HELICASE COMPLEX

DNA POLYMERASE-PRIMASE

SINGLE-STRANDED
DNA BINDING PROTEINS

SYNTHESIS OF LEADING STRAND

5'

B)

Dna A, *E. coli*; nuclear factor I, adenovirus;
T antigen, SV40; ARS-specific protein, yeast

— DnaB, mobile promoter-helicase, *E. coli*;
— DnaC, *E. coli*

— DnaG primase, *E. coli*; primase, other systems
— DNA polymerase

Figure 1a   Model of initiation of DNA replication.

The definitions of the replicon model are used here (133). The priming steps are as suggested by R. McMacken and Refs. 231 and 322. The initiator step is according to Refs. 69 and 90. The model is explained in the text.

Figure 1b   Known examples of the replication proteins in various systems.

actual synthesis of the RNA primer or transcriptional activation of the template to prepare for the priming event. A role in primer synthesis is disfavored since (*a*) in *Escherichia coli,* an oligonucleotide primase is both necessary and sufficient for primer synthesis, and in simian virus 40, the in vitro replication system does not require an α-amanitin-sensitive RNA polymerase (231, 322, 185, R. McMacken, personal communication); (*b*) primers that have been identified in vivo are more consistent with products of primases than of RNA polymerase both in bacteria and in SV40 (164, 114); and importantly, (*c*) the tight association of DNA polymerase with primase in eukaryotes eliminates the need for an additional protein to carry out de novo synthesis. The currently favored role for RNA polymerase is thus transcriptional activation of the template, as originally proposed for bacteriophage lambda and consistent with the requirement for the polyoma transcriptional enhancer for efficient polyoma virus replication in vivo (63). [Not all replicons use RNA polymerase this way. Bacteriophage T7 has the structural features that would allow primase initiation at the origin of replication, but in vitro results so far suggest that RNA polymerase synthesizes the primer on the leading strand (90a)].

After activation, a nucleoprotein complex is formed between the initiator, a replicon-specific protein, and the replicator, a DNA sequence that usually contains repeated elements important in recognition and binding by the initiator protein. It is inferred that the initiator wraps the DNA into a three-dimensional structure that serves as a nucleus for a series of protein-DNA and protein-protein interactions that lead to priming and DNA synthesis by a DNA polymerase-primase complex. While the details of these interactions are unclear, Figure 1 shows a scheme based on accumulated knowledge of bacterial chromosome replication and likely to be consistent with eukaryotic mechanisms. The prepriming proteins form a mobile promoter recognized by DNA polymerase and primase (231, 322). That one of the *E. coli* prepriming proteins is also a helicase suggests that the prepriming proteins may bind to each strand on both sides of the initiator binding site and thereby facilitate both priming and rapid movement of the newly created fork (R. McMacken, personal communication). Such a mobile priming apparatus has long been recognized for phage T7 and T4 (201), but has only recently been suggested for the chromosome. Finally, SSB, topoisomerases, and RNase Hs facilitate these processes but are not rate limiting.

While it is difficult to generalize, this model is meant to highlight several recent insights. First, the mechanism of priming on the leading strand at an origin of replication may be the same as the initiation of synthesis on the lagging strand at any point along the DNA. Second, DNA polymerase-primase plays a central role in formation of the initiation complex, implying a critical regulatory role for this protein, rather than a simple function in elongation. Third, replication may be somewhat simpler than previously realized, being guided by only five major classes of protein—RNA polymerase, initiator, prepriming proteins, DNA polymerase-primase, and helicase.

Our ability to propose such a detailed model for initiation in eukaryotic systems is due to numerous technical breakthroughs over the last three years. Three model systems have been especially useful: adenovirus, SV40, and yeast. In vitro replication systems for the viral DNAs have allowed identification of cellular replication proteins. In yeast, a more direct method combining modern and traditional genetics has allowed isolation not only of the proteins but also of the DNA sequences and genes involved in cellular replication. Continuing progress is further ensured by a new understanding of DNA polymerase α, the replicative DNA polymerase. These three model systems, DNA polymerase α, and the techniques used to study them will be reviewed here.

ADENOVIRUS

Genome Structure and Replication Mode

The adenovirus genome, Ad DNA-prot, is a linear, 35,000-nucleotide DNA containing an inverted terminal repetition of 102–162 nucleotides, depending on serotype, and with a 55-kd terminal protein (TP) covalently attached to each 5' end (for reviews see 38, 162, 280, 281). Replication initiates at either terminus (180), and is primed by a deoxynucleotide covalently linked to a preterminal protein (pTP) (35, 74, 131, 190, 238, 284, 301). Elongation is continuous and most likely occurs according to a simple strand-displacement mechanism (38, 162). By contrast, the chromosome's replication entails the complexities of internal initiation within a duplex DNA molecule and the existence of leading and lagging strand mechanisms to accomplish sequential replication of both strands. For this reason, adenovirus offers a somewhat limited model for chromosomal replication. Nevertheless, the simplicity of adenovirus replication facilitated the development of the first in vitro replication system for any eukaryotic DNA (36, 37), which has led to the most complete description to date of the proteins and DNA sequences required for any eukaryotic replicon.

In Vitro Replication

PROTEIN REQUIREMENTS FOR IN VITRO REPLICATION   Only five proteins are required to initiate and synthesize a full-length Ad DNA in vitro using the Ad DNA-prot as template (191, 227). The 59-kd DNA binding protein (Ad DBP) (39, 125, 155), an 80-kd preterminal protein (pTP) that serves as primer and is processed later in infection to the 55-kd terminal protein (TP) (35, 74, 131, 132, 189, 190, 238, 284, 301), and the 140-kd adenovirus DNA polymerase (Ad Pol) are the only viral gene products involved (79, 89, 189,

235, 286). All of these proteins are encoded by the viral E2A and E2B regions, and the mRNAs are produced from a common precursor RNA by differential splicing. The transcription of this unit is regulated by early region E1A gene products (280, 286). A requirement for at least two cellular factors, nuclear factors I and II, was demonstrated by attempts to reconstitute the reaction with purified proteins (107, 226–228, 251). Nuclear factor I is a 47-kd protein that binds to a 32-nucleotide DNA sequence near the termini of the Ad DNA (107, 251, 226, 228) and is required for initiation. Nuclear factor II is required for elongation, has type I topoisomerase activity, and can be replaced by HeLa and calf thymus topoisomerase I, but not by *E. coli* topoisomerase I (227).

INITIATION IN VITRO   The in vitro reaction has been separated into partial reactions. An early event in the initiation stage is the formation of a phosphodiester bond between the β-OH of a seryl residue of the 80-kd pTP and 5'-dCMP, the first nucleotide of the nascent strand (32, 39, 59, 190, 238, 284, 301). Formation of this pTP-dCMP complex is catalyzed by the 140-kd Ad Pol, which was first identified by in vitro complementation assay and which purifies as a stoichiometric complex with pTP (74, 132, 189, 235, 286). A DNA template is absolutely required for pTP-dCMP formation. The natural template is presumably Ad DNA-prot, and nuclear factor I stimulates formation of the pTP-dCMP on this template. Ad DBP inhibits; and in the presence of Ad DBP, nuclear factor I becomes essential and ATP stimulates (132, 226). The inhibitory effect of Ad DBP suggests that pTP-dCMP complex formation requires single-stranded DNA sites, or unwinding of the terminus. Supporting this model, Nagata et al (228) show that the pTP-Ad Pol complex binds to single- but not to double-stranded DNA at low ionic strength.

Is the presence of TP on the template essential for normal initiation? Formation of the pTP-dCMP complex can definitely occur on Ad DNA that has been deproteinated or plasmid DNA containing the left end of the Ad 5 DNA, provided the plasmid is linearized to place the Ad 5 terminus at the end of the molecule (40, 105–107, 251, 287, 301, 302, 321). On the other hand, the reaction on templates missing the terminal protein is different from that on the Ad DNA-prot complex. Synthesis is less efficient and the site of initiation on the linear plasmid is less precisely confined to the terminus (105, 106, 301, 321). Furthermore, in addition to the presence of the pTP, AD Pol, Ad DBP, nuclear factor I, and nuclear factor II, a protein from uninfected nuclei called pLP is required (105–107). This protein has no effect on Ad DNA-prot templates.

Single-stranded DNA can also serve as a cofactor for pTP-dCMP complex formation (40, 106, 132, 285, 301). There is no specific DNA sequence requirement and nuclear factor I, which is a sequence-specific DNA binding protein, is dispensable (105, 132, 226, 251).

ELONGATION   Elongation of the 3'-OH of the pTP-dCMP complex by Ad Pol extends to only 25–35% of the length of Ad DNA in the presence of Ad DNA-prot, ATP, Mg^{2+}, the four dNTPs, the pTP, Ad DBP, and nuclear factor I. Synthesis of full-length Ad DNA requires a topoisomerase I (227). Thus four of the five proteins required for Ad DNA replication are involved in initiation and only one additional one seems to be specific for elongation.

DNA SEQUENCE REQUIREMENTS FOR IN VITRO REPLICATION   Investigation of mutant DNA templates has revealed an organization of elements that suggests a possible mechanism of initiation by these proteins. Deletions and base substitutions have been introduced into plasmids containing the appropriate Ad DNA sequences and analysis of the effects of these mutations on pTP-dCMP complex formation and on DNA synthesis indicates that the first 50 base pairs of the Ad genome define the Ad origin of replication. Furthermore, there are two distinct domains within the 50-base-pair region.

The first domain consists of base pairs 1–18 of the Ad 5 genome and is sufficient to support low levels of initiation and elongation (40, 302). Within this region, base pairs 9–18, which are perfectly conserved in all human adenovirus DNAs that have been sequenced, are essential (40, 107, 251, 287, 301, 302, 321). There is also a requirement for the conserved 5' terminal dG : dC base pair. Though point mutations at position 4 from the terminus have no effect on either initiation or elongation, the nucleotides between 1 and 9 may serve as a spacer between the terminus and the conserved 10-base-pair core, since a one-base-pair deletion, four bases removed from the core sequence, inhibits formation of the pTP-dCMP complex (40, 107). Rijnders et al (255) reported that pTP-Ad Pol from Ad 5 binds specifically to restriction fragments containing the region of 9–22 in the Ad 2 DNA, equivalent to 9–18 in Ad 5, and on this basis have suggested that this sequence may play a role in the binding of pTP-Ad Pol to the origin. On the other hand, Nagata et al (228) have reported that pTP-Ad Pol binds only to single-stranded DNA. Thus, there is not yet a direct correlation between defects in binding pTP and defects in synthesis.

While the terminal 20 base pairs of Ad DNA will support both initiation and elongation, these reactions are inefficient in the absence of a second domain, consisting of a 32-base-pair region spanning nucleotides 17–48. This second domain contains a strong binding site for nuclear factor I (K_D=2 × 10^{-11} M) (107, 302, 251), and the bound protein protects the DNA segment between base pairs 19 and 43 from digestion with DNase I. There is a strict correlation between mutants deficient in binding and those deficient in initiation assays (64, 107, 251). These results support the idea that nuclear factor I is a specificity factor for initiation. It is not clear if nuclear factor I plays a role in elongation; nor is it clear if it participates in chromosomal replication.

MODEL OF THE IN VITRO REACTION   One model proposes that pTP-Ad Pol binds to duplex DNA at the terminus, the positioning or binding perhaps facilitated by the 55-kd TP, since in vitro synthesis on templates lacking TP is less efficient than on templates containing the protein (106). Nuclear factor I then binds and facilitates unwinding of the DNA, in conjunction with the 55-kd TP and ATP. Ad DBP next binds to and stabilizes the single-stranded DNA. Initiation follows and the action of nuclear factor II allows elongation. A second model, though it does not take into consideration the role of terminal protein, is similar in that it proposes that nuclear factor I acts as a recognition factor that binds its recognition sequence and thereby facilitates unwinding of the adjacent DNA. Initiation can then occur on the single-stranded DNA (251). Alternatively, the binding of nuclear factor I could directly facilitate binding of the pTP and polymerase by protein-protein interaction (251).

IMPLICATIONS   The adenovirus system, of course, may serve as a model for other viral systems or telomere replication where protein priming has been proposed to occur (93, 201, 252). A potentially more important result with respect to chromosomal replication is the unexpected finding that two of the five proteins required for viral replication are encoded by the host. In particular, nuclear factor I is highly conserved among eukaryotes, suggesting an important cellular function. Sequences that bind nuclear factor I (FIB sites) have been cloned, sequenced, and been shown to replace the adenovirus nuclear factor I site in supporting in vitro replication when present at the termini of linear templates (2, 100, 102). FIB sites have also been identified in the regulatory region of several viral and cellular genes, however (see e.g. Ref. 273). Thus it is not clear if they are involved in transcription or DNA replication.

SV40 DNA REPLICATION

Our detailed understanding of both the initiator (T antigen) and the replicator of the SV40 (simian virus 40) replicon combined with the recent development of efficient in vitro replication systems make this virus the most tractable system currently available for studying the enzymology of initiation at internal sites within DNA molecules. The structure, genetic organization, gene expression, and DNA replication of SV40 have been well studied and well reviewed (38, 57). Thus, work before 1982 will be discussed here only when needed for context.

The genome of SV40 is a circular duplex molecule of about 5243 base pairs, whose entire nucleotide sequence has been determined. The DNA is packaged in a chromatin structure similar to the chromosome, although 10–25% of the molecules are histone-free around the origin of replication. Replication begins at a unique site and proceeds bidirectionally. The boundaries of the origin of replication of SV40, which includes about 65 base pairs, have been deduced from studies of deletion mutants, evolutionary variants, and point mutations. Termination occurs when the two forks collide, without apparent sequence specificity. Completion of DNA synthesis and separation of two daughter molecules can occur by two mechanisms, different only in the extent to which the parental DNA strands are unwound before the two replication forks meet (8, 296, 333).

The Role of T Antigen in Initiation

Temperature-sensitive mutants mapping in the A gene fail to initiate DNA replication at the nonpermissive temperature. These mutations affect the SV40 T antigen, a 708-amino-acid protein which is the only viral gene product required for DNA replication (for an excellent, critical review see 254). T antigen is also required for numerous other functions, including initiation and maintenance of transformation, regulation of gene expression, stimulation of cellular DNA synthesis, stimulation of rDNA transcription, and the ability to allow adenovirus type 2 to grow in monkey cells, and possibly plays a role in the assembly of infectious virions.

An important advance was the purification of a biologically active, structurally related hybrid protein from the adenovirus SV40 hybrid, AD2+D2, called D2 T antigen, because this allowed a detailed analysis of the biochemical activities associated with T antigen and ultimately the correlation of these with the biological activities. T antigen is a multifunctional protein. It binds specifically to SV40 DNA or chromatin at the origin of DNA replication and both genetic and biochemical analyses suggest that initiation of replication involves a direct interaction of T antigen with the origin of replication (272, 340). In addition to DNA binding activity, it contains a DNA-dependent ATPase activity (94, 312), a nucleotide-binding activity (48), and sequences that target it to the nucleus (153, 178, 279), and it can be modified posttranslationally by adenylation (23), phosphorylation (15, 16), acylation, and ADP-ribosylation. T antigen can oligomerize (22, 225, 292), and T antigen binds to a cellular protein, p53, whose function is unknown (see citations in 254).

The study of T antigen constitutes an example of the most extensive use to date of in vitro mutagenesis combined with genetic and biochemical analysis carried out on a single replication protein (47, 95, 153, 154, 183, 197–199, 239, 248, 249, 290, 293, 314, 315, 340). The various strategies and their relative strengths provide a paradigm for future studies on similar proteins.

(I)DELETION ANALYSIS   Truncated proteins have been produced by tryptic digestion (222), in SV40-adenovirus hybrid constructions (248), or by deletion mutants created by in vitro mutagenesis (239). Analysis of these proteins shows that T antigen consists of multiple domains with different activities. The DNA binding domain lies in the N-terminal region between amino acids 83 and 214. A more C-terminal portion contains the ATPase, though the C-terminal 49 amino acids can be deleted (47). Furthermore, the DNA binding domain is distinct from that part of the protein required for stimulation of cell DNA synthesis and rDNA transcription (95, 277).

(II)SECOND SITE SUPPRESSORS   Studies of deletion mutants have the drawback that deletions may affect the overall folding of the protein rather than specific amino acids at an active site. A more powerful approach has been to isolate second site revertants that suppress cis-acting mutations in the origin of replication (272). Two such suppressors are mutations at amino acids 166 and 157 of T antigen, lying within the DNA binding region identified by deletion analysis (271). This strategy for identifying proteins that interact functionally with specific regulatory sequences should find wide applicability in systems like yeast, where the genes identified by the second site mutations can be cloned easily.

(III)POINT MUTANTS THAT SEPARATE REPLICATION AND TRANSFORMATION DOMAINS   Replication-defective and transformation-competent mutants have been selected after UV mutagenesis of cells containing SV40, and the DNAs have been cloned into plasmids (197, 198). Five mutant T antigens, the C series, have been particularly useful. Two mutants (amino acids 153 and 204) are defective in binding. One mutant (amino acid 522) causes defects in the ATPase. Two other mutants (amino acids 512 and 224) are neither deficient in ATPase nor in binding and yet are replication defective, suggesting that binding and ATPase are necessary but not sufficient for replication (198). Recently, the genes for these mutant proteins have been incorporated into an adenovirus vector that is capable of producing large amounts of T antigen when infected into human 293 cells. Since these cells are permissive to SV40, they can be expected to carry out essential posttranslational modifications and produce useful proteins for study (282). Immunoaffinity purification protocols that yield biologically active T antigen provide milligram quantities of T antigen in a single day from these strains, which will allow more direct physicochemical studies of T antigen function (68, 274).

Second site suppressors of mutations within the origin of replication (ori mutations), replication-defective–transformation-competent mutants, and evolutionary variants affecting binding (SV80) were detected only between the N terminus and amino acid 214 and from amino acid 566 to 660. The central portion of the molecule, where tsA mutants map, gave rise to no such mutants, suggesting this region may contain functions required for both replication and transformation or for the overall stability of the protein.

While these findings establish convincingly that T antigen consists of multiple domains and that binding and ATPase are necessary for replication, we have no molecular picture of how T antigen actually activates the origin. Insight into this mechanism should come from reconstitution by purified proteins of the reactions that occur in the cell-free systems described below, and the mutant collections should be invaluable aids in such experiments.

DNA Sequences Required for Replication

Another attempt to gain insight into the activation of the origin of replication before in vitro systems were available was characterization of the DNA sequences with which T antigen interacts. Methylation protection and footprinting experiments have defined three distinct T antigen binding sites overlapping the SV40 origin of replication (see 254). The highest affinity site, site I, spans about 30 base pairs (5185–5215) (Figure 2) (56, 144, 307, 308). Binding site II, a region of about 45 base pairs, starts approximately from the boundary of site I at base pair 5215 and extends through base pair 15 (56, 144, 307, 308). This region contains a 27-base-pair palindrome, the early RNA start sites, and borders on a 17-bp (15–31) AT-rich region containing the early promoter "TATA" sequence. (The region 5209–30 has been defined as approximating the functional origin of replication in vivo). The weakest site, site III, falls on the late transcription side of the origin. The 21-base-pair repeats that bind transcription factor SP1 and the 72-base-pair repeats of the transcriptional enhancer are also found here.

The interaction of wild-type (or D2) T antigen with these sites has been investigated using a variety of assays that include filter binding, methylation protection, protection from DNase digestion, immunoprecipitation, DNase footprinting, and alkylation interference (49, 56, 144, 145, 184, 209, 306–309, 340; see also 29 and 254). Alkylation interference experiments limit areas of close protein contact to a 15-base-pair subset of binding site I and a site of similar size in site II, and demonstrate the importance of a penta or hexanucleotide sequence (G)GCCTC, that is repeated in sites I and II (145). A slightly

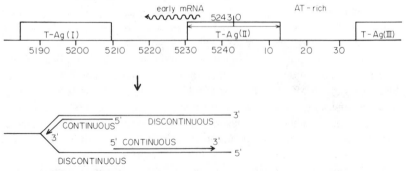

Figure 2   The replication origin region of SV40.

different view of the contact sites that leads to similar conclusions about the functional sequences has been reported (56, 184, 306). Since chemical modification protocols may themselves affect binding, either by steric hindrance, distortion of the template, or charge alteration, further physiochemical analysis is necessary to establish actual contacts.

The interaction of T antigen with binding site II has been shown to be central to the replication function. As mentioned above, the Nathans group demonstrated that site II point mutations constructed by site-directed mutagenesis resulted in cold-sensitive defects in viral DNA synthesis and that second site revertants of the mutations resulted in the production of altered T antigens with relaxed specificity for origin sequences (199, 271, 272).

A complementary approach has been to compare the in vivo effects of cis-acting ori mutations on viral DNA replication or replication of plasmids containing the SV40 origin in COS cells, that constitutively express T antigen, with the ability of D2 T and wild-type T antigen to bind in vitro (49, 82, 144, 184, 209, 307–309). In vivo, the distal third of binding site I can be removed without affecting origin function. However, mutations that completely remove site I (144), that remove both site I and the sequences between site I and II (5205–5215), and that separate sites I and II by an inserted bacterial sequence (65 bp), result in defective replication (144, 282). Small insertions and deletions in site II are defective.

In vitro, T antigen can bind to site II in the absence of site I; that is, deletions up to 5210 still bind T antigen (65, 66, 144, 184, 307–309). Such deletions are defective in replication, demonstrating that T antigen binding to site II is not sufficient for replication. (The previously held misconception that site I binding was required for binding to site II has been shown to be due to the fact that site II/T antigen complexes are not retained on nitrocellulose filters, either because the protein denatures and releases the DNA upon adsorption to nitrocellulose or because not enough protein binds to cause retention of DNA.) A 33-bp substitution between 5220 and 9, within site II, eliminates binding. A single-base-pair insertion in the center of the 27-base-pair palindrome that eliminates replication also drastically reduces binding (49).

However, there are sequences in site II that are essential for replication but not essential for T antigen binding in vitro. This could either be due to the binding to the mutant being nonproductive or suggest that additional types of signals essential for replication are found here (49, 144, 307). First, a four-base-pair deletion from 5239 to 5242 and an eight-base-pair substitution from 5239 to 5246 do not affect binding. This is interesting since the nucleotides contacted by T antigen appear to be 5232 to 5237 and 5243 to 5 (144). These deletions destroy the symmetry of the 27-base-pair palindrome and alter the spacing between hexameric sequences, though they do not eliminate these sequences and in one case create a new one. A possible role for the sequences between 5239 and 5246 is suggested by the existence of major initiation sites for RNA primers between 5239 and 12 (114, 115). Second, a deletion of nearly one half of the palindrome on the late transcription side still binds T antigen normally (49, 144, 307), consistent with the existence of recognition or contact sites in both halves of site II. Third, mutants in the sequence from 5205 to 5215, between sites I and II, are defective in replication but are binding competent. A possible role for this region in replication is to direct the transition from leading to lagging strand synthesis, which Hay & DePamphilis (114) have shown maps in the region 5210 to 5211. A role for these sequences independent of T antigen interaction can be further investigated by a genetic approach based on suppressor isolation (272), or carrying out replication in a defined replication system.

In Vitro Replication Systems

Several laboratories have now reported cell-free systems for the T antigen–dependent replication of SV40 DNA. Ariga & Sugano (8) demonstrated T antigen–dependent, semiconservative replication of exogenously added SV40 DNA by a mixture of soluble extracts of HeLa cell nuclei and the cytoplasm of SV40-infected COS1 cells (6–8). A much more efficient system was subsequently described by Li & Kelly (185), who demonstrated that cytoplasmic extracts of SV40-infected or uninfected monkey cells supplemented with purified SV40 T antigen could carry out all stages of SV40 DNA replication on exogenously added DNA. Synthesis after four hours amounts to 2000–4000 molecules replicated per cell equivalent of extract. Other systems based on that of Li & Kelly (185) have been described using human cells grown in suspension (186, 282, 283, 342).

By numerous criteria, synthesis in vitro corresponds precisely to DNA replication in vivo. A functional SV40 origin of DNA replication is required, and synthesis is entirely dependent on T antigen. Since almost all the synthesis in vitro is dependent on SV40 DNA and T antigen, there can be very little DNA repair synthesis in these systems. Synthesis occurs only in extracts of cells normally permissive to SV40 in vivo (simian, COS1, human HeLa and 293, Chinese hamster ovary), but not in those that are nonpermissive, such as BALB/3T3 mice (283). Initiation occurs in vitro since the product is not covalently attached to the template and synthesis begins at the origin of replication and proceeds bidirectionally primarily through theta form intermediates (6, 186, 283, 342). Replication is semiconservative and continues for several rounds. The products are similar to the replication termination intermediates observed by Sundin & Varshavsky (296) in vivo (186). Thus, all stages of replication, initiation, elongation, and termination have been reproduced.

Proteins Required for SV40 DNA Replication In Vitro

INITIATION   This system is truly exciting. Since the initiator, T antigen, is available in large quantity, it should soon be possible to reproduce the entire initiation process using purified proteins. This will be particularly interesting because the proteins required besides T antigen are almost certainly the same as those that act on the chromosome. The only limitation at the moment will be the difficulty in obtaining mutants affecting cellular proteins to confirm their roles.

While it is not yet clear how initiation occurs, it seems reasonable that T antigen somehow activates the origin of replication by altering its structure such that polymerase-primase can initiate synthesis. There is no evidence for a role for any of the cellular RNA polymerases. The reaction is dependent on ATP and Mg^{2+} but independent of the other three rNTPs. Furthermore, synthesis is not inhibited by α-amanitin, an inhibitor of RNA polymerases I, II, and III (8, 185). Synthesis is sensitive to aphidicolin, an inhibitor of polymerase-primase. Thus, a well-characterized DNA polymerase alpha will be helpful to successful resolution and reconstitution (8, 185, 283, 342).

T ANTIGEN   The in vitro system should prove a suitable means not only for defining what proteins interact with T antigen, but also for answering questions raised by previous studies of whether there are different subspecies of T antigen in the cell (22, 262, 263, 279, 292) and what the role of the multiple posttranslational modifications associated with T antigen is. Stillman et al (282) have studied the five T antigens encoded by the C series mutants described above (198). As in vivo, all are defective in in vitro replication. The new observation is that all five proteins reduce the amount of DNA replication in vitro even in the presence of wild-type T antigen. This dominance suggests that an additional function of T antigen besides binding and ATPase activity may be required for the initiation of DNA replication and that this additional activity is either interaction between multiple molecules of T antigen or interaction with other proteins involved in initiation (95, 197, 198, 282). T antigen has been shown to oligomerize (225) and to bind to a cellular protein, p53, and probably interacts with other proteins in replication.

TERMINATION   Replication in cytoplasmic extracts yields monomeric, relaxed products, and addition of nuclear extracts is required to introduce negative supercoiling (283). Since neither purified topoisomerase I nor topoisomerase II can supercoil the DNA when added to the reaction, additional factors must be present in the nuclei. Replication actually consists of three steps, disassembly of chromatin, DNA synthesis, and reassembly; and these results may indicate that reassembly is occurring in vitro.

DNA Sequences Required

It has been proposed that the SV40 origin may have signals similar to cellular initiation sites since, except for T antigen, all the same proteins are used in viral and cellular initiation. Furthermore, T antigen also has effects on cellular DNA synthesis. Chromosomal sequences isolated by homology with the SV40 origin, however, seem more likely to have effects on transcription than on replication (52, 208, 250). In any case, it would obviously be very useful if these extracts could replicate DNA containing chromosomal replicators as well as SV40 DNA. One study suggests that Blur8 sequences, the human Alu family sequences with some homology to the SV40 origin, cloned mouse DNAs that replicate as plasmids in COS monkey cells, and an intron sequence from a human *RAS* gene supports replication in vitro (7; H. Ariga, personal communication). However, this has not been reproduced in the system of Li & Kelly (186), and to date there is convincing evidence only for replication of SV40 and the genomes of the closely related papovaviruses BKV and JCV in vitro (186).

Minimal Origins Used In Vitro

Because the SV40 origin region contains signals important for both transcription and replication, it has sometimes been difficult to decide whether certain sequences affect replication, transcription, or both. For instance, Bergsma et al (17) conclude that the 21-bp repeats flanking T antigen site II, which serve as transcription factor SP1 binding sites, are required for replication. However, Myers & Tjian (see 224, 254) conclude they are only important for transcription. Replication in vitro, which does not require transcription by RNA polymerase II, can be used to discriminate. The origin sequences required for initiation have been systematically investigated by Stillman et al (282), and for the most part, the effect of the mutations is identical to that seen in vivo. Importantly, the 17-base-pair AT-rich sequence between nucleotides 15 and 31 plays a crucial role in in vitro initiation. It may or may not be relevant that this region also contains the "TATA" box for early gene transcription. On the other hand, mutations that remove either the early promoter elements such as the SP1 binding sites or the SV40 enhancer are not deficient in vitro.

Further study of the effect of the SP1 binding sites and the enhancer is of interest, however. Recent quantitative studies now demonstrate clearly that the polyoma transcriptional enhancer is required for polyoma DNA replication (62, 63), and that the presence of either the SV40 21-base-pair repeats or the transcriptional enhancer markedly increases SV40 DNA replication in vivo (186a). It is not clear whether these are required for transcriptional activation of the origin or if binding of specific proteins directly affects replication.

Regulation

While the SV40 in vitro replication system clearly represents a powerful tool for the investigation of the enzymatic events involved in establishing a replication fork, it is not clear to what extent the virus actually resembles a chromosomal replicon. SV40, by using T antigen as an initiator, bypasses the chromosomal initiation machinery. It is possible that the viral mechanism is similar to the chromosomal one, and that only the identity of the protein and specific DNA sequences differ. Alternatively, since the regulation of SV40 replication is different from that of the chromosome, in that more than one round of initiation can occur per cell cycle, it has been proposed that the SV40 system is a means of circumventing normal cell cycle regulation, rather than a model of normal regulation. To support the first possibility, the argument has been made that T antigen stimulates cellular functions such as DNA synthesis and rDNA transcription and that therefore T antigen may bind to chromosomal sequences. However, the domains of T antigen that stimulate cellular synthesis can be separated from those that stimulate viral replication, and therefore the stimulation is probably indirect. Until a chromosomal replicon is defined it will remain uncertain how far the analogy can be drawn. In the meantime, important information will be obtained anyway.

YEAST

Yeast is a useful model because its chromosomal organization and replication are typical of higher eukaryotes. Its rapid growth allows economical isolation of large quantities of replication proteins; its classical genetic system is powerful, and the yeast transformation system allows a union between recombinant DNA technology and genetics. The ability to use proteins to isolate genes, to mutagenize genes, and to reintroduce them into the chromosome has provided a new approach to eukaryotic cell biology. Yeast has been uniquely successful in contributing to our knowledge of eukaryotic DNA replication in at least two ways: (*a*) isolation of the DNA sequences essential for chromosomal transmission: centromeres, telomeres, and origins of replication; (*b*) isolation of genes and mutants that have defined several of the proteins involved in replication. Yeast is the only system currently available in which it is possible to study directly the molecular basis for the integration of S phase and the events of chromosomal replication into the eukaryotic cell cycle. Yeast DNA replication has been reviewed twice recently (31, 230).

DNA Sequences Required

The yeast genome is 1.35×10^4 kb in size and is divided into 17 linear chromosomes (150–2500 kb), mitochondrial DNA (15%), and 2 μm circle plasmid DNA (4%). The yeast chromosomes, though small, are organized like those of other eukaryotes, and the nuclear DNA is packaged in a typical chromatin structure. While the small size of the chromosomes makes classical cytology difficult, it has two advantages. First, orthogonal pulsed-field gel electrophoresis (OFAGE) can separate all of the yeast chromosomes (32, 266), facilitating mapping of cloned genes and the analysis of the behavior of whole chromosomes in replication, repair, or recombination. Second, the relatively high concentration of origins of replication (ARS), centromeres (CEN), and telomeres has allowed isolation of all these sequences. These probes should aid in definition of the roles of these elements in chromosome transmission in yeast and perhaps be useful for isolation of similar sequences from higher cell chromosomes. For instance, the ability to stabilize certain plasmids in yeast has been used to identify *Caenorhabditis elegans* sequences that may function in chromosome segregation in that organism (288).

CENTROMERES AND TELOMERES   Since little is yet known about the special features of the replication of centromeres and telomeres and since these elements have recently been reviewed (19, 122a), comments here are limited to two special aspects of telomere replication. The ends of eukaryotic chromosomes, called telomeres, are organized to provide stability and to allow complete replication of both the leading (3') and lagging (5') strands. The telomeres of yeast chromosomes end in sequences 100–300 base pairs in length, that consist of repeats of $C_{1-3}A$ units and that contain several single-strand breaks. Located just internal to the repeats there are a variable number of copies (0–4) of a 6.7-kb telomere-associated sequence called Y', another $C_{1-3}A$ region, and then a region called X. Duplication of the $C_{1-3}A$ repeats is an integral part of replication and though numerous models exist explaining how this occurs without loss of information from the ends, the mechanism is ill-defined. Several observations suggest that both recombination and nontemplated addition of sequences by polymerization are part of the process. That recombinational mechanisms may participate is suggested by the fact that a circular molecule containing telomere sequences is resolved to a linear one with two telomeres in yeast (71, 298, 325). That addition mechanisms may play a role in telomere replication is suggested by two findings. First, telomeres of individual chromosomes have variable numbers of $C_{1-3}A$ repeats from strain to strain (325–327). Second, $C_{1-3}A$ repeat units are added in an apparently non–template directed manner to the ends of *Tetrahymena* (269) and *Oxytricha* (245) DNAs during their replication as linear plasmids in yeast. One possible mechanism for such addition is a reaction like the reiterative, autocatalytic synthesis carried out by DNA polymerase I of *E. coli*. The DNA polymerase-primase complex of DNA polymerase I of yeast synthesizes repetitive units (see below) of six to ten nucleotides up to ~100 bases in length. Mutants that affect the length of telomeres have recently been isolated in yeast (32a), and it will be interesting to see if these are deficient in any of the yeast polymerases or if they define a new terminal transferase as proposed in other systems (99b).

ORIGINS OF REPLICATION   In eukaryotic cells, the multiple linear chromosomes are made up of multiple replicons, whose organization is conserved in all eukaryotes, including yeast (see 31 or 230 for complete references). The center-to-center distance of replication bubbles is about 12 μm or 36 kb, making an estimated 400 initiation sites per haploid genome. Replication proceeds bidirectionally at a rate of fork movement of 2–6 μm per minute, about 40 times slower than in *E. coli*. Termination appears to occur by fusion of replicons when two forks meet; there is no evidence for specific termination sequences. Replication occurs during only part of the cell cycle, S phase. Adjacent replicons initiate at about the same time, forming larger replication units. In higher cells, replicons seem to be activated in a specific temporal order, some initiating early in S and some late (30, 31, 96, 110, 230). In yeast, it has been difficult to demonstrate this unequivocally, but some evidence suggests that whole chromosomes may form temporal units. Finally, initiation occurs only once per S phase in any given replicon, and reinitiation is rarely observed. Reinitiation has been observed, however, after inhibition of replication with hydroxyurea (200) or freezing of forks with araCTP (344). Reinitiation has also been documented in the chorion genes of *Drosophila melanogaster* during oogenesis (54, 233).

Yeast chromosomal sequences that may be involved in regulating this pattern of replicon usage have been isolated. DNA segments that mediate autonomous replication of colinear DNA as plasmids in yeast have been called *ARS*s, autonomously replicating sequences (291). *ARS*s are thought to be origins of replication because: (*a*) origins have analogous properties in *E. coli*, (*b*) there are 400 *ARS*s per cell, corresponding to the average replicon size, (*c*) in contrast to mitochondria, they replicate under the control of genes that regulate the cell cycle and only during S phase, (*d*) they replicate only once per cell cycle, and (*e*) synthesis in several in vitro systems initiates specifically at an *ARS* sequence (see 31). Despite this, some doubt remains as to the role of *ARS*s within chromosomes because it has not been possible to precisely map initiation sites in chromosomes. There is good evidence of replication bubbles centered at the 2 μm *ARS* sequence, however, which is structurally and functionally equivalent to chromosomal *ARS*s. Furthermore, initiation does occur within the nontranscribed spacer of the rDNA repeats and one such repeat contains an *ARS*. While initiation in adjacent repeats has been observed, not all repeats are simultaneously activated (259). Similar observations were made in the rDNA repeat of sea urchin (20). How can we prove the elusive congruence between

*ARS*s and origins of replication? One approach is to identify the proteins that interact with the *ARS* (see below).

The structure of *ARS*s has been studied by analysis of the effects of mutations on the stability of *ARS*-containing plasmids (31, 34, 122, 161, 170, 278). *ARS1* and the *ARS* at the *HO* gene have been studied in the most detail, and the organization of *ARS1* is summarized in Figure 3 (31, 34, 230, 278). A current proposal is that domain A, a 14-base-pair sequence containing an 11-base-pair consensus (A/TTTTATPuTTTA/T) found at all *ARS* sequences and essential for *ARS* function, is a recognition site for a protein like T antigen that activates the neighboring sequences for replication. A 19-bp sequence containing the *ARS1* consensus allows autonomous replication (278). The role of domain B, which consists of more or less unique sequence, is unknown. Deletions of domain B have large effects on plasmid stability, though inefficient replication is possible in its absence (278). In assays that measure plasmid segregation, where 1:1 segregation is normal, 1:0 segregation is a replication failure, and 2:0 segregation is a result of nondisjunction, different *ARS*s show values of 1:0 segregation varying over 10-fold (122, 170; C. Newlon, personal communication). Most origins in other system are characterized by having repeated elements important for function and there are several 7–9-base-pair matches of the consensus sequence in domains B and C. Newlon has noted a correlation between the number of repeats and the strength of the *ARS*. Domain C deletions have only minor effects and are seen only in the absence of Domain B or in the presence of a centromere.

A role for *ARS*s in gene expression in addition to replication has been postulated. The *HMRE* sequence, the transcriptional "silencer" upon which the *SIR* system acts to keep *HMR* silent, has been localized to a 260-base-pair region containing *ARS* activity (1, 25, 78, 214). At the histone H2A-H2B locus there is an *ARS* at the 3'-end of the H2B gene that appeared to be involved in the

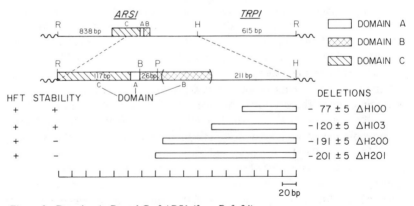

Figure 3   Domains A, B, and C of ARS1 (from Ref. 34).

S phase regulation of H2A and H2B gene expression (234). Recently, it has been shown that the *ARS* is not involved in differential cell cycle regulation, as previously thought, but that removal of the *ARS* leads instead to a lowering of the constitutive levels of H2A and H2B (M. Osley, L. Hereford, personal communication). Since transcriptional regulatory sequences are usually small, further refinement of the mapping will be necessary to confirm that the regulatory sequences and the *ARS* coincide.

DO EUKARYOTIC CHROMOSOMES HAVE SPECIFIC ORIGINS OF REPLICATION IN EACH REPLICON? Evidence, in addition to yeast *ARS*s, certainly suggests initiation is not completely random. First, all viral, plasmid, and higher cell mitochondrial DNAs have origins (57). Second, analysis of amplified genes suggests specific origins of replication (116, 213). Third, random DNAs when injected into mouse embryos do not replicate (341). Only when the polyoma or SV40 origins and corresponding T antigen genes are present on injected DNAs has replication been observed, suggesting the requirement for specific origins in mammalian embryos (341).

Assuming specific origins exist, it is unlikely that they are all the same. Kriegstein & Hogness (173) have shown that the pattern of initiation can change at different stages of development. Initiation-site spacing observed in *Drosophila* embryos, where rates of synthesis are very high, was only ~8 kb, while spacing in more slowly dividing tissue culture cells was ~40 kb. Furthermore, random injected DNA sequences replicate under cell cycle control in *Xenopus* and sea urchin eggs (88, 111, 210–212, but see also 41, 123, 331). Initiation proteins may be stored in the eggs or embryos and only after multiple cleavages is the concentration reduced such that only high-affinity sites for the initiator function (179). Alternatively, different classes of origins may be activated at different times by specific proteins.

Rates of replication also change depending on nutritional environment, and it has been proposed that this regulation depends on the number of initiations within the genome (73, 110). This again suggests that there may be different classes of origins used under different circumstances. In yeast, however, there is good evidence that the rate of fork movement also contributes to controlling the length of S phase (256, 257).

Finally, certain SV40 viruses containing mutations in T antigen replicate efficiently but do not initiate synthesis at the primary origin (202). This is similar to bacteriophage (299) and the *E. coli* chromosome itself, where specific primary origins of replication have been thoroughly demonstrated, but secondary origins are often used. In sum, none of these observations of flexible or aberrant initiation rules out specific origins of replication.

One approach to isolating origins of replication of higher organisms is to select for sequences that can replicate autonomously in yeast. While sequences from other organisms function as *ARS*s in yeast, the same sequences have not been shown to function efficiently in the organism from which the DNA was derived (9, 221, 258, 289). Other approaches are being tried but have not yet been successful (115, 213), being thwarted by the inability to demonstrate plasmid replication in higher eukaryotes. Replacement of sequences on bovine papilloma virus (BPV) (193) or Epstein-Barr virus (EBV) (355, 356), which can replicate as plasmids, may be useful, if higher cell centromeres cannot be isolated.

2 μm Circle DNA Replication

In addition to the chromosomes, yeast contains 50–100 copies of a 2 μm circle plasmid (see 26, 31, 230). The 2 μm *ARS* is similar to chromosomal *ARS*s both structurally and functionally, and under steady-state growth 2 μm plasmid replicates under the same regulation as a chromosomal replicon—the majority of plasmids replicate only during S phase and only once per cell cycle. Because the coding capacity of the plasmid is small, it is likely that the same proteins involved in chromosomal replication also are required for the plasmid. In these respects, 2 μm is a good model for a chromosomal replicon. However, certain adaptations of the plasmid are not characteristic of most chromosomal replicons (134). Under certain conditions the copy number of 2 μm circle can fall below the normal level. Under these circumstances, the plasmid is able to overcome the normal cell cycle regulation and to amplify its DNA to a stable high copy number in one cell cycle. This amplification requires the 2 μm circle–encoded site-specific recombination system (134; J. Broach, personal communication), which gives rise to multiple replication forks through recombination, without the need for multiple initiations. (This model could also account for higher cell gene amplification, if an extrachromosomal circular intermediate is proposed).

The segregation machinery of 2 μm plasmids is also different from the chromosome. 2 μm plasmids are more stable than *ARS* chimeric plasmids. This stability is mediated by one cis-acting plasmid locus, *REP3*, and two trans-acting plasmid loci, *REP1* and *REP2*. The *REP* system is now thought to function primarily in segregation, since *REP3* enhances stability of *ARS1* plasmids without changing copy number (163).

Strategies for Identifying Proteins Required for Replication

Significant progress has been made in isolating the proteins of the yeast replication apparatus and the genes that encode them.

IN VITRO REPLICATION   A number of extracts capable of replicating added 2 μm plasmid DNA or chimeric plasmids containing the chromosomal *ARS* sequences, *ARS1* or *ARS2*, have been described (see 230). While duplex plasmid DNAs provide models for all stages of replication, one of these extracts also efficiently copies single-stranded circular DNAs, providing a model similar to the ØX174 system of *E. coli* for lagging strand synthesis (33).

The replication of duplex plasmids requires DNA, ATP, and Mg^{2+}. There is less of a requirement for the dNTPs or other rNTPs. Aphidicolin and antibody against DNA polymerase I inhibit the reaction, implicating DNA polymerase I. α-amanitin does not inhibit. Antibody to yeast SSB-1, a single-stranded nucleic acid binding protein, also inhibits the reaction. The replication of single-stranded DNA, which is much more efficient, requires all four rNTPs, but is resistant to α-amanitin, suggesting that polymerase-primase is catalyzing the reaction.

Initiation has been shown to occur specifically at the *ARS* region in both 2 μm plasmids and *ARS1* plasmids, but to occur randomly on templates that do not contain a functional *ARS* (33, 136). Electron microscopic examination of plasmid DNA isolated from the extracts reveals dense protein "knobs" located specifically at the *ARS* regions (135–137, 139; see also 295). These do not appear to be transcription complexes, since α-amanitin does not inhibit their formation. They may be replication initiation complexes, perhaps similar to those observed in the oriC (90) system or with the lambda O protein (69) or T antigen (225).

Although these systems initiate replication in vitro, they have not yet proved very useful in dissecting the replication apparatus. The first problem is the inefficiency of initiation and the large contribution of repair synthesis observed. In one system, much of the synthesis does not represent initiation, but is due to preexisting "primers" in the exogenous template (5, 149, 167). It is likely that this inefficiency is because the initiator proteins for yeast replication are present in limiting quantities, as is the dnaA protein of *E. coli*. Thus there is great incentive to identify the yeast initiator by other means.

The second reason for lack of progress has been the absence of a collection of mutants that affect replication, that could be used for in vitro complementation. A fact that has been realized by workers in the yeast field but not appreciated in general is the following. While there are a number of mutants that affect the yeast cell cycle, none of them has yet been shown to affect directly the replication fork, and many others have been shown not to be required for replication per se.

REPLICATION MUTANTS   In bacteria, specific replication mutants were easily obtained from collections of random temperature-sensitive mutants by BUdR or [^{3}H]TdR suicide selections. This was not possible in yeast because the organism lacks thymidine kinase. The first attempt to isolate yeast replication mutants was the construction of the *cdc* (cell division cycle) collection of Hartwell and colleagues (53, 112, 113). A random population of temperature-sensitive mutants was screened for those having defects at specific points in the cell cycle, as observed by arrest at high temperature with a specific morphology, called the terminal phenotype. *cdc7* prevents entry into S phase, but it is not clear if it interacts directly with DNA to initate replication (118). Recent sequence analysis suggests that the *CDC7* protein has homology to protein kinases and it may act at some level regulating the G1 to S transition (M. Rosamond, personal communication). Among mutants that appear to have defects within S phase, *cdc8* is defective in thymidylate kinase (117, 118, 146, 147, 174, 175, 237, 267). *CDC21* encodes thymidylate synthetase; and *CDC9* encodes DNA ligase (31, 230). The remaining S phase mutants, *cdc2, 6, 40,* and *16,* have been shown by subsequent analyses to synthesize substantial amounts of DNA at the nonpermissive temperature in vivo and their gene products have not been identified. *cdc40* seems to be deficient in DNA repair. *cdc2* and *16* are deficient in permeabilized cells that carry out replication at replication forks present before permeabilization (176), but they are not deficient in synthesis in vivo or in soluble in vitro replication systems. A similar collection of cold-sensitive *cdc* mutants has also been described (219, 220).

Why are so few of the *cdc* mutants involved directly in replication? One reason might be that the terminal phenotypes used to select the mutants were very broad and can include many types of mutants other than replication mutants. For instance, nuclear transport mutants have the same terminal phenotype as DNA polymerase mutants. Other possible explanations are discussed in Ref. 230.

Among attempts to refine methods of isolating replication mutants to provide more useful collections, Dumas and colleagues isolated 60 complementation groups deficient in incorporation of precursors into DNA in vivo (70). Only six have primary defects in DNA synthesis (L. Dumas, personal communication). Johnston and colleagues used a primary terminal phenotype screen and a secondary screen for deficiencies in DNA but not in RNA synthesis in vivo (141–143). Screening random ts mutants for defects in DNA synthesis in permeable cells has identified an additional 14 complementation groups (176). Herpes simplex virus thymidine kinase gene has been expressed in yeast, allowing BUdR selections, and a number of DNA negative complementation groups have been identified (M. Engler, personal communication).

Mutants that show increased frequency of loss of plasmids containing both an *ARS* and *CEN* have also been identified, and may be replication mutants (122, 194). None of these mutants has been shown to encode a replication protein, however, and the uncertainty as to whether they really represent replication mutants makes it difficult to use them to identify replication proteins.

REVERSE GENETICS   The goal of isolating useful replication mutants has been achieved first, not through classical genetics, but through the use of "brute force," a biochemical approach, and "reverse genetics," a combination of biochemistry, recombinant DNA technology, and yeast genetics. "Brute force" involves assaying extracts of mutants for a specific defect; this was used to isolate the mutants affecting topoisomerases I and II (67, 310, 311, 319).

"Reverse genetics" involves purification of proteins likely to be involved in replication on the basis of what we know in other replication systems and use of the protein to produce antibody or an oligonucleotide probe based on the amino acid sequence of the protein. These probes can then be used to isolate the coresponding gene from a genomic or cDNA expression library (276, 360, 361). The first DNA metabolism gene to be isolated in this way was that encoding topoisomerase II, and *TOP2* has been shown by gene disruption to be essential (98, 99, 124). Temperature-sensitive mutants prepared by in vitro mutagenesis of the cloned gene and gene replacement (124, 264), along with the ts mutants of Sternglanz and colleagues (67), have shown that topoisomerase II is required at mitosis and thus is implicated in the segregation of chromosomes, as expected from the models of Sundin & Varshavsky (296), and not in the initiation of replication, as proposed in prokaryotes. Confirmation of this, however, requires that the double mutants missing both topoisomerases I and II be investigated, since it is known that such topoisomerases can compensate for each other in other systems (99a). In the *Schizosaccharomyces pombe* system, *TOP1* and *TOP2* mutants have also been isolated by direct assay and their phenotype is the same as those in *Saccharomyces* (319).

The genes for DNA polymerase I and SSB-1 have also been isolated in this way. Their usefulness will be described below.

Polymerase-Primase, Proteins, and Gene

YEAST DNA POLYMERASE-PRIMASE   Table 1 lists the DNA polymerases of eukaryotic cells, all of which, except DNA polymerase β, have specific counterparts in yeast.

The enzymatic properties of DNA polymerases I and II of yeast have been reviewed with the major difference being that DNA polymerase I uses polyribonucleotide primers more efficiently than does DNA polymerase II (31). Recent studies of yeast polymerases have focused on DNA polymerase I, since its abundance, its ability to use RNA primers, and its similarity to DNA polymerase alpha suggested it was the polymerase essential for replication.

Both conventional purifications carried out on fresh extracts in the presence of protease inhibitors with multiple specificities (11, 140) and preparations prepared by rapid immunoaffinity chromatography using mouse monoclonal antibody (240, 241) yield a DNA polymerase with the polypeptide composition given in Table 2. A core catalytic subunit has been identified by separation of subunits in SDS-containing polyacrylamide gels, renaturation and assay of polymerase activity in situ, the so-called activity gel assay (11, 13, 130, 268), and by analysis on immunoblots (11, 140). This subunit exhibits a heterogeneous composition with antigenically related peptides ranging from 140 to 180 kd in mass. There is also a 70-kd subunit, the gene for which has recently been cloned (D. Hinkle, personal communication) and two subunits of 48 and 58 kd, that may be associated with a primase activity (see below).

The catalytic subunit (140-kd form) has been purified to homogeneity (140; and J. Campbell, unpublished results). Both aphidicolin and antibody prepared against this purified subunit inhibit DNA synthesis confirming that this subunit contains the catalytic site and establishing it as a target of inhibition by aphidicolin (140).

ISOLATION OF THE YEAST DNA POLYMERASE I GENE   A mutant resistant to aphidicolin has been isolated in yeast (294), and DNA polymerase I purified from the mutant shows a 20-fold greater resistance to the drug than wild-type. DNA polymerase II shows no change in response to the drug (294). This supports a role for DNA polymerase I in replication. In order to investigate this

Table 1 Eukaryotic DNA polymerases

Higher cell polymerase (169)	Properties	Equivalent in yeast
DNA polymerase α	Major replicative polymerase[a]; contains an associated primase; inhibited by aphidicolin[b]	DNA polymerase I
DNA polymerase β[c]	Small; not found in fungi, free living protozoa, or some plants	None
DNA polymerase γ	Only polymerase found in mitochondria	Mitochondrial DNA polymerase (poorly characterized)
DNA polymerase δ	Large; contains an associated 3'-exonuclease[d]	DNA polymerase II (?)

[a]Refs. 108, 109, 168, 188, 192, 216, 223, 334.
[b]A drug that inhibits cell growth and that inhibits DNA polymerase α, but not β (127).
[c]Gene recently cloned (L. M. S. Chang, personal communication; 45, 304, 354).
[d]Until recently it was held that a polymerase with an associated 3'-exonuclease was prokaryotic in nature, whereas one without was eukaryotic, since polymerase III of *E. coli* contained 3'-exonuclease activity and polymerase α did not (355). That this distinction is essentially meaningless is proved by several recent findings. DNA polymerase III of *E. coli* has been shown to contain its proofreading exonuclease as a separate protein subunit, the ε subunit encoded by $dnaQ^+$ ($mutD^+$) (265). In fact, *E. coli* DNA polymerase III α subunit has recently been shown to be the catalytic core, having polymerase activity in itself, just as the large subunit of DNA polymerase I of yeast and polymerase α (195, 196).

further, a yeast genomic DNA expression library in λgtll (276, 360, 361) and an antibody prepared against yeast DNA polymerase I were used to isolate the gene encoding DNA polymerase I (140). The identity of the DNA polymerase I gene was confirmed by showing that cells containing the gene cloned in a high copy number plasmid overproduce the polymerase activity four- to fivefold and that insertion of the gene downstream from a bacteriophage T7 promoter allows synthesis of catalytically active yeast DNA polymerase I in *E. coli*.

Gene disruption and Southern hybridization experiments show that the polymerase is encoded by an essential, single-copy gene (140). Examination of germinated spores containing the disrupted gene reveals a defect in nuclear division and a terminal phenotype typical of replication mutants. Thus, DNA polymerase I is required for replication and DNA polymerase II cannot compensate.

The availability of the gene has allowed isolation of temperature-sensitive mutants (M. Budd, J. Campbell, unpublished) that will be useful for analyzing the remaining polymerases in yeast, for genetic studies to identify proteins that interact with polymerase (see for example 21, 206), and assessing the role of polymerase I in recombination and repair. Use of the gene as probe may allow cloning of other eukaryotic polymerases and should give information about the cell-cycle regulation of synthesis and subcellular localization of the protein to complement monoclonal antibody probes used in higher cells to date (55, 171, 205, 229, 305). Availability of the gene should also allow identification of functional domains through isolation of deletion mutants. Analysis of the DNA sequence will provide a definitive molecular weight of the primary translation product, and, by comparison with the recently described crystal structure of the *E. coli* DNA polymerase I protein (232, 232a), further structural information.

YEAST PRIMASE   Prokaryotic DNA polymerases and eukaryotic DNA polymerases β and γ cannot initiate DNA synthesis de novo, but instead require a primer hybridized to the template. In contrast, a number of workers observed that even highly purified preparations of DNA polymerase α could use unprimed single-stranded DNA as a template. DNA polyerase αs from many different sources have been shown to contain an associated primase activity (4, 50, 51, 152, 159, 172, 253, 270, 348–351). Yeast DNA polymerase I was also shown to contain a primase (240, 241, 275). Although tightly associated with polymerase, the primase of yeast is catalytically and structurally distinct from the core catalytic polymerase, as has also been demonstrated for *Drosophila*, mouse, and chick (75, 151, 297, 316, 317). Active yeast polymerase and primase can be separated by differential elution from an immunoaffinity column (240) or urea gradient analysis (D. Hinkle, personal communication). Primase is associated with one or both of the 48- and 58-kd subunits, as also

Table 2 Comparison of yeast DNA polymerase I and DNA polymerase alpha from various sources

Source	Purification scheme	Subunit[a] Composition Mass (kd)	Subunit[a] Function	Remarks	Refs.[c]
Yeast	Immunoaffinity	140–180	Catalytic core (activity gel)	Gene isolated	(11, 140)
		74	?		
		58> 48	Primase		(240, 241)
Drosophila melanagaster (embryo)	Conventional (calf thymus DNA template)	182	Catalytic core (by sedimentation)	First holoenzyme; size in vivo determined with subunit-specific antibodies	(150–152, 323)
		73			
		60> 50	Primase		(151)
Human-KB cell	Immunoaffinity[b]	185	Catalytic core	Mechanism of substrate binding and recognition consistent with other polymerase (28, 60, 61, 207, 216, 329, 330)	(328)
	Protein A elution gives a preparation containing polymerase plus IgG	77 55 49	At least one subunit contains primase		(305, 328) (328, 126)
Human-HeLa	Conventional (poly(dT) template)	155–145> 130–135	(major)> Catalytic (minor) core	Purified as polymerase-primase	(101)
		64			
		52			
		45			

Calf thymus	Immunoaffinity—Monoclonal antibody covalently attached to protein A, 3.2 M MgCl$_2$ elution(46)	160 68 55 > 48	Catalytic core Proposed anchor of core to primase Primase?	Copies poly(dT), SS to RF, Binds tightly to template, K_m=0.36 µg/ml	(46, 204, 324) (see also 42, 43, 45, 127–129, 336)
Monkey	Immunoprecipitation with monoclonal against calf thymus polymerase α followed by gel electrophoresis	190	Catalytic core	Activity demonstrated by elution from gel Active peptides of 40 kd, 70 kd, and 115 kd also seen in minor amounts.	(157, 158)
Monkey CV-1	Conventional	176–118 62 57 53 30			(354)
Human-HeLa and monkey	Conventional (activated salmon sperm DNA template)	αC$_1$C$_2$ C$_1$=25,000 C$_2$=96,000		Cofactors C$_1$C$_2$ stimulate α by eliminating nonproductive binding to ssDNA α subassembly not characterized	(246, 247)

[a] Stoichiometry of subunits uncertain for all tested.
[b] For all monoclonals thus far, epitopes lie in catalytic core but cross-reaction between polymerase α and β has been observed.
[c] Plant and viral polymerases are not discussed (58, 91, 217, 218).

observed in mouse (317). A 60–65-kd yeast primase with activities like those of the polymerase-primase complex has been purified from extracts free of polymerase, although it has not yet been strictly shown that the free primase is the same polypeptide that copurifies with the polymerase (138, 338). Yeast DNA polymerase I catalytic subunit has also been purified free of primase (140).

The catalytic properties of the primases—either free or in association with polymerase—are similar in yeast and higher cells and are those expected for an activity involved in synthesizing the RNA primers found in replicative intermediates in eukaryotes in vivo (see 57 for review). Primase synthesizes 5'-terminal oligoribonucleotides of discrete length, about 10 nucleotides, and of random sequence. Two aspects of the primase mechanism characteristic of eukaryotes and not found in prokaryotic primase are the synthesis of primers in quantized increments of unit length and the synthesis of alternating blocks of ribo- and deoxyribonucleotide tracts, instead of randomly mixed ribo and deoxyribo primers (126, 138, 275, 328, 352, 353). Importantly, there is no evidence yet for sequence specificity in priming. ATP is required and most primers begin with a purine. Studies of primer length and composition, though numerous, are not yet definitive (11, 12, 126, 138, 152, 172, 317, 318, 328, 338, 348–350, 352, 353, 357, 358, 359). One study suggests pppApN(prN)$_{4-5}$ is the priming sequence in *Drosophila* embryos (165, 357). Specific sites mapped on a minute virus of mouse using purified murine DNA polymerase α-primase suggest a specific termination sequence (76; see also 10). Using either mouse primase or a simian DNA polymerase α, two groups have attempted to map primer initiation sites at the SV40 origin of replication in vitro (318, 353, 354). An advantage of yeast is that chromosomal origins of replication are available, and mapping of primer starts in the *ARS* regions using purified enzymes may give insight into the chromosomal initiation process and sequence specificity of priming.

The next few years should see the answers to a number of questions raised by these findings. Are primers encoded by a specific sequence as in prokaryotes? Is the tight association of polymerase and primase physiologically significant? What is the enzymatic mechanism of the transition from priming to polymerization in the complex? Polymerase can extend preexisting primers, but is the transition from initiation to elongation continuous in the polymerase-primase complex? Sometimes? Always? Do primase and polymerase exist in free as well as in complexed states in vivo, allowing, for instance, for different mechanisms of initiation on the leading and lagging strands? [One other priming mechanism has been proposed in which Ap$_4$A serves as primer (14).]

COMPARISON OF YEAST DNA POLYMERASE I TO HIGHER CELL DNA POLYMERASE α The last three years have yielded unprecedented advances in our understanding of the structure of DNA polymerase α and its role in the cell. Thanks to three approaches—conventional purifications carried out rapidly and in the presence of multiple protease inhibitors, rapid immunoaffinity purifications, and characterization by activity gels and immunological reagents—a reproducible picture of the subunit structure of a core polymerase α assembly has finally emerged. Analysis of the data in Table 2 suggests a consensus core structure identical to yeast DNA polymerase I.

The chief advance over previous work is the ability to purify high-molecular-weight enzymes of reproducible composition. Much of the previously confusing heterogeneity has been eliminated by avoiding proteolysis. However, since even in immunoprecipitations of extracts, the catalytic core subunit is present as a mixture of polypeptides ranging from 145 to 190 kd, there may be different active forms of the enzyme in vivo. It should also be noted that more complicated forms of polymerase α than the core holoenzyme shown in Table 2 have also been reported (14, 160, 236, 246, 247). Additional proteins may modulate the core holoenzyme activity in replication. One preparation, for instance, contains an ATP-dependent DNA helicase (130a, 236). Another contains an associated 3'-exonuclease.

While studies on the previously available degraded forms of the enzyme may have been informative, further studies along two lines are in order with the new preparations. First, the conservation of the structure among many systems suggests it is biologically significant. Therefore the extent to which these structures represent a holoenzyme and the roles of the individual subunits must now be investigated directly by assays on more natural templates than have been used to date, such as those used to study DNA polymerase III of *E. coli* (169, 335). In a few studies, rates of synthesis, processivity, and fidelity have been compared to similar processes in vivo or to other replicative polymerases, and these are good criteria for how closely these enzymes resemble what one expects of a replicative polymerase (27, 103, 104, 129, 150, 187, 323, 332). Kaguni et al (150) have shown processivity is greater than for degraded forms and about like that of the *E. coli* polymerase III missing the beta subunit. What influences the increased processivity (150, 328) and increased length of product synthesized by the new forms of the enzyme? Is there an anchor protein like the beta subunit of DNA polymerase III of *E. coli* (77), the products of genes 44/62 and 45 of phage T4 (3), or the thioredoxin accessory protein of phage T7 polymerase (300)?

A second use for the new preparations is to study the detailed mechanism of polymerization in eukaryotes. A number of elegant mechanistic studies carried out with the degraded enzymes offer an important set of parameters against which the activity of the new preparations can be assessed. Extensive kinetic analyses of DNA polymerase α have led to (*a*) models of order of substrate recognition and binding (60, 61, 80, 85, 86, 303), (*b*) models predicting the existence of two physically separate but interactive template-primer binding sites (60, 81), (c) models of structural requirements for template recognition and binding suggesting specificity according to base composition (83, 85, 339), (d) models of the structural requirements for primer recognition and binding (14, 83, 84, 87, 215, 246, 247, 329, 362). A number of the findings, such as the order of substrate recognition and binding, are also true of DNA polymerase I of *E. coli* (28, 207) and DNA polymerase α (330), and can be expected to be general properties of all polymerases that will probably hold true for the newly available undegraded forms. Other aspects of mechanism, however, may have to be reevaluated using the new holoenzyme preparations. For instance, the presence of an associated primase can be expected to have profound effects on mechanism. Furthermore, little is known of the mechanism of translocation along a template and how this bears on the regulation of fidelity by enzymes that do not catalyze reactions normally associated with proofreading (150).

While advantages of the new preparations are clear, there are still problems. The quantities that can be isolated are small and the immunoaffinity purifications require harsh elution procedures that may lead to denaturation or disassembly and reassortment of subunits. In one case, with KB cells, IgG remains bound to the polymerase, which may affect the activity. The yeast system, with its large quantities of protein and genetics, will aid in clarifying the function of individual subunits. Also, with respect to mechanism, large quantities should enable physical studies to complement the more indirect kinetic studies that have been possible to date with limited amounts of protein.

Other Proteins

Over the last ten years, a number of yeast proteins representing some but not all of the classes of replication proteins defined in Figure 1 have been purified and characterized biochemically. These studies serve as an important resource now in view of the ability to use proteins to obtain the genes encoding them and to construct mutants. How these proteins and genes are regulated should provide insight into regulation of cell proliferation in eukaryotic cells.

Among these proteins are polymerase accessory proteins (44, 243, 244), ATPases (242), RNase H's (156, 345–347), and topoisomerases (Type I, 12, 72, 97; Type II, 12, 72, 97–99). Interestingly, one RNase H copurifies with and stimulates DNA polymerase I. The gene for one RNase H has been cloned and shown to be dispensable.

SINGLE-STRANDED NUCLEIC ACID BINDING PROTEINS   Numerous species of cellular proteins that bind tightly to single-stranded DNA columns have been described in eukaryotes (see review, this volume, Chase & Williams). Two important points are, first, that none of the SSBs has been shown to have a role in replication equivalent to that played by the *E. coli* SSB. The ability of an SSB to stimulate DNA polymerase has been used, by analogy with bacteria, to implicate a given SSB, such as calf thymus UP1 (119–121, 166), in replication, but such involvement is not proved by the data. Second, the DNA binding proteins such as calf thymus UP1 and the RNA binding proteins such as hnRNPs, though isolated independently and previously thought to have distinct functions, have recently been shown to be antigenically related (320). [hnRNP proteins are involved in the stability and processing of RNA (18, 182).] Thus, the eukaryotic SSBs may be encoded by a gene family or be related to each other by RNA or protein processing (260, 320, 337).

Recently, the major, abundant single-stranded nucleic acid binding proteins in yeast have been purified. There are at least five such proteins, 50-kd, 45-kd, 31-kd, 22.5-kd, 20 kd, and some smaller species (148). The 20-kd protein, SSB-m, is found only in mitochondria. The 45-kd species corresponds to SSB-1, originally described by LaBonne & Dumas (177). The 50-kd species has been designated SSB-2. SSB-m, SSB-2, and SSB-1 have been shown to be antigenically distinct.

Since antibody to SSB-1 inhibits in vitro DNA replication and since this protein stimulates DNA polymerase I, it has been thought to play a role in replication. Recent evidence suggests that SSB-1 may play a role in RNA metabolism as well. Reverse genetics has again been useful here. The gene for SSB-1 has been cloned and, surprisingly, gene disruptions show that it is not essential for yeast viability. However, two immunologically related, but genetically distinct proteins of 55 and 75 kd have been found in extracts and may provide compensating activities (148a). A related gene has also been found. Numerous other abundant, essential proteins are encoded by gene pairs in yeast.

THE YEAST INITIATOR PROTEIN   Yeast analogues of several proteins in Figure 1, notably prepriming and initiator proteins, remain to be identified. Several groups have recently had success in purifying proteins on the basis of their binding to specific DNA sequences (24, 313). Using DNase I footprinting as an assay, a protein has been purified that binds specifically to DNA containing *ARS1* sequences (92, K. Sweder, J. Campbell, unpublished results). It is not yet known, however, if the protein is involved in DNA replication. At the same time, mutants that suppress the *ARS⁻* phenotype of plasmids containing mutations in the consensus sequence have been isolated and can be examined for defects in this protein (S. Kearsey, unpublished results).

CONCLUSIONS

It is clear that eukaryotic DNA replication has become amenable to molecular approaches and that the complete enzymatic mechanism of initiation in various model systems will be worked out in the very near future. Furthermore, in yeast, the ability to isolate genes, mutagenize them, and reintroduce them into the chromosome, as well as having the chromosomal sequences that regulate the copying and segregation of chromosomes in hand, should allow formation of a clear picture of how the events of chromosomal replication are regulated and integrated into the eukaryotic cell cycle.

ACKNOWLEDGMENTS

I thank Elizabeth Bertani, Ambrose Jong, Elliot Meyerowitz, Bruce Stillman, and Barbara Wold for helpful critical reading of the manuscript. Special thanks to Linda Cusimano for organization and typing of the manuscript.

Literature Cited

1. Abraham, J., Nasmyth, K. A., Strathern, J. N., Klar, A. J. S., Hicks, J. B. 1984. *J. Mol. Biol.* 176:307–31
2. Adhya, S., Schneidman, P. S., Hurwitz, J. 1985. In press
3. Alberts, B. M., Barry, J., Bedinger, P., Burke, R. L., Hibner, U., Liu, C.-C., Sheridan, R. 1980. *ICN-UCLA Symp.* 19:449–73
4. Albert, W., Grummt, F., Hübscher, U., Wilson, S. H. 1982. *Nucleic Acids Res.* 10:935–46
5. Arendes, J., Kim, K. C., Sugino, A. 1983. *Proc. Natl. Acad. Sci. USA* 80:673–77
6. Ariga, H. 198. *Nucleic Acids Res.* 12:6053–62
7. Ariga, H. 1984. *Mol. Cell. Biol.* 4:1476–82
8. Ariga, H., Sugano, S. 1983. *J. Virol* 48:481–91
9. Ariga, H., Tsuchihashi, Z., Naruto, M., Yamada, M. 1985. *Mol. Cell. Biol.* 5:563–68
10. Badaracco, G., Bianchi, M., Valsasnini, P., Magni, G., Plevani, P. 1985. *EMBO J.* 4:1313–17
11. Badaracco, G., Capucci, L., Plevani, P., Chang, L. M. S. 1983. *J. Biol. Chem.* 258:10720–26
12. Badaracco, G., Plevani, P., Ruyechan, W. T., Chang, L. M. S. 1983. *J. Biol. Chem.* 258:2022–26
13. Banks, G. R., Spanos, A. 1975. *J. Mol. Biol.* 93:63–77
14. Baril, E., Bonin, P., Burstein, D., Mara, K., Zamecnik, P. 1983. *Proc. Natl. Acad. Sci. USA* 80:4931–35
15. Baumann, E. A. 1985. *Eur. J. Biochem.* 147:495–501
16. Baumann, E. A., Hand, R. 1982. *J. Virol.* 44:78–87
17. Bergsma, D. J., Olive, D. M., Hartzell, S. W., Subramanian, K. N. 1982. *Proc. Natl. Acad. Sci. USA* 79:381–85
18. Beyer, A. J., Christensen, M. E., Walker, B. W., LeStourgeon, W. M. 1977. *Cell* 11:127–38
19. Blackburn, E. H., Szostak, J. W. 1984. *Ann. Rev. Biochem.* 53:163–94
20. Botchan, P. M., Dayton, A. 1982. *Nature* 299:453–56
21. Botstein, D., Maurer, R. 1982. *Ann. Rev. Genet.* 16:61–83
22. Bradley, M. K., Griffin, J. D., Livingston, D. M. 1982. *Cell* 28:125–34
23. Bradley, M. K., Hudson, J., Villanueva, M. S., Livingston, D. M. 1984. *Proc. Natl. Acad. Sci. USA* 81:6973–77
24. Bram, R. J., Kornberg, R. D. 1985. *Proc. Natl. Acad. Sci. USA* 82:43–47
25. Brand, A. H., Breeden, L., Abraham, J., Sternglanz, R., Nasmyth, K. 1985. *Cell* 41:41–48
26. Broach, J. R. 1981. *The Molecular Biology of the Yeast Saccharomyces,* pp. 445–70. New York: Cold Spring Harbor Lab.
27. Brosius, S., Grosse, F., Krauss, G. 1983. *Nucleic Acids Res.* 11:193–202
28. Bryant, F. R., Johnson, K. A., Benkovic, S. J. 1983. *Biochemistry* 22:3537–46
29. Buchman, A. R., Berg, P. 1984. *Mol. Cell. Biol.* 4:1915–28
30. Calza, R. E., Eckhardt, D. T., Schildkraut, C. L. 1984. *Cell* 36:689–96
31. Campbell, J. L. 1983. *Genetic Engineering,* Vol. 5, pp. 109–55. New York: Plenum
32. Carle, G. F., Olsen, M. V. 1984. *Nucleic Acids Res.* 12:5647–64
32a. Carson, M., Hartwell, L. H. 1985. *Cell* 42:249–57

33. Celniker, S. E., Campbell, J. L. 1982. *Cell* 31:563–73
34. Celniker, S. E., Sweder, K. S., Srienc, F., Bailey, J. E., Campbell, J. L. 1984. *Mol. Cell. Biol.* 4:2455–66
35. Challberg, M. D., Desiderio, S. V., Kelly, T. J. Jr. 1980. *Proc. Natl. Acad. Sci. USA* 77:5105–9
36. Challberg, M. D., Kelly, T. J. Jr. 1979. *J. Mol. Biol.* 135:999–1012
37. Challberg, M. D., Kelly, T. J. Jr. 1979. *Proc. Natl. Acad. Sci. USA* 76:655–59
38. Challberg, M. D., Kelly, T. J. Jr. 1982. *Ann. Rev. Biochem.* 51:901–34
39. Challberg, M. D., Ostrove, J. M., Kelly, T. J. Jr. 1982. *J. Virol.* 41:265–70
40. Challberg, M. D., Rawlins, D. R. 1984. *Proc. Natl. Acad. Sci. USA* 81:100–4
41. Chambers, J. C., Watanabe, S., Taylor, J. H. 1982. *Proc. Natl. Acad. Sci. USA* 79:5572–76
42. Chang, L. M. S., Bollum, F. J. 1972. *Science* 175:1116–17
43. Chang, L. M. S., Bollum, F. J. 1981. *J. Biol. Chem.* 256:494–98
44. Chang, L. M. S., Lurie, K., Plevani, P. 1978. *Cold Spring Harbor Symp. Quant. Biol.* 43:587–95
45. Chang, L. M. S., Plevani, P., Bollum, F. J. 1982. *Proc. Natl. Acad. Sci. USA* 79:758–61
46. Chang, L. M. S., Rafter, E., Augl, C., Bollum, F. J. 1984. *J. Biol. Chem.* 259:14679–87
47. Clark, R., Peden, K. W. C., Pipas, J. M., Nathans, D., Tjian, R. 1983. *Mol. Cell. Biol.* 3:220–28
48. Clertante, P., Gandray, P., May, E., Cuzin, F. 1984. *J. Biol. Chem.* 259:15196–203
49. Cohen, G. L., Wright, P. C., DeLucia, A. L., Lewton, B. A., Anderson, M. E., Tegtmeyer, P. 1984. *J. Virol* 51:91–96
50. Conaway, R. C., Lehman, I. R. 1982. *Proc. Natl. Acad. Sci. USA* 79:2523–27
51. Conaway, R. C., Lehman, I. R. 1982. *Proc. Natl. Acad. Sci. USA* 79:4585–88
52. Conrad, S. E., Botchan, M. R. 1982. *Mol. Cell. Biol.* 2:949–65
53. Culotti, J., Hartwell, L. H. 1971. *Exp. Cell Res.* 67:389–401
54. de Cicco, D. V., Spradling, A. C. 1984. *Cell* 38:45–54
55. Delfini, C., Alfani, E., Venezia, V. D., Oberholtzer, G., Tomasello, C., et al. 1985. *Proc. Natl. Acad. Sci. USA* 82:2220–24
56. DeLucia, A. L., Lewton, B. A., Tjian, R., Tegtmeyer, P. 1983. *J. Virol.* 46:143–50
57. DePamphilis, M. L., Wassarman, P. M.

58. 1982. In *Organization and Replication of Viral DNA*, pp. 37–114. Boca Raton, Fla: CRC
58. Derse, D., Bastow, K. F., Cheng, Y.-C. 1982. *J. Biol. Chem.* 257:10251–60
59. Desiderio, S. V., Kelly, T. J. Jr. 1981. *J. Mol. Biol.* 145:319–37
60. Detera, S. D., Becerra, S. P., Swack, J. A., Wilson, S. H. 1981. *J. Biol. Chem.* 256:6933–43
61. Detera, S. D., Wilson, S. H. 1982. *J. Biol. Chem.* 257:9770–80
62. de Villiers, J., Olsen, L., Banerji, J., Schaffner, W. 1983. *Cold Spring Harbor Symp. Quant. Biol.* 47:911–19
63. de Villiers, J., Schaffner, W., Tyndall, C., Lupton, S., Kamen, R. 1984. *Nature* 312:212–46
64. de Vries, E., van Driel, W., Tromp, M., van Boom, J., van der Vliet, P. C. 1985. *Nucleic Acids Res.* 13:4935–52
65. DiMaio, D., Nathans, D. 1980. *J. Mol. Biol.* 140:129–42
66. DiMaio, D., Nathans, D. 1982. *J. Mol. Biol.* 156:531–48
67. DiNardo, S., Voelkel, K., Sternglanz, R. 1984. *Proc. Natl. Acad. Sci. USA* 81:2616–20
68. Dixon, R. A., Nathans, D. 1985. *J. Virol* 53:1001–4
69. Dodson, M., Roberts, J., McMacken, R., Echols, H. 1985. *Proc. Natl. Acad. Sci. USA* 82:4678–82
70. Dumas, L. B., Lussky, J. P., McFarland, E. J., Shampay, J. 1982. *Mol. Gen. Genet.* 187:42–46
71. Dunn, B., Szauter, P., Pardue, M. L., Szostak, J. W. 1984. *Cell* 39:191–201
72. Durnford, J. M., Champoux, J. J. 1978. *J. Biol. Chem.* 253:1086–89
73. Edenberg, H. J., Huberman, J. 1975. *Ann. Rev. Genet.* 9:245–84
74. Enomoto, T., Lichy, J. H., Ikeda, J.-E., Hurwitz, J. 1981. *Proc. Natl. Acad. Sci. USA* 78:6779–83
75. Enomoto, T., Suzuki, M., Takahashi, M., Kawasaki, K., Watanabe, Y., et al. 1985. *Cell Struct. Funct.* 10:161–71
76. Faust, E. A., Nagy, R., Davey, S. K. 1985. *Proc. Natl. Acad. Sci. USA* 82:4023–27
77. Fay, P. J., Johanson, K. O., McHenry, C. S., Bambara, R. 1982. *J. Biol. Chem.* 257:5692–99
78. Feldman, J. B., Hicks, J. B., Broach, J. R. 1984. *J. Mol. Biol.* 178:815–34
79. Field, J., Gronostajski, R. M., Hurwitz, J. 1984. *J. Biol. Chem.* 259:9487–95
80. Filpula, D., Fisher, P. A., Korn, D. 1982. *J. Biol. Chem.* 257:2029–40
81. Fisher, P. A., Chen, J. T., Korn, D. 1981. *J. Biol. Chem.* 256:133–41
82. Fisher, E. F., Feist, P. L., Beaucage, S.

L., Mayers, R. M., Tjian, R., et al 1984. *Biochemistry* 23:5938–44

83. Fisher, P. A., Korn, D. 1979. *J. Biol. Chem.* 254:11033–39
84. Fisher, P. A., Korn, D. 1979. *J. Biol. Chem.* 254:11040–46
85. Fisher, P. A., Korn, D. 1981. *Biochemistry* 20:4560–69
86. Fisher, P. A., Korn, D. 1981. *Biochemistry* 20:4570–78
87. Fisher, P. A., Wang, T. S.-F., Korn, D. 1979. *J. Biol. Chem.* 254:6128–37
88. Flytzanis, C. N., McMahon, A. P., Hough-Evans, B. R., Britten, R. J., Davidson, E. H. 1985. *Dev. Biol.* 108:431–42
89. Friefeld, B. R., Lichy, J. H., Hurwitz, J., Horwitz, M. S. 1983. *Proc. Natl. Acad. Sci. USA* 80:1589–93
90. Fuller, R. S., Funnell, B. E., Kornberg, A. 1984. *Cell* 38:889–900
90a. Fuller, C. W., Richardson, C. C. 1985. *J. Biol. Chem.* 260:3197–206
91. Furman, P. A., St. Clair, M. H., Spector, T. 1984. *J. Biol. Chem.* 259:9575–79
92. Galas, D. J., Schmitz, A. 1978. *Nucleic Acids Res.* 5:3157–70
93. Gerlich, W. H., Robinson, W. S. 1980. *Cell* 21:801–9
94. Giacherio, D., Hager, L. P. 1979. *J. Biol. Chem.* 254:8113–16
95. Gluzman, Y., Ahrens, B. 1982. *Virology* 123:78–92
96. Goldman, M. A., Holmquist, G. P., Gray, M. C., Caston, L. A., Ng, A. 1984. *Science* 224:686–92
97. Goto, T., Laipis, P., Wang, J. C. 1984. *J. Biol. Chem.* 259:10422–29
98. Goto, T., Wang, J. C. 1982. *J. Biol. Chem.* 257:5866–72
99. Goto, T., Wang, J. C. 1984. *Cell* 36: 1073–80
99a. Goto, T., Wang, J. C. 1985. *Proc. Natl. Acad. Sci. USA* 82:7178–82
99b. Greider, C. W., Blackburn, G. H. 1985. *Cell* 43:405–13
100. Gronostajski, R. M., Adhya, S., Nagata, K., Guggenheimer, R. A., Hurwitz, J. 1985. *Mol. Cell. Biol.* 5:964–71
101. Gronostajski, R. M., Field, J., Hurwitz, J. 1984. *J. Biol. Chem.* 259:9479–86
102. Gronostajski, R. M., Nagata, K., Hurwitz, J. 1984. *Proc. Natl. Acad. Sci. USA* 81:4013–17
103. Grosse, F., Krauss, G. 1981. *Biochemistry* 20:5470–75
104. Grosse, F., Krauss, G., Knill-Jones, J. W., Fersht, A. R. 1983. *EMBO J.* 2: 1515–19
105. Guggenheimer, R. A., Nagata, K., Kenny, M., Lindenbaum, J., Hurwitz, J. 1984. *J. Biol. Chem.* 259:7807–14

106. Guggenheimer, R. A., Nagata, K., Kenny, M., Hurwitz, J. 1984. *J. Biol. Chem.* 259:7815–25
107. Guggenheimer, R. A., Stillman, B. W., Nagata, K., Tamanoi, F., Hurwitz, J. 1984. *Proc. Natl. Acad. Sci. USA* 81: 3069–73
108. Hanaoka, F., Murakimi, Y., Hori, T., Yamada, M. 1985. *Mol. Biol. Med.* In press
109. Hanaoka, F., Tandai, M., Miyazawa, H., Hori, T., Yamada, M. 1985. *Jpn. J. Cancer Res. (Gann)* 76:441–44
110. Hand, R. 1978. *Cell* 15:317–25
111. Harland, R. M., Laskey, R. A. 1980. *Cell* 21:761–71
112. Hartwell, L. H. 1971. *J. Mol. Biol.* 59:183–94
113. Hartwell, L. H. 1973. *J. Bacteriol.* 115: 966–74
114. Hay, R. T., DePamphilis, M. L. 1982. *Cell* 28:767–79
115. Hay, R. T., Hendrickson, E. A., DePamphilis, M. L. 1984. *J. Mol. Biol.* 175:131–57
116. Heintz, N. H., Milbrandt, J. D., Greisen, K. S., Hamlin, J. L. 1983. *Nature* 302:439
117. Hereford, L. M., Hartwell, L. H. 1971. *Nature* 234:171–72
118. Hereford, L. M., Hartwell, L. H. 1974. *J. Mol. Biol.* 84:445–61
119. Herrick, G., Alberts, B. 1976. *J. Biol. Chem.* 251:2124–32
120. Herrick, G., Alberts, B. 1976. *J. Biol. Chem.* 251:2133–41
121. Herrick, G., Alberts, B. 1976. *J. Biol. Chem.* 251:2142–46
122. Hieter, P., Mann, C., Snyder, M., Davis, R. W. 1985. *Cell* 40:381–92
122a. Hieter, P., Pridmore, D., Hegemann, J. H., Thomas, M., Davis, R., Philippsen, P. 1985. *Cell* 42:913–21
123. Hines, P. J., Benbow, R. M. 1982. *Cell* 30:459–68
124. Holm, C., Goto, T., Wang, J. C., Botstein, D., 1985. *Cell* 41:553–63
125. Horwitz, M. S. 1978. *Proc. Natl. Acad. Sci. USA* 75:4291–95
126. Hu, S. Z., Wang, T. S. F., Korn, D. 1984. *J. Biol. Chem.* 259:2602–9
127. Huberman, J. 1981. *Cell* 23:647–48
128. Hübscher, U. 1983. *EMBO J.* 2:133–36
129. Hübscher, U., Gershwiler, P., McMaster, G. K. 1982. *EMBO J.* 1:1513–19
130. Hübscher, U., Spanos, A., Albert, W., Grummt, F., Banks, G. R. 1981. *Proc. Natl. Acad. Sci. USA* 78:6771–75
130a. Hübscher, U., Stalder, H.-P. 1985. *Nucleic Acids Res.* 13:5471–83
131. Ikeda, J. E., Enomoto, T., Hurwitz, J. 1981. *Proc. Natl. Acad. Sci. USA* 78:884–88

132. Ikeda, J. E., Enomoto, T., Hurwitz, J. 1982. *Proc. Natl. Acad. Sci. USA* 79: 2442–46
133. Jacob, F., Brenner, S., Cuzin, F. 1963. *Cold Spring Harbor Symp. Quant. Biol.* 28:329–48
134. Jayaram, M., Sutton, A., Broach, J. R. 1985. *Mol. Cell Biol.* 5:2466–75
135. Jazwinski, S. M. 1982. *Acta Biochim. Pol.* 29:159
136. Jazwinski, S. M., Edelman, G. M. 1982. *Proc. Natl. Acad. Sci. USA* 79:3428–32
137. Jazwinski, S. M., Edelman, G. M. 1984. *J. Biol. Chem.* 259:6852–57
138. Jazwinski, S. M., Edelman, G. M. 1985. *J. Biol. Chem.* 260:4995–5002
139. Jazwinski, S. M., Niedzwiecka, A., Edelman, G. M. 1983. *J. Biol. Chem.* 258:2754–57
140. Johnson, L. M., Snyder, M., Chang, L. M. S., Davis, R. W., Campbell, J. L. 1985. *Cell.* 43:369–77
141. Johnston, L. H., Thomas, A. P. 1982. *Mol. Gen. Genet.* 186:439–44
142. Johnston, L. H., Thomas, A. P. 1982. *Mol. Gen. Genet.* 186:445–48
143. Johnston, L. H., Thomas, A. P. 1982. *Mol. Gen. Genet.* 187:42–46
144. Jones, K. A., Myers, R. M., Tjian, R. 1984. *EMBO J.* 3:3247–55
145. Jones, K. A., Tjian, R. 1984. *Cell* 36:155–62
146. Jong, A. Y.-S., Campbell, J. L. 1984. *J. Biol. Chem.* 259:14394–98
147. Jong, A. Y.-S., Kuo, C.-L., Campbell, J. L. 1984. *J. Biol. Chem.* 259:11052–59
148. Jong, A. Y.-S., Aebersold, R., Campbell, J. L. 1985. *J. Biol. Chem.* 260: 16367–74
148a. Jong, A. Y.-S., Campbell, J. L. 1986. *Proc. Natl. Acad. Sci. USA.* In press
149. Jong, A. Y.-S., Scott, J. 1985. *Nucleic Acids Res.* 13:2943–58
150. Kaguni, L. S., DiFrancesco, A. A., Lehman, I. R. 1984. *J. Biol. Chem.* 259: 9314–19
151. Kaguni, L. S., Rossignol, J. M., Conaway, R. C., Banks, G. R., Lehman, I. R. 1983. *J. Biol. Chem.* 258:9037–39
152. Kaguni, L. S., Rossignol, J.-M., Conaway, R., Lehman, I. R. 1983. *Proc. Natl. Acad. Sci.* 80:2221–25
153. Kalderon, D., Richardson, W. D., Markham, A. F., Smith, A. E. 1984. *Nature* 311:33–37
154. Kalderon, D., Smith, A. E. 1984. *Virology* 139:109–37
155. Kaplan, L. M., Ariga, H., Hurwitz, J., Horwitz, M. S. 1979. *Proc. Natl. Acad. Sci. USA* 76:5534–38
156. Karwan, R., Blutsch, H., Wintersberger, U. 1984. *Biochemistry* 22:5500
157. Karawya, E., Swack, J., Albert, W., Fedorko, J., Minna, J. D., Wilson, S. H. 1984. *Proc. Natl. Acad. Sci. USA* 81: 7777–81
158. Karawya, E. M., Wilson, S. H. 1982. *J. Biol. Chem.* 257:13129–34
159. Kaufmann, G., Falk, H. H. 1982. *Nucleic Acids Res.* 10:2309–21
160. Kawasaki, K., Nagata, K., Enomoto, T., Hanaoka, F., Yamada, M. 1984. *J. Biochem.* 95:485–93
161. Kearsey, S. 1984. *Cell* 37:299–307
162. Kelly, T. J. Jr. 1984. *The Adenoviruses,* pp. 271–307. New York: Plenum
163. Kikuchi, Y. 1983. *Cell* 35:487–93
164. Kitani, T., Yoda, K.-Y., Ogawa, T., Okazaki, T. 1985. *J. Mol. Biol.* 184:45–52
165. Kitani, T., Yoda, K.-Y., Okazaki, T. 1984. *Mol. Cell. Biol.* 4:1591–96
166. Koerner, T. J., Meyer, R. R. 1983. *J. Biol. Chem.* 258:3126–33
167. Kojo, H., Greenberg, B. D., Sugino, A. 1981. *Proc. Natl. Acad. Sci. USA* 78: 7261–65
168. Kornberg, A. 1980. *DNA Replication.* San Francisco: Freeman
169. Kornberg, A. 1981. *DNA Replication—Supplement.* San Francisco: Freeman
170. Koshland, D., Kent, J. C., Hartwell, L. H. 1985. *Cell* 40:393–403
171. Kozu, T., Yagura, T., Seno, T. 1982. *Cell Struct. Funct.* 7:9–19
172. Kozu, T., Yagura, T., Seno, T. 1982. *Nature* 298:180–82
173. Kriegstein, M. S., Hogness, D. S. 1974. *Proc. Natl. Acad. Sci. USA* 71:135–39
174. Kuo, C.-L., Campbell, J. L. 1982. *Proc. Natl. Acad. Sci. USA* 79:4243–47
175. Kuo, C.-L., Campbell, J. L. 1983. *Mol. Cell. Biol.* 3:1730–37
176. Kuo, C.-L., Huang, N.-H., Campbell, J. L. 1983. *Proc. Natl. Acad. Sci. USA* 80:6465–69
177. LaBonne, S. G., Dumas, L. B. 1983. *Biochemistry* 22:3214–19
178. Lanford, R. E., Butel, J. S. 1984. *Cell* 37:801–13
179. Laskey, R. A., Gurdon, J. B., Trendelenburg, M. 1979. *Br. Soc. Devel. Biol. Symp.* 4:65–80
180. Lechner, R. L., Kelly, T. J. Jr. 1977. *Cell* 12:1007–20
181. Deleted in proof
182. LeStourgeon, W. M., Beyer, A. L., Christensen, M. E., Walker, B. W., Poupore, S. M., et al. 1978. *Cold Spring Harbor Symp. Quant. Biol.* 42:885–98
183. Lewis, E. D., Chen, S., Kumar, A., Blanck, G., Pollack, R. E., et al. 1983. *Proc. Natl. Acad. Sci. USA* 80:7065–69
184. Lewton, B. A., DeLucia, A. L., Tegtmeyer, P. 1984. *J. Virol.* 49:9–13

185. Li, J. J., Kelly, T. J. 1984. *Proc. Natl. Acad. Sci. USA* 81:6973–77
186. Li, J. J., Kelly, T. J. 1985. *Mol. Cell. Biol.* 5:1238–46
186a. Li, J. J., Peden, K. W. C., Dixon, R. A. F., Kelly, T. 1986. In press
187. Liu, P. K., Chang, C. C., Trosko, J. E., Dube, D. K., Martin, G. M., Loeb, L. A. 1983. *Prod. Natl. Acad. Sci. USA* 80:797–801
188. Liu, P. K., Loeb, L. A. 1984. *Science* 226:833–35
189. Lichy, J. H., Field, J., Horwitz, M. S., Hurwitz, J. 1982. *Prod. Natl. Acad. Sci. USA* 79:5225–29
190. Lichy, J. H., Horwitz, M. S., Hurwitz, J. 1981. *Proc. Natl. Acad. Sci. USA* 78:2678–82
191. Lichy, J. H., Nagata, K., Friefeld, B. R., Enomoto, T., Field, J., et al 1983. *Cold Spring Harbor Symp. Quant. Biol.* 47: 731–40
192. Lönn, U., Lönn, S. 1983. *Proc. Natl. Acad. Sci. USA* 80:3996–99
193. Lusky, M., Botchan, M. R. 1984. *Cell* 36:391–401
194. Maine, G. T., Sinha, P., Tye, B.-K. 1984. *Genetics* 106:365–85
195. Maki, H., Horiuchi, T., Kornberg, A. 1985. *J. Biol. Chem.* 260:12982–86
196. Maki, H., Kornberg, A. 1985. *J. Biol. Chem.* 260:12987–92
197. Manos, M. M., Gluzman, Y. 1984. *Mol. Cell. Biol.* 4:317–23
198. Manos, M. M., Gluzman, Y. 1985. *J. Virol.* 53:120–27
199. Margolskee, R. F., Nathans, D. 1984. *J. Virol.* 49:386–93
200. Mariani, B. D., Schimke, R. T. 1984. *J. Biol. Chem.* 259:1901–10
201. Marians, K. J. 1984. *CRC Crit. Rev. Biochem.* 17:153–215
202. Martin, R. G., Setlow, V. P. 1980. *Cell* 20:381–91
203. Masaki, S., Koiwai, O., Yoshida, S. 1982. *J. Biol. Chem.* 257:7172–77
204. Masaki, S., Tanabe, K., Yoshida, S. 1984. *Nucleic Acids Res.* 12:4455–67
205. Matsukage, A., Yamaguchi, M., Tanabe, K., Nishizawa, M., Takahashi, T., et al. 1982. *Gann* 73:850–53
206. Maurer, R., Osmond, B. C., Botstein, D. 1981. *ICN-UCLA Symp. Mol. Cell. Biol.* 12:375–86
207. McClure, W. R., Jovin, T. M. 1975. *J. Biol. Chem.* 250:4073–80
208. McCutchen, T. F., Singer, M. F. 1981. *Proc. Natl. Acad. Sci. USA* 78:95–99
209. McKay, R., DiMaio, D. 1981. *Nature* 289:810–13
210. McMahon, A. P., Flytzanis, C. N., Hough-Evans, B. R., Katula, K. S., Britten, R. J., et al. 1985. *Dev. Biol.* 108:420–30
211. Mechali, M., Kearsey, S. 1984. *Cell* 38:55–64
212. McTiernan, C. F., Stambrook, P. J. 1980. *J. Cell. Biol.* 87:45a
213. Milbrandt, J. D., Heintz, N. H., White, W. C., Rothman, S. M., Hamlin, J. L. 1981. *Proc. Natl. Acad. Sci. USA* 78: 6043–47
214. Miller, A. M., Nasmyth, K. A. 1984. *Nature* 312:247–51
215. Miller, P. S., Annon, N. D., McFarland, K. B., Pulford, S. M. 1982. *Biochemistry* 21:2507–12
216. Miller, M. R., Ulrich, R. G., Wang, T. S.-F., Korn, D. 1985. *J. Biol. Chem.* 260:134–38
217. Misumi, M., Weissbach, A. 1982. *J. Biol. Chem.* 257:2323–29
218. Mizutani, S., Temin, H. H. 1974. *J. Virol.* 13:1020–29
219. Moir, D., Botstein, D. 1982. *Genetics* 100:565
220. Moir, D., Stewart, S. E., Osmond, B. C., Botstein, D. 1982. *Genetics* 100: 547–64
221. Montiel, J. C., Norbury, M., Tuite, M., Dodson, J., Mills, A. 1984. *Nucleic Acids Res.* 12:1049–68
222. Morrison, B., Kress, M., Khoury, G., Jay, G. 1983. *J. Virol.* 47:106–14
223. Murakami, Y., Yasuda, H., Miyazawa, H., Hanaoka, F., Yamada, M. 1985. *Proc. Natl. Acad. Sci. USA* 82:1761–65
224. Myers, R. M., Tjian, R. 1980. *Proc. Natl. Acad. Sci. USA* 77:6491–95
225. Myers, R. M., Williams, R. C., Tjian, R. 1981. *J. Mol. Biol.* 148:347–53
226. Nagata, K., Guggenheimer, R. A., Enomoto, T., Lichy, J. H., Hurwitz, J. 1982. *Proc. Natl. Acad. Sci. USA* 79:6438–42
227. Nagata, K., Guggenheimer, R. A., Hurwitz, J. 1983. *Proc. Natl. Acad. Sci. USA* 80:4266–70
228. Nagata, K., Guggenheimer, R. A., Hurwitz, J. 1983. *Proc. Natl. Acad. Sci. USA* 80:6177–81
229. Nakamura, H., Morita, T., Masaki, S., Yoshida, S. 1984. *Exp. Cell. Res.* 151: 123–33
230. Newlon, C. S. 1984. *The Yeasts*, Vol. 3. Academic
231. Ogawa, T., Baker, T. A., van der Ende, A., Kornberg, A. 1985. *Proc. Natl. Acad. Sci. USA* 82:3562–66
232. Ollis, D. L., Brick, P., Hamlin, R., Xuong, N. G., Steitz, T. A. 1985. *Nature* 313:762–66
232a. Ollis, D. L., Kline, C., Steitz, T. A. 1985. *Nature* 313:818–19
233. Orr, W., Konitopolous, K., Kafatos, F.

C. 1984. *Proc. Natl. Acad. Sci. USA* 81:3773–77

234. Osley, M., Hereford, L. 1982. *Proc. Natl. Acad. Sci. USA* 79:7689–93

235. Ostrove, J. M., Rosenfeld, P., Williams, J., Kelly, T. J. Jr. 1983. *Proc. Natl. Acad. Sci. USA* 80:935–39

236. Ottiger, H.-P., Hubscher, U. 1984. *Proc. Natl. Acad. Sci. USA* 81:3993–97

237. Petes, T. D., Newlon, C. S. 1974. *Nature* 251:637–39

238. Pincus, S., Robertson, W., Rekosh, D. M. K. 1981. *Nucleic Acids Res.* 9:4919–38

239. Pipas, J. M., Peden, K. W. C., Nathans, D. 1983. *Mol. Cell. Biol.* 3:203–13

240. Plevani, P., Foiari, M., Valsasnini, P., Badaracco, G., Cheriathundam, E., Chang, L. M. S. 1985. *J. Biol. Chem.* 260:7102–7

241. Plevani, P., Badaracco, G., Augl, C., Chang, L. M. S. 1984. *J. Biol. Chem.* 259:7532–39

242. Plevani, P., Badaracco, G., Chang, L. M. S. 1980. *J. Biol. Chem.* 255:4957–63

243. Plevani, P., Chang, L. M. S. 1977. *Proc. Natl. Acad. Sci. USA* 74:1937–41

244. Plevani, P., Chang, L. M. S. 1978. *Biochemistry* 17:2530–36

245. Pluta, A. F., Dani, G. M., Spear, B. B., Zakian, V. A. 1984. *Proc. Natl. Acad. Sci. USA* 81:1475–79

246. Pritchard, C. G., DePamphilis, M. L. 1983. *J. Biol. Chem.* 258:9801–9

247. Pritchard, C. G., Weaver, D. T., Baril, E. F., DePamphilis, M. L. 1983. *J. Biol. Chem.* 258:9810–19

248. Prives, C., Barnet, B., Scheller, A., Khoury, G., Jay, G. 1982. *J. Virol.* 43:73–82

249. Prives, C., Covey, L., Scheller, A., Gluzman, Y. 1983. *Mol. Cell. Biol.* 3:1958–66

250. Queen, C., Lord, S. T., McCutchan, T. F., Singer, M. F. 1981. *Mol. Cell. Biol.* 1:1061–68

251. Rawlins, D. R., Rosenfeld, P. J., Wides, R. J., Challberg, M. D., Kelly, T. J. Jr. 1984. *Cell* 37:309–19

252. Revie, D., Tseng, B. Y., Grafstrom, R. H., Goulian, M. 1979. *Proc. Natl. Acad. Sci. USA* 76:5539–43

253. Riedel, H. D., König, H., Stahl, H., Knippers, R. 1982. *Nucleic Acids Res.* 10:5621–35

254. Rigby, P. J., Lane, D. P. 1983. In *Advances in Viral Oncology*, Vol. 3, pp. 31–57. New York: Raven

255. Rijnders, A. W. M., van Bergen B. G. M., van der Vliet, P. C., Sussenbach, J. S. 1983. *Nucleic Acids Res.* 24:8777–89

256. Rivin, C. J., Fangman, W. L. 1980. *J. Cell Biol.* 85:96

257. Rivin, C. J., Fangman, W. L. 1980. *J. Cell Biol.* 85:108

258. Roth, G. E., Blanton, H. M., Hager, L., Zakian, V. A. 1983. *Mol. Cell. Biol.* 3:1898–1908

259. Saffer, L. D., Miller, O. L. Jr. 1983. *J. Cell Biol.* 97:111a

260. Sapp, M., König, H., Riedel, H. D., Richter, A., Knippers, R. 1985. *J. Biol. Chem.* 260:1550–56

261. Sauer, B., Lehman, I. R. 1982. *J. Biol. Chem.* 257:12394–98

262. Scheidtmann, K. H., Hardung, M., Echle, B., Walter, G. 1984. *J. Virol.* 50:1–12

263. Scheller, A., Covey, L., Barnet, B., Prives, C. 1982. *Cell* 29:375–83

264. Scherer, S., Davis, R. W. 1979. *Proc. Natl. Acad. Sci. USA* 76:4951–55

265. Scheuermann, R. H., Echols, H. 1984. *Proc. Natl. Acad. Sci. USA* 81:7747–51

266. Schwartz, D. C., Cantor, C. R. 1984. *Cell* 37:67–75

267. Sclafani, R. A., Fangman, W. L. 1984. *Proc. Natl. Acad. Sci. USA* 81:5821–28

268. Scovassi, A., Torsello, S., Plevani, P., Badaracco, G. F., Bertazzoni, U. 1982. *EMBO J.* 1:1161–65

269. Shampay, J., Szostak, J. W., Blackburn, E. H. 1984. *Nature* 310:154–57

270. Shioda, M., Nelson, E. M., Bayne, M. L., Benbow, R. M. 1982. *Proc. Natl. Acad. Sci. USA* 79:7209–13

271. Shortle, D. R., Margolskee, R. F., Nathans, D. 1979. *Proc. Natl. Acad. Sci. USA* 76:6128–31

272. Shortle, D., Nathans, D. 1979. *J. Mol. Biol.* 131:801–37

273. Siebenlist, U., Hennighausen, L., Battey, J., Leder, P. 1984. *Cell* 37:381–91

274. Simanis, V., Lane, D. P. 1983. *Virology* 144:88–100

275. Singh, H., Dumas, L. B. 1984. *J. Biol. Chem.* 259:7936–40

276. Snyder, M., Davis, R. W. 1985. *Hybridomes in Biotechnology and Medicine.* New York: Plenum. In press

277. Soprano, K. J., Galanti, N., Jonak, G., McKercher, S., Pipas, J. M., et al. 1983. *Mol. Cell. Biol.* 3:214–19

278. Srienc, F., Bailey, J. E., Campbell, J. L. 1985. *Mol. Cell. Biol.* 5:1676–84

279. Staufenbiel, M., Deppet, W. 1983. *Cell* 33:173–81

280. Stillman, B. W. 1983. *Cell* 35:7–9

281. Stillman, B. W. 1985. *Genetic Engineering,* Vol. 7, pp. 1–27, New York: Plenum

282. Stillman, B. W., Gerard, R. D., Gug-

genheimer, R. A., Gluzman, Y. 1985. *EMBO J.* In press

283. Stillman, B. W., Gluzman, Y. 1985. *Mol. Cell. Biol.* 5:2051–60

284. Stillman, B. W., Lewis, J. B., Chow, L. T., Mathews, M. B., Smart, J. E. 1981. *Cell* 23:497–508

285. Stillman, B. W., Tamanoi, F. 1983. *Cold Spring Harbor Symp. Quant. Biol.* 47:741–50

286. Stillman, B. W., Tamanoi, F., Mathews, M. B. 1982. *Cell* 31:613–23

287. Stillman, B. W., Topp, W. C., Engler, J. A. 1982. *J. Virol.* 44:530–37

288. Stinchcomb, D. T., Mello, C., Hirsh, D. 1985. *Proc. Natl. Acad. Sci. USA* 82: 4167–71

289. Stinchcomb, D. T., Thomas, M., Kelly, J., Selker, E., Davis, R. 1980. *Proc. Natl. Acad. Sci. USA* 77:4559–63

290. Stringer, J. R. 1982. *J. Virol.* 42:854–64

291. Struhl, K., Stinchcomb, D., Scherer, S., Davis, R. W. 1979. *Proc. Natl. Acad. Sci. USA* 76:1035–39

292. Sturzbecher, H. W., Morike, M., Montenarh, M., Henning, R. 1985. *FEBS Lett.* 180:285–90

293. Sugano, S., Yamaguchi, N. 1984. *J. Virol.* 52:884–91

294. Sugino, A., Kojo, H., Greenberg, B. D., Brown, P. O., Kim, K. C. 1981. *ICN-UCLA Symp. Mol. Cell. Biol.* 11:529–53

295. Sugino, A., Sakai, A., Wilson-Coleman, F., Arendes, J., Kim, K. C. 1983. *Mechanisms of DNA Replication and Recombination,* pp. 527–52. New York: Liss.

296. Sundin, O., Varshavsky, A. 1980. *Cell* 21:103–14

297. Suzuki, M., Enomoto, T., Hanaoka, F., Yamada, M. 1985. *J. Biochem.* In press

298. Szostak, J. W. 1983. *Cold Spring Harbor Symp. Quant. Biol.* 47:1187–94

299. Tabor, S., Engler, M. J., Fuller, C. W., Lechner, R. L., Matson, S. W., et al. 1981. *ICN-UCLA Symp. Mol. Cell. Biol.* 11:387–468.

300. Tabor, S., Huber, H. E., Richardson, C. C. 1986. In *Ninthe Karaolinska Institute Nobel Conference.* New York: Raven. In press

301. Tamanoi, F., Stillman, B. W. 1982. *Proc. Natl. Acad. Sci. USA* 79:2221–25

302. Tamanoi, F., Stillman, B. W. 1983. *Proc. Natl. Acad. Sci. USA* 80:6446–50

303. Tanabe, K., Bohn, E. W., Wilson, S. H. 1979. *Biochemistry* 18:3401–6

304. Tanabe, K., Yamaguchi, M., Matsukage, A., Takahashi, T. 1981. *J. Biol. Chem.* 256:3098–102

305. Tanaka, S., Hu, S.-Z., Wang, S.-F., Korn, D. 1982. *J. Biol. Chem.* 257: 8386–90

306. Tegtmeyer, P., Lewton, B. A., DeLucia, A. L., Wilson, V. G., Ryder, K. 1983. *J. Virol.* 46:151–61

307. Tenen, D. G., Haines, L. L., Livingston, D. M. 1982. *J. Mol. Biol.* 157:473–92

308. Tenen, D. G., Livingston, D. M., Wang, S. S. 1983. *Cell* 34:629–39

309. Tenen, D. G., Taylor, T. S., Haines, L. L., Bradley, M. K., Martin, R. G., et al. 1983. *J. Mol. Biol.* 168:791–808

310. Thrash, C., Bankier, A. T., Barrell, B. G., Sternglanz, R. 1985. *Proc. Natl. Acad. Sci. USA* 82:4374–78

311. Thrash, C., Voelkl, K., DiNardo, S., Sternglanz, R. 1984. *J. Biol. Chem.* 259:1375–77

312. Tjian, R., Robbins, A. 1979. *Proc. Natl. Acad. Sci. USA* 76:610–17

313. Topol, J., Parker, C. S. 1984. *Cell* 36:357–69

314. Tornow, J., Cole, C. N. 1983. *J. Virol.* 47:487–94

315. Tornow, J., Polvino-Bodnor, M., Santangelo, G., Cole, C. N. 1985. *J. Virol.* 53:415–24

316. Tseng, B. Y., Ahlem, C. N. 1982. *J. Biol. Chem.* 257:7280–83

317. Tseng, B. Y., Ahlem, C. N. 1983. *J. Biol. Chem.* 258:9845–49

318. Tseng, B. Y., Ahlem, C. N. 1984. *Proc. Natl. Acad. Sci. USA* 81:2342–46

319. Uemura, T., Yanagida, M. 1984. *EMBO J.* 3:1737–44

320. Valentini, O., Biamonti, G., Pandolfo, M., Morandi, C., Riva, S. 1985. *Nucleic Acids Res.* 13:337–46

321. van Bergen, B. G. M., van der Ley, P. A., van Driel, W., van Mansfield, A. D. M., van der Vliet, P. C. 1983. *Nucleic Acids Res.* 11:1975–89

322. van der Ende, A., Baker, T. A., Ogawa, T., Kornberg, A. 1985. *Proc. Natl. Acad. Sci. USA* 82:3954–58

323. Villani, G., Fay, P. J., Bambara, R. A., Lehman, I. R. 1981. *J. Biol. Chem.* 256:8202–7

324. Wahl, A. F., Kowalski, S. P., Harwell, L. W., Lord, E. M., Bambera, R. A. 1984. *Biochemistry* 23:1895–99

325. Walmsley, R. W., Chan, C. S., Tye, B. K., Petes, T. D. 1984. *Nature* 310:157–60

326. Walmsley, R. W., Petes, T. D. 1985. *Proc. Natl. Acad. Sci. USA* 82:506–10

327. Walmsley, R. W., Szostak, J. W., Petes, T. D. 1983. *Nature* 302:84–86

328. Wang, T. S.-F., Hu, S.-Z., Korn, D. 1984. *J. Biol. Chem.* 259:1854–65

329. Wang, T. S.-F., Korn, D. 1980. *Biochemistry* 19:1782–90

330. Wang, T. S.-F., Korn, D. 1982. *Biochemistry* 21:1597–608

331. Watanabe, S., Taylor, J. H. 1980. *Proc. Natl. Acad. Sci. USA* 77:5292–96
332. Weaver, D. T., DePamphilis, M. L. 1982. *J. Biol. Chem.* 257:2075–86
333. Weaver, D. T., Fields-Berry, S. C., De-Pamphilis, M. L. 1985. *Cell* 41:565–75
334. Weissbach, A. 1977. *Ann. Rev. Biochem.* 46:25–47
335. Wickner, S. H. 1978. *Ann. Rev. Biochem.* 47:1163–91
336. Wierowski, J. V., Lawton, K. G., Hockensmith, J. W., Bambara, R. A. 1983. *J. Biol. Chem.* 258:6250–54
337. Williams, K. R., Stone, K. L., LoPresti, M. B., Merrill, B. M., Planck, S. R. 1985. *Proc. Natl. Acad. Sci. USA* 82: In press
338. Wilson, F. E., Sugino, A. 1985. *J. Biol. Chem.* 260:8173–81
339. Wilson, S. H., Matsukage, A., Bohn, E. W., Chen, Y.-C., Siverajan, M. 1977. *Nucleic Acids Res.* 4:3981–96
340. Wilson, V. G., Tevethia, M. J., Lewton, B. A., Tegtmeyer, P. 1982. *J. Virol.* 44:458–66
341. Wirak, D. O., Chalifour, L. E., Wassarman, P. M., Muller, W. F., Hassell, J. A., et al. 1985. *Mol. Cell. Biol.* In press
342. Wobbe, C. R., Dean, F., Weissbach, L., Hurwitz, J. 1985. *Proc. Natl. Acad. Sci. USA* 82:5710–14
343. Deleted in proof
344. Woodcock, D. M., Cooper, I. A. 1981. *Cancer Res.* 41:2483–90
345. Wyers, F., Huet, J., Sentenac, A., Fromageot, P. 1976. *Eur. J. Biochem.* 69:385–95
346. Wyers, F., Sentenac, A., Fromageot, P. 1973. *Eur. J. Biochem.* 35:270–81
347. Wyers, F., Sentenac, A., Fromageot, P. 1976. *Eur. J. Biochem.* 69:377–83
348. Yagura, T., Kozu, T., Seno, T. 1982. *J. Biochem.* 91:607–18
349. Yagura, T., Kozu, T., Seno, T. 1982. *J. Biol. Chem.* 257:11121–27
350. Yagura, T., Kozu, T., Seno, T., Saneyosh, M., Hiraga, S., et al. 1983. *J. Biol. Chem.* 258:13070–75
351. Yagura, T., Tanaka, S., Tomoko, K., Seno, K., Korn, D. 1983. *J. Biol. Chem.* 258:6698–700
352. Yamaguchi, M., Hendrickson, E. A., DePamphilis, M. L. 1985. *J. Biol. Chem.* 260:6254–63
353. Yamaguchi, M., Hendrickson, E. A., DePamphilis, M. L. 1985. *Mol. Cell. Biol.* 5:1170–83
354. Yamaguchi, M., Matsukage, A., Takahashi, T., Takahashi, T. 1985. *J. Biol. Chem.* 257:3932–36
355. Yates, J., Warren, N., Reisman, D., Sugden, B. 1984. *Proc. Natl. Acad. Sci. USA* 81:3806–10
356. Yates, J. L., Warren, N., Sugden, B. 1985. *Nature* 313:812–15
357. Yoda, K., Okazaki, T. 1983. *Nucleic Acids Res.* 11:3433–56
358. Yoshida, S., Suzuki, R., Masaki, S., Koiwai, O. 1983. *Biochim. Biophys. Acta* 741:348–57
359. Yoshida, S., Suzuki, R., Masaki, S., Koiwai, O. 1985. *J. Biochem.* 98:427–33
360. Young, R. A., Davis, R. W. 1983. *Proc. Natl. Acad. Sci. USA* 80:1194–98
361. Young, R. A., Davis, R. W. 1984. *Science* 222:778–82
362. Zamecnik, P. C., Rapaport, E., Baril, E. F. 1982. *Proc. Natl. Acad. Sci. USA* 79:1791–94

Ann. Rev. Biochem. 1986. 55:773–99
Copyright © 1986 by Annual Reviews Inc. All rights reserved

NEUROPEPTIDES: MULTIPLE MOLECULAR FORMS, METABOLIC PATHWAYS, AND RECEPTORS[1]

David R. Lynch and Solomon H. Snyder[2]

Departments of Neuroscience, Pharmacology and Experimental Therapeutics, Psychiatry and Behavioral Sciences, The Johns Hopkins University School of Medicine, 725 N. Wolfe Street, Baltimore, Maryland 21205

CONTENTS

[1]Abbreviations used: CNS, Central Nervous System; MSH, Melanocyte Stimulating Hormone; CPH, Carboxypeptidase H; CRF, Corticotropin Releasing Factor; LHRH, Luteinizing Hormone Releasing Factor.

[2]To whom all correspondence and reprint requests should be addressed.

0066-4154/86/0701-0773$02.00

PERSPECTIVES AND SUMMARY

In the past 15 years the number of polypeptides proposed as neurotransmitters has grown from several to more than 50. While it is not clear exactly how many peptide neurotransmitters there are, and the evidence that many of those discovered are transmitters is not complete, it is clear that the discovery of the large number of putative peptide neurotransmitters greatly increases our understanding of synaptic transmission. Neuropeptides and receptors for them have been found in all regions of the brain and in animals in all areas of the evolutionary scale.

Perhaps the most interesting observation about peptide neurotransmitters is that each one may exist in a variety of forms. The opioid peptide enkephalins were first characterized in two forms, leu-enkephalin (tyr-gly-gly-phe-leu) and met-enkephalin (tyr-gly-gly-phe-met) (1). Now there are at least nine endogenous peptides with opiate activity, all of which contain a met-enkephalin or leu-enkephalin sequence at their amino terminus. Cholecystokinin has been isolated in at least five different forms, all with the same sequence at the carboxyl terminus (2). Similarly, angiotensin exists in at least three and probably four forms produced from each other by extracellular cleavages (3). This suggests that families of neuropeptides exist within the CNS.

This review examines the explosion in neuropeptide research, focusing on the multiple forms of neuropeptides and how they influence neural transmission. For a substance to be accurately classified as a neurotransmitter, it must meet several criteria. First, it must be produced and stored presynaptically. In addition, it must be released into the synapse upon depolarization of the neuron. Finally, when released into the synaptic cleft, it must act postsynaptically to change the properties of the postsynaptic cell and must be inactivated in a specific way. This review concentrates on the first and last of these criteria and specifically on how these criteria relate to the multiplicity of forms of peptide neurotransmitters.

HISTORICAL BACKGROUND

To appreciate the greatly enhanced knowledge on neuropeptides in recent years, it is necessary to review how neurotransmitters have been traditionally identified. Classical neurotransmitters have been discovered mainly through their physiological effects, and the most familiar neuropeptides were found in the same way. Acetylcholine was first described as a substance that could slow the rate of isolated hearts (4). Substance P was a vasodilator in intestine and brain (5), and angiotensin was first observed as a vasoconstrictive peptide in plasma (6). Neurotensin, although first purified from brain tissue, was also recognized because of its vasoactive effects (7). The opioid peptide enkephalins were characterized and isolated by assaying their ability to act as morphinelike substances in bioassays and in receptor-binding assays (1, 8, 9).

These findings illustrate a major principle in the distribution of neuropeptides. Most neuropeptides are found peripherally as circulating hormones, and it is these forms that classically have been most easily characterized. Because of the heterogeneity of the different regions of the brain, most neurotransmitters are present in low overall levels within whole brain extracts. Consequently, it has proven more reasonable to characterize bioactive peptides peripherally and then extend the observations to the brain. Through the use of specific antibodies for such peptides, one can examine the brain by immunocytological methods or with immunoassays for their presence. Localizing neuropeptides by immunohistochemistry has been especially valuable for clarifying peptide function within the brain.

Identifying new neuropeptides has benefited greatly from improved biochemical techniques for peptide isolation and sequencing and genetic cloning approaches to characterize peptide precursors. While substance P was first described physiologically in 1931 (5), it was not sequenced until 1970 (10) and its precursor was not cloned until 1982 (11). In contrast, atrial natiuretic factors or atriopeptins have been suspected to exist since the 1960s (12), but were first demonstrated physiologically in the early 1980s (13, 14) and were isolated from atrial extracts in 1983 (15–20). Since that time their precursor (21–24) sequence and gene sequence (25–27) have been determined, and specific antibodies have been made against peptides synthesized from the sequences (28–30). These antibodies have permitted the detection of these peptides in the CNS by both immunocytochemistry and radioimmunoassay.

Molecular cloning has also allowed detection of putative peptide neurotransmitters simply on the basis of mRNA or gene structures. As will be discussed later, neuropeptides are commonly contained within long precursors in which the active peptide is flanked by basic amino acids. By examining the precursor sequence predicted by cloned mRNAs one can predict that certain peptides will be produced. Antibodies can then be generated against synthetic peptides made based on the sequence and used to seek the postulated peptides in the brain.

SYNTHESIS OF NEUROPEPTIDES

Classical neurotransmitters such as acetylcholine and norepinephrine are made by the action of specific enzymes. Acetylcholine is synthesized from choline and acetyl CoA by the action of choline acetyl transferase (31). The acetyl CoA is derived from the mitochondria and choline comes from a high-affinity neuronal uptake of choline from the extracellular fluid (32).

The possibility that peptide neurotransmitters are made nonribosomally by specific enzymes at the synapse has also been considered. Carnosine, a putative neurotransmitter in primary olfactory neurons, consists of B-alaninyl-histidine linked through the gamma carboxyl and is thought to be synthesized in this manner by the enzyme carnosine synthetase (33). However, most putative peptide neurotransmitters are considerably longer than two amino acids so that enzymatic linkage would not prove efficient. Also, the variety of similar but not identical forms of the various neuropeptides is difficult to explain by synthesis from individual amino acids.

Protein precursors of neuropeptides and hormones were first characterized for insulin in the pancreatic islets (34), and a similar pattern occurs for most if not all peptide neurotransmitters. The biologically active peptides are usually flanked in the precursor by two basic amino acids so that the sequential action of a trypsinlike enzyme followed by a carboxypeptidase B–like enzyme is capable of producing the active peptide from the precursor (35). The precursors also contain sites for other posttranslational modifications including acetylation of the N terminal residue, glycosylation, and amidation of the carboxyl terminal residue. These modifications are often critical in activating or inactivating the precursors.

The structures of the various precursors have been established by both gene cloning experiments and classical protein sequencing methods. These structures have been reviewed recently in this series (36), so that the present review concentrates on a few examples. These precursors often produce multiple peptides with distinct biological activities. Processing of the precursors varies in a tissue-specific manner for many of these precursors. Thus, only by characterizing both the expression of the precursor gene and its processing can one know which peptides derived from a given precursor may act as neurotransmitters. Immunocytochemistry and radioimmunoassays have permitted localization of peptides to discrete regions of the brain. Although the technique is not yet perfected, in situ hybridization may be employed to localize peptide-specific mRNAs to specific brain regions and individual cells. These themes are well illustrated by the proopiomelanocortin (POMC), pro-enkephalin A and B, procholecystokinin, and substance P precursors.

Proopiomelanocortin (POMC, Pro-ACTH-Endorphin)

POMC was the first precursor to be shown to be differentially processed in a tissue-specific manner to yield multiple active peptides. Processing of POMC in the anterior and intermediate lobes of the pituitary, recently reviewed by Douglass et al (36), will not be treated in detail here. Briefly, POMC contains an N-terminal portion incorporating gamma-MSH, a middle portion with ACTH and alpha-MSH, and a C-terminal portion with beta-lipotrophin (beta-LPH) and beta-endorphin (36–39). In the anterior lobe, cleavage of POMC yields a large N-terminal fragment, ACTH, and B-LPH (36, 40). In the neurointermediate lobe, processing proceeds farther so that the smaller peptides, gamma-MSH, alpha-MSH, and beta-endorphin, are the major products (36–40). All of these cleavages occur at dibasic amino acids, and the majority of these sites are lys-arg sites suggesting that this sequence is specifically favored in the cleavage of POMC. The sequence of met-enkephalin is also contained in POMC but is not flanked by dibasic amino acids on its carboxyl side so that it is not thought to be produced. POMC also undergoes several other posttranslational modifications, including phosphorylation, sulfation, glycosylation, and acetylation. N-terminal acetylation of beta-endorphin destroys its opiate activity (41, 42).

In the brain POMC occurs in discrete neuronal cell populations in the arcuate nucleus of the hypothalamus (43–46) and the nucleus tractus solitarius in the medulla (47, 48). The presence of POMC mRNA in the cells of the arcuate nucleus has been confirmed using in situ hybridization (49, 50). The cells in the arcuate nucleus were characterized first and shown to be distinct from those neurons that contain enkephalinlike immunoreactivity, again showing that POMC is not processed to met-enkephalin (43). These cells send axons to a wide variety of locations in the brain. The cells of the nucleus tractus solitarius were discovered more recently so their projections are not yet fully known.

Processing of POMC in the hypothalamus resembles the neurointermediate lobe in that the major forms of immunoreactivity in size resemble alpha MSH and beta-endorphin rather than ACTH and beta-LPH (51–53). However, unlike the intermediate lobe, most of the beta-endorphin is nonacetylated and the sequence of cleavages seems to vary (53–56). Interestingly, the medulla oblongata presents yet another pattern of processing in favoring acetylated forms of beta-endorphin, like the intermediate pituitary, but has lower amounts of C-terminally truncated forms of beta-endorphin than the intermediate lobe (57).

POMC processing in the brain proceeds farther than in peripheral tissues. While this might reflect a longer time before secretion resulting from axonal transport, this seems unlikely in view of the specificity of the cleavages involved. More extensive processing in the brain also occurs with proenkephalin A and procholecystokinin.

Proenkephalin A and B

The structure of two putative precursors of the enkephalins has been determined from cloned cDNAs. Bovine preproenkephalin A (Figure 1) contains 268 amino acids and begins with a 20-amino-acid putative signal peptide sequence (58–60). The amino terminal one third of the precursor is sometimes called synenkephalin and contains no enkephalin sequences. The carboxyl terminal two thirds of proenkephalin A contains four copies of met-enkephalin and one copy of leu-enkephalin, each flanked by two basic amino acids. This portion of the molecule also contains one copy of met-enkephalin-arg-gly-leu and the peptide met-enkephalin-arg-phe at the carboxyl terminus.

While the simple precursor structure suggests that enkephalin would be the major end product, this is apparently not always the case. In the brain, antibodies to met-enkephalin or met-enkephalin-arg-phe recognize mainly enkephalin pentapeptides (61). In the adrenal, most enkephalinlike immunoreactivity is found as higher-molecular-mass peptides of 5,000–10,000 daltons (61, 62). Thus processing in the adrenal is incomplete and does not occur at all dibasic amino acids. In the brain processing seems to occur at all dibasic amino acids leading to enkephalin pentapeptides. This suggests that the major neurotransmitters produced from proenkephalin A are the enkephalin pentapeptides. Other experiments suggest that the routes of processing may differ between these two tissues. In the adrenal most of the high-molecular-mass enkephalin-containing peptides have synenkephalin immunoreactivity associated with them, suggesting that the synenkephalin is cleaved from the

PROCESSING OF PREPROENKEPHALIN A

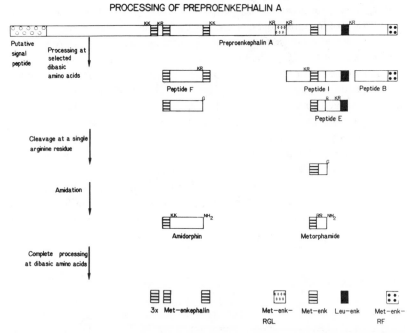

Figure 1   Processing of preproenkephalin A. Preproenkephalin A contains four copies of met-enkephalin, one copy of leu-enkephalin, one copy of met-enkephalin-arg-gly-leu, and one copy of met-enkephalin-arg-phe flanked by dibasic amino acids (R = arginine, K = lysine). Processing at selected dibasic sites yields the peptides found in the adrenal gland: Peptide F, Peptide E, Peptide B, and Peptide I. Cleavage at a single arginine residue followed by amidation can yield amidorphin or metorphamide (G = glycine, NH₂ = amide). Complete processing at dibasic amino acids yields the penta peptide enkephalins, met-enkephalin-arg-gly-leu and met-enkephalin-arg-phe.

enkephalin-containing portions later in the sequence of cleavages (63). In the supraoptic nucleus of the bovine hypothalamus higher-molecular-mass precursors have met-enkephalin-arg-phe immunoreactivity. This suggests that proenkephalin processing in the adrenal begins in the carboxyl terminal region while processing in the supraoptic nucleus begins in the amino-terminal region (63).

Recently two amidated peptides derived from proenkephalin A have been described. Metorphamide consists of met-enkephalin-arg-arg-val-NH$_2$ and is derived by cleavage at a single arginine residue followed by amidation (64, 65). Another peptide called amidorphin is made from residues 104–128 in proenkephalin A followed by C-terminal amidation (66). These two peptides are also found in the brain and consequently may act as neurotransmitters. However, their presence alone does not establish them as neurotransmitters. While amidorphin levels in the adrenal are quite high, so that amidorphin may be released as a hormone, most of the amidorphinlike immunoreactivity in the brain is found as smaller peptides (66). This suggests that in the brain amidorphin is cleaved to make other peptides such as met-enkephalin or possibly the putative peptides in the carboxyl-terminal half of amidorphin. Future studies on the synthesis of these peptides must involve the examination of the brain immunocytochemically for these peptides to determine if they are produced in selected neuronal populations. Initial immunocytochemical studies have identified a subset of neurons within the brain that produce metorphamide (67). Consequently it is possible that different neurons within the brain may produce different neurotransmitters from proenkephalin A.

The enkephalins can also arise from a second precursor designated proenkephalin B or prodynorphin. Proenkephalin B (Figure 2) contains three copies of leu-enkephalin each flanked by dibasic amino acids (68, 69). Each of these is contained within the sequences of the longer opioid peptides dynorphin, neoendorphin, and rimorphin. By selective processing at lys-arg sequences, proenkephalin B would yield dynorphin and neoendorphin, while cleavage at a single arginine can make dynorphin (1–8) (another opiatelike peptide) and rimorphin. Complete processing at dibasic amino acids can yield leu-enkephalin.

Given the possibilities for variations in processing of proenkephalin B, which peptides represent true neurotransmitters? Immunocytochemical studies and detailed radioimmunoassays have proved valuable in determining which peptides are made from this precursor. All of the proenkephalin B–derived peptides colocalize to the same neurons by immunocytochemistry (70). However, the ratios of these peptides vary in different brain regions (71–77). For example, while the level of neoendorphins generally exceeds that of dynorphin, a few nuclei within the brain do contain more dynorphin than neoendorphin (71, 74–76). In addition the levels of alpha-neoendorphin greatly

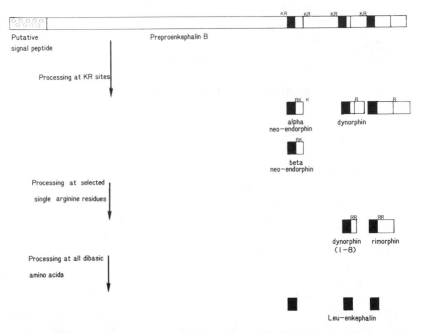

Figure 2   Processing of preproenkephalin B. Preproenkephalin B contains the sequences of the opioid peptides alpha and beta neoendorphin, dynorphin, dynorphin (1–8), and rimorphin. Leu-enkephalin is the amino terminal portion of all of these peptides. Processing at only lys-arg sites produces the neoendorphins and dynorphin, while rimorphin and dynorphin (1–8) can be made by further cleavage at selected single basic amino acids. Leu-enkephalin is made by cleavage at all dibasic amino acids.

exceed those of beta-neoendorphin in a few regions of the brain, suggesting a heterogeneity in the processing of the neoendorphins (71, 74–76). While the levels of leu-enkephalin cannot be used here to determine how much leu-enkephalin is derived from proenkephalin B, it seems that in at least one region leu-enkephalin is made from proenkephalin B. Lesions of the enkephalinergic striatonigral pathway decrease leu-enkephalin and other proenkephalin B–derived peptides but do not affect levels of proenkephalin A–derived peptides (77). While some argue that the leu-enkephalin remaining is restricted to a small region of the substantia nigra where the striatonigral pathway does not terminate (78), it still seems likely that enkephalin is made from proenkephalin B in at least some instances. The absolute levels of dynorphin and rimorphin are substantially less than those of the neoendorphins. This would suggest that in many cases in the brain dynorphin and rimorphin serve as precursors of smaller peptides, including leu-enkephalin. Still, the synaptic connections of the brain are so diverse that this may vary regionally and even within regions in the brain.

Procholecystokinin (CCK)

Originally discovered for its ability to cause gallbladder contraction, CCK is present in the nervous system as well. CCK and gastrin share carboxyl terminal homology, and CCK within the nervous system was first detected with anti-gastrin antibodies (79–81). Chemically authentic gastrin is only made within the central nervous system in the magnocellular nuclei of the hypothalamus, and all other gastrinlike immunoreactivity in the central nervous system is actually CCK (82).

The precursor for CCK has been cloned and contains the sequences of all the active forms of CCK ranging in size from 58 to 4 amino acids (Figure 3; 83, 84). However, while the carboxyl terminus of CCK is flanked by a gly-arg-arg sequence, production of several CCK species, including CCK-8, would require processing at single basic amino acids. While this step is unusual, it is also required for production of the opioid peptides dynorphin (1–8), rimorphin, and metorphamide (63, 64, 68).

Like proenkephalin A and POMC, proCCK undergoes tissue-specific processing. The intestine produces mainly CCK-33 (85), while the brain produces CCK-8 and smaller fragments (82). Nerves innervating the pancreas may use an even smaller fragment, CCK-4 (86). The distribution in the brain of the

PREPROCHOLECYSTOKININ

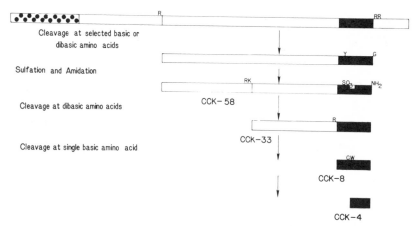

Figure 3   Processing of preprocholecystokinin. Preprocholecystokinin contains the sequences of all forms of cholecystokinin. The common carboxyl terminal of these peptides is flanked in the precursor by a glycine (G) and two basic amino acids so that the amidated carboxyl terminus (NH$_2$) can arise from the glycine residue. This region also contains the tyrosine that is sulfated in CCK. The amino terminus of the various forms is produced by cleavage at a single basic amino acid in the case of CCK-58 and CCK-8, at dibasic amino acids for CCK-33, and between a glycine (G) and a tryptophan (W) for CCK-4.

different forms of CCK has not yet been mapped in detail by im-munohistochemistry or detailed radioimmunoassay studies. While the levels of CCK in the brain are greater than those of perhaps any other neurotransmitter, it is unclear as yet how the levels of the different forms vary between regions. Such information is necessary to clarify which CCK-related peptides act as true neurotransmitters and which are precursors or degradation products.

Protachykinins

The precursors for substance P have also been cloned and sequenced from the cDNA (Figure 4; 10). Interestingly, two different putative precursor structures, called alpha- and beta-preprotachykinin, are obtained in this way, each of which contains the structure of substance P flanked by appropriate processing sites. However, B-PPT also contains a second sequence flanked by appropriate dibasic processing sites that is very similar to an amphibian peptide known as kassinin. This peptide, designated substance K for its homology to kassinin, contains the carboxyl terminal sequence phe-x-gly-leu-met-NH$_2$, which is common to all tachykinins including substance P. Consequently, this precursor sequence suggests the presence of a new putative peptide neurotransmitter that had not been identified by physiological means.

PROCESSING OF PREPROTACHYKININ GENE

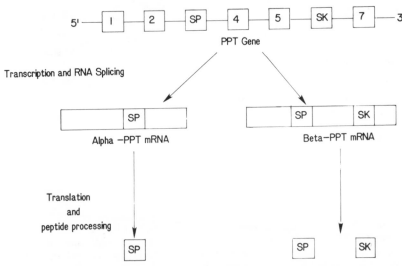

Figure 4 Processing of the preprotachykinin gene. One preprotachykinin gene can produce two mRNAs by differential RNA splicing. Following translation and peptide processing, alpha-preprotachykinin mRNA produces substance P while beta-preprotachykinin produces substance P and substance K.

The two distinct precursors of substance P appear to arise from a single gene (87). This gene is organized in the form of seven exons with the last six encoding portions of the protein. The sixth exon, which encodes substance K, is missing in the smaller precursor of substance P. Presumably, there is differential splicing of RNA transcribed from a single gene.

Interestingly, the alternate splicing does not seem to be a random event, as levels of the two precursor RNAs vary between tissues (87, 88). The central nervous system produces more of the alpha-PPT mRNA while the peripheral tissues produce more of the beta-PPT with the thyroid gland producing essentially no alpha-PPT. Thus the alternate splicing of the PPT gene offers another mechanism by which various forms of a given neuropeptide can be produced.

The examples of POMC, proenkephalin, and CCK precursors have demonstrated that even the mRNA sequence does not reveal exactly which peptides are formed because of the possibility of the differential processing of the precursors. Fortunately, concurrent with the cloning of the precursors, two new tachykinins were isolated from porcine spinal cord (89), one of which (termed neuromedin L or neurokinin alpha) is identical to substance K (90). The second novel tachykinin, termed neurokinin beta or neuromedin K (91), has not been found within any presently known precursor.

Which of these tachykinins are neurotransmitters in particular regions of the brain? Unfortunately, most of the original studies on the immunocytochemical localization of substance P within the brain used antisera which may cross-react with substance K or neurokinin beta. One recent study demonstrated the presence of a new N-terminally extended form of substance K in porcine brain but found relatively little authentic substance K in the brain (92). Antisera specific for each of these peptides may reveal whether they are differentially distributed in the brain and how the brain levels of these peptides correspond to the levels of the two mRNAs.

The two tachykinin precursors provide an example of how differential splicing of mRNAs, like tissue-specific processing, can give rise to different neuropeptides. By a similar alternate splicing event, the gene for calcitonin gives rise to two peptides, calcitonin and calcitonin gene–related polypeptide (CGRP) (93, 94). Calcitonin seems to be the major peptide made in the C cells of the thyroid, while CGRP is the major gene product in the nervous system. Both immunocytochemical (94, 95) and physiological experiments (96) suggest that CGRP is a transmitter in primary sensory cells.

Substance K and CGRP thus represent two putative neurotransmitters identified by an understanding of precursors and peptide processing rather than their physiological effects. Other experimenters have sought to take this approach further with recombinant DNA technology. By cloning the mRNA expressed in different regions of the brain and then screening for clones expressed in the brain but not in "non-neural" tissue like the liver, one might isolate clones that code for potential neuropeptides. Then by sequencing the clones and determining the amino acid sequence of the coded protein, one can see if the structure contains internal dibasic sites that could be processed to smaller bioactive peptides. The immunocytochemistry and physiology of such peptides could determine whether it did act as a transmitter. Sutcliffe and colleagues (97) used such an approach and predicted the presence of one precursor with three peptides that may act as transmitters. The immunocytochemistry shows that the three peptides are found in the same neurons and are distributed discretely in the brain. While this pattern is consistent with that of many neuropeptides, more physiological tests are necessary to show that such peptides are formed and that they are physiologically significant. Still, the strategy represents an interesting approach to the discovery of neuropeptides.

PROCESSING ENZYMES

The differential processing of neuropeptide precursors illustrates the importance of understanding the enzymes that produce such cleavages. In vitro a variety of enzymes can produce the proper cleavages; for example, trypsin and carboxypeptidase B are capable of producing enkephalins from proenkephalin A (61). However, the exact identity of enzymes that carry out such reactions in vivo is uncertain. A major question is whether these enzymes act only in the processing of propeptides and whether such enzymes act only on selected propeptides or groups of propeptides.

Enkephalin Convertase

Perhaps the best-characterized putative propeptide processing enzyme is a carboxypeptidase B–like enzyme designated carboxypeptidase H, carboxypeptidase E, or enkephalin convertase (EC 3.4.17.10) (98, 99). It is a novel carboxypeptidase whose properties suggest that it may be physiologically involved in processing precursors of enkephalins and other neuropeptides. It is exclusively carboxypeptidase B–like in its specificity as it removes arginine or lysine from the carboxyl end of peptides but does not cleave other amino acids. Mechanistically, it is classified as a metallopeptidase and is inhibited by EDTA and phenanthroline. It has an optimum pH of 5.5 and has been colocalized to secretory granules with enkephalins in the adrenal gland (98, 100) and with POMC-derived peptides in the anterior pituitary (101). It also has been shown to be cosecreted with enkephalins in adrenal chromaffin cells (102) and POMC-derived peptides in AtT 20 cells (103). Interestingly, it is present in both membrane and soluble forms, which differ in molecular mass by 2000 daltons (104). Another neurotransmitter-synthesizing enzyme, dopamine B-hydroxylase, which produces norepinephrine, is also found in membrane and soluble forms in the nervous system (105).

These characteristics all suggest that enkephalin convertase may be important in the synthesis of neuropeptides. More recent mapping of its microscopic localization within the brain and other parts of the body supports this conclusion. Guanidinoethylmercaptosuccinic acid (GEMSA) is a selective inhibitor of enkephalin convertase which has a K_i of about 10 nM for CPH but greater than 1 μM for other carboxypeptidase B–like enzymes (106). In tissue sections and homogenates the tritiated form of GEMSA binds selectively to enkephalin convertase (107). Consequently, the labeled GEMSA can be used to localize the membrane form of enkephalin convertase by light microscopic autoradiography (108). In the brain enkephalin convertase levels are highest in the median eminence and in other areas known to contain high levels of neuropeptides. While its overall distribution in general resembles that of enkephalins, it probably processes other peptides in the brain as well. Studies using lesioning techniques have shown that enkephalin convertase must exist in neurons that do not contain enkephalins (109).

The highest concentrations of enkephalin convertase in the body are found in the pituitary gland (99, 100). (3H)-GEMSA autoradiography localizes enkephalin convertase here to all three lobes with the highest levels of membrane-bound convertase being found in the intermediate lobe in the rat (100). An enkephalin convertase–like enzyme has also been characterized in insulinoma tissue, where it likely is important in the processing of insulin precursors (110). Consequently it seems likely that enkephalin convertase may play a role in the processing of a variety of neuropeptide precursors (111). How many precursors enkephalin convertase processes can only be speculated on at this point. The precursor structures reveal no need for the presence of multiple carboxypeptidase B–like enzymes. However, since at present not all neuropeptides have been discovered, it is hard to say that enkephalin convertase is involved in all the carboxypeptidase processing steps. In addition, in some areas of the brain such as the caudate nucleus and globus pallidus, the levels of enkephalin convertase are much lower than would be expected based on peptide mRNA levels. This could be accounted for by the presence of a distinct carboxypeptidase or by other mechanisms. Thus it is not clear exactly how many precursors enkephalin convertase acts on in vivo.

The regulation of enkephalin convertase levels has also been investigated. In cell culture of AtT 20 cells enkephalin convertase secretion is increased by administration of corticotropin-releasing factor (CRF), consistent with the packaging of enkephalin convertase in the same granules as POMC-derived peptides (103). However, enkephalin convertase levels are unaffected by long-term administration of glucocorticoids while POMC-derived peptide levels decrease, suggesting that enkephalin convertase levels are not regulated coordinately with peptide levels (103). This has been confirmed by in vivo studies as well, which show that treatments that raise adrenal enkephalin levels do not affect adrenal medullary enkephalin convertase levels (100). Similarly, haloperidol treatment, which increases anterior lobe POMC levels, does not affect anterior lobe enkephalin convertase levels (100). One study suggests that enkephalin convertase activity is increased by reserpine treatment of adrenal medullary chromaffin granules but the number of enkephalin convertase molecules did not change (112). Overall, these results suggest that enkephalin convertase levels are not coregulated with neuropeptide synthesis, possibly because enkephalin convertase seems to be present in excess in comparison to the levels of peptides in most tissues and thus may not catalyze a rate-limiting step in neuropeptide synthesis. These observations suggest that enkephalin convertase levels are not regulated in coordination with the neuropeptide levels.

Alpha Amidation Activity

Many neuropeptides contain carboxyl terminals that are alpha-amidated and in many cases the amidated end is required for biological activity. Neuropeptide Y was initially characterized on the basis of its amidated carboxyl terminal (113). In precursors of amidated peptides, the active peptide is followed by glycine and two basic amino acids. This suggests that amidation occurs by oxidative removal of glyoxylate from glycine leaving a carboxyl terminal amide (114–116).

The enzyme carrying out this reaction has been referred to as peptidyl glycine alpha-amidating monooxygenase (PAM) in view of its oxidative capabilities. It has been characterized from pig pituitary (114), all lobes of the rat pituitary (115–117), AtT 20 pituitary tumor cells (118), rat hypothalamus (119), and thyroid gland (120), and is similar in all tissues. It can be assayed by monitoring the conversion of D-tyr-val-gly to D-tyr-val-NH_2 (114). The enzyme requires molecular oxygen and copper and uses ascorbate as a cofactor (116). In these ways it resembles dopamine beta-hydoxylase, which synthesizes norepinephrine from dopamine (121). A variety of peptides with carboxyl terminal glycines can inhibit activity suggesting that the substrate specificty of this enzyme is broad. High levels of the enzyme are generally found in tissues known to produce large amounts of neuropeptides or peptide hormones such as all three lobes of the pituitary, the hypothalamus, and the submandibular gland (122). It is also released with POMC-derived peptides in AtT 20 cells showing that it is contained in secretory granules there (103). Unfortunately, the enzyme has not yet been mapped in detail in the nervous system so that one cannot yet assess which neuropeptide precursors it acts on. The biochemical data suggest that it works on many.

The regulation of PAM has been investigated in cell culture in AtT 20 cells and in whole animal experiments. Like enkephalin convertase it is released with POMC-derived peptides and release is stimulated by CRF (103). However, in contrast to enkephalin convertase, levels of PAM decrease with long-term administration of glucocorticoids, showing that PAM activity is coregulated with peptide levels (103). In rat intermediate lobe, PAM activity is regulated coordinately with POMC-derived peptides as levels increase with treatment with haloperidol, a dopaminergic receptor antagonist, and decline with treatment with bromocriptine, a dopaminergic agonist (123). In the rat anterior lobe PAM activity increases with adrenalectomy and with castration, and decreases with thyroidectomy and haloperidol treatment (123). This is consistent with coregulation of PAM and POMC-derived peptides in corticotrophs, but the heterogeneous nature of the anterior pituitary makes interpretation of the other changes difficult (123). Still, the overall scheme suggests that PAM is regulated with neuropeptide levels in many cases.

Other Processing Enzymes

The potentially most important processing enzymes, the endopeptidases that act at dibasic amino acids, have yet to be fully characterized. Since the distribution of these enzymes is likely responsible for the tissue-specific processing differences, there are probably multiple such enzymes whose activity is tightly controlled. Several candidate enzymes have been characterized (124–131). They have a variety of substrate specificities and pH optima. However, until we know more about the biochemistry and the subcellular localization of these enzymes, it is impossible to assess their role in peptide processing. The search for such enzymes still represents the most important challenge in the field of processing enzymes.

INACTIVATION OF NEUROPEPTIDES

If neuropeptides serve as neurotransmitters, there must be specific ways of terminating their actions at the synapse. Classical transmitters are inactivated enzymatically, as with the hydrolysis of acetylcholine by acetylcholinesterase (132), or by reuptake, as for norepinephrine and dopamine (133). For peptides the major proposed mechanism of inactivation is enzymatic degradation by peptidases. Numerous proteases capable of inactivating peptide neurotransmitters are present in the brain, but it is likely that relatively few physiologically inactivate peptide transmitters. To demonstrate conclusively that a specific protease is important in the synaptic inactivation of a neuropeptide several criteria must be met. Such peptidases must have the proper biochemical specificity to inactivate the peptide and must act under conditions present at the synapse. They also must be localized to the appropriate regions of the brain and at the proper synapses to carry out such functions. Finally, interference with the function of such putative degradative enzymes in vivo should potentiate the physiological or behavioral actions of the peptide. By determining the answers to these questions one can ascertain whether such peptidases act on one transmitter, on many transmitters, or if they function in the role of inactivating neuropeptides in vivo at all.

The questions have been best addressed for the peptidases that degrade the enkephalins. The enkephalins can be inactivated by cleavage at the tyr-gly bond by aminopeptidases, at the gly-gly bond by dipeptidyl aminopeptidases, or at the gly-phe bond by dipeptidyl carboxypeptidases (134). One dipeptidyl aminopeptidase capable of inactivating enkephalins in vitro has been characterized. Its distribution, when mapped histochemically, is unique but does not correspond to that of enkephalin or to that of other neurotransmitters (135–137). Thus, if it does act as an inactivating enzyme, it must certainly act on a variety of substrates.

The aminopeptidases capable of inactivating enkephalins in vitro have been better characterized. Two aminopeptidases have been characterized in rat brain membrane fractions, one of which is sensitive to the inhibitors puromycin and bestatin and the other of which is sensitive to bestatin but insensitive to puromycin (138, 139). The bestatin-sensitive, puromycin-insensitive enzyme appears to be similar if not identical to the kidney enzyme aminopeptidase M (138, 140). Biochemically these enzymes may cleave a variety of substrates so that their potential role in enkephalin metabolism is unclear based solely on in vitro criteria.

Similarly, at least two dipeptidyl carboxypeptidaselike enzymes can degrade the enkephalins. Angiotensin converting enzyme (ACE) cleaves enkephalin but with a K_m greater than that of other enzymes (141–143). ACE is selectively inhibited by the drug captopril. A second enzyme designated enkephalinase or endopeptidase 24.11 has also been implicated in the metabolism of enkephalins. This enzyme was first thought to be a dipeptidylcarboxypeptidase (144–147), but was subsequently found to be identical to a phosphoramidon-sensitive endopeptidase characterized from rabbit kidney (148). It is also inhibited with nanomolar potency by thiorphan (149), and can cleave numerous substrates in vitro including substance P (150, 151), CCK (152), and insulin (148).

In vitro biochemical specificity determines which peptidases are capable of cleaving enkephalins, but does not demonstrate which peptidases perform this function physiologically. If one adds (3H)-tyr-enkephalins to brain slices, the products are (3H)-tyr (60%), (3H)-tyr-gly (2%), and (3H)-tyr-gly-gly (18%) (153, 154). These results suggest that at least for exogenously added enkephalins the aminopeptidase and dipeptidylcarboxypeptidase activities are more important in enkephalin degradation. Studies with selective inhibitors have proved useful in determining which specific enzymes carry out these reactions for endogenous enkephalins. Bestatin and thiorphan prevent the degradation of endogenously released enkephalins suggesting that one of the bestatin-sensitive aminopeptidases and enkephalinase act in the degradation of endogenously released enkephalins (153). However, neither puromycin nor captopril prevent degradation of exogenously added or endogenously released enkephalins, suggesting that neither the puromycin-sensitive aminopeptidase nor ACE physiologically degrade enkephalins. These results conflict with the in vitro results for the aminopeptidases, which suggest that enkephalin is more easily degraded by the puromycin-sensitive enzyme (139), emphasizing that peptides may be metabolized differently physiologically than in vitro.

Other tests have examined the ability of thiorphan and bestatin to potentiate the behavioral effects of enkephalins. Both of these inhibitors exhibit antinociceptive effects in selected in vivo experiments while captopril has no such effect (153–155). The antinociceptive effects are reversed by naloxone, showing that these effects are mediated through endogenous opioid peptides. These results suggest that a bestatin-sensitive aminopeptidase and enkephalinase act as enkephalin-degrading enzymes in vivo, while ACE does not.

The results of these experiments should be interpreted cautiously. Thiorphan itself is not absolutely specific for enkephalinase so that thiorphan-induced analgesia might involve enzymes other than enkephalinase. The precise localization of these enzymes in the brain is particularly important in assessing their physiological significance. It is quite possible that they exert their antinociceptive effects at locations remote from enkephalinergic synapses. In the kidney both enkephalinase (156) and aminopeptidase M (141) are found on the brush borders of the proximal tubules, and in the gastrointestinal tract enkephalinase occurs on the luminal surface of the mucosa (156). Such a localization is inconsistent with a role in processing any specific peptide hormone as these enzymes will cleave a variety of peptides with which they are presented. Thus in peripheral tissues the localization of these enzymes does not suggest any specific function. In the brain the detailed microscopic localization of such enzymes has recently been characterized. The regional distribution of enkephalinase assayed in homogenates generally parallels that of enkephalins and opiate receptors (144, 145). Recently, Roques and colleagues have used a tritiated inhibitor of enkephalinase to localize it in the brain by light microscopic autoradiography (157), and Gorenstein and colleagues (158) have localized enkephalinase using histochemical techniques. Its localization closely parallels that of opiate receptors in the brain, suggesting that enkephalinase is selectively associated with enkephalin degradation. While more physiological experiments are needed to confirm this, the localization of enkephalinase supports its role as a physiological inactivating enzyme for the enkephalins.

While physiological experiments suggest that ACE is not involved in inactivating enkephalins, knowledge on its localization does demonstrate how the combination of biochemical and anatomical properties can help determine the function of a putative degradative enzyme. ACE was originally discovered by its ability to convert angiotensin I to angiotensin II, a potent vasoconstrictor, in the lung (159). It also inactivates the vasodilator bradykinin so that peripherally it may play a role in the maintenance of blood pressure (159). It acts on other substrates including the amidated peptides substance P (160–163) and LHRH (164). ACE has been best characterized from the lung (159), the kidney, and the testis (165), but it is abundant in the brain as well (166–174).

In the brain ACE has been localized by dissection to the choroid plexus and to the corpus striatum (171, 172). More recently, the rat brain isozyme has been localized using in vitro autoradiograpthy with (3H)-captopril, a selective and potent inhibitor of ACE (173). ACE is found in a striatonigral pathway, the subfornical organ, the choroid plexus, and the magnocellular nuclei of the hypothalamus (173). In the magnocellular nuclei and the subfornical organ, there are high levels of angiotensin so that ACE found there may be related to angiotensin metabolism (166). These two regions of the brain have been implicated in fluid balance and blood pressure regulation so that ACE here may subserve the same physiological functions as in the periphery. However, in the corpus striatum and substantia nigra angiotensin levels are low so that there may be another endogenous neuropeptide substrate for ACE.

Recently, the striatal isozyme of ACE has been isolated (174). Though its molecular weight differs from that of the lung isozyme, its catalytic properties are identical to those of the lung isozyme except for the ability of the striatal isozyme to cleave amidated substrates such as substance P and substance K (160, 174). The striatal isozyme cleaves amidated peptides much faster than the lung enzyme and the peptide substance K is cleaved by striatal ACE but not by lung ACE. Interestingly, in the striatonigral pathway, substance K is thought to be a major neurotransmitter. This suggests that the ACE here metabolizes substance K.

While this hypothesis is quite attractive, it still requires in vivo confirmation. In addition the physiological specificity of ACE for substance K and substance P in vivo must be confirmed. Still this provides an excellent example of how the anatomical and biochemical properties of an enzyme can combine to produce a potentially specific metabolic enzyme.

POSTSYNAPTIC EFFECTS

The variety of posttranslational modifications and degradative schemes demonstrate how the multiple forms of neuropeptides can arise. But why are there different forms and what is their physiological significance? One possibility is that each form may serve a separate function or interact with a subgroup of postsynaptic receptors. Here again it is instructive to examine actions of classical neurotransmitters. Acetylcholine acts on at least two types of receptors: nicotinic receptors and muscarinic receptors. Here, there is one endogenous transmitter for two receptors. The case of circulating catecholamines is different. At least four different receptors can bind adrenergic hormones peripherally with different selectivities for epinephrine and norepinephrine (175). The various receptors appear to select differentially for various ligands.

Such would also seem to be the case for some of the peptide receptors that have been characterized. Opiate receptors and tachykinin receptors provide useful examples.

Opiate Receptors

Multiple types of opiate receptors, designated mu receptors, delta receptors, and kappa receptors, have been identified in the brain. They differ in the physiological effects they mediate, in their pharmacology, and in their localization within the brain. Mu receptors are most closely associated with the traditional analgesic properties of morphine (176), while kappa receptors are associated with the sedative effects of some opiates (177). Delta receptors seem to mediate some limbic system functions including reward behaviors and limbic seizures (178).

These receptors also differ in their binding pharmacology. Mu receptors display selective high affinity for morphine and its derivatives and are most specifically labeled by Tyr-D-ala-gly-Mephe-gly-ol (DAGO, DAMGE) (179–181). Delta receptors are selectively labeled by derivatives of the enkephalins such as D-pen2-D-pen5-enkephalin (182). Kappa receptors have high affinity for ethylketocyclazocine and bromazocine (181) and the selective kappa agonist U50488 (183). This differential pharmacology has permitted localization of these receptors within the brain using autoradiography. Mu receptors are most concentrated in the thalamus, hypothalamus, hippocampus, and the periaqueductal gray (184). Delta receptors are enriched in the amygdala and nucleus accumbens while the substantia gelatinosa of the spinal cord and the nucleus of the solitary tract contain both delta and mu receptors (184). Kappa receptors are concentrated in the deep layers of the cerebral cortex and in lower levels in the cerebellum in the guinea pig (185, 186), while in the rat the highest levels seem to be in the striatum (187). These localizations correlate relatively well with their proposed functions, with mu receptors being high in areas important for pain transmission like the periaqueductal grey and delta receptors being concentrated in the limbic system, which mediates emotional responses.

POMC and proenkephalin A– and B–derived peptides differ in affinity for the subtypes of opiate receptors. Enkephalins bind with high affinity to mu and delta receptors but have a low affinity for kappa receptors while B-endorphin binds better to mu than to delta receptors (188). The proenkephalin B–derived peptides, the neoendorphins, rimorphins, and dynorphins, have selectively high affinity for the kappa receptor (189–193). The larger proenkephalin A–derived peptides have lower affinity for delta receptors than the enkephalins but higher affinity for the kappa receptor (194). Interestingly, metorphamide has high affinity for mu and kappa receptors, a different profile than either the enkephalins or the proenkephalin B–derived peptides (64).

However, high affinity of a peptide for a receptor does not establish the peptide as the endogenous ligand for that receptor. The distribution of the receptors must be compared to that of the putative endogenous ligand. In the case of the enkephalins, the distribution of enkephalins generally matches that of the mu and delta receptors, indicating that the enkephalins are endogenous ligands for the mu and delta receptors. However, the distribution of kappa receptors does not match that of the proenkephalin A– or B–derived peptides. Kappa receptors are found in high concentration in the deep layers of cerebral cortex where levels of proenkephalin B–derived peptides are low (76). Consequently perhaps some other ligand exists for the kappa receptor here. Alternatively, these receptors may be expressed for no apparent reason. High levels of catecholamine receptors occur in the rat striatum even though epinephrine and norepinephrine concentrations are low here (195, 196). Thus, the exact endogenous ligand or ligands for the kappa receptor have not yet been determined.

Tachykinin Receptors

A second family of peptides, the tachykinins, also provides an opportunity to correlate multiple forms of neuropeptides with multiple receptors. This family includes substance P, the amphibian peptide kassinin, and the molluscan peptide physalaemin as well as the more recently discovered peptides, neuromedin K and substance K. The presence of multiple receptors for these peptides is suggested by bioassays (197–200). Substance P and physalaemin are relatively potent compared to kassinin at contracting smooth muscle in the guinea pig ileum but far less potent in their actions on the vas deferens or bladder. The receptor found in the guinea pig ileum assay has been designated the substance P-P (SP-P) receptor while the receptor of the bladder and vas deferens has been designated the substance P-E (SP-E) receptor. The more recently discovered peptides, substance K (201) and neuromedin K (202), resemble kassinin in their preference for the SP-E receptor.

The finding of differential responses in bioassays has been complemented by studies with binding assays. I(125)-substance P binds with high affinity to membranes from brain and peripheral tissues (203–206). Binding is inhibited by substance P and its fragments while kassinin and eledoisin are substantially less potent. The different potencies of tachykinins roughly match the responses of the guinea pig ileum to these compounds, suggesting that the labeled site is the substance P-P receptor. Perhaps another receptor might mediate the activities of the SP-E response. This is confirmed through the binding of I(125)-eledoisin to rat brains (207, 208). This binding site has a pharmacology like the response of the rat vas deferens with kassinin and eledoisin being relatively potent and substance P less potent. The two sites are also differentially localized within rat brain (209). The septum has the highest levels of SP-P receptors while SP-E receptors are highest in the cerebral cortex. In the spinal cord the two types have been differentially localized by in vitro autoradiography (210, 211). While I(125)-substance P binds to both the dorsal and ventral horns of the spinal cord, I(125)-eledoisin binding is discretely localized to the dorsal most portion of the dorsal horn of the spinal cord.

While the high affinity of substance P for the I(125)-substance P–binding site suggests that it might be the endogenous ligand for this receptor, the ligand for the eledoisin site might be some other peptide. The recently discovered peptides, substance K and neuromedin K (212–214), seem to act most like kassinin, and neuromedin K is more potent at displacing I(125)-eledoisin than I(125)-substance P. Substance K seems to be potent at both binding sites (212–214). Interestingly, I(125)-substance K is displaced most potently by itself, suggesting perhaps a third class of binding sites (214). By autoradiography, substance K–binding sites differ from substance P–binding sites in localization (210). Thus binding studies support the idea that neuromedin K may be the endogenous ligand for the SP-E receptor, while substance K may act through its own receptor or through the other receptors.

How do the levels of tachykinins and receptor subtypes match anatomically within the central nervous system? Substance P immunoreactivity has been mapped and matches that of substance P receptors in selected regions of the nervous system (208–214). However, the mapping has used antibodies that may cross-react with substance K or neuromedin K. Immunocytochemical mapping of these new tachykinins has not yet been reported. In vitro autoradiography localizes substance P–, eledoisin-, and substance K–binding sites to distinct regions of the nervous system (208–210). In the substantia nigra there are high levels of tachykinins and high levels of substance K binding but low levels of substance P binding. To fully understand what tachykinins act at what receptors it will be necessary to localize the individual peptides specifically with immunocytochemistry.

CONCLUSIONS: FUNCTIONAL RELEVANCE

What are the behavioral functions of neuropeptides? Some peptides have been linked to specific functions while others have not. In general neuropeptides play similar roles in the CNS and periphery. In the gastrointestinal tract CCK promotes gallbladder contraction, while in the brain it seems to play a role in satiety (215, 216). More interesting still are the various peptides that regulate blood pressure and fluid balance. Angiotensin is a potent vasoconstrictor which is produced by extracellular cleavage of the serum protein angiotensinogen. In the brain angiotensin and its receptors have been localized to a few sites, many of which are important in fluid balance. Angiotensin is found in the supraoptic and paraventricular nuclei of the hypothalamus (166), which contain the cells that produce and release vasopressin into the blood from the posterior pituitary. Vasopressin acts on the kidney to promote water reabsorption. Angiotensin and its receptors are also found in the subfornical organ, which regulates drinking behavior. This portion of the brain is outside the blood-brain barrier so that circulating angiotensin can interact with receptors here. Angiotensin II is also found in the parabrachial nucleus and the nucleus of the solitary tract, two regions involved in blood pressure control. In spontaneously hypertensive rats angiotensin II turnover is increased, thus further associating the central and peripheral functions of this peptide (217).

Atrial natriuretic factor acts peripherally to lower blood pressure by causing vasodilation and natriuresis. In the central nervous system, it has been localized to the dorsal tegmental nucleus, which modulates blood pressure, and the organum vasculosum of the lamina terminalis (28–30, 218). Receptors for it are found in the subfornical organ and the nucleus of the solitary tract, two regions that are thought to be important in controlling fluid balance and blood pressure (219, 220).

That some peptides mediate particular functions both peripherally and in the CNS is important in considering therapeutic interventions. Different neuropeptides share a great many characteristics, such as pathways, methods of synthesis, and inactivation. Future research may clarify differences between neuropeptides, such as tissue-specific and peptide-specific synthesizing and degrading enzymes as well as specific receptors.

ACKNOWLEDGMENTS

We wish to thank our colleagues who have kindly supplied us with reprints to aid us in the research for this manuscript. These people include Huda Akil, Betty Eipper, Ron Emeson, Detlev Ganten, Jacques Glowinski, Ed Herbert, Leslie Iversen, A. John Kenny, Dick Mains, Shigetada Nakanishi, Shosaku Numa, Jens Rehfeld, Bernard Roques, Jean-Charles Schwartz, Howard Tager, Yvette Torrens, Anthony J. Turner, Stanley Watson, and Nadav Zamir.

Literature Cited

1. Hughes, J., Smith, T. W., Kosterlitz, H. W., Fothergill, L. H., Morgan, B. A., Morris, H. R. 1975. *Nature* 258:577–579
2. Rehfeld, J. F. 1985. *J. Neurochem.* 44: 1–10
3. Ganten, D., Lang, R. E., Lehmann, E., Unger, T. 1984. *Biochem. Pharmacol.* 33:3523–28
4. Loewi, O. 1921. *Pluegers Arch. Ges. Physiol.* 189:239–42
5. Von Euler, U. S., Gaddum, J. H. 1931. *J. Physiol.* 72:74–87
6. Page, I. H., Helmer, O. M. 1940. *J. Exp. Med.* 71:29–42
7. Carraway, R., Leeman, S. E., Niall, H. D. 1971. *J. Biol. Chem.* 248:6854–61
8. Terenius, L., Wahlstrom, A. 1974. *Acta Pharmacol. Toxicol.* 35 (Suppl. 1): 55 (Abstr.)
9. Simantov, R., Snyder, S. H. 1976. *Proc. Natl. Acad. Sci. USA* 73:2515–19
10. Chang, M. M., Leeman, S. E., Niall, H. D. 1971. *Nature New Biol.* 232:86–87
11. Nawa, H., Hirose, T., Takashima, H.,

Inayama, S., Nakanishi, S. 1983. *Nature* 306:32–36

12. Gauer, O. H., Henry, J. P., Sieker, H. O. 1961. *Prog. Cardiovasc. Dis.* 4:1–26
13. deBold, A. J., Borenstein, H. B., Veress, A. T., Sonnenberg, H. 1981. *Life Sci.* 28:89–94
14. Trippodo, N. C., Macphee, A. A., Cole, F. E., Blakesley, H. L. 1982. *Proc. Soc. Exp. Biol. Med.* 170:502–8
15. Flynn, T. G., deBold, M. L., deBold, A. J. 1983. *Biochem. Biophys. Res. Commun.* 117:859–65
16. Currie, M. G., Geller, D. M., Cole, B. R., Siegel, N. R., Fok, K. F., et al. 1984. *Science* 223:67–69
17. Kangawa, K., Fukuda, A., Minamino, N., Matsuo, H. 1984. *Biochem. Biophys. Res. Commun.* 119:933–40
18. Misono, K. S., Fukumi, H., Grammer, R. T., Inagami, T. 1984. *Biochem. Biophys. Res. Commun.* 119:524–29
19. Seidah, N. G., Lazure, C., Chretien, M., Thibault, G., Garcia, R., et al. 1984. *Proc. Natl. Acad. Sci. USA* 81:2640–44
20. Atlas, S. A., Kleiner, H. D., Camargo, M. J., Januszewicz, A., Sealey, J. E., et al. 1984. *Nature* 309:717–19
21. Yamanaka, M., Greenberg, B., Johnson, L., Seilhamer, J., Brewer, M., et al. 1984. *Nature* 309:719–22
22. Maki, M., Takayanagi, R., Misono, K. S., Pandey, K. N., Tibbetts, C., Inagami, T. 1984. *Nature* 309:722–24
23. Kakayama, K., Ohkubo, H., Hirose, T., Inayama, S., Nakanishi, S. 1984. *Nature* 310:699–701
24. Oikawa, S., Imai, M., Ueno, A., Tanaka, S., Noguchi, T., et al. 1984. *Nature* 309:724–26
25. Nemer, M., Chamberland, M., Sirois, D., Argentin, S., Drouin, J., et al. 1984. *Nature* 312:654–56
26. Greenberg, B. D., Bencen, G. H., Seilhamer, J. J., Lewicki, J. A., Fiddes, J. C. 1984. *Nature* 312:656–58
27. Argentin, S., Nemer, M., Drouin, J., Scott, G. K., Kennedy, B. P., Davies, P. L. 1985. *J. Biol. Chem.* 260:4568–71
28. Jacobowitz, D. M., Skofitsch, G., Keiser, H. R., Eskay, R. L., Zamir, N. 1985. *Neuroendocrinology* 40:92–95
29. Zamir, N., Skofitsch, G., Eskay, R. L., Jacobowitz, D. M. 1986. *Brain Res.* In press
30. Morii, N., Nakao, K., Sugawara, A., Sakamoto, M., Suda, M., et al. 1985. *Biochem. Biophys. Res. Commun.* 127:413–19
31. Hebb, C. 1972. *Physiol. Rev.* 52:918–55
32. Freeman, J. J., Jenden, D. J. 1976. *Life Sci.* 19:949–61
33. Horinishi, H., Grillo, M., Margolis, F. 1978. *J. Neurochem.* 31:909–19

34. Nolan, C., Margoliash, E., Peterson, J. D., Steiner, D. F. 1971. *J. Biol. Chem.* 246:2780–95
35. Docherty, K., Steiner, D. F. 1982. *Ann. Rev. Physiol.* 44:625–38
36. Douglass, J., Civelli, O., Herbert, E. 1984. *Ann. Rev. Biochem.* 53:665–715
37. Eipper, B. A., Mains, R. E. 1980. *Endocrine Rev.* 1:1–27
38. Drouin, J. P., Goodman, H. M. 1980. *Nature* 288:610–12
39. Chretien, M., Seidah, N. G. 1981. *Mol. Cell. Endocrinol.* 21:101–27
40. Eipper, B. A., Mains, R. E. 1978. *J. Biol. Chem.* 253:5732–44
41. Akil, H., Young, E., Watson, S. J., Coy, D. H. 1981. *Peptides* 2:289–92
42. Deakin, J. F., Dostrovsky, J. O., Smyth, D. G. 1980. *Nature* 279:74–75
43. Bloom, F., Battenberg, E., Rossier, J., Ling, N., Guillemin, R. 1978. *Proc. Natl. Acad. Sci. USA* 75:1591–95
44. Watson, S. J., Richard, C. W. III, Barchas, J. D. 1978. *Science* 200:1180–82
45. Oliver, C., Porter, J. C. 1978. *Endocrinology* 102:697–705
46. Joseph, S. A. 1980. *Am. J. Anat.* 158:533–648
47. Schwartzberg, D. G., Nakane, P. K. 1983. *Brain Res.* 276:351–56
48. Romagnano, M. A., Joseph, S. A. 1983. *Brain Res.* 276:1–16
49. Gee, C. E., Chen, C. L., Roberts, J. L., Thompson, R., Watson, S. J. 1983. *Nature* 306:374–76
50. Kelsey, J., Watson, S. J., Burke, S., Akil, H., Roberts, J. L. 1986. *J. Neurosci.* In press
51. Gramsch, C., Kleber, G., Hollt, V., Pasi, A., Mehraein, P., Herz, A. 1980. *Brain Res.* 192:109–19
52. Barnea, A., Cho, G., Porter, J. C. 1981. *J. Neurochem.* 36:1083–92
53. Emeson, R., Eipper, B. A. 1986. *J. Neurosci.* 6:In press
54. Zakarian, S., Smyth, D. 1979. *Proc. Natl. Acad. Sci. USA* 76:5972–76
55. Zakarian, S., Smyth, D. G. 1982. *Nature* 296:250–52
56. Liotta, A. S., Advis, J. P., Krause, J. E., McKelvy, J. F., Krieger, D. T. 1984. *J. Neurosci.* 4:956–65
57. Dores, R. M., Jain, M., Akil, H. 1986. *Brain Res.* In press
58. Noda, M., Furutani, Y., Takahashi, H., Toyosato, M., Hirose, T., et al. 1982. *Nature* 295:202–6
59. Gubler, U., Seeburg, P., Hoffman, B. J., Gage, L. P., Udenfriend, S. 1982. *Nature* 295:206–8
60. Comb, M., Seeburg, P. H., Adelman, J., Eiden, L., Herbert, E. 1982. *Nature* 295:663–66
61. Boarder, M. R., Lockfeld, A. J., Bar-

chas, J. D. 1982. *J. Neurochem.* 39:149–54

62. Stern, A. S., Lewis, R. V., Kimura, S., Rossier, J., Stein, S., Udenfriend, S. 1980. *Arch. Biochem. Biophys.* 205:606–13

63. Liston, D., Patey, G., Rossier, J., Verbanck, P., Vanderhaeghen, J. J. 1984. *Science* 225:734–37

64. Weber, E., Esch, F. S., Bohlen, P., Paterson, S., Corbett, A. D., et al. 1983. *Proc. Natl. Acad. Sci. USA* 80:7362–66

65. Matsuo, H., Miyata, A., Mizuno, K. 1984. *Nature* 305:721–23

66. Seizinger, B. R., Liebisch, D. C., Gramsch, C., Herz, A., Weber, E., et al. 1985. *Nature* 313:57–59

67. Merchenthaler, I., Maderdrut, J. L., Dockray, G. J., Altschuler, R. A., Petrusz, P. 1986. *Neuroscience.* In press

68. Kakidani, H., Furutani, Y., Takahashi, H., Noda, M., Morimoto, Y., et al. 1982. *Nature* 298:245–49

69. Civelli, O., Douglass, J., Goldstein, A., Herbert, E. 1985. *Proc. Natl. Acad. Sci. USA* 82:4291–95

70. Weber, E., Barchas, J. D. 1983. *Proc. Natl. Acad. Sci. USA* 80:1125–29

71. Zamir, N., Palkovits, M., Brownstein, M. J. 1984. *J. Neurosci.* 4:1240–47

72. Zamir, N., Palkovits, M., Brownstein, M. J. 1984. *J. Neurosci.* 4:1248–52

73. Weber, E., Evans, C. J., Chang, J. K., Barchas, J. D. 1982. *Biochem. Biophys. Res. Commun.* 108:81–88

74. Weber, E., Evans, C. J., Barchas, J. D. 1982. *Nature* 299:77–79

75. Seizinger, B. R., Grimm, C., Hollt, V., Herz, A. 1984. *J. Neurochem.* 42:447–57

76. Zamir, N., Weber, E., Palkovits, M., Brownstein, M. J. 1984. *Proc. Natl. Acad. Sci. USA* 81:6886–89

77. Zamir, N., Palkovits, M., Mezey, E., Brownstein, M. J. 1984. *Nature* 307:642–45

78. Khachaturian, H., Lewis, M. E., Schafer, M. K. H., Watson, S. J. 1985. *Trends. Neurosci.* 8:111–19

79. Vanderhaeghen, J. J., Signeau, J. C., Gepts, W. 1975. *Nature* 257:604–5

80. Dockray, G. J. 1976. *Nature* 264:568–70

81. Rehfeld, J. F. 1978. *J. Biol. Chem.* 253:4022–30

82. Marley, P. D., Rehfeld, J. F., Emson, P. C. 1984. *J. Neurochem.* 42:1523–35

83. Gubler, U., Chua, A. O., Hoffman, B. J., Collier, K. J., Eng, J. 1984. *Proc. Natl. Acad. Sci. USA* 81:4307–10

84. Deschenes, R. J., Lorenz, L. J., Haun, R. S., Roos, B. A., Collier, K. J., Dixon, J. E. 1984. *Proc. Natl. Acad. Sci. USA* 81:726–30

85. Jorpes, E., Mutt, V. 1966. *Acta Physiol. Scand.* 66:196–202

86. Rehfeld, J. F., Larsson, L. I., Goltermann, N. R., Schwartz, T. W., Holst, J. J., et al. 1980. *Nature* 284:33–38

87. Nawa, H., Kotani, H., Nakanishi, S. 1984. *Nature* 312:729–34

88. Nakanishi, S. 1985. *Trends Neurosci.* In press

89. Kimura, S., Okada, M., Sugita, Y., Kanazawa, I., Munekata, E. 1983. *Proc. Jpn. Acad.* 59:101–4

90. Maggio, J. E., Sandberg, B. E. B., Bradley, C. V., Iversen, L. L., Santikarn, S., Williams, D. H., et al. 1983. *Ir. J. Med. Sci.* 152 (Suppl. 1):20–21

91. Kangawa, K., Minamino, N., Fukuda, A., Matsuo, H. 1983. *Biochem. Biophys. Res. Commun.* 114:533–40

92. Tatemoto, K., Lunberg, J. M., Jornvall, H., Mutt, V. 1985. *Biochem. Biophys. Res. Commun.* 128:947–53

93. Amara, S. G., Jonas, V., Rosenfeld, M. G., Ong, E. S., Evans, R. M. 1982. *Nature* 298:240–44

94. Rosenfeld, M. G., Mermod, J. J., Amara, S. G., Swanson, L. W., Sawchenko, P. E., et al. 1983. *Nature* 304:129–35

95. Gibson, S. J., Polak, J. M., Bloom, S. R., Sabate, I. M., Mulderry, P. M., et al. 1984. *J. Neurosci.* 4:3101–11

96. Mason, R. T., Peterfreund, R. A., Sawchenko, P. E., Corrigan, A. Z., Rivier, J. E., Vale, W. W. 1984. *Nature* 308:653–55

97. Bloom, F. E., Battenberg, E. L. F., Milner, R. J., Sutcliffe, J. G. 1985. *J. Neurosci.* 5:1781–802

98. Fricker, L. D., Snyder, S. H. 1982. *Proc. Natl. Acad. Sci. USA* 79:3886–90

99. Fricker, L. D., Snyder, S. H. 1983. *J. Biol. Chem.* 258:10950–55

100. Strittmatter, S. M., Lynch, D. R., Desouza, E. B., Snyder, S. H. 1985. *Endocrinology.* 117:1667–74

101. Hook, V. Y. H., Loh, Y. P. 1984. *Proc. Natl. Acad. Sci. USA* 81:2776–80

102. Hook, V. Y. H., Eiden, L. E. 1985. *Biochem. Biophys. Res. Commun.* 128:563–70

103. Mains, R. E., Eipper, B. A. 1984. *Endocrinology* 115:1683–90

104. Supattapone, S., Fricker, L. D., Snyder, S. H. 1984. *J. Neurochem.* 42:1017–23

105. Molinoff, P. B., Axelrod, J. 1971. *Ann. Rev. Biochem.* 40:465–500

106. Fricker, L. D., Plummer, T. H., Snyder, S. H. 1983. *Biochem. Biophys. Res. Commun.* 111:994–1000

107. Strittmatter, S. M., Lynch, D. R., Snyder, S. H. 1984. *J. Biol. Chem.* 259:11812–17

108. Lynch, D. R., Strittmatter, S. M., Snyder, S. H. 1984. *Proc. Natl. Acad. Sci. USA* 81:6543–47
109. Lynch, D. R., Strittmatter, S. M., Venable, J. C., Snyder, S. H. 1986. *J. Neurosci.* In press
110. Docherty, K., Hutton, J. C. 1983. *FEBS Lett.* 162:137–41
111. Fricker, L. D. 1985. *Trends Neurosci.* 8:210–14
112. Hook, V. Y. H., Eiden, L. E., Pruss, R. M. 1985. *J. Biol. Chem.* 260:5991–97
113. Tatemoto, K., Carlquist, M., Mutt, V. 1982. *Nature* 296:659–60
114. Bradbury, A. F., Finnie, M. D. A., Smyth, D. G. 1982. *Nature* 298:686–88
115. Eipper, B. A., Mains, R. E., Glembotski, C. C. 1983. *Proc. Natl. Acad. Sci. USA* 80:5144–48
116. Glembotski, C. C., Eipper, B. A., Mains, R. E. 1984. *J. Biol. Chem.* 259:6385–92
117. Eipper, B. A., Glembotski, C. C., Mains, R. E. 1983. *Peptides* 4:921–28
118. Mains, R. E., Glembotski, C. C., Eipper, B. A. 1984. *Endocrinology* 114:1522–30
119. Emeson, R. B. 1984. *J. Neurosci.* 4:2604–13
120. Husain, I., Tate, S. S. 1983. *FEBS Lett.* 152:277
121. Diliberto, E. J., Allen, P. L. 1981. *J. Biol. Chem.* 256:3385–93
122. Eipper, B. A., Myers, A. C., Mains, R. E. 1985. *Endocrinology* 116:2497–504
123. Mains, R. E., Myers, A. C., Eipper, B. A. 1985. *Endocrinology* 116:2505–15
124. Fletcher, D. J., Quigley, J. P., Bauer, G. E., Noe, B. D. 1981. *J. Cell Biol.* 90:312–22
125. Loh, Y. P., Gainer, H. 1982. *Proc. Natl. Acad. Sci. USA* 79:108–12
126. Evangelista, R., Ray, P., Lewis, R. V. 1982. *Biochem. Biophys. Res. Commun.* 106:895–902
127. Mizuno, K., Miyata, A., Kangawa, K., Matsuo, H. 1982. *Biochem. Biophys. Res. Commun.* 108:1235–42
128. Mizuno, K., Matsuo, H. 1984. *Nature* 309:558–60
129. Lindberg, I., Yang, H. Y. T., Costa, E. 1982. *Biochem. Biophys. Res. Commun.* 106:186–93
130. Lindberg, I., Yang, H. Y. T., Costa, E. 1984. *J. Neurochem.* 42:1411–19
131. Mizuno, K., Kojima, M., Matsuo, H. 1985. *Biochem. Biophys. Res. Commun.* 128:884–91
132. Dudai, Y., Herzberg, M., Silman, I. 1973. *Proc. Natl. Acad. Sci. USA* 70:2473–77
133. Iversen, L. L. 1971. *Br. J. Pharmacol.* 41:571–91
134. Schwartz, J. C., De la Baume, S., Yi, C. C., Chaillet, P., Marcais-Collado, H., Costentin, J. 1982. *Prog. Neuro-Psychopharmacol. Biol. Psychiatr.* 6:665–71
135. Gorenstein, C., Snyder, S. H. 1979. *Life Sci.* 25:2065–70
136. Gorenstein, C., Tran, V. T., Snyder, S. H. 1981. *J. Neurosci.* 1:1096–102
137. Lee, C. M., Snyder, S. H. 1982. *J. Biol. Chem.* 257:12043–50
138. Gros, C., Giros, B., Schwartz, J. C. 1985. *Neuropeptides* 5:485–88
139. Hersh, L. 1985. *J. Neurochem.* 44:1427–35
140. Gros, C., Giros, B., Schwartz, J. C. 1985. *Biochemistry* 24:2179–85
141. Erdos, E. G., Johnson, A. L., Boyden, N. T. 1978. *Biochem. Pharmacol.* 27:843–48
142. Erdos, E. G., Skidgel, R. A. 1985. *Biochem. Soc. Trans.* 13:35–54
143. Lentzen, H., Simon, R., Reinsch, I., Busse, M. 1985. *Int. Narcotics Res. Conf. Abstr.* 75 (Abstr.)
144. Malfroy, B., Swerts, J. P., Guyon, A., Roques, B. P., Schwartz, J. C. 1978. *Nature* 276:523–26
145. Gorenstein, C., Snyder, S. H. 1980. *Proc. R. Soc. London* 210:123–32
146. Llorens, C., Malfroy, B., Schwartz, J. C., Gacel, G., Roques, B., et al. 1982. *J. Neurochem.* 39:1081–89
147. Malfroy, B., Schwartz, J. C. 1984. *J. Biol. Chem.* 259:14365–70
148. Kerr, M. A., Kenny, A. J. 1974. *Biochem. J.* 137:477–88
149. Roques, B. P., Fournie-Zaluski, M. C., Soroca, E., Lecomte, J. M., Malfroy, B., et al. 1980. *Nature* 288:286–88
150. Matsas, R., Fulcher, I. S., Kenny, A. J., Turner, A. J. 1983. *Proc. Natl. Acad. Sci. USA* 80:3111–15
151. Matsas, R., Kenny, A. J., Turner, A. J. 1984. *Biochem. J.* 223:433–40
152. Matsas, R., Turner, A. J., Kenny, A. J. 1984. *FEBS Lett.* 175:124–28
153. De la Baume, S., Yi, C. C., Schwartz, J. C., Chaillet, P., Marcais-Collado, H., Costentin, J. 1983. *Neuroscience* 8:143–51
154. Chaillet, P., Marcais-Collado, H., Costentin, J., Yi, C. C., De la Baume, S., Schwartz, J. C. 1983. *Eur. Jr. Pharmacol.* 86:329–36
155. Roques, B. P., Lucas-Soroca, E., Chaillet, P., Costentin, J., Fournie-Zaluski, M. C. 1983. *Proc. Natl. Acad. Sci. USA* 80:3178–82
156. Gee, N. S., Matsas, R., Kenny, A. J. 1983. *Biochem. J.* 214:377–86
157. Waksman, G., Hamel, E., Fournie-

Zaluski, M. C., Roques, B. P. 1986. *Proc. Natl. Acad. Sci. USA.* In press

158. Back, S. A., Gorenstein, C. 1985. *Soc. Neurosci. Abstr.* 15:389 (Abstr.)

159. Soffer, R. L. 1976. *Ann. Rev. Biochem.* 45:73–94

160. Thiele, E. A., Strittmatter, S. M., Snyder, S. H. 1985. *Biochem. Biophys. Res. Commun.* 128:317–24

161. Cascieri, M. A., Bull, H. G., Mumford, R. A., Patchett, A. A., Thornberry, N. A., Liang, T. 1984. *Mol. Pharmacol.* 25:287–93

162. Yokosawa, H., Endo, S., Ogura, Y., Ishii, S. 1983. *Biochem. Biophys. Res. Commun.* 116:735–42

163. Skidgel, R. A., Engelbrecht, S., Johnson, A. R., Erdos, E. G. 1984. *Peptides* 5:769–76

164. Skidgel, R. A., Erdos, E. G. 1985. *Proc. Natl. Acad. Sci. USA* 82:1025–29

165. Velletri, P. A. 1985. *Life Sci.* 36:1597–608

166. Brownfield, M. S., Reid, I. A., Ganten, D., Ganong, W. F. 1982. *Neuroscience* 7:1759–769

167. Rix, E., Ganten, D., Schull, B., Unger, T., Taugner, R. 1981. *Neurosci. Lett.* 22:125–30

168. Yang, H. Y. T., Neff, N. H. 1972. *J. Neurochem.* 19:2443–50

169. Wigger, H. J., Stalcup, S. A. 1978. *Lab. Invest.* 38:581–85

170. Defendini, R., Zimmerman, E. A., Weare, J. A., Alhenc-Gelans, F., Erdos, E. G. 1983. *Neuroendocrinology* 37:32–40

171. Saavedra, J. M., Fernandez-Pardol, J., Chevillard, C. 1982. *Brain Res.* 245:317–25

172. Chevillard, C., Saavedra, J. M. 1982. *J. Neurochem.* 38:281–84

173. Strittmatter, S. M., Lo, M. M. S., Javitch, J. A., Snyder, S. H. 1984. *Proc. Natl. Acad. Sci. USA* 81:1599–603

174. Strittmatter, S. M., Thiele, E. A., Snyder, S. H. 1985. *J. Biol. Chem.* 260:9825–32

175. Reisine, T. 1981. *Neuroscience* 6:1471–502

176. Herz, A., Blasig, J., Emrich, H. M., Cording, C., Piree, S., et al. 1978. *Adv. Biochem. Psychopharmacol.* 18:333–39

177. Martin, W. R., Eades, C. G., Thompson, J. A., Huppler, R. E., Gilbert, P. E. 1976. *J. Pharmacol. Exp. Ther.* 197:517–32

178. Urca, G., Frenk, H., Liebeskind, J. C., Taylor, A. N. 1977. *Science* 197:83–86

179. Chang, K. J., Cooper, B. R., Hazum, E., Cuatrecasas, P. 1979. *Mol. Pharmacol.* 16:91–104

180. Kosterlitz, H. W., Paterson, S. J. 1980. *Proc. R. Soc. London Ser. B* 210:113–22

181. Kosterlitz, H. W., Paterson, S. J., Robson, L. E. 1981. *Br. J. Pharmacol.* 73:939–49

182. Mosberg, H. I., Hust, R., Hruby, V. J., Gee, K., Yamamura, H. I. 1983. *Proc. Natl. Acad. Sci. USA* 80:5871–74

183. Von Voigtlander, P. F., Lahti, R. A., Ludens, J. H. 1983. *J. Pharmacol. Exp. Ther.* 224:7–12

184. Goodman, R. R., Snyder, S. H., Kuhar, M. J., Young, W. S. 1980. *Proc. Natl. Acad. Sci. USA* 77:6239–43

185. Goodman, R. R., Snyder, S. H. 1982. *Proc. Natl. Acad. Sci. USA* 79:5703–7

186. Robson, L. E., Foote, R. W., Maurer, R., Kosterlitz, H. W. 1984. *Neuroscience* 12:621–27

187. Quirion, R., Weiss, A. S., Pert, C. B. 1983. *Life Sci.* 33(Suppl. 1):183–86

188. Lord, J. A. H., Waterfield, A. A., Hughes, J., Kosterlitz, H. W. 1977. *Nature* 267:495–99

189. Chavkin, C., Goldstein, A. 1981. *Proc. Natl. Acad. Sci. USA* 78:6543–47

190. Chavkin, C., Goldstein, A. 1981. *Nature* 291:591–93

191. Corbett, A. D., Paterson, S. J., McKnight, A. T., Magnan, J., Kosterlitz, H. W. 1982. *Nature* 299:79–81

192. Oka, T., Negishi, K., Kajiwara, M., Watanabe, Y., Ishizuka, Y., Matsumiya, T. 1982. *Eur. J. Pharmacol.* 79:301–5

193. Suda, M., Nakao, K., Yoshimasa, T., Ikeda, Y., Sakamoto, M., et al. 1983. *Life Sci.* 33(Suppl. 1):275–78

194. Rezvani, A., Hollt, V., Way, E. L. 1983. *Life Sci.* 33(Suppl. 1):271–74

195. Lindvall, O., Bjorklund, A. 1983. In *Chemical Neuroanatomy,* ed. P. C. Emson, pp. 229–55. New York: Raven

196. Rainbow, T. C., Parsons, B., Wolfe, B. B. 1984. *Proc. Natl. Acad. Sci. USA* 81:1585–89

197. Lee, C. M., Iversen, L. L., Hanley, M. R. 1982. *Arch. Pharmacol.* 318:281–87

198. Erspamer, G. F., Erspamer, V., Pigginelli, D. 1980. *Arch. Pharmacol.* 311:61–65

199. Erspamer, V. 1981. *Trends Neurosci.* 4:267–69

200. Hunter, J. C., Maggio, J. E. 1984. *Eur. J. Pharmacol.* 97:159–60

201. Nawa, H., Doteuchi, M., Igano, K., Inouye, K., Nakanishi, S. 1984. *Life Sci.* 34:1153–60

202. Kimura, S., Goto, K., Ogawa, T., Sugita, Y., Kanazawa, I. 1984. *Neurosci. Res.* 2:97–104

203. Hanley, M. R., Sandberg, B. E. B., Lee,

C. M., Iversen, L. L., Brundish, D. E., Wade, R. 1980. *Nature* 268:810–12

204. Viger, A., Beaujouan, J. C., Torrens, Y., Glowinski, J. 1983. *J. Neurochem.* 40:1030–39

205. Torrens, Y., Beaujouan, J. C., Viger, A., Glowinski, J. 1983. *Arch. Pharmacol.* 324:134–39

206. Lee, C. M., Javitch, J. A., Snyder, S. H. 1983. *Mol. Pharmacol.* 23:563–69

207. Torrens, Y., Beaujouan, J. C., Glowinski, J. 1985. *Neuropeptides* 6:59–70

208. Cascieri, M. A., Liang, T. 1984. *Life Sci.* 35:179–84

209. Rothman, R. B., Danks, J. A., Herkenham, M., Cascieri, M. A., Chicchi, G. G., et al. 1984. *Neuropeptides* 4:343–49

210. Mantyh, P. W., Maggio, J. E., Hunt, S. P. 1984. *Eur. J. Pharmacol.* 102:361–64

211. Ninkovic, M., Beaujouan, J. C., Torrens, Y., Saffroy, M., Hall, M. D., Glowinski, J. 1985. *Eur. J. Pharmacol.* 106:463–64

212. Torrens, Y., Lavielle, S., Chassaing, G., Marquet, A., Glowinski, J., Beaujouan, J. C. 1984. *Eur. J. Pharmacol.* 102:381–82

213. Cascieri, M. A., Chicchi, G. G., Liang, T. 1985. *J. Biol. Chem.* 260:1501–7

214. Buck, S. H., Maurin, Y., Burks, T. F., Yamamura, H. I. 1984. *Life Sci.* 34:497–507

215. Smith, G. P., Gibbs, J., Young, R. C. 1974. *Fed. Proc.* 33:1146–49

216. Della-Ferra, M. A., Baile, C. A. 1979. *Science* 206:471–73

217. Lind, R. W., Swanson, L. W., Ganten, D. 1984. *Brain Res.* 321:209–15

218. Saper, C. B., Standaert, D. G., Currie, M. G., Schwartz, D. G., Geller, D. M., Needleman, P. 1985. *Science* 227:1047–49

219. Hermann, K., McDonald, W., Unger, T., Lang, R. E., Ganten, D. 1984. *J. Physiol.* 79:471–80

220. Quirion, R., Dalpe, M., De Lean, A., Gutkowska, J., Cantin, M., Genest, J. 1984. *Peptides* 5:1167–72

Ann. Rev. Biochem. 1986. 55:801-30

TRANSCRIPTION OF CLONED EUKARYOTIC RIBOSOMAL RNA GENES

Barbara Sollner-Webb and John Tower

Department of Biological Chemistry, The Johns Hopkins University, School of Medicine, Baltimore, Maryland 21205

CONTENTS

PERSPECTIVES AND SUMMARY

Although the typical eukaryotic cell produces about 10,000 different RNA species, nearly half of its transcriptional capacity is devoted to the synthesis of only one kind of RNA. This is the ~35–47S ribosomal RNA (rRNA), precursor to the mature ~18S, ~28S, and 5.8S RNAs of the ribosome. It is the sole product of synthesis by RNA polymerase I. The transcription of ribosomal

801

DNA (rDNA) is not only very efficient but also highly regulated, in large part reflecting the cellular need to produce more than a million new ribosomes per generation which in turn will be required to support protein synthesis in the daughter cells. Ribosomal RNA genes were the earliest characterized, and are still among the most thoroughly studied eukaryotic genes.

In the last few years great progress has been made in understanding the transcription of eukaryotic rDNA at the molecular level. Methods have been developed to accurately transcribe cloned rRNA genes, and this transcription has been found to be quite species-specific. The multiple domains of rDNA promoters have been delineated and ribosomal enhancer sequences have been described and characterized. Furthermore, relevant transcription factors are being identified and purified, their interactions with the promoter are being characterized, and the bases of a number of rRNA regulatory events are being discerned. Finally, new aspects of rRNA processing have been described, and progress is being made toward understanding polymerase I transcriptional termination. The older (pre-1981) literature is comprehensively reviewed in a number of works that stress the structural organization of rRNA genes (1–6), conditions that modulate rRNA accumulation (1, 4–8), characterization of RNA polymerase I (9–12b), rRNA processing (13, 14), the organization of the nucleolus (15–17), and the history of the field (1, 16, 17). Information concerning transcription of rRNA in cytoplasmic organelles and in prokaryotes, and transcription of 5S RNA by RNA polymerase III, may be found in Refs. 18, 19, and 20, respectively. This article focuses on the recent advances in understanding the transcription of eukaryotic nuclear rRNA genes that are outlined above, advances which have resulted almost entirely from studies using cloned ribosomal templates.

BACKGROUND ON rDNA STRUCTURE AND TRANSCRIPTION

The rRNA gene has played a central role in the history of molecular biology. In fact, the basic patterns of rDNA organization and rRNA production were already delineated prior to the recombinant DNA revolution (1, 17). This was possible because rRNA genes are very actively transcribed, and because amphibian rRNA genes are very GC-rich (21) and highly amplified in oocytes (22, 23). The *Xenopus* rRNA gene was thus the first eukaryotic gene to be isolated (21) and the first eukaryotic gene to be cloned (24). Amphibian rRNA genes were also the first genes whose transcriptional pattern was visualized in the electron microscope (25, 26) and rRNA was the first eukaryotic transcript whose basic processing pattern was discerned (14, 27–29). From early studies (reviewed in 1, 2), substantiated and greatly extended by more recent results from many species (largely reviewed in 3, 5, 6), a clear picture of the overall architecture of these genes has emerged.

In most eukaryotic species, rRNA genes are repeated ~100–5,000 times per haploid genome (3), although a few species contain a larger (3) or smaller (30) number. Indeed, for a rapidly growing mammalian cell to produce its requisite ~2 × 10⁶ ribosomes per ~15 hour generation time, with a ~30 nucleotides/ second polymerase I elongation rate (7) and a ~100 bp polymerase I spacing (25), at least 50 active rRNA genes/haploid chromosome complement would be required, consistent with measured values of 50–250 copies. The rRNA genes are located at one or a few chromosomal sites (3) called nucleolar organizers. Nucleoli (15, 16) are densely staining organelles which form at these positions during interphase and are the site of rDNA transcription, rRNA processing, and much of ribosome assembly.

The primary rRNA transcript is a large 35–47S (or 6–15 kb) precursor, which is processed through a series of endonucleolytic (and possibly exonucleolytic) cleavages (14, 31) to yield the mature rRNA species. The pre-rRNA begins with a transcribed spacer and extends through the 18S, 5.8S, and 28S rRNA regions, each separated by additional segments of transcribed spacer (2, 3, 27–29, 32, 33). The transcription of rDNA is catalyzed by RNA polymerase I (10, 34–36). In almost all species, synthesis by polymerase I can be distinguished from that of polymerases II and III by virtue of its resistance to high concentrations of the fungal toxin α-Amanitin (36). Transcription of rDNA is evidently the sole function of polymerase I, for the small nuclear RNAs whose synthesis was reported to be α-Amanitin-resistant (37, 38) now appear to be cleavage products of pre-rRNA (39, 40).

The rRNA coding regions are organized as head-to-tail repeats in the genomes of virtually all organisms, with the transcribed regions separated by segments of 'nontranscribed spacer' (1, 2, 41). This is beautifully illustrated in electron micrographs of ribosomal chromatin (or 'Miller spreads'), where transcribed regions bearing closely packed nascent ribonucleoprotein fibrils of increasing length alternate with nontranscribed spacer regions (25, 26). In different species there is considerable variation in the sizes of the transcribed spacers, and especially of the nontranscribed spacers. Typical rRNA repeating units of *Xenopus* and mouse, the first species used to effect 'in vivo' and in vitro transcription of cloned rDNA, are shown in Figure 1. In *Xenopus,* as well as in *Drosophila,* the rDNA promoter is duplicated at a number of positions in the

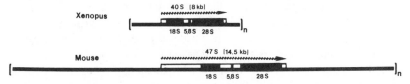

Figure 1   The repeating unit of *X. laevis* and mouse rDNA: The nontranscribed spacer (single line), transcribed spacer (open boxes), and regions coding for the mature rRNAs (solid boxes), as well as the primary rRNA transcript (wavy line), are shown.

'nontranscribed spacer' (42–44; 45–48) and these spacer promoters evidently give rise to transcription units at a variable, but generally very low, frequency (41, 44, 47–58).

METHODS OF CATALYZING AND ASSAYING TRANSCRIPTION

The vast majority of the recent progress in understanding rDNA transcription can be directly attributed to the development of systems in which cloned rRNA genes are faithfully transcribed. Establishing such systems required: (*a*) developing methods to synthesize RNA from a ribosomal template, (*b*) locating the initiation site of this RNA, and (*c*) confirming that this is at the in vivo initiation site of the cellular gene.

(A) TRANSCRIPTION REACTION   Crude extracts from a variety of cell types have been shown to catalyze transcription by all three classes of eukaryotic RNA polymerase. The most common extracts are the S-100, an isotonic, ~100,000 × *g* supernatant of hypotonically lysed tissue culture cells (59–61), first applied to rRNA transcription by Grummt (62), and the 'whole cell extract,' a 0.4 M $(NH_4)_2SO_4$ extract of tissue culture cells subsequently precipitated by 2.5 M $(NH_4)_2SO_4$ (63), first applied to rRNA transcription by Learned & Tjian (64). In certain cases, extracts prepared at different salt concentrations (65, 66), nuclear extracts (67), and extracts of cells from intact animals (68, 69) have also been used to synthesize rRNA. These extracts transcribe both linearized and closed circular rDNA templates of the homologous species.

Ribosomal RNA synthesis has also been accomplished 'in vivo' upon reintroduction of closed circular rDNA into intact cells. This was first shown with microinjected *Xenopus* oocytes (70–73), but transfected cultured cells that contain extrachromosomal (74–79) or integrated (77, 80a, 80b) copies of cloned rDNA also catalyze accurate transcription.

(B) MAPPING THE RESULTANT TRANSCRIPT   In vitro transcription reactions are generally performed with radiolabeled ribonucleoside triphosphates and a template that is truncated a few hundred bp beyond the initiation site. The length of the labeled 'run-off' RNA thus serves to localize the initiation site. In contrast, RNA transcribed from closed circular templates is generally analyzed by hybridization to an excess of a radiolabeled DNA probe followed by S1 nuclease mapping (81, 82) or primer extension analysis (83). (These latter two techniques require that there not be appreciable amounts of endogenous RNA of the same sequence.) Since all these assays measure steady-state RNA levels, differential RNA stability can greatly affect the amount of transcript that is detected, especially in intact cells (84, 85), and rapid specific processing of the transcript can also occur, in vitro as well as in vivo (66, 86). By S1 or primer extension analysis, the presumptive initiation site can be mapped to ± 1 bp on sequencing gels (43, 87). In addition, the initiating nucleotide of in vitro synthesized rRNA can be definitively identified by dinucleotide analysis coupled with apparent K_M measurements for each rXTP (64, 87), or by analysis of the polyphosphorylated nucleotide at the 5' terminus of a radiolabeled transcript (88).

(C) THE IN VIVO INITIATION SITE The 5' terminus of the primary rRNA transcript can be identified by virtue of its retention of a di- or triphosphate moiety. Although 5' polyphosphorylated residues have been directly demonstrated using in vivo labeled rRNA (89–91), these residues are more commonly identified by specific labeling of isolated RNA using capping enzymes (92, 93). The in vivo initiation site can then be determined from sequence analysis of this radioactively capped RNA (43). Alternatively, if a substantial fraction of the pre-rRNA is polyphosphorylated, the in vivo initiation site can be identified by S1 mapping or primer extension analysis of the bulk rRNA. Such analyses have allowed determination of the rRNA initiation site in *Xenopus laevis* (43), *Drosophila* (89, 94), rat (95, 96), human (97), and yeast (98). The latter S1 mapping procedure has also been used for pre-rRNA with a low or undetermined content of polyphosphorylated molecules (86, 99–101), but in this application it is necessary 1. to search for potential larger precursor species that might be rapidly processed to the major pre-rRNA using a probe that diverges in sequence from the template a short distance upstream of the presumptive initiation site and 2. to rule out artifactual cleavage in S1-sensitive rU:dA regions (86, 102).

rDNA TRANSCRIPTION SYSTEMS

INITIAL STUDIES Beginning in the late 1960s, many attempts were made to reproduce accurate transcription on purified eukaryotic rRNA genes using prokaryotic RNA polymerase, isolated RNA polymerase I, various cell extracts, nucleoli, and nuclei (reviewed in 4, 5, 7). In retrospect, it is clear that the early (pre ~1978) studies were greatly hindered by the lack of an assay for accurate transcription, especially if it occurred amid a background of nonspecific synthesis. The accumulated data nonetheless indicated that appreciable levels of accurate initiation were not occurring, and that components in addition to RNA polymerase I would be required to transcribe purified rRNA genes. From parallel studies using the endogenous chromatin template of nuclei, however, it was concluded that rRNA synthesis could be made to initiate in vitro (103–105).

XENOPUS   The first cloned rRNA gene that gave evidence of directing correct transcription was that of *Xenopus laevis*. In 1978, Trendelenburg & Gurdon (71) microinjected a cloned repeating unit of *X. laevis* rDNA into the nuclei of *Xenopus* oocytes and visualized the resultant chromatin by electron microscopy. A limited number of plasmid molecules with packed transcription units of approximately the same length as those of the endogenous rRNA genes were detected (71, 106), suggesting that cloned *Xenopus* rRNA genes were able to direct faithful transcription. Moss (72) and Sollner-Webb & McKnight (73) subsequently used S1 nuclease mapping to demonstrate that cloned *X. laevis* rDNA did indeed direct faithful in vivo transcription. These studies involved microinjecting *X. laevis* rDNA into *X. borealis* oocytes so that endogenous rRNA of the oocyte would not interfere with the S1 assay. However, transcription of a laevis rDNA 'pseudogene' can also be detected after injection into *X. laevis* oocytes using primer extension analysis (107). Transcription in oocytes requires that the rDNA plasmid be closed circular upon injection (73) and throughout the synthesis period (108). Curiously, the rRNA gene and spacer promoters appear to have differential sensitivities to altered degrees of superhelical tension (109). Cloned *Xenopus* rDNA has also been transcribed after injection into one-cell *Xenopus* embryos (110). As with the endogenous genes, transcription commences at the thirteenth cleavage of the developing embryo and persists through many cell divisions.

The first demonstration of accurate in vitro transcription of cloned *Xenopus* rDNA (111) utilized an extract of oocyte nuclei, but specific transcription of *Xenopus* rDNA has also been effected in S-100 extracts of *Xenopus* (99) and mouse (87, 111) tissue culture cells. All of these in vitro systems utilize linear as well as closed circular templates.

MOUSE   Advances in 1981 resulted in the establishment of the first faithful in vitro transcription system for cloned rDNA. Grummt (62) initially found that a specific RNA was produced when mouse rDNA was incubated in an S-100 extract of mouse cells. Miller & Sollner-Webb (86) then demonstrated that two distinct RNA species were actually produced in vitro; the larger reflected de novo initiation while the smaller, which corresponded to the previously identified transcript, resulted from in vitro processing of the larger RNA. Both the in vitro initiation (86, 88, 114) and processing (86) occur at the same position in vitro as in vivo, and thus the S-100 extract supports accurate transcription and primary rRNA processing. Whole cell extracts (115) and nuclear extracts (67) of mouse tissue culture cells were also found to catalyze specific synthesis of mouse rRNA. A number of studies suggest that active template molecules can direct several rounds of initiation in vitro (116, 117, E. A. Thompson, personal communication).

Recently, transcription of cloned mouse rDNA has also been achieved 'in vivo,' using transiently (74, 76; R. Baserga, personal communication) and stably (80a, 80b) transfected rodent cells. Transfections into mouse cells (74) were assayed by primer extension analysis, while transfections into Chinese hamster ovary (76, 80b) and rat (80a) cells were assayed by S1 mapping.

RAT   Rat rDNA is accurately transcribed in vitro using S-100 extracts of mouse (118) or rat (95) cells. Extracts can also be prepared from a solid tumor, and extract components of lesser activity have been isolated from liver (69).

HUMAN   In 1982, the laboratories of Grummt (119), Arnheim (120), Muramatsu (118), and Tjian (64) demonstrated faithful transcription of cloned human rDNA in S-100 extracts (118–120) or whole cell extracts (64) of cultured human cells. Human rDNA-containing plasmids have also been transcribed accurately 'in vivo' using transiently transfected primate cells (75).

DROSOPHILA   Also in 1982, Rae's laboratory (58) reported that cloned *Drosophila melanogaster* rRNA genes are faithfully transcribed in an S-100 extract of *D. melanogaster* cells. Notably, the genes with insertions in the 28S coding region, which appear virtually inactive in vivo (121), as well as the duplicated spacer promoters, are active in vitro (45, 58).

ACANTHAMOEBA   Using additional RNA polymerase I to supplement an S-100 extract from cells of the protozoa *Acanthamoeba castellanii*, Paule's laboratory has demonstrated faithful in vitro transcription of cloned *Acanthamoeba* rRNA genes (119, 101).

YEAST   Warner's (77) and Planta's (78) laboratories and Quincey & Arnold (79) have demonstrated that yeast rRNA genes direct transcription when reintroduced into intact yeast cells, either on multi-copy (77–79) or single-copy (77) plasmids, or when integrated into the genome (77). The template in these studies was a yeast minigene into which a foreign segment of DNA had been inserted, allowing its transcript to be analyzed by Northern blotting and by primer extension.

When reacted in an S-100 extract, yeast rDNA only yields a transcript that initiates 2 kb further upstream than does the normal 37S pre-rRNA (122). Although the presence of this upstream transcript in vivo is still in question (122, 123), it does not simply arise from the known preference (124) of yeast polymerase I to initiate nonspecifically in AT-rich sequences (125). It is striking that this in vitro promoter region has enhancer-like activity for rRNA initiation in vivo (77; see below).

NEUROSPORA   Using a 0.9 M $(NH_4)_2SO_4$ extract of a *Neurospora* lysate, Tyler & Giles (66) have recently developed a transcription system for cloned *Neurospora* rDNA. In contrast to most polymerase I transcription systems, reinitiation is unequivocally demonstrated in this case, for ~25 rRNAs are transcribed per input rDNA molecule. Moreover, the in vitro synthesized transcript is very efficiently processed, apparently reproducing an early in vivo rRNA processing event.

TETRAHYMENA   Cloned *Tetrahymena* rRNA genes also appear to be faithfully transcribed when reacted in a homologous S-100 extract supplemented with a high salt macronuclear extract (65).

SPECIES AND POLYMERASE SPECIFICITY

In contrast to transcription by RNA polymerases II and III, transcription of rRNA genes by polymerase I appears to be quite species-selective. As first shown by Grummt, Roth, and Paule (119), mouse rRNA genes do not direct transcription in a human cell extract and vice versa (64, 118, 126), a restriction that is also observed in vivo (74, 75). Moreover, mouse rRNA genes are not transcribed in *Acanthamoeba* extracts and *Acanthamoeba, Drosophila melanogaster,* and *Physarum* rRNA genes are evidently not transcribed in a mouse cell extract (119). *D. virilis* genes are also not transcribed in a *D. melanogaster* extract (58). An obvious explanation for why the polymerase I transcriptional machinery can evolve rapidly relative to that involving polymerases II and III is that polymerase I acts on the promoter of only one kind of gene and this gene family undergoes relatively rapid recombination (2, 127). However, polymerase I transcriptional barriers appear to evolve more slowly than do the species themselves, for rRNA synthetic machinery is compatible across many closely related pairs of species, such as *X. laevis/X. borealis* (72, 73, 128), human/monkey (64), mouse/rat (80a, 118), and mouse/hamster (76, 80b, R. Baserga, personal communication).

Two results indicate that the interactions between rDNA promoters and transcription factors may be considerably more conserved across species than is generally assumed. First, nucleolar DNA from the water beetle *Dytiscus* directs the formation of normal length transcription units when injected into *Xenopus* oocyte nuclei (70), suggesting that frog factors can transcribe insect rDNA. Second, contrary to initial reports (119), extracts derived from mouse cells efficiently transcribe cloned frog rDNA (111) and they utilize exactly the same large promoter sequence as does the homologous frog extract (129). Further study of cross-species reactions and rDNA promoter organization will be needed in order to draw general conclusions about the conserved and diverged features of rRNA initiation.

Several laboratories (74–76, 130, R. Baserga, personal communication) have investigated another aspect of polymerase specificity in eukaryotic transcription, namely, whether RNA polymerase I can catalyze synthesis of a 'pre-mRNA' which can be correctly processed and used for translation. To this end, RNA polymerase I promoter regions were cloned adjacent to the body of protein-coding genes and the resultant plasmids were introduced into cultured cells. Initial reports suggested that the encoded protein could indeed be translated from the polymerase I–synthesized RNA (74, 130). However, it now appears that RNA polymerase II initiates fortuitously within the rDNA region of these transfected constructs to produce transcripts that in turn direct the detected translation (75, 76).

THE rDNA PROMOTER

To determine the extent of the rDNA promoter—the region that directs transcriptional initiation—systematic series of deletion mutants approaching the initiation site from the 5' and 3' directions have been constructed from the rDNA of frog (72, 85), mouse (115, 131, 132), human (133), *Drosophila* (134, 135), and *Acanthamoeba* (136, 137). The template capacity of these mutants has been assessed, and an extensive array of data evidently reflecting both species and assay differences has been accumulated, the highlights of which are summarized below. (Nucleotide positions are designated relative to the site of initiation at $+1$.) From analysis in injected *Xenopus* oocytes, the *X. laevis* promoter was reported to extend upstream of position -35 by one laboratory (72), but to be between positions -7 and $+6$ by another (85). From in vitro analysis, it was observed to extend between residues ~ -140 and $+6$ (85). The in vitro mouse rDNA promoter has been reported to encompass the region from position ~ -39 (115, 131) on the 5' side to position $\sim +23$ (115) or $\sim +5$ (132) on the 3' side, although these borders seem to be gradual rather than sharp. In different experiments, a 5' border of the mouse rDNA promoter was reported to be upstream of residue ~ -80 (131) and at position ~ -140 (132). In analogous in vitro experiments mapping the human rDNA promoter, maximal transcription required the sequences extending between positions ~ -140 and $+18$, but a reduced level of initiation was directed by the sequences between positions ~ -50 and $\sim +5$ (133). Also using in vitro transcription, the *Drosophila* rDNA promoter was found to extend from position ~ -40 to $+4$ (134, 135), and the *Acanthamoeba* rDNA promoter between position ~ -29 and $\sim +5$, although sequences 10–15 nucleotides further upstream in the latter rDNA can also augment initiation (136). Finally, for another lower eukaryote, yeast, preliminary in vivo analysis shows that the rDNA promoter is contained within the regions -200 (77–79) to $+15$, with quantitatively important sequences extending upstream of position -130 (J. Klootwijk, A. Kempers-Veenstra, and R. Planta, personal communication). For the in vitro transcription that initiates at position ~ -2.2 kb in the yeast rDNA, a 22-bp promoter has been defined (125).

Recent results from the author's laboratory have demonstrated that the conditions of the transcription reaction can greatly affect the size of the rDNA region that is detected as the promoter (85, 132, 138a, 138b, J. Windle, S. Henderson, unpublished observations). A small rDNA region surrounding the initiation site is sufficient to direct maximal levels of transcription when 5' and 3' deletion mutants of mouse or frog rDNA are assayed under the most efficient synthesis conditions (85, 132). However, when reaction conditions are made more stringent by any of several different means (including alteration of ionic, factor, or template conditions), additional promoter domains that extend considerably further upstream become necessary to obtain a transcriptional signal equivalent to that from the wild-type gene. Both with *Xenopus* (85, 138b) and mouse (132; S. Henderson, B. Sollner-Webb, unpublished observations), as well as with human rDNA analyzed in Tjian's laboratory (75, 133), in vitro analyses detect promoter regions that extend in the 5' direction to position ~ -140 and in vivo analyses demonstrate promoter domains that extend to or beyond position ~ -165.

The accumulated data suggest a general model for the eukaryotic rDNA promoter that may serve to unify results from many diverse species and experiments. There is a 'proximal promoter domain' contained within the region ~ -40 to $\sim +5$ that productively binds the necessary transcription factors and is sufficient to direct efficient initiation in most optimized in vitro reactions. In addition, there are 'upstream promoter domains' that extend in the 5' direction for an additional ~ 120 bp. While their role in initiating transcription of cloned rDNA is only detected under more stringent reaction conditions, it is most likely that the entire large promoter region functions in initiating transcription of cellular rRNA genes in vivo.

To obtain additional information about the internal organization of the rDNA promoter, two other kinds of mutants are being studied. First are 'linker scanner' mutants (139) in which the promoter is traversed by a systematic series of clustered point mutations. Preliminary results with human (R. Tjian, personal communication) and *Xenopus* (J. Windle, B. Sollner-Webb, unpublished observations; D. Pennock, R. Reeder, personal communication) linker scanners indicate that regions from ~ -150 to ~ -110 and from ~ -40 to $\sim +10$ have major importance, while mutations in the intervening segment generally have less effect. These data suggest that the proximal and upstream promoter domains are separated by a region where the sequence is of lesser importance.

The final kind of promoter mapping involves point mutations. These have been constructed within the proximal domain of the mouse rDNA promoter in Grummt's (140) and Muramatsu's (141) laboratories. Of the sixteen positions examined, only changes at positions -7 (141), -16^{1} (140, 141), and -25

---

[1]The G residue at position -16 appears to be highly conserved in evolution (140). Other regions of the promoter generally show only a rather limited degree of sequence homology except in more closely related species (66, 97, 100, 142).

(141) reduce the transcriptional signal. The ~10-nucleotide spacing of these important positions is consistent with a model in which a transcription factor recognizes the proximal promoter principally from one side of the rDNA helix (141). Thirty nine point mutations have also been created within the *Acanthamoeba* rDNA promoter region. Transcription is depressed by the 10 mutations between positions −45 and −39, −30 and −28, and −17 and −11 (M. Paule, personal communication).

STABLE rDNA TRANSCRIPTION COMPLEXES

When one rDNA template is preincubated with S-100 or whole cell extract prior to addition of a different template, and synthesis is then initiated by addition of rXTPs, rRNA is only transcribed from the first template (116, 117). This indicates that the initial template became stably activated for synthesis by associating with transcription factors of the cell extract, a phenomenon first observed with polymerase III transcription factors and 5S DNA (143). The rDNA stable complex forms rapidly and remains intact during prolonged incubation with competitor DNA (116, 117, 144) and evidently through multiple rounds of transcription (116, 117, E. A. Thompson, personal communication). This association of rDNA and transcription factors is sequence-specific, for prior incubation with plasmids lacking the homologous species' rDNA promoter does not reduce the extract's ability to transcribe a subsequently added rDNA template (117, 126). Using the template commitment assay, stable transcription complexes have been demonstrated with rDNA from mouse (116, 117, 126, 144), human (126, 145), *Acanthamoeba* (137), and *Xenopus* (146, 147). One striking feature of rDNA pre-initiation complexes is that they are selectively sedimentable (145); this is evidently caused by their associating into large structures in the in vitro reaction. While this sedimentable association is indeed dependent upon the formation of the stable complex (147), it remains to be determined whether this reflects aspects of the higher order organization of active cellular rRNA genes.

To discern which sequences are required for stable complex formation, mutant rDNAs have been pre-incubated with cell extract prior to challenging with a competitor template. Using systematic series of 5' and 3' deletion mutants, the minimal region required for stable complex formation of mouse (147, 156), *Xenopus* (147), and *Acanthamoeba* (137) rDNA have been determined to be upstream of the initiation site. With mouse and *Xenopus* rDNA, this region was seen to coincide with the upstream portion of the minimal promoter region detected under the same conditions (residues −39 to −15 and −150 to ~−75, respectively). Nonetheless, mouse sequences extending over the entire promoter region (~−140 to +10) did augment complex stability. The importance of proximal promoter sequences in forming the mouse stable complex was also demonstrated by experiments showing that point mutations at mouse rDNA positions −25 (141) and −16 (140, 141), which markedly reduce transcription, similarly interfere with stable complex formation. In contrast, when mapping experiments were performed with *Acanthamoeba* rDNA deletions (137), sequences upstream of and partially overlapping the minimal promoter region (positions −45 to ~−25) were found to be necessary and sufficient for complex formation. Corroborating this data, point mutations between positions −31 and −24 markedly reduce the ability of *Acanthamoeba* rDNA to form a stable complex (M. Paule, personal communication). In human rDNA, the sequences both upstream and downstream of residue −30 appear to be required for stable complex formation (126). Further studies should resolve whether the apparent differences in complex binding regions noted above truly result from species differences or whether they instead reflect differences in assay conditions.

A complementary approach to studying complex formation uses DNase protection studies to map the rDNA regions that are closely associated with proteins. Using the partly purified sequence-specific transcription factor from *Acanthamoeba* (149), the region of the homologous species' rDNA that is protected from cleavage by DNase I (positions −68 to −13) contains the entire region identified above by the sequential addition assay. However, when both necessary *Acanthamoeba* transcription factors are used, the DNase-protected region is extended downstream to position +20 (149), indicating additional protein contacts in the segment of the promoter that flanks the initiation site. Related results have also been reported with *Xenopus* (148), in which the DNase I sensitivity of the entire rDNA promoter domain was reduced using crude oocyte extracts.

In summary, it appears as if the primary essential protein contacts occur in the upstream segments of the promoter, but that transcription components also bind to the other domains of the promoter region.

ENHANCER SEQUENCES IN THE rDNA

Sequences quite distant from the rRNA initiation site have also been shown to greatly augment the level of rRNA synthesized in intact cells. The effect of distant sequences was first reported by Moss (57), who injected *Xenopus* oocytes with pairs of marked *X. laevis* rRNA genes. These constructs contained different amounts of upstream nontranscribed spacer bearing 0, 1, or 2 copies of a 1-kb segment that consists of a duplicated promoter and 60/81-bp repetitive elements (see Figure 2). The genes with the larger spacers were transcribed with about a 10-fold preference over the genes with the smaller spacers, although when injected individually each gene was efficiently transcribed. Since the second spacer repeat starts more than 1 kb upstream from the promoter, this segment exerts its effect over a large distance (57). The stimulatory effect of

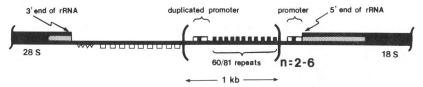

Figure 2 The *X. laevis* rDNA 'nontranscribed spacer': The 3' end of one *X. laevis* rRNA gene, the 5' end of the adjacent gene, and the intervening spacer are depicted. The spacer contains from 2 to ~6 copies of a ~1-kb region that consists of a duplicated promoter (>90% homologous to position −147 to +4 of the gene promoter) and ~10 copies of a 60 or 81 bp repeated sequence (■, that is 80% homologous to position −119 to −72 of the gene promoter). Further upstream, the spacer contains two other kinds of repeated sequences (designated by triangles and open boxes).

these spacer regions was also reported by Busby & Reeder (110), who observed increased transcription of genes bearing larger spacers in microinjected *Xenopus* embryos.

To examine which part of the *Xenopus* rDNA spacer is responsible for the preferential promoter utilization, Reeder's laboratory (107, 146) constructed a plasmid that contained the gene promoter and the 60/81-bp elements but not the duplicated spacer promoters. When injected into oocytes, this construct was transcribed with about a tenfold preference relative to a gene on a coinjected control plasmid lacking the 60/81-bp elements (146). Thus, the spacer effect can be largely mimicked by the 60/81-bp elements and does not require the presence of the upstream duplicated promoters. Moreover, Reeder's lab has demonstrated that the 60/81-bp elements are similarly effective when inserted in the normal and reverse orientation (146) and when positioned several kb from the 5' end of the transcription unit (107). Due to their ability to act bidirectionally to stimulate transcription from distant promoters in cis, the 60/81-bp repeats have been termed polymerase I enhancer elements (107, 150).

While it is not clear how this (or any) enhancer functions, a number of insights have been obtained. First, a segment of each 60/81-bp element is ~90% homologous to positions −119 to −72 of the rRNA gene promoter (43, 150, 151a), and this promoter region is involved in (but not sufficient for) stable binding of polymerase I transcription factors in vitro (147, 148). Second, the rDNA spacer elements have a cumulative effect, with larger regions conferring increasing activity to juxtaposed promoters (110, 146). Third, a subcloned block of 60/81-bp repeats can compete with a coinjected rRNA gene for necessary transcription components (107). Fourth, this rDNA enhancement appears to be polymerase-specific, for the 60/81-bp elements do not augment synthesis from a juxtaposed polymerase II or polymerase III promoter (138a) and the MSV enhancer does not augment synthesis from a juxtaposed rDNA promoter (75, 76). Finally, the 60/81-bp elements evidently act in concert with the entire upstream domain of the rDNA promoter, for they only augment transcription from genes with complete upstream domains and only under conditions where the stimulatory effect of these domains is exhibited (138a;

J. Windle, B. Sollner-Webb, unpublished observations). Taken together, these data suggest that the 60/81-bp enhancer elements may act by initially binding a polymerase I transcription factor, which in turn augments productive binding to the upstream domain of an adjoining gene promoter by favoring either sliding or cooperative interaction.

Other recent experiments have demonstrated that the duplicated promoters in the *X. laevis* rDNA spacer can also have a marked stimulatory effect on the transcription of a cis located rRNA gene (151b). Further studies should help to determine whether yet additional regions of the *X. laevis* nontranscribed spacer also have a major effect on transcription and how these regions act.

In studies involving yeast rDNA, Elion & Warner (77; personal communication) have demonstrated that the spacer that separates adjacent rRNA genes also has a polymerase I enhancer activity. They found that a 200-bp region located 2.2 kb upstream of the pre-rRNA initiation site stimulates transcription of an adjoining rDNA gene ~15-fold relative to a similar plasmid construct lacking this segment. This occurs whether the plasmids are maintained in yeast cells either in single- or multi-copy form or inserted into the genome (77). This 200-bp region also stimulates transcription when positioned at other sites relative to the gene promoter (77) and in either orientation (J. Warner, personal communication). In contrast to the frog enhancer, however, the yeast enhancer does not appear to have an additive effect and its action can be blocked by the termination, and certain other, rDNA regions (77). Recent results from M. Holland's laboratory (personal communication) demonstrate that the 22-bp segment of the yeast enhancer that acts as an in vitro promoter (125) is required for enhancement in vivo. Furthermore, Planta and colleagues (J. Klootwijk, personal communication) have recently obtained the intriguing result that transcription of yeast rDNA is also stimulated ~10-fold when two copies of a minigene are cloned in tandem, but the relationship of this effect to the action of the upstream enhancer sequence is not yet clear.

The extent to which other species possess rDNA enhancers also remains to be determined, but it might be noted that sequences that reside more than 250 bp upstream from the initiation site in the rDNA of human (75) and mouse (M. Lopata, D. Cleveland, B. Sollner-Webb, unpublished observations) also stimulate the level of transcription several fold in vivo, suggesting the possibility of mammalian polymerase I enhancer-like sequences as well.

FACTORS INVOLVED IN rDNA TRANSCRIPTION

Studies of the factors that catalyze rDNA transcription have involved both chromatographic fractionation techniques and stable complex analysis. Using phosphocellulose chromatography of mouse and human S-100 cell extracts, Muramatsu's laboratory reported the first resolution of polymerase I transcription factors (118). They found that components in the 0.1 M KCl flowthrough

('A'), the 0.4–0.6 M KCl elution ('C'), and the 0.6–1.0 M KCl elution ('D') were all necessary for efficient transcription, and that the 0.1–0.4 M KCl eluting material ('B') suppressed nonspecific synthesis. Such a background-reducing activity was subsequently shown to be poly(ADP-ribose) polymerase (152, 153), an enzyme that exhibits analogous activity in polymerase II transcription systems (154). Moreover, phosphocellulose fraction A has been found to act by preventing rapid RNA degradation and can be fully replaced by the RNase inhibitor, RNasin (155). The essential polymerase I transcription factors are therefore contained in fractions C and D.[2]

The purification of rDNA transcription factors has been pursued in a number of laboratories, using extracts of rodent (67–69, 126, 129, 144, 147, 155–158), human (126, 157, 159), and *Acanthamoeba* (149) cells. The results through the summer of 1985 are summarized below. Arnheim and colleagues (126, 157) have resolved mouse and human A, C, and D activities on phosphocellulose and further purified human D on Biorex 70, where it elutes with ~1/5,000 of the extract protein. Grummt (67) has used step elution from heparin columns to separate mouse factor D (designated 'TFIB') from factor C (designated 'TFIA'). The latter was found to co-elute with RNA polymerase I upon step elution from DEAE, heparin, and Cibacron Blue columns, but to be in part resolved from polymerase I by Biorex 70 chromatography. Thompson's laboratory (144, 158) has reported the separation of mouse C from 99% of the polymerase I in the S-100 extract by step elution from DEAE and phosphocellulose resins, although in more recent studies they find that ~10% of the polymerase I activity copurifies with factor C (E. A. Thompson, personal communication). Tower & Sollner-Webb (155, 156) have purified mouse factor D ~5000-fold by pelleting the stable transcription complex followed by gradient elution from phosphocellulose. They have also purified mouse factor C using DNA-cellulose, gradient elution from DEAE and Heparin resins, and sucrose gradient centrifugation to obtain a preparation of virtually homogeneous polymerase I polypeptides. Furthermore, they find that C and polymerase I also copurify upon gradient elution from phosphocellulose, carboxymethyl, and Cibacron Blue resins. Muramatsu's laboratory (152) has also reported that a 'purified polymerase I' contains C activity. Jacob and colleagues (68) have used DEAE and Heparin chromatography to obtain a substantially purified polymerase I fraction that retains both C and D activity. In a final study in a mammalian system, Tjians's laboratory (159) has used gradient elution from Heparin-agarose and gel filtration to purify human 'SL1' ~100,000-fold. Most likely this activity is the same as factor D. They have also partly purified a 'polymerase I fraction' by step elution from heparin and

---

[2]For the following discussion, C and D are defined as activities that participate in accurate rDNA transcription; RNA polymerase I is defined by its ability to catalyze α-Amanitin-resistant nonspecific transcription.

DEAE matrixes. In parallel studies using the unicellular eukaryote *Acantha-moeba*, Paule's laboratory has resolved two activities that are necessary to catalyze rDNA transcription. 'TIF-1,' which is most likely the *Acanthamoeba* analogue of mammalian D, has been purified ~20,000-fold using Cibacron Blue and gradient elution from phosphocellulose and DEAE matrixes (149). The second *Acanthamoeba* factor copurifies with polymerase I upon gradient elution from DEAE, phosphocellulose, and Heparin matrixes, and through glycerol gradient centrifugation, to virtual homogeneity of the polymerase polypeptides (160, 161) and is evidently analogous to factor C. To simplify the following discussion we shall follow Muramatsu's original nomenclature (118) and denote all the C-like factors by 'C' and all the D-like factors by 'D.' [We fear that conversion to the TFIA (=C) and TFIB (=D) nomenclature (67) might cause confusion with the originally defined (118) A and B activities.]

C activity copurifies with polymerase I (67, 68, 118, 129, 152, 155, 159, 161) to virtual homogeneity of the polymerase in two different systems (161; Tower, Sollner-Webb, unpublished observations). It therefore appears to be an RNA polymerase I, or at least very tightly associated with one. However, C activity (which is involved in catalyzing specific rDNA transcription) is clearly different from bulk polymerase I (which only catalyzes nonspecific RNA synthesis). In the first line of data that distinguishes C from bulk polymerase I (described in the section on regulation below) inactive extracts prepared from cells in which rDNA transcription was inhibited were found to contain normal amounts of polymerase I activity but virtually no C activity (144, 155, 161). Second, mammalian C can be resolved from a very large part of the polymerase I activity of the S-100 extract. In addition to the separations noted above from Thompson's (144, 158) and Grummt's (167b) laboratories, this has also been shown by pelleting studies in Sollner-Webb's laboratory in which ~90% of C activity cosedimented with the stable preinitiation complex, while ~90% of total polymerase I activity remained soluble (147, 156). Taken together, these data indicate that C activity defines a specific activation of RNA polymerase I that allows these molecules to participate in accurate rDNA transcription.

What still remains unresolved is the basis for this activation of polymerase. One possibility is that C is a component or subunit that is generally very closely associated with polymerase (155, 156, 161) but can be separated from it under appropriate conditions (158, 167b) and subsequently reconstituted to regain activity. This notion is consistent with heat denaturation studies (158, 167b). However, rigorous testing of this model would require a demonstration that factor C, which has been chromatographically freed from the bulk RNA polymerase I (158, 167b), only supports transcription with D when it is also supplemented with purified polymerase that lacks the C activity. The other possibility is that C is a form of polymerase that is activated by a covalent or other nondissociable modification. In the *Acanthamoeba* system studied in Paule's laboratory (161), differential heat stability studies have indicated that the vast majority of the polymerase I activity in log phase cells is of the activated form and SDS-PAGE analysis of this activated polymerase revealed a subunit pattern identical to that of nonactivated polymerase from nongrowing cells. This suggests that the activation may be due to a covalent modification. While there are a number of obvious possibilities for a covalent modification that could confer polymerase activation/inactivation, two deserve mention. First, RNA polymerase I is a highly phosphorylated enzyme (11, 162–165); the phosphorylation state appears to affect enzyme activity (162, 166) and therefore could possibly activate the enzyme for specific synthesis. Second, the C-terminal region of the large subunit of RNA polymerase II can be rapidly cleaved off, and the resultant enzyme is no longer able to participate in specific transcription although it can still catalyze nonspecific synthesis (167a; J. Corden, personal communication); conceivably an analogous phenomenon could occur with polymerase I. Through future studies the identity of C will undoubtedly be resolved.

Fraction D has been found to bestow the species-specific nature to rDNA transcription. By cross-combining phosphocellulose-derived fractions from mouse and human cell extracts, Muramatsu's laboratory (118) observed that all fractions except D could cross the mouse-human barrier. This result, which has been confirmed elsewhere (126), suggested that D may recognize species-specific rDNA sequences. Support for this hypothesis has been provided by studies showing that isolated factor D binds stably to the promoter region of homologous rDNA and thereby precludes synthesis from subsequently added competitor templates (126, 137, 156). In the mouse system, the region that is involved in D binding (residues ~-39 to ~-15) is the segment that confers stable complex formation using the template commitment assay (156). Parallel studies using Exonuclease III footprinting have identified an overlapping rDNA segment (residues -21 to -6) that is protected from digestion by factor D (67). Furthermore, as noted above in the section on stable complexes, the *Acanthamoeba* analogue of D protects a region that contains the upstream portion of the *Acanthamoeba* promoter from digestion by DNase I (149). Evidently C does not form a stable complex in the absence of D but does stably bind to the D/rDNA preinitiation complex (149, 156). In the *Acanthamoeba* system, association of C with the D/rDNA complex extends the region of DNase protection in the 3' direction from position ~-14 to position $\sim+20$ (149). These data suggest that the association of D with the rDNA promoter directs C to bind and initiate transcription. However, it remains to be determined why these results differ from those of Tjian and coworkers (159), who have been unable to detect a stable association of human rDNA with their 'SL1' fraction which otherwise seems analogous to factor D.

Analysis of the physical characteristics of D may be hampered by its estimated low abundance of <500 molecules per cell (159). However, native molecular masses of ~300 kd for *Acanthamoeba* D (149) and of ~100 kd for human D (159) have been deduced from gel filtration studies. In addition, Tjian's laboratory has prepared a polyclonal antibody to a greatly enriched preparation of human D (159). This antibody specifically inhibits human rDNA transcription in vitro and selectively binds to primate nucleoli in situ.

Finally, the possibility remains that cellular factors in addition to D and activated polymerase may have a role in rDNA transcription. Such putative additional factors could be masked under the currently detected activities, could have a stimulatory or inhibitory action, or could be involved in enhancement or other processes that have not yet been reproduced in vitro.

REGULATION OF rDNA TRANSCRIPTION

During the past two decades, considerable effort has focused on the regulation of eukaryotic rRNA gene expression. Numerous experiments using whole animals, intact cells, nuclei, and nucleoli have shown that rRNA synthesis is regulated in response to a variety of treatments, principally ones that alter the rate of protein synthesis or cell growth. Moreover, the amount of run-on transcription obtained in nuclei and nucleoli has been found to reflect the rate of rRNA production of the cells from which they were isolated. Ribosomal RNA synthesis is stimulated when cells are induced to proliferate by partial removal of an organ, refeeding after starvation, treatment with various chemical agents, or infection with SV40. Conversely, rRNA synthesis is decreased upon treatment with protein synthesis inhibitors, starvation for essential nutrients, infection with certain viruses, or upon entry into stationary phase. This vast literature has been extensively reveiwed in a number of articles (4, 5, 7, 8, 16, 168, 169) to which the reader is referred for specific references.

Several studies have shown that the amount of polymerase I that is tightly bound to the chromatin of nuclei and nucleoli—evidently molecules that are 'engaged' in synthesizing nascent transcripts (170)—correlates closely with the in vivo rate of rRNA synthesis (171; reviewed in 5, 7, 8). This result suggests that alterations in the rate of rRNA production are mediated largely by changes in the rate of transcriptional initiation. The alternative possibilities, that rRNA production may be regulated primarily by changes in rRNA stability or in the rate of elongation, appear unlikely although these may be contributing factors in certain situations (172–175).

When protein synthesis is inhibited by withdrawal of essential amino acids (176), by treatment with cycloheximide or puromycin (171), or through the use of temperature-sensitive mutants (177), the resultant inhibition of rRNA synthesis occurs rapidly ($t_{1/2} \sim 30$ min), while total polymerase I activity does not change significantly during a several hour period. From this result many workers have concluded that rDNA transcription requires a short-lived protein ancillary to the polymerase (178; reviewed in 5, 7, 16).

CELL EXTRACTS MIMIC THE IN VIVO TRANSCRIPTIONAL ACTIVITY   Studies using cell-free extracts and cloned rDNA have shown that the transcriptional activity of these extracts reflects the rate of rRNA synthesis in the cells from which they were prepared. Grummt first reported that transcription reactions containing extracts of stationary-phase or serum-starved cells do not accumulate a processed rDNA transcript (62), and Miller & Sollner-Webb found that extracts of stationary phase cells also do not produce the primary transcript (179). These results demonstrated that the inhibition of synthesis is at the level of transcription rather than processing, and mixing experiments indicated that this was due to the lack of a necessary component rather than to the presence of an additional inhibitor (62, 179). Extracts prepared from cells in which protein synthesis had been inhibited by cycloheximide treatment (129, 155; E. A. Thompson, personal communication) or by amino acid (67) or serum (167b) starvation are also deficient in the ability to catalyze transcription of cloned rDNA. Similarly, Thompson's laboratory has reported that the inhibition of rRNA synthesis observed in glucocorticoid-treated P1798 cells (180) is mimicked by S-100 extracts prepared from these cells (158), and Paule's laboratory has found that the decrease in rRNA synthesis that occurs during *Acanthamoeba* encystment is closely reflected in a reduced rRNA transcriptional activity of the S-100 extracts (161).

Transcriptional activity can be restored in all of these inactive extracts by addition of a component from extracts of log phase cells. This component is factor C, which was suggested above to be an activation of RNA polymerase I that allows it to participate in accurate rDNA transcription (see section on factors above). Supplementing with factor C allows these inactive extracts to transcribe cloned rDNA, whether C is added as a partially purified fraction (67, 129, 144, 155, 158, 167b) or as a preparation that had been purified to virtual homogeneity of the polymerase I polypeptides (161; J. Tower, B. Sollner-Webb, in preparation). These inactive extracts have approximately normal amounts of D activity and bulk polymerase I activity, and addition of these components does not restore the specific transcriptional activity (155, 161, 180). Thus, these extracts are inactive for rDNA transcription because they lack factor C. By inference, the decreased in vivo rRNA synthesis in amino acid– or serum-starved cells and in cycloheximide-treated cells, as well as in glucocorticoid treated cells and cells undergoing encystment, may also be due to low amounts of the activated form of polymerase I. However, it should be noted that while the in vitro data clearly demonstrate a correlation between the amount of available C and the in vivo rate of rRNA synthesis, no causal relationship has yet been established.

The rRNA transcriptional inhibition that occurs ~12–18 hours after cells are infected with Adenovirus (181–183) has been investigated in Tjian's laboratory using the in vitro reaction. As with the regulatory events described above, extracts prepared from Adenovirus-infected cells fail to initiate transcription of cloned rDNA (133). Mixing experiments have demonstrated that this is due to the lack of a necessary component, but its identity has not yet been investigated. However, since Adenovirus infection also causes a parallel shutdown of host protein synthesis (181, 184), it seems likely that the reduced transcriptional capacity of extracts of Adenovirus-infected cells is also due to reduced C activity.

In contrast to the case with Adenovirus, cells that have been infected with SV40 exhibit a 2–3-fold elevated rate of rRNA synthesis in vivo (185, 186). This stimulation is evidently due to an effect of T-antigen (187–189). Tjian's laboratory has investigated this phenomenon using in vitro assays, and found that the transcriptional activity of extracts prepared from normal cells is reproducibly increased 2–3-fold by addition of purified T-antigen (133). While it is not yet clear how T-antigen increases the activity of the extract to which it is added, preliminary results indicate that it involves a mechanism other than direct binding of T-antigen to the rDNA template (133).

An additional regulatory treatment that has been investigated using transcription of cloned rDNA is the selective inhibition of rRNA synthesis that occurs when cells are administered low doses of Actinomycin D (AMD). In this case, neither transient transfection systems (75) nor S-100 extracts of the treated cells (129) mimic the in vivo transcriptional inactivity. Instead, they catalyze 3–5-fold more rDNA transcription than do control extracts. The increased in vitro transcription results from a 3–5-fold elevation of C activity in these extracts, while the D level is not appreciably altered (J. Tower, B. Sollner-Webb, unpublished observations). Earlier electron microscopic observations of ribosomal chromatin from AMD-treated cells had indicated that this drug causes the nascent transcripts to be prematurely released from the ribosomal template (190a, 190b). Since this RNA release may well be accompanied by release of the transcribing polymerases, this result may readily explain the increased levels of C (activated polymerase) found in the extracts of AMD-treated cells. It is not yet clear why AMD causes a selective repression of rRNA synthesis in vivo, but since the repression is not reproduced on cloned rDNA promoters, it is evidently not inherent to the polymerase I transcriptional machinery. Most likely it is due to AMD intercalation into the large, very GC-rich rDNA transcribed region, but possibly characteristics special to transcription of genomic rDNA or to transcription occurring in nucleoli may also be involved.

NUCLEOLAR DOMINANCE   Nucleolar dominance is the transcriptional regulation observed in cross-species hybrids, where rRNA is only transcribed from the genes of one of the parental species. First reported for the plant *Crepis* (191), this phenomenon has been studied in a number of other cases, including *X. laevis-X. borealis* hybrid frogs (192) and mouse-human hybrid cells (193, 194). Recent advances in understanding the factors and the nucleotide sequences that direct rDNA transcription have allowed the biological bases for two kinds of nucleolar dominance to be understood.

In the case of mouse-human hybrid cells, most lines shed some of their human chromosomes and express only the mouse and not the human rRNA gene (M>H; 194–197), although certain lines exhibit the reverse dominance (H>M; 187, 194, 198). Reeder's (199) laboratory and Arnheim and collaborators (157) have shown that extracts of M>H cells mimic the in vivo situation by transcribing mouse but not human cloned rDNA, and the latter group has demonstrated that this is because the extract contains only the mouse but not the human species-specific transcription factor, D. Addition of human factor D to these transcription reactions allows cloned human rDNA to be transcribed (157). Thus, mouse-human nucleolar dominance appears to result from the lack of factor D specific for the recessive species of the hybrid cells. It is intriguing that a number of reports have indicated that introduction of SV40 T-antigen into mouse-human hybrid cells reactivates transcription of the recessive species' rDNA (187–189, and references therein), suggesting that mouse-human nucleolar dominance may be due to a reversible silencing, rather than a loss, of the recessive species' gene for factor D. Alternatively, T-antigen could act by altering the specificity of the dominant species' factor D.

In contrast, the dominance of *X. laevis* over *X. borealis* rDNA (192) can not be due to species-specific factors because each species' rDNA can be transcribed in oocytes of the other species (73, 128). Instead, Reeder's laboratory has shown that *Xenopus* nucleolar dominance is attributable to enhancer effects (200). Since *X. laevis* rDNA has ~20–60 copies of the 60/81-bp enhancer elements per gene, while *X. borealis* has only a small number of these sequences (201), results from studies of the *Xenopus* rDNA enhancer (see above) would suggest that when the two species' genes are simultaneously present in a *Xenopus* cell, laevis rDNA would be transcribed and borealis rDNA would be repressed. By coinjection studies, Reeder & Roan (200) have shown that this is indeed the case. In addition, by coinjecting hybrid plasmids containing a laevis spacer joined to a borealis gene and vice versa, they demonstrated that the dominance is determined by the enhancer-containing spacer (200). Thus, nucleolar dominance in *Xenopus* results from a competition between two species' rDNA for common transcription components, while the nucleolar dominance in mammalian cross-species hybrids is due to the presence of the specific transcription factor of only one species.

RAPID 5' AND 3' rRNA PROCESSING AND TRANSCRIPTIONAL TERMINATION

In 1971, Tiollais et al (202) used double-label experiments and sucrose gradient analysis to obtain strong evidence that 45S RNA was not the primary transcript of mammalian rDNA, but that it resulted from rapid processing of a 47S transcript through a 46S intermediate. Due to the lack of subsequent corroborating data this result was largely ignored, and for the next decade interest in rRNA processing focused on the boundaries between the transcribed spacers and the 18S, 5.8S, and 28S regions (13, 14, 27–29). Then in 1981, Miller & Sollner-Webb (86) showed that ~90% of the pre-rRNA of mouse was a processed species, the 5' end of which lacked the first 650 nucleotides of the primary transcript. This processing appears to involve an endonucleolytic cleavage followed by rapid degradation of the upstream fragment (86). In the intact cell it occurs quite rapidly after completion of the primary transcript (86, 203a, 204), and the processing is efficiently reproduced in vitro in S-100 extracts (86). It appears that human pre-rRNA may undergo an analogous processing at nucleotides +414 of the primary transcript (97), suggesting that 5' processing may be a common step in maturation of mammalian rRNA. Recently, *Neurospora* rRNA has also been shown to be very rapidly cleaved at nucleotide ~120 both in vivo and in vitro (66), demonstrating that efficient 5' processing is not restricted to mammalian species. However, 5' processing of the primary rRNA transcript to yield a major species of pre-rRNA is certainly not ubiquitous in eukaryotes, for the vast majority of the pre-rRNAs of *Drosophila* (89) and *Xenopus* (43) retain the 5' terminal residue of the primary transcript.

With regard to the 3' end of the transcript, S1 mapping data accumulated through the early 1980s suggested that the cellular pre-rRNA of mouse (205), *X. laevis* (43, 57), *Drosophila* (206), yeast (207), and *Tetrahymena* (208) ended coincident with the 3' end of the 28S coding region (43, 206), or ~7 (207), ~15 (57, 208), or ~30 (205) nucleotides beyond this position. In apparent agreement, Miller spreads of transcribing chromatin which had been either cleaved a short distance upstream from the 3' end of the 28S coding region (33) or deleted for rDNA sequences downstream of this site (71,106) suggested that transcription of *Xenopus* rDNA indeed terminated at the end of the 28S region (although other studies indicated that polymerases lacking transcripts remain attached to the gene for a short additional distance; 209). These results, in conjunction with earlier secondary structure analysis of pre-rRNA and 28S RNA (27, 28), have led to the standard textbook model in which the 3' ends of the primary rRNA transcript and 28S RNA are virtually coincident. However, new data has demonstrated that significant processing can also occur at the 3' end of the pre-rRNA. Using S1 mapping, Grummt (210) found that mouse rRNA transcripts extend at least several hundred bp beyond the end of the 28S region. In a striking extension of this work, Gurney (203a) has electrophoretically resolved three species of pre-rRNA from mouse cells that differ in length by ~600–700 nucleotides and presumably correspond to the 47S, 46S, and 45S species reported earlier by Tiollais et al (202). Using hybrid selection to 5' and 3' regions of the rDNA, the largest RNA was shown to contain the first 650 nucleotides of the transcript and a similar length of sequence 3' to the 28S region, the second contained the 3' trailer but not the 5' segment, and the third retained neither the 5' nor 3' regions (203a). Very recent studies from Grummt's laboratory (203b) indicate that a site ~560 bp 3' to the 28S region is an in vivo and in vitro termination site, the action of which involves a trans-acting factor. Thus, mouse rRNA appears to be sequentially processed at sites ~600 nucleotides in from each end of the presumptive primary transcript. At the 3' end of *Xenopus laevis* rRNA gene, P. Labhart and R. Reeder (personal communication) have observed a similar situation in which transcription continues for at least several hundred bp beyond the 28S region. In fact, their data indicate the revolutionary idea that *Xenopus* rDNA transcription actually extends through virtually the entire 'nontranscribed spacer' to yield a very unstable RNA segment. It remains to be determined how many other species' primary rRNA transcripts also have 3' trailers and the extent of these additional regions of transcription.

CONCLUDING REMARKS

This is a very exciting time in the study of rDNA transcription. Within the last five years, in vitro and in vivo systems that accurately transcribe cloned rRNA genes have been developed, and studies involving these systems are beginning to come to fruition. As was described in this article, great progress has been made in understanding the rDNA sequences, promoters as well as enhancers, and the protein factors that direct faithful and efficient transcription. Our knowledge of the functional role of these components, the mechanisms by which rDNA transcription is regulated, and the termination and processing of the rRNA transcript are also advancing rapidly. However, many fundamental questions still remain unanswered in virtually all of these areas, and other important aspects of rDNA transcription have barely been approached. A sampling of some of the intriguing questions that await resolution are enumerated below:

1. How do rDNA enhancer sequences act to augment transcription, and how general is their occurrence and mode of action across different species? The possible relevance of spacer transcription clearly merits investigation. An in vitro system that reproduces enhancement would certainly be of great value in these studies.
2. What are the nature and mode of action of the polymerase I transcription factors? It will be important to identify the physical modification (or association) that distinguishes factor C from bulk polymerase I. The purification of factor D and studies of how it interacts with the ribosomal template and factor C are also of high priority. It also remains to be determined whether there are specific polymerase I transcription factors in addition to C and D, possibly components that are not detectable in the current cell-free transcription systems.

3. How is rRNA gene transcription regulated in vivo? If many different rDNA regulatory conditions are indeed mediated by a rapid reduction in the availability of factor C, how is this accomplished and controlled? On the other hand, it is possible that C levels are reduced as a consequence of there being less polymerase I engaged in transcription, and in that eventuality it would be important to learn what is the primary cause of the decreased rDNA transcription. Modes of regulation that are not currently reproduced in vitro, for instance, ones that might involve enhancers, topological constraints, or nucleolar attachment sites, must be considered. The question of how SV40 T-antigen acts to stimulate transcription, especially how it may reactivate the recessive species' rRNA genes in hybrid cell lines, is also intriguing. Furthermore, the possibility exists that eukaryotic rDNA transcription might be controlled by a feedback regulation mechanism, at least formally analogous to the situation in prokaryotes (19, 211, 212). Models proposed to explain rDNA transcriptional regulation must also take into account electron microscopic data which suggests that each rRNA gene is generally either transcribed at maximal levels or not at all (e.g. 5, 17, 213). Ultimately, of course, one wants to learn how the coordinate production of the various components of the ribosome, which involves transcription by each of the three polymerase classes, is regulated.

4. How is the termination of rDNA transcription and the processing of the primary rRNA transcript effected? It remains to be determined where most rRNA genes actually terminate transcription, how the processing events that generate the major pre-rRNA species occur, and what (if any) functional significance they have.

5. How is rDNA organized in chromatin and within the nucleolar framework? Many studies involving nuclease digestion and electron microscopy have focused on ribosomal chromatin (reviewed in 213–217). Although these have been more than sufficient to demonstrate that it has a structure different than inactive chromatin, further investigations using additional techniques (218–220) appear to be needed to determine what the structure really is. Most likely, actively transcribing rDNA regions are not packaged into normal nucleosomes, a result that is hardly surprising considering that every ~100 bp along the gene there is a polymerase (which is ~3.5 times the size of a nucleosome) as well as a nascent rRNA transcript. It is striking, however, that the RNA gene region also appears to be in an extended conformation prior to the onset of transcription (see e.g. 221), but the structural basis of this altered chromatin configuration is not yet known. In various studies the nontranscribed spacer regions have appeared either to contain normal nucleosomes, to contain an extended chromatin structure, or even to be free of protein (e.g. 71, 106, 214, 217–222), but the bulk of the data suggest that the nontranscribed spacer is generally in an altered chromatin organization. The nature of the chromatin of this region and its potential

relation to the transcriptional process still await resolution. In addition, particular rDNA sites appear to be nuclease hypersensitive (223–226) but their relevance to rDNA transcription is still unclear. Further study of the intriguing result that such hypersensitive positions can be binding sites for topoisomerase I (227) deserves special attention, especially since topo I is highly enriched in the nucleolus (228, 229) and has been implicated in the establishment of active chromatin domains (230). Furthermore, the issue of the attachment of rDNA to the nucleolar framework (231–234) also awaits resolution. And finally, although the nucleolus has been the subject of literally thousands of studies over the last >200 years (16, 17), little is actually known about its precise organization and role in catalyzing transcription. At one extreme, the nucleolus might be essential to provide a requisite concentration or spatial arrangement of factors for rDNA transcription, rRNA processing, and ribosome assembly; at the other extreme, the nucleolus might form quite passively and merely reflect an aggregation of ribosomal precursors. Further study, presumably in part involving cloned rDNA that has been reintroduced into cells (233), will be needed to resolve these issues.

In conclusion, the last five years have been a time of major advances in the field of eukaryotic rDNA transcription and there is every reason to believe that this marked progress will continue and that a much clearer understanding of rDNA transcription will emerge.

ACKNOWLEDGMENTS

For making unpublished data available to us and for helpful and interesting discussions we thank many of our colleagues, especially R. Baserga, J. Corden, V. Culotta, I. Grummt, M. Holland, J. Klootwijk, M. Paule, R. Reeder, A. Thompson, R. Tjian, M. Trendelenburg, J. Warner, and J. Windle. We would also like to apologize to those whose experiments were not cited. Acknowledgment is also made for research support from NIH grants 27720 and 43231.

Literature Cited

1. Reeder, R. H. 1974. In *Ribosomes*, ed. M. Nomura, et al., pp. 489–519. Cold Spring Harbor, NY: Cold Spring Harbor Lab.
2. Fedoroff, N. 1979. *Cell* 16:697–701
3. Long, E. O., Dawid, I. B. 1980. *Ann. Rev. Biochem.* 49:727–64
4. Sollner-Webb, B., Wilkinson, J. K., Miller, K. G. 1982. *Cell Nucleus* 12:31–67
5. Miller, K. G., Sollner-Webb, B. 1982. *Cell Nucleus* 12:69–100
6. Mandal, R. K. 1984. *Prog. Nucleic Acid Res. Mol. Biol.* 31:115–60
7. Grummt, I. 1968. *Cell Nucleus* 5:373–414
8. Muramatsu, M., Matsui, M., Onishi, T., Mishima, Y. 1979. *Cell Nucleus* 7:123–61
9. Chambon, P. 1975. *Ann. Rev. Biochem.* 44:613–38
10. Roeder, R. 1976. In *RNA Polymerase*, ed. R. Losick, M. Chamberlin, pp. 258–

329. Cold Spring Harbor, NY: Cold Spring Harbor Lab.

11. Jacob, S. T., Rose, K. M. 1978. *Methods Cancer Res.* 14:191–241

12a. Paule, M. R. 1981. *Trends Biochem. Sci.* 6:128–31

12b. Sentenac, A. 1985. In *Critical Reviews in Biochemistry* 18:31–90

13. Crouch, R. 1984. In *Processing of RNA,* ed. D. Apirion, pp. 214–26. Cleveland, Ohio: CRC

14. Perry, R. P. 1976. *Ann. Rev. Biochem.* 45:605–29

15. Busch, H., Smetana, K. 1970. *The Nucleolus.* New York: Academic

16. Hadjiolov, A. A. 1985. *The Nucleolus and Ribosome Biogenesis.* New York: Springer-Verlag

17. Miller, O. L. 1981. *J. Cell. Biol.* 91:515–527

18. Clayton, D. A. 1984. *Ann. Rev. Biochem.* 53:573–94

19. Nomura, M., Gourse, R., Baughman, G. 1984. *Ann. Rev. Biochem.* 53:75–117

20. Korn, L., Bogenhagen, D. 1982. *Cell Nucleus* 12:1–29

21. Birnstiel, M., Speirs, J., Purdom, I., Jones, K., Loening, U. E. 1968. *Nature* 219:454–59

22. Brown, D. D., Dawid, I. B. 1968. *Science* 160:272–80

23. Gall, J. 1968. *Proc. Natl. Acad. Sci. USA* 60:553–57

24. Morrow, J., Cohen, S., Chang, A., Boyer, H., Goodman, H., Hellings, R. 1974. *Proc. Natl. Acad. Sci. USA* 71:1743–47

25. Miller, O. L., Beatty, B. R. 1969. *Science* 164:955–57

26. Miller, O. L., Bakken, A. H. 1972. *Acta Endocrinol. Suppl.* 168:155–77

27. Wellauer, P. K., Dawid, I. B. 1973. *Proc. Natl. Acad. Sci. USA* 70:2827–31

28. Wellauer, P. K., Dawid, I. B. 1974. *J. Mol. Biol.* 89:379–95

29. Wellauer, P. K., Dawid, I. B., Kelley, D. E., Perry, R. P. 1974. *J. Mol. Biol.* 89:379–91

30. Dame, J. B., McCutchan, T. F. 1983. *J. Biol. Chem.* 258:6984–90

31. Bowman, L. H., Goldman, W. E., Goldberg, G. I., Hebert, M. B., Schlessinger, D. 1983. *Mol. Cell. Biol.* 3:1501–10

32. Dawid, I. B., Wellauer, P. K. 1976. *Cell* 8:443–48

33. Reeder, R. H., Higashinakagawa, T., Miller, O. 1976. *Cell* 8:449–54

34. Blatti, S. P., Ingles, C. J., Lindell, T. J., Morris, P. W., Weaver, R. F. et al. 1970. *Cold Spring Harbor Symp. Quant. Biol.* 35:649–63

35. Roeder, R., Rutter, W. 1970. *Proc. Natl. Acad. Sci. USA* 65:675–769

36. Kedinger, C., Gniazdowski, M., Mandel, J. L., Chambon, P. 1970. *Biochem. Biophys. Res. Commun.* 38:165–73

37. Benecke, B. J., Penman, S. 1977. *Cell* 12:939–46

38. Benecke, B. J., Penman, S. 1979. *J. Cell. Biol.* 80:778–83

39. Reichel, R., Monstein, H. J., Jansen, H. W., Phillipson, L., Benecke, B. J. 1982. *Proc. Natl. Acad. Sci. USA* 79:3106–10

40. Reichel, R., Benecke, B. J. 1984. *EMBO J.* 3:473–79

41. Rungger, D., Crippa, M. 1976. *Prog. Biophys. Mol. Biol.* 31:247–69

42. Boseley, P., Moss, T., Machler, M., Portmann, R., Birnstiel, M. 1979. *Cell* 17:19–31

43. Sollner-Webb, B., Reeder, R. H. 1979. *Cell* 18:485–99

44. Moss, T., Birnstiel, M. L. 1979. *Nucleic Acids Res.* 6:3733–43

45. Kohorn, B. D., Rae, P. M. M. 1982. *Nucleic Acids Res.* 10:6879–86

46. Coen, E. S., Dover, G. A. 1982. *Nucleic Acids Res.* 10:7017–26

47. Miller, J. R., Hayward, D. C., Glover, D. M. 1983. *Nucleic Acids Res.* 11:11–19

48. Murtif, V. L., Rae, P. M. M. 1985. *Nucleic Acids Res.* 13:3221–39

49. Treco, D., Brownell, E., Arnheim, N. 1982. *Cell Nucleus* 12:101–26

50. Scheer, U., Trendelenburg, M. F., Franke, W. W. 1973. *Exp. Cell Res.* 80:175–90

51. Trendelenburg, M. F., Franke, W. W. 1973. *Nature* 245:167–70

52. Williams, M. A., Trendelenburg, M. F., Franke, W. W. 1981. *Differentiation* 20:36–44

53. Trendelenburg, M. F. 1981. *Biol. Cell* 42:1–12

54. Trendelenburg, M. F. 1982. *Prog. Clin. Biol. Res.* 85A:199–210

55. Morgan, G. T., Reeder, R. H., Bakken, A. H. 1983. *Proc. Natl. Acad. Sci. USA* 80:6490–94

56. Morgan, G. T., Roan, J. G., Bakken, A. H., Reeder, R. H. 1984. *Nucleic Acids Res.* 12:6043–52

57. Moss, T. 1983. *Nature* 302:223–28

58. Kohorn, B. D., Rae, P. M. M. 1982. *Proc. Natl. Acad. Sci. USA* 79:1501–1505

59. Wu, G., Zubay, G. 1974. *Proc. Natl. Acad. Sci. USA* 71:1803–1807

60. Wu, G. 1978. *Proc. Natl. Acad. Sci. USA* 75:2175–79

61. Weil, A., Luse, D. S., Segall, J., Roeder, R. G. 1979. *Cell* 18:469–84

62. Grummt, I. 1981. *Proc. Natl. Acad. Sci. USA* 78:727–31

63. Manley, J. L., Fire, A., Cano, A.,

Sharp, P. A., Gefter, M. L. 1980. *Proc. Natl. Acad. Sci. USA* 77:3855–59
64. Learned, R. M., Tjian, R. 1982. *J. Mol. Appl. Genet.* 1:575–84
65. Sutiphong, J., Matzura, C., Niles, E. G. 1984. *Biochemistry* 23:6319–26
66. Tyler, B. M., Giles, N. H. 1985. *Nucleic Acids Res.* 13:4311–32
67. Grummt, I. 1986. In *Molecular Biology of Development, UCLA Symp.*, Vol. 20. New York: Liss. In press
68. Kurl, R. N., Rothblum, L. I., Jacob, S. T. 1984. *Proc. Natl. Acad. Sci. USA* 81:6672–75
69. Kurl, R. N., Jacob, S. T. 1985. *Proc. Natl. Acad. Sci. USA* 82:1059–63
70. Trendelenburg, M. F., Zentgraf, H., Franke, W. W., Gurdon, J. B. 1978. *Proc. Natl. Acad. Sci. USA* 75:3791–95
71. Trendelenburg, M. F., Gurdon, J. B. 1978. *Nature* 276:292–94
72. Moss, T. 1982. *Cell* 30:835–42
73. Sollner-Webb, B., McKnight, S. L. 1982. *Nucleic Acids Res.* 10:3391–405
74. Grummt, I., Skinner, J. A. 1985. *Proc. Natl. Acad. Sci. USA* 82:722–26
75. Smale, S. T., Tjian, R. 1985. *Mol. Cell. Biol.* 5:352–62
76. Lopata, M., Sollner-Webb, B., Cleveland, D. 1985. *Mol. Cell. Biol.* 5:2842–46
77. Elion, E. A., Warner, J. R. 1984. *Cell* 39:663–73
78. Kempers-Veenstra, A. E., van Heerikhuizen, H., Musters, W., Klootwijk, J., Planta, R. J. 1984. *EMBO J.* 3:1377–82
79. Quincey, R. V., Arnold, R. E. 1984. *Biochem. J.* 224:497–503
80a. Vance, V. B., Thompson, E. A., Bowman, L. H. 1985. *Nucleic Acids Res.* 13:7499–513
80b. Dhar, V., Miller, D., Miller, C. J. 1985. *Mol. Cell. Biol.* 5:2943–50
81. Berk, A. J., Sharp, P. A. 1977. *Cell* 12:721–32
82. Sollner-Webb, B., Felsenfeld, G. 1977. *Cell* 10:537–47
83. Bina-Stein, M., Thoren, M., Salzman, N., Thompson, J. A. 1979. *Proc. Natl. Acad. Sci. USA* 76:731–35
84. Green, M. R., Maniatis, T., Melton, D. A. 1983. *Cell* 32:681–94
85. Sollner-Webb, B., Wilkinson, J. A., Roan, J., Reeder, R. H. 1983. *Cell* 35:199–206
86. Miller, K. G., Sollner-Webb, B. 1981. *Cell* 27:165–74
87. Wilkinson, J. A. K., Miller, K. G., Sollner-Webb, B. 1983. *J. Biol. Chem.* 258:13919–28
88. Grummt, I. 1981. *Nucleic Acids Res.* 9:6093–102
89. Levis, R., Penman, S. 1978. *J. Mol. Biol.* 121:219–38
90. Nikolaev, N., Georgiev, O. I., Venkov, P. V., Hadjiolov, A. A. 1979. *J. Mol. Biol.* 127:297–308
91. Klootwijk, J., deJonge, P., Planta, R. J. 1979. *Nucleic Acids Res.* 6:27–39
92. Moss, B. 1977. *Biochem. Biophys. Res. Commun.* 74:374–83
93. Reeder, R. H., Sollner-Webb, B., Wahn, H. L. 1977. *Proc. Natl. Acad. Sci. USA* 74:5402–5406
94. Long, E. O., Rebbert, M. L., Dawid, I. B. 1979. *Proc. Natl. Acad. Sci. USA* 78:1513–17
95. Rothblum, L. I., Reddy, R., Cassidy, B. 1982. *Nucleic Acids Res.* 10:7345–62
96. Financsek, I., Mizumoto, K., Muramatsu, M. 1982. *Gene* 18:115–22
97. Financsek, I., Mizumoto, K., Mishima, Y., Muramatsu, M. 1982. *Proc. Natl. Acad. Sci. USA* 79:3092–96
98. Klemenz, R., Geiduschek, E. P. 1980. *Nucleic Acids Res.* 8:2679–89
99. McStay, B., Bird, A. 1983. *Nucleic Acids Res.* 11:8167–81
100. Bach, R., Allet, B., Crippa, M. 1981. *Nucleic Acids Res.* 9:5311–30
101. Paule, M. R., Iida, C. T., Perna, P. J., Harris, G. H., Shimer, S. B., Kownin, P. 1984. *Biochemistry* 23:4167–72
102. Martin, F. H., Tinoco, I. 1980. *Nucleic Acids Res.* 8:2295–305
103. Crampton, J. M., Woodland, H. R. 1979. *Dev. Biol.* 70:467–78
104. Hipskind, R. A., Reeder, R. H. 1980. *J. Biol. Chem.* 255:7896–906
105. McKnight, S. L., Hipskind, R. A., Reeder, R. H. 1980. *J. Biol. Chem.* 255:7907–11
106. Bakken, A., Morgan, G., Sollner-Webb, B., Roan, J., Busby, S., Reeder, R. H. 1982. *Proc. Natl. Acad. Sci. USA* 79:56–60
107. Labhart, P., Reeder, R. H. 1984. *Cell* 37:285–89
108. Pruitt, S. C., Reeder, R. H. 1984. *J. Mol. Biol.* 174:121–39
109. Pruitt, S. C., Reeder, R. H. 1984. *Mol. Cell. Biol.* 4:2851–57
110. Busby, S. J., Reeder, R. H. 1983. *Cell* 34:989–96
111. Wilkinson, J. K., Sollner-Webb, B. 1982. *J. Biol. Chem.* 257:14375–83
112. Deleted in proof
113. Deleted in proof
114. Mishima, Y., Yamamoto, O., Kominami, R., Muramatsu, M. 1981. *Nucleic Acids Res.* 9:6773–85
115. Yamamoto, O., Takakusa, N., Mishima,

Y., Kominami, R., Muramatsu, M. 1984. *Proc. Natl. Acad. Sci. USA* 81: 299–303

116. Wandelt, C., Grummt, I. 1983. *Nucleic Acids Res.* 11:3795–809

117. Cizewski, V., Sollner-Webb, B. 1983. *Nucleic Acids Res.* 11:7043–56

118. Mishima, Y., Financsek, I., Kominami, R., Muramatsu, M. 1982. *Nucleic Acids Res.* 10:6659–70

119. Grummt, I., Roth, E., Paule, M. R. 1982. *Nature* 296:173–74

120. Miesfeld, R., Arnheim, N. 1982. *Nucleic Acids Res.* 10:3933–49

121. Long, E. O., Rebbert, M. L., Dawid, I. B. 1981. *Proc. Natl. Acad. Sci. USA* 78:1513–17

122. Swanson, M. E., Holland, M. J. 1983. *J. Biol. Chem.* 258:3242–50

123. Klootwijk, J., Verbeet, M. P., Veldman, G. M., deRegt, V. C., van Heerikhuizen, H., et al. 1984. *Nucleic Acids Res.* 12:1377–90

124. Gabrielsen, O. S., Oyen, T. B. 1982. *Nucleic Acids Res.* 10:5893–904

125. Swanson, M. E., Yip, M., Holland, M. J. 1985. *J. Biol. Chem.* 260:9905–15

126. Miesfeld, R., Arnheim, N. 1984. *Mol. Cell. Biol.* 4:221–27

127. Dover, G. A., Flavell, R. B. 1984. *Cell* 38:622–23

128. Morgan, G. T., Bakken, A. H., Reeder, R. H. 1982. *Dev. Biol.* 93:484–87

129. Sollner-Webb, B., Tower, J., Culotta, V. C., Windle, J. 1985. In *Genetic Engineering*, ed. J. Setlow, A. Hollaender, 7:309–32. New York: Plenum

130. Fleischer, S., Grummt, I. 1983. *EMBO J.* 2:2319–22

131. Grummt, I. 1982. *Proc. Natl. Acad. Sci. USA* 79:6908–11

132. Miller, K. G., Tower, J., Sollner-Webb, B. 1985. *Mol. Cell. Biol.* 5:554–62

133. Learned, R. M., Smale, S. T., Haltiner, M. M., Tjian, R. 1983. *Proc. Natl. Acad. Sci. USA* 80:3558–62

134. Kohorn, B. D., Rae, P. M. M. 1983. *Proc. Natl. Acad. Sci. USA* 80:3265–69

135. Kohorn, B. D., Rae, P. M. M. 1983. *Nature* 304:179–81

136. Kownin, P., Iida, C. T., Brown-Shimer, C., Paule, M. R. 1985. *Nucleic Acids Res.* 13:6237–48

137. Iida, C. T., Kownin, P., Paule, M. R. 1985. *Proc. Natl. Acad. Sci. USA* 82: 1668–72

138a. Sollner-Webb, B., Miller, K. G., Tower, J., Culotta, V. C., Windle, J. 1985. *J. Cell. Biochem. B* 9:174

138b. Windle, J., Sollner-Webb, B. 1986. *Mol. Cell. Biol.* 6:1228–34

139. McKnight, S. L., Kingsbury, R. 1982. *Science* 217:316–24

140. Skinner, J. A., Ohrlein, A., Grummt, I. 1984. *Proc. Natl. Acad. Sci. USA* 81: 2137–41

141. Kisimoto, T., Nagamine, M., Saski, T., Takakusa, N., Miwa, T., et al. 1985. *Nucleic Acids Res.* 13:3515–32

142. Sommerville, J. 1984. *Nature* 310:189–90

143. Bogenhagen, D. F., Wormington, W. M., Brown, D. D. 1982. *Cell* 28:413–21

144. Cavanaugh, A. H., Thompson, E. A. 1985. *Nucleic Acids Res.* 13:3357–69

145. Culotta, V. C., Wides, R. J., Sollner-Webb, B. 1985. *Mol. Cell. Biol.* 5:1582–90

146. Reeder, R. H., Roan, J., Dunaway, M. 1983. *Cell* 35:449–56

147. Culotta, V. C., Sollner-Webb, B. 1985. *J. Cell. Biochem. B* 9:156

148. Dunaway, M., Reeder, R. H. 1985. *Mol. Cell. Biol.* 5:313–9

149. Bateman, E., Iida, C., Kowin, P., Paule, M. 1985. *Proc. Natl. Acad. Sci. USA* 82:8004–8

150. Reeder, R. H. 1984. *Cell* 38:349–51

151a. Boseley, P., Moss, T., Machler, M., Portmann, R., Birnstiel, M. 1979. *Cell* 17:19–31

151b. deWinter, R., Moss, T. 1986. *Cell* In press

152. Muramatsu, M., Mishima, Y., Yamamoto, O., Nagamine, M., Takakusa, N., et al. 1983. *Nucle(ol)ar Workshop Abstr.*, p. 48

153. Kurl, R. N., Jacob, S. T. 1985. *Nucleic Acids Res.* 13:89–101

154. Slattery, E., Dignam, J. D., Matsui, T., Roeder, R. G. 1983. *J. Biol. Chem.* 258:5955–59

155. Tower, J., Culotta, V. C., Sollner-Webb, B. 1985. *J. Cell. Biochem. B* 9:206

156. Tower, J., Culotta, V. C., Sollner-Webb, B. 1986. *Mol. Cell. Biol.* In press

157. Miesfeld, R., Sollner-Webb, B., Croce, C., Arnheim, N. 1984. *Mol. Cell. Biol.* 4:1306–12

158. Cavanaugh, A. H., Gokal, P. K., Lawther, R. P., Thompson, E. A. 1984. *Proc. Natl. Acad. Sci. USA* 81:718–21

159. Learned, R. M., Cordes, S., Tjian, R. 1985. *Mol. Cell. Biol.* 5:1358–69

160. Spindler, S. R., Duester, G. L., D'Alessio, J. M., Paule, M. R. 1978. *J. Biol. Chem.* 253:4669–75

161. Paule, M. R., Iida, C. T., Perna, P. J., Harris, G. H., Knoll, D. A., D'Alessio, J. M. 1984. *Nucleic Acids Res.* 12:8161–80

162. Hirsch, J., Martelo, O. J. 1976. *J. Biol. Chem.* 251:5408–13

163. Breant, B., Buhler, J. M., Sentenac, A.,

Fromageot, P. 1983. *Eur. J. Biochem.* 130:247–51
164. Rose, K. M., Stetler, D. A., Jacob, S. T. 1981. *Proc. Natl. Acad. Sci. USA* 78:2833–37
165. Rose, K. M., Duceman, B. W., Jacob, S. T. 1981. *Curr. Top. Biol. Med. Res.* 5:115–41
166. Duceman, B. W., Rose, K. M., Jacob, S. T. 1981. *J. Biol. Chem.* 256:10755–58
167a. Dahmus, M., Kedinger, C. 1983. *J. Biol. Chem.* 258:2303–7
167b. Buttgereit, D., Pflugfelder, G., Grummt, I. 1985. *Nucleic Acids Res.* 13:8165–80
168. Hadjiolov, A., Nikolaev, N. 1976. *Prog. Biophys. Mol. Biol.* 31:349–51
169. Warner, J. 1974. See Ref. 1, pp. 461–83
170. Yu, F. L., Feigelson, P. 1971. *Proc. Natl. Acad. Sci. USA* 68:2177–80
171. Mishima, Y., Matsui, T., Muramatsu, M. 1979. *J. Biochem.* 85:807–18
172. Cooper, H. 1969. *J. Biol. Chem.* 244:5590–96
173. Wolf, S., Sameshima, M., Liebhaber, S., Schlessinger, D. 1980. *Biochemistry* 19:3484–90
174. Dauphinais, C. 1981. *Eur. J. Biochem.* 114:487–92
175. Dembinski, T. C., Bell, P. A. 1984. *J. Steroid Biochem.* 21:497–504
176. Grummt, I., Smith, V. A., Grummt, F. 1976. *Cell* 7:439–45
177. Gross, K. J., Pogo, O. A. 1976. *Biochemistry* 15:2070–81
178. Benecke, B. J., Ferenez, A., Seifart, K. H. 1973. *FEBS Lett.* 31:53–58
179. Miller, K. G., Sollner-Webb, B. 1982. *Cell Nucleus* 12:95
180. Cavanaugh, A. H., Thompson, E. A. 1983. *J. Biol. Chem.* 258:9768–73
181. Castiglia, C., Flint, J. 1983. *Mol. Cell. Biol.* 3:662–71
182. Raskas, H. J., Thomas, D. C., Green, M. 1970. *Virology* 40:893–902
183. Todaro, G., Takemoto, K. 1969. *Proc. Natl. Acad. Sci. USA* 62:1031–37
184. Ginsberg, H., Bello, L., Levine, A. 1976. In *Molecular Biology of Viruses,* ed. J. Colter, W. Paranolych, pp. 547–62. New York: Academic
185. Salomon, C., Turler, H., Weil, R. 1977. *Nucleic Acids Res.* 4:1483–1503
186. Ide, T., Whelly, S., Baserga, R. 1977. *Proc. Natl. Acad. Sci. USA* 74:3189–92
187. Soprano, K. J., Dev, V. G., Croce, C. M., Baserga, R. 1979. *Proc. Natl. Acad. Sci. USA* 76:3885–89
188. Soprano, K. J., Jonak, G. J., Galanti, N., Floros, J., Baserga, R. 1981. *Virology* 109:127–36
189. Soprano, K. J., Galanti, N., Jonak, G.

J., McKercher, S., Pipas, J. M., et al. 1983. *Mol. Cell. Biol.* 3:214–19
190a. Scheer, U., Trendelenburg, M. F., Franke, W. W. 1975. *J. Cell Biol.* 65:163–79
190b. Puvion-Dutilleul, F., Bachellerie, J. P. 1979. *J. Ultrastruct. Res.* 66:190–99
191. Navashin, M. 1934. *Cytologia* 5:169–75
192. Honjo, T., Reeder, R. H. 1973. *J. Mol. Biol.* 80:217–28
193. Eliceiri, G. L., Green, H. 1969. *J. Mol. Biol.* 41:253–60
194. Perry, R. P., Kelley, D. E., Schibler, U., Huebner, K., Croce, C. M. 1979. *J. Cell. Physiol.* 98:553–60
195. Marshall, C. J., Handmaker, S. D., Bramwell, M. E. 1975. *J. Cell. Sci.* 17:307–25
196. Miller, D. A., Dev, V. G., Tantravahi, R., Miller, O. J. 1976. *Exp. Cell. Res.* 101:235–42
197. Dev, V. G., Miller, D. A., Rechsteiner, M., Miller, O. J. 1979. *Exp. Cell. Res.* 123:47–54
198. Croce, C. M., Talavera, A., Basilico, C., Miller, O. J. 1977. *Proc. Natl. Acad. Sci. USA* 74:694–97
199. Onishi, T., Berglund, C., Reeder, R. H. 1984. *Proc. Natl. Acad. Sci. USA* 81:484–87
200. Reeder, R. H., Roan, J. G. 1984. *Cell* 38:39–44
201. LaVolpe, A., Taggart, M., MacLeod, D., Bird, A. 1982. *Cold Spring Harbor Symp. Quant. Biol.* 47:585–92
202. Tiollais, P., Galibert, F., Boiron, M. 1971. *Proc. Natl. Acad. Sci. USA* 68:1117–21
203a. Gurney, T. 1985. *Nucleic Acids Res.* 13:4905–19
203b. Grummt, I., Maier, U., Ohrlein, A., Hassouna, N., Bachellerie, J. 1985. *Cell* 43:801–10
204. Fetherston, J., Werner, E., Patterson, R. 1984. *Nucleic Acids Res.* 12:7187–98
205. Kominami, R., Mishima, Y., Urano, Y., Sakai, M., Muramatsu, M. 1982. *Nucleic Acids Res.* 10:1963–79
206. Mandal, R. K., Dawid, I. B. 1981. *Nucleic Acids Res.* 9:1801–11
207. Veldman, G. M., Klootwijk, J., de-Jonge, P., Leer, R. J., Planta, R. J. 1980. *Nucleic Acids Res.* 8:5179–92
208. Din, N., Engberg, J., Gall, J. G. 1982. *Nucleic Acids Res.* 10:1503–13
209. Trendelenburg, M. F. 1982. *Chromosoma* 86:703–15
210. Grummt, I., Sorbaz, H., Hofmann, A., Roth, E. 1985. *Nucleic Acids Res.* 13:2293–304
211. Jinks-Robertson, S., Gourse, R. L., Nomura, M. 1983. *Cell* 33:865–76
212. Gourse, R. L., Takebe, Y., Sharrock, R.

A., Nomura, M. 1985. *Proc. Natl. Acad. Sci. USA* 82:1069–73

213. Franke, W. W., Scheer, U., Spring, H., Trendelenburg, M. F., Zentgraf, H. 1979. *Cell Nucleus* 7:49–95

214. McKnight, S. L., Martin, K. A., Beyer, A. L., Miller, O. L. Jr. 1979. *Cell Nucleus* 7:97–122

215. Igo-Kemenes, T., Horz, W., Zachau, H. G. 1982. *Ann. Rev. Biochem.* 51:89–121

216. Scheer, U., Zentgraf, H. 1982. *Cell Nucleus* 11:143–76

217. Reeves, R. 1984. *Biochim. Biophys. Acta* 782:343–93

218. Sogo, J. M., Ness, P. J., Widmer, R. M., Parish, R. W., Koller, T. 1984. *J. Mol. Biol.* 178:897–919

219. Prior, C. P., Cantor, C. R., Johnson, E. M., Littau, V. C., Allfrey, V. G. 1983. *Cell* 34:1033–42

220. Davis, A. H., Reudelhuber, T. L., Garrard, W. T. 1983. *J. Mol. Biol.* 167:133–55

221. Pruitt, S. C., Grainger, R. M. 1981. *Cell* 23:711–20

222. Labhart, P., Koller, T. 1982. *Cell* 28:279–92

223. LaVolpe, A., Taggart, M., McStay, B., Bird, A. 1983. *Nucleic Acids Res.* 11:5361–80

224. Udvardy, A., Louis, C., Han, S., Schedl, P. 1984. *J. Mol. Biol.* 175:113–30

225. Bonven, B., Westergaard, O. 1982. *Nucleic Acids Res.* 10:7593–608

226. Franz, G., Tautz, D., Dover, G. A. 1985. *J. Mol. Biol.* 183: In press

227. Bonven, B., Gocke, E., Westergaard, O. 1985. *Cell* 41:541–51

228. Fleischmann, G., Pflugfelder, G., Steiner, E. K., Javaherian, K., Howard, G. C., et al. 1984. *Proc. Natl. Acad. Sci. USA* 81:6958–62

229. Muller, M. T., Pfund, W. P., Mehta, V. B., Trask, D. K. 1985. *EMBO J.* 4:1237–43

230. Ryoji, M., Worcel, A. 1985. *Cell* 40:923–32

231. Vogelstein, B., Pardoll, D. M. 1980. *Exp. Cell Res.* 128:467

232. Jackson, D. A., Cook, P. R., Patel, S. B. 1984. *Nucleic Acids Res.* 12:6709–26

233. Bolla, R. I., Bratten, D. C., Shiomi, Y., Hebert, M. B., Schlessinger, D. 1985. *Mol. Cell. Biol.* 5:1287–94

234. Ciejek, E. M., Nordstrom, J. L., Tsai, M.-J., O'Malley, B. W. 1982. *Biochemistry* 21:4945–53

Ann. Rev. Biochem. 1986. 55:831–54

DNA POLYMORPHISM AND HUMAN DISEASE

J. F. Gusella

Neurogenetics Laboratory, Massachusetts General Hospital, Department of Genetics, Harvard Medical School, Boston, Massachusetts 02114

CONTENTS

PERSPECTIVES AND SUMMARY

The traditional genetic definition of a polymorphism is a heritable variation for which the proportion of chromosomes carrying the most frequent allele (encoding the most common morph) does not exceed 99%. Polymorphisms have usually been detected as electrophoretic or antigenic protein variants, but all are ultimately due to a difference in the primary sequence of genomic DNA

831

0066-4154/86/0701-0831$02.00

between the pair of chromosomes encoding the characteristic. More recently, recombinant DNA techniques have permitted direct analysis of the coding DNA, and delineation of the specific base sequence alterations leading to particular altered protein molecules. Moreover, the same techniques can be used to scan nonexpressed DNA—intervening sequences, flanking sequences, regulatory sites, individual representatives of repeat families, and all segments of unknown function—for variation in sequence.

The utility of genetic polymorphisms is primarily that they enable the investigator to distinguish between the two copies of a particular locus in the human genome. This ability to identify a specific segment of a gene or chromosomal region despite the presence of a homologous segment on the other chromosome can be used in a number of ways to investigate human genetic diseases. The use of DNA sequence differences as polymorphisms has dramatically increased the potential for these investigations and has provided new approaches for many genetic disorders. This review will survey recent representative developments in the application of DNA polymorphism to human genetic disease to emphasize the rapid advances in this field.

RESTRICTION FRAGMENT LENGTH POLYMORPHISM

Definition and Detection

DNA polymorphism can potentially be detected in a number of ways including direct sequence analysis, or measurement of physical parameters of the molecule (1). In practice, however, restriction endonuclease digestion is the most straightforward approach to screening for variation, although only a subset of the total sequence is scanned. Since restriction enzymes have specific recognition sequences in DNA, sequence changes in genomic DNA can lead to the appearance or disappearance of particular cleavage sites, thereby altering the sizes of fragments generated from a given region (Figure 1A). Similarly, insertion or deletion of DNA between two recognition sites for an enzyme will alter the size of the restriction fragment produced by digestion with the enzyme. Differences in the sizes of fragments resulting from the digestion of the corresponding region of DNA from homologous chromosomes have been termed restriction fragment length polymorphisms or "RFLPs" (2). These are generally detected directly in genomic DNA by restriction enzyme digestion, followed by gel fractionation of the resulting fragments, transfer of the DNA to a solid support (Southern blotting), and hybridization to a specific cloned probe for the locus in question (Figure 1B). Alternative approaches such as hybridization directly to DNA in gels, or the comparison by agarose gel electrophoresis of cloned homologous sequences have been used less frequently to monitor RFLPs (3, 4).

A   Map of Enzyme B sites at Hypothetical Locus A

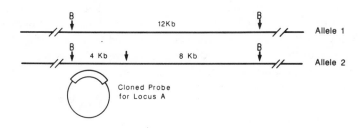

B   Detection of RFLP with Probe for Locus A

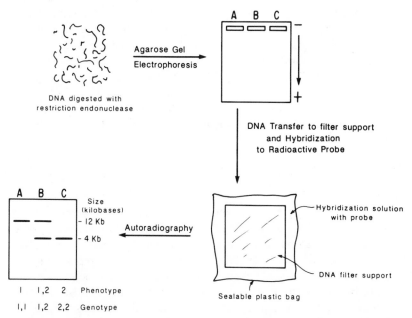

Figure 1A   Map of enzyme B sites at hypothetical locus A. A hypothetical restriction fragment length polymorphism is revealed at an arbitrary locus (A) by digestion with endonuclease B. Alleles 1 and 2 differ by the presence of an additional restriction enzyme site in the latter that reduces from 12 kilobases to 4 kilobases the size of fragment detected by a probe for the locus cloned in a plasmid vector.

Figure 1B   Detection of RFLP with probe for locus A. This flowchart depicts the steps taken to determine the phenotype and corresponding genotype at locus A in genomic DNA of three individuals using hybridization to the cloned probe from Figure 1A. The resultant autoradiograph indicates that the three individuals each have a different phenotype, corresponding to one homozygote for the 1 allele, one heterozygote, and one homozygote for the 2 allele.

Gene Probes

A major contribution to the total pool of RFLPs so far described has come from the use of cDNA clones or genomic clones from genes of known function. In fact, the first human RFLPs characterized were at the globin and insulin loci (5–10). Many RFLPs at known gene loci have been found by chance during the characterization of the genomic DNA by restriction mapping or sequencing. Increasingly, as the usefulness of RFLPs has become more apparent (2, 11, 12), investigators have specifically screened for the ability of particular cDNA clones to detect polymorphism. Table 1 presents a selected list of many genes for which RFLPs have been detected. At least one polymorphic gene has been identified on each chromosome except the Y (13). In many cases, these genes have been cloned and characterized primarily because they result in well-known inherited disorders when defective. With probes for some coding sequences, such as metallothionein and argininosuccinate synthetase, RFLPs have been detected at loci that are actually processed pseudogenes, distant from the site of the genuine expressed loci which are on chromosomes 16 and 9 respectively. The list presented in Table 1 is by no means a complete compilation of human polymorphic gene loci, which can be obtained in (13). It should be noted, however, that the 50 loci on this partial list already far outnumber the roughly 30 traditional expressed polymorphisms commonly employed by human geneticists.

Anonymous DNA Probes

The second major class of probes used to discover RFLPs is arbitrarily chosen single-copy cloned segments of genomic DNA. These are picked at random from genomic libraries based only on their ability to detect single-copy fragments on genomic blots and are subsequently screened for their ability to detect RFLPs. The sequences used are therefore "anonymous," not representing any known function. They might contain coding sequence, noncoding parts of a gene, or be unassociated with any gene. To the human geneticist, it is their ability to detect polymorphism, not their function, that is important.

Anonymous probes have typically been derived from genomic libraries after first identifying and then discarding those recombinants containing significant repetitive sequences by hybridization to human repetitive DNA (14). In some cases, probes containing both single-copy and repeat DNA have been used directly against genomic blots by competing the repeat sequence hybridization with excess unlabeled human repetitive DNA (15). Total genomic libraries of large and small insert sizes have both been used to generate useful polymorphic loci (16–18). These have then been assigned to chromosomes by hybridization to genomic DNA from panels of human-rodent somatic cell hybrid lines, or by in situ hybridization of labeled probe directly to metaphase chromosomes.

Occasionally anonymous cDNA clones have also been used rather than genomic sequences (19).

Increasingly, chromosome-specific libraries have been employed to target the discovery of RFLPs on particular chromosomes. Many libraries have been constructed from DNA of specific chromosomes purified by fluorescence-activated flow-sorting (20–24). Chromosome-specific sequences have also been obtained by making genomic libraries from somatic cell hybrid lines containing single human chromosomes and identifying clones containing human inserts by the species-specific hybridization of repeat sequences (14, 25–27).

Once a probe has been prepared, its ability to detect useful RFLPs is assessed by hybridization to genomic DNA from unrelated individuals. The probability that an RFLP will be found is, in part, a function of how many enzymes, and how many unrelated individuals, are used in this analysis. Since polymorphisms are generally most useful when a significant proportion of the population is heterozygous, screening panels of only 4–6 individuals have often been employed to favor the detection of frequent RFLPs (28). The number of enzymes tested is highly variable from laboratory to laboratory, and the cost of certain restriction enzymes is often the major factor precluding their use. A number of studies aimed at defining new RFLPs have indicated that the level of human DNA polymorphism is remarkably high (25, 29–31). Approximately one base in 250–500 differs between any two chromosomes chosen at random, although it has been suggested that the level of heterozygosity is lower for the X chromosome than for autosomes (30). Most of these sequence differences are presumed to represent neutral mutations conferring no selective advantage, or disadvantage.

The vast majority of RFLPs detected thus far alter single restriction enzyme sites and therefore appear to be point mutations. It has been proposed that these differences may occur preferentially at CpG dinucleotides because of the increased mutation rate at 5-methylcytosine (16, 30). This view has evolved from the observed frequency with which RFLPs are detected with the enzymes TaqI and MspI, both of which contain CpG in their recognition sequence. Recently, a model has been developed by Wijsman (32) for estimating the relative efficiency with which restriction enzymes can be used to detect RFLPs based on observed frequencies of dinucleotides in the human genome. This model accurately predicts the relative frequency with which RFLPs are found with a number of enzymes, including TaqI and MspI, without assuming any bias in mutation rate for different bases. The model also makes the prediction, which had already become evident to investigators in the field, that large probe molecules are more likely to detect RFLPs than small probes.

The first RFLP detected by an anonymous DNA probe was reported by Wyman & White in 1980 (33). This DNA polymorphism does not result from a

Table 1   Selected gene loci displaying RFLPs[a]

CHR[a]	Gene name	Disease
1	Anti-thrombin III	AT3 deficiency
	Beta nerve growth factor	
2	Carbamyl phosphate synthetase	
	Proopiomelanocortin	
	Apolipoprotein B	
3	Somatostatin	
	Transferrin	
4	Albumin	analbuminemia
	Alpha-fetoprotein	hereditary persistence of AFP
	Metallothionein pseudogene	
	Fibrinogen cluster	
5	Fms oncogene	
6	Complement C4	
	HLA genes	
	Plasminogen	
7	T-cell receptor beta chain	
	Argininosuccinate synthetase pseudogene	
8	Mos oncogene	
	Tissue plasminogen activator	
	Thyroglobulin	
	Carbonic anhydrase II	
9	Fibroblast interferon	
10	Urokinase	
11	Beta globin complex	sickle cell anemia
		beta thalassemia
	Harvey ras oncogene	
	Insulin	
	Insulin-like growth factor 2	
	Parathyroid hormone	
	Apolipoprotein A1	
12	Kirsten ras oncogene	
	Phenylalanine hydroxylase	phenylketonuria
13	Alpha propionyl CoA carboxylase	
14	Immunoglobulin	
	Heavy chain cluster	
	Alpha-1 antitrypsin	AAT deficiency
16	Alpha globin complex	alpha thalassemia
	Adenine phosphoribosyl transferase	
17	Growth hormone complex	GH deficiency
18	Prealbumin	hereditary amyloidosis
19	Complement C3	linked to myotonic dystrophy
	Apolipoprotein CII	linked to myotonic dystrophy
	Low-density lipoprotein receptor	familial hypercholesterolemia
20	Adenosine deaminase	combined immune deficiency
21	Superoxide dismutase 1	
22	sis oncogene	

Table 1   *(continued)*

CHR[a]	Gene name	Disease
	Immunoglobulin lambda chain	
	Myoglobin	
X	Factor VIII	hemophilia A
	Factor IX	hemophilia B
	Hypoxanthine phosphoribosyl transferase	Lesch-Nyhan Syndrome
	Ornithine transcarbamylase	OTC deficiency

[a]Extensive references for these as well as many additional polymorphic gene loci have been compiled in (13, 38). CHR = Chromosomal map location.

point mutation, but from the insertion or deletion of DNA that changes the size of a restriction fragment. Polymorphisms of this type are therefore not limited to two alleles. In fact, the DNA marker reported by Wyman & White has at least eight alleles. Similar insertion-deletion polymorphisms due to hypervariable stretches of DNA have been detected near the insulin (6–9), Harvey ras (34), and alpha globin loci (10, 35), as well as by additional anonymous probes (13), but are far less frequent than RFLPs detected as point mutations.

Subsequent to the initial report of an anonymous sequence RFLP, the number of DNA polymorphisms detected by arbitrary probes has increased steadily. These DNA markers are compiled every two years at Human Gene Mapping Workshops that bring together the majority of investigators in the field (36–38). In order to facilitate information exchange in an area of ever increasing complexity, a system of nomenclature has been developed for naming the loci identified only by anonymous DNA probes. The locus name begins with "D" for DNA segment, followed by a number or letter representing the chromosome to which the DNA segment maps. The next symbol in the name is "S" if the sequence is single-copy or "NF" if numerous fragments hybridizing to the same probe are present in the genome. The final component of the locus name is a number assigned in sequential order at the Human Gene Mapping Workshops. The locus detected by Wyman & White is therefore D14S1, since it is the first segment mapped to chromosome 14. Figure 2 displays a cumulative histogram of the number of RFLPs detected by arbitrary probes for each chromosome after Human Gene Mapping Workshops 6, 7, and 8 in 1981, 1983, and 1985, respectively. Each chromosome currently contains at least one anonymous sequence RFLP except chromosome 8. Certain chromosomes, in which there is more intense interest, such as chromosomes 4, 21, and X, have a disproportionate share of the total number of RFLPs reported. This is due primarily to the intensive use of chromosome-specific libraries for these regions (20–23, 25, 28). In addition to DNA markers on the autosomes and X chromosome, there are RFLPs exclusive to the Y chromosome, as well as RFLPs for loci found on both the X and Y chromosomes (13). The total number of anonymous DNA markers to which "D" numbers had been assigned as of HGM8 in 1985 was approximately 250. It should be noted, however, that this represents a considerable underestimate since there are large private collections of DNA markers for which "D" numbers and map positions have not yet been assigned (39). Still, when added to the polymorphic gene loci discussed above, the catalogued number of polymorphic loci that could be used as genetic markers now exceeds by more than tenfold the classical expressed polymorphic loci.

Repetitive Probes

Probes for the detection of RFLPs have also been obtained by using repetitive sequence clones. In some cases, such as the alpha satellite, the repetitive sequence probes represent families where all repeat units are clustered in a particular region of the genome such that RFLPs detected can be used to trace the inheritance of the region (40, 41). In other cases, the repetitive probes detect copies present in multiple regions of the genome. When the polymorphism attributable to any particular copy of the repeat family is easily distinguished, that locus becomes a potential genetic marker for the region in which it resides. Sometimes, multiple loci can be monitored at once in this way since each is detected by the same repetitive sequence probe. The most extreme, and potentially useful example of this type of probe is the hypervariable "minisatellite" described by Jeffreys and his colleagues (42, 43). These investigators isolated a

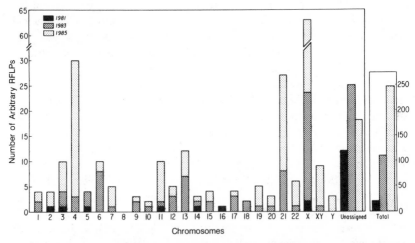

Figure 2 Number of arbitrary polymorphic DNA loci by chromosome. A cumulative histogram of the number of anonymous DNA probes detecting RFLPs in the human genome is presented. The data was compiled from Human Gene Mapping Workshops 6, 7, and 8, held in the summers of 1981, 1983, and 1985 respectively (13, 37, 38). The rapid increase in the number of arbitrary RFLP loci is indicative of the intense activity in the field of human DNA polymorphism.

repeat sequence probe from the region of the myoglobin gene on chromosome 22. Sequence analysis showed significant homology with the Chi recombination signal in *Escherichia coli*. This probe can be used to simultaneously reveal many polymorphic loci from throughout the genome. The degree of polymorphism of the cross-hybridizing loci presumably results from the tandem repeat structure of the core repeat sequence in the minisatellite which may frequently lead to unequal exchanges during recombination, and thereby generate altered restriction fragments. The level of polymorphism is so high, in fact, that it may be possible to "fingerprint" an individual since the probability that two unrelated individuals share all fragments at all cross-hybridizing loci is exceedingly low. One drawback of this type of RFLP marker, however, is that it is often possible only to determine the presence or absence of a given allele without being able to identify which of the many fragments on the blot represents the alternate allele.

LOCALIZATION OF HUMAN DISEASE GENES

Disease Categories

Human genetic diseases can be divided into two categories based on our knowledge of their etiology. For a relatively small number of monogenic Mendelian disorders, a defective protein has been identified as the cause of the disease phenotype. In these cases, it is relatively straightforward to use recombinant DNA techniques to isolate and characterize the culprit gene. This was the motivation for isolating many of the genes in Table 1. For the vast majority of inherited diseases, however, no specific defect has yet been uncovered. RFLPs are having their most significant impact on the investigation of these latter disorders.

For a genetic disease whose cause is not known, it is often still possible to determine that the root problem lies in a single primary deficit based on a monogenic pattern of inheritance for the disorder. One method of approaching such diseases is to pursue genetic linkage studies to determine the chromosomal location of the defective gene. This is done by searching for another genetic characteristic (a genetic marker) that displays a correlated pattern of inheritance with the disease allele as it is transmitted in families. If the disease allele and a particular allele for some other genetic polymorphism reside on different chromosomes, they will be transmitted independently and will show no correlation in their pattern of inheritance. Similarly, if the two are far apart on the same chromosome, multiple recombination events between them will eliminate any correlation. If, on the other hand, a high degree of coinheritance is observed for these two, it suggests that the disease gene and the gene encoding the other genetic characteristic are located close to each other in the same chromosomal region (Figure 3). Knowledge of the map location of the genetic marker locus thereby permits the inferred assignment of the disease gene to the same region. To perform a genetic linkage study, two components are necessary. First, the investigator must have access to families in which the disease gene is segregating and must characterize the pattern of inheritance of the disease allele. Second, he must have polymorphic genetic markers with which to test for coinheritance of the disease allele.

Monogenic disorders in humans are often subdivided into those displaying a dominant phenotype and those displaying a recessive phenotype. In humans, those disorders referred to as "dominant" require only a single copy of the disease allele to elicit disease symptoms in an individual. "Recessive" diseases require the presence of two disease alleles, and therefore the absence of a normal allele to cause the disease. Recessive diseases often represent enzyme

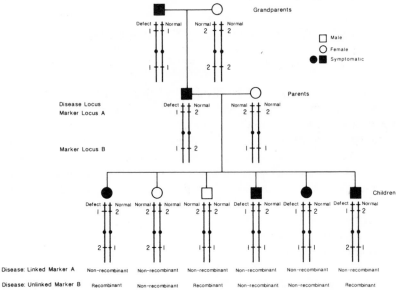

Figure 3   Family study of 2 marker loci: One linked and one unlinked to a dominant disease locus. The inheritance of two DNA marker loci from the same chromosome, A and B, each possessing two alleles, 1 and 2, has been traced in a disease family using the methods outlined in Figure 1 to demonstrate the principles of a genetic linkage study. The transmission of the dominant disease locus from the paternal grandfather who is homozygous at both marker loci indicates that the defect must be cis to 1 alleles at both A and B. Each transmission of the chromosome from the father to a child represents an independent test of the hypothesis of linkage to the disease locus for both marker A and marker B. No recombination events were detected in 6 chances between the disease and marker A while 3 out of 6 were seen for marker B. These results suggest that marker locus A maps close to the disease locus while marker locus B is distant from it, and shows independent segregation. Considerable additional data would have to be gathered, however, for either of these results to achieve statistical significance.

deficiencies (44). While far fewer dominant diseases have been characterized at the molecular level, they are commonly thought to involve alterations in structural proteins, receptors, or regulatory loci although enzyme deficiency has been seen. In fact, the distinction between dominant and recessive diseases is somewhat artificial since individuals homozygous for dominant disease alleles usually have far more severe symptoms than victims with only a single copy of the disease allele (45). Furthermore, in recessive diseases, the heterozygote can sometimes be distinguished from both the disease homozygote and the normal individual by biochemical tests or other signs of the disease allele. The pattern of inheritance in both of these cases would better be termed "codominant" since the phenotypes provide evidence for the presence of both alleles in heterozygotes. Similarly, the mode of inheritance of most RFLPs is codominant since the fragment sizes for both alleles are usually visible. Precise delineation of the mode of inheritance of the phenotypes associated with a disease allele is extremely important for genetic linkage analysis, where unambiguously tracing the disease allele through as many individual meioses as possible is paramount. Since this is achieved by analyzing the progeny of a given mating in which the disease allele was involved, it is clearly much easier in a dominant disorder than in a recessive disease in which the heterozygote cannot be distinguished from a normal individual. Consequently, the genetic linkage approach has initially been more successful for disorders displaying a dominant phenotype, and for X-linked disorders where the allele at the disease locus in males is unambiguous.

RFLPs as Genetic Markers

Once an investigator has identified and sampled disease pedigrees, it is possible to type them with polymorphic markers in an effort to detect correlated inheritance of a marker and the disease gene. A good genetic marker is one that is relatively easily assayed on a readily available tissue, and that shows a high degree of polymorphism. Until the characterization of RFLPs, the genetic markers available were relatively few in number and scattered haphazardly about the genome. These consist of a collection of plasma proteins and red blood cell antigens that display sufficient polymorphism to permit genetic linkage studies. In particular, the HLA locus encoding transplantation antigens has so many different alleles that virtually any two unrelated individuals have different genotypes. It is almost always possible, therefore, to unambiguously trace the parent of origin of each HLA allele in a given child. This makes the HLA locus an ideal genetic marker from the point of view of level of polymorphism. Unfortunately, its excessive cost and stringent sample collection requirements have prevented the use of HLA typing in many studies. The other expressed markers all have lower numbers of alleles and are therefore less "informative" genetic markers. It is not possible in many matings to unambiguously identify all four chromosomes using these markers. In particular, for two allele systems, it is never possible to follow all four parental chromosomes. In a heterozygote by homozygote mating, only the chromosomes of the heterozygote can be followed. Those of the homozygote are indistinguishable at the marker locus. In a heterozygote by heterozygote mating, approximately one half of the children will also be heterozygous for the two-allele marker, and in these it is impossible to determine which parent contributed which allele. Botstein and associates (2) have used the polymorphism information content or PIC value of a marker to express its potential in genetic linkage studies. This parameter is a function of the number of alleles at a given marker locus and of their frequencies in the population. It represents the probability that a child from a mating chosen at random will yield information on the inheritance of the marker locus. The PIC provides an estimate of how many matings will need to be typed before one is identified that can be used to test for cosegregation of a genetic marker and a disease gene. The PIC of HLA exceeds 0.9, but overall less than 25 of the commonly used expressed polymorphisms have a PIC higher than 0.15 (2).

Most RFLPs are detected as point mutations and consequently, they might be expected to make poor genetic markers since they will only yield two-allele systems. This is not the case for two reasons. First, for most RFLPs detected in a screen of only a few unrelated individuals, both alleles show a significant frequency in the population. The PIC values for most of these RFLPs range from 0.15 to 0.37, the maximum PIC that can be achieved with a two-allele system (2). Second, DNA markers are not limited to two alleles because a single probe will often detect multiple RFLPs in the same vicinity. Since the sequence differences underlying each of these RFLPs all reside within a few thousand base pairs of DNA, the probability of a recombination event occurring between them is vanishingly small. On average, it is estimated that recombination between two markers separated by 1,000,000 base pairs will only occur with a frequency of 1%. The information from each independent RFLP from a given locus can therefore be combined to construct haplotypes for that locus. This is analogous to the construction of haplotypes at the HLA locus by combining information on different antigenic determinants. Each haplotype can then be used as an allele for that locus. A probe that identifies four independent RFLPs, whether they are revealed by one restriction enzyme or by four, has the potential to generate a DNA marker with 16 different alleles. Each allele would consist of a different arrangement of cutting or noncutting at each of the four sites. Not all alleles may necessarily be seen in the population since their frequencies depend in part on the relative frequencies of each of the individual RFLPs. Furthermore, some haplotypes may occur less or more frequently than expected because of nonrandom associations between the sites. The latter phenomenon, known as linkage disequilibrium, can result if there has been insufficient time since introduction of one or another of the polymorphisms into the population for recombination to randomize its two alleles with respect to the other RFLPs (46). It must be considered when estimating the PIC value of a DNA marker for which there are multiple individual RFLPs.

Insertion-deletion polymorphisms often have many alleles that can be detected with any of a number of enzymes and therefore afford a certain degree of flexibility while also having high PIC values. Through the construction of haplotypes, DNA markers representing point mutations can clearly be made more informative and their PIC values accordingly increased. Even if a search for additional RFLPs fails with a given probe, the probe can be used to screen a genomic library and isolate neighboring fragments for use in expanding the degree of polymorphism. As a result, DNA markers not only surpass the expressed markers in number, but by and large in informativeness as well. Although typing expressed markers generally involves less expense, they only cover 10–15% of the genome, and for other areas, DNA markers represent the only option. Furthermore, although an individual RFLP is expensive to type, the ease of sample collection and storage, as well as the reusability of blots with multiple probes reduces cost considerations. Perhaps the most useful aspect of DNA markers relative to the expressed markers, however, is that given an interesting result, the investigator can expand the informativeness of the locus virtually at will, thereby enabling him to gain the maximum information possible from a set of pedigrees without resorting to costly and difficult collection of additional families.

Linkage Analysis for Mapping Disease Loci

With an adequate resource of family material, the ease of generating and typing DNA markers has made it almost certain that, with reasonable effort, an investigator can locate the map position of a genetic disease locus of interest. What constitutes the necessary family material will vary for every different disease depending on its mode of inheritance, age of onset, penetrance of the disease gene, etc. A general rule, however, is that it is preferable to have a large pedigree sufficient to achieve a significant score for coinheritance of the marker and disease locus than to attempt to sum data across several smaller families. The reason is that no matter how consistent the disease phenotype, there is no a priori guarantee that different families have defects at the same locus. If nonallelic heterogeneity were present, the data from individual pedigrees might cancel each other, and a significant linkage score might never be achieved. The use of only very large kindreds, when they are available, permits many tests of coinheritance of the disease gene and the genetic marker locus within a single pedigree where all affected individuals must carry the same defect.

Once a disease pedigree has been typed with a DNA marker, the data must be analyzed for coinheritance of the marker and the disease gene (47). One meiotic event is chosen and the allele of the marker traveling with the disease allele in that mating is assumed to be physically linked to the gene defect. Every additional meiosis in the pedigree then forms an independent test of the hypothesis. Calculations of likelihood, assuming various degrees of recombination, that a marker and disease gene are linked can often be performed by hand, but it is easiest for complex pedigrees to analyze the data with the aid of a computer. Although several programs are available to test for linkage and calculate the maximum likelihood estimate for the distance separating two loci, the LIPED program of Ott is most commonly used (48).

It should be noted that genetic linkage is an extraordinarily powerful method of eliminating a particular gene as the candidate for causing a disease (49–51). Often, the phenotype caused by a disease allele will lead the investigator to conclude that the defect might lie in some particular gene. An example would be the possibility that the defects underlying certain familial cancers might be in the primary structure of one of the known oncogenes (50). To establish that a particular gene is not responsible for causing the disease might take years of biochemical investigation involving studies at the DNA, RNA, and protein levels in both normal and tumor tissue. The genetic approach, however, can eliminate the gene as a candidate in a single experiment. If the gene causes the disorder, the investigator would expect to observe no recombination events between the candidate locus and the disease gene in a family study. Therefore, if a probe for the candidate gene detects an RFLP, and a single recombinant is found in the family study, the gene is effectively eliminated as the primary cause even though it might be involved at a secondary level in the pathogenesis of the disorder.

Distances along the chromosome can be measured using recombination rather than physical length. The unit of measure most commonly employed, the centiMorgan (cM) represents 1% recombination. The entire human genome consists of approximately 3000 cM. With adequate family material, an informative DNA marker can be expected to detect the presence of a disease gene by linkage analysis if the defect lies within 10 cM of the RFLP locus. Thus, each DNA marker can be used to exclude (or include) the disease gene from about 20 cM. The probability of detecting a linkage to an autosomal disease locus with a randomly chosen probe is 1/150. Thus, the investigator has a reasonably good chance of locating a disease gene by simply typing random DNA markers. This approach has been successful in mapping the gene causing Huntington's disease to chromosome 4 and is currently being applied to a wide variety of physical and behavioral disorders (52, 53). The chances of identifying a DNA marker linked to a disease gene are increased tremendously if there is some foreknowledge of the chromosome to which the defect maps. Sex-linked inheritance has provided the clue necessary to obtain DNA markers for such disorders as Duchenne muscular dystrophy and Becker muscular dystro hy (which might be allelic) (54–57), Alport syndrome (58), X-linked Charcot Marie Tooth disease (59, 60), choroideremia (61), ocular albinism (62), X-linked retinitis pigmentosa (63), Norrie disease (64), fragile X–mental retardation syndrome (65), adrenoleukodystrophy (66), etc. Linkage with expressed markers has allowed human geneticists to undertake efficient strategies for locating DNA markers close to the loci for such autosomal diseases as myotonic muscular dystrophy on chromosome 19 (67), and Wilson's disease on chromosome 13 (68).

Construction of a Human Linkage Map

As DNA markers are typed in disease families, knowledge is gained concerning linkages of RFLPs' loci to each other even if no linkage to the disease gene is found. It has been proposed that information on the cosegregation of DNA markers could be used to construct a complete linkage map of the human genome (2). This would then permit investigators to systematically search for the location of disease loci using a standard battery of equally spaced polymorphic markers. Two major benefits would ensue from such an approach. First, the duplication of effort involved in testing DNA markers covering the same region of the genome could be eliminated. Second, it would be possible to employ more efficient multipoint methods of analyzing linkage data. For example, the recently developed LINKAGE package of computer programs allows more information to be gained from typing data by simultaneously analyzing multiple linked markers for linkage to the disease locus (69–71).

Theoretically 150–200 equally spaced markers would provide a map that would guarantee the presence of an RFLP locus within 10 cM of any disease gene (2, 72–75). In practice, it will undoubtedly take several hundred markers more than those currently available to achieve this goal since randomly chosen markers are not equally spaced and not all optimally informative. Construction of the genetic linkage map will take a number of years, and in the meantime, much information necessary to build it will continue to come from disease pedigrees. These are not generally ideal in structure, however, for defining the linkage relationships of individual DNA markers with each other. As a result, many investigators interested in preparing linkage maps of individual chromosomes have concentrated their efforts on nondisease families. Some workers have concentrated on nuclear families with all four grandparents available while others have developed large pedigrees with many interrelated sibships (76, 77). The common feature in all cases, however, is that the families chosen have very large sibships to provide maximum information on cosegregation of any pair of markers for which one of the parents is doubly heterozygous. Permanent lymphoblastoid cell lines have also been established to provide a permanent source of DNA from these "reference" pedigrees. A worldwide collaboration for the investigation of some of these has been established through the efforts of the Centre de L'Etude du Polymorphisme Humain in Paris under the direction of Dr. Jean Dausset. In addition, many of these and other families have been deposited with the Human Genetic Mutant Cell Repository, Camden, N.J., and are available to all interested investigators (78). Linkage maps for all chromosomes will likely be developed within the next few years. The efforts of a number of groups have already resulted in the construction of preliminary linkage maps for large regions of chromosomes 1, 6, 11, 12, 13, 19, 21, and 22 as well as for the X-chromosome (79–87). Investigation in these regions now centers on firmly establishing the order of markers, and on generating additional highly polymorphic loci with which to construct a complete map with evenly spaced informative markers. The use of standard sets of reference pedigrees is particularly helpful in this regard since the delineation of recombination events detected during construction of the preliminary map allows rapid assignment of any new markers. The availability of highly informative genetic markers covering the entire genome should eventually lead to the mapping of all monogenic diseases for which adequate families are available. Of perhaps greater ultimate significance, it may be possible using the map to develop analytical tools for approaching disorders with multifactorial inheritance (genetically increased risk of cancer, heart disease, diabetes, etc.) and to demonstrate the genetic component of certain complex behavioral disorders such as schizophrenia.

ADDITIONAL APPLICATIONS OF RFLPS

Diagnosis of Genetic Disease

Many of the genes for which RFLPs have been found, presented in Table 1, were originally isolated to explore their roles in causing specific genetic disorders. The result in most cases has been improved potential for prenatal diagnosis of these diseases since it is possible to obtain and type DNA from the fetus by aminocentesis or chorion villus biopsy. In some cases, such as sickle cell anemia and certain thalassemias, the sequence alteration responsible for the generation of a disease allele also affects a restriction enzyme site and thereby produces an RFLP (88, 89). The disease allele in these cases gives a distinctive restriction fragment whose presence can be monitored directly. Direct determination of the disease allele is also possible for those families in which the defect is caused by the deletion of a significant stretch of DNA as can occur for alpha thalassemia or antithrombin III deficiency (90, 91). For the majority of diseases and disease families, however, it is only possible to detect the disease allele indirectly, by linkage to an RFLP which may also be present on normal chromosomes. This requires the availability of additional family members in whom the inheritance of the disease gene can be traced, but can potentially be done for any of the cloned genes listed in Table 1, or any other cloned gene that displays RFLPs. Linkage testing, unlike direct testing, is not applicable to all cases, but can only be used when the RFLPs used are segregating in an informative fashion in a pedigree of adequate size and structure. For example, in a recessive disease, such as phenylketonuria, prenatal diagnosis is limited to those families in which there is already an affected child whose DNA can be used to determine the phase of the marker alleles (cis or trans) with respect to the defective allele (92).

Linkage tests are also possible even if a gene has not been isolated if a linked genetic marker has been found. When the gene itself is the probe, the RFLP used will likely be located within a few thousand bases of the defect; hence, the possibility of recombination between the two is very low. When an arbitrary DNA probe provides the linked marker, however, significant recombination will be expected to contribute a measurable error rate in diagnostic testing. In order to reduce this problem, it is helpful to identify a second linked DNA marker on the opposite side of the disease gene from the first marker. A simple crossover between the defect and either of the two markers will also be detected as a recombination event between the marker loci. In these cases, no test result would be given because of the possibility of error. Misdiagnosis would only result in those rare cases where two recombination events occurred between the markers, one on either side of the defect. The applicability of linkage testing depends on the informativeness of both markers in the family, while the accuracy depends on their location with respect to the defect. In practice therefore, multiple closely linked markers are sought once the location of a disease gene has been determined. This has been achieved in the case of Duchenne muscular dystrophy (93), and similar efforts are under way for Huntington's disease, myotonic dystrophy, and other mapped disorders.

Detection of genetic diseases through the use of DNA polymorphism has improved prenatal diagnostic capabilities, but also has the potential to improve diagnosis in adult individuals for late onset disorders. Huntington's disease, for example, is a dominant neurodegenerative disorder that normally begins in mid-life (94). The victim of this disease suffers a slow deterioration, both physically and mentally, over a period of 10–20 years. The availability of a linked DNA marker makes it theoretically possible to diagnose the presence of the Huntington's disease allele in healthy individuals long before the onset of symptoms (52, 95–97). This raises a number of new medical, social, legal, and ethical issues which must be addressed for this disorder but also have more widespread implications for the future of genetic diagnosis (97, 98). These revolve around such issues as: the advisability of informing certain psychologically unstable individuals whether they carry the disease allele for a disorder such as Huntington's disease with its psychiatric effects and high suicide rate; potential deleterious effects of presymptomatic test results on marital and family relationships; the difficulty of imparting linkage-based results in a comprehensible fashion to a patient with no knowledge of genetic mechanisms; confidentiality of test information with regard to family members, employers, and insurers; etc (97, 98). Similar considerations will be raised for other late onset diseases such as Familial Alzheimer's disease and for persons with behavioral disorders such as bipolar affective disorder, panic disorder, etc, when the genetic causes of these are located on the map and linked DNA markers are available. The most important scientific question to settle for any genetic disease to be diagnosed by linkage testing will be that of nonallelic heterogeneity. The possibility of a second locus that causes similar symptoms must be eliminated before testing is performed on a population-wide basis. This issue is currently being addressed in Huntington's disease before presymptomatic testing is undertaken (95, 99, 100).

Ultimately, it is hoped that the approach of mapping gene defects by linkage to DNA markers will not only improve diagnosis, but will provide new avenues for isolating the genes responsible. These strategies, unlike the traditional approaches based on a knowledge of the defective protein, will be based on the chromosomal locations of the disease genes and will not be limited by assumptions concerning the biochemical nature of the defect (53). The knowledge gained by cloning and characterization of the genes causing such lethal diseases as Duchenne muscular dystrophy and cystic fibrosis may provide some clear means for treating victims of these disorders. In particular, the preclinical diagnostic capability afforded by recombinant DNA technology in late-onset disorders like Huntington's disease increases for many the immediacy with which therapies to prevent or delay the degenerative process are required. The further application of recombinant DNA techniques to identifying the primary defect seems the most hopeful route for fulfilling this need.

Detection of Abnormal Meioses

The importance of DNA polymorphisms to the human geneticist lies in their ability to distinguish between otherwise homologous chromosomal regions. Much of the research emphasis has been on applying them to trace particular chromosomal regions in family studies. They are also applicable, however, to identifying particular chromosomal regions in a number of circumstances other than traditional linkage studies. DNA polymorphisms can be used to determine the origin of various chromosomal imbalances and abnormalities such as trisomies, monosomies, deletions, and translocations by comparison of the restriction digestion pattern from the proband with that from parents and normal sibs for probes known to detect RFLPs. For example, Down syndrome or trisomy 21 is usually caused by nondysjunction during meiosis in one of the parents. RFLPs have been used to determine which parent contributed the extra chromosome. Similar studies have been performed to study many different sex chromosome and autosomal abnormalities (22, 65, 101).

Detection of Abnormal Mitoses

In addition to the application of RFLPs to abnormal meioses, it is also possible to investigate abnormal mitoses. The most dramatic application of this approach has been the study of pediatric malignancies including retinoblastoma, a tumor of the developing retina, and Wilms' tumor, which arises in the kidney. An association between each of these and deletion of a particular chromosomal band, 13q14 for retinoblastoma and 11p13 for Wilms' tumor, had previously been demonstrated in some cases by cytogenetic analysis. The mechanisms leading to tumor formation have now been investigated for both cancers using RFLPs (102–106). The approach has been to compare the allele pattern for markers on the chromosome of interest with the pattern seen in DNA prepared from normal white blood cells of the same individual. The results have indicated that several types of abnormal mitotic events can potentially lead to tumor formation. These include deletion, mitotic nondysjunction leading to loss of the chromosome, and mitotic recombination. The former two mechanisms result in hemizygosity in the tumor for part or all of the chromosome in question. The latter leaves part of the chromosome homozygous for all genes. It has been suggested that for each tumor, a critical region of the chromosome contains a gene that can exist in two allelic forms, one that prevents tumor formation, and one that does not. The former would encode a normal gene product and lead to normal development when it is present regardless of which allele is on the other homologous chromosome. The second allelic form would result from mutation at the locus, probably to inactivity, and be unable to prevent tumor formation. Under normal circumstances, a cell must undergo mutation of the normal locus on one chromosome and then the normal allele on the other chromosome must be eliminated by deletion, nondysjunction, mitotic recombination, gene conversion, or independent mutation before a tumor forms. The frequency with which each of these events occurs dictates that one should never observe more than one tumor in the same individual.

In fact, bilateral retinoblastomas, tumors arising independently in each eye, are seen as are cases of bilateral Wilms' tumor. These occur in individuals predisposed by family history to developing the tumor in question. For these cases, it has been proposed that one allele at the critical locus has already been inactivated, either by mutation or deletion, and the inactive locus is passed through the germ line. Tumor formation would then require only a single mutational event to eliminate the normal allele on the other chromosome. This model, originally put forward by Knudson (107–109) many years ago, has also received support from recent work with chromosome 13 RFLPs in familial retinoblastoma (110). In addition, evidence is increasing that other tumors may arise by similar mechanisms (111).

Other Applications

Several other applications of RFLPs to medicine have already been put into practice or are being explored. RFLPs are discovered by screening for differences between unrelated individuals. Experience has demonstrated that RFLPs can be found in the human genome in very high numbers and that if they are not located close to each other on the same chromosome they will segregate independently. It follows that no two unrelated people will have exactly the same set of alleles at all polymorphic loci. Thus, it should be possible to combine data from a number of highly informative polymorphic markers to provide an individual identification of the source of a particular tissue or chromosome. DNA markers could therefore be applied to paternity testing in the same way that HLA is used now with even greater accuracy. In fact polymorphism at the HLA locus is detectable using restriction enzymes but can be assayed in some cases where antigen determination is not possible (112, 113).

The same considerations suggest that RFLPs could become valuable tools in the forensic sciences. For example, if DNA can be recovered from semen samples in rape cases, identification of the rapist might be possible. Similarly, blood samples from the scene of a violent crime might potentially be compared by DNA analysis to those on the clothes of the suspect in order to prove his complicity.

RFLPs also have potential in transplant surgery which has only partially been realized. DNA polymorphism has been used to distinguish between donor and recipient cells in bone marrow transplants for leukemia victims, and to determine the origin of tumor cells in these patients in relapse (114). RFLPs might eventually be applied, however, to augment tissue typing in all forms of transplant surgery for improving the matching of the recipients and potential donors.

Finally, highly informative RFLP loci should increase the potential for association studies in diseases that show a familial nature without clear-cut Mendelian inheritance. These investigations are not based on demonstrating coinheritance of a marker and disease locus, but rather on demonstrating the coincidence of a particular marker allele and the disease in a population survey. In the past, many such investigations have been carried out with the HLA locus, and certain haplotypes have been found more frequently than expected in individuals with certain diseases (115). The major factor limiting the potential for such investigations in other regions of the genome has been the lack of multi-allele genetic markers. This limitation has now been eliminated by DNA polymorphisms and such studies are possible for many regions. One marker-disease system that has been explored is the association between a highly informative insertion-deletion polymorphism beside the insulin gene and non–insulin-dependent diabetes. Two studies have shown that a particular allele at the insulin locus is found more frequently in disease victims than is expected based on its level in the general population (116, 117). Although this is a heterogeneous disorder, two family studies with a variant of the disorder showing apparent autosomal dominant inheritance have been negative (118, 119). The positive association may be the result of linkage disequilibrium between the insulin allele and another gene on the short arm of chromosome 11 that is the primary cause of one form of non–insulin-dependent diabetes. Alternatively, the insulin gene itself might be one of several loci having a role in the etiology of the disorder and different alleles at this locus might differ in their effects. It is difficult with association studies to distinguish between these possibilities. Similar investigations have been conducted with insulin-dependent diabetes and hyperlipidemia, and it is likely that similar approaches will be undertaken with other complex disorders (120, 121). If the result for any marker-disease system is a very strong association, the marker could prove quite useful in assessing genetic risks of individuals with a family history of the disorder.

CONCLUDING REMARKS

DNA polymorphism has taken a leading role in human genetic research as investigators have realized the potential power and relative ease of recombinant DNA technology. Within only a few years, this tool has already had a major impact on several important genetic diseases, most notably Duchenne muscular dystrophy and Huntington's disease. These are only the forerunners, however, of a large number of genetic defects whose map positions will be found within the next few years. Cystic fibrosis, von Recklinghausen's neurofibromatosis, and familial Alzheimer disease are just a few of the important disorders that are currently the subjects of extensive investigation using genetic linkage to DNA markers. Improved diagnostic capability is the first benefit derived from this approach, but by far the most important result will be the new avenues it provides for isolating disease genes and exploring the mechanisms underlying their defects.

ACKNOWLEDGMENTS

This work was supported by NINCDS grants NS16367 (Huntington's Disease Center Without Walls) and NS20012, and by grants form the McKnight Foundation and Hereditary Disease Foundation. The author is supported by the Searle Scholar Program/Chicago Community Trust.

Literature Cited

1. Lerman, L. S., Fischer, S. G., Hurley, I., Silverstein, K., Lumelsky, N. 1984. *Ann. Rev. Biophys. Bioeng.* 13:399–423
2. Botstein, D., White, R. L., Skolnick, M., Davis, R. W. 1980. *Am. J. Hum. Genet.* 32:314–31
3. Purrello, M., Balazs, I. 1983. *Anal. Biochem.* 128:393–97
4. Gusella, J. 1982. *Adv. Exp. Med. Biol.* 154:153–64
5. Jeffreys, A. J. 1979. *Cell* 18:1–10
6. Ullrich, A., Dull, T. J., Gray, A., Brosius, J., Sures, I. 1980. *Science* 209:612–15
7. Rotwein, P., Chyn, R., Chirgwin, J., Cordell, B., Goodman, H. M., Permutt, M. A. 1981. *Science* 213:1117–20
8. Bell, G. I., Karam, J. H., Rutter, W. J. 1981. *Proc. Natl. Acad. Sci. USA* 1981. 78:5759–63
9. Owerbach, D., Nerup, J. 1982. *Diabetes* 31:275–77
10. Higgs, D. R., Goodbourn, S. E. Y., Wainscoat, J. S., Clegg, J. B., Weatherall, D. J. 1981. *Nucleic Acids Res.* 9:4213–24
11. Solomon, E., Bodmer, W. F. 1979. *Lancet* 1:923
12. Kurnit, D. M., Hoehn, H. 1979. *Ann. Rev. Genet.* 13:235–58
13. Willard, H. F., Skolnick, M. H., Pearson, P. L., Mandel, J. L. 1985. *Cytogenet. Cell Genet.* 40:360–489
14. Gusella, J. F., Keys, C., Varsanyi-Breiner, A., Kao, F.-T., Jones, C., et al. 1980. *Proc. Natl. Acad. Sci. USA* 77:2820–33
15. Litt, M., White, R. 1984. *Am. J. Hum. Genet.* 36:430A
16. Barker, D., Schafer, M., White, R. 1984. *Cell* 36:131–38
17. Gusella, J. F., Tanzi, R., Anderson, M., Ottina, K., Watkins, P. 1983. In *Banbury Report 14: Recombinant DNA Applications to Human Disease,* ed. C. T., Caskey, R. L., White, pp. 261–66. New York: Cold Spring Harbor Lab.
18. Feder, J., Yen, L., Wijsman, E., Wang, L., Wilkins, L., et al. 1985. *Am. J. Hum. Genet.* 37:635–50
19. Balazs, I., Purrello, M., Alhadeff, B., Grzeschik, K. H., Szabo, P. 1984. *Hum. Genet.* 68:57–61
20. Davies, K. E., Young, B. D., Elles, R. G., Hill, M. E., Williamson, R. 1981. *Nature* 293:374–76
21. Kunkel, L., Tantravahi, U., Eisenhard, M., Latt, S. 1982. *Nucleic Acids Res.* 10:1557–78
22. Stewart, G. D., Harris, P., Galt, J., Ferguson-Smith, M. A. 1985. *Nucleic Acids Res.* 13:4125–32
23. Gilliam, T. C., Healey, S., Tanzi, R., Stewart, G., Gusella, J. 1985. *Cytogenet. Cell Genet.* 40:641
24. Van Dilla, M., Deaven, L. L. 1985. *Am. J. Hum. Genet.* 37:531A
25. Watkins, P. C., Tanzi, R. E., Gibbons, K. T., Tricoli, J. V., Landes, G., et al. 1985. *Nucleic Acids Res.* 13:6075–88
26. Cavenee, W., Leach, R., Mohandas, T., Pearson, P., White, R. 1984. *Am. J. Hum. Genet.* 36:10–24
27. Bruns, G., Gusella, J., Keys, C., Leary, A., Housman, D., Gerald, P. 1982. *Adv. Exp. Med. Biol.* 154:60–72
28. Aldridge, J., Kunkel, L., Bruns, G., Tantravahi, U., Lalande, M., et al. 1984. *Am. J. Hum. Genet.* 36:546–64
29. Murray, J. C., Mills, K. A., Demopulos, C. M., Hornung, S., Motulsky, A. G. 1984. *Proc. Natl. Acad. Sci. USA* 81:3486–90
30. Cooper, D. N., Schmidtke, J. 1984. *Hum. Genet.* 66:1–16
31. Cooper, D. N., Smith, B. A., Cooke, H. J., Niemann, S., Schmidtke, J. 1985. *Hum. Genet.* 69:201–5
32. Wijsman, E. M. 1984. *Nucleic Acids Res.* 12:9209–26
33. Wyman, A. R., White, R. 1980. *Proc. Natl. Acad. Sci. USA* 77:6754–58
34. Capon, D. J., Chen, E. Y., Levinson, A. D., Seeburg, P. H., Goeddel, D. V. 1983. *Nature* 302:33–37
35. Goodbourn, S. E. Y., Higgs, D. R., Clegg, J. B., Weatherall, D. J. 1983. *Proc. Natl. Acad. Sci. USA* 80:5022–26
36. Human Gene Mapping Workshop: Human Gene Mapping 6: 6 Int. Workshop in Human Gene Mapping. 1982. *Cytogenet. Cell. Genet.* 32:1–324
37. Human Gene Mapping Workshop: Human Gene Mapping 7: 7th Int. Workshop in Human Gene Mapping. 1984. *Cytogenet. Cell. Genet.* 37:1–666
38. Human Gene Mapping Workshop: Human Gene Mapping 8: 8th Int. Workshop in Human Gene Mapping. 1985. *Cytogenet. Cell. Genet.* 40:1–823
39. Schumm, J., Knowlton, R., Braman, J., Barker, D., Vovis, G., et al. 1985. *Cytogenet. Cell Genet.* 40:739
40. Jabs, E. W., Wolf, S. F., Migeon, B. R. 1984. *Proc. Natl. Acad. Sci. USA* 81:4884–88
41. Willard, H. F., Waye, J. S., Schwartz, C., Skolnick, M., Phillips, J. A. 1985. *Cytogenet. Cell Genet.* 40:778–79

42. Jeffreys, A. J., Wilson, V., Thein, S. L. 1985. *Nature* 314:67–73
43. Jeffreys, A. J., Wilson, V., Thein, S. L. 1985. *Nature* 316:77–79
44. McKusick, V. 1983. *Mendelian Inheritance in Man.* Baltimore, MD: Johns Hopkins Univ. Press
45. Pauli, R. M. 1983. *Am. J. Med. Genet.* 16:455–58
46. Chakravarti, A., Buetow, K. H., Antonarakis, S. E., Waber, P. G., Boehm, C. D., Kazazian, H. H. 1984. *Am. J. Hum. Genet.* 36:1239–58
47. Ott, J. 1984. *Analysis of Human Genetic Linkage.* Baltimore, MD: Johns Hopkins Univ. Press
48. Ott, J. 1974. *Am. J. Hum. Genet.* 26:588–97
49. Phillips, J. A., Parks, J. S., Hjelle, B. L., Herd, J. E., Plotnick, P. 1982. *J. Clin. Invest.* 70:489–95
50. Barker, D., McCoy, M., Weinberg, R., Goldfarb, M., Wigler, M., et al. 1983. *Mol. Biol. Med.* 1:199–206
51. Seizinger, B. R., Tanzi, R. E., Gilliam, T. C., Bader, J., Perry, D., et al. 1986. *Ann. NY Acad. Sci.* In press
52. Gusella, J. F., Wexler, N. S., Conneally, P. M., Naylor, S. L., Anderson, M. A., et al. 1983. *Nature* 306:234–38
53. Gusella, J. F., Tanzi, R. E., Anderson, M. A., Hobbs, W., Gibbons, K., et al. 1984. *Science* 225:1320–26
54. Murray, J. M., Davies, K. E., Harper, P. S., Meredith, L., Mueller, C. R., Williamson, R. 1982. *Nature* 300:69–71
55. Davies, K. E., Pearson, P. L., Harper, P. S., Murray, J. M., O'Brien, T. O., et al. 1983. *Nucleic Acids Res.* 11:2303–12
56. Kingston, H. M., Thomas, N. S. T., Pearson, P. L., Sarfarazi, M., Harper, P. S. 1983. *J. Med. Genet.* 20:255–58
57. Roncuzzi, L., Fadda, S., Mochi, M., Prosperi, L., Sangiorgi, S., et al. 1985. *Am. J. Hum. Genet.* 37:407–17
58. Menlove, L., Kirschner, N., Nguyen, K., Morrison, T., Aldridge, J., et al. 1985. *Cytogenet. Cell Genet.* 40:697–98
59. Beckett, J., White, B. N., Simpson, N. E., Ebers, G. C., Holden, J., MacLeod, P. M. 1985. *Cytogenet. Cell Genet.* 40:579
60. Gal, A., Mucke, J., Theile, H., Bernhard, A., Wieacker, P., et al. 1985. *Cytogenet. Cell Genet.* 40:633
61. Nussbaum, R. L., Lewis, R. A., Lesko, J. G., Ferrell, R. 1985. *Am. J. Hum. Genet.* 37:473–82
62. Kidd, J. R., Castiglione, C. M., Davies, K., Pakstis, A. J., Gusella, J., et al. 1985. *Am. J. Hum. Genet.* 37:475A
63. Bhattacharya, S. S., Wright, A. F., Clayton, J. F., Price, W. H., Phillips, C. I., et al. 1984. *Nature* 309:253–55
64. Gal, A., Bleeker-Wagemakers, L., Wienker, T. F., Warburg, M., Ropers, H. H. 1985. *Cytogenet. Cell Genet.* 40:633
65. Goodfellow, P., Davies, K., Ropers, H. H. 1985. *Cytogenet. Cell Genet.* 40:296–352
66. Boue, J., Oberle, I., Heilig, R., Mandel, J. L., Moser, A., et al. 1985. *Hum. Genet.* 69:272–74
67. Naylor, S., Lalouel, J. M., Shaw, D. 1985. *Cytogenet. Cell Genet.* 40:242–67
68. Frydman, M., Bonne-Tamir, B., Farrer, L. A., Conneally, P. M., Magazanik, A., et al. 1985. *Proc. Natl. Acad. Sci. USA* 82:1819–21
69. Lathrop, G. M., Lalouel, J. M., Julier, C., Ott, J. 1984. *Proc. Natl. Acad. Sci. USA* 81:3443–46
70. Lathrop, G. M., Lalouel, J. M. 1984. *Am. J. Hum. Genet.* 36:460–65
71. Lathrop, G. M., Lalouel, J. M., Julier, C., Ott, J. 1985. *Am. J. Hum. Genet.* 37:482–99
72. Skolnick, M. H., White, R. 1982. *Cytogenet. Cell Genet.* 32:58–67
73. Bishop, D. T., Skolnick, M. H. 1983. In *Banbury Report 14: Recombinant DNA Applications to Human Disease,* ed. C. T. Caskey, R. L. White, pp. 251–60. New York: Cold Spring Harbor Lab.
74. Lange, K., Boehnke, M. 1982. *Am. J. Hum. Genet.* 34:842–45
75. Bishop, D. T., Williamson, J. A., Skolnick, M. H. 1983. *Am. J. Hum. Genet.* 35:795–815
76. White, R., Leppert, M., Bishop, D. T., Barker, D., Berkowitz, J., et al. 1985. *Nature* 313:101–5
77. Gusella, J. 1985. In *Genetic Engineering,* ed. J. K. Setlow, A., Hollaender, 7:333–47. New York: Plenum
78. NIGMS Human Genetic Mutant Cell Repository Catalogue. 1983. US Dept. Health and Human Services Publ. 83–2011
79. Povey, S., Morton, N. E. 1985. *Cytogenet. Cell Genet.* 40:67–106
80. Gerhard, D., Kidd, K., Kidd, J., Gusella, J. F., Housman, D. 1984. *ICSU Short Rep.* 1:172–73
81. Demars, R., Hasstedt, S., White, R. 1984. *Am. J. Hum. Genet.* 36:144S
82. Kittur, S. D., Hoppener, J. W. M., Antonarakis, S. E., Daniels, J. D. J., Meyers, D. A., et al. 1985. *Proc. Natl. Acad. Sci. USA* 82:5064–67
83. Grzeschik, K. H., Kazazian, H. H. 1985. *Cytogenet. Cell Genet.* 40:179–205
84. White, R., Leppert, M., O'Connell, P.,

Holm, T., Cavenee, W., et al. 1984. *Am. J. Hum. Genet.* 36:158S

85. Gusella, J. F., Tanzi, R. E., Watkins, P. C., Gibbons, K. T., Hobbs, W. J., et al. 1984. *Ann. NY Acad. Sci.* 450: 25–32

86. Julier, C., Lathrop, M., Lalouel, J. M., Kaplan, J. C. 1985. *Cytogenet. Cell Genet.* 40:663–64

87. Drayna, D., Davies, K., Hartley, D., Mandel, J.-L., Camerino, G., et al. 1984. *Proc. Natl. Acad. Sci. USA* 81: 2836–39

88. Orkin, S. H., Little, P. F. R., Kazazian, H. H., Boehm, C. 1982. *N. Engl. J. Med.* 307:32–36

89. Orkin, S. 1984. *Blood* 63:249–53

90. Antonarakis, S. E., Kazazian, H. H., Orkin, S. H. 1985. *Hum. Genet.* 69:1–14

91. Prochownik, E. V., Antonarakis, S., Bauer, K. A., Rosenberg, R. D., Fearon, E. R., Orkin, S. H. 1983. *N. Engl. J. Med.* 308:1549–52

92. Woo, S. L. C., Lidsky, A. S., Guttler, F., Thirumalachary, C., Robson, K. J. H. 1984. *J. Am. Med. Assoc.* 251:1998–2001

93. Bakker, E., Goor, N., Wrogemann, K., Kunkel, L. M., Fenton, W. A., et al. 1985. *Lancet* 1:655–58

94. Havden, M. R. 1981. *Huntington's Chorea.* New York: Springer-Verlag

95. Martin, J., Gusella, J. F. 1986. *N. Engl. J. Med.* In press

96. Harper, P. S. 1984. *Practical Genetic Counseling.* Bristol: Wright

97. Wexler, N. S., Conneally, P. M., Housman, D., Gusella, J. F. 1985. *Arch. Neurol.* 42:20–24

98. Bird, S. J. 1985. *J. Am. Med. Assoc.* 253:3286–91

99. Folstein, S. E., Phillips, J. A., Meyers, D. A., Chase, G. A., Abbott, M. H., et al. 1985. *Science* 229:776–79

100. Harper, P. S., Youngman, S., Anderson, M. A., Sarfarazi, M., Quarrell, O., et al. 1985. *J. Med. Genet.* 22:447–50

101. Gusella, J. F., Tanzi, R. E., Bader, P. I., Phelan, M. C., Stevenson, R., et al. *Nature.* 318:75–78

102. Cavenee, W. K., Dryja, T. P., Phillips, R. A., Benedict, W. F., Godbout, R., et al. 1983. *Nature* 305:779–83

103. Koufos, A., Hansen, M. F., Lampkin, B. C., Workman, M. L., Copeland, N. G., et al. 1984. *Nature* 309:172–74

104. Orkin, S. H., Goldman, D. S., Sallan, S. E. 1984. *Nature* 309:170–72

105. Reeve, A. E., Housiaux, P. J., Gardner, R. J. M., Chewings, W. E., Grindley, R. M., Millow, L. J. 1984. *Nature* 309: 174–76

106. Fearon, E. R., Vogelstein, B., Feinberg, A. P. 1984. *Nature* 309:176–78

107. Knudsen, A. G. 1971. *Proc. Natl. Acad. Sci. USA* 68:820–23

108. Knudsen, A. G., Strong, L. C., Anderson, D. E. 1973. *Prog. Med. Genet.* 9:113–58

109. Knudsen, A. G., Hethcote, H. W., Brown, B. W. 1975. *Proc. Natl. Acad. Sci. USA* 72:5116–20

110. Cavenee, W. K., Hansen, M. F., Nordenskjold, M., Maumenee, I., Squire, J. A., et al. 1985. *Science* 228:501–3

111. Koufos, A., Hansen, M. F., Copeland, N. G., Jenkins, N. A., Lampkin, B. C., Cavenee, W. K. 1985. *Nature* 316:330–34

112. Lamm, L. U., Olaisen, B. 1985. *Cytogenet. Cell Genet.* 40:128–55

113. Bosch, M. L., Schreuder, G. M. T., Spits, H., Termijtelen, A., Tilanus, M. G. J., Giphart, M. J. 1985. *J. Immunol.* 134:3212–17

114. Minden, M. D., Messner, H. A., Belch, A. 1985. *J. Clin. Invest.* 75:91–93

115. Ryder, L. P., Svejgaard, A., Dausset, J. 1981. *Ann. Rev. Genet.* 15:169–87

116. Rotwein, P. S., Chirgwin, J., Province, M., Knowler, W. C., Pettit, D. J., et al. 1983. *N. Engl. J. Med.* 308:65–71

117. Owerbach, D., Nerup, J. 1982. *Diabetes* 31:275–77

118. Bell, J. I., Wainscoat, J. S., Old, J. M., Chlouverakis, C., Keen, H., Turner, R. C., Weatherall, D. J. 1983. *Br. Med. J.* 286:590–92

119. Andreone, T., Fajans, S., Rotwein, P., Skolnick, M., Permutt, M. A. 1985. *Diabetes* 34:108–14

120. Owerbach, D., Hagglof, B., Lernmark, A., Holmgren, G. 1984. *Diabetes* 33: 958–65

121. Vella, M., Kessling, A., Jowett, N., Rees, A., Stocks, J., et al. 1985. *Hum. Genet.* 69:275–76

Ann. Rev. Biochem. 1986. 55:855–78
Copyright © 1986 by Annual Reviews Inc. All rights reserved

β-AMINO ACIDS: MAMMALIAN METABOLISM AND UTILITY AS α-AMINO ACID ANALOGUES

Owen W. Griffith

Department of Biochemistry, Cornell University Medical College, New York, NY 10021

CONTENTS

PERSPECTIVES AND SUMMARY

Four β-aminocarboxylic acids occur naturally in mammals. β-Alanine and R-β-aminoisobutyrate (R-β-AiB) are catabolites of uracil and thymine, respectively. S-β-Aminoisobutyrate (S-β-AiB) is formed by transamination of S-methylmalonate semialdehyde, a catabolite of valine. β-Leucine, only recently

855

discovered in mammals, is probably a precursor of L-leucine; valine and isofatty acids are β-leucine precursors. β-Amino acids do not occur in proteins; much of the β-alanine pool is, however, present as the dipeptide carnosine (β-alanyl-L-histidine). Both β-alanine and carnosine are putative neurotransmitters.

Several aspects of β-amino acid metabolism are of current clinical and basic research interest. Since rates of pyrimidine catabolism are altered in neoplasia, R-β-AiB formation and excretion reflect the response of some cancers to chemotherapy. Pharmacological inhibition of the enzymes initiating pyrimidine catabolism extends the half-lives of several antineoplastic nucleosides. Recent studies continue to elucidate the pathways of β-amino acid catabolism. Whereas early studies established that β-alanine, R-β-AiB, and S-β-AiB are initially transaminated to the corresponding aldehydes, the further metabolism of the aldehydes was poorly defined. Although the mammalian enzymes responsible for the metabolism are still not purified or even clearly identified, isotopic studies suggest that malonate semialdehyde and S-methylmalonate semialdehyde are oxidatively decarboxylated directly to acetyl-CoA and propionyl-CoA, respectively. Data in support of these and possible minor catabolic pathways are reviewed.

β-Amino acids have received considerable attention as analogues of the common protein amino acids. New syntheses and methods of resolution have been developed. β-Amino acids have been incorporated into synthetic peptides or employed as alternative substrates or inhibitors of enzymes. Recent applications are reviewed.

INTRODUCTION

Chemical synthesis of β-alanine, the simplest of the β-aminocarboxylic acids, was reported by Heintz in 1870 (1) only 20 years after Strecker's classic, albeit inadvertent, synthesis of α-alanine (2). Within 30 years six additional syntheses of β-alanine were reported (cf 3), and β-amino acids were often included in early synthetic peptides. Occurrence of β-alanine as a microbial catabolite of aspartate was reported as early as 1911 (4) [eliciting some controversy (5)], and carnosine (β-alanylhistidine) was identified in skeletal muscle in 1900 (6). Notwithstanding this initial interest, by the early 1900s it was clear that proteins were composed exclusively of α-amino acids, and β-amino acids were increasingly perceived as lying outside the mainstream of mammalian biochemistry.

This perception was strongly supported by the scientific developments that followed. Whereas dozens of novel and frequently complex β-amino acids were isolated from plants and microorganisms as free molecules and components of peptides and antibiotics [for reviews see (7, 8)], few β-amino acids were found in mammals and their metabolism appeared, at least initially, to be simple. β-Alanine and R-β-aminoisobutyrate (R-β-AiB)[1] were identified in the early 1950s as catabolites of uracil-cytosine and thymine, respectively (9, 10); a few years later S-β-aminoisobutyrate (S-β-AiB) was shown to be a catabolite of valine (11). More recently, β-aminoisocaproate (β-leucine) was reported to be a precursor of a portion of the L-leucine pool (12). Only these four β-aminocarboxylic acids have been shown to occur naturally in mammals. Interestingly, the metabolism of these four amino acids, particularly their catabolism, is not yet fully defined; a discussion of the available data is one focus of the present review.

Beyond their limited metabolic role, β-amino acids hold interest as analogues of α-amino acids. β-Amino acids have, for example, been used to probe the structural specificity and topology of several α-amino acid–binding sites and have found use as enzyme-specific alternative substrates. The limited normal metabolism of β-amino acids suggests that many β-amino analogues of α-amino acids will be metabolically stable in vivo; enzyme inhibitors may thus have extended biological half-lives when designed as β- rather than α-amino acids. Similarly, incorporation of β-amino acids into peptides of pharmacological interest has in some cases been found advantageous in terms of biological activity, metabolic stability, or both. Selected examples of the use of β-amino acids as α-amino acid analogues are a second focus of the present review.

CHEMISTRY OF β-AMINO ACIDS

Nomenclature

As illustrated in Figure 1, three different β-amino acids are plausible analogues of an α-amino acid. In type I analogues the α-amino group is "moved" to the β-carbon whereas in type II or III analogues an extra methylene group is inserted between the original carboxyl and amino groups. Type I and II analogues are named systematically as 3-amino derivatives of the parent acid but are frequently referred to by the nomenclature shown. Note that there is some redundancy (β-leucine = β-homovaline, etc); in such cases the type I name is preferred. α-Substituted-β-alanines (type III analogues) are named systematically as substituted propanoic acids or, less commonly, as 2-aminomethyl derivatives of the parent acid (e.g. 2-aminomethyl-4-methylpentanoic acid in the example shown). With the exception of β-proline (3-carboxypyrrolidine), α-substituted-β-alanines are not named by reference to the homologous α-amino acid. The prefix "iso" is, however, used with some α-substituted-β-alanines. Isoserine and isocysteine are α-hydroxy- and α-

---

[1]Abbreviations used: β-AiB; β-aminoisobutyrate; MS, malonate semialdehyde; MMS, methylmalonate semialdehyde; GABA, γ-aminobutyrate.

mercapto-β-alanine, respectively; isothreonine is β-amino-α-hydroxybutanoic acid. Many di- and trisubstituted-β-alanines bear trivial names related to the plant, microorganism, or antibiotic in which they were discovered (7, 8).

Both α- and β-substituted-β-alanines generally have at least one chiral carbon and thus occur in R- and S-configurations. Although all of the protein L-α-amino acids except cysteine are S-enantiomers, the configurationally analogous β-amino acids may be R or S. The enantiomers shown in Figure 1, all analogues of L-α-leucine, are R-β-leucine, S-β-homoleucine, and R-3-amino-2-(2-methyl)propyl-propanoic acid. The D and L designations common to α-amino acids are frequently applied to type I and II analogues by evaluating the configuration at the β-carbon as if it were the α-carbon. The D and L designations are less clearly assigned with type III analogues. R-β-AiB, the type III analogue of L-alanine, is, for example, referred to as D-β-AiB. Correspondingly, in the sense illustrated in Figure 1, S- or L-β-AiB is an analogue of D-alanine (13).

Synthesis

Since only β-alanine and a few racemic β-aminocarboxylic acids are available commercially, most β-amino acids and all pure enantiomers of these compounds must be prepared chemically or isolated from natural sources. Balenovic (14) and Drey (7, 8) have discussed this subject in detail and only procedures of general utility are described here.

α-Substituted-β-alanines in which R is alkyl or arylalkyl are obtained in good yields by routes 1 (11, 13) and 2 (15) (Figure 2). N-Chloromethylphthalimide and many of the required bromoesters and substituted malonates are now commercially available. The utility of route 3 for α-substituted-β-alanines (R = H; R' ≠ H) is limited by the availability of α-substituted acrylates and acrylonitriles, but aminomethylsuccinate (16) and aminomethylglutarate (17), type III analogues of Asp and Glu, respectively, are easily prepared. The synthesis of β-amino acids by reduction of pyrimidines (route 4) is suited to the preparation of both α- and β-substituted-β-alanines (18, 19). β-Amino[^{14}C-

Nomenclature

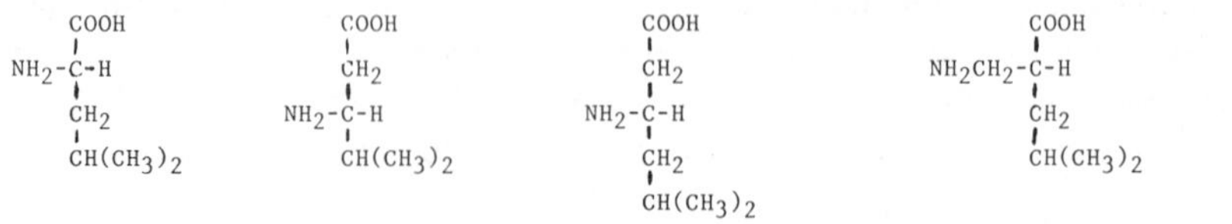

COOH	COOH	COOH	COOH
NH₂-C-H	CH₂	CH₂	NH₂CH₂-C-H
CH₂	NH₂-C-H	NH₂-C-H	CH₂
CH(CH₃)₂	CH(CH₃)₂	CH₂	CH(CH₃)₂
		CH(CH₃)₂	
	I	II	III
L-Leucine	β-Leucine	β-Homoleucine	3-Amino-2-(2-methylpropyl)-propanoic acid

Figure 1   β-Amino acid analogues of the α-amino acid L-leucine.

CH$_3$]isobutyrate (R = ^{14}CH$_3$, R' = H) is obtained in good yield from [^{14}C-CH$_3$]thymine.

Racemic β-substituted-β-amino acids are commonly prepared by routes 3 or 4 (7, 8). Use of chiral reagents in route 3 produces β-amino acids with modest enantiomeric enrichment (20–22). Use of urea (route 4) avoids the imine formation possible in route 3. Hydroxylamine has been used in place of NH$_3$ in route 3 to yield β-hydroxyamino acids (23); the products, of interest as enzyme inhibitors (24), can be reduced to β-amino acids (23). α-Amino acids are converted to type II analogues with retention of configuration by the reactions shown in route 5 (14). Stereoselective procedures for converting serine and threonine to isoserine (25) and isothreonine (25, 26), respectively, are available. Racemic isocysteine has been prepared (27, 28).

Preparative and Analytical Resolution of Enantiomers

β-Amino acid racemates are resolved by the same methods used with α-amino acids. Acylated S-β-amino-n-butyrate, S-β-aminovalerate, and S-β-phenyl-β-alanine are, for example, selectively hydrolyzed by *Escherichia coli* ben-

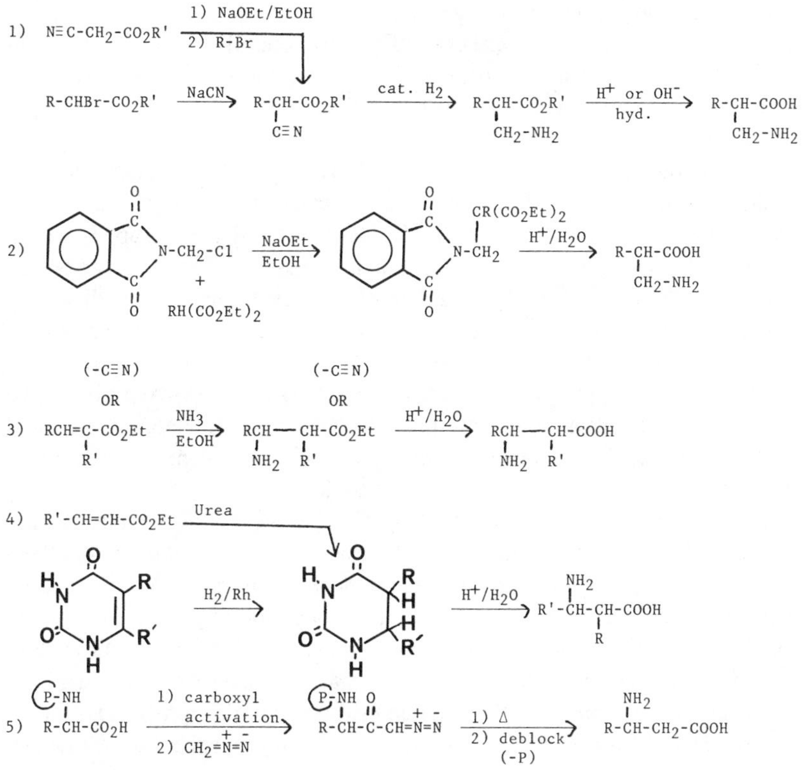

Figure 2 Synthesis of β-amino acids.

zylpenicillinacylase, an enzyme active toward single enantiomers of a wide variety of β-substituted-β-alanines (29). The enzyme is apparently not useful in resolving α-substituted-β-alanines. Chiral monoesters and monoamides of β-glutamate have also been obtained enzymatically. Glutamine synthetase, for example, forms only D-β-glutamine from β-glutamate and NH_3 (30) and chymotrypsin cleaves diethyl N-acetyl-β-glutamate to L-3-acetamido-4-carbethoxybutyric acid (31). Hog liver esterase forms the same enantiomer from dimethyl N-acetyl-β-glutamate, but has the opposite stereo-specificity with the N-carbobenzoxy, N-t-butyloxycarbonyl, and N-benzyl derivatives of β-glutamate diesters (32).

Intact animals and, more commonly, microorganisms have also been used to resolve β-amino acids. Organisms that selectively cleave one ester of dialkyl N-acyl-β-glutamates have been described (33) and β-amino-n-butyrate is resolved by a strain of *Pseudomonas putida* that selectively catabolizes the L-enantiomer (34). Baker's yeast selectively catabolizes S-β-AiB allowing R-β-AiB to be obtained from its racemate (35). The same enantiomer is obtained from the urine of individuals deficient in R-β-AiB transaminase (see below) (13, 36).

Although the procedures are generally tedious (14), many β-amino acids have been resolved by fractional crystallization of their diastereomeric salts. In the context of mammalian biochemistry, the resolution of RS-β-AiB was a major advance in that it allowed the metabolism of the individual enantiomers to be studied (13).

Chromatographic resolution of RS-β-AiB has been achieved analytically by gas chromatography of diastereomeric peptides or esters (35, 37). Similarly, the Manning-Moore diastereomeric dipeptide method (38) is applicable to β-substituted-β-alanines [e.g. L,(S)-glutamyl-RS-β-amino-n-butyrate yields two dipeptide peaks (34)]. The latter approach was unsuccessful with several α-substituted-β-alanines in which the β-amino group is not attached to the chiral carbon (O. Griffith, unpublished). α-Substituted-β-alanines are resolved if the dipeptide sequence is reversed; thus RS-β-AiB-L,(S)-Asp yields two well-resolved dipeptide peaks by amino acid analysis (39). Notably, the chiral centers of the resolvable dipeptides are separated by three bonds whereas in the unresolvable dipeptides the separation is by four bonds.

α-Amino acids are easily resolved by HPLC in the presence of either chiral solvents (40–43) or chiral stationary phases (44–46). β-Amino acids are less easily resolved by chiral solvent systems presumably because the increased separation between the amino and carboxyl groups allows greater flexibility in the complex with the solvent component. Use of RP-C18 columns with solvents containing N,N-dipropyl-L-alanine (43) did, however, resolve many α-substituted-β-alanines and β-substituted-β-alanines; RS-β-AiB was not resolved (39a, 47, O. Griffith, unpublished). N-Dinitrobenzoyl derivatives of

RS-β-AiB esters and similar derivatives of all other β-amino acids examined (>15 compounds) are resolved on chiral stationary phases developed by W. H. Pirkle and his associates (39a, 47). In some cases the order of enantiomer elution is constant within a series and predictable on the basis of expected solute-sorbant interactions; assignment of β-amino acid configurations from chromatographic data may thus be possible.

β-Amino acid configurations have been correlated with optical properties (48, 49). The most general rule applies to *N*-dithioethoxycarbonyl derivatives of β-substituted-β-alanines and correlates configuration with the sign of the CD Cotton effect (50).

Racemization

There is little published information on the resistance of β-amino acids to chemical racemization. Crumpler showed that *R*-β-AiB in urine could be racemized by boiling for 12 hr in 1 N NaOH (36). We find that the $t_{1/2}$ for racemization of enantiomers of α-ethyl-β-alanine at 110°C in 1 N NaOH and 6 N HCl are 200 min and 72 min, respectively. These times are significantly shorter than those observed with L-α-aminobutyrate ($t_{1/2}$ in NaOH = 309 min; $t_{1/2}$ in HCl = 249 min) (47). β-Substituted-β-alanines in which the chiral carbon is not α- to the carboxyl group are expected to be stable to racemization.

METABOLISM OF ENDOGENOUS β-AMINOCARBOXYLIC ACIDS

Pyrimidine Catabolism

As shown in Figure 3, β-alanine is a catabolite of uracil and cytosine; dihydrouracil and β-ureidopropionate are intermediates (9, 51–53). Although bacteria hydrolyze cytosine (R = H) directly to uracil, in mammals reaction 1 occurs at the nucleoside level (R = ribosyl, deoxyribosyl, arabinosyl, etc) (54). Uridine phosphorylase then converts the uracil nucleoside to uracil and a pentose-1-phosphate. The enzymes of uracil catabolism are also active toward thymine, metabolizing it via dihydrothymine and β-ureidoisobutyrate to *R*-β-aminoisobutyrate (Figure 4) (10, 53, 55, 56). The enzymes of pyrimidine degradation are abundant in liver, and in rodents that tissue can account quantitatively for the pyrimidine catabolism of the whole animal (52, 57–59). In fact, since β-ureidopropionase is apparently found only in liver, biosynthesis of free β-alanine and β-aminoisobutyrate from pyrimidines may be restricted to that site. However, since dihydropyrimidine dehydrogenase (reaction 2, Figure 3 or reaction 1, Figure 4) is irreversible in vivo, it is the committing step of pyrimidine catabolism (51, 52). Since the dehydrogenase is widely distributed (60, 61), it is probable that the partitioning of uracil and thymine between catabolism and salvage (i.e. resynthesis into nucleotides) is determined locally

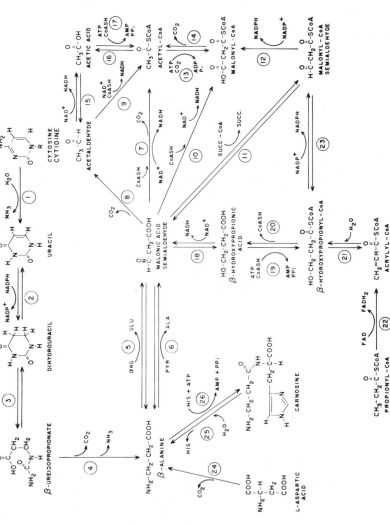

Figure 3 Formation and metabolism of β-alanine. The circled numbers refer to the following enzymes or metabolic processes: 1, cytosine ring deamination (see text); 2, dihydrouracil dehydrogenase; 3, dihydropyrimidinase; 4, β-ureidopropionase; 5, β-alanine-α-ketoglutarate transaminase; 6, β-alanine-pyruvate transaminase; 7, malonate semialdehyde dehydrogenase (acetylating); 8, nonenzymatic or (?) malonate semialdehyde decarboxylase; 9, acetaldehyde dehydrogenase; 10, as 9 (?); 11, succinyl-acetoacetyl-CoA transferase; 12, malonate semialdehyde dehydrogenase; 13, acetyl-CoA carboxylase; 14, malonyl-CoA carboxylase; 15, aldehyde dehydrogenase; 16, acetyl-CoA hydrolase; 17, acetyl-CoA synthetase; 18, 3-hydroxypropionate dehydrogenase; 19, [...]

by the enzymology of specific tissues (62, 63). Dihydropyrimidines or β-ureidoacids formed in extrahepatic tissues may be transported to the liver and metabolized or, less importantly, excreted in the urine (52, 58). As would be expected, the ratio between pyrimidine catabolism and salvage is high in mitotically quiescent tissues such as adult liver (64) but is low in proliferating tissues such as newborn liver, regenerating liver, hepatomas, and other neoplasms (65–67). Dietary pyrimidines, whether free or in nucleic acids, are extensively catabolized by the liver and gastrointestinal tissues (62, 68).

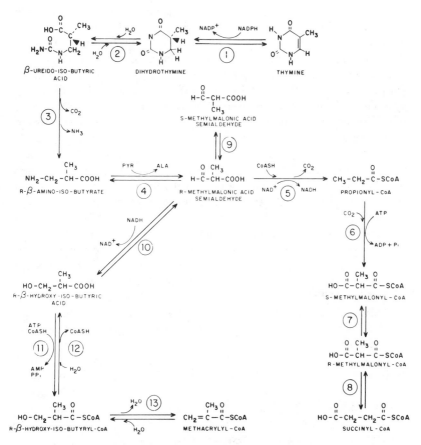

Figure 4 Formation and metabolism of R-β-aminoisobutyrate. The circled numbers refer to the following enzymes or metabolic processes: 1, dihydrouracil dehydrogenase; 2, dihydropyrimidinase; 3, β-ureidopropionase; 4, β-aminoisobutyrate-pyruvate transaminase (see text); 5, methylmalonate semialdehyde dehydrogenase (acetylating) (?); 6, propionyl-CoA carboxylase; 7, methylmalonyl-CoA racemase; 8, methylmalonyl-CoA mutase; 9, nonenzymatic (see text); 10, 3-hydroxyisobutyrate dehydrogenase; 11, an acyl-CoA synthetase (?); 12, 3-hydroxyisobutyryl-CoA hydrolase; 13, nonenzymatic for R-isomers (?).

The enzymology of pyrimidine metabolism has been reviewed previously (62, 63, 69, 70). In liver, where the catabolic pathway is complete, the dihydropyrimidine dehydrogenase reaction is rate-limiting in vitro (51, 71) and in vivo (52) [see, however, (72)]. Although early studies suggested that the enzymes catalyzing reactions 2–4 (Figure 3) might form a complex in vivo (58, 73), later studies have not sustained this view (71, 72). Since cytidine deaminase and dihydropyrimidine dehydrogenase deactivate and/or detoxify many antiviral and antineoplastic pyrimidine analogues (e.g. 5-fluorouracil, Ara-C, etc) (33, 54, 74–76), the purification, properties, and inhibition of these enzymes has received particular attention. Cytidine deaminase, widely distributed in the tissues of many but not all species (54, 76, 77), has been partially purified from human leukocytes (78), leukemic mouse spleen (75), human (79–81) and sheep (82) liver, and mouse kidney (76, 83). An affinity resin recently used to purify the *E. coli* enzyme to homogeneity may facilitate further purification of the mammalian enzyme (74). The deaminase is active toward cytidine and 5-substituted cytidines and deoxycytidines; it thus accounts for the initial catabolism of unusual nucleosides such as 5-hydroxymethylcytidine (83). Several potent inhibitors of cytidine deaminase are known (79, 84–86). Dihydropyrimidine dehydrogenase has been partially purified from several mammalian sources (59, 71) and has been purified to homogeneity from rat liver cytoplasm (87). The enzyme contains FAD and iron (87), is NADPH-specific (71, 87), and is active toward uracil, thymine, and some related pyrimidines (58, 88). The absolute stereochemistry of the reduction has recently been established (88a). Several inhibitors are known (58, 89–91). Yamada and coworkers report that the dihydrouracil dehydrogenase activity of rat liver is heterogeneous, that uracil reduction is catalyzed in vivo by a particulate-bound enzyme capable of using both NADH and NADPH, and that cytoplasm contains a NADPH-specific enzyme that preferentially catalyzes dihydrouracil oxidation (88, 92). The basis for the discrepancy between these findings and those of other investigators is not yet apparent. All studies indicate, however, that measurable dihydrouracil oxidation does not occur in vivo (52, 89, 93, 94).

Dihydropyrimidinase, which catalyzes reaction 3 (Figure 3), is detected only in liver and kidney (95) and has been partially purified from calf (96) and rat (71, 72) liver. The enzyme hydrolyzes dihydrouracil (96), *R*-dihydrothymine (97), hydantoins (96, 98) and *R*-5-alkylhydantoins, *R*-5-phenylhydantoin, *R*-α-phenylsuccinimide (97), glutarimide, and adipimide (99); the broad specificity accounts for the metabolism of several drugs and drug catabolites (93, 97, 99). Notably, the stereospecificity with respect to substituted dihydropyrimidines, hydantoins, and cyclic imides suggests that these compounds are bound dissimilarly in the active site (99). β-Ureidopropionase has been purified from rat (71, 100) and mouse (72) liver and several nonmammalian sources (94). A single cytoplasmic enzyme catalyzes the hydrolysis of β-ureidopropionate, β-ureidoisobutyrate, and β-ureido-α-fluoropropionate (from fluorouracil) (72, 100). Studies of pyrimidine catabolism have been greatly facilitated by advances in chromatographic (71, 100) and assay methodologies (101).

Although disorders of pyrimidine catabolism were until recently unknown, congenital deficiency of dihydropyrimidine dehydrogenase is now established in several individuals (102–104). The disorder is characterized by greatly increased plasma and urine levels of uracil, thymine (102–104), and 5-hydroxymethyluracil (102). Dihydrouracil and dihydrothymine are not elevated in urine (102–104), and are metabolized normally (102, 104). Severe deficiency of dihydropyrimidine dehydrogenase (0–2% of control) has been shown in leukocytes (102) and fibroblasts (104). In three infant or juvenile patients the disorder is associated with developmental and neurological problems (102, 104), but the disorder was suspected in an adult only following an adverse reaction to fluorouracil, which was detoxified poorly (103). A sibling of the adult patient excretes high levels of pyrimidines but is otherwise normal (103). Notably, urinary levels of pyrimidines are 2–10-fold higher in the younger, more severely affected patients (102–104). Wadman et al have also described a child with persistent, isolated uraciluria; the underlying metabolic defect is unknown (104). Disorders of pyrimidine catabolism are also rare in animals, but studies by Dagg et al established that pyrimidines are catabolized abnormally in the C5781/6 mouse which is β-ureidopropionase deficient (73).

Carnosine Metabolism and Other Sources of β-Alanine

β-Alanine is required for carnosine biosynthesis (reaction 26) and is formed by carnosine hydrolysis (reaction 25) and aspartate decarboxylation (reaction 24) (Figure 3). β-Alanine synthesis from aspartate is catalyzed by a specific aspartate decarboxylase in some bacteria (105) but in mammals results from a minor activity of cysteinesulfinate decarboxylase (106) and glutamate decarboxylase (107). The importance of these reactions in vivo is unknown but probably small.

Carnosine represents a large endogenous store of β-alanine; in rat skeletal muscle, carnosine and β-alanine levels are 1.5–4, and 0.13 $\mu mol \cdot gm^{-1}$, respectively (108, 109). Although the physiological role of carnosine is not yet fully established, the dipeptide affects muscle contraction (110) and has been implicated as a neurotransmitter in the olfactory pathway (111) [see, however, (112, 113)]. Since carnosine also represents a large store of histidine, its catabolism delays the development of histidine deficiencies (109, 114) and supports histamine biosynthesis (108, 115, 116). The latter effect may account for the beneficial effect of carnosine on wound healing (115) and for its depletion during infection (108). Dietary carnosine is partially degraded, but is also absorbed intact. The intestinal transport of carnosine and β-alanine have been studied (117, 118).

Carnosine is hydrolyzed to β-alanine and histidine by carnosine, a widely distributed, heterogeneous peptidase. In mouse and rat, carnosinase levels are high in kidney, uterus, and olfactory bulb mucosa and are considerably lower in skeletal muscle. Moderate activity was found in most tissues examined (119, 120). The enzyme has been purified from swine (121) and mouse (111) kidney. Margolis et al find that antisera prepared to the homogeneous mouse kidney enzyme cross-reacts with and inhibits carnosinase activity in mouse kidney, olfactory mucosa, and uterus but does not react with the activity of mouse heart, liver, or central nervous system (119). These and other studies suggest the existence of two distinct tissue forms of carnosinase (119). Tissue carnosinase is active toward carnosine and anserine (β-alanyl-L-(1-methyl)histidine) but does not hydrolyze homocarnosine (γ-aminobutyryl-L-histidine) (111, 119, 120). Human serum, however, contains a distinct carnosinase that hydrolyzes homocarnosine at 5% of the rate at which carnosine is split (122). Inherited deficiency of serum carnosinase is associated with progressive neurologic disease and developmental retardation in several male children (117, 123) but has produced no clinical problems in a female child (123). Postmortem tissue carnosine levels were normal in affected males (117, 123), and physiological explanations for the clinical findings remain elusive.

Carnosine synthetase catalyzes the ATP-dependent synthesis of carnosine from β-alanine and histidine (124, 125); AMP and PPi are coproducts (124, 126). Although not ubiquitous (127), the enzyme is widely distributed (128); activity in olfactory bulb and epithelium is particularly high (128) and may be neuronally located (128–130). The latter finding supports a role for the peptide as a neurotransmitter. However, K. Bauer et al find carnosine synthetase activity in cultured glia cells (131, 132). Carnosine synthetase has been partially purified from chick red blood cells (133) and muscle (125, 134), rabbit muscle (124), rat brain (126), and mouse olfactory bulbs (129), the last source yielding the highest specific activity. As substrate, β-alanine can be replaced by γ-aminobutyrate (yielding homocarnosine) and β-amino-n-butyrate (129); β-AiB is active with chick muscle but not with mouse olfactory bulb enzyme (124, 129). 3-Aminopropanesulfonate and 5-aminovalerate are weak inhibitors (135). Histidine can be replaced by L-ornithine and, less effectively, by L-lysine, D,L-1,4 diaminobutyrate, and 1-methylhistidine (129); some of the resulting dipeptides occur in rat brain and cultured glia cells (131). Carnosine synthetase from several mouse and rat tissues is reportedly stimulated by NAD^+ (126, 128), but this observation was not confirmed with the highly purified mouse olfactory bulb enzyme (129).

β-Alanine Catabolism

Early studies in rats established that β-[1-^{14}C]alanine and β-[3-^{14}C]alanine are both catabolized to $^{14}CO_2$ in >70% yield (52, 136). Whereas the rate and extent of $^{14}CO_2$ production are greatest with β-[1-^{14}C]alanine, a portion of the radiolabel from β-[2-^{14}C]- and β-[3-^{14}C]alanine (but not from β-[1-^{14}C]alanine) is trapped in products of the acetyl pool (e.g. fatty acids, cholesterol, acetyl-*p*-aminobenzoate) (52, 136). Although the initial step of β-alanine catabolism is clearly transamination to form malonic acid semialdehyde (see below), the further metabolism of this intermediate is poorly defined. Several possible pathways, some precedented only in bacteria, are discussed briefly below.

Liver contains two or three enzymes (isozymes?) catalyzing the α-ketoglutarate-dependent transamination of β-alanine and γ-aminobutyrate (GABA) (137, 138); the enzymes may be identical to (137) or similar to (138) the well-characterized brain GABA transaminases (139, 140). The homogeneous rabbit and hog liver enzymes also show activity toward δ-aminovalerate and S-β-aminoisobutyrate; of several α-ketoacids tested only α-ketoglutarate is a substrate (137, 138, 141). Inhibitors of GABA transaminase are of pharmacological interest for their ability to increase brain levels of GABA, an inhibitory neurotransmitter [reviews (142, 143)]. β-Alanine levels are increased modestly in physiological fluids following administration of some inhibitors (144, 145). Since a single infant with inherited hyper-β-alaninemia died with severe neurological problems (117), the clinical safety of compounds inhibiting β-alanine transamination has been questioned (145) and responded to (146).

β-Alanine is also transaminated by a mitochondrial pyruvate-dependent enzyme (reaction 6, Figure 3). The same enzyme transaminates *R*-β-AiB to *R*-methylmalonic acid semialdehyde (*R*-MMS) (reaction 4, Figure 4) and is thus *R*-β-AiB-pyruvate transaminase (147). Enzyme activity is highest in liver, much lower in kidney and intestine, and virtually undetectable in several other guinea pig tissues. In man, activity is present in normal human liver but is low or absent in liver of individuals having β-aminoisobutyric aciduria, a benign inherited condition in which up to 400 μg β-AiB·mg creatinine is excreted. Preliminary studies with mice and humans suggest a sex difference in the transamination of *R*-β-AiB (148). *R*-β-AiB-pyruvate transaminase has been partially purified from hog (147), human (149), and rat (150) liver. It is potently inhibited by D-aminooxyalanine, a metabolite of D-cycloserine (151), and by several α-hydrazinoalkanoic acids (150).

The catabolism of malonate semialdehyde (MS) is poorly defined. Of the pathways shown in Figure 3, those beginning with reactions 11 or 18 do not account for the finding that CO_2 is derived primarily from the carboxyl of MS; the flux through these pathways, if any, must therefore be small. Recent studies by Scholem & Brown with liver homogenates and fibroblasts also indicate malonyl-CoA is not a quantitatively significant metabolite of β-alanine in man (152); their findings suggest that MS is either metabolized directly to acetyl-

CoA (reaction 7) or is decarboxylated to acetaldehyde (reaction 8) and then further metabolized. Although not yet identified in mammals, an MS-specific, NAD^+-dependent enzyme catalyzing reaction 7 had been purified from *Pseudomonas fluorescens* (153, 154). In early studies Pihl & Fritzon suggested acetaldehyde might be formed from MS (reaction 8) (135). Although no enzymes catalyzing this reaction are known, the spontaneous decarboxylation of MS is well-documented (135). In mammals acetaldehyde is metabolized to CoASAc via acetate; reaction 9 is known in bacteria (155).

There is considerable evidence that mammals can carry out some of the additional metabolisms shown in Figure 3. Studies by Coon and coworkers establish that propionate can be metabolized to MS via reactions 18 to 22 (156, 157); these reactions account in part for the formation of CO_2 from propionate in patients affected by inherited or acquired disorders of the methylmalonyl-CoA pathway (158). Formation of malonyl-CoA semialdehyde from MS (reaction 11) is catalyzed by pig heart succinyl-acetoacetyl-CoA transferase; oxidation of β-hydroxypropionyl-CoA to malonyl-CoA semialdehyde (reaction 23) occurs in *Clostridia* (159) but has not yet been demonstrated in mammals. Oxidation of malonyl-CoA semialdehyde to malonyl-CoA (reaction 12) is also reported in *Clostridia* (159). The interconversions of malonyl-CoA and acetyl-CoA (reactions 13 and 14) are well-characterized in mammals. The formation of malonyl-CoA directly from malonate semialdehyde (reaction 10) has not yet been demonstrated but is analogous to reaction 9.

R-β-*Aminoisobutyrate Catabolism*

Early studies by Fink et al established that rat liver slices convert $[CH_3-^{14}C]$-thymine to radiolabeled glucose and alanine in addition to the expected intermediates of the thymine → β-AiB pathway (55). R-β-AiB is thus a glycogenic amino acid. The catabolism of R-MMS to propionyl-CoA and then to succinyl-CoA (Figure 4) is consistent with these findings but the pathway shown is based on studies of S-MMS catabolism and may not apply to the R-enantiomer. The catabolism of S-MMS is discussed below in the context of S-β-AiB.

If methylmalonate semialdehyde dehydrogenase, the enzyme catalyzing reaction 5 (Figure 4), is specific for the S-enantiomer, two processes may possibly serve to racemize R-MMS. The first, reaction 9, is the nonenzymatic racemization of MMS. The facile enolization of β-oxoacids and the relatively high acidity of their α-protons is well-established (160, 161), but the rate of racemization of MMS has apparently not been determined. A similar racemization is invoked to account for the R pathway of isoleucine metabolism (162). A second pathway interconverting R- and S-MMS is possible if both can be reversibly metabolized to methacrylyl-CoA, a compound in which chirality is lost (reactions 10–13, Figure 4). In the context of the present discussion (i.e.

R-MMS → *S*-MMS), the metabolism of methacrylyl-CoA to *S*-MMS is well-established and central to the catabolism of valine (see below); the metabolism of *R*-MMS to methacrylyl-CoA is less clear. Although 3-hydroxy-isobutyrate dehydrogenase (reaction 10) is active toward both *R*- and *S*-MMS and reaction 11 is probably catalyzed by synthetases active toward branched-chain or β-hydroxy acids, enzymes catalyzing the dehydration of *R*-β-hydroxyisobutyryl-CoA have not been characterized. Nonenzymatic hydration and dehydration reportedly occur (163), although the rate may be slow.

Valine Catabolism and S-β-Aminoisobutyrate Metabolism

S-β-AiB is reversibly formed by transamination of *S*-MMS, a catabolite of L-valine (Figure 5). As noted, the α-ketoglutarate-dependent transamination is catalyzed by an enzyme that also transaminates GABA and β-alanine (141). Although *S*-β-AiB transaminase has been considered a separate activity, its distinction from GABA transaminase is not established.

Since valine catabolism has been recently reviewed (164), only studies relating to the stereochemistry of *S*-MMS formation and the mechanism of its conversion to *S*-methylmalonyl-CoA are discussed here. The stereochemistry of MMS is fixed at reaction 4 catalyzed by enoyl-CoA hydratase. Hydration of methacrylyl-CoA occurs by addition of H- to the re face of the α-carbon of methacrylyl-CoA to give *S*-β-hydroxyisobutyryl-CoA (165–167). As noted, nonenzymatic hydration to *RS*-β-hydroxy-isobutyryl-CoA may also occur; the extent of this reaction in vivo is unknown but probably small. For example, rats given isobutyrate form and then hydrate methacrylyl-CoA (only the last step is shown in Figure 5); the finding that urinary β-hydroxyisobutyrate is at least 95% *S* on the basis of optical rotation (166) suggests that enzymatic hydration predominates in vivo. (Formation of small amounts of *R*-isomer followed by its efficient catabolism or renal absorption are not excluded by available data.)

The conversion of *S*-MMS to *S*-methylmalonyl-CoA could, in principle, occur by direct aldehyde oxidation or by various pathways in which propionyl-CoA is an intermediate (propionyl-CoA is readily carboxylated to *S*-methylmalonyl-CoA by propionyl-CoA carboxylase). If propionyl-CoA is an intermediate, free propionaldehyde and/or propionic acid might also be intermediates. Recent studies by Tanaka and coworkers demonstrate that [2-^{14}C]valine and [1-^{14}C]isobutyrate are metabolized to [1-^{14}C]methylmalonate semialdehyde but not to radiolabeled methylmalonate (168). This finding strongly supports the intermediacy of propionyl-CoA (Figure 5). A bacterial enzyme catalyzing reaction 8 forms propionyl-CoA from both MMS and propionaldehyde, but the K_m value for propionaldehyde is quite high. It is thus unlikely that propionaldehyde is formed as a free intermediate (169). Since the stereospecificity of the bacterial enzyme has not been reported, it is unknown if R-MMS is catabolized.

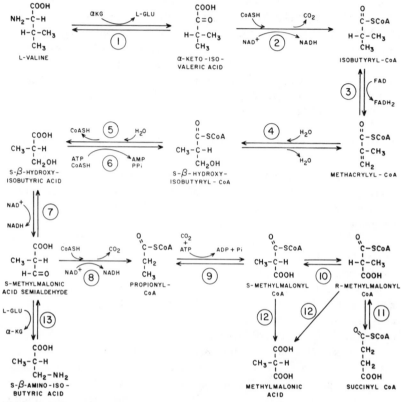

Figure 5 Relationship between valine catabolism and S-β-aminoisobutyrate metabolism. The circled numbers correspond to the following enzymes: 1, branched-chain amino acid transaminase(s); 2, branched-chain α-keto acid dehydrogenase; 3, isobutyryl-CoA dehydrogenase; 4, enoyl-CoA hydratase; 5, 3-hydroxyisobutyryl-CoA hydrolase; 6, an acyl-CoA synthetase (?); 7, 3-hydroxyisobutyryl-CoA dehydrogenase; 8, methylmalonate semialdehyde dehydrogenase (acylating); 9, propionyl-CoA carboxylase; 10, methylmalonyl-CoA racemase; 11, methylmalonyl-CoA mutase; 12, thioester hydrolase (?); 13, S-β-aminoisobutyrate-α-ketoglutarate transaminase (see text).

β-Leucine Metabolism

In 1976 Poston reported that extracts of *Clostridia* and of several mammalian tissues (e.g. rat liver and kidney, monkey liver, human leukocytes) catalyze the interconversion of L-leucine and β-leucine; β-leucine was further catabolized to isobutyrate and acetate via β-ketoisocaproate (Figure 6) (170). The interconversion of α-and β-leucine is attributed to leucine 2,3-aminomutase, a previously undescribed enzyme which is stimulated by adenosylcobalamine and inhibited by intrinsic factor (170). Subsequent studies suggested that β-leucine may be formed in vivo from L-leucine, L-valine, and terminally

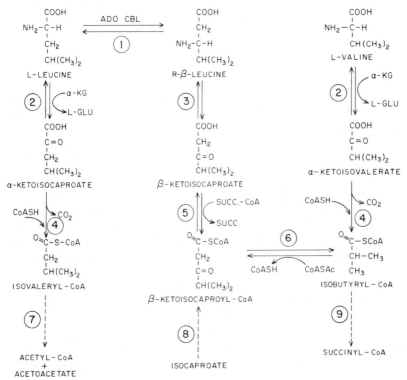

Figure 6 Metabolism of β-leucine. The circled numbers correspond to the following enzymes: 1, leucine 2,3-aminomutase; 2, branched-chain amino acid transaminase; 3, β-leucine transaminase (?); 4, branched-chain keto acid dehydrogenase; 5, succinyl-CoA: 3-oxo-acid CoA-transferase; 6, ketothiolase; 7, several enzymes of leucine catabolism; 8, several enzymes of fatty acid oxidation; 9, several enzymes of valine catabolism (see Figure 5).

branched (i.e. iso-) fatty acids (Figure 5) (12, 171–174). Although neither leucine 2,3-aminomutase nor the β-leucine tranasaminase has been fully characterized in vitro, there is substantial evidence in support of the pathway shown. Thus α-leucine is formed by rat liver homogenates in reaction mixtures containing valine, α-ketovalerate, isobutyrate, or a longer isofatty acid (isocaproate, isomyristate, or isostearate). In the presence of intrinsic factor α-leucine formation is reduced and in some (but not all) cases β-leucine formation is then observed (174). β-Leucine occurs in normal human serum (4.8 ± 3.1 μmol·L^{-1}) and is elevated in the serum of patients with pernicious anemia (24.7 ± 12.4 μmol·L^{-1}). The latter finding, in patients presumed to be leucine 2,3-aminomutase deficient, suggests that β-leucine is normally an α-leucine precursor (172, 174). As a catabolic route in vitro, the relative importance of the β-leucine pathway was high in testis (40%) but low in most other tissues

(0.4–5%) (173). Although the configuration of β-leucine in mammals is unknown, the leucine 2,3-aminomutase of *Andrographis paniculata* forms R-β-leucine (L-β-leucine) (175); that metabolism is, however, vitamin B_{12}-independent.

β-*Amino Acids in Physiological Fluids*

β-Alanine and β-AiB are minor constituents of human plasma (concentration <10 μM); S-β-AiB accounts for about 80% of the circulating β-AiB pool (176). Whereas most persons excrete little β-AiB in their urine (0–1.7 μg·mg creatinine^{-1}), some clinically normal persons are congenitally deficient in R-β-AiB-pyruvate transaminase and excrete much larger amounts of R-β-AiB (36, 177). Although rare in Caucasians (<10%), the high excretion trait is common in some Oriental populations (~40%) where a bimodal distribution has been shown (177). An acquired and generally transient increase in β-AiB excretion is sometimes associated with radiation exposure, surgery, glucose infusion, iron-copper deficiency anemia, growth retardation, thalassemia major, pregnancy, starvation, Down's syndrome, cancer, and chemotherapy of cancer [review (178), also (179, 180)]. In cancer patients excretion of the R-enantiomer of β-AiB is selectively elevated (36); it is probable but not yet demonstrated that R-β-AiB accounts for the increased excretion in the other conditions as well. Since in man the capacity to catabolize R-β-AiB is limited and the renal clearance is high (63), conditions or therapies that increase tissue destruction or turnover (or, more specifically, pyrimidine turnover) are likely to significantly increase β-AiB excretion. It is also clear, however, that the causes and sources of increased β-AiB formation are complex (178, 179). The increased excretion following surgery, for example, cannot reasonably be accounted for on the basis of tissue destruction alone (181).

The potential utility of β-AiB as a cancer marker is exemplified by the finding that β-AiB excretion increases 2–10 months prior to clinically detectable recurrences of urothelial tumors (179). It is also reported that the incidence of abnormally high β-AiB excretion increases with the grade of tumor cell dysplasia (182). Studies in which the thymine of tRNA and DNA are independently radiolabeled indicate that in both normal and tumor-bearing rats the largest portion of the urinary β-AiB is derived from tRNA (63).

β-AMINO ACIDS AS ANALOGUES OF α-AMINO ACIDS

β-*Amino Acid–Containing Peptides*

Early studies by Abderhalden suggested that peptide bonds involving β-amino acids are resistant to enzymatic hydrolysis (183, 184). Although later studies tempered this view (185), interest in the synthesis, hydrolytic stability, and conformational characteristics of β-amino acid–containing peptides continues.

As reviewed in depth by Drey (7, 8), several peptide antibiotics contain β-amino acid residues, and analogues of these peptides and others (e.g. enkephalins) have been prepared in an effort to achieve greater stability or specificity of action. Since pure enantiomers of β-amino acids are not readily available, β-alanine is most frequently used to replace glycine, or less commonly, another α-amino acid. Several analogues have substantial or full biological activity, whereas in other cases the activity spectrum is modified. Replacement of the C-terminal $GlyNH_2$ of vasopressin with β-$AlaNH_2$, for example, yields an analogue with antidiuretic and uterotonic activity but without pressor activity (186). On the other hand, 25 analogues of the sweetener Aspartame (L-Asp-L-Phe-OCH_3) were synthesized with the Phe residue replaced by any of several α- or β-substituted-β-alanines; none approached the sweetness of the parent molecule (187). Limited information on the topology of binding sites and on the conformational effects of inserting β-amino acids into peptides precludes reliable prediction of the biological activities of analogues.

β-Amino Acids as Substrate Analogues

Enzymes that normally act on α-amino acids sometimes bind β-amino acids as inhibitors or alternative substrates. As noted earlier, glutamine synthetase forms D-β-glutamine from β-glutamate (30). That finding was exploited in preparing and using β-glutamylphosphate to demonstrate the intermediacy of an enzyme-bound acylphosphate; γ-glutamylphosphate, the presumed true intermediate, is too unstable to isolate (188). β-Glutamate is also a substrate of γ-glutamylcysteine synthetase, which catalyzes a mechanistically related reaction; the configuration of the product is not yet known (189). Both glutamine and γ-glutamylcysteine synthetases utilize α-aminomethylglutarate, the type III analogue of glutamate, as substrate. Although glutamine synthetase uses only one enantiomer, γ-glutamylcysteine synthetase forms a dipeptide with both enantiomers, albeit at different rates (17). In the presence of L-Glu and RS-β-AiB, γ-glutamylcysteine synthetase forms both diastereomers of L-γ-glutamyl-β-aminoisobutyrate. β-AiB is, in fact, a better substrate than L-Ala, the α-amino acid it presumably mimics (190). L-α-Aminobutyrate, the Ala homologue, is a much better substrate than Ala, but its type III analogue, α-ethyl-β-alanine, is almost without activity (O. Griffith, unpublished). D-α-Amino acids and β-substituted-β-alanines are not substrates. Together the findings suggest that the extra methylene group in β-AiB forces the α-carbon far enough out of its normal position that a methyl group but not an ethyl group can be accommodated on either remaining α-carbon valence. Glutathione synthetase catalyzes the ATP-dependent formation of glutathione from γ-glutamylcysteine and glycine; although neither glycine nor the cysteine residue can be replaced by β-amino acids, β-glutamyl analogues are active (191). γ-Glutamylcyclotransferase is strongly inhibited by β-glutamyl-L-α-aminobutyrate, a substrate analogue in which the normal enzyme reaction would require the formation of a strained four-member ring (192). These several examples suggest that both α- and β-substituted-β-alanines deserve attention as α-amino acid analogues. It is also noted that carnitine acyltransferases are inhibited by aminocarnitine (3-amino-4-trimethyl-aminobutyrate) and its acylated derivatives (193, 194). The β-amino analogues mimic carnitine, a β-hydroxy acid. Aminocarnitine and its derivatives are metabolically stable, a property not shown by most α-amino acids (193). There is little information on the metabolic stability of other nonmammalian β-amino acids.

ACKNOWLEDGMENT

Studies from the author's laboratory were supported in part by NIH grant GM 32907. The author is a recipient of an Irma T. Hirschl Career-Scientist Award.

Literature Cited

1. Heintz, W. 1870. *Ann. Chem.* 156:25–51
2. Strecker, A. 1850. *Ann. Chem.* 75:27–45
3. Clarke, H. T., Behr, L. D. 1943. *Org. Synth.* 2:19–21
4. Ackermann, D. 1911. *Z. Biol.* 56:87–90
5. Abderhalden, E., Fodor, A. 1913. *Z. Physiol. Chem.* 85:112–30
6. Gulewitsch, W., Amiradzibi, S. 1900. *Ber. Dtsch. Chem. Ges.* 33:1902–3
7. Drey, C. N. C. 1976. *The Chemistry and Biochemistry of Amino Acids,* ed. B. Weinstein, 4:241–99. New York: Dekker
8. Drey, C. N. C. 1984. *Chemistry and Biochemistry of the Amino Acids,* ed. G. C. Barrett, pp. 25–54. London: Chapman & Hall
9. Fritzon, P., Nakken, K. F. 1956. *Acta Chem. Scand.* 10:161
10. Fink, K., Henderson, R. B., Fink, R. M. 1952. *J. Biol. Chem.* 197:441–52
11. Kupiecki, F. P., Coon, M. J. 1957. *J. Biol. Chem.* 229:743–54
12. Poston, J. M. 1986. *Adv. Enzymol.* 58:173–89
13. Kakimoto, Y., Armstrong, M. D. 1961. *J. Biol. Chem.* 236:3283–86
14. Balenovic, K. 1958. *Ciba Found. Symp. Amino Acids and Peptides with Antimetabolic Activity,* ed. G. E. W. Wolstenholme, C. M. O'Connor, pp. 5–16. Boston: Little, Brown
15. Bohme, H., Broese, R., Eiden, F. 1959. *Chem. Ber.* 92:1258–62
16. Bauce, L. G., Goren, H. J. 1979. *Int. J. Peptide Protein Res.* 14:216–26
17. Sekura, R., Hochreiter, M., Meister, A. 1976. *J. Biol. Chem.* 251:2263–70
18. Birkofer, L., Storch, I. 1953. *Chem. Ber.* 86:529–34
19. Dietrich, R. F., Sakurai, T., Kenyon, G. L. 1979. *J. Org. Chem.* 44:1894–95
20. Furukawa, M., Okawara, T., Noguchi, Y., Terawaki, Y. 1978. *Chem. Pharm. Bull.* 26:260–63
21. Bellassoued, M., Arous-Chtara, R., Gaudemar, M. 1982. *J. Organ. Chem.* 231:185–89
22. Furukawa, M., Okawara, T., Terawaki, Y. 1977. *Chem. Pharm. Bull.* 25:1319–25
23. Basheeruddin, K., Siddiqui, A. A., Khan, N. H., Saleha, S. 1979. *Synth. Commun.* 9:705–12
24. Cooper, A. J. L., Griffith, O. W. 1979. *J. Biol. Chem.* 254:2748–53
25. Shimohigashi, Y., Waki, M., Izumiya, N. 1979. *Mem. Fac. Sci. Kyushu Univ., Ser. C* 11:217–24. (Chem. Abstr. 91: 39804u)
26. Shimohigashi, Y., Waki, M., Izumiya, N. 1979. *Bull. Chem. Soc. Jpn.* 52:949–50
27. Wingo, W. J., Lewis, H. B. 1946. *J. Biol. Chem.* 165:339–46
28. Gundermann, K. D. 1954. *Ann. Chem.* 588:167–81
29. Rossi, D., Lucente, G., Romeo, A. 1977. *Experientia* 33:1557–59
30. Khedouri, E., Meister, A. 1965. *J. Biol. Chem.* 240:3357–60
31. Cohen, S. G., Khedouri, E. 1961. *J. Am. Chem. Soc.* 83:1093–96
32. Ohno, M., Kobayashi, S., Iimori, T., Wang, Y.-F., Izawa, T. 1981. *J. Am. Chem. Soc.* 103:2405–6
33. Kotani, H., Kuze, Y., Uchida, S., Miyabe, T., Iimori, T., Okano, K., et al. 1983. *Agric. Biol. Chem.* 47:1363–65

34. Winnacker, E. L., Herbst, M. M., Barker, H. A. 1971. *Biochim. Biophys. Acta* 237:280–83
35. Pollock, G. 1974. *Anal. Biochem.* 57:82–88
36. Crumpler, H. R., Dent, C. E., Harris, H., Westall, R. G. 1951. *Nature* 167:307–8
37. Solem, E. 1974. *Clin. Chim. Acta* 53:183–90
38. Manning, J. M., Moore, S. 1968. *J. Biol. Chem.* 243:5591–97
39. Griffith, O. W. 1983. *Fed. Proc.* 42:2235
39a. Griffith, O. W., Campbell, E. C., Pirkle, W. H., Tsipouras, A., Hyun, M. H. 1986. *J. Chromatogr.* In press
40. Gil-Av, E., Tishbee, A., Hare, P. E. 1980. *J. Am. Chem. Soc.* 102:5115–17
41. Lam, S. K. 1982. *J. Chromatogr.* 234:485–88
42. Gilon, C., Leshem, R., Grushka, E. 1980. *Anal. Chem.* 52:1206–9
43. Weinstein, S., Engel, M. H., Hare, P. E. 1982. *Anal. Biochem.* 121:370–77
44. Pirkle, W. H., Hyun, M. H., Tsipouras, A., Hamper, B. C., Banks, B. 1984. *J. Pharmacol. Biomed. Anal.* 2:173–81
45. Gubitz, G., Jellenz, W., Santi, W. 1981. *J. Chromatogr.* 203:377–84
46. Pirkle, W. H., Finn, J. M., Schreiner, J. L., Hamper, B. C. 1981. *J. Am. Chem. Soc.* 103:3964–66
47. Griffith, O. W., Campbell, E. B., Pirkle, W. H. 1985. *Fed. Proc.* 44:1213
48. Sjoberg, B., Hansson, B., Dahlbom, R. 1962. *Acta Chem. Scand.* 16:1057–59
49. Fritzon, P. 1957. *J. Biol. Chem.* 226:223–28
50. Furukawa, M., Okawara, T., Noguchi, Y., Terawaki, Y. 1979. *Chem. Pharm. Bull.* 27:2223–26
51. Fritzon, P. 1957. *J. Biol. Chem.* 226:223–268
52. Fritzon, P., Pihl, A. 1957. *J. Biol. Chem.* 226:229–35
53. Canellakis, E. S. 1956. *J. Biol. Chem.* 221:315–22
54. Camiener, G. W., Smith, C. G. 1965. *Biochem. Pharmacol.* 14:1405–15
55. Fink, K., Cline, R. E., Henderson, R. B., Fink, R. M. 1956. *J. Biol. Chem.* 221:425–33
56. Roberts, E., Bregoff, H. M. 1953. *J. Biol. Chem.* 201:393–98
57. Rutman, R. J., Cantarow, A., Paschkis, K. E. 1954. *J. Biol. Chem.* 210:321–29
58. Barrett, H. W., Munavalli, S. N., Newmark, P. 1964. *Biochim. Biophys. Acta* 91:199–204
59. Fritzon, P. 1960. *J. Biol. Chem.* 235:719–25
60. Moyer, J. D., Malinowski, N., Ayers, O. 1985. *J. Biol. Chem.* 260:2812–18
61. Marsh, J. C., Perry, S. 1964. *Arch. Biochem. Biophys.* 104:146–49
62. Levine, R. L., Hoogenraad, N. J., Kretchmer, N. 1974. *Pediatr. Res.* 8:724–34
63. Baliga, B. S., Borek, E. 1975. *Adv. Enzyme Regul.* 13:27–36
64. Plentl, A. A., Schoenheimer, R. 1944. *J. Biol. Chem.* 153:203–17
65. Ferdinandus, J. A., Morris, H. P., Weber, G. 1971. *Cancer Res.* 31:550–56
66. Queener, S. F., Morris, H. P., Weber, G. 1971. *Cancer Res.* 31:1004–9
67. Fritzon, P. 1962. *J. Biol. Chem.* 237:150–56
68. Sonoda, T., Tatibana, M. 1978. *Biochem. Biophys. Acta* 521:55–66
69. Wasternack, C. 1980. *Pharmacol. Ther.* 8:629–51
70. Levine, R. L., Hoogenraad, N. J., Kretchmer, N. 1974. *Pediatr. Res.* 8:629–51
71. Traut, T. W., Loechel, S. 1984. *Biochemistry* 23:2533–39
72. Sanno, Y., Holzer, M., Schimke, R. T. 1970. *J. Biol. Chem.* 245:5668–76
73. Dagg, C. P., Coleman, D. L., Fraser, G. M. 1964. *Genetics* 49:979–89
74. Ashley, G. W., Bartlett, P. A. 1984. *J. Biol. Chem.* 259:13615–20
75. Rothman, I. K., Malathi, V. G., Silber, R. 1978. *Methods Enzymol.* 51:408–12
76. Tomchick, R., Saslaw, L. D., Waravdekar, V. S. 1968. *J. Biol. Chem.* 243:2534–37
77. Maley, G. F., Maley, F. J. 1964. *J. Biol. Chem.* 239:1168–76
78. Chabner, B. A., Johns, D. G., Coleman, C. N., Drake, J. C., Evans, W. H. 1974. *J. Clin. Invest.* 53:922–31
79. Stoller, R. G., Myers, C. E., Chabner, B. A. 1977. *Biochem. Pharmacol.* 27:53–59
80. Wentworth, D. F., Wolfenden, R. 1978. *Methods Enzymol.* 51:401–7
81. Mancini, W. R., Liu, T.-S. 1983. *Biochem. Pharmacol.* 32:2427–32
82. Wisdom, G. B., Orsi, B. A. 1969. *Eur. J. Biochem.* 7:223–30
83. Creasey, W. A. 1963. *J. Biol. Chem.* 238:1772–76
84. Ashley, G. W., Bartlett, P. A. 1984. *J. Biol. Chem.* 259:13621–27
85. Cohen, R. M., Wolfenden, R. 1971. *J. Biochem.* 246:7561–65
86. Marquez, V. E., Rao, K. V. B., Silverton, J. V., Kelley, J. A. 1984. *J. Org. Chem.* 49:912–19
87. Shiotani, T., Weber, G. 1981. *J. Biol. Chem.* 256:219–24
88. Newmark, P., Stephens, J. D., Barrett, H. W. 1962. *Biochim. Biophys. Acta* 62:414–16

88a. Gani, D., Hitchcock, P. B., Young, D. W. 1985. *J. Chem. Soc. Perkin Trans.* 1:1363–72

89. Hallock, R. O., Yamada, E. W. 1976. *Can. J. Biochem.* 54:178–84

90. Cooper, G. M., Greer, S. 1970. *Cancer Res.* 30:2937–41

91. Sebesta, K., Baverova, J., Sormova, Z. 1961. *Biochim. Biophys. Acta* 50:393–94

92. Smith, A. E., Yamada, E. W. 1971. *J. Biol. Chem.* 246:3610–17

93. Maguire, J. H., Dudley, K. H. 1978. *Drug. Metab. Dispos.* 6:601–5

94. Wasternack, C., Lippmann, G., Reinbotte, H. 1979. *Biochim. Biophys. Acta* 570:341–51

95. Dudley, K. H., Butler, T. C., Bius, D. L. 1974. *Drug Metab. Dispos.* 2:103–12

96. Wallach, D. P., Grisolia, S. 1957. *J. Biol. Chem.* 226:277–88

97. Dudley, K. H., Roberts, J. B. 1978. *Drug Metab. Dispos.* 6:133–44

98. Eadie, G. S., Bernheim, F., Bernheim, M. L. C. 1949. *J. Biol. Chem.* 181:449–58

99. Maguire, J. H., Dudley, K. H. 1978. *Drug Metab. Dispos.* 6:140–45

100. Sommadossi, J.-P., Gewirtz, D. A., Diasio, R. B., Aubert, C., Cano, J.-P., Goldman, I. D. 1982. *J. Biol. Chem.* 257:8171–76

101. West, T. P., Shanley, M. S., O'Donovan, G. A. 1982. *Anal. Biochem.* 122:345–47

102. Berger, R., Stoker-de Vries, S. A., Wadman, S. K., Duran, M., Beemer, F. A., et al. 1984. *Clin. Chim. Acta* 141:227–34

103. Tuchman, M., Stoeckeler, J. S., Kiang, D. T., O'Dea, R. F., Ramnaraine, M. L., Mirkin, B. L. 1985. *N. Engl. J. Med.* 313:245–49

104. Wadman, S. K., Beemer, F. A., de Bree, P. K., Duran, M., Van Gennip, A. H., et al. 1984. *Adv. Exp. Med. Biol.* 165(A):109–14

105. Williamson, J. M., Brown, G. M. 1979. *J. Biol. Chem.* 254:8074–78

106. Griffith, O. W. 1983. *J. Biol. Chem.* 258:1591–98

107. Wu, J. Y. 1976. *GABA in Nervous System Function*, ed. E. Roberts, D. B. Tower, pp. 7–55. New York: Raven

108. Fitzpatrick, D., Amend, J. F., Squibb, R. L., Fisher, H. 1980. *Proc. Soc. Exp. Biol. Med.* 165:404–8

109. Tamaki, N., Morioka, S., Ikeda, T., Harada, M., Hama, T. 1980. *J. Nutr. Sci. Vitaminol.* 26:127–39

110. Severin, S. E., Bocharnikova, I. M., Vul'fson, P. L., Grigorovich, Y. A., Solov'eva, G. A. 1963. *Biokhimiya* 28:510–16. 28:415–20 (Engl ed.)

111. Margolis, F. L., Grillo, M., Grannot-Reisfeld, N., Farbman, A. I. 1983. *Biochim. Biophys. Acta* 744:237–48

112. MacLeod, N. K., Stroughan, D. W. 1979. *Brain Res.* 34:183–88

113. Nicoll, R. A., Alger, B. E., Jahr, C.-E. 1980. *Proc. R. Soc. London, Ser. B* 210:133–49

114. Ousterhout, L. E. 1960. *J. Nutr.* 70:226–34

115. Fitzpatrick, D. W., Fisher, H. 1982. *Surgery* 91:56–60

116. Greene, S. M., Margolis, F. L., Grillo, M., Fisher, H. 1984. *Eur. J. Pharmacol.* 99:79–84

117. Scriver, C. R., Perry, T. L., Nutzenadel, W. 1983. *The Metabolic Basis of Inherited Disease*, ed. J. B. Stanbury, J. B. Wyngaarden, D. S. Fredrickson, J. L. Goldstein, M. S. Brown, pp. 570–85. New York: McGraw-Hill. 5th ed.

118. Navab, F., Beland, S. S., Cannon, D. J., Texter, E. C. Jr. 1984. *Am. J. Physiol.* 247:643–51

119. Margolis, F. L., Grillo, M., Brown, C. E., Williams, T. H., Pitcher, R. G., Elgar, G. J. 1979. *Biochim. Biophys. Acta* 570:311–23

120. Lenney, J. F. 1976. *Biochim. Biophys. Acta* 429:214–19

121. Rosenberg, A. 1960. *Arch. Biochem. Biophys.* 88:83–93

122. Lenney, J. F., George, R. P., Weiss, A. M., Kucera, C. M., Chan, P. W. H., Rinzler, G. S. 1982. *Clin. Chim. Acta* 123:221–31

123. Murphey, W. H., Lindmark, D. G., Patchen, L. I., Housler, M. E., Harrod, E. K., Mosovich, L. 1973. *Pediatr. Res.* 7:601–6

124. Kalyankar, G. D., Meister, A. 1959. *J. Biol. Chem.* 234:3210–18

125. Winnick, R. E., Winnick, T. 1959. *Biochim. Biophys. Acta* 31:47–55

126. Skaper, S. D., Das, S., Marshall, F. D. 1973. *J. Neurochem.* 21:1429–45

127. Harding, J., Margolis, F. L. 1976. *Brain Res.* 110:351–60

128. Ng, R. H., Marshall, F. D. 1976. *Comp. Biochem. Physiol. B* 54:519–21

129. Horinishi, H., Grillo, M., Margolis, F. L. 1978. *J. Neurochem.* 31:909–19

130. Ng, R. H., Marshall, F. D., Henn, F. A., Sellstrom, A. 1977. *J. Neurochem.* 28:449–52

131. Bauer, K., Salnikow, J., de Vitry, F., Tixier-Vidal, A., Kleinkauf, H. 1979. *J. Biol. Chem.* 254:6402–7

132. Bauer, K., Hallermayer, K., Salnikow, J., Kleinkauf, H., Hamprecht, B. 1982. *J. Biol. Chem.* 257:3593–97

133. Ng, R. H., Marshall, F. D. 1976. *Comp. Biochem. Physiol. B* 54:523–25

134. Stenesh, J. J., Winnick, T. 1960. *Biochem. J.* 77:575–81
135. Seely, J. E., Marshall, F. D. 1982. *Life Sci.* 30:1763–68
136. Pihl, A., Fritzon, P. 1955. *J. Biol. Chem.* 215:345–51
137. Tamaki, N., Aoyama, H., Kubo, K., Ikeda, T., Hama, T. 1982. *J. Biochem.* 92:1009–17
138. Buzenet, A. M., Fages, C., Bloch-Tardy, M., Gonnard, P. 1978. *Biochim. Biophys. Acta* 522:400–11
139. Maitre, M., Ciesielski, L., Cash, C., Mandel, P. 1975. *Eur. J. Biochem.* 52:157–69
140. Schousboe, A., Wu, J.-Y., Roberts, E. 1973. *Biochemistry* 12:2868–73
141. Kakimoto, Y., Kanazawa, A., Taniguchi, K., Sano, I. 1968. *Biochim. Biophys. Acta* 156:374–80
142. Metcalf, B. W. 1979. *Biochem. Pharmacol.* 28:1705–12
143. Palfreyman, M. G., Schechter, P. J., Buckett, W. R., Tell, G. P., Koch-Weser, J. 1981. *Biochem. Pharmacol.* 30:817–24
144. Grove, J., Schechter, P. J., Tell, G., Koch-Weser, J., Sjoerdsma, A., et al. 1981. *Life Sci.* 28:2431–39
145. Brandt, N. J., Christensen, E. 1984. *Lancet* 1:450–51
146. Schechter, P. J., Lewis, P. J., Newberne, J. W. 1984. *Lancet* 1:737–38
147. Kakimoto, Y., Taniguchin, K., Sano, I. 1969. *J. Biol. Chem.* 244:335–40
148. Miyake, M., Kanazawa, A., Kakimoto, Y. 1979. *Biochem. Genet.* 17:785–94
149. Taniguchi, K., Tsujio, T., Kakimoto, Y. 1972. *Biochim. Biophys. Acta* 279:475–80
150. Griffith, O. W. 1983. *Fed. Proc.* 42:2235
151. Yasumitsu, T., Takao, T., Kakimoto, Y. 1976. *Biochem. Pharmacol.* 25:253–58
152. Scholem, R. D., Brown, G. K. 1983. *Biochem. J.* 216:81–85
153. Yamada, E. W., Jakoby, W. B. 1960. *J. Biol. Chem.* 235:589–94
154. Hayaishi, O., Nishizuka, Y., Tatibana, M., Takeshita, M., Kuno, S. 1961. *J. Biol. Chem.* 236:781–90
155. Burton, R. M., Stadtman, E. R. 1953. *J. Biol. Chem.* 202:873–90
156. Den, H., Robinson, W. G., Coon, M. J. 1959. *J. Biol. Chem.* 234:1666–71
157. Rendina, G., Coon, M. J. 1957. *J. Biol. Chem.* 225:523–34
158. Ando, T., Rasmussen, K., Nyhan, W. L., Hull, D. 1972. *Proc. Natl. Acad. Sci. USA* 69:2807–11
159. Vagelos, P. R., Earl, J. M. 1959. *J. Biol. Chem.* 234:2272–80
160. Kabachnik, M. I., Ioffe, S. T., Vatsuro, K. V. 1957. *Tetrahedron* 1:317–27
161. Nierlich, F., Polansky, O. E. 1968. *Monatsh. Chem.* 99:1351–54
162. Mamer, O. A., Tjoa, S. S., Scriver, C. R., Klassen, G. A. 1976. *Biochem. J.* 160:417–26
163. Robinson, W. G., Nagle, R., Bachhawat, B. K., Kupiecki, F. P., Coon, M. J. 1957. *J. Biol. Chem.* 224:1–11
164. Tanaka, K., Rosenberg, L. E. 1983. See Ref. 117, pp. 440–73
165. Amster, J., Tanaka, K. 1980. *J. Biol. Chem.* 255:119–20
166. Amster, J., Tanaka, K. 1979. *Biochim. Biophys. Acta* 585:643–44
167. Aberhart, D. J. 1977. *Bioorg. Chem.* 6:191–201
168. Baretz, B. H., Tanaka, K. 1978. *J. Biol. Chem.* 253:4203–13
169. Bannerjee, D., Sanders, L. E., Sokatch, J. R. 1970. *J. Biol. Chem.* 245:1828–35
170. Poston, J. M. 1976. *J. Biol. Chem.* 251:1859–63
171. Poston, J. M. 1977. *Science* 195:301–2
172. Poston, J. M. 1981. *Dev. Biochem. (Metab. Clin. Implic. Branched Chain Amino Ketoacids)*, pp. 401–4
173. Poston, J. M. 1984. *J. Biol. Chem.* 259:2059–61
174. Poston, J. M. 1980. *J. Biol. Chem.* 255:10067–72
175. Freer, I., Pedrocchi-Fantoni, G., Picken, D. J., Overton, K. H. 1981. *J. Chem. Soc. Chem. Commun.* 1981:80–82
176. Solem, E., Jellum, E., Eldjarn, L. 1974. *Clin. Chem. Acta* 50:393–403
177. Yanai, J., Kakimoto, Y., Tsujio, T., Sano, I. 1968. *Am. J. Hum. Genet.* 21:115–32
178. Berry, H. K. 1960. *Metabolism* 9:373–76
179. Nielsen, H. R., Nyholm, K., Sjolin, K. E. 1971. *Rev. Eur. Etudes Clin. Biol.* 16:444–50
180. Doi, Y., Hotate, T., Taniai, T., Morimoto, K., Kataura, A. 1978. *Igaku No Ayumi* 106:539–41. (Chem. Abstr. 89:177621s)
181. Levey, S., Woods, T., Abbott, W. E. 1963. *Metabolism* 12:148–55
182. Nyholm, K. K., Sjolin, K. E., Hammer, M., Knudsen, J., Stahl, D., Nielsen, H. R. 1976. *Molecular Base of Malignancy*, ed. E. Deutsch, K. Moser, H. Rainer, A. Stacher, pp. 33–35. Stuttgart: Thieme
183. Abderhalden, E., Reich, F. 1928. *Fermentforschung* 10:173–94
184. Abderhalden, E., Fleischmann, R. 1928. *Fermentforschung* 10:195–206
185. Hanson, H. T., Smith, E. L. 1948. *J. Biol. Chem.* 175:833–48
186. Anagnostaras, P., Cordopatis, P., Theo-

doropoulos, D. 1981. *Eur. J. Med. Chem. Clin. Ther.* 16:171–73

187. Miyoshi, M., Nunami, K., Sugano, H., Fujii, T. 1978. *Bull. Chem. Soc. Jpn* 51:1433–40

188. Kledouri, E., Wellner, V. P., Meister, A. 1964. *Biochemistry* 3:824–28

189. Sekura, R., Meister, A. 1977. *J. Biol. Chem.* 252:2599–605

190. Rathbun, W. B. 1967. *Arch. Biochem. Biophys.* 122:73–84

191. Oppenheimer, L., Wellner, V. P., Griffith, O. W., Meister, A. 1979. *J. Biol. Chem.* 254:5184–90

192. Griffith, O. W., Meister, A. 1977. *Proc. Natl. Acad. Sci. USA* 74:3330–34

193. Jenkins, D. L., Griffith, O. W. 1985. *J. Biol. Chem.* 260:14748–55

194. Kanamaru, T., Shinagawa, S., Asai, M., Okazaki, H., Sugiyama, Y., et al. 1985. *Life Sci.* 37:217–23

Ann. Rev. Biochem. 1986. 55:879–912

THE TRANSPORT OF PROTEINS INTO CHLOROPLASTS

Gregory W. Schmidt and Michael L. Mishkind[1]

Botany Department, University of Georgia, Athens, Georgia 30602

[1]Present address: Department of Biochemistry and Microbiology, Cook College, Rutgers University, New Brunswick, New Jersey 08903

CONTENTS

PERSPECTIVES AND SUMMARY

The means by which proteins become compartmentalized is an interest of those concerned with organelle biogenesis and presents a formidable area for research on membrane-protein interactions. Chloroplast proteins that are synthesized by

879

cytoplasmic ribosomes typically are synthesized as higher-molecular-weight precursors. They contain NH_2-terminal extensions, termed transit sequences, that are both necessary and sufficient for transport. The structural properties of the precursors have not been resolved completely, but they possess a number of functional characteristics that distinguish them from mature chloroplast proteins. The precursors uniquely are able to interact with the outer envelope membranes of chloroplasts, apparently by associating with specific receptors. Next, by an ATP-dependent mechanism, the precursors pass through the two envelope membranes, the inner of which maintains a permeability barrier to low-molecular-weight solutes during this process. Upon transport into the chloroplast, the precursors undergo endoproteolytic maturation in a pathway that in some instances involves at least two steps. The maturation enzymes have been partially characterized but it is not established whether all precursors are processed by identical proteases. Many of the major proteins of chloroplasts reside in thylakoids, membranes that harbor the photosynthetic electron transport components. Since thylakoids are not connected with the envelope, the proteins of this compartment that are synthesized by cytoplasmic ribosomes penetrate the envelope membranes, traverse the soluble stromal compartment, and then integrate into the thylakoid or, in some cases, become localized in its lumen. This last process is unique to chloroplasts since in other organelles the site for stable membrane integration of proteins is not separated from their site of membrane transport.

It will become clear that there are many outstanding problems to be solved with regard to the transport of proteins into chloroplasts. Because these are also areas of current research in other membrane and organelle transport systems, we occasionally draw attention to relevant work on protein secretion and import into mitochondria. Our emphasis, however, will be to review what is currently understood concerning the characteristics of precursors, the mechanisms for their transport through membrane bilayers, the maturation pathways and, finally, translocation of the chloroplast proteins to their ultimate organellar compartment.

OVERVIEW OF TRANSPORT OF PROTEINS THROUGH MEMBRANES

Cotranslational Transport

Among the most important breakthroughs toward understanding the mechanisms by which proteins are directed to particular subcellular compartments were the findings that proteins that are transported through membranes are synthesized as higher-molecular-weight precursors. This was extensively documented for secretory proteins of animals (1–3) and bacteria (4) and, later, for the storage proteins of plants (5). In these systems, precursors usually could be detected only through the use of in vitro mRNA translation systems. In vivo, the precursor extensions, termed signal sequences, are generally removed cotranslationally in concert with the movement of nascent polypeptide chains through the membrane (6). Cotranslational transport has been demonstrated to be obligatory for many of these proteins (6, 7). As a means to prevent the wasteful production of precursors in a completely synthesized, nontransportable form, the signal sequence, emerging from ribosomes, serves as a prompt for the temporary arrest of translation (8). In eukaryotes, this is achieved through signal sequence binding by a complex consisting of 7S RNA and cytosolic proteins termed the signal recognition particle (SRP). The SRP-ribosome-mRNA-nascent polypeptide complex remains translationally dormant until bound to membrane "docking proteins" (also termed SRP receptors) on the surface of the endoplasmic reticulum (E.R.). After ribosome-binding components of the membranes stabilize the E.R. association of the translational apparatus, the SRP is released and translation resumes. The 15–30 amino acids of signal sequences, consisting of mostly hydrophobic amino acids, are presumed to interact with the membrane, initiate the transfer of the elongating protein through the membrane, and then are cleaved by membrane-bound "signal peptidases" located on the lumenal side of the E.R. The remainder of the secreted protein is injected into and through the membrane by a mechanism that is dependent upon polypeptide chain elongation. As evidenced from the number of gene loci in *Escherichia coli* that affect protein export, the machinery to regulate and facilitate transport of proteins through membranes is highly complex (9).

Posttranslational Transport

In the late 1970s, the question of how proteins become associated with chloroplasts, mitochondria, peroxisomes, and glyoxysomes began to be studied. Initial approaches involved the use of cell-free protein synthesis systems for characterizing the primary translation products of mRNAs that direct cytoplasmic synthesis of organelle polypeptides. This first was performed with mRNA purified from the unicellular green alga, *Chlamydomonas reinhardtii*. Dobberstein et al (10) discovered that the small subunit of ribulose 1,5-bisphosphate carboxylase/oxygenase (RuBPCase) is synthesized as a higher-molecular-weight precursor. As compared to the 2–3-kilodalton signal sequences of secretory proteins, the extension of the small subunit precursor appeared to be longer with an apparent molecular mass of 3500 daltons. When incubated with algal extracts, the small subunit precursor could be converted to its mature, 14-kilodalton form. Another important discovery was that small subunit precursor mRNA was localized in preparations of free polysomes, contrasting markedly with the mRNAs for secretory proteins which are enriched in membrane-associated polysome fractions. These findings indicated that chloroplast proteins are transported through the two envelope membranes of chloroplasts by a posttranslational mechanism, a hypothesis consistent with the virtual absence of cytoplasmic ribosomes intimately associated with the organelles as determined by transmission electron microscopy (10).

To test the hypothesis for posttranslational transport of proteins, methods to reconstitute the process in vitro were developed. This was first accomplished by simply adding in vitro translation products to chloroplasts isolated from vascular plant leaves. Although the initial studies differed in incubation buffers and method of chloroplast isolation, the results were in general agreement (11, 12). The precursors to the small subunits of pea and spinach were transported into isolated chloroplasts and processed to their mature forms. The use of postribosomal supernatants (11), measurements of isotope incorporation (12), and the use of protein synthesis inhibitors by both groups demonstrated unequivocally that import occurs posttranslationally. Evidence that transport across both plastid membranes had occurred included resistance of the imported small subunit to digestion by exogenously added proteases. Also, the matured protein was recovered from the organelle's soluble fraction as part of the RuBPCase holoenzyme; therefore, it had assembled with chloroplast-synthesized large subunit upon its localization in the stroma (11).

In subsequent years, higher-molecular-weight precursors of proteins of mitochondria and microbodies (peroxisomes and glyoxysomes) were identified and their posttranslational transport into their respective organelles was documented (13–15). Overall, the mechanisms for transport of proteins into organelles are similar. However, it is becoming apparent that chloroplasts, mitochondria, and microbodies utilize rather different modes of protein import due to the number of membranes that are traversed and the energetics of this process.

CHLOROPLAST TOPOLOGY: DIVERGENT PATHWAYS FOR PROTEIN IMPORT AND SORTING

Chloroplast proteins that are nuclear-encoded can reside in one of six suborganellar compartments. These are the outer and inner membranes of the envelope, the envelope intermembrane space, the soluble or stroma fraction, the thylakoid membrane and, finally, the lumen compartment of thylakoids. It has been proposed that translocation of cytoplasmic products to these specific sites is achieved by a variety of receptors which interact with specific determinants of the imported precursors (16). Whether this is the case has not been proven but this hypothesis is supported by circumstantial evidence.

Proteins of the Stromal Compartment

Plastids are responsible for many of the major biosynthetic activities of plant and algal cells. The principal pathways found, at least in part, in the stroma of plastids in all plant cell types include those for nitrogen assimilation (17) and synthesis of fatty acids (18), many amino acids (19), terpenoids (20), and porphyrins (21). As far as is known, all of the above processes are accomplished by enzymes that are nuclear gene products synthesized by cytoplasmic ribosomes. Other plastid functions require at least some products of the organelle's genome as well as a host of nuclear gene products (22). Prominent among these are components for protein synthesis: the organelle genome encodes ribosomal and tRNAs, a few ribosomal proteins, and elongation factors (23). The replication and transcription machinery, however, is derived from nuclear genes in most systems (but see 24). In the case of the highly specialized photosynthetic organelles, chloroplasts, the major soluble enzyme, RuBPCase, contains a polypeptide, the large subunit, encoded by the organelle. All other chloroplast enzymes of the reductive pentose phosphate cycle for carbon dioxide assimilation and starch biosynthesis are synthesized in the cytoplasm (22). Thus, the vast majority of stromal proteins must be transported into this compartment after they are synthesized by cytoplasmic ribosomes.

The small subunit of RuBPCase is one of the most abundantly synthesized proteins in photosynthetic cells. High levels of small subunit mRNA enabled the early protein transport studies, summarized above, to be accomplished in spite of suboptimal conditions. Also, the highly expressed small subunit genes have been especially amenable to isolation via cDNA cloning (25). Consequently, small subunit precursors from several vascular plant species have been characterized by transport and nucleotide sequence analyses (26–34). Although the amino acid sequences of the mature polypeptides are highly homologous, the NH_2-terminal transit sequences of small subunit precursors have only a few regions of homology (Figure 1). In photosynthetic prokaryotes, where membrane transport of the small subunit does not occur, the small subunit is not made as a precursor (35). Likewise, in some eukaryotic algae, such as the chromophyte *Olisthodiscus,* the small subunit is a chloroplast rather than a nuclear gene product and also is synthesized in its mature form (36). Thus, the occurrence of a precursor of the small subunit correlates with the necessity for posttranslational transport through envelope membranes.

A vast number of other stromal proteins are posttranslationally imported into chloroplasts in vitro, especially when more optimal incubation conditions are employed (37). However, partly because their mRNA levels are low, few of these proteins have been characterized extensively at the molecular level. As an example, attempts to characterize the primary translation product for fructose-1,6-bisphosphatase failed because of low mRNA abundance or low incorporation of the radioactive amino acids used for in vitro translation (37). In spite of hindrances of this kind, information concerning other nuclear-encoded chloroplast proteins is beginning to emerge.

RuBPCase Small Subunits

Ref.

```
                  oo     oo     +        o          +oo    o++   - oo   oo  + o
Lemna        H₂N-MASSMMA STAAVARVGP AQTNMVGPFN GLRSSVPFPA TRKANNDLST LPSSGGRVSC M    29
                  V            R         A    C      A

                  oo  o o    o+o              o    +o  o   o++   - oo    o    +
Tobacco          MASSVLS  SAAVATRSNV AQANMVAPFT GLKSAASFPV SRKQNLDITS IASNGGRVQC M   33
                                  R          M

               o  oo o  oo o+ o +   o          +o o    ++  o- oo   oo   + +
Pea            MASMISS SAVTTVSRAS RGQSAAVAPF GGLKSMTGFP VKKVNTDITS ITSNGGRVKC M   27

                  oo     oo     oo    +         o  +o   o++o  - oo   o    +
Soybean          MASSM ISSPAVTTVN RAGAGMVAPF TGLKSMAGFP TRKTNNDITS IASNGGRVQC M   28

                     oo oo          +oo       o +o oo   o o o    + +
Wheat            MAPAVMA SSATTVAPFQ GLKSTAGLPI SCRSGSTGLS SVSNGGRIRC M   34

                        +oo o    + +oo +    │ +  +            - -
Chlamydomonas       MAVI AKSSVSAAVA RPARSSVRPM AALKPAVKAA PVVAPAEAND M   63
                      S                                A
```

Ferredoxin

```
                 oo oo o  o o    +        oo  o        +    o+ + o
Silene          MASTLSTLS VSASLLPKQQ PMVASSLPTN MGQALFGLKA GSRGRVTA M A   75
```

LHCP II Polypeptides

```
                     0000 o   oo o    +  +    oo   -    + o  +
Pea              MAASSSS SMALSSPTLA GKQLKLNPSS QELGAARFTM R   128

                  oo     oo      o    + + +-- ++    o- + o  +
Lemna            MAASSAI QSSAFAGQTA LKQRDELVRK VGVSDGRFSM R   130

                    o    0000    +  + 00000 - o   + o  +
Petunia          MAAAT MALSSSSFAG KAVKLSSSSS EITGNGKVTM R   93
                      T         P           NV      Q  R   A

                    o o o 0000    +  +   oo     - +     +
Wheat            MAAT TMSLSSSFA GKAVKNLPSS ALIGDARVNM R   129
```

Plastocyanin

```
           o  oo o     o    + 0000+  o +    o  + o + o +-       o
Silene    MATVTS SAAVAIPSFA GLKASSTTRA ATVKVAVATP RMSIKASLKD VGVVVAATAA AGILAGNAMA A   133
```

```
          │          │          │          │          │          │          │
         -60        -50        -40        -30        -20        -10        +1
```

Figure 1 Transit sequences of proteins transported into chloroplasts. Amino acids with hydroxyl (o), negatively charged (−), and positively charged (+) chains are indicated. Underlined regions are conserved sequences that may serve for receptor and/or transit peptidase recognition.

One letter code is:

A alanine	C cysteine	D aspartate	E glutamate
F phenylalanine	G glycine	H histidine	I isoleucine
K lysine	L leucine	M methionine	N asparagine
P proline	Q glutamine	R arginine	S serine
T threonine	V valine	W tryptophan	Y tyrosine

Pyruvate orthophosphate dikinase (PPDK) is responsible for regeneration of phosphoenolpyruvate for CO_2 fixation in the mesophyll chloroplasts of C_4 plants, but is also found in the cytoplasm of other tissues and in C_3 plants. In maize, a C_4 plant, the chloroplast-localized form is synthesized as a precursor with an apparent molecular mass of 110 kilodaltons whereas the mature protein is 94–95 kilodaltons (38, 39). The size of the PPDK transit sequence therefore appears to be much larger (>100 amino acids) than that of the small subunit precursors even though both proteins are localized in the stroma. In nonphotosynthetic tissues, cytoplasmic PPDK, though related immunologically to the plastid form, is not synthesized as a precursor since the translation product of its mRNA is 94 kilodaltons (39). As suggested by Aoyagi & Bassham (39), it is possible that the only difference between the two mRNAs is the occurrence of coding sequences for the transit sequence. Thus, the primary transcript of a single structural gene, containing an exon encoding the transit sequence, could be processed in alternative pathways to give rise to mRNAs encoding for either the chloroplastic or cytosolic forms of the enzyme. An analogous process occurs in yeast for formation of secretory and cytosolic forms of invertase (eg. Ref 40) and for a tRNA methylase for both the cytosol and mitochondria (41).

Alternative transcript maturation pathways could also be a means of generating chloroplast and cytosolic forms of enzymes in the same cell. Plastid and cytoplasmic aldolases of wheat, except for their compartmentation, are indistinguishable (42). In contrast, different genes must encode cytosolic and chloroplast aldolases in spinach and corn because antibodies against these enzymes recognize completely different antigenic determinants (43, 44). Proplastids of castor bean endosperm appear to contain 6-phosphogluconate dehydrogenase identical in size and enzymatic properties to that of the cytosol although these differ slightly in net charge (45). Similarly, 1–2% of leaf calmodulin could become chloroplast localized (46, 47); however, chloroplast association of calmodulin is an unsettled matter and there is evidence that it functions at the level of the chloroplast (and mitochondrial) envelope membranes as opposed to having physiological functions inside the organelles (48–50). Studies on the structures of genes encoding bimodally localized proteins will determine if retention or removal of exons encoding transit sequence is a causative mechanism for differential sorting of some of the above enzymes, particularly those that are immunologically related. It should be noted that cytosolic and mitochondrial forms of rat liver fumarase share antigenic determinants and are enzymatically indistinguishable but, unexpectedly, differ in sequence at their amino termini (51). Thus, posttranscriptional processes may not necessarily account for the occurrence of immunologically related proteins found in different cellular compartments.

In most instances, cytosolic and chloroplastic forms of similar enzymes are indisputably products of different genes as has been shown for triose phosphate isomerases (52, 53) and glyceraldehyde-3-phosphate dehydrogenases (54). In the latter case, the primary translation products of the mRNAs for two subunits of the chloroplast forms were 4–12 kilodaltons larger than the mature (36–43-kilodalton) subunits, depending on the plant species. The great variability in the apparent length of the putative transit sequences, 36–110 amino acids, led to consideration that the precursor extension does not contain the functional determinant for chloroplast import. It was suggested that, instead, transit sequences merely serve as effectors for folding of the precursors in order to expose "true" recognition sites; precursor maturation would cause conformational changes that ensure that the proteins remain in the chloroplasts when the enzymes acquire their functional form (54). This notion, espoused in earlier reviews of protein transport through membranes (55, 56), is not supported by recent data.

In recent years, precursors of many other stromal proteins that are nuclear gene products have been identified. These include nitrite reductase (57), proteins of chloroplast ribosomes (58), phytoferritin (59), ferredoxin-NADP oxidoreductase (37), UDP-glucosyl transferase (waxy protein) of maize endosperm starch granules (60), the acyl carrier proteins for fatty acid synthesis (61), and a subset of stress proteins that become chloroplast localized (61a, 62). In some of these studies, import of the precursors into isolated chloroplasts has been accomplished, resulting in their maturation.

Transit Sequences of Stromal Protein Precursors

Until sequence analysis was available, it was a matter of speculation whether the amino acid extension of the small subunit precursor was at the amino or carboxy terminus. The early studies of Dobberstein et al indicated that there was a single entity for the precursor chain because a rather large fragment could be released from the *Chlamydomonas* small subunit precursor upon its maturation with algal extracts (10). Microsequencing techniques, which had been of enormous utility in the characterization of the precursors of secretory proteins (6), were equally useful in determining that small subunit precursors of *Chlamydomonas* contain 44 amino acids at their NH_2 termini that are not present in the mature protein (63). This extension was termed a "transit sequence" since its presumed function (posttranslational transport) as well as composition (mostly apolar and polar residues) differed significantly from the mostly hydrophobic amino acids characteristic of the signal sequences of secretory proteins. The transit sequence was precisely removed upon incubation with extracts from *Chlamydomonas*, yielding a correctly matured small subunit. Moreover, the 44 amino acids of the transit sequence account fully for the molecular weight differences observed upon gel electrophoresis of precursor and mature forms of the small subunit (63). Therefore, it seemed unlikely that any of the carboxy termini are responsible for the larger size of the precursors.

The precise roles of transit sequences in the import and maturation of precursors into organelles have not been resolved. Transit sequences of quite different lengths and amino acid sequence share functional properties: small subunit precursors of monocots and algae can be imported in vitro into chloroplasts from dicots (64, 65). Likewise, chloroplasts in plants transformed with plasmids containing small subunit genes from other species will import the heterologous proteins, process them to their mature size, and utilize them for formation of multisubunit complexes (66). Transport is dependent upon the presence of transit sequences because if they are partially or completely removed from precursors, the proteins cannot be transported into chloroplasts in vitro (65). Recombinant plasmids containing sequences for transit sequences have been constructed with foreign genes placed in frame for their in vivo expression: the resulting products also undergo chloroplast transport with concomitant maturation of the chimeric precursors (67, 68). This elegantly demonstrates that transit sequences are all that are required for the import of proteins into chloroplasts. Thus, genetic engineering techniques now can be used to modify transit sequences as a means of identifying their functional determinants. Also, high resolution analysis of protein transport has been achieved with homogeneous preparations of mRNA (69). Therefore, it is feasible to use plasmids containing mutated genes for their transcription and translation in vitro followed by assays of the transport activity of altered transit sequences. At this time, however, information concerning the effects of transit sequence mutations have not been published. Consequently, we can discuss their roles only on the basis that evolutionary preservation of transport activity results from conservation of at least some of the structural features of the transit sequence.

The lengths of small subunit precursor transit sequences vary from 44 amino acids *(Chlamydomonas)* to 57 amino acids in tobacco and *Lemna*. Figure 1 displays the published transit sequences of the precursors of chloroplast proteins, which, except for small subunit precursor of *Chlamydomonas,* have been deduced from nucleotide sequences of cDNA or genomic clones. The most striking features of small subunit precursor transit sequences are their net positive charge and, for the vascular plants, highly homologous primary structures near the maturation site. The net positive charge, concentrated in the maturation site, has been proposed to be important in interaction of the cytosolic forms of the precursors with the negatively charged envelope membranes of chloroplasts (16). A net positive charge appears to be essential also in the import and maturation of precursors of mitochondrial proteins; when the arginine analogue canavanine, which has a pK_a of 6.6, is used during labeling studies of HeLa cells transfected with high copy plasmids, precursor and "intermediate" processing forms of ornithine transcarbamylase accumulate (70).

Von Heijne has used comparative sequence analyses to identify functional determinants in the signal sequences of secretory proteins (71). Following his lead, we have identified three domains in small subunit transit sequences between which correlations of structure and function can be inferred (65). Structurally, Domain I consists of the first 12–25 NH$_2$-terminal amino acids of which 24–40% are threonine and serine and approximately 5% are arginine or lysine. The great variability in length and primary structure of Domain I among different species appears to preclude a role for it in specific interactions with chloroplast receptors. However, the hydrophilic character of Domain I may be essential for keeping transit sequences on the protein surface during synthesis of the precursors.

Domain II varies in length from 11 to 18 residues in the small subunit precursors and is underlined in the middle segments of transit sequences in Figure 1. This region exhibits evolutionary conservation with regard to the series of residues surrounding glycine-leucine-lysine. This site is where the small subunit precursors evidently experience the first of at least two maturation events upon chloroplast import (65), discussed in more detail below. Because this region is the only part of small subunit transit sequences that is well conserved in plants and algae, we have proposed it constitutes the binding site for envelope receptors (65). Domain II is also positively charged and contains one or two prolines which may allow for distinctive secondary structure characteristics in the middle of the transit sequence. Proline may also facilitate small subunit precursor conformational changes as it traverses the envelope membranes. This amino acid reversibly undergoes cis/trans isomerization at its nitrogen atom during protein folding (72, 73). Proline is also an abundant residue in the hinge region of immunoglobulins, presumably facilitating protein flexibility (74).

Except in the case of the *Chlamydomonas* small subunit precursor, Domain III is serine/threonine-rich, has a net positive charge of 2–3, and possesses cysteine at the maturation site. We have suggested that the conserved primary sequence in this region of vascular plant transit sequences signifies its importance for precursor maturation but is of minor importance for chloroplast transport (65). The small subunit precursor of *Chlamydomonas* can be imported by chloroplasts from spinach and pea even though it lacks the Domain III sequence of vascular plant small subunit precursors (65). On the basis that there is no structural-functional correlation for chloroplast import in this region of transit sequences, we conclude that Domain III does not contain a binding site for chloroplast receptors.

Ferredoxin, a polypeptide that is associated loosely with the stromal surface of thylakoid membranes, may be imported into chloroplasts by a similar mechanism as that for the small subunit precursor. The ferredoxin precursor is about 5 kilodaltons larger than the mature (22-kilodalton) protein (59, 75). Its transit sequence (75), 48 residues in length, contains a sequence, underlined in Figure 1, which resembles Domain II of the small subunit precursor transit sequence. Hence, pre-ferredoxin might use the same receptors as the small subunit precursors. As discussed below, it is probable that maturation of pre-ferredoxin involves the same enzyme as the small subunit because of homologies at the processing site. Aside from these limited similarities in primary structure, there are no other homologies in primary sequence or, as shown by Smeekens et al, in the hydropathicity profiles of the small subunit and ferredoxin transit sequences (75).

It is likely that the tertiary structure of precursors of chloroplast proteins is of major importance in their interactions with envelope membrane receptors. At present, only predictions of their secondary structures can be made and this has not been pursued extensively. In the analysis of the small subunit precursor transit sequence from *Chlamydomonas*, Chou-Fasman paradigms predict that the first 15 residues of Domain I have a propensity for α-helix formation (63). The remainder of the transit sequence appears to be a random coil bordered by β turns. Recently, Pongor & Szalay developed a quantitative method for assessing secondary structure homologies of peptides and applied this to analyses of the RuBPCase small subunit precursors from several vascular plant species (76). As compared to the NH_2 terminus of the small subunit of the cyanobacterium *Synechococcus* (35), which is not made as a precursor, transit sequences of vascular plants are disposed toward β-structure formation. Pongor & Szalay consider their calculations to be consistent with the general tendency of proteins that undergo membrane translocation to form NH_2-terminal β-structures (77). However, this rule is not universal because the signal sequence of the *E. coli* LamB protein (β-lactamase) must form α-helixes (78): some LamB mutations that block cotranslational transport concordantly affect the ability of signal sequences to assume α-helixes in hydrophobic environments (78, 79).

Proteins of the Thylakoid Compartment

Formation of thylakoids is a light-regulated process in vascular plants. Photoactivation of phytochrome leads to increased transcription of nuclear genes encoding chloroplast proteins, and photoreduction of protochlorophyllide initiates chlorophyll synthesis (80). Except for NADPH-protochlorophyllide reductase (NPCR), little is known about the synthesis and function of proteins in the prolamellar membranes of etioplasts in dark-grown plants. NPCR is a major product of cytoplasmic protein synthesis in plants kept in darkness for prolonged periods and appears to be responsible for the elaborate morphology of prolamellar bodies (81). The mRNA for NPCR yields a 44-kilodalton precursor of the 36-kilodalton polypeptide when translated in the wheat germ system (81–83). As for many photosynthetic proteins, accumulation of NPCR is light-regulated via phytochrome, but in this case illumination is followed by dramatic decreases of its mRNA levels, especially in monocots (82, 83). Shortly after the onset of illumination, the enzyme begins to be selectively degraded, a process that is enhanced in vitro when its substrates become depleted (84, 85). A membrane-associated protease appears to be the cause for NPCR degradation and disorganization of prolamellar bodies which occurs during the early stages of greening.

Meyer & Kloppstech (86) have suggested that NPCR proteolysis may be due to a nuclear gene product whose mRNA rises and falls during the first 5 hours of illumination of etiolated pea seedlings. In vitro translation of the mRNA produces a 24-kilodalton precursor which is processed to a 17-kilodalton thylakoid membrane polypeptide upon in vitro transport into chloroplasts. There is no direct evidence about this protein's role in developing chloroplasts and other suggestions were made concerning its possible function in mediating light-dependent formation of photosynthetic membranes (86).

In fully developed chloroplasts of light-grown plants, thylakoids contain several chloroplast-synthesized integral membrane proteins of the photosynthetic electron transport pathway (22, 23). These include the chlorophyll a-binding polypeptides of Photosystem I and II reaction centers, at least two quinone-binding polypeptides, cytochromes b_6, b_{559}, and f, and two of the three subunits (I and III) of the CF_0 moiety of the ATP synthase. The few extrinsic proteins of the thylakoid that are made in the organelle are the α, β, and ϵ subunits of the CF_1 ATP synthase. Thus, most of the extrinsic proteins of thylakoids are imported from the cytoplasm. The light-harvesting chlorophyll a/b-binding proteins that associate either primarily with Photosystem II (LHCP II) or Photosystem I (LHCP I) and subunit II of CF_0 ATP synthase are notable in that they are integral membrane proteins that are products of cytoplasmic ribosomes (22, 87).

When LHCP II complexes are purified from thylakoid membranes and subjected to electrophoresis under denaturing conditions, several discrete polypeptides with apparent molecular masses of 25–30 kilodaltons are detected (88–90). These are the most abundant proteins of photosynthetic membranes, are immunologically related (91, 92), and are encoded by highly homologous nuclear genes: in Petunia, the LHCP multigene family contains at least 16 loci (93). The LHCP II polypeptides noncovalently bind antenna pigments, chlorophyll a, chlorophyll b, lutein, neoxanthin, and violaxanthin, in approximate molar ratios of $6:6:3:1:1$ per mole of apoprotein (90, 94). In addition, the LHCP II proteins assemble with each other (95), probably to form hexameric complexes in thylakoids (96, 97). Finally, they function to transfer light energy to Photosystem II reaction centers primarily by interactions that appear to be dependent upon monogalactosyl diacylglyceride (98), one of the major thylakoid lipids. The LHCP II polypeptides are highly hydrophobic (99, 100), and are integrated into thylakoids with 1–2 kilodaltons of their NH_2 termini oriented toward the stroma (101). Phosphorylation of the exposed domains by a thylakoid-associated protein kinase causes dissociation of LHCP II complexes from Photosystem II reaction centers (102). The surface moieties of the LHCP II polypeptides are implicated also in thylakoid adhesion to form grana stacks (103). This summary is incomplete but is intended only to emphasize that the LHCP II apoproteins become able to interact with many membrane components as well as with each other upon their import into chloroplasts.

Genetic evidence that genes encoding the LHCP II polypeptides are localized in the nucleus was provided by Kung et al, who analyzed *Nicotiana tabacum, Nicotiana glauca,* and their hybrids. In these species chloroplast DNA is inherited maternally but inheritance of tryptic peptide patterns of the LHCP II polypeptides is biparental (104). Their synthesis by cytoplasmic ribosomes was established by in vivo labeling studies with inhibitors of cytoplasmic and chloroplast protein synthesis (89, 105–107), and their absence among the proteins made by isolated chloroplasts (108). Chloroplast-synthesized proteins are not involved in the import, maturation, or thylakoid insertion of the LHCP II polypeptides: long-term growth of *Chlamydomonas* in the presence of inhibitors of chloroplast protein synthesis has no effect on their biogenesis (87).

Accumulation of the LHCP II polypeptides is light-dependent and is subject to transcriptional control by a phytochrome-dependent process (109–113). This was determined initially by Apel & Kloppstech, who also provided the first data concerning synthesis of the LHCP polypeptides as higher-molecular-weight precursors (109). They found that antibodies against the 25-kilodalton LHCP II polypeptide of barley immunoprecipitate a 29.5-kilodalton polypeptide from in vitro translation mixtures. The precursor-product relationship was substantiated by partial peptide mapping. Clarification of whether the LHCP II precursors are imported into chloroplasts by a direct, cotranslational mode, or via a posttranslational pathway similar to that for the small subunit precursor, depended on improving the sensitivity of the in vitro transport system (37, 114). Enhanced resolution of LHCP II transport was also achieved by using plastids deficient in endogenous LHCP II, obtained from the inner leaves of Romaine lettuce (89). In the case of pea leaf mRNA, in vitro translation products contain two polypeptides of 33 and 32 kilodaltons that are immunoprecipitated with antibodies to the major LHCP II polypeptide of *Chlamydomonas* (89). The translation products were shown to be structurally related to LHCP II polypeptides of 27–28 kilodaltons by peptide mapping analysis. When recovered from postribosomal supernatants of in vitro translation mixtures and incubated with trypsin, the LHCP II precursors are completely degraded. This demonstrated that the precursors are soluble in aqueous solutions and do not become sequestered in membrane vesicles that conceivably occur in the wheat germ translation system. Therefore, one function of the precursors' amino acid extension is to affect the overall hydrophilic character of these polypeptides.

The amino acid extension of the LHCP II precursors also appears to function as a transit sequence as shown by import of the precursors into isolated lettuce or pea chloroplasts (89). During in vitro transport, the precursors are converted to their mature size, are integrated fully and correctly into the thylakoid membrane, and bind chlorophylls to form pigment-protein complexes. Pre-existing pools of chlorophylls support formation of LHCP II complexes since these were formed from precursors imported into chloroplasts in darkness (89). No precursors, processing intermediates, or mature forms of the in vitro synthesized LHCPs were recovered with either the chloroplast envelope or stroma fractions. Since lettuce chloroplasts mature and integrate the pea LHCP II polypeptides in the same manner as pea chloroplasts, it was apparent that the import and maturation pathways for these polypeptides have been evolutionarily conserved. Moreover, how they become topologically organized in thylakoids is an inherent property of these proteins (89).

The other chlorophyll binding proteins that are nuclear encoded are the LHCP I apoproteins (D. L. Herrin, F. G. Plumley, G. W. Schmidt, submitted). Although their precursors have not been identified, it has been shown that five polypeptides that associate with Photosystem I complexes are also transported into chloroplasts by a posttranslational process and are recovered in high-molecular-weight complexes with the Photosystem I reaction centers (115).

The intermediate steps in the translocation of the LHCP polypeptides to the thylakoid have been difficult to resolve. As indicated above, concomitant chlorophyll synthesis is not required for their import and thylakoid integration since these steps can be performed in darkness in vitro (89). One question that appears to be settled concerns the role of chlorophyll b in assembly of the proteins into thylakoids. In chlorophyll b–deficient mutants of vascular plants, and in phenocopies produced by growing seedlings under intermittent illumination, thylakoids are severely deficient in both LHCP I and LHCP II (116). However, these plants contain normal levels of mRNAs for the LHCP II polypeptides in association with cytoplasmic polysomes (109, 117–119). From pulse-labeling and in vitro transport studies, the LHCP II polypeptides can be synthesized in these plants whereupon they are transported into chloroplasts and associate with thylakoids (118, 120). Thus, chlorophyll b does not participate in protein translocation to the thylakoids, but it is necessary to confer protection of the LHCP II apoproteins from degradation by a highly specific, chloroplast-localized protease. The lack of a chlorophyll b requirement for integration of LHCP apoproteins is also evident in studies of a chlorophyll b–less mutant of *Chlamydomonas* which possesses thylakoids with normal levels of LHCP II apoproteins (121).

The final class of thylakoid membrane proteins are those that function in the thylakoid lumen. These proteins are extrinsic components and include three polypeptides of the water oxidation complex associated with Photosystem II (122) and plastocyanin (123), whose copper atom carries electrons from the cytochrome f/b_6 complex to Photosystem I reaction centers. Each of these proteins is separated from its cytoplasmic site of synthesis by three membranes.

In the initial studies of the primary translation products of plastocyanin mRNA from pea leaves (37), the precursor exhibited an electrophoretic mobility of 25 kilodaltons. This is more than twice the size of the mature protein which, on the basis of amino acid sequences (124), is 10 kilodaltons in a wide range of photosynthetic organisms. It was proposed that the transit sequence of this protein was considerably larger (130–140 amino acids) than that of other precursors because it must be transported through thylakoid as well as envelope membranes (37). It still is unclear how plastocyanin translocation into the thylakoid space is achieved, but the work of Bohner et al indicates that copper plays a role (125). They found that growth of the alga *Scenedesmus acutus* in copper-free media leads to accumulation of a 14-kilodalton polypeptide that is immunoprecipitable with plastocyanin antibody. In contrast, copper deficiency in *Chlamydomonas* results in the absence of detectable plastocyanin even though these cells have high levels of its mRNA (126). When immunoprecipitated from in vitro translation mixtures, the plastocyanin precursor from *Chlamydomonas* has an electrophoretic mobility of 17 kilodaltons while, in the same gels, the mature form migrates as a 6-kilodalton polypeptide (126). Presumably, copper is necessary for protection of either the precursor or mature forms of plastocyanin from degradation, assuming the mRNA is translated in vivo. Copper-deficient *Chlamydomonas* cells synthesize cytochrome c_{553} (<6 kilodaltons), a plastocyanin substitute in algae. In vitro translation of cytochrome c_{553} mRNA yields a product with an apparent size of 14 kilodaltons (126).

The work on plastocyanin indicates that transit sequences of nearly the size of the mature polypeptides may be required for their transport to the thylakoid lumen. However, studies of the water oxidation polypeptides indicate that large transit sequences are not obligatory for this process. In spinach, these polypeptides have molecular masses of approximately 34, 23, and 16 kilodaltons, while their precursors are, respectively, 40, 33, and 26 kilodaltons (127). Thus, their transit sequences vary from the size of RuBPCase small subunit precursors to that of NADPH-dependent glyceraldehyde 3-phosphate dehydrogenase of the stroma. Westhoff et al have followed the import of precursors to the water oxidation polypeptides into spinach chloroplasts; they become localized inside thylakoid membranes as mature proteins (127). These workers suggested that precursor maturation occurs at or in the thylakoids but were unable to demonstrate this directly.

Transit Sequences of Thylakoid Membrane Proteins

Genomic and cDNA clones for LHCP II polypeptides of pea (128), wheat (129), Petunia (93), and *Lemna* (130) have been isolated and sequenced. In the case of the RuBPCase small subunit, the transit sequences and two amino acids of the mature polypeptide generally are coded by exons (e.g. 28), consistent with hypotheses of the role of exons in encoding for different functional domains in eukaryotes (131, 132). However, the genes for preLHCPs, except for *Lemna,* contain no intervening sequences. The transit sequences of the LHCP II precursors are NH_2-terminal as determined by comparisons of the deduced sequences with that of a partial NH_2-terminal sequence of one of the major LHCP components of pea (101). The transit sequences of these proteins are 37–38 residues in length and, for the species that have been examined so far, exhibit a high degree of homology (Figure 1). Although they are serine/threonine-rich and possess an overall net positive charge, there is minimal homology with transit sequences of small subunit precursors. The main exception is the three-amino-acid segment near the processing site, indicating the LHCP II precursors and soluble proteins are matured by similar, if not identical, endoproteases. Since the LHCP II precursor transit sequences are well-conserved evolutionarily, it is difficult to identify functional determinants such as those for receptor binding. However, their hydrophilicity undoubtedly promotes the solubility of the membrane precursors upon their synthesis by free cytoplasmic polysomes.

A cDNA clone to plastocyanin mRNA has been obtained and sequenced (133). The precursor contains a smaller NH_2-terminal transit sequence than was estimated from gel electrophoresis analysis of the translation product of pea mRNA (37). Pre-plastocyanin is actually 16.6 kilodaltons due to a transit sequence of 66 amino acids (Figure 1). There are no homologies with any of the transit sequences of other precursors, although serine and threonine are prevalent in the pre-plastocyanin's NH_2 terminus. Smeekens et al observed that the processing site is preceded by a region of 20 uncharged amino acids which could form a membrane-spanning region and proposed this contributes to translocation of the precursor through the thylakoid membrane (133).

Envelope Membrane Proteins

Chloroplast envelope membranes serve many roles in plant cells besides maintenance of the photosynthetic compartment. Comprehensive reviews of their roles in metabolite and ion transport (134) and their biosynthetic activities (lipid and terpenoid synthesis) (135, 136) have been published recently. Like the outer membranes of gram-negative bacteria (137) and mitochondria (138), the outer membrane of chloroplast envelopes is permeable and allows unhindered passage of small solutes (134). Proteins smaller than 9–10 kilodaltons, including trypsin, also can diffuse easily into the intermembrane space; this is due to pores with a calculated diameter of 2.5–3 nm (139). Although both the inner and outer membranes have been characterized in terms of their composition (140–145), the pore polypeptides have not been identified. The only proteins of either envelope membrane that have assigned functions are the 30-kilodalton phosphate translocator (146) and, possibly, the dicarboxylic acid transporter (147).

Another similarity between the chloroplast envelopes, mitochondrial membranes, and the outer membranes of bacteria is the occurrence of zones of adhesion at intervals between the proximal membranes. So far, the characterization of adhesion zones has been mostly structural (16, 148–151), although in bacteria they have been implicated in the translocation of proteins from the inner to the outer membranes (149, 152), and in the case of chloroplasts the transport of proteins from the outer to inner membrane compartments (16). The physical and biochemical nature of the membrane connections is unclear but they do not allow for continuity of the two lipid bilayers: in chloroplasts, the inner and outer membranes have quite distinct lipid, enzyme, and pigment compositions (139, 142–145).

Most, if not all, chloroplast envelope membrane polypeptides are nuclear gene products (136), but only one study has appeared concerning their biogenetic pathways (153). Two inner envelope polypeptides, the 30-kilodalton phosphate translocator and a 36-kilodalton protein, are synthesized in wheat germ extracts as precursors with molecular masses 11 and 2 kilodaltons greater, respectively, than the mature forms (153). A major outer envelope protein of 22 kilodaltons is synthesized in vitro as a 32-kilodalton precursor (153). It is not yet known whether these higher-molecular-mass forms possess amino terminal extensions or if they are competent for cotranslational or posttranslational integration into chloroplasts. The occurrence of a higher-molecular-mass precursor for an outer envelope membrane polypeptide contrasts with the outer membrane polypeptides of mitochondria; these apparently do not experience proteolytic maturation during membrane assembly (154–156 but see 157).

ENVELOPE RECEPTORS

Studies with isolated intact chloroplasts have shown that precursors bind receptors exposed at the cytoplasmic surface of the outer envelope membrane as a first step in their import. Proteins bound to the outer surface of repurified plastids can be distinguished from those in internal compartments by their sensitivity to nonpenetrating proteases. Although trypsin and chymotrypsin have been employed for this purpose, thermolysin is more selective because it cleaves a subset of outer envelope membrane proteins without affecting those of the inner membrane (142, 158). Grossman et al (114) found that transport, but not binding, of precursors, is greatly reduced when incubations are performed at 4°C. At least some of the precursors that bind in the cold are transferred into the chloroplast when the temperature is raised. As another means of uncoupling binding from transport, Cline et al employed the $K+/H+$ ionophore nigericin in studies of small subunit and LHCP II polypeptide binding and import (69);

for this work they used translation products of mRNAs purified by hybridization to cDNA clones. As compared to untreated chloroplasts, the number of precursor polypeptides that accumulate at the organelle surface in the presence of nigericin increases two to fourfold for the small subunit precursor and up to twofold for the LHCP II precursors. When the ionophore is subsequently removed, 80% of the small subunit precursors and 25% of those for the LHCP II polypeptides are transported into the reisolated chloroplasts. Thus, at least some of the binding in the presence of the uncoupler is of functional significance to the transport process. From independent quantitative assays, it has been estimated that a chloroplast can bind between 3000 and 5000 molecules (69, 159). Pfisterer et al calculate that this binding capacity would enable rapid import of the large amounts of chloroplast proteins that are synthesized during light-induced chloroplast development (159).

Isolated envelopes also selectively bind precursors when incubated with in vitro translation products from total cytoplasmic mRNAs (22, 159). Pfisterer et al found that purified thylakoid membranes do not selectively bind organelle precursors (159). In contrast, Ellis reported that both envelopes and thylakoids were equal in their avidity for precursors while red blood cell membranes were without specificity (22). Cline et al point out that preferential binding of translation products by isolated envelopes is not due solely to receptors since the highly basic precursors are expected to interact ionically with the highly negatively charged outer envelope membrane (69); intact chloroplasts have a pI of 4.5 (160).

Neither precursor binding nor import occurs if isolated chloroplasts are pretreated with proteases, presumably because binding sites for transit sequences are destroyed (16, 69). Although further studies are in order, the protease studies provide circumstantial evidence that the LHCP II and small subunit precursors have different receptors. We note from the data of Cline et al that import of LHCP II polypeptides is inhibited about 90% when chloroplasts are pretreated with 200 μg/ml of thermolysin whereas a similar pretreatment reduces small subunit import by less than 50% (Figure 4 of 69). Since the components of the outer envelope membrane critical for LHCP II precursor import exhibit hypersensitivity to proteolysis, they may be distinct from the envelope determinants required by the small subunit. Even with untreated chloroplasts, import of the LHCP precursors is less efficient than that of the small subunit precursors, especially in suboptimal reconstitution systems (e.g. 11). This could be due to aggregation of the LHCP precursors in the in vitro translation systems, a deficiency or low avidity of transport factors specific for the LHCP precursors relative to those for the small subunit, or differences in the number and/or affinity of receptors for the two classes of precursors. Cline et al calculated that isolated chloroplasts bind 70% more small subunit precursor than LHCP II precursor when nigericin is present to block import (69). Of potential relevance to these differences, inclusion of 5 mM EDTA in transport incubation mixtures completely blocks LHCP II precursor uptake but has no effect on the level of small subunit accumulation (69, M. L. Mishkind, G. W. Schmidt, unpublished). Although the EDTA effect is consistent also with distinct import pathways for soluble and membrane proteins, other explanations are possible. EDTA may directly alter the conformation of the LHCP precursors since their mature forms avidly bind divalent cations. Also, selective effects on import may be misleading since EDTA might induce degradation of imported LHCP apoproteins in the isolated chloroplasts.

The mitochondrial outer membrane appears to be similar to the chloroplast envelope with regard to the occurrence of receptors with affinity for specific precursors. Precursors to cytochrome c [equivalent to apocytochrome c because the precursor is not proteolytically processed after transport (161)], the ADP/ATP carrier (162), cytochrome b_2, and citrate synthase (163) tightly bind to the outer mitochondrial membrane at sites from which transport may occur. Unlabeled apocytochrome c competes for transport of trace amounts of labeled apocytochrome c but does not affect transport of precursors to the adenine nucleotide translocator or subunit 9 of the ATP synthetase (162). As in chloroplasts, there is likely to be more than one class of receptors at the organelle surface.

ENERGETICS OF TRANSPORT

Unlike precursor binding to envelope membranes, the transport of both thylakoid and stromal proteins into chloroplasts requires energy. It is certain that ATP, rather than an electrochemical gradient, supplies the energy for this process. Grossman et al demonstrated that light stimulates four to sevenfold the uptake of soluble and membrane proteins by isolated pea chloroplasts (164), in contrast to earlier work (11). Blocking photosynthetic noncyclic electron flow with DCMU (dichlorophenyldimethylurea) has no effect, however, on protein import in the light because dithiothreitol, present in transport mixtures, presumably serves as a reductant for cyclic photophosphorylation under these conditions. Complete inhibition of proton gradient formation and, consequently, ATP synthesis can be achieved with the protonophores salicylanilide XIII (Sal) and 3,5-di-tert-butyl-4-hydroxybenzylidenemalononitrile (SF6847) (164) or the ionophore nigericin (69, 165). In the presence of these inhibitors, import of proteins into illuminated chloroplasts decreases to levels observed in darkness. Inhibition of transport by these reagents substantiates the requirement for energy but does not distinguish its direct source. Fortunately, exogenous ATP supports import of the small subunit in darkness and in the presence of the inhibitors at levels comparable to that of uninhibited, illuminated chloroplasts (69, 164). Although ATP restoration of LHCP II polypeptide import appears to be less complete than that of the small subunit, the differential effect on the two classes of precursors may not be significant: the degree of light stimulation (e.g. 11) and ionophore inhibition (69) of import of precursors fluctuates considerably. Moreover, endogenous ATP levels vary in chloroplasts from different plant species and in plants of different physiological or developmental states (164). The effectiveness of exogenously supplied ATP might vary also with the activity of the adenylate transporter of the envelope. This carrier exchanges stromal ADP for external ATP and presumably functions to supply the chloroplast with ATP in darkness (134).

Further understanding of the energetics of transport of proteins into chloroplasts requires resolution of the mode by which ATP-derived energy is coupled to the transfer of proteins across the envelope membranes. It is not known whether the ATP relevant to transport is employed outside the chloroplast, between the envelope membranes, or in an internal compartment. Analyses of transport under conditions that selectively deplete specific ATP pools, such as incubation of isolated chloroplast with hexokinase and glucose, a nonpenetrating ATP-consuming system (e.g. 166), are just beginning (U. I. Flugge, personal communication). An outer envelope membrane protein kinase that phosphorylates in vitro the RuBPCase small subunit has been suggested to be involved (167), but there are no other clues about the energetics of the intermediate steps of transport in chloroplasts.

In other systems, electrochemical gradients are the energy source for transport of proteins through lipid bilayers. This is the case for translocation of proteins through the mitochondrial inner membrane (15, 168–170). These structures possess the electron transport chain, ATP synthetase complexes, and a substantial proton motive force. The topologically analogous inner envelope membranes of chloroplasts, especially those of the bundle sheath cells of maize (171), do not support a large electrochemical gradient. In *E. coli*, electrochemical gradients also have been considered to be the energy source for protein export, based on the accumulation of precursors in cells treated with ionophores (172–174). It has been proposed that in bacteria this energy is utilized in a manner analogous to electrophoresis as the membrane polarity is positive outside (175, 176). Since mitochondrial protein import would be toward the negative side of the membrane, Schatz & Butow suggested that the electrochemical gradient in these organelles may serve to "labilize" lipid bilayers to induce formation of inverted micelles that conduct the transmembrane movement of proteins (177). For certain mitochondrial proteins, import as well as binding occurs in the absence of energy input (15, 168, 178). These polypeptides, however, are limited to the outer envelope membrane, and in the case of apocytochrome *c*, to the intermembrane space. Energy-independent transport has not been demonstrated for any chloroplast polypeptide.

In contrast to the above, secretory proteins in eukaryotes are vectorially discharged into the lumen of the endoplasmic reticulum by the ATP-derived force of protein chain elongation; consistent with the absence of sources for formation of a large membrane potential in these membranes, uncouplers reportedly do not impair cotranslational transport in reconstituted microsome preparations (179). Although the necessity for ATP in posttranslational transport has been thought to be unique to chloroplasts, there are recent indications that it is also the energy source for the posttranslational transport of proteins into plasma membrane vesicles from *E. coli* (166). While it is possible that bacteria and chloroplasts have similar energetic means for transporting proteins through membranes, electrochemical gradients in bacteria are also of apparent importance, enabling precursor maturation to occur with optimal efficiency (166).

TRANSIT PEPTIDASES

All cytoplasmically synthesized chloroplast proteins so far examined experience proteolytic maturation during or shortly after transport into the organelle. Toward elucidating where and how precursor processing occurs, current objectives are to define the number of maturation steps, to identify the amino acid determinants in transit sequences that constitute processing sites, and to characterize the number, specificity, and location of the maturation enzymes.

Of the proteases that remove transit sequences from chloroplast precursors, only enzymes (transit peptidases) from pea (180) and *Chlamydomonas* (10, 63, 181) have been partially purified and characterized. The transit peptidase from *Chlamydomonas* that correctly and completely matures the small subunit precursor is a soluble protease that is inhibited by sulfhydryl reagents such as N-ethyl maleimide and mercuric chloride. This enzyme is a nuclear gene product (16, 182) and is enriched in high-molecular-weight fractions of cell extracts (181). It processes the algal precursor in vitro (181) and in vivo (182) in the absence of large subunits of RuBPCase, but does not recognize the small subunit precursor of pea (181). An analogous protease from pea chloroplasts, although originally thought to be membrane associated (12), is a soluble enzyme (183) of about 180 kilodaltons as estimated by gel filtration (180). This protease has been purified 348-fold but, as yet, is not sufficiently homogeneous to enable identification of its polypeptide(s) (180). However, we note that enrichment of its activity by anion exchange chromatography closely coincides with the occurrence of a high-molecular-mass polypeptide (>100 kilodaltons) in silver-stained polyacrylamide gels of the column fractions (Figure 3 of 180). In contrast to the *Chlamydomonas* transit peptidase, the pea enzyme is not inhibited by sulfhydryl-modifying reagents (180). Similar to the matrix peptidase that processes mitochondrial precursors (184–186), the pea enzyme is a metallo-protease with sensitivity to 1, 10-*o*-phenanthroline and EDTA (180). The pea transit peptidase preparation also has been reported to mature the precursors of wheat and barley plastocyanin and ferredoxin-NADP-oxidore-ductase (180).

As indicated by the underlined segments near the carboxy termini of the transit sequences displayed in Figure 1, the precursors of the soluble and some of the membrane proteins of vascular plants possess a highly similar stretch of three amino acids beginning five residues before their maturation sites. If this amino acid series (small-basic-hydropathic residues) constitutes the transit peptidase recognition site, the enzyme hydrolyzes peptide bonds two residues toward the carboxy terminus from where it binds. Smeekens et al also noted that the sequence three residues from the maturation site of pre-ferredoxin (glycine-arginine-valine) is homologous to the processing region of the small subunit precursors of vascular plants (75). Maturation of pre-ferredoxin could actually occur two residues toward the carboxy terminus from the hydrophobic amino acid, as postulated for small subunit precursors, but the resulting NH_2-terminal methionine is subsequently removed, a common occurrence (187). It is signifi-cant that the small subunit precursor of *Chlamydomonas* lacks this sequence and is processed only to an intermediate form upon transport into vascular plant chloroplasts (65). In their analysis of the precursor of plastocyanin, Smeekens et al noted that there is no homology at the processing site with that of the small subunit precursors, indicating that these may have different maturation path-ways (133). However, the transit peptidase preparation obtained by Robinson & Ellis is active with pre-plastocyanin as well as precursors to the vascular plant small subunits and ferredoxin-NADP-oxidoreductase (180). More highly puri-fied transit peptidase preparations are required before strong conclusions can be drawn about their substrates and sequence specificity.

The sequence similarities adjacent to the processing sites in the LHCP II polypeptide and small subunit precursors of vascular plants indicate that the transit peptidases for these two classes of proteins could be identical. Un-fortunately, the activity of a partially purified transit peptidase from pea chloroplasts with precursors of the LHCP II polypeptides has not been de-scribed (180). Crude preparations of a soluble *Chlamydomonas* transit pepti-dase, although highly active with small subunit precursors of the alga, will convert only 60% of the LHCP II precursors to their mature size even when high concentrations of cell extracts are employed (188). Although the two classes of precursors are distinguishable by their susceptibility to processing, it is not necessarily due to distinct transit peptidases. As Marks et al (188) suggest, the soluble forms of the LHCP II precursors may be poor processing substrates in vitro because of their conformation. Upon becoming membrane-associated in vivo, efficient maturation of the LHCP II precursors may be achieved by the same enzyme responsible for small subunit maturation.

Nascent chains of the precursors to the LHCP polypeptides appear to be better substrates for transit peptidases than full-length polypeptides, supporting the notion that the tertiary structure of the LHCP II precursors affects their ability to be processed. This is seen in the work of Pfisterer et al, who employed a wheat germ readout system to translate polysomes containing mRNA for the RuBPCase small subunit and LHCP polypeptides (159). Apparently, transit peptidase present in polysome preparations generates discrete, immunoprecipitable polypeptides corresponding in size to the mature forms of both protein classes (159). Perhaps the conformation of incompletely synthesized precursors of the LHCP polypeptides resembles that which they assume when undergoing envelope membrane translocation in vivo. Association of transit peptidase activity with polysomes is precedented in the studies of Dobberstein et al (10), Roy et al (189), and Gray & Kekwick (190). Verification that polysome readout products are correctly matured, however, requires analysis of the polypeptides by amino acid sequencing or, at least, two-dimensional electrophoresis. There also have been incidental reports of the recovery of mature chloroplast polypeptides from mRNA translation products generated in the wheat germ (58, 159) and reticulocyte lysate (58) systems. There are no satisfactory explanations for these observations, especially in the case of the reticulocyte lysate translations.

Recent studies indicate that at least some imported polypeptides experience multiple proteolytic cleavages during their maturation in chloroplasts. In a kinetic analysis of processing by the partially purified pea transit peptidase, transient appearance of an intermediate form (18 kilodaltons) of the pea small subunit was detected (191). Toward assessing whether the intermediate arises by proteolysis within the transit sequence, the precursor was treated with iodoacetate. This carboxymethylates the cysteine residue located at the carboxyl end of the transit sequence, introducing charge and changing side chain length at the junction with the mature polypeptide. Upon reaction of the modified precursor with transit peptidase, the 18-kilodalton intermediate was the only proteolytic product (191). The carboxymethylated precursor retains competence for import into isolated chloroplasts but is recovered as a series of polypeptides ranging in molecular mass from about 12 to 18 kilodaltons. These results indicate that the pea small subunit precursor contains two processing sites, both of which are recognized by the partially purified processing enzyme. If carboxymethylated, the site for generation of the mature polypeptide is not recognized. To explain the apparent degradation of the 18-kilodalton intermediate upon import of the modified precursor in chloroplasts, it was proposed that the intermediate fails to assemble into RuBPCase holoenzyme and consequently becomes a substrate for other proteases (191). The amino-terminal sequence of the 18-kilodalton derivative has not been determined. Consequently, it is not known if the intermediate arises by cleavage within the transit sequence or, alternatively, at the carboxy terminus of the mature protein. As discussed for maturation of the plastocyanin precursor, whether a single protease mediates cleavage at both sites will be resolved only upon purification of the enzyme to homogeneity.

Two-step processing of the RuBPCase small subunit is also evident from heterologous in vitro transport studies performed with *Chlamydomonas* precursors and chloroplasts isolated from spinach or pea (65). The algal small subunit precursor is transported into these organelles but is recovered as a form intermediate in size between the mature polypeptide and its precursor. Although it is present in the stroma, the intermediate is not assembled with vascular plant large subunits to form RuBPCase holoenzyme. The inability of vascular plant transit peptidases to completely mature the algal precursor probably are due to transit sequence differences described above. The *Chlamydomonas*-specific maturation site is retained, however, since the algal transit peptidase processes the intermediate to its mature form. These data demonstrate that an enzyme in vascular plant chloroplasts specifically cleaves the algal small subunit within its transit sequence. From amino acid sequence analysis, the intermediate results from processing within a segment (Domain II) that is represented in the transit sequences of vascular plant small subunit precursor sequences and is indicated by the arrow in Figure 1. This region is characterized by a hydrophobic amino acid (leucine) followed by a positively charged residue (lysine or arginine) and preceded by a small amino acid (glycine or alanine). Invariably, proline and valine are located at positions -4 and -6, respectively, relative to leucine. A hydrophobic amino acid (phenylalanine or methionine) occurs at position -3. The structural conservation in this segment in all small subunit precursors so far characterized suggests that it mediates critical steps in transport such as the binding of precursors to envelope receptors or proteolytic maturation (65). Cleavage within this region may allow the release of receptor-bound precursors to the stroma.

Robinson & Ellis have concluded that both processing steps of the pea small subunit precursor are catalyzed by the high-molecular-weight soluble metalloendoprotease (191). Whether this is the enzyme that generates the algal small subunit intermediate is uncertain but seems unlikely. Since pea chloroplast lysates process small subunit precursors from pea (180, 183), but not from *Chlamydomonas,* (M. L. Mishkind, G. W. Schmidt, unpublished), it appears that there must be two different enzymes. Unlike the chelator-sensitive enzyme, the activity in pea chloroplasts that cleaves the algal precursor appears to be quite labile. A similarly labile activity has been detected in *Chlamydomonas:* chloroplasts isolated from the alga will transport and partially mature pea small subunit precursors whereas crude cell lysates exhibit no processing activity with the vascular plant precursor (M. L. Mishkind, G. W. Schmidt, unpublished).

Evidence that processing intermediates occur in vivo has been obtained for the 15.5-kilodalton chloroplast ribosomal protein, L-18, from *Chlamydomonas*

(192). Whereas the primary translation product synthesized in vitro is an 18.5-kilodalton polypeptide, an intermediate of 17 kilodaltons is detected by immunoprecipitating extracts of pulse-labeled cells. The intermediate's half-life normally is less than 5 min but this can be extended when chloroplast protein synthesis is inhibited; these conditions lead to degradation of the intermediate rather than processing to its mature size (192).

The L-18 intermediate can also be detected during in vitro processing of its precursor (192). Like the pea small subunit intermediate, the 17-kilodalton polypeptide appears transiently during incubation of the primary translation product with algal extracts (58, 192). It was suggested that, in vivo, the L-18 precursor is rapidly converted to the 17-kilodalton intermediate by a processing event closely coupled to transport. Because it is degraded rather than processed to its mature form when cells are treated with chloroplast protein synthesis inhibitors, further maturation of the intermediate appears to require a product of chloroplast protein synthesis. Assembly of L-18 into a ribosome is not a prerequisite for the second processing step since complete maturation occurs in vitro with postribosomal supernatant fractions (58, 192). Two-step processing has not been observed for other imported *Chlamydomonas* chloroplast ribosomal proteins that are products of cytoplasmic protein synthesis (192).

Several imported mitochondrial proteins are also processed to their mature forms in two steps. The primary processing step of these polypeptides, all of which are located in the intermembrane space or within the inner membrane, is performed by the soluble matrix metallo-endoprotease (193–196).

DIRECTIONS AND PROSPECTS

Translocation to Inner Compartments

When stromal proteins reach the inner envelope membrane, their transit sequences probably are removed by proteolysis, an event that may facilitate release from receptors. In the case of the RuBPCase small subunit, the L-18 ribosomal protein, and perhaps other stromal proteins, this step may occur via envelope-localized proteases to generate maturation intermediates. Subsequently, processing of intermediates is completed by soluble transit peptidases in the stroma. Translocation of thylakoid membrane proteins from the chloroplast envelope to the photosynthetic membrane undoubtedly is more complex than the steps for localization of the stromal proteins. At present, there are few clues about this process, partly because the site where maturation of precursors of thylakoid proteins occurs and whether it involves more than one processing step have not been determined.

Although it is plausible that preLHCPs undergo transit through the stroma as soluble precursors or processing intermediates, there is much merit to proposals that vesicles derived from the inner envelope membrane function as carriers for these polypeptides. Morphological evidence for envelope-derived vesicles has been reviewed by Douce et al (136). Carde et al (150) have provided convincing ultrastructural evidence for the evagination of the inner envelope membrane in spinach chloroplasts in vivo and in vitro. They noted that plastids in young etiolated seedlings are relatively devoid of inner envelope evaginations. However, during light-dependent chloroplast development, inner envelope evaginations become abundant and often appear as membrane tubules. Vesicles also have been visualized in young chloroplasts subjected to centrifugation; due to their low density, vesicles float to the end of the intact organelles opposite from pelleted thylakoids (197). Nonthylakoid membranes, when very abundant, form a "peripheral reticulum" of anastomizing tubules which is usually a characteristic of photosynthetic cells of plants with the C_4 dicarboxylic acid mode of carbon dioxide fixation. Laetsch has noted that these membranes are different in composition from the envelope membranes and thylakoids on the basis that the peripheral reticulum is preferentially destroyed when permanganate is used as a fixative (198).

Douce et al (136) envisage that transit peptides interact with the external surface of the inner membrane to induce the formation of vesicles for translocation to thylakoids of products of envelope biosynthetic pathways. Since the hydrophilic precursor extensions may be removed as soon as they become exposed in the stroma, we believe they cannot be very effective in causing membrane evagination. We propose that vesicles are formed, at least in part, from interactions among the mature portions of imported proteins that are integral components of thylakoids. The LHCP II polypeptides are suitable candidates partly because they are major products of cytoplasmic protein synthesis. When LHCP precursors reach the inner envelope membrane they are likely to be oriented, similar to their mature forms (101), with the NH_2-terminal transit sequences plus 2 kilodaltons of the mature proteins exposed in the stroma. Upon maturation, the LHCP apoproteins may then be able to interact with each other to form higher order complexes like those that they form in thylakoids (95–97). They also could associate with some of the components synthesized in the inner membrane that intimately associate with the mature pigment-protein complexes. These include xanthophylls and lipids which have been determined to be specific components of LHCP II (94, 98, 199). Other envelope products, including β-carotene, the remaining thylakoid-destined lipids, plastoquinone, and geranylgeranyl precursors for the phytol chain of chlorophyll (136), could subsequently become sequestered in the forming vesicles through less specific lipophilic interactions. Thus, when the membrane evaginates, several constituents of photosynthetic membranes would undergo bulk transport as proposed by Douce et al (136). This scheme presents a solution to the problem of the absence of stromal proteins for lipid transfer from the envelope to thylakoids (200). In addition, it is possible that proteins destined for the thylakoid lumen are sequestered inside these vesicles and thereby would not undergo transport through the lipid bilayers of either the thylakoid or inner envelope membranes.

Formation of vesicles from membranes is topologically difficult and must involve localized disruption of lipid bilayers. Although this might be accomplished locally by hydrophobic portions of proteins, "nonbilayer-forming" lipids, possessing "intrinsic curvature" properties, have been proposed to be essential for evagination/invagination processes (201, 202). The major non-bilayer forming lipid in both the chloroplast envelope membrane and in the thylakoid is monogalactosyldiacylglycerol (MGDG) (136). As a lipid of this class, novel procedures must be used for its reconstitution into bilayers (203). MGDG appears to associate rather specifically with LHCP II complexes or PS II reaction centers or both. Sonication of Triton X-100–solubilized thylakoids with MGDG, but not other thylakoid lipids, leads to restoration of exciton transfer from LHCP II to PS II, as measured spectrofluorometrically and by formation of particles that could be sedimented at $17,000 \times g$ (98).

Xanthophylls also may be necessary components for vesicle evagination since these are found in the mature LHCP II complexes and are important effectors of the conformation of the LHCP polypeptides in the thylakoid membrane (F. G. Plumley, G. W. Schmidt, in preparation). Consistent with this hypothesis, inhibitors of carotenogenesis and mutants blocked in their synthesis exhibit severe defects in thylakoid formation (204, 205). Moreover, a mutant of maize which is blocked in transcription of genes encoding the LHCP polypeptides is nearly nonpigmented and lacks thylakoids (205). It was suggested that products of the carotenogenic pathway are directly involved in the regulation of expression of the LHCP genes (205), a hypothesis that may be correct. However, we feel that the studies of the maize mutants also indicate that pigment accumulation in chloroplasts requires the continual synthesis of the LHCP polypeptides.

Although the above model emphasizes a role of LHCP II polypeptides in the formation of envelope vesicles, other imported membrane proteins might perform this role as well. NADPH-protochlorophyllide reductase, which also can form polymeric complexes as the major integral membrane component in prothylakoids of etioplasts (82–85), is possibly another vesicle-forming protein.

Transit Sequences

Transit sequences are potent effectors for the biogenesis of chloroplast proteins and appear to be all that are necessary for import of soluble proteins into the stromal compartment. Whether transit sequences of proteins of the envelope membranes, the thylakoid membrane, and the thylakoid lumen serve autonomously in the suborganellar localization of these classes remains to be examined through genetic engineering techniques. If integral membrane proteins of thylakoids, such as the LHCP polypeptides, are translocated as a function of their mature sequences as well as their transit sequences, identification of the functional determinants for this process may be difficult. However, determining where fusion proteins containing the transit sequences of precursors to the LHCP polypeptides or plastocyanin become localized is a worthwhile endeavor. Comparative sequence analysis of different classes of precursors, especially when evolutionarily divergent species are employed that retain the capacity for heterologous transport, can aid also in the further identification of functional determinants of transit sequences. In this manner, more information can be obtained about the specificity of envelope receptors and the enzymes that are involved in precursor maturation. Finally, site-directed mutagenesis of cloned precursors will be of enormous use in pinpointing the necessary amino acid sequences for chloroplast import and maturation.

Import Factors

Isolated intact chloroplasts and in vitro synthesized precursor proteins provide factors sufficient for supporting polypeptide uptake. Since precursors synthesized in nonplant cellfree translation systems such as *E. coli* (67) and reticulocyte lysates (206) are capable of transport, the translation system is unlikely to provide plant-specific components other than the precursor polypeptides. It is possible, however, that cytoplasmic factors associated with isolated chloroplasts are part of the transport machinery. In yeast, a soluble, trypsin-sensitive factor of about 40 kilodaltons has been shown to be required for the import of the β-subunit of the mitochondrial ATP synthetase (207).

Although large complexes of precursors and cytosolic factors appear to be important in the import of proteins into mitochondria, the occurrence of chloroplast precursors in high-molecular-weight complexes has not been observed. The small subunit precursor, for example, is not incorporated into a rapidly sedimenting particle (65). Evidence that the small subunit precursor is not sequestered completely in a high-molecular-weight complex is apparent also from the ability of the IgG fraction of affinity-purified antibodies to selectively and almost completely block import of the small subunit precursor into chloroplasts (M. L. Mishkind, G. W. Schmidt, unpublished). Precursors of the LHCP polypeptides of *Chlamydomonas* do form higher-molecular-weight complexes in cell-free translation systems (Figure 5C of 65) but these complexes are much smaller than those of precursors of animal mitochondrial proteins. Pre-ornithine transcarbamylase (40 kilodaltons) of rat liver associates with a ribonucleoprotein in reticulocyte lysates to form a 400-kilodalton complex that is essential for organelle import (208). Aggregates (or complexes with cytosolic factors) have been noted also for the 45-kilodalton precursors of rat mitochondrial fumarase which elutes as a 300–400-kilodalton aggregate of 6–8 subunits on gel filtration columns (51) and the precursor of subunit 9 of the mitochondrial ATP synthetase in *Neurospora* (209). The *Neurospora* ADP/ ATP carrier protein, which is not made as a larger precursor, has a molecular mass of 31 kilodaltons but elutes in gel filtration columns as two bands, one of 120 kilodaltons and one of 500 kilodaltons or higher (210, but also see 211). Except for the data showing that LHCP precursors synthesized in the wheat germ system sediment partially into sucrose gradients (65), possibly due to their self-associating properties, there is no evidence for a role of precursor complexes in chloroplast transport. However, cytosolic complexes and factors warrant more detailed study.

In the case of mitochondria, porphyrins have been documented to participate in the import of some cytoplasmically synthesized proteins. Import of the precursor of δ-aminolevulinate synthase, the first enzyme for porphyrin synthesis, is blocked in vivo and in vitro by the presence of hemin (212, 213). The precursor of cytochrome c_1 is imported into the yeast mitochondrial matrix where it is converted to a maturation intermediate, then is translocated back to the outer face of the inner mitochondrial membrane. If heme is present, a second maturation step takes place resulting in the functional association of cytochrome c_1 with other components of the respiratory chain in the intermembrane space (194). The evidence for this pathway was derived from studies of a heme-deficient mutant of yeast which accumulates the cytochrome c_1 intermediate.

So far, whether chlorophyll *a* and/or its precursors participate in the import and maturation of LHCP apoproteins is not resolved. However, Johanningmeier & Howell (214) have suggested that chlorophyll precursors may participate in this process, perhaps through the involvement of magnesium protoporphyrin methyl ester which is formed by chloroplast envelope enzymes (215, 216). Their proposal derives from studies of accumulation of mRNAs for the LHCP II polypeptides in *Chlamydomonas* using chlorophyll biosynthesis inhibitors and mutants that accumulate chlorophyll precursors. Under conditions where protoporphyrin IX, magnesium protoporphyrin methyl ester, or protochlorophyllide accumulate, LHCP mRNA levels are low relative to normal cells, indicating porphyrins somehow play a role in regulation of nuclear gene expression in photosynthetic cells. From in vivo pulse-labeling studies on chlorophyll biosynthesis mutants of *Chlamydomonas,* we find that the mRNAs for the LHCP II polypeptides, although at low levels, are translated and mature LHCP apoproteins transiently accumulate in membranes (G. W. Schmidt, unpublished). Since thylakoids are virtually absent in these strains, we suppose they are blocked in transport at the level of the envelope and work is in progress toward verifying this possibility. If this is the case, it is possible that LHCP precursors or their mature forms must interact with the chlorophyll precursors as an early event in the import process.

Transport Mutants

Further resolution of the pathway and components for the transport of proteins into chloroplasts could be gained from genetic approaches. As examples, yeast mutants blocked in the import of mitochondrial proteins have been isolated by screening temperature-sensitive mutants for the accumulation of precursors of the β-subunit of the F_1-ATPase (217). Two nonallelic nuclear mutants were characterized and found to accumulate precursor in the cytosol; other proteins also are imported at reduced rates in these mutants. The primary lesions for these phenotypes have not been defined, but molecular analyses of these mutants may lead to identification of important components for mitochondrial protein transport. Mutants also have provided invaluable insights into the complexities of protein export in bacteria (9). Although genetic analyses provide an alternative and useful approach to the study of protein uptake by chloroplasts, the development of selection and screening protocols for mutants must be carefully considered: defects in the import machinery are likely to be lethal or lead to slow growth. However, conditional mutants should be obtainable and these will use genetic approaches to dissect the steps involved in the transport of proteins to the six chloroplast compartments.

ACKNOWLEDGMENTS

We are grateful to F. Gerald Plumley for his many contributions to this review. Preparation of this manuscript and research cited from our laboratory was supported from grants from the National Science Foundation and the US Department of Energy.

Literature Cited

1. Milstein, C., Brownlee, G. G., Harrison, T. M., Mathews, M. B. 1972. *Nature* 239:117–120
2. Blobel, G., Dobberstein, B. 1975. *J. Cell Biol.* 67:835–51
3. Blobel, G., Dobberstein, B. 1975. *J. Cell Biol.* 67:852–62
4. Inouye, S., Wang, S., Sekizawa, J., Halegoua, S., Inouye, I. J. 1977. *Proc. Natl. Acad. Sci. USA* 74:1004–8
5. Burr, B., Burr, F. A., Rubenstein, I., Simon, M. N. 1978. *Proc. Natl. Acad. Sci. USA* 75:696–700
6. Blobel, G. 1980. *Proc. Natl. Acad. Sci. USA* 77:1496–500
7. Silhavy, T. J., Benson, S. A., Emr, S. D. 1983. *Microbiol. Rev.* 47:313–44
8. Walter, P., Gilmore, R., Blobel, G. 1984. *Cell* 38:5–8
9. Benson, S. A., Hall, M. N., Silhavy, T. J. 1985. *Ann. Rev. Biochem.* 54:101–34
10. Dobberstein, B., Blobel, G., Chua, N.-H. 1977. *Proc. Natl. Acad. Sci. USA* 74:1082–85
11. Chua, N.-H., Schmidt, G. W. 1978. *Proc. Natl. Acad. Sci. USA* 75:6110–14
12. Highfield, P. E., Ellis, R. J. 1978. *Nature* 271:420–24
13. Kindl, H. 1982. *Int. Rev. Cytol.* 80:193–229
14. Lord, J. M., Roberts, L. M. 1983. *Int. Rev. Cytol.* Suppl. 15:115–56
15. Hay, R., Bohni, P., Gasser, S. 1984. *Biochim. Biophys. Acta* 779:65–87
16. Chua, N.-H., Schmidt, G. W. 1979. *J. Cell Biol.* 81:461–83
17. Oaks, A., Hirel, B. 1985. *Ann. Rev. Plant Physiol.* 36:345–65
18. Roughan, G., Slack, R. 1984. *Trends Biochem. Sci.* 9:383–86
19. Miflin, B. J., Lea, P. J. 1977. *Ann. Rev. Plant Physiol.* 28:299–329

20. Camara, B. 1984. *Plant Physiol.* 74: 112–16
21. Castelfranco, P. A., Beale, S. I. 1983. *Ann. Rev. Plant Physiol.* 34:241–78
22. Ellis, R. J. 1983. *Subcell. Biochem.* 9:237–61
23. Whitfeld, P. R., Bottomley, W. 1983. *Ann. Rev. Plant Physiol.* 34:279–310
24. Watson, J. C., Surzycki, S. J. 1982. *Proc. Natl. Acad. Sci. USA* 79:2264–67
25. Cashmore, A. R. 1979. *Cell* 17:383–88
26. Bedbrook, J. R., Smith, S. M., Ellis, R. J. 1980. *Nature* 287:692–97
27. Coruzzi, G., Broglie, R., Cashmore, A. R., Chua, N.-H. 1983. *J. Biol. Chem.* 258:1399–402
28. Berry-Lowe, S. L., McKnight, T. D., Shah, D. M., Meagher, R. M. 1982. *J. Mol. Appl. Genet.* 1:482–98
29. Stiekema, W. J., Wimpee, C. F., Tobin, E. M. 1983. *Nucleic Acids Res.* 11:8051–61
30. Dunsmuir, P., Smith, S., Bedbrook, J. 1983. *Nucleic Acids Res.* 11:4177–83
31. Pinck, L., Fleck, J., Pinck, M., Hadidane, R., Hirth, L. 1983. *FEBS Lett.* 154:145–48
32. Smith, S. M., Bedbrook, J., Speirs, J. 1983. *Nucleic Acids Res.* 11:2719–34
33. Mazur, B. J., Chui, C.-F. 1985. *Nucleic Acids Res.* 13:2373–86
34. Broglie, R., Coruzzi, G., Lamppa, G., Keith, B., Chua, N.-H. 1983. *Bio/Technology* 1:55–61
35. Shinozaki, K., Sugiura, M. 1983. *Nucleic Acids Res.* 11:6957–64
36. Reith, M. E., Cattolico, R. A. 1985. *Biochemistry* 24:2556–61
37. Grossman, A. R., Barlett, S. G., Schmidt, G. W., Mullet, J. E., Chua, N.-H. 1982. *J. Biol. Chem.* 257:1558–63
38. Hague, D. R., Uhler, M., Collins, P. D. 1983. *Nucleic Acids Res.* 11:4853–65
39. Aoyagi, K., Bassham, J. A. 1984. *Plant Physiol.* 76:278–80
40. Perlman, D., Raney, P., Halvorson, H. O. 1984. *Mol. Cell. Biol.* 4:1682–88
41. Hopper, A. K., Furukawa, A. H., Pham, H. D., Martin, N. C. 1982. *Cell* 28:543–50
42. Murphy, D. J., Walker, D. A. 1981. *FEBS Lett.* 134:163–66
43. Kruger, I., Schnarrenberger, C. 1983. *Eur. J. Biochem.* 136:101–6
44. Lebherz, H. G., Leadbetter, M. M., Bradshaw, R. A. 1984. *J. Biol. Chem.* 259:1011–17
45. Simcox, P. D., Dennis, D. T. 1978. *Plant Physiol.* 62:287–90
46. Muta, S. 1982. *FEBS Lett.* 147:161–64
47. Biro, R. L., Daye, B. S., Serlin, M. E., Terry, N., Datta, N., et al. 1984. *Plant Physiol.* 75:382–86
48. Dieter, P., Marme, D. 1984. *J. Biol. Chem.* 259:184–89
49. Simon, P., Bonzon, M., Greppin, H., Marme, D. 1984. *FEBS Lett.* 167:332–38
50. Roberts, D. M., Zielinski, R. E., Schleicher, M., Watterson, D. M. 1983. *J. Cell. Biol.* 97:1644–47
51. Ono, H., Yoshimura, N., Sato, M., Tuboi, S. 1985. *J. Biol. Chem.* 260: 3402–7
52. Pichersky, E., Gottlieb, L. D., Higgins, R. C. 1984. *Mol. Gen. Genet.* 193:158–61
53. Kurzok, H.-G., Feierabend, J. 1984. *Biochim. Biophys. Acta* 788:222–33
54. Cerff, R., Kloppstech, K. 1982. *Proc. Natl. Acad. Sci. USA* 79:7624–28
55. Sabatini, D. D., Kreibich, G., Morimoto, T., Adesnik, M. 1982. *J. Cell. Biol.* 92:1–22
56. Neupert, W., Schatz, G. 1981. *Trends Biochem. Sci.* 6:1–4
57. Small, I. S., Gray, J. C. 1984. *Eur. J. Biochem.* 145:291–97
58. Schmidt, R. J., Myers, A. M., Gillham, N. W., Boynton, J. E. 1984. *J. Cell Biol.* 98:2011–18
59. Van der Mark, F., van den Briel, W., Huisman, H. G. 1983. *Biochem. J.* 214:943–50
60. Shure, M., Wessler, S., Federoff, N. 1983. *Cell* 35:225–33
61. Ohlrogge, J. B., Kuo, T.-M. 1984. *What's New in Plant Physiol.* 15:41–44
61a. Vierling, E., Mishkind, M. L., Schmidt, G. W., Key, J. L. 1986. *Proc. Natl. Acad. Sci. USA.* 83:361–65
62. Kloppstech, K., Meyer, G., Schuster, G., Ohad, I. 1985. *EMBO J.* 4:1901–9
63. Schmidt, G. W., Devillers-Thiery, A., Desruisseaux, H., Blobel, G., Chua, N.-H. 1979. *J. Cell. Biol.* 83:615–22
64. Coruzzi, G., Broglie, R., Lamppa, G., Chua, N.-H. 1983. In *Structure and Function of Plant Genomes*, ed. O. Ciferri, L. Dure III, pp. 47–59. New York: Plenum
65. Mishkind, M. L., Wessler, S. R., Schmidt, G. W. 1985. *J. Cell Biol.* 100:226–34
66. Broglie, R., Coruzzi, G., Fraley, R. T., Rogers, S. G., Horsch, R. B., et al. 1984. *Science* 224:838–43
67. Van den Broeck, G., Timko, M. P., Kausch, A. P., Cashmore, A. R., Van Montagu, M., Herrera-Estrella, L. 1985. *Nature* 313:358–63
68. Schreier, P. H., Seftor, E. A., Schell, J., Bohnert, H. J. 1985. *EMBO J.* 4:25–32
69. Cline, K., Werner-Washburne, M., Lubben, T. H., Keegstra, K. 1985. *J. Biol. Chem.* 260:3691–96

70. Horwick, A. L., Fenton, W. A., Firgaira, F. A., Fox, J. E., Kolansky, D., et al. 1985. *J. Cell. Biol.* 100:1515–21
71. Von Heijne, G. 1984. *J. Mol. Biol.* 173:243–51
72. Brandts, J. F., Halvorson, H. R., Brennan, M. 1975. *Biochemistry* 14:4953–63
73. Kim, P. S., Baldwin, R. L. 1982. *Ann. Rev. Biochem.* 51:459–89
74. Davies, D. R., Metzger, H. 1983. *Ann. Rev. Immunol.* 1:87–177
75. Smeekens, S., Binsbergen, V., Weisbeek, P. 1985. *Nucleic Acids Res.* 13:3179–94
76. Pongor, S., Szalay, A. A. 1985. *Proc. Natl. Acad. Sci. USA* 82:366–70
77. Garnier, F., Gaye, P., Mercier, F.-C., Robson, B. 1980. *Biochimie* 6:231–39
78. Emr, S. D., Silhavy, T. 1983. *Proc. Natl. Acad. Sci. USA* 80:4599–603
79. Briggs, M. S., Gierasch, L. M. 1984. *Biochemistry* 23:3111–13
80. Tobin, E. M., Silverthorne, J. 1985. *Ann. Rev. Plant Physiol.* 36:569–93
81. Apel, K. 1981. *Eur. J. Biochem.* 120: 89–93
82. Apel, K., Santel, H.-J., Redlinger, T. E., Falk, H. 1980. *Eur. J. Biochem.* 111:251–58
83. Meyer, G., Bliedung, H., Kloppstech, K. 1983. *Plant Cell Rep.* 2:26–29
84. Hauser, I., Dehesh, K., Apel, K. 1984. *Arch. Biochem. Biophys.* 228:577–86
85. Kay, S. A., Griffith, W. T. 1983. *Plant Physiol.* 72:229–36
86. Meyer, G., Kloppstech, K. 1984. *Eur. J. Biochem.* 138:201–7
87. Chua, N.-H., Gillham, N. W. 1977. *J. Cell Biol.* 74:441–52
88. Delepelaire, P., Chua, N.-H. 1981. *J. Biol. Chem.* 256:9300–7
89. Schmidt, G. W., Bartlett, S. G., Grossman, A. R., Cashmore, A. R., Chua, N.-H. 1981. *J. Cell Biol.* 91:468–78
90. Suss, K.-H. 1983. *Photobiochem. Photobiophys.* 5:317–24
91. Chua, N.-H., Blomberg, F. 1979. *J. Biol. Chem.* 254:215–23
92. Plumley, F. G., Schmidt, G. W. 1983. *Anal. Biochem.* 134:86–95
93. Dunsmuir, P. 1985. *Nucleic Acids Res.* 13:2504–18
94. Braumann, T., Weber, G., Grimme, L. H. 1982. *Photobiochem. Photobiophys.* 4:1–8
95. Kuhlbrandt, W. 1984. *Nature* 307: 478–80
96. McDonnel, A., Staehelin, L. A. 1980. *J. Cell Biol.* 84:40–56
97. Li, J. 1985. *Proc. Natl. Acad. Sci. USA* 82:386–90
98. Siefermann-Harms, D., Ross, J. W.,

Kaneshiro, K. H., Yamamoto, H. Y. 1982. *FEBS Lett.* 149:191–96
99. Henriques, F., Park, R. B. 1976. *Biochim. Biophys. Acta* 430:312–20
100. Chua, N.-H., Matlin, K., Bennoun, P. 1975. *J. Cell Biol.* 67:361–77
101. Mullett, J. E. 1983. *J. Biol. Chem.* 258:9941–48
102. Bennett, J. 1983. *Biochem. J.* 212:1–13
103. Staehelin, L. A., Arntzen, C. J. 1983. *J. Cell Biol.* 97:1327–37
104. Kung, S. D., Thornber, J. P., Wildman, S. G. 1972. *FEBS Lett.* 24:185–88
105. Hoober, J. K. 1970. *J. Biol. Chem.* 245:4327–33
106. Cashmore, A. R. 1976. *J. Biol. Chem.* 251:2848–53
107. Machold, O., Aurich, O. 1972. *Biochim. Biophys. Acta* 281:103–12
108. Ellis, R. J. 1983. *Subcell. Biochem.* 9:237–61
109. Apel, K., Kloppstech, K. 1978. *Eur. J. Biochem.* 85:581–88
110. Apel, K. 1979. *Eur. J. Biochem.* 97: 183–88
111. Cuming, A. C., Bennett, J. 1981. *Eur. J. Biochem.* 118:71–80
112. Tobin, E. M. 1981. *Plant Physiol.* 67:1078–83
113. Kaufmann, L. S., Thompson, W. F., Briggs, W. R. 1984. *Science* 226: 1447–49
114. Grossman, A. R., Bartlett, S., Schmidt, G. W., Chua, N.-H. 1979. *Ann. NY Acad. Sci.* 343:266–74
115. Mullet, J. E., Grossman, A. R., Chua, N.-H. 1982. *Cold Spring Harbor Symp. Quant. Biol.* 46:979–84
116. Mullet, J. E., Burke, J. J., Arntzen, C. J. 1980. *Plant Physiol.* 65:823–27
117. Schwarz, H. P., Kloppstech, K. 1982. *Planta* 155:116–23
118. Bellemare, G., Bartlett, S. G., Chua, N.-H. 1982. *J. Biol. Chem.* 257:7762–67
119. Viro, M., Kloppstech, K. 1982. *Planta* 154:18–23
120. Bennett, J. 1981. *Eur. J. Biochem.* 118:61–70
121. Michel, H., Tellenbach, M., Boschetti, A. 1983. *Biochim. Biophys. Acta* 725: 417–24
122. Akerlund, H. E., Jansson, C. 1981. *FEBS Lett.* 124:229–32
123. Haehnel, W., Berzborn, R. J., Andersson, B. 1981. *Biochim. Biophys. Acta* 637:389–99
124. Boulter, D., Haslett, B. G., Peacock, D., Ramshaw, J. A. M., Scawen, M. D. 1977. In *Plant Biochemistry*, ed. D. H. Northcote, 2:1–40. Baltimore: University Park Press

125. Bohner, H., Bohme, H., Boger, P. 1981. *FEBS Lett.* 131:386–89
126. Merchant, S., Bogorad, L. 1986. *Mol. Cell. Biol.* 6:462–69
127. Westhoff, P., Jansson, C., Klein-Hitpass, L., Berzborn, R., Larsson, C., Bartlett, S. G. 1985. *Plant Mol. Biol.* 4:137–46
128. Cashmore, A. R. 1984. *Proc. Natl. Acad. Sci. USA* 81:2960–64
129. Lamppa, G. K., Morelli, G., Chua, N.-H. 1985. *Mol. Cell. Biol.* 5:1370–78
130. Karlin-Neumann, G. A., Kohorn, B. D., Thornber, J. P., Tobin, E. M. 1985. *J. Mol. Appl. Genet.* 3:45–61
131. Gilbert, W. 1978. *Nature* 271:501
132. Mount, S. M. 1982. *Nucleic Acids Res.* 10:459–72
133. Smeekens, S., de Groot, M., van Binsbergen, J., Weisbeek, P. 1985. *Nature* 317:456–58
134. Heber, U., Heldt, H. W. 1981. *Ann. Rev. Plant Physiol.* 32:139–68
135. Keegstra, K., Werner-Washburne, M., Cline, K., Andrews, J. 1984. *J. Cell. Biochem.* 24:55–68
136. Douce, R., Block, M. A., Dorne, A.-J., Joyard, J. 1984. *Subcell. Biochem.* 10:1–84
137. Nikaido, H., Vaara, M. 1985. *Microbiol. Rev.* 49:1–32
138. Colombini, M. 1979. *Nature* 279:643–45
139. Flugge, U. I., Benz, R. 1984. *FEBS Lett.* 169:85–89
140. Joyard, J., Billecocq, A., Bartlett, S. G., Block, M. A., Chua, N.-H., Douce, R. 1983. *J. Biol. Chem.* 258:10000–6
141. Cline, K., Andrews, J., Mersey, B., Newcomb, E. H., Keegstra, K. 1981. *Proc. Natl. Acad. Sci. USA* 78:3595–99
142. Joyard, J., Grossman, A., Bartlett, S. G., Douce, R., Chua, N.-H. 1982. *J. Biol. Chem.* 257:1095–101
143. Block, M. A., Dorne, A. J., Joyard, J., Douce, R. 1983. *J. Biol. Chem.* 258:13273–80
144. Block, M. A., Dorne, A.-J., Joyard, J., Douce, R. 1983. *J. Biol. Chem.* 258:13281–86
145. Werner-Washburne, M., Cline, K., Keegstra, K. 1983. *Plant Physiol.* 73:569–75
146. Flugge, U. I., Heldt, H. W. 1981. *Biochim. Biophys. Acta* 638:296–304
147. Somerville, S. C., Somerville, C. R. 1985. *Plant Sci. Lett.* 37:217–22
148. Hackenbrock, C. R., Miller, K. J. 1975. *J. Cell Biol.* 65:615–30
149. Smit, J., Nikaido, N. L. 1978. *J. Bacteriol.* 135:687–702
150. Carde, J.-P., Joyard, J., Douce, R. 1982. *Biol. Cell* 44:315–24
151. Wanner, G., Formanek, H., Theimer, R. R. 1981. *Planta* 151:109–25
152. DiRienzo, J. M., Inouye, M. 1979. *Cell* 17:155–61
153. Flugge, U. I., Wessel, D. 1984. *FEBS Lett.* 168:255–59
154. Sagara, Y., Ito, A. 1982. *Biochem. Biophys. Res. Commun.* 109:1102–7
155. Gasser, S. M., Schatz, G. 1983. *J. Biol. Chem.* 258:3427–30
156. Freitag, H., Janes, M., Neupert, W. 1982. *Eur. J. Biochem.* 126:197–202
157. Shore, G. C., Power, F., Bendayan, M., Carignan, P. 1981. *J. Biol. Chem.* 256:8761–66
158. Cline, K., Werner-Washburne, M., Andrews, J., Keegstra, K. 1984. *Plant Physiol.* 75:675–78
159. Pfisterer, J., Lachmann, P., Kloppstech, K. 1982. *Eur. J. Biochem.* 126:143–48
160. Stocking, C. R., Franceschi, V. R. 1982. *Plant Physiol.* 70:1255–59
161. Hennig, B., Neupert, W. 1981. *Eur. J. Biochem.* 121:203–12
162. Zwizinski, C., Schleyer, M., Neupert, W. 1983. *J. Biol. Chem.* 258:4071–74
163. Reizman, H., Hay, R., Witte, C., Nelson, N., Schatz, G. 1983. *EMBO J.* 2:1113–18
164. Grossman, A., Bartlett, S., Chua, N.-H. 1980. *Nature* 285:625–28
165. Werner-Washburne, M., Keegstra, K. 1985. *Plant Physiol.* 78:221–27
166. Chen, L., Tai, P. C. 1985. *Proc. Natl. Acad. Sci. USA* 82:4384–88
167. Soll, J., Buchanan, B. B. 1983. *J. Biol. Chem.* 258:6686–89
168. Reid, G. A. 1985. *Curr. Top. Biomembr. Transp.* 34:295–336
169. Gasser, S. M., Daum, G., Schatz, G. 1982. *J. Biol. Chem.* 257:13034–41
170. Schleyer, M., Schmidt, B., Neupert, W. 1982. *Eur. J. Biochem.* 125:109–16
171. Walker, G. H., Izawa, S. 1979. *Plant Physiol.* 63:133–38
172. Von Heijne, G. 1985. *Curr. Top. Membr. Transp.* 24:151–79
173. Duffaude, G. D., Lehnhardt, S. K., March, P. E., Inouye, M. 1985. *Curr. Top. Membr. Transp.* 24:65–103
174. Chen, L., Rhoads, D., Tai, P. C. 1985. *J. Bacteriol.* 161:973–80
175. Inouye, S., Soberon, X., Franceschini, T., Nakamura, K., Itakura, K., Inouye, M. 1982. *Proc. Natl. Acad. Sci. USA* 79:3438–41
176. Copeland, B. R., Landick, R., Nazos, P. M., Oxender, D. L. 1984. *J. Cell. Biochem.* 24:345–56

177. Schatz, G., Butow, R. A. 1983. *Cell* 32:316–18
178. Freitag, H., Janes, M., Neupert, W. 1982. *Eur. J. Biochem.* 126:197–202
179. Walter, P., Gilmore, R., Blobel, G. 1984. *Cell* 38:5–8
180. Robinson, C., Ellis, R. J. 1984. *Eur. J. Biochem.* 142:337–42
181. Mishkind, M. L., Jensen, K. H., Branagan, A. J., Plumley, F. G., Schmidt, G. W. 1985. *Curr. Top. Plant Biochem. Physiol.* 4:34–50
182. Schmidt, G. W., Mishkind, M. L. 1983. *Proc. Natl. Acad. Sci. USA* 80:2632–36
183. Smith, S. M., Ellis, R. J. 1979. *Nature* 278:662–64
184. McAda, P. C., Douglas, M. G. 1982. *J. Biol. Chem.* 257:3177–82
185. Miura, S., Mori, M., Amaya, Y., Tatibana, M. 1982. *Eur. J. Biochem.* 122:641–47
186. Bohni, P. C., Daum, G., Schatz, G. 1983. *J. Biol. Chem.* 258:4937–43
187. Spencer, N. 1984. *J. Theor. Biol.* 107:405–15
188. Marks, D. B., Keller, B. J., Hoober, J. K. 1985. *Plant Physiol.* 79:108–13
189. Roy, H., Patterson, R., Jagendorf, A. T. 1976. *Arch. Biochem. Biophys.* 172:64–73
190. Gray, J. C., Kekwick, R. G. O. 1973. *FEBS Lett.* 38:67–69
191. Robinson, C., Ellis, R. J. 1984. *Eur. J. Biochem.* 142:343–46
192. Schmidt, R. J., Gillham, N. W., Boynton, J. E. 1985. *Mol. Cell. Biol.* 5:1093–99
193. Teintze, M., Slaughter, M., Weiss, H., Neupert, W. 1982. *J. Biol. Chem.* 257:10364–71
194. Gasser, S. M., Ohashi, A., Daum, G., Bohni, P. C., Gibson, J., et al. 1982. *Proc. Natl. Acad. Sci. USA* 79:267–71
195. Schmidt, B., Wachter, E., Sebald, W., Neupert, W. 1984. *Eur. J. Biochem.* 144:581–88
196. Ohashi, A., Gibson, J., Gregor, I., Schatz, G. 1982. *J. Biol. Chem.* 257:13042–47
197. Beams, H. W., Kessel, R. G., Shih, C. Y. 1979. *Biol. Cell.* 35:87–96
198. Laetsch, W. M. 1974. *Ann. Rev. Plant Physiol.* 25:27–52
199. Tremolieres, A., Dubacq, J. P., Ambard-Bretteville, F., Remy, R. 1981. *FEBS Lett.* 130:27–31
200. Schwitzguebel, J. P., Nguyen, T. D., Siegenthaler, P. 1984. In *Function and Metabolism of Plant Lipids,* ed. P. A. Siegenthaler, W. Eichenberger, pp. 299–302. Amsterdam: Elsevier
201. Verkleij, A. J. 1982. *Biochim. Biophys. Acta* 779:43–63
202. Gruner, S. M. 1985. *Proc. Natl. Acad. Sci. USA* 82:3665–69
203. Sprague, S. G., Staehelin, L. A. 1984. *Plant Physiol.* 502–4
204. Harpster, M. H., Mayfield, S. P., Taylor, W. C. 1984. *Plant Mol. Biol.* 3:59–71
205. Mayfield, S. P., Taylor, W. C. 1984. *Eur. J. Biochem.* 144:79–84
206. Mishkind, M. L., Greer, K. S., Schmidt, G. W. 1986. *Methods Enzymol.* In press
207. Ohta, S., Schatz, G. 1984. *EMBO J.* 3:651–57
208. Firgaira, F. A., Hendrick, J. P., Dalousek, F., Draus, J. P., Rosenberg, L. E. 1984. *Science* 226:1319–22
209. Schmidt, B., Henning, B., Zimmermann, R., Neupert, W. 1983. *J. Cell Biol.* 96:248–55
210. Zimmermann, R., Neupert, W. 1980. *Eur. J. Biochem.* 109:217–29
211. Miura, S., Mori, M., Amaya, Y., Tatibana, M., Cohen, P. P. 1981. *Biochem. Int.* 2:302–12
212. Ades, I. Z. 1983. *Biochem. Biophys. Res. Commun.* 110:42–47
213. Yamaoto, M., Watanabe, N., Kikuchi, G. 1983. *Biochem. Biophys. Res. Commun.* 115:700–6
214. Johanningmeier, U., Howell, S. H. 1984. *J. Biol. Chem.* 259:13541–49
215. Fuesler, T. P., Castelfranco, P. A., Wong, Y.-S. 1984. *Plant Physiol.* 74:928–33
216. Crawford, M. S., Wang, W.-Y., Jensen, K. G. 1982. *Mol. Gen. Genet.* 188:1–6
217. Yaffe, M. P., Schatz, G. 1984. *Proc. Natl. Acad. Sci. USA* 81:4819–23

Ann. Rev. Biochem. 1986. 55:913–51

METALLOTHIONEIN[1],[2]

Dean H. Hamer

Laboratory of Biochemistry, National Cancer Institute, National Institutes of Health, Bethesda, Maryland 20205

CONTENTS

PERSPECTIVES AND SUMMARY

The heavy metals zinc, copper, cadmium, silver, mercury, gold, nickel, and cobalt are minor yet ubiquitous components of the biosphere. Zinc and copper,

---

[1]The US Government has the right to retain a nonexclusive royalty-free license in and to any copyright covering this paper.

[2]Abbreviations used: MT, metallothionein; bp, base pairs; Kb, Kilo base pairs; tk, thymidine kinase; SV40, simian virus 40; BPV, bovine papilloma virus; cDNA, complimentary DNA.

which participate in a variety of enzymatic reactions, are essential trace nutrients for all life forms but are toxic when present at inappropriately high concentrations. The other ions serve no known essential function, are present in lower and more variable concentrations in the environment, and are generally more potent toxins. We know remarkably little about how living organisms handle and utilize these metals. None of the molecules involved in membrane transport, initial ion liganding, control of the valence state, or activation of metalloenzymes have been identified. The initial targets for heavy metal poisoning are also unknown. In fact, the only molecule for which a clear role in intracellular metal metabolism can be ascribed is the small, cysteine-rich metal binding protein known as metallothionein (MT).

MT was discovered as a cadmium and zinc protein in horse kidney (1, 2). Subsequently, similar molecules have been identified in a broad range of eukaryotic species and in many different tissues and cell types. MTs bind to heavy metals through clusters of thiolate bonds, and the strength of binding can vary by as much as six orders of magnitude depending on the ion. The synthesis of MT is homeostatically regulated in cells and organisms exposed to heavy metals. This is now known to result from increases in the rate of MT gene transcription mediated through interactions between upstream regulatory DNA sequences and unidentified cellular factors. MT gene transcription is also induced, in some mammalian cells, by glucocorticoids, interferon, and stress conditions. At more complex levels, MT expression can be altered by changes in gene structure, such as amplification and methylation, and by cellular differentiation and development.

The function of MT has been debated ever since its discovery. A role in metal metabolism or detoxification is strongly suggested by the ability of MT to both bind to and be induced by heavy metal ions. A large body of corroborative evidence for such roles has been gathered, and recently has it been possible to directly test their validity through gene transfer and replacement experiments. Other possible functions suggested for MT, but still untested, include control of the intracellular redox potential, activated oxygen detoxification, or sulfur metabolism.

Three major groups of scientists have contributed to our knowledge of MT: physiologists and toxicologists interested in its role in heavy metal metabolism and detoxification, protein chemists and spectrocopists intrigued by its unusual structural features, and most recently, molecular biologists interested in gene regulation and the use of MT promoter sequences for genetic engineering experiments. Because I belong to the last group, this review will stress recent studies of MT at the molecular biological level. Work on the occurrence, structure, and physiology of MT is also described, but in summary rather than comprehensive form. Two symposium volumes deal with the more classic aspects of MT research in detail (3, 4).

NOMENCLATURE, OCCURRENCE, AND DETECTION

Because of the lack of a known enzymatic function, MTs have traditionally been classified according to their structural features (3). The following characteristics are hallmarks of MT:

1. high content of heavy metals (typically 4–12 atoms/mole) bound exclusively by clusters of thiolate bonds;
2. high content of cysteine (typically 23–33 mole%) and paucity of aromatic and hydrophobic amino acid residues;
3. low molecular weight (typically less than 10,000);
4. demonstrated structural or functional homology to mammalian MT.

All vertebrates examined contain two or more distinct MT isoforms which are grouped into two classes, designated MT-I and MT-II, depending on the elution position from DEAE-cellulose. In many cases each class actually consists of several different isoproteins which are designated $MT-I_A$, $MT-I_B$, $MT-I_C$, etc. Classification of proteins as members of the MT-I or MT-II class is based solely on chromatographic behavior and does not necessarily imply structural or functional homology, especially when comparing proteins from divergent species such as rodents versus primates.

Table 1 lists the organisms and organs in which MT and MT-like proteins have been demonstrated. The current list contains a wide variety of eukaryotic

Table 1 Occurrence of MT and MT-like proteins

Species	Organs	General references[a]	Sequence references[b]
man	liver, kidney, heart, testis, cultured cells	5, 6	7–11
monkey	kidney, cultured cells	12	13
horse	liver, kidney, intestine	14	15
cow	liver, duodenum	16, 17	
sheep	liver, kidney	18	19
pig	liver, kidney	20–22	
dog	liver, kidney, spleen	23, 24	
rabbit	liver, kidney	25, 26	27
hamster	liver, kidney, cultured cells	28	29
rat	liver, kidney, intestine, spleen, pancreas, placenta, testis, brain	30–32	33–35
mouse	liver, kidney, pancreas, placenta, testis, blood, cultured cells	36	37a–39
seal	liver, kidney	40a	

[a]Because of space limitations, it was impossible to include an exhaustive list of references. More complete citations can be found in (3).
[b]Both amino acid and DNA sequences are referenced.

Table 1   *(continued)*

Species	Organs	General references[a]	Sequence references[b]
chicken	liver	40b	
duck	liver, kidney	40c	
quail	liver, kidney	40d	
frog	liver, kidney	41, 42	
salamander	liver	43	
water lizard	liver	44	
earthworm	soft tissue	45	
fruit fly	larva	47	
fleshfly	larva	48	
stonefly	larva	49	
cockroach	ileum	50	
midge	larva	51	
sea urchin	embryo	52	53
crab	hepatopancreas	54–56	57
plaice	liver	59	59
flounder	liver	60	
skipjack	liver	61	
goldfish	liver, kidney	62	
trout	liver	63	
carp	liver	64	
eel	liver, gills	65	
limpet	soft tissue	66	
oyster	soft tissue	67, 68	
mussel	mantle	68–71	
grass	root	72, 73	
cabbage	leaves, stems, roots	74	
tobacco	leaves	75	
tomato	root	76	
Ochromonas dancia		77	
Neurospora crassa		78	78
Saccharomyces cerevisiae		79a, b	80–82

[a]Because of space limitations, it was impossible to include an exhaustive list of references. More complete citations can be found in (3).
[b]Both amino acid and DNA sequences are referenced.

species including vertebrates, invertebrates, plants, and microorganisms. Most studies have focused on liver and kidney, the primary sites of heavy metal accumulation, but MT can also be detected in many other organs and in various cultured cell types. Small metal-binding proteins have also been reported in prokaryotes (83), but detailed analysis of the protein from *Pseudomonas putida* suggests a complex metal binding site that contains histidine and glutamine as well as thiolate ligands (84).

MT is usually detected by virtue of its high content of metals (detected by atomic absorption spectrophotometry or labeling with radionuclides) or of cysteine (detected by ^{35}S-labeling). Commonly used separation methods include gel filtration (3, 4), reversed phase high pressure liquid chromatography (6), and gel electrophoresis (85). Polyclonal and monoclonal antibodies have been raised against several mammalian MTs and can be used in immunodiffusion, immunoelectrophoresis, and radioimmunoassays (86–91). These methods are particularly useful for detecting low levels of MT (92) and for subcellular localization studies by immunofluorescence (93–95).

STRUCTURE AND HEAVY METAL BINDING

The structure of MT has been studied by a variety of biophysical and biochemical techniques including UV, CD, ESR, and NMR spectroscopy, amino acid sequencing and partial proteolysis and, most recently, X-ray crystallography. The experiments leading to our present view of MT structure are reviewed in detail elsewhere (6).

Mammalian MT is a 61- or 62-amino-acid peptide containing 20 cysteines, 6–8 lysines, 7–10 serines, a single acetylated methionine at the amino terminus, and no aromatics or histidines. A consensus sequence, derived by comparison of 14 molecules for which protein or DNA data are available, is shown in Figure 1 *(top)*. Note that the majority of cysteine residues are present in Cys-X-Cys and Cys-Cys sequences. The metal content of purified MT is highly variable and depends on organism, tissue, and history of heavy metal exposure. For example, MT isolated from human liver autopsy samples contains almost exclusively zinc whereas MT from kidney contains substantial levels of cadmium and copper. These differences probably reflect both the natural heavy metal exposure of the organs and the expression of different MT isoforms (6). MT isolated from cells or organisms that have been experimentally exposed to inducing levels of heavy metals contain predominantly, but not exclusively, the administered metal. For example, MT-II isolated from the liver of cadmium-treated rats contains 5 atoms of cadmium and 2 of zinc (96).

Metals are associated with MT exclusively through thiolate bonds to all 20 cysteine residues. The metals can be removed by exposure to low pH and the resulting apothionein can be reconstituted with 7 atoms of cadmium or zinc or 12 atoms of copper. Overall stability constants, estimated either by pH titration or by a ligand substitution method, range from 10^{19} to 10^{17} for copper, 10^{17} to 10^{15} for cadmium, and 10^{14} to 10^{11} for zinc (97, 98; D. Petering, personal communication). Mammalian MT can also bind to 7 atoms of mercury, cobalt, lead, and nickel or to 10–12 atoms of silver and gold (6, 99).

The metals in MT are contained in two distinct, polynuclear clusters whose existence was initially inferred from [113]Cd-NMR studies (100a,b). The A cluster contains 11 cysteines, binds four atoms of zinc or cadmium or five to six

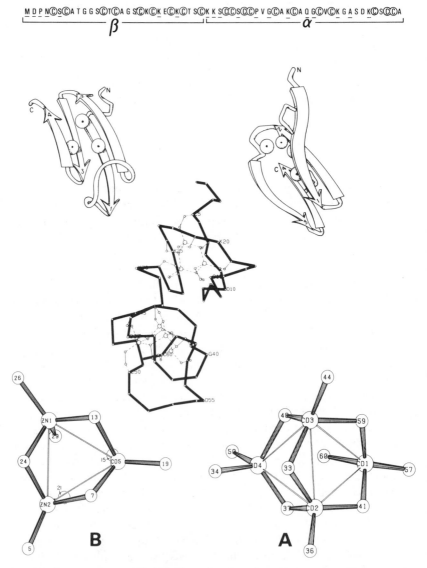

M D P N©S©A T G G S©T©A G S©K©K E©K©T S©K K S©©S©©P V G©A K©A Q G©V©K G A S D K©S©©A

β ————————————————————————— α

Figure 1   Structure of mammalian MT. The top line shows the consensus amino acid sequence of mammalian MT (see Table 1 for sequence references). The invariant cysteines are circled and other invariant residues are underlined. The rest of the figure shows the structure of rat cadmium, zinc-MT-II as determined by X-ray crystallography (W. Furey, A. Robbins, L. Clancy, D. Winge, B. Wang, C. Stout, manuscript submitted). *Top,* protein folding patterns in the β and α domains. *Middle,* overall structure. *Bottom,* coordination chemistry of the B and A clusters.

atoms of copper, and is contained within the carboxy-terminal α domain extending from amino acid 31 to 61. The B cluster contains nine cysteines, binds four atoms of zinc or cadmium or six atoms of copper, and is contained in the amino-terminal β cluster extending from amino acid 1 to 30 (101). Figure 1 *(bottom)* shows the cysteine sulfur coordination structures of the two metal clusters in rat cadmium, zinc-MT-II as deduced by X-ray crystallography at 2.3 Å resolution (W. Furey, A. Robbins, L. Clancy, D. Winge, B. Wang, C. Stout, manuscript submitted). All metal ions are tetrahedrally coordinated to four cysteine thiolate ligands. The A cluster contains four cadmium atoms of which two are bonded by three bridging and one terminal sulfurs and the other two by two bridging and two terminal sulfurs. The four cadmium atoms are arranged as a distorted tetrahedron or "butterfly" embedded within two overlapping six-atom rings. The B cluster contains two zinc and one cadmium atoms, all of which are bonded by two bridging and two terminal sulfurs. The metal ions form an equilateral triangle within a six-atom ring that adapts a chair configuration. The metal-metal distances in the two clusters range from 3.9 to 5.2 Å. The coordination chemistry of the metal clusters shown in Figure 1 is in agreement with an earlier model based on [113]Cd-NMR data (100a), but the actual assignments of the cysteine-metal contacts are quite different. [113]Cd-NMR has also been used to study the possibility that MT in low-ionic-strength solutions exists as multiple isomers (102).

The tentative overall X-ray structure of cadmium, zinc-MT-II has dramatically confirmed the two-domain nature of the protein (Figure 1, *center;* W. Furey et al, manuscript submitted). Both domains are globular, with diameters of 15–20 Å, and are linked by residues 30 and 31 to form a prolate ellipsoid. The protein-folding patterns in the α and β domains are topologically similar, but of opposite chirality, and are characterized by a high proportion of reverse turns. In both clusters, the polypeptide chain makes three turns to spiral around the metal atoms. Hence the structure is similar to a β sandwich except that the normal amino acid hydrogen bonds are replaced by sulfur-metal bonds. The highly disordered structure of apothionein (6) is expected from this model. Hydrophilic residues are clustered in two exposed loops (residues 55–56 in α and 8–12 in β). The majority of adjacent and closely spaced cysteine residues in the molecule are coordinated to a common metal atom, most frequently through one terminal and one bridging sulfer. Contacts between the two domains are largely limited to the α-β interface. The paucity of interdomain contacts is in good agreement with the observation that the separated α and β domains bind metals with the same stoichiometry as the intact molecule (103).

While the detailed structure of mammalian copper-MT has not yet been solved, it is likely to be quite different from that of the cadmium, zinc form as copper binds in the +1 rather than +2 valence state and the binding stoichiometry is 11–12 atoms of copper compared to 7 atoms of cadmium or zinc. This suggests that binding is primarily trigonal, rather than tetrahedral, which would in turn alter the folding and tertiary structure of the protein.

The order and cooperativity of metal binding to the two clusters has been investigated by taking advantage of the fact that MT and its metal-filled domains are much more resistant to proteases than are apothionein or unsaturated domains (101, 104, 105). It was found that zinc and cadmium first fill the A cluster then the B cluster whereas copper first fills the B cluster then the A cluster. The binding reaction is highly cooperative. For example, incubation of 1 mole of apothionein with 1 mole of cadmium gives a 25% (i.e. stoichiometric) yield of molecules with a fully saturated four-metal A cluster and no evidence for partially filled species. It is unclear whether the preferential and cooperative binding to the two clusters reflects kinetic or thermodynamic factors.

The copper MTs isolated from two fungi display interesting structural similarities and contrasts to mammalian MT. *Neurospora crassa* MT contains only 25 amino acids including seven cysteines that can be precisely aligned with those in the amino-terminal copper-binding domain of mammalian MT (78). The protein binds six copper atoms, suggestive of a linear cluster with five bridge and two terminal sulfurs, or in vitro to three atoms of zinc, cadmium, mercury, cobalt, or nickel (106). *Saccharomyces cerevisiae* MT, purified from a genetically selected overproducing strain, consists of 53 amino acids and contains eight copper atoms ligated to 12 cysteine residues (82). Surprisingly, the purified protein lacks the eight amino-terminal residues predicted by the DNA sequence of the gene. The removed sequence is unusual for MT in that it contains four hydrophobic residues including two aromatics. There is little obvious conservation of the positions of the cysteines or of the primary structure of yeast and mammalian MT except for the hexapeptide Lys-Lys-Ser-Cys-Cys-Ser. The binding stoichiometry of 12 cysteines to eight copper atoms suggests an octahedral cluster structure involving exclusively bridge sulfurs. The yeast protein can also bind in vitro to eight atoms of silver or to four atoms of zinc or cadmium. Hence the lower eukaryotic MTs resemble mammalian MT in their ability to bind group 1b and 2b metals in distinct configurations.

The possibility that MT is involved in transfer of metal ions to apometalloenzymes has led to several studies of MT metal exchange reactions. Mammalian zinc-MT can reactivate various zinc-dependent enzymes including carbonic anhydrase, aldolase, thermolysin, and alkaline phosphatase with approximately the same rates as inorganic zinc salts (98, 107a). The transfer of copper to the binuclear copper proteins tyrosinase and hemocyanin has been studied with *Neurospora* copper-MT (107b). The freshly isolated protein, containing Cu(I)-thiolate complexes, was incompetent for copper transfer but activity could be demonstrated in a derivative obtained by air oxidation followed by dithionite reduction. Small ligands, such as EDTA and nitrilotriacetate, and synthetic

MT-related oligopeptides have also been used as model compounds to study MT metal exchange and binding reactions (98, 107c; D. Petering, personal communication).

In summary, the structure of mammalian cadmium and zinc MT is now well understood but the structure of the copper forms and the possible role of MT in metal exchange reactions requires further study. Two major conclusions of the structural work are that MT is exquisitely designed to bind heavy metals in a tight and cooperative fashion and that the protein can adopt varied conformations in response to different classes of ions.

GENE STRUCTURE, ORGANIZATION, AND EVOLUTION

Molecular cloning is an important tool for understanding the regulation and function of MT. Cloned genes can be used as probes to determine gene copy number and mRNA levels, as substrates for mutagenesis to determine structure-function relationships, and as a means to understand the physiological role of MT through gene replacement and transduction experiments.

Cloned MT Genes

MAMMALIAN GENES   Complimentary DNA (cDNA) clones have been isolated for mouse MT-I (108), human MT-II (109), monkey MT-I and MT-II (13), chinese hamster MT-I and MT-II (29), rat MT-I (35), and sheep MT-I (19). Positive clones were identified by hybrid selected translation, "plus-minus" screening with probes from uninduced versus induced cells, or hybridization with previously cloned probes. The DNA sequences of the cloned MT cDNAs are colinear with the amino acid sequences of purified MT polypeptides, and codon utilization is nonrandom with a strong preference for C in the third position. As predicted from the conservation of mammalian MT protein structure, the coding regions of the cloned cDNAs are strongly homologous. The noncoding sequences are more divergent, although short homologous regions can be detected in both the 5' and 3' untranslated regions.

Genomic or "natural" MT genes have been cloned and sequenced from bacteriophage λ libraries of mouse and human DNA. To date, six functional genes have been reported: mouse MT-I (38, 108), mouse MT-II (39), human MT-II$_A$ (9), human MT-I$_A$ (10), human MT-I$_E$, and human MT-I$_F$ (11). Five human pseudogenes have also been described: ψMT-II$_B$ (9), ψMT-I$_C$ and ψMT-I$_D$ (10), ψMT-I$_G$ (ψMT-I in Ref. 11), and ψMT-I$_H$ (ψMT-I in Ref. 110). The criteria that have been used to distinguish functional from nonfunctional genes include ability to encode a typical MT polypeptide as deduced from the DNA sequence, ability to render transfected cells resistant to heavy metal toxicity, and expression in cultured cells or whole organisms as determined by hybridization to gene-specific probes derived from the 5' or 3' untranslated regions.

The functional mouse and human MT genes share a similar tripartite structure in which two introns interrupt three exons at precisely homologous positions (9–11, 38, 39). Exon 1 encodes the 5' untranslated region and amino acids 1-8-⅔, exon 2 encodes amino acids 9–31-⅓, and exon 3 encodes amino acids 31-⅔–62 and the 3' untranslated region. Intron lengths range from 250–590 bp for intron 1 to 140–350 bp for intron 2. The intron/exon boundaries conform to the GT . . .AG splicing rule (111), and all of the genes contain the typical polyadenylation signal AATAAA in their 3' untranslated regions. The mouse and human genes share little obvious homology in their intron and flanking sequences except in the promoter region.

The human ψMT-I$_G$ pseudogene (11) is similar to the functional MT genes in its tripartite structure but contains several mutations that would be expected to render the gene nonfunctional: an in phase termination codon at amino acid 40, replacement of the conserved cysteine at position 5 with tyrosine and of the conserved serine at position 35 with phenylalanine, and an alteration in the TATA box sequence in the 5' flanking region. Because ψMT-I$_G$ retains substantial homology to the functional human MT-I genes in its 5' and 3' flanking regions, it is likely that this gene has been inactivated by an accumulation of point mutations rather than a displacement in the genome. The ψMT-I$_C$ and ψMT-I$_D$ pseudogenes also retain introns and have in frame termination codons and/or deletions that would affect the reading frame (10). In contrast, the human ψMT-I$_H$ (110) and ψMT-II$_B$ (9) pseudogenes represent examples of processed genes in which no introns interrupt the coding sequences. Both genes terminate in poly(A) stretches, display no obvious homology to functional human MT genes in their 5' or 3' flanking regions, and are flanked by short direct repeats thought to represent duplications of the insertion site. These structural features are consistent with the notion that ψMT-II$_B$ and ψMT-I$_H$ arose by the insertion of reverse transcripts of MT mRNA into the genome of germ line cells (112).

The organization of rodent and primate MT gene families has been studied by gel transfer hybridization, cloning, and chromosomal localization experiments. Mice, which synthesize only two MT isoforms (37a, b), appear to contain only two functional MT genes, MT-I and MT-II. These genes are tightly linked with the MT-I gene lying 6 Kb downstream of the MT-II gene and oriented in the same direction of transcription (39). The mouse MT-I (and by inference MT-II) genes have been assigned to chromosome 8 by an analysis of somatic cell hybrids segregating mouse chromosomes (113). Rats are also thought to synthesize only two MT isoforms (33, 34), but gel transfer hybridization experiments with an MT-I cDNA probe demonstrated at least four MT-related sequences in the genome (35). Single-copy MT-I and MT-II genes, together with several weakly hybridizing MT-related sequences, have been identified in Chinese hamster. The MT-I and MT-II genes are located on chromosome 3 and are probably closely linked (114, 115).

The MT gene families of primates are more complex than those of rodents. Gel transfer hybridization of EcoRI-digested human DNA revealed a complex pattern of 12–14 bands, several of which contained more than one gene (9, 11). The MT-I_A, ψMT-I_D, ψMT-I_C, and MT-I_B genes are closely linked, in that order from 5' to 3', and are transcribed in the same direction (9). Physical linkage for the other human MT genes has not been established. The chromosomal localization of the human MT genes and pseudogenes was investigated by analysis of the DNA from human-mouse and human–Chinese hamster somatic cell hybrids (116). These experiments showed that the functional genes MT-II_A, MT-I_E, and MT-I_F and the nonprocessed pseudogene ψMT-I_G are linked on chromosome 16 between phosphoglycolate phosphatase and di-aphorase-4, probably on the proximal portion of the long arm (16 qcen16q21). Subsequent studies, utilizing transcript mapping of human-mouse hybrids, confirmed the chromosome 16 location of MT-II_A and showed that MT-I_A is also in this locus (117a). Hence, all of the known functional human MT genes are chromosomally linked. The remaining MT-related sequences are dispersed to at least four other autosomal sites (116). The ψMT-II_B processed pseudogene, which can exist as either of two allelic variants distinguishable by EcoRI digestion (9, 117b), lies on chromosome 4 in the p11-q21 region (116, 117c). Chromosomes 20 and 18 each contain a single locus while chromosome 1 contains two distinct loci, one within the distal two thirds of the short arm and the other probably on the long arm (116).

In summary, the mammalian MTs exhibit many of the features characteristic of multigene families in higher eukaryotes including the interruption of the coding sequences by introns, the chromosomal linkage of functional and nonprocessed pseudogenes, and the presence of processed pseudogenes dispersed throughout the genome. It remains to be seen whether the simplicity of the mouse family or the complexity of the human family is most characteristic.

Recently MT sequences have also been cloned from several nonmammalian sources. A sea urchin MT cDNA clone was obtained by using the RNA of zinc-animalized embryos to screen a library derived from normal mesenchyne blastula RNA (52). A cDNA for *Drosophila melanogaster* MT was identified by plus-minus screening of a library from copper-induced larvae (47). The MT gene from a lower eukaryotic, the yeast *Saccharomyces cerevisiae,* was cloned by virtue of its ability to render sensitive cells resistant to copper poisoning (118).

Evolution

The similar tripartite structure of human and mouse genomic MT genes suggests that they are common products of an ancestral gene that arose prior to the rodent-mammalian radiation. It is particularly interesting that the junction between exons 2 and 3 falls precisely at the boundary of the α and β domains, supporting the theory that exons correspond to protein domains. The coding sequences of the mammalian MTs are strongly conserved. For example, monkey MT-II and human MT-II$_a$ differ by 1.2% replacement and 4.8% silent changes whereas rat MT-I and human MT-I$_F$ differ by 9.9% replacement and 28% silent changes (11). A peculiarity of the mammalian MT genes is the lack of a clear relationship between the MT-I and MT-II classes in divergent species. Within primates, the MT-I genes of man and monkey are more closely related to one another than to the MT-II genes of either species. Similarly, within rodents the mouse, rat, and hamster MT-I genes are more closely related to one another than to the MT-II genes of these organisms. However, a comparison of primate to rodent MT genes shows that the primate MT-Is are less divergent from primate MT-IIs than from rodent MT-Is (11). One possible explanation of this paradox is that both rodents and primates independently acquired two MT isoform genes. A second possibility is that the two isoform genes arose prior to the divergence of rodents and primates but that early in primate evolution a gene conversion event made the two isoforms more similar. Finally, it is possible that the two isoforms arose prior to the divergence of primates and rodents but in primates the MT-I gene was inactivated and the functional genes all evolved from MT-II.

The MTs of *Drosophila* (47), plaice (59), crab (57), and *Neurospora* (78) are noticeably homologous to the mammalian MTs, and it has been proposed that these proteins be designated "class 1" MTs (6). In particular, the positions of the cysteine residues in these proteins is almost perfectly conserved. The main differences are in length; the two crab MTs contain 57 and 58 amino acids and correspond to the α plus β domains with several small deletions and insertions; *Drosophila* MT consists of 40 amino acids and corresponds to the β domain with an insertion; and *Neurospora* MT contains just 25 amino acids homologous to the β domain. A comparison of these proteins to mammalian MT gives an evolutionary rate of 7×10^{-10} substitutions/codon/year, in good agreement with the estimate from the mammalian DNA sequence data (6). This rate falls between the previously estimated rates for cytochrome c and globin.

The MTs of sea urchin (53) and yeast (82) are less obviously homologous to mammalian MT. Although both contain a high proportion of Cys-X-Cys and Cys-Cys sequences, there is no obvious linear alignment between the positions of these residues in the different molecules. It has been proposed that the carboxy-terminal half of sea urchin MT corresponds to the amino-terminal β domain of mammalian MT and, conversely, that the amino-terminal half of sea urchin corresponds to an α-like domain. A "central segment" in common between the sea urchin, yeast, and class 1 MTs has also been postulated (53). Further sampling of taxonomic groups will be required to test the validity of this modular divergent evolution model.

GENE AMPLIFICATION AND METHYLATION

Genetically selected and naturally occurring lines of both cultured mammalian cells and yeast exhibit substantial variations in MT synthesis and heavy metal resistance. Experiments utilizing cloned MT gene hybridization probes have demonstrated that such differences can be caused by changes in the copy number or methylation status of MT genes.

Amplification in Mammalian Cells

Stable cadmium-resistant lines of cultured mammalian cells can readily be isolated by continuous exposure of the cells to stepwise increases of cadmium in the culture medium (119, 120). Analysis of a cadmium-resistant mouse Friend leukemia cell line revealed 14-fold more stable MT-I mRNA, a sixfold higher rate of MT-I gene transcription, and a sixfold increase in the number of MT-I genes (121). The largest amplified MT-I DNA fragment observed was 55 Kb, and thus included the closely linked MT-II gene (39). The cadmium-resistant cells exhibited chromosomal aneuploidy and the presence of three small chromosomes absent from the parental line.

MT gene amplification has also been studied in Chinese hamster ovary cells, a line that is unusually sensitive to poisoning by low concentrations of cadmium. Lines resistant to successively higher cadmium concentrations exhibited coordinate 3- to 60-fold amplification of both the MT-I and MT-II genes and accumulated increased levels of MT-I and MT-II mRNA and polypeptides (114, 122). Cytogenic analyses of the highly cadmium-resistant variants consistently revealed breakages and rearrangements involving chromosome 3p, the locus of the MT-I and MT-II genes (115).

The mechanism of MT gene amplification is unknown. One hypothesis, suggested by the association between amplification and chromosomal aneuploidy and rearrangements, is that the first event is a random break or breaks in the chromosomal DNA around the MT locus. This could liberate MT gene fragments capable of autonomous replication or place the MT genes between DNA sequences capable of mispairing and unequal sister chromatid exchange. Fluctuation tests on cultured mouse cells have shown that the acquisition of low-level cadmium resistance is a random rather than induced event (J. P. Thirion, personal communication). It has also been shown that tumor-promoting agents can increase the frequency of the appearance of cadmium-resistant clones containing amplified MT genes (123). The maximal amplification levels for MT genes are considerably lower than for the dihydrofolate reductase gene (124), suggesting the possibility of a rate-limiting factor (cysteinyl tRNA?) for MT synthesis.

Methylation in Mammalian Cells

Most cultured mammalian cells synthesize MT and are resistant to moderate levels of heavy metal ions, but two exceptions are the W7 line of mouse thymoma cells and Chinese hamster ovary cells. The postulated inverse correlation between the extent of DNA methylation and higher eukaryotic gene expression (126) led to investigations of MT gene methylation in these lines.

Mouse W7 cells contain a normal MT-I gene, as determined by DNA gel transfer hybridization, yet synthesize no detectable MT-I mRNA and are sensitive to poisoning by low levels of cadmium (127). Treatment of exponentially growing W7 cells with 5-azacytidine, a cytidine analogue that is incorporated into nucleic acids but cannot be methylated, led to dramatic increases in both the basal and cadmium-induced levels of MT-I mRNA (128). The effect of 5-azacytidine was maximal within 8 hours, persisted for as long as 310 hours, and could be blocked by hydroxyurea, an inhibitor of DNA replication. Moreover, lines resistant to cadmium (10 μM) could be derived from mass cultures treated with the methylation inhibitor. The methylation status of the MT-I gene was determined by gel transfer hybridization analysis of chromosomal DNA digested with the methylation-sensitive restriction endonuclease enzymes HpaII versus MspI. Six HpaII sites in the vicinity of the mouse MT-I gene were heavily methylated in the parental W7 cells but not in cadmium-resistant lines derived by 5-azacytidine treatment. A low degree of MT-I gene amplification ($\sim$2-fold) was also observed in the resistant cells. These results were interpreted according to a model in which 5-azacytidine inhibits DNA methylation during the first round of DNA replication, thereby generating hemimethylated MT genes that are capable of normal transcription and induction. A second round of replication would generate 50% fully methylated inactive genes and 50% fully unmethylated active genes, the latter being capable of stable inheritance.

Analogous results have been obtained with Chinese hamster ovary cells. Lines resistant to moderate levels of cadmium (2 μM) show no obvious gene amplification (114) but are hypomethylated at both the MT-I and MT-II loci. Also, the frequency of moderately cadmium-resistant lines can be increased by 5-azacytidine treatment (C. Hildebrand, personal communication).

The mechanism by which methylation of C residues inhibits MT gene expression is unclear. Possibilities include interference with the binding of RNA polymerase or regulatory factors, alterations in nucleosome phasing, or changes in the higher-order packaging of chromatin. It is also unclear why or when W7 and Chinese hamster ovary cell MT genes became hypermethylated, since other thymoma and ovary cell lines exhibit normal MT gene expression.

Amplification in Yeast

The *Saccharomyces cerevisiae* copper-MT gene, *CUP1,* has provided an especially useful model for studying the mechanisms of MT gene amplification. *CUP1* was originally described as a Mendelianly inherited locus conferring resistance ($CUP1^R$) or sensitivity ($cup1^S$) to copper poisoning (129). $CUP1^R$ strains contain ten or more tandem copies of a 2-Kb segment of DNA that includes the copper-MT coding sequences, as well as a long open reading frame of unknown function, whereas $cup1^S$ strains contain only one copy of the reiteration unit (80, 118). Stepwise selection for $CUP1^R$ variants exhibiting increased copper resistance resulted in further increases in the $CUP1$ gene copy number, either through tandem amplification or chromosomal disomy (131). Amplification of the $CUP1$ locus has also been demonstrated in industrial yeast strains (132).

The amplification of the $CUP1$ gene in yeast is currently envisioned to involve two steps: an initial duplication of the repeat unit, followed by amplification (and contraction) events such as unequal crossing over, sister chromatid exchange, or gene conversion. The mechanism of the first step is unknown. An obvious possibility is recombination between related DNA sequences at the beginning and end of the repeat unit, but sequence analysis failed to reveal any obvious homologies at the expected positions (80, 81; J. Welch, personal communication). Moreover, attempts to experimentally observe MT gene amplification in single-copy $cup1^S$ strains have been unsuccessful (J. Welch, personal communication; D. Hamer, M. Walling, unpublished results).

The mechanism of the second step, amplification and contraction of a previously reiterated unit, has been approached by an elegant genetic analysis of $CUP1^R/CUP1^R$ and $CUP1^R/cup1^S$ diploids heterozygous at various linked and unlinked loci (135). A total of 136 random tetrads, each consisting of four meiotic progeny, were analyzed for $CUP1$ copy number by gel transfer hybridization. Remarkably, 17.6% of these showed aberrant segregations, the majority of which were nonreciprocal in character. This result indicates that gene conversion, rather than unequal crossing over or sister chromatid exchange, is the major driving force behind meiotic MT gene amplification in yeast. A model invoking misaligned synaptic pairing, a symmetric single-strand information transfer, and loops of unpaired nonhomologous sequences at the termini of the repeat unit was suggested to account for the high frequency of non-Mendelian, nonreciprocal segregation events.

TRANSCRIPTIONAL REGULATION

MT gene expression is rapidly and transiently induced by exposure to heavy metals and, in mammalian cells, various circulating factors such as hormones and interferon. In considering the mechanisms of these responses, two questions are paramount: what are the *cis*-acting control DNA sequences and what *trans*-acting cellular factors interact with them?

Gene Transfer Systems

The reintroduction of cloned MT genes into the cells and organisms from which they were derived has proven a critical methodology for studying *cis*-acting control sequences. Many of these experiments have utilized hybrid genes in which MT transcriptional regulatory sequences drive the expression of heterologous coding sequences. This facilitates distinction between expression of the reintroduced and endogenous genes.

STABLE TRANSFECTION   DNA can be stably introduced into cultured mammalian cells by cotransfection with linked or unlinked selectable marker genes (137). This approach has been used to study the expression of the mouse MT-I gene in human HeLa or mouse L cells (136) and of the human MT-I_A and MT-II_A genes in rat 2 cells (10, 138). Typically the reintroduced gene is present in 1 to 50 or more tandemly arranged copies integrated into a single but apparently random chromosomal site. In the case of the mouse MT-I gene, 75% of the transformants expressed MT mRNA with cadmium induction ratios ranging from 2- to 5-fold as compared to 10- to 20-fold for the endogenous genes; induction by dexamethasone was not detected (136). In the case of the human genes, roughly 60% of the clones were cadmium inducible and approximately half of these also responded to dexamethasone (10, 138). While an obvious advantage of stable transfection is that the cell lines can be studied over a long period of time, an important liability is the high degree of clone-to-clone variability in copy number, expression levels, and inducibility.

INFECTION AND ACUTE TRANSFECTION   In this method, cells are infected or transfected with encapsidated or naked DNA molecules and assayed for expression 12–72 hours later. Vectors based on the small DNA tumor virus simian virus 40 (SV40) have been used to study regulation of the mouse MT-I gene in infected or transfected monkey kidney cells (139–141) and human fibroblasts (142) or cervical carcinoma cells (143). In all cases the intact MT gene, and hybrid genes derived from it, was transcribed from the correct initiation site and responded to cadmium but not glucocorticoids. Plasmid vectors lacking any viral sequences have also been used to study metal regulation of the mouse MT-I gene in hamster cells (144) and of the human MT-II_A gene in mouse cells (145). Advantages of the acute methods are simplicity, rapidity, and reproducibility. Limitations are the inability to perform long-term experiments and the nonphysiologically high copy numbers of the recombinants.

BOVINE PAPILLOMA VIRUS   BPV is unusual among the DNA tumor viruses in that it is maintained in the nucleus of transformed cells as a free, autonomously replicating episome at 50–100 copies per cell. Introduction of a mouse MT-I-human growth hormone fusion gene into mouse C127 cells on a

BPV vector led to the isolation of transformed foci secreting mature growth hormone at various levels (146). Expression was inducible by cadmium, but not dexamethasone, whereas the endogenous genes in the same cells responded to both agents. BPV vectors have also been used to introduce the intact human MT-II$_A$ (147), MT-I$_A$ (10), MT-I$_E$, and MT-I$_F$ (11) genes into mouse cells. In all cases the transformed cells became resistant to cadmium toxicity due to the overproduction of MT mRNA and polypeptide. The high basal level transcription of MT genes in BPV has made this system useful for genetic engineering experiments.

EMBRYO INJECTION AND TRANSGENIC MICE   The ability to transfer cloned MT genes into whole organisms is critical for understanding their regulation during development and differentiation. Initial experiments demonstrated that a mouse MT-I-thymidine kinase (tk) fusion gene was appropriately regulated by cadmium following microinjection into the male pronucleus of fertilized one-cell mouse ova (148). To obtain "transgenic" mice, eggs microinjected with MT-tk (149), MT-rat growth hormone (150), or MT-human growth hormone (151) genes were reimplanted into the oviducts of pseudopregnant surrogate mothers. Approximately 25% of the offspring incorporated and expressed the hybrid genes. Typically the foreign DNA is present in 1 to 100 or more tandem head-to-tail copies, is incorporated into all tissues and organs including the germ line, and is inserted into a single, Mendelianly inherited site. This allows the construction of mouse lines homozygous for the injected DNA by mating experiments (reviewed in 152). Expression of the mouse MT-I fusion genes in transgenic mice is inducible by heavy metals and bacterial endotoxin but not glucocorticoids (150, 151, 153, 154).

YEAST SYSTEMS   Gene transfer systems are particularly well developed in the yeast *Saccharomyces cerevisiae* (155). The copper-MT gene, *CUP1,* and fusion genes derived from it, have been introduced into yeast cells on stable and unstable episomal vectors capable of either high or low copy number replication (81, 118, 156, 157). The ability of yeast to undergo efficient homologous recombination between chromosomal and transfected DNA sequences has also allowed integration of wild-type or mutated *CUP1* genes at various loci and the construction of strains lacking their endogenous *CUP1* coding sequences (118, 157).

Heavy Metals

The ability of heavy metals to induce MT synthesis was discovered by Piscator, who found increased MT levels in the livers of rabbits exposed to cadmium (25). Since then, this form of regulation has been recognized in all species that synthesize MT and in many different tissues and cell types (see Table 1 for references). MT synthesis in mammals and most other higher eukaryotes is inducible by a variety of heavy metal ions including cadmium, zinc, and copper, whereas the copper-binding MTs of fungi appear to be induced only by this ion. The optimal metal concentration for induction varies in different systems but is generally just below the level causing cell toxicity. The kinetics with which maximal MT synthesis is achieved also vary, ranging from days in whole mammals to minutes in yeast. Because metals appear to stimulate MT gene transcription almost immediately (see below), this probably reflects differences in mRNA and protein half-lives and the speed with which the metal reaches and penetrates the target cell.

The early observations that induction of mammalian MT synthesis is blocked by actinomycin D (158–160), and is accompanied by increases in translatable MT mRNA (161–165), suggested that heavy metals act at the level of mRNA synthesis, processing, or degradation. To distinguish between these possibilities, nuclei were isolated from cadmium-treated or uninduced mouse liver and kidney and incubated with radioactive UTP under conditions in which previously initiated RNA chains are elongated but no new chains are initiated (166). Synthesis of MT-I RNA increased 17- to 25-fold within 1 hour, well before the maximal accumulation of stable MT-I mRNA at 4–6 hours. Similar "run-off" data have been obtained for the mouse MT-II gene (39) and for human MT genes transfected into rat cells (138). These results, together with the inducibility of hybrid genes containing only upstream MT sequences (see below), demonstrate that heavy metal regulation occurs largely if not exclusively at the level of transcription initiation. Expression of the yeast copper-MT gene is also controlled at this level (81, 157).

CIS-ACTING CONTROL SEQUENCES   The upstream MT gene regulatory sequences involved in heavy metal induction have been studied by "reversed genetic" experiments in which mutated or hybrid control sequences are constructed by recombinant DNA techniques and reintroduced into living cells. This approach has been extensively used to analyze the control sequences of the mouse MT-I (136, 139, 144, 148, 167, 169) and human MT-II$_A$ (138, 145, 170) genes. The results obtained for the primate and rodent genes were similar and can be considered together (Figure 2, *top*).

Analysis of sequential 5' deletion mutants revealed that less than 60 bp of 5' flanking DNA were sufficient to bestow heavy metal inducibility on transcription. Surprisingly, though, 3' deletion or fusion mutants lacking all of the 5' flanking region to as far as position −126 also showed some heavy metal regulation (167, 170). (Nucleotides are numbered from position +1 at the initiation site, negative numbers for 5' flanking sequences and positive numbers for transcribed sequences.) These apparently contradictory results could be reconciled by postulating the existance of two or more functional heavy metal

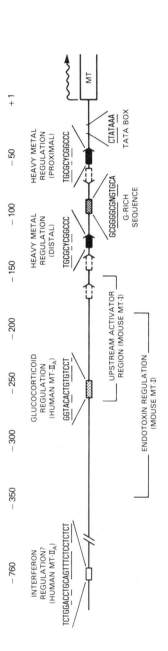

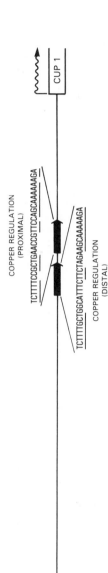

Figure 2   Transcriptional control sequences. *Top*, mammalian MT gene control elements. The solid arrows indicate the proximal and distal heavy metal control elements. A possible consensus sequence is shown with invariant nucleotides underlined. The hollow arrows indicate partial repeats of this sequence in the mouse MT-I gene (167, 169, 170). The dotted box indicates the glucocorticoid control element in the human MT-II$_A$ gene. Nucleotides homologous to mouse mammary tumor virus glucocorticoid control elements are underlined (170). The hollow box indicates the putative interferon control element. Nucleotides homologous to interferon-induced histocompatibility genes are underlined (190). The brackets indicate the maximum boundaries of the mouse MT-I sequences involved in the endotoxin response (154) and in transcriptional activation (B. Felber, D. Hamer, manuscript submitted). *Bottom*, yeast *CUP*1 gene control elements. The solid arrows indicate the copper-responsive elements. The sequences are shown with homologous nucleotides underlined (81, 157; D. Thiele, D. Hamer, manuscript submitted).

control sequences. Analysis of a series of small deletion and insertion mutants in the mouse MT-I gene indicated maximal boundaries of -15 to -84 for the more proximal region and -84 to -151 for the more distal region (169). While both regions render transcription sensitive to heavy metals, they differ quantitatively. The proximal region gives high induction ratios but low transcriptional efficiency, whereas the distal region gives low induction ratios but high transcriptional efficiency.

The apparent duplication of the information for heavy metal regulation led to a search for repeated DNA sequences that might be capable of interacting with identical or related cellular factors. It was found that both the mouse MT-I (167, 169) and human MT-II$_A$ (170) genes contain an imperfectly duplicated 12-nucleotide sequence present once in the proximal region (centered at position -46 for mouse or -49 for human) and once in the distal region (centered at position -114 or -145). Moreover, closely related sequences are present, at similar locations, in front of the genes for human MT-I$_A$ (10), MT-I$_E$, and MT-I$_F$ (11) and rat MT-I (R. Andersen, H. Herschman, personal communication). The mouse MT-I gene also contains partial repeats of the sequence centered at positions -160, -134, and -62 (inverse orientation). A consensus version of this heavy metal control sequence is shown in Figure 2. To test the functionality of this sequence, an oligonucleotide corresponding to positions -42 to -56 of the mouse MT-I gene was placed upstream of the heterologous herpes virus tk gene. The synthetic oligonucleotide conferred a low level of heavy metal inducibility when present in a single copy (167) and a much greater response when present in two or more copies (144). Induction was not dependent on the precise location of the synthetic oligonucleotide, although substantial variations in the levels of basal transcription were observed in different constructions (144).

A comparison of the induction of the mouse MT-I and MT-II genes by different metals showed identical dose-response curves for each ion tested both in whole mice (39) and cultured cells (171). Recently similar results have been obtained for the human MT-I$_E$ and MT-I$_F$ isoform genes (11; C. Schmidt, D. Hamer, manuscript submitted). In contrast, a comparison of human MT-II$_A$ to MT-I$_A$ showed that the MT-II gene responded equally well to cadmium and zinc whereas the MT-I gene was strongly inducible only by cadmium (10). Transfection experiments using MT-tk fusion constructs demonstrated that this differential regulation is controlled by the 5' flanking region, but the precise sequences involved have not been mapped (10).

Mutational analysis has revealed that sequences outside of the heavy metal control regions can also play an important role in determining the efficiency of transcription (Figure 2, *top*). In the case of the human MT-IIa gene, a G-rich sequence centered at position -79, between the two metal control elements, was implicated in regulating the level of basal transcription (170). Homologous sequences are found in other MT genes from both man and mouse (11, 39). In the case of the mouse MT-I gene, a region further upstream, between positions -155 and -268, has been found to play a role in determining the efficiency of both basal and induced transcription (148, 169). Placement of this region upstream of a truncated globin promoter activated transcription by 30-fold or more. Interestingly, this effect was independent of the orientation and precise location of the fragment and did not require heavy metal stimulation (B. Felber, D. Hamer, manuscript submitted). The possible relationship of these sequences to enhancer elements (172) is being actively investigated. The MT genes also contain typical TATA-box sequences in the -25 to -35 region. Mutations in these sequences, which are thought to act as recognition sites for RNA polymerase or other common transcription factors (111), reduce the efficiency of transcription without altering heavy metal inducibility (169).

Reverse genetics has also been used to study the *cis*-acting DNA sequences involved in copper induction of yeast *CUP1* gene transcription (81, 157). Analysis of deletion and fusion contructs showed the presence of two upstream copper control regions (Figure 2, *bottom*). Either of these regions is capable of conferring copper sensitivity, but both together are required to give normal transcription levels and induction ratios. The two upsteam regions contain an imperfect repeat of a 32–34-nucleotide sequence centered at positions -123 and -165. A synthetic oligonucleotide containing the proximal sequence conferred copper sensitivity to a heterologous promoter when present in two copies (D. Thiele, D. Hamer, manuscript submitted). The primary sequences of the yeast and mammalian heavy metal control sequences bear no obvious extended homology to one another.

A striking generalization from the above results is that the heavy metal regulatory elements of MT genes, like those of many other eukaryotic genes, are present in multiple copies (144, 172). One possible explanation for this sort of arrangement is that it allows cooperative binding of identical regulatory factors or differential binding of distinct but related factors. A second is that it prevents induction of genes that, by chance, contain a single copy of a sequence related to the regulatory element. Yet a third speculation is that binding of multiple factors affects chromatin structure in a manner not afforded by binding of a single factor.

TRANS-ACTING FACTORS   The search for cellular regulatory factors has proven less straightforward than the identification of the DNA sequences with which they interact. One approach has been to study the chromatin conformation of endogenous MT genes in uninduced versus induced cells. The mouse MT-I gene was found to contain a 250–300-bp region of DNA, near the 5' end of the gene, that was hypersensitive to digestion by various nucleases (173). The strongest digestion sites were centered around positions -225 within the upstream activator sequence and at -30 close to both the TATA box and proximal heavy metal control element. Chromatin from mouse W7 cells, which contain inactive hypermethylated MT genes (128), was less sensitive to nuclease digestion. Interestingly, cadmium induction led to increased nuclease hypersensitivity within a region between -30 and $+60$. It is not yet clear whether this results from increased transcription or from binding of regulatory factors. The chromatin structure of the mouse MT-I gene has also been studied by light micrococcal nuclease digestion (174).

Competition between transfected MT genes has provided a second method to detect cellular regulatory factors (141). The principle behind these experiments is that regulatory factors can be titrated out by introducing sufficient control sequence DNA into living cells. If the factors act as transcriptional activators in the presence of heavy metals, competition will lead to a decrease in the level of induced transcription. If they act as repressors in the absence of heavy metals, competition will lead to an increase in the basal level of transcription. In fact, it was found that addition of increasing amounts of a competitor plasmid containing mouse MT-I 5' flanking sequences led to a sequential decrease in the expression of a indicator plasmid in which the MT control sequences were fused to bacterial chlorampenicol acetyltransferase coding sequences. Competition was specific for MT regulatory factors, as compared to polymerase or other general factors, as shown by the inability of a strongly transcribed globin gene to affect MT expression. Analysis of deletion mutants showed that both heavy metal control elements are required for optimal competition; the proximal region alone, which gives only low transcriptional levels, showed no competition, whereas the distal region alone, which gives high transcription values, gave partial competition. There results suggested that a limiting cellular factor for MT gene induction is a positive activator that, in the presence of heavy metals, binds to the same sequences required for maximal transcription.

The molecular nature of the MT regulatory factors in unknown. It was originally speculated that apothionein itself might bind to the promoter region and repress transcription in the absence of heavy metals (175). This mechanism now seems unlikely since both the competition experiments and the mutational analyses indicate positive rather than negative regulation (see above). Moreover, an MT fusion gene was normally inducible in transfected Chinese hamster ovary cells that contain hypermethylated endogenous MT genes and produce little or no MT polypeptide (D. Hamer, unpublished results). A second possibility is that there exists an unidentified regulatory factor that contains two domains, one that binds to heavy metals and the other to the control sequences (176). Finally, regulation might occur through a cascade involving two or more factors. For example, there could be an activator that binds to the DNA and a metal-sensitive "anti-activator" that binds to the activator in the absence of metal ions. It will not be possible to formally distinguish between these speculations until the factors have been purified and studied in vitro.

Hormones and Other Circulating Factors

MT synthesis in mammals is inducible by a number of circulating factors including glucocorticoids, an unidentified product of the inflammatory response, and interferon. In general the response to these factors is less than to heavy metals, is more restricted in cell type, and is more variable among different isoform genes.

GLUCOCORTICOIDS AND OTHER HORMONES   It was observed early on that MT is induced by various stress conditions, such as exposure to hot or cold or starvation, that also raise circulating steroid hormone levels (177–179). The first direct evidence for a role of glucocorticoids in regulating MT levels was obtained in primary rat hepatocyte cultures (180). Subsequent studies in human cervical carcinoma cells demonstrated that glucocorticoid stimulation is a primary response in that it is not blocked by cycloheximide, is accompanied by increases in MT mRNA, and occurs prior to accumulation of zinc from the medium (181–183). Glucocorticoid induction has also been observed, at varying levels, in cultured fibroblast, hepatoma, sarcoma, and erythroid cells, but not in cells that contain inactive hypermethylated MT genes (127). In intact rats and mice, glucocorticoids stimulate MT synthesis primarily in the liver with smaller effects on kidney, skeletal muscle, and spleen (184).

The level at which glucocorticoids act has been investigated by "run-off" experiments on nuclei isolated from mouse liver (184). The results showed that induction occurs primarily at the level of transcription initiation with a minor effect (2- to 3-fold) on mRNA stability. In the mouse, which synthesizes only 2-MT isoforms, the MT-I and MT-II genes are equally inducible by dexamethasone (171). In contrast, various levels of regulation are observed from the multiple MT isoform genes of humans; the MT-II_A and MT-I_E genes are strongly inducible, the MT-I_A gene shows a partial response, and the MT-I_F gene is unaffected (10, 11; C. Schmidt, D. Hamer, manuscript submitted).

A popular model for glucocorticoid action is that the hormone binds to its receptor which in turn stimulates transcription by direct interaction with nuclear control DNA sequences. This model has been tested for the human MT-II_A gene by gene transfer and DNA binding experiments. Following stable transfection into cultured rat cells, the intact MT-II_A gene retained glucocorticoid sensitivity in approximately half the clones analyzed. Inducibility was also observed for fusion genes in which the MT 5' flanking sequences were placed as far as 630 bp upstream of the herpes virus tk gene (138). Analysis of a series of 5' deletion mutants revealed that sequences critical for the glucocorticoid response are located between positions −236 and −268 (170). The ability of partially purified glucocorticoid-receptor complexes to interact with DNA sequences in this region was demonstrated by filter binding, DNAse protection, and methylation experiments. Major binding occurred between nucleotides

−248 and −263, a region with significant homology to the glucocorticoid control elements of mouse mammary tumor virus (170).

Quite different results have been obtained for the mouse MT-I and MT-II genes. Although both genes are strongly inducible by glucocorticoids in their natural chromosomal state (171), they lose this property following gene transfer (136, 143, 153). Glucocorticoid regulation is also lost following amplification of the chromosomal MT-I gene in cadmium-resistant mouse cells (185). One interpretation of these results is that glucocorticoid induction of the mouse genes requires an epigenetic event, such as DNA methylation, or a commitment event that changes the chromatin structure early in development. An alternative possibility is that the glucocorticoid regulatory sequences are located distant (at least 25 Kb, the minimal size of the amplification unit) from the genes. In this context, it is notable that neither the mouse MT-I nor MT-II genes contain homologies to the glucocorticoid control consensus sequence (170) in their immediate 5' flanking DNA (38, 39).

MT synthesis in rat liver is also stimulated by the peptide hormones glucagon and angiotensin II and by both α and β adrenergic agonists (186a). These regulatory pathways have not yet been studied at the molecular level.

INFLAMMATORY RESPONSE   MT synthesis is also induced by tissue injury resulting from injection of agents such as turpentine, carbon tetrachloride, or bacterial endotoxin (177, 186b). The ability of turpentine to induce MT in adrenalectomized rats suggested that the response to these inflammatory agents was independent of the glucocorticoid-mediated pathway (187). The inflammatory response also appears to be independent of metal stimulation since injection of mice with bacterial endotoxin led to increased liver and kidney MT mRNA levels well before accumulation of zinc in these organs (154). The DNA sequences responsible for the inflammatory response have been investigated using transgenic mice carrying MT-tk fusion genes (154). A fusion gene with ~350 bp of upstream MT flanking sequences responded to both cadmium and endotoxin (but not glucocorticoids), whereas a gene with 185 bp of flanking sequences was induced only by cadmium. These results suggest that the inflammatory stress control element lies between positions −185 and −350 and confirm that the response is independent of both glucocorticoid hormones and heavy metals (Figure 2, *top*).

INTERFERON   The effect of interferon on MT gene transcription was serendipitously discovered by differential screening of a cDNA library from human neuroblastoma cells (189). The steady-state level of MT mRNA was increased 3- to 8-fold within 8 hours of interferon treatment and the rate of transcription increased 5-fold within 2 hours. Comparison of the 5' flanking sequences of the human MT-II$_A$ gene to those of three histocompatibility antigen genes that are induced by interferon revealed a homology region approximately 760 bp upstream of the MT gene (189). The consensus sequence of this region is shown in Figure 2, but it should be realized that the functionality of this putative control element has not been directly tested.

DEVELOPMENT AND DIFFERENTIATION

MT is generally considered a "housekeeping" protein because it is expressed in many different tissues and cell types. Nevertheless, there are substantial quantitative variations in levels of MT expression during embryogenesis and in different tissues of the adult. The development of cloned gene probes and gene transfer methods has allowed initial molecular analyses of these developmental pathways in mammals and the sea urchin.

MT gene expression in early mammalian development has been most thoroughly studied in rodents. The liver is the primary site of MT mRNA and polypeptide synthesis in fetal mice and rats (191–194). Levels of MT mRNA are detectable by day 12 of gestation, increase fivefold to reach a maximum by day 16, then remain approximately constant to parturition. Following birth there is a rapid drop in MT synthesis, although MT mRNA levels may remain constant throughout suckling (192). The intracellular localization of MT during pre- and post-natal development of the rat has been studied by immunochemical staining methods (93, 193). Surprisingly, intense nuclear staining was observed in the fetal and newborn liver up to day 9. By day 17 the staining was primarily cytoplasmic, similar to the pattern found in adult liver.

Substantial levels of MT mRNA are also found in the visceral and parietal yolk sac of the mouse fetus (194). In these organs, MT levels rise sevenfold from day 9 to 11 of gestation, remain constant until day 15, then drop 20-fold or more by parturition. Separation of the visceral yolk sac into endodermal and mesodermal layers revealed that essentially all of the MT mRNA is contained in the endodermal cells. Similarly high levels of MT mRNA are found in cultured embryonal carcinoma cells with primitive, parietal, or visceral endoderm characteristics, whereas low or undetectable levels of MT mRNA are present in the fetal brain, kidney, bowel, heart, placenta, and amnion and in undifferentiated embryonal carcinoma cells (191, 194).

In the adult mammal, the liver and kidney are well known to be the primary sites of MT and heavy metal accumulation (3, 4). Consistent with this, measurements of MT-I mRNA levels in cadmium-injected mice revealed the highest levels in the liver and kidney with lower levels in intestine, heart, testes, muscle, brain, and spleen (166). The accumulation of MT mRNA was inducible by cadmium in all tissues except the testes.

The development of transgenic mice has provided a unique opportunity to explore the molecular basis of the tissue-specific expression of MT genes.

Initial studies of mice carrying an MT-tk fusion gene revealed that enzyme levels were highest in liver, next highest in kidney, and lowest in brain (149). This pattern parallels the expression levels of the endogenous MT genes in these organs, suggesting that the cloned fusion gene contained all the information required for the establishment of tissue-specific regulation. However, more detailed studies of mice carrying an MT–growth hormone hybrid gene revealed substantial variations in the ratios of foreign to endogenous gene expression in different tissues (151). For example, the ratio of steady-state MT-growth hormone to MT mRNA was ~0.4 in liver and testes versus ~0.02 in kidney or intestine. It is unclear to what extent these variations reflect differences in transcription rate, mRNA stability, or integration site of the foreign gene. Transgenic mice have also been used to study the relationship between methylation and gene expression (153). The offspring of mice carrying an MT-tk fusion gene exhibited large variations in enzyme expression relative to the transgenic parent, and in two pedigrees a converse correlation could be demonstrated between the extent of methylation and enzyme expression. Evidence was also presented that hypermethylated DNA sequences were preferentially lost in some of the transgenic offspring.

The tissue specificity of human MT gene expression has been studied using cultured cell lines and liver autopsy samples (C. Schmidt, D. Hamer, manuscript submitted). The single MT-II$_A$ gene was ubiquitously expressed in all cell types examined whereas the MT-I$_E$ and MT-I$_F$ isoform genes exhibited differential patterns of activity. Kidney, chorion, and cervical carcinoma cells contained substantial levels of MT-I$_F$ mRNA but no detectable MT-I$_E$ RNA, whereas higher levels of MT-I$_E$ than of MT-I$_F$ RNA were found in liver and in cell lines derived from hepatocarcinoma, colon, and bladder. Only in teratocarcinoma cells, representing an early developmental stage, were the two MT-I isoform genes equally active. A possible relationship between methylation and the differential expression of the MT-I isoform genes was demonstrated by the ability of 5-azacytidine to reactivate MT-I$_E$ transcription in cervical carcinoma cells. An unexpected result of these studies was the effect of oncogenes on MT gene regulation. In particular, cells transformed by an activated cHa-*ras* oncogene expressed both the MT-II$_A$ and MT-I$_E$ genes at a high basal level that could be only slightly induced by cadmium. It is tempting to speculate that this reflects the effect of one of the normal cellular *ras* genes, which are thought to play a role in cell growth and differentiation, on MT gene transcription.

The sea urchin is particularly useful for studying early stages in gene regulation because of the wealth of classic embryological studies on this organism. A cloned cDNA probe has been used to follow the steady-state levels of MT mRNA during the development of both normal and zinc-treated embryos (52). The egg contains substantial levels of maternal mRNA that is polyadenylated shortly after fertilization. During the first 9 hours of development, which is the period of most rapid cell division, no new MT mRNA is synthesized. Accumulation of the mRNA begins around 12 hours and reaches a maximum at the 20-hour mesenchyme blastula stage. There is then a decrease in the amount of MT mRNA, until the gastrula stage at about 40 hours, followed by an increase during further development to the early pluteus stage. Thus MT mRNA accumulation reaches two discrete peaks, separated by a valley, during the course of embryogenesis.

The tissue specificity of MT gene expression in the sea urchin has been studied by comparing fractionated mesodermal and endodermal derivatives to ectodermal derivatives (52). The mesoderm-endoderm showed a low level of basal MT mRNA that was strongly induced by zinc treatment, whereas the ectoderm contained a high constitutive level of MT mRNA that was unaffected by zinc. These results were confirmed by analyzing gastrula, mesenchyine blastalae, and zinc "animalized" (ectoderm-predominant) embryos. A possible interpretation of these results is that the ectodermal cells, which are in constant contact with sea water, constitutively maintain MT at a high level to protect against sudden encounters with heavy metal ion, whereas the interior mesodermal and endodermal cells induce MT gene expression in response to metal stimulation as a second line of defense. It is not yet known whether these differences reflect the transcription of distinct MT genes in the different tissues.

These results show that MT gene expression can vary considerably during development. Of particular interest are the high basal MT levels in both mammalian and sea urchin embryos and their possible relationship to a fetal function of MT. At least two distinct mechanisms may be involved in the differential expression of MT genes during development. First, epigenetic changes, such as DNA methylation or chromatin modification, could alter the accessibility of MT genes to transcriptional and regulatory factors. Second, different cell types, perhaps in response to their role in heavy metal metabolism, might vary in their content of diffusable MT gene regulatory factors. Such hypothetical factors could in principle either act early (e.g. metal receptors) or late (e.g. DNA binding proteins) in the regulatory pathway.

GENETIC DISEASES

Genetic diseases provide useful "experiments of nature" for studying gene expression and function. Menkes' kinky hair syndrome is a severe, X-linked pediatric disease characterized by a maldistribution of bodily copper. Decreased levels of copper and of several copper-dependent enzymes are found in liver, brain, and serum whereas increased levels of copper and copper-MT are present in most nonhepatic tissues and in cultured cells (195–198). The over-accumulation of copper in cultured Menkes' fibroblasts is correlated with the ability of low concentrations of copper to induce MT-I$_E$ and MT-II$_A$ gene transcription in mutant but not normal cells (142). Transfection experiments with an MT fusion gene showed that this abnormal regulatory response is due to a defect in a diffusible or *trans*-acting factor. Low concentrations of copper are also unusually toxic to Menkes' cells and induce the expression of two heat-shock or stress proteins. These data suggest that the Menkes' mutation alters a factor that acts at an early stage of the metal regulatory pathway, most likely at the level of copper transport or metabolism, rather than a DNA-binding or transcription factor (142).

The X-linked mottled *(Mo)* mutants of mice may provide a useful animal model for Menkes' disease. Viable mutants share many features in common with Menkes' patients including maldistribution of bodily copper, deficiencies in copper enzymes, and overaccumulation of copper-MT in kidney and cultured cells (196, 199, 200).

Wilson's disease also affects copper metabolism in man but is clearly distinct from Menkes' disease in that it is a recessive autosomal trait that is expressed in adolescence or early adult life (196, 201, 202). The characteristic defects are overaccumulation of copper and copper-MT in the liver and brain, leading to hepatic damage and neurological defects, and deficiency in circulating cerulo-plasmin. A primary defect in MT gene regulation seems unlikely because cultured Wilson's cells show only minor abnormalities in the accumulation of MT polypeptide and mRNA (203, 204; A. Leone, D. Hamer, unpublished results). Thus Menkes' disease, Wilson's disease, and the mottled mouse mutants all appear to affect MT gene regulation indirectly by altering copper metabolism rather than directly by altering MT gene structure or DNA binding factors.

GENETIC ENGINEERING APPLICATIONS

The aim of genetic engineers is to construct organisms that overproduce desirable gene products or that display useful physiological traits. MT genes are well suited for such experiments because they are inducible, selectible, and expressed at high levels.

BPV recombinants carrying MT fusion genes have been used to produce a variety of mammalian gene products in cultured mouse cells, particularly proteins that must be posttranslationally modified in a mammalian host cell to obtain full biological activity. Expression of an MT-human growth hormone gene in BPV led to the synthesis and extracellular secretion of growth hormone from which the hydrophobic amino-terminal signal sequence was precisely processed. Production levels ranged from 5 mg/L ($\sim 10^8$ molecules/cell/day) under standard laboratory conditions up to 100 mg/L in a mammalian cell bioreactor (146; C. Henschel, personal communication). BPV recombinants carrying an MT-hepatitis B surface antigen fusion have been used to direct the synthesis of surface antigen that was appropriately secreted, glycosylated, and assembled into subviral particles and was immunogenic in hamsters. Expression levels were 1 mg/L and the transformed lines were capable of continuous production of surface antigen for as long as 2 months (205). BPV-MT vectors have also been used to produce influenza virus hemagglutinin and to study sequences involved in the processing and transport of this integral membrane protein (206). Other proteins under development in the BPV-MT system include tissue plasminogen activator, human chorionin gonadotropin, and several animal fertility hormones.

A particularly dramatic example of genetic engineering technology has come from the establishment of transgenic mice expressing MT-rat growth hormone (150) and MT-human growth hormone (151) genes. A high fraction of the progeny in these experiments efficiently expressed the fusion gene in the liver and other organs, accumulated circulating growth hormone, and grew up to twice the size of normal litter mates when fed a zinc-enriched diet. Potential applications of the transgenic technology include correcting genetic defects, accelerating the growth of livestock animals, and "farming" of useful gene products. Towards the first goal, it has been shown that the slow growth of *little* mice, which lack growth hormone, can be corrected by incorporation of an MT-growth hormone gene (207). The recent extension of the transgenic technology to sheep, pigs, and rabbits should facilitate achievement of the latter two goals (152).

In yeast, the ability of the *CUP1* copper-MT gene to confer resistance to copper toxicity has been exploited to sequentially amplify chromosomally integrated DNA sequences (M. Walling, D. Hamer, manuscript submitted). A plasmid carrying *CUP1* and a yeast-selectable marker was integrated into the *cup1^S* locus of a single copy strain to generate a duplication of *CUP1* sequences surrounding the foreign DNA. Selection for resistance to stepwise increases in copper concentration led to sequential 5- to 10-fold amplification of the entire repeat unit including the foreign DNA.

FUNCTION

In one of the first papers on MT, Kagi & Vallee speculated on its possible role in "... catalysis, storage, immune phenomena, or detoxification . . ." (2). Despite a quarter of a century of research, the function of MT remains enigmatic (209). It is generally agreed, in view of the ability of MT to both bind to and be induced by heavy metal ions, that its major role must be somehow related to metal metabolism. However, defining the exact nature of this role has proven difficult. One potentially powerful approach to this problem is to create mutants that underproduce or overproduce MT. Recent developments in gene transfer technology have made it possible to conduct such tests in single cells; studies in more complex systems, such as whole mammals, still depend on less direct methods.

Yeast Model System

The yeast *Saccharomyces cerevisiae* is particularly well suited for genetic studies because its genome can be manipulated through recombination with cloned DNA molecules. This method has been used to construct strains in which the *CUP1* sequences coding for copper-MT are completely replaced by a heterologous marker gene (157). Such strains, despite their inability to synthesize MT, grow with a normal doubling period on standard media containing low copper and are fully competent at mating, diplophase growth, sporulation, and germination. They also accumulate normal levels of total cell copper and of the copper-dependent form of superoxide dismutase. Therefore yeast MT is not essential for growth, development, copper accumulation, or the activation of a copper enzyme under normal laboratory conditions.

Yeast lacking MT do exhibit two abnormalities in heavy metal metabolism (157). First, they are hypersensitive to poisoning by high copper concentrations, thereby confirming the role of *CUP1* in copper detoxification (118, 129). Second, in low copper media they initiate transcription from the *CUP1* promoter at a high, constitutive rate. This autoregulatory defect could, in principle, be due to a role of MT as a transcriptional factor or to its ability to chelate free intracellular copper that would otherwise be available to activate transcription. The latter hypothesis is strongly supported by three observations: 1. the same upstream *CUP1* sequences are involved in copper induction and autoregulation (157); 2. purified yeast MT and apothionein do not bind to DNA (82; D. Hamer, unpublished observation); and 3. monkey MT downregulates *CUP1* transcription despite the lack of homology between the yeast and mammalian genes and proteins (see below).

These observations indicate that a major function of yeast MT is to maintain low levels of free intracellular copper. Under conditions of high environmental exposure, MT synthesis is induced so that excess free ions are chelated. Under normal conditions, basal MT expression is balanced such that it is sufficiently low to allow copper enzyme activation yet sufficiently high to prevent futile transcription from the *CUP1* promoter. This autoregulatory circuit provides a facile means of shutting down MT expression once excess copper has been chelated and may represent a particularly economical form of regulation for a gene that is frequently amplified.

The relevance of these results to higher eukaryotes was tested by transfecting yeast that lack their endogenous MT gene with plasmids in which monkey MT-I and MT-II coding sequences were placed under control of the *CUP1* promoter (210). Expression of the two monkey MT polypeptides protected the yeast against poisoning by high copper concentrations and restored normal transcriptional regulation to the *CUP1* promoter. Thus mammalian MT can complement both known functions of yeast MT despite the lack of obvious sequence homology.

Possible Roles in Higher Eukaryotes

MT was discovered by virtue of its ability to bind to cadmium, a nonessential and highly toxic element. Hence, it is not surprising that much of the research on the function of MT has focused on its possible role in the detoxification of this ion. The subsequent discoveries that MT is inducible by zinc and copper, and that its concentration changes during early development and physiological stress, led to increased interest in its role in the metabolism of the essential metals.

CADMIUM DETOXIFICATION   The ability of MT to act as a cadmium detoxifying agent in cultured mammalian cells has been demonstrated by three observations: 1. cell lines that fail to produce MT due to gene hypermethylation are unusually sensitive to cadmium poisoning (114, 128); 2. cell lines selected for cadmium resistance overproduce MT due to gene amplification (114, 121, 122); and 3. cell lines containing high copy number BPV-MT recombinants are also highly resistant to cadmium due to MT overproduction (11, 147). Studies of the amplified cell lines have shown that MT can also protect against poisoning by mercury, copper, and bismuth, but less efficiently than against cadmium. Results with zinc have been variable, ranging from no effect to a 2- to 3-fold protection (114, 176).

The possibility that MT can protect against acute caduium toxicity in whole mammals has also been investigated. The testicles are one of the organs most sensitive to cadmium poisoning. Pretreatment of mice with low doses of cadmium protected the animals from testicular damage and led to increased levels of a low-molecular-weight cadmium-binding protein (211). These results were interpreted to mean that metal pretreatment induces the synthesis of MT which is then capable of detoxifying high cadmium doses. The recent discovery that MT mRNA accumulation is not inducible by cadmium in mouse testes (166) suggests an alternative hypothesis, namely that the testes are hypersensitive to cadmium poisoning because they are incapable of producing increased MT polypeptide. Pretreatment of mice with low doses of cadmium also protected against subsequent liver damage and lethality (212, 213). Interestingly, pretreatment with copper had no effect on the lethality of a subsequent cadmium injection, while pretreatment with zinc actually increased mortality (214).

Interest in chronic cadmium toxicity, and the possible role of MT in this process, was sparked by the finding of environmental cadmium pollution as a causative factor in Itai-Itai disease in Japan (215). Both in man and mouse, chronic cadmium exposure leads to accumulation of the metal predominately in the renal cortex where it causes tubular damage, uremia, and excretion of cadmium into the urine. Similar symptoms are observed when purified cadmium-MT is injected into experimental animals. There is a good correlation between the levels of cadmium exposure, the extent of kidney damage, and the concentration of MT both in the kidney and the circulation (213, 216).

ZINC AND COPPER METABOLISM   The role of MT in the metabolism of these essential ions has been studied both in cultured cells and whole organisms. Cell lines that fail to synthesize MT are fully viable and grow with a normal doubling time (114, 128). Because many zinc and copper enzymes are involved in normal cell division, this strongly argues against any essential role of MT in the intracellular activation of metalloenzymes. On the other hand, cell lines that overproduce MT accumulate larger quantities of zinc and copper then do normal cells. Recently BPV-MT recombinants have been used to study the effect of individual human MT isoform proteins in essential metal metabolism (11; C. Schmidt, M. Wong, D. Hamer, D. Winge, manuscript in preparation). Surprisingly, cells overexpressing MT-I_F accumulated a much higher ratio of copper to zinc than cells expressing the closely related MT-I_E polypeptide. This correlates well with the tissue-specific expression of these genes in kidney and liver, respectively.

Studies in whole organisms have been less direct. Treatment of animals with glucocorticoids or with inflammatory agents leads to large increases in the hepatic levels of zinc and, to a lesser extent, copper (177–179, 186a). The accumulation of metals under these conditions appear to be an effect rather than a cause of MT gene induction. Unusually high levels of zinc and copper are also found in embryos during the periods of greatest MT synthesis. Finally, pretreatment of animals with low levels of cadmium results in several abnormalities in the absorption and circulation of zinc and copper, possibly due to induction of MT synthesis (reviewed in 3, 4).

OTHER POSSIBILITIES   The induction of MT under stress conditions has led to several suggestions concerning its possible role in cellular adaptation mechanisms. Interferon and interleukin 1, which induce MT synthesis, activate macrophages and neutrophils, which release active oxygen species. Recently it has been shown that zinc, cadmium-MT can scavenge free hydroxyl ions leading to the speculation that MT plays a direct role in the detoxification of this reactive species (217). It has also been observed that cultured mammalian cells that overexpress MT are unusually resistant to X-ray damage, leading to the suggestion that MT might protect cells against ionizing radiation (28, 209). The possibility that MT, by virtue of its high content of reduced cysteine, functions in sulfur metabolism or control of the intracellular redox potential, has also been mentioned (3), but at present there is no experimental evidence for such a role.

A Working Model

In attempting to synthesize the above observations, it might be useful to propose a working model for the function of MT. Any such model must bear in mind two key points. First, different heavy metal ions have different biological roles; cadmium is highly toxic and nonessential, copper is also highly toxic but is required for the activity of a limited number of enzymes, while zinc is relatively nontoxic and is essential for a large and diverse collection of enzymes. Second, MT binds to each of the ions with vastly different avidities; the stability constant for copper is ~100-fold greater than that for cadmium, which in turn is ~1000-fold greater than that for zinc. Given these facts, it seems reasonable to postulate that MT actually plays at least three distinct functions, one for each metal.

Cadmium is a natural component of the earth's minerals, and therefore it is inevitable that it will be absorbed by living organisms either directly or through food chains. Moreover, this element has an extremely long half-life, estimated as more than 20 years in humans. I propose that the main function of MT with regard to cadmium is protection against long-term toxicity, and further that MT is the major protein involved in this task. Under normal conditions of chronic low-level exposure, the basally synthesized MT is probably sufficient to chelate most or all of the cadmium ions. Under unusual conditions of sudden high-level exposure, the induction of MT gene transcription may provide an additional measure of safety. The high affinity of MT for cadmium, and the long half-life of the cadmium-containing protein, make it well suited for this detoxifying function.

Copper is much more abundant than cadmium in biological systems and plays an essential role in several enzymes involved in electron transfer and oxidation reactions. However, free copper is a potent toxin. I propose that the primary function of MT in regard to copper is to maintain low intracellular concentrations of the free ion while at the same time permitting the activation of copper enzymes. In contrast to cadmium, it seems likely that MT plays only a minor role in the metabolism of copper under normal conditions because the binding of copper to MT is too tight to allow ready transfer to enzymes. Presumably other binding proteins or ligands, still unidentified, are primarily responsible for copper accumulation and transport. Under conditions of high copper exposure, induced MT would provide a highly effective detoxifying agent because of its strong avidity for the free ion.

Zinc is also widely distributed in biological systems, but is considerably less reactive and toxic than copper. Also, zinc enzymes catalyze a much broader range of reactions than do copper enzymes including various steps in anabolism, catabolism, and the transfer of genetic information. I propose that MT acts as a zinc storage protein that plays a true homeostatic role in the metabolism of this ion. It is unlikely that MT is essential for direct intracellular zinc enzyme activation because cultured cells lacking MT are fully viable and because MT is no more competent at zinc enzyme activation than are inorganic zinc salts. Instead, I suggest that MT synthesized in the liver serves as a labile source of zinc that can be subsequently utilized in many different organs for the activation of still unidentified enzymes that are essential in early development and under stress conditions. The low stability constant of zinc-MT would make it ideally suited for such a purpose.

Testing this model, or any other related to the function of MT, will require the development of at least two critical experimental technologies. First, it will be necessary to engineer whole organisms with conditional mutations in MT synthesis. In principle this could be accomplished by generating transgenic animals from eggs submitted to gene replacement. Second, it will be essential to identify the metalloenzymes that are physiologically activated by zinc and copper during conditions such as early development and stress. A stumbling block is the lack of a sensitive, rapid, and high resolution method (e.g. two-dimensional electrophoresis) to detect metalloproteins. The development of such a methodology would revolutionize the field.

ACKNOWLEDGMENTS

This review is dedicated to my daughter, Addie, who was born midway through its preparation. I thank my many colleagues for communication of their published and unpublished results, Carl Schmidt for his assistance with the early stages of the manuscript, C. Stout for the illustrations used in Figure 1, and Gail Gray, Carolyn Ray, and Nadine Horne for their editorial assistance.

Literature Cited

1. Margoshes, M., Vallee, B. L. 1957. *J. Am. Chem. Soc.* 79:4813–14
2. Kagi, J. H. R., Vallee, B. L. 1960. *J. Biol. Chem.* 235:3460–65
3. Kagi, J. H. R., Nordberg, M., eds. 1979. *Metallothionein*. Basel: Birkhauser Verlag
4. Foulker, E. C., ed. 1982. *Biological Roles of Metallothionein*. New York: Elsevier
5. Pulido, P., Kagi, J. H. R., Vallee, B. L. 1965. *Biochemistry* 5:1768–77
6. Kagi, J. H. R., Vasak, M., Lerch, K., Gilg, D. E. O., Hunziker, P., et al. 1984. *Environ. Health Perspect.* 54:93–103
7. Kissling, M. M., Kagi, J. H. R. 1979. See Ref. 3, pp. 145–51
8. Kissling, M. M., Kagi, J. H. R. 1977. *FEBS Lett.* 82:247–50
9. Karin, M., Richards, R. I. 1982. *Nature* 299:797–802
10. Richards, R. I., Heguy, A., Karin, M. 1984. *Cell* 37:263–72
11. Schmidt, C. J., Jubier, M.-F., Hamer, D. H. 1985. *J. Biol. Chem.* 260:7731–37
12. Koizumi, S., Otaki, N., Kimura, M. 1985. *J. Biol. Chem.* 260:3672–75
13. Schmidt, C. J., Hamer, D. H. 1983. *Gene* 24:137–46
14. Kagi, J. H. R., Himmelhoch, S. R., Whanger, P. D., Bethune, J. L., Vallee, B. L. 1974. *J. Biol. Chem.* 249:3537–42
15. Kojima, Y., Berger, C., Kagi, J. H. R. 1979. See Ref. 3, pp. 153–61

16. Bremner, I., Marshall, R. B. 1974. *Br. J. Nutr.* 32:283–91
17. Hartman, H.-J., Weser, U. 1977. *Biochem. Biophys. Acta* 491:211–22
18. Bremner, I., Williams, R. B., Young, B. W. 1977. *Br. J. Nutr.* 38:87–92
19. Peterson, M. G., Lazdins, I., Danks, D. M., Mercer, J. F. B. 1984. *Eur. J. Biochem.* 143:507–11
20. Cousins, R. J., Barber, A. K., Trout, J. R. 1973. *J. Nutr.* 103:964–72
21. Webb, M., Daniel, M. 1975. *Chem. Biol. Interact.* 10:269–76
22. Bremner, I., Young, B. W. 1976. *Biochem. J.* 155:631–35
23. Vostal, J. J., Cherian, M. G. 1974. *Fed. Proc. Am. Soc. Exp. Biol.* 33:519
24. Amacher, D. E., Ewing, K. L. 1975. *Arch. Environ. Health* 30:510–13
25. Piscator, M. 1964. *Nord. Hyg. Tidskr.* 45:76–82
26. Nordberg, M., Trojanowska, B., Nordberg, G. F. 1974. *Environ. Physiol. Biochem.* 4:149–58
27. Kimura, M., Otaki, N., Imano, M. 1979. See Ref. 3, pp. 163–68
28. Bakka, A., Webb, M. 1981. *Biochem. Pharmacol.* 30:721–25
29. Griffith, B. B., Walters, R. A., Enger, M. D., Hildebrand, C. E., Griffith, J. K. 1983. *Nucleic Acids Res.* 11:901–10
30. Winge, D. R., Rajagopalan, K. V. 1972. *Arch. Biochem. Biophys.* 153:755–62
31. Shaikh, Z. A., Hirayama, K. 1979. *Environ. Health Perspect.* 28:267–71
32. Chen, R. W., Ganther, H. E. 1975. *Environ. Physiol. Biochem.* 5:378–88
33. Berger, C., Kissling, M. M., Andersen, R. D., Weser, U., Kagi, J. H. R. 1981. *Experientia* 37:619–25
34. Kissling, M. M., Berger, C., Kagi, J. H. R., Andersen, R. D., Weser, U. 1979. See Ref. 3, pp. 181–85
35. Andersen, R. D., Birren, B. W., Ganz, T., Piletz, J. E., Herschman, H. R. 1983. *DNA* 2:15–22
36. Nordberg, M., Nordberg, G. F., Piscator, M. 1975. *Environ. Physiol. Biochem.* 5:396–403
37a. Huang, I.-Y., Kimura, M., Hata, A., Tsunoo, H., Yoshida, A. 1981. *J. Biochem.* 89:1839–45
37b. Huang, I.-Y., Yoshida, A., Tsunoo, H., Nakajima, H. 1977. *J. Biol. Chem.* 252:8217–21
38. Glanville, N., Durnam, D. M., Palmiter, R. D. 1981. *Nature* 292:267–69
39. Searle, P. F., Davison, B. L., Stuart, G. W., Wilkie, T. M., Norstedt, G., Palmiter, R. D. 1984. *Mol. Cell. Biol.* 4: 1221–30
40a. Olafson, R. W., Thompson, J. A. J. 1974. *Mar. Biol.* 28:83–86
40b. Weser, U., Donay, F., Rupp, H. 1973. *FEBS Lett.* 32:171–74
40c. Brown, D. A., Bawden, C. A., Chatel, K. W., Parsons, T. R. 1977. *Environ. Conserv.* 4:213–16
40d. Yamamura, M., Suzuki, K. T. 1984. *Comp. Biochem. Physiol.* 77B:101–6
41. Suzuki, K. T., Tanaka, Y., Kawamura, R. 1983. *Comp. Biochem. Physiol.* 75C:33–37
42. Suzuki, K. T., Akitomi, H. 1983. *Comp. Biochem. Physiol. C* 75:211–15
43. Suzuki, K. T., Ebihara, Y. 1984. *Comp. Biochem. Physiol. C* 78:35–38
44. Suzuki, K. T., Akitomi, H., Kawamura, R. 1984. *Toxicol. Lett.* 21:179–84
45. Yamamura, M., Mori, T., Suzuki, K. T. 1981. *Experientia* 37:1187–89
46. Maroni, G., Watson, D. 1985. *Insect Biochem.* 15:55–63
47. Lastowski-Perry, D., Otto, E., Maroni, G. 1985. *J. Biol. Chem.* 260:1527–30
48. Aoki, Y., Suzuki, K. T., Kubota, K. 1984. *Comp. Biochem. Physiol. C* 77: 279–82
49. Everard, L. B., Swain, R. 1983. *Comp. Biochem. Physiol. C* 75:275–80
50. Bouquegneau, J. M., Ballan-Dufrancais, C., Jeantet, A. Y. 1985. *Comp. Biochem. Physiol. C* 80:95–98
51. Yamaura, M., Suzuki, T., Hatakeyama, S., Kubota, K. 1983. *Comp. Biochem. Physiol. C* 75:21–24
52. Nemer, M., Travaglini, E. C., Rondinelli, E., D'Alonzo, J. 1984. *Dev. Biol.* 102:471–82
53. Nemer, M., Wilkinson, D. G., Travaglini, E. C., Sternberg, E. J., Butt, T. R. 1985. *Proc. Natl. Acad. Sci. USA.* 82:4992–97
54. Olafson, R. W., Sim, R. G., Boto, K. G. 1979. *Comp. Biochem. Physiol.* 62: 407–16
55. Overnell, J., Trewhella, E. 1979. *Comp. Biochem. Physiol. C* 64:69–76
56. Brouwer, M., Brouwer-Hoexum, T., Engel, D. W. 1984. *Mar. Environ. Res.* 14:71–88
57. Lerch, K., Ammer, D., Olafson, R. W. 1982. *J. Biol. Chem.* 257:2420–26
58. Deleted in proof
59. Overnell, J., Berger, C., Wilson, K. J. 1981. *Biochem. Soc. Trans.* 9:217–18
60. Brown, D. A. 1977. *Mar. Biol.* 44:203–9
61. Takeda, H., Shimizu, C. 1982. *Nippon Suisan Gakkaishi* 48:717–23
62. Marafante, E. 1976. *Experientia* 32: 149–50
63. Bonham, K., Gedamu, L. 1984. *Biosci. Rep.* 4:633–42
64. Kito, H., Ose, Y., Mizuhira, V., Sato, T., Ishikawa, T., Tazawa, T. 1982. *Comp. Biochem. Physiol. C* 73:121–27

65. Noel-Lambot, F., Gerday, C., Disteche, A. 1978. *Comp. Biochem. Physiol. C* 61:177–87
66. Howard, A. G., Nickless, G. 1975. *J. Chromatogr.* 104:457–59
67. Ridlington, J. W., Fowler, B. A. 1979. *Chem. Biol. Interact.* 25:127–38
68. Engel, D. W., Brouwer, M. 1984. *Mar. Environ. Res.* 13:177–94
69. George, S. G., Carpene, E., Coombs, T. L., Overnell, J., Youngson, A. 1979. *Biochim. Biophys. Acta* 580:225–33
70. Carpene, E., Cortesi, P., Crisetig, G., Serranzanetti, G. P. 1980. *Thalassia Jugosl.* 16:317–23
71. Viarengo, A., Pertica, M., Mancinelli, G., Palmero, S., Zanicchi, G., Orunesu, M. 1981. *Mar. Pollut. Bull.* 12:347–50
72. Rauser, W. E. 1984. *Plant Physiol.* 74:1025–29
73. Rauser, W. E., Curvetto, N. R. 1980. *Nature* 287:563–64
74. Wagner, G. J. 1984. *Plant Physiol.* 76:797–805
75. Wagner, G. J., Trotter, M. M. 1982. *Plant Physiol.* 69:804–9
76. Bartolf, M., Brennan, E., Price, C. A. 1980. *Plant Physiol.* 66:438–41
77. Piccini, E., Coppelotti, O., Guidolin, L. 1985. *Comp. Biochem. Physiol.* In press
78. Lerch, K. 1980. *Nature* 284:368–70
79a. Premakuman, R., Winge, D. R., Wiley, R., Rajagopalan, K. V. 1975. *Arch. Biochem. Biophys.* 170:278–83
79b. Prinz, R., Weser, U. 1975. *FEBS Lett.* 54:224–29
80. Karin, M., Najarian, R., Haslinger, A., Valenzuela, P., Welch, J., Fogel, S. 1984. *Proc. Natl. Acad. Sci. USA* 81:337–41
81. Butt, T. R., Sternberg, E. J., Gorman, J. A., Clark, P., Hamer, D., et al. 1984. *Proc. Natl. Acad. Sci. USA* 81:3332–36
82. Winge, D. R., Nielson, K. B., Gray, N. R., Hamer, D. H. 1985. *J. Biol. Chem.* 260:14464–70
83. Higham, D. P., Sadler, P. J. 1983. *Inorg. Chim. Acta* 79:140–42
84. Higham, D. P., Sadler, P. J., Scawen, M. D. 1984. *Science* 225:1043–45
85. Koizumi, S., Otaki, N., Kimura, M. 1982. *Ind. Health* 20:101–8
86. Brady, F. O., Kafka, R. L. 1979. *Anal. Biochem.* 98:89–94
87. Zelazowski, A. J., Szymanska, A., Cierniewski, C. S. 1980. *Chem. Biol. Interact.* 33:115–25
88. Mehra, F. K., Bremner, I. 1983. *Biochem. J.* 213:459–65
89. Garvey, J. S., Vander Mallie, R. J., Chang, C. C., eds. 1982. *Methods Enzymol.* 84:121–39

90. Winge, D. R., Garvey, J. S. 1983. *Proc. Natl. Acad. Sci. USA* 80:2472–76
91. Masui, T., Utakoji, T., Kimura, M. 1983. *Experientia* 39:182–84
92. Garvey, J. S. 1984. *Environ. Health Perspect.* 54:117–27
93. Danielson, K. G., Ohi, S., Huang, P. C. 1982. *Proc. Natl. Acad. Sci. USA* 79:2301–4
94. Banerjee, D., Onosaka, S., Cherian, M. G. 1982. *Toxicology* 24:95–105
95. Clarkson, J. P., Elmes, M. E., Jasani, B., Webb, M., eds. 1984. *Trace Element–Analytical Chemistry in Medicine and Biology*, 3:292–97. New York
96. Winge, D. R., Nielson, K. B., Zeikus, R. D., Gray, W. R. 1984. *J. Biol. Chem.* 259:11419–25
97. Kagi, J. H. R., Vallee, B. L. 1961. *J. Biol. Chem.* 236:2435–42
98. Li, T.-Y., Kraker, A. J., Shaw, C. F. III, Petering, D. H. 1980. *Proc. Natl. Acad. Sci. USA* 77:6334–38
99. Nielson, K. B., Atkin, C. L., Winge, D. R. 1985. *J. Biol. Chem.* 260:5342–50
100a. Otvos, J. D., Armitage, I. M. 1980. *Proc. Natl. Acad. Sci. USA* 77:7094–98
100b. Boulanger, Y., Goodman, C. M., Forte, C. P., Fesik, S. W., Armitage, I. M. 1983. *Proc. Natl. Acad. Sci. USA* 80:1501–5
101. Winge, D. R., Miklossy, K. A. 1982. *J. Biol. Chem.* 257:3471–76
102. Vasak, M., Hawkes, G. E., Nicholson, J. K., Sadler, P. J. 1984. *Biochemistry* 24:740–47
103. Nielson, K. B., Winge, D. R. 1985. *J. Biol. Chem.* 260:5342–50
104. Nielson, K. B., Winge, D. R. 1983. *J. Biol. Chem.* 258:13063–69
105. Nielson, K. B., Winge, D. R. 1984. *J. Biol. Chem.* 259:4941–46
106. Beltramini, M., Lerch, K., Vasak, M. 1984. *Biochemistry* 23:3422–27
107a. Adam, A. O., Brady, F. O. 1980. *Biochem. J.* 187:329–35
107b. Beltramini, M., Lerch, K. 1982. *FEBS Lett.* 142:219–22
107c. Yoshida, A., Kaplan, B., Kimura, M. 1979. *Proc. Natl. Acad. Sci. USA* 76:486–90
108. Durnam, D. M., Perrin, F., Gannon, F., Palmiter, R. D. 1980. *Proc. Natl. Acad. Sci. USA* 77:6511–15
109. Karin, M., Richards, R. 1982. *Nucleic Acids Res.* 10:3165–73
110. Varshney, U., Gedamu, L. 1984. *Gene* 31:135–45
111. Breathnach, R., Chambon, P. 1981. *Ann. Rev. Biochem.* 50:349–83
112. Nishioka, Y., Leder, A., Leder, P. 1980. *Proc. Natl. Acad. Sci. USA* 77:2806–9

113. Cox, D. R., Palmiter, R. D. 1983. *Hum. Genet.* 64:61–64
114. Crawford, B. D., Enger, M. D., Griffith, B. B., Griffith, J. K., Hanners, J. L., et al. 1985. *Mol. Cell. Biol.* 5:320–29
115. Stallings, R. L., Munk, A. C., Longmire, J. L., Hildebrand, C. E., Crawford, B. D. 1984. *Mol. Cell. Biol.* 4:2932–36
116. Schmidt, C. J., Hamer, D. H., McBride, O. W. 1984. *Science* 224:1104–6
117a. Karin, M., Eddy, R. L., Henry, W. M., Haley, L. L., Byers, M. G., Shows, T. B. 1984. *Proc. Natl. Acad. Sci. USA* 81:5494–98
117b. Varshney, U., Hoar, D. I., Starozik, D., Gedamu, L. 1985. *Mol. Biol. Med.* 2:193–206
117c. Lieberman, H. B., Rabin, M., Barker, P. E., Ruddle, F. H., Varshney, U., Gedamu, L. 1985. *Cytogenet. Cell Genet.* 39:109–15
118. Fogel, S., Welch, J. S. 1982. *Proc. Natl. Acad. Sci. USA* 79:5342–46
119. Rugstad, H. E., Norseth, T. 1975. *Nature* 257:136–37
120. Hildebrand, C. E., Tobey, R. A., Campbell, E. W., Enger, M. D. 1979. *Exp. Cell Res.* 124:237–46
121. Beach, L. R., Palmiter, R. D. 1981. *Proc. Natl. Acad. Sci. USA* 78:2110–14
122. Gick, G. G., McCarty, K. S. 1982. *J. Biol. Chem.* 257:9049–53
123. Hayashi, K., Fujiki, H., Sugimura, T. 1983. *Cancer Res.* 43:5433–36
124. Schimke, R. T., ed. 1982. *Gene Amplification.* New York: Cold Spring Harbor
125. Deleted in proof
126. Holliday, R., Pugh, J. E. 1975. *Science* 187:226–32
127. Mayo, K., Palmiter, R. D. 1981. *J. Biol. Chem.* 256:2621–24
128. Compere, S. J., Palmiter, R. D. 1981. *Cell* 25:233–40
129. Brenes-Pomales, A., Lindegren, G., Lindegren, C. C. 1955. *Nature* 176:841–42
130. Deleted in proof
131. Fogel, S., Welch, J. W., Cathala, G., Karin, M. 1983. *Curr. Genet.* 7:347–55
132. Welch, J. W., Fogel, S., Cathala, G., Karin, M. 1983. *Mol. Cell. Biol.* 3:1353–61
133. Deleted in proof
134. Deleted in proof
135. Fogel, S., Welch, J. W., Louis, E. J. 1985. *Cold Spring Harbor Symp. Quant. Biol.* In press
136. Mayo, K. E., Warren, R., Palmiter, R. D. 1982. *Cell* 29:99–108
137. Perucho, M., Hanahan, D., Wigler, M. 1980. *Cell* 22:309–17
138. Karin, M., Haslinger, A., Holtgreve, H., Cathala, G., Slater, E., Baxter, J. D. 1984. *Cell* 36:371–79
139. Hamer, D. H., Walling, M. 1982. *J. Mol. Appl. Genet.* 1:273–88
140. Carter, A. D., Felber, B. K., Walling, M., Jubier, M. F., Schmidt, C. J., Hamer, D. H. 1984. *Proc. Natl. Acad. Sci. USA* 81:7392–96
141. Seguin, C., Felber, B. K., Carter, A. D., Hamer, D. H. 1984. *Nature* 312:781–85
142. Leone, A., Pavlakis, G. N., Hamer, D. H. 1985. *Cell* 40:301–9
143. Pavlakis, G. N., Hamer, D. H. 1983. *Recent Prog. Horm. Res.* 39:353–85
144. Searle, P., Stuart, G., Palmiter, R. 1985. *Mol. Cell. Biol.* 5:1480–89
145. Karin, M., Holtgreve, H. 1984. *DNA* 3:319–26
146. Pavlakis, G. N., Hamer, D. H. 1983. *Proc. Natl. Acad. Sci. USA* 80:397–401
147. Karin, M., Cathala, G., Nguyen-Huu, M. C. 1983. *Proc. Natl. Acad. Sci. USA* 80:4040–44
148. Brinster, R. L., Chen, H. Y., Warren, R., Sathy, A., Palmiter, R. D. 1982. *Nature* 296:39–42
149. Brinster, R. L., Chen, H. Y., Trumbauer, M., Senear, A. W., Warren, R., Palmiter, R. D. 1981. *Cell* 27:223–31
150. Palmiter, R. D., Brinster, R. L., Hammer, R. E., Trumbauer, M. E., Rosenfeld, M. G., et al. 1982. *Nature* 300:611–15
151. Palmiter, R. D., Norstedt, G., Gelinas, R. E., Hammer, R. E., Brinster, R. L. 1983. *Science* 222:809–14
152. Palmiter, R. D., Brinster, R. L. 1985. *Cell* 41:343–45
153. Palmiter, R. D., Chen, H. Y., Brinster, R. L. 1982. *Cell* 29:701–10
154. Durnam, D. M., Hoffman, J. S., Quaife, C. J., Benditt, E. P., Chen, H. Y., Palmiter, R. D. 1984. *Proc. Natl. Acad. Sci. USA* 81:1053–56
155. Strathern, J., Jones, E., Broach, J., eds. 1981. *The Molecular Biology of the Yeast Saccharomyces.* New York: Cold Spring Harbor
156. Butt, T. R., Sternberg, E., Herd, J., Crooke, S. T. 1984. *Gene* 27:23–33
157. Hamer, D. H., Thiele, D. J., Lemontt, J. E. 1985. *Science* 228:685–90
158. Squibb, K. S., Cousins, R. J. 1974. *Environ. Physiol. Biochem.* 4:24–30
159. Richards, M. P., Cousins, R. J. 1975. *Bioinorg. Chem.* 4:215–24
160. Squibb, K. S., Cousins, R. J., Feldman, S. L. 1977. *Biochem. J.* 164:223–28
161. Squibb, K. S., Cousins, R. J. 1977.

Biochem. Biophys. Res. Commun. 75: 806–12
162. Shapiro, S. G., Squibb, K. S., Markowitz, L. A., Cousins, R. J. 1978. *Biochemistry* 175:833–40
163. Andersen, R. D., Weser, U. 1978. *Biochem. J.* 175:841–52
164. Ohi, S., Cardenosa, G., Pine, R., Huang, P. C. 1981. *J. Biol. Chem.* 256:2180–84
165. Enger, M. D., Rall, L. B., Hildebrand, C. E. 1979. *Nucleic Acids Res.* 7:271–88
166. Durnam, D. M., Palmiter, R. D. 1981. *J. Biol. Chem.* 256:5712–6
167. Stuart, G., Searle, P., Chen, H., Brinster, R., Palmiter, R. D. 1984. *Proc. Natl. Acad. Sci. USA* 81:7318–22
168. Deleted in proof
169. Carter, A., Felber, B., Walling, M. J., Jubier, M.-F., Schmidt, C., Hamer, D. H. 1984. *Proc. Natl. Acad. Sci. USA* 81:7392–96
170. Karin, M., Haslinger, A., Holtgreve, H., Richards, R., Krauter, P., et al. 1984. *Nature* 308:513–19
171. Yagle, M. K., Palmiter, R. D. 1985. *Mol. Cell. Biol.* 5:291–94
172. Hamer, D. H., Khoury, G. 1983. In *Enhancers and Eukaryotic Gene Expression*, ed. Y. Gluzman, T. Shenk, pp. 1–15. New York: Cold Spring Harbor Lab.
173. Senear, A. W., Palmiter, R. D. 1982. *Cold Spring Harbor Symp. Quant. Biol.* 47:539–47
174. Koropatnick, J., Andrews, G., Duerksen, J. D., Varshney, U., Gedamu, L. 1983. *Nucleic Acids Res.* 11:3255–67
175. Karin, M., Richards, R. I. 1984. *Environ. Health Perspect.* 54:111–15
176. Durnam, D. M., Palmiter, R. D. 1984. *Mol. Cell. Biol.* 4:484–91
177. Oh, S. H., Deagen, J. T., Whanger, P. D., Weswig, P. H. 1978. *Am. J. Physiol.* 234:E282–85
178. Ryden, L., Deutsch, H. F. 1978. *J. Biol. Chem.* 253:519–24
179. Wong, K.-L., Klaassen, C. D. 1981. *Eur. J. Biochem.* 113:267–72
180. Failla, M. L., Cousins, R. J. 1978. *Biochim. Biophys. Acta* 538:435–44
181. Karin, M., Herschman, H. R. 1979. *Science* 204:176–77
182. Karin, M., Herschman, H. R. 1981. *Eur. J. Biochem.* 113:267–72
183. Karin, M., Andersen, R. D., Slater, E., Smith, K., Herschman, H. R. 1980. *Nature* 286:295–97
184. Hager, L. J., Palmiter, R. D. 1981. *Nature* 291:340–42
185. Mayo, K. E., Palmiter, R. D. 1982. *J. Biol. Chem.* 257:3061–67

186a. Cousins, R. J. 1985. *Physiol. Rev.* 65:238–309
186b. Sobocinski, P. Z., Canterbury, W. J., Mapes, C. A. 1977. *Fed. Proc.* 36:1100
187. Sobocinski, P. Z., Canterbury, W. J., Knutsen, G. L., Hauer, E. C. 1981. *Inflammation* 5:153–64
188. Deleted in proof
189. Friedman, R. L., Manly, S. P., McMahon, M., Kerr, I. M., Stark, G. R. 1984. *Cell* 38:745–55
190. Friedman, R. L., Stark, G. R. 1985. *Nature* 314:637–39
191. Ouellette, A. J. 1982. *Dev. Biol.* 92: 240–46
192. Andersen, R. D., Piletz, J. E., Birren, B. W., Herschman, H. R. 1983. *Eur. J. Biochem.* 131:497–500
193. Panemangalore, M., Banerjee, D., Onosaka, S., Cherian, M. G. 1983. *Dev. Biol.* 97:95–102
194. Andrews, G. K., Adamson, E. D., Gedamu, L. 1984. *Dev. Biol.* 103:294–303
195. Menkes, J. H., Alter, M., Steigleder, G., Weakley, D. R., Sung, J. H. 1962. *Pediatrics* 29:764–79
196. Danks, D. M., Camakaris, J. 1984. *Adv. Hum. Genet.* 13:149–216
197. Horn, N. 1984. In *Metabolism of Trace Metals in Man*, ed. O. M. Rennert, W. Y. Chan, 2:25–52. Boca Raton, FL:CRC
198. Riordan, J. R., Jolicoeur-Paquet, L. 1982. *J. Biol. Chem.* 257:4639–45
199. Prins, H. W., Van den Hamer, C. J. A. 1979. *J. Inorg. Biochem.* 10:19–27
200. Packman, S., Chin, P., O'Toole, C. 1984. *J. Inherited Metab. Dis.* 7:168–70
201. Bearn, A. G. 1960. *Ann. Hum. Genet.* 24:33–43
202. Danks, D. M. 1983. In *The Metabolic Basis of Inherited Disease*, ed. J. B. Stanbury, J. B. Wyngaadren, D. S. Frederickson, J. L. Goldstein, M. S. Brown, 5:1251–68. New York: McGraw-Hill
203. Camakaris, J., Ackland, L., Danks, D. M. 1980. *J. Inherited Metab. Dis.* 3: 155–57
204. Chan, W. Y., Cushing, W., Coffman, M. A., Rennert, O. M. 1980. *Science* 208:299–300
205. Hsiung, N., Fitts, R., Strath, W., Milne, A., Hamer, D. H. 1984. *J. Mol. Appl. Gen.* 2:497–506
206. Sambrook, J., Rodgers, L., White, J., Gething, M. J. 1985. *EMBO J.* 4:91–103
207. Hammer, R. E., Palmiter, R. D., Brinster, R. L. 1984. *Nature* 311:65–67
208. Deleted in proof
209. Karin, M. 1985. *Cell* 41:9–10

210. Thiele, D., Walling, M. J., Hamer, D. 1985. *Science.* 231:854–56
211. Nordberg, G. F. 1971. *Environ. Physiol.* 1:171–87
212. Leber, A. P., Miya, T. S. 1976. *Toxicol. Appl. Pharmacol.* 37:403–14
213. Friberg, L., Piscator, M., Nordberg, G. F. 1971. In *Cadmium in the Environment.* Chicago: CRC

214. Yoshikawa, H. 1970. *Ind. Health* 8:184–86
215. Tsuchiya, K. 1978. *Cadmium Studies in Japan: A Review.* New York: Elsevier
216. Suzuki, K. T. 1982. In *Biological Roles of Metallothionein,* ed. E. C. Foulkes, pp. 215–35. New York: Elsevier
217. Thornalley, P. J., Vasak, M. 1985. *Biochem. Biophys. Acta* 827:36–44

Ann. Rev. Biochem. 1986. 55:953–85

MOLECULAR PROPERTIES OF VOLTAGE-SENSITIVE SODIUM CHANNELS

William A. Catterall

Department of Pharmacology, SJ-30, University of Washington, Seattle, Washington 98195

CONTENTS

PERSPECTIVES AND SUMMARY

The voltage-sensitive sodium channel is responsible for the increase in sodium permeability during the initial rapidly rising phase of the action potential in nerve, skeletal muscle, and heart. The general functional characteristics of

0066-4154/86/0701-0953$02.00

sodium channels were first described in the classical work of Hodgkin & Huxley in 1952 (1), in which they used the method of voltage clamping to separate and describe the characteristics of the ionic currents underlying the action potential in the squid giant axon. These methods have been applied to vertebrate myelinated nerve and skeletal muscle with generally similar results. Since the general properties of sodium channels as determined by these techniques have been the subject of many reviews (2–5), I will give only a brief outline in the paragraphs that follow, without citation of the original work, in order to introduce methods, terminology, and concepts that are used throughout this review.

Like most cells, electrically excitable cells maintain a high intracellular K^+ concentration and a low intracellular Na^+ concentration relative to the extracellular fluid through the energy-dependent pumping of these cations by Na^+, K^+-ATPase. Excitable cells also maintain a resting membrane potential, inside negative, because their surface membranes are specifically permeable to K^+. Excitable cells are distinguished, however, by having voltage-sensitive ion channels in their surface membranes that respond to membrane potential changes with a large regenerative increase in permeability to specific ions on a time scale of milliseconds. Changes in ionic permeability are accurately measured only if the membrane potential of the cell is controlled experimentally. Thus, the voltage clamp technique was developed to allow rapid recording of ionic currents across cell membranes following known changes in membrane potential imposed by the investigator. At known voltage, ionic currents provide a direct measure of ion movement across the cell membrane and therefore of cell membrane ionic permeability. With this technique, changes in sodium permeability occurring on the millisecond time scale can be accurately recorded.

The sodium permeability increase resulting from depolarization of the squid giant axon is biphasic. When the axon is depolarized, sodium permeability first increases dramatically and then after 1 msec decreases to the base-line level. Hodgkin & Huxley (1) described this biphasic behavior in terms of two experimentally separable processes that control sodium channel function: activation, which controls the rate and voltage dependence of sodium permeability increase following depolarization, and inactivation, which controls the rate and voltage dependence of the subsequent return of sodium permeability to the resting level during a maintained depolarization. The sodium channel can therefore exist in three functionally distinct states or groups of states: resting, active, and inactivated. Both resting and inactivated states are nonconducting, but channels that have been inactivated by prolonged depolarization are refractory unless the cell is repolarized to allow them to return to the resting state.

The classical voltage clamp experiments define three essential functional properties of sodium channels: voltage-dependent activation, voltage-dependent inactivation, and selective ion transport. In this chapter, I shall review experiments of the past decade using a combination of biophysical, biochemical, and molecular genetic techniques that have given insight into the molecular basis of these three aspects of sodium channel function. A major motivation behind studies of the sodium channel has been the view that unique structural features of this ion channel may provide clues to mechanisms of electrical excitation and ion transport that will prove applicable to voltage-dependent ion channels in general and thus elucidate a common molecular basis for electrical excitability. While this goal remains unrealized, many of the features of sodium channels which must underlie their function have been inferred from detailed biophysical studies; the protein components of the channel have been identified, purified, and restored to functional form in isolation; the chemical nature of the channel subunits and their functional roles have been examined by both biochemical and molecular genetic approaches; and specific models for the structural basis of sodium channel function have been proposed. This review covers work that had appeared in the literature or was available to me in preprint form in August, 1985.

NEUROTOXINS AS PROBES OF SODIUM-CHANNEL STRUCTURE AND FUNCTION

The voltage clamp technique provides detailed information on sodium channel function, but it is only applicable to intact membranes. Biochemical studies of sodium channels have been greatly facilitated by the use of a number of neurotoxins that bind with high affinity and specificity to sodium channels and alter their properties. The actions of these neurotoxins have been well described in previous reviews (6–9). The presentation in the following paragraphs focuses on their use as molecular probes of sodium-channel function and cites only original literature that has appeared subsequent to a previous review of this area in the *Annual Reviews* series (9).

The neurotoxins that have been useful to date in revealing new aspects of sodium-channel structure and function act at four separate receptor sites as outlined in Table 1.

Neurotoxin receptor site 1 binds the water-soluble heterocyclic guanidines tetrodotoxin and saxitoxin. These toxins inhibit sodium-channel ion transport by binding to a common receptor site that is thought to be located near the extracellular opening of the ion-conducting pore of the sodium channel.

Neurotoxin receptor site 2 binds several lipid-soluble toxins including grayanotoxin and the alkaloids veratridine, aconitine, and batrachotoxin. The

Table 1   Neurotoxin Receptor Sites on the Sodium Channel

Site	Neurotoxins	Physiological effect
1	Tetrodotoxin Saxitoxin	Inhibit ion transport
2	Veratridine Batrachotoxin Grayanotoxin Aconitine	Cause persistent activation
3	North African α scorpion toxins	Slow inactivation
4	American β scorpion toxins	Enhance activation

competitive interactions of these four toxins at neurotoxin receptor site 2 that were first demonstrated in ion flux experiments have been confirmed by direct measurements of specific binding of ^{3}H-labeled batrachotoxinin A 20-α-benzoate to sodium channels (10). These toxins cause persistent activation of sodium channels at the resting membrane potential by blocking sodium channel inactivation and shifting the voltage dependence of channel activation to more negative membrane potentials. Therefore, neurotoxin receptor site 2 is likely to be localized on a region of the sodium channel involved in voltage-dependent activation and inactivation.

Neurotoxin receptor site 3 binds polypeptide toxins purified from North African scorpion venoms or sea anemone nematocysts. These toxins slow or block sodium-channel inactivation. They also enhance persistent activation of sodium channels by the lipid-soluble toxins acting at neurotoxin receptor site 2. The affinity for binding of ^{125}I-labeled derivatives of the polypeptide toxins to neurotoxin receptor site 3 is reduced by depolarization. These data indicate that neurotoxin receptor site 3 is located on part of the sodium channel that is involved in inactivation.

Neurotoxin receptor site 4 binds a new class of scorpion toxins that has also proved valuable in studies of sodium channels. Cahalan (11) showed that the venom of the American scorpion *Centruroides sculpturatus* modifies sodium-channel activation rather than inactivation. Pure toxins from several American scorpions have a similar action (12–14). These toxins bind to a new receptor site on the sodium channel (15, 16) and have been designated β scorpion toxins.

These several neurotoxins provide specific high-affinity probes for distinct regions of the sodium-channel structure. The probes have been used to study sodium-channel function in intact excitable membranes, to identify and purify the protein components of sodium channels that bind the toxins, and to analyze the structural and functional properties of these proteins.

MOLECULAR PROPERTIES OF SODIUM CHANNELS INFERRED FROM FUNCTIONAL STUDIES

Studies of sodium-channel function in intact excitable membranes using voltage clamp, ion flux, and neurotoxin binding methods have led to important inferences about sodium-channel structure that are an essential complement to current efforts to identify, purify, and determine the structure of the molecular components of the sodium channel directly. In this section, a few notable examples of functional studies that have led to valuable working models of the structure of functional elements of the sodium channel are briefly considered.

ION TRANSPORT AND THE ION SELECTIVITY FILTER   While the macromolecular structure responsible for the increase in sodium permeability during an action potential has been called the "sodium channel" for a long time, experimental evidence in favor of a channel or pore mechanism of transport is more recent and is based on estimates of the ion transport capacity of an individual sodium channel. These estimates have been made by three separate approaches with roughly comparable results. Comparison of voltage clamp currents with measurements of sodium-channel density by saxitoxin or tetrodotoxin binding (17, 18) in squid giant axons or frog muscle fibers indicates a unit conductance of 2.5–8.6 pS (2.5–$8.6 \times 10^{-12} \Omega^{-1}$). Analysis of voltage-dependent membrane current fluctuations due to sodium channel activation using Fourier transform methods yields an estimate of 4.1–8.8 pS (19–21) in squid giant axon and frog node of Ranvier. More recently, techniques have been developed to measure the activity of individual sodium channels in electrically isolated 1-μm^2 patches of excitable membrane. These techniques give an estimate of 12–18 pS (22, 23) at 20°C and 140 mM Na^+. All of these estimates imply physiological ion transport rates greater than 10^7 ions/sec. The limiting rates of transport by small antibiotic ion carriers such as valinomycin are in the range of 10^4 ions/sec (24). Thus, the rapid rate of ion translocation mediated by the sodium channel must represent the movement of sodium ions through a fixed pore or channel rather than the cyclical movement of a larger membrane macromolecule across the permeability barrier with each ion transported. These rates can be compared to turnover numbers for enzymes, which range up to $5 \times 10^5 \sec^{-1}$ for carbonic anhydrase. Ion transport mediated by the sodium channel is among the most rapid protein-mediated processes.

In addition to being unusually rapid, ion transport by the sodium channel is selective. Potassium is about 8% as permeable as sodium, and rubidium and cesium are even less permanent (25, 26). Hille (27, 28) extended these early measurements to additional metal cations and a large number of organic cations and developed a detailed model of the narrowest region of the ion-conducting pore of the sodium channel, the ion selectivity filter. On the basis of the selectivity of the pore for organic cations, the limiting region was proposed to be approximated by a 3.1 Å × 5.1 Å rectangular orifice lined by oxygen atoms, which act as hydrogen bond acceptors during transport of organic cations and hydrated metal cations, but exclude similarly sized ions having non-hydrogen-bonding substituents like methyl groups. The transport of ions through the activated sodium channel is blocked by protonation of an acid group with a pK_a of approximately 5.2 (29). On the basis of this finding, it was proposed that two of the oxygen atoms acting as hydrogen bond acceptors at the ion selectivity filter are the oxygens of a carboxylic acid group that is required in the deprotonated form for effective ion transport.

The alkali metal cations are all smaller than the proposed limiting size of the pore and yet their transport is selective. Evidently, the chemistry of their interaction with the selectivity filter determines the selectivity of transport with only a small contribution from steric exclusion. The selectivity of cation transport parallels the binding of these cations at a high-field-strength ion exchange site (25, 28). Since the carboxylate anion postulated to be a required constituent of the selectivity filter is a high-field-strength site, interaction of partially hydrated metal cations with this site is proposed to provide the basis of selectivity among metal cations (28). Ion selectivity also depends upon the nature and concentration of the intracellular cations, suggesting that the interaction of entering ions with the ion selectivity filter may be altered by the presence and nature of ions in its intracellular end (30).

Transport of cations by the sodium channel is saturable (27, 28, 31, 32). For sodium, measurements of voltage clamp currents over the concentration range from 15 to 240 mM are approximately fit by a hyperbolic saturation curve with an apparent K_D of 368 mM at 0 mV (31). Other measurably permeant cations have apparent K_D values ranging from 50 to 368 mM. It is noteworthy that sodium has a relatively low affinity for the sodium channel. The rate of movement of sodium through the channel is within an order of magnitude of the diffusion-controlled rate in free solution. Strong interactions with the pore would cause sodium binding in the pore for a significant time during transport and prevent achievement of such high transport rates. Thus, cations with high affinity for the ion-conducting pore of the sodium channel cannot be rapidly transported and act as competitive inhibitors of more rapidly transported species. Coordination of the transported cations by the pore must be favorable enough to offset the loss of part of the normal hydration sphere of the ion but not so favorable that the ion blocks the pore.

AN ESSENTIAL CARBOXYL GROUP AT THE SAXITOXIN/TETRODOTOXIN RE-
CEPTOR SITE   Several lines of evidence indicate that there is an essential
carboxyl group in the saxitoxin/tetrodotoxin receptor site on the sodium channel. Saxitoxin and tetrodotoxin binding are blocked by protonation of a group
with a pK_a of approximately 5.4 in a number of experimental systems (33–37).
Toxin binding is also blocked by treatment of excitable membranes with
carboxyl-modifying reagents such as carbodiimides followed by a nucleophile
(38, 39) or trialkyloxonium salts (36, 40). These chemical modifications seem
specific, since the irreversible block of toxin binding is prevented if the
reactions are carried out in the presence of saturating concentrations of tetrodotoxin. Sodium channels made tetrodotoxin-insensitive by these chemical reactions are still active in generating action potentials (38–40) and have normal
voltage dependence of activation and inactivation and normal ion selectivity in
voltage clamp studies (41). However, modified sodium channels having this
carboxyl group methylated by trimethyloxonium have 35% of the maximum
sodium transport rate of normal channels (42). Thus, this carboxyl group is not
essential for ion transport, but it is required for binding of tetrodotoxin and
saxitoxin and the charged form is necessary for achieving maximum transport
rate.

EVIDENCE FOR A VOLTAGE-DEPENDENT CONFORMATIONAL CHANGE
ASSOCIATED WITH SODIUM-CHANNEL ACTIVATION   Although there is no
direct evidence, it is now believed by most investigators that the changes of
functional state resulting in activation are due to voltage-dependent conformational changes in protein component(s) of the sodium channel. On theoretical grounds, a membrane protein that responds to a change in membrane
potential must have charged and/or dipolar amino acid residues located within
the membrane electrical field. Changes in the membrane potential then exert a
force on these protein-bound dipoles and charges. If the energy of the field-
charge interactions is great enough, the protein may be induced to undergo a
change in conformation to a new stable state in which the net charge or the
location of charge within the membrane electrical field has been altered. For
such a voltage-driven change of state, the steepness of the state function versus
membrane potential curve defines the number of charges that move according
to a Boltzmann distribution. On this basis, Hodgkin & Huxley (1) predicted that
activation of sodium channels would require the movement of six positive
charges from the intracellular to the extracellular side of the membrane. The
movement of a larger number of charges through a proportionately smaller
fraction of the membrane electrical field would be equivalent. Such a movement of membrane-bound charge gives rise to a capacitative current that can, in
principle, be detected using electrophysiological techniques.

Capacitative currents associated with activation of sodium channels (gating currents) were first detected by Armstrong & Bezanilla (43) in studies of the squid giant axon. They detected outward gating currents that were approximately 0.3% of the sodium current during an action potential and reached a maximum in 80 μsec as the inward sodium current began to increase. Detailed analysis of the voltage and time dependence of gating currents supports the conclusion that they represent charge movements associated with the change of sodium channel state from resting to active (44, 45). Inactivation of sodium channels during a depolarizing prepulse blocks gating currents with the same time and voltage dependence as sodium currents (43, 46). These experiments leave little doubt that the small capacitative currents measured are due to movements of charged groups on the sodium channel during activation. In all probability, these charged groups are amino acids whose position in the protein structure is altered in the conformational change leading to activation.

A second line of evidence supporting the concept of voltage-dependent conformational change as an important component of the mechanism of sodium-channel activation is derived from studies of voltage-dependent scorpion toxin binding (47–49). The K_D for scorpion toxin binding to neurotoxin receptor site 3 on the sodium channel increases on depolarization and is closely correlated with the voltage dependence of activation of sodium channels in both frog muscle and neuroblastoma cells (49). These results show that the scorpion toxin receptor site undergoes a voltage-dependent conformational change on activation of the sodium channel that results in reduced binding of scorpion toxin.

Studies of drug and toxin action in voltage clamp experiments have given additional indirect evidence of a voltage-dependent conformational change on activation. Local anesthetics and related drugs inhibit sodium currents more rapidly and with greater affinity if excitable membranes are stimulated to activate sodium channels repetitively (50–52). These effects are postulated to be due to voltage-dependent conformational changes in the local anesthetic receptor site on the sodium channel associated with activation (52). Similarly, the lipid-soluble toxins aconitine and batrachotoxin modify sodium-channel properties more rapidly and with higher affinity if sodium channels are activated by repetitive stimulation (53–55), and cause persistent activation by binding with high affinity to active states of sodium channels (10, 56, 57). These data show that the active state of the sodium channel generated by transient membrane depolarization has selective high affinity for a number of drugs and toxins. Presumably these effects are due to a voltage-dependent protein conformational change that alters the structure of the different binding sites for these agents located on various parts of the channel structure.

Taken together, these results support a view of sodium-channel activation as a major conformational change or sequence of conformational changes of the sodium-channel protein that is driven by the force of the membrane electric field acting on protein-bound charges. This conformational change both activates the ion channel and alters the conformation at drug and toxin receptor sites that are likely to be spatially separate from the ion-conducting pore of the sodium channel.

PROTEIN COMPONENTS INVOLVED IN SODIUM-CHANNEL INACTIVATION   A component of gating current associated with inactivation of sodium channels has not been detected (46). If inactivation is the result of a voltage-dependent conformational change, an inactivation gating current would be expected from the considerations discussed previously. Therefore, inactivation is considered to be a rapid, nearly voltage-independent reaction or conformational change of the active state of the sodium channel (46, 58). The apparent voltage dependence of inactivation results from the necessity for at least part of the sequence of voltage-dependent conformational changes associated with activation to occur before inactivation can proceed. Several lines of evidence have shown that protein components located on the intracellular aspect of the sodium channel are essential for the inactivation of the sodium channel during a maintained depolarization. Intracellular perfusion of the squid giant axon with pronase, a mixture of proteolytic enzymes, blocks inactivation of the sodium channel in a time-dependent manner (59, 60). Alkaline protease b, the most substrate-specific enzyme of pronase, is responsible for the block of inactivation during intracellular perfusion with pronase (61). Trypsin also mimics the effect of alkaline protease b. Both these enzymes are specific for cleavage at the carboxyl group of lysyl and arginyl residues, suggesting that a protease-sensitive amino acid sequence containing lysine or arginine and located on the intracellular surface of the sodium channel is required for inactivation.

These results led Eaton et al (62) to examine the effects of the arginine-specific reagents glyoxal, phenylglyoxal, and 2,3-butanedione on sodium-channel inactivation. When perfused inside squid giant axons under voltage clamp, each of these reagents irreversibly blocked sodium-channel inactivation. In addition, N-bromoacetamide and the tyrosine-specific reagents N-acetylimidazole (63) and tetranitromethane, or I-plus lactoperoxidase (64), also block sodium-channel inactivation when perfused inside the squid giant axon. A number of other amino-acid-modifying reagents have no effect (63). Thus, both arginine and tyrosine residues located on the intracellular aspect of the sodium channel are implicated in sodium-channel inactivation. The action of both pronase and N-bromoacetamide to block inactivation is slightly voltage dependent, indicating that the susceptible residues are less available when the sodium channel is in the inactivated conformation (65).

ALLOSTERIC INTERACTIONS AMONG FUNCTIONALLY DISTINCT SODIUM-CHANNEL COMPONENTS   The previous sections described experiments that define several distinct sodium-channel loci on the basis of their functional properties and/or their interaction with specific agents: the ion selectivity filter, the internal site of sodium-channel inactivation, and the four neurotoxin receptor sites. Numerous experiments have now indicated allosteric coupling among these distinct sites on the sodium channel. The first clear evidence of allosteric interactions among separate sodium-channel sites was derived from studies of persistent activation of sodium channels by lipid-soluble toxins and scorpion toxin (57, 66, 67). α-scorpion toxins and sea anemone toxins do not cause persistent activation of sodium channels by themselves but markedly enhance the ability of lipid-soluble toxins to do so (57, 66–69). These results show that neurotoxin receptor sites 2 and 3 are allosterically coupled. This allosteric coupling is also observed in direct studies of binding of ^{125}I-labeled scorpion and sea anemone toxins and [^{3}H]batrachotoxinin A 20-α-benzoate (10, 48, 70–73). In each case, toxin binding at one of the two neurotoxin receptor sites is markedly enhanced by occupancy of the other.

Direct evidence for protein conformational changes that mediate the allosteric interactions between neurotoxin receptor sites 2 and 3 has come from studies of spectral changes and energy transfer efficiency of specifically bound fluorescent toxin derivatives (74, 75). Binding of α-scorpion toxins at neurotoxin receptor site 3 causes a red shift in the spectrum of specifically bound batrachotoxinin A 20-α-anthranilate consistent with a conformational change that places the chromophore in a more hydrophilic environment (75). Binding of batrachotoxin at neurotoxin receptor site 2 alters the efficiency of fluorescence energy transfer between fluorescent tetrodotoxin and α-scorpion toxin derivatives specifically bound at sites 1 and 3 consistent with changes of intersite distance or orientation (74).

Studies of the ion selectivity of neurotoxin-activated sodium channels have revealed an important allosteric interaction between neurotoxin receptor site 2 and the ion selectivity filter. Batrachotoxin (54, 76), aconitine (77), grayanotoxin (78), and veratridine (79) all increase sodium-channel permeability to larger organic and metal cations as measured by either voltage clamp or ion flux techniques. These results indicate that, after activation of sodium channels by neurotoxins, the ion selectivity filter is likely to be somewhat larger and to have a lower-field-strength anionic site than after activation by depolarization. Evidently, binding of the lipid-soluble neurotoxins to neurotoxin receptor site 2 induces a conformational change at the ion selectivity filter.

Neurotoxin receptor site 3, the scorpion toxin/sea anemone toxin receptor site, is located on the extracellular surface of the sodium channel (reviewed in 9). In contrast, proteolytic enzymes and group-specific reagents that block sodium-channel inactivation act only from the intracellular side of the membrane (59–64). These results indicate that polypeptide toxin binding at a site on the extracellular side of the sodium channel structure can modulate inactivation that is thought to occur on the intracellular side.

Taken together, these allosteric interactions imply substantial conformational flexibility of the sodium channel, with strong interactions among different functional loci. For example, binding of batrachotoxin or other lipid-soluble toxins to a single site shifts the voltage dependence of activation, blocks inactivation, alters ion selectivity, and enhances polypeptide toxin binding. The various "partial reactions" of sodium channel function can be separately defined but are apparently highly interactive in operation.

MOLECULAR PROPERTIES OF SODIUM CHANNELS IN EXCITABLE MEMBRANES

COVALENT LABELING OF POLYPEPTIDE COMPONENTS   Direct chemical identification of Na^+ channel components in situ was first achieved by specific covalent labeling of neurotoxin receptor site 3 with a photoreactive azidonitrobenzoyl derivative of the scorpion toxin from the North African scorpion *Leiurus quinquestriatus* (80). Irradiation with UV light causes covalent attachment of the specifically bound toxin derivative. Analysis of covalently labeled synaptosomes by polyacrylamide el electrophoresis under denaturing conditions in SDS reveals specific covalent labeling of two polypeptides of 260 kd and 36 kd that have subsequently been designated the α- and β1-subunits of the Na^+ channel (80). The covalent labeling of these two polypeptides in synaptosomes was shown to be specific by inhibition with unlabeled scorpion toxin or by block of voltage-dependent binding of scorpion toxin by membrane depolarization. The β-scorpion toxins derived from American scorpion venoms have been used to label neurotoxin receptor site 4 on the Na^+ channel (81, 82). Toxin γ from *Tityus serrulatus* was covalently attached to its receptor site by covalent cross-linking with disuccinimidyl suberate. A single polypeptide of 270 kd was labeled in rat brain synaptosomes (81). In contrast, photoreactive derivatives of toxin II from *Centruroides suffusus suffusus* label two polypeptides with molecular weights similar to the α and β1 subunits of the channel (82). These results suggest that neurotoxin receptor site 4, like site 3, is located near the contact regions of the α and β1 subunits of the sodium channel.

All of these labeling methods result in covalent attachment of amino groups in the toxins to nearby components of the sodium channel. Since each toxin contains multiple amino groups, the differential labeling of the α and β1 subunits of the sodium channel that has been observed may result from reaction with amino groups in different regions of the toxin. Evidence in favor of this has been developed by separation of photoreactive derivatives of the α-scorpion toxin from *L. quinquestriatus* by ion exchange chromatography (83). One derivative labeled only the α subunit, while the other preferentially labeled the β1 subunit. Evidently, individual amino groups on the *Leiurus* toxin can be in close proximity to either the α or β1 subunit when the toxin is bound at neurotoxin receptor site 3. Rapid labeling of a freshly prepared brain homogenate under conditions designed to minimize proteolysis provided evidence that the α and β1 subunits are present in the intact brain as well as in isolated synaptosomes (83).

RADIATION INACTIVATION   The size of functional units of membrane proteins can be estimated by measuring the target size for inactivation of that function by irradiation with high-energy X-rays. Targets with larger sizes are inactivated more rapidly. A single hit anywhere within the covalently bonded structure of the target protein is considered to be sufficient to inactivate the entire molecule. Measurements of the target size for inactivation of tetrodotoxin binding to the sodium channel in electroplax or brain membranes show that a structure of 230 kd to 260 kd is required for this activity (84). This size corresponds approximately to that of the α subunit and suggests that this subunit is required for saxitoxin and tetrodotoxin binding. More recent radiation inactivation studies of other neurotoxin binding activities yield more complicated results. These data are considered below under FUNCTIONAL ROLES OF THE SODIUM-CHANNEL SUBUNITS.

ISOLATION OF THE SAXITOXIN/TETRODOTOXIN RECEPTOR

SOLUBILIZATION AND PURIFICATION   An alternative approach to identification and characterization of the protein components of the sodium channel is to solubilize neurotoxin-binding activity with detergents and to analyze and purify the solubilized sodium channel components using neurotoxin binding as a specific assay. Henderson & Wang (85) and Benzer & Raftery (86) first showed that the tetrodotoxin-binding component of sodium channels in garfish olfactory nerve could be solubilized by nonionic detergents with retention of high affinity and specificity of toxin binding. Subsequent work has extended these findings to sodium channels in eel electroplax (87), mammalian brain (88, 89), mammalian skeletal muscle (90), and chicken heart (91). In contrast to the ease of solubilization of neurotoxin receptor site 1 with retention of saxitoxin- and tetrodotoxin-binding activity, both neurotoxin receptor site 2 (W. Catterall unpublished observations) and neurotoxin receptor site 3 (88) lose high-affinity neurotoxin-binding activity on solubilization. Binding activity at neurotoxin receptor site 4 is retained after solubilization (91, 92).

Although the saxitoxin/tetrodotoxin receptor of the sodium channel was successfully solubilized in 1972, the marked instability of the toxin-binding activity prevented progress on the purification of the sodium channel until the discovery that addition of phospholipid or phospholipid and calcium to the detergent-solubilized sodium channel markedly stabilized the saxitoxin-binding activity (87, 88, 93). This stabilization is proposed to result from formation of mixed detergent-phospholipid micelles having energetically favorable interaction with the solubilized channel. The broad specificity of this phospholipid requirement is consistent with this interpretation (93).

The saxitoxin receptors of the sodium channels from electric eel electroplax (92, 94, 95), rat brain (96–98), rat and rabbit skeletal muscle (99, 100), and chicken heart (91) have been purified to near homogeneity using a combination of conventional purification methods: anion exchange chromatography, adsorption chromatography on hydroxylapatite, affinity chromatography on wheat germ agglutinin-Sepharose, and size fractionation by velocity sedimentation in sucrose gradients or by gel filtration chromatography. Preparations from each of these tissues bind from 0.4 to 0.9 mol of [³H]saxitoxin or [³H]tetrodotoxin per mol of protein, indicating that a high degree of purity has been obtained (Table 2). A denatured form of the saxitoxin receptor from electric eel has also been purified by immunoaffinity chromatography (101). These solubilized and purified channel channel preparations have provided the necessary material for analysis of the molecular properties of the sodium channel.

Table 2 Properties of Purified Sodium-Channel Preparations[a]

Source	Oligomeric molecular mass (kd)	Specific saxitoxin binding (mol/mol)	Subunit molecular masses	
			Native (kd)	Deglycosylated (kd)
Eel electroplax	260	0.25–0.35 (94, 118)	260 (92, 94, 95, 101)	208 (95, 105, 106)
Mammalian brain	316 (102)	0.9 (98)	α: 260 (80, 96–98)	200–220 (104, 107, 141)
			β1: 36 (80, 96, 97, 98)	23 (104)
			β2: 33 (96, 97, 104)	21 (104)
Mammalian skeletal muscle	314 (103)	0.85 (100)	260 (100)	
			38 (99, 100)	

[a]Numbers in parentheses are reference numbers.

MOLECULAR WEIGHT OF THE SOLUBILIZED SAXITOXIN RECEPTOR   The size of the solubilized saxitoxin receptor has been estimated by hydrodynamic studies. The solubilized saxitoxin receptor/Triton X-100 complex from rat brain has a Stokes radius of 80 Å, a sedimentation coefficient of 12 S, and a partial specific volume of 0.82 cm³ g (102). These data define a molecular weight of 601,000 for the receptor/detergent complex. The contribution of protein and bound detergent can be estimated by comparison of the partial specific volume of the complex (0.82 cm³ g) with values for Triton X-100/phosphatidylcholine (5:1) (0.92 cm³ g) and for typical proteins (0.73 cm³ g). This comparison indicates that the detergent/receptor complex contains 0.9 g Triton X-100/phosphatidylcholine per gram of protein. The molecular weight of the saxitoxin receptor protein is therefore 316,000. A similar analysis of the saxitoxin receptor from skeletal muscle solubilized in Lubrol PX arrived at an estimate of 314,000 daltons (103).

SUBUNIT COMPOSITION OF THE PURIFIED SAXITOXIN RECEPTOR   The initial identification of the subunits of the sodium channel from rat brain by photoaffinity labeling in situ revealed two classes of polypeptides of 260 kd and 36 kd (80). Analysis of the purified saxitoxin receptor from rat brain has shown that it consists of three polypeptides: α of 260 kd, β1 of 36 kd, and β2 of 33 kd (96–98, 104). The α and β1 subunits, but not the β2 subunit, are covalently labeled by photoreactive scorpion toxin derivatives (83, 97). The β2 subunit is covalently attached to the α subunit by disulfide bonds while the β1 subunit is associated noncovalently (97, 104). Limited proteolytic maps of the β1 and β2 subunits indicate that they are nonidentical (104). The subunits appear to be present in a 1 : 1 : 1 stoichiometry (98), and the sum of their molecular weights (329,000, Refs. 97, 104) agrees closely with the oligomeric molecular weight of the solubilized saxitoxin receptor (316,000, Ref. 102).

The saxitoxin receptor purified from rat skeletal muscle sarcolemma was initially reported to have a very different subunit composition (90). However, more recent work with highly purified preparations made under conditions that limit proteolysis shows that both sarcolemmal and transverse tubule sodium channels consist of a major subunit of 260 kd and one or two smaller subunits of 38 kd (99, 100).

Saxitoxin receptor preparations from eel electroplax (92, 94, 95) and chicken heart (91) contain only a single polypeptide of 260 kd. This polypeptide is analogous in size (80, 94, 96), extent of glycosylation (95, 104–106), and amino acid sequence (107, 108) to the α subunit of the sodium channel from brain (see below). The specific saxitoxin-binding activity, molecular weight, and subunit composition of the saxitoxin receptor preparations from these different sources are summarized in Table 2.

POSTTRANSLATIONAL MODIFICATION OF THE SUBUNITS OF THE SAXITOXIN
RECEPTOR   The solubilized saxitoxin receptor is specifically adsorbed to and
eluted from lectin-Sepharose columns indicating that it is a glycoprotein (90,
96, 108). The α subunits from electroplax (95), brain (104, 105), and skeletal
muscle (99) are glycosylated. Twenty nine percent of the mass of the electroplax protein is carbohydrate (95), while 15–20% of the rat brain α subunit is
carbohydrate (104, 105). The β1 and β2 subunits of the sodium channel from
mammalian brain specifically bind wheat germ agglutinin and are heavily
glycosylated with 30–36% of their mass as carbohydrate (104, 105). The small
subunits of the saxitoxin receptor from skeletal muscle do not bind wheat germ
agglutinin and therefore may not be glycosylated (99). The apparent sizes of the
deglycosylated subunits are: α, 200 kd for electroplax or 220 kd for rat brain
(104); β1, 23 kd for rat brain (104); and β2, 21 kd for rat brain (104).

The sodium channel is also a phosphoprotein (110, 111). The α subunit of the
purified saxitoxin receptor from rat brain is rapidly and specifically phosphorylated at three or four sites by cAMP-dependent protein kinase (110, 111) and
protein kinase C (112) in solubilized form and in synaptosomal surface membranes. In addition, rapid phosphorylation by endogenous cAMP-dependent
protein kinase is observed when intact synaptosomes or cultured brain neurons
are incubated with 8-BrcAMP or the adenylate cyclase activator forskolin
(111). Phosphorylation of the α subunit in synaptosomes is accompanied by a
16–26% reduction in neurotoxin-stimulated $^{22}Na^+$ influx mediated by the
sodium channel.

RECONSTITUTION OF SODIUM-CHANNEL FUNCTION
FROM PURIFIED COMPONENTS

Restoration of full sodium channel function by incorporation of purified components into phospholipid vesicles provides the only rigorous proof that the
proteins identified and purified on the basis of their neurotoxin binding activity
are indeed sufficient to form a functional voltage-sensitive ion channel. Thus,
the goal of reconstitution experiments is to show, as quantitatively as possible,
that purified saxitoxin receptors of known protein composition can mediate
normal sodium channel functions. In addition, successful reconstitution provides a valuable experimental preparation for biochemical analysis of the
structure and function of sodium channels.

ION FLUX   Several groups of investigators have successfully restored aspects
of sodium channel function that are lost on detergent solubilization of neuronal
membranes and thereby shown that detergent solubilization does not irreversibly destroy channel functions (113–116). More recently, purified saxitoxin

receptors from rat skeletal muscle, rat brain, and eel electroplax have been incorporated into phosphatidylcholine vesicles by mixture with the phospholipid dispersed in nonionic detergent, followed by slow removal of detergent by adsorption to polystyrene beads (117–121). Single-walled phosphatidylcholine vesicles containing sodium channels are formed by this method. The channels are oriented approximately randomly with half of the saxitoxin/tetrodotoxin binding sites directed outward as in intact cells and half directed inward in inside-out configuration. Activation of the reconstituted sodium channels by incubation with veratridine or batrachotoxin increases the initial rate of $^{22}Na^+$ influx into the reconstituted channels 4–15-fold. Outside-out and inside-out channels contribute to the measured $^{22}Na^+$ influx equivalently. Appropriate concentrations of the alkaloid toxins are effective in activating the purified channels, and they are inhibited by appropriate concentrations of tetrodotoxin and local anesthetics. In the case of reconstituted channels from rat brain and skeletal muscle, the measured ion selectivity is $Na^+ > K^+ > Rb^+ > Cs^+$ as for native sodium channels (120, 121). Thus, there is strong evidence that at least a fraction of the saxitoxin receptor molecules in these highly purified preparations retain a selective ion channel that can be activated and inhibited by neurotoxins and drugs.

It is important in reconstitution experiments to estimate the fraction of purified saxitoxin receptors whose ion transport function can be restored. Only if a large fraction of the purified protein molecules are active in ion transport can it be concluded that the major species characterized biochemically is functionally active. Such an estimate has been made for sodium channels purified from mammalian brain by comparing the initial rate of neurotoxin-stimulated $^{22}Na^+$ influx in reconstituted vesicles to that in intact cells and by preparing vesicles having a mean of less than one reconstituted sodium channel each (4–6 pmol sodium channel per mg lipid), estimating the fraction of vesicles whose internal volume is accessible to neurotoxin-activated sodium channels, and applying appropriate statistical procedures to calculate the fraction of functionally active channels. Both these approaches lead to the conclusion that at least 30% and perhaps as many as 70% of the purified channels are functionally active (121). Since the protein preparations were judged 90% pure and no single contaminant comprised more than 2%, it was concluded that a complex of α, β1, and β2 subunits was sufficient to mediate neurotoxin-activated ion flux (98, 121).

Similar experiments have not been carried out with reconstituted sodium channels from skeletal muscle or electroplax. For skeletal muscle, the most recent experiments with a nearly homogeneous sodium channel preparation reconstituted at approximately 5 pmol/mg phospholipid yield half-times for vesicle $^{22}Na^+$ equilibration of <50 msec in the absence of ion gradients or a membrane potential (100). These rapid rates achieved with a low density of sodium channels in the vesicle bilayer are consistent with the possibility that a large fraction of the purified skeletal muscle channels are functionally active. For electroplax, reconstitution of purified saxitoxin receptor preparations at a much greater density (40–100 pmol/mg phospholipid) yielded half-times for vesicle $^{22}Na^+$ equilibration of approximately 6 sec (118). The reported specific flux rates (1000–3500 nmol/min/mg) correspond to 1.5–5.3 × 10^3 Na^+/ reconstituted sodium channel/min at 150 mM Na^+. These values are 200–600-fold lower than reconstituted rat brain sodium channels studied under similar conditions (121). The large amount of purified channel protein used in these reconstitution experiments and the slow rates of ion flux observed leave open the possibility that only a small fraction of the purified saxitoxin receptors from electroplax can form functional sodium channels.

VOLTAGE-DEPENDENT NEUROTOXIN BINDING   Although reconstitution into phosphatidylcholine vesicles is sufficient to restore neurotoxin-stimulated ion transport activity of the purified sodium channel, high-affinity binding of the α-scorpion toxin from *L. quinquestriatus* is not restored under these conditions (121). Since binding of toxins at neurotoxin receptor site 3 is voltage- and state-dependent (reviewed in 9), the failure to recover this binding activity likely indicates that the channel is locked in a state with low affinity for scorpion toxin. Reconstitution in mixtures of phosphatidylcholine and mixed brain lipids restores scorpion toxin binding (121). The requirement for brain lipids can be satisfied by addition of brain phosphatidylethanolamine alone or in combination with phosphatidylserine (122). Other minor brain lipids did not have detectable effects. In vesicles of 65% phosphatidylcholine/35% phosphatidylethanolamine, high-affinity binding of *Leiurus* scorpion toxin is restored in good yield (50–75%) and is voltage dependent (122). Evidently, the native voltage-dependent functional properties of the sodium channel from rat brain require reconstitution in an appropriate phospholipid environment.

SINGLE-CHANNEL RECORDING   The ion conductance of the sodium channel is normally regulated by changes in membrane potential on the msec time scale. This aspect of function of purified and reconstituted sodium channels has been studied by two different single-channel recording methods. These techniques have excellent time resolution and voltage control and are sufficiently sensitive to detect the square pulses of ionic conductance mediated by a single sodium channel. They have the disadvantage that the fraction of saxitoxin receptors whose functional activity has been restored cannot be determined.

Rosenberg et al (124) used the freeze-thaw method of Tank et al (123) to prepare large (5–20 μm) reconstituted vesicles of mixed phospholipid composition (phosphatidylethanolamine/phosphatidylserine/phosphatidylcholine = 5:4:1) containing purified electroplax sodium channels and recorded from excised patches of membrane attached across the tip of a micropipette (123). In some patches, single-channel currents were observed which had the appropriate conductance (11 pS, 5×10^6 ions per sec), ion selectivity ($P_{Na}/P_K = 7$), and time and voltage dependence of opening and closing to represent the activity of individual sodium channels (124). Thus, at least a fraction of sodium channels purified from electroplax are activated by membrane depolarization and have an appropriate time course of opening and closing on the msec time scale.

An alternative approach is to incorporate reconstituted sodium channels into planar phospholipid bilayers (125). The high electrical resistance of such bilayers allows measurement of the ion conductance mediated by a single ion channel. After treatment with batrachotoxin, sodium channels in synaptosomal membrane vesicles can be successfully incorporated into planar phospholipid bilayers by fusion of the membrane vesicles with preformed bilayers (126). The functional properties of batrachotoxin-modified sodium channels in native membranes are retained (126). The presence of batrachotoxin is required to block sodium channel inactivation and allow measurement of steady-state opening and closing of sodium channels as a function of time and membrane voltage.

Hanke et al (127) used this method to study partially purified and reconstituted saxitoxin receptors from both electroplax and rat brain consisting primarily of a single glycoprotein of 260 kd, without prior treatment with batrachotoxin to block inactivation. In contrast to the expected behavior of sodium channels, spontaneous conductance increases were observed at steady negative holding potentials but not at 0 m V or at positive voltages. Two different single-channel conductance events were observed: frequent conductance increases of 150 pS with $P_{Na}/P_K = 2.2$ and less frequent increases of 25 pS with $P_{Na}/P_K = 10$. Both of these values of single-channel conductance are higher than expected for sodium channels in physiological Na^+ concentration and the sodium selectivity of the larger conductance increase is lower than expected for sodium channels. However, inhibition by a high concentration of tetrodotoxin and modulation of the conductance time course by sea anemone toxin raise the possibility that these conductance events are mediated by purified sodium channels whose functional properties have been markedly altered during isolation.

Hartshorne et al (128) applied a similar recording technique to purified and reconstituted sodium channels from rat brain that consisted of a stoichiometric complex of α, $\beta 1$, and $\beta 2$ subunits (97, 98). No spontaneous conductance events were observed. Fusion of reconstituted vesicles in the presence of batrachotoxin was accompanied by the appearance of single-channel conductance events of 25 pS at 500 mM Na^+. The probability of channel opening was voltage-dependent, increasing from near 0 at -120 mV to 96.8% at all voltages more positive than -60 mV. The channels were open half of the time at -91 mV and the steepness of their voltage dependence corresponded to an apparent gating charge of 3.8 mV. Their ion selectivity (P_{Na} : P_K : P_{Rb} = 1 : 0.13 : 0.04), block by tetrodotoxin (K_D = 8.3 nM at −50 mV, 135 nM at 70 mV), single-channel conductance, and voltage dependence all agree precisely with those of native sodium channels in the presence of batrachotoxin. The results show that the voltage-dependent gating function, high single-channel conductance, and ion selectivity of the purified sodium channels recorded in these planar bilayers remain quantitatively intact after purification and reconstitution.

Considered together, the results of the different reconstitution studies complement each other and establish that the purified sodium channels from various sources retain functional properties. The ion flux results for rat brain and skeletal muscle show that the isolated sodium channel complexes of three nonidentical glycoprotein subunits are sufficient to form a selective, neurotoxin-activated ion channel. Patch clamp recordings from reconstituted electroplax sodium channels show that some of these isolated proteins can be activated by depolarization and retain a high single-channel conductance with appropriate ion selectivity in the absence of neurotoxins. Planar bilayer recordings of reconstituted rat brain sodium channels in the presence of batrachotoxin show that the voltage dependence of channel gating and the rate and selectivity of ion transport are quantitatively retained.

PRIMARY STRUCTURE AND MEMBRANE TOPOLOGY OF THE SODIUM-CHANNEL SUBUNITS

PRIMARY STRUCTURE OF THE α SUBUNIT   The availability of highly purified, functional preparations of sodium channels from various tissues has provided the necessary tools to undertake the isolation of cDNA clones encoding the primary structure of the protein components of the sodium channel. Noda et al (107) screened cDNA expression libraries formed in a plasmid vector with both antisera against the sodium channel from eel electroplax and oligonucleotide probes encoding known short segments of the amino acid sequence of that sodium channel which, as purified, consists of a single glycoprotein of 260 kd (92, 94, 95, 101). Short cDNAs were isolated and used to identify long cDNA clones which were subjected to sequence analysis to deduce the complete amino acid sequence of the protein. The inferred sequence consists of 1820 residues with a total size of 208,321 daltons in reasonable agreement with the size of the purified, deglycosylated protein from electroplax [180 kd, (106)] and the deglycosylated α subunit of the sodium channel from rat brain [220 kd, (104)]. In a parallel series of experiments, Dunn et al (108) used antisera directed against the α subunit of the rat brain sodium channel to identify cDNA clones in a bacteriophage expression library, determined their nucleotide sequence, and deduced the corresponding amino acid sequence of the α subunit of the mammalian brain sodium channel. The regions of the α subunit from rat brain that have been sequenced to date have an overall homology of 60% to the electroplax sequence. However, the extent of homology varies considerably over the length of the polypeptide.

Homology matrix comparison of the deduced sequence reveals the presence of four internal repeated domains of approximately 250 residues having greater than 50% homology in both the sequences from electroplax (107) and rat brain (108) as illustrated in Figure 1 for the electroplax protein. This high degree of sequence homology is strong evidence that these four domains arose from a common ancestor by gene duplication and that they adopt similar secondary structures in pseudosymmetric orientations in the channel structure.

TOPOLOGY OF THE α SUBUNIT   A number of lines of evidence indicate that the α subunit of the sodium channel is exposed to the extracellular environment:

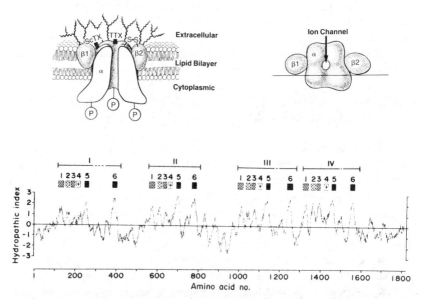

Figure 1   Topology and primary structure of the sodium channel. *Top.* A hypothetical model of sodium channel structure viewed in cross-section cut perpendicular to the plane of the membrane *(left)* or parallel to the plane of the membrane *(right).* The probable distribution of the three subunits of the mammalian channel is represented along with glycosylation, neurotoxin-binding sites, intersubunit disulfide bond, and sites of phosphorylation *(left).* The four homologous domains of the α subunit are shown forming a central transmembrane pore *(right).* Adapted from Ref. 152, with permission. *Bottom.* A plot of hydrophobicity vs amino acid sequence number for the electroplax sodium channel as adapted from Ref. 107. The four homologous domains (I, II, III, IV) and six of the helical segments within each domain (S1, S2, S3, S4, S5, S6) are noted. Adapted from Ref. 107 with permission.

it is glycosylated (95, 99, 104, 109) and it is covalently labeled by α- and β-scorpion toxins (80–83) and by tetrodotoxin derivatives (129) which act only on the extracellular side of the channel. The α subunit is also exposed on the intracellular side of the membrane since it is phosphorylated on four sites by cAMP-dependent protein kinase within intact synaptosomes (111) and intact brain neurons in cell culture (130). Thus, the α subunit is a transmembrane polypeptide as illustrated in the topographical model in Figure 1. Its size allows numerous membrane-spanning segments.

The amino acid sequence of the α subunit also supports the view that it is a transmembrane polypeptide with multiple membrane-spanning segments. Analysis of each domain for hydropathy (Figure 1) led Noda et al (107) to propose that each has at least four membrane-spanning regions as indicated by the cross-hatched and shaded areas. All of the predicted transmembrane segments (S1, S3, S5, and S6) are located in the four homologous domains (107, 108). Thus, these domains are likely to form four pseudosymmetric transmembrane units in the channel structure.

TOPOLOGY OF THE β1 AND β2 SUBUNITS   Both the β1 and β2 subunits of the sodium channel appear to be exposed at the extracellular surface. Each is heavily glycosylated with more than 30% of the apparent mass due to carbohydrate (104). The β1 subunit is covalently labeled by α- and β-scorpion toxins which act on the extracellular surface of the sodium channel (80, 82, 83). These two subunits have not been shown to be exposed at the intracellular surface. Since their protein mass is small (23 kd and 21 kd, respectively), and much of their mass must be extracellular to accommodate attachment of carbohydrate, they may not have extensive regions that protrude to the cytoplasmic surface of the membrane as illustrated in the topographical model of Figure 1.

FUNCTIONAL ROLES OF THE SODIUM-CHANNEL SUBUNITS

The functional role of individual sodium-channel subunits has been examined by four general approaches. The sites of neurotoxin binding have been identified by covalent labeling as described above. These data can be briefly summarized as follows: α- and β-scorpion toxins can specifically label both the α and β1 subunits of the sodium channel depending upon the method of covalent attachment and the derivative used (80–83). Tetrodotoxin derivatives can be covalently attached to the 260-kd component of the purified electroplax sodium channel (129), which is analogous to the α subunit of the mammalian channel. Thus neurotoxin receptor sites 1 and 3 are located on or near the α and β1 subunits, but not the β2 subunits. Further information on the functional roles of the sodium-channel subunits has been derived from radiation inactivation measurements, separation of the subunits under native conditions, and expression of mRNA encoding the channel subunits in *Xenopus* oocytes.

RADIATION INACTIVATION   The target size for inactivation of tetrodotoxin or saxitoxin binding to sodium channels in eel or mammalian brain membranes is 220–230-kd (84, 131). These results indicate that an intact α subunit is required for binding to neurotoxin receptor site 1 of sodium channels in native membranes. A requirement for subunits of the size of β1 and β2 would not be detected directly, since subunits with larger target size are inactivated more rapidly.

Binding of α-scorpion toxins to neurotoxin receptor site 3 is inactivated with a target size of 263 kd consistent with a requirement for an intact α-subunit (131). A parallel comparison of inactivation of the binding of saxitoxin and the α-scorpion toxin from *L. quinquestriatus* (131) detected a small, but possibly significant, difference in target size (220 kd vs 263 kd, respectively). These results were interpreted in terms of a requirement for only the α subunit for saxitoxin and tetrodotoxin binding, but both α and β1 subunits for α-scorpion toxin binding (131). This interpretation stretches the limits of resolution of the radiation inactivation method but agrees with earlier covalent labeling results (80, 82, 83) which place neurotoxin receptor site 3 at the contact regions between α and β1.

Two different target sizes have been reported for inactivation of binding β-scorpion toxins to neurotoxin receptor site 4 in synaptosomes: 266 kd (81) and 45 kd (131). This is a significant discrepancy since the former value indicates that the α subunit is required for toxin-binding at this site, while the latter suggests that the β1 subunit is both necessary and sufficient for this activity. While a number of methodological differences between the two studies have been discussed (131), it is important to resolve the inconsistency through additional experiments.

The target size for inactivation of the binding of [^{3}H]batrachotoxinin A 20-α-benzoate ([^{3}H]BTX-b) to neurotoxin receptor site 2 on sodium channels in synaptosomes has also been measured (131). Interpretation of these data is complicated by the fact that the standard BTX-b binding assay takes advantage of the allosteric interaction between α-scorpion toxins and BTX-b to increase the affinity and % specific binding by 20-fold (10). It would therefore seem that the target size could be no smaller than that for α-scorpion toxin binding (263 kd, 131) since destruction of each α-scorpion toxin-binding site should prevent high-affinity binding of BTX-b to that sodium channel. In contrast to this expectation, a curvilinear inactivation curve was observed which was fit to two exponentials corresponding to target sizes of 287 kd and 51 kd, suggesting roles for both α and β subunits in BTX-b binding. Further experiments are required to determine whether either of these values corresponds to the true functional target for neurotoxin binding at receptor site 2.

SUBUNIT SEPARATION   The functional saxitoxin/tetrodotoxin receptor purified from electroplax is composed of only a single glycoprotein of 260 kd analogous to the α subunit of the sodium channel from mammalian sources (92, 94, 95, 101). Thus, in electroplax, this polypeptide is autonomous in saxitoxin- and tetrodotoxin-binding in solubilized form. In addition, purified preparations can be incorporated into phospholipid vesicles (119) or planar bilayers (124) with recovery of normal ion transport properties. These results are consistent with the hypothesis that this single polypeptide is sufficient for physiological function (119, 124), but the low rates of ion flux measured (119) and the small number of functional single channels observed (124) do not yet allow a definitive conclusion to be drawn from these data as discussed above.

Selective removal of the β1 and β2 subunits from the purified mammalian sodium channel complex under mild conditions provides a useful approach to assessment of their functional role. The β1 subunit can be removed by treatment at high ionic strength (1.0 M $MgCl_2$) followed by sucrose gradient sedimentation (132). Dissociation of β1 is accompanied by stoichiometric loss of saxitoxin binding. Both dissociation of β1 and loss of saxitoxin binding are prevented in the presence of tetrodotoxin. Thus, the binding energy of tetrodotoxin is sufficient to stabilize the αβ1 complex against dissociation. This thermodynamic linkage between binding of tetrodotoxin and binding of the β1 subunit by the α subunit indicates that the conformation of the α subunit of the rat brain sodium channel having high affinity for saxitoxin and tetrodotoxin requires association with the β1 subunit, at least in detergent-solubilized form. In contrast to these results, which suggest an intimate association between the α and β1 subunits in maintaining a functional state of the channel, the disulfide-linked β2 subunit can be selectively removed by reduction of disulfide bonds in the presence of tetrodotoxin with no loss of saxitoxin binding (132). Moreover, the αβ1 complex is sufficient to mediate neurotoxin-activated ion flux and voltage-dependent scorpion toxin-binding when incorporated into phospholipid vesicles (133). Thus, these experiments indicate a requirement for β1, but not β2, in maintaining function of the purified and reconstituted sodium channel.

EXPRESSION OF mRNA ENCODING SODIUM CHANNEL SUBUNITS   Crude poly A+ mRNA fractions from mammalian brain and skeletal muscle direct the synthesis of functional sodium channels when injected into *Xenopus* oocytes (134, 135). Fractionation of these mRNA preparations by sucrose gradient sedimentation yields a single peak of RNA of undetermined size which directs sodium channel synthesis in oocytes (136). These results suggest that mRNA of a single size class is sufficient to encode a functional sodium channel. However, the low resolving power of the sucrose gradient separation method, the complex relationship between mRNA size and mobility in this separation technique, and the possibility of overlapping of mRNA species of different size do not allow a clear conclusion concerning which sodium channel subunits may be encoded by the injected mRNA. A definitive conclusion on this point awaits expression of mRNA synthesized in vitro from cloned cDNA probes encoding the individual subunits.

BIOSYNTHESIS OF THE SODIUM CHANNEL

In mammalian nerve and possibly also in skeletal muscle, the sodium channel is a heterotrimeric complex of three glycoprotein subunits and the α and $\beta2$ subunits are linked by disulfide bonds. The three subunits must therefore undergo several steps of posttranslational processing before a structurally mature sodium channel can be produced. Initial evidence for the importance of these posttranslational processing events has been derived from studies of the effects of tunicamycin on sodium channel levels in cultured neuroblastoma (137) and muscle (138) cells. Inhibition of N-linked protein glycosylation with tunicamycin reduces the level of functional sodium channels as measured by saxitoxin binding to less than 20% of normal in 48 hours of treatment. The number of saxitoxin-binding sites is reduced with a half-life of 18–22 hours, suggesting that the mature sodium channels present on the cell surface before treatment with tunicamycin have a half-life in this range.

More recently, the biosynthesis and precursor forms of the sodium channel in primary cultures of embryonic rat brain neurons have been studied with isotopic labeling methods (139, 140) and specific antisera directed primarily against the α subunit (111). Labeling of the entire sodium channel population of rat brain neurons with $[\gamma\text{-}^{32}P]ATP$ and cAMP-dependent protein kinase followed by immunoprecipitation reveals that approximately two-thirds of the α subunits in these cells, and in neonatal rat brain in vivo, are not linked to $\beta2$ subunits by disulfide bonds (139). Further analysis showed that these free α subunits are full-sized, membrane-associated, and have complex carbohydrate chains like mature α subunits. However, they are located in the intracellular compartment and are inactive in binding saxitoxin. Free α subunits are not observed in adult rat brain. Thus, it was proposed that they form an inactive reserve of sodium channels for incorporation into the cell surface during periods of rapid membrane assembly in development (139).

The first form of the newly synthesized α subunit, detected after a 5-min pulse of $[^{35}S]$methionine, has an apparent size of 224 kd (140). In the presence of tunicamycin to block cotranslational glycosylation, the newly synthesized α polypeptide is 203 kd (140). The amino acid sequence of the α subunit does not reveal a cleavable hydrophobic leader sequence (107, 108). Consistent with this, the size of this initial translation product increases progressively during further processing. Over two hours, this initial precursor undergoes a single-step increase in apparent size to 249 kd followed by a slow increase to the mature apparent size 260 kd. During this time, the carbohydrate chains are processed to give complex structures containing N-acetylglycosamine and sialic acid as indicated by binding to the carbohydrate-binding protein wheat germ agglutinin (140). After 1 hr, newly synthesized α subunits begin to be linked to β2 subunits by disulfide bonds and this process continues for approximately 8 hours until an average maximum of one-third of the newly synthesized α subunits have been linked to β2 subunits (140). α subunits disulfide-linked to β2 subunits are preferentially localized on the cell surface, suggesting that disulfide bond formation is a late event in sodium channel processing and assembly (139, 140). Since two-thirds of newly synthesized α subunits do not form a disulfide bond with β2 and remain in an inactive intracellular pool, it appears that posttranslational processing and assembly are rate-limiting steps in biosynthesis of functional cell-surface sodium channels in developing brain neurons.

The number of cell-surface sodium channels in cultured excitable cells is regulated by several different kinds of stimuli. The density of sodium channels in clonal pheochromocytoma (PC12) cells is increased 20-fold by nerve growth factor (154, 155). Electrical activity of the muscle cells causes a feedback down-regulation of sodium channel number which is mediated by changes in cytosolic calcium and cAMP (141, 142). These changes are observed without changes in channel degradation rate, implying that one or more of the steps in channel biosynthesis and processing must be the point of regulation (142). In contrast, persistent depolarization of cultured muscle cells by activation of sodium channels with batrachotoxin causes down regulation of channel number by increasing turnover rate as measured in the presence of cycloheximide (143). It will be of interest to determine which of the individual steps in the process of synthesis and degradation of sodium channels are affected by these stimuli.

TOWARD A STRUCTURAL MODEL OF SODIUM-CHANNEL FUNCTION

Information on the structure of the sodium channel is now sufficiently complete to warrant informed speculation on the structural basis of selective ion transport and voltage-dependent gating. The following two sections attempt to synthesize results from structural and functional studies in arriving at unified hypotheses.

IONIC CHANNEL STRUCTURE   The structures of two high-conductance ionic channels are known in more detail than the sodium channel and thus provide useful models for comparison. The gap junction channel which connects adjacent cells is formed by symmetrical hexagonal arrays of six identical subunits in the surface membrane of each cell. The subunits of each hexameric connexon are approximately cylindrical and are oriented perpendicular to the plane of the membrane (144). Electron diffraction and image reconstruction show that a transmembrane channel is formed at the center of each hexagonal array of subunits (144). The nicotinic acetylcholine receptor consists of a pseudosymmetric pentagonal array of five homologous subunits (reviewed in 145). As in the gap junction channel, a transmembrane channel is formed in the center of this symmetrical array of subunits with their long axes perpendicular to the membrane (146). These two examples suggest the generalization that high conductance ionic channels will form their transmembrane pore at the center of symmetrical arrays of homologous structural units.

The amino acid sequence of the α subunit of the sodium channel contains four internally homologous transmembrane domains as discussed above (107, 108). In light of the structure of the ionic channels of gap junctions and nicotinic acetylcholine receptors, Noda et al (107) proposed that the transmembrane pore of the sodium channel is formed at the center of a square array of the four transmembrane domains (Figure 1). The suggestion that this polypeptide can form a functional ion channel by itself is supported by its transmembrane orientation (107, 108, 111), by reconstitution studies on the purified sodium channel from electroplax (119, 124), and by expression studies of mRNA purified by size fractionation (136). Thus, the sodium channel may accomplish with homologous domains of a single polypeptide what the gap junction channel and the nicotinic acetylcholine receptor do with arrays of homologous subunits.

Analysis of the amino acid sequence of the subunits of the nicotinic acetylcholine receptor has revealed the presence of a segment that can form an amphipathic α-helix in each subunit (147, 148). These segments have repeated sequences with two or three hydrophobic residues followed by a charged residue. Thus, when in the α helix form, one side of the helix is hydrophobic and one is highly charged. These segments have now been shown to span the membrane and are proposed to form the walls of the ionic channel (149, 150).

In scanning the transmembrane segments of the sodium channel for similar amino acid sequences, some interesting features emerge. The segments S4 of each homologous domain contain sequences of lys or arg followed by two or three hydrophobic residues in a pattern repeated 4–6 times (107, 108). While the high concentration of positive charge in this structure is of interest in considerations of sodium channel gating (see below), it seems unlikely that these four segments by themselves could form the walls of a cation-selective ionic channel. The other α-helical segments do not show evidence for formation of amphipathic helixes that is as clear as the proposed components of the ionic channel of the acetylcholine receptor. However, each of these sets of helical segments contains charged and hydrophilic residues that may contribute to formation of the ionic channel. Further experimental work is required before the transmembrane orientation and role in formation of the transmembrane pore can be defined for any of these helical segments of the sodium-channel protein.

VOLTAGE-DEPENDENT ACTIVATION: THE SLIDING HELIX MODEL   As discussed above and in Ref. 4, activation of the sodium channel is considered to be a voltage-driven conformational change in which the equivalent of approximately six positive charges in the channel structure move outward across the membrane permeability barrier or a larger number of charges move a shorter distance across the membrane. The movement of protein-bound charge gives rise to gating current. The requirement for such large charge movements within the sodium channel protein focuses attention on the four positively charged S4 segments of the homologous domains of the sodium channel. Noda et al (107) suggested that these regions might be involved in channel gating. However, their topological model placed these segments on the cytoplasmic side of the membrane where they would be little affected by the membrane electrical field and no specific suggestions of gating mechanism were made. An alternative topological model places the S4 segments across the membrane but does not assign a clear role in channel gating (151). The purpose of this final section is to consider a specific mechanism whereby these charged elements might interact to mediate voltage-dependent gating: the *sliding helix model*.

The requirement for movement of six charges across the membrane places critical restraints on the underlying structural mechanisms. Mechanisms in which individual amino acid residues are translocated all the way across the permeability barrier seem excluded. Two other classes of mechanisms should be considered. 1. Diffuse gating charge mechanisms in which a large number of charges distributed widely in the protein structure move a fraction of the way across the membrane on activation are plausible. However, such mechanisms do not lead to specific and testable structural models since voltage-dependent gating is proposed to be a holistic property encompassing a large fraction of the molecule. 2. Localized gating charge mechanisms in which full charge transfer is postulated to result from simultaneous neutralization of one positive charge at the inner membrane surface and exposure of a different positive charge at the extracellular surface are also plausible. For these mechanisms, a process by which simultaneous charge exposure and neutralization is coupled to protein conformational change must be proposed. The structural features of the S4 segments of the sodium channel seem well designed for such a charge transfer model. The functional importance of these segments is supported by the finding that their amino acid sequence is essentially completely conserved from eel electroplax to rat brain (108). No other region of the channel structure is as highly conserved.

In considering how gating charge might be transferred across the membrane, Armstrong (4) suggested a model in which each gating element contains a row of paired positive and negative charges that extends across the membrane and acts as a charge transfer mechanism. Changes in electric field cause translocation of the negative charges relative to the positive charges resulting in loss of a positive charge on the intracellular side of the protein and appearance of a positive charge on the extracellular side. Thus, full charge translocation is achieved without movement of individual charged species fully across the bilayer. If the segments S4 are postulated to span the membrane, this helix forms a rigid structural element placing positive charges at regular intervals across the permeability barrier (Figure 2). Negative charges from other regions of the protein structure are postulated to form ion pairs with each of the positive charges of the helixes. Figure 2 illustrates a hypothetical voltage-sensing element. Depolarization alters the electrical force on the intramembranous charges causing negative charges to move inward and positive charges to move outward. An outward spiral movement (5 Å, 60° rotation) of the α-helix with respect to the array of negative charges causes an exchange of ion pair partners and creates an unneutralized positive charge in the extracellular compartment and an unneutralized negative charge in the intracellular compartment for a gating charge movement of +1. Movement of amino acid residues over similar distances have been observed in studies of conformational changes in much smaller proteins and therefore seem plausible. This movement of the S4 α-helix then initiates a conformational change in its domain as one step in channel activation.

Activation of the sodium channel follows a sigmoidal time course indicating that the channel protein must pass through multiple (at least three) nonconducting states before channel activation (1, 4). Analysis of gating current data suggests there are five transitions before activation occurs and the last requires movement of twice as much gating charge (4). If each of the four homologous domains contains a voltage-sensing element analogous to that in Figure 2 and undergoes an independent voltage-driven conformational change, a sequence of four changes of state is expected before activation. The sliding helix model of voltage-sensing therefore provides a mechanism to connect the sigmoid time course of channel activation with the presence of four homologous domains.

Movement of six positive charges outward through the membrane permeability barrier accompanies activation of the sodium channel (4). The voltage-sensing element depicted in Figure 2 mediates a movement of gating charge of +1. In fact, the S4 segments of domains I and II contain only four or five positive charges while those in domains III and IV contain six that would be membrane-associated (107, 108). Domains III and IV may undergo two cycles of ion pair exchange. The total gating charge of +6 may arise from progressive

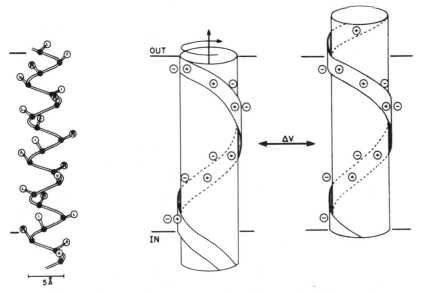

Figure 2   The sliding helix model of voltage-dependent gating. *Left.* A ball-and-stick, three-dimensional representation of the S4 helix of domain IV. Darkened circles represent the α-carbon of each amino acid residue. Open circles show the direction of projection of the side chain of each residue away from the core of the helix. Nonpolar residues are illustrated in thin letters by their single letter code: F, phe; A, ala; I, ileu; L, leu; V, val; G, glycine; T, threonine. Positively charged amino acids are illustrated in bold letters: R, arg. *Right.* The spiral path made by the positively charged arginine residues is illustrated in the form of a ribbon wrapped around the central core of the helix. *Right.* Movement of the S4 helix in response to membrane depolarization. The proposed transmembrane S4 helix is illustrated as a cylinder with a spiral ribbon of positive charge. At the resting membrane potential *(left),* all positively charged residues are paired with fixed negative charges on other transmembrane segments of the channel and the transmembrane segment is held in that position by the negative internal membrane potential. Depolarization reduces the force holding the positive charges in their inward position. The S4 helix is then proposed to undergo a screw-like motion through a rotation of approximately 60° and an outward displacement of approximately 5 Å. This movement leaves an unpaired negative charge on the inward surface of the membrane and reveals an unpaired positive charge on the outward surface to give a net charge transfer of +1. Adapted from Ref. 153 with permission.

voltage-driven transitions of the four domains yielding +1, +1, +2, and +2 charges transferred.

The sliding helix model attempts to define a workable structure for the voltage-sensing elements of the sodium channel. Each of these elements is proposed to initiate a conformational change in its own domain. The structure of each individual domain defines the energetic barrier to its voltage-driven conformational change and therefore the sequence of conformational transitions that occurs during the sigmoid time course of activational. Additional structural information is needed to construct a mechanism for coupling of the conformational changes in each domain to opening of a high conductance transmembrane pore and to eventual inactivation of the ion conductance mechanism.

Literature Cited

1. Hodgkin, A. L., Huxley, A. F. 1952. *J. Physiol.* 117:500–44
2. Hille, B. 1976. *Ann. Rev. Physiol.* 38:139–52
3. Ulbricht, W. 1977. *Ann. Rev. Biophys. Bioeng.* 6:7–31
4. Armstrong, C. M. 1981. *Physiol. Rev.* 61:644–83
5. Hille, B. 1984. *Ionic Channels of Excitable Membranes.* Sunderland, Mass: Sinauer
6. Narahashi, T. 1974. *Physiol. Rev.* 54: 813–89
7. Albuquerque, E. X., Daly, J. W. 1976. *Receptors and Recognition,* 1:299–336. London: Chapman & Hall
8. Lazdunski, M., Renand, J. F. 1982. *Ann. Rev. Physiol.* 44:463–73
9. Catterall, W. A. 1980. *Ann. Rev. Pharmacol. Toxicol.* 20:15–43
10. Catterall, W. A., Morrow, C. S., Daly, J. W., Brown, G. B. 1981. *J. Biol. Chem.* 256:8922–27
11. Cahalan, M. D. 1975. *J. Physiol.* 244: 511–34
12. Wang, G. K., Strichartz, G. 1982. *Biophys. J.* 40:175–79
13. Meves, H., Rubly, N., Watt, D. D. 1982. *Pfluegers Arch.* 393:56–62
14. Couraud, F., Jover, E., Dubois, J. M., Rochat, H. 1982. *Toxicon* 20:9–16
15. Jover, E., Couraud, F., Rochat, H. 1980. *Biochem. Biophys. Res. Commun.* 95: 1607–14
16. Barhanin, J., Giglio, J. R., Leopold, P., Schmid, A., Sampaio, S. V., Lazdunski, M. 1982. *J. Biol. Chem.* 257:12553–58
17. Levinson, S. R., Meves, H. 1975. *Proc. R. Soc. London Ser.* B 270:349–52
18. Almers, W., Levinson, S. R. 1975. *J. Physiol.* 247:483–509
19. Conti, F., DeFelice, L. J., Wanke, E. 1975. *J. Physiol.* 248:45–82
20. Conti, F., Hille, B., Neumcke, B., Nonner, W., Stampfi, R. 1976. *J. Physiol.* 262:729–42
21. Sigworth, F. J. 1980. *J. Physiol.* 307:97–129
22. Sigworth, F. J., Neher, E. 1980. *Nature* 287:447–49
23. Nagy, K., Kiss, T., Hof, D. 1983. *Pfluegers Arch.* 399:302–8
24. Stark, G., Benz, R., Lauger, P. 1971. *Biophys. J.* 11:981–94
25. Chandler, W. K., Meves, H. 1965. *J. Physiol.* 180:788–820
26. Moore, J. W., Anderson, N. C., Blaustein, M. P., Takata, M., Lettvin, J. Y., et al. 1966. *Ann. NY Acad. Sci.* 137:818–31
27. Hille, B. 1971. *J. Gen. Physiol.* 58: 599–619
28. Hille, B. 1972. *J. Gen. Physiol.* 59: 637–58
29. Hille, B. 1968. *J. Gen. Physiol.* 51: 199–219
30. Begenesich, T. B., Cahalan, M. D. 1980. *J. Physiol.* 307:217–42
31. Hille, B. 1975. *J. Gen. Physiol.* 66:535–60
32. Begenisich, T. B., Calahan, M. D. 1980. *J. Physiol.* 307:243–57
33. Henderson, R., Ritchie, J. M., Strichartz, G. R. 1973. *J. Physiol.* 235:783–804
34. Balerna, M., Fosset, M., Chicheportiche, R., Romey, G., Lazdunski, M. 1975. *Biochemistry* 14:5500–11
35. Henderson, R., Ritchie, J. M., Strichartz, G. R. 1974. *Proc. Natl. Acad. Sci. USA* 71:3936–40
36. Reed, J. K., Raftery, M. A. 1976. *Biochemistry* 15:944–53
37. Weigele, J. B., Barchi, R. L. 1978. *FEBS Lett.* 95:49–53
38. Shrager, P., Profera, C. 1973. *Biochim. Biophys. Acta* 318:141–46
39. Baker, P. F., Rubinson, K. A. 1975. *Nature* 257:412–14
40. Baker, P. F., Rubinson, K. A. 1976. *J. Physiol.* 266:3–4P
41. Spalding, B. C. 1979. *J. Physiol.* 305: 485–500
42. Sigworth, F. J., Spalding, B. C. 1980. *Nature* 283:293–95
43. Armstrong, C. M., Bezanilla, F. 1973. *Nature* 242:459–61
44. Bezanilla, F., Armstrong, C. M. 1974. *Science* 183:753–54
45. Keynes, R. D., Rojas, E. 1974. *J. Physiol.* 239:393–434

46. Armstrong, C. M., Bezanilla, F. 1977. *J. Gen. Physiol.* 70:567–90
47. Catterall, W. A., Ray, R., Morrow, C. S. 1976. *Proc. Natl. Acad. Sci. USA* 73:2683–86
48. Catterall, W. A. 1977. *J. Biol. Chem.* 252:8660–68
49. Catterall, W. A. 1979. *J. Gen. Physiol.* 74:375–91
50. Strichartz, G. R. 1973. *J. Gen. Physiol.* 62:37–57
51. Courtney, K. R. 1975. *J. Pharmacol. Exp. Ther.* 195:225–36
52. Hille, B. 1977. *J. Gen. Physiol.* 69:497–515
53. Bartels-Bernal, E., Rosenberry, T. L., Daly, J. W. 1977. *Proc. Natl. Acad. Sci. USA* 74:951–55
54. Khodorov, B. I. 1978. *Membrane Transport Processes*, 2:153–74. New York: Raven
55. Campbell, D. T. 1982. *J. Gen. Physiol.* 80:713–31
56. Catterall, W. A. 1975. *Proc. Natl. Acad. Sci. USA* 72:1782–86
57. Catterall, W. A. 1977. *J. Biol. Chem.* 252:8669–76
58. Aldrich, R. W., Corey, D. P., Stevens, C. F. 1983. *Nature* 306:436–41
59. Rojas, E., Armstrong, C. M. 1971. *Nature* 229:177–78
60. Armstrong, C. M., Bezanilla, F., Rojas, E. 1973. *J. Gen. Physiol.* 62:375–91
61. Rojas, E., Rudy, B. 1976. *J. Physiol.* 262:501–31
62. Eaton, D. C., Brodwick, M. S., Oxford, G. S., Rudy, B. 1978. *Nature* 271:473–76
63. Oxford, G. S., Wu, C. H., Narahashi, T. 1978. *J. Gen. Physiol.* 71:227–47
64. Brodwick, M. S., Eaton, D. C. 1978. *Science* 100:1494–96
65. Salgado, V. L., Yeh, J. Z., Narahashi, T. 1985. *Biophys. J.* 47:567–71
66. Catterall, W. A. 1975. *J. Biol. Chem.* 250:4053–59
67. Catterall, W. A. 1976. *J. Biol. Chem.* 251:5528–36
68. Catterall, W. A., Beress, L. 1978. *J. Biol. Chem.* 253:7393–96
69. Jacques, Y., Fosset, M., Lazdunski, M. 1978. *J. Biol. Chem.* 253:7383–92
70. Tamkun, M. M., Catterall, W. A. 1981. *Mol. Pharmacol.* 19:78–86
71. Ray, R., Morrow, C. S., Catterall, W. A. 1978. *J. Biol. Chem.* 253:7307–13
72. Lawrence, J. C., Catterall, W. A. 1981. *J. Biol. Chem.* 256:6223–29
73. Stengelin, S., Hucho, F. 1980. *Hoppe-Seyler's Z. Physiol. Chem.* 361:577–85
74. Angelides, K., Nutter, T. J. 1983. *J. Biol. Chem.* 258:11958–67
75. Angelides, K., Brown, G. B. 1984. *J. Biol. Chem.* 259:6117–26
76. Huang, L. Y. M., Catterall, W. A., Ehrenstein, G. 1979. *J. Gen. Physiol.* 73:839–54
77. Mozhayeva, G. N., Naumov, A. P., Negulyeva, Y. A., Nosyreva, E. D. 1977. *Biochim. Biophys. Acta* 466:461–73
78. Hironaka, T., Narahashi, T. 1977. *J. Membr. Biol.* 31:359–81
79. Frelin, C., Vigne, P., Lazdunski, M. 1981. *Eur. J. Biochem.* 119:437–42
80. Beneski, D. A., Catterall, W. A. 1980. *Proc. Natl. Acad. Sci. USA* 77:639–42
81. Barhanin, J., Schmid, A., Lombet, A., Wheeler, K. P., Lazdunski, M. 1983. *J. Biol. Chem.* 258:700–2
82. Darbon, H., Jover, E., Couraud, F., Rochat, H. 1983. *Biochem. Biophys. Res. Commun.* 115:415–22
83. Sharkey, R. G., Beneski, D. A., Catterall, W. A. 1984. *Biochemistry* 23:6078–86
84. Levinson, S. R., Ellory, J. C. 1973. *Nature* 245:122–23
85. Henderson, R., Wang, J. H. 1972. *Biochemistry* 11:4565–69
86. Benzer, T. I., Raftery, M. A. 1973. *Biochem. Biophys. Res. Commun.* 51:939–44
87. Agnew, W. S., Levinson, S. R., Brabson, J. S., Raftery, M. A. 1978. *Proc. Natl. Acad. Sci. USA* 75:2606–10
88. Catterall, W. A., Morrow, C. S., Hartshorne, R. P. 1979. *J. Biol. Chem.* 254:11379–87
89. Krueger, B. K., Ratzlaff, R. W., Strichartz, G. R., Blaustein, M. P. 1979. *J. Membr. Biol.* 50:287–310
90. Barchi, R., Cohen, S. A., Murphy, L. E. 1980. *Proc. Natl. Acad. Sci. USA* 77:1306–10
91. Lombet, A., Lazdunski, M. 1984. *Eur. J. Biochem.* 141:651–60
92. Norman, R. I., Schmid, A., Lombet, A., Barhanin, J., Lazdunski, M. 1983. *Proc. Natl. Acad. Sci. USA* 80:4164–68
93. Agnew, W. S., Raftery, M. A. 1979. *Biochemistry* 18:1912–19
94. Agnew, W. S., Moore, A. C., Levinson, S. R., Raftery, M. A. 1980. *Biochem. Biophys. Res. Commun.* 92:860–66
95. Miller, J. A., Agnew, W. S., Levinson, S. R. 1983. *Biochemistry* 22:462–70
96. Hartshorne, R. P., Catterall, W. A. 1981. *Proc. Natl. Acad. Sci. USA* 78:4620–24
97. Hartshorne, R. P., Messner, D. J., Coppersmith, J. C., Catterall, W. A. 1982. *J. Biol. Chem.* 257:13888–91
98. Hartshorne, R. P., Catterall, W. A. 1984. *J. Biol. Chem.* 259:1667–75

99. Barchi, R. L. 1983. *J. Neurochem.* 40: 1377–85
100. Kraner, S. D., Tanaka, J. C., Barchi, R. L. 1985. *J. Biol. Chem.* 260:6341–47
101. Nakayama, H., Withy, R., Raftery, M. A. 1982. *Proc. Natl. Acad. Sci. USA* 79:7575–79
102. Hartshorne, R. P., Coppersmith, J., Catterall, W. A. 1980. *J. Biol. Chem.* 255:10572–75
103. Barchi, R. L., Murphy, L. E. 1981. *J. Neurochem.* 36:2097–3000
104. Messner, D. J., Catterall, W. A. 1985. *J. Biol. Chem.* 260:10597–604
105. Grishin, E. V., Kovalenko, V. A., Pashkov, V. N., Shamotienko, O. G. 1984. *Biol. Membr.* 1:858–67
106. Agnew, W. S., Miller, J. A., Ellisman, M. H., Rosenberg, R. L., Tomiko, S. A., Levinson, S. R. 1983. *Cold Spring Harbor Symp. Quant. Biol.* 48:165–79
107. Noda, M., Shimizu, S., Tanabe, T., Takai, T., Kayano, T., et al. 1984. *Nature* 212:121–27
108. Dunn, R., Goldin, A., Dowsett, A., Downey, W., Catterall, W. A., Davidson, N. 1985. *J. Gen. Physiol.* 86:10a
109. Cohen, S. A., Barchi, R. L. 1981. *Biochim. Biophys. Acta* 645:253–61
110. Costa, M. R. C., Casnellie, J. E., Catterall, W. A. 1982. *J. Biol. Chem.* 257: 7918–21
111. Costa, M. R. C., Catterall, W. A. 1984. *J. Biol. Chem.* 259:8210–18
112. Costa, M. R. C., Catterall, W. A. 1984. *Cell. Mol. Neurobiol.* 4:291–97
113. Villegas, R., Villegas, G. M. 1981. *Ann. Rev. Biophys. Bioeng.* 10:387–419
114. Malysheva, M. K., Lishko, V. K., Chagovetz, A. M. 1980. *Biochim. Biophys. Acta* 602:70–77
115. Goldin, S. M., Rhoden, V., Hess, E. J. 1980. *Proc. Natl. Acad. Sci. USA* 77: 6884–88
116. Tamkun, M. M., Catterall, W. A. 1981. *J. Biol. Chem.* 256:11457–63
117. Weigele, J. B., Barchi, R. L. 1978. *FEBS Lett.* 95:49–53
118. Talvenheimo, J. A., Tamkun, M. M., Catterall, W. A. 1982. *J. Biol. Chem.* 252:1007–13
119. Rosenberg, R. L., Tomiko, S. A., Agnew, W. S. 1984. *Proc. Natl. Acad. Sci. USA* 81:1239–43
120. Tanaka, J. C., Eccleston, J. F., Barchi, R. L. 1983. *J. Biol. Chem.* 258:7519–26
121. Tamkun, M. M., Talvenheimo, J. A., Catterall, W. A. 1984. *J. Biol. Chem.* 259:1676–88
122. Feller, D. J., Talvenheimo, J. A., Catterall, W. A. 1985. *J. Biol. Chem.* 260:11542–47
123. Tank, D. W., Miller, C., Webb, W. W. 1982. *Proc. Natl. Acad. Sci. USA* 79:7749–53
124. Rosenberg, R. L., Tomiko, S. A., Agnew, W. S. 1984. *Proc. Natl. Acad. Sci. USA* 81:5594–98
125. Mueller, P., Rudin, D. O., Tien, H. T., Wescott, W. C. 1962. *Nature* 194: 979–80
126. Krueger, B. K., Worley, J. F., French, R. J. 1983. *Nature* 303:172–75
127. Hanke, W., Boheim, G., Barhanin, J., Pauron, D., Lazdunski, M. 1984. *EMBO J.* 3:509–15
128. Hartshorne, R. P., Keller, B. U., Talvenheimo, J. A., Catterall, W. A., Montal, M. 1985. *Proc. Natl. Acad. Sci. USA* 82:240–44
129. Lombet, A., Norman, R. I., Lazdunski, M. 1983. *Biochem. Biophys. Res. Commun.* 114:126–30
130. Rossie, S. R., Catterall, W. A. 1986. *Fed. Proc.* In press
131. Angelides, K. J., Nutter, T. J., Elmer, L. W., Kempner, E. S. 1985. *J. Biol. Chem.* 260:3431–39
132. Messner, D. J., Catterall, W. A. 1986. *J. Biol. Chem.* 261:211–15
133. Messner, D. J., Feller, D. J., Catterall, W. A. 1986. *J. Biol. Chem.* In press
134. Gundersen, C. B., Miledi, R., Parker, I. 1984. *Nature* 308:421–24
135. Gundersen, C. B., Miledi, R., Parker, I. 1984. *Proc. R. Soc. London Ser. B* 221:235–44
136. Sumikawa, K., Parker, I., Miledi, R. 1984. *Proc. Natl. Acad. Sci. USA* 81: 7994–98
137. Waechter, C. J., Schmidt, J. W., Catterall, W. A. 1982. *J. Biol. Chem.* 258: 5117–23
138. Bar-Sagi, D., Prives, J. 1983. *J. Cell. Physiol.* 114:77–81
139. Schmidt, J., Rossie, S., Catterall, W. A. 1985. *Proc. Natl. Acad. Sci. USA* 82:4847–51
140. Schmidt, J., Catterall, W. A. 1985. *Cell.* In press
141. Sherman, S. J., Catterall, W. A. 1984. *Proc. Natl. Acad. Sci. USA* 81:262–66
142. Sherman, S. J., Chrivia, J., Catterall, W. A. 1985. *J. Neurosci.* 6:1570–76
143. Bar-Sagi, D., Prives, J. 1985. *J. Biol. Chem.* 260:4740–44
144. Unwin, P. N. T., Zampighi, G. 1980. *Nature* 283:545–49
145. Popot, J. L., Changeux, J. P. 1984. *Physiol. Rev.* 64:1162–239
146. Kistler, J., Stroud, R. M., Klymkowski, M. W., Lalancette, R. A., Fairclough, R. H. 1982. *Biophys. J.* 37:371–83
147. Finer-Moore, J., Stroud, R. M. 1984. *Proc. Natl. Acad. Sci. USA* 81:155–59

148. Guy, H. R. 1984. *Biophys. J.* 45:249–61
149. Young, E. F., Ralston, E., Blake, J., Ramachandran, J., Hall, Z. W., Stroud, R. M. 1985. *Proc. Natl. Acad. Sci. USA* 82:626–30
150. Lindstrom, J., Criado, M., Hochschwender, S., Fox, L., Sarin, V. 1984. *Nature* 311:573–75
151. Kosower, E. M. 1985. *FEBS Lett.* 182:234–41

152. Catterall, W. A. 1986. *New Insight into Cell and Membrane Transport Processes,* ed. G. Poste, S. G. Crooke. New York: Plenum. In press
153. Catterall, W. A. 1986. *Trends Neurosci.* 9:7–10
154. Dichter, M. A., Tischler, A. S., Greene, L. A. 1977. *Nature* 268:501–4
155. Rudy, B., Kirschenbaum, B., Greene, L. A. 1982. *J. Neurosci.* 2:1405–11

Ann. Rev. Biochem. 1986. 55:987–1035
Copyright © 1986 by Annual Reviews Inc. All rights reserved

ACTIN AND ACTIN-BINDING PROTEINS. A CRITICAL EVALUATION OF MECHANISMS AND FUNCTIONS

Thomas D. Pollard

Department of Cell Biology and Anatomy, Johns Hopkins Medical School, Baltimore, Maryland 21205

John A. Cooper

Departments of Pathology and Biological Chemistry, Washington University School of Medicine, St. Louis, Missouri 63110

CONTENTS

987

0066-4154/86/0701-0987$02.00

PERSPECTIVES AND SUMMARY

Actin is an abundant, highly conserved protein that polymerizes into filaments that are essential for many forms of cellular motility, including muscle contraction, as well as the structure and mechanical properties of the cytoplasmic matrix. Owing to its importance in biology, actin has been studied in considerable detail, especially during the past 10 years. The field went through a somewhat euphoric stage in the early 1980s when the basic parameters of the polymerization process were elucidated and the major classes of actin-binding proteins were discovered. It seemed possible that the assembly and function of the actin system in cells might be explained by relatively simple mechanisms involving a small handful of proteins. However, since then it has become increasingly clear that complexities abound in this system and that the number of regulatory proteins may exceed 10 (even without considering isoforms) in most cells. Also, our perceptions of the mechanisms of action of actin-binding proteins have become increasingly complex. Even for apparently straightforward proteins, testing of simple mechanisms with extreme conditions, complex experiments, and new techniques has uncovered unexpected results that point to more complicated mechanisms. The simple picture that springs to mind at first glance is often not the entire story. This situation provides investigators with experimental and theoretical challenges, but promises fascinating insights as these mechanisms are elucidated.

At present there is a rather complete picture of how the primary structure of actin varies throughout the phylogenetic tree and the detailed three-dimensional structure of actin is nearly solved by X-ray crystallography. In contrast, there is a major controversy regarding the structure of the actin filament and how the actin molecule is oriented in the filament. Transient state kinetic experiments have recently provided a detailed quantitative description of the mechanism of polymerization. Although the approximate values of the rate constants for each major step are now available, more refined experiments continue to reveal important subtleties that are necessary to fully understand the assembly process. Much of the recent work in the field has been on proteins that bind to actin monomers or actin filaments to regulate polymerization or attach the filaments to other filaments, microtubules, membranes, or other structures in the cell. The mechanisms of some of these proteins are now understood in detail. Although most of these proteins have been localized in cells, there is almost no direct experimental evidence regarding the functions of these putative regulatory and structural proteins in living cells. This will be an important area of investigation in the future.

INTRODUCTION

The literature on actin and actin-binding proteins is enormous and a comprehensive summary of the field is beyond the scope of this review. Fortunately there are several excellent reviews. Korn (1) provides a comprehensive review of the literature on actin up to 1982. This is the best entry into the classical literature in the field. Frieden (2) covers work on the mechanism of polymerization up to mid-1984, emphasizing an evaluation of kinetic methods. Amos (3) reviews the studies up until mid-1984 on the structure of actin filaments and their complexes with tropomyosin and myosin. She provides an expert's evaluation of the technical problems that have led to the current uncertainties regarding the structure of these filaments. Stossel et al (4) cover the major classes of actin-binding proteins in detail and provide the most complete bibliography of references available through early 1985. Myosin is probably the most important molecule that binds to actin, and although it cannot be covered in our review, it is fortunate that two excellent reviews have recently been published: Harrington & Rodgers (5) cover the chemistry of muscle myosin; and Hibberd & Trentham (6) consider the relationship between chemical and mechanical events during muscle contraction.

Provided with these comprehensive reviews, we are able to focus our attention on the literature in the two-year period up to early 1986 and to evaluate the credibility of the major concepts in the field. We will pose a series of questions regarding molecular structure, reaction mechanisms, and cellular function and then summarize how well the available evidence provides answers. This, we hope, will clearly define the extent and limits of our knowledge and thereby the directions for future work. To keep the number of citations within reasonable limits we will usually refer to the most recent work on a topic, with the understanding that earlier important work will be referenced in these current papers and the reviews mentioned above.

THE ACTIN MOLECULE

Comparisons of primary structures show that the actin molecule has been highly conserved during evolution (see 1). Except for a series of variable acidic residues at the N terminus, there are only minimal ($<$5%) differences in the sequences of muscle and cytoplasmic actins of animals and protozoa. Plants have multiple actins that differ from each other and from animal actins by more than 10% (7). It is also well established that there are multiple actin genes that are differentially expressed in the various tissues of higher organisms (reviewed in 8). This survey of the phylogenetic tree makes it possible to ask:

How Has the Actin Molecule Evolved?

Based on amino acid sequence homologies, all metazoan cytoplasmic actins and the muscle actins of invertebrates appear to be direct descendants of the cytoplasmic actins found in protozoa and fungi, while the muscle actins of vertebrates form an easily distinguishable family that has subdivided more than twice to yield the distinctive actins found in the four types of muscle: skeletal muscle, cardiac muscle, vascular smooth muscle, and enteric smooth muscle (Figure 1 and Ref. 9). Plants diverged from animals 10^9 years ago, and their actin genes have evolved separately. In contrast to these conclusions based on coding sequences, comparisons based on intron positions indicate that vertebrate cytoplasmic actins are related more closely to vertebrate muscle actins than to the primitive actins in the muscles and nonmuscle cells of invertebrates (8). Presumably this means that there are different selective pressures on the protein sequence and the positioning of the introns.

How is the N terminus Acetylated?

All actins have the N-terminal amino acid blocked with an acetyl group that is added after synthesis (10). First the N-terminal methionine is acetylated and then removed by specific enzymes. For nonmuscle actins the next residue, an aspartic acid, is acetylated to complete the process. For muscle actins, the next residue, a cysteine, is acetylated and then cleaved to reveal an aspartic or glutamic acid residue. This acidic amino acid is then acetylated to complete the process. Nonmuscle cells contain the enzymes required to process muscle actin correctly (11).

What is the Structure of the Actin Molecule?

A low-resolution (i.e. 1.5-nm) model of the actin molecule obtained from electron microscopy and 3-D structure reconstruction of negatively stained crystalline sheets of pure actin (Figure 2A; 12, 13) showed that actin is bi-lobed, consisting of a large and a small domain. The overall dimensions are $3.3 \times 5.6 \times 5.0$ nm relative to the near-rectangular unit cell of $6.5 \times 5.5 \times 5.0$ nm that contains two actin molecules. However, the largest dimension of the molecule is closer to 6.5 nm along a diagonal.

There are 3-D co-crystals of 1:1 complexes of actin with DNase I (14) and with profilin (15) that have been subjected to X-ray diffraction analysis. The best-published structure (Figure 2B) is of the actin–DNase I complex at a

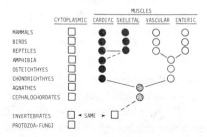

Figure 1 Evolution of the actin molecule in animals. Squares represent cytoplasmic actins that are as a group relatively invariant in animal cells. In vertebrates there are two major cytoplasmic actins and at least one minor cytoplasmic actin expressed. Note that invertebrates use a cytoplasmic actin in their muscles. Circles represent muscle actins. Plant actin genes have evolved separately into three classes that differ considerably from each other and from all of the actins shown here. Based on a diagram from Vandekerckhove & Weber (9).

resolution of 0.45 nm. Since the DNase I structure has been determined independently to 0.25-nm resolution, it has been possible to identify unambiguously the actin molecule in the composite map, even though the resolution is insufficent to trace its polypeptide chain. In this map the actin subunit has overall dimensions of $3.7 \times 4.0 \times 6.7$ nm and consists of a large and a small domain separated by a pronounced cleft. According to this model, the N-terminal ~ 150 amino acids are thought to be confined to the small domain, and the C-terminal $\sim 60\%$ of the polypeptide to the large domain. The highest electron density peak is found in the cleft and most likely represents the bound ATP. The small domain probably contains the high-affinity binding site for divalent cations and the large domain contains the nucleotide binding site (16, 16a). Both the domains contain a beta-pleated sheet, each surrounded by several alpha-helical stretches. There is good qualitative agreement between this actin model and the one derived by electron microscopy (cf Figures 2A, B). There is also an electron density map of the actin-profilin complex at 0.35 nm resolution, in which about 70% of the 375 amino acids of the actin subunit have already been tentatively identified (C. Shutt, personal communication), so there is hope that by the time this review is published, a near-atomic model of the actin molecule may finally be available.

What is the Molecular Structure of the Actin Filament?

For 30 years the helical parameters of the functionally important actin filament have been established by X-ray diffraction of oriented gels of filaments and of live muscle and by electron microscopy (Figure 2C). The actin subunits in the filament can be placed along a one-start, left-handed ('genetic') helix with a 5.9-nm pitch and a screw-angle of $-166°$. Another way to describe the filament is as a two-start, right-handed ('long-pitch') helix with a 71.5-nm pitch. The two long-pitch helixes are staggered by 2.75 nm and cross over every 35.8 nm. This latter representation is primarily gathered from electron micrographs of negatively stained (Figure 2C) or unidirectionally metal-shadowed actin filaments.

As summarized by Amos (3), the X-ray data are not adequate to solve the actin filament structure at molecular resolution, and when prepared for electron microscopy, the filaments are usually so disordered that the actin subunits cannot be resolved unambiguously. Attempts to improve the order by lateral aggregation of filaments into paracrystalline arrays have led to problems associated with 'superposition' and/or 'interdigitation' artifacts (3, 13).

What we consider the best presently available 3-D reconstruction of a single actin filament is shown in Figure 2D (16). According to the model, which has been contoured to include 100% mass, the filament is clearly built as a double-stranded helix with significant intersubunit contacts along the two long-pitch helixes. In contrast to earlier models, the filament appears narrowest

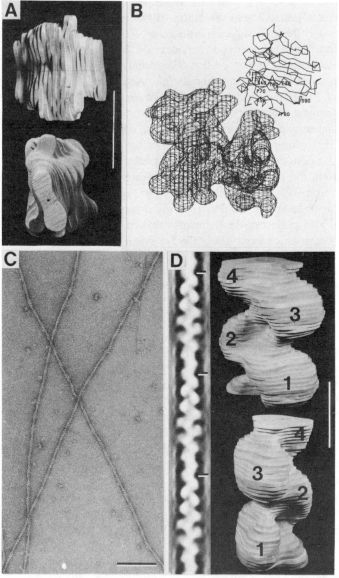

Figure 2   Actin molecules and filaments. *A*. A 1.5-nm resolution model of the actin molecule from a 3-D reconstruction from electron micrographs of negatively stained two-dimensional crystals of actin (12, 13). *B*. A 0.45-nm resolution model of the complex of actin-DNase-I from X-ray crystallography. The peptide chain of the DNase-I is shown on the upper right. The remainder of the model is actin (14). *C*. An electron micrograph of negatively stained actin filaments. The bar is 100 nm (17). *D*. A filtered image of a micrograph of a negatively stained actin filament and two views of a 3-D reconstruction of four subunits in such a filament. The numbers indicate the positions of four subunits. The bar is 5.5 nm (17).

(~6.5 nm) where the two strands come to lie side-by-side and widest (~8.2 nm) where the two strands lie on top of each other and are laterally staggered by 1.5–2 nm. In the model shown in Figure 2D polar actin subunits can be tentatively identified. They have overall dimensions of 3.3 × 5.5 × 6.0 nm with the 5.5-nm axis parallel to the filament axis. Thus the overall size and shape of the actin subunit deduced from the filament reconstruction (Figure 2D) is in good agreement with that obtained from the crystalline sheets (Figure 2A). There are several other proposals for how to orient the actin molecule in these filaments [see Amos (3)]. The correct model will become obvious once the molecular structure is available, since in filaments cysteine-374 is located less than 0.14 nm from lysine-191 of a neighboring subunit judging from chemical cross-linking experiments (Figure 3; 18, 19). Comparisons of the reactivity of lysines in monomers and polymers (20) may also help in the alignment of the subunits in the filament.

How do Nucleotides and Divalent Cations Bind to Actin?

Actin monomers bind both adenine nucleotides and divalent cations (Table 1; reviewed in Ref. 1). In low-salt buffers ATP binds to monomeric actin with a K_d in the nanomolar range. Assuming an invarient, diffusion-limited association rate constant of $6 \times 10^6 \, M^{-1}s^{-1}$, Ca^{2+}-actin binds ATP several times stronger than Mg^{2+}-actin (21). In 20-mM Tris with 0.8-mM Ca^{2+}, ATP binds almost 200 times stronger than ADP, but in buffers containing physiological concentrations of K^+ and Mg^{2+} the affinity for ATP is only slightly greater than for ADP (22). The absolute values of the rate and equilibrium constants under physiological conditions have not been reported. The situation is complicated, in part, because ATP binds Mg^{2+} and the reactivities of the free and bound species are not known. With regard to the mechanism of polymerization, the key step is the dissociation rate of each nucleotide, because it is presumably the rate-limiting step in nucleotide exchange.

Actin has a single high-affinity and multiple low-affinity binding sites for divalent cations (Table 1). Analysis of binding to the high-affinity site has been facilitated by the discovery that the fluorescence of actin labeled with N-iodoacetyl-N'-(5-sulfo-1-napthyl)ethylenediamine (IAEDANS) is sensitive to the binding of Mg^{2+}, due to a slow conformational change induced by the Mg^{2+} (23). Mg^{2+} initially binds weakly ($K_1 = 0.9$ mM at pH 8, 4.7 mM at pH 7)

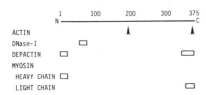

Figure 3   Linear map of the actin molecule with intermolecular contacts determined by chemical cross-linking indicated below. The actin-actin sites at lysine 191 and cys 374 are given by arrows. The other sites are given by bars corresponding to the cross-linked peptides (18, 19).

Table 1 Ligand binding constants

Ligand	Conditions	K_D (M)	k_+ $(M^{-1}s^{-1})$	k_- (s^{-1})	N	Ref.
ATP	0.8-mM CaCl$_2$ ⎱	1.3×10^{-10}	6×10^6	0.8×10^{-3}	1	244
	20-mM Tris, pH 8.2, 21° ⎰	7×10^{-10}	6×10^6	4×10^{-3}	1	21
	0.05-mM MgCl$_2$, 21°	23×10^{-10}	6×10^6	14×10^{-3}	1	21
	2-mM Tris, pH 8.0					
ADP	0.8-mM CaCl$_2$	2.2×10^{-8}	0.04×10^6	0.8×10^{-3}	1	244
	20-mM Tris, pH 8.2, 21°					
	1-mM MgCl$_2$, 0.1-mM CaCl$_2$			5×10^{-3}	1	62
	5-mM Tris, pH 7.8					
Mg^{2+}	10-µM ATP	0.045×10^{-3}			1	21
	2-mM Tris, pH 8					
	10-µM ATP	0.2×10^{-3}			1	21
	2-mM imidazole, pH 7					
Ca^{2+}	10-µM ATP	0.007×10^{-3}			1	21
	2-mM Tris, pH 8					
	10-µM ATP	0.01×10^{-3}			1	21
	2-mM imidazole, pH 7					
	2-mM HEPES, pH 7.6	0.2×10^{-3}			5	245

followed by isomerization of the actin ($k_{+2} \sim 0.2$ s^{-1}) that results in tight binding ($K_2 = 45$ µM at pH 8, 200 µM at pH 7).

$$A_1 + Mg^{2+} \underset{}{\overset{K_1}{\rightleftarrows}} A_1\text{-}Mg^{2+} \underset{k_{-2}}{\overset{k_{+2}}{\rightleftarrows}} A_1^*\text{-}Mg^{2+} \qquad \qquad 1.$$

In this and the following equations A_1 is actin monomer, K_d is a dissociation constant, k_+ is a forward rate constant, and k_- is a reverse rate constant. Ca^{2+} and Mg^{2+} compete for the same high-affinity site. Independent evidence for an Mg^{2+}-induced conformational change comes from the ability of Mg^{2+} to dissociate an antibody from actin; these antibodies bind to actin between residues 168 and 226 (24). See footnote on p. 1035.

MECHANISM OF POLYMERIZATION

The polymerization of actin can be conveniently divided into four steps: 1. activation (salt-binding and conformational changes in monomers); 2. nucleation (the formation of oligomers having a higher probability of growing into filaments than decomposing to monomers); 3. elongation (the bidirectional growth of polymers); and 4. annealing (the end-to-end joining of two filaments). All of these steps are reversible. For example, the reverse of annealing is the fragmentation of a filament into two shorter polymers. Because nucleation is slow and rate-limiting, the time course of spontaneous polymerization of monomers is sigmoidal with a lag at the beginning (Figure 4). In addition to these macromolecular steps, bound ATP is usually hydrolyzed during polymerization. Korn (1), Hill & Kirschner (25), Frieden (2), and Frieden & Goddette (26) provide good theoretical backgrounds for the following analysis. Frieden (2) also evaluates the various methods to measure polymerization.

Actin monomers are usually purified by cycles of polymerization and depolymerization, and monomers are usually separated from oligomers and minor contaminants by gel filtration in a low ionic strength buffer. Polymerization can be induced in many ways, usually by adding monovalent (KCl) or divalent ($MgCl_2$, $CaCl_2$) salts. Frieden (2) recommends using a single salt, while we usually use KCl, $MgCl_2$, and EGTA (to chelate Ca^{2+}) to mimic cytoplasmic conditions. Some apparent differences in the literature may be due to this lack of uniformity in conditions. In most ways muscle and cytoplasmic actins polymerize similarly, especially under conditions close to those in cells, but there are exceptions (see Ref. 1).

What is Monomer Activation?

Changes in sensitivity to trypsin (27) and UV absorption (28) first indicated that salt might induce a conformational change in actin prior to the association of actin molecules with each other to form a nucleus or longer polymer. The best evidence for such a conformational change is still the slow fluorescence change in IAEDANS-actin caused by binding of Mg^{2+} to the high-affinity site (Equation 1; Ref. 21). There is also a first order, rate-limiting step in the polymerization pathway when Ca^{2+}-actin is polymerized in Mg^{2+} (29–31). This first order reaction has a rate constant of about 0.05–0.33 s^{-1}, similar to the rate of the Mg^{2+}-induced conformational change in actin monomer (Equation 1, $k_{+2}=0.2$ s^{-1}; 21), so it seems reasonable to conclude that it is the monomer activation step. This step does not occur when Ca-actin is polymerized in Ca^{2+} and can be

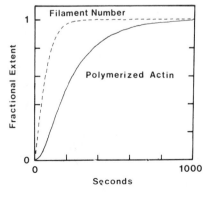

Figure 4 Time course of nucleation and polymerization of 10.9-μM actin in 100-mM KCl, 1-mM $MgCl_2$. Filament number and actin polymer concentration are normalized and calculated from a mathematical model for the processes. The time course of polymerization closely fits the experimental data used to select the values for the variables in the mathematical model (30).

circumvented by exchanging bound Ca^{2+} for Mg^{2+} prior to initiating polymerization. ADP-actin binds Mg^{2+} but does not change conformation or become activated like ATP-actin (32).

Activated Mg^{2+}-actin forms nuclei more than 10 times faster than unactivated actin (29, 31). It also elongates filaments more rapidly than Ca^{2+}-actin (33–35). Consequently, the overall polymerization process is much faster with Mg^{2+}-actin than Ca^{2+}-actin and with Mg^{2+}-ATP-actin than with Mg^{2+}ADP-actin.

What is Nucleation and How Can it be Evaluated?

The nucleus for polymerization is usually defined as the smallest actin oligomer that is more likely to grow into a filament than to dissociate into monomers. There is also a general consensus (29–31, 36) that the nucleus is most likely to be an actin trimer, formed by three actin molecules as follows:

$$A_1 + A_1 \underset{k_{-3}}{\overset{k_{+3}}{\rightleftarrows}} A_2 \qquad\qquad A_2 + A_1 \underset{k_{-4}}{\overset{k_{+4}}{\rightleftarrows}} A_3 \text{ (nucleus)} \qquad\qquad 2.$$

Nucleation is the rate-limiting step in spontaneous polymerization of actin monomers, because the reactions are so unfavorable, with dissociation constants in the range of 0.1 to 1 M (Table 2)! Formation of nuclei is responsible for the lag at the outset of spontaneous polymerization of actin (Figure 4).

In many ways nucleation is the most interesting and important step in polymerization, so it is unfortunate that it can only be studied indirectly, because assays for dimers or trimers, that could be used to evaluate the individual reactions, are not available. Instead, most of what we know about nucleation has been learned from kinetic modeling (reviewed in Ref. 2). This

Table 2   Nucleation constants

Conditions	$K_3 = \dfrac{k_{-3}}{k_3}$	$K_4 = \dfrac{k_{-4}}{k_{+4}}$	Nucleus size	Ref.
100-mM KCl, 1-mM $MgCl_2$, pH 8, 25°C	0.15	0.15	3	30[a]
100-mM KCl, 1-mM $MgCl_2$, pH 7, 25°C	0.07	0.07	3	29[b]
100-mM KCl, 0.1-mM $CaCl_2$, pH 8, 25°C	1.4	1.4	3	30[a]
100-mM KCl, 0.2-mM $CaCl_2$, pH 7, 25°C	1.3	1.3	3	29[b]
2-mM $MgCl_2$, pH 8, 25°C	0.17	0.17	3	30[a]
2-mM $MgCl_2$, pH 8, 20°C	0.08	5×10^{-6}	3	31[c]

[a]Absolute values calculated here from the relative complex constants in Ref. 30 using elongation association constants of 10^7, 5×10^6, and 10^7 $M^{-1}s^{-1}$. In this paper the nucleus is defined as one subunit larger than in other papers, and it has been redefined here to agree with others.
[b]Elongation rate constants measured.
[c]Elongation rate constants assumed to be 10^6 $M^{-1}s^{-1}$.

method has limitations, so one cannot be confident about the mechanism of nucleation given above.

The strategy of kinetic modeling is to use a computer to select a set of constants that will give the best fit of a mathematical model for all of the steps in polymerization to a set of experimental kinetic curves for the full time course of polymerization over a wide (6- to 10-fold) range of actin monomer concentrations. To obtain absolute values for the nucleation constants one must measure the elongation rate constants directly (29) or make some assumption about their value (31). Tobacman & Korn (30) measured relative elongation rate constants, so they obtained comparative, not absolute nucleation constants.

Most of the kinetic modeling work (29, 30) has employed a simplification proposed by Wegner & Engel (36), namely that all of the steps leading to the nucleus are rapidly reversible equilibrium reactions (a reasonable assumption) with the same equilibrium constants (a simple but undoubtedly incorrect assumption). A computer is used to solve a set of differential equations describing activation, nucleation, and elongation, varying the size of the nucleus and the equilibrium constant for nucleation until the theoretical curves simultaneously fit the whole data set (Table 2). Alternatively, using interactive computer simulation of the full time course of polymerization, Frieden (31) found the best fit using a much larger dissociation constant (~ 0.1 M) for dimers than for trimers (5 μM). This general approach is probably more realistic than the steady-state assumption of Wegner & Engel, but the constants obtained from the two methods are similar, lending some credence to both approaches. A third approach is to use a perturbation method that can simplify the mathematics of the theoretical analysis (37).

None of the modeling approaches have revealed the rate constants of the individual reactions or any details, such as the real possibility that there are separate steps for the formation of collision intermediates and for subsequent conformational changes in the oligomers. However, if the association reactions are diffusion limited (e.g. $k_{+3} = k_{+4} \approx 10^7 \text{M}^{-1}\text{s}^{-1}$), the dimer dissociation rate constants are in the range of 10^6 to 10^7s^{-1}! These large dissociation rates emphasize the instability of the intermediates on the nucleation pathway.

During spontaneous polymerization of 10–20-μM actin, the concentration of dimers and trimers is expected to be very low ($\ll 10^{-9}$M) and their lifetimes quite short. First the high rate of dissociation causes most dimers and trimers to disintegrate rapidly. Second, elongation (see next section) is so fast that trimers are consumed by growth into longer filaments that are much more stable. On the other hand, the total number of filaments continues to accumulate throughout the polymerization process, reaching a maximum when about half of the total actin is polymerized (Figure 4).

The nucleation rate depends on the solution conditions. Dimers and trimers are much more stable in Mg^{2+} than Ca^{2+}, presumably due to the conformation change induced by Mg^{2+} (23). ATP-actin forms nuclei very much faster than ADP-actin (32, 38). However, quite unexpectedly, it appears that mixtures of ATP-actin and ADP-actin form nuclei more rapidly in 1-mM $MgCl_2$, 0.1-mM $CaCl_2$ than ATP-actin alone (39), even though the nucleation rate of ADP-actin alone is negligible under these conditions. A plausible explanation is that the heterologous interaction is stronger than either homologous interaction.

A second approach for studying nucleation has been to test covalently cross-linked actin dimers and trimers for their ability to act as nuclei. As predicted by the kinetic modeling, trimers are much more effective nuclei than dimers (40, 41). However, the mechanism in these experiments appears to be more complex than anticipated. When Mg-ATP-actin monomers polymerize in 1-mM $MgCl_2$ with trimers as nuclei, there is a substantial lag at the outset of the reaction. Such a lag is not seen with Mg-ADP-actin monomers. Lal et al (41) proposed a two-step addition of a monomer to the trimer (equilibrium binding followed by a slow first order conformational change) to form an active nucleus. Another possibility suggested by Dr. U. Aebi (personal communication) is that the "trimers" are long-pitch helix trimers, not genetic helix trimers, and that the additional slow step is simply the unfavorable filling in of this structure to form a nucleus.

How do Filaments Elongate and Shorten?

Elongation refers to the association and dissociation of actin molecules at the ends of filaments. Since all of the subunits in a filament have the same polarity and the subunits at the ends of filaments are likely to differ in conformation from free monomers, the reactions of monomers must differ at the two ends. Based on the arrowhead pattern created when myosin heads bind to actin, one end is called the pointed (P) end and the other the barbed (B) end. Both ATP-actin and ADP-actin can react with the ends of filaments, so we must consider, at a minimum, eight different elongation reactions (Figure 5). In the nomenclature used here (42) k_+ represents an association rate constant and k_- is a dissociation constant. The superscripts B and P refer to the two ends of the filament. The superscripts T and D refer to the ATP and ADP.

Most of the rate constants have now been measured (Table 3), providing a full quantitative picture of the elongation process. Electron microscopy with three different types of morphologically identifiable nuclei has given consistent results in four different laboratories and is the only way to evaluate the reactions at the two ends of the filament independently. Spectroscopic assays with known concentrations of nuclei (e.g. actin trimers) give similar values for the rate constants, but they are, of course, the sum of the rate constants at the two ends. Since the spectroscopic constants are about half of the electron microscopic (EM) values (that are themselves minimal values), part of the trimers were probably not active in nucleating filaments. The only values that differ con-

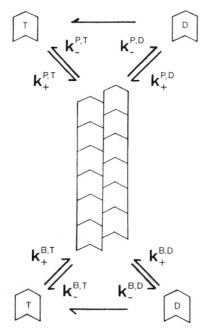

Figure 5   Reactions of actin monomers with the ends of actin filaments. T represents bound ATP and D represents bound ADP. The rate constants are defined in the text (44).

siderably from the average (Table 3), are the P-end rate constants obtained with brevin-capped actin filaments as nuclei. These values are lower than expected, so the number of active nuclei may have been overestimated.

For both ATP-actin and ADP-actin the growth is strongly biased toward the barbed end. For ATP-actin k_+^B is very large ($\sim 10^7$ $M^{-1}s^{-1}$ under optimal conditions), and the reaction is diffusion-limited judging from its dependence on viscosity and temperature (43). The association rate constant is smaller (~ 1–2×10^6 $M^{-1}s^{-1}$) at the pointed end. The dissociation rate constants are about 1 s^{-1}. ADP-actin has smaller association rate constants and larger dissociation rate constants than ATP-actin at the barbed end (38, 44). Preliminary EM experiments support the conclusion (44) that the rate constants are small for ADP-actin at the pointed end (D. Drenckhahn, T. Pollard, unpublished experiments).

At each end the rate of change of length of a filament is

$$\frac{dl}{dt} = k_+ A_1 - k_- \qquad\qquad 3.$$

When $k_+(A_1) = k_-$ the length of the filament is constant. This value of A_1 is called the *critical concentration*. The critical concentration can also be determined by measuring the minimum concentration of actin required to form

Table 3 Elongation rate constants

Method	Conditions	Barbed end			Pointed end			Ref.
		k_+ ($\mu M^{-1} s^{-1}$)	k_- (s^{-1})	$\bar{A}_1$ (μM)	k_+ ($\mu M^{-1} s^{-1}$)	k_- (s^{-1})	$\bar{A}_1$ (μM)	
ATP-actin								
EM (microvillar cores)	75-mM KCl, 5-mM MgSO$_4$	8.8	2.0	0.23	2.2	1.4	0.64	42
	20-mM KCl, 0.1-mM CaCl$_2$	5.9	6.0	1.0	0.8	0.7	0.9	42
EM (acrosomal processes)	75-mM KCl, 5-mM MgSO$_4$	12.3	2.0	0.16	1.5	0.7	0.5	103
	75-mM KCl, 0.2-mM CaCl$_2$			0.4			0.4	103
EM (decorated filaments)	150-mM KCl, 5-mM MgSO$_4$	3.0	0.3	0.1	1.2	1.4	1.2	131
EM (acrosomal processes)	50-mM KCl, 1-mM MgCl$_2$, 1-mM EGTA	8.4	0.5	0.15	1.1	0.5	0.5	43
EM (decorated filaments)	100-mM KCl, 0.2-mM CaCl$_2$	9.5	4.2	0.44	1.1	0.8	0.73	29
Fluorescence (trimer nuclei)	1-mM MgCl$_2$	1.8	0.6	0.34				41
	100-mM KCl, 1-mM MgCl$_2$	5.2	0.4	0.07				41
	1-mM CaCl$_2$	1.4	1.8	1.3				41
Fluorescence (trimer nuclei)	100-mM KCl, 1-mM CaCl$_2$	1.6	0.6	0.40				41
Fluorescence (brevin capped filaments)	1.5-mM MgCl$_2$, 0.05-mM CaCl$_2$				0.02	0.05	2.9	108
ADP-actin								
Fluorescence (filament nuclei)	50-mM KCl, 1-mM MgCl$_2$, 1-mM EGTA	1.3	6	5.0				44
Fluorescence (filament nuclei)	1-mM MgCl$_2$, 0.1-mM CaCl$_2$	0.7–1.8	5–12	8.0				59
Fluorescence (trimer nuclei)	1-mM MgCl$_2$	0.8	6.4	8.0				38
	100-mM KCl, 1-mM MgCl$_2$	0.9	1.8	2.0				38

any polymer either in solution or by EM. It is generally presumed that no polymer exists below the critical concentration, but laser light scattering or fluorescence photobleaching recovery (45, 46) detect species with diffusion coefficients larger than monomers at subcritical concentrations of actin in ~0.1-mM $MgCl_2$. Using neutron diffraction Goddette et al (47) find that these species are dimers. If these dimers participated in polymerization, one of the main assumptions in the field would be incorrect. However, the dimers may well be a dead-end side product that is not involved with polymerization (47).

For ATP-actin the critical concentrations at the two ends may be the same or different depending on the solution conditions (Table 3). In the presence of Mg^{2+}, the substantial difference at the two ends was well established by electron microscopy (Table 3) and has also been confirmed in solution studies with barbed-end capping proteins (48, 49). In Ca^{2+} and KCl, the critical concentrations at the two ends are the same judging from both EM (Table 3) and the capping protein experiments in solution.

Under those conditions where the critical concentrations differ at the two ends, the steady-state monomer concentration will fall between these two values with the result that there will be net addition of subunits at the barbed end balanced by net loss of subunits at the pointed end. This subunit flux at steady state was predicted theoretically by Wegner (50). Attempts to demonstrate this flux in bulk samples at steady state gave conflicting results (see Ref. 51 for summary), but the EM experiments are convincing. The quantitative analysis of this flux is complicated by the presence of ADP-actin reactions, that will be considered below after discussion of the hydrolysis of bound ATP.

Carlier et al (52) suggest that the scheme for elongation shown in Figure 5 is an oversimplification, because the nucleotide composition of the subunits near the ends of the filament may have an important effect on the elongation rate constants for ATP-actin. By two different indirect criteria, the critical concentration for ATP-actin addition to ATP-filaments appears to be considerably higher than for its addition to filaments with ADP-subunits near their ends, at least in 1-mM $MgCl_2$. These insightful observations may have other interpretations, so it will be important to test this hypothesis in the future.

Do Filaments Break or Anneal End to End?

It is well established that mechanical shearing such as sonic vibration (53) can break actin filaments into short pieces. With steady-state sonication the distribution of polymer sizes is constant, independent of polymer mass, and determined by the energy input (54).

A more interesting question is whether actin filaments fragment at the low shear rates expected in cells. The evidence is conflicting and its interpretation relies heavily on our incomplete knowledge of other steps in the polymerization process.

The best evidence favoring fragmentation in unstirred samples comes from mathematical modeling of the time course of spontaneous polymerization under conditions where the process is very slow (29, 55). Fragmentation has little effect on the lag phase but accelerates the late stages of polymerization. The kinetic curves can be modeled by a mechanism that includes random fragmentation at a rate proportional to the product of the number concentration of filaments and the square of the polymer length (29). This process increases the number of sites available for elongation over those expected from nucleation alone. The dependence on time, filament number concentration, and some power of filament length explains why fragmentation is detectable only late in the polymerization process and not at all when polymerization is fast.

The existence of annealing is more problematical. Judging from electron microscopy, actin filaments increase in length rapidly following the cessation of sonication. This suggested that short filaments bind end to end to form long filaments (53). Furthermore, mixtures of short bare filaments and filaments decorated with myosin heads form longer filaments that consist of alternating bare and decorated segments (56). These appear to form by end-to-end annealing. On the other hand, these EM studies provide little quantitative information on the frequency or rate of these apparent annealing reactions.

In contrast, quantitative measurements of filament number during the recovery of ADP-actin from sonication are inconsistent with annealing (54). The initial rate of decrease in filament number is directly proportional to the initial filament number, whereas annealing should give a rate proportional to the square of the initial filament concentration. The observed kinetics are consistent with a model where random fluctuations in the length of short filaments result in the disappearance of some filaments and provide subunits for the growth of other filaments. This process has not been analyzed in ATP, so more work will be required to learn whether annealing contributes substantially to length redistributions in ATP or under any other conditions.

Direct evidence for each of the processes that affect filament length is lacking owing to an absence of methods to measure filament length in solution. Preparation of samples for electron microscopy by negative staining can break long actin filaments. Fluorescence photobleaching and dynamic light scattering can yield an average filament length, but only if interactions between filaments are assumed not to exist. A rapid method to measure the distribution of lengths would be of great value.

When and Where is Bound ATP Hydrolyzed during Polymerization?

When ATP-actin polymerizes, the bound nucleotide is hydrolyzed so that the bulk of the subunits in filaments have bound ADP. Until recently the hydrolysis was considered to be tightly coupled to the incorporation of subunits, such that it would occur at the time each molecule binds to the end of the filament. If true, this would mean that the elongation process would consist exclusively of the association of ATP-actin monomers and the dissociation of ADP-actin (see for example Ref. 50).

This older model has now been discarded, because there is a delay between the incorporation of actin molecules into filaments and the hydrolysis of the bound ATP (57–59). This can be measured by release of inorganic phosphate or, perhaps better, an enhancement of the fluorescence of pyrene-labeled actin (59). The data are consistent with a mechanism where polymerized ATP-actin subunits hydrolyze bound ATP randomly, that is, by a first order reaction with a rate constant of about 0.02–0.06 s^{-1} in both Mg^{2+} and Ca^{2+} (58, 59). This is orders of magnitude faster than hydrolysis of ATP by unpolymerized actin. No other models for hydrolysis mechanism have been tested, but more complex mechanisms are, without doubt, also consistent with the data. One particularly interesting model is that hydrolysis occurs exclusively at the boundary between the central part of the filament consisting of ADP-subunits and the ends consisting of a variable number of ATP-subunits (39, 59a). In such a sequential mechanism a wave of hydrolysis would move from the center toward the ends and the rate constant would have to be much higher than 0.06 s^{-1}. In either case, there would be ATP-subunits at both ends of a rapidly growing filament, up to several hundred or more at high monomer concentrations. Some call these terminal regions "ATP-caps" (59), a term that regrettably may be confused with capping by other protein molecules.

What is the Behavior of Polymerized Actin at Steady State?

The situation at steady state is complex, because a minimum of nine reactions (Figure 5) will occur. If the nucleotide composition of the filament ends affects the rate constants (39) there will be more than nine reactions. Needless to say, analysis of the combined reactions at steady state is difficult or impossible, illustrating the value of separate determinations of each rate constant.

Using the constants in Tables 1 and 3 it is possible to make some rough guesses about events at steady state. In the presence of millimolar ATP and little or no ADP, ATP-actin will bind to the ends of filaments at rates of about 1 s^{-1} at both ends. (This is, of course, also the rate of subunit dissociation.) If ATP hydrolysis is a random process (58), it will occur infrequently on the terminal subunit and ATP-actin will be the main dissociating species (51). If the critical concentrations differ at the two ends, there will be net flux (or "treadmilling") of subunits from the barbed end towards the pointed end. The rates calculated from the rate constants in Table 3 are about 0.2 subunits per second with an efficiency of 10–20%.

On the other hand, if the mechanism of ATP hydrolysis is sequential (39), the hydrolysis rate constants will be higher than the exchange rate, hydrolysis will occur frequently on the terminal subunit, and ADP-actin will be the main dissociating species. Since k_-^D is greater than k_-^T, this mechanism predicts that the steady-state critical concentration should be greater than the critical concentration measured from initial rates of elongation. Since the critical concentrations are the same by both methods (34), this mechanism is, in our opinion, less likely to be correct.

Another consequence of $k_-^D > k_-^T$ is that there will be fluctuation in the length of individual filaments depending on the nucleotide composition of terminal subunits. When ADP-subunits are exposed at an end, the filament will shorten briefly by rapid dissociation of ADP-actin, until an ATP-subunit with a lower k_- binds to the end. These fluctuations in length are more dramatic for microtubules (see Ref. 60) since the difference in k_-^D and k_-^T may be more than 100-fold (61) rather than just fivefold as for actin (Table 3).

The predictions in the preceding paragraphs are based on the assumption that events at filament ends produce ADP-actin monomers relatively slowly so that exchange for ATP keeps the concentration of ADP-actin low. If the number of filament ends is very high, as during sonication, ADP-actin is produced more rapidly than the rate of nucleotide exchange and can become the main monomeric species. Then the critical concentration shifts to the value for ADP-actin (62).

MECHANICAL PROPERTIES

Because actin filaments are structural components of cells, it is essential to understand their mechanical properties, in terms of both individual filaments and solutions of monomers and filaments.

What are the Physical Properties of Individual Filaments?

Individual actin filaments are not rigid but can flex along their length and twist around their long axis. The twisting leads to disorder in the helical arrangement of subunits (63). Spectroscopic measurements with eosin-labeled actin are consistent with a root mean square twisting of about 4° between subunits, independent of temperature between 4° and 20°C (64). Compared with this torsional twisting, the filament is about 10 times stiffer with respect to longitudinal flexing (65). This flexibility may be necessary for actin filaments to form certain ordered structures (66).

Do Solutions of Filaments Form Networks?

Solutions of filaments formed from highly purified actin are complex materials exhibiting both viscous and elastic properties. Consequently they both resist flow (viscosity) and store mechanical energy like springs (elasticity). There is agreement that the viscosity measured either by continuous shearing (shear viscosity) or by nonperturbing oscillation methods (dynamic viscosity) is an inverse function of the shear rate, tending toward infinity at low shear rates (67–69).

There is some disagreement regarding the elasticity of solutions of actin filaments. Zaner & Stossel (68) concluded from creep experiments (where a constant stress is applied to the sample and the strain is followed with time) that actin filament solutions behave like relatively stiff rods that interfere with the motion of each other. Over short periods of time such entangled polymers could store some energy and act as an elastic solid, but over longer times the filaments might flow past each other like a liquid. They argued that no bonds between actin filaments are required to explain their data. Two other lines of evidence suggest that there are bonds between actin filaments in solution giving them properties of viscoelastic solids. First, during the polymerization of unstirred solutions of rhodamine-labeled actin the filaments become completely immobilized judging from fluorescence photobleaching (70). This shows that the filaments do not diffuse by tumbling end over end or even by snaking their way longitudinally among the other filaments. Second, in oscillation experiments the elastic modulus tends toward a constant value at low frequencies (10^{-3}–10^{-4} Hz) as expected for a viscoelastic solid (69). [The elastic modulus of a viscoelastic liquid, such as the model of actin filaments proposed by Zaner & Stossel (68), would go to zero at low frequencies]. These two completely different methods both suggest that pure actin filaments are physically attached to each other in solution. Jen et al (71) reached a similar conclusion from rheological measurements. The nature of the connections between the filaments is not known but they are weak. When subjected to a large strain, the continuous three-dimensional network of actin filaments is disrupted and the material flows like a liquid. The yield stress for 24-μM muscle actin filaments is about 10 dyn/cm^2 at a shear rate of 0.6 s^{-1}. It is presumed, but not proven, that weak connections between filaments are disrupted during flow, not the filaments themselves.

What about Nonfilamentous Actin?

A completely unexpected finding emerged from the rheological analysis of Sato et al (69)—that solutions of nonfilamentous actin also behave like viscoelastic solids both in low salt and in 0.6 M KI, providing that they are allowed to equilibrate without mixing for about 10 hours. Both the viscosity and elasticity are lower by a factor of 2 than for filamentous actin, but higher, by orders of magnitude, than for cytochrome-C or ovalbumin at least at low frequency. Like filamentous actin, the nonfilamentous actin is easily disrupted mechanically. These observations mean that actin molecules form nonfilamentous networks by binding weakly to each other. Perhaps similar interactions are responsible for the bonds between filaments.

ACTIN MONOMER BINDING PROTEINS

There are at least four different families of proteins that bind primarily to actin monomers (Table 4). Many of the capping proteins considered in the next section also have some affinity for actin monomers. Every one of the actin monomer binding proteins studied to date forms a complex with actin that does not polymerize as well as free actin monomers. It has generally been assumed that this inhibition is explained exclusively by the ability of these proteins to sequester monomers in a nonpolymerizable complex and that the fraction of actin in such a complex is determined by mass action (Figure 6). On the other hand, detailed experiments with both profilin and DNase-I raise questions about this simple mechanism.

Profilins

Profilins are small, soluble proteins that are probably present in very high concentrations throughout the cytoplasm, although this has been established for only one cell (72). Two profilins have been sequenced: a basic form of vertebrate profilin consists of 142 amino acids (73); and two neutral isoforms of *Acanthamoeba* profilin consist of 125 amino acids (including the rare amino acid trimethyllysine) (74). The N-terminal regions are homologous. The C-terminal 70% have similar patterns of hydrophilic and hydrophobic amino acids, but no sequence homology. *Acanthamoeba* has at least three different isoforms of profilin. The two neutral isoforms (profilin-IA and -IB) have nearly identical sequences (74). The basic isoform, profilin-II, has a much different sequence judging from tryptic peptide maps (75). Nonetheless, profilin-I and -II are indistinguishable in their effects on actin polymerization (75). Since both neutral (76) and basic (73) profilins have been purified by different methods from vertebrate cells, it is possible that both neutral and basic isoforms are

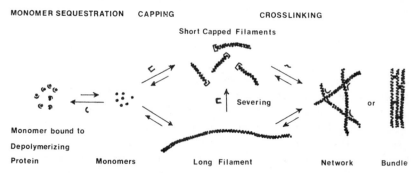

Figure 6 Actin-binding proteins. The actin-binding proteins are represented by the following symbols: "C" for monomer-binding proteins; bracket for capping and severing proteins; squiggle for cross-linking proteins (281).

Table 4 Actin monomer binding proteins

	M_r	IEP	Cellular concentration	Actin ATP exchange	Cytoplasmic actin monomer binding K_d (M)	Muscle actin monomer binding K_d (M)	Ref.
Profilins							
Acanthamoeba (P-I)	12,900	5.5–6.4	~80 μM	increases	$2\text{--}10\times10^{-6}$	$6\text{--}8\times10^{-5}$	78–81, 246
(P-II)	13,000	>9.0	~20 μM		10×10^{-6}		78
Physarum	13,000				$1\text{--}4\times10^{-6}$	$1\text{--}3\times10^{-5}$	247
Vertebrate							
Spleen	15,220	9.3					73, 83
Macrophage	15,500	7.8				3×10^{-6}	76
Brain	15,000			increases	$1\text{--}2\times10^{-6}$	6×10^{-6}	88, 248
DNase-I							
Vertebrate	33,000			decreases	10^{-10}	10^{-10}	89, 91, 92
Vitamin-D-binding protein							
Vertebrate blood	58,000		10 μM (serum)			$1\text{--}10\times10^{-8}$	96

present in many cells and that the reported differences in primary structure (73, 74) represent isoform, not species, differences.

Profilins do not bind to actin filaments judging from pelleting assays (77), but binding to monomers can be detected in several ways. Profilin increases the rate of exchange of ATP bound to actin (78), inhibits the actin monomer ATPase (79), increases the fluorescence of pyrene-labeled actin monomers (80), and inhibits actin polymerization (75–77, 79). Profilin has none of these effects on covalently cross-linked dimers of actin (81). From the profilin concentration dependence of these effects, one can determine apparent dissociation constants (K_d). To a first approximation, all of these methods give the same estimates for the K_d for any given combination of profilin and actin. Most profilins have a higher affinity for cytoplasmic than muscle actins. In the only case examined, *Acanthamoeba* profilin has a higher affinity for ADP-actin than ATP-actin (82). It has been suggested (82) that the absence of nucleotide from the complex of actin and profilin (profilactin) originally isolated from spleen (83) may account for the much greater stability of this complex compared with that reconstituted from purified components (Table 4). In some (82, 84) but not all (76) cases, chemical modification of cysteine 374 of actin drastically lowers the affinity for profilin as does removal of the C-terminal phenylalanine from actin (85). The enhancement of fluorescence of pyrene attached to cysteine-374 by profilin (80) and preliminary interpretation of the electron density maps of actin-profilin crystals are also consistent with binding of profilin near the C terminus of actin (C. Schutt, personal communication).

Since the inhibition of nucleation and elongation in bulk samples and the steady-state extent of polymerization can all be explained quantitatively by reduction of the free actin monomer concentration due to profilin binding to a single site on monomeric actin, it might seem that its mechanism of action is both simple and well established. On the other hand, a growing number of observations are incompatible with a simple monomer sequestration mechanism. First, *Acanthamoeba* profilin inhibits elongation only weakly at the barbed end of actin filament (75, 86). Judging from the dependence of the barbed end elongation rate on the concentrations of both profilin and actin, and assuming that only free actin can polymerize, the concentration of free actin appears to be determined by a K_d of about 50 μM. At the pointed end the K_d appears to be about 5 μM as expected from other assays. This discrepancy was originally observed for a profilin-like protein from echinoderm sperm (87). Since the actin-profilin complex cannot have two different K_d's, we proposed a more complex mechanism where the concentration of free actin monomers is established by binding of profilin to actin with a K_d of 5–10 μM and only free monomers can form nuclei or add to the pointed end. At the barbed end, not only free actin but also profilin and the actin-profilin complex can bind. The model proposes that profilin inhibits elongation at the barbed end of the filament by capping the end but does not prevent dissociation of the terminal subunit. Therefore, the apparent K_d of 50 μM is a measure of the affinity of profilin for the filament end, not its affinity for monomer. This model explains the electron microscopic data, but not why the effect of profilin on the steady-state monomer concentration is so much stronger (82).

Second, experiments with two different barbed-end capping proteins (76, 88) suggest that the effect of vertebrate profilins on elongation at the pointed end is much stronger than expected from their effect on the steady-state monomer concentration of actin alone. Third, similar experiments with the pointed-end capping protein, acumentin, indicate that vertebrate profilin has a much weaker effect on barbed-end elongation than expected (76). The simple monomer sequestration mechanism cannot explain these observations, so electron microscopic experiments are called for to learn whether the barbed-end capping mechanism may be operative for vertebrate profilins.

IS PROFILIN REGULATED?   Vertebrate profilin binds to phosphatidylinositol and several other anionic phospholipids. Since actin has a lower affinity for the profilin-lipid complex than free profilin, these lipids may regulate the activity of profilin (88a). For example, it is possible, though far from proven, that the turnover of phosphoinositols that follows the binding of ligands to cell surface receptors may release actin monomers bound to profilin. There is no experimental evidence for other modes of regulation of profilin, such as calcium sensitivity or phosphorylation. Although it is not a regulatory mechanism per se, it is clear that the critical concentration of actin is an important determinant of the concentration of actin-profilin complexes at steady state (79). In fact, profilin can amplify small changes in the critical concentration into substantial changes in the concentration of polymerized actin.

WHAT DOES PROFILIN DO IN THE CELL?   The high concentration of profilin in cells and its ability to bind to actin monomers have suggested that profilin may buffer the actin monomer concentration in the cell. As such, it is likely to be responsible for much of the unpolymerized actin found in cell extracts (see Ref. 1). On the other hand, profilin by itself cannot explain fully the high concentration of unpolymerized actin, because the K_d for the actin-profilin complex is so high that the free actin concentration in the cytoplasm would still be far above the critical concentration for polymerization. Consequently, other actin-binding proteins must contribute to the pool of unpolymerized actin. If the barbed-end capping mechanism is correct, profilin would suppress spontaneous nucleation more effectively than elongation at the barbed end of filaments, possibly allowing the cell to limit polymerization to the ends of existing filaments and other specific nucleating sites. Regardless of the mechanism, more information is required about the cellular concentrations and possible isoforms of profilins before their functions are understood. More work is also required on reconstituted mixtures of profilin with other actin-binding proteins.

DNase-I

The pancreatic secretory protein DNase-I binds with very high affinity (Table 4) near the N terminus (Figure 1) of most types of actin (89). One exception is actin from *Entamoeba histolytica* (90). The binding site is near the cleft in actin where ATP binds. Either directly or as a result of a conformational change of the actin, DNase-I inhibits the rate of nucleotide exchange (91).

The complex of actin and DNase-I cannot polymerize (92), so DNase-I inhibits the rate and extent of polymerization stoichiometrically. In addition to this sequestration of monomers, there is one report that incubation of actin filaments with very low concentrations of DNase-I can prevent their elongation by subsequent addition of actin monomers (92). The authors suggest that this is due to binding of DNase-I near the barbed end of the filaments with a K_d of $<10^{-8}$ M.

Since actin appears to be able to bind simultaneously to DNase-I and to either profilin (93), vitamin-D binding protein (94), cofilin (93), a depactin-like protein from vertebrates (93), or several of the capping proteins (see below), the DNase-I site (Figures 1 and 3) may not overlap any of these other sites. (One caveat is that the number of actin molecules in some of these complexes may be greater than one.) Spatial separation of these sites is also indicated by an enhancement of the fluorescence of pyrene-actin by vitamin-D binding protein (95) and profilin (80), but not by DNase-I (92).

IS THERE PHYSIOLOGICAL SIGNIFICANCE TO THE BINDING OF DNASE-I TO ACTIN? The function, if any, of DNase-I binding to actin remains uncertain. DNase-I is a secretory protein stored in zymogen granules separated from actin in the cytoplasm by an impermeable membrane, so the proteins are unlikely to meet inside the parotid or pancreatic acinar cell. Some actin is shed from the surface of cells and this may account for the presence of actin-DNase-I complexes in the pancreatic juice (95). Also, actin inhibits bovine, but not rat, DNase-I, so the interaction is not general (95a). Until more information is available, it is probably best to conclude that the binding of the two proteins is fortuitous.

Vitamin-D Binding Protein

Vitamin-D binding protein (DBP) is present in blood plasma at a concentration of about 10 μM and binds not only the vitamin, but also actin monomers (96). The K_d for actin is about 10^{-8} M judging from enhancement of the fluorescence of pyrene-labeled actin monomers and $2–10 \times 10^{-8}$ M based on the shift in the critical concentration for polymerization of muscle actin in 2-mM $MgCl_2$ (96).

The available evidence is consistent with a simple monomer sequestration mechanism of action, but quantitative studies of nucleation and growth at the two polymer ends need to be done to learn if the mechanism is more complex as suggested by the apparent difference in the affinity measured by the two different methods. DBP may complement serum gelsolin as an inhibitor of polymerization of any actin released from cells into the blood, for example, during tissue injury. No information is currently available on the influence of DBP on actin clearance from the blood.

CAPPING PROTEINS

Capping proteins bind to one of the ends of actin filaments and influence subunit reactions there (Figure 6). They were discovered using several different assays: inhibition of the binding of radiolabeled cytochalasins to actin filaments, decrease in viscosity of solutions of actin filaments, and affinity chromatography of crude cell extracts on immobilized DNase-I. Most of the proteins purified to date cap the barbed end of actin filaments and will be considered before those that cap the pointed end.

How are the Different Capping Proteins Related?

One can divide capping proteins into three families—gelsolin/villin, fragmin/severin, and a group termed simply "capping protein" (Table 5). The fragmin/severin and capping protein families may be universal in their distribution since they have been isolated from protozoa and vertebrates. The gelsolin/villin family has been found in vertebrates, but attempts to purify them from protozoa have been unsuccessful. The capping protein family includes heterodimers with subunits of 30–35 kd, which are not sensitive to calcium, and cap and nucleate but do not sever. The proteins in the fragmin/severin family consist of polypeptides of about 45 kd, which are often isolated as a 1:1 complex with actin. They require calcium to cap, nucleate, and sever. The proteins in the gelsolin/villin family are monomers of 90–95 kd that also require calcium and are sometimes isolated as 1:1 complexes with actin. They are present in different mammalian tissues. Brevin, found in plasma, is nearly identical to gelsolin, found in cytoplasm, except that brevin has 25 extra residues at the N terminus, based on limited amino acid sequence data and peptide maps. Therefore, brevin is sometimes called plasma gelsolin. A cell line synthesizes both, but secretes only brevin (97). Functionally, the proteins in this group cap, nucleate, and sever. Villin is unique among capping proteins in also possessing bundling activity.

The structural relationships among family members and between families, and the correlation of structure with function, are interesting issues. Sequence data reveal a limited area of homology between gelsolin and villin, which may

Table 5 Capping proteins

Protein name	Source	Subunits and M_r	Actin/protein molar ratio in cell	Ref.
Barbed-end capping proteins				
Capping protein	*Acanthamoeba*	28,000; 31,000	150	49
	Dictyostelium	32,000; 34,000		249
	Brain	31,000; 36,000		250
	Skeletal muscle			Footnote e
Fragmin/severin group				
Fragmin	*Physarum*	42,000		252
Severin	*Dictyostelium*	40,000	50[a]	109
45-kd protein	Sea urchin eggs	45,000	100[a]	253
				254
42-kd protein	Thyroid	42,000		255
Cap 42a+b	*Physarum*	42,000; 42,000		99
Gelsolin/villin group				
Gelsolin	Macrophage	90,000	9[a]	124
	Platelet	90,000	32[a]	115
	Smooth muscle	90,000		256
	Brain	90,000		200
	Cardiac muscle	90,000		257
	Various mammalian tissues			134
Brevin	Plasma (gelsolin)	93,000	2μM[c]	112
Villin	Intestinal brush border	95,000	10[b]	147
Pointed-end capping proteins				
Acumentin	Macrophages and neutrophils	65,000	2.6	127
Beta-actinin	Skeletal muscle	37,000; 35,000		128
Spectrin plus band 4.1	Mammalian tissues	260,000; 225,000; 78,000	2.5[d]	132

[a]Molar ratio in high-speed supernatant of extract.
[b]Molar ratio in microvilli.
[c]Absolute concentration of brevin in plasma.
[d]Molar ratio of complex to actin subunits in erythrocytes (166).
[e]J. F. Casella, S. Lin, submitted for publication.

have a common function (98). In *Physarum,* Cap 42a and Cap 42b subunits have extensive homology with actin and fragmin, respectively (99), but in *Acanthamoeba* no homology is found among actin, capping protein, profilin, and actophorin (100), and no homology is found by comparing the available sequences of profilin (73) or gelsolin (97) with that of actin. Most of this work has been with antibody cross-reactivity, peptide maps, and limited amino acid sequences, and DNA sequences, when available, will answer these questions in much greater detail.

How Much Capping Protein is in Cells?

Given the importance of absolute concentrations in deducing the physiologic role of a protein, it is regrettable that little such information is available for capping proteins. In general the cellular concentrations of capping proteins are low with a molar ratio to actin of about 1 : 100 (Table 5). This means there is one capping protein per filament even if the filaments are as short as 0.3 μm. Supernatants of high-speed centrifugations of cell homogenates appear to have higher ratios of capping protein to actin (Table 5), which is reasonable if part of the capping protein is free and some is bound to short, poorly sedimentable filaments. The physical nature of capping protein-actin complexes and of nonsedimentable actin in cells is unknown, but of great importance. In microvilli the molar ratio of villin to actin is unusually high (1 : 10) (101, 102), probably because it serves there primarily as a bundling protein rather than as a capping protein.

What is Capping and What is the Evidence that it Occurs?

EFFECTS ON ACTIN POLYMERIZATION   In theory, a protein that binds to the end of an actin filament can interfere with either the association of monomers, the dissociation of the terminal subunits, or both. Capping proteins and cytochalasins inhibit the addition of actin monomers to growing barbed ends of actin filaments, which is best evaluated by electron microscopy because the two ends can be distinguished. Effects on subunit dissociation can be tested by measurements of the critical concentration (below) or the kinetics of depolymerization. A full characterization should also include quantitation of the dependence of the effect on the concentration of capping protein.

Another line of evidence that capping proteins block barbed ends is their effect on the critical concentration. In Mg^{2+} and ATP, the critical concentration is different at the two ends of actin filaments (103). At steady state with both ends free, the critical concentration is near that of the barbed end, since the elongation rate constants of the barbed end are so much larger than those of the pointed end. If capping protein blocks the addition and loss of actin subunits at all of the barbed ends in a sample, the critical concentration should increase to that of the pointed end, as observed in one case (49). The quantitative dependence of the change in critical concentration on the capping protein concentration provides an inhibition constant, or apparent dissociation constant, of less than 0.01 nM for villin (105), less than 0.06 nM for *Acanthamoeba* capping protein (49), and about 0.5 nM for brain capping protein (106).

High concentrations of some capping proteins increase the critical concentration beyond that of the pointed end (48, 104). This unexpected behavior occurred with impure preparations of *Acanthamoeba* capping protein (49), but there are other possible explanations. Capping proteins might bind actin monomers or alter the behavior of the pointed end by causing actin filaments to be very short. Also, the production of a large number concentration of filaments may lead to a higher apparent critical concentration because of accumulation of ADP-actin monomers (39).

ACTUAL PHYSICAL BINDING   To support the model that capping proteins bind to the barbed end, experiments should show specific, saturable binding of one capping protein per actin filament with a dissociation constant equal to the inhibition constant measured in elongation and critical concentration experiments. This correlation has not yet been possible since the quantitation of binding is imprecise, in part because the capping proteins themselves alter the length of the actin filaments and the number of binding sites. Furthermore, these short filaments sediment poorly, complicating the separation of proteins bound to filaments from those free in solution. With prolonged ultracentrifugation, one brevin pellets per actin filament (107). The diffusion coefficient of brevin attached to actin filaments is the same as predicted for 1 : 1 binding with a dissociation constant of about 1 nM (108). Rapid gel filtration showed that *Acanthamoeba* capping protein binds to actin filaments (49). Fluorescence energy transfer was used to measure the binding of *Dictyostelium* severin to actin filaments and monomers (109). The dissociation constant of a 1 : 1 severin : actin complex was about 25 nM but may be lower because, curiously, dissociation of severin from the filaments was not observed. Capping proteins might be expected to compete with radiolabeled cytochalasin B for binding to barbed ends, but both brevin (96) and villin (110) increase the binding of cytochalasin, perhaps by creating more ends, which can bind both cytochalasin and the capping protein.

LOCALIZATION   Actin filaments grow with their barbed ends attached to particles or surfaces that are coated with capping proteins including villin (111), brevin (112), and severin (113). However, attempts to localize capping proteins on actin filaments with specific antibodies and electron microscopy have been fruitless. These studies show that binding to barbed ends is highly preferred, but the absence of interactions elsewhere on the filament, an important point, is unproven.

How do Capping Proteins Interact with Actin Monomers?

Capping proteins accelerate spontaneous polymerization of actin monomers by shortening the lag phase, and it is widely presumed that the nucleation rate is increased. Kinetic modeling studies show that the data are consistent with three different mechanisms where *Acanthamoeba* capping protein accelerates any one of the steps in nucleation process (114). Each model makes different testable predictions about the molecular interactions. Even though only the slow ends elongate, the overall polymerization rate is faster because the rate-limiting step of nucleation is faster and produces more filaments. The final result is numerous short filaments with capped barbed ends. The kinetic data are inconsistent with models where *Acanthamoeba* capping protein fragments actin filaments or simply acts like a pointed end. Similarly, plasma gelsolin and its complexes with actin accelerate but do not bypass nucleation (114a).

Some capping proteins, including gelsolin and fragmin, form stable complexes with one- or two-actin monomers even under conditions that do not support polymerization (108, 109, 115). The 1 : 1 complex is unusually stable, with a half time of dissociation of 30–35 days (Sedlar and J. Bryan, unpublished data). Cytochalasin D also forms complexes with actin dimers (116). In solutions of capping proteins and actin filaments, the capping proteins are probably bound both to filament ends and to monomers or dimers, with very little free. The proportions of these three forms still need to be measured.

How do Capping Proteins Shorten Actin Filaments?

One controversial activity of capping proteins has been their ability to shorten actin filaments. When actin monomers polymerize in the presence of a capping protein, the increased nucleation rate accounts for the large number of short filaments. However, addition of capping protein to actin filaments also results in short filaments. There are two theories to explain this finding. One theory is that capping proteins cap barbed ends and nucleate with monomers to form more filaments, which, combined with the usual rate of subunit exchange, leads to shorter, more numerous filaments. It seems reasonable that the final, steady-state length distribution should be the same whether capping protein is added to either actin monomers or filaments, since monomers and filaments are interconvertible physical states of the same molecule. This theory incorporates only processes for which independent evidence exists. A variation of this theory holds that filaments fragment and anneal at rapid rates, and capping proteins simply cap ends and inhibit annealing.

Another, very exciting, theory proposes that some capping proteins actively break the actin filaments, a process usually called severing. In this model a capping protein molecule binds to the side of an actin filament and intercalates between two adjacent actin subunits. Two new filament ends are created—the new pointed end is free and the new barbed end may have the capping protein bound to it. The impetus for this theory is that capping proteins shorten actin filaments very rapidly. In many cases the mixing process fragments filaments so that capping proteins may simply rapidly cap the newly created barbed ends. However, in some experiments fragmentation is minimized, and shortening is faster than expected from depolymerization. These sorts of experiments have been performed by mixing a capping protein either with actin filaments in solution (117), with filaments attached to *Limulus* bundles (112, 118), or with actin filaments attached to erythrocyte membranes (96). While the apparent depolymerization rates are very rapid, and severing seems to be the most likely explanation, we have some reservation about this conclusion simply because of uncertainties about actin polymerization, especially the rate constants for elongation, fragmentation, and annealing.

How are Capping Proteins Regulated?

Some capping proteins are regulated by factors that may be relevant to cell motility. The gelsolin/villin and fragmin/severin groups are sensitive to calcium. Cytoplasmic gelsolin requires calcium for activity (119) and binds calcium with a dissociation constant of 1 μM and stoichiometry of 1.7 calcium per gelsolin (120). Plasma gelsolin binds more than one calcium with a K_d of 20 μM and undergoes a conformational change on binding calcium (120a, 120b). The actin-binding and calcium-binding domains of plasma gelsolin can be separated by proteolysis (120c).

Nucleation and filament shortening by villin require calcium, but bundling does not. Villin has three binding sites for calcium and undergoes a conformational change upon binding calcium. One binding site has high affinity with little or no exchange, and the other two have K_d's of about 5 μM. The change in conformation may not, however, be related to the change in function since limited proteolysis eliminates the conformational change but not the calcium binding and calcium sensitivity of the functions (121, 122). The dependence of the capping activity on calcium concentration shows that one of the 5-μM sites may be involved, but the severing activity has a different calcium dependence that is not explained by the calcium-binding data (117).

Villin readily dissociates from actin on addition of EGTA, but gelsolin and severin remain bound, which casts doubt on the ability of calcium to regulate their activity in vivo (115). However, free gelsolin can be isolated from several sources under the proper conditions (123), and the gelsolin-actin complex is also functional and calcium-sensitive (124).

The calcium dependence of cap 42(a+b) from *Physarum* varies with phosphorylation of the protein by a specific kinase (125). This protein does not sever filaments. Fragmin, also from *Physarum,* is calcium-dependent and severs, but like gelsolin, can exist as a 1 : 1 complex with actin, which does not sever.

How are all these Various Functions Related?

By definition all these proteins can cap (inhibit subunit exchange at ends of actin filaments), but they vary in their ability to sever filaments, promote nucleation, bind to actin monomers, and bundle actin filaments. Are these activities attributable to different active sites on the protein? With limited proteolysis of villin, for example, the bundling activity is lost, but the other activities are not (126), so there must be two active sites. Regulatory factors sometimes distinguish various activities. For example, with villin, the capping and severing activities depend on different concentrations of calcium (117). Comparing the functional activities of the various proteins has been complicated because different techniques and conditions are employed in different laboratories and many assays do not have a unique quantitative interpretation. Owing to our limited knowledge about the mechanisms of nucleation and severing, it seems possible that all the activities (except bundling) may some day be explained by a unified mechanism with one type of interaction between the capping protein and actin.

What about the Pointed End?

The above discussion focused on barbed-end capping proteins. Few proteins have been found that cap pointed ends (Table 5), and their characterization is incomplete. Their binding has not been measured directly, they have not been localized on filaments, and some of their properties are not those expected of a pointed-end capping protein.

Acumentin, a 63.5-kd monomer from macrophages, inhibits elongation at the pointed end of filaments, as measured by electron microscopy, and probably shortens filaments and nucleates actin polymerization from monomers. Curiously, acumentin also markedly inhibits elongation of filaments in solution, a property not expected for a pointed-end capping protein (127).

Beta-actinin, from skeletal muscle, has some functional properties of capping proteins and is localized to the region of sarcomeres that includes pointed ends of actin filaments, but inhibits growth at both ends of actin filaments (128). Beta-actinin may explain why thin filaments in skeletal muscle have a defined length and why actin does not add to their pointed ends in isolated myofibrils (129). Some actin filaments grow from erythrocyte membranes with their barbed ends free (130, 131), and the complex of spectrin and band 4.1 has some properties expected for a pointed-end capper (132).

More detailed characterization of more pointed-end capping proteins is an area of great interest for the future, especially since these proteins may be useful in other studies on actin filaments.

What do Capping Proteins Do in Cells?

Their in vitro properties suggest that capping proteins may have several roles in cells: binding of capping proteins to filament ends would stop subunit addition and loss, stabilize filament length, and stop treadmilling. There are enough of the various capping proteins in cells so that every actin filament could be capped, and the ionic conditions in cells favor this binding. Recent evidence shows that treadmilling (i.e. subunit flux) may exist in cells (133), but much uncertainty still exists about the treadmilling process, the nature of ends of actin filaments, the length of actin filaments, and even the ionic conditions in cells.

Since capping proteins make actin filaments short, they may control filament length and thereby cytoplasmic viscosity. Local variations in viscosity may control the flow of cytoplasm and the movement of particles about the interior of cells. The notion is that intracellular organelles can flow into an area of a sol, but not a gel. Interestingly, gelsolin was localized by immunofluorescence microscopy to an area of cytoplasm under a particle being phagocytized, where breaking apart a filament network may be important (134), but gelsolin localization in general needs to be reconsidered (134a). One theory of cell movement holds that solution of the cytoplasm (that might be accomplished by severing actin filaments) is necessary for myosin-mediated contraction to occur (135).

Capping proteins are thought perhaps to bind barbed ends of actin filaments to membranes, especially plasma membranes, where this orientation is generally observed. Localization studies with the light microscope have not supported this idea (49), but studies at the electron microscopic level and biochemical studies of proteins binding actin to membranes may yet bear it out. In isolated brush borders, actin can add to the barbed ends of microvillar bundles (136), so perhaps the membrane-associated ends are not capped, despite the traditional electron microscope images which suggest that they are. Elongation of bundles of actin filaments at their barbed, membrane-associated ends does occur in *Thyone* sperm (137) and *Chlamydomonas* (138). In these cases the elongation of actin filaments creates a structure for the cell. The mechanism stopping the growth is not known, but in *Thyone* sperm it may simply be that all the monomeric actin released by the acrosomal reaction polymerizes.

LOW-MOLECULAR-WEIGHT SEVERING PROTEINS

These proteins, listed in Table 6, are monomers of 15–20 kd and have been found in protozoa, echinoderms, and vertebrates. They bind actin monomers and sever actin filaments. No structural comparisons, including antibody cross-reactivity, have been made, so it is not known how similar the proteins are. The functional properties of the proteins from different sources seem to be similar, but their mechanism is not yet clear, and the available data cannot be explained by a simple model. The proteins cause a stoichiometric increase in apparent critical concentration, can be chemically cross-linked into a 1 : 1 complex with actin, and decrease the sedimentability of actin (100, 139–141). These properties are characteristic of a monomer binder. However, the proteins also shorten actin filaments rapidly and increase the rate of actin polymerization from monomers, both in a manner consistent with severing. Electron micrographs show that mixtures of actin filaments with these proteins are short, clumped, and have an appearance that suggests decoration, but the proteins do not sediment with actin filaments (100, 139, 141). It is difficult to conceive of a simple model that explains all the properties. The binding site is probably

Table 6 Low-molecular-weight severing proteins

Protein name	Source	Subunits and M_r	Actin/protein molar ratio in cell	Ref.
Actin depolymeriz-ing factor	Brain, liver, kidney, hepa-toma	19,000		140 141 258
Depactin	Starfish oocyte	20,000	1.8[a]	139
Actophorin	*Acanthamoeba*	15,000	10	100

[a]Measured in high speed supernatant of cell extract, not whole cell.

different from that of DNase I, and the dissociation constant is estimated at 0.4 μM for depactin (139). Cells contain a moderate amount of these proteins [10 mol actin/mol actophorin in *Acanthamoeba* whole cell (100) and 1.8 actin/ depactin in starfish oocyte high-speed supernatant (139)], so they may account for part of the nonsedimentable actin. No factors that modulate their activity have been described.

PROTEINS THAT BIND TO THE SIDES OF ACTIN FILAMENTS

These proteins will be considered in three groups based on their functional properties. One group, including tropomyosin, binds to only one filament at a time, so they do not link filaments together. The proteins in the second group, called cross-linkers or gelation proteins, are divalent and act as spot welds to cross-link different filaments (Figure 6). A third group cause actin filaments to form bundles. The distinction between cross-linkers and bundlers is not sharp because under certain conditions cross-linkers can form coarse bundles, and bundlers can affect the mechanical properties of solutions, as cross-linkers do. A number of nonspecific agents, mainly polycations, can bundle actin filaments and gel solutions of actin filaments (142). The activities are inhibited by ionic strength and high pH. Some proteins, such as myosin subfragment-1 (143), aldolase (144), and nerve growth factor (145), bundle actin filaments in vitro, but further work will be required to learn whether they have such a role in vivo.

Bundling Proteins

The bundling proteins (Table 7) include the fascin group, fimbrin, villin, and band 4.9. The fascin group are 57-kd monomers that have been isolated from protozoa, echinoderms, *Limulus,* and mammals. Fimbrin is a 68-kd monomer found in intestinal brush border and a few other vertebrate tissues (146).

Table 7 Proteins that bundle actin filaments

Protein name	Source	Subunits and M_r	Actin/protein molar ratio in cell bundle	Ref.
Fascin group				
57-kd	*Limulus* sperm	57,000	1	156
Fascin	Sea urchin egg	57,000	5	149
	Sea urchin coelomo-cyte	57,000	10[a]	259
	Starfish sperm	57,000		260
55-kd	HeLa	57,000		148
55-kd	*Acanthamoeba*	2×55,000	6	261
53-kd	Porcine brain	53,000		262
50-kd	*Dictyostelium*	50,000		262a
Others				
Fimbrin	Brush border	68,000	8	151
	HeLa	68,000		148
Villin	Brush border	95,000	10	147
30-kd	*Dictyostelium*	31,700	150[b]	183
Band 4.9	Erythrocytes	3×48,000	30[c]	159

[a]Molar ratio in whole cells is 3:1.
[b]Estimate from yield of purification.
[c]Molar ratio in erythrocyte membranes.

Intestinal brush border has two bundling proteins, fimbrin and villin (147), and HeLa cells contain both fimbrin and fascin (148). Fimbrin is not related to either villin or fascin, based on a lack of antibody cross-reactivity. Villin, a 95-kd monomer, is unique because it is a capping protein (discussed above) and a bundler. Band 4.9, a trimer of 48-kd subunits, has only been described in erythrocyte membranes, and 30-kd protein, a dimer, has only been described in *Dictyostelium* (183).

HOW DO BUNDLING PROTEINS WORK?   Bundling proteins bind to the sides of actin filaments and lie between them in bundles. Fascin, fimbrin, and villin create bundles in which the filaments are very close together (about 10 nm apart laterally) and have the same polarity (148a, 150). Bundles with fascin are regularly spaced at intervals that agree with the stoichiometry of binding (one fascin per 4–5 actin) (149), and synthetic bundles resemble those isolated from cells (150). Fimbrin also sediments with the bundles of actin filaments and saturates at a ratio of about one fimbrin to 3.5 actins (151). Bundling proteins probably have two distinct binding sites for actin filaments on each monomer of bundler, an idea strongly supported by the high degree of order among the filaments and bundling proteins in structural studies of bundles. The biochemical identification and characterization of these sites, especially the comparison of different proteins, is an interesting area for future research.

HOW ARE BUNDLES USED IN CELLS?   Many cells contain tight bundles of parallel actin filaments with bundling proteins. Fascin is concentrated in bundles that form the cores of microvilli in sea urchin eggs and coelomocytes (152, 153). A 220K protein that can aggregate fascin/actin bundles but not actin filaments (149) is also present in microvilli of starfish oocytes. Bundles with less fascin have less order (154), and bundle formation can be separated from both filament and microvillus formation (155), so it seems likely that regulation of fascin controls bundle formation. No parameters have yet been found, however, that control the activity of fascin.

Limulus sperm contain a tight bundle of actin filaments arranged in a supercoil with straight segments and bends (66). A 57-kd protein of the fascin group is present at an unusually high molar ratio of $1:1$ with actin (156). The bundle shifts from a coiled to a straight configuration to produce the acrosomal process (157). This transition is remarkable because actin bundles in other places, such as microvilli and filopodia, are stable except for their assembly and disassembly. The role of actin-binding proteins in this transition is probably crucial but is not yet known. Fascin also inhibits the binding of myosin to *Limulus* bundles (156), a property that may be relevant in other systems, but probably not *Limulus* sperm.

Microvilli in intestinal brush border contain tight bundles of actin filaments, with two bundlers, fimbrin and villin (147). The relationship between the two and even the need for two bundlers is not clear. Actin filament bundles made with fimbrin and villin have a lateral spacing similar to that of native microvillar bundles, but the synthetic bundles do not have the hexagonal packing arrangement seen in the native ones (158). In other cells, fimbrin is localized in areas of cells with actin filaments and filament bundles (146). Curiously, the bundling activity of fimbrin in vitro is inhibited by concentrations of KCl and $MgCl_2$ that may be physiologic (151). Both fimbrin and villin bind calcium with dissociation constants in the micromolar range. Calcium decreases the activity of fimbrin (148a), but its effect on villin's bundling activity is unclear since the severing, nucleating, and capping activities of villin increase with calcium.

The 30-kd protein of *Dictyostelium* is a dimer that bundles and gels actin filaments, is inhibited by calcium, and is concentrated in filopodia (158a, 183).

The role of band 4.9 in erythrocytes (159) is unclear. Recent discoveries add tropomyosin (160) and myosin (161, 162) to band 4.1 and spectrin as erythrocyte proteins capable of binding to actin. Single short actin filaments with attached spectrin molecules are seen (163, 163a), but the location of these other actin-binding proteins remains to be determined. Also, a barbed-end capping protein has not been found in erythrocytes. Perhaps a side-binding protein like band 4.9 or tropomyosin can inhibit end interactions, or perhaps none is required. Determination of the in situ structure and its reconstitution with purified components is an exciting possibility for the future.

Cross-linker proteins are divalent molecules that presumably bind the sides of two different actin filaments, forming physical connections that create a three-dimensional network. They were discovered and characterized as factors that dramatically increase the viscosity of solutions of actin filaments. The different groups (Table 8) are actin-binding protein/filamin, alpha-actinin, spectrin, 120-kd proteins, low-molecular-weight proteins, and microtubule-associated proteins (MAPs).

Actin-binding protein and filamin are similar proteins with two long, flexible, 270-kd subunits that bind end-to-end to form dimers (164) that can aggregate (165). They have only been found in vertebrates. The spectrin group, recently reviewed by Bennett (166), are long flexible heterodimers of 240–265-kd subunits (167). Two subunits bind along their sides to form the dimer, and two dimers bind at their ends to form a tetramer. The tetramers can self-associate to form networks (168). Spectrins, originally described in erythrocyte membranes, have been found in many other vertebrate tissues, mainly associated with plasma membranes (166). In the terminal web of intestinal epithelial cells spectrin seems to cross-link actin filaments to each other but not to membranes (147). Two protozoan proteins resemble spectrins in physical and functional properties (169, 170), and the *Acanthamoeba* protein shows immunologic cross-reactivity with spectrin (169). These proteins also resemble actin-binding protein, so they may be evolutionary ancestors of both groups of vertebrate proteins.

Alpha-actinins have two subunits of 90–110 kd that appear to bind side-to-side to form a short rod. Alpha-actinins from different cells have similar physical properties, show immunologic cross-reactivity, and are inhibited by calcium (except the muscle variants). Alpha-actinin was first purified from skeletal muscle, where it is a component of the Z-line. Its presence in nonmuscle cells supports the idea that stress fibers in nonmuscle cells are analogues of sarcomeres.

HOW DO CROSS-LINKERS BIND TO FILAMENTS?   The widely accepted mechanism for these proteins is that the dimers contain two binding sites for filaments so they cross-link filaments into a three-dimensional network, which makes the solution a gel.

The proteins certainly bind to the sides of actin filaments. Brain alpha-actinin shows calcium-sensitive cooperative binding with a 1:10 stoichiometry (171), macrophage actin-binding protein has a K_d of 0.5 μM with a stoichiometry of 1:14 (164), and caldesmon binding saturates at a 1:6 ratio (172). These data suggest that the cross-linkers lie alongside actin filaments in a manner similar to tropomyosin in muscle. Macrophage alpha-actinin is an exception with a 1:1

Table 8   Cross-linker proteins

Protein name	Source	Subunits and M_r	Actin/protein molar ratio in cell	Ref.
ABP/Filamin group				
Actin-binding	Macrophage	270,000; 270,000	115	263
protein	Platelet	260,000; 260,000	46	264
Filamin	Smooth muscle	250,000; 250,000	165[d]	265
240-kd	*Dictyostelium*	240,000; 240,000		265a
Alpha-actinin group				
Alpha-actinin	Smooth muscle	110,000; 110,000	115[d]	266, 207
	Platelet	105,000; 105,000		182
	HeLa	105,000; 105,000		267
	Brain	105,000; 105,000		171
	Macrophage	103,000; 103,000	50	173
	Kidney	100,000; 100,000		207
	Skeletal I	100,000; 100,000		268
	muscle II	112,000; 112,000		
	Sea urchin egg	95,000; 95,000	15[a]	269
	Acanthamoeba	90,000; 90,000	50	270
	Dictyostelium	95,000; 95,000	30	271
Actinogelin	Ehrlich tumor	112,000; 112,000	25[a]	272
Spectrin group				
Spectrin	Erythrocytes	220,000; 240,000	2.5[c]	166
TW 260/240	Brush border	260,000; 240,000		273
Fodrin	Brain	235,000; 240,000		166
GP 260	*Acanthamoeba*	260,000; 260,000		169
HMW protein	*Physarum*	230,000; 230,000	350	170
Fodrin	*Dictyostelium*	235,000; 240,000		273a
120-kd proteins				
120-kd	*Dictyostelium*	120,000; 120,000	25	177
Caldesmon	Smooth muscle	120,000	40	172
Low-molecular-weight proteins				
36-kd	*Physarum*	36,000; 36,000	4	206
Gelactin I	*Acanthamoeba*	23,000	120[a]	274
Gelactin II	*Acanthamoeba*	28,000; 28,000	50[a]	274
Gelactin III	*Acanthamoeba*	32,000; 32,000	220[a]	274
Gelactin IV	*Acanthamoeba*	38,000; 38,000	600[a]	274
MAPs	Brain			181
				275

[a] Estimate from yield of purification.
[b] Based on monomer molecular weight.
[c] Based on dimers in erythrocyte membranes.
[d] (172)

stoichiometry of binding (173). Spectrin and band 4.1 bind synergistically to actin filaments, with maximum binding of one spectrin per actin protomer (174). One key point, which has never been shown by quantitative binding studies, is that one cross-linker molecule can bind to two actin filaments.

Morphology supports the idea that cross-linker proteins act as spot welds between actin filaments. With electron microscopy actin-binding protein is seen in shadowed specimens at intersections of actin filaments, and actin-binding protein is associated with an increased number of orthogonal crossovers between filaments (175). Actin-binding protein (176) and *Dictyostelium* 120K (177) are also associated with short actin filaments and "T" junctions between filaments, which implies that the proteins may nucleate actin polymerization.

HOW DOES CROSS-LINKING AFFECT SOLUTIONS OF ACTIN FILAMENTS? The cross-linking of the filaments presumably creates the gel observed in purification assays. This mechanism is very appealing and is generally accepted despite a paucity of evidence that quantitatively relates cross-linker binding, filament length, and the physical properties of filaments to mechanical properties of the solution. This topic was reviewed by Stossel and coworkers (4). The gel points of solutions of actin filaments of varying length mixed with either macrophage actin-binding protein (119) or smooth muscle filamin (179) are qualitatively predicted by Flory's theory (178) for network formation in chemical polymers with covalent bonds between subunits and between subunits and cross-links. Flory's theory considers the cross-links permanent and unbreakable, but rheology of solutions of actin filaments with *Acanthamoeba* alpha-actinin shows that cross-links may be labile (179a). Extension of this approach with more sophisticated rheology, a more complex theory, and incorporation of new knowledge about cross-linker binding and filament properties will provide ample opportunities for future research.

With higher concentrations of cross-linker proteins, the actin filaments form bundles that are generally coarse. The exact relationship between bundling and gelation is not understood, especially the effect of bundle formation on the mechanical properties of these materials. This is another question that suffers from the lack of techniques to determine the physical state of filaments, especially length and three-dimensional arrangement, without disruption and artefact. At high concentrations microtubule-associated-protein (MAP-2) links actin filaments into tight bundles with cross-striations (180), while lower concentrations form gels with actin filaments (181). Band 4.9, which bundles filaments, decreases the low shear viscosity of actin filaments at low molar ratios and increases it at high ratios (159). Macrophage alpha-actinin has less gelation activity at physiologic temperatures (173), so it may be primarily a bundling protein.

HOW ARE CROSS-LINKERS REGULATED? Calcium decreases the binding of nonmuscle, but not muscle, alpha-actinin to actin filaments (171, 182), and also decreases the activity of the 30-kd protein from *Dictyostelium* (183). Membrane lipids are associated with alpha-actinin in activated platelets (184), but the properties of that modified alpha-actinin are not known. Smooth muscle filamin is phosphorylated by a cAMP-dependent kinase, but the effect of phosphorylation on activity is not known (184a). Platelet actin-binding protein contains two phosphates per 260 kd polypeptide in resting platelets. Removal of one phosphate in vitro leads to a loss of activity (185). Platelet actin-binding protein is also degraded to an inactive form by a calcium-activated protease during platelet activation, but the physiologic role of the proteolysis is unclear (186). Phosphorylation of MAPs, including MAP-2 and tau, decreases their interaction with actin filaments (181).

Calmodulin and calcium regulate the binding of caldesmon to actin filaments. Calcium itself does not influence caldesmon binding to actin. However, calcium is required for calmodulin to bind to caldesmon, and the caldesmon-calmodulin-calcium complex does not bind actin (187). In separate experiments, caldesmon does not inhibit actomyosin ATPase when prepared in a phosphorylated form (188), and aggregation of caldesmon, which may or may not be physiologic, is required for its bundling activity (189).

HOW MIGHT CROSS-LINKERS ACT IN CELLS? Much of the interest in these proteins stems from the fact that cytoplasm has the mechanical properties of a non-Newtonian, viscoelastic material with a yield stress, meaning that it is a gel (190). Rheologic studies have been performed on solutions of actin filaments (68, 69) and mixtures of actin with spectrin (191) and alpha-actinin (179a), and studies with other cross-linker proteins should soon be available. Actin filaments alone become immobilized in solution (70) and have mechanical properties that indicate the existence of filament-filament interactions (69). The elastic modulus of actin filament solutions is greatly increased by the addition of spectrin, especially at high protein concentration (191). The experimental data are in need of theory to relate parameters such as filament length, filament flexibility, bundle formation, and cross-link density to the mechanical properties of solutions. Initial computer modeling studies show the possibility of relating biochemical parameters, such as calcium concentration, to mechanical properties and movement of cytoplasm (reviewed in Ref. 192).

One difficult question will be how to substantiate that one or more cross-linker proteins account for the mechanical properties of cytoplasm. Localization studies generally show that cross-linker proteins are in the same places as actin filaments in cells (193, 194), supporting the idea that physically continuous networks exist in cells. Some of these proteins may function more as bundlers than gelation factors. Alpha-actinin is associated with bundles of actin filaments at striated muscle Z-lines (195), smooth muscle dense bodies and attachment plaques (196), and epithelial cell zonula adherens (197). Actin filaments have their barbed ends at some of these sites, which prompted the idea that alpha-actinin was a capping protein. Although this is not the case, the original observation remains unexplained, along with the nature of barbed ends at such sites.

Some cross-linkers, such as filamin (205), *Dictyostelium* 120K (177), caldesmon (188), and *Physarum* 36K (206) inhibit the actin-activation of the ATPase of myosin. Interestingly, some cross-linkers, including *Dictyostelium* alpha-actinin (177), kidney alpha-actinin (207), TW 260/240 (207a), and filamin (205, 208), increase actomyosin ATPase at molar ratios far below stoichiometry. The reason is uncertain, but may involve holding filaments of actin and myosin close together (177) or a propagated change in structure of the actin filament (209).

Side-Binding Proteins that do not Cross-link

Tropomyosin, an elongated dimer of 35-kd subunits, binds to the sides of actin filaments in striated muscle to regulate contraction. It is present in a number of nonmuscle cells [reviewed in (4)] including erythrocytes (160). By itself it blocks actin-myosin interaction. Calcium and troponin relieve this inhibition in striated muscle, but troponin has not been found in nonmuscle cells, so the role of tropomyosin is unclear. Tropomyosin inhibits spontaneous fragmentation of actin filaments (198) and severing by some capping proteins (199, 117) but not others (200). There are several isoforms of tropomyosin in nonmuscle cells, and transformation, which changes cell shape, also changes the relative concentrations of these isoforms (201).

Cofilin, a 20-kd monomer found in procine brain and kidney, binds to actin filaments and inhibits tropomyosin binding and actomyosin ATPase (202). It probably also binds to monomers with a K_d of 0.2 μM, shown by an increased rate of nucleotide exchange (203), and it binds near the N terminus of actin, shown by chemical cross-linking (204). Its relationship to other actin-binding proteins is unknown.

MEMBRANE ATTACHMENT PROTEINS

Actin filaments often have their ends or sides close to plasma membranes, plasma membranes sometimes have unusual shapes that conform to the shape of actin bundles, and preparations of plasma membranes have actin filaments on their cytoplasmic surface. Proteins that may bind actin and actin-binding proteins to membranes have been described (Table 9). The sides of actin filaments are clearly attached to membranes, and most of these proteins bind to the sides of actin filaments, but few good candidates for end-binding attach-

Table 9 Proteins that attach actin to membranes

Protein name	Source	Subunits and Mr	Actin/protein molar ratio in cell	Ref.
110-kd	Intestinal brush border	110,000	9[a]	147, 280
Vinculin	Smooth muscle	130,000		218
	Striated muscle	130,000		216
	Platelet	130,000		219
	Fibroblast	130,000		229
Metavinculin	Smooth muscle	150,000		223, 224
Talin	Smooth muscle	215,000		227
24-kd	*Dictyostelium*	24,000		232
Protein I (85-kd)	Intestinal brush border	2×36,000; 10,000		231
GP IIb-III	Platelet	155,000; 114,000		276
GP Ib	Platelet	165,000		277
Cell surface actin	Lymphocyte	42,000		210
58-kd/CAG	Adenocarcinoma	58,000; 75,000	1[b]	278
140-kd	Fibroblast	140,000		225
130-kd	Fibroblasts, muscle	130,000		225a
Spectrin	Several mammalian tissues	265×2; 260×2		167
Band 4.1 (Synapsin)	Several mammalian tissues	75–80,000		279

[a]Molar ratio in microvilli (280).
[b]Molar ratio in Triton residue.

ment proteins exist. One novel notion is that actin or a modified actin inserts directly into the membrane (210). In general, this area is poorly defined, and only in the case of the erythrocyte (166) can molecular connections be traced from the actin filaments in the cytoplasm to integral membrane proteins.

Many cells have proteins like spectrin and band 4.1, which bind actin to erythrocyte membrane, and spectrin clearly associates with actin filaments as a cross-linker, so these molecules may bind filaments laterally to membranes [reviewed in (166)]. This mechanism may account for filament attachment in general, away from specialized areas. One striking piece of evidence against the importance of spectrin/actin interaction is that microinjection of antibodies to spectrin causes aggregation of spectrin, but no change in cell shape or actin filament distribution (211). Synapsin, which is related to band 4.1, is found on synaptic vesicles in brain and is phosphorylated in response to various stimuli (212). An alternative lateral connector may be actin-binding protein, which may link actin to plasma membranes in platelets (212a).

Brush-border microvilli have side arms that connect the actin bundles to the plasma membrane [reviewed in (147)]. These arms contain a complex of 110-kd protein and calmodulin, which binds actin in the absence of ATP (213) and exhibits actin-activated ATPase (214). There is some controversy over whether the 110-kd protein inserts into the plasma membrane (215) or attaches to another protein, but its extraction does cause the membrane to balloon away from the filament bundle (216).

Historically, vinculin has been the premier candidate for a protein linking actin filaments to membranes. It is defined as a 130-kd protein localized to attachment plaques in fibroblasts and smooth muscle cells, which are sites where actin filament bundles meet the plasma membrane. Vinculin is also associated with the zonula adherens of epithelial cells (197), the plasma membrane of striated muscle at sites adjacent to Z lines (217, 217a), and at intercalated disks (218). In all these places, actin filaments appear to attach to membranes, and alpha-actinin is seen on the cytoplasmic side of the vinculin. Some preparations of vinculin had capping activity, which fostered the idea that vinculin bound barbed ends of actin filaments to membranes, but recent work with purer material shows that the capping activity resides in contaminants and not the 130-kd vinculin (219, 220, 220a). The bundling activity of vinculin, which was also observed in crude preparations (221), is also being re-examined. In vitro vinculin does bind to alpha-actinin (221a). It is interesting that vinculin is not found in dense bodies (222), which are structures in the cytoplasm that contain alpha-actinin and bind actin filaments (196). Vinculin itself is a soluble protein, but metavinculin, a 152-kd analogue of vinculin, has solubility properties similar to membrane proteins and may represent the actual attachment molecule (223, 224). A 140-kd protein, of unknown relationship to vinculin, also localizes to attachment plaques, and extracellular antibodies against it inhibit attachment of cells to fibronectin surfaces (225). This protein may be identical to a 130-kd membrane surface glycoprotein that localizes at actin attachment sites (225a). Talin, a 215-kd protein of attachment plaques (226), binds vinculin but not actin (227).

Vinculin is of special interest because transformation with Rous sarcoma virus leads to phosphorylation of vinculin and dramatic alterations in the actin cytoskeleton (228). Tumor promoters also increase the phosphorylation of vinculin, although not on tyrosine (229). Platelet-derived growth factor leads to loss of vinculin from attachment plaques, loss of stress fibers, and then a reappearance of vinculin at attachment plaques, all with no effect on talin (230). Another protein phosphorylated by Rous sarcoma virus transformation, the 36-kd subunit of Protein I of brush border, is associated with membranes and bundles actin filaments (231).

A 24-kd protein from plasma membranes of *Dictyostelium* binds to actin (232), and fragments of plasma membrane bind to the sides of actin filaments

(233). Rod-shaped bridges connect the sides of actin filaments to plasma membranes, and myosin subfragment-1 disrupts these connections but not end attachments, good evidence for the existence of end-attachment proteins (234). Further studies with this system are promising because it is so amenable to biochemistry.

Some membrane-bound intracellular organelles, such as chromaffin granules (235) and lysosomes (236), bind actin, but the significance of the interaction is in some doubt (237). Interestingly, the gamma isoform of actin is selectively localized to mitochondria in muscle (238), which argues for specificity. Organelles from an alga move along actin filaments (239), and gelsolin inhibits movements of axoplasmic vesicles (240), but moving organelles in squid and bovine axoplasm are attached to microtubules (241).

FUTURE RESEARCH

During the past ten years many types of actin-binding proteins have been found (often in triplicate), and at first glance, they do just what one imagined that they would. Now the field will probably proceed in several directions. First, there will be a more critical evaluation of mechanisms, with assays and experimental designs that yield straightforward interpretations and quantitative results. Mechanisms will be tested by predictions for new experiments, especially ones that combine actin-binding proteins. Structure will be analyzed by peptide and sequence analysis to determine relationships among actin-binding proteins, and the structures of representative proteins will be solved by X-ray diffraction, so that functions will be ascribed to specific portions of these molecules. Site-directed mutagenesis will generate altered proteins for these biochemical experiments.

The physiologic role in cells of actin-binding proteins, and of actin itself, will be examined. Measurements of binding constants and absolute concentrations of actin-binding proteins and regulator molecules in cells are a start. More regulators will be discovered and related to physiologic stimuli and motility responses. Better microscopy will show us where actin-binding proteins and actin are, both in fixed and living cells. Some well-defined, quantitative assays of motility (242, 243) and the state of the actin-based cytoskeleton (243a) have now been developed, but others, based on living cells, cell models, and purified components, are still needed. An important application of such assays will be experiments that alter the activities of actin-binding proteins by addition of actin-binding proteins, fragments of proteins, and inhibitors. Although still in its infancy in this field, it seems possible that genetic methods will provide decisive tests for physiologic functions in this complex, multicomponent system (243b).

ACKNOWLEDGMENTS

We are grateful to Drs. Susan Craig, Detlev Drenckhahn, Elliot Elson, Carl Frieden, and Robin Michaels for reading the manuscript, and to Drs. Ueli Aebi, David DeRosier, and Paul Matsudaira for helpful conversations. J.A.C. is a Damon Runyon-Walter Winchell Cancer Fund postdoctoral fellow and is grateful to Dr. Elson for support. T.P. is supported by research grants from the N.I.H. and the Muscular Dystrophy Association.

Literature Cited

1. Korn, E. D. 1982. *Physiol. Rev.* 62:672–737
2. Frieden, C. 1985. *Ann. Rev. Biophys. Biophys. Chem.* 14:189–210
3. Amos, L. A. 1985. *Ann. Rev. Biophys. Biophys. Chem.* 14:291–313
4. Stossel, T. P., Chaponnier, C., Ezzell, R. M., Hartwig, J. H., Janmey, P. A., et al. 1985. *Ann. Rev. Cell Biol.* 1:353–402
5. Harrington, W. F., Rodgers, M. E. 1984. *Ann. Rev. Biochem.* 53:35–73
6. Hibberd, M. G., Trentham, D. R. 1986. *Ann. Rev. Biophys. Biophys. Chem.* 15:119–61
7. Hightower, R. C., Meagher, R. B. 1985. *EMBO J.* 4:1–8
8. Buckingham, M., Alonso, S., Bugaisky, G., Barton, P., Cohen, A., et al. 1985. *Adv. Exp. Med.* 182:333–44
9. Vandekerckhove, J., Weber, K. 1984. *J. Mol. Biol.* 179:391–413
10. Strauch, A. R., Rubenstein, P. A. 1984. *J. Biol. Chem.* 259:7224–29
11. Solomon, L. R., Rubenstein, P. A. 1985. *J. Biol. Chem.* 260:7659–64
12. Smith, P. R., Fowler, W. E., Pollard, T. D., Aebi, U. 1983. *J. Mol. Biol.* 167:641–60
13. Smith, P. R., Fowler, W. E., Aebi, U. 1984. *Ultramicroscopy* 13:113–23
14. Kabsch, W., Mannherz, H. G., Suck, D. 1985. *EMBO J.* 4:2113–18
15. Carlsson, L., Nystrom, L., Lindberg, U., Kannan, K. K., Cid-Dresdner, H., et al. 1976. *J. Mol. Biol.* 105:353–66
16. Mornet, D., Ue, K. 1984. *Proc. Natl. Acad. Sci. USA Biol.* 81:3680–84
16a. Hegyi, G., Szilagyi, L., Elzinga, M. 1985. *J. Muscle Res. Cell. Motil.* 6:74
17. Aebi, U., Millonig, R., Salvo, H. 1985. *Proc. 43rd Ann. Meet. Electron Microscopy Soc. Am.*, pp. 480–81. San Francisco Press
18. Sutoh, K. 1984. *Biochemistry* 23:1942–46
19. Elzinga, M., Phelan, J. J. 1984. *Proc. Natl. Acad. Sci. USA* 81:6599–602
20. Hitchcock-DeGregori, S. E., Mandala, S., Sachs, G. A. 1982. *J. Biol. Chem.* 257:12573–80
21. Frieden, C. 1982. *J. Biol. Chem.* 257:2882–86
22. Wanger, M., Wegner, A. 1983. *FEBS Lett.* 162:112–16
23. Frieden, C., Lieberman, D., Gilbert, H. R. 1980. *J. Biol. Chem.* 255:8991–93
24. Roustan, C., Benyamin, Y., Boyer, M., Bertrand, R., Audemard, E., Jauregui-Adell, J. 1985. *FEBS Lett.* 181:119–23
25. Hill, T. L., Kirschner, M. W. 1982. *Int. Rev. Cytol.* 78:1–125
26. Frieden, C., Goddette, D. W. 1983. *Biochemistry* 22:5836–43
27. Rich, S. A., Estes, J. E. 1976. *J. Mol. Biol.* 104:777–92
28. Rouayrenc, J. F., Travers, F. 1981. *Eur. J. Biochem.* 116:73–77
29. Cooper, J. A., Buhle, E. L. Jr., Walker, S. B., Tsong, T. Y., Pollard, T. D. 1983. *Biochemistry* 22:2193–202
30. Tobacman, L. S., Korn, E. D. 1983. *J. Biol. Chem.* 258:3207–14
31. Frieden, C. 1983. *Proc. Natl. Acad. Sci. USA* 80:6513–17
32. Frieden, C., Patane, K. 1985. *Biochemistry* 24:4192–96
33. Gershman, L. C., Newman, J., Selden, L. A., Estes, J. E. 1984. *Biochemistry* 23:2199–203
34. Pollard, T. D. 1983. *Anal. Biochem.* 134:406–12
35. Pollard, T. D. 1984. In *Calcium Regulation in Biological Systems. Proc. Takeda Sci. Found. Symp. Biosci.*, Japan: Academic
36. Wegner, A., Engel, J. 1975. *Biophys. Chem.* 3:215–25
37. Bishop, M. F., Ferrone, F. A. 1984. *Biophys. J.* 46:631–44
38. Lal, A. A., Brenner, S. L., Korn, E. D. 1984. *J. Biol. Chem.* 259:13061–65
39. Pantaloni, D., Carlier, M. F., Korn, E. D. 1985. *J. Biol. Chem.* 260:6572–78
40. Gilbert, H. R., Frieden, C. 1983. *Bio-

chem. Biophys. Res. Commun. 111:404–8

41. Lal, A. A., Korn, E. D., Brenner, S. L. 1984. *J. Biol. Chem.* 259:8794–800

42. Pollard, T. D., Mooseker, M. S. 1981. *J. Cell Biol.* 88:654–59

43. Drenckhahn, D., Pollard, T. D. 1986. *Biophys. J.* 49:4162

44. Pollard, T. D. 1984. *J. Cell Biol.* 99:769–77

45. Newman, J., Estes, J. E., Selden, L. A., Gershman, L. C. 1985. *Biochemistry* 24:1538–44

46. Mozo-Villarias, A., Ware, B. R. 1985. *Biochemistry* 24:1544–48

47. Goddette, D. W., Uberbacher, E. C., Bunick, G. J., Frieden, C. 1986. *J. Biol. Chem.* 261:2605–9

48. Wegner, A., Isenberg, G. 1983. *Proc. Natl. Acad. Sci. USA* 80:4922–25

49. Cooper, J. A., Blum, J. D., Pollard, T. D. 1984. *J. Cell Biol.* 99:217–25

50. Wegner, A. 1976. *J. Mol. Biol.* 108:139–40

51. Brenner, S. L., Korn, E. D. 1983. *J. Biol. Chem.* 258:5013–20

52. Carlier, M. F., Pantaloni, D., Korn, E. D. 1985. *J. Biol. Chem.* 260:6565–71

53. Nakaoka, Y., Kasai, M. 1969. *J. Mol. Biol.* 44:319–32

54. Carlier, M. F., Pantaloni, D., Korn, E. D. 1984. *J. Biol. Chem.* 259:9987–91

55. Wegner, A., Savko, P. 1982. *Biochemistry* 21:1909–13

56. Kondo, H., Ishiwata, S. 1976. *J. Biochem.* 79:159–71

57. Pardee, J. D., Spudich, J. A. 1982. *J. Cell Biol.* 93:648–54

58. Pollard, T. D., Weeds, A. G. 1984. *FEBS Lett.* 170:94–98

59. Carlier, M. F., Pantaloni, D., Korn, E. D. 1984. *J. Biol. Chem.* 259:9983–86

59a. Pantaloni, D., Hill, T. L., Carlier, M.-F., Korn, E. D. 1985. *Proc. Natl. Acad. Sci. USA* 82:7207–11

60. Mitchison, T., Kirschner, M. 1984. *Nature* 312:237–41

61. Carlier, M. F., Hill, T. L., Chen, Y. 1984. *Proc. Natl. Acad. Sci. USA* 81:771–75

62. Pantaloni, D., Carlier, M. F., Coue, M., Lal, A. A., Brenner, S. L., Korn, E. D. 1984. *J. Biol. Chem.* 259:6274–83

63. Egelman, E. H., Francis, N., DeRosier, D. J. 1983. *J. Mol. Biol.* 166:605–29

64. Yoshimura, H., Nishio, T., Mihashi, K., Kinosita, K., Ikegami, A. 1984. *J. Mol. Biol.* 179:453–67

65. Oosawa, F. 1980. *Biophys. Chem.* 11:443–46

66. DeRosier, D. J., Tilney, L. G. 1984. *J. Mol. Biol.* 175:57–73

67. Maruyama, K., Kaibara, M., Fukada, E. 1974. *Biochim. Biophys. Acta* 271:20–29

68. Zaner, K. S., Stossel, T. P. 1983. *J. Biol. Chem.* 258:1004–9

69. Sato, M., Leimbach, G., Schwarz, W. H., Pollard, T. D. 1985. *J. Biol. Chem.* 260:8585–92

70. Tait, J. F., Frieden, C. 1982. *Biochemistry* 21:3666–74

71. Jen, C. J., McIntire, L. V., Bryan, J. 1982. *Arch. Biochem.* 216:126–32

72. Tseng, P. C. H., Runge, M. S., Cooper, J. A., Williams, R. C., Pollard, T. D. 1984. *J. Cell Biol.* 98:214–21

73. Nystrom, L. E., Lindberg, U., Kendrick-Jones, J., Jakes, R. 1979. *FEBS Lett.* 101:161–65

74. Ampe, C., Vandekerckhove, J., Brenner, S. L., Tobacman, L., Korn, E. D. 1985. *J. Biol. Chem.* 260:834–40

75. Kaiser, D. A., Sato, M., Ebert, R., Pollard, T. D. 1986. *J. Cell Biol.* 102:221–26

76. Dinubile, M. J., Southwick, F. S. 1985. *J. Biol. Chem.* 260:7402–9

77. Reichstein, E., Korn, E. D. 1979. *J. Biol. Chem.* 254:6174–79

78. Mockrin, S. C., Korn, E. D. 1980. *Biochemistry* 19:5359–62

79. Tobacman, L. S., Korn, E. D. 1982. *J. Biol. Chem.* 257:4166–70

80. Lee, S., Cooper, J. A., Pollard, T. D. 1982. *J. Cell Biol.* 95:297a

81. Mockrin, S. C., Korn, E. D. 1983. *J. Biol. Chem.* 258:3215–21

82. Lal, A. A., Korn, E. D. 1985. *J. Biol. Chem.* 260:10132–38

83. Carlsson, L., Nystrom, L. E., Sundkvisk, I., Markey, F., Lindberg, U. 1977. *J. Mol. Biol.* 115:465–83

84. Malm, B. 1984. *FEBS Lett.* 173:399–402

85. Malm, B., Nystrom, L. E., Lindberg, U. 1980. *FEBS Lett.* 113:241–44

86. Pollard, T. D., Cooper, J. A. 1984. *Biochemistry* 23:6631–41

87. Tilney, L. G., Bonder, E. M., Coluccio, L. M., Mooseker, M. S. 1983. *J. Cell Biol.* 97:113–24

88. Nishida, E., Maekawa, S., Sakai, H. 1984. *J. Biochem.* 95:399–404

88a. Lassing, I., Lindberg, U. 1985. *Nature* 314:472–74

89. Mannherz, H. G., Barrington-Leigh, J., Leberman, R., Pfrang, H. 1975. *FEBS Lett.* 60:34–38

90. Gadasi, H. 1982. *Biochem. Biophys. Res. Commun.* 104:158–64

91. Hitchcock, S. E. 1980. *J. Biol. Chem.* 255:5668–73

92. Pinder, J. C., Gratzer, W. B. 1982. *Biochemistry* 21:4886–90

93. Maekawa, S., Nishida, E., Ohta, Y., Sakai, H. 1984. *J. Biochem.* 95:377–85
94. Goldschmidt-Clermont, P. J., Galbraith, R. M., Emerson, D. L., Marsot, F., Nel, A. E., Arnaud, P. 1985. *Biochem. J.* 228:471–77
95. Rohr, G., Mannherz, H. G. 1978. *Eur. J. Biochem.* 89:151–57
95a. Lacks, S. A. 1981. *J. Biol. Chem.* 256:2644–48
96. Lees, A., Haddad, J. G., Lin, S. 1984. *Biochemistry* 23:3038–47
97. Yin, H. L., Kwiatkowski, D. J., Mole, J. E., Cole, F. S. 1984. *J. Biol. Chem.* 259:5271–76
98. Matsudaira, P., Jakes, R., Walker, J. E. 1985. *Nature* 315:248–50
99. Maruta, H., Knoerzer, W., Hissen, H., Isenberg, G. 1984. *Nature* 312:424–26
100. Cooper, J. A., Blum, J. D., Williams, R. C. Jr., Pollard, T. D. 1986. *J. Biol. Chem.* 261:477–85
101. Craig, S. W., Powell, L. 1980. *Cell* 22:739–46
102. Bretscher, A., Weber, K. 1978. *J. Cell Biol.* 79:839–45
103. Bonder, E. M., Fishkind, D. J., Mooseker, M. S. 1983. *Cell* 34:491–501
104. Yamamoto, K., Pardee, J. D., Reidler, J., Stryer, L., Spudich, J. A. 1982. *J. Cell Biol.* 95:711–19
105. Walsh, T. P., Weber, A., Higgins, J., Bonder, E. M., Mooseker, M. S. 1984. *Biochemistry* 23:2613–21
106. Wanger, M., Wegner, A. 1985. *Biochemistry* 24:1035–40
107. Harris, D. A., Schwartz, J. H. 1981. *Proc. Natl. Acad. Sci. USA* 78:6798–802
108. Doi, Y., Frieden, C. 1984. *J. Biol. Chem.* 259:11868–75
109. Giffard, R. G., Weeds, A. G., Spudich, J. A. 1984. *J. Cell Biol.* 98:1796–803
110. Cribbs, D. H., Glenney, J. R., Kaulfus, P., Weber, K., Lin, S. 1982. *J. Biol. Chem.* 257:395–99
111. Glenney, J. R., Kaulfus, P., Weber, K. 1981. *Cell* 24:471–80
112. Harris, H., Weeds, A. G. 1984. *FEBS Lett.* 177:184–88
113. Spudich, J. A., Kron, S. J., Sheetz, M. P. 1985. *Nature* 315:584–85
114. Cooper, J. A., Pollard, T. D. 1985. *Biochemistry* 24:793–99
114a. Coue, M., Korn, E. D. 1985. *J. Biol. Chem.* 260:15033–41
115. Bryan, J., Kurth, M. C. 1984. *J. Biol. Chem.* 259:7480–87
116. Goddette, D., Frieden, C. 1985. *Biochem. Biophys. Res. Commun.* 128:1087–92
117. Walsh, T. P., Weber, A., Davis, K., Bonder, E., Mooseker, M. 1984. *Biochemistry* 23:6099–102
118. Bonder, E. M., Mooseker, M. S. 1983. *J. Cell Biol.* 96:1097–107
119. Yin, H. L., Zaner, K. S., Stossel, T. P. 1980. *J. Biol. Chem.* 255:9494–500
120. Yin, H. L., Stossel, T. P. 1980. *J. Biol. Chem.* 255:9490–93
120a. Kilhoffer, M. C., Gerard, D. 1985. *Biochemistry* 24:5653–60
120b. Hwo, S., Bryan, J. 1986. *J. Cell Biol.* 102:227–36
120c. Bryan, J., Hwo, S. 1986. *J. Cell Biol.* 102: In press
121. Hesterberg, L. K., Weber, K. 1983. *J. Biol. Chem.* 258:365–69
122. Hesterberg, L. K., Weber, K. 1983. *J. Biol. Chem.* 258:359–64
123. Kurth, M. D., Bryan, J. 1984. *J. Biol. Chem.* 259:7473–79
124. Janmey, P. A., Chaponnier, C., Lind, S. E., Zaner, K. S., Stossel, T. P., Yin, H. L. 1985. *Biochemistry* 24:3714–23
125. Maruta, H., Isenberg, G. 1984. *J. Biol. Chem.* 259:5208–13
126. Glenney, J. R., Geisler, N., Kaulfus, P., Weber, K. 1981. *J. Biol. Chem.* 256:8156–61
127. Southwick, F. S., Hartwig, J. H. 1982. *Nature* 297:303–7
128. Yokota, E., Maruyama, K. 1983. *J. Biochem.* 94:1897–900
129. Ishiwata, S., Funatsu, T. 1985. *J. Cell Biol.* 100:282–91
130. Cohen, C. M., Branton, D. 1979. *Nature* 279:163–65
131. Tsukita, S., Tsukita, S., Ishikawa, H. 1984. *J. Cell Biol.* 98:1102–10
132. Pinder, J. C., Ohanian, V., Gratzer, W. B. 1984. *FEBS Lett.* 169:161–64
133. Wang, Y.-L. 1985. *J. Cell Biol.* 101:597–602
134. Yin, H. L., Albrecht, J. H., Fattoum, A. 1981. *J. Cell Biol.* 91:901–6
134a. Carron, C. P., Hwo, S., Dingus, J., Benson, D. M., Meza, I., Bryan, J. 1986. *J. Cell Biol.* 102:237–45
135. Taylor, D. L., Condeelis, J. 1979. *Int. Rev. Cytol.* 56:57–144
136. Mooseker, M. S., Pollard, T. D., Wharton, K. A. 1982. *J. Cell Biol.* 95:223–33
137. Tilney, L. G., Inoue, S. 1985. *J. Cell Biol.* 100:1273–83
138. Detmers, P., Goodenough, U., Condeelis, J. 1983. *J. Cell Biol.* 97:522–32
139. Mabuchi, I. 1983. *J. Cell Biol.* 97:1612–21
140. Bamburg, J. R., Harris, H. E., Weeds, A. G. 1980. *FEBS Lett.* 121:178–82
141. Nishida, E., Maekawa, S., Muneyuki, E., Sakai, H. 1984. *J. Biochem.* 95:387–98
142. Griffith, L. M., Pollard, T. D. 1982. *J. Biol. Chem.* 257:9135–42

143. Ando, T., Scales, D. 1985. *J. Biol. Chem.* 260:2321–27
144. Clarke, F. M., Morton, D. J. 1976. *Biochem. J.* 159:797–98
145. Castellanii, L., O'Brien, E. J. 1981. *J. Mol. Biol.* 147:205–13
146. Bretscher, A., Weber, K. 1980. *J. Cell Biol.* 86:335–40
147. Mooseker, M. S. 1985. *Ann. Rev. Cell Biol.* 1:209–41
148. Yamashiro-Matsumura, S., Matsumura, F. 1985. *J. Biol. Chem.* 260:5087–97
148a. Glenney, J. R. Jr., Kaulfus, P., Matsudaira, P., Weber, K. 1981. *J. Biol. Chem.* 256:9283–88
149. Bryan, J., Kane, R. E. 1978. *J. Mol. Biol.* 125:207–24
150. DeRosier, D., Mandelkow, E., Silliman, A., Tilney, L., Kane, R. E. 1977. *J. Mol. Biol.* 113:679–95
151. Bretscher, A. 1981. *Proc. Natl. Acad. Sci. USA* 78:6849–53
152. Otto, J. J., Kane, R. E., Bryan, J. 1980. *Cell Motility* 1:31–40
153. Otto, J., Kane, R. E., Bryan, J. 1979. *Cell* 17:285–93
154. DeRosier, D. J., Edds, K. T. 1980. *Exp. Cell Res.* 126:490–94
155. Begg, D. A., Rebhun, L. I., Hyatt, H. 1982. *J. Cell Biol.* 93:24–32
156. Tilney, L. G. 1975. *J. Cell Biol.* 64:289–310
157. DeRosier, D. J., Tilney, L. G., Bonder, E. M., Frankl, P. 1982. *J. Cell Biol.* 93:324–37
158. Matsudaira, P., Mankelkow, E., Renner, W., Hesterberg, L. K., Weber, K. 1983. *Nature* 301:209–14
158a. Fechheimer, M. 1985. *J. Cell Biol.* 101:407a
159. Siegel, D. L., Branton, D. 1985. *J. Cell Biol.* 100:775–85
160. Fowler, V. M., Bennett, V. 1984. *J. Biol. Chem.* 259:5978–89
161. Fowler, V. M., Davis, J. Q., Bennett, V. 1985. *J. Cell Biol.* 100:47–56
162. Wong, A. J., Kiehart, P., Pollard, T. D. 1985. *J. Biol. Chem.* 260:46–49
163. Shen, B. W., Josephs, R., Steck, T. L. 1984. *J. Cell Biol.* 99:810–21
163a. Byers, T. J., Branton, D. 1985. *Proc. Natl. Acad. Sci. USA* 82:6153–57
164. Hartwig, J. H., Stossel, T. P. 1981. *J. Mol. Biol.* 145:563–81
165. Nunnally, M. H. 1981. *Actin-filamin gelation: a model for regulation of cytoplasmic consistency.* PhD thesis. Johns Hopkins Univ., Baltimore, MD
166. Bennett, V. 1985. *Ann. Rev. Biochem.* 54:273–304
167. Shotton, D. M., Burke, B. M., Branton, D. 1979. *J. Mol. Biol.* 131:303–29
168. Morrow, J. S., Marchesi, V. T. 1981. *J. Cell Biol.* 88:463–68
169. Pollard, T. D. 1984. *J. Cell Biol.* 99:1970–80
170. Sutoh, K., Iwane, M., Matsuzaki, F., Kikuchi, M., Ikai, A. 1984. *J. Cell Biol.* 98:1611–18
171. Duhaiman, A. S., Bamburg, J. R. 1984. *Biochemistry* 23:1600–8
172. Bretscher, A. 1984. *J. Biol. Chem.* 259:12873–80
173. Bennett, J. P., Zaner, K. S., Stossel, T. P. 1984. *Biochemistry* 23:5081–86
174. Cohen, C. M., Foley, S. F. 1984. *Biochemistry* 23:6091–98
175. Hartwig, J. H., Tyler, J., Stossel, T. P. 1980. *J. Cell Biol.* 87:841–48
176. Niederman, R., Amrein, P. C., Hartwig, J. H. 1983. *J. Cell Biol.* 96:1400–13
177. Condeelis, J., Vahey, M., Carboni, J. M., DeMey, J., Ogihara, S. 1984. *J. Cell Biol.* 99:119s–26s
178. Flory, P. J. 1953. *Principles of Polymer Chemistry.* Ithaca, NY: Cornell Univ. Press
179. Nunnally, M. H., Powell, L. D., Craig, S. W. 1981. *J. Biol. Chem.* 256:2083–86
179a. Sato, M., Schwartz, W. H., Pollard, T. D. 1985. *J. Cell Biol.* 101:408a
180. Sattilaro, R. F., Dentler, W. L., Lecluyse, E. L. 1981. *J. Cell Biol.* 90:467–73
181. Selden, S. C., Pollard, T. D. 1983. *J. Biol. Chem.* 258:7064–71
182. Rosenberg, S., Stracher, A., Burridge, K. 1981. *J. Biol. Chem.* 256:2986–91
183. Fechheimer, M., Taylor, D. L. 1984. *J. Biol. Chem.* 259:4514–20
184. Burn, P., Rotman, A., Meyer, R. K., Burger, M. M. 1985. *Nature* 314:469–71
184a. Wallach, D., Davies, P. J. A., Pastan, I. 1978. *J. Biol. Chem.* 253:4739–45
185. Zhuang, Q. Q., Rosenberg, S., Lawrence, J., Stracher, A. 1984. *Biochem. Biophys. Res. Commun.* 118:508–13
186. Fox, J. E. B., Goll, D. E., Reynolds, C. C., Phillips, D. R. 1985. *J. Biol. Chem.* 260:1060–66
187. Sobue, K., Muramoto, Y., Fujita, M., Kakiuchi, S. 1981. *Proc. Natl. Acad. Sci. USA* 78:5652–55
188. Ngai, P. K., Walsh, M. P. 1984. *J. Biol. Chem.* 259:13656–59
189. Sobue, K., Takahashi, K., Tanaka, T., Kanda, K., Ashino, N., et al. 1985. *FEBS Lett.* 182:201–4
190. Sato, M., Wang, T. Z., Allen, R. D. 1983. *J. Cell Biol.* 97:1089–97
191. Schanus, E., Booth, S., Hallaway, B., Rosenberg, A. 1985. *J. Biol. Chem.* 260:3724–30
192. Bray, D. 1985. *Nature* 313:738
193. Nunnally, M. H., D'Angelo, J. M.,

Craig, S. W. 1980. *J. Cell Biol.* 87: 219–26

194. Valerius, N. H., Stendahl, O. I., Hartwig, J. H., Stossel, T. P. 1981. *Cell* 24:195–202

195. Lazarides, E., Granger, B. 1978. *Proc. Natl. Acad. Sci. USA* 75:3683–87

196. Fay, F. S., Fujiwara, K., Rees, D. D., Fogarty, K. E. 1983. *J. Cell Biol.* 96:783–95

197. Geiger, B., Dutton, A. H., Tokuyasu, K. T., Singer, S. J. 1981. *J. Cell Biol.* 91:614–28

198. Wegner, A. 1982. *J. Mol. Biol.* 161: 217–27

199. Fattoum, A., Hartwig, J. H., Stossel, T. P. 1983. *Biochemistry* 22:1187–93

200. Verkhovsky, A. B., Surgucheva, I. G., Gelfand, V. I. 1984. *Biochem. Biophys. Res. Commun.* 123:596–603

201. Lin, J. J.-C., Helfman, D. M., Hughes, S. H., Chan, C.-S. 1985. *J. Cell Biol.* 100:692–703

202. Nishida, E., Maekawa, S., Sakai, H. 1984. *Biochemistry* 23:5307–13

203. Nishida, E. 1985. *Biochemistry* 24: 1160–64

204. Muneyuki, F., Nishida, E., Sutoh, K., Sakai, H. 1985. *J. Biochem.* 97:563–68

205. Davies, G., Bechtel, P., Pastan, I. 1977. *FEBS Lett.* 77:228–32

206. Ogihara, S., Tonomura, Y. 1982. *J. Cell Biol.* 93:604–14

207. Kobayashi, R., Tashima, Y. 1983. *Biochim. Biophys. Acta* 745:209–16

207a. Coleman, T. R., Mooseker, M. S. 1985. *J. Cell Biol.* 101:1850–57

208. Sosinski, J., Szpacenko, A., Dabrowska, R. 1984. *FEBS Lett.* 178:311–14

209. Craig-Schmidt, M. C., Robson, R. M., Goll, D. E., Stromer, M. H. 1981. *Biochim. Biophys. Acta* 670:9–16

210. Sanders, S. K., Craig, S. W. 1983. *J. Immunol.* 131:370–77

211. Mangeat, P. H., Burridge, K. 1984. *J. Cell Biol.* 98:1363–77

212. Huttner, W. B., Schiebler, W., Greengard, P., DeCamilli, P. 1983. *J. Cell Biol.* 96:1374–85

212a. Fox, J. E. B. 1985. *J. Biol. Chem.* 260:11970–77

213. Verner, K., Bretscher, A. 1985. *J. Cell Biol.* 100:1455–65

214. Collins, J. H., Borysenko, C. W. 1984. *J. Biol. Chem.* 259:14128–36

215. Glenney, J. R. Jr., Glenney, P. 1984. *Cell* 37:743–51

216. Howe, C. L., Mooseker, M. S. 1983. *J. Cell Biol.* 97:974–85

217. Pardo, J. V., Siliciano, J. D., Craig, S.

W. 1983. *Proc. Natl. Acad. Sci. USA* 80:1008–12

217a. Pardo, J. V., Siliciano, J. D., Craig, S. W. 1983. *J. Cell Biol.* 97:1081–88

218. Tokuyasu, K. T., Dutton, A. H., Geiger, B., Singer, S. J. 1981. *Proc. Natl. Acad. Sci. USA* 78:7619–23

219. Evans, R. R., Robson, R. M., Stromer, M. H. 1984. *J. Biol. Chem.* 259:3916–25

220. Rosenfeld, G. C., Hou, D. C., Dingus, J., Meza, I., Bryan, J. 1985. *J. Cell Biol.* 100:669–76

220a. Wilkins, J., Lin, S. 1986. *J. Cell Biol.* 102:1085–92

221. Jockusch, B. M., Isenberg, I. 1981. *Proc. Natl. Acad. Sci. USA* 78:3005–9

221a. Craig, S. W. 1985. *J. Cell Biol.* 101:136a

222. Small, J. V. 1985. *EMBO J.* 4:45–49

223. Siliciano, J. D., Craig, S. W. 1982. *Nature* 300:533–36

224. Feramisco, J. R., Smart, J. E., Burridge, K., Helfman, D. M., Thomas, G. P. 1982. *J. Biol. Chem.* 257:11024–31

225. Chen, W.-T., Hasegawa, E., Hasegawa, T., Weinstock, C., Yamada, K. M. 1985. *J. Cell Biol.* 100:1103–14

225a. Rogaski, A. A., Singer, S. J. 1985. *J. Cell Biol.* 101:785–801

226. Burridge, K., Connell, L. 1983. *J. Cell Biol.* 97:359–67

227. Burridge, K., Mangeat, P. 1984. *Nature* 308:744–46

228. Sefton, B. M., Hunter, T., Ball, E. H., Singer, S. J. 1981. *Cell* 24:165–74

229. Werth, D. K., Pastan, I. 1984. *J. Biol. Chem.* 259:5264–70

230. Herman, B., Pledger, W. J. 1985. *J. Cell Biol.* 100:1031–40

231. Gerke, V., Weber, K. 1984. *EMBO J.* 3:227–33

232. Stratford, C. A., Brown, S. S. 1985. *J. Cell Biol.* 100:727–35

233. Goodloe-Holland, C. M., Luna, E. J. 1984. *J. Cell Biol.* 99:71–78

234. Bennett, H., Condeelis, J. 1984. *J. Cell Biol.* 99:1434–40

235. Meyer, D. I., Burger, M. M. 1979. *FEBS Lett.* 101:129–33

236. Mehrabian, M., Bame, K. J., Rome, L. H. 1984. *J. Cell Biol.* 99:680–85

237. Kao, L. S., Westhead, E. W. 1984. *FEBS Lett.* 173:119–23

238. Pardo, J. V., Pittenger, M. F., Craig, S. W. 1983. *Cell* 32:1093–103

239. Kachar, B. 1985. *Science* 227:1355–56

240. Brady, S. T., Lasek, R. J., Allen, R. D., Yin, H. L., Stossel, T. P. 1984. *Nature* 310:56–58

241. Schnapp, B. J., Vale, R. D., Sheetz, M. P., Reese, T. S. 1985. *Cell* 40:455–62

242. Sheetz, M. P., Chasan, R., Spudich, J. A. 1984. *J. Cell Biol.*, 99:1867–71

243. Vale, R. D., Schnapp, B. J., Reese, T. S., Sheetz, M. P. 1985. *Cell* 40:559–69

243a. Pasternak, C., Elson, E. 1985. *J. Cell Biol.* 100:860–72

243b. Novik, P., Bostein, D. 1985. *Cell* 40:405–16

244. Neidl, C., Engel, J. 1979. *Eur. J. Biochem.* 101:163–69

245. Strzelecka-Golaszewska, H., Prochniewicz, E., Drabikowski, W. 1978. *Eur. J. Biochem.* 88:229–37

246. Tobacman, L. S., Brenner, S. L., Korn, E. D. 1983. *J. Biol. Chem.* 258:8806–12

247. Ozaka, K., Hatano, S. 1984. *J. Cell Biol.* 98:1919–25

248. Nishida, E. 1985. *Biochemistry* 24:1160–64

249. Schleicher, M., Gerisch, G., Isenberg, G. 1984. *EMBO J.* 3:2095–100

250. Kilimann, M. W., Isenberg, G. 1982. *EMBO J.* 1:889–94

251. Deleted in proof

252. Hasegawa, T., Takahashi, S., Hayashi, H., Hatano, S. 1980. *Biochemistry* 19:2677–83

253. Hosoya, H., Mabuchi, I. 1984. *J. Cell Biol.* 99:994–1001

254. Wang, L.-L., Spudich, J. A. 1984. *J. Cell Biol.* 99:844–51

255. Tawata, M., Kobayashi, R., Mace, M. L., Nielsen, T. B., Field, J. B. 1983. *Biochem. Biophys. Res. Commun.* 111:415–23

256. Kanno, K., Sasaki, Y., Hidaka, H. 1985. *FEBS Lett.* 184:202–6

257. Rouayrenc, J. F., Fattoum, A., Gabrion, J., Audemard, E., Kassab, R. 1984. *FEBS Lett.* 167:52–58

258. Ohta, Y., Endo, S., Nishida, E., Murofushi, H., Sakai, H. 1984. *J. Biochem.* 96:1547–58

259. Otto, J. J., Bryan, J. 1981. *Cell Motility* 1:179–92

260. Maekawa, S., Endo, S., Sakai, H. 1982. *J. Biochem.* 92:1959–72

261. Ueno, T., Korn, E. D. 1985. *Biophys. J.* 47:118a

262. Maekawa, S., Endo, S., Sakai, H. 1983. *J. Biochem.* 94:1329–37

262a. Hock, R. S., Condeelis, J. 1985. *J. Cell Biol.* 101:407a

263. Hartwig, J. H., Stossel, T. P. 1975. *J. Biol. Chem.* 250:5696–705

264. Rosenberg, S., Stracher, A. 1982. *J. Cell Biol.* 94:51–55

265. Wang, K. 1977. *Biochemistry* 16:1857–67

265a. Hock, R. S., Condeelis, J. 1985. *J. Cell Biol.* 101:406a

266. Feramisco, J. R., Burridge, K. 1980. *J. Biol. Chem.* 255:1194–99

267. Burridge, K., Feramisco, J. R. 1981. *Nature* 294:565–67

268. Kobayashi, R., Itoh, H., Tashima, Y. 1984. *Eur. J. Biochem.* 143:125–31

269. Mabuchi, I., Hamaguchi, Y., Kobayashi, T., Hosoya, H., Tsukita, S., Tsukita, S. 1985. *J. Cell Biol.* 100:375–83

270. Pollard, T. D. 1981. *J. Biol. Chem.* 256:7666–70

271. Brier, J., Fechheimer, M., Swanson, J., Taylor, D. L. 1983. *J. Cell Biol.* 97:178–85

272. Mimura, N., Asano, A. 1982. *J. Cell Biol.* 93:899–909

273. Pearl, M., Fishkind, D., Mooseker, M., Keene, D., Keller, T. III. 1984. *J. Cell Biol.* 98:66–78

273a. Bennett, H., Condeelis, J. 1985. *J. Cell Biol.* 101:285a

274. Maruta, H., Korn, E. D. 1977. *J. Biol. Chem.* 252:199–402

275. Arakawa, T., Frieden, C. 1984. *J. Biol. Chem.* 259:11730–34

276. Painter, R. G., Prodouz, K. N., Gaarde, W. 1985. *J. Cell Biol.* 100:652–57

277. Okita, J. R., Pidard, D., Newman, P. J., Montgomery, R. R., Kunicki, T. J. 1985. *J. Cell Biol.* 100:317–21

278. Carraway, C. A. C., Jung, G., Hinkley, R. E., Carraway, K. L. 1985. *Exp. Cell Res.* 157:71–82

279. Baines, A. J., Bennett, V. 1985. *Nature* 315:410–13

280. Matsudaira, P. T., Burgess, D. R. 1979. *J. Cell Biol.* 83:667–73

281. Craig, S. W., Pollard, T. D. 1982. *Trends Biochem. Sci.* 7:88–91

NOTE ADDED IN PROOF: New observations have shown that the dissociation constants for Mg^{2+} and Ca^{2+} binding to the high-affinity sites on actin are in the nanomolar range, not the micromolar range indicated in Table 1. Ref: Gershman, L. C., Selden, L. A., Estes, J. E. 1986. *Biophys. J.* 49:417A and *Biochem. Biophys. Res. Commun.* In press

Ann. Rev. Biochem. 1986. 55:1037–57

BIOCHEMICAL INTERACTIONS OF TUMOR CELLS WITH THE BASEMENT MEMBRANE

Lance A. Liotta, C. Nageswara Rao, and Ulla M. Wewer

Laboratory of Pathology, National Cancer Institute, Bethesda, Maryland 20205

CONTENTS

PERSPECTIVES AND SUMMARY

The vertebrate organism is divided into a series of tissue compartments containing specialized cell types supported by an extracellular matrix. The most common unit of tissue organization is an organ parenchymal cell attached to a basement membrane which is anchored to the interstitial stroma. The stroma

0066-4154/86/0701-1037$02.00

contains blood vessels, lymphatics, and nerves which support the parenchymal cells. The basement membrane and stroma are composed of specific extracellular matrix components including collagens, glycoproteins, elastin, and proteoglycans. These components form the structural scaffolding for the matrix, and their spacing, orientation, and charge influence selective filtration of soluble macromolecules through the matrix. In addition, the matrix components play an important role in cell morphology, mitogenesis, and differentiation. Regulatory signals are transmitted from components of the matrix through specific cell surface receptors. This in turn influences cell shape, cell polarity, and cell migration. The development of a malignant neoplasm is accompanied by profound alterations in the distribution and composition of the adjacent extracellular matrix. Normal cells may require specific interactions with the matrix for survival and normal functions. In contrast, the malignant tumor cells may lack such requirements or may respond in an abnormal way to the matrix. This may facilitate the ability of malignant tumor cells to locally invade host tissues and to metastasize to distant sites.

Recent work in the field of connective tissue biochemistry and cell biology has lead to significant insights into biochemical mechanisms that play a role in tumor invasion and metastases. The transition from in situ to invasive human carcinomas is accompanied by a marked disorganization and localized loss of the basement membrane components type IV collagen and laminin. Tumor cells have been found to utilize specific extracellular matrix receptors such as the laminin receptor for attachment to the basement membrane during invasion. Metastatic tumor cells produce proteases such as type IV collagenase and heparan sulfate proteoglycan–degrading enzymes which can participate in the lysis of the basement membrane. Locomotion of invading tumor cells through the extracellular matrix may be influenced by chemotactic factors. These biochemical findings have provided new strategies for tumor diagnosis and therapy of metastases. These strategies are being applied to routine surgical pathology and are being studied in animal models of metastases therapy.

TUMOR INVASION AND METASTASES

Tumor invasion is the first step in the complex multistep process that leads to the formation of metastases (1–5). After local invasion of adjacent host tissue barriers, the tumor cell must invade the vascular wall or lymphatic channels in order to disseminate. Tumor cells entering the circulation must be able to evade host defenses, survive the mechanical trauma of the blood flow, and arrest in the venous or capillary bed of the target organ. Once arrested, the tumor cells must again invade the vascular wall to enter the organ parenchyma. The extravasated tumor cell must be able to grow in a foreign "soil" different from the tissue of origin in order to initiate a metastatic colony (Figure 1). A metastatic tumor cell must, therefore, possess the capability of traversing all of the indicated steps.

The major cause of death in most human malignancies is metastasis (1, 4). About 50% of patients who develop a malignant tumor are cured by the various therapies now used. In the treatment failure group most patients die because of metastasis. Distant metastases are often too small to be detected at the time the primary tumor is treated. Widespread initiation of metastatic colonies usually occurs before clinical symptoms of metastatic disease are evident. Most patients suffer from multiple metastases. Since metastasis colony formation is a continuous process increasing over time, the identification of large metastases in a given organ frequently is accompanied by a larger number of occult micrometastases that were seeded more recently. The size and age variation in metastases, their dispersed anatomical location, and their heterogeneous composition are all factors that hinder surgical removal and limit the effective concentration of anticancer drugs that can be delivered to the metastatic colonies.

Consequently, a major challenge to cancer scientists is the development of improved methods to: (a) predict the metastatic aggressiveness of a patient's individual tumor; (b) prevent local invasion; and (c) identify and treat clinically silent micrometastases. Many laboratories are studying the fundamental mechanisms of invasion and metastases, with the hope of identifying specific

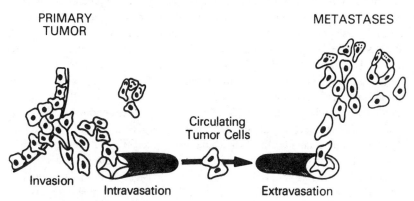

Figure 1 Diagram of the metastatic cascade. Tumor cells invade the epithelial basement membrane to reach the interstitial stroma where they gain access to blood vessels for further dissemination. Tumor cells invade the basement membrane and endothelial layer of the vessel wall and are dislodged into the circulation in single cells and clumps. Circulating tumor cells arrest in the precapillary venules of the target organ by adherence to endothelial cell lumenal surfaces, or exposed basement membranes. They must again invade the basement membrane to initiate a secondary tumor colony called a metastasis.

biochemical factors that can be the basis for such diagnostic or therapeutic strategies. Over the last several years, significant progress has been made toward this goal.

The complexity of the metastatic process has forced investigators to focus on one step at a time in order to reduce the number of variables to a reasonable level. In recent years, our laboratory has focused on the interaction of metastatic tumor cells with the extracellular matrix. The metastasizing tumor cell must interact with the matrix at many stages of tumor invasion and metastases. One particular type of matrix, the basement membrane, appears to play a crucial role during the progression of invasive tumors and during hematogenous dissemination.

MOLECULAR ORGANIZATION OF BASEMENT MEMBRANES

During organogenesis and tissue modeling, endothelia, parenchyma, mesothelia, myocytes, adipocytes, and nerves become separated from their underlying interstitial stroma by thin amorphous matrixes called basement membranes (6–12). Basement membranes act as an interface between histologically dissimilar tissues that arise from different primary germ layers. The basement membrane plays an important role in cell adhesion, selective filtration, morphogenesis, and mitogenesis. Prototype basement membranes are those located beneath capillary endothelium or at the dermal-epidermal junction. Based on electron microscopy studies, the basement membrane zone can be divided into four major segments: 1. The plasma membrane of the basal or endothelial cells apposed to the basement membrane through direct receptor-mediated contact and via hemidesmosomes; 2. The lamina lucida zone, an electron lucent zone 30–40 nm beneath the plasma membrane; 3. The lamina densa zone, an amorphous or finely fibrillar electron-dense layer lying beneath the lamina lucida and extending for a thickness of 50 nm or greater depending on the tissue; and 4. The basement membrane reticulum, a fibrillar area below the lamina densa containing anchoring fibrils, bundles of microfibril-like elements, and collagen fibers. The basement membrane of the kidney glomerulus does not contain a basement membrane reticulum zone. Instead, it has a laminina rara interna and a laminina rara externa on each side of the lamina densa.

Basement membranes are known to contain collagen, proteoglycans, and glycoproteins. The major components that have been identified and reasonably well characterized are a basement membrane–specific collagen; type IV collagen; laminin; a large complex glycoprotein; and a heparin sulfate proteoglycan specific to basement membranes (13–16). Additional components which are in the process of being characterized are entactin, nidogen, bullous phemphigoid antigen, and epidermolysis bullosa acquisita antigen (17, 18).

Type V collagen and type VII collagen may not be unique components of the basement membrane, but do extend from the interstitial side into the lamina densa. They may serve to anchor the basement membrane to the underlying stroma. The interstitial stroma contains its own set of collagens types I and III, and a different set of proteoglycans, glycoproteins, and elastins. Fibronectin is present throughout the stroma and is found in some basement membranes. Since fibronectin is present in plasma, it can be deposited in the basement membrane by simple filtration or it can be synthesized by the organ cells resting on the basement membrane.

DEFECTIVE INTERACTION OF INVASIVE TUMOR CELLS WITH BASEMENT MEMBRANES

The extracellular matrix is a dense latticework of collagen and elastin, embedded in a viscoelastic ground substance composed of proteoglycans and glycoproteins. It is a supporting scaffold that isolates tissue compartments, mediates cell attachment, and influences tissue architecture (6–9). The matrix acts as a selective macromolecule filter, and plays a role in mitogenesis and differentiation. Interactions between normal cells and the matrix may be altered in neoplasia, and this may influence tumor proliferation and invasion (10). The vertebrate organism is separated into tissue compartments bordered by the basement membrane and interstitial stroma (6). For most tissues, the organ parenchymal cells secrete and assemble the basement membrane. General and widespread changes occur in the distribution and quantity of the epithelial basement membrane during the transition from benign to undifferentiated invasive carcinomas (19–22). Benign pathologic disorders with epithelial disorganization or proliferation are usually characterized by a continuous basement membrane separating the epithelium from the stroma. In contrast, invasive carcinomas consistently exhibit a defective extracellular basement membrane adjacent to the invading tumor cells in the stroma. The basement membrane is also defective around tumor cells in lymph node and organ metastases (21). In certain regions of well-differentiated carcinoma, basement membrane formation by differentiated structures can be identified. Even in these locations the basement membrane is often abnormal because it is discontinuous or focally reduplicated. Electron microscopy reveals focal defects in the continuity of the basement membrane lamina densa of carcinoma in situ. These defects may be the earliest stages of progression to invasive carcinoma because in zones of actual microinvasion the basement membrane is markedly fragmented or absent altogether. Defective basement membrane organization and loss may be due to decreased synthesis, or abnormal assembly of secreted components. Alternatively the loss may be due to increased breakdown caused by tumor- or host-derived proteases. Normal epithelial cells and benign proliferating parenchymal cells are thought to require a basement membrane for anchorage and growth (6, 7, 9). Invasive tumor cells may lack such a requirement.

During the transition from in situ to invasive carcinoma, tumor cells penetrate the epithelial basement membrane and enter the underlying interstitial stroma. Once tumor cells enter the stroma, they gain access to lymphatics and blood vessels for further dissemination. Fibrosarcomas and angiosarcomas, developing from stroma cells, invade surrounding muscle basement membrane and destroy myocytes. Tumor cells must cross basement membranes to invade nerve and most types of organ parenchyma. During intrasvasation or extravasation, tumor cells of any histologic origin must penetrate the subendothelial basement membrane. In the distant organ where metastases colonies are initiated, extravasated tumor cells must migrate through the perivascular interstitial stroma before tumor colony growth occurs in the organ parenchyma. Therefore, tumor cell interaction with the extracellular matrix occurs at multiple stages in the metastatic cascade.

1. The tumor can modify the extracellular matrix in the following ways. Degradation of matrix components associated with invasion: such matrix degradation in a malignant tumor may be restricted to localized regions where active invasion is taking place. Marked disorganization and fragmentation of stromal elements and loss of the basement membrane is a general finding during the transition from benign to malignant epithelial neoplasms.

2. Stimulated accumulations of matrix components by host cells in response to the local presence of the tumor: *desmoplasia* is the term applied to the phenomenon of such excessive accumulation of connective tissue associated with malignant breast, bowel, and prostate cancers. A current working hypothesis is that tumor cells elaborate soluble host factors that diffuse away from the tumor cells and elicit a chemotactic or mitogenic response by host fibroblasts or myofibroblasts.

3. Synthesis of matrix components by tumor cells: matrix components such as collagen synthesized by a tumor cell are generally of the same type produced by the normal cell counterpart. The actual amount of matrix produced by the tumor cells is frequently much less than the normal counterpart and dependent on the differentiated state.

Tumors are histologically heterogeneous. Some regions may be anaplastic, poorly differentiated, and invasive, whereas other zones of the same tumor may comprise orderly uniform cells forming differentiated structures. Tumor cell subpopulations are well established to be heterogeneous with regard to various phenotypic properties, including metastatic propensity. Consequently, for any individual tumor one of the above three types of matrix modification may predominate, or all three may take place simultaneously in different regions of the same tumor.

THREE-STEP THEORY OF INVASION

The interstitial stroma of most tissues does not normally contain pre-existing passageways for cells. The basement membrane is an insoluble continuous but flexible structure which is impermeable to large proteins (11). These types of extracellular matrix become focally permeable to cell movement only during tissue healing and remodeling, inflammation, and neoplasia. Cell infiltration of the matrix undoubtedly depends on multiple factors including properties of both the infiltrating cell, associated host cells, as well as properties of the matrix itself.

We have proposed a three-step hypothesis (23) describing the sequence of biochemical events during tumor cell invasion of the extracellular matrix (Figure 2). The first step is tumor cell attachment via cell surface receptors that specifically bind to components of the matrix such as laminin (for the basement membrane) and fibronectin (for the stroma). The anchored tumor cell next secretes hydrolytic enzymes (or induces host cells to secrete enzymes) which can locally degrade the matrix (including degradation of the attachment components). Matrix lysis most likely takes place in a highly localized region close to the tumor cell surface. The third step is tumor cell locomotion into the region of the matrix modified by proteolysis. Continued invasion of the matrix may take place by cyclic repetition of these three steps.

LAMININ

Laminin is a large and complex glycoprotein synthesized by a variety of cell types and deposited in basement membranes (24–38). All types of basement membranes (i.e. subepithelial, subendothelial, epidermal, myocyte, etc) contain laminin as a major constituent. Laminin is the first extracellular matrix component synthesized by the mouse embryo, initially in the unfertilized egg and subsequently from the four-cell morula stage onward. By immunoelectron microscopy laminin is detected throughout the basement membrane, but may be present in a higher concentration in the lamina lucida. Laminin as visualized by rotary shadowing electron microscopy has a distinctive cruciform shape with three short arms (35 nm) and one long arm (75 nm) (Figure 3). All arms have globular end regions. The short arms may have two globular domains, while the long arm has one end globular domain of a size larger than the globular domains of the short arms (Figure 3). The large size of the laminin molecule is such that a single extended molecule could conceivably span the length of the basement membrane lamina lucida. The specialized structure of the laminin molecule may contribute to its multiple biologic functions. Laminin plays a role in cell attachment, cell spreading, mitogenesis, neurite outgrowth, morphogenesis, and cell movement (27). Laminin also plays a role in the architecture of the

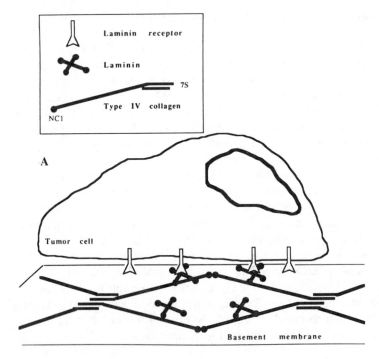

A

Step 1:   Attachment

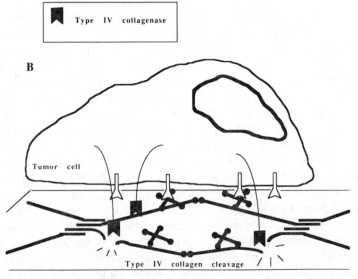

B

Step   2:   Dissolution

C

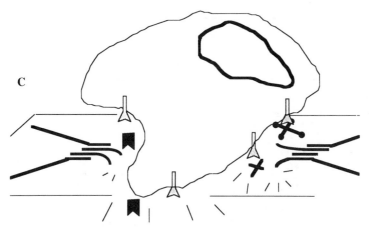

Step   3:   Locomotion

Figure 2   Three-step hypothesis of tumor cell invasion of extracellular matrix: schematic diagram (not to scale) of tumor cell invasion of the basement membrane. (*A*) The first step is tumor cell attachment to the matrix. This process may be mediated by specific attachment factors such as laminin, which form a bridge between the cell surface laminin receptor and type IV collagen. (*B*) The second step is local degradation of the matrix by tumor-cell–associated proteases. Such proteases may degrade both the attachment proteins as well as the structural collagenous proteins of the matrix. Type IV collagenase makes a single cleavage 25% of the distance from the N terminus of type IV collagen. Proteolysis may be localized at the tumor cell surface where the amount of active enzyme outbalances the natural protease inhibitors present in the matrix. (*C*) The third step is tumor cell locomotion into the region of the matrix modified by proteolysis. The direction of locomotion may be influenced by chemotactic factors. Continued invasion of the extracellular matrix may take place by cyclic repetition of these three steps.

basement membrane by virtue of its ability to bind to multiple matrix components including type IV collagen, heparin sulfate, proteoglycan, and entactin.

By gel electrophoresis, laminin consists of three major types of disulfide-bonded glycosylated polypeptide chains: two chains with an approximate molecular mass of 200,000 daltons (B1 and B2 chains), and a large chain of approximately 400,000 daltons (A chain). Translation of mRNA from parietal endoderm cells has provided evidence for two B1 chains migrating as a closely spaced doublet on SDS gel electrophoresis. It is unknown whether laminin from different basement membrane sources is composed of heteropolymers with different subunit compositions. It is further unclear how the polypeptide chains of laminin fit together to form the cross-shaped structure. The polypeptide chains are linked by 50 or 60 disulfide bonds present in the central protease-resistant part of the three short arms of the cross. Circular dichroism studies suggest that laminin may contain 20–30% alpha helix and 15% beta sheet

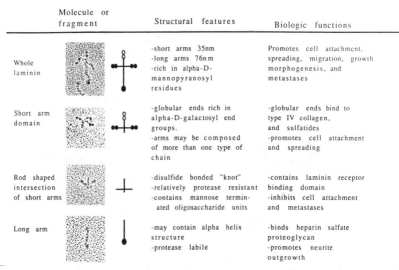

Molecule or fragment		Structural features	Biologic functions
Whole laminin		-short arms 35nm -long arms 76nm -rich in alpha-D-mannopyranosyl residues	Promotes cell attachment, spreading, migration, growth morphogenesis, and metastases
Short arm domain		-globular ends rich in alpha-D-galactosyl end groups. -arms may be composed of more than one type of chain	-globular ends bind to type IV collagen, and sulfatides -promotes cell attachment and spreading
Rod shaped intersection of short arms		-disulfide bonded "knot" -relatively protease resistant -contains mannose terminated oligosaccharide units	-contains laminin receptor binding domain -inhibits cell attachment and metastases
Long arm		-may contain alpha helix structure -protease labile	-binds heparin sulfate proteoglycan -promotes neurite outgrowth

Figure 3 Structural and functional domains of laminin. The rotary shadowing image of the molecule or fragment is shown adjacent to the diagram of the fragments.

structure (30). Protease digestion studies and Western blotting studies using monoclonal antibodies that recognize the B chains but not the A chains provide evidence that (*a*) at least a portion of the B chains reside in the cross-shaped intersection of the short arms, and (*b*) a portion or all of the A chain is embodied in the long arm. At this time it is unresolved whether the B or A polypeptide chains extend through the cross intersection to comprise a portion of both the long and the short arms.

Recombinant DNA techniques offer a means to answer questions about the primary sequences of the laminin polypeptide chains and the regulation of laminin gene expression. Barlow et al (24) have described the characterization and sequencing of cDNAs covering the 3' ends of both laminin B1 and B2 mRNAs. The amino acid sequences deduced from the cDNAs provide information about laminin structure and predict the existence of C-terminal coiled-coil alpha-helical domains. No significant homology was noted between the laminin B1 and B2 chains at the nucleotide level. The two polypeptides are coded for by mRNAs of very different size. One mRNA (B1) is close to the minimum size for the coding of a protein of molecular mass 200,000 daltons. The other mRNA (B2) has a considerable length of noncoding sequences. The question of whether laminin contains double or triple helixes is unanswered at this time. The length and mass of the arms of laminin are close to those expected for triplet or paired double helixes; however, double helixes are more common in complex proteins. The complete characterization and interrelationship of the polypeptide chains of laminin await isolation and sequencing of longer cDNAs and rotary shadowing mapping using monoclonal antibodies raised against either synthetic peptides or against recombinant fusion proteins made in bacteria.

The domains of laminin are heterogeneous in function and composition. The carbohydrate structure of the globular end regions of the molecule are different from the rod-shaped regions (32). The function of the carbohydrate groups is as yet undefined. The protease-resistant central region of laminin contains the binding site for the laminin receptor, presumably on one or more of the rod-shaped regions of the short arms. The long arm of laminin contains a heparin-binding site (35, 36) at the globular end region. The long arm also stimulates neurite outgrowth (28). One or more globular end regions of the short arms promote cell spreading, bind to plasma membrane sulfatides (35), and also bind to type IV collagen (34, 36–38). The type IV collagen–binding site for laminin is located approximately 125 nm from the C-terminal globular domain (37, 38).

LAMININ RECEPTOR

Many types of normal and neoplastic cells contain high-affinity cell surface–binding sites (laminin receptors) for laminin (39–43). Our laboratory found that laminin in solution would bind via such specific receptors to suspended or attached cells or to isolated cell membranes. In time course experiments the plateau of laminin binding is reached after 30 minutes at room temperature. With increasing concentrations of labeled laminin added to the cells or the membranes, the binding is saturable, and occurs with a high ratio of specific to nonspecific binding. Scatchard analysis is linear. The binding affinity constant is in the nanomolar range with 10,000–100,000 receptors per cell. We isolated the laminin receptor by laminin affinity chromatography (39, 40, 44). It has a molecular mass of slightly less than 70 kd, contains interchain disulfide bonds, and has an isoelectric pH value of 5.2. The isolated receptor retains the ability to bind laminin but not fibronectin or any type of collagen. A laminin receptor with a similar molecular mass and binding coefficient has been reported by the groups of Wicha & Von der Mark (41, 42).

In order to study mechanisms regulating the expression and function of the laminin receptor, a library of monoclonal antibodies (mAbs) was developed. The mAbs were prepared against the purified laminin receptor extracted from human breast carcinoma plasma membranes (44, 45). Two mAbs (LR1 and LR2) were found to differ in their effects on laminin binding to the receptor. By solid phase radioimmunoassay LR1 and LR2 bound with equal titer to the purified receptor. Using immunoblotting, both LR1 and LR2 recognize a single 67-Kd component among all the proteins extracted from the membranes of breast carcinoma tissue (Figure 4). The antibodies also bound with equal titer to isolated microsomal membranes or living breast carcinoma cells. No binding to

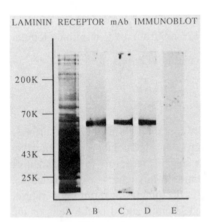

LAMININ RECEPTOR mAb IMMUNOBLOT

200K —

70K —

43K —

25K —

A   B   C   D   E

Figure 4   Immunoblot of human breast carcinoma plasma membrane extracts with anti-laminin receptor mAbs. Plasma membranes from human breast carcinoma were extracted with NP40 and run (reduced) onto a 7% Na Dod SO_4 polyacrylamide electrophoresis gel. The gel was blotted onto nitrocellulose (NC). The NC was incubated with the mAbs followed by rabbit anti-mouse second antibody. The bound antibodies were detected by ^{125}I protein A using fluorography.

Lane A: Total extracted proteins transferred to NC.

Lane B: Iodinated purified human breast carcinoma laminin receptor antigen.

Lane C: mAb LR2 blotted to total protein extract (replicate extract shown in Lane A). A single component is recognized among all the protein bands.

Lane D: mAb LR1 blotted to total protein extract (replicate extract shown in Lane A). A single component is recognized.

Lane E: Control immunoblot using human serum. No immunoreactivity is present.

serum components was evident. When added together with the labeled ligand, mAb LR1 produced a dose-dependent inhibition of specific laminin binding to human breast carcinoma cells (44). In contrast, mAb LR2 had no effect on laminin binding. The two classes of antibodies may therefore recognize different structural domains on the receptor molecule. We next studied the effect of these antibodies on the attachment of human melanoma and carcinoma cells to native human amnion basement membrane surfaces. Human amnion membranes consist of a single layer of epithelium, a continuous basement membrane, and an underlying nonvascular interstitial stroma. The epithelium of the amnion can be removed leaving a continuous laminin-rich basement membrane surface. The receptor binding fragment of laminin competitively inhibits attachment of cells to the amnion basement membrane surface (46). mAb LR1 markedly inhibits attachment of melanoma and carcinoma cells to the basement membrane surface, but not the stromal surface (which lacks laminin). Control antibodies fail to inhibit attachment. Thus, the orientation of laminin in the native basement membrane is such that the cross-shaped receptor binding domain of the short arms is exposed on the attachment surface.

Laminin receptors may be altered in number or degree of occupancy in human carcinomas. This may be the indirect result of defective basement membrane organization in the carcinomas. Breast carcinoma and colon carcinoma tissue contains a higher number of exposed (unoccupied) receptors compared to benign lesions (43, 45). The laminin receptors of normal epithelium may be polarized at the basal surface and occupied with laminin in the basement membrane. In contrast, the laminin receptors on invading carcinoma cells may be distributed over the entire surface of the cell. They may be unoccupied because of the loss of formed basement membrane associated with the invading cells (10). Using mAb LR1 we have isolated a human cDNA clone for the laminin receptor (U. Wewer et al, manuscript submitted). This should be useful for future studies of the genetic regulation of laminin receptor expression.

The laminin receptor can be shown to play a role in hematogenous metastases (46). In animal models, tumor cells selected for the ability to attach via laminin by Terranova et al (47) produced 10-fold more metastases following intravenous injection. Whole laminin on the tumor cell surface will stimulate hematogenous metastases (46, 48). This stimulatory effect requires the globular end regions of the molecule (Figure 3). Treating the cells with the receptor binding fragment of laminin markedly inhibits or abolishes lung metastases from hematogenously introduced tumor cells (46). Thus, the laminin receptor can play a role in hematogenous metastases through at least two mechanisms. If the receptor is unoccupied it can be used by the cell to bind directly to host laminin. If the receptor is occupied with laminin the cell can utilize the surface laminin as an attachment bridge through the globular end regions. The fragment of laminin which binds to the receptor, but lacks the laminin globular end regions, inhibits both of these mechanisms.

TYPE IV COLLAGENASE

In vitro studies of tumor cell invasion of the extracellular matrix have shown that cell proliferation is not absolutely required (49). However, protein synthesis and proteolysis do seem to play an important role. Inhibitors of protein synthesis or natural inhibitors of metallo-proteinases block invasion of the matrix (49, 50). Thus, invasion of the matrix is not merely due to passive growth pressure, but requires active biochemical mechanisms.

Many research groups have proposed that invasive tumor cells secrete matrix-degrading proteinases (51, 52). Collagen is an important substrate because it constitutes the structural scaffolding upon which the other components of the matrix are assembled (6). Tumor-derived collagenases that degrade interstitial collagen types I, II, and III have been characterized by a number of investigators (52–54). Tumor collagenases have properties similar to classic vertebrate collagenase first described by Gross and coworkers (55). They are metal ion (calcium and zinc)–dependent enzymes that function at neutral pH. Classic collagenase produces a single cleavage in the collagen molecule (interstitial collagen types I, II, and III) at 25°C producing 3/4 and 1/4 size fragments (75% of the distance from the N terminus). The molecular mass of classic collagenase ranges from 33,000 to 80,000, depending on the source. In some studies the amount of tumor collagenase can be correlated with the aggressive behavior of the tumor (53, 58–60). Our laboratory extended the hypothesis of interstitial collagen proteolysis by tumor cells to proteolysis of basement membranes. Detailed studies of tumor cell extravasation (56) or invasion of muscle (22) demonstrated local fragmentation of the host basement membrane adjacent to the tumor cell. Consequently, local proteolytic modification of the basement membrane may be necessary for (or at least augment) the migration of cells through this structure.

We found that tumor cells could degrade both collagenous and noncollagenous components of the basement membrane (23). Tumor cells collected from the venous effluent of a transplanted sarcoma exhibited higher basement membrane–degrading activity compared to the general population of cells in the tumor. Thus, tumor cells entering the circulation may be a subpopulation selected for the ability to degrade vascular basement membranes. The selection of aggressive tumor subpopulations is in keeping with the concept of tumor cell heterogeneity and selection of metastatic variants proposed by Fidler (2). A series of subsequent studies have extended the finding that metastatic tumor cells have the capacity to degrade basement membranes (62–64). Investigation into the proteases involved in tumor cell destruction of basement membranes revealed that basement membrane collagens type IV and V differed markedly from interstitual collagens I, II, and III with regard to proteolytic susceptibility (57, 65, 66). Collagen types IV and V are not susceptible to classic collagenase, which degrades collagen types I, II, and III. A separate family of collagenolytic enzymes were proposed to be required for catabolism of basement membranes. In support of this concept, a type IV collagenolytic metalloproteinase was identified in highly metastatic tumor cells and in endothelial cells (69). Separate metalloproteinases were found to degrade type V collagen (66–68). Type IV collagenase has a molecular mass of approximately 62,000 to 65,000 after activation (61). It is secreted in a latent form but may also exist on the cell surface. The level of type IV collagenase is augmented in many highly metastatic tumor cells. Antibodies prepared against type IV collagenase react with invading breast carcinoma cells and breast carcinoma lymph node metastases by immunohistology (20).

Type IV collagenase produces two sets of cleavage products as assayed by gel electrophoresis (59). The size of the fragments is consistent with a single major cleavage through both chains of type IV collagen. The substrate cleavage region for type IV collagenase has recently been identified by Fessler and coworkers (70). Procollagen IV monomers are triple helical molecules that contain a noncollagenous globular knob at one end and the "7S" bow-tie shaped domain at the other end. Four type IV monomers can associate by cross-links and disulfide bonds to form a tetramer. The individual molecules are joined at the 7S domain, and from this junction the knob-like termini extend, as shown by electron microscopy (11). The 7S domain is at the NH_2 termini, and the tumor cell type IV–specific collagenase cleaves at a site 25% of the distance from this end. The tumor enzyme, therefore, has the ability to break apart the tetramer and potentially dissociate the type IV network in the basement membrane lamina densa. Timpl et al (15) and Furthmayr's group (38) have hypothesized that type IV collagen molecules may be linked at their end regions, and possibly side to side, to form a uniform hexagonal network. Type IV collagenase can effectively break down this network by cleaving each triple helical molecule and breaking each side of the hexagon unit (Figure 2).

POSSIBLE GENETIC LINKAGE OF TYPE IV COLLAGENASE EXPRESSION WITH THE METASTATIC PHENOTYPE

A metastatic colony is the end result of a complex series of tumor host interactions (1–5). It is apparent that these interactions involve multiple gene products. A cascade or coordinated group of gene products expressed above a certain threshold level may be required for a tumor cell to successfully traverse the successive steps in the metastatic process. The crucial gene products may regulate host immune recognition of the tumor cell, cell growth, attachment, proteolysis, locomotion, and differentiation. The specific family of gene products necessary for metastases may be different for each histologic type of tumor.

We employed DNA transfection (71) and somatic cell hybridization (72) to investigate whether type IV collagenase may be a possible member of this hypothetical group of metastases-associated gene products. Transfection of tumor DNA into recipient NIH 3T3 cells has been shown by our laboratory and others to induce the metastatic phenotype assayed in nude mice (71–75). The metastatic phenotype was elicited in secondary and tertiary transfectants. Transfection of the ras[H] oncogene alone could induce the metastatic phenotype only in certain types of recipient cells (73) following both intravenous or subcutaneous injection into nude mice. After studying this phenomenon in a variety of different recipient cells including second generation diploid rat embryo fibroblasts, we have developed a working hypothesis (73): Induction of the metastatic phenotype requires at least two (and possibly more) gene complementation groups. In the correct recipient cells, one of these genes may be the activated but not the cellular (proto-oncogene) form of the ras^H oncogene. The exact DNA sequence and function of the other members of the gene complementation group are under intense study by many groups (73, 74, 76).

Among a series of NIH 3T3 cell transfectants that exhibit metastatic propensity, all secreted high levels of type IV collagenase compared to NIH 3T3 parent cells or spontaneously transformed (tumorigenic but nonmetastatic) NIH 3T3 cells.

A similar association of type IV collagenase with the metastatic phenotype was also observed following somatic cell hybridization. The results of somatic cell hybridization must be carefully analyzed because hybrid cells are unstable and the exact karyotypic of each hybrid clone will be different (77, 78). However, it is possible to interpret the data derived from an individual hybridization system as regards the correlation of a specific gene product with the metastatic phenotype. The metastatic phenotype of tumor-tumor cell and tumor cell–normal cell hybrids were compared with their type IV collagenase activity (72, 77). Fusion of metastatic tumor cells with nonmetastatic tumor cells resulted in maintenance or augmentation of the metastatic phenotype. However, when metastatic cells were fused with normal cells, in this series, the metastatic phenotype was suppressed. Furthermore, the hybrids retained the ability to produce tumors. Thus, tumorigenicity was shown to be distinct from metastatic propensity. The levels of type IV collagenase in the hybrid cells was altered in parallel with the metastatic behavior. Suppression of metastases resulted in suppression of type IV collagenase. In no case was a metastatic hybrid identified that had lost the ability to elaborate type IV collagenase. Thus, for this particular series of hybrids, type IV collagenase may be one of many gene products expressed concomitantly with other gene products necessary for metastases formation.

HEPARAN SULFATE PROTEOGLYCAN

Proteoglycans are a class of sulfated macromolecules that contain a core protein with at least one covalently bound glycosaminoglycan chain (82–84). The different groups of glycosaminoglycans include hyaluronic acid, chondroitin sulfate, keratan sulfate, dermatan sulfate, heparan sulfate, and heparin. Proteoglycans form the ground substance or filler material between other components of the extracellular matrix. They expand upon hydration and serve important structural and filtration functions. For example, the highly sulfated and negatively charged proteoglycan in glomerular basement membranes has an important role in the regulation of permeability and selective filtration (82). Heparan sulfate proteoglycan exists in a number of different forms and locations, such as associated with the cell surface, within the basement membrane, and in the interstitial stroma. Basement membrane heparan sulfate proteoglycan has been isolated from glomerular basement membranes, and a number of basement membrane–producing cell lines. The size of the high-molecular-mass, low-density heparan sulfate proteoglycan as purified from the EHS tumor is approximately 700,000 daltons, with heparan sulfate side chains of 70,000 daltons, and a core protein in the range of 400,000 daltons. Heparan sulfate proteoglycan comprises only a small percentage of the total basement membrane composition. Nevertheless, it is thought to play an important role in the organization of the basement membrane components and its filtration properties.

HEPARAN SULFATE ENDOGLYCOSIDASE

Nakajima et al, and Vlodovsky et al have examined the interactions of metastatic tumor cells with the in vitro-produced basement membrane–like matrix of cultured endothelial cells (85–87). These investigators found that the tumor cells cleaved the ^{35}S-sulfate labeled heparan sulfate to fragments that were approximately one third the size of the intact heparan sulfate chains. Highly invasive and metastatic tumor sublines degraded the sulfated glycosaminoglycans at a higher rate compared to tumor sublines that were tumorigenic but not metastatic. The time dependence of heparan sulfate degradation into particular fragments indicated that it was cleaved at specific intrachain sites (85). The terminal monosaccharides of the fragments were found to be L-gulonic acid. Therefore, the enzyme is an endoglucuronidase (heparanase). The heparanase has a pH optimum of 5 and has been partially purified by concanavalin A affinity chromatography. Full purification and characterization of the enzyme remain. Nevertheless, it may become an important marker to predict metastatic propensity.

Type IV collagenase and heparanase is only one member of a family of proteinases that participate in the physiologic turnover of basement membranes. A cascade of proteases including thiol proteases, heparanases, and serine proteases such as plasminogen activator all contribute to facilitating tumor invasion (79–82). Proteolysis regulation can take place at many levels including tumor cell–host cell interactions (62), and protease inhibitors (54) produced by the host or by the tumor cells themselves. Protease secretion or activation by invading cells may also be coupled to cell shape or locomotion (69). Expression of matrix-degrading enzymes is not tumor cell specific. The actively invading tumor cells may merely respond to different regulatory signals compared to their noninvasive counterparts.

FIBRONECTIN

The fibronectins are a group of large extracellular glycoproteins that are composed of structurally similar subunits varying in size between 210,000 and 250,000 daltons (7). One form of fibronectin found in blood plasma (termed plasma fibronectin) consists of four types of subunits that fall into two size classes. The fibronectin synthesized by fibroblasts (termed cellular fibronectin) contains subunits not found in plasma fibronectin. Schwarzbauer et al (88) have reported the isolation of cDNA and genomic clones encoding rat fibronectins, and have shown that several different mRNAs arise by alternate splicing of the transcript of a single fibronectin gene. Rat liver was found to contain three fibronectin mRNA species coding for serum fibronectin. The difference in size between the subunits of plasma fibronectin is caused by alternative RNA splicing within the coding sequence.

Fibronectin has been implicated in a variety of biologic phenomena relating to cell substrate interactions (7). It can mediate cell attachment, spreading, and phagocytosis, and has numerous effects on cell morphology. Fibronectin also binds a series of ligands including collagen, fibrin, heparin, hyaluronic acid, and actin. It stimulates cell movement based on haptotaxis and chemotaxis. Fibronectin is observed at the leading edge of closing epidermal wounds, and it promotes the healing of chronic corneal ulcers. Based on extensive studies with protease-derived fragments of fibronectin conducted by many groups, many of the ligand-binding domains of fibronectin have been localized to specific sites on the molecule. Plasma fibronectin is a dimer composed of two (A and B) polypeptide chains. The chains contain a heparin- and fibrin-binding domain at the amino terminus, a collagen-binding domain 1/4 of the distance from the amino terminus, and a cell-binding domain in the midregion of the chains. The primary amino acid sequence of at least one major cell-binding domain of fibronectin is an amazingly small region of only four amino acids: Arg-Gly-Asp-Ser (RGDS) (89, 90). A strong heparin-binding domain is located near the COOH terminus. The extreme COOH terminal domain includes a free sulfhydryl and interchain disulfide bonds.

Yamada's group has shown that fibronectin can have a biphasic function (7). Fibronectin or its proteolytic fragments can promote adhesion when coated on substrates; however, extremely high levels of the same material added in solution will inhibit the adhesion of Chinese hamster ovary (CHO) and baby hamster kidney (BHK) cells to fibronectin substrates. Synthetic peptides containing the tetrapeptide, RGDS, were shown to inhibit cell adhesion of CHO and BHK cells. These results were compatible with the concept that the RGDS peptide bound to a specific cell-surface fibronectin receptor. When this receptor was occupied by the peptide, the receptor was blocked from performing its function of initiating contact between the cell and the fibronectin substrate.

FIBRONECTIN RECEPTOR

Pytela et al (91) took a direct approach to identify a putative fibronectin receptor. In view of the large number of binding sites in the fibronectin molecule, they decided to limit their investigation to isolating a cell-surface component that recognized the synthetic peptide containing the RGDS sequence. The peptide was coupled to Sepharose and incubated with detergent extracts of cells. A 140,000-dalton protein was bound by the affinity matrix from octylglucoside extracts of MG-63 human osteosarcoma cells. The same protein was specifically eluted with the synthetic peptide Gly-Arg-Gly-Asp-Ser-Pro. The 140,000-dalton protein was labeled by cell-surface iodination and could be incorporated into liposomes with a high efficiency. The liposomes containing the protein showed a high affinity toward fibronectin-coated substrates, and this could be inhibited by the cell attachment peptide. These results support the concept of the 140,000-dalton protein functioning as a membrane-embedded cell surface fibronectin receptor. The binding coefficient of the receptor, the number of receptors per cell, and the specificity of the receptor for other matrix or serum components remain to be elucidated.

Literature Cited

1. Sugarbaker, E. V., Weingard, D. N., Roseman, J. M. 1982. *Cancer Invasion and Metastases,* ed. L. A. Liotta, I. R. Hart, pp. 427–65. Boston: Nijhoff
2. Fidler, I. J., Gersten, D. M., Hart, I. R. 1978. *Adv. Cancer Res.* 28:149–60
3. Nicolson, G. L. 1982. *Biochem. Biophys. Acta* 695:113–20
4. Liotta, L. A. 1985. In *Progress in Oncology,* ed. V. T. DeVita, S. Hellamn, S. A. Rosenberg, Vol. 1, pp. 28–42. Philadelphia: Lippincott
5. Poste, G., Fidler, I. 1980. *Nature* 283:139–46
6. Hay, E. D. 1982. *Cell Biology of Extracellular Matrix.* New York: Plenum
7. Yamada, K. M., Akiyama, S. K., Hasegawa, T., Hasegawa, E., Humphries, M. J., et al. 1985. *J. Cell. Biochem.* 28:79–97
8. Bernfield, M. R., Banerjee, S. D., Cohn, R. H. 1972. *J. Cell. Biol.* 52:674–80
9. Wicha, M. S., Liotta, L. A., Garbisa, S., Kidwell, W. R. 1980. *Exp. Cell. Res.* 124:181–90
10. Liotta, L. A., Rao, C. N., Barsky, S. H. 1983. *Lab. Invest.* 49:636–49
11. Vracko, R. 1974. *Am. J. Pathol.* 77:313
12. Martinez-Hernandez, A., Amenta, P. S. 1983. *Lab. Invest.* 48:656–80
13. Kefalides, N. A., Alper, R., Clark, C. C. 1979. *Int. Rev. Cytol.* 61:167–80
14. Madri, J. A., Furthmayr, H. 1979. *Am. J. Pathol.* 94:322–32
15. Timpl, R., Wiedemann, H., VanDelden, V., Furthmayr, H., Kuhn, K. 1981. *Eur. J. Biochem.* 120:203–12
16. Hogan, B. L. M. 1980. *Dev. Biol.* 76:275–81
17. Carlin, B., Jaffe, R., Bender, B., Chung, A. E. 1981. *J. Biol. Chem.* 256:5209–18
18. Stanley, J. R., Hawley-Nelson, P., Yuspa, S. H., Shevach, E. M., Katz, S. I. 1981. *Cell* 24:897–904
19. Siegal, G. P., Barsky, S. H., Terranova, V. P., Liotta, L. A. 1981. *Invasion Metastasis* 1:54–65
20. Barsky, S. H., Siegal, G. P., Jannotta, F., Liotta, L. A. 1983. *Lab. Invest.* 49:140–47
21. Burtin, P., Chavanel, G., Foidart, J. M., Martin, E. 1982. *Int. J. Cancer.* 30:13–20
22. Babai, F. 1976. *J. Ultrastruct. Res.* 56:287–97
23. Liotta, L. A., Kleinerman, J., Catanzara, P., Rynbrandt, D. 1977. *J. Natl. Cancer Inst.* 58:1427–39
24. Barlow, D. P., Green, N. M., Kurkinen, M., Hogan, B. L. M. 1984. *EMBO J.* 3:2355–62
25. Terranova, V. P., Rohrbach, D. H., Martin, G. R. 1980. *Cell* 22:719–26
26. Timpl, R., Rhode, H., Robey, P. G., Rennard, S. I., Foidart, J. M., Martin, G. R. 1979. *J. Biol. Chem.* 254:9933–41
27. Kleinman, H. K., Cannon, F. B., Laurie, G. W., Hassell, J. R., Aumailley, M., et al. 1985. *J. Cell Biol.* 27:317–25
28. Edgar, D., Timpl, R., Thoenen, H. 1984. *EMBO J.* 3:1463–68
29. McCarthy, J. B., Palm, S. L., Furcht, L. T. 1983. *J. Cell Biol.* 97:772–77
30. Engel, J., Odermatt, E., Engel, A., Mad-

1056 LIOTTA, RAO & WEWER

ri, J. A., Furthmayr, H., et al. 1981. *Mol. Biol.* 150:97–108

31. Liotta, L. A., Goldfarb, R. H., Brundage, R., Siegal, G. P., Terranova, V., Garbisa, S. 1981. *Cancer Res.* 41:4629–35

32. Rao, C. N., Goldstein, I. J., Liotta, L. A. 1983. *Arch. Biochem. Biophys.* 227:118–24

33. Rao, C. N., Margulies, I. M. K., Goldfarb, R. H., Madri, J. A., Woodley, D. T., Liotta, L. A. 1982. *Arch. Biochem. Biophys.* 219:65–70

34. Rao, C. N., Margulies, I. M. K., Tralka, T. S., Terranova, V. P., Madri, J. A., Liotta, L. A. 1982. *J. Biol. Chem.* 257:9740–50

35. Roberts, D. D., Rao, C. N., Magnani, J. L., Spitalnik, S. L., Liotta, L. A., Ginsburg, V. 1985. *Proc. Natl. Acad. Sci. USA* 82:1306–10

36. Timpl, R., Johansson, S., VanDelden, V., Oberbaumer, I., Hook, M. 1983. *J. Biol. Chem.* 258:8922–28

37. Rao, C. N., Margulies, I. M. K., Liotta, L. A. 1985. *Biochem. Biophys. Res. Commun.* 128:45–52

38. Charonis, A. S., Tsilibary, E. C., Yurchenko, P. D., Furthmayr, H. 1985. *J. Cell. Biol.* 100:1848–53

39. Terranova, V. P., Rao, C. N., Kalebic, T., Margulies, I. M. K., Liotta, L. A. 1983. *Proc. Natl. Acad. Sci. USA* 80:444–51

40. Rao, C. N., Barsky, S. H., Terranova, V. P., Liotta, L. A. 1983. *Biochem. Biophys. Res. Commun.* 111:804–8

41. Malinoff, H., Wicha, M. S. 1983. *J. Cell Biol.* 96:1475–80

42. Lesot, H., Kuhl, U., Von der Mark, K. 1983. *EMBO J.* 2:861–70

43. Barsky, S. H., Rao, C. N., Hyams, D., Liotta, L. A. 1984. *Breast Cancer Res. Treatment.* 4:181–88

44. Liotta, L. A., Horan Hand, P., Rao, C. N., Bryant, G., Barsky, S. H., Schlom, J. 1985. *Exp. Cell Res.* 156:177–86

45. Hand, P. H., Thor, A., Schlom, J., Rao, C. N., Liotta, L. 1985. *Cancer Res.* 45:2713–19

46. Barsky, S. H., Rao, C. N., Williams, J. E., Liotta, L. A. 1984. *J. Clin. Invest.* 74:843–48

47. Terranova, V. P., Liotta, L. A., Russo, R. G., Martin, G. R. 1982. *Cancer Res.* 42:2265–73

48. Varani, J., Lovett, E. J., McCoy, J. P., Shibata, S., Maddox, S., et al. 1983. *Am. J. Pathol.* 111:27–34

49. Thorgeirsson, U. P., Turpeenniemi-Hujanen, T., Neckers, L. M., Johnson, D. W., Liotta, L. A. 1984. *Invasion Metastasis* 4:73–84

50. Thorgeirsson, U. P., Liotta, L. A., Kalebic, T., Margulies, I. M. K., Thomas, K., et al. 1982. *J. Natl. Cancer Inst.* 69:1049–57

51. Strauli, P. 1980. In *Proteinases and Tumor Invasion, Monogr. Ser. Eur. Organ. Res. Treatment on Cancer,* ed. P. Strauli, A. J. Barrett, A. Baici, 6:215. New York: Raven

52. Liotta, L. A., Thorgeirsson, U. P., Garbisa, S. 1982. *Cancer Metastasis Rev.* 1:277–97

53. Wirl, G., Frick, J. 1979. *Urol. Res.* 7:103–8

54. Woolley, D. E., Tetlow, L. C., Mooney, C. J., Evanson, J. M. 1980. See Ref. 51, pp. 97–115

55. Gross, J., Nagai, Y. 1965. *Proc. Natl. Acad. Sci. USA* 54:1197–210

56. Wallace, A. C., Chew, E., Jones, D. S. 1978. In *Pulmonary Metastasis,* ed. L. Weiss, H. A. Gilbert, 3:26–32. Boston: Hall

57. Liotta, L. A., Abe, S., Gehron, P., Martin, G. R. 1979. *Proc. Natl. Acad. Sci. USA* 76:2268–76

58. Liotta, L. A., Tryggvason, K., Garbisa, S., Hart, I., Foltz, C. M., Shafie, S. 1980. *Nature* 284:67–68

59. Liotta, L. A., Tryggvason, K., Garbisa, S., Rohey, P. G., Abe, S. 1981. *Biochemistry* 20:100–8

60. Salo, T., Liotta, L. A., Keski-Oja, J., Turpeenniemi-Hujanen, T., Tryggvason, K. 1982. *Int. J. Cancer* 30:669–74

61. Salo, T., Liotta, L. A., Tryggvason, K. 1983. *J. Biol. Chem.* 258:2058–62

62. Henry, N., Eeckhout, Y., van Lamsweerde, A.-L., Vaes, G. 1983. *FEBS Lett.* 161:243–46

63. Nakajima, M., Custead, S. E., Welch, D. R., Nicolson, G. L. 1984. *AACR Proc.* 244:162

64. Starkey, J. R., Hosick, H. L., Stanford, D. R., Liggitt, H. D. 1984. *Cancer Res.* 44:1585–94

65. Welgus, H. G., Jeffrey, J. J., Eisen, A. Z. 1981. *J. Biol. Chem.* 256:9511–25

66. Mainardi, C. L., Seyer, J. M., Kang, A. H. 1980. *Biochem. Biophys. Res. Commun.* 97:1108–12

67. Liotta, L. A., Lanzer, W. L., Garbisa, S. 1981. *Biochem. Biophys. Res. Commun.* 98:184–89

68. Murphy, G., Cawston, T. E., Galloway, W. A., Barnes, M. J., Bunning, R. A. D., et al. *Biochem. J.* 199:807–11

69. Kalebic, T., Garbisa, S., Glaser, B., Liotta, L. A. 1983. *Science* 221:281–83

70. Fessler, L. I., Duncan, K., Fessler, J., Salo, T., Tryggvason, K. 1984. *J. Biol. Chem.* 259(15):9783–89

71. Thorgeirsson, U. P., Turpeenniemi-Huganen, T., Williams, J. E., Westin, E. H., Heilman, C. A., et al. 1985. *Mol. Cell. Biol.* 5:259–62

72. Turpeenniemi-Juganen, T., Thorgeirsson, U. P., Hart, I. R., Grant, S. S., Liotta, L. A. 1985. *J. Natl. Cancer Inst.* 74:99–103

73. Muschel, R. J., Williams, J. E., Lowy, D. R., Liotta, L. A. 1985. *Am. J. Pathol.* 121:1–8

74. Bernstein, S. C., Weinberg, R. A. 1985. *Proc. Natl. Acad. Sci. USA* 82:1726–30

75. Greig, R. G., Koestler, T. P., Trainer, D. L., Corwin, S. P., Miles, L., et al. 1985. *Proc. Natl. Acad. Sci. USA* 82:3698–701

76. Wallich, R., Bulbuc, N., Hammerlin, G. J., Katzav, S., Segal, S., Feldman, M. 1985. *Nature* 315:301–5

77. Hart, I. R. 1985. In *Cancer Invasion and Metastasis: Biologic and Therapeutic Aspects,* ed. G. L. Nicoloson, L. Milas, pp. 133–43. New York: Raven

78. Sidebottom, E., Clark, S. R. 1983. *Br. J. Cancer* 47:399–406

79. Ossowski, L., Reich, E. 1983. *Cell* 35:611–19

80. Poole, A. R., Recklies, A. D., Mort, J. S. 1980. See Ref. 51, pp. 81–95

81. Sloane, B. F., Dunn, J. R., Horn, K. V. 1981. *Science* 212:1151–52

82. Kanwar, Y. S., Farquhar, M. G. 1979. *Proc. Natl. Acad. Sci. USA* 76:4493–97

83. Hassell, J. R., Leyshon, H. K., Ledbetter, S. R., Tyree, B., Suzuki, S., et al. 1985. *J. Biol. Chem.* 260:8098–105

84. Wewer, U. M., Albrechtsen, R., Hassell, J. R. 1985. *Differentiation* 30:61–67

85. Nakajima, M., Irimura, T., Di Ferrante, N., Nicolson, G. L. 1984. *J. Biol. Chem.* 259(4):2283–90

86. Nakajima, M., Irimura, T., Di Ferrante, D., Di Ferrante, N., Nicolson, G. L. 1983. *Science* 220:611–12

87. Vlodovsky, I., Fuks, Z., Bar-Ner, M., Ariav, Y., Schirrmacher, V. 1983. *Cancer Res.* 43:2704–11

88. Schwarzbauer, J. E., Paul, J. I., Hynes, R. O. 1985. *Proc. Natl. Acad. Sci. USA* 82:1424–28

89. Pierschbacher, M. D., Ruoslahti, E. 1984. *Nature* 309:30–33

90. Pierschbacher, M. D., Hayman, E. G., Ruoslahti, E. 1981. *Cell* 26:259–67

91. Pytela, R., Pierschbacher, M. D., Ruoslahti, E. 1985. *Cell* 40:191–98

Ann. Rev. Biochem. 1986. 55:1059–89

HORMONAL REGULATION OF MAMMALIAN GLUCOSE TRANSPORT[1]

Ian A. Simpson and Samuel W. Cushman

Experimental Diabetes, Metabolism and Nutrition Section; Molecular, Cellular and Nutritional Endocrinology Branch; National Institute of Arthritis, Diabetes, and Digestive and Kidney Diseases; National Institutes of Health, Bethesda, Maryland 20892

CONTENTS

PERSPECTIVES AND SUMMARY

The condition now known as diabetes mellitus was first described in India circa 1500 B.C. (1). It is characterized by the triad of symptoms polyuria, polydipsia, and polyphagia, all of which result from excess circulating glucose. The hyperglycemia of type 1 diabetes can now be attributed to a primary failure of

---

[1]The US Government has the right to retain a nonexclusive royalty-free license in and to any copyright covering this paper.

the β cells of the pancreas to produce and secrete insulin while that of type 2 diabetes appears to be the consequence of a not yet completely resolved combination of defective insulin production/secretion and insulin resistance in peripheral target tissues. While a direct relationship between insulin and systemic glucose metabolism was established during the pioneering studies of Banting & Best (2–4), an understanding of its underlying mechanism did not begin to evolve until the late 1940s and 1950s when Levine and colleagues (5–7) proposed that insulin might stimulate the transport of glucose across the plasma membrane of target cells. Only in 1965 did Crofford & Renold (8, 9) provide the first direct evidence in vitro for this fundamental action of insulin at the cellular level.

In most mammalian tissues, glucose enters the cell by a carrier-mediated facilitated diffusion mechanism which exhibits no energy or counter-ion requirements and is not regulated by insulin. Notable exceptions are kidney and intestinal epithelia which contain a Na^+- and energy-dependent glucose transport system in addition to that of the facilitated diffusion type, and muscle and adipose tissue where glucose entry occurs by facilitated diffusion but is acutely regulated by insulin and other hormones. In the latter, the stimulation of glucose transport activity by insulin has been shown to occur through an increase in the maximum transport velocity (V_{max}) rather than a change in the apparent affinity (K_M) of the carrier for glucose (10), the latter being comparable in both insulin-sensitive and -insensitive tissues (5–10 mM).

The mechanism by which insulin achieves this change in glucose transport activity represents the central theme of this chapter. The discussion concentrates on the action of insulin on the rat adipose cell. By mass, adipose tissue is relatively unimportant in the total consumption of systemic glucose and the maintenance of glucose homeostasis as compared to muscle (11, 12). However, practical considerations such as the ease with which homogeneous cell suspensions can be obtained (13) and the exquisite sensitivity and responsiveness of these cells to hormones have made the isolated rat adipose cell the experimental preparation of choice for the study of insulin action.

Initially, because kinetic analyses could not distinguish between increases in glucose transporter intrinsic activity and number in response to insulin, a technique for assaying glucose transporter concentration in the plasma membrane fraction of rat adipose cells was developed in our laboratory employing cytochalasin B, a potent competitive inhibitor of glucose transport in many cell types (14). This technique was then used to demonstrate that insulin stimulates glucose transport primarily through an increase in glucose transporter number (14). More recently, this same technique was used in our laboratory to examine the subcellular distribution of glucose transporters and thus provide evidence for a novel concept for this action of insulin, namely, the translocation of glucose transporters to the plasma membrane from a large intracellular pool

(15–17). A similar hypothesis was independently proposed by Kono and colleagues (18) using a reconstitution technique (19) to assess the subcellular distribution of glucose transport activity.

Further studies have now extended this hypothesis to the stimulation of glucose transport by insulinomimetic agents in the rat adipose cell (20), and by insulin in the isolated guinea pig (21) and human (22) adipose cell and rat diaphragm (23, 24) and heart (25). They have also demonstrated in the rat adipose cell that the reduced glucose transport response to insulin in insulin-resistant metabolic states such as streptozotocin diabetes (26) and the potentiated glucose transport response to insulin in hyper-insulin-responsive states such as chronic experimental hyperinsulinemia (27) appear to be explained by corresponding changes in the number of glucose transporters in the intracellular pool in the basal state and translocation of glucose transporters to the plasma membrane in response to insulin. However, two hyper-insulin-responsive states have very recently been observed to be accompanied by increases in rat adipose cell glucose transporter intrinsic activity, as well as number and translocation (28, 29).

In addition, studies now in progress in our laboratory reveal striking counter-regulatory effects of adenylate cyclase stimulators and inhibitors on insulin-stimulated glucose transport activity in the rat adipose cell which appear to entail modulations of glucose transporter intrinsic activity and possibly the insulin-signaling mechanism itself, and occur through a cAMP-independent mechanism (30–32). Furthermore, the glucose transporter from a human hepatoma cell line, HepG2, has just been cloned by Mueckler et al (33), thus providing major new insights into the relationship between this integral membrane protein's structure and function. Finally, rabbit antisera prepared against the purified human erythrocyte glucose transporter (34, 35) and direct photolabeling of the glucose transporter with cytochalasin B (36–38) have provided confirmatory evidence for the translocation hypothesis of insulin action and are presently in use in studies of the structure of the rat adipose cell glucose transporter as it relates to the underlying molecular mechanisms (39).

MECHANISM OF INSULIN ACTION

Conceptually, two mechanisms can be envisaged that would give rise to an increase in the V_{max} for glucose transport in response to insulin. The first entails a change in plasma membrane glucose transporter intrinsic activity such that each carrier would transport glucose at a more rapid rate; the second comprises an increase in the number of functional glucose transporters in the plasma membrane exposed to the extracellular medium. In fact, it is the latter mechanism, shown schematically in Figure 1 (16), that is now considered to be the predominant mechanism by which insulin stimulates glucose transport in both

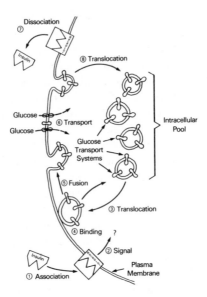

Figure 1   Hypothetical model of insulin's action on glucose transport [from Karnieli et al (16)].

adipose and muscle cells. This model was proposed independently by Kono and colleagues (18) at Vanderbilt University and by our own group (15) at the National Institutes of Health in 1981.

The translocation hypothesis of insulin action proposes the following sequence of events: (*a*) Insulin binds to its specific cell surface receptor inducing a signal the nature of which still remains unknown despite intense investigation [for review, see Kahn (40) and Czech (41)]. (*b*) In response to this signal, intracellular vesicles containing the glucose transporter are translocated by an exocyticlike mechanism to the plasma membrane in a manner comparable to the various secretory processes (42, 43). (*c*) Following fusion of these vesicles with the plasma membrane, glucose transporters are then exposed to the extracellular medium giving rise to the increase in glucose transport activity. (*d*) On removal of insulin from its receptor, glucose transporters are retranslocated back to the same intracellular pool by an endocyticlike mechanism similar to receptor-mediated endocytosis (44). Before discussing the evidence for this model and its more recent embellishments, however, a brief description of the current methodologies used to assess glucose transport activity and the subcellular distribution of glucose transporters is appropriate.

Methodology

Both muscle and adipose cells have a considerable capacity to metabolize glucose; thus, in order to measure glucose transport activity directly, it is necessary to use a nonmetabolizable glucose analogue. Of the many that have been studied, 3-*O*-methylglucose now appears to be the one of choice and its transport properties, together with protocols for its use, are comprehensively discussed in a review by Gliemann & Rees (45).

To determine the subcellular distribution of glucose transporters, two general approaches have been adopted for fractionating the cell. The first, used in our laboratory, is a differential ultracentrifugation approach based on the methodology originally described by McKeel & Jarett (46) in which membrane vesicles and/or subcellular organelles are separated into four fractions: a plasma membrane fraction enriched in marker enzymes such as hormone-sensitive adenylate cyclase and 5'-nucleotidase; a high-density microsomal membrane fraction enriched in endoplasmic reticulum marker enzyme activities such as rotenone-insensitive cytochrome *c* reductase and glucose-6-phosphate phosphatase; a low-density microsomal membrane fraction enriched in marker enzymes of the Golgi apparatus such as galactosyl and sialyltransferase; and a mitochondrial/nuclear fraction (17). The terms "high-density" and "low-density," used routinely in our laboratory to describe the two microsomal membrane fractions, are actually erroneous in that differential ultracentrifugation separates membrane vesicles by size rather than density: the plasma membranes represent the largest vesicles and the low-density microsomes, the smallest (17).

The other approach, originally developed by Kono et al (18, 47) to investigate the fate of internalized insulin, employs various continuous sucrose gradients to separate roughly equivalent subcellular fractions and uses the same marker enzyme activities to monitor the procedure. Application of either approach to several muscle preparations and cultured cell lines has proven considerably more difficult and is the primary reason for the paucity of data on glucose transporter subcellular distribution in the former (23–25) and a complete lack of data in the latter.

Until fairly recently, two methods were available for quantitating the distribution of glucose transporters among these various subcellular fractions. The one adopted in our laboratory is to measure the binding of cytochalasin B, first recognized as a potent competitive inhibitor of glucose transport in the human erythrocyte (48). The binding assay is performed in the presence of cytochalasin E, an analogue that does not inhibit binding to the glucose transporter at the concentrations used, but dramatically reduces specific binding to other membrane proteins such as actin (15). To further ensure specificity, only that component of binding that is inhibitable by a high concentration of D-glucose, but not L-glucose, is considered to reflect a quantitative estimate of the concentration of glucose transporters (15). The second method is to measure directly glucose transport activity following solubilization and reconstitution of the glucose transporter into artificial liposomes (18, 19). Glucose transporter distribution is then assessed by determining either the rate of D-glucose transport into the vesicles or the fraction of the total number of vesicles containing glucose transporter following reconstitution of less than one glucose transporter per vesicle (49).

Two more recent methodologies that have not only confirmed the observations made using the above techniques, but have also given further insight into the structure of the glucose transporter are: (a) photochemical cross-linking of [^{3}H]cytochalasin B to the glucose transporter (36–39), and (b) Western blotting with antisera prepared against the purified human erythrocyte glucose transporter that cross-react with the rat adipose cell glucose transporter (34, 35). Both techniques have revealed the rat adipose cell glucose transporter to be a 45–50-kd protein, the characteristics of which will be discussed in detail in a later section.

Steady-State Action of Insulin

Armed with these methods, let us now consider the action of insulin on glucose transport in the intact isolated rat adipose cell. Figure 2 compares the insulin concentration dependencies typical of insulin binding to its receptor and the stimulation of glucose transport activity. In cells exposed to a maximally stimulating concentration of insulin, the rate of 3-O-methylglucose transport is increased ~30-fold as compared to basal cells. The action of insulin is clearly positively cooperative with Hill coefficients of 2–3, and the range of insulin concentrations eliciting a half-maximal response is 0.05–0.20 nM (7–28 μU/

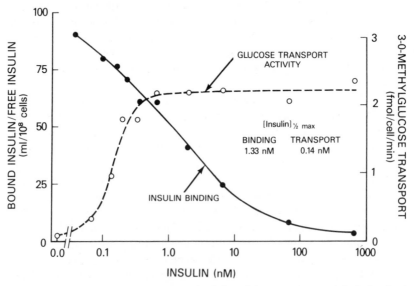

Figure 2   Steady-state insulin binding and stimulation of glucose transport activity by insulin at 37°C in the isolated rat adipose cell.

ml). This is in contrast to insulin's binding to its receptor where maximal binding is only achieved at concentrations > 100 nM (> 15,000 μU/ml), the binding curve is slightly negatively cooperative (Hill coefficient > 0.9), and half-maximal binding occurs at a concentration of ~1.33 nM (~200 μU/ml), a full order of magnitude greater than that required for half-maximal stimulation of glucose transport activity. This disparity, first recognized by Kono & Barham (50) in 1971, is now referred to as the "spare receptor" concept and still remains a major enigma in the field of insulin action.

Figure 3 illustrates the fundamental evidence that the increase in glucose transport activity in response to insulin results from a change in the subcellular distribution of glucose transporters (17); here, the data using the cytochalasin B binding technique are shown, although comparable data are obtained with the other methodologies. In membranes prepared from adipose cells not exposed to insulin (basal), by far the highest concentration of glucose transporters is observed in the low-density microsomal membrane fraction with relatively low concentrations detected in either the plasma or high-density microsomal membrane fractions; none are detected in the mitochondrial/nuclear fraction (not illustrated). Exposure of the cells to a maximally stimulating concentration of

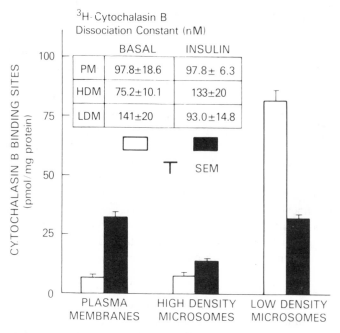

	BASAL	INSULIN
PM	97.8±18.6	97.8± 6.3
HDM	75.2±10.1	133±20
LDM	141±20	93.0±14.8

Figure 3   Steady-state subcellular distribution of glucose transporters in the isolated rat adipose cell incubated at 37°C in the absence (basal) or presence of 0.67-nM (100-μU/ml) insulin [from Simpson et al (17)].

insulin results in an ~6-fold increase in the concentration of glucose transporters in the plasma membranes, an ~2-fold increase in the high-density microsomes [almost certainly due to plasma membrane contamination (17)], and an ~60% decrease in the low-density microsomes. This steady-state redistribution of glucose transporters, induced by insulin, occurs under circumstances where no change in marker enzyme recovery or distribution is observed (17).

An assessment of the stoichiometry of this redistribution of glucose transporters is beset with several problems (17, 20). First, the direct measurement of absolute recoveries of glucose transporters is presently impossible since none of the current techniques are capable of determining the number of glucose transporters present in either the intact adipose cell or the initial homogenate. This has led to the use of marker enzyme activities to determine recovery. However, while this indirect technique is adequate for assessing the recovery of plasma membranes and consequently those glucose transporters associated with the plasma membranes, it cannot be applied with certainty in the case of the low-density microsomes because the subcellular distribution of galactosyltransferase does not correspond with that of the intracellular glucose transporters. No specific marker for the latter has yet been identified. Thus, while the best estimates suggest that the redistribution of glucose transporters in response to insulin is stoichiometric (17, 20), a rapid input of newly synthesized glucose transporters into the process cannot be ruled out (51).

Pertaining to this problem is the disparity between insulin's stimulatory action on glucose transport activity in the intact adipose cell, which is of the order of 20–40-fold, and the measured increase in the concentration of plasma membrane glucose transporters, which is of the order of 4–6 fold (17, 18). It remains uncertain whether this is due to an overestimate of the concentration of glucose transporters present in the plasma membranes from basal cells resulting from contamination with the highly enriched low-density microsomes, an additional effect of insulin on the intrinsic activity of the glucose transporter, or some other unknown factor. If the latter is indeed the case, this activation of the glucose transporter in the intact cell is not retained during isolation of the membrane fractions since assessment of glucose transport activity in both isolated plasma membrane vesicles (17, 52, 53) and following reconstitution of all membrane fractions (18, 49) reveals activities that are consistent with the membrane concentrations of glucose transporters measured by cytochalasin B binding.

In the insert to Figure 3 are shown the dissociation constants for specific D-glucose-inhibitable cytochalasin B binding to the glucose transporter in the various membrane fractions. Significant differences are clearly detectable between the low-density microsomal glucose transporters in basal adipose cells and those in the plasma membranes from either basal or insulin-stimulated cells, as well as those in the low-density microsomes from insulin-stimulated cells. These differences will be discussed later in the context of glucose transporter heterogeneity.

Kinetics of Action of Insulin

The data so far discussed represent the steady-state response to insulin. However, to further validate the translocation hypothesis, it is necessary to demonstrate the temporal relationship between the stimulation of glucose transport activity in the intact adipose cell and the translocation of glucose transporters, as shown in Figure 4 (16). In these experiments, the time courses for insulin's stimulation of glucose transport activity and its reversal with anti-insulin antibody are compared with the corresponding redistributions of glucose transporters between the plasma membranes and low-density microsomes. While the present illustration clearly indicates the close parallelism between these two processes, an evaluation of the half times not shown here reveals that the translocation of glucose transporters from the low-density microsomes to the plasma membranes in response to insulin actually precedes the onset of glucose transport activity by ~1.5 min (16). This is compatible with the lag times of up to 45 sec reported by others (54, 55) for the onset of insulin-stimulated glucose transport activity. On reversing the action of insulin, in this case with an excess of anti-insulin antibody or with collagenase (56), no

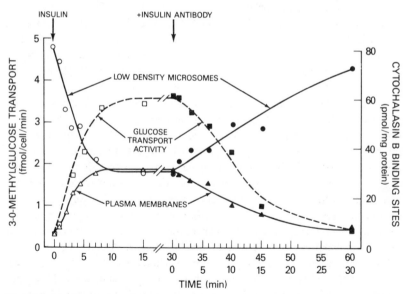

Figure 4 Stimulation of glucose transport activity and the subcellular redistribution of glucose transporters by insulin at 37°C in the isolated rat adipose cell: time course in response to 0.67-nM (100-μU/ml) insulin and reversal using a 300-fold excess of the IgG fraction of an anti-insulin antiserum [from Karnieli et al (16)].

corresponding lag is observed between the disappearance of glucose transporters and the loss of glucose transport activity.

Measurements of the absolute rates of reversal of insulin-stimulated glucose transport activity under these circumstances are complicated by the finite time required to remove insulin from its receptor. However, when the experiment is performed at 16°C as shown in Figure 5, a clear dissociation can be observed between the removal of insulin and the reversal of stimulated glucose transport activity (57). Furthermore, a markedly different temperature dependency for the stimulation and reversal of glucose transport activity is observed. While the half time for stimulation of glucose transport activity is only doubled at 16°C compared to 37°C, the half time for its reversal following the removal of insulin is increased by more than 10-fold. Indeed, stimulation of glucose transport activity can be measured even at 4°C, although an Arrhenius plot of the temperature dependency for insulin stimulation reveals a greater activation energy at temperatures below 16°C (4–16°C) than that seen above 16°C (16–37°C) (57). However, no such break in activation energy is observed for the intrinsic activity of the glucose transporter itself in either intact adipose cells or isolated plasma membranes over the range of 4–37°C (53, 57, 58).

Another consequence of lowering the incubation temperature is an elevation of the basal rate of glucose transport in both muscle (59–61) and adipose cells (62–65). When initially observed, this suggested different temperature dependencies for basal and insulin-stimulated glucose transport activity in intact cells which was lost in the preparation of plasma membranes. However, this dichotomy has been resolved in the adipose cell by the observation that a non-insulin-mediated translocation of glucose transporters to the plasma membrane occurs with reduction of the incubation temperature (58).

The ability to perform such temperature studies depends on the observations of Kono and colleagues (56, 66), who demonstrated that both insulin's stimulation of glucose transport and its reversal could be blocked by metabolic inhibitors such as KCN, NaN_3, and dinitrophenol, all of which cause a depletion of cellular ATP by inhibiting oxidative phosphorylation. This would suggest that while the transport of glucose itself does not require ATP, the action of insulin to induce translocation and the reversal of this process do exhibit an ATP dependence. It is therefore possible to perform experiments at lower incubation temperatures and, at particular times, halt the translocation of glucose transporters by addition of one of these metabolic inhibitors; all subsequent measurements of glucose transport activity can then be carried out at a constant incubation temperature such as 37°C. This same methodology has proven extremely useful in studies of the movement of other integral membrane proteins, such as the IGF-II receptor, which shuttle between the plasma membrane and intracellular sites (67).

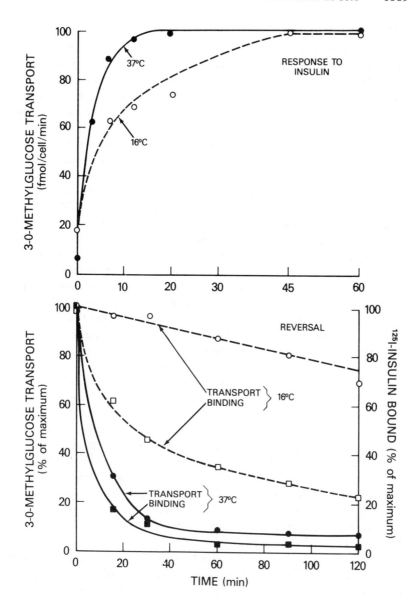

Figure 5   Effect of temperature on the stimulation of glucose transport activity by insulin in the isolated rat adipose cell: time course in response to 670-nM (100,000-μU/ml) insulin, and reversal of the response to 0.67-nM (100-μU/ml) insulin and removal of bound insulin using 3-mg/ml crude collagenase [from Simpson et al (57)].

Insulinomimetic Hormones/Agents

Over the course of several years, a wide and diverse range of agents including spermine, vitamin K_5, hydrogen peroxide, concanavilin A, p-chloromecuri-phenylsulfonate, and sodium vanadate has been shown to stimulate glucose transport activity in rat adipose cells by a mechanism that does not appear to require an intact insulin binding site (68); a role for some other portion of the insulin receptor in these responses, however, has not yet been ruled out. Every agent so far investigated appears to stimulate glucose transport by the same translocation process observed with insulin, strengthening the thesis that this is indeed the primary mechanism (20).

One class of hormones that could be considered insulinomimetic is that comprising the insulinlike growth factors types I and II (IGF-I and -II). Traditionally, their insulinlike actions on glucose transport and metabolism have been attributed to the ability of these hormones to interact with the insulin receptor [for review see (69)]. Recently, however, in the cultured BC_3H1 muscle cell line, IGF-I has been shown to be more potent than insulin, and IGF-II, equipotent with insulin in the stimulation of glucose oxidation, and by inference glucose transport, through interactions with their own specific recep-tors (70). These results suggest that a hormone receptor other than the insulin receptor is capable of eliciting an insulinlike modulation of glucose transport, the ramification of which, particularly in the course of tissue development, remains to be assessed.

Two further agents, which cannot be considered insulinomimetic in the strict sense but have been shown to modulate glucose transport activity, are the sulphonylureas and the tumor-promoting phorbol esters. For the past 20 years, sulphonylureas have been widely prescribed in the treatment of hyperglycemia associated with type II (non-insulin-dependent) diabetes mellitus (71). Howev-er, it was not until 1981 that their ability to augment insulin's stimulation of glucose transport, and not basal glucose transport, was recognized in the rat adipose cell (72). These initial observations have since been confirmed and the augmentation of insulin action has been attributed to an enhanced translocation of glucose transporters under circumstances where total glucose transporter number remains unaltered (73).

The phorbol esters have been shown to stimulate glucose uptake in a wide variety of cell types (74–76), although the mechanism by which this is achieved is still under investigation (76). One potentially related action of these agents is the stimulation of the Ca^{2+}-phospholipid-dependent protein kinase, protein kinase C. Recently, Witters et al (77) have shown that a partially purified preparation of rat brain protein kinase C brings about a serine-specific phosphorylation of the partially purified human erythrocyte glucose transporter in vitro. In addition, these investigators demonstrated that phorbol esters stimulate the phosphorylation of the glucose transporter in intact human red cells. In neither case, however, has a corresponding change in glucose transport activity been reported. Nevertheless, these observations raise the possibility that hormones such as epidermal growth factor, which activate protein kinase C as part of their normal mechanism of action, might exert a regulatory action on glucose transport by a similar mechanism (77, 78).

Summary

The studies just described regarding the translocation of glucose transporters in rat adipose cells in response to insulin have relied entirely on cell fractionation to demonstrate the subcellular distribution of glucose transporters. The question is thus raised as to the extent these observations might reflect an artifact of the homogenization procedure (79). In an elegant study using the essentially impermeant glucose analogue ethylidene glucose, Oka & Czech (80) have now demonstrated the translocation phenomenon in the intact cell. This glucose analogue is capable of blocking the photochemical cross-linking of [^{3}H]-cytochalasin B to those glucose transporters exposed to the extracellular medium, but not those inside the cell. Insulin can thus be shown to increase the number of cell surface glucose transporters.

The data so far discussed strongly suggest that the principle mechanism by which insulin stimulates glucose transport in the rat adipose cell is through the stoichiometric translocation of glucose transporters from a large, but as-yet-uncharacterized, intracellular pool to the plasma membrane. This process is rapid, reversible, and energy- and insulin concentration–dependent, and also occurs in response to insulinomimetic agents. In studies not described here, the translocation hypothesis has also been extended to the isolated guinea pig (21) and human (22) adipose cell, and rat diaphragm (23, 24) and heart (25). However, as will be discussed in the following sections, other levels of regulation are imposed on this underlying mechanism of insulin action. Indeed, contrary to earlier observations, Whitesell & Abumrad (81) have recently reported that insulin may influence both the apparent affinity (K_M) of the transporter for glucose and the maximum transport velocity (V_{max}) under certain circumstances.

ACUTE MODULATION OF INSULIN ACTION

Adenylate Cyclase Stimulators and Inhibitors

The regulation of lipolysis in the rat adipose cell by agents such as ACTH, glucagon, and β adrenergic agonists which increase cellular cAMP concentrations through the activation of adenylate cyclase, and by agents such as adenosine, prostaglandins, and nicotinic acid which decrease cellular cAMP concentrations through the inhibition of adenylate cyclase, has long been established. Similarly, the antilipolytic actions of insulin have long been recognized, although the locus of action still remains unclear. By contrast, the counterregulatory actions of these adenylate cyclase stimulators and inhibitors on the synthesis of triglyceride from glucose, a process stimulated by insulin, have only been observed relatively recently. This temporal disparity is almost certainly due to the failure to appreciate the observations of Schwabe et al (82), who indicated that concentrated suspensions of isolated rat adipose cells are bathed in very high endogenous concentrations of adenosine, a powerful inhibitor of adenylate cyclase.

Thus, it was not until 1976 that Halperin and coworkers (83, 84) demonstrated in the rat adipose cell that stimulators of lipolysis were inhibitors of insulin-stimulated glucose oxidation and by inference, inhibitors of glucose transport. This could only be demonstrated, however, when the endogenous adenosine was removed using adenosine deaminase and has since been confirmed by other investigators (30–32, 85–88). In fact, other studies with catecholamines in the presence of endogenous adenosine demonstrated a paradoxical stimulation of basal glucose transport activity (52, 88).

Table 1 illustrates that with the removal of endogenous adenosine, all of the ligands that stimulate adenylate cyclase in the rat adipose cell can now be shown to inhibit insulin-stimulated glucose transport activity (32). Similarly, the adenylate cyclase inhibitors such as prostaglandin E_1, nicotinic acid, and N^6-phenylisopropyladenosine, a nonmetabolizable adenosine analogue, prevent or reverse this inhibition. At submaximal levels, both the inhibitory actions of the lipolytic hormones/agents and the stimulatory effects of the antilipolytic hormones/agents are additive (not illustrated), suggesting that both

Table 1 Counterregulation of insulin-stimulated glucose transport activity by adenylate cyclase stimulators and inhibitors in the isolated rat adipose cell[a,b]

		Adenylate cyclase stimulators[c]		
Adenylate cyclase inhibitors[c]	Insulin (6.7-nM)	Isoproterenol (1000-nM)	ACTH (100-nM)	Glucagon (1000-nM)
		Relative 3-O-methylglucose transport, % ± SEM		
− Endogenous adenosine[d]	100[e]	45 ± 5	39 ± 2	51 ± 4
+ PIA[f] (1000-nM)	141 ± 3	124 ± 6		
+ Nicotinic acid (1000 nM)	140 ± 3	124 ± 7		
+ Prostaglandin E_1 (10 nM)	137 ± 3	94 ± 5		
+ Endogenous adenosine	143 ± 4	136 ± 4	133 ± 6	139 ± 2

[a]From Kuroda et al (31).
[b]Measured at steady state at 37°C.
[c]Added alone or in combination following achievement of the steady-state response to insulin.
[d]Achieved using 1-U/ml adenosine deaminase.
[e]Set to 100% with all other values expressed relative to that value in each experiment.
[f]N^6-phenylisopropyladenosine.

actions are mediated by mechanisms distal to the interactions of the ligands with their individual receptors, perhaps at the level of the guanine nucleotide regulatory subunits of adenylate cyclase, N_s and N_i (32). Finally, both the stimulatory action of adenosine and the inhibitory action of isoproterenol on glucose transport activity have been shown to occur through changes in the maximum transport velocity (V_{max}) (30).

However, unlike the action of insulin, Figure 6 demonstrates that this counterregulation of insulin-stimulated glucose transport activity is apparently not accompanied by corresponding changes in the number of glucose transporters in the plasma membrane (30–32). Indeed, neither the potentiation of insulin-stimulated glucose transport activity by adenosine, the inhibition by isoproterenol, nor the restitution of inhibited activity by N^6-phenylisopropyladenosine is accompanied by changes in the extent of translocation of glucose transporters induced by insulin. Thus, these ligands appear to exert their actions on the intrinsic activity of those glucose transporters present in the plasma membrane rather than through changes in the translocation process (31).

Circumstantial evidence has led to the proposal that the effects of these various adenylate cyclase stimulators and inhibitors are mediated by cAMP (30, 83–88). However, recent studies by Kuroda and coworkers (31, 32), illustrated in Figure 7, suggest that this might not be the case. In these studies, changes in the cellular concentrations of cAMP, as reflected in changes in the activity of cAMP-dependent protein kinase (protein kinase A), are compared with the corresponding changes in insulin-stimulated glucose transport activity. The enhancement of glucose transport activity by adenosine and the restitution of isoproterenol-inhibited glucose transport activity by N^6-phenylisopropyladenosine represent two examples where changes in glucose transport activity do not correlate with modulation of cAMP-dependent protein kinase. In the one situation where a correspondence is observed, the action of isoproterenol in the absence of adenosine, both the stimulation of protein kinase A activity and the inhibition of glucose transport activity exhibit similar isoproterenol concentration dependencies (not illustrated) suggesting that the same class of receptors is responsible for both effects.

In addition to their direct counterregulatory actions on glucose transporter intrinsic activity, adenylate cyclase stimulators and inhibitors also modulate the sensitivity of the glucose transport response to insulin in the rat adipose cell (32). As illustrated in Figure 8, adenosine increases the sensitivity to insulin by a factor of ~2 whereas isoproterenol decreases the sensitivity by ~3-fold. Furthermore, under these circumstances where the cells are exposed simultaneously to insulin and either adenosine or isoproterenol, the effects of the latter two agents appear to occur at the level of glucose transporter translocation (not illustrated), and may be mediated by a direct modulation of the insulin receptor signaling mechanism. Thus, for any given concentration of insulin to which the cell is exposed, the rate of glucose transport can be dramatically

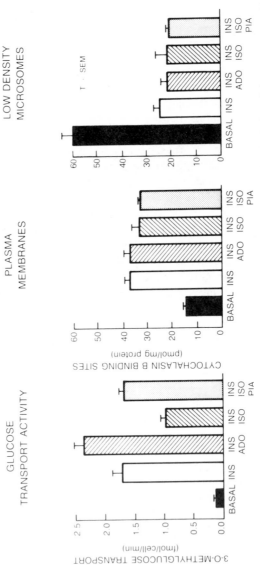

Figure 6 Counterregulation of insulin-stimulated glucose transport activity and the subcellular distribution of glucose transporters by adenylate cyclase stimulators and inhibitors at steady state at 37°C in the isolated rat adipose cell: basal, the responses to 6.7-nM (1000-μU/ml) insulin (INS) in the absence and presence of endogenous adenosine (ADO), the subsequent response to 200-nM isoproterenol (ISO) in the absence of endogenous adenosine, and the further subsequent response to 1-μM N^6-phenylisopropyladenosine (PIA) [from Kuroda et al (31)]. Removal of endogenous adenosine was achieved using 1-U/ml adenosine deaminase.

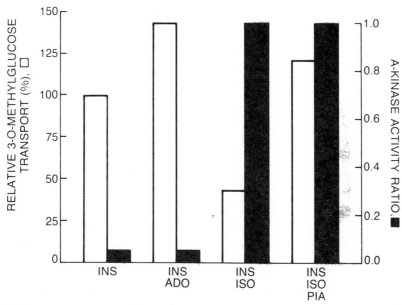

Figure 7   Relationship between the counterregulation of insulin-stimulated glucose transport activity and cAMP-dependent protein kinase activity by adenylate cyclase stimulators and inhibitors at steady state at 37°C in the isolated rat adipose cell: the responses to 6.7-nM (1000-μU/ml) insulin (INS) in the absence and presence of endogenous adenosine (ADO), the subsequent response to 200-nM isoproterenol (ISO) in the absence of endogenous adenosine, and the further subsequent response to 1-μM N^6-phenylisopropyladenosine (PIA) [from Kuroda et al (31)]. Removal of endogenous adenosine was achieved using 1-U/ml adenosine deaminase.

altered by the relative concentrations of these stimulatory and inhibitory hormones.

Other Hormones/Agents

Two additional hormones that are known to modulate the regulation of glucose transport by insulin in the rat adipose cell, albeit not as acutely as those just described, are the glucocorticoids and growth hormone. It has long been established that glucocorticoids, and more specifically the analogue dexamethasone, induce a decrease in both basal and insulin-stimulated glucose transport activity (89–96). The onset of these actions in vitro requires 1–2 hr and may be prevented by inhibitors of both mRNA and protein synthesis (95). The dependence on protein synthesis to express these inhibitory effects is clearly complex; Carter-Su & Okamoto (96) have recently indicated that the reduction in basal glucose transport activity results from an ~30% decrease in the concentration of glucose transporters in the plasma membranes without any apparent change in the concentration of glucose transporters residing in the

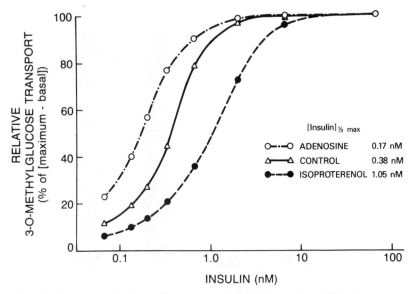

Figure 8   Steady-state stimulation of glucose transport activity by insulin at 37°C in the isolated rat adipose cell in the absence of endogenous adenosine (control), in the presence of endogenous adenosine, and in the presence of 200-nM isoproterenol and absence of endogenous adenosine [from Simpson et al (32)]. Removal of endogenous adenosine was achieved using 1-U/ml adenosine deaminase.

intracellular pool. These results suggest that the action of dexamethasone is to either promote an accelerated degradation of those glucose transporters present in the plasma membrane and/or induce a redistribution of glucose transporters in favor of the intracellular pool. Both mechanisms would require mediation by a protein(s) distinct from the glucose transporter.

An excess of circulating growth hormone in both rat and man appears to correlate with insulin resistance, while growth hormone–deficient conditions are associated with an increased response to insulin (97–99). However, somewhat paradoxically, adipose cells isolated from hypophysectomized rats exhibit enhanced basal rates of glucose oxidation/transport and a correspondingly diminished response to insulin that can subsequently be reversed by the administration of growth hormone (98). Further advances in the understanding of the action of growth hormone have been restricted by the apparent inability to mimic these effects of growth hormone in vitro. To date, Maloff et al (100), using a primary culture of adipose tissue, have observed the supression of basal glucose transport activity induced by growth hormone but have failed to observe any effect of growth hormone on either the sensitivity or responsiveness of the cells isolated from this tissue to insulin.

CHRONIC MODULATION OF INSULIN ACTION

Various pathophysiological conditions in both rat and man are accompanied by perturbations in insulin's ability to stimulate glucose transport at the cellular level in vitro. Conditions in which a diminished glucose transport response to insulin is observed in the adipose cell, so-called "insulin resistant states," include the streptozotocin diabetic rat (26), the high fat–fed rat (101), the aged, obese rat (102), the starved rat (28), and the near-term pregnant rat (103), and also in muscle, the streptozotocin diabetic rat (104, 105). Examples in which an augmented glucose transport response to insulin is observed in the adipose cell include the chronically hyperinsulinemic rat (27, 106–110), the young, genetically obese Zucker fatty rat (111), the fasted-refed rat (28), and the insulin-treated streptozotocin diabetic rat (29).

The streptozotocin diabetic rat model has been chosen to illustrate what appears now to be a perturbation common to all of the above-cited examples of insulin resistance so far examined (26). In this model, as illustrated in Figure 9, insulin-stimulated glucose transport activity is reduced by 67% in the adipose cells isolated from the diabetic rats as compared to the controls, correlating well with the 53% diminished glucose transporter concentration in the corresponding plasma membranes. The cause of this diminished response to insulin does not, however, appear to be due to an impairment in the translocation mechanism itself, but instead to a decreased concentration of glucose transporters in the intracellular pool in the basal state. In basal cells from the diabetic rats, the concentration of glucose transporters in the low-density microsomes is reduced by 46% as compared to that in the equivalent control cells, and it is this reduction in the intracellular pool in the basal state that is observed in the other insulin-resistant states. A relatively small size of the intracellular pool of glucose transporters would also appear to account for the relatively small response to insulin seen in isolated human (22) and guinea pig (21) adipose cells and rat heart (25) and diaphragm (23, 24).

The insulin hyperresponsive states can be divided into two categories. The first includes the chronically hyperinsulinemic rat model (27, 106–110) and the young, genetically obese Zucker fatty rat which is spontaneously hyperinsulinemic (111), and may be characterized as the complete antithesis to the insulin-resistant states. As seen in the preliminary experiment illustrated in Figure 10 (27), insulin-stimulated glucose transport activity in the adipose cells from rats made hyperinsulinemic using osmotic mini-pumps is increased by 43% as compared to the controls. This increase in glucose transport activity correlates with a 64% increase in the corresponding plasma membrane concentration of glucose transporters, but appears to be associated with an actual decrease in the concentration of glucose transporters in the low-density microsomes in the basal state. However, when a substantial increase in the in-

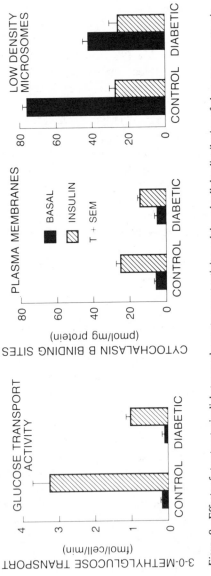

Figure 9 Effects of streptozotocin diabetes on glucose transport activity and the subcellular distribution of glucose transporters in the isolated rat adipose cell at steady state at 37°C in the absence (basal) or presence of 6.7-nM (1000-μU/ml) insulin [from Karnieli et al (26)].

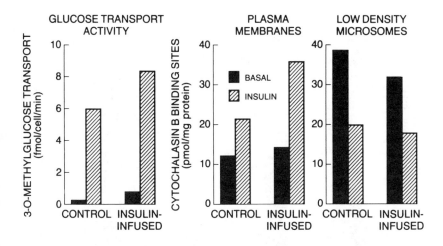

Figure 10 Effects of experimentally induced chronic hyperinsulinemia on glucose transport activity and the subcellular distribution of glucose transporters in the isolated rat adipose cell at steady state at 37°C in the absence (basal) or presence of 6.7-nM (1000-µU/ml) insulin [from Kahn et al (27)].

tracellular membrane protein of the cells from the hyperinsulinemic rats is taken into account, an increase in the total number of glucose transporters in the basal intracellular pool is observed (not illustrated) which is sufficient to explain the increased appearance of glucose transporters in the plasma membranes in response to insulin.

The second category of insulin hyperresponsive states includes the fasted-refed rat (28) and insulin-treated streptozotocin diabetic rat (29). Here, not only is there an increase in the total number of glucose transporters in the basal intracellular pool (not illustrated) as just described for the chronically hyperinsulinemic rat, but as seen in Figure 11 for diabetic rats treated with insulin for seven days using osmotic mini-pumps (29), there is a markedly greater increase in insulin-stimulated glucose transport activity in the intact adipose cell than can be explained by the much smaller increase in the concentration of glucose transporters in the corresponding plasma membranes. Thus, these two hyper-insulin-responsive states appear to be associated with both increased glucose transporter number and intrinsic activity.

Roles for plasma insulin and/or glucose are suggested in the altered glucose transport response to insulin in both the insulin-resistant and hyper-insulin-responsive states. For example, in the resistant states represented by fasting, high fat feeding, and diabetes, the levels of circulating insulin are decreased and the levels of circulating glucose unchanged or increased, whereas in all of the hyperresponsive states except fasting-refeeding, elevated levels of insulin and decreased levels of glucose are apparent. It is also potentially significant that in

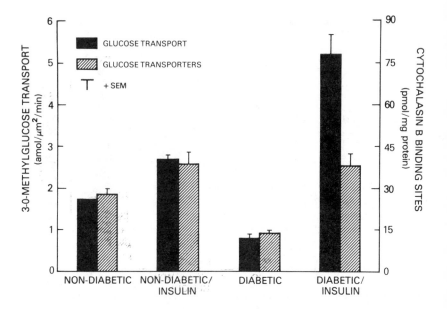

Figure 11 Comparison between glucose transport activity per unit cellular surface area in intact cells and the concentration of glucose transporters in the plasma membranes in the maximally insulin-stimulated isolated rat adipose cell with experimentally induced chronic hyperinsulinemia, streptozotocin diabetes, and insulin treatment of streptozotocin diabetes.

the hyperresponsive conditions where an enhanced glucose transporter intrinsic activity is observed, the animals have undergone a transition from a resistant state to a hyperresponsive state. However, in the aged, obese rat model and in some human type II diabetics, circulating insulin levels are elevated and glucose levels either unchanged or also elevated compared to controls, yet these often enlarged adipose cells exhibit a reduced glucose transport response to insulin. In addition, it is important to emphasize that changes in circulating insulin and prevailing glucose concentrations markedly influence the levels of many other hormones which may ultimately initiate the observed changes in the glucose transport response.

Situations somewhat analogous to the hyper-insulin-responsive states just described also appear to occur in other cell types. For example, differentiation of the mouse mammary gland from the virgin state through pregnancy to lactation is accompanied by a marked increase in glucose transport capacity which reverts on involution (C. G. Prosser, Y. J. Topper, personal communication). Similarly, transformation of cultured chick embryo fibroblasts with Rous sarcoma virus (113–115) and glucose deprivation of the same cells and mouse 3T3 fibroblast cell lines (116, 117) also appear to induce both more glucose transporters and potentially different forms of the glucose transporter.

The differentiation of cultured mouse 3T3-L1 cells from fibroblasts to adipocytelike cells occurring spontaneously at a slow rate or more rapidly in the presence of insulin, isobutylmethylxanthine (IBMX), and dexamethasone is accompanied by little change in basal glucose transport activity but a marked increase in the glucose transport response to insulin (118). The differentiated adipocyte form of this cell line is becoming of increasing importance as an alternative to the mature rat adipose cell for studies of insulin action in general, and specifically in studies of the turnover of the glucose transporter (51). As with the rat adipose cell in altered metabolic states, modulation of either the synthesis or degradation of glucose transporters appears to constitute another level of regulation of cellular glucose transport capacity.

GLUCOSE TRANSPORTER STRUCTURE AND FUNCTION

The human erythrocyte glucose transporter is the most extensively studied mammalian glucose transporter because of its presence in high concentrations in the plasma membrane and the ready availability of this cell type. It has been purified to homogeneity by several groups (51, 119–125) and both polyclonal and monoclonal antibodies have been produced against this integral membrane glycoprotein (51, 125–127). Determination of its molecular weight by Na dodecylsulfate/polyacrylamide gel electrophoresis (SDS-PAGE) yields a consensus relative molecular mass of ~55 kd, with the protein running as a very broad band reflecting the marked heterogeneity of its carbohydrate moiety (122). Removal of the carbohydrate with endoglycosidase F gives rise to a more discrete band of ~46 kd (128). Proteolysis studies reveal that this protein is transmembrane in nature, with an extracellular domain that is essentially resistant to proteolysis and a cytoplasmic domain that is susceptible to trypsin cleavage (129–131). A tryptic fragment of ~19 kd contains the cytochalasin B binding site and the epitopes for most of the available anti–human erythrocyte glucose transporter antibodies (51, 129, 130).

The antibodies generated against the purified human erythrocyte glucose transporter have been shown by Western blotting to cross-react extensively with avian, murine, rat, and human tissues, suggesting that these various glucose transporters have common determinants (34, 35, 51, 114). Indeed, these antibodies have now been used to study the biosynthesis of the glucose transporter in human heptatocarcinoma (HepG2) cells and fibroblasts and murine preadipocytes in culture, revealing the same primary translation product of ~38 kd and similar immediate precursors of the mature glucose transporter of ~42 kd (51). Similarly, a rabbit polyclonal antiserum prepared against the purified human erythrocyte glucose transporter has recently been used by Mueckler et al (33) to identify positive clones of the HepG2 glucose transporter, thus providing the most definitive information to date on the structure of the mammalian glucose transporter. These investigators have deduced the amino acid sequence of the HepG2 glucose transporter by analysis of a cDNA clone and confirmed its identity by fast-atom-bombardment mapping and gas phase Edman degradations. They further suggest that the protein moiety of the HepG2 glucose transporter is highly homologous, if not identical, to that of the human erythrocyte glucose transporter.

Figure 12 depicts the model of the HepG2 glucose transporter proposed by Mueckler et al (33). Analysis of its primary amino acid sequence suggests that this protein of 492 amino acids (54,117 daltons) contains 12 membrane-spanning domains, several of which could potentially form amphipathic α helixes containing numerous hydroxyl and amide side chains and thus provide either a glucose binding site and/or an appropriate lining for a pore through the membrane. The amino and carboxy termini and a relatively large central hydrophobic domain are predicted to exist on the cytoplasmic face of the membrane in at least partial agreement with previous predictions based on enzymatic digestion of the human erythrocyte glucose transporter (129–131). The N-linked glycosylation site has, however, been assigned to asparagine 45 which is close to the amino terminus.

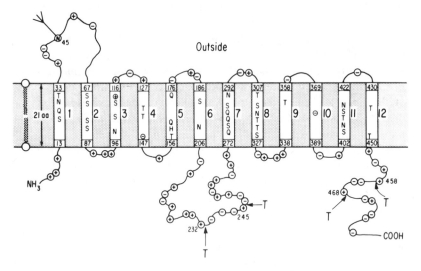

Figure 12   Proposed model for the orientation of the glucose transporter in the membrane [from Mueckler et al (33)]. The 12 putative membrane-spanning domains are numbered and shown as rectangles. The relative positions of acidic (Glu, Asp) and basic (Lys, Arg) amino acid residues are indicated by circled (+) and (−) signs, respectively. Uncharged polar residues within the membrane-spanning domains are indicated by their single-letter abbreviations: S, serine; T, threonine; H, histidine; N, asparagine; and Q, glutamine. The predicted position of the N-linked oligosaccharide at Asn 45 is shown. The arrows point to the positions of known tryptic cleavage sites in the native, membrane-bound erythrocyte glucose transporter.

The commonality of this structure to the glucose transporters in other tissues is further suggested by Mueckler et al (33) by the detection of mRNA species homologous to that from HepG2 cells in RNA extracts of K562 leukemic cells, HT29 colon adenocarcinoma cells, and human kidney tissue. These investigators are currently assessing whether this same glucose transporter structure is observed in insulin-responsive tissues and cells, and raise interesting future questions regarding the relationship between glucose transporters expressed constitutively in the plasma membrane in the basal state and those found in the intracellular pool, and the regions of the intracellular glucose transporter that might be involved in the translocation process.

Some information regarding the former of these two questions is now available. Initial studies comparing the affinities for cytochalasin B binding to the glucose transporter among the various subcellular membrane fractions prepared from basal and insulin-stimulated adipose cells suggest a heterogeneity of glucose transporters in this cell type (Figure 3) (17). This concept is further supported by the observation that certain, but not all, polyclonal antisera prepared against the purified human erythrocyte glucose transporter cross-react differentially with glucose transporters in the plasma membranes and low-density microsomes (34). However, the most compelling evidence for the presence of at least two distinct species of glucose transporter is obtained using the cytochalasin B cross-linking technique. As briefly described earlier in this report, high intensity UV light will induce the direct cross-linking of cytochalasin B to the glucose transporter in several cell types. This technique, again initially developed for the human erythrocyte glucose transporter (132–134), is now widely used to monitor changes in glucose transporter number in both whole cells (80) and isolated membranes in response to external stimuli such as glucose deprivation (117), dexamethasone (96), and insulin (37–39, 135).

A variation of this technique has recently been reported by Horuk et al (38, 39) in which the bifunctional reagent hydroxysuccinimdyl-4-azidobenzoate is used to covalently bind [^{3}H]cytochalasin B to the glucose transporter. Using this approach as illustrated in Figure 13 (39), glucose transporters in the plasma membranes and low-density microsomes prepared from basal and insulin-stimulated rat adipose cells were labeled with cytochalasin B, partially purified by SDS-PAGE, and then further characterized by isoelectric focusing. In the plasma membranes, only one isoform of the glucose transporter is observed with an isoelectric point of 5.6. In addition, the incorporation of [^{3}H]cytochalasin B into this pH 5.6 isoform is greater in the plasma membranes prepared from insulin-stimulated than basal cells, thus reconfirming the appearance of glucose transporters in this subcellular fraction in response to insulin. In contrast, two isoforms of the glucose transporter are evident in the low-density microsomes with isoelectric points of 5.6 and 6.4. However, only the pH 5.6

PLASMA MEMBRANES

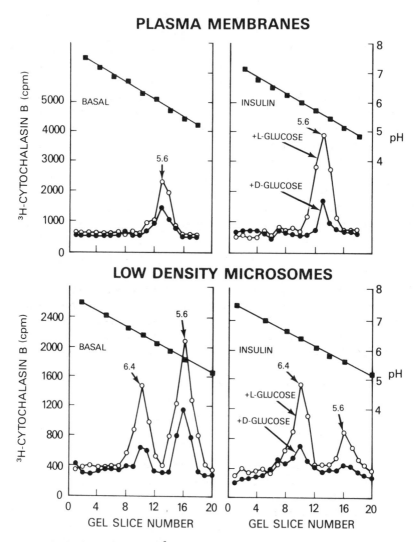

Figure 13   Isoelectric focusing of [³H]cytochalasin B-labeled glucose transporters in the plasma membranes and low-density microsomes of basal and maximally insulin-stimulated isolated rat adipose cells [from Horuk et al (39)]. Membranes were photoaffinity-labeled with [³H]cytochalasin B in the presence of 500-mM D- or L-glucose, the labeled membranes run on Na dodecylsulfate/ polyacrylamide gel electrophoresis, and the 40–55-kd proteins excised and eluted for isoelectric focusing.

isoform appears to decrease in response to insulin treatment of the intact cells, the same isoform that increases in the plasma membranes, while the level of the pH 6.4 isoform appears to remain unchanged.

The insulin-responsive rat adipose cell glucose transporter exhibits a lower relative molecular mass (~45 kd) on SDS-PAGE than the constitutive glucose transporters in other cell types (~55 kd), suggesting less extensive glycosylation. This raises the possibility that the pH 6.4 isoform may represent a biosynthetic precursor form which, unlike that in the mouse 3T3-L1 cell (51), cannot be resolved from the mature pH 5.6 isoform by SDS-PAGE. Some support for this hypothesis is provided by the observation that only the pH 5.6 isoform of the glucose transporter is susceptible to neuraminidase treatment, which removes sialic acid moieties from terminally glycosylated proteins, generating a pH 6.0 isoform; the pH 6.4 isoform is unaltered by neuraminidase suggesting incomplete glycosylation (39). Further identification of these two isoforms in the rat adipose cell and a comparison with the glucose transporters from non–insulin responsive tissues are currently under intensive investigation in several laboratories. Such studies will almost certainly provide molecular insight into the acute translocation of glucose transporters in response to insulin and the chronic modulation of glucose transport activity in altered metabolic states.

GENERAL CONSIDERATIONS

The translocation of integral membrane proteins from one subcellular compartment to another occurs in a variety of cellular activities including secretion, exocytosis, and receptor-mediated and fluid-phase endocytosis, and in the biosynthesis of various subcellular organelles. The acute, hormone-mediated movement of such proteins, however, is not as yet so extensively documented. Nevertheless, the translocation of glucose transporters in response to insulin is not unique and additional examples now include histamine-stimulated acid secretion (136, 137) and vasopressin-activated water permeability (138, 139), recently reviewed by Lienhard (140), and the action of glucose on somatostatin binding to the β cells of the pancreas during insulin secretion (43). Indeed, insulin itself appears to regulate the subcellular distribution of yet another membrane protein, the IGF-II receptor, in the rat adipose cell (67, 141, 142); however, while the characteristics of this process are qualitatively similar to those of the translocation of glucose transporters, they are sufficiently different to suggest that IGF-II receptors are present in distinct intracellular vesicles (143).

Considerable advances have been made in our understanding of how insulin regulates glucose transport in target tissues since this subject was last discussed in the *Annual Review of Biochemistry* (144). However, the nature of the signaling mechanism that is activated in response to insulin [see Kahn (40) in this series] and its subsequent translation into the translocation of glucose transporters have yet to be elucidated and should stimulate more than sufficient interest and endeavor for future reviews.

ACKNOWLEDGMENTS

The authors wish to thank their many colleagues, both former and current, for their indispensable contributions to the concepts and experimental results described here. These investigators include: Kenneth C. Appell, Deborah L. Baly, James E. Foley, Paul J. Hissin, Rupert C. Honnor, Richard Horuk, Hans-Georg Joost, Barbara B. Kahn, Eddy Karnieli, Masao Kuroda, Constantine Londos, Lester B. Salans, Ulf Smith, Lawrence J. Wardzala, Teresa M. Weber, Thomas J. Wheeler, Dena R. Yver, and Mary Jane Zarnowski. The authors also wish to thank Christin Carter-Su, Mike Mueckler, and Amira Klip for making manuscripts available prior to publication, D. B. Baly, H.-G. Joost, and B. B. Kahn for their critical comments regarding this report, and Betty J. Morris for her patience and expertise in typing the manuscript.

Literature Cited

1. Papaspyros, N. S. 1964. *The History of Diabetes Mellitus.* Stuttgart: George Thieme Verlag. 2nd ed.
2. Banting, F. G., Best, C. H. 1922. *J. Lab. Clin. Med.* 7:464–72
3. Best, C. H., Hoet, J. P., Marks, H. P. 1926. *Proc. R. Soc. London Ser. B* 100:32–54
4. Best, C. H., Dale, H. H., Hoet, J. P., Marks, H. P. 1926. *Proc. R. Soc. London Ser. B* 100:55–73
5. Levine, R., Goldstein, M. S., Klein, S., Huddlestun, B. 1950. *Am. J. Physiol.* 163:70–76
6. Goldstein, M. S., Henry, W. L., Huddlestun, B., Levine, R. 1953. *Am. J. Physiol.* 173:207–11
7. Levine, R. 1960. *Diabetes* 10:421–31
8. Crofford, O. B., Renold, A. E. 1965. *J. Biol. Chem.* 240:14–21
9. Crofford, O. B., Renold, A. E. 1965. *J. Biol. Chem.* 240:3237–44
10. Vinten, J., Gliemann, J., Østerlind, K. 1976. *J. Biol. Chem.* 251:794–800
11. DeFronzo, R. A., Ferrannini, E., Heudler, R., Wahren, J., Felig, P. 1981. *Diabetes* 30:1000–7
12. Ferrannini, E., Wahren, J., Felig, P., DeFronzo, R. A. 1981. *Metabolism* 29:28–36
13. Rodbell, M. 1964. *J. Biol. Chem.* 239:375–80
14. Wardzala, L. J., Cushman, S. W., Salans, L. B. 1978. *J. Biol. Chem.* 253:8002–5
15. Cushman, S. W., Wardzala, L. J. 1980. *J. Biol. Chem.* 255:4758–62
16. Karnieli, E., Zarnowski, M. J., Hissin, P. J., Simpson, I. A., Salans, L. B., et al. 1981. *J. Biol. Chem.* 256:4772–77
17. Simpson, I. A., Yver, D. R., Hissin, P. J., Wardzala, L. J., Karnieli, E., et al. 1983. *Biochim. Biophys. Acta* 763:393–407
18. Suzuki, K., Kono, T. 1980. *Proc. Natl. Acad. Sci. USA* 77:2542–45
19. Robinson, F. W., Blevins, T. L., Suzuki, K., Kono, T. 1982. *Anal. Biochem.* 122:10–19
20. Kono, T., Robinson, F. W., Blevins, T. L., Ezaki, O. 1982. *J. Biol. Chem.* 257:10942–47
21. Horuk, R., Rodbell, M., Cushman, S. W., Wardzala, L. J. 1983. *J. Biol. Chem.* 258:7425–29
22. Cushman, S. W., Karnieli, E., Foley, J. E., Hissin, P. J., Simpson, I. A., et al. 1982. *Clin. Res.* 30:388A (Abstr.)
23. Wardzala, L. J., Jeanrenaud, B. 1981. *J. Biol. Chem.* 256:7090–93
24. Wardzala, L. J., Jeanrenaud, B. 1983. *Biochim. Biophys. Acta* 730:49–56
25. Watanabe, T., Smith, M. M., Robinson, F. W., Kono, T. 1984. *J. Biol. Chem.* 259:13117–22
26. Karnieli, E., Hissin, P. J., Simpson, I. A., Salans, L. B., Cushman, S. W. 1981. *J. Clin. Invest.* 68:811–14
27. Kahn, B. B., Horton, E. S., Cushman, S. W. 1984. *Clin. Res.* 32:399A (Abstr.)
28. Kahn, B. B., Cushman, S. W. 1984. *Diabetes* 33(1):71A (Abstr.)
29. Kahn, B. B., Cushman, S. W. 1985. *Clin. Res.* 33:433A (Abstr.)
30. Smith, U., Kuroda, M., Simpson, I. A. 1984. *J. Biol. Chem.* 259:8758–63

31. Kuroda, M., Simpson, I. A., Honnor, R. C., Londos, C., Cushman, S. W. 1985. *Fed. Proc.* 44:480 (Abstr.)
32. Simpson, I. A., Kuroda, M., Appell, K. C., Honnor, R. C., Londos, C., et al. 1985. *Diabetes* 34(1):83A (Abstr.)
33. Mueckler, M., Caruso, C., Baldwin, S. A., Panico, M., Blench, I., et al. 1985. *Science* 229:941–45
34. Wheeler, T. J., Simpson, I. A., Sogin, D. C., Hinkle, P. C., Cushman, S. W. 1982. *Biochem. Biophys. Res. Commun.* 105:89–95
35. Lienhard, G. E., Kin, H. K., Ransome, K. J., Gorga, J. C. 1982. *Biochem. Biophys. Res. Commun.* 105:1150–56
36. Shanahan, M. F., Olson, S. A., Weber, M. J., Lienhard, G. E., Gorga, J. C. 1982. *Biochem. Biophys. Res. Commun.* 107:38–43
37. Pessin, J. E., Massague, J., Czech, M. P. 1984. *Investigation of Membrane Located Receptors*, pp. 295–302. London: Plenum
38. Horuk, R., Rodbell, M., Cushman, S. W., Simpson, I. A. 1983. *FEBS Lett.* 164:261–66
39. Horuk, R., Matthaei, S., Olefsky, J. M., Baly, D. L., Cushman, S. W., et al. 1986. *J. Biol. Chem.* 261:1823–28
40. Kahn, C. R. 1985. *Ann. Rev. Med.* 36:429–51
41. Czech, M. P. 1985. *Molecular Basis of Insulin Action*. New York: Plenum
42. Pollard, H. B., Pazoles, C. J., Creutz, C. E., Zinder, O. 1979. *Int. Rev. Cytol.* 58:158–97
43. Mehler, P. S., Sussman, A. L., Maman, A., Leitner, J. W., Sussman, K. E. 1980. *J. Clin. Invest.* 66:1334–38
44. Silverstein, S. C., Steinman, R. M., Cohn, Z. A. 1977. *Ann. Rev. Biochem.* 46:669–722
45. Gliemann, J., Rees, W. D. 1983. *Curr. Top. Membr. Transp.* 18:339–79
46. McKeel, D. W., Jarett, L. 1970. *J. Cell Biol.* 44:417–32
47. Suzuki, K., Kono, T. 1979. *J. Biol. Chem.* 254:9786–94
48. Lin, S., Spudich, J. A. 1974. *J. Biol. Chem.* 249:5778–83
49. Gorga, J. C., Lienhard, G. E. 1984. *Fed. Proc.* 43:2237–41
50. Kono, T., Barham, F. W. 1971. *J. Biol. Chem.* 246:6210–16
51. Haspel, H. C., Birnbaum, M. J., Wilk, E. M., Rosen, O. M. 1985. *J. Biol. Chem.* 260:7219–25
52. Ludvigsen, C., Jarett, L., McDonald, J. M. 1980. *Endocrinology* 106:786–90
53. Ludvigsen, C., Jarett, L. 1980. *Diabetes* 29:373–78
54. Ciaraldi, T. P., Olefsky, J. M. 1979. *Arch. Biochem. Biophys.* 193:221–31
55. Häring, H. U., Kemmler, W., Renner, R., Hepp, K. D. 1978. *FEBS Lett.* 95:177–80
56. Kono, T., Suzuki, K., Dansey, L. E., Robinson, F. W., Blevins, T. L. 1981. *J. Biol. Chem.* 256:6400–7
57. Simpson, I. A., Zarnowski, M. J., Cushman, S. W. 1983. *Fed. Proc.* 42:1790 (Abstr.)
58. Ezaki, O., Kono, T. 1982. *J. Biol. Chem.* 57:14306–10
59. Yu, K. T., Gould, M. K. 1981. *Diabetologia* 21:482–88
60. Brown, D. H., Park, C. R., Daughaday, W. H., Cornblath, M. 1952. *J. Biol. Chem.* 197:167–74
61. Kipnis, D. M., Cori, C. F. 1957. *J. Biol. Chem.* 244:681–93
62. Czech, M. P. 1976. *Mol. Cell. Biochem.* 11:51–63
63. Whitesell, R. R., Gliemann, J. 1979. *J. Biol. Chem.* 254:5276–83
64. Armatruda, J. M., Finch, E. D. 1979. *J. Biol. Chem.* 254:2619–25
65. Vega, F. V., Kono, T. 1979. *Arch. Biochem. Biophys.* 192:120–27
66. Kono, T., Robinson, F. W., Sarver, J. A., Vega, F. V., Pointer, R. H. 1977. *J. Biol. Chem.* 252:2226–33
67. Wardzala, L. J., Simpson, I. A., Rechler, M. M., Cushman, S. W. 1984. *J. Biol. Chem.* 259:8378–83
68. Czech, M. P. 1980. *Diabetes* 29:399–409
69. Rechler, M. M., Kasuga, M., Sasaki, N., De Vroede, M. A., Romanus, J. A., et al. 1982. *Insulin-like Growth Factors/Somatomedins: Basic Chemistry, Biology, and Clinical Importance*, pp. 459–90. Berlin: de Gruyter
70. De Vroede, M. A., Romanus, J. A., Standaert, M. L., Pollet, R. J., Nissley, S. P., et al. 1984. *Endocrinology* 114:1917–29
71. Joost, H. G. 1985. *Trends Pharm. Sci.* 6:239–41
72. Maloff, B. L., Lockwood, D. H. 1981. *J. Clin. Invest.* 68:85–90
73. Jacobs, D. B., Jung, C. Y. 1985. *J. Biol. Chem.* 260:2593–96
74. Driedger, P. E., Blumber, P. M. 1977. *Cancer Res.* 37:3257–65
75. Lee, L. S., Weinstein, I. B. 1979. *J. Cell Physiol.* 99:451–60
76. Klip, A., Rothstein, A., Mack, E. 1984. *Biochem. Biophys. Res. Commun.* 124:14–22
77. Witters, L. A., Vater, C. A., Lienhard, G. E. 1985. *Nature* 315:777–78

78. Barnes, D., Colowick, S. P. 1976. *J. Cell Physiol.* 89:633–40
79. Carter-Su, C., Czech, M. P. 1980. *J. Biol. Chem.* 255:10382–86
80. Oka, Y., Czech, M. P. 1984. *J. Biol. Chem.* 259:8125–33
81. Whitesell, R. R., Abumrad, N. A. 1985. *J. Biol. Chem.* 260:2894–99
82. Schwabe, U., Ebert, R., Erbler, H. C. 1975. *Adv. Cyclic Nucleotide Res.* 5:569–84
83. Taylor, W. M., Halperin, M. L. 1979. *Biochem. J.* 178:381–89
84. Taylor, W. M., Mak, M. L., Halperin, M. L. 1976. *Proc. Natl. Acad. Sci. USA* 73:4359–63
85. Green, A. 1983. *Biochem. J.* 212:189–95
86. Kashiwagi, A., Heucksteadt, T. P., Foley, J. E. 1983. *J. Biol. Chem.* 258:13685–92
87. Kirsch, D. M., Baumgarten, M., Deufel, I., Rinniger, F., Kemmler, W., et al. 1984. *Biochem. J.* 217:737–45
88. Kashiwagi, A., Foley, J. E. 1982. *Biochem. Biophys. Res. Commun.* 107:1151–57
89. Munck, A. 1962. *Biochim. Biophys. Acta* 57:318–26
90. Czech, M. P., Fain, J. N. 1972. *Endocrinology* 91:518–22
91. Olefsky, J. M. 1975. *J. Clin. Invest.* 56:1499–508
92. Livingston, J. N., Lockwood, D. H. 1975. *J. Biol. Chem.* 250:8353–60
93. Malchoff, D. M., Maloff, B. L., Livingston, J. N., Lockwood, D. H. 1982. *Endocrinology* 110:2081–87
94. Foley, J. E., Cushman, S. W., Salans, L. B. 1978. *Am. J. Physiol.* 234:E112–E119
95. Carter-Su, C., Okamoto, K. 1985. *Am. J. Physiol.* 248:E215–E223
96. Carter-Su, C., Okamoto, K. 1985. *J. Biol. Chem.* 260:11091–98
97. Goodman, H. M. 1966. *Endocrinology* 78:819–25
98. Schoenle, E., Zapf, J., Froesch, E. R. 1979. *Endocrinology* 105:1237–42
99. Schoenle, E., Zapf, J., Froesch, E. R. 1979. *Diabetologia* 16:41–46
100. Maloff, B. L., Levine, S. H., Lockwood, D. H. 1980. *Endocrinology* 107:538–44
101. Hissin, P. J., Karnieli, E., Simpson, I. A., Salans, L. B., Cushman, S. W. 1982. *Diabetes* 31:589–92
102. Hissin, P. J., Foley, J. E., Wardzala, L. J., Karnieli, E., Simpson, I. A., et al. 1982. *J. Clin. Invest.* 70:780–90
103. Toyoda, N., Murata, K., Sugiyama, Y. 1985. *Endocrinology* 116:998–1002
104. Nesher, R., Karl, I. E., Kipnis, D. M. 1985. *Am. J. Physiol.* 249:C226–C232
105. Wallberg-Henriksson, H., Holloszy, J. O. 1985. *Am. J. Physiol.* 249:C233–C237
106. Kobayashi, M., Olefsky, J. M. 1978. *Am. J. Physiol.* 235:E53–E62
107. Kobayashi, M., Olefsky, J. M. 1978. *J. Clin. Invest.* 62:73–81
108. Whittaker, J., Alberti, K. G., York, D. A., Singh, J. 1979. *Biochem. Soc. Trans.* 7:1055–66
109. Wardzala, L. J., Hirshman, M., Pofcher, E., Horton, E. D., Mead, P. M., et al. 1985. *J. Clin. Invest.* 76:460–69
110. Trimble, E. R., Weir, G. C., Gjinovci, A., Assimacopoulos-Jeannet, F., Benzi, R., et al. 1984. *Diabetes* 33:444–49
111. Guerre-Millo, M., Lavau, M., Horne, J. S., Wardzala, L. J. 1985. *J. Biol. Chem.* 260:2197–201
112. Deleted in proof
113. Inui, K. I., Moller, D. E., Tillotsen, L. G., Isselbacher, K. J. 1979. *Proc. Natl. Acad. Sci. USA* 76:3972–76
114. Salter, D. W., Baldwin, S. A., Lienhard, G. E., Weber, M. J. 1982. *Proc. Natl. Acad. Sci. USA* 79:1540–44
115. Salter, D. W., Weber, M. J. 1979. *J. Biol. Chem.* 254:3554–61
116. Ullrey, D. B., Franchi, A., Pouyssegur, J., Kalcker, H. M. 1982. *Proc. Natl. Acad. Sci. USA* 79:3777–79
117. VanPutten, J. P. M., Krans, H. M. J. 1985. *J. Biol. Chem.* 260:7996–8001
118. Resh, M. D. 1982. *J. Biol. Chem.* 257:6978–86
119. Kasahara, A., Hinkle, P. C. 1976. *Proc. Natl. Acad. Sci. USA* 73:396–400
120. Kasahara, A., Hinkle, P. C. 1977. *J. Biol. Chem.* 252:7384–90
121. Wheeler, T. J., Hinkle, P. C. 1981. *J. Biol. Chem.* 257:8907–14
122. Gorga, F. R., Baldwin, S. A., Lienhard, G. E. 1979. *Biochem. Biophys. Res. Commun.* 91:955–61
123. Baldwin, J. M., Gorga, J. C., Lienhard, G. E. 1981. *J. Biol. Chem.* 256:3685–89
124. Baldwin, J. M., Lienhard, G. E., Baldwin, S. A. 1980. *Biochim. Biophys. Acta* 599:699–714
125. Sogin, D. C., Hinkle, P. C. 1980. *Proc. Natl. Acad. Sci. USA* 77:5725–29
126. Baldwin, S. A., Lienhard, G. E. 1980. *Biochem. Biophys. Res. Commun.* 94:1401–8
127. Allard, W. J., Lienhard, G. E. 1985. *J. Biol. Chem.* 260:8668–75
128. Lienhard, G. E., Crabb, J. H., Ransome, K. J. 1984. *Biochim. Biophys. Acta* 769:404–10

129. Klip, A., Denziel, M., Walker, D. 1984. *Biochem. Biophys. Res. Commun.* 122:218–24
130. Cairns, M. T., Elliot, D. A., Scudder, P. R., Baldwin, S. A. 1984. *Biochem. J.* 221:179–88
131. Shanahan, M. F., D'Artel-Ellis, J. 1984. *J. Biol. Chem.* 259:13878–84
132. Shanahan, M. F. 1982. *J. Biol. Chem.* 257:7290–93
133. Shanahan, M. F. 1983. *Biochemistry* 22:2750–56
134. Carter-Su, C., Pessin, J. E., Gitomer, W., Czech, M. P. 1980. *J. Biol. Chem.* 257:5419–25
135. Klip, A., Walker, D., Ransome, K. J., Schoer, D. W., Lienhard, G. E. 1983. *Arch. Biochem. Biophys.* 226:198–205
136. Forte, J. G., Black, J. A., Forte, T. M., Machen, T. E., Wolosin, J. M. 1981. *Am. J. Physiol.* 241:G349–G358
137. Wolosin, J. M., Forte, J. G. 1981. *J. Biol. Chem.* 256:3149–52
138. Wade, J. B., Stetson, D. L., Lewis, S. A. 1981. *Ann. NY Acad. Sci.* 372:106–17
139. Gronowicz, G., Masur, S. K., Holtzman, E. 1980. *J. Membr. Biol.* 52:221–35
140. Lienhard, G. E. 1983. *Trends Biochem. Sci.* 8:125–27
141. Oppenheimer, C. L., Pessin, J. E., Massague, J., Gitomer, W., Czech, M. P. 1983. *J. Biol. Chem.* 258:4824–30
142. Oka, Y., Mottola, C., Oppenheimer, C. L., Czech, M. P. 1984. *Proc. Natl. Acad. Sci. USA* 81:4028–32
143. Appell, K. A., Simpson, I. A., Rechler, M. M., Cushman, S. W. 1985. *Fed. Proc.* 44:480 (Abstr.)
144. Czech, M. P. 1977. *Ann. Rev. Biochem.* 46:359–84

Ann. Rev. Biochem. 1986. 55:1091–117

COMPLEX TRANSCRIPTIONAL UNITS: DIVERSITY IN GENE EXPRESSION BY ALTERNATIVE RNA PROCESSING

Stuart E. Leff, Michael G. Rosenfeld

Howard Hughes Medical Institute Eukaryotic Regulatory Biology Program, School of Medicine, University of California, San Diego, La Jolla, California 92093

Ronald M. Evans

Howard Hughes Medical Institute Gene Expression Laboratory, The Salk Institute, 10010 North Torrey Pines Road, La Jolla, California 92037

CONTENTS

0066-4154/86/0701-1091$02.00

PERSPECTIVES AND SUMMARY

The expression of eukaryotic genes requires the activities of complex biochemical machinery to transcribe, process, and transport mature mRNA before it can be translated into a functional product. These activities relating to RNA production, maturation, transport, or stability have been shown to regulate the levels of specific gene products during development. Simple transcription units in eukaryotes can be defined as templates for RNA polymerase that encode information for the production of a single protein product. In contrast, complex transcription units give rise to multiple protein products, and since non–protein coding portions of mRNAs may contain sequences that are critical to the regulation of protein product levels, we extend this definition to transcription units that produce multiple mature mRNAs.

In this review we describe the role of alternative RNA processing as a strategy employed to generate diversity in the expression of complex transcription units. To do so we briefly outline the biochemical pathways and mechanisms that are involved in mRNA production and processing. We describe the different modes of alternative RNA processing that have been observed, and we review how the resulting diversity in some cases may be utilized in both specialized tissues and during development. Last, we speculate on how such discrete choices in RNA processing may be biochemically regulated.

EXPRESSION OF THE SIMPLE TRANSCRIPTION UNIT

A consideration of the factors and processes involved in expressing a simple transcription unit is critical to understanding the regulation of product formation from complex transcriptional units. The elementary schema of a eukaryotic RNA polymerase II transcription unit involves signals that specify the appropriate site of transcriptional initiation and termination; 3' end processing of the transcript by cleavage and polyadenylation; and removal of intervening sequences (introns) from the primary transcript. These features will only be briefly discussed as they have been reviewed more extensively in this volume by R. A. Padgett, P. Sharp, and colleagues (1) and in earlier articles by others (2–6).

Transcript Initiation and Promoter Signals

Much is now known about the promoter signals for specific initiation of RNA polymerase II–dependent transcriptional units. Early work on mapping transcription units and their start sites (7–16) as well as identifying important cis-acting signals was critical to this understanding. The mapping of specific initiation sites allowed the identification of specific promoter sequences such as the TATA Box (for reviews, see 2, 17). This sequence, while absent in a small percentage of eukaryotic and viral genes (18–23), is otherwise absolutely required for polymerase II transcription in vitro, and mutations or deletions in this sequence result in a loss of specificity in initiation sites in vivo (for review see 24, 25). Thus, the TATA sequence appears to specify accurate initiation about 25 nucleotides downstream at a site that serves to generate the 5' cap of the mature mRNA (26, 27). In addition, other upstream elements such as the CCAAT or GGGCGG homologies have been identified for many promoters (for review see 24, 28).

Termination of Transcription and 3' End Formation

TERMINATION   While relatively little is known about the signals that regulate termination of transcription in eukaryotes, it is clear that transcription proceeds and terminates beyond the site(s) of polyadenylation. Darnell and colleagues observed this first for the late adenovirus transcriptional units (29), where five poly(A) sites have been identified. Experiments examining transcription rates along the length of this transcription unit revealed that termination did not occur at a poly(A) site but rather near the end of the genome (29). This finding has now been confirmed for many other genes (12, 15, 30–38). Thus it may be generally hypothesized that transcription proceeds through the poly(A) site to a point hundreds to thousands of nucleotides downstream where several weak signals may direct termination (32, 33, 39). These signals operate in an orientation-dependent fashion exerting a negative *cis*-effect on transcription (40). During this time, or during transcription, cleavage and polyadenylation may occur at specific poly(A) addition site(s) upstream (29). However, whether usage of particular poly(A) sites in complex transcription units in vivo is solely specified by cleavage rather than early termination remains uncertain for some genes. Indeed both mechanisms may regulate the class (IgM vs IgD) of immunoglobulin heavy chain that is expressed during B-lymphocyte development (37).

POLYADENYLATION   As discussed above, after transcription has passed a putative poly(A) site, an endonucleolytic cleavage reaction may occur on the nascent RNA transcript, creating the 3' end to which a poly(A) tail of about 200 nucleotides is added (for review see 2, 5, 5b). A consensus sequence AAUAAA is found 10–30 nucleotides upstream from the poly(A) addition site for most transcripts (41). This sequence is highly conserved, although natural variations of its sequence can be found still active in some genes (5, 34, 42–45), and mutations that are at least partially active have been analyzed (46). Nevertheless, deletion of this sequence prevents formation of stable mRNA (47), and mutation to AAGAAA greatly reduces accurate cleavage although polyadenylation still occurs at normal and abnormal sites (48). Based on deletion analyses (43, 49–52) and sequence comparisons (5, 53–55), additional downstream sequences appear to be involved in specifying correct poly(A) site usage. These studies have led to the hypothesis that stemloop structures may direct correct 3' end cleavage (49, 52), possibly with the involvement of U4 snRNP particles (53) similar to the involvement of U7 snRNPs in the formation of correct 3' ends of unpolyadenylated sea urchin histone mRNAs (5, 56–58).

The kinetics for polyadenylation are rapid (within minutes of transcription) for some viral transcripts (29, 59) and for nuclear RNA analyzed from Chinese hamster ovary cells and HeLa cells (59a). Analyses of newly synthesized transcripts from viral genes (15, 29, 60, 60a, 60b, 61, 61a) demonstrated the presence of unspliced polyadenylated precursors. In most of these studies, spliced, unpolyadenylated transcripts were not detected although one exception has been reported for adenovirus (61b). Thus, for adenovirus and SV40 transcripts, polyadenylation appears to precede splicing. In the case of cellular genes, pulse labeling experiments demonstrate that polyadenylation can precede splicing (62, 62a, 62b, 62c, 62d), although it cannot be ruled out that they occur at the same time. Studies using cordycepin to block polyadenylation (63) and injecting unpolyadenylated β-globin transcripts into *Xenopus laevis* oocytes (64) show that polyadenylation is not a requirement for splicing to occur. Furthermore, the formation of heterogenous nuclear ribonucleoprotein particles (hnRNPs), which may represent, in part, pre-splicing complexes, occurs concomitantly with transcription (65–68). Thus, while polyadenylation occurs soon after transcription in most cases studied, the relative order of polyadenylation and splicing, or "pre-splicing," is not clear for all genes. Determination of this sequence of events may be important to understanding the regulation of mRNA production from complex transcription units that have multiple poly(A) sites and that utilize alternative splicing pathways.

mRNA Splicing

Since the discovery of intervening sequences (introns) in eukaryotic transcription units and the process of their removal by splicing (8, 69–72), much has been learned about the general mechanisms involved in this form of RNA processing. Intron removal and splicing must be a precise process. Single protein-coding exons are often interrupted. To ensure fidelity in splicing, short consensus sequences are found at intron boundaries (4, 73–77), and mutational analyses confirm that these sequences are important signals recognized in splicing (78, 79). However, since chimaeric introns comprised of 5' and 3' splice junctions from different genes can be spliced correctly (80), additional rules and mechanisms must be involved to ensure correct splice site choices, i.e. to prevent "exon skipping." Scanning mechanisms in which splicing factors would attach to the RNA and read sequences in search of proximal splice junction would prevent such exon skipping and simplify the process of splice site selection. Such models have been proposed (4, 81), and supported by experiments conducted with human β-globin gene recombinant vectors (81). In these experiments, the upstream copies of tandemly duplicated splice sites (both 5' donor and 3' acceptor) were used when these constructs were introduced into COS cells. Similar experiments conducted with rabbit β-globin recombinant vectors yielded different results (82). In this case, splice sites further removed from the intron were utilized when the vectors were introduced into HeLa cells. These latter results are not consistent with either the 5' or the 3' scanning model (82). In fact, recent experiments in vitro have indicated that exons from separate RNA precursors can be spliced in trans (83, 84), further suggesting that orderly splicing must be regulated by a mechanism other than simple linear scanning. In these studies, however, inter-molecular *trans*-splicing was facilitated by the formation of RNA-RNA hybrids. Thus, some mechanism that utilizes diffusion of a complex along the RNA molecule cannot be ruled out. Regardless, understanding these mechanisms and the factors involved in their regulation will be critically important to understanding the overall regulation of alternative mRNA production from complex transcription units.

Several RNP structures are involved in general splicing mechanisms. Sequence complementarity between U1 snRNAs (small nuclear RNA) and the splice junctions of pre-mRNAs suggest a role for U1 snRNPs in splicing (75, 76). This proposal has received support from experiments that show a requirement of U1 snRNPs in pre-mRNA splicing (85, 86), although other experiments indicate that U1 snRNPs are not required for the removal of the intron found in SV40 small t-antigen pre-mRNAs (87). Nevertheless, U1 snRNPs appear to be only one element of a pre-mRNA splicing complex found in vitro in both yeast (88) and in higher eukaryotic cells (89, 90). In addition, a related class of molecules, the U2 snRNPs, may be involved in recognizing sites upstream from the 3' splice acceptor region (91). Thus, the packaging of newly transcribed pre-mRNAs into hnRNP particles may involve the formation of complexes that act to correctly align exons for accurate splicing (92).

How does the splicing occur? Studies conducted in vitro (93–95) and in vivo (96–99) demonstrate the involvement of a two-step process. The first step is cleavage of the 5' splice site at its consensus guanosine residue. The free 5' guanosine at the end of the intron reacts with a residue, usually adenosine, upstream from the 3' splice acceptor forming a 5' to 2' phosphodiester bond. The transient product of this interaction is referred to as a lariat structure (see 6 for review). The second step involves the release of the intron concomitant with the ligation of the exons. Although these reactions are proposed to be initiated within the pre-mRNA splicing complex described above, the mechanisms bringing the correct exons together and promoting the second step remain unexplained (88–90). For some genes, secondary structure has been hypothesized to play a role in splice site choice (100), and the introductions of deletions or insertions distal from splice sites have been reported to affect the efficiency of their usage (101, 102). Furthermore, *Cyt*-4 mutants of *Neurospora* show defective splicing of their large mitochondrial rRNA as a secondary consequence of incorrect 3' end formation (102a). On the other hand, sizable intron deletions or insertions have had little effect on splice site choices in other genes (103). Whether RNA secondary structure plays a significant role in specifying alternative splicing pathways in complex transcriptional units remains speculative.

GENOMIC ARRANGEMENTS OF COMPLEX TRANSCRIPTIONAL UNITS

The production of multiple mRNAs from a single transcriptional unit can involve the usage of alternative signals for any of the steps of mRNA production or processing outlined above. In some cases the protein coding regions are unaffected and thus the consequences of alternative mRNA production remain obscure; in other cases, where the protein product is altered, the choice of alternative mRNA production appears stochastic. For genes in which alternative protein products are predicted to have an important consequence to cell phenotype, a means to selectively regulate RNA processing is required. In turn, such regulation may be subject to developmental or tissue-specific control. Individual examples of complex transcriptional units can be grouped according to their genomic structure (see Table 1, Figure 2). Thus, multiple mRNAs may arise from the use of multiple sites of initiation or polyadenylation, or from the use of alternative modes of exon splicing, or from any combination of these.

Table 1 Cellular genes that encode multiple mRNAs

Gene	Protein product affected?	Developmental or tissue-specific regulation?	Ref.
A. Single 5' end, multiple poly(A) sites, differential splicing			
1. IgM heavy chain	yes	yes	121–123
2. IgD heavy chain	yes	yes	124
3. IgE heavy chain	yes	yes	127
4. IgG heavy chain	yes	yes	221, 222
5. Rat calcitonin	yes	yes	131, 132
6. Bovine prekininogens	yes	?	177
7. *Drosophila* myosin heavy chain	?	yes	189
8. *Drosophila* myosin alkalai light chain	yes	yes	188
9. *Drosophila* glycinamide ribotidetransformylase	yes	?	223
10. Mouse major urinary proteins	no	?	224
11. Human fibrinogen	yes	no	158, 159

Table 1   *(continued)*

Gene	Protein product affected?	Developmental or tissue-specific regulation?	Ref.
B. Multiple 5' ends, single or multiple poly(A) sites, differential splicing			
1. Chicken myosin light chain	yes	yes	138
2. Mouse myosin light chain	yes	yes	139
3. *Drosophila* EH8	yes	yes	225
4. *Drosophila* alcohol dehydrogenase	no	yes	226
5. Mouse salivary & liver α-amylase	no	yes	45, 140
6. Hamster 3-hydroxy-3-methyl glutaryl coenzyme A	no	?	227
C. Single 5' end, single or multiple poly(A) sites, differential splicing			
1. Bovine substance P and K	yes	yes	145
2. Mouse α-crystallins	yes	no	228, 229
3. Human interleukin-2 receptor	yes	?	178, 179
4. Rat fibronectin	yes	?	148, 149
5. Human fibronectin	yes	yes	146, 201
6. Human arginosucinate synthetase	no	yes	230
7. Rat troponin T	yes	yes	150, 151
8. *Drosophila* tropomyosin	yes	yes	190
9. Rat γ-fibrinogens	yes	no	147
10. Human growth hormone	yes	no	174, 175
11. Rat prolactin	yes	no	176
12. H–2K antigens	yes	?	231–235
13. Qa/Tla antigens	yes	?	236
14. c-ki-*ras*2	yes	?	142–144
15. Myelin basic protein	yes	?	237
16. Ovomucoid	yes	no	238
D. Multiple 5' ends or multiple poly(A) sites, no splice changes			
1. Dihydrofolate reductase (5', pA)	no	?	164–166
2. Rat α2μ-globulin (pA)	no	?	239
3. Chicken lysozyme (5')	no	?	240
4. Yeast invertase (5')	yes	yes	160–162
5. Yeast alcohol dehydrogenase (5', pA)	no	?	241
6. Chicken vimentin (pA)	no	yes	242, 243
7. Chicken X gene (pA)	no	yes	244
8. Human N-*ras* (pA)	no	?	157

Genes Using Alternative Poly(A) Sites and Alternative Splicing

ADENOVIRUS 2 MAJOR LATE MESSAGES   The major late transcription unit of Adenovirus type 2 (Ad2) was the first characterized example of a complex transcription unit that uses multiple poly(A) sites and alternative modes of RNA splicing (29, 70, 104–106). The complex arrangement of this gene (see Figure 1) allows for a diverse group of proteins to be produced from a single transcription unit that is 25 kb long (see 3, 107 for review).

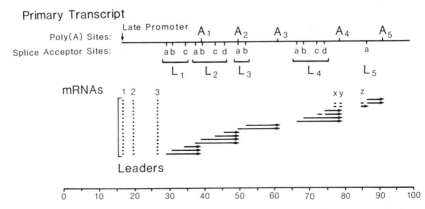

Figure 1   Five families of Adenovirus late mRNAs. A_{1-5} identify the poly(A) sites for the five families of 3' coterminal late messages L_{1-5} respectively. Lower case letters identify alternative splice acceptor sites used to generate families of mRNAs. Adapted with permission from Ref. 107.

During both early and late times of infection initiation of transcription occurs at a unique site termed the P16 promoter. The 3' ends of mRNAs from this transcription unit can occur at five separate poly(A) sites (105, 106, 108, 109). The relative usage of these poly(A) sites occurs at a fixed ratio of $1:2:3:2:2$ at late times of infection (29). During early times of infection an alternative termination site is active resulting in the selective use of only the first two poly(A) sites (110–112).

Figure 1 also shows that multiple mRNAs can be produced from transcripts ending at each of the five poly(A) sites. Each mRNA contains three 5' leader sequences that are spliced to each series of 3' coterminal messages. As the leader sequences appear to lack an initiator methionine AUG, coding sequences arise entirely from the body of the messages (107). Larger mRNAs that potentially encode more than one protein translate only the coding sequences found proximal to the 5' leader sequence (105, 109, 113). Thus, alternative splicing of these transcripts can regulate the production of different protein products from single 3' coterminal regions of the Ad2 late transcription unit (107). The mechanisms that regulate these choices remain unknown.

CELLULAR GENES   Several other eukaryotic genes have now been identified to use alternative polyadenylation sites along with differential patterns of RNA splicing (see Table 1). Two members of this class, the genes for rat calcitonin and immunoglobulin heavy chain, are shown in Figure 2A. The alternative mRNA products from these two genes are regulated in development or tissue specifically such that alternative protein products are produced.

Immunoglobulin heavy chain   There are two forms of regulation that affect immunoglobulin gene expression. One is gene rearrangement, which creates

A. SINGLE INTIATION SITE, MULTIPLE POLY (A) SITES, DIFFERENTIAL
   SPLICING

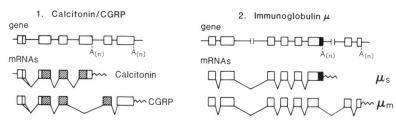

1. Calcitonin/CGRP 2. Immunoglobulin μ

B. MULTIPLE INTIATION SITES, SINGLE OR MULTIPLE POLY (A) SITES,
   DIFFERENTIAL SPLICING AT THE 5' END

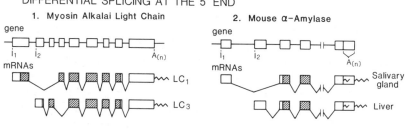

1. Myosin Alkalai Light Chain 2. Mouse α-Amylase

C. SINGLE INTIATION SITE, SINGLE POLY (A) SITE, VARIALBLE SPLICING

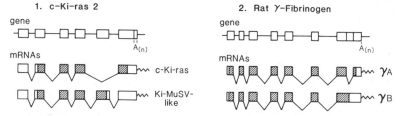

1. c-Ki-ras 2 2. Rat γ-Fibrinogen

D. MULTIPLE 5' ENDS OR MULTIPLE 3' ENDS, SPLICING UNAFFECTED

1. Heterogenous 5' end:
   Yeast Invertase

2. Heterogenous 5' ends and Poly (A) sites
   Mouse Dihydrofolate Reductase (DHFR)

Figure 2 Examples of cellular complex transcription units: classes of genomic arrangements generating multiple mRNAs. Coding regions (except for immunoglobulin mRNAs) are shaded to show origin of protein polymorphisms. For IgM (μ heavy chain)-encoding transcripts the filled region denotes sequences specific for secreted IgM heavy chains while the transmembrane coding region found in membrane-bound IgM is encoded by the downstream exons. A(n) denotes poly(A) sites and I identifies multiple initiation sites.

functional genes and allows for class switching (114–116). The other is RNA processing which includes poly(A) site selection, splice site selection, and transcription termination. Immunoglobulin heavy chain gene expression shows multiple changes during B-lymphocyte development (for review see 117, 118). Early immunoglobulin heavy chain expression is in the form of membrane-associated IgM monomer before a progressive appearance of IgD on the surface. Stimulation by antigen or mitogen and the appropriate lymphokines produces a loss of membrane-associated IgM and IgD followed by the production of secreted IgM pentamer (119). The switch from membrane-bound to secreted immunoglobulin is a consequence of a change in the carboxy terminus of the molecule. mRNA encoding membrane or secreted immunoglobulins differ in both their coding sequences and poly(A) sites. The coding change is a consequence of alternative splicing that substitutes the membrane-anchoring sequence exon for an exon that allows passage of the immunoglobulin through the membrane. Each exon is closely followed by its poly(A) site (Figure 2A; 120–123). This genomic organization is consistent with poly(A) site selection controlling the differential splicing, although no direct evidence has been provided. Similarly, when B-cells express two immunoglobulins such as IgM and IgD (or IgE), the expression of the isotypes results from alternative splicing of nuclear precursors that use different poly(A) sites (124–127). In such cases where two poly(A) sites are available for use, how are the relative levels of the mRNA products regulated? This problem is similar to that observed for Ad-2 late mRNA synthesis where multiple poly(A) sites are active. *Trans*-acting factors have been hypothesized to regulate poly(A) site selection and thus the relative levels of secreted vs membrane-bound IgM and IgD (37). Such factors may be hypothesized to act more generally on large classes of poly(A) site (118), or to act independently of sequence but rather due to the proximity to the promoter (128). In this model, the factor is an endonuclease that recognizes different poly(A) sites with different efficiencies and whose levels may vary in different tissues. In the case of IgM-secreting plasma cells, there appears to be an additional regulation at the level of transcriptional termination occurring between the exons encoding the constant regions for IgM and IgD.

Calcitonin/CGRP gene   The production of multiple mRNAs from a single calcitonin gene was first observed during studies of calcitonin gene expression in rat medullary thyroid carcinomas (MTCs). Observations that serially transplanted tumors could switch from high to low calcitonin production states were explained by the identification of a structurally distinct mature transcript (129, 130) referred to as calcitonin gene–related peptide (CGRP) mRNA (131). This second mRNA (1200 nucleotides in length) was identified by hybridization to calcitonin cDNA plasmid probes based on regions of shared homology, and it encoded a new 16,000-dalton protein product that could not be precipitated with calcitonin antisera. Figure 2A shows the structure of the rat calcitonin/CGRP gene and the alternative patterns of RNA processing that yield selectively calcitonin and CGRP mRNA. The first three exons used by both mRNAs are a 5' noncoding first exon and two exons that encode the region common to both mRNAs and protein precursors. The fourth exon contains the calcitonin peptide coding and 3' noncoding sequences while CGRP mRNAs alternatively contain the fifth (coding) and sixth (3' noncoding) exons. The two mRNAs diverge precisely at the third exon/intron junction indicating that alternative modes of splicing distinguish the two mRNAs (see Figure 3A). In addition, an alternative splice in the 5' untranslated region of the common first exon of both calcitonin and CGRP mRNA (34) is found. This splice has no apparent effects and it is not regulated.

In either calcitonin- or CGRP-producing MTC cell lines transcription begins at the same "cap" site and proceeds through the entire gene, followed by cleavage and polyadenylation occurring at the end of the fourth and/or sixth exons (34). Thus, calcitonin mRNA is produced by splicing of the first three exons to the fourth (polyadenylated) exon, while for CGRP mRNA production the first three exons are spliced to the fifth and sixth exons. During this splicing event(s), the calcitonin coding exon and poly(A) site are excised as an intron although no such excised intron species have been detected in vivo or in vitro.

The production of calcitonin and CGRP mRNAs appear to be regulated by changes in poly(A) site selection associated with alternative splicing reactions. Subsequent analyses have shown that the alternative production of these two mRNAs is regulated tissue specifically (131, 132). Using probes specific for calcitonin and CGRP sequences, mRNA homologous to CGRP was found in hypothalamus, mid-brain, lateral medulla, and trigeminal ganglia where calcitonin mRNA was undetectable (131, 132). Conversely, thyroid tissue contains predominantly (greater than 98%), although not exclusively, calcitonin mRNA (133). The organization of the human calcitonin/CGRP gene is very similar to that of rat (134, 135), and it can be predicted that the tissue-specific regulation of mRNA processing will be analogous.

Since transcription across the calcitonin/CGRP gene is constant through at least 0.64 kb downstream from the CGRP poly(A) site, selective RNA splicing and polyadenylation rather than alternative RNA transcriptional termination appear to be the regulated events. The similarities between the regulation of these RNA processing events and those for the regulation of the secreted μ_s vs μ_m heavy chain expression have led to experiments introducing the calcitonin gene into B-cells at various stages of development. Remarkably, terminally differentiated IgM-secreted plasma cells produce primarily mature calcitonin and not CGRP mRNA from the exogenous gene. Therefore, only a single poly(A) site and splicing pattern is used. However, early B-cells expressing predominantly μ_m mRNA showed similar behavior (137). We know that

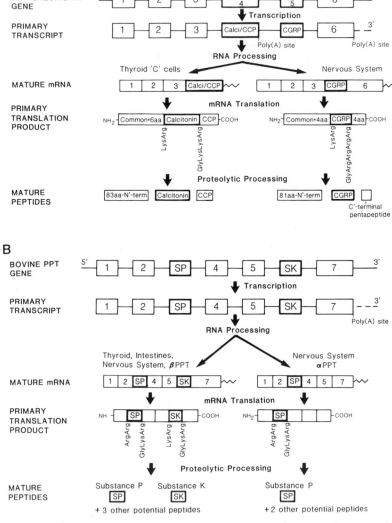

Figure 3 Pathways of transcription, mRNA processing, translation, and proteolytic processing used in the generation of calcitonin/CGRP and substance P/substance K peptides. CCP denotes a 16-amino-acid COOH-terminal calcitonin cleavage product.

calcitonin mRNA production is not a specific feature of these vectors, because other nonlymphoid cell lines can produce either intermediate or predominantly CGRP mRNA. For example, a mouse teratocarcinoma cell line that produces primarily CGRP mRNA from its endogenous calcitonin gene will preferentially synthesize CGRP mRNA from a transfected exogenous rat gene (S. Leff, unpublished observation). Thus, the "machinery" necessary to direct alternative RNA processing is operative in specific cell types, and these results would suggest that it is probably mediated in part by a *trans*-acting factor. However, there is no direct evidence as yet that such putative factors act on transcripts from multiple genes. Available evidence favors the hypothesis that such factors may act to regulate poly(A) site selection for the calcitonin gene (34). However, these factors appear to commit pre-mRNA splicing patterns to favor CGRP mRNA in CGRP-producing cells and thus indirectly regulate poly(A) site selection (S. Leff, M. G. Rosenfeld, in preparation).

Genes Using Multiple Initiation Sites and Alternative Splicing

Two genes that utilize alternative sites of initiation in the generation of heterogenous mRNAs, myosin alkalai light chain and α-amylase, are shown in Figure 2B. One difference between these two genes should be noted. For myosin alkalai light chain mRNAs differential splicing alters the coding sequence leading to the production of protein isoforms LC_1 and LC_3 (138, 139). α-amylase alternative mRNAs specific for the salivary gland or liver differ only in their 5' nontranslated sequences and thus the protein-coding sequences are not affected (45, 140). An additional heterogeneity at the 3' end of α-amylase mRNAs has been found in both liver and salivary gland where about 5% of the mRNAs are polyadenylated 237 nucleotides downstream from the major poly(A) site (141).

The regulating mechanisms of the alternative splicing pattern for these two genes are not known. However, the regulation of promoter site choice can be predicted to be involved. For the mouse α-amylase gene, a simple 5'-3' scanning model for RNA splicing would be consistent with the splicing patterns observed. This is not the case for myosin light chain transcripts where two internal exons are alternatively "skipped" in the LC_1 and LC_3 mRNAs. The observed pattern of alternative splicing for this gene predicts that secondary or tertiary structure of the pre-mRNA plays an important role in splice site selection (138, 139). This may be a common feature of other genes that produce mRNAs heterogenous at their 5' ends due to multiple start sites and differential splicing (see Table 1).

Genes Producing Differentially Spliced mRNAs that Have Common Ends

Several genes are known to use only alternative splicing to generate diversity in their mRNA products (Table 1, Figure 2C). This can occur by the selective inclusion or exclusion of a particular intron as in the case of the c-Kirstin-(Ki) *ras*2 protooncogene (Figure 2C; 142–144), the preprotachyinin gene (Figure 3B; 145), and the human fibronectin gene (Figure 4; 146). Other modes of splicing in this class include alternative removal of portions of exons to alter protein structure such as seen for rat γ-fibrinogen (Figure 2C; 147) and rat

A. TROPONIN mRNAs

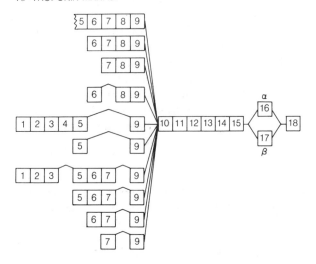

B. RAT FIBRONECTIN GENE- 3' EXONS

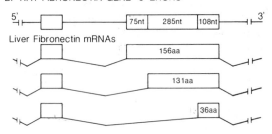

C. HUMAN FIBRONECTIN GENE- TYPE III HOMOLOGY REGION

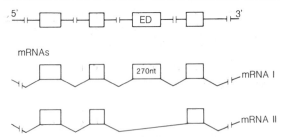

Figure 4 Alternatively spliced mRNAs produced by the troponin T(A), rat (B), and human (C) fibronectin genes. For human fibronectin mRNA I, inclusion of the 270-nucleotide extra domain (ED) adds a 90-amino-acid segment in a type III homology domain. (A) is adapted with permission from Ref. 151.

fibronectin (Figure 4*B*; 148, 149); and usage of alternative exons as seen for rat troponin T (Figure 4*A*; 150, 151).

Figure 2*C* shows the organization and splicing patterns for the c-Ki-*ras*2 gene. All *ras* genes including c-Ha-*ras*1 (Harvey ras) and N-*ras* encode 188–189-amino-acid p21 proteins (142–144, 152–155). c-Ki-*ras*2 encodes two major transcripts of 5.5 kb and 3.8 kb. The gene has five coding exons, although in 98% of the transcripts the fourth (IVa) exon is spliced out, and only the fifth (exon IVb) is used to translate the C-terminal portion of c-Ki-*ras* (144). Viral-Ki-*ras* produced by Murine Sarcoma Virus utilizes the fourth exon, and thus it encodes a p21 different from the c-Ki-*ras* in a portion of the carboxy terminus (142–144). mRNA polymorphism has been observed for at least one other member of the *ras* gene family, N-*ras* (154, 156, 157), although this appears to be due to alternative poly(A) site usage without a change in protein sequence (157).

Alternative RNA splicing appears to be the basis for the presence of a minor form (B) of γ-fibrinogen in both rat (147) and human (158, 159). The seventh intron in the rat γ-fibrinogen gene is not spliced out during the formation of γB-fibrinogen mRNA. These sequences in the seventh intron encode a unique 12-amino-acid carboxy terminus not found in rat γA-fibrinogen. The alternative carboxy terminus for γA-fibrinogen is 4 amino acids encoded by the eighth exon. A similar organization is seen for the human γ-fibrinogen gene except that the 3'-most intron sequences included in human γB-fibrinogen mRNA encode a poly(A) site that is distinct from that found in human γA-fibrinogen mRNA. If the mechanisms that regulate or dictate these alternative mRNA processing pathways are the same in both rat and human, then poly(A) site selection may not be the critical step in determining the splicing pathway for a given pre-mRNA. However, there is no evidence that alternative RNA processing of γ-fibrinogen transcripts is regulated developmentally. Although its functional significance is unknown, the fact that γB-fibrinogen is found across several species suggests that it may have a functional role in blood coagulation (158).

Like other complex transcription units, little is known about the mechanisms that regulate alternative pre-mRNA splicing. One may speculate that RNA secondary structure may play a role in determining the ratios of alternatively spliced mRNAs. For genes of this class that show tissue-specific or developmental regulation of their alternative mRNA products (see Table 1), both rates of transcription and levels of *trans*-acting factors could influence the formation of secondary structures specifying RNA splicing patterns.

Genes Producing mRNAs With 5' and/or 3' Terminal Heterogeneity Having Invariant Splicing

Several genes utilize only alternative sites of initiation or polyadenylation to generate multiple mRNAs (Table 1). In most of these cases, the multiple mRNAs encode identical proteins. One exception to this generality is the gene encoding yeast invertase (Figure 2D). Two sites of initiation can be used to generate either a secreted or cytoplasmic form of the enzyme (160, 161). Usage of the downstream start site produces a cytoplasmic form of invertase that is constitutively expressed (160, 161). Usage of the upstream start produces a secreted form of invertase that encodes a N-terminal signal peptide and that is subject to catabolite repression (160–162). These separate promoters appear to respond to a variety of cellular factors that permit their independent yet coordinate regulation (163).

An example of mRNA heterogeneity at the 3' end is shown in Figure 2D. As many as four mRNAs that differed in their 3' end were observed for the enzyme dihydrofolate reductase (DHFR) in mouse cells (164). In subsequent studies seven such mRNAs were observed (165), each having different lengths of 3' noncoding sequences. Further work has identified heterogenous cap sites for DHFR mRNAs at their 5' ends (166). It has not been determined whether specific 5' and 3' terminal sites are utilized coordinately, but the usage of different poly(A) sites appears to be regulated during growth (167).

What role might alternative flanking noncoding sequences have in regulating gene expression? Data from sequence analyses predict an important function. Examination of 3' untranslated sequences of isotypic mRNAs of different species shows a stronger than expected conservation of nucleotide sequences (168, 169) and exon length. This conservation was far greater than is seen between mRNAs for isoforms of proteins that are encoded by separate genes (169). Thus this conservation implies that 3' noncoding sequences are functionally important to gene expression. One possibility is that 3' noncoding sequences interact with specific factors that affect RNA transport or stability. In the case of the c-*fos* protooncogene, the transformation potential is inhibited by 3' noncoding sequences that appear to interact with the carboxy terminal coding sequences to reduce translational or transport efficiencies of the c-*fos* mRNA (169a). Indeed, such sequences that might mediate mRNA transport of stability may be differentially recognized during development or in different tissues.

ALTERNATIVE RNA PROCESSING: DEVELOPMENTAL AND BIOLOGICAL IMPLICATIONS

Neuroendocrine Peptides and Hormone Receptors

MULTIPLE PEPTIDE mRNAS ENCODED BY SINGLE GENES   Analysis of alternative mRNA products expressed by the calcitonin gene have predicted a novel family of neuropeptides (see 169b). Similar analysis has led to the discovery of an additional member of the tachykinin family as a constituent of the substance P gene. An outline of these events for both genes is shown in Figure 3. Nucleotide sequence analysis of cDNA clones from bovine brain determined that two separate mRNAs can encode substance P (170). These two mRNAs have common 5' and 3' sequences and differ by the inclusion of a region that encodes a peptide that shares considerable homology with substance P and an amphibian peptide kassinin. Based on the homology to kassinin (170), and a common immunoreactivity (171), this peptide was named substance K. The shorter mRNA encodes a 112-amino-acid precursor protein named α-preprotachykinin (α-PPT). This precursor contains an N-terminal signal sequence and an internal substance P sequence that is bounded by signals for excision and a carboxy terminal glycine required for amidation (Figure 3). The second mRNA encodes β-preprotachykinin (β-PPT), a 130-amino-acid precursor that differs from α-PPT only by the inclusion of sequences encoding an excisable amidated form of substance K at amino acids 98–107.

Subsequent analysis of the PPT gene has elucidated the role of alternative mRNA splicing in the generation of α- and β-PPT mRNAs (145). As shown in Figure 3, the PPT gene is comprised of seven exons; the third and sixth encode substance P and K, respectively. Exclusion of the sixth exon during mRNA splicing distinguishes α-PPT mRNA. While this event parallels the removal of the fourth (calcitonin-coding) exon in calcitonin/CGRP mRNA processing, PPT processing differs in that both α- and β-PPT mRNAs share the same 3' end poly(A) site. Thus, alternative RNA processing of PPT transcripts may be regulated entirely by splice site selection.

How might the regulation of mRNA expression for these two genes be related? Studies of the distribution of PPT and calcitonin/CGRP mRNAs and immunoreactivity indicate that they are expressed in distinct yet overlapping tissues and areas of the nervous system (132, 145, 169b, 172, 173). Analysis of the mRNA or peptide products expressed from these two genes in neurons of sensory ganglia and thyroid C-cells shows that α-PPT and CGRP mRNAs predominate over β-PPT and calcitonin mRNA levels in sensory neurons, while the reverse relationship is found in thyroid C-cells (132, 145). Thus, it is tempting to speculate that common factors or mechanisms may operate on transcripts from these two genes in thyroid C-cells to include the substance K and calcitonin-coding exons, while in sensory neurons mRNA processing favors the exclusion of these two exons and the greater production of α-PPT and CGRP mRNAs (Figure 3). However, it cannot be ruled out that cell- and sequence-specific mRNA stability mechanisms may be operating on these gene products to contribute to the selective and alternative pattern of mRNA expression that is observed.

Genes for several other endocrine peptides utilize alternative RNA processing events to generate diversity in their peptide products. In some cases these processing events appear to be stochastic, and alternative minor forms of the peptides, differing by only a few amino acids, are generated. Thus, use of an alternative downstream splice acceptor in human growth hormone transcripts leads to a 15-amino-acid deletion in about 15% of the growth hormone molecules contained in pituitary (174, 175). Analyses of rat prolactin cDNA and genomic DNA sequences reveal the presence of a second 3' splice acceptor for the second prolactin exon leading to an insertion or deletion of a single alanine residue in the N-terminal signal sequence (176).

Of potentially greater functional importance is the expression of multiple protein precursors from the bradykinin gene. This important vasoactive peptide is excised from both a high- and low-molecular-weight form of prekininogen. Analysis of the 3' end of this gene reveals the alternative splicing and polyadenylation pathways that generate these two forms of kininogen (177). The high-molecular-weight prekininogen mRNAs appear to use an upstream poly(A) site and coding sequences to generate a larger carboxy terminal light chain that may contribute to the production of a cofactor for contact activation of intrinsic coagulation and fibrinolysis (177). Low-molecular-weight prekininogen mRNAs appear to alternatively splice to a small downstream exon that uses a separate poly(A) site and that does not encode such a function.

HORMONE RECEPTORS   The recent elucidation of the structure of several peptide and steroid hormone receptors by recombinant DNA technology has uncovered another important functional class of complex transcription units. Initial studies of the receptor for the T-cell growth factor, interleukin-2 (IL-2), have identified the presence of two major sizes of mRNA (1.5 and 3.5 kb) that appear to arise from a single human gene (178, 179). More detailed analysis of the IL-2 receptor gene and its cDNAs to its messages revealed the presence of two poly(A) sites and an alternatively spliced exon such that four mRNAs can possibly arise from IL-2 receptor transcripts (179). Exclusion of the apparently alternatively spliced exon (216 nucleotides) yields an mRNA that appears to encode an altered membrane protein that does not bind IL-2.

mRNA heterogeneity has also been seen for the human insulin (180) and epidermal growth factor (EGF) receptors (181–183). In normal fibroblasts, two major human EGF receptor mRNAs appear having sizes of roughly 6 and 10 kb in length. Subsequent analyses of the 5' portion of the gene indicate that all of the mRNAs found in human A431 carcinoma cells initiate at one of several sites between 107 and 258 nucleotides upstream from the predicted translation start site (184). This microheterogeneity at the 5' end would not account for the size differences observed for the heterogenous mRNA. Other studies suggest that only a single major polypeptide is translated in *Xenopus* oocytes from the high-molecular-weight EGF receptor mRNAs, although the presence of both high-molecular-weight mRNAs was not proven (185). Thus it may be predicted that the two major EGF receptor mRNAs show a divergence in their 3' noncoding sequences.

Recent experiments suggest that alternative mRNA processing is operative in regulating the expression of isoforms of the human glucocorticoid receptor (185a). Both alternative splicing and poly(A) site selection appear to direct synthesis of mRNAs encoding a major 94-kd glucocorticoid receptor as well as a 90-kd variant that does not bind glucocorticoids in vitro. Additional analysis should elucidate the functional identity and regulation of these interesting variants.

Proteins of Muscle and Components of the Extracellular Matrix

Studies of the constituent proteins of muscle have elucidated the widespread existence of isoforms encoded by multigene families whose expression is tissue-specifically and developmentally regulated at the level of transcription (186). In addition, alternative RNA processing has been observed to contribute to this diversity by directing the production of multiple products from the genes for myosin light chain (138, 139, 187, 188); myosin heavy chain (189); tropomyosin (190–192); and troponin T (150, 151, 193). Troponin T serves as a remarkable example of alternative RNA processing generating extreme heterogeneity (Figure 4A) (151). Based on genomic and cDNA analyses, 18 small exons are alternatively spliced in an intricate pattern to generate a minimum of 10 and as many as 64 unique mRNAs. This heterogeneity is consistent with earlier observations of antigenically diverse isoforms of troponin T that appear in different types of muscle and at different stages of development (194–200). The regulation of these alternative splicing events remains unknown, but it can be predicted that *trans*-acting factors are involved (151).

Consistent with the diverse functions performed by plasma and extracellular matrix-associated fibronectins, this gene also utilizes RNA processing events to generate polymorphism (146, 148, 149, 201). The genomic structure and splicing pathways for this large gene have been only partially characterized in rat (148, 149) and human (146, 201), and at least two regions involving alternative mRNA processing have been identified (Figure 4B, C). cDNAs synthesized from rat liver RNA predicted three variants that differ by the inclusion of sequences encoding 95 or 120 extra amino acids 276 residues internal from the carboxy terminus. A separate region of polymorphism was found for human fibronectin mRNAs, whereby a 270-bp in frame internal fragment (named ED, extra domain) was found included in a proportion of cDNAs cloned from a human cell line. Based on observations that most human liver mRNAs lack this ED domain, and that the predicted amino acid sequence for this mRNA form (mRNAII) better matches the sequence of a peptide fragment of bovine plasma fibronectin, Kornblihtt et al (146, 201) suggest that ED-containing fibronectins are a cellular form and plasma fibronectin subunits lack the ED sequence. Future studies elucidating the structure of other fibronectin mRNAs can be expected to aid in the understanding of the molecular basis to the versatile functions and behavior of different fibronectins.

Other Transcription Units

While a complete detailed survey of all complex transcription units known to date would be impractical, several additional examples of such cellular genes can be found in Table 1. Other examples that are too numerous to cover in this review are complex transcription units found in a large number of animal viruses. Complex processing pathways have been observed for several transcription units in adenoviruses, SV40, polyoma and BK viruses, parvoviruses, herpes viruses, influenza viruses, RNA tumor viruses, and probably papilloma viruses (for reviews see 3, 107, 202–207). In fact, complex processing of viral transcripts appears to be the usual biological event rather than the exception. Thus, it may be hypothesized that alternative patterns of mRNA processing in viruses are used to derive greater functional capability out of a genome of limited capacity. The biological consequences and regulation of these patterns of RNA processing during viral infection include enabling and regulating the usage of overlapping open reading frames, affecting the efficiency of translation from particular AUG codons, assembling varied and alternative combinations of reading frames, and regulating the program of viral gene expression (for review see 202).

The importance of regulated RNA processing events in development has recently been reinforced by studies of developmentally regulated loci in *Drosophila* and other metazoans. Thus, multiple overlapping mRNA products or cDNAs have been attributed to the Antennapedia (208, 209) and Bithorax (210) loci in *Drosophila*. Remarkably, Hogness and colleagues suggest that a portion of this domain, *bxd,* encodes transcripts that act to *cis*-modulate alternative splicing pathways found for transcripts of the Ultra-bithorax *(Ubx)* domain of the Bithorax-complex (210). mRNA polymorphism that is developmentally modulated also arises from a locus in the *Xenopus* laevis genome that shares homology to the Antennapedia locus of *Drosophila* (211). It is tempting to speculate that recent observations of the different expression of U1 snRNA isoforms during early *Xenopus* development (212) may play some role in generating such developmentally regulated patterns of alternatively processed mRNAs.

Studies of transposable elements from yeast and *Drosophila* suggest that some of these elements exhibit characteristics similar to those of RNA retroviruses and therefore may utilize heterogenous RNA products to mediate their activities as transposons (see 213–216). Indeed the identification of multiple mRNAs from *copia* elements of *Drosophila* (217) and the detection of sequence homology between coding regions of avian retroviruses and several *copia*-like transposable elements (216) suggest that alternative RNA processing events like those found to be important in retroviruses may also be important to the function of these elements. A more intriguing example of regulated RNA processing events playing a role in *Drosophila* recombination comes from recent experiments with transposable elements conducted by G. Rubin and collaborators. These elements, which appear to mediate their own transposition (218) solely in germ-line tissue, can encode multiple apparently overlapping transcripts (219). Rubin and colleagues find that while transcription of P-elements occurs in both somatic and germ-line tissues, correct splicing of P-element transcripts to support transposition of the element is restricted to the germ line (219a). This system should allow the use of a genetic approach to identifying putative factors that regulate particular splicing events during development or in specific tissues.

CONCLUDING REMARKS

Both the general paradigms that operate in alternative RNA processing and specific examples have been reviewed in this chapter. Some of these examples appear to involve stochastic processes while others identify processes that appear highly regulated. Regardless, the mechanisms governing alternative RNA processing remain poorly understood even though considerable information has been gathered about the complex biochemical mechanisms of each individual step of mRNA formation. For transcription termination, polyadenylation and splicing *cis*-acting signal sequences have been identified. However, the considerable degeneracy found in these consensus sequences in metazoans suggests that the factors involved in the recognition of those sequences may exist as multiclass families, and they may utilize differential affinities for these signals to regulate these complex processing events. Indeed, U1 snRNA polymorphism has been identified (212) reflecting one potential level at which splicing diversity may be regulated. The involvement of factors in the recognition of different promoters is well recognized (see 28) and similarly, tissue-specific differences in the proteins that contribute to hnRNP particle formation can be expected (see 92).

It is not yet clear whether individual RNA processing events are regulated independently or are affected by distal processing events. If poly(A) site selection occurs as an independently regulated event, two general models may be considered. The first model states that poly(A) site selection is regulated by varying the intracellular levels of a poly(A) site-specific endonuclease (118). Such endonucleases would either recognize different sites with different affinities or select promoter-proximal over promoter-distal sites (128). It is unclear, however, whether such a nonspecific mechanism could operate to make such clear poly(A) site choices as those seen for the calcitonin/CGRP gene in brain versus thyroid (132). An alternative model favors the use of sequence-specific factors that may enhance or block the usage of a particular poly(A) site. This model, however, relies on either the existence of a separate factor for each gene or the proposal that poly(A) sites for different genes can be regulated coordinately. No specific data favoring this model have yet been presented.

Independent regulation of alternative splicing pathways can be hypothesized to involve either factors that affect the folding of the pre mRNA to present particular donor-acceptor pairs in proximity, or factors that block or favor the usage of a particular splicing junction. An extension of this model suggests that the overall processing of an individual transcript involves coordinate and simultaneous events. While data exist that favor the general procession of splicing to occur 5' to 3' for β-globin transcripts (81, 219b), more complex transcripts containing as many as eight introns, such as those for the chicken ovalbumin and ovomucoid gene, appear to show a different preferred order of splicing that is not processive (220). Since new transcripts that are to be processed rapidly become part of hnRNPs in the nucleus, it may be considered that each RNA adopts a preferred conformation that depends upon its sequence and the RNP-associated proteins (see 92). Thus the folding of the RNA in this RNP structure may determine the particular sequence of splicing and/or polyadenylation events, and therefore tissue- or developmentally specific patterns of RNA processing events could be determined by the presence of different RNP protein constituents in different tissues. A critical challenge remains to determine if such mechanisms occur to regulate complex RNA processing events and whether specific factors act on transcripts from different genes to coordinatedly regulate gene expression from some complex transcription units during development.

ACKNOWLEDGMENTS

We thank Dr. Andrew Russo for discussion and Marijke ter Horst for excellent secretarial assistance. S. Leff is a recipient of an American Cancer Society Postdoctoral Fellowship. Research was supported by grants from the American Cancer Society and NIH.

Literature Cited

1. Padgett, R. A., Grabowski, P. J., Konarska, M. M., Seiler, S., Sharp, P. A. 1986. *Ann. Rev. Biochem.* 55:1119–50
2. Nevins, J. R. 1983. *Ann. Rev. Biochem.* 52:441–66
3. Darnell, J. E. Jr. 1982. *Nature* 297:365–71
4. Sharp, P. A. 1981. *Cell* 23:643–46
5. Birnstiel, M. L., Busslinger, M., Strub, K. 1985. *Cell* 41:349–59
5a. Flint, S. J. 1984. In *Processing of RNA*, ed. D. Apirion, pp. 151–80. Boca Raton, Fla: CRC

5b. Nevins, J. R. 1984. See Ref. 5a, pp. 133–50
6. Keller, W. 1984. *Cell* 39:423–25
7. Bachenheimer, S. L., Darnell, J. E. 1975. *Proc. Natl. Acad. Sci. USA* 72:4445–49
8. Goldberg, S., Weber, J., Darnell, J. E. 1977. *Cell* 10:617–21
9. Weber, J., Jelinek, W., Darnell, J. E. 1977. *Cell* 10:611–16
10. Evans, R. M., Fraser, N. W., Ziff, E., Weber, J., Wilson, M., Darnell, J. E. 1977. *Cell* 12:733–39

11. Wilson, M. C., Fraser, N. W., Darnell, J. E. 1979. *Virology* 94:175–84
12. Nevins, J. R., Blanchard, J.-M., Darnell, J. E. 1980. *J. Mol. Biol.* 144:377–86
13. Berk, A. J., Sharp, P. A. 1977. *Cell* 12:45–55
14. Fraser, N. W., Sehgal, P. B., Darnell, J. E. 1978. *Nature* 272:590–93
15. Ford, J. P., Hsu, M.-T. 1978. *J. Virol.* 28:795–801
16. Ziff, E. B., Evans, R. M. 1978. *Cell* 15:1463–76
17. Chambon, P., Dierich, A., Gaub, M. P., Jackowlev, S., Jongstra, J., et al. 1984. *Recent Prog. Hormone Res.* 40:1–39
18. Baker, C. C., Herisse, J., Courtois, G., Galibert, F., Ziff, E. B. 1979. *Cell* 18:569–80
19. Reddy, V., Thimmappaya, B., Dhar, R., Subramanian, K., Zain, S., et al. 1978. *Science* 200:494–500
20. Soeda, E., Arrand, J. R., Smolar, N., Walsh, J. E., Griffin, B. E. 1980. *Nature* 283:445–53
21. McKnight, S. L., Kingsbury, R. 1982. *Science* 217:316–24
22. Melton, D. W., Konecki, D. S., Brennand, J., Caskey, C. T. 1984. *Proc. Natl. Acad. Sci. USA* 81:2147–51
23. Reynolds, G. A., Basu, S. K., Osborne, T. F., Chin, D. J., Gil, G., et al. 1984. *Cell* 38:275–85
24. Breathnach, R., Chambon, P. 1981. *Ann. Rev. Biochem.* 50:349–83
25. Shenk, T. 1981. *Curr. Top. Immunol.* 93:25–46
26. Weil, P. A., Luse, D. S., Segall, J., Roeder, R. G. 1979. *Cell* 18:469–84
27. Manley, J. L., Fire, A., Cano, A., Sharp, P. A., Gefter, M. L. 1980. *Proc. Natl. Acad. Sci. USA* 77:3855–59
28. Dynan, W. S., Tjian, R. 1985. *Nature* 316:774–78
29. Nevins, J. R., Darnell, J. E. 1978. *Cell* 15:1477–93
30. Hofer, E., Darnell, J. E. 1981. *Cell* 23:585–93
31. Hofer, E., Hofer-Warbinek, R., Darnell, J. E. 1982. *Cell* 29:887–93
32. LeMeur, M. A., Galliot, B., Gerlinger, P. 1984. *EMBO. J.* 3:2779–86
33. Hagenbuchle, O., Wellauer, P. K., Cribbs, D. L., Schibler, U. 1984. *Cell* 38:737–44
34. Amara, S. G., Evans, R. M., Rosenfeld, M. G. 1984. *Mol. Cell. Biol.* 4:2151–60
35. Sheffery, M., Marks, P. A., Rifkind, R. A. 1984. *J. Mol. Biol.* 172:417–36
36. Weintraub, H., Larsen, A., Groudine, M. 1981. *Cell* 24:333–44
37. Mather, E. L., Nelson, K. J., Haimovich, J., Perry, R. P. 1984. *Cell* 36:329–38
38. Frayne, E. G., Leys, E. J., Crouse, G. F., Hook, A. G., Kellems, R. E. 1984. *Mol. Cell. Biol.* 4:2921–24
39. Citron, B., Falck-Pedersen, E., Salditt-Georgieff, M., Darnell, J. E. 1984. *Nucleic Acids Res.* 12:8723–31
40. Falck-Pedersen, E., Logan, J., Shenk, T., Darnell, J. E. Jr. 1985. *Cell* 40:897–905
41. Proudfoot, N. J., Brownlee, G. G. 1976. *Nature* 263:211–14
42. Wickens, M., Stephenson, P. 1984. *Science* 226:1045–51
43. Simonsen, C. L., Levinson, A. D. 1983. *Mol. Cell. Biol.* 3:2250–58
44. Jung, A., Sippel, A. E., Grez, M., Schutz, G. 1980. *Proc. Natl. Acad. Sci. USA* 77:5759–63
45. Hagenbuchle, O., Tosi, M., Schibler, U., Bovey, R., Wellauer, P. K., et al. 1981. *Nature* 289:643–46
46. Mason, P. J., Jones, M. B., Ellington, J. A., Williams, J. G. 1985. *EMBO J.* 4:205–11
47. Fitzgerald, M., Shenk, T. 1981. *Cell* 24:251–60
48. Montell, C., Fisher, E. F., Caruthers, M. H., Berk, A. J. 1983. *Nature* 305:600–5
49. McDevitt, M. A., Imperiale, M. J., Ali, H., Nevins, J. R. 1984. *Cell* 37:993–99
50. Gil, A., Proudfoot, N. J. 1984. *Nature* 312:473–74
51. Sadofsky, M., Alwine, J. C. 1984. *Mol. Cell. Biol.* 4:1460–68
52. Woychik, R. P., Lyons, R. H., Post, L., Rottman, F. M. 1984. *Proc. Natl. Acad. Sci. USA* 81:3944–48
53. Berget, S. M. 1984. *Nature* 309:179–81
54. Lai, E. C., Stein, J. P., Catterall, J. F., Woo, S. L. C., Mace, M. L., et al. 1979. *Cell* 18:829–42
55. McLaughlan, J., Gaffney, D., Whitton, J. L., Clements, J. B. 1985. *Nucleic Acids Res.* 13:1347–68
56. Stunnenberg, H. G., Birnstiel, M. L. 1982. *Proc. Natl. Acad. Sci. USA* 79:6201–4
57. Galli, G., Hofstetter, H., Stunnenberg, H. G., Birnstiel, M. L. 1983. *Cell* 34:823–28
58. Strub, K., Galli, G., Busslinger, M., Birnstiel, M. L. 1984. *EMBO J.* 3:2801–7
59. Acheson, N. H. 1984. *Mol. Cell. Biol.* 4:722–29
59a. Salditt-Georgieff, M., Harpold, M., Sawicki, S., Nevins, J., Darnell, J. E. Jr. 1980. *J. Cell Biol.* 86:844–48
60. Lai, C. J., Dhar, R., Khoury, G. 1978. *Cell* 14:251–61
60a. Craig, E. A., Raskas, H. J. 1976. *Cell* 8:205–13

60b. Goldenberg, C., Raskas, H. J. 1979. *Cell* 16:131–38
61. Nevins, J. R. 1979. *J. Mol. Biol.* 130:493–506
61a. Weber, J., Blanchard, J.-M., Ginsberg, H., Darnell, J. E. Jr. 1980. *J. Virol.* 33:286–91
61b. Berget, S. M., Sharp, P. A. 1979. *J. Mol. Biol.* 129:547–65
62. Ross, J., Knecht, D. A. 1978. *J. Mol. Biol.* 119:1–20
62a. Schibler, U., Marku, K., Perry, R. P. 1979. *Cell* 15:1495–509
62b. Gilmore-Hebert, M., Wall, R. 1979. *J. Mol. Biol.* 135:879–91
62c. Harpold, M. M., Dobner, P. R., Evans, R., Bancroft, F. C., Darnell, J. E. Jr. 1979. *Nucleic Acids Res.* 6:3133–44
62d. Roop, D. R., Nordstrom, J. L., Tsai, S. Y., Tsai, M.-J., O'Malley, B. W. 1978. *Cell* 15:671–85
63. Zeevi, M., Nevins, J. R., Darnell, J. E. 1981. *Cell* 26:39–46
64. Green, M. R., Maniatis, T., Melton, D. A. 1983. *Cell* 32:681–94
65. Beyer, A., Osheim, Y. N. 1985. *J. Cell. Biochem.* Suppl. 9A:18
66. Malcolm, D. B., Sommerville, J. 1974. *Chromosoma* 48:137–58
67. McKnight, S. L., Miller, O. L. Jr. 1976. *Cell* 8:305–19
68. Patton, J. R., Ross, D. A., Chae, C. B. 1985. *Mol. Cell. Biol.* 5:1220–28
69. Berget, S. M., Moore, C., Sharp, P. A. 1977. *Proc. Natl. Acad. Sci. USA* 74:3171–75
70. Klessig, D. F. 1977. *Cell* 12:9–21
71. Chow, L. T., Gelinas, R. E., Broker, T. R., Roberts, R. J. 1977. *Cell* 12:1–8
72. Jeffreys, A. J., Flavell, R. A. 1977. *Cell* 12:1097–108
73. Breathnach, R., Benoist, C., O'Hare, K., Gannon, F., Chambon, P. 1978. *Proc. Natl. Acad. Sci. USA* 75:4853–57
74. Seif, I., Khoury, G., Dhar, R. 1979. *Nucleic Acids Res.* 6:3387–89
75. Rogers, J., Wall, R. 1980. *Proc. Natl. Acad. Sci. USA* 77:1877–79
76. Lerner, M. R., Boyle, J. A., Mount, S. M., Wolin, S. L., Steitz, J. A. 1980. *Nature* 283:220–24
77. Mount, S. M. 1982. *Nucleic Acids Res.* 10:459–72
78. Treisman, R., Proudfoot, N. J., Shander, M., Maniatis, T. 1982. *Cell* 29:903–11
79. Wieringa, B., Meyer, F., Reiser, J., Weissmann, C. 1983. *Nature* 301:38–43
80. Chu, G., Sharp, P. A. 1981. *Nature* 289:378–82
81. Lang, K. M., Spritz, R. A. 1983. *Science* 220:1351–55
82. Kuhne, T., Wieringa, B., Reiser, J., Weissmann, C. 1983. *EMBO J.* 2:727–33
83. Solnick, D. 1985. *Cell* 42:157–64
84. Konarska, M. M., Padgett, R. A., Sharp, P. A. 1985. *Cell* 42:165–71
85. Padgett, R. A., Mount, S. M., Steitz, J. A., Sharp, P. A. 1983. *Cell* 35:101–7
86. Kramer, A., Keller, W., Appel, B., Luhrmann, R. 1984. *Cell* 38:299–307
87. Fradin, A., Jove, R., Hemenway, C., Keiser, H. D., Manley, J. L., et al. 1984. *Cell* 37:927–36
88. Brody, E., Abelson, J. 1985. *Science* 228:963–67
89. Frendewey, D., Keller, W. 1985. *Cell* 42:355–67
90. Grabowski, P. J., Seiler, S. R., Sharp, P. A. 1985. *Cell* 42:345–53
91. Keller, E. B., Noon, W. A. 1984. *Proc. Natl. Acad. Sci. USA* 81:7417–20
92. Pederson, T. 1983. *J. Cell. Biol.* 97:1321–26
93. Ruskin, B., Krainer, A. R., Maniatis, T., Green, M. R. 1984. *Cell* 38:317–31
94. Padgett, R. A., Konarska, M. M., Grabowski, P. J., Hardy, S. F., Sharp, P. A. 1984. *Science* 225:898–903
95. Grabowski, P. J., Padgett, R. A., Sharp, P. A. 1984. *Cell* 37:415–27
96. Pikielny, C. W., Teem, J. L., Rosbash, M. 1983. *Cell* 34:395–403
97. Rodriguez, J. R., Pikielny, C. W., Rosbash, M. 1984. *Cell* 39:603–10
98. Domdey, H., Apostol, B., Lin, R. J., Newman, A., Brody, E., Abelson, J. 1984. *Cell* 39:611–21
99. Zeitlin, S., Efstratiadis, A. 1984. *Cell* 39:589–602
100. Munroe, S. H. 1984. *Nucleic Acids Res.* 12:8437–56
101. Rautmann, G., Matthes, H. W. D., Gait, M. J., Breathnach, R. 1984. *EMBO J.* 3:2021–28
102. Somasekhar, M. B., Mertz, J. E. 1985. *Nucleic Acids Res.* 13:5591–609
102a. Garriga, G., Bertrand, H., Lambowitz, A. M. 1984. *Cell* 36:623–34
103. Wieringa, B., Hofer, E., Weissmann, C. 1984. *Cell* 37:915–25
104. Wilson, M. C., Darnell, J. E. 1981. *J. Mol. Biol.* 148:231–51
105. Ziff, E., Fraser, N. W. 1978. *J. Virol.* 25:897–906
106. McGrogan, M., Raskas, H. J. 1978. *Proc. Natl. Acad. Sci. USA* 75:625–29
107. Ziff, E. B. 1980. *Nature* 287:491–99
108. Fraser, N., Ziff, E. 1978. *J. Mol. Biol.* 124:27–51
109. Nevins, J. R., Darnell, J. E. 1978. *J. Virol.* 25:811–23
110. Chow, L. T., Broker, T. R., Lewis, J. B. 1979. *J. Mol. Biol.* 134:265–303

111. Shaw, A. R., Ziff, E. B. 1980. *Cell* 22:905–16
112. Nevins, J. R., Wilson, M. C. 1981. *Nature* 290:113–18
113. Anderson, C. W., Lewis, J. B., Atkins, J. F., Gesteland, R. F. 1974. *Proc. Natl. Acad. Sci. USA* 71:2756–60
114. Honjo, T., Kataoka, T. 1978. *Proc. Natl. Acad. Sci. USA* 75:2140–44
115. Cory, S., Adams, J. M. 1980. *Cell* 19:37–51
116. Coleclough, C., Cooper, D., Perry, R. P. 1980. *Proc. Natl. Acad. Sci. USA* 77:1422–26
117. Joho, R., Nottenburg, C., Coffman, R. L., Weissman, I. L. 1983. *Curr. Top. Dev. Biol.* 18:16–58
118. Blattner, F. R., Tucker, P. W. 1984. *Nature* 307:417–22
119. Melchers, F., Anderson, J. 1974. *Eur. J. Immunol.* 4:181–88
120. Perry, R. P., Kelley, D. E. 1979. *Cell* 18:1333–39
121. Alt, F. W., Bothwell, A. L. M., Knapp, M., Siden, E., Mather, E., et al. 1980. *Cell* 20:293–301
122. Rogers, J., Early, P., Carter, C., Calame, K., Bond, M., et al. 1980. *Cell* 20:303–12
123. Early, P., Rogers, J., Davis, M., Calame, K., Bond, M., et al. 1980. *Cell* 20:313–19
124. Maki, R., Roeder, W., Traunecker, A., Sidman, C., Wabl, M., et al. 1981. *Cell* 24:353–65
125. Moore, K. W., Rogers, J., Hunkapiller, T., Early, P., Nottenburg, C., et al. 1981. *Proc. Natl. Acad. Sci. USA* 78:1800–4
126. Liu, C. P., Tucker, P. W., Mushinski, J. F., Blattner, F. R. 1980. *Science.* 209:1348–53
127. Yaoita, Y., Kumagai, Y., Okumura, K., Honjo, T. 1982. *Nature* 297:697–99
128. Nishikura, K., Vuocolo, G. A. 1984. *EMBO J.* 3:689–99
129. Rosenfeld, M. G., Amara, S. G., Roos, B. A., Ong, E. S., Evans, R. M. 1981. *Nature* 290:63–65
130. Rosenfeld, M. G., Lin, C. R., Amara, S. G., Stolarsky, L., Roos, B. A., et al. 1982. *Proc. Natl. Acad. Sci. USA* 79:1717–21
131. Amara, S. G., Jonas, V., Rosenfeld, M. G., Ong, E. S., Evans, R. M. 1982. *Nature* 298:240–44
132. Rosenfeld, M. G., Mermod, J. J., Amara, S. G., Swanson, L. W., Sawchenko, P. E., et al. 1983. *Nature* 304:129–33
133. Sabate, M. I., Stolarsky, L. S., Polak, J. M., Bloom, S. R., Varndell, I. M., et al. 1985. *J. Biol. Chem.* 260:2589–92
134. Jonas, V., Lin, C. R., Kawashima, E., Semon, D., Swanson, L. W., et al. 1985. *Proc. Natl. Acad. Sci. USA* 82:1994–98
135. Steenberg, P. H., Hoppener, J. W. M., Zandberg, J., Lips, C. J. M., Jansz, H. S. 1985. *FEBS Lett.* 183:403–7
136. Deleted in proof
137. Leff, S. E., Amara, S. G., Nelson, C., Evans, R. M., Rosenfeld, M. G. 1985. *J. Cell. Biochem.* Suppl. 9A:22
138. Nabeshima, Y., Fujii-Kuriyama, Y., Muramatsu, M., Ogata, K. 1984. *Nature* 308:333–38
139. Robert, B., Daubas, P., Akimenko, M. A., Cohen, A., Garner, I., et al. 1984. *Cell* 39:129–40
140. Young, R. A., Hagenbuchle, O., Schibler, U. 1981. *Cell* 23:451–58
141. Tosi, M., Young, R. A., Hagenbuchle, O., Schibler, U. 1981. *Nucleic Acids Res.* 9:2313–23
142. McGrath, J. P., Capon, D. J., Smith, D. H., Chen, E. Y., Seeburg, P. H., et al. 1983. *Nature* 304:501–7
143. Shimizu, K., Birnbaum, D., Ruley, M. A., Fasano, O., Suard, Y., et al. 1983. *Nature* 304:497–500
144. Capon, D. J., Seeburg, P. H., McGrath, J. P., Hayflick, J. S., Edman, V., et al. 1983. *Nature* 304:507–12
145. Nawa, H., Kotani, H., Nakanishi, S. 1984. *Nature* 312:729–34
146. Kornblihtt, A. R., Vibe-Pedersen, K., Baralle, F. E. 1984. *EMBO J.* 3:221–26
147. Crabtree, G. R., Kant, J. A. 1982. *Cell* 31:159–66
148. Schwarzbauer, J. E., Tamkun, J. W., Lemischka, I. R., Hynes, R. O. 1983. *Cell* 35:421–31
149. Tamkun, J. W., Schwarzbauer, J. E., Hynes, R. O. 1984. *Proc. Natl. Acad. Sci. USA* 81:5140–44
150. Medford, R. M., Nguyen, H. T., Destree, A. T., Summers, E., Nadal-Ginard, B. 1984. *Cell* 38:409–21
151. Breitbart, R. E., Nguyen, H. T., Medford, R. M., Destree, A. T., Mahdavi, V., et al. 1985. *Cell* 41:67–82
152. Chang, E. H., Gonda, M. A., Ellis, R. W., Scolnick, E. M., Lowy, D. R. 1982. *Proc. Natl. Acad. Sci. USA* 79:4848–52
153. Capon, D. J., Chen, E. Y., Levinson, A. D., Seeburg, P. H., Goeddel, D. V. 1983. *Nature* 302:33–37
154. Hall, A., Marshall, C. J., Spurr, N., Weiss, R. A. 1983. *Nature* 303:396–400
155. Taparowsky, E., Shimizu, K., Goldfarb, M., Wigler, M. 1983. *Cell* 34:581–86
156. Murray, M. J., Cunningham, J. M., Parada, L. F., Pautry, F., Lebowitz, P., et al. 1983. *Cell* 33:749–57
157. Hall, A., Brown, R. 1985. *Nucleic Acids Res.* 13:5255–68

158. Fornace, A. J., Cummings, D. E., Comeau, C. M., Kant, J. A., Crabtree, G. R. 1984. *J. Biol. Chem.* 259:12826–30

159. Chung, D. W., Davie, E. W. 1984. *Biochemistry* 23:4232–36

160. Perlman, D., Halvorson, H. O., Cannon, L. E. 1982. *Proc. Natl. Acad. Sci. USA* 79:781–85

161. Carlson, M., Botstein, D. 1982. *Cell* 28:145–54

162. Perlman, D., Halvorson, H. O. 1981. *Cell* 25:525–36

163. Perlman, D., Raney, P., Halvorson, H. O. 1984. *Mol. Cell. Biol.* 4:1682–88

164. Setzer, D. R., McGrogan, M., Nunberg, J. H., Schimke, R. T. 1980. *Cell* 22:361–70

165. Setzer, D. R., McGrogan, M., Schimke, R. T. 1982. *J. Biol. Chem.* 257:5143–47

166. McGrogan, M., Simonsen, C. C., Smouse, D. T., Farnham, P. J., Schimke, R. T. 1985. *J. Biol. Chem.* 260:2307–14

167. Kaufman, R. J., Sharp, P. A. 1983. *Mol. Cell. Biol.* 3:1598–608

168. Miyata, T., Yasunaga, T., Nishida, T. 1980. *Proc. Natl. Acad. Sci. USA* 77:7328–32

169. Yaffe, D., Nudel, U., Mayer, Y., Neuman, S. 1985. *Nucleic Acids Res.* 13:3723–37

169a. Miller, A. D., Curran, T., Verma, I. M. 1984. *Cell* 36:51–60

169b. Rosenfeld, M. G., Amara, S. G., Evans, R. M. 1984. *Science* 225:1315–20

170. Nawa, H., Hirose, T., Takashima, H., Inayama, S., Nakanishi, S. 1983. *Nature* 306:32–36

171. Maggio, J. E., Sandberg, B. E. B., Bradley, C. V., Iversen, L. L., Santikarn, S., et al. 1983. *Br. J. Med. Sci.* 152(Suppl. 1):20–21

172. Gibson, S. J., Polak, J. M., Bloom, S. R., Sabate, I. M., Mulderry, P. M., et al. 1984. *J. Neurosci.* 4:3101–11

173. Mulderry, P. K., Ghatei, M. A., Rodrigo, J., Allen, J. M., Rosenfeld, M. G., et al. 1985. *Neuroscience* 14:947–54

174. Wallis, M. 1980. *Nature* 284:512

175. DeNoto, F. M., Moore, D. D., Goodman, H. M. 1981. *Nucleic Acids Res.* 9:3719–30

176. Maurer, R. A., Erwin, C. R., Donelson, J. E. 1981. *J. Biol. Chem.* 256:10524–28

177. Kitamura, N., Takagaki, Y., Furuto, S., Tanaka, T., Nawa, H., et al. 1983. *Nature* 305:545–49

178. Nikaido, T., Shimizu, A., Ishida, N., Sabe, H., Teshigawara, K., et al. 1984. *Nature* 311:631–35

179. Leonard, W. J., Depper, J. M., Crabtree, G. R., Rudikoff, S., Pumphrey, J., et al. 1984. *Nature* 311:626–31

180. Ullrich, A., Bell, J. R., Chen, E. Y., Herrera, R., Petruzzelli, L. M., et al. 1985. *Nature* 313:756–61

181. Lin, C. R., Chen, W. S., Kruijer, W., Stolarsky, L. S., Weber, W., et al. 1984. *Science* 224:843–48

182. Ullrich, A., Coussens, L., Hayflick, J. S., Dull, T. J., Gray, A., et al. 1984. *Nature* 309:418–25

183. Xu, Y., Ishii, S., Clark, A. J. L., Sullivan, M., Wilson, R. K., et al. 1984. *Nature* 309:806–10

184. Ishii, S., Xu, Y., Stratton, R. H., Roe, B. A., Merlino, G. T., et al. 1985. *Proc. Natl. Acad. Sci. USA* 82:4920–24

185. Simmen, F. A., Gope, M. L., Schultz, T. Z., Wright, D. A., Carpenter, G., et al. 1984. *Biochem. Biophys. Res. Commun.* 124:125–32

185a. Hollenberg, S. M., Weinberger, C., Ong, E. S., Cerelli, G., Oro, A., et al. 1985. *Nature* 318:635–41

186. Pearson, M. L., Epstein, H. F., eds. 1982. *Molecular and Cellular Control of Muscle Development.* Cold Spring Harbor, NY: Cold Spring Harbor Lab.

187. Periasamy, M., Strehler, E. E., Garfinkel, L. I., Gubits, R. M., Ruiz-Opazo, N., et al. 1984. *J. Biol. Chem.* 259: 13595–604

188. Falkenthal, S., Parker, V. P., Davidson, N. 1985. *Proc. Natl. Acad. Sci. USA* 82:449–53

189. Rozek, C. E., Davidson, N. 1983. *Cell* 32:23–34

190. Basi, G. S., Boardman, M., Storti, R. V. 1984. *Mol. Cell. Biol.* 4:2828–31

191. Karlik, C. C., Mahaffey, J. W., Coutu, M. D., Fyrberg, E. A. 1984. *Cell* 37:469–81

192. Ruiz-Opazo, N., Weinberger, J., Nadal-Ginard, B. 1985. *Nature* 315:67–70

193. Cooper, T. A., Ordahl, C. P. 1984. *J. Cell. Biol.* 99:346a

194. Dhoot, G. K., Frearson, N., Perry, S. V. 1979. *Exp. Cell Res.* 122:339–50

195. Dhoot, G. K., Perry, S. V. 1979. *Nature* 278:714–18

196. Dhoot, G. K., Perry, S. V. 1980. *Exp. Cell Res.* 127:75–87

197. Matsuda, R., Obinata, T., Shimada, Y. 1981. *Dev. Biol.* 82:11–19

198. Toyoto, N., Shimada, Y. 1981. *J. Cell. Biol.* 91:497–504

199. Toyoto, N., Shimada, Y. 1983. *Cell* 33:297–304

200. Wilkinson, J. M., Moir, A. J. G., Waterfield, M. D. 1984. *Eur. J. Biochem.* 143:47–56

201. Kornblihtt, A. R., Vibe-Pedersen, K.,

Baralle, F. E. 1984. *Nucleic Acids Res.* 12:5853–67
202. Broker, T. R. 1984. See Ref. 5a, pp. 182–212
203. Nevins, J. R., Chen-Kiang, S. 1981. In *Advances in Virus Research,* 26:1–35. New York: Academic
204. Flint, S. J. 1982. *Biochem. Biophys. Acta* 651:175–208
205. Tooze, J., ed. 1981. *DNA Tumor Viruses, Molecular Biology of Tumor Viruses.* Cold Spring Harbor, NY: Cold Spring Harbor Lab. 1073 pp. 2nd ed.
206. Wagner, E. K. 1985. In *The Herpes Viruses,* ed. B. Roizman, pp. 45–104. New York: Plenum
207. Weiss, R., Teich, N., Varmus, H., Coffin, J., eds. 1982. *RNA Tumor Viruses, Molecular Biology of Tumor Viruses.* Cold Spring Harbor, NY: Cold Spring Harbor Lab. 1396 pp. 2nd ed.
208. Scott, M. P., Weiner, A. J., Hazelrigg, T. I., Polisky, B. A., Pirrotta, V., et al. 1983. *Cell* 35:763–76
209. Garber, R. L., Kuroiwa, A., Gehring, W. J. 1983. *EMBO J.* 2:2027–36
210. Hogness, D. S., Beachy, P. A., Lipshitz, H., Pealtie, D., Paro, R., et al. 1985. *Cold Spring Harbor Symp. Quant. Biol.* 50:24
211. Carrasco, A. E., McGinnis, W., Gehring, W. J., DeRobertis, E. M. 1984. *Cell* 37:409–14
212. Forbes, D. J., Kirschner, M. W., Caput, D., Dahlberg, J. E., Lund, E. 1984. *Cell* 38:681–89
213. Roeder, G. S., Fink, G. R. 1983. In *Mobile Genetic Elements,* ed. J. A. Shapiro, pp. 299–328. Orlando, Fla: Academic
214. Rubin, G. M. 1983. See Ref. 213, pp. 329–61
215. Varmus, H. E. 1983. See Ref. 213, pp. 411–503
216. Mount, S. M., Rubin, G. M. 1985. *Mol. Cell. Biol.* 5:1630–38
217. Flavell, A. J., Ruby, S. W., Toole, J. T., Roberts, B. E., Rubin, G. M. 1980. *Proc. Natl. Acad. Sci.* 77:7107–11
218. Spradling, A. C., Rubin, G. M. 1982. *Science* 218:341–47
219. Karess, R. E., Rubin, G. M. 1984. *Cell* 38:135–46
219a. Laski, F. A., Rio, D. C., Rubin, G. M. 1986. *Cell* 44:7–19
219b. Lang, K. M., van Santen, V. L., Spritz, R. A. 1985. *EMBO J.* 4:1991–96
220. Tsai, M.-J., Ting, A. C., Nordstrom, J. L., Zimmer, W., O'Malley, B. W. 1980. *Cell* 22:219–30
221. Rogers, J., Choi, E., Souza, L., Carter, C., Word, C., et al. 1981. *Cell* 26:19–27

222. Tyler, B. M., Cowman, A. F., Adams, J. M., Harris, A. W. 1981. *Nature* 293:406–8
223. Henikoff, S., Sloan, J. S., Kelly, J. D. 1983. *Cell* 34:405–14
224. Clark, A. J., Clissold, P. M., Alshawi, R., Beattie, P., Bishop, J. 1984. *EMBO J.* 3:1045–52
225. Vincent, A., O'Connell, P., Gray, M. R., Rosbash, M. 1984. *EMBO J.* 3:1003–13
226. Benyajati, C., Spoerel, S., Haymerle, H., Ashburner, M. 1983. *Cell* 33:125–33
227. Reynolds, G. A., Goldstein, J. L., Brown, M. S. 1985. *J. Biol. Chem.* 260:10369–77
228. King, C. R., Piatigorsky, J. 1983. *Cell* 32:707–12
229. King, C. R., Piatigorsky, J. 1984. *J. Biol. Chem.* 259:1822–26
230. Freytag, S. O., Beaudet, A. L., Bock, H. G. O., O'Brien, W. E. 1984. *Mol. Cell. Biol.* 4:1978–84
231. Steinmetz, M., Moore, K. W., Frelinger, J. G., Sher, B. T., Shen, F.-W., et al. 1981. *Cell* 25:683–92
232. Reyes, A. A., Schold, M., Itakura, K., Wallace, R. B. 1982. *Proc. Natl. Acad. Sci. USA* 79:3270–74
233. Kress, M., Glaros, D., Khoury, G., Jay, G. 1983. *Nature* 306:602–4
234. Lalanne, J.-L., Cochet, M., Kummer, A.-M., Gachelin, G., Kourilsky, P. 1983. *Proc. Natl. Acad. Sci. USA* 80:7561–65
235. Transy, K. C., Lalanne, J.-L., Kourilsky, P. 1984. *EMBO J.* 5:2383–86
236. Brickell, P. M., Latchman, D. S., Murphy, D., Willison, K., Rigby, P. W. J. 1983. *Nature* 306:756–60
237. Takahashi, N., Roach, A., Teplow, D. B., Prusiner, S. B., Hood, L. 1985. *Cell* 42:139–48
238. Stein, J. P., Catterall, J. F., Kristo, P., Means, A. R., O'Malley, B. W. 1980. *Cell* 21:681–87
239. Unterman, R. D., Lynch, K. R., Nakhashi, H. L., Dolan, K. P., Hamilton, J. W., et al. 1981. *Proc. Natl. Acad. Sci. USA* 78:3478–82
240. Grez, M., Land, H., Giesecke, K., Schutz, G. 1981. *Cell* 25:743–52
241. Bennetzen, J. L., Hall, B. D. 1982. *J. Biol. Chem.* 257:3018–25
242. Capetanaki, Y. G., Ngai, J., Flytzanis, C. N., Lazarides, E. 1983. *Cell* 35:411–20
243. Zehner, Z. E., Paterson, B. M. 1983. *Proc. Natl. Acad. Sci. USA* 80:911–15
244. Heilig, R., Perrin, F., Gannon, F., Mandel, J. L., Chambon, P. 1980. *Cell* 20:625–37

Ann. Rev. Biochem. 1986. 55:1119–150

SPLICING OF MESSENGER RNA PRECURSORS

Richard A. Padgett[1], Paula J. Grabowski, Maria M. Konarska, Sharon Seiler, and Phillip A. Sharp

Center for Cancer Research and Department of Biology, Massachusetts Institute of Technology, 77 Massachusetts Avenue, Cambridge, Massachusetts 02139

CONTENTS

Perspectives and Summary

Rapid advances have been made in the past two years in the analysis of the splicing of mRNA precursors. This has primarily occurred through the development of soluble reactions that accurately process exogenously added substrate RNA. Some features of the mRNA precursor splicing mechanism can now be

---

[1]Present Address: Department of Biochemistry, University of Texas, Health Sciences Center, Dallas, Texas 75235

1119

0066-4154/86/0701-1119$02.00

compared to the splicing of tRNA precursors and to the self-splicing of ribosomal and mitochondrial precursors. Surprisingly, the splicing mechanism of mRNA precursors has some features in common with the self-splicing or RNA-catalyzed splicing processes.

A novel RNA form, a lariat RNA, is generated during the splicing of mRNA precursors. These RNAs have a circular component with an extending tail and are formed by a branch where an adenosine residue is linked through a 2'-5' phosphodiester bond to the 5' end of the intervening sequence, and a 3'-5' phosphodiester bond to the remainder of the intervening sequence. The site of branch formation is typically 20–50 nucleotides upstream of the 3' splice site. The excised intervening sequence is released intact as a lariat RNA. A kinetic intermediate in the splicing of mRNA precursors is generated by cleavage at the 5' splice site. Thus the intermediate is composed of two RNAs, the 5' exon and the lariat form of the intervening sequences linked by a normal phosphodiester bond to the 3' exon. This intermediate structure is part of a multicomponent complex which has a sedimentation velocity coefficient of 40–60S. Such complexes have been coined spliceosomes and contain, in addition to yet undefined components, both small nuclear ribonucleoprotein particles and proteins previously identified as constituents of heterogeneous ribonucleoprotein particles. Formation of the spliceosome precedes generation of the intermediate RNA form and is probably the critical step in specifying the splicing reaction.

The specific sequences recognized during splicing of mRNA precursors in mammals are probably confined to the consensus sequences at the boundaries of the intron. Sequences encompassing the branch site can be deleted without altering the specificity. In contrast, sequences at the branch site are critical in yeast.

Small nuclear ribonucleoproteins containing U1 and U2 RNAs are essential for splicing. Evidence suggests that other types of snRNPs are also important. These snRNPs probably recognize sequences at both the 5' and 3' splice sites and organize the spliceosome. Mutant introns with altered 5' and 3' splice site sequences can become arrested at either of two stages: before formation of the spliceosome, or after formation of the intermediate lariat RNA. The mechanism of splicing of mRNA precursors does not preclude a *trans*-splicing reaction where exons from two separate RNA molecules are joined. The efficiency of *trans*-splicing increases significantly if the two RNA substrates have complementary sequences.

Related Reviews

Over the short period of time since the discovery of splicing, a number of excellent reviews have appeared. These are highlighted by Abelson's review (1) of "RNA Processing and the Intervening Sequence Problem," Breathnach &

Chambon's review (2) of "Eukaryotic Split Genes," and Nevins's review (3) on "The Pathway of Eukaryotic mRNA Formation." A recent extensive review on splicing was written by Rogers (4). This current volume contains a review covering RNA-catalyzed splicing reactions written by Cech & Bass (5).

Structure of mRNA Precursors

Synthesis of mRNA precursors in eukaryotic cells is the result of RNA polymerase II initiation at a promoter sequence. Shortly after initiation, the 5' end of the RNA is modified by addition of 7-methyl guanosine triphosphate and methylation to form the characteristic cap structure (7mGpppmXpmY . . .) (6). As the RNA chain grows, it is thought to become organized into a ribonucle-oprotein (RNP) complex by association with a family of highly conserved basic proteins (7–9). During or shortly after synthesis of the RNA, approximately 0.1% of adenosine residues are methylated at the 6 position, usually at the sequence N_1-(G/A)-m^6A-C-N_2 where N_1 is typically a purine and N_2 is rarely a guanosine (10, 11). While the functional significance of this methylation is unknown, it is worth noting that a sizable fraction of the methylated residues appear in the mature cytoplasmic mRNA (12). Transcription typically con-tinues beyond the 3' end of the mature mRNA and a processing reaction generates the site of poly(A) addition. Polyadenylation of the 3' end commonly precedes splicing; however, there are examples where nascent transcripts are spliced (13). This is direct evidence that the two processes are not mechanisti-cally coupled. Evidence from studies employing drugs to inhibit polyadenyl-ation is also consistent with independent processes (14).

As the process of splicing is kinetically slower than 3' cleavage and polyadenylation, the in vivo substrate for splicing is typically a capped and polyadenylated linear RNA. There does not appear to be a strict order for the excision of multiple introns from a precursor RNA (15–17); rather, the order of removal seems to be kinetically determined as if the excision of some introns was intrinsically faster than others. This results in partially spliced molecules with different patterns of introns. The half-life of introns within precursor RNAs can vary from a few seconds to 10–20 minutes. For example, the excision of the first intervening sequence of the major late transcription unit of adenovirus is half complete in 30 seconds (13), while complete splicing of a typical viral mRNA requires about 20 minutes. In contrast, the complete splicing of the rabbit β-globin precursor requires approximately 5 minutes (18). The aspects of a natural intron that determine its rate of removal are unknown. As reviewed below, most changes in introns decrease the rate of splicing. However, such decreases do not necessarily reduce the level of cytoplasmic mRNA; in many cases, the sole effect is an increase in the level of nuclear precursor RNA (19).

Splicing of mRNA precursors appears to occur exclusively in the nucleus

(20). Whether there exists a mechanism that retains unspliced RNA in the nucleus is unclear. In cases where the precursor cannot be properly spliced, the unspliced RNA remains in the nucleus and is degraded (21). However, if the RNA contains an intron that is part of an alternative splicing pattern, intron sequences can be transported to the cytoplasm. It is possible that sequences within the precursor RNA perhaps within the intron are recognized by components that are themselves confined to the nucleus. Some of these components could be part of the splicing machinery.

Another uncertainty is whether splicing is required for the efficient transport of RNA to the cytoplasm. Certainly mRNAs from genes that do not contain introns are transported without splicing. On the other hand, when the 16S mRNA transcription unit of SV40 was replaced with a cDNA copy of itself or a cDNA copy of β-globin (22), and expressed on a replicating SV40 vector, the inserted cDNA gene did not produce cytoplasmic mRNA. Reinsertion of an intron into the 16S transcription unit restored the production of cytoplasmic mRNA (23, 23a). This result is not universal; others have not seen such a strict requirement for splicing in the production of cytoplasmic mRNA (24, 25). Thus, while the processes of splicing and transport to the cytoplasm appear to be linked, the linkage is not absolute as it would be in a model where the splicing process releases the RNA from the nucleus.

The process of splicing requires that the sequences at the 5' and 3' splice sites be correctly cleaved and the two ends accurately joined. In the case of mRNA precursor splicing, evolutionarily conserved sequences at the two splice sites specify the reaction. These consensus sequences are best demonstrated in the form of a matrix that shows the percentage of occurrence of each base at each position relative to the virtually invariant GU and AG dinucleotides at the splice sites (2, 26). Table 1 shows such a matrix. In compiling the table, all cases that violated the GU or AG rules were excluded as these were primarily due to incorrect assignment of intron boundaries. However, there are two clear cases where the dinucleotide GC is used at the 5' splice site (27–30). The consensus region evident in the matrix extends from the last two nucleotides in the exon through the first six nucleotides of the intron at the 5' splice site AG:GU(A)AGU (where the colon denotes the site of cleavage and ligation). The consensus at the 3' splice site extends from at least the last 15 nucleotides of the intron through the first nucleotide of the exon to give a consensus of $(U/C)_{11}NCAG:G$. Examinations of intron sequences have not revealed other conserved sequences with the exception of the branch site sequences (see below). It is worth noting that the consensus sequences shown above are probably not the sole determinants of functional splice sites since they can be found within both exons and introns and yet are not used. As discussed later, additional factors must play a role in specifying splice sites.

The above discussion concerned splice sites found in higher eukaryotes. An

Table 1 Consensus sequences of vertebrates[a]

5' splice site

	exon			intron								
%G	23	14	13	77	100	0	32	12	84	18	30	22
%A	34	35	62	8	0	0	60	74	9	15	33	22
%U	15	12	13	8	0	100	5	7	3	50	17	25
%C	29	38	12	8	0	0	3	7	4	17	21	31
	—	—	A	G:	G	U	A	A	G	U	—	—

3' splice site

	intron																exon		
%G	18	13	15	10	10	7	7	10	9	5	5	5	24	0	0	100	55	27	24
%A	17	11	8	9	7	4	9	8	10	8	4	9	26	2	100	0	20	21	19
%U	37	44	46	46	56	59	43	49	41	46	42	46	23	19	0	0	8	32	28
%C	28	33	32	35	27	30	42	33	40	40	49	41	27	78	0	0	17	20	28
	—	—	Py	Py	Py	Py	Py	Py	Py	Py	Py	Py	N	C	A	G:	G	—	—

[a] A tabulation of the sequences at assigned 5' and 3' splice sites in approximately 400 vertebrate genes in the Gen Bank Data Base. All examples were included where intervening sequences began with a GU dinucleotide and terminated with an AG dinucleotide.

interesting difference is found when the splice sites of yeast are considered (Table 2). In yeast, the 5' splice site is almost invariant while the 3' splice site has a much less apparent polypyrimidine tract. The specificity at the 3' splice site seems to mainly reside in the invariant sequence at the branch site (see below). Since the typical intron in a yeast gene is short, these conserved sequences seem sufficient to specify the splice sites.

The length of introns in vertebrate genes ranges from approximately 50 to well over 10,000 bases with no obvious periodicity (31). This is consistent with the observations that introns in equivalent genes from two vertebrate species can vary extensively in length and sequence so as to be almost totally nonhomologous. Exon lengths, in contrast, seem to be clustered around means of 52, 140, 223, and 299 bases (31). Exons as small as 7 bases have been described (32); however, if the flanking introns were spliced at different times, the actual exon length used in the splicing reaction could be larger. The shortest introns reported in nonviral genes are approximately 60 bases long. An intron of only 31 bases is excised from a late mRNA species of SV40 and polyoma but this could be processed from a concatameric precursor (33).

Biochemistry of Splicing

Since the discovery of pre-mRNA splicing, considerable effort has been expended to determine how the specific sites are recognized, brought together, and spliced. This required the development of a cell-free system where conditions could be controlled and the components purified and reconstituted. Cell-free splicing systems are based on whole cell (34) or nuclear extracts (35) of cultured cells, usually HeLa. These extracts contain all the components that can be solubilized by moderate salt extraction of lysed cells or crude nuclei, respectively. Reactions formed with such extracts show accurate transcription (34), polyadenylation (36), and splicing (37).

The pre-mRNA substrates studied in the in vitro systems are produced either by endogenous transcription of DNA templates to produce a coupled transcription/splicing system (37) or by the addition of exogenous pre-mRNA

Table 2 Consensus sequences at splice sites and branch site in genes of yeast[a]

5' splice site	branch site	3' splice site
:GUAUGU _____	UACUAAC _____	^C_UAG:
% % % % % %	% % % % % % %	% % %

[a]The sequences were taken from a tabulation of introns in yeast ribosomal protein, actin, and mating type genes (208). The UACUAAC sequence is between 16 and 64 nucleotides upstream of the 3' splice site. The fraction listed under each position is the frequency of occurrence of the indicated base.

(38–40). The exogenous RNA is generally made in vitro by transcription of appropriate DNA templates by mammalian transcription reactions (39) or by purified prokaryotic RNA polymerases (41). In addition, the use of purified unspliced RNA isolated from cultured cells has also been reported (42). The most common system uses a nuclear extract of HeLa cells and pre-mRNA produced by in vitro transcription of DNA templates using the purified SP6 RNA polymerase (40). The nuclear extract appears to give higher efficiencies of splicing with many RNAs than does the whole cell extract. Transcription by the SP6 RNA polymerase can be primed with a cap dinucleotide to yield RNAs with capped 5' termini (43).

The reaction characteristics of these in vitro splicing systems are interesting (38–40). There is a requirement for ATP which could not be substituted by any other cofactor. There is also a requirement for Mg^{2+} at rather low levels while higher levels of Mg^{2+} inhibited splicing. Monovalent cation concentrations of about 60 mM and temperatures of 30°C seemed to be optimum. Perhaps the most interesting characteristics of the systems are their distinctly nonlinear kinetics. A lag period is observed at the start of the reaction before the onset of splicing. The length of the lag period varies between 15 and 45 minutes following which spliced RNA accumulates with approximately linear kinetics for 2–3 hours. Events occurring during this lag require ATP and substrate RNA. Recent results have shown that the substrate RNA is being assembled into a specific complex during this period (see below). An exception to the above conclusions is a system derived from MOPC 315 cells (44) which can use either ATP or GTP, does not require Mg^{2+}, does not show a lag period, is only active at high substrate concentrations, and has much faster kinetics. The differences between these two sets of results are not understood.

Investigations of the structural requirements of the substrate RNA showed that neither complete exons nor poly(A) at the 3' end were necessary. The requirement of a 5' 7mGpppX cap structure differed between the whole cell and nuclear extracts. In the whole cell extract, this structure was required for efficient splicing (43). Furthermore, addition of cap analogues of the general structure 7mGpppX could completely inhibit the splicing reaction at quite low levels (10–100 μM). Interestingly, this inhibition could be observed only if the analogue was added before the lag period, implying that the cap-dependent step occurred early in the reaction. In contrast, the nuclear extract system was only partially inhibited by the cap analogues, and splicing of uncapped RNA could occur although the efficiency and accuracy of the reaction was less than when capped RNA was used (40). Recently, it has been shown that cap analogue sensitivity can be demonstrated in the nuclear extract if it is pre-incubated prior to the addition of substrate RNA (45). It is possible that the cap requirement observed in vitro could reflect an in vivo process where recognition of the 5' end of the precursor might be important in the assembly of the splicing complex.

The above discussion is concerned only with in vitro systems based on mammalian cell extracts. It has recently been found that a whole cell extract of yeast can splice yeast pre-mRNA in vitro (46–48). The analysis of this reaction shows that it is very similar to the mammalian case in its requirements and in the intermediates and products as described below.

Biochemical fractionation of the HeLa nuclear extract in combination with inactivation and complementation studies have identified multiple factors, at least seven, required for mRNA splicing (49, 50). Some of these partially purified factors contain essential nucleic acid components, such as U1 and U2 small nuclear ribonucleoprotein particles (50; see below) in addition to other heat-labile and heat-stable components.

Reaction Intermediates and Products

A general scheme of the pre-mRNA splicing reaction derived from studies of the in vitro reaction is shown in Figure 1 (51–53). The first covalent modification of the substrate RNA is a cleavage at the precise 5' splice site. In either a kinetically rapid second reaction or mechanistically coupled reaction, the 5' end of the intervening sequence is joined via a 2'-5' phosphodiester bond to an adenosine residue near the 3' splice site to produce a lariat or tailed circular molecule containing the intervening sequence and the 3' exon. This bipartite intermediate is subsequently resolved by cleavage at the precise 3' splice site and ligation of the 5' and 3' exons via a 3'-5' phosphodiester bond. This produces two products: the spliced exons and the excised intervening sequence in the form of a lariat. While this pathway is derived from in vitro reactions, similar lariat form intermediates have been identified in vivo in the splicing of the rabbit β-globin second intervening sequence (54) and a yeast intron (55, 56).

The fate of the phosphate moieties at the splice sites has been determined and is shown in Figure 1 (48, 52, 57). The phosphate at the 5' splice site becomes incorporated into the 2'-5' phosphodiester bond in the branch structure. Consistent with this, the 5' exon intermediate RNA has been shown to end in a 3' hydroxyl group. The phosphate at the 3' splice site is used to form the 3'-5' phosphodiester bond linking the two exons in the product RNA. Consistent with this the excised intervening sequence ends in a 3' hydroxyl group. It is interesting to note that the entire reaction as well as the two subreactions conserve the number of phosphodiester bonds as well as conserving the phosphate moieties themselves. Thus although the reaction as a whole appears to require the input of energy in the form of ATP, the actual cleavage and ligation steps could occur by concerted transesterifications in a manner similar to that seen in the splicing of the *Tetrahymena* ribosomal RNA precursor (58).

Branch type structures in RNA were first suggested as an explanation of nuclease-resistant products isolated from nuclear poly(A)-containing RNA

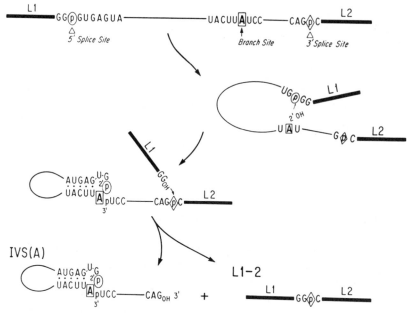

Figure 1   The nuclear mRNA splicing process. Complementarity between the sequences at the 5' splice site and branch site is indicated by base pairs (dots). Phosphate moieties from the 5' splice site Ⓟ and 3' splice site Ⓟ are retained during processing.

from HeLa cells (59). Analysis of the branch structures indicated that the base labeled X below was predominantly adenosine and that Y+Z was about 50% guanosine and 25% each cytosine and uridine.

$$\ldots \text{pX}^{2'5}_{3'5}{}^{\text{pYp}\ldots}_{\text{pZp}\ldots}$$

These identifications agree with the structure seen in the cases of individual in vitro splicing products where the X position is without exception an A residue, the Y position is a G in normal splices, and the Z position is usually a pyrimidine. Other bases can be found in the Y and Z positions in various mutant RNAs as described below.

The frequency of occurrence in vivo is about one branch for every 10,000 bases of poly(A) containing nuclear RNA (59). This suggests that the lariat form of the intervening sequence is degraded rapidly and/or that the branch structure is specifically hydrolyzed. It has been noted that HeLa cell extracts contain an activity that can cleave the 2'-5' phosphodiester bond in lariat RNAs to give the corresponding linear form (60). This has been used to identify branch-containing RNAs and oligonucleotides (43, 61, 62, 68). The role of this activity in vivo is not known.

Role of Branch Site in Splicing

The discovery that an intron sequence was used as a site of branch formation in mammalian cells suggested a role for the previously recognized conserved internal sequence found 20–40 bases upstream of the 3' splice site in all yeast nuclear mRNA introns (63, 64). This sequence, UACUAAC, is highly conserved in yeast and analysis of point mutants has shown that only small deviations from the canonical sequence are tolerated (65). When the pathway of yeast pre-mRNA splicing was investigated both in vivo (55, 56) and in vitro (47, 48), it was shown that the final A in the sequence was indeed the site of branch formation in a pathway identical to that of mammalian pre-mRNAs. An examination of mammalian intervening sequences for sequences resembling the yeast-conserved signal identified a weak consensus sequence located 20–50 bases upstream of the 3' splice site (53, 54, 66). As additional branch sites have been mapped in vitro, it is clear that this consensus sequence does predict the location of most branch sites (61). However, unlike the case in yeast where deletion or mutation of the conserved sequence abolishes splicing, such deletions of mammalian branch sites serve only to slow the reaction by a variable amount while other sequences are used as branch sites (62, 67, 68). In some cases, the new branch sites bear almost no relation to the consensus sequence. It was suggested first for yeast (64) and later for mammalian introns (57) that sequence complementarity between the 5' splice site and the branch site could play an important role in aligning and stabilizing a splicing complex. An experimental test of this idea in yeast by attempting to rescue a mutation in the branch site sequence with a compensating mutation in the 5' splice site sequence did not restore wild-type levels of spliced products (68a). The cryptic branch site sequences used in mammalian introns when the normal branch sites have been deleted or mutated show no complementarity to the 5' splice sites (other than that expected for two random sequences) (68). It appears from these findings that sequences at the branch site in mammalian pre-mRNAs do not play an essential role in directing the splicing reaction to a particular 3' splice site but may be important in the efficiency of the reaction. This is also supported by results obtained with point mutants discussed below.

Mutations Affecting Splicing

As with any sequence-specific reaction, mutational analysis should show what sequences are required and may suggest the role they play in the reaction. In the case of splicing, both naturally occurring and constructed mutations have been investigated.

The study of the defects in α and β-globin genes present in several forms of thalassemia has provided many examples of natural splicing mutants (reviewed in 69). There are two types of mutations in the splice site consensus sequences that affect splicing. The first is a class of mutations of the normal 5' or 3' splice sites which reduce the amount of normal mature mRNA and often activate nearby cryptic splice sites leading to the production of aberrant mRNAs. The activation of cryptic splice sites, many of which are poor matches with the consensus sequences, indicates that the splicing reaction has considerable flexibility in the use of splice site sequences. It also suggests that the availability of a single functional splice site creates an imperative to find a cognate sequence with which to react. The lack of reactivity of the cryptic sites in the presence of the normal site may simply reflect a much higher rate or affinity of the reaction components for the normal site versus the cryptic sites.

The second major class of mutants are those in which a new splice site is produced by mutation within the intervening sequence. As before, these mutations cause the production of aberrantly spliced mRNAs. In this case the new site must be able to out-compete the normal site in the reaction. In one example, the mutation created a new 5' splice site near the 3' end of the intron (70). Upstream of this new 5' splice site, a cryptic 3' splice site within the intron was activated to react with the normal 5' splice site. This produced an RNA with an additional exon derived from intron sequences.

To date, the first class of mutations have only been identified in the first 6 bases of 5' splice site sequences and the last 2 bases of 3' splice site sequences (4). This suggests that individual bases in the rest of the sequences in the exons and introns do not play a critical role in determining the site of splicing. The only examples of the second class of mutations occur within introns near the normal 3' splice sites. It is worth noting that none of the natural mutations of mammalian genes that have been analyzed affect the branch site sequence. This suggests a relative lack of sequence specificity in the branch formation reaction.

By far the largest group of splicing mutants have been produced in vitro and assayed either in vivo or in vitro or both. The earliest findings using modified introns were that most of the sequences within the intron were nonessential for splicing thus further emphasizing the importance of the splice site sequences. In addition, it was found that 5' and 3' splice sites from two different genes could be accurately spliced together (71, 72). This suggested that the two splice sites flanking an intron were not uniquely related to one another in terms of accurate splicing. In tests of models suggesting that the splicing entity scanned intervening sequences and joined the first two splice sites encountered, splice sites were duplicated in tandem to produce a situation where a single 5' splice site could react with two different 3' splice sites or vice versa. The results of these experiments varied. In some constructions the distal of the pairs of sites were used in vivo (73), while in others it was the proximal of the pair (74). The former result strongly suggests that splicing does not involve a scanning mechanism.

The most exhaustive analysis of the sequences necessary for accurate splicing is that of Wieringa et al (75). By progressively deleting the second intron of rabbit β-globin gene, they determined that only the 5' most 6 bases and the 3' most 24 bases were needed to obtain nearly wild-type levels of mature cytoplasmic mRNA in vivo. They further found that there was a minimum intron length of about 60 bases but that essentially random sequences could be inserted between the minimal 5' and 3' splice sites to restore activity. Although many of the mutants of Wieringa et al (75) lacked the wild-type branch site, the method of construction placed a cryptic branch site near the 3' splice sites (68).

It is interesting that mutations in either the 5' splice site or the 3' splice site consensus sequence can result in the accumulation of the intermediate RNA in lariat form (47, 61, 62, 75a). Mutants with altered 5' splice sequences yielded a lariat RNA with an abnormal sequence at the 2' position of the branch site. This suggests that such sequence changes do not block recognition of the 5' splice site, probably by U1 RNP and interactions between the branch site and 5' splice site. After cleavage at the 5' splice site, there appears to be a proof editing step which examines sequences at the branch site as well as the 3' splice site before executing the second step in the reaction. It is probable that these blocked intermediate RNAs with abnormal branch site sequences or mutant 3' splice site sequences are rapidly degraded in vivo.

Thus it appears that all the sequences necessary for splicing small, single introns are located near the splice sites. Mutations within introns away from the splice sites can block the system only by creating new splice sites. This is somewhat surprising since the limited set of consensus sequences at the 5' and 3' splice sites seem inadequate to specify the processing of very long introns. This lack of intron involvement in the splicing process distinguishes nuclear pre-mRNA splicing from some of the types of splicing described below where intron sequences play an intimate role in splicing even to the extent of catalyzing the reaction.

Role of snRNPs

The first suggestion that small nuclear ribonucleoprotein particles (snRNPs) may be involved in splicing comes from the hypothesis of Lerner et al (76) and Rogers & Wall (77) that the U1 small nuclear RNA (snRNA) recognized sequences at the 5' splice sites by virture of sequence complementarity. This hypothesis has had a considerable impact on the field.

The U class of snRNAs (reviewed in 78, 79) are a class of abundant small RNAs whose sequences show significant phylogenetic conservation. These RNAs are complexed with proteins to form ribonucleoprotein particles (snRNPs; reviewed in 80, 81). A subset of these RNAs, U1, U2, U4, U5, U6, U7, and the recently identified U8, U9, and U10 species (82) are complexed with a common set of proteins, which allows coimmunoprecipitation by antisera directed against a common determinant. In addition, U1 and U2 snRNPs contain one or more unique proteins that can be recognized by specific antisera. These antisera, mostly obtained from patients suffering from one of several autoimmune diseases, have proven to be very useful in establishing the requirement of snRNPs in splicing. Another distinguishing feature of these RNAs (with the exception of U6 snRNA) is that they have a unique 5' cap structure containing a 2,2,7-trimethyl guanosine to which antibodies have been raised (82a).

The hypothesis that U1 RNP acts during recognition in the splicing process was one of several proposals for the function of snRNPs. Oshima et al (83) proposed that U2 RNA could form base pairs with the exon sequences at either end of an intron and thus contribute to the specificity of the splicing reaction. A more recent hypothesis by Keller & Noon (66) is that U2 snRNA, by virtue of complementarity to the consensus branch site sequence, provides 3' splice site specificity.

A variety of prior observations were consistent with a role for U snRNPs, especially U1 and U2 snRNPs, in the processing of mRNA precursors (reviewed in 78, 79). Sedimentation analyses of the ribonucleoprotein complexes of mammalian nuclei had shown snRNAs sedimenting with large (30–60S) structures containing high molecular weight RNA as well as soluble 10S snRNP particles (84–86). At monovalent salt concentrations below 150 mM, at least half of the U1 snRNPs are associated with the fundamental 30S subunit that packages the high-molecular-weight precursors of messenger RNAs (76, 84, 86). Strikingly, a degradation form of U1 RNA (designated U1*) that is missing only its first six nucleotides sediments exclusively as a 10S snRNP particle, as if the absence of this short region complementary to 5' splice sites prevents the association of U1*-containing snRNPs with the 30S particles (76). In addition, experiments with aminomethyltrioxsalen, a chemical reagent that reversibly cross-links base-paired RNAs, indicated that both U1 and U2 RNAs can be found base-paired to high molecular weight RNAs in vivo (87–89).

A substantial body of data now supports the hypothesis that U1 and U2 snRNPs are required for the splicing of mRNA precursors. Splicing of messenger RNA precursors in in vitro systems is inhibited by pretreatment of soluble mammalian splicing extracts with anti-Sm, an antibody specific for a determinant common to all U snRNPs (90, 91), anti-(U1)RNP, specific for U1 RNP (51, 90) or anti-trimethylguanosine antibodies which react with the 5' cap structure (91). In an entirely analogous fashion, antisera specific for either U1 snRNPs or the Sm antigen block splicing and thus cause the accumulation of unprocessed primary transcripts in *Xenopus* oocytes (92, 93). It has recently been found that pretreatment of mammalian splicing extracts with anti-(U1)RNP antiserum prevents the formation of the spliceosome, the multicomponent RNA-protein complex that is specific to the splicing reaction (see below).

Partially purified U1 snRNPs selectively bind in vitro and protect from RNase digestion a 15–17 nucleotide region containing the 5' splice site of a β-globin pre-mRNA (94). This association of U1 RNA and precursor RNA can also be detected in splicing extracts. Anti-Sm antibodies immunoprecipitate the precursor, intermediate, and spliced product RNA species from total in vitro splicing reactions and from the spliceosome-containing region of gradient-fractionated reactions (95). Anti-(U1)RNP antisera likewise immunoprecipi-tate all the RNA species of the splicing reaction from spliceosome-containing gradient fractions (95). However, the same anti-(U1)RNP antisera will only immunoprecipitate the precursor RNA from total splicing reactions containing all of the above species and this coimmunoprecipitation of precursor with U1 snRNP is diminished upon incubation with ATP (95, 96). These seemingly anomalous observations are most probably due to the sequestration of (U1)RNP determinants during the splicing reaction and their exposure during gradient sedimentation of such reactions. The binding of U1 snRNP to precursor RNA in in vitro splicing systems can be localized by RNase digestion to the same 15 nucleotides at the 5' splice site region that is protected when purified U1 snRNPs are mixed with the isolated transcript (96). RNase digestion of the immunoprecipitates formed with antisera to U2 RNP reveals that U2 snRNP is associated with a region of the intron that includes the branch site (96). Consistent with this, antibodies to U2 snRNP immunoprecipitate the precursor, intermediate, and product RNAs from in vitro splicing reactions (96; S. Seiler, unpublished).

Specific hydrolysis of the 5' eight to fourteen nucleotides of U1 RNA present in either crude HeLa nuclear extracts or partially fractionated extracts is sufficient to abolish splicing in vitro (50, 91, 96). Specific hydrolysis of the 14–15 nucleotides from the 5' terminus of U2 RNA (50, 96) or of a 15-nucleotide loop region near the 5' end of the molecule (96) also prevents splicing in vitro. The 5' terminus of U2 RNA is complementary to a 10-nucleotide region near the 5' terminus (nucleotides 11–20) of U1 RNA. Stretches within the first loop of U2 RNA also exhibit complementarity to a loop in U1 RNA, to 3 nucleotides of sequence in the 3' splice site consensus, and by bulging one nucleotide to the branch site consensus sequence (97). This latter region, which was specifically proposed by Keller & Noon (66) to recognize the branch site, is inaccessible to cleavage by RNase H digestion in the presence of complementary deoxyoligonucleotides (96). Thus while it seems likely that the 5' terminus of the RNA moiety of U1 snRNP binds the 5' splice site via the formation of intermolecular base pairs and so specifies the 5' cleavage of the splicing reaction, the manner in which U2 snRNP protects the branch site and its exact role in the splicing reaction is less clearly understood. In a recent study by Chabot et al (98) a third snRNP has been identified as a factor involved in mRNA splicing. This snRNP, probably U5 snRNP, selectively binds to the 3' splice site of a human β-globin pre-mRNA substrate. Interestingly, this factor has been shown to possess Sm determinants in addition to a 5'-trimethyl guanosine cap structure characteristic of snRNPs of the Sm class, yet its binding to 3' splice sites is micrococcal nuclease–resistant. In binding to the 3' splice site of a pre-mRNA this factor may play a role in positioning U2 snRNP at the branch site (98). It remains an intriguing possibility that yet other snRNPs, in addition to U1, U2, and U5, may be required for splicing.

It is important to note that the experiments described above do not prove the mechanism proposed by Lerner et al (76) and Rogers & Wall (77). There is no evidence that actual base pairing between U1 RNA and pre-mRNA is involved. Indeed, while the original version of the models included U1 complementarity to the 3' splice site, no evidence in support of this hypothesis has been found. The association of U2 RNA sequences with branch site sequences may play a role in the reaction but considering the variety of sequences that can serve as branch sites, this interaction almost certainly does not represent the major specificity determinant at the 3' splice site.

Role of 40S RNP Structure

Heterogeneous nuclear RNA (hnRNA) is closely associated with a defined set of nuclear proteins to form ribonucleoprotein (RNP) particles or complexes (hnRNP). Recent reviews of hnRNP structure are suggested to the reader for details that are beyond the scope of this article (7, 8, 99). HnRNP is probably assembled during transcription on the nascent pre-mRNA chain and this form is thought to be retained until the fully processed mRNP is exported to the cytoplasm. Thus, the hnRNP structure may be important in RNA processing events (100). In analogy with chromatin, these structures may also function to condense the RNA chain and maintain it in an untangled conformation in the nucleus (7, 101).

Constituents of the 40S RNP Structure

The hnRNP as visualized by electron microscopy appears as a "beads-on-a-string" structure (for review see 7, 8) containing repeating monomer units approximately 200–300 Å in diameter linked by ribonuclease-sensitive strands. The monomer RNP particles, released by RNase treatment of isolated nuclei, sediment in sucrose gradients at approximately 30–50S (101–103). The constituents of the particles are approximately 500 nucleotides of rapidly labeled nonribosomal RNA and a set of 8 or more proteins (7, 104). Six of these proteins constitute a high proportion of the 40S particle protein mass and are present at a fixed stoichiometry relative to one another (102). These proteins behave as 3 pairs of closely related polypeptides, group A (A1 and A2), group B (B1 and B2), and group C (C1 and C2). The A, B, and C group proteins exhibit pairwise behavior in that they migrate in SDS gels as closely spaced doublets (A1=32,000, A2=34,000; B1=36,000, B2=37,000; C1=42,000, C2=44,000 daltons), and they dissociate from the 40S monomer in a pairwise manner with increasing ionic strength. The A group proteins are the first to dissociate from the RNP complex at 0.2 M NaCl; at 0.4 M NaCl both the A and B group proteins are dissociated. Under these conditions (0.4 M NaCl), the C group proteins remain bound but are completely dissociated at 0.6 M salt. Not surprisingly, the A and B group proteins are positively charged at physiological pH and probably bind RNA principally by electrostatic interactions. Interestingly, the C group proteins, which bind the hnRNA most tightly, are weakly acidic. Thus, it is likely that the C group proteins have at least one binding surface specific for RNA. The high abundance and conserved nature of the A, B, and C group proteins suggests that they have an important structural role in pre-mRNA packaging. In addition, the high glycine content of these proteins is reminiscent of structural proteins such as collagen, silk, and chorion.

In the lower eukaryote, *Artemia salina,* a single protein constituent of pre-mRNP structures has been identified as a 40,000-dalton polypeptide species, termed helix-destabilizing protein (HD-40, Ref. 105). This protein has biochemical and immunological characteristics that are strikingly similar to a composite of the A, B, and C group proteins isolated from mammalian sources and therefore is likely to be a functional analogue in lower eukaryotes. The unique property of HD-40 to unwind the RNA chain while at the same time condensing it may be important in permitting sequence recognition by splicing factors.

Monoclonal antibodies have been particularly useful in isolating, by immunoprecipitation, apparently pure forms of hnRNP (7, 106, 107). Analysis of isolated hnRNP structure has confirmed the presence of the A, B, and C group proteins and has also identified weakly associated components of larger molecular mass, the 68,000- and 120,000-dalton proteins (107). The same set of hnRNP proteins were also detected by UV-cross-linking of protein and RNA in intact HeLa cells, suggesting that these proteins are in direct contact with hnRNA (107). This same set of hnRNP proteins (in a fixed stoichiometry) were immunoprecipitated using independent monoclonal antibodies, two specific for the C group proteins and others specific for either the 68,000- or 120,000-dalton protein. By this analysis a minimum of 8–10 proteins have been identified as genuine constituents of the hnRNP. However, due to the limitation of the technique, the possibility cannot be excluded that there are additional protein and RNA components of the hnRNP structure. Furthermore, it is not known if the 40S monomer structure is homogeneous (108, 109). Without a functional assay for an intact form of hnRNP, it will be difficult to resolve these uncertainties.

The interactions that maintain the hnRNP structure appear to be due primarily to protein-protein interactions. That is, RNase digestion experiments have shown that although approximately 600 nt of RNA is found in a given 40S RNP structure, further RNase treatment does not disrupt the structure even if the RNA component is nicked into 50–100 nt segments (107, 110). These results show that the great majority of the hnRNA chain is accessible to RNase. Maintaining a large hnRNA in a compact, untangled conformation while permitting the recognition of specific signals on the RNA is a nontrivial problem exemplified by the CAD (carbamoyl-phosphate synthetase, aspartate transcarbamylase, dihydro-orotase) RNA precursor which is 25 kilobases long and contains 37 or more intervening sequences (111). This RNA precursor has been detected in 200S complexes that contain fragments of CAD pre-mRNA (112).

Recent evidence suggests that some of the proteins involved in 40S RNP structure are important for splicing. Preincubation of an extract with a monoclonal antibody specific for C1 and C2 polypeptides of the core hnRNP particle completely inhibits splicing (113). Surprisingly, a second monoclonal antibody reactive to a different epitope on the same two polypeptides does not result in inhibition. This suggests that binding to a particular region of the polypeptide is essential in order to block the reaction. Monoclonal antibodies to the 68,000 and 120,000 K proteins do not interfere with splicing in the extract; however, it is difficult to interpret a negative result in these types of experiments. At this preliminary stage of analysis, it seems likely that one type of hnRNP protein is an important component in the splicing process.

Spliceosome

There are many reasons to believe that the assembly of a multicomponent RNA-protein complex is important in specifying the splicing process (114). Recent experiments have provided direct evidence for and a preliminary characterization of such a multicomponent splicing complex, termed the spliceosome, both in yeast (46) and mammalian (95, 115) systems.

The spliceosome was first identified by the presence of the two RNA intermediates of splicing, the 5' exon and the lariat intron-3' exon RNAs. It is probably essential that these two RNA species remain juxtaposed until the second stage of the reaction joins the exons. Complexes containing these two RNAs have been identified by velocity sedimentation in sucrose or glycerol gradients and range in size from 40S in yeast to 50–60S in mammalian extracts. It is not clear if these size differences are due to different conditions of preparation or differences in composition. Characterization of the components of the complexes is obviously needed.

The complexes require ATP for their formation, and nonhydrolyzable analogues of ATP will not substitute. In mammalian extracts, the spliceosome forms during the lag period of the reaction and contains almost exclusively intact pre-mRNA. Later in the reaction, the two intermediate RNAs are found exclusively in the complex. When supplemented with fresh nuclear extract and ATP the RNA within the 60S complex can be chased into spliced RNA with accelerated kinetics and higher efficiencies than newly added precursor RNA (95). These results strongly suggest that the 60S complex formed in vitro is a functional intermediate in the splicing process.

The formation of the spliceosome also requires U snRNPs as evidenced by inhibition of complex formation by either direct addition of anti-(U1)snRNP antiserum to the extract (95) or by use of an extract depleted in U snRNPs by pre-treatment with an anti-Sm antiserum (115). As anticipated, sequences important for splicing also play an important role in the formation of the spliceosome. In yeast extracts, a mutation at the conserved branch site sequence leads to inefficient complex formation (46). In vivo, this mutant precursor is deficient in splicing. In mammalian extracts, an examination of a variety of mutants has revealed the importance of the 5' splice site and the polypyrimidine tract in splicing complex formation (115). In addition, an RNA substrate consisting entirely of bacterial sequences and therefore lacking authentic 5' and 3' splice sites does not form a splicing complex in vitro (95).

Although the first covalent reaction in the splicing pathway involves only the 5' splice site and the branch site, there is evidence that events at the 3' splice site have a major effect on this reaction. When mutants at the 3' splice site were examined in vitro, it was found that they reduced the efficiency of the 5' cleavage reaction (61, 116). This is consistent with the suggestion that sequences at both splice sites are recognized in the assembly of an active RNA-protein complex. It is of interest that during the splicing reaction, one or more nuclear factors protect the branchpoint region (96, 117) and 3' splice site (98) of pre-mRNA from nuclease digestion. The sequence specificity and ATP requirement for protection suggests that these interactions are specific for splicing and perhaps involve initial complex formation events.

The formation of a spliceosome complex in mammalian extracts that sediments at 50–60S suggests that a large number of cellular components must bind to the substrate RNA. For comparison, a 60S ribosomal subunit contains approximately 5300 nucleotides of RNA and 45 proteins, a total molecular mass of 3×10^6 daltons. The typical substrate RNAs used for spliceosome formation vary from approximately 200–500 nucleotides in length. Immunoprecipitation results provide evidence that spliceosomes contain both U1 and U2 snRNP (95; S. Seiler et al, unpublished) as well as antigens related to the C1 or C2 polypeptides of hnRNP (113). The discovery of a third required snRNP, which is likely to be U5 snRNP (98), suggests that the spliceosome may contain U5 in addition to U1 and U2 snRNPs. Given its apparent mass, it is possible that a 60S spliceosome will contain additional non-snRNP components and/or multiples of the above cellular factors.

Regulation of Splicing

Expression of a single gene may in some cases result in the production of multiple forms of mature RNA transcripts. This can be achieved either by alternative splicing of a primary RNA, utilization of different promoters or polyadenylation sites, or a combination of the above. This mode of gene expression can generate a set of related polypeptides and may be viewed as a mechanism for increasing the coding capacity of a genome. More interestingly, it also opens the possibility that splicing might be involved in regulation of gene expression. For example, unique mRNAs, specific for different cells or tissues or for certain stages of development, could be generated from a single gene by utilization of different 5' or 3' splice sites or by skipping of entire exons. In some cases, the alternative splicing may itself be a consequence of other processes influencing the structure of pre-RNA as in cases of differential transcription initiation or polyadenylation. Thus, the ideal example of the regulatory function of alternative splicing is one where a common primary transcript with common 5' and 3' termini is differentially spliced in a cell type or developmental stage-specific manner.

Many of the characterized examples of alternatively spliced pre-mRNAs do not fulfill the above requirements. Differential splicing of μ, δ, and γ immunoglobulin genes (119–122), giving rise to secreted or membrane-bound forms, could be controlled by either regulation of splicing or by the selection of RNA termination and polyadenylation sites. A similar situation occurs in the case of calcitonin (123–125), prekininogen (126, 127), γ-fibrinogen (128, 129), *Drosophia* GAR transformylase (130), or mouse major urinary protein genes (131). In addition, the utilization of different sites of initiation of transcription may influence alternate pathways of splicing, as in the cases of mouse α-amylase (132, 133) or *Drosophila* alcohol dehydrogenase (134), where the selection of a downstream promoter removes the first potential 5' splice site. A similar but more complicated pattern of RNA processing is observed in the expression of the myosin light chain gene (135–138) where the selection of one of two possible transcription initiation sites correlates with the combination of exons included in the mature mRNA.

The earliest reports of alternative splicing of a primary transcript with constant 5' and 3' termini came from the study of viral genomes. The differential choice of the 5' splice site that is subsequently joined to a common 3' splice site regulates production of small and large T antigens in SV40 and polyoma viruses (reviewed by Ziff, 139). A similar mode of gene expression is found in many adenovirus transcription units (reviewed in 3, 139). Alternative selection of different 3' splice sites joining to a common 5' splice site has also been observed. Again, viral genomes provide the best-studied examples; the whole family of adenovirus late transcripts, adenovirus early EIV RNAs, and late SV40 and polyoma transcripts are generated in this way. Cellular genes also utilize alternate 3' splice sites; the known examples include fibronectin (140, 141), and mouse class I transplantation antigens (142–144). Splicing of mRNAs for pro-opiomelanocortin (145), possibly human insulin (146), insulin-like growth factor (147), or human growth factor (148) may also be processed in a similar fashion; however, the physiological relevance and specificity of these cases have not been determined.

Alternative patterns of splicing of a common primary transcript that result in deletion of entire exons, as in the case of αA-crystallin (30, 149), interleukin-2 receptors (150), argininosuccinate synthetase (151), or tropomyosin (152), have been documented. Expression of troponin T (153, 154), *Drosophila* myosin heavy and light chains (155, 156), fibronectin (141), or pre-protachykinin (157), results in formation of multiple proteins produced in specific tissues or at different stages of development. This mode of RNA splicing allows for complicated combinatorial patterns of exon selection and sometimes involves choices of mutually exclusive splicing pathways (154).

Finally, both unspliced and spliced versions of the same pre-mRNA may be active, generating different, specific products. This mode of alternative expression is used in yeast mitochondria to produce maturases and subsequently other mitochondrial proteins from the same primary transcript (158; reviewed by Grivell & Borst, 159). A similar mechanism operates in expression of retroviral subgenomic RNAs (review by Bishop, 160). Alternative splicing has also been proposed to be involved in expression of development-controlling ANT-C and BX-C gene complexes in *Drosophila* (161, 162); however, further studies are required to verify this possibility.

The involvement of alternative splicing in the expression of a gene does not necessarily imply regulation at the level of splicing. In fact, in most of the systems listed above the alternative splicing could reflect the random utilization of alternative splice sites based on their relative strengths. Demonstration of regulation of gene expression at the level of splicing would follow from the finding that the relative efficiency of splicing of alternative pathways was different in different physiological states, cell types, or developmental stages. In a strict sense, this type of regulation has not been documented in any system with the possible exception of an example of alternative splicing of adenovirus mRNAs since it is very difficult to directly measure in vivo the relative efficiencies of splicing reactions. However, assuming that the steady-state levels of mRNAs reflect the relative efficiency of splicing, the examples discussed above of viral genes (for review see 3, 139) as well as troponin T (153, 154), *Drosophila* myosin heavy (155) and light (156) chains, fibronectin (141), and preprotachykinin (157), suggest that regulation of splicing occurs.

What mechanisms might be responsible for the regulation of splicing? Since the pattern of splicing of an identical precursor varies between different cellular states, *trans*-acting factors affecting the process must vary between these states.

Such *trans*-acting factors could be different levels of specific snRNPs that stimulate the activity of different splice sites. In this regard, it is interesting that regulation of expression of subsets of genes encoding U1 RNAs has been observed during oogenesis of *Xenopus laevis* (163). Expression of other small nuclear RNAs may be similarly regulated. Another class of *trans*-acting factors that could affect the pattern of splicing would be those that alter the gross conformation of the precursor RNA, perhaps by altering the secondary structure. Factors of this type could bind at distal sites within exons or introns and influence the processing at particular splice sites. A third class of *trans*-acting factors would be the presence or absence of another RNA complementary to one or more regions of the precursor. This RNA could bind to the precursor and affect the splicing pattern. The recent findings that secondary structures within introns are compatible with splicing both in vitro (164, 165) and in vivo (165a) suggests that interactions of a precursor with a complementary RNA are not precluded by the mechanism of mRNA splicing.

Trans-splicing

Splicing is typically pictured as the joining of two exons from a covalent precursor. However, the *trans*-splicing of exons from two separate precursor molecules (intermolecular splicing) might be important in certain biological systems and in particular circumstances. It was somewhat surprising to find that two such separate RNA substrates each containing a single exon segment with the adjacent portion of the intervening sequence (i.e. a single 5' or 3' splice site) can be joined together in an in vitro trans-splicing process. The efficiency of the reaction was enhanced when the termini of the intervening sequences of the two RNA molecules were paired in a short RNA duplex (164, 165). However, formation of this secondary structure was not strictly required since pairs of RNA substrates with no significant complementarity were also *trans*-spliced, albeit at lower efficiency (164). These findings support the notion that secondary structures within intervening sequences can influence splicing of flanking exons (see above). In addition, the presence of the *trans*-splicing activity precludes a simple scanning mechanism for mRNA splicing that threads along the phosphodiester bonds of the precursor.

The mechanism of processing of most mRNA precursors in the nucleus of mammalian cells must prevent a *trans*-splicing reaction (164). This could be achieved by formation of complexes on nascent RNA transcripts, sequestering newly synthesized sequences from interactions in *trans* with other precursors with complementary sequences. However, some biological systems might utilize *trans*-splicing as a natural step in mRNA synthesis. The 5' ends of many mRNAs in trypanosomes contain a common "leader" segment (166–168). In many cases, this short RNA fragment is encoded by sequences on a separate chromosome from the sequences comprising the mRNA body. The manner by which these short leader sequences are distributed during trypanosome mRNA synthesis is unknown but *trans*-splicing has been proposed as a possibility.

It is interesting to speculate that *trans*-splicing in mammalian extracts might occur by formation of partial complexes on RNAs containing the 5' splice sites and 3' splice sites followed by the coalescence of these partial structures into a complete complex. Partial complexes have been described for RNAs containing either 5' or 3' splice sites (95, 115).

Other Splicing Mechanisms

Introns have been described in ribosomal, tRNA, and protein-encoding genes in nuclear, mitochondrial, and chloroplast genomes. It is parsimonious to consider that RNA splicing only arose once and that introns in these various types of genes might have a common evolutionary origin and the splicing processes that excise them might be related.

Many of the nuclear tRNA genes of eukaryotes contain a short intron located one residue downstream of the anticodon (1). Such introns are excised from precursor tRNAs by two distinct enzymes, which have been partially or completely purified from yeast: an endonuclease that cleaves at the two splice sites to yield 2', 3'-cyclic phosphate and 5' hydroxyl termini (169), and an RNA ligase with a number of unusual activities. The mechanism for the reaction shown is in Figure 2. The ligation step in the reaction requires a series of activities: polynucleotide kinase for addition of phosphate to the 5' terminus, adenylylation of the 5' terminus, 3'-phosphodiesterase which resolves the 2'-3' cyclic phosphate to a 2' phosphate, and ligase activity which forms the 3'-5' phosphodiester linkage (170). At the moment, all of these activities are thought to be associated with one polypeptide. This polypeptide is apparently adenylated as an intermediate in the reaction. In this tRNA splicing mechanism the phosphate group forming the 3'-5' phosphodiester bond between the two exons originated from the γ ATP cofactor. The mechanism of splicing of tRNA precursors in vertebrate cells must differ as it has been shown that the phosphate group at the 3' end of the upstream exon is retained in the phosphodiester bond between the two exons (171–173).

An intermediate in the splicing of tRNA in yeast is the formation of a 2'-phosphomonoester, 3',5'-phosphodiester linkage. A 2'-phosphatase activity probably generates the 2'-hydroxyl group of the final product (170). Formation of the 2'-phosphomonoester, 3',5'-phosphodiester linkage is a general feature of RNA ligation in plants (174, 175) and, in these systems, the process responsible for ligation of the two tRNA exons may not be specific for tRNA precursors. In contrast, the yeast ligation activity shows specificity for tRNA halves (170). Several aspects of the tRNA ligation reaction are similar to reactions previously described for the RNA ligase encoded by T4 bacteriophage (176, 177). In fact, the combination of T4 polynucleotide kinase and T4 RNA ligase will join the two tRNA halves (170). Ligation activities have been

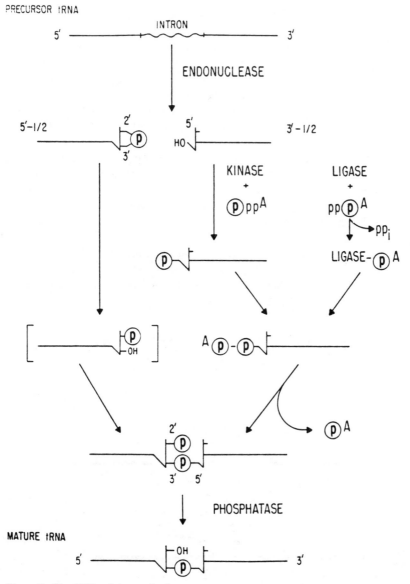

Figure 2   The tRNA splicing mechanism.

identified in mammalian extracts that join 2'-3' cyclic phosphate termini to 5'
hydroxyl termini retaining the phosphate in the newly formed 3'-5'
phosphodiester bond (178). This ligation activity in vitro is not specific for
tRNA substrates. A cyclase activity, converting 3' terminal phosphate to 2'-3'
cyclic phosphate, has also been characterized in mammalian extracts (178,

179). This enzyme may be responsible for maintaining the 2'-3' cyclic termini generated by endonuclease cleavage.

The endonuclease responsible for cleaving the pre-tRNA substrate at the correct sites probably recognizes the pre-tRNA tertiary structure. Interestingly, the endonuclease seems to be associated with the nuclear membrane suggesting that processing and transport may be related for tRNAs (170). Another indication that pre-tRNA structure is important in the reaction is that the two halves of a given tRNA will associate and undergo ligation, while 5' and 3' halves from two heterologous tRNAs will not (180).

At the superficial level where they can now be compared, the mechanism for splicing of tRNA precursors seems unrelated to the mRNA splicing process described above. In particular, the tRNA splicing process generates 2' phosphates, depends upon complementary secondary structures between exons, and yields linear intervening sequences which are not characteristic of the mRNA process.

The introns of mitochondria have been separated into two groups on the basis of sequence and secondary structure. Group I introns contain four sets of consensus sequences that are highly conserved and, as pairs, show partial complementarity (181, 182). These consensus sequences are also present in the

Splicing of Tetrahymena Ribosomal RNA

self-splicing intron from the *Tetrahymena* pre-rRNA (183–185), and thus all the group I introns are thought to possess the catalytic RNA structure. Some of these introns have been shown to have self-splicing activity (186). Self-splicing of RNA is the subject of other reviews (5, 187, 188) and will not be discussed here in detail.

A comparison of the overall schemes for the group I self-splicing mechanism (58) and the splicing of nuclear mRNA precursors (57) is of interest (Figures 3*A* and 3*B*). In both schemes, the first biochemical modification of the precursor

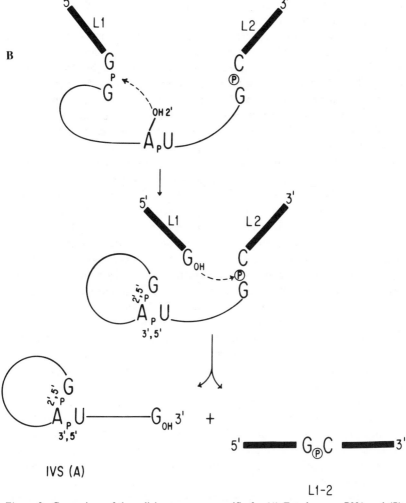

Figure 3 Comparison of the splicing processes specific for (*A*) *Tetrahymena* rRNA and (*B*) nuclear mRNA.

RNA is cleavage at the 5' splice site. In the self-splicing process, a 3' hydroxyl group of the guanosine cofactor attacks the phosphodiester bond and displaces the 3' hydroxyl group of the upstream exon. This is a transesterification reaction. In the mRNA splicing process, the 2' hydroxyl group at the branch site might have the equivalent role as the 3' hydroxyl group of the cofactor. The second step in the two mechanisms is also similar, the 3' hydroxyl group of the upstream exon is involved in a transesterification reaction with the 3' hydroxyl group of the intervening sequences. Consistent with a transesterification mechanism, the phosphate moieties at splice sites in both reactions are conserved. These similarities in steps between these two processes could be fortuitous or may reflect a common mechanism, or an ancestral process that was the antecedent of both current processes (189).

Group II-type introns, which have also been found in chloroplasts, in addition to the mitochondria, possess a common pattern of secondary structures as well as a set of conserved sequences (182, 185, 190, 191). In addition, most, if not all, of these introns can be aligned with a *GU* at the 5' terminus of a pyrimidine tract with A_C^T at the 3' terminus (187). Characterization of RNA from yeast mitochondria has shown that group II introns accumulate as circular RNAs (192, 193) with an unusual structure near the site where the termini are joined (194). This unusual structure blocks the elongation of reverse transcriptase at a specific site, a feature typical of a branch. Recently, Peebles et al (195) and van der Veen et al (196) have reported that the last intron in the yeast mitochondrial oxi 3 gene, a group II intron, will self-splice in an in vitro system. More interestingly, the intron is excised as a lariat RNA containing a branch. Consistent with a self-splicing mechanism using a 2' hydroxyl group, splicing of substrate from this oxi 3 gene intron only requires Mg^{2+} and not a nucleotide cofactor as do group I introns. This strongly suggests that the 2' hydroxyl group at the branch site participates in a transesterification reaction resulting in cleavage at the 5' splice site. This cleavage reaction and the subsequent ligation reaction are catalyzed by RNA sequences in the precursor RNA, probably in the intron.

The formation of branches during the self-splicing of group II introns and the overall similarity of the splicing processes of mRNA precursors and self-splicing RNAs supports the hypothesis that the current mRNA process is derived from a self-splicing process. It is possible that the mRNA process could also be RNA-catalyzed. The sequences responsible for the catalysis could not be part of each intron but instead would have to be part of a *trans*-acting factor perhaps as part of one or more snRNP. Alternatively, the biochemical reaction in the mRNA precursor splicing process could now be catalyzed by protein enzymes having subsumed this role from the original RNA-catalyzed process. If the mRNA splicing process is related to the self-splicing process, it would suggest that the introns found in genes encoding protein could be descendants of self-splicing introns (189).

Origin of Introns

The evolutionary origin of introns is a fascinating question but one that will be difficult to answer with any degree of confidence. Several hypotheses have been made which fall into two groups. One proposal holds that genes originally evolved with a discontinuous structure, i.e. with introns (197, 198). In this scenario, the genes of vertebrate cells would be more typical of those of primordial cells than the genes of prokaryotic cells. The alternative proposal is that genes originally evolved as continuous sets of sequences, similar to the gene structure in prokaryotic cells, and that introns arose by insertion of transposons (199, 200). The major difficulty with this latter hypothesis is that a splicing system must either be present or coevolved that can excise the intervening sequences (201). The sequence specificity of the transposable element would need to be related to the specificity of the splicing process. The nuclear transposable elements that have been characterized to date do not insert with boundary sequences that generate splice sites. The insertion of an intron by a gene conversion process has been described for yeast mitochondria (202, 203). This process is promoted by a protein encoded by the intron. Insertion of introns by this process can be considered a special case of a general process where nonhomologous recombination near the boundaries of that intron would place an intron at a new position. Comparison of the intron positions in a family of genes encoding serine proteases suggests that introns have been inserted within genes since the divergence of vertebrate species (204). However, it is difficult to establish whether the presence of introns at different positions in closely related genes reflects the acquisition or loss of introns during evolution.

An interesting hypothesis for the origin of introns can be developed from the general belief that RNA-specified processes may have developed very early in evolution. The RNA-catalyzed cleavage and ligation reactions associated with self-splicing of RNA have been proposed as examples of such primordial processes (188, 205). If this was the case, then as genes evolved in primitive systems, the self-splicing process may have been present both for the excision of introns as well as the insertion of introns. Reversal of the excision reaction, an event that may occur on a thermodynamic basis, would insert introns in new positions in RNA (205). Then the introns in nuclear genes encoding proteins may have evolved from self-splicing introns, a suggestion strengthened by the similarities in the two splicing processes as discussed above (189).

It is likely that most of the introns in protein-encoding genes have no function in the contemporary life of the organism. Their presence must reflect events in the evolution of the organism and the lack of sufficient selective advantage to select for individuals that have deleted an intron. At one time, some positive selective advantage must have been associated with the presence of introns to account for their persistence in biological systems. In early biological systems, splicing of RNA and introns might have played a role in the assembly of small tracts of protein-encoding sequences into larger and more complicated genes

(197, 206). During the evolution of contemporary organisms, the incorporation of exons that specify functional domains of proteins has been used to generate new genes with new functions (207). This process is probably facilitated by the presence of introns that can be used as sites for nonhomologous recombinations. As reviewed above, the ability to join various combinations of exons at the processing level can be used to regulate the expression of genetic material. This could also have been important for the evolution of multicellular organisms.

ACKNOWLEDGMENTS

We thank Robert Marciniak for the consensus sequence analysis.

Preparation of this article was supported by NIH grant RO1-GM34277 and in part by NCI grant P30-CA14051 (to P.A.S.). R.A.P. was supported by a Myron A. Bantrell Charitable Trust Fellowship, M.M.K. by a Jane Coffin Childs Memorial Fund Fellowship, and P.J.G. by a Helen Hay Whitney Foundation Fellowship.

Literature Cited

1. Abelson, J. 1979. *Ann. Rev. Biochem.* 48:1035–69
2. Breathnach, R., Chambon, P. 1981. *Ann. Rev. Biochem.* 50:349–83
3. Nevins, J. R. 1983. *Ann. Rev. Biochem.* 52:441–66
4. Rogers, J. H. 1984. *Int. Rev. Cytol. Suppl.* 17:1–71
5. Cech, T. R., Bass, B. L. 1986. *Ann. Rev. Biochem.* 55:599–629
6. Salditt-Georgieff, M., Harpold, M., Chen-Kiang, S., Darnell, J. E. Jr. 1980. *Cell* 19:69–78
7. Martin, T. E., Pullman, J. M., McMullen, M. D. 1980. In *Cell Biology: A Comprehensive Treatise,* ed. D. M. Prescott, L. Goldstein, 4:137–74. New York: Academic
8. LeStourgeon, W. M., Lothstein, L., Walker, B. W., Beyer, A. L. 1981. In *Cell Nucleus,* ed. H. Busch, 9:49–87. New York: Academic
9. Choi, Y. D., Dreyfuss, G. 1984. *J. Cell Biol.* 99:1997–2004
10. Schibler, U., Kelly, D. E., Perry, R. P. 1977. *J. Mol. Biol.* 115:695–714
11. Chen-Kiang, S., Nevins, J. R., Darnell, J. E. Jr. 1979. *J. Mol. Biol.* 135:733–52
12. Revel, M., Groner, Y. 1978. *Ann. Rev. Biochem.* 47:1079–126
13. Keohavong, P., Gattoni, R., LeMoullea, J. M., Jacob, M., Stevenin, J. 1982. *Nucleic Acids Res.* 10:1215–29
14. Zeevi, M., Nevins, J. R., Darnell, J. E. Jr. 1981. *Cell* 26:39–46
15. Nordstrom, J. L., Roop, D. R., Tsai, M. J., O'Malley, B. W. 1979. *Nature* 278:328–30
16. Roop, D. R., Nordstrom, J. L., Tsai, S.-Y., Tsai, M. J., O'Malley, B. W. 1978. *Cell* 15:651–85
17. Berget, S. M., Sharp, P. A. 1979. *J. Mol. Biol.* 129:547–65
18. Curtis, P. J., Montei, N., Weissmann, C. 1977. *Cold Spring Harbor Symp. Quant. Biol.* 42:971–84
19. Pikielny, C. W., Rosbash, M. 1985. *Cell* 41:119–26
20. Nevins, J. R. 1979. *J. Mol. Biol.* 130:493–506
21. Khoury, G., Gruss, P., Dhar, R., Lai, C.-J. 1979. *Cell* 18:85–92
22. Hamer, D. H., Leder, P. 1979. *Cell* 18:1299–302
23. Gruss, P., Lai, C.-J., Dhar, R., Khoury, G. 1979. *Proc. Natl. Acad. Sci. USA* 76:4317–21
23a. Gruss, P., Khoury, G. 1980. *Nature* 286:634–37
24. Treisman, R., Novak, U., Favaloro, J., Kamen, R. 1981. *Nature* 292:595–698
25. Gething, M.-J., Sambrook, J. 1981. *Nature* 293:620–25
26. Breathnach, R., Benoist, C., O'Hare, K., Gannon, F., Chambon, P. 1978. *Proc. Natl. Acad. Sci. USA* 75:4853–57
27. Dodgson, J. B., Engel, J. D. 1983. *J. Biol. Chem.* 258:4623–29
28. Erbil, C., Niessing, J. 1983. *EMBO J.* 2:1339–43

29. Wieringa, B., Meyer, F., Reiser, J., Weissmann, C. 1983. *Nature* 301:38–43
30. King, C. R., Piatigorsky, J. 1983. *Cell* 32:707–12
31. Naora, H., Deacon, N. J. 1982. *Proc. Natl. Acad. Sci. USA* 79:6196–200
32. Baldwin, A. S., Kittlek, E. W., Emerson, C. P. 1985. *Proc. Natl. Acad. Sci. USA.* 82:8080–84
33. Treisman, R., Kamen, R. 1981. *J. Mol. Biol.* 148:273–301
34. Manley, J. L., Fire, A., Cano, A., Sharp, P. A., Gefter, M. L. 1980. *Proc. Natl. Acad. Sci. USA* 77:3855–59
35. Dignam, J. D., Lebowitz, R. M., Roeder, R. G. 1983. *Nucleic Acids Res.* 11:1475–89
36. Moore, C. L., Sharp, P. A. 1984. *Cell* 36:581–91
37. Padgett, R. A., Hardy, S. F., Sharp, P. A. 1983. *Proc. Natl. Acad. Sci. USA* 80:5230–34
38. Hernandez, N., Keller, W. 1983. *Cell* 35:89–99
39. Hardy, S. F., Grabowski, P. J., Padgett, R. A., Sharp, P. A. 1984. *Nature* 308:375–77
40. Krainer, A. R., Maniatis, T., Ruskin, B., Green, M. R. 1984. *Cell* 36:993–1005
41. Melton, D. A., Krieg, P. A., Rebagliati, M. R., Maniatis, T., Zinn, K., Green, M. R. 1984. *Nucleic Acids Res.* 12:7035–56
42. Goldenberg, C. J., Raskas, H. J. 1981. *Proc. Natl. Acad. Sci. USA* 78:5430–34
43. Konarska, M. M., Padgett, R. A., Sharp, P. A. 1984. *Cell* 38:731–36
44. Goldenberg, C. J., Hauser, S. D. 1983. *Nucleic Acids Res.* 11:1337–48
45. Edery, I., Sonenberg, N. 1985. *Proc. Natl. Acad. Sci. USA.* 82:7590–94
46. Brody, E., Abelson, J. 1985. *Science* 228:963–67
47. Newman, A. J., Lin, R.-J., Cheng, S.-C., Abelson, J. 1985. *Cell* 42:335–44
48. Lin, R.-J., Newman, A. J., Cheng, S.-C., Abelson, J. 1985. *J. Biol. Chem.* 260:14780–92
49. Furneaux, H. M., Perkins, K. K., Freyer, G. A., Arenas, J., Hurwitz, J. 1985. *Proc. Natl. Acad. Sci. USA* 82:684–88
50. Krainer, A. R., Maniatis, T. 1985. *Cell* 42:725–36
51. Grabowski, P. J., Padgett, R. A., Sharp, P. A. 1984. *Cell* 37:415–27
52. Padgett, R. A., Konarska, M. M., Grabowski, P. J., Hardy, S. F., Sharp, P. A. 1984. *Science* 225:898–903
53. Ruskin, B., Krainer, A. R., Maniatis, T., Green, M. R. 1984. *Cell* 38:317–31
54. Zeitlin, S., Efstratiadis, A. 1984. *Cell* 39:589–602
55. Rodriguez, J. R., Pikielny, C. W., Rosbash, M. 1984. *Cell* 39:603–10
56. Domdey, H., Apostol, B., Lin, R.-J., Newman, A., Brody, E., Abelson, J. 1984. *Cell* 39:611–21
57. Konarska, M. M., Grabowski, P. J., Padgett, R. A., Sharp, P. A. 1985. *Nature* 313:552–57
58. Cech, T. R., Zaug, A. J., Grabowski, P. J. 1981. *Cell* 27:487–96
59. Wallace, J. C., Edmonds, M. 1983. *Proc. Natl. Acad. Sci. USA* 80:950–54
60. Ruskin, B., Green, M. R. 1985. *Science* 229:135–40
61. Reed, R., Maniatis, T. 1985. *Cell* 41:95–105
62. Ruskin, B., Greene, J. M., Green, M. R. 1985. *Cell* 41:833–44
63. Langford, C. J., Gallwitz, D. 1983. *Cell* 33:519–27
64. Pikielny, C. W., Teem, J. L., Rosbash, M. 1983. *Cell* 34:395–403
65. Langford, C. J., Klinz, F.-J., Donath, C., Gallwitz, D. 1984. *Cell* 36:645–53
66. Keller, E. B., Noon, W. A. 1984. *Proc. Natl. Acad. Sci. USA* 81:7417–20
67. Rautmann, G., Breathnach, R. 1985. *Nature* 315:430–32
68. Padgett, R. A., Konarska, M. M., Aebi, M., Hornig, H., Weissmann, C., Sharp, P. A. 1985. *Proc. Natl. Acad. Sci. USA.* 82:8349–53
68a. Jacquier, A., Rodriguez, J. R., Rosbash, M. 1985. *Cell* 43:423–30
69. Treisman, R., Orkin, S. H., Maniatis, T. 1983. In *Globin Gene Expression and Hematopoietic Differentiation*, ed. G. Stamatoyannopoulos, A. W. Nienhuis, pp. 99–121. New York: Liss
70. Treisman, R., Orkin, S. H., Maniatis, T. 1983. *Nature* 302:591–96
71. Chu, G., Sharp, P. A. 1981. *Nature* 289:378–82
72. Horowitz, M., Cepko, C. L., Sharp, P. A. 1983. *J. Mol. Appl. Gen.* 2:147–59
73. Kuhne, T., Wieringa, B., Reiser, J., Weissmann, C. 1983. *EMBO J.* 2:727–33
74. Lang, K. M., Spritz, R. A. 1983. *Science* 220:1351–55
75. Wieringa, B., Hofer, E., Weissmann, C. 1983. *Cell* 37:915–25
75a. Parker, R., Guthrie, C. 1985. *Cell* 41:107–18
76. Lerner, M. R., Boyle, J. A., Mount, S. M., Wolin, S. L., Steitz, J. A. 1980. *Nature* 283:220–24
77. Rogers, J., Wall, R. 1980. *Proc. Natl. Acad. Sci. USA* 77:1877–79
78. Reddy, R., Busch, H. 1981. In *Cell Nu-*

1148   PADGETT, GRABOWSKI, KONARSKA ET AL

cleus, ed. H. Busch, 8:261–306. New York: Academic

79. Zieve, G. W. 1981. *Cell* 25:296–97
80. Lerner, M. R., Steitz, J. A. 1981. *Cell* 25:298–300
81. Busch, H., Reddy, R., Rothblum, L., Choi, Y. C. 1982. *Ann. Rev. Biochem.* 51:617–54
82. Reddy, R., Henning, D., Busch, H. 1985. *J. Biol. Chem.* 260:10930–35
82a. Bringmann, P., Rinke, J., Appel, B., Reuter, R., Lührmann, R. 1983. *EMBO J.* 2:1129–35
83. Oshima, Y., Itoh, M., Okada, N., Miyata, T. 1981. *Proc. Natl. Acad. Sci. USA* 78:4471–74
84. Deimel, B., Louis, C., Sekeris, C. E. 1977. *FEBS Lett.* 73:80–84
85. Howard, E. F. 1978. *Biochemistry* 17:3228–36
86. Zieve, G., Penman, S. 1981. *J. Mol. Biol.* 145:501–23
87. Calvet, J. P., Pederson, T. 1981. *Cell* 26:363–70
88. Calvet, J. P., Meyer, L. M., Pederson, T. 1982. *Science* 217:456–58
89. Setyono, B., Pederson, T. 1984. *J. Mol. Biol.* 174:285–95
90. Padgett, R. A., Mount, S. M., Steitz, J. A., Sharp, P. A. 1983. *Cell* 35:101–7
91. Kramer, A., Keller, W., Appel, B., Luhrmann, R. 1984. *Cell* 38:299–307
92. Bozzoni, I., Annesi, F., Beccari, E., Fragapane, P., Pierandrei-Amaldi, P., Amaldi, F. 1984. *J. Mol. Biol.* 180:1173–78
93. Fradin, A., Jove, R., Hemenway, C., Keiser, H. D., Manley, J. L., Prives, C. 1984. *Cell* 37:927–36
94. Mount, S. M., Pettersson, I., Hinterberger, M., Karmas, A., Steitz, J. A. 1983. *Cell* 33:509–19
95. Grabowski, P. J., Seiler, S. R., Sharp, P. A. 1985. *Cell* 42:345–53
96. Black, D. L., Chabot, B., Steitz, J. A. 1985. *Cell* 42:737–50
97. Rinke, J., Appel, B., Blocker, H., Frank, R., Luhrmann, R. 1984. *Nucleic Acids Res.* 12:4111–26
98. Chabot, B., Black, D., LeMaster, D. M., Steitz, J. A. 1985. *Science* 230:1344–49
99. Pederson, T. 1983. *J. Cell. Biol.* 97:1321–26
100. Osheim, Y. N., Miller, O. L., Beyer, A. L. 1985. *Cell* 43:143–51
101. Samarina, O. P., Lukanidin, E. M., Molar, J., Georgiev, G. P. 1968. *J. Mol. Biol.* 33:251–63
102. Beyer, A. L., Christensen, M. E., Walker, B. W., LeStourgeon, W. M. 1977. *Cell* 11:127–38
103. Karn, J., Vidali, G., Boffa, L. C.,

Allfrey, V. G. 1977. *J. Biol. Chem.* 252:7307–22
104. LeStourgeon, W. M., Beyer, A. L., Christensen, M. E., Walker, B. W., Poupore, S. M., Daniels, L. P. 1977. *Cold Spring Harbor Symp. Quant. Biol.* 42:885–98
105. Thomas, J. D., Glowacka, S. K., Szer, W. 1983. *J. Mol. Biol.* 171:439–55
106. Dreyfuss, G., Choi, Y. D., Adam, S. A. 1984. *Mol. Cell. Biol.* 4:1104–14
107. Choi, Y. D., Dreyfuss, G. 1984. *Proc. Natl. Acad. Sci. USA* 81:7471–75
108. Gattoni, R., Stevenin, J., Devilliers, G., Jacob, M. 1978. *FEBS Lett.* 90:318–23
109. Stevenin, J., Gallinaro-Matringe, H., Gattoni, R., Jacob, M. 1977. *Eur. J. Biochem.* 74:589–602
110. Kinniburgh, A. J., Martin, T. E. 1976. *Proc. Natl. Acad. Sci. USA* 73:2725–29
111. Padgett, R. A., Wahl, G. M., Stark, G. R. 1982. *Mol. Cell. Biol.* 2:293–301
112. Sperling, R., Sperling, J., Levine, A. D., Spann, P., Stark, G. R., Kornberg, R. D. 1985. *Mol. Cell. Biol.* 5:569–75
113. Choi, Y. D., Grabowski, P. J., Sharp, P. A., Dreyfuss, G. 1986. *Science* 231:1534–39
114. Sharp, P. A. 1981. *Cell* 23:643–46
115. Frendewey, D., Keller, W. 1985. *Cell* 42:355–67
116. Ruskin, B., Green, M. R. 1985. *Nature* 317:732–34
117. Ruskin, B., Green, M. R. 1985. *Cell.* 43:131–42
118. Deleted in proof
119. Alt, F. W., Bothwell, A. L. M., Knapp, M., Siden, E., Mather, E., et al. 1980. *Cell* 20:293–301
120. Rogers, J., Early, P., Carter, C., Calame, K., Bond, M., et al. 1980. *Cell* 20:303–12
121. Early, P., Rogers, J., Davis, M., Calame, K., Bond, M., et al. 1980. *Cell* 20:313–19
122. Maki, R., Roeder, W., Traunecker, A., Sidman, C., Wabl, M., et al. 1981. *Cell* 24:353–65
123. Rosenfeld, M. G., Lin, C. R., Amara, S. G., Stolarsky, L., Roos, B. A., et al. 1982. *Proc. Natl. Acad. Sci. USA* 79:1717–21
124. Amara, S. G., Jonas, V., Rosenfeld, M. G., Ong, E. S., Evans, R. M. 1982. *Nature* 298:240–44
125. Rosenfeld, M. G., Amara, S. G., Evans, R. M. 1984. *Science* 225:1315–20
126. Kitamura, N., Takagaki, Y., Furuto, S., Tanaka, T., Nawa, H., Nakanishi, S. 1983. *Nature* 305:545–49
127. Kitamura, N., Kitagawa, H., Fukushima, D., Takagaki, Y., Miyata, T., Nakanishi, S. 1985. *J. Biol. Chem.* 260:8610–17

128. Crabtree, G. R., Kant, J. A. 1982. *Cell* 31:159–66
129. Fornace, A. J. Jr., Cummings, D. E., Comeau, C. M., Kant, J. A., Crabtree, G. R. 1984. *J. Biol. Chem.* 259:12826–30
130. Henikoff, S., Sloan, J. S., Kelly, J. D. 1983. *Cell* 34:405–14
131. Clark, A. J., Clissold, P. M., Alshawi, R., Beattie, P., Bishop, J. 1984. *EMBO J.* 3:1045–52
132. Young, R. A., Hagenbüchle, O., Schibler, U. 1981. *Cell* 23:451–58
133. Schibler, U., Hagenbüchle, O., Wellauer, K., Pittet, A. C. 1983. *Cell* 33:501–8
134. Benyajati, C., Spoerel, N., Haymerle, H., Ashburner, M. 1983. *Cell* 33:125–33
135. Robert, B., Daubas, P., Akimenko, M.-A., Cohen, A., Garner, I., et al. 1984. *Cell* 39:129–40
136. Nabeshima, Y. I., Fujii-Kuriyama, Y., Muramatsu, M., Ogata, K. 1984. *Nature* 308:333–38
137. Periasamy, M., Strehler, E. E., Granfinkel, L. I., Gubits, R. M., Ruiz-Opazo, N., Nadal-Ginard, B. 1984. *J. Biol. Chem.* 259:13595–604
138. Barton, P. J. R., Cohen, A., Robert, B., Fiszman, M. Y., Bonhomme, F., et al. 1985. *J. Biol. Chem.* 260:8578–84
139. Ziff, E. B. 1980. *Nature* 287:491–99
140. Schwarzbauer, J. E., Tamkun, J. W., Lemischka, I. R., Hynes, R. O. 1983. *Cell* 35:421–31
141. Kornblihtt, A. R., Vibe-Pedersen, K., Barelle, F. E. 1984. *EMBO J.* 3:221–26
142. Kress, M., Glavos, D., Khoury, G., Jay, G. 1983. *Nature* 306:602–4
143. Transy, C., Lalanne, J.-L., Kourilsky, P. 1984. *EMBO J.* 3:2383–86
144. Sher, B. T., Navin, R., Coligan, J. E., Hood, L. E. 1985. *Proc. Natl. Acad. Sci. USA* 82:1175–79
145. Oates, E., Herbert, E. 1984. *J. Biol. Chem.* 259:7421–25
146. Laub, O., Rutter, W. J. 1983. *J. Biol. Chem.* 258:6043–50
147. Jansen, M., van Schaik, F. M. A., van Tol, H., Van den Brande, J. L., Sussenbach, J. S. 1985. *FEBS Lett.* 179:243
148. DeNoto, F. M., Moore, D. D., Goodman, H. M. 1981. *Nucleic Acids Res.* 9:3719–30
149. King, C. R., Piatigorsky, J. 1984. *J. Biol. Chem.* 259:1822–26
150. Leonard, W. J., Depper, J. M., Crabtree, G. R., Rudikoff, S., Pumphrey, J., et al. 1984. *Nature* 311:626–31
151. Freytag, S. O., Beaudet, A. L., Bock, H. G. O., O'Brien, W. E. 1984. *Mol. Cell. Biol.* 4:1978–84

152. Basi, G. S., Boardman, M., Storki, R. V. 1984. *Mol. Cell. Biol.* 4:2828–31
153. Medford, R. M., Nguyen, H. T., Destree, A. T., Summers, E., Nadal-Ginard, B. 1984. *Cell* 38:409–21
154. Breibart, R. E., Nguyen, H. T., Medford, R. M., Destree, A. T., Mahdavi, V., Nadal-Ginard, B. 1985. *Cell* 41:67–82
155. Rozek, C. E., Davidson, N. 1983. *Cell* 32:23–34
156. Falkenthal, S., Parker, V. P., Davidson, N. 1985. *Proc. Natl. Acad. Sci. USA* 82:449–53
157. Nawa, H., Kotani, H., Nakanishi, S. 1984. *Nature* 312:729–34
158. Lazowska, J., Jacq, C., Slonimski, P. P. 1980. *Cell* 22:333–48
159. Grivell, L. A., Borst, P. 1982. *Nature* 298:703–4
160. Bishop, J. M. 1983. *Ann. Rev. Biochem.* 52:301–54
161. Scott, M. P., Weiner, A. J., Hazelrigg, T. I., Polisky, B. A., Pirrotta, V., et al. 1983. *Cell* 35:763–76
162. Beachy, P. A., Helfand, S. L., Hogness, D. S. 1985. *Nature* 313:545–51
163. Forbes, D. J., Kirschner, M. W., Caput, D., Dahlberg, J. E., Lund, E. 1984. *Cell* 38:681–89
164. Konarska, M. M., Padgett, R. A., Sharp, P. A. 1985. *Cell* 42:157–64
165. Solnick, D. 1985. *Cell* 42:157–64
165a. Solnick, D. 1985. *Cell* 43:667–76
166. Boothroyd, J. C., Cross, G. A. M. 1982. *Gene* 20:281–89
167. Parsons, M., Nelson, R. G., Watkins, K. P., Agabian, N. 1984. *Cell* 38:309–16
168. Guyaux, M., Cornelissen, A. W. C. A., Pays, E., Steinert, M., Borst, P. 1985. *EMBO J.* 4:995–98
169. Peebles, C. L., Gegenheimer, P., Abelson, J. 1983. *Cell* 32:525–36
170. Greer, C. L., Peebles, C. L., Gegenheimer, P., Abelson, J. 1983. *Cell* 32:537–46
171. Nishikura, K., DeRobertis, E. M. 1981. *J. Mol. Biol.* 145:405–20
172. Filipowicz, W., Shatkin, A. J. 1983. *Cell* 32:547–57
173. Laski, F. A., Fire, A. Z., RajBhandary, U. L., Sharp, P. A. 1983. *J. Biol. Chem.* 258:11974–80
174. Konarska, M. M., Filipowicz, W., Domdey, H., Gross, J. H. 1981. *Nature* 293:112–6
175. Konarska, M. M., Filipowicz, W., Gross, J. H. 1982. *Proc. Natl. Acad. Sci. USA* 79:1474–78
176. Cranston, J. W., Silber, R., Malathi, V. G., Hurwitz, J. 1984. *J. Biol. Chem.* 249:7447–56

177. Uhlenbeck, O. C., Gumport, R. I. 1981. In *The Enzymes*, ed. P. Boyer, 15:31–58. New York: Academic

178. Filipowicz, W., Konarska, M., Shatkin, A. J. 1983. *Nucleic Acids Res.* 11:1405–18

179. Reinberg, D., Arenas, J., Hurwitz, J. 1985. *J. Biol. Chem.* 260:6088–97

180. Peebles, C. L., Ogden, R. C., Knapp, G., Abelson, J. 1979. *Cell* 18:27–35

181. Davies, R. W., Waring, R. B., Ray, J. A., Brown, T. A., Scazzocchio, C. 1982. *Nature* 300:719–24

182. Michel, F., Jacquier, A., Dujon, B. 1981. *Biochemie* 64:867–81

183. Cech, T. R., Tanner, N. K., Tinoco, I., Weir, B. R., Zuker, M., et al. 1983. *Proc. Natl. Acad. Sci. USA* 80:3903–7

184. Waring, R. B., Scazzocchio, C., Brown, T. A., Davies, R. W. 1983. *J. Mol. Biol.* 167:595–605

185. Michel, F., Dujon, B. 1983. *EMBO J.* 2:33–38

186. Tabak, H. F., van der Horst, G., Osinga, K. A., Arnberg, A. C. 1984. *Cell* 39:623–29

187. Cech, T. R. 1985. *Cell* 34:713–16

188. Cech, T. R. 1985. *Int. Rev. Cytol.* 93:3–22

189. Sharp, P. A. 1985. Cell. 42:397–400

190. Koller, B., Gingrich, J. C., Stiegler, G. L., Farley, M. A., Delius, H., et al. 1984. *Cell* 36:545–53

191. Sugita, M., Shinozaki, K., Sugiura, M. 1985. *Proc. Natl. Acad. Sci. USA* 82:3557–61

192. Arnberg, A. C., van Ommen, G.-J. B., Grivell, L. A., van Bruggen, E. F. J.,

Borst, P. 1980. *Cell* 19:313–19

193. Halbreich, A., Pajot, P., Foucher, M., Grandchamp, C., Slonimski, P. 1980. *Cell* 19:321–29

194. Grivell, L. A., Bonen, L., Borst, P. 1983. In *Genes: Structure and Expression,* ed. A. M. Kroon, pp. 279–306. Chichester: Wiley

195. Peebles, C. L., Perlman, P. S., Mecklenburg, K. L., Petrillo, M. L., Tabor, J. H., et al. 1986. *Cell* 44:213–23

196. van der Veen, R., Arnberg, A. C., van der Horst, G., Bonen, L., Tabak, H. F., et al. 1986. *Cell* 44:225–34

197. Doolittle, W. F. 1978. *Nature* 272:581–82

198. Darnell, J. E. Jr. 1978. *Science* 202:1257–60

199. Cavalier-Smith, T. J. 1978. *J. Cell. Sci.* 34:247–78

200. Crick, F. H. C. 1979. *Science* 204:264–71

201. Cavalier-Smith, T. J. 1985. *Nature* 315:283–84

202. Jacquier, A., Dujon, B. 1985. *Cell* 41:383–94

203. Macreadie, I. G., Scott, R. M., Zinn, A. R., Buton, R. A. 1985. *Cell* 41:395–402

204. Rogers, J. 1985. *Nature* 315:458–59

205. Kruger, K., Grabowski, P. J., Zaug, A. J., Sands, J., Gottschling, D. E., et al. 1982. *Cell* 31:147–57

206. Gilbert, W. 1978. *Nature* 271:501

207. Gilbert, W. 1985. *Nature* 228:823–24

208. Teem, J. L., Abovich, N., Kauter, N. F., Schwindinger, W. F., Warner, J. R., et al. 1984. *Nucleic Acids Res.* 22:8295–311

Ann. Rev. Biochem. 1986. 55:1151–91

THE HEAT-SHOCK RESPONSE

Susan Lindquist

Department of Molecular Genetics and Cell Biology, The University of Chicago, Chicago, Illinois 60637

CONTENTS

PERSPECTIVES AND SUMMARY

When cultured cells or whole organisms are exposed to elevated temperatures, they respond by synthesizing a small number of highly conserved proteins, the heat-shock proteins, or hsps. This response is universal. It has been observed in every organism in which it has been sought, from eubacteria to archebacteria, from mice to soybeans. It is found in nearly every cell- and tissue-type of multicellular organisms, in explanted tissues, and in cultured cells. It may be

0066-4154/86/0701-1151$02.00

that some creature living in the depths of the ocean does not have a heat-shock response, but that is doubtful. The proteins are induced by a wide variety of other stresses, seem to have very general protective functions, and may well play a role in normal growth and development.

Man has long studied the effects of heat on himself and other living things, but studies of the heat-shock response per se began in 1962 with the publication of a little-noticed paper describing a new set of puffs on the salivary gland chromosomes of a fruit fly, *Drosophila busckii*, puffs induced by heat, dinitrophenol, or sodium salicylate (1). For the next decade the response was studied solely at the cytological level, but several important observations were made. Most notably, it was shown that the puffs were (*a*) induced by several other stress treatments (1–4), (*b*) produced within a few minutes (2, 3), (*c*) associated with newly synthesized RNA (1, 4), (*d*) found in other *Drosophila* species and in many different tissues (5, 6), and (*e*) accompanied by the disappearance of previously active puffs (2–4). In 1973, Tissières & Mitchell inaugurated the molecular analysis of the response by reporting that the induction of these puffs coincided with the synthesis of a small number of new proteins (7). Investigations quickly shifted to *Drosophila* tissue-culture cells (8, 9), far more amenable to biochemical analysis, and the pace accelerated.

From the first, the response was hailed as a model system for investigating gene structure and regulation. In this vein, investigations have proven extremely successful. The genes for the *Drosophila* hsps were among the first eukaryotic genes to be cloned (10–14), to have their organization within the genome defined (10, 13–22), to have their chromatin structure determined before and after activation (23–26), to have their regulatory sequences identified (27, 28), and to have the transcription factors interacting with these sequences characterized (29–33). Genes for one of the proteins, hsp70, provided one of the first convincing examples of gene conversion acting on an evolutionary time scale (34). The response also provided one of the first examples of selective gene expression operating at the level of translation (35, 36).

It was not until 1978–1979 that investigators, working on other organisms, discovered that heat and many other types of stress could induce the synthesis of similar proteins in cultured avian cells (37), in yeast (38, 39), and in Tetrahymena (40). Within a few years similar responses had been reported in an extraordinary variety of organisms. Investigations of gene structure and regulation in these organisms are proceeding apace and several interesting lessons have already been learned. We now know that some heat-shock genes have been very highly conserved during evolution, not only in their protein-coding sequences (41–45), but also in their regulatory sequences (27, 41–48). Furthermore, it is now clear that the responses of different cells and organisms are regulated in different ways, reflecting their specific biological characteristics and providing a marvelous demonstration of biological versatility and adaptation (49). Studies in *Escherichia coli* have led to the identification of a new sigma factor, suggesting that alternate sigma factors play a more general role in prokaryotic regulation than previously supposed (50, 51).

Unfortunately, with respect to the question of function, investigations have met with less success. The ubiquity of the response and the remarkable conservation of some of the genes attest to their importance. Accordingly, much attention has recently been given to determining what the proteins do. There is now convincing, though by no means incontrovertible, evidence that the proteins provide protection from the toxic effects of stress. The specific protective mechanisms, however, have proven elusive quarry. The functions of only a few of the bacterial proteins have been established and these, for the most part, only in the context of their requirements for the growth of bacteriophage. Will this information prove relevant to eukaryotic cells? The answer is not clear.

Of course, the picture is far from bleak. Combining data from the genetic analyses of yeast and *E. coli* with data from the biochemical and immunological analyses of higher eukaryotes is beginning to yield important insights. Furthermore, awareness is growing that aspects of the response are directly pertinent to important human problems: the enhancement of agricultural productivity, the control of virulent dimorphic pathogens, the effective use of hyperthermia in cancer therapy, and the prevention of developmental anomalies caused by teratogenic agents. The third decade of research holds great promise.

CHARACTERIZATION OF THE RESPONSE

Comparison: Different Organisms and Stages of Development

Drosophila cells are normally grown at 25°C. Hsps are induced when the temperature is raised to 29–38°C, but the maximum response is observed at 36–37°C (52). At such temperatures, heat-shock messages are produced within four minutes. Within an hour several thousand transcripts are present per cell (53). At the same time, both the transcription of previously active genes (2, 9, 54, 55) and the translation of pre-existing messages are repressed (36, 37). Heat-shock messages appear in the cytoplasm within a few minutes of temperature elevation and are immediately translated with very high efficiency (53). As long as cells are maintained at high temperatures, hsps continue to be the primary products of protein synthesis. When cells are returned to normal temperatures, normal protein synthesis gradually resumes, the timing a reflection of the severity of the preceding heat shock (56, 57).

The induction of hsps in other organisms is equally rapid, but the maximum induction temperature varies, correlating with the normal range of environmental exposure. In salmon and trout, the maximum response is observed at about

28°C (58), in slime molds at 30°C (59, 60), in sea urchins at 30–32°C (61), in yeast at 39–40°C (62), in corn plants at 40–45°C (63, 64), in cyanobacteria (65) and *E. coli* (66–68) at 45–50°C, and in Halobacteria at 60°C (69). In organisms that grow over a broad range of temperatures, the maximum response is usually achieved at 10–15°C above the optimum growth temperature. In organisms that grow over a more restricted range, maximum response occurs at about 5°C above the optimum.

The response appears to be transient in some organisms and sustained in others. The distinction may be largely artificial, based on the limited number of experimental parameters tested. In *E. coli,* the response is transient when cells raised at 30–37°C are shifted to 42°C, but sustained when they are shifted to 45–50°C (66–68). In salmon the response is transient at 24°C and sustained at 28°C (70). In corn plants it is transient at 40°C and sustained at 45°C (71). Such findings suggest the response will be transient in most organisms at moderate temperatures, with growth resuming after a temporary pause, and sustained at higher temperatures until cells slowly begin to die. The temperature range for a transient response may be very narrow in some organisms, and very broad in others.

There are several sources of confusion in the literature. (*a*) Transient responses are not completely transient. As hsps accumulate, their synthesis declines but plateaus at a higher level (68). (*b*) In eukaryotes, close relatives of the hsps are synthesized at normal temperatures. These have often been confused with the hsps themselves (43, 72–76). (*c*) Hsps have characteristic and relatively stable breakdown products. Hsp70, for example, is proteolytically reduced to doublets of ~40–44 kd (kilodaltons) and ~18–22 kd (77; S. Lindquist, unpublished observations). The breakdown products have sometimes been reported as novel hsps. (*d*) In order to achieve a rapid response, mammalian and avian cells are often given very short exposures to extreme temperatures, with the result that hsps accumulate only when cells are returned to normal temperatures. (*e*) The response is influenced by the metabolic state of the cell. In the yeast *Saccharomyces cerevisiae,* for example, the response of fermenting cells grown at 25°C is transient at 36°C and sustained at 40°C; in respiring cells, the response is transient at 34°C and sustained at 36°C (62). (*f*) Synthesis is influenced by previous incubation temperatures. When *Drosophila* cells are shifted directly to high temperatures, the maximum response occurs at 37°C and very little synthesis is observed at 39°C. When the temperature is increased gradually (a regimen more likely to reflect natural environmental exposure), the response is extended by several degrees (52). Similar results have been obtained in several other organisms.

In multicellular organisms the overwhelming majority of cells respond to heat in like fashion. In *Drosophila,* imaginal wing discs, the brain, Malpighian tubules, salivary glands, and tissue-culture cells respond identically (7). In the rat, heart, lung, thymus, kidney, brain, liver, and adrenal gland produce a similar spectrum of proteins (78). In maize, plumules, mesocotyls, radicles, and young leaves respond alike (71). There are, however, important exceptions to this general rule. For example, in the rat, hsp70 cannot be induced in the brain until three weeks postpartum (78). In chickens, lymphocytes produce the full spectrum of hsps, but reticulocytes produce only hsp70 (79). In quails, hsp25 is not produced in myotubes, but is produced in undifferentiated myoblasts (80, 81).

Development and differentiation have still another important effect on the response. Early embryos are incapable of producing hsps in fruit flies, sea urchins, mice, and frogs, species of such evolutionary divergence as to suggest a universal phenomenon. The inability to respond begins with oocyte development or at the time of fertilization. In *Drosophila,* competence is achieved at the time of cellular blastoderm formation (82–84); in the sea urchin, at the hatched-blastula stage (85, 86); in the frog, at the late blastula/early gastrula stage (87, 88); and in mammalian embryos, at the morula/blastocyst stage (88a). Furthermore, mouse teratocarcinoma stem cells, equivalent in many ways to the pluripotent inner mass cells of the blastocyst, do not synthesize hsps in response to heat, but do so after being induced to differentiate with retinoic acid in vitro (89) or by subcutaneous implantation into mice in vivo (90). Germinating pollen tubules are also unable to induce the synthesis of hsps in response to heat (91). Paradoxically, many early embryos appear to synthesize a subset of the hsps at normal temperatures. (For more on this subject, see *Developmental Inductions* below.)

The Proteins Induced by Heat

Heat-shock proteins are generally defined as those whose synthesis is sharply and dramatically induced at high temperatures. Such proteins will be considered in this section. At present, much of the literature remains disconnected phenomenology, but in several cases functional patterns are emerging. Throughout this discussion it is important to remember that the cellular response to heat is complex. Several reports have identified minor proteins whose synthesis is increased by heat (92–94). Since very little is known about them, they will not be discussed here. Also, heat treatments at moderate temperatures, in which cells continue to grow and develop, produce complex readjustments of metabolism, with several pre-existing proteins increasing and others decreasing. The nature of these adjustments has been little studied.

HSP70   This is the most highly conserved of the hsps and, as such, has aroused the greatest interest. The complete amino acid sequence of hsp70, from four eukaryotes (41, 43–45) and one prokaryote (95), is presented in Figure 1. The human protein is 73% identical to the *Drosophila* protein and 50% identical to the *E. coli dna*K product. Note that many differences, especially among the eukaryotic proteins, are homologous substitutions—(R-K), (D-E), (A-G), etc. There are regions of extraordinary conservation. For example, from amino acids 143 to 187, there are only two differences between the human and the *Drosophila* proteins, and only four between these and the *E. coli* protein. An interesting feature of the conservation of these genes is that there appears to have been a constraint on third codon substitutions. Several stretches of identity in the nucleotide sequence link the human and *Drosophila* genes (45) and suggest that aspects of their nucleic acid structure may be subject to selection. Concerning the diverged regions in the amino acid sequence, an unanswered question is what changes are due to flexibility in the sequence and what changes reflect selected biological adaptations in different organisms.

   *E. coli dna*K was originally identified as a host gene required for replication of bacteriophage lambda (96, 97). It codes for an abundant protein in uninfected cells even at normal temperatures (98). Synthesis accelerates immediately upon shift to high temperature, and then declines, after 20 minutes or so, to a new steady-state level that is proportional to growth temperature. Purified dnaK protein has a weak, DNA-independent ATPase activity (turnover rate: one ATP per molecule per minute), and is autophosphorylated at threonine (99). Less than 5% of the protein is phosphorylated in vivo at normal temperatures.

   Biochemical and genetic evidence indicates that dnaK is required for replication of lambda DNA. In in vitro assays the protein is required in very high concentrations and probably does not function catalytically. It appears to interact with both the lambda O and P proteins, the former stimulating, the latter inhibiting its ATPase activity (99). The fact that it has not been possible to obtain knockout mutations in the gene suggests it is essential for growth in *E. coli* (C. Gross, personal communication), but its normal cellular function is not known. When ts mutants of *dna*K are placed at restrictive temperatures, RNA synthesis appears to stop before DNA synthesis (100).

   As demonstrated by Craig and her colleagues, the *dna*K gene is unique (95), but in *Drosophila* (72, 73) and *Saccharomyces cerevisiae* (43), hsp70 genes belong to a multigene family whose members respond to temperature in different ways. Evidence now suggests that such is the case in all eukaryotes. A monoclonal antibody, produced against *Drosophila* hsp70, reacts with a group of proteins of approximately 70 kd in *Drosophila*, sea urchin, nematode, chicken, and human cells (S. Lindquist, R. Morimoto, unpublished observations). Some of these proteins are induced by heat, others are not. Two 70-kd heat-induced polypeptides in mammalian cells have distinct proteolytic cleavage patterns and induction profiles (101–103). Several members of the *hsp70* gene family have recently been cloned from mammalian and

```
HUMAN        MA-KAAAUGI DLGTTYSCUG UFQHGKUEII AMDQGMATTP SYUAFT-DTE ALIGDAAKMQ
XENOPUS      ..T.GU...  .....T....  .......... ........... .......-... ..........
DROSOPHILA   .P----.I.. .......... .Y........M .Y........ ........-.S. ..M.EP....
YEAST        .S-.--...  .........A H.AMDA.D..  .......... .F....-...  ..........
E.COLI       .G-.--II.. .....M...A IMDGTTPRUL E.AE.D.... .II.Y.Q.G. T.U.QP..A.
                                                                           60
UALMPQMTUF DAKRLIGRKF GDPUUQSDMK HWPFQUIMDG DKPKUQUSYK GETKAFYPEE ISSMULTKMK
..M.......  .......... M.....C.L. A.....US.E G...K.E.. ..E.S.F...  ..........
..M..R...  ..AK.....Y D..KIAE... ....K.US.. G...IG.E.. ..S.R.A...  ..........
A.M..S...  .......M. M..E..A... .F..KL.DUD G..QI..EF. ....M.T..Q ......G...
AUT.....L.  AI......A. Q.EE..R.US IM..KI.AAD -M6DAW.EU. .--QKMA.PQ ..AE..K...
                                                                           130
EIAEAYLGYP UTMAUITUPA YFMDSQRQAT KDAGUIAGLM ULRIIMEPTA ARIAYGLDAT GKGEAMULIF
.T......H.  .......... .......... .....L....  I......... .....KG AA..Q......
.T......ES I.D......  .......... ...H......  .......... ..L.....KM L.........
.T..S...AK .MD..U....  .......... ....T......  .......... .....KK ..E.H7....
KT..D...E.  ..E.......  .......A...  .......A...E .K........  ..L.....K- .T.M.TIAUY
                                                                           200
DL6G6TFDUS ILTIDDG--- -IFEUKATAG DTHLGGEDFD MALUMHFUEE FKRKHKKDIS QMKRAURALR
..........  .......---  -......... .......... .M........  ......G ......L...
..........  .....E.S-- -L...AS...  .......... ....T.LA. ....Y...LR S.P..L....
..........  L.F.E.---  -.........  .......... ...IQ..... ...M...L. T.Q..L....
.......I.  ..IE..EUDGE KT...L..M.  .......... S..I.YL... ..KDQGI.LR MDPL.MQ..K
                                                                           270
TACERAKATL SSSTGASLEI DSLFEGIDFY TSI----TRA AFEELCSDLF ASTLEPUEKA LADAKLDKAQ
...D......  ...SQ..I..  .......... .A.----..  .......... .G........  .........S.
..A.......  ...E.TI..  .A....Q...  .KU----S..  ......AM. .M..Q.....  .M...M..GQ
..........  ...AQT.U..  .......... ...----...  .......A.. ...D....U  .........S.
E.A.K..IE.  ..AQQTDUML PUITADATGP KHMMIKUT.. KL.S.UE..U MASI..LKU.  .Q..G.SUSD
                                                                           340
IHDLULUGGS TRIPKUQKLL QDFFMGADLM KSIMPDEAUG YGAAUQRAIL MGDKSEMUQD LLLLDUAPLS
..EI......  .......... .......E..  .......... A.........  .......... ..........
...I......  .......S..  .E..H.KM. L........A  .......... SC.Q.GKI.. U..U......
UDEI......  .........U T.Y...KEP. A........A  .......... T..E.SKT..  ..........
.D.UI...Q  ..M.M...KU AE..-.KEPA .DU..E...A I.....GGU. T..----.K. U.....T...
                                                                           410
LGLETAGGUM TALIKAMSTI PTKQTQIFTT YSDMQPGULI QUYEGERAMT KDMMLLGRFE LSCIPPAP-G
..........  .U.....T.. .......S..  .......... ..F....... .K.. ..G....A.
..I.......  .K..E..CA. .C...KT.S.  .........S. .......... ...A.T.D ..G....A.
..I.......  .K..P.....  S..KFE..S. .A........  ..F....K. .......K. ..G....A.
..I..M....  .T..AK.T..  ...HS.U.S. AE...SA.T. H.LQ...KRA A..KS..Q.M .DG.M...A.
                                                                           480
UPQIEUTFDI DAMGILMUTA TKDSTGKAMK ITITMDKGAL SKEEIERMUQ EAEKYKAEDE UQREAUSAKM
.......... .......S. UEK.S..Q..  .......... ..D..K....  .D.D A....D...
.......L  .......S. KEM.....KM ...K......  .QA..D...M .....AD... KMGQ.ITSA.
.......U  .S......S. UEKG...S..  .......... ..D..K..A  ...F.E.. KESQ.IAS..
M.......... ..D...H.S. KDKMS...EQ. ...KASS.-. MED..QK..A D..AMAEA.A KFE.L.QTR.
                                                                           550
ALESYAFMMK SAUEDEGLKG KISEADKKKU LDKCQEUISW LDAMTLAEKD EFEHKRKELE QUCMPIISGL
.......L.  .M....MU..  ...DE..ATI SE..TQ.... .EM.Q....E .YAFQQ.D. K..Q...TK.
.....UY.U.  QS..QAPA-.  .LD....MS. ...M.T.A. ..S..T...E ..D..ME..T RH.S..MTKM
Q...IAYSL.  MTISEA.D.L EQADK.TUTK KAEET--...  ..S..T.S.E ..DD.L...Q DIA...M.K.
Q6DHLLHSTA KQ...EA.D-- ..LPAD..TAI ESALTALETA .KGEDK.AIE A----.MQ..A ..SQKLMEIA
                                                                           620
YQ-6A---6G -P6------P 6GF-----GA Q6PK66SG-S GPTIEEU-D   HUMAN       640aa
YQ------6G UP66U----P 6GMPGSSC6A QARQ66--MS GPTIEEU-D   XENOPUS     647aa
HQQGA6AAG6 -PGAMCGQQA 6GF------- -----6--YS GPTUEEU-D   DROSOPHILA  641aa
YQ--A6--6A -P6GAAG6AP 6GFPG---6A PPAPEAE--- GPTUEEU-D   YEAST       643aa
QQQHAQQQTA 6ADASAMMAK DDDUUD---- --- -AE-F- ----EEUKDKK E.COLI      638aa
```

Figure 1 The complete amino acid sequence of five *hsp*70 genes: human (heat-inducible and serum regulated), *Xenopus laevis* (heat-inducible, oocyte constitutive), *Drosophila melanogaster* (chromosomal locus 87C, clone G3), yeast (*Saccharomyces cerevisiae*, *YG*100), and *E. coli* (*dna*K). A dot (·) indicates the same amino acid as in the human sequence. In the last row, where varied homologies can be drawn, the single-letter code is given for each species. Data from Refs. 41, 43–45, and M. Slater and E. Craig, personal communication.

avian cells (E. Hickey, R. Morimoto, S. Munroe, J. Nevins, H. Pelham, P. Thomas, R. Vollemy, L. Weber, personal communications).

Sufficient sequence data are available for the yeast *hsp*70 gene family to produce an approximate picture of their relatedness (Figure 2) (104). Note that some of these genes are as distantly related to each other as they are to the *E. coli* protein. Transcription of YG104 and YG105 is not affected by temperature. YG102 is induced only slightly by heat, if at all. YG100 is induced moderately.

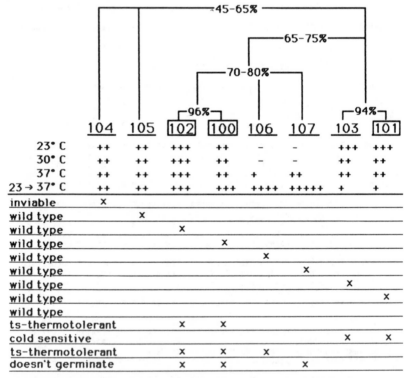

Figure 2 Sequence homology, RNA expression patterns, and mutant phenotypes of the *hsp*70 gene family in *Saccharomyces cerevisiae*. *Top:* Estimated homologies given as percent identical nucleotides. The boxed genes have been sequenced in their entirety, the others, only in part. *Middle:* mRNA levels during steady-state growth at 23, 30, and 37°C and in response to shift from 23 to 37°C. *Bottom:* Phenotypes of strains carrying disruption mutations in genes indicated by X. Note that these disruptions are in the middle of the gene and peptide fragments with some biological activity may be produced. Data from Refs. 105, 106, and personal communication, E. Craig.

YG106 and YG107 are induced very strongly, but only at high temperatures. YG101 and YG103, on the other hand, are repressed by heat shock (Figure 2) (104).

The development of site-directed mutation methods in yeast has permitted the placement of mutations in individual members of the family. Mutations in the closely related *YG*101 and *YG*103 genes produce no phenotypes individually, but the double mutant is cold-sensitive (the cells grow well at 37°C, but not at 23°C) (105). Similarly, cells carrying individual mutations in YG100 or its close relative, YG102, have no phenotypes, but the double mutant is temperature sensitive (ts) (the cells grow well at 23°C, but not at 37°C) (106). This double mutant has another curious phenotype: although ts for growth, the cells are more tolerant than wild-type cells to short exposure at extreme temperatures. That is, they are thermotolerant. Significantly, YG107 and several other heat-shock proteins are overproduced in these cells and may be responsible for thermotolerance. An artificial gene placing the YG100 coding region under the control of the YG101 promoter is unable to rescue the cold-sensitive phenotype of the YG101/YG103 double mutant. A gene placing the YG101 coding region under the control of the YG100 promoter is unable to rescue the temperature-sensitive phenotype of the YG100/YG102 double mutant (105). These results demonstrate that closely related members of the yeast hsp70 family are functionally homologous while distantly related members are functionally distinct. Perhaps the functions of all four proteins are

PROTEIN & PHOSPHORYLATED NUCLEOTIDE INDUCTION IN E. COLI

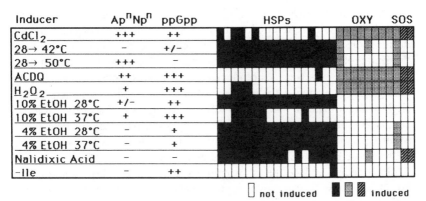

Inducer	Ap^nNp^n	ppGpp	HSPs	OXY	SOS
CdCl₂	+++	++			
28→ 42°C	−	+/−			
28→ 50°C	+++	−			
ACDQ	++	+++			
H₂O₂	+	+++			
10% EtOH 28°C	+/−	++			
10% EtOH 37°C	+	+++			
4% EtOH 28°C	−	+			
4% EtOH 37°C	−	+			
Nalidixic Acid	−	−			
−Ile	−	++			

☐ not induced   ■ ▨ induced

Figure 3   The induction of unusual nucleotides and stress proteins in *E. coli*. Proteins listed by their alphanumeric designations. Hsps: B25.3, B56.5 (groEL), B66.0 (dnaK), B83.0 (rpoD), C14.7, C15.4 (groES), C62.5, D33.4, D48.5, D60.5 (lysU), F10.1, F21.5, F84.1, G13.5, G21.0, H94.0 (lon), H26.5 (dnaJ). Oxidative stress proteins: B15.5, B20.9, B44.8, D78.0, D78.1, F35.6, F50.6, G15.4, I21.3 (Super Oxide Dismutase). SOS proteins: recA, single-strand binding protein. ACDQ: 6-amino-7-chloro-5,8, dioxoquinoline. Data provided by R. VanBogelen, P. Kelly, and F. Neidhardt.

related, the YG101/103 protein pair having evolved to function at low temperatures, the YG100/102 pair at high temperatures.

Mutations have also been made in other members of the hsp70 family (E. Craig, personal communication). Cells carrying a mutation in YG104 are inviable. Apparently, this protein has a unique and indispensible function. In contrast, cells carrying individual mutations in YG105, YG106, or YG107 are apparently wild type. Double mutants of YG106 and YG107 are also wild type. Triple mutants of YG100, YG102, and YG106 have the same phenotype as the double mutant YG100/YG102, i.e. they are temperature sensitive for growth, but thermotolerant. Triple mutants of YG100, YG102, and YG107 fail to germinate. When other mutant combinations are analyzed and the proteins that are superinduced in them determined, a clearer picture of their functional relationships will emerge.

The *Drosophila hsp70* gene family contains five or six members. Two, hsp70 and hsp68, are induced by heat shock, hsp70 several fold higher than hsp68 (14). At normal temperatures, hsp70 is not detectable, but during incubation at high temperatures it becomes one of the major proteins stainable on polyacrylamide gels (107). During heat shock it is concentrated primarily within nuclei and, secondarily, at cell membranes. During recovery from heat shock, the protein disappears from the nucleus and is found in the cytoplasm (108). Evidence suggests that it is capable of self-degradation in vitro (77). At normal temperatures other members of the family are expressed at varying levels, and one appears to be developmentally regulated (75; 75a; E. Craig, K. Palter, personal communication).

Analysis of the *hsp70* gene family of vertebrate cells is complicated. Different investigators have referred to the proteins by different molecular weight designations. Furthermore, the heat-inducible and the constitutively synthesized proteins have sometimes been confused. These problems should soon be alleviated by the analysis of cloned genes. Interesting stories are already emerging. For example, the sequence of one clone predicts a protein that is slightly larger than the others (~75 kd) containing a signal sequence at its 5' end, typical of proteins inserted into the endoplasmic reticulum (H. Pelham, S. Munro, personal communication). It may be related to a 76-kd protein of plant cells, whose message is not translated in cell-free lysates (109).

Most species produce at least two prominent heat-inducible proteins of ~70 kd. One of these is also growth-regulated (110, 110a). When serum-starved cells are refed, synthesis of the protein is induced 12–14 hrs later, during S phase. It is a prominent protein in HeLa cells at normal temperatures, but not in WI38 cells (101, 112). The activity of this gene is also induced by viral and cellular oncogenes (111–113), perhaps through their effects on growth regulation. Monoclonal antibodies specific for this protein produce diffuse cytoplasmic and punctate nuclear staining at normal temperatures, with exponentially growing cells showing intense nuclear staining and quiescent cells very little. Nuclear staining is also intensified by heat shock (114).

Polyclonal antibodies specific for the other heat-inducible 70-kd protein of mammalian cells also show both cytoplasmic and nuclear staining, but during heat shock and for a period immediately following it, the protein concentrates intensely in nucleoli (115). When the *Drosophila hsp70* gene is introduced into mammalian cells, the protein shows a very similar pattern of localization (116). The functional significance of this localization was demonstrated by placing the *Drosophila* gene under the control of another promoter, so that it would be produced in the absence of heat shock. When cells containing this gene were treated with actinomycin to inhibit the production of endogenous hsps, then heat-shocked and allowed to recover, nucleolar morphology recovered much more rapidly than in cells not producing the *Drosophila* protein (117). Taken together with observations on the effect of heat on hnRNAs (118, 119, 282a; see REGULATION OF THE RESPONSE, *RNA Processing,* below), the results suggest that the protein may catalyze the reassembly of RNP structures damaged by heat.

Cell fractionation experiments indicate that a portion of the nuclear hsp70, in *Drosophila* as well as in mammalian cells, is in a form that is resistant to salt extraction, and may be associated with the nuclear matrix (115, 120–122). On return to normal temperatures the protein relocalizes to the cytoplasm. Different laboratories report it associating with different cytoskeletal networks. It may be that hsp70 is promiscuous in its associations with other proteins or that different members of the family have different associations. Alternatively, variance in the results may simply stem from differences in experimental technique.

Members of the hsp70 family have acidic isoelectric points and similar tryptic peptide patterns in all species examined. Unfortunately, data are not yet available for a systematic comparison of other physical properties. The mammalian proteins bind ATP very tightly, a property that has recently been exploited to purify them (101). They are methylated at lysine and arginine residues (123) and bind the fatty acids palmitate and stearate in a 1:1 ratio (124). *Dictyostelium* hsp70 is reported to be phosphorylated (125), but attempts to detect phosphorylation of the chicken, mammalian, and *Drosophila* proteins have not been successful (126; S. Lindquist, unpublished results). The proteins have variously been reported to purify as monomers, dimers, or in higher-order structures. Again, these differences may stem from several factors: confusion between the different 70-kd proteins, the variety of species analyzed, the different physiological states examined, or the multiplicity of techniques employed.

Recently, clathrin-uncoating ATPase (an enzyme that removes clathrin cages from coated vesicles) was identified as a constitutively synthesized member of the *hsp*70 gene family by (*a*) copurification with the HeLa 73-kd heat-inducible protein on ATP-agarose affinity columns, (*b*) partial proteolytic peptide mapping, (*c*) two-dimensional gel analysis and, most convincingly, (*d*) by antigenic cross-reactivity (127). The immunogen was a synthetic peptide from a region of the protein that is conserved across species and between heat-inducible and constitutively synthesized members (positions 8–18 of the sequence in Figure 1). Recalling the ATPase activity of the *E. coli dna*K product, the postulated nucleolar function of hsp70 (above), and hypotheses on the complementary roles of ubiquitin and heat-shock proteins (below), it is possible to envisage various members of the hsp70 family adapted to use the energy of nucleotide triphosphates to disaggregate or disassemble a variety of protein structures or aggregates in the cell. At present, however, this notion must be considered highly speculative.

HSP83   This is the second-most highly conserved heat-shock protein. The *Drosophila* and yeast proteins are 83 kd, and the related *E. coli* protein has an apparent molecular mass of 62.5 kd. Sequence comparisons of the *Drosophila* and yeast genes indicate 60% identity of amino acids (128, 129). The *Drosophila* and *E. coli* genes have 36% amino-acid identity, but selected regions have conservation as high as 90% (J. Bardwell, E. Craig, personal communication).

All eukaryotic cells produce a prominent heat-shock protein in the range of 83 to 90 kd. Sequence data are not yet available for other organisms, but antibodies for chicken hsp89 cross-react with the human, mouse, frog, and *Drosophila* proteins (126). In all cases, the protein has a moderately acidic pI (5.1–5.4). Biochemical characterization of the protein is more extensive in vertebrates than in other organisms. It is methylated at lysine and arginine residues (123) and is phosphorylated (128). Cell-fractionation studies indicate that it is a soluble, cytoplasmic protein (121, 130), and immunocytological localizations show diffuse cytoplasmic staining with no concentration in mitochondria or other organelles (132). In most cells, it is abundant at normal temperatures and induced by heat, but abundance and inducibility vary in a tissue-specific way (134). Rat cardiac cells and quail myotubes have very little of the protein, either at normal temperatures or during heat shock (80, 132). Since it is also glucose-regulated, such tissue-specific differences in expression may relate to differences in energy metabolism (135). In yeast, separate genes produce the constitutive and inducible forms (D. Finkelstein, personal communication), but the proteins are functionally homologous. Individual mutations in either gene are wild-type, but the double mutation is lethal. In higher eukaryotes it is not known whether the constitutively synthesized and the heat-inducible proteins are produced by separate genes.

To date, the most interesting property reported for this protein is its transient association with retroviral transforming proteins and steroid hormone receptor complexes. Immediately after synthesis, the transforming protein of Rous Sarcoma virus, pp60^{src}, is quantitatively associated with hsp89 and another protein, of 50 kd (136–138). The complex is cytosolic, and in this form pp60^{src} is phosphorylated at serine residues, but not at tyrosines. In pulse-chase experiments, pp60^{src} simultaneously loses its association with hsp89, is activated as a kinase, is phosphorylated at tyrosine residues, and becomes associated with membranes (139, 140). Similar results have been obtained with the tyrosine-specific protein kinases of the FSV and Y73 viruses (136). Notably, the hsp90-pp50-pp60^{src} complex normally has a half-life of only 15 minutes. In various transformation-defective mutants of pp60^{src}, however, the complex is stable, and the protein neither associates with membranes nor becomes phosphorylated at tyrosines (139).

Recently, monoclonal antibodies prepared against the 8S progesterone receptor revealed that hsp90 is a major component of the complex (141, 141a). Again, the association with hsp90 is transient. When the receptor is converted to the 4S form, it loses its association with hsp90, binds hormone more tightly, and acquires the ability to bind to DNA. Associations with a glucocorticoid receptor (141b), a membrane ATPase (142), and with hsp100, a Golgi protein (143), have also been described, but not characterized in detail. These findings suggest that the protein may serve to shuttle important membrane proteins around the cell, holding them in an inactive form as it does so. Remember, however, that hsp90 is an abundant protein in most cells, and only a small portion of it is found in such complexes. The bulk of the protein sediments as a monomer, and immunoprecipitation reveals little association with other cellular proteins (131, 132). Thus, a question arises: is it present in excess to ensure that all of its target proteins will be bound, or does it have some other purpose?

SMALL HSPS   All organisms produce one or more small hsps. They are a heterogeneous group. Even among different *Drosophila* species the size and number of the proteins vary (144). Nevertheless, they can be considered homologous on the basis of (*a*) limited sequence relatedness, (*b*) similarities in predicted protein structure, (*c*) similarities in intracellular distribution, and (*d*) the tendency to form particles of similar size and structure. In *Drosophila melanogaster*, four proteins of this class have been extensively characterized: hsp28, 26, 23, and 22 (19–22, 145, 146). Recently, though, additional proteins have been identified. They are regulated in a tissue- and stage-specific manner (146a). Yeast cells produce a single prominent protein of 26 kd (38, 62), and nematodes, three proteins of 25, 18, and 16 kd (147). *Xenopus* cells produce a single protein of 30 kd with heat shock, but the hybrid selection of messenger RNAs indicates that there may be a larger class of related proteins (44). The small hsps are particularly prominent and heterogeneous in plant cells. In soybeans, messenger RNAs for 20 different proteins, ranging from 14 to 27 kd, accumulate to 20,000 copies per cell within two hours of heat shock (148).

The small hsps of any given species show greater homology to each other than to the proteins of other species, suggesting a continuing process of homogenization through evolution, either by gene conversion or by repeated duplication and deletion. The soybean proteins sequenced to date have 90% identity (149). The four *Drosophila* proteins have an overall homology of ~50% (145, 146). However, the *Drosophila* proteins have little homology with the soybean proteins, the most convincing being the sequence (asn or asp)-gly-val-leu-thr found in the most hydrophobic domain near the carboxyl terminus. Intriguingly, the small hsps of most species show homology to α-crystallin lens protein. A region of 75 amino acids in lens crystallin has 50% identity with the *Drosophila* proteins (146). The nematode hsps have greater homology with α-crystallin than with the *Drosophila* hsps (147). Antibodies produced against hsp24 of chicken cells cross-react with lens protein purified from embryonic chick (150).

The homology with crystallin may relate to the capacity of the small hsps to form higher-order structures. Cytoplasmic particles containing the hsps complexed with small RNAs have been isolated from *Drosophila* and yeast during recovery from heat shock (151; A.-P. Arrigo, personal communication). During heat shock itself, the proteins concentrate in nuclei, where they are also associated with RNA (118, 122, 152–159). The particles are similar in size and structure to ones existing prior to heat shock, containing a similar, but not identical, group of proteins. They have a distinctive, circular, hollow-core morphology, identical to the prosome RNP particles of mouse and duck cells (151, 155). Similar particles containing the small hsps and RNA have been observed in tomato and corn cells, but these are larger and form both during heat shock and recovery (156, 157).

The small hsps are modified, but, as with other proteins, there has been no systematic comparison of different species. The rat proteins, synthesized at normal temperatures and induced by heat shock, are phosphorylated exclusively at serine residues. Phosphorylation patterns differ in cells exposed to heat, sodium arsenite, calcium ionophores, phorbol esters, or fresh serum (153, 158). The plant proteins are both methylated and phosphorylated (159).

Deletion and disruption mutations have been created in the *Saccharomyces cerevisiae* gene for hsp26, the major small hsp of this organism (159a). Surprisingly, they have no affect on growth rates at any temperature, in rich or poor media, in respiratory or fermentative metabolism. Nor do they affect acquired thermotolerance in log or stationary phase. Moreover, although the protein accumulates at high concentrations in sporulating cells (75a, 160), the mutations do not influence sporulation or germination at any temperature. It may be that yeast cells contain a related gene, which covers the hsp26 lesions, but, if so, it is not related enough to be detected by low-stringency DNA hybridization or by antigenic cross-reaction. Furthermore no other protein appears to compensate for it by increased synthesis. The function of this protein may be far more subtle than previously supposed.

UBIQUITIN   In both chickens and yeast, ubiquitin has recently been found to be heat-inducible (161; D. Findley, A. Varshavsky, personal communication). Considering the evolutionary divergence of these species, it is likely to be a more general phenomenon. Presumably, it was missed in other studies because proteins of such small size, 7–8 kd, are not usually resolved on gels. The potential significance of this finding is underscored by two others. First, mammalian cells containing a temperature-sensitive, ubiquitin-activating enzyme induce hsps at the nonpermissive temperature, a temperature which does not induce hsps in the parental cell line (162, 103). Second, several *E. coli* hsps either are proteases themselves or affect the activity of proteases (see below). It has been suggested that ubiquitin and the hsps are complementary methods of dealing with a common problem, i.e. the production of denatured protein aggregates in heat-shocked cells. Ubiquitin would remove them from the cell through its protease-targeting activity. Other hsps would either prevent the aggregates from forming or disaggregate them, once formed (162).

OTHER HSPS OF EUKARYOTIC CELLS   Mammalian cells produce proteins of 100 and 110 kd which do not appear to have counterparts in *Drosophila*. Yeast and plant cells produce 96-kd proteins, but their relationship to the larger mammalian proteins is not clear. Immunological localizations of the 100-kd protein demonstrate concentration in the Golgi (163). Antibodies against the 110-kd protein reveal localization in or around nucleoli (164). Treatment with DNase disrupts this staining, while treatment with ribonuclease eliminates it altogether. Both the 100- and 110-kd proteins are constituents of normal cells and are glucose regulated. Their induction patterns are complex: under various conditions they and the so-called glucose-regulated proteins are induced together, independently, or reciprocally (165–169).

Several other proteins that are induced by heat have been identified in individual organisms. In *Drosophila*, histone H2b is induced (170, 171). In mammalian cells, γ-interferon is induced (172). And in yeast, enolase (173) and glyceraldehyde-3-phosphate dehydrogenase (L. Petko, S. Lindquist, unpublished information) are induced by heat.

OTHER HSPS IN *E. COLI*   Several of the *E. coli* hsps have been identified. Certain of their properties are similar to those of eukaryotic hsps, but no direct homologies have yet been demonstrated.

The groE proteins were first identified as essential for the morphogenesis of bacteriophage. On the basis of mutant phenotypes and the mapping of bypass mutations, these proteins appear to be required for essential proteolytic cleavages occurring at different assembly steps in different phage. In lambda assembly, they interact with B protein, which forms the precollar ring connecting head and tail (174–177). In T4, the interaction is with the gene 31 protein at an early step in coat-protein assembly for the head (174, 178). In T5 morphogenesis, they are required for tail assembly (179).

It is now known that the groE operon encodes two proteins, groES and groEL (15 and 65 kd, respectively), but, in the earlier literature, mutations mapping to the two genes were not distinguished as such (180, 181). GroES and groEL interact both in vivo and in vitro. Mutations in groEL can suppress multiple phenotypic effects of mutations in groES. And mutations in either gene yield similar phenotypic effects on phage assembly (182, 183). In vitro, the two proteins cosediment in the presence of ATP, and purified groES protein inhibits the weak ATPase activity of the groEL protein.

As was the case for dnaK, the cellular functions of the groEL and groES proteins are not known. Ts-lethal mutations in these genes demonstrate that they are essential for growth at high temperatures, but, in the absence of deletion mutations, it is not certain that they are required for growth at normal temperatures. At high temperatures they account for an astounding 15% of total cellular protein (98, 184). Even at normal temperatures, they are among the most abundant proteins in the cell. Mutations have pleiotropic effects, altering permeability, causing filamentation, and disrupting normal patterns of DNA, RNA, and protein synthesis at nonpermissive temperatures (185–187). GroEL forms a cylindrical 25S particle with sevenfold rotational symmetry composed of 14 subunits (188, 189). In electron micrographs these particles appear remarkably similar to those formed by the small hsps of eukaryotic cells.

Two other *E. coli* proteins were originally identified as essential for the growth of phage lambda, dnaJ and grpE. DnaJ is a 37-kd protein that migrates as a dimer in nondenaturing gels (183). It binds with no apparent specificity to both double- and single-stranded DNA. It is required for lambda DNA replication, and appears to interact with lambda P protein and with the host proteins, dnaB and dnaK. DnaJ has a basic pI of 8.5, and is quantitatively associated with membranes in cell-fractionation experiments. The grpE protein has an apparent molecular mass of 23.5 kd. When affinity columns of dnaK are washed with high salt and eluted with ATP, highly purified grpE protein is released (M. Zylicz, personal communication). Again, the cellular functions of these proteins are not clear.

The ATP-dependent lon protease is a heat-shock protein of 94 kd (190). It is believed to play a major role in the degradation of abnormal proteins and in regulating the turnover rates of certain normal cellular proteins (191–193). Recall that ubiquitin, a recently identified eukaryotic hsp, has a similar cellular function. Lon mutants are extremely pleiotropic. Among other things, they accumulate large quantities of mucopolysaccharides, have an abnormal SOS response, and have a decreased ability to lysogenize phage lambda (194–196).

One of two E. coli lysyl-tRNA synthetases, lysU, is also a heat-shock protein (197, 198). This is the only amino acyl-tRNA synthetase activity in E. coli known to be encoded by two different genes. Its reputed ability to participate in the formation of the various phosphorylated nucleotides known as alarmones suggests it may participate in the regulation of a variety of stress responses.

Synthesis of the sigma^{70} subunit of RNA polymerase, rpoD, is also induced by heat shock. It is a 70-kd protein that associates with core polymerase, conferring upon it the ability to bind to normal cellular promoters and initiate RNA synthesis (199–201). Why is this subunit induced by heat shock, but not the other polymerase subunits? Evidence suggests that rpoD and rpoH, the heat-shock sigma factor, compete for polymerase and thereby determine the specificity of transcription. Thus, it may be that transcription of sigma^{70} is required both to turn off transcription of heat-shock promoters and to re-establish the transcription of normal promoters in transient heat-shock responses.

RNAs Induced by Heat

In most cells, the heat-shock response is transcriptionally regulated, and messages coding for the hsps are induced 10- to 1000-fold. There are, in addition, a few RNAs of unknown function that accumulate in heat-shocked cells at high levels. Such RNAs have been characterized only in a few organisms, but they may well be a common phenomenon.

The chromosomal locus 93D is one of the largest heat-shock puffs in D. melanogaster (2). Similar large puffs, distinguished by the presence of unusually large RNP particles, have been found in other Drosophila species (202, 203). Clones have been characterized from 93D in D. melanogaster (204) and from the analogous locus 2-48C in D. hydei (205). In both cases, the transcribed region is very long. It is made up of short, imperfect repeats of 280 and 115 bp respectively, with considerable sequence divergence between the two species. The transcribed region includes sequences adjacent to and interspersed with the repeats. Curiously, much of the RNA produced by these puffs is confined to the nucleus, heterogeneous in size, and not polyadenylated (206). Its function is unknown. Fruit flies carrying overlapping deficiencies of 93D are viable, and have a normal heat-shock response (207). That it has some function is conjectured from its common occurrence. A locus of similar morphology exists in Chironomus as well (208).

Another heat-inducible RNA, but one which is presumed not to have a cellular function, is the $\alpha\beta$ sequence. It, too, is found in D. melanogaster, where it is transcribed from tandemly reiterated genes at the chromosomal locus

87C, one of two loci containing genes for hsp70. Transcripts of different lengths, containing various numbers of repeats, are produced (209). The sequence is also found at the chromocenter, but here it is not transcribed. Since the sibling species *D. simulans* contains the chromocenter sequence but not the heat-inducible sequence, the latter is judged not to to be functional. Presumably, the element transposed into the heat-shock locus at 87C in *D. melanogaster*, where it picked up heat-shock promoter sequences (210).

Another transposon with heat-shock promoter sequences has been found in *Dictyostelium* (211, 212). The element is 4.7 kb long, with 330-bp terminal repeats and three long open-reading frames, a structure similar to the transposable elements of many organisms. It is repeated 40–50 times in the *Dictyostelium* genome. That transposon-like sequences are heat-inducible in two such divergent organisms suggests it may be a more common phenomenon, perhaps of evolutionary significance. However, it is certainly not universal. The yeast transposable element, TY-1, is not associated with heat-inducible promoters and is repressed by heat shock (36).

Several small nuclear RNAs that are induced by heat, or at least not repressed, have been observed in *Drosophila*. One of these is 5S ribosomal RNA, whose processing is disrupted by heat (213). Another is U1 RNA (214).

OTHER INDUCTIONS OF HSPs

Developmental Inductions

The best documented examples of heat-shock gene expression during the normal course of development are found in *Drosophila*. (*a*) Messages for hsp26, hsp 28, and hsp83 are induced at high levels in ovarian nurse cells and passed into the developing oocyte (215). Hsp70, 68, 22, and 23 messages are not produced and cannot be induced at this time, even with a heat shock. (*b*) Hsp22, 23, 26, and 28, are also induced at the late third instar larva/prepupa stage of development (216, 217). Relative levels of induction differ from those in heat-shocked cells but, again, the induction is most notably distinguished from heat shock in that hsp70 is not produced. The trigger for this induction is believed to be the molting hormone ecdysone, which reaches a peak in late third instar larvae. Ecdysone induces synthesis of the same proteins in tissue-culture cells (218).

Ecdysone-resistant cell lines do not produce the proteins when exposed to the hormone, but are fully capable of producing all of the hsps in response to heat (219). This suggests the two forms of induction are independently regulated. That this is indeed the case has recently been demonstrated for both the ovarian induction and the prepupal induction by analysis of genes containing varying amounts of 5' sequence transformed into the *Drosophila* germ line. Sequences responsible for ovarian induction of hsp26 lie in a region from 522 to 352 bases upstream of the start site of transcription, while sequences responsible for heat-shock induction are located between 341 and 14 (220). Similar experiments have been performed with the hsp28 gene. Here, sequences responsible for 80% of heat-induced synthesis map more than 1.1 kb upstream of the gene, while sequences responsible for prepupal induction map within 227 bases of the gene (221).

Certain hsps appear to be constitutively produced in the early embryos of other organisms. For example, in eight-day mouse ectoderm and in embryonal carcinoma cells two prominent proteins of 89 and 70 kd appear, which have the same electrophoretic mobilty on 2-D gels as hsps. Unfortunately, it is not clear whether these are identical to the heat-inducible proteins or other members of the same families (222). While hsps are not inducible in germinating pollen tubules, two proteins of 70 and 72 kd, with mobilities identical to two hsps, are present in high concentrations (223). (A remarkable developmental induction of hsp70 mRNA in *Xenopus* oocytes will be discussed in the REGULATION OF THE RESPONSE section below.)

Reminiscent of the *Drosophila* oocyte induction, mRNAs for hsp26 and 83 accumulate to high levels as yeast cells approach stationary phase and enter sporulation (75a, 160, 224). Hsp26 accumulates to become a major Coomassie-stainable protein in these cells. Furthermore, two members of the hsp70 family, detected with a cross-reacting monoclonal antibody, accumulate to high levels. This, too, is paralleled in *Drosophila,* as a protein detected by the same antibody accumulates in ovaries (75a; K. Palter, E. Craig, personal communication).

Particularly interesting cases of developmentally regulated hsp expression have been described in two types of dimorphic pathogens, the trypanosomatid protozoans *Leishmania* and *Trypanosoma,* and the fungus *Histoplasma capsulatum.* The nonparasitic forms of these organisms grow at ~25°C. They are induced to differentiate into their pathogenic forms at high temperatures. The first change in gene expression detected during this transition is the synthesis of hsps (225–228). Virulent strains of *H. capsulata* induce hsps and differentiate into pathogenic yeast forms at higher temperatures than avirulent strains (G. Medoff, B. Maresca, personal communication).

Other Inducers

The earliest studies of heat-induced puffs in *Drosophila* demonstrated that they could be induced by a variety of other stress treatments. Most of these interfered with oxidative phosphorylation or electron transport, so it was inferred that the induction served to protect cells from respiratory stress. Heat shock was presumed to increase rates of metabolism and act through the same mechanism (230). Several lines of evidence now indicate that this simple notion is incorrect. First, a large number of other agents, with no specificity for respiration, also induce the response. Second, yeast mutants that are respiration-deficient nevertheless produce hsps in response to high temperatures (62). Third, hsps do not concentrate in mitochondria. In fact, electron-microscopic autoradiograms of *Drosophila* cells reveal mitochondria as clear islands in a black sea of autoradiographic grains (231). This is not to say that mitochondrial respiration is irrelevant. Respiration-deficient yeast cells do not recover from heat shock at high temperatures. In other fungi, heat shock transiently inhibits respiration. Furthermore, some of the small hsps of plant cells appear to localize to mitochondria (234). A connection must exist, but its nature is unclear at present.

Many inducers of hsps are effective across a broad range of species. Ethanol induces hsps in mammalian cells (232), yeast (233), and *E. coli* (68). Sodium arsenite induces the proteins in *Drosophila* (154), mammalian cells (232), trout (58), and soybeans (234). Cadmium induces in *Drosophila* (235), mammalian cells (236), and soybeans (234). However, there are many species-specific differences. Ethanol, sodium arsenite, and cadmium do not induce hsps in *Dictyostelium* (239). Anaerobiosis, and recovery from anaerobiosis, induce hsps in many organisms, but in plants anaerobiosis induces a different group of proteins (238). Many inducers—zinc, copper and mercury ions, sulfhydryl reagents, calcium ionophores, steroid hormones, chelating agents, pyridoxine, methylene blue, glucosamine, deoxyglucose, and a variety of DNA and RNA viruses—have only been tested on a few cell types. (See 237 for a recent compilation.)

Is There a Common Mechanism?

As we have already seen, the developmental and heat-shock inductions of hsps are independently regulated. What about the inductions by various forms of stress? Do they all work via the same mechanism, and, if so, what is it? From the earliest studies to the present, the plethora of inducing agents has presented a challenge to advocates of a universal mechanism. Many of the agents are so diverse in their effects on cell physiology that it is difficult to envisage them all provoking the same trigger. Over the years several theories have been advanced, but, as new data have accumulated, these have all broken down. A recent, particularly promising example will serve to illustrate. An unusual class of phosphorylated dinucleotides, ApppA, ApppGpp, ApppG, and AppppG and AppppA, together termed alarmones, have been shown to accumulate in *E. coli* and *Salmonella typhimurium* within minutes of heat shock (240). The same nucleotides accumulate in response to ethanol and hydrogen peroxide, both of which induce hsps, and it has therefore been proposed that the alarmones are the trigger for the heat-shock response. However, these experiments were done with a severe heat shock, from 28°C to 50°C. Less severe heat shocks, 28°C to 42°C, and treatment with other agents, such as nalidixic acid, induce hsps, but do not appreciably induce the alarmones (R. VanBogelen, F. Neidhardt, personal communication; see Figure 3). Furthermore, the kinetics of the induction do not fit the model. The heat-shock response is transcriptionally activated in *E. coli* in less than one minute; within five minutes, synthesis of heat-shock messenger RNAs has already begun to decline. Accumulation of the alarmones appears to peak between 5 and 15 minutes. Variations in the model can be evoked to fit these circumstances. In its simplest form, however, it is beginning to appear less probable.

It may be that the challenge will not be answered, that there will be no universal inducer. It has long been noted that many agents induce only a selected subset of the hsps. Recent data from Neidhardt and VanBogelen, in Figure 3, illustrate for *E. coli*. While this points to independent regulation, in and of itself it is not compelling. It might be that different genes have different affinities for the "master regulator." Thus, if some agents caused only a moderate increase in its activity, only those genes with a particularly high affinity for it would be activated. Furthermore, heat-shock messages and proteins might have different turnover rates under different conditions. All the genes might be transcriptionally activated, but differential turnover could produce different patterns of accumulation. However, recent experiments indicate that, in a few cases at least, the heat-shock genes are subject to independent systems of activation by stress.

A case in point is the oxidative stress regulation system of *S. typhimurium*, discussed above. Dominant mutations, selected for resistance to killing by hydrogen peroxide, map to the *oxyR* locus and constitutively overexpress nine of twelve oxidative stress proteins and three hsps (241). Two of the three, D64a and E89, can be further induced by heat in these strains; the third, F52a, cannot. On the other hand, recessive deletion mutants of *oxyR* are noninducible for the nine oxidative stress proteins, yet all of the hsps, excepting only F52a, are still inducible by heat. Thus, *oxyR* is a positive regulator for nine oxidative stress proteins and for three hsps, D64a, E89, and F52a. It may be the only regulator for F52a, but D64a and E89 must have an additional mechanism of induction that operates during heat shock, probably through an *rpoH* system (see below).

Another case is presented by one of the hsp70 genes of human cells that is induced by heat, cadmium, the adenovirus *E1a* protein, and by the addition of serum to serum-starved cells. Through analysis of recombinant DNA clones (110), sequences responsible for cadmium and heat-shock induction have been mapped 107 to 68 bases upstream of the transcription start site. Sequences responsible for serum stimulation (and possibly E1a induction, as well) have been mapped 58 to 47 bases upstream of transcription. Although the cadmium and heat-shock inductions were not separated in this study, it is noteworthy that the region delineated as sufficient for their induction contains both a heat-shock consensus element (see below) and, separately, a metallothionein consensus element.

It is too early to tell how many inducers will act via the same mechanism as heat shock itself. It is quite possible, even probable, that several of them will. The question remains: what is the trigger for this induction?

GENOME ORGANIZATION

The heat-shock genes of both eukaryotic and prokaryotic organisms are scattered at various chromosomal locations. Some related genes are clustered, some are interspersed with independently regulated genes. Fairly complete pictures are available for *D. melanogaster* and *E. coli*. Data from other organisms are still very incomplete and will not be presented here. A remarkable feature of the heat-shock genes, one common to all organisms, is an absence of intervening sequences. This will be considered again below (See REGULATION OF THE RESPONSE, *RNA Processing*).

The location of the heat-shock genes in *Drosophila* was apparent from their initial description as puffs on polytene chromosomes (3). Individual coding sequences were first mapped to these sites by in situ hybridization of heat-induced messenger RNAs (8, 9, 242), and the locations confirmed and extended by the analysis of cloned genes and deletion mutations (10–18).

Genes for hsp70 are present in five copies per haploid genome, two at chromosomal site 87A, and three at 87C on the right arm of chromosome three. The genes at 87A are arranged head-to-head, with ~1.7 kb of spacer DNA between them. Two of the genes at 87C are in tandem, separated from the third, in divergent orientation, by ~38 kb of DNA containing the repetitious $\alpha\beta$ sequences described earlier (11–13, 16, 17, 209). The nucleic acid sequences of genes at the two loci indicate 97% identity in the protein-coding sequence and additional homology extending for ~400 bp upstream of transcription initiation (41, 243). Divergence appears to be kept in check by gene conversion. Restriction-enzyme polymorphisms at the two loci indicate that the *hsp*70 genes within a species are more similar to each other than would be expected if they were evolving independently (34). The gene for hsp68 has ~85% homology with the *hsp*70 genes and is located at 95D (15).

Genes for the other hsps are located on the left arm of the third chromosome. All four of the small hsps are encoded at locus 67B, within a 7 kb-stretch of DNA in the order hsp28, hsp23, hsp26, hsp22 (20–22). Each gene has its own promoter, with hsp26 transcribed in the opposite direction to the other three. At least three other genes are encoded at this locus, one between the genes for hsp23 and hsp26, and two others on the far side of the hsp22 gene. These are expressed at normal temperatures and are developmentally regulated (145, 217). Hsp83 coding sequences are found at 63BC.

Seven of the *E. coli* hsps have been mapped. The *groE* genes are part of the same operon, promoter-*groES-groEL,* at 94 minutes on the circular 100-minute *E. coli* chromosome (181, 183). The genes for dnaK and dnaJ form an operon with two heat-inducible promoters located 115 and 40 bp upstream of the *dnaK* transcription start-site (201, 244, 245). The gene for sigma factor, *rpoD,* is part of an operon that also contains the genes for ribosomal protein S21, rpsU, and DNA primase, dnaG, in the order *rpsU-dnaG-rpoD* at 67 minutes (246). Heat-inducible transcripts of *rpoD* are initiated from a promoter located within the *dnaG* gene (247). The heat-inducible isoform of lysyl-tRNA synthetase, *lysU,* is in close proximity to the gene for lysine decarboxylase, *cadA,* at 93 minutes (198). Lysine decarboxylase is not heat-inducible, but the two genes are coinduced under other conditions. The *lon* gene is found at 10 minutes.

REGULATION OF THE RESPONSE

A striking feature of heat-shock gene expression is that the responses of different organisms and, indeed, of different cell types within an organism are regulated in different ways (49). In *E. coli* (248) and in yeast (36, 249), the response is controlled primarily at the level of transcription. In *Drosophila,* regulation is exerted on both transcription and translation (8, 9, 35, 36). In *Xenopus,* the response of somatic cells is controlled primarily at the transcriptional level, the response of oocytes at the translational level (250). These differences make perfect biological sense. Messenger RNAs have half-lives on the order of 6–9 hours in *Drosophila.* Thus, to effect a rapid change in protein synthesis, *Drosophila* cells must block the translation of pre-existing messages. Yeast and *E. coli* cells do not have this problem. Given the larger size of the organism, it is unlikely that *Xenopus* would need to respond as rapidly as *Drosophila* to an increase in environmental temperature. Accordingly, somatic cells rely primarily on transcriptional mechanisms (87). Oocytes, however, face a special problem. Given the enormous size of these cells, with their large store of messenger RNAs and their single tetraploid nucleus, it is estimated that they would require 10–100 days to synthesize an effective quantity of heat-shock messenger RNA (250). The *hsp*70 gene is therefore constitutively transcribed but translationally repressed, ready to be activated in the event of exposure to heat.

Transcription

EUKARYOTES A common regulatory mechanism appears to control the transcription of all eukaryotic heat-shock genes. When the *Drosophila* genes and their 5' flanking regions are introduced into mammalian cells and *Xenopus* or sea urchin oocytes, they are activated at temperatures appropriate to those organisms (27, 28, 251, 252). Thus, the genes carry the information required for their activation, but this information is not acted upon by temperature in a self-sufficient way. It depends upon *trans*-activating mechanisms that are induced by stress in the host cell. Both elements in this regulation have now been identified.

Sequence analysis of the *Drosophila* heat-shock genes early revealed striking homologies in their 5' upstream sequences, homologies postulated to serve in transcriptional activation (41, 41a, 46). Deletion analysis mapped the activation site more precisely, to a consensus element located (on the *hsp*70 genes) between 47 and 66 bp upstream of transcription initiation (27). A synthetic promoter sequence, derived from this consensus region, is sufficient for heat-inducible transcription in heterologous systems (28). The heat-shock genes of all organisms analyzed to date have similar sequences in their 5' regions. The sequence of the consensus element (HSE) is rotationally symmetric, C—GAA—TTC—G, and is generally located 1.5 helical turns upstream of the TATA box (253). HSEs from several organisms are presented in the top of Figure 4.

EUKARYOTIC PROMOTERS

PROTEIN	
D. hsp70 (215 & 144)	ATAAAGAATATTCTAGAA------CTCGAGAAATTTCTCGG
(36 & 15)	TTCTCGTTGCTTCGAGAGAGCGCGCCTCGAATGTTCGCGA
D. hsp83 (35,25,15)	ATCCAGAAGCCTCTAGAAGTTTCTAGAGACTTCCAGTT
D. hsp23 (97)	CCCGAGAAGTTTCGTGTC
D. hsp26 (13)	TTCCGGACTCTTCTAGAA
X. hsp70 (72)	CTCGGGAAACTTCGGGTC
X. hsp30 (14)	CTCGGGAACGTCCCAGAA
S. hsp17 (28)	CCCAAGGACTTTCTCGAA
H. hsp70 (80)	CCCTGGAATATTCCCGAC
consensus	C--GAA--TTC--G

PROKARYOTIC PROMOTERS

PROTEIN	-35 REGION	-10 REGION
groE	ITTCCCCCTTGAAGGGGCGAAGCCATCCCCATTTCTCTGGTCAC	
dnaK P1	TCTCCCCCTTGATGACGTCCTTTACGACCCCATTTAGTAGTCAA	
dnaK P2	TTGGGCAGTTGAAACCAGACGTTTCGCCCCTATTACAGACTCAC	
C62.5	GCTCTCGCTTGAAATTATTCTCCCTTGTCCCCATCTCTCCCACATC	
rpoD PHS	TGCCACCCTTGAAAAACTGTCGATGTGGGACGATAGCAGATA	
lon	TCTCGGCGTTGAATGTGGGGGAAACATCCCCATATACTGACGTAC	
consensus	T tC CcCTTGAA (13–15 bp) CCCCATtTa	

Figure 4   Heat-shock promoter sequences: *Drosophila, Xenopus,* soybean, human, and *E. coli*. Locations of the eukaryotic consensus sequences are given in parentheses. Data from Refs. 41, 43–48, 201.

Chromatin-mapping studies of the heat-shock genes indicate that they are found in an open configuration at normal temperatures, as if poised for immediate activation, with multiple hypersensitive sites at their 5' ends (23–26, 254–256). The hsp70 and hsp83 genes have two prominent hypersensitive sites; the hsp22, 23, 26, and 28 genes have four or five sites. In all of the genes, hypersensitive sites bracket, and sometimes extend into, the HSE. Upon heat shock, the chromatin structure changes profoundly. First, the coding region becomes more accessible to nucleolytic attack. Second, the position and number of hypersensitive sites changes (25, 254, 256). Most notably, regions centered around the HSE become refractory to digestion, denoting the binding of protein. Certain hypersensitive sites in front of the small heat-shock genes, however, are not affected by heat. Presumably, these are related to developmental regulation. For hsp26, they are upstream of the HSE; for hsp28, downstream. These are the relative positions of the developmental-regulation sequences mapped by deletion analysis (220, 221; see *Developmental Inductions,* above).

A heat-shock transcription factor (HSTF) that binds to the HSE in vivo and in vitro has been isolated and characterized (29–33). In in vitro transcription assays it promotes the transcription of heat-shock genes, but does not affect the activity of other genes. In DNA-footprinting studies, it protects the consensus element and surrounding sequences from nucleolytic digestion, consistent with the above-mentioned chromatin digestion experiments. HSTF can be isolated both from heat-shocked cells and from cells grown at normal temperatures. Though more active when isolated from heat-shocked cells, the difference in its activity is 10- to 50-fold less than the difference in in vivo transcription activity. It may be that HSTF is partially activated during purification, or that it requires either additional factors or a particular type of chromatin structure for full activity. Evidence for at least one additional factor, which binds the TATA box, has been obtained (32).

A single consensus element is sufficient to confer heat-inducible transcription on exogenous genes in heterologous, high–copy number systems (27, 28). It is striking, however, that most heat-shock genes characterized to date contain multiple consensus elements. There are a total of four HSEs in front of one of the *Drosophila hsp70* genes. As transformation experiments demonstrate, full transcriptional activity in *Drosophila* cells requires at least the two most proximal sites (257, 258). An explanation is found in the work of Parker and colleagues, which shows that the binding of HSTF to these sites is cooperative (259, 260). Alkylation and protection experiments map the HSTF contact sites to HSE residues, and indicate that contact points in site 1 change when site 2 is filled. In vitro transcription assays confirm the importance of cooperative binding. DNAs containing sites 1 and 2 were transcribed much more efficiently than those containing either site alone.

The transcription of heat-shock genes is also negatively regulated (56, 57, 261). When hsps have accumulated to a specific level, a level that is proportional to the the severity of the heat treatment, further transcription is repressed. In *Drosophila,* the evidence indicates that hsp70 is the critical element in this regulation (57), but it is not known whether it acts directly or indirectly. Findings from two other organisms suggest it may be a universal feature of heat-shock regulation: (*a*) deletion of two *hsp*70 genes in yeast, *YG*100 and *YG*102, results in constitutive synthesis of other hsps (104, 106), and (*b*) *dnaK,* the *hsp*70 analogue of *E. coli,* is required to repress the heat-shock response in this organism (262).

PROKARYOTES   The promoter sequences of several *E. coli* heat-shock genes have been identified by mapping the 5′ ends of their messenger RNAs, the presumed start sites of transcription (201). Comparison of the sequences reveals two consensus elements, in the -35 and -10 regions: TNtCNCcCTTGAA and CCCCATtT, respectively (See Figure 4). Different consensus sequences in the -35 and -10 regions of other *E. coli* genes, TTGA-CA and TATAAT, respectively, are crucial for transcriptional activation by sigma^{70}-RNA polymerase. The sequence TTGA is shared by both the heat shock and the sigma^{70} promoters in the -35 region, but there is nothing in common in the -10 region (201).

The transcription activating factor that operates on the heat-shock consensus sequences was first identified by an amber mutation, *htpR165* or *hin165,* carried in a strain with a temperature-sensitive suppressor (263, 264). At high temperatures, hsps were not induced and the cells died. The gene was cloned, mapped to 76 minutes on the *E. coli* chromosome, and shown to encode a 33-kd protein (51). Remarkably, sequence analysis revealed it has homology to the polymerase subunit sigma^{70}, although it is only half the size of that protein. That htpR is an alternate sigma factor, the first such identified in *E. coli,* was demonstrated by isolation of the protein and characterization by in vitro transcription assays. The protein has therefore been renamed sigma^{32}, the gene, *rpoH* (50).

The regulation of sigma^{32} is complex. Changes in both the concentration of the protein and its activity appear to be important (265). Furthermore, the concentration can itself be affected by at least two mechanisms, that is, by a change in synthesis or by a change in stability. The concentration of sigma^{32} mRNA is increased by heat shock (C. Georgopoulus, personal communication). How this is effected, however, is unclear, since *rpoH* apparently does not have a heat-shock promoter (C. Gross, C. Georgopoulus, personal communication). The protein normally has a very short half-life, on the order of a few minutes, but this is prolonged in *dnaK* mutations (C. Georgopoulus, C. Gross, personal communication). Negative regulation of the response by *dnaK* (262) may work by sigma factor degradation. The activity of sigma^{32} is also affected by changes in the concentration of sigma^{70}, presumably through competition for core polymerase or holoenzyme. Competition between these factors might explain why increased transcription of heat-shock genes is accompanied by decreased transcription of normal messages.

Translation

Heat treatment of *Drosophila* cells produces the most dramatic changes in translational specificity yet described. Within ten minutes of a shift from 25 to 37°C, normal polysomes virtually disappear (8). This is not brought about by a general inhibition in protein synthesis. When heat-shock mRNAs appear, they are translated with very high efficiencies (53). Pre-existing messages are not degraded in this process, but are retained in the cell, and reactivated during recovery (35, 36, 56, 57, 266, 267). The existence of similar translational mechanisms has been reported in many organisms, including sea urchins (85), slime molds (59, 239), nematodes (268), soybeans (91), corn plants (71), and tomatoes (237), but the *Drosophila* response is the only one that has been characterized in detail.

High temperatures are important in establishing this response. Many inducers can activate the change in transcription, few the change in translation. A simple explanation, then, would be that temperature itself is the mechanism. If messages required a certain secondary structure for translation, heat might melt this structure in normal messages, but not in heat-shock messages. That this is not the case has been demonstrated in several ways. First, when heat-shock and control messages are mixed and translated in reticulocyte lysates, no difference in specificity is observed at various temperatures. Second, when heat-shocked cells are returned to normal temperatures, they do not immediately revert to normal patterns of protein synthesis (36). Rather, hsps are synthesized exclusively until a specific quantity of protein has accumulated (57). Most notably, translation lysates prepared from heat-shocked *Drosophila* cells retain the ability to discriminate between the two classes of messages, while lysates prepared from control cells do not (35, 267, 269). Thus, high temperatures act as the trigger, not the effector.

It has been known for some time that heat-shock messages have several distinctive structural features that might be used for recognition by the translational machinery of heat-shocked cells. They have unusually long 5' untranslated leader sequences that are rich in adenosine residues, have little secondary structure, and have conserved sequences in the middle and at their 5' ends (46). Recently, it has been demonstrated that these leaders do contain the signal for recognition, but its nature is unclear. Deletions and insertions in the leaders can convert the messages, so that they are translated only at low temperatures. However, simple deletions of the conserved elements have only small effects (270, 271).

The mechanism by which ribosomes are able to use the information in the leader to discriminate between heat-shock and normal cellular messages is also unclear. Fractionation of cell-free lysates produces conflicting results. In some cases discrimination activity fractionates with ribosomes, while in others with the supernatant (269, M. Saunders personal communication). It may be a loosely bound factor. Changes in ribosome-protein composition and modification have been reported (272–274), but no experiments have demonstrated that these changes are effective in altering translational specificity. In other systems, evidence suggests that the cytoskeleton is associated with the translational machinery (275). Elements of the cytoskeleton are disrupted by heat-shock (276) but the importance of this phenomenon in translational regulation has been questioned, at least for mammalian cells (277). It is not even clear what step in protein synthesis is effected. The rapid disappearance of polysomes suggests initiation is the crucial step (8, 278), but there is also evidence to suggest that control is exerted at the level of elongation (279, 280).

A further mystery is, what governs the return to normal patterns of translation after heat shock? When hsps have accumulated in sufficient quantities, the translation of pre-existing messages is restored and the translation of heat-shock messages is repressed. Unlike heat-shock induction, recovery is a very gradual process (56, 57, 281, 282). There is no sudden change in specificity. The synthesis of hsps (hsp70 appears to be particularly important) is required to restore normal protein synthesis (57, 281, 282). Since the rates of protein synthesis are high in heat-shocked *Drosophila* cells (53), hsps are probably required to effect a change in specificity, rather than to repair heat-induced damage to the translational machinery (57).

RNA Processing

Of the 70 or so heat-shock genes that have been cloned and analyzed, only two have been found to have introns, the hsp83 gene of *Drosophila* (46, 128) and one of the small hsp genes of nematodes (147). Considering how rare it is for the genes of higher organisms not to have such interruptions, it is almost certainly of functional significance. An obvious possibility, and one in keeping with the emergency nature of the response, is that this feature has been deployed to ensure that the proteins will be made as rapidly as possible. But this appears not to be the only reason.

Hsp83 has a pattern of expression that is distinct from that of other hsps: it is produced at moderately elevated temperatures, but not at high temperatures (52). The block in synthesis at high temperatures is not due to a failure in translation. When cells are heat shocked at moderate temperatures, and then shifted to high temperatures in the presence of actinomycin, hsp83 mRNA is translated with high efficiency (52). Rather, the block appears to be due to a failure in RNA processing. The *hsp*83 gene is abundantly transcribed at high temperatures, but intervening sequences are not removed from its RNA. Furthermore, when the intron-containing gene for alcohol dehydrogenase is placed under the control of a heat-shock promoter, this RNA too accumulates as an unspliced precursor (282a). Thus, at high temperatures, there is a general block in intron processing. The absence of introns in most heat-shock genes allows their messages to circumvent this block.

What causes the block in processing is not known. Heat shock has been reported to disrupt hnRNA structure (119). However, conflicting results have also been published (118, 257). The discrepancies between these reports may originate from the fact that the block in processing occurs only at extreme temperatures. Interestingly, mild heat treatments, which induce the synthesis of hsps, potentiate processing under otherwise restrictive conditions (282a). At present it is not clear whether hsps play a direct role in RNA processing or a more general role in preserving nuclear structure. Hsps have been reported to associate with hnRNA (118). However, this may simply be a consequence of their very high nuclear concentrations (108, 151, 152).

TOLERANCE TO HEAT AND OTHER FORMS OF STRESS

It is a natural and widely held assumption that the purpose of the heat-shock response is to protect organisms from the toxic effects of heat and other forms of stress. There is a great deal of evidence to support this notion. The proteins are induced in different organisms at different temperatures, but in each case, the temperatures correspond to the upper portion of the organism's natural growth range. More compellingly, the induction of the proteins coincides with the acquisition of tolerance to more extreme temperatures. Evidence in conflict with this notion does exist, however. Interpretations are complicated by the fact that some of the proteins are likely to play important roles in normal physiology as well as during heat shock. Moreover, resistance to high temperatures is probably mediated not only by the synthesis of hsps, but by other mechanisms as well.

Heat-Induced Lethality and Thermotolerance

A simple design is common to many experiments testing the hypothesis that hsps provide protection from high temperatures. A group of cells or organisms is split into two identical sets. One is shifted directly to an extreme temperature, and the kinetics of lethality are then measured. The other is incubated at a more moderate elevated temperature, thereby inducing the synthesis of hsps, and is then exposed to the same extreme temperature as was the first set. A dramatic increase in survival, often of at least two orders of magnitude, is observed in the second set.

Thermotolerance experiments of this type have been performed on a broad range of organisms at various stages of development and on cultured cells grown under many different conditions. Examples include mammalian tissue-culture cells (232, 236, 283–286), whole mice (287), *Drosophila* embryos, larvae, pupae, and adults (82, 83, 108, 288), slime molds (59), sea urchin embryos and plutei (85, 88), and yeast cells (264, 289).

Of course, many protective accommodations in cell physiology and structure might be made during the conditioning incubation at moderate temperature, accommodations that may be only coincidentally related to the induction of the hsps. But several lines of evidence suggest that the synthesis of hsps is the crucial element.

1. Conditioning heat treatments can be replaced by treatments with a wide variety of agents having the common property of inducing hsps at normal temperatures: ethanol, hypoxia, sodium arsenite, and cadmium chloride, to name a few. Such agents do not induce heat-shock proteins in all organisms, but, in those in which they do, they also induce thermotolerance (108, 232–234). (The converse is also true. Heat shock induces tolerance to ethanol, anoxia, and several other forms of stress.) On the other hand, when hsps are induced by amino-acid analogues that become incorporated into proteins, thermotolerance is not observed (236). This is an exception that supports the rule. Hsps produced under these conditions do not assume their characteristic intracellular distributions. Their functions have been disrupted by incorporation of the analogue (57).

2. The kinetics of thermotolerance induction, under many different conditions, are tightly correlated with the kinetics of heat-shock protein synthesis and reach a maximum when the accumulation of hsps plateaus (285, 286, 290). Li has carefully quantitated the data from several different types of experiments. The concentration of hsp70 shows the best correlation with thermotolerance (290).

3. The decay of thermotolerance coincides with the degradation of hsps. When heat-shocked cells are incubated for extended periods at normal temperatures, they gradually lose the ability to survive exposure to extreme temperatures. At the same time, and with similar kinetics, hsps are degraded (285, 290).

4. At the stages in development in which hsps cannot be induced, organisms are extremely sensitive to heat, and thermotolerance cannot be induced. This is most carefully documented during the embryonic development of fruit flies (82, 83), sea urchins (85, 88a), frogs (88), and mammals (88a, 89, 90, 222, 291). (See *Comparison: Different Organisms and Stages of Development,* above.) It is also noteworthy that sperm development in many organisms is extremely sensitive to high temperatures. In *Drosophila,* it has recently been found that primary spermatocytes are one of the very few cell types in the adult that do not respond to high temperatures by inducing hsps (292).

5. Mammalian cell lines and yeast mutants selected for their ability to survive exposure to high temperatures, without conditioning pretreatments, constitutively produce hsps. A Chinese hamster fibroblast cell line constitutively produces hsp70 at much higher levels than the parental line (293). A yeast mutant, *hsr1*, synthesizes two 48-kd hsps (enolase isoforms) and two other proteins of 73 and 56 kd (173, 294).

6. Mutants, initially characterized as unable to acquire thermotolerance, are defective in hsp synthesis. In *E. coli*, cells carrying amber mutations in the positive regulatory factor sigma[32] (variously known as *hin, htpr,* and *rpoD*) are unable to produce hsps at nonpermissive temperatures and die. In *Dictyostelium*, a mutant that is unable to synthesize the small hsps at high temperatures is also unable to acquire thermotolerance (239).

7. When cycloheximide is added to cells immediately before heat shock, blocking the synthesis of hsps, the acquisition of thermotolerance is also blocked (249, 295).

There are a number of contradictions in the literature. One investigator has reported that the addition of cycloheximide to yeast cells did not prevent the induction of thermotolerance (296). However, it was not clear how effective the drug had been in blocking protein synthesis. Other contradictions concern which proteins are essential in establishing thermotolerance. Certain experiments support the notion that the small hsps are important, as in the case of the *Dictyostelium* mutant described above. Also, ecdysone induces both thermotolerance and small hsps in *Drosophila* (300). On the other hand, yeast cells that carry deletions of *hsp*26, the major small hsp gene, have normal thermotolerance (159a), and *Drosophila* embryos that constitutively synthesize small hsps are, nonetheless, extremely thermosensitive (82, 215). Another case in point is the isolation of two melanoma cell lines with widely different levels of thermotolerance but no qualitative or quantitative differences in hsp synthesis (297).

A particularly interesting line of research concerns the change in thermotolerance that accompanies transformation. Hyperthermia, administered in conjunction with ionizing radiation, is becoming an important tool in cancer therapy (see 298 for review). The basis of the method is that transformed cells are more susceptible to the toxic effects of heat than are their untransformed counterparts, both in tumors in vivo and in cultured cells in vitro. It has recently been found that transformation of mouse embryo cells with SV40, while increasing their sensitivity to heat, also causes them to produce higher levels of hsps both at normal temperatures and after heat shock (299). The implication is that hsps are not providing these cells with thermotolerance; the question is, would they be even more sensitive to heat without the hsps?

Overall, a great many experiments support the notion that hsps are crucial to the induction of thermotolerance. It seems almost certain that they are. However, the physiological effects of heat are complex. Cells in different states of metabolism and in different stages of differentiation may be killed by different mechanisms. Furthermore, the causes of lethality by short-term exposures to extreme temperatures may be different from those caused by long-term incubations at moderately high temperatures. Different proteins may be involved in protecting cells from these different toxicities. Finally, cells have diverse mechanisms for coping with high temperatures; the adjustment of membrane fluidity is a salient example. If the state of metabolism or differentiation should block these other mechanisms from coming into play, the synthesis of hsps would be irrelevant.

Phenocopies

Determining the specific causes of heat-induced lethality and elucidating the molecular mechanisms of protection will not be easy, especially because these phenomena are likely to vary from cell to cell. The molecular mechanisms of phenocopy induction and protection are more amenable to analysis. Dead cells, like dead men, tell no tales. But tissues involved in phenocopy induction continue growth and development, and can be analyzed in considerable detail.

PHENOCOPY INDUCTION   Phenocopies are nonheritable, environmentally induced anomalies of development that closely mimic mutant phenotypes. The term was coined by Goldschmidt in 1935 to describe a broad range of defects observable in adult *Drosophila* when their pupae had been subjected to high temperatures at certain stages of development. Such abnormalities can be induced by treatment with chemicals such as ethanol and ether as well as by heat. The phenomenon is by no means limited to insects. Developmental anomalies induced by heat and chemicals have been extensively studied in avian (301) and mammalian embryos (302–304). Agents that are teratogenic in mammalian cells are also teratogenic in *Drosophila* and furthermore induce the synthesis of hsps (305). Thus, there is a direct connection between the two phenomena.

A distinctive feature of phenocopy induction is that it is stage-specific. Stress treatments characteristically lead to developmental defects during periods of rapid growth and morphogenesis—in vertebrates, during embryogenesis, or, in insects, during embryogenesis or pupal metamorphosis. During the susceptible period, narrow windows of development determine the particular type of abnormality later to be manifested. A remarkable example is the production of bithorax phenocopies (four-winged flies) in *Drosophila*. The stress-sensitive period occurs early in embryonic development, four hours after fertilization, and lasts for only ten minutes (306).

PHENOCOPY PREVENTION   The relevance of these studies to the synthesis of hsps stems from Milkman's discovery that phenocopies (and heat-induced lethality) could be prevented in *Drosophila* by giving pupae a mild heat shock prior to the more severe inducing treatment (307). Mitchell & Petersen later determined that such protective treatments induced the synthesis of hsps. Of the many phenocopies that they have described (288, 308), those affecting hair and bristle morphology have received the most study. The multihair phenocopy is illustrated in Figure 5. During normal development at 25°C, each hair cell

MULTIHAIR PHENOCOPIES

CRITICAL DEVELOPMENTAL PERIOD - THORAX

25° C → 40.8° C x 40 min.

| 40 hrs. | 41 - 45 hrs. | 47 hrs. |

PHENOCOPY PROTECTION BY MILD PRE HS

Wing - critical period 37 to 40 hrs.

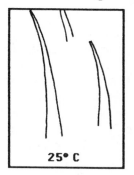

| 25° C | 40.8°C x 40' | pre hs 35°C x 30' |

Figure 5 Multihair phenocopies in adult flies of *Drosophila melanogaster*. *Top:* multihairs appear on the thorax when pupae are exposed to 40.8°C for 40 minutes any time between 41 and 45 hours after puparium formation. *Bottom:* if pupae are exposed to 35°C for 30 minutes before the 40.8°C treatment, the phenocopy is prevented. Tracings of micrographs provided by N. Petersen and H. Mitchell.

produces a single unbranched hair. When pupae are briefly exposed to extreme temperature—40.8°C for 35–40 minutes—multiple and branched hairs are produced. Treatments at different periods in development affect hairs on different portions of the fly, reflecting the fact that different portions of the organism reach the stage of hair morphogenesis at different times. When pupae are exposed to 35°C for 30 minutes prior to the 40.8°C treatment, the pheno-copy is prevented and normal hairs are formed.

The sensitive period for the production of multihair occurs at a time of rapid changes in protein and RNA synthesis, when the projection of the hair cell produces its outer coat of cuticulin (309, 310). Shortly thereafter, cytoplasm is pushed into the projection to raise the hair. Severe heat treatments disrupt protein synthesis for as long as 16 hours. When it is restored, the normal pattern of gene expression is discoordinate, and cytoplasm is pushed into the hair before gaps in the cuticulin are closed. In animals that have been given a mild pre-shock before the 40.8°C treatment, protein synthesis recovers much more rapidly, and coordination between the two processes is restored.

A recent development is the extension of such studies to mammals. In humans, maternal temperatures greater than 38.9°C occurring during the 21st to 28th day of gestation are associated with a variety of neural tube defects, including anencephaly, microcephaly, and spina bifida. Raising the tempera-ture of pregnant mice or explanted rat embryos to 42–43°C for short periods produces a similar range of defects (302–304). Mild pre-heat shocks, adminis-tered before the higher teratogenic exposure, prevent the anomalies (311; G. Chernoff, personal communication). In rat embryos, this pre-treatment has been shown to induce the synthesis of hsp71 and hsp88 in cells of the neural plate and to greatly enhance the recovery of normal protein synthesis (311).

CONCLUDING REMARKS

A great deal of progress has been made in the decade since molecular studies were initiated. Much remains to be done. The two most important questions, what is the trigger for the induction and what is the function of the proteins, remain. With the extension of investigations to so many different organisms the questions have multiplied. How are different patterns of induction established in different cell types and what are their particular purposes? The answers, it appears, will be not only intellectually satisfying, but also directly relevant to human biology and medicine.

ACKNOWLEDGMENTS

I am extremely grateful to E. Buckbee and E. Gordon for reviewing the manuscript. Special thanks go to several colleagues for providing me with manuscripts in advance of publication and/or allowing me to cite their unpublished data: A.-P. Arrigo, J. Bardwell, I. Cartwright, T. Chappel, G. Chernoff, R. Cohen, V. Corces, E. Craig, S. Elgin, J. Feramisco, D. Findley, C. Georgopoulos, C. Gross, L. Hightower, E. Hoffman, C. Hunt, K. Jacobsen, R. Kingston, J. Key, S. Kurtz, G. Li, M. Meselson, H. Mitchell, T. McGarry, R. Morimoto, S. Munroe, R. Nagao, F. Neidhardt, K. Palter, C. Parker, H. Pelham, N. Petersen, L. Petko, J. Rossi, J. Rothman, D. Ruden, M. Saunders, D. Schlossman, M. Schlesinger, F. Schofel, D. Shuly, M. Slater, J. Topol, D. Walsh, W. Welch, S. Wheelan, B. Wu, R. VanBogelen, A. Varshavsky, and J. Yost.

Literature Cited

1. Ritossa, F. 1962. *Experientia* 18:571–73
2. Berendes, H. D. 1968. *Chromosoma* 24:418–37
3. Ashburner, M. 1970. *Chromosoma* 31:356–76
4. Leenders, H. J., Berendes, H. D. 1972. *Chromosoma* 37:434–44
5. Ritossa, F. M. 1964. *Exp. Cell Res.* 36:515–23
6. Berendes, H. D. 1965. *Dev. Biol.* 11:371–84
7. Tissières, A., Mitchell, H. K., Tracy, U. M. 1974. *J. Mol. Biol.* 84:389–98
8. Lindquist McKenzie, S., Henikoff, S., Meselson, M. 1975. *Proc. Natl. Acad. Sci. USA* 72:1117–21
9. Spradling, A., Penman, S., Pardue, M. L. 1975. *Cell* 4:395–404
10. Livak, K. J., Freund, R., Schweber, M., Wensink, P. C., Meselson, M. 1978. *Proc. Natl. Acad. Sci. USA* 75:5613
11. Schedl, P., Artavanis-Tsakonas, S., Steward, R., Gehring, W. J., Mirault, M. E., et al. 1978. *Cell* 14:921–29
12. Craig, E. A., McCarthy, B. J., Wadsworth, S. C. 1979. *Cell* 16:575–88
13. Moran, L., Mirault, M. E., Lis, J., Schedl, P., Artavanis-Tsakonas, S., Gehring, W. J. 1979. *Cell* 17:1–8
14. Artavanis-Tsakonas, S., Schedl, P., Mirault, M. E., Moran, L., Lis, J. 1979. *Cell* 17:9–18
15. Holmgren, R. K., Livak, K., Morimoto, R., Freund, R., Meselson, M. 1979. *Cell* 18:1359–70
16. Ish-Horowicz, D., Holden, J., Gehring, W. J. 1977. *Cell* 12:643–52
17. Ish-Horowicz, D., Pinchin, S. M., Schedl, P., Artavanis-Tsakonas, S., Mirault, M. E. 1979. *Cell* 18:1351–58
18. Mirault, M. E., Goldschmidt-Clermont, M., Artavanis-Tsakonas, S., Schedl, P. 1979. *Proc. Natl. Acad. Sci. USA* 76:5254–58
19. Wadsworth, S., Craig, E. A., McCarthy, B. J. 1980. *Proc. Natl. Acad. Sci. USA* 77:2134–37
20. Craig, E. A., McCarthy, B. J. 1980. *Nucleic Acids Res.* 8:4441–57
21. Corces, V., Holmgren, R., Freund, R., Morimoto, R., Meselson, M. 1980. *Proc. Natl. Acad. Sci. USA* 77:5390–93
22. Voellmy, R., Goldschmidt-Clermont, M., Southgate, R., Tissières, A., Levis, R., Gehring, W. J. 1981. *Cell* 23:261–70
23. Wu, C., Bingham, P. M., Livak, K. J., Holmgren, R., Elgin, S. C. R. 1979. *Cell* 16:797–806
24. Wu, C., Wong, Y.-C., Elgin, S. C. R. 1979. *Cell* 16:807–14
25. Wu, C. 1980. *Nature* 286:854–60
26. Keene, M. A., Corces, V., Lowenhaupt, K., Elgin, S. C. R. 1981. *Proc. Natl. Acad. Sci. USA* 143:46
27. Pelham, H. R. 1982. *Cell* 30:517–28
28. Pelham, H. R., Bienz, M. 1982. *EMBO J.* 1:1473–77
29. Parker, C. S., Topol, J. 1984. *Cell* 36:357–69
30. Parker, C. S., Topol, J. 1984. *Cell* 37:253–62
31. Wu, C. 1980. *Nature* 286:854–60
32. Wu, C. 1984. *Nature* 309:229–41
33. Wu, C. 1985. *Nature* 317:84–87
34. Brown, A. J. L., Ish-Horowicz, D. 1981. *Nature* 290:677–82
35. Storti, R. V., Scott, M. P., Rich, A., Pardue, M. L. 1980. *Cell* 4:395–404
36. Lindquist, S. 1981. *Nature* 293:311–14
37. Kelly, P. M., Schlesinger, M. J. 1982. *Cell* 15:1277–86
38. McAlister, L., Strausberg, A., Kulaga, A., Finkelstein, D. B. 1979. *Curr. Genet.* 1:63–74
39. Miller, M. J., Xuong, N. H., Geiduschek, E. P. 1979. *Proc. Natl. Acad. Sci. USA* 76:5222–26
40. Guttman, S. D., Gorovsky, M. A. 1979. *Cell* 17:305–17
41. Ingolia, T. D., Craig, E. A., McCarthy, B. 1980. *Cell* 21:669–76
41a. Karch, F., Torok, I., Tissières, A. 1981. *J. Mol. Biol.* 148:219–30
42. Craig, E. A., Ingolia, T. D., Slater, M.

J., Manseau, L. J., Bardwell, J. C. A. 1982. In *Heat Shock from Bacteria to Man*, pp. 11–16. Cold Spring Harbor, NY: Cold Spring Harbor Lab. 440 pp.

43. Ingolia, T. D., Slater, M. J., Craig, E. A. 1982. *Mol. Cell. Biol.* 2:1388–98
44. Bienz, M. 1984. *EMBO J.* 3:2477–83
45. Hunt, C., Morimoto, R. 1985. *Proc. Natl. Acad. Sci. USA* 82:6455–59
45a. Vollemy, R., Ahmed, A., Schiller, P., Bromley, P., Rungger, D. 1985. *Proc. Natl. Acad. Sci. USA* 82:4949–53
46. Holmgren, R. V., Corces, V., Morimoto, R., Blackman, R., Meselson, M. 1981. *Proc. Natl. Acad. Sci. USA* 78:3775–78
47. Cohen, S. M., Capello, J., Lodish, H. F. 1984. *Mol. Cell. Biol.* 4:2332–40
48. Schoffl, F., Raschke, E., Nagao, R. T. 1984. *EMBO J.* 3:2491–97
49. Lindquist, S. 1985. See Ref. 183, pp. 232–35
50. Grossman, A. D., Erickson, J. W., Gross, C. A. 1984. *Cell* 38:383–90
51. Landick, R., Vaughn, V., Lau, E. T., VanBogelen, R. A., Erickson, J. W., Neidhardt, F. C. 1984. *Cell* 38:175–82
52. Lindquist, S. 1980. *Dev. Biol.* 77: 463–79
53. Lindquist, S. 1980. *J. Mol. Biol.* 137: 151–58
54. Jamrich, M., Greenleaf, A. L., Bautz, E. K. F. 1977. *Proc. Natl. Acad. Sci. USA* 79:2079–83
55. Findly, R. C., Pederson, T. 1981. *J. Cell Biol.* 88:223–28
56. DiDomenico, B. J., Bugaisky, G. E., Lindquist, S. 1982. *Proc. Natl. Acad. Sci. USA* 79:6181–85
57. DiDomenico, B. J., Bugaisky, G. E., Lindquist, S. 1982. *Cell* 31:593–603
58. Kothary, R. K., Candido, E. P. 1982. *Can. J. Biochem.* 60:347–55
59. Loomis, W. F., Wheeler, S. 1980. *Dev. Biol.* 79:399–408
60. Francis, D., Lin, L. 1980. *Dev. Biol.* 79:238–42
61. Giudice, G., Roccheri, M. C., Di-Bernardo, M. G. 1980. *Cell Biol. Int. Rep.* 4:69–74
62. Lindquist, S., DiDomenico, B. J., Bugaisky, G., Kurtz, S., Petko, L., Sonoda, S. 1982. See Ref. 42, pp. 167–76
63. Key, J. L., Lin, C. Y., Chen, Y. M. 1981. *Proc. Natl. Acad. Sci. USA* 78:3526–30
64. Baszczynski, C. L., Walden, D. B., Atkinson, B. G. 1982. *Can. J. Biochem.* 60:569–79
65. Borbely, G., Suranyi, G., Korcz, A., Palfi, Z. 1985. *J. Bacteriol.* 161:1125–30

66. Yamamori, T., Ito, K., Nakamura, Y., Yura, T. 1978. *J. Bacteriol.* 134:1133–40
67. Lemaux, P. G., Herendeen, S. L., Bloch, P. L., Neidhardt, F. C. 1978. *Cell* 13:427–34
68. Neidhardt, F. C., VanBogelen, R. A., Vaughn, V. 1984. *Ann. Rev. Genet.* 18:295–329
69. Daniels, C. J., McKee, A. H. Z., Doolittle, W. F. 1984. *EMBO J.* 3:745–49
70. Gedamu, L., Culham, B., Heikkila, J. J. 1983. *Bioscience* 3:647–58
71. Baszczynski, C. L., Walden, D. B., Atkinson, B. G. 1985. In *Changes in Eukaryotic Gene Expression in Response to Environmental Stress*, pp. 349–71. London: Academic. 379 pp.
72. Ingolia, T. D., Craig, E. A. 1982. *Proc. Natl. Acad. Sci. USA* 79:525–29
73. Wadsworth, S. C. 1982. *Mol. Cell. Biol.* 2:286–92
74. Ellwood, M. S., Craig, E. A. 1984. *Mol. Cell. Biol.* 4:1454–59
75. Craig, E. A., Ingolia, T. D., Manseau, L. J. 1983. *Dev. Biol.* 99:418–26
75a. Kurtz, S., Rossi, J., Petko, L., Lindquist, S. 1986. *Science* 231:1154–57
76. Lowe, D. G., Moran, L. A. 1984. *Proc. Natl. Acad. Sci. USA* 81:2317–21
77. Mitchell, H. K., Petersen, N. S., Buzin, C. H. 1985. *Proc. Natl. Acad. Sci. USA* 82:4969–73
78. White, F. P., Currie, W. R. 1982. See Ref. 42, pp. 379–86
79. Morimoto, R., Fodor, E. 1984. *J. Cell Biol.* 99:1316–23
80. Atkinson, B. G. 1981. *J. Cell Biol.* 89:666–73
81. Voellmy, R., Bromley, P. A. 1982. *Mol. Cell. Biol.* 2:479–83
82. Graziosi, G., Micali, F., Marzari, R., de Cristini, F., Savoini, A. 1980. *J. Exp. Zool.* 214:141–45
83. Dura, J.-M. 1981. *Mol. Gen. Genet.* 184:381–85
84. Bergh, S., Arking, R. 1984. *J. Exp. Zool.* 231:379–91
85. Roccheri, M. C., Di Bernardo, M. G., Giudice, G. 1981. *Dev. Biol.* 83:173–77
86. Howlett, S., Miller, J., Schultz, G. A. 1983. *Biol. Bull.* 165:500
87. Bienz, M. 1984. *Proc. Natl. Acad. Sci. USA* 81:3138–42
88. Heikkila, J. J., Kloc, M., Bury, J., Schultz, G. A., Browder, L. W. 1985. *Dev. Biol.* 107:483–89
88a. Heikkila, J. J., Miller, J. G. O., Schultz, G. A., Kloc, M., Browder, L. W. 1985. See Ref. 71, pp. 135–58
89. Morange, M., Diu, A., Bensaude, O., Babinet, C. 1984. *Mol. Cell. Biol.* 4:730–35

90. Wittig, S., Hensse, S., Keitel, C., Elsner, C., Wittig, B. 1983. *Dev. Biol.* 96:507–14
91. Mascarenhas, J. P., Altschuler, M., 1985. See Ref. 71, pp. 321–26
92. Anderson, N. L., Giomettim, C. S., Gemmell, M. A., Nance, S. L., Anderson, N. G. 1982. *Clin. Chem.* 28:1084–92
93. Reiter, T., Penman, S. 1983. *Proc. Natl. Acad. Sci. USA* 80:4737–41
94. Maytin, E. V., Colbert, R. A., Young, D. A. 1985. *J. Biol. Chem.* 260:2384–92
95. Bardwell, J. C., Craig, E. A. 1984. *Proc. Natl. Acad. Sci. USA* 81:848–52
96. Georgopoulos, C. P. 1977. *Mol. Gen. Genet.* 151:35–39
97. Sunshine, M., Feiss, M., Stuart, J., Yochem, J. 1977. *Mol. Gen. Genet.* 151:27–34
98. Herendeen, S. L., VanBogelen, R. A., Neidhardt, F. C. 1979. *Bacteriology* 139:185–94
99. Zylicz, M., LeBowitz, J. H., McMacken, R., Georgopoulos, C. 1983. *Proc. Natl. Acad. Sci. USA* 80:6431–35
100. Saito, H., Uchida, H. 1977. *J. Mol. Biol.* 113:1–25
101. Welch, W. J., Feramisco, J. R. 1985. *Mol. Cell. Biol.* 5:1229–37
102. Lowe, D. G., Moran, L. A. 1984. *Proc. Natl. Acad. Sci. USA* 81:2317–21
103. Subjeck, J. R., Sciandra, J. J., Shyy, T. T. 1985. *Int. J. Radiat. Biol.* 47:275–84
104. Craig, E. A., Slater, M. R., Boorstein, W. R., Palter, K. 1985. *UCLA Symp. Mol. Cell. Biol.* (NS), 30:659–68. New York: Liss.
105. Craig, E. A., Jacobsen, K. *Mol. Cell. Biol.* In press
106. Craig, E. A., Jacobsen, K. 1984. *Cell* 38:841–49
107. Velazquez, J. M., Sonoda, S., Bugaisky, G., Lindquist, S. 1983. *J. Cell Biol.* 96:286–90
108. Velazquez, J. M., Lindquist, S. 1984. *Cell* 36:655–62
109. Baszczynski, C. L., Walden, D. B., Atkinson, B. G. 1983. *Can. J. Biochem. Cell Biol.* 61:395–403
110. Wu, B. J., Morimoto, R. I. 1985. *Proc. Natl. Acad. Sci. USA* 82:6070–74
110a. Kao, H. T., Capasso, O., Heintz, N., Nevins, J. R. 1985. *Mol. Cell. Biol.* 5:628–33
111. Imperiale, M. J., Kao, H. T., Feldman, L. T., Nevins, J. R., Strickland, S. 1984. *Mol. Cell. Biol.* 4:867–74
112. Nevins, J. R. 1982. *Cell* 29:913–19
113. Kingston, R. E., Baldwin, A. S. Jr., Sharp, P. A. 1984. *Nature* 312:280–82
114. La Thangue, N. B. 1984. *EMBO J.* 3:1871–79
115. Welch, W. J., Feramisco, J. R. 1984. *J. Biol. Chem.* 259:4501–13
116. Pelham, H., Lewis, M., Lindquist, S. 1984. *Phil. Trans. R. Soc. London Ser. B* 307:301–7
117. Pelham, H. R. B. 1984. *EMBO J.* 3: 3095–100
118. Kloetzel, P. M., Bautz, E. K. F. 1983. *EMBO J.* 2:705–10
119. Mayrand, S., Pederson, T. 1983. *Mol. Cell Biol.* 3:161–71
120. Sinibaldi, R. M., Morris, P. W. 1981. *J. Biol. Chem.* 256:10735–38
121. Levinger, L., Varahavsky, A. 1981. *J. Cell Biol.* 90:793–96
122. Tanguay, R. M. 1985. See Ref. 71, pp. 91–113
123. Wang, C., Lazarides, E. 1984. *Biochem. Biophys. Res. Commun.* 119:735–43
124. Guidon, P. T. Jr., Hightower, L. E. 1985. *J. Cell. Biol.* 99:454a (Part 2)
125. Loomis, W. F., Wheeler, S., Schmidt, S. 1982. *Mol. Cell. Biol.* 2:484–89
126. Schlesinger, M. J., Kelley, P. M., Aliperti, G., Malfer, C. 1982. See Ref. 42, pp. 243–50
127. Chappell, T. G., Welch, W. J., Schlossman, D. M., Palter, K. B., Schlesinger, M. J., Rothman, J. E. *Cell.* In press
128. Hackett, R. W., Lis, J. T. 1983. *Nucleic Acids Res.* 11:7011–30
129. Farrelly, F. W., Finkelstein, D. B. 1984. *J. Biol. Chem.* 259:5745–51
130. Carlsson, L., Lazarides, E. 1983. *Proc. Natl. Acad. Sci. USA* 80:4664–68
131. Lanks, K. W., Kasambalides, E. J. 1979. *Biochim. Biophys. Acta* 578:1–12
132. Lai, B.-T., Chin, N. W., Stanek, A. E., Keh, W., Lanks, K. W. 1984. *Mol. Cell. Biol.* 4:2802–10
133. Deleted in proof
134. Morange, M., Diu, A., Bensaude, O., Babinet, C. 1984. *Mol. Cell. Biol.* 4: 730–35
135. Kasambalides, E. J., Lanks, K. W. 1983. *J. Cell Physiol.* 114:93–98
136. Yonemoto, W., Lipsich, L. A., Darrow, D., Brugge, J. S. 1982. See Ref. 42, pp. 289–98
137. Brugge, J., Erikson, E., Erikson, R. L. 1981. *Cell* 25:363–72
138. Oppermann, H., Levinson, W., Bishop, J. M. 1981. *Proc. Natl. Acad. Sci. USA* 78:1067–71
139. Brugge, J., Yonemoto, W., Darrow, D. 1983. *Mol. Cell. Biol.* 3:9–19
140. Courtneidge, S. A., Bishop, J. M. 1982. *Proc. Natl. Acad. Sci. USA* 79:7117–21
141. Catelli, M. G., Binart, N., Jung-Testas, I., Renoir, J.-M., Baulieu, E.-E., et al. 1986. *EMBO J.* In press

141a. Schuh, S., Yonemoto, W., Brugge, J., Bauer, V. J., Riehl, R. M., et al. 1985. *J. Biol. Chem.* 260:14292–96

141b. Sanchez, E. R., Toft, D. O., Schlessinger, M. J., Pratt, W. B. 1985. *J. Biol. Chem.* 260:12398–401

142. Burdon, R. H., Cutmore, C. M. 1982. *FEBS Lett.* 140:45–48

143. Welch, W. J., Garrels, J. I., Feramisco, J. R. 1982. See Ref. 42, pp. 257–66

144. Sinibaldi, R. M., Storti, R. V. 1982. *Biochem. Genet.* 20:791–807

145. Southgate, R., Ayme, A., Voellmy, R. 1983. *J. Mol. Biol.* 165:35–57

146. Ingolia, T. D., Craig, E. A. 1982. *Proc. Natl. Acad. Sci. USA* 79:2360–64

146a. Ayme, A., Tissières, A. 1985. *EMBO J.* 4:2949–54

147. Russnak, R. H., Jones, D., Candido, E. P. M. 1983. *Nucleic Acids Res.* 11:3187–205

148. Schoffl, F., Key, J. L. 1982. *J. Mol. Appl. Genet.* 1:301–14

149. Nagao, R. T., Czarnecka, E., Gurley, W. B., Schoffl, F., Key, J. L. 1985. *Mol. Cell. Biol.* 5:3417–25

150. Schlesinger, M. J. 1985. See Ref. 71, pp. 183–95

151. Arrigo, A.-P., Darlix, J.-L., Khandjian, E. W., Simon, M., Spahr, P.-F. 1985. *EMBO J.* 4:399–406

151a. Schuldt, C., Kloetzel, P. M. 1985. *Dev. Biol.* 110:65–74

152. Arrigo, A.-P., Ahmad-Zadeh, C. 1981. *Mol. Gen. Genet.* 184:74–79

153. Kim, Y.-J., Shuman, J., Sette, M., Przybyla, A. 1984. *Mol. Cell. Biol.* 4:468–74

154. Tanguay, R. M., Vincent, M. 1982. *Can. J. Biochem.* 60:306–15

155. Schmid, H. P., Akhayat, O., Martin de Sa, C., Puvion, F., Koehler, K., Scherrer, K., 1984. *EMBO J.* 3:29–34

156. Nover, L., Scharf, K.-D., Neumann, D. 1983. *Mol. Cell. Biol.* 3:1648–55

157. Fransolet, S., Deltour, R., Bronchart, R., Van de Walle, C. 1979. *Planta* 146:7–18

158. Welch, W. J. 1985. *J. Biol. Chem.* 260:3058–62

159. Nover, L., Scharf, K.-D. 1984. *Eur. J. Biochem.* 139:303–13

159a. Petko, L., Lindquist, S. 1986. *Cell* In press

160. Kurtz, S., Lindquist, S. 1984. *Proc. Natl. Acad. Sci. USA* 81:7323–27

161. Bond, U., Schlesinger, M. J. 1985. *Mol. Cell. Biol.* 5:949–56

162. Finley, D., Ciechanover, A., Varshavsky, A. 1984. *Cell* 37:43–55

163. Lin, J. J.-C., Welch, W. J., Garrels, J. I., Feramisco, J. R. 1982. See Ref. 42, pp. 267–73

164. Subjeck, J. R., Shyy, T., Shen, J., Johnson, R. J. 1983. *J. Cell Biol.* 97:1389–95

165. Welch, W. J., Garrels, J. I., Thomas, G. P., Lin, J.J.-C., Feramisco, J. R. 1983. *J. Biol. Chem.* 258:7102–11

166. Hightower, L. E., White, F. P. 1982. See Ref. 42, pp. 369–77

167. Sciandra, J. J., Subjeck, J. R., Hughes, C. S. 1984. *Proc. Natl. Acad. Sci. USA* 81:4843–47

168. Kasambalides, E. J., Lanks, K. W. 1985. *J. Cell Physiol.* 123:283–87

169. Whelan, S. A., Hightower, L. E. 1985. *J. Cell. Physiol.* 125:251–58

170. Saunders, M. M. 1981. *J. Cell Biol.* 91:579–83

171. Tanguay, R. M., Camato, R., Lettre, F., Vincent, M. 1983. *Can. J. Biochem. Cell Biol.* 61:414–20

172. Taylor, M. W., Long, T., Martinez-Valdez, H., Downing, J., Zeige, G. 1984. *Proc. Natl. Acad. Sci. USA* 81:4033–36

173. Iida, H., Yahara, I. 1985. *Nature* 315:688–91

174. Georgopoulos, C. P., Hendrix, R. W., Casjens, S. R., Kaiser, A. D. 1973. *J. Mol. Biol.* 76:45–60

175. Murialdo, H. 1979. *Virology* 96:341–67

176. Sternberg, N. 1973. *J. Mol. Biol.* 76:1–23

177. Sternberg, N. 1973. *J. Mol. Biol.* 76:25–44

178. Coppo, A., Manzi, A., Pulitzer, J. F., Takahashi, H. 1973. *J. Mol. Biol.* 76:61–87

179. Zweig, M., Cummings, D. J. 1973. *J. Mol. Biol.* 80:505–18

180. Tilly, K., Murialdo, H., Georgopoulos C. 1981. *Proc. Natl. Acad. Sci. USA* 78:1629–33

181. Tilly, K., VanBogelen, R. A., Georgopoulos, C., Neidhardt, F. C. 1983. *J. Bacteriol.* 154:1505–7

182. Tilly, K., Georgopoulos, C. 1982. *J. Bacteriol.* 149:1082–88

183. Tilly, K., Chandrasekhar, G. N., Zylicz, M., Georgopoulos, C. 1985. *Microbiology—1985*, ed. L. Leive, D. Schlessinger, pp. 322–26. Washington, DC: Am. Soc. Microbiol.

184. Neidhardt, F. C., Phillips, T. A., VanBogelen, R. A., Smith, M. W., Georgalis, Y., Subramanian, A. R. 1981. *J. Bacteriol.* 145:513–20

185. Georgopoulos, C. P., Eisen, H. 1974. *J. Supramol. Struct.* 2:349–59

186. Takano, R., Kakefuda, T. 1972. *Nature New Biol.* 239:34–37

187. Wada, M., Itikawa, H. 1984. *J. Bacteriol.* 157:694–96

188. Hendrix, R. W., Tsui, L. 1978. *Proc. Natl. Acad. Sci. USA* 75:136–39

189. Hohn, T., Hohn, B., Engel, A., Wurtz, M., Smith, P. R. 1979. *J. Mol. Biol.* 129:359–73
190. Phillips, T. A., VanBogelen, R. A., Neidhardt, F. C. 1984. *J. Bacteriol.* 159:283–87
191. Charette, M. F., Henderson, G. W., Markovitz, A. 1981. *Proc. Natl. Acad. Sci. USA* 78:4728–32
192. Chung, C. H., Goldberg, A. L. 1981. *Proc. Natl. Acad. Sci. USA* 78:4931–35
193. Larimore, F. S., Waxman, L., Goldberg, A. L. 1982. *J. Biol. Chem.* 257:4187–95
194. Mackie, G., Wilson, D. B. 1972. *J. Biol. Chem.* 247:2973–78
195. Hua, S.-S., Markovitz, A. 1972. *J. Bacteriol.* 110:1089–99
196. Gottesman, S., Gottesman, M., Shaw, J. E., Pearson, M. L. 1981. *Cell* 24:225–33
197. Hirshfield, I. N., Bloch, P. L., VanBogelen, R. A., Neidhardt, F. C. 1981. *J. Bacteriol.* 146:345–51
198. VanBogelen, R. A., Vaughn, V., Neidhardt, F. C. 1983. *J. Bacteriol.* 153:1066–68
199. Chamberlin, M. J. 1974. *Ann. Rev. Biochem.* 43:721–75
200. Gross, C. A., Blattner, F. R., Taylor, W. E., Lowe, P. A., Burgess, R. R. 1979. *Proc. Natl. Acad. Sci. USA* 76:5789–93
201. Cowing, D. W., Bardwell, J. C., Craig, E. A., Woolford, C., Hendrix, R. W., Gross, C. A. 1985. *Proc. Natl. Acad. Sci. USA* 82:2679–83
202. Peters, F. P. A. M. N., Lubsen, N. H., Walldorf, U., Moormann, R. J., Hovemann, B. 1984. *Mol. Gen. Genet.* 197:392–98
203. Dangli, A., Grond, C., Kloetzel, P., Bautz, E. K. F. 1983. *EMBO J.* 2:1747–51
204. Walldorf, U., Richter, S., Ryseck, R. P., Steller, H., Edstrom, J. E., et al 1984. *EMBO J.* 3:2499–504
205. Peters, F. P. A. M. N., Grond, C. J., Sondermeijer, P. J., Lubsen, N. H. 1982. *Chromosoma* 85:237–49
206. Lengyel, J. A., Ransom, L. J., Graham, M. L., Pardue, M. L. 1980. *Chromosoma* 80:237–52
207. Mohler, J., Pardue, M. L. 1982. *Chromosoma* 86:457–67
208. Santa-Cruz, M. C., Morcillo, G., Diez, J. L. 1984. *Biol. Cell.* 52:205–11
209. Lis, J. T., Prestidge, L., Hogness, D. S. 1978. *Cell* 14:901–19
210. Mason, P. J., Torok, I., Kiss, I., Karch, F., Udvardy, A. 1982. *J. Mol. Biol.* 156:21–35
211. Zuker, C., Cappello, J., Lodish, H. F., George, P., Chung, S. 1984. *Proc. Natl. Acad. Sci. USA* 81:2660–64
212. Rosen, E., Sivertsen, A., Firtel, R. A. 1983. *Cell* 35:243–51
213. Rubin, G. M., Hogness, D. S. 1975. *Cell* 6:207–13
214. Wieben, E. D., Pederson, T. 1982. *Mol. Cell. Biol.* 2:914–20
215. Zimmerman, J. L., Petri, W., Meselson, M. 1983. *Cell* 32:1161–70
216. Cheney, C. M., Shearn, A. 1983. *Dev. Biol.* 95:325–30
217. Sirotkin, K., Davidson, N. 1982. *Dev. Biol.* 89:196–210
218. Ireland, R. C., Berger, E. M. 1982. *Proc. Natl. Acad. Sci. USA* 79:855–59
219. Berger, E., Vitek, M., Morgenelli, C. M. 1984. In *Invertebrate Systems in Vitro.* 11. Amsterdam
220. Cohen, R. S., Meselson, M. 1985. *Cell* 43:737–46
221. Hoffman, E., Corces, V. 1986. *Mol. Cell. Biol.* 6:663–73
222. Bensaude, O., Morange, M. 1983. *EMBO J.* 2:173–78
223. Cooper, P., Ho, T.-H.D., Hauptmann, R. 1984. *Plant Physiol.* 75:431–41
224. Lindquist, S. 1984. *J. Embryol. Exp. Morphol.* 83(Suppl):147–61
225. Hunter, K. W., Cook, C. L., Hayunga, E. G. 1984. *Biochem. Biophys. Res. Commun.* 125:755–60
226. Lawrence, F., Robert-Gero, M. 1985. *Proc. Natl. Acad. Sci. USA* 82:4414–17
227. Van-der-Ploeg, L. H., Giannini, S. H., Cantor, C. R. 1985. *Science* 228:1443–36
228. Lambowitz, A. M., Kobayashi, G. S., Painter, A., Medoff, G. 1983. *Nature* 303:806–8
229. Deleted in proof
230. Leenders, H. J., Berendes, H. D., Helmsing, P. J., Derksen, J., Koninkx, J. F. J. G. 1974. *Sub-Cell. Biochem.* 3:119–47
231. Velazquez, J. M., DiDomenico, B. J., Lindquist, S. 1980. *Cell* 20:679–89
232. Li, G. C. 1983. *J. Cell Physiol.* 115:116–22
233. Plesset, J., Palm, C., McLaughlin, C. S. 1982. *Biochem. Biophys. Res. Commun.* 108:1340–45
234. Key, J. L., Kimpel, J., Vierling, E., Lin, C.-Y., Nagao, R. T., et al. 1985. See Ref. 71, pp. 327–48
235. Courgeon, A. M., Maisonhaute, C., Best-Belpomme, M. 1984. *Exp. Cell. Res.* 153:515–21
236. Li, G. C., Laszlo, A. 1985. *J. Cell. Physiol.* 122:91–97
237. Nover, L., ed. 1984. *Heat Shock Response of Eukaryotic Cells.* Leipzig: VEB Georg Thieme. 82 pp.
238. Sachs, M. M., Freeling, M., Okimoto, R. 1980. *Cell* 20:761–67

239. Rosen, E., Sivertsen, A., Firtel, R. A., Wheeler, S., Loomis, W. F. 1985. See Ref. 71, pp. 257–78

240. Lee, P. C., Bochner, B. R., Ames, B. N. 1983. Proc. Natl. Acad. Sci. USA 80:7496–500

241. Christman, M. F., Morgan, R. W., Jacobson, F. S., Ames, B. N. 1985. Cell 41:753–62

242. McKenzie, S. L., Meselson, M. 1977. J. Mol. Biol. 117:279–83

243. Karch, F., Torok, I., Tissieres, A. 1981. J. Mol. Biol. 148:219–30

244. Saito, H., Uchida, H. 1978. Mol. Gen. Genet. 164:1–8

245. Yochem, J., Uchida, H., Sunshine, M., Saito, H., Georgopoulos, C. P., et al. 1978. Mol. Gen. Genet. 14:9–14

246. Burton, Z. F., Gross, C. A., Watanabe, K. K., Burgess, R. R. 1983. Cell 32:335–49

247. Taylor, W. E., Straus, D. B., Grossman, A. D., Burton, Z. F., Gross, C. A., Burgess, R. R. 1984. Cell 38:371–81

248. Yamamori, T., Yura, T. 1980. J. Bacteriol. 142:843–51

249. McAlister, L., Finkelstein, D. B. 1980. J. Bacteriol. 143:606–19

250. Bienz, M., Gurdon, J. B. 1982. Cell 29:811–19

251. Corces, V., Pellicer, A., Axel, R., Meselson, M. 1981. Proc. Natl. Acad. Sci. USA 78:7038–42

252. McMahon, A. P., Novak, T. J., Britten, R. J., Davidson, E. H. 1984. Proc. Natl. Acad. Sci. USA 81:7490–94

253. Bienz, M. 1985. Trends Biochem. Sci. 10:157–61

254. Cartwright, I., Elgin, S. C. R. 1986. Mol. Cell. Biol. In press

255. Costlow, N., Lis, J. T. 1984. Mol. Cell. Biol. 4:1853–63

256. Udvardy, A., Schedl, P. 1984. J. Mol. Biol. 172:385–403

257. Dudler, R., Travers, A. A. 1984. Cell 38:391–98

258. Amin, J., Mestril, R., Lawson, R., Klapper, H., Voellmy, R. 1985. Mol. Cell. Biol. 5:197–203

259. Topol, J., Ruden, D. M., Parker, C. S. 1985. Cell 42:527–37

260. Shuey, D. J., Parker, C. S. 1986. J. Biol. Chem. In press

261. Corces, V., Pellicer, A. 1984. J. Biol. Chem. 259:14812–17

262. Tilly, K., McKittrick, N., Zylicz, M., Georgopoulos, C. 1983. Cell 34:641–46

263. Neidhardt, F. C., VanBogelen, R. A. 1981. Biochem. Biophys. Res. Commun. 100:894–900

264. Yamamori, T., Yura, T. 1982. Proc. Natl. Acad. Sci. USA 79:860–64

265. Grossman, A. D., Cowing, D., Erickson, J., Baker, T., Zhou, Y. N., Gross, C. 1985. See Ref. 183, pp. 327–31

266. Mirault, M. E., Goldschmidt-Clermont, M., Moran, L., Arrigo, A. P., Tissières, A. 1978. Cold Spring Harbor Symp. Quant. Biol. 42:819–27

267. Kruger, C., Benecke, B.-J. 1981. Cell 23:595–603

268. Snutch, T. P., Baillie, D. L. 1983. Can. J. Biochem. Cell Biol. 61:480–87

269. Scott, M., Pardue, M. L. 1981. Proc. Natl. Acad. Sci. USA 78:3353–58

270. Klemenz, R., Hultmark, D., Gehring, W. J. 1985. EMBO J. 4:2053–60

271. McGarry, T. J., Lindquist, S. 1985. Cell 42:903–11

272. Glover, C. V. 1982. Proc. Natl. Acad. Sci. USA 79:1781–85

273. Olsen, A. S., Triemer, D. F., Sanders, M. M. 1983. Mol. Cell. Biol. 3:2017–27

274. Scharf, K. D., Nover, L. 1982. Cell 30:427–37

275. Cerveza, M., Dreyfuss, G., Penman, S. 1981. Cell 23:113–20

276. Biessmann, H., Falkner, F. G., Saumweber, H., Walter, M. F. 1982. See Ref. 42, pp. 275–81

277. Welch, W. J., Feramisco, J. R. 1985. Mol. Cell. Biol. 5:1571–81

278. Sirkin, E., Lindquist, S. 1985. UCLA Symp. Mol. Cell. Biol. 30:669–80

279. Ballinger, D. G., Pardue, M. L. 1983. Cell 33:103–13

280. Thomas, G. P., Mathews, M. B. 1982. See Ref. 42, pp. 207–13

281. Lindquist, S., DiDomenico, B. 1985. See Ref. 71, pp. 71–90

282. Petersen, N. S., Mitchell, H. K. 1981. Proc. Natl. Acad. Sci. USA 78:1708–11

282a. Yost, H. J., Lindquist, S. 1986. Cell In press

283. Hahn, G. M., Li, G. C. 1982. Radiat. Res. 92:452–57

284. Li, G. C., Werb, Z. 1982. Proc. Natl. Acad. Sci. USA 79:3218–22

285. Landry, J., Bernier, D., Cretien, P., Nicole, L. M., Tanguay, R. M., Marceau, N. 1982. Cancer Res. 42:2457–61

286. Subjeck, J. R., Sciandra, J. J., Johnson, R. J. 1982. Br. J. Radiol. 55:579–84

287. Li, G. C., Meyer, J., Mak, Y. K., Hahn, G. M. 1983. Cancer Res. 43:5758–60

288. Mitchell, H. K., Moller, G., Petersen, N. S., Lipps-Sarmiento, L. 1979. Dev. Genet. 1:181–92

289. Tobe, T., Ito, K., Yura, T. 1984. Mol. Gen. Genet. 195:10–16

290. Li, G. C., Laszlo, A. 1985. See Ref. 71, pp. 227–54

291. Muller, W. U., Li, G. C., Goldstein, L. S. 1985. Int. J. Hyperthermia 1:97–102

292. Bonner, J. J., Parks, C., Parker-Thornburg, J., Mortin, M. A., Pelham, H. R. 1984. *Cell* 37:979–91
293. Li, G. C. 1985. *Int. J. Radiation Oncology Biol. Phys.* 11:165–77
294. Iida, H., Yahara, I. 1984. *J. Cell Biol.* 99:1441–50
295. Plesofsky-Vig, N., Brambl, R. 1985. *J. Bacteriol.* 162:1083–91
296. Hall, B. G. 1983. *J. Bacteriol.* 156:1363–65
297. Ferrini, U., Falcioni, R., Delpino, A., Cavaliere, R., Zupi, G., Natali, P. G. 1984. *Int. J. Cancer* 34:651–55
298. Dewey, W. C., Holahan, E. V. 1984. *Prog. Exp. Tumor Res.* 28:198–219
299. Omar, R. A., Lanks, K. W. 1984. *Cancer Res.* 44:3976–82
300. Berger, E. M., Woodward, M. P. 1983. *Exp. Cell Res.* 147:437–42
301. Landauer, W. 1958. *Ann. Nat.* 92:201–13
302. Pleet, H., Graham, J. M. Jr., Smith, D. W. 1981. *Pediatrics* 67:785–89
303. Bellve, A. R. 1973. *J. Reprod. Fertil.* 35:393–403
304. Webster, W. S., Germian, M. A., Edwards, M. J. 1985. *Teratology* 31:73–82
305. Buzin, C. H., Bournias-Vardiabasis, N. 1984. *Proc. Natl. Acad. Sci. USA* 81:4075
306. Capdevila, M. P., Garcia-Bellido, A. 1974. *Nature* 250:500–2
307. Milkman, R. 1966. *Biol. Bull.* 131:331–45
308. Mitchell, H. K., Petersen, N. S. 1982. *Dev. Genet.* 3:91–102
309. Mitchell, H. K., Petersen, N. S. 1981. *Dev. Biol.* 85:233–42
310. Petersen, N. S., Mitchell, H. K. 1982. See Ref. 42, pp. 345–52
311. Walsh, D. A., Klein, N. W., Hightower, L. E., Edwards, M. J. 1985. *Teratology* 31(3):A30–A31

AUTHOR INDEX

(Names appearing in capital letters indicate authors of chapters in this volume.)

SUBJECT INDEX

A

Acanthamoeba
profilins of, 1006-8
rDNA promoter in, 809
rDNA transcription in, 807
rDNA transcription complexes
in, 811-12
Acetaldehyde
metabolism of, 868
Acetate
conversion to butyrate and
caproate, 11
lactate fermentation and, 11
Acetylation
proopiomelanocortin and,
777
Acetylcholine
platelet-activating factor and,
503
receptors for, 790
synthesis of, 775
Acetyl CoA
acetylcholine synthesis and,
775
I-cell fibroblasts and, 189
N-Acetylgalactosamine
lectins and, 49
N-Acetylglucosamine
glycoprotein lectins and,
40
isolectins and, 42
N-Acetylimidazole
sodium channels and, 961
Acetylkinase
periplasmic transport systems
and, 417
N-Acetylneuraminic acid
M. gallisepticum and, 53
Acetylphosphate
periplasmic transport systems
and, 416-17
Acid phosphatase
I-cell fibroblasts and, 188-
89
phosphate in, 173
Acids
See specific type
Aconitine
sodium channels and, 955,
960
ACTH
proopiomelanocortin and, 776-
77
Actin, 987-1030
divalent cations and, 993-94
DNase-I and, 1010
low-molecular-weight severing
proteins and, 1018-19
nonfilamentous, 1005
nucleotides and, 993-94
polymerization of, 994-1004

vitamin-D binding protein
and, 1010-11
Actin-binding protein, 1006-11
calcium-dependent proteinase
and, 466
Actin filaments
bundling proteins and, 1019-
26
capping proteins and, 1011-18
elongation of, 998-1001
low-molecular-weight severing
proteins and, 1018-19
membrane attachment proteins
and, 1026-29
molecular structure of, 991-93
properties of, 1004-5
α-Actinin
actin filaments and, 1022
calcium-dependent proteinase
and, 466
β-Actinin
actin filaments and, 1017
Actin molecule, 989-94
evolution of, 989-90
structure of, 990-91
Actinomyces naeslundii
fimbriae of, 53
Actinomyces viscosus
fimbriae of, 53
Actinomycin D
rRNA synthesis and, 820
Active transport
sugar:ion cotransport and,
226, 229-31
Acumentin
actin filaments and, 1017
Adenine
diphtheria toxin and, 200
lectins and, 48
Adenocarcinoma cells
taurine chloramine and, 448
Adenosine
glucose transport and, 1071
Adenosine triphosphate
actin filaments and, 998-1001
arginine-binding protein and,
408
cellular energy transfer and,
287-88
chloroplast protein transport
and, 897-99
diphtheria toxin and, 204
endocytic/exocytic pathways
and, 664
lysosomal acidification and,
673
periplasmic transport systems
and, 415-16
phosphorylation and, 15
platelet-activating factor and,
491
protein degradation and, 456

RNA transcription and, 351
synthesis of
photosynthetic electron
transport and, 29
Adenosine triphosphate/ADP carrier
transport of, 275
Adenoviruses
DNA-binding proteins and,
125-27
DNA replication in, 736-39
Adenylate cyclase
glucose transport and, 1071-75
prostacyclin and, 73
Adipocytes
organogenesis and, 1040
β-Adrenergic receptor
affinity chromatography and,
57
Adrenoleukodystrophy
DNA markers and, 845
Aerobacter aerogenes
fructokinase mutants of, 319
Affinity chromatography
β-adrenergic receptors and,
57
glycoproteins and, 57
laminin and, 1047
lectins and, 36, 43
prostacyclin synthase and, 72
Affinity electrophoresis
lectins and, 43
Agaricus bisporus
lectins of, 43
Agglutinin
See specific type
Aging
lectin binding patterns and, 58
β-Alanine, 855-56
carnosine biosynthesis and,
865-66
catabolism of, 866-68
synthesis of, 856-59
transamination of, 867
Albumin
colocalization of, 183
1-deoxynojirimycin and, 170
prostaglandin D₂ and, 74
Aldolase
actin filaments and, 1019
cathepsin M and, 460-61
E. coli and, 325-26
yeasts and, 326
Algae
glucose uptake in, 235
plastids and, 882-83
phosphorus turnover in, 15
sugar:ion cotransport in, 226
Alkaline phosphatase
lysosomal enzymes and, 173
Alkaline protease b
sodium channels and, 961

1263

CUMULATIVE INDEXES

CONTRIBUTING AUTHORS, VOLUMES 51–55

CHAPTER TITLES, VOLUMES 51–55

1290   CHAPTER TITLES

Annual Reviews Inc. |ORDER FORM|

A NONPROFIT SCIENTIFIC PUBLISHER

4139 El Camino Way, Palo Alto, CA 94306-9981, USA • (415) 493-4400

al Reviews Inc. publications are available directly from our office by mail or telephone (paid by credit card or ase order), through booksellers and subscription agents, worldwide, and through participating professional ties. Prices subject to change without notice.

ividuals: Prepayment required on new accounts by check or money order (in U.S. dollars, check drawn on . bank) or charge to credit card — American Express, VISA, MasterCard.

titutional buyers: Please include purchase order number.

idents: $10.00 discount from retail price, per volume. Prepayment required. Proof of student status must be vided (photocopy of student I.D. or signature of department secretary is acceptable). Students must send ers direct to Annual Reviews. Orders received through bookstores and institutions requesting student rates be returned.

fessional Society Members: Members of professional societies that have a contractual arrangement with tual Reviews may order books through their society at a reduced rate. Check with your society for infor- tion.

lar orders: Please list the volumes you wish to order by volume number.

ding orders: New volume in the series will be sent to you automatically each year upon publication. Cancel- may be made at any time. Please indicate volume number to begin standing order.

ublication orders: Volumes not yet published will be shipped in month and year indicated.

ornia orders: Add applicable sales tax.

age paid (4th class bookrate/surface mail) **by Annual Reviews Inc.** Airmail postage extra.

NNUAL REVIEWS SERIES		Prices Postpaid per volume USA/elsewhere	Regular Order Please send: Vol. number	Standing Order Begin with: Vol. number
al Review of ANTHROPOLOGY (Prices of Volumes in brackets effective until 12/31/85)				
Vols. 1-10	(1972-1981)	$20.00/$21.00]		
Vol. 11	(1982)	$22.00/$25.00]		
Vols. 12-14	(1983-1985)	$27.00/$30.00]		
Vols. 1-14	(1972-1985)	$27.00/$30.00		
Vol. 15	(avail. Oct. 1986)	$31.00/$34.00	Vol(s). _____	Vol. _____
al Review of ASTRONOMY AND ASTROPHYSICS (Prices of Volumes in brackets effective until 12/31/85)				
Vols. 1-2, 4-19	(1963-1964; 1966-1981)	$20.00/$21.00]		
Vol. 20	(1982)	$22.00/$25.00]		
Vols. 21-23	(1983-1985)	$44.00/$47.00]		
Vols. 1-2, 4-20	(1963-1964; 1966-1982)	$27.00/$30.00		
Vols. 21-23	(1983-1985)	$44.00/$47.00		
Vol. 24	(avail. Sept. 1986)	$44.00/$47.00	Vol(s). _____	Vol. _____
al Review of BIOCHEMISTRY (Prices of Volumes in brackets effective until 12/31/85)				
Vols. 30-34, 36-50	(1961-1965; 1967-1981)	$21.00/$22.00]		
Vol. 51	(1982)	$23.00/$26.00]		
Vols. 52-54	(1983-1985)	$29.00/$32.00]		
Vols. 30-34, 36-54	(1961-1965; 1967-1985)	$29.00/$32.00		
Vol. 55	(avail. July 1986)	$33.00/$36.00	Vol(s). _____	Vol. _____
al Review of BIOPHYSICS AND BIOPHYSICAL CHEMISTRY (Prices of Vols. in brackets effective until 12/31/85)				
rmerly Annual Review of Biophysics and Bioengineering)				
Vols. 1-10	(1972-1981)	$20.00/$21.00]		
Vol. 11	(1982)	$22.00/$25.00]		
Vols. 12-14	(1983-1985)	$47.00/$50.00]		
Vols. 1-11	(1972-1982)	$27.00/$30.00		
Vols. 12-14	(1983-1985)	$47.00/$50.00		
Vol. 15	(avail. June 1986)	$47.00/$50.00	Vol(s). _____	Vol. _____
al Review of CELL BIOLOGY				
Vol. 1	(1985)	$27.00/$30.00		
Vol. 2	(avail. Nov. 1986)	$31.00/$34.00	Vol(s). _____	Vol. _____
al Review of COMPUTER SCIENCE				
Vol. 1	(avail. late 1986) . . . . **Price not yet established**		Vol. _____	Vol. _____
al Review of EARTH AND PLANETARY SCIENCES (Prices of Volumes in brackets effective until 12/31/85)				
Vols. 1-9	(1973-1981)	$20.00/$21.00]		
Vol. 10	(1982)	$22.00/$25.00]		
Vols. 11-13	(1983-1985)	$44.00/$47.00]		
Vols. 1-10	(1973-1982)	$27.00/$30.00		
Vols. 11-13	(1983-1985)	$44.00/$47.00		
Vol. 14	(avail. May 1986)	$44.00/$47.00	Vol(s). _____	Vol. _____

ANNUAL REVIEWS SERIES	Prices Postpaid per volume USA/elsewhere	Regular Order Please send:	Standing O Begin wi

Annual Review of ECOLOGY AND SYSTEMATICS (Prices of Volumes in brackets effective until 12/31/85)

[Vols. 1-12	(1970-1981). . . . . $20.00/$21.00]		
[Vol. 13	(1982) . . . . $22.00/$25.00]		
[Vols. 14-16	(1983-1985). . . . . $27.00/$30.00]		
Vols. 1-16	(1970-1985). . . . . $27.00/$30.00		
Vol. 17	(avail. Nov. 1986). . . . . $31.00/$34.00	Vol(s). _____	Vol. _____

Annual Review of ENERGY (Prices of Volumes in brackets effective until 12/31/85)

[Vols. 1-6	(1976-1981). . . . . $20.00/$21.00]		
[Vol. 7	(1982) . . . . $22.00/$25.00]		
[Vols. 8-10	(1983-1985). . . . . $56.00/$59.00]		
Vols. 1-7	(1976-1982). . . . . $27.00/$30.00		
Vols. 8-10	(1983-1985). . . . . $56.00/$59.00		
Vol. 11	(avail. Oct. 1986). . . . . $56.00/$59.00	Vol(s). _____	Vol. _____

Annual Review of ENTOMOLOGY (Prices of Volumes in brackets effective until 12/31/85)

[Vols. 9-16, 18-26	(1964-1971; 1973-1981). . . . . $20.00/$21.00]		
[Vol. 27	(1982) . . . . $22.00/$25.00]		
[Vols. 28-30	(1983-1985). . . . . $27.00/$30.00]		
Vols. 9-16, 18-30	(1964-1971; 1973-1985). . . . . $27.00/$30.00		
Vol. 31	(avail. Jan. 1986). . . . . $31.00/$34.00	Vol(s). _____	Vol. _____

Annual Review of FLUID MECHANICS (Prices of Volumes in brackets effective until 12/31/85)

[Vols. 1-5, 7-13	(1969-1973; 1975-1981). . . . . $20.00/$21.00]		
[Vol. 14	(1982) . . . . $22.00/$25.00]		
[Vols. 15-17	(1983-1985). . . . . $28.00/$31.00]		
Vols. 1-5, 7-17	(1969-1973; 1975-1985). . . . . $28.00/$31.00		
Vol. 18	(avail. Jan. 1986). . . . . $32.00/$35.00	Vol(s). _____	Vol. _____

Annual Review of GENETICS (Prices of Volumes in brackets effective until 12/31/85)

[Vols. 1-15	(1967-1981). . . . . $20.00/$21.00]		
[Vol. 16	(1982) . . . . $22.00/$25.00]		
[Vols. 17-19	(1983-1985). . . . . $27.00/$30.00]		
Vols. 1-19	(1967-1985). . . . . $27.00/$30.00		
Vol. 20	(avail. Dec. 1986). . . . . $31.00/$34.00	Vol(s). _____	Vol. _____

Annual Review of IMMUNOLOGY

Vols. 1-3	(1983-1985). . . . . $27.00/$30.00		
Vol. 4	(avail. April 1986). . . . . $31.00/$34.00	Vol(s). _____	Vol. _____

Annual Review of MATERIALS SCIENCE (Prices of Volumes in brackets effective until 12/31/85)

[Vols. 1-11	(1971-1981). . . . . $20.00/$21.00]		
[Vol. 12	(1982) . . . . $22.00/$25.00]		
[Vols. 13-15	(1983-1985). . . . . $64.00/$67.00]		
Vols. 1-12	(1971-1982). . . . . $27.00/$30.00		
Vols. 13-15	(1983-1985). . . . . $64.00/$67.00		
Vol. 16	(avail. August 1986). . . . . $64.00/$67.00	Vol(s). _____	Vol. _____

Annual Review of MEDICINE (Prices of Volumes in brackets effective until 12/31/85)

[Vols. 1-3, 5-15, 17-32	(1950-52; 1954-64; 1966-81). . . . . $20.00/$21.00]		
[Vol. 33	(1982) . . . . $22.00/$25.00]		
[Vols. 34-36	(1983-1985). . . . . $27.00/$30.00]		
Vols. 1-3, 5-15, 17-36	(1950-52; 1954-64; 1966-85). . . . . $27.00/$30.00		
Vol. 37	(avail. April 1986). . . . . $31.00/$34.00	Vol(s). _____	Vol. _____

Annual Review of MICROBIOLOGY (Prices of Volumes in brackets effective until 12/31/85)

[Vols. 18-35	(1964-1981). . . . . $20.00/$21.00]		
[Vol. 36	(1982) . . . . $22.00/$25.00]		
[Vols. 37-39	(1983-1985). . . . . $27.00/$30.00]		
Vols. 18-39	(1964-1985). . . . . $27.00/$30.00		
Vol. 40	(avail. Oct. 1986). . . . . $31.00/$34.00	Vol(s). _____	Vol. _____

Annual Review of NEUROSCIENCE (Prices of Volumes in brackets effective until 12/31/85)

[Vols. 1-4	(1978-1981). . . . . $20.00/$21.00]		
[Vol. 5	(1982) . . . . $22.00/$25.00]		
[Vols. 6-8	(1983-1985). . . . . $27.00/$30.00]		
Vols. 1-8	(1978-1985). . . . . $27.00/$30.00		
Vol. 9	(avail. March 1986). . . . . $31.00/$34.00	Vol(s). _____	Vol. _____